Table of Atomic Masses and Numbers

Based on the 1991 Report of the Commission on Atomic Weights and Isotopic Abundances of the International Union of Pure and Applied Chemistry and for the elements as they exist naturally on earth. Scaled to the relative atomic mass of carbon-12. The estimated uncertainties in values, between ±1 and ±9 units in the last digit of an atomic mass, are in parentheses after the atomic mass. (From *Journal of Physical and Chemical Reference Data*, Vol. 22(1993), pp. 1571–1584. © 1993 IUPAC)

Element	Symbol	Atomic Number	Atomic Mass		Element	Symbol	Atomic Number	Atomic Mass	
Actinium	Ac	89	227.0278	(L)	Molybdenum	Mo	42	95.94(1)	(g)
Aluminum	Al	13	26.981539(5)		Neodymium	Nd	60	144.24(3)	(g)
Americium	Am	95	243.0614	(L)	Neon	Ne	10	20.1797(6)	(g, m)
Antimony	Sb	51	121.757(3)		Neptunium	Np	93	237.0482	(L)
Argon	Ar	18	39.948(1)	(g, r)	Nickel	Ni	28	58.6934(2)	
Arsenic	As	33	74.92159(2)		Nielsbohrium	Ns	107	262.12	(L, n)
Astatine	At	85	209.9871	(L)	Niobium	Nb	41	92.90638(2)	
Barium	Ba	56	137.327(7)		Nitrogen	N	7	14.00674(7)	(g, r)
Berkelium	Bk	97	247.0703	(L)	Nobelium	No	102	259.1009	(L)
Beryllium	Be	4	9.012182(3)		Osmium	Os	76	190.23(3)	(g)
Bismuth	Bi	83	208.98037(3)		Oxygen	O	8	15.9994(3)	(g, r)
Boron	B	5	10.811(5)	(g, m, r)	Palladium	Pd	46	106.42(1)	(g)
Bromine	Br	35	79.904(1)		Phosphorus	P	15	30.973762(4)	
Cadmium	Cd	48	112.411(8)	(g)	Platinum	Pt	78	195.08(3)	
Calcium	Ca	20	40.078(4)	(g)	Plutonium	Pu	94	244.0642	(L)
Californium	Cf	98	251.0796	(L)	Polonium	Po	84	208.9824	(L)
Carbon	C	6	12.011(1)	(g, r)	Potassium	K	19	39.0983(1)	(g)
Cerium	Ce	58	140.115(4)	(g)	Praseodymium	Pr	59	140.90765(3)	
Cesium	Cs	55	132.90543(5)		Promethium	Pm	61	144.9127	(L)
Chlorine	Cl	17	35.4527(9)	(m)	Protactinium	Pa	91	231.03588(2)	
Chromium	Cr	24	51.9961(6)		Radium	Ra	88	226.0254	(L)
Cobalt	Co	27	58.93320(1)		Radon	Rn	86	222.0176	(L)
Copper	Cu	29	63.546(3)	(r)	Rhenium	Re	75	186.207(1)	
Curium	Cm	96	247.07003	(L)	Rhodium	Rh	45	102.90550(3)	
Dysprosium	Dy	66	162.50(3)	(g)	Rubidium	Rb	37	85.4678(3)	(g)
Einsteinium	Es	99	252.083	(L)	Ruthenium	Ru	44	101.07(2)	(g)
Erbium	Er	68	167.26(3)	(g)	Rutherfordium	Rf	104	261.11	(L, n)
Europium	Eu	63	151.965(9)	(g)	Samarium	Sm	62	150.36(3)	(g)
Fermium	Fm	100	257.0951	(L)	Scandium	Sc	21	44.955910(9)	
Fluorine	F	9	18.9984032(9)		Seaborgium	Sg	106	263.118	(L, n)
Francium	Fr	87	223.0197	(L)	Selenium	Se	34	78.96(3)	
Gadolinium	Gd	64	157.25(3)	(g)	Silicon	Si	14	28.0855(3)	(r)
Gallium	Ga	31	69.723(4)		Silver	Ag	47	107.8682(2)	(g)
Germanium	Ge	32	72.61(2)		Sodium	Na	11	22.989768(6)	
Gold	Au	79	196.96654(3)		Strontium	Sr	38	87.62(1)	(g, r)
Hafnium	Hf	72	178.49(2)		Sulfur	S	16	32.066(6)	(g, r)
Hahnium	Ha	105	262.114	(L, n)	Tantalum	Ta	73	180.9479(1)	
Hassium	Hs	108	265	(n)	Technetium	Tc	43	98.9072	(L)
Helium	He	2	4.002602(2)	(g, r)	Tellurium	Te	52	127.60(3)	(g)
Holmium	Ho	67	164.93032(3)		Terbium	Tb	65	158.92534(3)	
Hydrogen	H	1	1.00794(7)	(g, m, r)	Thallium	Tl	81	204.3833(2)	
Indium	In	49	114.818(3)		Thorium	Th	90	232.0381(1)	(g)
Iodine	I	53	126.90447(3)		Thulium	Tm	69	168.93421(3)	
Iridium	Ir	77	192.22(3)		Tin	Sn	50	118.710(7)	(g)
Iron	Fe	26	55.847(3)		Titanium	Ti	22	47.88(3)	
Krypton	Kr	36	83.80(1)	(g, m)	Tungsten	W	74	183.84(1)	
Lanthanum	La	57	138.9055(2)	(g)	Ununnilium	Uun	110	269	L
Lawrencium	Lr	103	262.11	(L)	Unununium	Uuu	111	272	L
Lead	Pb	82	207.2(1)	(g, r)	Uranium	U	92	238.0289(1)	(g, m)
Lithium	Li	3	6.941(2)	(g, m, r)	Vanadium	V	23	50.9415(1)	
Lutetium	Lu	71	174.967(1)	(g)	Xenon	Xe	54	131.29(2)	(g, m)
Magnesium	Mg	12	24.3050(6)		Ytterbium	Yb	70	173.04(3)	(g)
Manganese	Mn	25	54.93805(1)		Yttrium	Y	39	88.90585(2)	
Meitnerium	Mt	109	266	(n)	Zinc	Zn	30	65.39(2)	
Mendelevium	Md	101	258.10	(L)	Zirconium	Zr	40	91.224(2)	(g)
Mercury	Hg	80	200.59(2)						

(g) Geologically exceptional specimens of this element are known that have different isotopic compositions. For such samples, the atomic mass given here may not apply as precisely as indicated.

(L) This atomic mass is the relative mass of the isotope of longest half-life. The element has no stable isotopes.

(m) Modified isotopic compositions can occur in commercially available materials that have been processed in undisclosed ways, and the atomic mass given here might be quite different for such samples.

(n) Name and symbol approved in 1995 for use in the United States by the Nomenclature Committee of the American Chemical Society.

(r) Ranges in isotopic compositions of normal samples obtained on earth do not permit a more precise atomic mass for this element, but the tabulated value should apply to any normal sample of the element.

Chemistry

The Study of Matter and Its Changes
Second Edition

Chemistry

The Study of Matter and Its Changes
Second Edition

James E. Brady
St. John's University, New York

John R. Holum
Augsburg College (Emeritus), Minnesota

John Wiley & Sons, Inc.
New York / Chichester / Brisbane / Toronto / Singapore

ACQUISITIONS EDITOR   Nedah Rose
DEVELOPMENTAL EDITOR   Kathleen Dolan
MARKETING MANAGER   Catherine Faduska
PRODUCTION EDITOR   Suzanne Magida
DESIGNER   Kevin Murphy
MANUFACTURING MANAGER   Mark Cirillo
PHOTO EDITOR   Lisa Passmore
ILLUSTRATION EDITOR   Sigmund Malinowski

This book was set in 10/12 New Baskerville by Progressive Information Technologies and
printed and bound by Von Hoffman Press. The cover was printed by Phoenix Color.

Electronic Illustrations were provided by Fine Line.

Library of Congress Cataloging-in-Publication Data
Brady, James E., 1938–
    Chemistry: the study of matter and its changes/James E. Brady,
John R. Holum. —2nd ed.
        p.   cm.
    Includes index.
    ISBN 0-471-10042-0 (cloth : alk. paper)
    1. Chemistry.   I. Holum, John R.   II. Title.
QD33.B82   1996
540—dc20   95-45557
                                                                              CIP

Printed in the United States of America

10 9 8 7 6 5 4 3 2 1

Preface

We were gratified by the enthusiastic reception this text received in its previous edition, and users continue to tell us they are pleased with the writing style, student-friendly attitude, and the clarity and thoroughness of our explanations of difficult concepts. Therefore, in approaching this revision our broadest goal was to retain these features while addressing suggestions that would make the text more useful still to both the student and the teacher. The result, we believe, is a textbook well suited to the needs of the mainstream general chemistry course for science and engineering students.

OVERALL PHILOSOPHY AND GOALS

The philosophy of the text continues to be based on our conviction that a general chemistry course serves several goals in the education of a student. First, of course, it must provide the student with a sound foundation in the basic facts and concepts of chemistry. In doing so, the course should build a solid footing of factual knowledge upon which the theoretical models of chemistry can be constructed. The general chemistry course should also give the student an appreciation of the importance of chemistry in society and day-to-day living, and it should enable the student to develop skills in analytical thinking and problem solving.

FEATURES THAT SERVE THE PHILOSOPHY AND GOALS

To fulfill the goals described above, we have focused on a number of major areas, taking into account during the revision process both reviewers' comments and suggestions made by users of the text. Significant changes in the second edition are highlighted in italics.

Organization

Our aim from the outset has been to provide a logical progression of topics arranged to provide the maximum flexibility for the teacher in organizing his or her course. To improve the usefulness of the text, both for the teacher and for the student who will be the ultimate consumer, we have made some significant changes in organization as well as revisions of individual chapters.

Overall Structure

There is a consensus among those teaching general chemistry that textbooks have become too large and contain more information than can reasonably be taught in two semesters. For practical reasons (primarily time constraints), most instructors skip the discrete end-of-book chapters dealing with the descriptive chemistry of the elements—topics of interest principally to chemistry majors who do not make up the bulk of their students. *To provide a shorter textbook tailored to the* **realistic** *needs of teachers and students, we have elected to place the classical descriptive chemistry of the metals and nonmetals in a separate supplement.* We have, however, retained within the main textbook (Chapter 19) discussions of complex ions of the transition elements, including structure, bonding, and nomenclature, and continue to integrate substantial amounts of descriptive chemistry throughout the discussions of principles.

Concept Development

The development of concepts in the text is revealed by the content of chapters. Chapters 1 through 5 develop a foundation in reaction chemistry, stoichiometry, and thermochemistry, with a basic introduction to the structure of matter and the periodic table.

To enable students to get an early start on understanding energy concepts, particularly the difference between heat and temperature, *we have introduced some basic notions of kinetic theory in Chapter 1* when we first discuss the concept of energy. This is expanded upon further in Chapter 5 (Energy and Thermochemistry).

Because so many of the substances we encounter on a daily basis are organic compounds, *we provide an early but* brief *overview of the kinds of compounds formed by carbon. Some simple hydrocarbons are introduced in Chapter 2, and the structural features of the most common functional groups are defined in Chapter 7 (Bonding I).* This will make it easier for students to understand later discussions of weak acids and bases as well as the physical properties of liquids, where organic compounds are often used as examples during discussions of basic principles.

New Chapter 4 (Chemical Reactions in Solution) incorporates portions of previous Chapters 11 and 12 that dealt with acid–base and redox chemistry. This new chapter provides students with a foundation in the basic descriptive chemistry of solution reactions, including acid–base, metathesis, and redox reactions. These topics give students a knowledge base that serves as a foundation for theoretical concepts developed in Chapters 6–8 (which deal with atomic structure and bonding). Chapter 4 also provides a thorough discussion of the stoichiometry of reactions in solution and prepares students for meaningful experiments in the laboratory.

Chapters 6 through 8 cover electronic structures of atoms and bonding in compounds. *In Chapter 6 (Atomic and Electronic Structure), discussions of irregularities in periodic trends in ionization energy and electron affinity now appear in a Special Topic.* This enables teachers who do not wish to dwell on these finer details to omit them easily. Yet they are available for teachers who wish to explore them.

The discussion of bonding is divided between two chapters. The first treats the topic at a relatively elementary level, describing the principal features of ionic and covalent bonds using Lewis structures. The second bonding chapter deals with molecular structure (VSEPR theory) and the valence bond and molecular orbital theories. *In this edition we have moved the discussion of bond energies and their measurement to Chapter 13 (Thermodynamics), where it appears as a Special Topic. Lewis acids and bases are now introduced in Chapter 8* when the concept of coordinate covalent bonds is presented.

We have added a new chapter (Chapter 9) entitled Chemical Reactions: Periodic Correlations. The focus here is on Brønsted acid–base chemistry and redox reactions, with particular emphasis on properties that can be correlated with an element's position in the periodic table.

Chapters 10 through 12 focus on the physical properties of the states of matter and solutions. *In the chapter on gases (Chapter 10), the discussion of manometers has been expanded slightly. The treatment of the individual gas laws has been trimmed, with emphasis now on the combined gas law. The discussion of the van der Waals equation has been shortened. We have dropped the detailed treatment of fractional crystallization that appeared in the first edition in the chapter on Solutions.*

Chapters 13 through 15 take the student into the study of the fundamental factors that determine the spontaneity of change and equilibrium. Chapter 13 (Thermodynamics) seeks to answer the question "Is a change possible?" Chapter 14 (Kinetics) examines the question "If a change is possible, how fast does it occur?" And Chapter 15 (Equilibrium) addresses the question "What is a system like when it ceases to change?"

Chapters 16 and 17 examine equilibrium in greater detail as they apply to acid–base chemistry and solubility equilibria. *Chapter 16 on acid–base equilibrium has been completely rewritten to make it more concise.* Emphasis is on situations where simplifying approximations apply. Discussion of problems requiring the quadratic equation or the method of successive approximations are placed in a separate section.

Chapter 18, which covers electrochemistry, brings together concepts of thermodynamics and equilibrium as well as practical applications of electrolysis and galvanic cells.

Chapter 19 provides a discussion of the structure, nomenclature, bonding, and chemical equilibria involving complex ions, particularly those of the transition metals.

Chapter 20 presents an overview of nuclear reactions and the role they play in chemistry and society. *This chapter, which could actually be presented earlier in the course should the teacher elect to do so, has been revised to keep it up to date.*

Chapter 21, the final chapter of the text, serves as an introduction to organic and biochemistry, and includes the elements of polymer chemistry.

Level

We have developed discussions carefully to provide students with clear, easily understood explanations of difficult topics while maintaining a light and student-friendly writing style. As in the previous edition, students are not assumed to have had a previous course in chemistry, and mastery of only basic algebra is expected. The level of the discussions in this book is probably best judged by the topic coverage in Chapters 13–16, those that deal with the driving forces for reactions, with chemical equilibria, and with exercises calling for a combination of thinking and computational skills. We believe that the level is right for the mainstream general college chemistry course.

PEDAGOGICAL FEATURES

In response to student difficulties in problem solving, we have adopted a unique approach to developing thinking skills. We distinguish three types of learning aids: those that further comprehension and learning; those that enhance problem-solving skills; and those that extend the breadth and knowledge of the student.

Features That Further Comprehension and Learning

Margin comments make it easy to enrich a discussion, without carrying the aura of being essential. Some margin comments jog the student's memory concerning a definition of a term.

Periodic table correlations. One of our goals in this edition has been to call particular attention to the usefulness of the periodic table in correlating chemical and physical properties of the elements. To emphasize this, *a special icon, shown at the left, is placed in the margin to highlight particular periodic correlations when they are introduced.* We also include a table inside the rear cover of the text that serves as an index to these periodic correlations.

Boldface terms alert the student to "must-learn" items.

Chapter summaries use the boldface terms to show how the terms fit into statements that summarize concepts.

Features that Enhance Problem-Solving Abilities

Many students entering college today lack experience in analytical thinking. A course in chemistry should provide an ideal opportunity to help students sharpen their skills because problem solving in chemistry operates on two levels. Because of the nature of the subject, in addition to mathematics many problems also involve the application of theoretical concepts. Students have difficulty at both levels, and one of the goals of this text has been to develop a unified approach that addresses each level.

Chemical Tools Approach to Problem Analysis. In studying chemistry, students learn many simple skills, such as finding the number of grams in a mole of a substance or writing the Lewis structure of a molecule. Problem solving involves bringing together a sequence of such simple tasks. Therefore, if we are to teach problem solving, we must teach students how to seek out the necessary relationships required to obtain solutions to problems.

In the previous edition, we introduced a new and innovative approach to problem solving that makes use of an analogy between the abstract tools of chemistry and the concrete tools of a mechanic. Students are taught to think of simple skills as tools that can be used to solve more complex problems. When faced with a new problem, the student is encouraged to examine the tools that have been taught and to select those that bear on the problem at hand.

To foster this approach to thinking through problems, we present a comprehensive program of reinforcement and review. *The **icon** in the margin calls attention to each chemical tool when it is first introduced and is accompanied now by a brief statement that identifies the tool.* Following the Summary at the end of the chapter, the tools are reviewed under the heading **Tools You Have Learned,** preparing students for the exercises that follow.

Worked Examples that involve an assemblage of concepts include a section entitled *Analysis* that describes the thought processes involved in the identification of the tools needed to solve the problem. *Examples usually conclude with a check for the reasonableness of the answer.*

Practice Exercises follow the worked examples to enable the student to apply what has just been studied to a similar problem.

Thinking it Through is an innovative selection of end-of-chapter problems that takes the emphasis of problem solving away from the answer itself and focuses it instead on the information and method needed to solve the problem. The intent of the Thinking It Through problems is to further accustom the student to analyzing a problem before trying to carry out any computations. *We now divide these problems into two categories. Students who have developed reasonably good thinking skills can skip the Level 1 problems and proceed directly to the more challenging problems in Level 2.*

Review Exercises provide routine practice in the use of basic tools as well as opportunities to incorporate these tools into the solution of more complex problems. Most Review Exercises are classified according to topic type, but several **Additional Exercises** at the end of most problem sets are unclassified. Many problems are cumulative, requiring two or more concepts, and several in later chapters require skills learned in earlier chapters.

Tests of Facts and Concepts provide students with an opportunity to review concepts from the preceding chapters, and they include many problems that require the student to use concepts from more than one chapter.

Features that Extend the Breadth of Knowledge of the Student

Chemicals in Use. These are two-page, illustrated essays placed between most chapters. The essays offer students the opportunity to learn about some practical, real-world applications of chemistry in industry, medicine, and the environment. The first one, following Chapter 1, focuses on Linus Pauling and his enormous contributions to chemistry and reveals the human side of chemists and chemistry. Many of the essays are new and updated essay topics include "Lasers in Chemistry," "Superconductivity," "Synthetic Diamonds and Diamond Coatings," "Buffers and Breathing," and "Ozone in the Stratosphere."

Special Topics. The Special Topics sometimes resemble Chemicals in Use essays but have lengths unsuited to the two-page format. Some Special Topics provide historical background; others concern chemistry and the environment. A list of special topics is included before the Table of Contents.

Illustrations. Illustrations, which are distributed liberally throughout the text, have been carefully crafted using modern computer techniques to provide accurate, eye-appealing complements to discussions. Color is used constructively rather than for its own sake. For example, a consistent set of colors is used to identify atoms of the elements in drawings that illustrate molecular structure. These are shown in the margin.

Photographs. The large number of striking photographs in the book serve two purposes. One is to provide a sense of reality and color to the chemical and physical phenomena described in the text. Toward this end, many photographs of chemicals and chemical reactions are included. The other purpose of the photographs is to illustrate how chemistry relates to the world outside the laboratory. The chapter-opening photos, for example, call the students'

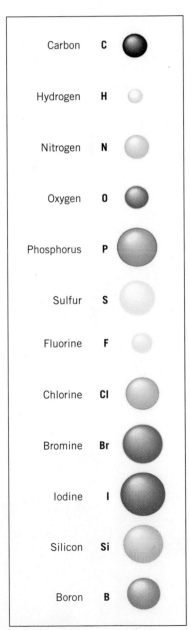

Carbon	C
Hydrogen	H
Nitrogen	N
Oxygen	O
Phosphorus	P
Sulfur	S
Fluorine	F
Chlorine	Cl
Bromine	Br
Iodine	I
Silicon	Si
Boron	B

attention to the relationship between the chapter's content and common (and often not-so-common) things. Similar photos within the chapters illustrate common, practical examples and applications of chemical reactions and physical phenomena.

SUPPLEMENTS

A comprehensive package of supplements has been created to assist both the teacher and the student and includes the following:

Descriptive Chemistry of the Elements by James E. Brady and John R. Holum. This full color, paperback supplement contains three chapters covering the classical descriptive chemistry of the metals and nonmentals.

A Student's Companion for Chemistry by Thomas J. Greenbowe and Kathy Burke, both of Iowa State University and Jeffrey Pribyl, Mankato State University. This small, portable quick reference to the essential concepts and equations in general chemistry is intended for use in lecture or for review purposes. It also makes an ideal reference source in "open-book" exams.

Study Guide by James E. Brady and John R. Holum. This Guide has been written to further enhance understanding of concepts. It is an invaluable tool for students and contains chapter overviews, additional worked-out problems, and alternate problem-solving approaches as well as extensive review exercises.

Solutions Manual by Michael Kenney of Michigan State University and Paul Gaus of Wooster College. The Manual contains worked-out solutions for text problems whose answers appear in Appendix D.

Chemistry in the Laboratory: A Study of Chemical and Physical Changes by Jo A. Beran of Texas A & M University—Kingsville. This volume contains 35 experiments, including significant microscale experiments designed to reduce the generation of waste. Each experiment begins with an introduction providing the "big picture" relevance of the experiment. A comprehensive Instructor's Manual accompanies this laboratory.

Instructor's Manual by Mark A. Benvenuto of University of Detroit—Mercy. This rich teacher's resource manual contains learning objectives, chapter overviews, lecture lead-ins and transitions, class demonstrations and group activities, discussion and critical thinking questions, teaching tips for first-time lecturers, and helpful software and multimedia resources. The manual also contains complete worked-out solutions to all text problems and practice exercises.

Test Bank by Raymond F. X. Williams of Howard University. Contains over 1,800 test items, including multiple-choice, true–false, short answer questions, and critical thinking problems.

Computer Test Bank. IBM and Macintosh versions of the entire Test Bank are available with full editing features to help you customize tests.

Chemistry: *An Experimental Science* **Videotape or Videodisk** developed by George M. Bodner and Paul E. Smith, Purdue University. Both video resources contain the same 38 chemical demonstrations. Each demonstration is 1–4 minutes in length, with narration that explains the demonstration, its techniques, and its underlying principles. Proper safety precautions are observed and emphasized in all of the demonstrations.

Four-Color Overhead Transparencies. Over 125 four-color illustrations from the text are provided in a form suitable for projection in the classroom.

Chemistry Teaching Graphics: General Chemistry CD-ROM by Darrell J. Woodman, Chemistry Teaching Computer Graphics Project, The University of Washington. The dual platform CD-ROM contains 30 general chemistry concept presentations that have been developed for classroom or student viewing to help explain chemical concepts involving: 1) the particulate view of matter, 2) complex spatial relationships and structures, and 3) dynamic processes. The units included have been found useful in teaching at test sites over the last three years. The CD-ROM includes broadcast-quality graphics, true 3D perspective and frequent use of molecular dynamics animations and quantum mechanical calculation of geometries and/or energies. These demonstrations are designed not merely to be attractive and to clarify concepts forcefully, but also to be scientifically accurate representations of molecular phenomena based on extensive consultation with subject-matter experts and other experienced teachers.

ACKNOWLEDGMENTS

We begin with fond words of thanks to our wives, Mary Holum and June Brady, and to our children Liz, Ann, and Kathryn Holum and Mark and Karen Brady who have given us constant support and encouragement. Our appreciation is also extended to our colleagues, Earl Alton and Sandra Olmsted of Augsburg College, and Ernest Birnbaum, Neil Jespersen, and William Pasfield of St. John's University for their helpful discussions and stimulating ideas. We also recognize the support of our colleagues at the administrative level—President Charles S. Anderson of Augsburg College and Rev. David O'Connell, C. M., of St. John's University. It is with particular pleasure that we thank the staff at Wiley for their careful work, encouragement, and sense of humor, particularly our editor, Nedah Rose, our developmental editor, Kathleen Dolan, our marketing manager, Catherine Faduska, our photo editor, Lisa Passmore, our designer, Kevin Murphy, our illustration editor, Sigmund Malinowski, and our production editor, Suzanne Magida. We also extend our most sincere appreciation to Connie Parks for her diligence in her help with the proofreading, to Michael Kenney of Michigan State University for his assistance with the development of the problem sets and answers, and to Suzane Holladay for her careful checking of the answers for accuracy, and to the colleagues listed below, whose careful reviews, helpful suggestions, and thoughtful criticisms of the manuscript have been of such great value to us in developing this book. We especially appreciate the suggestions for end-of-chapter Exercises by some of the reviewers and we have cited their specific contributions within the text.

Reviewers of Second Edition

Wesley Bentz
Alfred University

Mark Benvenuto
University of Detroit—Mercy

C. Eugene Burchill
University of Manitoba

Kathleen Crago
Loyola University

David Dobberpuhl
Creighton University

Joseph Dreisbach
University of Scranton

Barbara Drescher
Middlesex County College

Thomas Greenbowe
Iowa State University

Peter Hambright
Howard University

Henry Harris
Armstrong State College

Ernest Kho
University of Hawaii—Hilo

Ken Loach
SUNY—Plattsburgh

Glen Loppnow
University of Alberta

David Marten
Westmont College

Jeanette Medina
SUNY—Geneseo

Robert Nakon
West Virginia University

Brian Nordstrom
Embry–Riddle Aeronautical University

Sabrina Godfrey Novick
Hofstra University

Les Pesterfield
Western Kentucky University

Ronald See
St. Louis University

Anton Shurpik
U.S. Merchant Marine Academy

Reuben Simoyi
West Viginia University

Agnes Tenney
University of Portland

Roselin Wagner
Hofstra University

Reviewers of First Edition

Wendy Elcesser
Indiana University of Pennsylvania

Mark Amman
Alfred State College

Andrew Jorgensen
University of Toledo

Henry Daley
Bridgewater State College

David Frank
Ferris State University

Reynold Kero
Saddleback College

William Davies
Emporia State University

Nina Klein
Montana Tech

Patricia Moyer
Phoenix College

Janice Kelland
Memorial University of Newfoundland

David F. Dever
Macon College

Henry C. Kelly
Texas Christian University

Ralph W. Sheets
Southwest Missouri State University

James Niewahner
Northern Kentucky University

Diana Daniel
General Motors Institute

Warren Yeakel
Henry Ford Community College

Donald Brandvold
New Mexico Institute of Mining and Technology

Dale Arrington
South Dakota School of Mines and Technology

Larry Krannich
University of Alabama-Birmingham

Judith Burstyn
University of Wisconsin

Edward Senkbeil
Salisbury State University

James Farrar
University of Rochester

Russell D. Larsen
Texas Technical University

Robert M. Kren
University of Michigan-Flint

Keith O. Berry
Oklahoma State University

William Deese
Louisiana Tech University

Donna Friedman
Florrisant Valley Community College

Simon Bott
University of North Texas

Mary Sohn
Florida Institute of Technology

Wendy L. Keeney-Kennicutt
Texas A & M University

Darel Straub
University of Pittsburgh

Robert F. Bryan
University of Virginia

William Bitner
Alvin Community College

Paul Gaus
College of Wooster

Karl Seff
University of Hawaii

DonnaJean A. Fredeen
Southern Connecticut State University

Barbara A. Burke
California State Polytechnic University-Pomona

A Student Guide: How to Use This Book

You are about to begin what could be one of the most exciting courses that you will undertake in college. It offers you the opportunity to learn what makes our world "tick" and to gain insight into the roles that natural and synthetic chemicals play in nature and in our society. This knowledge will not come without some effort, however, and this book has been carefully designed and written with an awareness of the kinds of difficulties you may face. Therefore, before you begin, we would like to outline some of the key features of the book that will aid you in your studies. Please take some time now to familiarize yourself with them.

- *Inside front and back covers*   We have used the space inside the front and back covers of the book to provide useful data you will need often. Notice the periodic table inside the front cover, and the tables of conversion factors and other tables inside the back cover.
- *Table of Contents*   Take a few moments to study the Table of Contents. It will give you an overview of the subject you are about to undertake.
- *Appendices*   There are five lettered appendices located at the back of the book. We call your attention to two or them in particular. Appendix A gives a brief review in the basic math concepts needed in the course as well as some instruction in using your calculator. You may find it helpful to refer to this appendix from time to time. Appendix D gives answers to all the Practice Exercises and selected Review Exercises.
- *Glossary*   A glossary of chemical terms is also found at the back of the book. Notice that with each glossary entry there is a number in parentheses. This is the section number in the text where the term is introduced and discussed.
- *Index*   A well-constructed index is a key part of a useful textbook. Learn to use the index whenever you wish to find a given topic in the book. It can save you a lot of time.

In addition to these general features of the book, each chapter contains a variety of learning aids that you should be sure to make full use of. To help you recognize them, we present the following "pages in miniature."

On prolonged exposure to air and moisture, bronze statues such as this Fisherman's Memorial in Gloucester, MA, gradually corrode and become coated with a green layer of $Cu_2(OH)_2CO_3$. The corrosion of metals involves a process called oxidation, *which is one of the chemical terms introduced in this chapter.*

Chapter 4

Chemical Reactions in Solution

4.1
SOLUTIONS AND CHEMICAL REACTIONS

When two or more reactants are involved in a chemical reaction, their particles—atoms, ions, or molecules—must make physical contact. The particles need freedom of motion for this, which exists in gases and liquids but not in solids. Whenever possible, therefore, reactions are carried out with all of the reactants in one fluid phase, liquid or gas. When necessary, a liquid is used to dissolve solid reactants so their particles can move about. A *solid* mixture of sodium carbonate and citric acid, for example, can be made without any reaction occurring. At the moment water is added, however, and these substances start to dissolve, a furious fizzing occurs as gaseous carbon dioxide forms and leaves the mixture. Representing citric acid as H_3Cit, the reaction in solution is

$$3Na_2CO_3(aq) + 2H_3Cit(aq) \longrightarrow 2Na_3Cit(aq) + 3CO_2(g) + 3H_2O$$

You've seen this fizzing if you've use a medication like Alka-Seltzer, which also contains aspirin (Figure 4.1).

In this chapter we discuss the kinds of reactions that occur primarily in aqueous solutions and how to deal with these reactions quantitatively in the laboratory.

Some Definitions

When a solution forms, at least two substances are involved. One is called the *solvent*, and all of the others are called *solutes*. The solvent is the medium into which the solutes are mixed or dissolved. Water is a typical and very common solvent, but the solvent can actually be in any physical state, a solid, a liquid, or a gas. Unless stated otherwise, we will assume that any solutions we mention are *aqueous solutions*, so that liquid water is understood to be the solvent.

FIGURE 4.1
Carbon dioxide forms when sodium bicarbonate reacts with citric acid in water, as is occurring here to a tablet of a carbonated medication, Alka-Seltzer.

3.2
MEASURING MOLES OF ELEMENTS AND COMPOUNDS

Diamond is one form of pure carbon. A 12 g diamond would be a 60 carat stone.

FIGURE 3.1

Each quantity of these elements contains the same number of atoms, Avogadro's number.

Atomic mass [T]

We ignore the estimated uncertainties in the atomic mass values given in the Table of Atomic Masses and Numbers. These are numbers between 1 and 9 set in parentheses immediately after most of the atomic masses.

EXAMPLE 3.2
Converting Grams to Moles

[▪] The rounding off rules were given in footnote 6, page 14.

According to the SI definition, we have 1 mol of carbon-12 when we have exactly 12 g of this isotope. However, naturally occurring carbon is a mixture of isotopes, and the "average" carbon atom has a mass of 12.011 u. Avogadro's number of these "average atoms" therefore has a mass of 12.011 g. In other words, for naturally occurring carbon: 1 mol C = 12.011 g C. The extension to all the elements is important and worth emphasizing.

One mole of any element has a mass in grams that is numerically equal to the element's atomic mass.

The sizes *in grams* of the 1-mol samples of four elements shown in Figure 3.1 depend on the individual atomic masses, but each sample contains the *same* number of atoms (Avogadro's number).

In our chemical thinking we use *moles* to express the quantitative relationships between formula un... ...measure mole amounts, a... we weigh out 12.011 g of c... The next two examples ill... between grams and moles

Rounding Atomic Ma...

We will often obtain data... from the Periodic Table... known to several places be... the Tables. *The data in the*... *mass*. We round an atomi... accommodate the numbe... lation. Often we simply r... appear in the data given i...

How many moles of silicon...

ANALYSIS This is our first ve... tions, finding the number... restate the question as follo...

The atomic mass of silicon... tool, because we know fro... about silicon.

Like any equivalency, this o... the tool we need for our gr...

$$\frac{28...}{1...}$$

Key Concepts
Key concepts and equations are set off between blue rules. These are especially important and you should be sure to know them and understand how to apply them.

STEP 9   Combine H⁺ and OH⁻ to form H₂O. The left side has $2OH^-$ and $2H^+$, which become $2H_2O$. So in place of $2OH^- + 2H^+$ we write $2H_2O$.

$$2H_2O + 3SO_3^{2-} + 2MnO_4^- \longrightarrow 3SO_4^{2-} + 2MnO_2 + H_2O + 2OH^-$$

STEP 10   Cancel any H₂O that you can. In this equation, one H_2O can be canceled from both sides. The final equation, balanced for basic solution, is

$$H_2O + 3SO_3^{2-} + 2MnO_4^- \longrightarrow 3SO_4^{2-} + 2MnO_2 + 2OH^-$$

■ **Practice Exercise 13** Balance the following equation for a basic solution.

$$MnO_4^- + C_2O_4^{2-} \longrightarrow MnO_2 + CO_3^{2-}$$

The result of Steps 8 and 9 is to replace H⁺ with an equivalent number of H₂O, and the addition of that number of OH⁻ to the other side of the equation.

When we carry out reactions in solution, we normally dissolve the solutes first and then mix their solutions. To work quantitatively with the stoichiometry of these solutions, we must have ways of expressing their concentrations. The expression of concentration most suited to this purpose is called the **molar concentration** or **molarity** (abbreviated *M*). *The molarity of a solution is the number of moles of solute per liter of solution.* We can write this as a ratio of the moles of solute to the volume of the solution expressed in liters.

$$\text{Molarity } (M) = \frac{\text{moles of solute}}{\text{liters of solution}} \qquad (4.1)$$

4.10
MOLAR CONCENTRATION

[T] Molar concentration

Thus, a solution that contains 0.100 mol of NaCl in 1.00 L has a molarity of 0.100 *M*, and we would refer to this solution as 0.100 *molar* NaCl or as 0.100 *M* NaCl. The same concentration would result if we dissolved 0.0100 mol of NaCl in 0.100 L (100 mL) of solution, because the *ratio* of moles of solute to liters of solution is the same.

$$\frac{0.100 \text{ mol NaCl}}{1.00 \text{ L NaCl soln}} = \frac{0.0100 \text{ mol NaCl}}{0.100 \text{ L NaCl soln}} = 0.100 \text{ } M \text{ NaCl}$$

Molarity is particularly useful because it lets us obtain chemicals by moles simply by measuring volumes of solutions, and this is easy to do in the lab. If we had a 0.100 *M* NaCl solution, for example, and we needed 0.100 mol of NaCl for a reaction, we would simply measure out 1.00 L of the solution because in this volume there is 0.100 mol of NaCl.

Figure 4.11 (p. 162) illustrates the preparation of a standard solution of the yellow compound sodium chromate, Na_2CrO_4.

Molarity as a Conversion Factor

A solution's molarity is a source of two conversion factors that relate moles of solute to volume of solution. When a solution is 0.100 *M* NaCl, for example, we can construct the following conversion factors.

$$\frac{0.100 \text{ mol NaCl}}{1.00 \text{ L NaCl soln}} \qquad \text{and} \qquad \frac{1.00 \text{ L NaCl soln}}{0.100 \text{ mol NaCl}}$$

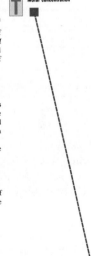

Margin Notes
You will find many of these throughout the text. Sometimes they help clarify discussions, sometimes they offer supporting data or interesting asides, and sometimes they are intended to jog your memory about something you learned earlier.

[T] **Tool Icon**
Learning to solve problems is learning how to think and how to incorporate information into the solution. To help you develop your problem-solving skills, we find it useful to compare solving chemistry problems with repairing a car or a toaster. The repair job requires the use of certain tools, one at a time, until the task is completed. Solving chemistry problems likewise involves the use of "tools," they are just *different* tools. This icon in the margin, with its brief message, calls your attention to such a problem-solving tool, which you will be taught to apply. (Additional explanation of the Chemical Tools approach to problem solving is given in the Study Guide, which is available from your bookstore or the publisher.)

Photographs

You will find numerous striking photographs throughout the book. Some provide support for discussions of chemical events, while others relate chemical phenomena to the world outside the laboratory.

(a)   (b)   (c)   (d)

If a solution contains less solute than required for saturation, we call it an **unsaturated solution.** Unsaturated solutions are able to dissolve more solute.

Usually, the solubility of a solute increases with increasing temperature, so as the temperature of the mixture is raised, more solute dissolves. If the temperature of the solution is subsequently lowered, we would expect this additional solute to separate from the solution, and indeed, this tends to happen. However, sometimes the solute doesn't separate and we obtain a **supersaturated solution,** a solution that contains more solute than required for saturation at a given temperature.

Supersaturated solutions are unstable and can be prepared only if there are no traces of undissolved solute left in contact with the solution. If even a tiny crystal of the solute is added, the extra solute will crystallize and separate from the solution. Figure 4.2 shows what happens when a tiny crystal of sodium acetate is added to a supersaturated solution of this salt in water.

FIGURE 4.2

When a small seed crystal of sodium acetate is added to a supersaturated solution of the salt, excess solute crystallizes rapidly until the solution is just saturated. The crystallization shown in this sequence took less than 10 seconds.

In Chapter 2 you learned that pure water does not conduct electricity. This is because water consists of uncharged molecules that are incapable of transporting electrical charge. However, when an ionic compound such as $CuSO_4$ or NaCl is dissolved in the water, an electrically conducting solution is formed.

Solutes such as $CuSO_4$ or NaCl that yield electrically conducting aqueous solutions are called **electrolytes** (Figure 4.3). These ionic compounds are electrolytes because their solid crystals consist of already existing ions. When they dissolve in water, the ions separate from each other and enter the solution

4.2

ELECTROLYTES AND NONELECTROLYTES

For ionic compounds, the ions are present both in the solid and in aqueous solutions.

(a)   (b)

FIGURE 4.3

Electrical conductivity of solutions of electrolytes *vs* nonelectrolytes. (a) The copper sulfate solution is a strong conductor, and $CuSO_4$ is a strong electrolyte. (b) Neither sugar nor water is an electrolyte, and this sugar solution is a nonconductor.

FIGURE 2.22

An ionic crystal shatters when struck. If ions of like charge come face to face, the repulsions between them can force parts of the crystal apart.

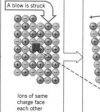

A blow is struck

Crystal is stable because ions of opposite charge face each other.

Ions of same charge face each other

Crystal shatters because of repulsions

In a crystal at room temperature, the ions jiggle about within a cage of other ions. They are held too tightly to move away. When heat is added to raise the temperature of the crystal, the ions have more kinetic energy and neighboring ions bounce off each other more violently. Eventually a temperature is reached at which the violent motions overcome the attractions between the ions and the crystal collapses—the compound melts. However, because the net attractions between the ions are so large, the temperature required is very high. This is why all ionic compounds are solids at room temperature and tend to have high melting points.

You have probably never seen an ionic substance melt. Most of them melt well above room temperature. For example, ordinary table salt, NaCl, melts at 801 °C. Some ionic substances melt only at extremely high temperatures. For instance, aluminum oxide, Al_2O_3, melts at about 2000 °C, and for this reason it is made into special bricks that are used to line the inside walls of furnaces.

Another property of ionic compounds is that their solids are generally relatively hard and quite brittle. Molecular substances, such as paraffin wax or car wax, tend to be soft and easily crushed, but a crystal of rock salt is much harder. When struck by a hammer, however, the salt crystal shatters. The slight movement of a layer of ions within an ionic crystal suddenly places ions of the *same* charge next to each other, and for that instant there are large repulsive forces that split the solid, as illustrated in Figure 2.22.

Electrical Properties

In the solid state, ionic compounds do not conduct electricity. This is because electrical conductivity requires the movement of electrical charges, and in the solid the attractive forces prevent the movement of ions through the crystal. When the solid is melted, however, the ions become free to move about and the liquid conducts electricity well.

This can be demonstrated experimentally with the help of the apparatus depicted in Figure 2.23, which consists of a pair of metal electrodes that can be dipped into a container holding a substance whose electrical conductivity we wish to test. One of the electrodes is wired to an electric light bulb that glows if electricity is able to pass between the two electrodes. When this apparatus is used to test the electrical conductivity of solid salt crystals, the bulb fails to light. However, when a flame is applied, the bulb lights brightly as soon as the salt melts.

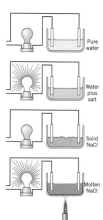

Pure water

Water plus salt

Solid NaCl

Molten NaCl

FIGURE 2.23

An apparatus to test for electrical conductivity. The electrodes are dipped into the substances to be tested. If the lightbulb glows when electricity is applied, the sample is an electrical conductor. Here we see that neither pure water (a molecular substance) nor solid sodium chloride conduct. Salt does conduct, however, if it is melted or dissolved in water.

Computer-Crafted Drawings

Illustrations such as these will help you understand abstract concepts. Study them carefully. You may find questions on exams based on the illustrations.

When we use our general symbol ⇔ in stoichiometry, it will mean "is *stoichiometrically* equivalent to." The relationships in P_4O_{10} can thus be expressed in each of the following ways.

$$1 \text{ mol } P_4O_{10} \Leftrightarrow 4 \text{ mol P}$$
$$1 \text{ mol } P_4O_{10} \Leftrightarrow 10 \text{ mol O}$$
$$4 \text{ mol P} \Leftrightarrow 10 \text{ mol O}$$

We can use these equivalencies to devise conversion factors for solving problems in stoichiometry. The following conversion factors, for example, express the equivalencies inherent in P_4O_{10}.

$$\frac{1 \text{ mol } P_4O_{10}}{4 \text{ mol P}} \quad \text{or} \quad \frac{4 \text{ mol P}}{1 \text{ mol } P_4O_{10}}$$

$$\frac{1 \text{ mol } P_4O_{10}}{10 \text{ mol O}} \quad \text{or} \quad \frac{10 \text{ mol O}}{1 \text{ mol } P_4O_{10}}$$

$$\frac{4 \text{ mol P}}{10 \text{ mol O}} \quad \text{or} \quad \frac{10 \text{ mol O}}{4 \text{ mol P}}$$

The reaction of phosphorus with oxygen that gives P_4O_{10} produces a brilliant light, often used in fireworks displays.

EXAMPLE 3.1
Stoichiometry of Chemical Formulas

Cl_2O_7 is an oily and very explosive liquid.

How many moles of O atoms are combined with 4.20 moles of Cl atoms in Cl_2O_7?

ANALYSIS Note the words *How many moles . . . combined with . . . moles.* These words are the clue that we have a stoichiometric equivalency question. We can restate the problem as:

$$4.20 \text{ mol Cl} \Leftrightarrow ? \text{ mol O}$$

What we need therefore is a moles-to-moles conversion factor. The subscripts in Cl_2O_7 supply it, because the formula, Cl_2O_7, actually means

$$2 \text{ mol Cl} \Leftrightarrow 7 \text{ mol O}$$

So we have the following conversion factors.

$$\frac{2 \text{ mol Cl}}{7 \text{ mol O}} \quad \text{or} \quad \frac{7 \text{ mol O}}{2 \text{ mol Cl}}$$

SOLUTION To find the moles of O atoms stoichiometrically equivalent to 4.20 moles of Cl atoms in Cl_2O_7, we multiply 4.20 mol Cl by the first factor. Notice how the units *mol Cl* (not just *mol*) cancel properly. (Draw in the cancel lines yourself.)

$$4.20 \text{ mol Cl} \times \frac{7 \text{ mol O}}{2 \text{ mol Cl}} = 14.7 \text{ mol O}$$

Thus 14.7 mol of O is combined with 4.20 mol of Cl in Cl_2O_7. (We leave three significant figures in the answer because the least precise number, 4.20, has three significant figures. The 7 and the 2 in the conversion factor, which come from the formula's subscripts, are *exact* numbers of atoms, and so they have an infinite number of significant figures.)

CHECK Does the *size* of the answer make sense? When working problems, make simple mental checks to see that the *size* of the answer makes sense in relationship to the given data. Notice here, for example, that the answer, 14.7 mol O, is numerically larger than 4.20 mol Cl. It *must* be larger because of the 7 to 2 ratio of O to Cl in Cl_2O_7, so the answer's size "makes sense."

■ **Practice Exercise 1** How many moles of nitrogen atoms are combined with 8.60 mol of oxygen atoms in dinitrogen pentoxide, N_2O_5?

Worked Examples
Be sure to study the many worked examples you will find throughout the text. They will aid you in applying concepts and they will teach you how to use the problem-solving tools. The strategy used to solve the problem is divided into several steps.

Analysis
Here we explain the reasoning that's applied to solving the problem. This is the most important step in finding the solution, because it is where we decide which of the problem-solving tools to use. Don't just read through the analysis quickly. Study it and make every effort to understand it thoroughly.

Solution
After figuring out *how* to solve the problem, we show you how the answer is obtained. This is really the easiest part, because the path to the answer has already been decided.

Check
Whenever you work a numerical problem you should ask yourself "Does the answer make sense?" Learn to develop a feel for the expected size of an answer, so you can judge whether the answer is reasonable. What we do here will help you develop this sense.

Practical Exercises
These follow most examples and give you the opportunity to immediately test what you have just learned. Be sure to make working the practice exercises part of your study routine.

FIGURE 2.19

The heart of this tiny electronic device is the microcircuit in the center, which is set into the surface of a tiny silicon chip. Silicon's semiconductor properties make possible microelectronic devices such as this.

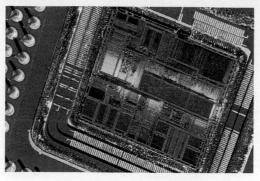

The heart of these devices is a microcircuit embedded in the surface of a tiny silicon chip (Figure 2.19).

Trends in the Periodic Table

The symbol below, shaped like the periodic table, will be placed in the margin when we wish to call your attention to a correlation between the location of an element in the periodic table and some property of the element or its compounds. Here we note the correlation between metallic character and the location of the elements in the periodic table.

The occurrence the metalloids between the metals and the nonmetals is our first example of trends in properties within the periodic table. We will see frequently that as we move from position to position across a period or down a group, chemical and physical properties change in a more or less regular fashion. There are few abrupt changes in the characteristics of the elements as we scan across a period or down a group. The location of the metalloids can be seen, then, as an example of the gradual transition between metallic and nonmetallic properties. F_____ f_____ i_____ _____ _____ s_____ f_____ _____ _____ num, an element that ha_____ _____ ductor; to phosphorus, a_____ similar gradual change is a nonmetal, silicon and metals. Trends such as the properties.

2.6
REACTIONS OF THE ELEMENTS: MOLECULAR AND IONIC COMPOUNDS

A property possessed by other elements to form c_____ ble, however. For examp_____ sodium chloride, but no_____ When we examine the c_____ find that certain generali_____ for a number of reasons. _____ we can study chemical be_____

Special Topics

These short essays serve several purposes. Some, such as this one, relate issues that face society to the subject being studied in the chapter. Others provide interesting explanations of common applications of chemistry. Reading these essays will give you a better understanding of how chemistry relates to the world around you. There is a Table of Contents for the Special Topics on page xxii.

SPECIAL TOPIC 6.2 / ELECTROMAGNETIC FIELDS AND THEIR POSSIBLE PHYSIOLOGICAL EFFECTS

Over the past few years you may have heard reports of possible dangers associated with living near high-voltage power lines or operating electrical equipment. What has caused public concern is the fear that 60 Hz electromagnetic radiation emitted by electricity passing through wires might be affecting the health of those nearby. Such fears have been fueled by news media reports of increased cancer rates among groups receiving strong exposure to this radiation, although the actual evidence supporting a relationship between exposure and cancer is weak and even partly contradictory.

When an oscillating electric current with a certain frequency passes through a wire, it emits radiation with that same frequency. In fact, that's how radio and TV stations broadcast their signals—by pulsing an electric current through a transmitting antenna. Ordinary household AC (alternating current) electricity has a frequency of 60 Hz, and weak electromagnetic signals are emitted by all wires that carry it. This radiation is most intense when the voltage is highest, as in the lines that carry electricity over long distances between power plants and cities.

The energy possessed by photons of 60 Hz electromagnetic radiation is extremely small (you might try the calculation), so small that it cannot affect the bonds that hold molecules together or even cause heating effects the way microwaves do. How, then, can they affect the activities of cells? The answer might be in the weak pulsating electric and magnetic fields induced in the body by this radiation.

Strong electromagnetic fields have been shown to affect the rate of bone growth as well as the amounts of various proteins produced in cells. Experiments have also revealed that cells exposed to an electric field hold onto calcium ions more than cells not exposed. Other experiments have demonstrated that

High-voltage power lines such as these emit low levels of 60 Hz electromagnetic radiation.

weak magnetic fields increase the uptake of calcium ions in cells that have been exposed to a substance that triggers cell division. It is believed that this additional calcium increases the tendency of these cells to divide. Since cancer growth depends on the rate of cell division, these results suggest one way cancers could be promoted by such radiation.

Despite laboratory evidence, the connection between electromagnetic radiation and cancer remains inconclusive. Although it is at least *possible* that cancer can be induced by electromagnetic fields, a lot of additional research will be necessary to pin down the answer.

Periodic Table Icon

Icons such as this one call your attention to properties that vary in a systematic way with an element's location in the periodic table. Knowing such "periodic trends" is an important part of knowing chemistry. A table on the inside rear cover of the book lists important periodic trends discussed in this text.

determined by the "energy levels" of the various steps of the staircase. If the ball is raised to a higher step, its potential energy is increased. When it drops to a lower step, its potential energy decreases. But each time the ball stops, it stops on one of the steps, never in between. Therefore, the energy changes for the ball are restricted to the differences in potential energy between the steps.

So it is with an electron in an atom. The electron can only have energies corresponding to the set of electron energy levels in the atom. When the atom is supplied with energy by an electric discharge, an electron is raised from a low-energy level to a higher one. When the electron drops back, energy equal to the difference between the two levels is released and emitted as a photon. Because only certain energy jumps can occur, only certain frequencies can appear in the spectrum.

The existence of specific energy levels in atoms, as implied by atomic spectra, forms the foundation of all theories about electronic structure. Any model of the atom that attempts to describe the positions or motions of electrons must also account for atomic spectra.

The potential energy of the ball at rest is quantized.

Summary

Each chapter contains a brief summary that provides an overview of the chapter and incorporates the bold-faced terms. Use the Summary as a review; *don't* rely on it as a way to avoid reading the chapter.

■ **SUMMARY**

Solution Vocabulary   A considerable vocabulary has developed for describing **solutions.** Some terms are qualitative—**dilute, concentrated, saturated, unsaturated,** and **supersaturated,** for example. The **solubility** of a substance is often stated in grams of **solute** per 100 g of **solvent.**

Electrolytes   Substances that **dissociate** or **ionize** in water to produce cations and anions are **electrolytes;** those that do not are called **nonelectrolytes.** Electrolytes include salts and metal hydroxides as well as molecular acids and bases that ionize by reaction with water.

Acids and Bases as Electrolytes   The modern version of the **Arrhenius definition of acids and bases** is that an **acid** is a substance that produces hydronium ions, H_3O^+, when dissolved in water, and a **base** one that produces hydroxide ions, OH^-, when dissolved in water.

The oxides of nonmetals are generally **acidic anhydrides** and react with water to give acids. Metal oxides are usually **basic anhydrides** because they tend to react with water to give metal hydroxides or bases.

Strong acids and bases are **strong electrolytes; weak acids and bases** are **weak electrolytes.** In a solution of a weak electrolyte there is a chemical equilibrium between the nonionized molecules of the solute and the ions formed by the reaction of the solute with water.

Acid–Base Neutralization and Ionic Reactions   Acids react with bases in neutralization reactions to produce a salt and water. Equations for these reactions can be written in three different ways. In **molecular equations,** complete formulas for all reactants and products are used. In an **ionic equation,** soluble strong electrolytes are written in dissociated (ionized) form; "molecular" formulas are used for solids and weak electrolytes. A **net ionic equation** is obtained by eliminating **spectator ions** from the ionic equation. An ionic or net ionic equation is balanced only if both atoms *and* net charge are balanced.

duce gases in metathesis reactions, which are found in Table 4.2.

Oxidation–Reduction   **Oxidation** is the loss of electrons or an increase in oxidation number; **reduction** is the gain of electrons or a decrease in oxidation number. Both always occur together in **redox** reactions. The substance oxidized is the **reducing agent;** the substance reduced is the **oxidizing agent. Oxidation numbers** are a bookkeeping device that we use to follow changes in redox reactions. They are assigned according to the rules on page 152. The term **oxidation state** is equivalent to oxidation number.

Ion–Electron Method   In a balanced redox equation, the number of electrons gained by one substance is always equal to the number lost by another substance. The **ion–electron method** divides a *skeleton* net ionic equation into two **half-reactions,** which are balanced separately before being recombined to give the final balanced net ionic equation. For reactions in basic solution, the equation is balanced as if it occurred in an acidic solution, and then the balanced equation is converted to its proper form for basic solution by adding an appropriate number of OH^-.

Solution Stoichiometry   **Molar concentration (molarity)** equals the number of moles of solute per liter of solution. This concentration unit makes available the two conversion factors,

$$\frac{\text{mol solute}}{1\ \text{L soln}} \quad \text{and} \quad \frac{1\ \text{L soln}}{\text{mol solute}}$$

Solutions of known molarity are made in two ways. One is to dissolve a known number of moles of the solute into a solvent until the final volume of the solution reaches a known value. The other is to take a specific volume of a relatively concentrated solution and dilute it with the solvent until the final volume is reached. For dilution problems, remember the relationship

$$V_{\text{dil}} \cdot M_{\text{dil}} = V_{\text{concd}} \cdot M_{\text{concd}}$$

Figure 4.16 gives an overview of how to use stoichiometric equivalencies to move from moles to the units commonly used in measurements—grams or volumes of solutions of known molarities.

Titration is a technique used to make quantitative measurements of the amounts of solutions needed to obtain a complete reaction. In an acid–base titration, the **end point** is normally detected visually using an **acid–base indicator.** For redox reactions, $KMnO_4$ is a useful titrant because it serves as its own indicator in acidic solutions.

Tools You Have Learned

Here we have compiled a list of the problem-solving tools introduced in the chapter. The nature of each tool is described and its function in solving problems is given. You will find it especially helpful to refer to this table when you get stuck on a problem.

■ **TOOLS YOU HAVE LEARNED**

The table below lists the tools you have learned in this chapter that are applicable to problem solving. Review them if necessary, and refer to them when working on the Thinking-It-Through problems and the Review Exercises that follow.

Tool	Function
Chemical formula (page 90)	Subscripts in a formula establish atom ratios and mole ratios between the elements in the substance. These ratios serve as stoichiometric equivalencies that are used to solve problems.
Atomic mass (page 92)	Used to form a conversion factor to calculate mass from moles of an element, or moles from the mass of an element.
Formula mass; molecular mass (page 95)	Used to form a conversion factor to calculate mass from moles of a compound, or moles from the mass of a compound.
Avogadro's number (page 90)	Relates macroscopic lab-sized quantities (e.g., mass) to numbers of individual atomic-sized particles such as atoms, molecules, or ions.
Percentage composition (page 97)	To represent the composition of a compound and be the basis for computing the empirical formula. Comparing experimental and theoretical percentage compositions can help establish the identity of a compound.
Balanced chemical equation (page 107)	The coefficients are used to establish stoichiometric equivalencies that relate moles of one substance to moles of another in a chemical reaction. They also establish the ratios by formula units among the reactants and products.
Theoretical, actual, and percentage yields (page 114)	To estimate the efficiency of a reaction. Remember that the theoretical yield is calculated from the limiting reactant using the balanced chemical equation.

THINKING IT THROUGH

Remember, you are not asked to obtain answers for the following problems. Instead, assemble the data necessary to solve the problems and describe how you would use the data to obtain the answers. For numerical problems, set up the calculations using appropriate conversion factors.

The problems are divided into two groups. Those in Level 2 are significantly more challenging than those in Level 1 and provide an opportunity to really hone your problem solving skills.

Level 1 Problems

1. Carbon and hydrogen atoms form a compound called *ethane.* If you were asked to state the *ratio by atoms* of carbon and hydrogen in ethane, what additional information would you need?

2. Sulfur and oxygen atoms are present in sulfur trioxide, SO_3. How many atoms of S and how many of O are needed to make 10^{20} molecules of SO_3?

3. In a certain compound of tin and chlorine, the ratio of the atoms is described as 1 mol of tin atoms to 4 mol of

Thinking It Through

These questions ask you to *think* about how to solve problems. In fact, here the *answer* isn't the answer. Instead, your goal is to describe *how* you would obtain the answer. These problems are divided into two groups. If you already have well-developed problem-solving skills, you can skip to the Level 2 Problems, which are more difficult than those in Level 1.

H^+ concentration in the mixture is 0.400 *M*. Dichromate ion oxidizes Fe^{2+} to Fe^{3+} and is reduced to Cr^{3+}. After the reaction in the mixture has ceased, how many milliliters of 0.0100 *M* NaOH will be required to neutralize the remaining H^+? (Set up the solution to the problem.)

12. An organic compound contains carbon, hydrogen, and sulfur. A sample of it with a mass of 1.045 g was burned in oxygen to give gaseous CO_2, H_2O, and SO_2. These gases were passed through 500 mL of an acidified 0.0200 *M* $KMnO_4$ solution, which caused the SO_2 to be oxidized to SO_4^{2-} and part of the available $KMnO_4$ to be reduced to Mn^{2+}. Next, 50.00 mL of 0.0300 *M* $SnCl_2$ was added to a 50.00 mL portion of the partially reduced $KMnO_4$ solution. All the MnO_4^- in the 50 mL portion was reduced and the excess Sn^{2+} that remained after reaction was titrated with 0.0100 *M* $KMnO_4$, requiring 27.28 mL of the titrant to reach an end point. Explain in detail how you would determine the percentage of sulfur in the original sample of the organic compound. Set up the solution to the problem.

13. A solution contained a mixture of SO_3^{2-} and $S_2O_3^{2-}$. A 100.0 mL portion of the solution was found to react with 80.00 mL of 0.0500 *M* CrO_4^{2-} in a basic solution to give CrO_2^-. The only sulfur-containing product was SO_4^{2-}. After the reaction, the solution was treated with excess $BaCl_2$ solution, which precipitated $BaSO_4$. This solid was filtered from the solution, dried, and found to weigh 0.9336 g. Explain in detail how you can determine the molar concentrations of SO_3^{2-} and $S_2O_3^{2-}$ in the original solution.

REVIEW EXERCISES   *Answers to questions whose numbers are printed in color are given in Appendix D. Challenging questions are marked with asterisks.*

Electrolytes

4.1 Define the following: (a) solvent, (b) solute, (c) concentration.

4.2 Define the following: (a) concentrated, (b) dilute, (c) saturated, (d) unsaturated, (e) supersaturated.

4.3 Why are chemical reactions often carried out in solutions?

4.4 What is an electrolyte? What is a nonelectrolyte?

4.5 Define dissociation as it applies to ionic compounds that dissolve in water.

4.6 Write equations for the dissociation of the following ionic compounds in water: (a) LiCl, (b) $BaCl_2$, (c) $Al(C_2H_3O_2)_3$, (d) $(NH_4)_2CO_3$, (e) $FeCl_3$

4.7 Write equations for the dissociation of the following ionic compounds in water: (a) $CuSO_4$, (b) $Al_2(SO_4)_3$, (c) $CrCl_3$, (d) $(NH_4)_2HPO_4$, (e) $KMnO_4$

Acids and Bases as Electrolytes

4.8 Give two general properties of an acid. Give two general properties of a base.

4.9 How did Arrhenius define an acid and a base?

4.10 What is the difference between *dissociation* and *ionization* as applied to the discussion of electrolytes?

4.11 Pure $HClO_4$ is molecular. Write an equation for its reaction with water to give H_3O^+ and ClO_4^- ions.

4.12 Hydrazine is a toxic substance that can form when household ammonia is mixed with a bleach such as Clorox. Its formula is N_2H_4, and it is a weak base. Write a chemical equation showing its reaction with water.

4.13 Which of the following would yield an acidic solution when dissolved in water? (a) P_4O_{10}, (b) K_2O, (c) SeO_3, (d) Cl_2O_7

4.14 What is the difference between a strong electrolyte and a weak electrolyte?

4.15 If a substance is a weak electrolyte, what does this mean in terms of the tendency of the ions to react to reform the molecular compound? How does this compare with strong electrolytes?

4.16 Nitrous acid, HNO_2, is a weak acid. Write an equation showing its reaction with water.

4.17 Why don't we use double arrows in the equation for the reaction of a strong acid with water?

4.18 Write the formulas for the strong acids discussed in Section 4.3. (If you must

4.19 $HClO_3$ is a strong ... tion with water.

4.20 When diprotic acid... release their H^+, they do... dynamic equilibrium. W... trate this for H_2CO_3.

4.21 Phosphoric acid, ... goes ionization in three s... each of these reactions.

Acid–Base Neutralizatio...

4.22 What are the diff... and net ionic equations?

4.23 What two conditi... ionic equation?

4.24 The following eq... wrong with it?

$$NO_2 + H_2...$$

4.25 Complete and ba... each, write the molecula... (All the products are sol...

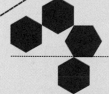

Review Exercises
Each chapter has a complete set of exercises to aid you in learning the chapter contents. Those with numbers printed in blue have answers in Appendix D at the back of the book. Most of the exercises are categorized according to subject to aid you in selecting study questions. As you gain confidence in problem solving, you should attempt to solve the Additional Exercises as well as the more difficult problems that are identified by asterisks.

TEST OF FACTS AND CONCEPTS: CHAPTERS 1–4

Many of the fundamental concepts and problem-solving skills developed in the preceding chapters will carry forward into the rest of this book. Therefore, we recommend that you pause here to see how well you have grasped the concepts, how familiar you are with important terms, and how able you are at working chemical problems.

Some of the problems here require data or other information found in tables in this book, including those inside the covers. Freely use these tables as needed. For problems that require mathematical solutions, we recommend that you first assemble the necessary information in the form of equivalencies and then use them to set up appropriate con-

(a) sulfurous acid   (d) phosphoric acid
(b) nitric acid   (e) carbonic acid
(c) hypochlorous acid

6. The formula mass of a substance is 60.2. Therefore, what is the mass in grams of one of its molecules?

7. A sample of a compound with a mass of 204 g consists of 1.00×10^{23} molecules. What is its formula mass?

8. Calculate the formula mass of $Fe_4[Fe(CN)_6]_3$.

9. How many grams of copper(II) nitrate trihydrate, $Cu(NO_3)_2 \cdot 3H_2O$, are present in 0.118 mol of this compound?

*... iodide hexahydrate,
... of this compound?*

*... was analyzed and
...3 g H, and 0.1013 g
... of nicotine.*

*... following percent-
...0%; P, 29.79%. The
... cal formula of this
... bols in the order*

*...hol, C_2H_5OH, are in
...sity of ethyl alcohol*

*...by a sample of ethyl-
...00 \times 10^{24} molecules?
...g/mL.*

*...be used to prepare
...any moles and how*

...O

*...INO_3, are needed to
...ving reaction?*

...+ 2NO + 4H_2O

*...amonia can be con-
...wing reaction?*

...+ 6H_2O

*...s of O_2 are needed to
...eaction?*

*...nL. of 18.0 M H_2SO_4
...atory bench top, solid
...tainer of sodium bi-
...g before this use and*

CHEMICALS IN USE 4
The Greenhouse Effect

The Earth's average temperature would be over 30 degrees colder except for certain gases in the atmosphere, particularly carbon dioxide, methane, dinitrogen monoxide (N_2O), water vapor, and some synthetics. These "greenhouse gases" have an Earth-insulating or **greenhouse effect** because their molecules are able to absorb part of the energy that the Earth continually radiates to outer space and then reradiate some of the energy back to Earth, thus keeping the Earth warmer than otherwise. (A real greenhouse works differently; it prevents air movements from taking heat away.)

EARTH'S ENERGY BALANCE
A nearly perfect balance exists between the energy received at the Earth's surface, mostly from the sun but some from the Earth's interior, and the energy that the Earth loses, mostly by radiation. Without such a balance, the Earth's average temperature would either increase or decrease drastically.

As shown in Figure 4a, for every 100 units of energy incoming from the sun, roughly 30 units are reflected either by clouds or by the Earth's surface. The rest is absorbed, 25 units by clouds and 45 units by the ground. A net of 70 energy units leaves the planet as infrared radiation ("heat rays"). Many processes in the lower atmosphere contribute to the 70 units, including the heating caused by updrafts of air and the formation of clouds as well as by radiation from the ground. Of the 70 units that leave the earth for outer space, a net of 16 units is the difference between the energy radiated by the ground and that trapped by the greenhouse effect. For every 104 units radiated

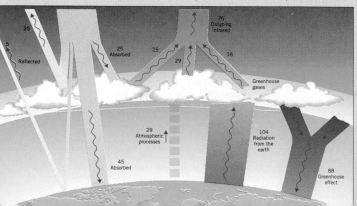

Brief Table of Contents

Special Topics Table of Contents

Contents

page 19

page 61

page 93

page 188

Chapter 6

Atomic and Electronic Structure   221

Chapter 7

Chemical Bonding I   269

page 366

Chapter 11

Intermolecular Attractions and the Properties of Liquids and Solids   441

Chapter 12

Solutions   493

Chapter 13

Thermodynamics   541

page 549

Chapter 14

Kinetics: The Study of Rates of Reaction   585

Chapter 15

Chemical Equilibrium—General Concepts   635

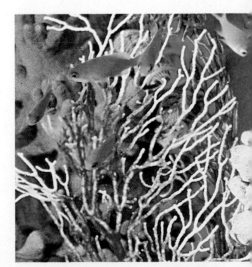

page 653

Chapter 16

Acid–Base Equilibria   679

Chapter 17

Solubility and Simultaneous Equilibria   733

page 732

Chapter 18

Electrochemistry   763

page 811

Chapter 19

Metal Complexes and Their Equilibria   811

Chapter 20

Nuclear Reactions and Their Role in Chemistry   847

Chemistry impacts our lives in many ways. In the late 1980s, the discovery by chemists of materials able to conduct electricity without resistance at relatively high temperatures led to the vision of trains, suspended above their tracks by magnetic levitation, able to travel at high speeds with no frictional drag other than that offered by the air. Scientists have been working to develop these superconducting materials to the point where they can be put to such practical uses. The photo here is of Japan's experimental "Mag-Lev" train, expected to travel at speeds over 250 mph!

Chapter 1

Building a Foundation

As the authors of your textbook, we would like to welcome you to your course in general chemistry. We hope the knowledge you gain here will help you better understand how nature operates, and that it will be valuable to you in your career. While expressing these sentiments, we realize that you may not plan to major in chemistry. If you are not a chemistry major, you are probably taking chemistry because it's a required course, and perhaps you've wondered why.

Chemistry, physics, biology, and other related sciences have in common the central fact that they all study the behavior of material things—things composed of chemical substances. Only their scientific points of view differ. For example, a biologist may study how a cricket metabolizes nutrients and how it derives energy from the process. A physicist, on the other hand, may be interested in the mechanics of motion of the limbs of the cricket, which allow it to jump. Both scientists study the same creature, but their interests in the cricket differ.

Today we understand that living organisms such as crickets are fascinating combinations of chemicals that possess a unique quality we call life. Chemistry is the science that studies chemicals, and the term *chemical* can be applied to *all* the substances that exist in the world around us, whether alive in crickets or lifeless, whether manufactured or of natural origin.

If chemicals are to be considered the realm of chemistry, then it is no wonder that those who wish to study biology or physics find it helpful, and perhaps even necessary, to spend some time looking at the world through the eyes of chemists. By seeking to understand the value of chemistry as it pertains to your chosen major, you will be better able to appreciate both your chemistry course and the career that awaits you.

1.1
CHEMISTRY: WHERE IT FITS AMONG THE SCIENCES

If you've had a previous course in chemistry, much of what is discussed in this chapter may be familiar. Nevertheless, be sure to review it thoroughly because the concepts developed here will be used in later chapters.

1

1.2
CHEMISTRY AS A SCIENCE

Let's begin our study of chemistry by defining exactly what the subject is about. **Chemistry** is a science that seeks to discover what substances are made of and to learn how the properties of substances are related to their compositions. One obvious goal is to be able to make new matcrials with new and useful properties that satisfy particular needs. Beyond this, chemistry seeks to provide us with a better understanding of the underlying workings of nature.

Chemical Reactions

Chemistry is especially concerned with learning and understanding the *changes* that take place between chemicals, changes that we call chemical reactions. In a **chemical reaction,** chemicals interact with each other to form entirely *different* substances with different properties. Sometimes these changes can be quite dramatic, as illustrated by the reation of sodium with chlorine.

Sodium, shown in Figure 1.1*a*, is a metal. Like other metals, it is shiny and conducts electricity well. Unlike other metals, though, it is very soft and is easily cut with a knife. Notice that the outside of the bar of sodium is coated with a white film. This was formed by the reaction of sodium with oxygen and moisture in the air. The tendency of sodium to react rapidly with oxygen and water makes it a dangerous chemical with which to work. Sodium reacts violently in water, producing a lot of heat and liberating the flammable gas hydrogen. In this same reaction sodium also forms a substance called sodium hydroxide, commonly known as lye, which is very corrosive toward flesh. Contact of sodium with your skin can cause severe chemical burns.

Chlorine, shown in the flask in Figure 1.1*b*, is different from sodium in many ways. It is a pale, yellow-green gas. If you've ever smelled a liquid laundry bleach such as Clorox, you have smelled chlorine, which escapes from the bleach in small amounts. In concentrated form chlorine is especially dangerous to inhale, causing severe lung damage that can easily lead to death. In fact, chlorine gas has been used as a weapon of war.

When metallic sodium and gaseous chlorine come together they react violently, as shown in Figure 1.1*c*. The substance formed in this reaction is a white powder, quite different in appearance from either sodium or chlorine. Its chemical name is sodium chloride, although it's known to most people by its common name, *salt.*

FIGURE 1.1

(*a*) A freshly cut piece of sodium reveals a shiny metallic surface. The metal reacts with oxygen and moisture, so it cannot be touched with bare fingers.
(*b*) Chlorine is a pale green gas.
(*c*) When a small piece of sodium is melted in a metal spoon and thrust into the flask of chlorine, it burns brightly as the two elements react to form sodium chloride. The smoke coming from the flask is composed of fine crystals of salt.

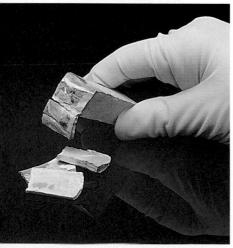

(*a*)

(*b*)

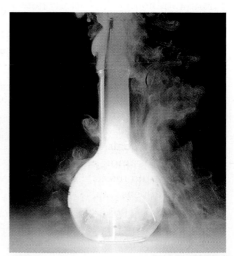

(*c*)

If you think about this reaction for a moment, perhaps it will strike you as really quite amazing, and even a bit like magic. Here we have two chemicals, sodium and chlorine, whose ingestion can produce severe medical problems or even death. Yet, when they react with each other they form a substance our bodies cannot do without! Such startling events help make chemistry fascinating. As you study this course, perhaps you will discover for yourself that same sense of magic that has caught the imagination of others and produced generations of chemists.

Most chemical reactions are not quite as spectacular as the one between sodium and chlorine. Nevertheless, they take place in us and around us all the time. We metabolize the foods we consume, while photosynthesis creates their replacements. Chemical reactions in batteries supply us with energy as does the combustion of fuels. Chemical reactions in the air lead to smog, while in the upper atmosphere sunlight interacting with gases released from aerosol cans and damaged air conditioners degrades the natural ozone layer that shields the Earth from harmful ultraviolet radiation. Understanding these reactions and seeking to control the ones that can be harmful to us is an important role for chemistry in modern science and in our society.

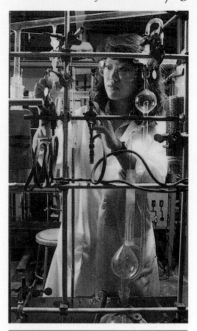

A scientist working in a chemical research laboratory.

Chemistry is said to be a **natural science,** which means that it is concerned with the workings of the natural world.

The Scientific Method and How It Relates to Chemistry

As a science, chemistry is a dynamic subject, constantly changing as new discoveries are made by scientists who work in university and industrial laboratories. The general approach that these scientists bring to their work is called the **scientific method.** It is, quite simply, a common sense approach to developing an understanding of natural phenomena.

A scientific study normally begins with some observation that fires our curiosity and raises some questions about the behavior of nature. Usually, we begin to search for answers in the work of others who have published in scientific journals. As our knowledge grows, we begin to plan experiments that permit us to make our own observations. Generally, these experiments are performed in a laboratory under controlled conditions so the observations are *reproducible.* In fact, the ability to obtain the same results when experiments are repeated is what separates a true science from a pseudoscience such as astrology.

Empirical Facts and Scientific Laws

The observations we make in the course of performing our experiments provide us with **empirical facts**—so named because we learn them by *observing* some physical, chemical, or biological system. These facts are referred to as our **data.** For example, if we study the behavior of gases, such as the air we breathe, we soon discover that the volume of a gas depends on a number of factors, including the mass of the gas, its temperature, and its pressure. The bits of information we record relating these factors are our data.

One of the goals of science is to organize facts so that relationships or generalizations among the data can be established. For instance, one generalization we would make from our observations is that when the temperature of a gas rises, the gas tends to expand and occupy a larger volume. If we were to repeat our experiments many times with numerous different gases, we would find that this generalization is uniformly applicable to them all. Such a broad generalization, based on the results of many experiments, is called a **law** or **scientific law.**

Webster's defines *empirical* as "pertaining to, or founded upon, experiment or experience."

Hypotheses and Theories

As useful as they may be in summarizing the results of experiments, laws can only state what happens. They do not explain *why* substances behave the way they do. Human beings are curious creatures, though, and we seek explanations. Therefore, after we've collected and studied our data we begin to speculate about the reasons for the results of our experiments. *The tentative explanation we formulate is called a* **hypothesis.** We use a hypothesis to make predictions of new behavior and then design experiments to test our predictions. If the results of these new experiments prove us wrong, we must discard the hypothesis and seek a new one. However, if our explanation survives repeated testing, it is gradually accepted and may ultimately achieve the status of a theory. A **theory** *is a tested explanation of the behavior of nature*. Most useful theories are broad, with many far-reaching and subtle implications. It is impossible to perform every test that may show such a theory to be wrong, so we can never be *absolutely* sure the theory is correct.

The sequence of steps just described—observation, explanation, and the testing of an explanation by additional experiments—constitute the **scientific method.** Despite its name, this method is not used only by those who call themselves scientists. An auto mechanic follows the same steps when fixing your car. First, tests are performed (*observation*) that enable the mechanic to suggest the probable cause of the problem (*a hypothesis*). Then parts are replaced and the car is checked to see whether the problem has been solved (*testing of the hypothesis by experiment*). In short, we all use the scientific method as much by instinct as by design.

From the preceding discussion you may get the impression that scientific progress always proceeds in a dull, orderly, and stepwise fashion. This isn't true; science is exciting and provides a rewarding outlet for cleverness and creativity. Luck, too, sometimes plays an important role. For example, in 1828 Frederick Wöhler, a German chemist, was heating a substance called ammonium cyanate in an attempt to add support to one of his hypotheses. His experiment, however, produced an unexpected substance, which out of curiosity he analyzed and found to be urea (a constituent of urine). This was an exciting discovery, because it was the first time anyone had ever knowingly made a substance produced only by living creatures from a chemical not having a life origin. The fact that this could be done led to the beginning of a whole branch of chemistry called *organic chemistry*. Yet, had it not been for Wöhler's curiosity and his application of the scientific method to his unexpected results, the significance of his experiment might have gone unnoticed.

As a final note, it is significant that the most spectacular and dramatic changes in science occur when major theories are proven wrong. This happens only rarely, but when it does occur, scientists are sent scrambling to develop new theories, and exciting new frontiers are opened.

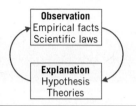

| Observation |
| Empirical facts |
| Scientific laws |

| Explanation |
| Hypothesis |
| Theories |

The scientific method is cyclical. Observations suggest explanations, which suggest new experiments, which suggest new explanations, and so on.

Many breakthrough discoveries in science have come about by accident.

1.3
CHEMISTRY IN THE LABORATORY

The Importance of Measurements

In our description of the reaction of sodium with chlorine, we mentioned that sodium is a metal with the ability to conduct electricity and that chlorine is a pale green gas. Observations of this kind are said to be **qualitative** because they do not involve numbers.

However interesting, qualitative observations are of limited usefulness in any science because they simply do not provide enough information. To make significant progress, every science must resort to **quantitative observations,** and these involve making numerical measurements.

A necessary aspect of all measurements is a uniform and generally accepted system of units. In the United States most people are accustomed to using the *English system* of units in which we measure distances in inches, feet, and miles; volume in ounces, quarts, and gallons; and mass in ounces and pounds. However, in the sciences and increasingly for consumer products (Figure 1.2) a metric-based system is used. Its chief advantage is that larger and smaller units are related to each other by multiples of 10.

The International System of Units

In 1960, a modification of the original metric system was adopted by the General Conference on Weights and Measures (an international body). It is called the **International System of Units,** abbreviated **SI** from the French name, *Le Système International d'Unités.*

For the most part, scientists have willingly adopted the SI units, although there is still some usage of the older metric units. The primary emphasis throughout this book will be on the SI, but there will be times where we will use some older units because of their convenience in dealing with certain laboratory measurements.

SI Base Units

The SI has as its foundation a set **base units** for seven basic measured quantities. These are given in Table 1.1. The size of each base unit is very precisely defined. For example, the international standard for the SI unit of mass, the kilogram, is a carefully preserved platinum-iridium alloy block stored at the International Bureau of Weights and Measures in France (Figure 1.3). This metal block is *defined* as having a mass of *exactly* one kilogram (1 kg). Most countries keep a carefully calibrated replica of the international standard kilogram in some central location. For example, the National Institute of Standards and Technology (formerly, the National Bureau of Standards) located near Washington, DC, has its own standard kilogram, which as near as possible is a duplicate of the one in France. This U.S. kilogram serves indirectly as the calibrating standard for all "weights" used for scales and balances in the United States. Thus, the masses you measure on the balances in your general chemistry laboratory can be traced to the U.S. standard, and ultimately to the international standard.

Objects created by human hands, of course, can be destroyed or lost and so are not the most desirable choices for standards. In the SI, only the kilogram is

FIGURE 1.2

Metric units are becoming commonplace on many consumer products.

FIGURE 1.3

The international standard kilogram, made of a platinum-iridium alloy, which is kept at the International Bureau of Weights and Measures in France. Other nations, such as the United States, maintain their own standard masses that have been carefully calibrated against this international standard.

TABLE 1.1   The SI Base Units

Measurement	Unit	Symbol
Length	meter	m
Mass	kilogram	kg
Time	second	s
Electric current	ampere	A
Temperature	kelvin	K
Amount of substance	mole	mol
Luminous intensity	candela	cd

Not all of these units will have meaning for you now, but we will encounter most of them later in this book and their meanings will be made clear when they are needed.

defined by using an object. The other base units are established in terms of reproducible physical phenomena. For instance, the meter is defined as exactly the distance light travels in a vacuum in $1/299{,}792{,}458$ of a second. Everyone has access to this standard because light and a vacuum are available to all.

Derived Units

The base units, which form the foundation of the SI, are used to define additional **derived units.** For example, there is no SI base unit for area, but we know that to calculate the area of a rectangular room we multiply its length by its width. Therefore, the *unit* for area is derived by multiplying the *unit* for length by the *unit* for width.

$$\text{length} \times \text{width} = \text{area}$$

$$(\text{meter}) \times (\text{meter}) = (\text{meter})^2$$

$$\text{m} \times \text{m} = \text{m}^2$$

The units we attach to numbers are sometimes referred to as dimensions.

In the English system we measure area in square feet (ft^2) or square yards (yd^2). Carpeting, for example, is priced by the square yard.

The SI unit for area is therefore m^2 (read as *meter squared,* or *square meter*). In obtaining this unit we employ a very important concept that we will use repeatedly throughout this book when we perform calculations: *Units undergo the same kinds of mathematical operations that numbers do.* This is a concept we discuss at greater length in Section 1.4.

Decimal Multipliers

When making measurements, we sometimes find that the basic units are of an awkward size. For instance, the meter (slightly larger than a yard) is not convenient for expressing the sizes of very small objects such as bacteria. Neither is it convenient for expressing the very large distances between planets or stars. In the SI, we form units that more closely suit our needs by modifying the basic units with **decimal multipliers.** These modifiers and the prefixes we use to identify them are given in Table 1.2. Those we shall use most frequently are given in bold type. Be sure to learn them.

Exponential notation and powers of 10 are reviewed in Appendix A.

When the name of a unit is preceded by one of these prefixes, the size of the unit is modified by the corresponding decimal multiplier. For instance, the prefix *kilo-* indicates a multiplying factor of 10^3, or 1000. Therefore, a kilometer is a unit of length equal to 1000 meters. The symbol for kilometer (km) is formed by applying the symbol meaning kilo (k) as a prefix to the symbol for

SI prefixes

TABLE 1.2   Decimal Multipliers That Serve as SI Prefixes

Prefix	Symbol	Multiplication Factor	Prefix	Symbol	Multiplication Factor
exa	E	10^{18}	**deci**	d	10^{-1}
peta	P	10^{15}	**centi**	c	10^{-2}
tera	T	10^{12}	**milli**	m	10^{-3}
giga	g	10^{9}	**micro**	μ	10^{-6}
mega	M	10^{6}	**nano**	n	10^{-9}
kilo	k	10^{3}	**pico**	p	10^{-12}
hecto	h	10^{2}	femto	f	10^{-15}
deka	da	10^{1}	atto	a	10^{-18}

meter (m). Thus 1 km = 1000 m (or alternatively, 1 km = 10³ m). Similarly a decimeter (dm) is 1/10 of a meter, so 1 dm = 0.1 m. The following practice exercise will give you some experience applying these decimal multipliers. Refer to Table 1.2 if necessary.

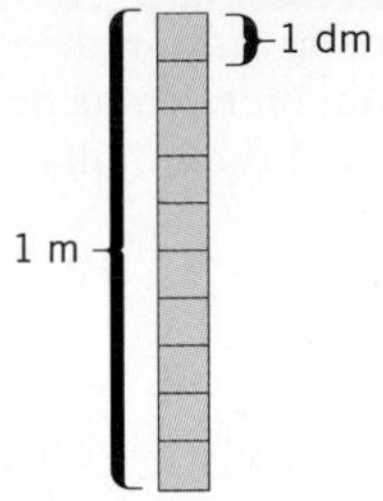

■ **Practice Exercise 1** Give the abbreviation for (a) milligram, (b) micrometer, (c) picosecond. How many meters are in (a) 1 nm, (b) 1 cm, (c) 1 pm? What symbol is used to mean (a) 10^{-2} g, (b) 10^{6} m, (c) 10^{-9} s?

Units for laboratory measurements

Units for Laboratory Measurements

The most common measurements you will make in the laboratory will be those of length, volume, mass, and temperature. Although you will nearly always use SI units, it is wise to be aware of how they relate in size to the perhaps more familiar English units. Some common conversions between the English system and the SI are given in Table 1.3.

Length   The SI base unit for length, the **meter (m),** is too large for most laboratory purposes. More convenient units are the **centimeter (cm)** and the **millimeter (mm).** They are related to the meter as follows.

$$1 \text{ cm} = 10^{-2} \text{ m} = 0.01 \text{ m}$$

$$1 \text{ mm} = 10^{-3} \text{ m} = 0.001 \text{ m}$$

It is also useful to know the relationships

$$1 \text{ m} = 100 \text{ cm} = 1000 \text{ mm}$$

$$1 \text{ cm} = 10 \text{ mm}$$

Many 12-in. rulers have one side marked in centimeters and millimeters (Figure 1.4). It is helpful to remember that 1 in. equals approximately 2.5 cm or about 25 mm (1 in. = 2.54 cm, *exactly*).

FIGURE 1.4

Centimeters and millimeters are conveniently sized units for most laboratory measurements of length. Here is a common ruler that is marked in both English and SI units.

TABLE 1.3   Some Useful Conversions

Measurement	English to Metric	Metric to English
Length	1 in. = 2.54 cm	1 m = 39.37 in.
	1 yd = 0.9144 m	1 km = 0.6215 mile
	1 mile = 1.609 km	
Mass	1 lb = 453.6 g	1 kg = 2.205 lb
	1 oz = 28.35 g	
Volume	1 gal = 3.785 L	1 L = 1.057 qt
	1 qt = 946.4 mL	
	1 oz (fluid) = 29.6 mL	

[1] Originally, these conversions were established by measurement. For example, if a metric ruler is used to measure the length of an inch, it is found that 1 in. equals 2.54 cm. Later, to avoid confusion about the accuracy of such measurements, it was agreed that these relationships would be taken to be exact. For instance, 1 in. is now defined as *exactly* 2.54 cm. Exact relationships also exist for the other quantities in Table 1.3, but for simplicity many have been rounded off. For example, 1 lb = 453.59237 g, *exactly*.

Comparison among the cubic meter, the liter (cubic decimeter), and the milliliter (cubic centimeter).

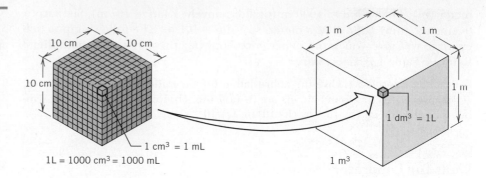

$1 \text{ cm}^3 = 1 \text{ mL}$

$1 \text{ L} = 1000 \text{ cm}^3 = 1000 \text{ mL}$

$1 \text{ dm}^3 = 1 \text{ L}$

1 m^3

One cubic meter is somewhat larger than one cubic yard, which is much too large to conveniently express most volumes measured in the laboratory.

$1 \text{ dm}^3 = (10 \text{ cm})^3 = 1000 \text{ cm}^3$

FIGURE 1.5

Common laboratory glassware used for measuring volumes includes graduated cylinders, volumetric flasks, pipets, and a buret.

Paper clip   0.4 g

Penny   3.1 g

The masses of some common things.

Volume   Volume has the dimensions of (distance)^3. For example, the volume of a room is calculated by multiplying its length times its width times its height. With these dimensions expressed in meters, the derived SI unit for volume is the **cubic meter, m^3.**

In chemistry, measurements of volume usually arise when we measure amounts of liquids. The traditional metric unit of volume used for this is the **liter (L).** In SI terms, a liter is defined as exactly 1 cubic decimeter.

$$1 \text{ L} = 1 \text{ dm}^3$$

However, even the liter is too large to conveniently express most volumes measured in the laboratory. The glassware we normally use, such as that illustrated in Figure 1.5, is marked in **milliliters (mL).**[2]

$$1 \text{ L} = 1000 \text{ mL}$$

Because 1 dm = 10 cm, then $1 \text{ dm}^3 = 1000 \text{ cm}^3$. Therefore, 1 mL is exactly the same as 1 cm^3.

$$1 \text{ cm}^3 = 1 \text{ mL}$$

$$1 \text{ L} = 1000 \text{ cm}^3 = 1000 \text{ mL}$$

Sometimes you may see cm^3 abbreviated cc (especially in medical applications), although this symbol is frowned on in the SI.

Mass   In the SI, the base unit for the measurement of mass is the kilogram (kg), although the gram (g) is a more conveniently sized unit for most laboratory measurements. One gram, of course, is 1/1000 of a kilogram (0.001 kg).

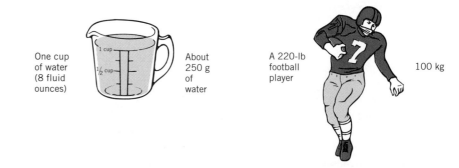

One cup of water (8 fluid ounces)

About 250 g of water

A 220-lb football player   100 kg

[2] Use of the abbreviations L for liter and mL for milliliter is rather recent. Confusion between the printed letter l and the number 1 prompted the change from l for liter and ml for milliliter. You may encounter the abbreviation ml in other books or on older laboratory glassware.

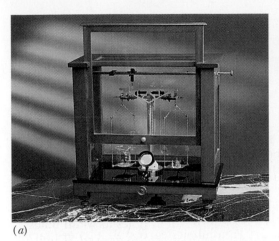

(*a*)   (*b*)   (*c*)

FIGURE 1.6

Typical laboratory balances.
(*a*) A traditional two-pan analytical balance capable of measurements to the nearest 0.0001 g.
(*b*) A modern top-loading balance capable of mass measurements to the nearest 0.001 g (fitted with a cover to reduce the effects of air currents and thereby improve precision).
(*c*) A modern analytical balance capable of measurements to the nearest 0.0001 g.

Mass and weight are not the same. We will discuss this further in Section 1.5.

Mass is measured by comparing the weight of a sample with the weights of known standard masses. The apparatus used is called a **balance** (some examples are shown in Figure 1.6). To use the balance in Figure 1.6*a*, we place our sample on the left pan and then add standard masses to the other. When the weight of the sample and the total weight of the standards are in balance (when they match), their masses are then equal.

Notice that two of the balances in Figure 1.6 have only one pan. This is true of most modern balances. For some, the standard masses supplied by the manufacturer are located within the balance case where they are protected from dust (and the probing fingers of curious people). In many modern balances, such as the one shown in Figure 1.6*b*, there are no standard masses at all. These electronic balances work on an altogether different principle, which is too complex to discuss here.

Temperature   To measure temperature we use a thermometer, which consists of a long glass tube with a very thin bore connected to a reservoir containing a liquid, usually mercury (Figure 1.7). As the temperature rises, the liquid in the reservoir expands and its length in the column increases.

FIGURE 1.7

A typical laboratory thermometer.

Thermometers are graduated in *degrees* according to one of two temperature scales. Both scales use as reference points the temperature at which water freezes[3] and the temperature at which it boils. On the **Fahrenheit scale** water freezes at 32 °F and boils at 212 °F. If you've been raised in the United States, this is probably the scale you are most familiar with. In recent times, however, you have probably noticed an increased use of the Celsius scale, especially in

---

[3] Water freezes and ice melts at the same temperature, and a mixture of ice and water will maintain a constant temperature of 32 °F or 0 °C. If heat is added, some ice melts; if heat is removed, some liquid water freezes, but the temperature doesn't change. This constancy of temperature is what makes the "ice point" convenient for calibrating thermometers.

FIGURE 1.8

Comparison of the Celsius and
Fahrenheit temperature scales.

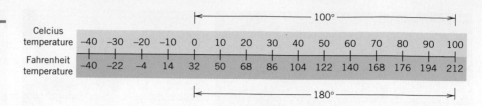

Biologists who study microorganisms often carry out their experiments at 37 °C because that is normal human body temperature.

Temperature conversions

Notice that the name of the temperature scale, the Kelvin scale, is capitalized, but the name of the unit, the kelvin, is not. However, the symbol for the kelvin is the capital letter K.

weather broadcasts. This is the scale we use in the sciences. On the **Celsius scale** water freezes at 0 °C and boils at 100 °C (see Figure 1.8).

As you can see in Figure 1.8, on the Celsius scale there are 100 degree units between the freezing point of water and its boiling point, while on the Fahrenheit scale this same temperature range is spanned by 180 degree units. Therefore, each Celsius degree is nearly twice as large as a Fahrenheit degree (actually, 5 Celsius degrees is the same as 9 Fahrenheit degrees). If it is necessary to convert between these temperature scales, we can use the equation[4]

$$t_F = \left(\frac{9 \ °\text{F}}{5 \ °\text{C}} \right) t_C + 32 \ °\text{F} \tag{1.1}$$

where t_F is the Fahrenheit temperature and t_C is the Celsius temperature.

The SI unit of temperature is the **kelvin (K),** which is the degree unit on the **Kelvin temperature scale.** Notice that the temperature unit is K, not °K (the degree symbol, °, is omitted). Kelvin temperatures must be used in many equations in which the temperature enters directly into the calculations. We will come across this situation many times throughout the book.

Figure 1.9 shows how the Kelvin, Celsius, and Fahrenheit temperature scales relate to each other. Notice that the kelvin is **exactly** the same size as the Celsius degree. The only difference between these two temperature scales is the zero point. The zero point on the Kelvin scale is called **absolute zero** and corresponds to nature's coldest temperature. It is 273.15 degree units below the zero point on the Celsius scale, which means that 0 °C equals 273.15 K, and 0 K equals −273.15 °C. Thermometers are never marked with the Kelvin

FIGURE 1.9

Comparison among Kelvin, Celsius, and Fahrenheit temperature scales.

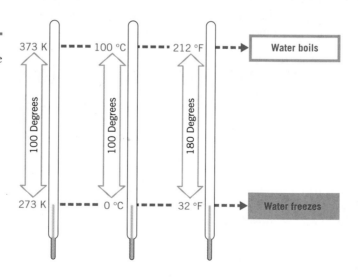

[4] Many equations will be numbered to make it easy to refer to them later on in the book or in class discussions.

scale, so when we need to express a temperature in kelvins, we have to do some arithmetic. The relationship between the Kelvin temperature, T_K, and the Celsius temperature, t_C, is

$$T_K = t_C + 273.15 \qquad (1.2)$$

Temperature conversions

In our calculations, we will nearly always round this to the nearest degree and use the equation

$$T_K = t_C + 273 \qquad (1.3)$$

EXAMPLE 1.1
Converting Among Temperature Scales

Thermal pollution, the release of large amounts of heat into rivers and other bodies of water, is a serious problem near power plants and can affect the survival of some species of fish. For example, trout die if the temperature of the water rises above approximately 25 °C. (a) What is this temperature in °F? (b) What is this temperature in kelvins?

ANALYSIS: This is your first opportunity in this text to study the principles of problem solving. If you have not already done so, we suggest you read the section titled "To the Student," located at the beginning of the book, where we examine in a general way the strategy we will use to tackle problems of all sorts. Here the problem is relatively simple, but we will use the same approach in future examples.

 Our first job in solving a problem is determining the kinds of tools required to do the work. Both parts of the problem here deal with temperature conversions. Therefore, we ask ourselves, "What relationships do we have that relate temperature scales to each other?" Let's write them:

Equation 1.1 $\qquad t_F = \left(\dfrac{9\ ^\circ F}{5\ ^\circ C} \right) t_C + 32\ ^\circ F$

Equation 1.3 $\qquad T_K = t_C + 273$

Equation 1.1 relates Fahrenheit temperatures to Celsius temperatures, so this is the relationship we need to answer part (a). Equation 1.3 relates Kelvin temperatures to Celsius temperatures, which is what we need for part (b). Now that we have what we need, the rest follows.

We will use a capital *T* to stand for the Kelvin temperature and a lowercase *t* (as in t_C) to stand for the Celsius temperature. If we wished to be very precise about the cancellation of units, we could express Equation 1.3 as

$$T_K = \left(\dfrac{1\ K}{1\ ^\circ C} \right) t_C + 273.15\ K$$

SOLUTION TO (a): We substitute the value of the Celsius temperature (25 °C) for t_C.

$$t_F = \left(\frac{9\ ^\circ F}{5\ ^\circ C} \right) (25\ ^\circ C) + 32\ ^\circ F$$

$$= 77\ ^\circ F$$

Therefore, 25 °C = 77 °F. (Notice that we have canceled the unit °C in the equation above. As noted earlier, units behave the same as numbers do in calculations.)

SOLUTION TO (b): Once again, we have a simple substitution. Since $t_C = 25$ °C, the Kelvin temperature is

$$T_K = (25) + 273$$

$$= 298\ K$$

Thus, 25 °C = 298 K.

■ **Practice Exercise 2** What Celsius temperature corresponds to 50 °F? What Kelvin temperature corresponds to 68 °F (expressed to the nearest whole kelvin unit)?

1.4
SIGNIFICANT FIGURES AND SCIENTIFIC CALCULATIONS

In the preceding section, we discussed the importance of measurements in science. When a scientist writes down a number obtained from a measurement, two kinds of information are being given at the same time. One, of course, is the magnitude of the measurement. *The other is the extent of its reliability.* Let's look at an example.

Suppose you asked two different people to measure the width of your room. The first person reports that the room is 11.2 ft wide. The second tells you that he measured the room at the same place and obtained 11.13 ft. Obviously the room can have only one true width at any given place. What, then, do these two numbers tell us?

The first number, 11.2 ft, implies that the width of the room was measured with a tape measure that required the person using it to estimate the tenths place. Figure 1.10*a* illustrates how a tape measure such as this would be marked. Because the tenths place must be *estimated*, different people measuring the width of the room might report distances that differ by ± 0.1 ft. In other words, we might expect another person's estimate of the distance to be 0.1 ft larger or smaller than the reported value of 11.2 ft. Therefore, we view this measurement as being *uncertain* by ± 0.1 ft.

FIGURE 1.10

(*a*) Measurement of length with a tape measure marked only every 1 ft. An estimate must be made of the tenths place. (*b*) A portion of the scale of another tape measure that is marked every 0.1 ft. This scale permits estimation of the hundredths place.

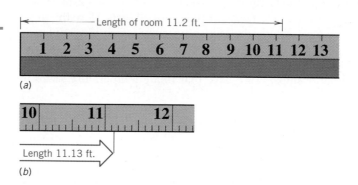

The second measurement, 11.13 ft, suggests that the tape measure used to obtain it was more finely divided, as shown in Figure 1.10*b*. The markings on this scale allow the user to be certain of the tenths place but require that the hundredths place be estimated. Therefore, we can expect the second measurement to be uncertain by about ± 0.01 ft.

We would certainly expect the second measurement to be more reliable than the first because it has more digits and a smaller amount of uncertainty. *The reliability of a piece of data is indicated by the number of digits used to represent it.* This is so important we have a special terminology to describe numbers that come from measurement.

Digits that result from measurement such that only the digit farthest to the right is not known with certainty are called **significant figures.**

The number of significant figures in a measured value is equal to the number of digits known for sure *plus* one that is uncertain. The first measurement of 11.2 ft has 3 significant figures; the second, more reliable value of 11.13 ft has 4 significant figures.

Counting Significant Figures

Usually, it is simple to determine the number of significant figures in a number; we just count the digits. Thus, 3.25 has three significant figures and 56.205 has five of them. When zeros come at the beginning or the end of a number, however, they sometimes cause confusion.

*When zeros appear at the end of a number **and** to the right of the decimal point, they are always counted as significant figures.* Thus, 4.500 m has four significant figures because the zeros would not be written unless those digits were known to be zeros.

Any zeros that precede the first nonzero digit are never counted as significant figures. For instance, a length of 2.3 mm is the same as 0.0023 m. Since we are dealing with the same measured value, its number of significant figures cannot change when we change the units. Therefore, both quantities have two significant figures and in 0.0023 m we don't count the zeros that come before the 2. They are needed, however, to locate the position of the decimal point.

With large numbers, zeros are also often needed to locate the decimal point, and this can lead to questions as to the number of significant figures in a reported value. For instance, suppose you were told that at a football game there were 45,000 fans. If this were just a rough estimate, it might be uncertain by as much as several thousand, in which case the value 45,000 represents just two significant figures and none of the zeros count as a significant figure. On the other hand, suppose the official count of tickets collected was reported to be 45,000 "give or take about 10." In this case, the value represents 45,000 ± 10 fans and contains four significant figures, two of which are zeros. We can see, therefore, that a simple statement such as "There were 45,000 fans at the game" is ambiguous. We can't tell whether any of the zeros should count as significant figures unless the statement is accompanied by a description of how uncertain the value is.

To avoid this type of confusion, so that we can indicate the proper number of significant figures as well as the location of the decimal, scientific notation comes in handy. Thus, we can write the rough estimate of 45,000 as 4.5×10^4. The 4.5 shows the number of significant figures and the 10^4 tells us the location of the decimal. The value obtained from the ticket count, on the other hand, can be expressed as 4.500×10^4. This time the 4.500 shows four significant figures and an uncertainty of ± 10 people.

4.5×10^4 two significant figures

4.500×10^4 four significant figures

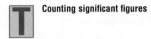
Counting significant figures

For the rough estimate, the number of fans present is 45,000 ± 1000.

Scientific notation is reviewed in Appendix A.

Accuracy and Precision

These are terms often used to describe measurements. **Accuracy** *refers to how close a measurement is to the actual, true value.* The closer it is to the true value, the more accurate it is. Accurate measurements depend on careful calibration of the measuring device to be sure it gives us correct values when used correctly.

Precision *relates to how close measurements are to each other.* Measurements with the rule in Figure 1.10*a* agree to the nearest 0.1 ft but those obtained with the rule in Figure 1.10*b* agree to the nearest 0.01 ft. The latter measurements are of greater precision because they have a smaller uncertainty.

We usually assume that a very precise measurement is also of high accuracy. We can be wrong, however, if our instruments are improperly calibrated. For

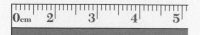

How accurate would measurements be with this ruler?

FIGURE 1.11

This improperly marked ruler will yield measurements that are each wrong by one whole unit. The measurements might be precise, but the accuracy would be very poor.

example, the poorly marked ruler in Figure 1.11 might yield measurements with a precision of ± 0.01 cm, but all the measurements would be too large by 1 cm, a case of good precision but poor accuracy.

Combining Numbers in Calculations

When several numbers are obtained in an experiment they are usually combined in some way to calculate a desired quantity. For example, to determine the area of a rectangular carpet we require two measurements, length and width, which are then multiplied to give the answer we want. If one of these measurements is very precise and the other is not, we can't expect too much precision in the calculated area; the large uncertainty in the less precise measurement carries through to give a large uncertainty in the area. Therefore, to get some idea of how precise the area really is, we need a way to take into account the precision of the various values used in the calculation. To make sure this happens, we follow certain rules according to the kinds of arithmetic being performed.

Rules for significant figures in calculations

Multiplication and Division   *For multiplication and division, the number of significant figures in the answer should not be greater than the number of significant figures in the least precise measurement.* Let's look at a typical problem involving some measured quantities.

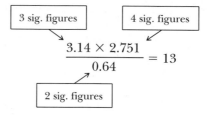

The answer to this calculation displayed on a calculator[5] is 13.497093. However, the rule says that the answer should have only as many significant figures as the least precise factor. Because the least precise factor, 0.64, has only two significant figures, the answer should have only two. The correct answer, 13, is then obtained by rounding off the calculator answer.[6]

Rules for significant figures in calculations

Addition and Subtraction   *For addition and subtraction, the answer should have the same number of decimal places as the quantity with the fewest number of decimal places.* As an example, consider the following addition of measured quantities.

$$
\begin{array}{r}
3.247 \\
41.36 \\
+\ 125.2 \\
\hline
169.8
\end{array}
$$

← This number has only 1 decimal place.
← The answer has been rounded to 1 decimal place.

---

[5] Calculators usually give too many significant figures. An exception is when the answer has zeros at the right that are significant figures. For example, an answer of 1.200 would be displayed on most calculators as 1.2. If the zeros belong in the answer, be sure to write them down.

[6] When we wish to round off a number at a certain point, we simply drop the digits that follow if the first of them is less than 5. Thus, 8.1634 rounds to 8.16, if we wish to have only two decimal places. If the first digit after the point of round off is larger than 5, or if it is 5 followed by other nonzero digits, then we add 1 to the preceding digit. Thus 8.167 and 8.1653 both round to 8.17. Finally, when the digit after the point of round off is a 5 and no other digits follow the 5, then we drop the five if the preceding digit is even and add one if it is odd. Thus, 8.165 rounds to 8.16 and 8.175 rounds to 8.18.

In this calculation, the digits beneath the 6 and the 7 are unknown; they could be anything. (They're not necessarily zeros because if we *knew* they were zeros, then zeros would have been written there.) Adding an unknown digit to the 6 or the 7 will give an answer that's also unknown, so for this sum we are not justified in writing digits in the second and third places after the decimal point. Therefore, we round the answer to the nearest tenth.

Exact Numbers   Not all the numbers we use come from measurement. Those that come from definitions, such as 12 in. = 1 ft, and those that come from a direct count, such as the number of people in a room, have no uncertainty and are called **exact numbers**. When exact numbers are used in a calculation, we assume them to have an infinite number of significant figures and we don't take them into account when we apply the rules described above.

Unit Conversions: The Factor-Label Method

Factor label method

As you saw in Example 1.1, solving a mathematical problem involves two steps. The first is assembling the necessary information and the second is using the information correctly to obtain the answer. Our goal is to teach you both, but our principal focus at this point is on a system called the **factor-label method** (also called **dimensional analysis**), which scientists use to help them perform the correct arithmetic to solve a problem.

To solve a problem, you first have to collect all the information you need. The factor label method will then help you set up the arithmetic correctly.

In the factor-label method we treat a numerical problem as one involving a conversion of units from one kind to another. To do this we use one or more *conversion factors* to change the units of the given quantity to the units of the answer.

$$\begin{pmatrix} \text{given} \\ \text{quantity} \end{pmatrix} \times \begin{pmatrix} \text{conversion} \\ \text{factor} \end{pmatrix} = \begin{pmatrix} \text{desired} \\ \text{quantity} \end{pmatrix}$$

A **conversion factor** *is a fraction formed from a valid relationship or equality between units that is used to switch from one system of measurement and units to another.* To illustrate, suppose we want to express a person's height of 72.0 in. in centimeters. To do this we need the relationship between the inch and the centimeter. We can obtain this from Table 1.3.

$$2.54 \text{ cm} = 1 \text{ in. (exactly)} \tag{1.4}$$

To construct a valid conversion factor, the relationship between the units must be true. For example, the statement

3 ft = 41 in.

is false. Although you might make a conversion factor out of it, any answers you would calculate are sure to be incorrect. *Correct answers require correct relationships between units.*

If we divide both sides of this equation by 1 in., we obtain a conversion factor.

$$\frac{2.54 \text{ cm}}{1 \text{ in.}} = \frac{1 \text{ in.}}{1 \text{ in.}} = 1$$

Notice that we have canceled the units from both the numerator and the denominator of the center fraction. *Units behave just as numbers do in mathematical operations,* which is a key part of the factor-label method. This leaves the first fraction equaling 1. Let's see what happens if we multiply 72.0 in., the height that we mentioned, by this fraction.

$$72.0 \text{ in.} \times \frac{2.54 \text{ cm}}{1 \text{ in.}} = 183 \text{ cm}$$

$$\begin{pmatrix} \text{given} \\ \text{quantity} \end{pmatrix} \times \begin{pmatrix} \text{conversion} \\ \text{factor} \end{pmatrix} = \begin{pmatrix} \text{desired} \\ \text{quantity} \end{pmatrix}$$

The relationship between the inch and the centimeter is exact, so the numbers in 1 in. = 2.54 cm have an infinite number of significant figures.

Because we have multiplied 72.0 in. by something that is equal to 1, we know we haven't changed the magnitude of the person's height. We have, however, changed the units. Notice that we have canceled the unit "inches." The only unit left is "centimeters," which is the unit we want for the answer. The result, therefore, is the person's height in centimeters.

One of the benefits of the factor-label method is that it lets you know when you have done the *wrong* arithmetic. From the relationship in Equation 1.4, we can actually construct two conversion factors:

$$\frac{2.54 \text{ cm}}{1 \text{ in.}} \quad \text{and} \quad \frac{1 \text{ in.}}{2.54 \text{ cm}}$$

We used the first one correctly, but what would've happened if we had used the second by mistake?

$$72.0 \text{ in.} \times \frac{1 \text{ in.}}{2.54 \text{ cm}} = 28.3 \text{ in.}^2/\text{cm}$$

In this case, none of the units cancels. We get units of in.²/cm because in. × in. = in.² Even though our calculator may be very good at arithmetic, we've got the wrong answer. *The factor-label method let's us know we have the wrong answer because the units are wrong!*

We will use the factor-label method extensively throughout this book to aid us in setting up the proper arithmetic in problems. In fact, we will see that it also helps us assemble the information we need to solve the problem. The following examples illustrate the method.

EXAMPLE 1.2
Applying the Factor-Label Method

Convert 3.25 m to millimeters (mm).

ANALYSIS: A good way to begin a problem is to restate the question in the form of an equation. We will do this by writing the given quantity (with its units) on the left and the *units* of the desired answer on the right.

$$3.25 \text{ m} = ? \text{ mm}$$

To solve this by the factor-label method we need a conversion factor that can be used to change the unit "meter" to the unit "millimeter," and this requires a relationship between the two units. Reviewing what has been covered so far we realize that the tool we need to solve this problem comes from the table of decimal multipliers. The prefex *milli* means $\times 10^{-3}$, so we can write

$$1 \text{ mm} = 10^{-3} \text{ m}$$

Notice that this relationship connects the units given to the units desired

Now we have the information we need to solve the problem, so let's proceed with the solution.

SOLUTION: From the relationship above, we can make two conversion factors.

$$\frac{1 \text{ mm}}{10^{-3} \text{ m}} \quad \text{and} \quad \frac{10^{-3} \text{ m}}{1 \text{ mm}}$$

We know we have to cancel the unit "meter," so we need to multiply by a conversion factor with this unit in the denominator. Therefore, we select the one on the left. This gives

$$3.25 \text{ m} \times \frac{1 \text{ mm}}{10^{-3} \text{ m}} = 3.25 \times 10^3 \text{ mm}$$

Notice we have expressed the answer to three significant figures because that is how many there are in the given quantity, 3.25 m.

Most chemists and physicists use the factor-label method. There must be a good reason why.

CHECKING THE ANSWER FOR REASONABLENESS Before we leave this problem, let's take a moment to be sure our answer is *reasonable,* so we can be confident we have performed the arithmetic correctly. Such a check should always be the last step in your problem solving.

We know that millimeters are much smaller than meters, so 3.25 m must represent a lot of millimeters. Our answer, therefore, makes sense.

A liter, which is slightly larger than a quart, is defined as 1 cubic decimeter ($1\ dm^3$). How many liters are there in 1 cubic meter ($1\ m^3$)?

ANALYSIS: Let's begin once again by stating the problem in equation form.

$$1\ m^3 = ?\ L$$

Next, we assemble the tools. What relationships do we know that relate these various units? We are given the relationship between liters and cubic decimeters,

$$1\ L = 1\ dm^3 \qquad (1.5)$$

From the table of decimal multipliers, we also know the relationship between decimeters and meters,

$$1\ dm = 0.1\ m$$

but we need a relationship between cubic units. Since units undergo the same kinds of operations numbers do, we simply cube each side of this equation (being careful to cube *both* the numbers and the units).

$$(1\ dm)^3 = (0.1\ m)^3$$
$$1\ dm^3 = 0.001\ m^3 \qquad (1.6)$$

Notice how Equations 1.5 and 1.6 provide a path from the given units to those we seek. Such a path is always a necessary condition when we apply the factor-label method.

$$m^3 \xrightarrow{\text{Equation 1.6}} dm^3 \xrightarrow[\text{Equation 1.5}]{} L$$

Now we are ready to solve the problem.

SOLUTION: The first step is to eliminate the units m^3. We use Equation 1.6.

$$1\ m^3 \times \frac{1\ dm^3}{0.001\ m^3} = 1000\ dm^3$$

Then we use Equation 1.5 to take us from dm^3 to L.

$$1000\ dm^3 \times \frac{1\ L}{1\ dm^3} = 1000\ L$$

Thus, $1\ m^3 = 1000\ L$.

Usually, when a problem involves the use of two or more conversion factors, they can be "strung together" to avoid having to compute intermediate results. For example, this problem can be set up as follows.

$$1\ m^3 \times \frac{1\ dm^3}{0.001\ m^3} \times \frac{1\ L}{1\ dm^3} = 1000\ L$$

CHECKING THE ANSWER FOR REASONABLENESS One liter is about a quart. A cubic meter is about a cubic yard. Therefore, we expect a large number of liters in a cubic meter, so our answer seems reasonable. (Notice here that in our analysis we

EXAMPLE 1.3
Applying the Factor-Label Method

have approximated the quantities in the calculation in units of quarts and cubic yards, which are more familiar than liters and m³ if you've been raised in the U.S. We get a feel for the approximate magnitude of the answer using our familiar units and then relate this to the actual units of the problem.)

EXAMPLE 1.4
Applying the Factor-Label Method

In 1975, the world record for the long jump was 29.21 ft. What is this distance in meters?

ANALYSIS: The problem can be stated as

$$29.21 \text{ ft} = ? \text{ m}$$

One of several sets of relationships we can use is

$$1 \text{ ft} = 12 \text{ in.}$$

$$1 \text{ in.} = 2.54 \text{ cm} \qquad \text{(From Table 1.3)}$$

$$1 \text{ cm} = 10^{-2} \text{ m} \qquad \text{(From Table 1.2)}$$

Notice how they take us from feet to inches to centimeters to meters.

SOLUTION: Now we apply the factor-label method by eliminating unwanted units to bring us to the units of the answer.

$$29.21 \text{ ft} \times \frac{12 \text{ in.}}{1 \text{ ft}} \times \frac{2.54 \text{ cm}}{1 \text{ in.}} \times \frac{10^{-2} \text{ m}}{1 \text{ cm}} = 8.903 \text{ m}$$

ft to in.
ft to cm
ft to m

Notice that if we were to stop after the first conversion factor, the units of the answer would be inches; if we stop after the second, the units would be centimeters, and after the third we get meters—the units we want. This time the answer has been rounded to four significant figures because that's how many there were in the measured distance. Notice that the numbers 12 and 2.54 do not affect the number of significant figures in the answer because they are exact numbers derived from definitions.

This is not the only way we could have solved this problem. Other sets of conversion factors could have been chosen. For example, we could have used

$$3 \text{ ft} = 1 \text{ yd}$$

$$1 \text{ yd} = 0.9144 \text{ m}$$

Then the problem would have been set up as

$$29.21 \text{ ft} \times \frac{1 \text{ yd}}{3 \text{ ft}} \times \frac{0.9144 \text{ m}}{1 \text{ yd}} = 8.903 \text{ m}$$

Many problems that you meet, just like this one, have more than one path to the answer. There isn't any *one* correct way to set up the solution. *The important thing is for you to be able to reason your way through a problem and find some set of relationships that can take you from the given information to the answer.* The factor-label method can help you search for these relationships if you keep in mind the units that must be eliminated by cancellation.

CHECKING THE ANSWER FOR REASONABLENESS A meter is slightly longer than a yard. In 29 ft, there are slightly less than 10 yards, so the answer of 8.903 m seems to be reasonable.

■ **Practice Exercise 3** Use the factor-label method to perform the following conversions: (a) 3.00 yd to inches, (b) 1.25 km to centimeters, (c) 3.27 mm to feet, (d) 20.2 miles/gallon to kilometers/liter.

1.5 MATTER AND ENERGY

Having discussed the importance of measurement and the kinds of observations we make in the laboratory, it is now time to turn our attention to the meat of our subject. The two terms that form the title to this section basically define the focus of chemistry, and they are intimately related. All the materials we see around us are examples of *matter,* and when samples of matter react chemically, *energy* is almost always absorbed or given off (as heat or light, for instance). Thus, the combustion reaction between gasoline and oxygen liberates energy in the form of heat that we harness to power vehicles. In fact, one of the principal uses of chemical reactions is to provide the energy needs of our society.

Matter

Matter is defined as *anything that occupies space and has mass.* It is the stuff our universe is made of and includes all tangible things, from rocks to pizza to people. Notice that in setting out this definition we have used the term *mass* rather than *weight.* The words *mass* and *weight* are often used interchangeably even though they refer to different things. **Mass** *refers to how much matter there is in a given object;* **weight** *refers to the force with which the object is attracted by gravity.* For example, a golf ball contains a certain amount of matter and has a certain mass, which is the same regardless of the golf ball's location. However, the weight of the golf ball can vary. On earth its weight is approximately six times larger than the weight it would have on the moon because the gravitational attraction of the earth is six times that of the moon. Because mass does not vary from place to place, we use mass rather than weight when we specify the amount of matter in an object.

Mass is measured by a procedure called **weighing,** even though we do not really measure the sample's weight. As discussed in the Section 1.3, we use a balance to *compare* the sample's weight with the weights of standard masses. When both weigh the same, their masses must be the same.

When we observe a sample of matter, it can be in one (or more) of three different **physical states:** solid, liquid, and gas. Water, for instance, can exist as solid ice, as liquid water, and as gaseous steam, depending on the temperature. Even though the physical appearances of these three states are quite different, the chemical makeup of water is the same in all three states.

Energy

When chemical reactions occur they are almost always accompanied by an absorption or release of energy, so the study of energy is an integral part of the study of chemistry. The concept of energy is more difficult to grasp than that

Mass is also related to an object's inertia, or resistance to a change in its motion. An object of large mass has a large inertia and is difficult to move.

Solid, liquid, and gas are referred to as *states of matter.*

The iceberg and the sea that surrounds it are both composed of the same substance, water. They represent the solid and liquid states of this substance.

Work is done by an object when it causes something to move. A moving car has energy because it can move another car in a collision.

of matter because energy is intangible; you can't hold it in your hand to study it and you can't put it in a bottle. **Energy** *is something an object has if the object is able to do work.* It can be possessed by the object in two different ways, as kinetic energy and as potential energy.

Kinetic energy *is the energy an object has when it is moving.* It depends on the object's mass and velocity; the larger its mass and the greater its velocity, the more kinetic energy it has. A simple equation relates kinetic energy (K.E.) to these quantities:

$$K.E. = \tfrac{1}{2}mv^2 \tag{1.7}$$

where m is the mass and v is the velocity.

Potential energy *is energy an object has that can be changed to kinetic energy; it can be thought of as **stored energy**.* For example, when you wind an alarm clock, you give the spring potential energy (stored energy). This potential energy is then gradually changed into kinetic energy by the clock's mechanism as the timepiece operates. Water, high in the mountains, also has potential energy because of its attraction by gravity. When the water falls to a lower altitude we can use this energy to turn turbines that generate electricity or operate machinery. Chemicals also possess potential energy that is sometimes called **chemical energy.** This is energy stored in chemicals that can be liberated during chemical reactions. For example, the foods we eat have potential energy that is changed to kinetic energy by the process we call metabolism. This kinetic energy ultimately appears as movements of muscles and as body heat.

The conversion of energy from one form to another is controlled by one of nature's most important physical laws, the **law of conservation of energy,** *which states that energy can be neither created nor destroyed; it can only be changed from one form to another.* You've experienced this law if you have ever tossed a ball in the air. You give the ball some initial amount of kinetic energy when you throw it. As it rises, its potential energy increases. Because energy cannot come from nothing, the potential energy increase comes at the expense of the ball's kinetic energy, and the ball slows down. When all the kinetic energy has changed to potential energy the ball can go no higher; it has stopped moving and its potential energy is at a maximum. The ball then begins to fall and its potential energy is changed back to kinetic energy.

Units of Energy

The SI unit of energy is called the **joule (J),** named after James Joule, a British scientist who discovered that the energy of moving objects could be converted entirely into heat. The joule is defined as the amount of kinetic energy possessed by a 2-kg object moving at a speed of 1 meter per second (1 m/s). The relationship comes from the equation for kinetic energy given in Equation 1.7.

$$1\,J = \frac{1}{2}\,(2\text{ kg})\,(1\text{ m/s})^2 = 1\text{ kg m}^2/s^2$$

Thus, the joule is a derived unit formed directly from the SI base units. It really represents a rather small amount of energy. If you drop a 4.4 lb mass from a height of about 4 in. onto your foot, you would receive an amount of energy of about 1 joule.

An older and perhaps more familiar-sounding unit of energy is the **calorie (cal).** Originally, the calorie was defined as the amount of heat needed to raise the temperature of 1 g of water by 1 °C. With the introduction of the SI, the calorie was redefined more precisely as

$$1 \text{ cal} = 4.184 \text{ J (exact)}$$

Both the joule and calorie represent small amounts of energy. Better suited for measuring energies associated with chemical reactions are the kilojoule (kJ) and kilocalorie (kcal). These represent 1000 J and 1000 cal, respectively. You should also learn the relationship: 1 kcal = 4.184 kJ.

In this text we shall report energies in kilojoules or joules. However, you should be able to deal with energies in either set of units.

The nutritional Calorie (written with a capital letter C) is really a kilocalorie.

Heat and Temperature

Energy makes its presence known in a variety of ways. It can appear entirely as *mechanical energy,* which is simply energy of motion or kinetic energy. We can observe this energy when it is transferred in a collision between one object and another. However, perhaps the most common way we observe the transfer of energy is as heat.

Mention of heat immediately brings to mind the notion of temperature. **Heat** (also called **thermal energy**) is actually a form of kinetic energy. Specifically, it is the *total* kinetic energy of all the tiny particles (atoms) that make up an object. These tiny particles have kinetic energy because they are constantly moving and jiggling around. The **temperature** of an object is related to the *average* kinetic energy of the jiggling and moving atoms. Stated another way, the temperature of an object is related to the kinetic energy of an "average" atom within it, whereas the heat an object contains equals the *total* kinetic energy of all of its atoms.

Energy and Its Relationship to Chemical Change

The concepts discussed in the preceding paragraph form the foundation of one of science's most important theories, which we will discuss in much greater detail in Chapter 5. For now, however, let's see how these concepts allow us to explain how changes in chemical energy during chemical reactions make themselves known.

In a gasoline engine, the combustion of gasoline produces hot gases from which heat is extracted to perform work. What is the origin of the heat that appears in this reaction, and why do the substances formed in the reaction have such high temperatures?

When gasoline burns, the gases that form have a lower potential energy than the starting materials. Therefore, during this reaction, the chemicals

Recall that chemical energy is the potential energy stored in chemicals.

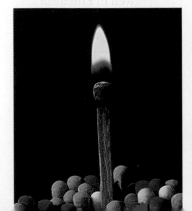

The difference between heat and temperature. Both fires could be at the same temperature if the *average* kinetic energies of the atoms in the flames are the same. The forest fire has many more atoms with high kinetic energies, so its *total* kinetic energy is greater than that of the match flame. Therefore, the forest fire has more heat in it.

undergo a net decrease in potential energy. This energy cannot just disappear, however. Instead, it is transformed into kinetic energy (heat energy). Because the total kinetic energy has increased, the average kinetic energy per atom must also have increased, and that means that the temperature has risen.

As you can see, the analysis of temperature changes in chemical reactions can tell us much about the energy changes that occur. And as we shall see, the energy changes that accompany reactions reveal a great deal about the chemicals themselves.

1.6
PROPERTIES OF MATTER

If you were searching through a pile of books for your chemistry book, you would no doubt recognize it by its size, color, and the printing on the cover. These are the characteristics that books have that help you distinguish among them. Similarly, in chemistry we use the characteristics, or *properties*, of matter to distinguish one kind from another. To help organize our thinking we classify these properties into different types.

Physical and Chemical Properties

A **physical property** *is one that can be observed without changing the chemical makeup of a substance.* For example, you know that when you stir sugar in water the solid sugar disappears as it dissolves. This ability of sugar to dissolve in water is a physical property because the act of dissolving the sugar doesn't alter its chemical composition. We can show this by evaporating the water; the sugar is recovered chemically unchanged. Because the sugar isn't altered chemically as it dissolves, forming the solution is said to be a **physical change.**

A **chemical property** *describes a chemical change (chemical reaction) that a substance undergoes.* If we melt sugar in a pan and heat it to a high temperature, the color of the sugar darkens as it begins to decompose into carbon and water. The decomposition of sugar at high temperatures is a chemical property, and the act of observing this chemical property leads to a chemical reaction—a permanent change in chemical composition that is not reversed by cooling the sugar to room temperature.

Intensive and Extensive Properties

Some properties that we observe depend on the size of the sample under study and others do not. For example, the masses and volumes of two different pieces of gold can be different, but both have the same characteristic shiny yellow color and both conduct electricity. Mass and volume are **extensive properties,** *or properties that depend on sample size.* Color and electrical conductivity are **intensive properties,** *or properties that are independent of sample size.*[7]

1.7
DENSITY AND SPECIFIC GRAVITY

One of the interesting things about extensive properties is that if you take the ratio of two of them, the resulting quantity is usually independent of sample size. In effect, the sample size cancels out and the calculated quantity becomes

[7] Color can often be a useful property for identification, but there are instances where you can be fooled, particularly when particle size is very small. For example, silver is a white metal with a high luster; however, in a very finely divided state, as in the image on black-and-white photographic film or paper, metallic silver appears black.

an intensive property. One useful property obtained this way is **density,** *which is defined as the ratio of an object's mass to its volume.* Using the symbols *d* for density, *m* for mass, and *V* for volume, we can express this mathematically as

$$d = \frac{m}{V} \qquad (1.8)$$

 Density

Notice that to determine an object's density we make two measurements, mass and volume.

A student measured the volume of an iron nail to be 0.880 cm³. She found that its mass was 6.92 g. What is the density of iron?

EXAMPLE 1.5
Calculating Density

ANALYSIS: This is a very straightforward calculation that requires that you remember the definition of density given by Equation 1.8. Don't neglect to learn such definitions, because they are tools you will need in situations like this.

SOLUTION: To determine the density we simply take the ratio of mass to volume.

$$\text{Density} = \frac{6.92 \text{ g}}{0.880 \text{ cm}^3}$$

$$= 7.86 \text{ g/cm}^3$$

This could also be written as

$$\text{Density} = 7.86 \text{ g/mL}$$

because 1 cm³ = 1 mL.

TABLE 1.4 Densities of Some Common Substances in g/cm³ at Room Temperature

Water	1.00
Aluminum	2.70
Iron	7.86
Silver	10.5
Gold	19.3
Glass	2.2
Air	0.0012

CHECKING THE ANSWER FOR REASONABLENESS In calculating the density, we are dividing 6.92 by a number slightly smaller than 1. The answer should therefore be slightly larger than 6.92. Our answer of 7.86 seems to be okay.

Each pure substance has its own characteristic density (Table 1.4). Gold, for instance, is much more dense than iron. Each cubic centimeter of gold has a mass of 19.3 g, so its density is 19.3 g/cm³. By comparison, the density of water is 1.00 g/cm³ and the density of air at room temperature is about 0.0012 g/cm³.

There is more mass in 1 cm³ of gold than in 1 cm³ of iron.

Most substances, such as the mercury in the bulb of a thermometer, expand slightly when they are heated. The same amount of matter occupies a larger volume at a higher temperature, so the amount of matter packed into each cubic centimeter is less. Therefore, density decreases slightly with increasing temperature.[8] For solids and liquids the size of this change is small, as you can see from the data for water in Table 1.5. Except when very precise calculations are involved we can generally ignore the variation of density with temperature.

TABLE 1.5 Density of Water as a Function of Temperature

Temperature (°C)	Density (g/cm³)
10	0.999700
15	0.999099
20	0.998203
25	0.997044
30	0.995646

Using Density

One reason why density is a useful property is that it provides a way to convert between the mass and volume of a substance. In Example 1.5, we found that iron has a density of 7.86 g/cm³. This density defines a relationship, which we

---

[8] Liquid water behaves oddly. Its maximum density is at 4 °C, so when water at 0 °C is warmed, its density increases until the temperature reaches 4 °C. As the temperature is increased further the density of water gradually decreases.

will call an **equivalence,** between the amount of mass and its volume. In words, we could express this by saying that 7.86 g of iron are equivalent to 1.00 cm³ of iron, or alternatively that 1.00 cm³ of iron is equivalent to 7.86 g of iron. We express this relationship symbolically as

$$7.86 \text{ g iron} \Leftrightarrow 1.00 \text{ cm}^3 \text{ iron}$$

where we have used the symbol ⟺ to mean "is equivalent to." (We can't really use an equal sign in this expression because grams can't *equal* cubic centimeters; one is a unit of mass and the other is a unit of volume.)

In setting up calculations by the factor-label method, an equivalence can be used to construct conversion factors just as equalities can. From the equivalence we have just written we can form two conversion factors:

$$\frac{7.86 \text{ g iron}}{1.00 \text{ cm}^3 \text{ iron}} \quad \text{and} \quad \frac{1.00 \text{ cm}^3 \text{ iron}}{7.86 \text{ g iron}}$$

The following example illustrates how we use density in calculations.

EXAMPLE 1.6
Calculations Using Density

A sample of vegetable oil has a density of 0.916 g/mL. (a) What is the mass of 225 mL of the oil? (b) How many milliliters are occupied by 45.0 g of the oil?

ANALYSIS: What we need here is a tool that lets us convert between the mass and volume of a substance, and that is precisely what the density is. For this oil the density tells us that

$$1.00 \text{ mL oil} \Leftrightarrow 0.916 \text{ g oil}$$

From this relationship we can construct two conversion factors.

$$\frac{1.00 \text{ mL oil}}{0.916 \text{ g oil}} \quad \text{and} \quad \frac{0.916 \text{ g oil}}{1.00 \text{ mL oil}}$$

SOLUTION TO (a): The question can be restated as: 225 mL oil = ? g oil. We need to eliminate the unit "mL oil," so we choose the conversion factor on the right.

$$225 \text{ mL oil} \times \frac{0.916 \text{ g oil}}{1.00 \text{ mL oil}} = 206 \text{ g oil}$$

Therefore, 225 mL of the oil has a mass of 206 g.

SOLUTION TO (b): The question is: 45.0 g oil = ? mL oil. This time we need to eliminate the unit "g oil," so we use the conversion factor on the left.

$$45.0 \text{ g oil} \times \frac{1.00 \text{ mL oil}}{0.916 \text{ g oil}} = 49.1 \text{ mL}$$

Thus, 45.0 g of the oil has a volume of 49.1 mL.

CHECKING THE ANSWERS FOR REASONABLENESS Notice that the density tells us that 1 mL of oil has a mass of slightly less than 1 g. Therefore, for part (a) we might expect 225 mL of oil to have a mass slightly less than 225 g. Our answer, 206 g, is reasonable. For part (b), 45 g of oil should have a volume not too far from 45 mL, so our answer of 49.1 mL is also in the "right ball park."

■ **Practice Exercise 4** A bar of aluminum has a volume of 1.45 mL. Its mass is 3.92 g. What is its density?

■ **Practice Exercise 5** The density of silver is 10.5 g/cm³. (a) What volume would 2.86 g of silver occupy? (b) What is the mass of 16.3 cm³ of silver?

Specific Gravity

The numerical value for the density of a substance depends on the units used for mass and volume. For example, if we express mass in grams and volume in milliliters, the density of water is 1.00 g/mL. However, if mass is given in pounds and volume in gallons, the density of water is 8.34 lb/gal; and if mass is in pounds and volume is in cubic feet, water's density is 62.4 lb/ft³. In a similar way, the densities of other substances have different numerical values for different units. One way to avoid having to tabulate densities in all sorts of different units is to tabulate specific gravities instead. *The* **specific gravity** *of a substance is defined as the ratio of the density of the substance to the density of water.*

$$\text{sp. gr.} = \frac{d_\text{substance}}{d_\text{water}} \qquad (1.9)$$

 Specific gravity

The specific gravity tells us how much denser than water a substance is. For instance, if the specific gravity of a substance is 2.00, then it is twice as dense as water; if its specific gravity is 0.50, then it is only half as dense as water. This means that if we know the density of water in a particular set of units, we can multiply it by a substance's specific gravity to obtain the substance's density in these same units. Rearranging Equation 1.9 gives

$$d_\text{substance} = (\text{sp. gr.})_\text{substance} \times d_\text{water} \qquad (1.10)$$

Let's look at an example.

Methanol, a liquid fuel that can be made from coal, has a specific gravity of 0.792. What is the density of methanol in units of g/mL, lb/gal, and lb/ft³?

ANALYSIS: The question asks us to relate specific gravity to density, so Equation 1.9 (or its equivalent, Equation 1.10) is the tool we need to get the answer. We also need some data—the densities of water in the requested units. Ordinarily we would look for them in a table of densities of water, although this time the necessary data were given in the preceding discussion.

$$\begin{aligned} d_\text{water} &= 1.00 \text{ g/mL} \\ &= 8.34 \text{ lb/gal} \\ &= 62.4 \text{ lb/ft}^3 \end{aligned}$$

SOLUTION: Now that we have all the necessary tools and data, we bring them together to obtain the answers.

$$\begin{aligned} d_\text{methanol} &= (\text{sp. gr.})_\text{methanol} \times d_\text{water} \\ &= 0.792 \times 1.00 \text{ g/mL} \\ &= 0.792 \text{ g/mL} \end{aligned}$$

Similarly, for the other units,

$$\begin{aligned} d_\text{methanol} &= 0.792 \times 8.34 \text{ lb/gal} \\ &= 6.61 \text{ lb/gal} \end{aligned}$$

EXAMPLE 1.7
Using Specific Gravity

Although the density of water varies slightly with temperature, it is useful to remember the value 1.00 g/cm³. It can be used if the water is near room temperature and only three (or fewer) significant figures are required.

and

$$d_{\text{methanol}} = 0.792 \times 62.4 \text{ lb/ft}^3$$
$$= 49.4 \text{ lb/ft}^3$$

Notice that when the density is expressed in grams per milliliter it is numerically the same as the specific gravity.

Finally, take a moment to assure yourself that the answers we have obtained are reasonable.

■ **Practice Exercise 6** The density of aluminum is 2.70 g/mL. What is its specific gravity? What is the density of aluminum in units of lb/ft³?

■ **Practice Exercise 7** Ethyl acetate is a clear, colorless solvent with a fruity odor and is often used in the manufacture of plastics. Its specific gravity is 0.902. What is its density in both g/mL and lb/gal?

1.8
IDENTIFICATION OF SUBSTANCES BY THEIR PROPERTIES

A job chemists are often called upon to perform is chemical analysis. What is it that a particular sample is composed of? To answer such a question, the chemist relies on the properties of the chemicals that make up the sample.

For identification purposes, intensive properties are more useful than extensive ones because every sample of a given substance exhibits the same set of intensive properties. For instance, if you were asked to decide whether a particular liquid sample were water, finding either its mass or volume wouldn't help in your identification because mass and volume are not unvarying properties of a substance. By taking appropriate amounts, it is possible for *any* liquid to have a given mass or a given volume. However, if we measure both the mass *and* volume of the sample, we can calculate the density of the liquid, and density is an intensive property. Since water has a density of 1.00 g/mL, we will know for sure that the liquid is not water if its density differs from this value. If the measured density is 1.00 g/mL, there is a reasonable possibility the liquid is water, so we might determine some additional properties such as melting point and boiling point. If after doing our experiments we find that the liquid in question is colorless and has a density of 1.00 g/mL, a freezing point of 0 °C, and a boiling point of 100 °C, we can feel quite certain that it *is* water, because these are properties that all samples of water have in common. It is also unlikely that another liquid would have these same identical properties.

Color, density, freezing point, and boiling point are examples of physical properties that can help us identify substances. Chemical properties are also intensive properties and also can be used for identification. For example, gold miners were able to distinguish between real gold and fool's gold (a mineral also called pyrite; Figure 1.12) by heating the material in a flame. Nothing happens to the gold, but the pyrite sputters, smokes, and releases bad smelling fumes because of its ability when heated to react chemically with oxygen in the air.

FIGURE 1.12

The color of the mineral pyrite, also called iron pyrite, accounts for its nickname, "fool's gold."

The Importance of Accuracy and Precision of Measurements

If we are to rely on properties such as density for identification of substances, it is very important that our measurements be reliable. The terms *accuracy* and *precision* address this question.

The importance of accuracy is obvious. If we have no confidence that our measured values are close to the true values, we certainly cannot trust the results of our experiments.

Precision of measurement can be equally important. For example, suppose we had two samples of a substance and believed them to be of the same mass. We could check our suspicion by weighing them. Suppose we selected an inexpensive balance capable of measurements to the nearest ± 0.1 g and obtained values of 12.3 g and 12.4 g. Are the masses really different? With an uncertainty of ± 0.1 g, we cannot tell whether the difference between the values is the result of an actual difference in mass or just a manifestation of the uncertainty in the measurements.

Suppose we now measure the masses with a higher quality balance, capable of measurements to the nearest ± 0.0001 g, and obtain values of 12.3115 g and 12.3667 g. These values differ by 0.0552 g. This difference is much larger than the uncertainty in the measured masses (± 0.0001 g), so we can be sure there is a real difference in the masses of the two samples.

The point of the preceding discussion is that to trust the results of our experiments, we must be sure our measurements are accurate and that they are of sufficient precision to be meaningful. This has a lot to do with how we plan experiments in the laboratory.

SUMMARY

Chemistry and Chemical Reactions Scientists from all fields must be concerned with chemistry because the material things they study are composed of chemicals. Today, as in the past, chemistry plays a crucial role in fulfilling the needs of society. **Chemistry** is a **natural science** that studies the properties and composition of matter and the way substances interact with each other in chemical reactions. When substances undergo a **chemical reaction** (or **chemical change**) they change into new substances that have different properties from the starting materials. Before the reaction begins we observe the properties of the starting materials; after the reaction is over we observe the properties of the new substances that are formed.

Scientific Method This is the general procedure by which science advances. **Observations** are made and the **empirical facts** or **data** that are collected from many experiments are often summarized in **scientific laws. Hypotheses** and **theories** that explain the data are tested by other experiments. Although the general pattern in the development of science is the cycle of observation–explanation–observation–explanation . . . , many discoveries are made accidentally by people who have learned to be observant through scientific training.

Units of Measurement **Qualitative observations** lack numerical information, whereas **quantitative observations** require numerical measurements. The units used for measurements are based on the set of seven **SI base units,** which can be combined to give various **derived units.** These all can be scaled to various sizes by applying **decimal multiplying factors.** In the laboratory we routinely measure length, volume, mass, and temperature. Convenient units for these purposes are, respectively, **centimeters** or **millimeters, liters** or **milliliters, grams,** and **degrees Celsius** or **kelvins.**

Significant Figures The **precision** of a measured quantity is revealed by the number of **significant figures** that it contains, which equals the number of digits known with certainty plus the first one that possesses some uncertainty. Measured values are **precise** if they contain many significant figures and therefore differ from each other by small amounts. A measurement is **accurate** if its value lies very close to the true value. When measurements are combined by multiplication or division, the answer should not contain more significant figures than the least precise factor. When addition or subtraction is used, the answer is rounded to the same number of decimal places as the quantity having the fewest number of decimal places. **Exact numbers** do not enter into determining the number of significant figures in a calculated quantity because they have no uncertainty. **Scientific notation** is useful for writing large or small numbers in compact form and for expressing unambiguously the number of significant figures in a number.

Factor-Label Method The **factor-label method** is based on the ability of units to undergo the same mathematical operations as numbers. **Conversion factors** are constructed from *valid relationships* between units. These relationships can be either equalities or equivalencies between units. Unit cancellation serves as a guide to the use of conversion factors and aids us in correctly setting up the arithmetic for a problem.

Matter and Energy   **Matter** is anything that has mass and occupies space. **Mass** is proportional to the amount of matter in a substance and differs from **weight,** which is determined by how mass is attracted to the Earth by gravity. **Energy** is the capacity to do work. An object can have energy as **kinetic energy** (because of its motion) or as **potential energy** (stored energy). All forms of energy can be converted to heat. Heat and temperature are different; **heat** is the *total* kinetic energy of the atoms in a substance; **temperature** is a measure of the *average* kinetic energy of the atoms.

Properties of Matter   When studying matter we are concerned with its physical and chemical properties. **Physical properties** can be measured without changing the chemical makeup of the sample. **Chemical properties** relate to how substances change into other substances in chemical reac-

tions. **Intensive properties** are independent of sample size; **extensive properties** depend on sample size.

Density and Specific Gravity   **Density** is a useful intensive property that is defined as the ratio of a sample's mass to its volume. Besides serving as a means for identifying substances, density is a conversion factor that relates mass to volume. **Specific gravity** is the ratio of a sample's density to that of water. The numerical values of specific gravity and density are the same if density is expressed in the units g/mL.

Identification of Substances by Their Properties   Intensive properties are more useful than extensive properties for identification of substances. Accuracy and precision of measurement affect our ability to make positive identifications.

TOOLS YOU HAVE LEARNED

The table below lists the tools you have learned in this chapter that are applicable to problem solving. Review them if necessary, and refer to them when working on the Thinking-It-Through problems and the Review Exercises that follow.

Tool	Function
SI prefixes (Table 1.2, page 6)	Create larger and smaller units. To form conversion factors for converting between differently sized units.
Units in laboratory measurements (page 7)	To convert between commonly used laboratory units of measurement.
Temperature Conversions (pages 10 and 11)	To convert among temperature units. Be especially sure you can convert between Celsius and Kelvin temperatures, because that will be needed most in this course.
Rules for counting significant figures. (page 13)	Determining the number of significant figures in a number.
Rules for arithmetic and significant figures (page 14)	Rounding the answers to the correct number of significant figures.
Factor-label method (page 15)	To set up the arithmetic in numerical problems by assembling necessary relationships into conversion factors and applying them to obtain the desired units of the answer.
Density (page 23)	To calculate the density of a substance from measurements of mass and volume. To convert between mass and volume for a substance.
Specific gravity (page 25)	Convert density from one set of units to a different set of units.

THINKING IT THROUGH

The aim of the following problems is to give you practice in thinking your way through solving problems. The goal is not to obtain the answer itself. Instead, you are asked to assemble the information needed to obtain the answer, state what additional data (if any) are needed, and describe how you would use the data to answer the question. For problems involving unit conversions, list the relationships among the units needed to carry out the conversions, construct appropriate conversion factors and use them to set up the solution.

The problems are divided into two groups. Those in Level 2 are significantly more challenging than those in Level 1 and provide an opportunity to really hone your problem solving skills.

Level 1 Problems

1. How many nanometers are there in 14.6 cm?

2. Suppose a projectile is fired at a velocity of 1450 m/s. What is this speed expressed in kilometers per hour?

3. A substance has a density of 6.85 g/cm^3. What is this density expressed in units of μg/μm^3?

4. How many cubic millimeters are there in 2.20 cm^3?

5. A person is 5 ft 9 in. tall. What is this height expressed in centimeters?

6. Normal human body temperature is 98.6 °F. What is this temperature expressed in °C and in kelvins?

7. Suppose a car weighing 1.50 tons (1 ton = 2000 lb) is traveling at a speed of 30.0 miles per hour. How much kinetic energy, expressed in joules, does the car have? What is this energy expressed in calories?

8. In one year (365 days), light travels a distance of 9.46×10^{12} km. What is the speed of light expressed in *meters per second*?

9. Light travels through a vacuum at a speed of 3.0×10^{10} cm s^{-1}. What is the speed of light in units of *miles per hour*.

10. A cylindrical metal bar has a diameter of 0.753 cm and a length of 2.33 cm. It has a mass of 8.423 g. What is the density of the metal in the units g/cm^3?

11. What is the specific gravity of the metal described in the preceding question? What is its density in pounds per cubic foot?

12. What is the volume (in cubic centimeters) of a bar of lead that has a mass of 256.4 g? The density of lead is 11.34 g/cm^3.

13. For the preceding question, how many significant figures should be reported in the answer?

14. A liquid has a specific gravity of 1.12. What is the mass in pounds of 3.45 gallons of the liquid?

Level 2 Problems

15. A liquid has a specific gravity of 0.824. What is the mass in pounds of 146 in.^{3} of it?

16. Because of the serious consequences of lead poisoning, the Federal Centers for Disease Control in Atlanta has set a threshold of concern for lead levels in children's blood. This threshold was based on a study that suggested that lead levels in blood as low as *10 micrograms of lead per deciliter of blood* can result in subtle effects of lead toxicity. Suppose a child had a lead level in her blood of 2.5×10^{-4} g of lead per liter of blood. Is this person in danger of exhibiting the effects of lead poisoning?

17. Gold has a density of 19.31 g/cm^3. How many grams of gold are required to provide a gold coating 0.50 mm thick on a ball bearing having a diameter of 0.500 in.?

18. A truck with a mass of 4000 kg is traveling at a speed of 80 km per hour. If the driver slams on the brakes, how many calories of heat will be liberated in the brake linings as the truck comes to a stop?

19. Methanol has a specific gravity of 0.792. What is its density in units of pounds per cubic meter?

REVIEW EXERCISES

Answers to questions whose numbers are printed in color are given in Appendix D. More challenging questions are marked with asterisks.

General

1.1 After some thought, give two reasons why a course in chemistry will benefit *you* in the pursuit of your particular major.

1.2 What does the science of chemistry seek to study?

1.3 Look around and list 10 items you see that are made of synthetic materials, that is, materials not found in nature.

1.4   In answering Exercise 1.3, what have you learned about the contribution of chemistry to modern civilization?

Chemical Reactions

1.5   What is a chemical reaction?

1.6   Lye is a common name for a substance called sodium hydroxide. Muriatic acid is the common name for hydrochloric acid. Either of these chemicals can cause severe burns if left in contact with the skin, but when water solutions of them are mixed in just the right proportions, the resulting solution contains only sodium chloride. From this description, how do you know that a chemical reaction occurs between sodium hydroxide and hydrochloric acid?

1.7   If you swallow a water solution of baking soda, a gas (carbon dioxide) forms in your stomach, which causes you to burp. Has a chemical reaction occurred? Explain your answer.

Scientific Method

1.8   What steps are involved in the scientific method?

1.9   What is the function of a laboratory?

1.10   Define: (a) data, (b) hypothesis, (c) law, (d) theory.

1.11   What role does luck play in the advancement of science?

Units of Measurement

1.12   Several commercial products that use metric units are pictured in Figure 1.2. Can you find any others among the items with which you come in contact each day?

1.13   What does the abbreviation SI stand for?

1.14   Which SI base unit is defined in terms of a physical object?

1.15   On page 20 we saw that we could calculate kinetic energy by using the equation K.E. = $\frac{1}{2}mv^2$. The SI unit for mass is the kilogram (kg), and the derived unit for velocity or speed is meter/second (m/s). What is the SI derived unit for energy?

1.16   What is the meaning of these prefixes? (a) centi-, (b) milli-, (c) kilo-, (d) micro-, (e) nano-, (f) pico-, (g) mega-

1.17   What abbreviation is used for each of the prefixes named in Exercise 1.16?

1.18   What number should replace the question mark in each of the following?
(a) 1 cm = ? m
(b) 1 km = ? m
(c) 1 m = ? pm
(d) 1 dm = ? m
(e) 1 g = ? kg
(f) 1 cg = ? g

1.19   What numbers should replace the question marks below?
(a) 1 nm = ? m
(b) 1 μg = ? g
(c) 1 kg = ? g
(d) 1 Mg = ? g
(e) 1 mg = ? g
(f) 1 dg = ? g

1.20   What units are most useful in the laboratory for measuring (a) length, (b) volume, and (c) mass?

1.21   How is mass measured? What is the difference between a balance and a spring scale found in a market?

1.22   What reference points do we use in calibrating the scale of a thermometer? What temperature on the Celsius scale do we assign to each of these reference points?

1.23   In each pair, which is larger:
(a) A Fahrenheit degree or a Celsius degree?
(b) A Celsius degree or a kelvin?
(c) A Fahrenheit degree or a kelvin?

1.24   Perform the following conversions.
(a) 60 °C to °F
(b) 20 °C to °F
(c) 45.5 °F to °C
(d) 59 °F to °C
(e) 40 °C to K
(f) − 20 °C to K

1.25   Perform the following conversions.
(a) 86 °F to °C
(b) − 4 °F to °C
(c) − 25 °C to °F
(d) 373 K to °C
(e) 298 K to °C
(f) 30 °C to K

1.26   A clinical thermometer registers a patient's temperature to be 37.46 °C. What is this temperature in °F?

1.27   The coldest permanently inhabited place on earth is the Siberian village of Oymyakon in Russia. In 1964 the temperature reached a shivering − 96 °F! What is this temperature in °C?

1.28   Helium has the lowest boiling point of any liquid. It boils at 4 K. What is its boiling point in °C and °F?

1.29   Liquid nitrogen is used as a commercial refrigerant to flash freeze foods. Nitrogen boils at − 196 °C. What is this temperature on the Kelvin temperature scale?

1.30   Liquid oxygen is used as an oxidizer in the launch of the space shuttle. It boils at − 183 °C. What is this temperature on the Kelvin temperature scale?

Significant Figures

1.31   Define the term *significant figures*.

1.32   What is *accuracy*? What is *precision*?

1.33   How many significant figures do the following measured quantities have?
(a) 27.53 cm
(b) 39.240 cm
(c) 102.0 g
(d) 0.00021 kg
(e) 0.06080 m
(f) 0.0002 L

1.34   How many significant figures do the following measured quantities have?
(a) 0.0240 g
(b) 101.303 m
(c) 0.008 kg
(d) 615.00 mg
(e) 1.00050 L
(f) 3.5105 mm

1.35   Perform the following arithmetic and round off the answers to the correct number of significant figures. Assume that all of the numbers were obtained by measurement.
(a) 0.0022×315
(b) $83.25 - 0.01075$
(c) $(84.45 \times 0.02)/(31.2 \times 9.8)$

(d) $(22.4 + 102.7 + 0.005)/(5.478)$
(e) $(315.44 - 208.1) \times 7.2234$

1.36 Perform the following arithmetic and round off the answers to the correct number of significant figures. Assume that all of the numbers were obtained by measurement.
(a) $3.58/1.739$
(b) $4.02 + 0.001$
(c) $(22.4 \times 8.3)/(1.142 \times 0.002)$
(d) $(1.345 + 0.022)/(13.36 \times 8.4115)$
(e) $(74.335 - 74.332)/(4.75 \times 1.114)$

Review of Scientific Notation (see Appendix A.1)

1.37 Express the following numbers in scientific notation. Assume three significant figures in each number.
(a) 2340   (d) 45,000
(b) 31,000,000   (e) 0.00000400
(c) 0.000287   (f) 324,000

1.38 Express the following numbers in scientific notation. Assume, in this problem, that only the nonzero digits are significant figures.
(a) 389   (d) 0.00225
(b) 0.00075   (e) 2.33
(c) 81,300   (f) 28,320

1.39 Write the following numbers in standard, nonexponential form.
(a) 2.1×10^5 (d) 4.6×10^{-12}
(b) 3.35×10^{-6} (e) 34.6×10^{-7}
(c) 3.8×10^3 (f) 8.5×10^8

1.40 Write the following numbers in standard, nonexponential form.
(a) 4.27×10^{-8} (d) 2.85×10^{-2}
(b) 7.11×10^5 (e) 5.0000×10^7
(c) 33.5×10^{-9} (f) 17.2×10^2

1.41 Perform the following arithmetic and express the answers in scientific notation.
(a) $\dfrac{(1.0 \times 10^7) \times (4.0 \times 10^5)}{(2.0 \times 10^8)}$
(b) $\dfrac{(4.0 \times 10^{-5}) \times (6.0 \times 10^{10})}{(3.0 \times 10^{-2})}$
(c) $\dfrac{(5.0 \times 10^{-4}) \times (2.0 \times 10^{-6})}{(1.0 \times 10^{-12})}$
(d) $(3.0 \times 10^4) + (2.1 \times 10^5)$
(e) $(8.0 \times 10^{12}) \div (2.0 \times 10^{-3})^2$

1.42 Perform the following arithmetic and express the answers in scientific notation.
(a) $\dfrac{(8.0 \times 10^6)}{(2.0 \times 10^5) \times (1.0 \times 10^3)}$
(b) $\dfrac{(1.6 \times 10^{15}) \times (1.0 \times 10^{-5})}{(8.0 \times 10^4)}$
(c) $\dfrac{(4.5 \times 10^{28})}{(3.0 \times 10^{-6})^2}$
(d) $(1.4 \times 10^5) - (3.0 \times 10^4)$
(e) $(3.3 \times 10^{-4}) + (2.52 \times 10^{-2})$

Factor-Label Method

1.43 Suppose someone suggested using the fraction 1 yd / 2 ft as a conversion factor to change a length expressed in feet to its equivalent in yards. What is wrong with this conversion factor? Can we construct a valid conversion factor relating centimeters to meters from the equation 1 cm = 1000 m? Explain your answer.

1.44 In 1 hour there are 3600 seconds. By what conversion factor would you multiply 250 seconds to convert it to hours? By what conversion factor would you multiply 3.84 hours to convert it to seconds?

1.45 If you were to convert the measured length 4.165 ft to yards by multiplying by the conversion factor (1 yd /3 ft), how many significant figures should the answer contain? Why?

Unit Conversions by the Factor-Label Method

1.46 Perform the following conversions.
(a) 32.0 dm to km   (d) 137.5 mL to L
(b) 8.2 mg to μg   (e) 0.025 L to mL
(c) 75.3 mg to kg   (f) 342 pm to dm

1.47 Perform the following conversions; express your answers in scientific notation.
(a) 183 nm to cm   (d) 33 dm to mm
(b) 3.55 g to dg   (e) 0.55 dm to km
(c) 6.22 km to nm   (f) 53.8 ng to pg

1.48 Perform the following conversions. If necessary, refer to Tables 1.2 and 1.3.
(a) 36 in. to cm   (d) 1 cup (8 oz) to mL
(b) 5.0 lb to kg   (e) 55 mi /hr to km /hr
(c) 3.0 qt to mL   (f) 50.0 mi to km

1.49 Perform the following conversions. If necessary, refer to Tables 1.2 and 1.3.
(a) 250 mL to qt   (d) 1.75 L to fluid oz
(b) 3.0 ft to m   (e) 35 km /hr to mi /hr
(c) 1.62 kg to lb   (f) 80.0 km to mi

1.50 Cola is often sold in 12 oz cans. How many milliliters are in each can?

1.51 How many fluid ounces are in a 2.00 L bottle of cola?

1.52 A metric ton is 1000 kg. How many pounds is this?

1.53 In the United States, a "long ton" is 2240 lb. What is this mass expressed in metric tons (1 metric ton = 1000 kg)?

1.54 If a person is 6 ft 2 in. tall, what is the person's height in centimeters?

1.55 If a person weighs 81.6 kg, what is the person's weight in pounds?

1.56 Perform the following conversions. (a) 6.2 yd^2 to m^2, (b) 4.8 in.^{2} to mm^2, (c) 3.7 ft^3 to L

1.57 Perform the following conversions. (a) 9.5×10^2 mm^2 to in.^{2}, (b) 7.3 m^3 to ft^3, (c) 288 in.^{2} to m^2

1.58 A bullet is fired at a speed of 2235 ft/s. What is this speed expressed in kilometers per hour?

*1.59 Distances between stars are often expressed in units of *light-years*. A light-year is the distance that light travels during 1 year. Use the factor-label method to calculate the number of miles in one light-year, given that light travels at a speed of 3.0×10^8 meters per second. (*Hint:* Begin with *1 year* (365 days) and then use conversion factors to change that to a distance traveled by light.)

1.60 In the United States, the speed limit on some highways is 65 miles per hour. What is this speed expressed in kilometers per hour?

*1.61 Radio waves travel at the speed of light, 3.0×10^8 meters per second. If you were to broadcast a question to an astronaut on the moon, which is 239,000 miles from Earth, what is the minimum time that you would have to wait to receive a reply?

Matter and Energy

1.62 Define *matter* and *energy*. Which of the following are examples of matter? (a) air, (b) a pencil, (c) a cheese sandwich, (d) a squirrel, (e) your mother

1.63 What are the three states of matter?

1.64 How do mass and weight differ?

1.65 Define (a) *kinetic energy* and (b) *potential energy*. What two things determine how much kinetic energy an object possesses?

1.66 State the equation used to calculate an object's kinetic energy. Define the symbols used in the equation.

1.67 How do *temperature* and *heat* differ?

1.68 State the law of conservation of energy. When a ball rolls down a hill, what happens to its potential energy? What happens to its kinetic energy? Are these observations consistent with the law of conservation of energy? Explain.

1.69 When gasoline burns, it reacts with oxygen in the air and forms carbon dioxide and water vapor. How does the potential energy of the gasoline and oxygen compare with the potential energy of the carbon dioxide and water vapor?

1.70 What is *mechanical energy*?

Properties of Matter

1.71 What is *a physical property*? What is *a chemical property*? What is the chief distinction between physical and chemical properties?

1.72 How does a physical change differ from a chemical change? When a substance changes from one state of matter to another (for example, from solid to liquid), is the change a physical or chemical change?

1.73 Define the terms *intensive property* and *extensive property*. Give two examples of each.

1.74 Choose an object on your desk and list as many of its properties as you can. Which are physical properties and which are chemical properties? Which properties are independent of the size of the object?

Density and Specific Gravity

1.75 Gasoline has a density of about 0.65 g/mL. How much do 18 gal weigh in kilograms? In pounds?

1.76 A sample of kerosene weighs 36.4 g. Its volume was measured to be 45.6 mL. What is the density of the kerosene?

1.77 Acetone, the solvent in nail polish remover, has a density of 0.791 g/mL. What is the volume of 25.0 g of acetone?

1.78 A glass apparatus contains 26.223 g of water when filled at 25 °C. At this temperature, water has a density of 0.99704 g/mL. What is the volume of this apparatus?

1.79 A graduated cylinder was filled with water to the 15.0 mL mark and weighed on a balance. Its mass was 27.35 g. An object made of silver was placed in the cylinder and completely submerged in the water. The water level rose to 18.3 mL. When reweighed, the cylinder, water, and silver object had a total mass of 62.00 g. Calculate the density of silver.

1.80 Ethyl ether has been used as an anesthetic. Its density is 0.715 g/mL. What is its specific gravity?

1.81 Propylene glycol is used as a food additive and in cosmetics. Its density is 8.65 lb/gal. The density of water is 8.34 lb/gal. Calculate the specific gravity of propylene glycol.

1.82 Trichloroethylene is a common dry-cleaning solvent. Its specific gravity is 1.47 measured at 20 °C. What is its density in grams per milliliter?

1.83 Gold has a specific gravity of 19.3. Water has a density of 62.4 lb/ft³. Calculate the mass, in pounds, of one cubic foot of gold.

1.84 A liquid known to be either ethanol (ethyl alcohol) or methanol (methyl alcohol) was found to have a density of 0.798 ± 0.001 g/mL. Consult the *Handbook of Chemistry and Physics* to determine which liquid it is. What other measurements could help to confirm the identity of the liquid?

1.85 An unknown liquid was found to have a density of 69.22 lb/ft³. The density of ethylene glycol (the liquid used in antifreeze) is 1.1088 g/mL. Is the unknown liquid ethylene glycol?

Identification of Substances by Their Properties

1.86 Suppose you were told that behind a screen there were two samples of liquid, one of them water and the other gasoline. You are told that Sample 1 occupies 3 fluid ounces and Sample 2 occupies 7 fluid ounces.
 (a) What kind of property (intensive/extensive) is volume?
 (b) Can you use the information given to you in this question to determine which sample is water and which is gasoline (explain)?

1.87 Name two intensive properties that you *could* use to distinguish between water and gasoline. Give one chemical property you could use.

1.88 Many reference books list physical properties that can aid in the identification of substances. One such book is the *Handbook of Chemistry and Physics,* which no doubt is available in your school library.

(a) Use the Table of Physical Constants of Inorganic Compounds in the *Handbook of Chemistry and Physics* to tabulate the melting points, boiling points, and colors of cadmium iodide, lead iodide, and bismuth tribromide.

(b) Which of these tabulated properties could you use to distinguish among these three substances?

(c) A chemist who was asked to analyze a sample was able to isolate a yellow solid from an experiment. This substance was found to melt at 402 °C and boil at 954 °C. Which of the chemicals described in part (a) of this question could the chemist have obtained from the experiment?

Additional Exercises

*1.89 A pycnometer is a glass apparatus used for accurately determining the density of a liquid. When dry and empty, a certain pycnometer had a mass of 27.314 g. When filled with distilled water at 25.0 °C, it weighed 36.842 g. When filled with chloroform (a liquid once used as an anesthetic before its toxic properties were known), the apparatus weighed 41.428 g. At 25.0 °C, the density of water is 0.99704 g/mL. (a) What is the volume of the pycnometer? (b) What is the density of chloroform?

*1.90 The strange Ydarb tribe of a former French colony thrives on cabbage and canned potatoes! Cabbage costs 42 francs per head and potatoes cost 26 francs per can. The average Ydarb consumes 3 cans of potatoes per head of cabbage. If an Ydarb spends 186 francs on cabbage, how much money will be spent on potatoes?

1.91 Suppose you have a job in which you earn $2.50 for each 30 minutes that you work.

(a) Express this information in the form of an equivalence between dollars and minutes worked.

(b) Use the equivalence defined in (a) to calculate the number of dollars earned in 1 hr 45 min.

(c) Use the equivalence defined in (a) to calculate the number of minutes you would have to work to earn $7.35.

1.92 Carbon tetrachloride and water do not dissolve in each other. When mixed, they form two separate layers. Carbon tetrachloride has a density of 13.3 lb/gal and water has a density of 62.4 lb/ft³. Which liquid will float on top of the other when the two are mixed?

1.93 If you were given only the masses of samples of water and methanol, or if you were given only the volume of each liquid, you could not tell them apart. However, if you were given both their masses *and* their volumes, you could determine which was the methanol and which was the water. How?

1.94 The density of propylene glycol, a substance used as a nontoxic antifreeze, is 1.040 g/mL at 20 °C. What is the density of this substance at 20 °C expressed in units of (a) grams per liter, and (b) kilograms per cubic meter?

1.95 The density of isopropyl alcohol, used in rubbing alcohol, is 785 kg/m³ at 20 °C. What is the density of this alcohol at 20 °C expressed in units of (a) pounds per gallon and (b) grams per liter?

*1.96 A standard LP phonograph record has a diameter of 12 in. and rotates at $33\frac{1}{3}$ rpm (revolutions per minute) while it is being played. How fast is a point on the edge of the record moving, expressed in miles per hour?

1.97 When an object is heated to a high temperature, it glows and gives off light. The color balance of this light depends on the temperature of the glowing object. Photographic lighting is described, in terms of its color balance, as a temperature in kelvins. For example, a certain electronic flash gives a color balance (called color temperature) rated at 5800 K. What is this temperature expressed in °C?

1.98 Suppose that when a chemical reaction occurs in a solution in an insulated container, the solution becomes cool.

(a) What has happened to the average kinetic energy of the atoms in the liquid?

(b) What has happened to the total kinetic energy of the atoms in the liquid?

(c) What has happened to the total potential energy of the substances in the solution?

*1.99 There exists a single temperature at which the value reported in °F is the same as the value reported in °C. What is this temperature?

*1.100 In the text, the Kelvin scale of temperature is defined as an absolute scale in which one Kelvin degree unit is equivalent to one Celsius degree unit. A second absolute temperature scale exists called the Rankine scale. On this scale, one Rankine degree unit is equivalent in size to one Fahrenheit degree unit. (a) What is the only temperature at which the Kelvin and Rankine scales possess the same numeric value? Explain your answer. (Hint: No calculations are necessary to answer this question.) (b) What is the boiling point of water expressed in °R?

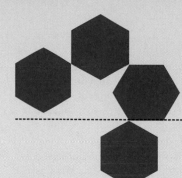

A TRIBUTE
TO A CHEMIST
Linus Pauling

"Professor Pauling, I'd like to know what goes through your mind when you decide that you understand something?" asked an undergraduate chemistry major at an informal reception held for Pauling at Augsburg College in the 1960s.

There was no hesitation, no searching for a lofty philosophical principle. With a face wreathed in good humor and exuberance, Pauling replied, "When I say 'Aha! So that's how it works!' " Then he went on, "But it isn't long before I realize that what I thought I had gotten right begged a number of questions. I realize that I hadn't really understood; that there's more to the problem than I had thought. So I puzzle over the problem some more until I again say, 'Aha! So that's how it *really* works.' Of course, that isn't the end of it, because still more questions arise, still more 'Aha's!' "

Linus Pauling (1901–1994), one of the greatest scientists of all time, stands in the company of Albert Einstein, Charles Darwin, Isaac Newton, and Galileo Galilei. Pauling was born on February 28, 1901, in Portland, Oregon, and was educated at Oregon State College (Corvallis) and the California Institute of Technology (Pasadena), where he joined the faculty at the age of 21. Normally in this "between chapters"

FIGURE 1a

Linus Pauling (1901–1994)

space we would have a unit on "Chemicals in Use," but it seemed fitting to us this time to open the series with a tribute to Pauling instead, an essay that might be called "A Chemist in Use."

Pauling and the Nature of the Chemical Bond

One of the traits of genius is the ability to see connections among seemingly unrelated observations and then to develop a general principle that applies even to more observations. Pauling developed connections out of an immense knowledge of chemical properties and a thorough understanding of a seemingly unrelated field, quantum physics. But, at bottom, both physics and chemistry deal with matter and energy, so numerous relationships and applications of one field to the other have developed. Pauling's intimate knowledge of the two fields led him to realize that quantum physics most helpfully intersects that of chemical properties at the points of the chemical bond and molecular structure. Structure determines properties, and bonds determine structure.

Among a long list of exceptional accomplishments, Pauling regarded his work on the nature of the chemical bond as his most significant. In an interview of Pauling conducted in 1994 by George B. Kauffman and Laurie M. Kauffman,[1] Pauling speculated on the problem that Arthur Becket Lamb, editor of the *Journal of the American Chemical Society,* must have had when Pauling submitted his now famous 1931 paper on the nature of the chemical bond. In scientific publications, the articles are called *papers,* and they are not published until they have been scrutinized and approved by one to three referees, scientists who are peers of the author in technical expertise and are knowledgeable in the field of the paper (but who remain anonymous to the author). It's called *peer review,* and all publications that report original research use the process. Pauling's paper focused quantum *physics*

[1] G. B. Kauffman and L. M. Kauffman, "Linus Pauling: Reflections," *American Scientist,* November–December, 1994, page 522. We gratefully acknowledge this article as one source of information for this essay. The personal experiences of one of us (JH) is another source.

on a fundamental *chemical* problem, and Lamb's dilemma was that there was no one in the world with the knowledge that Pauling had in *both* major fields. There existed, in other words, no qualified peers to referee the paper. So Lamb, who had already had confidence-building experiences with Pauling, broke the rule and went ahead and published the paper anyway. He was not disappointed nor, of course, were the worlds of chemistry and physics.

Eight years later, Cornell University Press published Pauling's book on the subject, *The Nature of the Chemical Bond and the Structure of Molecules and Crystals: An Introduction to Modern Structural Chemistry.* This book and succeeding editions, the latest an abridgment published in 1967 as the *The Chemical Bond: A Brief Introduction to Modern Structural Chemistry,* has been studied at the graduate level by thousands of chemists around the world.

Pauling and the Shapes of Protein Molecules

Pauling's work on the chemical bond and molecular structure won him his first of two unshared Nobel Prizes, the 1954 Nobel Prize in chemistry. His output of ideas continued unabated after the Prize. His knowledge of another area of physics, that of X-ray diffraction, was brought to bear on fundamental problems in biochemistry, one being the structures of proteins. Using both X-ray data and the building of molecular models, Pauling discovered two of the ways, the α-helix and the β-pleated sheet, by which huge protein molecules develop unique, overall shapes (to be discussed in Section 21.10).

The breadth of interest displayed by Pauling was immense. He once heard an anesthesiologist say that the noble gas argon is a good anesthetic. But argon is an element having no chemical reactions with any substance found in living systems, so how could it work? This led Pauling to ideas about how general anesthetics probably work. No doubt he said "Aha!" more than once when musing in this field.

One of the characteristics of many geniuses is a disarming modesty. In a conversation that one of us (JH) had with Pauling in the early 1960s, Pauling commented on the fact that he was very good at what he did. It was said matter of factly, as if to say that he couldn't help it; it just was true. Yet there was no evidence anywhere on the mantels or on the walls of his office of the coveted gold disc, the Nobel Prize medal itself. When asked about this, Pauling went over to a filing cabinet, rummaged among the thin boxes that filled the top drawer (all containing prizes and medals), and brought out *the* medal itself. Then he put it back. He was clearly more interested in ideas.

He would hope for all beginning students of chemistry that they also have a profound interest in how things work. Without that interest, there are no "Aha's!" and life is poorer.

Pauling and the Idea of Molecular Diseases

Pauling's interest in protein structure eventually led him to the concept of a *molecular disease,* that is, a disease attributed not to a germ (bacterium or virus), nor to a lack of a vitamin, nor to a poison, but simply to a defect in the structure of a molecule needed by cells, a defect originating in the molecular structure of a gene, a fundamental hereditary unit. Sickle cell anemia, but one example, is a molecular disease of hemoglobin, the red substance in blood that transports oxygen. In 1945, Pauling had first predicted that sickle cell anemia is the result of a mutant form of hemoglobin, and, in 1956, Vernon Ingram found the specific molecular flaw. The hemoglobin molecule is made in the body by piecing together portions of hundreds of small "building blocks" called amino acids, very roughly somewhat in the way that a poem might be made from "piecing" together letters of the alphabet. Taking the pieces of the right kind and in the right order are everything of course. Hamlet began his soliloquy with "To be or not to be?" He didn't say, "To be or now to be." In the hemoglobin of sickle cell anemia, two of the building block pieces or "letters" are wrong. Pauling's idea thus opened a huge door to the investigation of hundreds of genetic diseases. What marks an idea as particularly outstanding is its influence on a wide range of problems. As we have just seen, Pauling demonstrated his genius for such ideas over and over again.

Pauling and World Peace

Pauling's lively interests extended far beyond chemistry and physics. He was an American citizen who saw himself also as a world citizen, this during a time, the "cold war," when such an interest could get you into a lot of trouble. In 1962 he also won his second unshared Nobel Prize, the Nobel Prize for Peace.

Pauling and Vitamin C

Pauling left the California Institute of Technology in 1962. Already he was onto something involving vitamin C and the common cold. The medical community pooh-poohed the idea for years, but Pauling and his private research foundation, the Linus Pauling Institute of Science and Medicine, has moved considerable rethinking of conventional medical wisdom.

Archaeologists excavate a site at East Turkana in northern Kenya where the exceptionally well-preserved fossil remains of an elephant have been found. The fact that there are different kinds of atoms of an element (called isotopes), will enable the age of the fossils to be determined. In this chapter you will learn about the internal structures of atoms and what makes isotopes of elements possible. We will discuss the dating of archaeological objects in more detail in Chapter 20.

Chapter 2

The Structure of Matter: Atoms, Molecules, and Ions

One of the goals of chemistry is to organize information about chemical substances so that similarities and differences are easy to spot. Organization is the theme of this chapter, and we begin here with a discussion of three principal classes of matter—elements, compounds, and mixtures. We will also study in this chapter the basic structures of atoms, as well as the periodic table, the chemist's chief tool for organization.

2.1
ELEMENTS, COMPOUNDS, AND MIXTURES

Elements

If a chemical reaction changes one substance into two or more others, a **decomposition** has occurred. For example, if we pass electricity through molten sodium chloride (salt), the silvery metal sodium and the pale green gas chlorine are formed. In this example, we have decomposed sodium chloride into two simpler substances. No matter how we try, however, sodium and chlorine cannot be decomposed further by chemical reactions into still simpler substances that can be stored and studied.

In chemistry, *substances that cannot be decomposed into simpler materials by chemical reactions are called* **elements.** Sodium and chlorine are two examples. Others you may be familiar with are iron, chromium, lead, copper, aluminum, and carbon (as in charcoal). Some elements are gases at room temperature, including chlorine, oxygen, hydrogen, nitrogen, and helium.

So far, scientists have discovered 90 existing elements in nature and have made 21 more, for a total of 111. The names of the elements are given on the inside front cover of the book. The list may seem long, but many elements are rare; we will be most interested in only a relatively small number of them.

Sodium and chlorine are the same substances that we observed reacting with each other to form sodium chloride in Chapter 1.

These 111 elements, by themselves and in chemical combination, account for all matter in all its enormous variety everywhere in the known universe. That out of the seemingly infinite variety of things there are only just over a hundred elementary substances has greatly simplified the chemist's goal to understand the workings of nature.

37

Compounds

By means of chemical reactions, elements combine in various *specific proportions* to give all the more complex substances in nature. For instance, hydrogen and oxygen combine to form water, and sodium and chlorine combine to form sodium chloride. Water and sodium chloride are examples of compounds. A **compound** *is a substance formed from two or more elements in which the elements are always combined in the same fixed (i.e., constant) proportions by mass.* For example, if any sample of pure water is decomposed, the mass of oxygen obtained is *always* eight times the mass of hydrogen. Similarly, when hydrogen and oxygen react to form water, the mass of oxygen consumed is always eight times the mass of hydrogen, never more and never less. Thus, if 1.0 g of hydrogen is mixed with 10.0 g of oxygen, the reaction that produces water uses only 8.0 g of oxygen. The other 2.0 g of oxygen is left over, unreacted.

In our description of the makeup of a compound, there is a subtle but significant point to be understood. For example, we say that the compound sodium chloride is *composed of* sodium and chlorine. However, in this compound these two elements are not present in the same form as we find them in their pure states. As noted earlier, the pure element sodium is a metal, but salt certainly has no metallic characteristics. Similarly, pure chlorine is a pale green gas, but salt is not green and there is nothing gaseous about it. *When these or other elements react to form a compound, their individual properties are lost and in their place we find the properties of the compound.*

Mixtures

Elements and compounds are examples of **pure substances.**[1] The composition of a pure substance is always the same, regardless of its source. All samples of table salt, for example, contain sodium and chlorine combined in the same proportions by mass, and all samples of water contain the same proportions by mass of hydrogen and oxygen. Pure substances are rare, however. Usually we encounter mixtures of compounds or elements. Unlike elements and compounds, **mixtures** *can have variable compositions*. Sugar and water are examples of compounds because each consists of elements chemically combined in definite proportions. But, when we dissolve sugar in water, we form a mixture in which the proportions can be varied almost as much as we wish. (Some people like their coffee sweeter than others!)

Mixtures are either homogeneous or heterogeneous. A **homogeneous mixture** *has the same properties throughout the sample.* Examples are thoroughly stirred mixtures of salt in water or sugar in water. We call a homogeneous mixture such as this a **solution.** Solutions need not be liquids, just homogeneous. Brass, for example, is a solid solution of copper and zinc.

A **heterogeneous mixture** *consists of two or more regions called* **phases** *that differ in properties.* A mixture of gasoline and water, for example, is a two-phase mixture in which the gasoline floats on the water as a separate layer (Figure 2.1). The phases in a mixture don't have to be chemically different substances like gasoline and water, however. A mixture of ice and liquid water, for example, is a two-phase heterogeneous mixture in which the phases have the same chemical composition but occur in different physical states.

Water and sodium chloride must be more complex than elements because they can be decomposed to elements.

Diet cola | Regular soft drink | Syrup

Mixtures containing sugar can have variable compositions

Solid solutions of metals, such as brass and stainless steel, are called *alloys.*

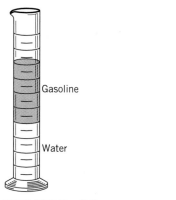

Gasoline

Water

FIGURE 2.1

A two-phase heterogeneous mixture.

[1] We have used the term *substance* rather loosely until now. Strictly speaking, **substance** really means *pure substance*. Each unique chemical element and compound is a substance; a mixture consists of two or more substances.

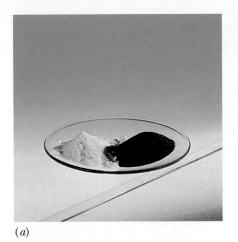

(*a*)

(*b*)

FIGURE 2.2

(*a*) Samples of powdered sulfur and powdered iron. (*b*) A mixture of sulfur and iron is made by stirring the two powders together.

FIGURE 2.3

Formation of the mixture is a physical change; it hasn't changed the iron and sulfur into a compound of these two elements. The mixture can be separated by pulling the iron out with a magnet.

An important way that mixtures differ from compounds is in the changes that occur when they form. Consider, for example, the elements iron and sulfur, which are pictured in powdered form in Figure 2.2*a*. We can make a mixture simply by dumping them together and stirring them. In the mixture (Figure 2.2*b*), both elements retain their original properties. The process we use to create this mixture involves a physical change, rather than a chemical change, because no new chemical substances form. To separate the mixture, we could similarly use just physical changes. For example, we could remove the iron by stirring the mixture with a magnet—a physical operation. The iron powder sticks to the magnet as we pull it out, leaving the sulfur behind (Figure 2.3). The mixture could also be separated by treating it with a liquid called carbon disulfide, which is able to dissolve the sulfur but not the iron. Filtering the sulfur solution from the solid iron, followed by evaporation of the liquid carbon disulfide from the sulfur solution, gives the original components, iron and sulfur, separated from each other.

As we noted earlier, formation of a compound from its elements involves a chemical reaction. Iron and sulfur, for example, combine to form a compound often called "fool's gold" because of its appearance (Figure 2.4). In this compound the elements no longer have the same properties they had before they were combined and they cannot be separated by physical means. Just as the formation of a compound involves a chemical reaction, so also does its decomposition. The decomposition of fool's gold into iron and sulfur is therefore a chemical change.

The relationships among elements, compounds, and mixtures are shown in Figure 2.5 on page 40.

FIGURE 2.4

A sample of iron pyrite, commonly called "fool's gold."

Chemical Symbols

In chemistry we use a system of symbols to write the formulas of compounds and to describe chemical reactions by chemical equations. In this system, each element is assigned its own unique **chemical symbol.** In most cases, the symbol is formed from one or two letters of the English name for the element. For instance, the symbol for carbon is C, for bromine it is Br, and for silicon it is Si.

FIGURE 2.5

Classification of matter.

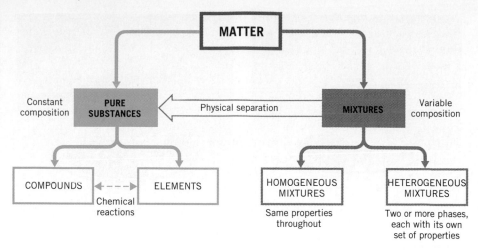

Latin was the universal language of science in the early days of chemistry.

For some elements, the symbols are derived from the Latin names given to those elements long ago. Table 2.1 contains a list of elements whose symbols come to us in this way.[2]

Regardless of the origin of the symbol, the first letter is always capitalized and the second letter, if there is one, is always written lowercase. Thus, the symbol for copper is Cu, not CU. Be careful to follow this rule so you can avoid confusion between such symbols as Co (cobalt) and CO (carbon monoxide).

Sometimes we use the chemical symbol to stand for the name of an element. We also use the symbol to stand for an **atom** of the element, which is the smallest bit of the element that retains its chemical identity. Thus, Co can be used in place of the name cobalt and also to mean one atom of cobalt.

Chemical Formulas

In one sense a chemical formula is a shorthand way of writing the name for a compound, just as chemical symbols are shorthand notations for the names of elements. The most important characteristic of a formula, however, is that it specifies the composition of a substance.

TABLE 2.1 Elements That Have Symbols Derived from Their Latin Names

Element	Symbol	Latin Name
Sodium	Na	Natrium
Potassium	K	Kalium
Iron	Fe	Ferrum
Copper	Cu	Cuprum
Silver	Ag	Argentum
Gold	Au	Aurum
Mercury	Hg	Hydrargyrum
Antimony	Sb	Stibium
Tin	Sn	Stannum
Lead	Pb	Plumbum

[2] The symbol for tungsten is W, from the German name *wolfram*. This is the only element whose symbol is neither related to its English name nor derived from its Latin name.

Atoms come together in definite proportions when they combine chemically. Often, the atoms join to produce larger stable particles called **molecules.** Sometimes the combining atoms acquire electrical charges and cling together because of the attraction between opposite charges.

Regardless of how the atoms are held to each other, in the formula for the compound each atom is identified by its chemical symbol. When more than one atom of an element is present, the number of atoms is given by a subscript. For example, the iron oxide in rust has the formula Fe_2O_3, which tells us that the compound is composed of iron and oxygen and that in this compound there are two atoms of iron for every three atoms of oxygen. When no subscript is written, we assume it to be one. For instance, the formula H_2O tells us that in water there are two H atoms for every one O atom. Similarly, the formula for chloroform, $CHCl_3$, indicates that one atom of carbon, one atom of hydrogen, and three atoms of chlorine have combined.

For more complicated compounds, we sometimes find formulas containing parentheses. An example is the formula for urea, $CO(NH_2)_2$, which tells us that the group of atoms within the parentheses, NH_2, occurs twice. (The formula for urea also could be written as CON_2H_4, but there are good reasons for writing certain formulas with parentheses, as you will see later.)

In addition to describing compounds, chemical formulas also are used for elements that are found in nature as molecules. Examples are oxygen and nitrogen, the major gases found in the atmosphere. When these elements are not combined with other elements in compounds, we always find them as *diatomic* (two atom) molecules having the formulas O_2 and N_2. A number of important elements occur as diatomic molecules and are listed in the table in the margin. Be sure to learn them; you will encounter them often throughout the course.

Hydrates   Certain compounds form crystals that contain water molecules. An example is ordinary plaster—the material often used to coat the interior walls of buildings. Plaster consists of crystals of calcium sulfate, $CaSO_4$, that contain two molecules of water for each $CaSO_4$. These water molecules are not held very tightly and can be driven off by heating the crystals. The dried crystals absorb water again if exposed to moisture, and the amount of water absorbed always gives crystals in which the H_2O to $CaSO_4$ ratio is 2 to 1. Compounds whose crystals contain water molecules in fixed ratios are quite common and are called **hydrates.** The formula for this hydrate of calcium sulfate is written $CaSO_4 \cdot 2H_2O$ to show that there are two molecules of water per $CaSO_4$. The raised dot is used to indicate that the water molecules are not bound too tightly in the crystal and can be removed.

Sometimes the *dehydration* (removal of water) of hydrate crystals produces changes in color. An example is copper sulfate, which is sometimes used as an agricultural fungicide. Copper sulfate forms blue crystals with the formula $CuSO_4 \cdot 5H_2O$ in which there are five water molecules for each $CuSO_4$. When these blue crystals are heated, most of the water is driven off and the solid that remains, now nearly pure $CuSO_4$, is almost white (Figure 2.6). If left exposed to the air, the $CuSO_4$ will absorb moisture and form blue $CuSO_4 \cdot 5H_2O$ again.

Counting Atoms in Formulas   Counting the number of atoms of the elements in a chemical formula is an operation you will have to perform many times, so let's look at an example.

Later in the book, when we wish to refer to individual atoms or molecules in general, we will simply use the term *particle.*

 Subscripts in a chemical formula.

Elements That Occur Naturally as Diatomic Molecules

Hydrogen	H_2	Fluorine	F_2
Nitrogen	N_2	Chlorine	Cl_2
Oxygen	O_2	Bromine	Br_2
		Iodine	I_2

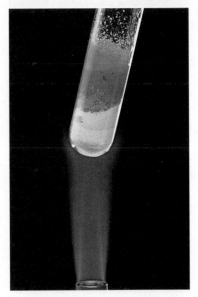

FIGURE 2.6

When blue crystals of copper sulfate hydrate ($CuSO_4 \cdot 5H_2O$) are heated, they readily lose water. The white solid that forms is pure $CuSO_4$.

When all the water is removed, the solid is said to be anhydrous, meaning *without water.*

EXAMPLE 2.1
Counting Atoms in Formulas

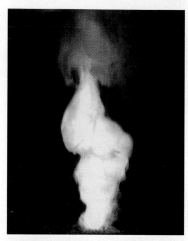

FIGURE 2.7

A mixture of zinc (Zn) and sulfur (S) reacts violently when ignited and forms the compound zinc sulfide, ZnS.

FIGURE 2.8

The combustion of butane, C_4H_{10}.

Coefficients in chemical equations are also called **stoichiometric coefficients.** As you will learn in Chapter 3, *stoichiometry* is a word associated with amounts of substances involved in chemical reactions.

How many atoms of each element are in the formulas (a) $Al_2(SO_4)_3$ and (b) $CoCl_3 \cdot 6H_2O$?

SOLUTION:

(a) Here we must recognize that all the atoms within the parentheses occur three times. Therefore, the formula $Al_2(SO_4)_3$ shows

$$2\ Al \qquad 3\ S \qquad 12\ O$$

(b) This is a formula for a hydrate, as indicated by the raised dot. It contains six water molecules, each with two H and one O, for every $CoCl_3$. Therefore, the formula $CoCl_3 \cdot 6H_2O$ represents:

$$1\ Co \qquad 3\ Cl \qquad 12\ H \qquad 6\ O$$

■ **Practice Exercise 1** How many atoms of each element are expressed by the formulas: (a) $NiCl_2$, (b) $FeSO_4$, (c) $Ca_3(PO_4)_2$, (d) $Co(NO_3)_2 \cdot 6H_2O$?

Chemical Equations

A **chemical equation** *describes what happens when a chemical reaction occurs.* It uses chemical formulas to provide a before-and-after picture of the chemical substances involved. Consider, for example, the reaction that occurs when a mixture of powdered zinc (Zn) and sulfur (S) is ignited. The reaction, shown in Figure 2.7, is quite violent and produces the compound zinc sulfide, which has the formula ZnS. This reaction is represented by the chemical equation

$$Zn + S \longrightarrow ZnS$$

The two substances that appear to the left of the arrow, Zn and S, are called the **reactants.** These are the substances that exist before the reaction occurs. To the right of the arrow is the formula for zinc sulfide, ZnS, which is the **product** of the reaction. In this example, only one substance is formed in the reaction, so there is only one product. As we will see, however, in most chemical reactions there is more than one product. The products are the substances that are formed and that exist after the reaction is over. The arrow means "reacts to yield." Thus, this equation tells us that *zinc and sulfur react to yield zinc sulfide.*

Coefficients

Many reactions are more complex than the one we've just examined. An example is the burning of butane, C_4H_{10}, the fluid in disposable cigarette lighters (Figure 2.8). The chemical equation for this reaction is

$$2C_4H_{10} + 13O_2 \longrightarrow 8CO_2 + 10H_2O$$

In this equation there are numbers, called **coefficients,** in front of the formulas. These are present to balance the equation. *An equation is* **balanced** *if there is the same number of atoms of each element indicated on both sides of the arrow.* The 2 before the C_4H_{10} tells us that 2 molecules of butane react. This involves a total of 8 carbon atoms and 20 hydrogen atoms. (Notice we have multiplied the numbers of atoms of C and H in one molecule of C_4H_{10} by the coefficient 2.) On the right we find 8 molecules of CO_2, which contain a total of 8 carbon atoms. Similarly, 10 water molecules contain 20 hydrogen atoms. Finally, we can count 26 oxygen atoms on both sides of the equation. You will learn to balance equations such as this in Chapter 3.

In a chemical equation we sometimes specify the physical states of the reactants and products, that is, whether they are solids, liquids, or gases. This is done by writing *s, l,* or *g* in parentheses after the chemical formulas. For example, the equation for the combustion of the carbon in a charcoal briquette can be written

$$C(s) + O_2(g) \longrightarrow CO_2(g)$$

At times we will also find it useful to indicate that a particular substance is dissolved in water. We do this by writing *aq*, meaning *aqueous* solution, in parentheses after the formula. For instance, the reaction between stomach acid (HCl) and the active ingredient in TUMS, $CaCO_3$, is

$$2HCl(aq) + CaCO_3(s) \longrightarrow CaCl_2(aq) + H_2O(l) + CO_2(g)$$

■ **Practice Exercise 2** How many atoms of each element appear on each side of the arrow in the equation

$$Mg(OH)_2 + 2HCl \longrightarrow MgCl_2 + 2H_2O$$

■ **Practice Exercise 3** Rewrite the equation in Practice Exercise 2 to show that $Mg(OH)_2$ is a solid, HCl and $MgCl_2$ are dissolved in water, and H_2O is a liquid.

s = solid
l = liquid
g = gas

Aqueous means dissolved in water.

2.2 ATOMS AND ATOMIC MASSES

In modern science, we have come to take for granted the existence of atoms. Even in everyday conversation we hear talk of *atomic energy* and *nuclear energy*. However, scientific evidence for the existence of atoms is relatively recent, and chemistry did not progress very far until that evidence was found.

The concept of atoms began nearly 2500 years ago when certain Greek philosophers expressed the belief that matter is ultimately composed of tiny indivisible particles, and it is from the Greek word *atomos,* meaning "not cut," that our word atom is derived. Their conclusions, however, were not based on any evidence; they were derived simply from philosophical reasoning.

Laws of Chemical Combination

The concept of atoms remained a philosophical belief, having limited scientific usefulness, until the discovery of two quantitative laws of chemical combination—the *law of conservation of mass* and the *law of definite proportions*. The evidence that led to the discovery of these laws came from the work of many scientists in the eighteenth and early nineteenth centuries.

Law of Conservation of Mass

No detectable gain or loss of mass occurs in chemical reactions. Mass is *conserved*.

Law of Definite Proportions

In a given chemical compound, the elements are always combined in the same proportions by mass.

The law of conservation of mass means that if we place chemical reactants in a sealed vessel that permits no matter to enter or escape, the mass of the vessel and its contents after a reaction will be identical to its mass before. Although this may seem quite obvious to us now, it wasn't quite so clear in the early history of modern chemistry. Not until scientists made sure that *all reactants* and *all products,* including any that were gases, were included when masses were measured could the law of conservation of mass be truly tested.

The law of definite proportions is illustrated by the compound zinc sulfide, whose formation from the elements is illustrated in Figure 2.7. For *any* sample of zinc sulfide that's decomposed, we obtain 1.000 g of Zn for each 0.490 g of S. In this compound, therefore, the zinc-to-sulfur ratio is 1.000 g Zn /0.490 g S. Similarly, it is always found that when 1.000 g of Zn combines with sulfur, exactly 0.490 g of S reacts. If a mixture of 1.000 g of Zn and 1.000 g of S is ignited, only 0.490 g of the sulfur reacts and 0.510 g of S is left over, unreacted. It is impossible to cause 1.000 g of zinc to combine with more than 0.490 g of S to form zinc sulfide.

This constancy of composition of compounds is one of the most useful fundamental laws of chemistry. In fact, on page 38 we *defined* a compound as a substance in which two or more elements are chemically combined in a *definite fixed proportion by mass.*

The Atomic Theory

The law of definite proportions brings up a question also raised by the law of conservation of mass: "What must be true about the nature of matter, given the truth of these laws?" In other words, what is matter made of?

At the beginning of the nineteenth century, John Dalton (1766–1844), an English scientist, used the Greek concept of atoms to make sense out of the emerging law of definite proportions. Dalton reasoned that if atoms really existed, they must have certain properties to account for the two laws of chemical combination. He described such properties, and the list constitutes what we now call **Dalton's atomic theory.**

Dalton's Atomic Theory

1. Matter consists of tiny particles called atoms.

2. Atoms are indestructible. In chemical reactions, the atoms rearrange but they do not themselves break apart.

3. The atoms of one particular element are all identical in mass and other properties.

4. The atoms of different elements differ in mass and other properties.

5. When atoms of different elements combine to form compounds, new and more complex particles form. However, in a given compound the constituent atoms are always present in the same fixed numerical ratio.

Dalton's theory easily explained the law of conservation of mass. According to the theory, a chemical reaction is simply a reordering of atoms from one combination to another. If no atoms are gained or lost, then the mass after the reaction must be the same as the mass before. This explanation of the law of conservation of mass works so well that it serves as the reason for balancing chemical equations.

The law of definite proportions is also easy to explain. According to the theory a given compound always has atoms of the same elements in the same numerical ratio. For example, a compound might contain two elements, say *A* and *B*, in a numerical ratio of 1 to 1 (so the formula is *AB*). If the mass of *B* is twice that of *A*, then every time we encounter a molecule of *AB* the mass ratio (*A* to *B*) is 1 to 2. This same mass ratio would exist regardless of how many molecules we had in our sample, so in samples of this compound the elements *A* and *B* are always present in the same proportion by mass.

Law of Multiple Proportions

Strong support for Dalton's theory came when Dalton and other scientists studied elements that are able to combine to give two (or more) compounds. For example, sulfur and oxygen form two different compounds, which we call sulfur dioxide and sulfur trioxide. If we decompose a 2.00 g sample of sulfur dioxide, we find it contains 1.00 g of S and 1.00 g of O. If we decompose a 2.50 g sample of sulfur trioxide, we find it also contains 1.00 g of S, but 1.50 g of O. This is summarized in the following table.

Compound	Sample Size	Mass of Sulfur	Mass of Oxygen
Sulfur dioxide	2.00 g	1.00 g	1.00 g
Sulfur trioxide	2.50 g	1.00 g	1.50 g

First, notice that sample sizes were chosen so that each had the *same mass of sulfur*. Second, the ratio of the masses of oxygen in the two samples is one of small whole numbers.

$$\frac{\text{mass of oxygen in sulfur trioxide}}{\text{mass of oxygen in sulfur dioxide}} = \frac{1.50 \text{ g}}{1.00 \text{ g}} = \frac{3}{2}$$

Similar observations are made when we study other elements that form more than one compound with each other and these observations form the basis of the **law of multiple proportions.**

Law of Multiple Proportions

Whenever two elements form more than one compound, the different masses of one element that combine with the same mass of the other element are in the ratio of small whole numbers.

Dalton's theory explains the law of multiple proportions in a very simple way. Suppose sulfur trioxide has the formula SO_3 and sulfur dioxide has the formula SO_2 (Figure 2.9). If we had just one molecule of each, then our samples each would have one sulfur atom and therefore the same mass of sulfur. Then, comparing the oxygen atoms, we find they are in a numerical ratio of 3 to 2. But because oxygen atoms all have the same mass, the mass ratio must also be 3 to 2.

The law of multiple proportions was not known before Dalton presented his theory. It was discovered because the theory suggested its existence. Although this law is not of much practical use in chemistry today, for many years it was one of the strongest arguments in favor of the existence of atoms.

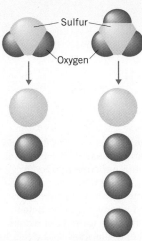

FIGURE 2.9

Molecules of sulfur trioxide, SO_3, and sulfur dioxide, SO_2. Each has one sulfur atom and, therefore, the same mass of sulfur. The oxygen ratio is 3 to 2, both by atoms and by mass.

Dividing both 1.50 and 1.00 by 0.50 gives a whole-number 3 to 2 ratio.

The discovery of the law of multiple proportions is a fine illustration of the scientific method at work.

Atomic Masses

One of the most useful concepts to come from Dalton's atomic theory is that atoms of an element have a constant, characteristic **atomic mass** (or **atomic weight**). This concept opened the door to the determination of chemical formulas and ultimately to one of the most useful devices chemists have for organizing chemical information, the periodic table of the elements. But how can the masses of atoms be measured?

Individual atoms are much too small to weigh on a balance, so their masses in grams can't be measured directly. However, the *relative masses* of the atoms of elements can be determined *provided that the formula of a compound containing them is known*. Let's look at an example to see how this could work.

Hydrogen (H) combines with the element fluorine (F) to form the compound hydrogen fluoride. The formula of this compound is HF, which means that in *any* sample of this substance the fluorine-to-hydrogen *atom ratio* is always 1 to 1. It is found that when a sample of HF is decomposed, the mass of fluorine obtained is always 19.0 times larger than the mass of hydrogen, so the fluorine-to-hydrogen *mass ratio* is always 19.0 to 1.00.

F to H atom ratio: 1 to 1
F to H mass ratio: 19.0 to 1.00

How could a 1 to 1 atom ratio give a 19.0 to 1.00 mass ratio? *Only if each fluorine atom is 19.0 times heavier than each H atom.*

Notice that even though we haven't found the actual masses of F and H atoms, we do now know how their masses compare (i.e., we know their *relative masses*). Similar procedures with other elements in other compounds are able to establish relative-mass relationships among the other elements as well. What we need next is a way to place all these masses on the same mass scale, but before we study this, let's take a closer look at one of Dalton's hypotheses.

Isotopes   The cornerstone of Dalton's theory was the idea that all of the atoms of an element have identical masses. Actually, most elements occur in nature as uniform mixtures of two or more kinds of atoms that have slightly different masses, which we call **isotopes.** An iron nail, for example, is made up of a mixture of four isotopes, and the chlorine in salt is a mixture of two isotopes.

The existence of isotopes did not affect the development of Dalton's theory for two reasons. First, all the isotopes of a given element have virtually identical *chemical* properties—all give the same kinds of chemical reactions. Second, the relative proportions of its different isotopes are essentially constant, regardless of where on the Earth or in the atmosphere the element is found. As a result, every sample of a particular element has the same isotopic composition, and the *average* mass per atom is the same from sample to sample. Therefore, in the laboratory elements *behave* as though their atoms have masses equal to the average.

The Carbon-12 Atomic Mass Scale

To establish a uniform mass scale for atoms it is necessary to define a standard against which the relative masses can be compared. Currently, the agreed-upon reference is the most abundant isotope of carbon, called carbon-12 and symbolized ^{12}C. An atom of this isotope is assigned *exactly* 12 units of mass.

To establish the first table of relative atomic masses, the formulas for compounds needed to be known. At the time Dalton presented his atomic theory no method was available to determine what the formula of a compound actually was. This difficulty was not overcome for nearly 50 years, so a reliable set of relative atomic masses was not established until the mid-1800s.

Even the smallest laboratory sample of an element has so many atoms that the relative proportions of the isotopes is constant.

These are called **atomic mass units,** symbolized in the SI by the letter **u**. Notice that this assignment establishes the size of the atomic mass unit to be 1/12th of the mass of a single carbon-12 atom, or stated another way:

$$1 \text{ atom } {}^{12}\text{C} \Longleftrightarrow 12 \text{ u (exactly)}$$

In modern terms, then, the atomic mass of an element is the average mass of the element's atoms (as they occur in nature) relative to an atom of carbon-12, which is assigned a mass of 12 units. Thus, if an average atom of an element has a mass twice that of a ${}^{12}\text{C}$ atom, its atomic mass would be 24.

The definition of the size of the atomic mass unit is really quite arbitrary. It could just as easily have been selected to be 1/24th of the mass of a carbon atom, or 1/10th the mass of an iron atom, or any other value. Why 1/12th the mass of a ${}^{12}\text{C}$ atom? First, carbon is a very common element, available to any scientist. Second, and most important, by choosing the atomic mass unit of this size, the atomic masses of nearly all the other elements are almost whole numbers, with the lightest atom (hydrogen) having a mass of approximately 1 u.

The atomic mass unit is sometimes called a dalton:

$$1 \text{ u} = 1 \text{ dalton}$$

Chemists generally work with whatever *mixture* of isotopes comes with a given element as it occurs naturally. Because the composition of this isotopic mixture is very nearly constant regardless of the source of the element, we can speak of an *average atom* of the element—average in terms of mass. For example, naturally occurring hydrogen is a mixture of two isotopes in the relative proportions given in the margin. The "average atom" of the element hydrogen, as it occurs in nature, has a mass that is 0.083992 that of a ${}^{12}\text{C}$ atom. Since $0.083992 \times 12.000 \text{ u} = 1.0079 \text{ u}$, the average atomic mass of hydrogen is 1.0079 u. Notice that this average value is just a little larger than the atomic mass of ${}^{1}\text{H}$ because naturally occurring hydrogen also contains a little ${}^{2}\text{H}$.

Hydrogen Isotope	Mass	Percentage Abundance
${}^{1}\text{H}$	1.007825 u	99.985
${}^{2}\text{H}$	2.0140 u	0.015

Originally, the relative atomic masses of the elements were determined in a way similar to that described for hydrogen and fluorine in our earlier discussion. A sample of a compound was analyzed and from the formula of the substance the relative atomic masses were calculated. These were then adjusted to place them on the unified atomic mass scale. In modern times, methods have been developed to measure very precisely both the relative abundances of the isotopes of the elements and their atomic masses. This kind of information has permitted the calculation of more precise values of the average atomic masses, which are found in the table on the inside front cover of the book. Example 2.2 illustrates how this calculation is done.

Naturally occurring chlorine is a mixture of two isotopes. In every sample of this element 75.77% of the atoms are ${}^{35}\text{Cl}$ and 24.23% are atoms of ${}^{37}\text{Cl}$. The accurately measured atomic mass of ${}^{35}\text{Cl}$ is 34.9689 u and that of ${}^{37}\text{Cl}$ is 36.9659 u. From these data, calculate the average atomic mass of chlorine.

ANALYSIS: In a sample containing many atoms of chlorine, 75.77% of the mass is contributed by atoms of ${}^{35}\text{Cl}$ and 24.23% is contributed by atoms of ${}^{37}\text{Cl}$. This means that when we calculate the mass of the "average atom," we have to weight it according to both the masses of the isotopes and their relative abundances. To do this it is convenient to imagine an "average atom" to be 75.77% ${}^{35}\text{Cl}$ and 24.23% ${}^{37}\text{Cl}$. (Of course, such an atom doesn't really exist, but this is a simple way to see how we can calculate the average atomic mass of this element.)

EXAMPLE 2.2
Calculating Average Atomic Masses from Isotopic Abundances

SOLUTION: We will calculate 75.77% of the mass of an atom of ^{35}Cl, which is the contribution of this isotope to the "average atom." Then we will calculate 24.23% of the mass of an atom of ^{37}Cl, which is the contribution of this isotope. Adding these contributions gives the total mass of the "average atom."

$$0.7577 \times 34.9689 \text{ u} = 26.50 \text{ u} \qquad \text{(for } ^{35}Cl\text{)}$$
$$\underline{0.2423 \times 36.9659 \text{ u} = 8.957 \text{ u}} \qquad \text{(for } ^{37}Cl\text{)}$$
$$\text{Total mass of average atom} = 35.46 \text{ u} \qquad \text{(rounded)}$$

The average atomic mass of chlorine is therefore 35.46 u.

■ **Practice Exercise 4** Aluminum atoms have a mass that is 2.24845 times that of an atom of ^{12}C. What is the atomic mass of aluminum?

■ **Practice Exercise 5** How much heavier than an atom of ^{12}C is the average atom of naturally occurring copper? Refer to the table inside the front cover of the book for the necessary data.

■ **Practice Exercise 6** Naturally occurring boron is composed of 19.8% of ^{10}B and 80.2% of ^{11}B. Atoms of ^{10}B have a mass of 10.0129 u and those of ^{11}B have a mass of 11.0093 u. Calculate the average atomic mass of boron.

2.3
A MODERN VIEW OF ATOMIC STRUCTURE

As you probably know, atoms are not quite as indestructible as Dalton had thought. They can be broken into simpler parts that we call **subatomic particles** and a knowledge of the internal structure of atoms is important to understanding the chemical and physical properties of the elements. In this chapter we will examine atomic structure in general terms, reserving a more detailed study for Chapter 6.

Subatomic Particles: Protons, Neutrons, and Electrons

There are three principal subatomic particles: **protons, neutrons,** and **electrons.** Experiments have shown that at the center of an atom there exists a very tiny, extremely dense core called the **nucleus,** which is where an atom's protons and neutrons are found. Because they are found in nuclei, protons and neutrons are sometimes called **nucleons.** The electrons in an atom surround the nucleus and fill the remainder of the volume of the atom. (How the electrons are distributed around the nucleus is the subject of Chapter 6.) The properties of the subatomic particles are summarized in Table 2.2 and the general structure of the atom is illustrated in Figure 2.10.

Notice that two of the subatomic particles carry electrical charges. Protons carry a single unit of positive charge, which is one type of electrical charge. Electrons carry one unit of the opposite charge, a negative charge. These electrical charges are such that like charges repel each other and opposite charges attract, so electrons are attracted to protons. In fact, it is this attraction that holds the electrons around the nucleus.

Protons are in *all* nuclei. Except for ordinary hydrogen, all nuclei also contain neutrons.

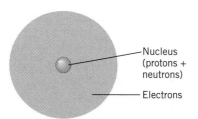

FIGURE 2.10

An atom is composed of a tiny nucleus that holds all the protons and neutrons, plus electrons that fill the space outside the nucleus.

TABLE 2.2 Properties of Subatomic Particles

Particle	Mass (g)	Mass (u)	Electrical Charge	Symbol
Electron	$9.1093897 \times 10^{-28}$	0.0005485712	1−	$^{0}_{-1}e$
Proton	$1.6726430 \times 10^{-24}$	1.007277252	1+	$^{1}_{1}H, ^{1}_{1}p$
Neutron	1.674954×10^{-24}	1.008665	0	$^{1}_{0}n$

SPECIAL TOPIC 2.1 / THE DISCOVERY OF SUBATOMIC PARTICLES

Our current knowledge of atomic structure has been pieced together from facts obtained by scientists from experiments that began in the nineteenth century. In 1834, Michael Faraday discovered that the passage of electricity through aqueous solutions could cause chemical changes, which was the first hint that matter was electrical in nature. Later in that century scientists began to experiment with *gas discharge tubes,* in which a high voltage electric current was passed through a gas at low pressure in a glass tube (Figure 1). Such a tube is fitted with a pair of metal *electrodes* and when the electricity begins to flow between them the gas in the tube glows. This flow of electricity is called an electric discharge, which is how the tubes got their names. (Modern neon signs work this way.)

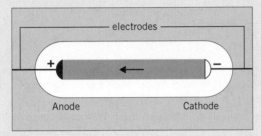

FIGURE 1

A gas discharge tube. Cathode rays flow from the negatively charged cathode to the positively charged anode.

The physicists who first studied this phenomenon did not know what caused the tube to glow, but tests soon revealed that negatively charged particles were moving from the negative electrode (the *cathode*) to the positive electrode (the *anode*). The physicists called these emissions *rays,* and because the rays came from the cathode, they called them *cathode rays.*

In 1897 the British physicist J. J. Thomson constructed a special gas discharge tube to make quantitative measurements of the properties of cathode rays. In some ways, the *cathode ray tube* he used was similar to a television picture tube, as Figure 2 shows. In Thomson's tube, a beam of cathode rays was focused on a glass surface coated with a phosphor that glowed when the cathode rays struck it (point 1). The cathode ray beam passed between the poles of a magnet and between a pair of metal electrodes that could be given electrical charges. The magnetic field tends to bend the beam in one direction (to point 2), whereas the charged electrodes bend the beam in the opposite direction (to point 3). By adjusting the charge on the electrodes, the two effects can be made to cancel, and from the amount of charge on the electrodes required to balance the effect of the magnetic field, Thomson was able to calculate the first bit of

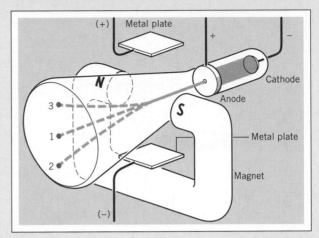

FIGURE 2

Thomson's cathode ray tube, which was used to measure the charge-to-mass ratio for the electron.

quantitative information about a cathode ray particle—the ratio of its charge to its mass (often expressed as e/m, where e stands for charge and m stands for mass). The charge-to-mass ratio has a value of -1.76×10^8 coulombs per gram, where the coulomb (C) is the SI unit of electrical charge, and the negative sign reflects the negative charge on the particle.

Many experiments were performed using the cathode ray tube and they demonstrated that cathode ray particles are in all matter. They are, in fact, *electrons.*

MEASURING THE CHARGE AND MASS OF THE ELECTRON

In 1909 a researcher at the University of Chicago, Robert Millikan, designed a clever experiment that enabled him to measure the electron's charge (Figure 3). During an experiment he would spray a fine mist of oil droplets above a pair of parallel metal plates, the top one of which had a small hole in it. As the oil drops settled, some would pass through this hole into the space between the plates, where he would irradiate them briefly with X rays. The X rays knocked electrons off molecules in the air, and the electrons became attached to the oil drops, which thereby were given an electrical charge. By observing the rate of fall of the charged drops both when the metal plates were electrically charged and when they were not, Millikan was able to calculate the amount of charge carried by each drop. When he examined his results, he found that all the values he obtained were whole-number multiples of -1.60×10^{-19} C. He reasoned that since a drop could only pick up whole numbers of electrons, this value must be the charge carried by each individual electron.

SPECIAL TOPIC 2.1 / Continued

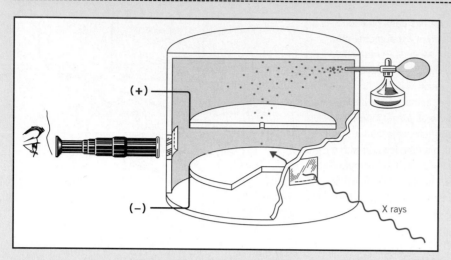

FIGURE 3

Millikan's oil drop experiment. Electrons, which are ejected from air molecules by the X rays, are picked up by very small drops of oil falling through the tiny hole in the upper metal plate. By observing the rate of fall of the charged oil drops, with and without electrical charges on the metal plates, Millikan was able to calculate the charge carried by an electron.

Once Millikan had measured the electron's charge, its mass could then be calculated from Thomson's charge-to-mass ratio. This mass was calculated to be 9.09×10^{-28} g. More precise measurements have since been made, and the mass of the electron is currently reported to be $9.1093897 \times 10^{-28}$ g.

DISCOVERY OF THE PROTON

The removal of electrons from an atom gives a positively charged particle (called an *ion*). To study these, a modification was made in the construction of the cathode ray tube to give a new device called a *mass spectrometer.* This apparatus is described in Special Topic 2.3 and was used to measure the charge-to-mass ratios of positive ions. These ratios were found to vary, depending on the chemical nature of the gas in the discharge tube, showing that their masses also varied. The lightest positive particle observed was produced when hydrogen was in the tube, and its mass was about 1800 times as heavy as an electron. When other gases were used, their masses always seemed to be whole-number multiples of the mass observed for hydrogen atoms. This suggested the possibility that clusters of the positively charged particles made from hydrogen atoms made up the positively charged particles of other gases. The hydrogen atom, minus an electron, thus seemed to be a fundamental particle in all matter and was named the *proton,* after the Greek *proteios,* meaning "of first importance."

DISCOVERY OF THE ATOMIC NUCLEUS

Early in this century, Hans Geiger and Ernest Marsden, working under Ernest Rutherford at Great Britain's Manchester University, studied what happened when *alpha rays* hit thin metal foils.

Because of their identical charges, electrons repel each other and protons repel each other. The repulsions between the electrons keep them spread out throughout the volume of the atom and it is the balance between the attractions and repulsions felt by the electrons that controls the sizes of atoms. The repulsions between protons are apparently offset by nuclear forces that involve other subatomic particles, which we will not study.

Matter as we generally find it in nature appears to be electrically neutral, which means that it contains equal numbers of positive and negative charges. Therefore, *in a neutral atom, the number of electrons equals the number of protons.*

The proton and neutron are much more massive than the electron, so in any atom almost all of the atomic mass is contributed by the particles that are found in the nucleus. It is also interesting to note, however, that the diameter of the atom is approximately 10,000 times the diameter of its nucleus, so almost all the *volume* of an atom is occupied by its electrons, which fill the space around the nucleus. (To place this on a more meaningful scale, if the

The electron has a mass that is only about 1/1836 that of a proton or neutron.

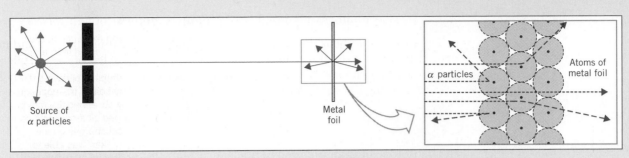

FIGURE 4

Alpha particles are scattered in all directions by a thin metal foil. Some hit something very massive head on and are deflected backward. Many sail through. Some, making near misses with the massive "cores" (nuclei), are still deflected, because alpha particles have the same kind of charge (+) as those cores.

Alpha rays are composed of particles having masses four times those of the proton and bearing two positive charges; they are emitted by certain unstable atoms in a phenomenon called radioactivity. Most of the alpha particles sailed right on through as though the foils were virtually empty space (Figure 4). A significant number of alpha particles, however, were deflected at very large angles. Some were even deflected backward, as if they had hit stone walls. Rutherford was so astounded that he compared the effect to that of firing a 15 in. artillery shell at a piece of tissue paper and having it come back and hit the gunner. From studying the angles of deflection of the particles, Rutherford reasoned that only something extraordinarily massive and positively charged could cause such an occurrence. Since most of the alpha particles went straight through, he further reasoned that the metal atoms in the foils must be mostly empty space. Rutherford's ultimate conclusion was that virtually all of the mass of an atom must be concentrated in a particle having a very small volume located in the center of the atom. He called this massive particle the atom's *nucleus*.

DISCOVERY OF THE NEUTRON

From the way alpha particles were scattered by a metal foil, Rutherford and his students were able to estimate the number of positive charges on the nucleus of an atom of the metal. This had to be equal to the number of protons in the nucleus, of course. But when they computed the nuclear mass based on this number of protons, the value always fell short of the actual mass. In fact, Rutherford found that only about half of the nuclear mass could be accounted for by protons. This led him to suggest that there were other particles in the nucleus that had a mass close to or equal to that of a proton, but with no electrical charge. This suggestion initiated a search that finally ended in 1932 with the discovery of the *neutron* by Sir James Chadwick, a British physicist.

nucleus were 1 ft in diameter, it would lie at the center of an atom with a diameter of approximately 1.9 miles!)

Atomic Numbers, Mass Numbers, and Isotopes

What distinguishes one element from another is the number of protons in the nuclei of its atoms, because *all the atoms of a particular element have an identical number of protons.* This means that each element has associated with it a unique number, which we call its **atomic number** (Z), that equals the number of protons in the nuclei of any of its atoms.

It is the atomic number of an atom that identifies which element it is.

$$\text{Atomic number, } Z = \text{number of protons}$$

What makes isotopes of the same element different are the numbers of neutrons in their nuclei. The **isotopes** *of a given element have atoms with the same number of protons but different numbers of neutrons.* The numerical sum of the

We will use the symbol Z to stand for atomic number frequently in later discussions.

The mass number is simply the number of nucleons and has no units.

protons and neutrons in the atoms of a particular isotope is called the **mass number** *(A)* of the isotope.

$$\text{Mass number, } A = \text{number of protons} + \text{number of neutrons}$$

Thus, every isotope is fully defined by two numbers, its atomic number and its mass number. Sometimes these numbers are added to the left of the chemical symbol of an element as a subscript and a superscript, respectively. Thus, if *X* stands for the chemical symbol for the element, an isotope of *X* is represented by

$$_Z^A X$$

The isotope of uranium used in nuclear reactors, for example, can be symbolized as follows.

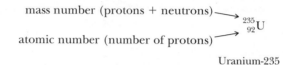

$$\text{mass number (protons + neutrons)} \longrightarrow {}_{92}^{235}\text{U}$$
$$\text{atomic number (number of protons)} \longrightarrow$$

Uranium-235

It is useful to remember that for a neutral atom, the atomic number equals both the number of protons and the number of electrons.

As indicated, the name of this isotope is uranium-235. Each neutral atom contains 92 protons and $(235 - 92) = 143$ neutrons as well as 92 electrons. In writing the symbol for the isotope, the atomic number is often omitted because it is really redundant. Every atom of uranium has 92 protons, and every atom that has 92 protons is an atom of uranium. Therefore, this uranium isotope can be represented simply as ${}^{235}\text{U}$.

In naturally occurring uranium, a more abundant isotope is ${}^{238}\text{U}$. Atoms of this isotope also have 92 protons, but the number of neutrons is 146. Therefore, atoms of ${}^{235}\text{U}$ and ${}^{238}\text{U}$ have the identical number of protons, but differ in the numbers of neutrons.

In general, the mass number of an isotope differs slightly from the atomic mass of the isotope. For instance, the isotope ${}^{35}\text{Cl}$ has an atomic mass of 34.968852 u. In fact, the only isotope that has an atomic mass equal to its mass number is ${}^{12}\text{C}$; *by definition* the mass of this atom is exactly 12 u.

■ **Practice Exercise 7** Write the symbol for the isotope of plutonium (Pu) that contains 146 neutrons.

■ **Practice Exercise 8** How many protons, neutrons, and electrons are in each atom of ${}_{17}^{35}\text{Cl}$?

2.4
THE PERIODIC TABLE

So far, we have discussed several different kinds of substances including elements and compounds. Among compounds we noted, at least briefly, that some are composed of discrete molecules while others are made up of atoms that have acquired electrical charges. For elements such as sodium and chlorine, we mentioned metallic and nonmetallic properties. If we were to continue on this way, without attempting to build our subject around some central organizing structure, it would not be long before we became buried beneath a mountain of information in the form of seemingly unconnected facts.

The need for organization was recognized by many early chemists, and there were numerous attempts to discover relationships among the chemical and physical properties of the elements. A number of different sequences of elements were tried in search of some sort of order or pattern. A few of these arrangements came quite close, at least in some respects, to our current peri-

SPECIAL TOPIC 2.2 / THE MASS SPECTROMETER AND THE MEASUREMENT OF ATOMIC MASSES

When a spark is passed through a gas, electrons are knocked off the gas molecules. Because electrons are negatively charged, the particles left behind carry positive charges; they are called positive ions. These positive ions have different masses, depending on the masses of the molecules from which they are formed. Thus, some molecules have large masses and give heavy ions, while some have small masses and give light ions.

The device that is used to study the positive ions produced from gas molecules is called a *mass spectrometer* (illustrated in the figure at the right). In a mass spectrometer, positive ions are created by passing an electrical spark (called an electric discharge) through a sample of the particular gas being studied. As the positive ions are formed, they are attracted to a negatively charged metal plate that has a small hole in its center. Some of the positive ions pass through this hole and travel onward through a tube that passes between the poles of a powerful magnet.

One of the properties of charged particles, both positive and negative, is that their paths become curved as they pass through a magnetic field. This is exactly what happens to the positive ions in the mass spectrometer as they pass between the poles of the magnet. However, the extent to which their paths are bent depends on the masses of the ions. This is because the path of a heavy ion, like that of a speeding cement truck, is difficult to change, but the path of a light ion, like that of a motorcycle, is influenced more easily. As a result, heavy ions emerge from between the magnet's poles along different lines from the lighter ions. In effect, an entering beam containing ions of different mass is sorted by the magnet into a number of

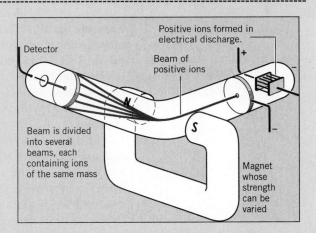

beams, each containing ions of the same mass. This spreading out of the ion beam thus produces an array of different beams called a *mass spectrum*.

In practice, the strength of the magnetic field is gradually changed, which sweeps the beams of ions across a detector located at the end of the tube. As a beam of ions strikes the detector, its intensity is measured and the masses of the particles in the beam are computed based on the strength of the magnetic field and the geometry of the apparatus.

Among the benefits derived from measurements using the mass spectrometer are very accurate isotopic masses and relative isotopic abundances. These serve as the basis for the very precise values of the atomic masses that you find in the modern periodic table (Figure 2.12 on page 56).

odic table, but either they were flawed in some way or they were presented to the scientific community in a manner that did not lead to their acceptance.

Mendeleev's Periodic Table

The periodic table we use today is based primarily on the efforts of a Russian chemist, Dmitri Ivanovich Mendeleev (1834–1907), and a German physicist, Julius Lothar Meyer (1830–1895). Working independently, these scientists developed similar periodic tables only a few months apart in 1869. Mendeleev is usually given the credit, however, because he had the good fortune to publish first.

Mendeleev was preparing a chemistry textbook for his students at the University of St. Petersburg. He found that when he arranged the elements in order of increasing atomic mass, similar chemical properties occurred over and over again at regular intervals. For instance, the elements lithium (Li), sodium (Na), potassium (K), rubidium (Rb), and cesium (Cs) have similar chemical properties. Each forms a water-soluble chlorine compound with the general formula *M*Cl: LiCl, NaCl, KCl, RbCl, and CsCl. Moreover, the elements that immediately follow each of these elements also constitute a set with similar chemical properties. Beryllium (Be) follows lithium, magnesium (Mg) follows sodium, calcium (Ca) follows potassium, strontium (Sr) follows rubidium, and barium (Ba) follows cesium. All of these elements form a water soluble chlorine compound with the general formula MCl_2: $BeCl_2$, $MgCl_2$, $CaCl_2$, $SrCl_2$, and $BaCl_2$. Based on extensive observations of this type, Mendeleev devised the original form of the **periodic law.** It stated that *the chemical and physical properties of the elements vary in a periodic way with their atomic weights.* Mendeleev used this law to construct his periodic table, which is illustrated in Figure 2.11.

The elements in Mendeleev's table are arranged in rows, called **periods,** in order of increasing atomic mass. When the rows are broken at the right places and stacked, the elements fall naturally into columns, called **groups,** in which the elements of a given group have similar chemical properties.

Mendeleev's genius rested on his placing elements with similar properties in the same group even when this left occasional gaps in the table. For example, he placed arsenic (As) in Group V under phosphorus because its chemical properties were similar to those of phosphorus, even though this left gaps in Groups III and IV. Mendeleev reasoned, correctly, that the elements that belonged in these gaps had simply not yet been discovered. In fact, on the basis of the location of these gaps Mendeleev was able to predict with remarkable accuracy the properties of these yet-to-be-found substances. His predictions helped serve as a guide in the search for the missing elements. Table 2.3 compares Mendeleev's predicted properties of "eka-silicon" and some of its compounds with the actual properties measured for the element germanium after it was discovered.

Periodic refers to the recurrence of properties at regular intervals.

FIGURE 2.11

Mendeleev's periodic table roughly as it appeared in 1871. The numbers next to the symbols are atomic masses.

	Group I	Group II	Group III	Group IV	Group V	Group VI	Group VII	Group VIII
1	H 1							
2	Li 7	Be 9.4	B 11	C 12	N 14	O 16	F 19	
3	Na 23	Mg 24	Al 27.3	Si 28	P 31	S 32	Cl 35.5	
4	K 39	Ca 40	— 44	Ti 48	V 51	Cr 52	Mn 55	Fe 56, Co 59 Ni 59, Cu 63
5	(Cu 63)	Zn 65	— 68	— 72	As 75	Se 78	Br 80	
6	Rb 85	Sr 87	?Yt 88	Zr 90	Nb 94	Mo 96	— 100	Ru 104, Rh 104 Pd 105, Ag 108
7	(Ag 108)	Cd 112	In 113	Sn 118	Sb 122	Te 128	I 127	
8	Cs 133	Ba 137	?Di 138	?Ce 140	—	—	—	———
9	—	—	—	—	—	—	—	
10	—	—	?Er 178	?La 180	Ta 182	W 184	—	Os 195, Ir 197 Pt 198, Au 199
11	(Au 199)	Hg 200	Tl 204	Pb 207	Bi 208	—		
12	—	—	—	Th 231	—	U 240	—	——— ——

TABLE 2.3   Comparison of Some Predicted Properties of Eka-Silicon and Observed Properties of Germanium

Property	Observed for Silicon (Si)	Predicted for Eka-Silicon (Es)	Observed for Tin (Sn)	Found for Germanium (Ge)
Atomic weight	28	72	118	72.59
Melting point (°C)	1410	High	232	947
Density (g/cm^3)	2.33	5.5	7.28	5.35
Formula of oxide	SiO_2	EsO_2	SnO_2	GeO_2
Density of oxide (g/cm^3)	2.66	4.7	6.95	4.23
Formula of chloride	$SiCl_4$	$EsCl_4$	$SnCl_2$	$GeCl_2$
Boiling point of chloride (°C)	57.6	100	114	84

Two elements, tellurium (Te) and iodine (I), caused Mendeleev some problems. According to the best estimates at that time, the atomic mass of tellurium was greater than that of iodine. Yet if these elements were placed in the table according to their atomic masses, they would not fall into the proper groups required by their properties. Therefore, Mendeleev switched their order and in so doing violated his own periodic law. (Actually, he believed that the atomic mass of tellurium had been incorrectly measured, but this wasn't so.)

The table that Mendeleev developed is in many ways similar to the one we use today. One of the main differences, though, is that Mendeleev's table lacks the column containing the elements helium (He) through radon (Rn). In Mendeleev's time, none of these elements had yet been found because they are relatively rare and because they have virtually no tendency to undergo chemical reactions. When these elements were finally discovered, beginning in 1894, another problem arose. Two more elements, argon (Ar) and potassium (K), did not fall into the groups required by their properties if they were placed in the table in the order required by their atomic masses. Another switch was necessary and another exception to Mendeleev's periodic law had been found. Apparently, then, atomic mass was not the true basis for the periodic repetition of the properties of the elements. To determine what the true basis was, however, scientists had to await the discoveries of the atomic nucleus, the proton, and atomic numbers.

The Modern Periodic Table

One of the gratifying dividends of the discovery of atomic numbers was that the elements in Mendeleev's table turned out to be arranged in precisely the order of increasing atomic number. In other words, if we take atomic numbers as the basis for arranging the elements in sequence, no annoying switches are required and the elements Te and I or Ar and K are no longer a problem. This leads to our present statement of the **periodic law.**

The Periodic Law

The chemical and physical properties of the elements vary in a periodic way with their atomic *numbers*.

SPECIAL TOPIC 2.3 / DMITRI MENDELEEV

Dmitri Ivanovich Mendeleev was one of Russia's most famous and versatile chemists. While at the University of St. Petersburg he taught organic chemistry and wrote a textbook on this subject, published in 1863, which was very popular. In 1867 he was appointed professor of general chemistry and when he began teaching that course he felt the need for a new textbook for his students. While preparing this book he wrote down the properties of the elements on cards. Sorting through these cards he came upon the periodic repetition of properties that was the cornerstone of his periodic table. Mendeleev's great textbook, *General Chemistry,* went through eight Russian editions and numerous translations into English, German, and French.

The fact that it is the atomic number—the number of protons in the nucleus of an atom—that determines the order of elements in the table is very significant. We will see later that this has important implications with regard to the relationship between the number of electrons in an atom and the atom's chemical properties.

The modern periodic table is shown in Figure 2.12 and also appears on the inside front cover of the book. We will refer to the table frequently, so it is important for you to become familiar with it and with some of the terminology applied to it.

FIGURE 2.12

The modern periodic table.

	IA (1)	IIA (2)	IIIB (3)	IVB (4)	VB (5)	VIB (6)	VIIB (7)	VIII (8)	VIII (9)	VIII (10)	IB (11)	IIB (12)	IIIA (13)	IVA (14)	VA (15)	VIA (16)	VIIA (17)	Noble gases 0 (18)
1	1 H 1.00794																	2 He 4.00260
2	3 Li 6.941	4 Be 9.01218											5 B 10.81	6 C 12.011	7 N 14.00674	8 O 15.9994	9 F 18.99840	10 Ne 20.1797
3	11 Na 22.98977	12 Mg 24.3050											13 Al 26.98154	14 Si 28.0855	15 P 30.97376	16 S 32.066	17 Cl 35.4527	18 Ar 39.948
4	19 K 39.0983	20 Ca 40.078	21 Sc 44.9559	22 Ti 47.88	23 V 50.9415	24 Cr 51.9961	25 Mn 54.9380	26 Fe 55.847	27 Co 58.93320	28 Ni 58.69	29 Cu 63.546	30 Zn 65.39	31 Ga 69.723	32 Ge 72.61	33 As 74.92159	34 Se 78.96	35 Br 79.904	36 Kr 83.80
5	37 Rb 85.4678	38 Sr 87.62	39 Y 88.90585	40 Zr 91.224	41 Nb 92.90638	42 Mo 95.94	43 Tc 98.9072	44 Ru 101.07	45 Rh 102.90550	46 Pd 106.42	47 Ag 107.8682	48 Cd 112.411	49 In 114.82	50 Sn 118.710	51 Sb 121.75	52 Te 127.60	53 I 126.90447	54 Xe 131.29
6	55 Cs 132.90543	56 Ba 137.327	57 *La 138.9055	72 Hf 178.49	73 Ta 180.9479	74 W 183.85	75 Re 186.207	76 Os 190.2	77 Ir 192.22	78 Pt 195.08	79 Au 196.96654	80 Hg 200.59	81 Tl 204.3833	82 Pb 207.2	83 Bi 208.98037	84 Po 208.9824	85 At 209.9871	86 Rn 222.0176
7	87 Fr 223.0197	88 Ra 226.0254	89 †Ac 227.0278	104 Rf 261.11	105 Ha 262.114	106 Sg 263.118	107 Ns 262.12	108 Hs (265)	109 Mt (266)	110 Uun (269)	111 Uuu (272)							

Periods

Atomic number — 1 H 1.00794 — Atomic mass

*	58 Ce 140.115	59 Pr 140.90765	60 Nd 144.24	61 Pm 144.9127	62 Sm 150.36	63 Eu 151.965	64 Gd 157.25	65 Tb 158.92534	66 Dy 162.50	67 Ho 164.93032	68 Er 167.26	69 Tm 168.93421	70 Yb 173.04	71 Lu 174.967

†	90 Th 232.0381	91 Pa 231.0359	92 U 238.0289	93 Np 237.0482	94 Pu 244.0642	95 Am 243.0614	96 Cm 247.07003	97 Bk 247.0703	98 Cf 251.0796	99 Es 252.083	100 Fm 257.0951	101 Md 258.10	102 No 259.1009	103 Lr 260.105

Terminology Associated with the Periodic Table As in Mendeleev's table, the elements are arranged in rows called **periods,** but here they are arranged in order of increasing atomic number. For identification purposes the periods are numbered. We will find these numbers useful later on. Below the main body of the table are two long rows of 14 elements each. These actually belong in the main body of the table following La ($Z = 57$) and Ac ($Z = 89$), as shown in Figure 2.13. They are almost always placed below the table simply to conserve space. Fully spread out, the table is difficult to read if printed on one page. Notice that in the fully extended form of the table, with all the elements arranged in their proper locations, there is a great deal of empty space. An important requirement of a detailed atomic theory, which we will get to in Chapter 6, is that it must explain not only the repetition of properties, but also why there is so much empty space in the table.

Again, as in Mendeleev's table, the vertical columns are called **groups.** However, there is not uniform agreement among chemists on how they should be numbered. In the past, the groups were labeled with Roman numerals and divided into A groups and B groups, separated by the three short columns headed by Fe, Co, and Ni (Group VIII), as indicated in Figure 2.12. This corresponds quite closely to Mendeleev's original designations and is preferred by many chemists in the United States. However, another version of the table, popular in Europe, has the first seven groups from left to right labeled A, followed by the three short columns headed by Fe, Co, and Ni, and then the next seven groups labeled B. In an attempt to standardize the table, the International Union of Pure and Applied Chemistry (the IUPAC), an international body of scientists responsible for setting standards in chemistry, has officially adopted a third system in which the groups are simply numbered sequentially from left to right using Arabic numerals. Thus, Group IA in the old system is Group 1 in the IUPAC table, and Group VIIA in the old system is Group 17 in the IUPAC table. In Figure 2.12 and on the inside front cover of the book, we have used both the older U.S. labels as well as those preferred by the IUPAC. Because of the lack of uniform agreement among chemists on how the groups should be specified, and because many chemists still prefer the more traditional A group/B group designations in Figure 2.12, we will use just the latter when we wish to specify a particular group.

As we have already noted, the elements in a given group bear similarities to each other. Because of such similarities, groups are sometimes referred to as **families of elements.** The elements in the longer columns (the A groups and Group 0) are known as the **representative elements.** Those that fall into the B groups in the center of the table are called **transition elements.** The elements in the two long rows below the main body of the table are called the **inner**

Recall that the symbol *Z* stands for atomic number.

If you continue in chemistry you will certainly need to know the customary U.S. system because so many references use it.

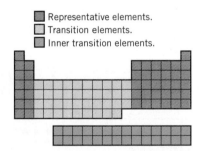

■ Representative elements.
□ Transition elements.
■ Inner transition elements.

FIGURE 2.13

Extended form of the periodic table. The two long rows of elements below the main body of the table in Figure 2.12 are placed in their proper places in this table.

1 H																	2 He
3 Li	4 Be											5 B	6 C	7 N	8 O	9 F	10 Ne
11 Na	12 Mg											13 Al	14 Si	15 P	16 S	17 Cl	18 Ar
19 K	20 Ca	21 Sc	22 Ti	23 V	24 Cr	25 Mn	26 Fe	27 Co	28 Ni	29 Cu	30 Zn	31 Ga	32 Ge	33 As	34 Se	35 Br	36 Kr
37 Rb	38 Sr	39 Y	40 Zr	41 Nb	42 Mo	43 Tc	44 Ru	45 Rh	46 Pd	47 Ag	48 Cd	49 In	50 Sn	51 Sb	52 Te	53 I	54 Xe
55 Cs	56 Ba	57 La	72 Hf	73 Ta	74 W	75 Re	76 Os	77 Ir	78 Pt	79 Au	80 Hg	81 Tl	82 Pb	83 Bi	84 Po	85 At	86 Rn
87 Fr	88 Ra	89 Ac	104 Rf	105 Ha	106 Sg	107 Ns	108 Hs	109 Mt	110 Uun	111 Uuu							

58 Ce	59 Pr	60 Nd	61 Pm	62 Sm	63 Eu	64 Gd	65 Tb	66 Dy	67 Ho	68 Er	69 Tm	70 Yb	71 Lu
90 Th	91 Pa	92 U	93 Np	94 Pu	95 Am	96 Cm	97 Bk	98 Cf	99 Es	100 Fm	101 Md	102 No	103 Lr

The IUPAC scheme also recognizes the groupings called the lanthanide and actinide series.

transition elements, and each row is named after the element that it follows in the main body of the table. For example, elements 58–71 are called the **lanthanide elements** because they follow lanthanum ($Z = 57$) and elements 90–103 are called the **actinide elements** because they follow actinium ($Z = 89$).

Some of the groups have acquired common names. For example, except for hydrogen, the Group IA elements are metals. They form compounds with oxygen that dissolve in water to give solutions that are strongly alkaline, or caustic. As a result, they are called the **alkali metals,** or simply the alkalis. The Group IIA elements are also metals. Their oxygen compounds are alkaline, too, but many compounds of the Group IIA elements are unable to dissolve in water and are found in deposits in the ground. Because of their properties and where they occur in nature, the Group IIA elements became known as the **alkaline earth metals.**

On the right side of the table, in Group 0, are the **noble gases.** They used to be called the inert gases until it was discovered that the heavier members of the group show a small degree of reactivity. The term *noble* is used when we wish to suggest a very limited degree of chemical reactivity. Gold, for instance, is often referred to as a noble metal because so few chemicals are capable of reacting with it.

Finally, the elements of Group VIIA are called the **halogens,** derived from the Greek words meaning sea or salt. Chlorine (Cl), for example, is found in familiar table salt, NaCl, a compound that accounts in large measure for the salty taste of sea water.

■ **Practice Exercise 9** Circle the correct choices.
(a) Representative elements are: K, Cr, Pr, Ar, Al.
(b) A halogen is: Na, Fe, O, Cl, Cu.
(c) An alkaline earth metal is: Rb, Ba, La, As, Kr.
(d) A noble gas is: H, Ne, F, S, N.
(e) An alkali metal is: Zn, Ag, Br, Ca, Li.
(f) An inner transition element is: Ce, Pb, Ru, Xe, Mg.

2.5 METALS, NONMETALS, AND METALLOIDS

The periodic table organizes all sorts of chemical and physical information about the elements and their compounds. It allows us to study systematically the way properties vary with an element's position within the table and, in turn, makes the similarities and differences among the elements easier to understand and remember.

Even a casual inspection of samples of the elements reveals that some are familiar metals and that others, equally familiar, are not metals. Most of us are already familiar with metals such as lead, iron, or gold and nonmetals such as oxygen or nitrogen. A closer look at the nonmetallic elements, though, reveals that some of them, silicon and arsenic to name two, have properties that lie between those of true metals and true nonmetals. These elements are called **metalloids.** Division of the elements into the categories of metals, nonmetals, and metalloids is not even, however (see Figure 2.14). Most elements are metals, slightly over a dozen are nonmetals, and only a handful are metalloids.

Notice that the metalloids are grouped around the bold stair-step line that is drawn diagonally from boron (B) to astatine (At).

Metals

You probably know a metal when you see one. Metals tend to have a shine so unique that it's called a *metallic luster.* For example, the silvery sheen of the freshly exposed surface of sodium in Figure 1.1 (page 2) would most likely

	IA (1)	IIA (2)	IIIB (3)	IVB (4)	VB (5)	VIB (6)	VIIB (7)	VIII (8)	VIII (9)	VIII (10)	IB (11)	IIB (12)	IIIA (13)	IVA (14)	VA (15)	VIA (16)	VIIA (17)	0 (18) Noble gases
1	H																	He
2	Li	Be											B	C	N	O	F	Ne
3	Na	Mg											Al	Si	P	S	Cl	Ar
4	K	Ca	Sc	Ti	V	Cr	Mn	Fe	Co	Ni	Cu	Zn	Ga	Ge	As	Se	Br	Kr
5	Rb	Sr	Y	Zr	Nb	Mo	Tc	Ru	Rh	Pd	Ag	Cd	In	Sn	Sb	Te	I	Xe
6	Cs	Ba	*La	Hf	Ta	W	Re	Os	Ir	Pt	Au	Hg	Ti	Pb	Bi	Po	At	Rn
7	Fr	Ra	†Ac	Rf	Ha	Sg	Ns	Hs	Mt	Uun	Uuu							

Periods

*	Ce	Pr	Nd	Pm	Sm	Eu	Gd	Tb	Dy	Ho	Er	Tm	Yb	Lu
†	Th	Pa	U	Np	Pu	Am	Cm	Bk	Cf	Es	Fm	Md	No	Lr

☐ Metals
☐ Nonmetals
☐ Metalloids

FIGURE 2.14

Distribution of metals, nonmetals, and metalloids among the elements in the periodic table.

lead you to identify sodium as a metal even if you had never seen or heard of it before. We also know that metals conduct electricity. Few of us would hold an iron nail in our hand and poke it into an electrical outlet. In addition, we know that metals conduct heat very well. On a cool day, metals always feel colder to the touch than do neighboring nonmetallic objects because metals conduct heat away from your hand very rapidly. Nonmetals seem less cold because they can't conduct heat away as quickly and therefore their surfaces warm up faster.

Other properties that metals possess, to varying degrees, are **malleability**—the ability to be hammered or rolled into thin sheets—and **ductility**—the ability to be drawn into wire. The production of sheet steel (Figure 2.15) for automobiles and household appliances depends on the malleability of iron

Gold is so malleable that it can be hammered into sheets that are only 1/280,000 of an inch thick.

FIGURE 2.15

The production of sheet steel for use in automobiles and household appliances begins with hot bars of steel that are rolled thinner and thinner by powerful rollers. The entire process relies on the malleability of iron.

FIGURE 2.16

The ductility of copper allows it to be drawn into wire. Here copper wire passes through one die after another as it is drawn into thinner and thinner wire.

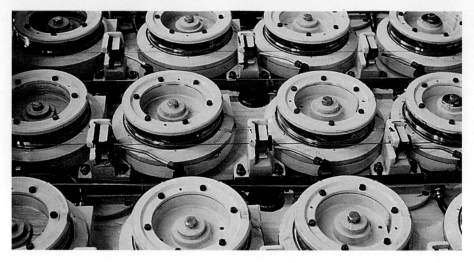

Thin lead sheets are used for sound deadening because the easily deformed lead absorbs the sound's vibrations.

FIGURE 2.17

A bead of liquid mercury on a porcelain surface.

We use the term *free element* to mean an element that is not chemically combined with any other element.

and steel, and the manufacture of electrical wire (Figure 2.16) is based on the ductility of copper.

Another important physical property that we usually think of for metals is hardness. Some metals, such as chromium or iron, are indeed quite hard; but others, like copper and lead, are rather soft. The alkali metals are so soft they can be cut with a knife, but they are also so chemically reactive that we rarely get to see them as free elements.

All the metallic elements, except mercury, are solids at room temperature (Figure 2.17). Mercury's low freezing point (− 39 °C) and fairly high boiling point (357 °C) make it useful as a fluid in thermometers. Most of the other metals have much higher melting points, and some are used primarily because of this. Tungsten, for example, has the highest melting point of any metal (3400 °C, or 6150 °F), which explains its use as filaments that glow white hot in electric light bulbs.

The chemical properties of metals vary tremendously. Some, such as gold and platinum, are very unreactive toward almost all chemical agents. This property, plus their natural beauty and rarity, makes them highly prized for use in jewelry. Other metals, however, are so reactive that few people except chemistry students ever get to see them in their "free" states. For instance, in Chapter 1 you learned that the metal sodium reacts very quickly with oxygen or moisture in the air, and its bright metallic surface tarnishes almost immediately. By contrast, compounds of sodium are quite stable and very common. Examples are table salt ($NaCl$), baking soda ($NaHCO_3$), lye ($NaOH$), and bleach ($NaOCl$). We will have more to say about the chemical properties of metals in Section 2.7.

Nonmetals

Substances such as plastics, wood, and glass that lack the properties of metals are said to be *nonmetallic,* and an element that has nonmetallic properties is called a **nonmetal.** Most often, we encounter the nonmetals in the form of compounds or mixtures of compounds. There are some, however, that are very important to us in their elemental forms. The air we breathe, for instance, contains mostly nitrogen, N_2, and oxygen, O_2. Both are gaseous, colorless, and odorless nonmetals. Since we can't see, taste, or smell them, however, it's difficult to experience their existence. (Although if you step into an atmosphere without oxygen, your body will very quickly tell you that something is missing!) Probably the most commonly *observed* nonmetallic element is carbon. We find it as the graphite in pencils, as coal, and as the charcoal used for barbecues. It also occurs in a more valuable form as diamond. Although diamond and graphite differ in appearance, each is a form of elemental carbon.

Photographs of some of the nonmetallic elements appear in Figure 2.18. Their properties are almost completely opposite those of metals. Each of these elements lacks the characteristic appearance of a metal. They are poor conductors of heat and, with the exception of the graphite form of carbon, are also poor conductors of electricity. The electrical conductivity of graphite appears to be an accident of molecular structure, since the structures of metals and graphite are completely different.

Many of the nonmetals are solids at room temperature and atmospheric pressure, while many others are gases. All of the Group 0 elements are gases in which the particles consist of single atoms. The other gaseous elements—hydrogen, oxygen, nitrogen, fluorine, and chlorine—are composed of diatomic molecules. Their formulas are H_2, O_2, N_2, F_2, and Cl_2. Bromine and iodine are also diatomic, but bromine is a liquid and iodine is a solid at room temperature. (The formulas of the diatomic nonmetals were mentioned on page 41.)

The nonmetallic elements lack the malleability and ductility of metals. A lump of sulfur crumbles when hammered and breaks apart when pulled on. Diamond cutters rely on the brittle nature of carbon when they split a gem-quality stone by carefully striking a quick blow with a sharp blade.

As with metals, nonmetals exhibit a broad range of chemical reactivity. Fluorine, for instance, is extremely reactive. It reacts readily with almost all the other elements. At the other extreme is helium, the gas used to inflate children's balloons and the Goodyear blimp. This element does not react with anything. Chemists find helium useful when they want to provide a totally *inert* (unreactive) atmosphere inside some apparatus.

Metalloids

The properties of metalloids lie between those of metals and nonmetals. This shouldn't surprise us since the metalloids are located between the metals and the nonmetals in the periodic table. In most respects, metalloids behave as nonmetals, both chemically and physically. However, in their most important physical property, electrical conductivity, they somewhat resemble metals. Metalloids tend to be **semiconductors;** they conduct electricity, but not nearly so well as metals. This property, particularly as found in silicon and germanium, is responsible for the remarkable progress made during the last two decades in the field of solid-state electronics. Virtually every stereo system, television receiver, and AM–FM radio relies heavily on transistors made from semiconductors. Perhaps the most amazing advance of all has been the fantastic reduction in the size of electronic components that semiconductors have allowed. To semiconductors we owe the development of small and versatile TV cameras, compact disc players, hand-held calculators, and microcomputers.

Diamonds such as these are simply another form of the element carbon.

Mercury and bromine are the only two liquid elements at room temperature and pressure.

FIGURE 2.18

Some nonmetallic elements. In the bottle on the left is dark red liquid bromine, which vaporizes easily to give a deeply colored orange vapor. Pale green chlorine fills the round flask in the center. Solid iodine lines the bottom of the flask on the right and gives off a violet vapor. Powdered red phosphorus occupies the dish in front of the flask of chlorine, and black-powdered graphite is in the watch glass. Also shown are lumps of yellow sulfur.

FIGURE 2.19

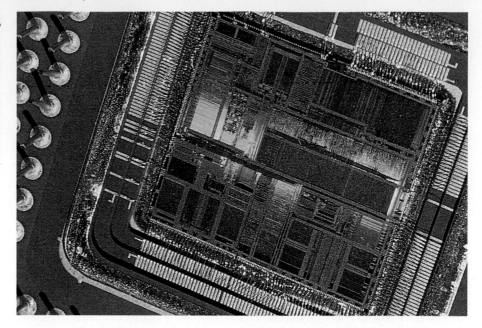

The heart of this tiny electronic device is the microcircuit in the center, which is set into the surface of a tiny silicon chip. Silicon's semiconductor properties make possible microelectronic devices such as this.

The heart of these devices is a microcircuit embedded in the surface of a tiny silicon chip (Figure 2.19).

Trends in the Periodic Table

The symbol below, shaped like the periodic table, will be placed in the margin when we wish to call your attention to a correlation between the location of an element in the periodic table and some property of the element or its compounds. Here we note the correlation between metallic character and the location of the elements in the periodic table.

The occurrence the metalloids between the metals and the nonmetals is our first example of trends in properties within the periodic table. We will see frequently that as we move from position to position across a period or down a group, chemical and physical properties change in a more or less regular fashion. There are few abrupt changes in the characteristics of the elements as we scan across a period or down a group. The location of the metalloids can be seen, then, as an example of the gradual transition between metallic and nonmetallic properties. From left to right across period 3, we go from aluminum, an element that has every appearance of a metal; to silicon, a semiconductor; to phosphorus, an element with clearly nonmetallic properties. A similar gradual change is seen going down Group IVA. Carbon is certainly a nonmetal, silicon and germanium are metalloids, and tin and lead are metals. Trends such as these are useful to spot, because they help us remember properties.

2.6
REACTIONS OF THE ELEMENTS: MOLECULAR AND IONIC COMPOUNDS

A property possessed by nearly every element is the ability to combine with other elements to form compounds. Not all combinations appear to be possible, however. For example, sodium reacts vigorously with chlorine to form sodium chloride, but no compound is formed between sodium and iron. When we examine the chemical properties of the elements, though, we do find that certain generalizations are possible. It is worth looking at these now for a number of reasons. First, they provide a general framework within which we can study chemical behavior in greater detail later. Second, they illustrate how the periodic table can make it easier to organize and remember chemical properties. And finally, they show us that to *understand* the chemical behavior of the elements, we need a theoretical model that explains what actually occurs between atoms when they react.

Molecular and Ionic Compounds

Atoms combine with each other in two broad general ways: either by the sharing of electrons between atoms or by the transfer of one or more electrons from one atom to another. Electron sharing gives discrete electrically neutral particles that we call **molecules,** whereas electron transfer produces electrically charged particles that we call **ions.** Compounds composed of molecules are called *molecular compounds* and those composed of ions are called *ionic compounds.*

Molecular Compounds   Some molecular compounds are made up of molecules that contain as few as two atoms. As you learned earlier, they are referred to as diatomic molecules. An example is the compound carbon monoxide, CO, a poisonous gas. (As you probably know, it is one of the substances in the exhaust of gasoline engines and in cigarette smoke.) Most molecules are more complex, however, and contain more atoms. Molecules of water, for example, have the formula H_2O and those of ordinary table sugar have the formula $C_{12}H_{22}O_{11}$. There also are molecules that are very large, such as those that occur in plastics and in living organisms, some of which contain millions of atoms. The formulas that describe the compositions of molecules are called **molecular formulas.**

On page 41 you learned that some elements occur as diatomic molecules.

The attractions that hold atoms to each other in compounds are called **chemical bonds,** and in molecules these attractions are strong enough that the group of atoms that make up a molecule move about together and behave as a single particle.

Molecular Compounds of Nonmetals

The kinds of compounds that form when elements react with each other often can be correlated with the locations of the combining elements in the periodic table. For example, when the elements that combine are nonmetals, located in the upper right corner of the periodic table (plus hydrogen, in Group IA), molecular substances are produced. Nonmetals combine with each other in a variety of ways to give molecules of varying degrees of complexity, which reaches a maximum with compounds in which carbon is combined with a handful of other elements such as hydrogen, oxygen, and nitrogen. There are so many of these compounds, in fact, that their study encompasses the chemical specialties called organic chemistry and biochemistry.

Hydrogen is a special case. It is not a metal like the other Group IA elements.

Molecular Formulas of the Nonmetals

	Group Number		
IVA	VA	VIA	VIIA
C^a	N_2	O_2	F_2
	P_4	S_8	Cl_2
	As_4	Se_8	Br_2
			I_2

a Carbon forms crystals of graphite and diamond, which contain enormous numbers of atoms linked in either a two- or three-dimensional interlocking network, and large cage-like molecules called fullerenes, with formulas such as C_{60}.

Even in their elemental states, most nonmetals exist as molecules. The only ones that do not are the noble gases, which occur as simple uncombined atoms. The formulas of the nonmetals in Groups IVA through VIIA are shown in the table in the margin.

At this early stage we can only begin to look for signs of order among the many nonmetal–nonmetal compounds, so we will look only briefly at some simple compounds that the nonmetals form with hydrogen and oxygen, as well as some simple compounds of carbon.

TABLE 2.4 Simple Hydrogen Compounds of the Nonmetallic Elements

Period	IVA	VA	VIA	VIIA
		Group		
2	CH_4	NH_3	H_2O	HF
3	SiH_4	PH_3	H_2S	HCl
4	GeH_4	AsH_3	H_2Se	HBr
5		SbH_3	H_2Te	HI

TABLE 2.5 Simplest Formulas of Some Oxides of the Nonmetallic Elements

IVA	VA	VIA
	Group	
Carbon	*Nitrogen*	
CO_2	N_2O_3	
	N_2O_5	
Silicon	*Phosphorus*	*Sulfur*
SiO_2	P_2O_3	SO_2
	P_2O_5	SO_3
Germanium	*Arsenic*	*Selenium*
GeO_2	As_2O_3	SeO_2
	As_2O_5	SeO_3
	Antimony	*Tellurium*
	Sb_2O_3	TeO_2
	Sb_2O_5	TeO_3

Many of the nonmetals form more complex compounds with hydrogen, but we will not discuss them here.

Compounds of Nonmetals with Hydrogen and Oxygen Compounds that elements form with hydrogen are called *hydrides*. The formulas of the simple hydrides of the nonmetals are given in Table 2.4.[3] Notice that the formulas are similar for nonmetals within a given group of the periodic table. If you can remember the formula for the hydride of the top member of the group, then you know the formulas of all of them in that group.

The oxygen compounds of the nonmetals, which we call *oxides*, are more complex than the hydrides and many nonmetals form more than one of them. Sulfur, for example, forms two oxides, SO_2 and SO_3, both of which are industrially important and both of which are also serious air pollutants. Nitrogen oxides are even more numerous: NO, NO_2, N_2O, N_2O_3, N_2O_4, and N_2O_5. Despite this complexity there is still some order to be found, as we can see in Table 2.5, in which we have included formulas of some of the oxides of the nonmetals and metalloids. Patterns such as those shown in the table also helped Mendeleev fix the locations of the nonmetallic elements in their respective groups in the periodic table.

Our goal at this time is to acquaint you with some of the important kinds of organic compounds we encounter regularly, so our discussion here is brief.

Compounds of Carbon Among all the elements, carbon is unique in its ability to form strong bonds both to itself and to other elements such as hydrogen, oxygen, and nitrogen. As a consequence, the number and complexity of such compounds is enormous, and their study constitutes the major specialty called **organic chemistry.** The term *organic* here comes from an early belief that these compounds could only be made by living organisms. We now know this isn't true, but the name organic chemistry persists nonetheless.

[3] This table shows how the formulas are normally written. The order in which the hydrogens appear in the formula is not of concern to us now. Instead, we are interested in the *number* of hydrogens that combine with a given nonmetal.

TABLE 2.6 Hydrocarbons Belonging to the Alkane Series

Compound	Name	Boiling Point (°C)
CH_4	Methane[a]	−161.5
C_2H_6	Ethane[a]	−88.6
C_3H_8	Propane[a]	−42.1
C_4H_{10}	Butane[a]	−0.5
C_5H_{12}	Pentane	36.1
C_6H_{14}	Hexane	68.7

[a] Gases at room temperature (25 °C) and atmospheric pressure.

The study of organic chemistry begins with **hydrocarbons** (compounds of carbon and hydrogen). The simplest hydrocarbon is methane, CH_4, which is a member of a series of hydrocarbons with the general formula C_nH_{2n+2}, where n is a whole number. The first six members of this series, called the **alkane** series, are given in Table 2.6 along with their boiling points. Molecules of methane, ethane, and propane are illustrated in Figure 2.20.

The alkanes are common compounds. They are the principal constituents of petroleum and most of our useful fuels are produced from petroleum. Methane itself is the major component of natural gas. Gas-fired barbecues use propane as a fuel, and butane is the fuel in inexpensive cigarette lighters. Hydrocarbons with higher boiling points are found in gasoline, kerosene, paint thinners, diesel fuel, and even candle wax.

Alkanes are not the only class of hydrocarbons. For example, there are three two-carbon hydrocarbons. In addition to ethane, C_2H_6, there are also ethylene, C_2H_4 (from which polyethylene is made), and acetylene, C_2H_2 (the fuel used in *acetylene* torches).

The hydrocarbons serve as the foundation for organic chemistry. Derived from hydrocarbons are various other classes of organic compounds. An example is the class of compounds called alcohols, in which the atoms OH replace a hydrogen in the hydrocarbon. For example, *methanol*, CH_3OH (also called *methyl alcohol*), can be seen to be related to methane, CH_4, if one H is removed and replaced with OH. Methanol is used as a fuel and as a raw material for making other organic chemicals. Another familiar alcohol is *ethanol* (also called *ethyl alcohol*), C_2H_5OH. Ethanol, also known as grain alcohol because it is obtained from the fermentation of grains, is in alcoholic beverages. It is also mixed with gasoline to give a fuel mixture known as gasohol.

Alcohols constitute just one kind of compound derived from hydrocarbons. We will discuss some others when you have learned more about how atoms bond to each other and about the structures of molecules.

The chemically correct names for ethylene and acetylene are ethene and ethyne, respectively. Naming organic compounds is discussed in Chapter 21.

Methanol is also known as wood alcohol. It is quite poisonous. Ethanol in high doses is also a poison.

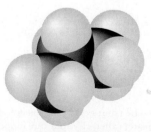

Methane, CH_4 Ethane, C_2H_6 Propane, C_3H_8

FIGURE 2.20

The first three members of the alkane series of hydrocarbons. White atoms represent hydrogen.

2.7
IONIC COMPOUNDS

Here we are concentrating on what happens to the individual atoms, so we have not shown chlorine as diatomic Cl_2 molecules.

Na = sodium atom
Na^+ = sodium ion

A neutral sodium atom has 11 protons and 11 electrons. A sodium ion has 11 protons and 10 electrons.

An important property of metals is their ability to combine chemically with nonmetals to form ionic compounds. In this reaction, one or more electrons are transferred from the metal atom to nonmetal atoms. This is what happens, for example, when sodium comes in contact with chlorine. Each sodium atom gives up one electron to a chlorine atom, which thereby gains one. We can diagram these changes in equation form by using the symbol e^- to stand for an electron.

$$\overset{\overset{\displaystyle e^-}{\frown}}{Na + Cl} \longrightarrow Na^+ + Cl^-$$

The electrically charged particles formed in this reaction are called **ions,** specifically a sodium ion (Na^+) and a chloride ion (Cl^-). The sodium ion has a positive 1+ charge, indicated by the superscript plus sign, because it now has one more proton in its nucleus than there are electrons outside. Similarly, by gaining one electron the chlorine atom has added one more negative charge, so the chloride ion has a single negative charge indicated by the minus sign. In referring to these particles we will frequently call a positively charged ion a **cation** (pronounced *CAT-ion*) and a negatively charged ion an **anion** (pronounced *AN-ion*).[4] Solid sodium chloride is composed of these charged sodium and chloride ions and is said to be an **ionic compound.**

Figure 2.21 illustrates the structures of water and sodium chloride and demonstrates an important difference between molecular and ionic compounds. In water it is safe to say that two hydrogen atoms "belong" to each oxygen atom in a particle having the formula H_2O. However, in NaCl it is impossible to say that a particular Na^+ ion belongs to a particular Cl^- ion. The ions in a crystal of NaCl are simply stacked in the most efficient way, so that positive ions and negative ions can be as close to each other as possible. In this

FIGURE 2.21

(*a*) In water there are discrete molecules that each consist of one atom of oxygen and two atoms of hydrogen. Each particle has the formula H_2O. (*b*) In sodium chloride, ions are packed in the most efficient way. Each Na^+ is surrounded by six Cl^-, and each Cl^- is surrounded by six Na^+. Because individual molecules do not exist, we simply specify the ratio of ions as NaCl.

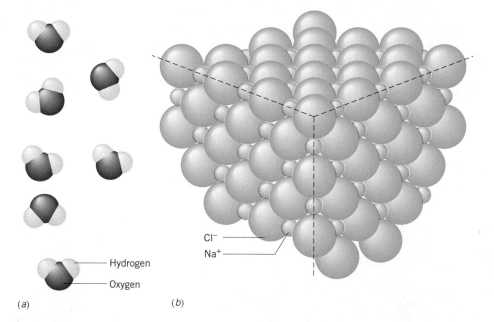

Hydrogen

Oxygen

Cl^-
Na^+

(*a*)   (*b*)

[4] The names *cation* and *anion* come from the way the ions behave when electrically charged metal plates called electrodes are dipped into a solution that contains them. We will discuss this in detail in Chapter 18.

way, the attractions between oppositely charged ions, which are responsible for holding the compound together, can be as strong as possible.

Because molecules don't exist in ionic compounds, the subscripts in their formulas are always chosen to specify the smallest whole-number ratio of the ions. This is why the formula of sodium chloride is given as NaCl rather than Na_2Cl_2 or Na_3Cl_3. Although the smallest unit of an ionic compound can't be called a molecule, the idea of "smallest unit" is still quite often useful. Therefore, we take the smallest unit of an ionic compound to be whatever is represented in its formula and call this unit a **formula unit.** Thus, one formula unit of NaCl consists of one Na^+ and one Cl^-, whereas one formula unit of the ionic compound $CaCl_2$ consists of one Ca^{2+} and two Cl^- ions. (In a broader sense, we can use the term *formula unit* to refer to whatever is represented by a formula. Sometimes the formula specifies a set of ions, as in NaCl; sometimes it is a molecule, as in O_2 or H_2O; sometimes it can be just an ion, as in Cl^- or Ca^{2+}; and sometimes it may be just an atom, as in Na.)

Notice that the charges on the ions are omitted in writing the formula.

Ions Formed by Representative Metals and Nonmetals

We will have to wait until a later chapter to study the reasons why certain atoms gain or lose one electron each, whereas other atoms gain or lose two or more electrons. For now, however, we can use the periodic table as the basis for some generalizations that can help us remember the kinds of ions formed by many of the representative elements. For example, except for hydrogen, the neutral atoms of each of the Group IA elements always lose one electron when they react, thereby becoming ions with a charge of 1+. Similarly, atoms of the Group IIA elements always lose two electrons when they react; so these elements always form ions with a charge of 2+. In Group IIIA, the only important positive ion we need consider now is that of aluminum, Al^{3+}; an aluminum atom loses three electrons when it reacts to form this ion.

Positive ions are formed by metals.

All these ions are listed in Table 2.7. *Notice that the number of positive charges on each of these cations is the same as the group number when we use the traditional numbering of groups in the periodic table.* Thus, sodium is in Group IA and forms an ion with a 1+ charge, barium (Ba) is in Group IIA and forms an ion with a 2+ charge, and aluminum is in Group IIIA and forms an ion with a 3+ charge. Although this generalization doesn't work for all the metallic elements (it doesn't work for the transition elements, for instance), it does help us remember what happens to the metallic elements of Groups IA and IIA and aluminum when they react.

Among the nonmetals on the right side of the periodic table we also find some useful generalizations. For example, when they combine with metals, the halogens (Group VIIA) form ions with a 1− charge and the nonmetals in

TABLE 2.7 Some Ions Formed from the Representative Elements

Group Number						
IA	IIA	IIIA	IVA	VA	VIA	VIIA
Li^+	Be^{2+}		C^{4-}	N^{3-}	O^{2-}	F^-
Na^+	Mg^{2+}	Al^{3+}	Si^{4-}	P^{3-}	S^{2-}	Cl^-
K^+	Ca^{2+}				Se^{2-}	Br^-
Rb^+	Sr^{2+}				Te^{2-}	I^-
Cs^+	Ba^{2+}					

Formulas of cations and anions.

Negative ions are formed by the nonmetals when they combine with metals.

Group VIA form ions with a 2− charge. Notice that *the number of negative charges on the anion is equal to the number of spaces to the right that we have to move in the periodic table to get to a noble gas.*

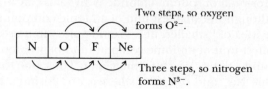

Two steps, so oxygen forms O^{2-}.

Three steps, so nitrogen forms N^{3-}.

Writing Formulas for Ionic Compounds

Because all chemical compounds are electrically neutral, the ions in an ionic compound always occur in a ratio such that the total positive charge is equal to the total negative charge. This is why the formula for sodium chloride is NaCl; the 1 to 1 ratio of Na^+ to Cl^- gives electrical neutrality. In addition, as we've already mentioned, discrete molecules do not exist in ionic compounds, so we always use the smallest set of subscripts that specify the correct ratio of the ions. The following, therefore, are the rules we use in writing the formulas of ionic compounds.

A substance is electrically neutral, with a net charge of zero, if the total positive charge equals the total negative charge.

Rules for writing formulas of ionic compounds.

Rules for Writing Formulas of Ionic Compounds

1. The positive ion is given first in the formula. (This isn't required by nature, but it is a custom we always follow.)

2. The subscripts in the formula must produce an electrically neutral formula unit. (Nature does require electrical neutrality.)

3. The subscripts should be the smallest set of whole numbers possible.

EXAMPLE 2.3
Writing Formulas for Ionic Compounds

Write the formulas for the ionic compounds formed from (a) Ba and S, (b) Al and Cl, and (c) Al and O.

SOLUTION: In each case, the ions must be combined in a ratio that produces an electrically neutral formula unit with the smallest set of whole-number subscripts.

(a) The ions here are Ba^{2+} and S^{2-}. Since the charges are equal but opposite, a 1 to 1 ratio will give a neutral formula unit. Therefore, the formula is BaS.

(b) For these elements, the ions are Al^{3+} and Cl^-. We can obtain a neutral formula unit by combining one Al^{3+} with three Cl^-. (The charge on Cl is 1−; the 1 is understood.)

$$1(3+) + 3(1-) = 0$$

The formula is $AlCl_3$.

(c) For these elements, the ions are Al^{3+} and O^{2-}. In the formula we seek there must be the same number of positive charges as negative charges. This number must be a whole-number multiple of both 3 and 2. The smallest number that satisfies this condition is 6, so there must be two Al^{3+} and three O^{2-} in the formula.

$$2(3+) = 6+$$
$$3(2-) = 6-$$
$$\overline{\text{Sum} = 0}$$

The formula is Al_2O_3.

■ **Practice Exercise 10** Write formulas for ionic compounds formed from (a) Na and F, (b) Na and O, (c) Mg and F, and (d) Al and C.

There is another rather simple way to obtain the formulas of the compounds in Example 2.3. The procedure is to make the subscript for one ion equal to the number of charges on the other. For example, for Al^{3+} and Cl^-, we can write

$$Al^{③+} \times Cl^{①-}$$

which gives Al_1Cl_3 or simply $AlCl_3$.

For the ions Al^{3+} and O^{2-} we write

$$Al^{③+} \times O^{②-}$$

This gives the formula Al_2O_3.

If you are not careful, you can be fooled in following this procedure. For example, if you apply this method to the compound formed from the ions Ba^{2+} and S^{2-}, it gives the formula Ba_2S_2. However, by convention we always choose the smallest whole-number ratio of ions (Rule 3 above). Notice in Ba_2S_2 that both subscripts are divisible by 2. Therefore, to obtain the correct formula we reduce the subscripts to the smallest set of whole numbers, which gives BaS. Exercising appropriate care, you might go back and try this method on Practice Exercise 10.

Many of our most important chemicals are ionic compounds. We have mentioned NaCl, common table salt, and $CaCl_2$, a substance used to melt ice on walkways in the winter and to keep dust down on dirt roads in the summer. Other examples are sodium fluoride, NaF, used by dentists to give fluoride treatments to teeth, and calcium oxide, CaO, an important ingredient in cement.

Transition Metals and Post-transition Metals

The transition elements are located in the center of the periodic table, from Group IIIB on the left to Group IIB on the right. All of them lie to the left of the metalloids, and they all are metals. Included here are some of our most familiar metals, including iron, chromium, copper, silver, and gold.

Most of the transition metals are much less reactive than the metals of Groups IA and IIA, but when they react they also transfer electrons to nonmetal atoms to form ionic compounds. However, the charges on the ions of the transition metals do not follow as straightforward a pattern as do those of the alkali and alkaline earth metals. One of the characteristic features of the transition metals is the ability of many of them to form more than one positive ion. Iron, for example, can form two different ions, Fe^{2+} and Fe^{3+}. This means that iron can form more than one compound with a given nonmetal. For

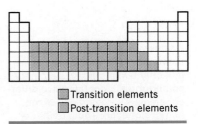

☐ Transition elements
☐ Post-transition elements

Distribution of transition and post-transition metals in the periodic table.

**TABLE 2.8   Ions of Some Transition Metals
and Post-transition Metals**

Transition Metals

Chromium	Cr^{2+}, Cr^{3+}
Manganese	Mn^{2+}, Mn^{3+}
Iron	Fe^{2+}, Fe^{3+}
Cobalt	Co^{2+}, Co^{3+}
Nickel	Ni^{2+}
Copper	Cu^{+}, Cu^{2+}
Zinc	Zn^{2+}
Silver	Ag^{+}
Cadmium	Cd^{2+}
Gold	Au^{3+}
Mercury	Hg_2^{2+}, Hg^{2+}

Post-transition Metals

Tin	Sn^{2+}, Sn^{4+}
Lead	Pb^{2+}, Pb^{4+}
Bismuth	Bi^{3+}

example, with chloride ion, Cl^-, iron forms two compounds, with the formulas $FeCl_2$ and $FeCl_3$. With oxygen, we find the compounds FeO and Fe_2O_3. As usual, we see that the formulas contain the ions in a ratio that gives electrical neutrality. Some of the most common ions of the transition metals are given in Table 2.8.

■ **Practice Exercise 11**   Write formulas for the chlorides and oxides formed by (a) chromium and (b) copper.

The prefix *post-* means "after."

The **post-transition metals** are those metals that occur in the periodic table immediately following a row of transition metals. The two most common and important ones are tin (Sn) and lead (Pb). These post-transition metals are quite different from the metals that precede the transition metals. One of the most significant differences is their ability to form two different ions, and therefore to form two different compounds with a given nonmetal. For example, tin forms two oxides, SnO and SnO_2. Lead also forms two oxides that have similar formulas (PbO and PbO_2). The ions that these metals form are also included in Table 2.8.

Compounds Containing Ions Composed of More than One Element

A substance is **diatomic** if it is composed of molecules that contain only two atoms. It is a **binary compound** if it contains two different elements, regardless of the number of each. Thus, BrCl is a binary compound and is also diatomic; CH_4 is a binary compound, but is not diatomic.

The metal compounds that we have discussed so far have been **binary compounds**—compounds formed between *two different elements*. There are many other ionic compounds that contain more than two elements. These substances usually contain **polyatomic ions,** which are ions that are themselves composed of two or more atoms linked by the same kinds of bonds that hold molecules together. Polyatomic ions differ from molecules, however, in that they contain either too many or too few electrons to make them electrically neutral. Table 2.9 lists some important polyatomic ions. The formulas of ionic compounds formed from them are determined in the same way as are those of binary ionic compounds; the ratio of the ions must be such that the formula unit is electrically neutral, and the smallest set of whole-number subscripts is used.

TABLE 2.9   Formulas and Names of Some Polyatomic Ions

Ion	Name (Alternate Name in Parentheses)	Ion	Name (Alternate Name in Parentheses)
NH_4^+	ammonium ion	CO_3^{2-}	carbonate ion
H_3O^+	hydronium ion[a]	HCO_3^-	hydrogen carbonate ion (bicarbonate ion)[b]
OH^-	hydroxide ion	SO_3^{2-}	sulfite ion
CN^-	cyanide ion	HSO_3^-	hydrogen sulfite ion (bisulfite ion)[b]
NO_2^-	nitrite ion	SO_4^{2-}	sulfate ion
NO_3^-	nitrate ion	HSO_4^-	hydrogen sulfate ion (bisulfate ion)[b]
ClO^-	hypochlorite ion (often written OCl^-)	SCN^-	thiocyanate ion
ClO_2^-	chlorite ion	$S_2O_3^{2-}$	thiosulfate
ClO_3^-	chlorate ion	CrO_4^{2-}	chromate ion
ClO_4^-	perchlorate ion	$Cr_2O_7^{2-}$	dichromate ion
MnO_4^-	permanganate ion	PO_4^{3-}	phosphate ion
$C_2H_3O_2^-$	acetate ion	HPO_4^{2-}	monohydrogen phosphate ion
$C_2O_4^{2-}$	oxalate ion	$H_2PO_4^-$	dihydrogen phosphate ion

[a] You will only encounter this ion in aqueous solutions.   [b] You will often see and hear the alternate names for these ions.

Write the formula for the ionic compound calcium phosphate, which is formed from Ca^{2+} and PO_4^{3-}.

SOLUTION: As before, we write the positive ion first and then make the number of charges on one ion equal to the subscript for the other.

$$Ca^{2+} \qquad PO_4^{3-}$$

The formula is written with parentheses to show that the PO_4^{3-} ion occurs three times in the formula unit.

$$Ca_3(PO_4)_2$$

EXAMPLE 2.4
Formulas That Contain Polyatomic Ions

■ **Practice Exercise 12** Write formulas for the ionic compound formed from (a) Na^+ and CO_3^{2-}, (b) NH_4^+ and SO_4^{2-}, (c) potassium ion and acetate ion, (d) strontium ion and nitrate ion, and (e) Fe^{3+} and acetate ion.

Polyatomic ions are found in a large number of very important compounds. Some common ones are $CaSO_4$ (calcium sulfate, in plaster of Paris), $NaHCO_3$ (sodium bicarbonate, also called baking soda), $NaOCl$ (sodium hypochlorite, in liquid laundry bleach), $NaNO_2$ (sodium nitrite, a meat preservative), $MgSO_4$ (magnesium sulfate, also known as Epsom salts), and $NH_4H_2PO_4$ (ammonium dihydrogen phosphate, a fertilizer).

The properties of ionic compounds reflect the way ions interact with each other. The attractive force between ions of opposite charge is very large, as is the repelling force between ions of like charge. Therefore, in an ionic compound the ions arrange themselves so that the attractions between oppositely charged ions are at a maximum and the repulsions between like-charged ions are at a minimum.

2.8
PROPERTIES OF IONIC AND MOLECULAR COMPOUNDS

FIGURE 2.22

An ionic crystal shatters when struck. If ions of like charge come face to face, the repulsions between them can force parts of the crystal apart.

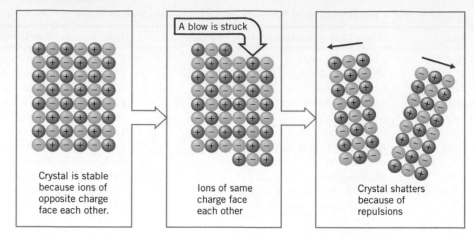

A blow is struck

Crystal is stable because ions of opposite charge face each other.

Ions of same charge face each other

Crystal shatters because of repulsions

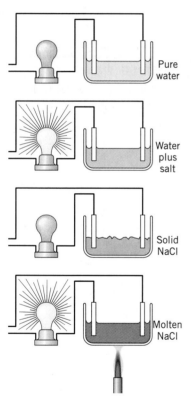

Pure water

Water plus salt

Solid NaCl

Molten NaCl

FIGURE 2.23

An apparatus to test for electrical conductivity. The electrodes are dipped into the substances to be tested. If the lightbulb glows when electricity is applied, the sample is an electrical conductor. Here we see that neither pure water (a molecular substance) nor solid sodium chloride conduct. Salt does conduct, however, if it is melted or dissolved in water.

In a crystal at room temperature, the ions jiggle about within a cage of other ions. They are held too tightly to move away. When heat is added to raise the temperature of the crystal, the ions have more kinetic energy and neighboring ions bounce off each other more violently. Eventually a temperature is reached at which the violent motions overcome the attractions between the ions and the crystal collapses—the compound melts. However, because the net attractions between the ions are so large, the temperature required is very high. This is why all ionic compounds are solids at room temperature and tend to have high melting points.

You have probably never seen an ionic substance melt. Most of them melt well above room temperature. For example, ordinary table salt, NaCl, melts at 801 °C. Some ionic substances melt only at extremely high temperatures. For instance, aluminum oxide, Al_2O_3, melts at about 2000 °C, and for this reason it is made into special bricks that are used to line the inside walls of furnaces.

Another property of ionic compounds is that their solids are generally relatively hard and quite brittle. Molecular substances, such as paraffin wax or car wax, tend to be soft and easily crushed, but a crystal of rock salt is much harder. When struck by a hammer, however, the salt crystal shatters. The slight movement of a layer of ions within an ionic crystal suddenly places ions of the *same* charge next to each other, and for that instant there are large repulsive forces that split the solid, as illustrated in Figure 2.22.

Electrical Properties

In the solid state, ionic compounds do not conduct electricity. This is because electrical conductivity requires the movement of electrical charges, and in the solid the attractive forces prevent the movement of ions through the crystal. When the solid is melted, however, the ions become free to move about and the liquid conducts electricity well.

This can be demonstrated experimentally with the help of the apparatus depicted in Figure 2.23, which consists of a pair of metal electrodes that can be dipped into a container holding a substance whose electrical conductivity we wish to test. One of the electrodes is wired to an electric light bulb that glows if electricity is able to pass between the two electrodes. When this apparatus is used to test the electrical conductivity of solid salt crystals, the bulb fails to light. However, when a flame is applied, the bulb lights brightly as soon as the salt melts.

The apparatus in Figure 2.23 can also be used to show that water solutions of ionic compounds conduct electricity. If the electrodes are dipped into pure distilled water, the bulb remains dark; pure water is not a conductor of electricity. If salt crystals are then added to the water and the mixture stirred, the solution that's formed conducts electricity well. The reason it conducts is that the ions become separated when the salt dissolves, and they are therefore free to move about. This freedom of movement of the charged ions permits the conduction of electricity[5] and has a profound influence on the reactions of ionic compounds.

Properties of Molecular Compounds

The properties of molecular substances usually differ markedly from those of ionic compounds. Within the individual molecules the atoms are held to each other very strongly, but between neighboring molecules the attractions are very weak. These weak attractions are responsible for many of the properties of molecular substances, just as the strong attractions between ions are responsible for many of the properties of ionic compounds. For example, molecular compounds such as water and candle wax tend to have low melting points. The molecules of these substances don't have to bounce around inside their crystals as violently as do the ions in an ionic crystal to overcome the attractive forces and become a liquid. Crystals of molecular compounds are also usually soft because the molecules easily slide past each other.

Molecular substances differ from ionic compounds in their electrical characteristics, too. Molecules are uncharged particles, so they do not conduct electricity in the solid state or when melted. Pure water, for example, does not conduct electricity. Most molecular substances also will not conduct electricity when dissolved in water. For example, if the electrodes of the conductivity apparatus shown in Figure 2.23 are dipped into a solution of sugar in water, the light bulb doesn't glow. Sugar molecules carry no electrical charge, so there are no electrical charges in the solution to provide conduction. As we will see later, however, there are certain kinds of molecules that react with water to give ions, and their solutions do conduct electricity.

> Notice how the notion of temperature being related to the average kinetic energy of the particles within a substance helps us explain the high melting points of ionic compounds.

In conversation, chemists rarely use formulas to describe compounds. Instead, names are used. For example, you already know that water is the name for the compound having the formula H_2O and that sodium chloride is the name of NaCl.

At one time there was no uniform procedure for assigning names to compounds, and those who discovered compounds used whatever method they wished. Without some sort of system, however, remembering names for the rapidly increasing number of compounds soon became impossible. The search for a solution led chemists around the world to agree on a systematic method for naming substances. As a result, we are able to write the correct formula for any given compound given the correct name, and vice versa.

In this section we discuss the **nomenclature** (naming) of simple **inorganic compounds.** In general, these are substances that would *not* be considered

2.9
INORGANIC CHEMICAL NOMENCLATURE

 Rules for naming compounds.

---

[5] This kind of conduction takes place by a different means than the conduction in metals and is described in more detail in Chapter 18.

to be derived from hydrocarbons such as methane (CH_4), ethane (C_2H_6), and other carbon–hydrogen compounds. As we discussed earlier, the hydrocarbons and compounds that can be thought of as coming from them are called organic compounds. We will have more to say about naming them in Chapter 21.

Binary Compounds Containing a Metal and a Nonmetal

For ionic compounds made from a metal and a nonmetal, the name of the cation is given first, followed by the name of the anion formed from the nonmetal. The latter is created by adding the suffix *-ide* to the stem of the name for the nonmetal. A familiar example is NaCl, sodium chloride. Table 2.10 lists some common *monatomic* (one-atom) negative ions and their names. Other examples of compounds formed from them are

CaO	calcium oxide
ZnS	zinc sulfide
Mg_3N_2	magnesium nitride

The *-ide* suffix is usually used only for monatomic ions, although there are two common exceptions—*hydroxide ion* (OH^-) and *cyanide ion* (CN^-).

TABLE 2.10 Monatomic Negative Ions

H^-	hydride	N^{3-}	nitride	O^{2-}	oxide	F^-	fluoride
C^{4-}	carbide	P^{3-}	phosphide	S^{2-}	sulfide	Cl^-	chloride
Si^{4-}	silicide	As^{3-}	arsenide	Se^{2-}	selenide	Br^-	bromide
				Te^{2-}	telluride	I^-	iodide

EXAMPLE 2.5
Naming Compounds and
Writing Formulas

(a) What is the name of $SrBr_2$? (b) What is the formula for aluminum selenide?

SOLUTION:

(a) The compound is composed of the ions Sr^{2+} and Br^-. The cation simply takes the name of the metal, which is strontium. The anion's name is derived from bromine by replacing *-ine* with *-ide;* it is the bromide ion. The name of the compound is strontium bromide.

(b) The name tells us that the cation is the aluminum ion, Al^{3+}, and the anion is the selenide ion, which is formed from selenium. This ion is Se^{2-}. The correct formula must represent an electrically neutral formula unit, so the formula is Al_2Se_3.

Many of the transition metals and post-transition metals are able to form more than one positive ion. Iron, a typical example, forms ions with either a 2+ or a 3+ charge (Fe^{2+} or Fe^{3+}). Compounds that contain these different iron ions have different formulas, so in their names it is necessary to specify which iron ion is present. There are two ways of doing this. In the old system, the suffix *-ous* is used to specify the ion with the lower charge and the suffix *-ic* is used to specify the ion with the higher charge. With this method, we use the Latin stem for elements whose symbols are derived from their Latin names.

Some examples are

Fe^{2+}	ferrous ion	$FeCl_2$	ferrous chloride
Fe^{3+}	ferric ion	$FeCl_3$	ferric chloride
Cu^+	cuprous ion	$CuCl$	cuprous chloride
Cu^{2+}	cupric ion	$CuCl_2$	cupric chloride

Additional examples are given in Table 2.11. Notice that mercury is an exception; we use the English stem when naming its ions.

TABLE 2.11   Metals That Form More than One Ion

Cr^{2+}	chromous	Mn^{2+}	manganous	Fe^{2+}	ferrous	Cu^+	cuprous
Cr^{3+}	chromic	Mn^{3+}	manganic	Fe^{3+}	ferric	Cu^{2+}	cupric
Hg_2^{2+}	mercurous	Sn^{2+}	stannous	Pb^{2+}	plumbous	Co^{2+}	cobaltous
Hg^{2+}	mercuric	Sn^{4+}	stannic	Pb^{4+}	plumbic	Co^{3+}	cobaltic

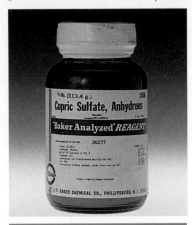

The older system of nomenclature is still found on the labels of many laboratory chemicals. This bottle contains copper(II) sulfate, which according to the older system is called cupric sulfate.

The currently preferred method for naming ions of metals that can have more than one charge in compounds is called the **Stock system.** Here we use the English name followed, *without a space,* by the numerical value of the charge written as a Roman numeral in parentheses.

Fe^{2+}	iron(II)	$FeCl_2$	iron(II) chloride
Fe^{3+}	iron(III)	$FeCl_3$	iron(III) chloride
Cr^{2+}	chromium(II)	CrS	chromium(II) sulfide
Cr^{3+}	chromium(III)	Cr_2S_3	chromium(III) sulfide

Alfred Stock (1876–1946), a German inorganic chemist, was one of the first scientists to warn the public of the dangers of mercury poisoning.

Copper(I) sulfate is Cu_2SO_4.
Copper(II) sulfate is $CuSO_4$.

Remember that the Roman numeral is the positive charge on the metal ion; it is not necessarily a subscript in the formula. You must figure out the formula from the ionic charges, as discussed in Section 2.7 and illustrated in the preceding example. Even though the Stock system is now preferred, some chemical companies still label bottles of chemicals using the old system. These old names also appear in the older scientific literature, which still holds much excellent data. Unfortunately, this means that you must know both systems.

■ **Practice Exercise 13** Name the compounds K_2S, Mg_3P_2, $NiCl_2$, and Fe_2O_3. Use the Stock system where appropriate.

■ **Practice Exercise 14** Write formulas for (a) aluminum sulfide, (b) strontium fluoride, (c) titanium(IV) oxide, and (d) chromous bromide.

Binary Compounds Between Two Nonmetals

In naming binary compounds containing two nonmetals, we usually use a method that specifies the actual numbers of atoms in a molecule. This system makes use of the following Greek prefixes.

mono-	= 1 (often omitted)	hexa-	= 6
di-	= 2	hepta-	= 7
tri-	= 3	octa-	= 8
tetra-	= 4	nona-	= 9
penta-	= 5	deca-	= 10

For example, NO_2 is nitrogen *di*oxide and N_2O_4 is *di*nitrogen *tetra*oxide (Sometimes we drop the *a* before an *o* for ease of pronunciation. N_2O_4 would then be named dinitrogen tetroxide.) Some other examples are:

HCl   hydrogen chloride (mono- $AsCl_3$ arsenic trichloride
              omitted)
CO   carbon monoxide SF_6 sulfur hexafluoride

■ **Practice Exercise 15** Name the following compounds using Greek prefixes when needed: PCl_3, SO_2, Cl_2O_7.

Binary Acids and Their Salts

Acids are substances that react with water to yield hydronium ions, H_3O^+, and anions. An example is hydrogen chloride, HCl. In the pure state, this compound is a gas and consists of molecules. When it dissolves in water, however, it reacts to yield ions according to the equation

$$HCl(g) + H_2O \longrightarrow H_3O^+(aq) + Cl^-(aq)$$

Thus, the acid transfers an H^+ ion to a water molecule to form H_3O^+.

The binary compounds of hydrogen with many of the nonmetals are acidic, and in their water solutions they are referred to as **binary acids.** Some other examples are HBr and H_2S. In naming these substances as acids, we add the prefix *hydro-* and the suffix *-ic* to the stem of the nonmetal name, followed by the word *acid*. For example, water solutions of hydrogen chloride and hydrogen sulfide are named as follows:

HCl(*g*)   hydrogen chloride   HCl(*aq*)   *hydro*chlor*ic acid*
H_2S(*g*)   hydrogen sulfide H_2S(*aq*)   *hydro*sulfur*ic acid*

Notice that the gaseous molecular substances are named in the usual way as binary compounds. *It is their aqueous solutions that are named acids.*

One of the important reactions that acids undergo is called **neutralization.** This occurs when an acid reacts with a substance called a **base,** often a metal hydroxide such as NaOH. For example, aqueous HCl reacts with aqueous NaOH as follows:

$$HCl(aq) + NaOH(aq) \longrightarrow NaCl(aq) + H_2O$$
A neutralization reaction

In general, the reaction of an acid with a base produces water and an ionic compound, in this case, sodium chloride, or salt. This reaction is so general, in fact, that the term salt has taken on a broader meaning than just NaCl. We will use the word **salt** to mean any ionic compound that doesn't contain either the OH^- or the O^{2-} ion. (Compounds that contain these ions are bases.)

When acids are neutralized, the salt that is produced contains the anion formed by removing a hydrogen ion, H^+, from the acid molecule. Thus HCl yields salts containing the chloride ion, Cl^-. Similarly, HBr gives salts containing the bromide ion, Br^-.

■ **Practice Exercise 16** Name the water solutions of the following acids: HF, HBr. Name the sodium salts formed by neutralizing these acids with NaOH.

Oxoacids and Their Salts

Acids that contain hydrogen, oxygen, and another element are called **oxoacids.** Examples are H_2SO_4 and HNO_3. These acids do not take the prefix *hydro-*. Many nonmetals form two or more oxoacids that differ in the number of oxygen atoms in their formulas, and they are named according to which one has the larger or smaller number of oxygens. The acid with the larger number of oxygens takes the suffix *-ic* and the one with the fewer number of oxygens takes the suffix *-ous.*

H_2SO_4	sulfur*ic acid*	HNO_3	nitr*ic acid*
H_2SO_3	sulfur*ous acid*	HNO_2	nitr*ous acid*

The halogens can occur in as many as four different oxoacids. The oxoacid with the most oxygens has the prefix *per-*, and the one with the least has the prefix *hypo-*.

$HClO$	*hypo*chlor*ous acid* (usually written HOCl)	$HClO_3$	chlor*ic acid*
$HClO_2$	chlor*ous acid*	$HClO_4$	*per*chlor*ic acid*

The neutralization of oxoacids produces negative polyatomic ions. There is a very simple relationship between the name of the polyatomic ion and that of its parent acid.

1. *-ic* acids give *-ate* anions: HNO_3 (nitr*ic acid*) $\rightarrow NO_3^-$ (nitr*ate* ion)

2. *-ous* acids give *-ite* anions: H_2SO_3 (sulfur*ous acid*) $\rightarrow SO_3^{2-}$ (sulf*ite* ion)

In naming polyatomic anions, the prefixes *per-* and *hypo-* carry over from the name of the parent acid. Thus perchloric acid, $HClO_4$, gives perchlorate ion, ClO_4^-, and hypochlorous acid, $HClO$, gives hypochlorite ion, ClO^-.

■ **Practice Exercise 17** The formula for arsenic acid is H_3AsO_4. What is the name of the salt Na_3AsO_4?

Acid Salts

In the formula of an acid such as HCl or H_2SO_4 the hydrogens that can be removed by neutralization are almost always written first. When the acid has just one hydrogen that can be neutralized, it is called a **monoprotic acid,** but if it contains more than one such hydrogen it is called a **polyprotic acid.** Neutralization of polyprotic acids occurs stepwise and can be halted before all the hydrogens have been removed. For example, if sulfuric acid is combined with sodium hydroxide in a ratio of one formula unit of acid to two formula units of base, then complete neutralization of the acid occurs and the SO_4^{2-} ion is formed.

$$H_2SO_4 + 2NaOH \longrightarrow Na_2SO_4 + 2H_2O$$

However, if the acid and base are combined in a one to one ratio, only half of the available hydrogens are neutralized.

$$H_2SO_4 + NaOH \longrightarrow NaHSO_4 + H_2O$$

The salt $NaHSO_4$, which can be isolated as crystals by evaporating the reaction mixture, is referred to as an **acid salt** because it contains the HSO_4^- ion, which is still capable of furnishing additional H^+.

The hydrogen ion (H^+) given up by an acid is just a proton. A proton is all that's left after the single electron of hydrogen is removed from a neutral hydrogen atom.

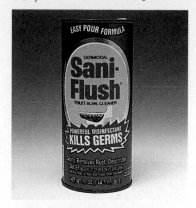

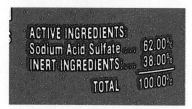

Many acid salts have useful applications. As its active ingredient, this familiar product contains sodium hydrogen sulfate (sodium bisulfate), which the manufacturer calls "sodium acid sulfate."

In naming acid salts formed by ions such as HSO_4^-, we specify the number of hydrogens that can still be neutralized if the salt were to be treated with additional base. For example,

NaHSO₄   sodium hydrogen sulfate
NaH₂PO₄   sodium dihydrogen phosphate

For acid salts of diprotic acids, the prefix *bi-* is still often used.

NaHCO₃   sodium bicarbonate
or
sodium hydrogen carbonate

Notice that the prefix bi- does *not* mean "two"; it means that there is an acidic hydrogen in the compound.

■ **Practice Exercise 18** What is the formula for sodium bisulfite? What is the chemically correct name for this compound?

Common Names

Not every compound is named according to the systematic procedure described above. Many familiar substances were discovered long before a systematic method for naming them had been developed and they acquired common names that are so well known that no attempt has been made to rename them. For example, following the scheme described above we might expect H_2O to have the name dihydrogen oxide. Although this isn't wrong, the common name water is so well known that it is always used. Another example is ammonia, NH_3, whose odor you have no doubt experienced while using household ammonia solutions or the glass cleaner Windex.

Another class of compounds for which common names are often used is very complex substances. A common example is sucrose, which is the chemical name for table sugar, $C_{12}H_{22}O_{11}$. The structure of this compound is pretty complex, and its name assigned following the systematic method is equally complex. It is much easier to say the simple name sucrose, and be understood, than to struggle with the cumbersome systematic name for this common compound.

SUMMARY

Elements, Compounds, and Mixtures An **element** is a substance that cannot be decomposed into something simpler by a chemical reaction. In more precise terms, it is a substance whose atoms all have the same number of protons in their nuclei. Elements combine in fixed proportions to form **compounds.** Elements and compounds are **pure substances** that may be combined in varying proportions to give **mixtures.** A one-phase mixture is called a **solution** and is **homogeneous.** If a mixture consists of two or more phases it is **heterogeneous.** Formation or separation of a mixture into its components can be accomplished by a **physical change** in which the chemical properties of the components do not change. Formation or decomposition of a compound takes place by a **chemical change.**

Symbols, Formulas, and Equations Each element has an internationally agreed upon **chemical symbol.** These symbols are used to write **formulas** for chemical compounds in which the symbol stands for an atom of the element. Subscripts are used to specify how many atoms of each kind are present. Some compounds form crystals, called **hydrates,** that contain water molecules in definite proportions. Heating a hydrate usually can drive off the water. **Chemical equations** present before-and-after descriptions of chemical reactions. When **balanced,** an equation contains **coefficients** that make the number of atoms of each kind the same among the **reactants** and the **products.**

Laws of Chemical Combination When accurate masses of all the reactants and products in a reaction are measured and compared, no observable changes in mass accompany chemical reactions (the **law of conservation of mass**). The mass ratios of the elements in any compound are constant regardless of the source of the compound or how it is prepared (the **law of definite proportions**). Whenever two elements form more than one compound, then the different masses of one element that combine with a fixed mass of the other are in a ratio of small whole numbers (the **law of multiple proportions,** discovered after Dalton had proposed his theory).

Dalton's Atomic Theory Dalton explained the laws of chemical combination by proposing that matter consists of indestructible atoms with masses that do not change during chemical reactions. Actually, most elements consist of a small number of **isotopes** with slightly differing masses. However, all isotopes of an element have very nearly identical chemical properties and the percentages of the isotopes that make up an element are generally so constant throughout the world that we can say that the average mass of their atoms is a constant.

Atoms, Isotopes, and Atomic Masses Atoms can be split into **subatomic particles** such as **electrons, protons,** and **neutrons. Nucleons** are particles that make up the atomic **nucleus** and include the protons (charge $= 1+$) and neutrons (no charge). The number of protons is called the **atomic number (Z)** of the element. Each element has a different atomic number. The electrons (each with a charge of $1-$) are found outside the nucleus; their number equals the atomic number in a neutral atom. Isotopes of an element have identical atomic numbers but different numbers of neutrons. An element's **atomic mass (atomic weight)** is the relative mass of its atoms on a scale in which atoms of carbon-12 have a mass of exactly 12 u **(atomic mass units).**

The Periodic Table The search for similarities and differences among the properties of the elements led Mendeleev to discover that when the elements are placed in (approximate) order of increasing atomic mass, similar properties recur at regular, repeating intervals. In the modern **periodic table** the elements are arranged in rows, called **periods,** in order of increasing atomic number. The rows are stacked so that elements in the columns, called **groups** or **families,** have similar chemical and physical properties. The A-group elements (IUPAC Groups 1, 2, and 13–18) are called **representative elements;** the B-group elements (IUPAC Groups 3–12) are called **transition elements.** The two long rows of **inner transition elements** located below the main body of the table consist of the **lanthanides,** which follow La ($Z = 57$), and the **actinides,** which follow Ac ($Z = 89$). Certain groups are given family names: Group IA (Group 1), except for hydrogen, constitutes the **alkali metals** (the alkalis); Group IIA (Group 2), the **alkaline earth metals;** Group VIIA (Group 17), the **halogens;** Group 0 (Group 18), the **noble gases.**

Metals, Nonmetals, and Metalloids. Most elements are **metals;** they occupy the lower left-hand region of the periodic table (to the left of a line drawn approximately from boron, B, to astatine, At). **Nonmetals** are found in the upper right-hand region of the table. **Metalloids** occupy a narrow band between the metals and nonmetals.

Metals exhibit a **metallic luster,** tend to be **ductile** and **malleable,** and conduct electricity. They react with nonmetals to form ionic compounds called **salts.** Nonmetals tend to be brittle, lack metallic luster, and are nonconductors of electricity. Many nonmetals are gases. Besides combining with metals, nonmetals combine with each other to form molecules without undergoing electron transfer. Bromine (a nonmetal) and mercury (a metal) are the two elements that are liquids at ordinary room temperature. **Metalloids** have properties intermediate between those of metals and nonmetals.

Molecular and Ionic Compounds When atoms chemically combine to form compounds, the reactions produce either ions of opposite charge or neutral molecules, depending on the elements involved. When molecules form, the atoms are linked by the sharing of electrons. Molecules in **molecular compounds** carry no electrical charge. Compounds of nonmetals with hydrogen are called **hydrides;** those with oxygen are **oxides. Organic compounds** are **hydrocarbons** or compounds considered to be derived from hydrocarbons by replacing H atoms with other atoms.

When ionic **binary compounds** are formed, electrons are transferred from a metal to a nonmetal. The metal atom becomes a positive ion (a **cation**); the nonmetal atom becomes a negative ion (an **anion**). The formulas of ionic compounds are controlled by the requirement that the compounds must be electrically neutral. Many ionic compounds also contain **polyatomic ions**—ions that are composed of two or more atoms. Ionic compounds tend to be brittle, high-melting, nonconducting solids. When melted or dissolved in water, however, they do conduct electricity. Most molecular compounds tend to be soft and low melting.

Naming Compounds International agreement between chemists provides a system that allows us to write a single formula from a compound's name. For **salts** (ionic compounds not containing OH^- or O^{2-}), the **Stock system** is preferred, but the older system must also be learned. Compounds between nonmetals use Greek prefixes to specify number. **Acids** are substances that react with water to give H_3O^+ (hydronium ions). They react with **bases** such as NaOH in **neutralization** reactions that yield water and salts. Binary acids are *hydro . . . ic acids* and give *-ide* anions. **Oxoacids** and their anions are related: *-ic acid* produces *-ate* ion; *-ous acid* produces *-ite* ion. **Acid salts,** formed by the partial neutralization of **polyprotic acids,** contain acidic hydrogens and when named use the prefix *bi-* or contain the name hydrogen.

TOOLS YOU HAVE LEARNED

The table below lists the tools that you have learned in this chapter. You will need them to answer chemistry questions. Review them if necessary, and refer to the tools when working on the Thinking-It-Through problems and the Review Exercises that follow.

Tool	Function
Subscripts in a formula (page 41)	Specifies the number of atoms of each element in a formula unit of a substance. You must be able to interpret the formula in terms of the numbers of atoms of each element expressed by the formula.
Rules for writing formulas of ionic compounds (page 68)	Permits us to write correct chemical formulas for ionic compounds. You will need to learn to use the periodic table to remember the charges on the cations and anions of the representative metals and nonmetals. You also should learn the ions formed by the transition and post-transition metals in Table 2.8, and it is essential that you learn the names and formulas of the polyatomic ions in Table 2.9.
Rules for naming compounds (page 73)	Allows us to write a formula from a name and a name from a formula.

THINKING IT THROUGH

Remember, you are not asked to obtain answers for the following problems. Instead, assemble the data necessary to solve the problems and describe how you would use the data to obtain the answers. For numerical problems, set up the calculations using appropriate conversion factors.

The problems are divided into two groups. Those in Level 2 are significantly more challenging than those in Level 1 and provide an opportunity to really hone your problem-solving skills.

Level 1 Problems

1. How many times heavier than an atom of ^{12}C is the average atom of iron? (Explain how you can obtain the answer.)

2. Suppose the atomic mass unit had been defined as $\frac{1}{10}$th of the mass of an atom of phosphorus. What would the atomic mass of carbon be on this scale?

3. How do we find the chemical symbol for a particle that has 43 electrons outside a nucleus that contains 45 protons and 58 neutrons?

4. How can you tell whether the element that has a nucleus that contains 49 protons is a metal, a nonmetal, or a metalloid?

5. Describe how you would determine the chemical formula for the compound formed between the element with atomic number 56 and the element with atomic number 35.

6. A certain compound formed by two elements is a gas at room temperature. When cooled to a very low temperature it liquefies and then freezes to form a solid that is soft and easily crushed. How can you tell whether this compound is more likely to be molecular or ionic?

7. How do we know that the name *hydroselenic acid* refers to a binary acid?

8. What reasoning is involved in determining the formula for sodium bioxalate? Oxalic acid is $H_2C_2O_4$.

Level 2 Problems

9. Carbon forms a compound with element X that has the formula CX_4. A sample of this compound was analyzed and found to contain 1.50 g of C and 39.95 g of X. Explain how you would calculate the atomic mass of X.

10. In an experiment it was found that 3.50 g of phosphorus combines with 12.0 g of chlorine to form the compound

PCl_3. How many grams of chlorine would combine with 8.50 g of phosphorus to form this same compound? (Set up the calculation).

11. Aluminum combines with oxygen to form aluminum oxide. In an experiment a chemist found that 15.0 g of aluminum combined with 13.3 g of oxygen. If this experiment was repeated with 25.0 g of aluminum, how many grams of aluminum oxide would be formed? (Set up the calculations.)

12. When the element phosphorus burns in air it is found that 2.40 g of phosphorus combines with 2.89 g of oxygen to form the oxide P_2O_5. If this experiment were repeated with a mixture of 5.60 g of phosphorus and 8.00 g of oxygen, how many grams of P_2O_5 would be formed and how many grams of oxygen would remain unreacted? (Set up the calculations.)

13. Manganese forms several compounds with oxygen. In one of them it was found that 1.45 g of Mn was combined with 0.845 g of oxygen. In a sample of a different compound it was found that 3.26 g of Mn was combined with 1.42 g of oxygen. Set up the calculations that would show that these two compounds obey the law of multiple proportions.

14. Arsenic forms two compounds with sulfur. In 6.00 g of one of these compounds there was 3.62 g of As. In 8.00 g of the other compound there was 3.86 g of As. Explain in detail how you would show that these compounds exhibit the law of multiple proportions.

REVIEW EXERCISES

Answers to questions whose numbers are printed in color are given in Appendix D. More challenging questions are marked with asterisks.

Elements, Compounds, and Mixtures

2.1 Define: (a) element, (b) compound, (c) mixture, (d) homogeneous, (e) heterogeneous, (f) phase, (g) solution, (h) physical change

2.2 What kind of change is needed to separate a compound into its elements?

2.3 In places like Saudi Arabia fresh water is scarce and is recovered from seawater. When seawater is boiled, the water evaporates and the steam can be condensed to give pure water that people can drink. If all the water is evaporated, solid salt is left behind. Are the changes described here chemical or physical?

Chemical Symbols

2.4 What is the chemical symbol for each of the following elements? (a) chlorine, (b) sulfur, (c) iron, (d) silver, (e) sodium, (f) phosphorus, (g) iodine, (h) copper, (i) mercury, (j) calcium

2.5 What is the name of each of the following elements? (a) K, (b) Zn, (c) Si, (d) Sn, (e) Mn, (f) Mg, (g) Ni, (h) Al, (i) C, (j) N

2.6 What are two ways to interpret a chemical symbol?

Chemical Formulas

2.7 What is the difference between an atom and a molecule?

2.8 How many atoms of each element are represented in each of the following formulas? (a) $K_2C_2O_4$, (b) H_2SO_3, (c) $C_{12}H_{26}$, (d) $HC_2H_3O_2$, (e) $(NH_4)_2HPO_4$, (f) $(CH_3)_3COH$

2.9 How many atoms of each kind are represented in the following formulas? (a) Na_3PO_4, (b) $Ca(H_2PO_4)_2$, (c) C_4H_{10}, (d) $Fe_3(AsO_4)_2$, (e) $Cu(NO_3)_2$, (f) $MgSO_4 \cdot 7H_2O$

2.10 How many atoms of each kind are represented in the following formulas? (a) $Ni(ClO_4)_2$, (b) $CuCO_3$, (c) $K_2Cr_2O_7$, (d) CH_3CO_2H, (e) $NH_4H_2PO_4$, (f) $C_3H_5(OH)_3$

2.11 Asbestos, a known cancer-causing agent, has as a typical formula, $Ca_3Mg_5(Si_4O_{11})_2(OH)_2$. How many atoms of each element are given in this formula?

2.12 Write the formulas and names of the elements that exist in nature as diatomic molecules.

Chemical Equations

2.13 The combustion of a thin wire of magnesium metal (Mg) in an atmosphere of pure oxygen produces the brilliant light of a flashbulb. After the reaction, a thin film of magnesium oxide is seen on the inside of the bulb. The equation for the reaction is

$$2Mg + O_2 \longrightarrow 2MgO$$

(a) State in words how this equation is read.
(b) Give the formula(s) of the reactants.
(c) Give the formula(s) of the products.
(d) Rewrite the equation to show that Mg and MgO are solids and O_2 is a gas.

2.14 What are the numbers called that are written in front of the formulas in a balanced chemical equation?

2.15 How many atoms of each element are represented in each of the following expressions? (a) $3N_2O$, (b) $7CH_3CO_2H$ (c) $4NaHCO_3$, (d) $2(NH_2)_2CO$, (e) $2CuSO_4 \cdot 5H_2O$, (f) $5K_2Cr_2O_7$

2.16 Consider the balanced equation

$$2Fe(NO_3)_3 + 3Na_2CO_3 \longrightarrow Fe_2(CO_3)_3 + 6NaNO_3$$

(a) How many atoms of Na are on each side of the equation?
(b) How many atoms of C are on each side of the equation?
(c) How many atoms of O are on each side of the equation?

2.17 Consider the balanced equation for the combustion of octane, a component of gasoline:

$$2C_8H_{18} + 25O_2 \longrightarrow 16CO_2 + 18H_2O$$

(a) How many atoms of C are on each side of the equation?

(b) How many atoms of H are on each side of the equation?

(c) How many atoms of O are on each side of the equation?

2.18 Rewrite the chemical equation in the preceding question so that it specifies gasoline and water as liquids and oxygen and carbon dioxide (CO_2) as gases.

Laws of Chemical Combination and Dalton's Theory

2.19 Name and state the three laws of chemical combination. Which law was not discovered until after Dalton had proposed his theory?

2.20 Which postulate of Dalton's theory is based on the law of conservation of mass? Which is based on the law of definite proportions?

2.21 Why didn't the existence of isotopes affect the apparent validity of the atomic theory?

2.22 In your own words, describe how Dalton's theory explains the law of conservation of mass and the law of definite proportions.

2.23 Which of the laws of chemical combination is used to define the term *compound?*

2.24 Laughing gas is a compound formed from nitrogen and oxygen in which there are 1.75 g of nitrogen to 1.00 g of oxygen. Below are given the compositions of several nitrogen–oxygen compounds. Which of these is laughing gas?

(a) 6.35 g nitrogen, 7.26 g oxygen

(b) 4.63 g nitrogen, 10.58 g oxygen

(c) 8.84 g nitrogen, 5.05 g oxygen

(d) 9.62 g nitrogen, 16.5 g oxygen

(e) 14.3 g nitrogen, 40.9 g oxygen

2.25 A compound of nitrogen and oxygen has the formula NO. In this compound there are 1.143 g of oxygen for each 1.000 g of nitrogen. A different compound of nitrogen and oxygen has the formula NO_2. How many grams of oxygen would be combined with each 1.000 g of nitrogen in NO_2?

2.26 Tin forms two compounds with chlorine, $SnCl_2$ and $SnCl_4$.

(a) When combined with the same mass of tin, what would be the ratio of the masses of chlorine in the two compounds?

(b) In the compound $SnCl_2$, 0.597 g of chlorine is combined with each 1.000 g of tin. In $SnCl_4$, how many grams of chlorine would be combined with 1.000 g of tin?

Atomic Masses

2.27 Write the symbol for the isotope that forms the basis of the atomic mass scale. What is the mass of this atom expressed in atomic mass units?

2.28 The actual mass of the atomic mass unit is $1.66056520 \times 10^{-24}$ g. Using this value, calculate the mass in grams of one atom of carbon-12.

2.29 In the compound CH_4, 0.33597 g of hydrogen is combined with 1.0000 g of carbon-12. Use this information to calculate the atomic mass of the element hydrogen.

2.30 If an atom of carbon-12 had been assigned a relative mass of 24.0000 u, what would be the atomic weight of hydrogen relative to this mass?

2.31 Naturally occurring copper is composed of 69.17% of ^{63}Cu, with an atomic mass of 62.9396 u, and 30.83% of ^{65}Cu, with an atomic mass of 64.9278 u. Use these data to calculate the average atomic mass of copper.

2.32 Naturally occurring magnesium is composed of 78.99% of ^{24}Mg (atomic mass, 23.9850 u), 10.00% of ^{25}Mg (atomic mass, 24.9858 u), and 11.01% of ^{26}Mg (atomic mass, 25.9826 u). Use these data to calculate the average atomic mass of magnesium.

Atomic Structure

2.33 What are the names, symbols, electrical charges, and masses (expressed in u) of the three subatomic particles introduced in this chapter?

2.34 Where is nearly all of the mass of an atom located? Explain your answer in terms of the particles that contribute to this mass.

2.35 What is a *nucleon?* Which ones have we studied?

2.36 Define the terms *atomic number* and *mass number.*

2.37 Which is better related to the chemistry of an element, its mass number or its atomic number? Give a brief explanation in terms of the basis for the periodic table.

2.38 In terms of the structures of atoms, how are isotopes of the same element alike? How do they differ?

2.39 Name one property that atoms of two *different* elements might possibly have in common?

2.40 Consider the symbol $_b^aX_d^c$, where X stands for the chemical symbol for an element. What information is given in locations: (a) a, (b) b, (c) c, and (d) d?

2.41 Write the symbols of the isotopes that contain the following. (Use the table of atomic masses and numbers printed inside the front cover for additional information, as needed.)

(a) An isotope of iodine whose atoms have 78 neutrons.

(b) An isotope of strontium whose atoms have 52 neutrons.

(c) An isotope of cesium whose atoms have 82 neutrons.

(d) An isotope of fluorine whose atoms have 9 neutrons.

2.42 Give the numbers of neutrons, protons, and electrons in the atoms of each of the following isotopes. (Use the table of atomic masses and numbers printed inside the front cover for additional information, as needed.)

(a) radium-226   (c) $_{82}^{206}Pb$

(b) carbon-14   (d) $_{23}^{11}Na$

2.43 Give the numbers of electrons, protons, and neutrons in the atoms of each of the following isotopes. (As necessary, consult the table of atomic masses and numbers printed inside the front cover.)

(a) cesium-137   (c) $_{92}^{238}U$

(b) iodine-131   (d) $_{79}^{197}Au$

2.44 How many electrons, protons, and neutrons are in each of the following particles?
(a) $^{81}_{35}Br^-$ (b) $^{58}_{26}Fe^{3+}$ (c) $^{63}_{29}Cu^{2+}$ (d) $^{87}_{37}Rb^+$

The Periodic Table

2.45 What is the general formula for the compounds formed by Li, Na, K, Rb, and Cs with chlorine? What is the general formula for the compounds formed by Be, Mg, Ca, Sr, and Ba with chlorine? How did this kind of information lead Mendeleev to develop his periodic table?

2.46 On what basis did Mendeleev construct his periodic table? On what basis are the elements arranged in the modern periodic table?

2.47 In the periodic table, what is a "period"? What is a "group"?

2.48 Why were there gaps in Mendeleev's periodic table?

2.49 In the text, we identified two places in the modern periodic table where the atomic mass order was reversed. Using the table on the inside front cover, locate two other places where this occurs.

2.50 Below are some data for the elements sulfur and tellurium.

	Sulfur	Tellurium
Atomic mass	32.06	127.60
Melting point (°C)	112.8	449.5
Boiling point (°C)	445	990
Formula of oxide	SO_2	TeO_2
Melting point of oxide (°C)	−72.7	733
Density (g/cm³)	2.07	6.25

(a) Had the element selenium not been known in the time of Mendeleev, he would have called it eka-sulfur. Estimate its properties as an average of those of sulfur and tellurium.
(b) In the *Handbook of Chemistry and Physics* (located in your school library), look up the values for the properties of selenium and its oxide and compare them with the values you obtained in (a).

2.51 On the basis of their positions in the periodic table, why is it not surprising that strontium-90, a dangerous radioactive isotope of strontium, replaces calcium in newly formed bones?

2.52 In the refining of copper, sizable amounts of silver and gold are recovered. Why is this not surprising?

2.53 Why would you reasonably expect cadmium to be a contaminant in zinc but not in silver?

2.54 Make a rough sketch of the periodic table and mark off those areas where you would find (a) the representative elements, (b) the transition elements, and (c) the inner transition elements.

2.55 What group numbers are used to designate the representative elements following (a) the traditional U.S. system for designating groups and (b) the IUPAC system?

2.56 Supply the IUPAC group numbers that correspond to the following customary U.S. designations: (a) Group IA, (b) Group VIIA, (c) Group IIIB, (d) Group IB, (e) Group IVA

2.57 Based on discussions in this chapter, explain why it is unlikely that scientists will discover a new element, never before observed, having an atomic weight of approximately 73.

2.58 Which of the following is an alkali metal? Ca, Cu, In, Li, S.

2.59 Which of the following is a halogen? Ce, Hg, Si, O, I.

2.60 Which of the following is a transition element? Pb, W, Ca, Cs, P.

2.61 Which of the following is a noble gas? Xe, Se, H, Sr, Zr.

2.62 Which of the following is a lanthanide element? Th, Sm, Ba, F, Sb.

2.63 Which of the following is an actinide element? Ho, Mn, Pu, At, Na.

2.64 Which of the following is an alkaline earth metal? Mg, Fe, K, Cl, Ni.

Physical Properties of Metals, Nonmetals, and Metalloids

2.65 Give five physical properties that we usually observe for metals.

2.66 Why is mercury used in thermometers? Why is tungsten used in light bulbs?

2.67 What property of metals allows them to be drawn into wire?

2.68 Gold can be hammered into sheets so thin that some light can pass through them. What property of gold allows such thin sheets to be made?

2.69 Only two metals are colored (the rest are "white," like iron or lead). You have surely seen both of them. Which metals are they?

2.70 Which nonmetals occur as monatomic gases (gases whose particles consist of single atoms)?

2.71 Which nonmetals occur in nature as diatomic molecules? Which of these are gases?

2.72 Which two elements exist as liquids at room temperature and pressure?

2.73 Which physical property of metalloids distinguishes them from metals and nonmetals?

2.74 Sketch the shape of the periodic table and mark off those areas where we find (a) metals, (b) nonmetals, and (c) metalloids.

2.75 Which metals can you think of that are commonly used to make jewelry? Why isn't iron used to make jewelry?

Molecular Compounds of Nonmetals

2.76 Describe what kind of event must occur (involving electrons) if the atoms of two different elements are to react to form (a) an ionic compound or (b) a molecular compound.

2.77 Which are the only elements that exist as free, individual atoms when not chemically combined with other elements?

2.78 What are the chemical formulas for the molecules formed by the following nonmetals in their elemental states? (a) chlorine, (b) sulfur, (c) phosphorus, (d) nitrogen, (e) oxygen, (f) hydrogen

2.79 Without referring to Table 2.4 but using the periodic table, write chemical formulas for the simplest hydrogen compounds of: (a) carbon, (b) nitrogen, (c) tellurium, (d) iodine.

2.80 Astatine forms a compound with hydrogen. Predict its chemical formula.

2.81 Under appropriate conditions, tin can be made to form a simple molecular compound with hydrogen. Predict its formula.

2.82 Based on what you've learned in Section 2.6, what are likely formulas of the two oxides of bismuth? (If necessary, refer to Table 2.5.)

2.83 What are the formulas of (a) methanol and (b) ethanol?

2.84 What is the formula for the alkane that has 10 carbon atoms?

2.85 Candle wax is a mixture of hydrocarbons, one of which is an alkane with 23 carbon atoms. What is the formula for this hydrocarbon?

Ionic Compounds

2.86 With what kind of elements do metals react?

2.87 In what major way do compounds formed between two nonmetals differ from those formed between a metal and a nonmetal?

2.88 Why are nonmetals found in more compounds than are metals, even though there are fewer nonmetals than metals?

2.89 What is an ion? How does it differ from an atom or a molecule?

2.90 Consider the sodium atom and the sodium ion.
(a) Write the chemical symbol of each.
(b) Do these particles have the same number of nuclei?
(c) Do they have the same number of protons?
(d) Could they have different numbers of neutrons?
(e) Do they have the same number of electrons?

2.91 What holds an ionic compound together? Can we identify individual molecules in an ionic compound?

2.92 Why do we use the term *formula unit* for ionic compounds instead of the term *molecule*?

2.93 Define *cation* and *anion*.

2.94 When calcium reacts with chlorine, each calcium atom loses two electrons and each chlorine atom gains one electron. What are the formulas of the ions created in this reaction?

2.95 How many electrons has a titanium atom lost if it has formed the ion Ti^{4+}? What are the total numbers of protons and electrons in a Ti^{4+} ion?

2.96 If an atom gains an electron to become an ion, what kind of electrical charge does the ion have?

2.97 How many electrons has a nitrogen atom gained if it has formed the ion N^{3-}? How many protons and electrons are in a N^{3-} ion?

2.98 What three rules do we apply when we write the formulas for ionic compounds?

2.99 What is wrong with the formula $RbCl_3$? What is wrong with the formula SNa_2?

2.100 A student wrote the formula for an ionic compound of titanium as Ti_2O_4. What is wrong with this formula? What should the formula be?

2.101 The formula for a compound is correctly given as C_4H_{10}. How do we know that this is a molecular compound and not an ionic compound?

2.102 Use the periodic table, but not Table 2.7, to write the symbols for the ions of: (a) K, (b) Br, (c) Mg, (d) S, (e) Al.

2.103 Use the periodic table, but not Table 2.7, to write the symbols for ions of: (a) barium, (b) oxygen, (c) fluorine, (d) strontium, (e) rubidium.

2.104 Which are the post-transition metals? Write their symbols.

2.105 What are the formulas of the ions formed by: (a) iron, (b) cobalt, (c) mercury, (d) chromium, (e) tin, (f) lead, (g) copper?

2.106 Write formulas for ionic compounds formed between: (a) Na and Br, (b) K and I, (c) Ba and O, (d) Mg and Br, (e) Ba and F.

2.107 Which of the following formulas are incorrect? (a) NaO_2, (b) $RbCl$, (c) K_2S, (d) Al_2Cl_3, (e) MgO_2.

2.108 What are the formulas for: (a) cyanide ion, (b) ammonium ion, (c) nitrate ion, (d) sulfite ion, (e) chlorate ion?

2.109 What are the formulas for: (a) hypochlorite ion, (b) bisulfate ion, (c) phosphate ion, (d) dihydrogen phosphate ion, (e) permanganate ion.

2.110 What are the names of the following ions? (a) $Cr_2O_7^{2-}$, (b) OH^-, (c) $C_2H_3O_2^-$, (d) CO_3^{2-}, (e) CN^-, (f) ClO_4^-?

2.111 Write formulas for the ionic compounds formed from: (a) K^+ and nitrate ion, (b) Ca^{2+} and acetate ion, (c) ammonium ion and Cl^-, (d) Fe^{3+} and carbonate ion, (e) Mg^{2+} and phosphate ion.

2.112 Write formulas for the ionic compounds formed from: (a) Zn^{2+} and hydroxide ion, (b) Ag^+ and chromate ion, (c) Ba^{2+} and sulfite ion, (d) Rb^+ and sulfate ion, (e) Li^+ and bicarbonate ion.

2.113 Write formulas for two compounds formed between O^{2-} and: (a) lead, (b) tin, (c) manganese, (d) iron, (e) copper.

2.114 Write formulas for the ionic compounds formed from Cl^- and: (a) cadmium ion, (b) silver ion, (c) zinc ion, (d) nickel ion.

*2.115 From what you have learned in Sections 2.6 and 2.7, write correct balanced equations for the reactions between (a) calcium and chlorine, (b) magnesium and oxygen, (c) aluminum and oxygen, and (d) sodium and sulfur.

Properties of Ionic and Molecular Compounds

2.116 Compare the properties of ionic and molecular compounds.

2.117 Why does molten KCl conduct electricity even though solid KCl does not?

2.118 What happens when an ionic compound dissolves in water that allows its solutions to conduct electricity?

2.119 Why are ionic compounds so brittle?

2.120 Why do most ionic compounds have high melting points? Naphthalene (moth flakes) consists of soft crystals that melt at a relatively low temperature of 80 °C. Does this substance have characteristics of an ionic or a molecular compound?

Naming Compounds

2.121 What is the difference between a binary compound and one that is diatomic? Give examples that illustrate this difference.

2.122 Name the following. (a) CaS, (b) NaF, (c) $AlBr_3$, (d) Mg_2C, (e) Na_3P, (f) Li_3N, (g) Ba_3As_2, (h) Al_2O_3.

2.123 Name the following, using both the old nomenclature system and the Stock system. (a) $CrCl_3$, (b) Mn_2O_3, (c) CuO, (d) Hg_2Cl_2, (e) SnO_2, (f) PbS, (g) $CoCl_2$, (h) FeS.

2.124 Name the following. (a) SiO_2, (b) ClF_3, (c) XeF_4, (d) S_2Cl_2, (e) P_4O_{10}, (f) N_2O_5, (g) Cl_2O_7, (h) $AsCl_5$.

2.125 Define *acid, base, neutralization,* and *salt.* What is the name of the ion H_3O^+?

2.126 Iodine, like chlorine, forms several acids. What are the names of the following? (a) HIO_4, (b) HIO_3, (c) HIO_2, (d) HIO, (e) HI. *Contributed by Prof. David Dobberpuhl, Creighton University*

2.127 For the acids in the preceding question, (a) write the formulas and (b) name the ions formed by removing a hydrogen ion (H^+) from each acid. *Contributed by Prof. David Dobberpuhl, Creighton University*

2.128 Name the following. If necessary, refer to Table 2.9 on page 71. (a) $NaNO_2$, (b) K_3PO_4, (c) $KMnO_4$, (d) $NH_4C_2H_3O_2$, (e) $BaSO_4$, (f) $Fe_2(CO_3)_3$, (g) KSCN, (h) $Na_2S_2O_3$.

2.129 Write the formula for (a) chromic acid, (b) carbonic acid, and (c) oxalic acid. (Hint: Check the table of polyatomic ions.)

2.130 Name the following: (a) $CrCl_2$, (b) $NH_4C_2H_3O_2$, (c) KIO_3, (d) $HClO_2$, (e) $CaSO_3$, (f) AgCn, (g) $ZnBr_2$, (h) $H_2Se(g)$, (i) $H_2Se(aq)$, (j) $V(NO_3)_3$, (k) $Co(C_2H_3O_2)_2$, (l) Au_2S_3, (m) Au_2S, (n) $GeBr_4$, (o) K_2CrO_4, (p) $Fe(OH)_2$, (q) I_2O_4, (r) I_4O_9, (s) P_4Se_3.

2.131 Write formulas for the following. (a) sodium monohydrogen phosphate, (b) lithium selenide, (c) sodium hydride, (d) chromic acetate, (e) nickel(II) cyanide, (f) iron(III) oxide, (g) stannic sulfide, (h) antimony pentafluoride, (i) dialuminum hexachloride, (j) tetraarsenic decaoxide, (k) magnesium hydroxide, (l) cupric bisulfate, (m) ammonium thiocyanate, (n) potassium thiosulfate.

2.132 Write the formulas for the following. (a) ammonium sulfide, (b) chromium(III) sulfate, (c) molybdenum(IV) sulfide, (d) tin(IV) chloride, (e) iron(III) oxide, (f) calcium bromate, (g) mercury(II) acetate, (h) barium bisulfite, (i) silicon tetrafluoride, (j) boron trichloride, (k) stannous sulfide, (l) calcium phosphide, (m) magnesium dihydrogen phosphate, (n) calcium oxalate.

2.133 Write formulas for the following: (a) gold(III) nitrate, (b) copper(II) sulfate, (c) ammonium bromate, (d) nickel(II) iodate, (e) lead(IV) oxide, (f) antimony(III) sulfide, (g) platinum(II) chloride, (h) cadmium sulfide, (i) barium acetate, (j) mercury(I) chloride, (k) mercury(II) chloride, (l) strontium phosphate, (m) barium arsenide, (n) cobaltous hydroxide, (o) aluminum sulfite, (p) ammonium dichromate, (q) diiodine pentaoxide, (r) tetraphosphorus heptasulfide, (s) disulfur decafluoride.

2.134 Define: *polyprotic acid, monoprotic acid.*

2.135 Name the following acid salts: (a) $NaHCO_3$, (b) KH_2PO_4, (c) $(NH_4)_2HPO_4$.

Additional Exercises

2.136 An element has 28 protons in its nucleus. When this element reacts with chlorine, is the compound formed likely to be ionic or molecular? Justify your answer.

***2.137** Elements *X* and *Y* form the compound XY_4. When these elements react, it is found that 1.00 g of *X* combines with 5.07 g of *Y*. When *X* combines with oxygen, it forms the compound XO_2 in which 1.00 g of *X* combines with 1.14 g of O.
(a) What is the atomic mass of *X*?
(b) What is the atomic mass of *Y*?

2.138 An iron nail is composed of four isotopes with the percentage abundances and atomic masses given in the table below. Calculate the average atomic mass of iron.

Isotope	Percentage abundance	Atomic mass (u)
^{54}Fe	5.80	53.9396
^{56}Fe	91.72	55.9349
^{57}Fe	2.20	56.9354
^{58}Fe	0.28	57.9333

***2.139** Write balanced chemical equations for the complete neutralization of H_2SO_4 by (a) $Al(OH)_3$ and (b) $Ca(OH)_2$.

2.140 Write the formulas for all the acid salts that could be formed from the reaction of NaOH with the acid H_3PO_4.

2.141 Name the following oxoacids and give the names and formulas of the salts formed from them by neutralization with NaOH: (a) HOCl, (b) HIO_2, (c) $HBrO_3$, (d) $HClO_4$.

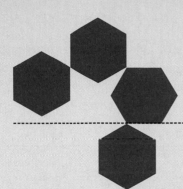

CHEMICALS IN USE 2
Fossil Fuels

Long ago, nature used photosynthesis to transform energy from the sun into the chemical energy of ancient plants and then locked up this energy in their fossilized remains. Today called the *fossil fuels*—principally petroleum, coal, and natural gas—they are a legacy from the past now being rapidly consumed. For the first time in history we are concerned about running out of them and about the impact on civilization if their disappearance occurs too suddenly for us to adapt. What few people realize is that the fossil fuels are also a major source of raw materials for making the chemical compounds needed to manufacture virtually all medicinal agents, dyes, plastics, coatings, and synthetic fibers.

When the fossil fuels burn in sufficient oxygen, the products are carbon dioxide and water. The release of carbon dioxide into the atmosphere accelerated so much during the twentieth century that climatologists are deeply concerned about the effect on the heat balance of the planet (see *Chemicals in Use 4*, "The Greenhouse Effect.") Dealing with the atmospheric carbon dioxide problem might someday require a greater reliance on energy from the sun in all of its forms, including windmills (the winds being a product of the uneven solar heating of the earth and the Earth's rotation.)

PETROLEUM AND CRUDE OIL

The fossil fuels formed over a span of hundreds of thousands of years during the Carboniferous Period in geologic history, roughly 280 to 345 million years ago. Vast areas of the continents, little more than monotonous plains, then basked in sunlight near sea level. In the oceans, countless tiny, photosynthesizing plants like the diatoms—tons of them per acre of ocean surface during the early spring—soaked up solar energy to power their fugitive lives. Then they died. According to one theory, the death of each such plant released a tiny droplet of oily material that eventually settled into the bottom muds. The muds grew in thickness and compacted, sometimes into a rock called shale, sometimes into limestone and sandstone deposits. Under the pressure and heat of com-

pacting, the oily matter changed into petroleum, a mixture of crude oil, water, and natural gas. (*Petroleum* is from *petra,* "rock" and *oleum,* "oil.") In some parts of the world, the petroleum managed to move slowly through porous rock and collect into vast underground pools to form the great petroleum reserves of our planet. In other regions, this movement could not occur, and the oily substances remain to this day locked in enormous deposits of oil shales and oil sands.

The story is told of a gentleman in a western state who built a new home and made the fireplace out of a local shale. When he lit the first fire, both the fireplace and his house burned up! His shale was rich in oil. In the Green River region of the United States, where Utah, Colorado, and Idaho meet, there is a shale formation estimated to contain 2000 billion barrels of *shale oil* (see Figure 2a). (The United States uses roughly 5 billion barrels of oil per year.) When the oil shale is crushed and heated to about 260 °C, a substance essentially the same as petroleum is released. Rock qualifies as oil shale if it holds an average of 10 gallons of petroleum per ton of rock. The cost of wringing the petroleum from the rock, however, is presently too high for commercial exploitation, even if the management of the associated environmental problems is excluded from the cost.

Near Lake Atabasca in the province of Alberta, Canada, in a land area about the size of Lake Michigan, there are huge reserves of a material much like crude oil but intermixed with sand, not shale (see

FIGURE 2a

Oil shale in the Mahogany zone near Parachute, Colorado.

FIGURE 2b

Syncrude's tar sands project in Alberta, Canada.

Figure 2b). Each ton of this *oil sand* holds about two thirds of a barrel of oil, and the total deposit is estimated to contain over 600 billion barrels of oil, which is equivalent to about half of the entire petroleum reserves of the world.

COAL

On the marshy lands bordering the ancient oceans, lush vegetation flourished and died in a moist and sunny setting. The rate of decay of the remains of these plants, covered by stagnant, oxygen-poor, often acidic water, was slower than the rate at which the plants died. The slowly rotting mass accumulated to huge depths, became fibrous and turned to *peat,* a woody material used as fuel in some regions of the world.

Where peat layers became thick enough or were compressed by later deposits of sedimentary rock like limestone and sandstone, the peat changed into lignite ("brown coal"). Although lignite is over 40% water, it is still an important fuel. Many lignite deposits became thick enough for further compaction to occur, and most of the water was squeezed out. In this way bituminous coal ("soft coal") formed, which is less than 5% water but contains considerable quantities of volatile matter. (*Volatile* means "easily evaporated.") When still further compaction took place and nearly all of the volatile matter was forced out, anthracite ("hard coal") formed, which is over 95% carbon.

The energy content in the coal reserves of the world exceeds that of the known petroleum reserves by a wide margin. In estimates that have the available supplies of petroleum lasting less than a century, the coal reserves are considered to have a lifetime of up to three centuries. Coal, however, contains very small quantities of radioactive elements, as well as compounds of mercury and sulfur. Their concentrations are very small, as we said, but because of the enormous quantities of coal annually burned, mostly at electric power plants, air and water pollution problems are caused. So much coal is burned that, according to the calculations of a physicist at the Oak Ridge National Laboratory, the combustion of coal at power plants produces an annual release of radioactive contaminants (uranium and thorium) exceeding the entire annual U. S. consumption of nuclear fuels. The mercury pollution in many lakes north of the industrial regions of the United States stems in part from what is released by burning coal.

NATURAL GAS

In both soft coal and petroleum, the most volatile hydrocarbons, methane and ethane, also accumulated. Natural gas, the gas used for bunsen burners in most labs, is largely methane. Natural gas also contains ethylene, which is removed before the remainder of the gas is sent on for combustion purposes. Ethylene is the raw material for making many chemical compounds as well as polyethylene plastics.

USEFUL SUBSTANCES FROM CRUDE OIL BY REFINING

Crude oil is a complex mixture of compounds, but nearly all are hydrocarbons (compounds made up exclusively of carbon and hydrogen atoms). Small but vexing amounts of sulfur-containing compounds are also present. The object of refining crude oil is to separate it into products of varying uses, such as solvents, gasoline, kerosene, diesel oil, fuel oil, paraffin wax, and asphalt (see Section 21.2).

Questions
----
1. What are the names of the three principal fossil fuels used today?

2. What are the differences among petroleum, crude oil, and natural gas?

3. In general terms, describe what happened to change peat into lignite, lignite into soft coal, and soft coal into hard coal.

In some respects, a chemical reaction is like a camel race. There's a lot of chaos, but things are moving in a particular direction and the operative ratio is expressible in whole numbers, like 1 camel to 1 jockey.

Chapter 3

Stoichiometry: Quantitative Chemical Relationships

Stoichiometry is the branch of chemistry concerned with the ratios *by atoms* of the elements in compounds and with the ratios *by atoms, molecules,* or *formula units* of the substances in chemical reactions.

The ratios of atoms in chemical formulas are expressed in *whole*, not fractional numbers, because it isn't possible to have a fraction of an atom. In Section 2.2 we noted that hydrogen combines with fluorine to form HF in the ratio of 1 H atom to 1 F atom. The formula of water, H_2O, tells us that the combining ratio is 2 atoms of H to 1 atom of O. To make only *one molecule* of H_2O with no atoms of H or O left over, we would be bound by the formula of water to take *two* atoms of hydrogen and *one* of oxygen.

$$2 \text{ atoms H} + 1 \text{ atom O} \longrightarrow 1 \text{ molecule } H_2O$$

To make a dozen (12) molecules of H_2O, we must take 2 dozen atoms of H and 1 dozen atoms of O, still a *ratio* of 2 : 1.

$$2 \text{ doz H atoms} + 1 \text{ doz O atoms} \longrightarrow 1 \text{ doz } H_2O \text{ molecules}$$

So whether we express the ratio of H to O present in H_2O in atoms or dozens of atoms, the important ratio is the same, 2 : 1. This ratio is the stoichiometric "law" for water and is simply a fact about our world. *Every pure substance has a stoichiometric "law" given by its formula.*

Avogadro's Number and the Mole

Because atoms and molecules are so small, we need far more than dozens of them to have samples we can actually see and manipulate in the laboratory. It is convenient, therefore, to define a quantity similar to the dozen but repre-

3.1
THE MOLE CONCEPT

"Stoy-kee-*ah*-meh-tree" is from the Greek *stoicheion,* meaning "element," and *metron,* "measure."

Recall that a formula unit, a term used instead of "molecule" when dealing with ionic compounds, consists of whatever is represented by the ionic compound's formula (see page 67).

Elemental hydrogen and oxygen occur ordinarily as diatomic molecules, H_2 and O_2, but their ratio by *atoms* in water is still 2 : 1.

senting many more things. We call this quantity the **mole** (abbreviated **mol**), which is the SI base unit for the *amount of chemical substance.*

One **mole** of any element or compound has the same number of formula units as there are atoms in exactly 12 g of carbon-12.

Amedeo Avogadro (1776–1856) was an Italian scientist who did important early work in stoichiometry.

The number of formula units in a mole of a substance is enormous. It has been determined experimentally to be 6.0221367×10^{23} and is called **Avogadro's number** in honor of Amedeo Avogadro.[1] When we need this number in calculations, we will usually round it to three or four significant figures, namely to 6.02×10^{23} or 6.022×10^{23}. Thus, to four significant figures, we can write

$$1 \text{ mol} = 6.022 \times 10^{23} \text{ things}$$

Avogadro's number

The "things" of importance in chemistry, of course, are atoms, ions, molecules, and formula units. Thus, 1 mol of carbon has Avogadro's number of C atoms, 1 mol of water has the same number of H_2O molecules, and 1 mol of NaCl has Avogadro's number of formula units. In stoichiometry there is no more significant generalization than the following: *Equal numbers of moles contain equal numbers of fundamental particles, be they atoms, ions, molecules, or formula units.* This simple idea is at the heart of all stoichiometry. It is the foundation of all quantitative reasoning in chemistry.

We can now use the mole concept to reexpress the ratios of the atoms in molecules of water, H_2O.

Ratio by atoms	Ratio by dozens of atoms	Ratio by moles
$\dfrac{2 \text{ H atoms}}{1 \text{ O atom}}$	$\dfrac{2 \text{ doz H atoms}}{1 \text{ doz O atoms}}$	$\dfrac{2 \text{ mol H atoms}}{1 \text{ mol O atoms}}$

In all, the key stoichiometric ratio is $2:1$, and one of the most important ideas in all of chemistry is that *moles of atoms always combine in the same ratio as the individual atoms themselves.*

Stoichiometric Equivalencies

The **stoichiometric equivalence** of two elements in a chemical formula is the ratio (by atoms or by moles) in which they occur in the formula. For example, the subscripts in P_4O_{10} give us the ratio of P to O in this substance, both by atoms and by moles of atoms:

P_4O_{10} is tetraphosphorus decaoxide.

By atoms: 4 P atoms to 10 O atoms
By moles: 4 mol P to 10 mol O

Chemical formula

Notice how these relationships are immediately available from the formula P_4O_{10}. Thus, from here on, we will view a chemical formula as the tool that establishes the stoichiometric equivalencies used in solving problems. The formula P_4O_{10} gives a specific *stoichiometric* equivalence between P and O; 4 mol of P is stoichiometrically equivalent to 10 mol of O.

---

[1] Avogadro's number is so huge that it is beyond easy comprehension. A mole of dollars, for example, would be enough to guarantee every person in the world, roughly 6 billion people, an income of over \$3 million *per second* for a hundred years. Yet Avogadro's number of carbon-12 atoms has a mass of only 12 g. This shows how extremely small atoms are.

When we use our general symbol $\Leftrightarrow$ in stoichiometry, it will mean "is *stoichiometrically* equivalent to." The relationships in P_4O_{10} can thus be expressed in each of the following ways.

$$1 \text{ mol } P_4O_{10} \quad \Leftrightarrow \quad 4 \text{ mol } P$$
$$1 \text{ mol } P_4O_{10} \quad \Leftrightarrow \quad 10 \text{ mol } O$$
$$4 \text{ mol } P \quad \Leftrightarrow \quad 10 \text{ mol } O$$

We can use these equivalencies to devise conversion factors for solving problems in stoichiometry. The following conversion factors, for example, express the equivalencies inherent in P_4O_{10}.

$$\frac{1 \text{ mol } P_4O_{10}}{4 \text{ mol } P} \quad \text{or} \quad \frac{4 \text{ mol } P}{1 \text{ mol } P_4O_{10}}$$

$$\frac{1 \text{ mol } P_4O_{10}}{10 \text{ mol } O} \quad \text{or} \quad \frac{10 \text{ mol } O}{1 \text{ mol } P_4O_{10}}$$

$$\frac{4 \text{ mol } P}{10 \text{ mol } O} \quad \text{or} \quad \frac{10 \text{ mol } O}{4 \text{ mol } P}$$

The reaction of phosphorus with oxygen that gives P_4O_{10} produces a brilliant light, often used in fireworks displays.

How many moles of O atoms are combined with 4.20 moles of Cl atoms in Cl_2O_7?

ANALYSIS Note the words *How many moles . . . combined with . . . moles*. These words are the clue that we have a stoichiometric equivalency question. We can restate the problem as:

$$4.20 \text{ mol Cl} \Leftrightarrow ? \text{ mol O}$$

What we need therefore is a moles-to-moles conversion factor. The subscripts in Cl_2O_7 supply it, because the formula, Cl_2O_7, actually means

$$2 \text{ mol Cl} \Leftrightarrow 7 \text{ mol O}$$

So we have the following conversion factors.

$$\frac{2 \text{ mol Cl}}{7 \text{ mol O}} \quad \text{or} \quad \frac{7 \text{ mol O}}{2 \text{ mol Cl}}$$

SOLUTION To find the moles of O atoms stoichiometrically equivalent to 4.20 moles of Cl atoms in Cl_2O_7, we multiply 4.20 mol Cl by the first factor. Notice how the units *mol Cl* (not just *mol*) cancel properly. (Draw in the cancel lines yourself.)

$$4.20 \text{ mol Cl} \times \frac{7 \text{ mol O}}{2 \text{ mol Cl}} = 14.7 \text{ mol O}$$

Thus 14.7 mol of O is combined with 4.20 mol of Cl in Cl_2O_7. (We leave three significant figures in the answer because the least precise number, 4.20, has three significant figures. The 7 and the 2 in the conversion factor, which come from the formula's subscripts, are *exact* numbers of atoms, and so they have an infinite number of significant figures.)

CHECK Does the *size* of the answer make sense? When working problems, make simple mental checks to see that the *size* of the answer makes sense in relationship to the given data. Notice here, for example, that the answer, 14.7 mol O, is numerically larger than 4.20 mol Cl. It *must* be larger because of the 7 to 2 ratio of O to Cl in Cl_2O_7, so the answer's size "makes sense."

■ **Practice Exercise 1** How many moles of nitrogen atoms are combined with 8.60 mol of oxygen atoms in dinitrogen pentoxide, N_2O_5?

EXAMPLE 3.1
Stoichiometry of Chemical Formulas

Cl_2O_7 is an oily and very explosive liquid.

3.2

MEASURING MOLES OF ELEMENTS AND COMPOUNDS

Diamond is one form of pure carbon. A 12 g diamond would be a 60 carat stone.

FIGURE 3.1

Each quantity of these elements contains the same number of atoms, Avogadro's number.

Atomic mass

We ignore the estimated uncertainties in the atomic mass values given in the Table of Atomic Masses and Numbers. These are numbers between 1 and 9 set in parentheses immediately after most of the atomic masses.

According to the SI definition, we have 1 mol of carbon-12 when we have exactly 12 g of this isotope. However, naturally occurring carbon is a mixture of isotopes, and the "average" carbon atom has a mass of 12.011 u. Avogadro's number of these "average atoms" therefore has a mass of 12.011 g. In other words, for naturally occurring carbon: 1 mol C = 12.011 g C. The extension to all the elements is important and worth emphasizing.

One mole of any element has a mass in grams that is numerically equal to the element's atomic mass.

The sizes *in grams* of the 1-mol samples of four elements shown in Figure 3.1 depend on the individual atomic masses, but each sample contains the *same number* of atoms (Avogadro's number).

In our chemical thinking we use *moles* to express the quantitative relationships between formula units. For this to be useful, however, we must be able to measure mole amounts, and in the lab we use the balance to do this. Thus, if we weigh out 12.011 g of carbon, we know we have taken one mole of carbon. The next two examples illustrate how we use atomic masses as tools to covert between grams and moles for an element.

Rounding Atomic Masses

We will often obtain data from the Table of Atomic Numbers and Masses or from the Periodic Table for various calculations. Many atomic masses are known to several places beyond the decimal point, as you can readily see from the Tables. *The data in the given problem are what dictate how far to round an atomic mass.* We round an atomic mass so as to have enough significant figures to accommodate the number of significant figures of the other data in the calculation. Often we simply retain as many decimal places in an atomic mass as appear in the data given in the problem.

EXAMPLE 3.2
Converting Grams to Moles

The rounding off rules were given in footnote 6, page 14.

How many moles of silicon are in 4.60 g of Si?

ANALYSIS This is our first venture into one of the most common chemical calculations, finding the number of *moles* present in a given *mass* of substance. Let's restate the question as follows.

$$4.60 \text{ g Si} \Longleftrightarrow ? \text{ mol Si}$$

The atomic mass of silicon, 28.09 (rounded from 28.0855), gives us the needed tool, because we know from the definition of the mole that the following is true about silicon.

$$28.09 \text{ g Si} \Longleftrightarrow 1 \text{ mol Si}$$

Like any equivalency, this one makes available two conversion factors. One will be the tool we need for our grams-to-moles conversion.

$$\frac{28.09 \text{ g Si}}{1 \text{ mol Si}} \quad \text{and} \quad \frac{1 \text{ mol Si}}{28.09 \text{ g Si}}$$

Thus if we multiply the given, 4.60 g Si, by the second conversion factor, we will obtain the answer.

SOLUTION We carry out the following calculation. Notice how the units, g Si, cancel to leave the correct unit, mol Si. (Draw in the cancel lines yourself.)

$$4.60 \text{ g Si} \times \frac{1 \text{ mol Si}}{28.09 \text{ g Si}} = 0.164 \text{ mol Si}$$

In other words,

$$4.60 \text{ g Si} \Longleftrightarrow 0.164 \text{ mol Si}$$

CHECK Does the answer "make sense?" Is its *numerical size* at least in the right range? Yes, because we began with 4.60 g Si, clearly *less* than a whole mole of silicon, so the answer has to reflect this obvious fact.

■ **Practice Exercise 2** How many moles of sulfur are present in 35.6 g of sulfur?

Circular silicon wafers are being loaded into a high-temperature furnace during one step in the manufacture of silicon chips.

How many grams of gold are in 0.150 mol of Au?

ANALYSIS The question can be restated as follows:

$$0.150 \text{ mol Au} \Longleftrightarrow ? \text{ g Au}$$

The atomic mass of gold is 196.97, so the following is true about gold.

$$1 \text{ mol Au} \Longleftrightarrow 196.97 \text{ g Au}$$

The conversion factor that we need to convert the number of moles of Au to the number of grams of Au comes from this equivalency.

SOLUTION The given is 0.150 mol Au, so we multiply it by the ratio of grams of Au to moles of Au obtained from the atomic mass of gold.

$$0.150 \text{ mol Au} \times \frac{196.97 \text{ g Au}}{1 \text{ mol Au}} = 29.5 \text{ g Au}$$

Thus, we can write 0.150 mol Au ⟺ 29.5 g Au.

CHECK We have much less than a whole mole of gold, so the numerical part of the answer has to be much less than 196.97, gold's atomic mass, which it is.

■ **Practice Exercise 3** How many grams of silver are in 0.263 mol of Ag?

EXAMPLE 3.3
Converting Moles to Grams

Very often chemists round atomic masses to the second decimal place even when other data, like 0.150 mol, indicate that fewer numbers of significant figures are required in the rounded atomic masses.

Using Avogadro's Number in Calculations

Avogadro's number is rarely used in solving ordinary stoichiometry problems. Only when we need to relate some macroscopic lab-size property, like mass, to numbers of individual atomic-size particles do we require this number. For example, we might want to know how many atoms are in a sample of an element with a certain mass, or we might want to know the mass in grams of one or more atoms of an element. The following example illustrates how we carry out such a calculation, and it provides further practice in using a basic definition as a specific tool for a calculation.

Gold nugget.

EXAMPLE 3.4
Converting Mass to Number of Atoms

How many atoms are in a sample of uranium with a mass of 1.000×10^{-6} g (1.000 μg)?

ANALYSIS Notice the wording of the question: How many *atoms . . . in* 1.000×10^{-6} *g*. This is the key that tells us we are going to have to use Avogadro's number. Because the atomic mass of uranium is 238.03, we may write

$$1 \text{ mol U} \Leftrightarrow 238.03 \text{ g U}$$

Using Avogadro's number, we may write

$$1 \text{ mol U} \Leftrightarrow 6.022 \times 10^{23} \text{ atoms U}$$

Therefore,

$$238.03 \text{ g U} \Leftrightarrow 6.022 \times 10^{23} \text{ atoms U}$$

The last equivalency provides the necessary conversion factor.

SOLUTION The given is 1.000×10^{-6} g U, so we multiply it by the ratio of atoms U to g U given in the last equivalency.

$$1.000 \times 10^{-6} \text{ g U} \times \frac{6.022 \times 10^{23} \text{ atoms U}}{238.03 \text{ g U}} = 2.530 \times 10^{15} \text{ atoms U}$$

Thus a 1-microgram sample of uranium, a tiny speck, contains over 2.5 million billion ($10^6 \times 10^9$) atoms.

■ **Practice Exercise 4** How many atoms are in 1.000×10^{-9} g of lead?

FIGURE 3.2

One mole of four different compounds. Each sample contains the identical number of formula units or molecules, Avogadro's number.

Formula Masses and Molecular Masses

Just as each element has an atomic mass, each compound has a **formula mass,** the sum of the atomic masses of the atoms shown in the formula. The mole concept works with compounds exactly as it does with elements. Variations on the term *formula mass* are common. When a substance is known to be molecular, we refer to its **molecular mass,** reserving *formula mass* for ionic compounds where, instead of a molecule, we think of a formula unit.

Figure 3.2 shows 1.00 mole samples of four compounds. Their sizes in *grams* vary with their formula (molecular) masses, but they have identical numbers of molecules or formula units. *For every chemical substance, its formula (molecular) mass in grams equals one mole of it.* This makes the calculation of a formula (molecular) mass a common operation.

EXAMPLE 3.5
Calculating the Formula Mass of a Compound

What is the formula mass of calcium phosphate, $Ca_3(PO_4)_2$, expressed to the second decimal place?

ANALYSIS We simply add up the atomic masses, remembering that a subscript *outside* parentheses is a multiplier for all atoms inside.

SOLUTION Using atomic masses of 40.08 for Ca, 30.97 for P, and 16.00 for O,

$$Ca_3(PO_4)_2 = \quad 3 \text{ Ca} \quad + \quad 2 \text{ P} \quad + \quad 8 \text{ O}$$

Formula mass of $Ca_3(PO_4)_2 = (3 \times 40.08) + (2 \times 30.97) + (8 \times 16.00)$

$$= \quad 120.24 \quad + \quad 61.94 \quad + \quad 128.00$$

$$= \quad 310.18$$

Many chemists use "molecular weight" or "formula weight" for molecular (or formula) mass.

To have 1 mol of calcium phosphate, the sample's mass must be 310.18 g. In other words, 1 mol $Ca_3(PO_4)_2 \Leftrightarrow 310.18$ g of $Ca_3(PO_4)_2$.

Calcium phosphate, called in the agricultural world "white gold" and mined as a mineral, is a major source of phosphate fertilizer.

■ **Practice Exercise 5** What is the mass in grams of 1 mol of Na_2CO_3, expressed to the second decimal place?

We can calculate grams from moles or moles from grams using the formula masses of compounds just as we did for elements in Examples 3.1 and 3.3.

 Formula mass; molecular mass

A student has worked out the details for an experiment that will use 0.115 mol of $Ca_3(PO_4)_2$, calcium phosphate, as a starting material. How many grams of $Ca_3(PO_4)_2$ should be weighed out?

EXAMPLE 3.6
Calculating Grams from Moles for Compounds

ANALYSIS The problem can be restated as follows:

$$0.115 \text{ mol } Ca_3(PO_4)_2 \Longleftrightarrow ? \text{ g } Ca_3(PO_4)_2$$

The tool we need is the formula mass, and in Example 3.5, we found that for $Ca_3(PO_4)_2$ the formula mass is 310.18. This gives us the following equivalency.

$$1 \text{ mol } Ca_3(PO_4)_2 \Longleftrightarrow 310.18 \text{ g } Ca_3(PO_4)_2$$

SOLUTION The given is $0.115 \text{ mol } Ca_3(PO_4)_2$, so we multiply it by the ratio of g $Ca_3(PO_4)_2$ to mol $Ca_3(PO_4)_2$ that we obtain from the formula mass of $Ca_3(PO_4)_2$.

$$0.115 \text{ mol } Ca_3(PO_4)_2 \times \frac{310.18 \text{ g } Ca_3(PO_4)_2}{1 \text{ mol } Ca_3(PO_4)_2} = 35.7 \text{ g } Ca_3(PO_4)_2$$

In other words,

$$0.115 \text{ mol } Ca_3(PO_4)_2 \Longleftrightarrow 35.7 \text{ g } Ca_3(PO_4)_2$$

CHECK Does the *size* of the answer make sense? Yes. The student requires a little over one-tenth of a mole of $Ca_3(PO_4)_2$, so a bit over one-tenth of 310 g is needed.

■ **Practice Exercise 6** How many grams of sodium carbonate, Na_2CO_3, must be taken to obtain 0.125 mol of Na_2CO_3?

■ **Practice Exercise 7** A sample of 45.8 g of H_2SO_4 contains how many moles of H_2SO_4?

Applications of Stoichiometry

One of the most common uses of stoichiometry in the lab occurs when we must relate the mass of one substance needed to make a compound to the *mass* of another that is also required.

EXAMPLE 3.7
Reaction Stoichiometry

$SiCl_4$ is used to make transistor-grade elemental silicon.

How many grams of Cl are needed to combine with 24.4 g of Si to make silicon tetrachloride, $SiCl_4$?

ANALYSIS This is our first opportunity to combine some of the concepts that we've studied and to begin to develop *in you* a feel for how to approach chemistry problems. Let's begin by restating the problem as follows.

$$24.4 \text{ g Si} \Leftrightarrow ? \text{ g Cl (for the compound } SiCl_4)$$

We have not learned of a *direct* way to go from the number of *grams* of Si to the number of *grams* of Cl, so to solve this problem we will need to combine more than one step. Where do we begin? Let's start by examining what we *do* know to attempt to find the tools that will take us to the answer.

We do know a formula, $SiCl_4$, and it is the key to the problem's solution. Think of this formula as a tool that permits us to write the following mole equivalency.

$$1 \text{ mol Si} \Leftrightarrow 4 \text{ mol Cl}$$

We also know the mass of Si. We must go from 24.4 g of Si to *moles* of Si. Then we can use the number of moles of Si and the previous equivalency (1 mol Si ⇔ 4 mol Cl) to calculate the number of *moles* of Cl. Finally, we can use the number of moles of Cl to calculate the corresponding number of grams of Cl. Our successive moves can be diagrammed as follows. The starting point is at 24.4 g Si. The tools we need are by the arrows.

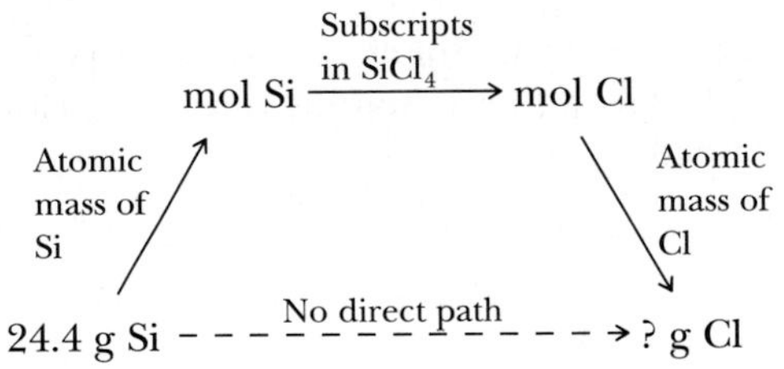

The top level, the mole level, is where the stoichiometric information in the formula $SiCl_4$ comes into play. *In general, the key calculation in a stoichiometry problem occurs at the mole level.*

SOLUTION First we move from "g Si" to "mol Si" using the atomic mass of Si.

$$1 \text{ mol Si} \Leftrightarrow 28.09 \text{ g Si}$$

So,

$$24.4 \text{ g Si} \times \frac{1 \text{ mol Si}}{28.09 \text{ g Si}} = 0.869 \text{ mol Si}$$

Next we use the stoichiometric equivalency we earlier found in the formula $SiCl_4$ to find the number of moles of Cl.

$$0.869 \text{ mol Si} \times \frac{4 \text{ mol Cl}}{1 \text{ mol Si}} = 3.48 \text{ mol Cl}$$

Finally, we change the number of moles of Cl to the number of grams of Cl.

$$1 \text{ mol Cl} \Leftrightarrow 35.45 \text{ g Cl}$$

So

$$3.48 \text{ mol Cl} \times \frac{35.45 \text{ g Cl}}{1 \text{ mol Cl}} = 1.23 \times 10^2 \text{ g Cl}$$

The answer is that 1.23×10^2 g of Cl combines with 24.4 g of Si to make $SiCl_4$.

A more efficient way to solve this problem is to string together or "chain" the conversion factors. Then it isn't necessary to write down the intermediate answers. (Insert the cancel lines yourself to show that the answer is in the correct units.)

$$24.4 \text{ g Si} \times \frac{1 \text{ mol Si}}{28.09 \text{ g Si}} \times \frac{4 \text{ mol Cl}}{1 \text{ mol Si}} \times \frac{35.45 \text{ g Cl}}{1 \text{ mol Cl}} = 1.23 \times 10^2 \text{ g Cl}$$

Grams Si $\longrightarrow$ mole Si $\longrightarrow$ mole Cl $\longrightarrow$ grams Cl

CHECK Remember to ask, "Does the answer make sense?" It seems like a great deal of Cl, 123 g, for the 24.4 g of Si, particularly since the atomic masses of Si and Cl are not too different, 28.09 versus 35.45. But notice that *in moles* the formula, $SiCl_4$, demands four times as much Cl as Si, and 4 times 24.4 is approaching the numerical part of the answer.

■ **Practice Exercise 8** How many grams of iron are needed to combine with 25.6 g of O to make Fe_2O_3?

Among the most exciting experimental activities in chemistry is making an entirely new compound or isolating from a natural source a compound that has never been reported before. Immediate questions are then, What is it? And what is its formula? To find out, the research begins with **qualitative analysis,** a series of procedures to identify all of the elements making up the substance. Next come procedures for **quantitative analysis,** which determine the mass of each element in a weighed sample of the substance. The usual form for describing the relative masses of the elements in a compound is a list of *percentages by mass* called the compound's **percentage composition.** The **percentage by mass** of an element is the number of grams of the element present in 100 g of the compound. In general, a percentage by mass is found by using the following equation.

$$\% \text{ by mass of element} = \frac{\text{mass of element}}{\text{mass of whole sample}} \times 100\%$$

3.3
PERCENTAGE COMPOSITION

 Percentage composition

EXAMPLE 3.8
Calculating a Percentage Composition

A sample of a liquid with a mass of 8.657 g was decomposed into its elements and gave 5.217 g of carbon, 0.9620 g of hydrogen, and 2.478 g of oxygen. What is the percentage composition of this compound?

ANALYSIS We must apply the equation for the percent by mass of an element, given above, to each element. The "mass of whole sample" here is 8.657 g, so we take each element in turn and perform the calculations. The "check" is that the percentages must add up to 100%, allowing for small differences caused by rounding.

SOLUTION

For C: $\dfrac{5.217 \text{ g}}{8.657 \text{ g}} \times 100\% = 60.26\% \text{ C}$

For H: $\dfrac{0.9620 \text{ g}}{8.657 \text{ g}} \times 100\% = 11.11\% \text{ H}$

For O: $\dfrac{2.478 \text{ g}}{8.657 \text{ g}} \times 100\% = \underline{28.62\% \text{ O}}$

Sum of percentages: 99.99%

The results tell us that in 100.00 g of the liquid there are 60.26 g of carbon, 11.11 g of hydrogen, and 28.62 g of oxygen.

■ **Practice Exercise 9** From 0.5462 g of a compound there was isolated 0.1417 g of nitrogen and 0.4045 g of oxygen. What is the percentage composition of this compound? Are any other elements present?

Theoretical Percentage Composition

Another kind of calculation is to find the *theoretical percentage composition,* which is calculated from the formula. This is needed, for example, when the same elements can combine in two or more ways. Nitrogen and oxygen, for example, form all of the following compounds: N_2O, NO, NO_2, N_2O_3, N_2O_4, and N_2O_5. By comparing the percentage composition found by experiment for some unknown compound of nitrogen and oxygen to the theoretical percentages for each possible formula, a choice can be made among the candidates. Which formula, for example, fits the percentage composition calculated in Practice Exercise 9? We will work this as an example.

NO_2 causes the red-brown color of smog.

EXAMPLE 3.9
Calculating a Theoretical Percentage Composition

Do the mass percentages of 25.94% N and 74.06% O match the formula N_2O_5?

ANALYSIS To calculate the theoretical percentages by mass of N and O in N_2O_5, we need the masses of N and O in a specific sample of N_2O_5. *If we choose 1 mol of the given compound to be this sample, the computations to find the rest of the data will be simpler.*

SOLUTION We know that 1 mol of N_2O_5 must contain 2 mol N and 5 mol O. The corresponding number of grams of N and O are found as follows.

$$2 \text{ mol N} = 2 \times 14.01 \text{ g N} = 28.02 \text{ g N}$$
$$5 \text{ mol O} = 5 \times 16.00 \text{ g O} = \underline{80.00 \text{ g O}}$$
$$1 \text{ mol } N_2O_5 = 108.02 \text{ g } N_2O_5$$

Now we can calculate the percentages:

$$\text{For \% N: } \frac{28.02 \text{ g}}{108.02 \text{ g}} \times 100\% = 25.94\% \text{ N in } N_2O_5$$

$$\text{For \% O: } \frac{80.00 \text{ g}}{108.02 \text{ g}} \times 100\% = 74.06\% \text{ O in } N_2O_5$$

Thus the theoretical percentages match the experimental values, so the correct formula is N_2O_5.

■ **Practice Exercise 10** Calculate the theoretical percentage composition of N_2O_3.

3.4
EMPIRICAL AND MOLECULAR FORMULAS

Empirical is from the Latin *empiricus,* "pertaining to experience."

The compound that forms when phosphorus burns in oxygen consists of molecules with the formula P_4O_{10}. When a formula gives the composition of one *molecule,* it is called a **molecular formula.** Notice, however, that both the subscripts 4 and 10 are divisible by 2, so the smallest numbers that tell us the *ratio* of P to O are 2 and 5. In a simpler (but less informative) kind of formula, the **empirical formula,** the subscripts are the *smallest* whole numbers that describe the ratios of the atoms in the substance. The empirical formula of P_4O_{10} is thus P_2O_5. It is an *empirical* formula that is calculated from the mass data obtained by quantitative analysis, a calculation illustrated in the next example.

A sample of a compound between tin and chlorine with a mass of 2.571 g was found to contain 1.171 g of tin. What is its empirical formula?

ANALYSIS The empirical formula must specify the ratio *by atoms* of Sn and Cl, but how do we obtain this from the *mass* data given? The answer lies in the mole concept. Recall that for any compound, *moles of atoms always combine in the same ratio as the individual atoms themselves.* Thus if we can find the *mole* ratio of Sn to Cl, then we know it will be identical to the atom ratio and we will have the empirical formula. The first step, therefore, is to convert the numbers of grams of Sn and Cl to the numbers of moles of Sn and Cl. Then we convert these numbers into their simplest *whole-number* ratio.

You no doubt noticed that the problem did not give the mass of chlorine in the 2.571 g sample. But the mass of Cl is simply the difference between 2.571 g and the mass of Sn present in the sample, namely, 1.171 g of Sn. The problem actually illustrates a common rule in chemical analysis: *the quantity of one of the elements in a sample of a compound generally is not directly determined by analysis but is calculated by difference.*

SOLUTION First, we find the mass of Cl in 2.571 g of compound:

$$\text{Mass of Cl} = 2.571 \text{ g} - 1.171 \text{ g} = 1.400 \text{ g Cl}$$

Now we use the atomic mass tool for each element to convert the mass data for tin and chlorine into moles. We'll carry three decimal places for the atomic masses, using 118.710 for Sn and 35.453 for Cl. (This is often done in quantitative analysis so that the atomic masses have at least as many decimal places as are in the mass data.)

$$1.171 \text{ g Sn} \times \frac{1 \text{ mol Sn}}{118.710 \text{ g Sn}} = 0.009864 \text{ mol Sn}$$

$$1.400 \text{ g Cl} \times \frac{1 \text{ mol Cl}}{35.453 \text{ g Cl}} = 0.03949 \text{ mol Cl}$$

We could now write a formula: $Sn_{0.009864}Cl_{0.03949}$, but we want whole numbers as subscripts for a proper empirical formula. To change the subscripts in $Sn_{0.009864}Cl_{0.03949}$ into a set of whole numbers *that are in the same ratio,* we divide each by a common divisor. The easiest common divisor is the smaller number in the set. *This is always the way to begin the search for whole-number subscripts; pick the smallest number of the set as the divisor.* It's guaranteed to make at least one subscript a whole number, namely, 1. Here, we divide both numbers by 0.009864.

$$Sn_{\frac{0.009864}{0.009864}}Cl_{\frac{0.03949}{0.009864}} = Sn_{1.000}Cl_{4.003}$$

As a rule, if a calculated subscript, expressed to the proper number of significant figures, differs from a whole number by less than a few units in the last significant digit, we can safely round to the nearest whole number. We may round 4.003 to 4, so we can write the empirical formula as $SnCl_4$.

■ **Practice Exercise 11** A 1.5246 g sample of a compound between nitrogen and oxygen contains 0.7117 g of nitrogen. Calculate its empirical formula. (Write the symbol of nitrogen first.)

Sometimes our strategy of using the lowest common divisor does not give whole numbers. Let's see how to handle such a situation.

EXAMPLE 3.10
Calculating an Empirical Formula from Mass Data

An empirical formula is also called a compound's *simplest formula.*

EXAMPLE 3.11
Calculating an Empirical
Formula from Mass
Composition

Magnetite is so named because it
has magnetic properties.

One of the compounds of iron and oxygen, "black iron oxide," occurs naturally in the mineral magnetite. When a 2.4480 g sample was analyzed it was found to have 1.7714 g of Fe. Calculate the empirical formula of this compound.

ANALYSIS When we calculate the mass of O in the sample by difference, we'll have the masses in grams of both elements, but we need the numbers of *moles* for a formula. So we'll use the atomic masses of Fe and O as tools for converting their respective masses to numbers of moles. Then we'll write a trial formula and see whether we can adjust the coefficients to their smallest whole numbers by the strategy learned in Example 3.10.

SOLUTION The mass of O is, as we said, found by difference.

$$2.4480 \text{ g} - 1.7714 \text{ g} = 0.6766 \text{ g}$$

Using three decimal places, the atomic masses of Fe and O give us the following equivalencies.

$$55.847 \text{ g Fe} \Leftrightarrow 1 \text{ mol Fe}$$

$$15.999 \text{ g O} \Leftrightarrow 1 \text{ mol O}$$

The moles of Fe and O in the sample can now be calculated.

$$1.7714 \text{ g Fe} \times \frac{1 \text{ mol Fe}}{55.847 \text{ g Fe}} = 0.031719 \text{ mol Fe}$$

$$0.6766 \text{ g O} \times \frac{1 \text{ mol O}}{15.999 \text{ g O}} = 0.04229 \text{ mol O}$$

These results let us write the formula as $Fe_{0.031719}O_{0.04229}$.

Our first effort to change the ratio of 0.031719 to 0.04229 into whole numbers is to divide both by the smaller, 0.031719.

$$Fe_{\frac{0.031719}{0.031719}}O_{\frac{0.04229}{0.031719}} = Fe_{1.0000}O_{1.333}$$

The numbers before the *last* place are known with certainty, according to our definition of significant figures (page 13).

This time the procedure seems to fail. Given the fact that our number of significant figures is four, we can be sure that the digits 1.33 in 1.333 are known with certainty. Therefore the subscript for O, 1.333, is much too far from a whole number to round off and retain the precision allowed. In a *mole* sense, the ratio of 1 to 1.333 is correct; we just have not yet found a way to state the ratio in whole numbers. Let's multiply each subscript in $Fe_{1.0000}O_{1.333}$ by a whole number, 2. *This does not change the ratio;* it changes only the size of the numbers used to state it.

$$Fe_{(1.0000 \times 2)}O_{(1.333 \times 2)} = Fe_{2.0000}O_{2.666}$$

This didn't work either; 2.666 is also too far from a whole number for the allowed precision to be rounded off. Let's try using 3 instead of 2 *on the first ratio of 1.0000 to 1.333.*

$$Fe_{(1.0000 \times 3)}O_{(1.333 \times 3)} = Fe_{3.0000}O_{3.999}$$

We are now justified in rounding; 3.999 is acceptably close to 4. The empirical formula of the oxide of iron is Fe_3O_4.

■ **Practice Exercise 12** A 2.012 g sample of a compound of nitrogen and oxygen has 0.5219 g of nitrogen. Calculate its empirical formula.

Empirical Formulas from Percentage Compositions

You have seen that to calculate an empirical formula *we need the number of grams of each of the elements in a sample of the compound.* Only rarely is it possible, however, to use the *same* sample of a compound to obtain the masses of *all* of its elements. Normally, two or more analyses are carried out on separate samples. The results are converted to percentages by mass *in order to put the data obtained from different samples on the same basis, namely, a 100 gram sample.* Each mass percentage can then be thought of as a certain number of grams and be converted into the number of moles of the element. Then the mole proportions are converted to whole numbers, which are used as subscripts to write the empirical formulas.

A white powder used in paints, enamels, and ceramics has the following percentage composition: Ba, 69.58%, C, 6.090%; and O, 24.32%. What is its empirical formula?

ANALYSIS The percentages are numerically equal to *masses* if we choose a 100 g sample of the compound, but subscripts express *moles*. So we have to convert masses to moles using the atomic mass tool in a grams-to-moles conversion. We can't expect that the moles found this way will be simple whole numbers. We expect that we'll have to divide the numbers of moles by their lowest value as the first step to getting whole-number coefficients.

SOLUTION Using three decimal places, the atomic masses are 137.327 for Ba, 12.011 for C, and 15.999 for O. Each gives the usual type of conversion factor, so the grams-to-moles operations are

$$\text{Ba:} \quad 69.58 \text{ g Ba} \times \frac{1 \text{ mol Ba}}{137.327 \text{ g Ba}} = 0.5067 \text{ mol Ba}$$

$$\text{C:} \quad 6.090 \text{ g C} \times \frac{1 \text{ mol C}}{12.011 \text{ g C}} = 0.5070 \text{ mol C}$$

$$\text{O:} \quad 24.32 \text{ g O} \times \frac{1 \text{ mol O}}{15.999 \text{ g O}} = 1.520 \text{ mol O}$$

Our preliminary empirical formula is then

$$Ba_{0.5067}C_{0.5070}O_{1.520}$$

We next divide each subscript by the smallest, 0.5067.

$$Ba_{\frac{0.5067}{0.5067}}C_{\frac{0.5070}{0.5067}}O_{\frac{1.520}{0.5063}} = Ba_1C_{1.001}O_{3.000}$$

The coefficients are acceptably close to whole numbers, so the empirical formula is $BaCO_3$, representing barium carbonate.

■ **Practice Exercise 13** A white solid used to whiten paper has the following percentage composition: Na, 32.37%; S, 22.57%. The unanalyzed element is oxygen. What is the compound's empirical formula?

EXAMPLE 3.12
Calculating an Empirical
Formula from Percentage
Composition

Indirect Analysis

A compound is seldom broken down completely to its *elements* in a quantitative analysis, as our examples may lead you to think. Instead, the compound is changed into other *compounds*. The reactions separate the elements by capturing each one entirely (quantitatively) in a *separate* compound *whose formula is known*.

We illustrate how to use the resulting data by an example of a compound made entirely of carbon, hydrogen, and oxygen. Such compounds burn completely in pure oxygen—the reaction is called *combustion*—and the sole products are carbon dioxide and water. (This kind of indirect analysis is sometimes called *combustion analysis.*) The complete combustion of methyl alcohol (CH_3OH), for example, occurs according to the following equation.

$$2CH_3OH + 3O_2 \longrightarrow 2CO_2 + 4H_2O$$

The carbon dioxide and water can be separated and are individually weighed. Notice that all of the carbon atoms in the original compound end up as CO_2 molecules, and all of the hydrogen atoms are in H_2O molecules. In this way at least two of the original elements, C and H, are entirely separated. From the masses of CO_2 and H_2O and the mass of the original sample, the percentages of C and H in the original compound can be calculated. They will not add up to 100% when oxygen is also present in the original compound, but the percentage of O is simply the difference between 100.00% and the sum of the C and H percentages.

%C + %H + %O = 100%

EXAMPLE 3.13
Empirical Formula from Indirect Analysis

A 0.5438 g sample of a liquid consisting of only C, H, and O was burned in pure oxygen, and 1.039 g of CO_2 and 0.6369 g of H_2O were obtained. What is the empirical formula of the compound?

ANALYSIS There are two broad parts to this problem. We first have to calculate the *percentage composition* of the original compound. Then we use these results to calculate the *empirical formula*.

For the first part, we start with the numbers of *grams* of CO_2 and H_2O. We convert these to the numbers of *moles* of CO_2 and H_2O so that we can use the stoichiometric equivalences given by the subscripts in their formulas to find the numbers of moles of C and H.

$$1 \text{ mol } CO_2 \Leftrightarrow 1 \text{ mol C}$$

$$1 \text{ mol } H_2O \Leftrightarrow 2 \text{ mol H}$$

The numbers of moles of C and H must next be changed to the numbers of *grams* of C and H in order to calculate their *mass percentages* in the original sample. In short, we have the following series of calculations.

$$\text{Grams } CO_2 \longrightarrow \text{moles } CO_2 \longrightarrow \text{moles C} \longrightarrow \text{grams C} \longrightarrow \%C$$

$$\text{Grams } H_2O \longrightarrow \text{moles } H_2O \longrightarrow \text{moles H} \longrightarrow \text{grams H} \longrightarrow \%H$$

We find the percentage oxygen by difference.

$$100.00\% - (\%C + \%H) = \%O$$

In the second half of the solution, we use the percentage composition to calculate the empirical formula as in Example 3.12.

SOLUTION For the first half, we will use a chain calculation to handle all four steps in each series. Be sure that you see how each conversion factor changes the original units. The "flow" of the calculation is given beneath each chain. (We've figured the molecular masses from atomic masses taken to their third decimal places: CO_2, 44.009; H_2O, 18.015.) For the number of grams of C in 1.039 g of CO_2,

$$1.039 \text{ g } CO_2 \times \frac{1 \text{ mol } CO_2}{44.009 \text{ g } CO_2} \times \frac{1 \text{ mol C}}{1 \text{ mol } CO_2} \times \frac{12.011 \text{ g C}}{1 \text{ mol C}} = 0.2836 \text{ g C}$$

$$\text{Grams } CO_2 \longrightarrow \text{moles } CO_2 \longrightarrow \text{moles C} \longrightarrow \text{grams C}$$

For the number of grams of H in 0.6369 g of H_2O,

$$0.6369 \text{ g } H_2O \times \frac{1 \text{ mol } H_2O}{18.015 \text{ g } H_2O} \times \frac{2 \text{ mol } H}{1 \text{ mol } H_2O} \times \frac{1.008 \text{ g } H}{1 \text{ mol } H} = 0.07127 \text{ g } H$$

Grams H_2O → moles H_2O → moles H → grams H

Now we use the calculated masses of C and H and the mass of the sample, 0.5438 g, to calculate the mass percentages of C and H.

$$\text{For C, } \frac{0.2836 \text{ g}}{0.5438 \text{ g}} \times 100 = 52.15\% \text{ C}$$

$$\text{For H, } \frac{0.07127 \text{ g}}{0.5438 \text{ g}} \times 100 = \underline{13.11\% \text{ H}}$$
$$65.26\% \text{ (for C + H)}$$

The difference between this total and 100.00% is the percentage oxygen (the only other element). We find that $100.00 - 65.26 = 34.74$; the percentage by mass of O is thus 34.74%.

Now we can convert the mass percentages to an empirical formula. Taking the percentages as numbers of grams, we find the numbers of moles by a grams-to-moles calculation using atomic masses.

$$\text{For C: } \quad 52.15 \text{ g C} \times \frac{1 \text{ mol C}}{12.011 \text{ g C}} = 4.342 \text{ mol C}$$

$$\text{For H: } \quad 13.11 \text{ g H} \times \frac{1 \text{ mol H}}{1.008 \text{ g H}} = 13.01 \text{ mol H}$$

$$\text{For O: } \quad 34.74 \text{ g O} \times \frac{1 \text{ mol O}}{15.999 \text{ g O}} = 2.171 \text{ mol O}$$

Our preliminary molecular formula is thus $C_{4.342}H_{13.01}O_{2.171}$. We divide all of these subscripts by the smallest number, 2.171.

$$C_{\frac{4.342}{2.171}}H_{\frac{13.01}{2.171}}O_{\frac{2.171}{2.171}} = C_{2.000}H_{5.993}O_1$$

The results are acceptably close to C_2H_6O, the answer.

■ **Practice Exercise 14** The combustion of a 5.048 g sample of a compound of C, H, and O gave 7.406 g CO_2 and 3.027 g H_2O. Calculate the empirical formula of the compound.

Molecular Formulas from Empirical Formulas and Molecular Masses

What we obtain by using mass data to calculate a formula is an *empirical* formula, as we have said, and the empirical formula is all that we can get for ionic compounds. For molecular compounds, however, chemists prefer *molecular* formulas because they contain more information, such as the exact composition of one molecule.

Recall that the subscripts of a molecular formula are whole-number multiples of those in the empirical formula. The subscripts of the molecular formula, P_4O_{10}, for example, are each two times those in the empirical formula, P_2O_5. The molecular mass of P_4O_{10} is likewise two times the formula mass of P_2O_5. You can see that to find out whether a formula calculated from the percentage composition of a molecular compound is only an *empirical* for-

For many substances, the empirical and molecular formulas are identical. Examples are water, H_2O, and ammonia, NH_3.

mula, we need to determine one more fact about the compound, its molecular mass. If the experimental molecular mass *equals* the calculated empirical formula mass, the empirical formula itself is a molecular formula. Otherwise, the experimental molecular mass will be some whole number multiple of the value calculated from the empirical formula. Whatever the whole number is, it's a common multiplier for the subscripts of the empirical formula.

EXAMPLE 3.14
Calculating a Molecular Formula

Styrene, the raw material for polystyrene foam plastics, has an empirical formula of CH. Its molecular mass is 104. What is its molecular formula?

ANALYSIS The molecular mass of styrene, 104, is some simple multiple of the formula mass calculated for its empirical formula, CH. So we have to compute the latter, and then see how many times larger 104 is.

SOLUTION For CH, the formula mass is

$$12.01 + 1.01 = 13.01$$

To find how much larger 104 is than 13.01, we divide.

$$\frac{104}{13.01} = 7.99$$

Rounding this to 8, we see that 104 is 8 times larger than 13.01, so the correct molecular formula of styrene must have subscripts 8 times those in CH. Styrene, therefore, is C_8H_8.

■ **Practice Exercise 15** The empirical formula of hydrazine is NH_2, and its molecular mass is 32.0. What is its molecular formula?

3.5
WRITING AND BALANCING CHEMICAL EQUATIONS

In experimental work, chemists spend much of their time dealing with the *chemical* properties of substances, which means that they work with chemical reactions. Chemists therefore nearly always plan experiments around chemical equations.

Chemical Equations

We learned in Chapter 2 that a *chemical equation* is a shorthand, quantitative description of a chemical reaction. An equation is *balanced* when all atoms present among the reactants are also somewhere among the products. *Coefficients,* we learned, are multiplier numbers for respective formulas and their values determine whether an equation is balanced.

Balancing Equations

Always approach the balancing of an equation as a two-step process.

STEP 1 *Write the unbalanced "equation."* Organize the formulas in the pattern of an equation with plus signs and an arrow. Use *correct* formulas. (You learned to write many of them in Chapter 2, but until we have studied more chemistry, you will usually be given formulas.)

STEP 2 *Adjust the coefficients to get equal numbers of each kind of atom on both sides of the arrow.*

Coefficient
↓
$3CO_2$
↑
Subscript

When doing step 2, make no changes in the formulas, either in the atomic symbols or their subscripts. If you do, the equation will involve different substances from those intended. You may still be able to balance it, but the equation will not be for the reaction you want.

We'll begin with simple equations that can be balanced easily by inspection. An example is the reaction of zinc metal with hydrochloric acid (margin photo). First, we need the correct formulas, and this time we'll include the physical states because they are different. The reactants are zinc, $Zn(s)$, and hydrochloric acid, an aqueous solution of the gas, hydrogen chloride, HCl, and so symbolized as $HCl(aq)$. We also need formulas for the products. Zn changes to a water-soluble compound, zinc chloride, $ZnCl_2(aq)$, and hydrogen gas, $H_2(g)$, bubbles out as the other product. (Recall that hydrogen occurs naturally as a *diatomic molecule*, not as atoms.)

STEP 1   Write an unbalanced equation.

$$Zn(s) + HCl(aq) \longrightarrow ZnCl_2(aq) + H_2(g) \qquad \text{(unbalanced)}$$

STEP 2   Adjust the coefficients to get equal numbers of each kind of atom on both sides of the arrow.

There is no simple set of rules for adjusting coefficients. Experience is the greatest help. You can begin almost anywhere, with any element obviously out of balance. We see two out of balance here, H and Cl. Because there are two Cl to the right of the arrow but only one to the left, put a 2 in front of the HCl on the left side. The result is

$$Zn(s) + 2HCl(aq) \longrightarrow ZnCl_2(aq) + H_2(g)$$

Everything is now balanced. On each side we find 1 Zn, 2 H, and 2 Cl. Equations that can be balanced by inspection do not get much harder than this.

One complication is that an infinite number of *balanced* equations can be written for any given reaction! We might, for example, have adjusted the coefficients so that our equation came out as follows.

$$2Zn(s) + 4HCl(aq) \longrightarrow 2ZnCl_2(aq) + 2H_2(g)$$

This equation is also balanced. So we add one further convention to the rules about balanced equations, and it is only a *convention*, not a law of nature. The *smallest* whole-number coefficients are generally used to write balanced equations. Occasionally we ignore this convention, but only for a special reason.

Zinc metal reacts with hydrochloric acid.

EXAMPLE 3.15
Writing a Balanced Equation

Sodium hydroxide, NaOH, and phosphoric acid, H_3PO_4, react as aqueous solutions to give sodium phosphate, Na_3PO_4, and water. The sodium phosphate remains in solution. Write the balanced equation for this reaction.

SOLUTION We first write an unbalanced equation that includes all of the formulas arranged according to the conventions for a chemical equation. We include the designation (*aq*) for all substances dissolved in water (except H_2O itself, which we'll not give any designation when it is in its liquid state).

$$NaOH(aq) + H_3PO_4(aq) \longrightarrow Na_3PO_4(aq) + H_2O \qquad \text{(unbalanced)}$$

Among the several things not in balance, perhaps the easiest to follow across the arrow is Na, so we begin with it. There are 3 Na on the right side, so we put a 3 in front of NaOH on the left, as a trial.

$$3NaOH(aq) + H_3PO_4(aq) \longrightarrow Na_3PO_4(aq) + H_2O \qquad \text{(unbalanced)}$$

In deciding where to start, try leaving the balancing of H and O atoms (when they are involved) to the last.

"PO₄" is actually the phosphate
ion, PO₄³⁻.

Now the Na are in balance, and we leave the unit of PO_4 alone. It appears to be a unit, because it occurs as such on both sides of the arrow. Not counting the PO_4 unit, we have on the left 3 O and 3 H in 3 NaOH plus 3 H in H_3PO_4, for a net of 3 O and 6 H on the left. On the right, in H_2O, we have 1 O and 2 H. The ratio of 3 O to 6 H on the left is equivalent to the ratio of 1 O to 2 H on the right, so we write the multiplier (coefficient) 3 in front of H_2O.

$$3NaOH(aq) + H_3PO_4(aq) \longrightarrow Na_3PO_4(aq) + 3H_2O$$

We now have a balanced equation. On each side we have 3 Na, 1 PO_4, 6 H, and 3 O besides those in PO_4, and it is obvious that our coefficients cannot be reduced to smaller whole numbers.

■ **Practice Exercise 16** When aqueous solutions of calcium chloride, $CaCl_2$, and potassium phosphate, K_3PO_4, are mixed, a reaction occurs in which solid calcium phosphate, $Ca_3(PO_4)_2$, separates from the solution. The other product is $KCl(aq)$. Write the balanced equation.

The strategy in Example 3.15 of leaving whole units of atoms, like PO_4, unchanged is extremely useful. These are usually polyatomic ions that go through reactions unaffected. If we leave them alone, we have less atom counting to do.

Another suggestion for picking trial coefficients for balancing an equation is to use subscripts to suggest them. The reaction of aluminum with oxygen illustrates how this works.

EXAMPLE 3.16
Balancing Chemical
Equations

Although they are bright and shiny, aluminum objects are actually covered with a tight, invisible coating of solid aluminum oxide, Al_2O_3. This coating forms instantly when fresh aluminum metal is exposed to oxygen in the air. Write the balanced equation for the formation of this oxide. Remember that oxygen occurs in air as the diatomic molecule, $O_2(g)$.

SOLUTION The unbalanced equation comes first.

$$Al(s) + O_2(g) \longrightarrow Al_2O_3(s)$$

It doesn't matter where we start, but chemists know that starting with the *least simple* adjustment can actually work best over all. So let's begin with oxygen. More O is on the right than on the left, so we try its *subscript* in Al_2O_3 as the *coefficient* of O_2 on the left. We write 3 in front of O_2.

Subscript of O suggests
coefficient of O_2

$$Al(s) + 3O_2(g) \longrightarrow Al_2O_3(s)$$

Now we have (3×2) O on the left yet only 3 O on the right, but now the *subscript* in O_2 suggests a *coefficient* for Al_2O_3. So we place a 2 before Al_2O_3.

Subscript of O_2 suggests
coefficient of Al_2O_3

$$Al(s) + 3O_2(g) \longrightarrow 2Al_2O_3(s)$$

This leaves (2×2) Al on the right, so we write a 4 in front of Al on the left, and the equation is then balanced.

$$4Al(s) + 3O_2(g) \longrightarrow 2Al_2O_3(s)$$

■ **Practice Exercise 17** Iron reacts with chlorine, a gaseous, diatomic element, to form the solid, $FeCl_3$. Write the balanced equation.

The coefficients in an equation give us the ratios *by moles* of the reactants and products. In the following reaction, for example,

$$3NaOH(aq) + H_3PO_4(aq) \longrightarrow Na_3PO_4(aq) + 3H_2O$$

the coefficients tell us that to make 1 mol of Na_3PO_4 from 1 mol of H_3PO_4, we must also use 3 mol of NaOH. We're not compelled, of course, to carry out the reaction with these actual numbers of moles. The coefficients tell us only that we must take whatever quantities we choose in the *proportions* set by the coefficients. *The chemist makes the decision about the actual quantities, the magnitude, or the scale of the reaction. The natural world tells us what must be the combining proportions in moles for whatever the scale may be.* Regardless of the scale of a reaction, the coefficients of a chemical equation give the ratio in which the *moles* of one substance react with or produce *moles* of another. When you fully understand this fact, you will have mastered the heart of the stoichiometry of chemical reactions.

Look upon a balanced equation as a calculating tool, because its coefficients give us *stoichiometric equivalencies* between the substances involved. We have the following equivalencies, for example, in the reaction shown above between sodium hydroxide and phosphoric acid.

$$3 \text{ mol NaOH} \Leftrightarrow 1 \text{ mol H}_3\text{PO}_4$$

$$3 \text{ mol NaOH} \Leftrightarrow 1 \text{ mol Na}_3\text{PO}_4$$

$$3 \text{ mol NaOH} \Leftrightarrow 3 \text{ mol H}_2\text{O}$$

$$1 \text{ mol H}_3\text{PO}_4 \Leftrightarrow 1 \text{ mol Na}_3\text{PO}_4$$

$$1 \text{ mol H}_3\text{PO}_4 \Leftrightarrow 3 \text{ mol H}_2\text{O}$$

Any of these can be used to construct conversion factors for stoichiometric calculations.

3.6
USING CHEMICAL EQUATIONS IN CALCULATIONS

A balanced equation serves as a tool for relating moles of substances that react or are produced.

 Use of a balanced equation

How many moles of sodium phosphate, Na_3PO_4, can be made from 0.240 mol of NaOH by the following reaction?

$$3NaOH(aq) + H_3PO_4(aq) \longrightarrow Na_3PO_4(aq) + 3H_2O$$

ANALYSIS The question asks about a *moles-to-moles relationship* within a given reaction. We can restate the question as follows.

$$0.240 \text{ mol NaOH} \Leftrightarrow ? \text{ mol Na}_3\text{PO}_4$$

The balanced equation is our tool because it gives the coefficients we need. From the coefficients, we know the following.

$$3 \text{ mol NaOH} \Leftrightarrow 1 \text{ mol Na}_3\text{PO}_4$$

This enables us to prepare the conversion factor that we need.

EXAMPLE 3.17
Stoichiometry of Chemical Reactions

SOLUTION We convert 0.240 mol NaOH to the Na_3PO_4 equivalent to it in the reaction as follows.

$$0.240 \text{ mol NaOH} \times \frac{1 \text{ mol } Na_3PO_4}{3 \text{ mol NaOH}} = 0.0800 \text{ mol } Na_3PO_4$$

Thus, in this reaction,

$$0.240 \text{ mol NaOH} \Leftrightarrow 0.0800 \text{ mol } Na_3PO_4$$

CHECK Don't forget to ask, "Does the size of the answer make sense?" Yes. The equation told us that 3 mol NaOH $\Leftrightarrow$ 1 mol Na_3PO_4, so the actual number of moles of Na_3PO_4 should be one-third the actual number of moles of NaOH (0.240 mol), and it is.

■ **Practice Exercise 18** How many moles of sulfuric acid, H_2SO_4, are needed to react with 0.366 mol of NaOH by the following reaction?

$$2NaOH(aq) + H_2SO_4(aq) \longrightarrow Na_2SO_4(aq) + 2H_2O$$

EXAMPLE 3.18
Reaction Stoichiometry

Amateur gold seekers, unaware of the extremely poisonous nature of sodium cyanide, sometimes died using it carelessly.

Gold particles can be separated from crushed rock by passing air and an aqueous solution of sodium cyanide, NaCN, through the mixture. The gold dissolves by the following reaction in which oxygen in the air is also a reactant.

$$2Au(s) + 4NaCN(aq) + O_2(g) + 2H_2O \longrightarrow$$
$$2NaAu(CN)_2(aq) + H_2O_2(aq) + 2NaOH(aq)$$

How many moles of NaCN are required to dissolve 0.136 mol of Au?

ANALYSIS The question can be restated as follows.

$$0.136 \text{ mol Au} \Leftrightarrow ? \text{ mol NaCN}$$

The coefficients tell us:

$$2 \text{ mol Au} \Leftrightarrow 4 \text{ mol NaCN}$$

We will use this equivalency to make a conversion factor.

Gold metal is obtained from $NaAu(CN)_2$ by a reaction with zinc metal:

$$Zn(s) + 2 NaAu(CN)_2(aq) \longrightarrow$$
$$2 Au(s) + Na_2Zn(CN)_4(aq)$$

SOLUTION We multiply the given, 0.136 mol Au, by the appropriate conversion factor

$$0.136 \text{ mol Au} \times \frac{4 \text{ mol NaCN}}{2 \text{ mol Au}} = 0.272 \text{ mol NaCN}$$

CHECK Does the *size* of the answer make sense? Of course. The number of moles of NaCN has to be twice the number of moles of Au because the ratio of coefficients in 4 NaCN and 2 Au is 4:2 or 2:1.

■ **Practice Exercise 19** If 0.575 mol of CO_2 is produced by the combustion of propane, C_3H_8, how many moles of oxygen are consumed? The balanced equation is as follows (omitting physical states):

$$C_3H_8 + 5O_2 \longrightarrow 3CO_2 + 4H_2O$$

Before continuing, please notice a point of major importance. *Stoichiometric equivalencies obtained from an equation's coefficients apply only to the given reaction.* It is possible in the lab, for example, to arrange a reaction between sodium hydroxide and phosphoric acid according to any of the following equations.

$$NaOH(aq) + H_3PO_4(aq) \longrightarrow NaH_2PO_4(aq) + H_2O$$

where   1 mol NaOH $\Leftrightarrow$ 1 mol H_3PO_4

$$2NaOH(aq) + H_3PO_4(aq) \longrightarrow Na_2HPO_4(aq) + 2H_2O$$

where   2 mol NaOH $\Leftrightarrow$ 1 mol H_3PO_4

$$3NaOH(aq) + H_3PO_4(aq) \longrightarrow Na_3PO_4(aq) + 3H_2O$$

where   3 mol NaOH $\Leftrightarrow$ 1 mol H_3PO_4

The stoichiometric equivalencies between NaOH and H_3PO_4 are different in all three reactions. Stoichiometric equivalencies between substances are therefore always a function of the particular reaction being studied.

Stoichiometric Mass Calculations

In practical work, a chemist is often confronted by a question such as the following. "If I start with so many grams of reactant *A*, how many grams of reactant *B* ought I use, and how many grams of a particular product should be produced?" Notice that the question concerns *grams*, not moles, for the practical reason that masses in *grams* are delivered by laboratory balances. The coefficients of the desired reaction, however, know nothing about grams, only about relative numbers of moles. This is why the solutions to such questions flow from a given mass to the corresponding moles, then through stoichiometric equivalencies to the moles of something else, and finally to the equivalent in mass of the latter. Figure 3.3 outlines this flow. We emphasize again that when we know two facts, namely, the *balanced equation* and the *mass* of any substance in it, we can calculate the required or expected mass of *any* other substance in the equation. Example 3.19 illustrates how it works.

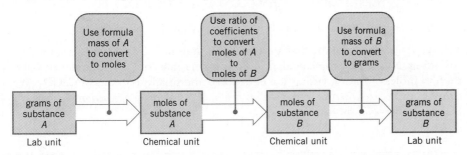

FIGURE 3.3

The flow of the calculation for solving stoichiometry problems. The flow applies to the calculation of the required or the expected mass of *any* reactant or product.

Portland cement is a mixture of the oxides of calcium, aluminum, and silicon. The raw material for its calcium oxide is calcium carbonate, which occurs as the chief component of a natural rock, limestone. When calcium carbonate is strongly heated it decomposes by the following reaction. One product, CO_2, is driven off to leave the desired CaO as the only other product.

$$CaCO_3(s) \xrightarrow{\text{heat}} CaO(s) + CO_2(g)$$

A chemistry student is to prepare 1.50×10^2 g of CaO in order to test a particular "recipe" for portland cement. How many grams of $CaCO_3$ should be used, assuming that all will be converted?

EXAMPLE 3.19
Stoichiometric Mass
Calculations

CaO is commonly called "quick lime" and it is an ingredient in portland cement.

Portland cement gets its name from the color of stone formations near Portland, England.

ANALYSIS We're given the number of *grams* of one substance, but we know that the fundamental chemical reasoning must be done in terms of *moles*. So we must first translate the desired mass of product into moles. Then we can employ the balanced equation as a tool, using its coefficients and the calculated number of moles of CaO to find the equivalent number of moles of the reactant, $CaCO_3$. Finally, we convert the calculated moles of the reactant into grams.

SOLUTION Making a conversion factor out of the formula mass of CaO, 56.08, we translate 1.50×10^2 g CaO into moles as follows.

$$1.50 \times 10^2 \text{ g CaO} \times \frac{1 \text{ mol CaO}}{56.08 \text{ g CaO}} = 2.67 \text{ mol CaO}$$

The coefficients of the equation tell us that the stoichiometric equivalency between CaO and $CaCO_3$ is

$$1 \text{ mol CaCO}_3 \Leftrightarrow 1 \text{ mol CaO}$$

So the 2.67 mol of CaO requires 2.67 mol of $CaCO_3$. To weigh out this much $CaCO_3$, the student must convert the number of moles of $CaCO_3$ into grams. The formula mass of $CaCO_3$ is 100.09, so we do the following calculation.

$$2.67 \text{ mol CaCO}_3 \times \frac{100.09 \text{ g CaCO}_3}{1 \text{ mol CaCO}_3} = 2.67 \times 10^2 \text{ g CaCO}_3$$

Thus to prepare 1.50×10^2 g of CaO, 2.67×10^2 g of $CaCO_3$ has to be used.

CHECK Does the *size* of the answer, which is numerically larger than the grams of CaO to be made, make sense? It does, because the formula mass of $CaCO_3$ is larger than that of CaO.

CHAIN CALCULATION ROUTE Had we solved the problem by a chain calculation, not leaving out the obvious stoichiometric step of the 1:1 mole ratio of mol $CaCO_3$ to mol CaO, the set up would have looked like

$$150 \text{ g CaO} \times \frac{1 \text{ mol CaO}}{56.08 \text{ g CaO}} \times \frac{1 \text{ mol CaCO}_3}{1 \text{ mol CaO}} \times \frac{100.09 \text{ g CaCO}_3}{1 \text{ mol CaCO}_3} = 268 \text{ g CaCO}_3$$

Grams CaO → moles CaO → moles $CaCO_3$ → grams $CaCO_3$

The difference between the first answer, 267 g, and the second, 268 g, is caused by rounding. Notice how the calculation flows from grams of CaO to moles of CaO, then to moles of $CaCO_3$ (using the equation), and finally to grams of $CaCO_3$. We cannot emphasize too much that *the key step in all calculations of reaction stoichiometry is the use of the balanced equation.*

Concrete is a hardened mixture of cement and sand, gravel, or rock.

EXAMPLE 3.20
Stoichiometric Mass Calculations

One of the most spectacular reactions of aluminum, the thermite reaction, is with iron oxide, Fe_2O_3, by which metallic iron is made. So much heat is generated that the iron forms in the liquid state (Figure 3.4). The equation is

$$2Al(s) + Fe_2O_3(s) \longrightarrow Al_2O_3(s) + 2Fe(l)$$

A certain welding operation, used over and over, requires that each time at least 86.0 g of Fe be produced. What is the minimum mass in grams of Fe_2O_3 that must be used for each operation? (Do a chain calculation.) Calculate also how many grams of aluminum are needed.

ANALYSIS Remember that all problems in reaction stoichiometry must be solved at the mole level because an equation's coefficients disclose *mole* ratios, not mass ratios. So we have to convert the number of grams of Fe to moles. Then we can use the stoichiometric equivalency given by the coefficients in the balanced equation,

$$1 \text{ mol Fe}_2\text{O}_3 \Leftrightarrow 2 \text{ mol Fe}$$

to see how many *moles* of Fe_2O_3 are needed. We finally convert this answer into grams of Fe_2O_3. The other calculations follow the same pattern.

SOLUTION We'll set up the first calculation as a chain.

$$86.0 \text{ g Fe} \times \frac{1 \text{ mol Fe}}{55.85 \text{ g Fe}} \times \frac{1 \text{ mol Fe}_2\text{O}_3}{2 \text{ mol Fe}} \times \frac{159.70 \text{ g Fe}_2\text{O}_3}{1 \text{ mol Fe}_2\text{O}_3} = 123 \text{ g Fe}_2\text{O}_3$$

Grams Fe → moles Fe → moles Fe_2O_3 → grams Fe_2O_3

A minimum of 123 g of Fe_2O_3 is required to make 86.0 g of Fe.

Next, we calculate the number of grams of Al needed, but we know that we must first find the number of *moles* of Al required. Only from this can the grams of Al be calculated. The relevant stoichiometric equivalency, again using the balanced equation, is

$$2 \text{ mol Al} \Leftrightarrow 2 \text{ mol Fe}$$

Expressing this in the smallest whole numbers, we have

$$1 \text{ mol Al} \Leftrightarrow 1 \text{ mol Fe}$$

Employing another chain calculation to find the mass of Al needed to make 86.0 g of Fe, we have (using 26.98 as the atomic mass of Al)

$$86.0 \text{ g Fe} \times \frac{1 \text{ mol Fe}}{55.85 \text{ g Fe}} \times \frac{1 \text{ mol Al}}{1 \text{ mol Fe}} \times \frac{26.98 \text{ g Al}}{1 \text{ mol Al}} = 41.5 \text{ g Al}$$

Grams Fe → moles Fe → moles Al → grams Al

CHECK Think about the size of the answer, 41.5 g of Al, in relationship to the following: the mass of Fe needed, the stoichiometric equivalency in this reaction of Fe and Al (1 mol:1 mol), and the relative atomic masses of Fe (55.85) and Al (26.98). Does the answer make sense? (Should it take *less* mass of Al than the mass of Fe made?)

■ **Practice Exercise 20** How many grams of aluminum oxide are also produced by the reaction described in Example 3.20?

FIGURE 3.4

The thermite reaction for using aluminum to convert the iron in iron oxide to molten iron. Pictured here is a device for making white hot iron by this reaction and letting it run down into a mold between the ends of two steel railroad rails. The rails are thereby welded together.

3.7
LIMITING REACTANT CALCULATIONS

The industrial synthesis of ammonia is a major operation worldwide, because ammonia is applied to fields as a source of nitrogen that crops can use.

In many situations a chemist deliberately mixes reactants in a mole ratio that does not agree with the coefficients of the equation. Some reactions proceed better when one reactant is in stoichiometric excess, for example. One such situation is the preparation of ammonia, NH_3, from its elements. The equation is

$$N_2(g) + 3H_2(g) \longrightarrow 2NH_3(g)$$

Suppose that a chemist mixed 1.00 mol of N_2 with 5.00 mol of H_2. What is the maximum number of moles of product that could form? We first note that the coefficients tell us that 1 mol of N_2 consumes 3 mol of H_2.

$$1 \text{ mol N}_2 \Leftrightarrow 3 \text{ mol H}_2$$

But 5 mol of H_2 was used, not 3, so there will be 2 mol of H_2 left over. Once the 1 mol of N_2 taken is consumed, no additional NH_3 can form. In this sense, the reactant that is *completely* consumed limits the amount of product that forms, so it is called the **limiting reactant.** *We must always base the calculation of the maximum yield of product on the stoichiometric equivalency between it and the limiting reactant.* Because N_2 is the limiting reactant in our example, the equivalency we must use is:

$$1 \text{ mol } N_2 \Leftrightarrow 2 \text{ mol } NH_3$$

This tells us that the 1.00 mol of N_2 can be converted to a maximum of 2.00 mol of NH_3.

The identity of the limiting reactant depends on the actual composition of the mixture of starting materials. Suppose, for example, that we had taken a mixture of 2.00 mol N_2 and 5.00 mol H_2. We can see (from 1 mol $N_2 \Leftrightarrow 3$ mol H_2) that 2.00 mol of N_2 will require 6.00 mol of H_2, but we took only 5.00 mol of H_2. Therefore all of the H_2 will be used up, and so now H_2 is the limiting reactant. The amount of product that forms (based on 3 mol $H_2 \Leftrightarrow 2$ mol NH_3) is found as follows.

$$5.00 \text{ mol } H_2 \times \frac{2.00 \text{ mol } NH_3}{3.00 \text{ mol } H_2} = 3.33 \text{ mol } NH_3$$

In Example 3.21 we'll see how to solve limiting reactant problems when the amounts of the reactants are given in mass units.

EXAMPLE 3.21
Limiting Reactant

Gold(III) hydroxide, $Au(OH)_3$, is used for electroplating gold onto other metals. It can be made by the following reaction.

$$2KAuCl_4(aq) + 3Na_2CO_3(aq) + 3H_2O \longrightarrow 2Au(OH)_3(aq) + 6NaCl(aq) + 2KCl + 3CO_2(g)$$

To prepare a fresh supply of $Au(OH)_3$, a chemist at an electroplating plant has mixed 20.00 g of $KAuCl_4$ with 25.00 g of Na_2CO_3 (both dissolved in a large excess of water). What is the maximum number of grams of $Au(OH)_3$ that can form?

ANALYSIS The clue that tells us to *begin* the solution by figuring out what is the limiting reactant is that *the quantities of two reactants are given.* Once we determine which reactant limits the product, we use its mass in our calculation.

To find the limiting reactant, we arbitrarily pick one of the candidates ($KAuCl_4$ or Na_2CO_3) — there is no general rule for making the choice — and calculate whether it would all be used up. If so, we've found the limiting reactant. If not, the other reactant limits. (We were told that water is in excess, so we know that it does not limit the reaction.)

SOLUTION We'll arbitrarily work with the first candidate, $KAuCl_4$, and calculate how many grams of the second, Na_2CO_3, *should* be provided to react with 20.00 g of $KAuCl_4$. The formula masses are 377.88 for $KAuCl_4$, and 105.99 for Na_2CO_3. We'll set up a chain calculation using the stoichiometric equivalency:

$$2 \text{ mol } KAuCl_4 \Leftrightarrow 3 \text{ mol } Na_2CO_3$$

$$20.00 \text{ g } KAuCl_4 \times \frac{1 \text{ mol } KAuCl_4}{377.88 \text{ g } KAuCl_4} \times \frac{3 \text{ mol } Na_2CO_3}{2 \text{ mol } KAuCl_4} \times \frac{105.99 \text{ g } Na_2CO_3}{1 \text{ mol } Na_2CO_3} = 8.415 \text{ g } Na_2CO_3$$

Grams $KAuCl_4 \rightarrow$ moles $KAuCl_4 \rightarrow$ moles $Na_2CO_3 \rightarrow$ grams Na_2CO_3

We find that 20.00 g of $KAuCl_4$ needs 8.415 g of Na_2CO_3. The 25.00 g of Na_2CO_3 taken is therefore more than enough to let the $KAuCl_4$ react completely. $KAuCl_4$ is limiting.

A gold-plated table place setting.

Had we tested Na_2CO_3 instead of $KAuCl_4$ as the limiting reactant, we would have calculated how many grams of $KAuCl_4$ are required by 25.00 g of Na_2CO_3.

$$25.00 \text{ g } Na_2CO_3 \times \frac{1 \text{ mol } Na_2CO_3}{105.99 \text{ g } Na_2CO_3} \times \frac{2 \text{ mol } KAuCl_4}{3 \text{ mol } Na_2CO_3} \times \frac{377.88 \text{ g } KAuCl_4}{1 \text{ mol } KAuCl_4} = 59.42 \text{ g } KAuCl_4$$

Grams $Na_2CO_3 \rightarrow$ moles $Na_2CO_3 \rightarrow$ moles $KAuCl_4 \rightarrow$ grams $KAuCl_4$

The given mass of Na_2CO_3 would require much more $KAuCl_4$ than provided, so we confirm that $KAuCl_4$ is the limiting reactant.

From here on, we have a routine calculation going from mass of reactant, $KAuCl_4$, to mass of product, $Au(OH)_3$. We know from the equation's coefficients that

$$1 \text{ mol } KAuCl_4 \Leftrightarrow 1 \text{ mol } Au(OH)_3$$

Using this, plus formula masses, we do the following chain calculation.

$$20.00 \text{ g } KAuCl_4 \times \frac{1 \text{ mol } KAuCl_4}{377.88 \text{ g } KAuCl_4} \times \frac{1 \text{ mol } Au(OH)_3}{1 \text{ mol } KAuCl_4} \times \frac{247.99 \text{ g } Au(OH)_3}{1 \text{ mol } Au(OH)_3} = 13.13 \text{ g } Au(OH)_3$$

Grams $KAuCl_4 \rightarrow$ moles $KAuCl_4 \rightarrow$ moles $Au(OH)_3 \rightarrow$ grams $Au(OH)_3$

Thus from 20.00 g of $KAuCl_4$ we can make a maximum of 13.13 g of $Au(OH)_3$.

In this synthesis, some of the initial 25.00 g of Na_2CO_3 is left over; 20.00 g of $KAuCl_4$ requires only 8.415 g of Na_2CO_3 out of 25.00 g Na_2CO_3. The difference, (25.00 g − 8.415 g) = 16.58 g of Na_2CO_3, remains unreacted. It is possible that the chemist used an excess to ensure that every last bit of the very expensive $KAuCl_4$ would be changed to $Au(OH)_3$.

■ **Practice Exercise 21** In an industrial process for making nitric acid, the first step is the reaction of ammonia with oxygen at high temperature in the presence of a platinum gauze. Nitrogen monoxide forms as follows:

$$4NH_3 + 5O_2 \longrightarrow 4NO + 6H_2O$$

How many grams of nitrogen monoxide can form if a mixture of 30.00 g of NH_3 and 40.00 g of O_2 is taken initially?

In most experiments designed for chemical synthesis, the amount of a product actually isolated falls short of the calculated maximum amount. Losses occur for several reasons. Some are mechanical, such as materials sticking to glassware. In some reactions, losses occur by the evaporation of a *volatile* product. In others, a product is a solid that comes out of solution as it forms because it is largely but not entirely insoluble. The solid is removed by filtration, and what stays in solution, although relatively small, contributes to some loss of product.

One of the common causes of obtaining less than the stoichiometric amount of a product is the occurrence of a **competing reaction.** It produces a **by product,** a substance made by a reaction that competes with the **main reaction.** The synthesis of phosphorus trichloride, for example, gives some phosphorus pentachloride as well, because PCl_3 can react further with Cl_2.

Main reaction: $\quad 2P(s) + 3Cl_2(g) \longrightarrow 2PCl_3(l)$

Competing reaction: $\quad PCl_3(l) + Cl_2(g) \longrightarrow PCl_5(s)$

The **actual yield** of desired product is simply how much is isolated, stated in mass units or moles. The **theoretical yield** of the product is what must be obtained if no losses occur. When less than the theoretical yield of product is

3.8
THEORETICAL YIELD AND PERCENTAGE YIELD

A **volatile** liquid is one with a low boiling point that evaporates quickly, like acetone (bp 56 °C), nail polish remover.

The competition is between newly forming PCl_3 and still unreacted P for unchanged Cl_2. A competing reaction is often referred to as a **side reaction.**

Both actual and theoretical yields must, of course, be in the same units when a percentage yield is calculated.

obtained, chemists generally calculate the *percentage yield* of product to describe how well the preparation went. The **percentage yield** is the actual yield calculated as a percentage of the theoretical yield.

$$\text{Percentage yield} = \frac{\text{actual yield}}{\text{theoretical yield}} \times 100\%$$

Theoretical, actual, and percentage yields

We'll work an example that combines the determination of the limiting reactant with a calculation of percentage yield.

EXAMPLE 3.22
Calculating a Percentage Yield

A chemist set up a synthesis of phosphorus trichloride by mixing 12.0 g P with 35.0 g Cl_2 and obtained 42.4 g of PCl_3. Calculate the percentage yield of this compound. The equation for the main reaction is, as we noted earlier,

$$2P(s) + 3Cl_2(g) \longrightarrow 2PCl_3(l)$$

ANALYSIS Notice that the masses of *both* reactants are given, so we know at the outset that we have a limiting reactant problem. We must first figure out which reactant, P or Cl_2, is the limiting reactant, because we have to base the calculation of the theoretical yield of PCl_3 on it.

SOLUTION Recall that in any limiting reactant problem, we arbitrarily pick one reactant and do a calculation to see whether it can be entirely used up. We'll choose phosphorus and see whether there is enough available to react with 35.0 g of chlorine. The following chain calculation gives us the answer.

$$12.0 \text{ g P} \times \frac{1 \text{ mol P}}{30.97 \text{ g P}} \times \frac{3 \text{ mol } Cl_2}{2 \text{ mol P}} \times \frac{70.90 \text{ g } Cl_2}{1 \text{ mol } Cl_2} = 41.2 \text{ g } Cl_2$$

Thus, with 35.0 g of Cl_2 provided but 41.2 g of Cl_2 needed, there is not enough Cl_2 for the 12.0 g of P. The Cl_2 can be all used up, so Cl_2 is the limiting reactant. We therefore base the calculation of the theoretical yield of PCl_3 on Cl_2.

To find the *theoretical yield* of PCl_3, we calculate how many grams of PCl_3 could be made from 35.0 g of Cl_2 if everything went perfectly according to the equation given.

$$35.0 \text{ g } Cl_2 \times \frac{1 \text{ mol } Cl_2}{70.90 \text{ g } Cl_2} \times \frac{2 \text{ mol } PCl_3}{3 \text{ mol } Cl_2} \times \frac{137.32 \text{ g } PCl_3}{1 \text{ mol } PCl_3} = 45.2 \text{ g } PCl_3$$

Grams Cl_2 → moles Cl_2 → moles PCl_3 → grams PCl_3

The actual yield was 42.4 g of PCl_3, not 45.2 g, so the percentage yield is calculated as follows.

$$\text{Percentage yield} = \frac{42.4 \text{ g } PCl_3}{45.2 \text{ g } PCl_3} \times 100\% = 93.8\%$$

Thus 93.8% of the theoretical yield of PCl_3 was obtained.

■ **Practice Exercise 22** Ethyl alcohol, C_2H_6O, can be converted to acetic acid (the acid in vinegar), $C_2H_4O_2$, by the action of sodium dichromate in aqueous sulfuric acid according to the following equation.

$$3C_2H_6O(aq) + 2Na_2Cr_2O_7(aq) + 8H_2SO_4(aq) \longrightarrow$$

$$3C_2H_4O_2(aq) + 2Cr_2(SO_4)_3(aq) + 2Na_2SO_4(aq) + 11H_2O$$

In one experiment, 24.0 g of C_2H_6O, 90.0 g of $Na_2Cr_2O_7$, and a known excess of sulfuric acid were mixed, and 26.6 g of acetic acid ($C_2H_4O_2$) was isolated. Calculate the percentage yield of $C_2H_4O_2$.

SUMMARY

Stoichiometric Equivalencies The ratios by *particles* in chemical compounds and in reactions are given by the subscripts within formulas and by the coefficients within **balanced equations.** These ratios describe the **stoichiometric equivalencies** within compounds or between substances in reactions.

Mole Concept and Formula Mass In the SI *definition,* one **mole** of any substance is an amount with the same number, **Avogadro's number** (6.022×10^{23}), of atoms, molecules, or formula units as there are atoms in 12 g (exactly) of carbon-12. The sum of the atomic masses of all of the atoms appearing in a chemical formula gives the **formula mass** or the **molecular mass.**

Formula (molecular) mass (g) $\Leftrightarrow$ 1 mol of substance

Like an atomic mass, a formula or molecular mass is a tool for grams-to-moles or moles-to-grams conversions.

Chemical Formulas The actual composition of a molecule is given by its **molecular formula.** An **empirical formula** gives the ratio of atoms, but in the smallest whole numbers, and it is generally the *only* formula we write for ionic compounds. In the case of a molecular compound, the molecular mass is equal either to that calculated from the compound's empirical formula or to a simple multiple of it. When the two calculated masses are the same, then the molecular formula is the same as the empirical formula.

An empirical formula can be found from the masses of the elements obtained by the quantitative analysis of a known sample of the compound or it can be calculated from the **percentage composition.** The percentage of an element in a compound is the same as the number of grams of the element in 100 g of the compound. If there is $A\%$ of X in X_xZ_z, then 100.00 g of X_xZ_z contains A g of X. From the grams of X and Z, the moles of X and Z can be calculated. When these are adjusted to their corresponding whole-number ratios, the subscripts in the empirical formula are obtained.

Formula Stoichiometry A chemical formula is a tool for stoichiometric calculations, because its subscripts tell us the mole ratios in which the various elements are combined. In X_xZ_z, the stoichiometric equivalency between X and Z is simply

$$z \text{ mol } Z \Leftrightarrow x \text{ mol } X$$

The conversion factors available from this equivalency are

$$\frac{z \text{ mol } Z}{x \text{ mol } X} \quad \text{and} \quad \frac{x \text{ mol } X}{z \text{ mol } Z}$$

Balanced Equations and Reaction Stoichiometry A balanced equation is a tool for reaction stoichiometry because its coefficients disclose the stoichiometric equivalencies. When balancing an equation, only the coefficients can be adjusted, never the subscripts. In a generalized reaction:

$$xA + mW \longrightarrow yB + nZ$$

we know that

$$x \text{ mol } A \Leftrightarrow m \text{ mol } W$$
$$x \text{ mol } A \Leftrightarrow y \text{ mol } B$$
$$x \text{ mol } A \Leftrightarrow n \text{ mol } Z$$

and so on.

All problems of reaction stoichiometry must be solved at the mole level of the substances involved. If grams are given, they must first be changed to moles.

Yields of Products A reactant taken in a quantity less than required by another reactant, as determined by the reaction's stoichiometry, is called the **limiting reactant.** The **theoretical yield** of a product can be no more than permitted by the limiting reactant. Sometimes **side reactions** producing by-products reduce the **actual yield.** The ratio of the actual to the theoretical yield, expressed as a percentage, is the **percentage yield.**

TOOLS YOU HAVE LEARNED

The table below lists the tools you have learned in this chapter that are applicable to problem solving. Review them if necessary, and refer to them when working on the Thinking-It-Through problems and the Review Exercises that follow.

Tool	Function
Chemical formula (page 90)	Subscripts in a formula establish atom ratios and mole ratios between the elements in the substance. These ratios serve as stoichiometric equivalencies that are used to solve problems.
Atomic mass (page 92)	Used to form a conversion factor to calculate mass from moles of an element, or moles from the mass of an element.
Formula mass; molecular mass (page 95)	Used to form a conversion factor to calculate mass from moles of a compound, or moles from the mass of a compound.
Avogadro's number (page 90)	Relates macroscopic lab-sized quantities (e.g., mass) to numbers of individual atomic-sized particles such as atoms, molecules, or ions.
Percentage composition (page 97)	To represent the composition of a compound and be the basis for computing the empirical formula. Comparing experimental and theoretical percentage compositions can help establish the identity of a compound.
Balanced chemical equation (page 107)	The coefficients are used to establish stoichiometric equivalencies that relate moles of one substance to moles of another in a chemical reaction. They also establish the ratios by formula units among the reactants and products.
Theoretical, actual, and percentage yields (page 114)	To estimate the efficiency of a reaction. Remember that the theoretical yield is calculated from the limiting reactant using the balanced chemical equation.

THINKING IT THROUGH

Remember, you are not asked to obtain answers for the following problems. Instead, assemble the data necessary to solve the problems and describe how you would use the data to obtain the answers. For numerical problems, set up the calculations using appropriate conversion factors.

The problems are divided into two groups. Those in Level 2 are significantly more challenging than those in Level 1 and provide an opportunity to really hone your problem solving skills.

Level 1 Problems

1. Carbon and hydrogen atoms form a compound called *ethane*. If you were asked to state the *ratio by atoms* of carbon and hydrogen in ethane, what additional information would you need?

2. Sulfur and oxygen atoms are present in sulfur trioxide, SO_3. How many atoms of S and how many of O are needed to make 10^{20} molecules of SO_3?

3. In a certain compound of tin and chlorine, the ratio of the atoms is described as 1 mol of tin atoms to 4 mol of chlorine atoms. What is the empirical formula of the compound?

4. The following information applies to a certain compound of carbon and hydrogen: In 4 mol of the compound there are 12 mol of C. In 6 molecules of the compound, there are 48 atoms of H. Describe how you would use this information to obtain the empirical formula for the compound.

5. Concerning a certain compound of sodium, sulfur, and oxygen, you are given the following stoichiometric equivalencies:

$$4 \text{ mol Na} \Leftrightarrow 10 \text{ mol O}$$

$$5 \text{ mol O} \Leftrightarrow 2 \text{ mol S}$$

Explain how you would determine the empirical formula. Set up the calculation to determine the number of moles of sulfur combined with 2.7 mol Na.

6. Chloroform, a chemical once used as an anesthetic, has the formula $CHCl_3$. (a) How many molecules of $CHCl_3$ would have to be decomposed to give 18 atoms of Cl? (b) How many moles of $CHCl_3$ would have to be decomposed to give enough chlorine atoms to make 0.36 mol of Cl_2?

7. A fuel that has been used in rocket engines is a compound called dimethylhydrazine. Its formula can be written $(CH_3)_2N_2H_2$. How many moles of H_2 gas could be made from all the hydrogen in 0.56 mol of $(CH_3)_2N_2H_2$?

8. If during combustion in a rocket engine all the carbon in 0.40 mol of $(CH_3)_2N_2H_2$ is changed to CO_2, how many moles of CO_2 form?

9. How many moles of copper are in an old copper penny that has a mass of 3.14 g? (New pennies are made of zinc with only a thin surface coating of copper.)

10. How many moles of carbon are combined with 0.60 mol of hydrogen atoms in $(CH_3)_2N_2H_2$?

11. For element *X*, 59.8 g corresponds to 2.6 moles. What is the atomic mass of the element?

12. What additional information, if any, do you need to calculate the number of moles of sodium, Na, in 22.99 grams of Na?

13. What must be the atomic mass of an element if 1.500 mol weighs 24.00 g?

14. To answer the following question, state what additional information, if any, is needed. How many atoms of fluorine make up 9.50 g of fluorine?

15. If the stoichiometric equivalency of sulfur and chlorine in one of their compounds can be expressed as follows, then what is the *empirical* formula and what is the formula mass corresponding to the empirical formula of the compound?

$$2 \text{ S} \Leftrightarrow 8 \text{ Cl}$$

16. For a grams-to-moles conversion involving CH_4, what conversion factor must be used (including specific units)? Specify three significant figures where this is relevant.

17. The formula mass of compound *X* is 122.0. What conversion factor is needed to calculate the number of moles of *X* in a given number of grams of *X*?

18. How many moles of Ca are combined with 1.34 mol of Br in $CaBr_2$?

19. What additional information available in tables is needed to answer the following question? How many grams of Ca are combined with 1.34 g of Br in $CaBr_2$?

20. From a 1.015 g sample of a binary compound of calcium and bromine, there was obtained 0.2035 g of Ca. Outline the steps needed to calculate the percentage composition of the compound and its empirical formula. Show relevant conversion factors.

21. A compound whose molecules consist only of C, H, and O was found to have the following percentage composition: C, 62.04%; H, 10.41%. Outline the steps needed to calculate an empirical formula, showing all conversion factors. Can a molecular formula be calculated?

22. A sample of the hydrate $CuSO_4 \cdot 5H_2O$ has a mass of 12.5 g. If it is completely dehydrated, how many grams of $CuSO_4$ would remain? How many moles of water would be released during the dehydration process?

23. Calcium hydroxide and hydrobromic acid react as follows:

$$Ca(OH)_2(s) + 2HBr(aq) \longrightarrow CaBr_2(aq) + 2H_2O$$

(a) Is the equation balanced?
(b) If the reaction is carried out to produce 0.125 mol of $CaBr_2$, how many moles of HBr are needed (or is sufficient information given)?
(c) If the reaction is carried out with 1.24 g of $Ca(OH)_2$, how many moles and how many grams of HBr are needed?

24. A reaction between 0.100 mol of CH_4 and 0.100 mol of Cl_2 is planned. The sought-for result is given in the following balanced equation:

$$CH_4(g) + Cl_2(g) \longrightarrow CH_3Cl(g) + HCl(g)$$

There was isolated 3.02 g of CH_3Cl. What is the theoretical yield of this compound (in grams), and what is the percentage yield in this experiment?

Level 2 Problems

25. In the binary compound *XY*, the percentage composition was found to be 50.00% *X* and 50.00% *Y*. This is a hypothetical situation, but what would have to be true about the atomic masses of *X* and *Y* for this to be true?

26. A sample of iron ore in which the iron occurs in the mineral Fe_3O_4 is found to contain 12.5% (by mass) of iron. What is the percentage by mass of iron oxide Fe_3O_4 in the ore?

27. How many grams of carbon are combined with 0.85 g H in a sample of $(CH_3)_2N_2H_2$?

28. A white powder was known to be either Na_3PO_4, Na_2HPO_4, or NaH_2PO_4. A sample of it was analyzed and found to contain 21.82% phosphorus by mass. What was the percentage by mass of sodium in the sample?

29. If compound X and compound Y react with each other in a 1 to 1 ratio by moles, and if 25.0 g of X combines exactly with 29.0 g of Y, which compound has the higher formula mass?

30. Al_2O_3 reacts with HI as follows:

$$Al_2O_3(s) + 6HI(aq) \longrightarrow 2AlI_3(aq) + 3H_2O$$

A mixture containing 16.0 g of Al_2O_3 and 140 g of HI produces 127 g of AlI_3. How many grams of which reactant were present in excess in the original reaction mixture.

REVIEW EXERCISES

Answers to questions whose numbers are printed in color are given in Appendix D. Challenging questions are marked with asterisks.

The Mole Concept and Stoichiometric Equivalencies

3.1 The use of *whole* numbers for the coefficients in a chemical equation recognizes what chemical truth about atoms, ions, and molecules?

3.2 In what ratio must N and O combine to make dinitrogen tetroxide, N_2O_4? (Express this ratio in the smallest whole numbers.)

3.3 How many atoms are in 6 g (exactly) of carbon-12?

3.4 How many atoms are in 1.5 mol (exactly) of carbon-12? How many grams does this much carbon-12 weigh?

3.5 What stoichiometric equivalence exists between the atoms in each of the following compounds? (a) SO_2, (b) As_2O_3, (c) PbN_6, (d) Al_2Cl_6.

3.6 Give the stoichiometric equivalencies that exist between the atoms in the following compounds. (a) Mn_3O_4, (b) Sb_2S_5, (c) P_4O_6, (d) Hg_2Cl_2

3.7 How many moles of Al atoms are needed to combine with 1.58 mol of O atoms to make aluminum oxide, Al_2O_3?

3.8 How many moles of vanadium atoms, V, are needed to combine with 0.565 mol of O atoms to make vanadium pentoxide, V_2O_5?

3.9 How many moles of F atoms are needed to combine with 0.0230 mol of U atoms in UF_6?

Measuring Moles

3.10 The atomic mass of aluminum is 26.98. What specific conversion factors does this value make available for relating a mass of aluminum (in grams) and a quantity of aluminum given in moles?

3.11 On the average (and to four significant figures), how much heavier is a gold atom (as typically found in nature) than an atom of carbon-12?

3.12 How many atoms are in a sample of lead with a mass of 1.000×10^{-12} g (one-trillionth gram)?

3.13 What is the mass in grams of 1.000×10^{10} (10 billion) atoms of mercury, the silvery liquid in common thermometers?

3.14 Calculate the formula masses of the following to two decimal places.
(a) $NaHCO_3$ (d) $Al_2(SO_4)_3$
(b) $K_2Cr_2O_7$ (e) $CuSO_4 \cdot 5H_2O$
(c) $(NH_4)_2CO_3$

3.15 Calculate the formula masses of the following to two decimal places.
(a) $Ca(NO_3)_2$ (d) $Fe_4[Fe(CN)_6]_3$
(b) $Pb(C_2H_5)_4$ (e) $Na_2SO_4 \cdot 10H_2O$
(c) $Mg_3(PO_4)_2$

3.16 Which has a larger mass, 0.5 mol of H_2O or 2 mol of He?

3.17 Calculate the mass in grams of the following.
(a) 1.25 mol Ca   (d) 1.45 mol $(NH_4)_2CO_3$
(b) 0.625 mol Fe   (e) 2.15×10^{-3} mol $KMnO_4$
(c) 0.600 mol C_4H_{10}

3.18 What is the mass in grams of the following?
(a) 0.754 mol Zn   (c) 0.322 mol $POCl_3$
(b) 0.194 mol I_2 (d) 4.31×10^{-5} mol $(NH_4)_2HPO_4$

3.19 Calculate the number of moles in each of the following samples.
(a) 21.5 g $CaCO_3$ (c) 16.8 g $Sr(NO_3)_2$
(b) 1.56 g NH_3 (d) 6.98 μg Na_2CrO_4

3.20 What are the number of moles in the following samples?
(a) 9.36 g $Ca(OH)_2$ (c) 4.29 g H_2O_2
(b) 38.2 g $PbSO_4$ (d) 4.65 mg $NaAuCl_4$

3.21 How many grams of carbon are in each of the following?
(a) 0.200 mol Na_2CO_3 (d) 14.5 g of CO_2
(b) 0.200 mol C_3H_8 (e) 25.0 g of C_6H_{14}
(c) 25.0 g of $Fe_2(CO_3)_3$

3.22 How many grams of chlorine are in each of the following samples?
(a) 0.146 mol $CaCl_2$ (c) 18.5 g Al_2Cl_6
(b) 0.558 mol $PbCl_4$ (d) 33.2 g NO_2Cl

3.23 If a sample of borazon, BN, contains 0.0500 mol of boron, how many moles and how many grams of nitrogen are also in the sample? (Borazon is the hardest of all known substances.)

3.24 One sample of CaC_2 contains 0.150 mol of carbon. How many moles and how many grams of calcium are also in the sample? [Calcium carbide, CaC_2, was once used to make signal flares for ships. Water dripped onto CaC_2 reacts to give acetylene (C_2H_2), which burns brightly.]

3.25 How many moles of iodine, I, are in 0.500 mol of $Ca(IO_3)_2$? How many grams of calcium iodate, $Ca(IO_3)_2$, are needed to supply this much iodine? [Iodized salt contains a trace amount of calcium iodate to help prevent a thyroid condition called goiter.]

3.26 How many moles of nitrogen, N, are in 0.650 mol of $(NH_4)_2CO_3$? How many grams of this compound supply this much nitrogen?

3.27 How many moles of nitrogen, N, are in 0.556 mol of NH_4NO_3? How many grams of NH_4NO_3 supply this much nitrogen?

*3.28 One recipe for making soap calls for mixing 13 oz of sodium hydroxide, NaOH, 5.0 cups of water, and 6.0 lb of tallow. Assume that the formula of tallow is $C_{57}H_{110}O_6$. The mole ratio should theoretically be 3.0 mol of NaOH to 1.0 mol of tallow. What is the actual mole ratio of NaOH to tallow in the recipe?

*3.29 One recipe for hydroponic plant food calls for 1.50 oz of KNO_3, 1.00 oz of $CaSO_4$, 0.750 oz of $MgSO_4$, 0.500 oz of $CaHPO_4$, and 0.250 oz of $(NH_4)_2SO_4$—all dissolved in 10.0 gal of water. How many grams and how many moles of each of these substances are in the quantities given? (Hydroponics is a method of raising plants by using a balanced nutrient medium, with or without any mechanical support such as soil, sand, or gravel.)

3.30 In the early 1990s, the annual United States' production of ammonia, NH_3, was about 34 billion pounds (34×10^9 lb). How many moles per year of NH_3 does this represent?

3.31 In the early 1990s the annual United States' production of sulfuric acid, H_2SO_4, was roughly 89 billion pounds (87×10^9 lb). How many moles of H_2SO_4 is this?

3.32 How many grams of oxygen are needed to combine with 24.00 g of sulfur to make (a) SO_2 and (b) SO_3?

3.33 How many grams of nitrogen are needed to combine with 12.00 g of oxygen to make each of the following oxides of nitrogen? (a) N_2O, (b) NO, (c) NO_2, (d) N_2O_5.

3.34 If a sample of sulfuric acid contains 12.64 g of S, how many grams of O and of H are also present?

*3.35 If a sample of $Na_2CO_3 \cdot 10H_2O$ contains 2.34 g of C, how many grams of O are also present? How many grams of sodium, Na? How many grams of H?

Percentage Composition

3.36 The percentage by mass of carbon in carbon dioxide is C, 27.29%.
(a) What is the percentage of oxygen?
(b) How many grams of carbon are present in 50.00 g of carbon dioxide?

3.37 If 100.00 g of copper(I) oxide, Cu_2O, contains 88.82 g of Cu, what is the percentage of oxygen in the compound?

3.38 Calculate the percentage of nitrogen in ammonia, NH_3, and urea, $CO(NH_2)_2$. (These are the two major nitrogen fertilizers used in agriculture worldwide.)

3.39 Calculate the percentage composition for each of the following: (a) NaH_2PO_4, (b) $NH_4H_2PO_4$, (c) $(CH_3)_2CO$, (d) $CaSO_4$, (e) $CaSO_4 \cdot 2H_2O$

3.40 Phencyclidine ("angel dust") is $C_{17}H_{25}N$. A sample suspected of being this illicit drug was found to have a percentage composition of 83.71% C, 10.42% H, and 5.61% N. Do these data acceptably match the theoretical data for phencyclidine?

3.41 The hallucinogenic drug LSD has the molecular formula $C_{20}H_{25}N_3O$. One suspected sample contained 74.07% C, 7.95% H, and 9.99% N.
(a) What is the percentage O?
(b) Are these data consistent for LSD?

Empirical and Molecular Formulas

3.42 Explain why calculations based on percentage composition give only empirical formulas. (How do we interpret percentage data?)

3.43 What additional information besides the empirical formula must be obtained for a new compound to write its molecular formula?

3.44 Write empirical formulas for the following compounds. (a) S_2Cl_2, (b) $C_6H_{12}O_6$, (c) C_4H_{10}, (d) As_2O_6, (e) H_2O_2

3.45 What are the empirical formulas of the following compounds? (a) NH_4NO_2, (b) $Na_2S_2O_6 \cdot 2H_2O$, (c) $H_4P_2O_6 \cdot 2H_2O$, (d) $Ti_2(C_2O_4)_3 \cdot 2H_2O$, (e) $(NH_4)_2S_2O_8$

3.46 A 0.3148 g sample of a compound between phosphorus and oxygen was found to contain 0.1774 g P. What is its empirical formula?

3.47 A sample with a mass of 0.3886 g of a compound of mercury and bromine was found to contain 0.1107 g bromine. Its molecular mass was found to be 561. What are its empirical and molecular formulas?

3.48 A 0.6662 g sample of "antimonal saffron" was found to contain 0.4017 g of antimony. The remainder was sulfur. The formula mass of this compound is 404. What are the empirical and molecular formulas of this pigment, arranging their atomic symbols in the order SbS? (This compound is a red pigment used in painting.)

3.49 Quantitative analysis of a sample of sodium pertechnetate with a mass of 0.8961 g found 0.1114 g of sodium and 0.4748 g of technetium. The remainder was oxygen. Calculate the empirical formula of sodium pertechnetate, arranging the atomic symbols in the formula in the order of NaTcO. (A radioactive form of sodium pertechnetate is used as a brain-scanning agent in medicine.)

3.50 One compound of mercury with a formula mass of 519 contains 77.26% Hg, 9.25% C, and 1.17% H (with the balance being O). Calculate the empirical and molecular formulas, arranging the symbols in the order HgCHO.

3.51 A sample of hydrocarbon weighing 0.5992 g was found to contain 0.5040 g of carbon and 0.09515 g of hydrogen. Its molecular mass is 114. Write the empirical and molecular formulas of the compound, arranging the symbols in the order CH.

*3.52 Sodium reacts with oxygen to form sodium peroxide.
(a) What is its empirical formula if it can be made by combining 0.4681 g Na with 0.3258 g O?
(b) On this basis, how many grams of sodium can combine with 1.000 g of oxygen?
(c) Another known oxide of sodium is called sodium oxide. *Using only your answer to part b,* calculate some possible values for the grams of sodium combined with 1.000 g of oxygen in sodium oxide.
(d) A sample of sodium oxide was found to contain 1.145 g of Na and 0.3983 g of O. How many grams of Na can combine with 1.000 g of O to make sodium oxide?
(e) Compare the answers to parts c and d. What is the ratio of the two masses of sodium that are able to combine with 1.000 g of O as they form the two different oxides of sodium? What law of chemical combination is illustrated?
(f) Calculate the empirical formula of sodium oxide.

*3.53 Potassium reacts with oxygen to form a superoxide.
(a) What is the empirical formula of this compound, if it can be made by combining 0.5634 g of potassium with 0.4611 g of oxygen?
(b) How many grams of oxygen can combine with 1.000 g of potassium to make potassium superoxide?
(c) Another known oxide of potassium is called potassium oxide. *Using only your answer to part b,* calculate some possible values for the grams of oxygen that are present with 1.000 g of potassium in this oxide.
(d) Potassium oxide contains its elements in a ratio of 0.8298 g of K to 0.1698 g of O. What is the empirical formula of potassium oxide?
(e) How many grams of oxygen can combine with 1.000 g of potassium to form potassium oxide?

(f) In what ratio do the grams of oxygen calculated in parts b and e stand? Which law of chemical combination is illustrated by these examples?

3.54 A 0.1246 g sample of a compound of chromium and chlorine was dissolved in water. All of the chloride ion was then captured by silver ion in the form of AgCl. A mass of 0.3383 g of AgCl was obtained. Calculate the empirical formula of the compound of Cr and Cl.

*3.55 A compound of Ca, C, N, and S was subjected to quantitative analysis and formula mass determination, and the following data were obtained. A 0.250 g sample was mixed with Na_2CO_3 to convert all of the Ca to 0.160 g of $CaCO_3$. A 0.115 g sample of the compound was carried through a series of reactions until all of its S was changed to 0.0802 g of $BaSO_4$. A 0.712 g sample was processed to liberate all of its N as NH_3, and 0.155 g NH_3 was obtained. The formula mass was found to be 156. Determine the empirical and molecular formulas of this compound.

3.56 When 6.853 mg of a sex hormone was burned in a combustion analysis, 19.73 mg of CO_2 and 6.391 mg of H_2O were obtained. The formula mass was found to be 290. What is the molecular formula? (The compound contains only C, H, and O.)

3.57 When a sample of a compound in the vitamin D family with a formula mass of 399 was burned in a combustion analysis, 5.982 mg of the compound gave 18.490 mg of CO_2 and 6.232 mg of H_2O. What is the molecular formula of this compound? (The compound contains only C, H, and O.)

Balanced Equations

3.58 When given the *unbalanced* equation
$$Na(s) + Cl_2(g) \longrightarrow NaCl(s)$$
and asked to balance it, student *A* wrote
$$Na(s) + Cl_2(g) \longrightarrow NaCl_2(s)$$
and student *B* wrote
$$2Na(s) + Cl_2(g) \longrightarrow 2NaCl(s)$$
(a) Both equations are balanced, but which student is correct?
(b) Explain why the other student's answer is incorrect.

3.59 How many moles of iron are part of the expression $2Fe_4[Fe(CN)_6]_3$, taken from a balanced equation?

3.60 How many moles of oxygen are part of the expression, $3Fe(H_2PO_4)_3$, taken from a balanced equation?

3.61 Why is the changing of subscripts not allowed when balancing a chemical equation?

3.62 Write the equation that expresses in acceptable chemical shorthand the following statement: "Iron can be made to react with molecular oxygen (O_2) to give iron oxide with the formula Fe_2O_3."

3.63 The conversion of one air pollutant, NO, produced in vehicle engines, into another, NO_2, occurs when NO reacts with molecular oxygen in the air. Write the balanced equation for this reaction.

3.64 Balance the following equations.
(a) $Ca(OH)_2 + HCl \rightarrow CaCl_2 + H_2O$
(b) $AgNO_3 + CaCl_2 \rightarrow Ca(NO_3)_2 + AgCl$
(c) $Fe_2O_3 + C \rightarrow Fe + CO_2$
(d) $NaHCO_3 + H_2SO_4 \rightarrow Na_2SO_4 + H_2O + CO_2$
(e) $C_4H_{10} + O_2 \rightarrow CO_2 + H_2O$

3.65 Balance the following equations.
(a) $SO_2 + O_2 \rightarrow SO_3$
(b) $P_4O_{10} + H_2O \rightarrow H_3PO_4$
(c) $Pb(NO_3)_2 + Na_2SO_4 \rightarrow PbSO_4 + NaNO_3$
(d) $Fe_2O_3 + H_2 \rightarrow Fe + H_2O$
(e) $Al + H_2SO_4 \rightarrow Al_2(SO_4)_3 + H_2$

Stoichiometry

3.66 If two substances react completely in a 1 : 1 ratio *both* by mass and by moles, what must be true about these substances?

3.67 A mixture of 0.020 mol of Mg and 0.020 mol of Cl_2 reacted completely to form $MgCl_2$ according to the equation

$$Mg + Cl_2 \longrightarrow MgCl_2$$

(a) What information describes the *stoichiometry* of this reaction?
(b) What information gives the size of the reaction?

3.68 In a report to a supervisor, a chemist described an experiment in the following way: "0.0800 mol of H_2O_2 decomposed into 0.0800 mol of H_2O and 0.0400 mol of O_2." Express the chemistry and stoichiometry of this reaction by a conventional chemical equation.

3.69 Chlorine is used by textile manufacturers to bleach cloth. Excess chlorine is destroyed by its reaction with sodium thiosulfate, $Na_2S_2O_3$, as follows.

$Na_2S_2O_3(aq) + 4Cl_2(g) + 5H_2O \longrightarrow$
$$2NaHSO_4(aq) + 8HCl(aq)$$

(a) How many moles of $Na_2S_2O_3$ are needed to react with 0.12 mol of Cl_2?
(b) How many moles of HCl can form from 0.12 mol of Cl_2?
(c) How many moles of H_2O are required for the reaction of 0.12 mol of Cl_2?

3.70 The octane in gasoline burns according to the following equation.

$$2C_8H_{18} + 25O_2 \longrightarrow 16CO_2 + 18H_2O$$

(a) How many moles of O_2 are needed to react fully with 6 mol of octane?
(b) How many moles of CO_2 can form from 0.5 mol of octane?
(c) How many moles of water are produced by the combustion of 8 mol of octane?
(d) If this reaction is used to synthesize 6.00 mol of CO_2, how many moles of oxygen are needed? How many moles of octane?

3.71 The incandescent white of a fireworks display is caused by the reaction of phosphorus with O_2 that produces P_4O_{10}.
(a) Write the balanced chemical equation for the reaction.
(b) How many moles of O_2 are needed to combine with 0.221 mol of P?
(c) How many moles of P_4O_{10} can be made from 0.250 mol of O_2?
(d) How many moles of P are needed to combine with 0.114 mol of O_2?

3.72 The combustion of a sample of butane, C_4H_{10}, produced 3.46 g of water as well as CO_2.
(a) Write the balanced chemical equation for the reaction.
(b) How many moles of water formed? How many millimoles? (Remember, 1 mol = 1000 mmol.)
(c) How many moles of butane burned?
(d) How many grams of butane burned?
(e) How much oxygen was used up, in moles? In grams?

3.73 One way to change the iron mineral, Fe_2O_3, into metallic iron is to heat it with H_2. Besides Fe, water also forms.
(a) Write the balanced chemical equation for the reaction.
(b) How many moles of iron are made from 22 mol of Fe_2O_3?
(c) How many moles of H_2 are needed to make 24 mol of Fe?
(d) If 95 mol of H_2O forms, how many grams of Fe_2O_3 were used up?

3.74 Ammonium nitrate will detonate if ignited in the presence of certain impurities. The equation for this reaction at a high temperature is

$$2NH_4NO_3(s) \xrightarrow{\text{>300 °C}} 2N_2(g) + O_2(g) + 4H_2O(g)$$

Notice that all of the products are gases and so must occupy a vastly greater volume than the solid reactant.
(a) How many moles of *all* gases are produced from 1 mol of NH_4NO_3?
(b) If 1.00 ton of NH_4NO_3 exploded according to this equation, how many moles of *all* gases would be produced? (1 ton = 2000 lb)

(In April 1947 a ship cargo of ammonium nitrate, NH_4NO_3, blew up in the harbor of Texas City, Texas, with a loss of nearly 600 lives. In 1995, The Federal building in Oklahoma City was also destroyed by an explosion of NH_4NO_3.)

3.75 In *dilute* nitric acid, HNO_3, copper metal dissolves according to the following equation.

$3Cu(s) + 8HNO_3(aq) \longrightarrow$
$$3Cu(NO_3)_2(aq) + 2NO(g) + 4H_2O$$

How many moles of HNO_3 are needed to dissolve 11.45 g of Cu according to this equation?

3.76 The Solvay process for the manufacture of sodium carbonate begins by passing ammonia and carbon dioxide through a solution of sodium chloride to make sodium bicarbonate and ammonium chloride. The equation for the overall reaction is

$$H_2O + NaCl + NH_3 + CO_2 \longrightarrow NH_4Cl + NaHCO_3$$

In the next step, sodium bicarbonate is heated to give sodium carbonate and two gases, carbon dioxide and steam.

$$2NaHCO_3 \longrightarrow Na_2CO_3 + CO_2 + H_2O$$

(a) How many moles of sodium carbonate could be made from 120 mol of NaCl?

(b) What is the theoretical yield of sodium carbonate from 456 g of NaCl in moles? In grams?

Limiting Reactants

3.77 Zinc and sulfur react to form zinc sulfide according to the equation

$$Zn + S \longrightarrow ZnS$$

In an experiment, 30.0 g of zinc and 36.0 g of sulfur are mixed.

(a) Which chemical is the limiting reactant?

(b) How many grams of ZnS can form?

(c) How many grams of the excess reactant will be left over after the reaction?

3.78 Powdered aluminum and iron(III) oxide react with a large evolution of heat (per mole) according to the following equation.

$$2Al + Fe_2O_3 \longrightarrow Al_2O_3 + 2Fe$$

In one experiment, 4.20 mol of Al was mixed with 1.75 mol of Fe_2O_3.

(a) Which reactant, if either, was the limiting reactant?

(b) Calculate the theoretical yield (in moles) of iron.

3.79 Phosphorus(V) chloride reacts with water to give phosphoric acid and hydrogen chloride according to the following equation.

$$PCl_5 + 4H_2O \longrightarrow H_3PO_4 + 5HCl$$

In one experiment, 0.360 mol of PCl_5 was slowly added to 2.88 mol of water.

(a) Which reactant, if either, was the limiting reactant?

(b) Calculate the theoretical yields (in moles) of H_3PO_4 and HCl.

*3.80 Silver nitrate, $AgNO_3$, reacts with iron(III) chloride, $FeCl_3$, to give silver chloride, AgCl, and iron(III) nitrate, $Fe(NO_3)_3$. A solution containing 18.0 g of $AgNO_3$ was mixed with a solution containing 32.4 g of $FeCl_3$.

(a) Write the equation for this reaction.

(b) Which reactant is the limiting reactant?

(c) What is the maximum number of moles of AgCl obtainable from this experiment?

(d) What is the maximum number of grams of AgCl obtainable from the experiment?

(e) How many grams of the reactant in excess remains after the reaction is over?

Percentage Yield

3.81 Barium sulfate, $BaSO_4$, is made by the following equation.

$$Ba(NO_3)_2(aq) + Na_2SO_4(aq) \longrightarrow$$
$$BaSO_4(s) + 2NaNO_3(aq)$$

A chemist began with 75.00 g of $Ba(NO_3)_2$.

(a) How many grams of Na_2SO_4 should be taken?

(b) After collecting and drying the product, 63.45 g of $BaSO_4$ was obtained. Calculate the percentage yield of the product.

3.82 Aluminum sulfate can be made by the following reaction.

$$2AlCl_3(aq) + 3H_2SO_4(aq) \longrightarrow Al_2(SO_4)_3(aq) + 6HCl(aq)$$
$$\text{aluminum sulfate}$$

It is quite soluble in water, so to isolate it the solution has to be evaporated to dryness. This drives off the volatile HCl, but the residual solid has to be heated to a little over 200 °C to drive off all of the water. In one experiment, 25.00 g of $AlCl_3$ was used.

(a) How many grams of H_2SO_4 are needed?

(b) There was eventually isolated 28.36 g of pure $Al_2(SO_4)_3$. Calculate the percentage yield.

*3.83 For a research project, a student decided to test the effect of the lead(II) ion (Pb^{2+}) on the ability of salmon eggs to hatch. This ion was obtainable from the water-soluble salt, lead(II) nitrate, $Pb(NO_3)_2$, which the student decided to make by the following reaction. (The desired product was to be isolated by the slow evaporation of the water.)

$$PbO(s) + 2HNO_3(aq) \longrightarrow Pb(NO_3)_2(aq) + H_2O$$

Losses of product for various reasons were expected, and a yield of 86.0% was expected. In order to have 5.00 g of product at this yield, how many grams of PbO should be taken? (Assume that sufficient nitric acid, HNO_3, would be used.)

*3.84 The potassium salt of benzoic acid, potassium benzoate ($KC_7H_5O_2$), can be made by the action of potassium permanganate on toluene (C_7H_8), as follows.

$$C_7H_8 + 2KMnO_4 \longrightarrow$$
$$KC_7H_5O_2 + 2MnO_2 + KOH + H_2O$$

If the yield of potassium benzoate cannot realistically be expected to be more than 71%, what is the minimum number of grams of toluene needed to achieve this yield while producing 11.5 g of potassium benzoate?

*3.85 Manganese trifluoride, MnF_3, can be prepared by the following reaction.

$$2MnI_2(s) + 13F_2(g) \longrightarrow 2MnF_3(s) + 4IF_5(l)$$

What is the minimum number of grams of F_2 that must be used to react with 12.0 g of MnI_2 if the overall yield of MnF_3 is no more than 75% of theory?

Additional Exercises

3.86 A lawn fertilizer is rated as 6.00% nitrogen, meaning 6 g of N in 100 g of fertilizer. The nitrogen is present in the form of urea, $(NH_2)_2CO$. How many grams of urea are present in 100 g of the fertilizer to supply the rated amount of nitrogen?

3.87 Based solely on the amount of available carbon, how many grams of sodium oxalate, $Na_2C_2O_4$, could be obtained from 125 g of C_6H_6? (Assume that no loss of carbon occurs in any of the reactions needed in the synthesis.)

3.88 Balance the following equations.

(a) $Mg(OH)_2 + HBr \longrightarrow MgBr_2 + H_2O$

(b) $HCl + Ca(OH)_2 \longrightarrow CaCl_2 + H_2O$

(c) $Al_2O_3 + H_2SO_4 \longrightarrow Al_2(SO_4)_3 + H_2O$

(d) $KHCO_3 + H_3PO_4 \longrightarrow K_2HPO_4 + H_2O + CO_2$

(e) $C_9H_{20} + O_2 \longrightarrow CO_2 + H_2O$

3.89 Balance the following equations.

(a) $CaO + HNO_3 \longrightarrow Ca(NO_3)_2 + H_2O$

(b) $Na_2CO_3 + Mg(NO_3)_2 \longrightarrow MgCO_3 + NaNO_3$

(c) $(NH_4)_3PO_4 + NaOH \longrightarrow Na_3PO_4 + NH_3 + H_2O$

(d) $LiHCO_3 + H_2SO_4 \longrightarrow Li_2SO_4 + H_2O + CO_2$

(e) $C_4H_{10}O + O_2 \longrightarrow CO_2 + H_2O$

3.90 A sample of a compound of C, H, N, and O with a mass of 0.6216 g was found to contain 0.1735 g C, 0.01455 g H, and 0.2024 g N. Its molecular mass is 129. Calculate its empirical and molecular formulas, arranging their symbols in alphabetical order.

3.91 Strychnine, a deadly poison, has a molecular mass of 334 and a percentage composition of 75.42% C, 6.63% H, 8.38% N, and the balance oxygen. Calculate the empirical and molecular formulas of strychnine (arranging the symbols alphabetically).

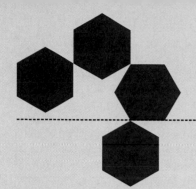

Almost everybody in the world is familiar with sodium chloride, NaCl, and English-speaking people call it *salt*. It is as vital to our lives as air, water, and food. The migrations and patterns of settlement of whole tribes and peoples in ancient times were governed as much by the availability of salt as that of anything else. Even in early U. S. history, according to the historian Frederick Jackson Turner, little westward migration could occur until sources of salt were found beyond the Allegheny mountains in what are now western parts of Virginia, West Virginia, New York, and Kentucky.

OUR NEED FOR SALT

Every cell of the body contains the ions of salt, Na^+ and Cl^-. Altogether, the average adult human body contains about 160 to 175 g of salt, enough to fill six 1-oz salt shakers. To compensate for losses by perspiration and excretion in the urine, we need about 200 mg a day—roughly one tenth of a teaspoonful.

Meat contains enough salt so that meat eaters can satisfy all of their daily requirements without supplementary salt. The need for extra salt arose when ancient tribes discovered and relied more and more on crops for their diets. Vegetables and cereal grains cannot alone replace the daily losses of salt from the body.

Another use of salt discovered by ancient peoples was that meat does not spoil when coated heavily with crystalline salt or soaked in concentrated salt brine. Decay-causing microorganisms cannot live in a high concentration of salt. Until the development of modern refrigeration, meat packers did exactly that, they packed meat in salt. Without salting, only deep freezing can indefinitely preserve meat. Fishermen who go to sea for weeks to harvest fish must carry tons of salt to preserve the catch, unless they are equipped with good deep-freezing systems. No wonder that sailors have long been called "old salts."

So essential has salt been both to the body and for food preservation that in some salt-poor regions it has enjoyed the status of money itself. When Marco Polo visited Tibet on his way to China in the thirteenth century, he found coins made of rock salt bearing the seal of the ruler. Ancient Rome had to use armed troops to guard the salt supplies obtained from salt-making operations near Ostia, southwest of Rome. The soldiers who protected and transported the salt along the Via Salarium (the Salt Road) were originally paid in salt. In time, they were given money with which to buy the salt, and this payment was called their *salarium*, the Latin root of the English word, *salary*. We still speak of hard workers as people "worth their salt."

MANUFACTURE OF SALT

The salt available to us in grocery stores has a purity that we take for granted, but only in certain deposits of solid salt found beneath the Earth's surface is the purity truly high. Thus, people of high moral purity and integrity came to be called "the salt of the earth." Quite often, salt obtained from solid deposits contains claylike solids plus ions of calcium, magnesium, and sulfate. The salt sold to poor people in some countries is often of inferior quality because it is contaminated by so much earthy matter. The poorest quality salt is "salt that has lost its taste" and is merely the earthy residue left after the sodium chloride leached out.

Salt isn't actually "made" in the way that aspirin is made from substances that aren't aspirin. The "manufacture" of salt refers instead to harvesting it from its natural sources and purifying it. Since the most ancient of times, people near the ocean have let sunlight evaporate seawater to obtain salt. The salt content of the oceans isn't high—3.3% on the average (meaning 3.3 g NaCl per 100 g of seawater). Despite this, seawater is still the principal source of salt for most countries. Taken together, the world's oceans have enough salt—over 500,000 cubic miles of it—to cover the entire area of the United States to a depth of nearly 1.5 miles.

To obtain salt from the oceans, seawater is allowed to flow into a salt pan, a low-lying area of several acres surrounded by dikes (see Figure 3a). As sunlight works on the still water, the undissolved matter (sand and clay) settles. As more and more water evaporates, the least soluble substances (the sulfates and carbonates of calcium) begin to deposit from the solution. Now the concentrated brine is allowed to flow successively into crystallizing pans where further evaporation leaves "sea salt" of varying grades, the highest being 96% NaCl.

In the Middle East and many other regions, salt is obtained from the natural brines of salt marshes and swamps. The brines are frequently home to the few microorganisms that can thrive in salt water, and they include some that lend a striking red color to the salt pan (see Figure 3b). Water that is both red and salty —characteristics of blood—might be confused as blood. Evidently, the army of Moab made this mistake (2 Kings 3:22–23). Looking at the sun glinting red off the salt pans near the camp of the combined armies of Judah, Israel, and Edom, the Moabites thought that the enemies had slaughtered each other. These salt pans were probably those of Sodom, a city whose name may be from the Hebrew words for "field" (sade) and "red" (adom).

Rock salt is the other principal source of salt. It can be obtained by conventional methods of underground mining or by injecting water into the deposits, pumping out the brine, and then processing the brine in various ways.

FIGURE 3b

The red color of certain microorganisms that can live in very salty water gives these salt pans in California the appearance of blood.

INDUSTRIAL USES OF SALT

Most of the annual United States' salt production is used to make other chemicals, chiefly chlorine (Cl_2), hydrochloric acid (HCl), and sodium hydroxide (NaOH). Chlorine is essential to the manufacture of plastics such as Saran Wrap and PVC (polyvinyl chloride, used in making clear bottles), drugs like aspirin, many solvents, antifreeze compounds, and a large number of other commercially important substances. Millions of tons of chlorine are made from salt each year in the United States for these purposes and as well as to furnish municipal water systems with chlorine for killing disease-causing organisms. Thus, salt not only supports life at its most physical level, it supports the standards of living of all industrialized countries.

FIGURE 3a

Most of the world's supply of salt for human use is obtained by the evaporation of water from seawater spread out in large salt pans. Here we see a worker in Brazil harvesting salt by hand.

Questions

1. What are the chief sources of sodium chloride?

2. In what ways was salt vital to ancient peoples?

3. What is the average percentage concentration of salt in the ocean?

4. Which three chemicals of industrial importance are manufactured from salt?

On prolonged exposure to air and moisture, bronze statues such as this Fisherman's Memorial in Gloucester, MA, gradually corrode and become coated with a green layer of $Cu_2(OH)_2CO_3$*. The corrosion of metals involves a process called* oxidation, *which is one of the chemical terms introduced in this chapter.*

Chapter 4

Chemical Reactions in Solution

4.1

SOLUTIONS AND CHEMICAL REACTIONS

When two or more reactants are involved in a chemical reaction, their particles—atoms, ions, or molecules—must make physical contact. The particles need freedom of motion for this, which exists in gases and liquids but not in solids. Whenever possible, therefore, reactions are carried out with all of the reactants in one fluid phase, liquid or gas. When necessary, a liquid is used to dissolve solid reactants so their particles can move about. A *solid* mixture of sodium carbonate and citric acid, for example, can be made without any reaction occurring. At the moment water is added, however, and these substances start to dissolve, a furious fizzing occurs as gaseous carbon dioxide forms and leaves the mixture. Representing citric acid as H_3Cit, the reaction in solution is

$$3Na_2CO_3(aq) + 2H_3Cit(aq) \longrightarrow 2Na_3Cit(aq) + 3CO_2(g) + 3H_2O$$

You've seen this fizzing if you've use a medication like Alka-Seltzer, which also contains aspirin (Figure 4.1).

In this chapter we discuss the kinds of reactions that occur primarily in aqueous solutions and how to deal with these reactions quantitatively in the laboratory.

Some Definitions

When a solution forms, at least two substances are involved. One is called the *solvent,* and all of the others are called *solutes.* The **solvent** is the medium into which the solutes are mixed or dissolved. Water is a typical and very common solvent, but the solvent can actually be in any physical state, a solid, a liquid, or a gas. Unless stated otherwise, we will assume that any solutions we mention are *aqueous solutions,* so that liquid water is understood to be the solvent.

FIGURE 4.1

Carbon dioxide forms when sodium bicarbonate reacts with citric acid in water, as is occurring here to a tablet of a carbonated medication, Alka-Seltzer.

127

A **solution** is a homogeneous mixture in which the molecules or ions of the components freely intermingle.

Concentrations described as grams of solute per 100 g of *solution* are called **percentage concentrations.**

A **solute** is any substance dissolved in the solvent. It might be a gas, like the carbon dioxide dissolved in carbonated beverages. Some solutes are liquids, like ethylene glycol dissolved in water to protect a vehicle's radiator against freezing. Solids, of course, can be solutes, like the sugar dissolved in lemonade.

We often find it necessary to describe the composition of a solution in terms of the amounts of solute and solvent. We use the term **concentration** to describe the *ratio* of the amount of solute to either the amount of solvent or the amount of solution. For example, we might have a solution of salt in water in which the concentration is 2 g salt/100 g of water. This expression gives the solute-to-solvent ratio. The concentration of salt in seawater is commonly given as 3 g salt/100 g seawater, which expresses the solute-to-solution ratio. There are other ways of expressing concentration, as we will see later.

Several terms are used to describe the *relative* amounts of solute and solvent without specifying actual quantities. In a **dilute solution,** for example, the ratio of solute to solvent is small, sometimes very small. A few crystals of salt in a glass of water make a very dilute salt solution. In a **concentrated solution,** the ratio of solute to solvent is large. Syrup, for example, is a very concentrated solution of sugar in water.

Concentrated and *dilute* are relative terms. For example, a solution of 100 g of sugar in 100 mL of water is concentrated compared to one with just 10 g of sugar in 100 mL of water, but the latter solution is more concentrated than one that has 1 g of sugar in 100 mL of water.

Usually there is a limit to the amount of a solute that can dissolve in a given solvent. For example, at 20 °C only 36.0 g of sodium chloride is able to dissolve in 100 g of water. If more is added, it simply rests at the bottom of the solution. We say this is a **saturated solution** because at its present temperature it cannot dissolve any more solute. To make sure that a solution is saturated, extra solid solute is sometimes put into its container. The presence of this solid at the bottom of the container over a period of time is visible evidence that the solution is saturated.

The **solubility** of a solute is usually described by the number of grams that dissolve in 100 g of solvent at a given temperature to make a saturated solution. The temperature must be specified because solubilities vary with temperature. Table 4.1 gives the solubilities of several substances in water, and you can see that they vary widely. Some saturated solutions, like sodium hydroxide, are concentrated, but others, like copper sulfide, are extremely dilute.

TABLE 4.1   Solubilities of Some Common Compounds in Water

Substance	Formula	Solubility (g/100 g water)[a]
Ammonium chloride	NH_4Cl	29.7 (0 °C)
Boric acid	H_3BO_3	6.35 (30 °C)
Calcium carbonate	$CaCO_3$	0.0015 (25 °C)
Calcium chloride	$CaCl_2$	74.5 (20 °C)
Copper sulfide	CuS	3.3×10^{-5} (18 °C)
Lead sulfate	$PbSO_4$	4.3×10^{-3} (25 °C)
Sodium hydroxide	$NaOH$	42 (0 °C)
		347 (100 °C)
Sodium chloride	$NaCl$	35.7 (0 °C)
		39.12 (100 °C)

[a] At the temperature given in parentheses.

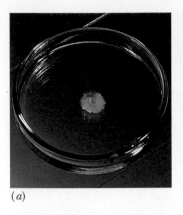
(a)

(b)

(c)

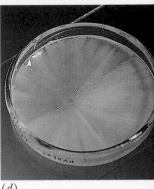
(d)

If a solution contains less solute than required for saturation, we call it an **unsaturated solution.** Unsaturated solutions are able to dissolve more solute.

Usually, the solubility of a solute increases with increasing temperature, so as the temperature of the mixture is raised, more solute dissolves. If the temperature of the solution is subsequently lowered, we would expect this additional solute to separate from the solution, and indeed, this tends to happen. However, sometimes the solute doesn't separate and we obtain a **supersaturated solution,** a solution that contains more solute than required for saturation at a given temperature.

Supersaturated solutions are unstable and can be prepared only if there are no traces of undissolved solute left in contact with the solution. If even a tiny crystal of the solute is added, the extra solute will crystallize and separate from the solution. Figure 4.2 shows what happens when a tiny crystal of sodium acetate is added to a supersaturated solution of this salt in water.

FIGURE 4.2

When a small seed crystal of sodium acetate is added to a supersaturated solution of the salt, excess solute crystallizes rapidly until the solution is just saturated. The crystallization shown in this sequence took less than 10 seconds.

In Chapter 2 you learned that pure water does not conduct electricity. This is because water consists of uncharged molecules that are incapable of transporting electrical charge. However, when an ionic compound such as $CuSO_4$ or NaCl is dissolved in the water, an electrically conducting solution is formed.

Solutes such as $CuSO_4$ or NaCl that yield electrically conducting aqueous solutions are called **electrolytes** (Figure 4.3). These ionic compounds are electrolytes because their solid crystals consist of already existing ions. When they dissolve in water, the ions separate from each other and enter the solution

4.2
ELECTROLYTES AND NONELECTROLYTES

For ionic compounds, the ions are present both in the solid and in aqueous solutions.

(a)

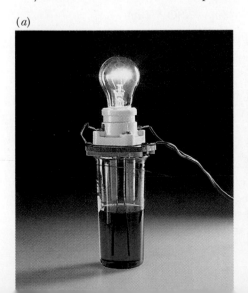

(b)

FIGURE 4.3

Electrical conductivity of solutions of electrolytes *vs* nonelectrolytes. (*a*) The copper sulfate solution is a strong conductor, and $CuSO_4$ is a strong electrolyte. (*b*) Neither sugar nor water is an electrolyte, and this sugar solution is a nonconductor.

as more or less independent particles. This change is called the **dissociation** of the ionic compound.

Aqueous solutions of most molecular compounds do not conduct electricity, and such solutes are called **nonelectrolytes.** Examples are sugar and ethylene glycol (mentioned above as the solute in antifreeze solutions). Both of these solutes consist of molecules that stay intact when they dissolve. They simply intermingle with water molecules when their solutions form. The solution contains no charged particles, so it is unable to conduct electricity.

Ethylene glycol, $C_2H_4(OH)_2$, is a type of alcohol. Other alcohols, such as ethanol and methanol, are also nonelectrolytes.

Equations for Dissociation Reactions

A convenient way to describe the dissociation of an ionic compound is with a chemical equation.

$$NaCl(s) \longrightarrow Na^+(aq) + Cl^-(aq)$$

We use the symbol *aq* after a charged particle to mean that it is dissolved in water and surrounded by water molecules. When an ion is surrounded by water molecules, we say it is hydrated. By writing the formulas of the ions separately, we mean that they are essentially independent of each other in the solution.

We can write a similar equation for the dissociation of calcium chloride, $CaCl_2$.

$$CaCl_2(s) \longrightarrow Ca^{2+}(aq) + 2Cl^-(aq)$$

This shows that each formula unit of $CaCl_2(s)$ releases three ions, one $Ca^{2+}(aq)$ and two $Cl^-(aq)$. The symbols (*s*) and (*aq*) once again indicate that the change is the dissociation of the ionic compound. Quite often, however, these symbols are omitted. When the context makes it clear that the system is aqueous, these symbols can be "understood." You should not be fooled, therefore, when you see an equation such as

$$CaCl_2 \longrightarrow Ca^{2+} + 2Cl^-$$

Unless something is said to the contrary, it means

$$CaCl_2(s) \longrightarrow Ca^{2+}(aq) + 2Cl^-(aq)$$

Polyatomic ions generally remain intact as dissociation occurs. When sodium sulfate, Na_2SO_4, dissolves in water, for example, the solution contains both sodium ions and intact sulfate ions, released as follows.

$$Na_2SO_4(s) \longrightarrow 2Na^+(aq) + SO_4^{2-}(aq)$$

To write equations like this properly, you must know both the formulas and the charges of the polyatomic ions. If necessary, refer to Table 2.9 for a review of the formulas of polyatomic ions.

■ **Practice Exercise 1** Write equations that show what happens when the following crystalline ionic compounds dissolve in water: (a) $MgCl_2$, (b) $Al(NO_3)_3$, (c) Na_2CO_3, (d) $(NH_4)_2SO_4$.

4.3
ACIDS AND BASES AS ELECTROLYTES

Some of our most familiar chemicals, as well as many important laboratory reagents, are acids and bases. The vinegar in a salad dressing, the sour juice of a lemon, the gastric juice that aids digestion in the stomach, vitamin C, and the liquid in the battery of an automobile are similar in at least one respect—they

Some common acids.

Some common bases.

all contain acids. The white crystals of lye in certain drain cleaners, the white substance that makes milk of magnesia opaque, and household ammonia are all bases.

Usually, we use the term **reagent** to mean a chemical commonly kept on hand to be used in chemical reactions.

Common Properties of Acids and Bases

Long ago, people had simple tests to decide whether something was an acid or a base. All acids, for example, have a tart (sour) taste. This isn't a very safe test, and you are strongly urged not to apply it to anything without more knowledge and experience. (Battery acid, for example, would destroy your tongue!)

Acids generally turn blue litmus red. Litmus is a dye, and one remarkable property about this natural dye is that is has a red color in acids and a blue color in bases.[1] Experimentally, **acids** turn litmus red and their aqueous solutions usually have a tart taste.

Many substances classified as acids by these tests also corrode some metals. (Perhaps you have seen this effect following the spill of battery acid somewhere on a metal surface under the hood of a car.)

Needless to say, anything called an "acid" should be treated carefully until you are sure about it. Battery acid—somewhat concentrated sulfuric acid—is one thing; but the citric acid in lemon juice, or ascorbic acid (vitamin C) are something else when it comes to potential harm to yourself. Obviously, we want to learn why substances posing such wide differences in harm or benefit to us can all be so narrowly placed in the same family—acids.

The old tests to decide whether something is a base also used taste and behavior to litmus. Aqueous solutions of **bases** generally have a somewhat bitter taste, and they turn litmus blue. They also have a soapy "feel." Like acids, bases pose varying risks. The base in milk of magnesia, $Mg(OH)_2$, can be taken internally as a laxative or as something to soothe the stomach when too much gastric acid (hydrochloric acid) is present. But another base, sodium

Litmus paper, a strip of paper impregnated with the dye litmus, becomes blue in aqueous ammonia (a base) and red in lemon juice (which contains citric acid).

---

[1] Litmus paper, commonly found among the items in a locker in the general chemistry lab, consists of strips of absorbent paper that have been soaked in a solution of litmus and dried. Red litmus paper is used to test if a solution is basic. A basic solution turns red litmus blue. To test if the solution is acidic, blue litmus paper is used. Acidic solutions turn blue litmus red.

hydroxide (lye), is *extremely* hazardous, particularly to the eyes, but even to the skin. (The hazards posed by acids and bases are good reasons for wearing eye protective devices whenever you do laboratory work or are observing it.)

Arrhenius Definition of Acids and Bases

Arrhenius was nearly failed by his examining committee for proposing such a bizarre idea that electrically charged particles exist in solution. In 1903, this view won him the Nobel Prize.

The first comprehensive theory concerning acids, bases, and electrically conducting solutions appeared in 1884 in the Ph.D. thesis of a Swedish chemist, Svante Arrhenius. He proposed that ions form directly when salts dissolve in water, which was a radical notion at the time. This is why *all* aqueous solutions of salts conduct electricity, according to Arrhenius.

Arrhenius also theorized that all acids release hydrogen ions, H^+, in water, and all bases release hydroxide ions, OH^-. This explains why acids have many properties in common and why bases behave similarly to each other.

One of the properties of acids and bases is their reaction with each other. For example, if solutions of hydrochloric acid, $HCl(aq)$, and sodium hydroxide, $NaOH(aq)$, are mixed in a 1 mol to 1 mol ratio, the resulting solution has no effect on litmus paper because the following reaction occurs.

$$HCl(aq) + NaOH(aq) \longrightarrow NaCl(aq) + H_2O$$

The acidic and basic properties of the solutes disappear; the solution of the products is neither acidic nor basic. We say that an *acid–base neutralization* has occurred. According to Arrhenius, acid–base neutralization is simply the combination of a hydrogen ion with a hydroxide ion to produce a water molecule, thus making H^+ ions and OH^- ions disappear.

Ionization Reactions for Acids and Bases in Water

Arrhenius did not know about hydronium ions, only hydrogen ions, H^+. But today, we know that H^+ can exist in water only attached to something, as in H_3O^+. We view H_3O^+ as a species that carries H^+ in water. However, for the sake of convenience, we often use the term *hydrogen ion* as a substitute for *hydronium ion,* and in many equations, we use $H^+(aq)$ to stand for $H_3O^+(aq)$. In fact, whenever you see the symbol $H^+(aq)$, think of $H_3O^+(aq)$.

For most purposes, we find that the following modified versions of Arrhenius's definitions work satisfactorily when we deal with aqueous solutions.

Arrhenius Definition of Acids and Bases

An **acid** is a substance that reacts with water to produce hydronium ion, H_3O^+.

A **base** is a substance that produces hydroxide ion in water or is able to react with hydronium ion.

Substances That Are Acids

If gaseous HCl is cooled to about $-85\ °C$, it condenses to a liquid that doesn't conduct electricity. No ions are present in pure liquid HCl.

In general, **acids** are molecular substances that react with water to produce ions, one of which is the hydronium ion. For example, pure hydrogen chloride is a gas and is molecular, not ionic. As it dissolves in water, however, the following chemical reaction occurs, which produces ions that did not preexist in $HCl(g)$.

$$HCl(g) + H_2O \longrightarrow H_3O^+(aq) + Cl^-(aq)$$

A reaction like this, in which ions form where none have existed before, is called an **ionization reaction.** One result of such a reaction in water is a solution that conducts electricity, so acids can be classified as electrolytes.

Similar ionization reactions occur for other acids as well. For example, nitric acid reacts with water according to the equation

$$HNO_3(l) + H_2O \longrightarrow H_3O^+(aq) + NO_3^-(aq)$$

Notice that in the ionization reactions of HCl and HNO_3 described above, the hydronium ion is formed by the transfer of an H^+ ion from the acid molecule to the water molecule. The particle left after loss of the H^+ is an anion.

In a sense, the active ingredient in the hydronium ion is H^+, which is why $H^+(aq)$ is often used in place of $H_3O^+(aq)$ in equations. The ionization of HCl and HNO_3 in water can be represented as

$$HCl(g) \xrightarrow{H_2O} H^+(aq) + Cl^-(aq)$$

and

$$HNO_3(l) \xrightarrow{H_2O} H^+(aq) + NO_3^-(aq)$$

Sometimes an acid also contains hydrogen atoms that are not able to transfer to water molecules. An example is acetic acid, $HC_2H_3O_2$, the acid that gives vinegar its sour taste. This acid reacts with water as follows.

$$HC_2H_3O_2(l) + H_2O \longrightarrow H_3O^+(aq) + C_2H_3O_2^-(aq)$$

Notice that only the hydrogen written first in the formula is able to transfer to H_2O to give hydronium ions. (We will discuss why the hydrogens are different in a later section.)

Polyprotic acids, which can furnish more than one H^+ per molecule of acid, undergo reactions similar to those of HCl and HNO_3, except that the loss of H^+ by the acid occurs in two or more steps. Thus, the ionization of sulfuric acid, a *diprotic acid,* takes place by two successive steps.

$$H_2SO_4(aq) + H_2O \longrightarrow H_3O^+(aq) + HSO_4^-(aq)$$

$$HSO_4^-(aq) + H_2O \longrightarrow H_3O^+(aq) + SO_4^{2-}(aq)$$

■ **Practice Exercise 2** Write equations for (a) the ionization of $HCHO_2$ (formic acid) in water, and (b) the stepwise ionization of the triprotic acid H_3PO_4 in water.

Nonmetal Oxides as Acids   The acids we've discussed so far have been molecules that contain hydrogen atoms that can be transferred to water molecules. Another class of compounds that fits our description of acids is *nonmetal oxides.* Here we have oxides such as SO_3, CO_2, and N_2O_5, whose aqueous solutions contain H_3O^+ and turn litmus red. These oxides are called **acidic anhydrides,** where *anhydride* means "without water." They react with water to form molecular acids containing hydrogen, which are then able to undergo reaction with water to yield H_3O^+.

$$SO_3(g) + H_2O \longrightarrow H_2SO_4(aq) \qquad \text{sulfuric acid}$$

$$N_2O_5(g) + H_2O \longrightarrow 2HNO_3(aq) \qquad \text{nitric acid}$$

$$CO_2(g) + H_2O \longrightarrow H_2CO_3(aq) \qquad \text{carbonic acid}$$

Although carbonic acid is too unstable to isolate as a pure compound, its solution in water is quite common. Carbon dioxide from the atmosphere dissolves in groundwater, for example, where it exists partly as carbonic acid and its ions. In carbonated beverages, carbonic acid also is present in the solution.

Not all nonmetal oxides are acidic anhydrides, only those that are able to react with water. For example, carbon monoxide doesn't react with water, its solutions in water are not acidic, and therefore CO is not classified as an acidic anhydride.

Substances That Are Bases

Bases fall into two categories, ionic compounds that contain OH^- or O^{2-}, and molecular compounds that react with water to give hydroxide ion.

Ionic bases include metal hydroxides such as NaOH and $Ca(OH)_2$. When dissolved in water, they dissociate.

$$NaOH(s) \longrightarrow Na^+(aq) + OH^-(aq)$$

$$Ca(OH)_2(s) \longrightarrow Ca^{2+}(aq) + 2OH^-(aq)$$

Soluble metal oxides are **basic anhydrides** because they react with water to form the hydroxide ion as one of the products. Calcium oxide is typical.

$$CaO + H_2O \longrightarrow Ca(OH)_2$$

Continual contact of your hands with fresh Portland cement can lead to irritation because the mixture is quite basic.

This reaction occurs when cement is made because calcium oxide or "quick lime" is an ingredient in this material. In this case it is the oxide ion, O^{2-}, that actually forms the OH^-.

$$O^{2-} + H_2O \longrightarrow 2OH^-$$

Molecular Bases   The most common molecular base is the gas ammonia, NH_3, which dissolves in water and reacts to give a basic solution.

$$NH_3(aq) + H_2O \longrightarrow NH_4^+(aq) + OH^-(aq)$$

Since bases yield solutions that contain ions, they are also electrolytes.

This is also an ionization reaction because ions have been formed where none had previously existed. Notice that when a molecular base reacts with water an H^+ is lost by the water molecule and gained by the base. One product is a cation that has one more H and one more positive charge than the reactant base. Loss of H^+ by the water gives the other product, the OH^- ion, which is why the solution is basic.

Strong and Weak Acids and Bases

Sodium chloride and calcium chloride are both examples of **strong electrolytes**—electrolytes that break up essentially 100% into ions in water. No molecules of either NaCl or $CaCl_2$ are detectable in their aqueous solutions; they exist entirely as ions. This behavior is typical of all ionic compounds, which gives us a useful generalization.

All ionic compounds are strong electrolytes, even those of very limited solubility.

Solid $CaCO_3$ (chalk) illustrates an ionic compound that is nearly insoluble in water; just a trace of it dissolves. Yet it is a strong electrolyte because what does dissolve dissociates 100% into the ions, Ca^{2+} and CO_3^{2-}.

Metal hydroxides are ionic compounds and so are also strong electrolytes. These include all of the hydroxides of the Group IA and IIA metals. Some common examples are

Group IA	NaOH	Sodium hydroxide
	KOH	Potassium hydroxide
Group IIA	$Mg(OH)_2$	Magnesium hydroxide
	$Ca(OH)_2$	Calcium hydroxide

Those of Group IIA are very slightly soluble in water; yet what does dissolve breaks up 100% into the metal ions and hydroxide ions. Because these compounds are completely dissociated, they are said to be **strong bases.** Any base that is a strong electrolyte is also a strong base.

Hydrochloric acid is also a strong electrolyte. Its ionization in water is essentially complete and no molecules of HCl can be detected in its solutions. Acids, like hydrochloric acid, that are strong electrolytes are called **strong acids.** All strong acids are strong electrolytes, but there are very few. *These should be memorized;* the most common are

$HClO_4(aq)$	Perchloric acid
$HCl(aq)$	Hydrochloric acid
$HBr(aq)$	Hydrobromic acid
$HI(aq)$	Hydroiodic acid
$HNO_3(aq)$	Nitric acid
$H_2SO_4(aq)$	Sulfuric acid

 List of strong acids

Weak Acids and Bases

Most acids are not completely ionized in water. For instance, a solution of acetic acid, $HC_2H_3O_2$, is a relatively poor conductor of electricity compared to a solution of HCl with the same concentration (Figure 4.4), so acetic acid is classified as a **weak electrolyte.** This is because in the acetic acid solution only a small fraction (less than 1%) of the molecules of the acid actually exist as H_3O^+ and $C_2H_3O_2^-$ ions. The rest of the acetic acid is present as molecules of $HC_2H_3O_2$. For now, let's represent this weak ionization as follows.

$$HC_2H_3O_2(aq) + H_2O \xrightarrow{\text{small percentage}} H_3O^+(aq) + C_2H_3O_2^-(aq)$$

FIGURE 4.4

Electrical conductivity of solutions of strong and weak acids and bases. (*a*) 1 *M* HCl is 100% ionized and it's a strong conductor, enabling the light to glow brightly. (1 *M* means a concentration of 1 mol of solute per liter of solution.) (*b*) 1 *M* $HC_2H_3O_2$ is a weaker conductor than 1 *M* HCl because the extent of its ionization is far less, so the light is dimmer. (*c*) 1 *M* NH_3 also is a weaker conductor than 1 *M* HCl because the extent of its ionization is low.

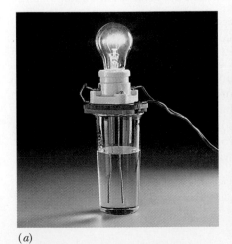

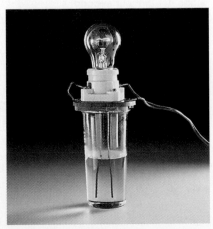

(*a*)　　　　　(*b*)　　　　　(*c*)

Formic acid, HCHO$_2$, is not among the list of strong acids, so we would conclude it is a weak acid. This acid is in the venom of stinging insects, such as fire ants.

Strong bases are ionic metal hydroxides, such as NaOH; weak bases are molecular bases, such as NH$_3$.

Solutions of acetic acid do not have high concentrations of H$_3$O$^+$ and are not strongly acidic. Acetic acid is therefore said to be a **weak acid.** All weak acids are weak electrolytes. Other examples are carbonic acid, H$_2$CO$_3$, and nitrous acid, HNO$_2$. In fact, *if an acid is not one of the strong acids listed above, you can assume it to be a weak acid.*

Certain bases are also weak electrolytes and have a low percentage ionization. They are classified as **weak bases.** Ammonia is perhaps the most common example (see Figure 4.4*c*). In a solution of this base only a small fraction of the solute is ionized to give NH$_4$$^+$ and OH$^-$. Most of the base is present as ammonia molecules.

$$NH_3(aq) + H_2O \xrightarrow{\text{small percentage}} NH_4^+(aq) + OH^-(aq)$$

It is interesting to note that *virtually all molecular bases are weak bases.*

Let's briefly summarize the results of our discussion.

Weak acids and bases are weak electrolytes.

Strong acids and bases are strong electrolytes.

Always bear in mind that whether a given compound is strong or weak in any category—acids, bases, or electrolytes—hinges on its *percentage ionization,* not its solubility. As we have said, many bases are quite insoluble in water, but they are still strong bases (and strong electrolytes). Ammonia, on the other hand, is very soluble in water, yet it is a weak base.

Dynamic Equilibria in Solutions of Weak Acids and Bases

A question that might have occurred to you during the discussion of weak acids and bases is, "If some of the solute molecules in aqueous ammonia or aqueous acetic acid can react with water, why can't all of them?" Actually, all do have the potential to react, but we meet here examples of a very important chemical phenomenon— *dynamic equilibrium.* Let's mentally go inside an acetic acid solution to see what is happening.

As soon as the acetic acid is mixed with water, solute and solvent molecules start to collide with each other. Because of the chemical nature of acetic acid, however, only a small percentage of such collisions produces ions. As a result, the rate (or speed) at which the ions form is quite small compared with the rate of collision. There is a *low* probability of the following reaction occurring. When it happens, we call it the *forward reaction* in the acetic acid–water system.

For now, consider *rate* to mean the number of events per unit of volume per unit of time, like collisions per cm^3 per second.

$$HC_2H_3O_2(aq) + H_2O \xrightarrow[\text{low probability}]{\text{small percentage}} H_3O^+(aq) + C_2H_3O_2^-(aq)$$

Once H$_3$O$^+$ ions and C$_2$H$_3$O$_2$$^-$ form, they wander around in the solution and experience collisions with solvent molecules. Occasionally, of course, an H$_3$O$^+$ ion and a C$_2$H$_3$O$_2$$^-$ ion meet and collide. Because of the chemical nature of these ions, there is a *high* probability that such a collision transfers H$^+$ from H$_3$O$^+$ back to C$_2$H$_3$O$_2$$^-$. We also represent this reaction by an equation, but one in which the arrow goes from right to left. We call it the *reverse reaction* because it tends to undo the forward reaction.

The probability of ions forming or disappearing as a result of collisions varies from solute to solute.

$$HC_2H_3O_2(aq) + H_2O \xleftarrow[\text{high probability}]{\text{high percentage}} H_3O^+(aq) + C_2H_3O_2^-(aq)$$

Let's go over the dynamics of this again. Just after $HC_2H_3O_2$ is added to water, the concentrations of H_3O^+ and $C_2H_3O_2^-$ build up because of the ionization reaction, the *forward reaction*. As their concentrations increase however, they naturally encounter each other more frequently. So the *reverse reaction* occurs with increasing frequency. Eventually the concentrations of the ions become large enough *to make the rate of the reverse reaction equal the rate of the forward reaction*. At this point, ions disappear as rapidly as they form, so their concentrations remain constant from this moment on. The entire chemical system consisting of H_2O, $HC_2H_3O_2$, H_3O^+, and $C_2H_3O_2^-$ has reached a state of balance called **chemical equilibrium.** It is also said to be a **dynamic equilibrium** because the forward and reverse reactions don't cease; they continue to occur even after the concentrations have stopped changing.

We indicate a dynamic equilibrium by using double arrows ($\rightleftharpoons$) in the chemical equation. The ionization of acetic acid is thus represented as follows.

$$HC_2H_3O_2(aq) + H_2O \rightleftharpoons H_3O^+(aq) + C_2H_3O_2^-(aq)$$

In discussing an equilibrium like this, we will often talk about the *extent of completion* of a forward or a reverse reaction. To call acetic acid a *weak* acid, for example, is just another way of saying that the forward reaction in this equilibrium is far from completion. For any weak electrolyte, only a very small percentage of the solute is actually ionized at any instant after equilibrium is reached.

With molecular compounds that are strong electrolytes, the tendency of the forward ionization reaction to occur is very high, while the tendency of the reverse reaction to occur is extremely small. In aqueous HCl, for example, there is little tendency during a collision between Cl^- and H_3O^+ for neutral molecules of HCl and H_2O to form. As a result, the reverse reaction has essentially no tendency to occur, so in a very brief time all of the HCl is converted to ions—it becomes 100% ionized. For this reason, *we do not use double arrows in describing what happens when HCl(g) or any other strong electrolyte undergoes ionization or dissociation.*

■ **Practice Exercise 3** Nitrous acid, HNO_2, is a weak acid. Write the chemical equation that represents the equilibrium for this reaction.

For this juggler, balls are going up at the same rate as they are coming down, so the number of balls in the air stays constant. A somewhat similar situation exists in a dynamic equilibrium, where products form at the same rate as they disappear and the amount of products stays constant.

4.4 ACID–BASE NEUTRALIZATION

Acid–base neutralizations generally produce water and a **salt** (which we defined in Chapter 2 as any ionic compound that does not contain H^+, OH^-, or O^{2-} ions). The salt is composed of the anion of the acid and the cation of the base. An example is the reaction of nitric acid with potassium hydroxide.

$$HNO_3(aq) + KOH(aq) \longrightarrow KNO_3(aq) + H_2O$$

The products are water and a salt, KNO_3, which is made of K^+ and NO_3^- ions.

There are some salts, the *acid salts,* that include an ion still capable of providing H^+. Sodium hydrogen sulfate, $NaHSO_4$, for example, contains the HSO_4^- ion. This enables the salt to neutralize sodium hydroxide as follows and to make still another member of the salt family, sodium sulfate, Na_2SO_4.

$$NaHSO_4(aq) + NaOH(aq) \longrightarrow Na_2SO_4(aq) + H_2O$$

Molecular, Ionic, and Net Ionic Equations

The equations we've written above for these neutralization reactions are sometimes referred to as molecular equations, even though in each case both reactants and one of the products are strong electrolytes and are present as ions in the reaction mixture. In a **molecular equation,** *complete formulas are written for all the reactants and products; the formulas of the compounds do not indicate the presence of ions.*

To provide a more accurate description of the actual reaction, we can write an **ionic equation** *in which the formulas of all soluble strong electrolytes are shown in their dissociated (or ionized) forms.* Let's do this for the reaction of nitric acid with potassium hydroxide.

We begin with the reactants, which are both fully dissociated (ionized) in water. In place of HNO_3 we write the ions formed by its ionization, and in place of KOH we write the ions formed in its dissociation. On the right side of the equation only one product, KNO_3, is a strong electrolyte, so we can write it in "dissociated form." This gives the ionic equation

$$H^+(aq) + NO_3^-(aq) + K^+(aq) + OH^-(aq) \longrightarrow K^+(aq) + NO_3^-(aq) + H_2O$$

Notice that the H^+ and OH^- ions on the left disappear and are replaced by water molecules on the right; the neutralization actually involves the reaction of H^+ with OH^- ions. The K^+ and NO_3^- ions don't change. They are present before the reaction, and they are present afterwards. *Ions that do not actually take part in a reaction are sometimes called* **spectator ions;** in a sense they just stand by and watch what's going on.

To emphasize the actual reaction that occurs, we write the **net ionic equation,** *which is obtained by eliminating the spectator ions from the ionic equation.* Let's cross out the K^+ and NO_3^- ions.

$$H^+(aq) + \cancel{NO_3^-(aq)} + \cancel{K^+(aq)} + OH^-(aq) \longrightarrow \cancel{K^+(aq)} + \cancel{NO_3^-(aq)} + H_2O$$

What remains is the net ionic equation,

$$H^+(aq) + OH^-(aq) \longrightarrow H_2O$$

> The net ionic equation tells us that the reaction between HNO_3 and KOH is actually a reaction between H^+ and OH^-.

Let's briefly look at another example, the reaction of $HCl(aq)$ with $Ca(OH)_2(aq)$ to give a solution of $CaCl_2$.

$$2HCl(aq) + Ca(OH)_2(aq) \longrightarrow CaCl_2(aq) + 2H_2O$$

In writing the ionic equation, we have to be careful to give the correct number of ions of each type. One $Ca(OH)_2$ gives one Ca^{2+} and two OH^-, and two HCl must give two $H^+(aq)$ and two Cl^-. The ionic equation is therefore

$$2H^+(aq) + 2Cl^-(aq) + Ca^{2+}(aq) + 2OH^-(aq) \longrightarrow$$
$$Ca^{2+}(aq) + 2Cl^-(aq) + 2H_2O$$

and after eliminating spectator ions, the net ionic equation is

$$2H^+(aq) + 2OH^-(aq) \longrightarrow 2H_2O$$

Reducing the coefficients to the smallest set of whole numbers gives

$$H^+(aq) + OH^-(aq) \longrightarrow H_2O$$

Notice that we obtain the same net ionic equation for both reactions. In fact, this is the net reaction that occurs *whenever a strong acid and a strong base are combined.*

Criteria for a Balanced Ionic or Net Ionic Equation

In the ionic and net ionic equations we've written, not only are the atoms in balance, but so is the net electrical charge, which is the same on both sides of the equation. Thus, in the ionic equation above, the sum of the charges on $2H^+$, $2Cl^-$, Ca^{2+}, and $2OH^-$ is zero, which matches the sum of the charges on the Ca^{2+}, $2Cl^-$, and $2H_2O$. In the net ionic equation, on the left we have H^+ and OH^-, with a net charge of zero; on the right we have H_2O also with a charge of zero. This is an additional requirement for an ionic equation or net ionic equation to be balanced: *The net electrical charge on both sides of the equation must be the same.*

Criteria for balanced ionic equations

Criteria for Balanced Ionic and Net Ionic Equations

1. *Material balance*   All atoms on one side of the arrow must appear somewhere on the other side.

2. *Electrical balance*   The net electrical charge on the left must equal the net electrical charge on the right (although this charge does not necessarily have to be zero).

Neutralization When One Reactant Is a Weak Electrolyte

Let's look at the ionic and net ionic equations when acetic acid ($HC_2H_3O_2$), a weak acid, and sodium hydroxide, a strong base, react to give a salt, sodium acetate ($NaC_2H_3O_2$), and water. The molecular equation is

$$HC_2H_3O_2(aq) + NaOH(aq) \longrightarrow NaC_2H_3O_2(aq) + H_2O$$

In this reaction, only two substances are strong electrolytes, NaOH and $NaC_2H_3O_2$. Acetic acid is a weak electrolyte and is present in solution mostly in the form of molecules. Therefore, when we write the ionic equation, we do not express the acid in ionic form. We simply write the formula for molecular acetic acid. The ionic equation for this reaction is therefore,

In writing an ionic equation, we always write the formulas of weak electrolytes in "molecular form."

$$HC_2H_3O_2(aq) + Na^+(aq) + OH^-(aq) \longrightarrow Na^+(aq) + C_2H_3O_2^-(aq) + H_2O$$

This time there is only one spectator ion, Na^+, and when we eliminate it from the equation, we obtain the net ionic equation,

$$HC_2H_3O_2(aq) + OH^-(aq) \longrightarrow C_2H_3O_2^-(aq) + H_2O$$

The net ionic equation tells us this time that the reaction in the solution is actually one between hydroxide ions and molecules of acetic acid.

A slightly different situation exists in the reaction of hydrochloric acid with aqueous ammonia, a weak base, to give the salt ammonium chloride (NH_4Cl). The molecular equation for the reaction is

$$NH_3(aq) + HCl(aq) \longrightarrow NH_4Cl(aq)$$

Notice that no water is formed in this reaction. This is because the molecular base does not contain hydroxide ions. A solution of ammonia is basic because of the equilibrium

$$NH_3(aq) + H_2O \rightleftharpoons NH_4^+(aq) + OH^-(aq)$$

but the reaction occurs to an extent of less than 1%, so in a solution of ammonia nearly all of the base is present as molecules. In the neutralization reaction, therefore, ammonia is the principal reactant. The net ionic equation for the neutralization is

$$NH_3(aq) + H^+(aq) \longrightarrow NH_4^+(aq)$$

■ **Practice Exercise 4** Write molecular, ionic, and net ionic equations for the reaction of (a) HCl with KOH, (b) $HCHO_2$ with LiOH, and (c) N_2H_4 with HCl.

Reactions of Acids with Insoluble Hydroxides and Oxides

The tendency for water to be formed in neutralization reactions is so strong that it serves to drive reactions of acids (both strong and weak) with insoluble hydroxides and oxides. An example is the reaction of magnesium hydroxide with hydrochloric acid, pictured in Figure 4.5. This is the reaction that occurs when milk of magnesia neutralizes stomach acid.

Magnesium hydroxide has a very low solubility in water and is the solid that makes milk of magnesia white. The molecular equation for its reaction with $HCl(aq)$ is

$$Mg(OH)_2(s) + 2HCl(aq) \longrightarrow MgCl_2(aq) + 2H_2O$$

FIGURE 4.5

A solution of hydrochloric acid is neutralized by magnesium hydroxide, $Mg(OH)_2$, in milk of magnesia. The mixture is clear where the solid $Mg(OH)_2$ has dissolved.

When we write the ionic equation, we do not write the $Mg(OH)_2$ in dissociated form, even though it is an ionic compound. This is because it is not dissolved in water in the reaction mixture, so the ions are not free to roam about in the solution. Magnesium chloride, on the other hand, is water soluble and is fully dissociated. (Notice in Figure 4.5 that the reaction mixture is not cloudy where the reaction is complete, indicating that all the solid has dissolved. This tells us that the $MgCl_2$ is soluble in water.) The ionic equation is

$$Mg(OH)_2(s) + 2H^+(aq) + 2Cl^-(aq) \longrightarrow Mg^{2+}(aq) + 2Cl^-(aq) + 2H_2O$$

The net ionic equation is obtained by eliminating the Cl^- spectator ions.

$$Mg(OH)_2(s) + 2H^+(aq) \longrightarrow Mg^{2+}(aq) + 2H_2O$$

Many metal oxides are similarly dissolved by acids. For example, before steel can be given a protective coating of zinc (a process called galvanizing), rust must first be stripped from the steel. This is usually done by dissolving the rust, mostly Fe_2O_3, in hydrochloric acid. The molecular, ionic, and net ionic equations are

FIGURE 4.6

A precipitation reaction. When a colorless solution of $Cd(NO_3)_2$ is poured into a colorless solution of Na_2S, a bright orange-yellow precipitate of cadmium sulfide, CdS, appears. In this reaction, Cd^{2+} replaces Na^+ in the sulfide and Na^+ replaces Cd^{2+} in the nitrate; hence the name *double replacement*.

$$Fe_2O_3(s) + 6HCl(aq) \longrightarrow 2FeCl_3(aq) + 3H_2O$$

$$Fe_2O_3(s) + 6H^+(aq) + 6Cl^-(aq) \longrightarrow 2Fe^{3+}(aq) + 6Cl^-(aq) + 3H_2O$$

$$Fe_2O_3(s) + 6H^+(aq) \longrightarrow 2Fe^{3+}(aq) + 3H_2O$$

■ **Practice Exercise 5** Write molecular, ionic, and net ionic equations for the reaction of $Al(OH)_3(s)$ with $HCl(aq)$. (Aluminum hydroxide is an ingredient in the antacid Digel.)

4.5
IONIC REACTIONS THAT PRODUCE PRECIPITATES

Acid–base neutralization reactions of the type discussed in the preceding section are not the only ionic reactions that are observed. When a solution of cadmium nitrate, $Cd(NO_3)_2$, is added to a solution of sodium sulfide, Na_2S, a bright orange-yellow solid forms immediately, as shown in Figure 4.6. A solid such as this which is formed in a solution is called a **precipitate.** In this case the precipitate is cadmium sulfide, CdS, which can be recovered by filtering the solid from the rest of the mixture, as pictured in Figure 4.7. The remaining solution that passes through the filter paper is called the *filtrate*. By evaporating this to dryness, we obtain the other product, sodium nitrate, $NaNO_3$. From this information, we can write the following equation.

$$Cd(NO_3)_2(aq) + Na_2S(aq) \longrightarrow CdS(s) + 2NaNO_3(aq)$$

This kind of reaction, in which ions change partners, is called **metathesis** (pronounced *me-TATH-e-sis*) or **double replacement.** In this example, cadmium ions exchange places with sodium ions.

As with acid–base neutralization, it is useful to write the ionic and net ionic equations for metathesis reactions. The soluble strong electrolytes are written in "ionic form" to show they are dissociated. The precipitate, however, is written in "molecular form" because the ions that make up the solid are not free to wander through the solution. The ionic equation is therefore

$$Cd^{2+}(aq) + 2NO_3^-(aq) + 2Na^+(aq) + S^{2-}(aq) \longrightarrow$$
$$CdS(s) + 2Na^+(aq) + 2NO_3^-(aq)$$

and by eliminating the spectator ions, NO_3^- and Na^+, we obtain the net ionic equation,

$$Cd^{2+}(aq) + S^{2-}(aq) \longrightarrow CdS(s)$$

Thus, the reaction is actually between Cd^{2+} and S^{2-}. The formation of the precipitate from the ions in solution is illustrated by Figure 4.8.

The usefulness of a net ionic equation for metathesis reactions is that it allows us to *generalize;* it allows us to extend what we have learned about one reaction to similar reactions. In our example, it calls to our attention the fact

FIGURE 4.7

Filtration is a procedure that separates a solid precipitate from a solution.

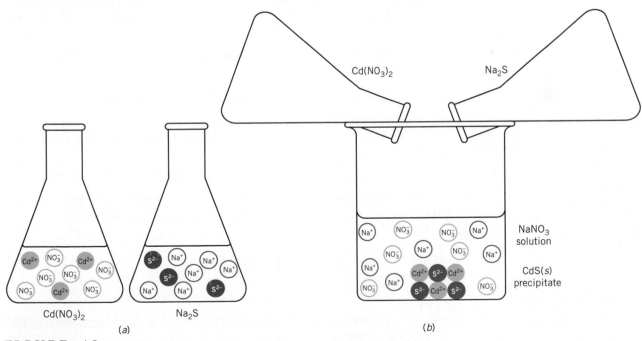

FIGURE 4.8

The reaction of $Cd(NO_3)_2$ with Na_2S. (*a*) These ionic compounds exist as separated ions in their aqueous solutions. (*b*) When the two solutions are combined, the cadmium ions, Cd^{2+}, and sulfide ions, S^{2-}, collect together and form an insoluble precipitate of cadmium sulfide, CdS. However, the sodium ions, Na^+, and nitrate ions, NO_3^-, remain free.

that *any* soluble compound that provides Cd^{2+} will react in solution with any soluble compound that furnishes S^{2-} to form a precipitate of CdS. This is very helpful; like any generalization, it greatly reduces the number of facts about the chemical properties of individual compounds that have to be learned. If we have a solution containing, for example, $CdCl_2$ instead of $Cd(NO_3)_2$, and another containing K_2S instead of Na_2S, we now can confidently know that a precipitate of CdS will form when the solutions are mixed. This is because the solution of $CdCl_2$ contains Cd^{2+} ions and that of K_2S has S^{2-} ions, and these ions — *regardless of their origins* — always form an insoluble precipitate of CdS in water.

$CdCl_2$ and K_2S are both water-soluble, strong electrolytes.

EXAMPLE 4.1
Writing Equations for Ionic Reactions

When solutions of sodium sulfate, Na_2SO_4, and barium nitrate, $Ba(NO_3)_2$, are mixed, a precipitate of barium sulfate forms. The other product is sodium nitrate, which is soluble in water. Write molecular, ionic, and net ionic equations for the reaction.

ANALYSIS We will start with the molecular equation, *being sure to use correct chemical formulas* (and their states). Then, to write the ionic equation, separate the formulas of the water-soluble, strong electrolytes. Finally, eliminate the spectator ions and rewrite what is left to give the net ionic equation. *Always be sure that any two species that you cancel are in the same state.* Both must not only have the same formula but also the same designation of state, like (*aq*) or (*s*).

SOLUTION *The Molecular Equation* Among the reactants the ions are Na^+ and SO_4^{2-} from the Na_2SO_4, and Ba^{2+} and NO_3^- from the $Ba(NO_3)_2$. In assembling the products, the Na^+ goes with the NO_3^-, so the correct formula is $NaNO_3$. Similarly, the Ba^{2+} goes with the SO_4^{2-}, so the correct formula is $BaSO_4$. We now arrange these in the proper equation form:

$$Na_2SO_4(aq) + Ba(NO_3)_2(aq) \longrightarrow BaSO_4(s) + NaNO_3(aq) \qquad \text{(unbalanced)}$$

To balance, we need a coefficient of 2 in front of $NaNO_3$.

$$Na_2SO_4(aq) + Ba(NO_3)_2(aq) \longrightarrow BaSO_4(s) + 2NaNO_3(aq)$$

The Ionic Equation All of the formulas except $BaSO_4$ represent water-soluble, strong electrolytes. So we split them up as follows to obtain the ionic equation.

Notice that *one* $Ba(NO_3)_2$ gives *one* Ba^{2+} and *two* NO_3^-. Also, *two* $NaNO_3$ give *two* Na^+ and *two* NO_3^-.

$$2Na^+(aq) + SO_4^{2-}(aq) + Ba^{2+}(aq) + 2NO_3^-(aq) \longrightarrow$$
$$BaSO_4(s) + 2Na^+(aq) + 2NO_3^-(aq)$$

The Net Ionic Equation On each side of the ionic equation are $2Na^+(aq)$ and $2NO_3^-(aq)$, so we eliminate them. They are spectator ions. This leaves

$$SO_4^{2-}(aq) + Ba^{2+}(aq) \longrightarrow BaSO_4(s)$$

Let's rewrite this with the cation first, just to match the order of the ions in the formula of the solid.

$$Ba^{2+}(aq) + SO_4^{2-}(aq) \longrightarrow BaSO_4(s)$$

Notice that the equation satisfies the requirements for both a material balance and an electrical balance.

■ **Practice Exercise 6** As discussed earlier, milk of magnesia is a suspension of finely divided solid magnesium hydroxide, $Mg(OH)_2$, in water. It can be made by adding a solution of sodium hydroxide, NaOH, to a solution of $MgCl_2$. The $Mg(OH)_2$ precipitates and sodium chloride is left in solution. Write molecular, ionic, and net ionic equations for this reaction.

The Usefulness of Each Kind of Equation   Perhaps at this point you are wondering which is *the* correct way to write the equation of an ionic reaction. The answer is that no one kind of equation is more correct than any other, nor is one better than the other. Each has its own usefulness.

We usually want the *molecular equation* for planning an experiment, because it identifies the actual reactants and lets us work out the amounts to use.

We use the *ionic equation* to visualize all that is happening in the solution when the reaction is taking place.

We need the *net ionic equation* when we want to generalize the reaction — to appreciate that more than one set of reactants can lead to the same net reaction. This equation focuses our attention most sharply on the chemical change that is actually taking place in the solution.

Sometimes a product of a metathesis reaction is a substance that normally is a gas at room temperature and is not very soluble in water. An example is hydrogen sulfide, H_2S, the compound that gives rotten eggs their foul odor. It is a weak electrolyte and is formed when a strong acid like HCl is added to a metal sulfide like sodium sulfide, Na_2S. Hydrogen sulfide has a low solubility in water, so it tends to bubble out of the solution as it forms and to escape as a gas. Once it has escaped, there is no possible way for it to participate in any sort of reverse reaction, so its departure drives the reaction to completion. (The departure of H_2S removes ions, $2H^+$ and S^{2-}.) The molecular, ionic, and net ionic equations for the reaction are

4.6 IONIC REACTIONS THAT PRODUCE GASES

H_2S can be detected by odor when its concentration is only 0.15 ppb (parts per billion) or 21 μg H_2S/m^3 air.

Molecular Equation:
$$2HCl(aq) + Na_2S(aq) \longrightarrow 2NaCl(aq) + H_2S(g)$$

Ionic Equation:
$$2H^+(aq) + 2Cl^-(aq) + 2Na^+(aq) + S^{2-}(aq) \longrightarrow 2Na^+(aq) + 2Cl^-(aq) + H_2S(g)$$

Net Ionic Equation:
$$2H^+(aq) + S^{2-}(aq) \longrightarrow H_2S(g)$$

Reactions of Carbonates and Bicarbonates with Acids

Carbonic acid (H_2CO_3) forms when an acid reacts with either a bicarbonate or a carbonate. For example, consider the reaction of sodium bicarbonate ($NaHCO_3$) with hydrochloric acid, which is pictured in Figure 4.9. As the sodium bicarbonate solution is added to the hydrochloric acid, bubbles of carbon dioxide are released. This is the same reaction that occurs if you take sodium bicarbonate to soothe an upset stomach. Stomach acid is HCl and its reaction with the $NaHCO_3$ both neutralizes the acid and produces CO_2 gas (burp!). The molecular equation for the reaction is

$$HCl(aq) + NaHCO_3(aq) \longrightarrow NaCl(aq) + H_2CO_3(aq)$$

Earlier it was mentioned that carbonic acid is too unstable to be isolated in pure form. When it forms in appreciable amounts as the product of metathesis reactions, it decomposes into water and its anhydride, the gas CO_2. Carbon dioxide is only slightly soluble in water, so most of the CO_2 bubbles out of the

FIGURE 4.9

The reaction of sodium bicarbonate with hydrochloric acid. The bubbles contain the gas carbon dioxide.

Carbonic acid formed and decomposed when sodium carbonate, pumped by a highway snowblower, was used to neutralize concentrated nitric acid spilling from a ruptured tankcar in a switching yard in Denver, Colorado (April, 1983).

solution. The decomposition reaction is

$$H_2CO_3(aq) \longrightarrow H_2O + CO_2(g)$$

Therefore, the overall equation for the reaction is

$$HCl(aq) + NaHCO_3(aq) \longrightarrow NaCl(aq) + H_2O + CO_2(g)$$

The ionic equation is

$$H^+(aq) + Cl^-(aq) + Na^+(aq) + HCO_3^-(aq) \longrightarrow$$
$$Na^+(aq) + Cl^-(aq) + H_2O + CO_2(g)$$

and the net ionic equation is

$$H^+(aq) + HCO_3^-(aq) \longrightarrow H_2O + CO_2(g)$$

Similar results are obtained if we begin with a carbonate instead of a bicarbonate. The reaction of sodium carbonate with hydrochloric acid follows the molecular equation

$$Na_2CO_3(aq) + 2HCl(aq) \longrightarrow 2NaCl(aq) + CO_2(g) + H_2O$$

In the net ionic equation derived from this, we first have the carbonate ion reacting with $H^+(aq)$ to give carbonic acid, followed by its decomposition to CO_2 and H_2O.

$$2H^+(aq) + CO_3^{2-}(aq) \longrightarrow H_2CO_3(aq)$$
$$H_2CO_3(aq) \longrightarrow H_2O + CO_2(g)$$

Overall, the net reaction is

$$2H^+(aq) + CO_3^{2-}(aq) \longrightarrow H_2O + CO_2(g)$$

Reactions of Insoluble Carbonates with Acid

The release of CO_2 by the reaction of a carbonate with an acid is such a strong driving force for reaction that it enables insoluble carbonates to dissolve in acids. The reaction of limestone, $CaCO_3$, with hydrochloric acid is shown in Figure 4.10. The molecular, ionic, and net ionic equations for the reaction are as follows.

$$CaCO_3(s) + 2HCl(aq) \longrightarrow CaCl_2(aq) + CO_2(g) + H_2O$$
$$CaCO_3(s) + 2H^+(aq) + 2Cl^-(aq) \longrightarrow Ca^{2+}(aq) + 2Cl^-(aq) + CO_2(g) + H_2O$$
$$CaCO_3(s) + 2H^+(aq) \longrightarrow Ca^{2+}(aq) + CO_2(g) + H_2O$$

The ability of calcium carbonate to dissolve in acid is responsible for one of nature's most marvelous wonders, limestone caverns, as described in Chemicals in Use 14.

Reactions of Acids with Sulfites

Another unstable weak acid that decomposes when formed in large amounts is sulfurous acid, H_2SO_3. The reaction is

$$H_2SO_3(aq) \longrightarrow H_2O + SO_2(g)$$

Whenever a solution contains the ions necessary to form H_2SO_3, we can expect a release of the gas SO_2. For example, the net ionic equations for the

FIGURE 4.10

Bubbles of CO_2 are formed in the reaction of limestone ($CaCO_3$) with hydrochloric acid.

reaction of bisulfite and sulfite ions with acid are

$$HSO_3^-(aq) + H^+(aq) \longrightarrow H_2O + SO_2(g)$$

$$SO_3^{2-}(aq) + 2H^+(aq) \longrightarrow H_2O + SO_2(g)$$

Reactions of Bases with Ammonium Salts

In writing an equation for the metathesis reaction between NH_4Cl and $NaOH$, one of the products we are tempted to write is NH_4OH—so-called "ammonium hydroxide." However, NH_4OH has never been detected, either in pure form or in an aqueous solution. Instead, "ammonium hydroxide" is simply a solution of NH_3 in water. (Note the equivalence in terms of atoms: $NH_4OH \Leftrightarrow NH_3 + H_2O$.) Therefore, in place of NH_4OH, we write $NH_3 + H_2O$.

$$NaOH(aq) + NH_4Cl(aq) \longrightarrow NaCl(aq) + NH_3(aq) + H_2O$$

The net ionic equation, after eliminating Na^+ and Cl^-, is

$$NH_4^+(aq) + OH^-(aq) \longrightarrow NH_3(g) + H_2O$$

This equation shows that *any* ammonium salt will react with a strong base to yield ammonia. Even though ammonia is very soluble in water, some escapes as a gas and its odor can be easily detected over the reaction mixture.

Table 4.2 contains a more comprehensive list of substances that are released as gases in metathesis reactions. You should learn these reactions so that you can use them in writing ionic and net ionic equations.

In the preceding sections you've learned that certain combinations of reactants yield predictable products. In neutralization, for example, a strong acid and a strong base always yield a salt and water. In metathesis, if we know which products are insoluble, we can write correct net ionic equations for the reactions. Because of this predictability, we can often anticipate the outcome of an

4.7
PREDICTING METATHESIS REACTIONS

 Gases formed in metathesis reactions

TABLE 4.2 Gases Formed in Metathesis Reactions

Gas	Formed by Reactions of Acids with:	Equations for Formation[a]
H_2S	Sulfides	$2H^+ + S^{2-} \rightarrow H_2S$
CO_2	Carbonates	$2H^+ + CO_3^{2-} \rightarrow (H_2CO_3) \rightarrow H_2O + CO_2$
	Bicarbonates	$H^+ + HCO_3^- \rightarrow (H_2CO_3) \rightarrow H_2O + CO_2$
SO_2	Sulfites	$2H^+ + SO_3^{2-} \rightarrow (H_2SO_3) \rightarrow H_2O + SO_2$
	Hydrogen sulfites	$H^+ + HSO_3^- \rightarrow (H_2SO_3) \rightarrow H_2O + SO_2$
HCN	Cyanides	$H^+ + CN^- \rightarrow HCN$

Gas	Formed by Reaction of Bases with:	Equation for Formation[b]
NH_3	Ammonium salts	$NH_4^+ + OH^- \rightarrow NH_3 + H_2O$

[a] Formulas in parentheses are of unstable compounds that break down according to the continuation of the sequence.

[b] In writing a metathesis reaction, you might be tempted to write NH_4OH as a formula for "ammonium hydroxide." This compound does not exist. In water, it is nothing more than a solution of NH_3.

Solubility rules

TABLE 4.3 Solubility Rules for Ionic Compounds in Water

Soluble Compounds

1. All compounds of the alkali metals (Group IA) are soluble.
2. All salts containing NH_4^+, NO_3^-, ClO_4^-, ClO_3^-, and $C_2H_3O_2^-$ are soluble.
3. All chlorides, bromides, and iodides (salts containing Cl^-, Br^-, or I^-) are soluble *except* when combined with Ag^+, Pb^{2+}, and Hg_2^{2+} (note the subscript "2").
4. All sulfates (salts containing SO_4^{2-}) are soluble *except* those of Pb^{2+}, Ca^{2+}, Sr^{2+}, Hg_2^{2+}, and Ba^{2+}.

Insoluble Compounds

5. All hydroxides (OH^- compounds) and all metal oxides (O^{2-} compounds) are insoluble *except* those of Group IA and of Ca^{2+}, Sr^{2+}, and Ba^{2+}.
   Note: When metal oxides do dissolve, they react with water to form hydroxides. The oxide ion, O^{2-}, does not exist in water. For example:

   $$Na_2O(s) + H_2O \longrightarrow 2NaOH(aq)$$

6. All compounds that contain PO_4^{3-}, CO_3^{2-}, SO_3^{2-}, and S^{2-} are insoluble, *except* those of Group IA and NH_4^+.

ionic reaction without actually carrying out the experiment. However, we do need certain information at our fingertips to do this.

One of the most important things to know is solubility. Fortunately there is a set of rules that we can use to determine, in many cases, whether an ionic compound is soluble or insoluble. These **solubility rules** are given in Table 4.3. Let's see what they tell us.

Rule 1 states that all compounds of the alkali metals are soluble in water. This means that you can expect any salt containing Na^+ or K^+, or any of the Group IA metal ions, *regardless of the anion,* to be soluble. If one of the reactants in a metathesis is Na_3PO_4, you now know from this rule that it is soluble. Therefore, you would write it in *dissociated* form in the ionic equation. Similarly, Rule 6 states, in part, that all carbonate salts are *insoluble* except those of the alkali metals and the ammonium ion. Thus, if one of the products in a metathesis reaction is $CaCO_3$, you expect it to be insoluble, because the cation is not an alkali metal or NH_4^+. Therefore, you would write its formula in its undissociated form when you construct the ionic equation. Special Topic 4.1 describes how the insolubility of $CaCO_3$ bears on the softening of "hard water."

EXAMPLE 4.2
Predicting Reactions and Writing Their Equations

Predict whether a reaction could occur when aqueous solutions of $Pb(NO_3)_2$ and $Fe_2(SO_4)_3$ are mixed. If a reaction is possible, write molecular, ionic, and net ionic equations for it.

ANALYSIS We first have to make trial predictions of possible products. The reactants are ionic compounds, and we are told that they are in solution and so are water soluble. Solubility rules 1 and 4 tell us this also. We begin, therefore, by seeing what a double replacement (metathesis) *might* produce. If even one new combination is *insoluble* in water, according to the solubility rules, then we can predict that a reaction occurs.

SPECIAL TOPIC 4.1 / BOILER SCALE AND HARD WATER

Precipitation reactions occur around us all the time, and we hardly ever take notice until they cause a problem. One common problem is caused by **hard water**—groundwater that contains the "hardness ions," Ca^{2+}, Mg^{2+}, Fe^{2+}, or Fe^{3+}, in concentrations high enough to form precipitates with ordinary soap. Soap normally consists of the sodium salts of organic acids derived from animal fats or oils (so-called *fatty acids*). An example is sodium stearate, $NaC_{17}H_{35}O_2$. The negative ion of the soap forms an insoluble "scum" with hardness ions, which reduces the effectiveness of the soap for removing dirt and grease.

Hardness ions can be removed from water in a number of ways. One way is to add hydrated sodium carbonate, $Na_2CO_3 \cdot 10H_2O$, often called washing soda, to the water. The carbonate ion forms insoluble precipitates with the hardness ions; an example is $CaCO_3$.

$$Ca^{2+}(aq) + CO_3^{2-}(aq) \longrightarrow CaCO_3(s)$$

Once precipitated, the hardness ions are not available to interfere with the soap.

Another problem when the hard water of a particular locality is rich in bicarbonate ion is the precipitation of insoluble carbonates on the inner walls of hot water pipes. When solutions containing HCO_3^- are heated, the ion decomposes as follows.

$$2HCO_3^-(aq) \longrightarrow H_2O + CO_2(g) + CO_3^{2-}(aq)$$

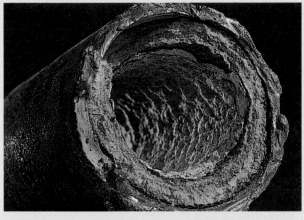

Boiler scale built up on the inside of a water pipe.

Like most gases, carbon dioxide becomes less soluble as the temperature is raised, so CO_2 is driven from the hot solution and the HCO_3^- is gradually converted to CO_3^{2-}. As the carbonate ions form, they are able to precipitate the hardness ions. This precipitate, which sticks to the inner walls of pipes and hot water boilers, is called *boiler scale*. In locations that have high concentrations of Ca^{2+} and HCO_3^- in the water supply, boiler scale is a very serious problem, as illustrated in the accompanying photograph.

SOLUTION Our reactants are $Pb(NO_3)_2$ and $Fe_2(SO_4)_3$, which contain the ions Pb^{2+} and NO_3^-, and Fe^{3+} and SO_4^{2-}, respectively. To write the formulas of the possible products, we interchange cations. We combine Pb^{2+} with SO_4^{2-}, and for electrical neutrality, we must use one ion of each. Therefore, we write $PbSO_4$ as one possible product. For the other product, we combine Fe^{3+} with NO_3^-. Electrical neutrality now demands that we use *three* NO_3^- to *one* Fe^{3+} to make $Fe(NO_3)_3$. The correct formulas of the possible products,[2] then, are $PbSO_4$ and $Fe(NO_3)_3$.

Is either of these possible products insoluble? Rule 2 says that all nitrates, like $Fe(NO_3)_3$, are soluble, but Rule 4 tells us that the sulfate of Pb^{2+} is *insoluble*. A precipitation reaction thus occurs, so we start by writing the unbalanced molecular equation:

$$Fe_2(SO_4)_3(aq) + Pb(NO_3)_2(aq) \longrightarrow Fe(NO_3)_3(aq) + PbSO_4(s) \quad \text{(unbalanced)}$$

When it is balanced, we obtain the *molecular equation*.

$$Fe_2(SO_4)_3(aq) + 3Pb(NO_3)_2(aq) \longrightarrow 2Fe(NO_3)_3(aq) + 3PbSO_4(s)$$

Next, we expand this to give the ionic equation in which soluble compounds are written in dissociated (separated) form as ions, and insoluble compounds are written in "molecular" form. From the wording of the problem as well as from the

Always write equations in two steps:
1. Write the correct formulas.
2. Balance the equation.

[2] Some students might be tempted (without thinking) to write $Pb(SO_4)_2$ and $Fe_2(NO_3)_3$, or even $Pb(SO_4)_3$ and $Fe_2(NO_3)_2$. This is a common error. Always be careful to figure out the charges of the ions that must be combined in the formula. Then take the ions in a ratio that gives a neutral formula unit.

solubility rules, we can place the following labels beneath the formulas of the molecular equation.

$$Fe_2(SO_4)_3(aq) + 3Pb(NO_3)_2(aq) \longrightarrow 2Fe(NO_3)_3(aq) + 3PbSO_4(s)$$

soluble	soluble	soluble	insoluble
(Rule 4)	(Rule 2)	(Rule 2)	(Rule 4)

We now separate the ions of the water-soluble species to obtain the *ionic equation*.

$$2Fe^{3+}(aq) + 3SO_4^{2-}(aq) + 3Pb^{2+}(aq) + 6NO_3^-(aq) \longrightarrow$$
$$2Fe^{3+}(aq) + 6NO_3^-(aq) + 3PbSO_4(s)$$

By removing spectator ions, we obtain:

$$3Pb^{2+}(aq) + 3SO_4^{2-}(aq) \longrightarrow 3PbSO_4(s)$$

Finally, we reduce the coefficients to give us the correct *net ionic equation*.

$$Pb^{2+}(aq) + SO_4^{2-}(aq) \longrightarrow PbSO_4(s)$$

EXAMPLE 4.3
Predicting Reactions and Writing Their Equations

What reaction (if any) occurs in water between KNO_3 and NH_4Cl?

ANALYSIS Let's proceed by writing molecular, ionic, and net ionic equations as in the preceding example.

SOLUTION Solubility Rule 1 tells us that both KNO_3 and NH_4Cl are soluble. So at the instant of mixing solutions of these compounds, we'll have the following ions intermingling: K^+, NO_3^-, NH_4^+, and Cl^-. By exchange of partners, the new combinations link K^+ with Cl^- as KCl, and NH_4^+ with NO_3^- as NH_4NO_3. By Solubility Rule 1, both products are also soluble in water. The anticipated molecular equation is therefore

$$KNO_3(aq) + NH_4Cl(aq) \longrightarrow KCl(aq) + NH_4NO_3(aq)$$

and the ionic equation is

$$K^+(aq) + NO_3^-(aq) + NH_4^+(aq) + Cl^-(aq) \longrightarrow$$
$$K^+(aq) + Cl^-(aq) + NH_4^+(aq) + NO_3^-(aq)$$

Notice that the right side of the equation is the same as the left side except for the order in which the ions are written. When we eliminate spectator ions, everything goes. *There is no net ionic equation, which means there is no net reaction.* In fact, when we mix solutions of salts whose potential products in a metathesis reaction are all soluble ionic compounds, we can anticipate in advance that no reaction will occur. It really isn't necessary to actually write the chemical equations.

■ **Practice Exercise 7** Predict what reaction (if any) would occur upon mixing the following solutions. Write net ionic equations for any reactions that can take place. (a) $AgNO_3$ and NH_4Cl, (b) Na_2S and $Pb(C_2H_3O_2)_2$, (c) $BaCl_2$ and NH_4NO_3.

Reactions in Which a Weak Electrolyte Forms

Our preceding examples illustrated a major generalization: *Expect a reaction to occur if ions can be removed from a solution.* The formation of an insoluble compound is not the only way this can occur. The formation of a *soluble* compound will serve just as well, *provided that it is a nonelectrolyte, like water, or a sufficiently weak electrolyte, like some weak acid.*

We have already discussed neutralization reactions in which water is a product. However, a similar reaction also occurs when we mix a strong acid with a salt that contains an anion of a weak acid. An example is the reaction between $HCl(aq)$ and $NaC_2H_3O_2(aq)$. The products of a metathesis reaction between these solutes are $NaCl$ and $HC_2H_3O_2$. The molecular equation is

$$HCl(aq) + NaC_2H_3O_2(aq) \longrightarrow NaCl(aq) + HC_2H_3O_2(aq)$$

In writing the ionic equation, we see that all the solutes are strong electrolytes except $HC_2H_3O_2$, which is a weak acid (and, therefore, a weak electrolyte). The ionic equation, therefore, is

$$H^+(aq) + Cl^-(aq) + Na^+(aq) + C_2H_3O_2^-(aq) \longrightarrow$$
$$Na^+(aq) + Cl^-(aq) + HC_2H_3O_2(aq)$$

Dropping out the spectator ions gives the net ionic equation,

$$H^+(aq) + C_2H_3O_2^-(aq) \longrightarrow HC_2H_3O_2(aq)$$

Summary

We have now seen several factors that drive ionic reactions, including the formation of a gas discussed in Section 4.6. These are summarized below.

An ionic reaction occurs when:

1. A precipitate forms from a solution of soluble reactants.
2. Water forms in the reaction of an acid and a base.
3. A weak electrolyte forms from a solution of strong electrolytes.
4. A gas forms that escapes from the reaction mixture.

Criteria for predicting ionic reactions

Here's one more example that illustrates how we can use these criteria to establish the outcome of a reaction.

EXAMPLE 4.4
Predicting a Reaction

What reaction, if any, occurs between $Ba(OH)_2$ and $(NH_4)_2SO_4$ in an aqueous system?

ANALYSIS We will consider the possible products of a metathesis reaction and apply the criteria established above to determine what actually happens.

SOLUTION Exchanging cations between the two reactants gives as possible products $BaSO_4$ and NH_4OH. However, earlier you have learned that NH_4OH doesn't really exist and that we should write instead $NH_3(g) + H_2O$. Let's assemble this into a balanced molecular equation and apply the solubility rules next.

$$Ba(OH)_2 + (NH_4)_2SO_4 \longrightarrow BaSO_4 + 2NH_3(g) + 2H_2O$$

soluble	soluble	insoluble
(Rule 5)	(Rule 2)	(Rule 4)

Now we write the ionic equation, expressing soluble ionic compounds in dissociated form.

$$Ba^{2+}(aq) + 2OH^-(aq) + 2NH_4^+(aq) + SO_4^{2-}(aq) \longrightarrow$$
$$BaSO_4(s) + 2NH_3(g) + 2H_2O$$

Notice that there are no ions that appear the same on both sides; there are no spectator ions to eliminate, so the ionic equation is also the net ionic equation.

4.8
OXIDATION–REDUCTION REACTIONS

Among the first reactions studied by early scientists were those that involved oxygen. The combustion of fuels and the reactions of metals with oxygen to give oxides were described by the word *oxidation*. The removal of oxygen from metal oxides to give the metals in their elemental forms was described as *reduction*.

As time went by, scientists realized that reactions involving oxygen were actually special cases of a much more general phenomenon, one in which electrons are transferred from one substance to another. Collectively, electron transfer reactions came to be called **oxidation–reduction reactions,** or simply **redox reactions.** The term **oxidation** was applied to the loss of electrons by one reactant, and **reduction** to the gain of electrons by another. For example, the reaction between sodium and chlorine involves a loss of electrons by sodium (*oxidation* of sodium) and a gain of electrons by chlorine (*reduction* of chlorine). We can write these changes in equation form including the electrons.

$$Na \longrightarrow Na^+ + e^- \qquad \text{(oxidation)}$$

$$Cl_2 + 2e^- \longrightarrow 2Cl^- \qquad \text{(reduction)}$$

We say that sodium is oxidized and chlorine is reduced.

Oxidation and reduction always occur together. No substance is ever oxidized unless something else is reduced. Otherwise, electrons would appear as a product of the reaction, and this is never observed. During a redox reaction, then, some substance must accept the electrons that another substance loses. This electron-accepting substance is called the **oxidizing agent** because it helps something else to be oxidized. The substance that supplies the electrons is called the **reducing agent** because it helps something else to be reduced. Sodium is a reducing agent, for example, when it supplies electrons to chlorine. In the process, sodium is oxidized. Chlorine is an oxidizing agent when it accepts electrons from the sodium, and when that happens, chlorine is reduced to chloride ion. One way to remember this is by the following summary:

The *oxidizing agent* causes oxidation to occur by accepting electrons. The *reducing agent* causes reduction by supplying electrons.

The substance that is oxidized is the reducing agent.

The substance that is reduced is the oxidizing agent.

Redox reactions are very common. They occur in batteries, which are built so that the electrons transferred pass through some external circuit that lights a flashlight or operates a pocket calculator. The metabolism of foods, which supplies our bodies with energy, also occurs by a series of redox reactions. And ordinary household bleach works by oxidizing substances that stain fabrics, making them colorless or easier to remove from the fabric. (See Special Topic 4.2.)

Liquid bleach contains hypochlorite ion, OCl⁻, as the oxidizing agent.

EXAMPLE 4.5
Identifying Oxidation–Reduction

Calcium and oxygen react to form calcium oxide, CaO, an ionic compound.

$$2Ca + O_2 \longrightarrow 2CaO$$

Which element is oxidized and which is reduced? What are the oxidizing and reducing agents?

SPECIAL TOPIC 4.2 / CHLORINE AND LAUNDRY BLEACH

Although the combustion of fuels is probably the best known example of a practical application of redox reactions, there are other less well-known reactions that are almost as important. Each year huge amounts of chlorine, manufactured from salt, are used in purifying municipal drinking water supplies and treating wastewater. Chlorine is a powerful oxidizing agent, and its function in water purification is to kill bacteria. It does this by oxidizing the bacteria and causing their death.

Another use for chlorine and the chlorine-containing anion ClO^- is as a bleach. Ordinary chlorine bleaches such as Clorox consist of a dilute solution of sodium hypochlorite, $NaClO$. The hypochlorite ion is a strong oxidizing agent and performs its bleaching action by destroying dyes and other colored matter. In the paper and cotton processing industries, this bleaching function has been accomplished by using chlorine directly.

Today, industry is moving away from the use of chlorine as a bleach because of environmental concerns. It is feared that the products of the reaction of chlorine with organic substances may be toxic, and in fact, many well-known insecticides are chlorine-containing compounds.

"Chlorine" bleach, $NaOCl(aq)$, destroys dyes by oxidizing them to colorless compounds.

SOLUTION When calcium reacts with oxygen, its atoms change to Ca^{2+} ions, which means that the calcium atoms must lose electrons.

$$Ca \longrightarrow Ca^{2+} + 2e^-$$

Because Ca loses electrons, it is oxidized. Since Ca is oxidized, it must be the reducing agent.

When oxygen reacts with calcium, it forms O^{2-} ions, which means that the atoms in O_2 must gain electrons.

$$O_2 + 4e^- \longrightarrow 2O^{2-}$$

Because O_2 gains electrons, it is reduced and must be the oxidizing agent.

■ **Practice Exercise 8** Identify the substances oxidized and reduced and the oxidizing and reducing agents in the reaction of calcium and chlorine to form calcium chloride.

Oxidation Numbers

The reaction of calcium with oxygen in the preceding example is clearly a redox reaction. However, not all reactions with oxygen produce ionic products. For example, sulfur reacts with oxygen to give sulfur dioxide, SO_2, which is molecular. Nevertheless, it is convenient to also view this as a redox reaction, but to do so we must make some changes in the way we define oxidation and reduction. To do this, chemists have developed a bookkeeping system called **oxidation numbers,** which provide a way to keep tabs on electron transfers.

The oxidation number of an element in a particular compound is assigned according to a set of rules, which are described below. For simple, monatomic ions in a compound such as NaCl, the oxidation numbers are the same as the charges on the ions, so in NaCl the oxidation number of Na^+ is $+1$ and the oxidation number of Cl^- is -1. The real value of oxidation numbers is that they can also be assigned to atoms in molecular compounds. In such cases, it is important to realize that the oxidation number does not actually equal a charge on an atom. To be sure not to confuse oxidation numbers with actual electrical charges, we will specify the sign *before* the number when writing oxidation numbers, and *after* the number when writing electrical charges. Thus, a sodium ion has a charge of $1+$ and an oxidation number of $+1$.

A term that is frequently used interchangeably with oxidation number is **oxidation state.** In NaCl, for example, sodium has an oxidation number of $+1$ and is said to be "in the $+1$ oxidation state." Similarly, the chlorine in NaCl is said to be in the -1 oxidation state. There are times when it is convenient to specify the oxidation state of an element when its name is written out. This is done by writing the oxidation number as a Roman numeral in parentheses after the name of the element. For example, "iron(III)" means iron in the $+3$ oxidation state.

We can use these new terms now to redefine a redox reaction.

A **redox reaction** is a chemical reaction in which changes in oxidation numbers occur.

We will find it easy to follow redox reactions by taking note of the changes in oxidation numbers. To do this, however, we must be able to assign oxidation numbers to atoms in a quick and simple way.

Assignment of Oxidation Numbers

We can use some basic knowledge learned earlier plus the set of rules below to determine the oxidation numbers in almost any compound.

Rules for assigning oxidation numbers

Rule 1 means that O in O_2, and P in P_4, and S in S_8 all have oxidation numbers of zero.

Rules for Assigning Oxidation Numbers

1. The oxidation number of any free element (an element not combined chemically with a different element) is zero, regardless of how complex its molecules might be.

2. The oxidation number for any simple, monatomic ion (e.g., Na^+ or Cl^-) is equal to the charge on the ion.

3. The sum of all the oxidation numbers of the atoms in a molecule or polyatomic ion must equal the charge on the particle.

4. In its compounds, fluorine has an oxidation number of -1.

5. In its compounds, hydrogen has an oxidation number of $+1$.

6. In its compounds, oxygen has an oxidation number of -2.

As you will see shortly, occasionally, there is a conflict between two rules. When this happens, we apply the rule with the lower number and ignore the conflicting rule.

In addition to these basic rules, there is some other chemical knowledge you will need to remember. Recall that we can use the periodic table to help us remember the charges on certain ions of the elements. For instance, all the metals in Group IA form ions with a 1+ charge, and all those in Group IIA form ions with a 2+ charge. This means that when we find sodium in a compound, we can assign it an oxidation number of +1 because its simple ion, Na^+, has a charge of 1+. (We have applied Rule 2.) Similarly, calcium in a compound exists as Ca^{2+} and has an oxidation number of +2.

In binary ionic compounds with metals, the nonmetals have oxidation numbers equal to the charges on their anions. For example, the compound Fe_2O_3 contains the oxide ion, O^{2-}, which is assigned an oxidation number of −2. Similarly, Mg_3P_2 contains the phosphide ion, P^{3-}, which has an oxidation number of −3.

With this as background, let's look at some examples that illustrate how we apply the rules, especially when they don't explicitly cover all the atoms in the formula.

Molybdenum disulfide, MoS_2, has a structure that allows it to be used as a dry lubricant, much like graphite. What are the oxidation numbers of the atoms in MoS_2?

EXAMPLE 4.6
Assigning Oxidation Numbers

SOLUTION This is a binary compound of a metal and a nonmetal, so for the purposes of assigning oxidation numbers we assume it to contain ions. Molybdenum is a transition metal, so there is no simple rule to tell what its ions are. However, sulfur is a nonmetal in Group VIA, and its ion is S^{2-}. (If necessary, you might review Table 2.7 on page 67, which summarizes the ions formed by the representative elements.) Once we know the oxidation number of sulfur, we can use Rule 3 (the summation rule) to determine the oxidation number of molybdenum.

$$
\begin{array}{lll}
S & (2 \text{ atoms}) \times (-2) = -4 & (\text{Rule 2}) \\
\underline{Mo} & \underline{(1 \text{ atom}) \times \quad (x) = \quad x} & \\
& \text{Sum} = \quad 0 & (\text{Rule 3})
\end{array}
$$

The value of x must be +4 for the sum to be zero. Therefore, the oxidation numbers are

$$Mo = +4 \qquad \text{and} \qquad S = -2$$

Chlorite ion, ClO_2^-, has been shown to be a potent disinfectant, and solutions of it are sometimes used to disinfect air conditioning systems in cars. What are the oxidation numbers of chlorine and oxygen in this ion?

EXAMPLE 4.7
Assigning Oxidation Numbers

SOLUTION Examining our rules, we see we have one for oxygen, so let's use it and the summation rule to figure out the oxidation number of chlorine.

$$
\begin{array}{lll}
O & (2 \text{ atoms}) \times (-2) = -4 & (\text{Rule 6}) \\
\underline{Cl} & \underline{(1 \text{ atom}) \times \quad (x) = \quad x} & \\
& \text{Sum} = -1 & (\text{Rule 3})
\end{array}
$$

For the sum to be −1, the value of x must be +3, which is the oxidation number of the chlorine.

$$Cl = +3 \qquad \text{and} \qquad O = -2$$

EXAMPLE 4.8
Assigning Oxidation Numbers

What are the oxidation numbers of the atoms in hydrogen peroxide, H_2O_2, a common antiseptic purchased in drug stores.

SOLUTION This time we are going to have a conflict between two of the rules. Rule 5 tells us to assign an oxidation number of $+1$ to hydrogen and Rule 6 tells us to give oxygen an oxidation number of -2. Both of these can't be correct because the sum must be zero (Rule 3).

$$
\begin{array}{llll}
H & (2 \text{ atoms}) \times (+1) = +2 & (\text{Rule 5}) \\
O & (2 \text{ atoms}) \times (-2) = -4 & (\text{Rule 6}) \\
\hline
& \qquad\qquad\quad \text{Sum} \neq \quad 0
\end{array}
$$

As mentioned above, when there is a conflict between the rules, we ignore the one with the higher number that causes the conflict. Rule 6 is the one with the higher number, so we will ignore it this time and just apply Rules 3 and 5.

$$
\begin{array}{llll}
H & (2 \text{ atoms}) \times (+1) = +2 & (\text{Rule 5}) \\
O & (2 \text{ atoms}) \times \quad (x) = \; 2x \\
\hline
& \qquad\qquad\quad \text{Sum} = \quad 0 & (\text{Rule 3})
\end{array}
$$

For the sum to be zero, $2x = -2$, so $x = -1$. Therefore, in this compound, oxygen has an oxidation number of -1.

$$H = +1 \qquad O = -1$$

Sometimes, oxidation numbers calculated by the rules have fractional values, as illustrated in the next example.

EXAMPLE 4.9
Assigning Oxidation Numbers

The air bags used as safety devices in modern autos are inflated by the very rapid decomposition of the ionic compound sodium azide, NaN_3. The reaction gives elemental sodium and gaseous nitrogen. What is the average oxidation number of the nitrogen in sodium azide.

SOLUTION In ionic compounds, sodium exists as the ion Na^+. Therefore, azide ion (N_3^-) must have a charge of $1-$.

$$
\begin{array}{llll}
Na & (1 \text{ atom}) \times (+1) = +1 & (\text{Rule 5}) \\
N & (3 \text{ atoms}) \times \quad (x) = \; 3x \\
\hline
& \qquad\qquad\quad \text{Sum} = \quad 0 & (\text{Rule 3})
\end{array}
$$

The sum of the oxidation numbers of the three nitrogen atoms in this ion must add up to -1, so each nitrogen must have an oxidation number of $-\frac{1}{3}$.

The charge on the polyatomic ion equals the sum of the oxidation numbers of its atoms.

Many compounds contain familiar polyatomic ions such as ammonium (NH_4^+), sulfate (SO_4^{2-}), nitrate (NO_3^-), and acetate ($C_2H_3O_2^-$). The charges on these ions can be considered to represent the *net oxidation number* of the atoms that make up the ion. We can use this sometimes to help assign oxidation numbers, as shown in Example 4.10.

What is the oxidation number of chromium in the compound $Cr(NO_3)_3$?

SOLUTION The nitrate ion has a charge of 1−, which we can take to be its net oxidation number. Then we apply the summation rule.

$$
\begin{array}{rcr}
Cr \text{ (1 atom)} \times & (x) = & x \\
NO_3^- \text{ (3 ions)} \times & (-1) = & -3 \\
\hline
& Sum = & 0 \quad (\text{Rule 3})
\end{array}
$$

Obviously, the oxidation number of chromium is $+3$.

■ **Practice Exercise 9** Assign oxidation numbers to each atom in (a) $NiCl_2$, (b) Mg_2TiO_4, (c) $K_2Cr_2O_7$, (d) HPO_4^{2-}, and (e) $V(C_2H_3O_2)_3$.

■ **Practice Exercise 10** Iron forms an oxide with the formula Fe_3O_4 that contains both Fe^{2+} and Fe^{3+} ions. What is the *average* oxidation number of iron in this oxide?

EXAMPLE 4.10
Assigning Oxidation Numbers

Oxidation Numbers and Redox Reactions

Oxidation numbers can be used in several ways. One is to define oxidation and reduction in the most comprehensive manner, as follows.

Oxidation is an increase in oxidation number.

Reduction is a decrease in oxidation number.

Let's see how these definitions apply to the reaction of hydrogen with chlorine. To avoid ever confusing oxidation numbers with actual electrical charges, we will write oxidation numbers directly above the chemical symbols of the elements.

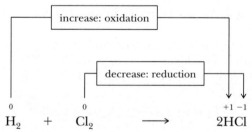

$$
\overset{0}{H_2} \; + \; \overset{0}{Cl_2} \; \longrightarrow \; \overset{+1 \; -1}{2HCl}
$$

Notice we have assigned the atoms in H_2 and Cl_2 oxidation numbers of zero, in accord with Rule 1. The changes in oxidation number tell us that hydrogen is oxidized and chlorine is reduced.

Identify the substance oxidized and the substance reduced, as well as the oxidizing and reducing agents, in the reaction

$$2KCl + MnO_2 + 2H_2SO_4 \longrightarrow K_2SO_4 + MnSO_4 + Cl_2 + 2H_2O$$

SOLUTION To identify redox changes using oxidation numbers, we first must assign oxidation numbers to each atom on both sides of the equation. Following the rules we get

$$
\overset{+1\,-1}{2KCl} + \overset{+4\;-2}{MnO_2} + \overset{+1\,+6\,-2}{2H_2SO_4} \longrightarrow \overset{+1\,+6\,-2}{K_2SO_4} + \overset{+2\,+6\,-2}{MnSO_4} + \overset{0}{Cl_2} + \overset{+1\,-2}{2H_2O}
$$

EXAMPLE 4.11
Using Oxidation Numbers to follow Redox Reactions

Next we look for changes, keeping in mind that an increase in oxidation number is oxidation and a decrease is reduction.

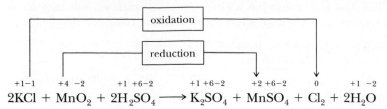

$$+1-1 \quad +4 \;-2 \quad +1 +6 -2 \qquad +1 +6 -2 \quad +2 +6 -2 \quad 0 \quad +1 \;-2$$
$$2KCl + MnO_2 + 2H_2SO_4 \longrightarrow K_2SO_4 + MnSO_4 + Cl_2 + 2H_2O$$

Thus the Cl in KCl is oxidized and the Mn in MnO_2 is reduced. The reducing agent is KCl, and the oxidizing agent is MnO_2.

4.9 BALANCING REDOX EQUATIONS BY THE ION–ELECTRON METHOD

Equations for redox reactions are often more complex than those for metathesis and can be difficult to balance by inspection. The ion–electron method is a systematic procedure for balancing redox equations.

Both $FeCl_3$ and $SnCl_2$ are soluble salts.

Each half-reaction must have the same elements on both sides. If Sn is on one side, it must also be on the other side. Keep this in mind when dividing a skeleton equation into half-reactions.

Many redox reactions take place in aqueous solution and many of these involve ions; they are ionic reactions. An example is the reaction of laundry bleach with substances in the wash water. The active ingredient in the bleach is hypochlorite ion, OCl^-, which is the oxidizing agent in these reactions. To study redox reactions, it is often helpful to write ionic and net ionic equations, just as we did in our analysis of metathesis reactions. Balancing net ionic equations for redox reactions is especially easy if we follow a procedure called the ion–electron method.

In the **ion–electron method,** we divide the oxidation and reduction processes into individual equations called **half-reactions** that are balanced separately. Then we combine the balanced half-reactions to obtain the fully balanced net ionic equation.

To illustrate the method, let's balance the net ionic equation for the reaction of iron(III) chloride, $FeCl_3$, with tin(II) chloride, $SnCl_2$, which changes the iron to Fe^{2+} and the tin to Sn^{4+}. In the reaction the chloride ion is unaffected—it's a spectator ion.

We begin by writing a **skeleton equation,** which shows only the ions (or sometimes molecules, too) involved in the reaction. The reactants are Fe^{3+} and Sn^{2+}, and the products are Fe^{2+} and Sn^{4+}. The skeleton equation is therefore

$$Fe^{3+} + Sn^{2+} \longrightarrow Fe^{2+} + Sn^{4+}$$

The next step is to divide the equation into two half-reactions. In this case, we have one for tin and one for iron. We choose one of the reactants, let's say Sn^{2+}, and write it at the left of an arrow. On the right, we write what Sn^{2+} changes to, which is Sn^{4+}. This gives us the beginnings of one half-reaction. For the second half-reaction, we write the other reactant, Fe^{3+}, on the left and the other product, Fe^{2+}, on the right.

$$Sn^{2+} \longrightarrow Sn^{4+}$$

$$Fe^{3+} \longrightarrow Fe^{2+}$$

Next, we balance the half-reactions so that *each* obeys *both* criteria for a balanced ionic equation: *both atoms and charge have to balance.* Obviously, the atoms are already balanced in each equation. The charge, however, is not. *To bring the charge into balance, we add electrons.* For the first half-reaction, two electrons are added to the right, so the net charge on both sides will be 2 +. In the second half-reaction, one electron is added to the left, which makes the net charge on both sides equal to 2 +.

$$Sn^{2+} \longrightarrow Sn^{4+} + 2e^-$$

$$Fe^{3+} + e^- \longrightarrow Fe^{2+}$$

Electrons must be added to whichever side of the half-reaction is more positive (or less negative).

In any redox reaction, the number of electrons lost always equals the number gained. To make the electron gain equal to the electron loss, the second reaction has to occur twice each time the first reaction occurs once. We make this so by multiplying each of the coefficients in the second equation by 2.

$$Sn^{2+} \longrightarrow Sn^{4+} + 2e^-$$

$$2(Fe^{3+} + e^- \longrightarrow Fe^{2+})$$

This gives

$$Sn^{2+} \longrightarrow Sn^{4+} + 2e^-$$

$$2Fe^{3+} + 2e^- \longrightarrow 2Fe^{2+}$$

In an oxidation the electrons appear as a product in the half-reaction, and in reduction they appear as a reactant. That's the way we identify which is oxidation and which is reduction when we balance an equation this way. In this example, Sn^{2+} is oxidized and Fe^{3+} is reduced.

Finally, we recombine the balanced half-reactions by adding them together. The result is

$$Sn^{2+} + 2Fe^{3+} + 2e^- \longrightarrow Sn^{4+} + 2Fe^{2+} + 2e^-$$

Dropping $2e^-$ from both sides gives us the final balanced equation.

$$Sn^{2+} + 2Fe^{3+} \longrightarrow Sn^{4+} + 2Fe^{2+}$$

Notice that the charge *and* the number of atoms are now balanced.

■ **Practice Exercise 11** Balance the following equation by the ion–electron method.

$$SnCl_3^-(aq) + HgCl_2(aq) \longrightarrow SnCl_6^{2-}(aq) + Hg_2Cl_2(s)$$

(Hint: In this equation, you will have to use Cl^- to help balance the half-reactions.)

Reactions Involving H⁺ and OH⁻

In many redox reactions in aqueous solutions, H^+ or OH^- ions play an important role, as do water molecules. For example, when solutions of $K_2Cr_2O_7$ and $FeSO_4$ are mixed, we find that the acidity of the mixture decreases as dichromate ion, $Cr_2O_7^{2-}$, oxidizes iron(II). This is because the reaction uses up H^+ as a reactant and produces H_2O as a product. In other reactions, OH^- is consumed, while in still others H_2O is a reactant. Another fact is that in many cases the products (or even the reactants) of a redox reaction will differ depending on the acidity of the solution. For example, in an acidic solution MnO_4^- is reduced to Mn^{2+} ion, but in a neutral or slightly basic solution, the reduction product is insoluble MnO_2.

Because of these factors, redox reactions are generally carried out in solutions containing a substantial excess of either acid or base, so before we can apply the ion–electron method, we have to know whether the reaction occurs in an acidic or a basic solution. (This information will always be given to you in this book.)

A solution of $K_2Cr_2O_7$ being added to an acidic solution containing Fe^{2+}.

Balancing Redox Equations for Acidic Solutions

As you have just learned, $Cr_2O_7^{2-}$ reacts with Fe^{2+} in an acidic solution, giving Cr^{3+} and Fe^{3+} as products. This information gives the skeleton equation,

$$Cr_2O_7^{2-} + Fe^{2+} \longrightarrow Cr^{3+} + Fe^{3+}$$

We now proceed through the following steps to find the balanced equation.

> As we balance the equation, the ion−electron method will tell us how H^+ and H_2O are involved in the reaction. We don't need to know this information in advance.

S T E P 1   Divide the skeleton equation into half-reactions. We create the beginnings of two half-reactions. Except for hydrogen and oxygen, the same elements must appear on both sides of a given half-reaction.

$$Cr_2O_7^{2-} \longrightarrow Cr^{3+}$$
$$Fe^{2+} \longrightarrow Fe^{3+}$$

S T E P 2   Balance atoms other than H and O. There are two Cr atoms on the left and only one on the right, so we place a coefficient of 2 in front of Cr^{3+}. The second half-reaction is already balanced in terms of atoms, so nothing need be done to it.

$$Cr_2O_7^{2-} \longrightarrow 2Cr^{3+}$$
$$Fe^{2+} \longrightarrow Fe^{3+}$$

> We use H_2O, not O or O_2 to balance oxygen atoms, because H_2O is what is actually present in the solution.

S T E P 3   Balance oxygen by adding H_2O to the side that needs O. There are seven oxygen atoms on the left of the first half-reaction and none on the right. Therefore, we add $7H_2O$ to the right side of the first half-reaction to balance the oxygens.

$$Cr_2O_7^{2-} \longrightarrow 2Cr^{3+} + 7H_2O$$
$$Fe^{2+} \longrightarrow Fe^{3+}$$

> By balancing oxygen with H_2O, we have created an imbalance in H. We are allowed to correct this by adding H^+ because the solution is acidic.

S T E P 4   Balance hydrogen by adding H^+ to the side that needs H. After adding the water, we see that the first half-reaction has 14 hydrogens on the right and none on the left. To balance these, we add $14H^+$ to the left side of this equation. When you do this step (or others) *be careful to write the charges on the ions.* If they are omitted, you will not obtain a balanced equation in the end.

$$14H^+ + Cr_2O_7^{2-} \longrightarrow 2Cr^{3+} + 7H_2O$$
$$Fe^{2+} \longrightarrow Fe^{3+}$$

Now each half-reaction is balanced in terms of atoms. Next we will balance the charge.

> This is a critical step. Be sure you understand how to perform it.

S T E P 5   Balance the charge by adding electrons. First we compute the net electrical charge on each side. For the first half-reaction we have

$$\underbrace{14H^+ + Cr_2O_7^{2-}}_{\text{Net charge} = (14+) + (2-) = 12+} \longrightarrow \underbrace{2Cr^{3+} + 7H_2O}_{\text{Net charge} = 2(3+) + 0 = 6+}$$

The algebraic difference between the net charges on the two sides equals the number of electrons that must be added to the more positive (less negative) side. In this instance, we must add $6e^-$ to the left side of the equation.

$$6e^- + 14H^+ + Cr_2O_7^{2-} \longrightarrow 2Cr^{3+} + 7H_2O$$

> Check the net charge after adding e^-; it must be the same on both sides.

This half-reaction is now complete; it is balanced in terms of both atoms and charge.

To balance the other half-reaction, we add one electron to the right.

$$Fe^{2+} \longrightarrow Fe^{3+} + e^-$$

STEP 6 Make the number of electrons gained equal to the number lost and then add the two half-reactions. At this point we have the two balanced half-reactions

$$6e^- + 14H^+ + Cr_2O_7{}^{2-} \longrightarrow 2Cr^{3+} + 7H_2O$$
$$Fe^{2+} \longrightarrow Fe^{3+} + e^-$$

Six electrons are gained in the first, but only one is lost in the second. Therefore, before combining the two equations, we multiply all of the coefficients of the second one by 6.

$$6e^- + 14H^+ + Cr_2O_7{}^{2-} \longrightarrow 2Cr^{3+} + 7H_2O$$
$$6(Fe^{2+} \longrightarrow Fe^{3+} + e^-)$$

(Sum) $6e^- + 14H^+ + Cr_2O_7{}^{2-} + 6Fe^{2+} \longrightarrow 2Cr^{3+} + 7H_2O + 6Fe^{3+} + 6e^-$

STEP 7 Cancel anything that is the same on both sides. This is the final step. Six electrons cancel from both sides to give the final balanced equation.

$$14H^+ + Cr_2O_7{}^{2-} + 6Fe^{2+} \longrightarrow 2Cr^{3+} + 7H_2O + 6Fe^{3+}$$

Notice that both the charge and the atoms balance.

In some reactions, you may have H_2O or H^+ on both sides—for example, $6H_2O$ on the left and $2H_2O$ on the right. Cancel as many as you can.

$$\dots + 6H_2O \dots \longrightarrow \dots + 2H_2O \dots$$

gives

$$\dots + 4H_2O \dots \longrightarrow \dots$$

The following is a summary of the steps we've followed for balancing an equation for a redox reaction in an acidic solution.

Ion–Electron Method—Acidic Solution

Ion–electron method: acid solutions

STEP 1 Divide the equation into two half-reactions.

STEP 2 Balance atoms other than H and O.

STEP 3 Balance O by adding H_2O.

STEP 4 Balance H by adding H^+.

STEP 5 Balance net charge by adding e^-.

STEP 6 Make e^- gain equal e^- loss; then add half-reactions.

STEP 7 Cancel anything that's the same on both sides.

If you don't skip any steps and you perform them in the order given, you will always obtain a properly balanced equation.

Balance the following equation. The reaction occurs in an acidic solution.

$$MnO_4{}^- + H_2SO_3 \longrightarrow SO_4{}^{2-} + Mn^{2+}$$

SOLUTION We follow the steps given above.

STEP 1 $MnO_4{}^- \longrightarrow Mn^{2+}$

$$H_2SO_3 \longrightarrow SO_4{}^{2-}$$

EXAMPLE 4.12
Using The Ion–Electron
Method

STEP 2 There is nothing to do for this step. All the atoms except H and O are already in balance.

STEP 3
$$MnO_4^- \longrightarrow Mn^{2+} + 4H_2O$$
$$H_2O + H_2SO_3 \longrightarrow SO_4^{2-}$$

STEP 4
$$8H^+ + MnO_4^- \longrightarrow Mn^{2+} + 4H_2O$$
$$H_2O + H_2SO_3 \longrightarrow SO_4^{2-} + 4H^+$$

STEP 5
$$5e^- + 8H^+ + MnO_4^- \longrightarrow Mn^{2+} + 4H_2O$$
$$H_2O + H_2SO_3 \longrightarrow SO_4^{2-} + 4H^+ + 2e^-$$

STEP 6
$$2(5e^- + 8H^+ + MnO_4^- \longrightarrow Mn^{2+} + 4H_2O)$$
$$5(H_2O + H_2SO_3 \longrightarrow SO_4^{2-} + 4H^+ + 2e^-)$$

$$10e^- + 16H^+ + 2MnO_4^- + 5H_2O + 5H_2SO_3 \longrightarrow$$
$$2Mn^{2+} + 8H_2O + 5SO_4^{2-} + 20H^+ + 10e^-$$

STEP 7 Cancel $10e^-$, $16H^+$, and $5H_2O$ from both sides. The final equation is

$$2MnO_4^- + 5H_2SO_3 \longrightarrow 2Mn^{2+} + 3H_2O + 5SO_4^{2-} + 4H^+$$

Once again, notice that each side of the equation has the same number of atoms of each element and the same net charge. This is what makes it a balanced equation.

■ **Practice Exercise 12** What is the balanced equation for the following reaction in an acidic solution?

$$Cu + NO_3^- \longrightarrow Cu^{2+} + NO$$

Balancing Redox Equations for Basic Solutions

In basic solutions, the concentration of H^+ is very small; the dominant species are H_2O and OH^-. Strictly speaking, these should be used to balance the half-reactions. However, the simplest way to obtain a balanced equation for a basic solution is to first *pretend* that the solution is acidic. We balance the equation using the seven steps just described, and then we use a simple three-step procedure to convert the equation to the correct form for a basic solution.

As an example, suppose we wanted to balance the following equation for a basic solution.

$$SO_3^{2-} + MnO_4^- \longrightarrow SO_4^{2-} + MnO_2$$

Following Steps 1 through 7 for acidic solutions gives

$$2H^+ + 3SO_3^{2-} + 2MnO_4^- \longrightarrow 3SO_4^{2-} + 2MnO_2 + H_2O$$

Conversion of this equation to one appropriate for a basic solution uses the fact that H^+ and OH^- react in a 1-to-1 ratio to give H_2O. The following three steps are involved.

Ion−electron method: basic solutions

STEP 8 **Add to both sides of the equation the same number of OH^- as there are H^+.** The equation for acidic solution has $2H^+$ on the left, so we add $2OH^-$ to *each* side. This gives

$$2OH^- + 2H^+ + 3SO_3^{2-} + 2MnO_4^- \longrightarrow 3SO_4^{2-} + 2MnO_2 + H_2O + 2OH^-$$

STEP 9   Combine H$^+$ and OH$^-$ to form H$_2$O.   The left side has 2OH$^-$ and 2H$^+$, which become 2H$_2$O. So in place of 2OH$^-$ + 2H$^+$ we write 2H$_2$O.

$$2H_2O + 3SO_3^{2-} + 2MnO_4^- \longrightarrow 3SO_4^{2-} + 2MnO_2 + H_2O + 2OH^-$$

STEP 10   Cancel any H$_2$O that you can. In this equation, one H$_2$O can be canceled from both sides. The final equation, balanced for basic solution, is

$$H_2O + 3SO_3^{2-} + 2MnO_4^- \longrightarrow 3SO_4^{2-} + 2MnO_2 + 2OH^-$$

■ **Practice Exercise 13** Balance the following equation for a basic solution.

$$MnO_4^- + C_2O_4^{2-} \longrightarrow MnO_2 + CO_3^{2-}$$

The result of Steps 8 and 9 is to replace H$^+$ with an equivalent number of H$_2$O, and the addition of that number of OH$^-$ to the other side of the equation.

When we carry out reactions in solution, we normally dissolve the solutes first and then mix their solutions. To work quantitatively with the stoichiometry of these solutions, we must have ways of expressing their concentrations. The expression of concentration most suited to this is called the **molar concentration** or **molarity** (abbreviated **M**). *The molarity of a solution is the number of moles of solute per liter of solution.* We can write this as a ratio of the moles of solute to the volume of the solution expressed in liters.

4.10 MOLAR CONCENTRATION

 Molar concentration

$$\text{Molarity } (M) = \frac{\text{moles of solute}}{\text{liters of solution}} \qquad (4.1)$$

Thus, a solution that contains 0.100 mol of NaCl in 1.00 L has a molarity of 0.100 *M*, and we would refer to this solution as 0.100 *molar* NaCl or as 0.100 *M* NaCl. The same concentration would result if we dissolved 0.0100 mol of NaCl in 0.100 L (100 mL) of solution, because the *ratio* of moles of solute to liters of solution is the same.

$$\frac{0.100 \text{ mol NaCl}}{1.00 \text{ L NaCl soln}} = \frac{0.0100 \text{ mol NaCl}}{0.100 \text{ L NaCl soln}} = 0.100 \text{ } M \text{ NaCl}$$

Molarity is particularly useful because it lets us obtain chemicals by moles simply by measuring volumes of solutions, and this is easy to do in the lab. If we had a 0.100 *M* NaCl solution, for example, and we needed 0.100 mol of NaCl for a reaction, we would simply measure out 1.00 L of the solution because in this volume there is 0.100 mol of NaCl.

Figure 4.11 (p. 162) illustrates the preparation of a standard solution of the yellow compound sodium chromate, Na$_2$CrO$_4$.

Molarity as a Conversion Factor

A solution's molarity is a source of two conversion factors that relate moles of solute to volume of solution. When a solution is 0.100 *M* NaCl, for example, we can construct the following conversion factors.

$$\frac{0.100 \text{ mol NaCl}}{1.00 \text{ L NaCl soln}} \quad \text{and} \quad \frac{1.00 \text{ L NaCl soln}}{0.100 \text{ mol NaCl}}$$

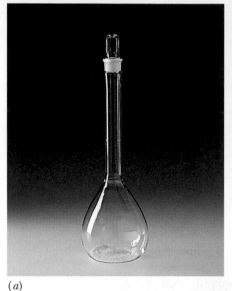

(a)

(b)

(c)

FIGURE 4.11

The preparation of a solution having a known molarity. (*a*) A 250 mL volumetric flask, one of a number of sizes available for preparing solutions. When filled to the line etched around its neck, this flask contains exactly 250 mL of solution. The flask here already contains a weighed amount of solute. (*b*) Water is being added. (*c*) The solute is brought completely into solution before the level is brought up to the narrow neck of the flask. (*d*) More water is added to bring the level of the solution to the etched line. (*e*) The flask is stoppered and then tipped over several times to mix its contents thoroughly.

(d)

(e)

Generally, the volumes used in experiments are better expressed in milliliters rather than liters. Many students find it easier to translate the "1.00 L" part of these factors into the equivalent 1000 mL here rather than convert between liters and milliliters at some other stage of the calculation. Factors such as these, therefore, can also be rewritten as follows whenever it is convenient:

$$\frac{0.100 \text{ mol NaCl}}{1000 \text{ mL NaCl soln}} \quad \text{and} \quad \frac{1000 \text{ mL NaCl soln}}{0.100 \text{ mol NaCl}}$$

EXAMPLE 4.13
Calculating the Molarity of a
Solution

To study the effect of dissolved salt on the rusting of an iron sample, a student prepared a solution of NaCl by dissolving 1.461 g of NaCl in a 250.0 mL volumetric flask. What is the molarity of this solution?

ANALYSIS To work with molarity, it is imperative that you know its definition: Molarity is the ratio of *moles of solute* to *liters of solution*. To calculate this ratio from the given data, we first have to convert 1.461 g of NaCl into moles of NaCl, and we have to convert 250.0 mL into liters. Then we simply divide the number of moles by the number of liters to find the molarity.

SOLUTION The number of moles of NaCl are found using the formula mass of NaCl, 58.44.

$$1.461 \text{ g NaCl} \times \frac{1 \text{ mol NaCl}}{58.44 \text{ g NaCl}} = 0.02500 \text{ mol NaCl}$$

The volume of the solution in liters equivalent to 250.0 mL is 0.2500 L. The ratio of moles to liters, therefore, is

$$\frac{0.02500 \text{ mol NaCl}}{0.2500 \text{ L}} = 0.1000 \, M \text{ NaCl}$$

CHECKING THE ANSWER FOR REASONABLENESS The mass of NaCl taken, 1.461 g, is a small fraction of 1 mol NaCl (58.44 g), but the given volume is by no means so small a fraction of 1 L. Therefore, the ratio of the two has to be less than 1 *M*, which it is.

■ **Practice Exercise 14** As part of a study to determine whether just the chloride ion in NaCl accelerates the rusting of iron, a student decided to see whether iron rusted as rapidly when exposed to a solution of Na_2SO_4. This solution was made by dissolving 3.550 g of Na_2SO_4 in water using a 125.0 mL volumetric flask to adjust the final volume. What is the molarity of this solution?

EXAMPLE 4.14
Using Molar Concentrations

How many milliliters of 0.250 *M* NaCl solution must be measured to obtain 0.100 mol of NaCl?

ANALYSIS Once again we use the given molarity as a tool to construct a conversion factor. From the molarity we have the following options:

$$\frac{0.250 \text{ mol NaCl}}{1.00 \text{ L NaCl soln}} \quad \text{or} \quad \frac{1.00 \text{ L NaCl soln}}{0.250 \text{ mol NaCl}}$$

SOLUTION We operate with the second factor on 0.100 mol NaCl:

$$0.100 \text{ mol NaCl} \times \frac{1.00 \text{ L NaCl soln}}{0.250 \text{ mol NaCl}} = 0.400 \text{ L of } 0.250 \, M \text{ NaCl}$$

Because 0.400 L corresponds to 400 mL, 400 mL of 0.250 *M* NaCl provides 0.100 mol of NaCl.

CHECKING THE ANSWER FOR REASONABLENESS A whole liter provides 0.250 mol, so we need a fraction of a liter (1000 mL) to provide just 0.100 mol. The answer, 400 mL, is reasonable.

■ **Practice Exercise 15** To continue the experiment on rusting (Practice Exercise 14), a student decided to see how rapidly iron rusted when chloride ion but not sodium ion was present. How many milliliters of 0.150 *M* KCl solution must be taken to provide 0.0500 mol of KCl?

In the laboratory, we often must prepare a solution with a specific molarity. Example 4.15 illustrates the kind of calculation required, and also demonstrates a useful relationship: *Whenever you know **both** the volume and molarity of a solution, you can calculate the number of moles of solute in it.* This is because the product of molarity and volume (expressed in liters) equals the moles of solute.

molarity × volume = moles

$$\text{molarity} \times \text{volume (L)} = \text{moles of solute} \qquad (4.2)$$

$$\frac{\text{mol solute}}{\text{L soln}} \times \text{L soln} = \text{mol solute}$$

This relationship is a valuable tool for solving problems in solution stoichiometry.

EXAMPLE 4.15
Preparing a Solution with a Known Molarity

A student needs to prepare 250 mL of 0.100 M Cd(NO$_3$)$_2$ solution. How many grams of cadmium nitrate are required?

ANALYSIS The key to solving this problem is recognizing that because we know both the molarity and the volume of the solution, we can calculate the number of moles of Cd(NO$_3$)$_2$ that will be in solution after it's prepared. Once we know the number of moles of Cd(NO$_3$)$_2$, we can calculate the mass using the formula mass of the salt.

SOLUTION The stated molarity, 0.100 M Cd(NO$_3$)$_2$, provides the following conversion factors.

$$\frac{0.100 \text{ mol Cd(NO}_3)_2}{1.00 \text{ L Cd(NO}_3)_2 \text{ soln}} \quad \text{or} \quad \frac{1.00 \text{ L Cd(NO}_3)_2 \text{ soln}}{0.100 \text{ mol Cd(NO}_3)_2}$$

We have 250 mL, or 0.250 L, Cd(NO$_3$)$_2$ solution. To find moles of Cd(NO$_3$)$_2$ we multiply by the first factor.

$$0.250 \text{ L Cd(NO}_3)_2 \text{ soln} \times \frac{0.100 \text{ mol Cd(NO}_3)_2}{1.00 \text{ L Cd(NO}_3)_2 \text{ soln}} = 0.0250 \text{ mol Cd(NO}_3)_2$$

We next convert from moles to grams using the formula mass of Cd(NO$_3$)$_2$, which is 236.43.

$$0.0250 \text{ mol Cd(NO}_3)_2 \times \frac{236.43 \text{ g Cd(NO}_3)_2}{1 \text{ mol Cd(NO}_3)_2} = 5.91 \text{ g Cd(NO}_3)_2$$

Thus, to prepare the solution, 5.91 g of Cd(NO$_3$)$_2$ would be weighed out, placed in a 250.0 mL volumetric flask, and then dissolved in water. (This kind of flask is pictured in Figure 4.11.) Then more water would be added to make the solution reach the flask's etched mark. Now the flask can carry the label 0.100 M Cd(NO$_3$)$_2$.

We could have set this up as a chain calculation as follows:

$$0.250 \text{ L Cd(NO}_3)_2 \text{ soln} \times \frac{0.100 \text{ mol Cd(NO}_3)_2}{1.00 \text{ L Cd(NO}_3)_2 \text{ soln}} \times \frac{236.43 \text{ g Cd(NO}_3)_2}{1 \text{ mol Cd(NO}_3)_2} = 5.91 \text{ g Cd(NO}_3)_2$$

CHECKING THE ANSWER FOR REASONABLENESS If we were working with a full liter of this solution, it would contain 0.1 mol of Cd(NO$_3$)$_2$. The formula mass of the salt is 236.43, so 0.1 mole is about 24 g. However, we are working with just a quarter of a liter (250 mL), so the amount of Cd(NO$_3$)$_2$ needed is about a quarter of 24 g, or

6 g. The answer, 5.91 g, is very close to this so it makes sense. (The more you try to work these rough calculations in your head, the quicker and surer you will become at understanding and solving stoichiometry problems.)

■ **Practice Exercise 16** How many grams of $CaCl_2$ are needed to prepare 250 mL of 0.125 M $CaCl_2$ solution?

Diluting Solutions

It is not always necessary to begin with a solute in pure form to prepare a solution with a known molarity. The solute can already be dissolved in a solution of relatively high concentration, and this can be *diluted* to make one of the desired lower concentration. The choice of apparatus used depends on the precision required. If high precision is necessary, pipets (Figure 4.12) and volumetric flasks are used. If we can do with less precision, we might use graduated cylinders instead.

The process of diluting a solution involves distributing a given amount of solute throughout a larger volume of solution. The amount of solute remains constant, however. Since the product of molarity and volume equals the number of moles of solute, it must therefore be true that

$$\begin{pmatrix} \text{volume of} \\ \text{dilute solution} \\ \text{to be prepared} \end{pmatrix} \times M_{\text{dilute}} = \begin{pmatrix} \text{volume of} \\ \text{concentrated solution} \\ \text{to be used} \end{pmatrix} \times M_{\text{concd}}$$

Or

$$V_{\text{dil}} \cdot M_{\text{dil}} = V_{\text{concd}} \cdot M_{\text{concd}} \tag{4.3}$$

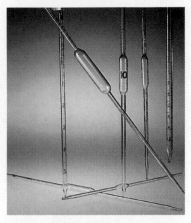

FIGURE 4.12

A selection of volumetric pipets, each marked for a particular volume and each with an etched line to designate where the filling of the pipet should stop.

 Diluting solutions of known molarity

The units used for volume in Equation 4.3 can be any we please, provided that they are the same on both sides of the equation. We thus normally solve dilution problems using *milliliters* directly in Equation 4.3.

How can we prepare 100 mL of 0.0400 M $K_2Cr_2O_7$ from 0.200 M $K_2Cr_2O_7$?

ANALYSIS This is the way such a question comes up in the lab, but what is really asked is: How many milliliters of 0.200 M $K_2Cr_2O_7$ (the more concentrated solution) must be diluted to give a solution with a final volume of 100 mL and a final molarity of 0.0400 M? Once we see the question this way, we see that Equation 4.3 applies.

SOLUTION It's a good idea to assemble the data first, noting what is missing (and therefore what has to be calculated).

$$V_{\text{dil}} = 100 \text{ mL} \qquad M_{\text{dil}} = 0.0400 \text{ } M$$

$$V_{\text{concd}} = ? \qquad M_{\text{concd}} = 0.200 \text{ } M$$

Next, we use Equation 4.3

$$100 \text{ mL} \times 0.0400 \text{ } M = V_{\text{concd}} \times 0.200 \text{ } M$$

EXAMPLE 14.16
Preparing a Solution of
Known Molarity by Dilution

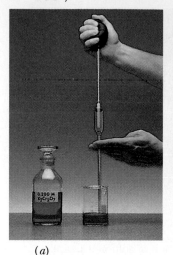

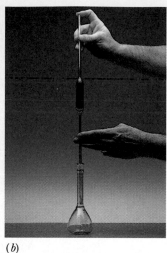

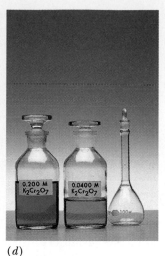

(a)   (b)   (c)   (d)

FIGURE 4.13

Preparing a solution by dilution. (a) The calculated volume of the more concentrated solution is withdrawn from the stock solution by means of a volumetric pipet. (b) The solution is allowed to drain entirely from the pipet into the volumetric flask. (c) Water is added to the flask, the contents are mixed, and the final volume is brought up to the etch mark on the narrow neck of the flask. (d) The new solution is put into a labeled container.

Solving for V_{concd} gives

$$V_{concd} = \frac{100 \text{ mL} \times 0.0400 \text{ } M}{0.200 \text{ } M}$$

$$= 20.0 \text{ mL}$$

This answer means that we would withdraw 20.0 mL of 0.200 M $K_2Cr_2O_7$, place it in a 100 mL volumetric flask, and then add water until the final volume is 100 mL (Figure 4.13).

CHECKING THE ANSWER FOR REASONABLENESS Think about the *magnitude* of the dilution, from 0.2 M to 0.04 M. The concentrated solution is 5 times as concentrated as the dilute solution ($5 \times 0.04 = 0.2$). So we should make a 1:5 dilution, and we see that 100 mL is 5 times 20 mL.

■ **Practice Exercise 17** How could 100 mL of 0.125 M H_2SO_4 solution be made from 0.500 M H_2SO_4 solution?

4.11
STOICHIOMETRY OF REACTIONS IN SOLUTION

In this section we work examples that illustrate how to do stoichiometry problems involving solutions of known molar concentrations. As you will see, we can work either with molecular equations or ionic equations. We begin with a simple problem using a molecular equation.

EXAMPLE 4.17
Stoichiometry Involving Reactions in Solution

One of the solids present in photographic film is silver bromide, AgBr. Suppose we wished to prepare this compound from aqueous solutions of silver nitrate and calcium bromide by the following precipitation reaction.

$$2AgNO_3(aq) + CaBr_2(aq) \longrightarrow 2AgBr(s) + Ca(NO_3)_2(aq)$$

How many milliliters of 0.125 M $CaBr_2$ solution must be used to react with the solute in 50.0 mL of 0.115 M $AgNO_3$?

ANALYSIS The key equivalency is seen in the coefficients of the equation

$$2 \text{ mol AgNO}_3 \Leftrightarrow 1 \text{ mol CaBr}_2$$

We can calculate the moles of $AgNO_3$ from the data given, using the volume and the molarity of the $AgNO_3$ solution. We then use the coefficients in the equation to translate to moles of $CaBr_2$. Finally, we use the molarity of the $CaBr_2$ solution as a conversion factor to find the volume of the solution needed. The calculation flow will look like the following.

$$V_{AgNO_3} \times M_{AgNO_3} \nearrow \text{mol AgNO}_3 \xrightarrow{\times \frac{1 \text{ mol CaBr}_2}{2 \text{ mol AgNO}_3}} \text{mol CaBr}_2 \searrow {\times \frac{1.00 \text{ L CaBr}_2 \text{ soln}}{0.125 \text{ mol CaBr}_2}}$$

AgNO₃ soln
Given: 50.0 mL of
0.115 M AgNO₃

CaBr₂ soln
To find: ? mL of
0.125 M CaBr₂

Silver bromide, AgBr, precipitates when solutions of CaBr₂ and AgNO₃ are mixed.

SOLUTION First, we find the moles of $AgNO_3$ taken.

$$0.0500 \text{ L AgNO}_3 \text{ soln} \times \frac{0.115 \text{ mol AgNO}_3}{1.00 \text{ L AgNO}_3 \text{ soln}} = 5.75 \times 10^{-3} \text{ mol AgNO}_3$$

Next, we use the coefficients of the equation to calculate the amount of $CaBr_2$ required.

$$5.75 \times 10^{-3} \text{ mol AgNO}_3 \times \frac{1 \text{ mol CaBr}_2}{2 \text{ mol AgNO}_3} = 2.88 \times 10^{-3} \text{ mol CaBr}_2$$

Finally we calculate the volume (mL) of 0.125 M CaBr₂ that contains this many moles of $CaBr_2$.

$$2.88 \times 10^{-3} \text{ mol CaBr}_2 \times \frac{1.00 \text{ L CaBr}_2 \text{ soln}}{0.125 \text{ mol CaBr}_2} = 0.0230 \text{ L CaBr}_2 \text{ soln}$$

Thus 0.0230 L, or 23.0 mL, of 0.125 M CaBr₂ has enough solute to combine with the $AgNO_3$ in 50.0 mL of 0.115 M AgNO₃.

CHECKING THE ANSWER FOR REASONABLENESS The molarities of the two solutions are about the same, but only 1 mol of $CaBr_2$ is needed for each 2 mol $AgNO_3$. Therefore, the volume of $CaBr_2$ solution needed (23.0 mL) should be about half the volume of $AgNO_3$ solution taken (50.0 mL), which it is.

■ **Practice Exercise 18** How many milliliters of 0.124 M NaOH contain enough NaOH to react with the H_2SO_4 in 15.4 mL of 0.108 M H₂SO₄ according to the following equation?

$$2NaOH(aq) + H_2SO_4(aq) \longrightarrow Na_2SO_4(aq) + 2H_2O$$

Calculating Concentrations of Ions in Solutions of Electrolytes

In working problems involving solutions of electrolytes, it is often necessary to know the concentrations of the ions in a solution. This information is easy to calculate from the molar concentration of the electrolyte solute. For example, suppose we are working with a 0.10 M solution of $CaCl_2$. In 1.0 L of this solution there is dissolved 0.10 mol of $CaCl_2$, which is fully dissociated into Ca^{2+} and Cl^- ions.

$$CaCl_2 \longrightarrow Ca^{2+} + 2Cl^-$$

From the stoichiometry of the dissociation, we see that 1 mol Ca^{2+} and 2 mol Cl^- are formed from each 1 mol $CaCl_2$. Therefore, 0.10 mol $CaCl_2$ will yield 0.10 mol Ca^{2+} and 0.20 mol Cl^-. In 0.10 M $CaCl_2$, then, the concentration of Ca^{2+} is 0.10 M and the concentration of Cl^- is 0.20 M. Notice that the concentration of a particular ion equals the concentration of the salt multiplied by the number of ions of that kind in one formula unit of the salt.

EXAMPLE 4.18
Calculating the
Concentrations of Ions in a
Solution

What are the molar concentrations of the ions in 0.20 M $Al_2(SO_4)_3$?

ANALYSIS The concentrations of the ions are determined by the stoichiometry of the salt. Therefore, we determine the number of ions of each kind formed from one formula unit of $Al_2(SO_4)_3$. These values are then used along with the given concentration of the salt to calculate the ion concentrations.

SOLUTION Each formula unit of $Al_2(SO_4)_3$ yields two Al^{3+} ions and three SO_4^{2-} ions, so 0.20 mol $Al_2(SO_4)_3$ yields 0.40 mol Al^{3+} and 0.60 mol SO_4^{2-}. Therefore, the solution contains 0.40 M Al^{3+} and 0.60 M SO_4^{2-}.

The answers here have been obtained by simple mole reasoning. We could have found the answers in a more formal manner using the factor label method. We can start with the given concentration and proceed as follows:

$$\frac{0.20 \text{ mol } Al_2(SO_4)_3}{1.0 \text{ L soln}} \times \frac{2 \text{ mol } Al^{3+}}{1 \text{ mol } Al_2(SO_4)_3} = \frac{0.40 \text{ mol } Al^{3+}}{1.0 \text{ L soln}} = 0.40 \ M \ Al^{3+}$$

$$\frac{0.20 \text{ mol } Al_2(SO_4)_3}{1.0 \text{ L soln}} \times \frac{3 \text{ mol } SO_4^{2-}}{1 \text{ mol } Al_2(SO_4)_3} = \frac{0.60 \text{ mol } SO_4^{2-}}{1.0 \text{ L soln}} = 0.60 \ M \ SO_4^{2-}$$

Study both methods. With just a little practice, you will have little difficulty with the reasoning approach that we used first.

EXAMPLE 4.19
Calculating the Concentration
of a Salt from the
Concentration of One of Its
Ions

A student found that the sulfate ion concentration in a solution of $Al_2(SO_4)_3$ was 0.90 M. What was the concentration of $Al_2(SO_4)_3$ in the solution?

ANALYSIS Once again, we use the formula of the salt to determine the number of ions released when it dissociates. This time we use the information to work backwards to find the salt concentration.

SOLUTION Let's set up the problem using the factor label method to be sure of our procedure. We will use the fact that 1 mol $Al_2(SO_4)_3$ yields 3 mol SO_4^{2-} in solution.

$$\frac{0.90 \text{ mol } SO_4^{2-}}{1.0 \text{ L soln}} \times \frac{1 \text{ mol } Al_2(SO_4)_3}{3 \text{ mol } SO_4^{2-}} = \frac{0.30 \text{ mol } Al_2(SO_4)_3}{1.0 \text{ L soln}} = 0.30 \ M \ Al_2(SO_4)_3$$

The reasoning approach here would have worked as follows. We know that 1 mol $Al_2(SO_4)_3$ yields 3 mol SO_4^{2-} in solution. Thus, the number of moles of $Al_2(SO_4)_3$ is only $\frac{1}{3}$ the number of moles of SO_4^{2-}. So the concentration of $Al_2(SO_4)_3$ must be $\frac{1}{3}$ of 0.90 M, or 0.30 M.

■ **Practice Exercise 19** What are the molar concentrations of the ions in 0.40 M $FeCl_3$?

■ **Practice Exercise 20** In a solution of Na_3PO_4, the PO_4^{3-} concentration was determined to be 0.250 M. What was the sodium ion concentration in the solution?

Using Net Ionic Equations in Solution Stoichiometry Calculations

You have seen that a net ionic equation is convenient for focusing on the net chemical change in an ionic reaction. In fact, for redox reactions, the balanced equation obtained by the ion–electron method *is* the net ionic equation. Let's study some examples that illustrate how such equations can be used in stoichiometric calculations.

How many milliliters of 0.100 M AgNO$_3$ solution are needed to react completely with 25.0 mL of 0.400 M CaCl$_2$ solution? The net ionic equation for the reaction is

$$Ag^+(aq) + Cl^-(aq) \longrightarrow AgCl(s)$$

EXAMPLE 4.20
Stoichiometric Calculations Using a Net Ionic Equation

ANALYSIS The problem describes solutions of salts with formulas expressed in "molecular form," but the equation relates a reaction between the ions of the salts. Therefore, the first thing we will do is calculate the concentrations of the ions in the solutions being mixed. Then, using the volume and molarity of the Cl$^-$ solution, we calculate the moles of Cl$^-$ available. This value equals the moles of Ag$^+$ that react because Ag$^+$ and Cl$^-$ combine in a one-to-one mole ratio. Finally we use the molarity of the Ag$^+$ solution to determine the volume of the 0.100 M AgNO$_3$ solution needed.

SOLUTION In 0.100 M AgNO$_3$, the Ag$^+$ concentration is 0.100 M. In 0.400 M CaCl$_2$, the Cl$^-$ concentration is 0.800 M. Having these values, we can now restate the problem: How many milliliters of 0.100 M Ag$^+$ solution are needed to react completely with 25.0 mL of 0.800 M Cl$^-$ solution?

We have the volume and molarity of the Cl$^-$ solution, so we can calculate the number of moles of Cl$^-$ available for reaction

$$0.0250 \text{ L Cl}^- \text{ soln} \times \frac{0.800 \text{ mol Cl}^-}{1.00 \text{ L Cl}^- \text{ soln}} = 0.0200 \text{ mol Cl}^-$$

(Notice we have been careful to express the volume of the chloride solution in liters.)

Since 0.0200 mol Cl$^-$ is available, the amount of Ag$^+$ required for reaction is also 0.0200 mol. Now we calculate the volume of the Ag$^+$ solution using its molarity as a conversion factor.

$$0.0200 \text{ mol Ag}^+ \times \frac{1.00 \text{ L Ag}^+ \text{ soln}}{0.100 \text{ mol Ag}^+} = 0.200 \text{ L Ag}^+ \text{ soln}$$

Our calculations tell us that we must use 0.200 L or 200 mL of the AgNO$_3$ solution. We could have used the following chain calculation, of course.

$$0.0250 \text{ L Cl}^- \text{ soln} \times \frac{0.800 \text{ mol Cl}^-}{1.00 \text{ L Cl}^- \text{ soln}} \times \frac{1 \text{ mol Ag}^+}{1 \text{ mol Cl}^-} \times \frac{1.00 \text{ L Ag}^+ \text{ soln}}{0.100 \text{ mol Ag}^+} = 0.200 \text{ L Ag}^+ \text{ soln}$$

CHECKING THE ANSWER FOR REASONABLENESS The chloride concentration is eight times the silver ion concentration. Since the ions react one for one, we will need only one eighth as much chloride solution as silver solution. One eighth of our answer, 200 mL, is 25 mL, which is the amount of chloride solution we used. Therefore, the answer appears to be correct.

■ **Practice Exercise 21** How many milliliters of 0.500 M KOH are needed to react completely with 60.0 mL of 0.250 M FeCl$_2$ solution to precipitate Fe(OH)$_2$?

EXAMPLE 4.21
Calculation Involving the Stoichiometry of an Ionic Reaction

Milk of magnesia is a suspension of $Mg(OH)_2$ in water. It can be made by adding a base to a solution containing Mg^{2+}. Suppose that 40.0 mL of 0.200 *M* NaOH solution is added to 25.0 mL of 0.300 *M* $MgCl_2$ solution. What mass of $Mg(OH)_2$ will be formed, and what will be the concentrations of the ions in the solution after the reaction is complete? The net ionic equation for the reaction is

$$Mg^{2+}(aq) + 2OH^-(aq) \longrightarrow Mg(OH)_2(s)$$

ANALYSIS The first thing we should notice here is that we've been given the volume and molarity for both solutions. By now, you will realize that volume and molarity give us moles, so in effect, we have been given the number of moles of each reactant. This means we have a limiting reactant problem (similar to the kind we studied on page 111).

To answer all the parts of this question, the best approach is to begin by tabulating the number of moles of each ion supplied by their respective solutions. Then we'll use the ionic equation to see which ions are removed and calculate which, if any, is limiting. Then we can go after the concentrations of the remaining ions and the mass of $Mg(OH)_2$ that forms.

Actually, the reaction occurs as the solutions are being poured together. For purposes of the calulations, however, we may pretend that the solutes can be mixed first and then allowed to react. The initial and final states will still be the same.

Milk of magnesia is a home remedy for acid indigestion.

SOLUTION Let's begin by determining the number of moles of $MgCl_2$ and NaOH supplied by the volumes of their solutions used. The conversion factors are taken from their molarities—0.200 *M* NaOH and 0.300 *M* $MgCl_2$.

$$0.0400 \text{ L NaOH soln} \times \frac{0.200 \text{ mol NaOH}}{1.00 \text{ L NaOH soln}} = 8.00 \times 10^{-3} \text{ mol NaOH}$$

$$0.0250 \text{ L MgCl}_2 \text{ soln} \times \frac{0.300 \text{ mol MgCl}_2}{1.00 \text{ L MgCl}_2 \text{ soln}} = 7.50 \times 10^{-3} \text{ mol MgCl}_2$$

From this information, we obtain the number of moles of each ion present *before* any reaction occurs. In doing this, notice we take into account that 1 mol of $MgCl_2$ gives 2 mol Cl^-.

Moles of Ions Before Reaction			
Mg^{2+}	7.50×10^{-3} mol	Cl^-	15.0×10^{-3} mol
Na^+	8.00×10^{-3} mol	OH^-	8.00×10^{-3} mol

Now we refer to the net ionic equation, where we see that only Mg^{2+} and OH^- react. From the coefficients of the equation,

$$1 \text{ mol Mg}^{2+} \Leftrightarrow 2 \text{ mol OH}^-$$

This means that 7.50×10^{-3} mol Mg^{2+} would require 15.0×10^{-3} mol OH^-, but we have only 8.00×10^{-3} mol OH^-. Insufficient OH^- is provided for the Mg^{2+}, so OH^- must be the limiting reactant. Therefore, all of the OH^- will be used up and some Mg^{2+} will be unreacted. The amount of Mg^{2+} that *does* react to form $Mg(OH)_2$ can be found as follows.

$$8.00 \times 10^{-3} \text{ mol OH}^- \times \frac{1 \text{ mol Mg}^{2+}}{2 \text{ mol OH}^-} = 4.00 \times 10^{-3} \text{ mol Mg}^{2+}$$
$$\text{(this amount reacts)}$$

Now we can tabulate the number of moles of each ion left in the mixture *after* the reaction is complete. For Mg^{2+}, the amount remaining equals the initial amount minus the amount that reacts.

Moles of Ions After Reaction

Mg^{2+}	$(7.50 \times 10^{-3} \text{ mol}) - (4.00 \times 10^{-3} \text{ mol}) = 3.50 \times 10^{-3} \text{ mol}$
Cl^-	$15.0 \times 10^{-3} \text{ mol (no change)}$
Na^+	$8.00 \times 10^{-3} \text{ mol (no change)}$
OH^-	$0.00 \text{ mol (all used up)}$

The question asks for the concentrations of the ions in the reaction mixture, so we must now divide each of these numbers of moles by the *total volume of the final solution* (40.0 mL + 25.0 mL = 65.0 mL), expressed in liters (0.0650 L). For example, for Mg^{2+}, its concentration is

$$\frac{3.50 \times 10^{-3} \text{ mol } Mg^{2+}}{0.0650 \text{ L soln}} = 0.0538 \text{ } M \text{ } Mg^{2+}$$

Performing similar calculations for the other ions gives

Concentrations of Ions After Reaction

Mg^{2+}	$0.0538 \text{ } M$	Cl^-	$0.231 \text{ } M$
Na^+	$0.123 \text{ } M$	OH^-	$0.0 \text{ } M$

Now we can turn our attention to the mass of $Mg(OH)_2$ that's formed. If 4.00×10^{-3} mol Mg^{2+} reacts, then 4.00×10^{-3} mol $Mg(OH)_2$ must be produced. The formula mass of $Mg(OH)_2$ is 58.3, so the mass of $Mg(OH)_2$ formed is

> Stoichiometrically,
> 1 mol Mg^{2+} $\Leftrightarrow$ 1 mol $Mg(OH)_2$

$$4.00 \times 10^{-3} \text{ mol } Mg(OH)_2 \times \frac{58.3 \text{ g } Mg(OH)_2}{1 \text{ mol } Mg(OH)_2} = 0.233 \text{ g } Mg(OH)_2$$

■ **Practice Exercise 22** How many moles of $BaSO_4$ will form if 20.0 mL of 0.600 M $BaCl_2$ is mixed with 30.0 mL of 0.500 M $MgSO_4$? What will the concentrations of each ion be in the final reaction mixture?

Chemical Analyses

Chemical analyses fall into two categories. In a **qualitative analysis** we simply determine which substances are present in a sample without measuring their amounts. In a **quantitative analysis,** our goal is to measure the amounts of the various substances in a sample.

When chemical reactions are used in a quantitative analysis, a useful strategy is to convert *all* of an element present in a sample into a substance with a known formula. From the amount of this substance obtained, we can determine how much of the element was present in the original sample.

EXAMPLE 4.22
Calculation Involving a
Quantitative Analysis

A certain insecticide is known to contain carbon, hydrogen, and chlorine. Reactions are carried out on a 1.000 g sample of the compound that convert all of its chlorine to chloride ion dissolved in water. This solution is treated with an excess amount of $AgNO_3$ solution, and the AgCl precipitate is collected and weighed. Its mass is 2.022 g. What is the percentage by mass of Cl in the original insecticide sample?

ANALYSIS The strategy is to determine the mass of Cl in 2.022 g of AgCl. We then assume that all this chlorine was in the original 1.000 g sample, and calculate the percentage Cl.

SOLUTION The formula mass of AgCl is 143.3, and the atomic mass of Cl is 35.45. In 1 mol of AgCl there must be 1 mol of Cl, so in 143.3 g AgCl there must be 35.45 g Cl. With this information, we can now calculate the mass of Cl in 2.022 g AgCl.

$$2.022 \text{ g AgCl} \times \frac{35.45 \text{ g Cl}}{143.3 \text{ g AgCl}} = 0.5002 \text{ g Cl}$$

The percentage Cl in the sample can be calculated as

$$\% \text{ Cl} = \frac{\text{mass of Cl}}{\text{mass of sample}} \times 100\%$$

$$= \frac{0.5002 \text{ g Cl}}{1.000 \text{ g sample}} \times 100\%$$

$$= 50.02\%$$

The insecticide was 50.02% Cl.

■ **Practice Exercise 23** A sample of a mixture containing $CaCl_2$ and $MgCl_2$ weighed 2.000 g. The sample was dissolved in water, and H_2SO_4 was added until the precipitation of $CaSO_4$ was complete. The $CaSO_4$ was filtered, dried completely, and weighed. A total of 0.736 g of $CaSO_4$ was obtained.

(a) How many moles of Ca^{2+} were in the $CaSO_4$?

(b) How many moles of Ca^{2+} were in the original 2.000 g sample?

(c) How many moles of $CaCl_2$ were in the 2.000 g sample?

(d) How many grams of $CaCl_2$ were in the 2.000 g sample?

(e) What was the percentage of $CaCl_2$ in the mixture?

Titrations

Titration is an important laboratory procedure used in performing chemical analyses. The apparatus is shown in Figure 4.14. The long tube is called a **buret,** and it is marked for volumes, usually in increments of 0.10 mL. In a typical analysis, a solution containing an unknown amount of the substance to be analyzed is placed in the receiving flask. Precisely measured volumes of a solution of known concentration—a **standard solution**—are then added from the buret. This addition is continued until some visual effect, like a color change, signals that the two reactants have been combined in just the right ratio to give a complete reaction.

The valve at the bottom of the buret is called a **stopcock,** and it permits the analyst to control the amount of **titrant** (the solution in the buret) that is delivered to the receiving flask. This permits the addition of the titrant to be stopped once the reaction is complete.

Titrations are used for both acid–base and redox reactions. Prior to the start of an acid–base titration, a drop or two of an *indicator* solution is added to the solution in the receiving flask. An **acid–base indicator** is a dye that has one color in an acidic solution and a different color in a basic solution. We have already mentioned litmus. Another, phenolphthalein, is more commonly used in acid–base titrations. It is colorless in acid and pink in base. (The theory of acid–base indicators is discussed in Chapter 16.)

Acid–Base Titrations

During an acid–base titration, the analyst looks for a change in the indicator color. This signals that the solution has changed from acidic to basic (or basic

FIGURE 4.14

Apparatus for titration. The rate of addition of the titrant from the buret is controlled by manipulating the stopcock. Seen here is the use of the apparatus in an acid–base titration employing phenolphthalein as the indicator.

The titration procedure is used for many kinds of reactions.

to acidic, depending on the nature of the reactants in the buret and flask). Usually this color change is very abrupt. As the end of the reaction is approached, it normally occurs with the addition of only one final drop of the titrant. At this moment, the titration is stopped. The **end point** has been reached, and the total volume of the titrant added to the receiving flask is recorded.

A student prepares a solution of hydrochloric acid that is approximately 0.1 *M* and wishes to determine its precise concentration. A 25.00 mL portion of the HCl solution is transferred to a flask, and after a few drops of indicator are added, the HCl solution is titrated with 0.0775 *M* NaOH solution. The titration requires exactly 37.46 mL of the standard NaOH solution. What is the molarity of the HCl solution?

ANALYSIS We'll need the balanced equation to learn the stoichiometric equivalency between HCl and NaOH. Then we calculate the moles of NaOH used (from the volume and molarity of the NaOH solution) to find the equivalent moles of HCl and use this and the volume of the HCl solution to calculate the desired molarity.

SOLUTION The chemical equation for the neutralization reaction is

$$HCl(aq) + NaOH(aq) \longrightarrow NaCl(aq) + H_2O$$

From the molarity and volume of the NaOH solution, we calculate the number of moles of NaOH consumed in the titration.

$$0.03746 \ \cancel{L \ NaOH \ soln} \times \frac{0.0775 \ mol \ NaOH}{1.00 \ \cancel{L \ NaOH \ soln}} = 2.90 \times 10^{-3} \ mol \ NaOH$$

The coefficients in the equation tell us that NaOH and HCl react in a one-to-one mole ratio, so in this titration, 2.90×10^{-3} mol HCl was in the flask. To calculate the molarity of the HCl, we simply take the ratio of the number of moles of HCl that reacted (2.90×10^{-3} mol HCl) to the volume (in liters) of the HCl solution used (0.02500 L).

$$\text{Molarity of HCl soln} = \frac{2.90 \times 10^{-3} \ mol \ HCl}{0.02500 \ L \ HCl \ soln}$$

$$= 0.116 \ M \ HCl$$

The concentration of the hydrochloric acid is 0.116 *M*.

CHECKING THE ANSWER FOR REASONABLENESS The volume of NaOH solution used is larger than the volume of HCl solution. Because the reactants combine in a one-to-one ratio, this must mean that the HCl solution is more concentrated than the NaOH solution. The value we obtained, 0.116 *M* is larger than 0.0775 *M*, so our answer seems reasonable.

■ **Practice Exercise 24** In a titration, a sample of H_2SO_4 solution having a volume of 15.00 mL required 36.42 mL of 0.147 *M* NaOH solution for *complete* neutralization. What is the molarity of the H_2SO_4 solution?

■ **Practice Exercise 25** "Stomach acid" is hydrochloric acid. A sample of gastric juice having a volume of 5.00 mL required 11.00 mL of 0.0100 *M* KOH solution for neutralization in a titration. What was the molar concentration of HCl in this fluid?

EXAMPLE 4.23
Calculation Involving Acid–Base Titration

We could also use the following conversion factor:

$$\frac{0.0775 \ mol \ NaOH}{1000 \ mL \ NaOH \ soln}$$

Stoichiometrically,

1 mol NaOH ⇔ 1 mol HCl

In concentrated solutions, MnO_4^- is purple, but dilute solutions of the ion appear pink.

FIGURE 4.15

A solution of $KMnO_4$ being added to a stirred acidic solution containing Fe^{2+}. The reaction oxidizes the pale blue-green Fe^{2+} to Fe^{3+} while the MnO_4^- is reduced to the almost colorless Mn^{2+} ion. The purple color of the permanganate will continue to be destroyed until all of the Fe^{2+} has reacted. Only then will the iron-containing solution take on a pink or purple color. This ability of MnO_4^- to signal the completion of the reaction makes it especially useful in redox titrations.

Redox Titrations

Unlike in acid–base titrations, there are no simple indicators that can be used to conveniently detect the end points in redox titrations. One of the most useful reactants for redox titrations is potassium permanganate, $KMnO_4$, especially when the reaction can be carried out in an acidic solution. Permanganate ion is a powerful oxidizing agent, so it oxidizes most substances that are capable of being oxidized. That's one reason why it is used. Especially important, though, is the fact that the MnO_4^- ion has a deep purple color and its reduction product in acidic solution is the almost colorless Mn^{2+} ion. Therefore, when a solution of $KMnO_4$ is added from a buret to a solution of a reducing agent, the chemical reaction that occurs forms a nearly colorless product. As the $KMnO_4$ solution is added, the purple color continues to be destroyed as long as there is any reducing agent left (Figure 4.15). However, after the last trace of the reducing agent has been consumed, the MnO_4^- ion in the next drop of titrant has nothing to react with, so it colors the solution pink. This signals the end of the titration. In this way, permanganate ion serves as its own indicator in redox titrations. Example 4.24 illustrates a typical analysis using $KMnO_4$ in a redox titration.

EXAMPLE 4.24
Redox Titrations in Chemical Analysis

All the iron in a 2.000 g sample of an iron ore was dissolved in an acidic solution and converted to Fe^{2+}, which was then titrated with 0.1000 M $KMnO_4$ solution. In the titration the iron was oxidized to Fe^{3+}. The titration required 27.45 mL of the $KMnO_4$ solution.

(a) How many grams of iron were in the sample?

(b)   What was the percentage of iron in the sample?

(c) If the iron was present in the sample as Fe_2O_3, what was the percentage by weight of Fe_2O_3 in the sample?

ANALYSIS The first step, of course, will be to write a balanced equation. Let's analyze the problem after that point. From the volume of the $KMnO_4$ solution and its concentration we can determine the number of moles of titrant used. The coefficients of the equation then allow us to compute the number of moles of iron(II) that reacted. Since all the iron had been previously changed to iron-(II), this is the number of moles of iron in the sample. Converting moles of iron to grams of iron gives the mass of iron in the sample, from which the percentage of iron can be calculated easily. Let's tackle the first two parts of the problem now, and then return to the last part afterward.

SOLUTION The skeleton equation for the reaction is

$$Fe^{2+} + MnO_4^- \longrightarrow Fe^{3+} + Mn^{2+}$$

Balancing it by the ion–electron method for acidic solutions gives

$$5Fe^{2+} + MnO_4^- + 8H^+ \longrightarrow 5Fe^{3+} + Mn^{2+} + 4H_2O$$

The number of moles of $KMnO_4$ consumed is calculated from the volume of the solution used in the titration and its concentration.

$$0.02745 \text{ L KMnO}_4 \times \frac{0.1000 \text{ mol KMnO}_4}{1.00 \text{ L KMnO}_4} = 0.002745 \text{ mol KMnO}_4$$

The chemical equation tells us five moles of iron react per mole of permanganate consumed. The number of moles of iron that reacted is

$$0.002745 \text{ mol KMnO}_4 \times \frac{5 \text{ mol Fe}}{1 \text{ mol KMnO}_4} = 0.01372 \text{ mol Fe}$$

and the mass of iron in the sample is

$$0.01372 \text{ mol Fe} \times \frac{55.85 \text{ g Fe}}{1 \text{ mol Fe}} = 0.7663 \text{ g Fe}$$

This is the answer to part (a) of the problem. Next we calculate the percentage of iron in the sample, which is the mass of iron divided by the mass of the sample, all multiplied by 100%.

$$\% \text{ Fe} = \frac{\text{mass of Fe}}{\text{mass of sample}} \times 100\%$$

Substituting gives

$$\% \text{ Fe} = \frac{0.7663 \text{ g Fe}}{2.000 \text{ g sample}} \times 100\% = 38.32\% \text{ Fe}$$

ANALYSIS CONTINUED Now we can work on the last part of the question. Earlier in the problem we determined the number of moles of iron that reacted, 0.01372 mol Fe. How many moles of Fe_2O_3 would have contained this number of moles of iron? That's the critical question we have to answer. Once we know this, we can calculate the mass of the Fe_2O_3 and the percentage of Fe_2O_3 in the original sample.

SOLUTION CONTINUED The chemical formula for the iron oxide gives us

$$1 \text{ mol Fe}_2O_3 \Leftrightarrow 2 \text{ mol Fe}$$

This provides the conversion factor we need to determine how many moles of Fe_2O_3 were present in the sample. Working with the number of moles of Fe,

$$0.01372 \text{ mol Fe} \times \frac{1 \text{ mol Fe}_2O_3}{2 \text{ mol Fe}} = 0.006860 \text{ mol Fe}_2O_3$$

This is the number of moles of Fe_2O_3 in the sample. The formula mass of Fe_2O_3 is 159.7 g mol^{-1}, so the mass of Fe_2O_3 in the sample was

$$0.006860 \text{ mol Fe}_2O_3 \times \frac{159.7 \text{ g Fe}_2O_3}{1 \text{ mol Fe}_2O_3} = 1.096 \text{ g Fe}_2O_3$$

Finally, the percentage of Fe_2O_3 in the sample was

$$\% \text{ Fe}_2O_3 = \frac{1.096 \text{ g Fe}_2O_3}{2.000 \text{ g sample}} \times 100\% = 54.80\% \text{ Fe}_2O_3$$

The ore sample contained 54.80% Fe_2O_3.

■ **Practice Exercise 26** A sample of a tin ore weighing 0.3000 g was dissolved in an acid solution, and all the tin in the sample was changed to tin(II). The solution was titrated with 8.08 mL of 0.0500 M $KMnO_4$ solution, which oxidized the tin(II) to tin(IV). (a) What is the balanced equation for the reaction in the titration? (b) How many grams of tin were in the sample? (c) What was the percentage by weight of tin in the sample? (d) If the tin in the sample had been present in the compound SnO_2, what would have been the percentage by weight of SnO_2 in the sample?

SUMMARY

Solution Vocabulary A considerable vocabulary has developed for describing **solutions.** Some terms are qualitative —**dilute, concentrated, saturated, unsaturated,** and **supersaturated,** for example. The **solubility** of a substance is often stated in grams of **solute** per 100 g of **solvent.**

Electrolytes Substances that **dissociate** or **ionize** in water to produce cations and anions are **electrolytes;** those that do not are called **nonelectrolytes.** Electrolytes include salts and metal hydroxides as well as molecular acids and bases that ionize by reaction with water.

Acids and Bases as Electrolytes The modern version of the **Arrhenius definition of acids and bases** is that an **acid** is a substance that produces hydronium ions, H_3O^+, when dissolved in water, and a **base** one that produces hydroxide ions, OH^-, when dissolved in water.

The oxides of nonmetals are generally **acidic anhydrides** and react with water to give acids. Metal oxides are usually **basic anhydrides** because they tend to react with water to give metal hydroxides or bases.

Strong acids and bases are **strong electrolytes; weak acids and bases** are **weak electrolytes.** In a solution of a weak electrolyte there is a chemical equilibrium between the nonionized molecules of the solute and the ions formed by the reaction of the solute with water.

Acid–Base Neutralization and Ionic Reactions Acids react with bases in neutralization reactions to produce a salt and water. Equations for these reactions can be written in three different ways. In **molecular equations,** complete formulas for all reactants and products are used. In an **ionic equation,** soluble strong electrolytes are written in dissociated (ionized) form; ''molecular'' formulas are used for solids and weak electrolytes. A **net ionic equation** is obtained by eliminating **spectator ions** from the ionic equation. An ionic or net ionic equation is balanced only if both atoms *and* net charge are balanced.

Metathesis Reactions **Metathesis** or **double replacement** reactions occur between the ions in solutions of soluble salts and produce precipitates. Writing net ionic equations allows us to generalize. Metathesis can also produce gaseous compounds with low solubility in water or compounds that are unstable and decompose to give slightly soluble gases.

Predicting Metathesis Reactions You should learn the **solubility rules,** and remember that all salts are strong electrolytes. Remember that all strong acids and bases are strong electrolytes, too. Everything else is a weak electrolyte or a nonelectrolyte. Be sure to learn the reactions that produce gases in metathesis reactions, which are found in Table 4.2.

Oxidation–Reduction **Oxidation** is the loss of electrons or an increase in oxidation number; **reduction** is the gain of electrons or a decrease in oxidation number. Both always occur together in **redox** reactions. The substance oxidized is the **reducing agent;** the substance reduced is the **oxidizing agent. Oxidation numbers** are a bookkeeping device that we use to follow changes in redox reactions. They are assigned according to the rules on page 152. The term **oxidation state** is equivalent to oxidation number.

Ion–Electron Method In a balanced redox equation, the number of electrons gained by one substance is always equal to the number lost by another substance. The **ion–electron method** divides a *skeleton* net ionic equation into two **half-reactions,** which are balanced separately before being recombined to give the final balanced net ionic equation. For reactions in basic solution, the equation is balanced as if it occurred in an acidic solution, and then the balanced equation is converted to its proper form for basic solution by adding an appropriate number of OH^-.

Solution Stoichiometry **Molar concentration (molarity)** equals the number of moles of solute per liter of solution. This concentration unit makes available the two conversion factors,

$$\frac{\text{mol solute}}{\text{1 L soln}} \quad \text{and} \quad \frac{\text{1 L soln}}{\text{mol solute}}$$

Solutions of known molarity are made in two ways. One is to dissolve a known number of moles of the solute into a solvent until the final volume of the solution reaches a known value. The other is to take a specific volume of a relatively concentrated solution and dilute it with the solvent until the final volume is reached. For dilution problems, remember the relationship

$$V_{dil} \cdot M_{dil} = V_{concd} \cdot M_{concd}$$

Figure 4.16 gives an overview of how to use stoichiometric equivalencies to move from moles to the units commonly used in measurements—grams or volumes of solutions of known molarities.

Titration is a technique used to make quantitative measurements of the amounts of solutions needed to obtain a complete reaction. In an acid–base titration, the **end point** is normally detected visually using an **acid–base indicator.** For redox reactions, $KMnO_4$ is a useful titrant because it serves as its own indicator in acidic solutions.

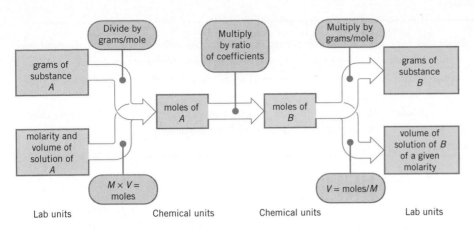

FIGURE 4.16

An overview of the paths that are followed in working stoichiometry problems involving solutions. The central feature is the conversion between moles of one reactant and moles of another, which makes use of the coefficients of the balanced chemical equation. Moles are thus the central fundamental chemical units used to construct chemical equivalencies. We can obtain moles from more than one laboratory unit, from grams or from volumes of solutions of known molarities.

Tools You Have Learned

The table below lists the tools that you have learned in this chapter. You will need them to answer chemistry questions. Review them if necessary, and refer to the tools when working on the Thinking It Through problems and the Review Exercises that follow.

Tool	Function
List of strong acids (page 135)	To enable the recognition of any other acids as weak.
	To decide which formula should represent the acid in a net ionic equation.
Criteria for balanced ionic equations (page 139)	To obtain balanced ionic and net ionic equations.
Types of substances that react to give gases either with acids or bases (page 145)	To predict when to expect CO_2, SO_2, HCN, H_2S, or NH_3 to form from a reaction.
Solubility rules (page 146)	To tell whether a given salt is soluble in water and dissociates fully.
	To enable the prediction of metathesis reactions.
Criteria for predicting ionic reactions (page 149)	To determine when a net ionic reaction will occur in an aqueous solution.
Rules for assigning oxidation numbers (page 152)	To assign oxidation numbers to the atoms in a chemical formula. Oxidation numbers enable you to identify oxidation and reduction.
Ion-electron method (pages 159 and 160)	Enables you to balance net ionic equations for redox reactions.
Molar concentration (pages 161, 164, and 165)	To relate the amount of solute in a solution to the volume of the solution. Used to construct conversion factors based on the relationship
	Moles of solute ⟺ liters of solution

THINKING IT THROUGH

Remember, you are not asked to obtain answers for the following problems. Instead, assemble the data necessary to solve the problems and describe how you would use the data to obtain the answers. For numerical problems, set up the calculation using appropriate conversion factors.

The problems are divided into two groups. Those in Level 2 are significantly more challenging than those in Level 1 and provide an opportunity to really hone your problem-solving skills.

Level 1 Problems

1. Calcium hydroxide and hydrobromic acid react as follows:

$$Ca(OH)_2(s) + 2HBr(aq) \longrightarrow CaBr_2(aq) + 2H_2O$$

If 24.50 mL of a certain solution of HBr are able to combine exactly with 0.2067 g of $Ca(OH)_2$, what is the molar concentration of the HBr solution? (Set up the solution to the problem.)

2. In analytical work of the highest precision possible involving solutions, all volumetric glassware, like pipets and volumetric flasks, have to be *calibrated*. This simply means that their volumes, when filled to the etched marks, are actually determined. To do this for a volumetric flask, the flask can first be dried and its mass measured with high precision. The flask is then filled to the etch mark with pure water while the entire system—flask and contents—is kept at a precisely known temperature. Once the temperature has stabilized and its contents are as close to the etched mark as humanly discernible, the system's mass is measured. Suppose that a flask nominally labeled "250 mL" is calibrated in this way, that the initial mass is 142.67 g, the final mass is 396.29 g, and the temperature of the system is 25.00 °C. At this temperature the density of water is 0.99707 g/mL. Describe how you can use this information to determine the true volume of the flask.

3. You are planning an experiment that requires the presence of the hydroxide ion at a concentration of at least 0.2 *M*. Which of the following could you use as the solute: NaOH, KOH, $Ca(OH)_2$, or $Mg(OH)_2$? Describe the information you need to know to answer the question.

4. Using both molecular and net ionic equations, describe the reaction of solid $MgCO_3$ with an aqueous solution of hydrobromic acid by means of which the carbonate dissolves. Suppose that you mix 100.0 mL of 0.1250 *M* HBr with a 10.54 g sample of solid $MgCO_3$ in a large beaker.

(a) What do you *see* as you carry this out?
(b) When the reaction has subsided, what do you *see*? (Describe the possibilities and how you could predict which to expect.)
(c) Describe how you can determine the identities and the molarities of the principal components *dissolved* in the solution after the reaction is complete.
(d) Explain how you can determine whether any $MgCO_3$ remains undissolved. How would you determine how many grams of $MgCO_3$ there are left over?

5. An unknown compound was either NaCl, NaBr, or NaI. When 0.546 g of the compound was dissolved in water, it took 22.1 mL of 0.248 *M* $AgNO_3$ to combine with all of the halide ion in the solution. Describe how you would use this information to determine which compound the unknown was.

6. Is the reaction $HIO_4 + 2H_2O \rightarrow H_5IO_6$ a redox reaction? (Describe how you go about answering this question.)

7. Iodate ion, IO_3^-, oxidizes $NO_2(g)$ to give iodide ion and nitrate ion as products. How many milliliters of 0.0200 *M* IO_3^- solution will react with 0.230 g of NO_2? (Set up the solution to the problem.)

8. A certain reaction is expected to evolve Cl_2 gas as a product, and it is desired to keep it from escaping into the lab. Therefore, the gases formed in the reaction will be trapped and passed through a solution of $Na_2S_2O_3$. The Cl_2 reacts with $S_2O_3^{2-}$ to give Cl^- and SO_4^{2-}. If 0.020 mol of Cl_2 is expected to be formed, what is the minimum volume (in milliliters) of 0.500 *M* $Na_2S_2O_3$ that will be required to react with all the Cl_2? (Set up the solution to the problem.)

9. How many milliliters of 0.200 *M* $Na_2S_2O_3$ solution are required to react completely with 0.020 mol of I_3^-? In the reaction, I_3^- is reduced to I^- and $S_2O_3^{2-}$ is oxidized to $S_4O_6^{2-}$. (Set up the solution to the problem.)

Level 2 Problems

10. Concentrated sulfuric acid contains 96% H_2SO_4 by mass and has a specific gravity at room temperature of 1.839. You are assigned the preparation of 250 mL of 0.10 *M* H_2SO_4. How would you proceed to prepare the more dilute solution in the volume specified? Describe any necessary calculations.

11. A mixture is made by combining 300 mL of 0.0200 *M* $Na_2Cr_2O_7$ with 400 mL of 0.060 *M* $Fe(NO_3)_2$. Initially, the

H^+ concentration in the mixture is 0.400 *M*. Dichromate ion oxidizes Fe^{2+} to Fe^{3+} and is reduced to Cr^{3+}. After the reaction in the mixture has ceased, how many milliliters of 0.0100 *M* NaOH will be required to neutralize the remaining H^+? (Set up the solution to the problem.)

12. An organic compound contains carbon, hydrogen, and sulfur. A sample of it with a mass of 1.045 g was burned in oxygen to give gaseous CO_2, H_2O, and SO_2. These gases were passed through 500 mL of an acidified 0.0200 *M* $KMnO_4$ solution, which caused the SO_2 to be oxidized to SO_4^{2-} and part of the available $KMnO_4$ to be reduced to Mn^{2+}. Next, 50.00 mL of 0.0300 *M* $SnCl_2$ was added to a 50.00 mL portion of the partially reduced $KMnO_4$ solution. All the MnO_4^- in the 50 mL portion was reduced and the

excess Sn^{2+} that remained after reaction was titrated with 0.0100 *M* $KMnO_4$, requiring 27.28 mL of the titrant to reach an end point. Explain in detail how you would determine the percentage of sulfur in the original sample of the organic compound. Set up the solution to the problem.

13. A solution contained a mixture of SO_3^{2-} and $S_2O_3^{2-}$. A 100.0 mL portion of the solution was found to react with 80.00 mL of 0.0500 *M* CrO_4^{2-} in a basic solution to give CrO_2^-. The only sulfur-containing product was SO_4^{2-}. After the reaction, the solution was treated with excess $BaCl_2$ solution, which precipitated $BaSO_4$. This solid was filtered from the solution, dried, and found to weigh 0.9336 g. Explain in detail how you can determine the molar concentrations of SO_3^{2-} and $S_2O_3^{2-}$ in the original solution.

REVIEW EXERCISES

Answers to questions whose numbers are printed in color are given in Appendix D. Challenging questions are marked with asterisks.

Electrolytes

4.1 Define the following: (a) solvent, (b) solute, (c) concentration.

4.2 Define the following: (a) concentrated, (b) dilute, (c) saturated, (d) unsaturated, (e) supersaturated.

4.3 Why are chemical reactions often carried out in solutions?

4.4 What is an electrolyte? What is a nonelectrolyte?

4.5 Define dissociation as it applies to ionic compounds that dissolve in water.

4.6 Write equations for the dissociation of the following ionic compounds in water: (a) LiCl, (b) $BaCl_2$, (c) $Al(C_2H_3O_2)_3$, (d) $(NH_4)_2CO_3$, (e) $FeCl_3$

4.7 Write equations for the dissociation of the following ionic compounds in water: (a) $CuSO_4$, (b) $Al_2(SO_4)_3$, (c) $CrCl_3$, (d) $(NH_4)_2HPO_4$, (e) $KMnO_4$

Acids and Bases as Electrolytes

4.8 Give two general properties of an acid. Give two general properties of a base.

4.9 How did Arrhenius define an acid and a base?

4.10 What is the difference between *dissociation* and *ionization* as applied to the discussion of electrolytes?

4.11 Pure $HClO_4$ is molecular. Write an equation for its reaction with water to give H_3O^+ and ClO_4^- ions.

4.12 Hydrazine is a toxic substance that can form when household ammonia is mixed with a bleach such as Clorox. Its formula is N_2H_4, and it is a weak base. Write a chemical equation showing its reaction with water.

4.13 Which of the following would yield an acidic solution when dissolved in water? (a) P_4O_{10}, (b) K_2O, (c) SeO_3, (d) Cl_2O_7

4.14 What is the difference between a strong electrolyte and a weak electrolyte?

4.15 If a substance is a weak electrolyte, what does this mean in terms of the tendency of the ions to react to reform the molecular compound? How does this compare with strong electrolytes?

4.16 Nitrous acid, HNO_2, is a weak acid. Write an equation showing its reaction with water.

4.17 Why don't we use double arrows in the equation for the reaction of a strong acid with water?

4.18 Write the formulas for the strong acids discussed in Section 4.3. (If you must peek, they are listed on page 135.)

4.19 $HClO_3$ is a strong acid. Write an equation for its reaction with water.

4.20 When diprotic acids, like H_2CO_3, react with water to release their H^+, they do so in two steps, each of which is a dynamic equilibrium. Write chemical equations that illustrate this for H_2CO_3.

4.21 Phosphoric acid, H_3PO_4, is a weak acid that undergoes ionization in three steps. Write chemical equations for each of these reactions.

Acid–Base Neutralization

4.22 What are the differences among molecular, ionic, and net ionic equations? What are spectator ions?

4.23 What two conditions must be fulfilled by a balanced ionic equation?

4.24 The following equation is not balanced. What is wrong with it?

$$NO_2 + H_2O \longrightarrow NO_3^- + 2H^+$$

4.25 Complete and balance the following equations. For each, write the molecular, ionic, and net ionic equations. (All the products are soluble in water.)
(a) $Ca(OH)_2(aq) + HNO_3(aq) \rightarrow$
(b) $Al_2O_3(s) + HCl(aq) \rightarrow$
(c) $Zn(OH)_2(s) + H_2SO_4(aq) \rightarrow$

4.26 Complete and balance the following equations. For each, write the molecular, ionic, and net ionic equations. (All the products are soluble in water.)
(a) $HC_2H_3O_2(aq) + Mg(OH)_2(s) \rightarrow$
(b) $HClO_4(aq) + NH_3(aq) \rightarrow$
(c) $H_2CO_3(aq) + NH_3(aq) \rightarrow$

Ionic Reactions That Produce Precipitates

4.27 What is another name for *metathesis reaction*? What is a precipitate?

4.28 Write ionic and net ionic equations for these reactions.
(a) $(NH_4)_2CO_3(aq) + MgCl_2(aq) \rightarrow 2NH_4Cl(aq) + MgCO_3(s)$
(b) $CuCl_2(aq) + 2NaOH(aq) \rightarrow Cu(OH)_2(s) + 2NaCl(aq)$
(c) $3FeSO_4(aq) + 2Na_3PO_4(aq) \rightarrow Fe_3(PO_4)_2(s) + 3Na_2SO_4(aq)$
(d) $2AgC_2H_3O_2(aq) + NiCl_2(aq) \rightarrow 2AgCl(s) + Ni(C_2H_3O_2)_2(aq)$

4.29 Write balanced ionic and net ionic equations for these reactions.
(a) $CuSO_4(aq) + BaCl_2(aq) \rightarrow CuCl_2(aq) + BaSO_4(s)$
(b) $Fe(NO_3)_3(aq) + LiOH(aq) \rightarrow LiNO_3(aq) + Fe(OH)_3(s)$
(c) $Na_3PO_4(aq) + CaCl_2(aq) \rightarrow Ca_3(PO_4)_2(s) + NaCl(aq)$
(d) $Na_2S(aq) + AgC_2H_3O_2(aq) \rightarrow NaC_2H_3O_2(aq) + Ag_2S(s)$

4.30 Aqueous solutions of sodium sulfide, Na_2S, and copper nitrate, $Cu(NO_3)_2$, are mixed. A precipitate of copper sulfide, CuS, forms at once. Left behind is a solution of sodium nitrate, $NaNO_3$. Write the net ionic equation for this reaction.

4.31 Silver bromide is "insoluble." What does this mean about the concentrations of Ag^+ and Br^- in a saturated solution of AgBr? Explain why a precipitate of AgBr forms when solutions of the soluble salts $AgNO_3$ and NaBr are mixed.

4.32 Silver bromide (mentioned in the preceding question) is the chief light-sensitive substance used in the manufacture of photographic film. It can be prepared by mixing solutions of $AgNO_3$ and NaBr. Taking into account that $NaNO_3$ is soluble, write molecular, ionic, and net ionic equations for this reaction.

4.33 Trisodium phosphate, Na_3PO_4, is a useful cleaning agent, but it must be handled with care because its solutions are quite caustic. If a solution of Na_3PO_4 is added to one containing a calcium salt such as $CaCl_2$, a precipitate of calcium phosphate is formed. Write molecular, ionic, and net ionic equations for the reaction.

Ionic Reactions That Produce Gases

4.34 What gas is formed if HCl is added to (a) $NaHCO_3$, (b) Na_2S, (c) K_2SO_3?

4.35 Suppose you suspected that a certain solution contained ammonium ion. What simple chemical test could

you perform that would tell you whether your suspicion was correct?

4.36 Write balanced net ionic equations for the reaction between:
(a) $HNO_3(aq) + K_2CO_3(aq)$
(b) $Ca(OH)_2(aq) + NH_4NO_3(aq)$
(c) $H_2SO_4(aq) + NaHSO_3(aq)$

Predicting Metathesis Reactions

4.37 Use the solubility rules to decide which of the following compounds are *soluble* in water.
(a) $Ca(NO_3)_2$ (d) $AgNO_3$
(b) $FeCl_2$ (e) $BaSO_4$
(c) $Ni(OH)_2$ (f) $CuCO_3$

4.38 Use the solubility rules to decide which of the following compounds are *insoluble* in water.
(a) $AgCl$ (d) $Ca_3(PO_4)_2$
(b) $Cr_2(SO_4)_3$ (e) $Al(C_2H_3O_2)_3$
(c) $(NH_4)_2CO_3$ (f) ZnO

4.39 Predict which of the following compounds are *soluble* in water.
(a) $HgBr_2$ (d) $Sr(NO_3)_2$
(b) Hg_2Br_2 (e) $(NH_4)_3PO_4$
(c) PbI_2 (f) $Pb(C_2H_3O_2)_2$

4.40 Predict which of the following compounds are *insoluble* in water.
(a) $Ca(OH)_2$ (d) NiS (g) $PbCl_2$
(b) $CsOH$ (e) $MgSO_3$ (h) Fe_2O_3
(c) K_3PO_4 (f) $Mg(ClO_4)_2$ (i) $CaCO_3$

4.41 Complete and balance the following reactions and then write the ionic and net ionic equations. If all ions cancel, indicate that no reaction (N.R.) takes place.
(a) $Na_2SO_3 + Ba(NO_3)_2 \rightarrow$
(b) $K_2S + ZnCl_2 \rightarrow$
(c) $NH_4Br + Pb(C_2H_3O_2)_2 \rightarrow$
(d) $NH_4ClO_4 + Cu(NO_3)_2 \rightarrow$

4.42 Complete and balance the following reactions and then write the ionic and net ionic equations. If all ions cancel, indicate that no reaction (N.R.) takes place.
(a) $(NH_4)_2S + NaCl \rightarrow$
(b) $Cr_2(SO_4)_3 + K_2CO_3 \rightarrow$
(c) $Sr(OH)_2 + MgCl_2 \rightarrow$
(d) $BaCl_2 + Na_2SO_3 \rightarrow$

4.43 Complete and balance the molecular, ionic, and net ionic equations for the following reactions.
(a) $HNO_3 + Cr(OH)_3 \rightarrow$
(b) $HClO_4 + NaOH \rightarrow$
(c) $Cu(OH)_2 + HC_2H_3O_2 \rightarrow$
(d) $ZnO + HBr \rightarrow$

4.44 Complete and balance molecular, ionic, and net ionic equations for the following reactions.
(a) $NaHSO_3 + HBr \rightarrow$
(b) $(NH_4)_2CO_3 + NaOH \rightarrow$
(c) $(NH_4)_2CO_3 + Ba(OH)_2 \rightarrow$
(d) $FeS + HCl \rightarrow$

4.45 How would the electrical conductivity of a solution of $Ba(OH)_2$ change as a solution of H_2SO_4 is added slowly to it? Use a net ionic equation to justify your answer.

4.46 Washing soda is $Na_2CO_3 \cdot 10H_2O$. Explain, using chemical equations, how this substance is able to remove Ca^{2+} ions from "hard water."

4.47 If a solution of trisodium phosphate, Na_3PO_4, is poured into seawater, precipitates of calcium phosphate and magnesium phosphate are formed. (Magnesium and calcium ions are among the principal ions found in seawater.) Write net ionic equations for these reactions.

*4.48 Choose reactants that would yield the following net ionic equations. Write molecular equations for each.
(a) $HCO_3^- + H^+ \rightarrow H_2O + CO_2(g)$
(b) $Fe^{2+} + 2OH^- \rightarrow Fe(OH)_2(s)$
(c) $Ba^{2+} + SO_3^{2-} \rightarrow BaSO_3(s)$
(d) $2Ag^+ + S^{2-} \rightarrow Ag_2S(s)$
(e) $ZnO(s) + 2H^+ \rightarrow Zn^{2+} + H_2O$

*4.49 Suppose that you wished to prepare copper(II) carbonate by a precipitation reaction involving Cu^{2+} and CO_3^{2-}. Which of the following pairs of reactants could you use as solutes?
(a) $Cu(OH)_2 + Na_2CO_3$
(b) $CuSO_4 + (NH_4)_2CO_3$
(c) $Cu(NO_3)_2 + CaCO_3$
(d) $CuCl_2 + K_2CO_3$
(e) $CuS + NiCO_3$

4.50 Explain why the following reactions take place.
(a) $CrCl_3 + 3NaOH \rightarrow Cr(OH)_3 + 3NaCl$
(b) $ZnO + 2HBr \rightarrow ZnBr_2 + H_2O$
(c) $MnCO_3 + H_2SO_4 \rightarrow MnSO_4 + H_2O + CO_2$
(d) $Na_2C_2O_4 + 2HNO_3 \rightarrow 2NaNO_3 + H_2C_2O_4$

Oxidation–Reduction, Oxidation Numbers

4.51 Define oxidation and reduction (a) in terms of electron transfer and (b) in terms of oxidation numbers.

4.52 In the reaction $2Mg + O_2 \rightarrow 2MgO$, which substance is the oxidizing agent and which is the reducing agent? Which substance is oxidized and which is reduced?

4.53 Why must both oxidation and reduction occur simultaneously during a redox reaction?

4.54 In the compound As_4O_6, arsenic has an oxidation number of $+3$. What is the oxidation state of the arsenic in this compound?

4.55 Assign oxidation numbers to the atoms indicated by boldface type: (a) $\mathbf{S}^{2-}$, (b) $\mathbf{S}O_2$, (c) $\mathbf{P}_4$, (d) $\mathbf{P}H_3$.

4.56 Assign oxidation numbers to the atoms indicated by boldface type: (a) $\mathbf{Cl}O_4^-$, (b) $\mathbf{Cr}Cl_3$, (c) $\mathbf{Sn}S_2$, (d) $\mathbf{Au}(NO_3)_3$.

4.57 Assign oxidation numbers to all of the atoms in the following compounds: (a) Na_2HPO_4, (b) $BaMnO_4$, (c) $Na_2S_4O_6$, (d) ClF_3.

4.58 Assign oxidation numbers to all the atoms in the following ions: (a) NO_3^-, (b) SO_3^{2-}, (c) NO^+, (d) $Cr_2O_7^{2-}$.

4.59 Assign oxidation numbers to nitrogen in the following.

(a) NO
(b) NO_2
(c) N_2O_3
(d) N_2O_5
(e) N_2H_4
(f) N_2
(g) NH_2OH
(h) NH_3
(i) NaN_3

4.60 Is the following a redox reaction? Explain.
$$2NO_2 \longrightarrow N_2O_4$$

4.61 Is the following a redox reaction? Explain.
$$2CrO_4^{2-} + 2H^+ \longrightarrow Cr_2O_7^{2-} + H_2O$$

4.62 Assign oxidation numbers to each atom in the following: (a) $NaOCl$, (b) $NaClO_2$, (c) $NaClO_3$, (d) $NaClO_4$.

4.63 Assign oxidation numbers to the elements in the following. (a) $Ca(VO_3)_2$, (b) $SnCl_4$, (c) MnO_4^{2-}, (d) MnO_2.

4.64 Assign oxidation numbers to the elements in the following. (a) PbS, (b) $TiCl_4$, (c) $Sr(IO_3)_2$, (d) Cr_2S_3.

4.65 Assign oxidation numbers to the elements in the following. (a) OF_2, (b) HOF, (c) CsO_2, (d) O_2F_2.

4.66 What is the average oxidation number of carbon in (a) C_2H_5OH (grain alcohol), (b) $C_{12}H_{22}O_{11}$ (sucrose—table sugar), (c) $CaCO_3$ (limestone), (d) $NaHCO_3$ (baking soda)?

4.67 If the oxidation number of nitrogen in a certain molecule changes from $+3$ to -2 during a reaction, is the nitrogen oxidized or reduced? How many electrons are gained (or lost) by each nitrogen atom?

4.68 For the following reactions, identify the substance oxidized, the substance reduced, the oxidizing agent, and the reducing agent.
(a) $2HNO_3 + 3H_3AsO_3 \rightarrow 2NO + 3H_3AsO_4 + H_2O$
(b) $NaI + 3HOCl \rightarrow NaIO_3 + 3HCl$
(c) $2KMnO_4 + 5H_2C_2O_4 + 3H_2SO_4 \rightarrow 10CO_2 + K_2SO_4 + 2MnSO_4 + 8H_2O$
(d) $6H_2SO_4 + 2Al \rightarrow Al_2(SO_4)_3 + 3SO_2 + 6H_2O$

4.69 For the following reactions, identify the substance oxidized, the substance reduced, the oxidizing agent, and the reducing agent.
(a) $Cu + 2H_2SO_4 \rightarrow CuSO_4 + SO_2 + 2H_2O$
(b) $3SO_2 + 2HNO_3 + 2H_2O \rightarrow 3H_2SO_4 + 2NO$
(c) $5H_2SO_4 + 4Zn \rightarrow 4ZnSO_4 + H_2S + 4H_2O$
(d) $I_2 + 10HNO_3 \rightarrow 2HIO_3 + 10NO_2 + 4H_2O$

The Ion–Electron Method

4.70 The following equation is not balanced. Why? Use the ion–electron method to balance it.
$$Ag^+ + Fe \longrightarrow Ag + Fe^{2+}$$

4.71 Balance the following half-reactions occurring in an acidic solution. Indicate whether each is an oxidation or a reduction.
(a) $BiO_3^- \rightarrow Bi^{3+}$
(b) $Pb^{2+} \rightarrow PbO_2$
(c) $NO_3^- \rightarrow NH_4^+$
(d) $Cl_2 \rightarrow ClO_3^-$

4.72 Balance the following half-reactions occurring in a basic solution. Indicate whether each is an oxidation or a reduction.
(a) $Fe \rightarrow Fe(OH)_2$
(b) $SO_2Cl_2 \rightarrow SO_3^{2-} + Cl^-$
(c) $Mn(OH)_2 \rightarrow MnO_4^{2-}$
(d) $H_4IO_6^- \rightarrow I_2$

4.73 Balance the following equations for reactions occurring in an acidic solution.
(a) $S_2O_3^{2-} + OCl^- \rightarrow Cl^- + S_4O_6^{2-}$
(b) $NO_3^- + Cu \rightarrow NO_2 + Cu^{2+}$
(c) $IO_3^- + AsO_3^{3-} \rightarrow I^- + AsO_4^{3-}$
(d) $SO_4^{2-} + Zn \rightarrow Zn^{2+} + SO_2$
(e) $NO_3^- + Zn \rightarrow NH_4^+ + Zn^{2+}$
(f) $Cr^{3+} + BiO_3^- \rightarrow Cr_2O_7^{2-} + Bi^{3+}$
(g) $I_2 + OCl^- \rightarrow IO_3^- + Cl^-$
(h) $Mn^{2+} + BiO_3^- \rightarrow MnO_4^- + Bi^{3+}$
(i) $H_3AsO_3 + Cr_2O_7^{2-} \rightarrow H_3AsO_4 + Cr^{3+}$
(j) $I^- + HSO_4^- \rightarrow I_2 + SO_2$

4.74 Balance these equations for reactions occurring in an acidic solution.
(a) $Sn + NO_3^- \rightarrow SnO_2 + NO$
(b) $PbO_2 + Cl^- \rightarrow PbCl_2 + Cl_2$
(c) $Ag + NO_3^- \rightarrow NO_2 + Ag^+$
(d) $Fe^{3+} + NH_3OH^+ \rightarrow Fe^{2+} + N_2O$
(e) $HNO_2 + I^- \rightarrow I_2 + NO$
(f) $C_2O_4^{2-} + HNO_2 \rightarrow CO_2 + NO$
(g) $HNO_2 + MnO_4^- \rightarrow Mn^{2+} + NO_3^-$
(h) $H_3PO_2 + Cr_2O_7^{2-} \rightarrow H_3PO_4 + Cr^{3+}$
(i) $VO_2^+ + Sn^{2+} \rightarrow VO^{2+} + Sn^{4+}$
(j) $XeF_2 + Cl^- \rightarrow Xe + F^- + Cl_2$

4.75 Balance equations for these reactions occurring in a basic solution.
(a) $CrO_4^{2-} + S^{2-} \rightarrow S + CrO_2^-$
(b) $MnO_4^- + C_2O_4^{2-} \rightarrow CO_2 + MnO_2$
(c) $ClO_3^- + N_2H_4 \rightarrow NO + Cl^-$
(d) $NiO_2 + Mn(OH)_2 \rightarrow Mn_2O_3 + Ni(OH)_2$
(e) $SO_3^{2-} + MnO_4^- \rightarrow SO_4^{2-} + MnO_2$
(f) $CrO_2^- + S_2O_8^{2-} \rightarrow CrO_4^{2-} + SO_4^{2-}$
(g) $SO_3^{2-} + CrO_4^{2-} \rightarrow SO_4^{2-} + CrO_2^-$
(h) $O_2 + N_2H_4 \rightarrow H_2O_2 + N_2$
(i) $Fe(OH)_2 + O_2 \rightarrow Fe(OH)_3 + OH^-$
(j) $Au + CN^- + O_2 \rightarrow Au(CN)_4^- + OH^-$

*4.76 Balance the following equations by the ion–electron method.
(a) $NBr_3 \rightarrow N_2 + Br^- + HOBr$ (basic solution)
(b) $Cl_2 \rightarrow Cl^- + ClO_3^-$ (basic solution)
(c) $H_2SeO_3 + H_2S \rightarrow S + Se$ (acidic solution)
(d) $MnO_2 + SO_3^{2-} \rightarrow Mn^{2+} + S_2O_6^{2-}$ (acidic solution)
(e) $XeO_3 + I^- \rightarrow Xe + I_2$ (acidic solution)
(f) $HXeO_4^- \rightarrow XeO_6^{4-} + Xe + O_2$ (basic solution)
(g) $(CN)_2 \rightarrow CN^- + OCN^-$ (basic solution)

4.77 Write a balanced net ionic equation for the reaction of OCl^- with $S_2O_3^{2-}$. The OCl^- is reduced to chloride ion, and the $S_2O_3^{2-}$ is oxidized to sulfate ion.

4.78 Write a balanced net ionic equation for the oxidation of $H_2C_2O_4$ by $K_2Cr_2O_7$ in an acidic solution. The reaction yields Cr^{3+} and CO_2 among the products.

Molar Concentration

4.79 Define *molar concentration*.

4.80 A solution is labeled 0.25 M HCl. Use this information to construct two conversion factors that relate moles of

solute to volume of solution expressed in liters. Rewrite the conversion factors relating moles of solute to milliliters of solution.

4.81 Calculate the molarity of a solution prepared by dissolving
(a) 4.00 g of NaOH in 100.0 mL of solution.
(b) 16.0 g of $CaCl_2$ in 250.0 mL of solution.
(c) 14.0 g of KOH in 75.0 mL of solution.
(d) 6.75 g of $H_2C_2O_4$ in 500 mL of solution.

4.82 Calculate the molarity of a solution that contains
(a) 3.60 g of H_2SO_4 in 450 mL of solution.
(b) 2.00×10^{-3} mol $Fe(NO_3)_2$ in 12.0 mL of solution.
(c) 1.65 mol HCl in 2.16 L of solution.
(d) 18.0 g HCl in 0.375 L of solution.

4.83 Calculate the number of grams of each solute that has to be taken to make each of the following solutions.
(a) 125 mL of 0.200 M NaCl
(b) 250 mL of 0.360 M $C_6H_{12}O_6$ (glucose)
(c) 250 mL of 0.250 M H_2SO_4

4.84 How many grams of solute are needed to make each of the following solutions?
(a) 250 mL of 0.100 M K_2SO_4
(b) 100 mL of 0.250 M K_2CO_3
(c) 500 mL of 0.400 M KOH

4.85 Calculate the number of moles of each of the ions in the following solutions.
(a) 35.0 mL of 1.25 M KOH
(b) 32.3 mL of 0.45 M $CaCl_2$
(c) 18.5 mL of 0.40 M $(NH_4)_2CO_3$
(d) 30.0 mL of 0.35 M $Al_2(SO_4)_3$

4.86 Calculate the concentrations of each of the ions in
(a) 0.25 M $Cr(NO_3)_2$, (b) 0.10 M $CuSO_4$, (c) 0.16 M Na_3PO_4, (d) 0.075 M $Al_2(SO_4)_3$

4.87 Calculate the concentrations of each of the ions in
(a) 0.060 M $Ca(OH)_2$, (b) 0.15 M $FeCl_3$, (c) 0.22 M $Cr_2(SO_4)_3$, (d) 0.60 M $(NH_4)_2SO_4$

4.88 A solution contains only the solute Na_3PO_4. What is the molar concentration of this salt if the sodium concentration is 0.21 M?

4.89 A solution contains $Fe_2(SO_4)_3$ as the only solute. The Fe^{3+} concentration in the solution is 0.48 M. What is the molar concentration of the sulfate ion?

Dilution of Solutions

4.90 If 25.0 mL of 0.56 M H_2SO_4 is diluted to a volume of 125 mL, what is the molarity of the resulting solution?

4.91 A 150 mL sample of 0.45 M HNO_3 is diluted to 450 mL. What is the molarity of the resulting solution?

4.92 To what volume must 25.0 mL of 18.0 M H_2SO_4 be diluted to produce 1.50 M H_2SO_4?

4.93 How many milliliters of water must be added to 150 mL of 2.5 M KOH to give a 1.0 M solution? (Assume volumes are additive.)

*4.94 How many milliliters of 0.10 M HCl must be added to 50.0 mL of 0.40 M HCl to give a final solution that has a molarity of 0.25 M?

Solution Stoichiometry

4.95 What is the molarity of an aqueous solution of potassium hydroxide if 21.34 mL is exactly neutralized by 20.78 mL of 0.116 M HCl by the following reaction?

$$KOH(aq) + HCl(aq) \longrightarrow KCl(aq) + H_2O$$

4.96 What is the molarity of an aqueous sulfuric acid solution if 12.88 mL is neutralized by 26.04 mL of 0.1024 M NaOH? The reaction is

$$2NaOH(aq) + H_2SO_4(aq) \longrightarrow Na_2SO_4(aq) + 2H_2O$$

4.97 How many milliliters of 0.25 M $NiCl_2$ solution are needed to react completely with 20.0 mL of 0.15 M Na_2CO_3 solution? How many grams of $NiCO_3$ will be formed?

4.98 Epsom salts is $MgSO_4 \cdot 7H_2O$. How many grams of this compound are needed to react completely with 50.0 mL of 0.125 M $BaCl_2$ solution?

4.99 How many milliliters of 0.100 M NaOH are needed to completely neutralize 25.0 mL of 0.250 M H_3PO_4?

4.100 How many grams of baking soda, $NaHCO_3$, are needed to react with 162 mL of stomach acid having an HCl concentration of 0.052 M?

4.101 Consider the reaction of calcium chloride with silver nitrate.
(a) Write the molecular equation for this reaction.
(b) How many milliliters of 0.250 M $CaCl_2$ would be needed to react completely with 20.0 mL of 0.500 M $AgNO_3$ solution?

4.102 How many milliliters of ammonium sulfate solution having a concentration of 0.250 M are needed to react completely with 50.0 mL of 1.00 M NaOH solution? The net ionic equation for the reaction is

$$NH_4^+(aq) + OH^-(aq) \longrightarrow NH_3(g) + H_2O$$

*4.103 Suppose that 4.00 g of solid Fe_2O_3 is added to 25.0 mL of 0.500 M HCl solution. What will the concentration of the Fe^{3+} be when all the HCl has reacted? What mass of Fe_2O_3 will not have reacted?

*4.104 Suppose that 25.0 mL of 0.440 M NaCl is added to 25.0 mL of 0.320 M $AgNO_3$.
(a) How many moles of AgCl would precipitate?
(b) What would be the concentrations of each of the ions in the reaction mixture after the reaction?

*4.105 A mixture is prepared by adding 25.0 mL of 0.185 M Na_3PO_4 to 34.0 mL of 0.140 M $Ca(NO_3)_2$.
(a) What weight of $Ca_3(PO_4)_2$ will be formed?
(b) What will be the concentrations of each of the ions in the mixture after reaction?

4.106 How many milliliters of 0.230 M $KMnO_4$ solution are needed to react completely with 40.0 mL of 0.250 M $SnCl_2$ solution, given the following chemical equation for the reaction?

$$16H^+ + 2MnO_4^- + 5Sn^{2+} \longrightarrow 5Sn^{4+} + 2Mn^{2+} + 8H_2O$$

4.107 In an acidic solution, HSO_3^- ion reacts with ClO_3^- ion to give SO_4^{2-} ion and Cl^- ion.
(a) Write a balanced net ionic equation for the reaction.
(b) How many milliliters of 0.150 M $NaClO_3$ solution are needed to react completely with 30.0 mL of 0.450 M $NaHSO_3$ solution?

4.108 Lead(IV) oxide reacts with hydrochloric acid to give chlorine. The equation for the reaction is

$$PbO_2 + 4Cl^- + 4H^+ \longrightarrow PbCl_2 + 2H_2O + Cl_2$$

How many grams of PbO_2 must react to give 15.0 g of Cl_2?

4.109 How many grams of aluminum must react to displace all the silver from 25.0 g of silver nitrate? The reaction occurs in aqueous solution.

4.110 Manganese(II) ion is oxidized to permanganate ion by bismuthate ion, BiO_3^-, in an acidic solution. In the reaction, BiO_3^- is reduced to Bi^{3+}.
(a) Write a balanced net ionic equation for the reaction.
(b) How many grams of $NaBiO_3$ are needed to oxidize the manganese in 18.5 g of $Mn(NO_3)_2$?

4.111 Sodium iodate reacts with sodium sulfite according to the equation

$$NaIO_3 + 3Na_2SO_3 \longrightarrow 3Na_2SO_4 + NaI$$

(a) In this reaction, which substance is the reducing agent?
(b) How many grams of Na_2SO_3 are needed to react with 5.00 g of $NaIO_3$?

Titrations and Chemical Analyses

4.112 Describe the following.
(a) buret   (d) end point
(b) titrant   (e) stopcock
(c) indicator   (f) standard solution

4.113 In a titration, 23.25 mL of 0.105 M NaOH was needed to react with 21.45 mL of HCl solution. What is the molarity of the acid?

4.114 A 12.5 mL sample of vinegar, containing acetic acid, $HC_2H_3O_2$, was titrated using 0.504 M NaOH solution. The titration required 20.65 mL of the base.
(a) What was the molar concentration of acetic acid in the vinegar?
(b) Assuming the density of the vinegar to be 1.0 g/mL, what was the percentage (by mass) of acetic acid in the vinegar?

4.115 Lactic aid, $HC_3H_5O_3$, is a monoprotic acid that forms when milk sours. A 18.5 mL sample of a solution of lactic acid required 17.25 mL of 0.155 M NaOH to reach an end point in a titration. How many moles of lactic acid were in the sample?

*4.116 Magnesium sulfate forms a hydrate known as Epsom salts. A student dissolved 1.24 g of this hydrate in water and added a $BaCl_2$ solution until the precipitation of $BaSO_4$ was complete. The precipitate was filtered, dried, and found to weigh 1.174 g.
(a) How many moles of $BaSO_4$ were formed?
(b) How many moles of $MgSO_4$ were in the original 1.24 g sample?
(c) How many grams of H_2O were in the original 1.24 g sample?
(d) Determine the formula of Epsom salts.

*4.117 A sample of iron chloride weighing 0.300 g was dissolved in water, and the solution was treated with $AgNO_3$ solution to precipitate the chloride as $AgCl$. After precipitation was complete, the $AgCl$ was filtered, dried, and found to weigh 0.678 g.
(a) How many grams of Cl were in the iron chloride sample?
(b) Determine the empirical formula of the iron chloride.

*4.118 A 1.500 g sample of a mixture of limestone, $CaCO_3$, and rock was pulverized and then treated with 50.0 mL of 0.200 M HCl. The mixture was warmed to expel the last traces of CO_2, and the unreacted HCl was then titrated with 0.0500 M NaOH. The volume of base required was 34.60 mL.
(a) How many moles of NaOH were used in the titration?
(b) How many moles of HCl remained after reaction with the $CaCO_3$?
(c) How many moles of $CaCO_3$ had reacted?
(d) What was the percentage by weight of $CaCO_3$ in the original sample?

*4.119 Aspirin is a monoprotic acid called acetylsalicylic acid. Its formula is $HC_9H_7O_4$. A certain pain reliever was analyzed for aspirin by dissolving 0.250 g of it in water and titrating it with 0.0300 M KOH solution. The titration required 29.40 mL of base. What is the percentage by weight of aspirin in the drug?

4.120 Hydrogen peroxide (H_2O_2) can be purchased in drug stores for use as an antiseptic. A sample of such a solution weighing 1.000 g was acidified with H_2SO_4 and titrated with a 0.02000 M solution of $KMnO_4$. The net ionic equation for the reaction is

$$6H^+ + 5H_2O_2 + 2MnO_4^- \longrightarrow 5O_2 + 2Mn^{2+} + 8H_2O$$

The titration required 17.60 mL of $KMnO_4$ solution.
(a) How many grams of H_2O_2 reacted?
(b) What is the percentage by weight of the H_2O_2 in the original antiseptic solution?

4.121 A sample of a chromium-containing alloy weighing 3.450 g was dissolved in acid, and all the chromium in the sample was oxidized to CrO_4^{2-}. It was then found that 3.18 g of Na_2SO_3 was required to reduce the CrO_4^{2-} to CrO_2^- in a basic solution, with the SO_3^{2-} being oxidized to SO_4^{2-}.
(a) Write a balanced equation for the reaction of CrO_4^{2-} with SO_3^{2-} in a basic solution.
(b) How many moles of CrO_4^{2-} reacted with the Na_2SO_3?
(c) How many grams of chromium were in the alloy sample?
(d) What was the percentage by mass of chromium in the alloy?

4.122 Solder is an alloy containing the metals tin and lead. A particular sample of this alloy, weighing 1.50 g, was dis-

solved in acid. All the tin was then converted to the +2 oxidation state. Next, it was found that 0.368 g of $Na_2Cr_2O_7$ was required to oxidize the Sn^{2+} to Sn^{4+} in an acidic solution. In the reaction the chromium was reduced to Cr^{3+} ion.
(a) Write a balanced net ionic equation for the reaction between the Sn^{2+} and $Cr_2O_7^{2-}$ in an acidic solution.
(b) Calculate the number of grams of tin that were in the sample of solder.
(c) What was the percentage by mass of tin in the solder?

4.123 Sodium nitrite, $NaNO_2$, is used as a preservative in meat products such as frankfurters and bologna. In an acidic solution, nitrite ion is converted to nitrous acid, HNO_2, which reacts with permanganate ion according to the equation.

$$H^+ + 5HNO_2 + 2MnO_4^- \longrightarrow 5NO_3^- + 2Mn^{2+} + 3H_2O$$

A 1.000 g sample of a water-soluble solid containing $NaNO_2$ was dissolved in dilute H_2SO_4 and titrated with 0.01000 M $KMnO_4$ solution. The titration required 12.15 mL of the $KMnO_4$ solution. What was the percentage of $NaNO_2$ in the original 1.000 g sample?

4.124 Both calcium chloride, $CaCl_2$, and sodium chloride are used to melt ice and snow on roads in the winter. A certain company was marketing a mixture of these two compounds for this purpose. A chemist, wishing to analyze the mixture, dissolved 2.463 g of it in water and precipitated the calcium by adding sodium oxalate, $Na_2C_2O_4$.

$$Ca^{2+} + C_2O_4^{2-} \longrightarrow CaC_2O_4(s)$$

The calcium oxalate was then carefully filtered from the solution, dissolved in sulfuric acid, and titrated with 0.1000 M $KMnO_4$ solution. The reaction that occurred was

$$6H^+ + 5H_2C_2O_4 + 2MnO_4^- \longrightarrow 10CO_2 + 2Mn^{2+} + 8H_2O$$

The titration required 21.62 mL of the $KMnO_4$ solution.
(a) How many moles of $C_2O_4^{2-}$ were present in the CaC_2O_4 precipitate?
(b) How many grams of $CaCl_2$ were in the original 2.463 g sample?
(c) What was the percentage of $CaCl_2$ in the sample?

4.125 A way to analyze a sample for nitrite ion is to acidify a solution containing NO_2^- and then allow the HNO_2 that is formed to react with iodide ion in the presence of excess I^-. The reaction is

$$2HNO_2 + 2H^+ + 3I^- \longrightarrow 2NO + 2H_2O + I_3^-$$

Then the I_3^- is titrated with $Na_2S_2O_3$ solution using starch as an indicator.

$$I_3^- + 2S_2O_3^{2-} \longrightarrow 3I^- + S_4O_6^{2-}$$

In a typical analysis, a 1.104 g sample that was known to contain $NaNO_2$ was treated as described above. The titration required 29.25 mL of 0.3000 M $Na_2S_2O_3$ solution.
(a) How many moles of I_3^- had been produced in the first reaction?

(b) How many moles of NO_2^- had been in the original 1.104 g sample?

(c) What was the percentage of $NaNO_2$ in the original sample?

4.126 A solution containing 0.1244 g of $K_2C_2O_4$ was acidified, changing the $C_2O_4^{2-}$ ions to $H_2C_2O_4$. The solution was then titrated with 13.93 mL of a $KMnO_4$ solution to reach a faint pink end point.

(a) Write the net ionic equation for the reaction involved in the titration.

(b) What causes the solution to become slightly pink at the end point in this titration?

(c) What was the molarity of the $KMnO_4$ solution used in the titration?

4.127 A sample of a copper ore with a mass of 0.4225 g was dissolved in acid. A solution of potassium iodide was added, which caused the reaction

$$2Cu^{2+}(aq) + 5I^-(aq) \longrightarrow I_3^-(aq) + 2CuI(s)$$

The I_3^- that was formed required 29.96 mL of 0.02100 M $Na_2S_2O_3$ in a titration in which the reaction was

$$I_3^-(aq) + 2S_2O_3^{2-}(aq) \longrightarrow 3I^-(aq) + S_4O_6^{2-}(aq)$$

(a) What was the percentage by weight of copper in the ore?

(b) If the ore contained $CuCO_3$, what was the percentage by weight of $CuCO_3$ in the ore?

Additional Problems

4.128 Classify each of the following as a strong electrolyte, weak electrolyte, or nonelectrolyte. *Contributed by Prof. David Dobberpuhl, Creighton Univ.*

(a) KCl
(b) glycerin
(c) NaOH
(d) $C_{12}H_{22}O_{11}$ (sucrose, or table sugar)
(e) $HC_2H_3O_2$ (acetic acid)
(f) CH_3OH (methyl alcohol)
(g) H_2SO_4
(h) NH_3

4.129 Write the net ionic equations for the following reactions. Include the designations for the states—solid, liquid, gas, or aqueous solution. (For any substance soluble in water, assume that it is used as its aqueous solution.)

(a) $CaCO_3 + HNO_3 \rightarrow Ca(NO_3)_2 + CO_2 + H_2O$
(b) $CaCO_3 + H_2SO_4 \rightarrow CaSO_4 + CO_2 + H_2O$
(c) $FeS + HBr \rightarrow H_2S + FeBr_2$
(d) $KOH + SnCl_2 \rightarrow KCl + Sn(OH)_2$

4.130 Examine each pair of substances, determine whether they will react together, and then write the molecular and net ionic equations for any reactions that you predict.

(a) $Na_2S(aq)$ and $H_2SO_4(aq)$
(b) $LiHCO_3(aq)$ and $HNO_3(aq)$
(c) $(NH_4)_3PO_4(aq)$ and $KOH(aq)$
(d) $K_2SO_3(aq)$ and $HCl(aq)$
(e) $BaCO_3(s)$ and $HBr(aq)$
(f) $CaCl_2(aq)$ and $HNO_3(aq)$

*4.131 A 0.113 g sample of a mixture of sodium chloride and sodium nitrate was dissolved in water, and enough silver nitrate solution was added to precipitate all of the chloride ion. The mass of silver chloride obtained was 0.277 g.

(a) What mass of sodium chloride was in the sample?

(b) What mass percentage of the sample was sodium chloride?

*4.132 A mixture of sodium carbonate and sodium chloride with a mass of 0.326 g was dissolved in water and titrated with 0.116 M HCl until no more CO_2 could be removed from the solution. By then, 43.6 mL of the standard HCl had been added. Calculate the percentage by mass of the sodium carbonate in the mixture.

*4.133 Qualitative analysis of an unknown acid found only carbon, hydrogen, and oxygen. Quantitative analysis gave the following data. A 10.46 mg sample, when burned in oxygen, gave 22.17 mg CO_2 and 3.40 mg H_2O. Its molecular weight was determined to be 166. When a 0.1680 g sample of the acid was titrated with 0.1250 M NaOH, the end point was reached after 16.18 mL of the base had been added.

(a) Calculate the percentage composition of the acid.
(b) What is its empirical formula?
(c) What is its molecular formula?
(d) Is the acid mono-, di-, or triprotic?

*4.134 A solution contains $Ce(SO_4)_3^{2-}$ at a concentration of 0.0150 M. It was found that in a titration, 25.00 mL of this solution reacted completely with 23.44 mL of 0.032 M $FeSO_4$ solution. The reaction gave Fe^{3+} as a product in the solution. In this reaction, what is the final oxidation state of the Ce?

*4.135 Sulfur dioxide is a component of air pollution that can be determined using an acid–base titration. An air sample containing SO_2 is bubbled through a solution containing hydrogen peroxide, H_2O_2. The SO_2 reacts with the H_2O_2 to produce H_2SO_4, $SO_2 + H_2O_2 \rightarrow H_2SO_4$. The sulfuric acid is then titrated against a standardized solution of sodium hydroxide. A 2.2 m³ sample of polluted air having a density of 1.19 g/L is bubbled through a solution containing 0.3% H_2O_2. The resulting solution requires 32.48 mL of 0.01042 M NaOH to completely neutralize the resulting sulfuric acid. Calculate the number of milligrams of SO_2 in the initial air sample. Also report the concentration of SO_2 in the air sample using units of parts per million (ppm = mg SO_2/kg air.)

TEST OF FACTS AND CONCEPTS: CHAPTERS 1–4

Many of the fundamental concepts and problem-solving skills developed in the preceding chapters will carry forward into the rest of this book. Therefore, we recommend that you pause here to see how well you have grasped the concepts, how familiar you are with important terms, and how able you are at working chemical problems.

Some of the problems here require data or other information found in tables in this book, including those inside the covers. Freely use these tables as needed. For problems that require mathematical solutions, we recommend that you first assemble the necessary information in the form of equivalencies and then use them to set up appropriate conversion factors needed to obtain the answers.

1. A rectangular box was found to be 24.6 cm wide, 0.35140 m high, and 7424 mm deep.
(a) How many significant figures are in each measurement?
(b) Calculate the volume of the box in units of cm^3. Be sure to express your answer to the the correct number of significant figures.
(c) Use the answer in part (b) to calculate the volume of the box in cubic feet.
(d) Suppose the box were solid and composed entirely of zinc, which has a specific gravity of 7.140. What would be the mass of the box in kilograms?

2. The atoms of an isotope of plutonium, Pu, each contain 94 protons, 110 neutrons, and 94 electrons. Write a symbol for this element that incorporates its mass number and atomic number.

3. Give chemical formulas for the following.
(a) potassium nitrate
(b) calcium carbonate
(c) cobalt(II) phosphate
(d) lithium hydrogen sulfate
(e) magnesium sulfite
(f) iron(III) bromide
(g) magnesium nitride
(h) aluminum selenide
(i) cupric perchlorate
(j) sulfurous acid
(k) nitrous acid
(l) hypoiodous acid
(m) barium bisulfite
(n) bromine pentafluoride
(o) dinitrogen pentaoxide
(p) strontium acetate
(q) ammonium dichromate
(r) copper(I) sulfider

4. Give chemical names for the following.
(a) $NaClO_3$
(b) HIO_3
(c) $Ca(H_2PO_4)_2$
(d) $NaMnO_4$
(e) AlP
(f) ICl_3
(g) PCl_3
(h) $HC_2H_3O_2$
(i) K_2CrO_4
(j) $HOCl$
(k) $Ca(CN)_2$
(l) $MnCl_2$
(m) $NaNO_2$
(n) $Fe(HSO_4)_3$

5. Write the formulas of any acid salts that could form by the reaction of the following acids with potassium hydroxide.

(a) sulfurous acid
(b) nitric acid
(c) hypochlorous acid
(d) phosphoric acid
(e) carbonic acid

6. The formula mass of a substance is 60.2. Therefore, what is the mass in grams of one of its molecules?

7. A sample of a compound with a mass of 204 g consists of 1.00×10^{23} molecules. What is its formula mass?

8. Calculate the formula mass of $Fe_4[Fe(CN)_6]_3$.

9. How many grams of copper(II) nitrate trihydrate, $Cu(NO_3)_2 \cdot 3H_2O$, are present in 0.118 mol of this compound?

10. How many moles of nickel(II) iodide hexahydrate, $NiI_2 \cdot 6H_2O$, are in a sample of 15.7 g of this compound?

11. A sample of 0.5866 g of nicotine was analyzed and found to consist of 0.4343 g C, 0.05103 g H, and 0.1013 g N. Calculate the percentage composition of nicotine.

12. A compound of potassium had the following percentage composition: K, 37.56%; H, 1.940%; P, 29.79%. The rest was oxygen. Calculate the empirical formula of this compound (arranging the atomic symbols in the order K H P O.)

13. How many molecules of ethyl alcohol, C_2H_5OH, are in 1.00 fluid ounce of the liquid? The density of ethyl alcohol is 0.798 g/mL (1 oz = 29.6 mL).

14. What volume in liters is occupied by a sample of ethylene glycol, $C_2H_6O_2$, that consists of 5.00×10^{24} molecules? The density of ethylene glycol is 1.11 g/mL.

15. If 2.56 g of chlorine, Cl_2, are to be used to prepare dichlorine heptoxide, Cl_2O_7, how many moles and how many grams of oxygen, O_2, are needed?

16. Balance the following equations.
(a) $Fe_2O_3 + HNO_3 \rightarrow Fe(NO_3)_3 + H_2O$
(b) $C_{21}H_{30}O_2 + O_2 \rightarrow CO_2 + H_2O$

17. How many moles of nitric acid, HNO_3, are needed to react with 2.56 mol of Cu in the following reaction?

$$3Cu + 8HNO_3 \longrightarrow 3Cu(NO_3)_2 + 2NO + 4H_2O$$

18. Under the right conditions, ammonia can be converted to nitric oxide, NO, by the following reaction.

$$4NH_3 + 5O_2 \longrightarrow 4NO + 6H_2O$$

How many moles and how many grams of O_2 are needed to react with 56.8 g of ammonia by this reaction?

19. To neutralize the acid in 10.0 mL of 18.0 M H_2SO_4 that was accidentally spilled on a laboratory bench top, solid sodium bicarbonate was used. The container of sodium bicarbonate was known to weigh 155.0 g before this use and

out of curiosity its mass was measured as 144.5 g afterwards. The reaction is

$$H_2SO_4 + 2NaHCO_3 \longrightarrow Na_2SO_4 + 2CO_2 + 2H_2O$$

Was sufficient sodium bicarbonate used? Determine the limiting reactant and calculate the maximum yield in grams of sodium sulfate.

20. How many milliliters of concentrated sulfuric acid (18.0 M) are needed to prepare 125 mL of 0.144 M H_2SO_4?

21. The density of concentrated phosphoric acid solution is 1.689 g solution/mL solution at 20°C. It contains 144 g H_3PO_4 per 1.00×10^2 mL of solution.
(a) Calculate the molarity of H_3PO_4 in this solution.
(b) Calculate the number of grams of this solution required to hold 50.0 g H_3PO_4.

22. A mixture consists of lithium carbonate (Li_2CO_3) and potassium carbonate (K_2CO_3). These react with hydrochloric acid as follows.

$$Li_2CO_3(s) + 2HCl(aq) \longrightarrow 2LiCl(aq) + H_2O + CO_2(g)$$
$$K_2CO_3(s) + 2HCl(aq) \longrightarrow 2KCl(aq) + H_2O + CO_2(g)$$

When 4.43 g of this mixture was analyzed, it consumed 53.2 mL of 1.48 M HCl. Calculate the number of grams of each carbonate and their percentages.

23. Dolomite is a mineral consisting of calcium carbonate ($CaCO_3$) and magnesium carbonate ($MgCO_3$). When dolomite is strongly heated, its carbonates decompose to their oxides (CaO and MgO) and carbon dioxide is expelled.
(a) Write the separate equations for these decompositions of calcium carbonate and magnesium carbonate.
(b) When a dolomite sample with a mass of 5.78 g was heated strongly, the residue had a mass of 3.02 g. Calculate the masses in grams and the percentages of calcium carbonate and magnesium carbonate in this sample of dolomite.

24. Adipic acid, $C_6H_{10}O_4$, is a raw material for making nylon, and it can be prepared in the laboratory by the following reaction between cyclohexene, C_6H_{10}, and sodium dichromate, $Na_2Cr_2O_7$, in sulfuric acid.

$$3C_6H_{10}(l) + 4Na_2Cr_2O_7(aq) + 16H_2SO_4(aq) \longrightarrow$$
$$3C_6H_{10}O_4(s) + 4Cr_2(SO_4)_3(aq) + 4Na_2SO_4(aq) + 16H_2O$$

There are side reactions. These plus losses of product during its purification reduce the overall yield. A typical yield of purified adipic acid is 68.6%.
(a) To prepare 12.5 g of adipic acid in 68.6% yield requires how many grams of cyclohexene?
(b) The only available supply of sodium dichromate is its dihydrate, $Na_2Cr_2O_7 \cdot 2H_2O$. (Since the reaction occurs in an aqueous medium, the water in the dihydrate causes no problems, but it does contribute to the mass of what is taken of this reactant.) How many grams of this dihydrate are also required in the preparation of 12.5 g of adipic acid in a yield of 68.6%?

25. When 6.584 g of one of the hydrates of sodium sulfate

(Na_2SO_4) was heated so as to drive off all of its water of hydration, the residue of anhydrous sodium sulfate had a mass of 2.889 g. What is the formula of the hydrate?

26. One way to prepare iodine is to let sodium iodate, $NaIO_3$, react with hydroiodic acid, HI. The following reaction occurs.

$$NaIO_3 + 6HI \longrightarrow 3I_2 + NaI + 3H_2O$$

Calculate the number of moles and the number of grams of iodine that can be made this way from 16.4 g of $NaIO_3$.

27. *C.I. Pigment Yellow 45* ("sideran yellow") is a pigment used in ceramics, glass, and enamel. When analyzed, a 2.164 g sample of this substance was found to contain 0.5259 g of Fe and 0.7345 g of Cr. The remainder was oxygen. Calculate the empirical formula of this pigment.

28. How many grams of 4.00% solution of KOH in water are needed to neutralize completely the acid in 10.0 mL of 0.256 M H_2SO_4?

29. Calculate the molar concentration of 15.00% Na_2CO_3 solution at 20.0°C given that its density is 1.160 g/mL.

30. What is the difference between a strong electrolyte and a weak electrolyte? Formic acid, $HCHO_2$, is a weak acid. Write a chemical equation showing its reaction with water.

31. Methylamine, CH_3NH_2, is a weak base. Write a chemical equation showing its reaction with water.

32. A certain toilet cleaner uses $NaHSO_4$ as its active ingredient. In an analysis, 0.500 g of the cleaner was dissolved in 30.0 mL of distilled water and required 24.60 mL of 0.105 M NaOH for complete neutralization in a titration. The net ionic equation for the reaction is

$$HSO_4^- + OH^- \longrightarrow H_2O + SO_4^{2-}$$

What was the percentage by weight of $NaHSO_4$ in the cleaner?

33. A volume of 28.50 mL of a freshly prepared solution of KOH was required to titrate 50.00 mL of 0.0922 M HCl solution. What was the molarity of the KOH solution?

34. Assign oxidation numbers to the atoms in the following formulas: (a) As_4, (b) $HClO_2$, (c) $MnCl_2$, (d) $V_2(SO_3)_3$.

35. For the following unbalanced equations, write the reactants and products in the form they should appear in an ionic equation. Then write balanced net ionic equations by applying the ion–electron method.
(a) $K_2Cr_2O_7 + HCl \rightarrow KCl + Cl_2 + H_2O + CrCl_3$
(b) $KOH + SO_2(aq) + KMnO_4 \rightarrow K_2SO_4 + MnO_2 + H_2O$

36. Balance the following equations by the ion–electron method for *acidic solutions*.
(a) $Cr_2O_7^{2-} + Br^- \rightarrow Br_2 + Cr^{3+}$
(b) $H_3AsO_3 + MnO_4^- \rightarrow H_2AsO_4^- + Mn^{2+}$

37. Balance the following equations by the ion–electron method for *basic solutions*.
(a) $I^- + CrO_4^{2-} \rightarrow CrO_2^- + IO_3^-$
(b) $SO_2 + MnO_4^- \rightarrow MnO_2 + SO_4^{2-}$

38. In the previous two questions, identify the oxidizing agents and reducing agents.

It takes all of a one-mile run to work off the calories in only one slice of unbuttered bread. Nature, even at Mesa Arch, Canyonlands National Park, Utah, is unyielding about such matters, which fall in the realm of thermochemistry, the major topic of this chapter.

Chapter 5

Energy and Thermochemistry

5.1
KINETIC AND POTENTIAL ENERGY REVISITED

Matter and energy define the fundamental domain of chemistry, and the two are intimately intertwined. Virtually all chemical reactions give off or absorb energy (usually heat) as they occur, but we do not often indicate the energy in an equation. You might see the equation for the combustion of carbon written as follows.

$$C(s) + O_2(g) \longrightarrow CO_2(g)$$

But the equation would be more informative if we had some way to indicate in the equation that heat is evolved. For example, the equation could show heat as a *product*.

$$C(s) + O_2(g) \longrightarrow CO_2(g) + \text{heat}$$

The decomposition of potassium chlorate ($KClO_3$), on the other hand, does not proceed without a steady supply of heat from the outside. The equation for this reaction, therefore, could include heat as a *reactant*.

$$2KClO_3(s) + \text{heat} \longrightarrow 2KCl(s) + 3O_2(g)$$

Just including ''heat'' somewhere in an equation, however, isn't very satisfying; we want to include the quantity of heat. One of the questions we must consider in this chapter is how to do this.

The study of the energies given off or absorbed by chemical reactions, a field called **thermochemistry,** has not only given us many useful applications but also important clues concerning the nature of matter itself. Because thermochemical concepts rely heavily on those of kinetic and potential energy, introduced in Chapter 1, a brief review of the latter is warranted before we move into thermochemistry itself.

Kinetic and Potential Energy

As we learned in Section 1.5, energy is not a *physical* thing to be weighed and bottled. Rather it's an *ability* possessed by things, the ability to do work. We

189

Kinetic is from the Greek *kinetikos,* meaning "of motion."

learned that objects can have energy in two fundamental ways, kinetic and potential. *Kinetic energy* or KE, the energy of motion associated with mechanical work, is calculated from a moving object's mass (m) and velocity (v) by the equation $KE = \frac{1}{2}mv^2$. *Potential energy,* or PE, is energy "in storage," existing *because there are in the universe natural attractions or repulsions that objects or their parts experience for each other.* When you lift a book off your desk, for example, you convert kinetic energy (associated with your motions) into stored or potential energy. The book and the Earth have a natural force of attraction for each other called the gravitational force, so you have to use some of your own kinetic energy to increase the distance between the Earth and the book. Your energy, now transformed into potential energy possessed by the book, can be recovered as kinetic energy by dropping the book. The higher you move the book from the floor, the more potential energy you give it, and the more kinetic energy released when you drop the book. Thus, *one very important property of potential energy is that it can be converted into kinetic energy.*

Springs store energy when they are compressed, to give another example, and the stored energy is put into the spring by the person or machine compressing it. The kinetic energy of the compressing work is converted into the potential energy of the spring. When the spring is released, it might push on something and thus transform its potential energy back into kinetic energy and mechanical work.

Winding a watch puts some of your energy into storage in the mainspring. This energy then changes to the kinetic energy of the moving hands.

These two examples illustrate the two basic mechanisms for increasing the supply of energy stored as potential energy. *Potential energy increases whenever objects that attract each other are pulled apart,* as in lifting a book to a higher level above the Earth. *Potential energy also increases whenever objects that naturally experience a repelling force are pushed together,* as in compressing a spring.

The examples also suggest the two mechanisms by which potential energy can decrease. Potential energy decreases in the fall of the book, because objects that attract each other are allowed to act. It also decreases in the release of the compressed spring, because things that repel are allowed to do so.

Another important aspect of energy introduced in Chapter 1 was the *law of conservation of energy:* energy can be neither created nor destroyed, but only changed from one form to another.[1] An implication of this concept is that *the total energy of the universe is constant.*

$$\text{Total energy} = KE + PE$$

Regardless of the kinds of changes and transformations of KE into PE or PE into KE, the total energy remains unchanged.

Chemical Bonds and Chemical Energy

The relationship of potential energy to attractions and repulsions is extremely important to the study of matter and its changes, because matter is made of

---

[1] The famous physicist Albert Einstein showed that matter and energy are interconvertible, and in some circumstances matter can actually be changed to energy and vice versa. This happens, for example, in nuclear reactions, which are discussed in Chapter 20. The law of conservation of energy, therefore, should really be stated as the *law of conservation of mass–energy:* The total mass and energy of the universe is a constant. However, for chemical reactions under normal conditions the quantity of such interconversion, mass to energy or energy to mass, is entirely negligible.

things that have natural attractions and repulsions for each other. As we mentioned in Section 2.3, the pieces held together in compounds are atomic *nuclei* and *electrons*. The nuclei are positively charged because they carry an atom's protons, and *electrons* are negatively charged. Because oppositely charged particles attract each other and like-charged particles repel each other, compounds made up of two or more nuclei and electrons have built into them both attractive and repulsive forces. The electrons and nuclei within atoms and molecules, for example, *attract* each other. Electrons, however, *repel* other electrons, and atomic nuclei *repel* other atomic nuclei. Therefore, to hold together the pieces of atoms present in chemical compounds, some net attractive force must be at work. The term *chemical bond* is simply the name we give to this net attractive force.

We'll have much more to say about bonds in Chapters 7 and 8, but remember that forces are related to energy in one form or another. What is important, therefore, about chemical bonds for the present chapter is simply that they give rise to a compound's potential energy. The potential energy that resides in chemical bonds is called *chemical energy*. Thus, the chemical energy in things like firecrackers or gasoline is associated with the chemical bonds within them. The chemical energy in gasoline contributes much to the overall potential energy of a parked car.

5.2 KINETIC THEORY OF MATTER

The concept introduced in Section 1.5 that *heat* is actually a form of kinetic energy is part of a major understanding of matter called the **kinetic theory of matter.** Its central idea, both simple and useful for the current chapter, is that atoms and molecules are in constant motion of some sort, either moving from one place to another or simply jiggling and vibrating in place. Atoms and molecules, we learned, have *molecular kinetic energy.*

Physical evidence for the motions of atoms and molecules comes from several sources and observations, some of which you have experienced yourself. The fragrance of a drop of perfume or cologne, when placed at one corner of a room, even in still air, will eventually be noticed by someone at the opposite corner. The only way that the molecules responsible for the fragrance can make this trip from corner to corner in a room without drafts is by actually being in motion themselves. Because of collisions between moving molecules, these motions are quite *random,* going in every conceivable direction. A given molecule does not go far before it hits another and careens off on a different course. So it may take a given molecule a long time to make much net progress in one direction. With time, however, the random motions and collisions of the perfume's molecules result in their being uniformly mixed with all of the air in the room.

Random motions of molecules occur in liquids, too. When you add sugar without stirring to hot coffee, the sugar molecules eventually become equally distributed among the other molecules in the cup as they move randomly about. The reason that the whole cup of coffee does not rattle around on the table is because the effects of the random molecular motions within the coffee cancel each other out entirely. On the average, for every molecule moving upward there are some moving in every other possible direction.

The particles in solids do not move from place to place, but they do vibrate about their positions within the solid.

Between collisions, molecules in air at room temperature move with an average speed of about 400 m s^{-1} (900 mph)!

Not all collisions are head-on collisions. It's like a demolition derby, or bumper cars at an amusement park.

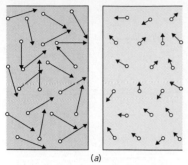

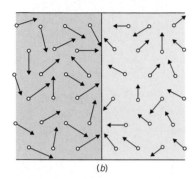

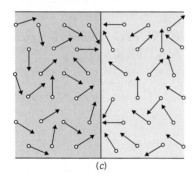

FIGURE 5.1

Energy transfer from a warmer to a cooler object. (*a*) The longer arrows on the left denote higher kinetic energies and so something warmer, like hot water just before it's poured into a cooler coffee cup. (*b*) The hot water and the cup's inner surface are now in thermal contact. Energetic water molecules are slowed down by collisions with molecules in the cup's material. Thus energy transfers from the water to the material of the cup. (*c*) Thermal equilibrium; the average kinetic energies of the molecules of both materials are now the same, so their temperatures are now equal.

Temperature Change and Average Molecular Kinetic Energy

One of the most important measurements taken in thermochemistry involves the initial and final temperatures of a mixture undergoing a chemical reaction. What a temperature change means for the energy change harkens back to the meaning of temperature, introduced in Section 1.5.

The particles making up a given sample of matter, like a gas, do not all move with identical speeds. We can imagine, however, an *average speed* of the particles, and we can therefore think about their *average kinetic energy*. As mentioned in Section 1.5, *the absolute or Kelvin temperature of any object is directly proportional to the constituent particles' average molecular kinetic energy*. In a very cold object, the average molecular kinetic energy is small. At absolute zero, 0 K, the average *kinetic* energy is actually zero, because the molecules have no velocities; they're not moving around at all. At higher and higher temperatures, the average molecular kinetic energies become correspondingly greater. We understand a temperature *change,* therefore, as a change in the *average* molecular kinetic energy of the particles that make up the object. This is an important relationship. A temperature change signals an energy change, specifically one involving the transfer of *heat.*

The Kinetic Theory and the Transfer of Heat

Earlier, energy was defined as the ability to do work. It is thus the ability to cause a change of some sort. Anyone who has done any cooking knows that heat can cause chemical changes, but our most basic understanding of what heat is considers the kinds of *physical* changes that involve heat. **Heat** is able to cause two kinds of physical change, namely, changes in temperature and changes in the physical states of matter (like melting a solid to a liquid and boiling a liquid to a gas). Thus, an object can possess heat energy because of its ability to change the temperature of another, *colder* object when the two are brought into some kind of contact. Although it seems as if something flows from the warmer to the colder object, we do not think of heat as a "thing." Instead, we understand "heat" as another term for *molecular kinetic energy.* The heat possessed by an object is thus the total of all of the individual values of the molecular kinetic energies possessed by its constituent particles. The temperature of the object, remember, is a measure of the *average* molecular kinetic energy of its constituent particles.

These concepts let us understand better how heat can transfer. When you pour boiling water into a cool cup, for example, the cup will soon be too hot to hold (except by its handle). Figure 5.1 shows how the transfer of energy occurs at the molecular level. High-energy water molecules strike the lower energy particles in the cup's material. These collisions give some of the molecular kinetic energy of the water molecules to the cup's molecules, which vibrate more energetically. The water molecules slow down as a result, and have less average molecular kinetic energy. By this mechanism, some of the molecular kinetic energy in the hot water transfers to the material of the cup. The temperature of the water decreases and the temperature of the cup itself increases. Of course, in the air next to the cup's outer surface, air molecules are constantly striking the cup. Following most of such collisions with the warmer cup, the air molecules rebound with a higher average molecular kinetic energy than they had. The temperature of the air immediately around the cup, therefore, increases.

The Energy Consequences of Bond Making and Breaking

Chemical reactions generally involve *both* the breaking and making of chemical bonds (Section 2.6). When bonds *form,* things that attract each other are enabled to do so, which tends to decrease the potential energy of the reacting system. When bonds *break,* on the other hand, things that normally attract each other are forced apart, which increases the potential energy of the reacting system. Each reaction, therefore, has a certain net overall energy change, a net balance between the "costs" of breaking bonds and the "profits" from making them.

In many reactions, the products have *less* chemical energy than the reactants. When the gasoline in an engine burns, for example, molecules of relatively large chemical energy change into products with less carbon dioxide and water. The difference in chemical energy between reactants and products appears partly as the mechanical (kinetic) energy that drives the car and partly as waste heat released largely via the cooling system.

In some reactions the products have *more* chemical energy than the reactants. An example is the production of glucose, $C_6H_{12}O_6$, and oxygen in plants from carbon dioxide and water by the process called *photosynthesis.* The energy increase is supplied by solar energy, which is absorbed by the plant's green pigment, chlorophyll.

$$6CO_2 + 6H_2O + \text{solar energy} \xrightarrow[\text{(several steps)}]{\text{chlorophyll}} C_6H_{12}O_6 + 6O_2$$

Heat, Temperature Change, and Spontaneous Chemical Reactions

Many reactions, like the combustion of gasoline, give off heat. To study further the origin of this heat, we need to define three terms that we will use frequently: *system, surroundings,* and *boundary.* By **system** we mean the particular part of the universe that we have decided to study, such as a mixture undergoing a chemical reaction in a beaker. Literally everything else in the universe besides the system constitutes the system's **surroundings.** Separating the system from its surroundings is a **boundary,** which might be real (like the walls of a beaker) or imaginary (like the imaginary envelope between the atmosphere's troposphere and the stratosphere). We have an **insulated system** or a *closed system* when energy cannot cross the system's boundary in any direction.

The total energy of an insulated system, the sum of its molecular kinetic and potential energies, is constant (law of conservation of energy). Transformations of one energy form into the other, however, occur. When a chemical reaction, like the combustion of gasoline, occurs in an *insulated* system, the lowering of the system's potential energy must be balanced by an increase in its molecular kinetic energy. (Otherwise, the total energy would not be constant.) Increasing the average molecular kinetic energy means increasing the system's temperature. Hence when a spontaneous reaction in an *insulated* system produces an increase in the average molecular kinetic energy by a lowering of the potential energy, the temperature of the system increases. *The only way that an insulated system's temperature can increase spontaneously is by a decrease in the system's potential energy and an increase in its molecular kinetic energy.*

When the same reaction is carried out in an *uninsulated* container, the buildup of molecular kinetic energy (heat) can be offset by a flow of heat *out of*

5.3
ENERGY CHANGES IN CHEMICAL REACTIONS

When we study energy changes in chemical reactions, our concern is with changes in *potential energy* (chemical energy), not kinetic energy.

Solar energy is light energy from the sun. Light energy is studied in Section 6.1.

The troposphere is the zone of the atmosphere nearest the Earth.

Spontaneous means happening without sustained human intervention.

exo means "out"
endo means "in"
therm means "heat"

If it "flows" at all, heat always flows spontaneously from a region of higher temperature to one of lower temperature; always.

the system into the cooler surroundings. Because of the direction of this heat flow, the reaction is said to be **exothermic;** heat is flowing *out* of the system.

Similarly, systems undergoing reactions in *insulated* vessels leading to an increase in potential energy at the expense of molecular kinetic energy become cooler when they occur. *The only way that a closed or insulated system's temperature can decrease spontaneously is by an increase in the system's potential energy and a decrease in its molecular kinetic energy.* In an open or *uninsulated* vessel where cooling is occurring, heat will tend to flow from the warmer surroundings into the container. Such reactions are described as being **endothermic;** heat is flowing *into* the system. Let's summarize these important relationships.

1. During an **exothermic** reaction, the system becomes warmer and a net decrease of the potential energy of the substances occurs. If uninsulated, the system gives off heat to the surroundings.

2. During an **endothermic** reaction, the system becomes cooler and a net increase in the potential energy of the substances occurs. If uninsulated, the system receives heat from the surroundings.

Usually, chemical reactions are not well-insulated from their surroundings, and heat flows into or out of such systems. Whenever the temperature of a reaction mixture increases above that of the surroundings, heat flows out. When a reaction causes the temperature to become lower than that of the surroundings, heat flows in.

5.4 HEATS OF REACTION; CALORIMETRY

The Special Place of Heat as a Form of Energy

One of the important facts about our world is that *all forms of energy can be converted quantitatively into heat.* The mechanical energy of a moving car, for example, is converted into heat by the frictional action of the brakes, and the brake shoes and drums can become very hot. When a current of electrons is forced into something with high electrical resistance, like the heating element in a toaster, the electrical energy of the current changes entirely into heat.

The opposite is not true; heat cannot be converted *quantitatively* into another form of energy. The heat generated, for example, by burning coal at an electrical power plant cannot *even in principle* be changed 100% into electrical energy. We must leave to Chapter 13 any further discussion of this major limitation set by nature on what we can do. Our interest here is limited. It concerns only the fact that any or all forms of energy can be changed entirely into heat *and so measured this way.*

One way that specialists in thermochemistry measure the heat generated by an exothermic reaction is to let the heat flow into a known mass of cooler water and measure the increase in the temperature of the water. To calculate the heat that transfers from the temperature change of the water, we need to know something about the capacity of water to absorb heat. This capacity is one of the *thermal properties of water.* The **thermal properties** of a substance are those that describe its ability to absorb or release heat without changing chemically. They include *heat capacity, specific heat,* and *molar heat capacity.*

Specific Heat and Heat Capacity

Every object has a **heat capacity,** defined as *the amount of heat energy required to raise its temperature one degree Celsius.* Heat capacity is the ratio of the amount of heat absorbed by the object to the change in temperature, Δt. (The symbol Δ indicates a *change* in a quantity. Recall that we use the lowercase t for Celsius temperature reserving T for kelvins.)

$$\text{Heat capacity} = \frac{\text{heat absorbed}}{\Delta t}$$

The units of heat capacity are joules per degree Celsius or $J\,°C^{-1}$ (which can be written as $J/°C$). For larger amounts of heat, we think in terms of kilojoules per degree Celsius, $kJ\,°C^{-1}$ (or $kJ/°C$).

When we generalize from "heat *absorbed*" to heat flow in any direction, whether absorbed or released, we have the following simple relationship:

$$\text{Heat capacity} \times \Delta t = \text{heat energy} \qquad (5.1)$$

Heat capacity is an extensive property, meaning that the heat capacity of an object depends not only on its *materials* but also on its *mass*. An object with a large mass of water, like a lake, for example, obviously needs more heat to warm up by 1 °C than does a cup of lake water. One gram of water has a heat capacity of roughly $4\,J\,°C^{-1}$, whether the water is in a cup or a lake, but 250 g of water (about 1 cup) has a heat capacity 250 times as great as that of 1 g of water.

Each substance has an *intensive* thermal property related to heat capacity, an *inherent* ability to accept or to release heat, and which is called the substance's *specific heat*. The **specific heat** of a substance is the heat needed to raise the temperature of specifically *one gram* by one degree Celsius.[2] In other words, the specific heat of a substance is its *heat capacity per gram*.

$$\text{Specific heat} = \frac{\text{heat capacity}}{\text{mass (g)}} \qquad (5.2)$$

By dividing heat capacity in $J\,°C^{-1}$ by the mass in grams, we obtain the units of specific heat, $J\,g^{-1}\,°C^{-1}$ (joules per gram per degree Celsius). The specific heat of water, for example, is $4.184\,J\,g^{-1}\,°C^{-1}$ for the one specific degree between 14.5 and 15.5 °C. When we express water's specific heat in calories instead of joules, its value is $1\,cal\,g^{-1}\,°C^{-1}$. In fact, one calorie was originally *defined* this way (see page 20). The value of specific heat varies somewhat with temperature, but not by much. At 25 °C, the value for water is very close to $4.184\,J\,g^{-1}\,°C^{-1}$, being $4.1796\,J\,g^{-1}\,°C^{-1}$. To two significant figures, liquid water's specific heat is $4.2\,J\,g^{-1}\,°C^{-1}$ over the whole range between its freezing and boiling points. Relative to water, metals have low specific heats (see Table 5.1).

By rearranging Equation 5.2, we obtain another expression for heat capacity.

$$\text{Heat capacity} = \text{specific heat} \times \text{mass}$$

When we substitute this expression for heat capacity into Equation 5.1, we have another useful equation for the amount of an energy change. (Notice how the units on the left reduce to J.)

$$\text{Specific heat} \times \text{mass} \times \Delta t = \text{heat energy} \qquad (5.3)$$

$$\frac{J}{g\,°C} \times \quad g \quad \times °C = J$$

The symbol Δ, pronounced "delta" denotes a *change between some initial and final state* in whatever follows the symbol. Thus, $\Delta t = t_{\text{final}} - t_{\text{initial}}$.

The SI unit of energy, the joule (J), and its common multiple, the kilojoule (kJ), were defined in Section 1.5 on page 20. $1\,J = 10^3\,kJ$.

An analysis of the units in Equation 5.1 shows that the product on the left, heat capacity $\times \Delta t$, has the same net unit as energy, J.

$$\frac{J}{°C} \times °C = J$$

Because the kelvin and the Celsius degrees have the same *size*, so do Δt (in °C) and ΔT (in K). Hence, *with no change in the numerical values,* the units of specific heat can be expressed as either as $J\,g^{-1}\,K^{-1}$ or as $kJ\,kg^{-1}\,K^{-1}$. Similarly, heat capacities can be given, without numerical changes, either in $J\,K^{-1}$ (for $J\,°C^{-1}$) or $kJ\,K^{-1}$ (for $kJ\,°C^{-1}$).

The "calorie" used in nutrition is actually the kilocalorie.

 Heat capacity equation

---

[2] When one extensive quantity, like heat capacity, is divided by another extensive quantity, like mass, the result is an intensive quantity, in this example, the specific heat.

It is important to remember the distinction: *heat capacity* is an extensive property, but *specific heat* is an intensive property. Specific heat is heat capacity *per gram*.

EXAMPLE 5.1
Calculating an Energy Change from a Temperature Change, Mass, and Specific Heat

If a gold ring with a mass of 5.5 g changes in temperature from 25.0 to 28.0 °C, how much energy (in joules) has it absorbed?

ANALYSIS The question actually asks for the *change* in heat energy that occurs in the gold ring as a result of the temperature change, Δt. Equation 5.3 gives the relationship among these factors, and we will use the specific heat of gold at 25 °C, 0.129 J g^{-1} °C^{-1} (Table 5.1).

SOLUTION The temperature increases from 25.0 to 28.0 °C, so Δt is 3.0 °C. The mass of the ring is 5.5 g. We substitute our values of specific heat, temperature change, and grams into Equation 5.3 and solve for the *energy change*.

$$\text{Specific heat} \times \text{mass} \times \Delta t = \text{heat energy} \qquad \text{(Equation 5.3)}$$

$$\frac{J}{g \, °C} \times \quad g \quad \times °C = J$$
$$= (0.129 \, J \, g^{-1} \, °C^{-1}) \times (5.5 \, g) \times (3.0 \, °C)$$
$$= 2.1 \, J$$

Notice that g and g^{-1} cancel; °C and °C^{-1} also cancel. Only J remains. Thus, only 2.1 J raises the temperature of 5.5 g of gold by 3.0 °C.

■ **Practice Exercise 1** The temperature of 250 g of water is changed from 25.0 to 30.0 °C. How much energy was transferred into the water? Calculate your answer in joules, kilojoules, calories, and kilocalories.

TABLE 5.1   Specific Heats

Substance	Specific Heat J g^{-1} °C^{-1} (at 25 °C)
Carbon (graphite)	0.711
Copper	0.387
Ethyl alcohol	2.45
Gold	0.129
Granite	0.803
Iron	0.4998
Lead	0.128
Olive oil	2.0
Silver	0.235
Water (liquid)	4.1796

We have defined thermal properties in terms of the heat that a substance must *absorb* for a given *increase* in temperature. Specific heat is also the heat that 1 g of a substance must *lose* if its temperature is to *decrease* by 1 °C. Specific heat, thus, tells us not only how relatively hard it is to increase the temperature of a specific amount of substance, but also how relatively difficult it is to cool it.

Molar Heat Capacity

When dealing with the thermal properties of a pure substance, it is often more useful to work with one mole instead of one gram. We therefore define the **molar heat capacity** of a substance as the heat needed to raise the temperature of one mole of it by one degree Celsius. In terms of the joule, the units of molar heat capacity are J mol^{-1} °C^{-1}.

Because the molar mass of water is 18.0 g mol^{-1}, the molar heat capacity of water is 18.0 times its specific heat, namely, 75.3 J mol^{-1} °C^{-1} or 18.0 cal mol^{-1} °C^{-1}.

The Unusual Thermal Properties of Water

Water has an unusually high specific heat, higher than the values for almost all known materials (Table 5.1). It is roughly 10 times that of iron, for example, and nearly 33 times that of gold. In other words, the heat that raises the

SPECIAL TOPIC 5.1 / HYPOTHERMIA—WHEN THE BODY'S "THERMAL CUSHION" IS OVERTAXED

All of the reactions in the body, collectively called *metabolism,* have rates sensitive to temperature changes. Metabolic reactions generally are faster at higher temperatures and slower at lower temperatures. We will discuss here a consequence of overusing the body's thermal cushion to thwart a decrease in body temperature.

Some of the critical functions sustained by the energy of metabolism are the operation of the central nervous system, including the brain, and the working of the heart. *Hypothermia* is a decrease in the temperature of the inner core of the body. As the inner core cools, rates of metabolism slow down and the symptoms given in the table progressively occur. When a newspaper states that someone died of "exposure," nearly always it is a case of hypothermia.

The onset of hypothermia is a serious medical emergency requiring prompt action by others. The victim may either not recognize the problem or be confused by it. First aid consists of getting the victim out of the wind, dried off, wrapped in warm, dry blankets, and fed warm fluids or sweet food. Contrary to the legend about the St. Bernard dogs, brandy should never be given because, once in the bloodstream, the alcohol in brandy causes the blood capillaries to dilate. When capillaries near the cold skin suddenly release extra loads of chilled blood toward the body's core, the core cooling accelerates.

Core Temperature (°C)	Symptoms
37	Normal body temperature
35–36	Intense, uncontrollable shivering
33–35	Amnesia, confusion
30–32	Muscular rigidity, reduced shivering
26–30	Unconsciousness, erratic heartbeat
Below 26	Death by pulmonary edema and heart failure

temperature of 1 g of water only 1 °C raises the temperature of 1 g of iron by roughly 10 °C or 1 g of gold by 33 °C. In cooling terms, if we remove a calorie of heat from a gram of iron, its temperature falls by 10 °C. But the same loss of heat from a gram of water makes its temperature fall by only 1 °C. *Water's high specific heat means that it is relatively difficult to make the temperature of water swing widely.*

The adult body is about 60% water by mass so it has a high heat capacity. It is thus relatively easy for the human body to maintain a steady temperature of 37 °C, which is vital to survival. In other words, the body can exchange considerable energy with the environment but experience only a small change in temperature. With this large thermal "cushion," the body can adjust to large and sudden changes in outside termperature and experience little fluctuation of its core temperature. *Hypothermia* is an emergency brought on when the body cannot prevent its core temperature from decreasing (see Special Topic 5.1).

An infant's body is about 80% water.

Because of water's thermal "cushion," cities by oceans or other very large bodies of water, like one of the Great Lakes, have relatively mild climates compared to cities several miles inland.

Calorimetry

We discussed the thermal properties of matter because they are needed to calculate the heat of reaction from a temperature change caused by the reaction. Most heats of reaction are determined by measuring the change in temperature caused by the heat absorbed by a certain mass of water. The apparatus used is called a **calorimeter** (*a calorie meter*). Calorimeter design is not standard, varying according to the kind of reaction and the precision desired. The science of using a calorimeter and determining heats of reaction is called **calorimetry.** Specialists in calorimetry have developed many sophisticated methods to measure temperature changes caused by reactions, which are the data used to calculate the heat evolved or absorbed. We'll look only at a very simple calorimeter, one that you might use in the lab.

Heat of reaction

The Coffee Cup Calorimeter

The *coffee cup calorimeter,* a simple, inexpensive device used in many general chemistry labs, is made of two nested and capped cups made of styrofoam, a very good insulator. When a reaction occurs in such a calorimeter, little heat is lost to the surroundings (or gained from it). If the reaction evolves heat, for example, nearly all of the heat stays within the solution in the calorimeter, where it causes an easily measured temperature increase. From the predetermined heat capacity of the calorimeter's contents before the reaction, the heat absorbed or evolved by the reaction is found by Equation 5.1. (In our calculations involving a coffee cup calorimeter, we will ignore the negligible heat absorbed by the styrofoam. This works acceptably well when the maximum temperature measurement is reached soon after the reaction is initiated.)

A calorimeter does not measure joules *directly;* its thermometer is the only part of a calorimeter to provide raw data, the temperature change. The *energy* associated with this must be calculated. There is no instrument that directly measures energy.

EXAMPLE 5.2
Coffee Cup Calorimetry

The reaction of hydrochloric acid and sodium hydroxide is exothermic. The equation is

$$HCl(aq) + NaOH(aq) \longrightarrow NaCl(aq) + H_2O$$

In one experiment, a student placed 50.0 mL of 1.00 M HCl at 25.5 °C in a coffee cup calorimeter. To this was added 50.0 mL of 1.00 M NaOH solution also at 25.5 °C. The mixture was stirred, and the temperature quickly increased to a maximum of 32.4 °C. What is the energy evolved in joules per mole of HCl? Because the solutions are relatively dilute, we can assume that their specific heats are close to that of water, 4.18 J g^{-1} °C^{-1}, and that their densities are 1.00 g mL^{-1}. (We also assume that no heat is lost to the coffee cup itself or to the surrounding air.)

The actual densities at 25 °C are 1.02 g mL^{-1} for 1.0 M HCl and 1.04 g mL^{-1} for 1.0 M NaOH.

ANALYSIS Before we can use Equation 5.1 to find joules evolved, we have to calculate two things: the system's heat capacity and the temperature change. To calculate the system's heat capacity using the given specific heats we need Equation 5.2.

$$\text{Specific heat} = \frac{\text{heat capacity}}{\text{mass (g)}} \qquad \text{(Equation 5.2)}$$

By rearranging, we have

$$\text{Heat capacity} = \text{mass (g)} \times \text{specific heat}$$

The mass in this equation refers to the *total* grams of the combined solutions, but we were given volumes. So we have to use their densities to calculate mass:

$$\text{Density} = \frac{\text{mass}}{\text{volume}}$$

Because the densities for the solutions are both 1.00 g mL^{-1}, the milliliters are *numerically* the same as the grams of the solutions. Each solution mixed has a mass of 50.0 g, so the total mass of the final solution is 100.0 g.

The calculation flow for Example 5.2 is

Total volume
  | Density
  ↓ calculation
Total mass
  | Specific heat
  ↓ calculation
Heat capacity of cup contents
  | Equation 5.1
  | calculation
  ↓ using Δ*t*
Total joules evolved
  | Moles of acid
  | from final
  | solution's
  | molarity and
  ↓ volume
Joules per mole

SOLUTION We can now calculate the heat capacity of the final solution at the instant of mixing and immediately before any reaction occurs to increase its temperature. It was given that the solution's specific heat is 4.18 J g^{-1} °C^{-1}, so

$$
\begin{aligned}
\text{Heat capacity} &= \text{mass} \times \text{specific heat} \\
&= 100.0 \text{ g} \times 4.18 \text{ J g}^{-1}\,°\text{C}^{-1} \\
&= 481 \text{ J }°\text{C}^{-1}
\end{aligned}
$$

Thus the heat capacity of the solution in the calorimeter is 418 J °C^{-1}.

The reaction increases the system's temperature by $\Delta t = 32.4\ °C - 25.5\ °C = 6.9\ °C$, so the energy evolved by the reaction can be calculated from Equation 5.1.

$$\text{Heat capacity} \times \Delta t = \text{heat energy (evolved)}$$

Or

$$\text{Heat evolved} = 418\ \text{J}\ °C^{-1} \times 6.9\ °C$$
$$= 2.9 \times 10^3\ \text{J}$$

This is the heat evolved for the specific mixture prepared, but the problem calls for joules *per mole* of HCl. We calculate the number of moles of HCl used with the molarity-to-moles tool given by the definition of molarity. The molarity of the acid is 1.00 *M*, so in 50.0 mL of HCl solution we have 0.0500 mol HCl.

$$\text{Molarity} = M = \frac{\text{moles solute}}{1000\ \text{mL soln}}$$

$$50.0\ \text{mL HCl soln} \times \frac{1.00\ \text{mol HCl}}{1000\ \text{mL HCl soln}} = 0.0500\ \text{mol HCl}$$

The neutralization of 0.0500 mol of acid released 2.9×10^3 J. So, properly put, we should express the heat released as the ratio:

$$\text{Energy evolved} = \frac{2.9 \times 10^3\ \text{J}}{0.0500\ \text{mol HCl}} = 58 \times 10^3\ \text{J mol}^{-1}$$
$$= 58\ \text{kJ mol}^{-1}$$

■ **Practice Exercise 2** When pure sulfuric acid dissolves in water, a great deal of heat is given off. To measure it, 175 g of water was placed in a coffee cup calorimeter and chilled to 10.0 °C. Then 4.90 g of sulfuric acid (H_2SO_4), also at 10.0 °C, was added, and the mixture was quickly stirred with a thermometer. The temperature rose rapidly to 14.9 °C. Assume that the value of the specific heat of the solution is known to be 4.18 J g^{-1} °C^{-1}, and that all of the heat evolved is absorbed by the solution. Calculate the heat evolved in kilojoules by the formation of this solution. (Remember to use the *total* mass of the solution, the water plus the solute.) Calculate also the heat evolved *per mole* of sulfuric acid.

Although the total energy of the universe is believed to be constant, the energy possessed by any given *system* within the universe can vary widely. The energy in the *surroundings* simply adjusts as transfers of energy occur into or out of systems. We will now focus our attention just on *systems,* and on the energy content they can have.

5.5 ENTHALPY CHANGES: HEATS OF REACTION AT CONSTANT PRESSURE

State Functions

The *amount* of energy possessed by a system depends neither on how the system received it nor on what might become of it at some future time. A system's energy is a function only of the condition or the *state of the system*. We have specified or defined the **state of a system** when we have stated certain of its properties, particularly its chemical composition and its pressure, temperature, and volume. Any property that does not depend on a system's history or future is called a **state function.** When we recognize that some property is a *state* function, many calculations are much easier.

It's easy to understand why several common physical properties of a system, like volume, pressure, and temperature, are state functions. A system's temperature, for example, does not depend on what it was yesterday. Nor does it matter *how* the system reached it. If it is now 25 °C, for example, we know all we

Note carefully that the term *state* in thermochemistry has a broader meaning than it has in such expressions as "liquid state" or "solid state."

can or need to know about its temperature. We do not have to specify how it got there. Also, if the temperature were to increase to 35 °C, the change in temperature, Δt, is simply the difference between the final and the initial temperatures, $t_{final} - t_{initial}$. To make this calculation, we do not have to know what caused the temperature change—exposure to sunlight, heating by an open flame, or any other mechanism. All we need are the initial and final values. *This independence from the method or mechanism by which a change occurs is the important feature of all state functions.*

Energy as a State Function

The energy of a system, E, like its temperature, is also a state function. Unlike a system's temperature, however, which can be measured by a thermometer, a system's energy cannot be measured with any instrument. We must calculate energy from other raw data. Of course, what we calculate is a *change* in energy brought about by the reaction itself.

Remarkably, a system's *total* energy, whether initial or final, *cannot, in fact, be known!* Think for a moment about a calorimeter charged with its load of reactants ready to react. The calorimeter and its contents do have a *total* energy, namely, the sum of their combined kinetic and potential energy. But how can we ever know, for example, what the total kinetic energy actually is? We can be sure of one fact; the kinetic energy is not zero simply because the system is *resting* at one location in the lab. The calorimeter has velocity and so has kinetic energy from several sources. It is on a spinning planet, for example, which is orbiting the sun in a galaxy also in motion. We have no way of knowing how all of these motions affect the kinetic energy of the calorimeter, so we cannot know what its *total* energy is either. However, as long as the galactic and planetary motions remain steady, we do not need to know this. Our interest is solely in the *change* in energy brought about in the reacting system itself. Only from this *change* in energy might we get something useful from the system. Thus, even though we can *imagine* total values of initial and final energy, $E_{initial}$ and E_{final}, we can obtain by experiment only a *change* in energy, ΔE (read "delta E"). This is defined as follows.

> The *total* potential energy, which depends both on the atomic nuclei present as well as on the location of the system in the universe, is likewise unknowable.

$$\Delta E = E_{final} - E_{initial}$$

E and ΔE are state functions.

Enthalpy

Most reactions carried out in the lab are in open vessels, in glassware open to the atmosphere and so at constant (atmospheric) pressure. Only in special circumstances is the vessel sealed, permitting no matter (like a gas) to escape. For reasons to be discussed in Chapter 13, we have a special term for energy when the system is open to the atmosphere, the **enthalpy** of the system, symbolized by H. Like E, H is a state function.

> Another name for enthalpy is heat content.

When a system reacts at constant pressure and absorbs or evolves energy, we say that it experiences an **enthalpy change, ΔH**, defined by the equation

> H = enthalpy
> H = hydrogen atom

$$\Delta H = H_{final} - H_{initial}$$

H_{final} is the enthalpy of the system in its final state and $H_{initial}$ is the enthalpy of the system in its initial state. Like H, ΔH *is a state function*. Its value depends solely on the difference in enthalpy between the initial and the final states and not on the mechanism by which the system undergoes this change.

For a chemical reaction, the initial state refers to the reactants and the final state to the products, so for a chemical reaction the equation for ΔH can be rewritten as follows.

$$\Delta H = H_{\text{products}} - H_{\text{reactants}} \qquad (5.4)$$

We will encounter other quantities that are defined as a difference between two absolute values. In each situation, *we will subtract the value for the initial state from the value for the final state.* Simply put, this always means "products minus reactants."

Our formal definition of H in an absolute sense serves only one purpose — to define the meaning of positive and negative values of ΔH. We said that when a system absorbs energy from its surroundings, the change is endothermic, so the system's final enthalpy has to be larger than its initial enthalpy.

In endothermic changes: $\qquad H_{\text{final}} > H_{\text{initial}}$

Therefore, if we could calculate ΔH from H_{final} and H_{initial}, we would subtract a smaller value from a larger one, so the difference (ΔH) would have a positive sign. **In endothermic changes: ΔH is positive.**

On the other hand, if a change is exothermic, the system loses energy and its final enthalpy must be less than its initial enthalpy.

In exothermic changes: $\qquad H_{\text{final}} < H_{\text{initial}}$

Now, to calculate ΔH from H_{initial} and H_{final} (if we could know absolute values of H), we would subtract a larger number from one that is smaller, so the value of ΔH for all exothermic changes is negative. **In exothermic changes: ΔH is negative.**

> Unless we specify otherwise, whenever we write ΔH, we mean ΔH for the *system,* not the surroundings.

Enthalpy Changes and Heats of Reaction at Constant Pressure

The enthalpy change for a chemical reaction can make itself known either as work or as heat or as some of each. It depends on how we let it occur. If the reaction occurs in a battery, for example, at least part of the enthalpy change can appear as electrical work in the external circuit. If, however, the battery's terminals are wired together, the chemical reaction would still occur but all of the enthalpy change would appear as heat and none as electrical work. When *all* of the enthalpy change appears as heat, ΔH is equal to the **heat of reaction at constant pressure,** and is given the symbol q.

$$\Delta H = q \qquad \text{(at constant pressure)}$$

The sign of q, thus, is the sign of ΔH. **For exothermic changes, q is negative** because ΔH is negative. **For endothermic changes, q is positive** because ΔH is positive.

> Sometimes you'll see the symbol q written as q_p, where the subscript signifies constant pressure.

Standard Conditions for Enthalpy Changes

To facilitate comparisons of enthalpy changes among different systems, chemists have agreed to a set of reference conditions. The standard reference temperature is 25 °C, and the reference pressure is approximately the pressure exerted by our atmosphere at sea level. (In Chapter 10 we will put atmospheric pressure on a precise basis.) The standard pressure is called one **standard atmosphere** and its symbol is **atm.**

> We can easily keep a reaction at 25 °C by immersing the reaction vessel in a vat of water kept at 25 °C by thermostatic control.

> The pressure of the atmosphere is caused by the gravitational pull of Earth on the surrounding air.

Standard Heats of Reaction

The **standard heat of reaction** is the value of ΔH for a reaction occurring under standard conditions and involving the actual numbers of *moles* specified by the coefficients of the equation. To show that ΔH is for *standard* conditions, we add a superscript to ΔH, the degree sign, to make $\Delta H°$. The units of $\Delta H°$ are normally kilojoules.

To illustrate clearly what we mean by $\Delta H°$, let us use the reaction between gaseous nitrogen and hydrogen that produces gaseous ammonia.

$$N_2(g) + 3H_2(g) \longrightarrow 2NH_3(g)$$

The values of $\Delta H°$ are taken from available reference sources.

When specifically 1.000 mol of N_2 and 3.000 mol of H_2 react to form 2.000 mol of NH_3 at 25 °C and 1 atm, the reaction releases 92.38 kJ. Hence, for the reaction *as given by the above equation,* $\Delta H° = -92.38$ kJ. Often the enthalpy change is made a part of the equation; for example:

$$N_2(g) + 3H_2(g) \longrightarrow 2NH_3(g) \qquad \Delta H° = -92.38 \text{ kJ}$$

Thermochemical equations

An equation that includes the value of $\Delta H°$ is called a **thermochemical equation.** It always shows the physical states of the reactants and products, and *its $\Delta H°$ value is true only when the given coefficients are taken to mean the number of actual moles they express.* The above equation, for example, shows a release of 92.38 kJ if *two* moles of NH_3 form. If we were to make twice as much or 4.000 mol of NH_3 (from 2.000 mol of N_2 and 6.000 mol of H_2), then twice as much heat (184.8 kJ) would be released. On the other hand, if only 0.5000 mol of N_2 and 1.500 mol of H_2 were to react to form only 1.000 mole of NH_3, then only half as much heat (46.19 kJ) would be released. For the various sizes of the reactions just described, for example, we would have the following thermochemical equations.

$$N_2(g) + 3H_2(g) \longrightarrow 2NH_3(g) \qquad \Delta H° = -92.38 \text{ kJ}$$

$$2N_2(g) + 6H_2(g) \longrightarrow 4NH_3(g) \qquad \Delta H° = -184.8 \text{ kJ}$$

$$\tfrac{1}{2}N_2(g) + \tfrac{3}{2}H_2(g) \longrightarrow NH_3(g) \qquad \Delta H° = -46.19 \text{ kJ}$$

Because the coefficients of a thermochemical equation always mean *moles,* we can use fractional coefficients. Normally we avoid them because we cannot have fractions of *molecules,* but we can have fractions of moles.

EXAMPLE 5.3
Writing a Thermochemical Equation

The following thermochemical equation is for the exothermic reaction of hydrogen and oxygen that produces water.

$$2H_2(g) + O_2(g) \longrightarrow 2H_2O(g) \qquad \Delta H° = -483.6 \text{ kJ}$$

What is the thermochemical equation for this reaction when it is conducted to produce 1.000 mol H_2O?

ANALYSIS The given equation is for 2.000 mol of H_2O, and any changes in the coefficient for water must be made identically to all other coefficients, as well as to the value of $\Delta H°$.

SPECIAL TOPIC 5.2 / PERPETUAL MOTION MACHINES—IMPOSSIBLE

A perpetual motion machine is one that creates more energy than it uses and so theoretically could run forever. The U.S. Patent Office, wearied by claims for these impossibilities, now requires that all patent applications be accompanied by "reductions to practice," that is, *working* models. Because working models of perpetual motion machines are impossible, the Patent Office is no longer bothered with such applications.

Here's how such a machine would work in principle. Suppose that it takes only 200 kJ to break 1 mol of CO_2 back to C plus O_2. If we get 394 kJ by burning a mole of carbon to CO_2 but only 200 kJ is required to recycle the system back to C plus O_2,

our profit is 194 kJ per mole. Over and over, we could let the elements burn, get 394 kJ per mole, save 200 kJ out of this, and get 194 kJ left over. We could use this energy to swat flies, bake cookies, repair worn-out parts of the machinery, or run tractors while we took it easy. Very nice—but impossible. We haven't explained *why* it's impossible; we don't know why, except that this is the way of nature. All we have done is *discover* that it is impossible. Out of this discovery, often repeated by would-be inventors of perpetual motion machines, came the law of conservation of energy.

SOLUTION We divide everything by 2, to obtain

$$H_2(g) + \tfrac{1}{2}O_2(g) \longrightarrow H_2O(g) \qquad \Delta H^\circ = -241.8 \text{ kJ}$$

■ **Practice Exercise 3** What is the thermochemical equation for the formation of 2.500 mol of H_2O?

Thermochemical Equations for Experimentally Difficult Reactions

Once we have the thermochemical equation for a given reaction, we can write one for the reverse reaction, regardless of how hard it might actually be to make the reverse reaction occur. The thermochemical equation, for example, for the combustion of carbon to give carbon dioxide is

$$C(s) + O_2(g) \longrightarrow CO_2(g) \qquad \Delta H^\circ = -393.5 \text{ kJ}$$

The reverse reaction, which is extremely difficult to carry out, would be the decomposition of carbon dioxide to carbon and oxygen.

$$CO_2(g) \longrightarrow C(s) + O_2(g)$$

Despite how hard it would be to cause this reaction, we can still know what its ΔH° *must be;* it must be $+393.5$ kJ. It must be numerically equal but opposite in sign to the ΔH° for the forward reaction. The law of conservation of energy *requires* this remarkable result. If the values of ΔH° for the forward and the reverse reactions were not equal but opposite in sign, then perpetual motion machines would be possible. But they are not (see Special Topic 5.2). Regardless of the difficulty of directly decomposing CO_2 into its elements, we can still write a thermochemical equation for it:

$$CO_2(g) \longrightarrow C(s) + O_2(g) \qquad \Delta H^\circ = +393.5 \text{ kJ}$$

Thus, if we know ΔH° for a given reaction, then we also know ΔH° for the reverse reaction; it has the same numerical value, but the opposite algebraic sign. This extremely useful fact makes thermochemical data available that would otherwise be impossible to measure.

Multiple versus Single Path Routes and Enthalpy Changes

We can imagine two paths leading from 1 mole each of carbon and oxygen to 1 mol of carbon dioxide.

One-Step Path.   Let C and O_2 react to give CO_2 directly.

$$C(s) + O_2(g) \longrightarrow CO_2(g) \qquad \Delta H^\circ = -393.5 \text{ kJ}$$

Two-Step Path.   Let C and O_2 react to give CO, and then let CO react with O_2 to give CO_2.

$$\text{Step 1: } C(s) + \tfrac{1}{2}O_2(g) \longrightarrow CO(g) \qquad \Delta H^\circ = -110.5 \text{ kJ}$$

$$\text{Step 2: } CO(g) + \tfrac{1}{2}O_2(g) \longrightarrow CO_2(g) \qquad \Delta H^\circ = -283.0 \text{ kJ}$$

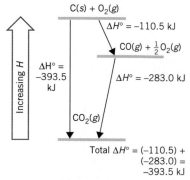

C(s) + O_2(g)

$\Delta H^\circ = -110.5$ kJ

CO(g) + $\tfrac{1}{2}O_2$(g)

$\Delta H^\circ = -393.5$ kJ

$\Delta H^\circ = -283.0$ kJ

Increasing H

CO_2(g)

Total $\Delta H^\circ = (-110.5) + (-283.0) = -393.5$ kJ

Enthalpy diagram for C(s) + O_2 → CO_2(g)

FIGURE 5.2

An enthalpy diagram for the formation of $CO_2(g)$ from its elements by two different paths; read *down* from the topmost line. On the left is path 1, the direct conversion of C(s) and $O_2(g)$ to $CO_2(g)$. On the right, showing two shorter, downward-pointing arrows, is path 2. The first step of path 2 takes the elements to $CO(g)$, and the second step takes $CO(g)$ to $CO_2(g)$. The overall enthalpy change is identical for both paths, as it must be, because enthalpy is a state function.

Overall, the two-step path consumes 1 mol each of C and O_2 to make 1 mol of CO_2, just like the one-step path. The initial and final states for the two routes to CO_2, in other words, are identical.

Because ΔH° is a state function dependent only on the initial and final states and is independent of path, then the values of ΔH° for both routes should be identical. We can see that this is exactly true simply by adding the equations for the two-step path and comparing the result with the equation for the single step.

$$\text{Step 1: } C(s) + \tfrac{1}{2}O_2(g) \longrightarrow CO(g) \qquad \Delta H^\circ = -110.5 \text{ kJ}$$

$$\text{Step 2: } CO(g) + \tfrac{1}{2}O_2(g) \longrightarrow CO_2(g) \qquad \Delta H^\circ = -283.0 \text{ kJ}$$

$$CO(g) + C(s) + O_2(g) \longrightarrow CO_2(g) + CO(g) \qquad \Delta H^\circ = -110.5 \text{ kJ} + (-283.0 \text{ kJ})$$
$$\Delta H^\circ = -393.5 \text{ kJ}$$

The equation resulting from adding steps 1 and 2 has "$CO(g)$" appearing *identically* on opposite sides of the arrow. The two may therefore be canceled to obtain the net equation. As we learned in Section 4.4, such a cancellation is permitted only when both the formula and the physical state are identically given on opposite sides of the arrow. The net thermochemical equation for the two-step process, therefore, is

$$C(s) + O_2(g) \longrightarrow CO_2(g) \qquad \Delta H^\circ = -393.5 \text{ kJ}$$

The results, chemically and thermochemically, are thus identical for both routes to CO_2, demonstrating that ΔH° is a state function.

Enthalpy diagrams

Enthalpy Diagrams

We can show the energy relationships among the alternative pathways for the same overall reaction by an *enthalpy diagram,* as illustrated in Figure 5.2 for the formation of CO_2 from C and O_2. Horizontal lines correspond to different *absolute* values of enthalpy, *H,* which we cannot know. However, an upper line represents a larger value of *H* than a lower line. *Changes* in enthalpy, ΔH°, which we can know, are represented by the vertical distances between these lines. Arrows pointing upward represent increases in enthalpy and have positive values of ΔH. Arrows pointing downward represent decreases in enthalpy and have negative values of ΔH.

For an exothermic reaction, the reactants are always on a higher line than the products, because the reactants have more potential energy and so more enthalpy than the products. Notice in Figure 5.2 that the line for the absolute enthalpy of $C(s) + O_2(g)$, taken as the sum, is above the line for the final product, CO_2. Thus we think of a horizontal line as representing the *sum* of the enthalpies of all of the substances on the line *in the physical states specified*. A downward-pointing arrow signifies an exothermic reaction with a negative ΔH.

To the left in Figure 5.2, we see a long vertical arrow connecting the enthalpy levels for the reactants, $C(s) + O_2(g)$, and the final product, $CO_2(g)$. This corresponds to the one-step path, the direct path. On the right we have the two-step path. Here, the overall change stops at an intermediate enthalpy level corresponding to the intermediate products, $CO(g) + \frac{1}{2}O_2(g)$, which are those made or left over in the first step. Then Step 2 occurs to give the final product. The total decrease in enthalpy, however, is the same regardless of the path.

Hydrogen peroxide, H_2O_2, decomposes into water and oxygen by the following equation.

$$H_2O_2(l) \longrightarrow H_2O(l) + \tfrac{1}{2}O_2(g)$$

Construct an enthalpy diagram for the following two reactions of hydrogen and oxygen, and use the diagram to determine the value of $\Delta H°$ for the decomposition of hydrogen peroxide.

$$H_2(g) + O_2(g) \longrightarrow H_2O_2(l) \qquad \Delta H° = -188 \text{ kJ}$$

$$H_2(g) + \tfrac{1}{2}O_2(g) \longrightarrow H_2O(l) \qquad \Delta H° = -286 \text{ kJ}$$

EXAMPLE 5.4
Preparing an Enthalpy Diagram

ANALYSIS The two given reactions are exothermic, so their values of $\Delta H°$ will be associated with downward-pointing arrows. The highest enthalpy level, therefore, must be for the elements themselves. The lowest level must be for the product of the reaction with the largest negative $\Delta H°$, the formation of water by the second equation. The enthalpy level for the H_2O_2, formed by a reaction with the less negative $\Delta H°$, must be in between.

SOLUTION We diagram the two reactions having known values of $\Delta H°$ and then think about how the answer to the problem might be determined.

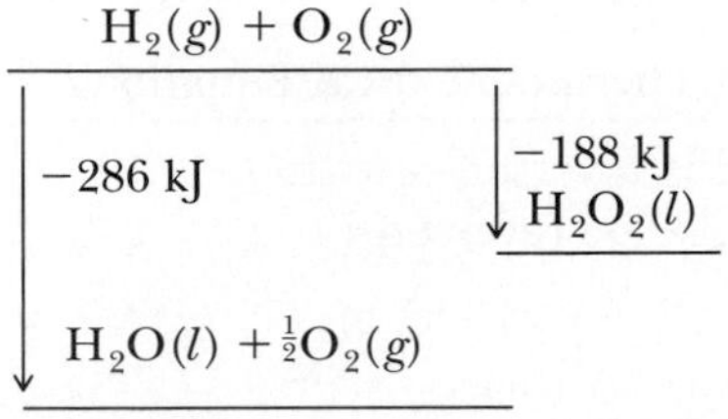

$$H_2(g) + O_2(g)$$

-286 kJ $\qquad\qquad$ -188 kJ
$H_2O_2(l)$

$H_2O(l) + \tfrac{1}{2}O_2(g)$

Notice that the gap on the right side corresponds exactly to the change of $H_2O_2(l)$ into $H_2O(l) + \frac{1}{2}O_2(g)$, the reaction for which $\Delta H°$ is sought. This enthalpy separation corresponds to the difference between the enthalpy levels, -286 kJ and -188 kJ. Thus for the decomposition of hydrogen peroxide by the equation given,

$$\Delta H^\circ = [-286 \text{ kJ} - (-188 \text{ kJ})] = -98 \text{ kJ}$$

Our completed enthalpy diagram looks like this:

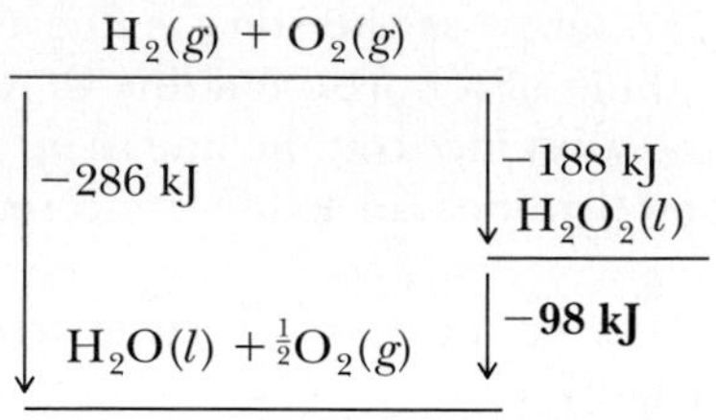

$$H_2(g) + O_2(g)$$

-286 kJ

-188 kJ
$H_2O_2(l)$

$H_2O(l) + \frac{1}{2}O_2(g)$

-98 kJ

■ **Practice Exercise 4** Two oxides of copper can be made from copper by the following reactions.

$$2Cu(s) + O_2(g) \longrightarrow 2CuO(s) \qquad \Delta H^\circ = -310 \text{ kJ}$$

$$2Cu(s) + \tfrac{1}{2}O_2(g) \longrightarrow Cu_2O(s) \qquad \Delta H^\circ = -169 \text{ kJ}$$

Using these data, construct an enthalpy diagram that can be used to find ΔH° for the following reaction: $Cu_2O(s) + \tfrac{1}{2}O_2(g) \longrightarrow 2CuO(s)$

Hess's Law

Although enthalpy diagrams are instructive, you may have already considered that they are not necessary to calculate ΔH° for a reaction from known thermochemical equations. Something like algebraic summing should work, and **Hess's law of heat summation** gives us this tool.

Hess's Law of Heat Summation

For any reaction that can be written in steps, ΔH° is the same as the sum of the values of ΔH° for the individual steps.

The chief use of Hess's law is to calculate the enthalpy change for a reaction for which such data cannot be determined experimentally or are otherwise unavailable. This usually requires that we manipulate equations, as we did when we added those for the two-step process to CO_2, described on page 204, or the process shown in Example 5.4. A few rules that govern the manipulations of equations follow.

We reverse an equation by interchanging reactants and products. This leaves the arrow pointing from left to right.

Rules for Manipulating Thermochemical Equations

1. When an equation is reversed—written in the opposite direction—the sign of ΔH° must also be reversed.[3]

[3] To illustrate, the reverse of the equation:
$$C(s) + O_2(g) \longrightarrow CO_2(g) \qquad \Delta H^\circ = -394 \text{ kJ}$$
is the equation:
$$CO_2(g) \longrightarrow C(s) + O_2(g) \qquad \Delta H^\circ = +394 \text{ kJ}$$

2. Formulas canceled from both sides of an equation must be for the substance in identical physical states.

3. If all the coefficients of an equation are multiplied or divided by the same factor, the value of $\Delta H°$ must likewise be changed.

Carbon monoxide is often used in metallurgy to remove oxygen from metal oxides and thereby give the free metal. The thermochemical equation for the reaction of CO with iron(III) oxide, Fe_2O_3, is

$$Fe_2O_3(s) + 3CO(g) \longrightarrow 2Fe(s) + 3CO_2(g) \qquad \Delta H° = -26.7 \text{ kJ}$$

Use this equation and the equation for the combustion of CO,

$$CO(g) + \tfrac{1}{2}O_2(g) \longrightarrow CO_2(g) \qquad \Delta H° = -283.0 \text{ kJ}$$

to calculate the value of $\Delta H°$ for the following reaction.

$$2Fe(s) + \tfrac{3}{2}O_2(g) \longrightarrow Fe_2O_3(s)$$

EXAMPLE 5.5
Using Hess's Law

ANALYSIS We cannot simply add the two given equations, because this will not produce the equation we want. We first have to manipulate these equations so that when we add them we will get the target equation.

SOLUTION We can manipulate the two given equations using the following reasoning.

STEP 1 We begin by trying to get the iron atoms to come out right. The target equation must have 2Fe on the *left*, but the first equation above has 2Fe to the *right* of the arrow. To move it to the left, we must reverse the *entire* equation, remembering also to reverse the sign of $\Delta H°$. We also are pleased to note that this manipulation places Fe_2O_3 to the right of the arrow, which is where it has to be after we add our adjusted equations. After these manipulations, and reversing the sign of $\Delta H°$, we have:

$$2Fe(s) + 3CO_2(g) \longrightarrow Fe_2O_3(s) + 3CO(g) \qquad \Delta H° = +26.7 \text{ kJ}$$

STEP 2 There must be $\tfrac{3}{2}O_2$ on the left, and we must be able to cancel *three* CO and *three* CO_2 when the equations are added. If we multiply the second of the equations given above by 3, we will obtain the necessary coefficients. We must also multiply the value of $\Delta H°$ of this equation by 3, because three times as much chemicals are now involved in the reaction. When we have done this, we have:

$$3CO(g) + \tfrac{3}{2}O_2(g) \longrightarrow 3CO_2(g) \qquad \Delta H° = 3 \times (-283.0 \text{ kJ}) = -849.0 \text{ kJ}$$

Let's now put our two equations together and find the answer.

$$2Fe(s) + 3CO_2(g) \longrightarrow Fe_2O_3(s) + 3CO(g) \qquad \Delta H° = +26.7 \text{ kJ}$$
$$3CO(g) + \tfrac{3}{2}O_2(g) \longrightarrow 3CO_2(g) \qquad \Delta H° = -849.0 \text{ kJ}$$

Sum: $\qquad Fe(s) + \tfrac{3}{2}O_2(g) \longrightarrow Fe_2O_3(s) \qquad \Delta H° = -822.3 \text{ kJ}$

■ **Practice Exercise 5** Ethanol, C_2H_5OH, is made industrially by the reaction of water with ethylene, C_2H_4. Calculate the value of $\Delta H°$ for the reaction

$$C_2H_4(g) + H_2O(l) \longrightarrow C_2H_5OH(l)$$

given the following thermochemical equations:

$$C_2H_4(g) + 3O_2(g) \longrightarrow 2CO_2(g) + 2H_2O(l) \qquad \Delta H° = -1411.1 \text{ kJ}$$

$$C_2H_5OH(l) + 3O_2(g) \longrightarrow 2CO_2(g) + 3H_2O(l) \qquad \Delta H° = -1367.1 \text{ kJ}$$

5.6
STANDARD HEATS OF FORMATION AND HESS'S LAW

Appendix E.1 gives a much larger table of standard enthalpies of formation.

The **standard enthalpy of formation, $\Delta H_f°$,** of a substance, also called its **standard heat of formation,** is the amount of heat absorbed or evolved when *one mole* of the substance is formed at 25 °C and 1 atm from its elements in their *standard states*. An element is in its **standard state** when it is at 25 °C and 1 atm and in its most stable form and physical state (solid, liquid, or gas). Oxygen, for example, is in its standard state only as a gas at 25 °C and 1 atm and only as O_2 molecules, not as O atoms or O_3 (ozone) molecules. Carbon must be in the form of graphite, not diamond, to be in its standard state, because the graphite form of carbon is the more stable form under standard conditions.

Extensive tables for standard enthalpies of formation, like Table 5.2, are available. All values of $\Delta H_f°$ for elements (in their standard states) are zero. (Forming an element *from itself*, of course, has no meaning.)

It is important to keep in mind the significance of the subscript f in the symbol $\Delta H_f°$. This subscript is applied to a value of $\Delta H°$ only when *one mole* of the substance is formed *from its elements in their standard states*. Below are listed four thermochemical equations and their corresponding values of $\Delta H°$.

$$H_2(g) + \tfrac{1}{2}O_2(g) \longrightarrow H_2O(l) \qquad \Delta H_f° = -285.9 \text{ kJ/mol}$$

$$2H_2(g) + O_2(g) \longrightarrow 2H_2O(l) \qquad \Delta H° = -571.8 \text{ kJ}$$

$$CO(g) + \tfrac{1}{2}O_2(g) \longrightarrow CO_2(g) \qquad \Delta H° = -283.0 \text{ kJ}$$

$$2H(g) + O(g) \longrightarrow H_2O(l) \qquad \Delta H° = -971.1 \text{ kJ}$$

Only in the first equation can $\Delta H°$ have the subscript f. It is the only reaction that satisfies the conditions specified above. The second equation shows the formation of *two* moles of water, not one. The third involves a *compound* as one of the reactants. The fourth involves the elements as *atoms,* which are *not standard states* for these elements. Also notice that the units of $\Delta H_f°$ are kilojoules *per mole,* not just kilojoules. We can obtain the enthalpy of formation of two moles of water ($\Delta H°$ for the second equation) simply by multiplying the $\Delta H_f°$ value for 1 mol of H_2O by the factor 2.

$$\left(\frac{-285.9 \text{ kJ}}{\text{mol } H_2O(l)}\right) \times 2 \text{ mol } H_2O(l) = -571.8 \text{ kJ}$$

EXAMPLE 5.6
Writing an Equation for a Standard Heat of Formation

What equation must be used to represent the formation of nitric acid, $HNO_3(l)$, when we want to include its value of $\Delta H_f°$?

ANALYSIS Because we are allowed to show only *one mole* of the product, we begin with its formula and take whatever fractions of moles of the elements needed to

TABLE 5.2 Standard Enthalpies of Formation of Typical Substances

Substance	ΔH_f° (kJ mol^{-1})	Substance	ΔH_f° (kJ mol^{-1})
Ag(s)	0.00	$H_2O_2(l)$	-187.6
AgBr(s)	-100.4	HBr(g)	-36
AgCl(s)	-127.0	HCl(g)	-92.30
Al(s)	0.00	HI(g)	26.6
$Al_2O_3(s)$	-1669.8	$HNO_3(l)$	-173.2
C(s, graphite)	0.00	$H_2SO_4(l)$	-811.32
CO(g)	-110.5	$HC_2H_3O_2(l)$	-487.0
$CO_2(g)$	-393.5	Hg(l)	0.00
$CH_4(g)$	-74.848	Hg(g)	60.84
$CH_3Cl(g)$	-82.0	$I_2(s)$	0.00
$CH_3I(g)$	14.2	K(s)	0.00
$CH_3OH(l)$	-238.6	KCl(s)	-435.89
$CO(NH_2)_2(s)$ (urea)	-333.19	$K_2SO_4(s)$	-1433.7
$CO(NH_2)_2(aq)$	-319.2	$N_2(g)$	0.00
$C_2H_2(g)$	226.75	$NH_3(g)$	-46.19
$C_2H_4(g)$	52.284	$NH_4Cl(s)$	-315.4
$C_2H_6(g)$	-84.667	NO(g)	90.37
$C_2H_5OH(l)$	-277.63	$NO_2(g)$	33.8
Ca(s)	0.00	$N_2O(g)$	81.57
$CaBr_2(s)$	-682.8	$N_2O_4(g)$	9.67
$CaCO_3(s)$	-1207	$N_2O_5(g)$	11
$CaCl_2(s)$	-795.0	Na(s)	0.00
CaO(s)	-635.5	$NaHCO_3(s)$	-947.7
$Ca(OH)_2(s)$	-986.59	$Na_2CO_3(s)$	-1131
$CaSO_4(s)$	-1432.7	NaCl(s)	-411.0
$CaSO_4 \cdot \frac{1}{2}H_2O(s)$	-1575.2	NaOH(s)	-426.8
$CaSO_4 \cdot 2H_2O(s)$	-2021.1	$Na_2SO_4(s)$	-1384.5
$Cl_2(g)$	0.00	$O_2(g)$	0.00
Fe(s)	0.00	Pb(s)	0.00
$Fe_2O_3(s)$	-822.2	PbO(s)	-219.2
$H_2(g)$	0.00	S(s)	0.00
$H_2O(g)$	-241.8	$SO_2(g)$	-296.9
$H_2O(l)$	-285.9	$SO_3(g)$	-395.2

make it. We also remember to include the physical states. Table 5.2 gives the value of ΔH_f° for $HNO_3(l)$, -173.2 kJ mol^{-1}.

SOLUTION The three elements all occur as diatomic molecules in the gaseous state, so the following fractions of moles supply exactly enough to make one mole of HNO_3.

$$\tfrac{1}{2}H_2(g) + \tfrac{1}{2}N_2(g) + \tfrac{3}{2}O_2(g) \longrightarrow HNO_3(l) \qquad \Delta H_f^\circ = -173.2 \text{ kJ mol}^{-1}$$

■ **Practice Exercise 6** Write the thermochemical equation that would be used to represent the standard heat of formation of sodium bicarbonate, $NaHCO_3(s)$.

Standard enthalpies of formation are useful because they provide a convenient method for applying Hess's law without the need for manipulating thermochemical equations. To see how this works, let's consider the reaction

$$SO_3(g) \longrightarrow SO_2(g) + \tfrac{1}{2}O_2(g) \qquad \Delta H° = ?$$

for which we wish to determine the heat of reaction.

To apply standard enthalpies of formation, we need to imagine a special path from the reactant to the products, one that involves first decomposing SO_3 into its elements in their standard states and then recombining the elements to form the products. This path is shown in Figure 5.3. The first step, whose enthalpy change is indicated as $\Delta H_1°$, corresponds to the decomposition of SO_3 into sulfur and oxygen. This is just the *reverse* of the formation of SO_3, so we can write

$$\Delta H_1° = -\Delta H_{fSO_3(g)}°$$

The minus sign is written because when we reverse a process, we change the sign of its ΔH.

The second step in Figure 5.3, whose enthalpy change is indicated as $\Delta H_2°$, is for the formation of SO_2 plus half a mole of O_2 from sulfur and oxygen. Therefore, we can write

$$\Delta H_2° = \Delta H_{fSO_2(g)}° + \tfrac{1}{2}\Delta H_{fO_2(g)}°$$

The sum of these two steps gives the net change we want, so the sum of $\Delta H_1°$ and $\Delta H_2°$ must equal the desired $\Delta H°$.

$$\Delta H° = \Delta H_1° + \Delta H_2°$$

By substitution

$$\Delta H° = [-\Delta H_{fSO_3(g)}°] + [\Delta H_{fSO_2(g)}° + \tfrac{1}{2}\Delta H_{fO_2(g)}°]$$

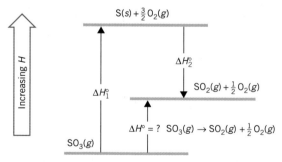

FIGURE 5.3

Enthalpy diagram for the reaction

$$SO_3(g) \longrightarrow SO_2(g) + \tfrac{1}{2}O_2(g)$$

The path of the reaction in this diagram involves the elements in their standard states as the intermediate state. Here we see the reactant being decomposed into its elements (longer upward-pointing arrow); then the elements are recombined to form the products (short downward-pointing arrow). The difference in the lengths of these two arrows is proportional to the net enthalpy change (shorter upward-pointing arrow).

This can be rewritten as

$$\Delta H^\circ = [\Delta H^\circ_{\mathrm{fSO}_2(g)} + \tfrac{1}{2}\Delta H^\circ_{\mathrm{fO}_2(g)}] + [-\Delta H^\circ_{\mathrm{fSO}_3(g)}]$$

The change of sign gives

$$\Delta H^\circ = [\Delta H^\circ_{\mathrm{fSO}_2(g)} + \tfrac{1}{2}\Delta H^\circ_{\mathrm{fO}_2(g)}] - [\Delta H^\circ_{\mathrm{fSO}_3(g)}]$$

The net result is that ΔH° equals the sum of the heats of formation of the products of the reaction minus the sum of the heats of formation of the reactant(s), each multiplied by the appropriate coefficients. The general form of Hess's law is

$$\Delta H_{\mathrm{reaction}} = \left[\begin{array}{c}\text{Sum of } \Delta H_f \text{ of all} \\ \text{of the products}\end{array}\right] - \left[\begin{array}{c}\text{Sum of } \Delta H_f \text{ of all} \\ \text{of the reactants}\end{array}\right] \qquad (5.5)$$

 Hess's law equation

Some chefs keep baking soda, $NaHCO_3$, handy to put out grease fires. When thrown on the fire, baking soda partly smothers the fire and the heat decomposes it to give CO_2, which further smothers the flame. The equation for the decomposition of $NaHCO_3$ is

$$2NaHCO_3(s) \longrightarrow Na_2CO_3(s) + H_2O(l) + CO_2(g)$$

Use the data in Table 5.2 to calculate the ΔH° for this reaction in kilojoules.

ANALYSIS Hess's law equation (Equation 5.5) is now our basic tool for computing values of ΔH°. So we calculate the values of ΔH° for the products, taken as a set, and the values of ΔH° for the reactants, also taken as a set. Then we subtract the latter from the former to calculate ΔH°. We compute values of ΔH° using ΔH°_f data from Table 5.2 and the coefficients of the chemical equation.

SOLUTION

$$\begin{aligned}\Delta H^\circ = &[1 \text{ mol } Na_2CO_3(s) \times \Delta H^\circ_{\mathrm{fNa}_2\mathrm{CO}_3(s)} + 1 \text{ mol } H_2O(l) \times \Delta H^\circ_{\mathrm{fH}_2\mathrm{O}(l)} \\ &+ 1 \text{ mol } CO_2(g) \times \Delta H^\circ_{\mathrm{fCO}_2(g)}] - [2 \text{ mol } NaHCO_3(s) \times \Delta H^\circ_{\mathrm{fNaHCO}_3(s)}]\end{aligned}$$

We now use Table 5.2 to find the values of ΔH°_f for each substance in its proper physical state.

$$\begin{aligned}\Delta H^\circ = &[1 \text{ mol } Na_2CO_3 \times (-1131 \text{ kJ/mol } Na_3CO_3) + 1 \text{ mol } H_2O \\ &\times (-285.9 \text{ kJ/mol } H_2O) + 1 \text{ mol } CO_2 \times (-393.5 \text{ kJ/mol } CO_2)] \\ &- [2 \text{ mol } NaHCO_3 \times (-947.7 \text{ kJ/mol } NaHCO_3)]\end{aligned}$$

$$\begin{aligned}\Delta H^\circ &= (-1810 \text{ kJ}) - (-1895 \text{ kJ}) \\ &= +85 \text{ kJ}\end{aligned}$$

Thus, under standard conditions, the reaction is endothermic by 85 kJ. (Notice that we did not have to manipulate any equations.)

■ **Practice Exercise 7** Calculate ΔH° for the following reactions.

(a) $2NO(g) + O_2(g) \rightarrow 2NO_2(g)$

(b) $NaOH(s) + HCl(g) \rightarrow NaCl(s) + H_2O(l)$

EXAMPLE 5.7
Using Hess's Law and Standard Enthalpies of Formation

SUMMARY

Energy Sources and Units Because electrical attractions and repulsions occur within atoms, molecules, and ions, substances have **potential energy,** often called **chemical energy.** As these particles of substances rearrange during chemical reactions, some potential energy might be converted into heat that evolves; such reactions are **exothermic.** In **endothermic** reactions, the particles absorb energy from their surroundings which becomes part of the potential energy of the products. **Photosynthesis** in plants converts solar energy into the chemical energy of the compounds made by the plants from simple molecules. The study of energy changes in reactions is called **thermochemistry.**

Specific Heat and Heat Capacity Two thermal properties that all substances have are *specific heat* and *heat capacity*. Heat capacity, like mass, is a function of the size of the sample. Hence, it is more significant to define a quantity for each substance that gives us the heat capacity per gram (or per mole). **Heat capacity** is the number of joules needed to change the temperature of the entire sample by one degree Celsius. The **specific heat** is the heat capacity per gram; if calculated on a per mole basis, the quantity is called the **molar heat capacity.** Water has an unusually high specific heat.

Thermochemistry The **heat of reaction,** q, can be calculated from the temperature change caused when known quantities of reactants undergo the reaction in a **system** of known heat capacity. The system might give off heat to its **surroundings,** or it might take heat from the surroundings. The **law of conservation of energy** ensures that the energy which crosses the **boundary** from a system is identical in magnitude to the energy that enters its surroundings. The value of q depends not just on the particular reaction but also on its size or scale, the temperature of its system, the pressure, and other variables that define the **state of the system.**

When the system is under constant pressure, the heat of the reaction is called its **enthalpy change,** ΔH. Exothermic reactions have negative values of ΔH; endothermic changes have positive values. (Only enthalpy *changes* can be determined, not absolute values of enthalpy.)

An enthalpy change is a **state function,** because its value depends only on the initial and final states of the system and not on the path taken between these states.

The law of conservation of energy ensures that an enthalpy change for a forward reaction has the same magnitude as but an opposite sign to the enthalpy change for the reverse reaction (assuming that the conditions are otherwise constant).

Enthalpy changes can be determined by **calorimetry.** A "coffee cup" system works well when low precision is acceptable.

Standard Heats of Reaction The reference conditions for thermochemistry, called the **standard conditions,** are 25 °C and 1 atm of pressure. An enthalpy change measured under these conditions is called the **standard enthalpy of reaction** or the **standard heat of reaction.** Its symbol is $\Delta H°$, where the degree sign signifies that standard conditions are involved. The value of $\Delta H°$ is a function of the amounts of substances involved in the reaction. If the number of moles of reactants is doubled, the standard heat of the reaction is twice as large. The units for $\Delta H°$ are generally joules or kilojoules.

When the enthalpy change is for the formation of *one* mole of a substance under standard conditions from its elements *in their standard states,* it is called the **standard heat of formation** of the compound, symbolized as $\Delta H_f°$, with units of kilojoules per mole (kJ mol^{-1}).

A balanced chemical equation that includes the enthalpy change and that specifies the physical states of the substances is called a **thermochemical equation.** Thermochemical equations can be added, reversed (reversing also the sign of the enthalpy change), multiplied by a constant multiplier (doing the same to the enthalpy change), and similarly manipulated. If formulas are canceled or added, they must be of substances in identical physical states.

Hess's Law **Hess's law of heat summation** is possible because ΔH is a state function and does not depend on the kinds or numbers of steps between the initial and final states. The ΔH of a reaction can be calculated as the sum of the ΔH_f values of the products minus the sum of the ΔH_f values of the reactants. The coefficients in the equation and the values of $\Delta H_f°$ from tables are used to calculate values of ΔH_f for products and reactants. Otherwise, values of ΔH can be determined by the manipulation of any combination of thermochemical equations that add up to the final equation.

Tools you have learned

The table below lists the tools you have learned in this chapter that are applicable to problem solving. Review them if necessary, and refer to them when working on the Thinking-It-Through problems and the Review Exercises that follow.

Tool	Function
Heat capacity equation (page 195)	To use specific heat data, mass, and a temperature change to calculate a heat capacity.
Heat of reaction (page 198)	To use a heat capacity and a temperature change to calculate the heat of a reaction or to use specific heat data and a mass together with a temperature change to calculate the heat of a reaction.
Thermochemical equations (page 202)	To use thermochemical equations for one set of reactions to write a thermochemical equation for some net reaction.
Enthalpy diagrams (page 204)	To visualize enthalpy changes in multistep reactions.
Hess's law (page 211)	To use standard heats of formation to calculate the enthalpy of a reaction.

THINKING IT THROUGH

The goal for each of the following problems is to give you practice in thinking your way through problems. The goal is not to find the answer itself; instead, you are only asked to assemble the available information needed to obtain the answer, state what additional data (if any) are needed, and describe how you would use the data to answer the question. For problems involving unit conversions, list the relationships among the units that are needed to carry out the conversions. Construct the conversion factors that can be formed from these relationships. Then set up the solution to the problem by arranging the conversion factors so the units cancel correctly to give the desired units of the answer.

The problems are divided into two groups. Those in Level 2 are more challenging than those in Level 1 and provide an opportunity to really hone your problem-solving skills.

Level 1 Problems

1. Which vehicle possesses more kinetic energy and why, vehicle *X* with a mass of 2000 kg and a velocity of 50 km hr^{-1} or vehicle *Y* with a mass of 1000 kg and a velocity of 100 km hr^{-1}?

2. To calculate the heat absorbed by an object, what information is needed, other than the value of its heat capacity?

3. If you know the specific heat of an object, what else besides the temperature change has to be known before you can calculate the energy flow caused by the temperature change?

4. Which will involve the larger numerical value for 250 g of water, its specific heat or its heat capacity? Explain.

5. Which will involve the larger number, a value for a substance's specific heat in J g^{-1} °C^{-1} or cal g^{-1} °C^{-1}?

6. If you want to calculate how much energy is absorbed by a bar of iron when it has been kept in a fireplace with a blazing log fire, which of the following information is needed: the initial temperature of the iron before it went into the fire, the length of time it is in the fire, the final temperature of the iron, the mass of the iron bar, the volume of the iron bar, the temperature of the burning material, the specific heat of iron?

7. You have been told that for a given reaction the *heat of reaction* has a value of 500 kJ mol^{-1}. For this information to be meaningful, what else (besides the identity of the reaction itself) should be known?

Level 2 Problems

8. A piece of metallic lead with a mass of 51.36 g was heated to 100.0 °C in boiling water. It was quickly dried and immersed in 10.0 g of liquid water that was at a temperature of 24.6 °C. The system (water plus lead) came to a temperature of 35.2 °C. What is the specific heat of lead?

9. Suppose a truck with a mass of 14.0 tons (1 ton = 2000 lb) is traveling at a speed of 45.0 mi/h. If the truck driver slams on the brakes, the kinetic energy of the truck is changed to heat as the brakes slow the truck to a stop. How much would the temperature of 5.00 gallons of water increase if all this heat could be absorbed by the water?

10. If you want to calculate how much energy is released by a reaction in a coffee cup calorimeter, which of the following data are needed, or has something been omitted from the list? The mass of the reactant in the cup, the specific heat of the water in the cup, the length of time needed for the reaction to go to completion, the mass of the cup, the total mass of the water in the cup, the initial temperature of the system, including the reactants in the cup, the final temperature of the system.

11. How many kilojoules of heat are evolved in the combustion of 35.5 g of methyl alcohol. The *unbalanced equation* for the reaction is

$$CH_3OH(l) + O_2(g) \longrightarrow CO_2(g) + H_2O(g)$$

12. In a thermochemical equation for which the value of $\Delta H°$ is given, the units of $\Delta H°$ are simply energy units (e.g., kJ) and not energy units *per mole*. Why?

13. Would it be possible to write a thermochemical equation for the following reaction entirely from the data given after it, or would additional information be needed?

$$C_2H_2(g) + 2H_2(g) \longrightarrow C_2H_6(g) \qquad \Delta H° = ?$$
acetylene   ethane

Given:

$$C_2H_2(g) + H_2(g) \longrightarrow C_2H_4(g) \qquad \Delta H° = -175.1 \text{ kJ}$$
acetylene   ethylene

$$C_2H_4(g) + H_2(g) \longrightarrow C_2H_6(g) \qquad \Delta H° = -136.4 \text{ kJ}$$
ethylene   ethane

14. The thermochemical equation for the formation of acetylene in its standard state from the elements in their standard states is as follows.

$$2C(s, \text{graphite}) + H_2(g) \longrightarrow C_2H_2(g)$$
$$\Delta H_f° = +227 \text{ kJ mol}^{-1}$$

(a) Why are the units for $\Delta H_f°$ in kJ mol^{-1} but the units for $\Delta H°$ are simply kJ?
(b) Is the decomposition of acetylene into its elements an exothermic or an endothermic process?
(c) As you compare the standard heat of formation of acetylene with the values for most other compounds in the tables provided, do you see anything unusual about it?

REVIEW EXERCSES

Answers to questions whose numbers are printed in color are given in Appendix D. Challenging questions are marked with asterisks.

Kinetic and Potential Energy

5.1 In terms of natural attractions and repulsions between things, and in general terms only, explain how the potential energy of a system can be increased.

5.2 Do the following pairs of particles attract or repel each other?
(a) two electrons
(b) two atomic nuclei
(c) one electron and one atomic nucleus

5.3 What is meant by the term *chemical energy*?

5.4 How are the following understood in terms of the kinetic theory of matter?
(a) the fundamental *model* of matter
(b) temperature
(c) heat content of matter

5.5 What do the following terms mean?
(a) system
(b) surroundings
(c) boundary

5.6 What must be true about a system to say that it is *insulated*?

5.7 What descriptive adjective do we use when a reaction liberates heat to its surroundings? When it receives heat from its surroundings?

Thermal Properties

5.8 What is the name of the thermal property whose values can have the following units?
(a) J g^{-1} °C^{-1} (b) J mol^{-1} °C^{-1} (c) J °C^{-1}

5.9 Which kind of substance needs more energy to undergo an increase of 5 °C, something with a *high* or with a *low* specific heat? Explain.

5.10 Which kind of substance experiences the larger increase in temperature when it absorbs 100 J, something with a *high* or a *low* specific heat?

5.11 If the specific heat values in Table 5.1 were in units of kJ, namely, kJ kg^{-1}K^{-1}, would the values be *numerically* different? Explain.

5.12 How much heat in kilojoules must be removed from 175 g of water to lower its temperature from 25.0 to 15.0 °C (which would be like cooling a glass of lemonade)?

5.13 How much heat in kilojoules is needed to bring 1.0 kg of water from 25 to 99 °C (comparable to making four cups of coffee)?

5.14 How many joules are needed to increase the temperature of 15.0 g of Fe from 20.0 to 40.0 °C?

5.15 The addition of 250 J to 30.0 g of copper initially at 22 °C will change its temperature to what final value?

5.16 When a 20.0 g sample of an unknown substance, but one listed in Table 5.1, received 47.0 J of energy, its temperature changed from 25.0 to 35.0 °C. What was this substance?

5.17 If 500 mL of olive oil, initially at 25.0 °C, receives 1.25 kJ of heat energy, what is its final temperature? (The density of olive oil is 0.91 g mL^{-1}.)

5.18 A 5.00 g mass of a metal was heated to 100 °C and then plunged into 100 g of water at 24.0 °C. The temperature of the resulting mixture became 28.0 °C.
(a) How many joules did the water absorb?
(b) How many joules did the metal lose?
(c) What is the heat capacity of the metal sample?
(d) What is the specific heat of the metal?

5.19 A sample of copper was heated to 120 °C and then thrust into 200 g of water at 25.00 °C. The temperature of the mixture became 26.50 °C.
(a) How much heat in joules was absorbed by the water?
(b) The copper sample lost how many joules?
(c) What was the mass in grams of the copper sample?

5.20 Fat tissue is 85% fat and 15% water. The complete breakdown of the fat itself converts it to CO_2 and H_2O, which releases 9.0 kcal per gram of the fat in the tissue.
(a) How many kilocalories are released by the loss of 1.0 lb of fat *tissue* in a weight-reduction program?
(b) A person running 8.0 miles hr^{-1} expends about 5.0×10^2 kcal hr^{-1} of extra energy. How far does a person have to run to burn off 1.0 lb of fat *tissue* by this means alone?

5.21 A well-nourished person adds about 0.50 lb of fat tissue for each 3.5×10^2 kcal of food energy taken in over and above that needed. Suppose you decided to reduce your weight simply by omitting butter but keeping every other aspect of your diet and your activities the same. How many days would be needed to lose 1.0 kg of fat *tissue* by this strategy alone? The caloric content of dietary fats and oils is 9.0 kcal g^{-1}. Suppose that you have been eating 0.25 lb of salad oil and butter per day.

5.22 Calculate the molar heat capacity of iron in J mol^{-1} °C^{-1}. Its specific heat is 0.4498 J g^{-1} °C^{-1}.

5.23 What is the molar heat capacity of ethyl alcohol, C_2H_5OH, in units of J mol^{-1}, if its specific heat is 0.586 cal g^{-1} °C^{-1}?

5.24 A vat of 4.54 kg of water underwent a decrease in temperature from 60.25 to 58.65 °C. How much energy in kilojoules left the water? (For this range of temperature, use a value of 4.18 J g^{-1} °C^{-1} for the specific heat of water.)

Calorimetry

5.25 Nitric acid, HNO_3, reacts with potassium hydroxide, KOH, as follows: $HNO_3(aq) + KOH(aq) \rightarrow KNO_3(aq) + H_2O(l)$. A student placed 55.0 mL of 1.3 M HNO_3 in a coffee cup calorimeter, noted that the temperature was 23.5 °C, and added 55.0 mL of 1.3 M KOH, also at 23.5 °C. After quickly stirring the mixture with a thermometer, its temperature rose to 31.8 °C^{-1}. Calculate the heat evolved in joules. Assume that the specific heats of all solutions are 4.18 J g^{-1} °C^{-1} and that all densities are 1.00 g mL^{-1}. Calculate the heat evolved per mole of acid (in units of kJ mol^{-1}).

*5.26 A dilute solution of hydrochloric acid with a mass of 610.29 g and containing 0.33183 mol of HCl was exactly neutralized in a calorimeter by the sodium hydroxide in 615.31 g of a comparably dilute solution. The temperature increased from 16.784 to 20.610 °C. The specific heat of the hydrochloric acid solution was 4.031 J g^{-1} °C^{-1}; that of the sodium hydroxide solution was 4.046 J g^{-1} °C^{-1}. The heat capacity of the calorimeter was 77.99 J °C^{-1}. Use these data to calculate heat evolved, which occurs by the following equation: $HCl(aq) + NaOH(aq) \rightarrow NaCl(aq) + H_2O(l)$. Assume that the original solutions made independent contributions to the total heat capacity of the system following their mixing.

*5.27 The hydrochloric acid in a dilute solution with a mass of 610.28 g, a specific heat of 4.031 J g^{-1} °C^{-1}, and containing 0.33143 mol of HCl reacted exactly with the KOH in 619.69 g of KOH solution, specific heat 4.003 J g^{-1} °C^{-1}. The equation is

$$HCl(aq) + KOH(aq) \longrightarrow KCl(aq) + H_2O(l)$$

The calorimeter's heat capacity was 77.99 J g^{-1}. The reaction caused the system's temperature to change from 15.533 to 19.410 °C. Use these data to calculate the heat evolved in units of kJ mol^{-1}. (Assume that the individual solutions make independent contributions to the total heat capacity of the system when they are mixed.)

Enthalpy and Heats of Reaction at Constant Pressure

5.28 What are some facts about a system that we should specify to describe the *state* of the system. What is meant by the term *state function*? Give some examples.

5.29 What is meant by the *enthalpy* of a system?

5.30 What equation defines ΔH in general terms? When the system involves a chemical reaction?

5.31 If the enthalpy of a system increases by 100 kJ, what must be true about the enthalpy of the surroundings? Why?

5.32 What is the sign of ΔH for an exothermic change?

5.33 Why do standard reference values for temperature and pressure have to be selected when we consider and compare heats of reaction for various reactions? What are the values for the standard temperature and standard pressure?

5.34 What distinguishes a *thermochemical* equation from an ordinary chemical equation?

5.35 If a thermochemical equation has a coefficient of 1/2 for a formula, what *unit* does it signify?

5.36 One thermochemical equation for the reaction of carbon monoxide with oxygen is

$$3CO(g) + \tfrac{3}{2}O_2(g) \longrightarrow 3CO_2(g) \qquad \Delta H° = -849 \text{ kJ}$$

(a) Write the thermochemical equation using 2 mol of CO.
(b) What is $\Delta H°$ for the formation of 1 mol of CO_2 by this reaction?

5.37 Ammonia reacts with oxygen as follows.

$$4NH_3(g) + 7O_2(g) \longrightarrow 4NO_2(g) + 6H_2O(g)$$
$$\Delta H° = -1132 \text{ kJ}$$

(a) Calculate the enthalpy change for the reaction of 1 mol of NH_3.

(b) Write the thermochemical equation for the reaction of 1 mol of O_2.

5.38 Aluminum and iron(III) oxide, Fe_2O_3, react to form aluminum oxide, Al_2O_3, and iron. For each mole of aluminum used, 426.9 kJ of energy is released under standard conditions. Write the thermochemical equation that shows the consumption of 4 mol of Al. (All of the substances are solids.)

5.39 Benzene, C_6H_6, burns according to the following equation.

$$2C_6H_6(l) + 15O_2(g) \longrightarrow 12CO_2(g) + 6H_2O(l)$$
$$\Delta H° = -6542 \text{ kJ}$$

What is $\Delta H°$ for the combusion of 1.500 mol of benzene?

5.40 The following equation represents the dehydration of calcium hydroxide to make "quick lime," CaO, an ingredient in cement.

$$Ca(OH)_2(s) \longrightarrow CaO(s) + H_2O(l) \qquad \Delta H° = +65.3 \text{ kJ}$$

When water is mixed with cement, one of the reactions is the reverse of this dehydration. Write the thermochemical equation for the reaction of 10 mol of CaO with water under standard conditions.

Hess's Law

5.41 Construct an enthalpy diagram that shows the enthalpy changes for a one-step conversion of germanium, Ge(s), into its dioxide, $GeO_2(s)$. On the same diagram, show the two-step process, first to the monoxide, GeO(s), and then its conversion to the dioxide. The relevant thermochemical equations are the following.

$$Ge(s) + \tfrac{1}{2}O_2(g) \longrightarrow GeO(s) \qquad \Delta H° = -255 \text{ kJ}$$
$$Ge(s) + O_2(g) \longrightarrow GeO_2(s) \qquad \Delta H° = -534.7 \text{ kJ}$$

Using this diagram, determine $\Delta H°$ for the following reaction.

$$GeO(s) + \tfrac{1}{2}O_2(g) \longrightarrow GeO_2(s)$$

5.42 Construct an enthalpy diagram for the formation of $NO_2(g)$ from its elements by two pathways. First, from its elements, and second by a two-step process, also from the elements. The relevant thermochemical equations are:

$$\tfrac{1}{2}N_2(g) + O_2(g) \longrightarrow NO_2(g) \qquad \Delta H° = +33.8 \text{ kJ}$$
$$\tfrac{1}{2}N_2(g) + \tfrac{1}{2}O_2(g) \longrightarrow NO(g) \qquad \Delta H° = +90.37 \text{ kJ}$$
$$NO(g) + \tfrac{1}{2}O_2(g) \longrightarrow NO_2(g) \qquad \Delta H° = ?$$

Be sure to note the signs of the values of $\Delta H°$, because in an enthalpy diagram downward-pointing arrows signify exothermic steps and upward-pointing arrows are for endothermic steps. Using the diagram, determine the value of $\Delta H°$ for the third equation.

5.43 What is Hess's law of heat summation? What important property of enthalpy makes Hess's law possible?

5.44 Show how the equations:

$$N_2O_4(g) \longrightarrow 2NO_2(g) \qquad \Delta H° = +57.93 \text{ kJ}$$
$$2NO(g) + O_2(g) \longrightarrow 2NO_2(g) \qquad \Delta H° = -113.14 \text{ kJ}$$

can be manipulated to give $\Delta H°$ for the following reaction.

$$2NO(g) + O_2(g) \longrightarrow N_2O_4(g)$$

5.45 Show how the thermochemical equations of Review Exercise 5.44 can be manipulated to give the thermochemical equation for the conversion of 1 mol of NO(g) into $NO_2(g)$.

5.46 We can generate hydrogen chloride by heating a mixture of sulfuric acid and potassium chloride according to the equation

$$2KCl(s) + H_2SO_4(l) \longrightarrow 2HCl(g) + K_2SO_4(s)$$

Calculate $\Delta H°$ in kilojoules for this reaction from the following thermochemical equations.

$$HCl(g) + KOH(s) \longrightarrow KCl(s) + H_2O(l)$$
$$\Delta H° = -203.6 \text{ kJ}$$
$$H_2SO_4(l) + 2KOH(s) \longrightarrow K_2SO_4(s) + 2H_2O(l)$$
$$\Delta H° = -342.4 \text{ kJ}$$

5.47 Calculate $\Delta H°$ in kilojoules for the following reaction, the preparation of the unstable acid nitrous acid, HNO_2.

$$HCl(g) + NaNO_2(s) \longrightarrow HNO_2(l) + NaCl(s)$$

Use the following thermochemical equations.

$$2NaCl(s) + H_2O(l) \longrightarrow 2HCl(g) + Na_2O(s)$$
$$\Delta H° = +507.31 \text{ kJ}$$
$$NO(g) + NO_2(g) + Na_2O(s) \longrightarrow 2NaNO_2(s)$$
$$\Delta H° = -427.14 \text{ kJ}$$
$$NO(g) + NO_2(g) \longrightarrow N_2O(g) + O_2(g)$$
$$\Delta H° = -42.68 \text{ kJ}$$
$$2HNO_2(l) \longrightarrow N_2O(g) + O_2(g) + H_2O(l)$$
$$\Delta H° = +34.35 \text{ kJ}$$

5.48 Barium oxide reacts with sulfuric acid as follows.

$$BaO(s) + H_2SO_4(l) \longrightarrow BaSO_4(s) + H_2O(l)$$

Calculate $\Delta H°$ in kilojoules for this reaction. The following thermochemical equations can be used.

$$SO_3(g) + H_2O(l) \longrightarrow H_2SO_4(l) \qquad \Delta H° = -78.2 \text{ kJ}$$
$$BaO(s) + SO_3(g) \longrightarrow BaSO_4(s) \qquad \Delta H° = -213 \text{ kJ}$$

5.49 Exothermic reactions occur spontaneously more frequently than do endothermic changes. In which direction will the following system most likely react, left to right or right to left? Calculate $\Delta H°$ for the reaction and then answer the question.

$$2Ag(s) + Zn(NO_3)_2(aq) \longrightarrow Zn(s) + 2AgNO_3(aq)$$

The following thermochemical equations can be used.

$$Cu(NO_3)_2(aq) + Zn(s) \longrightarrow Zn(NO_3)_2(aq) + Cu(s)$$
$$\Delta H° = -258 \text{ kJ}$$
$$2AgNO_3(aq) + Cu(s) \longrightarrow Cu(NO_3)_2(aq) + 2Ag(s)$$
$$\Delta H° = -106 \text{ kJ}$$

5.50 Copper metal can be obtained by heating copper oxide, CuO, in the presence of carbon monoxide, CO, according to the following reaction.

$$CuO(s) + CO(g) \longrightarrow Cu(s) + CO_2(g)$$

Calculate $\Delta H°$ in kilojoules, using the following thermochemical equations.

$$2CO(g) + O_2(g) \longrightarrow 2CO_2(g) \qquad \Delta H° = -566.1 \text{ kJ}$$
$$2Cu(s) + O_2(g) \longrightarrow 2CuO(s) \qquad \Delta H° = -310.5 \text{ kJ}$$

5.51 Calcium hydroxide reacts with hydrochloric acid by the following equation.

$$Ca(OH)_2(aq) + 2HCl(aq) \longrightarrow CaCl_2(aq) + 2H_2O(l)$$

Calculate $\Delta H°$ in kilojoules for this reaction, using the following equations as needed.

$$CaO(s) + 2HCl(aq) \longrightarrow CaCl_2(aq) + H_2O(l)$$
$$\Delta H° = -186 \text{ kJ}$$
$$CaO(s) + H_2O(l) \longrightarrow Ca(OH)_2(s) \qquad \Delta H° = -62.3 \text{ kJ}$$
$$Ca(OH)_2(s) \xrightarrow{\text{dissolving in water}} Ca(OH)_2(aq)$$
$$\Delta H° = -12.6 \text{ kJ}$$

5.52 Calculate the $\Delta H°$ for the following reaction using the given thermochemical equations and others, as needed, that use data in tables in this chapter.

$$LiOH(aq) + HCl(aq) \longrightarrow LiCl(aq) + H_2O(l)$$
$$Li(s) + \tfrac{1}{2}O_2(g) + \tfrac{1}{2}H_2(g) \longrightarrow LiOH(s) \quad \Delta H° = -487.0 \text{ kJ}$$
$$2Li(s) + Cl_2(g) \longrightarrow 2LiCl(s) \qquad \Delta H° = -815.0 \text{ kJ}$$
$$LiOH(s) \xrightarrow{\text{dissolving in water}} LiOH(aq) \qquad \Delta H° = -19.2 \text{ kJ}$$
$$HCl(g) \xrightarrow{\text{dissolving in water}} HCl(aq) \qquad \Delta H° = -77.0 \text{ kJ}$$
$$LiCl(s) \xrightarrow{\text{dissolving in water}} LiCl(aq) \qquad \Delta H° = -36.0 \text{ kJ}$$

Hess's Law and Standard Heats of Formation

5.53 Which of the following equations has a value of $\Delta H°$ that would properly be labeled as $\Delta H_f°$?
(a) $CaCO_3(s) \rightarrow CaO(s) + CO_2(g)$
(b) $Ca(s) + \tfrac{1}{2}O_2(g) \rightarrow CaO(s)$
(c) $2Fe(s) + O_2(g) \rightarrow 2FeO(s)$
(d) $SO_2(g) + \tfrac{1}{2}O_2(g) \rightarrow SO_3(g)$

5.54 Write the thermochemical equations, including values of $\Delta H_f°$ in kilojoules per mole (from Table 5.2), for the formation of each of the following compounds from their elements, everything in standard states.
(a) $HC_2H_3O_2(l)$, acetic acid
(b) $NaHCO_3(s)$, sodium bicarbonate
(c) $CaSO_4 \cdot 2H_2O(s)$, gypsum
(d) $CaSO_4 \cdot \tfrac{1}{2}H_2O(s)$, plaster of paris
(e) $CH_3OH(l)$, methyl alcohol

5.55 Using data in Table 5.2, calculate $\Delta H°$ in kilojoules for the following reactions.
(a) $2H_2O_2(l) \rightarrow 2H_2O(l) + O_2(g)$
(b) $HCl(g) + NaOH(s) \rightarrow NaCl(s) + H_2O(l)$
(c) $CH_4(g) + Cl_2(g) \rightarrow CH_3Cl(g) + HCl(g)$
(d) $2NH_3(g) + CO_2(g) \rightarrow CO(NH_2)_2(s) + H_2O(l)$

5.56 Write the thermochemical equations that would be associated with the standard heats of formation of the following compounds.
(a) $HCl(g)$
(b) $NH_4Cl(s)$
(c) $CO(NH_2)_2(s)$ (urea)
(d) $Na_2CO_3(s)$

Additional Problems

5.57 Iron metal can be obtained from iron ore, which we can assume for this problem to be Fe_2O_3, by its reaction with hot carbon.

$$2Fe_2O_3(s) + 3C(s) \longrightarrow 4Fe(s) + 3CO_2(g)$$

What is $\Delta H°$ (in kJ) for this reaction? Is the reaction exothermic or endothermic?

5.58 The combustion of acetylene, $C_2H_2(g)$, gives $CO_2(g)$ and $H_2O(l)$. The standard enthalpy change, $\Delta H°$, is $-1299.61 \text{ kJ mol}^{-1} C_2H_2$. Write the thermochemical equation for this reaction which involves 1 mol of $C_2H_2(g)$.

5.59 Phosphorus burns in air to give tetraphosphorus decaoxide.

$$4P(s) + 5O_2(g) \longrightarrow P_4O_{10}(s) \qquad \Delta H° = -3062 \text{ kJ}$$

The product combines with water to give phosphoric acid, H_3PO_4.

$$P_4O_{10}(s) + 6H_2O(l) \longrightarrow 4H_3PO_4(l) \qquad \Delta H° = -257.2 \text{ kJ}$$

Using these equations (and any others in the chapter, as needed), write the thermochemical equation for the formation of 1 mol of $H_3PO_4(l)$ from the elements. Include $\Delta H_f°$.

5.60 The amino acid glycine, $C_2H_5NO_2$, is one of the compounds used by the body to make proteins. The equation for its combustion is

$$4C_2H_5NO_2(s) + 9O_2(g) \longrightarrow$$
$$8CO_2(g) + 10H_2O(l) + 2N_2(g)$$

For each mole of glycine that burns, 973.49 kJ of heat is liberated. Use this information plus values of $\Delta H_f°$ for the products of combustion to calculate $\Delta H_f°$ for glycine.

5.61 The value of $\Delta H_f°$ for $HBr(g)$ was first evaluated using the following standard enthalpy values obtained experimentally. Write the thermochemical equation for the formation of 1 mol of $HBr(g)$ from its elements, including the value of $\Delta H_f°$.

$$Cl_2(g) + 2KBr(aq) \longrightarrow Br_2(aq) + 2KCl(aq)$$
$$\Delta H° = -96.2 \text{ kJ}$$
$$H_2(g) + Cl_2(g) \longrightarrow 2HCl(g) \qquad \Delta H° = -184 \text{ kJ}$$
$$HCl(aq) + KOH(aq) \longrightarrow KCl(aq) + H_2O(l)$$
$$\Delta H° = -57.3 \text{ kJ}$$
$$HBr(aq) + KOH(aq) \longrightarrow KBr(aq) + H_2O(l)$$
$$\Delta H° = -57.3 \text{ kJ}$$
$$HCl(g) \xrightarrow{\text{dissolving in water}} HCl(aq) \qquad \Delta H° = -77.0 \text{ kJ}$$
$$Br_2(l) \xrightarrow{\text{dissolving in water}} Br_2(aq) \qquad \Delta H° = -4.2 \text{ kJ}$$
$$HBr(g) \xrightarrow{\text{dissolving in water}} HBr(aq) \qquad \Delta H° = -79.9 \text{ kJ}$$

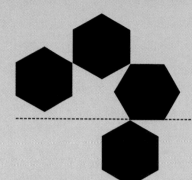

CHEMICALS IN USE 4
The Greenhouse Effect

The Earth's average temperature would be over 30 degrees colder except for certain gases in the atmosphere, particularly carbon dioxide, methane, dinitrogen monoxide (N_2O), water vapor, and some synthetics. These "greenhouse gases" have an Earth-insulating or **greenhouse effect** because their molecules are able to absorb part of the energy that the Earth continually radiates to outer space and then reradiate some of the energy back to Earth, thus keeping the Earth warmer than otherwise. (A real greenhouse works differently; it prevents air movements from taking heat away.)

EARTH'S ENERGY BALANCE

A nearly perfect balance exists between the energy received at the Earth's surface, mostly from the sun but some from the Earth's interior, and the energy that the Earth loses, mostly by radiation. Without such a balance, the Earth's average temperature would either increase or decrease drastically.

As shown in Figure 4a, for every 100 units of energy incoming from the sun, roughly 30 units are reflected either by clouds or by the Earth's surface. The rest is absorbed, 25 units by clouds and 45 units by the ground. A net of 70 energy units leaves the planet as infrared radiation ("heat rays"). Many processes in the lower atmosphere contribute to the 70 units, including the heating caused by updrafts of air and the formation of clouds as well as by radiation from the ground. Of the 70 units that leave the earth for outer space, a net of 16 units is the difference between the energy radiated by the ground and that trapped by the greenhouse effect. For every 104 units radiated

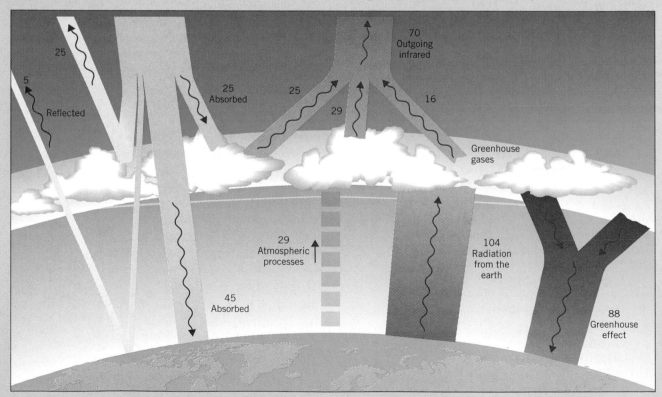

FIGURE 4a

Greenhouse gases insulate the Earth and help to prevent a reduction in global temperature. (Adapted by permission from S. H. Schneider, "The Changing Climate," *Scientific American*, September 1989, page 70. All rights reserved.)

from the ground, 88 units are trapped and returned to Earth.

THE GREENHOUSE EFFECT AND CLIMATE

When news services speak of the "greenhouse effect," they refer not to the normal phenomenon just described but to a possible *accelerated* warming of the Earth caused by increasing levels of the greenhouse gases. The concern is not about the presence of greenhouse gases—they are nearly all of natural origin and perform an essential service—but about too high a concentration of them.

Small changes in *average* air temperatures could have major changes in patterns of world climate and ecology. Since about 1900, the average global air temperature has risen by 0.4 °C. The average *annual* temperature of the Earth rose more than this in 1990 alone, but there have always been yearly fluctuations. If the increase of the 1990s were to continue, however, huge ecosystems, like the Arctic Ocean, the nearby Arctic tundra, mangrove swamps, and other coastal wetlands would be under severe pressure. Tropical storms and hurricanes would intensify and perhaps destroy the coral reefs that protect coastlines. It is not that the Earth has never before undergone large changes in climate with successful adaptations by most living species, but the greenhouse warming would entail an unprecedented *rate* of change with which vulnerable species could not cope.

CARBON DIOXIDE, THE CHIEF GREENHOUSE GAS

The atmosphere receives each year an estimated 9×10^{15} mol of CO_2, largely from three sources: the respiration of plants and animals (over 93%), forest fires and the burning of other vegetable matter (2%), and the burning of fossil fuels like oil, coal, and natural gas (5%). Two natural "sinks" take away over 95% of the released CO_2, each contributing about equally. One is photosynthesis by which plants, using solar energy, change CO_2, water, and certain soil or aquatic substances into complex plant compounds. The other sink is the uptake by the oceans as CO_2 reacts with ions of calcium (and magnesium) to form limestone deposits. The CO_2 uptake thus leaves roughly 5% of the CO_2 in the air, a small but important net amount.

The chief contributor of the extra CO_2 is the accelerated burning of fossil fuels, particularly throughout the last 100 years. This burning annually adds more to the greenhouse gas level than all of the other greenhouse gases combined, about 55% of the total additions.

Since 1900, the average atmospheric CO_2 level has risen from 280 to 345 ppm,[1] and some specialists see this going to 600 ppm in another 50 to 75 years, maybe in less time.

OTHER GREENHOUSE GASES

Methane—according to one estimate, about 15% of the greenhouse gases—originates from several sources, chiefly the action of bacteria in soil, in wetlands and rice paddies, and in the digestive tracts of animals, but also by some releases during oil and gas production. The earth has several "sinks" for methane, but they do not remove it as rapidly as it is being generated. The atmosphere's concentration of methane has risen over the last two centuries from 750 ppb to about 1700 ppb.

Dinitrogen monoxide, N_2O, which contributes a little over 5% of the total greenhouse gas concentration, arises from the action of soil bacteria on the increased amounts of fertilizers being used worldwide.

A group of synthetic compounds called the *chlorofluorocarbons* or CFCs—examples are $CFCl_3$ and CF_2Cl_2—contribute nearly 25% of the *increase* in the concentration of the greenhouse gases. Their use is being phased out, however. (The CFCs are discussed in *Chemicals in Use* 7 "Ozone in the Stratosphere," page 362.)

Current Status as Estimated by Climatologists
Has the greenhouse effect arrived? Is accelerated global warming now a certainty? The United Nations Intergovernmental Panel on Climate Change (IPCC) is a group of scientists from 25 countries seeking answers. The panel has taken into account certain Earth *cooling* effects, particularly local increases in particles hurled into the atmosphere by major volcanic eruptions. Such particles are Earth cooling because they increase the percentage of incoming sunlight reflected directly back to space. The IPCC consensus in the early 1990s was that the Earth's average temperature will increase by roughly 0.3 °C per decade so that by 2100, the average temperature of the Earth will have increased by 3 °C.

Questions

1. Name the greenhouse gases and describe their principal origins.

2. Explain briefly how the greenhouse effect works. What happens that tends to promote global warming?

3. What "sinks" are at work to reduce the CO_2 level of the atmosphere?

[1] The concentrations of trace contaminants in air are commonly expressed in *parts per million (ppm)* or *parts per billion (ppb)*. One part per million means, for example, 1 mL of gas in 1×10^6 mL of air; one part per billion means 1 mL of gas in 1×10^9 mL of air.

Navigational aids such as this, which are located along waterways to warn boaters of hazards, use solar energy to charge batteries so they can display flashing warning beacons at night. Light is one way that energy is transmitted through space, and the way atoms emit energy as light is an important topic discussed in this chapter.

Chapter 6

Atomic and Electronic Structure

In Chapter 2 you learned that an atom consists of a small dense nucleus, which contains all of the atom's neutrons and protons, surrounded by electrons that fill the remaining volume of the atom. The nucleus serves to identify which element the atom is, and it determines the number of electrons the atom must have to be electrically neutral. However, when two or more atoms join to form a compound, the nuclei of the atoms stay relatively far apart. Only the atoms' outer reaches—the regions inhabited by electrons—come in close contact. Therefore, the chemical properties of the elements must be determined by the electrons of the various atoms, and the similarities and differences in these properties must have something to do with the way these electrons are distributed around the particular nuclei.

In this chapter we will study the way electrons are arranged around the nucleus of an atom, which is called the atom's **electronic structure.** The basic clue to electronic structure comes from the study of the light emitted when atoms of the elements are *excited,* or energized. To learn about this, however, we must first learn a little about light itself.

Electromagnetic Radiation

You've learned that objects can have energy in only two ways, as kinetic energy and as potential energy. You also learned that energy can be transferred between things, and in Chapter 5 our principal focus was on the transfer of heat (actually the transfer of molecular kinetic energy through collisions between atom-sized particles). You may recall, however, that energy can also be transferred between things as light energy (more properly called *electromagnetic*

6.1
ELECTROMAGNETIC RADIATION AND ATOMIC SPECTRA

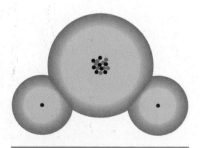

Atoms, not drawn to scale, as they are joined in a molecule of water. The nuclei stay far apart and only the outer parts of the atoms touch.

221

(a)

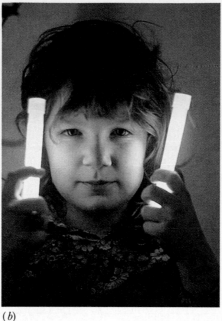

(b)

(c)

FIGURE 6.1

Light is given off in a variety of chemical reactions. (*a*) Combustion. (*b*) Cyalume light sticks. (*c*) A lightning bug.

energy). This is a very important form of energy in chemistry. For example, many chemical systems emit visible light as they react (Figure 6.1).

Electromagnetic energy is energy carried through space or matter by means of wavelike oscillations. These oscillations are systematic fluctuations in the intensities of very tiny electrical and magnetic forces. Each changes rhythmically with time, and the successive series of these oscillations that travel through space is called **electromagnetic radiation** (popularly, a light wave).

Figure 6.2 shows how the **amplitude** or intensity of the wave varies with time and with distance as the wave travels through space. In Figure 6.2*a*, we see two complete oscillations or *cycles* of the wave during a 1 second interval. The number of cycles per second is called the **frequency** of the electromagnetic radiation, and its symbol is ν (a Greek letter pronounced "new").

The concept of frequency extends beyond just electromagnetic radiation to other events that recur at regular intervals. For example, you go to school 5

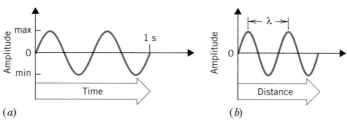

(a)   (b)

FIGURE 6.2

Two views of electromagnetic radiation. (*a*) The frequency, ν, of a light wave is the number of complete oscillations each second. Here two cycles span a 1-second time interval, so the frequency is 2 cycles per second, or 2 Hz. (*b*) Electromagnetic radiation frozen in time. This curve shows how the amplitude varies along the direction of travel. The distance between two peaks is the wavelength, λ, of the electromagnetic radiation.

days *per week,* or you pay your bills once *each month.* Each of these statements describes how frequently an event occurs, and they have in common the notion of *per unit of time,* or 1/time. In the SI, the unit of time is the second, so frequency is given the unit "per second," which is 1/second, or (second)$^{-1}$. This unit is given the special name **hertz (Hz).**

The SI symbol for the second is s.

$$s^{-1} = \frac{1}{s}$$

$$1 \text{ Hz} = 1 \text{ s}^{-1}$$

As electromagnetic radiation moves away from its source, the positions of maximum and minimum amplitude are regularly spaced. The distance separating maximum values (or minimum values) is called the radiation's **wavelength,** symbolized by λ (the Greek letter, *lambda*) (Figure 6.2*b*). Because wavelength is a distance, it has distance units (for example, meters).

If we multiply a wave's wavelength by its frequency, the result is the speed of the wave. We can see this if we analyze the units.

$$\text{meters} \times \frac{1}{\text{second}} = \frac{\text{meters}}{\text{second}} = \text{speed}$$

For any wave, the product of its wavelength and its frequency equals the speed of the wave.

$$\text{m} \times \frac{1}{\text{s}} = \frac{\text{m}}{\text{s}} = \text{m s}^{-1}$$

The speed of electromagnetic radiation in a vacuum is a constant and is commonly called the *speed of light.* Its value to three significant figures is 3.00×10^8 m/s (or m s^{-1}). This important physical constant is given the symbol c.

The speed of light is one of our most precisely measured constants, and a more precise value is given in a table on the inside rear cover of this book.

$$c = 3.00 \times 10^8 \text{ m s}^{-1}$$

From the preceding discussion we obtain a very important relationship that allows us to convert between λ and ν.

 Wavelength-frequency relationship

$$\lambda \times \nu = c = 3.00 \times 10^8 \text{ m s}^{-1} \qquad (6.1)$$

PROBLEM What is the frequency in hertz of yellow light that has a wavelength of 625 nm?

ANALYSIS To convert between wavelength and frequency, we use Equation 6.1. However, we must be careful about the units.

SOLUTION To calculate the frequency, we solve Equation 6.1 for ν.

$$\nu = \frac{c}{\lambda}$$

Next, we substitute for c (3.00×10^8 m s^{-1}) and for the wavelength. However, to cancel units correctly, we must have the wavelength in meters. Recall from Chapter 1 that nm means nanometer and the prefix nano implies the factor "$\times 10^{-9}$." Therefore, 625 nm equals 625×10^{-9} m. Substituting gives

$$\nu = \frac{3.00 \times 10^8 \text{ m s}^{-1}}{625 \times 10^{-9} \text{ m}}$$

$$= 4.80 \times 10^{14} \text{ s}^{-1}$$
$$= 4.80 \times 10^{14} \text{ Hz}$$

EXAMPLE 6.1
Calculating Frequency from Wavelength

EXAMPLE 6.2
Calculating Wavelength from
Frequency

Radio station WGBB on Long Island, New York, broadcasts its AM signal, a form of electromagnetic radiation, at a frequency of 1240 kHz. What is the wavelength of these radio waves expressed in meters?

SOLUTION This time we solve Equation 6.1 for the wavelength.

$$\lambda = \frac{c}{\nu}$$

Now we must cancel the unit s^{-1} in 3.00×10^8 m s^{-1}. The prefix "k" in kHz means kilo and stands for "$\times 10^3$," and Hz means s^{-1}. The frequency is therefore 1240×10^3 s^{-1}. Substituting gives

$$\lambda = \frac{3.00 \times 10^8 \text{ m s}^{-1}}{1240 \times 10^3 \text{ s}^{-1}}$$

$$= 242 \text{ m}$$

■ **Practice Exercise 1** A certain shade of green light has a wavelength of 550 nm. What is the frequency of this light in hertz?

■ **Practice Exercise 2** An FM radio station in West Palm Beach, Florida, broadcasts electromagnetic radiation at a frequency of 104.3 MHz (megahertz). What is the wavelength of these radio waves, expressed in meters?

The Electromagnetic Spectrum

Electromagnetic radiation comes in a large range of frequencies called the **electromagnetic spectrum,** illustrated in Figure 6.3. Each portion of the spectrum has a popular name. For example, radio waves are electromagnetic radiations having very low frequencies (and therefore very long wavelengths). Microwaves, which also have low frequencies, are emitted by radar instruments such as those the police use to monitor the speeds of cars. Microwaves are also absorbed by molecules in food, and the energy that these molecules take on raises their temperature. This is why foods cook quickly in a microwave oven. Infrared radiation is emitted by hot objects and consists of the range of frequencies that can make molecules of most substances vibrate internally. You can't see infrared radiation, but you can feel how it is absorbed by your body by holding your hand near a hot radiator; the absorbed radiation makes your hand warm. Each substance absorbs a uniquely different set of infrared frequencies. A plot of the frequencies absorbed versus the intensities of absorption is called an infrared absorption spectrum. It can be used to identify a compound, because each infrared spectrum is as unique as a set of fingerprints (Figure 6.4). Many substances absorb visible and ultraviolet radiations in unique ways, too, and they have visible and ultraviolet spectra. Gamma rays are at the high-frequency end of the electromagnetic spectrum. They are produced by certain elements that are radioactive. X rays are very much like

Remember that there is an inverse relationship between wavelength and frequency. The lower the frequency, the longer the wavelength.

F I G U R E   6.3

The electromagnetic spectrum.

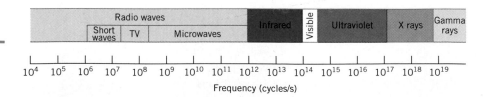

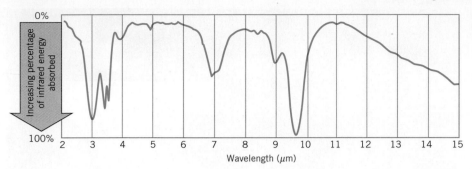

FIGURE 6.4
Infrared absorption spectrum of methyl alcohol, wood alcohol. (Courtesy Sadtler Research Laboratories, Inc., Philadelphia, PA.)

gamma rays, but they are usually made by special equipment. Both X rays and gamma rays penetrate living things easily.

Most of the time, you are bombarded with electromagnetic radiations from all portions of the electromagnetic spectrum. Radio and TV signals pass through you; you feel infrared radiation when you sense the warmth of a radiator; X rays and gamma rays fall on you from space; and light from a lamp reflects into your eyes from the page you're reading. Of all these radiations, your eyes are able to sense only a very narrow band of wavelengths ranging from about 400 to 700 nm. This band is called the **visible spectrum** and consists of all the colors you can see, from red through orange, yellow, green, blue, and violet. White light is composed of all these colors in roughly equal amounts and can be separated into them by focusing a beam of white light through a prism, which spreads the various wavelengths apart. This is illustrated in Figure 6.5a (see page 226). A photograph showing the production of a visible spectrum is shown in Figure 6.5b.

The Energy of Electromagnetic Radiation

In 1900 a German physicist named Max Planck (1858–1947) launched one of the greatest upheavals in the history of science when he proposed that electromagnetic radiation is emitted only in tiny packets or **quanta** of energy, which were later called **photons.** Each photon pulses with a frequency, ν, and travels with the speed of light. Planck proposed and Albert Einstein (1879–1955) confirmed that *the energy of a photon of electromagnetic radiation is proportional to its frequency,* not to its intensity or brightness, as had been believed up to that time. (See Special Topic 6.1.)

The energy of one photon is called one *quantum* of energy.

$$\text{Energy of a photon} = E = h\nu \qquad (6.2)$$

Energy of a photon

In this expression, h is a proportionality constant that we now call **Planck's constant.**

Planck's and Einstein's discovery was really quite surprising. If a particular event requiring energy, such as photosynthesis in green plants, is initiated by the absorption of light, it is the frequency of the light that is important, not its intensity or brightness. An analogy would be a group of pole vaulters trying to get over a wall. If they have long enough poles, each can clear the wall. If the poles are too short, however, they can batter the wall with as much intensity as they want, but they'll never clear the top. Not even doubling or tripling the number of vaulters with short poles will get anyone across.

The value of Planck's constant is $h = 6.63 \times 10^{-34}$ J s. It has units of energy (joules) multiplied by time (seconds).

FIGURE 6.5

(*a*) A diagram showing how white light is refracted by a glass prism, which spreads out the colors of the visible spectrum. (*b*) In this color photograph we see the continuous rainbow of colors formed from white light.

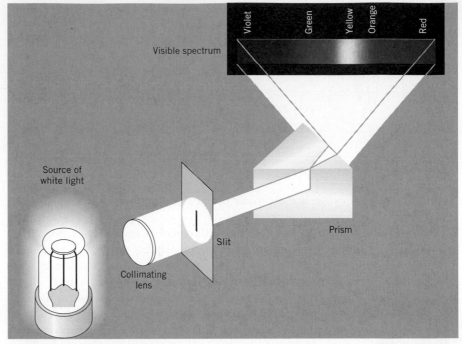

(*a*)

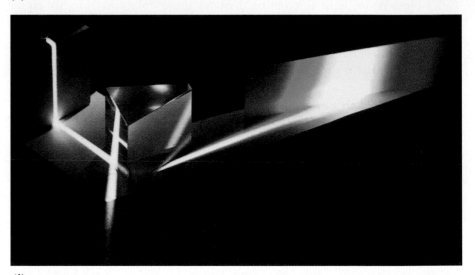

(*b*)

6.2
THE BOHR MODEL OF THE HYDROGEN ATOM

The spectrum described in Figure 6.5 is called a **continuous spectrum** because it contains light of *all* colors. It is formed when the light from the sun, or any other object that's been heated to a very high temperature (such as the filament in an electric light bulb), is split by a prism and displayed on a screen. A rainbow after a summer shower is a continuous spectrum that most people have seen. In this case, the colors contained in sunlight are spread out by tiny water droplets in the air.

A rather different kind of spectrum is observed if we examine the light that is given off when an *electric discharge,* or spark, passes through a gas such as

SPECIAL TOPIC 6.1 / PHOTOELECTRICITY AND ITS APPLICATIONS

One of the earliest clues to the relationship between the frequency of light and its energy was the discovery of the photoelectric effect. In the latter part of the nineteenth century, it was found that certain metals acquired a positive charge when they were illuminated by light. Apparently, light is capable of kicking electrons out of the surface of the metal.

When this phenomenon was studied in detail, it was discovered that electrons could be made to leave a metal's surface only if the frequency of the incident radiation was above some certain value, which was named the threshold frequency. This threshold frequency differs for different metals, depending on how tightly the metal atom holds onto electrons. Above the threshold frequency, the kinetic energy of the emitted electron increases with increasing frequency, but its kinetic energy does not depend on the intensity of the light. In fact, if the frequency of the light is below the minimum frequency, no electrons are observed at all, no matter how bright the light is. To physicists of that time, this was very perplexing. The explanation of the phenomenon was finally given by Albert Einstein in the form of a very simple equation.

$$KE = h\nu - w$$

where KE is the kinetic energy of the electron that is emitted, $h\nu$ is the energy of the photon of frequency ν, and w is the minimum energy needed to eject the electron from the metal's surface. Stated another way, part of the energy of the photon is needed just to get the electron off the surface of the metal. This amount is w. Any energy left over ($h\nu - w$) appears as the electron's kinetic energy.

Besides its important theoretical implications, the photoelectric effect has many practical applications. For example, automatic "electric eye" door openers use this phenomenon by sensing the interruption of a light beam caused by the person wishing to use the door. The phenomenon is also responsible for photoconduction by certain substances that are used in light meters in cameras and other devices. The production of sound in motion pictures makes use of a strip along the edge of the film (called the sound track) that causes the light passing through it to fluctuate in intensity according to the frequency of the sound that's been recorded. A photocell converts this light to a varying electric current that is amplified and played through speakers in the theater. Even the sensitivity of photographic film to light is related to the release of photoelectrons within tiny grains of silver bromide that are suspended in a coating on the film.

hydrogen. The electric discharge is an electric current that *excites,* or energizes, the atoms of the gas. More specifically, the electric current excites and gives energy to the electrons in the atom. The atoms then emit the absorbed energy in the form of light as the electrons return to a lower energy state. When a narrow beam of this light is passed through a prism, as shown in Figure 6.6, we do *not* see a continuous spectrum. Instead, only a few colors are observed, displayed as a series of individual lines. This series of lines is called the

Atoms of an element can also be excited by adding them to the flame of a Bunsen burner.

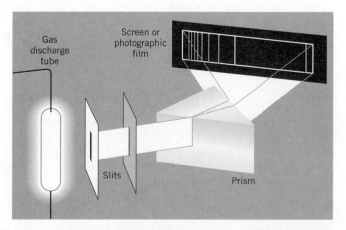

FIGURE 6.6

Production of a line spectrum. The light emitted by excited atoms is formed into a narrow beam and passed through a prism. This light beam is divided into relatively few narrow beams with frequencies that are characteristic of the particular element that is emitting the light.

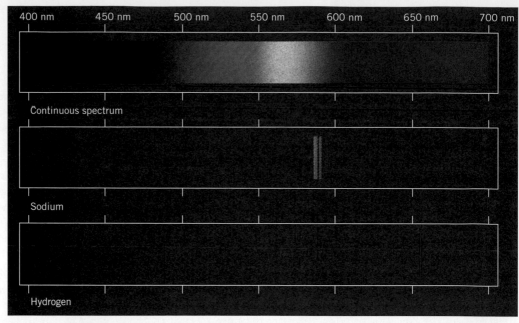

FIGURE 6.7

(*Top*) The continuous visible spectrum produced by an incandescent lamp. (*Center*) The atomic spectrum (line spectrum) produced by sodium. (*Bottom*) The atomic spectrum (line spectrum) produced by hydrogen.

An emission spectrum is also called a *line spectrum* because the light corresponding to the individual emissions appears as lines on the screen.

element's **atomic spectrum** or **emission spectrum.** Figure 6.7 shows the visible portions of the atomic spectra of two common elements, sodium and hydrogen, and how they compare with a continuous spectrum. Notice that the spectra of these elements are quite different. In fact, each element has its own unique atomic spectrum that is as characteristic as a fingerprint.

The Significance of Atomic Spectra

Earlier you saw that there is a simple relationship between the frequency of light and its energy, $E = h\nu$. Because excited atoms emit light of only certain specific frequencies, it must be true that only certain characteristic energy changes are able to take place within the atoms. For instance, in the spectrum of hydrogen there is a red line (see Figure 6.7) that has a wavelength of 656.4 nm and a frequency of 4.567×10^{14} Hz. As shown in the margin, the energy of a photon of this light is 3.026×10^{-19} J. Whenever a hydrogen atom emits red light, the frequency of the light is always precisely 4.567×10^{14} Hz and the energy of the atom decreases by *exactly* 3.026×10^{-19} J, never more and never less. Atomic spectra, then, tell us that *when an excited atom loses energy, not just any arbitrary amount can be lost.*

$E = h\nu$

$h = 6.626 \times 10^{-34}$ J s

$E = (6.626 \times 10^{-34}$ J s)

$\quad \times (4.567 \times 10^{14}$ s$^{-1})$

$\quad = 3.026 \times 10^{-19}$ J

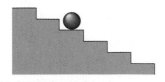

FIGURE 6.8

A ball on a staircase. The ball can have only certain amounts of potential energy when at rest.

How is it that atoms of a given element always undergo exactly the same specific energy changes? The answer seems to be that in an atom an electron can have only certain definite amounts of energy and no others. In the words of science, we say that the electron is restricted to certain **energy levels,** and that the energy of the electron is **quantized.**

The energy of an electron in an atom might be compared to the potential energy of a ball on a staircase (Figure 6.8). The ball can only come to rest on a step, so at rest it can only have certain specific amounts of potential energy as

SPECIAL TOPIC 6.2 / ELECTROMAGNETIC FIELDS AND THEIR POSSIBLE PHYSIOLOGICAL EFFECTS

Over the past few years you may have heard reports of possible dangers associated with living near high-voltage power lines or operating electrical equipment. What has caused public concern is the fear that 60 Hz electromagnetic radiation emitted by electricity passing through wires might be affecting the health of those nearby. Such fears have been fueled by news media reports of increased cancer rates among groups receiving strong exposure to this radiation, although the actual evidence supporting a relationship between exposure and cancer is weak and even partly contradictory.

When an oscillating electric current with a certain frequency passes through a wire, it emits radiation with that same frequency. In fact, that's how radio and TV stations broadcast their signals—by pulsing an electric current through a transmitting antenna. Ordinary household AC (alternating current) electricity has a frequency of 60 Hz, and weak electromagnetic signals are emitted by all wires that carry it. This radiation is most intense when the voltage is highest, as in the lines that carry electricity over long distances between power plants and cities.

The energy possessed by photons of 60 Hz electromagnetic radiation is extremely small (you might try the calculation), so small that it cannot affect the bonds that hold molecules together or even cause heating effects the way microwaves do. How, then, can they affect the activities of cells? The answer might be in the weak pulsating electric and magnetic fields induced in the body by this radiation.

Strong electromagnetic fields have been shown to affect the rate of bone growth as well as the amounts of various proteins produced in cells. Experiments have also revealed that cells exposed to an electric field hold onto calcium ions more than cells not exposed. Other experiments have demonstrated that

High-voltage power lines such as these emit low levels of 60 Hz electromagnetic radiation.

weak magnetic fields increase the uptake of calcium ions in cells that have been exposed to a substance that triggers cell division. It is believed that this additional calcium increases the tendency of these cells to divide. Since cancer growth depends on the rate of cell division, these results suggest one way cancers could be promoted by such radiation.

Despite laboratory evidence, the connection between electromagnetic radiation and cancer remains inconclusive. Although it is at least *possible* that cancer can be induced by electromagnetic fields, a lot of additional research will be necessary to pin down the answer.

---

determined by the "energy levels" of the various steps of the staircase. If the ball is raised to a higher step, its potential energy is increased. When it drops to a lower step, its potential energy decreases. But each time the ball stops, it stops on one of the steps, never in between. Therefore, the energy changes for the ball are restricted to the differences in potential energy between the steps.

The potential energy of the ball at rest is quantized.

So it is with an electron in an atom. The electron can only have energies corresponding to the set of electron energy levels in the atom. When the atom is supplied with energy by an electric discharge, an electron is raised from a low-energy level to a higher one. When the electron drops back, energy equal to the difference between the two levels is released and emitted as a photon. Because only certain energy jumps can occur, only certain frequencies can appear in the spectrum.

The existence of specific energy levels in atoms, as implied by atomic spectra, forms the foundation of all theories about electronic structure. Any model of the atom that attempts to describe the positions or motions of electrons must also account for atomic spectra.

The Atomic Spectrum of Hydrogen

The first success in explaining atomic spectra quantitatively came with the study of the spectrum of hydrogen. This is the simplest element, since its atoms have only one electron, and it produces the simplest spectrum with the fewest lines.

The atomic spectrum of hydrogen actually consists of several series of lines. One is in the visible region of the electromagnetic spectrum and is shown in Figure 6.7. Another series is in the ultraviolet region, and the rest are in the infrared. In 1885, J. J. Balmer found an equation that was able to give the wavelengths of the lines in the visible portion of the spectrum. This was soon extended to a more general equation, called the **Rydberg equation,** that could be used to calculate the wavelengths of *all* the spectral lines of hydrogen.

> It's really rather amazing that an equation as simple as this, involving only a constant and a set of integers, can be used to calculate the energies that an electron can have in the hydrogen atom.

$$\text{Rydberg equation:} \qquad \frac{1}{\lambda} = R_{\text{H}} \left(\frac{1}{n_1^2} - \frac{1}{n_2^2} \right)$$

The symbol λ stands for the wavelength, R_{H} is a constant (109,678 cm^{-1}), and n_1 and n_2 are variables whose values are whole numbers that range from 1 to ∞. The only restriction is that the value of n_2 must be larger than n_1. (This assures that the calculated wavelength has a positive value.) Thus, if $n_1 = 1$, acceptable values of n_2 are 2, 3, 4, . . . , ∞. The Rydberg constant, R_{H}, is an *empirical constant,* which means its value was chosen so that the equation gives values for λ that match the ones determined experimentally. The use of the Rydberg equation is straightforward, as illustrated in the following example.

EXAMPLE 6.3
Calculating the Wavelength of a Line in the Hydrogen Spectrum

PROBLEM The lines in the visible portion of the hydrogen spectrum are called the Balmer series, for which $n_1 = 2$. Calculate, to four significant figures, the wavelength in nanometers of the spectral line in this series for which $n_2 = 4$.

SOLUTION To solve this problem, we substitute values into the Rydberg equation, which will give us $1/\lambda$. Taking the reciprocal will then give the wavelength. As usual, we must be careful with the units.

Substituting $n_1 = 2$ and $n_2 = 4$ into the Rydberg equation gives

$$\frac{1}{\lambda} = 109678 \text{ cm}^{-1} \left(\frac{1}{2^2} - \frac{1}{4^2} \right)$$

$$= 109678 \text{ cm}^{-1} \left(\frac{1}{4} - \frac{1}{16} \right)$$

$$= 109678 \text{ cm}^{-1} (0.2500 - 0.0625)$$

$$= 109678 \text{ cm}^{-1} (0.1875)$$

$$= 2.056 \times 10^4 \text{ cm}^{-1}$$

Taking the reciprocal gives the wavelength in centimeters.

$$\lambda = \frac{1}{2.056 \times 10^4 \text{ cm}^{-1}}$$

$$= 4.864 \times 10^{-5} \text{ cm}$$

Finally, we convert to nanometers.

$$\lambda = 4.864 \times 10^{-5} \text{ cm} \times \frac{10^{-2} \text{ m}}{1 \text{ cm}} \times \frac{1 \text{ nm}}{10^{-9} \text{ m}}$$

$$= 486.4 \text{ nm}$$

This is the green line in the hydrogen spectrum in Figure 6.7.

■ **Practice Exercise 3** Calculate the wavelength in nanometers of the line in the visible spectrum of hydrogen for which $n_1 = 2$ and $n_2 = 3$. What color is this line?

The discovery of the Rydberg equation was both exciting and perplexing. The fact that the wavelength of any line in the hydrogen spectrum can be calculated by a simple equation involving just one constant and the reciprocals of the squares of two whole numbers is remarkable. What is there about the behavior of the electron in the atom that can account for such simplicity?

The Bohr Theory of the Hydrogen Atom

The first theoretical model of the hydrogen atom that successfully accounted for the Rydberg equation was proposed in 1913 by Niels Bohr (1885–1962), a Danish physicist. In his model, Bohr likened the electron moving around the nucleus to a planet circling the sun. He suggested that the electron moves around the nucleus along fixed paths, or orbits. His model broke with the classical laws of physics by placing restrictions on the sizes of the orbits and the energy that the electron could have in a given orbit. This ultimately led Bohr to an equation that described the energy of the electron in the atom. The equation includes a number of physical constants such as the mass of the electron, its charge, and Planck's constant. It also contains an integer, n, that Bohr called a **quantum number.** Each of the orbits is identified by its value of n. When all the constants are combined, Bohr's equation becomes

$$E = \frac{-b}{n^2} \qquad (6.3)$$

where E is the energy of the electron and b is the combined constant (its value is 2.18×10^{-18} J). The allowed values of n are whole numbers that range from 1 to ∞ (i.e., n can equal 1, 2, 3, 4, . . . , ∞). From this equation the energy of the electron in any particular orbit can be calculated.

Because of the negative sign in Equation 6.3, the lowest (most negative) energy value occurs when $n = 1$, which corresponds to the *first Bohr orbit.* This is the most stable energy state for the atom and is called the **ground state.** According to Bohr's theory, this orbit brings the electron closest to the nucleus.

When a hydrogen atom absorbs energy, as it does when an electric discharge passes through it, the electron is raised from the orbit having $n = 1$ to a higher orbit, to $n = 2$ or $n = 3$ or even higher. These higher orbits are less stable than the lower ones, so the electron quickly drops to a lower orbit. When this happens, energy is emitted in the form of light (Figure 6.9). Since

Niels Bohr won the 1922 Nobel Prize in physics for his work on atomic structure.

Classical physical laws place no restrictions on the sizes or energies of orbits.

Bohr's equation for the energy actually is $E = -\dfrac{2\pi^2 me^4}{n^2 h^2}$ where m is the mass of the electron, e is the charge on the electron, n is the quantum number, and h is Planck's constant. Therefore, in Equation 6.3, $b = \dfrac{2\pi^2 me^4}{h^2}$.

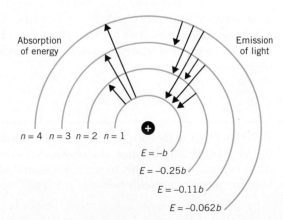

Absorption of energy

Emission of light

$n = 4$ $n = 3$ $n = 2$ $n = 1$

$E = -b$

$E = -0.25b$

$E = -0.11b$

$E = -0.062b$

FIGURE 6.9

Absorption of energy and emission of light by the hydrogen atom. When the atom absorbs energy, the electron is raised to a higher energy level. When the electron falls to a lower energy level, light of a particular energy and frequency is emitted.

the energy of the electron in a given orbit is fixed, a drop from one particular orbit to another, say from $n = 2$ to $n = 1$, always releases the same amount of energy, and the frequency of the light emitted because of this change is always precisely the same.

The success of Bohr's theory was in its ability to account for the Rydberg equation. When the atom emits a photon, an electron drops from a higher initial energy E_h to a lower final energy E_l. If the initial quantum number of the electron is n_h and the final quantum number is n_l, then the energy change, calculated as a positive quantity, is

$$\Delta E = E_h - E_l$$

$$= \left(\frac{-b}{n_h^2}\right) - \left(\frac{-b}{n_l^2}\right)$$

This can be rearranged to give

$$\Delta E = b\left(\frac{1}{n_l^2} - \frac{1}{n_h^2}\right) \qquad \text{with } n_h > n_l$$

By combining Equations 6.1 and 6.2, the relationship between the energy "ΔE" of a photon and its wavelength λ is

$$\Delta E = \frac{hc}{\lambda} = hc\left(\frac{1}{\lambda}\right)$$

Substituting and solving for $1/\lambda$ gives

$$\frac{1}{\lambda} = \frac{b}{hc}\left(\frac{1}{n_l^2} - \frac{1}{n_h^2}\right) \qquad \text{with } n_h > n_l$$

Notice how closely this equation derived from Bohr's theory matches the Rydberg equation, which was obtained solely from the experimentally measured atomic spectrum of hydrogen. Equally satisfying is that the combination of constants, b/hc, has a value of 109,730 cm^{-1}, which differs by only 0.05% from the experimentally derived value of R_H in the Rydberg equation.

Why the Bohr Theory Had to Be Abandoned

Bohr's model of the atom was both a success and a failure. By calculating the energy changes that occur between energy levels, Bohr was able to account for the Rydberg equation and, therefore, for the atomic spectrum of hydrogen. However, the theory was not able to explain quantitatively the spectra of atoms more complex than hydrogen, and all attempts to modify the theory to make it work met with failure. Gradually, it became clear that Bohr's picture of the atom was flawed and that another theory would have to be found. Nevertheless, the concepts of quantum numbers and fixed energy levels were important steps forward.

6.3
WAVE PROPERTIES OF MATTER AND WAVE MECHANICS

Bohr's efforts to develop a theory of electronic structure were doomed from the very beginning because the classical laws of physics—those known in his day—simply do not apply to particles as tiny as the electron. Classical physics fails for atomic particles because matter is not really as our physical senses perceive it. Under appropriate circumstances, small particles such as an electron behave not like solid particles, but instead like waves. This idea was proposed in 1924 by a young French graduate student, Louis de Broglie.

In Section 6.1 you learned that light waves are characterized by their wavelengths and their frequencies. The same is true of matter waves. De Broglie suggested that the wavelength of a matter wave, λ, is given by the equation.

$$\lambda = \frac{h}{mv} \qquad (6.4)$$

where h is Planck's constant, m is the particle's mass, and the v is its velocity.

When first encountered, the concept of a particle of matter behaving as a wave rather than as a solid object is difficult to comprehend. This book certainly seems solid enough, especially if you drop it on your toe. The reason for the book's apparent solidity is that in de Broglie's equation (Equation 6.4) the mass appears in the denominator. This means that heavy objects have extremely short wavelengths. The peaks of the matter waves for heavy objects are so close together that the wave properties go unnoticed and can't even be measured experimentally. But tiny particles with very small masses have much longer wavelengths; therefore, their wave properties become an important part of their overall behavior.

Perhaps by now you've begun to wonder if there is any way to *prove* that matter has wave properties. Actually, these properties can be demonstrated by a phenomenon that you have probably witnessed. When raindrops fall on a quiet pond, ripples spread out from where the drops strike the water, as shown in Figure 6.10. When two sets of ripples cross, there are places where the waves are *in phase,* which means that the peak of one wave coincides with the peak of the other. At these points the intensities of the waves add and the height of the water is equal to the sum of the heights of the two crossing waves. At other places the crossing waves are *out of phase,* which means the peak of one wave occurs at the trough of the other. In these places the intensities of the waves cancel. This reinforcement and cancellation of wave intensities, referred to,

All the objects that had been studied by scientists until the time of Bohr were large and massive in comparison with the electron, so no one had detected the limits of classical physics.

De Broglie was awarded a Nobel Prize in 1929.

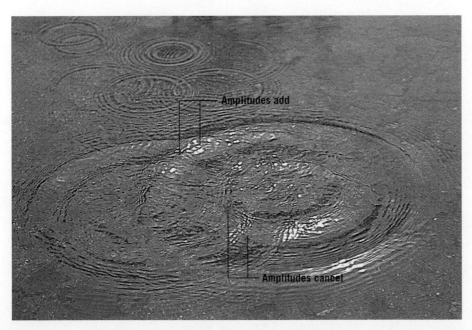

Amplitudes add

Amplitudes cancel

FIGURE 6.10

Diffraction of water waves on the surface of a pond. As the waves cross, the amplitudes increase where the waves are in phase and cancel where they are out of phase.

FIGURE 6.11

(*a*) Waves in phase produce constructive interference and an increase in intensity. (*b*) Waves out of phase produce destructive interference and yield cancellation of intensity.

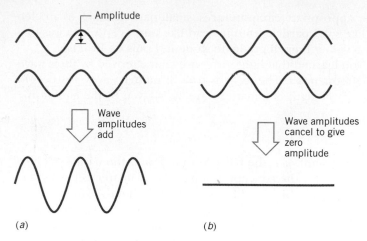

(*a*) (*b*)

respectively, as *constructive* and *destructive interference,* is a phenomenon called **diffraction.** It is examined more closely in Figure 6.11.

If you've ever noticed the rainbow of colors that shines from the surface of a compact disk (Figure 6.12), you have seen the diffraction of light waves. When white light that contains radiations of all wavelengths is reflected from the closely spaced "grooves" on the CD, it is divided into many individual light beams. The light waves in these beams experience interference with each other, and for a given angle between the incoming and reflected light, all colors cancel except for one, which reinforces. Because we see only colors that reinforce, as the angle changes, so do the colors of the light we see.

Diffraction is a phenomenon that can only be explained as a property of waves, and we have seen how it can be demonstrated with water waves and light waves. Experiments can also be done to show that electrons, protons, and neutrons experience diffraction, which demonstrates their wave nature. In fact, diffraction is the principle on which the electron microscope is based (Special Topic 6.3).

FIGURE 6.12

A rainbow of colors is produced by the diffraction of reflected light from the closely spaced grooves on the surface of a compact disk.

Electron Waves in Atoms

The wave properties of matter form the foundation for a theory called **wave mechanics,** which serves as the basis of all current theories of electronic structure. The term **quantum mechanics** is also used because wave mechanics predicts quantized energy levels. In 1926 Erwin Schrödinger (1887–1961), an Austrian physicist, became the first scientist to successfully apply the concept of the wave nature of matter to an explanation of electronic structure. His work and the theory that developed from it are highly mathematical. Fortunately, we need only a qualitative understanding of electronic structure, and the main points of the theory can be understood without all the math.

Schrödinger won a Nobel Prize in 1933 for his work.

Properties of Waves

Before we discuss electron waves, we need to know a little more about waves in general. There are basically two kinds of waves, *traveling waves* and *standing waves.* On a lake or ocean the wind produces waves whose crests and troughs move across the water's surface, as shown in Figure 6.13. The water moves up and down while the crests and troughs travel horizontally in the direction of the wind. These are examples of **traveling waves.**

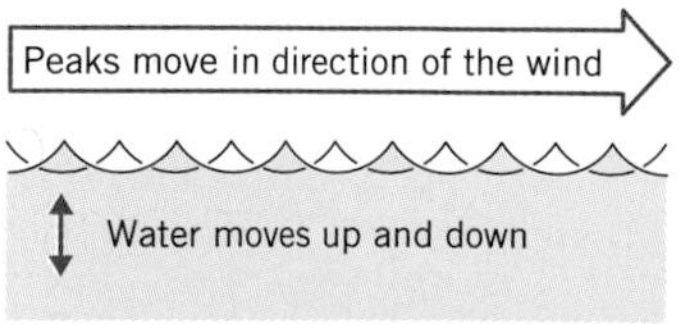

FIGURE 6.13

Traveling waves.

SPECIAL TOPIC 6.3 / THE ELECTRON MICROSCOPE

The usefulness of a microscope in studying small specimens is limited by its ability to distinguish between closely spaced objects. We call this ability the *resolving power* of the microscope. Through optics, it is possible to increase the magnification and thereby increase the resolving power, but only within limits. These limits depend on the wavelength of the light that is used. Objects with diameters less than the wavelength of the light cannot be seen in detail. Since visible light has wavelengths of the order of about 200 to 800 nm, objects smaller than this can't be examined with a microscope that uses visible light.

The electron microscope uses electron waves to "see" very small objects. De Broglie's equation, $\lambda = h/mv$, suggests that if an electron, proton, or neutron has a very high velocity, its wavelength will be very small. In the electron microscope, electrons are accelerated to high speeds across high-voltage electrodes. This gives electron waves with typical wavelengths of about 0.006 to 0.001 nm that strike the sample and are then focused magnetically onto a fluorescent screen where they form a visible image. Because of other difficulties, the actual resolving power is considerably less than this—generally on the order of 6 to 1 nm. Some high-resolution electron microscopes, however, are able to reveal the shadows of individual atoms in very thin specimens through which the electron beam passes.

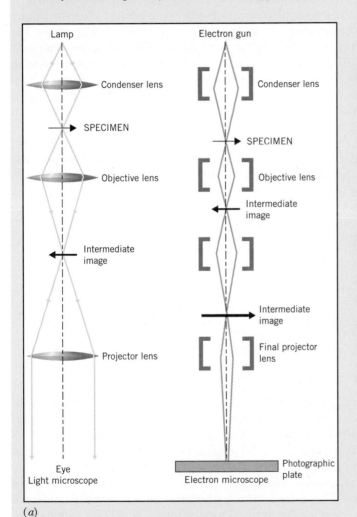

(a)

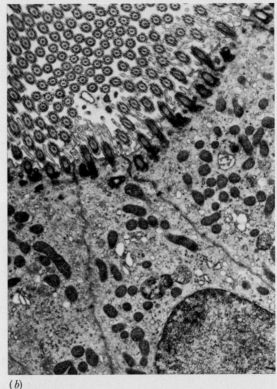

(b)

(*a*) Schematics of both the light microscope and the electron microscope (from A. G. Marshall, *Biophysical Chemistry,* 1978, John Wiley & Sons, NY, used by permission). (*b*) Electron micrograph of the retina of a rabbit. Magnification: 8400×.

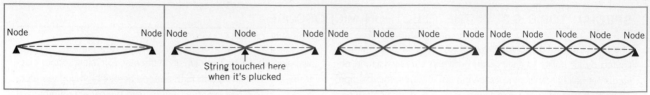

String touched here
when it's plucked

FIGURE 6.14

Standing waves on a guitar string.

The wavelength of this wave is actually twice the length of the string.

Notes played this way are called harmonics.

A more important kind of wave for us is the standing wave. An example is the vibrating string of a guitar. When the string is plucked, its center vibrates up and down while the ends, of course, remain fixed. The crest, or point of maximum amplitude of the wave, occurs at one position. At the ends of the string are points of zero amplitude, called **nodes,** and their positions are also fixed. A **standing wave,** then, is one in which the crests and nodes do not change position.

As you know, many notes can be played on a guitar string by shortening its effective length with a finger placed at frets along the neck of the instrument. But even without shortening the string, we can play a variety of notes. For instance, if the string is touched momentarily at its midpoint at the same time it is plucked, the string vibrates as shown in Figure 6.14 and produces a tone an octave higher. The wave that produces this higher tone has a wavelength exactly half of that formed when the untouched string is plucked. In Figure 6.14 we see that other wavelengths are possible, too, and each gives a different note.

If you examine Figure 6.14, you will see that there are some restrictions on the wavelengths that can exist. Not just any wavelength is possible because the nodes at either end of the string are in fixed positions. The only waves that can occur are those for which a half-wavelength is repeated *exactly* a whole number of times. Expressed another way, the length of the string is a whole-number multiple of half-wavelengths. In a mathematical form we could write this as

$$L = n\left(\frac{\lambda}{2}\right)$$

where L is the length of the string, λ is a wavelength (therefore, $\lambda/2$ is half the wavelength), and n is an integer. We see that whole numbers (similar to quantum numbers) appear quite naturally in determining which vibrations can occur.

Standing Electron Waves in Atoms

Knowing something about standing waves, we can now look at matter waves, concentrating on electron waves in particular. To study this topic, you may find it best to read through the entire discussion rather quickly to get an overview of the subject, and then read it again more slowly. All of the information fits together in the end, somewhat like the pieces of a puzzle.

The theory of quantum mechanics tells us that in the atom, electron waves are standing waves. Like guitar strings, electrons can have many different waveforms or wave patterns. Each of these waveforms, which are called **orbitals,** has a characteristic energy. Not all of the energies are different, but most are. *Energy changes within an atom are simply the result of an electron changing from a wave pattern with one energy to a wave pattern with a different energy.*

Electron waves are described by the term *orbital* to differentiate them from the notion of *orbits,* which was part of the Bohr model of the atom.

We will be interested in two properties of orbitals, their energies and their shapes. Their energies are important because we normally find atoms in their most stable states, referred to as their **ground states,** which occur when the electrons have waveforms with the lowest possible energies. The shapes of the wave patterns (i.e., where their amplitudes are large and where they are small) are important because the theory tells us that the amplitude of a wave at any particular place is related to the likelihood of finding the electron there. This will be important when we study how and why atoms form chemical bonds to each other.

Most stable **almost always means lowest energy.**

In much the same way that the characteristics of a wave on a one-dimensional guitar string can be related to a single integer, wave mechanics tells us that the three-dimensional electron waves (orbitals) can be characterized by a set of *three* integer quantum numbers, n, ℓ, and m_ℓ. In discussing the energies of the orbitals, it is usually most convenient to sort the orbitals into groups according to these quantum numbers.

The Principal Quantum Number, n   The quantum number n is called the **principal quantum number,** and all orbitals that have the same value of n are said to be in the same **shell.** The values of n can range from $n = 1$ to $n = \infty$. The shell with $n = 1$ is called the *first shell,* the shell with $n = 2$ is the *second shell,* and so forth. The various shells are also sometimes identified by letters, beginning (for no significant reason) with K for the first shell ($n = 1$).

The term *shell* **comes from an early notion that atoms could be thought of as similar to onions, with the electrons being arranged in layers around the nucleus.**

The principal quantum number is related to the size of the electron wave (i.e., how far the wave effectively extends from the nucleus). The higher the value of n, the larger is the electron's average distance from the nucleus. This quantum number is also related to the energy of the orbital. As n increases, the energies of the orbitals also increase.

Bohr's theory took into account only the principal quantum number, n. His theory worked fine for hydrogen because it just happens to be the one element in which all orbitals having the same value of n also have the same energy. Bohr's theory failed for atoms other than hydrogen, however, because orbitals with the same value of n can have different energies when the atom has more than one electron.

The Secondary Quantum Number, ℓ   The secondary quantum number, ℓ, divides the shells into smaller groups of orbitals called **subshells.** The value of n determines which values of ℓ are allowed. For a given n, ℓ can range from $\ell = 0$ to $\ell = (n - 1)$. Thus, when $n = 1$, ℓ can have only one value, 0. This means that when $n = 1$, there is only one subshell (the shell and subshell are really identical). When $n = 2$, ℓ can have values of 0 or 1. (The maximum value of $\ell = n - 1 = 2 - 1 = 1$.) This means that when $n = 2$, there are two subshells. One has $n = 2$ and $\ell = 0$, and the other has $n = 2$ and $\ell = 1$. The relationship between n and the allowed values of ℓ are summarized in the table in the margin.

ℓ **is also called the** *azimuthal quantum number.*

Subshells could be identified by their value of ℓ. However, to avoid confusing numerical values of n with those of ℓ, a letter code is normally used to specify the value of ℓ.

Relationship between n and ℓ

Value of n	Values of ℓ
1	0
2	0, 1
3	0, 1, 2
4	0, 1, 2, 3
5	0, 1, 2, 3, 4

value of ℓ	0	1	2	3	4	5	. . .
letter designation	*s*	*p*	*d*	*f*	*g*	*h*	. . .

To designate a particular subshell, we write the value of its principal quantum number followed by the letter code for the subshell. For example, the subshell with $n = 2$ and $\ell = 1$ is the $2p$ subshell; the subshell with $n = 4$ and $\ell = 0$ is

The number of subshells in a given shell equals the value of *n* for that shell.

the 4*s* subshell. Notice that because of the relationship between *n* and ℓ, every shell has an *s* subshell (l*s*, 2*s*, 3*s*, etc.); all the shells except the first have a *p* subshell (2*p*, 3*p*, 4*p*, etc.); and all but the first and second shells have a *d* subshell (3*d*, 4*d*, etc.); and so forth.

■ **Practice Exercise 4** What subshells would be found in the shells with *n* = 3 and *n* = 4?

The secondary quantum number determines the shape of the orbital, which we will examine more closely later. Except for hydrogen, the value of ℓ also affects the energy; the subshells within a given shell differ slightly in energy, with the energy of the subshell increasing with increasing ℓ. This means that within a given shell, the *s* subshell is lowest in energy, *p* is the next lowest, followed by *d*, then *f*, and so on. For example,

$$4s < 4p < 4d < 4f$$
— increasing energy →

Spectroscopists used m_ℓ to explain additional lines in the spectra of atoms when they emit light while in a magnetic field, which explains how this quantum number got its name.

The Magnetic Quantum Number, m_ℓ The third quantum number, m_ℓ, is known as the **magnetic quantum number.** It divides the subshells into individual orbitals and is related to the way the individual orbitals are oriented relative to each other in space. As with ℓ, there are restrictions as to the possible values of m_ℓ, which can range from $+\ell$ to $-\ell$. When $\ell = 0$, m_ℓ can have only the value 0 because $+0$ and -0 are the same. An *s* subshell, then, has just a single orbital. When $\ell = 1$, the possible values of m_ℓ are $+1$, 0, and -1. A *p* subshell therefore has three orbitals: one with $\ell = 1$ and $m_\ell = 1$, another with $\ell = 1$ and $m_\ell = 0$, and a third with $\ell = 1$ and $m_\ell = -1$. Similarly, we find that a *d* subshell has five orbitals and an *f* subshell has seven orbitals. The numbers of orbitals in the subshells are easy to remember because they follow a simple arithmetic progression.

$$s \quad p \quad d \quad f \ldots$$
$$1 \quad 3 \quad 5 \quad 7 \ldots$$

■ **Practice Exercise 5** How many orbitals are there in a *g* subshell?

The Whole Picture

The relationships among all three quantum numbers are summarized in Table 6.1. In addition, the relative energies of the subshells in an atom containing two or more electrons are depicted in Figure 6.15. Several important features should be noted. First, observe that each orbital on this energy diagram is indicated by a separate circle—one for an *s* subshell, three for a *p* subshell, and so forth. Second, notice that all the orbitals of a given subshell have the *same* energy. Third, note that, in going upward on the energy scale, the spacing between successive shells decreases as the number of subshells increases. This leads to the overlapping of shells having different values of *n*. For instance, the 4*s* subshell is lower in energy than the 3*d* subshell, 5*s* is lower than 4*d*, and 6*s* lower than 5*d*. In addition, the 4*f* subshell is below the 5*d* subshell and 5*f* is below 6*d*.

We will see shortly that Figure 6.15 is very useful for predicting the electronic structures of atoms. Before discussing this, however, we must study another very important property of the electron, called spin.

TABLE 6.1   Summary of Relationships Among n, ℓ, and m_ℓ

Value of n	Value of ℓ	Values of m_ℓ	Subshell	Number of Orbitals
1	0	0	$1s$	1
2	0	0	$2s$	1
	1	$-1, 0, 1$	$2p$	3
3	0	0	$3s$	1
	1	$-1, 0, 1$	$3p$	3
	2	$-2, -1, 0, 1, 2$	$3d$	5
4	0	0	$4s$	1
	1	$-1, 0, 1$	$4p$	3
	2	$-2, -1, 0, 1, 2$	$4d$	5
	3	$-3, -2, -1, 0, 1, 2, 3$	$4f$	7

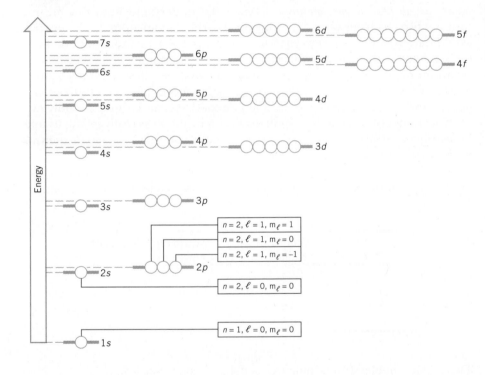

FIGURE 6.15

Approximate energy level diagram for atoms with two or more electrons. The quantum numbers associated with the orbitals in the first two shells are also shown.

Earlier it was stated that an atom is in its most stable state (its ground state) when its electrons have the lowest possible energies. This occurs when the electrons "occupy" the lowest energy orbitals that are available. But what determines how the electrons "fill" these orbitals? Fortunately, there are some simple rules that can help. These govern both the maximum number of electrons that can be in a particular orbital and how orbitals with the same energy become filled. One important factor that influences the distribution of electrons is the phenomenon known as *electron spin*.

6.4
ELECTRON SPIN AND THE PAULI EXCLUSION PRINCIPLE

Electrons moving in curved paths through wires in the large electromagnet suspended from a crane produce a magnetic field that is strong enough to pick up large portions of scrap steel.

Pauli received the 1945 Nobel Prize in physics for his discovery of the exclusion principle.

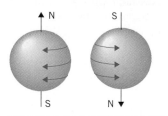

The electron can spin in either of two directions.

Magnetic Properties of the Electron

The concept of electron spin is based on the fact that electrons behave as tiny magnets. This can be explained by imagining that an electron spins around its axis, like a toy top. The revolving electrical charge of the electron creates its own magnetic field. The same effect is used to make electric motors work. The passage of electrical charge through the curved windings of an electric motor sets up magnetic forces that push and pull, thereby causing the rotor within the device to turn.

Electron spin gives us a fourth quantum number for the electron, called the **spin quantum number, m_s**. Like a toy top, the electron can spin in either of two directions, so the spin quantum number can take on either of two possible values; they are $m_s = +\frac{1}{2}$ or $m_s = -\frac{1}{2}$. The actual values of m_s and the reason they are not integers aren't very important to us, but the fact that there are *only* two values is very significant.

In 1925 an Austrian physicist, Wolfgang Pauli (1900–1958), expressed the importance of electron spin in determining electronic structure. The **Pauli exclusion principle** states that *no two electrons in the same atom can have identical values for all four of their quantum numbers.* To understand what this means, suppose two electrons were to occupy the $1s$ orbital of an atom. Each electron would have $n = 1$, $\ell = 0$, and $m_\ell = 0$. Since these three quantum numbers are the same for both electrons, the exclusion principle requires that their fourth quantum numbers (their spin quantum numbers) be different; one electron must have $m_s = +\frac{1}{2}$ and the other, $m_s = -\frac{1}{2}$. No more than two electrons can occupy the $1s$ orbital of the atom simultaneously because there are only two possible values of m_s. Thus the Pauli exclusion principle is really telling us that *the maximum number of electrons in any orbital is two,* and that *when two electrons are in the same orbital, they must have opposite spins.*

The limit of two electrons per orbital also limits the maximum electron populations of the shells and subshells. For the subshells we have

Subshell	Number of Orbitals	Maximum Number of Electrons
s	1	2
p	3	6
d	5	10
f	7	14

The maximum electron population per shell is shown below.

Shell	Subshells	Maximum Shell Population
1	$1s$	2
2	$2s\ 2p$	8 (2 + 6)
3	$3s\ 3p\ 3d$	18 (2 + 6 + 10)
4	$4s\ 4p\ 4d\ 4f$	32 (2 + 6 + 10 + 14)

In general, the maximum electron population of a shell is $2n^2$.

Magnetic Properties of Atoms

We have seen that two electrons occupying the same orbital must have different values of m_s. When this occurs, we say that the spins of the electrons are *paired*, or simply that the electrons are *paired*. Such pairing leads to the cancellation of the electrons' magnetic effects because the north pole of one electron magnet is opposite the south pole of the other. Atoms with more electrons that spin in one direction than in the other are said to contain *unpaired* electrons. For these atoms, the magnetic effects do not cancel and the atoms themselves become tiny magnets that can be attracted to an external magnetic field. This weak attraction to a magnet of a substance containing unpaired electrons is called **paramagnetism.** Substances in which all the electrons are paired are not attracted to a magnet and are said to be **diamagnetic.**

Paramagnetism and diamagnetism are measurable properties that provide experimental verification of the presence or absence of unpaired electrons in substances. In addition, the quantitative measurement of the strength of the attraction of a paramagnetic substance toward a magnetic field permits the calculation of the number of unpaired electrons in its atoms, molecules, or ions.

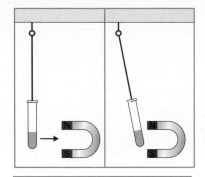

A paramagnetic substance is attracted to a magnetic field.

Diamagnetic substances are actually weakly repelled by a magnetic field.

6.5
ELECTRONIC STRUCTURES OF MULTIELECTRON ATOMS

The distribution of electrons among the orbitals of an atom is called the atom's **electronic structure** or **electron configuration.** This is something very useful to know about an element because the arrangement of electrons in the outer parts of an atom, which is determined by its electron configuration, controls the chemical properties of the element.

We shall be interested in the ground state electron configurations of the elements. This is the configuration that yields the lowest energy for an atom and can be predicted for many of the elements by the use of the energy level diagram in Figure 6.15 and application of the Pauli exclusion principle. To see how we go about this, let's begin with the simplest atom of all, hydrogen.

Hydrogen has an atomic number, Z, equal to 1, so a neutral hydrogen atom has one electron. In its ground state this electron occupies the lowest energy orbital that's available, which is the $1s$ orbital. To indicate symbolically the electron configuration, we list the subshells that contain electrons and indicate their electron populations by appropriate superscripts. Thus the electron configuration of hydrogen is written as

$$\text{H} \qquad 1s^1$$

Another way of expressing electron configurations that we will sometimes find useful is the **orbital diagram.** In it, each orbital will be represented by a circle and arrows will be used to indicate the individual electrons, head up for spin in one direction and head down for spin in the other. The orbital diagram for hydrogen is simply

$$\text{H} \qquad \textcircled{\uparrow}$$
$$1s$$

It doesn't matter whether the arrow points up or down. The energy of the electron is the same whether it has one spin or the other. For consistency, when an orbital is half-filled, we will show the electron with an arrow that points up.

The Aufbau Principle

To arrive at the electron configuration of an atom of another element, we imagine that we begin with a hydrogen atom and then add enough protons (plus whatever neutrons are also needed) to obtain the nucleus of the atom. As we proceed, we also add electrons, one at a time, to the lowest available orbital, until we have added enough electrons to give the neutral atom of the element in question. This imaginary process for obtaining the electronic structure of an atom is known as the **aufbau principle,** the word *aufbau* being German for *building up*. Let's look at the way this works for helium, which has $Z = 2$. This atom has two electrons, both of which are permitted to occupy the $1s$ orbital. The electron configuration of helium can therefore be written as

$$\text{He} \quad 1s^2 \qquad \text{or} \qquad \text{He} \quad \text{(⭡⭣)}$$
$$1s$$

Notice that the orbital diagram shows that both electrons in the $1s$ orbital are paired.

We can proceed in the same fashion to predict successfully the electron configurations of most of the elements in the periodic table. For example, the next elements in the table are lithium, Li ($Z = 3$), and beryllium, Be ($Z = 4$), which have three and four electrons, respectively. The Pauli exclusion principle tells us that the $1s$ subshell is filled with two electrons, and Figure 6.15 shows that the orbital of next lowest energy is the $2s$, which can also hold up to two electrons. Therefore, we can represent the electronic structures of lithium and beryllium as

$$\text{Li} \quad 1s^22s^1 \qquad \text{or} \qquad \text{Li} \quad \text{(⭡⭣)} \quad \text{(⭡)}$$
$$1s \quad 2s$$

$$\text{Be} \quad 1s^22s^2 \qquad \text{or} \qquad \text{Be} \quad \text{(⭡⭣)} \quad \text{(⭡⭣)}$$
$$1s \quad 2s$$

After beryllium comes boron, B ($Z = 5$). Referring to Figure 6.15, we see that the first four electrons of this atom complete the $1s$ and $2s$ subshells, so the fifth electron must be placed into the $2p$ subshell.

$$\text{B} \quad 1s^22s^22p^1$$

In the orbital diagram for boron, the fifth electron can be put into any one of the $2p$ orbitals—which one doesn't matter because they are all of equal energy.

$$\text{B} \quad \text{(⭡⭣)} \quad \text{(⭡⭣)} \quad \text{(⭡)(○)(○)}$$
$$1s \quad 2s \quad 2p$$

Notice, however, that when we give this orbital diagram we show *all* of the orbitals of the $2p$ subshell even though two of them are empty.

Next we come to carbon, which has six electrons. As before, the first four electrons complete the $1s$ and $2s$ orbitals. The remaining two electrons go in the $2p$ subshell to give

$$\text{C} \quad 1s^22s^22p^2$$

Now, however, to write the orbital diagram we have to make a decision as to where to put the two p electrons. (At this point you may have an unprintable suggestion! But try to bear up. It's really not all that bad.) To make this decision, we apply **Hund's rule,** which states that *when electrons are placed in a set of orbitals of equal energy, they are spread out as much as possible to give as few paired electrons as possible.* Both theory and experiment have shown that following this rule leads to the electron configuration with the lowest energy. For carbon, it means that the two *p* electrons are in separate orbitals and their spins are in the same direction.

C (↑↓) (↑↓) (↑)(↑)( )
　　1s　2s　　2p

Applying the Pauli exclusion principle and Hund's rule, we can now complete the electron configurations and orbital diagrams for the rest of the elements of the second period.

　　　　　　1s　2s　　2p　　　　　　　　　　　1s　2s　　2p
N $1s^2 2s^2 2p^3$ (↑↓) (↑↓) (↑)(↑)(↑)　　F $1s^2 2s^2 2p^5$ (↑↓) (↑↓) (↑↓)(↑↓)(↑)

O $1s^2 2s^2 2p^4$ (↑↓) (↑↓) (↑↓)(↑)(↑)　　Ne $1s^2 2s^2 2p^6$ (↑↓) (↑↓) (↑↓)(↑↓)(↑↓)

We can continue to predict electron configurations in this way, using Figure 6.15 as a guide to tell us which subshells become occupied and in what order. For instance, after completing the 2*p* subshell at neon, Figure 6.15 predicts that the next two electrons enter the 3*s*, followed by the filling of the 3*p*. Then we find the 4*s* lower in energy than the 3*d*, so it is filled first. Next, the 3*d* is completed before we go on to fill the 4*p*, and so forth.

On the basis of this discussion, it would seem that Figure 6.15 must be available to us whenever we wished to write the electron configuration of an element. However, as you will see in the next section, all the information contained in this figure is also contained in the periodic table.

In Chapter 2 you learned that when the periodic table was constructed, elements with similar chemical properties were arranged in vertical columns called groups. The basic structure of the periodic table is one of the strongest empirical supports for the quantum theory that we have been using to predict electron configurations, and it also permits us to use the periodic table itself as a device for predicting electron configurations.

Consider, for example, the way the table is laid out (Figure 6.16). On the left there is a block of *two* columns of elements, on the right there is a block of *six* columns, in the center there is a block of *ten* columns, and below the table there are two rows consisting of *fourteen* elements each. These numbers—2, 6, 10, and 14—are *precisely* the numbers of electrons that the quantum theory tells us can occupy *s*, *p*, *d*, and *f* subshells, respectively. In fact, *we can use the structure of the periodic table to predict the filling order of the subshells when we write the electron configuration of an element.*

To use the periodic table to predict electron configurations, we follow the aufbau principle as before. We start with hydrogen and then move through

6.6
ELECTRON CONFIGURATIONS AND THE PERIODIC TABLE

This would be an amazing coincidence if the theory were wrong.

FIGURE 6.16

The overall structure of the periodic table.

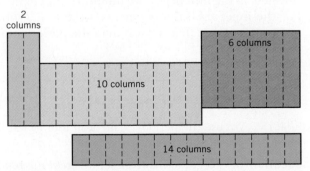

FIGURE 6.17

Subshells that become filled as we cross periods.

Periodic table yields electron configurations

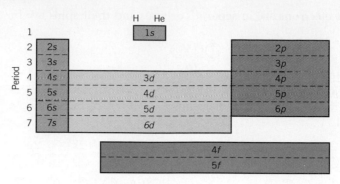

the table row after row until we reach the element of interest, noting as we go along which regions of the table we pass through. For example, consider the element calcium. Using Figure 6.15, we obtain the electron configuration.

$$\text{Ca} \quad 1s^2 2s^2 2p^6 3s^2 3p^6 4s^2$$

To see how this configuration relates to the periodic table, refer to the inside front cover of the book. Notice that the first period has just two elements, H and He. Starting with hydrogen and passing through this period, two electrons are added to the atom. These enter the $1s$ subshell. Next we move across the second period, where the first two elements, Li and Be, are in the block of two columns. As you saw above, the last electron to enter each of these atoms goes into the $2s$ subshell. Then we enter the block of six columns, and as we move across this region in the second row we gradually fill the $2p$ subshell. Now we go to the third period where we first pass through the block of two columns, filling the $3s$ subshell, and then through the block of six columns, filling the $3p$. Finally, we move to the fourth period. To get to calcium we step through two elements in the two-column block, filling the $4s$ subshell.

Now let's see what we can learn from this. Notice that every time we move across the block of two columns we add electrons to an s subshell that has a principal quantum number equal to the period number. Also note that every time we move across the block of six columns we add electrons to a p subshell that has a principal quantum number equal to the period number. This is summarized in Figure 6.17.

Figure 6.17 also includes the subshells that are filled when we cross through rows of transition elements and through rows of inner transition elements. For the transition elements, we simply subtract 1 from the period number to find the principal quantum number of the d subshell that's filled. For instance, as we move through the transition elements of the fourth period, we fill the $3d$ subshell, in the fifth period we fill the $4d$, and in the sixth period we fill the $5d$. For the inner transition elements, we fill an f subshell that has a principal quantum number that is 2 less than the period number of the element.

EXAMPLE 6.4
Predicting Electron Configurations

What is the electron configuration of (a) Mn and (b) Bi?

ANALYSIS We use the periodic table as a tool to tell us which subshells become filled. It is best if you can do this without referring to Figure 6.17.

SOLUTION (a) To get to manganese, we cross the following regions of the table, with the results indicated.

Period 1   Fill the 1s subshell

Period 2   Fill the 2s and 2p subshells

Period 3   Fill the 3s and 3p subshells

Period 4   Fill the 4s and then move 5 places in the 3d region

The electron configuration of Mn is therefore

$$\text{Mn}\quad 1s^22s^22p^63s^23p^64s^23d^5$$

Often it is desirable to group all subshells of the same shell together. For manganese, this gives

$$\text{Mn}\quad 1s^22s^22p^63s^23p^63d^54s^2$$

(b) To get to bismuth, we fill the following subshells:

Period 1   Fill the 1s

Period 2   Fill the 2s and 2p

Period 3   Fill the 3s and 3p

Period 4   Fill the 4s, 3d, and 4p

Period 5   Fill the 5s, 4d, and 5p

Period 6   Fill the 6s, 4f, 5d, and then add 3 electrons to the 6p

This gives

$$\text{Bi}\quad 1s^22s^22p^63s^23p^64s^23d^{10}4p^65s^24d^{10}5p^66s^24f^{14}5d^{10}6p^3$$

Grouping subshells with the same value of n gives

$$\text{Bi}\quad 1s^22s^22p^63s^23p^63d^{10}4s^24p^64d^{10}4f^{14}5s^25p^65d^{10}6s^26p^3$$

■ **Practice Exercise 6** Use the periodic table to predict the electron configurations of (a) Mg, (b) Ge, (c) Cd, (d) Gd. Group subshells of the same shell together.

■ **Practice Exercise 7** Draw orbital diagrams for (a) Na, (b) S, (c) Fe.

Electronic Basis for the Periodic Recurrence of Properties

You learned that Mendeleev constructed the periodic table by placing elements with similar properties in the same group. We are now ready to understand the reason for these similarities in terms of the electronic structures of atoms.

When we consider the chemical reactions of atoms, our attention is usually focused on the distribution of electrons in the **outer shell** of the atom (the occupied shell with the largest value of n). This is because the outer electrons (those in the outer shell) are the ones that are exposed to other atoms when the atoms react. The inner electrons of an atom, called the **core electrons,** are buried deep within the atom and normally do not play a role when chemical bonds are formed. It seems reasonable, therefore, that elements with similar properties should have similar outer shell electron configurations. This is, in fact, exactly what we observe. For example, let's look at the alkali metals of Group IA. Going by our rules, we obtain the following configurations.

Li $1s^22s^1$
Na $1s^22s^22p^63s^1$
K $1s^22s^22p^63s^23p^64s^1$
Rb $1s^22s^22p^63s^23p^63d^{10}4s^24p^64d^{10}5s^1$
Cs $1s^22s^22p^63s^23p^63d^{10}4s^24p^64d^{10}5s^25p^66s^1$

Each of these elements has just one outer-shell electron which is in an *s* sub-shell. We know that when they react, the alkali metals each lose one electron to form ions with a charge of $1+$. For each, the electron that is lost is this outer *s* electron, and the electron configuration of the ion that is formed is the same as that of the preceding noble gas.

Li^+	$1s^2$		He	$1s^2$
Na^+	$1s^2 2s^2 2p^6$		Ne	$1s^2 2s^2 2p^6$
K^+	$1s^2 2s^2 2p^6 3s^2 3p^6$		Ar	$1s^2 2s^2 2p^6 3s^2 3p^6$

etc.

If you write the electron configurations of the members of any of the groups in the periodic table, you will find the same kind of similarity among the configurations of the outer-shell electrons. The differences are in the value of the principal quantum number of these outer electrons.

Abbreviated Electron Configurations

Because we are interested primarily in the electrons of the outer shell, we often write electron configurations in an abbreviated, or shorthand form. To illustrate, let's consider the elements sodium and magnesium. Their full electron configurations are

$$Na \quad 1s^2 2s^2 2p^6 3s^1$$
$$Mg \quad 1s^2 2s^2 2p^6 3s^2$$

The outer electrons of both atoms are in the $3s$ subshell and each has the core configuration $1s^2 2s^2 2p^6$, which is the same as that of the noble gas neon.

To write the shorthand configuration for an element, we indicate what the core is by placing in brackets the symbol of the noble gas whose electron configuration is the same as the core configuration. This is followed by the configuration of the outer electrons for the particular element. The noble gas used is almost always the one that occurs at the end of the period preceding the period containing the element whose configuration we wish to represent. Thus, for sodium and magnesium we write

$$Na \quad [Ne] \; 3s^1$$
$$Mg \quad [Ne] \; 3s^2$$

Even with the transition elements, we need only concern ourselves with the outermost shell and the *d* subshell just below. For example, iron has the configuration

$$Fe \quad 1s^2 2s^2 2p^6 3s^2 3p^6 3d^6 4s^2$$

Only the $4s$ and $3d$ electrons play a role in the chemistry of iron. In general, the electrons below the outer *s* and *d* subshells of a transition element—the *core* electrons—are relatively unimportant. In every case these core electrons have the electron configuration of a noble gas. Therefore, for iron the core is $1s^2 2s^2 2p^6 3s^2 3p^6$, which is the same as the electron configuration of argon. The abbreviated configuration of iron is therefore written as

$$Fe \quad [Ar] \; 3d^6 4s^2$$

What is the shorthand electron configuration of manganese? Draw the orbital diagram for manganese using this shorthand configuration.

SOLUTION Manganese is in period 4. The preceding noble gas is argon, Ar, so to write the abbreviated configuration, we write the symbol for argon in brackets followed by the electron configuration that exists beyond the argon core. We can obtain this by noting that to get to Mn in period 4, we first cross the "s region" by adding two electrons to the $4s$ subshell, and then go five steps into the "d region," where we add five electrons to the $3d$ subshell. Therefore, the shorthand electron configuration for Mn is

$$\text{Mn} \quad [\text{Ar}] \ 4s^2 3d^5$$

Placing the electrons that are in the highest shell farthest to the right gives

$$\text{Mn} \quad [\text{Ar}] \ 3d^5 4s^2$$

To draw the orbital diagram, we simply distribute the electrons in the $3d$ and $4s$ orbitals following Hund's rule. This gives

$$\text{Mn} \quad [\text{Ar}] \quad \uparrow \ \uparrow \ \uparrow \ \uparrow \ \uparrow \qquad \uparrow\downarrow$$
$$\qquad\qquad\qquad\quad 3d \qquad\qquad 4s$$

Notice that each of the $3d$ orbitals is half-filled.

■ **Practice Exercise 8** Write shorthand configurations and abbreviated orbital diagrams for (a) P and (b) Sn. Where appropriate, place the electrons that are in the highest shell farthest to the right.

EXAMPLE 6.5
Writing Shorthand Electron Configurations

Valence Shell Electron Configurations

For the representative elements (those in the longer columns), the only electrons that are normally important in controlling chemical properties are the ones in the outer shell. This outer shell is known as the **valence shell** and it is always the occupied shell with the largest value of n. The electrons in the valence shell are called **valence electrons.** (The term *valence* comes from the study of chemical bonding and relates to the combining capacity of an element, but that's not important here.)

For the representative elements it is very easy to determine the electron configuration of the valence shell by using the periodic table. The valence shell always consists of just the s and p subshells that we encounter crossing the period that contains the element in question. Thus, to determine the valence shell configuration of sulfur, a period-3 element, we note that to reach sulfur in period 3 we place two electrons into the $3s$ and four electrons into the $3p$. The valence shell configuration of sulfur is therefore

$$\text{S} \quad 3s^2 3p^4$$

Predict the electron configuration of the valence shell of arsenic ($Z = 33$).

SOLUTION To reach arsenic in period 4, we add electrons to the $4s$, $3d$, and $4p$ subshells. But the $3d$ is not part of the fourth shell and therefore not part of the

EXAMPLE 6.6
Writing Valence Shell Configurations

valence shell, so all we need be concerned with are the electrons in the $4s$ and $4p$ subshells. This gives as the valence shell configuration of arsenic,

$$\text{As} \quad 4s^2 4p^3$$

■ **Practice Exercise 9** What is the valence shell electron configuration of (a) Se, (b) Sn, and (c) I?

6.7 SOME UNEXPECTED ELECTRON CONFIGURATIONS

The rules you've learned for predicting electron configurations work most of the time, but not always. Appendix B gives the electron configurations of all of the elements as determined experimentally. Close examination reveals that there are quite a few exceptions to the rules. Some of these exceptions are important to us because they occur with common elements.

Two important exceptions are for chromium and copper. Following the rules, we would expect the configurations to be

$$\text{Cr} \quad [\text{Ar}] \; 3d^4 4s^2$$
$$\text{Cu} \quad [\text{Ar}] \; 3d^9 4s^2$$

However, the actual electron configurations, determined experimentally, are

$$\text{Cr} \quad [\text{Ar}] \; 3d^5 4s^1$$
$$\text{Cu} \quad [\text{Ar}] \; 3d^{10} 4s^1$$

The corresponding orbital diagrams are

Cr [Ar] ⇡⇡⇡⇡⇡   ⇡
Cu [Ar] ⇅⇅⇅⇅⇅   ⇡
       $3d$ $4s$

Notice that for chromium, an electron is "borrowed" from the $4s$ subshell to give a $3d$ subshell that is exactly half-filled. For copper the $4s$ electron is borrowed to give a completely filled $3d$ subshell. A similar thing happens with silver and gold, which have filled $4d$ and $5d$ subshells, respectively.

$$\text{Ag} \quad [\text{Kr}] \; 4d^{10} 5s^1$$
$$\text{Au} \quad [\text{Xe}] \; 4f^{14} 5d^{10} 6s^1$$

Apparently, half-filled and filled subshells (particularly the latter) have some special stability that makes such borrowing energetically favorable. This subtle, but nevertheless important phenomenon affects not only the ground state configurations of atoms but also the relative stabilities of some of the ions formed by the transition elements.

6.8 SHAPES OF ATOMIC ORBITALS

The same theory that tells us of the energies of atomic orbitals also describes their shapes. How do we know these shapes are correct? We don't for sure, but many of the predictions that have been made using the theory seem to be borne out by experiments. This gives the theoretical explanations strength and support. For example, the fact that the results of wave mechanics account for the shape of the periodic table so very well gives the theory a good deal of credibility.

The difficulty in describing where electrons are in an atom stems from the basic problem that we face when we attempt to picture a particle as a wave. There is nothing in our worldly experience that is comparable. The way we get around this perplexing conceptual problem, so that we can still think of the electron as a particle in the usual sense, is to speak in terms of the statistical probability of the electron being found at a particular place.

Describing the electron's position in terms of statistical probability is based on more than simple convenience. The German physicist Werner Heisenberg showed mathematically that it is impossible to measure with complete precision both a particle's velocity and position at the same instant, and that such measurements are subject to a minimum uncertainty. This is Heisenberg's famous **uncertainty principle.** The theoretical limitations on measuring speed and position are not significant for large objects. However, for small particles such as the electron, these limitations prevent us from ever knowing or predicting where in an atom an electron will be at a particular instant, so we speak of probabilities instead.

Wave mechanics views the probability of finding an electron at a given point in space as equal to the square of the amplitude of the electron wave at that point. It seems quite reasonable to relate probability to amplitude, or intensity, because where a wave is intense its presence is strongly felt. The amplitude is squared because, mathematically, the amplitude can be either positive or negative, but probability only makes sense if it is positive. Squaring the amplitude assures us that the probabilities will be positive. We need not be very concerned about this point, however.

The notion of electron probability leads to two very important and frequently used concepts. One is that an electron behaves as if it were spread out around the nucleus in a sort of **electron cloud.** Figure 6.18 is a *dot-density diagram* that illustrates the way the probability of finding the electron varies in space for a 1*s* orbital. In those places where the dot density is large (i.e., where there are large numbers of dots per unit volume), the amplitude of the wave is large and the probability of finding the electron is also large.

The other important concept is **electron density,** which relates to how much of the electron's charge is packed into a given volume. Because of its wave nature, the electron (and its charge) is spread out around the nucleus. In regions of high probability there is a high concentration of electrical charge and the electron density is large; in regions of low probability, the electron density is small. In looking at the way the electron density distributes itself in atomic orbitals, we are interested in three things—the *shape* of the orbital, its *size,* and its *orientation* in space relative to other orbitals.

Shapes and Sizes of *s* and *p* Orbitals

Electron density doesn't end abruptly at some particular distance from the nucleus. It gradually fades away. To define the size and shape of an orbital, it is useful to picture some imaginary surface enclosing, say, 90% of the electron density of the orbital, and on which the probability of finding the electron is everywhere the same. For the 1*s* orbital in Figure 6.18, we find that if we go out a given distance from the nucleus in *any* direction, the probability of finding the electron is the same. This means that all the points of equal probability lie on the surface of a sphere, so we can say that the shape of the orbital is spherical. In fact, all *s* orbitals are spherical. As suggested earlier, their size

Heisenberg won the 1932 Nobel Prize in physics for his work.

The amplitude of an electron wave is described by a *wave function,* which is usually given the symbol ψ (the Greek letter psi). The probability of finding the electron in a given location is given by ψ^2.

FIGURE 6.18

Electron probability distribution for a 1*s* electron.

FIGURE 6.19

Size variations among *s* orbitals. The orbitals become larger as the principal quantum number, *n*, becomes larger. The diagrams shown here represent cross sections of the spherical electron density distributions.

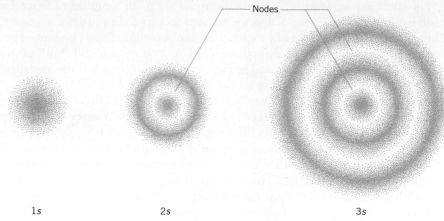

1*s*   2*s*   3*s*

FIGURE 6.20

(*a*) Cross section of the probability distribution in a 2*p* orbital.
(*b*) Cross section of a 3*p* orbital.

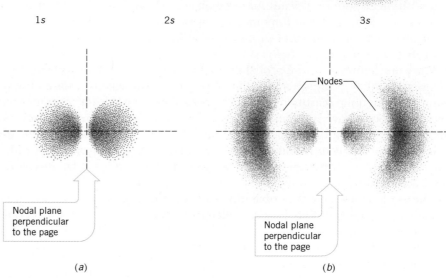

Nodal plane perpendicular to the page

Nodal plane perpendicular to the page

(*a*)   (*b*)

increases with increasing *n*. This is illustrated in Figure 6.19. Notice that beginning with the 2*s* orbital, there are certain places where the electron density drops to zero. These are the nodes of the electron wave. It is interesting that electron waves have nodes just like the waves on a guitar string. For electron waves, however, the nodes consist of imaginary *surfaces* on which the electron density is zero.

The *p* orbitals are quite different from *s* orbitals, as shown in Figure 6.20. Notice that the electron density is equally distributed in two regions on opposite sides of the nucleus. Figure 6.20*a* shows the two "lobes" of *one* 2*p* orbital. The electron density of each lobe is concentrated around an imaginary line that passes through the nucleus. At the nucleus and perpendicular to the line passing through the centers of the two lobes there is a *nodal plane*—an imaginary flat surface on which every point has an electron density of zero. The size of the *p* orbitals also increases with increasing *n* as illustrated by the cross section of a 3*p* orbital in Figure 6.20*b*. The 3*p* and higher *p* orbitals have additional nodes besides the nodal plane that passes through the nucleus.

Orientations of Orbitals in *p* and *d* Subshells

A *p* subshell consists of three orbitals of equal energy whose directions lie at 90° to each other along the axes of an imaginary *xyz* coordinate system (Figure 6.21). For convenience in referring to the individual *p* orbitals, they are often

FIGURE 6.21

The orientations of the three *p* orbitals in a *p* subshell.

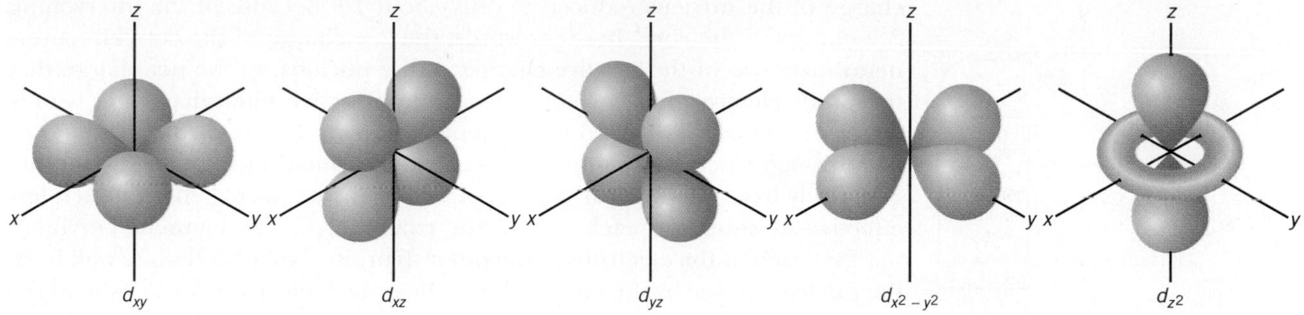

FIGURE 6.22

The shapes and directional properties of the five d orbitals of a d subshell.

labeled according to the axis along which they lie. The p orbital concentrated along the x axis is labeled p_x, and so forth.

The shapes of the d orbitals, illustrated in Figure 6.22, are a bit more complex than are those of the p orbitals. Because of this, and because there are five orbitals in a d subshell, we haven't attempted to draw all of them at the same time on the same set of coordinate axes. Notice that four of the five d orbitals have the same shape and consist of four lobes of electron density. These orbitals differ only in their orientations around the nucleus (their labels come from the mathematics of wave mechanics). The fifth d orbital, labeled d_{z^2}, has two lobes that point in opposite directions along the z axis plus a doughnut-shaped ring of electron density around the center that lies in the x–y plane. We will see that the d orbitals are important in the formation of chemical bonds in certain molecules, and that their shapes and orientations are important in understanding the properties of the transition metals, which will be discussed in Chapter 19.

The f orbitals are even more complex than the d orbitals, but we will have no need to discuss their shapes.

6.9 VARIATION OF ATOMIC PROPERTIES WITH ELECTRONIC STRUCTURE

There are many chemical and physical properties that vary in a more or less systematic way according to an element's position in the periodic table. For example, in Chapter 2 we noted that the metallic character of the elements increases from top to bottom in a group and decreases from left to right across a period. In this section we discuss several physical properties of the elements that have an important influence on chemical properties. We will see how these properties correlate with an atom's electron configuration, and because electron configuration is also related to the location of an element in the periodic table, we will study their periodic variations as well.

Effective Nuclear Charge

Many of an atom's properties are determined by the amount of positive charge felt by the atom's outer electrons. Except for hydrogen, this positive charge is always *less* than the full nuclear charge, because the negative charge of the electrons in inner shells partially offsets, or "neutralizes," the positive charge of the nucleus.

To gain a better understanding of this, consider the element lithium, which has the electron configuration $1s^2 2s^1$. The electrons beneath the valence shell (those in the $1s^2$ core) are tightly packed around the nucleus and for the most part lie between the nucleus and the electron in the outer shell. This core has a charge of $2-$, and it surrounds a nucleus that has a charge of $3+$. When the outer $2s$ electron "looks toward" the center of the atom, it "sees" the $3+$

The inner electrons partially shield the outer electrons from the nucleus, so the outer electrons "feel" only a fraction of the full nuclear charge.

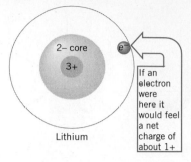

Lithium

If an electron were here it would feel a net charge of about 1+

FIGURE 6.23

If the 2 − charge of the $1s^2$ core of lithium were 100% effective at shielding the $2s$ electron from the nucleus, the valence electron would feel an effective nuclear charge of only about 1+.

An electron spends very little time between the nucleus and another electron in the same shell, so it shields that other electron poorly.

The C—H distance in most hydrocarbons is about 110 pm (110 × 10⁻¹² m).

The angstrom is named after Anders Jonas Ångström (1814–1874), a Swedish physicist who was the first to measure the wavelengths of the four most prominent lines of the hydrogen spectrum.

charge of the nucleus reduced to only about 1+ because of the intervening 2 − charge of the core. In other words, the 2 − charge of the core effectively neutralizes two of the positive charges of the nucleus, so the net charge that the outer electron feels, which we call the **effective nuclear charge,** is only about 1+. This is illustrated rather simplistically in Figure 6.23.

Although electrons in inner shells shield the electrons in outer shells quite effectively from the nuclear charge, electrons in the *same* shell are much less effective at shielding each other. For example, in the element beryllium ($1s^2 2s^2$) each of the electrons in the outer $2s$ orbital is shielded quite well from the nuclear charge by the inner $1s^2$ core, but one $2s$ electron doesn't shield the other $2s$ electron very well at all. This is because electrons in the same shell are at about the same average distance from the nucleus, and in attempting to stay away from each other they only spend a very small amount of time one below the other, which is what's needed to provide shielding. Since electrons in the same shell hardly shield each other at all from the nuclear charge, *the effective nuclear charge felt by the outer electrons is determined primarily by the difference between the charge on the nucleus and the charge on the core*. With this as background, let's examine some properties controlled by the effective nuclear charge.

Sizes of Atoms and Ions

The wave nature of the electron makes it difficult to define exactly what we mean by the "size" of an atom or ion. As we've seen, the electron cloud doesn't simply stop at some particular distance from the nucleus; instead it gradually fades away. Nevertheless, atoms and ions do behave in many ways as though they have characteristic sizes. For example, in a whole host of hydrocarbons, ranging from methane (CH_4, natural gas) to octane (C_8H_{18}, in gasoline) to many others, the distance between the nuclei of carbon and hydrogen atoms is virtually the same. This would suggest that carbon and hydrogen have the same relative sizes in each of these compounds.

Experimental measurements reveal that the diameters of atoms range from about 1.4×10^{-10} to 5.7×10^{-10} m. Their radii, which is the usual way that size is specified, range from about 7.0×10^{-11} to 2.9×10^{-10} m. Such small numbers are difficult to comprehend. A million carbon atoms placed side by side in a line would extend a little less than 0.2 mm, or about the diameter of the period at the end of this sentence.

The sizes of atoms and ions are rarely expressed in meters because the numbers are so cumbersome. Instead, a unit is chosen that makes the values easier to comprehend. A unit that scientists have traditionally used is called the **angstrom** (symbolized **Å**), which is defined as

$$1 \text{ Å} = 1 \times 10^{-10} \text{ m}$$

However, the angstrom is not an SI unit, and in many current scientific journals, atomic dimensions are given in picometers, or sometimes in nanometers (1 pm = 10^{-12} m and 1 nm = 10^{-9} m). In this book, we will normally express atomic dimensions in picometers, but because much of the scientific literature has these quantities in angstroms, you may someday find it useful to remember the conversions:

$$1 \text{ Å} = 100 \text{ pm}$$

$$1 \text{ Å} = 0.1 \text{ nm}$$

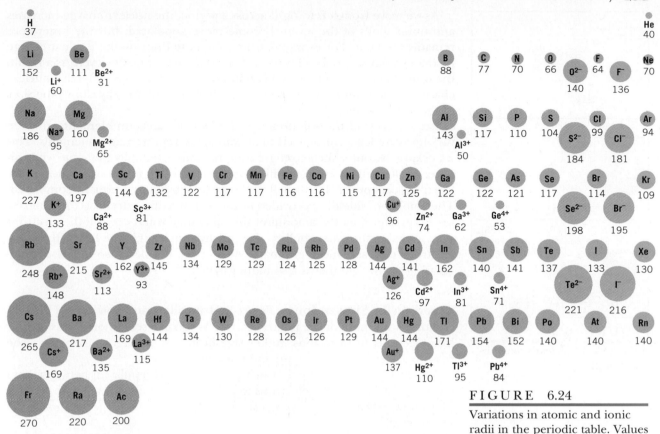

FIGURE 6.24

Variations in atomic and ionic radii in the periodic table. Values are in picometers.

Periodic Variations in Atomic Size

The variations in atomic radii within the periodic table are illustrated in Figure 6.24. Here we see that atoms generally become larger going from top to bottom in a group, and they become smaller going from left to right across a period. To understand these variations we must consider two factors. One is the value of the principal quantum number of the valence electrons, and the other is the effective nuclear charge felt by the valence electrons.

Going from top to bottom within a group, the effective nuclear charge felt by the outer electrons remains nearly constant, while the principal quantum number of the valence shell increases. For example, consider the elements of Group IA. For lithium, the valence shell configuration is $2s^1$; for sodium, it is $3s^1$; for potassium, it is $4s^1$; and so forth. For each of these elements, the core has a negative charge that is one less than the nuclear charge, so the valence electron of each experiences nearly the same effective nuclear charge of about $1+$. However, as we descend the group, the value of n for the valence shell increases, and as you learned earlier, the larger the value of n, the larger the orbital. Therefore, the atoms become larger as we go down a group simply because the orbitals containing the valence electrons become larger. This same argument applies whether the valence shell orbitals are s or p.

Moving from left to right across a period, electrons are added to the same shell. The orbitals holding the valence electrons all have the *same* value of n. In this case we have to examine the variation in the effective nuclear charge felt by the valence electrons.

Variations in atomic size

Large atoms are found in the lower left of the periodic table, and small atoms are found in the upper right.

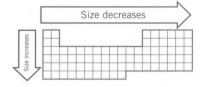

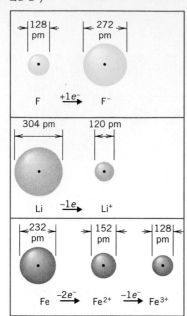

FIGURE 6.25

Adding electrons leads to an increase in the size of the particle; removing electrons leads to a decrease in the size of the particle.

Adding electrons creates an ion that is larger than the neutral atom; removing electrons produces an ion that is smaller than the neutral atom.

Fe   [Ar] $3d^6 4s^2$
Fe^{2+} [Ar] $3d^6$
Fe^{3+} [Ar] $3d^5$

As we move from left to right across a period, the nuclear charge increases and outer shells of the atoms become more populated, but the inner core remains the same. For example, from lithium to fluorine the nuclear charge increases from 3+ to 9+. The core ($1s^2$) stays the same, however. As a result, the outer electrons feel an increase in positive charge (i.e., effective nuclear charge), which causes them to be drawn inward, and thereby causes the sizes of the atoms to decrease.

Across a row of transition elements or inner transition elements, the size variations are less pronounced than among the representative elements. This is because the outer-shell configuration remains essentially the same while an inner shell is filled. From atomic numbers 21 to 30, for example, the outer electrons occupy the $4s$ subshell while the $3d$ subshell is gradually completed. The amount of shielding provided by the addition of electrons to this inner $3d$ level is greater than the amount of shielding that would occur if the electrons were added to the outer shell, so the effective nuclear charge felt by the outer electrons increases more gradually. As a result, the decrease in size with increasing atomic number is also more gradual.

Trends in the Sizes of Ions   Figure 6.24 also illustrates how sizes of the ions compare with those of the neutral atoms. As you can see, when atoms gain or lose electrons to form ions, rather significant size changes take place. The reasons are easy to understand and remember.

When electrons are added to an atom, the mutual repulsions between them increase. This causes the electrons to push apart and occupy a larger volume. Therefore, *negative ions are always larger than the atoms from which they are formed* (Figure 6.25).

When electrons are removed from the valence shell, the electron–electron repulsions decrease, which allows the remaining electrons to be pulled closer together around the nucleus. Therefore, *positive ions are always smaller than the atoms from which they are formed.*

For atoms of the representative elements, formation of positive ions generally involves emptying the outer shell, which exposes the smaller core that lies beneath. For transition elements such as iron, the first electrons lost come from the outer s subshell, which exposes the d subshell underneath. Further electron loss comes from this d subshell. For example, the radius of an iron atom is 116 pm, whereas the radius of the Fe^{2+} ion is 76 pm. Removing yet another electron to give Fe^{3+} decreases electron–electron repulsions in the d subshell and gives the Fe^{3+} ion a radius of 64 pm.

■ **Practice Exercise 10**   Use the periodic table to choose the largest atom or ion in each set. (a) Ge, Te, Se, Sn, (b) C, F, Br, Ga, (c) Fe, Fe^{2+}, Fe^{3+}, (d) O, O^{2-}, S, S^{2-}

Ionization Energy

The **ionization energy** (abbreviated **IE**) *is the energy required to remove an electron from an isolated, gaseous atom or ion in its ground state.* For an element X, it is the increase in potential energy associated with the change

$$X(g) \longrightarrow X^+(g) + e^-$$

In effect, the ionization energy is a measure of how much work is required to remove an electron, so it reflects how tightly the electron is held by the atom.

TABLE 6.2 Successive Ionization Energies in kJ/mol for Hydrogen Through Magnesium

	1	2	3	4	5	6	7	8
H	1,312							
He	2,372	5,250						
Li	520	7,297	11,810					
Be	899	1,757	14,845	21,000				
B	800	2,426	3,659	25,020	32,820			
C	1,086	2,352	4,619	6,221	37,820	47,260		
N	1,402	2,855	4,576	7,473	9,442	53,250	64,340	
O	1,314	3,388	5,296	7,467	10,987	13,320	71,320	84,070
F	1,680	3,375	6,045	8,408	11,020	15,160	17,860	92,010
Ne	2,080	3,963	6,130	9,361	12,180	15,240	—	—
Na	496	4,563	6,913	9,541	13,350	16,600	20,113	25,666
Mg	737	1,450	7,731	10,545	13,627	17,995	21,700	25,662

Usually, the ionization energy is expressed in units of kilojoules per mole (kJ/mol), so it is really the energy needed to remove one electron from each atom in 1 mol of atoms.

Table 6.2 gives the ionization energies of the first 12 elements. As you can see, atoms with more than one electron have more than one ionization energy. These correspond to the stepwise removal of electrons, one after the other. Lithium, for example, has three ionization energies because it has three electrons. To remove the outer $2s$ electrons from one mole of isolated lithium atoms to give one mole of gaseous lithium ions, Li^+, requires 520 kJ; so the *first ionization energy* of lithium is 520 kJ/mol. The second IE of lithium is 7297 kJ/mol, and corresponds to the process

$$Li^+(g) \longrightarrow Li^{2+}(g) + e^-$$

This involves the removal of an electron from the now exposed $1s$ core of lithium. Removal of the third (and last) electron requires the third IE, which is 11,810 kJ/mol. In general, successive ionization energies always increase because each subsequent electron is being pulled away from an increasingly more positive ion, and that requires more work.

Periodic Trends in Ionization Energies Within the periodic table there are trends in the way IE varies that are useful to know and to which we will refer in later discussions. We can see these by examining a graph that shows how the first ionization energy varies with an element's position in the table, which is shown in Figure 6.26, page 256. Notice that the elements with the largest ionization energies are the nonmetals in the upper right of the periodic table, and that those with the smallest ionization energies are the metals in the lower left of the table. In general, then, the following trends are observed.

Variations in ionization energy

Ionization energy generally increases from bottom to top within a group and increases from left to right within a period.

FIGURE 6.26

Variation in first ionization energy with location in the periodic table. Elements with the largest ionization energies are in the upper right of the periodic table. Those with the smallest ionization energies are at the lower left.

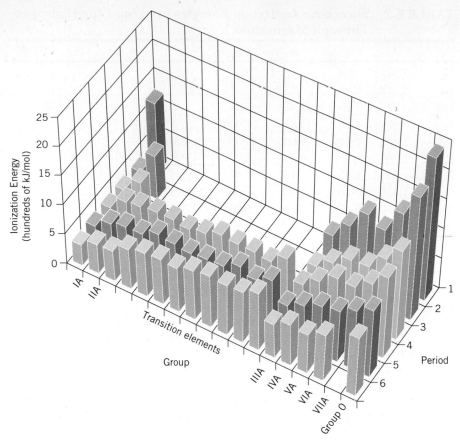

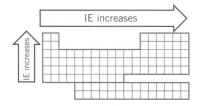

It is often helpful to remember that the trends in IE are just the opposite of the trends in atomic size within the periodic table; when size increases, IE decreases.

The same factors that affect atomic size also affect ionization energy. As the value of n increases going down a group, the orbitals become larger and the outer electrons are farther from the nucleus. Electrons farther from the nucleus are bound less tightly, so IE decreases from top to bottom. Of course, this is just the same as saying that it increases from bottom to top.

As you can see, there is a gradual overall increase in IE as we move from left to right across a period, although the horizontal variation of IE is somewhat irregular (see Special Topic 6.4). The reason for the overall trend is the increase in effective nuclear charge felt by the valence electrons across a period. As we've seen, this draws the valence electrons closer to the nucleus and leads to a decrease in atomic size from left to right. But the increasing effective nuclear charge also causes the valence electrons to be held more tightly, which makes it more difficult to remove them.

The results of these trends place elements with the largest IE in the upper right-hand corner of the periodic table. It is very difficult to cause these atoms to lose electrons. In the lower left-hand corner of the table are elements that have loosely held valence electrons. These elements form positive ions relatively easily, as you learned in Chapter 2.

Stability of the Noble Gas Configuration Table 6.2 shows that for a given element successive ionization energies increase gradually until the valence shell is emptied. Then a very much larger increase in IE occurs as the core is broken into. This is illustrated graphically in Figure 6.27 for the period-2

FIGURE 6.27

Variations in successive ionization energies for the elements
lithium through fluorine.

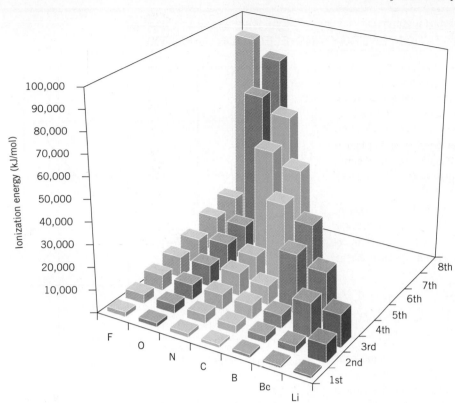

elements lithium through fluorine. For lithium we see that the first electron
(the $2s$ electron) is removed rather easily, but the second and third electrons,
which come from the $1s$ core, are much more difficult to dislodge. For beryllium, the large jump in IE occurs after two electrons (the two $2s$ electrons) are
removed. In fact, for all of these elements, the big jump in IE happens when
the core is broken into.

The data displayed in Figure 6.27 suggest that although it may be moderately difficult to empty the valence shell of an atom, it is *extremely* difficult to
break into the noble gas configuration of the core electrons. As you will learn,
this is one of the factors that influences the number of positive charges on ions
formed by the representative metals.

■ **Practice Exercise 11** Use the periodic table to select the atom with the largest
IE: (a) Na, Sr, Be, Rb, (b) B, Al, C, Si.

Electron Affinity

The **electron affinity** (abbreviated **EA**) *is the potential energy change associated
with the addition of an electron to a gaseous atom or ion in its ground state.* For an
element X, it is the change in potential energy associated with the process

$$X(g) + e^- \longrightarrow X^-(g)$$

For nearly all the elements, the addition of one electron to the neutral atom
is exothermic, and the EA is given as a negative value. This is because the
incoming electron experiences an attraction to the nucleus, which causes the
potential energy to be lowered as the electron approaches the atom. However,

SPECIAL TOPIC 6.4 / IRREGULARITIES IN THE PERIODIC VARIATIONS IN IONIZATION ENERGY AND ELECTRON AFFINITY

The variation in first ionization energy across a period is not a smooth one, as seen in the graph at the right for the elements in period 2. The first irregularity occurs between Be and B, where the IE increases from Li to Be, but then decreases from Be to B. This happens because there is a change in the nature of the subshell from which the electron is being removed. For Li and Be, the electron is removed from the $2s$ subshell, but at boron the first electron comes from the higher-energy $2p$ subshell where it is not bound as tightly.

Another irregularity occurs between nitrogen and oxygen. For nitrogen, the electron that's removed comes from a singly occupied orbital. For oxygen, the electron is taken from an orbital that already contains an electron. We can diagram this as follows:

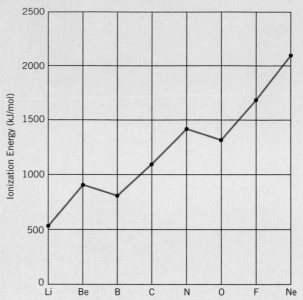

Variation in IE for the period 2 elements Li through F.

For oxygen, repulsions between the two electrons in the p orbital that's about to lose an electron helps the electron leave. This "help" is absent for the electron that's about to leave the p orbital of nitrogen. As a result, it is not as difficult to remove one electron from an oxygen atom as it is to remove one electron from a nitrogen atom.

As with ionization energy, there are irregularities in the periodic trends for electron affinity. Across a period, for example, the Group IIA elements have little tendency to acquire electrons because their outer-shell s orbitals are filled. The incoming electron must enter a higher-energy p orbital. We also see that the EA for elements in Group VA are either endothermic or only slightly exothermic. This is because the incoming electron must enter an orbital already occupied by an electron.

One of the most interesting irregularities occurs between periods 2 and 3 among the nonmetals. In any group, the element in period 2 has a less exothermic electron affinity than the element below it. The reason seems to be the small size of the nonmetal atoms of period 2, which are among the smallest elements in the periodic table. Repulsions between the many electrons in the small valence shells of these atoms leads to a lower-than-expected attraction for an incoming electron and a less exothermic electron affinity than the element below in period 3.

---

when a second electron must be added, as in the formation of the oxide ion, O^{2-}, work must be done to force the electron into an already negative ion.

Notice that the sign convention used here agrees with the one that we followed in determining the sign of ΔH in Chapter 5.

Change	EA (kJ/mol)
$O(g) + e^- \longrightarrow O^-(g)$	-141
$O^-(g) + e^- \longrightarrow O^{2-}(g)$	$+844$
$O(g) + 2e^- \longrightarrow O^{2-}(g)$	$+703$ (net)

Notice that more energy is absorbed adding an electron to the O^- ion than is released by adding an electron to the O atom. Overall, the formation of an isolated oxide ion is endothermic. The same applies to the formation of any negative ion with a charge larger than $1-$.

TABLE 6.3 Electron Affinities of the Representative Elements (kJ/mol)

IA	IIA	IIIA	IVA	VA	VIA	VIIA
H −73						
Li −60	Be +238	B −27	C −122	N ~+9	O −141	F −328
Na −53	Mg +230	Al −44	Si −134	P −72	S −200	Cl −348
K −48	Ca +155	Ga −30	Ge −120	As −77	Se −195	Br −325
Rb −47	Sr +167	In −30	Sn −121	Sb −101	Te −190	I −295
Cs −45	Ba +50	Tl −30	Pb −110	Bi −110	Po −183	At −270

The electron affinities of the representative elements are given in Table 6.3, and we see that periodic trends in electron affinity roughly parallel those for ionization energy.

Overall, electron affinity becomes more *exothermic* from left to right across a period and from bottom to top in a group (Figure 6.28).

This shouldn't be surprising, because a valence shell that loses electrons easily (low IE) will have little attraction for additional electrons (small EA). On the other hand, a valence shell that holds its electrons tightly will also tend to bind an additional electron tightly.

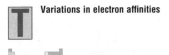
Variations in electron affinities

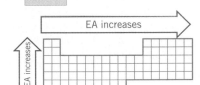

FIGURE 6.28

Variation of electron affinity within the periodic table.

SUMMARY

Electromagnetic Energy Electromagnetic energy, or light energy, travels through space at a constant speed of 3.00×10^8 m s^{-1} in the form of waves. The **wavelength,** λ, and **frequency,** ν, of the wave are related by the equation $\lambda\nu = c$, where c is the **speed of light.** The SI unit for frequency is the **hertz (Hz).** Light also behaves as if it consists of small packets of energy called **photons** or **quanta.** The energy delivered by a photon is proportional to the frequency of the light, and is given by the equation $E = h\nu$, where h is **Planck's constant.** White light is composed of all the frequencies visible to the eye and can be split into a **continuous spectrum.** Visible light represents only a small portion of the entire **electromagnetic spectrum,** which also includes **X rays, ultraviolet, infrared, microwaves,** and **radio** and **TV** waves.

Atomic Spectra The occurrence of **line spectra** tells us that atoms can emit energy only in discrete amounts and suggests that the energy of the electron is **quantized;** that is, the electron is restricted to certain specific **energy levels** in an atom. Niels Bohr recognized this and, although his theory was later shown to be incorrect, he was the first to propose a model that was able to account for the **Rydberg equation.** Bohr was the first to introduce the idea of **quantum numbers.**

Matter Waves The wave behavior of electrons and other tiny particles, which can be demonstrated by **diffraction** experiments, was suggested by de Broglie. Schrödinger applied wave theory to the atom and launched the theory we call **wave mechanics** or **quantum mechanics.** This theory tells us that electron waves in atoms are **standing waves** whose crests and **nodes** are stationary. Each standing wave, or **orbital**, is characterized by three quantum numbers, n, ℓ, and m_ℓ. **Shells** are designated by n (which can range from 1 to ∞), **subshells** by ℓ (which can range from 0 to $n - 1$), and orbitals within subshells by m_ℓ (which can range from $-\ell$ to $+\ell$).

Electron Configurations The electron has magnetic properties that are explained in terms of spin. The **spin quantum number**, m_s, can have values of $+\frac{1}{2}$ or $-\frac{1}{2}$. The **Pauli exclusion principle** limits orbitals to a maximum population of two electrons with **paired spins.** Substances with unpaired electrons are **paramagnetic** and are attracted weakly to a magnetic field. Substances with only paired electrons are **diamagnetic** and are slightly repelled by a magnetic field. The **electron configuration** of an element in its **ground state** is obtained by filling orbitals beginning with the $1s$ subshell and following the Pauli exclusion principle and **Hund's rule** (which states that electrons spread out as much as possible in orbitals of equal energy). The periodic table serves as a guide in predicting electron configurations. **Abbreviated configurations** show subshell populations outside a noble gas **core. Valence shell configurations** show the populations of subshells in the **outer shell** of an atom of the representative elements. Sometimes we represent electron configurations using **orbital diagrams.** Unexpected configurations occur for chromium and copper because of the extra stability of half-filled and filled subshells.

Orbital Shapes All s orbitals are spherical; each p orbital consists of two lobes with a nodal plane between them. A p subshell has three p orbitals whose axes are mutually perpendicular and point along the x, y, and z axes of an imaginary coordinate system centered at the nucleus. In each orbital the electron is conveniently viewed as an **electron cloud** with a varying **electron density.** Four of the five d orbitals in a d subshell have the same shape, with four lobes of electron density each. The fifth has two lobes of electron density pointing in opposite directions along the z axis and a ring of electron density in the x–y plane.

Atomic Properties The amount of positive charge felt by the valence electrons of an atom is the **effective nuclear charge.** This is less than the actual nuclear charge because core electrons partially shield the valence electrons from the full positive charge of the nucleus. **Atomic radii** depend on the value of n of the valence shell orbitals and the effective nuclear charge experienced by the valence electrons. These radii are expressed in units of picometers or nanometers, or an older unit called the **angstrom (Å).** 1 Å = 100 pm = 0.1 nm. Atomic radii decrease from left to right in a period and from bottom to top in a group in the periodic table. Negative ions are larger than the atoms from which they are formed; positive ions are smaller than the atoms from which they are formed.

Ionization energy (IE) is the energy needed to remove an electron from an isolated gaseous atom, molecule, or ion in its ground state. It is endothermic and the first ionization energy increases from left to right in a group and from bottom to top in a period. [Irregularities occur in a period when the nature of the orbital from which the electron is removed changes and when the electron removed is first taken from a doubly occupied p orbital.] Successive ionization energies become larger, but there is a very large jump when the next electron must come from the noble gas core beneath the valence shell.

Electron affinity (EA) is the potential energy change associated with the addition of an electron to a gaseous atom or ion in its ground state. For atoms, the first EA is usually exothermic. When more than one electron is added to an atom, the overall energy change is endothermic. In general, electron affinity becomes more exothermic from left to right in a period and from bottom to top in a group. [However, the EA of second period nonmetals is less exothermic than for the nonmetals of the third period. Irregularities across a period occur when the electron being added must enter the next higher energy subshell and when it must enter a half-filled p subshell.]

Tools You Have Learned

The table below lists the tools you have learned in this chapter. Notice that only two of them are related to numerical calculations. The others are conceptual tools that we use in analyzing properties of substances in terms of the underlying structure of matter. Review all these tools and refer to them, if necessary, when working on the Thinking-It-Through problems and the Review Exercises that follow.

Tool	Function
Wavelength–frequency relationship (page 223) $\lambda\nu = c$	To convert between wavelength and frequency.

Energy of a photon (page 225) $E = h\nu$	To calculate the energy carried by a photon of frequency ν. Also, ν can be calculated if E is known.
Periodic Table (page 244)	We will use the periodic table as a tool for many purposes. In this chapter you learned to use the periodic table as an aid in writing electron configurations of the elements and as a tool to correlate an element's location in the table to properties such as atomic radius, ionization energy, and electron affinity.
Periodic trends in atomic and ionic size (page 253) Atomic size increases from top to bottom in a group and decreases from left to right in a period.	To compare the sizes of atoms and ions.
Periodic trends in ionization energy (page 255) IE becomes larger from left to right in a group and from bottom to top in a group. For a given element, successive IE's become larger.	To compare the ease with which atoms of the elements lose electrons.
Periodic trends in electron affinity (page 259) EA becomes more exothermic from left to right in a period and from bottom to top in a group.	To compare the tendency of atoms or ions to gain electrons.

THINKING IT THROUGH

Remember, you are not asked to obtain answers for the following problems. Instead, assemble the data necessary to solve the problems and describe how you would use the data to obtain the answers. For numberical problems, set up the calculation using appropriate factors.

The problems are divided into two groups. Those in Level 2 are significantly more challenging than those in Level 1 and provide an opportunity to really hone your problem solving skills.

Level 1 Problems

1. An FM radio station broadcasts at a frequency of 101.9 MHz. What is the wavelength of these radio waves expressed in meters? How many peaks of these waves pass a given point each second? (Set up the calculation.)

2. Ultraviolet (UV) radiation can cause sunburn and has been implicated as a causative factor in the formation of certain skin cancers. A typical photon of UV light has a wavelength of 150 nm. How can you calculate the energy of a photon of this light? How many times larger is this than the energy of a photon emitted by a high-voltage wire carrying 60 Hz alternating current?

3. For a hydrogen atom, explain how to calculate the frequency of the photon that is emitted when the electron drops from an energy level that has $n = 6$ to one that has $n = 2$. Describe how you would calculate the wavelength of this light. How can you determine what color the light is?

4. Atoms of which of the following elements are paramagnetic? (a) Mg, (b) Zn, (c) As, (d) Ar, (e) Ag

5. The atoms of which element have valence electrons that experience the greatest effective nuclear charge, silicon or sulfur? Explain your reasoning.

6. State the specific relationship that allows you to determine

(a) which element has the larger atoms, Ge or S.
(b) which element is likely to have the larger first ionization energy, phosphorus or sulfur.
(c) which element is likely to have the more exothermic electron affinity, phosphorus or sulfur.

Level 2 Problems

7. Explain in detail how to calculate the ionization energy of the hydrogen atom from information provided in this chapter.

8. Magnesium forms the ion Mg^{2+} when a magnesium atom loses two electrons. Show how you would calculate the energy required to remove two electrons from a mole of magnesium atoms. How much water (in grams) could have its temperature raised from room temperature (25 °C) to the boiling point (100 °C) by the energy needed to remove two electrons from a mole of magnesium atoms? Set up the calculation.

9. Explain how you can easily determine the ionization energy of the fluoride ion.

10. How can you determine the electron affinity of an Al^{3+} ion?

11. How can you show that $Na(g) + Cl(g)$ is more stable (of lower energy) than $Na^+(g) + Cl^-(g)$? The potential energy of the Na^+ and Cl^- ions is inversely proportional to the distance between the ions. How can you determine the minimum distance at which the energy of the Na^+ and Cl^- ions is lower than the energy of the neutral atoms?

REVIEW EXERCISES

Answers to questions whose numbers are printed in color are given in Appendix D. More challenging questions are marked with asterisks.

Electromagnetic Radiation

6.1 In general terms, why do we call light *electromagnetic radiation?*

6.2 In general, what does the term *frequency* imply? What is meant by the term *frequency of light?* What symbol is used for it, and what is the SI unit (and symbol) for frequency?

6.3 What is meant by the term *wavelength* of light? What symbol is used for it?

6.4 Sketch a picture of a wave and label its wavelength and its amplitude.

6.5 Arrange the following regions of the electromagnetic spectrum in order of increasing wavelength (i.e., shortest wavelength → longest wavelength): microwave, TV, X ray, ultraviolet, visible, infrared, gamma rays.

6.6 What wavelength range is covered by the *visible spectrum?*

6.7 Arrange the following colors of visible light in order of increasing wavelength: orange, green, blue, yellow, violet, red.

6.8 Write the equation that relates the wavelength and frequency of a light wave.

6.9 What is the frequency in hertz of blue light having a wavelength of 430 nm?

6.10 A certain substance strongly absorbs infrared light having a wavelength of 6.85 μm. What is the frequency of this light in hertz?

6.11 Ozone protects the Earth's inhabitants from the harmful effects of ultraviolet light arriving from the sun. This shielding is a maximum for UV light having a wavelength of 295 nm. What is the frequency in hertz of this light?

6.12 Radar signals are electromagnetic radiations in the microwave region of the spectrum. A typical radar signal has a wavelength of 3.19 cm. What is its frequency in hertz?

6.13 In New York City, radio station WCBS broadcasts its FM signal at a frequency of 101.1 megahertz (MHz). What is the wavelength of this signal in meters?

6.14 Sodium vapor lamps are often used in residential street lighting. They give off a yellow light having a frequency of 5.09×10^{14} Hz. What is the wavelength of this light in nanometers?

6.15 There has been some concern in recent times about possible hazards to people who live very close to high-voltage electric power lines. The electricity in these wires oscillates at a frequency of 60 Hz, which is the frequency of any electromagnetic radiation that they emit. What is the wavelength of this radiation in meters? What is it in kilometers?

6.16 An X-ray beam has a frequency of 1.50×10^{18} Hz. What is the wavelength of this light in nanometers and in picometers?

6.17 How is the frequency of a particular type of radiation related to the energy associated with it? (Give an equation, defining all symbols.)

6.18 What is a photon?

6.19 Show that the energy of a photon is given by the equation $E = hc/\lambda$.

6.20 Examine each of the following pairs and state which of the two has the higher *energy.* (a) Microwaves and infrared. (b) Visible light and infrared. (c) Ultraviolet light and X rays. (d) Visible light and ultraviolet light.

6.21 What is a quantum of energy?

6.22 Calculate the energy in joules of a photon of red light having a frequency of 4.0×10^{14} Hz. What is the energy of one mole of these photons?

6.23 Calculate the energy in joules of a photon of green light having a wavelength of 560 nm.

6.24 Microwaves are used to heat food in microwave

ovens. The microwave radiation is absorbed by moisture in the food. This heats the water, and as the water becomes hot, so does the food. How many photons having a wavelength of 3.00 mm would have to be absorbed by 1.00 g of water to raise its temperature by 1.00 °C?

Atomic Spectra

6.25 What is an atomic spectrum? How does it differ from a continuous spectrum?

6.26 What fundamental fact is implied by the existence of atomic spectra?

6.27 In the spectrum of hydrogen, there is a line with a wavelength of 410.3 nm. (a) What color is this line? (b) What is its frequency? (c) What is the energy of each of its photons?

Bohr Atom and the Hydrogen Spectrum

6.28 Use the Rydberg equation to calculate the wavelength in nanometers of the spectral line of hydrogen for which $n_2 = 6$ and $n_1 = 3$. Would we be expected to see the light corresponding to this spectral line? Explain your answer.

*6.29 Calculate the wavelength in nanometers of the shortest wavelength of light emitted by a hydrogen atom.

6.30 Describe Niels Bohr's model of the structure of the hydrogen atom.

6.31 In qualitative terms, how did Bohr's model account for the atomic spectrum of hydrogen?

6.32 Calculate the energy in joules and the wavelength in nanometers of the spectral line produced in the hydrogen spectrum when an electron falls from the fourth Bohr orbit to the first. In which region of the electromagnetic spectrum (UV, visible, or infrared) is the line?

6.33 Calculate the wavelength of the spectral line produced in the hydrogen spectrum when an electron falls from the tenth Bohr orbit to the fourth. In which region of the electromagnetic spectrum (UV, visible, or infrared) is the line?

6.34 What is the term used to describe the lowest energy state of an atom?

6.35 In what way was Bohr's theory a success? How was it a failure?

Wave Nature of Matter

6.36 How does the behavior of very small particles differ from that of the larger, more massive objects that we meet in everyday life? Why don't we notice this same behavior for the larger, more massive objects?

6.37 Describe the phenomenon called diffraction. How can this be used to demonstrate that de Broglie's theory was correct?

6.38 What is the difference between a *traveling wave* and a *standing wave?*

Electron Waves in Atoms

6.39 What are the names used to refer to the theories that apply the matter–wave concept to electrons in atoms?

6.40 What is the term used to describe a particular waveform of a standing wave for an electron?

6.41 What are the two properties of orbitals in which we are most interested? Why?

Quantum Numbers

6.42 What are the allowed values of the principal quantum number?

6.43 What is the value for n for (a) the K shell and (b) the M shell?

6.44 What is the letter code for a subshell with (a) $\ell = 1$, (b) $\ell = 3$, (c) $\ell = 5$?

6.45 Give the values of n and ℓ for the following subshells: (a) 3s, (b) 5d, (c) 4f.

6.46 For the shell with $n = 6$, what are the possible values of ℓ?

6.47 Why does every shell contain an s subshell?

6.48 What are the possible values of m_ℓ for a subshell with (a) $\ell = 1$ and (b) $\ell = 3$?

6.49 How many orbitals are found in (a) an s subshell, (b) a p subshell, (c) a d subshell, and (d) an f subshell?

6.50 If the value of ℓ for an electron in an atom is 5, could this electron be in the 4th shell?

6.51 If the value of m_ℓ for an electron in an atom is 2, could another electron in the same subshell have a value of -3?

6.52 If the value of m_ℓ for an electron in an atom is -4, what is the smallest value of ℓ that the electron could have? What is the smallest value of n that the electron could have?

6.53 Suppose an electron in an atom has the following set of quantum numbers: $n = 2$, $\ell = 1$, $m_\ell = 1$, $m_s = +\frac{1}{2}$. What set of quantum numbers is impossible for another electron in this same atom?

6.54 How many orbitals are there in an h subshell ($\ell = 5$)?

Electron Spin

6.55 What physical property of electrons leads us to propose that they spin like toy tops?

6.56 What is the name of the magnetic property exhibited by atoms that contain unpaired electrons?

6.57 What is the Pauli exclusion principle? What effect does it have on the populating of orbitals by electrons?

6.58 Give the complete set of quantum numbers for all of the electrons that could populate the 2p subshell of an atom.

6.59 What are the possible values of the spin quantum number?

Electron Configuration of Atoms

6.60 What do we mean by the term *electronic structure?*

6.61 Within any given shell, how do the energies of the s, p, d, and f subshells compare?

6.62 What fact about the energies of subshells was responsible for the apparent success of Bohr's theory about electronic structure?

6.63 How do the energies of the orbitals belonging to a given subshell compare?

6.64 Give the electron configurations of the elements in period 2 of the periodic table.

6.65 Predict the electron configurations of (a) S, (b) K, (c) Ti, and (d) Sn.

6.66 Predict the electron configurations of (a) As, (b) Cl, (c) Ni, and (d) Si.

6.67 Give the correct electron configurations of (a) Cr and (b) Cu.

6.68 What is the correct electron configuration of silver?

6.69 Draw complete orbital diagrams for (a) Mg and (b) Ti.

6.70 Draw complete orbital diagrams for (a) As and (b) Ni.

6.71 How many unpaired electrons would be found in the ground state of (a) Mg, (b) P, and (c) V?

6.72 Write the shorthand electron configurations for (a) Ni, (b) Cs, (c) Ge, and (d) Br.

6.73 Draw orbital diagrams for the shorthand configurations of (a) Ni, (b) Cs, (c) Ge, and (d) Br.

6.74 Write the shorthand electron configurations for (a) Al, (b) Se, (c) Ba, and (d) Sb.

6.75 Draw orbital diagrams for the shorthand configurations of (a) Al, (b) Sc, (c) Ba, and (d) Sb.

6.76 How are the electron configurations of the elements in a given group similar? Illustrate your answer by writing shorthand configurations for the elements in Group VIA.

6.77 Define the terms *valence shell* and *valence electrons*.

6.78 What is the value of n for the valence shells of (a) Sn, (b) K, (c) Br, and (d) Bi?

6.79 Give the configuration of the valence shell for (a) Na, (b) Al, (c) Ge, and (d) P.

6.80 Draw the orbital diagram for the valence shell for (a) Na, (b) Al, (c) Ge, and (d) P.

6.81 Give the configuration of the valence shell for (a) Mg, (b) Br, (c) Ga, and (d) Pb.

6.82 Draw the orbital diagram for the valence shell of (a) Mg, (b) Br, (c) Ga, and (d) Pb.

6.83 How many unpaired electrons are expected to be found in the ground states of the following atoms? (a) Mn, (b) As, (c) S, (d) Sr, and (e) Ar.

6.84 Which of the following atoms in their ground states are expected to be diamagnetic? (a) Ba, (b) Se, (c) Zn, and (d) Si.

Shapes of Atomic Orbitals

6.85 In what general terms do we describe an electron's location in an atom?

6.86 Sketch the approximate shape of (a) a $1s$ orbital and (b) a $2p$ orbital.

6.87 How does the size of a given type of orbital vary with n?

6.88 How are the p orbitals of a given p subshell oriented relative to each other?

6.89 What is a *nodal plane*?

6.90 On appropriate coordinate axes, sketch the shape of the following d orbitals: (a) d_{xy}, (b) $d_{x^2-y^2}$, (c) d_{z^2}.

Atomic and Ionic Size

6.91 What is the meaning of *effective nuclear charge*? How does the effective nuclear charge felt by the outer electrons vary going down a group? How does it change as we go from left to right across a period?

6.92 If the core electrons were 100% effective at shielding the valence electrons from the nuclear charge and the valence electrons provided no shielding for each other, what would be the effective nuclear charge felt by a valence electron in (a) Na, (b) S, (c) Cl?

6.93 Choose the larger atom in each pair: (a) Na or Si, (b) P or Sb.

6.94 Choose the larger atom in each pair: (a) Al or Cl, (b) Al or In.

6.95 Choose the largest atom among the following: Ge, As, Sn, Sb.

6.96 In what region of the periodic table are the largest atoms found? Where are the smallest atoms found?

6.97 Place the following in order of increasing size: N^{3-}, Mg^{2+}, Na^+, Ne, F^-, O^{2-}. (Hint: Count the number of electrons and the nuclear charge in each particle.)

6.98 Why are the size changes among the transition elements more gradual than those among the representative elements going left to right in the periodic table?

6.99 Choose the larger particle in each pair: (a) Na or Na^+, (b) Co^{3+} or Co^{2+}, (c) Cl or Cl^-.

Ionization Energy

6.100 Define ionization energy. Why are ionization energies of atoms and positive ions endothermic quantities?

6.101 For oxygen, write an equation for the change associated with (a) its first ionization energy and (b) its third ionization energy.

6.102 Choose the atom with the larger ionization energy in each pair: (a) B or C, (b) O or S, (c) Cl or As.

6.103 Explain why ionization energy increases from left to right in a period and decreases from top to bottom in a group.

6.104 Why is an atom's second ionization energy always larger than its first ionization energy?

6.105 Why is the fifth ionization energy of carbon so much larger than its fourth?

6.106 Why is the first ionization energy of aluminum less than the first ionization energy of magnesium?

6.107 Why does phosphorus have a larger first ionization energy than sulfur?

Electron Affinity

6.108 Define *electron affinity*.

6.109 For sulfur, write an equation for the change associated with (a) its first electron affinity and (b) its second electron affinity. How should they compare?

6.110 Why does Cl have a more exothermic electron affinity than F? Why does Br have a less exothermic electron affinity than Cl?

6.111 Choose the atom with the more exothermic electron affinity in each pair: (a) Cl or Br, (b) Se or Br, (c) Si or Ga.

6.112 Why is the second electron affinity of an atom always endothermic?

6.113 How is electron affinity related to effective nuclear charge? On this basis, explain the relative magnitudes of the electron affinities of oxygen and fluorine.

Additional Problems

*6.114 How many grams of water could have its temperature raised by 5.0 °C by a mole of photons that have a wavelength of (a) 600 nm and (b) 300 nm?

*6.115 It has been found that when the chemical bond between chlorine atoms in Cl_2 is formed, 328 kJ is released per mole of Cl_2 formed. What is the wavelength of light that would be required to break chemical bonds between chlorine atoms?

6.116 Use the Rydberg equation to calculate the wavelengths of the lines in the spectrum of hydrogen that result when an electron falls from a Bohr orbit with (a) $n = 5$ to $n = 2$, (b) $n = 4$ to $n = 1$, and (c) $n = 6$ to $n = 4$. In which regions of the electromagnetic spectrum are these lines?

6.117 What, if anything, is wrong with the following electron configurations for atoms in their ground states?
(a) $1s^2 2s^1 2p^3$ (c) $[Kr]\ 3d^7 4s^2$
(b) $1s^2 2s^2 2p^4$ (d) $[Xe]\ 4f^{14} 5d^8 6s^1$

6.118 What, if anything, is wrong with the following orbital diagrams?
(a) ○ ⊕⊕⊕
(b) ⊕ ⊕○○
(c) ⊕ ⊕⊕⊕

6.119 How many electrons are in p orbitals in an atom of germanium?

6.120 What are the quantum numbers of the electrons that are lost by an atom of iron when it forms the ion Fe^{2+}?

*6.121 The removal of an electron from the hydrogen atom corresponds to raising the electron to the Bohr orbit that has $n = \infty$. On the basis of this statement, calculate the ionization energy of hydrogen in units of (a) joules per atom and (b) kilojoules per mole.

6.122 Use orbital diagrams to illustrate what happens when an oxygen atom gains two electrons. On the basis of what you have learned about electron affinities and electron configurations, why is it extremely difficult to place a third electron on the oxygen atom?

6.123 You learned that the $2s$ and $3s$ orbitals have nodes at certain distances from the nucleus. These were depicted in Figure 6.19. Similar nodes occur in $3p$, $4p$, etc., orbitals. On this basis, make a sketch of a dot-density diagram for a $3p$ orbital.

*6.124 From the data available in this chapter, determine the ionization energy of (a) F^-, (b) O^-, and (c) O^{2-}. Are any of these energies exothermic?

6.125 For an oxygen atom, which requires more energy, the addition of two electrons or the removal of one electron?

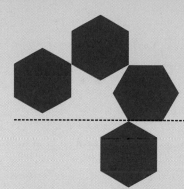

CHEMICALS IN USE 5
Lasers in Chemistry

A cry of "Fantastic!" is not an unusual reaction of someone watching a laser light show. Such entertainment is just one of the many common applications of lasers in our technological society. Compact disk players use a beam of laser light to read data from a CD so it can be changed to music. Laser printers produce printed pages that rival in quality those that come from a printing press, and laser surgery has enabled medical procedures impossible only a few decades ago.

The word **laser** is an acronym for *light amplification* by *stimulated emission* of *radiation.* In a laser, electrons are raised to a higher energy state by the absorption of energy in one form or another. If conditions are right, the number of excited atoms exceeds the number in the ground state and we say a **population inversion** exists. These are the conditions necessary to initiate laser emission. (A population inversion is only possible in certain systems, which is why not all substances are able to work as lasers.) The laser process begins when one atom emits a photon. This strikes another excited atom and stimulates it to emit a photon. These two then initiate further emissions, and so on until a cascade of photons is produced. In this way, the intensity of the original one-photon emission is amplified enormously.

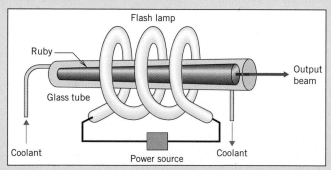

FIGURE 5a

Diagram of a ruby laser. Light from the flash lamp excites atoms in the ruby rod, which has a totally reflecting mirror on one end and a partially reflecting mirror on the other. The laser beam emerges from the partially mirrored end. Because of the large amount of energy absorbed by the ruby rod, it must be cooled to prevent overheating.

The ruby laser was one of the earliest and is illustrated in Figure 5a. A very bright flash lamp, similar to the kind used in the electronic flash lamp in a modern camera, wraps around a ruby rod and provides the energy to *pump* the laser into its excited state. The ends of the ruby rod are mirrored, one end being totally reflecting while the other end reflects only part of the light produced. The purpose of these mirrors is to contain the light within the ruby rod long enough to achieve maximum intensity. The laser beam then emerges from the ruby through the partially reflecting end.

Some lasers are solids; examples are ruby and yttrium aluminum garnet ($Y_3Al_2O_{15}$, commonly referred to as YAG). Others consist of dyes in solution, and still others are gases that are excited by passing an electric discharge (spark) through them. The first gas laser was a mixture of helium and neon which produces an intense beam of red light. One of the most powerful and efficient gas lasers uses CO_2 mixed with He and N_2. It produces laser light with a wavelength in the infrared region of the spectrum.

APPLICATIONS

The light from a laser has some unique properties. One is that it is extremely intense. The light from a He–Ne or CO_2 laser, for example, is many orders of magnitude brighter than sunlight. Another is that laser light is *coherent,* which means that all the waves of the individual photons are in phase. Because of this, the laser beam doesn't spread out much as it travels through space and is easily focused to a very tiny spot. These properties combine to permit the concentration of an enormous amount of energy into a very small area, which can produce very high localized temperatures. This is what makes laser surgery possible. Industrially, lasers are used to cut materials that range from silk to diamonds (Figure 5b), and robot-driven lasers are used now to weld entire automobile frames in a few seconds (Figure 5c).

One of the greatest advantages of a laser is that its radiation is concentrated in a very narrow band of wavelengths, so the light is much more nearly *monochromatic* (of *one color*) than the light from any other source. This property is especially valuable in chemistry because each substance, whether atomic or molecular, has its own characteristic *absorption spectrum*—wavelengths of light that are selectively absorbed by the substance as determined by the substance's unique set of energy levels. By choosing a laser with the proper wavelength, the chemist can selectively probe a system for individual chemical constituents. This ability has led to some amazing achievements.

In analytical chemistry, laser-based methods have proved to be so selective and sensitive that in some instances single atoms have been detected in the gas phase and single molecules in solution.

In the chemical industry, laser radiation of a particular wavelength is used to selectively destroy H_2S molecules in mixtures of CO and H_2 that are used to synthesize petrochemicals. The removal of the H_2S is essential because even trace amounts of less than one part per million can interfere with the synthesis.

It has been found that in the chemical synthesis of a compound used to make vitamin D, the use of carefully selected wavelengths of laser light can discriminate against unwanted side reactions and increase the percentage yield from about 30% to as much as 80%.

The electronics industry especially has found many uses for lasers, both because of their ability to direct their energy to specific chemical species as well as their ability to be focused to an extremely narrow beam. Pulses of laser radiation with a frequency chosen so as to be absorbed selectively by the epoxy resin in printed circuit boards permits this material to be vaporized without overheating the surroundings. Gas phase laser-induced dissociation of molecules such as $Sn(CH_3)_4$ are also used to deposit extremely thin (about 1 μm, or 0.00004 in.) metallic lines for electronic circuits such as those in memory chips in computers.

One of the most interesting applications is in the separation of isotopes from each other. Although they are atoms of the same element, their slight differences in mass cause slight differences in the wavelengths at which they absorb light. Laser light is so nearly monochromatic that its frequency can be tuned to excite or ionize only the atoms of a specific isotope. Once these ions are formed, they can be separated from the atoms of the other isotopes present. The enrichment of nuclear fuels for reactors is one potential application for the procedure.

The space available here is not sufficient to describe all the various ways lasers have been used in science and industry, but we hope you have become aware of what versatile tools lasers are. In chemistry today we have barely begun to explore the ways this marvelous phenomenon will help in preparing new and exciting materials for the future.

Questions

1. What is the origin of the word *laser*?

2. What is a *population inversion*?

3. How does the intensity of laser light compare to that from other kinds of light sources?

4. What does it mean when we say that laser light is *coherent*?

5. How is energy pumped into a ruby laser? How is a He–Ne laser pumped?

6. In what part of the spectrum do we find the laser emission from a carbon dioxide laser?

7. What property of laser light permits its use in the separation of the isotopes of an element from each other?

8. Define the term *monochromatic*.

FIGURE 5*b*

A laser is used to cut saw blades from a sheet of metal.

FIGURE 5*c*

Robots use lasers to weld an auto body on an assembly line.

Today's golfers, such as LPGA professional Debbi Miho Koyama, owe their better scores to golf clubs whose shafts are made of composite materials that include graphite fibers. Graphite is a form of carbon in which chemical attractions, called covalent bonds, hold atoms to each other in large molecules. Covalent bonding is one of the kinds of chemical bonding we will examine in this chapter.

Chapter 7

Chemical Bonding I

Atoms of nearly every element have the ability to combine with other atoms to form more complex substances. The forces of attraction that bind atoms to each other in such combinations are called **chemical bonds.** Understanding the nature and origin of chemical bonds is an important part of understanding chemistry, because changes in these bonding forces as old bonds break and new ones form constitute the underlying basis for all chemical reactions.

There are two principal classes of bonding forces. One occurs in molecules and involves the sharing of electrons; it is called *covalent bonding.* The other involves the transfer of electrons between atoms; this produces ions and is called *ionic bonding.* Ionic bonding is simpler to understand, so we will discuss it first.

7.1
ELECTRON TRANSFER AND THE FORMATION OF IONIC COMPOUNDS

Conditions That Control the Formation of an Ionic Compound

As you know, ionic compounds are formed when metals react with nonmetals. Among the examples discussed earlier was sodium chloride, table salt. You learned that when this compound is formed from its elements, each sodium atom loses one electron to form a sodium ion, Na^+, and each chlorine atom gains one electron to become a chloride ion, Cl^-. Once formed, these ions become tightly packed together, as illustrated in the margin, because their opposite charges attract. *This attraction between positive and negative ions in an ionic compound is what we call an* **ionic bond.**

The reason Na^+ and Cl^- ions attract each other is easy to understand. But *why* are electrons transferred between these and other atoms? *Why* does sodium form Na^+ and not Na^- or Na^{2+}? And *why* does chlorine form Cl^- instead of Cl^+ or Cl^{2-}? To answer these questions we must consider a number of factors that are related to the potential energy of the system of reactants and

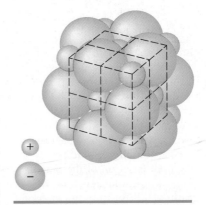

Packing of ions in NaCl.

269

Keep in mind the relationship between potential energy changes and endothermic and exothermic processes:

Endothermic ⇔ increase in PE
Exothermic ⇔ decrease in PE

products. This is because *for any stable compound to form from its elements, there must be a net lowering of the potential energy.* In other words, the reaction must be exothermic.

Importance of the Lattice Energy

In Chapter 5 you learned that we can use the law of conservation of energy (Hess's law) to combine the energy values associated with a series of changes when calculating the net energy change for some overall reaction. We can use the same principles here to analyze the factors that contribute to the net energy change when atoms react to form an ionic compound.

Figure 7.1 illustrates two paths from the elements, Na(s) and $\frac{1}{2}$Cl$_2$(g), to the product, NaCl(s). Hess's law demands that the energy change must be the same along each of them.

The direct path along the bottom has as its energy change the heat of formation of NaCl, ΔH_f°. The alternative path is divided into a number of steps. The first two steps, both of which are endothermic, change Na(s) and Cl$_2$(g) into gaseous atoms, Na(g) and Cl(g). The next two steps change these atoms to ions, first by the endothermic ionization energy (IE) of Na followed by the exothermic electron affinity (EA) of Cl. Notice that at this point, if we add all the energy changes, the ions are at a considerably higher energy than the reactants. If these were the only energy terms involved in the formation of NaCl, the heat of formation would be endothermic and the compound would be unstable; it could not be formed by direct combination of the elements.

The name for this energy comes from the term *lattice,* which is used to describe the regular pattern of ions that exists in crystals of the salt.

Ionic compounds such as NaCl are stable because of the final energy term in this alternative path. It is an energy called the **lattice energy,** defined as *the amount that the potential energy of the system is lowered when the ions in one mole of the compound are brought from a gaseous state to the positions the ions occupy in a crystal of the compound.* The potential energy is lowered because ions that have a net attraction for each other are being brought closer together; they change from isolated ions to the tightly packed arrangement found in the solid. As you learned earlier, when the potential energy decreases, the change is exothermic. This is why the lattice energy appears with a negative sign in Figure 7.1.

As you can see in Figure 7.1, the lattice energy is the largest of all the contributing energy factors in the formation of NaCl. It is able to overcome the net increase in potential energy associated with the formation of the isolated ions and cause the formation of the compound to be exothermic overall, which it must be for the compound to be stable.

What we have seen here for the formation of NaCl applies to other ionic compounds as well. For an ionic compound to be formed from the elements, the exothermic lattice energy must be larger than the endothermic combina-

FIGURE 7.1

Energy changes in the formation of NaCl, followed along two different paths. The lower path leads directly to NaCl. The upper path involves the formation of gaseous atoms from the elements, formation of ions from the gaseous atoms, and finally the condensation of the gaseous ions to give solid NaCl. Both paths yield the same net energy change.

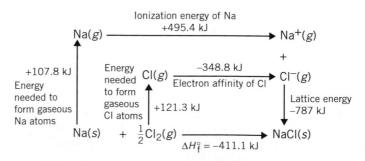

tion of factors involved in the formation of the ions themselves, which primarily involve the IE of the metal and the EA of the nonmetal.

Why Metals Form Cations and Nonmetals Form Anions

Now we can understand why metals tend to form positive ions and nonmetals tend to form negative ions. At the left of the periodic table are the metals—elements with small IE and EA. Relatively little energy is needed to remove electrons from them to produce positive ions. At the upper right of the periodic table are the nonmetals—elements with large IE and EA. It is quite difficult to remove electrons from these elements, but sizable amounts of energy are released when they gain electrons. On an energy basis, therefore, it is least "expensive" to form a cation from a metal and an anion from a nonmetal, so it is relatively easy for the energy-lowering effect of the lattice energy to exceed the net energy-raising effect of the IE and EA. In fact, metals combine with nonmetals to form ionic compounds simply because ionic bonding is favored energetically over other types whenever atoms of small IE combine with atoms of large EA.

Changes in Electron Configurations When Ions Form: The Octet Rule

Let's look now at how the electronic structures of the elements affect the kinds of ions they form. We begin by examining what happens when sodium loses an electron. The electron configuration of Na is

$$Na \quad 1s^2 2s^2 2p^6 3s^1$$

The electron that is lost is the one least tightly held, which is the single outer $3s$ electron. The electronic structure of the Na$^+$ ion, then, is

$$Na^+ \quad 1s^2 2s^2 2p^6$$

Na$^+$ has the same electron configuration as the noble gas Ne.

The removal of the first electron from Na does not require much energy because the first IE of sodium is small. Therefore, an input of energy equal to the first IE can be easily recovered by the exothermic lattice energy of ionic compounds that contain the Na$^+$ ion. However, removal of a second electron from sodium is very difficult because it involves breaking into the $2s^2 2p^6$ core. The second IE of Na is enormous, as indicated in the margin. As a result, the amount of energy required to create a Na^{2+} ion is *much* larger than the amount of energy that can be recovered by the lattice energy, so overall the formation of a compound containing Na^{2+} is energetically unfavorable. This is why we never observe compounds that contain this ion, and why sodium stops losing electrons once it has achieved a noble gas configuration.

For sodium:

1st IE = 496 kJ/mol
2nd IE = 4563 kJ/mol

A somewhat similar situation exists for other metals, too. Consider calcium, for example. We know that this metal forms ions with a 2+ charge. This means that when a calcium atom reacts, it loses its two outermost electrons.

$$Ca \quad 1s^2 2s^2 2p^6 3s^2 3p^6 4s^2$$
$$Ca^{2+} \quad 1s^2 2s^2 2p^6 3s^2 3p^6$$

Ca^{2+} has the same electron configuration as the noble gas Ar.

The two $4s$ electrons of Ca are not held too tightly, so the total amount of energy that must be invested to remove them (the sum of the first and second IE) can be recovered easily by the large lattice energy of a Ca^{2+} compound.

For calcium:

> 1st IE = 590 kJ/mol
> 2nd IE = 1140 kJ/mol
> 3rd IE = 4940 kJ/mol

However, the removal of yet another electron from calcium to form Ca^{3+} requires breaking into the noble gas core. As in the case of sodium, a tremendous amount of energy is needed to accomplish this, much more than can be regained by the lattice energy of a Ca^{3+} compound. Therefore, a calcium atom loses just two electrons when it reacts.

In both sodium and calcium, we find that the stability of the noble gas core that lies below the outer shell of electrons of these metals effectively limits the number of electrons that they lose, and that the ions that are formed have noble gas electron configurations.

Nonmetals also tend to have noble gas configurations when they form anions. For example, when a chlorine atom reacts, it gains one electron. For chlorine we have

$$\text{Cl} \quad 1s^2 2s^2 2p^6 3s^2 3p^5$$

and when an electron is gained, its configuration becomes

$$\text{Cl}^- \quad 1s^2 2s^2 2p^6 3s^2 3p^6$$

At this point, electron gain ceases, because if another electron were to be added, it would have to enter an orbital in the next higher shell. With oxygen, a similar situation exists. The formation of the oxide ion, O^{2-}, is endothermic, as you learned in the previous chapter.

> **Both Cl^- and O^{2-} have noble gas electron configurations.**

$$\text{O } (1s^2 2s^2 2p^4) + 2e^- \longrightarrow \text{O}^{2-} (1s^2 2s^2 2p^6) \qquad EA(\text{net}) = +703 \text{ kJ/mol}$$

However, the energy input is modest and can be overcome without much difficulty by the large lattice energies of ionic metal oxides. But we never observe the formation of O^{3-} because, once again, the third electron would have to enter an orbital in the next higher shell, and this is *very* energetically unfavorable.

Summary

> **Electron configurations of ions of the representative elements**

What we see here is that a balance of energy factors causes many atoms to form ions that have a noble gas electron configuration. Historically, this is expressed in the form of a generalization: *When they form ions, atoms of most of the representative elements tend to gain or lose electrons until they have obtained a configuration identical to that of the nearest noble gas.* Because all the noble gases except helium have outer shells with eight electrons, this rule has become known as the **octet rule,** which can be stated as follows: *Atoms tend to gain or lose electrons until they have achieved an outer shell that contains an octet of electrons (eight electrons).* Sodium and calcium achieve an octet by emptying their valence shells; chlorine and oxygen achieve an octet by gaining enough electrons to reach a noble gas configuration.

> **Hydrogen cannot obey the octet rule because its outer shell can contain just two electrons.**

Exceptions to the Octet Rule

The octet rule as applied to ionic compounds really works well only for the cations of the Group IA and IIA metals and for the anions of the nonmetals. It does not work as well for the transition metals and post-transition metals. For example, tin (a post-transition metal) forms two ions, Sn^{2+} and Sn^{4+}. The electron configurations are

$$
\begin{array}{ll}
\text{Sn} & [\text{Kr}]4d^{10}5s^2 5p^2 \\
\text{Sn}^{2+} & [\text{Kr}]4d^{10}5s^2 \\
\text{Sn}^{4+} & [\text{Kr}]4d^{10}
\end{array}
$$

Notice that the Sn^{2+} ion is formed by the loss of the higher energy $5p$ electrons first. Then, loss of the $5s$ electrons gives the Sn^{4+} ion. However, neither of these ions has a noble gas configuration.

For the transition elements, the first electrons lost are the s electrons of the outer shell. Then, if additional electrons are lost, they come from the underlying d subshell. An example is iron, which has the electron configuration

$$Fe \quad [Ar]3d^6 4s^2$$

When iron reacts, it loses its $4s$ electrons fairly easily to give Fe^{2+}. But because the $3d$ subshell is close in energy to the $4s$, it is not very difficult to remove still another electron to give Fe^{3+}.

$$Fe^{3+} \quad [Ar]3d^5$$

Notice here that the first electrons to be removed come from the shell with the largest value of n. Then, after this shell is emptied, the next electrons are removed from the shell below.

Because so many of the transition elements are able to form ions in a similar way, the ability to form more than one positive ion is usually cited as one of the characteristic properties of the transition elements. Frequently, one of the ions formed has a 2+ charge, which arises from the loss of the two outer s electrons. Ions with larger positive charges result when additional d electrons are lost. Unfortunately, it is not easy to predict exactly which ions can form for a given transition metal, nor is it simple to predict their relative stabilities. However, transition metals do tend to form ions with filled or half-filled outer electron subshells, as in Mn^{2+} and Fe^{3+}.

When an atom loses electrons, the first to go always comes from the shell with largest n. Within a given shell, electrons are always removed first from the highest energy subshell. This means that d is emptied before p, which is emptied before s.

 Electron configurations of ions of transition and post-transition

---

How do the electron configurations change (a) when a nitrogen atom forms the N^{3-} ion and (b) when an antimony atom forms the Sb^{3+} ion?

ANALYSIS For the nonmetals, you've learned that the octet rule does work, so the ion that is formed by nitrogen will have a noble gas configuration.

For antimony, a post-transition element, we must keep in mind that the p subshell of the valence shell loses electrons before the s subshell.

SOLUTION (a) The electron configuration for nitrogen is

$$N \quad [He]2s^2 2p^3$$

To form N^{3-}, three electrons are gained. These enter the $2p$ subshell because it is the lowest available energy level. The configuration for the ion is therefore

$$N^{3-} \quad [He]2s^2 2p^6$$

(b) Let's begin with the ground state electron configuration for antimony.

$$Sb \quad [Kr]4d^{10} 5s^2 5p^3$$

To form the Sb^{3+} ion, three electrons must be removed. These will come from the outer shell, which has $n = 5$. Within this shell, the energies of the subshells increase in the order $s < p < d < f$. Therefore, the $5p$ subshell is higher in energy than the $5s$, so the electrons are removed from the $5p$. This gives

$$Sb^{3+} \quad [Kr]4d^{10} 5s^2$$

■ **Practice Exercise 1** How does the electron configuration change when a chromium atom forms the following ions: (a) Cr^{2+}, (b) Cr^{3+}, (c) Cr^{6+}?

EXAMPLE 7.1
Writing Electron
Configurations of Ions

7.2
ELECTRON BOOKKEEPING: LEWIS SYMBOLS

In the last section you saw how the valence shells of atoms change when electrons are transferred during the formation of ions. We will soon see that many atoms share their valence electrons with each other when they form covalent bonds. In these discussions of bonding it is useful to be able to keep track of valence electrons. To help us do this, we use a simple bookkeeping device called Lewis symbols, named after their inventor, a famous American chemist, G. N. Lewis (1875–1946).

To draw the **Lewis symbol** for an element, we write its chemical symbol surrounded by a number of dots (or some other similar mark), which represent the atom's valence electrons. For example, the element lithium, which has one valence electron in its $2s$ subshell, has the Lewis symbol

$$\text{Li} \cdot$$

In fact, each element in Group IA has a similar Lewis symbol, because each has only one valence electron. The Lewis symbols for all of the Group IA metals are

$$\text{Li}\cdot \quad \text{Na}\cdot \quad \text{K}\cdot \quad \text{Rb}\cdot \quad \text{Cs}\cdot \quad \text{Fr}\cdot$$

The Lewis symbols for the eight A-group elements of period 2 are[1]

Lewis symbols

Group	IA	IIA	IIIA	IVA	VA	VIA	VIIA	0
Symbol	Li·	·Be·	·Ḃ·	·Ċ·	·N̈·	·Ö:	·F̈:	:N̈e:

The elements below each of these in their respective groups have identical Lewis symbols except, of course, for the chemical symbol of the element. Notice that when an atom has more than four valence electrons, the additional electrons are shown to be paired with others. Also notice that *for the representative elements, the group number is equal to the number of valence electrons* when the U.S. convention for numbering groups in the periodic table is followed.

This is one of the advantages of the U.S. system for numbering groups in the periodic table.

EXAMPLE 7.2
Writing Lewis Symbols

What is the Lewis symbol for arsenic, As?

SOLUTION Arsenic is in Group VA and therefore has five valence electrons. The first four are placed around the symbol for arsenic as follows:

$$\cdot \overset{\cdot}{\underset{\cdot}{\text{As}}} \cdot$$

The fifth electron is paired with one of the first four. This gives

$$\cdot \overset{\cdot}{\text{As}} :$$

The location of the fifth electron doesn't really matter, so equally valid Lewis symbols are:

$$\cdot \overset{\cdot \cdot}{\text{As}} \cdot \quad \text{or} \quad : \overset{\cdot}{\text{As}} \cdot \quad \text{or} \quad \cdot \underset{\cdot \cdot}{\text{As}} \cdot$$

■ **Practice Exercise 2** Write Lewis symbols for (a) Se, (b) I, and (c) Ca.

---

[1] For beryllium, boron, and carbon, the number of unpaired electrons in the Lewis symbol doesn't agree with the number predicted from the atom's electron configuration. Boron, for example, has two electrons paired in its $2s$ orbital and a third electron in one of its $2p$ orbitals; therefore, there is actually only one unpaired electron in a boron atom. The Lewis symbols are drawn as shown, however, because when beryllium, boron, and carbon form bonds, they *behave* as if they have two, three, and four unpaired electrons, respectively.

Although we will use Lewis symbols mostly to follow the fate of valence electrons in covalent bonds, they can also be used to describe what happens during the formation of ions. For example, when a sodium atom reacts with a chlorine atom, the electron transfer can be depicted as

$$Na \overset{\curvearrowright}{+} \cdot \ddot{\underset{..}{Cl}}: \longrightarrow Na^+ + \left[:\ddot{\underset{..}{Cl}}: \right]^-$$

The valence shell of the sodium atom is emptied, so no dots remain. The outer shell of chlorine, which formerly had seven electrons, gains one to give a total of eight. The brackets were drawn around the chloride ion to show that all eight electrons are the exclusive property of the Cl^- ion.

We can diagram a similar reaction between calcium and chlorine atoms.

$$:\ddot{\underset{..}{Cl}} \cdot \overset{\curvearrowleft}{} {}_{\circ}Ca \overset{\curvearrowright}{} \cdot \ddot{\underset{..}{Cl}}: \longrightarrow Ca^{2+} + 2\left[:\ddot{\underset{..}{Cl}}: \right]^-$$

EXAMPLE 7.3
Using Lewis Symbols

Use Lewis symbols to diagram the reaction that occurs between sodium and oxygen atoms to give Na^+ and O^{2-} ions.

SOLUTION First let's draw the Lewis symbols for Na and O.

$$Na \cdot \qquad \cdot \ddot{\underset{..}{O}}:$$

It takes two electrons to complete the octet around oxygen. Each Na can supply only one, so we need two Na atoms. Therefore,

$$Na \overset{\curvearrowright}{} \cdot \ddot{\underset{..}{O}}: \overset{\curvearrowleft}{} {}_{\circ}Na \longrightarrow 2Na^+ + \left[:\ddot{\underset{..}{O}}: \right]^{2-}$$

Notice that we have put brackets around the oxide ion.

■ **Practice Exercise 3** Diagram the reaction between magnesium and oxygen atoms to give Mg^{2+} and O^{2-} ions.

Most of the substances we encounter in our daily lives are not ionic. Rather than existing as collections of electrically charged particles (ions), they occur as electrically neutral combinations of atoms called **molecules.** Water, as you already know, consists of molecules made from two hydrogen atoms and one oxygen atom, and the formula for one particle of this compound is H_2O. Most substances consist of much larger molecules. For example, you've learned that the formula for table sugar is $C_{12}H_{22}O_{11}$.

Energy Changes in the Formation of a Covalent Bond

Earlier we saw that for ionic bonding to occur, the energy-lowering effect of the lattice energy must be greater than the combined energy-raising effects of the ionization energy (IE) and electron affinity (EA). Many times this is not possible, particularly when the ionization energies of all the atoms involved are large. This happens, for example, when nonmetals combine with each other. In such cases, nature uses a different way to lower the energy—electron sharing.

7.3
ELECTRON SHARING: THE FORMATION OF COVALENT BONDS

An ionic substance

A molecular substance

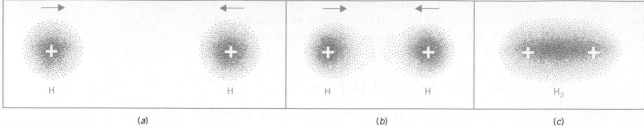

(a)   (b)   (c)

FIGURE 7.2

Formation of a bond between two hydrogen atoms. (*a*) Two H atoms separated by a large distance. (*b*) As the atoms approach each other, their electron densities begin to shift to the region between the two nuclei. (*c*) The electron density becomes concentrated between the nuclei.

Let's look at what happens when two hydrogen atoms join to form an H_2 molecule (Figure 7.2). As the two atoms approach each other, the electron of each atom begins to feel the attraction of both nuclei. This causes the electron density around each nucleus to shift toward the region between the two atoms. Therefore, as the distance between the nuclei decreases, there is an increase in the probability of finding either electron near either nucleus. In effect, as the molecule is formed, each of the hydrogen atoms in this H_2 molecule acquires a share of two electrons.

When the electron density shifts to the region between the two hydrogen atoms, it attracts both nuclei and pulls them together. Being of the same charge, however, the two nuclei also repel each other, as do the two electrons. In the molecule that forms, therefore, the atoms are held at a distance at which all these attractions and repulsions are balanced. Overall, the nuclei are kept from separating, and the net force of attraction produced by the sharing of the pair of electrons is called a **covalent bond.**

Every covalent bond is characterized by two quantities, the average distance between the nuclei held together by the bond, and the energy needed to separate the two atoms to produce neutral atoms again. In the hydrogen molecule, the attractive forces pull the nuclei to a distance of 75 pm, and this distance is called the **bond length** (or sometimes, the **bond distance**). Because a covalent bond holds atoms together, work must be done (energy must be supplied) to separate them. When the bond is *formed,* an equivalent amount of energy is released as the potential energies of the atoms are lowered. The amount of energy released when the bond is formed (or the amount of energy needed to "break" the bond) is called the **bond energy.**

Figure 7.3 shows how the potential energy changes when two hydrogen atoms form H_2. We see that the minimum potential energy occurs at a bond

As the distance between the two nuclei and the electron cloud that lies between them decreases, the potential energy decreases.

FIGURE 7.3

Changes in the potential energy of two hydrogen atoms as they approach each other to form the H_2 molecule.

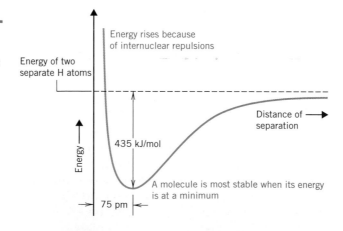

distance of 75 pm, and that 1 mol of hydrogen molecules is more stable than 2 mol of hydrogen atoms by 435 kJ. In other words, the bond energy of H_2 is 435 kJ/mol.

Pairing of Electrons

Before joining, each of the separate hydrogen atoms has one electron in its $1s$ orbital. When these electrons are shared, the $1s$ orbital of each atom is, in a sense, filled. Because the electrons now share the same space, they become paired as required by the Pauli exclusion principle; that is, m_s is $+\frac{1}{2}$ for one of the electrons and $-\frac{1}{2}$ for the other. In general, we almost always find that the electrons involved become paired when atoms form covalent bonds. In fact, a covalent bond is sometimes referred to as an **electron pair bond.**

Lewis symbols are often used to keep track of electrons in covalent bonds. The electrons that are shared between two atoms are shown as a pair of dots placed between the symbols for the bonded atoms. The formation of H_2 from hydrogen atoms, for example, can be depicted as

$$\text{H}\cdot \; + \; \text{H}\cdot \; \longrightarrow \; \text{H:H}$$

Because the electrons are shared, each H atom is considered to have two electrons.

(Two electrons can be counted around each of the H atoms.)

For simplicity, the electron pair in the covalent bond is usually represented as a dash. Thus, the hydrogen molecule is represented as

$$\text{H}\!-\!\text{H}$$

A formula such as this, which is drawn with Lewis symbols, is called a **Lewis formula** or **Lewis structure.** It is also called a **structural formula** because it shows which atoms are present in the molecule *and* how they are attached to each other.

Covalent Bonding and the Octet Rule

We have seen that when a nonmetal atom forms an anion, electrons are gained until the s and p subshells of its valence shell are completed. This tendency to finish with a completed valence shell, usually consisting of eight electrons, also influences the number of electrons an atom tends to acquire by sharing, and it thereby controls the number of covalent bonds that an atom forms.

Hydrogen, with just one electron in its $1s$ orbital, can complete its valence shell by obtaining a share of just one electron from another atom. When this other atom is hydrogen, the H_2 molecule is formed. Since hydrogen obtains a stable valence shell configuration when it shares just one pair of electrons with another atom, hydrogen atoms form only one covalent bond.

Many atoms form covalent bonds by sharing enough electrons to give them complete s and p subshells in their outer shells. This is the noble gas configuration mentioned earlier and is the basis of the octet rule described in Section 7.1. As applied to covalent bonding, the **octet rule** can be stated as follows: *When atoms form covalent bonds, they tend to share sufficient electrons so as to achieve an outer shell having eight electrons.*

In Chapter 6 you learned that when two electrons occupy the same orbital, and therefore share the same space, their spins must be paired. The pairing of electrons is an important part of the formation of a covalent bond.

One dash stands for two electrons.

As you will see, it is useful to remember that hydrogen atoms only form one covalent bond.

Often, the octet rule can be used to explain the number of covalent bonds an atom forms. This number normally equals the number of electrons the atom needs to have a total of eight (an octet) in its outer shell. For example, the halogens (Group VIIA) all have seven valence electrons. The Lewis symbol for a typical member of this group, chlorine, is

$$\cdot \ddot{\underset{\cdot\cdot}{\text{Cl}}} :$$

We can see that only one electron is needed to complete an octet. Of course, chlorine can actually gain this electron and become a chloride ion. This is what it does when it forms ionic compounds such as sodium chloride (NaCl). But when chlorine combines with another nonmetal, the transfer of electrons is not energetically favorable. Therefore, in forming such compounds as HCl or Cl_2, chlorine gets the one electron it needs by forming a covalent bond.

$$\text{H}\cdot \;+\; \cdot\ddot{\underset{\cdot\cdot}{\text{Cl}}}: \;\longrightarrow\; \text{H}:\ddot{\underset{\cdot\cdot}{\text{Cl}}}:$$

$$:\ddot{\underset{\cdot\cdot}{\text{Cl}}}\cdot \;+\; \cdot\ddot{\underset{\cdot\cdot}{\text{Cl}}}: \;\longrightarrow\; :\ddot{\underset{\cdot\cdot}{\text{Cl}}}:\ddot{\underset{\cdot\cdot}{\text{Cl}}}:$$

The HCl and Cl_2 molecules can also be represented using dashes for the bonds.

$$\text{H}-\ddot{\underset{\cdot\cdot}{\text{Cl}}}: \qquad \text{and} \qquad :\ddot{\underset{\cdot\cdot}{\text{Cl}}}-\ddot{\underset{\cdot\cdot}{\text{Cl}}}:$$

There are many nonmetals that form more than one covalent bond. For example, the three most important elements in biochemical systems are carbon, nitrogen, and oxygen.

$$\cdot\dot{\underset{\cdot}{\text{C}}}\cdot \qquad\qquad \cdot\dot{\underset{\cdot}{\text{N}}}\cdot \qquad\qquad \cdot\ddot{\underset{\cdot\cdot}{\text{O}}}:$$

In most of the compounds in which they occur, carbon forms four covalent bonds, nitrogen forms three, and oxygen forms two.

The simplest hydrogen compounds of these elements are methane, CH_4; ammonia, NH_3; and water, H_2O. Their Lewis structures are

$$\begin{array}{ccc}
\text{H} & \text{H} & \text{H} \\
\text{H}:\overset{\cdot}{\underset{\cdot}{\text{C}}}:\text{H} & \text{H}:\overset{\cdot\cdot}{\text{N}}:\text{H} & \text{H}:\ddot{\underset{\cdot\cdot}{\text{O}}}: \\
\text{H} & & \\
\end{array}$$

or   or   or

$$\begin{array}{ccc}
\text{H} & \text{H} & \text{H} \\
| & | & | \\
\text{H}-\text{C}-\text{H} & \text{H}-\underset{\cdot\cdot}{\text{N}}-\text{H} & \text{H}-\ddot{\underset{\cdot\cdot}{\text{O}}}: \\
| & & \\
\text{H} & & \\
\text{methane} & \text{ammonia} & \text{water}
\end{array}$$

Multiple Bonds

The molecules CH_4, NH_3, and H_2O have only single bonds.

The bond produced by the sharing of one pair of electrons between two atoms is called a **single bond.** So far, these have been the only kind we have discussed. There are, however, many molecules in which more than one pair of electrons are shared between two atoms. For example, we know that nitrogen, the most abundant gas in the atmosphere, occurs in the form of diatomic molecules, N_2. As we've just seen, the Lewis symbol for nitrogen is

$$\cdot\dot{\underset{\cdot}{\text{N}}}:$$

and each nitrogen atom needs three electrons to complete its octet. When the

N_2 molecule is formed, each of the nitrogen atoms shares three electrons with the other.

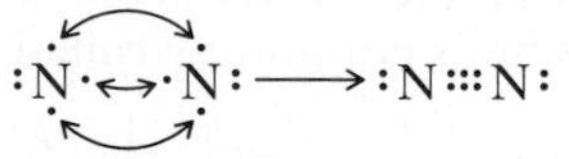

$$:\ddot{N}\cdot \longleftrightarrow \cdot \ddot{N}: \longrightarrow :N:::N:$$

The result is called a **triple bond.** Notice that in the Lewis formula for the molecule, the three shared pairs of electrons are placed between the two atoms. We count all of these electrons as though they belong to both of the atoms. Each nitrogen therefore has an octet.

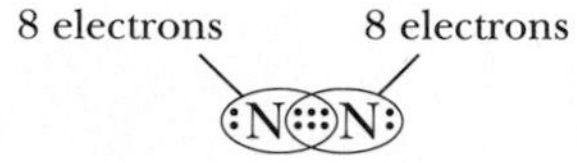

8 electrons   8 electrons

$$:N \textcircled{:::} N:$$

The triple bond is usually represented by three dashes, so the bonding in the N_2 molecule is normally shown as

$$:N \equiv N:$$

Double bonds also occur in molecules. An example is the CO_2 molecule, which contains a carbon atom that is bonded covalently to two separate oxygen atoms. We can diagram the formation of the bonds in CO_2 as follows.

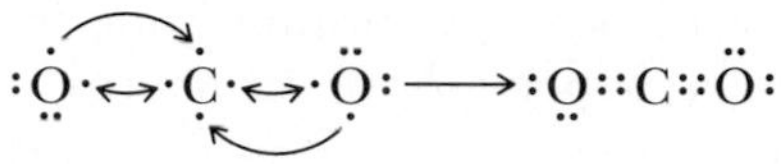

$$:\ddot{O}\cdot \longleftrightarrow \cdot \ddot{C}\cdot \longleftrightarrow \cdot \ddot{O}: \longrightarrow :\ddot{O}::C::\ddot{O}:$$

The central carbon atom shares two of its electrons with each of the oxygen atoms, and each oxygen shares two electrons with carbon. The result is the formation of two **double bonds.** Once again, if we circle the valence shell electrons that "belong" to each atom, we see that each has an octet.

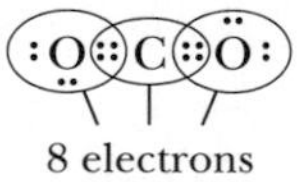

$$:O \textcircled{::} C \textcircled{::} \ddot{O}:$$

8 electrons

The Lewis structure for CO_2, using dashes, is

$$:\ddot{O}=C=\ddot{O}:$$

The location of the unshared pairs of electrons around the oxygen is unimportant. Two equally valid Lewis structures for CO_2 are $\ddot{O}=C=\ddot{O}$ and $:\ddot{O}=C=\ddot{O}:$.

7.4 SOME IMPORTANT COMPOUNDS OF CARBON

Carbon, hydrogen, oxygen, and nitrogen are the elements most commonly found in organic compounds. In Chapter 2 you learned that these compounds are substances that can be considered to be derived from hydrocarbons—compounds in which the basic molecular "backbones" are composed of carbon atoms linked to one another in a chainlike fashion. (Hydrocarbons themselves are the principal constituents of petroleum.)

One of the chief features of organic compounds is the tendency of carbon to complete its octet by forming four covalent bonds. It can accomplish this by bonding to atoms of other nonmetallic elements or to other carbon atoms. For example, the structures of methane, ethane, and propane are

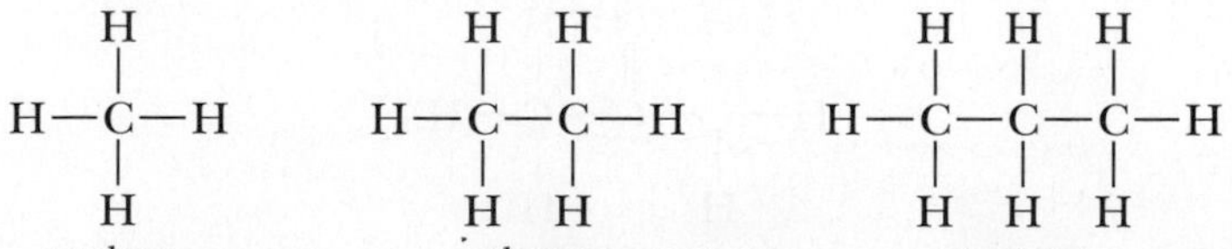

A more comprehensive discussion of organic compounds is found in Chapter 21.

The shapes of these molecules are illustrated in Figure 2.20 on page 65. Their structures can be abbreviated as

CH_4

CH_3CH_3

$CH_3CH_2CH_3$

methane   ethane   propane

The phenomenon seen here is one of the reasons there are so many compounds of carbon. For example, the formula $C_{20}H_{42}$ is found for 366,319 different compounds that differ in the way the carbon atoms are attached to each other.

Butane and isobutane are said to be isomers of each other.

When more than four carbon atoms are present, matters become more complex because there is more than one way to arrange the atoms. For example, butane has the formula C_4H_{10}, but there are two ways to arrange the carbon atoms. They occur in compounds commonly called butane and isobutane.

$$
\begin{array}{c}
\text{H} \\
\text{H}\,\text{—}\,\text{C}\,\text{—}\,\text{H} \\
\end{array}
$$

butane
C_4H_{10}
bp = -0.5 °C

isobutane
C_4H_{10}
bp = -11.7 °C

Even though they have the same molecular formula, these are actually different compounds with different properties, as you can see from the boiling points listed below their structures. The ability of atoms involved in the same molecular formula to arrange themselves in more than one way to give different compounds is called *isomerism* and is discussed more fully in Chapters 19 and 21.

Carbon can complete its octet by forming four single bonds, or it can form double or triple bonds. The structures of ethylene, C_2H_4, and acetylene, C_2H_2, are

$$
\text{H}\,\text{—}\,\text{C}\,\text{=}\,\text{C}\,\text{—}\,\text{H} \qquad \text{H}\,\text{—}\,\text{C}\,\text{≡}\,\text{C}\,\text{—}\,\text{H}
$$

Methyl alcohol is the fuel in this can of Sterno.

Ethyl alcohol is one of the chemicals added to gasoline to reduce pollution from automobiles during cold weather.

Compounds That Also Contain Oxygen and Nitrogen

In Chapter 2 it was noted that alcohols are organic compounds in which one of the hydrogen atoms of a hydrocarbon is replaced by OH. Examples are methyl alcohol and ethyl alcohol. The structures of these compounds are

methyl alcohol   ethyl alcohol

These can be written CH_3OH and CH_3CH_2OH

Ketones are important solvents.

Notice that the oxygen forms two bonds to complete its octet, just as in water. In alcohols these are single bonds, but oxygen can also form double bonds, as you saw for CO_2. One family of compounds in which a doubly bonded oxygen replaces a pair of hydrogen atoms is called *ketones*. The simplest example is acetone, the solvent in nail polish remover.

Acetone can also be represented by the formula

$$
CH_3\,\text{—}\,\overset{\displaystyle O}{\underset{}{C}}\,\text{—}\,CH_3
$$

acetone

Notice that the carbon bonded to the oxygen is also attached to *two* other carbon atoms. If at least one of the atoms attached to the C=O group (called a *carbonyl group,* pronounced *car-bon-EEL*) is a hydrogen, a different family of compounds is formed called *aldehydes.* An example is formaldehyde, which is used to preserve biological specimens, for embalming, and to make plastics.

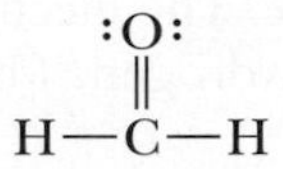

$$\begin{array}{c} :\ddot{O}: \\ \| \\ H-C-H \end{array}$$
formaldehyde

Another important family of oxygen-containing organic compounds are the organic acids. An example is acetic acid

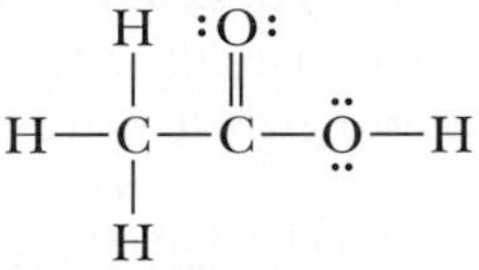

$$\begin{array}{c} H \quad :\ddot{O}: \\ |\quad\quad \| \\ H-C-C-\ddot{O}-H \\ | \\ H \end{array}$$
acetic acid

Notice that the acid has a doubly bonded oxygen and an OH group attached

to the end carbon atom. In general, molecules that contain the $\begin{array}{c} O \\ \| \\ -C-O-H \end{array}$ group (called a *carboxyl group*) are acids. When they react with water, the hydrogen that's transferred to the water to give H_3O^+ is the one bonded to the oxygen.

$$\begin{array}{c} H \quad :\ddot{O}: \\ |\quad\quad \| \\ H-C-C-\ddot{O}-H \\ | \\ H \end{array} + H_2O \rightleftharpoons H_3O^+ + \left[\begin{array}{c} H \quad :\ddot{O}: \\ |\quad\quad \| \\ H-C-C-\ddot{O}: \\ | \\ H \end{array}\right]^-$$

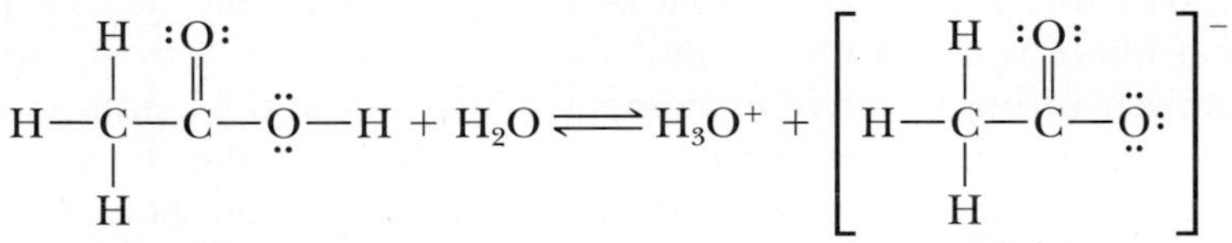

acetic acid   acetate ion

Nitrogen atoms need three electrons to complete an octet and in most of its compounds, nitrogen forms three bonds. The common nitrogen-containing organic compounds can be imagined as being derived from ammonia by replacing one or more of the hydrogens of NH_3 with hydrocarbon groups. They're called amines, and an example is methylamine, CH_3NH_2.

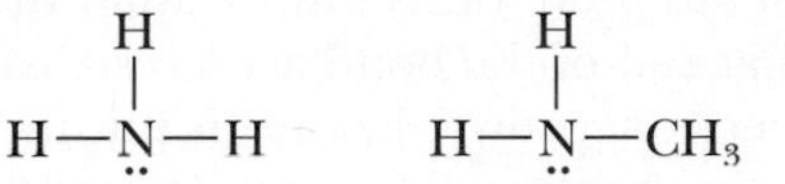

$$\begin{array}{c} H \\ | \\ H-\underset{\cdot\cdot}{N}-H \end{array} \qquad \begin{array}{c} H \\ | \\ H-\underset{\cdot\cdot}{N}-CH_3 \end{array}$$
ammonia   methylamine

Amines are strong-smelling compounds and often have a "fishy" odor. Like ammonia, they're basic.

$$CH_3NH_2(aq) + H_2O \rightleftharpoons CH_3NH_3^+(aq) + OH^-(aq)$$

We have just touched the surface of the subject of organic chemistry here, but we have introduced you to some of the kinds of organic compounds you will encounter in chapters ahead.

Acetaldehyde, shown below, is used in the manufacture of perfumes, dyes, plastics, and other products.

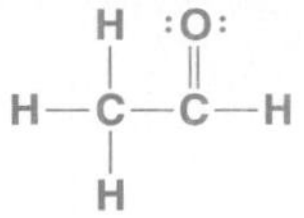

$$\begin{array}{c} H \quad\;\; :\ddot{O}: \\ |\quad\quad\;\; \| \\ H-C-C-H \\ | \\ H \end{array}$$
acetaldehyde

In general, compounds with the carboxyl group are weak acids.

When two identical atoms form a covalent bond, as in H_2 or Cl_2, each has an equal share of the electron pair in the bond. The electron density at both ends of the bond is the same, because the electrons are equally attracted to both nuclei. However, when different kinds of atoms combine, as in HCl, one nucleus usually attracts the electrons in the bond more strongly than the other.

7.5
ELECTRONEGATIVITY AND THE POLARITY OF BONDS

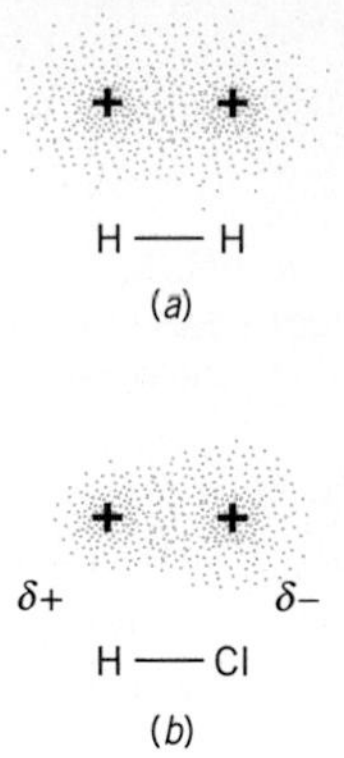

H —— H
(a)

$\delta+$ $\quad$ $\delta-$

H —— Cl
(b)

FIGURE 7.4

(a) The electron density of the pair of bonding electrons is distributed evenly in the bond between two H atoms. This gives a nonpolar covalent bond. (b) The electron density of the bonding pair between H and Cl is distributed unevenly in HCl, with more of the charge around the chlorine end of the bond. This causes the bond to be polar.

Experimental measurement of dipole moments is one way to check our theoretical models.

The noble gases have been assigned electronegativities of zero and are omitted from the table.

The effect of unequal attractions for the bonding electrons is an unbalanced distribution of electron density within the bond. For example, it has been found that a chlorine atom attracts electrons more strongly than a hydrogen atom does. In the HCl molecule, therefore, the electron cloud is pulled more tightly around the Cl, and that end of the molecule experiences a slight buildup of negative charge. The electron density that shifts toward the chlorine is removed from the hydrogen, which causes the hydrogen end to acquire a slight positive charge.

In HCl, electron transfer is incomplete. The electrons are still shared, but unequally. The charges on either end of the molecule are less than full 1+ and 1− charges. In HCl, for example, the hydrogen carries a charge of + 0.17 and the chlorine a charge of − 0.17. These are called **partial charges** and are usually indicated by the lowercase Greek letter delta, δ (Figure 7.4). Partial charges can also be indicated on Lewis structures. For example,

$$\text{H—}\overset{..}{\underset{..}{\text{Cl}}}\text{:}$$
$\delta+$ $\quad$ $\delta-$

A bond that carries partial positive and negative charges on opposite ends is called a **polar covalent bond,** or often simply a **polar bond** (the word *covalent* is understood). The term *polar* comes from the notion of *poles* of opposite charge at either end of the bond. Because there are *two poles* of charge involved, the bond is said to be a **dipole.**

The polar bond in HCl causes the molecule as a whole to have opposite charges on either end, so we say that HCl is a **polar molecule.** The HCl molecule as a whole is also a dipole, and the extent of its polarity is expressed quantitatively through its **dipole moment,** which is found by multiplying the amount of charge on either end by the distance between the charges.

The degree to which a covalent bond is polar depends on the difference in the abilities of the bonded atoms to attract electrons. The greater the difference, the more polar the bond and the more the electron density is shifted toward the atom that attracts electrons more.

The term that we use to describe the relative attraction of an atom for the electrons in a bond is called the **electronegativity** of the atom. In HCl, for example, chlorine is *more electronegative* than hydrogen. The electron pair of the covalent bond spends more of its time around the more electronegative atom, which is why that end of the bond acquires a partial negative charge.

The concept of electronegativity has been put on a quantitative basis—numerical values have been assigned for each element, as shown in Figure 7.5. This information is useful because the *difference* in electronegativity provides an estimate of the degree of polarity of a bond. In addition, the relative magnitudes of the electronegativities indicate which end of the bond carries the partial negative charge. For instance, fluorine is more electronegative than chlorine. Therefore, we expect an HF molecule to be more polar than an HCl molecule. In addition, hydrogen is less electronegative than either fluorine or chlorine, so in both of these molecules the hydrogen bears the partial positive charge.

$$\text{H—}\overset{..}{\underset{..}{\text{F}}}\text{:} \qquad \text{H—}\overset{..}{\underset{..}{\text{Cl}}}\text{:}$$
$\delta+$ $\;$ $\delta-$ $\qquad$ $\delta+$ $\;$ $\delta-$

■ **Practice Exercise 4** For each of the following bonds, choose the atom that carries the partial negative charge: (a) P—Br, (b) Si—C, (c) S—Cl.

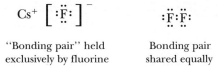

H 2.1																	
Li 1.0	Be 1.5											B 2.0	C 2.5	N 3.1	O 3.5	F 4.1
Na 1.0	Mg 1.3											Al 1.5	Si 1.8	P 2.1	S 2.4	Cl 2.9
K 0.9	Ca 1.1	Sc 1.2	Ti 1.3	V 1.5	Cr 1.6	Mn 1.6	Fe 1.7	Co 1.7	Ni 1.8	Cu 1.8	Zn 1.7	Ga 1.8	Ge 2.0	As 2.2	Se 2.5	Br 2.8
Rb 0.9	Sr 1.0	Y 1.1	Zr 1.2	Nb 1.3	Mo 1.3	Tc 1.4	Ru 1.4	Rh 1.5	Pd 1.4	Ag 1.4	Cd 1.5	In 1.5	Sn 1.7	Sb 1.8	Te 2.0	I 2.2
Cs 0.9	Ba 0.9	La 1.1	Hf 1.2	Ta 1.4	W 1.4	Re 1.5	Os 1.5	Ir 1.6	Pt 1.5	Au 1.4	Hg 1.5	Tl 1.5	Pb 1.6	Bi 1.7	Po 1.8	At 2.0
Fr 0.9	Ra 0.9	Ac 1.0														

Lanthanides: 1.0 – 1.2
Actinides: 1.0 – 1.2

FIGURE 7.5

The electronegativities of the elements.

Examination of electronegativity values and their differences reveals that there is no sharp dividing line between ionic and covalent bonding. Ionic bonding and nonpolar covalent bonding simply represent the two extremes. A bond is mostly ionic when the difference in electronegativity between two atoms is very large; the more electronegative atom acquires essentially complete control of the bonding electrons. In a **nonpolar covalent bond,** there is no difference in electronegativity, so the pair of bonding electrons is shared equally.

$$Cs^+ \left[\; :\!\ddot{\ddot{F}}\!: \; \right]^- \qquad\qquad :\!\ddot{\ddot{F}}\!:\!\ddot{\ddot{F}}\!:$$

"Bonding pair" held   Bonding pair
exclusively by fluorine   shared equally

The degree to which the bond is polar, which we might think of as the amount of **ionic character of the bond,** varies in a continuous way with changes in the electronegativity difference (Figure 7.6). The bond becomes more than 50% ionic when the electronegativity difference exceeds approximately 1.7.

Trends in Electronegativity in the Periodic Table

The figure in the margin illustrates the trends in electronegativity within the periodic table: *Electronegativity increases from bottom to top in a group, and from left to right in a period.* Notice that the trends follow those for ionization energy. This is because an atom that has a small IE will give away an electron more easily than an atom with a large IE, just as an atom with a small electronegativity will lose its share of an electron pair more readily than an atom with a large electronegativity.

Elements in the same region of the table (for example, the nonmetals) have similar electronegativities, which means that if they form bonds with each other, the electronegativity differences will be small and the bonds will be more covalent than ionic. On the other hand, if elements from widely separated regions of the table combine, large electronegativity differences will occur and the bonds will be predominantly ionic. This is what occurs, for example, when an element from Group IA or Group IIA reacts with an element from the upper-right-hand corner of the periodic table.

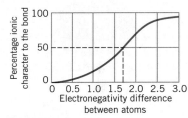

FIGURE 7.6

Variation in the percentage ionic character of a bond with electronegativity difference.

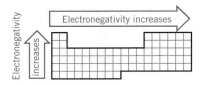

Periodic trends in electronegativity

It is found that the electronegativity is proportional to the average of the ionization energy and the electron affinity of an element.

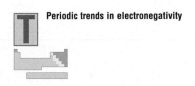

7.6
DRAWING LEWIS STRUCTURES

Keep in mind that Lewis structures are not meant to depict the shapes of molecules. They just describe which atoms are bonded to each other and the kinds of bonds involved. Thus, the Lewis structure for water can be drawn as H—Ö—H, but it does not mean the water molecule is linear, with all the atoms in a straight line. The actual shape of a water molecule is depicted below.

Lewis structures are very useful in chemistry. They are used to describe the structures of molecules, and much complex chemical reasoning has been based on them. Also, as you will learn in the next chapter, we can use the Lewis structure to make reasonably accurate predictions about the shape of a molecule.

In Sections 7.3 and 7.4 you saw Lewis structures for a variety of molecules that obey the octet rule. Examples included CO_2, Cl_2, N_2, and a variety of compounds of carbon. The octet rule is not always obeyed, however. For instance, there are some molecules in which one or more atoms must have more than an octet in the valence shell. Examples are PCl_5 and SF_6, whose Lewis structures are

In these molecules the formation of more than four bonds to the central atom requires that the central atom have a share of more than eight electrons. With the exception of period-2 elements like carbon and nitrogen, most nonmetals can have more than an octet of electrons in the outer shell.

There are also some molecules (but not many) in which the central atom behaves as though it has less than an octet. The most common examples involve compounds of beryllium and boron.

Four electrons around Be

Six electrons around B

Although Be and B sometimes have less than an octet, the elements in period 2 never exceed an octet. This is because their valence shells, having $n = 2$, can hold a maximum of only 8 electrons. (This explains why the octet rule works so well for atoms of carbon, nitrogen, and oxygen.) However, elements in periods below period 2, such as phosphorus and sulfur, sometimes do exceed an octet, because their valence shells can hold more than 8 electrons. For example, the valence shell for elements in period 3, for which $n = 3$, can hold a maximum of 18 electrons, and the valence shell for period 4 elements can hold as many as 32 electrons.

A Method for Drawing Lewis Structures

Drawing Lewis structures

We have used the Lewis structures of a variety of compounds in our discussions of covalent bonds. Such structures are also drawn for polyatomic ions, which are held together by covalent bonds, too. We now turn our attention to how these structures can be obtained in a systematic way.

The method of writing Lewis structures can be broken down into a number of steps, which are summarized in Figure 7.7. The first is to decide which atoms are bonded to each other, so that we know where to put the dots or dashes. This is not always a simple matter. Many times the formula suggests the way the atoms are arranged because the central atom, which is usually the least electronegative one, is usually written first. Examples are CO_2 and ClO_4^-, which have the *skeletal structures* (i.e., arrangements of atoms)

Sometimes, obtaining the skeletal structure is not quite so simple, especially when more than two elements are present. There are some generalizations possible, however. For example, the skeletal structure of nitric acid, HNO_3, is

$$\begin{matrix} & & & O & \\ H & O & N & O & \text{(correct)} \end{matrix}$$

rather than one of the following

$$\begin{matrix} O & & & & & & \\ O & N & O & \quad \text{or} \quad & H & O & O & N & O & \text{(incorrect)} \\ H & & & & & & \end{matrix}$$

Nitric acid is an oxoacid (Section 2.9) and it happens that the hydrogen atoms that can be released from molecules of oxoacids are always bonded to oxygen atoms, which are in turn bonded to the third nonmetal atom. Therefore, recognizing HNO_3 as the formula of an oxoacid allows us to predict that the three oxygen atoms are bonded to the nitrogen, and the hydrogen is bonded to one of the oxygens. (It is also useful to remember that hydrogen forms only one bond, so we would not choose it to be a central atom.)

There are times when no reasonable basis can be found for choosing a particular skeletal structure. If you must make a guess, choose the most symmetrical arrangement of atoms, because it has the greatest chance of being correct.

■ **Practice Exercise 5** Predict reasonable skeletal structures for SO_2, NO_3^-, $HClO_3$, and H_3PO_4.

After you've decided on the skeletal structure, the next step is to count all of the *valence electrons* to find out how many dots must appear in the final formula. Using the periodic table, locate the groups in which the elements in the formula occur to determine the number of valence electrons contributed by each atom. If the structure you wish to draw is that of an ion, *add one additional valence electron for each negative charge or remove a valence electron for each positive charge.* Some examples are:

SO_3	Sulfur (Group VIA) contributes $6e^-$	$1 \times 6 = 6$
	Oxygen (Group VIA) contributes $6e^-$	$\underline{3 \times 6 = 18}$
		Total $24e^-$
ClO_4^-	Chlorine (Group VIIA) contributes $7e^-$	$1 \times 7 = 7$
	Oxygen (Group VIA) contributes $6e^-$	$4 \times 6 = 24$
	Add one e^- for the $1-$ charge	$\underline{ +1}$
		Total $32e^-$

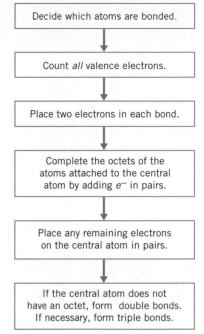

FIGURE 7.7

Summary of steps in the writing of Lewis structures. These rules lead to Lewis structures in which the octet rule is obeyed by the maximum number of atoms.

NH_4^+ Nitrogen (Group VA) contributes $5e^-$ $1 \times 5 = \quad 5$
Hydrogen (Group IA) contributes $1e^-$ $4 \times 1 = \quad 4$
Subtract one e^- for the 1+ charge $\underline{\qquad -1}$
Total $8e^-$

■ **Practice Exercise 6** How many valence electrons should appear in the Lewis structures of SO_2, PO_4^{3-}, and NO^+?

After we have determined the number of valence electrons, we place them into the skeletal structure in pairs following the steps outlined in Figure 7.7. Let's look at some examples of how we go about this.

EXAMPLE 7.4
Drawing Lewis Structures

What is the Lewis structure of the chloric acid molecule, $HClO_3$?

SOLUTION First, we select a reasonable skeletal structure. Since the substance is an oxoacid, we expect the hydrogen to be bonded to an oxygen, which in turn is bonded to the chlorine. The other two oxygens would also be bonded to the chlorine. This gives

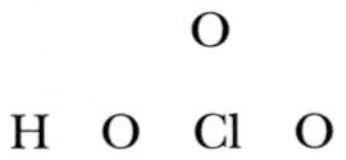

O

H O Cl O

The total number of electrons is 26 ($1e^-$ from H, $6e^-$ from each O, and $7e^-$ from Cl). Now we follow the steps in Figure 7.7. We begin by placing a pair of electrons in each bond.

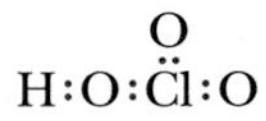

O
H:O:C̈l:O

This has used $8e^-$, so we still have $18e^-$ to go. Next, we work on the atoms surrounding the chlorine (which is the central atom in this structure). No additional electrons are needed around the H, because $2e^-$ are all that can occupy its valence shell. Therefore, we next complete the octets of the oxygens.

:Ö:
H:Ö:C̈l:Ö:

As you will see, it is useful to remember that hydrogen atoms form only one covalent bond.

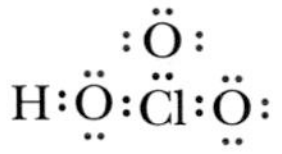

We have now used a total of $24e^-$, so there are two electrons left. These are placed on the Cl (the central atom) to give

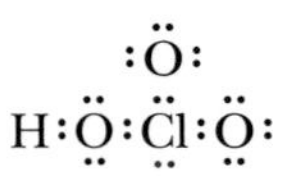

:Ö:
H:Ö:C̈l:Ö:

which we can also write as follows, using dashes for the electron pairs in the bonds.

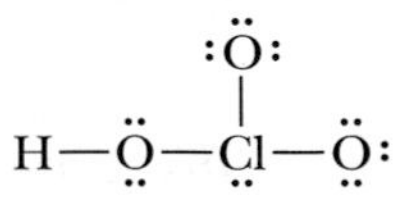

:Ö:
|
H—Ö—Cl—Ö:

The chlorine and the three oxygens have octets, and the hydrogen is completed with $2e^-$, so we are finished.

Write the Lewis structure for the SO₃ molecule.

SOLUTION We expect the skeletal structure to be

O

O S O

EXAMPLE 7.5
Drawing Lewis Structures

The total number of electrons in the formula is 24 ($6e^-$ from the sulfur, plus $6e^-$ from each oxygen). We begin to distribute the electrons by placing a pair in each bond. This gives

$$\text{O}$$
$$\text{O:S:O}$$

We have used $6e^-$, so there are $18e^-$ left. We next complete the octets around the oxygens, which uses the remaining electrons.

$$:\overset{..}{\text{O}}:$$
$$:\overset{..}{\text{O}}:\overset{..}{\text{S}}:\overset{..}{\text{O}}:$$

At this point all the electrons have been placed into the structure, but we see that the sulfur still lacks an octet. We cannot simply add more dots because the total must be 24. Therefore, according to the last step of the procedure in Figure 7.7, we have to create a multiple bond. To do this we move a pair of electrons that we have shown to belong solely to an oxygen into a sulfur–oxygen bond so that it can be counted as belonging to both the oxygen *and* the sulfur. In other words, we place a double bond between sulfur and one of the oxygens. It doesn't matter which oxygen we choose for this honor.

$$:\overset{..}{\text{O}}: \qquad\qquad :\overset{..}{\text{O}}: \qquad\qquad :\overset{..}{\text{O}}:$$
$$:\overset{..}{\text{O}}:\text{S}:\overset{..}{\text{O}}: \quad \text{gives} \quad :\overset{..}{\text{O}}::\text{S}:\overset{..}{\text{O}}: \quad \text{or} \quad :\overset{..}{\text{O}}=\text{S}-\overset{..}{\text{O}}:$$

Notice that each atom has an octet.

What is the Lewis structure of CO?

SOLUTION The skeletal structure is simply

C O

EXAMPLE 7.6
Drawing Lewis Structures

The total number of valence electrons is 10, and we begin distributing them by placing a pair into the bond.

C:O

Now we try to complete the octets with the remaining 8 electrons. To do this, we arbitrarily select one atom, say carbon, to be the "central atom" and complete the valence shell of the other. The two electrons that remain are then placed on the carbon to give

$$:\text{C}:\overset{..}{\text{O}}:$$

Carbon has only four electrons at this point, so we must move two unshared pairs of electrons that are on the oxygen into the bond. This gives a triple bond.

$$:\text{C}:\overset{..}{\text{O}}: \quad \text{gives} \quad :\text{C}:::\text{O}: \quad \text{or} \quad :\text{C}\equiv\text{O}:$$

EXAMPLE 7.7
Drawing Lewis Structures

What is the Lewis structure for SF_4?

SOLUTION The skeletal structure is

$$
\begin{array}{c}
\text{F} \\
\text{F} \quad \text{S} \quad \text{F} \\
\text{F}
\end{array}
$$

and there must be $6 + 28 = 34$ electrons in the structure. First we place $2e^-$ into each bond, and then complete the octets of the fluorine atoms. This uses 32 electrons.

$$
:\overset{..}{\underset{..}{F}}: \qquad :\overset{..}{\underset{..}{F}}:
$$

There are two electrons left, and according to Step 5 in Figure 7.7 they are placed on the central atom. We will redraw the structure to make room for them.

We are now finished. Each fluorine atom has an octet and the central atom does not have less than an octet. In fact, in this molecule, the sulfur violates the octet rule by having five pairs of electrons in its valence shell.

■ **Practice Exercise 7** Draw Lewis structures for OF_2, NH_4^+, SO_2, NO_3^-, ClF_3, and $HClO_4$.

7.7
FORMAL CHARGE AND THE SELECTION OF LEWIS STRUCTURES

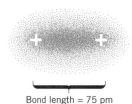

Bond length = 75 pm

The hydrogen molecule has a bond length of 75 pm.

Lewis structures are meant to describe how atoms share electrons in chemical bonds. Such descriptions are theoretical explanations or predictions that relate to the forces that hold molecules and polyatomic ions together. But, as you learned in Chapter 1, a theory is only as good as the observations on which it is based, so to have confidence in a theory about chemical bonding, we need to have a way to check it. We need experimental observations that relate to the description of bonding.

Two properties that are related to the number of electron pairs shared between two atoms are **bond length,** the distance between the nuclei of the bonded atoms, and **bond energy,** the energy required to separate the bonded atoms to give neutral particles. For example, we mentioned in Section 7.3 that measurement has shown the H_2 molecule to have a bond length of 75 pm and a bond energy of 435 kJ/mol, which means that it takes 435 kJ to break the bonds of 1 mol of H_2 molecules to give 2 mol of hydrogen atoms.

For bonds between the same elements, the bond length and energy depend on the **bond order,** which is defined as *the number of pairs of electrons shared between two atoms.* The bond order is a measure of the amount of electron density in the bond, and the greater the electron density, the more tightly the

TABLE 7.1   Average Bond Lengths and Bond Energies Measured for Carbon–Carbon Bonds

Bond	Bond Length (pm)	Bond Energy (kJ/mol)
C—C	154	348
C=C	134	615
C≡C	120	812

nuclei are held and the more closely they are drawn together. This is illustrated by the data in Table 7.1, which gives typical bond lengths and bond energies for single, double, and triple bonds between carbon atoms. In summary:

As the bond order increases, the bond length decreases and the bond energy increases, provided we are comparing bonds between the same elements.

A single bond has a bond order of 1, a double bond a bond order of 2, and a triple bond a bond order of 3.

Correlating bond properties with bond order

With this as background, let's examine the Lewis structure of sulfuric acid, drawn following the rules given in the preceding section.

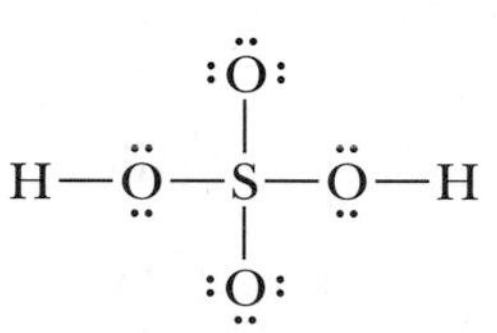

H—Ö—S—Ö—H   (Structure I)

It obeys the octet rule, and there doesn't seem to be any need to write any other structures for it. But a problem arises if we compare the predicted bond lengths with those found experimentally. In our Lewis structure, all four sulfur–oxygen bonds are single bonds, which means they should have about the same bond lengths. However, experimentally it has been found that the bonds are not of equal length, as illustrated in Figure 7.8. The S—O bonds are shorter than the S—OH bonds, which means they must have a larger bond order. Therefore, we need to modify our Lewis structure to make it conform to reality.

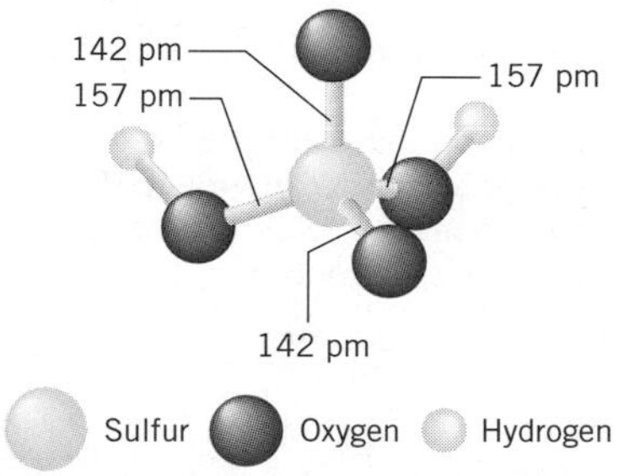

Because sulfur is in period 3, its valence shell has 3s, 3p, and 3d subshells, which together can accommodate more than eight electrons. Therefore, we are able to have more than four bonds to sulfur and we are able to increase the bond order in the S—O bonds by moving electron pairs to create sulfur–oxygen double bonds as shown below.

142 pm —
157 pm —
— 157 pm
142 pm

Sulfur   Oxygen   Hydrogen

FIGURE   7.8

The structure of sulfuric acid in the vapor state. Notice the differences in the sulfur–oxygen bond lengths.

H—Ö—S—Ö—H   gives   H—Ö—S—Ö—H   (Structure II)

Now we have a Lewis structure that better fits experimental observations because the sulfur–oxygen double bonds are expected to be shorter than the sulfur–oxygen single bonds. Because this second Lewis structure agrees better with the actual structure of the molecule, it is the *preferred* Lewis structure, even though it violates the octet rule.

Formal Charges

Are there any criteria that we could have applied that would have allowed us to predict that the second Lewis structure for H_2SO_4 is better than the one with only single bonds, even though it seems to violate the octet rule unnecessarily? To answer this question, let's take a closer look at the two Lewis structures we've drawn.

In Structure I, there are only single bonds between the sulfur and oxygen atoms. If the electrons in the bonds are shared equally by S and O, then each atom "owns" half of the electron pair, or the equivalent of one electron. In other words, the four single bonds place the equivalent of four electrons in the valence shell of the sulfur. A single atom of sulfur by itself, however, has six valence electrons. This means that in this Lewis structure, the sulfur has two electrons *less* than it does as just an isolated atom. Thus, at least in a bookkeeping sense, it would appear that if sulfur obeyed the octet rule in H_2SO_4, it would have a charge of 2+. This *apparent charge* on the sulfur atom is called its **formal charge.**

The formal charge on any atom in a Lewis structure can be calculated as follows:

Calculating formal charges

$$\begin{bmatrix} \text{Formal} \\ \text{charge} \end{bmatrix} = \begin{bmatrix} \text{Number of } e^- \text{ in valence} \\ \text{shell of the isolated atom} \end{bmatrix}$$
$$- \left[\begin{pmatrix} \text{Number of bonds} \\ \text{to the atom} \end{pmatrix} + \begin{pmatrix} \text{Number of} \\ \text{unshared } e^- \end{pmatrix} \right] \quad (7.1)$$

For example, for the sulfur in Structure I, we get

$$\text{Formal charge on S} = 6 - (4 + 0) = 2$$

> We add up the number of bonds and the number of unshared electrons and then subtract this total from the number of valence electrons possessed by a neutral atom of the element.

Let's also calculate the formal charges on the hydrogen and oxygen atoms in Structure I. An isolated H atom has one electron. In Structure I each H has one bond and no unshared electrons. Therefore,

$$\text{Formal charge on H} = 1 - (1 + 0) = 0$$

The hydrogens in this structure have no formal charge.

In Structure I there are two kinds of oxygen to consider. An isolated oxygen atom has six electrons, so we have, for the oxygens also bonded to hydrogen

$$\text{Formal charge} = 6 - (2 + 4) = 0$$

> Each unshared electron contributes one e^- to the total an atom has in its valence shell.

and for the oxygens not bonded to hydrogen

$$\text{Formal charge} = 6 - (1 + 6) = -1$$

Formal charges are indicated in a Lewis structure by placing them in circles alongside the atoms, as shown below

$$\begin{array}{c} :\overset{\cdot\cdot}{\underset{\cdot\cdot}{O}}:^{\ominus} \\ | \\ H-\overset{\cdot\cdot}{\underset{\cdot\cdot}{O}}-S^{(2+)}-\overset{\cdot\cdot}{\underset{\cdot\cdot}{O}}-H \\ | \\ :\overset{\cdot\cdot}{\underset{\cdot\cdot}{O}}:_{\ominus} \end{array}$$

Notice that the sum of the formal charges in the molecule is zero. In general, *the sum of the formal charges in any Lewis structure equals the charge on the species.*

Now let's look at the formal charges in Structure II. For sulfur we have

$$\text{Formal charge on S} = 6 - (6 + 0) = 0$$

so the sulfur has no formal charge. The hydrogens and the oxygens that are also bonded to H are the same in this structure as before, so they have no formal charges. And finally, the oxygens that are not bonded to hydrogen have

$$\text{Formal charge} = 6 - (2 + 4) = 0$$

These oxygens also have no formal charges.

Now let's compare the two structures side by side.

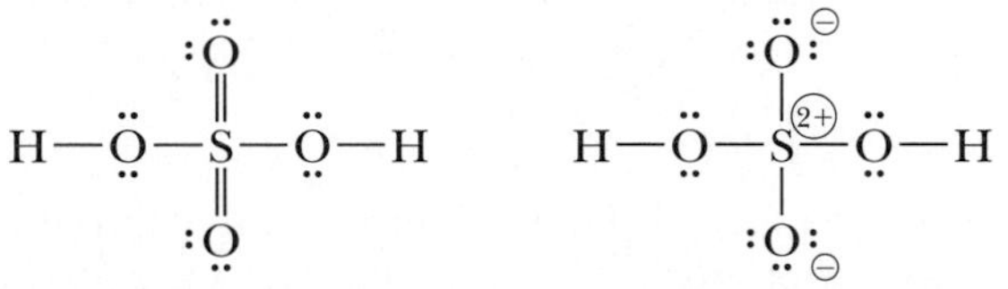

Imagine changing the one with the double bonds to the one with the single bonds. This involves creating two pairs of positive–negative charge from something electrically neutral. Stated another way, it involves separating negative charges from positive charges, and this requires an increase in the potential energy. Our conclusion is that the singly bonded structure on the right has a higher potential energy than the one with the double bonds. In general, the lower the potential energy of a molecule, the more stable it is. Therefore, the lower energy structure with the double bonds is preferred over the one with only single bonds, and it gives us a rule that we can use in selecting the best Lewis structures for a molecule or ion:

When several Lewis structures are possible, those with the smallest formal charges are the most stable and are preferred.

A student drew three Lewis structures for the nitric acid molecule:

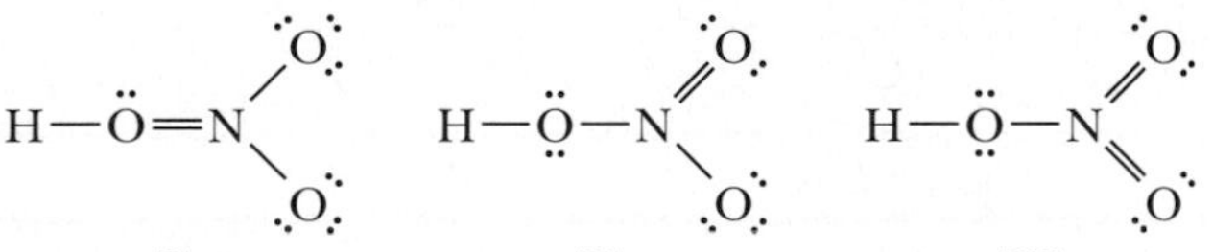

EXAMPLE 7.8
Selecting Lewis Structures
Based on Formal Charges

Which one is preferred?

SOLUTION Structure III can be eliminated at the very beginning. This structure is not acceptable because nitrogen has 10 electrons in its valence shell. Period-2 elements *never* exceed an octet because their valence shells have only *s* and *p* subshells and can accommodate a maximum of eight electrons. Let's now calculate formal charges on the atoms in each of the other two structures.

Structure I:

Formal charge on N $= 5 - (4 + 0) = +1$

Formal charge on O attached to H $= 6 - (3 + 2) = +1$

Formal charge on O not attached to H $= 6 - (1 + 6) = -1$

Structure II:

Formal charge on N = +1

Formal charge on O attached to H = 6 − (2 + 4) = 0

Formal charge on singly bonded O not attached to H = − 1

Formal charge on doubly bonded O = 6 − (2 + 4) = 0

Next, we assign these formal charges to the atoms in the structures.

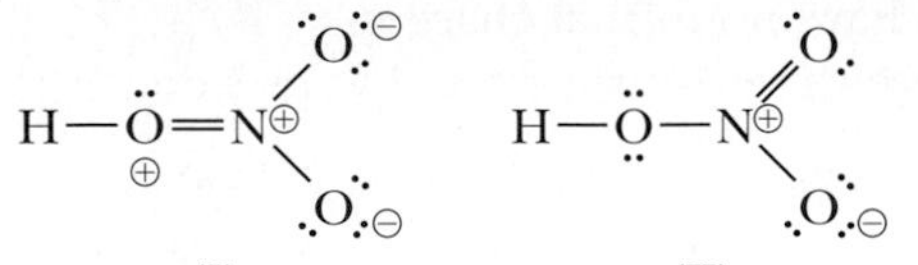

(I)　　　　　　　(II)

In Structure III, the formal charges are zero on each of the atoms, but this cannot be the "preferred structure" because the nitrogen atom has too many electrons in its valence shell.

Since Structure II has fewer formal charges than Structure I, it is the lower energy, preferred Lewis structure for HNO_3.

EXAMPLE 7.9
Selecting Lewis Structures Based on Formal Charges

What is the preferred Lewis structure for the molecule $POCl_3$, in which phosphorus is bonded to one oxygen and three chlorine atoms?

ANALYSIS First we must draw a Lewis structure following the rules in the previous section. Then we can assign formal charges. If there is a way to reduce these formal charges by moving electrons, we will do so to obtain a better Lewis structure.

SOLUTION By following the rules from the previous section, we obtain the following Lewis structure.

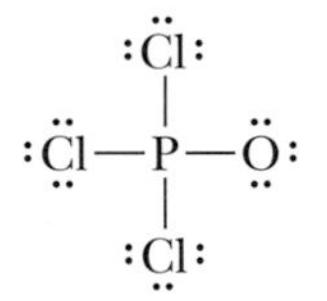

Now we assign formal charges.

	Phosphorus	Oxygen	Chlorine
Formal charge =	5 − (4 + 0) = +1	6 − (1 + 6) = − 1	7 − (1 + 6) = 0

Therefore, our Lewis structure looks like this

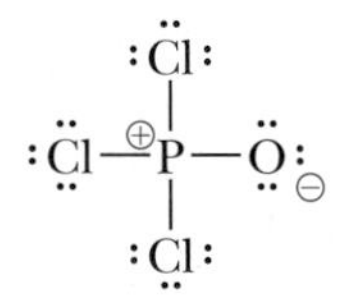

When an atom with a negative formal charge is bonded to an atom with a positive formal charge, the atom with the negative formal charge donates a pair of electrons to form a double bond.

We can reduce the formal charges if we can place another electron on phosphorus and remove one from the oxygen. This can be accomplished if we make one of the unshared pairs of electrons on the oxygen into a bond between P and O (which we are permitted to do because phosphorus is a period-3 element; its valence shell can hold more than eight electrons).

Since this structure contains no formal charges, it is preferred over the previous structure.

EXAMPLE 7.10
Selecting Lewis Structures
Based on Formal Charges

Two structures can be drawn for BCl_3, as shown below

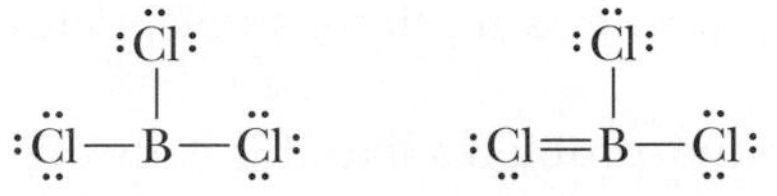

(I)   (II)

Why is the one that violates the octet rule preferred?

SOLUTION Let's assign formal charges.

(I)   (II)

In Structure I all the formal charges are zero. In Structure II, two of the atoms have formal charges, so this alone would argue in favor of Structure I. There is another argument in favor as well. Notice that the formal charges in Structure II place the positive charge on the more electronegative chlorine atom and the negative charge on the less electronegative boron. If charges could form in this molecule, they certainly would not be expected to form in this way. Therefore, there are two factors that make the structure with the double bond unfavorable, so we usually write the Lewis structure for BCl_3, as shown in (I).

■ **Practice Exercise 8** Assign formal charges to the atoms in the following Lewis structures:

(a) 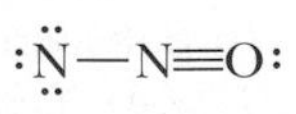 $:\ddot{N}—N≡O:$ (b) 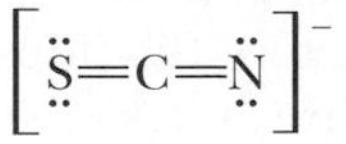 $\left[\ddot{S}=C=\ddot{N}\right]^-$

■ **Practice Exercise 9** Select the preferred Lewis structures for: (a) SO_2, (b) $HClO_3$, (c) H_3PO_4.

There are some molecules and ions for which we cannot write Lewis structures that agree with experimental measurements of bond length and bond energy. One example is the formate ion, CHO_2^-, which is produced by neutralizing formic acid, $HCHO_2$ (the substance that causes the stinging sensation in bites from fire ants). The skeletal structure for this ion is

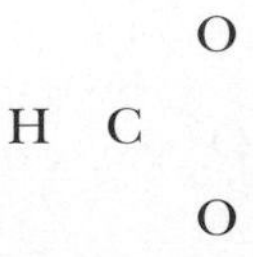

```
        O
   H   C
        O
```

7.8 RESONANCE: WHEN A SINGLE LEWIS STRUCTURE FAILS

Formic acid is an organic acid. It has the structure

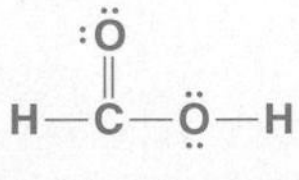

$$H—C(=\ddot{O})—\ddot{O}—H$$

and, following the usual steps, we would write its Lewis structure as

$$\left[\text{H}-\overset{\displaystyle \text{:O:}}{\underset{\displaystyle \text{:O:}}{\text{C}}} \right]^{-}$$

This structure suggests that one carbon–oxygen bond should be longer than the other, but experiment shows that they are identical. In fact, the C–O bond lengths are about halfway between the expected values for a single bond and a double bond. The Lewis structure doesn't match the experimental evidence, and there's no way to write one that does. It would require showing all of the electrons in pairs and, at the same time, showing 1.5 pairs of electrons per bond.

The way we get around problems like this is through the use of a concept called **resonance.** We view the actual structure of the molecule or ion, which we cannot draw satisfactorily, as a composite, or average, of a number of Lewis structures that we can draw. For example, for formate we write

No atoms have been moved; the electrons have just been redistributed.

$$\left[\text{H}-\overset{\displaystyle \text{:O:}}{\underset{\displaystyle \text{:O:}}{\text{C}}} \right]^{-} \longleftrightarrow \left[\text{H}-\overset{\displaystyle \text{:O:}}{\underset{\displaystyle \text{:O:}}{\text{C}}} \right]^{-}$$

where we have simply shifted electrons around in going from one structure to the other. The bond between the carbon and a particular oxygen is depicted as a single bond in one structure and as a double bond in the other. The average of these is 1.5 bonds (halfway between a single and a double bond), which is in agreement with the experimental bond lengths. These two Lewis structures are called **resonance structures** or **contributing structures,** and the actual structure of the ion is said to be a **resonance hybrid** of the two resonance structures that we've drawn. The double-headed arrow is used to show that we are drawing resonance structures and implies that the true hybrid structure is a composite of the two resonance structures.

The term *resonance* is often misleading to the beginning student. The word itself suggests that the actual structure flip-flops back and forth between the two structures shown. This is *not* the case! A mule, which is the *hybrid* offspring of a donkey and a horse, isn't a donkey one minute and a horse the next. Although it may have characteristics of both parents, a mule is a mule. A *resonance hybrid* also has characteristics of its "parents," but it never has the exact structure of any of them.

Method for determining resonance structures

There is a simple way to determine when resonance should be applied to Lewis structures. If you find that you must move electrons to create one or more double bonds while following the procedures developed in the previous sections, the number of resonance structures is equal to the number of equivalent choices for the locations of the double bonds. For example, in drawing the Lewis structure for the NO_3^- ion, we reach the stage

$$\overset{\displaystyle \text{:O:}}{\underset{\displaystyle \text{:O:}\quad\quad\text{:O:}}{\overset{|}{\text{N}}}}$$

A double bond must be created to give the nitrogen an octet. Since it can be formed in any one of three locations, there are three resonance structures for this ion.

$$
\begin{bmatrix} \ddot{\mathrm{O}} \\ \ddot{\mathrm{O}}^{\diagdown}\mathrm{N}^{\diagup}\ddot{\mathrm{O}} \end{bmatrix}^{-} \longleftrightarrow \begin{bmatrix} \ddot{\mathrm{O}} \\ \ddot{\mathrm{O}}^{\diagdown}\mathrm{N}{=}\ddot{\mathrm{O}} \end{bmatrix}^{-} \longleftrightarrow \begin{bmatrix} \mathrm{O} \\ {\parallel} \\ \ddot{\mathrm{O}}^{\diagdown}\mathrm{N}^{\diagup}\ddot{\mathrm{O}} \end{bmatrix}^{-}
$$

Notice that each structure is the same, except for the location of the double bond.

In the nitrate ion, the extra bond that is placed differently from one structure to another is actually distributed among all three positions. Therefore, the average bond order in the N—O bonds is expected to be $1\frac{1}{3}$, or 1.33.

The three oxygens in NO_3^- are said to be equivalent; that is, they are all alike in their chemical environment. The only thing that each oxygen is bonded to is the nitrogen atom.

Draw resonance structures for the sulfite ion, SO_3^{2-}.

SOLUTION Following our usual procedure we obtain the Lewis structure

$$
\begin{bmatrix} \ddot{\mathrm{O}} \\ {|} \\ \ddot{\mathrm{O}}{-}\mathrm{S}{-}\ddot{\mathrm{O}} \end{bmatrix}^{2-}
$$

and it doesn't seem that we need the concept of resonance. However, if we assign formal charges, we get

$$
\begin{bmatrix} \overset{\ominus}{\ddot{\mathrm{O}}} \\ {|} \\ \overset{\ominus}{\ddot{\mathrm{O}}}{-}\overset{\oplus}{\mathrm{S}}{-}\overset{\ominus}{\ddot{\mathrm{O}}} \end{bmatrix}^{2-}
$$

We can obtain a better Lewis structure by reducing the number of formal charges, which can be accomplished by forming a double bond between one of the oxygen atoms and the sulfur. Since there are three choices for the location of this double bond, there are three resonance structures.

$$
\begin{bmatrix} \ddot{\mathrm{O}} \\ {|} \\ \ddot{\mathrm{O}}{=}\mathrm{S}{-}\ddot{\mathrm{O}} \end{bmatrix}^{2-} \longleftrightarrow \begin{bmatrix} \mathrm{O} \\ {\parallel} \\ \ddot{\mathrm{O}}{-}\mathrm{S}{-}\ddot{\mathrm{O}} \end{bmatrix}^{2-} \longleftrightarrow \begin{bmatrix} \ddot{\mathrm{O}} \\ {|} \\ \ddot{\mathrm{O}}{-}\mathrm{S}{=}\ddot{\mathrm{O}} \end{bmatrix}^{2-}
$$

As with the nitrate ion, we expect an average bond order of 1.33.

■ **Practice Exercise 10** Draw the resonance structures for: (a) NO_2^-, (b) PO_4^{3-}.

EXAMPLE 7.11
Drawing Resonance Structures

Effect of Resonance on the Stability of Molecules and Ions

One of the benefits that a molecule or ion derives from existing as a resonance hybrid is that its total energy is lower than any one of its resonance structures. A particularly important example of this occurs with the compound benzene, C_6H_6. This is a ring-shaped molecule whose basic structure appears in many important organic molecules, ranging from plastics to amino acids.

Two resonance structures are drawn for benzene.

These are often represented as hexagons with dashes showing the locations of the double bonds. It is assumed that at each apex of the hexagon there is a carbon bonded to a hydrogen as well as to the adjacent carbon atoms.

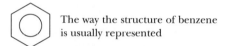

Usually, the actual structure of benzene—that of its resonance hybrid—is represented as a hexagon with a circle in the center. This is intended to show that the electron density of the three extra bonds is evenly distributed around the ring.

The way the structure of benzene is usually represented

Although the individual resonance structures for benzene show double bonds, the molecule does not react like other organic molecules that have true carbon–carbon double bonds. The reason appears to be that the resonance hybrid is considerably more stable than either of these resonance forms. In fact, it has been calculated that the actual structure of the benzene molecule is more stable than either of its resonance structures by approximately 146 kJ/mol. This extra stability achieved through resonance is called the **resonance energy.**

7.9
COORDINATE COVALENT BONDS: LEWIS ACIDS AND BASES

When a hydrogen atom forms a covalent bond with a chlorine atom, the pair of electrons that they share is made up of one electron from each of the two atoms.

$$\mathrm{H\cdot \; + \; \cdot \ddot{\underset{\cdot\cdot}{C}l}\colon \longrightarrow H\colon\ddot{\underset{\cdot\cdot}{C}l}\colon}$$

As we have seen, similar bonds are formed between a pair of chlorine atoms and when hydrogen forms bonds to carbon, nitrogen, and oxygen. Sometimes, however, bonds are formed in which both electrons of the shared pair come from the same atom. Let's look at an example.

When ammonia, NH_3, is placed into an acidic solution, it picks up a hydrogen ion, H^+, and becomes NH_4^+. Let's keep tabs on all of the electrons when this happens by using red dots for the hydrogen electrons and black dots for nitrogen electrons.

$$\text{H} \ :\!\overset{\cdot\cdot}{\underset{\cdot\cdot}{\text{N}}}\!: \ + \ \text{H}^+ \longrightarrow \left[\text{H}\!:\!\overset{\cdot\cdot}{\underset{\cdot\cdot}{\text{N}}}\!:\!\text{H}\right]^+$$

All electrons are alike, of course. We are using different colors for them so we can see where the electrons in the bond came from.

The H^+ ion has a vacant $1s$ orbital in its valence shell that can accommodate two electrons. When the H^+ becomes bonded to the nitrogen of NH_3, the nitrogen donates both of the electrons to the bond. *This type of bond, in which both electrons of the shared pair come from one of the two atoms, is called a* **coordinate covalent bond.**

It is important to understand that even though we make a distinction as to the *origin* of the electrons, once the bond is formed it is really the same as any other covalent bond. In other words, we can't tell where the electrons in the bond came from after the bond has been formed. In NH_4^+, for instance, all four of the N—H bonds are equivalent once they've been formed, and no distinction is made among them. We usually write the structure of NH_4^+ as

The notion of coordinate covalent bonds is used to explain how bonds are formed, rather than what the bonds are like after they have been formed.

$$\left[\begin{array}{c} \text{H} \\ | \\ \text{H}-\text{N}-\text{H} \\ | \\ \text{H} \end{array}\right]^+$$

Lewis Acids and Bases

One reason the coordinate covalent bond is a useful concept is that it allows us to extend our view of acids and bases to systems that do not involve hydrogen ions. To see why we might want to do this, consider the reactions of sulfur trioxide and calcium oxide with water, followed by the neutralization of the resulting acid and base.

$$SO_3(g) + H_2O \longrightarrow H_2SO_4(aq)$$

$$+$$

$$CaO(s) + H_2O \longrightarrow Ca(OH)_2(aq)$$

$$\downarrow \text{ (acid–base neutralization)}$$

$$CaSO_4(s) + 2H_2O$$

The reaction of CaO with water is really the reaction of oxide ion with water to give hydroxide ion.

Notice that two molecules of water go into the system when the acidic and basic anhydrides react, and two molecules of water come out in the subsequent neutralization that forms $CaSO_4$. The *net* change is simply the conversion of SO_3 and CaO into $CaSO_4$, and one might wonder whether the water molecules are actually necessary to accomplish this. As it turns out, they are not.

If SO_3 gas is passed over solid CaO in the absence of water, the two anhydrides react to yield $CaSO_4$. In fact, the reaction is actually between SO_3 and the oxide ion, O^{2-}.

$$SO_3 + O^{2-} \longrightarrow SO_4^{2-}$$

Since the beginning reactants (SO_3 and CaO) are the same, and the ultimate product ($CaSO_4$) is the same, whether or not water is present, it would be useful to be able to explain both reactions as acid–base reactions. To do this for the reaction in the absence of water, however, requires a broader definition of acids and bases. This is just what G. N. Lewis provided when he developed his acid–base theory. **Lewis acids and bases** are defined as follows.

Lewis Definitions of Acids and Bases
- -

1. An **acid** is any ionic or molecular species that can accept a pair of electrons in the formation of a coordinate covalent bond.

2. A **base** is any ionic or molecular species that can donate a pair of electrons in the formation of a coordinate covalent bond.

3. **Neutralization** is the formation of a coordinate covalent bond between the donor (base) and the acceptor (acid).

Examples of Lewis Acid–Base Reactions

An example of a Lewis acid–base neutralization is the exothermic reaction between BCl_3 and NH_3 to give a compound with the formula BCl_3NH_3. The reaction is exothermic because a bond is formed between the nitrogen and the boron.

Compounds like BCl_3NH_3, which are formed by simply joining two smaller molecules, are called *addition compounds.*

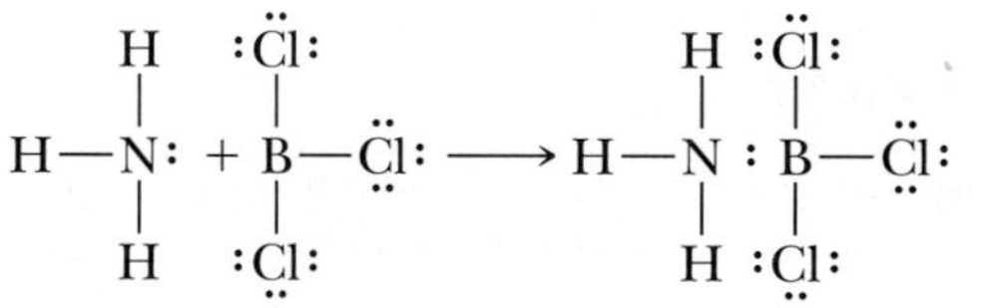

An arrow sometimes is used to represent the donated pair of electrons in the coordinate covalent bond. The direction of the arrow indicates the direction in which the electron pair is donated, in this case from the nitrogen to the boron.

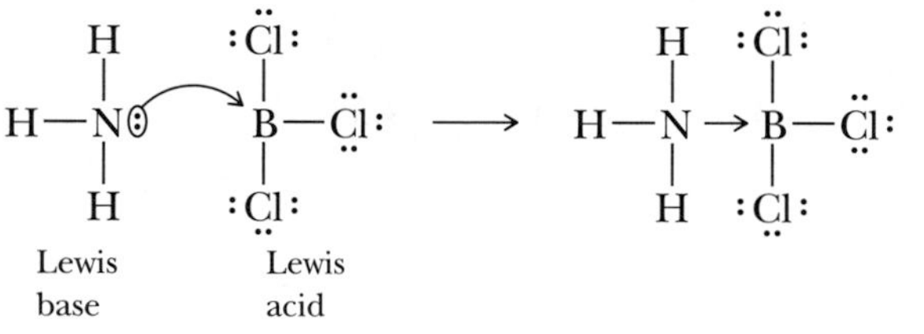

The ammonia molecule has an unshared pair of electrons and is a *Lewis base.* The boron atom in BCl_3 has only six electrons in its valence shell and seeks two more to achieve an octet. By accepting the electrons from the ammonia molecule, BCl_3 is a *Lewis acid.*

The Lewis concept includes the Arrhenius acids and bases as special cases. When hydronium ion reacts with hydroxide ion, for example, a proton is transferred. In the Lewis view, this proton is an acid that leaves one particle and becomes joined to another. When it becomes attached to the hydroxide ion, a Lewis acid–base neutralization takes place.

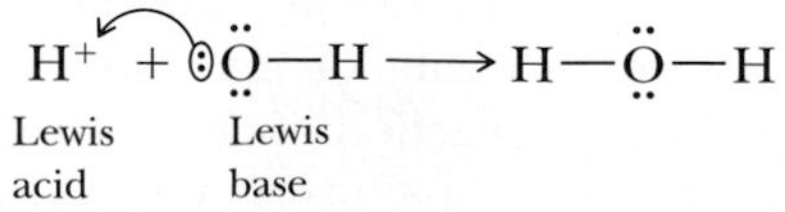

Earlier in this section we commented that the Lewis approach frees us from proton transfer as a criterion for an acid–base reaction. We can illustrate this by analyzing the reaction between sulfur trioxide and oxide ion in the following way.

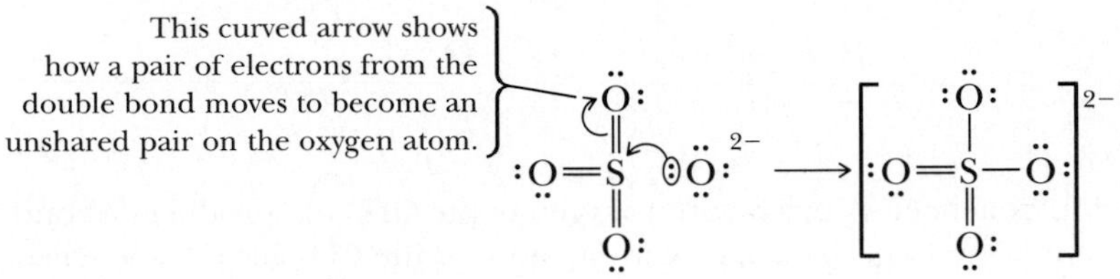

This curved arrow shows how a pair of electrons from the double bond moves to become an unshared pair on the oxygen atom.

Notice that the oxide ion supplies the electron pair for the covalent bond and that the electrons shift around in the SO_3 molecule to make room for the additional bond. The result is the formation of the sulfate ion, SO_4^{2-}.

Here is another example that involves the relocation of an atom as well. It is the reaction of sulfur trioxide with water, which occurs when SO_3 dissolves in water and sulfuric acid, H_2SO_4, forms.

$$H_2O \ + \ SO_3 \longrightarrow H_2SO_4$$

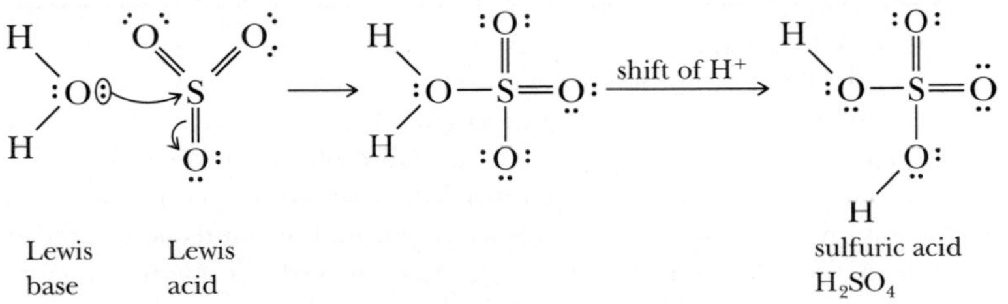

Lewis base   Lewis acid   sulfuric acid H_2SO_4

Identifying Substances as Lewis Acids and Bases

In the examples above, we can see some simple rules that help us identify substances that might be Lewis acids and bases. Notice that the Lewis bases (e.g., NH_3, H_2O, and O^{2-}) are substances that have completed valence shells and unshared pairs of electrons. Lewis acids, on the other hand, are substances with incomplete valence shells, such as BCl_3, or double bonds that by the shifting of electrons to other atoms can make room for an incoming pair of electrons from a Lewis base.

When carbon dioxide is bubbled into aqueous sodium hydroxide, the gas is instantly trapped as the bicarbonate ion. (In fact, this reaction makes it hard to store standard NaOH solutions in contact with air, in which traces of CO_2 are always present.) The reaction can be written as follows.

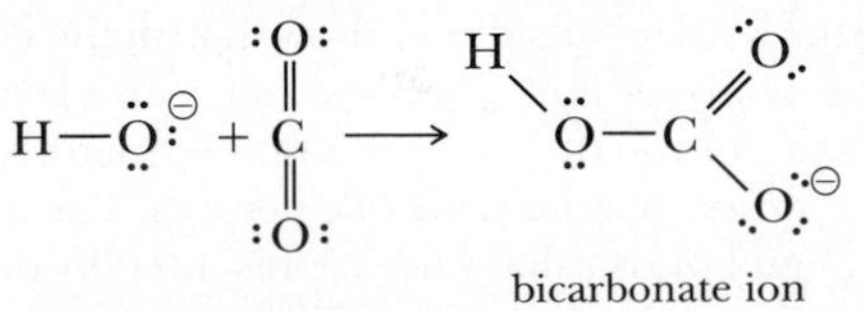

bicarbonate ion

Which is the Lewis acid and which is the Lewis base?

EXAMPLE 7.12

Identifying Lewis Acids and Bases

SOLUTION Notice that in this reaction a bond is formed from the oxygen of the OH^- ion to the carbon of the CO_2. In addition, a double bond is changed to a single bond. Let's follow the movement of electrons in this reaction.

The donation of electrons *from* the oxygen of the OH^- ion produces a bond, so the OH^- ion is the Lewis base. The carbon atom of the CO_2 accepts the electrons, so we identify the CO_2 as the Lewis acid.

■ **Practice Exercise 11** (a) Is the fluoride ion more likely to behave as a Lewis acid or a Lewis base? Explain. (b) Is the $BeCl_2$ molecule more likely to behave as a Lewis acid or a Lewis base? Explain.

SUMMARY

Ionic Bonding The forces of attraction between positive and negative ions in an ionic compound are called **ionic bonds.** The formation of ionic compounds by electron transfer is favored when atoms of low ionization energy react with atoms of high electron affinity. The chief stabilizing influence in the formation of ionic compounds is the **lattice energy**—the lowering of the potential energy that occurs when gaseous ions are brought together to form the crystal of the ionic compound. When atoms of the elements in Groups IA and IIA as well as the nonmetals form ions, they usually gain or lose enough electrons to achieve a noble gas electron configuration. Transition elements lose their outer *s* electrons first, followed by loss of *d* electrons from the shell below the outer shell. Post-transition metals lose electrons from their outer *p* subshell first, followed by loss of electrons from the *s* subshell in the outer shell.

Covalent Bonding Electron sharing between atoms occurs when electron transfer is energetically too "expensive." Shared electrons attract the positive nuclei and this leads to a lowering of the potential energy of the atoms as a covalent bond forms. Electrons generally become paired when they are shared. An atom tends to share enough electrons to complete its valence shell. Except for hydrogen, the valence shell usually holds eight electrons, which forms the basis of the octet rule. The **octet rule** states that atoms of the representative elements tend to acquire eight electrons in their outer shells when they form bonds. **Single, double,** and **triple bonds** involve the sharing of one, two, and three pairs of electrons, respectively, between two atoms. Compounds of B and Be often have less than an octet around the central atom. Atoms of the elements of period 2 cannot have more than an octet because their outer shells can hold only eight electrons. Elements in period 3 and below can exceed an octet if they form more than four bonds.

Bond energy (the energy needed to separate the bonded atoms) and **bond length** (the distance between the nuclei of the atoms connected by the bond) are two experimentally measurable quantities that can be related to the number of pairs of electrons in the bond. For bonds between the same atoms, bond energy increases and bond length decreases as the **bond order** increases.

Organic Compounds Carbon normally forms four covalent bonds, either to atoms of other elements or to other carbon atoms. Oxygen-containing organic compounds include **alcohols** (which have an O—H group bonded to carbon), **ketones** (which have a doubly bonded oxygen atom attached to a carbon that is also bonded to two other carbon atoms), **aldehydes** (in which the C=O group is attached to a hydrogen), and **organic acids,** which contain the carboxyl group. Amines are nitrogen-containing compounds derived from ammonia in which a hydrogen of NH_3 is replaced by a hydrocarbon group.

Lewis Symbols and Lewis Structures Lewis symbols are a bookkeeping device used to keep track of valence electrons in ionic and covalent bonds. The **Lewis symbol** of an element consists of the element's chemical symbol surrounded by a number of dots equal to the number of valence electrons. In the **Lewis structure** for an ionic compound, the Lewis symbol for the anion is enclosed in brackets (with the charge written outside) to show that all the electrons belong entirely to the ion. The Lewis structure for a molecule or polyatomic ion uses pairs of dots between chemical symbols to represent shared pairs of electrons. The electron pairs in covalent bonds usually are represented by dashes; one dash equals two electrons. The following procedure is used to draw the Lewis structure: (1) decide on the skeletal structure (remember that the central atom is usually first in the formula); (2) count all the valence electrons, taking into account the charge, if any; (3) place a pair of electrons in each bond; (4) complete the octets of atoms other than the central atom (but remember that hydrogen can only have two electrons); (5) place any leftover electrons on the central atom in pairs; (6) if the central atom still lacks an octet, move electron pairs to make double or triple bonds.

Electronegativity   The attraction an atom has for the electrons in a bond is called the atom's **electronegativity.** When atoms of different electronegativities form a bond the electrons are shared unequally and the bond is **polar,** with **partial positive** and **partial negative** charges at opposite ends. When the two atoms have the same electronegativity, the bond is **nonpolar.** The extent of polarity of the bond depends on the electronegativity difference between the two bonded atoms. When the electronegativity difference is very large, ionic bonding results. A bond is approximately 50% ionic when the electronegativity difference is 1.7. In the periodic table, electronegativity increases from left to right across a period and from bottom to top in a group.

Formal Charges   The **formal charge** assigned to an atom in a Lewis structure is the difference between the number of valence electrons of an isolated atom of the element and the number of electrons that "belong" to the atom because of its bonds to other atoms and its unshared valence electrons. The sum of the formal charges always equals the net charge on the molecule or ion. The most stable (lowest energy) Lewis structure for a molecule or ion is the one with the fewest formal charges. This is the preferred Lewis structure for the particle.

Resonance   Two or more atoms in a molecule or polyatomic ion are chemically equivalent if they are attached to the same kinds of atoms or groups of atoms. Bonds to chemically equivalent atoms must be the same; they must have the same bond length and the same bond energy, which means they must involve the sharing of the same number of electron pairs. Sometimes the Lewis structures we draw suggest that the bonds to chemically equivalent atoms are not the same. Typically, this occurs when it is necessary to form multiple bonds during the drawing of a Lewis structure. When alternatives exist for the location of a multiple bond among two or more equivalent atoms, then each possible Lewis structure is actually a **resonance structure** or **contributing structure,** and we draw them all. In drawing resonance structures, the relative locations of the nuclei must be identical in all. Remember that none of the resonance structures corresponds to a real molecule, but their composite—the **resonance hybrid**—does approximate the actual structure of the molecule or ion.

Coordinate Covalent Bonding   For bookkeeping purposes, we sometimes single out a covalent bond whose electron pair originated from one of the two bonded atoms. An arrow is sometimes used to indicate the donated pair of electrons. Once formed, a coordinate covalent bond is no different than any other covalent bond.

Lewis Acids and Bases   A **Lewis acid** accepts a pair of electrons from a **Lewis base** in the formation of a coordinate covalent bond. Lewis bases often have filled valence shells and must have at least one unshared electron pair. Lewis acids can have an incomplete valence shell that can accept an electron pair, or double bonds that allow electron pairs to be moved to make room for an incoming electron pair from a Lewis base.

Tools You Have Learned

The table below lists the tools that you have learned in this chapter. You will need them to answer chemistry questions. Review them if necessary, and refer to the tools when working on the Thinking-It-Through problems and the Review Exercises that follow.

Tool	Function
Rules for electron configurations of ions (page 272 and 273)	To be able to write correct electron configurations for cations and anions.
Lewis symbols (page 274)	A bookkeeping device for keeping track of valence electrons in atoms and ions. Learn to use the periodic table to construct the Lewis symbol.
Periodic trends in electronegativity (page 283)	To compare the relative polarities of chemical bonds and to identify the most (or least) electronegative element in a compound.

Method for drawing Lewis structures (page 284)	A systematic method for constructing the Lewis structure for a molecule or polyatomic ion.
Correlation between bond properties and bond order (page 289)	To compare experimental properties that are related to covalent bonds with those predicted by theory.
Formal charges (page 290)	To select the best Lewis structure for a molecule or polyatomic ion.
Method for determining resonance structures (page 294)	Resonance structures provide a way of expressing the structures of molecules for which a single satisfactory Lewis structure cannot be drawn.

THINKING IT THROUGH

By now you know that the goal here is not to find the answer itself; instead, you are asked to assemble the available information needed to obtain the answer, state what additional data (if any) are needed, and describe how you would use the data to answer the question.

The problems are divided into two groups. Those in Level 2 are significantly more challenging than those in Level 1 and provide an opportunity to really hone your problem solving skills.

Level 1 Problems

1. How can you determine how much energy would be absorbed or given off in the formation of the gaseous ions that could eventually come together to form one mole of Na_2O?

2. Describe how we could use Lewis symbols to follow the formation of K_2S from potassium and sulfur atoms.

3. Explain how you would figure out the changes in electron configuration that accompany the following:

$$Pb \longrightarrow Pb^{2+} \longrightarrow Pb^{4+}$$

4. Why is K_2S ionic but H_2S is molecular? What factors must you consider in answering this question?

5. How would you determine which molecule is likely to have the more polar bonds, PCl_3 or SCl_2? In these molecules, how can you tell which atoms carry the partial positive and partial negative charges?

Level 2 Problems

6. How many double bonds are in the Lewis structure for $HBrO_3$? Explain in detail how you would answer this question.

7. What is the preferred Lewis structure for H_3AsO_4? In detail, explain what knowledge you must have to answer this question?

8. Are the following Lewis structures considered to be resonance structures? Explain. How can you predict which is the more likely structure for $POCl_3$?

9. How can you make predictions about the relative S—O bond lengths in the SO_3^{2-} and SO_4^{2-} ions?

REVIEW EXERCISES

Answers to questions whose numbers are printed in color are given in Appendix D. More challenging questions are marked with asterisks.

Ionic Bonding

7.1 What must be true about the total energy of a collection of atoms for a stable compound to be formed?

7.2 What is an *ionic bond?*

7.3 How is the tendency to form ionic bonds related to the IE and EA of the atoms involved?

7.4 What is the lattice energy? In what ways does it contribute to the stability of ionic compounds?

7.5 Explain what happens to the electron configurations of Mg and Br when they react to form the compound $MgBr_2$.

7.6 Describe what happens to the electron configuration of a nitrogen atom when it forms the nitride ion, N^{3-}.

7.7 Magnesium forms the ion Mg^{2+}, but not the ion Mg^{3+}. Why?

7.8 Why doesn't chlorine form the ion Cl^{2-}?

7.9 What are the electron configurations of the Pb^{2+} and Pb^{4+} ions?

7.10 Why do many of the transition elements in period 4 form ions with a 2+ charge?

7.11 What are the abbreviated electron configurations of the following ions: (a) Ni^{2+}, (b) Ti^{2+}, (c) Ti^{4+}, (d) Cu^{2+}, (e) Au^{3+}?

7.12 What are the electron configurations of the following ions: (a) S^{2-}, (b) P^{3-}, (c) I^-, (d) Al^{3+}, (e) Zn^{2+}? Which obey the octet rule?

Lewis Symbols

7.13 Write Lewis symbols for the following atoms: (a) Si, (b) Sb, (c) Ba, (d) Al, (e) S.

7.14 Write Lewis symbols for the following atoms: (a) K, (b) Ge, (c) As, (d) Br, (e) Se.

7.15 Write Lewis symbols for the following ions: (a) K^+, (b) Al^{3+}, (c) S^{2-}, (d) Si^{4-}, (e) Mg^{2+}.

7.16 Write Lewis symbols for the following ions: (a) Br^-, (b) Se^{2-}, (c) Li^+, (d) C^{4-}, (e) As^{3-}.

7.17 Use Lewis symbols to diagram the reactions between: (a) Ca and Br, (b) Al and O, and (c) K and S.

7.18 Use Lewis symbols to diagram the reactions between: (a) Mg and S, (b) Mg and Cl, and (c) Mg and N.

Electron Sharing

7.19 In terms of the potential energy change, why doesn't ionic bonding occur when two nonmetals react with each other?

7.20 Describe what happens to the electron density around two hydrogen atoms as they come together to form an H_2 molecule.

7.21 What holds the nuclei together in a covalent bond?

7.22 What happens to the energy of two hydrogen atoms as they approach each other? What happens to the spins of the electrons?

7.23 What factors control the bond length in a covalent bond?

7.24 Why is a covalent bond sometimes called an *electron pair bond?*

7.25 Is bond formation endothermic or exothermic?

7.26 How much energy, in joules, is required to break the bond in *one* hydrogen molecule? The bond energy of H_2 is 435 kJ/mol.

7.27 How many grams of water could have their temperature raised from 25 °C (room temperature) to 100 °C (the boiling point of water) by the amount of energy released in the formation of one mole of H_2 from hydrogen atoms?

Covalent Bonding and the Octet Rule

7.28 Use Lewis structures to diagram the formation of (a) Br_2, (b) H_2O, and (c) NH_3 from neutral atoms.

7.29 What is the *octet rule?* What is responsible for it?

7.30 How many covalent bonds are normally formed by: (a) H, (b) C, (c) O, (d) N, (e) Cl?

7.31 Chlorine tends to form only one covalent bond because it needs just one electron to complete its octet. What is the Lewis structure for the simplest compound formed by chlorine with: (a) nitrogen, (b) carbon, (c) sulfur, (d) bromine?

7.32 Use the octet rule to predict the formula of the simplest compound formed from hydrogen and (a) Se, (b) As, and (c) Si. (Remember, however, that the valence shell of hydrogen can hold only two electrons.)

7.33 What would be the formula for the simplest compound formed from: (a) P and Cl, (b) C and F, and (c) I and Cl?

7.34 Why do period-2 elements never form more than four covalent bonds? Why are period-3 elements able to exceed an octet when they form bonds?

7.35 Define (a) *single bond,* (b) *double bond,* and (c) *triple bond.*

7.36 What is a structural formula?

7.37 $H—C\equiv N$: is the Lewis structure for hydrogen cyanide. Draw circles enclosing electrons to show that carbon and nitrogen obey the octet rule.

7.38 Why doesn't hydrogen obey the octet rule? How many covalent bonds does a hydrogen atom form?

Compounds of Carbon

7.39 What special "talent" does carbon have that allows it to form so many compounds?

7.40 Sketch the Lewis structures for: (a) methane, (b) ethane, (c) propane.

7.41 Draw the structure for a hydrocarbon that has a chain of five carbon atoms linked by single bonds. How many hydrogen atoms does the molecule have? What is the molecular formula for the compound?

7.42 How many *different* molecules have the formula C_5H_{12}? Sketch their structures.

7.43 Match the compounds on the left with the family types on the right.

$$CH_3—CH_2—\overset{\overset{\displaystyle O}{\|}}{C}—OH \qquad \text{hydrocarbon}$$

$$CH_3—\overset{\overset{\displaystyle O}{\|}}{C}—CH_2—CH_3 \qquad \text{aldehyde}$$

$$CH_3—CH_2—\overset{\overset{\displaystyle OH}{|}}{CH}—CH_3 \qquad \text{amine}$$

$$CH_3—CH_2—\overset{\overset{\displaystyle O}{\|}}{C}—H \qquad \text{acid}$$

$$CH_3—\overset{\overset{\displaystyle H}{|}}{N}—CH_2—CH_3 \qquad \text{alcohol}$$

$$CH_3—CH_2—\underset{\underset{\displaystyle CH_3}{|}}{CH}—CH_3 \qquad \text{ketone}$$

7.44 Write a chemical equation for the ionization in water of the acid in Exercise 7.43. Is the acid strong or weak?

7.45 Write a chemical equation for the reaction of the amine in Exercise 7.43 with water.

7.46 Write the net ionic equation for the reaction of the acid and amine in Exercise 7.43 with each other.

7.47 Draw the structures of (a) ethylene, (b) acetylene.

Polar Bonds and Electronegativity

7.48 What is a polar covalent bond? Define dipole moment.

7.49 Define electronegativity.

7.50 Which element has the highest electronegativity? Which is the second most electronegative element?

7.51 Which elements are assigned electronegativities of zero? Why?

7.52 Use Figure 7.5 to choose the atom in each of the following bonds that carries the partial positive charge: (a) N—S, (b) Si—I, (c) N—Br.

7.53 Which bond in Exercise 7.52 is most polar?

7.54 Which of the following bonds is most polar? For each bond, choose the atom that carries the partial negative charge. (a) Hg—I, (b) P—I, (c) Si—F, (d) Mg—N.

7.55 Are any of the covalent bonds in the preceding exercise more ionic than covalent?

7.56 If an element has a low electronegativity, is it likely to be a metal or a nonmetal? Explain your answer.

Failure of the Octet Rule

7.57 How many electrons are in the valence shells of: (a) Be in $BeCl_2$, (b) B in BCl_3, and (c) H in H_2O?

7.58 What is the minimum number of electrons that would be expected to be in the valence shell of As in $AsCl_5$?

Drawing Lewis Structures

7.59 What are the expected skeletal structures for: (a) $SiCl_4$, (b) PF_3, (c) PH_3, and (d) SCl_2?

7.60 What are the expected skeletal structures for: (a) HIO_3, (b) H_2CO_3, (c) HCO_3^-, (d) PCl_4^+?

7.61 How many dots must appear in the Lewis structures of: (a) $SiCl_4$, (b) PF_3, (c) PH_3, and (d) SCl_2?

7.62 How many dots must appear in the Lewis structures of: (a) HIO_3, (b) H_2CO_3, (c) HCO_3^-, (d) PCl_4^+?

7.63 Draw Lewis structures for: (a) $SiCl_4$, (b) PF_3, (c) PH_3, and (d) SCl_2.

7.64 Draw Lewis structures for: (a) HIO_3, (b) H_2CO_3, (c) HCO_3^-, (d) PCl_4^+.

7.65 Draw Lewis structures for: (a) CS_2, (b) CN^-, (c) SeO_3, and (d) SeO_2.

7.66 Draw Lewis structures for: (a) HNO_2, (b) $HClO_2$, and (c) H_2SeO_3.

7.67 Draw Lewis structures for (a) NO^+, (b) NO_2^-, (c) $SbCl_6^-$, and (d) IO_3^-. Which contain multiple bonds?

7.68 Draw Lewis structures for: (a) TeF_4, (b) ClF_5, (c) XeF_2, and (d) XeF_4.

7.69 Draw Lewis structures for: (a) ClO_4^-, (b) AsH_3, (c) $AsCl_4^+$, and (d) PCl_6^-.

7.70 Draw the Lewis structure for formaldehyde, CH_2O.

The central atom in the molecule is carbon, which is attached to two hydrogens and an oxygen.

7.71 Draw Lewis structures for: (a) $GeCl_4$, (b) CO_3^{2-}, (c) PO_4^{3-}, and (d) O_2^{2-}.

Bond Length and Bond Energy

7.72 Define *bond length* and *bond energy*.

7.73 Define bond order. How are bond energy and bond length related to bond order? Why are there these relationships?

7.74 The energy required to break the H—Cl bond to give H^+ and Cl^- ions would not be called the H—Cl bond energy. Why?

Formal Charge

7.75 What is the definition of formal charge?

7.76 How are formal charges used to select the best Lewis structure for a molecule? What is the basis for this method of selection?

7.77 Assign formal charges to each atom in

(a) $H-\ddot{O}-\overset{..}{\underset{..}{Cl}}-\ddot{O}:$

(b) $\ddot{O}=\overset{:\ddot{O}:}{\underset{..}{S}}-\ddot{O}:$

(c) $\ddot{O}=\ddot{S}-\ddot{O}:$

(d) $:\ddot{Cl}-\overset{:O:}{\underset{}{N}}-\ddot{O}:$ (with N double-bonded to top O)

(e) $:\ddot{F}-\overset{:\ddot{F}:}{\underset{:\ddot{F}:}{N}}-\ddot{O}:$

7.78 Draw the Lewis structure for $HClO_4$ according to the rules in Section 7.6. Assign formal charges to each atom in the formula. Select the preferred Lewis structure for this compound.

7.79 Below are two structures for $BeCl_2$. Why is the one on the left the preferred structure?

$$:\ddot{Cl}-Be-\ddot{Cl}: \qquad :\ddot{Cl}-Be=\ddot{Cl}:$$

7.80 Assign formal charges to all the atoms in

$$\overset{\textstyle :\ddot{O}-H}{\underset{\textstyle H-\ddot{O}:}{:\ddot{O}-As-\ddot{O}-H}}$$

7.81 Assign formal charges to all the atoms in

$$H-N\equiv N-\ddot{N}:$$

Resonance

7.82 What is the need for the concept of resonance?

7.83 What is a resonance hybrid? How does it differ from the resonance structures drawn for a molecule?

7.84 Draw the resonance structures for CO_3^{2-}.

7.85 How should the N—O bond lengths compare in the NO_3^- and NO_2^- ions?

7.86 Arrange the following in order of increasing C—O bond length: CO, CO_3^{2-}, CO_2, HCO_2^- (formate ion).

7.87 How many resonance structures could be written for the N_2O_4 molecule? Its skeletal structure is

O   O

N   N

O   O

7.88 The Lewis structure of CO_2 was given as

$$\ddot{O}=C=\ddot{O}$$

but two other resonance structures can also be drawn for it. What are they? On the basis of formal charges, why are they not preferred structures?

7.89 Write Lewis structures for the nitrate ion, NO_3^-, and for nitric acid, HNO_3. How many equivalent oxygens are there in each structure? Compare the number of resonance structures for each.

7.90 Following the rules in Figure 7.7, draw the Lewis structure for the sulfate ion. Assign formal charges to each atom in this structure. Reduce formal charges to obtain the best Lewis structure for the ion. Draw the resonance structures for the ion. What is the average sulfur–oxygen bond order in the ion?

Coordinate Covalent Bonds

7.91 What is a *coordinate covalent bond*?

7.92 Once formed, how (if at all) does a coordinate covalent bond differ from an ordinary covalent bond?

7.93 BCl_3 has an incomplete valence shell. Show how it could form a coordinate covalent bond with H_2O.

7.94 Use Lewis structures to show that the hydronium ion, H_3O^+, can be considered to be formed by the creation of a coordinate covalent bond between H_2O and H^+.

7.95 Use Lewis structures to show that the reaction

$$BF_3 + F^- \longrightarrow BF_4^-$$

involves the formation of a coordinate covalent bond.

Lewis Acids and Bases

7.96 Define Lewis acid and Lewis base.

7.97 Explain why the addition of a proton to a water molecule to give H_3O^+ is a Lewis acid–base reaction.

7.98 Methylamine, CH_3NH_2, has the structure

```
      H   H
      |   |
  H—C—N:
      |   |
      H   H
```

Use Lewis structures to illustrate the reaction of methylamine with boron trifluoride, BF_3.

7.99 Aluminum chloride, $AlCl_3$, forms molecules with itself with the formula Al_2Cl_6. Its structure is

```
  :Cl:      :Cl:      :Cl:
      \\    /    \\    /
       Al        Al
      /    \\    /    \\
  :Cl:   :Cl:     :Cl:
```

Use Lewis structures to show how the reaction $2AlCl_3 \rightarrow Al_2Cl_6$ is a Lewis acid–base reaction.

7.100 Use Lewis structures to show the Lewis acid–base reaction between SeO_2 and H_2O to give H_2SeO_3. Identify the Lewis acid and the Lewis base in the reaction.

7.101 Explain why the oxide ion, O^{2-}, can function as a Lewis base but not as a Lewis acid.

7.102 Use Lewis structures to diagram the reaction

$$CO_2 + O^{2-} \longrightarrow CO_3^{2-}$$

Identify the Lewis acid and Lewis base in this reaction.

7.103 Use Lewis structures to diagram the reaction

$$CO_2 + H_2O \longrightarrow H_2CO_3$$

Identify the Lewis acid and Lewis base in this reaction.

Additional Exercises

*7.104 To change 1 mol of $Mg(s)$ and $\frac{1}{2}$ mol of $O_2(g)$ to gaseous atoms requires a total of approximately 150 kJ of energy. The first and second ionization energies of magnesium are 737 and 1450 kJ/mol; the first and second electron affinities of oxygen are -141 and $+844$ kJ/mol, and the standard heat of formation of $MgO(s)$ is -602 kJ/mol. Construct an enthalpy diagram similar to the one in Figure 7.1 and use it to calculate the lattice energy of magnesium oxide. How does the lattice energy of MgO compare with that of NaCl? What might account for the difference?

7.105 In many ways, tin(IV) chloride behaves more like a covalent molecular species than as a typical ionic chloride. Draw the Lewis structure for the tin(IV) chloride molecule.

7.106 In each pair, choose the one with more polar bonds. (Use the periodic table to answer the question.)

(a) PCl_3 or $AsCl_3$ (b) $SiCl_4$ or SCl_2
(c) SF_2 or GeF_4 (d) SrO or SnO

7.107 How many electrons are in the outer shell of the Zn^{2+} ion?

7.108 The following are two Lewis structures that can be drawn for phosgene, a substance that has been used as a war gas.

```
      :Ö:                   Ö:
       |                    ||
 :Cl—C=Cl:        :Cl—C—Cl:
```

Which is the better Lewis structure? Why?

7.109 The following resonance structures were drawn for the azide ion, N_3^-. Identify the best and worst of them.

$$\left[:\ddot{N}=N=\ddot{N}:\right]^- \longleftrightarrow \left[\ddot{N}\equiv N-\ddot{N}:\right]^-$$

$$\left[:N\equiv N=\ddot{N}:\right]^- \longleftrightarrow \left[:\ddot{N}-N-\ddot{N}:\right]^-$$

7.110 How should the sulfur–oxygen bond lengths compare for the species SO_3, SO_2, SO_3^{2-}, and SO_4^{2-}?

7.111 What other resonance structures are there for the molecule described in Exercise 7.81?

7.112 What is the most reasonable Lewis structure for S_2Cl_2?

7.113 What is the Lewis structure for $SbCl_6^-$?

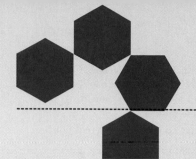

CHEMICALS IN USE 6
Superconductivity

ENERGY LOSS FROM ELECTRICAL RESISTANCE

At ordinary temperatures, metals like copper have a small but significant resistance to the flow of electricity. Moving electrons either bump into jiggling metal atoms or encounter defects in the metal structure. Such resistance means that a percentage of the electrical energy is converted to heat. In the long-distance transmission of electricity from power stations to users, such resistance requires, for compensation, extremely high voltages in the overhead power cables and it wastes energy. Resistance also limits the capacity of computer chips. Although the individual components on these chips use little power, when thousands of them are crowded together, much heat is generated and the computer chips must be cooled to keep them from burning out. This stands in the way of designing the more densely packed circuits needed to make computers smaller. It is not at all surprising, therefore, that scientists and engineers have for years intensely studied the phenomenon of superconductivity.

THE SUPERCONDUCTING STATE

A material in a state in which it offers no resistance to electricity and counteracts a surrounding magnetic field is called a *superconductor*. Prior to 1987, only a few materials, nearly all of them metals and metal alloys, could be put into such a state and, then, only if their temperatures were lowered to just a few degrees above absolute zero. H. Kamerlingh-Onnes, a Dutch physicist, won the 1913 Nobel prize in physics for his discovery that mercury is superconducting at 4.1 K. This required the immersion of the mercury in a bath of liquid helium, which boils at 4.2 K. The temperature at and below which a material is superconducting is given the symbol T_c. An intermetallic compound of niobium and germanium, Nb_3Ge, was discovered in 1975 for which T_c equals 23.2 K. The goal is to find materials of much higher values of T_c so that the many advantages of superconductivity will be more readily available.

In 1986, K. A. Müller and J. G. Bednorz of IBM in Zürich, Switzerland, discovered that a previously known ceramic material, a mixed oxide of lanthanum, barium, and copper with the formula $LaBa_2CuO_4$, has a value of T_c equal to 30 K. For their work, Müller and Bednorz shared the 1987 Nobel prize in physics. To keep the material in a superconducting state at 30 K, however, still requires refrigeration technology too expensive and too cumbersome to permit any important practical applications.

HIGH-TEMPERATURE SUPERCONDUCTIVITY

To make superconductivity practical, materials had to be discovered that could pass over to the superconducting state at temperatures much higher than the boiling point of liquid helium. Specifically, a material with T_c above the boiling point of liquid nitrogen, 77.4 K, was sought. The refrigeration technology for producing and handling liquid nitrogen is well-known and relatively simple, and liquid nitrogen costs about as much as milk or beer. Today, materials that have values of T_c above the boiling point of nitrogen are called *high-temperature superconductors*.

With the opening made by Müller and Bednorz toward ceramic-like substances, mixed metal oxides, the search continued among them. Chemists Maw-Kuen Wu and J. Ashburn (University of Alabama) were the first to synthesize a material superconducting above the boiling point of nitrogen. They substituted yttrium for lanthanum in the Müller-Bednorz compound and prepared a mixed oxide, $YBa_2Cu_3O_7$, with T_c equal to 93 K. Chemist Paul Chu (University of Houston) was on the same yttrium track, and the Chu and Wu teams made a joint announcement in early 1987. $YBa_2Cu_3O_7$ is sometimes called the "1-2-3 compound," because of the $1:2:3$ ratio of $Y:Ba:Cu$. The significant structural feature of this substance is the layering of copper and oxygen atoms (see Figure 1). The 1-2-3 compound is just one of a small family of the mixed oxides of these elements that are superconducting above 77 K.

Experimentally, the ceramic superconductors are easy to make. All it takes are the parent oxides and a high-temperature furnace. Teams around the world worked arduously to find materials that pushed T_c to still higher values. By 1988, a mixed oxide of thallium, calcium, barium, and copper—discovered by A. M. Hermann and Z. Sheng (University of Arkansas)—was found that had zero electrical resistance at 107 K. They ground a

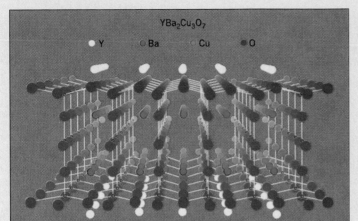

FIGURE 1

The 1-2-3 superconductor. (Adapted from A. W. Sleight, "Chemistry of High-Temperature Superconductors," *Science*, December 16, 1988, page 1521. Used by permission.

mixture of $BaCu_3O_4$ with CaO and thallium oxide, pressed it into a pellet, baked it in a stream of oxygen at 900 °C for a few minutes, and cooled it to room temperature. An IBM team (San Jose, CA), working with the same elements, found a material for which $T_c = 122$ K. Since then, materials with values of T_c above 130 K have been found.

DRAWBACKS

Problems exist with the ceramic superconductors. They are brittle, they tend to lose their superconducting states in the presence of magnetic fields, and the density of the electrical current they can carry is not yet as high as desired. Progress has been made, however, in depositing superconducting films on metallic supports first coated with zirconium dioxide. In 1995, for example, the Los Alamos National Laboratories reported the successful preparation of a superconducting tape that could be formed into a coil, a configuration needed for making small electrical motors.

APPLICATIONS

Levitation means "floating on air"; the verb is "to leviate." One of the most fascinating properties of materi-

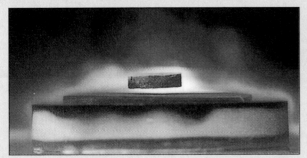

FIGURE 2

Levitation of a magnet above a supercooled sample of the 1-2-3 compound.

tals in their superconducting states is that they can be levitated by a magnetic field. Besides zero electrical resistance, a substance in its superconducting state permits no magnetic field within itself. A weak magnetic field is actually *repelled* by a superconductor, and this is what makes the levitation of superconductors possible (Figure 2). After the discovery of high-temperature superconductors, people envisioned that someday we might levitate entire trains and send them at high speed with no more frictional resistance than offered to airplanes by the surrounding air; prototypes of such "mag-lev trains are currently being tested by Japan and Germany.

Another possible application is the development of superconducting magnetic energy storage. Large rings of superconducting cables, immersed in liquid nitrogen or liquid helium, could store direct electric current indefinitely if there were no electrical resistance in the cables. This electrical energy could then be tapped during high-power needs.

Earlier we mentioned the heat-generating electrical resistance to current that puts a limitation on computer technology. Perhaps superconductivity might be used to overcome this and enable the design of even more powerful, yet modestly sized computer.

Just as electrical resistance is the enemy of high-efficiency power transmission, mechanical friction is the enemy of moving parts. Today, much interest in the practical applications of superconductivity is focused on small devices, such as those that must be able to run unattended for very long periods of time. Tiny rotors and solenoids for instruments on long space flights might be operated without friction, for example.

Questions

- -

1. What two properties must a material possess to be called a superconductor?

2. What is the economic significance of finding a superconducting material with a value of T_c greater than 77K?

Many people would easily recognize the fragrance given off by this large display of roses. Our ability to smell an aroma depends on the way molecules of a substance fit together with others that make up our olfactory system, and this in turn depends on the shapes of molecules. In this chapter we explore the various shapes that molecules have and how chemical bonds control them.

Chapter 8

Chemical Bonding II

The internal structure of an ionic compound, such as NaCl, is controlled primarily by the sizes of the ions and their charges. The attractions between the ions have no preferred directional properties, so if an ionic compound is melted, this structure is lost as the array of ions collapses into a jumbled liquid state. Molecular substances are quite different, however. Molecules have three-dimensional shapes that are determined by the relative orientations of their covalent bonds, and this structure is maintained regardless of whether the substance is a solid, a liquid, or a gas.

Many of the properties of a molecule depend on the three-dimensional arrangement of its atoms. For example, the functioning of enzymes, which are substances that affect how fast biochemical reactions occur, requires that there be a very precise fit between one molecule and another. Even slight alterations in molecular geometry can destroy this fit and deactivate the enzyme, which in turn prevents the biochemical reaction involved from occurring.

In this chapter we will explore molecular geometry and study theoretical models that allow us to predict the shapes of molecules. We will also examine theories that explain, in terms of wave mechanics and the electronic structures of atoms, why covalent bonds form and why they are so highly directional in nature. We begin with a discussion of the kinds of shapes that are observed for molecules.

Basic Molecular Shapes

Molecular shape only becomes a question when there are at least three atoms present. If there are only two, there is no doubt as to how they are arranged; one is just alongside the other. But when there are three or more atoms in a molecule we find that its shape can usually be considered derived from one or another of five basic geometrical structures, which are described on the pages

8.1
SOME COMMON MOLECULAR STRUCTURES

We will study the structures of ionic solids in Chapter 11.

Basic molecular shapes

309

Describing the three-dimensional structure of a molecule is a tool you will use later in analyzing the physical properties of substances. It will also be useful to you in other chemistry-related courses you might take.

ahead. As you study these structures, you should try hard to visualize them in three dimensions. You should also learn how to sketch them in a way that conveys the three-dimensional information.

Linear Molecules   In a **linear molecule** the atoms lie in a straight line. All diatomic molecules could therefore be considered to be linear. When there are three atoms present the angle formed by the covalent bonds, which we call the **bond angle,** equals 180° as illustrated below.

Planar Triangular Molecules   A **planar triangular molecule** is one in which three atoms are located at the corners of a triangle and are bonded to a fourth atom that lies in the center of the triangle.

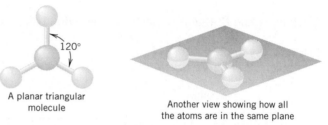

A planar triangular
molecule

Another view showing how all
the atoms are in the same plane

In this molecule, all four atoms lie in the same plane and the bond angles are all equal to 120°.

Tetrahedral Molecules   A **tetrahedron** is a four-sided geometric figure shaped like a pyramid with triangular faces. A **tetrahedral molecule** is one in which four atoms, located at the vertices of a tetrahedron, are bonded to a fifth atom in the center of the structure.

A tetrahedron

A tetrahedral molecule

All the bond angles in a tetrahedral molecule are the same and are equal to 109.5°.

This is the first molecular shape that you've encountered that can't be set down entirely on a two-dimensional surface, so you may need some practice in drawing the structure. It is really quite simple, as illustrated in Figure 8.1.

Drawing molecular shapes is a tool you will find useful in your study of chemistry, so learning to do so is worth the time spent.

FIGURE 8.1

Steps in drawing a tetrahedral molecule.

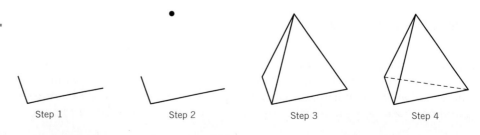

Step 1   Step 2   Step 3   Step 4

Trigonal Bipyramidal Molecules   A **trigonal bipyramid** consists of two *trigonal pyramids* (pyramids with triangular faces) that share a common base.

A trigonal bipyramid

In a **trigonal bipyramidal molecule,** the central atom is located in the middle of the triangular plane shared by the upper and lower trigonal pyramids and is bonded to five atoms that are at the vertices of the figure.

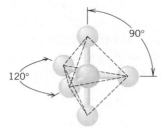

A trigonal bipyramidal molecule

In this structure, not all the bonds are equivalent. If we imagine the trigonal bipyramid centered inside a sphere, the atoms in the triangular plane are located around the equator and the bonds to these atoms are therefore called **equatorial bonds.** The angle between two equatorial bonds is 120°. The two vertical bonds pointing north and south are 180° apart and are called **axial bonds.** The bond angle between an axial bond and an equatorial bond is 90°.

A simplified representation of a trigonal bipyramid is illustrated in the margin. The equatorial triangular plane is sketched as it would look tilted forward, so we are looking at it from its edge. The axial bonds are represented as lines pointing up and down. Notice that the bond pointing down appears to be partially hidden by the triangular plane in the center.

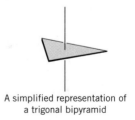

A simplified representation of a trigonal bipyramid

Octahedral Molecules   An **octahedron** is an eight-sided figure that you might think of as two *square pyramids* sharing a common square base. The octahedron has only six vertices, and in an **octahedral molecule** we find an atom in the center of the octahedron bonded to six other atoms at these vertices.

An octahedral

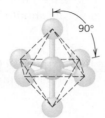

An octahedral molecule

All the bonds in an octahedral molecule are equivalent, with angles between adjacent bonds equal to 90°.

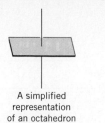

A simplified
representation
of an octahedron

A simplified representation of an octahedron is shown in the margin. The square plane in the center of the octahedron looks like a parallelogram when drawn in perspective and viewed from its edge. The bonds to the top and bottom of the octahedron are shown as vertical lines. Once again, the bond pointing down is drawn so it appears to be partially hidden by the square plane in the center.

8.2
PREDICTING THE SHAPES OF MOLECULES: VSEPR THEORY

VSEPR theory **T**

For a theoretical model to be useful, it should explain known facts and it should be capable of making accurate predictions. A theory that is remarkably successful at both of these, and that is also conceptually simple, has come to be called the **valence shell electron pair repulsion theory,** or **VSEPR theory** for short. The theory is based on the concept that *valence shell electron pairs, being negatively charged, stay as far apart as possible so that the repulsions between them are at a minimum.* Let's look at an example that illustrates how this simple concept enables us to predict the shape of a molecule.

Consider the $BeCl_2$ molecule. We've seen that its Lewis structure is

$$:\ddot{C}l-Be-\ddot{C}l:$$

But how are these atoms arranged? Is $BeCl_2$ linear or is it nonlinear; that is, do the atoms lie in a straight line, or do they form some angle less than 180°?

Cl—Be—Cl   or

180°

Be
Cl   Cl
<180°

According to VSEPR theory, the shape of a molecule is determined by how the electron pairs in the valence shell of the central atom minimize their mutual repulsions. In $BeCl_2$, there are two pairs of electrons around the beryllium atom. To minimize their repulsions the two electron pairs will seek positions as far apart as possible, which occurs when the electron pairs are on opposite sides of the nucleus. We can represent this as

to suggest the approximate locations of the electron clouds of the valence shell electron pairs. In order for the electrons to be in the Be—Cl bonds, the Cl atoms must be placed where the electrons are; the result is that we predict that a $BeCl_2$ molecule should be linear.

Cl—Be—Cl

In fact, this is the shape of $BeCl_2$ molecules in the vapor state.

Let's look at another example, the BCl_3 molecule. Its Lewis structure is

$$\begin{array}{c}:\ddot{C}l:\\ |\\ :\ddot{C}l-B-\ddot{C}l:\end{array}$$

Here the central atom has three electron pairs. What arrangement will lead to minimum repulsions? As you may have guessed, the electron pairs will be as far apart as possible when they are located at the corners of a triangle with the boron in the center. This places the three electron pairs and the boron nucleus all in the same plane.

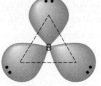

When we attach the Cl atoms, we obtain a planar triangular molecule.

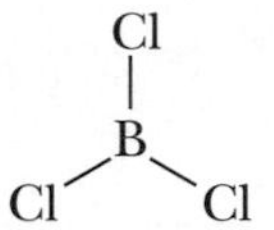

$$\text{Cl} - \overset{\displaystyle \text{Cl}}{\underset{\displaystyle \text{Cl}}{\text{B}}} $$

Experimentally, it has been proved that this is the shape of BCl_3.

Figure 8.2 shows the orientations expected for different numbers of electron pairs around the central atom. It can be shown that these provide the minimum repulsions between the electron pairs. Notice that they yield the same five shapes discussed in Section 8.1.

FIGURE 8.2

Shapes expected for different numbers of electron pairs.

Number of Electron Pairs	Shape	Example
Two pairs	 linear	$BeCl_2$ — 180° (Cl–Be–Cl) 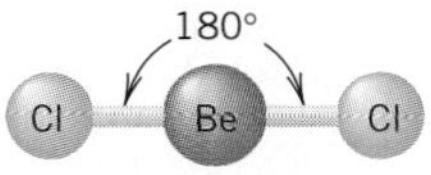
Three pairs	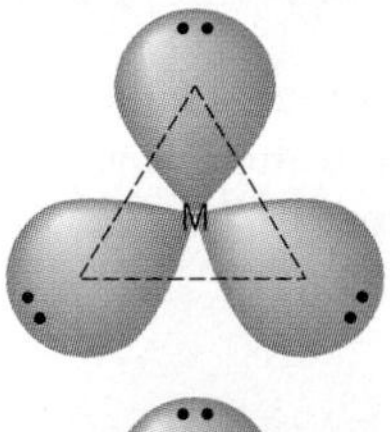 planar triangular	BCl_3 — 120° 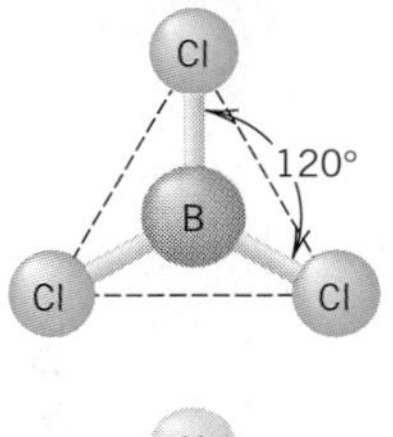
Four pairs	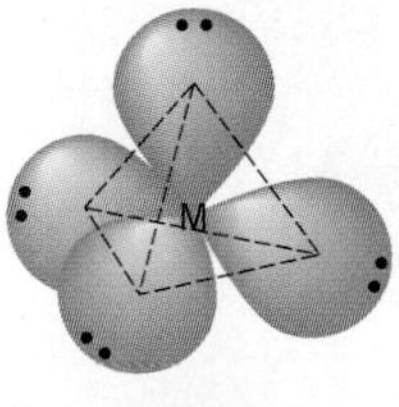 tetrahedral (A tetrahedron is pyramid shaped. It has four triangular faces and four corners.)	CH_4 — 109.5° 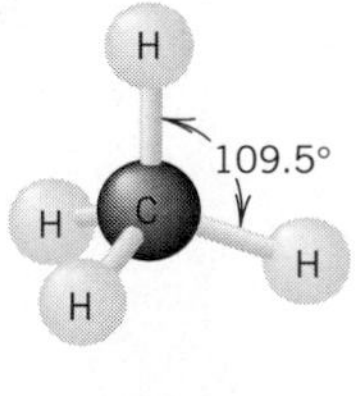
Five pairs	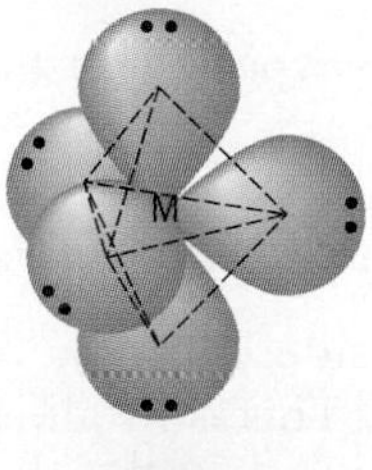 trigonal bipyramidal (This figure consists of two three-sided pyramids joined by sharing a common face–the triangular plane through the center.)	PCl_5 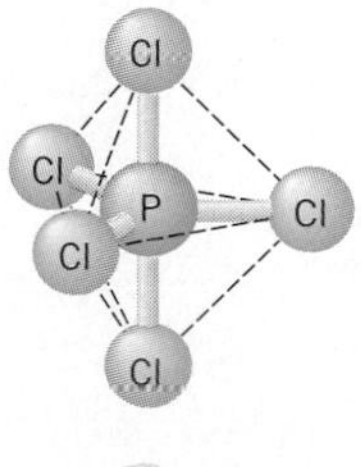
Six pairs	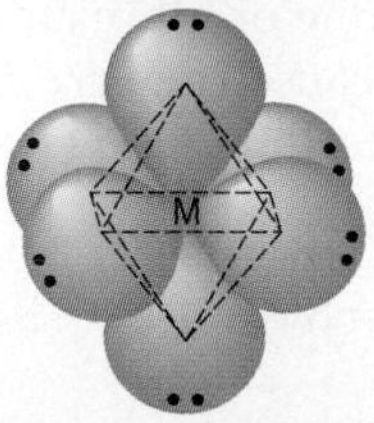 octahedral (An octahedron is an eight-sided figure with *six* corners. It consists of two square pyramids that share a common square base.)	SF_6 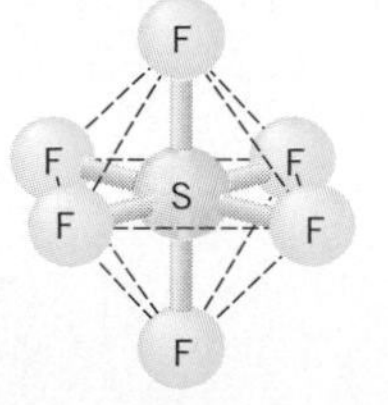

Using Lewis Structures to Predict Molecular Shapes

To predict the shape of a molecule or ion, we need to know how many sets of electron pairs surround the central atom. We can obtain this information by drawing the Lewis structure. For this purpose, we just need the Lewis structure constructed following the rules in Figure 7.7. It is not necessary to assign formal charges and go through the procedure of trying to select "preferred" Lewis structures. Nor is it necessary to draw multiple resonance structures when they might otherwise be appropriate. Just one simple Lewis structure is all we need.

EXAMPLE 8.1
Predicting Molecular Shapes

Carbon tetrachloride was once used as a cleaning fluid until it was discovered that it causes liver damage if absorbed by the body. What is the shape of the CCl_4 molecule?

ANALYSIS We will begin by drawing a Lewis structure for CCl_4. Then we will count the number of electron pairs in the valence shell of the central atom. From this, we will decide the structure of the molecule.

SOLUTION Following the rules, the Lewis structure of CCl_4 is

$$
\begin{array}{c}
\ddot{\mathrm{C}}\mathrm{l}{:} \\
|\\
{:}\ddot{\mathrm{C}}\mathrm{l}{-}\mathrm{C}{-}\ddot{\mathrm{C}}\mathrm{l}{:} \\
|\\
{:}\ddot{\mathrm{C}}\mathrm{l}{:}
\end{array}
$$

There are four electron pairs in the valence shell of carbon. These would have minimum repulsions when arranged tetrahedrally, so the molecule is expected to be tetrahedral. (This is, in fact, its structure.)

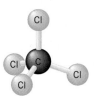

■ **Practice Exercise 1** What shape is expected for the $SbCl_5$ molecule?

Molecular Shapes When Some Electron Pairs Are Not in Bonds

In the molecules that we have considered so far, all the electron pairs of the central atom have been in bonds. This isn't always the case, though. Some molecules have a central atom with one or more pairs of electrons that are not shared with another atom. Still, because they are in the valence shell, these unshared electron pairs—also called **lone pairs**—affect the geometry of the molecule. An example is $SnCl_2$.

$$
{:}\ddot{\mathrm{C}}\mathrm{l}{-}\ddot{\mathrm{S}}\mathrm{n}{-}\ddot{\mathrm{C}}\mathrm{l}{:}
$$

There are three electron pairs around the tin atom, two in bonds plus the lone pair. As in BCl_3, mutual repulsions will direct them toward the corners of a triangle.

Placing the two chlorine atoms where two of the electron pairs are gives

We can't describe this molecule as triangular, even though that is how the electron pairs are arranged. *Molecular shape describes the arrangement of atoms, not the arrangement of electron pairs.* Therefore, we describe the shape of the $SnCl_2$ molecule as being **nonlinear** or **bent** or **V-shaped.**

Notice, now, that when there are three electron pairs around the central atom, *two* different molecular shapes are possible. If all three electron pairs are in bonds, as in BCl_3, a molecule with a planar triangular shape is formed. If one of the electron pairs is a lone pair, as in $SnCl_2$, the arrangement of the atoms in the molecule is said to be nonlinear. The predicted shapes of both, however, are *derived* by first noting the triangular arrangement of electron pairs around the central atom and *then* adding the necessary number of atoms.

We use the orientations of the electron pairs in the valence shell of the central atom to determine how the atoms in the structure are arranged. The arrangement of the atoms is what we mean by the *shape of the molecule.*

Molecules with Four Electron Pairs in the Valence Shell The most common type of molecule or ion has four electron pairs (an octet) in the valence shell of the central atom. When these electron pairs are used to form four single bonds, as in methane (CH_4), the resulting molecule is tetrahedral, as we've seen. There are many examples, however, where some of the pairs are lone pairs. For example,

$$H-\overset{\displaystyle\cdot\cdot}{N}-H \qquad H-\overset{\displaystyle\cdot\cdot}{\underset{\displaystyle\cdot\cdot}{O}}-H$$
$$\quad\;\;|$$
$$\quad\;\;H$$

One lone pair   Two lone pairs

Figure 8.3 shows how the lone pairs affect the shapes of molecules of this type.

Molecules with Five Electron Pairs in the Valence Shell When five electron pairs are present around the central atom, they are directed toward the vertices of a trigonal bipyramid. Molecules such as PCl_5 have this geometry, as we saw in Figure 8.2.

When there are lone pairs in the valence shell of the central atom, their locations are determined by which arrangement gives the minimum overall repulsions. To determine this, we must compare the sizes of lone pairs and

FIGURE 8.3

Molecular shapes with four electron pairs around the central atom.

Number of Pairs in Bonds	Number of Lone Pairs	Structure	
4	0	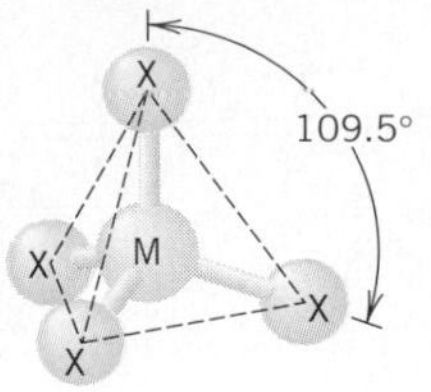	Tetrahedral (Example, CH_4) All bond angles are 109.5°.
3	1	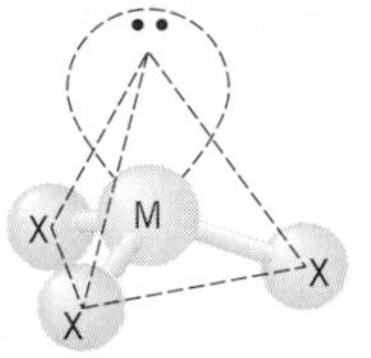	Trigonal pyramidal (Pyramid-shaped) (Example, NH_3)
2	2	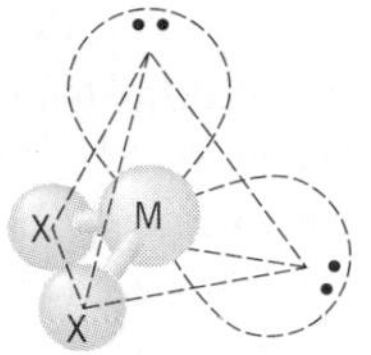	Nonlinear, bent (Example, H_2O)

bonding pairs. A bond pair (BP) has a nucleus at both ends, so its effective size is smaller than a lone pair (LP), which has a nucleus only at one end. Because of their greater size, lone pairs repel others more strongly than do bond pairs, so we can say that LP–LP repulsions are greater than LP–BP repulsions, which in turn are greater than BP–BP repulsions.

In the trigonal bipyramid, not all the electron pairs have the same nearest-neighbor environments. Those in axial positions each have three close neighbors at angles of 90° and one at a much larger angle of 180°. Those in the equatorial plane (the triangular plane through the center of the molecule) each have two neighbors at 90° and two at a much larger angle of 120°. The repulsions that affect the geometry of molecules depend primarily on the nearest-neighbor interactions, so when lone pairs occur in this structure, they *always* are located in the triangular plane, because this yields the fewest strong nearest-neighbor repulsions. Figure 8.4 shows the kinds of geometries that we find for different numbers of lone pairs in the trigonal bipyramid.

If the lone pair is an axial position, there are three strong LP–BP interactions. If the lone pair is in an equatorial position, there are only two strong LP–BP interactions.

Molecules with Six Electron Pairs in the Valence Shell Finally, we come to those molecules or ions that have six electron pairs around the central atom. When all are in bonds, as in SF_6, the molecule is octahedral. When one lone pair is present, the molecule or ion has the shape of a **square pyramid,** and when two lone pairs are present they take positions on opposite sides of the nucleus and the molecule or ion has a **square planar** structure. These shapes are shown in Figure 8.5.

No common molecule or ion with six electron pairs around the central atom has more than two of them as lone pairs.

Number of Pairs in Bonds	Number of Lone Pairs	Structure	
5	0	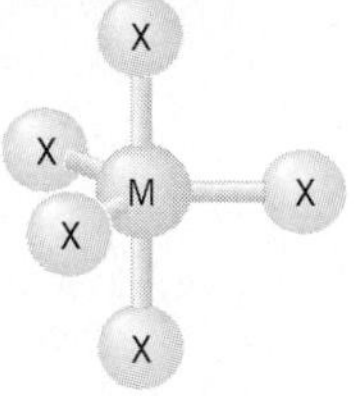	Trigonal bipyramidal (Example, PCl_5)
4	1	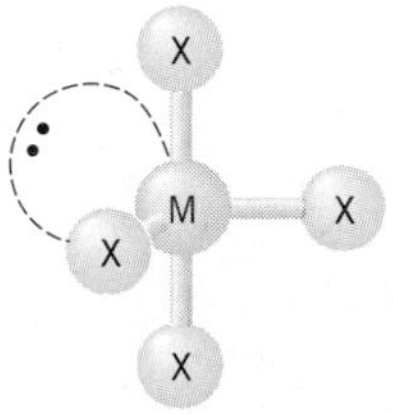	Unsymmetrical tetrahedron 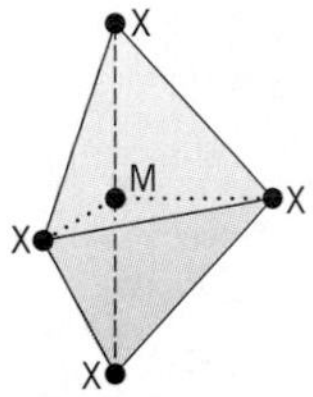 (Example, SF_4)
3	2	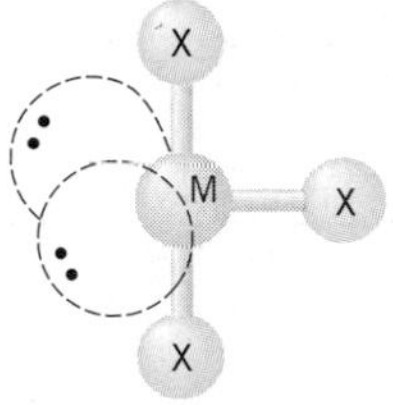	T-Shaped 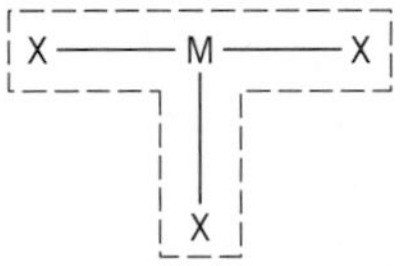 (Example, ClF_3)
2	3	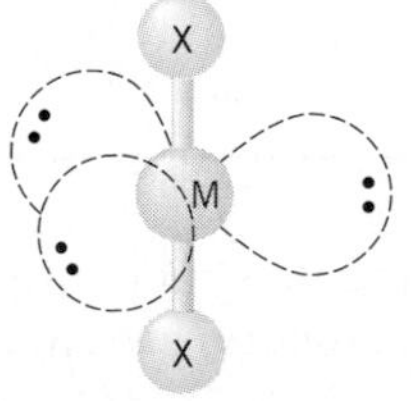	Linear (Example, I_3^-)

FIGURE 8.4

Molecular shapes with five electron pairs around the central atom.

Do we expect the ClO_2^- ion to be linear?

ANALYSIS To apply VSEPR theory, we first need the Lewis structure for the ClO_2^- ion, so drawing the Lewis structure is the first step in solving the problem. Then we will count the number of electron pairs around the central atom. This will tell us which of the basic geometries the molecular shape is based on. We sketch this shape and then attach the two oxygens. Finally, we ignore any lone pairs to determine the description of the shape of the molecule. In other words, in this last step we note how the *atoms* are arranged, not how the electron pairs are arranged.

SOLUTION Following our procedure, we obtain

$$\left[\ddot{\underset{..}{O}} - \ddot{\underset{..}{Cl}} - \ddot{\underset{..}{O}} \colon \right]^-$$

EXAMPLE 8.2
Predicting the Shapes of Molecules and Ions

FIGURE 8.5

Molecular shapes with six electron pairs around the central atom.

Number of Pairs in Bonds	Number of Lone Pairs	Structure	
6	0	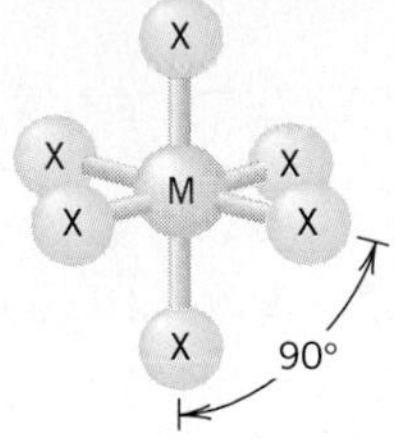	Octahedral (Example, SF$_6$) All bond angles are 90°.
5	1	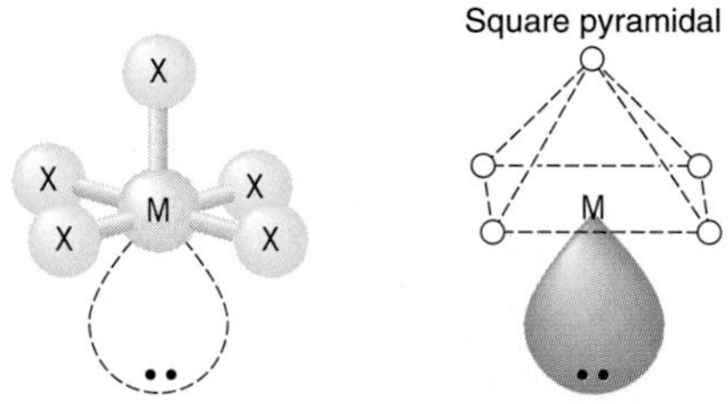	Square pyramidal (Example, BrF$_5$)
4	2	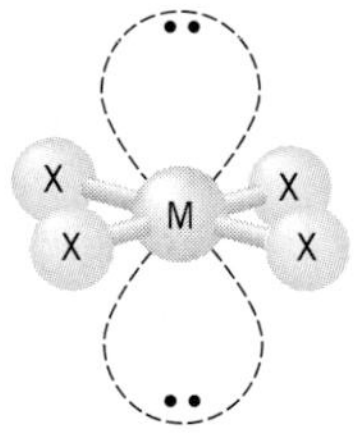	Square planar (Example, XeF$_4$)

Counting electron pairs, we see that there are four around the chlorine. Four electron pairs (according to the theory) are always arranged tetrahedrally. For the electron pairs, this gives

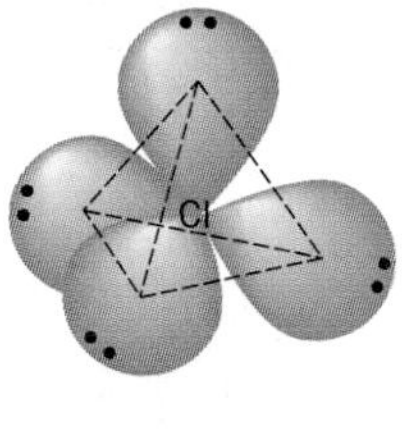

Now we add the two oxygens. It doesn't matter which locations in the tetrahedron we choose because all the bond angles are equal.

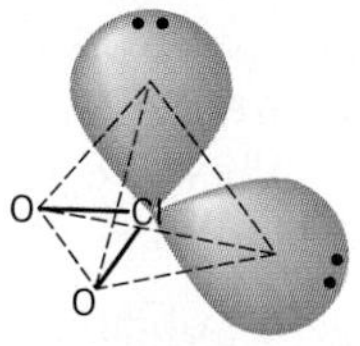

We see that the O—Cl—O angle is less than 180°, so the ion is expected to be nonlinear.

Xenon is one of the noble gases, and is generally quite unreactive. In fact, it was long believed that all the noble gases were totally unable to form compounds. It came as quite a surprise, therefore, when it was discovered that some compounds could be made. One of these is xenon difluoride, XeF_2. What would we expect the geometry of XeF_2 to be?

ANALYSIS To apply VSEPR theory, we need a Lewis structure. Then we count electron pairs around the central atom to determine which of the basic geometries forms the basis for the structure. We sketch the structure and attach the fluorine atoms. Finally, we ignore the electron pairs and observe how the atoms are arranged to describe the shape of the molecule.

SOLUTION The outer shell of xenon, of course, has a noble gas configuration, which contains 8 electrons. Each fluorine has 7 valence electrons, so the total number of electrons that must be placed in the structure is 22. These are distributed following our usual procedure, taking Xe as the central atom. The result is

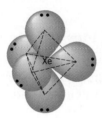

Next we count electron pairs around xenon; there are five of them. When there are five electron pairs, they are arranged in a trigonal bipyramid.

Now we must add the fluorine atoms. In a trigonal bipyramid, the lone pairs always occur in the equatorial plane through the center, so the fluorines go on the top and bottom. This gives

The three atoms, F—Xe—F, are arranged in a straight line, so the molecule is linear.

What is the probable geometry of the $XeOF_4$ molecule, which contains an oxygen atom and four fluorine atoms each bonded to xenon?

SOLUTION By now you should know the routine. First we draw the Lewis structure.

There are six electron pairs around xenon, and they must be arranged octahedrally.

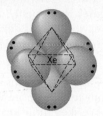

Next we attach the oxygen and four fluorines. The most symmetrical structure is

which we would describe as a square pyramid. Although the oxygen might, in principle, be placed in one of the positions at the corners of the square base (in place of one of the fluorines), the structure shown is the actual structure of the molecule.

Shapes of Molecules and Ions with Double or Triple Bonds

Our discussion to this point has dealt only with the shapes of molecules or ions having single bonds. Luckily, the presence of double or triple bonds does not complicate matters at all. In a double bond, both electron pairs must stay together between the two atoms; they can't wander off to different locations in the valence shell. This is also true for the three pairs of electons in a triple bond. *For the purposes of predicting molecular geometry, then, we can treat double and triple bonds just as we do single bonds.* For example, the Lewis formula for CO_2 is

$$:\ddot{O}\!\!=\!\!C\!\!=\!\!\ddot{O}:$$

The carbon atom has no lone pairs. Therefore, the two groups of electron pairs that make up the double bonds are located on opposite sides of the nucleus and a linear molecule is formed. Similarly, we would predict the following shapes for NO_2^- and NO_3^-.

We only need to consider one of the resonance structures to reach a conclusion about molecular structure.

nonlinear (bent)   planar triangular

EXAMPLE 8.5
Predicting the Shapes of
Molecules and Ions

The Lewis structure for the very poisonous gas hydrogen cyanide, HCN, is

$$H\!\!-\!\!C\!\!\equiv\!\!N:$$

Is the HCN molecule linear or nonlinear?

SOLUTION The triple bond behaves like a single bond for the purposes of predicting the overall geometry of a molecule. Therefore, we expect the triple and single bonds to locate themselves 180° apart, yielding a linear HCN molecule.

■ **Practice Exercise 2** Predict the shapes of SO_3^{2-}, XeO_4, and OF_2.

■ **Practice Exercise 3** Predict the geometry of CO_3^{2-}.

8.3 MOLECULAR SHAPE AND MOLECULAR POLARITY

One of the main reasons we are concerned about whether a molecule is polar or not is that many physical properties, such as melting point and boiling point, are affected by it. This is because polar molecules attract each other. The positive end of one polar molecule attracts the negative end of another (Figure 8.6). The strength of the attraction depends both on the amount of charge on either end of the molecule and on the distance between the charges; in other words, it depends on the molecule's dipole moment.

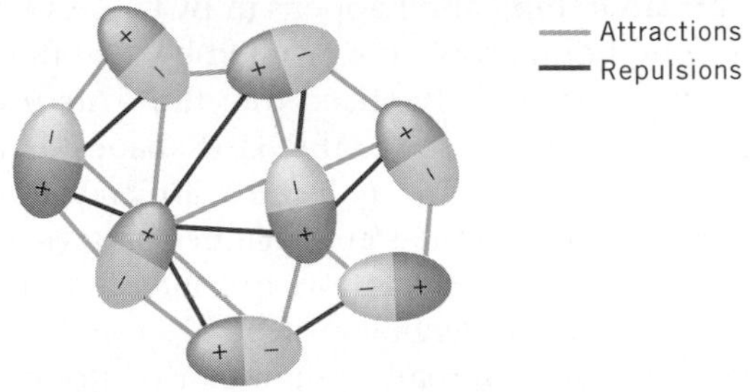

—— Attractions
—— Repulsions

FIGURE 8.6

Attractions between dipoles. Molecules tend to orient themselves so that the positive end of one molecular dipole is near the negative end of another.

The dipole moment of a molecule is a property that can be determined experimentally, and when this is done, an interesting observation is made. There are many molecules that have no dipole moment even though they contain bonds that are polar. Stated differently, they are nonpolar molecules, even though they have polar bonds. The reason for this can be seen if we examine the key role that molecular structure plays in determining molecular polarity.

Let's begin with the H—Cl molecule, which has only two atoms and therefore only one bond. As you've learned, this bond is polar because the two atoms differ in electronegativity. This means that the atoms at opposite ends of the bond carry partial charges of opposite sign. A molecule with equal but opposite charges on opposite ends is polar, so HCl is a polar molecule. In fact, any molecule composed of just two atoms that differ in electronegativity must be polar.

For molecules that contain more than two atoms, we have to consider the combined effects of all the polar bonds. Sometimes, when all the atoms attached to the central atom are the same, the effects of the individual polar bonds cancel and the molecule as a whole is nonpolar. Some examples are shown in Figure 8.7. In this figure the dipoles associated with the bonds themselves—the **bond dipoles**—are shown as arrows crossed at one end, ↦.

Molecular shape and molecular polarity

FIGURE 8.7

In symmetric molecules such as these, the bond dipoles cancel to give nonpolar molecules.

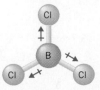

FIGURE 8.8

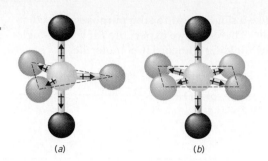

(*a*) A molecule with the general formula AX_5 and a trigonal bipyramidal structure. The set of three bond dipoles in the triangular plane in the center (in blue) cancel, as do the linear set of dipoles (red). Overall, the molecule is nonpolar. (*b*) An octahedral molecule AX_6 consists of three linear sets of bond dipoles. Cancellation occurs for each set, so the molecule is nonpolar overall.

(*a*)   (*b*)

Bond dipoles can be treated as vectors, and the polarity of a molecule is predicted by taking the vector sum of the bond dipoles. If you've studied vectors before, this may help your understanding.

The arrowhead indicates the negative end of the bond dipole and the crossed end corresponds to the positive end.

In CO_2 both bonds are identical, so each bond dipole is of the same magnitude. Because CO_2 is a linear molecule, these bond dipoles point in opposite directions and work against each other. The net result is that their effects cancel, and CO_2 is a nonpolar molecule. Although it isn't as easy to visualize, the same thing also happens in BCl_3 and CCl_4. In each of these, the influence of one bond dipole is cancelled by the effects of the others.

Perhaps you've noticed that the structures of the molecules in Figure 8.7 correspond to three of the basic shapes that we used in the VSEPR theory to derive the shapes of molecules. Molecules with the remaining two structures, trigonal bipyramidal and octahedral, also are nonpolar if all the atoms attached to the central atom are the same. Examples are shown in Figure 8.8. The trigonal bipyramidal structure can be viewed as a planar triangular set of atoms (shown in blue) plus a pair of atoms arranged linearly (shown in red). All the bond dipoles in the planar triangle cancel, as do the two dipoles of the bonds arranged linearly, so the molecule is nonpolar overall. Similarly, we can look at the octahedral molecule as consisting of three linear sets of bond dipoles. Cancellation of bond dipoles occurs in each set, so overall the octahedral molecule is also nonpolar.

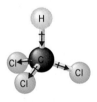

FIGURE 8.9

Bond dipoles in the chloroform molecule, $CHCl_3$. Because C is slightly more electronegative than H, the C—H bond dipole points toward the carbon. The small C—H bond dipole actually adds to the effects of the C—Cl bond dipoles. All the bond dipoles are additive and this causes $CHCl_3$ to be a polar molecule.

Polar Molecules

If all the atoms attached to the central atom are not the same, or if there are lone pairs in the valence shell of the central atom, the molecule is usually polar. For example, in $CHCl_3$, one of the atoms in the tetrahedral structure is different from the others. The C—H bond is less polar than the C—Cl bonds, and the bond dipoles do not cancel (Figure 8.9).

Two familiar molecules that have lone pairs in the valence shell of the central atom are shown in Figure 8.10. Here the bond dipoles are oriented in

FIGURE 8.10

When lone pairs occur on the central atom, the bond dipoles usually do not cancel, and polar molecules result.

Noncancellation gives

Noncancellation gives

such a way that their effects do not cancel. In water, for example, each bond dipole points partially in the same direction, toward the oxygen atom, and because of this the bond dipoles partially add to give a net dipole moment for the molecule. The same thing happens in ammonia, where three bond dipoles point partially in the same direction and add to give a polar NH_3 molecule.

Not every structure that contains lone pairs on the central atom produces polar molecules. The following are two exceptions.

The lone pairs also influence the polarity of a molecule, but we will not explore this any further here.

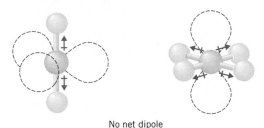

No net dipole

In the first case, we have a pair of bond dipoles arranged linearly, just as in CO_2. In the second, the bonded atoms lie at the corners of a square, which can be viewed as two linear sets of bond dipoles. If the atoms attached to the central atom are the same, cancellation of bond dipoles is bound to occur and produce nonpolar molecules. This means that molecules such as linear XeF_2 and square planar XeF_4 are nonpolar.

Based on the preceding discussions, let's see how we can use VSEPR theory to determine whether molecules are expected to be polar or nonpolar.

Do we expect the PCl_3 molecule to be polar or nonpolar?

ANALYSIS First, we need to know whether the bonds in the molecule are polar. If not, the molecule will be nonpolar regardless of its structure. If the bonds are polar, then we need to determine its structure. Based on the structure, we can then decide whether or not the bond dipoles cancel.

SOLUTION The electronegativities of the atoms (P = 2.1, Cl = 2.9) tell us that the individual P—Cl bonds will be polar. Therefore, we need to know the structure of the molecule to predict whether or not the molecule is polar. First we draw the Lewis structure following our usual procedure.

$$:\overset{\displaystyle ..}{\underset{\displaystyle ..}{Cl}}:$$
$$:\overset{\displaystyle ..}{\underset{\displaystyle ..}{Cl}}-P-\overset{\displaystyle ..}{\underset{\displaystyle ..}{Cl}}:$$

There are four electron pairs around the phosphorus, so according to VSEPR theory, they should be arranged tetrahedrally. This means that the PCl_3 molecule should have a trigonal pyramidal shape, as shown below.

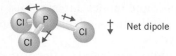

Net dipole

Because of the structure, the bond dipoles do not cancel, and we expect the molecule to be polar.

EXAMPLE 8.6
Predicting Molecular Polarity

EXAMPLE 8.7
Predicting Molecular Polarity

Would you expect the molecule SO_3 to be polar or nonpolar?

SOLUTION Oxygen is more electronegative then sulfur, so we expect the S—O bonds to be polar. Let's look at the molecular structure to see whether the bond dipoles cancel. The simple Lewis structure for SO_3 is

$$:\ddot{O}:$$
$$|$$
$$:\ddot{O}=S-\ddot{O}:$$

VSEPR theory tells us that the molecule should have a planar triangular shape. We saw in Figure 8.7 that such a molecule is nonpolar when all the attached atoms are the same, because the bond dipoles cancel.

EXAMPLE 8.8
Predicting Molecular Polarity

Would you expect the molecule HCN to be polar or nonpolar?

SOLUTION Once again we have polar bonds because carbon is slightly more electronegative than hydrogen and nitrogen is slightly more electronegative than carbon. The Lewis structure of HCN is

$$H-C\equiv N:$$

VSEPR theory predicts a linear shape, but the two bond dipoles do not cancel. One reason is that they are not of equal magnitude, which we know because the difference in electronegativity between C and H is 0.4, whereas the difference in electronegativity between C and N is 0.6. The other reason is because both bond dipoles point in the same direction, from the atom of low electronegativity to the one of high electronegativity. This is illustrated below.

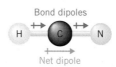

Notice that the bond dipoles add to give a net dipole moment for the molecule.

■ **Practice Exercise 4** Which of the following molecules would you expect to be polar? (a) SF_6, (b) SO_2, (c) $BrCl$, (d) AsH_3, (e) CF_2Cl_2.

8.4 ■
WAVE MECHANICS AND COVALENT BONDING: VALENCE BOND THEORY

So far, we have taken a very simple view of covalent bonding based primarily on the use of Lewis structures for electron bookkeeping. Lewis structures, however, tell us nothing about *why* covalent bonds are formed or *how* electrons manage to be shared between atoms. Nor does the VSEPR theory, as useful and accurate as it generally is at predicting molecular geometry, explain *how* the electron pairs in the valence shell of an atom manage to avoid each other. Thus, as helpful as these simple models are, we must look beyond them if we wish to understand more fully the covalent bond and the factors that determine molecular geometry.

Modern theories of bonding are based on the principles of wave mechanics. This is the theory, you recall, that gives us the electron configurations of atoms and the description of the shapes of atomic orbitals. When wave mechanics is applied to molecules, it considers how the orbitals of the atoms that come together to form a covalent bond interact with each other, and in so doing it attempts to explain in detail how atoms share electrons.

There are fundamentally two theories of covalent bonding that have evolved, and in many ways, they complement one another. They are called the **valence bond theory** (or **VB theory,** for short) and the **molecular orbital theory (MO theory).** They differ principally in the way they construct a theoretical model of the bonding in a molecule. The valence bond theory imagines individual atoms, each with its own orbitals and electrons, coming together to form the covalent bonds of the molecule. On the other hand, the molecular orbital theory doesn't concern itself with *how* the molecule is formed. It just views a molecule as a collection of positively charged nuclei surrounded in some way by electrons that occupy a set of *molecular orbitals,* in much the same way that the electrons in an atom occupy *atomic orbitals.* (In a sense, this theory would look at an atom as though it were a special case—a molecule having only one positive center, instead of many.)

In their simplest forms, the VB and MO theories appear to be rather different. However, it has been found that both theories can be extended and refined to give the same results. This is as it should be, of course, since they both attempt to explain the same set of facts—the structures and shapes of molecules, and the strengths of chemical bonds. We will examine both theories in an elementary way, but the greater emphasis will be on the VB theory because it is easier to understand.

Basic Postulates of the Valence Bond Theory

According to VB theory, *a bond between two atoms is formed when a pair of electrons with their spins paired is shared by two* **overlapping** *atomic orbitals, one orbital from each of the atoms joined by the bond.* By **overlap of orbitals** we mean that portions of two atomic orbitals from different atoms share the same space.

An important part of the theory is that only *one* pair of electrons, with their spins paired, can be shared by two overlapping orbitals. This electron pair becomes concentrated in the region of overlap and helps "cement" the nuclei together, so the amount that the potential energy is lowered when the bond is formed is determined in part by the extent to which the orbitals overlap. Therefore, atoms tend to position themselves so that the maximum amount of orbital overlap occurs because this yields the minimum potential energy and therefore the strongest bonds. As we will see, this is one of the major factors that controls the shapes of molecules.

The way VB theory views the formation of a hydrogen molecule is shown in Figure 8.11. As the two atoms approach each other, their 1s orbitals overlap, thereby giving the H—H bond. The description of the bond in this molecule provided by VB theory is essentially the same as that discussed in Section 7.3.

Now let's look at the HF molecule, which is a bit more complex than H_2. Following the usual rules we can write its Lewis structure as

$$H-\ddot{\underset{..}{F}}:$$

Remember, just two electrons with paired spins can be shared between two overlapping atomic orbitals.

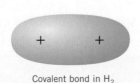

Separated H atoms   Overlapping of orbitals   Covalent bond in H_2

FIGURE 8.11

The formation of the hydrogen molecule according to valence bond theory.

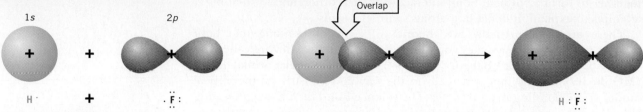

1s   2p   Overlap

H·   +   ·F̈:   H:F̈:

FIGURE 8.12

The formation of the hydrogen fluoride molecule according to valence bond theory. Only the half-filled 2p orbital of fluorine is shown.

and we can diagram the formation of the bond as

$$H\cdot + \cdot\ddot{F}: \longrightarrow H{-}\ddot{F}:$$

Our Lewis symbols suggest that the H—F bond is formed by the pairing of electrons, one from hydrogen and one from fluorine. To explain this according to VB theory, we must have two half-filled orbitals, one from each atom, that can be joined by overlap. (They must be half-filled, because we can't place more than two electrons into the bond.) To see clearly what must happen, it is best to look at the orbital diagrams of the valence shells of hydrogen and fluorine.

H   ⬆
        1s

F   ⬆⬇   ⬆⬇ ⬆⬇ ⬆
       2s   2p

The requirements for bond formation are met by overlapping the half-filled 1s orbital of hydrogen with the half-filled 2p orbital of fluorine; there are then two orbitals plus two electrons whose spins can adjust so they are paired. The formation of the bond is illustrated in Figure 8.12.

The overlap of orbitals provides a means for sharing electrons, thereby allowing each atom to complete its valence shell. It is sometimes convenient to indicate this using orbital diagrams. For example, the diagram below shows how the fluorine atom completes its 2p subshell by acquiring a share of an electron from hydrogen.

F (in HF)   ⬆⬇   ⬆⬇ ⬆⬇ ⬆⬇   (colored arrow is the H electron)
            2s   2p

Notice that both the Lewis and VB descriptions of the formation of the H—F bond account for the completion of the atoms' valence shells. Therefore, *a Lewis structure can be viewed, in a very qualitative sense, as a shorthand notation for the valence bond description of a molecule.*

Let's look now at a still more complex molecule, hydrogen sulfide, H_2S. Experiments have shown that this is a nonlinear molecule in which the H—S—H bond angle is about 92°.

H₂S is the compound that gives rotten eggs their foul odor.

S

H   H

92°

Valence bond theory explains this structure as follows.

The orbital diagram for the valence shell of sulfur is

S   ⬆⬇   ⬆⬇ ⬆ ⬆
      3s   3p

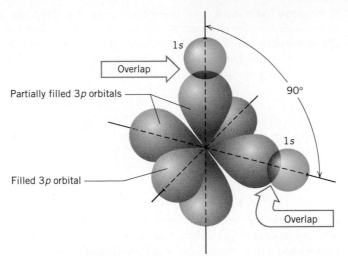

FIGURE 8.13

Bonding in H₂S. The hydrogen $1s$ orbitals must position themselves so that they can best overlap with the two partially filled $3p$ orbitals of sulfur.

Sulfur has two p orbitals that each contain only one electron. Each of these can overlap with the $1s$ orbital of a hydrogen atom, as shown in Figure 8.13. This overlap completes the $2p$ subshell of sulfur because each hydrogen provides one electron.

S (in H₂S) (colored arrows are H electrons)

$3s$ $3p$

In Figure 8.13, notice that when the $1s$ orbital of a hydrogen atom overlaps with a p orbital of sulfur, the best overlap occurs when the hydrogen atom lies along the axes of the p orbital. Because p orbitals are oriented at 90° angles to each other, the H—S bonds are expected to be at this angle, too. Therefore, the predicted bond angle is 90°. This is very close to the actual bond angle of 92° found by experiment. Thus, the VB theory requirement for maximum overlap quite nicely explains the geometry of the hydrogen sulfide molecule.

The overlap of p orbitals with each other is also possible. For example, according to VB theory the bonding in the fluorine molecule, F₂, occurs by the overlap of two $2p$ orbitals, as shown in Figure 8.14. The formation of the other diatomic molecules of the halogens, all of which are held together by single bonds, could be similarly described.

Two orbitals from different atoms never overlap simultaneously with opposite ends of the same p orbital.

■ **Practice Exercise 5** Use the principles of VB theory to explain the bonding in HCl. Give the orbital diagram for chlorine in the HCl molecule and indicate the orbital that shares its electron with one from hydrogen. Sketch the orbital overlap that gives rise to the H—Cl bond.

FIGURE 8.14

Bonding in the fluorine molecule according to valence bond theory. The two completely filled p orbitals on each fluorine atom are omitted, for clarity.

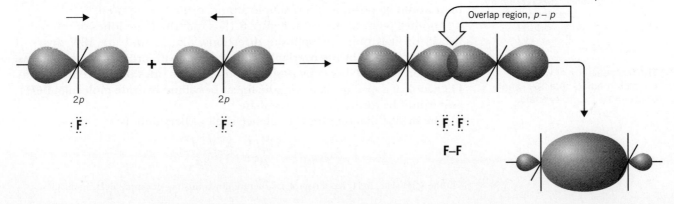

■ **Practice Exercise 6** The phosphine molecule, PH_3, has a trigonal pyramidal shape with H—P—H bond angles equal to 93.7°. Give the orbital diagram for phosphorus in the PH_3 molecule and indicate the orbitals that share electrons with those from hydrogen. On a set of *xyz* coordinate axes, sketch the orbital overlaps that give rise to the P—H bonds.

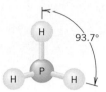

8.5
HYBRID ORBITALS

The approach that we have taken so far has worked well with some simple molecules. Their shapes are explained quite nicely by the overlap of simple atomic orbitals. However, there are many molecules with shapes and bond angles that fail to fit the model that we have developed. For example, methane, CH_4, has a shape that the VSEPR theory predicts (correctly) to be tetrahedral. The H—C—H bond angles in this molecule are 109.5°. But no simple atomic orbitals are oriented at this angle with respect to each other. Therefore, to explain the bonds in molecules such as CH_4 we must study the way atomic orbitals *of the same atom* can interact with each other when bonds are formed.

Formation of *sp*, *sp²*, and *sp³* Hybrid Orbitals

Theoreticians tell us that when atoms form bonds, their simple atomic orbitals often mix to form new orbitals that we call **hybrid orbitals.** These new orbitals have new shapes and new directional properties. The reason this mixing occurs can be seen if we look at the shapes of hybrid orbitals.

One kind of hybrid atomic orbital is formed by mixing an *s* orbital and a *p* orbital. This creates *two* new orbitals called *sp* **hybrid orbitals.** The label *sp* identifies the kinds of pure atomic orbitals from which the hybrids are formed, in this case, one *s* and one *p* orbital. The shapes and directional properties of the *sp* hybrid orbitals are illustrated in Figure 8.15. Notice that each of the hybrid orbitals has the same shape; each has one large lobe and another much smaller one. The large lobe extends farther from the nucleus than either the *s* or the *p* orbital from which the hybrid was formed. This allows the hybrid orbital to overlap more effectively with an orbital on another atom when a bond is formed. Therefore, hybrid orbitals form stronger, more stable bonds than would be possible if just simple atomic orbitals were used.

Another point to notice in Figure 8.15 is that the large lobes of the two *sp* hybrid orbitals point in opposite directions; that is, they are 180° apart. If bonds are formed by the overlap of these hybrids with orbitals of other atoms, the other atoms will occupy positions on opposite sides of this central atom. Let's look at a specific example, the linear beryllium hydride molecule, BeH_2, as it would be formed in the gas phase.[1]

The orbital diagram for the valence shell of beryllium is

In general, the greater the overlap of two orbitals, the stronger the bond. At a given internuclear distance, the greater "reach" of a hybrid orbital gives better overlap than either an *s* or *p* orbital.

The mathematics of wave mechanics predicts this 180° angle between *sp* hybrid orbitals.

$$\text{Be} \quad \underset{2s}{\text{Ⓝ}} \quad \underset{2p}{\text{◯◯◯}}$$

[1] In the solid state, BeH_2 has a complex structure not consisting of simple BeH_2 molecules.

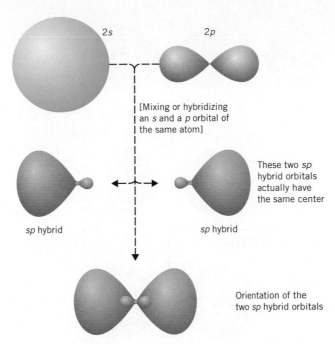

FIGURE 8.15

The formation of *sp* hybrid orbitals.

2s

2p

[Mixing or hybridizing an *s* and a *p* orbital of the same atom]

These two *sp* hybrid orbitals actually have the same center

sp hybrid

sp hybrid

Orientation of the two *sp* hybrid orbitals

Notice that the $2s$ orbital is filled and the three $2p$ orbitals are empty. For bonds to form at a $180°$ angle between beryllium and the two hydrogen atoms, two conditions must be met: (1) the two orbitals that beryllium uses to form the Be—H bonds must point in opposite directions, and (2) each of the beryllium orbitals must contain only one electron. The reason for the first requirement is obvious; the overlap of Be and H orbitals must give the correct experimentally measured bond angle. The reason for the second is that each bond must contain just two electrons. Since a hydrogen $1s$ orbital supplies one electron, the beryllium orbital with which it overlaps must also contain one electron. A filled Be orbital won't do, because overlap of a half-filled hydrogen $1s$ with a filled orbital on Be would produce a "bond" with three electrons in it—a situation that's forbidden by VB theory. An empty Be orbital won't do either, because overlap with a hydrogen $1s$ orbital would give a bond with only one electron in it. Although such a bond isn't forbidden, it would be much weaker than a bond with two electrons, so electron pair bonds are definitely preferred. The net effect of all this is that when the Be—H bonds form, the electrons of the beryllium atom become unpaired, and the resulting half-filled s and p atomic orbitals become hybridized.

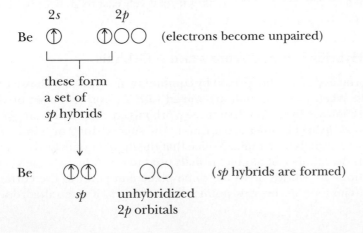

2s   2p

Be   (electrons become unpaired)

these form
a set of
sp hybrids

Be   (*sp* hybrids are formed)

sp   unhybridized
2p orbitals

These *sp* hybrid orbitals are said to be *equivalent* because they have the same size, shape, and energy.

FIGURE 8.16

Bonding in BeH₂ according to valence bond theory. Only the larger lobe of each *sp* hybrid orbital is shown.

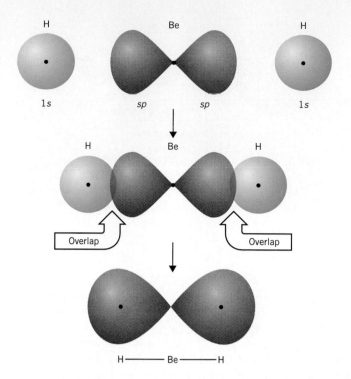

Now the 1*s* orbitals of the hydrogen atoms can overlap with the *sp* hybrids of beryllium as shown in Figure 8.16. Because the two *sp* hybrid orbitals of beryllium are identical in shape, the two Be—H bonds are identical except for the directions in which they point and we say that the bonds are *equivalent*. Since the bonds point in opposite directions, the linear geometry of the molecule is also explained. The orbital diagram for beryllium in this molecule is

Be (in BeH₂) ⟨↿⇂⟩⟨↿⇂⟩   ○○   (colored arrows are H electrons)

          sp   unhybridized
              2*p* orbitals

Even if we had not known the shape of the BeH₂ molecule, we could have obtained the same bonding picture by applying the VSEPR theory first. The Lewis structure for BeH₂ is H:Be:H, and VSEPR theory predicts that the molecule is linear. Once the shape is known, we can apply the VB theory to explain the bonding in terms of orbital overlaps. Thus, the VB and VSEPR theories complement each other well. VSEPR theory allows us to predict geometry in a simple way, and this makes it relatively easy to analyze the bonding in terms of VB theory.

Other Hybrids Formed from *s* and *p* Orbitals

Hybrid orbitals can also be formed by combining an *s* orbital with two or three *p* orbitals. When two *p* orbitals are mixed with an *s* orbital, a set of *three **sp²** hybrid orbitals* are formed. When three *p* orbitals are mixed with an *s* orbital, a set of *four **sp³** hybrid orbitals* are formed. The superscript 2 or 3 indicates the number of *p* orbitals in the mix. Notice that the number of hybrid orbitals in a set equals the number of atomic orbitals used to form them.

The three *sp²* hybrids lie in the same plane and point to the corners of a triangle. The four *sp³* hybrids point to the corners of a tetrahedron. Their

There is a conservation of orbitals during hybrid orbital formation. The number of hybrid orbitals of a given kind is always equal to the number of atomic orbitals that are mixed.

180°

two *sp* hybrids

linear

three *sp²* hybrids
(all orbitals are in the
plane of the paper)

120°

all angles = 120°   *planar triangular*

four *sp³* hybrids

109.5°

all angles = 109.5°   *tetrahedral*

directional properties are summarized in Figure 8.17. The following examples illustrate how they can be used to explain the bonding in molecules.

The BCl_3 molecule has a planar triangular shape. How is this explained in terms of valence bond theory?

EXAMPLE 8.9
Explaining Bonding with Hybrid Orbitals

ANALYSIS Since we know the shape of the molecule, we can anticipate the kinds of hybrid orbitals used to form the bonds. A planar triangular shape is consistent with sp^2 hybridization in the valence shell of the central atom, so we proceed from that point to determine whether the electronic structures of the atoms involved fit with the use of these kinds of orbitals.

SOLUTION First, let's examine the orbital diagram for the valence shell of boron.

B   ⊕   ⊕○○
    2*s*   2*p*

To form the three bonds to chlorine, boron needs three half-filled orbitals. These can be obtained by unpairing the electrons in the 2*s* orbital and placing one of them in the 2*p*. Then the 2*s* orbital and two of the 2*p* orbitals can be combined to give the set of sp^2 hybrids.

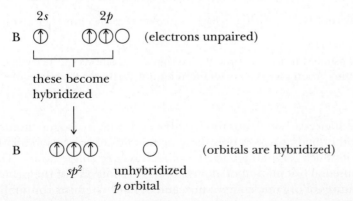

2*s*   2*p*

B   ⊕   ⊕⊕○   (electrons unpaired)

these become
hybridized

B   ⊕⊕⊕   ○   (orbitals are hybridized)
    sp²   unhybridized
            p orbital

Now let's look at the valence shell of chlorine.

Cl   ⊕   ⊕⊕⊕
      3s   3p

We see that the half-filled $3p$ orbital of a chlorine atom can overlap with a hybrid sp^2 orbital of boron to give a B—Cl bond.

B (in BCl₃)   ⊕⊕⊕   ◯   (colored arrows
               sp^2 unhybridized   are Cl electrons)
                        p orbital

Figure 8.18 illustrates the overlap of the orbitals to give the bonding in the molecule. Notice that the arrangement of atoms yields a planar triangular molecule, which is in agreement with the structure predicted by the VSEPR theory.

EXAMPLE 8.10
Explaining Bonding with Hybrid Orbitals

Methane, CH_4, is a tetrahedral molecule. How is this explained in terms of valence bond theory?

SOLUTION The tetrahedral structure of the molecule suggests that sp^3 hybrid orbitals are involved in bonding. Let's examine the valence shell of carbon.

C   ⊕   ⊕⊕◯
     2s   2p

To form four C—H bonds, we need four half-filled orbitals. Unpairing the electrons in the $2s$ and moving one to the vacant $2p$ orbital satisfies this requirement. Then we can hybridize all the orbitals to give the desired sp^3 set.

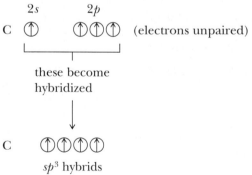

                2s   2p
C   ⊕   ⊕⊕⊕   (electrons unpaired)

                these become
                hybridized

                    ↓

C   ⊕⊕⊕⊕
          sp^3 hybrids

Then we form the four bonds to hydrogen 1s orbitals.

C (in CH₄)   ⊕⊕⊕⊕   (colored arrows are H electrons)
               sp^3

This is illustrated in the margin. Notice that the positions of the hydrogen atoms around the carbon give the correct tetrahedral shape for the molecule.

In methane, carbon forms four single bonds with hydrogen atoms by using sp^3 hybrid orbitals. Carbon uses these same kinds of orbitals in all of its compounds in which it is bonded to four other atoms by single bonds. This makes the tetrahedral orientation of atoms around carbon one of the primary structural features of organic compounds, and organic chemists routinely think in terms of "tetrahedral carbon," shown in Figure 8.19.

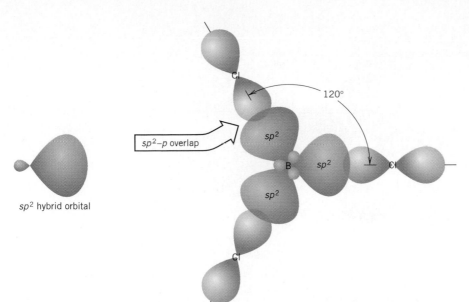

FIGURE 8.18

Valence bond description of the bonding in BCl_3. Only the half-filled p orbital of each chlorine is shown.

In the alkane hydrocarbons, carbon atoms are bonded to other carbon atoms. An example is ethane, C_2H_6.

$$
\begin{array}{ccc}
 & H & H \\
 & | & | \\
H - & C - C & - H \\
 & | & | \\
 & H & H \\
\end{array}
$$

In this molecule, the carbons are bonded together by the overlap of sp^3 hybrid orbitals (Figure 8.20). One of the most important characteristics of this bond is that the overlap of the orbitals in the C—C bond is hardly affected at all if one portion of the molecule rotates relative to the other around the bond axis. Such rotation, therefore, is said to occur freely and permits different possible relative orientations of the atoms in the molecule. These different relative orientations are called **conformations.** With complex molecules, the number of possible conformations is enormous. For example, Figure 8.21 illustrates three of the enormous number of possible conformations of the pentane molecule, C_5H_{12}, one of the low molecular weight organic compounds in gasoline.

$$
\begin{array}{ccccc}
H & H & H & H & H \\
| & | & | & | & | \\
H-C-&C-&C-&C-&C-H \\
| & | & | & | & | \\
H & H & H & H & H \\
\end{array}
$$

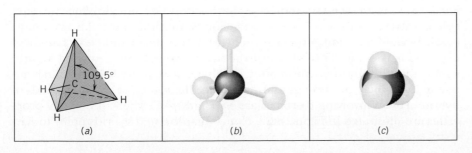

FIGURE 8.19

The "tetrahedral carbon." (*a*) The heavy lines are the axes of the bonds. (*b*) A ball-and-stick model of the CH_4 molecule. (*c*) A scale model of CH_4 that indicates the relative volumes occupied by the electron clouds.

FIGURE 8.20

The bonds in the ethane molecule. (*a*) Overlap of orbitals. (*b*) The degree of overlap of the sp^3 orbitals in the carbon–carbon bond is not appreciably affected by the rotation of the two CH_3 groups relative to each other around the bond.

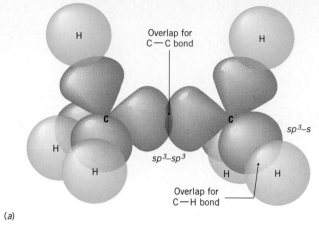

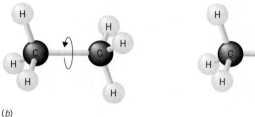

(*b*)

FIGURE 8.21

Three of the many conformations of the atoms in the pentane molecule, C_5H_{12}. Free rotation around single bonds makes these different conformations possible.

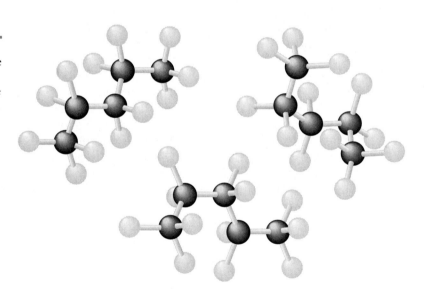

Hybridization When the Central Atom Has More Than an Octet

Earlier we saw that certain molecules have atoms that must violate the octet rule because they form more than four bonds. In these cases, the atom must reach beyond its *s* and *p* valence shell orbitals to form sufficient half-filled orbitals for bonding. This is because the *s* and *p* orbitals can be mixed to form a maximum of only four hybrid orbitals. When five or more hybrid orbitals are needed, *d* orbitals are brought into the mix. The two most common kinds of hybrid orbitals involving *d* orbitals are sp^3d and sp^3d^2. Their directional properties are illustrated in Figure 8.22. Notice that the sp^3d hybrids point toward

FIGURE 8.22

Orientations of hybrid orbitals that involve d atomic orbitals. (*a*) sp^3d hybrid orbitals. (*b*) sp^3d^2 hybrid orbitals.

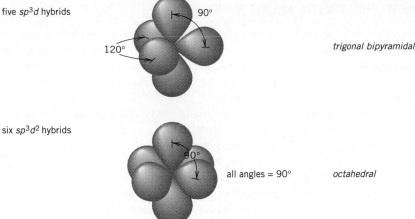

five sp^3d hybrids

90°

120°

trigonal bipyramidal

six sp^3d^2 hybrids

90°

all angles = 90°   octahedral

the corners of a trigonal bipyramid and the sp^3d^2 hybrids point toward the corners of an octahedron.

The sulfur hexafluoride molecule has an octahedral shape. Describe the bonding in this molecule in terms of valence bond theory.

SOLUTION As before, let's examine the valence shell of sulfur.

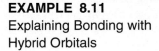

S ⇅ ⇅↑↑
  3s   3p

To form six bonds to fluorine atoms we need six half-filled orbitals, but we see only four orbitals all together. The structure of the molecule, however, gives us a clue to where we can obtain the others. An octahedral structure suggests the use of sp^3d^2 hybrid orbitals, so we need to find some d orbitals to form the hybrids.

An isolated sulfur atom has electrons only in its $3s$ and $3p$ subshells, so these are the only ones we usually show in the orbital diagram. But the third shell also has a d subshell, which is empty in a sulfur atom. Therefore, let's rewrite the orbital diagram to show the vacant $3d$ subshell.

S ⇅ ⇅↑↑ ○○○○○
  3s   3p   3d

Unpairing all of the electrons to give six half-filled orbitals, followed by hybridization, gives the required set of half-filled sp^3d^2 orbitals.

    3s   3p   3d
S ↑ ↑↑↑ ↑↑○○○ (electrons unpaired)
  └──────────────┘
      these become
      hybridized
          ↓
S ↑↑↑↑↑↑   ○○○ (sp^3d^2 hybrids formed)
   sp^3d^2 unhybridized
              $3d$ orbitals

EXAMPLE 8.11
Explaining Bonding with Hybrid Orbitals

sp^3d^2 hybrid orbitals of sulfur in SF_6

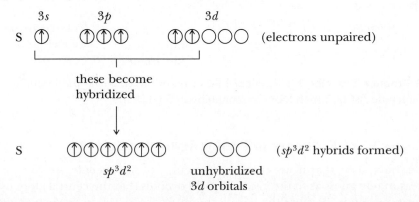

SPECIAL TOPIC 8.1 / WHAT'S THE TRUTH ABOUT WATER?

Over the years there's been quite a bit of controversy about the nature of the bonding in the water molecule. Specifically, the question has been: What kind of orbitals does oxygen really use in forming its bonds to the hydrogen atoms?

In our discussion of this molecule, we've given what has become a standard explanation for the H—O—H bond angle, the use by oxygen of sp^3 hybrid orbitals to overlap with the $1s$ orbitals of the hydrogen atoms. Based on bond angle data alone, this is plausible because the angles between these kinds of hybrids is quite close to the bond angle in H_2O. It also fits with the general rule of thumb that we've followed in the remainder of the discussion of VB theory—that atoms generally use hybrid orbitals to form bonds to neighboring atoms. By following this rule, we obtain reasonable explanations of molecular structure. Nevertheless, as useful as this rule may be for other molecules, there is considerable evidence that it doesn't really work well for water.

In our discussion of hybridization, we've ignored the fact that forming hybrid orbitals requires energy, which is usually more than paid back by the formation of strong covalent bonds. In oxygen and other period-2 elements, however, it is especially "expensive" to form the hybrids, because the energy of the $2s$ orbital is so much lower than the energy of the $2p$ orbitals. This makes it less attractive for these elements to use hybrid orbitals if unhybridized orbitals can also do the job, and this does appear to be the case with water, based on detailed calculations and some experimental evidence that are beyond the scope of this discussion. But, if oxygen uses its pure $2p$ orbitals to form the O—H bonds, how come the bond angle isn't close to 90°, as it is in H_2S (page 327) where sulfur uses its $3p$ orbitals to form the S—H bonds?

To explain the spreading of the H—O—H angle in water from 90° to 104°, we have to consider the small size of the oxygen atom. As you can see in the figure below, if the bond angle were 90°, the two hydrogen atoms would interpenetrate each other, which would cause their completed valence shells to overlap. This can't happen, however, because two pairs of electrons can never occupy the same space, so the angle spreads to relieve this problem. In H_2S, the bond angle doesn't have to increase appreciably from 90° because the sulfur atom is larger and the hydrogens don't interfere with each other.

The point of this discussion is that often there are alternative theoretical explanations of the same phenomenon (in this instance, the bonding in water). Some are simple and others are more complex. Which one we use depends on how precise we must be. The VSEPR theory, for example, is very simple, but it doesn't attempt to explain bonding in terms of the atomic orbitals involved. The model we've provided in this chapter using hybrid orbitals also works well most of the time, so we find it useful. But even this model is not completely accurate all the time, so when it really matters, many factors have to be weighed in deciding what is as close to the "truth" as possible.

  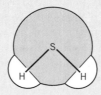

(*Left*) If the bond angle in H_2O were 90°, the H atoms would overlap severely. (*Center*) At a bond angle of 104°, the hydrogens don't interfere with each other very much. (*Right*) Because S is larger than O, the bond angle in H_2S can be much closer to 90°.

Finally, the six S—F bonds are formed by overlap of the half-filled $2p$ orbitals of fluorine with these half-filled sp^3d^2 hybrids.

S (in SF$_6$)   ⊕⊕⊕⊕⊕⊕   ○○○   (colored arrows
               sp^3d^2 unhybridized   are F electrons)
                                   $3d$ orbitals

■ **Practice Exercise 7** Use valence bond theory to describe the bonding in the molecule AsCl$_5$, which has a trigonal bipyramidal shape.

Using VSEPR Theory to Predict Hybridization

We have seen that if we know the structure of a molecule, we can make a reasonable guess as to the kind of hybrid orbitals that the central atom uses to form its bonds. Because the VSEPR theory works so well in predicting geometry, we can use it to help us obtain VB descriptions of bonding. For example, the Lewis structures of CH_4 and SF_6 are

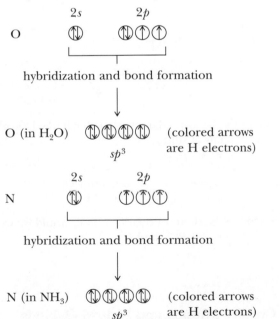

In CH_4, there are four electron pairs around carbon. The VSEPR model tells us that they should be arranged tetrahedrally. The only hybrid orbitals that are tetrahedral are sp^3 hybrids, and we have seen that they explain the structure of this molecule well. Similarly, VSEPR theory tells us that the six electron pairs around sulfur should be arranged octahedrally. The only octahedrally oriented hybrids are sp^3d^2, so the sulfur in the SF_6 molecule must use these hybrids.

■ **Practice Exercise 8** What kind of hybrid orbitals are expected to be used by the central atom in (a) SiH_4 and (b) PCl_5?

Hybridization in Molecules That Have Lone Pairs of Electrons

Methane is a tetrahedral molecule with sp^3 hybridization of the orbitals of carbon and H—C—H bond angles that are each equal to 109.5°. In ammonia, NH_3, the H—N—H bond angles are 107°, and in water the H—O—H bond angle is 104.5°. Both NH_3 and H_2O have H—X—H bond angles that are close to the bond angles expected for a molecule whose central atom has sp^3 hybrids. The use of sp^3 hybrids by oxygen and nitrogen, therefore, is often used to explain the geometry of H_2O and NH_3.

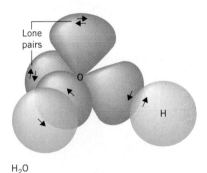

H_2O

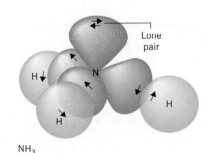

NH_3

According to these descriptions, not all of the hybrid orbitals of the central atom must be used for bonding. Lone pairs of electrons can be accommodated in them too. In fact, there is evidence to suggest that the lone pair on the nitrogen in ammonia does in fact reside in an sp^3 hybrid orbital. (The case is less convincing for water, however, as noted in Special Topic 8.1.)

EXAMPLE 8.12
Explaining Bonding with
Hybrid Orbitals

Use valence bond theory to explain the bonding in the SF_4 molecule.

SOLUTION Let's begin by constructing the Lewis formula for the molecule. Following the usual procedure, we obtain

$$\overset{\displaystyle :\!\ddot{F}\!: \quad :\!\ddot{F}\!:}{\underset{\displaystyle :\!\ddot{F}\!-\!S\!-\!\ddot{F}\!:}{\diagdown \diagup}}$$

The VSEPR theory predicts that the electron pairs around the sulfur should be in a trigonal bipyramidal arrangement, and the only hybrids that fit this geometry are sp^3d. To see how they are formed, we look at the valence shell of sulfur, including the vacant $3d$ subshell.

S   ⊕   ⊕ ↑ ↑   ○ ○ ○ ○ ○
    3s   3p   3d

To form the four bonds to fluorine atoms, we need four half-filled orbitals, so we unpair the electrons in one of the filled orbitals. This gives

        3s   3p   3d
S   ⊕   ↑ ↑ ↑   ↑ ○ ○ ○ ○

Next, we form the hybrid orbitals. In doing this, we use all the valence shell orbitals that have electrons in them.

         3s   3p   3d
S   ⊕   ↑ ↑ ↑   ↑ ○ ○ ○ ○
        └──────────┬──────────┘
              these become
               hybridized
                   │
                   ↓

S   ⊕ ↑ ↑ ↑ ↑   ○ ○ ○ ○
            sp^3d unhybridized
                           $3d$ orbitals

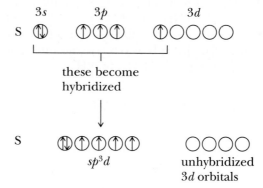

Lone
pair

The structure of SF_4.

Now four S—F bonds can be formed by overlap of half-filled $2p$ orbitals of fluorine with the sp^3d hybrid orbitals of sulfur.

S (in SF_4)   ⊕ ⊕ ⊕ ⊕ ⊕   ○ ○ ○ ○   (colored arrows
                  sp^3d unhybridized   are F electrons)
                                 $3d$ orbitals

■ **Practice Exercise 9** What kind of hybrid orbitals would we expect the central atom to use for bonding in (a) PCl_3 and (b) ClF_3?

Coordinate Covalent Bonds and Hybrid Orbitals

In Section 7.9 we defined a coordinate covalent bond as one in which both of the shared electrons are provided by just one of the joined atoms. For example, boron trifluoride, BF_3, can combine with an additional fluoride ion to form the tetrafluoroborate ion, BF_4^-, according to the equation

$$BF_3 + F^- \longrightarrow BF_4^-$$

tetrafluoroborate
ion

We can diagram this reaction as follows:

As we mentioned previously, the coordinate covalent bond is really no different from any other covalent bond once it has been formed. The distinction between them is made *only* for bookkeeping purposes. One place where such bookkeeping is useful is in keeping track of the orbitals and electrons used when atoms bond together.

The VB theory requirements for bond formation—two overlapping orbitals sharing two electrons—can be satisfied in two ways. One, as we have seen, is by the overlapping of two half-filled orbitals. This gives an "ordinary" covalent bond. The other is the overlapping of one filled orbital with one empty orbital. The shared pair of electrons is donated by the atom with the filled orbital and a coordinate covalent bond is formed.

The structure of the BF_4^- ion can therefore be explained as follows. First we examine the orbital diagram for boron.

To form four bonds, we need four hybrid orbitals. Since VSEPR theory predicts that the ion will have a tetrahedral shape, the boron will use sp^3 hybrids. Notice we spread the electrons out over the hybrid orbitals as much as possible.

Boron forms three ordinary covalent bonds with fluorine atoms plus one coordinate covalent bond with a fluoride ion.

Coordinate covalent bond—both electrons are from F^- (colored arrows are F electrons).

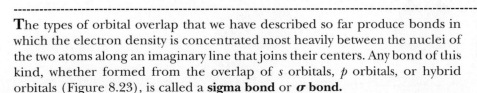

■ **Practice Exercise 10** What hybrid orbitals are used by phosphorus in PCl_6^-? Draw the orbital diagram for phosphorus in PCl_6^-. What is the shape of the PCl_6^- ion?

The types of orbital overlap that we have described so far produce bonds in which the electron density is concentrated most heavily between the nuclei of the two atoms along an imaginary line that joins their centers. Any bond of this kind, whether formed from the overlap of *s* orbitals, *p* orbitals, or hybrid orbitals (Figure 8.23), is called a **sigma bond** or **σ bond**.

8.6

DOUBLE AND TRIPLE BONDS

FIGURE 8.23

Sigma bonds. (*a*) From the overlap of *s* orbitals. (*b*) From the end-to-end overlap of *p* orbitals. (*c*) From the overlap of hybrid orbitals.

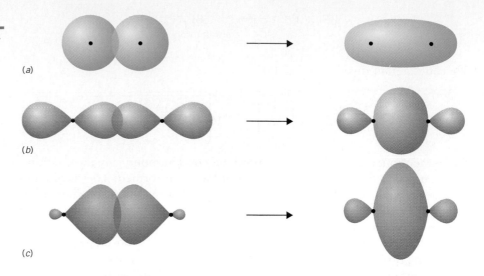

Another way that *p* orbitals can overlap is shown in Figure 8.24. This produces a bond in which the electron density is divided between two separate regions that lie on opposite sides of an imaginary line joining the two nuclei. This kind of bond is called a **pi bond** (or **π bond**). Notice that a π bond, like a *p* orbital, consists of two parts, and each part makes up just half of the π bond; it takes *both* of them to equal *one* π bond.

The formation of π bonds allows atoms to form double and triple bonds. To see how this occurs, let's look at the bonding in the compound ethene, C_2H_4, which has the Lewis structure

$$
\begin{array}{c}
\text{H} \qquad\qquad \text{H} \\
\diagdown \qquad\qquad \diagup \\
\text{C} = \text{C} \\
\diagup \qquad\qquad \diagdown \\
\text{H} \qquad\qquad \text{H}
\end{array}
$$

ethene

Ethene is also called ethylene. Polyethylene, a common plastic, is made from C_2H_4.

The molecule is planar and each carbon atom lies in the center of a triangle surrounded by three other atoms (two H and one C atom). As we've seen, this structure suggests that carbon uses sp^2 hybrid orbitals to form its bonds. Therefore, let's look at the distribution of electrons among the orbitals that carbon has available in its valence shell, assuming sp^2 hybridization.

FIGURE 8.24

Formation of a π bond. Two *p* orbitals overlap sideways instead of end to end. The electron density is concentrated in two regions on opposite sides of the bond axis.

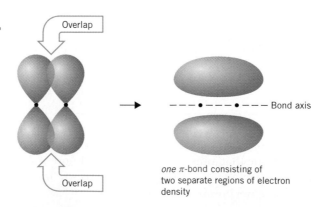

Overlap

Overlap

Bond axis

one π-bond consisting of two separate regions of electron density

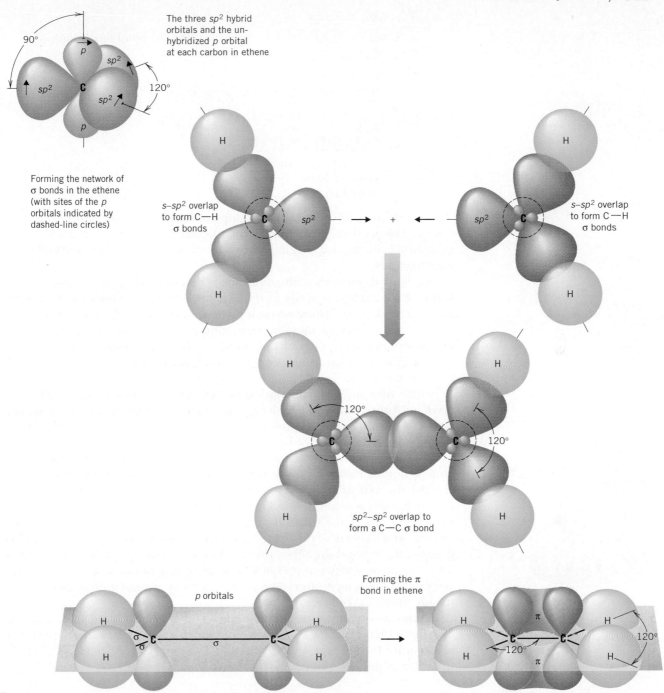

The three sp^2 hybrid orbitals and the un-hybridized p orbital at each carbon in ethene

90°

p

sp^2

sp^2

sp^2

p

120°

Forming the network of σ bonds in the ethene (with sites of the p orbitals indicated by dashed-line circles)

$s–sp^2$ overlap to form C—H σ bonds

sp^2

H

H

sp^2

$s–sp^2$ overlap to form C—H σ bonds

H

H

120°

120°

$sp^2–sp^2$ overlap to form a C—C σ bond

H

H

H

H

p orbitals

Forming the π bond in ethene

H

H

σ

σ

C

σ

C

H

H

H

H

C

C

H

H

π

π

120°

120°

FIGURE 8.25

The carbon–carbon double bond. (Adapted from J. R. Holum, *Organic and Biological Chemistry,* 2nd edition, 1986, John Wiley & Sons, New York.)

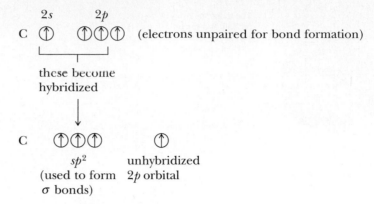

Notice that the carbon atom has an unpaired electron in an unhybridized $2p$ orbital. This p orbital is oriented perpendicular to the triangular plane of the sp^2 hybrid orbitals, as shown in Figure 8.25. Now we can see how the molecule goes together.

The basic framework of the molecule is determined by the formation of σ bonds. Each carbon uses two of its sp^2 hybrids to form σ bonds to hydrogen atoms. The third sp^2 hybrid on each carbon is used to form a σ bond between the two carbon atoms, thereby accounting for one of the two bonds of the double bond. Finally, the remaining unhybridized $2p$ orbitals, one from each carbon atom, overlap to produce a π bond, which accounts for the second bond of the double bond.

Notice how well this description of bonding fits with the observed (and predicted) structure of the molecule. The use of sp^2 hybrids by carbon makes available the unpaired electrons in the unhybridized $2p$ orbitals, which in turn makes it possible for the carbon atoms to form the extra bond between them.

This explanation also accounts for one of the most important properties of double bonds: rotation of one portion of the molecule relative to the rest around the axis of the double bond occurs only with great difficulty. The reason for this is illustrated in Figure 8.26. We see that as one CH_2 group is rotated relative to the other around the carbon–carbon bond, the unhybridized p orbitals become misaligned and can no longer overlap effectively. This destroys the π bond. In effect, then, rotation around a double bond involves bond breaking, which requires more energy than is normally available to molecules at room temperature. As a result, rotation around a double-bond axis usually doesn't take place.

In almost every instance, a double bond consists of a σ bond and a π bond. Another example is the compound formaldehyde (the substance used as a preservative for biological specimens and as an embalming fluid). The Lewis structure of this compound is

Restricted rotation around double bonds has important consequences in organic chemistry.

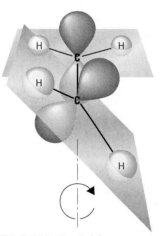

FIGURE 8.26

Restricted rotation around a double bond. Rotation of the CH_2 group closest to us relative to the one at the rear causes the unhybridized p orbitals to become misaligned, thereby destroying the π bond.

$$\begin{array}{c} H \\ \diagdown \\ C{=}\ddot{\overset{..}{O}} \\ \diagup \\ H \end{array}$$

formaldehyde

As with ethene, the carbon forms sp^2 hybrids, leaving an unpaired electron in an unhybridized p orbital.

The oxygen can also form sp^2 hybrids, with electron pairs in two of them and an unpaired electron in the third. This means that the remaining unhybridized p orbital also has an unpaired electron.

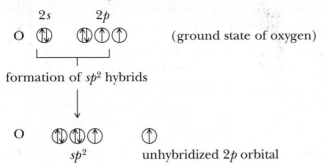

(ground state of oxygen)

formation of sp^2 hybrids

O sp^2 unhybridized $2p$ orbital

Figure 8.27 shows how the carbon, hydrogen, and oxygen atoms come together to form the molecule. As before, the basic framework of the molecule is

FIGURE 8.27

Bonding in formaldehyde. The carbon–oxygen double bond consists of a σ bond and a π bond.

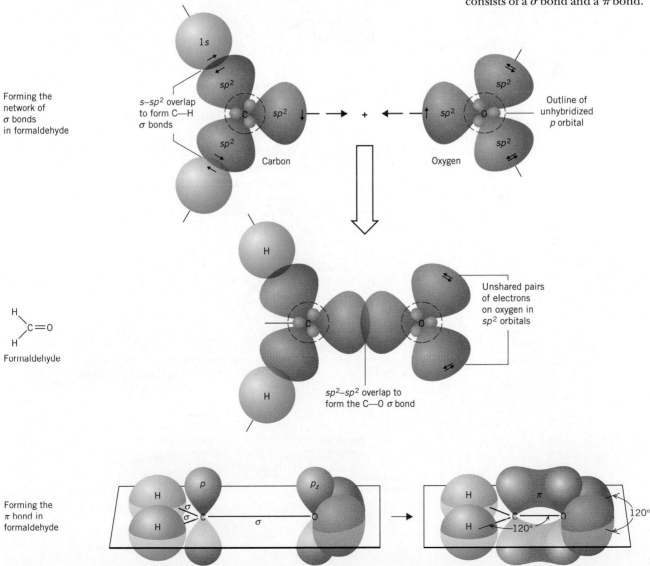

Forming the network of σ bonds in formaldehyde

Formaldehyde

Forming the π bond in formaldehyde

The two filled *sp²* hybrids on the oxygen become lone pairs on the oxygen atom in the molecule.

formed by the σ bonds. These determine the molecular shape. The carbon–oxygen double bond also contains a π bond formed by the overlap of the unhybridized *p* orbitals.

Now let's look at a molecule containing a triple bond. An example is acetylene, C_2H_2 (a gas used as a fuel for welding torches).

$$H—C≡C—H$$
acetylene

In the linear acetylene molecule, each carbon needs two hybrid orbitals to form two σ bonds—one to a hydrogen atom and one to the other carbon atom. These can be provided by mixing the 2s and one of the 2p orbitals to form *sp* hybrids. To help us visualize the bonding, we will imagine that there is an *xyz* coordinate system centered at each carbon and that it is the $2p_x$ orbital that becomes mixed in the hybrid orbitals.

We label the orbitals p_x, p_y, and p_z just for convenience; they are really all equivalent.

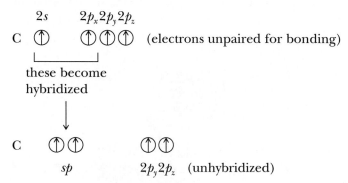

Figure 8.28 shows how the molecule is formed. The *sp* orbitals point in opposite directions and are used to form the σ bonds. The unhybridized $2p_y$ and $2p_z$ orbitals are perpendicular to the C—C bond axis and overlap sideways to form

FIGURE 8.28

The carbon–carbon triple bond in acetylene.

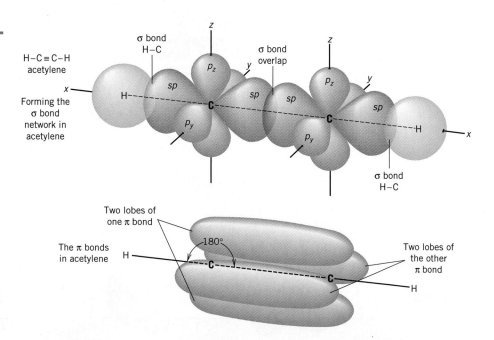

FIGURE 8.29

The triple bond in nitrogen, N_2.

:N≡N:

Nitrogen

The σ bond in N_2 is made by the overlap of two *sp* hybrid orbitals. Each nitrogen has an unshared pair of electrons in its other *sp* orbital.

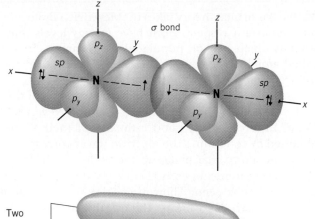

The π bonds in N_2

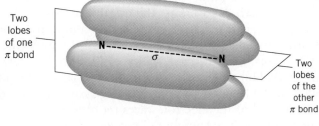

Two lobes of one π bond

Two lobes of the other π bond

two separate π bonds that surround the C—C σ bond. Notice that we now have three pairs of electrons in three bonds—one σ bond and two π bonds—whose electron densities are concentrated in different places. The three electron pairs therefore manage to avoid each other as much as possible. Also notice that the use of *sp* hybrid orbitals for the σ bonds allows us to explain the linear arrangement of atoms in the molecule.

Similar descriptions can be used to explain the bonding in other molecules that have triple bonds. Figure 8.29, for example, shows how the nitrogen molecule, N_2, is formed. In it, too, the triple bond is composed of one σ bond and two π bonds.

A Brief Summary

On the basis of the preceding discussion, we can make some observations that are helpful in applying the valence bond theory to a variety of molecules.

1. The basic molecular framework of a molecule is determined by the arrangement of its σ bonds.

2. Hybrid orbitals are used by an atom to form its σ bonds and to hold lone pairs of electrons.

3. The number of hybrid orbitals needed by an atom in a structure equals the number of atoms to which it is bonded plus the number of lone pairs of electrons in its valence shell.

4. When there is a double bond in a molecule, it consists of one σ bond and one π bond.

5. When there is a triple bond in a molecule, it consists of one σ bond and two π bonds.

8.7
MOLECULAR ORBITAL THEORY

Molecular orbital theory takes the view that a molecule is similar to an atom in one important respect. Both have energy levels that correspond to various orbitals which can be populated by electrons. In atoms, these orbitals are called atomic orbitals; in molecules, they are called **molecular orbitals.** (We will frequently call them MOs.)

In most cases, the actual shapes and energies of the molecular orbitals of a molecule cannot be determined exactly. Nevertheless, theoreticians have found that reasonably good estimates of their shapes and energies can be obtained by combining the electron waves corresponding to the atomic orbitals of the atoms that make up the molecule. In forming molecular orbitals, these waves interact by constructive and destructive interference just like other waves that we've seen. Their intensities are either added or subtracted when the atomic orbitals overlap. The way this occurs can be seen if we look at the overlap of a pair of 1s orbitals from two atoms in a molecule like H_2 (Figure 8.30). The *two* 1s orbitals combine when the molecule is formed to give *two* MOs. In one MO, the electron waves add together between the nuclei, which gives a buildup of electron density that helps hold the nuclei near each other. Such an MO is said to be a **bonding molecular orbital.** In the other MO, cancellation of the electron waves reduces the electron density between the nuclei. The reduced amount of negative charge between the nuclei allows the nuclei to repel each other strongly, so this MO is called an **antibonding molecular orbital.** Antibonding MOs tend to destabilize a molecule when occupied by electrons.

> The number of MOs formed is always equal to the number of atomic orbitals that are combined.

Both the bonding and antibonding MOs formed by the overlap of s orbitals have their maximum electron density on an imaginary line that passes through the two nuclei. Earlier, we called bonds that have this property sigma bonds. Molecular orbitals like this are also designated sigma (σ), an asterisk is used to indicate the MOs that are antibonding, and a subscript is written to report which atomic orbitals make up the MO. For example, the bonding MO formed by the overlap of 1s orbitals is symbolized as σ_{1s} and the antibonding MO is written as σ_{1s}^*.

Bonding MOs are lower in energy than antibonding MOs formed from the same atomic orbitals. This is also depicted in Figure 8.30. When electrons populate molecular orbitals, they fill the lower energy, bonding MOs first. The rules that apply to filling MOs are the same as those for filling atomic orbitals: *electrons spread out over orbitals of equal energy, and two electrons can only occupy the same orbital if their spins are paired.*

FIGURE 8.30

Interaction of 1s atomic orbitals to produce bonding and antibonding molecular orbitals. These are σ-type orbitals because the electron density is concentrated along the imaginary line that passes through both nuclei.

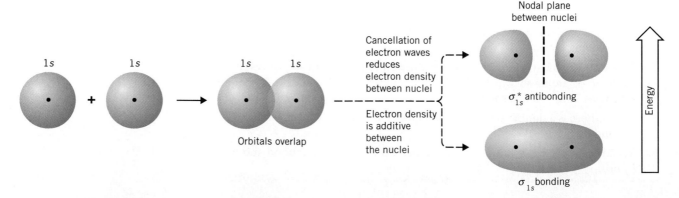

Why Some Molecules Exist and Others Do Not

Let's see how molecular orbital theory can be used to account for the existence of certain molecules, as well as the nonexistence of others. Figure 8.31 is an MO energy level diagram for H_2. The energies of the separate $1s$ atomic orbitals are indicated at the left and right; those of the molecular orbitals are shown in the center. The H_2 molecule has two electrons, and both can be placed in the σ_{1s} orbital. The shape of this bonding orbital, shown in Figure 8.30, should be familiar. It's the same as the shape of the electron cloud that we described using the valence bond theory.

Next, let's consider what happens when two helium atoms come together. Why can't a stable molecule of He_2 be formed? Figure 8.32 is the energy diagram for He_2. Notice that both bonding and antibonding orbitals are filled. In situations such as this there is a net destabilization because the antibonding MO is raised in energy more than the bonding MO is lowered, relative to the orbitals of the separated atoms. This means the total energy of He_2 is larger than that of two separate He atoms, so the "molecule" is unstable and immediately comes apart.

In general, the effects of *antibonding electrons* (those in antibonding MOs) cancel the effects of an equal number of bonding electrons, and molecules with equal numbers of bonding and antibonding electrons are unstable. If we remove an antibonding electron from He_2 to give He_2^+, there is a net excess of bonding electrons, and the ion should be capable of existence. In fact, the emission spectrum of He_2^+ can be observed when an electric discharge is passed through a helium-filled tube, which shows that He_2^+ is present during the electric discharge. However, the ion is not very stable and cannot be isolated.

He_2^+ has two bonding electrons and one antibonding electron.

Bond Orders

The concept of bond order was introduced in Chapter 7. Recall that it was defined as the number of pairs of electrons shared between two atoms. Thus, the sharing of one pair gives a single bond with a bond order of 1, two pairs give a double bond and a bond order of 2, and three pairs give a triple bond

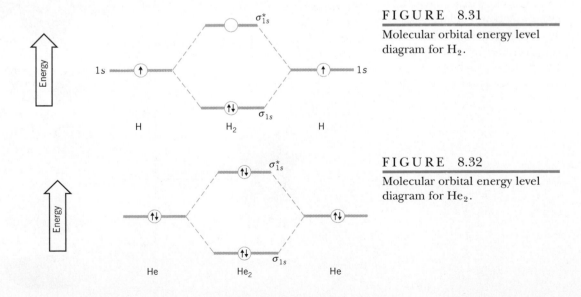

FIGURE 8.31

Molecular orbital energy level diagram for H_2.

FIGURE 8.32

Molecular orbital energy level diagram for He_2.

with a bond order of 3. To translate the MO description into these terms, we compute the bond order as

$$\text{Bond order} = \frac{(\text{number of bonding } e^-) - (\text{number of antibonding } e^-)}{2}$$

For the H_2 molecule, we have

$$\text{Bond order} = \frac{2 - 0}{2} = 1$$

A bond order of 1 corresponds to a single bond. For He_2 we have

$$\text{Bond order} = \frac{2 - 2}{2} = 0$$

A bond order of zero means no bond exists, so the He_2 molecule is unable to exist. However, the He_2^+ ion does form, and its calculated bond order is

$$\text{Bond order} = \frac{2 - 1}{2} = 0.5$$

Notice that the bond order does not have to be a whole number.

Bonding in Diatomic Molecules of Period 2

The outer shell of a period-2 element consists of $2s$ and $2p$ subshells. When atoms of this period bond to each other, the atomic orbitals of these subshells

FIGURE 8.33

Formation of molecular orbitals by the overlap of p orbitals. (*a*) Two p_x orbitals that point at each other give bonding and anti-bonding σ-type MOs. (*b*) Perpendicular to the $2p_x$ orbitals are $2p_y$ and $2p_z$ orbitals that overlap to give two sets of bonding and antibonding π-type MOs.

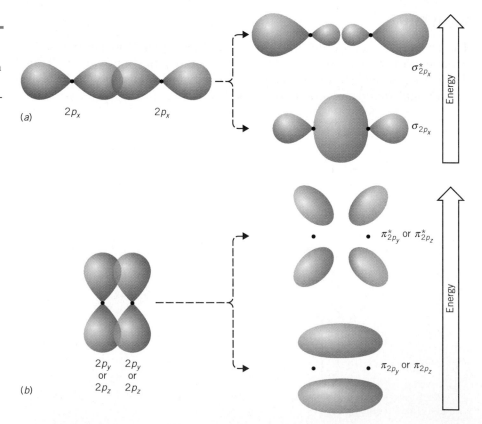

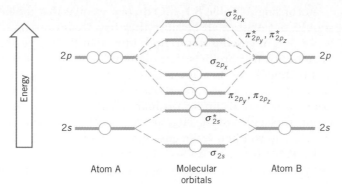

FIGURE 8.34

Approximate relative energies of molecular orbitals in second period diatomic molecules.

interact strongly to produce molecular orbitals. The $2s$ orbitals, for example, overlap to form σ_{2s} and σ_{2s}^{*} molecular orbitals having essentially the same shapes as the σ_{1s} and σ_{1s}^{*} MOs, respectively. Figure 8.33 shows the shapes of the bonding and antibonding MOs produced when the $2p$ orbitals overlap. If we label those that point toward each other $2p_x$, a set of bonding and antibonding MOs are formed that we can label as σ_{2p_x} and $\sigma_{2p_x}^{*}$. The $2p_y$ and $2p_z$ orbitals, which are perpendicular to the $2p_x$ orbitals, then overlap sideways to give π-type molecular orbitals. They are labeled π_{2p_y} and $\pi_{2p_y}^{*}$, and π_{2p_z} and $\pi_{2p_z}^{*}$.

The approximate relative energies of the MOs formed from the second shell atomic orbitals are shown in Figure 8.34. Using this energy diagram, we can predict the electronic structures of diatomic molecules of period 2. These *MO electron configurations* are obtained using the same rules that are applied to the filling of atomic orbitals in atoms.

1. Electrons fill the lowest energy orbitals that are available.

2. No more than two electrons, with spins paired, can occupy any orbital.

3. Electrons spread out as much as possible, with spins unpaired, over orbitals that have the same energy.

Applying these rules to the valence electrons of period-2 atoms gives the MO electron configurations shown in Table 8.1. Let's see how well MO theory performs by examining data that are available for these molecules.

TABLE 8.1 Molecular Orbital Populations and Bond Orders for Period-2 Diatomic Molecules[a]

		Li$_2$	Be$_2$	B$_2$	C$_2$	N$_2$	O$_2$	F$_2$	Ne$_2$
	$\sigma_{2p_x}^{*}$	◯	◯	◯	◯	◯	◯	◯	⇅
$\pi_{2p_y}^{*}$, $\pi_{2p_z}^{*}$		◯◯	◯◯	◯◯	◯◯	◯◯	↑ ↑	⇅ ⇅	⇅ ⇅
	σ_{2p_x}	◯	◯	◯	◯	⇅	⇅	⇅	⇅
π_{2p_y}, π_{2p_z}		◯◯	◯◯	↑ ↑	⇅ ⇅	⇅ ⇅	⇅ ⇅	⇅ ⇅	⇅ ⇅
	σ_{2s}^{*}	◯	⇅	⇅	⇅	⇅	⇅	⇅	⇅
	σ_{2s}	⇅	⇅	⇅	⇅	⇅	⇅	⇅	⇅
Number of bonding electrons		2	2	4	6	8	8	8	8
Number of antibonding electrons		0	2	2	2	2	4	6	8
Bond order		1	0	1	2	3	2	1	0
Bond energy (kJ/mol)		110	—	300	612	953	501	129	—
Bond length (pm)		267	—	158	124	109	121	144	—

(Energy axis at left, increasing upward.)

[a] σ_{2p_x} is thought to be lower in energy than π_{2p_y} and π_{2p_z} in O$_2$ and F$_2$, but this does not affect conclusions about the net number of bonds in these molecules or the number of unpaired electrons that they have. Therefore, we have used the same energy diagram for all period 2 diatomic molecules to minimize confusion.

According to Table 8.1, MO theory predicts that molecules of Be_2 and Ne_2 should not exist at all because they have bond orders of zero. In beryllium vapor and in gaseous neon, no evidence of Be_2 or Ne_2 has even been found. MO theory also predicts that diatomic molecules of the other period-2 elements should exist because they all have bond orders greater than zero. These molecules have, in fact, been observed. Although lithium, boron, and carbon are complex solids under ordinary conditions, they can be vaporized. In the vapor, molecules of Li_2, B_2, and C_2 can be detected. Nitrogen, oxygen, and fluorine, as you know, are gaseous elements that exist as N_2, O_2, and F_2.

In Table 8.1, we also see that the predicted bond order increases from boron to carbon to nitrogen and then decreases from nitrogen to oxygen to fluorine. As the bond order increases, the *net* number of bonding electrons increases, which means that more electron density is concentrated between the nuclei. This greater concentration of negative charge binds the nuclei more tightly and therefore gives a stronger bond. The attraction between the increased electron density and the positive nuclei also draws the nuclei closer to the center of the bond, thereby decreasing the bond length. The *experimentally measured* bond energies and bond lengths given in Table 8.1 follow these predictions quite nicely.

Molecular orbital theory is particularly successful in explaining the electronic structure of the oxygen molecule. Experiments show that O_2 is paramagnetic; the molecule contains two unpaired electrons. In addition, the bond length in O_2 is about what is expected for an oxygen–oxygen double bond. These data are not explained by valence bond theory. For example, if we write a Lewis structure for O_2 that shows a double bond and also obeys the octet rule, all the electrons appear in pairs.

$:\ddot{O}::\ddot{O}:$ (Not acceptable based on experimental evidence because all electrons are paired)

On the other hand, if we show the unpaired electrons, the structure has only a single bond and doesn't obey the octet rule.

$:\dot{\ddot{O}}:\dot{\ddot{O}}:$ (Not acceptable based on experimental evidence because of the O—O single bond)

With MO theory, we don't have any of these difficulties. When Hund's rule is applied, the two electrons in the π^* orbitals of O_2 spread out over these orbitals with their spins unpaired because both orbitals have the same energy. The electrons in the two antibonding π^* orbitals cancel the effects of two electrons in the two bonding π orbitals, so the net bond order is 2 and the bond is effectively a double bond.

■ **Practice Exercise 11** The MO energy level diagram for the nitric oxide molecule, NO, is essentially the same as that shown in Table 8.1 for the period-2 diatomic molecules. Indicate which MOs are populated in NO and calculate the bond order for the molecule.

Molecular oxygen is attracted weakly by a magnet.

Although MO theory handles easily the things that VB theory has trouble with, MO theory loses the simplicity of VB theory. For even quite simple molecules, MO theory is too complicated to make predictions without extensive calculations.

8.8 DELOCALIZED MOLECULAR ORBITALS

One of the least satisfying aspects of the way valence bond theory explains chemical bonding is the need to write resonance structures for certain molecules and ions. In Chapter 7 we discussed this in terms of Lewis structures, which we have since learned can be equated to shorthand notations for valence bond descriptions of molecules. As an example, let's look at the formate

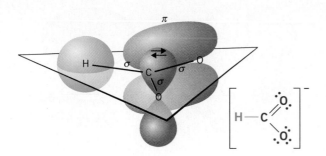

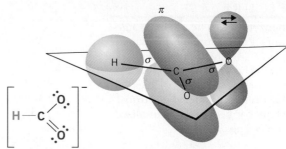

FIGURE 8.35

Valence bond descriptions of the resonance structures of the formate ion, CHO_2^-. Only the p orbitals of oxygen that can form π bonds to carbon are shown. The ion is planar because carbon uses sp^2 hybrid orbitals to form the σ bond framework.

ion, CHO_2^-, which we described in Chapter 7 as a resonance hybrid of the following two structures.

$$\left[H-C \begin{matrix} \overset{..}{O} \\ \\ \overset{..}{O} \end{matrix} \right]^- \longleftrightarrow \left[H-C \begin{matrix} \overset{..}{O} \\ \\ \overset{..}{O} \end{matrix} \right]^-$$

In terms of the overlap of orbitals, VB theory would picture the formate ion as shown in Figure 8.35. The σ-bond framework is formed by overlap of the sp^2 hybrid orbitals of carbon with the $1s$ orbital of hydrogen and the p orbitals of the oxygen atoms. As you can see in Figure 8.35, the π bond can be formed with either oxygen atom, which is how we obtain the two resonance structures.

Molecular orbital theory avoids the problem of resonance by recognizing that electron pairs can sometimes be shared among overlapping orbitals from three or more atoms. For the CHO_2^- ion, it allows all *three p* orbitals to overlap simultaneously to form one large π-type molecular orbital that spreads over all three nuclei, as shown in Figure 8.36. Since the π electrons in this MO are not required to stay "localized" between just two nuclei, we say that the bond is **delocalized.** Delocalized π-type molecular orbitals permit single descriptions of the electronic structures of molecules and ions.

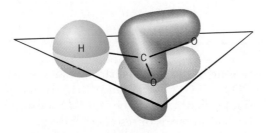

FIGURE 8.36

Molecular orbital theory allows the electron pair in the π-type bond to be spread out, or delocalized, over all three atoms of the $-CO_2^-$ unit in CHO_2^-.

The delocalized nature of π bonds that extend over three or more nuclei can be indicated with dotted lines rather than dashes when Lewis-type structural formulas are drawn. The formate ion, for example, can be drawn as

$$\left[H-C \begin{matrix} O \\ \\ O \end{matrix} \right]^-$$

A localized bond is one in which the electrons spend all their time shared between just two atoms. The electrons in a delocalized bond become shared among more than two atoms.

FIGURE 8.37

Benzene. (*a*) The σ-bond framework. All atoms lie in the same
plane. (*b*) The unhybridized *p*
orbitals at each carbon prior to
side-to-side overlap. (*c*) The double doughnut-shaped electron
cloud formed by the delocalized
π electrons.

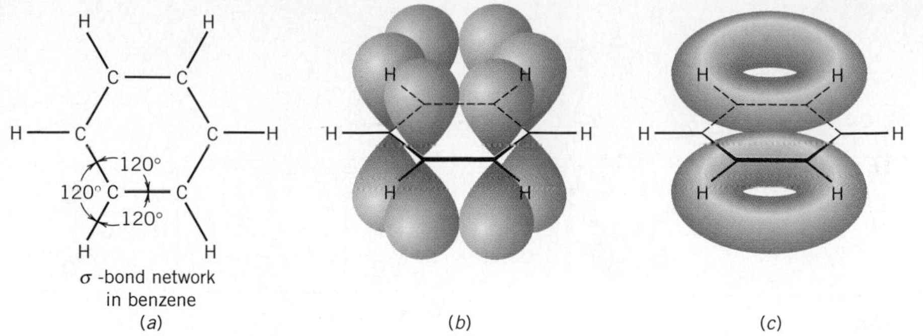

σ-bond network
in benzene
(*a*)

(*b*)

(*c*)

Delocalized π bonds are quite common in many kinds of molecules and
ions. Some other examples are

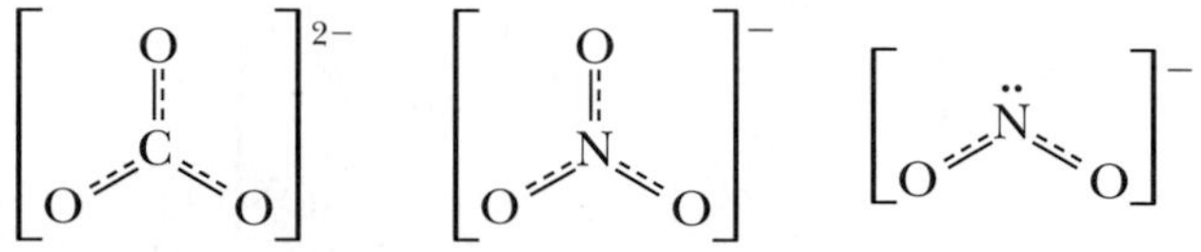

$$\left[\begin{array}{c} O \\ \| \\ O \overset{C}{\diagup \diagdown} O \end{array}\right]^{2-} \qquad \left[\begin{array}{c} O \\ \| \\ O = N = O \end{array}\right]^{-} \qquad \left[O = \overset{\cdot\cdot}{N} = O\right]^{-}$$

$$CO_3{}^{2-} \qquad\qquad NO_3{}^{-} \qquad\qquad NO_2{}^{-}$$

**The bonding in each of these ions
or molecules is explained by just
one structure in the MO theory.**

An important example from the realm of organic chemistry is benzene, C_6H_6.
As you learned in Chapter 7, this molecule has a ring structure whose resonance structures can be written as

**Benzene is an important industrial
solvent, but it is also quite toxic.**

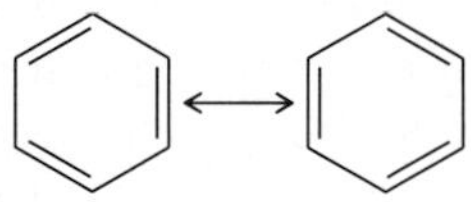

In benzene, the sigma bond framework requires that the carbon atoms use sp^2
hybrid orbitals because each carbon forms three σ bonds. This leaves each
carbon atom with a half-filled unhybridized *p* orbital perpendicular to the
plane of the ring. These *p* orbitals overlap to give a delocalized π-electron
cloud that looks something like two doughnuts with the sigma bond framework sandwiched between them (Figure 8.37). The delocalized nature of the
π electrons is the reason we usually represent the structure of benzene as

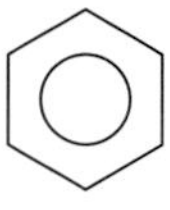

One of the special characteristics of delocalized bonds, like those found in
$CO_3{}^{2-}$, $NO_3{}^{-}$, $NO_2{}^{-}$, and C_6H_6, is that they make a molecule or ion more
stable than it would be if it had localized bonds. In Chapter 7 this was described in terms of resonance energy. In the molecular orbital theory, we no
longer speak of resonance; instead, we refer to the electrons as being delocalized. The extra stability that is associated with this delocalization is therefore
described, in the language of MO theory, as the **delocalization energy.**

**Functionally, the terms *resonance
energy* and *delocalization energy*
are the same; they just come from
different approaches to bonding
theory.**

8.9
BONDING IN SOLIDS

In Chapter 2 we described the division of the elements into three broad
classes: metals, nonmetals, and metalloids. One of the significant properties of
these substances is their differing abilities to conduct electricity. To explain

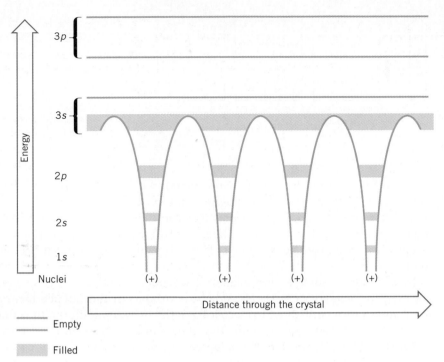

FIGURE 8.38

Energy bands in solid sodium. The delocalized 3s valence band extends throughout the entire solid, as do bands such as the 3p band formed from higher energy orbitals of the sodium atoms.

these differences adequately, a theory of bonding in solids was developed called the **band theory of solids.**

In a solid, an **energy band** is composed of a very large number of closely spaced energy levels that are formed by combining atomic orbitals of similar energy from each of the atoms within the substance. For example, in sodium the 1s atomic orbitals, one from each atom, combine to form a single 1s band. The number of energy levels in the band equals the number of 1s orbitals supplied by the entire collection of sodium atoms. The same thing occurs with the 2s, 2p, etc., orbitals, so that we also have 2s, 2p, etc., bands within the solid.

Figure 8.38 illustrates the energy bands in solid sodium. Notice that the electron density in the 1s, 2s, and 2p bands does not extend far from each individual nucleus, so these bands produce effectively localized energy levels in the solid. However, the 3s band is delocalized and extends continuously through the solid. The same applies to energy bands formed by higher energy orbitals.

Sodium atoms have filled 1s, 2s, and 2p orbitals, so the corresponding bands in the solid are also filled. The 3s orbital of sodium, however, is only half-filled, which leads to a half-filled 3s band. The 3p and higher energy bands in sodium are completely empty. When a voltage is applied across sodium, electrons in filled bands cannot move through the solid because orbitals in the same band on neighboring atoms are already filled and cannot accept an additional electron. In the 3s band, which is half-filled, an electron can hop from atom to atom with ease, and this allows sodium to conduct electricity well.

We refer to the band containing the outer-shell (valence-shell) electrons as the **valence band.** Any band that is either vacant or partially filled and uninterrupted throughout the solid is called a **conduction band,** because electrons in it are able to move through the solid and thereby serve to carry electricity.

The formation of delocalized energy bands in the solid is like the formation of the delocalized bonds discussed in the preceding section.

FIGURE 8.39

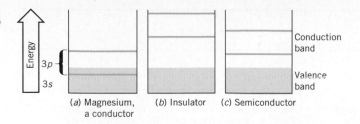

(a) Magnesium, a conductor (b) Insulator (c) Semiconductor

(*a*) In magnesium, the empty 3*p* conduction band overlaps the filled 3*s* valence band and provides a way for this metal to conduct electricity. (*b*) In an insulator, the energy gap between the filled valence band and the empty conduction band prevents electrons from populating the conduction band. (*c*) In a semiconductor, there is a small band gap and thermal energy can promote some electrons from the filled valence band to the empty conduction band. This enables the solid to conduct electricity weakly.

Recall that at any given temperature there is a distribution of kinetic energy among the particles of a substance. *Thermal energy is kinetic energy a particle possesses because of its temperature. The higher the temperature, the larger the thermal energy.*

In metallic sodium the valence band and conduction band are the same, so sodium is a good conductor. In magnesium the 3*s* valence band is filled and therefore cannot be used to transport electrons. However, the vacant 3*p* conduction band actually overlaps the valence band and can easily be populated by electrons when voltage is applied (Figure 8.39*a*). This permits magnesium to be a conductor.

In an insulator such as glass, diamond, or rubber, all the valence electrons are used to form covalent bonds, so all the orbitals of the valence band are filled and cannot contribute to electrical conductivity. In addition, the energy separation, or **band gap,** between the filled valence band and the nearest conduction band (empty band) is large. As a result, electrons cannot populate the conduction band, so these substances are unable to conduct electricity (Figure 8.39*b*).

In a semiconductor such as silicon or germanium, the valence band is also filled, but the band gap between the filled valence band and the nearest conduction band is small (Figure 8.39*c*). At room temperature, thermal energy possessed by the electrons is sufficient to promote some electrons to the conduction band, and a small degree of electrical conductivity is observed. One of the interesting properties of semiconductors is that their electrical conductivity increases with increasing temperature. This is because as the temperature rises, the number of electrons with enough energy to populate the conduction band also increases.

Transistors and Other Electronic Devices   One of the most significant discoveries in this century has been the way that the electrical characteristics of semiconductors can be modified by the controlled introduction of carefully selected impurities. This has led to the discovery of transistors, which have made possible all the marvelous electronic devices we now take for granted, such as portable TVs, CD players, radios, calculators, and microcomputers. In fact, almost everything we do today is influenced in some way by electronic circuits etched into tiny silicon chips.

In a semiconductor such as silicon, all the valence electrons are used to form covalent bonds to other atoms. If a small amount of a Group IIIA element such as boron is added to silicon, it can replace silicon atoms in the structure of the solid. (We say the silicon has been **doped** with boron.) However, for each boron added, one of the covalent bonds in the structure will be deficient in electrons because boron has only three valence electrons instead of the four that silicon has. Under an applied voltage an electron from a neighboring atom can move to fill this deficiency and thereby leave a positive "hole" behind. When this "hole" becomes filled, another is created elsewhere, and electrical conduction results from the migration of this positive "hole" through the solid. Because of the positive nature of the moving charge carrier, the substance is said to be a **p-type semiconductor.**

If the impurity added to the silicon is a Group VA element such as arsenic, it has one more electron in its valence shell than silicon. When the bonds are formed in the solid, there will be an electron left over that's not used in bonding. The extra electrons supplied by the impurity can enter the conduction band and move through the solid under an applied voltage, and because the moving charge now consists of negatively charged electrons, the solid is said to be an **n-type semiconductor.**

Transistors are made from n- and p-type semiconductors and can be formed directly on the surface of a silicon chip, which has made possible the microcircuits in computers and calculators. Another interesting application is in solar batteries, which are easier to understand.

A silicon **solar battery** (also called a **solar cell**) is composed of a silicon wafer doped with arsenic (giving an n-type semiconductor) over which is placed a thin layer of silicon doped with boron (a p-type semiconductor). This is illustrated in Figure 8.40. In the dark there is an equilibrium between electrons and holes at the interface between the two layers, which is called a **p–n junction.** Some electrons from the n-type layer diffuse into the holes in the p-layer and are trapped. This leaves some positive holes in the n-layer. Equilibrium is achieved when the positive holes in the n-layer prevent further movement of electrons into the p-layer.

When light falls on the surface of the cell, the equilibrium is upset. Energy is absorbed, which permits electrons that were trapped in the p-layer to return to the n-layer. As these electrons move across the p–n junction into the n-layer, other electrons leave the n-layer through the wire, pass through the electrical circuit, and enter the p-layer. Thus an electric current flows when light falls on the cell and the external circuit is completed. This electric current can be used to run a motor, power a hand-held calculator, or perform whatever other task we wish.

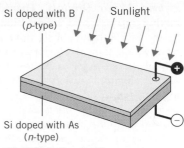

FIGURE 8.40

Construction of a silicon solar battery.

A hand-held solar calculator. The solar panel is located above the display.

SUMMARY

Molecular Shapes and VSEPR Theory The structures of most molecules can be described in terms of one or another of five basic geometries: **linear, planar triangular, tetrahedral, trigonal bipyramidal,** and **octahedral.** The **VSEPR theory** predicts molecular geometry by assuming that the electron pairs in the valence shell of an atom stay as far apart as possible so the repulsions between them are minimized. Figures 8.2 to 8.5 illustrate the structures obtained with different numbers of groups of electrons in the valence shell of the central atom in a molecule or ion and with different numbers of lone pairs and attached atoms. The correct shape of a molecule or polyatomic ion can usually be predicted from the Lewis structure.

Molecular Shape and Molecular Polarity A molecule that contains identical atoms attached to a central atom will be nonpolar if there are no lone pairs of electrons in the central atom's valence shell. It will be polar if lone pairs are present, except in two cases: (1) when there are three lone pairs and two attached atoms, and (2) when there are two lone pairs and four attached atoms. If all the atoms attached to the central atom are not alike, the molecule will usually be polar.

Valence Bond (VB) Theory This is one of two theories of chemical bonding based on wave mechanics discussed in this chapter. According to VB theory, a covalent bond is formed between two atoms when an atomic orbital on one atom **overlaps** with an atomic orbital on the other and a pair of electrons with paired spins is shared between the overlapping orbitals. In general, the better the overlap of the orbitals, the stronger the bond. A given atomic orbital can overlap with only one other orbital on a different atom, so a given atomic orbital can only form one bond with an orbital on the other atom.

Hybrid Atomic Orbitals   **Hybrid orbitals** are formed by mixing pure *s*, *p*, and/or *d* orbitals. Hybrid orbitals overlap better with other orbitals than the pure atomic orbitals from which they are formed, so bonds formed by hybrid orbitals are stronger than those formed by ordinary atomic orbitals. **Sigma bonds** (σ bonds) are formed by the following kinds of orbital overlap: *s*–*s*, *s*–*p*, end-to-end *p*–*p*, and overlap of hybrid orbitals. Sigma bonds allow free rotation around the bond axis. The side-by-side overlap of *p* orbitals produces a pi bond (π bond). **Pi bonds** do not permit free rotation around the bond axis because rotation around the axis of a π bond involves bond breaking. In complex molecules, the basic molecular framework is built with σ bonds. A double bond consists of one σ bond and one π bond. A triple bond consists of one σ bond and two π bonds.

Molecular Orbital (MO) Theory   This theory begins with the supposition that molecules are similar to atoms, except they have more than one positive center. They are treated as collections of nuclei and electrons, and the electrons of the molecule are in **molecular orbitals** of different energies. Molecular orbitals can spread over two or more nuclei, and can be considered to be formed by the constructive and destructive interference of the overlapping electron waves corresponding to the atomic orbitals of the atoms in the molecule. **Bonding MOs** concentrate electron density be-tween nuclei; **antibonding MOs** remove electron density from between nuclei. The rules for the filling of MOs are the same as those for atomic orbitals. The ability of MO theory to describe **delocalized orbitals** avoids the need for resonance theory. Delocalization of bonds leads to a lowering of the energy by an amount called the **delocalization energy** and produces more stable molecular structures.

Bonding in Solids   In solids, atomic orbitals of the atoms combine to yield **energy bands** that consist of many energy levels. The **valence band** is formed by orbitals of the valence shells of the atoms in the solid. A **conduction band** is a partially filled or empty band. In an **electrical conductor** the conduction band is either partially filled or is empty and overlaps a filled band. In an **insulator,** the **band gap** between the filled valence band and the empty conduction band is large, so no electrons populate the conduction band. In a **semiconductor,** the band gap between the filled valence band and the conduction band is small and thermal energy can promote some electrons to the conduction band. Silicon becomes a **p-type semiconductor,** in which the charge is carried by positive "holes," if it is doped with a Group IIIA element such as boron. It becomes an **n-type semiconductor,** in which the charge is carried by electrons, if it is doped with a Group VA element such as arsenic.

Tools You Have Learned

The table below lists the tools you have learned in this chapter that are applicable to problem solving. Review them if necessary, and refer to them when working on the Thinking-It-Through problems and the Review Exercises that follow.

Tool	Function
Basic molecular shapes (page 309)	Knowing how to draw them enables you to sketch the shapes of most molecules.
VSEPR theory (page 312)	This theory enables you to predict the shape of a molecule or polyatomic ion when its Lewis structure is known. It also enables you to determine the kind of hybrid orbitals used by the central atom in a molecule or polyatomic ion.
Molecular shape (page 321)	In this chapter you learned how we can use the shape of a molecule as a tool to determine whether the molecule is polar.

THINKING IT THROUGH

Remember, you are not asked to obtain answers for the following problems. Instead, assemble the data necessary to solve the problems and describe how you would use the data to obtain the answers. For numerical problems, set up the calculation using appropriate conversion factors.

The problems are divided into two groups. Those in Level 2 are significantly more challenging than those in Level 1 and provide an opportunity to really hone your problem solving skills.

Level 1 Problems

1. The BrO_3^- ion is pyramidal. Without drawing the Lewis structure for the ion, explain how we know that there is at least one lone pair of electrons on the bromine atom.

2. The molecule SF_2 is nonlinear whereas the molecule XeF_2 is linear. Without drawing their Lewis structures, what differences could there be between the valence shells of S and Xe in these compounds?

3. The molecule SF_5Cl is octahedral. Is it polar or nonpolar? Explain.

4. Antimony forms a compound with hydrogen that is called stibine. Its formula is SbH_3 and the H—Sb—H bond angles are 91.3°. Which kinds of orbitals does Sb most likely use to form the Sb—H bonds, pure p orbitals or hybrid orbitals? Explain your reasoning.

5. Describe in detail how you would predict whether the SF_2 molecule is polar or nonpolar.

6. The sequence of energy levels in the molecular orbital energy level diagram for nitrogen monoxide, NO, is essentially the same as for N_2 and O_2. How would you determine the average bond order in the NO molecule?

7. All of the s orbitals in the $3s$ energy band of magnesium are filled by electrons. Nevertheless, magnesium is a good conductor of electricity. Explain this on the basis of the band theory of solids.

Level 2 Problems

8. The molecule XCl_3 is pyramidal. In which group in the periodic table is element X found? If the molecule were planar triangular, in which group would X be found? If the molecule were T shaped, in which group would X be found. Why is it unlikely that element X is in Group VI?

9. Consider the molecule $COCl_2$, whose Lewis structure is

The VSEPR theory predicts this is a planar triangular molecule with bond angles of 120°. However, in this molecule the O—C—Cl bond angles are not the same as the Cl—C—Cl bond angle. Suggest two reasons why all the bond angles are not the same.

REVIEW EXERCISES

Answers to questions whose numbers are printed in color are given in Appendix D. More challenging questions are marked with asterisks.

Shapes of Molecules

8.1 Sketch the following molecular shapes and give the various bond angles in the structure: (a) planar triangular, (b) tetrahedral, (c) octahedral.

8.2 Sketch the following molecular shapes and give the bond angles in the structure: (a) linear, (b) trigonal bipyramidal.

VSEPR Theory

8.3 What is the underlying principle upon which the VSEPR theory is based?

8.4 What arrangements of electron pairs are expected when the valence shell of the central atom contains (a) three pairs, (b) six pairs, (c) four pairs, or (d) five pairs of electrons?

8.5 What molecular shapes are expected when the valence shell of the central atom has (a) three bonding electron pairs and one lone pair, (b) four bonding electron pairs and two lone pairs, and (c) two bonding electron pairs and one lone pair? Give the name for each structure and also sketch its shape.

8.6 Predict the shapes of: (a) FCl_2^+, (b) AsF_5, (c) AsF_3, and (d) SeO_2.

8.7 Predict the shapes of: (a) TeF_4, (b) $SbCl_6^-$, (c) NO_2^-, and (d) PO_4^{3-}.

8.8 Predict the shapes of: (a) IO_4^-, (b) ICl_4^-, (c) TeF_6, and (d) ICl_2^-.

8.9 Predict the shapes of: (a) CS_2, (b) BrF_4^-, (c) ICl_3, and (d) SeO_3.

8.10 Predict the shapes of: (a) SbH_3, (b) PCl_4^+, (c) ClO_3^-, and (d) SiO_4^{4-}.

8.11 Acetylene, a gas used in welding torches, has the Lewis structure $H—C\equiv C—H$. What would you expect the $H—C—C$ bond angle to be in this molecule?

8.12 Ethylene, a gas used to ripen tomatoes artificially, has the Lewis structure

$$\begin{array}{cc} H & H \\ | & | \\ H—C & =C—H \end{array}$$

What would you expect the $H—C—H$ and $H—C—C$ bond angles to be in this molecule? (Caution: Don't be fooled by the way the structure is drawn here.)

8.13 Formaldehyde has the Lewis structure

$$\begin{array}{c} H \\ | \\ H—C=\overset{..}{\underset{..}{O}} \end{array}$$

What would you predict its shape to be?

Predicting Molecular Polarity

8.14 Why is it useful to know the polarities of molecules?

8.15 How do we indicate a bond dipole when we draw the structure of a molecule?

8.16 What condition must be met if a molecule having polar bonds is to be nonpolar?

8.17 Use a drawing to show why the SO_2 molecule is polar.

8.18 Which of the following molecules would be expected to be polar? (a) HBr, (b) $POCl_3$, (c) CH_2O, (d) $SnCl_4$, (e) $SbCl_5$

8.19 Which of the following molecules would be expected to be polar? (a) PBr_3, (b) SO_3, (c) $AsCl_3$, (d) ClF_3, (e) BCl_3

Modern Bonding Theories

8.20 What is the theoretical basis of both valence bond (VB) theory and molecular orbital (MO) theory?

8.21 What shortcomings of Lewis structures and VSEPR theory do VB and MO theories attempt to overcome?

8.22 What is the main difference in the way VB and MO theories view the bonds in a molecule?

Valence Bond Theory

8.23 What is meant by *orbital overlap*?

8.24 What are the principal postulates of the valence bond theory?

8.25 Use sketches of orbitals to describe how the covalent bond in H_2 is formed.

8.26 Use sketches of orbitals to describe how VB theory would explain the formation of the H—Br bond in hydrogen bromide.

8.27 Hydrogen selenide is one of nature's most foul-smelling substances. Molecules of H_2Se have H—Se—H bond angles very close to 90°. How would VB theory explain the bonding in H_2Se? Illustrate with appropriate orbital diagrams.

8.28 Use sketches of orbitals to show how VB theory explains the bonding in the F_2 molecule. Illustrate with appropriate orbital diagrams as well.

Hybrid Orbitals

8.29 What term is used to describe the mixing of atomic orbitals of the same atom?

8.30 Why do atoms usually prefer to use hybrid orbitals for bonding rather than plain atomic orbitals?

8.31 Sketch figures that illustrate the directional properties of the following hybrid orbitals: (a) sp, (b) sp^2, (c) sp^3, (d) sp^3d, (e) sp^3d^2.

8.32 Why do period-2 elements never use sp^3d or sp^3d^2 hybrid orbitals for bond formation?

8.33 What relationship is there, if any, between Lewis structures and the valence bond descriptions of molecules?

8.34 Use orbital diagrams to explain how the $BeCl_2$ molecule is formed. What kind of hybrid orbitals does beryllium use in $BeCl_2$?

8.35 Draw Lewis structures for the following and use the geometry predicted by VSEPR theory for the electron pairs to determine what kind of hybrid orbitals the central atom uses in bond formation: (a) ClO_3^-, (b) SO_3, (c) OF_2, (d) $SbCl_6^-$, (e) $BrCl_3$, (f) XeF_4.

8.36 Use orbital diagrams to describe the bonding in (a) $SnCl_4$ and (b) $SbCl_5$.

8.37 Use VSEPR theory to help you describe the bonding in the following molecules according to VB theory: (a) $AsCl_3$, (b) ClF_3, (c) $SbCl_5$, (d) $SeCl_2$.

8.38 Using orbital diagrams, describe how sp^3 hybridization occurs in (a) carbon, (b) nitrogen, and (c) oxygen. If these elements use sp^3 hybrid orbitals to form bonds, how many lone pairs of electrons would be found on each?

8.39 Sketch the way the orbitals overlap to form the bonds in each of the following: (a) CH_4, (b) NH_3, (c) H_2O.

8.40 We explained the bond angles of 107° in NH_3 by using sp^3 hybridization of the central nitrogen atom. Had the original unhybridized p orbitals of the nitrogen been used to overlap with $1s$ orbitals of each hydrogen, what would have been the H—N—H bond angles? Explain.

8.41 Cyclopropane is a triangular molecule with C—C—C bond angles of 60°. Explain why the σ bonds joining carbon atoms in cyclopropane are weaker than the carbon–carbon σ bonds in the noncyclic propane.

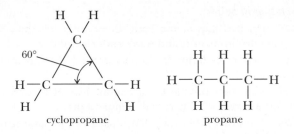

cyclopropane   propane

8.42 What facts about boron strongly suggested the need to consider sp^2 hybridization of its second shell atomic orbitals?

8.43 Using sketches of orbitals and orbital diagrams, describe sp^2 hybridization at (a) boron and (b) carbon.

8.44 Phosphorus trifluoride, PF_3, has F—P—F bond angles of 97.8°.

(a) How would VB theory use hybrid orbitals to explain these data?

(b) How would VB theory use unhybridized orbitals to account for these data?

(c) Do either of these models work very well?

8.45 If the central oxygen in the water molecule did not use sp^3 hybridized orbitals (or orbitals of any kind of hybridization), what would be the expected bond angle in H_2O (assuming no angle-spreading force)?

Coordinate Covalent Bonds and VB Theory

8.46 The ammonia molecule, NH_3, can combine with a hydrogen ion, H^+ (which has an empty $1s$ orbital), to form the ammonium ion, NH_4^+. (This is how ammonia can neutralize acid and therefore function as a base.) Sketch the geometry of the ammonium ion, indicating the bond angles.

8.47 Use orbital diagrams to show the formation of a coordinate covalent bond in H_3O^+ when H_2O reacts with H^+.

8.48 What kind of hybrid orbitals are used by tin in $SnCl_6^{2-}$? Draw the orbital diagram for Sn in $SnCl_6^{2-}$. What is the geometry of $SnCl_6^{2-}$?

8.49 Use orbital diagrams to show that the bonding in SbF_6^- involves the formation of a coordinate covalent bond.

Multiple Bonds and Hybrid Orbitals

8.50 How do σ and π bonds differ?

8.51 Why can free rotation occur easily around a σ-bond axis but not around a π-bond axis?

8.52 Using sketches, describe the bonds and bond angles in ethylene, C_2H_4.

8.53 Sketch the way the bonds form in acetylene, C_2H_2.

8.54 A six-membered ring of carbons can hold a double bond but not a triple bond. Explain.

cyclohexene (exists)   cyclohexyne (unknown)

8.55 A nitrogen atom can undergo sp^2 hybridization when it becomes part of a carbon–nitrogen double bond, as in H_2C=NH.

(a) Using a sketch, show the electron configuration of sp^2 hybridized nitrogen just before the overlapping occurs to make this double bond.

(b) Using sketches (and the analogy to the double bond in C_2H_4), describe the two bonds of the carbon–nitrogen double bond.

(c) Describe the geometry of H_2C=NH (using a sketch that shows all expected bond angles).

8.56 A nitrogen atom can undergo sp hybridization and then become joined to carbon by a triple bond to give the structural unit —C≡N:. This triple bond consists of one σ bond and two π bonds.

(a) Write the orbital diagram for sp hybridized nitrogen as it would look before any bonds form.

(b) Using the carbon–carbon triple bond as the analogy, and drawing pictures to show which atomic orbitals overlap with which, show how the three bonds of the triple bond in —C≡N: form.

(c) Again using sketches, describe all the bonds in hydrogen cyanide, H—C≡N:.

(d) What is the likeliest H—C—N bond angle in HCN?

8.57 How does VB theory treat the benzene molecule? (Draw sketches describing the orbital overlaps and the bond angles.)

8.58 Tetrachloroethylene, a common dry-cleaning solvent, has the formula C_2Cl_4. Its structure is

Use the VSEPR and VB theories to describe the bonding in this molecule. What are the expected bond angles in the molecule?

8.59 Phosgene, $COCl_2$, was used as a war gas during World War I. It reacts with moisture in the lungs of its victims to form CO_2 and gaseous HCl, which cause the lungs to fill with fluid. Phosgene is a simple molecule having the structure

$$\ddot{O}=C\overset{\displaystyle :\ddot{Cl}:}{\underset{\displaystyle :\ddot{Cl}:}{\big<}}$$

Describe the bonding in this molecule using VB theory.

Molecular Orbital Theory

8.60 Why is the higher energy MO in H_2 called an antibonding orbital?

8.61 Using a sketch, describe the two lowest energy MOs of H_2 and their relationship to their parent atomic orbitals.

8.62 Explain why He_2 does not exist but H_2 does.

8.63 How does MO theory account for the paramagnetism of O_2?

8.64 On the basis of MO theory, explain why Li_2 molecules can exist but Be_2 molecules cannot. Could the ion Be_2^+ exist?

8.65 Use the MO energy diagram to predict which in each pair has the greater bond energy: (a) O_2 or O_2^+, (b) O_2 or O_2^-, (c) N_2 or N_2^+.

8.66 What are the bond orders in: (a) O_2^+, (b) O_2^-, and (c) C_2^+?

8.67 What relationship is there between bond order and bond energy?

8.68 Sketch the shapes of the π_{2p_y} and $\pi_{2p_y}^*$ MOs.

8.69 What is a delocalized MO?

8.70 Use a Lewis-type structure to indicate delocalized bonding in the nitrate ion.

*8.71 There exists a hydrocarbon called butadiene, which has the molecular formula C_4H_6 and the structure

The C=C bond lengths are 134 pm (about what is expected for a carbon–carbon double bond), but the C—C bond length in this molecule is 147 pm, which is shorter than a normal C—C single bond. The molecule is planar (i.e., all the atoms lie in the same plane).

(a) What kind of hybrid orbitals do the carbon atoms use in this molecule to form the carbon–carbon bonds?

(b) Between which pairs of carbon atoms do we expect to find sideways overlap of p orbitals (i.e., π-type p–p overlap)?

(c) On the basis of your answer to part (b), do you expect to find localized or delocalized π bonding in the carbon chain in this molecule?

(d) Based on your answer to part (c), explain why the center carbon–carbon bond is shorter than a carbon–carbon single bond.

8.72 What problem encountered by VB theory does MO theory avoid by delocalized bonding?

8.73 Draw the representation of the benzene molecule that indicates its delocalized π system.

8.74 What effect does delocalization have on the stability of the electronic structure of a molecule?

8.75 What is delocalization energy? How is it related to resonance energy?

Bonding in Solids

8.76 On the basis of the band theory of solids, how do conductors, insulators, and semiconductors differ?

8.77 Define the terms (a) valence band and (b) conduction band.

8.78 Why does the electrical conductivity of a semiconductor increase with increasing temperature?

8.79 Construct a diagram that illustrates the band structure of potassium.

8.80 Construct a diagram that illustrates the band structure of calcium.

8.81 In calcium, why can't electrical conduction take place by movement of electrons through the $2s$ energy band? How does calcium conduct electricity?

8.82 What is a p-type semiconductor? What is an n-type semiconductor?

8.83 Name two elements that would make germanium a p-type semiconductor when added in small amounts.

8.84 Name two elements that would make silicon an n-type semiconductor when added in small amounts.

8.85 How does a solar cell work?

Additional Exercises

8.86 Draw the structure of the molecule CCl_2F_2 (the carbon is the central atom). Is the molecule polar or nonpolar? Explain.

*8.87 *A lone pair of electrons in the valence shell of an atom has a larger effective volume than a bonding electron pair. Lone pairs therefore repel other electron pairs more strongly than do bonding pairs.* On the basis of these statements, describe how the bond angles in TeF_4 and BrF_4^- deviate from those found in a trigonal bipyramid and an octahedron, respectively. Sketch the molecular shapes of TeF_4 and BrF_4^- and indicate these deviations on your drawing.

*8.88 *The two electron pairs in a double bond repel other electron pairs more than the single pair of electrons in a single bond.* On the basis of this statement, which bond angles should be larger in SO_2Cl_2, the O—S—O bond angles or the Cl—S—Cl bond angles? (In the molecule, sulfur is bonded to two oxygen atoms and two chlorine atoms. Hint: Assign formal charges and work with the best Lewis structure for the molecule.)

*8.89 *A hybrid orbital does not distribute electron density symmetrically around the nucleus of an atom. Therefore, a lone pair in a hybrid orbital contributes to the overall polarity of a molecule.* On the basis of these statements and the fact that NH_3 is a very polar molecule and NF_3 is a nearly nonpolar molecule, justify the notion that the lone pair of electrons in each of these molecules is held in an sp^3 hybrid orbital.

8.90 What kind of hybrid orbitals do the numbered atoms use in the following molecule?

$$\text{H} \quad \overset{②}{} \quad :\overset{..}{O}: \overset{④}{\longleftarrow} \quad \text{H}$$

H—C—C≡C—C—C=C
　①　H　③　H　H

8.91 What kinds of bonds (σ or π) are found in the numbered bonds in the following molecule?

$$\text{H} \quad \overset{②}{} \quad :\overset{..}{O}: \overset{④}{} \quad \text{H}$$

H—C—C≡C—C—C=C
　①　H　③　H　H

8.92 In a certain molecule, a *p* orbital overlaps with a *d* orbital as shown below. What kind of bond is formed, σ or π? Explain your choice.

*8.93 If we take the internuclear axis in a diatomic molecule to be the *x* axis, what kind of *p* orbital (p_x, p_y, or p_z) on one atom would have to overlap with a d_{xy} orbital on the other atom to give a sigma bond?

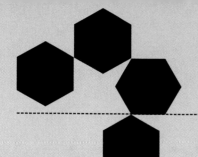

CHEMICALS IN USE 7
Ozone in the Stratosphere

STRATOSPHERIC OZONE CYCLE

The *stratosphere,* an envelope of space surrounding the Earth between altitudes of 10 and 50 km (6 to 31 miles), is host to a number of reactive chemical species at very low concentrations. It is bathed in solar radiation with about 7% in the ultraviolet (UV) region of the spectrum. UV radiation with wavelengths between 280 nm and 315 nm causes sunburn and can lead to one of two kinds of skin cancer.

Fortunately for us, nearly all UV radiation below 320 nm is absorbed in the stratosphere by the *stratospheric ozone cycle,* a typical *chemical chain reaction* (see Special Topic 14.1). The region of the stratosphere where the ozone cycle occurs is called the *ozone layer.* The cycle is initiated by the action of short UV wavelengths, those below 242 nm, which are able to split oxygen molecules.

$$O_2 \xrightarrow[\text{(λ = 242 nm or lower)}]{\text{UV radiation}} 2O^*$$

The reaction makes oxygen atoms, identified as O* to indicate that they are *electronically excited.* (They carry an outer-level electron in a higher than normal orbital.)

The cycle itself, now initiated, consists of two steps that repeat over and over. In the first step, ozone forms when O* collides with O_2 at the surface of a neutral particle, *M,* which might be a molecule of N_2 or O_2.

$$O^* + O_2 + M \longrightarrow O_3 + M + \text{heat} \qquad (1)$$

The function of *M* is to absorb some of the energy of the collision. Otherwise the new O_3 molecule has enough vibrational energy to split apart as soon as it forms. *What is desired, instead, is that UV energy split O_3.*

In the second step, ozone molecules absorb UV energy and break up.

$$O_3 + \text{UV energy} \xrightarrow[\text{(λ = 240–320 nm)}]{} O_2 + O^* \qquad (2)$$

Notice that the O* *product* of the second step is a necessary *reactant* for the first step (1). Thus, reactions 1 and 2 together constitute a chemical chain reaction, the stratospheric ozone cycle. Each "turn" of the cycle absorbs some UV energy. If we add equations 1 and 2, the net result is simply UV energy → heat.

The conversion of UV energy into heat is the net effect of the ozone cycle in the stratosphere. For each initial O*, the cycle can repeat up to hundreds of times before other events, such as a chance combination of two atoms of O to give back O_2, terminate a chain.

The small, net concentration or "level" of stratospheric ozone varies rather widely with latitude, altitude, and month of the year, so no single figure can be cited to describe it. At latitudes reaching 60° north or south from the equator, the ozone level in 1974 varied roughly between 260 and 360 Dobson units, reaching as high as 500 Dobson units. The *Dobson unit,* named for English scientist Gordon Dobson, equals 2.7×10^{16} molecules of O_3 in a column of air 1 cm^2 in area at its base (on the Earth's surface) and extending through the stratosphere.

If any reaction consumes O*, besides the reaction of O* with O_2 (Equation 1), an ozone cycle is broken and the means to regenerate fresh ozone is diminished. Increasing amounts of UV radiation would then pass through to where we live and cause increased incidence of skin cancer.

Atomic chlorine is one of the many species able to combine with O*, and Cl atoms have been indirectly supplied to the ozone layer for decades by the widespread use of chlorofluorocarbons or CFCs, like CCl_3F and CCl_2F_2, as air conditioner fluids and aerosol gases. UV radiation breaks C— Cl bonds, making the CFCs sources of atomic chlorine. For example,

$$CCl_2F_2 \xrightarrow{\text{UV radiation}} CClF_2 + Cl$$

Atomic chlorine is able to destroy ozone and so disrupt the ozone cycle by the following steps.

$$Cl \ + O_3 \longrightarrow ClO + O_2 \qquad (3)$$
$$ClO \ + O \longrightarrow Cl \ + O_2 \qquad (4)$$
$$\text{Net:} \quad O_3 + O \longrightarrow 2O_2$$

Because this ozone-destroying pair of steps is a chemical chain reaction, the breakup of only one CFC molecule is able to initiate the destruction of thousands of ozone molecules. (Other Cl-based cycles also occur.)

THE ANTARCTIC OZONE "HOLE"

Since the 1970s, a decline in the ozone level has occurred over the Antarctic during each Antarctic spring[1] reaching its lowest levels in October each year. By 1984, the loss was 30%; by 1989, 70%. The result is that a continent-broad column of the atmosphere over the Antarctic, the *Antarctic ozone hole*, annually becomes more transparent to UV radiation. A significant recovery in the ozone level in the region of the ozone hole occurs between November and March. (Seasonally, the Arctic has a similar "hole" over the North Pole, but so far it has not been as severe.)

WHY OVER THE ANTARCTIC? WHY IN THE ANTARCTIC SPRING?

A circulating wind pattern called the *Antarctic vortex*, generated by the rotation of the Earth and the sharply uneven heating of the atmosphere, develops over the South Pole during the Antarctic winter, June–August. The chemical processes that destroy ozone are confined largely to this vortex (see Figure 7*a*).

During the total darkness of the Antarctic winter, the stratospheric temperature drops below 195 K ($-78°C$). Vast, thin clouds, called the polar stratospheric clouds (PSCs), form. They consist of crystals both of water and of nitric acid trihydrate, which forms by the reaction of water with nitrogen dioxide. NO_2 occurs both naturally and as a product of pollution. Molecular chlorine accumulates during this time and becomes adsorbed by the cloud crystals. Cl_2 is made as follows from chlorine nitrate ($ClONO_2$) and hydrogen chloride.

$$HCl + ClONO_2 \longrightarrow Cl_2 + HONO_2$$

Chlorine nitrate itself forms when ClO, made during Antarctic sunshine by equation 3, reacts with NO_2.

$$ClO + NO_2 \longrightarrow ClONO_2$$

Hydrogen chloride forms when atomic chlorine, available from the ozone-destroying cycle (Equations 3–4), reacts with methane (CH_4) to give CH_3 and HCl. (Methane is a pollutant produced both naturally and by pipeline leaks.)

When the Antarctic winter is over and the sun reappears, chlorine molecules are released from the polar stratospheric clouds and are soon broken apart by UV radiation in the new sunshine ($Cl_2 + UV \rightarrow$ 2Cl). Therefore, *just as sunlight returns and the Antarctic*

FIGURE 7*a*

The Antarctic polar vortex and the ozone hole on October 7, 1989. The wind speeds are indicated by color code in the upper part of the figure (the cylinder missing a wedge). The darkest blue indicates essentially zero wind velocity and the red areas correspond to velocities up to 180 mph. Beneath the cylinder, the surface plot is color coded for total ozone, where the darkest blue colors are around the South Pole and signify values below 200 Dobson units. Latitudes are drawn in at 30° and 60° south. Extending to the lower right is the Andes chain of mountains of South America. Extending to the lower left is the tip of Africa. (Used by permission from M. R. Schoeberl and D. L. Hartmann, *Science,* January 4, 1991, page 47.)

spring commences, a large supply of chlorine atoms is released. They quickly destroy ozone by the cycle of Equations 3–4, and the Antarctic ozone hole once again appears. In October 1994, the ozone level in the ozone hole dropped to a low of 100 Dobson units. (In the Octobers prior to the mid 1970s, the ozone levels were about 300 Dobson units.)

Between November and March, the polar vortex breaks up and air richer in ozone migrates into the Antarctic stratosphere from mid-latitudes. The result is an overall decrease in the stratospheric ozone level over the southern hemisphere.

Paul Crutzen (Germany/Netherlands), Mario Molina (USA/Mexico), and F. S. Rowland (USA) shared the 1995 Nobel prize in chemistry for their research into the effects of chlorofluorocarbons on the stratospheric ozone layer.

Questions

1. Write the equations by means of which ozone cycles are initiated.

2. What two equations represent the ozone cycle?

3. What happens to the UV radiation as a result of the ozone cycle?

4. By means of equations, explain how CFCs can reduce the stratospheric ozone level.

5. What is the polar vortex? The Antarctic ozone hole?

6. Polar stratospheric clouds provide reservoirs for the subsequent release of chlorine atoms. How is this done? (Include equations.)

7. How does the reappearance of the sun at the onset of the Antarctic spring initiate the decline in ozone levels in the ozone hole?

---

[1] The times of the seasons in the southern hemisphere are opposite those of the northern hemisphere. Spring in the Antarctic occurs during the months of fall in the northern hemisphere.

380
YRS
OLD

1776 US REV
1800

1889 WA. BECAME A STATE

The knowledge that some things in nature are periodic gives access to other data. As soon as scientists knew that the number of tree rings is a periodic function of time—one ring equals one year—they had the power to figure out the age of any tree, such as that of this 380-year old fir tree in the state of Washington. We'll learn of other, more important examples of periodic relationships in this chapter.

Chapter 9

Chemical Reactions: Periodic Correlations

Two of the most common kinds of reactions in all of nature involve the transfers from one species to another of nature's smallest particles, the proton and the electron. Much of acid–base chemistry concerns proton transfers. All of redox chemistry is about electron transfers. This chapter focuses on the qualitative aspects of such reactions. We'll ask, What species are most likely to give up protons? What species accept them? If substances vary in their proton donating or proton accepting abilities, are there trends that correlate with the periodic table? The same kinds of questions can be asked of electron donating and electron accepting species.

We will be using the concept and the vocabulary of a *chemical equilibrium* in a qualitative way throughout the chapter so, if you feel the need, review now what was said on this topic in Section 4.3. There, we discussed the Arrhenius view of acids and bases, namely, that *acids* give hydrogen ions (H^+), *bases* give hydroxide ions (OH^-), and an *acid–base neutralization* reaction is one between these two ions to give the solvent, H_2O. The Arrhenius concept, however, is needlessly restrictive because too many reactions resemble acid–base neutralization without involving either H^+ or OH^-. Even the reaction between $SO_3(g)$ and $CaO(s)$ to give $CaSO_4(s)$, for example, which we studied in Section 7.9 and which occurs in the absence of water, is so similar to neutralization that we ought to use the same terms, namely, *acid* (for SO_3), *base* (for CaO), and *neutralization*. Yet, neither hydrogen ions nor hydroxide ions are involved. Out of this example and many others, the *Lewis concept of acids and bases* was born. It is the most general view of acids and bases that we'll study: acids are electron pair acceptors; bases are electron pair donors.

9.1
BRØNSTED ACIDS AND BASES

Today, when water is the solvent, we use H^+ as our stand-in symbol for H_3O^+, but this was not how Arrhenius saw the acidic species in water.

The Lewis concept of acids and bases was discussed in Section 7.9.

FIGURE 9.1

The reaction of gaseous HCl and gaseous NH₃ to form NH₄Cl. The HCl and NH₃ are escaping from their concentrated aqueous solutions and are forming micro-crystals of NH₄Cl, which cause the cloud.

Johannes Brønsted (1879–1947) was a Danish chemist.

Chemists often use the terms *proton* and *hydrogen ion* interchangeably when dealing with acid–base chemistry.

There are many other reactions that strongly resemble acid–base neutralization that are not between H^+ and OH^-. For example, a white cloud of tiny crystals of solid ammonium chloride, $NH_4Cl(s)$, forms when ammonia and hydrogen chloride mix in air (see Figure 9.1).

$$NH_3(g) + HCl(g) \longrightarrow NH_4Cl(s)$$

This reaction also "looks" like an acid–base neutralization, yet all that has happened is that H^+ has moved from HCl to NH_3.

Recall that another name for H^+ is *proton*, because H^+ is a hydrogen atom minus its electron and so is simply the *nucleus* of a hydrogen atom, namely, one proton. Thus the transfer of a *proton* occurs when ammonia and hydrogen chloride mix in air. When their molecules are properly oriented during a collision, a proton transfers from a molecule of HCl and becomes attached (more strongly) to the N atom of a molecule of NH_3. As ammonium ions and chloride ions form, they attract each other, gather, and settle into places in crystals of ammonium chloride.

Proton transfers from Cl to N

Brønsted Concept

Johannes Brønsted[1] is credited with being the first to see clearly that the important event in most acid–base reactions is simply the transfer of a proton from one species to another. Brønsted proposed that *acids* be defined as species that can donate protons and *bases* as species that can accept protons. The heart of the Brønsted theory of acids and bases is that *acid–base reactions are simply proton-transfer reactions*. Brønsted's definitions of acids and bases are therefore very simple.

Brønsted's Definitions of Acids and Bases

An **acid** is a proton donor.

A **base** is a proton acceptor.

According to these definitions, even in the gaseous state HCl is an acid in its reaction with NH_3 because it donates a proton to the NH_3 molecule. Similarly, ammonia is a base because it accepts a proton from the HCl molecule.[2]

Even when water is the solvent, chemists use the Brønsted definitions more often than those of Arrhenius. Thus, the reaction between hydrogen chloride and water to form hydrochloric acid, which is another proton transfer reac-

---

[1] Thomas Lowry (1874–1936), a British scientist working independently of Brønsted, developed many of the same ideas but did not carry them as far. Some references call this view of acids and bases the Brønsted–Lowry theory.

[2] Notice that NH_3 is also a *Lewis base*; in its acceptance of H^+, NH_3 provides an electron pair, as the equation shows. The H^+ part from $HCl(g)$ is the Lewis acid; it accepts the electron pair in forming the bond to N in NH_4^+.

tion, is clearly a Brønsted acid–base reaction. Molecules of HCl are the acid in this reaction, and water molecules are the base.

$$H_2O + HCl(g) \longrightarrow H_3O^+(aq) + Cl^-(aq)$$
$$\text{base} \qquad \text{acid}$$

Conjugate Acids and Bases

Under the Brønsted view, it is useful to consider any acid–base reaction as a chemical equilibrium, having both a forward and a reverse reaction. We first encountered the chemical equilibrium in aqueous acetic acid on page 136, but now let's reevaluate it in the light of Brønsted's definitions. Acetic acid is a weak acid, so we represent its ionization as a chemical equilibrium in which water is not just a solvent but also a chemical reactant.

$$HC_2H_3O_2(aq) + H_2O \rightleftharpoons H_3O^+(aq) + C_2H_3O_2^-(aq)$$

At equilibrium both the forward and reverse reactions occur at equal rates. Because of the chemical nature of $HC_2H_3O_2$ —it is a *weak* acid—the equilibrium strongly favors the *reactants;* the percentage ionization is small, meaning that the percentage of the molecules of $HC_2H_3O_2$ in solution that have changed to $C_2H_3O_2^-$ ions is small.

In the forward reaction, the $HC_2H_3O_2$ molecule donates its proton to the water molecule and changes to an acetate ion. Thus $HC_2H_3O_2$ behaves as a Brønsted acid, a proton donor. Because water accepts this proton from $HC_2H_3O_2$, water behaves as a Brønsted base, a proton acceptor.

Now let's look at the reverse reaction, the one read from right to left. In the reverse reaction, it is H_3O^+ that behaves as a Brønsted acid; it donates a proton to the $C_2H_3O_2^-$ ion. The $C_2H_3O_2^-$ ion behaves as a Brønsted base by accepting the proton in the reverse reaction.

The reaction between $HC_2H_3O_2$ and H_2O is typical of proton-transfer reactions in general. In them we can always identify *two* acids (e.g., $HC_2H_3O_2$ and H_3O^+) and *two* bases (e.g., H_2O and $C_2H_3O_2^-$). Notice, too, that the acid on the right of the double arrows (H_3O^+) is formed from the base on the left (H_2O), and the base on the right ($C_2H_3O_2^-$) is formed from the acid on the left ($HC_2H_3O_2$). Two substances, like H_3O^+ and H_2O, that differ from each other by only *one* proton are referred to as a **conjugate acid–base pair.** Hydronium ion and water are thus a conjugate acid–base pair. One member of any such pair is called the **conjugate acid** because it is the proton donor of the two. The other member is called the **conjugate base,** because it is the pair's proton acceptor. Thus for the pair H_3O^+/H_2O, we say that H_3O^+ is the conjugate acid of H_2O, and H_2O is the conjugate base of H_3O^+.

We have another conjugate acid–base pair in the equilibrium involving $HC_2H_3O_2$ and water, namely, the pair $HC_2H_3O_2$ and $C_2H_3O_2^-$. $HC_2H_3O_2$ has one more H^+ than $C_2H_3O_2^-$, so the conjugate acid of $C_2H_3O_2^-$ is $HC_2H_3O_2$; the conjugate base of $HC_2H_3O_2$ is $C_2H_3O_2^-$. One of the common ways to designate the two members of a conjugate acid–base pair in an equilibrium is to connect them by a line.

Acetic acid ($HC_2H_3O_2$)

Only the H in red is available as a proton in an acid–base neutralization

Acetate ion ($C_2H_3O_2^-$)

Notice how H_3O^+ and H_2O differ by just one H^+ and that H_3O^+, the conjugate *acid*, has the extra H^+.

In any acid–base equilibrium, there are invariably *two* conjugate acid–base pairs. We must learn how to pick them out of an equation by inspection and to write them from formulas.

EXAMPLE 9.1
Determining the Conjugate
Bases of Brønsted Acids

What are the conjugate bases of nitric acid, HNO_3, and the hydrogen sulfate ion, HSO_4^-?

ANALYSIS A conjugate base is always found by removing one H^+ from a given acid.

SOLUTION Removing one H^+ (both the atom and the charge) from HNO_3 leaves NO_3^-. The nitrate ion, NO_3^-, is the conjugate base of HNO_3. Deleting an H^+ from HSO_4^- leaves its conjugate base, SO_4^{2-}. [Notice that the charge goes from $1-$ to $2-$ because $(1-) - (1+) = (2-)$.]

■ **Practice Exercise 1** Write the formula of the conjugate base for each of the following Brønsted acids. (a) H_2O, (b) HI, (c) HNO_2, (d) H_3PO_4, (e) $H_2PO_4^-$, (f) HPO_4^{2-}, (g) H_2, (h) NH_4^+

EXAMPLE 9.2
Determining the Conjugate
Acids of Brønsted Bases

What are the formulas of the conjugate acids of OH^- and PO_4^{3-}?

ANALYSIS To find a conjugate acid we add one H^+ to the formula of the given base.

SOLUTION Adding H^+ to OH^- gives H_2O, the conjugate acid of the hydroxide ion. Adding H^+ to the formula PO_4^{3-} gives HPO_4^{2-}, the conjugate acid of PO_4^{3-}. Always be sure that the electrical charge is correctly shown.

■ **Practice Exercise 2** Write the formula of the conjugate acid for each of the following Brønsted bases. (a) HO_2^-, (b) SO_4^{2-}, (c) CO_3^{2-}, (d) CN^-, (e) NH_2^-, (f) NH_3, (g) $H_2PO_4^-$, (h) HPO_4^{2-}

EXAMPLE 9.3
Identifying Conjugate Acid–
Base Pairs in a Brønsted
Acid–Base Reaction

Sodium hydrogen sulfate is used
in the manufacture of certain
kinds of cement and to clean
oxide coatings from metals.

The anion of sodium hydrogen sulfate, HSO_4^-, reacts as follows with the phosphate anion, PO_4^{3-}.

$$HSO_4^-(aq) + PO_4^{3-}(aq) \longrightarrow SO_4^{2-}(aq) + HPO_4^{2-}(aq)$$

Identify the two Brønsted acids and the two Brønsted bases in this reaction.

ANALYSIS We look for pairs that differ by only one H^+. *The members of each pair must be on opposite sides of the arrow.*

SOLUTION Two of the formulas in the equation contain "PO_4," so they must belong to the same conjugate pair. The one with the greater number of hydrogens, HPO_4^{2-}, must be the Brønsted acid, and the other, PO_4^{3-}, must be the Brønsted base. Therefore, one conjugate acid–base pair is HPO_4^{2-} and PO_4^{3-}. The other two ions, HSO_4^- and SO_4^{2-}, belong to the second conjugate acid–base pair; HSO_4^- is the conjugate acid and SO_4^{2-} is the conjugate base.

conjugate pair

$$HSO_4^-(aq) + PO_4^{3-}(aq) \longrightarrow SO_4^{2-}(aq) + HPO_4^{2-}(aq)$$

acid   base   base   acid

conjugate pair

Identify the conjugate acid–base pairs in the following reaction.

$$CN^-(aq) + H_3O^+(aq) \longrightarrow HCN(aq) + H_2O$$

ANALYSIS To find the acid–base pairs, we look for formulas that differ only by one H^+.

SOLUTION We see that one pair must be CN^- and HCN, and the other pair must be H_3O^+ and H_2O. In each pair, the acid is the substance with one more hydrogen than the other.

■ Practice Exercise 3 When aqueous solutions of sodium bicarbonate and sodium dihydrogen phosphate are mixed in the proper mole ratios of solutes, a reaction occurs by the following net ionic equation.

$$HCO_3^-(aq) + H_2PO_4^-(aq) \longrightarrow H_2CO_3(aq) + HPO_4^{2-}(aq)$$

Identify the two Brønsted acids and the two Brønsted bases in this reaction. Rewrite the equation, showing the two conjugate acid–base pairs.

■ Practice Exercise 4 When some of the strong cleaning agent trisodium phosphate is mixed with household vinegar, which contains acetic acid, the following equilibrium is established. (The products are favored.) Identify the pairs of conjugate acids and bases.

$$PO_4^{3-}(aq) + HC_2H_3O_2(aq) \rightleftharpoons HPO_4^{2-}(aq) + C_2H_3O_2^-(aq)$$

Amphoteric Substances

Some molecules or ions are able to function as either an acid or a base, depending on the kind of substance mixed with them. The most common example is water. Consider again its reaction with $HCl(g)$.

$$\underset{\text{base}}{H_2O} + \underset{\text{acid}}{HCl(g)} \longrightarrow H_3O^+(aq) + Cl^-(aq)$$

Notice that H_2O behaves here as a *base* because it accepts a proton from the HCl molecule. Water, however, interacts as an *acid* with ammonia in the following equilibrium.

$$\underset{\text{acid}}{H_2O} + \underset{\text{base}}{NH_3(aq)} \rightleftharpoons NH_4^+(aq) + OH^-(aq)$$

Here it donates a proton to NH_3 in the forward reaction. The fact that the forward reaction is not the *favored* reaction simply reflects the chemical natures of both H_2O and NH_3; H_2O is a (very) weak acid and NH_3 is a weak base. Nonetheless, the forward reaction shows H_2O acting as a Brønsted acid.

Substances that can be either acids or bases depending on the other substance present are said to be **amphoteric.** Another term that is sometimes used is **amphiprotic** to stress that *proton* donating/accepting ability is of central concern.

EXAMPLE 9.4
Identifying Conjugate Acid–
Base Pairs

HCN(*g*) is hydrogen cyanide, a dangerous poison used by some states for executions. HCN(*aq*) is called hydrocyanic acid.

From the Greek *amphoteros,*
"partly one and partly the other."

Acid salts, like NaHCO$_3$, NaHSO$_3$, and NaH$_2$PO$_4$, are so named because they can neutralize OH$^-$.

As suggested above, amphoteric or amphiprotic substances need not be molecules. Some ions also have this property, especially the anions found in acid salts. For example, the bicarbonate ion found in baking soda can either donate a proton to a base or accept a proton from an acid. Toward hydroxide ion, HCO_3^- is an acid, the net ionic equation being

$$\underset{\text{acid}}{HCO_3^-(aq)} + \underset{\text{base}}{OH^-(aq)} \longrightarrow CO_3^{2-}(aq) + H_2O$$

Toward hydronium ion, however, HCO_3^- is a base.

$$\underset{\text{base}}{HCO_3^-(aq)} + \underset{\text{acid}}{H_3O^+(aq)} \longrightarrow H_2CO_3(aq) + H_2O$$

As it forms, the H$_2$CO$_3$ almost entirely decomposes to CO$_2$(*g*) and water.

■ **Practice Exercise 5** Using net ionic equations, illustrate how the anion in Na$_2$HPO$_4$ is amphoteric.

9.2 ■

STRENGTHS OF BRØNSTED ACIDS AND BASES AND PERIODIC TRENDS

Strong Brønsted acids are defined with reference specifically to water as both the solvent and the Brønsted base. Because, like strong electrolytes, *strong acids are essentially 100% ionized in water,* such acids are strong donors of protons toward the base, water. Weak acids are poor donors of protons toward water. A strong Brønsted base, like the hydroxide ion, is one that strongly accepts and binds protons from donors. A weak Brønsted base, like the water molecule, is a weak proton binder. The OH$^-$ ion and the H$_2$O molecule, of course, make up a conjugate acid–base pair, and one of the more illuminating comparisons of acid/base strengths involves just such pairs.

Comparing the Acid–Base Strengths of Conjugate Pairs

Consider once again the equilibrium in aqueous acetic acid. When we say that acetic acid is a weak acid, we are saying that the *position of the equilibrium* in *aqueous* acetic acid lies on the left. The reactants are favored; most HC$_2$H$_3$O$_2$ molecules exist in water most of the time as un-ionized molecules.

It is the *chemical nature* of acetic acid not to ionize much.

$$\underset{\text{acid}}{HC_2H_3O_2(aq)} + \underset{\text{base}}{H_2O} \rightleftharpoons \underset{\text{acid}}{H_3O^+(aq)} + \underset{\text{base}}{C_2H_3O_2^-(aq)}$$

What we now add to the discussion is that the *position* of an acid–base equilibrium tells us something about the relative strengths both of the two Brønsted acids present as well as of the two Brønsted bases.

In the aqueous acetic acid equilibrium, for example, there are two Brønsted acids, HC$_2$H$_3$O$_2$ and H$_3$O$^+$. Each acid is competing in an attempt to donate its proton to a base. The fact that nearly all the protons end up in the HC$_2$H$_3$O$_2$ molecules, and only relatively few spend their time in the H$_3$O$^+$ ions, means that the *hydronium ion is a better proton donor than the acetic acid molecule.*

The acetic acid equilibrium also has two bases, C$_2$H$_3$O$_2^-$ and H$_2$O, and we can conclude that C$_2$H$_3$O$_2^-$ is the stronger base. Both C$_2$H$_3$O$_2^-$ and H$_2$O compete as bases for the available protons. At equilibrium, however, because of the chemical nature of the aqueous HC$_2$H$_3$O$_2$ system, most of the protons originally provided by HC$_2$H$_3$O$_2$ are still found in HC$_2$H$_3$O$_2$ molecules rather than joined to H_2O in H_3O^+ ions. This means that *acetate ions must be more effective than water molecules at winning protons from donors.* This is the same as saying that the acetate ion is a stronger base than the water molecule. Thus, *the position of equilibrium favors the weaker acid and base,* which are $HC_2H_3O_2$ and H_2O in aqueous acetic acid. Such an outcome is generally true: *In an equilibrium involving conjugate pairs of Brønsted acids and bases, the position of equilibrium always favors the weaker species.* Stated another way, *stronger acids and bases tend to react with each other to produce their weaker conjugates.*

Another useful generalization concerning the relative strengths of conjugate acid–base pairs is that *the stronger a Brønsted acid is, the weaker is its own conjugate base.* To illustrate, recall that $HCl(g)$ is a very strong Brønsted acid. We know this because $HCl(g)$ is 100% ionized in water—$HCl(g)$ is strong enough a proton donor to give away all of its protons even to such weak acceptors as water molecules. (We don't even write double equilibrium arrows.)

$$HCl(g) + H_2O \xrightarrow{100\%} H_3O^+(aq) + Cl^-(aq)$$

At the same time, we can also see that the chloride ion, the conjugate base of $HCl(g)$, is a very weak Brønsted base. Even in the presence of a very strong proton donor, namely, H_3O^+, the chloride ion isn't able to win any protons. So the strong acid, HCl, has a particularly weak conjugate base, Cl^-.

An extension of the previous generalization is that *the weaker a Brønsted acid is, the stronger is its conjugate base.* As an example, let's compare the conjugate pair OH^- and O^{2-}. Of the two, the hydroxide ion is the conjugate acid(!)— only OH^- of this pair has a proton to donate—and the oxide ion is the conjugate base. The hydroxide ion must be exceedingly weak as a Brønsted acid; in fact, we've known it so far only as a *base.* Given that OH^- must be an extremely weak acid, its conjugate base, the oxide ion, must be (and is) a *very* strong base. In water, oxide ions are such powerful proton acceptors that they are able to win protons quantitatively from H_2O molecules, which are very weak proton donors as we know. (Again, we don't write the double equilibrium arrows.)

$$\underset{\text{base}}{O^{2-}(aq)} + \underset{\text{acid}}{H_2O} \xrightarrow{100\%} 2OH^-(aq)$$

By now you may have realized that considerable variations in weakness must exist among acids or bases described here only qualitatively as weak. In Chapter 16 we'll develop a quantitative description of relative weakness, so we'll continue the discussion here on a qualitative level. There is much to be learned about periodic relationships and acid–base strengths.

Strengths of Binary Acids and Periodic Relationships

One of the chief goals of our study throughout this book is to acquaint you with the periodic table and to give you a feel for the way chemists use the table to correlate, explain, and remember chemical facts. Let's now learn how the periodic table can be used to help us understand and remember the strong and weak acids.

Many of the binary compounds between hydrogen and nonmetals, any of which we can represent by HX, H_2X, H_3X, etc., are acidic and are called **binary acids.** HCl is an example already mentioned. Table 9.1 lists the binary hydro-

The hydronium ion is the strongest proton donor that can exist in the presence of water. Any stronger proton donor, like $HCl(g)$, reacts with water to give H_3O^+.

The hydroxide ion is the strongest proton acceptor that can exist in the presence of water. Any stronger proton acceptor added to water reacts with water to generate OH^-.

Periodic table and binary acids strengths

TABLE 9.1   Acidic Binary Compounds of Hydrogen and Nonmetals[a]

Group VI		Group VII	
(H_2O)		HF	Hydrofluoric acid
H_2S	Hydrosulfuric acid	*HCl	Hydrochloric acid
H_2Se	Hydroselenic acid	*HBr	Hydrobromic acid
H_2Te	Hydrotelluric acid	*HI	Hydriodic acid

[a] The *names* are for the aqueous solutions of these compounds. Strong acids are marked with asterisks.

gen compounds that are acids in water. Those that are strong are marked by asterisks. If you have not already done so, *you must memorize the names and formulas of the strong binary acids.*

Pure water is neutral in an acid–base sense.

Water, a very weak acid as we have mentioned, is included in the list. Its reaction with itself, however, produces H_3O^+ and OH^- ions in equal numbers.

Periodic Trends in the Strengths of Binary Acids   The relative strengths of the binary acids can be correlated with the periodic table in two ways.

The strengths of the binary acids increase from left to right in a period.

Electronegativity was discussed on page 282.

These horizontal trends in acid strength are in parallel with trends in electronegativity. As we go left to right from S to Cl in period 3 or from O to F in period 2, the electronegativity increases, so the H—X bond becomes more polar, making the partial positive charge on H greater. Hence the H separates from an H—X bond more and more easily as H^+, and the binary hydride becomes a stronger acid as we go from left to right in the same period. Thus, HCl is a stronger acid than H_2S, and HF is a stronger acid than H_2O.

Cl is to the *right* of S in period 3; F is to the *right* of O in period 2.

The second correlation will seem at first to violate what we just emphasized, namely, the importance of relative electronegativities.

Within a group, the strengths of binary acids increase from top to bottom.

Among the binary acids of the halogens, for example, the following is the order of relative acidity in water.

$$HF < HCl < HBr < HI$$

Relative to HCl, of course, H_2S, HF, and H_2O are all weak acids.

HF is the weakest acid in the series, and HI is the strongest. How can this be, since F is the most electronegative of the halogens, and I is the least? Shouldn't the H—F bond be more polar than the H—I bond? In fact it is; the H—F bond is more polar than the H—I bond. Therefore, why isn't H—F the most acidic of the binary acids of the halogens? Evidently, in our theory about the acid strengths of H—X acids, the relative electronegativity of X must be only one factor.

Another factor concerns the strength of the bond from the nonmetal atom to H. As we proceed down a group in the periodic table, the H—X bond involves electrons farther and farther from the nucleus of X. The farther that the electron pair of the bond is from the X nucleus, the weaker is the bond. With the increasing size of X, the ability of X to hold the H atom becomes progressively weaker. As we descend the halogen group, this loss in H—X bond strength tends to make the acids stronger. The loss in bond strength more than compensates for the lessening size of the partial positive charge on H as X becomes less electronegative. The net effect of these two opposing factors is an increase in the strengths of the binary acids of the halogens as we descend the group.

In the series of the binary acids of Group VIA elements we see the identical trend. Their formulas are of the general type H_2X, and the farther X is from the top of the group the stronger H_2X is as an acid.

■ **Practice Exercise 6** Using *only* the periodic table, choose the stronger acid of each pair: (a) H_2Se or HBr, (b) H_2Se or H_2Te, (c) CH_3OH or CH_3SH.

Oxoacids

Acids made of hydrogen, oxygen, and some other element are called **oxoacids,** and they include many common and important acids (see Table 9.2). Those that are strong are marked by asterisks in the table. If you have not already done so, *you should memorize the names and formulas of these strong acids* because, when added to the strong binary acids of Table 9.1, there are relatively few strong acids. Once you know which acids are strong, you will know that any others (there are hundreds) are weak.

The concept of electronegativity helps us understand the relative strengths of oxoacids acids—why some, like H_2SO_4, are strong but very similar acids, like H_2SO_3, arc much weaker. The ionizable hydrogens of oxoacids are attached to oxygen atoms and these, in turn, are held by some other atom, usually that of a nonmetal or metalloid found at or near the center of the molecule. Oxoacids, in other words, have one or more OH groups attached to a central atom. Here, for example, are the Lewis structures of two oxoacids of Group VIA elements. The central atom in sulfuric acid is S, and in selenic acid it is Se.

Strong acids

Relative acidities of some acids. The weaker the acid is, the lower it is on this list. (Only a few weak acids are given here.)

$HClO_4$	
HI	
HBr	**STRONG**
H_2SO_4	**ACIDS**
HCl	
HNO_3	
H_2SO_3	**MODERATE**
H_3PO_4	**ACIDS**
HNO_2	
HF	
$HC_2H_3O_2$	**WEAK**
H_2CO_3	**ACIDS**
HCN	

$$
\begin{array}{ccc}
& O & \\
& \parallel & \\
H-O-S-O-H & & H-O-Se-O-H \\
& \parallel & \\
& O & \\
\end{array}
$$

H_2SO_4
sulfuric acid

H_2SeO_4
selenic acid

TABLE 9.2 Some Oxoacids of Nonmetals and Metalloids[a]

Group IVA		Group VA		Group VIA		Group VIIA	
H_2CO_3	Carbonic acid	*HNO_3	Nitric acid			HFO	Hypofluorous acid
		HNO_2	Nitrous acid				
		H_3PO_4	Phosphoric acid	*H_2SO_4	Sulfuric acid	*$HClO_4$	Perchloric acid
		H_3PO_3	Phosphorous acid[b]	H_2SO_3	Sulfurous acid	*$HClO_3$	Chloric acid
						$HClO_2$	Chlorous acid
						$HClO$	Hypochlorous acid
		H_3AsO_4	Arsenic acid	*H_2SeO_4	Selenic acid	*$HBrO_4$	Perbromic acid[c]
		H_3AsO_3	Arsenous acid	H_2SeO_3	Selenous acid	*$HBrO_3$	Bromic acid
						HIO_4	Periodic acid
						(H_5IO_6)[d]	
						HIO_3	Iodic acid

[a] Strong acids are marked with asterisks.

[b] Phosphorous acid, despite its formula, is only a diprotic acid.

[c] Pure perbromic acid is unstable; a dihydrate is known.

[d] H_5IO_6 is formed from $HIO_4 + 2H_2O$.

Sulfuric acid is a stronger acid than selenic acid, and the question is, "Why?" But we might first ask a prior question, "Why should either be acidic at all?"

To answer the last question, we know that to be acidic means that H$^+$ can transfer to H$_2$O in an ionization. This H atom in the oxoacids is bound to an atom of the second most electronegative element, oxygen, so the O—H bond is a very polar bond. Even in the un-ionized molecular acid, the H atom already is partway toward having a positive charge. It is partway, in other words, toward separating as H$^+$.

$\overset{\delta-}{\text{O}}\!\!-\!\!\overset{\delta+}{\text{H}}$

Polarity of O—H
bond in oxoacids

> The largest that $\delta+$ could be is 1 +, which would mean a detached hydrogen ion.

Thus, one reason why oxoacids with O—H groups are acidic in the first place is because the O—H bond is already partly ionic.

Periodic table and oxoacids strengths

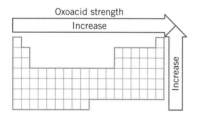

Oxoacid strength
Increase

Increase

Oxoacids with Different Central Atoms but the Same Number of Oxygens—The Effect of the Location of the Central Atom in the Periodic Table

When comparing oxoacids of the same structure except for the central atom, the size of the $\delta+$ on H in the O—H unit varies in a systematic way with the location of the central atom in the periodic table. Be sure to notice that our attention here is only on the oxoacids with central atoms holding identical numbers of oxygens. This keeps the influence of the oxygens themselves constant as we change the central atom. Oxoacids of this type vary in strength as the central atom is changed within the same vertical group in the periodic table in the following way.

When the central atoms of oxoacids are in the same *group* and they hold the same number of oxygen atoms, the acid strength increases from bottom to top within the group.

The reason for this is now easy to understand. The electronegativities of the central atoms in oxoacids with *X*—O—H systems have the same influence as they have in the binary acids with H—*X* bonds. Recall that central atoms of the same group in the periodic table have electronegativities that increase from bottom to top in their group. As we move up a group, *X* becomes more electronegative and thus draws electron density more strongly toward itself in bonds. The central atom in *X*—O—H systems can thus make O even more able to draw electron density away from H. An increase in the electronegativity of *X*, therefore, increases the shift of electron density down the entire *X*—O—H chain toward the *X* and away from the H. As a result, there is a larger $\delta+$ on the H. As with the binary acids, the larger this $\delta+$ on H is, the more able is the H to transfer as H$^+$.

We can now understand why H$_2$SO$_4$ is a stronger acid than H$_2$SeO$_4$. Their central atoms, S and Se, are in the same group, with S being nearer the top. Electronegativities increase from bottom to top in a group, so we know that S is more electronegative than Se. An S atom, therefore, can cause a larger $\delta+$ on the H of an adjoining OH group than can an Se atom. So sulfuric acid is a stronger acid than selenic acid.

The same trend holds among the Group VA elements. Phosphoric acid, H_3PO_4, is a stronger acid than arsenic acid, H_3AsO_4, because P is more electronegative than As, which is below P in Group VA.

Another generalization about the relative strengths of oxoacids also stems from the relative electronegativities of central atoms. To keep the effects of the O atoms constant, we again have to let their number on the central atom be constant. Such oxoacids vary in strength as the central atom's location changes in the same *period* in the following way.

When the central atoms of oxoacids are in the same *period* and they hold the same number of oxygen atoms, the acid strength increases from left to right within the period.

The reason for this is that electronegativities also increase from left to right within a period. Phosphorus, for example, is just to the left of sulfur in the third period and is less electronegative than sulfur. The P atom in H_3PO_4, therefore, cannot contribute to the size of $\delta+$ on the H of an O—H group as much as S, so H_3PO_4 is a weaker acid than H_2SO_4.

The percentage ionization in 1 *M* H_3PO_4 is far less than in 1 *M* H_2SO_4.

$$H-O-\overset{\displaystyle O}{\underset{\displaystyle O}{\overset{\|}{\underset{\|}{S}}}}-O-H \qquad H-O-\overset{\displaystyle O}{\overset{\|}{\underset{\displaystyle O-H}{\underset{|}{P}}}}-O-H$$

H_2SO_4 H_3PO_4
sulfuric acid   phosphoric acid

Oxoacids and the Effect of the Dispersal of Negative Charge to the Lone Oxygens

Still another factor that contributes to the relative acidities of oxoacids is their number of *lone oxygens*, those joined to the central atom but not also bonded to any H atoms. As you can see in their structures, H_2SO_4 has two lone oxygens and H_3PO_4 has only one. Following the ionizations of their first protons, three lone oxygens are in the resulting HSO_4^- anion but only two are in the $H_2PO_4^-$ anion.

In the oxoacids that have them, the *lone* oxygens, not the oxygens still bonded to hydrogens, carry most of the negative charge. This charge actually spreads uniformly to the lone oxygens. In HSO_4^-, the single $(1-)$ charge spreads over *three* lone oxygens; in $H_2PO_4^-$, it spreads just to *two*. In HSO_4^-, therefore, less negative charge occurs at any one place than in $H_2PO_4^-$. Less negative charge means, of course, less ability to attract something positively charged, like H^+. As a result, any given lone oxygen in HSO_4^- is less able than any lone oxygen in $H_2PO_4^-$ to recapture H^+ from H_3O^+ in the solution. In other words, HSO_4^- cannot become H_2SO_4 again as readily as $H_2PO_4^-$ can become H_3PO_4. So if we compare solutions of H_2SO_4 and H_3PO_4 of equal concentrations, there will be a larger percentage of HSO_4^- ions than of $H_2PO_4^-$ ions. Sulfuric acid, in other words, will be more fully ionized. Of course, this is how we define acid strength—in terms of the *percentage* of the acid molecules that become ionized in solution—so we say that sulfuric acid is a stronger acid than phosphoric acid.

$$H-O-\overset{\displaystyle O}{\underset{\displaystyle O}{\overset{\|}{\underset{\|}{S}}}}-O^-$$

HSO_4^-
Three lone O atoms

$$H-O-\overset{\displaystyle O}{\overset{\|}{\underset{\displaystyle O-H}{\underset{|}{P}}}}-O^-$$

$H_2PO_4^-$
Two lone O atoms

The HSO_4^- ion is also a stronger acid than the $H_2PO_4^-$ ion.

■ **Practice Exercise 7** Predict the stronger acid in each pair: (a) $HClO_4$ or $HBrO_4$, (b) H_3AsO_4 or H_2SeO_4.

Oxoacids with the Same Central Atom but Different Numbers of Oxygens In the two previous major generalizations, we stated an important condition—the number of oxygens held by the central atom must be the same in the formulas of the acids being compared. This is because acid strength also varies systematically with the number of such oxygens, provided we compare acids with the same central atom.

For a given central atom, the acid strength of an oxoacid increases with the number of oxygens it holds.

Sulfuric acid (H_2SO_4), for example, is a stronger acid than sulfurous acid (H_2SO_3). Nitric acid (HNO_3) is a stronger acid than nitrous acid (HNO_2).

$$
\begin{array}{cccc}
\text{H---O---}\overset{\overset{\displaystyle O}{\|}}{\underset{\underset{\displaystyle O}{\|}}{S}}\text{---O---H} & \text{H---O---}\overset{\overset{\displaystyle O}{\|}}{S}\text{---O---H} & \text{H---O---}\overset{\overset{\displaystyle O}{|}}{N}\!\!=\!\!O & \text{H---O---N}\!\!=\!\!O
\end{array}
$$

H_2SO_4		H_2SO_3		HNO_3		HNO_2
sulfuric acid	>	sulfurous acid		nitric acid	>	nitrous acid

The extra oxygen on S in sulfuric acid contributes to an additional withdrawal of electron density from each O—H bond and thus makes each O—H bond more polar in this acid than in sulfurous acid. So the size of $\delta+$ on H in sulfuric acid is greater than it is in sulfurous acid, making sulfuric acid the stronger acid of the two. The same argument explains why nitric acid is a stronger acid than nitrous acid.

Also at work here is the effect of the number of lone oxygens. In the HSO_4^- ion there are three lone oxygens and in HSO_3^- there are two. The greater number of lone oxygens in HSO_4^- leads to a greater spreading out of the negative charge than occurs in HSO_3^-. With a greater spreading out of negative charge, each site in HSO_4^- that shares in carrying this charge has a relatively low negative charge density compared to similar sites in HSO_3^-. Consequently, HSO_4^- has a lesser tendency than HSO_3^- to attract and so to recapture H^+. In solutions of equal concentrations of sulfuric acid and sulfurous acid, therefore, the percentage ionization in H_2SO_4 is higher, and sulfuric acid is the stronger acid. The same factors help to make nitric acid a stronger acid than nitrous acid.

> The HSO_4^- ion is also a stronger acid than the HSO_3^- ion.

■ **Practice Exercise 8** Predict the order of acid strength for the four oxoacids of chlorine in Table 9.2.

9.3
ACID–BASE PROPERTIES OF THE ELEMENTS AND THEIR OXIDES

You have probably already noticed that the elements most likely to form acids are the nonmetals in the upper right-hand corner of the periodic table. The elements most likely to form basic hydroxides are similarly grouped in one general location in the table, among the metals, particularly those in Groups IA and IIA along the left side. Thus elements can themselves be classified according to their abilities to be involved in acids or bases.

Few nonmetals form acids *directly* from the elements, however, and no metals form hydroxides directly from their elements. So the experimental basis for classifying the elements according to their abilities to be involved as acids or bases depends on how their *oxides* behave toward water. In general, *metal oxides react with water to form bases, and nonmetal oxides react with water to form acids.* Typical metal oxides that give hydroxides in water are sodium oxide, Na_2O, and calcium oxide, CaO

$$Na_2O + H_2O \longrightarrow 2NaOH \qquad \text{sodium hydroxide}$$

$$CaO + H_2O \longrightarrow Ca(OH)_2 \qquad \text{calcium hydroxide}$$

Metal oxides like Na_2O and CaO are called **basic anhydrides** (*anhydride* means "without water") because they can neutralize acids, and their aqueous mixtures test basic to litmus. Basic anhydrides are generally metal oxides that react with water to give metal hydroxides, as discussed in Chapter 4.

Metal oxides are invariably solids, and the oxide ions (O^{2-}) in many metal oxides are so tightly bound in the crystals that when such oxides are stirred with water their oxide ions are unable to win H^+ ions from surrounding water molecules. When a strong acid is present in the surrounding medium, however, then even some otherwise water-insoluble metal oxides dissolve. Iron(III) oxide, for example, dissolves in water by reacting as follows with acid.

$$Fe_2O_3(s) + 6H^+(aq) \longrightarrow 2Fe^{3+}(aq) + 3H_2O$$

Hydrates of Metal Ions with High Charge Densities as Weak Acids The partial negative sites on O in water molecules are attracted to metal ions in water. Water molecules thus gather about a cation to give a species called a *hydrated ion*. Hydrated metal ions are acidic in solution *when the central ion has a sufficiently high density of positive charge,* the ratio of the cation's charge to its ionic volume. To have a high charge density on a cation usually means that the charge must be more than 2 +.

The small, triply charged cation, Al^{3+}, is an example of a metal ion with sufficient positive charge density to form a hydrated ion in water and to make this hydrate a proton donor. The hexahydrate of aluminum forms, and it sets up the following equilibrium in water to generate hydronium ions.

$$[Al(H_2O)_6]^{3+}(aq) + H_2O \rightleftharpoons [Al(H_2O)_5(OH^-)]^{2+}(aq) + H_3O^+(aq)$$

Behind this behavior is the ability of the high positive charge density on Al^{3+} in $[Al(H_2O)_6]^{3+}$ to draw considerable electron density from the O—H bonds of the water molecules. We could even say that Al^{3+} is powerfully electronegative. This reduction of electron density in the O—H bonds weakens them enough to make an H_2O molecule in $[Al(H_2O)_6]^{3+}(aq)$ a proton donor. We saw exactly the same effect of electronegativity in strong oxoacids. Among them, the relatively higher electronegativity of a central atom in an acid like H_2SO_4 helps to weaken a neighboring O—H bond and to make the system an acid.

It's the charge *density* that matters, not just the size of the charge, so if the cation is very small, like Be^{2+}, the *ratio* of its smaller 2+ charge to its small size is yet large enough to make the hydrated Be^{2+} ion a weak acid.

Metal Ions with Small Charges as Nonacids None of the singly charged cations of the Group IA metals—Li^+, Na^+, K^+, Rb^+, Cs^+—generates extra hydrogen ions in water. Except for Be^{2+}, neither do any of the doubly charged cations of the Group IIA metals—Mg^{2+}, Ca^{2+}, Sr^{2+}, or Ba^{2+}. Apparently, in none of these cations is the ratio of charge to size large enough.

Calcium oxide, or *quicklime,* is a component of cement, and the reaction of CaO with water occurs when cement "sets."

The reactions of insoluble metal oxides with acids was discussed in Chapter 4.

The force of attraction between an ion and water molecules is often strong enough to persist in crystals when solid *hydrates* separate from solution, for example, $CuSO_4 \cdot 5H_2O$ (page 41).

The hydrated aluminum ion, $Al(H_2O)_6^{3+}$

Recall from Chapter 4 that not all nonmetal oxides react with water.

Nonmetal Oxides as Acids Nonmetal oxides are usually **acidic anhydrides;** they react with water to give acids. Typical examples of the formation of acids from nonmetal oxides are the following reactions.

$$SO_3(g) + H_2O \longrightarrow H_2SO_4(aq) \qquad \text{sulfuric acid}$$

$$N_2O_5(g) + H_2O \longrightarrow 2HNO_3(aq) \qquad \text{nitric acid}$$

$$CO_2(g) + H_2O \longrightarrow H_2CO_3(aq) \qquad \text{carbonic acid}$$

9.4 REACTIONS OF METALS WITH ACIDS

As we said, the transfers of subatomic particles characterize two major families of chemical reactions. We've just studied one family, *Brønsted acid–base reactions,* which feature proton transfers. We now turn to the second family, namely, *redox reactions,* which involve electron transfers. Such reactions were introduced in Sections 4.8 and 4.9, where we learned how to recognize **redox reactions** (oxidation–reduction reactions) as those involving changes in oxidation numbers. You may wish to review the vocabulary of these sections now as well as how to assign *oxidation numbers,* studied in Section 4.8. Recall that anything that undergoes an increase in oxidation number (becomes more positive or less negative) is said to be oxidized. In the same reaction mixture, something else necessarily experiences a decrease in oxidation number (becomes less positive or more negative) and so is reduced.

Recall that oxidizing agents are reduced and reducing agents are oxidized.

Oxidation: oxidation number becomes more positive or less negative.

Reduction: oxidation number becomes less positive or more negative.

We now turn our attention to some especially important kinds of redox reactions. One is the reaction of a metal with an acid.

An important property of acids is their tendency to react with certain metals. Battery acid (which is sulfuric acid, H_2SO_4) will begin to dissolve unprotected iron or steel parts of an automobile if it's spilled and not flushed away with water. A similar reaction occurs even faster, accompanied by vigorous bubbling, if the acid is spilled on the zinc surface of galvanized steel (see Figure 9.2).

In general, the reaction of an acid with a metal is a redox reaction in which the metal is oxidized and the acid is reduced. But in these reactions, the part of the acid that is reduced depends on the composition of the acid itself as well as on the metal.

When sulfuric acid reacts with the zinc coating on a galvanized garbage pail or a galvanized nail, the reaction is

In many cases,

$$\text{metal} + \text{acid} \longrightarrow H_2 + \text{salt}$$

$$Zn(s) + H_2SO_4(aq) \longrightarrow ZnSO_4(aq) + H_2(g)$$

The observed bubbling is caused by the release of hydrogen gas. If we assign oxidation numbers, we can analyze the oxidation–reduction changes that occur. In the diagram below, we use the symbol Δ to stand for the change in oxidation number (positive, as we said, for an increase in oxidation number, negative for a decrease).

$$\Delta = (-1 \text{ per H}) \times (2\text{H}) = -2 \text{ units}$$

$$\underset{0}{Zn(s)} + \underset{+1+6-2}{H_2SO_4(aq)} \longrightarrow \underset{+2+6-2}{ZnSO_4(aq)} + \underset{0}{H_2(g)}$$

$$\Delta = +2 \text{ units per Zn}$$

The oxidation number of zinc increases from 0 to $+2$, so zinc is oxidized. The oxidation number of hydrogen decreases from $+1$ to 0, so the hydrogen of the acid is reduced.

An even clearer picture of what occurs is seen if we write the net ionic equation. Sulfuric acid, you recall, is a strong acid, so in aqueous solution we can write it in ionized form. Similarly, zinc sulfate is a salt and is therefore a strong electrolyte; we write it in fully dissociated form as well. The ionic equation for the reaction is therefore

$$Zn(s) + 2H^+(aq) + SO_4{}^{2-}(aq) \longrightarrow Zn^{2+}(aq) + H_2(g) + SO_4{}^{2-}(aq)$$

After spectator ions are cancelled, the net ionic equation for the reaction is

$$Zn(s) + 2H^+(aq) \longrightarrow Zn^{2+}(aq) + H_2(g)$$

Once again, assigning oxidation numbers reveals that the zinc is oxidized and the H^+ of the acid is reduced.

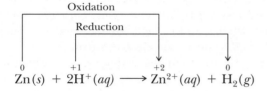

Stated another way, the H^+ of the acid is the oxidizing agent.

Many metals react with acids just as zinc does—by being oxidized by hydrogen ions. In these reactions a metal salt and gaseous hydrogen are the products.

■ **Practice Exercise 9** Write balanced molecular, ionic, and net ionic equations for the reaction of hydrochloric acid with (a) magnesium and (b) aluminum. (Both are oxidized by hydrogen ions.)

The ease with which metals lose electrons varies considerably from one metal to another. Some metals are easily oxidized and others are not. For many metals, like iron and zinc, the hydrogen ion is a sufficiently strong oxidizing agent to do the job, so when such metals are placed in a solution of an acid they are oxidized while the hydrogen ions are reduced. Such metals are said to be *more active* than hydrogen (H_2), and they dissolve in acids like HCl and H_2SO_4 to give hydrogen gas and a salt that contains the anion of the acid. Hydrogen ions, however, are not powerful enough to cause the oxidation of some other metals. Copper, for example, is significantly less reactive than zinc or iron, and H^+ cannot oxidize it. If a copper penny is dropped into sulfuric or hydrochloric acid, it just sits there. No reaction occurs. Copper is an example of a metal that is *less active* than H_2.

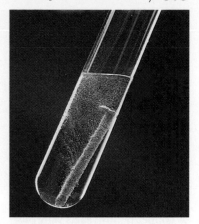

FIGURE 9.2

A galvanized (zinc-coated) iron nail in a dilute solution of sulfuric acid. The acid dissolves the zinc and liberates hydrogen gas, which bubbles to the surface.

Oxidizing Power of Acids

Hydrochloric acid contains H_3O^+ ions (which we abbreviate as H^+) and Cl^- ions. The hydrogen ion in hydrochloric acid is able to be an oxidizing agent by being reduced to H_2. However, the Cl^- ion in the solution has no tendency at all to be an oxidizing agent, because it would have to become a Cl^{2-} ion to gain an electron. As you learned in Chapter 7, this is just not feasible—too much energy would be required. In a solution of HCl, therefore, the only oxidizing agent is H^+.

FIGURE 9.3

A copper penny reacts vigorously with concentrated nitric acid, as this sequence of photographs shows. The dark red-brown vapors are nitrogen dioxide, the same gas that gives smog its characteristic color.

The strongest oxidizing agent in a solution of a "nonoxidizing" acid is H^+.

In an aqueous solution of sulfuric acid, we find a similar situation. In water this acid ionizes to give hydrogen ions and sulfate ions. Although the sulfate ion can be reduced (to SO_3^{2-}, for example), hydrogen ion is more easily reduced. Therefore, when we add a metal such as zinc to sulfuric acid, it is H^+, rather than SO_4^{2-}, that removes the electrons from Zn.

Compared with many other chemicals that we will study, the hydrogen ion in water is really a rather poor oxidizing agent, so hydrochloric acid and sulfuric acid have rather poor oxidizing abilities. For this reason, they are often called **nonoxidizing acids,** even though their hydrogen ions are able to oxidize certain metals. Actually, when we call something a nonoxidizing acid we are saying that the *anion* of the acid is a weaker oxidizing agent than H^+.

There are also **oxidizing acids,** acids whose anions are stronger oxidizing agents than H^+. An example is nitric acid, HNO_3. When dissolved in water, nitric acid ionizes to give H^+ and NO_3^- ions. However, in this solution the nitrate ion is a more powerful oxidizing agent than the hydrogen ion. In the competition for electrons, it is the nitrate ion that generally wins, and when nitric acid reacts with a metal, it is the nitrate ion that is reduced.

Because the NO_3^- ion is a stronger oxidizing agent than H^+, it is able to oxidize metals that H^+ cannot. For example, if a copper penny is dropped into concentrated nitric acid, it reacts violently (see Figure 9.3). The reddish brown gas is nitrogen dioxide, NO_2, which is formed by the reduction of the NO_3^- ion. The molecular equation for the reaction is

$$4HNO_3(aq) + Cu(s) \longrightarrow Cu(NO_3)_2(aq) + 2NO_2(g) + 2H_2O$$

If we change this to a net ionic equation and then assign oxidation numbers, we can see easily which is the oxidizing agent and which is the reducing agent.

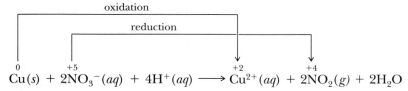

Notice that even though it is the nitrogen whose oxidation number is decreasing, we identify the substance that contains this nitrogen, the NO_3^- ion, as the oxidizing agent.

Nitrate ion is reduced and copper is oxidized. Therefore, nitrate ion is the oxidizing agent and copper is the reducing agent. Notice that in this reaction, *no hydrogen gas is formed.* The H^+ ions of the HNO_3 are an essential part of the reaction, but they just become part of water molecules without a change in oxidation number.

When oxidizing acids react with metals, it is often difficult to predict the products. Reduction of the nitrate ion, for example, can produce all sorts of compounds having different oxidation states for the nitrogen, depending on the reducing power of the metal and the concentration of the acid. When *concentrated* nitric acid reacts with a metal, it often produces nitrogen dioxide, NO_2, as the reduction product. When *dilute* nitric acid is used to dissolve a metal, the product is often nitric oxide, NO, instead. Copper, for example, can display either behavior toward nitric acid. The molecular and net ionic equations for the reactions are

Concentrated HNO₃

$$Cu(s) + 4HNO_3(aq) \longrightarrow Cu(NO_3)_2(aq) + 2NO_2(g) + 2H_2O$$

$$Cu(s) + 4H^+(aq) + 2NO_3^-(aq) \longrightarrow Cu^{2+}(aq) + 2NO_2(g) + 2H_2O$$

Dilute HNO₃

$$3Cu(s) + 8HNO_3(aq) \longrightarrow 3Cu(NO_3)_2(aq) + 2NO(g) + 4H_2O$$

$$3Cu(s) + 8H^+(aq) + 2NO_3^-(aq) \longrightarrow 3Cu^{2+}(aq) + 2NO(g) + 4H_2O$$

If very dilute nitric acid reacts with a metal that is a particularly strong reducing agent, like zinc, the nitrogen can be reduced all the way down to the -3 oxidation state that it has in NH_4^+ (or NH_3). The net ionic equation is

$$4Zn(s) + 10H^+(aq) + NO_3^-(aq) \longrightarrow 4Zn^{2+}(aq) + NH_4^+(aq) + 3H_2O$$

The nitrate ion in the presence of hydrogen ions makes nitric acid quite a powerful oxidizing acid. All metals except the very unreactive ones, such as platinum and gold, are attacked by it. Nitric acid also does a good job of oxidizing organic substances, so it is wise to be especially careful when working with this acid in the laboratory. Very serious accidents have occurred when inexperienced people have used concentrated nitric acid around organic substances.

The oxidation number of N in NO_2 is $+4$. In NO, the oxidation number is $+2$.

Nitric acid causes severe skin burns, so be careful when you work with it in the laboratory. If you spill any on your skin, wash it off immediately and seek the help of your lab teacher.

The evolution of hydrogen in the reaction of a metal with an acid is a special case of a more general phenomenon—one element displacing (pushing out) another element from a compound by means of a redox reaction. In the case of a metal–acid reaction, it is the metal that displaces hydrogen from the acid, changing $2H^+$ to H_2.

Another reaction of this same general type occurs when one metal displaces another metal from its compounds, and is illustrated by the experiment shown in Figure 9.4. Here we see a brightly polished strip of metallic zinc that is dipped into a solution of copper sulfate. After the zinc has been in the solution for a while, a reddish brown deposit of metallic copper has formed on it, and if the solution were analyzed, we would find that it now contains zinc ions, as well as some remaining unreacted copper ions.

The results of this experiment can be summarized by the equation

$$Zn(s) + CuSO_4(aq) \longrightarrow Cu(s) + ZnSO_4(aq)$$

A reaction such as this, in which one element replaces another in a compound, is sometimes called a **single replacement reaction.**

The redox changes in the reaction above become clear if we write the net ionic equation. Both copper sulfate and zinc sulfate are soluble salts, so they

9.5
DISPLACEMENT OF ONE METAL BY ANOTHER FROM COMPOUNDS

FIGURE 9.4

The reaction of zinc with copper ion. (*Left*) A piece of shiny zinc next to a beaker containing a copper sulfate solution. (*Center*) When the zinc is placed in the solution, copper ions are reduced to the free metal while the zinc dissolves. (*Right*) After a while the zinc becomes coated with a red-brown layer of copper. Notice that the solution is a lighter blue than before.

are completely dissociated. Sulfate ion is a spectator ion, and the net ionic equation is

$$Zn(s) + Cu^{2+}(aq) \longrightarrow Cu(s) + Zn^{2+}(aq)$$

We see that zinc has reduced the copper ion to metallic copper, and the copper ion has oxidized metallic zinc to the zinc ion. In the process, zinc ions have taken the place of the copper ions in the solution.

The reaction of zinc with copper ion is quite similar to the reaction of zinc with sulfuric acid. The more "active" zinc replaces another less "active" element in a compound. Furthermore, it is easy to show that the reverse reaction doesn't occur. Nothing happens if a brightly polished piece of copper is dipped into a solution of zinc sulfate (see Figure 9.5). No matter how long the copper is in the solution, its surface remains untarnished. This means that the reaction of copper with zinc sulfate doesn't occur.

$$Cu(s) + ZnSO_4(aq) \longrightarrow \text{no reaction}$$

In other words, the less "active" copper is unable to displace the more "active" zinc from the solution.

Activity series of metals

The Activity Series

Throughout the discussion in the preceding paragraph we used the word *active* to mean "easily oxidized." In other words, an element that is more easily oxidized will displace one that is less easily oxidized from its compounds. The relative ease of oxidation of two metals can be established in experiments just as simple as the ones pictured in Figures 9.4 and 9.5. After such comparisons are made for many pairs, the metals can be arranged in order of their ease of oxidation to give what is often called an **activity series** (see Table 9.3). According to the way the metals have been arranged in Table 9.3, those at the bottom are more easily oxidized (are more active) than those at the top. *This means that a given element will be displaced from its compounds by any metal below it in the table.*

Notice that we have included hydrogen in the activity series. Metals below hydrogen in the series can displace hydrogen from solutions containing H^+.

 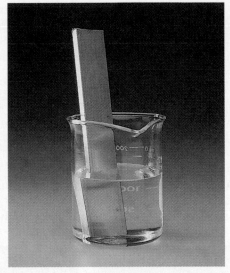 

FIGURE 9.5

Although metallic zinc will displace copper from a solution containing Cu^{2+} ion, metallic copper will not displace Zn^{2+} from its solutions. Here we see that the copper bar is unaffected by being dipped into a solution of zinc sulfate.

These are the metals that are capable of reacting with the nonoxidizing acids discussed in the previous section. On the other hand, metals above hydrogen in the table do not react with acids having H^+ as the strongest oxidizing agent.

> If a metal is above hydrogen in the activity series (Table 9.3), it will not react with nonoxidizing acids.

TABLE 9.3   Activity Series for Hydrogen and Some Metals

	Element	Oxidation Product
Least Active	Gold	Au^{3+}
	Mercury	Hg^{2+}
	Silver	Ag^+
	Copper	Cu^{2+}
	HYDROGEN	H^+
	Lead	Pb^{2+}
	Tin	Sn^{2+}
	Cobalt	Co^{2+}
	Cadmium	Cd^{2+}
	Iron	Fe^{2+}
	Chromium	Cr^{3+}
	Zinc	Zn^{2+}
	Manganese	Mn^{2+}
	Aluminum	Al^{3+}
	Magnesium	Mg^{2+}
	Sodium	Na^+
	Calcium	Ca^{2+}
	Strontium	Sr^{2+}
	Barium	Ba^{2+}
	Potassium	K^+
	Rubidium	Rb^+
Most Active	Cesium	Cs^+

Increasing Ease of Oxidation of the Metal

FIGURE 9.6

Metallic sodium reacts violently with water. Reduction of water molecules liberates hydrogen. The heat of the reaction ignites the sodium metal, which can be seen burning and sending sparks from the surface of the water.

Metals at the very bottom of the table are very easily oxidized and are extremely strong reducing agents. They are so reactive, in fact, that they are able to reduce the hydrogen in water. Sodium, for example, reacts vigorously with water to liberate H_2 according to the following equation (see Figure 9.6).

$$2Na(s) + H_2O \longrightarrow H_2(g) + 2NaOH(aq)$$

For metals below hydrogen in the activity series, an interesting parallel exists between the ease of oxidation of the metal and the speed with which it reacts with H^+. For example, in Figure 9.7, we see samples of iron, zinc, and magnesium reacting with solutions of hydrochloric acid. In each test tube the initial HCl concentration is the same, but we see that the magnesium reacts more rapidly than zinc, which reacts more rapidly than iron. You can see that the order of reactivity in Table 9.3 is the same; magnesium is more easily oxidized than zinc, which is more easily oxidized than iron.

FIGURE 9.7

The relative ease of oxidation of metals parallels their rates of reaction with hydrogen ions of an acid. The products are hydrogen gas and the metal ion in solution. All three test tubes contain HCl(*aq*) at the same concentration. The first also contains pieces of iron, the second pieces of zinc, and the third pieces of magnesium. Among these three metals, the ease of oxidation increases from iron to zinc to magnesium.

Using the Activity Series

The activity series in Table 9.3 permits us to make predictions of the outcome of single replacement redox reactions, as illustrated in the following examples.

EXAMPLE 9.5
Using the Activity Series

What will happen if an iron nail is dipped into a solution containing copper sulfate? If a reaction occurs, write its chemical equation.

ANALYSIS Because of the discussion above, it is obvious we will need to refer to the activity series. However, if this question had been asked in another setting, how would you know what to do?

Reading the question, we have to ask, what *could* happen? If a chemical reaction were to occur, iron would have to react with the copper sulfate. A *metal* possibly reacting with the *salt of another metal* should suggest the possibility of a single displacement reaction. In turn, you should recognize that such reactions can be predicted from the locations of the potential reactants in the activity series. According to the activity series in Table 9.3, iron is more easily oxidized than copper, so the iron will displace the copper from the copper sulfate. A reaction will occur. To write an equation for the reaction, we have to know the final oxidation state of the iron. In the table, this is indicated as $+2$, so the Fe atoms change to Fe^{2+} ions and pair with SO_4^{2-} ions to give $FeSO_4$. Copper(II) ions are reduced to copper atoms.

SOLUTION The analysis gave us the products, so the equation is

$$Fe(s) + CuSO_4(aq) \longrightarrow Cu(s) + FeSO_4(aq)$$

CHECK The equation is balanced as it stands; the question is answered.

In a *double replacement* reaction such as

$$CdCl_2(aq) + Na_2S(aq) \longrightarrow$$
$$CdS(s) + 2NaCl(s)$$

two anions (Cl^- and S^{2+}) acquire different cations. In a *single replacement* reaction, *one* anion acquires a different cation. Single replacement reactions are redox reactions; double replacement reactions are not.

What happens if an iron nail is dipped into a solution of aluminum sulfate?

ANALYSIS Scanning the activity series, we see that aluminum metal is *more* easily oxidized than iron metal. This means that aluminum atoms would be able to displace iron ions from an iron compound. But it also means that iron atoms cannot displace aluminum ions from its compounds, and iron *atoms* plus aluminum *ions* are what we're given.

SOLUTION If nature does not permit electrons to transfer from iron atoms to aluminum ions, we must conclude that no reaction can occur.

$$Fe(s) + Al_2(SO_4)_3(aq) \longrightarrow \text{no reaction}$$

■ **Practice Exercise 10** Write a chemical equation for the reaction that will occur, if any, when (a) aluminum metal is added to a solution of copper chloride and (b) silver metal is added to a solution of magnesium sulfate. If no reaction will occur, write "no reaction" in place of the products.

EXAMPLE 9.6
Using the Activity Series

If you were an engineer designing parts for a machine, you would consider several factors in choosing the right metal for the job. Certainly, some of them would be physical properties—strength and hardness, for example. However, you would also be concerned about the chemical properties of the metal. What is the tendency of the metal to react with air and moisture? Will the metal react with any chemicals that come into contact with it while the machine is operating? The answers to these chemical questions are very important to the life span of your machine.

When we raise questions about the reactivity of a metal, we are concerned with how easily the metal is oxidized. This is because in nearly every compound containing a metal, the metal exists in a positive oxidation state, and when a free metal reacts to form a compound, it is oxidized. As a result, a metal like sodium, which is very easily oxidized, is said to be very reactive, whereas a metal like platinum, which is very difficult to oxidize, is said to be unreactive.

There are several ways to compare how easily metals are oxidized. In the last section, we saw that by comparing the abilities of metals to displace each other from compounds we are able to establish their relative ease of oxidation. This was the basis for the activity series (Table 9.3).

9.6
PERIODIC TRENDS IN THE REACTIVITIES OF METALS

Reactivity refers in general to the tendency of a substance to react with something. The *reactivity of a metal* refers to its tendency specifically to undergo *oxidation*.

 Periodic table and metal reactivities

FIGURE 9.8

The variation in the ease of oxidation of metals with position in the periodic table.

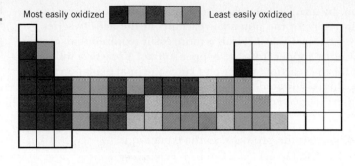

Ionization energy is the energy per mole required to remove single electrons from isolated gaseous atoms.

You should study Figure 9.8 so that you know where the very reactive elements are found and where the very unreactive ones are found.

As useful as the activity series is in predicting the outcome of certain redox reactions, it is difficult to remember in detail. Furthermore, often it is sufficient just to know approximately where an element stands in relation to others in a broad range of reactivity. This is where the periodic table can be especially useful to us once again, because there are trends and variations in reactivity within the periodic table that are simple to identify and remember.

Figure 9.8 illustrates how the ease of oxidation (the reactivity) of metals varies in the periodic table. In general, these trends roughly parallel the variations in ionization energy. You might expect this, since the ionization energy is a measure of how difficult it is to remove an electron from an atom, as we studied in Section 6.9. The relationship between reactivity and ionization energy (IE) is only approximate, however, because the IE refers to isolated gaseous atoms, and when we consider ease of oxidation, we are concerned with the removal of electrons from the metal in its usual solid form as well as the hydration of the cation that forms.

By examining Figure 9.8, we see that the metals that are most easily oxidized are found at the far left in the periodic table. The metals in Group IA, for example, are so easily oxidized that all of them react with water to liberate hydrogen. Because of their reactivity toward moisture and oxygen, they have no useful applications that require exposure to the atmosphere, so we rarely encounter them as free metals. The same is true of the heavier metals in Group IIA, calcium through barium. These elements also react with water to liberate hydrogen.

In Section 6.9, you learned how the ionization energies of the elements vary within the periodic table. Recall that *the ionization energy increases as we move from left to right in a period.* This places metals with the lowest ionization energies at the left of the table—in other words, in Groups IA and IIA. It should be no surprise, therefore, that these elements are the most easy to oxidize. You also learned that *the ionization energy decreases as we go down a group.* This explains why the heavier elements in Group IIA are especially easy to oxidize compared with those at the top of the group.

In Figure 9.8 we can also locate the metals that are the most difficult to oxidize. They occur for the most part among the heavier transition elements in the center of the periodic table, where we find the very unreactive elements platinum and gold—metals used to make fine jewelry. Their bright luster and lack of any tendency to corrode in air or water combine to make them particularly attractive for this purpose. This same lack of reactivity also is responsible for their industrial uses. Gold, for example, is used to coat the electrical contacts in low-voltage circuits found in microcomputers, because even small amounts of corrosion on more reactive metals would be sufficient to impede the flow of electricity so much as to make the devices unreliable.

The reactivity of a metal is determined by its ease of oxidation, and therefore its ability to serve as a reducing agent. *For a nonmetal, reactivity is usually gauged by its ability to serve as an oxidizing agent.* This ability varies according to the nonmetal's electronegativity. Nonmetals with high electronegativities have strong tendencies to acquire electrons and are therefore strong oxidizing agents. In parallel with changes in electronegativities in the periodic table, *the oxidizing abilities of nonmetals increase from left to right across a period and from bottom to top in a group.* Thus, the most powerful oxidizing agent is fluorine, followed closely by oxygen, both in the upper right-hand corner of the periodic table.

Single replacement reactions occur among the nonmetals, just as with the metals. For example, heating a metal sulfide in oxygen causes the sulfur to be replaced by oxygen. The sulfur then combines with oxygen to give sulfur dioxide. The equation for a typical reaction is

$$CuS(s) + \tfrac{3}{2}O_2(g) \longrightarrow CuO(s) + SO_2(g)$$

Displacement reactions are especially evident among the halogens, where a particular halogen, as an element, will oxidize the *anion* of any halogen below it in Group VIIA. Thus, F_2 will oxidize Cl^-, Br^-, and I^-. However, Cl_2 will only oxidize Br^- and I^-, and Br_2 will only oxidize I^-. Thus, the following reactions are observed.

Fluorine:

$$F_2 + 2Cl^- \longrightarrow 2F^- + Cl_2$$

$$F_2 + 2Br^- \longrightarrow 2F^- + Br_2$$

$$F_2 + 2I^- \longrightarrow 2F^- + I_2$$

Chlorine:

$$Cl_2 + 2Br^- \longrightarrow 2Cl^- + Br_2$$

$$Cl_2 + 2I^- \longrightarrow 2Cl^- + I_2$$

Bromine:

$$Br_2 + 2I^- \longrightarrow 2Br^- + I_2$$

In 1789, the French chemist Antoine Lavoisier discovered that *combustion* involves the reaction of chemicals in various fuels, like wood and coal, not just with air but specifically with the oxygen in air. In fact, the term *oxidation* was invented to describe reactions in which one of the reactants is elemental oxygen, O_2. Oxidation originally meant the reaction of a substance with oxygen. Only much later did chemists realize that the reactions of oxygen constitute a special case of a much broader class of reactions. This realization led to the extension of the meaning of "oxidation" to cover the kinds of reactions discussed in the previous sections.

Oxygen is a plentiful chemical; it's in the air and available to anyone who wants to use it, chemist or not. Furthermore, O_2 is a very reactive oxidizing agent, so its reactions have been well studied. When substances combine with oxygen, the products are generally oxides, *molecular oxides* when the oxygen reacts with nonmetals and *ionic oxides* when the oxygen reacts with metals.

9.7
PERIODIC TRENDS IN THE REACTIVITIES OF NONMETALS

F and O are the two most electronegative of all elements.

 Periodic table and nonmetal reactivities

The relative oxidizing abilities of the halogens is something you should remember.

9.8
MOLECULAR OXYGEN AS AN OXIDIZING AGENT

Antoine Laurent Lavoisier (1742–1794), a French chemist, is deservedly called the "father of modern chemistry." He was the first to insist on *quantitative* data from chemical research; he systematized chemical nomenclature; and his text *Traité élémentaire de chimie* was to modern chemistry what Newton's *Principia* was to the development of physics.

Combustion of Organic Compounds

Combustion is normally taken to mean a particularly *rapid* reaction of a substance with oxygen in which both heat and light are given off.

The general, unbalanced equation is

Hydrocarbon + $O_2 \longrightarrow CO_2$ + H_2O

Combustion of Hydrocarbons Fuels such as natural gas, gasoline, kerosene, heating oil, and diesel fuel are examples of *hydrocarbons*—compounds containing only the elements carbon and hydrogen. Natural gas is composed primarily of methane, CH_4. Gasoline is a mixture of hydrocarbons, the most familiar of which is octane, C_8H_{18}. Kerosene, heating oil, and diesel fuel are mixtures of hydrocarbons in which the molecules contain even more atoms of carbon and hydrogen.

When hydrocarbons burn in a *plentiful* supply of oxygen, the products of combustion are always carbon dioxide and water. Thus, methane and octane combine with oxygen according to the equations.

$$CH_4 + 2O_2 \longrightarrow CO_2 + 2H_2O$$

$$2C_8H_{18} + 25O_2 \longrightarrow 16CO_2 + 18H_2O$$

The complete combustion of all other organic compounds containing only carbon, hydrogen, and oxygen produces the same products, CO_2 and H_2O, and follows similar equations.

■ **Practice Exercise 11** Write a balanced equation for the combustion of butane, C_4H_{10}, in an abundant supply of oxygen. Butane is the fuel used in disposable cigarette lighters.

■ **Practice Exercise 12** Ethanol, C_2H_5OH, is now mixed with gasoline, and the mixture is sold under the name *gasohol*. Write a chemical equation for the combustion of ethanol.

When there is less than an abundant supply of oxygen during the combustion of a hydrocarbon, not all of the carbon is converted to carbon dioxide. Instead, some of it forms carbon monoxide. Its formation is a pollution problem associated with the use of gasoline engines, as you know.

$$CH_4 + \tfrac{3}{2}O_2 \longrightarrow CO + 2H_2O \qquad \text{(in a limited oxygen supply)}$$

When the oxygen supply is extremely limited, only the hydrogen of a hydrocarbon mixture is converted to the oxide (water). The carbon atoms emerge as elemental carbon. For example, the combustion of methane in a very limited oxygen supply follows the equation.

$$CH_4 + O_2 \longrightarrow C + 2H_2O \qquad \text{(in a very limited oxygen supply)}$$

This finely divided form of carbon is also called *carbon black*.

The carbon that forms is very finely divided and would be called *soot* by almost anyone observing the reaction. Nevertheless, such soot has considerable commercial value when collected and marketed under the name *lampblack*. This sooty form of carbon is used to manufacture inks and much of it is used in the production of rubber tires, where it serves as a binder and a filler. When soot from incomplete combustion is released into air, its tiny particles constitute one component of air pollution, namely, *particulates*, which contribute to the haziness of smog.

Combustion of Organic Compounds That Contain Sulfur A major pollution problem in industrialized countries is caused by the release into the atmosphere of sulfur dioxide formed by the combustion of fuels that contain sulfur or its compounds. The products of the combustion of organic compounds of sulfur are carbon dioxide, water, and sulfur dioxide. A typical reaction is

$$2C_2H_5SH + 9O_2 \longrightarrow 4CO_2 + 6H_2O + 2SO_2$$

A solution of sulfur dioxide in water is acidic, and dissolved SO_2 is one solute that converts rainwater into "acid rain" (see *Chemicals in Use 10*).

Reactions of Metals with Oxygen

We don't often think of metals as undergoing combustion, but have you ever seen an old-fashioned flashbulb fired to take a photograph? The source of light is the reaction of the metal magnesium with oxygen (see Figure 9.9). A close look at a fresh flashbulb reveals a fine web of thin magnesium wire within the glass envelope. The wire is surrounded by an atmosphere of oxygen, a clear colorless gas. When the flashbulb is used, a small electric current surges through the thin wire, causing it to become hot enough to ignite, and it burns rapidly in the oxygen atmosphere. The equation for the reaction is

$$2Mg + O_2 \longrightarrow 2MgO$$

Most metals react directly with oxygen, although not as spectacularly, and usually we refer to the reaction as **corrosion** or tarnishing because the oxidation products dull the shiny metal surface. Iron, for example, is oxidized fairly easily, especially in the presence of moisture. As you know, under these conditions the iron corrodes—it rusts. Rust is a form of iron(III) oxide, Fe_2O_3, that also contains an appreciable amount of absorbed water. The formula for rust is therefore normally given as $Fe_2O_3 \cdot xH_2O$ to indicate its somewhat variable composition. Although the rusting of iron is a slow reaction, the combination of iron with oxygen can be speeded up if the metal is heated to a very high temperature (see Figure 9.10).

An aluminum surface, unlike that of iron, is not noticeably dulled by the reaction of aluminum with oxygen. Aluminum is a common metal found around the home in uses ranging from aluminum foil to aluminum window frames, and it surely appears shiny. Yet, aluminum is a rather easily oxidized metal, as can be seen from its position in the activity series (Table 9.3). A *freshly* exposed surface of the metal does react very quickly with oxygen and becomes coated with a very thin film of aluminum oxide, Al_2O_3, so thin that it does not obscure the shininess of the metal beneath. Fortunately, the oxide coating adheres very tightly to the surface of the metal and makes it very difficult for additional oxygen to combine with the aluminum. Therefore, further oxidation of aluminum occurs very slowly.

■ **Practice Exercise 13** The oxide formed in the reaction shown in Figure 9.10 is Fe_2O_3. Write a balanced chemical equation for the reaction.

Reactions of Nonmetals with Oxygen

Most nonmetals combine as readily with oxygen as do the metals, and their reactions usually occur rapidly enough to be described as combustion. To most people, the most important nonmetal combustion is that of carbon,

FIGURE 9.9

Fine magnesium wire in an atmosphere of oxygen fills the flashbulb at the left. After being used (right), the interior of the bulb is coated with a white film of magnesium oxide.

FIGURE 9.10

Steel is cut by a stream of pure oxygen whose reaction with the red-hot metal produces enough heat to melt the steel and send a shower of burning steel sparks flying.

Nitrogen is one nonmetal that doesn't react readily with oxygen, especially at normal temperatures. But nitrogen oxides formed at high temperatures in automobile engines are a serious source of air pollution.

Most scientists are convinced that SO_2 released by coal-burning power plants in the Midwest is responsible for acid rain in the northeastern United States and Canada.

because the reaction is a source of heat. Coal and charcoal, for example, are common carbon fuels. Coal is used worldwide in large amounts to generate electricity, and charcoal is a popular barbecue fuel for broiling hamburgers. If plenty of oxygen is available, the combustion of carbon gives CO_2, but when the supply of oxygen is limited, some CO also forms. Manufacturers that package charcoal briquettes, therefore, print a warning on the bag that the charcoal shouldn't be used indoors for cooking or heating.

Sulfur is another nonmetal that readily burns in oxygen. In the manufacture of sulfuric acid, the first step is the combustion of sulfur to produce sulfur dioxide. As mentioned earlier, sulfur dioxide also forms when sulfur compounds burn, and the presence of both sulfur and sulfur compounds as impurities in coal and petroleum is a major source of air pollution. Combustion of these fuels releases their sulfur content in the form of SO_2, which enters the atmosphere, where it drifts on the wind until it finally dissolves in rainwater. Then, as a dilute solution of sulfurous acid, it falls to earth as one component of acid rain.

■ **Practice Exercise 14** Elemental phosphorus can exist in a white, waxy form that consists of molecules having the formula P_4. When phosphorus burns in an abundant supply of oxygen, it forms an oxide composed of molecules of P_4O_{10}. Write an equation for the reaction.

SUMMARY

Brønsted Acids and Bases. A **Brønsted acid** is a proton donor; a **Brønsted base** is a proton acceptor. Acid–base neutralization in the Brønsted theory is a proton transfer event. In an aqueous equilibrium involving a Brønsted acid or base, there are two **conjugate acid–base pairs** that differ from each other by only one H^+. A substance that can be either an acid or a base, depending on the nature of the other reactant, is **amphoteric** or, with proton-transfer reactions, **amphiprotic**. We may summarize the three views of acids and bases as follows.

Arrhenius view   Acids give H^+ in water; bases provide OH^- in water.

Brønsted view   Acids are proton donors; bases are proton acceptors.

Lewis view   Acids are electron pair acceptors; bases are electron pair donors.

Relative Acidities and the Periodic Table   **Binary acids** contain only hydrogen and another nonmetal. Their strengths increase from top to bottom within a group and left to right across a period. **Oxoacids,** which contain oxygen atoms in addition to hydrogen and another element, increase in strength as the number of oxygen atoms on the same central atom increases. Oxoacids having the same number of oxygens increase in strength as the central atom moves from bottom to top within a group and from left to right across a period.

Metal–Acid Reactions   Both **oxidation,** the loss of electrons or an increase in oxidation number, and **reduction,** the gain of electrons or a decrease in oxidation number, occur together in **redox** reactions. The substance oxidized is the **reducing agent;** the substance reduced is the **oxidizing agent.**

In **nonoxidizing acids,** the strongest oxidizing agent is H^+. The reaction of a metal with a nonoxidizing acid gives hydrogen and a salt of the acid. Only metals more active than hydrogen react this way. These are metals that are located below hydrogen in the activity series (Table 9.3). **Oxidizing acids,** like HNO_3, contain an anion that is a stronger oxidizing agent than H^+ and are able to oxidize many metals that nonoxidizing acids cannot.

Metal Displacement Reactions   If one metal is more easily oxidized than another, it can displace the other metal from its compounds by a redox reaction. Atoms of the more active metal become ions; ions of the less active metal generally become atoms. In this manner, any metal in the activity series can displace any of the others above it in the series from their compounds. Within the periodic table, the most reactive metals (those that are most easily oxidized) are located at the left in Groups IA and IIA. The least reactive metals are located in the center of the periodic table among the heavier transition metals.

Nonmetal Displacement Reactions The reactivities of nonmetals depend on their relative abilities to be oxidizing agents. The most reactive nonmetals, fluorine and oxygen, are thus the strongest electron acceptors, have the highest electronegativities, and are the strongest oxidizing agents. Within the same period, nonmetal reactivity increases from left to right. Within the same group, this reactivity increases from bottom to top. Oxygen thus displaces sulfur from metals sulfides, forming metal oxides and sulfur oxides. Among the Group VII elements, fluorine oxidizes other halide ions to halogens; chlorine oxidizes Br^- and I^-, and bromine oxidizes only I^-.

Oxidations by Molecular Oxygen **Combustion,** the rapid reaction of a fuel or a nonmetal with oxygen, is accompa-

nied by the evolution of heat and light. Combustion of a hydrocarbon in the presence of excess oxygen gives CO_2 and H_2O, two molecular oxides. When the supply of oxygen is limited, some CO also forms, and in a very limited supply of oxygen the products are H_2O and very finely divided, elemental carbon (as soot or lampblack). The combustion of organic compounds containing only carbon, hydrogen, and oxygen also gives the same products, CO_2 and H_2O. Sulfur burns to give SO_2, which also forms when sulfur-containing fuels burn. Most nonmetals also burn in oxygen to give molecular oxides.

Metals also combine with oxygen, but only sometimes is the reaction violent enough to be considered combustion. The products are ionic metal oxides.

TOOLS YOU HAVE LEARNED

The table below lists the tools you have learned in this chapter that are applicable to problem solving. Review them if necessary, and refer to them when working on the Thinking-It-Through Problems and the Review Exercises that follow.

Tool	Function
Periodic table	To help understand and recall the following: The relative strengths of the binary acids (page 371). The relative strengths of the oxoacids (page 374). The relative reactivities of the metals in redox reactions (page 385). The relative reactivities of the nonmetals in redox reactions (page 387).
List of strong acids (page 373)	To be able quickly to recognize that any acid not on the list is a weak acid.
Activity series of metals (page 382)	To help predict (or recall) single displacement reactions.

THINKING IT THROUGH

Remember, you are not asked to obtain answers for the following problems. Instead, assemble the data necessary to solve the problems and describe how *you would use the data to obtain the answers. For numerical problems, set up the calculation using appropriate conversion factors.*

The problems are divided into two groups. Those in Level 2 are significantly more challenging than those in Level 1 and provide an opportunity to really hone your problem solving skills.

Level 1 Problems

1. You intend to use a particular acid in an experiment, and it is important to you to know whether the acid is "strong" or "weak." All you have is the name of the acid. How can you make this judgment without taking measurements?

2. What is the conjugate base of CH_4? Is it a strong or a weak base? How can you tell without looking at a table of strong and weak bases?

3. What is the conjugate acid of $HSO_4{}^-$? Is it a weak acid? How can you tell without a table of weak acids?

4. Which of the following are amphoteric: H_3O^+, H_2O, $HSO_4{}^-$, $NO_3{}^-$? For those that are amphoteric, write the formulas of both their conjugate acids and bases.

5. Potassium is more easily oxidized than calcium. Use a property of the atoms of these elements to explain why this observation is not unexpected.

6. Will cadmium metal react spontaneously with Ni^{2+} ion to give Cd^{2+} and metallic nickel?

7. A solution contains Ga^{3+} and Pt^{2+} ions. Into the solution are placed strips of metallic Ga and Pt. What reaction will occur?

8. Why is it reasonable to expect that Cl_2 is a stronger oxidizing agent than Br_2?

9. A few minutes after you first start your car, drops of a colorless liquid can be seen falling from the exhaust pipe. After a while, this stops. Explain these observations in terms of the chemistry involved in the running of the engine and the physical changes that occur.

Level 2 Problems

10. Identify the conjugate acid–base pairs in the following equation, which lacks an arrowhead on the arrow. Place the arrowhead at the correct end to show what reaction occurs.

$$H_2O + HSO_3{}^- \text{---} H_3O^+ + SO_3{}^{2-}$$

11. Aqueous solutions of which ion, Al^{3+} or Ga^{3+}, should be more acidic? Explain your answer.

12. What is the formula for the acidic anhydride of chromic acid, H_2CrO_4?

13. What products would be expected in the reaction of CuSe with oxygen? Explain your reasoning.

14. Suppose you wished to determine the placement of vanadium in the activity series. Describe the experiments that you would have to perform to find where this element fits.

15. The element scandium, Sc, reacts with water with the evolution of hydrogen.

$$2Sc(s) + 6H_2O \longrightarrow 2Sc^{3+}(aq) + 6OH^-(aq) + 3H_2(g)$$

16. Where does Sc stand relative to zinc in the activity series? (b) How would you expect the reactivity of lanthanum, La, to compare with scandium? Explain your answers to both parts.

17. A mixture of the following was prepared: $Br_2(s)$, $At_2(s)$, $NaBr(aq)$, and $NaAt(aq)$. What reaction, if any, would we expect to observe? Explain your reasoning.

18. In the reaction

$$HCHO_2(aq) + C_2H_3O_2{}^-(aq) \rightleftharpoons CHO_2{}^-(aq) + HC_2H_3O_2(aq)$$

The position of equilibrium favors the products. Which is the stronger Brønsted base, $C_2H_3O_2{}^-$ or $CHO_2{}^-$? Which is the stronger Brønsted acid, $HC_2H_3O_2$ or $HCHO_2$? Explain your reasoning.

REVIEW EXERCISES

Answers to questions whose numbers are printed in color are given in Appendix D. Challenging questions are marked with asterisks.

Brønsted Acids and Bases

9.1 How is a Brønsted acid defined? How is a Brønsted base defined?

9.2 Give the conjugate acids of the following.
(a) F^- (d) $O_2{}^{2-}$
(b) N_2H_4 (e) $HCrO_4{}^-$
(c) C_5H_5N

9.3 Give the conjugate bases of the following.
(a) NH_3 (d) H_5IO_6
(b) $HCO_3{}^-$ (e) HNO_3
(c) HCN

9.4 Identify the conjugate acid–base pairs in the following reactions.
(a) $HNO_3 + N_2H_4 \rightleftharpoons NO_3{}^- + N_2H_5{}^+$
(b) $NH_3 + N_2H_5{}^+ \rightleftharpoons NH_4{}^+ + N_2H_4$
(c) $H_2PO_4{}^- + CO_3{}^{2-} \rightleftharpoons HPO_4{}^{2-} + HCO_3{}^-$
(d) $HIO_3 + HC_2O_4{}^- \rightleftharpoons IO_3{}^- + H_2C_2O_4$

9.5 Identify the conjugate acid–base pairs in the following reactions.
(a) $HSO_4{}^- + SO_3{}^{2-} \rightleftharpoons HSO_3{}^- + SO_4{}^{2-}$
(b) $S^{2-} + H_2O \rightleftharpoons HS^- + OH^-$
(c) $CN^- + H_3O^+ \rightleftharpoons HCN + H_2O$
(d) $H_2Se + H_2O \rightleftharpoons HSe^- + H_3O^+$

9.6 What is meant by the term *amphoteric?* Give two chemical equations that illustrate the amphoteric nature of water.

9.7 The position of equilibrium in the equation below lies far to the left. Identify the conjugate acid–base pairs. Which of the two acids is stronger?

$$HOCl(aq) + H_2O \rightleftharpoons H_3O^+(aq) + OCl^-(aq)$$

9.8 Consider the following: CO_3^{2-} is a weaker base than hydroxide ion, and HCO_3^- is a stronger acid than water. In the equation below, would the position of equilibrium lie to the left or to the right? Justify your answer.

$$CO_3^{2-}(aq) + H_2O \rightleftharpoons HCO_3^-(aq) + OH^-(aq)$$

9.9 Acetic acid, $HC_2H_3O_2$, is a weaker acid than nitrous acid, HNO_2. How do the strengths of the bases $C_2H_3O_2^-$ and NO_2^- compare?

9.10 Nitric acid, HNO_3, is a very strong acid. It is 100% ionized in water. In the reaction below, would the position of equilibrium lie to the left or to the right?

$$NO_3^-(aq) + H_2O \rightleftharpoons HNO_3(aq) + OH^-(aq)$$

Trends in Acid–Base Strength

9.11 Choose the stronger acid: (a) H_2S or H_2Se, (b) H_2Te or HI, (c) HIO_3 or HIO_4, (d) H_2SeO_4 or $HClO_4$

9.12 Choose the stronger acid: (a) HBr or HCl, (b) H_2O or HF, (c) H_3AsO_3 or H_3AsO_4, (d) H_2S or HBr

9.13 Suppose that a new element was discovered. Based on the discussions in this chapter, what properties (both physical and chemical) might be used to classify the element as a metal or a nonmetal?

9.14 Astatine, atomic number 85, is radioactive and does not occur in appreciable amounts in nature. On the basis of what you have learned in this chapter, answer the following. (a) How would the acidity of HAt compare to HI? (b) What might be a formula for an oxoacid of At?

9.15 Explain why nitric acid is a stronger acid than nitrous acid.

$$\begin{array}{cc} HO{-}NO_2 & HO{-}NO \\ \text{nitric acid} & \text{nitrous acid} \end{array}$$

9.16 Explain why H_2S is a stronger acid than H_2O.

9.17 Explain why $HClO_4$ is a stronger acid than H_2SeO_4.

Reactions of Metals with Acids

9.18 Write balanced ionic and net ionic equations for the reactions of the following metals with hydrochloric acid to give hydrogen plus the metal ion in solution.
(a) manganese (gives Mn^{2+})
(b) cadmium (gives Cd^{2+})
(c) tin (gives Sn^{2+})
(d) nickel (gives Ni^{2+})
(e) chromium (gives Cr^{3+})

9.19 Write balanced ionic and net ionic equations for the reaction of each metal in Review Exercise 9.18 with dilute sulfuric acid.

9.20 What is a nonoxidizing acid? Give two examples. What is the oxidizing agent in a nonoxidizing acid?

9.21 On the basis of the discussion in this chapter, suggest chemical equations for the oxidation of metallic silver to silver(I) ion with (a) dilute HNO_3 and (b) concentrated HNO_3.

9.22 What is the strongest oxidizing agent in an aqueous solution of nitric acid?

9.23 When hot and concentrated, sulfuric acid is a fairly strong oxidizing agent. Write a balanced net ionic equation for the oxidation of metallic copper to copper(II) ion by hot, concentrated H_2SO_4, in which the sulfur is reduced to SO_2. Write a balanced molecular equation for the reaction.

Displacement Reactions and the Activity Series

9.24 The following chemical reactions are *observed to occur* in aqueous solution.

$$2Al + 3Cu^{2+} \longrightarrow 2Al^{3+} + 3Cu$$
$$2Al + 3Fe^{2+} \longrightarrow 3Fe + 2Al^{3+}$$
$$Pb^{2+} + Fe \longrightarrow Pb + Fe^{2+}$$
$$Fe + Cu^{2+} \longrightarrow Fe^{2+} + Cu$$
$$2Al + 3Pb^{2+} \longrightarrow 3Pb + 2Al^{3+}$$
$$Pb + Cu^{2+} \longrightarrow Pb^{2+} + Cu$$

Arrange the metals Al, Pb, Fe, and Cu in order of increasing ease of oxidation.

9.25 In the preceding question, were all the experiments described actually necessary to establish the order?

9.26 According to the activity series in Table 9.3, which of the following metals react with nonoxidizing acids? (a) silver, (b) gold, (c) zinc, (d) magnesium

9.27 In each pair below, choose the metal that would most likely react more rapidly with a nonoxidizing acid such as HCl. (a) Aluminum or iron (b) Zinc or nickel, (c) Cadmium or magnesium

9.28 Use Table 9.3 to predict the outcome of the following reactions. If no reaction occurs, write N.R. If a reaction occurs, write a balanced net ionic equation for it.
(a) $Fe + Mg^{2+} \rightarrow$ (c) $Ag^+ + Fe \rightarrow$
(b) $Cr + Pb^{2+} \rightarrow$ (d) $Ag + Au^{3+} \rightarrow$

9.29 Use Table 9.3 to predict the outcome of the following displacement reactions. If no reaction occurs, write N.R. If a reaction occurs, write a balanced net ionic equation for it.
(a) $Mg + Fe^{2+} \rightarrow$ (d) $Mn + Co^{2+} \rightarrow$
(b) $Au + Ag^+ \rightarrow$ (e) $Cr + Sn^{2+} \rightarrow$
(c) $Cd + Zn^{2+} \rightarrow$

9.30 Use Table 9.3 to predict whether the following displacement reactions should occur. If no reaction occurs, write N.R. If a reaction does occur, write a balanced net ionic equation for it.
(a) $Zn + Sn^{2+} \rightarrow$ (d) $Mn + Pb^{2+} \rightarrow$
(b) $Cr + H^+ \rightarrow$ (e) $Zn + Co^{2+} \rightarrow$
(c) $Pb + Cd^{2+} \rightarrow$

9.31 Which metals in Table 9.3 will not react with nonoxidizing acids?

9.32 Which metals in Table 9.3 will react with water? Write chemical equations for each of these reactions.

Trends in Reactivity of Metals

9.33 When we say that aluminum is more reactive than iron, which property of these elements are we concerned with?

9.34 In what groups in the periodic table are the most reactive metals found? Where do we find the least reactive metals?

9.35 How is the ionization energy of a metal related to its reactivity?

9.36 Arrange the following metals in their approximate order of reactivity (most reactive first, least reactive last) based on their locations in the periodic table: (a) iridium, (b) silver, (c) calcium, (d) iron.

Oxidation by O_2

9.37 Define combustion.

9.38 Why is "loss of electrons" described as oxidation?

9.39 Are covalent oxides more likely to be formed by metals or by nonmetals? Explain your answer based on periodic trends in electronegativity, which you learned in Chapter 7.

9.40 Write balanced chemical equations for the complete combustion (in the presence of excess oxygen) of the following:
(a) C_6H_6 (benzene, an important industrial chemical)
(b) C_3H_8 (propane, a gaseous fuel used in many stoves)
(c) $C_{21}H_{44}$ (a component of paraffin wax)
(d) $C_{12}H_{26}$ (a component of kerosene)
(e) $C_{18}H_{36}$ (a component of diesel fuel)

9.41 Write balanced equations for the combustion of the hydrocarbons in Review Exercise 9.40 in (a) a slightly limited supply of oxygen and (b) a very limited supply of oxygen.

9.42 Methanol, CH_3OH, has been suggested as an alternative to gasoline as an automotive fuel. Write a balanced chemical equation for its complete combustion.

9.43 Metabolism of carbohydrates such as glucose, $C_6H_{12}O_6$, produces the same products as complete combustion. Write a chemical equation representing the metabolism (combustion) of glucose.

9.44 Sucrose, $C_{12}H_{22}O_{11}$, is ordinary table sugar. Write a balanced chemical equation representing the metabolism of sucrose. (See Review Exercise 9.43.)

9.45 What products are produced in the combustion of

$C_{10}H_{22}$ (a) if there is an excess of oxygen available? (b) If there is a slightly limited oxygen supply? (c) If there is a very limited supply of oxygen?

9.46 If one of the impurities in diesel fuel has the formula C_2H_6S, what products will be formed when it burns? Write a balanced chemical equation for the reaction.

9.47 Burning ammonia in an atmosphere of oxygen produces stable N_2 molecules as one of the products. What is the other product? Write the balanced equation for the reaction.

9.48 Write chemical equations for the reaction of oxygen with: (a) zinc, (b) aluminum, (c) magnesium, (d) iron, (e) calcium.

9.49 Complete and balance equations for the following. If no reaction occurs, write N.R.
(a) $KCl + Br_2 \rightarrow$ (d) $CaBr_2 + Cl_2 \rightarrow$
(b) $NaI + Cl_2 \rightarrow$ (e) $AlBr_3 + F_2 \rightarrow$
(c) $KCl + F_2 \rightarrow$ (f) $ZnBr_2 + I_2 \rightarrow$

9.50 In each pair, choose the better oxidizing agent.
(a) O_2 or F_2 (d) P_4 or S_8
(b) As_4 or P_4 (e) Se_8 or Cl_2
(c) Br_2 or I_2 (f) As_4 or S_8

Additional Exercises

*9.51 A sample of citric acid, a triprotic acid, weighing 0.200 g required 31.25 mL of 0.100 M NaOH for complete neutralization. What is the molecular mass of citric acid?

9.52 The standard heat of formation of sucrose, $C_{12}H_{22}O_{11}$, is -2230 kJ/mol. Use data in Table 5.2 (page 209) to compute the amount of energy (in kJ) released by metabolizing 1 oz (28.4 g) of sucrose. (See Review Exercise 9.43.)

9.53 For ethanol, C_2H_5OH, which is mixed with gasoline to make the fuel gasohol, $\Delta H_f^\circ = -277.63$ kJ/mol. Calculate the number of kilojoules released by burning completely 1 gal (6.56 kg) of ethanol. Use data in Table 5.2 (page 209) to help in the computation.

*9.54 A copper bar with a mass of 12.340 g is dipped into 255 mL of 0.125 M $AgNO_3$ solution. When the reaction that occurs has finally ceased, what will be the mass of unreacted copper in the bar? If all the silver that forms adheres to the copper bar, what will be the total mass of the bar after the reaction?

9.55 Why are gold artifacts often found with very little corrosion in archaeological digs on sites as old as 5000 years, but copper artifacts are often badly corroded at much younger sites?

TEST OF FACTS AND CONCEPTS: CHAPTERS 5–9

We pause again for you to test your understanding of concepts, your knowledge of scientific terms, and your skills at solving chemistry problems. Read through the following questions carefully and answer each as fully as possible. When necessary, review topics you are uncertain of. If you can answer these questions correctly, you are ready to go on to the next group of chapters.

1. The specific heat of helium is 5.19 J/g °C and of nitrogen is 1.04 J/g °C. How many joules can 1 mole of each gas absorb when its temperature increases by 1.00 °C?

2. When we say that the value of ΔH for a chemical reaction is a *state function,* what do we mean?

3. Label the following thermal properties as intensive or extensive. (a) specific heat, (b) heat capacity, (c) ΔH_f°, (d) ΔH°, (e) molar heat capacity

4. The thermochemical equation for the combustion reaction of half a mole of carbon monoxide is as follows.

$$\tfrac{1}{2}CO(g) + \tfrac{1}{4}O_2(g) \longrightarrow \tfrac{1}{2}CO_2(g) \quad \Delta H^\circ = -141.49 \text{ kJ}$$

Write the thermochemical equation for
(a) The combustion of 2 mol of $CO(g)$.
(b) The decomposition of 1 mol of $CO_2(g)$ to $O_2(g)$ and $CO(g)$.

5. The standard heat of combustion of eicosane, $C_{20}H_{42}(s)$, a typical component of candle wax, is 1.332×10^4 kJ/mol, when it burns in pure oxygen, and the products are cooled to 25 °C. The only products are $CO_2(g)$ and $H_2O(l)$. Calculate the value of the standard heat of formation of eicosane (in kJ/mol) and write the corresponding thermochemical equation.

6. Using data in Table 5.2 calculate values for the standard heats of reaction (in kilojoules) for the following reactions.
(a) $H_2SO_4(l) \rightarrow SO_3(g) + H_2O(l)$
(b) $C_2H_6(g) \rightarrow C_2H_4(g) + H_2(g)$

7. Calculate the standard heat of formation of calcium carbide, $CaC_2(s)$, in kJ/mol using the following thermochemical equations.

$$Ca(s) + 2H_2O(l) \longrightarrow Ca(OH)_2(s) + H_2(g)$$
$$\Delta H^\circ = -414.79 \text{ kJ}$$
$$2C(s) + O_2(g) \longrightarrow 2CO(g) \qquad \Delta H^\circ = -221.0 \text{ kJ}$$
$$CaO(s) + H_2O(l) \longrightarrow Ca(OH)_2(s) \quad \Delta H^\circ = -65.19 \text{ kJ}$$

$$2H_2(g) + O_2(g) \longrightarrow 2H_2O(l) \qquad \Delta H^\circ = -571.8 \text{ kJ}$$
$$CaO(s) + 3C(s) \longrightarrow CaC_2(s) + CO(g)$$
$$\Delta H^\circ = +462.3 \text{ kJ}$$

8. If a given shell has $n = 4$, which kinds of subshells (s, p, etc.) does it have? What is the maximum number of electrons that could populate this shell?

9. A beam of green light has a wavelength of 500 nm. What is the frequency of this light? What is the energy, in joules, of one photon of this light? What is the energy, in joules, of one mole of photons of this light? Would blue light have more or less energy per photon than this light?

10. Arrange the following kinds of electromagnetic radiation in order of increasing frequency: X rays, blue light, radio waves, gamma rays, microwaves, red light, infrared light, ultraviolet light.

11. What experimental evidence is there that matter has wavelike properties?

12. Use the periodic table to predict the electron configurations of (a) tin, (b) germanium, (c) silicon, (d) lead, and (e) nickel.

13. Give the electron configurations of the following ions. (a) Pb^{2+}, (b) Pb^{4+}, (c) S^{2-}, (d) Fe^{3+}, (e) Zn^{2+}

14. What causes an atom, molecule, or ion to be paramagnetic? Which of the ions in the preceding question are paramagnetic? What term describes the magnetic properties of the others?

15. Give the shorthand electron configurations of (a) Ni, (b) Cr, (c) Sr, (d) Sb, (e) Po

16. Which of the following elements has the largest difference between its second and third ionization energy? Explain your choice.
(a) Li   (b) Be   (c) B   (d) C

17. Which of the following processes are endothermic?
(a) $P^-(g) + e^- \rightarrow P^{2-}(g)$
(b) $Fe^{3+}(g) + e^- \rightarrow Fe^{2+}(g)$
(c) $Cl(g) + e^- \rightarrow Cl^-(g)$
(d) $S(g) + 2e^- \rightarrow S^{2-}(g)$

18. Give orbital diagrams for the valence shells of selenium and thallium.

19. Which ion would be larger: (a) Fe^{2+} or Fe^{3+}, (b) O^- or O^{2-}?

20. Which of the following pairs of elements would be expected to form ionic compounds: (a) Br and F, (b) H and P, (c) Ca and F?

21. Use Lewis symbols to diagram the reaction of calcium with sulfur to form CaS.

22. Draw Lewis structures for (a) SbH_3, (b) IF_3, (c) $HClO_2$, (d) C_2^{2-}, (e) AsF_5, (f) O_2^{2-}, (g) HCO_3^-, (h) TeF_6, (i) HNO_3.

23. Use the VSEPR theory to predict the shapes of (a) $SbCl_3$, (b) IF_5, (c) AsH_3, (d) BrF_2^-, (e) OF_2.

24. What kinds of hybrid orbitals are used by the central atom in each of the species in the preceding question?

25. Referring to your answers to questions 22 and 23, which of the following molecules would be nonpolar: SbH_3, IF_3, AsF_5, $SbCl_3$, OF_2?

26. The oxalate ion has the following arrangement of atoms.

$$\begin{array}{cc} O & O \\ C & C \\ O & O \end{array}$$

Draw all of its resonance structures.

27. Why, on the basis of formal charges and relative electronegativities, is it more reasonable to expect the structure of $POCl_3$ to be the one on the left rather than the one on the right?

28. A certain element X was found to form three compounds with chlorine having the formulas XCl_2, XCl_4, and XCl_6. One of its oxides has the formula XO_3, and X reacts with sodium to form the compound Na_2X.

(a) Is X a metal or a nonmetal?

(b) In which group in the periodic table is X located?

(c) In which periods in the periodic table could X possibly be located?

(d) Draw Lewis structures for XCl_2, XCl_4, XCl_6, and XO_3. (Where possible, follow the octet rule.) Which has multiple bonding?

(e) What do we expect the molecular structures of XCl_2, XCl_4, XCl_6, and XO_3 to be? Which are polar molecules?

(f) The element X also forms the oxide XO_2. Draw a Lewis structure for XO_2 that obeys the octet rule.

(g) Assign formal charges to the atoms in the Lewis structures for XO_2 and XO_3 drawn for parts (d) and (f).

(h) What kinds of hybrid orbitals would X use for bonding in XCl_4 and XCl_6?

(i) If X were to form a compound with aluminum, what would be its formula?

(j) Which compound of X would have the more ionic bonds, Na_2X or MgX?

(k) If X were in period 5, what would be the electron configuration of its valence shell?

29. Describe how molecular orbital theory explains the bonding in the oxygen molecule.

30. Draw diagrams that illustrate the band structures in typical conductors, nonconductors, and semiconductors.

31. If the element gallium, Ga, is added to germanium, which kind of semiconductor (n or p) will result? Explain your answer.

32. What are the conjugate acids of (a) HSO_3^-, (b) N_2H_4? What are the conjugate bases of (a) HSO_3^-, (b) N_2H_4, (c) $C_5H_5NH^+$?

33. Identify the conjugate acid–base pairs in the reaction

$$CH_3NH_2 + NH_4^+ \rightleftharpoons CH_3NH_3^+ + NH_3$$

34. Which is the stronger acid, H_3PO_3 or H_3PO_4? How can one tell without a table of weak and strong acids?

35. Which is the stronger acid, H_2S or H_2Te? How can one tell without a table of weak and strong acids?

36. X, Y, and Z are all nonmetallic elements in the same period of the periodic table where they occur, left to right, in the order given. Which would be a stronger binary acid than the binary acid of Y, the binary acid of X or the binary acid of Z? Explain.

37. Complete and balance the following equations if a reaction occurs.

(a) $Sn(s) + HCl(aq) \rightarrow$

(b) $Cu(s) + HNO_3(concd) \rightarrow$

(c) $Zn(s) + Cu^{2+}(aq) \rightarrow$

(d) $Ag(s) + Cu^{2+}(aq) \rightarrow$

38. Stearic acid, $C_{17}H_{35}CO_2H$, is derived from animal fat. Write a balanced chemical equation for the combustion of stearic acid in an abundant supply of oxygen.

Hot, expanding gases exert enough pressure to hurl clots of liquid lava from this erupting vent in the Alaid volcano in the Kuril island chain north of Japan. Among the simple natural laws about gases that we'll study in this chapter is one that relates gas pressure to temperature.

Chapter 10

Properties of Gases

Having studied the structure of matter, we begin here a systematic study of the *states* of matter—solids, liquids, and gases. We study gases first, because they are the easiest to understand and their behavior will help in the next chapter to explain some of the properties of liquids and solids. Many gases are important substances in their own right, of course, like oxygen and nitrogen in air and the carbon dioxide in carbonated beverages.

Air is roughly 21% O_2 and 79% N_2, but it has traces of several other gases.

One of the remarkable facts about gases is that despite wide differences in *chemical* properties, they all more or less obey the same set of *physical* laws—the gas laws. One question that we want to raise, in fact, is why there are general laws for gases but not similar laws for liquids and solids.

You no doubt are familiar with some of the ways gases behave. Blowing up a balloon, for example, gives most people some experience with gas pressure, and the "feel" of a balloon gives us the sense that this pressure acts equally in all directions.

The warning on all aerosol cans, "Do Not Incinerate," suggests how gas temperature affects its pressure. An increase in temperature might make the pressure go high enough to blow the can apart.

To study such matters as the effect of pressure and temperature on gases, we need to learn how four physical properties of gases are related—pressure *(P)*, volume *(V)*, temperature *(T)*, and the amount or moles *(n)*.

10.1
PROPERTIES COMMON TO ALL GASES

10.2
PRESSURE

Pressure is force *per unit area,* calculated by dividing the force by the area on which the force acts.

$$\text{Pressure} = \frac{\text{force}}{\text{area}}$$

The force normally acting on a pile of books, for example, is caused by the gravitational force of Earth acting on all objects nearby. Our measure of this gravitational force is called the *weight* of the object, so we use weight for force in the equation for pressure when the gravitational force is at work.

The distinction between force and pressure is important. You no doubt have experienced the extra comfort that comes from redistributing a constant force over a broader area. A number of books in a backpack causes less shoulder pain, for example, when the pack has wide straps rather than straps of string. The force (the weight) is the same in both case, but the pressure, or force per unit area, on your shoulder is less when the straps are wide. They distribute the force over a larger area, so the *ratio* of force to area, the pressure, is everywhere less. A book weighing 4.5 lb, for example, exerts a remarkably different pressure when it rests cover-side-down on your hand than if you balance it at its edge on a fingertip (see Figure 10.1). If your hand has an area of 19.5 in^2, the pressure in the first situation is only 4.5 lb ÷ 19.5 in.^{2} or 0.23 lb in.$^{-2}$. But if the book is resting on, say, only 0.0312 in.^{2}, an area 1/2 in. by 1/16 in., the pressure is 4.5 lb ÷ 0.0312 in.^{2} or 1.4×10^2 lb in.$^{-2}$, over 600 times as much.

Remember the distinction between *mass* and *weight* (page 19). We're dealing with *weight* here as the measure of force; on the moon the force (and so the pressure) exerted by the books would be less because their *weight* is less there, not their mass.

Atmospheric Pressure and the Barometer

Earth's gravity acts on the molecules in air to create a force, namely, that of the air pushing on the earth's surface. The resulting force *per unit area* is called **atmospheric pressure.** Atmospheric pressure can be demonstrated by the effects of pumping air from a steel can (see Figure 10.2). Before air is removed, the walls of the can experience atmospheric pressure identically inside and out. When air is pumped out, however, the pressure inside decreases. The atmospheric pressure outside is now greater, giving a net inward pressure sufficiently great to cause the can to collapse.

A **barometer** is an instrument used to measure atmospheric pressure, the simplest type being the *Torricelli barometer* (see Figure 10.3). It consists of a tube sealed at one end, 80 cm or more in length, and filled with mercury. When the tube is inverted over a dish of mercury, some mercury runs out, but not all. Atmospheric pressure, pushing on the surface of the mercury in the dish, acts to retain some mercury in the tube. Opposing the atmosphere is the downward pressure caused by the weight of the mercury still inside the tube. When the two pressures become equal, the mercury level stops falling. The space inside the tube above the mercury level has essentially no atmosphere; it's a *vacuum.*

The height of the mercury column, measured from the surface of the mercury in the dish, is directly proportional to atmospheric pressure. On days when the pressure is high, more mercury is forced into the tube and the height of the column increases. When the atmospheric pressure drops (during an approaching storm, for example), some mercury flows out of the tube and the height of the column decreases. Most people live where the height of the column fluctuates between 730 and 760 mm.

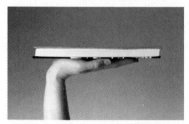

(a)

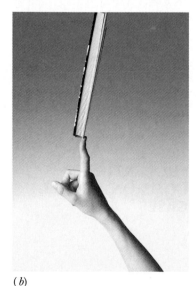

(b)

FIGURE 10.1

With the force distributed over a large area (*a*), the pressure is small. When the force acts on a small area (*b*), the pressure is large.

(*a*)   (*b*)

FIGURE 10.2

The effect of an unbalanced pressure. (*a*) The pressure inside the can is the same as outside. The pressures are balanced. (*b*) When a vacuum pump reduces the pressure inside the can, the unbalanced outside pressure quickly and violently makes the can collapse.

Units of Pressure

The height of the mercury in a barometer varies with altitude. On top of Earth's highest peak, Mt. Everest, the thin atmosphere can hold the mercury in a barometer to a height of only about 250 mm. At sea level, however, the height of the mercury column fluctuates around a value of 760 mm. Some days it is a little higher, some a little lower, depending on the weather. The average pressure at sea level has long been used by scientists as a unit of pressure and is called the **standard atmosphere** of pressure (abbreviated **atm**).[1]

In the SI, the unit of pressure is the **pascal,** symbolized **Pa.** The pascal is a very small pressure unit; 1 Pa is approximately the pressure exerted by the weight of a lemon spread over an area of 1 m².

To bring the standard atmosphere unit in line with other SI units, it has been redefined in terms of the pascal as follows.

$$1 \text{ atm} = 101{,}325 \text{ Pa (exactly)}$$

In Canada, atmospheric pressures are given in kilopascals (kPa) in weather reports; 1 atm is about 101 kPa.

For ordinary laboratory work, the *atmosphere* is inconveniently large, and chemists often use a smaller unit, the **torr** (named after Torricelli). The torr is defined as 1/760 of 1 atm.

$$1 \text{ torr} = \frac{1}{760} \text{ atm}$$

$$1 \text{ atm} = 760 \text{ torr (exactly)}$$

Mercury, a shiny, metallic element, is a liquid above − 39 °C.

$1 \text{ Pa} = 1 \text{ N m}^{-2}$, where N is the symbol for the *newton,* the SI unit of force.

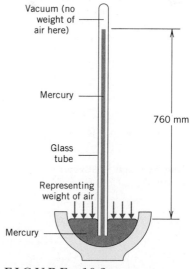

FIGURE 10.3

The mercurial or Torricelli barometer.

---

[1] Because any metal, including mercury, expands or contracts as the temperature increases or decreases, the height of the mercury column in the barometer varies a little with temperature. Therefore, the standard atmosphere was originally defined as the pressure able to support a column of mercury 760 mm high, measured at 0 °C and at sea level.

FIGURE 10.4

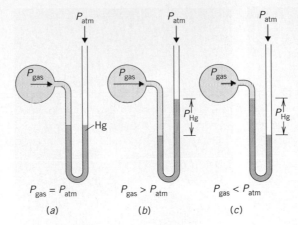

An open-end manometer. (*a*) The pressure of the entrapped gas, P_{gas}, equals P_{atm}, the atmospheric pressure. (Note that the arrows representing each pressure are here of *equal* length). (*b*) The entrapped gas pressure exceeds that of the atmosphere by an amount equal to the value of P_{Hg} when expressed in the unit of millimeters of mercury or mm Hg (but remember that 1 mm Hg – 1 torr). In other words, $P_{gas} = P_{atm} + P_{Hg}$. (*c*) The gas pressure is less than that of the atmosphere by an amount equal to P_{Hg}, so $P_{gas} = P_{atm} - P_{Hg}$.

The objection to "mm Hg" as a *pressure* unit is that a unit of length is not a good way to describe "force per unit area."

In the coming years, in all sciences, the kilopascal and the pascal will almost certainly replace other pressure units in the United States.

Advantages of mercury over other liquids are its low reactivity, low melting point, and particularly its very high density, which permits short manometer tubes.

Use Equation 10.1 when P_{gas} exceeds P_{atm}.

Thus, on most days near sea level, the torr is very close to the pressure that is able to support a column of mercury 1 mm high. In fact, the *millimeter of mercury* (abbreviated *mm Hg*) is often itself used as a pressure unit. Except when the most exacting measurements are being made, it is safe to use the relationship

$$1 \text{ torr} = 1 \text{ mm Hg}$$

The torr and atm are the units that we will use throughout this book, but those going into medicine in the United States will often see "mm Hg" used in connection with the gases of respiration or anesthesia.

Manometers

Gases produced by chemical reactions can be kept from escaping by using closed glassware. A **manometer** is used to measure pressures inside such vessels, and two types are common.

The **open-end manometer** consists of a U-tube partly filled with a liquid, usually mercury (see Figure 10.4). One arm of the U-tube is open to the atmosphere. The other end is exposed to the entrapped gas. When the pressure of the gas, P_{gas}, equals atmospheric pressure, P_{atm}, the two mercury levels are equal (Figure 10.4*a*). When the gas pressure exceeds that of the atmosphere (Figure 10.4*b*), the gas pushes the mercury to make its level in the open tube stand higher than its level nearest the gas. The difference in the two heights, P_{Hg}, represents the millimeters of mercury pressure (or number of torr) over and above the atmospheric pressure that day. Thus, the pressure of the gas in torr when the gas pressure *exceeds* atmospheric pressure is found by *adding* P_{Hg} to P_{atm}.

$$P_{gas} = P_{atm} + P_{Hg} \qquad \text{(used when } P_{gas} > P_{atm}\text{)} \qquad (10.1)$$

The value of P_{atm} in torr is measured by a laboratory barometer at the time when P_{Hg} is determined. If the difference in the mercury heights[2] in the two arms is, for example, 120 mm on a day when P_{atm} is 752 torr, the total gas pressure is 872 torr, by Equation 10.1.

$$P_{gas} = 752 \text{ torr} + 120 \text{ torr} = 872 \text{ torr}$$

---

[2] In this chapter, when we have trailing zeros *before* the decimal point (the 0 in 120, for example), we intend them to be counted as significant. Thus, in 120 torr there are three significant figures. In 760 torr, there are likewise three.

When the pressure on the entrapped gas is *less* than atmospheric pressure, the higher outside pressure is able to force mercury in the direction of the gas bulb (see Figure 10.4*c*). Now to find the gas pressure we must do a *subtraction*.

$$P_{gas} = P_{atm} - P_{Hg} \qquad \text{(used when } P_{gas} < P_{atm}) \qquad (10.2)$$

Use Equation 10.2 when P_{gas} is less than P_{atm}.

The level near the gas sample might, for example, become 200 mm *higher* than the level in the open arm, giving a pressure difference of 200 torr. If the atmospheric pressure is 752 torr, for example, then the pressure in the bulb is 552 torr, by Equation 10.2.

$$P_{gas} = 752 \text{ torr} - 200 \text{ torr} = 552 \text{ torr}$$

In a problem or exercise, if you can't remember whether to add or subtract, draw a picture of the manometer and roughly indicate the levels according to the data given (and always remember to ask whether the *size* of the answer makes sense).

The second common type of manometer, a **closed-end manometer,** is particularly convenient when the gas pressures to be measured are less than atmospheric pressure (see Figure 10.5). When a closed-end manometer is manufactured, the arm farthest from the gas sample is sealed, not open. The tube is then pumped free of air, and mercury is allowed to enter and fill the tube. Then the system is opened to atmospheric pressure, and some mercury drains out until the pressure of the atmosphere is able to retain the remaining mercury at a natural level in the arm nearest the gas container. No atmospheric pressure now exists above the mercury in the sealed arm.

When a closed-end manometer is connected to a gas sample having a pressure *less* than atmospheric, the level in the closed arm cannot now be maintained, and it falls. Some mercury is pulled into the arm nearest the gas sample, opening a space above the mercury in the closed arm, where a vacuum now exists just as in a Torricelli barometer. Because P_{atm} now actually equals 0, the difference in height of the two mercury levels, P_{Hg}, corresponds *directly* to the actual pressure of the gas sample. Thus a closed-end manometer gives us a pressure value directly, without the need of a calculation. One other advantage of a closed-end mercury manometer is that the tubes need not be as long as they must be for the open-end manometer.

Liquids Other than Mercury in Manometers and Barometers

Mercury is so dense (13.6 g mL^{-1}) that when a gas pressure differs very little from atmospheric pressure, the difference in mercury levels is correspondingly small. The precision of measurement therefore suffers. By using a liquid less dense than mercury, the difference in liquid levels is larger. If water ($d = 1.0$ g mL^{-1}) is used, for example, a pressure difference of 1 mm Hg would show up as a difference in heights of liquid levels 13.6 times as much or 13.6 mm (of water). In other words, a pressure that can sustain a 1 mm difference in height when mercury is the liquid can hold a 13.6 mm difference in height when the much less dense water is the liquid. A simple relationship exists between the two heights, h_A and h_B, when liquids A and B, having densities of d_A and d_B, are compared in a manometer or barometer measuring the same actual gas pressure.

$$h_B \times d_B = h_A \times d_A$$

Dividing by d_B gives

$$h_B = h_A \times \frac{d_A}{d_B} \qquad (10.3)$$

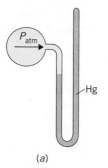

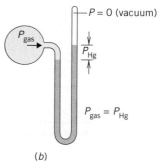

FIGURE 10.5

A closed-end manometer for measuring gas pressures less than 1 atm or 760 torr. (*a*) When constructed, the tube is fully evacuated and then mercury is allowed to enter the tube so that the closed arm fills completely. (*b*) When connected to a bulb containing a gas at low pressure, the gas pressure is insufficient to keep the mercury level at the top of the closed arm, and mercury flows out of that arm and into the other arm. The system comes to rest when the difference in mercury levels, P_{Hg}, represents the pressure of the confined gas; $P_{gas} = P_{Hg}$.

EXAMPLE 10.1
Using a Fluid Other than
Mercury in a Manometer

PROBLEM A liquid with a density of 0.854 g mL^{-1} is used in an open-end manometer to measure a gas pressure slightly greater than atmospheric pressure, which is
745.0 mm Hg at the time of measurement. The liquid level is 17.5 mm higher in
the open arm than in the arm nearest the gas sample (see Figure 10.4*b*). What is
the gas pressure in torr?

ANALYSIS The unit "torr" is a millimeter of *mercury* unit, so we need to calculate
what the given height would be if mercury had been the liquid instead. Only then
can we calculate the gas pressure in *torr*. Equation 10.3 is used, where liquid *B* is
mercury and liquid *A* is the one with a density of 0.854 g mL^{-1}.

SOLUTION

$$h_{Hg} = 17.5 \text{ mm} \times \frac{0.854 \text{ g mL}^{-1}}{13.6 \text{ g mL}^{-1}} = 1.10 \text{ mm Hg}$$

The height of the corresponding column of mercury, 1.10 mm, translates into 1.10
torr, so the pressure of the gas in torr is

$$P_{gas} = 745.0 \text{ torr} + 1.10 \text{ torr}$$
$$= 746.1 \text{ torr}$$

■ **Practice Exercise 1** If water were used instead of mercury to construct a Torricelli barometer (Figure 10.3), the tube would have to be at least how long in
millimeters and in feet?

10.3
PRESSURE–VOLUME–TEMPERATURE RELATIONSHIPS FOR A FIXED AMOUNT OF GAS

Now that we have all of the physical units of pressure, volume, and temperature needed to describe gases, we can study how fixed masses of gases respond
to changes in pressure, volume, and temperature.

Boyle's Law

Robert Boyle discovered how the volume of a fixed amount of gas varies with
the gas pressure at a constant temperature (see Figure 10.6). Boyle's discovery,
now called the **pressure–volume law** or **Boyle's law,** is that *the volume of a given
amount of gas held at constant temperature varies inversely with the applied pressure.*

$$V \propto \frac{1}{P} \qquad (T \text{ and mass are constant})$$

What is remarkable and is actually the heart of Boyle's discovery is that *this
relationship is essentially the same for all gases at ordinary temperatures and pressures.*
Neither liquids nor solids have a comparably general law.

We can remove the proportionality sign, $\propto$, by introducing a proportionality constant, *C.*

$$V = \frac{1}{P} \times C$$

Rearranging gives

$$PV = C$$

This relationship tells us that if the pressure is doubled the volume is cut in
half (at constant temperature).

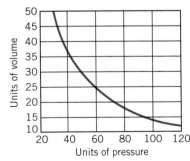

FIGURE 10.6

The relationship of volume to
pressure—Boyle's law. The unit
of pressure here is the same unit
that Boyle used, *inches of mercury.*
The volume varies reciprocally
with the pressure. Thus, the gas
volume is reduced by half, for example, when the pressure doubles from 30 to 60 units.

The Concept of an *Ideal Gas*

When very precise measurements are made, Boyle's law doesn't quite work. No gas obeys Boyle's law exactly over a wide range of pressures, but most gases come very close to it at ordinary temperatures and pressures. It's when a gas is under high pressure (10 atm or more) or at a low temperature (below 200 K) or some combination of these *and is therefore on the verge of changing to a liquid* that the gas most poorly fits Boyle's law. We'll see why later in the chapter.

By "ordinary" we mean near room temperature and atmospheric pressure.

Although real gases do not *exactly* obey Boyle's law, or any of the other gas laws that we'll study, it is often useful to imagine a hypothetical gas that would. Such a gas is called an *ideal gas*. An **ideal gas** obeys the gas laws exactly over all temperatures and pressures. A real gas behaves more and more like an ideal gas as its pressure decreases and its temperature increases.

Robert Boyle (1627–1691), an English scientist, stood at the threshold of modern science, living to see the publication of Newton's *Principia* but dying well before the time of Lavoisier (page 387).

Charles' Law

Jacques Charles discovered how the volume of a fixed amount of gas varies as the temperature is changed at constant pressure (see Figure 10.7). The gas used for obtaining the data in Figure 10.7 is one that happens to liquefy at − 100 °C. Each plot is a straight line, however, so it is easy to extrapolate (meaning, *reasonably extend*) lines to temperatures below − 100 °C. Such extrapolations, shown by the dashed lines in Figure 10.7, are simply reasonable predictions of how the gas would behave if it never changed to a liquid. Because other gases behave in the same way, we have another general law. According to the **temperature–volume law,** or **Charles' law,** *the volume of a fixed amount of gas is directly proportional to its Kelvin temperature when the gas pressure is held constant.* In other words,

Jacques Alexandre César Charles (1746–1823), a French scientist and more a mathematician than a chemist, had a keen interest in hot air balloons. He was the first to inflate a balloon with hydrogen.

$$V \propto T \qquad (P \text{ and mass are constant})$$

Using a constant of proportionality, C', we can write

$$V = C'T \qquad (P \text{ and mass are constant})$$

A *straight line* plot (Figure 10.7) signifies a *direct* proportionality, like Charles' law. Thus, if the Kelvin temperature is doubled, the volume is doubled (at constant pressure).

The temperature must be expressed in kelvins if the Charles' law relationship is to be expressible in this simple form. Thus, in calculations involving gas temperatures, we always translate degrees Celsius into kelvins.

Notice that the extrapolated lines in Figure 10.7 all converge on − 273.15 °C. The lines would all reach this temperature if the corresponding

The value of C' depends on the size and pressure of the gas sample.

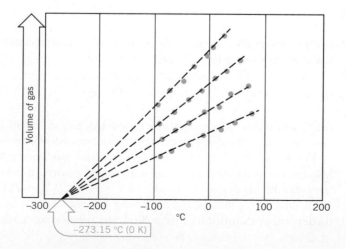

FIGURE 10.7

Charles' law plots. Each line shows how the gas volume changes with the Kelvin temperature for a different mass of the same gas.

gas volumes became zero. For a real gas, however, a zero volume is impossible. Yet this extrapolation is how scientists recognized that $-273.15\ °C$ must represent the coldest attainable temperature. A lower temperature would imply a gas with a negative volume. So $-273.15\ °C$ came to be called **absolute zero** and was designated zero on the Kelvin scale of temperatures.

Gay-Lussac's Law

Although observations of Jacques Charles helped, Joseph Gay-Lussac is credited with discovering how the pressure and temperature of a fixed amount of gas at constant volume are related. The specified conditions exist, for example, for a gas confined in a vessel with rigid walls, like an aerosol can. The relationship is today called the **pressure–temperature law,** or **Gay-Lussac's law:** *the pressure of a fixed amount of gas held at constant volume is directly proportional to the Kelvin temperature.* In other words,

$$P \propto T \qquad (V \text{ and mass are constant})$$

Using still another constant of proportionality, Gay-Lussac's law becomes

$$P = C''T \qquad (V \text{ and mass are constant})$$

The Combined Gas Law

In many experiments with gases, keeping either the temperature or the pressure of a gas constant is not attempted. Doing either is not necessary because of a more general law that relates the temperature, pressure, and volume of a fixed amount of gas. According to the **combined gas law,** *the ratio PV/T is a constant for a fixed amount of gas.*

$$\frac{PV}{T} = \text{constant (for a fixed amount of gas)}$$

If an initial set of conditions for a given gas sample is given the label "1," then

$$\frac{P_1 V_1}{T_1} = C'''$$

where C''' is the constant. If any or all of the conditions are changed to create a new set, which we may label "2," then

$$\frac{P_2 V_2}{T_2} = C'''$$

Because both PV/T expressions equal the same constant, the combined gas law may be written as

$$\frac{P_1 V_1}{T_1} = \frac{P_2 V_2}{T_2} \qquad (10.4)$$

Although temperatures must be in kelvins, the values of pressure and volume can be in any units *as long as they are used consistently for both sides.*

Equation 10.4 contains each of the other gas laws as special cases. Boyle's law, for example, requires a constant temperature; in other words, $T_1 = T_2$, so they would cancel in Equation 10.4 leaving $P_1 V_1 = P_2 V_2$. This signifies the heart of Boyle's law, namely, that the product PV is the same constant value under either conditions 1 or 2. Similarly, under constant pressure, P_1 cancels

P_2, and Equation 10.4 reduces to $V_1/T_1 = V_2/T_2$. This, of course, is another way of saying that the ratio V/T is a constant when P is constant, the heart of Charles' law. Under constant volume, Gay-Lussac's condition, V_1 cancels V_2, and Equation 10.4 reduces to $P_1/T_1 = P_2/T_2$. This signifies that at constant volume, the ratio P/T is a constant (Gay-Lussac's law). You can see why the gas law represented by Equation 10.4 is called the *combined* gas law.

A sample of argon is trapped in a gas bulb at a pressure of 760 torr when the volume is 100 mL and the temperature is 35.0 °C. What must its temperature be if its pressure becomes 720 torr and its volume 200 mL?

ANALYSIS Notice that the mass of the sample is *constant*. This is our clue that the combined gas law equation applies.

SOLUTION It's a good practice in gas law calculations to collect the data first, remembering to convert degrees Celsius into kelvins.

$V_1 = 100$ mL $\qquad P_1 = 760$ torr $\qquad T_1 = 308$ K $(35.0 + 273)$

$V_2 = 200$ mL $\qquad P_2 = 720$ torr $\qquad T_2 = ?$

The combined gas law equation,

$$\frac{P_1 V_1}{T_1} = \frac{P_2 V_2}{T_2}$$

is used, and after inserting the known values and rearranging we have

$$T_2 = 308 \text{ K} \times \frac{200 \text{ mL}}{100 \text{ mL}} \times \frac{720 \text{ torr}}{760 \text{ torr}}$$

$$= 584 \text{ K, or } 311 \text{ °C (from } 584 - 273)$$

CHECK The net effect is that a large *increase* in temperature must occur. Is this reasonable? Yes; notice that the volume is doubled (200 mL/100 mL), and Charles' law tells us that the temperature would have to be *increased* to achieve this. In fact, the kelvin temperature would have to be doubled (to 616 K) had it not been for the pressure change. The pressure decreases, and Gay-Lussac's law tells us that the temperature must be *decreased* to have this happen. But notice that the change in pressure is relatively small (720 torr/760 torr), not enough to offset the large increase in temperature mandated by Charles' law. The temperature, therefore, is not quite doubled.

■ **Practice Exercise 2** What will be the final pressure of a sample of nitrogen with a volume of 950 m³ at 745 torr and 25.0 °C if it is heated to 60.0 °C and given a final volume of 1150 m³?

As we said, the combined gas law can be used even when pressure, volume, or temperature is held constant. The next worked example illustrates this important fact and shows why separate equations for the laws of Boyle, Charles, and Gay-Lussac are not necessary.

Anesthetic gas is normally given to a patient when the room temperature is 20.0 °C and the patient's body temperature is 37.0 °C. What would this temperature change do to 1.60 L of gas if the pressure and mass stay constant?

EXAMPLE 10.2
Using the Combined Gas Law

Argon, a noble gas, is used for the low pressure atmosphere inside light bulbs because it won't chemically corrode the hot, glowing filament of wire in the bulb.

The combined gas law equation rearranges to

$$T_2 = T_1 \times \frac{V_2}{V_1} \times \frac{P_2}{P_1}$$

EXAMPLE 10.3
Using the Combined Gas Law

ANALYSIS This is a simple application of Equation 10.4. Because the pressure is constant, we can cancel P_1 and P_2 from the equation, to leave

$$\frac{V_1}{T_1} = \frac{V_2}{T_2}$$

SOLUTION Let's collect the data, changing degrees Celsius to kelvins:

$$V_1 = 1.60 \text{ L} \qquad T_1 = 293 \text{ K } (20.0 + 273)$$

$$V_2 = ? \qquad T_2 = 310 \text{ K } (37.0 + 273)$$

All that's left is the arithmetic. Inserting the data into the equation gives

$$\frac{1.60 \text{ L}}{293 \text{ K}} = \frac{V_2}{310 \text{ K}}$$

Solving for V_2 gives

$$V_2 = 1.60 \text{ L} \times \frac{310 \text{ K}}{293 \text{ K}}$$

$$= 1.69 \text{ L}$$

The volume increases by 0.09 L, about 90 mL. Although this is only a 6% change, it is a factor that anesthesiologists have to consider in administering anesthetic gases.

CHECK The answer is reasonable because the volume must increase with the temperature, but not by much because the volume-raising ratio of kelvin temperatures is small, only 310 K/293 K.

■ **Practice Exercise 3** A sample of nitrogen has a volume of 880 mL and a pressure of 740 torr. What pressure will change the volume to 870 mL at the same temperature?

10.4 THE IDEAL GAS LAW

Relationships of Gas Volumes in Gas Phase Reactions Many reactions occur in which gases constitute both reactants and products. Provided that comparisons are made under identical conditions of temperature and pressure, some very simple relationships have been discovered between the *volumes* of the gases in these experiments. Hydrogen gas reacts with chlorine gas, for example, to give gaseous hydrogen chloride.

$$\text{Hydrogen} + \text{chlorine} \longrightarrow \text{hydrogen chloride}$$
$$\text{1 vol} \qquad \text{1 vol} \qquad \text{2 vol}$$

The relationship of gas volumes, expressible in simple, whole number ratios, was discovered by Gay-Lussac.

Beneath the names are the measured *volumes* with which these gases interact when they are at the same pressure and temperature. Similar simple ratios by volume are observed when hydrogen combines with oxygen to give water, which is a gas above 100 °C.

$$\text{Hydrogen} + \text{oxygen} \longrightarrow \text{water (gaseous state)}$$
$$\text{2 vol} \qquad \text{1 vol} \qquad \text{2 vol}$$

The reacting volumes, measured under identical temperatures and pressures, given beneath the names, are in ratios of simple, whole numbers.

Avogadro's Principle

Amedeo Avogadro (1776–1856), an Italian scientist, helped to put chemistry on a quantitative basis.

When Amedeo Avogadro thought about how the volume ratios in the reactions of gases are in *whole* numbers, he was driven to conclude that equal volumes of gases must have identical numbers of molecules (at the same temperature and pressure). Today, we know that "equal numbers of *molecules*" is the same as "equal numbers of *moles*," so Avogadro's insight, now called **Avogadro's principle,** is expressed as follows. *When measured at the same temperature and pressure, equal volumes of gases contain equal numbers of moles.* In the equation for the reaction of hydrogen and chlorine, for example, we can see how the coefficients all stand in the same $1:1:2$ ratio as the gas volumes and the numbers of moles.

	$H_2(g)$	+	$Cl_2(g)$	$\longrightarrow$	$2HCl(g)$
Coefficients	1		1		2
Volumes	1 vol		1 vol		2 vol (experimental)
Molecules (or moles)	1		1		2 (Avogadro's principle)

A corollary to Avogadro's principle is that *the volume of a gas is directly proportional to its number of moles, n.*

$$V \propto n \qquad \text{(at constant } T \text{ and } P\text{)}$$

Avogadro's principle was a remarkable advance in our understanding of gases. His insight enabled chemists for the first time to determine the formulas of gaseous elements.[3]

Standard Molar Volume

A major implication of Avogadro's principle is that the volume occupied by one mole of *any* gas—its *molar volume*—must be identical for all gases under the same pressure and temperature. To compare experimental molar volumes, scientists have agreed to use 1 atm and 273.15 K (0 °C) as the **standard conditions of temperature and pressure,** or **STP,** for short. Under STP, the volume of one mole of a gas is 22.4 L, or very close to this. Table 10.1 gives the experimental data supporting this value. The value of 22.4 L mol^{-1} at STP is called the **standard molar volume** of a gas.

Formulation of the Ideal Gas Law

The constant C''' in the combined gas law, $PV/T = C'''$, is a constant only for a fixed number of moles of gas. C''', in fact, is directly proportional to n. To

TABLE 10.1   Standard Molar Volumes of Some Gases

Gas	Formula	Standard Molar Volume (L)
Helium	He	22.398
Argon	Ar	22.401
Hydrogen	H_2	22.410
Nitrogen	N_2	22.413
Oxygen	O_2	22.414
Carbon dioxide	CO_2	22.414

[3] Suppose that hydrogen chloride, for example, is correctly formulated as HCl, not as H_2Cl_2 or H_3Cl_3 or higher, and certainly not as $H_{0.5}Cl_{0.5}$. Then the only way that *two* volumes of hydrogen chloride could come from just *one* volume of hydrogen and *one* of chlorine is if each particle of hydrogen and each particle of chlorine consist of *two* atoms of H and Cl, respectively, H_2 and Cl_2. If these particles were single-atom particles, H and Cl, then one volume of H and one volume of Cl could give only *one* volume of HCl, not two. Of course, if the initial assumption were incorrect so that hydrogen chloride is, say, H_2Cl_2 instead of HCl, then hydrogen would be H_4 and chlorine would be Cl_4. The extension to larger subscripts is obvious.

create an equation even more general than the combined gas law, we split C''' into the product of the number of moles, n, and still another constant.

$$\frac{PV}{T} = C''' = n \times \text{new constant}$$

This "new constant" is given the symbol R and is called the **universal gas constant.** We can now write the combined gas law in a still more general form called the **ideal gas law.**

$$\frac{PV}{T} = nR$$

An ideal gas would obey this law exactly over all ranges of the gas variables. This equation, sometimes called the **equation of state for an ideal gas,** is usually rearranged and written as follows.

In some references, it is called the universal gas law.

Ideal gas law

Ideal Gas Law (Equation of State for an Ideal Gas)

$$PV = nRT \tag{10.5}$$

Equation 10.5 tells us how the four important variables for a gas, P, V, n, and T, are related. If we know the values of three, we can calculate the fourth. In fact, Equation 10.5 tells us that if three of the four variables are fixed for a given gas, *the fourth can only have one value.* This is a remarkable fact, and *it applies to all gases.* We can define the *state* of a given gas simply by specifying any three of the four variables. Nothing like the ideal gas law exists for liquids and solids.

To use the ideal gas law, we have to know the value of the universal gas constant, R. To calculate a value of R, we use the standard conditions of pressure and temperature, the standard molar volume, which sets 1 mol as n, the number of moles of the sample. We still have to decide what units to use for pressure and volume, so the value of R differs with these choices. Our choices, which reflect what is likely common practice among chemists, is to express volumes in *liters* and pressures in *atmospheres*. Thus for one molar volume at STP, $n = 1$ mol, $V = 22.4$ L, $P = 1.00$ atm, and $T = 273$ K. Using these values lets us calculate R as follows.

If n, P, and T in Equation 10.5 are known, for example, then V can have only one value.

$$R = \frac{PV}{nT} = \frac{(1.00 \text{ atm})(22.4 \text{ L})}{(1.00 \text{ mol})(273 \text{ K})}$$

$$= 0.0821 \frac{\text{atm L}}{\text{mol K}}$$

Or, arranging the units in a commonly used order,

$$R = 0.0821 \text{ L atm mol}^{-1} \text{ K}^{-1}$$

When a problem gives the values of pressure, volume, or temperature in any units other than, respectively, atmospheres, liters, or kelvins, unit conversions must be carried out before using the value of R in the ideal gas law.

EXAMPLE 10.4
Using the Ideal Gas Law

A sample of oxygen at 24.0 °C and 745 torr was found to have a volume of 455 mL. How many grams of O_2 were in the sample?

ANALYSIS In this question we are given the pressure, volume, and temperature of a gas and then asked to determine the number of *grams* of gas. The ideal gas law equation, $PV = nRT$, lets us calculate the number of *moles, n*, from *P, V*, and *T*, but an easy moles-to-grams conversion for O_2, using its molecular mass, will give us the mass of O_2 in grams.

SOLUTION To use $R = 0.0821$ L atm mol^{-1} K^{-1}, we must have the volume in liters, the pressure in atmospheres, and the temperature in kelvins. Gathering the data and making the necessary unit conversions as we go, we have

$$P = 0.980 \text{ atm} \qquad \text{From: } 745 \text{ torr} \times \frac{1 \text{ atm}}{760 \text{ torr}}$$

$$V = 0.455 \text{ L} \qquad \text{From: } 455 \text{ mL}$$

$$T = 297 \text{ K} \qquad \text{From: } 24.0 + 273$$

Solving the ideal gas law for *n* gives us

$$n = \frac{PV}{RT}$$

Substituting the proper values of *P, V*, and *T* into this equation gives

$$n = \frac{(0.980 \text{ atm})(0.455 \text{ L})}{(0.0821 \text{ L atm mol}^{-1} \text{ K}^{-1})(297 \text{ K})}$$

$$= 0.0183 \text{ mol of } O_2$$

The molecular mass of O_2 is 32.00, so 1 mol O_2 = 32.00 g O_2. A conversion factor from this relationship lets us convert "mol O_2" into "g O_2."

$$0.0183 \text{ mol } O_2 \times \frac{32.00 \text{ g } O_2}{1 \text{ mol } O_2} = 0.586 \text{ g } O_2$$

Thus, the sample of oxygen had a mass of 0.586 g.

■ **Practice Exercise 4** What volume in milliliters does a sample of nitrogen with a mass of 0.245 g occupy at 21 °C and 750 torr?

Determining the Molecular Mass of a Gas

When a chemist makes a new compound, its molecular mass must be determined in order to establish its chemical identity. When the compound is a gas, this determination can be made with the help of the ideal gas law.

Suppose that a gas has been trapped in a sealed bulb at a known pressure, volume, and temperature. Its mass in *grams* is also known. We use the ideal gas law equation and the *P, V*, and *T* data to compute the number of *moles* in the sample. Then we compute the ratio of *grams to moles* to find the molecular mass.

A student collected a sample of a gas in a 0.220 L gas bulb until its pressure reached 0.757 atm at a temperature of 25.0 °C. The sample weighed 0.299 g. What is the molecular mass of the gas?

ANALYSIS To find the molecular mass, we calculate the ratio of the number of grams of the sample to the number of moles. We've been given the number of grams, and we can use the ideal gas equation to compute the number of moles.

EXAMPLE 10.5
Determining the Molecular
Mass of a Gas

SOLUTION Pressure is already in atmospheres and the volume is in liters, but we must convert degrees Celsius into kelvins. Gathering our data, we have

$$P = 0.757 \text{ atm} \qquad V = 0.220 \text{ L} \qquad T = (25.0 + 273) = 298 \text{ K}$$

We substitute the data into the ideal gas law equation.

$$PV = nRT$$

$$(0.757 \text{ atm})(0.220 \text{ L}) = (n)(0.0821 \text{ L atm mol}^{-1} \text{ K}^{-1})(298 \text{ K})$$

Rearranging to solve for n gives

$$n = \frac{(0.757 \text{ atm})(0.220 \text{ L})}{(0.0821 \text{ L atm mol}^{-1} \text{ K}^{-1})(298 \text{ K})}$$

$$= 6.81 \times 10^{-3} \text{ mol}$$

The ratio of grams to moles, therefore, is

$$\frac{0.299 \text{ g}}{6.81 \times 10^{-3} \text{ mol}} = 43.9 \text{ g mol}^{-1}$$

When the molecular mass of a substance is expressed in units of grams per mole, the quantity is sometimes called the **molar mass** and has the units g mol⁻¹.

Because there are 43.9 grams per mole, the molecular mass must be 43.9. (The gas used for this example is CO_2, with a molecular mass of 44.0.)

■ **Practice Exercise 5** The label on a cylinder of an inert gas became illegible, so a student allowed some of the gas to flow into a 300 mL gas bulb until the pressure was 685 torr. The sample now weighed 1.45 g; its temperature was 27.0 °C. What is the molecular mass of this gas? Which of the Group 0 gases (the inert gases) was it?

EXAMPLE 10.6
Calculating the Molecular Mass from Gas Density

A gaseous compound of phosphorus and fluorine was found to have a density of 3.50 g L⁻¹ at a temperature of 25 °C and a pressure of 740 torr. It was also found to consist of 35.2% P and 64.8% F. What is its empirical formula, its molecular mass, and its molecular formula?

ANALYSIS To calculate the *empirical formula*, we use the percentage composition just as we studied in Example 3.12, page 101).

To calculate the *molecular mass*, we need both the number of moles, n, of the gas sample and the sample's mass in grams. We calculate the number of moles from $PV = nRT$, but the given data include only P, T, and the gas density, not V. Remember, however, that the units of gas density are grams per *liter*, so we can simplify our calculation by basing it on a volume of $V = 1.00$ L. At this volume the sample must have a mass of 3.50 g, because the gas density is 3.50 g *per liter*. So when we compute n, we can take the ratio 3.50 g/n to give us the molecular mass in grams per mole.

To determine the *molecular formula*, we compare the molecular mass just found with that calculated from the empirical formula and make any needed multiplication of subscripts in the empirical formula. Because this is a complicated calculation, let's summarize the steps.

STEP 1 Calculate the *empirical formula* from the percentage composition.

STEP 2 Calculate the *molecular mass* by relating the grams per liter (density) data to the number of grams per mole, using the ideal gas law to find the number of moles in a 1 L sample.

STEP 3 Calculate the *molecular formula* by comparing the experimental molecular mass (step 2) to that calculated from the empirical formula (step 1).

SOLUTION *STEP 1* To calculate the empirical formula from the percentages by mass of P (35.2%) and F (64.8%), we assume a sample with a mass of 100.0 g, which therefore holds the following number of moles of P and F.

$$35.2 \text{ g P} \times \frac{1 \text{ mol P}}{30.97 \text{ g P}} = 1.14 \text{ mol P}$$

and

$$64.8 \text{ g F} \times \frac{1 \text{ mol F}}{19.00 \text{ g F}} = 3.41 \text{ mol F}$$

The formula, then, could be written as $P_{1.14}F_{3.41}$, but we divide both of the subscripts by the smaller, 1.14. Because $3.41 \div 1.14 = 2.99$, close enough to 3, we can write the empirical formula as PF_3.

STEP 2 To calculate the number of moles of gas in 1.00 L, we use the ideal gas law, making unit conversions as needed.

$T = 298 \text{ K}$ From: $25 + 273$

$P = 0.974 \text{ atm}$ From: $740 \text{ torr} \times \dfrac{1 \text{ atm}}{760 \text{ torr}}$

$V = 1.00 \text{ L}$

$$n = \frac{PV}{RT} = \frac{(0.974 \text{ atm})(1.00 \text{ L})}{(0.0821 \text{ L atm mol}^{-1} \text{ K}^{-1})(298 \text{ K})}$$

$$= 3.98 \times 10^{-2} \text{ mol}$$

Because the mass of the gas in 1.00 L is 3.50 g (from the given density in g L^{-1}), the ratio of grams to moles is

$$\frac{3.50 \text{ g}}{3.98 \times 10^{-2} \text{ mol}} = 87.9 \text{ g mol}^{-1}$$

The molecular mass of the compound is thus 87.9.

STEP 3 The molecular mass that we can easily calculate for the empirical formula, PF_3, is 87.97. It is essentially identical to that found in Step 2, so PF_3 is the molecular formula of the compound as well as its empirical formula.

■ **Practice Exercise 6** A gaseous compound of phosphorus and fluorine with an empirical formula of PF_2 was found to have a density of 5.60 g L^{-1} at 23.0 °C and 750 torr. Calculate its molecular mass and its molecular formula.

For reactions involving gases, Avogadro's principle let us use a new kind of stoichiometric equivalency, one between *volumes* of gases. Earlier, for example, we noted the following reaction and its gas volume relationships.

$$2H_2(g) + O_2(g) \longrightarrow 2H_2O(g)$$
     2 vol   1 vol   2 vol

We can now write the following stoichiometric equivalencies, provided we understand that the pressure and temperature are constant.

2 vol $H_2(g) \Leftrightarrow$ 1 vol $O_2(g)$ just as 2 mol $H_2 \Leftrightarrow$ 1 mol O_2

2 vol $H_2(g) \Leftrightarrow$ 2 vol $H_2O(g)$ just as 2 mol $H_2 \Leftrightarrow$ 2 mol H_2O

1 vol $O_2(g) \Leftrightarrow$ 2 vol $H_2O(g)$ just as 1 mol $O_2 \Leftrightarrow$ 2 mol H_2O

10.5
THE STOICHIOMETRY OF REACTIONS BETWEEN GASES

The recognition that equivalencies in gas *volumes* are numerically the same as those for numbers of moles of gas in reactions involving gases simplifies many calculations.

EXAMPLE 10.7
Stoichiometry of Reactions of
Gases

How many liters of oxygen at STP are needed to combine exactly with 1.50 L of hydrogen at STP?

ANALYSIS We need the chemical equation

$$2H_2(g) + O_2(g) \longrightarrow 2H_2O(g)$$

and we need Avogadro's principle, which tells us that for this reaction of gases, 2 vol $H_2(g) \Leftrightarrow 1$ vol $O_2(g)$. We can write this equivalency because the *same* conditions of temperature and pressure, STP, are stipulated.

SOLUTION We multiply the given, 1.50 L H_2, by a conversion factor made from the equivalency 2 vol $H_2(g) \Leftrightarrow 1$ vol $O_2(g)$. Using liters as a volume unit,

$$\text{Liters of } O_2 = 1.50 \text{ L } H_2 \times \frac{1 \text{ L } O_2}{2 \text{ L } H_2}$$

$$= 0.750 \text{ L } O_2$$

CHECK The calculated fact that 0.750 L of O_2 combines with 1.50 L of H_2 is a reasonable answer because the coefficients and Avogadro's principle tell us that we need half as much O_2 by volume as H_2.

■ **Practice Exercise 7** Methane burns according to the following equation.

$$CH_4(g) + 2O_2(g) \longrightarrow CO_2(g) + 2H_2O(g)$$

The combustion of 4.50 L of CH_4 consumes how many liters of O_2, both volumes measured at STP?

Methane is the chief constituent of Bunsen burner gas.

In experimental work involving the generation of gases, questions arise such as, *What size flask is needed* to trap the gas that is going to be generated by a reaction when the gas is to be at some desired pressure and temperature? Or, *How many grams of some starting material* can generate enough gas to fill a given gas bulb at some value of pressure and temperature. The stoichiometric equivalencies between gas volumes and moles help us find the answers to such questions.

EXAMPLE 10.8
Calculating the Volume of a
Gas Product Using
Stoichiometric Equivalencies

A student needs to prepare some CO_2 and intends to use the following reaction in which $CaCO_3$ is heated strongly.

$$CaCO_3(s) \longrightarrow CO_2(g) + CaO(s)$$

The question concerns the size of the flask needed to accept the gas. How large should the container be to hold the CO_2 if 1.25 g of $CaCO_3$ is to be used? The pressure is 740 torr. The final temperature is to be 25.0 °C.

ANALYSIS The balanced equation tells us:

$$1 \text{ mol } CaCO_3 \Leftrightarrow 1 \text{ mol } CO_2$$

Once we calculate the *moles* of $CaCO_3$, we can use this value for n in the ideal gas law equation to find the gas volume.

SOLUTION The formula mass of $CaCO_3$ is 100.09, so

$$\text{moles of } CaCO_3 = 1.25 \text{ g } CaCO_3 \times \frac{1 \text{ mol } CaCO_3}{100.09 \text{ g } CaCO_3}$$

$$= 1.25 \times 10^{-2} \text{ mol } CaCO_3$$

Because 1 mol $CaCO_3 \Leftrightarrow$ 1 mol CO_2,

$$n = 1.25 \times 10^{-2} \text{ mol } CO_2$$

Before we use n in the ideal gas law equation, we must convert the given pressure and temperature into the units required by R.

$$P = 740 \text{ torr} \times \frac{1 \text{ atm}}{760 \text{ torr}} = 0.974 \text{ atm} \qquad T = (25.0 + 273) = 298 \text{ K}$$

By rearranging the ideal gas law equation we obtain

$$V = \frac{nRT}{P}$$

$$= \frac{(1.25 \times 10^{-2} \text{ mol})(0.0821 \text{ L atm mol}^{-1} \text{ K}^{-1})(298 \text{ K})}{0.974 \text{ atm}}$$

$$= 0.314 \text{ L} = 314 \text{ mL}$$

The container must therefore have a volume of 314 mL.

CHECK It's hard to check the "reasonableness" of an answer like this, but at least be sure that the unit conversions and cancellations are right.

■ **Practice Exercise 8** In one lab, the gas collecting apparatus used a gas bulb with a volume of 250 mL. How many grams of $Na_2CO_3(s)$ would be needed to prepare enough $CO_2(g)$ to fill this bulb when the pressure is 738 torr and the temperature is 23 °C? The equation is

$$Na_2CO_3(s) + 2HCl(aq) \longrightarrow 2NaCl(aq) + CO_2(g) + H_2O$$

10.6 DALTON'S LAW OF PARTIAL PRESSURES

In many applications, mixtures of nonreacting gases are used, and our question in this section is, How do the gas laws apply to gas mixtures? John Dalton discovered the answer.

In a mixture of nonreacting gases, each gas contributes to the total pressure in proportion to the fraction (by volume) in which it is present. Dalton called this contribution the **partial pressure** of the gas. It is the pressure that the gas would exert all by itself if it were the only gas in the same container at the same temperature. The general symbol for the partial pressure of gas a is P_a. For a particular gas, the formula of the gas may be put into the subscript, as in P_{O_2}. What Dalton discovered about partial pressures is now called **Dalton's law of partial pressures:** *The total pressure of a mixture of nonreacting gases is the sum of their individual partial pressures.* In equation form, the law is

$$P_{total} = P_a + P_b + P_c + \cdots \tag{10.6}$$

In dry CO_2-free air at STP, for example, P_{O_2} is 159.12 torr, P_{N_2} is 593.44 torr, and P_{Ar} is 7.10 torr. These partial pressures add up to 759.66 torr, just 0.34 torr less than 760 torr or 1.00 atm. The remaining 0.34 torr is contributed by several trace gases, including other noble gases.

The air we breathe is a common mixture of nonreacting gases. At least under ordinary temperatures and pressures, nitrogen and oxygen do not react.

 Dalton's law of partial pressures

Suppose you want to fill a pressurized tank with a volume of 4.00 L with oxygen-enriched air for use in diving, and you want the tank to contain 50.0 g of O_2 and 150 g of N_2. What is the total gas pressure in the tank at 25 °C?

EXAMPLE 10.9
Using Dalton's Law of Partial Pressures

ANALYSIS If we calculate the pressure that each gas would have if it were *alone* in the tank and then add the partial pressures, we'll have the answer. To find the individual partial pressures, we apply the ideal gas law using a temperature of 298 K (for 25 °C) and a volume of 4.00 L. To use this law, however, we must first convert the amounts of the gases in grams to the numbers of moles, a routine grams-to-moles conversion.

SOLUTION For moles of oxygen,

$$n = 50.0 \text{ g } O_2 \times \frac{1 \text{ mol } O_2}{32.00 \text{ g } O_2} = 1.56 \text{ mol } O_2$$

For moles of nitrogen,

$$n = 150 \text{ g } N_2 \times \frac{1 \text{ mol } N_2}{28.00 \text{ g } N_2} = 5.36 \text{ mol } N_2$$

Next, we calculate the partial pressures. For each, $P_a = nRT/V$, where V is the volume of the tank.

$$P_{O_2} = \frac{1.56 \text{ mol} \times 0.0821 \text{ L atm K}^{-1} \text{ mol}^{-1} \times 298 \text{ K}}{4.00 \text{ L}}$$

$$= 9.54 \text{ atm}$$

$$P_{N_2} = \frac{5.36 \text{ mol} \times 0.0821 \text{ L atm K}^{-1} \text{ mol}^{-1} \times 298 \text{ K}}{4.00 \text{ L}}$$

$$= 32.8 \text{ atm}$$

For the total pressure, using Dalton's law, we have

$$P_{\text{total}} = 9.54 \text{ atm} + 32.8 \text{ atm}$$
$$= 42.3 \text{ atm}$$

■ **Practice Exercise 9** How many grams of oxygen are present at 25 °C in a 5.00 L tank of oxygen-enriched air under a total pressure of 30.0 atm when the only other gas is nitrogen at a partial pressure of 15.0 atm?

Collecting Gases over Water

When gases that do not react with water are prepared in the laboratory, they can be trapped over water by an apparatus such as that illustrated in Figure 10.8. This method unavoidably produces a wet gas, one with as much water vapor as it can hold at the temperature of the water.

When water vapor is present in a mixture of gases, it has a partial pressure like any other gas. The space above *any* liquid always contains some of the

FIGURE 10.8

Collecting a gas over water. As the gas bubbles through the water, water vapor goes into the gas, so the total pressure inside the bottle includes the partial pressure of the water vapor at the temperature of the water.

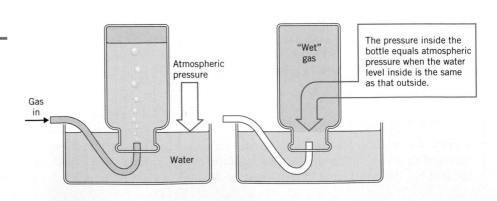

TABLE 10.2   Vapor Pressure of Water at Various Temperatures

Temperature (°C)	Vapor Pressure (torr)	Temperature (°C)	Vapor Pressure (torr)
0	4.579	50	92.51
5	6.543	55	118.0
10	9.209	60	149.4
15	12.79	65	187.5
20	17.54	70	233.7
25	23.76	75	289.1
30	31.82	80	355.1
35	42.18	85	433.6
37[a]	47.07	90	527.8
40	55.32	95	633.0
45	71.88	100	760.0

[a] Human body temperature.

vapor of the liquid. The vapor exerts its own pressure, called the **vapor pressure.** Its value for any given substance depends only on the temperature. The vapor pressures of water at different temperatures, for example, are given in Table 10.2.

When we collect a (wet) gas over water, we usually want to know how much *dry* gas this corresponds to, and we can use data for the wet gas to find out. From Dalton's law we can write

$$P_{total} = P_{gas} + P_{water}$$

The pressure of the gas is then, by rearrangement,

$$P_{gas} = P_{total} - P_{water}$$

We find values for P_{water} at various temperatures from a table, like Table 10.2. P_{total} is obtained from the laboratory barometer, provided that the water level inside the gas collecting bottle is the same as the water level outside. We thus calculate P_{gas}, and this pressure is what the gas would exert if it were dry and inside the same volume that was used to collect it.

Even the mercury in a barometer has a tiny vapor pressure—0.0011 torr at 20 °C—which is much too small to affect readings of barometers and the manometers studied in this chapter.

EXAMPLE 10.10
Calculation Using Data from the Collection of a Gas over Water

A sample of oxygen is collected over water at 20 °C and a pressure of 738 torr. Its volume is 310 mL. (a) What is the partial pressure of the oxygen? (b) What would be its volume when dry at STP?

ANALYSIS The initial pressure of 738 torr has to be corrected by a Dalton's law calculation, using the vapor pressure of water at 20 °C, which is 17.5 torr (from Table 10.2). The result will be the answer to the first question. We will then have *initial* values of P (here meaning P_{O_2}), V, and T.

The second part of the problem asks for a final value of V at new values of P and T representing standard temperature and pressure. In other words, after answering the first question we will have values for P_1, V_1, and T_1, as well as values for P_2 and T_2. We're asked to calculate V_2. The combined gas law tells us how these values are all related.

SOLUTION First, we find the partial pressure of the oxygen, using Dalton's law.

$$P_{O_2} = P_{total} - P_{water}$$
$$= 738 \text{ torr} - 17.5 \text{ torr} = 721 \text{ torr}$$

Second, we assemble the data.

$$P_1 = 721 \text{ torr (now } P_{O_2}) \qquad P_2 = 760 \text{ torr (standard pressure)}$$

$$V_1 = 310 \text{ mL} \qquad V_2 = ?$$

$$T_1 = (20 + 273) = 293 \text{ K} \qquad T_2 = 273 \text{ K (standard temperature)}$$

We use these in the combined gas law equation:

$$\frac{P_1 V_1}{T_1} = \frac{P_2 V_2}{T_2}$$

Or

$$\frac{(721 \text{ torr})(310 \text{ mL})}{293 \text{ K}} = \frac{(760 \text{ torr})(V_2)}{273 \text{ K}}$$

Solving for V_2 gives us

$$V_2 = \frac{(721 \text{ torr})(310 \text{ mL})(273 \text{ K})}{(760 \text{ torr})(293 \text{ K})}$$

$$= 274 \text{ mL}$$

Thus, when the water vapor is removed from the gas sample, the dry oxygen will occupy a volume of 274 mL at STP.

CHECK The dry volume (274 mL) must be *less* than the wet volume (310 mL), but not much less, so the size of the answer is reasonable.

■ **Practice Exercise 10** Suppose you prepared a sample of nitrogen and collected it over water at 15 °C at a total pressure of 745 torr and a volume of 310 mL. Find the partial pressure of the nitrogen and the volume it would occupy at STP.

Mole Fractions and Mole Percents for Gas Mixtures from Partial Pressures

The concept of mole fraction applies to any uniform mixture in any physical state, gas, liquid, or solid.

One of the useful ways of describing the composition of a mixture of gases is in terms of the *mole fractions* of the components. The **mole fraction** of any component in a mixture is the ratio of the number of its moles to the total number of moles of all components. Expressed mathematically, the mole fraction of substance *A* in a mixture of *A, B, C,* . . ., *Z* substances is

Mole fractions

$$X_A = \frac{n_A}{n_A + n_B + n_C + n_D + \cdots + n_Z} \qquad (10.7)$$

where X_A is the mole fraction of component *A*, and $n_A, n_B, n_C,$ . . ., n_Z, are the numbers of moles of each component, *A, B, C,* . . ., *Z*, respectively. The sum of all mole fractions for a mixture must always equal 1.

You can see in Equation 10.7 that the units (moles) cancel and that a mole fraction has no units. Nevertheless, always remember that a mole fraction stands for the ratio of *moles* of one component to the total number of *moles* of all components.

Sometimes a mole fraction is multiplied by 100 to give the **mole percent** of the component, symbolized **mole%.**

Partial pressure data can be used to calculate the mole fractions of individual gases in a gas mixture because the number of *moles* of each gas is directly proportional to its partial pressure. We can demonstrate this as follows. The partial pressure, P_A, for any one gas, *A*, in a gas mixture with a total volume *V* at a temperature *T* is found by the ideal gas law equation, $PV = nRT$. So to calculate the number of moles of *A* present, we have

$$n_A = \frac{P_A V}{RT}$$

For any particular gas mixture at a given temperature, the values of V, R, and T are all constants, making the ratio V/RT a constant, too. We can therefore simplify the previous equation by using C to stand for V/RT. In other words, we can write

$$n_A = P_A C$$

The result is the same as saying that the number of moles of a gas in a mixture of gases is directly proportional to the partial pressure of the gas. The constant C is the same for all gases in the mixture. So by using different letters to identify individual gases in the usual way, we can use this expression for each component in Equation 10.7 for gas mixtures. Thus, by Equation 10.7,

$$X_A = \frac{P_A C}{P_A C + P_B C + P_B C + \cdots + P_Z C}$$

The constants, C, can be factored out and canceled, so

$$X_A = \frac{P_A}{P_A + P_B + P_B + \cdots + P_Z}$$

The denominator is the sum of the partial pressures of all the gases in the mixture, but this sum equals the total pressure of the mixture (Dalton's law of partial pressures). Therefore, the previous equation simplifies to

$$X_A = \frac{P_A}{P_{total}} \qquad (10.8)$$

Equation 10.8 gives us a simple way to calculate the partial pressure of a gas in a gas mixture when we know its mole fraction or vice versa.

What are the mole fractions and mole percents of nitrogen and oxygen in air when their partial pressures are 160 torr for oxygen and 600 torr for nitrogen? Assume no other gases are present.

ANALYSIS This calls for Equation 10.8, which gives us the mole fraction of a gas from its partial pressure and the total pressure. But we must first calculate P_{total}, which is just the sum of the partial pressures (Equation 10.6). When we have P_{total}, we can apply Equation 10.8 to each gas.

SOLUTION

$$P_{total} = 600 \text{ torr} + 160 \text{ torr} = 760 \text{ torr}$$

Using Equation 10.8 first for the mole fraction of N_2:

$$X_{N_2} = \frac{P_{N_2}}{P_{total}} = \frac{600 \text{ torr}}{760 \text{ torr}}$$
$$= 0.789, \text{ or } 78.9 \text{ mole percent of } N_2$$

We do the same for oxygen.

$$X_{O_2} = \frac{P_{O_2}}{P_{total}} = \frac{160 \text{ torr}}{760 \text{ torr}}$$
$$= 0.211, \text{ or } 21.1 \text{ mole percent of } O_2$$

CHECK The two mole percentages must add up to 100%, and they do.

EXAMPLE 10.11
Using Partial Pressure Data to Calculate Mole Fractions

■ **Practice Exercise 11** What is the mole fraction and the mole percent of oxygen in exhaled air if P_{O_2} is 116 torr and P_{total} is 760 torr?

Partial Pressures from Mole Fractions

If we can calculate mole fractions from partial pressures, it must be equally simple to calculate partial pressures from mole fractions (both calculations using the value of the total pressure). This is a useful calculation for those, like mountain climbers, who are interested in working at high altitudes and wonder uneasily about the partial pressure of oxygen at such altitudes. We breath normally at or near sea level because our lungs and the hemoglobin concentration of our blood are adapted to the partial pressure of oxygen in air at low elevations, roughly 160 torr. At the summit of Mt. Everest (elevation 8848 m), however, the total atmospheric pressure is only about 250 torr. *The mole fractions of oxygen and nitrogen in air are essentially constant at all altitudes in the lower atmosphere,* so the mole fraction of oxygen, 0.211, is the same at sea level as at the top of Everest. What changes with altitude is the partial pressure of each gas because proportionately fewer numbers of moles of the gases in air are present in a given volume. For oxygen on top of Mt. Everest, the value of P_{O_2} must be the mole fraction of O_2 times the partial pressure there.

$$P_{O_2} = 0.211 \times 250 \text{ torr} = 52.8 \text{ torr (at Everest's summit)}$$

In other words, the value of P_{O_2} at any altitude is some fraction, the mole fraction, of the air pressure there. A value of P_{O_2} of only about 50 torr is too low for nearly all people to survive, but Nepalese porters (Sherpas) and several other well-trained peoples have cardiovascular systems adapted to high altitudes.

> Hemoglobin is the compound in our red blood cells that chemically picks up oxygen in the lungs and releases it where tissues, having done oxygen-consuming chemical work, need oxygen.

10.7
GRAHAM'S LAW OF EFFUSION

Diffusion is the complete spreading out and intermingling of the molecules of one gas, like those of perfume or cologne, among those of another gas, like the air in a room. We can think of this process as the movement of a perfume's molecules from a region of the perfume's higher partial pressure into the space where its partial pressure is lower.

A similar process, called **effusion,** is the movement of gas molecules through an extremely tiny opening into a region of lower pressure (see Figure 10.9). Helium-filled rubber balloons go limp overnight because helium atoms effuse through invisible pores in the balloon. Why *air*-filled balloons go limp more slowly is answered in this section.

Thomas Graham studied the rates of effusion of a series of gases through the same tiny hole and found that the more dense a gas is, the slower if effuses.

> Thomas Graham (1805–1869) was a Scottish scientist.

FIGURE 10.9

Effusion.

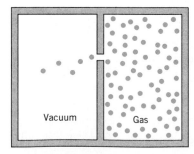

The relationship is now called **Graham's law:** *The rates of effusion of gases are inversely proportional to the square roots of their densities, d, when compared at identical pressures and temperatures.*

$$\text{Effusion rate} \propto \frac{1}{\sqrt{d}} \qquad (\text{constant } P \text{ and } T)$$

Two balloons newly filled with air (*left*) and helium (*right*).

This proportionality can be changed to an equation in the usual way, using a constant which we will call C^{**}.

$$\text{Effusion rate} \times \sqrt{d} = C^{**} \qquad (\text{constant } P \text{ and } T)$$

What is remarkable about C^{**} is that *it is virtually identical for all gases.* Thus if we have two gases, *A* and *B*, we can write:

$$\text{Effusion rate } (A) \times \sqrt{d_A} = \text{effusion rate } (B) \times \sqrt{d_B} = C^{**}$$

By rearrangement, we obtain:

$$\frac{\text{effusion rate } (A)}{\text{effusion rate } (B)} = \frac{\sqrt{d_B}}{\sqrt{d_A}} = \sqrt{\frac{d_B}{d_A}} \qquad (10.9)$$

Same balloons, one day later. Helium has escaped by effusion from its balloon.

It can be shown[4] that the density of a gas is directly proportional to its molecular mass. Thus, we can reexpress Equation 10.9 as follows:

$$\frac{\text{effusion rate } (A)}{\text{effusion rate } (B)} = \sqrt{\frac{d_B}{d_A}} = \sqrt{\frac{M_B}{M_A}} \qquad (10.10)$$

 Graham's law of effusion

where M_A and M_B are the molecular masses of gases *A* and *B*. Now we can see why helium effuses more rapidly than air from a rubber balloon. Helium's molecular mass is only 4.00, but the chief gases in air have much larger molecular masses. Oxygen's is 32.00 and nitrogen's is 28.00. By Equation 10.10, helium effuses $\sqrt{32.00/4.00}$ or 2.83 times more rapidly than O_2. Thus, helium atoms effuse out of a balloon faster than air molecules can effuse in. The net result is a fairly rapid loss of gas from the helium balloon.

Which effuses more rapidly and by what factor, ammonia or hydrogen chloride?

ANALYSIS A gas effusion problem requires the use of Graham's law, which says that the gas with the smaller molecular mass will effuse more rapidly. So we need to find the molecular masses. Then the relative rates of effusion are found by using these in Equation 10.10.

SOLUTION The molecular masses are 17.03 for NH_3 and 36.46 for HCl, so we know that NH_3, with its smaller value, effuses more rapidly than HCl. The ratio of the effusion rates are given by

$$\frac{\text{effusion rate } (NH_3)}{\text{effusion rate } (HCl)} = \sqrt{\frac{36.46}{17.03}}$$
$$= 1.463$$

Ammonia effuses 1.463 times more rapidly than HCl under the same conditions.

EXAMPLE 10.12
Using Graham's Law

---

[4] Density = mass/volume, so mass = (density)(volume).
Number of moles = n = mass/formula mass, so mass = (n)(formula mass)
Therefore, (density)(volume) = (n)(formula mass), from which it can be seen that (density) = (formula mass)(n)/(volume). So for fixed values of volume and number of moles (n), the density of a gas is proportional to its formula mass.

■ **Practice Exercise 12** The hydrogen halide gases all have the same general formula, H*X*, where *X* can be Cl, Br, or I. If HCl(*g*) effuses 1.88 times more rapidly than one of the others, which hydrogen halide is the other, HBr or HI?

10.8
KINETIC THEORY AND THE GAS LAWS

Back in the nineteenth century, scientists who already knew the gas laws couldn't help but ask, "How do gases 'work'?" They wondered what had to be true about all gases to explain their conformity to a *common set* of gas laws. The **kinetic theory of gases** was the answer. We introduced some of its ideas in Chapter 5, and we continue this study here.

The strategy for answering the question about the nature of a gas began with postulating a *model* or picture of an ideal gas, and it continued by using the laws of physics and statistics to see whether the model would correctly predict the gas laws. The results were so spectacularly successful that some specialists in the history of human thought have called it one of greatest triumphs of the human mind!

Postulates of the Kinetic Theory of Gases

The particles are assumed to be so small that they have no dimensions at all.

1. A gas consists of an extremely large number of very tiny particles that are in constant, random motion.
2. The gas particles themselves occupy a net volume so small in relation to the volume of their container that their contribution to the total volume can be ignored.
3. The particles often collide in perfectly elastic collisions[5] with themselves and with the walls of the container, and they move in straight lines between collisions neither attracting nor repelling each other.

Billiard players master the applications of the laws of mechanics, if not their mathematical forms.

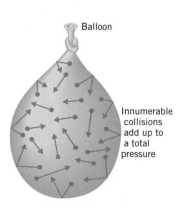
Balloon

Innumerable collisions add up to a total pressure

Physics, the study of matter, energy, and physical changes, has a branch called *mechanics* that deals with the laws of behavior of moving objects. In effect, the kinetic theory treats gases as if they were supersmall, constantly moving billiard balls that bounce off each other and the walls of their container. The gas particles are assumed to be so small that their individual volumes can be ignored.

According to the model, gases are mostly empty space. This explains why gases, unlike liquids and solids, can be compressed so much (squeezed to smaller volumes). It also explains why we have gas laws for gases, and *the same laws for all gases,* but not comparable laws for liquids or solids. The chemical identity of the gas does not matter, because gas molecules do not touch each other except when they collide and there are extremely weak interactions, if any, between them.

The theoretical calculations based on the model gave results that agreed splendidly with the known behavior of gases, as described by the gas laws. Scientists, therefore, concluded that the model itself must be very close to the

[5] In *perfectly elastic* collisions, no energy is lost by friction as the colliding objects deform momentarily.

truth about gases. When the behavior of real gases departs somewhat from the ideal, the model even helps to explain this, as we'll see later in this section.

The Gas Laws Explained by the Kinetic Theory

We cannot go over the mathematical details of the kinetic theory of gases, but we can describe some of the ways in which the model helps us understand the gas laws.

Kinetic Theory and Gas Temperature Revisited   The greatest triumph of the kinetic theory came with its explanation of gas temperature, which we discussed in Section 5.2. A gas sample has a huge number of particles moving randomly, colliding constantly, changing velocities, and so experiencing changes in their kinetic energies. Figure 10.10 shows how these different values of kinetic energy, KE, are distributed among the gas molecules at two different temperatures. The peaks of the curves represent the most frequently experienced values of KE. The *average* value of kinetic energy lies slightly to the right of the peak for each curve, because the curves are not symmetrical.

Just 1 mL of a gas at STP has over 2.5×10^{19} molecules.

The plots in Figure 10.10 are called *Maxwell–Boltzmann* distributions.

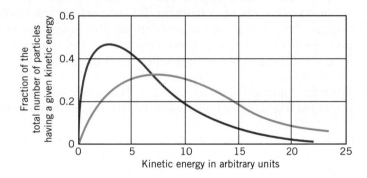

FIGURE 10.10

The distribution of kinetic energies among gas particles. The curve with the higher maximum shows how at room temperature (300 K) the relative number of particles having a particular value of kinetic energy changes with changes in the kinetic energy. The flatter curve shows what happens when the temperature is raised.

What the calculations based on the model of an ideal gas showed was that the product of pressure and volume, *PV*, is proportional to the average kinetic energy.

$$PV \propto \text{average KE}$$

But from the experimental study of gases, culminating in the equation of state for an ideal gas, we have another term to which *PV* is proportional, the Kelvin temperature of the gas.

$$PV \propto T$$

We know that the proportionality constant here is *nR*, because by the ideal gas law, $PV = nRT$. With *PV* proportional *both* to *T* and to "average KE," it must be true that the temperature of a gas is proportional to the average KE of its particles.

$$T \propto \text{average KE}$$

Kinetic Theory and the Pressure–Volume Law (Boyle's Law)   Using the model of an ideal gas, physicists were able to show that gas pressure is the net effect of innumerable collisions made by gas particles on the walls. Suppose that one wall of the gas container is a movable piston that we can push in

FIGURE 10.11

The kinetic theory and the pressure–volume law. When the gas volume is made smaller in going from (*a*) to (*b*), the frequency of the collisions per unit area of the container's walls increases. Therefore, the pressure increases.

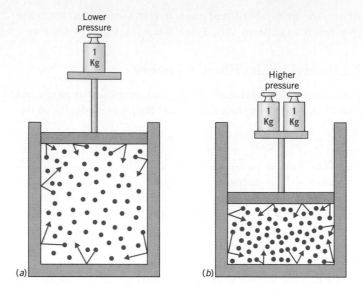

(or pull out) and so change the gas volume (see Figure 10.11). If we make the volume smaller without changing the temperature, no change occurs in the average kinetic energy of the molecules. But now there would be more gas particles per unit volume. If we reduce the volume by one-half, for example, we double the number of molecules per unit volume. This would, therefore, double the number of collisions per second with a unit area and so double the pressure. Thus, to cut the volume in half requires that we double the pressure, not triple it or quadruple it, which is exactly what Boyle discovered:

$$P \propto \frac{1}{V} \qquad \text{or} \qquad V \propto \frac{1}{P}$$

FIGURE 10.12

The kinetic theory and the pressure–temperature law. (*a*) The gas exerts only a small pressure, P_1, over and above the atmospheric pressure. (*b*) Raising the gas temperature raises the pressure to P_2, the extra mercury being added to keep the gas from expanding its volume. At the higher temperature the gas particles move with greater average velocity (as suggested by the longer motion arrows), so normally they would take up more space. The extra mercury is a measure of the increased pressure needed to prevent this.

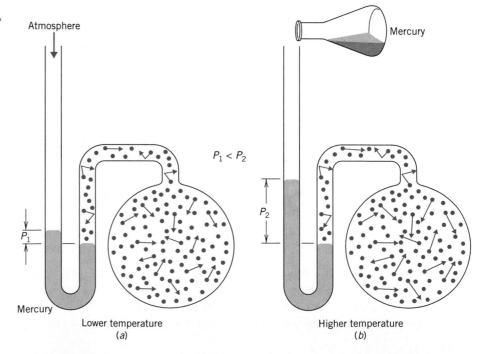

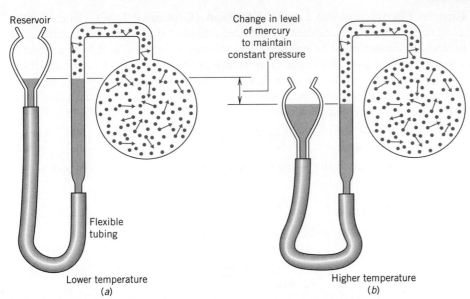

Reservoir

Change in level
of mercury
to maintain
constant pressure

Flexible
tubing

Lower temperature
(a)

Higher temperature
(b)

FIGURE 10.13

The kinetic theory and the temperature–volume law. The pressure is the same in both (a) and (b), as indicated by the mercury levels. However, at the lower temperature (a), the gas takes up less volume. At the higher temperature (b), the gas particles have a higher average energy and velocity. To hold the pressure constant (a condition of Charles' law), the gas has to be given more room by letting some mercury flow out of the tube and back into the reservoir.

Kinetic Theory and the Pressure–Temperature Law (Gay-Lussac's Law) The kinetic theory, as we have said, tells us that an increase in gas temperature increases the average velocity of gas particles. At higher velocities, the particles must strike the container's walls more frequently and with greater force. But at constant volume, the *area* being struck is still the same, so the force per unit area—the pressure—must therefore increase (see Figure 10.12). In this way the kinetic theory explains how gas pressure is proportional to gas temperature (under constant volume and mass), which is the pressure–temperature law of Gay-Lussac.

Kinetic Theory and the Temperature–Volume Law (Charles' Law) The kinetic theory tells us that increasing the temperature of a gas increases the average kinetic energy of its particles, which tends to increase the pressure of the gas. The only way we can keep P constant is to allow the gas to expand (see Figure 10.13). Therefore, a gas expands with increasing T in order to keep P constant, which is another way of saying that $V \propto T$ at constant P. Thus, the kinetic theory explains Charles' law.

Kinetic Theory and Dalton's Law of Partial Pressures The law of partial pressures is actually evidence for the part of the third postulate in the kinetic theory that pictures gas particles moving in straight lines between collisions neither attracting nor repelling each other. They act *independently*, in other words (see Figure 10.14). Only if the particles of each gas do act independently can the partial pressures of the gases add up in a simple way to give the total pressure.

FIGURE 10.14

Gas molecules would not travel in straight lines if they attracted each other as in (a) or repelled each other as in (b). This would influence the length of time between collisions with the walls and would therefore affect the collision frequency with the walls. In turn, this would affect the pressure each gas would exert. Only if the molecules traveled in straight lines with no attraction or repulsions, as in (c), would their individual pressures not be influenced by near misses or by collisions between the molecules.

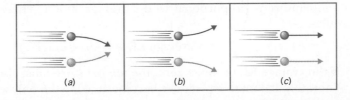

(a)   (b)   (c)

Kinetic Theory and the Law of Effusion (Graham's Law) The conditions of Graham's law are that the rates of effusion of two gases with different molecular masses must be compared at the same pressure and temperature. Thus when two gases have the same temperature, *their particles have identical average kinetic energies.* Remember that KE $= \frac{1}{2}mv^2$. Using subscripts 1 and 2 to identify two gases with molecules having different masses m_1 and m_2, we can write that at a given temperature

$$(\text{KE})_1 = (\text{KE})_2$$

Taking v_1 and v_2 to be the *average* velocities of the molecules of the gases, we have

$$\tfrac{1}{2}m_1v_1{}^2 = \tfrac{1}{2}m_2v_2{}^2$$

Rearranging to get the ratio of v^2 terms, we have

$$\frac{v_1{}^2}{v_2{}^2} = \frac{m_2}{m_1}$$

The ratio of velocities, then, is found by taking the square roots of both sides.

$$\frac{v_1}{v_2} = \sqrt{\frac{m_2}{m_1}}$$

For any substance, the individual mass of a molecule is proportional to the molecular mass. Representing a molecular mass of a gas by M, we can restate this as $m \propto M$. The proportionality constant is the same for all gases. (It's in grams per atomic mass unit when we express m in atomic mass units.) When we take a ratio of two molecular masses, the constant cancels anyway, so we can write

$$\frac{v_1}{v_2} = \sqrt{\frac{m_2}{m_1}} = \sqrt{\frac{M_2}{M_1}}$$

The rate of effusion of a gas, of course, has to be proportional to the average speed of its molecules.

$$\text{Effusion rate} \propto v$$

$$\text{Effusion rate} = C^{\#}v$$

Taking a ratio of speeds causes $C^{\#}$ to cancel, so we can now write

$$\frac{\text{effusion rate gas 1}}{\text{effusion rate gas 2}} = \sqrt{\frac{M_2}{M_1}}$$

This is the way we expressed Graham's law in Equation 10.10, page 421.

Kinetic Theory and Absolute Zero The kinetic theory found that gas temperature is proportional to the average kinetic energy of the gas molecules.

$$T \propto \text{average KE} \propto \tfrac{1}{2}(m)(\text{average } v)^2$$

If the average KE becomes zero, the temperature must also become zero. We know that the mass (m) cannot become zero, so the only way that the average

KE can be zero is if v goes to zero. A gas particle cannot move any slower than it does at a dead standstill, so if the gas molecules stop moving entirely, the gas is as cold as anything can get. It's at absolute zero.[6]

The combined gas law states that the ratio PV/T is a constant. According to the ideal gas law, PV/T actually equals a product of constants, nR. We know experimentally, however, that PV/T is not quite a constant for real gases. When we use experimental values of P, V, and T for a real gas, like O_2, to plot actual values of PV/T as a function of P, we get the curve shown in Figure 10.15. The *horizontal* plot in Figure 10.15 is what we should see if PV/T were truly constant over all values of P, as it would be for an ideal gas.

A real gas, like oxygen, deviates from ideal behavior for two important reasons. First, the model of an ideal gas assumes that gas molecules individually have no volume, but of course they do. (If all of the kinetic motions of the gas molecules ceased and the molecules settled, you could imagine the net space that the molecules would occupy in and of themselves.) Therefore, the actual space left for the kinetic motions of the molecules of the real gas is not as large as the volume V of the container. Fortunately, the volume taken up by the molecules themselves is a very tiny fraction of the container's volume at ordinary pressures. This is why real gases fit the gas laws so well at ordinary pressures. The fraction of space taken up by the molecules themselves, however, would increase if we forced more and more molecules into the container and so increased the pressure. It would be like jamming more and more people into a small room. The fraction of space available for each person to move around diminishes even though the volume of the room is unchanged. The increasing fraction of space taken up by the individual molecular volumes is one reason why real gases deviate more and more from ideal behavior at increasing pressures. At the higher values of P in the plot for a real gas in Figure 10.15, the volume occupied by the molecules themselves becomes too large in relationship to the container volume, V. Thus, the value of V used in a gas law calculation is larger than warranted, which tends to make the ratio PV/T greater for a real gas than that for the ideal gas, particularly at high pressures.

The second fact concerning real gases is that their particles do attract each other somewhat, unlike one assumption about an ideal gas. Any tendency for the gas particles to cling together, however slight, and so strike the container's walls less often would cause smaller values of P than expected from the model of an ideal gas. Thus, the ratio PV/T is *less* than that for an ideal gas, and the consequences show up particularly where the problem of particle volume is least, at lower pressures. The curve of Figure 10.15, therefore, dips at lower pressures.

Many attempts have been made to modify the equation of state of an ideal gas to get an equation that better fits the experimental data for individual real gases. One of the more successful attempts was that of J. D. van der Waals. He found ways to correct the measured values of P and V to give values that fit a

10.9
REAL GASES: DEVIATIONS FROM THE IDEAL GAS LAW

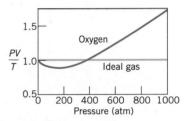

FIGURE 10.15

A graph of PV/T versus P for an ideal gas is like plotting a constant versus P. The graph must be a straight line, as shown, because the ideal gas law equation tells us that $PV/T = nR$ (a product of constants). The same plot for oxygen, however, is not a straight line, showing that O_2 is not "ideal."

J.D. van der Waals (1837–1923), a Dutch scientist, won the 1910 Nobel Prize in physics.

---

[6] Even at 0 K, there is slight motion required by the Heisenberg uncertainty principle. If the molecules were actually dead still, then there would be no uncertainty in their speed and thus the uncertainty in their position would be infinitely great. We would not know where they were! But we do know; they're in this or that container. Thus some uncertainty in speed must exist to have less uncertainty in position and so locate the sample.

TABLE 10.3   Van der Waals Constants

Substance	a (L^2 atm mol^{-2})	b (L mol^{-1})
Noble Gases		
Helium, He	0.03421	0.02370
Neon, Ne	0.2107	0.01709
Argon, Ar	1.345	0.03219
Krypton, Kr	2.318	0.03978
Xenon, Xe	4.194	0.05105
Other Gases		
Hydrogen, H_2	0.02444	0.02661
Oxygen, O_2	1.360	0.03183
Nitrogen, N_2	1.390	0.03913
Methane, CH_4	2.253	0.04278
Carbon dioxide, CO_2	3.592	0.04267
Ammonia, NH_3	4.170	0.03707
Water, H_2O	5.464	0.03049
Ethyl alcohol, C_2H_5OH	12.02	0.08407

general gas law equation. The result of his derivation, a derivation that we will not describe, is called the *van der Waals' equation of state for a real gas.*

$$\left(P + \frac{n^2 a}{V^2} \right)(V - nb) = nRT \qquad (10.11)$$

The constants a and b are called the *van der Waals constants* (see Table 10.3). They are determined for each real gas by carefully measuring P, V, and T under different conditions. Then trial calculations are made to figure out what values of the van der Waals constants give the best match between the observed data and van der Waals' equation.

Notice that the constant a involves a correction to the pressure term of the ideal gas law, so the size of a would indicate something about attractions between molecules. The constant b helps to correct for the volume term, so the size of b would indicate something about the sizes of particles in the gas. Larger values of a mean stronger attractive forces between molecules; larger values of b mean larger molecular sizes. Thus, the most easily liquefied substances, like water and ethyl alcohol, have the largest values of van der Waals' constant a, suggesting attractive forces between their molecules. In the next chapter we'll continue the study of factors that control the physical state of a substance, particularly attractive forces and their origins.

SUMMARY

Gas Laws An **ideal gas** is a hypothetical gas that obeys the gas laws exactly over all ranges of pressure and temperature. Real gases exhibit ideal gas behavior most closely at low pressures and high temperatures, which are conditions remote from those that liquefy a gas.

Boyle's Law (Pressure–Volume Law). Volume varies inversely with pressure at constant temperature and mass. $V \propto 1/P$.

Charles' Law (Temperature–Volume Law). Volume varies directly with the Kelvin temperature at constant pressure and mass. $V \propto T$.

Gay-Lussac's Law (Temperature–Pressure Law). Pressure varies directly with Kelvin temperature at constant volume and mass. $P \propto T$.

Avogadro's Principle. Equal volumes of gases contain equal numbers of moles when compared at the same temperature and pressure.

Combined Gas Law. *PV* divided by *T* for a given gas sample is a constant.

Ideal Gas Law. $PV = nRT$. When *P* is in atm and *V* is in L, the value of *R* is 0.0821 L atm mol^{-1} K^{-1} (*T* being, as usual, in kelvins).

Dalton's Law of Partial Pressures. The total pressure of a mixture of gases is the sum of the partial pressures of the individual gases.

$$P_{\text{total}} = P_a + P_b + P_c + \cdots$$

Graham's Law of Effusion. The rate of effusion of a gas varies inversely with the square root of its density (or the square root of its molecular mass) at constant pressure and temperature.

Kinetic Theory of Gases An ideal gas consists of innumerable very hard particles, with individual volumes so small they can be ignored, between which no forces of attraction or repulsion act. They are in constant, chaotic, random motion. When the laws of physics and statistics are applied to this model, and the results compared with the ideal gas law, the Kelvin temperature of a gas is proportional to the average kinetic energy of the gas particles. Pressure is the result of forces of collisions of the particles with the container's walls.

Real Gases Because individual gas particles do have real volumes and because small forces of attraction do exist between them, real gases do not exactly obey the gas laws. The van der Waals equation of state for a real gas makes corrections for the volume of the gas molecules and for the attractive force between gas molecules. The measure of the attractive force is given by *a*, a van der Waals constant, and the other constant, *b*, is a measure of the relative size of the gas molecules.

Reaction Stoichiometry: A Summary With the study in this chapter of the stoichiometry of reactions involving gases, we have completed our study of the tools needed for the calculations of all variations of reaction stoichiometry. The fact of central importance is that all such calculations must funnel through *moles*. Whether we start with grams of some compound in a reaction, or the molarity of its solution plus a volume, or *P–V–T* data for a gas in the reaction, *we must get the essential calculation into moles*. The equation's coefficients then give us the stoichiometric equivalencies needed to convert from the number of moles of one substance into the numbers of moles of any of the others in the reaction. Then we can move back to any other kind of unit we wish. The following flow chart summarizes what we have been doing.

The labels on the arrows of the flow chart suggest the basic tools. Formula masses or molecular masses get us from grams to moles or from moles to grams. Molarity and volume data move us from concentration to moles or back. With *P–V–T* data we can find moles or, knowing moles of a gas, we can find any one of *P, V,* or *T*, given the other two.

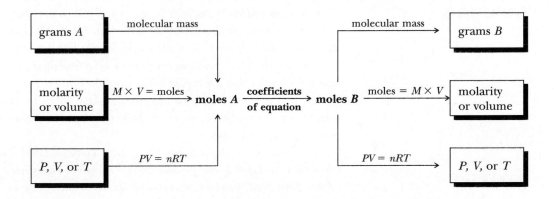

Tools You Have Learned

The table below lists the tools you have learned in this chapter that are applicable to problem solving. Review them if necessary, and refer to them when working on the Thinking-It-Through problems and the Review Exercises that follow.

Tool	Function
Combined gas law (page 406)	To be used to calculate a particular value of P, V, or T, given other values and the quantity of gas as fixed. To be used in applications of Boyle's, Charles', or Gay-Lussac's law.
Ideal gas law (page 410)	To be used when any three of the four variables of the physical state of a gas, P, V, T, or n, are known to calculate the value of the fourth.
Dalton's law of partial pressures (page 415)	To calculate the partial pressure of one gas in a mixture of gases from the total pressure and either the partial pressures of the other gases or their mole fractions. Given the volume and temperature of a gas collected over water and the atmospheric pressure, to calculate the volume the gas would have when dry.
Mole fractions (page 418)	To calculate the mole fraction or mole percent of one component of a mixture from other data about the mixture. To calculate the partial pressure of one gas in a mixture from its mole fraction (page 419).
Graham's law of effusion (page 421)	To calculate relative rates of effusion of gases. To calculate molecular masses of gases from relative rates of effusion.

THINKING IT THROUGH

The goal for each of the following problems is to give you practice in thinking your way through problems. The goal is not to find the answer itself; instead, you are only asked to assemble the available information needed to obtain the answer, state what additional data (if any) are needed, and describe how you would use the data to answer the question. For problems involving unit conversions, list the relationships among the units that are needed to carry out the conversions. Construct the conversion factors that can be formed from these relationships. Then set up the solution to the problem by arranging the conversion factors so the units cancel correctly to give the desired units of the answer.

The problems are divided into two groups. Those in Level 2 are more challenging than those in Level 1 and provide an opportunity to really hone your problem-solving skills. Both levels, however, may include questions that draw on material studied in earlier chapters.

Level 1 Problems

1. The plunger of a bicycle tire pump is to be pushed in so that the pressure of the entrapped volume of air changes from 1 atm to 2.5 atm. The air hose has been closed off; no air can escape. No temperature change occurs. If the initial volume of air is 175 mL, how can you calculate the expected final volume of the air?

2. How can one calculate how much change in temperature is required to increase the volume of a gas sample, initially at 24.0 °C, from 500 to 550 mL if the pressure is kept constant?

3. A sample of (dry) methane, CH_4, was collected at atmospheric pressure and a temperature of 25 °C in a sealed glass flask which the manufacturer said could withstand an internal pressure of 1.50 atm. Assuming the claim to be accurate, could this flask be safely heated to 50 °C? What additional data are needed (if any)?

4. A glass vessel with a volume of 1.50 L, fitted with a closed-end manometer and a valve, is filled with dry nitrogen. The initial manometer reading at 24.0 °C is 742 torr, the same as the atmospheric pressure on that day. The valve is left open and the bulb is heated to 610.0 °C.

(a) How can you predict what the manometer will read at the higher temperature?

(b) The valve is shut, and the flask is allowed to cool back to 24.0 °C. How can you predict what the manometer will now read?

5. Hydrogen gas is purchased in high-pressure, steel cylinders fitted with pressure gauges and valves. For an experiment using hydrogen gas, a student in an advanced chemistry class allowed hydrogen to transfer out of such a steel cylinder into another sealed vessel. The vessel had a volume of 235 mL. The pressure in the vessel changed from 740 torr to 9.50 atm. (The initial pressure was caused by the presence of argon, a noble gas in Group 0.) Are enough data provided to enable the student, without using a weighing balance, to calculate how much hydrogen (in moles) was now in the vessel? What additional data are needed? How would the calculation be made?

6. A sample of gas was collected over water at 25 °C and a pressure of 745 torr in a 275 mL container. How could the number of moles of the gas collected be calculated?

Level 2 Problems

7. Helium can be purchased as a gas under high pressure in heavy-walled steel cylinders fitted with pressure gauges and valves. They are certainly too heavy to weigh using balances ordinarily available in chemistry laboratories. In the United States, the gauges usually read in lb in.$^{-2}$. One such gauge gave a reading of 2.5×10^3 lb in.$^{-2}$. A chemist wanted to know how many moles of helium were still in the tank. How could this be calculated? If insufficient data are given here, what additional data must be obtained?

8. A dry sample of an unknown gas was collected in a 225 mL flask attached to an open-end manometer with a valve arrangement that would allow the chemist to add water to the flask. Enough gas was added to the flask to make the final pressure equal to 865 torr at 25 °C.

(a) What additional data, if any, are needed to calculate the number of moles of gas added to the flask? To calculate the number of grams added (without using a weighing balance)?

(b) Suppose that 100 mL of water, also at 25 °C, was allowed to enter the flask without allowing any of the gas to leave. What could be inferred about the gas if the pressure does *not* increase when this is done?

9. A 20.0 g sample of dilute, aqueous hydrogen peroxide was placed into a 1.00 L sealed vessel fitted to an open-end manometer with an initial reading of 740 torr. Then a valve was closed to prevent any gas from entering or leaving the vessel. The system was maintained at 25 °C. Hydrogen peroxide decomposes according the following equation.

$$2H_2O_2(aq) \longrightarrow O_2(g) + 2H_2O(l)$$

After all of the hydrogen peroxide had thus decomposed, the manometer had a reading of 907 torr. How could you calculate the number of grams of hydrogen peroxide in the initial sample? (Ignore the very small *change* in the gas space available in the flask caused by the formation of some additional water. Assume that oxygen is insoluble in water.) Is it necessary to do a Dalton's law calculation to correct for the vapor pressure of water? Explain.

10. A 1.710 g sample of a gas at 100.0 °C and a pressure of 740.0 torr occupies a volume of 1.250 L.

(a) If the gas is an ideal gas, how can its molecular mass be found from the given data?

(b) Suppose the gas consists of ethyl alcohol vapor, $C_2H_5OH(g)$, whose molecules are relatively large and between which exist relatively strong forces of attraction. How could one calculate the expected pressure of the same sample in the same vessel at 100.0 °C?

11. For as long as measurements have been taken, the percentage composition of oxygen and nitrogen in clean, dry air has been constant to four significant figures. Air thus appears to obey the law of constant composition (law of definite proportions). Yet, we call air a mixture, not a compound. Based on information in this chapter, what kind of experiment might be performed to show that clean, dry air is indeed a mixture?

12. The research team that you supervise has succeeded in preparing a binary compound of nitrogen and fluorine. It is a gas that boils below − 100 °C. You have yet to discover its molecular formula. A member of the team reports that a sample with a volume of 265 mL has a mass of 101 mg. Do you have sufficient data to calculate a molecular mass? If not, what additional data do you need? How would you use the data to calculate the molecular mass?

13. Both CO_2 and SF_6 are colorless and odorless gases that are chemically quite stable in the atmosphere. If you were to manufacture tennis balls, which gas would you use to give internal pressure to the balls? Explain. Assume that costs are not what matter the most, only quality.

REVIEW EXERCISES   *Answers to questions whose numbers are printed in color are given in Appendix D.*
Challenging questions are marked by asterisks.

Concept of Pressure

10.1 How are the following related?
(a) force and pressure   (c) torr and atm
(b) torr and mm Hg   (d) torr and pascal

10.2 Why is the high density of mercury an advantage in a Torricelli barometer?

10.3 Mercury has a low vapor pressure. What advantage does this give to the use of mercury in a Torricelli barometer?

10.4 At 20 °C the density of mercury is 13.5 g mL^{-1} and that of water is 1.00 g mL^{-1}. At 20 °C, the vapor pressure of mercury is 0.0012 torr and that of water is 18 torr. Give and explain two reasons why water would be an inconvenient fluid to use in a Torricelli barometer.

10.5 Which exerts the higher pressure, a force of 100 kg acting on 25 cm^2 or a force of 25 kg acting on 5 cm^2?

10.6 Carry out the following unit conversions.
(a) 735 mm Hg to torr   (c) 738 torr to mm Hg
(b) 740 torr to atm   (d) 1.45 × 10^3 Pa to torr

10.7 What is the pressure in torr of each of the following?
(a) 0.329 atm (summit of Mt. Everest, world's highest mountain)
(b) 0.460 atm (summit of Mt. Denali, highest mountain in the United States)

10.8 What is the pressure in kPa of each of the following? (These are the values of the pressures exerted individually by N$_2$, O$_2$, and CO$_2$, respectively, in typical inhaled air.)
(a) 595 torr   (b) 160 torr   (c) 0.300 torr

10.9 A suction pump works by creating a vacuum into which some fluid will rise. Suppose that you work for a construction company and are told to use a suction pump to remove water from a pit. The water level is over 35 ft below the site where you must place the pump. Can *any* suction pump raise water to this height? (The density of water is 1.00 g mL^{-1}; that of mercury is 13.6 g mL^{-1}.)

10.10 One of the oldest units for atmospheric pressure is lb in.$^{-2}$. Calculate the standard atmosphere in these units to three significant figures. Calculate the mass in pounds of a uniform column of water 33.9 ft high having an area of 1.00 in.^{2} at its base. (The density of water may be taken as 1.00 g mL^{-1}; 1 mL = 1 cm^3; 1 lb = 454 g; 1 in. = 2.54 cm.)

10.11 When the mercury level in the *closed* end of a closed-end manometer stands 12.4 cm above the level in the other arm, what is the pressure (in torr) in the bulb attached to the manometer? What would be the pressure in the bulb if the mercury level in the closed end were 12.4 cm *below* the level in the arm nearest the bulb?

10.12 A student carried out a reaction using a bulb connected to an open-end manometer on a day when the temperature was 24.0 °C and the pressure was 746 torr. Before

the reaction, the level of the mercury in both arms was at 12.50 cm, measured by a meter stick placed between the two arms of the U-shaped tube. After the reaction and after cooling the system back to 24.0 °C, the level in the arm nearest the bulb was 8.50 cm. What was the pressure in the bulb (in torr)?

10.13 In another experiment, using the same apparatus as described in Review Exercise 10.12, the mercury level in the arm nearest the bulb stood at 10.20 cm after the reaction. Before the reaction the level was 12.50 cm. The pressure in the lab that day was 741 torr. The temperature in the bulb was 23.0 °C when the mercury level readings were taken. What was the pressure in the bulb (in torr)?

10.14 A liquid with a density of 1.18 g mL^{-1} is used in an open-arm manometer instead of mercury on a day when the atmospheric pressure is 751.5 torr. A gas sample causes the open-arm liquid level to stand 18.5 mm higher than the liquid level in the arm nearest the gas sample. What is the pressure of the gas sample in torr?

Specific Gas Laws and the Combined Gas Law

10.15 State the following laws in words. By what other name is each law known?
(a) temperature–volume law
(b) law of partial pressures
(c) pressure–volume law
(d) law of partial pressures
(e) law of gas effusion
(f) Avogadro's princple

10.16 Which of the four important variables in the study of the physical properties of gases are assumed to be held constant in each of the following laws?
(a) Boyle's law   (d) Dalton's law
(b) Charles' law   (e) Graham's law
(c) Gay-Lussac's law   (f) Avogadro's principle

10.17 A sample of helium at a pressure of 740 torr and in a volume of 2.58 L was heated from 24.0 to 75.0 °C. The volume of the container expanded to 2.81 L. What was the final pressure (in torr) of the helium?

10.18 When a sample of neon with a volume of 648 mL and a pressure of 0.985 atm was heated from 16.0 to 63.0 °C, its volume became 689 mL. What was its final pressure (in atm)?

10.19 What must be the new volume of a sample of nitrogen (in L) if 2.68 L at 745 torr and 24.0 °C is heated to 375.0 °C under conditions that let the pressure change to 760 torr?

10.20 When 280 mL of oxygen at 741 torr and 18.0 °C was warmed to 33.0 °C, the pressure became 760 torr. What was the final volume (in mL)?

10.21 A sample of argon with a volume of 6.18 L, a pressure of 761 torr, and a temperature of 20.0 °C expanded to a volume of 9.45 L and a pressure of 373 torr. What was its final temperature in °C?

10.22 A sample of a refrigeration gas in a volume of 455 mL, at a pressure of 1.51 atm, and at a temperature of 25.0 °C was compressed into a volume of 220 mL with a pressure of 2.00 atm. To what temperature (in °C) did it have to change?

10.23 After a sample of xenon with a volume of 532 mL was heated from 22.0 to 86.0 °C, its volume changed to 587 mL and its pressure became 789 torr. What must have been its initial pressure in torr?

10.24 After a sample of nitrogen with a volume of 668 mL and a pressure of 1.00 atm was compressed to a volume of 332 mL and a pressure of 2.00 atm, its temperature was 27.0 °C. What must have been its initial temperature (in °C)?

10.25 To compress nitrogen at 755 torr from 740 mL to 525 mL, what must the new pressure be if the temperature is held constant?

10.26 If 625 mL of oxygen at 925 torr is allowed to expand at constant temperature until its pressure is 748 torr, what will be the volume?

10.27 Helium in a 100 mL container at a pressure of 525 torr is transferred to a container with a volume of 250 mL. What is the new pressure (a) if no change in temperature occurs? (b) If its temperature changes from 24 to 15 °C?

10.28 What will have to happen to the temperature of a sample of methane if 1.00 L at 735 torr and 25.0 °C is given a pressure of 809 torr and a volume of 0.900 L?

*10.29 In a diesel engine, the fuel is ignited when it is injected into hot compressed air, heated by the compression itself. In a typical high-speed diesel engine, the chamber in the cylinder has a diameter of 10.7 cm and a length of 13.4 cm. On compression, the length of the chamber is shortened by 12.7 cm (a "5-inch stroke"). The compression of the air changes its pressure from 1.00 to 34.0 atm. The temperature of the air before compression is 364 K. As a result of the compression, what will be the final air temperature (in K and °C) just before the fuel injection?

*10.30 Early one cool (60.0 °F) morning you start on a bike ride with the air pressure at 14.7 lb in.$^{-2}$ and the tire pressure at 50.0 lb in.$^{-2}$. (This means that the actual pressure in the tire is 64.7 lb in.$^{-2}$, 14.7 + 50.0 lb in.$^{-2}$. Tire gauges tell us how much *over* the atmospheric pressure the tire pressure is.) By late afternoon, the air had warmed up considerably, and this plus the heat generated by tire friction sent the temperature inside the tire to 104 °F. What will the tire gauge now read, assuming that the volume of the air in the tire and the atmospheric pressure have not changed?

Ideal Gas Law

10.31 The SI system generally uses its base units to compute constants involving derived units. The SI unit for volume, for example, is the cubic meter, called the *stere*, because the meter is the base unit of length. We learned about

the SI unit of pressure, the pascal, in this chapter. And the temperature unit is the kelvin.
(a) Calculate the value of the gas constant using SI units of pascals for pressure and (meter)^2 for volume.
(b) Calculate the value of the gas constant when we express the standard pressure as 760 torr and the standard molar volume as 22.4×10^3 mL.

10.32 Using oxygen as an example, carefully distinguish between 1 *molecule*, 1 *mole*, 1 *molar mass*, and 1 *molar volume*.

10.33 At STP how many molecules of H_2 are in 22.4 L?

10.34 What volume in liters does 1.00 mol of O_2 occupy at 20.0 °C and 760 torr?

10.35 A sample of 1.00 mol of N_2 at 50.0 °C and 745 torr occupies what volume, in liters?

10.36 A sample of 4.18 mol of H_2 at 18.0 °C occupies a volume of 24.0 L. Under what pressure, in atmospheres, is this sample?

10.37 If a steel cylinder with a volume of 1.60 L contains 10.0 mol of oxygen, under what pressure (in atm) is the oxygen if the temperature is 25.0 °C?

10.38 A steel cylinder containing 79.8 mol of helium with a volume of 10.0 L is under what pressure (in atm) at a temperature of 24 °C?

10.39 When the pressure in a certain gas cylinder with a volume of 4.50 L reaches 500 atm, the cylinder is likely to explode. If the cylinder contains 42.0 mol of argon at 24.0 °C, is it on the verge of exploding? (Calculate the pressure in atm.)

10.40 A steel cylinder with a volume of 25.0 L contains nitrogen under a pressure of 148 atm and a temperature of 25.0 °C. How many moles of nitrogen does the cylinder contain?

10.41 After the contents of a steel cylinder of pressurized neon with a volume of 18.5 L have been depleted until the pressure in the cylinder is 1.00 atm, the cylinder is called "empty." How many moles of neon remain, if the temperature is 24.0 °C?

*10.42 When a steel cylinder of oxygen with a volume of 14.5 L was used to supply oxygen to an oxyacetylene torch, the pressure in the cylinder changed from 208.0 to 201.0 atm. The temperature of the oxygen was 25.0 °C at the times of both pressure readings.
(a) How many moles and how many grams of oxygen had been used up?
(b) How many moles of acetylene, C_2H_2, were burned? (Assume that the sole products are CO_2 and H_2O.)

10.43 A small cylinder of H_2 with a volume of 850 mL and a pressure of 112 atm at 24.0 °C was used to supply hydrogen for a reaction. After the experiment, the pressure in the cylinder was half as much, and still at 24.0 °C. How many moles and how many grams of hydrogen were taken?

10.44 To three significant figures, what is the density in g L^{-1} of the following gases at STP?
(a) C_2H_6 (ethane)   (c) Cl_2
(b) N_2 (d) argon

10.45   Calculate the densities in mg mL^{-1} of the following gases at STP. Carry the calculations to three significant figures.
(a) neon   (c) CH_4 (methane)
(b) O_2 (d) CF_4

10.46   What density (in g L^{-1}) does oxygen have at 24.0 °C and 742 torr?

10.47   At 748.0 torr and 20.65 °C, what is the density of argon (in g L^{-1})?

10.48   At 22.0 °C and a pressure of 755 torr, a gas was found to have a density of 1.13 g L^{-1}. Calculate its molecular mass.

10.49   A gas was found to have a density of 0.08747 mg mL^{-1} at 17.0 °C and a pressure of 760 torr. What is its molecular mass? Can you tell what the gas most likely is?

10.50   A chemist isolated a gas in a glass bulb with a volume of 255 mL at a temperature of 25.0 °C and a pressure (in the bulb) of 10.0 torr. The gas weighed 12.1 mg. What is the molecular mass of this gas?

10.51   A chemist isolated 6.3 mg of one of the many boron hydrides in a glass bulb with a volume of 385 mL at 25.0 °C and a bulb pressure of 11 torr.
 (a) What is the molecular mass of this hydride?
 (b) What is its molecular formula, BH_3, B_2H_6, or B_4H_{10}?

10.52   To isolate a sample of the gas B_2H_6, a chemist used a glass bulb with a volume of 455 mL. At a bulb temperature of 25.0 °C, the pressure in the bulb was 20.0 torr. How much did the sample weigh, in milligrams?

Stoichiometry of Reactions of Gases

10.53   How many liters of F_2 at STP are needed to react with 4.00 L of H_2, also at STP, in the following reaction?
$$H_2(g) + F_2(g) \longrightarrow 2HF(g)$$

10.54   In the Haber process for the synthesis of ammonia:
$$N_2(g) + 3H_2(g) \longrightarrow 2NH_3(g)$$
how many liters of N_2 are needed to react completely with 45.0 L of H_2, if the volumes of both gases are measured at STP?

*10.55**   In the first step for one industrial preparation of nitric acid, HNO_3, ammonia reacts with oxygen at 650 °C and 1 atm. The reaction is
$$4NH_3(g) + 5O_2(g) \longrightarrow 4NO(g) + 6H_2O(g)$$
How many liters of oxygen at 650 °C and 1.00 atm are needed to react with 36.0 L of NH_3, at 400 °C and 6.00 atm?

10.56   Propylene, C_3H_6, reacts with hydrogen under pressure to give propane, C_3H_8:
$$C_3H_6(g) + H_2(g) \longrightarrow C_3H_8(g)$$
A sample of 18.0 g of propylene requires how many liters of hydrogen when measured at 740 torr and 24 °C?

*10.57**   Ammonia is converted to ammonium sulfate, an important fertilizer, by the reaction
$$2NH_3 + H_2SO_4 \longrightarrow (NH_4)_2SO_4$$
In one batch, 214 kg of ammonium sulfate is to be made.

 (a) How many liters of ammonia are needed when the volume is measured at 24 °C and 100 kPa?
 (b) How many moles of H_2SO_4 are required?
 (c) If the H_2SO_4 is in the form of a 8.00 M solution, what volume of this solution (in liters) is needed?

*10.58**   One industrial synthesis of acetylene, C_2H_2, a gas used as a raw material for making many synthetic drugs, dyes, and plastics, is the addition of water to calcium carbide, CaC_2.
$$CaC_2(s) + 2H_2O(l) \longrightarrow Ca(OH)_2(s) + C_2H_2(g)$$
 (a) In a small-scale test to improve efficiency, 50.0 g of CaC_2 is converted to acetylene. What is the theoretical yield of acetylene in moles? In liters (at 24 °C and 745 torr)?
 (b) To make 1.00×10^6 L of acetylene (at 24 °C and 745 torr) by this method requires how much calcium carbide, in moles? In kilograms?

*10.59**   A sample of an unknown gas with a mass of 1.620 g occupied a volume of 941 mL at a pressure of 748 torr and a temperature of 20.00 °C. When made to decompose into its elements, 1.389 g of carbon and 0.2314 g of hydrogen were obtained.
 (a) What is the percentage composition of this compound?
 (b) What is its empirical formula?
 (c) What is its molecular formula?

Dalton's Law

10.60   Calculate the partial pressures of oxygen and nitrogen in air in kilopascals when the pressures are 160.00 torr for O_2 and 594.70 torr for N_2. What discrepancy exists between these data and the fact that the total pressure of the atmosphere at the time they were obtained was 760.00 torr? What might cause this discrepancy?

10.61   When nitrogen is prepared and collected over water at 30 °C and a total pressure of 742 torr, what is its partial pressure in torr?

10.62   If you were to prepare oxygen and collect it over water at 10 °C and a total pressure of 748 torr, what would be its partial pressure in torr?

10.63   A sample of carbon monoxide was prepared and collected over water at a temperature of 20 °C and a total pressure of 754 torr. It occupied a volume of 268 mL. Calculate the partial pressure of the CO in torr as well as its dry volume (in mL) under a pressure of 1.00 atm at 20°C.

10.64   A sample of hydrogen was prepared and collected over water at 25 °C and a total pressure of 742 torr. It occupied a volume of 288 mL. Calculate its partial pressure (in torr) and what its dry volume would be (in mL) under a pressure of 1.00 atm at 25°C.

10.65   What volume of ''wet'' methane would you have to collect at 20.0 °C and 742 torr to be sure that the sample contains 244 mL of dry methane (at 742 torr and 20°C)?

10.66   What volume of ''wet'' oxygen would you have to collect if you need the equivalent of 275 mL of dry oxygen at 1.00 atm? (The atmospheric pressure in the lab is 746 torr.) The oxygen is to be collected over water at 15.0 °C.

10.67 What are the mole percentages of the components of air inside the lungs when they have the following partial pressures? For N_2, 570 torr; O_2, 103 torr; CO_2, 40 torr; and water vapor, 47 torr.

10.68 Assuming no other components are present, what are the mole percentages of oxygen and nitrogen (a) in the air at the top of Mt. Everest (elevation 8.8 km) on a day when $P_{N_2} = 197$ torr and $P_{O_2} = 53$ torr. (b) In air at sea level and 1 atm pressure, with $P_{N_2} = 593$ torr and $P_{O_2} = 159$ torr? (c) Comparing the answers calculated for parts (a) and (b), what has to be the reason it is hard to breathe without supplemental oxygen at high altitudes?

Graham's Law

10.69 Under conditions in which the density of CO_2 is 1.96 g L^{-1} and that of N_2 is 1.25 g L^{-1}, which gas will effuse more rapidly? What will be the ratio of the rates of effusion of N_2 to CO_2?

10.70 Ammonia effuses at a rate that is 2.93 times larger than that of an unknown gas. What is the molecular mass of the unknown?

10.71 Uranium hexafluoride is a white solid that readily passes directly into the vapor state. (Its vapor pressure at 20.0 °C is 120 torr.) A trace of the uranium in this compound—about 0.7%—is uranium-235, which can be used in a nuclear power plant. The rest of the uranium is essentially uranium-238, and its presence interferes with these applications for uranium-235. Gas effusion of UF_6 can be used to separate the fluoride made from ^{235}U and the fluoride made from ^{238}U. Which hexafluoride effuses more rapidly? By how much? (The very small difference that you will find is actually enough, although repeated effusions are necessary.)

Kinetic Theory of Gases

10.72 What model of a gas was proposed by the kinetic theory of gases?

10.73 To what properties of an ideal gas is its temperature proportional (a) according to the calculations of the kinetic theory? (b) According to the combined gas law?

10.74 What aspects of the kinetic theory of gases explain why two gases, given the freedom to mix, will always mix *entirely*?

10.75 If the molecules of a gas at constant volume are somehow given a lower average kinetic energy, what physical property of the gas will change and in what direction?

10.76 Explain *how* heating makes a gas expand at constant pressure. (Describe how the model of an ideal gas connects the increase in temperature to the gas expansion.)

10.77 Explain in terms of the kinetic theory *how* heating a confined gas makes its pressure increase.

10.78 How does the kinetic theory explain the existence of a minimum temperature, 0 K?

10.79 How does the kinetic theory explain the greater effusion rate of a gas with a low molecular mass compared to one with a higher molecular mass?

10.80 In what way does Dalton's law of partial pressures provide evidence for one postulate of the kinetic theory?

10.81 What postulates of the kinetic theory are not strictly true, and why?

10.82 If a given gas A has a larger value of the van der Waals constant b than gas B, what does this suggest about gas A?

10.83 A small value for the van der Waals constant a suggests something about the molecules of the gas. What?

10.84 What equation does van der Waals' equation become more and more like when the volume of a gas becomes larger and larger without any change in temperature or in the number of moles?

Additional Questions

*10.85 The range of temperatures over which an automobile tire must be able to withstand pressure changes is roughly -50 to 120 °F. If a tire is filled to 35 lb in.$^{-2}$ at -50 °F, what will be the pressure in the tire (in the same pressure units) at 120 °F? (Assume that neither the outside atmospheric pressure of 14.7 lb/in.^{2} nor the volume of the tire changes.)

10.86 What must be the new pressure of 450 mL of nitrogen if, at the same temperature, its volume is made 350 mL?

10.87 What new temperature must be used to change the pressure of 400 mL of argon, initially at 25 °C, from 1.00 atm to 2.50 atm?

*10.88 If 0.250 mol of H_2 is to be taken for a reaction from a cylinder at 120.0 atm and 25.0 °C, the valve on the cylinder should remain open as H_2 transfers into the reaction vessel until the gauge reads what pressure (in atm)? Assume no temperature change occurs.

*10.89 A common laboratory preparation of hydrogen on a small scale uses the reaction of zinc with hydrochloric acid. Zinc chloride is the other product.
 (a) Write the balanced equation for the reaction.
 (b) If 12.0 L of H_2 at 760 torr and 20.0 °C is wanted, how much zinc (in moles and grams) is needed, in theory?
 (c) How many moles of HCl are needed for part b?
 (d) If the acid is available as 8.00 M HCl, how many milliliters of this solution are sufficient?

*10.90 In an experiment designed to prepare a small amount of hydrogen by the method described in Review Exercise 10.89, a student was limited to using a gas collecting bottle with a maximum capacity of 335 mL. The method involved collecting the hydrogen over water. What are the minimum number of grams of Zn and the minimum number of milliliters of 6.00 M HCl needed to produce the *wet* hydrogen that can exactly fill this collecting bottle at 740 torr and 25.0 °C?

10.91   Carbon dioxide can be made in the lab by the reaction of hydrochloric acid with calcium carbonate.

$$CaCO_3(s) + 2HCl(aq) \rightarrow CaCl_2(aq) + H_2O(l) + CO_2(g)$$

How many grams of calcium carbonate and how many milliliters of 8.00 M HCl are needed to prepare 465 mL of dry CO_2 if it is to be collected at 20.0 °C and 745 torr?

10.92   In many countries, the fertilizer ammonia is made using methane as the source of hydrogen (and energy). The overall process—and there are several steps—is by the following equation, where N_4O is used as an approximate formula for air, another raw material.

$$7CH_4 + 10H_2O + 4N_4O \longrightarrow 16NH_3 + 7CO_2$$

Methane is sold in units of *tcf*, where 1 tcf = 1×10^3 ft³ = 28.3×10^3 L at STP.

(a)   What volume of NH_3 (in liters at STP) can be made by this process from 1.00 tcf of methane (also at STP)?

(b)   One thousand cubic feet of methane represents how many kilograms of ammonia?

10.93   A sample of an unknown gas with a mass of 3.620 g was made to decompose into 2.172 g of O_2 and 1.448 g of S.

Prior to the decomposition, this sample occupied a volume of 1120 mL at 750 torr and 25.0 °C.

(a)   What is the percentage composition of the elements in this gas?

(b)   What is the empirical formula of the gas?

(c)   What is its molecular formula?

*10.94   A sample of a new antimalarial drug with a mass of 0.2394 g was made to undergo a series of reactions that changed all of the nitrogen in the compound into N_2. This gas had a volume of 18.90 mL when collected over water at 23.80 °C and a pressure of 746.0 torr. At 23.80 °C, the vapor pressure of water is 22.110 torr.

(a)   Calculate the percentage of nitrogen in the sample.

(b)   When 6.478 mg of the compound was burned in pure oxygen, 17.57 mg of CO_2 and 4.319 mg of H_2O were obtained. What are the percentages of C and H in this compound? Assuming that any undetermined element is oxygen, write an empirical formula for the compound.

(c)   The molecular mass of the compound was found to be 324. What is its molecular formula?

10.95 In one analytical procedure for determining the percentage of nitrogen in unknown compounds, a weighed sample is made to decompose to N_2, which is collected over water at known temperatures and pressures. The volumes of N_2 are then translated into grams and then into percentages.

(a) Show that the following equation can be used to calculate the percentage of nitrogen in a sample having a mass of W grams when the N_2 has a volume of V mL and is collected over water at t_c °C at a total pressure of P torr. The vapor pressure of water occurs in the equation as $P^\circ_{H_2O}$.

$$\text{Percentage N} = 0.04489 \times \frac{V(P - P^\circ_{H_2O})}{W(273 + t_c)}$$

(b) Use this equation to calculate the percentage of nitrogen in the sample of Review Exercise 10.94.

*10.96** The "rotten-egg" odor is caused by hydrogen sulfide, H_2S. Most people can detect it at a concentration of 0.15 ppb (parts per billion), meaning 0.15 L of H_2S in 10^9 L of space. A typical student lab is $40 \times 20 \times 8$ ft.

(a) At STP, how many liters of H_2S could be present to have a concentration of 0.15 ppb?

(b) How many milliliters of 0.100 M Na_2S would be needed to generate the amount of H_2S in part a by the following reaction with hydrochloric acid?

$$Na_2S(aq) + 2HCl(aq) \longrightarrow H_2S(g) + 2NaCl(aq)$$

*10.97** A mixture was prepared in 500 mL reaction vessel from 300 mL of O_2 (measured at 25 °C and 740 torr) and 400 mL of H_2 (measured at 45 °C and 1250 torr). The mixture was ignited and the H_2 and O_2 reacted to form water. What was the final pressure inside the reaction vessel after the reaction was over if the temperature was held at 120 °C?

*10.98** A student collected 18.45 mL of H_2 over water at at 24 °C. The water level inside the collection apparatus was 8.5 cm higher than the water level outside. The barometric pressure in the lab was 746 torr. How many grams of zinc had to react with $HCl(aq)$ to produce the H_2 that was collected?

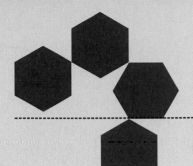

CHEMICALS IN USE 8
Ozone in Smog

Ozone, O_3, is a powerful oxidizing agent, being able to split molecules apart that have carbon–carbon double bonds. Because such double bonds occur widely in all living things, ozone is dangerous to plants and animals. The U.S. National Ambient Air Quality Standard for ozone is a daily maximum 1-hour average ozone concentration of only 0.12 ppm (120 ppb).[1] Dozens of U.S. urban areas exceed this at least once per year.

How does ozone originate in the air where we live? The *direct* and only source of ozone in the lower atmosphere is the combination of oxygen atoms with oxygen molecules when they collide at the surface of some particle, *M*.

$$O + O_2 + M \longrightarrow O_3 + M \qquad (1)$$

M is any molecule, like N_2 or O_2, that can absorb some of the kinetic energy involved in the collision. Thus the earlier question, "How does ozone originate?" becomes a new question, "How are oxygen *atoms* generated?" Once oxygen atoms are generated, there will be ozone. The chief source of oxygen atoms is the breakup of molecules of nitrogen dioxide, an air pollutant, a breakup made possible by the energy of sunlight.

$$NO_2 + \text{solar energy} \longrightarrow NO + O \qquad (2)$$

Oxides of Nitrogen
The NO_2 for Equation 2 forms in air from nitrogen monoxide, NO, which is produced inside vehicle engine cylinders where the high temperature and pressure enables the direct combination of nitrogen and oxygen.

$$N_2 + O_2 \xrightarrow[\text{Pressure}]{\text{High temperature}} 2NO \qquad (3)$$

As soon as newly made NO, now in the exhaust gas, hits the cooler outside air, it reacts further with oxygen to give nitrogen dioxide, NO_2, which gives smog its reddish-brown color (see Figure 8).

$$2NO + O_2 \longrightarrow 2NO_2 \qquad (4)$$

Thus, NO_2 forms in air made smoggy by vehicle exhaust. But Reaction 2 converts NO_2 back to nitrogen monoxide. Because Reactions 2 and 4 do not occur at identical rates, some NO is always present in smoggy air. The significance of this is that *nitrogen monoxide is able to destroy ozone.*

$$NO + O_3 \longrightarrow NO_2 + O_2 \qquad (5)$$

The reactions of Equations 1, 2, and 5, when added together, give us *no net chemical effect!* (Try it.) How then does atmospheric ozone develop at all?

The reactions of Equations 1–5 proceed at their own rates. The rates are unequal, however, so the reactions actually do not exactly cancel each other. The net effect is that a small, steady concentration of ozone develops even in clean air. The range of ozone concentration in clean air, however, is very low, between 20 and 50 ppb. In the more polluted urban areas, levels as high as 400 ppb commonly occur for brief periods of time each year.

Unburned Hydrocarbons and the Ozone in Smog
Equation 5 represents the reaction that can make most ozone disappear, *but other substances are able to remove the* NO *needed for this reaction before the* NO *is used to destroy ozone.* It is the removal of NO other than by Reaction 5 that enables a buildup in the ozone level of the lower atmosphere.

Net destroyers of nitrogen monoxide are themselves present in vehicle exhaust, namely, the unburned or partially oxidized hydrocarbons remaining from the incomplete combustion of the fuel. In sunlight and oxygen, some unburned hydrocarbon molecules are changed into organic derivatives of hydrogen peroxide called *peroxy radicals* usually symbolized

---

[1] *Ambient* means "all surrounding, all encompassing." The symbol *ppm* stands for "parts per million," and *ppb* means "parts per billion." Thus 0.12 ppm means 0.12 mL ozone in 10^6 mL of air.

FIGURE 8a

The effect of smog on visibility. (*a*) A clear day. (*b*) A day of heavy smog.

as R—O—O (or simply RO_2).[2] Peroxy radicals originate chiefly by the reaction of unburned hydrocarbons with hydroxyl radicals, HO. These arise by a number of mechanisms in polluted air, but we will not go into the details of how HO radicals form. Suffice it to say, HO radicals and hydrocarbons lead to ROO radicals, and these destroy NO molecules as follows.

$$ROO + NO \rightarrow RO + NO_2$$

This reaction reduces the supply of ozone-destroying NO.

The development of ozone in smog thus depends on the formation of ROO radicals whose formation, in turn, depends on hydrocarbon emissions. Specialists generally agree that emissions of hydrocarbons must be significantly reduced before the ozone problem will be appreciably solved. Even the loss of hydrocarbons in the vapors from fuel tanks or at gas pumps has to be reduced.

[2] The term "radical" is used for any species with an unpaired electron. Sometimes an electron dot is added to the formula of a radical, as in $RO_2\cdot$. The chlorine *atom*, for example, is a radical and sometimes symbolized as $Cl\cdot$. It has seven valence electrons, six occurring as three pairs plus a "lone" unpaired electron.

Reading

J. H. Seinfeld, Urban Air Pollution: State of the Science, *Science*, February 10, 1989, page 745.

Questions

1. What chemical reaction inside the cylinder of a vehicle engine launches the production of ozone in smog?
2. How does the NO_2 in smog form? (Write an equation.)
3. How is NO_2 involved in the production of ozone in smog? (Write equations.)
4. What chemical property of ozone makes it dangerous?
5. What is the connection between the presence of unburned hydrocarbons in smog and the ozone in smog? (Use equations as part of the answer.)

Intermolecular forces of attraction, the major topic of this chapter, affect a number of physical properties, like viscosity —resistance to a deformation of shape. We usually don't think of solids as having viscosity, but under enough pressure, even they "flow," as is evident in this view of the Ferris glacier in the Fairweather Range on the border of Glacier Bay National Park, Alaska.

Chapter 11

Intermolecular Attractions and the Properties of Liquids and Solids

Although much of the study of chemistry is devoted to *chemical* properties and reactions, the *physical* properties of substances often concern us more on a day-to-day basis. This is especially true when substances are in their *condensed states,* namely, as liquids or as solids. One physical property of most liquids, for example, is that they contract in volume when they freeze, but water is an exception. It *expands* when it freezes. If you are unaware of this property and also live where temperatures drop below 0 °C, the expansion of freezing water can lead to an expensive auto repair bill, unless you put antifreeze in your car's radiator in winter. The expansion of water as it solidifies not only cracks engine blocks and radiators, it splits huge boulders when water has seeped into cracks.

Another physical property of liquids is that their boiling points decrease with increasing altitude. The traditional "three-minute egg" needs to be cooked in boiling water a little longer in Denver, which is at an elevation of about 1 mile.

Still another physical property is that liquids vary widely in their resistance to flowing—their *viscosity*. Moreover, this resistance decreases sharply with increasing temperature. Knowing this, we generally use a relatively high viscosity motor oil in the hot months of the year. Oil that becomes too "thin" when hot is less effective as a lubricant, and it seeps around engine pistons and

11.1
WHY GASES DIFFER FROM LIQUIDS AND SOLIDS

The boiling point of water in Denver (elev 1610 m) is roughly 95 °C, so the flow of heat in joules per minute into an egg is slower.

The expression "as slow as molasses in January" reflects how viscosity increases as the temperature is lowered.

is burned up in the combustion chambers. All of these observations concern physical, not chemical properties.

One of the goals of chemistry has been to understand how chemical composition and molecular structure determine the physical properties of matter. From such studies have come a more complete knowledge of the structure of matter and an ability (at least to some degree) to design materials that have the kinds of physical properties we want. Examples of success include materials for transistors and computer chips and the many plastics that have come to replace traditional materials such as wood, steel, and natural rubber.

We began our study of the physical properties of substances in the last chapter where we discussed the properties of gases. In this chapter we will go on to investigate the properties of the other two states of matter—liquids and solids—as well as what happens when substances change from one state to another.

Gases, Liquids, and Solids

As we learned in Chapter 10, gases obey the gas laws rather well, regardless of chemical composition, but liquids and solids have no comparably general laws. There are two reasons. One is that gas molecules collectively, in and of themselves, occupy an extremely small volume compared to the total volume occupied by a gas, so *a gas is almost entirely empty space.* This fact alone accounts for the *compressibility* of gases, one of the most obvious differences between them and liquids or solids. In liquids and solids, the molecules are packed together very tightly with little empty space between them. Liquids and solids, therefore, are virtually incompressible.

The other reason for the existence of *general* laws for gases is that forces of attraction between gas molecules are uniformly so weak that they have hardly any influence on the physical properties of gases. We don't find similar "liquid laws" or "solid laws" because incompressiblilty and forces of attraction can't be ignored in liquids and solids; they play a very important role in determining the properties of these "condensed" states of matter.

11.2 INTERMOLECULAR ATTRACTIONS

The force between two electrical charges is called the *electrostatic force.* Other forces in nature are the magnetic, gravitational, nuclear strong, and nuclear weak forces.

Most of the *physical* properties of gases, liquids, and solids are actually controlled by the strengths of **intermolecular attractions,** the electrical forces of attraction between neighboring particles. In liquids and solids, these forces are much stronger than in gases. But why? What factors determine the *size* of an intermolecular attraction? We can name two.

One factor is distance, because an electrical force falls off rapidly with distance. If you have ever played with magnets you've noticed how a magnetic force diminishes rapidly as the distance between the magnets increases. Thus, when two magnets are near each other, the attraction can be quite strong, but if they are far apart, hardly any attraction is felt at all. Molecules aren't magnets, but electrical forces similarly diminish with distance, decreasing with the *square* of the distance between the charges. When the distance doubles, the force is reduced by $1/2^2$ or by one-fourth. In the liquid and solid states, distances between molecules are as small as they can be, so intermolecular attractions are stronger in the condensed states of matter than in gases.

It is important to realize that attractions *between* molecules (*inter*molecular) are always much weaker than attractions *within* molecules (*intra*molecular). In a molecule of HCl, for example, the H and Cl are held very tightly to each

Many common substances, such as synthetic fibers, were developed simply because they have desirable physical properties as well as chemical properties.

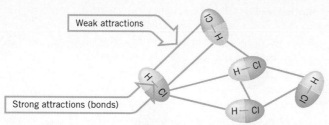

FIGURE 11.1

Strong attractions exist between H and Cl atoms within HCl molecules. Weaker attractions exist between neighboring HCl molecules.

other by a covalent bond. When a particular chlorine atom moves, the hydrogen atom bonded to it is forced to follow along, and the HCl molecule remains intact as it moves about (see Figure 11.1). Attractions between neighboring HCl molecules, in contrast, are much weaker.

Chemical composition is the second factor influencing intermolecular attractions, and it's of major importance for liquids and solids. In gases, the molecules are so far apart that attractive forces are almost negligible. Therefore, differences between two gases caused by chemical composition hardly matter at all to their physical properties. When chemical composition enables substances to be liquids or solids, with molecules close together, intermolecular attractions are relatively strong. Moreover, differences among such attractions, caused by differences in chemical makeup, are amplified, so the physical properties of liquids and solids depend considerably on chemical structure.

From the preceding discussion, it should be obvious that to understand and explain the physical properties of liquids and solids, we need to learn about the kinds and relative strengths of intermolecular attractions. Although they are all electrical in nature, it's useful to distinguish among several kinds.

Intermolecular attractions and structure

The strengths of the attractions within molecules (the chemical bonds) determine chemical properties. The attractions between molecules determine the physical properties of substances.

Dipole–Dipole Attractions

In Chapter 7 we saw that the HCl molecule is an example of a *polar molecule,* one with a partial positive charge at one end and a partial negative charge at the other. Because unlike charges attract, polar molecules tend to line up so that the positive end of one dipole is near the negative end of another. Thermal energy (molecular kinetic energy), however, causes the molecules to jiggle about, so the alignment isn't perfect. Nevertheless, there is still a net attraction between polar molecules (see Figure 11.2). We call this kind of intermolecular attractive force a **dipole–dipole attraction.** It is generally a much weaker force than a covalent bond, being only about 1% as strong.

Two reasons are behind the relative weakness of a dipole–dipole attraction. First, the charges associated with dipoles are only *partial* charges, not full charges. Second, the distance separating any two (partial) charges experiencing a dipole–dipole attraction is usually greater than a covalent bond distance. Therefore, the decrease in attractive force with distance is magnified when the

$$\overset{\delta+}{\text{H}}\!-\!\overset{\delta-}{\text{Cl}}$$

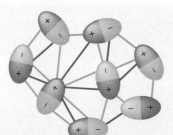

 Attractions (——) are greater than repulsions (——), so the molecules feel a net attraction to each other.

FIGURE 11.2

Attractions between polar molecules occur because the molecules tend to align themselves so that opposite charges are near each other.

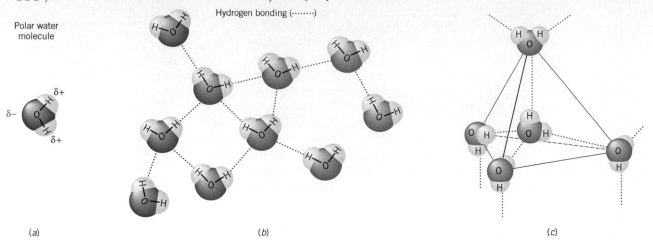

Polar water molecule

Hydrogen bonding (·······)

(a)

(b)

(c)

FIGURE 11.3

Hydrogen bonding in water. (*a*) The polar water molecule. (*b*) Hydrogen bonding produces strong attractions between water molecules in the liquid. (*c*) Hydrogen bonding (dotted lines) between water molecules in ice. Each water molecule is held by four hydrogen bonds in a tetrahedral environment.

Relative Electronegativities	
F	4.1
O	3.5
N	3.1
H	2.1

opposite partial charges of two dipoles are attracting each other. Nevertheless, dipole–dipole attraction is a significant factor in understanding many physical properties of liquids and solids, particularly boiling points, melting points, viscosities, and solubilities.

Hydrogen Bonds

A particularly important kind of dipole–dipole attraction occurs when hydrogen is covalently bonded to any one of three very small, highly electronegative atoms: fluorine, oxygen, and nitrogen. In this situation, unusually strong dipole–dipole attractions are often observed for which there are two reasons. First, because of the particularly large electronegativity differences, the covalent bonds in the F—H, O—H, or N—H systems are very polar. The ends of these dipoles thus carry substantial fractions of full charges. Second, the partial charges at the ends of these dipoles are highly concentrated because of the small sizes of the atoms involved. These two factors, partial charge size and charge concentration, combine to produce particularly effective attractions between molecules bearing F—H, O—H, or N—H systems. They are so effective that the attraction has a special name, **hydrogen bond.** Typically, a hydrogen bond is about five times stronger than other dipole–dipole attractions. Many molecules in biological systems, like proteins and nucleic acids, contain N—H and O—H bonds, and hydrogen bonding in these substances is one factor that determines their overall structures and biological functions.

Hydrogen Bonds in Water   The strength of the hydrogen bond gives water some very unusual and special properties. Most liquids become more dense

Hydrogen bonding between water molecules in ice causes its molecules to be farther apart in the solid than in liquid water. This makes ice less dense than liquid water, which is why icebergs float.

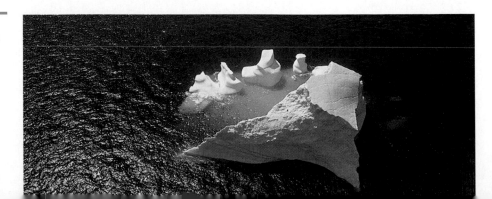

when they change to their solid forms. Not so with water, as we noted earlier. In liquid water, the molecules experience hydrogen bonds that constantly break and reform as the molecules move around (see Figure 11.3*b*). As water freezes, however, the molecules must take up definite locations (Figure 11.3*c*); they cannot move around anymore. Because of the shape and size of the water molecule, *hydrogen bond forces of attraction in crystalline water can be at their collective maximum only in a more expanded "open" arrangement of molecules.* The expansion thus *necessitated* if water is to freeze means that the volume occupied by a given mass of water, when frozen, is greater than the volume of the same mass when a liquid. The result is that ice is less dense than liquid water. Thus, ice floats on the more dense liquid (to the satisfaction of polar bears and hockey players, one might add).

Density (*d*) is the ratio of mass (*m*) to volume (*V*):

$$d = \frac{m}{V}$$

When *V* increases while *m* remains constant, *d* decreases.

London Forces

The attractions between polar molecules are fairly easy to understand. But attractive forces occur even between the particles of nonpolar substances, such as the atoms of the noble gases and the nonpolar molecules of Cl_2 and CH_4. These nonpolar substances can also be condensed to liquids and even to solids if they are cooled to low enough temperatures. Therefore, attractions between their particles, although weak, must exist to cause them to cling together. Electronegativity differences cannot be the origin of these attractions, so what is?

Let's first get a sense of how weak or strong the attractions are. We can use boiling point data for this purpose because, as we'll explain in more detail later in this chapter, the higher the boiling point, the stronger are the attractions between molecules in the liquid. Table 11.1 gives boiling point data and molecular masses for two polar substances, hydrogen chloride and hydrogen sulfide (H_2S), and two nonpolar substances, fluorine (F_2) and argon (Ar). All are gases at room temperature, but all exist in the liquid state at sufficiently low temperatures. These four are selected for comparison because they are somewhat similar in molecular mass and their particles have the same number of electrons. As much as possible, we want the principal variable in our comparisons to be molecular polarity or lack of it. Notice that the two nonpolar substances, F_2 and Ar, have very similar boiling points. The two polar compounds, HCl and H_2S, also have similar boiling points, but they are at least 100 degrees higher than those of the nonpolar substances. This tells us that intermolecular attractions in polar liquids are considerably stronger than those in nonpolar

Intermolecular attractions and boiling points

TABLE 11.1 Boiling Points of Some Polar and Nonpolar Substances

Substance	Boiling Point (°C)	Formula Mass	Number of Electrons
Polar			
HCl	−84.9	36	18
H_2S	−60.7	34	18
Nonpolar			
F_2	−188.1	38	18
Ar	−185.7	40	18

FIGURE 11.4

Instantaneous "frozen" views of the electron densities in two neighboring particles. Attractions occur between the instantaneous dipoles while they exist.

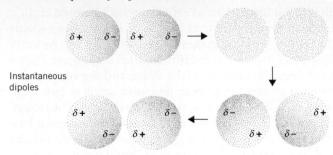

Instantaneous dipoles

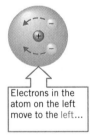

Electrons in the atom on the left move to the left...

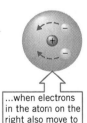

...when electrons in the atom on the right also move to the left.

liquids when substances with nearly the same molecular mass and equal numbers of electrons are compared.

In 1930 Fritz London, a German physicist, explained how the particles in even nonpolar substances can experience intermolecular attractions. He noted that in any atom or molecule the electrons are constantly moving. If we could examine such motions in two neighboring particles, we would find that the movement of electrons in one influences the movement of electrons in the other. This is because electrons repel each other and tend to stay as far apart as they can. Therefore, as an electron of one particle gets near the other particle, electrons on the second particle are pushed away. This happens continually as the electrons move around, so to some extent, the electron density in both particles flickers back and forth in a synchronous fashion. This is illustrated in Figure 11.4, which depicts a series of instantaneous "frozen" views of the electron density. Notice that *at any given moment the electron density of a particle can be unsymmetrical,* with more negative charge on one side than on the other. For that particular instant, the particle is a dipole, and we call it a momentary dipole or **instantaneous dipole.**

As an instantaneous dipole forms in one particle, it causes the electron density in its neighbor to become unsymmetrical, too. As a result, this second particle also becomes a dipole. We call it an **induced dipole** because it is caused by, or *induced* by, the formation of the first dipole. Because of the way the dipoles are formed, they always have the positive end of one near the negative end of the other, so there is a dipole–dipole attraction between them. It is a very short-lived attraction, however, because the electrons keep moving; the dipoles vanish as quickly as they form. But, in another moment, the dipoles will reappear in a different orientation and there will be another brief dipole–dipole attraction. In this way the short-lived dipoles cause momentary tugs between the particles. When averaged over a period of time, there is a net, overall attraction. It tends to be quite weak, however, because the attractive forces are only "turned on" part of the time.

The momentary dipole–dipole attractions that we've just discussed are called *instantaneous dipole–induced dipole attractions,* or **London forces,** to distinguish them from the kind of permanent dipole–dipole attractions that exist continuously in polar substances like HCl.

London forces constitute the only kind of attraction possible between nonpolar molecules. London forces also contribute to the total intermolecular attraction between polar molecules, being in addition to the regular dipole–dipole attractions. London forces even occur between oppositely charged ions, but their effects are relatively weak and contribute little to the overall attractions between ions.

TABLE 11.2 Boiling Points of the Halogens and Noble Gases

Group VIIA	Boiling Point (°C)	Group 0	Boiling Point (°C)
F_2	−188.1	He	−268.6
Cl_2	−34.6	Ne	−245.9
Br_2	58.8	Ar	−185.7
I_2	184.4	Kr	−152.3
		Xe	−107.1
		Rn	−61.8

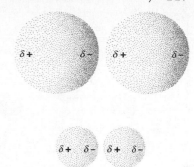

FIGURE   11.5

A large electron cloud is more easily deformed than a small one. Large molecules therefore experience stronger London forces than small molecules.

The Strengths of London Forces   The strengths of London forces depend chiefly on two factors. One is the size of the electron cloud of the particles. When the electron cloud is large, the outer electrons are generally not held very tightly by the nucleus (or nuclei, if the particle is a molecule). This makes the electron cloud "mushy" and rather easily deformed (see Figure 11.5), which favors the formation of an instantaneous dipole or the creation of an induced dipole. Particles with large electron clouds therefore form short-lived dipoles more easily and experience stronger London forces than do similar particles with small electron clouds. The effects of size can be seen if we compare the boiling points of the halogens or the noble gases (see Table 11.2). Going down Group 0, for example, the noble gas atoms become larger and the boiling points increase, reflecting stronger intermolecular attractions (stronger London forces) with the particles having larger electron clouds.

Another factor that affects the strength of London forces is the number of atoms in a molecule. For molecules containing the same elements, London forces increase with the number of atoms, as illustrated by the hydrocarbons (see Table 11.3). Their molecules consist of chains of carbon atoms with hydrogen atoms bonded to them all along the chain. If we compare two hydrocarbon molecules of different chain length—for example, C_3H_8 and C_6H_{14} —the one having the longer chain also has the higher boiling point. This means that the molecule with the longer chain length experiences the stronger intermolecular attractive forces. They are strong between the longer molecules because more places occur along their lengths where instantaneous

Intermolecular attractions and boiling points

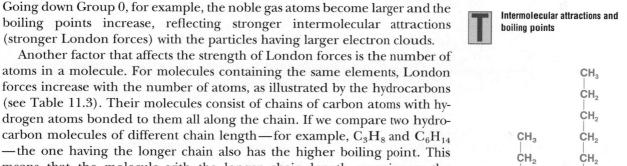

TABLE 11.3 Boiling Points of Some Hydrocarbons[a]

Molecular Formula	Boiling Point at 1 atm (°C)
CH_4	−161.5
C_2H_6	−88.6
C_3H_8	−42.1
C_4H_{10}	−0.5
C_5H_{12}	36.1
C_6H_{14}	68.7
⋮	⋮
$C_{10}H_{22}$	174.1
⋮	⋮
$C_{22}H_{46}$	327

[a] The molecules of each hydrocarbon in this table have carbon chains of the type C—C—C—C—etc.; that is, one carbon follows another.

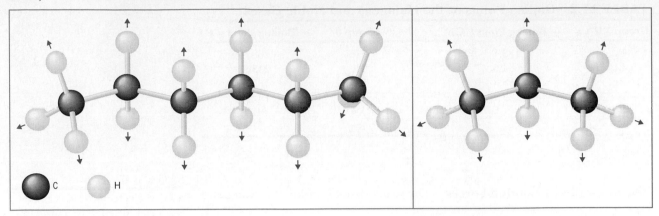

FIGURE 11.6

The C_6H_{14} molecule, left, has more places along its chain where it can be attracted to other molecules nearby than does the shorter C_3H_8 molecule, right.

dipoles can develop and lead to London attractions to other molecules (see Figure 11.6). Even if the strength of attraction at each location is about the same, the *total* attraction experienced between the longer C_6H_{14} molecules is greater than that felt between shorter C_3H_8 molecules. In Table 11.3 you can see that some of the hydrocarbons have boiling points that are quite high, which shows that the cumulative effects of London forces can be very strong attractions.

Ion–Induced Dipole Force of Attraction

The creation of a dipole in a neighbor is not restricted to instantaneous irregularities in the electron densities of neutral atoms and molecules. Ions, with full positive and negative charges, also induce dipoles in molecules nearby (like those of a solvent), leading to **ion–induced dipole** attractions. These can be quite strong because the charge on the ion does not flicker on and off like the instantaneous charges responsible for ordinary London forces.

11.3
SOME GENERAL PROPERTIES OF LIQUIDS AND SOLIDS

We'll begin our study of the general physical properties of liquids and solids by examining two properties that depend mostly on how tightly packed molecules are, namely, *compressibility* and *diffusion*. Other properties depend much more on the strengths of intermolecular attractive forces, properties such as *retention of volume or shape, surface tension*, the ability of a liquid to *wet* a surface, the *viscosity* of a liquid, and a solid's or liquid's *tendency to evaporate*.

Properties That Depend Primarily on Tightness of Packing

Compressibility In Section 11.1, we noted that in a liquid or solid most of the space is taken up by the molecules, and that there is very little empty space into which to crowd other molecules. As a result, it is very difficult to compress liquids or solids to a smaller volume by applying pressure, so we say that these states of matter are nearly **incompressible.** This is a property that we often make use of. When you "step on the brakes" of a car, for example, you rely on the incompressibility of the brake fluid to transmit the pressure you apply with your foot to the brake shoes on the wheels. If some air should get into the brake lines, you're in big trouble, because when you apply pressure to the brakes it simply compresses the air and doesn't force the brake shoes to stop the car. A similar application of the incompressibility of liquids is the foundation of the engineering science of *hydraulics*, which uses fluids to transmit forces that lift or move heavy objects.

Diffusion   As we indicated in Sections 5.2 and 10.7, diffusion occurs much more rapidly in gases than in liquids, and hardly at all in solids. In gases, molecules diffuse rapidly because they travel relatively long distances between collisions. Their movements from one place to another are not greatly hindered (see Figure 11.7). In liquids, however, a given molecule suffers many collisions as it moves about, so it takes longer to move from place to place and diffusion is much slower. Diffusion in solids is almost nonexistent at room temperature because the particles of a solid are held tightly in place. At high temperatures, though, the particles of a solid sometimes have enough kinetic energy to jiggle their way past each other, and diffusion can slowly occur. Such high-temperature solid-state diffusion is used to make electronic devices, like transistors (described in Section 8.9). Solid-state diffusion allows for small, carefully controlled amounts of some impurity—arsenic, for example—to be added to a semiconductor such as silicon or germanium. This process, called "doping," allows the conductivity of these materials to be modified in desirable ways.

Properties That Depend Primarily on the Strengths of Intermolecular Attractions

Retention of Volume and Shape   Volume and shape are obvious physical properties. You know that a solid such as an ice cube keeps both its shape and volume when transferred from one container to another. The same amount of water as a *liquid*, however, retains only its volume, not its shape, when transferred to a container with another shape. A gas does not necessarily retain either its volume or shape when transferred; a gas spontaneously adapts its shape and volume to any container. The underlying reason for the differences is simply intermolecular attractive forces.

In gases, intermolecular attractive forces are very weak. In liquids and solids, however, the attractions are much stronger and are able to hold the particles closely together, thereby preventing liquids and solids from expanding in volume in the manner of a gas. As a result, liquids and solids keep the same volume regardless of the size of their container. In a solid, the attractions are even stronger than in a liquid. They hold the particles more or less rigidly in place so a solid retains its shape when moved from one container to another.

Surface Tension   One characteristic of the liquid state is the phenomenon of **surface tension,** the property of a liquid's surface (in contact with its own vapor or with air) that tends to bring the liquid's contained volume into a form having a minimum surface area. The smallest surface area *for a given volume* is a sphere—it's a principle of solid geometry. The tendency of a liquid to spontaneously assume a form with a minimum surface area explains many common observations. Surface tension, for example, is responsible for the tendency of rain drops to be little spheres. Surface tension also causes the sharp edges of glass tubing to become rounded when the glass is softened in a flame, an

Hydraulic machinery such as this earth mover uses the incompressibility of liquids to transmit forces that accomplish work.

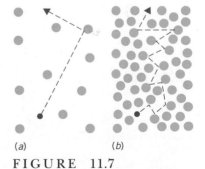

(*a*)   (*b*)

F I G U R E   11.7

(*a*) Diffusion in a gas is rapid because relatively few collisions occur between widely spaced molecules. (*b*) Diffusion in a liquid is slow because of many collisions between closely spaced particles.

FIGURE 11.8

A glass carefully filled with water above the rim.

Something like surface tension exists at the boundary between two immiscible liquids like water and gasoline. The phenomenon is here called *interfacial tension*.

The surface tension of water is roughly two to three times larger than the surface tension of any common organic solvent.

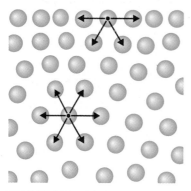

FIGURE 11.9

Molecules at the surface of a liquid are attracted toward the interior. Those within the liquid are uniformly attracted in all directions.

operation called "fire polishing." Surface tension is also what allows us to fill a water glass above the rim, giving the surface a rounded appearance (see Figure 11.8). The surface behaves as if it has a thin, invisible "skin" that lets the water in the glass pile up, trying to assume a spherical shape. Gravity, of course, works in opposition, tending to pull the water down. If too much water is added to the glass, the gravitational force finally wins and the skin breaks; the water overflows.

To understand surface tension we must ask two questions. What causes a liquid's molecules to set up a skinlike surface in the first place? And what causes the preference for a rounded or spherical shape?

The development of a skinlike surface is caused by the difference between the attractions felt by molecules at a liquid's surface and those felt by molecules within the liquid (see Figure 11.9). A molecule *within* the liquid is surrounded by densely packed molecules on all sides. A molecule at the *surface* however, has neighbors beside and below it, but relatively few above. Thus, the attractions experienced by a surface molecule are not uniform; instead, they occur mainly in the direction of the bulk of the liquid. As a result, the entire surface of the liquid experiences a pull toward the center. This causes a jamming together of surface molecules, making the surface behave as if it possessed a skin. The more strongly surface molecules jam together, the greater is the liquid's surface tension. Thus, the magnitude of the surface tension of a liquid is related to the strengths of the intermolecular attractive forces within it. Not surprisingly, water's surface tension is among the highest known (comparisons being at the same temperature); its intermolecular forces are hydrogen bonds, the strongest kind of dipole–dipole attraction.

For the second question, that concerning the *shape* of a liquid droplet, we have to consider how potential energy is related to stability. In general, *a system becomes more stable when its potential energy decreases.* Remember what we said on page 190: when things that naturally attract each other are free to act, the potential energy of the system is lowered. That's why, when the gravitational force is unhindered, a yardstick standing on end falls over; by doing so its potential energy decreases and it reaches a lower energy, more stable position. For a liquid, when intermolecular forces of attraction act to reduce its surface area, the system's potential energy decreases. *The lowest energy (most stable state) is achieved when the liquid has the smallest surface area possible.* For a given volume, the shape with the smallest surface area is a sphere, as we said, so liquids spontaneously tend to assume spherical shapes if they can.

To *increase* the surface area of a liquid it would be necessary to pull molecules away from the interior and make them part of the surface where fewer attractive forces would be acting. In other words, we'd be expecting things that naturally attract each other to pull apart. Thus, increasing the surface of a given volume of liquid would *increase* the liquid's potential energy, making the system *less* stable. The surface tension of a liquid is thus related to this potential energy increase—the more this energy tends to increase, the larger becomes the surface tension. When intermolecular attractions are strong, as they are in water, much energy would have to be expended to overcome them, making it difficult to move a molecule to the surface. Thus, the surface tension of water is large, roughly three times that of gasoline, which consists of relatively nonpolar hydrocarbon molecules able to experience only London forces.

Wetting of a Surface by a Liquid A property we associate with liquids, especially water, is their ability to wet things. **Wetting** is the spreading of a

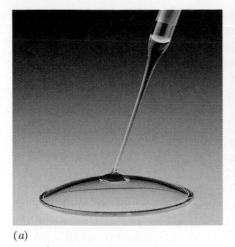
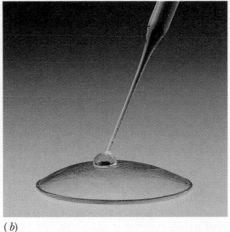

(*a*)   (*b*)

FIGURE 11.10

(*a*) Water wets a clean glass surface. (*b*) If the surface is greasy, the water doesn't wet it. The water resists spreading and forms a bead instead.

liquid across a surface to form a thin film. Water wets clean glass, such as the windshield of a car, by forming a thin film over the surface of the glass (see Figure 11.10*a*). Water won't wet a greasy windshield, however. Instead, on greasy glass water forms tiny beads (see Figure 11.10*b*).

For wetting to occur, the intermolecular attractive forces between the liquid and the surface must be of about the same strength as the forces within the liquid itself. Such a rough equality of forces exists when water touches clean glass. Water, despite its high surface tension and tendency to form beads, is yet able to wet clean glass because the glass surface contains lots of oxygen atoms. Water molecules can therefore form hydrogen bonds to the glass nearly as well as they can to each other. Thus, part of the energy needed to expand the water's surface area as wetting occurs is recovered by the formation of hydrogen bonds to the glass surface.

When the glass is coated by a film of oil or grease, however, the conditions are quite different (Figure 11.10*b*). The surface now is actually no longer glass, but oil and grease instead. Their molecules are relatively nonpolar, so their attractions to other molecules are largely limited to London forces. These are weak compared with hydrogen bonds, so the attractions *within* liquid water are much stronger than the attractions *between* water molecules and the greasy surface. The weak water-to-greasy glass attractions can't overcome the attractions within water that give it surface tension, so the water doesn't spread out; it forms beads instead.

One of the reasons that detergents are used for such chores as doing laundry or washing floors is that detergents contain chemicals called **surfactants** that drastically lower the surface tension of water. This makes the water "wetter," which allows the detergent solution to spread more easily across the surface to be cleaned.

When a liquid has a low surface tension, like gasoline, we know that it has weak intermolecular attractions, and such a liquid easily wets solid surfaces. The weak attractions between molecules in gasoline, for example, are readily overcome by attractions to almost any surface, so gasoline easily spreads to a thin film. If you've ever spilled a little gasoline, you have experienced firsthand that it doesn't bead.

Glass is characterized by a vast network of silicon-oxygen bonds (see *Chemicals in Use* 9.)

Viscosity   As everybody knows, syrup flows less readily or is more resistant to flow than water (both at the same temperature). Flowing is a change in the *form* of the liquid, and such resistance to a change in form is called the liquid's **viscosity.** We say that syrup is more *viscous* than water. The concept of viscosity is not confined to liquids, however, although it is with liquids that the property is most commonly associated. Solid things, even rock, also yield to forces acting to change their shapes but normally do so only gradually and imperceptibly. Gases also have viscosity, but they respond almost instantly to form-changing forces.

Viscosity has been called the "internal friction" of a material. It is caused largely by intermolecular attractions, including London forces; the stronger such attractions are, the more viscous the material is. Both permanent dipole–dipole and London forces affect viscosity. The ability of molecules to tangle with each other is also a factor. The long, floppy, entangling molecules in heavy machine oil, almost entirely a mixture of long-chain, nonpolar hydrocarbons, plus the London forces in the material, give it a viscosity roughly 600 times that of water at 15 °C. Vegetable oils, like the olive oil or corn oil used to prepare salad dressings, consist of molecules that are also large but generally nonpolar. Olive oil is roughly 100 times more viscous than water.

Viscosity also depends on temperature, increasing as the temperature decreases. When water, for example, is cooled from its boiling point to room temperature, its viscosity increases by over a factor of three. The increase in viscosity with cooling is why operators of vehicles use a "light," thin (meaning less viscous) motor oil during subzero weather. It's much easier to start an automobile at subzero temperatures when the cold crankcase oil flows readily and is not thick and syrupy. During very hot months, a "heavy" (more viscous) motor oil is used so that it cannot slip as readily between cylinders and pistons and so be lost even as it is doing its lubricating work.

> Powerful compression forces far below the Earth's crust act on solid rock of one form to change it to another.

> Intermolecular attractions and structure

Evaporation Finally, we come to one of the most important physical properties of liquids and solids—their tendency to undergo **evaporation,** the change of state from liquid to gas or from solid to gas. Everyone has seen liquids evaporate; it's what happens to water, for example, when streets, wet from a rain shower, gradually dry. Solids can also, without going through the liquid state, change directly to the gaseous state by evaporation, except that for a solid-to-gas change of state we use a special term, **sublimation.** Solid carbon dioxide, commonly called *dry ice,* is an example. It is "dry" ice because it doesn't melt; instead, at atmospheric pressure it *sublimes;* it changes directly to gaseous CO_2. Naphthalene, the ingredient in some brands of moth flakes, is a substance that can sublime and seemingly disappear.

To understand evaporation, we have to examine the motions of molecules in liquids and solids. Within these two states the molecules are not motionless. They bounce around against their neighbors, and at a given temperature, there is exactly the same kind of distribution of kinetic energies in a liquid or a solid as there is in a gas. This means that Figure 10.10 on page 423 applies to liquids and solids as well as gases. Some molecules have low kinetic energies and move slowly, while others with higher kinetic energies move faster. A small fraction of the molecules have very large kinetic energies and therefore very high velocities. If one of these high velocity molecules reaches the surface and is moving outward fast enough, it can escape the attractions of its neighbors and enter the vapor state. When this happens, we say the molecule has left by evaporation.

> A change of state is also called a *phase change.*

> Naphthalene sublimes when heated, and the vapor condenses directly to a solid when it encounters a cool surface. Here beautiful flaky naphthalene crystals have been formed on the bottom of a flask filled with ice water.

Evaporation and Cooling   One of the things we notice about the evaporation of a liquid is that it produces a cooling effect. Have you ever come out of the water after swimming and been chilled by a breeze? The evaporation of water from your body produced this effect. In fact, our bodies use the evaporation of water to maintain a constant body temperature. During warm weather or vigorous exercise we perspire, and the evaporation of our perspiration cools our skin. Evaporative cooling also occurs within the lungs, which puts water vapor into position to be exhaled, carrying heat out of the body with it. You've probably also used the cooling effect caused by evaporation by blowing gently on the surface of a hot bowl of soup or a cup of coffee. The stream of air stirs the liquid, bringing more hot liquid to the surface to be cooled as it evaporates.

We can see why liquids become cool during evaporation by examining Figure 11.11, which illustrates the kinetic energy distribution in a liquid at a particular temperature. A marker along the horizontal axis shows the minimum kinetic energy needed by a molecule to escape the attractions of its neighbors. Only molecules with kinetic energies equal to or greater than the minimum can leave the liquid. Others with less kinetic energy may begin to leave, but before they can escape they slow to a stop and then fall back. The situation is somewhat like attempting to launch a space ship using a cannon. If the velocity of the projectile is not very great, it will not rise very far before it slows to a halt and falls to Earth. However, if it is going fast enough (if it has its "escape velocity"), it can overcome Earth's gravity and leave.

Notice in Figure 11.11 that the minimum kinetic energy needed to escape is much larger than the average kinetic energy. This means that the molecules that evaporate are those with the largest kinetic energies. It also means that the average kinetic energy of those that remain decreases. (You might think of this as similar to removing all rowdy, noisy people from a large gathering. When done, the remaining people generate a lower average noise level.) Because the Kelvin temperature of what remains is directly proportional to the now lower average kinetic energy, the temperature of the liquid must also decrease; in other words, evaporation causes the liquid that remains to be cooler.

When a canteen of water is covered with a water-absorbing fabric that is kept wet, the evaporation of the water helps to keep the water in the canteen cool.

If the container is open, the molecules that evaporate wander away. This continues until all the liquid is gone.

Factors That Control the Rate of Evaporation

One of the important things we are going to be concerned about later in this chapter is the *rate of evaporation* of a liquid. There are several factors that control this. You are probably already aware of one of them—the surface area of the liquid. Because evaporation occurs from the liquid's surface and not

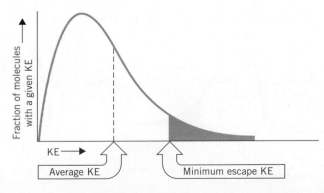

FIGURE 11.11

The minimum kinetic energy needed by a molecule to escape from the surface of the liquid is much larger than the average kinetic energy. The shaded area represents the total fraction of molecules that have the "escape KE."

FIGURE 11.12

Increasing the temperature increases the total fraction of molecules with enough kinetic energy to escape from the liquid.

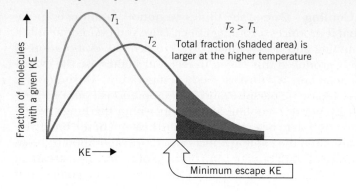

The evaporation rate refers to the rate at which material leaves the liquid (or solid) state *per unit of surface area.*

from within, it makes sense that as the surface area is increased, more molecules are able to escape and the liquid evaporates more quickly. For liquids having the same surface area, the rate of evaporation depends on two factors, namely, temperature and the strengths of intermolecular attractions. Let's examine each of them separately.

The influence of temperature on evaporation rate is no surprise; you already know that hot water evaporates faster than cold water. The underlying reason is brought out in Figure 11.12, which shows the kinetic energy distributions for the *same* liquid at two temperatures. Notice two important features of the figure. First, the same minimum kinetic energy is needed for the escape of molecules at both temperatures. This minimum is determined by the kinds of attractive forces between the molecules, and is independent of temperature. Second, the shaded area of the curve represents the total fraction of molecules having kinetic energies equal to or greater than the minimum. At the higher temperature, the total fraction is larger, which means that at the higher temperature a greater total fraction has the ability to evaporate. As you might expect, when more molecules have the needed energy, more evaporate in a unit of time, and the rate of evaporation of a given liquid is greater at a higher temperature.

The effect of intermolecular attractions on evaporation rate can be seen by studying Figure 11.13. Here we have kinetic energy distributions for two *different* liquids—call them A and B—both at the same temperature. In liquid A, the attractive forces are weak; they might be of the London type, for example. As we see, the minimum kinetic energy needed by A molecules to escape is not very large because they are not attracted very strongly to each other. In liquid

FIGURE 11.13

Kinetic energy distribution in two liquids, A and B, at the same temperature.

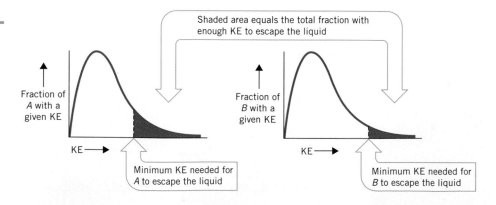

B, the intermolecular attractive forces are much stronger; they might be hydrogen bonds, for instance. Molecules of *B*, therefore, are held more tightly within the liquid and must have a higher kinetic energy to evaporate. As you can see from the figure, the fraction of molecules with enough energy to evaporate is greater for *A* than for *B*, which means that *A* evaporates faster than *B*. In general, then, *the weaker the intermolecular attractive forces, the faster is the rate of evaporation at a given temperature.* You are probably also aware of this phenomenon. Lighter fluid (butane, C_4H_{10}), whose molecules experience weak London forces of attraction, evaporates faster than water, whose molecules feel the effects of much stronger hydrogen bonds.

A change of state occurs when a substance is transformed from one physical state to another. The evaporation of a liquid and the sublimation of a solid, described in the preceding section, are two examples. Others are the melting of a solid such as ice and the freezing of a liquid such as water.

One of the important features about changes of state is that, at any particular temperature, they always tend toward physical equilibrium. We introduced the concept of *equilibrium* on page 136 with an example of a system at chemical equilibrium. The same general principles apply to a physical equilibrium. Let's examine in some detail what happens when a liquid evaporates in a sealed container and what is involved when equilibrium is reached.

When a liquid is added to an empty container, molecules begin to evaporate and collect in the space above the liquid (see Figure 11.14*a*). As they fly around in the vapor, the molecules collide with each other, with the walls of the container, and with the surface of the liquid itself. Those that strike the liquid's surface tend to stick because their kinetic energy becomes scattered among the surface molecules. It is somewhat like throwing a Ping-Pong ball into a large box of Ping-Pong balls. The incoming ball knocks others around, and by giving them kinetic energy it loses some of its own. Because its kinetic energy has been reduced, there is a high probability that the incoming ball won't bounce out.

The change of a vapor to its liquid state is called **condensation;** we say that the vapor *condenses* when it changes to a liquid. The rate at which vapor molecules collide with a liquid's surface and condense depends on the concentration of molecules in the vapor. When there are only a few in a given volume,

11.4
CHANGES OF STATE AND DYNAMIC EQUILIBRIUM

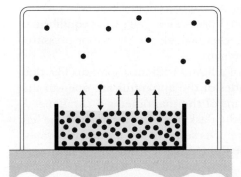
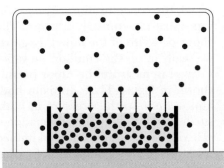

(*a*) (*b*)

FIGURE 11.14

(*a*) A liquid begins to evaporate into a closed container. (*b*) Equilibrium is reached when the rate of evaporation equals the rate of condensation.

the number of collisions per second with a unit area of the liquid's surface is small, and the rate of condensation is slow. When there are many molecules per unit of volume in the vapor, many such collisions occur each second with a unit of surface area, so the rate of condensation is higher.

Consider now a liquid as it is first introduced into an empty chamber; with virtually none of its molecules yet in the vapor state, the liquid's rate of evaporation is much larger than its vapor's rate of condensation. But as molecules accumulate in the vapor, the condensation rate increases and continues to increase until the rates of condensation and evaporation are the same (Figure 11.14*b*). From that moment on, the number of molecules in the vapor will remain constant, because over a given period of time the number that enters the vapor is the same as the number that leaves. At this point we have a condition of *dynamic equilibrium,* one in which two opposing effects, evaporation and condensation, are occurring at equal rates and their effects cancel. It is an *equilibrium* because there is no apparent change; the *number* of vapor molecules remains constant as does the number of liquid molecules. It is a *dynamic* equilibrium because activity has not ceased. Molecules continue to evaporate and condense; they just do so at equal rates.

Similar equilibria are also reached in melting and sublimation. When a solid is heated to make it melt, a temperature is reached, called the **melting point** of the solid, at which dynamic equilibrium exists between molecules in the solid and those in the liquid state. Molecules leave the solid and enter the liquid at the same rate as molecules leave the liquid and join the solid. As long as no heat is added or removed from a solid–liquid equilibrium mixture, melting and freezing occur at equal rates. For sublimation, the situation is exactly the same as in the evaporation of a liquid into a sealed container (see Figure 11.15). After a few moments, the rates of sublimation and condensation become the same and equilibrium is established. All these various equilibria have very profound effects on the properties of solids and liquids, as we shall soon see.

In chemistry (unless otherwise indicated), when we use the term *equilibrium,* we always mean *dynamic equilibrium.*

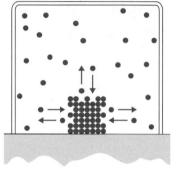

FIGURE 11.15

A solid–vapor equilibrium. Molecules evaporate and condense on the crystal at equal rates when equilibrium is reached.

11.5
VAPOR PRESSURES OF LIQUIDS AND SOLIDS

The vapor-pressure equilibrium is possible only in a closed container. When the container is open, vapor molecules drift away and the liquid might completely evaporate.

When a liquid evaporates, the molecules that enter the vapor exert a pressure called the **vapor pressure.** From the very moment a liquid begins to evaporate into the vapor space above it, there is a vapor pressure. If the evaporation is taking place inside a sealed container, this pressure grows until finally equilibrium is reached. Once the rates of evaporation and condensation become equal, the concentration of molecules in the vapor remains constant and the vapor exerts a constant pressure. This final pressure is called the **equilibrium vapor pressure** of the liquid. In general, when we refer to the vapor pressure, we really mean the equilibrium vapor pressure.

We can measure the vapor pressure of a liquid with an apparatus like that shown in Figure 11.16. Initially, both sides of the apparatus are open to the atmosphere, so the pressure in both arms of the manometer is the same. A small amount of liquid is then added to the flask from the funnel and the left side of the apparatus is sealed immediately. As some of the liquid evaporates, the pressure in the flask rises, forcing the fluid in the left side of the manometer downward. The increase in pressure, as measured by the difference in the levels in the manometer, is equal to the pressure exerted by the vapor, namely, the vapor pressure of the liquid.

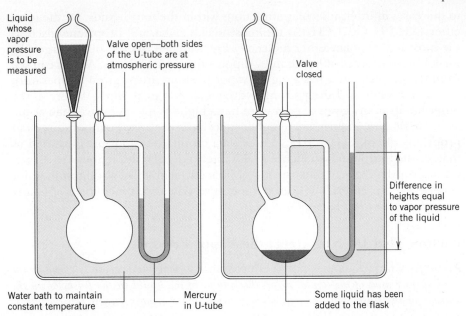

FIGURE 11.16

Measuring the vapor pressure of a liquid.

Factors That Affect Equilibrium Vapor Pressure

Figure 11.17 shows plots of equilibrium vapor pressure versus temperature for a few liquids. It's obvious that both a liquid's temperature and its chemical composition are the major factors affecting vapor pressure. Once we have selected a particular liquid, however, only the temperature matters. The reason is that the vapor pressure of a given liquid is a function solely of its rate of evaporation *per unit area of the liquid's surface*. When this rate is large, a large concentration of molecules in the vapor state is necessary to establish equilibrium, which is another way of saying that the vapor pressure is relatively high when the evaporation rate is high. As the temperature of a given liquid increases, so does its rate of evaporation and so does its vapor pressure.

As chemical composition changes in going from one liquid to another, the strengths of intermolecular attractions change. If the attractions increase, the rates of evaporation decrease, and the vapor pressures decrease. The vapor pressure data for the four liquids in Figure 11.17 illustrate these relationships. From left to right, the plots of vapor pressure versus temperature correspond

Intermolecular attractions and vapor pressures

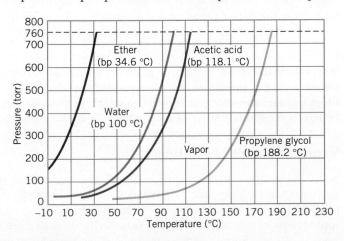

FIGURE 11.17

Variation of vapor pressure with temperature for some common liquids.

to increases in intermolecular attractions within the four liquids. Of the four, ether ($CH_3CH_2OCH_2CH_3$) is the most weakly polar and propylene glycol is the most highly polar. Ether molecules are attracted to each other largely by weak London forces rather than by dipole–dipole attractions. As you can see from the graphs, ether has a higher vapor pressure at any given temperature than water, where relatively strong hydrogen bonds are at work. Water, on the other hand, at any given temperature has a higher vapor pressure than acetic acid. Acetic acid, in turn, has a higher vapor pressure at any temperature than propylene glycol, $C_3H_6(OH)_2$. These data on relative vapor pressures tell us that, of the four liquids in Figure 11.17, intermolecular attractions are strongest in propylene glycol, next strongest in acetic acid, third strongest in water and weakest in ether. Thus, *we can use vapor pressures as indications of relative strengths of the attractive forces in liquids.*

Factors That Do Not Affect the Vapor Pressure

An important fact about vapor pressure is that *its magnitude doesn't depend on the total surface area of the liquid, or on the volume of the liquid in the flask, or on the volume of the flask itself, just as long as some liquid remains when equilibrium is reached.* The reason is that none of these factors affects the rate of evaporation per unit surface area.

Increasing the total *surface area* does increase the total rate of evaporation, but the larger area is also available for condensation; so the rate at which molecules return to the liquid also increases. The rates of both evaporation and condensation are thus affected equally, and no change occurs to the equilibrium vapor pressure.

Adding more liquid to the container can't affect the equilibrium either because evaporation occurs from the *surface*. Having more molecules in the bulk of the liquid does not change what is going on at the surface.

To understand why the vapor pressure doesn't depend on the size of the vapor space, imagine a liquid and its vapor in a cylinder with a movable piston. Withdrawing the piston increases the volume of the vapor space and momentarily lowers the vapor pressure. This means that fewer molecules of the vapor strike a given surface area of the liquid each instant, and the rate of condensation per unit area will have decreased. The rate of evaporation hasn't changed, however, so for a moment the substance evaporates faster than it condenses. This condition prevails until the concentration of molecules in the vapor is high enough to make the condensation rate again equal to the evaporation rate. Now the vapor pressure has returned to its original value. Similarly, we expect that reducing the volume of the vapor space above the liquid will also not affect the equilibrium vapor pressure.

■ **Practice Exercise 1** Suppose a liquid is in equilibrium with its vapor in a piston–cylinder apparatus like that just described. If the piston is withdrawn a short way and the system is allowed to return to equilibrium, what will have happened to the total number of molecules in both the liquid and the vapor?

Vapor Pressures of Solids

Solids have vapor pressures just as liquids do. In a crystal, the particles are not stationary, as we have said. They vibrate back and forth about their equilibrium positions. At a given temperature there is a distribution of kinetic energies, so some particles vibrate slowly while others vibrate with a great deal of molecular kinetic energy. Some particles at the surface have large enough kinetic energies to break away from their neighbors and enter the vapor state. When particles in the vapor collide with the crystal, they can be recaptured, so condensation can occur too. Eventually, the concentration of particles in the vapor reaches a point where the rate of sublimation equals the rate of condensation, and a dynamic equilibrium is established. The pressure of the vapor that is in equilibrium with the solid is called the *equilibrium vapor pressure of the solid*. As with liquids, this equilibrium vapor pressure is usually referred to simply as the vapor pressure. Like that of a liquid, the vapor pressure of a solid is determined by the strengths of the attractive forces between the particles and by the temperature.

In many solids, such as NaCl, the attractive forces are so strong that virtually no particles have enough kinetic energy at room temperature to escape, and essentially no evaporation occurs.

11.6 BOILING POINTS OF LIQUIDS

If you were asked to check whether a pot of water were boiling, what would you look for? The answer, of course, is *bubbles*. When a liquid boils, large bubbles usually form at many places on the inner surface of the container and rise to the top. If you were to place a thermometer into the boiling water, you would find that the temperature remains constant, regardless of how you adjust the flame under the pot. A hotter flame just makes the water bubble faster, but it doesn't raise the temperature. *Any pure liquid remains at a constant temperature while it is boiling,* a temperature called the liquid's **boiling point.**

If you measure the boiling point of water in Philadelphia, New York, or any place else that is nearly at sea level, your thermometer will read 100 °C or very close to it. However, if you try this experiment in Denver, Colorado, you will find that water boils at about 95 °C. Denver, at a mile above sea level, has a lower atmospheric pressure, so we find that the boiling point depends on the atmospheric pressure.

These observations raise some interesting questions. Why do liquids boil? And why does the boiling point depend on the pressure of the atmosphere? The answers become apparent when we realize that inside the bubbles of a boiling liquid is the *liquid's vapor,* not air. When water boils, the bubbles contain water vapor (steam); when alcohol boils, the bubbles contain alcohol vapor. As a bubble grows, liquid evaporates into it, and the pressure of the vapor pushes the liquid aside, making the size of the bubble increase (see Figure 11.18). Opposing the bubble's internal vapor pressure, however, is the pressure of the atmosphere pushing down on the top of the liquid, attempting to collapse the bubble. The only way the bubble can exist and grow is for the vapor pressure within it to equal (maybe just slightly exceed) the pressure exerted by the atmosphere. In other words, there cannot even be bubbles of vapor until the temperature of the liquid rises to a point at which the liquid's vapor pressure equals the atmospheric pressure. Thus, in scientific terms, the **boiling point** is defined as *the temperature at which the vapor pressure of the liquid is equal to the prevailing atmospheric pressure.*

Now we can easily understand why water boils at a lower temperature in Denver than it does in New York City. Because the atmospheric pressure is lower in Denver, the water there doesn't have to be heated to as high a temperature to make its vapor pressure equal to the atmospheric pressure. The lower temperature of boiling water at places with high altitudes, like Denver, makes it necessary to cook foods longer. At the other extreme, a pressure cooker is a device that increases the pressure over the boiling water and thereby raises the boiling point. At the higher temperature, the rate at which foods cook is faster.

We know that the water in this pot of vegetables is boiling because we can see bubbles of steam rising to the surface. Because the temperature of boiling water cannot rise above its boiling point, foods cooked in water can't become too hot and burn.

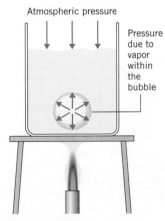

FIGURE 11.18

The pressure of the vapor within a bubble pushes the liquid aside against the opposing pressure of the atmosphere.

On the top of Mt. Everest, the world's tallest peak, water boils at only 69 °C.

To make it possible to compare the boiling points of different liquids, chemists have chosen 1 atm as the reference pressure. The boiling point of a liquid at 1 atm is called its **normal boiling point.** (If a boiling point is reported without also mentioning the pressure at which it was measured, we assume it to be the normal boiling point.) Notice in Figure 11.17, page 457, that we can find the normal boiling points of ether, water, acetic acid, and propylene glycol by noting the temperatures at which their vapor pressure curves cross the 1 atm pressure line.

Earlier we mentioned that the boiling point is a property whose value depends on the strengths of the intermolecular attractions in a liquid. When the attractive forces are strong, the liquid has a low vapor pressure at a given temperature, so it must be heated to a high temperature to bring its vapor pressure up to atmospheric pressure. High boiling points therefore result from strong intermolecular attractions, so we often use normal boiling point data to assess relative intermolecular attractions among different liquids.

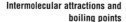
Intermolecular attractions and boiling points

The effects of intermolecular attractions on boiling point are easily seen by examining Figure 11.19, which gives the plots of the boiling points versus period numbers for some families of binary hydrides, compounds between hydrogen and another element. Notice, first, the gradual increase in boiling point for the hydrides of the Group IVA elements (CH_4 through GeH_4). These hydrides are composed of nonpolar tetrahedral molecules, and experience relatively weak London forces. The boiling points increase from CH_4 to GeH_4 simply because the molecules become larger.

When we look at the hydrides of the other nonmetals, we find the same trend from period 3 through period 5. Thus, for three hydrides of the Group VA series, PH_3, AsH_3, and SbH_3, there is a gradual increase in boiling point, corresponding again to the increasing strengths of London forces. Similar increases occur for three Group VIA hydrides—H_2S, H_2Se, and H_2Te—and for three Group VIIA hydrides—HCl, HBr, and HI. Significantly, however, the period 2 members of each of these series—NH_3, H_2O, and HF—have much higher boiling points than might otherwise be expected. The reason is that each is involved in hydrogen bonding, which is a much stronger attraction than London forces.

One of the most interesting and far-reaching consequences of hydrogen bonding is that it causes water to be a liquid, rather than a gas, at temperatures near 25 °C. If it were not for hydrogen bonding, water would have a boiling point somewhere near − 80 °C and could not exist as a liquid except at still

FIGURE 11.19

Boiling points of the hydrides of elements of Groups IVA, VA, VIA, and VIIA of the periodic table. The actual boiling points of HF, H_2O, and NH_3 are higher than expected, based on the trends shown by the other hydrides, because of hydrogen bonding.

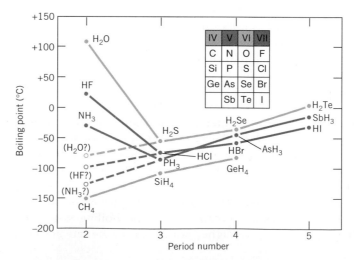

lower temperatures. At such low temperatures it is unlikely that life as we know it could have developed.

■ **Practice Exercise 2** The atmospheric pressure at the top of Mt. McKinley in Alaska, 3.85 miles above sea level, is 300 torr. Use Figure 11.17 to estimate the boiling point of water at the top of this mountain.

When a liquid or solid evaporates or a solid melts, there are increases in the distances between the particles of the substance. Particles that normally attract each other are forced apart, increasing the potential energies of the systems. Such energy changes affect our daily lives in many ways, especially the energy changes associated with the changes in state of water, changes which even control the weather on our planet. To study these energy changes, let's begin by examining how the temperature of a substance varies as it is heated.

11.7
ENERGY CHANGES DURING CHANGES OF STATE

Heating Curves and Cooling Curves

Figure 11.20*a* illustrates the way the temperature of a substance changes as we add heat to it *at a constant rate,* starting with the solid and finishing with the gaseous state of the substance. The graph is sometimes called a **heating curve** for the substance.

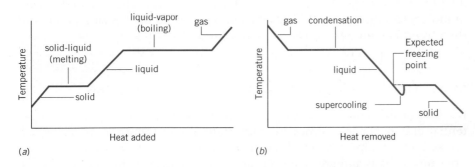

(a) (b)

FIGURE 11.20

(*a*) A heating curve in which heat is added to a substance at a constant rate. The temperatures corresponding to the flat portions of the curve occur at the melting point and boiling point. (*b*) A cooling curve in which heat is removed from a substance at a constant rate. Condensation of vapor to a liquid occurs at the same temperature as the liquid boils. Supercooling is seen here as the temperature of the liquid dips below its freezing point (the same temperature as its melting point). Once a tiny crystal forms, the temperature rises to the freezing point.

Notice that as heat is added to the solid, the temperature increases until the solid begins to melt. The temperature then remains constant as long as both solid and liquid phases coexist. When all the solid has melted, the continued addition of heat to what is now a liquid causes the temperature once again to climb. This continues until the boiling point is reached, at which point the temperature levels off again. As we noted, the temperature of the boiling liquid stays the same until all of it has boiled away and changed to a gas. Then, finally, more heat just raises the temperature of the gas. Let's analyze these changes in terms of changes in the kinetic energy and potential energy of the particles.

First, let's look at the portions of the graph that slope upward. These occur where we are increasing the temperature of the solid, the liquid, or the gas phases. Because temperature is related to average kinetic energy, any heat we add in these regions of the heating curve goes primarily to increasing the average kinetic energies of the particles. In other words, the added heat makes the particles go faster and so collide with each other with more force.

In those portions of the heating curve where the temperature remains constant, the average kinetic energy of the particles is not changing. This means that all the heat being added must go to increase the *potential energies* of the particles. During melting, the particles held rigidly in the solid begin to

When a solid or liquid is heated, the volume expands only slightly, so there are only small changes in the average distance between the particles. This means that very small changes in potential energy take place, so almost all the heat added goes to increasing the kinetic energy.

separate slightly as they form the mobile liquid phase. This separation of the particles gives rise to a potential energy increase, which is what we measure by following the quantity of heat input during the melting process. During boiling, an even greater increase occurs in the distances between the molecules. Here they go from the relatively tight packing in the liquid to the widely spaced distribution of molecules in the gas. This gives rise to an even larger increase in the potential energy, which we see as a longer flat region on the heating curve during the boiling of the liquid.

The opposite of a heating curve is a **cooling curve** (see Figure 11.20*b*). Here we start with a gas and gradually cool it—remove heat from it at a constant rate—until we have reached a solid. It looks very much like the opposite of a heating curve, except for what often happens when the temperature of the liquid approaches the freezing point. We might expect that when the freezing point is reached, removal of more heat would cause the immediate formation of some solid. Often, however, the liquid continues to cool *below its freezing point,* a phenomenon called **supercooling.**

Supercooling can happen when the freezing point is reached, because the molecules of the liquid are still in a jumbled disordered arrangement, not in the ordered structure characteristic of a solid (as we'll soon describe). While the molecules move about, waiting for a few to form the beginnings of a crystal, the temperature continues to drop because heat is continually being removed. Finally, a tiny crystal forms and other molecules quickly join it, losing potential energy, which is changed to kinetic energy. The sudden increase in molecular kinetic energy raises the temperature to the freezing point, and then freezing continues normally as more heat is taken away.

A heating curve can be used to measure accurately the melting point and boiling point of a liquid.

Supercooling of a vapor is also possible.

Molar Heats of Fusion, Vaporization, and Sublimation

The concept of enthalpy was introduced in Section 5.5.

These are also called enthalpies of fusion, vaporization, and sublimation.

Because phase changes occur at constant temperature and pressure, the potential energy changes associated with melting and vaporization can be expressed as enthalpy changes. Usually, enthalpy changes are expressed on a "per mole" basis and are given special names to identify the kind of change involved. For example, using the word **fusion** instead of *melting,* the **molar heat of fusion,** ΔH_{fusion}, is the heat absorbed by one mole of a solid when it melts to give a liquid at the same temperature and pressure. Similarly, the **molar heat of vaporization,** $\Delta H_{vaporization}$, is the heat absorbed when one mole of a liquid is changed to one mole of vapor at a constant temperature and pressure. Finally, the **molar heat of sublimation, $\Delta H_{sublimation}$**, is the heat absorbed by one mole of a solid when it sublimes to give one mole of vapor, once again at a constant temperature and pressure. The values of ΔH for fusion, vaporization, and sublimation are all positive because the phase change in each case is accompanied by a net increase in potential energy.

You have probably made use of these enthalpy changes at one time or another. For example, you've added ice to a drink to keep it cool because as the ice melts, it absorbs heat (its heat of fusion). Your body uses the heat of vaporization of water to cool itself through the evaporation of perspiration. During the summer, ice cream trucks carry dry ice because the sublimation of CO_2 absorbs its heat of sublimation and keeps the ice cream cold. Refrigerators and air conditioners work on the basis of making the heat unwanted in one space function as the heat of vaporization for the system's operating fluid (see Special Topic 11.1).

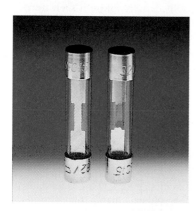

Fusion means melting. The thin metal band in an electrical fuse protects a circuit by melting if too much current is passed through it. On the right we see a fuse that has done its job.

SPECIAL TOPIC 11.1 / REFRIGERATION AND AIR CONDITIONING

The fact that the evaporation of a liquid is endothermic and the condensation of a gas is exothermic forms the basis for the operation of refrigerators and air conditioners. These devices use a gas that has a high critical temperature—ammonia or a halogenated hydrocarbon—to pump heat from one place to another.

In a refrigerator, a compressor squeezes the gas into a small volume, which causes the pressure to increase. At the same time, the work done compressing the gas becomes stored as heat—the gas becomes warm. The warm gas is then circulated through cooling coils that are usually located on the back of the refrigerator. (Did you ever notice that these coils are warm?) When the gas cools under pressure, it liquefies when its temperature drops below its critical temperature, and this releases

the substance's heat of vaporization, which is also given off to the surroundings. After being cooled, the pressurized liquid moves back into the refrigerator where it passes through a nozzle into a region of low pressure. At this low pressure, the liquid evaporates, and heat equal to the heat of vaporization is absorbed, which cools the inside of the refrigerator. Then the gas moves back to the compressor, and the cycle is repeated.

Through condensation and evaporation, the compressor uses the gas as the fluid in a heat pump that absorbs heat from inside the refrigerator and dumps it outside. An air conditioner works in essentially the same way, except that a room becomes the equivalent of the inside of a refrigerator, and the heat pump dumps the heat into the air outside.

Heats of Vaporization and Vapor Pressures

The increase of vapor pressure with temperature noted in Section 11.5 is related to the heat of vaporization of the substance. The relationship, however, is not a simple proportionality. It involves **natural logarithms,** which are logarithms to the base e, a nonrepeating decimal number (an irrational number) having the value 2.71828. . . . The natural logarithm of a quantity x, symbolized as $\ln x$, is the power to which e must be raised to give x.

Both natural logarithms and common (base 10) logarithms are discussed in detail in Appendix A.

The equation relating vapor pressure (P), heat of vaporization (ΔH_{vap}), and temperature is called the **Clausius–Clapeyron equation**[1]:

$$\ln P = \frac{-\Delta H_{vap}}{RT} + C \qquad (11.1)$$

The Clausius–Clapeyron equation can be derived from the principles of thermodynamics, a subject that is discussed in Chapter 13.

The quantity $\ln P$ is the natural logarithm of the vapor pressure, R is the gas constant expressed in energy units $(R = 8.314 \text{ J mol}^{-1} \text{ K}^{-1})$, T is the absolute temperature, and C is a constant.

The Clausius–Clapeyron equation provides a convenient graphical method for determining heats of vaporization from experimentally measured vapor pressure–temperature data. To see this, let's rewrite Equation 11.1 as follows.

 Clausius–Clapeyron equation

$$\ln P = \left(\frac{-\Delta H_{vap}}{R}\right)\frac{1}{T} + C$$

Recall from algebra that a straight line is represented by the general equation

$$y = mx + b$$

where x and y are variables, m is the slope, and b is the intercept of the line with the y axis. In this case, we can make the substitutions

$$y = \ln P \qquad x = \frac{1}{T} \qquad m = \left(\frac{-\Delta H_{vap}}{R}\right) \qquad b = C$$

---

[1] Rudolf Clausius (1822–1888) was a German physicist. Benoit Clapeyron (1799–1864) was a French engineer. Both made significant contributions to thermodynamics.

FIGURE 11.21

A graph showing plots of ln P versus $1/T$ for three liquids. (The numbers along the horizontal axis are equal to $1/T$ values multiplied by 1000 to make the axis easier to label.) Notice how well the data fit straight lines.

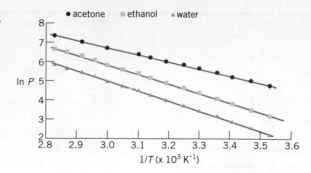

Therefore, we have

$$\ln P = \left(\frac{-\Delta H_{vap}}{R}\right)\frac{1}{T} + C$$

$$y = m \quad x + b$$

Thus, a graph of ln P versus $1/T$ should give a straight line that has a slope equal to $-\Delta H_{vap}/R$. Such straight line relationships are illustrated in Figure 11.21 in which experimental data are plotted for water, acetone, and ethanol.

Sometimes we wish to calculate the vapor pressure at some temperature given the vapor pressure at another temperature. Provided the value of ΔH_{vap} is known, we can use the Clausius–Clapeyron equation.

Suppose P_1 is the vapor pressure of the substance at a temperature T_1 and P_2 is the vapor pressure at temperature T_2. Then, we can write

$$\ln P_1 = \left(\frac{-\Delta H_{vap}}{R}\right)\frac{1}{T_1} + C$$

$$\ln P_2 = \left(\frac{-\Delta H_{vap}}{R}\right)\frac{1}{T_2} + C$$

Subtracting the second equation from the first gives

$$\ln P_1 - \ln P_2 = \left(\frac{-\Delta H_{vap}}{R}\right)\frac{1}{T_1} - \left(\frac{-\Delta H_{vap}}{R}\right)\frac{1}{T_2} + C - C$$

For logarithms,

$$\log a - \log b = \log (a/b)$$
$$\ln a - \ln b = \ln (a/b)$$

Combining logarithm terms on the left and factoring out $\Delta H_{vap}/R$ on the right yields

$$\ln \frac{P_1}{P_2} = \frac{\Delta H_{vap}}{R}\left(\frac{1}{T_2} - \frac{1}{T_1}\right) \tag{11.2}$$

Example 11.1 illustrates how we can use this "two-point" form of the Clausius–Clapeyron equation.

EXAMPLE 11.1
Using the Clausius–Clapeyron Equation

Hexane, C_6H_{14}, a component of paint thinners, has $\Delta H_{vap} = 30.1$ kJ/mol. At room temperature (25 °C), the vapor pressure of hexane is 148 torr. What is the vapor pressure of hexane at 50 °C?

ANALYSIS The problem is straightforward; we just have to substitute quantities into Equation 11.2. To avoid errors, let's extract the pressure–temperature data from the problem and arrange it in a table. Arbitrarily taking "1" to be the data for the lower temperature, we have

	Vapor Pressure	Temperature
1	148 torr	25 °C = 298 K
2	P_2	50 °C = 323 K

The unknown quantity, P_2, is in the logarithm term. To obtain it, we will first solve for the ratio, P_1/P_2. Then we can substitute the value of P_1 and solve for P_2.

SOLUTION Let's begin by substituting values into the right side of Equation 11.2. We must be careful about the energy units, however. Because R has energy units of joules, we must change the units of ΔH_{vap} from kilojoules to joules; thus, the value of ΔH_{vap} must be expressed as 30.1×10^3 J/mol.

$$\ln \frac{P_1}{P_2} = \frac{30.1 \times 10^3 \, \text{J mol}^{-1}}{8.314 \, \text{J mol}^{-1} \, \text{K}^{-1}} \left(\frac{1}{323 \, \text{K}} - \frac{1}{298 \, \text{K}} \right)$$

$$= (3.62 \times 10^3 \, \text{K}) \times (-2.60 \times 10^{-4} \, \text{K}^{-1})$$

$$= -0.941$$

To obtain the ratio P_1/P_2 we have to take the antilogarithm of -0.941. This is done by raising e to the -0.941 power. (If necessary, refer to Appendix A for discussions of logarithms, antilogarithms, and the use of your scientific calculator for handling them.)

$$\frac{P_1}{P_2} = e^{-0.941}$$

$$= 0.390$$

Solving the pressure ratio for P_2 gives

$$P_2 = \frac{P_1}{0.390}$$

$$= \frac{148 \, \text{torr}}{0.390}$$

$$= 379 \, \text{torr}$$

Thus, the vapor pressure of hexane at 50 °C is calculated to be 379 torr.

■ **Practice Exercise 3** Using data in Example 11.1, what is the vapor pressure of hexane at 60 °C?

Energy Changes and Intermolecular Attractions

For a given substance, $\Delta H_{vaporization}$ is larger than ΔH_{fusion}. This is because the particles separate to greater distances during vaporization than during melting, so the changes in potential energy are greater. For the same substance, $\Delta H_{sublimation}$ is even larger than $\Delta H_{vaporization}$ because the attractive forces in the solid are larger than those in the liquid. As you would expect, overcoming these stronger forces requires more energy.

The stronger the attractions, the more the potential energy will increase when the molecules become separated, and the larger will be the value of ΔH.

When a liquid evaporates or a solid sublimes, the particles go from a situation in which the attractive forces are very strong to one in which the attractive forces are so small they can almost be ignored. Therefore, the values of $\Delta H_{vaporization}$ and $\Delta H_{sublimation}$ give us directly the energy needed to separate molecules from each other. We can examine such values to obtain reliable comparisons of the strengths of intermolecular attractions.

TABLE 11.4   Some Typical Heats of Vaporization

Substance	$\Delta H_{vaporization}$ (kJ/mol)	Type of Attractive Force
H_2O	+43.9	Hydrogen bonding
NH_3	+21.7	Hydrogen bonding
HCl	+15.6	Dipole–dipole
SO_2	+24.3	Dipole–dipole
F_2	+5.9	London
Cl_2	+10.0	London
Br_2	+15.0	London
I_2	+22.0	London
CH_4	+8.16	London
C_2H_6	+15.1	London
C_3H_8	+16.9	London
C_6H_{14}	+30.1	London

Intermolecular attractions and molar heats of phase changes

The Lewis structure of SO_2 tells us that the molecule is nonlinear, and this means that it must be polar.

In Table 11.4, notice that the heats of vaporization of water and ammonia are very large, which is just what we would expect for hydrogen bonded substances. By comparison, a nonpolar substance of similar molecular mass, CH_4, has a very small heat of vaporization. Note also that polar substances such as HCl and SO_2 have fairly large heats of vaporization compared with nonpolar substances with the same number of electrons. For example, HCl and F_2 both have 18 electrons, which means the strengths of their London forces should be about the same. The more polar HCl, however, has a much larger $\Delta H_{vaporization}$ than the nonpolar F_2. We can make a similar comparison between SO_2 (32 electrons) and Cl_2 (34 electrons). Once again the polar substance (SO_2) has a larger heat of vaporization than the nonpolar one (Cl_2).

Heats of vaporization also reflect the factors that control the strengths of London forces. For example, the data in Table 11.4 show the effect of chain length on the intermolecular attractions between hydrocarbons; as the chain length increases from one carbon in CH_4 to six carbons in C_6H_{14}, the heat of vaporization also increases. The increase is in agreement with an increase in the total strengths of the London forces. In our discussion of London forces, we also mentioned that their strengths increase as the electron clouds of the particles become larger, and we used the boiling points of the halogens to illustrate this. The heats of vaporization of the halogens in Table 11.4 are in agreement with this, too.

11.8
DYNAMIC EQUILIBRIUM AND LE CHÂTELIER'S PRINCIPLE

In the new equilibrium, there is less liquid and more vapor, but there is *equilibrium* nonetheless.

Throughout this chapter, we have studied various dynamic equilibria. One example was the equilibrium that exists between a liquid and its vapor in a closed container. You learned that when the temperature of a liquid is increased in this system, its vapor pressure also increases. Let's now analyze this effect in terms of what occurs at the molecular level.

Initially, the liquid is in equilibrium with its vapor, which exerts a certain pressure. When the temperature is increased, equilibrium no longer exists because evaporation occurs more rapidly than condensation. Eventually, as the concentration of molecules in the vapor increases, *the system reaches a new equilibrium* in which there is more vapor and a little less liquid. The greater concentration of the vapor causes a larger pressure.

The response of a liquid's equilibrium vapor pressure to a temperature change is an example of a general phenomenon. Whenever a dynamic equilibrium is upset by some disturbance, the system responds in whichever way restores equilibrium or at least "moves" in this direction. Although the values of volume, pressure, and temperature might be different in the new equilibrium, there is at least equilibrium once again. It's important to realize that *dynamic equilibrium* only means "no net change," despite much "coming and going"; that *restoring* equilibrium means only restoring this condition, not necessarily restoring the originally measured physical or chemical properties.

Throughout the remainder of this book, we will deal with many kinds of equilibria, both chemical and physical. It would be very time consuming and sometimes very difficult to carry out a detailed analysis each time we wish to know the effects of some disturbance on an equilibrium system. Fortunately, there is a relatively simple and fast method for predicting the effect of a disturbance, one based on a principle proposed in 1888 by a brilliant French chemist, Henry Le Châtelier (1850–1936).

Le Châtelier's Principle

 Le Châtelier's principle

When a dynamic equilibrium in a system is upset by a disturbance, the system responds in a *direction* that tends to counteract the disturbance and, if possible, restore equilibrium.

Let's see how we can apply Le Châtelier's principle to a liquid–vapor equilibrium that is subjected to a temperature increase. We cannot increase a temperature, of course, without adding heat. Thus, *the addition of heat is really the disturbing influence when a temperature is increased.* So let's incorporate "heat" as a term in the equation used to represent the liquid–vapor equilibrium:

$$\text{Heat} + \text{liquid} \rightleftharpoons \text{vapor} \qquad (11.3)$$

Recall that $\rightleftharpoons$ double arrows are the way we indicate a dynamic equilibrium in an equation. They imply opposing changes happening at equal rates. Evaporation is endothermic, so we must show that heat is absorbed by the liquid when it changes to the vapor in Equation 11.3 and that heat is released when the vapor condenses to a liquid.

Le Châtelier's principle tells us that when the temperature of the liquid–vapor equilibrium system is raised by adding heat, the system will try to adjust in a way that absorbs some of the added heat. This can happen if some liquid evaporates because, as we noted, vaporization is endothermic. When liquid evaporates, the amount of vapor increases and the pressure rises. Thus, we have reached the correct conclusion in a very simple way, namely, that heating a liquid must increase its vapor pressure.

Recall that we often use the term *position of equilibrium* to refer to the relative amounts of the substances on opposite sides of the double arrows in an equilibrium expression such as Equation 11.3. Thus, we can think of how a disturbance affects the position of equilibrium. For example, increasing the temperature increases the amount of vapor and decreases the amount of liquid, and we say *the position of equilibrium has shifted;* in this case, it has shifted in the direction of the vapor, or it has *shifted to the right.* In using Le Châtelier's principle, it is often convenient to think of a disturbance as "shifting the position of equilibrium" in one direction or another in the equilibrium equation.

■ **Practice Exercise 4** Use Le Châtelier's principle to predict how a temperature increase will affect the vapor pressure of a solid. (*Hint:* Solid + heat ⇌ vapor.)

11.9
PHASE DIAGRAMS

Phase diagrams

Sometimes it is useful to know under what combinations of temperature and pressure a substance will be a liquid, a solid, or a gas or will produce an equilibrium between any two phases. A simple way to determine this is to use a **phase diagram**—a graphical representation of the pressure–temperature relationships that apply to the equilibria between the phases of the substance.

Figure 11.22 is the phase diagram for water. On it, there are three lines that intersect at a common point. These lines give temperatures and pressures at which equilibria between phases can exist. For example, line *AB* is called the vapor pressure curve for the solid (ice). Every point on line *AB* gives a temperature and a pressure at which ice and its vapor are able to be *in equilibrium*. For instance, ice can be in equilibrium with water vapor at −10 °C only when the vapor pressure is 2.15 torr. This kind of information would be very useful to know if you wanted to design a system for making freeze-dried coffee (see Special Topic 11.2). It tells you that ice will sublime at −10 °C if a vacuum pump can reduce the pressure of water vapor above the ice to *less* than 2.15 torr. The lower pressure upsets the ice–vapor equilibrium—disturbs it—and ice cannot help but respond by subliming away.

The melting point and boiling point can be read directly from the phase diagram.

Line *BD* is called the vapor pressure curve for liquid water. It gives the temperatures and pressures at which the liquid and vapor are able to be in equilibrium. Notice that when the temperature is 100 °C, the vapor pressure is 760 torr. Therefore, this diagram also tells us that water boils at 100 °C when the pressure is 1 atm (760 torr), because that is the temperature at which the vapor pressure equals 1 atm. At the top of Mt. McKinley, in Alaska, the atmospheric pressure is only about 330 torr. The phase diagram tells us that the boiling point of water there will be 78 °C.

The solid–vapor equilibrium line, *AB*, and the liquid–vapor line, *BD*, intersect at a common point, *B*. Because this point is on both lines, there is equilibrium between all three phases at the same time.

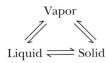

FIGURE 11.22

The phase diagram for water, distorted to emphasize certain features. Usually, a phase diagram is drawn properly to scale so that its temperature and pressure scales may be read accurately.

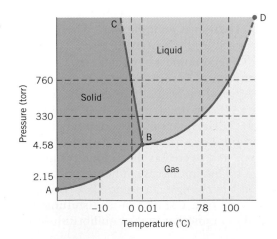

Anyone who has watched much television has certainly seen advertisements that praise the virtues of freeze-dried instant coffee. Campers routinely carry freeze-dried foods because they are lightweight (see Figure 1). In addition, bacteria cannot grow and reproduce in the complete absence of moisture, so freeze-dried foods need no refrigeration. Therefore, freeze-drying is clearly a useful way of preserving foods. It is accomplished by first freezing the food (or brewed coffee) and then placing it in a chamber that is connected to high-capacity vacuum pumps. These pumps lower the pressure in the chamber to below the vapor pressure of ice, which causes the ice crystals to sublime. Drying foods in this way has an important advantage over other methods—the delicate molecules responsible for the flavor of foods are not destroyed, as they would be if the food were heated. Freeze-dried foods may also be easily reconstituted simply by adding water.

Freeze-drying is also used by biologists to preserve tissue cultures and bacteria. Removing moisture from tissue and storing it at low temperatures allows the cells to survive in what

FIGURE 1

It's common to find freeze-dried foods among the supplies carried by campers.

amounts to a state of suspended animation for periods of at least several years.

The temperature and pressure at which this triple equilibrium occurs define the **triple point** of the substance. For water, the triple point occurs at 0.01 °C and 4.58 torr. Every known chemical substance except helium has its own characteristic triple point, which is controlled by the balance of intermolecular forces in the solid, liquid, and vapor.

Line *BC*, which extends upward from the triple point, is the solid–liquid equilibrium line or *melting point line*. It gives temperatures and pressures at which the solid and the liquid are able to be in equilibrium. At the triple point, the melting of ice occurs at + 0.01 °C (and 4.58 torr); at 760 torr, melting occurs very slightly lower, at 0 °C. Thus, we can tell that *increasing the pressure on ice lowers its melting point.*

The decrease in the melting point of ice that occurs when there is an increase in pressure can be predicted using Le Châtelier's principle and the knowledge that liquid water is more dense than ice. Let's reflect again on the equilibrium between ice and liquid water at 0 °C and 1 atm.

$$H_2O(s) \rightleftharpoons H_2O(l)$$

Imagine that the equilibrium is established in an apparatus like that shown in Figure 11.23 (and, again, keep in mind that ice is less dense than water). If the

In the SI, the triple point of water is used to define the Kelvin temperature of 273.16 K.

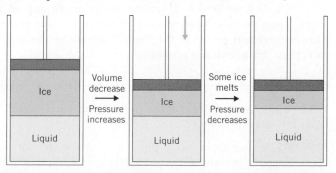

FIGURE 11.23

The effect of pressure on the equilibrium,

$$H_2O(s) \rightleftharpoons H_2O(l).$$

piston is pushed in slightly, the pressure increases, and this is a disturbance. According to Le Châtelier's principle, the system should respond in a way that tries to absorb the higher pressure, namely tries to reduce it. The only way that any pressure reduction can happen is for some of the ice to melt so that the ice–liquid mixture won't require as much space. Then the molecules won't push as hard against each other, and the pressure will drop. Thus, a pressure increasing disturbance to the system favors a volume decreasing change, the melting of some ice.

Next, suppose we have ice at a pressure just below the solid–liquid line, *BC*. If, *at constant temperature,* we raise the pressure to a point just above the line, the ice will melt and become a liquid. This can happen only if the melting point becomes lower as the pressure is raised.

Water is very unusual. Almost all other substances have melting points that increase with increasing pressure. Consider, for example, the phase diagram for carbon dioxide (see Figure 11.24). For CO_2 the solid–liquid line slants to the right (it slanted to the left for water). Also notice that solid carbon dioxide has a vapor pressure of 1 atm at -78 °C. This is the temperature of dry ice, which sublimes at atmospheric pressure at -78 °C.

Besides specifying phase equilibria, the three intersecting lines on a phase diagram serve to define regions of temperature and pressure at which only a single phase can exist. For example, between lines *BC* and *BD* in Figure 11.22 are temperatures and pressures at which water exists as a liquid without being in equilibrium with either vapor or ice. At 760 torr, water is a liquid anywhere between 0 °C and 100 °C. For instance, we are told by the diagram that we can't have ice with a temperature of 25 °C if the pressure is 760 torr (which, of course, you already knew; ice never has a temperature of 25 °C.) The diagram also says that we can't have water vapor with a pressure of 760 torr when the temperature is 25 °C (which, again, you already knew; the temperature has to be taken to 100 °C for the vapor pressure to reach 760 torr). Instead, we are told by the phase diagram that the *only* phase for pure water at 25 °C and 1 atm is the liquid. *Equilibrium* with other phases is impossible. Below 0 °C at 760 torr, water is a solid; it cannot be *in equilibrium* with any other phases. Above 100 °C at 760 torr, water is a vapor and cannot be *in equilibrium* with other phases. On the phase diagram of water, the phases that can exist in the different temperature–pressure regions are marked.

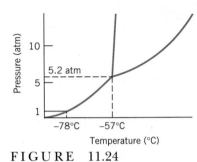

F I G U R E   11.24

The phase diagram for carbon dioxide.

EXAMPLE 11.2
Interpreting a Phase Diagram

What phase would we expect for water at 0 °C and 4.58 torr?

ANALYSIS The words, "What phase . . . " tell us to use the phase diagram of water (Figure 11.22).

SOLUTION First we find 0 °C on the temperature axis of the phase diagram of water. Then we move upward until we intersect a line corresponding to 4.58 torr. This intersection occurs in the "solid" region of the diagram. At 0 °C and 4.58 torr, then, water exists as a solid.

EXAMPLE 11.3
Interpreting a Phase Diagram

What phase changes occur if the pressure on water at 0 °C is gradually increased from 2.15 torr to 800 torr?

ANALYSIS Asking about "what phase changes occur" means that we must use the phase diagram of water (Figure 11.22).

SOLUTION At 0 °C and 2.15 torr, water exists as a gas. As we move upward along the 0 °C line, we first encounter the solid–vapor line. As we go above this line, the water will freeze and become ice. Continuing the climb at 0 °C, we next encounter the solid–liquid line at 760 torr. Above 760 torr the solid will melt. At 800 torr and 0 °C, the water will be liquid.

■ **Practice Exercise 5** What phase changes will occur if water at − 20 °C and 2.15 torr is heated to 50 °C under constant pressure?

■ **Practice Exercise 6** What phase will water be in if it is at a pressure of 330 torr and a temperature of 50 °C?

There is one final aspect of the phase diagram that has to be mentioned. The vapor pressure line for the liquid, which begins at point *B*, terminates at point *D*, which is known as the **critical point.** The temperature and pressure at *D* are called the **critical temperature,** T_c, and **critical pressure,** P_c. Above the critical temperature, a distinct liquid phase cannot exist, *regardless of the pressure.*

With the aid of Figure 11.25 we can see what happens to a substance as it approaches its critical point. In Figure 11.25*a*, we see a liquid in a container with some vapor above it. We can distinguish between the two phases because they have different densities, which causes them to bend light differently. This allows us to see the interface, or surface, between the more dense liquid and the less dense vapor. If this liquid is now heated, two things happen. First, more liquid evaporates. This causes an increase in the number of molecules per cubic centimeter of vapor, which, in turn, causes the density of the vapor to increase. Second, the liquid expands (exactly as mercury does in a thermometer). This means that a given mass of liquid occupies more volume, so its density decreases. As the temperature of the liquid and vapor continue to increase, the vapor density and the liquid density approach each other. Eventually these densities become equal, and a separate liquid phase no longer exists; everything is the same (see Figure 11.25*b*). The highest temperature at which a liquid phase still exists is the critical temperature, and the pressure of the vapor at this temperature is the critical pressure. A substance given a temperature above its critical temperature and a density near its liquid density is described as a **supercritical fluid.**

The values of the critical temperature and critical pressure are unique for every chemical substance. As with the other physical properties that we've discussed, the values T_c and P_c are controlled by the intermolecular attractions. Liquids with strong intermolecular attractions, like water, tend to have high critical temperatures. Substances with weak intermolecular attractions tend to have low critical temperatures.

More dense liquid at the bottom can be detected by the interface between the phases

(*a*)

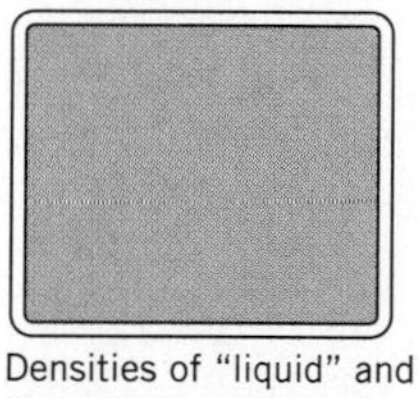

Densities of "liquid" and "vapor" have become the same - there is only one phase

(*b*)

FIGURE 11.25

Changes that are observed when a liquid is heated in a sealed container. (*a*) Below the critical temperature. (*b*) Above the critical temperature.

Above the critical temperature, a substance cannot be liquefied by the application of pressure.

Supercritical CO_2 has solvent properties that allow it to dissolve and remove caffeine from coffee beans. Some producers of decaffeinated coffee have switched from using potentially toxic solvents to the use of supercritical CO_2 because of the nontoxic nature of any traces of CO_2 that might remain in their decaffeinated product.

Solids have unique properties that are determined in large measure by the orderly packing of the particles within them. This order is revealed even when we examine the external features of crystals. For example, Figure 11.26 is a photograph of crystals of one of our most familiar chemicals, sodium chloride —table salt. Notice that each particle is very nearly a perfect little cube. You might think that the manufacturers went to a lot of trouble and expense to make such uniformly shaped crystals. Actually, they could hardly avoid it.

11.10
CRYSTALLINE SOLIDS

FIGURE 11.26

Crystals of table salt. The size of the tiny cubic sodium chloride crystals can be seen in comparison with a penny.

The way the ions are packed in NaCl is common to many compounds and is called the *rock salt structure*. (Rock salt is a common name for sodium chloride.)

The regular, symmetrical shapes of snowflakes are caused by the highly organized packing of water molecules within crystals of ice.

FIGURE 11.27

(*a*) A simple cubic unit cell showing the locations of the lattice positions. (*b*) A simple cubic unit cell with atoms having their nuclei at the corners. (*c*) A portion of a simple cubic lattice built by stacking simple cubic unit cells.

Whenever a solution of NaCl is evaporated, the crystals that form have edges that intersect at 90° angles. Cubes, then, are the norm, not the exception for NaCl.

When most substances freeze, or when they separate as a solid from a solution that is being evaporated, they normally form crystals that have highly regular features. The crystals have flat surfaces, for example, that meet at angles which are characteristic for the substance. The regularity of these surface features reflects the high degree of order among the particles that lie within the crystal. This is true whether the particles are atoms, molecules, or ions.

Crystal Lattices and Unit Cells

The particles in crystals are arranged in patterns that repeat over and over again in all directions. The resulting overall pattern is called a **crystal lattice.** Its high degree of regularity is the principal feature that makes solids different from liquids. A liquid lacks this long-range repetition of structure because the particles in a liquid are jumbled and disorganized as they move about.

Because there are millions of chemical compounds, it may seem that an enormous number of different kinds of lattices are possible. If this were true,

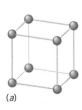

(*a*)

(*b*)

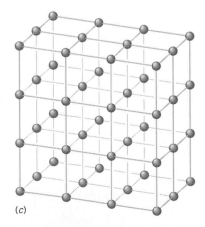

(*c*)

studying the crystal lattices of solids would be hopelessly complex. Fortunately, however, the number of *kinds* of lattices that are mathematically possible is quite limited. This has allowed for a great deal of progress in understanding the structures of solids.

To describe the structure of a crystal it is convenient to view it as being composed of a huge number of simple, basic units called **unit cells.** By repeating this simple structural unit up and down, back and forth, in all directions, we can build the entire lattice. Figure 11.27, for example, shows the simplest and most symmetrical unit cell of all, called the **simple cubic** unit cell. This unit cell is a cube having atoms (or molecules or ions) at each of its eight corners. Stacking these unit cells gives a **simple cubic lattice.**

Two other cubic unit cells are also possible: face-centered cubic and body-centered cubic. The **face-centered cubic** (abbreviated **fcc**) unit cell has identical particles at each of the corners plus another in the center of each face, as shown in Figure 11.28. Many common metals—copper, silver, gold, aluminum, and lead, for example—form crystals that have face-centered cubic lattices. Each of these metals has the same *kind* of lattice, but the *sizes* of their unit cells differ because the sizes of the atoms differ (see Figure 11.29).

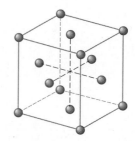

FIGURE 11.28

A face-centered cubic unit cell.

FIGURE 11.29

Two similar face-centered cubic unit cells. The atoms are arranged in the same way, but their unit cells have edges of different lengths because the atoms are of different sizes. (1Å = 1 × 10^{-10} m.)

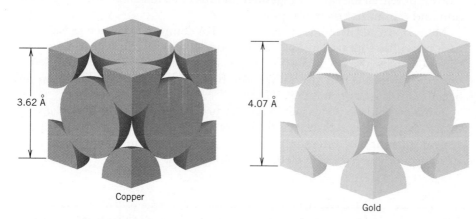

3.62 Å

Copper

4.07 Å

Gold

The **body-centered cubic (bcc)** unit cell has identical particles at each corner plus one in the center of the cell, as illustrated in Figure 11.30. The body-centered cubic lattice is also common among a number of metals—examples are chromium, iron, and platinum. Again, these are substances with the same *kinds* of lattice, but the dimensions of the lattices reflect the *different sizes* of the particular atoms.

Not all unit cells are cubic. Some have edges of different lengths or edges that intersect at angles other than 90°. Although you should be aware of the existence of other unit cells and lattices, we will limit the remainder of our discussion to cubic lattices and their unit cells.

We have seen that a number of metals have fcc or bcc lattices. The same is true for many compounds. Figure 11.31, for example, is a view of a portion of a sodium chloride crystal. The Cl^- ions (green) are located at the lattice positions that correspond to a face-centered cubic unit cell. The smaller gray spheres represent Na^+ ions. Notice that they fill the spaces between the Cl^- ions. Sodium chloride is said to have a face-centered cubic lattice, and the cubic shape of this lattice is behind the cubic shape of a sodium chloride crystal. Sodium chloride, in fact, cannot form a crystal of any other shape, given the sizes and charges of its ions and the requirement that the most stable

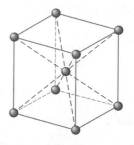

FIGURE 11.30

A body-centered cubic unit cell.

FIGURE 11.31

A face-centered cubic unit cell can be seen in the structure of NaCl.

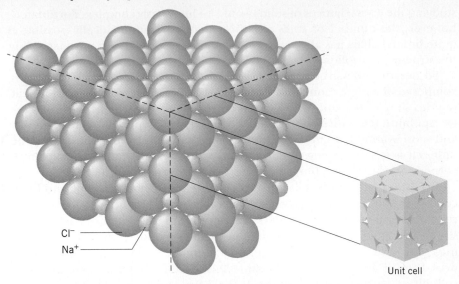

Cl⁻
Na⁺

Unit cell

(lowest potential energy) arrangement of ions must maximize forces of attraction and minimize forces of repulsion.

Many of the alkali halides (Group IA–VIIA compounds), such as NaCl and KCl, crystallize with fcc lattices. Because sodium chloride and potassium chloride both have the same kind of lattice, Figure 11.31 also could be used to describe the unit cell of KCl. The *sizes* of their unit cells are different, however, because K⁺ is a larger ion that Na⁺.

A major point to be learned from the preceding discussion is that *a single lattice type can be used to describe the structures of many different crystals.* For this reason, a handful of different lattice types is all we need to describe the crystal structures of every possible element or compound.

11.11
X-RAY DIFFRACTION

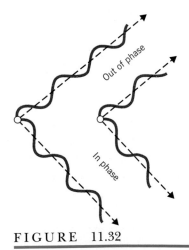

FIGURE 11.32

X rays emitted from atoms are in phase in some directions and out of phase in other directions.

When atoms are bathed in X rays, they absorb some of the radiation and then emit it again in all directions. In effect, each atom becomes a tiny X-ray source. If we look at radiation from two such atoms (see Figure 11.32), we find that the X rays emitted are in phase in some directions but out of phase in others. In Chapter 6 you learned that constructive (in-phase) and destructive (out-of-phase) interferences create a phenomenon called *diffraction.* X-Ray diffraction by crystals has enabled many scientists to win Nobel prizes by determining the structures of extremely complex compounds in a particularly elegant way.

In a crystal, there are enormous numbers of atoms, but *they are evenly spaced throughout the lattice.* When the crystal is bathed in X rays, intense beams are diffracted because of constructive interference, and they appear only in specific directions. In other directions, no X rays appear because of destructive interference. When the X rays coming from the crystal fall on photographic film, the diffracted beams form a **diffraction pattern** (see Figure 11.33). The film is darkened only where the X rays strike.[2]

In 1913, the British physicist William Henry Bragg and his son William Lawrence Bragg discovered that just a few variables control the appearance of

---

[2] Modern X-ray diffraction instruments use electronic devices to detect and measure the intensities of the diffracted X rays.

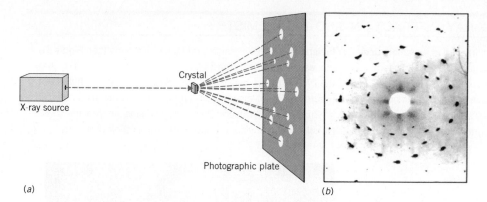

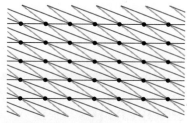

FIGURE 11.33

X-Ray diffraction. (*a*) The production of an X-ray diffraction pattern. (*b*) An X-ray diffraction pattern produced by sodium chloride.

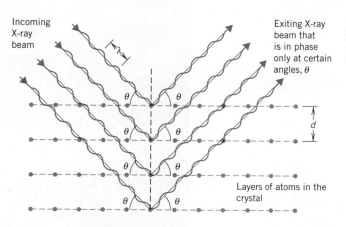

FIGURE 11.34

Diffraction of X rays from successive layers of atoms in a crystal.

an X-ray diffraction pattern. These are shown in Figure 11.34, which illustrates the conditions necessary to obtain constructive interference of the X rays from successive layers of atoms (planes of atoms) in a crystal. A beam of X rays having a wavelength λ strikes the layers at an angle θ. Constructive interference causes an intense diffracted beam to emerge at the same angle θ. The Braggs derived an equation, now called the **Bragg equation,** relating λ, θ, and the distance between the planes of atoms, d,

$$n\lambda = 2d \sin \theta \qquad (11.3)$$

where n is a whole number. The Bragg equation is the basic tool used by scientists in the study of solid structures. Let's briefly see how they use it.

In any crystal, many different sets of planes can be passed through the atoms. Figure 11.35 illustrates this idea in two dimensions for a simple pattern of points. When a crystal produces an X-ray diffraction pattern, many spots are observed because of the diffraction from the many sets of planes. The physical geometry of the apparatus used to record the diffraction pattern allows the measurement of the angles at which the diffracted beams emerge from each distinct set of planes. Because the wavelength of the X rays, the values of n (which can be determined), and the measured angles, θ, are known the distances between planes of atoms, d, can be computed. The next step is to use the calculated interplanar distances to work backward to deduce where the atoms in the crystal must be located so that layers of atoms are indeed separated by these distances. If this sounds like a difficult task, it is! Some sophisticated mathematics as well as computers are needed to accomplish it. The efforts, however, are well rewarded because the calculations give the locations of atoms within the unit cell and the distances between them. This informa-

The two Braggs shared the 1915 Nobel Prize in physics.

FIGURE 11.35

A two-dimensional pattern of points with many possible sets of parallel lines. In a three-dimensional crystal lattice there are many sets of parallel planes.

SPECIAL TOPIC 11.3 / X-RAY DIFFRACTION AND BIOCHEMISTRY

Biochemists have found that X-ray diffraction is extremely useful for studying the structures of large molecules. The most famous example of this use was the experimental determination of the molecular structure of DNA. DNA is found in the nuclei of cells and serves to carry an organism's genetic information. In 1953, using X-ray diffraction photographs of DNA fibers obtained by Rosalind Franklin and Maurice Wilkins (see Figure 1), James Watson and Francis Crick came to the conclusion that the DNA structure consists of the now-famous double helix (see Figure 2). We will examine this important structure in more detail in Chapter 21. Watson, Crick, and Wilkins shared the 1962 Nobel Prize in physiology and medicine for their discovery.

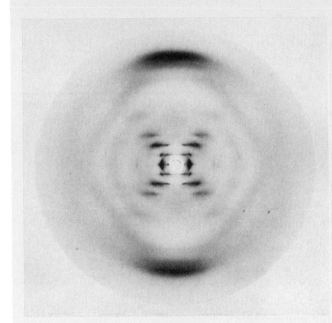

FIGURE 1
X-Ray diffraction photograph of DNA.

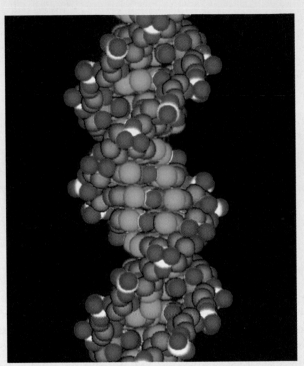

FIGURE 2
A computer-generated model of the DNA double helix.

tion, plus a lot of chemical "common sense," is used by chemists to arrive at the shapes and sizes of the molecules in the crystal. The general shape of DNA molecules, the chemicals of genes, was deduced by using X-ray diffraction (see Special Topic 11.3).

11.12 PHYSICAL PROPERTIES AND CRYSTAL TYPES

Properties of solids and crystal structure types

You know from personal experience that solids exhibit a wide range of properties. Some solids, like diamond, are very hard. Others, such as naphthalene (moth flakes) or ice, are soft by comparison, and are easily crushed. Some solids, like salt crystals or iron, have high melting points, whereas others, such as candle wax, melt at low temperatures. Some solids conduct electricity well, but others are nonconducting.

The physical properties just described depend on the kinds of particles in the solid as well as on the strengths of attractive forces holding the solid together. Even though we can't make exact predictions about such properties, some generalizations do exist. In discussing them, it is convenient to divide crystals into four types: ionic, molecular, covalent, and metallic.

Ionic Crystals

We already discussed some of the properties of **ionic crystals** in Chapter 2. They are relatively hard, have high melting points, and are brittle, properties reflecting strong attractive forces between ions of opposite charge as well as repulsions that occur when ions of like charge are near each other. You also learned that ionic compounds do not conduct electricity in their solid states but in their molten states conduct electricity well. This behavior is consistent with the mobility of ions in the liquid state, but not in the solid state.

Molecular Crystals

Molecular crystals are solids in which the lattice sites are occupied either by atoms (as in solid argon or krypton) or by molecules (as in solid CO_2, SO_2, or H_2O). Such solids tend to be soft and have low melting points because the particles in the solid experience relatively weak intermolecular attractions. In crystals of argon, for example, the attractive forces are exclusively London forces. In SO_2, which is composed of polar molecules, there are dipole–dipole attractions as well. And in water crystals (ice) the molecules are held in place by hydrogen bonds.

Covalent Crystals

Covalent crystals are solids in which lattice positions are occupied by atoms that are covalently bonded to other atoms at neighboring lattice sites. The result is a crystal that is essentially one gigantic molecule. These solids are sometimes called *network solids* because of the interlocking network of covalent bonds extending throughout the crystal in all directions. A typical example is diamond (see Figure 11.36). Covalent crystals tend to be very hard and to have very high melting points because of the strong attractions between covalently bonded atoms. Other examples of covalent crystals are quartz (SiO_2, found in typical grains of sand) and silicon carbide (SiC, a common abrasive used in sandpaper).

Metallic Crystals

Metallic crystals have properties that are quite different from those of the other three types. Metallic crystals conduct heat and electricity well, and they have the luster characteristically associated with metals. A number of different models have been developed to explain metallic crystals, and in Chapter 8 the bonding in metals is discussed in terms of the *band theory of solids*. A simpler model, however, views the lattice positions of a metallic crystal as being occupied by *positive ions* surrounded by electrons in a cloud that spreads throughout the entire solid (see Figure 11.37). The electrons in this cloud belong to no single positive ion, but rather to the crystal as a whole. Because the electrons aren't localized on any one atom, they are free to move easily, which accounts for the high electrical conductivity of metals. By their movement, the electrons can also transmit kinetic energy rapidly through the solid, so metals

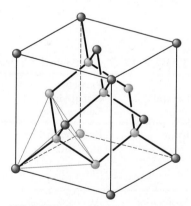

FIGURE 11.36

The structure of diamond. Notice that each carbon atom is covalently bonded to four others at the corners of a tetrahedron. This structure extends throughout an entire diamond crystal. (In diamond, of course, all the atoms are identical. They are different shades of gray here to make it easier to visualize the structure.)

FIGURE 11.37

The "electron sea" model of a metallic crystal.

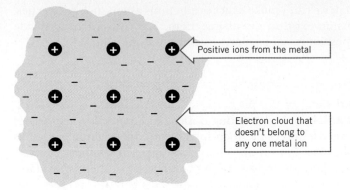

Positive ions from the metal

Electron cloud that doesn't belong to any one metal ion

are also good conductors of heat. This model explains the luster of metals, too. When light shines on the metal, the loosely held electrons vibrate easily and readily re-emit the light with essentially the same frequency and intensity.

It is not possible to make many simple generalizations about the melting points of metals. We've seen before that some, like tungsten, have very high melting points, whereas others, such as mercury, have quite low melting points. To some extent, the melting point depends on the charge of the positive ions in the metallic crystal. The atoms of the Group IA metals tend to exist as cations with a 1+ charge, and they are only weakly attracted to the "electron sea" that surrounds them. Atoms of the Group IIA metals, however, each lose two electrons to the metallic lattice and form ions with a 2+ charge. These more highly charged ions are attracted more strongly to the surrounding electron sea, so the Group IIA metals have higher melting points than their neighbors in Group IA. For example, magnesium melts at 650 °C whereas sodium, also in period 3, melts at only 98 °C. Those metals with very high melting points, like tungsten, must have very strong attractions between their

In terms of the band theory, the electrons in the "electron sea" model correspond to the electrons in the conduction band.

TABLE 11.5   Types of Crystals

Crystal Type	Particles Occupying Lattice Sites	Type of Attractive Force	Typical Examples	Typical Properties
Ionic	Positive and negative ions	Attractions between ions of opposite charge	$NaCl$, $CaCl_2$, $NaNO_3$	Relatively hard; brittle; high melting points; nonconductors of electricity as solids, but conduct when melted
Molecular	Atoms or molecules	Dipole–dipole attractions, London forces, hydrogen bonding	HCl, SO_2, N_2, Ar, CH_4, H_2O	Soft; low melting points; nonconductors of electricity in both solid and liquid states
Covalent (network)	Atoms	Covalent bonds between atoms	Diamond, SiC (silicon carbide), SiO_2 (sand, quartz)	Very hard; very high melting points; nonconductors of electricity
Metallic	Positive ions	Attractions between positive ions and an electron cloud that extends throughout the crystal	Cu, Ag, Fe, Na, Hg	Range from very hard to very soft; melting points range from high to low; conduct electricity in both solid and liquid states; have characteristic luster

atoms, which suggests that there probably is some covalent bonding between them as well.

The different ways of classifying crystals and a summary of their general properties are given in Table 11.5.

The elements of Group IIA are also harder than their neighbors in Group IA.

The metal osmium, Os, forms an oxide with the formula OsO_4. The soft crystals of OsO_4 melt at 40 °C, and the resulting liquid does not conduct electricity. In what form does OsO_4 probably exist in the solid?

ANALYSIS You might be tempted to suggest that the compound is ionic simply because it is formed from a metal and a nonmetal. However, this is not always the case, especially when the metal has a large positive charge. Therefore, we must examine the properties of the compound and find the kind of solid that fits the properties.

SOLUTION The characteristics of the OsO_4 crystals—softness and low melting point —suggest that solid OsO_4 exists as molecular crystals that contain molecules of OsO_4. This is further supported by the fact that liquid OsO_4 does not conduct electricity, which is evidence for the lack of ions in the liquid.

■ **Practice Exercise 7** Boron nitride, which has the empirical formula BN, melts under pressure at 3000 °C and is as hard as a diamond. What is the probable crystal type for this compound?

■ **Practice Exercise 8** Crystals of elemental sulfur are easily crushed and melt at 113 °C to give a clear yellow liquid that does not conduct electricity. What is the probable crystal type for solid sulfur?

EXAMPLE 11.4
Identifying Crystal Types from Physical Properties

FIGURE 11.38

Pieces of broken glass have sharp edges, but their surfaces are not flat planes.

If a cubic salt crystal is broken, the pieces still have flat faces that intersect at 90° angles. If you shatter a piece of glass, on the other hand, the pieces often have surfaces that are not flat. Instead, they tend to be smooth and curved (see Figure 11.38). This behavior illustrates a major difference between crystalline solids, like NaCl, and noncrystalline solids, or **amorphous solids,** such as glass.

The word *amorphous* is derived from the Greek word *amorphos,* which means "without form." Amorphous solids do not have the kinds of long-range repetitive internal structures that are found in crystals. In some ways their structures, being jumbled, are more like liquids than solids. Examples of amorphous solids are ordinary glass and many plastics. In fact, the word **glass** is often used as a general term to refer to any amorphous solid.

Substances that form amorphous solids often consist of long, chainlike molecules that are intertwined in the liquid state somewhat like long strands of cooked spaghetti (see Figure 11.39). To form a crystal from the melted material, these long molecules would have to become untangled and line up in specific patterns. But as the liquid is cooled, the molecules slow down. Unless the liquid is cooled extremely slowly, the molecular motion decreases too rapidly for the untangling to take place, and the substance solidifies with the molecules still intertwined (see Special Topic 11.4).

Compared with substances that produce amorphous solids, those that form crystalline solids behave quite differently when they are cooled. First, substances that form crystals solidify at a constant temperature, as you have seen

11.13
NONCRYSTALLINE SOLIDS

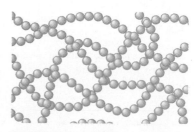

FIGURE 11.39

In an amorphous solid, long molecules are tangled and disorganized, so there is no long-range order characteristic of a crystal.

SPECIAL TOPIC 11.4 / CRYSTALLINE STRUCTURE AND THE PHYSICAL PROPERTIES OF PLASTICS

When you think of a plastic like polyethylene (the plastic used to make garbage bags and sandwich bags), physical strength and toughness are not properties that come to mind. In fact, most plastics we encounter are relatively soft, or they break easily. Yet, if formed and processed properly, some plastics, including polyethylene, can be as hard and strong as steel.

Polyethylene is formed by joining many ethylene molecules, C_2H_4, end to end to give long chains composed of CH_2 units that repeat thousands of times. Left to themselves, these chains fold and coil randomly to give an amorphous solid in which the individual polymer strands are jumbled together like a plate of cooked spaghetti. The resulting plastic is soft and lacks physical toughness.

In the early 1970s it was discovered by accident that by proper processing, polymer strands could be made to line up in a semicrystalline state. The result was called an oriented polymer, and these plastics have a whole new set of properties.

The first commercial oriented polymer, known as Kevlar, was produced by DuPont. Its molecules have structures that resemble stiff chains of connected carbon rings. Kevlar fibers are used to make bulletproof vests and helmets for the military as well as thin, strong, lightweight hulls for racing boats.

More recently, Allied-Signal Company produced an oriented polyethylene polymer they call Spectra. Weight for weight it is about 30 to 40% stronger than Kevlar. Spectra fibers are flexible

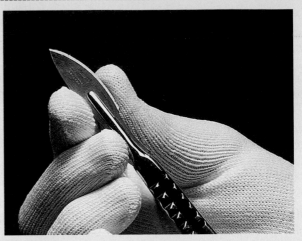

FIGURE 1

Surgical gloves made of Spectra fibers resist cuts by scalpels and protect a surgeon's hands.

and can be woven to produce a fabric that is very resistant to cuts by a sharp knife (see Figure 1). It is used to make thin, lightweight liners for surgical gloves, which resist cuts by scalpels, as well as industrial work gloves. Mixed with other plastics, it can be molded into strong, rigid forms such as helmets for military or sporting applications.

Some scientists prefer to reserve the term solid *for crystalline substances, so they refer to glass as a* supercooled liquid.

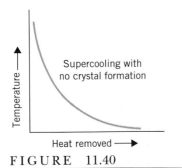

FIGURE 11.40

A cooling curve for a liquid that forms an amorphous solid. There is no sharply defined freezing point or melting point.

in Figure 11.20*b* on page 461. You also learned that sometimes the liquid can be cooled below its freezing point, and supercooling takes place. For most liquids, though, crystallization usually begins before long. But with substances that form amorphous solids, this never happens. The molecules cannot become untangled before they are frozen in place at a low temperature. Therefore, amorphous solids are sometimes described as "supercooled liquids," a term suggesting the kind of structural disorder found in liquids. The term also suggests that the material's constituent molecules retain some residual ability at least to flex their chains if not to diffuse throughout the material and, over a long period of time, achieve an improved degree of crystalline orderliness. But the rate at which such change occurs is typically so small that under ordinary conditions it cannot be observed and the material is fully rigid. Glass, for example, which is a typical amorphous solid, over a very long time will develop regions of higher and higher order, as revealed by X-ray diffraction patterns of old glass. Figure 11.40 shows a typical cooling curve for an amorphous solid. No sharp breaks occur in the curve as the material is cooled until it becomes rigid.

Amorphous solids also soften gradually when heated. This is the reason you can heat glass tubing in a flame to soften it so that you can bend it. By contrast, if you warm an ice cube (crystalline water), it won't become soft gradually; instead, at 0 °C it will suddenly melt and drip all over you!

SUMMARY

Physical Properties: Gases, Liquids, and Solids Although the chemical properties of substances are very important, physical properties often have a greater effect on our daily lives. Most physical properties depend primarily on intermolecular attractions. In gases, these attractions are weak because the molecules are so far apart. They are much stronger in liquids and solids, whose particles are packed together tightly.

Intermolecular Attractions Polar molecules attract each other primarily by **dipole–dipole attractions,** which arise because the positive end of one dipole attracts the negative end of another. Nonpolar molecules are attracted to each other by **London forces,** which are **instantaneous dipole–induced dipole attractions.** London forces are present between all particles, including atoms, polar and nonpolar molecules, and ions. London forces increase with increasing size of the particle's electron cloud; they also increase with increasing chain length among such molecules as the hydrocarbons. **Hydrogen bonding,** a special case of dipole–dipole attractions, occurs between molecules in which hydrogen is covalently bonded to a small, very electronegative atom—principally, nitrogen, oxygen, or fluorine. Hydrogen bonding is much stronger than the other types of intermolecular attractions.

General Properties of Liquids and Solids Properties that depend mostly on closeness of packing of particles are **compressibility** (or the opposite, incompressibility) and **diffusion.** Diffusion is slow in liquids and almost nonexistent in solids at room temperature. Properties that depend mostly on the strengths of intermolecular attractions are **retention of volume and shape, surface tension,** and **ease of evaporation.** Both solids and liquids retain volume; solids retain shape when transferred from one vessel to another. Surface tension is related to the energy needed to expand a liquid's surface area. A liquid can **wet** a surface if its molecules are attracted to the surface about as strongly as they are attracted to each other. Evaporation of liquids and solids is endothermic and produces a cooling effect. The overall rate of evaporation increases with increasing surface area. The rate of evaporation from a given surface area of a liquid increases with increasing temperature and decreasing intermolecular attractions. Evaporation of a solid is called **sublimation.**

The resistance that a liquid (or a solid) has to change in its physical form, its **viscosity,** is higher when the intermolecular attractions are greater or when the molecules entangle with each other. Viscosity decreases with increasing temperature.

Changes of State Changes from one physical state to another, such as melting, vaporization, or sublimation, can occur as dynamic equilibria. In a **dynamic equilibrium,** opposite processes occur at equal rates, so there is no apparent change in the composition of the system. For liquids and solids, equilibria are established when vaporization occurs in a sealed container. A solid is in equilibrium with its liquid at the melting point.

Vapor Pressures When the rates of evaporation and condensation of a liquid are equal, the vapor exerts a pressure called the **equilibrium vapor pressure** (or more commonly, just the **vapor pressure**). The vapor pressure is controlled by the rate of evaporation *per unit surface area.* When the intermolecular attractive forces are large, the rate of evaporation is small and the vapor pressure is small. Vapor pressure increases with increasing temperature because the rate of evaporation increases as the temperature rises. The vapor pressure is independent of the *total* surface area of the liquid because as the surface area increases, the rates of evaporation and condensation are both increased equally; there is no change in the number of molecules in the vapor, so there is no change in the vapor pressure. Solids have vapor pressures just as liquids do.

Boiling Point A substance boils when its vapor pressure equals the prevailing atmospheric pressure. The **normal boiling point** of a liquid is the temperature at which its vapor pressure equals 1 atm. Substances with high boiling points have strong intermolecular attractions.

Energy Changes Associated with Changes of State On a heating curve, flat portions correspond to phase changes in which the heat added changes the potential energies of the particles without changing their average kinetic energy. On a cooling curve, **supercooling** occurs sometimes when the temperature of the liquid drops below the freezing point of the substance. The enthalpy changes for melting, vaporization of a liquid, and sublimation are the **molar heat of fusion,** ΔH_{fusion}, the **molar heat of vaporization,** $\Delta H_{vaporization}$, and the **molar heat of sublimation,** $\Delta H_{sublimation}$, respectively. They are all endothermic and are related as follows: $\Delta H_{fusion} < \Delta H_{vaporization} < \Delta H_{sublimation}$. The sizes of these enthalpy changes are large for substances with strong intermolecular attractive forces. The heat of vaporization is related to the vapor pressure by the **Clausius–Clapeyron equation.**

Le Châtelier's Principle When the equilibrium in a system is upset by a disturbance, the system changes in a direction that minimizes the disturbance and, if possible, brings the system back to equilibrium. By this principle, we find that raising the temperature favors an endothermic change. Decreasing the volume favors a change toward a less dense phase.

Phase Diagrams Temperatures and pressures at which equilibria can exist between phases are given graphically in a **phase diagram.** The three equilibrium lines intersect at the **triple point.** The liquid–vapor line terminates at the **critical point.** At the **critical temperature,** a liquid has a vapor pressure equal to its **critical pressure.** Above the critical temperature a liquid phase cannot be formed; the single phase that exists is called a **supercritical fluid.** The equilibrium lines also divide a phase diagram into temperature–pressure regions in which a substance can exist in just a single phase. Water is different from most substances in that its melting point decreases with increasing pressure.

Crystalline Solids Crystalline solids have very regular features that are determined by the highly ordered arrangements of particles within their solids, which can be described in terms of repeating three-dimensional arrays called **lattices.** The simplest portion of the lattice is the **unit cell.** Three cubic unit cells are possible—**simple cubic, face-centered cubic,** and **body-centered cubic.** Many differ-

ent substances can have the same kind of lattice. Information about crystal structures are obtained experimentally from X-ray diffraction patterns produced when crystals are bathed in X rays. Distances between planes of atoms can be calculated by the Bragg equation, $n\lambda = 2d \sin \theta$, where n is a whole number, λ is the wavelength of the X rays, d is the distance between planes of atoms producing the diffracted beam, and θ is the angle at which the diffracted X ray beam emerges relative to the planes of atoms producing the diffracted beam.

Crystal Types Crystals can be divided into four general types: **ionic, molecular, covalent,** and **metallic.** Their properties depend on the kinds of particles within the lattice and on the attractions between the particles, as summarized in Table 11.5. A noncrystalline or **amorphous solid** is formed when a liquid is cooled without crystallization occurring. Such a solid has no sharply defined melting point and is called a **supercooled liquid.**

Tools you have learned

The table below lists the tools you have learned in this chapter that are applicable to problem solving. Review them if necessary, and refer to them when working on the Thinking-It-Through problems and the Review Exercises that follow.

Tool	Function
Chemical structure, boiling points, and vapor pressures	Assess relative strengths of intermolecular attractions: —from structure (pages 443 and 452) —from boiling points (pages 445, 447, and 460) —from vapor pressures (page 457).
Clausius–Clapeyron equation (page 463)	Determine heat of vaporization from vapor pressure–temperature data. Estimate vapor pressures.
Molar heats of fusion, vaporization, and sublimation	Assess relative strengths of intermolecular attractions (page 466).
Le Châtelier's principle (page 467)	Predict how an equilibrium will respond to a disturbance.
Phase diagrams (page 468)	Determine the combinations of temperature and pressure under which a substance can exist as a solid, a liquid, or a gas or under which two phases can be in equilibrium.
Physical properties of solids (page 476)	Judge type of crystal structure.

THINKING IT THROUGH

Remember, you are not asked to obtain answers for the following problems. Instead, assemble the data necessary to solve the problems and describe how you would use the data to obtain the answers. For numerical problems, set up the calculation using appropriate conversion factors.

The problems are divided into two groups. Those in Level 2 are significantly more challenging than those in Level 1 and provide an opportunity to really hone your problem solving skills.

Level 1 Problems

1. When hot coffee in an insulated cup gradually cools, the cooler liquid at the surface circulates to the bottom as fresh hot liquid rises to the top. Why?

2. Vigorous exercise on a hot muggy day produces more discomfort than the same exercise on an equally hot but dry day. Why?

3. If you pour out half of the contents of a filled can of gasoline and then seal the container, you will find that when the container is reopened sometime later there is a noticeable hiss as gases escape from the can. Why is there an increase in the pressure inside the can even if its temperature hasn't changed?

4. London forces are generally weaker than hydrogen bonds, yet water boils at 100 °C, whereas $C_{20}H_{42}$ (a component of candle wax) boils at 343 °C. Why?

5. At a given temperature, which of the liquids below is likely to have the greater vapor pressure?

(CH₃CH₂)₂O   (CH₃CH₂)₂NH

6. At a given temperature, which of the liquids below is likely to have the higher boiling point?

7. A flask is partially filled with water and fitted with a stopcock. With the stopcock open, the water is brought to a boil. After a few minutes heating is stopped and the stopcock is closed. The flask is allowed to cool for a short time and then ice water is poured over the top of it. This causes the water to boil vigorously! Why?

8. Why are indoor relative humidities generally very low in Minnesota in winter, even when it is snowing outside?

9. Why do clouds form when the humid air of a weather system known as a *warm front* encounters the cool, relatively dry air of a *cold front*?

10. When a liquid is spread over a larger surface area, it evaporates more rapidly. Why, then, does the surface area of a liquid in a sealed container not influence the equilibrium vapor pressure of the liquid?

11. Some liquid water is added to a large vessel. After a wait of a time period sufficient for the water to have reached equilibrium with its vapor it was discovered that the vapor pressure of the water was less than the equilibrium vapor pressure. What is the most probable reason for this?

12. When one mole of a liquid freezes, less energy is given off than when one mole of the substance condenses from a vapor to a liquid. Why?

13. Which of the following is likely to have the larger molar heat of vaporization?

14. Which of the following is likely to have the larger molar heat of vaporization?

15. Room temperature is above hydrogen's critical temperature. If H_2 at a pressure of 1 atm is compressed at room temperature, will a pressure eventually be reached at which it will condense to a liquid?

Level 2 Problems

16. If a gas such as Freon, CCl_2F_2, is allowed to expand freely into a vacuum, it cools slightly. Why does this cooling prove there are attractive forces between the CCl_2F_2 molecules?

17. Calcium oxide crystallizes with the same structure as NaCl. The radius of a Ca^{2+} ion is approximately 99 pm and that of an O^{2-} ion is approximately 140 pm. Using these values, estimate the density of CaO.

18. Melting point is sometimes used as an indication of the extent of covalent bonding in a compound—the higher the melting point, the more ionic the substance. On this basis, oxides of metals seem to become less ionic as the charge on the metal ion increases. Thus, Cr_2O_3 has a melting point of 2266 °C, whereas CrO_3 has a melting point of only 196 °C. The explanation often given is similar in some respects to explanations of the variations in the strengths of certain intermolecular attractions given in this chapter. The goal of this question, therefore, is to provide an explanation for the greater degree of electron sharing in CrO_3 as compared with Cr_2O_3.

REVIEW EXERCISES

Answers to questions whose numbers are printed in color are given in Appendix D. More challenging questions are marked with asterisks.

Comparisons among the States of Matter

11.1 List four items in your room whose physical properties account for their use in a particular application.

11.2 Under what conditions would we expect gases to obey the gas laws best?

11.3 Why is the behavior of a gas affected very little by its chemical composition?

11.4 Why are the intermolecular attractive forces stronger in liquids and solids than they are in gases?

Intermolecular Attractions

11.5 Describe dipole–dipole attractions.

11.6 What are London forces? How are they affected by molecular size?

11.7 What are hydrogen bonds?

11.8 Which nonmetals, besides hydrogen, are most often involved in hydrogen bonding? Why these and not others?

11.9 Which is expected to have the higher boiling point, C_8H_{18} or C_4H_{10}? Explain your choice.

11.10 Ethanol and dimethyl ether have the same molecular formula, C_2H_6O. Ethanol boils at 78.4°C, whereas dimethyl ether boils at -23.7 °C. Their structural formulas are

$$CH_3CH_2OH \qquad CH_3OCH_3$$
ethanol   dimethyl ether

Explain why the boiling point of the ether is so much lower than the boiling point of ethanol.

11.11 How do the strengths of covalent bonds and dipole–dipole attractions compare? How do the strengths of ordinary dipole–dipole attractions compare with the strengths of hydrogen bonds?

11.12 Explain why London forces are called instantaneous dipole–induced dipole forces.

11.13 What are *ion–induced dipole* attractions?

11.14 What kinds of intermolecular attractive forces (dipole–dipole, London, hydrogen bonding) are present in the following substances? (a) HF (b) CS_2 (c) PCl_3 (d) SF_6 (e) SO_2

General Properties of Liquids and Solids

11.15 Name two physical properties of liquids and solids that are controlled primarily by how tightly packed the particles are. Name three that are controlled mostly by the strengths of the intermolecular attractions.

11.16 Why does diffusion occur more slowly in liquids than in gases? Why does diffusion occur extremely slowly in solids?

11.17 Compare how shape and volume change when (a) a gas, (b) a liquid, and (c) a solid is transferred from one container to another.

11.18 Why are liquids and solids so difficult to compress?

11.19 On the basis of kinetic theory, would you expect the rate of diffusion in a liquid to increase or decrease as the temperature is increased? Explain your answer.

11.20 What is surface tension? Why do molecules at the surface of a liquid behave differently from those within the interior?

11.21 What kinds of observable effects are produced by the surface tension of a liquid?

11.22 What relationship is there between surface tension and the intermolecular attractions in the liquid?

11.23 Which liquid is expected to have the larger surface tension at a given temperature, CCl_4 or H_2O? Explain your answer.

11.24 What does *wetting* of a surface mean? What is a *surfactant*? What is its purpose and how does it function?

11.25 Polyethylene plastic consists of long chains of carbon atoms, each of which is also bonded to hydrogens, as shown below:

$$\cdots-\overset{\displaystyle H}{\underset{\displaystyle H}{\overset{|}{\underset{|}{C}}}}-\overset{\displaystyle H}{\underset{\displaystyle H}{\overset{|}{\underset{|}{C}}}}-\overset{\displaystyle H}{\underset{\displaystyle H}{\overset{|}{\underset{|}{C}}}}-\overset{\displaystyle H}{\underset{\displaystyle H}{\overset{|}{\underset{|}{C}}}}-\overset{\displaystyle H}{\underset{\displaystyle H}{\overset{|}{\underset{|}{C}}}}-\overset{\displaystyle H}{\underset{\displaystyle H}{\overset{|}{\underset{|}{C}}}}-\overset{\displaystyle H}{\underset{\displaystyle H}{\overset{|}{\underset{|}{C}}}}-\overset{\displaystyle H}{\underset{\displaystyle H}{\overset{|}{\underset{|}{C}}}}-\cdots$$

Water forms beads when placed on a polyethylene surface. Why?

11.26 The structural formula for glycerol is

$$H-\overset{\displaystyle H}{\underset{\displaystyle OH}{\overset{|}{\underset{|}{C}}}}-\overset{\displaystyle H}{\underset{\displaystyle OH}{\overset{|}{\underset{|}{C}}}}-\overset{\displaystyle H}{\underset{\displaystyle OH}{\overset{|}{\underset{|}{C}}}}-H$$

Would you expect this liquid to wet glass surfaces? Explain your answer.

11.27 On the basis of what happens on a molecular level, why does evaporation lower the temperature of a liquid?

11.28 On the basis of the distribution of kinetic energies of the molecules of a liquid, explain why increasing the liquid's temperature increases the rate of evaporation.

11.29 How is the rate of evaporation of a liquid affected by increasing the surface area of the liquid? How is the rate of evaporation affected by the strengths of intermolecular attractive forces?

11.30 Which liquid evaporates faster at 25 °C, butanol or diethyl ether? Both have the molecular formula, $C_4H_{10}O$, but their structural formulas are different, as shown below.

$$CH_3CH_2CH_2CH_2OH \qquad CH_3CH_2-O-CH_2CH_3$$
butanol diethyl ether

11.31 During the cold winter months, snow often disappears gradually without melting. How is this possible? What is name of the process responsible for this phenomenon?

11.32 How is freeze-drying accomplished? What are its advantages?

11.33 Syrup is much more viscous at room temperature than olive oil, yet syrup dissolves in water and olive oil does not. What kinds of intermolecular attractions are likely to be prevalent in syrup that are not present in olive oil? Explain.

Changes of State and Equilibrium

11.34 What is a ''change of state''?

11.35 Why do molecules of a vapor that collide with the surface of a liquid tend to be captured by the liquid?

11.36 Under what conditions is an equilibrium established in the evaporation of a liquid? Why do we call it a *dynamic equilibrium*?

11.37 Under what conditions is a dynamic equilibrium established between the liquid and solid forms of a substance?

11.38 What is the temperature called at which there is an equilibrium between a liquid and a solid?

11.39 Is it possible to establish an equilibrium between a solid and its vapor?

Vapor Pressure

11.40 Define *equilibrium vapor pressure*. Why do we call the equilibrium involved a *dynamic equilibrium*?

11.41 Explain why changing the volume of a container in which there is a liquid–vapor equilibrium has no effect on the vapor pressure.

11.42 What effect does increasing the temperature have on the vapor pressure of a liquid?

11.43 Which compound should have the higher vapor pressure at 25 °C, butanol or diethyl ether? Their structures are given in Review Exercise 11.30

11.44 Below are the vapor pressures of some relatively common chemicals measured at 20 °C. Arrange these substances in order of increasing intermolecular attractive forces.

Benzene, C_6H_6	80 torr
Acetic acid, $HC_2H_3O_2$	11.7 torr
Acetone, C_3H_6O	184.8 torr
Diethyl ether, $C_4H_{10}O$	442.2 torr
Water	17.5 torr

11.45 Why does humid air at 30 °C contain more moisture per cubic meter than humid air at 25 °C?

11.46 Why does rain often form when humid air is forced to rise over a mountain range?

11.47 Why does moisture condense on the outside of a cool glass of water in the summertime?

11.48 Why do we feel more uncomfortable in humid air at 90 °F than in dry air at 90 °F?

11.49 Why does the air in a heated building in the winter have such a low humidity?

11.50 Why doesn't increasing the surface area of a liquid cause an increase in the equilibrium vapor pressure?

Boiling Points of Liquids

11.51 Define *boiling point* and *normal boiling point*.

11.52 Why does the boiling point vary with atmospheric pressure?

11.53 Mt. Kilimanjaro in Tanzania is the tallest peak in Africa (19,340 ft). The normal barometric pressure at the top of this mountain is about 345 torr. At what Celsius temperature would water be expected to boil there? (See Figure 11.17.)

11.54 Why is the boiling point a useful property with which to identify liquids?

11.55 Which should have the higher boiling point, diethyl ether or butanol? Their structures are given in Review Exercise 11.30

11.56 The boiling points of some common substances are given here. Arrange these substances in order of increasing strengths of intermolecular attractions.

Ethanol, C_2H_5OH	78.4 °C
Ethylene glycol, $C_2H_4(OH)_2$	197.2 °C
Water	100 °C
Diethyl ether, $C_4H_{10}O$	34.5 °C

11.57 Explain how a pressure cooker works.

11.58 The radiator cap of an automobile engine is designed to maintain a pressure of approximately 15 lb/in.^{2} above normal atmospheric pressure. How does this help prevent the engine from ''boiling over'' in hot weather?

11.59 Butane, C_4H_{10}, has a boiling point of -0.5 °C (which is 31 °F). Despite this, liquid butane can be seen sloshing about inside a typical butane lighter, even at room temperature. Why isn't the butane boiling inside the lighter at room temperature?

11.60 Why does H_2S have a lower boiling point than H_2Se? Why does H_2O have a much higher boiling point than H_2S?

11.61 Explain why HF has a lower boiling point than H_2O, even though HF is more polar than H_2O and forms stronger hydrogen bonds.

Energy Changes That Accompany Changes of State

*11.62 Below is a cooling curve for one mole of a substance.

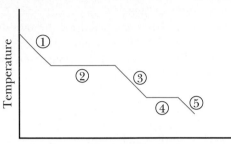

(a) On which portions of this graph do we find the average kinetic energy of the molecules of the substance changing?
(b) On which portions of this graph do we find the potential energy changing?
(c) Which portion of the graph corresponds to the release of the heat of vaporization?
(d) Which portion of the graph corresponds to the release of the heat of fusion?
(e) Which is larger, the heat of fusion or the heat of vaporization?
(f) On the graph, indicate the melting point of the solid.
(g) On the graph, indicate the boiling point of the liquid.
(h) On the drawing, indicate how supercooling would affect the graph.

11.63 What is the name and symbol given to the enthalpy change involved in (a) the conversion of one mole of liquid to one mole of vapor, (b) the conversion of one mole of solid to one mole of vapor, and (c) the conversion of one mole of solid to one mole of liquid?

11.64 The molar heat of vaporization of water at 25 °C is +43.9 kJ/mol. How many kilojoules of heat would be required to vaporize 125 mL (0.125 kg) of water?

11.65 The molar heat of vaporization of acetone, C_3H_6O, is 30.3 kJ/mol at its boiling point. How many kilojoules of heat would be liberated by the condensation of 5.00 g of acetone?

11.66 Why is $\Delta H_{vaporization}$ larger than ΔH_{fusion}? How does $\Delta H_{sublimation}$ compare with $\Delta H_{vaporization}$?

11.67 Would the "heat of condensation," $\Delta H_{condensation}$, be exothermic or endothermic?

11.68 Which would be expected to have a higher molar heat of vaporization, water (bp 100 °C) or ethyl alcohol (bp 78.4 °C)?

11.69 Rain forms when water vapor condenses in clouds. Explain why this is the source of energy for producing winds in storms.

11.70 Ethanol (grain alcohol) has a molar heat of vaporization of 39.3 kJ/mol. Ethyl acetate, a common solvent, has a molar heat of vaporization of 32.5 kJ/mol. Which of these substances has the larger intermolecular attractions?

11.71 Acetic acid has a heat of fusion of 10.8 kJ/mol and a heat of vaporization of 24.3 kJ/mol.

$$HC_2H_3O_2(s) \longrightarrow HC_2H_3O_2(l) \qquad \Delta H_{fusion} = 10.8 \text{ kJ/mol}$$
$$HC_2H_3O_2(l) \longrightarrow HC_2H_3O_2(g) \quad \Delta H_{vaporization} = 24.3 \text{ kJ/mol}$$

Use Hess's law to estimate the value for the heat of sublimation of acetic acid, in kilojoules per mole.

11.72 Suppose 45.0 g of water at 85 °C is added to 105.0 g of ice at 0 °C. The molar heat of fusion of water is 6.01 kJ/mol, and the specific heat of water is 4.18 J/g °C. Based on these data, (a) what will be the final temperature of the mixture and (b) how many grams of ice will melt?

11.73 A burn caused by steam is much more serious than one caused by the same amount of boiling water. Why?

11.74 Arrange the following substances in order of their increasing values of $\Delta H_{vaporization}$: (a) HF, (b) CH_4, (c) CF_4, (d) HCl.

Clausius–Clapeyron Equation

11.75 Ethyl acetate, a solvent for certain plastics, has a vapor pressure of 72.8 torr at 20 °C and a vapor pressure of 186.2 torr at 40 °C. Estimate the value of ΔH_{vap} for this solvent in units of kilojoules per mole.

11.76 Isopropyl alcohol, used in rubbing alcohol, has a heat of vaporization of 42.09 kJ/mol and a vapor pressure of 31.6 torr at 20 °C. What is the vapor pressure of this alcohol at 60 °C?

Le Châtelier's Principle

11.77 State Le Châtelier's principle in your own words.

11.78 What do we mean by the "position of equilibrium"?

11.79 Use Le Châtelier's principle to predict the effect of adding heat in the equilibrium: solid + heat $\rightleftharpoons$ liquid.

Phase Diagrams

11.80 Sketch the phase diagram for a substance that has a triple point at −15.0 °C and 0.30 atm, melts at −10.0 °C at 1 atm, and has a normal boiling point of 90 °C.

11.81 Based on the phase diagram of Review Exercise 11.80, below what pressure will the substance undergo sublimation?

11.82 Based on the phase diagram in Review Exercise 11.80, how does the density of the liquid compare with the density of the solid?

11.83 For most substances, the solid is more dense than the liquid. Use Le Châtelier's principle to explain why the melting point of such substances should *increase* with increasing pressure.

11.84 Define *critical temperature* and *critical pressure*.

11.85 What phases of a substance are in equilibrium at the triple point?

11.86 Why doesn't CO_2 have a normal boiling point?

11.87 According to Figure 11.24, what phase(s) should exist for CO_2 at: (a) -60 °C and 6 atm, (b) -60 °C and 2 atm, (c) -40 °C and 10 atm, and (d) -57 °C and 5.2 atm?

11.88 Describe what happens when CO_2 at 2 atm is warmed from -80 °C to 0 °C at constant pressure.

11.89 Describe what happens when the pressure on CO_2 is gradually raised from 1 atm to 20 atm at a constant temperature of -56.8 °C. What happens if the pressure is increased from 1 atm to 20 atm at a constant temperature of -58 °C?

11.90 Looking at the phase diagram for CO_2 (Figure 11.24), how can we tell that solid CO_2 is more dense than liquid CO_2?

11.91 If a substance exists as a liquid at room temperature and atmospheric pressure, what can we say about its triple point?

11.92 At room temperature, hydrogen can be compressed to very high pressures without liquefying. On the other hand, butane becomes a liquid at high pressure (at room temperature). What does this tell us about the critical points of hydrogen and butane?

Crystalline Solids and X-Ray Diffraction

11.93 What surface features do crystals have that suggest a high degree of order among the particles within them?

11.94 What is a *crystal lattice?* What is a *unit cell?* What relationship is there between a unit cell and a crystal lattice?

11.95 Describe simple cubic, face-centered cubic, and body-centered cubic unit cells. (Make a sketch of each.)

11.96 Make a sketch of a layer of sodium ions and chloride ions in a NaCl crystal. Indicate how the ions are arranged in a face-centered cubic pattern.

11.97 How do the crystal structures of copper and gold differ? In what way are they similar? Based on the location of the elements in the periodic table, what kind of crystal structure would you expect for silver?

11.98 Only 14 different kinds of crystal lattices are possible. How can this be true, considering the fact that there are millions of different chemical compounds that are able to form crystals?

11.99 Write the Bragg equation and define the symbols.

11.100 Explain, in general terms, how an X-ray diffraction pattern of a crystal and the Bragg equation provide information that allows chemists to figure out the structures of molecules.

11.101 How many atoms are contained within a simple cubic unit cell? (Hint: To answer the question, add up the parts of atoms shown in Figure 11.27*b*.)

11.102 How many copper atoms are within the face-centered cubic unit cell of copper? (Hint: See Figure 11.29, and add up all the *parts* of atoms in the fcc unit cell.)

11.103 Copper crystallizes with a face-centered cubic unit cell. The length of the edge of a unit cell is 362 pm. Sketch the face of a unit cell, showing the nuclei of the copper atoms at the lattice points. The atoms are in contact along the diagonal from one corner to another. The length of this diagonal is four times the radius of a copper atom. What is the atomic radius of copper?

11.104 Silver forms face-centered cubic crystals. The atomic radius of a silver atom is 144 pm. Draw the face of a unit cell with the nuclei of the silver atoms at the lattice points. The atoms are in contact along the diagonal. Calculate the length of an edge of this unit cell.

*11.105 Why can't $CaCl_2$ or $AlCl_3$ form crystals with the same structure as NaCl? (Hint: Count the number of ions of Na^+ and Cl^- in the unit cell of NaCl.)

Crystal Types

11.106 What kinds of particles are located at the lattice sites in a metallic crystal?

11.107 What kinds of attractive forces exist between particles in (a) molecular crystals, (b) ionic crystals, and (c) covalent crystals?

11.108 Why are covalent crystals sometimes called network solids?

11.109 Tin(IV) chloride, $SnCl_4$, has soft crystals with a melting point of -30.2 °C. The liquid is nonconducting. What type of crystal is formed by $SnCl_4$?

11.110 Magnesium chloride is ionic. What general physical properties are expected of its crystals?

11.111 Elemental boron is a semiconductor, is very hard, and has a melting point of about 2250 °C. What type of crystal is formed by boron?

11.112 Gallium crystals are shiny and conduct electricity. Gallium melts at 29.8 °C. What type of crystal is formed by gallium?

11.113 Titanium(IV) bromide forms soft orange-yellow crystals that melt at 39 °C to give a liquid that doesn't conduct electricity. The liquid boils at 230 °C. What type of crystals does $TiBr_4$ form?

11.114 Columbium is another name for one of the elements. This element is shiny, soft, and ductile. It melts at 2468 °C and the solid conducts electricity. What kind of solid does columbium form?

11.115 Elemental phosphorus consists of soft white "waxy" crystals that are easily crushed and melt at 44 °C. The solid does not conduct electricity. What type of crystal does phosphorus form?

Amorphous Solids

11.116 What does the word *amorphous* mean?

11.117 What is an amorphous solid? What happens when a substance that forms an amorphous solid is cooled from the liquid state to the solid state?

11.118 Why is glass sometimes called a supercooled liquid?

11.119 Compare what happens when glass is heated in a flame with what occurs when a crystal, such as ice, is heated.

Additional Exercises

11.120 Make a list of *all* of the attractive forces that exist in solid Na_2SO_3.

11.121 Should acetone molecules be attracted to water molecules more strongly than to other acetone molecules? Explain your answer. The structure of acetone is shown below.

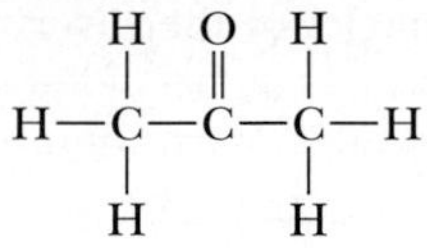

$$
\begin{array}{ccccc}
 & H & O & H & \\
 & | & \| & | & \\
H- & C- & C- & C & -H \\
 & | & & | & \\
 & H & & H &
\end{array}
$$

*11.122 The intermolecular forces of attraction in ethylene glycol are much stronger than in liquid acetone. Therefore, more energy is absorbed when a given amount of the glycol is converted to a liquid than when the same amount of acetone is changed to a liquid. Yet, if you spill acetone on your hand it produces a much greater cooling effect than if you spill an equivalent amount of the glycol on your hand. Why?

*11.123 When warm moist air sweeps in from the ocean and rises over a mountain range, it expands and cools. Explain how this cooling is related to the attractive forces between gas molecules. Why does this cause rain to form? When the air drops down the far side of the range, its pressure rises as it is compressed. Explain why this causes the air temperature to rise. How does the humidity of this air compare with the air that originally came in off the ocean? Now, explain why the coast of California is lush farmland, whereas the plains on the eastern side of the Rocky Mountains are arid and dry.

*11.124 If you shake a CO_2 fire extinguisher on a cool day, you can feel the liquid CO_2 sloshing around inside. However, if you shake the same fire extinguisher on a very hot summer day, you will not feel the liquid sloshing around. What is responsible for this difference?

11.125 The critical temperature of acetone is 236 °C, whereas the critical temperature of propane, C_3H_8, is 97 °C. Why are they so different?

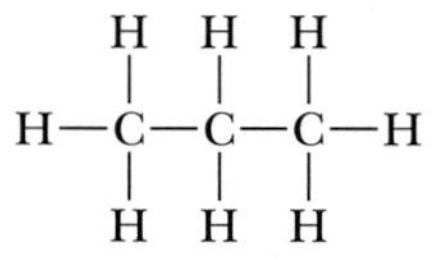

$$
\begin{array}{ccccc}
H & H & H & & \\
| & | & | & & \\
H-C- & C- & C & -H & \\
| & | & | & & \\
H & H & H & &
\end{array}
\qquad
\begin{array}{ccccc}
H & O & H & & \\
| & \| & | & & \\
H-C- & C- & C & -H & \\
| & & | & & \\
H & & H & &
\end{array}
$$

propane 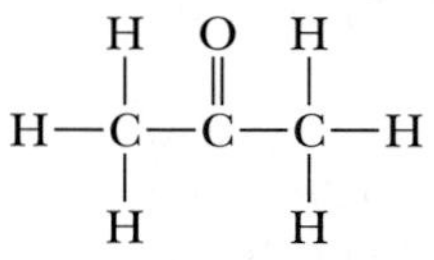   acetone

11.126 From the data in Exercise 11.76, estimate the boiling point of isopropyl alcohol.

11.127 Propane, the fuel used in gas barbecues, has the following vapor pressures:

Temp. (°C)	Vapor pressure (torr)
−92.4	40
−87.0	60
−79.6	100
−61.2	300
−47.3	600

Graphically determine the value of ΔH_{vap} for propane in units of kilojoules per mole.

*11.128 Potassium ions have a radius of 133 pm and bromide ions have a radius of 195 pm. The crystal structure of potassium bromide is the same as for sodium chloride. Estimate the length of the edge of the unit cell in potassium bromide.

*11.129 The unit cell edge in sodium chloride has a length of 564.0 pm. The sodium ion has a radius of 95 pm. What is the *diameter* of a chloride ion?

*11.130 Gold crystallizes in a face-centered cubic lattice. The edge of the unit cell has a length of 407.86 pm. The density of gold is 19.31 g/cm³. Use these data and the atomic mass of gold to calculate the value of Avogadro's number.

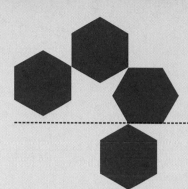

CHEMICALS IN USE 9
Glass

A *glass*, as we learned in Chapter 11, is any amorphous solid, one having no regularly repeating order to its constituent particles. Most chemists use the term this way, but in the public's mind, "glass" is understood in narrower terms. The American Society for Testing and Materials (ASTM), an organization that sets widely accepted standards for analytical procedures and for quality control, also prefers to limit the term. To the ASTM, *glass* is an inorganic material made by the thermal fusion of solids and cooled without crystallizing until it is rigid.

One substance that can be melted and then cooled in this manner is quartz sand, which is essentially all silicon dioxide—silica. The trouble is that it melts at a very high temperature—above 1700 °C—and becomes rigid so quickly when cooled that it is very hard to work into desired shapes.

SODA–LIME–SILICA GLASS

When sodium carbonate—soda ash—is mixed with silica sand, the melting point is lowered from a temperature of over 1700 °C (for pure silica) to the range of 700–850 °C. In other words, sodium carbonate acts as a *fluxing agent,* something that can help prepare a flux or a molten mass. When a molten mixture of silica and sodium carbonate cools, a glass forms called "water glass" because it is relatively soluble in water.

If limestone—calcium carbonate—is added along with the sodium carbonate to the silica sand, the melting-point lowering is still quite acceptable, but the glass that forms, called soda–lime–silica glass, is much less soluble in water. These facts were known to Egyptian glass workers at least 2000 years ago when this, now the most common kind of glass, began its continuous history of manufacture. If too much sodium carbonate is used, the resistance to water drops; if too much lime is used, the glass tends to "devitrify," meaning that crystalline regions develop. A typical batch is made up of 75% silica (of as uniform grain size as possible), 10% limestone, and 15% sodium carbonate.

Sometimes some of the limestone is replaced by magnesium carbonate and aluminum oxide. Because nearly all silica contains traces of iron compounds, the glass product is green. This kind of green glass is still the cheapest glass made. In time, glass workers found that by adding traces of oxides such as those of cobalt, arsenic, and selenium, the green color could be almost entirely removed.

Although the ingredients for soda–lime–silica glass are limestone and sodium carbonate along with the silica, the high temperatures involved decompose the carbonates to oxides.

$$\underset{\text{limestone}}{CaCO_3(s)} \longrightarrow \underset{\text{quicklime}}{CaO(s)} + CO_2(g)$$

$$Na_2CO_3(s) \longrightarrow Na_2O(s) + CO_2(g)$$

Thus, the melting of the initial batch generates considerable gas, and the mixture has to be maintained in the molten state for a few days to allow time for this gas to escape from the very thick, viscous fluid.

Glass can be given a variety of bright colors. The addition of finely divided copper selenide (CuSe), cadmium selenide (CdSe), or metallic gold produces a ruby red glass. When copper or cobalt compounds are dissolved in the melt, the glass is blue. Chromium compounds are used to make bright green glass. (See Figure 9a)

Soda–lime–silica glass is the "soft glass" familiar to chemistry students for making glass stirring rods and bent glass items fashioned from glass tubing. Sharp edges on cut soda–lime–silica glass soften readily in a strong Bunsen burner flame. The familiar yellow color that this procedure, called "fire polish-

FIGURE 9a

The presence of various transition metals in glass cause striking colors as seen in this photo of a stained glass window at the Palau de la Musica, Barcelona, Spain.

ing" glass, imparts to the flame is caused by the sodium ions in the glass.

OTHER KINDS OF GLASS

Borosilicate Glass

One of the problems with soda–lime–silica glass is that when heated it expands, and it contracts unevenly as it cools unevenly. This results in stress points in the cooled glass which make it break or shatter very readily. In other words, it is difficult to work with this kind of glass in situations involving wide changes in temperature. Table 9a shows the relative thermal expansions of various kinds of glasses, and soda–lime–silica glass shows poor stability to heat with 15 times the thermal expandability of pure silica glass.

If much of the sodium carbonate is replaced by boron oxide, B_2O_3, and some of the limestone is replaced by aluminum oxide, Al_2O_3, the resulting glass is much more thermally stable. It is called borosilicate glass, and the most common brand is Pyrex. Although it softens at a higher temperature—typically requiring oxygen–methane or oxygen–hydrogen flames—hot pieces of borosilicate glass can be fused together and annealed to give seals that withstand mechanical shocks very well.[1] Borosilicate glass is the material used to make nearly all glass kitchen ware and laboratory glassware (see Figure 9b). When glass of even lower thermal expandability is needed, then pure silica glass or a near relative, Vycor, is used (Table 9a).

Lead Crystal Glass

When lead oxide, PbO, is used as the fluxing agent instead of sodium carbonate, the resulting glass has a high refractive index (it can "bend" light rays well). When cut or polished to angular designs, lead crystal glass sparkles with a brilliance that has attracted glass artisans for over 300 years.

FIGURE 9b

Laboratory glassware is made from borosilicate glass.

Safety Glass

It is probably safe to say that the automobile industry would not have been possible without the invention of safety glass, glass that does not shatter and throw shards when struck. Safety glass can be made by two general methods. One is to seal a thin layer of clear plastic between two glass sheets. When such laminated glass shatters, its fragments are held by the plastic. If several alternating layers of glass and plastic are laminated to a thickness of 1.5 to 2 in., the product is bulletproof glass.

The safety glass commonly used today for vehicle windows other than windshields is made by a special process of heat treatment. A molded sheet of glass is heated until its temperature is just below the softening point and then a cold blast of air is spread over its surfaces. The glass nearest these surfaces becomes subject to great compressional strains that aid the glass in resisting forces of bending and twisting. If the glass receives a sharp, glass-breaking blow, it instantly develops cracks over its entire expanse. The glass particles, however, are small chunks and are relatively harmless because they do not fly about. (Laminated glass in the same circumstances develops cracks that radiate from the point of impact.)

Questions

1. How does the ASTM define glass?

2. Why is pure quartz seldom used to make ordinary glass items?

3. Sodium carbonate is called a fluxing agent for silica. What does this mean?

4. Why is calcium carbonate a part of the recipe for glass?

5. SiO_2 is an acidic anhydride. What basic anhydrides form during the manufacture of soda–lime–silica glass? (Give their names and formulas.)

6. What property of borosilicate glass suits it better to laboratory glassware than soda–lime–silica glass?

7. What is the fluxing agent (name and formula) in lead crystal glass, and what makes this glass attractive to glass artisans?

TABLE 9a   Thermal Expandability of Various Glasses

Type of glass	Relative expandability
Silica glass	1.0
Vycor	1.4
Borosilicate glass	6.1
Soda–lime–silica glass	15
Lead crystal glass	16

[1] *Annealing* means cooling the softened material slowly and uniformly and thus minimizing the development of stress points.

The brilliant colors of this sunset highlighting the "seastacks"
at Cannon Beach, Oregon, are caused by those frequencies of
sunlight that are not well scattered by dust and microdroplets
suspended in air. A Special Topic in this chapter explains why
the sky at midday is blue but at sunrise and sunset often is
beautifully colored.

Chapter 12

Solutions

A **solution** is a homogeneous mixture in which all of the particles have the sizes of atoms, small molecules, and small ions, those with average diameters in the range of 0.05 to 0.25 nm. As the particle sizes become larger, the mixtures take on the special properties of colloidal dispersions (Section 12.10) and suspensions. In Chapter 4, we introduced the common terms used for solutions, namely, *solute, solvent, solubility,* and others. If you aren't sure of them you should review them now. Our goal in this section and the next is to learn how chemical composition makes some solutes very soluble and others insoluble, particularly in water. To understand the effect of composition, however, we need a detailed study of the *process* by which solutions form.

A Tendency toward Randomness— One Driving Force for Solution Formation

Let's look more closely at how gases mix spontaneously with each other. Figure 12.1*a* shows a container holding two unmixed gases separated by a removable panel. When the panel is slid away (Figure 12.1*b*) the process of mixing begins spontaneously because of the random motions of the molecules. The gases mingle with no outside help. Once they have formed an homogeneous solution, their molecules will never spontaneously separate from one another and so restore the system to its original, unmixed state.

The spontaneous mixing of gases displays one of nature's strongest "driving forces" for change, namely, *the tendency of a system, left to itself, to become increasingly disordered.* At the instant the panel is slid away (Figure 12.1), we have two gases in the same container, but they are separated and on opposite sides of the container. This actually represents considerable *order*, like lines of boys and girls before a sixth grade dance. The ordered state suddenly becomes

12.1 FORMATION OF SOLUTIONS

Homogeneous means "everywhere alike."

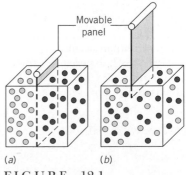

(a)   (b)

FIGURE 12.1

When two gases initially in separate compartments (*a*) suddenly find themselves in the same container, they mix spontaneously (*b*).

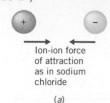

Ion-ion force
of attraction
as in sodium
chloride

(a)

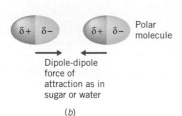

Dipole-dipole
force of
attraction as in
sugar or water

(b)

FIGURE 12.2

(*a*) Interionic force of attraction. (*b*) Intermolecular force of attraction.

Two liquids are *miscible* if they are completely soluble in all proportions. The opposite of miscible is *immiscible*.

```
    H   H
    |   |   δ-   δ+
H—C—C—O—H
    |   |
    H   H
```

Ethyl alcohol has a polar O—H group.

```
        H            H
        |            |
   H····O        H····O
  /              /     \
C₂H₅—O      H····O      H
  ⋮          /   |
 H—O      C₂H₅
     \
      H
```

FIGURE 12.3

Ethyl alcohol molecules, C₂H₅—O—H, experience hydrogen bonding (···) between themselves in pure alcohol.

```
        H
        |
        C
   H   / \\   H
    \ //   \ /
     C      C
     |      ||
     C      C
    / \\   / \
   H   \ //   H
        C
        |
        H
```

Benzene
C₆H₆

highly improbable as soon as the panel is removed. Now the vastly more probable distribution is one in which the molecules are thoroughly mixed. Nature's drive toward disorder is a powerful one. As the molecules mingle, the more disordered state is statistically so probable that it's a certainty for all practical purposes.

Attractions between Solute and Solvent— Another Driving Force for Solution Formation

To understand how *gaseous* solutions form, the only factor we have to consider is nature's driving force for disorder. Another possible factor, attractive forces between particles, plays a negligible role because they are so weak in a gas, and the particles are far apart, relative to their diameters.

When solids or liquids dissolve in liquid solvents, however, attractive forces between ions or polar molecules are very important and help to bring the particles close together (see Figure 12.2). Yet the separation of ions and polar molecules from each other must happen if liquid state solutions are to form. Success in preparing such a solution, therefore, depends not only on the drive toward disorder but also on the strengths of the intermolecular attractive forces in *both solvent and solute.*

Solutions of Liquids in Liquids

If we add ethyl alcohol to water, it dissolves completely in any proportion we want. We say that water and ethyl alcohol are completely *miscible.* Benzene, C₆H₆, on the other hand, is a liquid that is virtually insoluble in water. To understand why, let's look closely at what happens in each case when the liquids are combined. Overall, we know that the molecules of solute and solvent must be pushed apart to make room for those of the other if a solution is to form.

Ethyl alcohol (beverage alcohol) has a molecular structure that includes a polar O—H group. Its molecules, therefore, can form hydrogen bonds with water molecules, which also have O—H groups (see Figure 12.3). Water molecules can therefore attract alcohol molecules almost as strongly as they attract each other. Thus, the forces that must be overcome when water molecules are pushed apart to make room for alcohol molecules are compensated by similar attractions to the incoming alcohol molecules. The alcohol molecules are therefore able to move into the water and form the solution, and water molecules can similarly move into the alcohol. These two liquids, therefore, dissolve in each other. Because forces of attraction are easily accommodated in this system, nature's strong tendency toward the greater disorder of a solution can work its way.

Quite a different situation occurs if we try to dissolve benzene in water. Benzene molecules have no O—H groups and are otherwise nonpolar. Between them are only relatively weak London forces of attraction. Benzene molecules are therefore not attracted to water molecules very strongly and so cannot push water molecules aside. Water molecules, in other words, stick too tightly together to be pushed aside by the benzene molecules. Even though the tendency toward the greater disorder of a solution is present, the attractive forces between the water molecules are too strong to be overcome by benzene molecules. Thus, benzene is insoluble in water.

It is also easy to see why a solution of water in benzene cannot be made. Suppose that we did manage to disperse water molecules in benzene. As they move about, the water molecules would occasionally encounter each other. Because water molecules attract each other so much more strongly than they attract benzene molecules, they would stick together at each encounter. This would continue to happen until all the water was in a separate phase. A solution of water in benzene would thus not be stable, and it would not form spontaneously.

Although benzene cannot dissolve in water, it does dissolve quite well in nonpolar liquids, like carbon tetrachloride, CCl_4. The forces of attraction between CCl_4 molecules are about as weak as those between benzene molecules. Therefore CCl_4 molecules can easily leave their own kind and mingle with the benzene molecules. Nature's drive toward disorder easily overcomes what little resistance there is, and the solution readily forms.

In summary, we see that when the strengths of intermolecular attractions are similar in solute and solvent, solutions can form as illustrated by the easy formation of alcohol–water and benzene–carbon tetrachloride solutions. To summarize these and similar situations, chemists use the **"like dissolves like"** rule: when solute and solvent have molecules "like" each other in polarity, they tend to form a solution. When their molecules are quite different in polarity, solutions of any appreciable concentration do not form. The rule has long enabled chemists to use chemical composition and molecular structure to predict the likelihood of two substances dissolving in each other.

T   **"Like dissolves like" rule**

Solutions of Solids in Liquids

The situation for dissolving solids in liquids is only slightly different from that of dissolving liquids in liquids. Let us first examine what happens when a crystalline salt, like sodium chloride, dissolves in water.

Figure 12.4 shows a section of a crystal of NaCl in contact with water. Where water touches the crystal, the dipoles of water molecules orient themselves so that their negative ends point toward positive Na^+ ions and the positive ends of the dipoles point at negative Cl^- ions. In other words, *ion–dipole* attractions occur that tend to tug and pull ions from the crystal. Notice that ions at the corners and edges of a crystal are held by fewer neighbors and so are more readily dislodged than ions elsewhere on the crystal's surface. As water molecules dislodge these ions, new corners and edges are exposed, and the dissolution of the crystal can continue.

Water molecules collide everywhere along the crystal surface, but *successful* collisions—those that dislodge ions—are more likely to occur at corners and edges.

As they become free, the ions become completely surrounded by water molecules (also shown in Figure 12.4). The phenomenon is called the **hydration** of ions. The *general* term for the surrounding of a solute particle by solvent molecules is **solvation,** so hydration is just a special case of solvation. Ionic compounds are able to dissolve in water when the attractions between water dipoles and ions overcome the attractions of the ions for each other in the crystal. Then nature's tendency for disorder can work its dissolving ways.

We're clearly at the borderline between chemical and physical changes when we place the binding of water molecules to sodium or chloride ions in the realm of physical changes.

Similar events explain why solids composed of polar molecules, like those of sugar, dissolve in water (see Figure 12.5). Attractions between the solvent and solute dipoles help to dislodge molecules from the crystal and bring them into solution. Again we see that "like dissolves like"; a polar solute dissolves in a polar solvent.

The same reasoning explains why nonpolar solids like wax are soluble in nonpolar solvents such as benzene. Wax is a solid mixture of long chain hydrocarbons. Their molecules attract each other weakly by London forces, so the wax molecules easily slip away from the solid even though the attractions

FIGURE 12.4

Hydration of ions. Hydration involves a complex redirection of forces of attraction and repulsion. Before this solution forms, water molecules are attracted only to each other; and Na^+ and Cl^- ions have only each other in the crystal to be attracted to. In the solution, the ions have water molecules to take the places of their oppositely charged counterparts; in addition, water molecules find ions more attractive than even other water molecules.

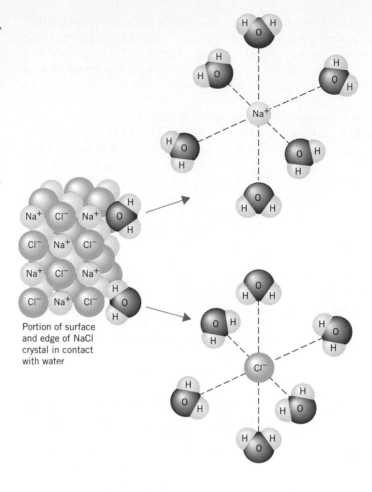

Portion of surface and edge of NaCl crystal in contact with water

FIGURE 12.5

Solvation of polar molecules. The molecules of polar compounds can trade forces of attraction to each other (in the crystal) for forces of attraction to the molecules of a polar solvent (in the solution).

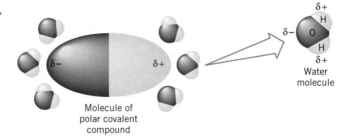

Molecule of polar covalent compound

Water molecule

between the molecules of the solvent (benzene) and the solute (wax) are weak themselves.

When intermolecular attractive forces within solute and solvent are sufficiently different, the two do not form a solution. For example, ionic solids or very polar molecular solids (like sugar) are insoluble in nonpolar solvents such as benzene, gasoline, or lighter fluid. The molecules of these solvents, all hydrocarbons, are unable to attract ions or very polar molecules with enough force to overcome the much stronger attractions that the ions or polar molecules experience within their own crystals. Nature's drive toward disorder can thus be stymied by interionic or intermolecular forces.

Because intermolecular attractive forces come into play, the formation of a solution is inevitably associated with energy exchanges. The energy exchanged between the system and its surroundings when one mole of a solute dissolves in a solvent at constant pressure to make a dilute solution is called the *molar enthalpy of solution,* or usually just the **heat of solution, ΔH_{soln}.**

It costs energy—it is *endothermic*—to separate the particles of solute and also those of the solvent and make them spread out to make room for each other. This change would *increase* the potential energy of the system (because things that naturally attract each other are being pulled apart). But once the spread-apart particles come back together as a *solution,* the attractive forces between approaching solute and solvent particles yield a decrease in the system's potential energy and is an *exothermic* change. The heat of solution, ΔH_{soln}, is simply the net result of these two opposing energy contributions. To look at this more closely, let us follow the potential energy changes that occur when a solid dissolves in a liquid.

Lattice Energies and Solvation Energies

In Chapter 5 you learned that we could use Hess's law to calculate heats of reaction from the heats of formation of the reactants and the products. This approach works because enthalpy is a *state function,* so the magnitude of a *change* in enthalpy, ΔH, does not depend on *how* the system changes from one state to another. The same property helps us analyze heats of solution and explain why some solutions form endothermically and others exothermically. We can devise any route or path we please for the *process* of forming a solution as long as we go from the same initial state—separated solute and solvent—to the same final state—the solution.

To take advantage of available energy data, we will devise a hypothetical two-step path and use an enthalpy diagram (see Figure 12.6). Overall, the two steps will take us from the solid solute and liquid solvent to the final solution,

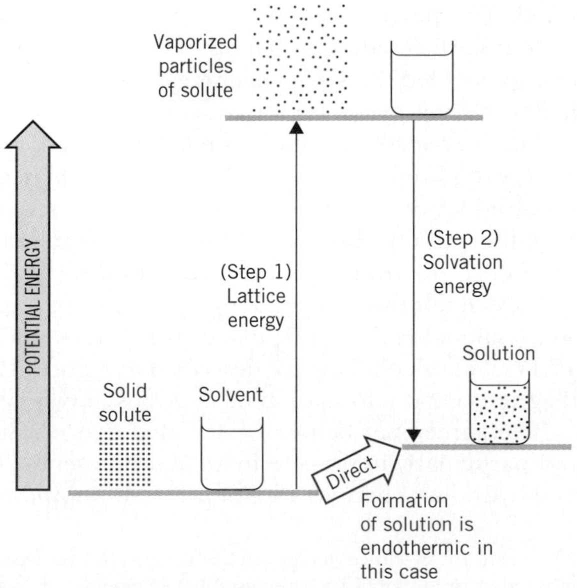

12.2
HEATS OF SOLUTION

$$\Delta H_{soln} = H_{soln} - H_{components}$$

The magnitude of ΔH_{soln} depends somewhat on the final concentration of the solution being made.

FIGURE 12.6

Ethalpy diagram for a solid dissolving in a liquid. In the real world, the solution is formed directly as indicated by the red arrow. We can analyze the energy change by imagining the two separate steps, because enthalpy changes are functions of state and are independent of path. The energy change along the direct path is the algebraic sum of Step 1 and Step 2.

TABLE 12.1 Lattice Energies, Hydration Energies, and Heats of Solution for Some Group IA Metal Halides

Compound	Lattice Energy (kJ mol^{-1})[a]	Hydration Energy (kJ mol^{-1})	ΔH_{soln}[b] Calculated ΔH_{soln} (kJ mol^{-1})	Measured ΔH_{soln} (kJ mol^{-1})
LiCl	+833	−883	−50	−37.0
NaCl	+766	−770	−4	+3.9
KCl	+690	−686	+4	+17.2
LiBr	+787	−854	−67	−49.0
NaBr	+728	−741	−13	−0.602
KBr	+665	−657	+8	+19.9
KI	+632	−619	+13	+20.33

[a] These values are the true lattice energies given opposite signs. True lattice energies have negative values, because they relate to the exothermic *formation* of crystalline lattices from their constituent, gaseous ions (or molecules).

[b] Heats of solution refer to the formation of extremely dilute solutions.

but you'll see that they do not represent the way a solution is actually made in the lab. In the first step, we imagine that the solid separates into its individual particles; in effect, we imagine that the solid is vaporized. In the second step, we imagine that the gaseous solute particles enter the solvent where they become solvated. As we said, we imagine this particular process to let us use existing experimental data, being permitted to do so only because the heat of solution is a state function.

As you can see from Figure 12.6, the first step is endothermic and *increases* the potential energy of the system. The particles in the solid attract each other, so energy must be supplied to separate them, energy called the *lattice energy*.[1] By receiving the requisite lattice energy, the solid is converted to its gaseous state.

In the second step, gaseous solute particles enter the solvent and are solvated. The potential energy of the system now *decreases*, because solute and solvent particles attract each other. This step is therefore exothermic, and its energy is called the **solvation energy,** which is a general term. The special term **hydration energy** is used for aqueous solutions.

The *heat of solution*, the net energy change, is the difference between the energy required for Step 1 and the energy released in Step 2. When the energy required for Step 1 exceeds the energy released in Step 2, the solution forms endothermically. But when more energy is released in Step 2 than is needed for Step 1, the solution forms exothermically.

We can test this analysis by comparing the experimental heats of solution of some salts with those calculated from lattice and hydration energies (Table 12.1). The calculations are described in Figures 12.7 and 12.8, using enthalpy diagrams for the formation of solutions of two salts, KI and NaBr.

The agreement between calculated and measured values in Table 12.1 is not particularly impressive in an absolute sense. This is partly because lattice and hydration energies are not precisely known and partly because the model

ΔH is positive for an endothermic change, so the arrow for Step 1 in Figure 12.6 points upward, the positive direction.

[1] The concept of lattice energy applies to any kind of crystalline solid, whether it is ionic or molecular. In Section 7.1 we discussed lattice energies as contributing to the stabilities of ionic compounds.

FIGURE 12.7

The formation of aqueous potassium iodide.

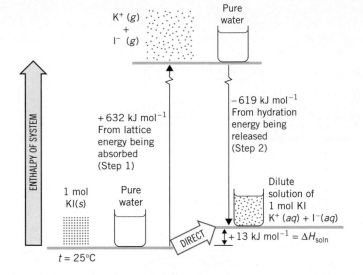

Step 1:	$KI(s) \rightarrow K^+(g) + I^-(g)$	$\Delta H = +632 \text{ kJ mol}^{-1}$
Step 2:	$K^+(g) + I^-(g) \rightarrow K^+(aq) + I^-(aq)$	$\Delta H = -619 \text{ kJ mol}^{-1}$
Net:	$KI(s) \rightarrow K^+(aq) + I^-(aq)$	$\Delta H_{soln} = +13 \text{ kJ mol}^{-1}$

FIGURE 12.8

The formation of aqueous sodium bromide.

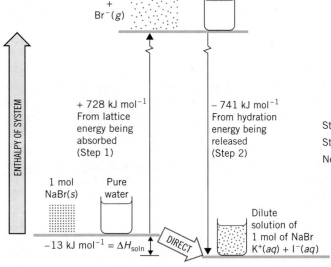

Step 1:	$NaBr(s) \rightarrow Na^+(g) + Br^-(g)$	$\Delta H = +728 \text{ kJ mol}^{-1}$
Step 2:	$Na^+(g) + Br^-(g) \rightarrow Na^+(aq) + Br^-(aq)$	$\Delta H = -741 \text{ kJ mol}^{-1}$
Net:	$NaBr(s) \rightarrow Na^+(aq) + Br^-(aq)$	$\Delta H_{soln} = -13 \text{ kJ mol}^{-1}$

used in our analysis is evidently too simple. Notice, however, that when "theory" predicts relatively large heats of solution, the experimental values are also relatively large, and that both values have the same sign (except for NaCl). Notice also that the changes in values show the same trends when we compare the three chloride salts—LiCl, NaCl, and KCl—or the three bromide salts—LiBr, NaBr, and KBr.

Small percentage errors in very large numbers can cause huge percentage changes in the *differences* between such numbers (as you can discover by working Review Exercise 12.18).

Solutions of Liquids in Liquids

To consider heats of solution when liquids dissolve in liquids, we will work with a three-step path to go from the initial to the final state (see Figure 12.9). First, we imagine that the molecules of one liquid are moved apart just far enough to make room for molecules of the other liquid. (We will designate one liquid as the solute.) Because we have to overcome forces of attraction, this step increases the system's potential energy and so is *endothermic*.

FIGURE 12.9

Hypothetical steps in the analysis of the enthalpy change for the formation of a solution of two liquids.

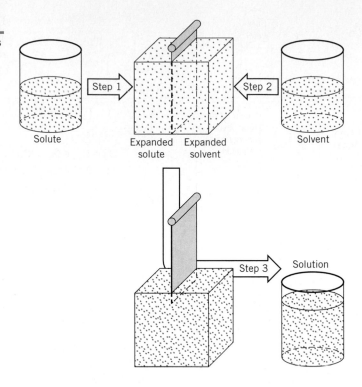

The second step is like the first, but is done to the other liquid (solvent). On an enthalpy diagram (Figure 12.10) we have climbed two energy steps and have both the solvent and the solute in their slightly "expanded" conditions.

The third step lets nature's drive for randomness take its course, as the molecules of expanded solvent and solute come together and intermingle. Because the molecules of the two liquids now experience mutual forces of attraction, the system's potential energy *decreases,* and Step 3 is exothermic. The value of ΔH_{soln} will, again, be the net energy change for these steps.

FIGURE 12.10

Enthalpy changes in the formation of an ideal solution. The three-step and direct-formation paths both start and end at the same place with the same enthalpy outcome.

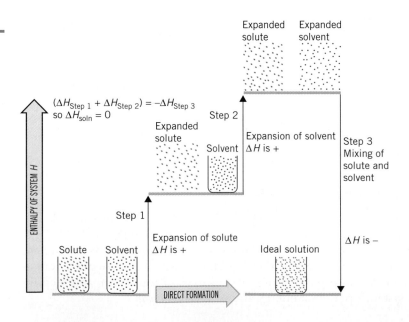

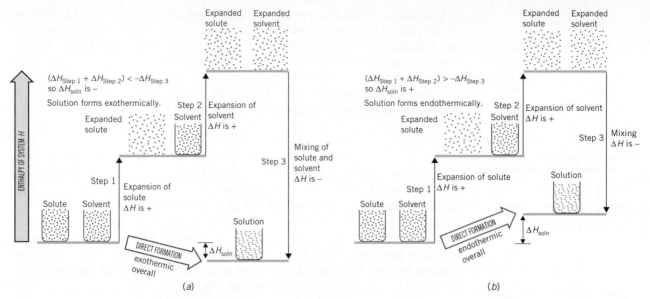

FIGURE 12.11

Enthalpy changes when a real solution forms. Because ΔH_{soln} is a state function, the multistep paths and the direct-formation paths have identical enthalpy outcomes. (*a*) The process is exothermic. (*b*) The process is endothermic.

The enthalpy diagram in Figure 12.10 shows the case when the sum of the energy inputs for Steps 1 and 2 is equal to the energy released in Step 3, so the overall value of ΔH_{soln} is zero. This is very nearly the case when we make a solution of benzene and carbon tetrachloride. Attractive forces between molecules of benzene are almost exactly the same as those between molecules of CCl_4, or between molecules of benzene and those of CCl_4. If all such intermolecular forces were identical, the net ΔH_{soln} would be exactly zero, and the resulting solution would be called an **ideal solution.** Be sure to notice the difference between an *ideal solution* and an *ideal gas*. In an ideal gas, there are no attractive forces. In an ideal solution, there are attractive forces, but they are all the same.

Acetone and water form a solution exothermically (see Figure 12.11*a*). With these liquids, the third step releases more energy than the sum of the first two chiefly because molecules of water and acetone attract each other more strongly than acetone molecules attract each other.

Ethyl alcohol and hexane form a solution endothermically (see Figure 12.11*b*). In this case, the release of energy in the third step is not enough to compensate for the energy demands of Steps 1 and 2, and the solution becomes cool as it forms. The problem is chiefly that ethyl alcohol molecules attract each other more strongly than they can attract hexane molecules. Hexane molecules cannot push their way in and among those of ethyl alcohol without breaking up some of the hydrogen bonding in the alcohol.

Solutions of Gases in Liquids

The formation of a solution of a gas in a liquid is usually exothermic, particularly when water is the solvent. The gas is already expanded, so there is no energy cost associated with "expanding the solute." The solvent, however, must be expanded slightly to accommodate the molecules of the gas, and this does require a small energy input—small because the attractive forces do not change very much when the solvent molecules are forced apart just a little bit. When the gas molecules finally fill the spaces thus made for them, the net ΔH_{soln} depends on how strongly the gas molecules become attracted to those of the solvent. If this attraction is strong enough, there is enough of a lowering of the potential energy as the solution forms to make the overall formation of the solution exothermic. This is because the gas molecules change from a condition in which the attractive forces are almost zero to a condition in which there are significant forces of attraction. The net energy change associated with the two steps of expanding the solvent slightly and then filling the spaces with solute molecules is simply the solvation energy.

12.3
EFFECT OF TEMPERATURE ON SOLUBILITY

By "solubility" we mean the mass of solute that forms a *saturated* solution with a given mass of solvent at a specified temperature. The units often used are grams of solute per 100 g of solvent. Because solubility refers to a *saturated* solution, if we try to make more solute dissolve, any extra solute we add to the system will just sit there as a separate phase. There will be a coming and going of solute particles between the dissolved and undissolved phases, of course, because in a saturated solution we have equilibrium between them (see Figure 12.12). We can write this equilibrium as follows.

$$\text{solute}_{\text{undissolved}} \rightleftharpoons \text{solute}_{\text{dissolved}}$$
[Solute contacts the saturated solution]   [Solute is in a saturated solution]

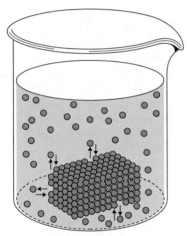

FIGURE 12.12

In a saturated solution dynamic equilibrium exists between the undissolved solute and the solute in the solution.

Effect of Temperature on the Solubility of Solids and Liquids in Other Liquids   As long as we keep the temperature constant, the equilibrium in a saturated solution holds. Only by subjecting the system to the specific stresses of heating or cooling it through changing its temperature can we shift its equilibrium. The equilibrium shifts in whichever direction most directly absorbs the stress (Le Châtelier's principle).

Let's consider the more common situation, one in which adding heat causes more solute to dissolve into a solution that is initially saturated. We may even place "heat" within the equilibrium expression, putting it on the left side, the undissolved solute side, because heat is absorbed when solute dissolves.

$$\text{Solute}_{\text{undissolved}} + \text{heat} \rightleftharpoons \text{solute}_{\text{dissolved}}$$

The stress of additional heat causes the equilibrium to shift to the *right,* a shift that "uses up the heat" and thus absorbs the stress *but also causes more solute to dissolve.* The individual solutes and solvents determine how much more dissolves, and there are large variations. Compare in Figure 12.13, for example, the widely different responses to increasing temperature in the solubilities of ammonium nitrate and sodium chloride.

Cerium(III) sulfate, $Ce_2(SO_4)_3$, represents a rare example of a substance that becomes *less* soluble with increasing temperature (see Figure 12.13). Heat must be *removed* from a saturated solution of cerium(III) sulfate to make more solute dissolve. For its equilibrium equation, we must show "heat" on the right side because heat is liberated when more of the solute dissolves into a saturated solution.

$$\text{Solute}_{\text{undissolved}} \rightleftharpoons \text{solute}_{\text{dissolved}} + \text{heat}$$

When we *increase* the temperature of this system, the equilibrium absorbs the stress, the added heat, by shifting to the left, and some of the dissolved solute comes out of solution. As we said, such systems are not common.

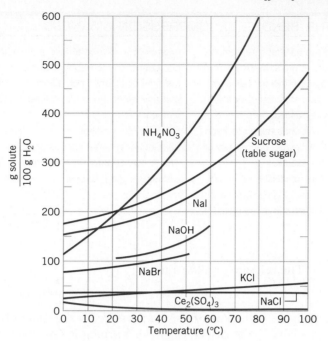

FIGURE 12.13

Solubility in water versus temperature for several substances.

Effect of Temperature on the Solubility of a Gas in a Liquid Table 12.2 gives data for the solubilities of several common gases in water at different temperatures but all under 1 atm of pressure. You can see that at constant pressure, the solubilities of gases in liquids always decrease with increasing temperature. This is true of *all* gases in *all* liquids, and the reasons are complex. We'll not study them; we're at a borderline here where the twin factors that affect the formation of a solution compete very closely—the drive to disorder and intermolecular attractive forces.

The solubility of a gas in a liquid is affected not only by temperature but also by pressure. The gases in air, for example, chiefly oxygen and nitrogen, are not very soluble in water under ordinary pressures, but at elevated pressures both become increasingly soluble (see Figure 12.14). Special Topic 12.1 describes the dangerous implications of these changes to those who must work under higher than normal air pressures.

12.4
EFFECT OF PRESSURE ON THE SOLUBILITIES OF GASES

TABLE 12.2   Solubilities of Common Gases in Water[a]

Gas	Temperature			
	0 °C	**20 °C**	**50 °C**	**100 °C**
Nitrogen, N_2	0.0029	0.0019	0.0012	0
Oxygen, O_2	0.0069	0.0043	0.0027	0
Carbon dioxide, CO_2	0.335	0.169	0.076	0
Sulfur dioxide, SO_2	22.8	10.6	4.3	1.8[b]
Ammonia, NH_3	89.9	51.8	28.4	7.4[c]

[a] Solubilities are in grams of solute per 100 g of water when the gaseous space over the liquid is saturated with the gas and the total pressure is 1 atm.

[b] Solubility at 90°C.   [c] Solubility at 96°C.

SPECIAL TOPIC 12.1 / THE BENDS—DECOMPRESSION SICKNESS

When people work in a space where the air pressure is much above normal, they have to be careful to return slowly to normal atmospheric pressure. Otherwise, they face the danger of the "bends"—severe pains in joints and muscles, fainting, possible deafness, paralysis, and death. Workers building deep tunnels, where higher than normal air pressure is maintained to keep water out, are at risk. So are deep-sea divers. Each 10 m of water depth adds roughly 1 atm to the pressure.

The bends can develop because the solubilities of both nitrogen and oxygen in blood are higher under higher pressure, as Henry's law says. Once the blood is enriched in these gases it must not be allowed to lose them suddenly. If microbubbles of nitrogen or oxygen appear at blood capillaries, they will block the flow of blood. Such a loss is particularly painful at joints, and any reduction in blood flow to the brain can be extremely serious.

For each atmosphere of pressure above the normal, about 20 minutes of careful decompression is usually recommended. This allows time for the respiratory system to gather and expel excess nitrogen. Excess oxygen is mostly used up by normal metabolism.

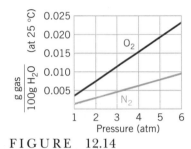

FIGURE 12.14

Solubility in water versus pressure for two gases.

The relevant equilibrium is

$$\text{Gas} + \text{solvent} \rightleftharpoons \text{solution} \qquad (12.1)$$

Suppose we now increase the pressure on the system by reducing the volume available to the gas. Le Châtelier's principle says that the system will change in a way to counteract this "stress," that is, the system will respond in a pressure-absorbing (pressure-reducing) way. To lower the pressure and reduce the stress requires a reduction in the amount of *gas,* so some gas molecules must leave the gas phase and dissolve in the solution. Thus, increasing the pressure of the gas shifts the equilibrium to the right, and the solubility of the gas increases. Conversely, if we make the pressure on the solution lower, the equilibrium must shift to the left, and some dissolved gas leaves the solution. Every time you pop open a carbonated beverage, dissolved carbon dioxide gas fizzes out in response to the sudden lowering of pressure.

To see more clearly the dynamics of how the effect of pressure on gas solubility works, imagine a closed container with a movable wall and partly filled with a solution of some gas in a liquid (see Figure 12.15). On the left, we begin with equilibrium; gas molecules come and go at equal rates between the dissolved and undissolved states (Figure 12.15a). The rate at which gas molecules enter the solution is proportional to the frequency with which they collide with the surface of the solution. This frequency increases when we increase the pressure of the gas (Figure 12.15b) because the gas molecules are squeezed closer together. More of them now collide with a given surface area of the solution in each second of time. Because liquids and liquid solutions are incompressible, the increased pressure alone has no effect on the frequency

FIGURE 12.15

How pressure increases the solubility of a gas in a liquid. (*a*) At some specific pressure, equilibrium exists between the vapor phase and the solution. (*b*) An increase in pressure puts stress on the equilibrium. (*c*) More gas dissolves and equilibrium is restored.

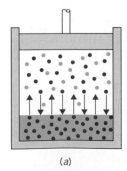

(a)

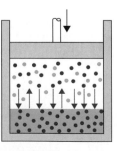

(b)

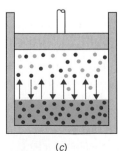

(c)

with which gas molecules can leave the solution. Therefore the increased pressure increases the rate of the forward reaction in our system (Equation 12.1) over the rate of the reverse reaction.

As more gas dissolves, however, the forward rate steadily slows and the reverse rate increases. It increases because at a higher concentration of dissolved gas, there are simply more gas molecules in solution at each unit of surface area. Their frequency of escape is proportional to this concentration. But the forward gas-dissolving reaction dominates until enough additional gas has dissolved to cause the opposing rates to equalize. We again have equilibrium (Figure 12.15c); in terms of what species are where, it is not the original equilibrium system but a new one because more gas molecules are in solution.

For gases *that do not react with the solvent,* there is a simple relationship between gas pressure and gas solubility—the **pressure–solubility law,** usually called **Henry's law,** after William Henry.

William Henry (1775–1836) was an English chemist and physician.

Henry's Law (Pressure–Solubility Law)

The concentration of a gas in a liquid at any given temperature is directly proportional to the partial pressure of the gas on the solution.

$$C_g = k_g P_g \qquad (T \text{ is constant})$$

Henry's law

In the equation, C_g is the concentration of the gas and P_g is the partial pressure of the gas above the solution. The proportionality constant, k_g, called the Henry's law constant, is unique to each gas. (Notice that constant temperature is assumed.) The equation is true only at low concentrations and pressures and, as we said, for gases that do not react with the solvent.

An alternate (and commonly used) expression of Henry's law is

$$\frac{C_1}{P_1} = \frac{C_2}{P_2} \tag{12.2}$$

Equation 12.2 is true because the two ratios, C_1/P_1 and C_2/P_2, equal the same Henry's law constant, k_g.

where C_1 and P_1 refer to initial conditions and C_2 and P_2 to final conditions.

At 20 °C the solubility of N_2 in water is 0.0150 g L^{-1} when the partial pressure of nitrogen is 580 torr. What will be the solubility of N_2 in water at 20 °C when its partial pressure is 800 torr?

EXAMPLE 12.1
Using Henry's Law

ANALYSIS This problem deals with the effect of gas pressure on gas solubility, so Henry's law applies. We use this law in its form given by Equation 12.2 because it lets us avoid having to know or calculate the Henry's law constant.

SOLUTION Let's gather the data first.

$$C_1 = 0.0150 \text{ g L}^{-1} \qquad C_2 = ?$$
$$P_1 = 580 \text{ torr} \qquad P_2 = 800 \text{ torr}$$

Using Equation 12.2, we have

$$\frac{0.0150 \text{ g L}^{-1}}{580 \text{ torr}} = \frac{C_2}{800 \text{ torr}}$$

Solving for C_2,

$$C_2 = 0.0207 \text{ g L}^{-1}$$

The solubility under the higher pressure is 0.0207 g L^{-1}.

CHECK In relationship to the initial concentration, the size of the answer makes sense because Henry's law tells us to expect a greater solubility at the higher pressure.

■ **Practice Exercise 1** How many grams of nitrogen and oxygen are dissolved in 100 g of water at 20 °C when the water is saturated with air? At 1 atm pressure, the solubility of oxygen in water is 0.00430 g O_2/100 g H_2O, and the solubility of nitrogen in water is 0.00190 g N_2/100 g H_2O. In pure, dry air, P_{N_2} equals 593 torr and P_{O_2} equals 150 torr.

Solubilities of Gases That Are Strongly Hydrated

The gases sulfur dioxide, ammonia, and, to a lesser extent, carbon dioxide are far more soluble in water than are oxygen or nitrogen (see Table 12.2). Part of the reason is that SO_2, NH_3, and CO_2 molecules have polar bonds and sites of partial charge that attract water molecules, forming hydrogen bonds to help hold the gases in solution. Ammonia molecules, in addition, not only can accept hydrogen bonds from water (O—H· · ·N) but also can donate them through their N—H bonds (N—H· · ·O).

The more soluble gases also react with water to some extent as the following chemical equilibria form.

$$CO_2(aq) + H_2O \rightleftharpoons H_2CO_3(aq) \rightleftharpoons H^+(aq) + HCO_3^-(aq)$$

$$SO_2(aq) + H_2O \rightleftharpoons H^+(aq) + HSO_3^-(aq)$$

$$NH_3(aq) + H_2O \rightleftharpoons NH_4^+(aq) + OH^-(aq)$$

The forward reactions contribute to the higher concentrations of the gases in solution, as compared to gases such as O_2 and N_2 that do not react with water at all. Gaseous sulfur trioxide is very soluble in water, reacting quantitatively with water to form sulfuric acid.[2]

$$SO_3(g) + H_2O \longrightarrow H_2SO_4(aq)$$

12.5 TEMPERATURE-INDEPENDENT CONCENTRATION UNITS

In experimental work, solutions are used for such a variety of purposes that it shouldn't be surprising that no single concentration expression could serve all needs. For the stoichiometry of chemical reactions in solution, *molar concentration* or *molarity,* mol/L, provides by far the best set of units. Molarity, however, varies with temperature because a solution's volume (but not its number of moles of solute) varies somewhat with temperature. Most solutions expand when heated and contract when cooled.

To describe and compare the *physical properties* of solutions, temperature-insensitive expressions for concentration work best. We have already studied

---

[2] The "concentrated sulfuric acid" of commerce has a concentration of 93 to 98% H_2SO_4, or roughly 18 M. This solution takes up water avidly and very exothermically, removing moisture even out of humid air itself (and so must be kept in stoppered bottles). It is dense (*d* about 1.8 g mL^{-1}), oily, sticky, and highly corrosive and must be handled with extreme care. Safety goggles and gloves must be worn when dispensing concentrated sulfuric acid. When making a more dilute solution, *always pour (slowly, with stirring) the concentrated acid into the water.* If (less dense) water is poured into concentrated sulfuric acid, it can layer on the acid's surface. At the interface, such intense heat can be generated that the steam thereby created could explode from the container, spattering acid around.

one, the *mole fraction,* or its near relative, the *mole percent* (see Section 10.6). Although it was introduced for mixtures of gases, the concept of mole fraction applies to any homogeneous mixture. We'll next study weight fractions (and weight percents) and molal concentrations, which are other temperature-insensitive concentration expressions.

In a mixture, the mole fraction of component A, X_A, is given by

$X_A =$

$$\frac{\text{no. of moles of } A}{\text{total no. of moles, all components}}$$

Weight Fraction and Weight Percent[3]

When we compute the ratio of the mass of one component to the total mass of the mixture or solution, we find the *weight fraction,* $w_{\text{component}}$. For a solution, we have

$$w_{\text{component}} = \frac{\text{mass of component}}{\text{mass of solution}}$$

For example, the weight fraction of NaCl in a solution consisting of 12.5 g of NaCl and 75.0 g of H_2O is

$$w_{\text{NaCl}} = \frac{12.5 \text{ g}}{(12.5 \text{ g} + 75.0 \text{ g})} = 0.143$$

Notice that the units cancel; weight fraction has no units. The weight fraction of water in this solution can be found in the same way or, because the fractions must add up to 1.000, we could find the weight fraction of water by difference: $(1.000 - 0.143) = 0.857$.

When a weight fraction is multiplied by 100, we have the *weight percent* or **percent by weight.** When we know the percent by weight of one solute in a solution, we know the *number of grams of solute per 100 grams of solution.* A solution labeled "0.9% NaCl," for example, is one in which the ratio of solute to solution is 0.9 g of NaCl to 100 g of NaCl solution.

T **Weight percents**

There are other ways to define *percent concentration,* but in most laboratories, when no specification of the kind of percentage is given, a *percent* concentration means the percent *by weight.* We'll follow this practice. Special Topic 12.2 gives some other concentration expressions: ppm (parts per million), ppb (parts per billion), and percent by volume.

■ **Practice Exercise 2** How many grams of NaOH are needed to prepare 250 g of 1.00% NaOH solution in water? How many grams of water are needed? How many milliliters are needed, given that the density of water at room temperature is 0.988 g mL^{-1}?

■ **Practice Exercise 3** Hydrochloric acid can be purchased from chemical supply houses as a solution that is 37% HCl. What mass of this solution contains 7.5 g of HCl?

Molal Concentration

The number of moles of solute per kilogram of *solvent* is called the **molal concentration** or the **molality** of a solution. The usual symbol for molality is *m.*

Notice that it's per kilogram of *solvent,* not kilogram of solution.

$$\text{Molality} = m = \frac{\text{mol of solute}}{\text{kg of solvent}}$$

 Molal concentration

---

[3] Strictly speaking, these terms should be *mass fraction* and *mass percentage,* but we bow to very widely accepted practice.

SPECIAL TOPIC 12.2 / OTHER CONCENTRATION UNITS

VOLUME–VOLUME PERCENT

Volume percent is similar to weight percent, but is often preferred when working with solutions where all components are liquids, like antifreeze solution in water. A **percent by volume**, v/v%, gives the number of volumes of one liquid that is needed to make up 100 units of volume of the solution. Thus, to prepare a 40% (v/v) antifreeze solution, one would mix, say 4 L of antifreeze with enough water to make the final volume equal to 10 L. Liquor alcohol is usually given a "proof" rating, where the proof is twice the volume percent of alcohol. Thus, 80 proof alcohol is 40% (v/v) in ethyl alcohol.

PARTS PER MILLION (ppm)

When solutions are very dilute, like those of pollutants in air, their concentrations are sometimes given in **parts per million**, or **ppm**. This means the number of parts in one million parts of the whole. The same unit must be used for "parts"; for example, 1 g in 10^6 grams is 1 ppm. For a sense of how small 1 ppm is, it's like one penny in $10,000 or 1 minute in 2 years. If the concentration of ozone in air reaches 1 ppm it refers to 1 gram of ozone in 1×10^6 grams of air. This ozone level is regarded as so hazardous to health that human activities should be curtailed so that as little as possible of the air is brought into the lungs.

PARTS PER BILLION (ppb)

To express the concentrations of extremely dilute solutions, *parts per billion* is sometimes used. It's similar to parts per million; 1 **part per billion**, or **ppb**, means one part per 10^9 parts. Any unit can be employed for "parts" provided its the same unit in both uses. One part per billion is like 1 drop of water in a swimming pool measuring 11 by 10 by 20 feet. Its roughly two drops of water in the largest tank car you've ever seen (about 33,000 gallons), or 1 minute in 2000 years. Some water pollutants are dangerous even at ppb levels.

It can be shown that

$$ppm = \text{weight fraction} \times 10^6$$
$$ppb = \text{weight fraction} \times 10^9$$

We can, of course, obtain any mass we need by taking the equivalent in volume, using the solvent's density to calculate the volume.

For example, if we dissolve 0.5000 mol of sugar in 1000 g (1.000 kg) of water, we have a 0.5000 molal solution of sugar. We would not even need a volumetric flask to prepare the solution, because we weigh the solvent. Some important physical properties of a solution are related in a simple way to its molality, as we'll soon see.

It is very important that you learn the distinction between *molarity* and *molality*. Molarity is useful in determining the number of moles of solute obtained when we pour a specified volume of solution *at the temperature for which the measuring device has been calibrated.*[4] The molarity of a solution, however, changes with temperature, as we pointed out, and molality does not.

The difference between molarity and molality is large when the solvent has a density considerably different from 1 g mL^{-1}, water's density. Consider bromine, for example; it is so corrosive that it is often dispensed as a dilute solution in carbon tetrachloride, a solvent of high density. If we had a 1 *molar* solution of Br_2 in CCl_4, the *volume* containing 1 mol of Br_2 would be 1000 mL. But if we had a 1 *molal* Br_2 solution in CCl_4, the volume containing 1 mol of Br_2 would be about 680 mL. Because of its considerable density, we obtain 1000 *grams* of CCl_4 with much less volume than 1000 mL.

For CCl_4 at 25 °C, $d = 1.589 \text{ g mL}^{-1}$.

It should also be pointed out that when water is the solvent, a solution's molarity approaches its molality as the solution becomes more dilute. In very dilute solutions, 1 L of *solution* is nearly 1 L of water, which has a mass close to 1 kg. Now the ratio of moles/liter (molarity) is very nearly the same as the ratio of moles/kilogram (molality).

[4] A volumetric flask, a pipet, or a buret is *calibrated* when we have experimentally determined precisely what volume is delivered for each unit of volume etched on the apparatus at the temperature value also etched.

An experiment calls for a 0.150 *m* solution of sodium chloride in water. How many grams of NaCl would have to be dissolved in 500 g of water to prepare a solution of this molality?

ANALYSIS We work with the basic definition; *molality* means moles of solute per kilogram of solvent. This is a *ratio,* so 0.150 *m* signifies the following ratio, where we substitute 1000 g for 1 kg.

$$\frac{0.150 \text{ mol NaCl}}{1000 \text{ g H}_2\text{O}}$$

To calculate the number of moles of NaCl we need for 500 g of H_2O, we use this ratio as a conversion factor.

SOLUTION

$$500 \text{ g H}_2\text{O} \times \frac{0.150 \text{ mol NaCl}}{1000 \text{ g H}_2\text{O}} = 0.0750 \text{ mol NaCl}$$

This gives us the number of *moles* of NaCl needed. We next use the formula mass of NaCl, 58.44, in the usual way as a tool to convert 0.0750 mol of NaCl to grams of NaCl.

$$0.0750 \text{ mol NaCl} \times \frac{58.44 \text{ g NaCl}}{1 \text{ mol NaCl}} = 4.38 \text{ g NaCl}$$

Thus, when 4.38 g of NaCl is dissolved in 500 g of H_2O, the concentration is 0.150 *m* NaCl.

■ **Practice Exercise 4** Water freezes at a lower temperature when it contains solutes. To study the effect of methyl alcohol on the freezing point of water, we might begin by preparing a series of solutions of known molalities. Calculate the number of grams of methyl alcohol (CH_3OH) needed to prepare a 0.250 *m* solution, using 2000 g of water.

■ **Practice Exercise 5** If you prepare a solution by dissolving 4.00 g of NaOH in 250 g of water, what is the molality of the solution?

Conversions among Concentration Units   Sometimes, in working with a solution whose weight percent is known, we run into a situation where we would like to know its molality. We'll show how to handle a problem like this to demonstrate how *the basic definitions of various concentration expressions are finally all that we need to work our way to the answer.*

EXAMPLE 12.2
Calculation to Prepare a
Solution of a Given Molality

What is the molality of 10.0% aqueous NaCl?

ANALYSIS Working from the basic definitions of molality and percent by weight, we translate the heart of the question into a form that displays them. In other words, the question asks us to go from

$$\frac{10.0 \text{ } grams \text{ } of \text{ } NaCl}{100.0 \text{ } grams \text{ } of \text{ } NaCl \text{ } soln} \quad \text{to} \quad \frac{? \text{ } mole \text{ } of \text{ } NaCl}{1 \text{ } kg \text{ } of \text{ } water}$$

The first ratio applies the definition of a weight percent, and the second uses that of molality. We can now see that we must carry out a grams-to-moles conversion for

EXAMPLE 12.3
Finding Molality from Weight
Percent

10.0 g of NaCl. Then we must calculate the number of grams of *water* in 100.0 grams of NaCl *solution* and change the result to kilograms. Finally, to obtain the molality we seek, we take the ratio of the number of moles of NaCl to the kilograms of water.

SOLUTION We calculate the number of moles of NaCl in 100 g of the solution in the usual way by a grams-to-mole conversion using the formula mass of NaCl, 58.44.

$$10.0 \text{ g NaCl} \times \frac{1 \text{ mol NaCl}}{58.44 \text{ g NaCl}} = 0.171 \text{ mol NaCl}$$

To calculate the number of kilograms of *water* in 100 g of the *solution*, we must first subtract out the mass of the solute in 100 g of the solution and change the result to kilograms.

$$100.0 \text{ g} - 10.0 \text{ g} = 90.0 \text{ g or } 0.0900 \text{ kg (of water)}$$

Now we compute the molality by taking the following ratio.

$$\text{Molality} = \frac{0.171 \text{ mol NaCl}}{0.0900 \text{ kg H}_2\text{O}}$$

$$= 1.90 \text{ } m \text{ NaCl}$$

Thus a 10.0% NaCl solution is also 1.90 molal.

■ **Practice Exercise 6** A certain sample of concentrated hydrochloric acid is 37.0% HCl. Calculate the molality of this solution.

Another kind of calculation is to find the molarity of a solution from its weight percent. This cannot be done without the density of the solution, as the next example shows.

EXAMPLE 12.4
Finding Molarity from Weight Percent

A certain supply of concentrated hydrochloric acid has a concentration of 36.0% HCl. The density of the solution is 1.19 g mL^{-1}. Calculate the molar concentration of HCl.

ANALYSIS Working again from basic definitions, here those of percent by weight and molarity, we see that our task is the conversion of

$$\frac{36.0 \text{ } grams \text{ HCl}}{100 \text{ } grams \text{ HCl soln}} \quad \text{to} \quad \frac{? \text{ } moles \text{ HCl}}{1 \text{ } liter \text{ HCl soln}}$$

The numerators show that we need a grams-to-moles conversion for the solute, HCl. The denominators tell us that we must carry out a grams-to-liters conversion for the HCl solution, which is why we need the solution's density. *Density is the tool that connects mass to volume.*

SOLUTION First, we find the number of moles of HCl in 100 g of 36.0% HCl solution in the usual way, using the molecular mass of HCl (36.46).

$$36.0 \text{ g HCl} \times \frac{1 \text{ mol HCl}}{36.46 \text{ g HCl}} = 0.987 \text{ mol HCl}$$

When the density of a *solution* is given, "g/mL" always means the ratio of grams of *solution* to milliliters of *solution*.

To find the number of liters of HCl solution in 100 g of HCl solution we use the density, 1.19 g mL^{-1}, as our tool. Because of the way we define density, *both* units in the given density refer to the *solution*. There are 1.19 g HCl *solution* per milliliter of HCl *solution*. The given density, for example, gives us the following conversion factors. (We *must* carry the complete units into the calculation if we intend to use the factor-label method.)

$$\frac{1 \text{ mL HCl soln}}{1.19 \text{ g HCl soln}} \quad \text{and} \quad \frac{1.19 \text{ g HCl soln}}{1 \text{ mL HCl soln}}$$

So to calculate the number of milliliters (and then, liters) of HCl solution in 100 g of HCl solution, we simply use the first conversion factor.

$$100 \text{ g HCl soln} \times \frac{1 \text{ mL HCl soln}}{1.19 \text{ g HCl soln}} = 84.0 \text{ mL or } 0.0840 \text{ L HCl soln}$$

Thus, the volume occupied by 100 g of 36.0% HCl is 0.0840 L. The molarity of the solution is then found by the ratio of the number of moles of HCl to liters of HCl solution:

$$\text{Molarity} = \frac{0.987 \text{ mol HCl}}{0.0840 \text{ L HCl soln}} = 11.8 \ M \text{ HCl}$$

Thus 36.0% HCl is also 11.8 M HCl.

■ **Practice Exercise 7** Hydrobromic acid can be purchased as 40.0% HBr. The density of this solution is 1.38 g mL^{-1}. What is the molar concentration of HBr in this solution?

Colligative Properties

The physical properties of solutions to be studied in this and succeeding sections are called **colligative properties,** because they depend mostly on the relative *populations* of particles in mixtures, not on their chemical identities. In this section, we examine the effect of a solute on the vapor pressure of a solvent in a liquid solution.

Solutions of a Nonvolatile Solute

All liquid solutions of nonvolatile solutes have lower vapor pressures than their pure solvents. How much lower is a function of the *mole fraction* of the solvent. Because a ratio by moles is the same as a ratio by molecules, the mole fraction of the solvent gives us the population of its molecules relative to the total population of all molecules.

Raoult's Law   A particularly simple law, **Raoult's law,** describes how a solution's vapor pressure is related to the mole fraction of *solvent*. Raoult's law says that the vapor pressure of a solution of a *nonvolatile* solute, P_{solution}, is the mole fraction of the *solvent*, X_{solvent}, times the vapor pressure of the pure solvent, $P^{\circ}_{\text{solvent}}$.

Raoult's Law
- - - - - - - - - - - - - - -

$$P_{\text{solution}} = X_{\text{solvent}}P^{\circ}_{\text{solvent}} \tag{12.3}$$

12.6
EFFECTS OF SOLUTES ON VAPOR PRESSURES OF SOLUTIONS

After the Greek *kolligativ,* depending on number and not on nature.

Nonvolatile means "cannot evaporate"; the substance has a negligible vapor pressure. Volatile means "can evaporate"; the substance has a high vapor pressure.

Francois Marie Raoult (1830–1901) was a French scientist.

A solution of sugar in water is an example of a solution of a nonvolatile solute.

T Raoult's law

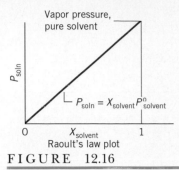

Vapor pressure,
pure solvent

$P_{soln} = X_{solvent}P^o_{solvent}$

$X_{solvent}$

Raoult's law plot

FIGURE 12.16

A Raoult's law plot of the vapor pressure of a solution versus the mole fracton of the solvent.

FIGURE 12.17

Lowering of vapor pressure by a nonvolatile solute. Because particles of the solute are present, the opportunities for the escape of the solvent particles are reduced. Therefore, the vapor pressure of this solution is less than the vapor pressure of the pure solvent (at the same temperature).

Because of the form of Equation 12.3, a plot of $P_{solution}$ versus $X_{solvent}$ should be *linear* at all concentrations if the system obeys Raoult's law (see Figure 12.16)

The reason for Raoult's law is easy to understand (see Figure 12.17). Although solute molecules cannot evaporate, they can interfere with the escape of solvent molecules into the vapor state, the forward change in the following equilibrium

Solvent molecules in solution $\rightleftharpoons$ solvent molecules in vapor state
(Their escape is impeded by solute)   (Their return is unimpeded by solute)

The reverse change, however, is not hindered by the presence of solute molecules. The equilibrium thus shifts to the left when the concentration of a nonvolatile solute is increased, and the vapor pressure decreases until the rates of the forward and reverse changes are again equal. How much lower the vapor pressure becomes depends on the fraction of the *solvent* molecules at the solution's surface. For example, in a solution with 75 mole % of *solvent*, 75% of its surface molecules are those of the solvent. The vapor pressure must therefore be 75% of its value for the pure solvent, which is exactly what Raoult's law says.

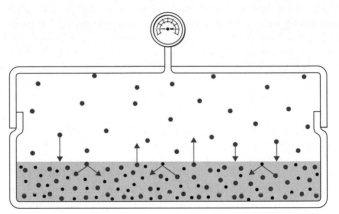

EXAMPLE 12.5

Calculating with Raoult's Law

Carbon tetrachloride has a vapor pressure of 100 torr at 23°C. This solvent can dissolve candle wax, which is essentially nonvolatile. Although candle wax is a mixture, we can take its molecular formula to be $C_{22}H_{46}$ (molecular mass 311). What is the vapor pressure at 23 °C of a solution prepared by dissolving 10.0 g of wax in 40.0 g of CCl_4 (molecular mass 154)?

ANALYSIS This problem involves the vapor pressure of a *solution* and so we need the Raoult's law equation (12.3). We must determine the mole fraction of the *solvent* before we can use this law, so we have to calculate the number of moles of solute and solvent separately and then add them.

SOLUTION

For CCl_4 $40.0 \text{ g } CCl_4 \times \dfrac{1 \text{ mol } CCl_4}{154 \text{ g } CCl_4} = 0.260 \text{ mol } CCl_4$

For $C_{22}H_{46}$ $10.0 \text{ g } C_{22}H_{46} \times \dfrac{1 \text{ mol } C_{22}H_{46}}{311 \text{ g } C_{22}H_{46}} = 0.0322 \text{ mol } C_{22}H_{46}$

The total number of moles = 0.292 mol

Now, we can calculate the mole fraction of the solvent, CCl_4.

$$X_{CCl_4} = \frac{0.260 \text{ mol}}{0.292 \text{ mol}} = 0.890$$

The vapor pressure of the solution will be this particular mole fraction, 0.890, of the vapor pressure of pure CCl_4 (100 torr), calculated by Raoult's law.

$$P_{\text{solution}} = 0.890 \times 100 \text{ torr} = 89.0 \text{ torr}$$

The presence of the wax in the CCl_4 lowers the vapor pressure of the CCl_4 from 100 to 89.0 torr.

■ **Practice Exercise 8** Dibutyl phthalate, $C_{16}H_{22}O_4$ (molecular mass 278), is an oil sometimes worked into plastic articles to give them softness. It has a negligible vapor pressure, having to be heated to 148 °C before its vapor pressure is even 1 torr. What is the vapor pressure, at 20 °C, of a solution of 20.0 g of dibutyl phthalate in 50.0 g of octane, C_8H_{18} (molecular mass, 114)? The vapor pressure of pure octane at 20 °C is 10.5 torr.

Solutions That Contain Two or More *Volatile* Components

When two (or more) components of a liquid solution can evaporate, the vapor contains molecules of each. Each volatile component contributes its own partial pressure to the total pressure. The partial pressure of a particular component is directly proportional to the component's mole fraction in the solution. This is reasonable because the rate of evaporation of each compound has to be lowered by molecules of the others in the same way we just described for nonvolatile solutes. So, by Dalton's law of partial pressures, the total vapor pressure will be the sum of the partial pressures. To calculate these partial pressures, we use Raoult's law for each component. When component A is present in a mole fraction of X_A, its partial pressure (P_A) is this fraction of its vapor pressure when pure, P_A°.

$$P_A = X_A P_A^\circ$$

And the partial pressure of component B is

$$P_B = X_B P_B^\circ$$

The total pressure of the solution of liquids A and B is then

$$P_{\text{total}} = P_A + P_B$$

Remember, P_A and P_B here are the *partial* pressures as calculated by Raoult's law.

EXAMPLE 12.6
Calculating the Vapor Pressure of a Solution of Two Liquids

Acetone is a solvent for both water and molecular liquids that do not dissolve in water, like benzene. At 20 °C acetone has a vapor pressure of 162 torr. The vapor pressure of water at 20 °C is 17.5 torr. What is the vapor pressure of a solution of acetone and water with 50.0 mole % of each?

ANALYSIS We assume that Raoult's law applies. To find P_{total} we need to calculate the individual partial pressures and then add them.

SOLUTION

$$P_{\text{acetone}} = 162 \text{ torr} \times 0.500 = 81.0 \text{ torr}$$

$$P_{\text{water}} = 17.5 \text{ torr} \times 0.500 = \underline{8.75 \text{ torr}}$$

$$P_{\text{total}} = 89.8 \text{ torr}$$

Thus the vapor pressure of the solution is much higher than that of pure water but much less than that of pure acetone.

■ **Practice Exercise 9** At 20 °C, the vapor pressure of cyclohexane, a hydrocarbon solvent, is 66.9 torr and that of toluene (another solvent) is 21.1 torr. What is the vapor pressure of a solution of the two at 20 °C when each is present at a mole fraction of 0.500?

Ideal Solutions and Raoult's Law

As we described in Section 12.2, an *ideal solution* is one for which ΔH_{soln} is zero, and therefore one in which forces of attraction between all molecules are identical. Only an ideal solution would obey Raoult's law exactly. Figure 12.18 shows how the total vapor pressure of a two-component, ideal solution of volatile liquids is related to the solution's composition. The two steeper lower lines show how the *partial* vapor pressures of the components change as their mole fractions change. These are straight lines, as required by the linear relationship of Equation 12.3 (Raoult's law). Any point on the top line in Figure 12.18 represents the total pressure at a particular pair of values of mole fractions. Each point on the top line is the sum of the partial pressures given by points on the lines just below it.

Deviations from Raoult's Law Few real, two-component solutions come close to being ideal. Figure 12.19 shows two of the typical ways in which real two-component solutions behave. When Raoult's law is not obeyed, the indi-

FIGURE 12.18

The vapor pressure of an ideal, two-component solution of volatile compounds.

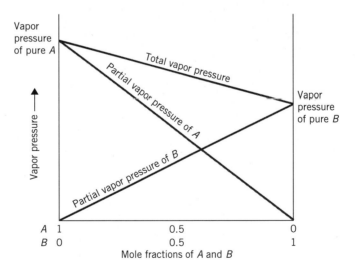

FIGURE 12.19

Typical deviations from ideal behavior of the total vapor pressures of real, two-component solutions of volatile substances.

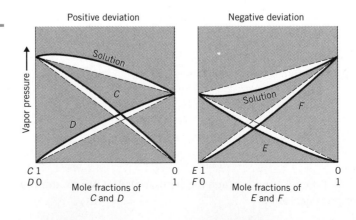

TABLE 12.3   Deviations from Raoult's Law

Relative Attractive Forces	ΔH_{soln}	Raoult's Law Plots
A to A or B to B forces are greater than those between A and B	Positive (endothermic)	Positive deviations
A to A and B to B forces are all equal and equal to those between A and B	Zero	Obeyed perfectly
A to A or B to B forces are less than those of A to B	Negative (exothermic)	Negative deviations

vidual partial pressures are not linear with mole fraction. Hence, the sums of the partial pressures are not linear either. The sums, represented by the upper plots of total pressure, show either upward or *positive deviations*, or they give downward or *negative deviations* from Raoult's law behavior. What causes this?

Negative deviations occur when solutions form exothermically. To see why, let's call the solute *A* and the solvent *B*. We have already explained that when a solution forms exothermically, the attractions between molecules of *A* and *B* in the *solution* are stronger than those between them when pure. *A* or *B* molecules, therefore, are held more strongly within the solution than as pure *A* or pure *B*. Hence they are less able to escape the solution than from their pure forms. As a result, their partial pressures above the solution are *lower* than calculated by Raoult's law, so the solution as a whole has a lower than expected vapor pressure. Negative deviations, in other words, imply stronger intermolecular forces between solute and solvent than exist in the pure components.

Just the opposite occurs when the intermolecular attractions within the solution of *A* and *B* are *weaker* than within their pure forms. It is thus easier for an *A* or *B* molecule to escape from the solution than from their pure liquids, so their partial pressures above the solution are *higher* than calculated by Raoult's law. Such solutions form endothermically and show positive overall deviations from Raoult's law.

Table 12.3 summarizes the relationships between Raoult's law predictions and the kinds of deviations just discussed.

Acetone–water solutions show negative deviations.

Ethyl alcohol–hexane solutions show positive deviations. (Hexane is a nonpolar component of gasoline.)

Solutes affect more than the vapor pressures of solutions. The boiling point of a solution, for example, is higher than that of the solvent itself, and the freezing point of a solution is lower than the pure solvent's freezing point. For aqueous solutions, we can understand why by returning to the phase diagram for water, first studied in Section 11.9.

12.7
EFFECTS OF SOLUTES ON FREEZING AND BOILING POINTS OF SOLUTIONS

Effect of a Solute on the Phase Diagram of Water

Figure 12.20*a* shows the phase diagram for pure water by the curves in blue. Remember that these are plots of the pressures and temperatures that have the only pairs of values at which the various phases are able to be *in equilibrium*. The *triple point* occurs where all three phases are able to be in equilibrium simultaneously. Two vertical dashed lines in Figure 12.20*a* extend from particular points on the solid–liquid and the vapor–liquid curves, points that correspond to a pressure of 760 torr (1 atm). These vertical lines thus intersect the temperature axis at the *normal* freezing point and the *normal* boiling point of pure water.

The word "normal" in *normal* boiling or freezing point means that the system is under 1 atm of pressure when boiling or freezing.

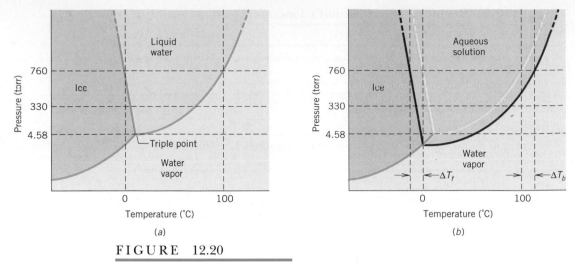

FIGURE 12.20

(*a*) Phase diagram for pure water. (*b*) Phase diagram for an aqueous solution.

Now let's replace the pure water in the phase diagram by an aqueous solution of a nonvolatile solute (see Figure 12.20*b*). *The solute stays entirely in the liquid phase.* None of its molecules are in the vapor phase, which is pure water vapor. None are in the solid (ice) phase, because the structure of the ice crystals does not admit alien molecules. Let's see now how all this affects the curves that separate the phases in the phase diagram.

First, the curved line that separates ice from water vapor is unaffected because neither ice nor water vapor contains solute molecules. However, the solute makes the vapor pressure of the *solution* less than that of the pure solvent (water) at every temperature. Therefore, the corresponding curved line (in red) between the solution and vapor phases *is everywhere lower* than the line (light blue) for pure water. The triple point is thus forced to a lower value. The line defining the equilibrium between the solid and liquid phases, which slants upward from the triple point, is therefore pushed to the left. The two new curves (in red) for an aqueous solution in Figure 12.20*b* are everywhere *below* the curves for pure water.

One technology for obtaining salt-free water from the ocean is to freeze seawater, separate the ice crystals (pure water, when washed free of brine), and let the ice melt.

Freezing Point Depression and Boiling Point Elevation

Find the points where the new (red) curves in Figure 12.20*b* and those for pure water (soft blue) intersect at the horizontal dashed line set at a pressure of 760 torr. The two intersections of the red curves with this horizontal line represent, left to right, the freezing point and the normal boiling point of the *solution.* As you can see, the freezing point of the solution is *lower* than that of pure water by an amount equal to ΔT_f, and the boiling point is *higher* by ΔT_b. The first phenomenon is called **freezing point depression** by a solute, and the latter is **boiling point elevation;** both are caused by the presence of a solute. We say that the solute *depresses* the freezing point and *elevates* the boiling point.

Freezing point depression and boiling point elevation are both *colligative properties.* The magnitudes of ΔT_f and ΔT_b are proportional to the relative populations of solute and solvent molecules. Molality (*m*) is the preferred concentration expression (rather than mole fraction or percent by weight) because of the resulting simplicity of the equations relating ΔT to concentra-

**TABLE 12.4 Molal Boiling Point Elevation and Freezing Point
Depression Constants**

Solvent	BP (°C)	K_b (°C m^{-1})	MP (°C)	K_f
Water	100	0.51	0	1.86
Acetic acid	118.3	3.07	16.6	3.57
Benzene	80.2	2.53	5.45	5.07
Chloroform	61.2	3.63	—	—
Camphor	—	—	178.4	37.7
Cyclohexane	80.7	2.69	6.5	20.0

tion. The following equations work reasonably well only for dilute solutions, however.

$$\Delta T_f = K_f m \qquad (12.4)$$

$$\Delta T_b = K_b m \qquad (12.5)$$

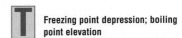 Freezing point depression; boiling point elevation

K_f and K_b are the proportionality constants and are called, respectively, the **molal freezing point depression constant** and the **molal boiling point elevation constant.** The values of both K_f and K_b are characteristic of each solvent (see Table 12.4). The units of each constant are °C m^{-1}. For example, the value of K_f for a given solvent corresponds to the number of degrees of freezing point lowering for each molal unit of concentration. The K_f for water is 1.86 °C m^{-1} and water freezes normally at 0 °C. Therefore, a 1.00 m solution in water freezes at 1.86 °C *below* the normal freezing point, or at −1.86 °C. Similarly, because K_b for water is 0.51 °C m^{-1}, a 1.00 m aqueous solution boils at (100 + 0.51) °C or 100.51 °C.

EXAMPLE 12.7
Estimating a Freezing Point
Using a Colligative Property

Estimate the freezing point of a permanent-type antifreeze solution made up of 100.0 g of ethylene glycol, $C_2H_6O_2$ (molecular mass, 62.1), and 100.0 g of water (molecular mass, 18.0).

ANALYSIS Equation 12.4 is clearly appropriate because the problem concerns freezing point depression. We must think only of an "estimate," however, because this solution is too concentrated for Equation 12.4 to be very reliable. To use Equation 12.4, we first calculate the molality of the solution, that is, the number of moles of ethylene glycol per kilogram of water.

HOCH₂CH₂OH
ethylene glycol

SOLUTION For the number of moles of ethylene glycol,

$$100.0 \text{ g } C_2H_6O_2 \times \frac{1 \text{ mol } C_2H_6O_2}{62.1 \text{ g } C_2H_6O_2} = 1.61 \text{ mol } C_2H_6O_2$$

This 1.61 mol of $C_2H_6O_2$ is contained in 100 g or 0.100 kg of water, so the molality is found by taking the following ratio.

$$\text{Molality} = \frac{1.61 \text{ mol}}{0.100 \text{ kg}} = 16.1 \ m$$

Now we use Equation 12.4 ($\Delta T_f = K_f m$), knowing from Table 12.4 that K_f for water is 1.86 °C m^{-1}.

$$\Delta T_f = (1.86 \text{ °C } m^{-1})(16.1 \ m)$$
$$= 29.9 \text{ °C}$$

The solution, therefore, freezes at $(0 - 29.9)$ °C or at -29.9 °C. (Actually, remember, the best we can say is that we *estimate* a freezing point of about -30 °C because of the uncertainty over the precision of Equation 12.4 at this concentration.)

■ **Practice Exercise 10** At what temperature will a 10% aqueous solution of sugar ($C_{12}H_{22}O_{11}$) boil?

Determination of Molecular Masses We have described the properties of freezing point depression and boiling point elevation as *colligative;* that is, they depend on the relative *numbers* of particles, not on their kinds. Because the effects are proportional to molal concentrations, experimental values of ΔT_f or ΔT_b should be useful for calculating the molecular masses of unknown solutes. We'll see how this works in the next example.

EXAMPLE 12.8
Calculating a Molecular Mass from Freezing Point Depression Data

A solution made by dissolving 5.65 g of an unknown molecular compound in exactly 100 g of benzene froze at 4.39 °C. What is the molecular mass of the solute?

ANALYSIS We can use Equation 12.4 to calculate the number of moles of solute in the given solution. Then we take the ratio of grams of solute to number of moles of solute to find the molecular mass. Our first task, therefore, is to calculate the number of moles of solute in 5.65 g of the unknown.

SOLUTION Table 12.4 tells us the following facts about benzene:

$$\text{Melting point} = 5.45 \text{ °C}$$

$$K_f = 5.07 \text{ °C } m^{-1}$$

The freezing point depression is

$$\Delta T_f = 5.45 \text{ °C} - 4.39 \text{ °C} = 1.06 \text{ °C}$$

We now use Equation 12.4 to find the molality of the solution

$$\Delta T_f = K_f m$$

$$\text{Molality} = \frac{\Delta T_f}{K_f}$$

$$= \frac{1.06 \text{ °C}}{5.07 \text{ °C } m^{-1}} = 0.209 \ m$$

This result means that there is a ratio of 0.209 mol of solute per 1 kg of benzene. But we do not have this much benzene, only 110 g or 0.110 kg. So the actual number of moles of solute in the given solution is found by

$$0.110 \text{ kg benzene} \times \frac{0.209 \text{ mol solute}}{1 \text{ kg benzene}} = 0.0230 \text{ mol solute}$$

With this result we know that 5.65 g of solute is the same as 0.0230 mol of solute, so the ratio of grams to moles is

$$\frac{5.65 \text{ g}}{0.0230 \text{ mol}} = 246 \text{ g mol}^{-1}$$

Thus, the molecular mass of the solute is 246.

■ **Practice Exercise 11** A solution made by dissolving 3.46 g of an unknown compound in 85.0 g of benzene froze at 4.13 °C. What is the molecular mass of the compound?

In living things, membranes of various kinds keep mixtures and solutions organized and separated. Yet some substances have to be able to pass through membranes in order that nutrients and products of chemical work can be distributed correctly. These membranes, in other words, must have a selective *permeability*. They must keep some substances from going through while letting others pass. Such membranes are said to be *semipermeable*.

The degree of permeability varies with the kind of membrane. Cellophane, for example, is permeable to water and small solute particles—ions or molecules—but impermeable to very large molecules, like those of starch or proteins. Special membranes can even be prepared that are permeable only to water, not to any solutes.

Depending on the kind of membrane separating solutions of different concentration, two similar phenomena, *dialysis* and *osmosis,* can be observed. Both are functions of the relative populations of the particles of the dissolved materials, so they are colligative properties. When a membrane is able to let both water and *small* solute particles through, such as the membranes in living systems, the process is called **dialysis,** and the membrane is called a *dialyzing membrane.* It does not permit extralarge molecules through. Artificial kidney machines use dialyzing membranes to help remove the smaller molecules of wastes from the blood while letting the blood retain its large protein molecules.

Osmosis

At its limit, a semipermeable membrane will let only solvent molecules get through. The phenomenon is now called *osmosis,* and the special membrane needed to observe it is called an *osmotic membrane.* Such membranes are rare, but they can be made.

One theory that explains osmosis is illustrated in Figure 12.21. An osmotic membrane separates two solutions from each other. One solution, B in Figure 12.21a, is more concentrated in solute than the other, A, which is water. The solvent in each has its own "escaping tendency," which is exactly like the vapor pressure of a solution. Like vapor pressure, the escaping tendency of the solvent is lowered by the presence of nonvolatile solutes. Solvent molecules are able to move through the membrane in both directions, but they cannot do so at equal rates. The pure water, A, has the larger mole fraction of *solvent,* so the escaping tendency of its solvent molecules is greater than that of solution B.

12.8
DIALYSIS AND OSMOSIS. OSMOTIC PRESSURE

Latin *permeare,* to go through.

The portability of this artificial kidney machine means that the user does not have to make trips to the dialysis clinic as often.

By Raoult's law:

$$P_{solution} = X_{solvent} \times P^\circ_{solvent}$$

FIGURE 12.21

Osmosis and osmotic pressure. (*a*) Initial conditions. A solution, B, is separated from pure water, A, by an osmotic membrane, but no change has yet occurred. (*b*) Enlarged view at the membrane. Water molecules (small circles) are moving into B more easily and hence more rapidly than they are leaving B for A. The solute molecules (larger dots) interfere with the movement of water from B to A. (*c*) After a while the volume of fluid in the tube has increased visibly. Osmosis has occurred. (*d*) A back pressure is needed to prevent osmosis. This is the osmotic pressure of the solution.

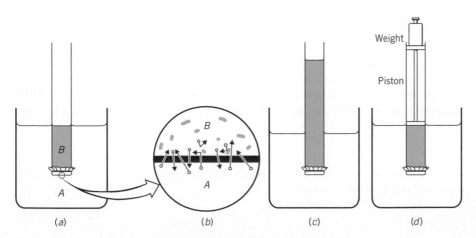

Weight			
Piston			
(a)	(b)	(c)	(d)

The direction of osmosis is here traditionally defined with reference to the concentration of the *solute,* not the solvent.

Thus, solvent passes more rapidly from *A* to *B* than from *B* to *A*, and a net transfer of solvent *from* the less concentrated side into the more concentrated solution occurs. The phenomenon is **osmosis,** a net shift of solvent (only) through the membrane from the side less concentrated in *solute* to the side more concentrated.

Osmotic Pressure

What can stop osmosis is the development of a back pressure. As illustrated in Figure 12.21*c*, a column of solution rises as solvent moves into the more concentrated solution. The extra weight of the liquid in the rising column creates a back pressure. We could actually prevent osmosis, or even force it to reverse, by applying an opposing pressure, as illustrated in Figure 12.21*d*. The exact back pressure needed to prevent any osmotic flow *when one of the liquids is pure solvent* is called the **osmotic pressure** of the solution.

Be careful to notice that the term "osmotic pressure" uses the word *pressure* in a novel way. A solution does not "have" a special pressure called osmotic pressure apart from osmosis. What the solution has is a *concentration* that can generate the occurrence of osmosis and the associated osmotic pressure in a special circumstance. Then, in proportion to the solution's concentration, it requires a specific back pressure to prevent osmosis. By *exceeding* this back pressure, osmosis can be reversed. The use of *reverse osmosis* to purify sea water is described in Special Topic 12.3.

It might be helpful in understanding osmosis to refer to Figure 12.22, where we have a beaker of water in an enclosed space *already saturated in water vapor* and we have just added a beaker of an aqueous solution of a nonvolatile solute. Water molecules escape from and return to both beakers. At equilibrium over the beaker of pure water, the rates of escape and return are equal, but a complication occurs above the solution. The rate of *return* of water molecules there is greater than the rate of escape because the solute molecules interfere with the evaporation (escape) of the solvent molecules. Thus, at the solution–vapor interface, more water molecules enter the solution than leave, and this takes water from the vapor state. To replace it and thus keep the vapor phase saturated, more water must evaporate from the pure water. The net effect is that water moves from the beaker of pure water into the vapor space and then down into the solution. In osmosis, an osmotic membrane instead of beakers separates the liquids, but the same principles are at work.

It is not hard to imagine that the higher the concentration of the *solution* is, the more water will transfer to it (referring to Figure 12.22). In other words, the osmotic pressure of a solution is proportional to the relative concentrations of its solute and solvent. Although *molality* would, in principle, serve better to express this proportionality by an equation, in any *dilute* aqueous solution molality and molarity are virtually identical, as we noted earlier. It simplifies matters to use *molarity*.

The symbol for osmotic pressure is the Greek capital pi, Π. In a dilute aqueous solution, Π is proportional both to temperature, *T*, and molar concentration, *M*. Therefore, we can write

$$\Pi \propto MT$$

The proportionality constant turns out to be the gas constant, *R*, so for a *dilute* aqueous solution we can write

$$\Pi = MRT \tag{12.6}$$

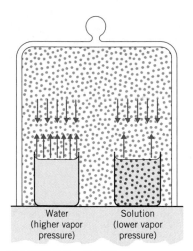

FIGURE 12.22

The vapor pressure of the pure solvent is greater than the vapor pressure of the solution made from this solvent. So liquid will transfer from the pure solvent to the solution.

The fact that the density of water is on the order of 1 gram per milliliter is what makes molality and molarity very nearly the same when the solution is dilute.

SPECIAL TOPIC 12.3 / DESALINATION OF SEAWATER BY REVERSE OSMOSIS

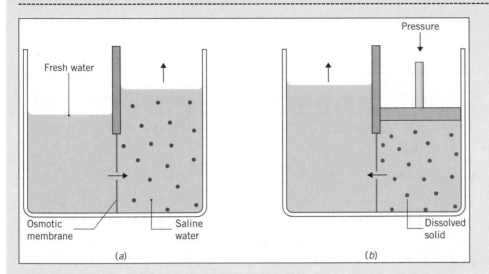

FIGURE 1
(*a*) Osmosis. (*b*) Reverse osmosis.

We defined *osmotic pressure* as the pressure needed to prevent osmosis from occurring in its natural direction, the direction that transfers water from pure water (or a dilute solution) to a concentrated solution. If we apply a pressure exceeding this, we not only prevent osmosis, we reverse it. We force water through an osmotic membrane from a concentrated solution into what can only be called pure water on the other side. The process is called **reverse osmosis** (Figure 1).

Pressure, of course, is not free. There have to be pressure pumps, suitable membranes, and the energy to run it all, but in many parts of the world energy is far more readily available than drinkable water. Membranes of cellulose acetate, a chemically modified form of plant cellulose, do not allow the ions or other impurities in seawater through. Reverse osmosis desalination plants are in operation today capable of producing 3 million gallons of fresh water daily. The U.S. military base at Guantánamo Bay, Cuba, has made fresh water by reverse osmosis since Fidel Castro became the Cuban dictator and threatened to cut the base off from its normal supply.

The technology for small-scale reverse osmosis also exists. Life rafts, for example, can be equipped with portable reverse osmosis desalination devices capable of generating enough

fresh water daily for 25 people (Figure 2). The devices have several internal chambers to enhance the pressures applied by the users.

FIGURE 2
Portable, hand-operated device for desalination of seawater by reverse osmosis.

Of course, *M* equals mol/L, which equals n/V, where n is the number of moles and V is the volume in liters. We can therefore substitute n/V for M in Equation 12.6. After rearranging terms, we have an equation for osmotic pressure identical in form to the ideal gas law.

$$\Pi V = nRT \qquad (12.7)$$

Osmotic pressure

Equation 12.7 is the *van't Hoff equation for osmotic pressure.*

Osmotic pressures can be measured by an instrument called an *osmometer*, illustrated and explained in Figure 12.23. Osmotic pressures can be very high, even in dilute solutions, as Example 12.9 shows.

EXAMPLE 12.9
Calculating an Osmotic Pressure

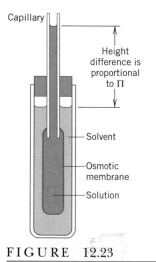

Capillary

Height difference is proportional to Π

— Solvent

— Osmotic membrane

— Solution

FIGURE 12.23

Simple osmometer.

When soil water is too concentrated in dissolved salts, the plant can't get the water it needs. In fact, the plant tends to *lose* water to the soil (as well as by evaporation) and it wilts.

A very dilute solution, 0.0010 *M* sugar in water, is separated from pure water by an osmotic membrane. What osmotic pressure in torr develops at 25 °C or 298 K?

ANALYSIS Equation 12.6 applies; recall that $R = 0.0821$ L atm mol^{-1} K^{-1}.

SOLUTION

$$\Pi = (0.0010 \text{ mol L}^{-1})(0.0821 \text{ L atm mol}^{-1} \text{ K}^{-1})(298 \text{ K})$$
$$= 0.024 \text{ atm}$$

The osmotic pressure is 0.024 atm. In torr we have

$$P = 0.024 \text{ atm} \times \frac{760 \text{ torr}}{1 \text{ atm}} = 18 \text{ torr}$$

The osmotic pressure of 0.0010 *M* sugar in water is 18 torr.

■ **Practice Exercise 12** What is the osmotic pressure in torr of a 0.0115 *M* glucose solution when the temperature is 25 °C?

In the preceding example, you saw that a 0.0010 *M* solution of sugar has an osmotic pressure of 18 torr. This pressure would support a column of the solution roughly 25 cm or 10 in. high. If the solution had been 100 times as concentrated, 0.100 *M* sugar—still relatively dilute—the height of the column supported would be roughly 25 m or over 80 ft! Thus osmotic pressure is a factor (not the only factor) in the rise of sap in trees.

Molecular Mass Determinations Using Osmotic Pressure An osmotic pressure measurement taken of a dilute solution can be used to determine the molar concentration (see Equation 12.6) and thus also the molecular mass of a molecular solute. The next example shows how.

EXAMPLE 12.10
Calculating Molecular Mass from Osmotic Pressure

An aqueous solution with a volume of 100 mL and containing 0.122 g of an unknown molecular compound has an osmotic pressure of 16.0 torr at 20.0 °C. What is the molecular mass of the solute?

ANALYSIS A molecular mass is the ratio of grams to moles. We're given the number of *grams* of the solute, so we need to find the number of *moles* equivalent to this mass in order to compute the ratio.

To find the number of *moles* of solute we must first use Equation 12.6 to find the moles per liter concentration of the solution, its *molarity*. Remember that to use the gas constant ($R = 0.0821$ L atm mol^{-1} K^{-1}), we need *P* in atm and *T* in kelvins. Then we can use the given volume of the solution (in liters) and the molarity to calculate the number of moles of solute that are represented by 0.122 g of solute. Finally, knowing both the grams and the number of moles of solute, we take the ratio of the two to find the molecular mass.

SOLUTION First, the solution's molarity, M, is calculated using the osmotic pressure. This pressure (Π), 16.0 torr, corresponds to 0.0211 atm. The temperature, 20.0°C, corresponds to 293 K (273 + 20.0). Using Equation 12.6,

$$\text{16.0 torr} \times \frac{\text{1 atm}}{\text{760 torr}} = \text{0.0211 atm}$$

$$\Pi = MRT$$

$$0.0211 \text{ atm} = (M)(0.0821 \text{ L atm mol}^{-1} \text{ K}^{-1})(293 \text{ K})$$

$$M = 8.77 \times 10^{-4} \frac{\text{mol solute}}{\text{L soln}}$$

The 0.122 g of solute is in 100 mL or 0.100 L of solution, not in a whole liter. So the numbers of moles of solute corresponding to 0.122 g is found by

$$\text{Moles of solute} = (0.100 \text{ L soln})\left(8.77 \times 10^{-4} \frac{\text{mol solute}}{\text{L soln}}\right)$$

$$= 8.77 \times 10^{-5} \text{ mol}$$

We have thus shown that 0.122 g of solute is the same as 8.77×10^{-5} mol. The ratio of grams to moles is calculated as follows.

$$\text{Molar mass} = \frac{0.122 \text{ g}}{8.77 \times 10^{-5} \text{ mol}} = 1.39 \times 10^3 \text{ g mol}^{-1}$$

Thus the molecular mass of the compound is 1.39×10^3.

■ **Practice Exercise 13** A solution of a carbohydrate prepared by dissolving 72.4 mg in 100 mL of solution has an osmotic pressure of 25.0 torr at 25.0 °C. What is the molecular mass of the compound?

Expected Effect of the Dissociation of a Solute

A 1.00 m solution of NaCl freezes at − 3.37 °C instead of − 1.86 °C, the expected freezing point of a *molecular* compound dissolved in water at the same molality, 1.00 m. This greater depression of the freezing point by the salt— almost twice as much—is not hard to understand if we remember that colligative properties depend on the concentrations of *particles*. We know that NaCl(s) dissociates into ions in water.

$$\text{NaCl}(s) \longrightarrow \text{Na}^+(aq) + \text{Cl}^-(aq)$$

If the ions are truly separated, 1.00 m NaCl actually has a concentration in dissolved solute particles of 2.00 m, twice the given concentration. Theoretically, then, 1.00 m NaCl should freeze at 2 × (−1.86 °C) or − 3.72 °C. (Why it actually freezes a little higher than this, at − 3.37 °C, will be discussed shortly.)

If we made up a solution of 1.00 m $(NH_4)_2SO_4$, we would have to consider the following dissociation.

$$(\text{NH}_4)_2\text{SO}_4(s) \longrightarrow 2\text{NH}_4^+(aq) + \text{SO}_4^{2-}(aq)$$

Thus, 1 mol of $(NH_4)_2SO_4$ can give a total of 3 mol of ions (2 mol of NH_4^+ and 1 mol of SO_4^{2-}). A 1 m solution of $(NH_4)_2SO_4$, therefore, has a calculated freezing point of 3 × (−1.86 °C) = − 5.58 °C. It actually freezes at − 3.6 °C. The relatively poor agreement between the experimental and calculated values suggest that the theory behind the calculation is flawed (see later).

When we want to roughly *estimate* a colligative property of a solution of an electrolyte, we recalculate the solution's molality using an assumption about the way the solute dissociates or ionizes.

12.9 COLLIGATIVE PROPERTIES OF SOLUTIONS OF ELECTROLYTES

We say "estimate" because the calculations can only give rough values when the simple assumption of 100% dissociation is used.

EXAMPLE 12.11
Estimating the Freezing Point
of a Salt Solution

Estimate the freezing point of aqueous 0.106 m MgCl$_2$, assuming that it dissociates completely.

ANALYSIS The equation that relates a change in freezing temperature to molality,

$$\Delta T_f = K_f m$$

requires a value of K_f from Table 12.4—it's 1.86 °C m^{-1} for water—and a recalculation of m to take into account the assumption that the solute dissociates entirely.

SOLUTION When MgCl$_2$ dissolves in water it breaks up as follows.

$$MgCl_2(s) \longrightarrow Mg^{2+}(aq) + 2Cl^-(aq)$$

Because 1 mol of MgCl$_2$ gives 3 mol of ions, the effective (assumed) molality of what we have called 0.106 m MgCl$_2$ is three times as great.

$$\text{Effective molality} = (3)(0.106 \ m) = 0.318 \ m$$

Because, $\Delta T_f = K_f m$

$$\Delta T_f = (1.86 \text{ °C } m^{-1})(0.318 \ m)$$
$$= 0.591 \text{ °C}$$

Thus, the freezing point is depressed below 0.000 °C by 0.591 °C, so we calculate that this solution freezes at -0.591 °C. (It actually freezes at -0.517 °C.)

■ **Practice Exercise 14** Calculate the freezing point of aqueous 0.237 m LiCl on the assumption that it is 100% dissociated. Calculate what its freezing point would be if the percent dissociation were 0%.

Effects of Interionic Attractions

Neither the 1.00 m NaCl nor the 1.00 m (NH$_4$)$_2$SO$_4$ solution described earlier in this section freezes quite as low as calculated. In Example 12.11 we saw a similar discrepancy. What is wrong, of course, is our assumption that an electrolyte separates *completely* into its ions, which is not quite correct. The ions are not really 100% free of each other. Some oppositely charged ions exist as very closely associated pairs, called *ion pairs*, which behave as single "molecules." Clusters larger than two ions probably also exist. The actual *particle* concentration in a 1.00 m NaCl solution, therefore, is not quite double the molality. It is less than 2.00 m because a small fraction of the Na$^+$ and Cl$^-$ ions are associated in the solution. As a result, the lowering of the freezing point of 1.00 m NaCl is not quite as great as calculated on the basis of 100% dissociation.

As solutions of electrolytes are made more and more *dilute,* the observed and calculated freezing points come closer and closer together. At greater dilutions, the association of ions is less and less a complication because the ions can be farther apart. So the solutes behave more and more as if they were 100% separated into their ions at ever higher dilutions.

Chemists compare the degrees of dissociation of electrolytes at different dilutions by a quantity called the **van't Hoff factor, i.** It is the ratio of the observed freezing point depression to the value calculated on the assumption that the solute dissolves as a nonelectrolyte.

$$i = \frac{(\Delta T)_{\text{measured}}}{(\Delta T)_{\text{calcd as a nonelectrolyte}}}$$

"Associated" means the coming together of particles to form larger particles by intermolecular attraction, without covalent (or ionic) bonds.

Jacobus van't Hoff (1852–1911), a Dutch scientist, won the first Nobel prize in chemistry (1901).

The hypothetical van't Hoff factor, i, is 2 for NaCl, KCl, and MgSO$_4$, which break up into two ions on 100% dissociation. For K$_2$SO$_4$, the theoretical value

TABLE 12.5   Van't Hoff Factors versus Concentration

	Van't Hoff Factor, i			
	Molal Concentration (mol salt/kg water)			If 100% Dissociation Occurred
Salt	0.1	0.01	0.001	
NaCl	1.87	1.94	1.97	2.00
KCl	1.85	1.94	1.98	2.00
K_2SO_4	2.32	2.70	2.84	3.00
$MgSO_4$	1.21	1.53	1.82	2.00

of i is 3 because one K_2SO_4 unit gives 3 ions. The actual van't Hoff factors for several electrolytes at different dilutions are given in Table 12.5. Notice that with decreasing concentration (that is, at higher dilutions) the experimental van't Hoff factors agree better with their corresponding hypothetical van't Hoff factors.

The increase in the percentage of complete dissociation that comes with greater dilution is not the same for all salts. In going from concentrations of 0.1 to 0.001 m, the increase in percentage dissociation of KCl, as measured by the change in i, is only about 7%. But for K_2SO_4, the increase for the same dilution is about 22%, a difference caused by the anion, SO_4^{2-}. It has twice the charge as the anion in KCl and so the SO_4^{2-} ion attracts K^+ more strongly than can Cl^-. Hence, letting an ion of $2-$ charge and an ion of $1+$ charge get farther apart by dilution has a greater effect on their acting independently than giving ions of $1-$ and $1+$ charge more room. When *both* cation and anion are doubly charged, the improvement in percent dissociation with dilution is even greater. Thus, there is an almost 50% increase in the value of i for $MgSO_4$ as we go from a 0.1 to a 0.001 m solution.

Estimation of Percent Dissociation from Freezing Point Depression

Data   In 1.00 m aqueous acetic acid, the following equilibrium exists.

Equilibria of this type are discussed more fully in Chapters 4 and 16.

$$HC_2H_3O_2(aq) \rightleftharpoons H^+(aq) + C_2H_3O_2^-(aq)$$

The solution freezes at $-1.90°C$, which is only a little lower than it would be expected to freeze ($-1.86\ °C$) if no ionization occurred. Evidently some ionization, but not much, has happened. We can estimate the percent ionization by the following calculation.

We first use the data to calculate the *apparent molality* of the solution, that is, the molality of all dissolved species, $HC_2H_3O_2 + C_2H_3O_2^- + H^+$. We again use Equation 12.4 ($\Delta T_f = K_f m$), only now letting m stand for the apparent molality; K_f is 1.86 °C m^{-1}.

$$\text{Apparent molality} = \frac{\Delta T_f}{K_f}$$

$$= \frac{1.90\ °C}{1.86\ °C\ m^{-1}}$$

$$= 1.02\ m$$

If there is 1.02 mol of all solute species in 1 kg of solvent, we have 0.02 mol of solute species more than we started with, because we began with 1.00 mol of acetic acid. Since we get just one *extra* particle for each acetic acid molecule that ionizes, the extra 0.02 mol of particles must have come from the ionization of 0.02 mol of acetic acid. So the percent ionization is

$$\text{Percent ionization} = \frac{\text{mol of acid ionized}}{\text{mol of acid available}} \times 100$$

$$= \frac{0.02}{1.00} \times 100 = 2\%$$

In other words, the percent ionization in 1.00 *m* acetic acid is estimated to be 2% by this procedure. (Other kinds of measurements offer better precision, and the percent ionization of acetic acid at this concentration is actually less than 1%.)

Association of Solute Molecules and Colligative Properties

Some solutes produce *weaker* colligative effects than their molal concentrations would lead us to predict. This is not often seen, but it points to another phenomenon that we should briefly note, namely, the phenomenon of the **association** of solute particles. For example, when dissolved in benzene, benzoic acid molecules associate as *dimers*. They are held together by hydrogen bonds, indicated by the dotted lines in the following.

Di- signifies "two," so a dimer is the result of the combination of two single molecules.

C_6H_5- means

$$2 \; C_6H_5-\overset{\overset{\displaystyle O}{\|}}{C}-O-H \;\rightleftharpoons\; C_6H_5-C\overset{O\cdots H-O}{\underset{O-H\cdots O}{\Big\langle}}C-C_6H_5$$

benzoic acid   benzoic acid dimer

Because of this association, the depression of the freezing point of a 1.00 *m* solution of benzoic acid in benzene is only about one-half the calculated value. By forming a dimer, the effective molecular mass of benzoic acid is twice as much as normally calculated. The larger effective molecular mass lessens the molal concentration of particles by half, and the effect on the freezing point is reduced by one-half.

12.10 COLLOIDAL DISPERSIONS

Importance of Particle Size: Solutions, Colloidal Dispersions, and Suspensions

As we mentioned at the beginning of this chapter, the *sizes* of particles, apart from their relative numbers or their identities, affect the physical properties of homogeneous mixtures. If we mix fine sand with water, for example, and shake the mixture constantly, it continues to be somewhat homogeneous, but the sand particles soon settle when we stop shaking it. Nothing like this occurs when we dissolve salt or sugar in water. Small ions and molecules do not settle out of solution, ever. They have weights too small for gravity to overcome the mixing influence of their kinetic motions.

The shaken mixture of sand and water is an example of a **suspension.** To form suspensions, particles must have at least one dimension larger than about 1000 nm (1 μm). Suspensions that don't settle at once, like powdered sand in water, can nearly always be separated by filtration using ordinary filter paper or an ordinary laboratory centrifuge. The high-speed rotor of a centrifuge creates a strong gravitation-like force sufficient to pull the suspended material down through the fluid.

Properties and Types of Colloidal Dispersions

A **colloidal dispersion**—often simply called a **colloid**—is a mixture in which the dispersed particles have at least one dimension in the range of 1 to 1000 nm (see Table 12.6). Colloidally dispersed particles are generally too small to be trapped by ordinary filter paper.

The tendency not to separate possessed by colloidal dispersions in a fluid state is aided by the collisions that the dispersed particles experience from the constantly moving molecules of the "solvent." Such erratic movement of colloidally dispersed particles is called **Brownian movement** after an English botanist, Robert Brown (1773–1858), who first noticed it when he used a microscope to study pollen grains in water. Particles in suspension, of course, experience the same bumping around, but they are too large to be kept in suspension without outside help, for example, by stirring or shaking the mixture.

Colloidal dispersions that do eventually separate are those in which the dispersed particles, over time, grow too large. For example, you can make a colloidal dispersion of olive oil in vinegar—a salad dressing—by vigorously shaking the mixture. Shaking breaks at least some of the oil into microdroplets having sizes in the right range for colloidal dispersions. However, the microdroplets eventually coalesce and grow in mass until gravity wins over Brownian movement. Evidently, to prepare a *stable* colloidal dispersion, we must not only make the dispersed particles initially small enough but must also keep them from joining together.

The dispersed particles will not coalesce if they carry the same kind of electrical charge—all positive or all negative (see Figure 12.24a). (Ions of the opposite charge would occur in the solvent to keep the system electrically neutral.) Some of the most stable dispersions form when the surfaces of their colloidal particles have preferentially attracted ions of just one kind of charge from a dissolved salt. The dispersed particles of most **sols**—colloidal dispersions of solids in a fluid—are examples. Alternatively, dispersions may form when extremely large, like-charged ions, such as those of proteins, are involved (see Figure 12.24b).

Emulsions—dispersions of one liquid in another—often are relatively stable colloidal dispersions, provided a third component called an **emulsifying agent** is also present. Mayonnaise, for example, is an oil-in-water emulsion (see

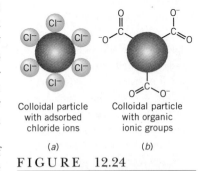

Colloidal particle with adsorbed chloride ions

Colloidal particle with organic ionic groups

(a)   (b)

FIGURE 12.24

Colloidal particles often bear electrical charges that stabilize the dispersion. (*a*). The colloidal particle has attracted chloride ions to itself. In both cases, these colloidal particles cannot join together, grow larger, and eventually separate from the dispersing medium. (*b*) A particle whose extremely large molecules carry negatively charged groups.

TABLE 12.6   Colloidal Systems

Type	Dispersed Phase[a]	Dispersing Medium[b]	Common Examples
Foam	Gas	Liquid	Soap suds, whipped cream
Solid foam	Gas	Solid	Pumice, marshmallow
Liquid aerosol	Liquid	Gas	Mist, fog, clouds, liquid air pollutants
Emulsion	Liquid	Liquid	Cream, mayonnaise, milk
Solid emulsion	Liquid	Solid	Butter, cheese
Smoke	Solid	Gas	Dust, particulates in smog, aerogels
Sol	Solid	Liquid	Starch in water, paint, jellies[c]
Solid sol	Solid	Solid	Alloys, pearls, opals

[a] The colloidal particles constitute the *dispersed phase.*

[b] The continuous matter into which the colloidal particles are dispersed is called the *dispersing medium.*

[c] Sols that become semisolid or semirigid, like gelatin dessert and fruit jellies, are called *gels.*

SPECIAL TOPIC 12.4 / SKY COLORS

When we look at a part of the sky well away from the sun on a reasonably clear day, we see the sky as blue. Why isn't it black? The answer is the occurrence of colloidal-sized particles in the air. No part of our sky is totally free from them, and they scatter sunlight back toward our eyes (Tyndall effect). Then why isn't the sky white, like the sun? Why is it blue? And why is the western sky after sunset often orange or red?

The answers lie in the fact that not all frequencies of light are scattered by colloidal particles with the same intensity. White light from the sun is a mixture of the frequencies of light from all colors of the spectrum—from the lowest visible frequency, red, through orange, yellow, green, blue, and indigo, to the highest frequency, violet. And the intensity of *scattered* light varies with the *fourth power of the frequency.* Thus, blue to violet frequencies are the most intensely scattered. So when the midday sunlight streams across the sky, these frequencies are preferentially scattered toward our vision, and we see a blue sky. When we're enjoying the midday sun, those far to the east or west of us are seeing either a sunset or a sunrise. As we see the scattered blue light, they see more of what isn't scattered so well— the oranges and reds (see Figure 1).

When the sky has a lot of colloidal particles—maybe from a forest fire or from heavy smog—so much blue light is scattered

FIGURE 1

Light scattering colors the sky at sunset.

as the setting sunlight streams toward us that the light that reaches us is left mostly vivid in reds and oranges.

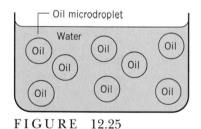

FIGURE 12.25

An oil-in-water emulsion.

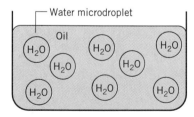

FIGURE 12.26

A water-in-oil emulsion.

Figure 12.25) of an edible oil in a dilute solution of an edible organic acid. The emulsifying agent is lecithin, a component of egg yolk. Its molecules act to give an electrically charged surface to each microdroplet of the oil, which keeps the microdroplets from coalescing. In homogenized milk, the microparticles of butterfat are coated with the milk protein casein. Water-in-oil emulsions are also possible (Figure 12.26); margarine is an example in which a soybean product is the emulsifying agent.

Starch has very large molecules and forms a colloidal dispersion in water. Usually, colloidal starch mixtures have a milky look or can even be opaque, because the colloidal particles are large enough to reflect and scatter visible light. Yet even when they are so dilute that the system appears as clear and transparent as water, the starch molecules are able to scatter enough light to show the light beam. When a beam of light is focused on a dilute and otherwise clear-looking dispersion of starch in water, the path of the light beam can be seen by the light that is scattered to the side (see Figure 12.27). Light scattering by colloidal dispersions is called the **Tyndall effect,** after John Tyndall (1820–1893), a British scientist. Solutes in true solutions, however, involve species too small to scatter light, so solutions do not give the Tyndall effect.

The Tyndall effect is quite common. Whenever you see sunlight "streaming" through openings in a forest canopy or between clouds, you are observing the Tyndall effect. Sunlight is scattered by colloidally dispersed particles of smoke or microdroplets of liquids exuded by trees. Not all wavelengths of visible light are equally well scattered. As a result, when the sun is near the horizon at sunrise or sunset and the air is polluted by smoke, brilliant sky colors are seen (see Special Topic 12.4).

SPECIAL TOPIC 12.5 / AEROGELS—"FROZEN SMOKE"

A *gel* is a colloidal dispersion of a solid in a liquid, and gelatin salads or desserts are familiar examples. They are prepared by adding to hot water a small amount of gelatin, a protein with huge, long molecules, and letting the mixture cool. The long protein molecules become randomly entangled like the brush in a brush pile, and water is caught within the small openings. When cool, the mixture holds the water in a nonfluid manner.

What would happen if the water in a gel could be displaced by air without seriously disturbing the "brush pile"? For one thing, you would have a mass of very low density, roughly 0.01 to 0.02 g mL^{-1}. (The density of air is 0.0012 g mL^{-1} at 20 °C and 1 atm.) Can a water-free gel be made?

The problem is that the surface tension of water is too high (see page 449). As water evaporates from the gel, intermolecular attractions in the residual water film exert internal forces that cause the "brush pile" to collapse. This is why the lungs collapse, for example, when there are insufficient amounts of natural surfactants (surface-active agents) coating the insides (the sides exposed to air) of the lung's small air sacs or alveoli. In the United States, over 60,000 premature infants are annually born having not yet developed enough of their own lung surfactant. Many develop severe respiratory distress syndrome and die. By delivering into the lungs of such infants a synthetic surfactant mixture, many lives are saved, but it requires prompt action.

SILICON AEROGELS

By preparing an aqueous gel of silica, a form of silicon dioxide with large filament-like molecules, and by using supercritical fluids to remove water, scientists have succeeded in preparing a solid, water-free gel so frothy in texture and so light in density that it's sometimes called "frozen smoke" (see Figure 1). Materials like this are called *aerogels,* and those made of silica are called *silica aerogels*. Samples of silica aerogels seem to be wisps of blue haze as seemingly insubstantial as puffs of smoke.

One way to prepare silica aerogels is to dilute a gelatinous mixture of silica and water and then treat it with a basic catalyst like ammonia. After the mixture is allowed to rest for several days the essential gel is made. It resembles gelatin. To avoid the problem of water's surface tension during the removal of the water, supercritical carbon dioxide is first used to replace the water and then is allowed to evaporate. This form of carbon dioxide is a liquid under pressure but has no surface tension. (Supercritical carbon dioxide is also used to dissolve caffeine out of coffee; see page 471.) It is therefore possible to remove the water without disturbing the filamentous structure of the silica gel itself. The filaments involve spherical clusters of several hundred silicon and oxygen atoms joined by connecting strands

FIGURE 1

The block of silica aerogel held by the balance pan on the left does not outweigh the several sunflower seeds on the pan to the right.

made of silicon and oxygen atoms. Synthetic organic polymers can also be used to prepare aerogels.

Silica aerogels have some amazing properties. They are outstanding thermal insulators, for example. They might be the much-needed replacement for refrigerator insulation material that, until recently, was based on the presence of chlorofluorocarbons (CFCs). *Chemicals in Use 7* ("Ozone in the Stratosphere") describes the effect of CFCs on the stratospheric ozone shield. If the aerogels can be made essentially transparent, they could be sealed into double-pane windows, making them even more useful for letting light through without losing heat (in winter) or gaining heat (in summer). The aerogels are also excellent sound insulators.

The aerogels are remarkably tough. After being compressed, they will bounce back to their original shape. Aerogels might be support systems for powdered metal catalysts, systems that would allow gaseous reaction mixtures to be pumped through, giving reactants contact with the catalysts. Perhaps someday abundant supplies of natural gas (methane) can be converted into less abundant liquid fuels by such a process, but a catalyst has yet to be found. Physicists have used aerogels to detect fast-moving subatomic particles. As they streak through an aerogel they cause the emission of bluish light (called Cerenkov radiation).

One drawback of the silica aerogels is that they are unstable in the presence of water. A sufficiently long exposure to humid air will let a film of water molecules form on the silica and cause its collapse. For many uses, however, this problem can be solved by sealing layers of aerogel between sheets of glass, plastic, or metal.

FIGURE 12.27

Tyndall effect. The tube on the left contains a colloidal starch dispersion, and the tube on the right has a colloidal dispersion of Fe_2O_3 in water. The middle tube has a solution of Na_2CrO_4, a colored solute. The thin red laser light is partly scattered in the two colloidal dispersions so it can be seen, but it passes through the middle solution unchanged.

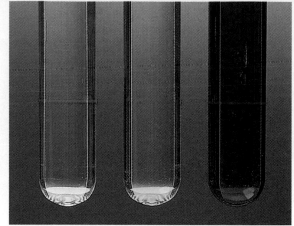

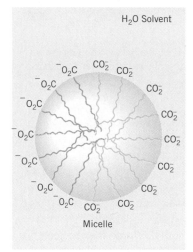

H₂O Solvent

Micelle

FIGURE 12.28

The formation of micelles in soap is driven by the water-avoiding properties of the hydrophobic tails of soap units and the water-attracting properties of their hydrophilic heads.

In synthetic detergents, the hydrophilic group is not:

$$-\overset{\displaystyle O}{\overset{\|}{C}}-O^- \quad \text{but} \quad -\overset{\displaystyle O}{\underset{\displaystyle O}{\overset{\|}{\underset{\|}{S}}}}-O^-.$$

A interesting *solid* colloidal dispersion is that of large molecules of silicon dioxide in air called *silicon aerogels* or "frozen smoke" (see Special Topic 12.5).

How Soaps and Detergents Work

The "like dissolves like" rule helps explain how soaps and detergents work. Oily and greasy matter constitute the "glue" that binds soil to fabrics or skin. Hot water alone does a poor job of removing such dirt-binding materials, but just a small amount of detergent makes a huge cleansing difference.

All detergents have a common feature, whether they are ionic or molecular. Their structures include a long, nonpolar, hydrocarbon "tail" holding a very polar or ionic "head." A typical system in ordinary soap, for example, has the following anion. (The charge-balancing cation is Na^+ in most soaps.)

$$CH_3CH_2CH_2CH_2CH_2CH_2CH_2CH_2CH_2CH_2CH_2CH_2CH_2CH_2CH_2CH_2CO^-$$

hydrocarbon tail 　　　　　　　　　　　　　　　　　anionic head

The tail of this species, like all hydrocarbons, is nonpolar and therefore has a strong tendency to dissolve in oil or grease but not in water. Its natural tendency is to avoid water; we say that the tail is **hydrophobic** (water fearing). The entire ion is dragged into solution in hot water only because of its ionic head, which can be hydrated. The ionic head is said to be **hydrophilic** (water loving). To minimize contacts between the hydrophobic tails and water and to maximize contacts between the hydrophilic heads and water, the soap's anions gather into colloidal-sized groups called soap **micelles** (see Figure 12.28).

When soap micelles encounter a greasy film, their anions find something else that can accommodate hydrophobic tails (see Figure 12.29). The tails of innumerable anions of soap work their way into the film; the tails, in a sense, dissolve in the grease or oil. But the hydrophilic heads of the anions will not be dragged into the grease; the heads stay out in the water. With a little agitating help from the washing machine, the greasy film breaks up into countless tiny droplets *each pincushioned by projecting, negatively charged heads.* Because the droplets, now colloidally dispersed, all have the same kind of charge, they repel each other and cannot again coalesce. For all practical purposes, the greasy film has been dissolved, and we send it on its way to be destroyed by bacteria at wastewater treatment plants or in the ground.

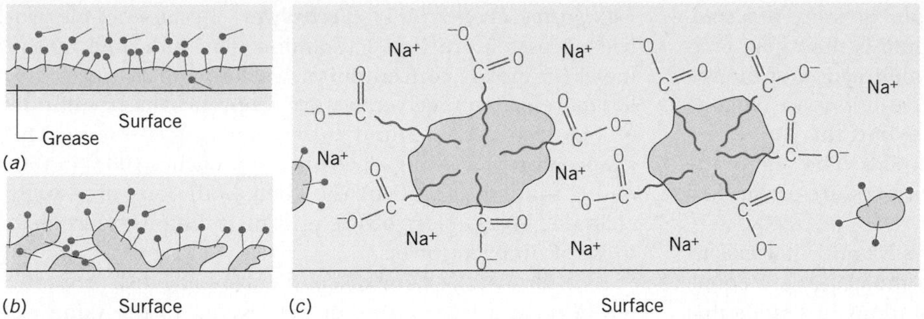

FIGURE 12.29

How soap loosens grease that might be coating a cloth surface. (*a*) The hydrophobic tails of soap anions work their way into the grease layer. The hydrophilic heads stay exposed to water. (*b*) The grease layer breaks up. (*c*) Microdroplets of grease form, each pincushioned and given like charges by soap anions. These are now easily poured out with the wash water.

SUMMARY

SOLUTIONS In **solutions** the dissolved particles have dimensions on the order of 0.05 to 0.25 nm (sometimes up to 1 nm). Solutions are fully homogeneous. They cannot be separated by filtration, and they remain physically stable indefinitely. (They do not display the **Tyndall effect** or **Brownian movement,** either.)

When liquid solutions form by physical means, generally polar or ionic solutes dissolve well in polar solvents, like water. Nonpolar, molecular solutes dissolve well in nonpolar solvents. These observations are behind the **"like dissolves like"** rule. Nature's driving force for establishing disorder and intermolecular attractions are the major factors in the formation of a solution. When both solute and solvent are nonpolar, nature's tendency toward more randomness dominates because intermolecular attractive forces are weak. Ion–dipole attractions and the **solvation** or **hydration** of dissolved species are major factors in forming solutions of ionic or polar solutes in something polar, like water.

Heats of Solution The molar enthalpy of solution, the **heat of solution,** is the net of the **lattice energy** and the **solvation energy** (or, when water is the solvent, the *hydration energy*). The lattice energy is the increase in the potential energy of the system required to separate the molecules or ions of the solute from each other. The solvation energy corresponds to the potential energy lowering that occurs in the system from the subsequent attractions of these particles to the solvent molecules as intermingling occurs.

When liquids dissolve in liquids, an **ideal solution** forms if the potential energy increase needed to separate the molecules equals the energy lowering as the separated molecules come together. Usually, some net energy exchange occurs, however, and few solutions are ideal.

When a gas dissolves in a gas, the energy cost to separate the particles is virtually zero, because they're already separated. They remain separated in the gas mixture, so the net heat of solution is very small if not zero.

Pressure and Gas Solubility At pressures not too much different from atmospheric pressure, the solubility of a gas in a liquid is directly proportional to the partial pressure of the gas—**Henry's law.**

Concentration Expressions To study and use colligative properties, a solution's concentration ideally is expressed either in mole fractions or as a **molal concentration,** or **molality.** Such expressions provide a temperature-independent description of a concentration that, in the final analysis, gives a ratio of solute to solvent particles. The **molality** of a solution, *m*, is the ratio of the number of moles of solute to the kilograms of *solvent* (not solution, but solvent).

Weight fractions and **weight percents** are concentration expressions used often for reagents when direct information about number of moles of solutes is unimportant.

Colligative Properties **Colligative properties** are those that depend on the ratio of the particles of the solute to the molecules of the solvent. These properties include the **lowering of the vapor pressure,** the **depression of a freezing point,** the **elevation of a boiling point,** and the **osmotic pressure** of a solution.

According to **Raoult's law,** the vapor pressure of a solution of a nonvolatile (molecular) solute is the vapor pressure of the pure solvent times the mole fraction of the solvent. When the components of a solution are volatile liquids, then Raoult's law calculations find the *partial pressures* of the vapors of the individual liquids. The sum of the partial pressures equals the total vapor pressure of the solution.

An **ideal solution** is one that obeys Raoult's law exactly and in which all attractions between molecules are equal. An ideal solution has $\Delta H_{soln} = $). Solutions of liquids that form exothermically usually display negative deviations from Raoult's law. Those that form endothermically have positive deviations.

In proportion to its molal concentration, a solute causes **freezing point depression** and **boiling point elevation.** The proportionality constants are the *freezing point depression constant* and the *boiling point elevation constant,* and they differ from solvent to solvent. Freezing point and boiling point data from a solution made from known masses of both solute and solvent can be used to calculate the molecular mass of a solute.

When a solution is separated from the pure solvent (or from a less concentrated solution) by an *osmotic membrane,* **osmosis** occurs, which is the net flow of solvent into the more concentrated solution. The back pressure is required to prevent osmosis is called the solution's *osmotic pressure,* which, in a dilute aqueous solution, is proportional to the product of the Kelvin temperature and the molar concentration. The proportionality constant is R, the ideal gas constant, and the equation relating these variables is $\Pi V = nRT$.

When the membrane separating two different solutions is a *dialyzing* membrane, it permits **dialysis,** the passage not only of solvent molecules but also of small solute molecules or ions. Only very large molecules are denied passage.

Colligative Properties of Electrolytes Because an electrolyte releases more ions in solution than indicated by the molal (or molar) concentration, a solution of an electrolyte has more pronounced colligative properties than a solution of a molecular compound at the same concentration. The dissociation of a strong electrolyte approaches 100% in very dilute solutions, particularly when both ions are singly charged. Weak electrolytes provide solutions more like those of nonelectrolytes.

The ratio of the value of a particular colligative property, like ΔT_f for a freezing point depression, to the value expected under no dissociation is the **van't Hoff factor, i.** A solute whose formula unit breaks into two ions, like NaCl or $MgSO_4$, would have a van't Hoff factor of 2 if it were 100% dissociated. Observed van't Hoff factors approach those corresponding to 100% dissociation only as the solutions are made more and more dilute.

Other Homogeneous Mixtures Relative particle sizes and homogeneity distinguish among the three simplest kinds of mixtures—**suspensions, colloidal dispersions,** and **solutions.** Suspensions have particles with at least one dimension larger than 1000 nm, they are not homogeneous, and they settle out spontaneously. They can be separated by filtration.

Colloidal Dispersions In colloidal dispersions, the dispersed particles have one or more dimensions in the range of 1 to 1000 nm. They are considered to be homogeneous, but we're at a borderline. They separate more slowly than suspensions in response to gravity, but many can be kept indefinitely if the dispersed particles bear like charges or are coated with an emulsifying agent. Colloidal particles can scatter light **(Tyndall effect),** can display **Brownian movement,** and cannot ordinarily be separated by filtration.

Tools you have learned

The table below lists the tools you have learned in this chapter that are applicable to problem solving. Review them if necessary, and refer to them when working on the Thinking-It-Through problems and the Review Exercises that follow.

Tool	Function
"Like dissolves like" rule (page 495)	To use chemical composition and structure to predict whether two substances can form a solution).
Henry's law (page 505)	To calculate the solubility of a gas at a given pressure from its solubility at another pressure.

Weight fraction; weight percent (page 507)	To calculate the mass of a solution that delivers a given mass of solute.
Molal concentration (page 507)	To have a temperature-independent concentration expression for use with colligative properties.
Raoult's law (page 511)	To calculate the effect of a solute on the vapor pressure of a solution.
Equations for freezing point depression and boiling point elevation (page 517)	To estimate freezing and boiling points or molecular masses. To estimate percent dissociation of a weak electrolyte (page 525).
Equation for osmotic pressure (page 521)	To estimate the osmotic pressure of a solution. To calculate molecular masses from osmotic pressure and concentration data (page 522).

THINKING IT THROUGH

The goal for each of the following problems is to give you practice in thinking your way through problems. The goal is not *to find the answer itself; instead, you are only asked to assemble the available information needed to obtain the answer, state what additional data (if any) are needed, and describe how you would use the data to answer the question. For problems involving unit conversions, list the relationships among the units that are needed to carry out the conversions. Construct the conversion factors that can be formed from these relationships. Then set up the solution to the problem by arranging the conversion factors so the units cancel correctly to give the desired units of the answer.*

The problems are divided into two groups. Those in Level 2 are more challenging than those in Level 1 and provide an opportunity to really hone your problem-solving skills. Both levels, however, may include questions that draw on material studied in earlier chapters.

Level 1 Problems

1. The solubility of a gas in water at 20 °C is 0.0176 g L^{-1} at 681 torr. What is its solubility at 0.989 atm at 20 °C?

2. You have been hired to be an assistant to a chemistry professor and are asked to prepare 500 g of 5.00% Na_2CO_3 solution. Describe how you would go about doing this.

3. A solution of sulfuric acid (H_2SO_4) is 10.0%. How many grams of this solution are needed to provide 0.100 mol of H_2SO_4? If the question had asked for the number of *milliliters* of the solution instead of the number of grams, what additional data about the solution would be needed, and how would they be used?

4. For an experiment involving osmosis, you need to know the molality of a 12.5% solution of sugar ($C_{12}H_{22}O_{11}$) in water. How can this be calculated?

5. The stockroom has a solution that is 0.125 m NH_4Cl. An experiment you want to carry out needs enough of this solution to provide 0.0575 mol of NH_4Cl. How much solution (supply the unit) is required? If the problem had called for the number of *milliliters* of the solution that would be necessary, what additional data would you need?

6. The stockroom has a solution labeled 50.00% H_2SO_4. Its specific gravity at 20 °C is 1.3977. What is the molarity of the solution?

7. A solution is prepared by dissolving 0.200 mol of glucose ($C_6H_{12}O_6$, a nonvolatile solute) in 1000 g of water. What is the vapor pressure of the solution at 25 °C, at which temperature the vapor pressure of pure water is 23.756 torr?

8. At 20 °C, the vapor pressure of pure water is 17.535 torr. How many grams of glucose ($C_6H_{12}O_6$, a nonvolatile solute) have to be dissolved in 100 g of water to have a solution with a vapor pressure of 17.000 torr?

9. A solution is made by dissolving 25.00 g of ethylene glycol ($C_2H_6O_2$, a nonvolatile solute) in water to make a solution with a mass of 125.00 g. At approximately what temperature will this solution freeze?

10. If the vapor pressure of ethylene glycol at 100 °C is negligible, at approximately what temperature will the solution of Question 9 boil?

11. What is the vapor pressure of a solution of tetrachloroethylene in methyl acetate at 40.0 °C in which the mole fraction of tetrachloroethylene is 0.450? At 40 °C, the vapor pressure of tetrachloroethylene is 40.0 torr and that of methyl acetate is 400 torr.

12. What is the osmotic pressure of a 0.0125 *M* solution of a nonvolatile solute in water at 20 °C?

13. If the $CaCl_2$ dissolved in a 0.075 *m* solution is 100% dissociated, at approximately what temperature does the solution freeze?

Level 2 Problems

14. A solution is prepared by mixing 150 g of table sugar (sucrose, $C_{12}H_{22}O_{11}$ with 350 mL of water at 20.0 °C. How would you go about estimating the freezing point of this solution by a calculation? What additional data (if any) are needed?

15. If you add 75.0 g of water to 150 g of a 10.0% solution of NaCl, what is the percent concentration of the new solution?

16. The mole fraction of NaCl in an aqueous solution (with no other component) is 0.0100. What is the weight percent of NaCl in the solution?

17. A solution of KNO_3 has a concentration that is 0.750 *m*. What is the mole fraction of KNO_3 in the solution?

18. If the van't Hoff factor for $CaCl_2$ in water at a concentration of 0.075 *m* is 2.30, what is the freezing point of the solution.

REVIEW EXERCISES

Answers to questions whose numbers are printed in color are given in Appendix D. The more challenging questions are marked with asterisks.

Why Solutions Form

12.1 Name three kinds of *homogeneous* mixtures. In which are the particle sizes the smallest?

12.2 What is the "driving force" behind the spontaneous dissolving of two gases in each other?

12.3 When substances form liquid solutions, what are the two driving forces involved?

12.4 Methyl alcohol, CH_3—O—H, and water are miscible in all proportions. What does this mean? Explain how the O—H unit in methyl alcohol contributes to this.

12.5 Gasoline and water are immiscible. What does this mean? Explain why, in terms of structural features of their molecules and forces of attraction between them.

12.6 Explain how ion–dipole forces help to bring potassium chloride into solution in water.

12.7 What does hydration refer to, and what kinds of substances, electrolytes or nonelectrolytes, does it tend to assist most in dissolving in water?

12.8 Explain why potassium chloride does not dissolve in benzene.

12.9 What kinds of substances are more soluble in water, hydrophilic compounds or hydrophobic compounds?

12.10 Explain why potassium chloride will not dissolve in carbon tetrachloride, CCl_4.

12.11 Iodine, I_2, is very slightly soluble in water but dissolves readily in carbon tetrachloride. Why won't it dissolve

very much in water, and what makes so easy the formation of the solution of iodine in carbon tetrachloride?

12.12 Iodine dissolves far better in ethyl alcohol (to form "tincture of iodine," an old antiseptic) than in water. What does this tell us about the alcohol molecules, as compared with water molecules?

Heats of Solution

12.13 How do we define *heat of solution*? Give both a narrative definition and an equation.

12.14 No solution actually forms by the two-step process that we used as a model for the formation of a solution of some ionic compound in water. When we are interested in the molar enthalpy of solution, however, we can get away with this model. Explain. What makes attractive the particular two steps we used for the purpose of estimating ΔH_{soln}?

12.15 The value of ΔH_{soln} for a soluble compound is, say, $+26$ kJ mol^{-1}, and a nearly saturated solution is prepared in an insulated container (e.g., a coffee-cup calorimeter). Will the system's temperature increase or decrease as the solute dissolves?

12.16 Referring to Review Exercise 12.15, which value for this compound would be numerically larger, its lattice energy or its hydration energy?

12.17 Consider the formation of aqueous potassium chloride.

(a) Write the thermochemical equations for the two steps in the formation of a solution of KCl in water. (We studied thermochemical equations in Section 5.5). The lattice energy of KCl is -690 kJ mol^{-1}, and the hydration energy is -686 kJ mol^{-1}.

(b) Write the sum of the two equations in the form of a thermochemical equation (showing also the net ΔH).

12.18 Referring to Table 12.1, suppose the lattice energy of KCl were just 2% greater than the value given. What then would be the calculated ΔH_{soln} (in kJ mol^{-1})? How would this new value compare with that given in the table? By what percentage is the new value different from the value in the table? (The point of this calculation is to show that small percentage errors in two large numbers, such as lattice energy and hydration energy, can cause large percentage changes in the difference between them.)

12.19 Which would be expected to have the larger hydration energy, Al^{3+} or Li^+? Why? (Both ions are about the same size.)

12.20 Suggest a reason why the value of ΔH_{soln} for a gas such as CO_2 is negative.

12.21 The value of ΔH_{soln} for the formation of an acetone–water solution is negative. Explain this in general terms that discuss intermolecular forces of attraction.

12.22 The value of ΔH_{soln} for the formation of an ethyl alcohol–hexane solution is positive. Explain this in general terms that involve intermolecular forces of attraction.

12.23 How can Le Châtelier's principle be applied to explain how the addition of heat causes undissolved solute to go into solution (in the cases of most saturated solutions)?

Temperature and Solubility

12.24 If a saturated solution of NH_4NO_3 at 70 °C is cooled to 10 °C, how many grams of solute will separate if the quantity of the solvent is 100 g? (Use data in Figure 12.13).

12.25 A hot concentrated solution in which equal masses of NaI and NaBr are dissolved is slowly cooled from 50 °C. Which salt precipitates first? (Use data in Figure 12.13).

Pressure and Solubility

12.26 What is Henry's law?

12.27 The solubility of methane, the chief component of natural gas, in water at 20 °C and 1.0 atm pressure is 0.025 g L^{-1}. What is its solubility in water at 1.5 atm and 20 °C?

*12.28 At 740 torr and 20 °C, nitrogen has a solubility in water of 0.018 g L^{-1}. At 620 torr and 20 °C, its solubility is 0.015 g L^{-1}. Does nitrogen obey Henry's law? (Do the necessary calculations to support your answer.)

12.29 What makes ammonia or carbon dioxide so much more soluble in water than nitrogen?

12.30 Sulfur dioxide is an air pollutant wherever sulfur-containing fuels have been used. Rainfall that washes sulfur dioxide out of the air and onto the land is called acid rain. How does sulfur dioxide contribute to acid rain? Write an equation as part of your answer.

Expressions of Concentration

12.31 What is the molal concentration of a solution made by dissolving 24.0 g of glucose (molecular mass 180) in 1.00 kg of water?

12.32 If you dissolved 11.5 g of NaCl in 1.00 kg of water, what would be its molal concentration? The volume of this solution is virtually identical to the original volume of the 1.00 kg of water. Therefore, what is the molar concentration of this solution? What would have to be true about any solvent for one of its dilute solutions to have essentially the same molar and molal concentrations?

*12.33 A solution of ethyl alcohol, CH_3CH_2OH, in water has a concentration of 1.25 m. Calculate the weight percent of ethyl alcohol.

*12.34 An aqueous solution of $NaNO_3$ has a concentration of 0.363 m. Its density is 1.0185 g mL^{-1}. Calculate the following.
(a) the weight percent concentration of $NaNO_3$
(b) the molar concentration of $NaNO_3$

*12.35 In an aqueous solution of sulfuric acid, the concentration is 1.89 mole % of acid. The density of the solution is 1.0645 g mL^{-1}. Calculate the following.
(a) the molal concentration of H_2SO_4
(b) the weight percent of the acid
(c) the molarity of the solution

12.36 A solution of NaCl in water has a concentration of 19.5%. Calculate the molality of the solution.

*12.37 A solution of NH_3 in water is at a concentration of 5.00%. Its density is 0.9787 g mL^{-1}. Calculate the following.
(a) the molarity of the solution
(b) the molality of the solution

Lowering of the Vapor Pressure

12.38 What specific fact about a physical property of a solution must be true to call it a colligative property?

12.39 What causes a solution with a nonvolatile solute to have a lower vapor pressure than the solvent at the same temperature?

12.40 How is Raoult's law used when we want to calculate the total vapor pressure of a liquid solution made up of two *volatile* liquids?

12.41 What kinds of data would have to be obtained to find out if a binary solution of two miscible liquids is almost exactly an ideal solution?

12.42 At 25 °C, the vapor pressure of water is 23.8 torr. Assuming that the solution described in Review Exercise 12.31 is ideal, calculate its vapor pressure.

12.43 The vapor pressure of water at 20 °C is 17.5 torr. A 20% solution of ethylene glycol in water is prepared. Assuming that the solute is nonvolatile, do a calculation to estimate the vapor pressure of the solution.

12.44 Pentane and heptane are two hydrocarbon liquids present in gasoline. At 20 °C, the vapor pressure of pentane is 420 torr and the vapor pressure of heptane is 36.0 torr. What is the total vapor pressure of a solution prepared with a concentration of 25.0 mole % pentane in heptane? (Calculate to two significant figures.)

12.45 Benzene and toluene help get good engine performance from lead-free gasoline. At 40 °C, the vapor pressure of benzene is 180 torr and that of toluene is 60 torr. To prepare a solution of these that will have a total vapor pressure of 96 torr at 40 °C requires what mole percent concentration of each?

12.46 At 21.0 °C, a solution of 18.26 g of a nonvolatile, nonpolar compound in 33.25 g of ethyl bromide, C_2H_5Br, had a vapor pressure of 336.0 torr. The vapor pressure of pure ethyl bromide at this temperature is 400.0 torr. Calculate the following.
(a) the mole fractions of solute and solvent
(b) the number of moles of solute present
(c) the molecular mass of the solute

12.47 An aqueous solution showed a *positive deviation* from the expected relationship between the vapor pressure of an ideal binary solution and its mole fraction composition. What does this mean? Explain the positive deviation in terms of intermolecular attractions.

Freezing Point Depression and Boiling Point Elevation

12.48 When an aqueous solution of sodium chloride starts to freeze, why don't the ice crystals contain ions of the salt?

12.49 Explain why a nonvolatile solute dissolved in water makes the system have (a) a higher boiling point than water, and (b) a lower freezing point than water.

*12.50 Ethylene glycol, $C_2H_6O_2$, is used in some antifreeze mixtures. Protection against freezing to as low as − 40 °F is sought.
(a) How many moles of solute are needed per kilogram of water to ensure this protection?
(b) The density of ethylene glycol is 1.11 g mL^{-1}. To how many milliliters of solute does your answer to part (a) correspond?
(c) Look up and use conversion factors to calculate how many quarts of ethylene glycol should be mixed with each quart of water to get the desired protection.

12.51 Calculate what would be the estimated boiling point of 2.00 *m* sugar in water. What would be its estimated freezing point? (It's largely the sugar in ice cream that makes it difficult to keep ice cream frozen hard.)

12.52 Glycerol, $C_3H_8O_3$ (molecular mass 92), is essentially a nonvolatile liquid that is very soluble in water. A solution is made by dissolving 46.0 g of glycerol in 250 g of water. By calculations, estimate the following:
(a) the boiling point of the solution at 1 atm
(b) its freezing point
(c) its vapor pressure at 25 °C (At this temperature, the vapor pressure of water is 23.8 torr.)

12.53 A solution of 12.00 g of an unknown molecular compound dissolved in 200.0 g of benzene froze at 3.45 °C. Calculate the molecular mass of the unknown.

12.54 A solution of 14 g of an unknown (molecular) compound in 1.0 kg of benzene boiled at 8.17 °C. Calculate the molecular mass of the unknown.

*12.55 Benzene reacts with hot concentrated nitric acid dissolved in sulfuric acid to give chiefly nitrobenzene, $C_6H_5NO_2$. A byproduct is often obtained, which consists of 42.86% C, 2.40% H, and 16.67% N, all weight percents.
(a) Calculate the empirical formula of the byproduct.
(b) The boiling point of a solution of 5.5 g of the byproduct in 45 g of benzene was 1.84 °C higher than that of benzene. Calculate a molecular mass of the byproduct, based on these data, and determine its molecular formula.

*12.56 An experiment calls for the use of the dichromate ion, $Cr_2O_7^{2-}$, in sulfuric acid as an oxidizing agent for isopropyl alcohol, C_3H_8O. The chief product is acetone, C_3H_6O, which forms according to the following equation.
$$3C_3H_8O + Na_2Cr_2O_7 + 4H_2SO_4 \longrightarrow$$
$$3C_3H_6O + Cr_2(SO_4)_3 + Na_2SO_4 + 7H_2O$$
(a) The oxidizing agent is available only as sodium dichromate dihydrate. In theory, how many grams of sodium dichromate dihydrate are needed to oxidize 21.4 g of isopropyl alcohol according to the balanced equation?
(b) The amount of acetone actually isolated was 12.4 g. Calculate the percentage yield of acetone.
(c) The reaction produces a volatile byproduct. When a sample of it with a mass of 8.654 mg was burned in oxygen, it was converted into 22.368 mg of carbon dioxide and 10.655 mg of water, the sole products. (Assume that any unaccounted for element is oxygen.) Calculate the percentage composition of the byproduct and determine its empirical formula.
(d) A solution prepared by dissolving 1.338 g of the byproduct in 115.0 g of benzene had a freezing point of 4.87 °C. Calculate the molecular mass of the byproduct and write its molecular formula.

Dialysis and Osmosis

12.57 Why do we call dialyzing and osmotic membranes *semipermeable*? What is the opposite of *permeable*?

12.58 What is the key difference between dialyzing and osmotic membranes?

12.59 What is meant by *dialysis*?

12.60 In osmosis, why *must* the net migration of solvent be from the side of the membrane less concentrated in solute to the side more concentrated in solute?

12.61 Two glucose solutions of unequal molarity are separated by an osmotic membrane. Which solution will *lose* water, the one with the higher or the one with the lower molarity?

12.62 Two glucose solutions of unequal molarity are separated by a dialyzing membrane. Given the various driving forces in nature, what change or changes should be expected?

12.63 Which aqueous solution has the higher osmotic pressure, 10% glucose, $C_6H_{12}O_6$, or 10% sucrose, $C_{12}H_{22}O_{11}$? (Both are molecular compounds.)

12.64 What is the osmotic pressure in torr of a 0.010 M aqueous solution of a molecular compound at 25 °C?

12.65 (a) Show that the following equation is true.

$$\text{Molar mass of solute} = \frac{(\text{grams of solute})RT}{\Pi V}$$

(b) An aqueous solution of a compound with a very high molecular mass was prepared in a concentration of 2.0 g L^{-1} at 298 K. Its osmotic pressure was 0.021 torr. Calculate the molecular mass of the compound.

12.66 When a solid is *associated* in a solution, what does this mean? What difference does it make to expected colligative properties?

Colligative Properties of Electrolytes

12.67 Why are colligative properties of solutions of ionic compounds usually more pronounced than those of solutions of molecular compounds of the same molalities?

12.68 The vapor pressure of water at 20 °C is 17.5 torr. If the solute in a solution made from 10.0 g of NaCl in 1.00 kg of water is 100% dissociated (and is an ideal solution), what is the vapor pressure of the solution at 20 °C?

12.69 Which aqueous solution, if either, has the lower freezing point, 10% NaCl or 10% NaI?

12.70 Which aqueous solution, if either, has the higher boiling point, 0.50 m NaI or 0.50 m Na$_2$CO$_3$?

12.71 Which aqueous solution, if either, has the higher osmotic pressure, 2.0% glucose (C$_6$H$_{12}$O$_6$) or 2.0% NaCl?

12.72 What is the van't Hoff factor? What is its calculated value for all molecular solutes? For all ionic solutes of the Na$_2$SO$_4$ type?

12.73 The van't Hoff factor for the solute in 0.100 m NiSO$_4$ is 1.19. What would this factor be if the solution behaved as if it were 100% dissociated?

12.74 The van't Hoff factor for the solute in 0.118 m LiCl is 1.89.

(a) Calculate the freezing point of the solution.

(b) This solution is roughly as dilute as the solution described in Review Exercise 12.73. Explain why the van't Hoff factor for LiCl is so much greater than that of NiSO$_4$.

12.75 A 1.00 m aqueous solution of HF freezes at −1.91 °C. According to these data, what is the percent ionization of HF in the solution?

12.76 An aqueous solution of a weak electrolyte, HX, with a concentration of 0.125 m has a freezing point of −0.261 °C. What is the percent ionization of the compound (to two significant figures)?

Kinds of Mixtures

12.77 What feature about mixtures distinguishes them from elements and compounds?

12.78 What single feature most distinguishes suspensions, colloidal dispersions, and solutions from each other?

Colloidal Dispersions

12.79 List three factors that can contribute to the stability of a colloidal dispersion.

12.80 What general name is given to a colloidal dispersion of one liquid in another?

12.81 What is an emulsifying agent? Give an example.

12.82 What is the Tyndall effect, and why do colloidal dispersions but not solutions show it?

12.83 What causes Brownian movement in fluid colloidal dispersions?

12.84 What is a sol, and how can one often be stabilized?

12.85 What simple test could be used to tell if a clear, aqueous fluid contains colloidally dispersed particles?

12.86 Soap, as a solute in water, spontaneously forms micelles. What are micelles, and what is the driving force for their formation?

Additional Problems

*12.87 A solution of HNO$_3$ in water has a concentration of 10.00%. Its density is 1.0543 g mL^{-1}. Calculate the following.

(a) the molality of this solution

(b) the molarity of the HNO$_3$

*12.88 A solution of ethyl alcohol, C$_2$H$_5$OH, in water has a concentration of 4.613 mol L^{-1}. At 20°C, its density is 0.9677 g mL^{-1}. Calculate the following.

(a) the molality of the solution

(b) the percent concentration of the alcohol

(c) The density of ethyl alcohol is 0.7893 g mL^{-1} and the density of water is 0.9982 g mL^{-1} at 20°C. Calculate the percent by *volume* of ethyl alcohol in this solution (see Special Topic 12.2).

12.89 Consider an aqueous 1.00 m solution of Na$_3$PO$_4$.

(a) Calculate the boiling point of the solution on the assumption that it does not ionize at all in solution.

(b) Do the same calculation assuming that the van't Hoff factor for Na$_3$PO$_4$ reflects 100% dissociation into its ions.

(c) The 1.00 m solution boils at 100.183 °C at 1 atm. Calculate the van't Hoff factor for the solute in this solution.

12.90 A clear, colorless aqueous solution gave a Tyndall effect and also conducted electricity.

(a) On the basis of only this information, what can be said about the system?

The solution was placed in a cellophane pouch and immersed in a beaker of pure water for several hours. A small sample of the water from the solution outside the pouch was then found to give an instantaneous precipitate with a few drops of 1% AgNO$_3$. Other small samples from the same solution in the beaker, individually tested, gave no precipitates with aqueous solutions of sodium sulfate, sodium carbonate, and sodium hydroxide.

(b) What additional conclusion(s) might now be drawn about the original solution?

12.91 Do a calculation to estimate the osmotic pressure, in atmospheres, of seawater, which may be taken to have a concentration of 3.6 g NaCl/100 mL solution. The estimated pressure is considerably more than a person can press, which is why portable reverse desalination devices—Special Topic 12.3—have multiple chambers. (Contributed by Prof. Mark Benvenuto, University of Detroit−Mercy.)

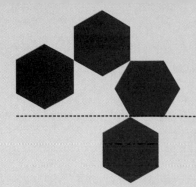

CHEMICALS IN USE 10
Concerning Pure Water

ACID RAIN

Normally, water from even the purest of natural sources is slightly acidic because of its dissolved carbon dioxide. The following equilibria occur, the last one producing hydrogen ion.

$$CO_2(g) \rightleftharpoons CO_2(aq)$$

$$CO_2(aq) + H_2O \rightleftharpoons H_2CO_3(aq)$$

$$H_2CO_3(aq) \rightleftharpoons H^+(aq) + HCO_3^-(aq)$$

Rain, however, is often more acidic than normal and then is called **acid rain.** Recorded extreme examples are rain as acidic as lemon juice, observed in 1964 in the northeastern section of the United States, and rain as acidic as vinegar, which fell at Pitlochry, Scotland, in 1974.

When rainwater becomes more acidic by falling through polluted air or by leaching the deposits of certain pollutants from soil, the waters of nearby lakes and streams also become more acidic, sometimes enough so to render them lethal habitats for the hatchlings of game fish. Moreover, acid rain leaches aluminum ions from soil, ions that also damage fish and forest life. Whole forests in parts of Europe have thus been severely harmed (see Figure 10a).

Certain building materials are corroded by acid rain. Limestone and marble are particularly sensitive, because they are chiefly calcium carbonate, and all carbonates are dissolved by acids.

$$CaCO_3(s) + 2H^+(aq) \longrightarrow$$
$$Ca^{2+}(aq) + CO_2(g) + H_2O$$

Several major cathedrals in Europe, built of limestone and marble, are under constant attack by acid rain. Acid is also corrosive to exposed metals such as railroad rails, vehicles, and machinery.

Dust particles settling on buildings, metals, and soil also carry acidic materials as solids, so specialists prefer the term **acid deposition** to *acid rain* to encompass all means whereby acids enter surface features of the Earth. Southern parts of the Scandinavian peninsula receive acid deposition from Germany's Ruhr valley, the English Midlands, and countries of eastern Europe. Parts of southern Canada and the northern United States receive acid deposition from the industrial belt curving from Boston to Chicago. Industrial areas of eastern Europe have released into their atmospheres enormous loads of the oxides of sulfur and nitrogen, the chief causes of acid rain. Let's see how human activities produce these gases and what might be done about them. Volcanic eruptions also inject huge quantities of sulfur oxides into the atmosphere; however, they are *relatively* small sources, and we can do nothing to prevent such natural events. Volcanic SO_2 releases have been estimated to be roughly 13×10^6 tons per year, which is only 5 to 10% of the quantity released by human activities.

Oxides of Sulfur

During the combustion of coal and oil, their sulfur impurities are oxidized to sulfur dioxide. Although relatively small quantities of sulfur are in the fossil fuels, seldom over 3%, often much less, such enormous quantities of the fuels are burned, that hundreds of millions of tons of SO_2 are generated worldwide each year.

Sulfur dioxide dissolves in water by forming hydrates, $SO_2 \cdot nH_2O$, where n varies with the SO_2 con-

FIGURE 10a

Acid rain has destroyed the forest on the Krusne Hory mountains near Teplice, Czech Republic.

centration and with the temperature.[1] The hydrates, which we could also represent as $SO_2(aq)$, are in equilibrium with hydrogen ion and hydrogen sulfite ion, HSO_3^-.

$$SO_2(aq) + H_2O \rightleftharpoons H^+(aq) + HSO_3^-(aq)$$

Although the percentage to which the forward reaction occurs is much less than 100%, when falling rain picks up gaseous SO_2 from the air, the rainwater reaching the ground is more acidic than normal. To make matters worse, both oxygen and the ozone (O_3) in smog convert some SO_2 to sulfur trioxide, particularly in sunlight when fine dust is present. When SO_3 reacts with water, sulfuric acid, a strong acid (one with a high percentage ionization), forms. Thus SO_3 also contributes to the acidity of rain where air pollution occurs.

Nitrogen Dioxide

Another contributor to acid rain, nitrogen dioxide (NO_2), forms from the nitrogen monoxide in vehicle exhausts, as was described in *Chemicals in Use 8,* page 438. In water, NO_2 reacts to gived
two acids, namely, nitric acid (HNO_3), a strong acid, and nitrous acid (HNO_2), a weak acid.

$$2NO_2(g) + H_2O \longrightarrow HNO_3(aq) + HNO_2(aq)$$

In this way, nitrogen dioxide contributes to acid rain.

Meeting the Acid Rain Problem

The removal of most of the sulfur impurities from coal and oil *before* use is prohibitively expensive. However, it is technically feasible to remove sulfur dioxide from the smokestack gases *following* fuel combustion. Sulfur dioxide, for example, reacts with wet calcium hydroxide according to the following equation.

$$SO_2(g) + Ca(OH)_2(s) \longrightarrow CaSO_3(s) + H_2O$$

The reaction, however, substitutes a solid waste disposal problem—the calcium sulfite—for an air pollution problem. Moreover, not all SO_2 is removed and, given the enormous quantities of coal and oil burned worldwide, emissions of SO_2 still occur. Thus, it's only in a *technological* sense that the SO_2 problem is controllable. Relatively few people are willing to accept as alternatives to the use of coal and oil a drastic

cutback in energy consumption made possible by a simpler life style or by a switch to a greater reliance on nuclear power. Nuclear power bears its own pollution ills, and it presently is costlier in every way than power obtained from the burning of coal or oil. More reliance on solar energy is urged by many, and considerable research is being done in this area.

MERCURY POLLUTION

Mercury is another pollutant that can get into the natural water supply, and once present, it remains for a long time. Some mercury arises from natural sources, and some is released by burning coal. The operators of sawmills situated by lakes and streams once used mercury compounds to retard the growths of molds on the sluiceways. Manufacturers of certain plastics once dumped mercury wastes into nearby streams. Such uses are now prohibited, but the mercury wastes already present will be recycled for decades by natural processes within bodies of water.

Biological processes in aquatic bacteria convert mercury to derivatives of the methylmercury ion, CH_3Hg^+, whose compounds are somewhat soluble in fat and muscle tissue. This ion binds to proteins in cells and renders them incapable of carrying out their functions. Fish in mercury-polluted waters are able to accumulate so much mercury that humans eating the fish endanger their brain and nerve tissue. Terrible physical deformations showed up in the 1960s in Japanese infants born to mothers living downstream from a plastics manufacturer near the Bay of Minamata and who ate fish contaminated by mercury. Today, it is routine for American states and Canadian provinces to issue mercury "advisories" that warn fishermen not to eat more than a designated amount of fish per week. Such warnings should be heeded.

Questions

1. What are the chief gases that are responsible for acid rain and how do they get into the Earth's atmosphere? (Write equations.) Why is *acid deposition* a better term?

2. How do sulfur dioxide and nitrogen dioxide contribute to an increase in the concentration of hydrogen ion in water? (Write the equilibria.)

3. Briefly describe some of the problems caused by acid rain in living systems and materials.

4. What form of mercury (give a formula) is a threat to humans?

5. In general terms, explain how mercury persists for so long in an aquatic environment.

---

[1] Traditionally, water with dissolved sulfur dioxide is said to contain sulfurous acid, H_2SO_3, but evidence for the presence of such a molecule in water does not exist. Nevertheless, it's convenient to represent the formation of hydrogen ions in aqueous sulfur dioxide as the ionization of H_2SO_3.

At a parade in New York City celebrating victory in Operation Desert Storm, multicolored confetti falls through the cheering crowd. Nature's laws are at work mixing the various confetti colors, and they also require that the confetti fall down, not up. The same laws that control the fate of confetti are also at work in influencing the outcome of chemical reactions, as you will learn in this chapter.

Chapter 13

Thermodynamics

13.1
INTRODUCTION

In our world, some events happen spontaneously and others do not. A river runs to the sea, a stone wall gradually crumbles, an abandoned car slowly rusts. These are spontaneous changes, which take place without assistance. On the other hand, climbing a mountain in a car or arranging the cards in a shuffled deck into a specific sequence are events that do not happen by themselves. Without outside assistance, these events are impossible.

In chemistry we also find that some changes are spontaneous and others are not. Hydrogen and oxygen combine quite spontaneously to produce water, but water does not decompose spontaneously to produce hydrogen and oxygen.

Observations such as these make us wonder about the factors that control chemical and physical changes, and in this and the next two chapters we will examine what these factors are. We begin here with *thermodynamics,* which explores the question, "What determines whether or not a change is possible?" To see something happen, of course, it must not only be *possible,* but it must also occur fast enough. Therefore, in Chapter 14 we will study the rates of chemical reactions to answer the query, "What determines how fast chemical changes take place?" Finally, in Chapter 15 we will study chemical equilibria, so we can answer the question, "What is a chemical system like when it gets where it's going?"

The answers to such questions also go a long way toward helping us understand the world in which we live, because we are constantly concerned with whether certain changes are possible, how fast they occur, and what the results will be.

Thermodynamics

The study of **thermodynamics** is concerned principally with energy changes and the flow of energy from one substance to another. Remarkably, a detailed study of energy transfers has led to an understanding of why certain changes are inevitable and why others are simply impossible. We can thus understand

Thermo implies heat, *dynamics* implies movement.

why rivers flow downhill instead of up, and *why* stone walls crumble instead of being formed during the passage of time from heaps of sand.

You were introduced to some of the principles of thermodynamics in Chapter 5 when you learned to calculate enthalpy changes (heats of reaction). You should recall that enthalpy changes deal with the exchange of heat between chemical systems and their surroundings. Such transfers of heat, however, represent only one aspect of thermodynamics. Another is nature's drive toward disorder (randomness), which we discussed briefly in Chapter 12. Both energy and disorder are prime topics in thermodynamics.

If necessary, review in Chapter 5 the meanings of such terms as *system, surroundings, state,* and *state function.*

13.2 ENERGY CHANGES IN CHEMICAL REACTIONS—A SECOND LOOK

Thermodynamics is organized around three fundamental, experimentally derived laws of nature. For ease of reference, they are identified by number and are called the first law, the second law, and the third law. The first law of thermodynamics deals with exchanges of energy between a system and its surroundings and is basically a statement of the law of conservation of energy. This law, you recall, serves as the foundation for Hess's law, which we used in our computations involving enthalpy changes in Chapter 5.

The First Law of Thermodynamics

In Chapter 5, the *total energy* of a system was given the symbol E, and in thermodynamics it is called the system's **internal energy.** Regardless of a system's composition, the internal energy is the sum of all the kinetic energies (KE) and potential energies (PE) of its particles.

$$E_{\text{system}} = (\text{KE})_{\text{system}} + (\text{PE})_{\text{system}}$$

Despite the importance of this definition, we can never actually know the value of E_{system}. As we commented in Chapter 5, the kinetic and potential energies of the particles of a system depend in part on velocities and attractions that we are unable to determine, so we can't measure E or calculate its value. This is no terrible crisis, however, because we are really concerned with energy changes, since these are what we can measure.

The value of E_{system} depends only on the state of the system, so internal energy is a state function. This means that for ΔE we can write

$$\Delta E = E_{\text{final}} - E_{\text{initial}}$$

or, for a chemical reaction,

$$\Delta E = E_{\text{products}} - E_{\text{reactants}}$$

Notice that we have followed the practice established in Chapter 5—"final minus initial" or "products minus reactants." This means that if a system absorbs energy from its surroundings during a change, its final energy is greater than its initial energy and ΔE is positive. This is what happens, for example, when the surroundings supply energy to charge a battery. As the system (the battery) absorbs the energy, its energy increases and becomes available for later use.

Heat and Work   Thermodynamics considers two ways by which a system can exchange energy with its surroundings, heat and work. When a system undergoes a change, it might absorb or lose heat. When heat is absorbed, the system's energy increases, and when heat is lost, the system's energy decreases.

Work is exchanged when a force pushing against the system moves through some distance. For example, the system does work if it pushes back an opposing force, as when the hot expanding gases in the cylinder of an engine push a piston. As the piston is pushed, the system of gases loses some if its energy. On the other hand, the system might have some work done on it, as in the compression of a gas, and thereby increase its amount of stored energy.

The **first law of thermodynamics** relates the change in the internal energy of a system to the exchanges of heat and work by the equation

$$\Delta E = q + w \qquad (13.1)$$

where q represents the amount of heat *absorbed by* the system, and w is the amount of work *done on* the system. In effect, Equation 13.1 states that the change in the internal energy of a system is the sum of the amount of energy it gains as heat plus the amount of energy it gains from having work done on it.

ΔE = (heat input) + (work input)

In the preceding paragraph we were very careful to define q and w precisely, because by doing so we establish the algebraic signs of these quantities. Thus, q is given a positive sign if heat flows into the system and a negative sign if heat flows out. (Notice that this is consistent with the sign conventions we established in Chapter 5 for endothermic and exothermic changes.) Similarly, w is given a positive sign if the system gains energy by having work done on it, as in compressing a spring. On the other hand, if the system performs work, its energy reserves are depleted somewhat (i.e., the system loses energy) and w is given a negative sign. These sign conventions are summarized as follows:

q is $(+)$	Heat is absorbed by the system.
q is $(-)$	Heat is released by the system.
w is $(+)$	Work is done on the system.
w is $(-)$	Work is done by the system.

Heat, Work, and State Functions

Although ΔE is independent of how a change takes place, the values of q and w are not. In other words, the path that's followed determines the form in which the energy appears. For example, let's consider the discharge of an automobile battery by two different paths, as illustrated in Figure 13.1 on page 544. Both paths take us between the same two states, one being the fully charged state and the other the fully discharged state. Because the same initial and final states are involved, ΔE *must be the same for both paths*. But how about q and w?

In Path 1, we simply short the battery by placing a heavy wrench across the terminals. If you have ever done this, even by accident, you know how violent the result can be. Sparks fly and the wrench becomes very hot as the battery quickly discharges. Lots of heat is given off, but the system does no work ($w = 0$). Thus, the entire ΔE appears as heat.

In Path 2, we discharge the battery more slowly by using it to operate a motor. Along this path, much of the energy represented by ΔE appears as work and only a relatively small amount appears as heat caused by friction within the motor and the electrical resistance of the wires.

There are two things to be learned from this example. The first is that q and w are not state functions. Their values depend entirely on how a change between two states occurs. However, it is interesting to note that their sum, ΔE, *is* a state function. The second lesson here is that the amount of energy that we

FIGURE 13.1

The complete discharge of a battery along two different paths yields the same total amount of energy, ΔE. However, if the battery is simply shorted with a heavy wrench, as shown in Path 1, this energy appears entirely as heat. Path 2 gives part of the total energy as heat, but much of the energy appears as work done by the motor.

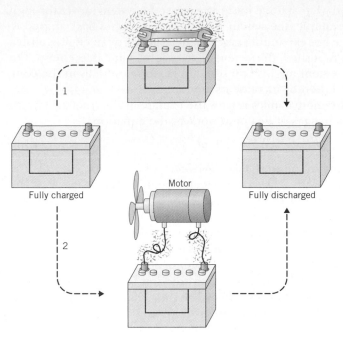

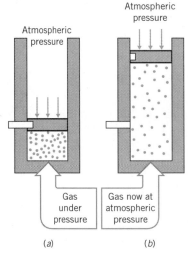

(a)   (b)

FIGURE 13.2

(a) A gas confined under pressure in a cylinder fitted with a piston that is held in place by a sliding pin. *(b)* When the piston is released, the gas inside the cylinder expands and pushes the piston upward against the opposing pressure of the atmosphere. As it does so, the gas does some work.

If a gas is *compressed* by an external pressure, *w* is positive because work is done on the gas.

can extract from a system in the form of work depends on how rapidly we withdraw the energy. We will have more to say about this later.

Pressure–Volume Work

Most of the reactions that we carry out in the lab do not produce electricity and therefore are not able to do electrical work. But there is another important kind of work these systems can be involved in that is related to expansion or contraction under the influence of an outside pressure. For example, consider a compressed gas in a cylinder fitted with a piston that is supported by the pressure of the atmosphere (Figure 13.2). When the pin is pulled out, thereby allowing the piston to move, the gas pushes the piston back against the opposing pressure of the atmosphere. In doing so, the gas does some work on the surroundings. The amount of this work can be calculated by the equation

$$w = -P\Delta V \qquad (13.2)$$

where P is the *opposing pressure* against which the piston pushes and ΔV is the change in the volume of the system (the gas) during the expansion. The negative sign is required by the convention that work done *by the system* is negative.

The origin of Equation 13.2 can be seen by analyzing the units. As noted earlier, work is accomplished by moving an opposing force, F, through some distance or length, L. The amount of work done is the product of this force and length.

$$\text{work} = F \times L$$

Pressure is defined as force per unit area. But an area is simply length squared, L^2. So we can write pressure as

$$P = \frac{F}{L^2}$$

Volume (or a volume change) has dimensions of length cubed, L^3, so pressure times volume change is

$$P\Delta V = \frac{F}{L^2} \times L^3$$
$$= F \times L$$

Thus, pressure times volume change is really the same as force times distance, which is work. A common example is the work done by the expanding gases in a cylinder of a car engine as they move a piston. Another example is the work we do compressing air with a tire pump. This kind of work is often called **pressure–volume work,** or P–V work.

Heats of Reaction at Constant Volume

Suppose that the only kind of work a system can do is P–V work. Then we can modify Equation 13.1 as follows.

$$\Delta E = q + w$$
$$= q + (-P\Delta V)$$
$$= q - P\Delta V$$

Let's apply this equation to a chemical reaction occurring under conditions of constant volume. An example is a reaction taking place in the apparatus illustrated in Figure 13.3, called a *bomb calorimeter* because of the resemblance of the reaction chamber to a small bomb. Because of the construction of the apparatus, the volume of the reaction mixture cannot change, so ΔV must be equal to zero when the reaction occurs. This means, of course, that $P\Delta V$ must also be zero. Therefore, under conditions of constant volume,

$$\Delta E = q - 0 = q$$

If there is no volume change, then $V_{final} = V_{initial}$ and ΔV must equal zero.

Thus, the heat of reaction measured in a bomb calorimeter is the *heat of reaction at constant volume,* often symbolized as q_V, and corresponds to ΔE for the reaction.

$$\Delta E = q_V$$

ΔE is the heat of reaction at constant volume.

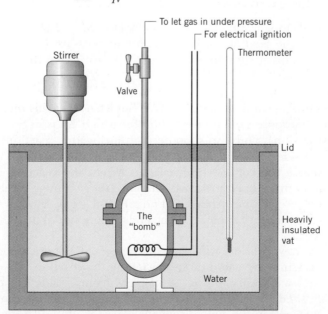

FIGURE   13.3

A bomb calorimeter. The water bath is usually equipped with devices for adding or removing heat from the water and thus keep its temperature constant up to the moment when the reaction occurs in the bomb. The reaction chamber is of fixed volume, so $P\Delta V$ must equal zero for reactions in this apparatus.

Stirrer

To let gas in under pressure
For electrical ignition
Thermometer

Valve

Lid

The "bomb"

Heavily insulated vat

Water

Enthalpy, Heats of Reaction at Constant Pressure

Most reactions that are of interest to us do not occur at constant volume. Instead, they are carried out in open containers where they experience the constant pressure of the atmosphere. Because of this, thermodynamicists invented the quantity we call **enthalpy.** It is defined mathematically by the equation.

$$H = E + PV$$

If we only consider changes that occur at constant pressure, but allow the volume to change, then the change in enthalpy, ΔH, is

$$\Delta H = \Delta E + P\Delta V \qquad (13.3)$$

Let's see how Equation 13.3 applies to a chemical reaction taking place at constant pressure and during which the system expands against the opposing pressure of the atmosphere. As we saw in Equation 13.2, the work involved for the expanding system is

$$w = -P\Delta V$$

and therefore

$$\Delta E = q - P\Delta V$$

Substituting this into Equation 13.3 gives

$$\Delta H = q - P\Delta V + P\Delta V$$

which reduces to

$$\Delta H = q_P$$

We use the subscript P to show that the heat involved this time is the heat of reaction at *constant pressure*. This equation states that the heat of reaction measured under conditions of constant pressure is equal to ΔH. Recall that this is exactly how we defined the enthalpy change in Chapter 5.

The Difference between ΔE and ΔH

Let's look again at the equation $\Delta H = \Delta E + P\Delta V$, because it permits us to see why the heat of reaction at constant pressure (ΔH) differs from the heat of reaction at constant volume (ΔE). Rearranging this equation gives

$$\Delta H - \Delta E = P\Delta V$$

Thus ΔH and ΔE differ by $P\Delta V$. But what does this mean, physically?

Suppose we have a reaction for which ΔE is negative (i.e., an exothermic reaction), and also suppose that during this reaction a gas is formed. Let's consider first what happens if we carry out the reaction in a bomb calorimeter, where the volume can't change. When the reaction occurs, a certain total amount of energy is given off. Since the volume is unable to change, ΔV must be zero, so the system can't do any P–V work. This means that *all* the energy released must appear as heat, and the amount of heat that we measure is equal to ΔE.

Now, consider what happens if we carry out the same reaction at a constant pressure. We could do this by letting the reaction take place inside a cylinder with a movable piston. As the gas forms inside the cylinder, the piston is pushed back against the opposing pressure of the atmosphere and the pressure within the system remains constant. But, as the gas moves the piston it

The pressure inside the reaction vessel increases, but the gas can't do any work because the volume can't change.

does some work, and the energy for this comes from the total energy given off by the reaction. Therefore, the amount of energy left over to be given off as the heat of reaction at constant pressure (ΔH) is equal to the total amount of energy released in the reaction (ΔE) minus the amount that was lost in pushing back the piston. Thus, the difference between ΔE and ΔH is the $P\Delta V$ work that the system does as it expands against the opposing pressure of the atmosphere.

Whenever a reaction system expands under the constant pressure of the atmosphere, the measured ΔH is a little less negative than ΔE. This is a small energy penalty that we pay for the convenience of carrying out the reaction at constant pressure. On the other hand, if there is a decrease in the volume of the reacting system at constant pressure, then the value of ΔH is actually a bit more negative than ΔE, and we get a bonus. In either case, though, the size of the $P\Delta V$ term is small for reactions at atmospheric pressure, so differences between ΔE and ΔH are small.

> The ΔV for reactions involving only solids and liquids are very tiny, so ΔE and ΔH for these reactions are nearly the same size.

Conversions between ΔE and ΔH

The only time ΔE and ΔH differ by a significant amount is when gases are formed or consumed in a reaction. If it is necessary to calculate one from the other, there is a simple relationship that can be used.

Recall that ΔE and ΔH differ by $P\Delta V$. If we assume ideal gas behavior, then

$$\Delta V = \Delta\left(\frac{nRT}{P}\right)$$

For a change at constant pressure and temperature we can rewrite this as

$$\Delta V = \Delta n\left(\frac{RT}{P}\right)$$

Thus, when the reaction occurs, the volume change is caused by a change in the number of moles of *gas*, Δn_{gas}, which is defined as

$$\Delta n_{gas} = (n_{gas})_{final} - (n_{gas})_{initial} \tag{13.4}$$

> Remember, in calculating Δn, we only count moles of *gases* among the reactants and products.

The P–V work is therefore,

$$P\Delta V = P \cdot \Delta n\left(\frac{RT}{P}\right)$$

$$= \Delta n_{gas} RT$$

 Converting between ΔE and ΔH

Substituting into the equation for ΔH gives

$$\Delta H = \Delta E + \Delta n_{gas} RT \tag{13.5}$$

The following example illustrates how small the difference is between ΔE and ΔH.

The decomposition of calcium carbonate in limestone is used industrially to make carbon dioxide.

$$CaCO_3(s) \longrightarrow CaO(s) + CO_2(g)$$

The reaction is endothermic and has $\Delta H° = +571$ kJ. What is the value of $\Delta E°$ for this reaction? What is the percentage difference between $\Delta E°$ and $\Delta H°$?

EXAMPLE 13.1
Conversion between ΔE and ΔH

ANALYSIS The problem asks us to convert between ΔE and ΔH, so we need to use Equation 13.5. The superscript ° on ΔH and ΔE tells us the temperature is 25 °C (standard temperature for measuring heats of reaction). It is also important to remember that in calculating Δn, we count just the numbers of moles of gas.

We also need a value for R. Since we want to calculate the term ΔnRT in units of kilojoules, we need to have R in appropriate energy units. The value of R we will use in this and other calculations that involve energies is found in a table of constants like the one inside the rear cover of this book. It is $R = 8.314\ \text{J mol}^{-1}\ \text{K}^{-1}$.

SOLUTION The equation we need is

$$\Delta H = \Delta E + \Delta n_{\text{gas}}RT$$

Solving for ΔE and applying the superscript ° gives

$$\Delta E° = \Delta H° - \Delta n_{\text{gas}}RT$$

For the reaction in this problem, taking coefficients to represent numbers of moles, there is one mole of gas among the products and no moles of gas among the reactants, so

$$\Delta n_{\text{gas}} = 1 - 0 = 1$$

The temperature is 298 K, and $R = 8.314\ \text{J mol}^{-1}\ \text{K}^{-1}$. Substituting,

$$\begin{aligned}\Delta E° &= +571\ \text{kJ} - (1\ \text{mol})(8.314\ \text{J mol}^{-1}\ \text{K}^{-1})(298\ \text{K}) \\ &= +571\ \text{kJ} - 2.48\ \text{kJ} \\ &= +569\ \text{kJ}\end{aligned}$$

The percentage difference is calculated as follows:

$$\begin{aligned}\%\ \text{difference} &= \frac{2.48\ \text{kJ}}{571\ \text{kJ}} \times 100\% \\ &= 0.434\%\end{aligned}$$

Notice how small the relative difference is between ΔE and ΔH.

■ **Practice Exercise 1** The reaction

$$CaO(s) + 2HCl(g) \longrightarrow CaCl_2(s) + H_2O(g)$$

has $\Delta H° = -217.1\ \text{kJ}$. Calculate $\Delta E°$ for this reaction. What is the percentage difference between $\Delta E°$ and $\Delta H°$?

13.3
ENTHALPY CHANGES AND SPONTANEITY

We take many spontaneous events for granted. What would you think if you dropped a book and it rose to the ceiling?

Now that you have a better understanding of energy changes, we turn our attention to one of the main goals of thermodynamics—finding relationships among the factors that control whether events are spontaneous. A **spontaneous change** is one that occurs without continuous outside assistance. Examples are water flowing over a waterfall and the melting of ice cubes in a cold drink on a warm day. These are events that proceed on their own.

Some spontaneous changes occur very rapidly. An example is the set of biochemical reactions that take place when you accidentally touch something that is very hot—they cause you to jerk your hand away quickly. Other spontaneous events, such as the gradual erosion of a mountain, occur slowly and many years pass before a change is noticed. Still others occur at such an extremely slow rate under ordinary conditions that they appear not to be spontaneous at all. Gasoline–oxygen mixtures appear perfectly stable indefinitely at room temperature because under these conditions they react very slowly.

However, if heated, their rate of reaction increases tremendously and they react explosively.

Each day we also witness events that are obviously not spontaneous. We may pass by a pile of bricks in the morning and later in the day find that they have become a brick wall. We know from experience that the wall didn't get there by itself. A pile of bricks becoming a brick wall is *not* spontaneous; it requires the intervention of a bricklayer. Similarly, the decomposition of water into hydrogen and oxygen is not spontaneous. We see water all the time and we know that it's stable. Nevertheless, we can cause water to decompose by passing an electric current through it in a process called *electrolysis* (Figure 13.4).

$$2H_2O(l) \xrightarrow{\text{Electrolysis}} 2H_2(g) + O_2(g)$$

This will continue, however, only as long as the electric current is maintained. As soon as the supply of electricity is cut off, the decomposition ceases. This example demonstrates the difference between spontaneous and nonspontaneous changes. Once a spontaneous event begins, it has a tendency to continue until it is finished. A nonspontaneous event, on the other hand, can continue only as long as it receives some sort of outside assistance.

Nonspontaneous changes have another common characteristic. They can occur only when accompanied by some spontaneous change. The bricklayer consumes food, and a series of spontaneous biochemical reactions then occur that supply the necessary muscle power to build the wall. Similarly, the nonspontaneous electrolysis of water requires some sort of spontaneous mechanical or chemical change to generate the needed electricity. In short, all nonspontaneous events occur at the expense of spontaneous ones. Everything that happens can be traced, either directly or indirectly, to spontaneous changes.

Spontaneity and Potential Energy Changes

Because spontaneous reactions are so important, it is necessary for us to understand the factors that favor spontaneity. Let's begin by examining some everyday events such as those depicted in Figure 13.5. One thing that each of these has in common is a lowering of the potential energy. Both the snowboard and the water lose potential energy as they move from a higher to a

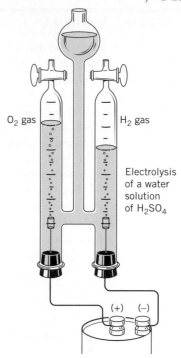

O₂ gas   H₂ gas

Electrolysis of a water solution of H_2SO_4

(+)   (−)

FIGURE 13.4

The electrolysis of water produces H_2 and O_2 gases. It is a nonspontaneous change that continues only as long as electricity is supplied.

FIGURE 13.5

Three common spontaneous events—a person rides a "snowboard" downhill, water cascades over a waterfall, and fuel burns.

SPECIAL TOPIC 13.1 / ENTHALPY CHANGES AND BOND ENERGIES

In Chapter 5 you learned that enthalpy changes for chemical reactions, obtained either by measurement or by calculation using Hess's law, give us information about the heat given off or absorbed when chemical reactions occur. Studying heats of reaction, and heats of formation in particular, also can yield fundamental information about the chemical bonds in the substances that react, because the origin of the energy changes in chemical reactions is changes in bond energies.

Recall that the **bond energy** is the amount of energy needed to break a chemical bond to give electrically neutral fragments. It is a useful quantity to know in the study of chemical properties, because during chemical reactions, bonds within the reactants are broken and new ones are formed as the products appear. The first step—bond breaking—is one of the factors that controls the reactivity of substances. Elemental nitrogen, for example, has a very low degree of reactivity, which is generally attributed to the very strong triple bond in N_2. Reactions that involve the breaking of this bond in a single step simply do not occur. When N_2 does react, it is by a stepwise involvement of its three bonds, one at a time.

MEASUREMENT OF BOND ENERGIES

The bond energies of simple molecules such as H_2, O_2, and Cl_2 are usually measured *spectroscopically*. That is, the light that is emitted by the molecules when they are energized (excited) by a flame or an electric arc is analyzed, and the amount of energy needed to break the bond is computed.

For more complex molecules, thermochemical data can be used to calculate bond energies in a Hess's law kind of calculation. We will use the standard heat of formation of methane to illustrate how this is accomplished. However, before we can attempt such a calculation, we must first define a thermochemical quantity that we will call the **atomization energy,** symbolized ΔH_{atom}. This is the amount of energy needed to rupture all the chemical bonds in one mole of gaseous molecules to give gaseous atoms as products. For example, the atomization of methane is

$$CH_4(g) \longrightarrow C(g) + 4H(g)$$

and the enthalpy change for the process is ΔH_{atom}. For this particular molecule, ΔH_{atom} corresponds to the total amount of energy needed to break all the C—H bonds in one mole of CH_4; therefore, division of ΔH_{atom} by 4 would give the average C—H bond energy in methane, expressed in kJ/mol.

Figure 1 shows how we can use the standard heat of formation, ΔH_f°, to calculate the atomization energy. Across the bottom we have the chemical equation for the formation of CH_4 from its elements. The enthalpy change for this reaction, of

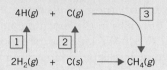

FIGURE 1

Two paths for the formation of methane from its elements in their standard states. Steps 1, 2, and 3 of the upper path involve the formation of gaseous atoms of the elements and the formation of the bonds in CH_4.

course, is ΔH_f°. In this figure we also can see an alternative three-step path that leads to $CH_4(g)$. One step is the breaking of H—H bonds in the H_2 molecules to give gaseous hydrogen atoms, another is the vaporization of carbon to give gaseous carbon atoms, and the third is the combination of the gaseous atoms to form CH_4 molecules. These changes are labeled 1, 2, and 3 in the figure.

Since ΔH is a state function, the net enthalpy change from one state to another is the same regardless of the path that we follow. This means that the sum of the enthalpy changes along the upper path must be the same as the enthalpy change along the lower path, ΔH_f°. Perhaps this can be more easily seen in terms of Hess's law if we write the changes along the upper path in the form of thermochemical equations.

Steps 1 and 2 have enthalpy changes that are called *standard heats of formation of gaseous atoms.* Values for these quantities have been measured for many of the elements, and some are given in Table E.2. Step 3 is the opposite of atomization, and its enthalpy change will therefore be the negative of ΔH_{atom} (recall that if we reverse a reaction, we change the sign of its ΔH).

Step 1	$2H_2(g) \longrightarrow 4H(g)$	$\Delta H_1^\circ = 4\Delta H_f^\circ[H(g)]$
Step 2	$C(s) \longrightarrow C(g)$	$\Delta H_2^\circ = \Delta H_f^\circ[C(g)]$
Step 3	$4H(g) + C(g) \longrightarrow CH_4(g)$	$\Delta H_3^\circ = -\Delta H_{atom}$
	$2H_2(g) + C(s) \longrightarrow CH_4(g)$	$\Delta H^\circ = \Delta H_f^\circ[CH_4(g)]$

Notice that by adding the first three equations, we get the equation for the formation of CH_4 from its elements in their standard states. This means that adding the ΔH° values of the first three equations should give ΔH_f° for CH_4.

$$\Delta H_1^\circ + \Delta H_2^\circ + \Delta H_3^\circ = \Delta H_f^\circ[CH_4(g)]$$

Let's substitute for ΔH_1°, ΔH_2°, and ΔH_3°, and then solve for ΔH_{atom}. First, we substitute for the ΔH° quantities.

$$4\Delta H_f^\circ[H(g)] + \Delta H_f^\circ[C(g)] + (-\Delta H_{atom}) = \Delta H_f^\circ[CH_4(g)]$$

Next, we solve for $(-\Delta H_{atom})$.

$$-\Delta H_{atom} = \Delta H_f^\circ[CH_4(g)] - 4\Delta H_f^\circ[H(g)] - \Delta H_f^\circ[C(g)]$$

Changing signs and rearranging the right side of the equation just a bit gives

$$\Delta H_{atom} = 4\Delta H_f^\circ[H(g)] + \Delta H_f^\circ[C(g)] - \Delta H_f^\circ[CH_4(g)]$$

Now all we need are values for the ΔH_f°'s on the right side. From Table E.2 we obtain $\Delta H_f^\circ[H(g)]$ and $\Delta H_f^\circ[C(g)]$, and the value of $\Delta H_f^\circ[CH_4(g)]$ is obtained from Table 5.2.

$$\Delta H_f^\circ[H(g)] = +217.9 \text{ kJ/mol}$$

$$\Delta H_f^\circ[C(g)] = +716.7 \text{ kJ/mol}$$

$$\Delta H_f^\circ[CH_4(g)] = -74.8 \text{ kJ/mol}$$

Substituting these values gives

$$\Delta H_{atom} = 1663.1 \text{ kJ/mol}$$

and division by 4 gives an estimate of the average C—H bond energy in this molecule.

$$\text{Bond energy} = \frac{1663.1 \text{ kJ/mol}}{4}$$
$$= 415.8 \text{ kJ/mol of C—H bonds}$$

This value is quite close to the one in Table E.3, which is an average of the C—H bond energies in different compounds. The other bond energies in Table E.3 are also based on thermochemical data and were obtained by similar calculations.

USES OF BOND ENERGIES

An amazing thing about covalent bond energies is that they are very nearly the same in many different compounds. This suggests, for example, that a C—H bond is very nearly the same in CH_4 as it is in a large number of other compounds that contain this kind of bond.

Because the bond energy doesn't vary much from compound to compound, we can use tabulated bond energies to estimate the heats of formation of substances. For example, let's calculate the standard heat of formation of methyl alcohol vapor, $CH_3OH(g)$. The structural formula for methanol is

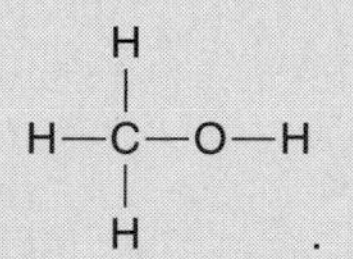

$$\begin{array}{c} \text{H} \\ | \\ \text{H—C—O—H} \\ | \\ \text{H} \end{array}$$

To perform this calculation, we set up two paths from the elements to the compound, as shown in Figure 2. The lower path has an enthalpy change corresponding to $\Delta H_f^\circ[CH_3OH(g)]$, while the upper path takes us to the gaseous elements and then

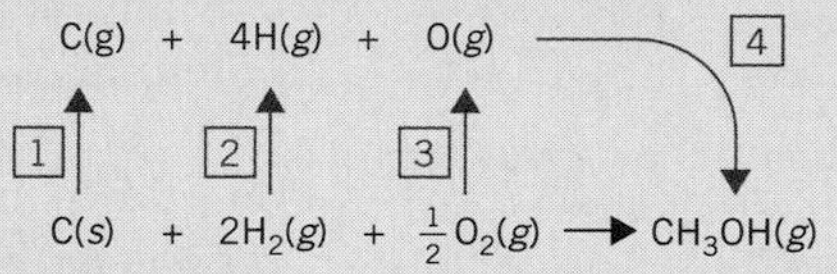

FIGURE 2

Two paths for the formation of methyl alcohol vapor from its elements. The numbered steps along the upper path are referred to in the discussion.

through the energy released when the bonds in the molecule are formed. This latter energy can be computed from the bond energies in Table E.3. As before, the sum of the energy changes along the upper path must be the same as the energy change along the lower path, and this permits us to compute $\Delta H_f^\circ[CH_3OH(g)]$.

Steps 1, 2, and 3 involve the formation of the gaseous atoms from the elements, and their enthalpy changes are obtained from Table E.2.

$$\Delta H_1^\circ = \Delta H_f^\circ[C(g)] = 1 \text{ mol} \times 716.7 \text{ kJ/mol} = 716.7 \text{ kJ}$$

$$\Delta H_2^\circ = 4\Delta H_f^\circ[H(g)] = 4 \text{ mol} \times (217.9 \text{ kJ/mol}) = 871.6 \text{ kJ}$$

$$\Delta H_3^\circ = \Delta H_f^\circ[O(g)] = 1 \text{ mol} \times (249.2 \text{ kJ/mol}) = 249.2 \text{ kJ}$$

Adding these values gives a total energy input for the first three steps of 1837.5 kJ. The net ΔH° for the first three steps is +1837.5 kJ.

The formation of the CH_3OH molecule from the gaseous atoms is exothermic because energy is always released when atoms become joined by covalent bonds. In this molecule we count three C—H bonds, one C—O bond, and one O—H bond. Their formation releases energy equal to their bond energies, which we obtain from Table E.3.

Bond		Energy (kJ)
3(C—H)	3 × (412 kJ/mol) = 1236	
C—O		360
O—H		463

Adding these together gives a total of 2059 kJ. Therefore, ΔH_4° is −2059 kJ (because it is exothermic). Now we can compute the total enthalpy change for the upper path.

$$\Delta H^\circ = (+1837.5 \text{ kJ}) + (-2059 \text{ kJ})$$
$$= -222 \text{ kJ}$$

The value just calculated should be equal to ΔH_f° for $CH_3OH(g)$. For comparison, it has been found experimentally that ΔH_f° for this molecule (in the vapor state) is −201 kJ/mol. At first glance, the agreement doesn't seem very good, but on a relative basis the calculated value (−222 kJ) differs from the experimental one by only about 10%.

lower altitude. Similarly, the chemical substances in the gasoline–oxygen mixture lose energy by evolving heat as they react to produce CO_2 and H_2O.

Many, but not all, spontaneous events are accompanied by a lowering of the potential energy, so an energy lowering is *one* of the factors that work in *favor* of spontaneity. Since a change that lowers the potential energy of a system can be said to be exothermic, we can state this factor another way — *exothermic changes have a tendency to proceed spontaneously.*

In Section 13.2 we saw that thermodynamics looks to ΔE and ΔH for the potential energy change in reacting systems. There you learned that it is ΔE that we observe when the reaction happens at constant volume, and it is ΔH that we measure when the reaction occurs at constant pressure. Since we are most interested in reactions at constant pressure, we will devote our attention to ΔH. This quantity, you recall, has a negative sign for an exothermic change, so we can say that reactions for which ΔH is negative *tend* to proceed spontaneously. The enthalpy change thus serves as *one* of the thermodynamic factors that influence whether or not a given process can occur by itself.

> Most, but not all, chemical reactions that are exothermic occur spontaneously.

> If we study systems at constant temperature, the kinetic energy remains constant and any energy changes we measure must be changes in potential energy.

13.4 ENTROPY AND SPONTANEOUS CHANGE

In Chapter 12 we discussed the solution process, and you learned that one of the principal driving forces in the formation of a solution is the increase in disorder, or randomness, that occurs when particles of the solute mix with those of the solvent. In fact, this increase in disorder can be so important that it can outweigh energy effects. For example, the dissolving of NaI in water is endothermic (ΔH_{soln} is positive); nevertheless, NaI crystals will dissolve quite spontaneously in water, even though the energy effect would seemingly make the process nonspontaneous.

The dissolving of salts such as NaI in water is just one example of a change that occurs spontaneously even though it is endothermic. Other examples are the melting of ice when the weather becomes warm and the evaporation of water from a puddle or a pond. Both of these changes are also endothermic and spontaneous, and they take place for the same reason that NaI dissolves in water. Each is accompanied by an increase in the randomness of the distribution of its particles. When ice melts, the water molecules leave the highly organized crystalline state and become jumbled and disorganized in the liquid. When water evaporates, the molecules are no longer confined to the region of other water molecules; instead, they are free to roam throughout the entire atmosphere.

It is a universal phenomenon that something that brings about randomness is more likely to occur than something that brings about order. We can see the reason for this if we examine the close relationship between randomness and statistical probability. Suppose we try a simple experiment with a new deck of playing cards. When first unwrapped, they are separated according to suit and arranged numerically. Now, suppose we toss the deck in the air and let the cards fall to the floor. When we sweep them up and restack them, we will almost certainly find they have become disordered. We expect this disordering to occur because *there are so many millions of ways for the cards to be disordered, but there is only one way for them to come together again in the original sequence.* As soon as the cards are tossed, a random sequence becomes much more probable than the ordered one with which we began. Therefore, the spontaneous change from the ordered arrangement to a disordered one takes place simply because of the laws of chance.

The cards are highly ordered

After they've been tossed they are disordered

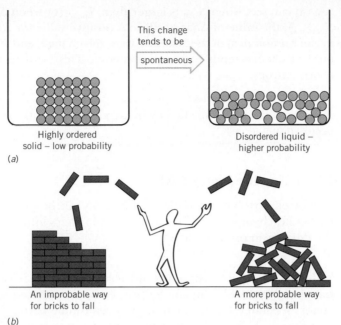

FIGURE 13.6

(a) If water molecules were dropped into the container, they would not be expected to land in just the right places to give the highly ordered solid. If the way they land is determined purely by chance, the disordered liquid is more likely to occur. *(b)* Tossing bricks into the air is more likely to give a pile of bricks than a brick wall.

The same laws of chance that apply to the ordering of a deck of playing cards also apply to the ordering within distributions of particles in chemical and physical systems. Consider, for example, the melting of an ice cube (Figure 13.6). Of all the possible ways of placing water molecules into a container, the highly ordered arrangement within the crystalline ice cube has a very low probability, while the disordered arrangement of the molecules found in the liquid is more probable. We can see this more clearly, perhaps, by analogy. Just imagine tossing bricks in the air and observing how they land. The chance that they will fall one on the other to form a brick wall is certainly very small; the brick wall is an improbable result. A much more likely arrangement is a jumbled, disordered pile. Thus, the collapse of a brick wall and the melting of an ice cube have something in common. In each case, the system passes spontaneously from a state of low probability to one of higher probability.

Entropy Changes and Spontaneity

Because statistical probability is so important in determining the outcome of chemical and physical events, thermodynamics defines a quantity, called **entropy** (symbol S), that describes the degree of randomness of a system. The larger the value of the entropy, the larger is the degree of randomness of the system and, therefore, the larger is its statistical probability.

The greater the statistical probability of a particular state, the greater the entropy of that state.

Like the enthalpy, entropy is a state function. It depends only on the state of the system, so a change in entropy, ΔS, is independent of the path from start to finish. As with other thermodynamic quantities, ΔS is defined as "final minus initial" or "products minus reactants." Thus

$$\Delta S = S_{final} - S_{initial}$$

or, for a chemical system,

$$\Delta S = S_{products} - S_{reactants}$$

As you can see, when S_{final} is larger than $S_{initial}$ (or when $S_{products}$ is larger than $S_{reactants}$), the value of ΔS is positive. A positive value for ΔS means an increase in the randomness of the system during the change, and we have seen that this kind of change tends to be spontaneous. This leads to a general statement about entropy:

Any event that is accompanied by an increase in the entropy of the system *tends* to occur spontaneously.

Predicting the Sign of ΔS

Predicting the sign of ΔS

It is often possible to predict whether ΔS will be positive or negative for a particular change. This is because several factors influence the magnitude of the entropy in predictable ways.

Volume For gases, the entropy increases with increasing volume, as illustrated in Figure 13.7. At the left we see a gas confined to one side of a container, separated from a vacuum by a removable partition. Let's suppose the partition could be pulled away in an instant, as shown in Figure 13.7*b*. Now we find a situation in which all the molecules of the gas are at one end of a larger container. This is a very improbable distribution of the molecules, so the gas expands spontaneously to achieve a more probable (higher entropy) particle distribution.

Temperature The entropy is also affected by the temperature; the larger the temperature the larger the entropy. For example, when a substance is a solid at absolute zero, its particles are essentially motionless. The atoms are at their equilibrium lattice positions and there is a minimum in the amount of disorder (and entropy) of the particles (Figure 13.8*a*). If some heat is added to the solid, the kinetic energy of the particles increases along with the temperature. This causes the particles to move and vibrate within the crystal, so at a particular moment (pictured in Figure 13.8*b*) the particles are not found exactly at their lattice sites. Thus, we have introduced some disorder and increased the entropy. At a still higher temperature the instantaneous disorder is even greater and the solid has a still higher entropy (Figure 13.8*c*).

Physical State One of the major factors that affects the entropy of a system is its physical state, which is demonstrated in Figure 13.9. We have already concluded that the distribution of particles in a liquid is more probable and of

FIGURE 13.7

The expansion of a gas into a vacuum. *(a)* A gas in a container separated from a vacuum by a partition. *(b)* The gas at the moment the partition is removed. *(c)* The gas expands to achieve a more probable (higher entropy) particle distribution.

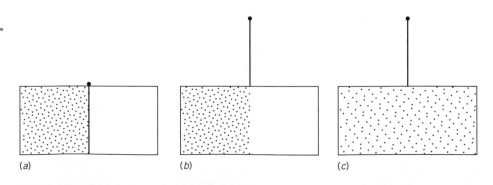

(a)   (b)   (c)

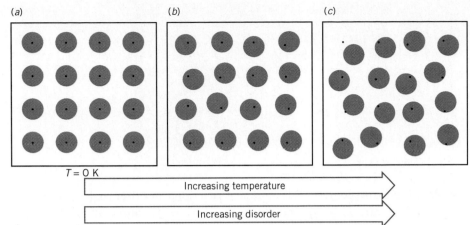

FIGURE 13.8

(a) At absolute zero the atoms, represented by colored circles, rest at their equilibrium lattice positions, represented by black dots. There is perfect order and minimum entropy. *(b)* At a higher temperature, the particles vibrate about their equilibrium positions and there is a greater disorder at any particular instant. *(c)* At a still higher temperature, vibration is more violent and at any instant the particles are found in even more disordered arrangements.

a higher entropy than that in a solid. A gas has an even more random, higher entropy particle distribution. In fact, a gas has such a large entropy compared with a liquid or solid that changes which produce gases from liquids or solids are almost always accompanied by increases in entropy.

$$S_{solid} < S_{liquid} \ll S_{gas}$$

Entropy Changes in Chemical Reactions

When a chemical reaction produces or consumes gases, the sign of its entropy change is usually easy to predict. This is because the entropy of a gas is so much larger than that of either a liquid or solid. For example, the thermal decomposition of sodium bicarbonate produces two gases, CO_2 and H_2O.

For reactions that involve gases, we can simply calculate the change in the number of moles of gas, Δn_{gas}, on going from reactants to products. When Δn_{gas} is positive, so is the entropy change.

$$2NaHCO_3(s) \xrightarrow{\text{Heat}} Na_2CO_3(s) + CO_2(g) + H_2O(g)$$

Because the amount of gaseous products is larger than the amount of gaseous reactants, we can predict that the entropy change for the reaction is positive. On the other hand, the reaction

$$CaO(s) + SO_2(g) \longrightarrow CaSO_3(s)$$

(which can be used to remove sulfur dioxide from a gas mixture) has a negative entropy change.

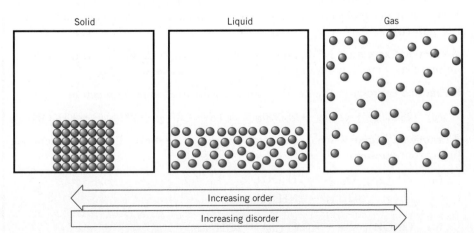

FIGURE 13.9

Comparison of the entropies of the solid, liquid, and gaseous states of a substance. The crystalline solid is highly ordered and has a very low entropy. The liquid has a higher entropy because it is less ordered, but all the particles are still found at one end of the container. The gas has the highest entropy because the particles are randomly distributed throughout the entire container.

FIGURE 13.10

Hydrogen atoms grouped in pairs in H_2 is a less probable (lower entropy) particle distribution than individual H atoms.

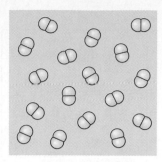 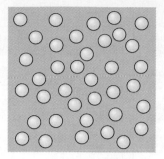

Lower entropy　　　　　　　Higher entropy

Molecular Complexity　For chemical reactions, another major factor that controls the sign of ΔS is changes in the degree of complexity of the molecules. For example, Figure 13.10 illustrates two different ways for atoms of hydrogen to be distributed in a container. If there were no attractions between the atoms, which is the more likely way for the atoms to find themselves, as individual H atoms drifting in the container or as pairs of atoms in H_2? Clearly, the more random distribution is that of individual atoms, and this is also the one of higher entropy. Therefore, the reaction

$$H_2(g) \longrightarrow 2H(g)$$

is accompanied by a positive ΔS. Notice that as the less complex particles form, the total number of particles increases. We conclude, then, that *when a reaction occurs with a decrease in the degree of molecular complexity and an increase in the number of particles, there tends to be an increase in entropy.*

For a given number of atoms, the more particles there are, the less complex each must be. Two moles of H has more particles than 1 mol of H_2; an atom of H is less complex than a molecule of H_2.

EXAMPLE 13.2
Predicting the Sign of ΔS

Predict the algebraic sign of ΔS for the reactions:

(a)　$2NO_2(g) \rightarrow N_2O_4(g)$

(b)　$C_3H_8(g) + 5O_2(g) \rightarrow 3CO_2(g) + 4H_2O(g)$

SOLUTION　In reaction (a) we are forming fewer, more complex molecules (N_2O_4) from simpler ones (NO_2). Since molecular complexity is increasing, entropy must be decreasing, so ΔS must be negative.

For reaction (b), we can count the number of molecules on both sides. On the left of the equation we have six molecules; on the right there are seven. The only way to form more molecules from the same number of atoms is to make them less complex. If the molecules are less complex, they are of higher entropy, so for reaction (b), we expect ΔS to be positive.

■ **Practice Exercise 2**　Predict the sign of the entropy change for (a) the condensation of steam to liquid water, and (b) the sublimation of a solid.

■ **Practice Exercise 3**　Predict the sign of ΔS for the following reactions:

(a)　$2SO_2(g) + O_2(g) \rightarrow 2SO_3(g)$　　(b)　$CO(g) + 2H_2(g) \rightarrow CH_3OH(g)$

■ **Practice Exercise 4**　What is the expected sign of ΔS for the following reactions? Justify your answers.

(a)　$2H_2(g) + O_2(g) \rightarrow 2H_2O(l)$

(b)　$N_2(g) + 3H_2(g) \rightarrow 2NH_3(g)$

(c)　$Ca(OH)_2(s) \xrightarrow{H_2O} Ca^{2+}(aq) + 2OH^-(aq)$

SPECIAL TOPIC 13.2 / POLLUTION AND THE SECOND LAW OF THERMODYNAMICS

A problem we hear about often is pollution. Environmentalists and politicians talk frequently about how to avoid polluting the environment and how to clean up the mess that already exists. Such discussions raise some interesting questions. Can we entirely avoid polluting our environment? Why are pollutants so easily spread and so difficult to eliminate? Many of the answers can be found in the second law of thermodynamics.

According to the second law, any spontaneous activity is accompanied by an increase in the total entropy of the universe. This means that whenever we create order somewhere, our activities must generate an even greater disorder in our surroundings.

Much of our time and effort is spent in creating order in the world around us. We clean up our desks and put out the garbage, mow the lawn and rake leaves in the autumn. Overall our activities are spontaneous, because the biochemical reactions driven by the foods we eat and digest are spontaneous and allow us to do these things. According to the second law, however, the total entropy of the universe must be increased by our spontaneous activities. Therefore, the increased order that we create for ourselves has to be balanced by an even larger increase in disorder somewhere in our surroundings—the environment.

Pollution involves the scattering of undesirable substances through our surroundings and is accompanied by an enormous increase in entropy. It is also a direct result of our efforts to create an orderly world. For example, we vacuum the floors in our homes to keep them clean, but the combustion of fuels to generate the electricity to power the vacuum cleaner releases polluting gases such as SO_2 into the atmosphere. Once released, such pollutants cannot be recovered effectively because of the emormous expense that would be required.

The entropy effect on pollution is important to understand, particularly when we consider the consequences of releasing harmful substances into the environment. Such materials include the highly toxic elements released in trace amounts when coal or nuclear fuels are used, as well as many toxic chemicals

Fires raging in the oil fields of Kuwait after the Mideast war of 1991 released large amounts of pollutants into the environment.

such as DDT and dioxin. When these chemicals are released, their spread is unavoidable because of the large entropy increase that occurs as they are scattered about. Eliminating them once they have had an opportunity to disperse is an almost impossible task, because doing so requires an enormous expenditure of energy and must generate even more disorder around us. Only if a pollutant is extremely hazardous does it warrant the effort necessary to reduce its concentration in the environment, and even if this is done, we can never eliminate the pollutant entirely. We can only reduce our risk of injury.

The surest way to overcome pollution, of course, is not to create it in the first place. For the benefits of activities that pollute, we try to reduce the risks as much as we judge possible. Thus, if we must use coal to generate electricity, it is worth the effort to remove as much sulfur as we can from the coal *before* it is burned. Beyond that, we must simply face the fact that we can never avoid pollution entirely. It's a no-win situation and a trade-off between risks and benefits.

The Second Law of Thermodynamics

One of the most far-reaching observations in science is incorporated into the **second law of thermodynamics,** which states, in effect, that *whenever a spontaneous event takes place in our universe, it is accompanied by an overall increase in entropy.* Notice that the increase in entropy here is for the *universe* (system *plus* surroundings), not just the system alone. This means that a system's entropy can decrease, just as long as there is a larger increase in the entropy of the surroundings so that the *net* entropy change is positive. Because everything that happens relies on spontaneous changes of some sort, the entropy of the universe is constantly increasing. As described in Special Topic 13.2, this has some serious implications regarding the safekeeping of our environment.

13.5
THE THIRD LAW OF THERMODYNAMICS

25 °C and 1 atm are the same standard conditions we used in our discussion of $\Delta H°$ in Chapter 5.

In the preceding section we described how the entropy of a substance depends on temperature, and we noted that at absolute zero the order within a crystal is a maximum and the entropy is a minimum. The **third law of thermodynamics** goes one step further by stating: *At absolute zero the entropy of a perfectly ordered pure crystalline substance is zero.*

$$S = 0 \quad \text{at} \quad T = 0 \text{ K}$$

Because we know the point at which entropy has a value of zero, it is possible by *experimental measurement* and calculation to determine the total amount of entropy that a substance has at temperatures above 0 K. If the entropy of 1 mol of a substance is determined at a temperature of 298 K (25 °C) and a pressure of 1 atm, we call it the **standard entropy, $S°$**. Table 13.1 lists the standard entropies for a number of substances.[1] Notice that entropy has the

TABLE 13.1 Standard Entropies of Some Typical Substances (25 °C, 1 atm)

Substance	$S°(\text{J mol}^{-1}\text{K}^{-1})$	Substance	$S°(\text{J mol}^{-1}\text{K}^{-1})$
$Ag(s)$	42.55	$H_2O(g)$	188.7
$AgCl(s)$	96.2	$H_2O(l)$	69.96
$Al(s)$	28.3	$HCl(g)$	186.7
$Al_2O_3(s)$	51.00	$HNO_3(l)$	155.6
$C(s, \text{graphite})$	5.69	$H_2SO_4(l)$	157
$CO(g)$	197.9	$HC_2H_3O_2(l)$	160
$CO_2(g)$	213.6	$Hg(l)$	76.1
$CH_4(g)$	186.2	$Hg(g)$	175
$CH_3Cl(g)$	234.2	$K(s)$	64.18
$CH_3OH(l)$	126.8	$KCl(s)$	82.59
$CO(NH_2)_2(s)$	104.6	$K_2SO_4(s)$	176
$CO(NH_2)_2(aq)$	173.8	$N_2(g)$	191.5
$C_2H_2(g)$	200.8	$NH_3(g)$	192.5
$C_2H_4(g)$	219.8	$NH_4Cl(s)$	94.6
$C_2H_6(g)$	229.5	$NO(g)$	210.6
$C_2H_5OH(l)$	161	$NO_2(g)$	240.5
$Ca(s)$	154.8	$N_2O(g)$	220.0
$CaCO_3(s)$	92.9	$N_2O_4(g)$	304
$CaCl_2(s)$	114	$Na(s)$	51.0
$CaO(s)$	40	$Na_2CO_3(s)$	136
$Ca(OH)_2(s)$	76.1	$NaHCO_3(s)$	102
$CaSO_4(s)$	107	$NaCl(s)$	72.38
$CaSO_4 \cdot \frac{1}{2}H_2O(s)$	131	$NaOH(s)$	64.18
$CaSO_4 \cdot 2H_2O(s)$	194.0	$Na_2SO_4(s)$	149.4
$Cl_2(g)$	223.0	$O_2(g)$	205.0
$Fe(s)$	27	$PbO(s)$	67.8
$Fe_2O_3(s)$	90.0	$S(s)$	31.9
$H_2(g)$	130.6	$SO_2(g)$	248.5
		$SO_3(g)$	256.2

[1] In our earlier discussions of standard states (Chapter 5) we defined the *standard pressure* as 1 atm. This was the original pressure unit used by thermodynamicists. However, in the SI, the recognized unit of pressure is the pascal (Pa), not the atmosphere. After considerable discussion, the SI adopted the **bar** as the standard pressure for thermodynamic quantities: 1 bar = 10^5 Pa. One bar differs from one atmosphere by only 1.3%, and for thermodynamic quantities that we deal with in this text, their values at 1 atm and at 1 bar differ by an insignificant amount. Since you are more familiar with atmospheres than pascals, we shall continue to refer to the standard pressure as 1 atm.

SPECIAL TOPIC 13.3 / WHY THE UNITS OF ENTROPY ARE ENERGY/TEMPERATURE

Entropy is a state function, just like enthalpy, so the value of ΔS doesn't depend on the "path" that is followed during a change. In other words, we can proceed from one state to another in any way we like and ΔS will be the same. If we choose a *reversible path* in which just a slight alteration in the system can change the direction of the process, we can measure ΔS directly. For example, at 25 °C an ice cube will melt and there is nothing we can do at that temperature to stop it. This change is *nonreversible* in the sense described above, so we couldn't use this path to measure ΔS. At 0 °C, however, it is simple to stop the melting process and reverse its direction. At 0 °C the melting of ice is a reversible process.

If we set up a change so that it is reversible, we can calculate the entropy change as $\Delta S = q/T$, where q is the heat added to the substance and T is the temperature at which the heat is added. We can understand this in the following way. In a system of molecules held together by attractive forces, disorder is introduced by adding heat, which makes the molecules move more

violently. In a crystal, for instance, the molecules vibrate about their equilibrium lattice positions, so that at any instant we really have a "not-quite-ordered" collection of particles. Adding heat increases molecular motion and leads to greater instantaneous disorder. In other words, as illustrated in Figure 13.8, if we could freeze the motion of the molecules the way a photographer's flash freezes action, we would see that after we have added some heat there is greater disorder and therefore greater entropy. This entropy increase is *directly proportional to the amount of heat added*. However, the added heat makes a more noticeable and significant difference if it is added at low temperature, where little disorder exists, rather than at high temperature where there is already substantial disorder. For a given quantity of heat the entropy change is therefore *inversely proportional to the temperature* at which the heat is added. Thus, $\Delta S = q/T$, and entropy has units of energy divided by temperature (e.g., J/K or cal/K).

dimensions of energy/temperature (i.e., joules per kelvin); the reason is explained in Special Topic 13.3.

Once we have the entropies of a variety of substances, we can calculate the standard **entropy change, $\Delta S°$,** for chemical reactions in much the same way as we calculated $\Delta H°$ in Chapter 5.

$$\Delta S° = (\text{sum of } S° \text{ of the products}) - (\text{sum of } S° \text{ of the reactants}) \quad (13.6)$$

 Calculating $\Delta S°$ from standard entropies

If the reaction we are working with happens to correspond to the formation of 1 mol of a compound from its elements, then the $\Delta S°$ that we calculate can be referred to as the **standard entropy of formation, $\Delta S_f°$.** Values of $\Delta S_f°$ are not tabulated, however; if we need them for some purpose, we must calculate them from tabulated values of $S°$.

This is simply a Hess's law type of calculation. Note, however, that elements have nonzero $S°$ values, which must be included in the bookkeeping.

Urea (from urine) reacts slowly with water to produce ammonia and carbon dioxide.

$$CO(NH_2)_2(aq) + H_2O(l) \longrightarrow CO_2(g) + 2NH_3(g)$$
$$\text{urea}$$

What is the standard entropy change when 1 mol of urea reacts with water?

SOLUTION The standard entropies of the reactants and products are found in Table 13.1. Let's first collect the data and assemble them as we have done at the right.

EXAMPLE 13.3
Calculating $\Delta S°$ from Standard Entropies

Substance	$S°$ (J/mol K)
$CO(NH_2)_2(aq)$	173.8
$H_2O(l)$	69.96
$CO_2(g)$	213.6
$NH_3(g)$	192.5

Applying Equation 13.6, we have

$$\Delta S^\circ = [S^\circ_{CO_2(g)} + 2S^\circ_{NH_3(g)}] - [S^\circ_{CO(NH_2)_2(aq)} + S^\circ_{H_2O(l)}]$$

$$= \left[1 \text{ mol} \times \left(\frac{213.6 \text{ J}}{\text{mol K}} \right) + 2 \text{ mol} \times \left(\frac{192.5 \text{ J}}{\text{mol K}} \right) \right]$$

$$- \left[1 \text{ mol} \times \left(\frac{173.8 \text{ J}}{\text{mol K}} \right) + 1 \text{ mol} \times \left(\frac{69.96 \text{ J}}{\text{mol K}} \right) \right]$$

$$= (598.6 \text{ J/K}) - (243.8 \text{ J/K})$$

$$= 354.8 \text{ J/K}$$

Notice that the unit *mol* cancels in each term, so the units of ΔS° are joules per kelvin.

Thus, the standard entropy change for this reaction is $+ 354.8$ J/K.

■ **Practice Exercise 5** Calculate the standard entropy change, ΔS°, in J/K for the following:

(a) $CaO(s) + 2HCl(g) \rightarrow CaCl_2(s) + H_2O(l)$

(b) $C_2H_4(g) + H_2(g) \rightarrow C_2H_6(g)$

13.6
THE GIBBS FREE ENERGY

We have studied two factors, enthalpy and entropy, that affect whether or not a physical or chemical event will be spontaneous. Sometimes these factors work together. For example, when a stone wall crumbles, its potential energy decreases and the stones become disordered. Therefore, the enthalpy decreases and entropy increases. Since both of these favor a spontaneous change, the two factors complement one another. In other situations, the effects of enthalpy and entropy are in opposition. Such is the case, as we have seen, in the melting of ice or the evaporation of water. The endothermic nature of these changes tends to make them nonspontaneous, while the increase in the randomness of molecules tends to make them spontaneous. In the reaction of H_2 with O_2 to form H_2O the enthalpy and entropy changes are also in opposition. In this case, the exothermic nature of the reaction is sufficient to overcome the negative value of ΔS and cause the reaction to be spontaneous.

When enthalpy and entropy oppose one another, their relative importance in determining spontaneity is far from obvious. In addition, temperature becomes a third factor that can influence the direction in which a change is spontaneous. For example, if we attempt to raise the temperature of an ice–water slush to 25 °C, all the solid will melt. At 25 °C the change *solid* → *liquid* is spontaneous. On the other hand, if we attempt to cool an ice–water slush to −25 °C, freezing occurs, so at −25 °C the opposite change is spontaneous. Thus, there are actually three factors that can influence spontaneity: the enthalpy change, the entropy change, and the temperature.

Specialists in thermodynamics, in studying the second law, defined a quantity called the **Gibbs free energy, G,** named to honor one of America's most important scientists, Josiah Willard Gibbs (1839–1903). (It's called *free energy*

The origin of *free* in *free energy* is discussed in Section 13.8.

because it is related, as we will see later, to the maximum energy in a change that is "free" or "available" to do useful work.) The Gibbs free energy is defined as a function of enthalpy, *H*, entropy, *S*, and temperature, *T*, in the following way.

$$G = H - TS \tag{13.7}$$

As with the enthalpy, *H*, we cannot determine the absolute amount of free energy a substance has. However, what affects us and what we need most to know about are *changes* in free energy, not absolute values. So we will consider *changes* in the Gibbs free energy, ΔG, and we will deal with systems at constant temperature and constant pressure. With these conditions in mind, Equation 13.7 becomes

$$\Delta G = \Delta H - T\Delta S \qquad\qquad (13.8)$$

Because *G* is defined entirely in terms of state functions, it is also a state function. This means that

$$\Delta G = G_{\text{final}} - G_{\text{initial}}$$

The special importance of the free energy change is summarized in the following statement.

At constant temperature and pressure, change can only be spontaneous if it is accompanied by a decrease in the free energy of the system.

In other words, for a change to be spontaneous, G_{final} must be less than G_{initial} and ΔG must be negative. With this in mind, we can now examine how ΔH, ΔS, and *T* are related in determining spontaneity.

> Reactions that occur with a free energy decrease are sometimes said to be **exergonic.** Those that occur with a free energy increase are sometimes said to be **endergonic.**

When ΔH Is Negative and ΔS Is Positive

When a change is exothermic and is also accompanied by an increase in entropy, both factors favor spontaneity.

$$\Delta H \text{ is negative } (-)$$

$$\Delta S \text{ is positive } (+)$$

$$\Delta G = \Delta H - T\Delta S$$
$$= (-) - [T(+)]$$

Notice that regardless of the absolute temperature, which must be a positive number, ΔG is negative. This means that regardless of the temperature, such a change must be spontaneous. An example of such an event is the collapse of a building (Figure 13.11 on the next page).

> The confetti shown falling in the chapter opening photo also experiences both a negative ΔH and a positive ΔS.

When ΔH Is Positive and ΔS Is Negative

When a change is endothermic and is accompanied by a lowering of the entropy, both factors work against spontaneity.

$$\Delta H \text{ is positive } (+)$$

$$\Delta S \text{ is negative } (-)$$

$$\Delta G = \Delta H - T\Delta S$$
$$= (+) - [T(-)]$$

Now, no matter what the temperature is, ΔG will be positive and the change must be nonspontaneous. An example is the formation of a building from a pile of bricks. From common experience you know it won't happen by itself, and when a cartoonist uses it in a story it is unexpected and amusing.

FIGURE 13.11

The collapse of a building such as this, once begun, will always be spontaneous because it is accompanied by a lowering of the potential energy and an increase in entropy.

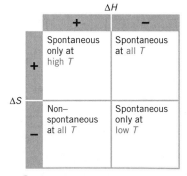

FIGURE 13.12

Summary of the effects of the signs of ΔH and ΔS on spontaneity.

When ΔH and ΔS Have the Same Sign

When ΔH and ΔS have the same algebraic sign, the temperature becomes the determining factor in controlling spontaneity. If ΔH and ΔS are both positive, then

$$\Delta G = (+) - [T(+)]$$

Thus, ΔG is the difference between two positive quantities, ΔH and $T\Delta S$. This difference will be negative only if the term $T\Delta S$ is larger in magnitude than ΔH, and this will be true only when the temperature is high. In other words, *when ΔH and ΔS are both positive, the change will be spontaneous at high temperature but not at low temperature.* A familiar example is the melting of ice.

$$H_2O(s) \longrightarrow H_2O(l)$$

This is endothermic and is accompanied by an increase in entropy. We know that at high temperature (above 0 °C) melting is spontaneous, but at low temperature (below 0 °C) it is not.

For similar reasons, when ΔH and ΔS are both negative, ΔG will be negative only when the temperature is low.

$$\Delta G = (-) - [T(-)]$$

Only when the negative value of ΔH is larger in magnitude than the negative value of $T\Delta S$ will be ΔG be negative. Such a change is only spontaneous at low temperature. An example is the freezing of water.

$$H_2O(l) \longrightarrow H_2O(s)$$

This is an exothermic change that is accompanied by a decrease in entropy; it is only spontaneous at low temperatures (i.e., below 0 °C).

Figure 13.12 summarizes the effects of the signs of ΔH and ΔS on ΔG, and hence on the spontaneity of physical and chemical events.

 Spontaneity determined by sign of ΔG

When ΔG is determined at 25 °C (298 K) and 1 atm, we call it the **standard free energy change, $\Delta G°$.** There are several ways of obtaining $\Delta G°$ for a reaction. One of them is to compute $\Delta G°$ from $\Delta H°$ and $\Delta S°$.

$$\Delta G° = \Delta H° - (298 \text{ K})\Delta S°$$

Experimental measurement of $\Delta G°$ is also possible, but we will discuss how this is done at a later time.

13.7
STANDARD FREE ENERGIES

Calculate $\Delta G°$ for the reaction of urea with water from values of $\Delta H°$ and $\Delta S°$.

$$CO(NH_2)_2(aq) + H_2O(l) \longrightarrow CO_2(g) + 2NH_3(g)$$

ANALYSIS We can calculate $\Delta G°$ with the equation

$$\Delta G° = \Delta H° - T\Delta S°$$

The data needed to calculate $\Delta H°$ come from Table 5.2 and require a Hess's law calculation. To obtain $\Delta S°$, we normally would need to do a similar calculation with data from Table 13.1. However, we already performed this calculation in Example 13.3 on page 559.

SOLUTION First we calculate $\Delta H°$ from data in Table 5.2.

$$\Delta H° = [\Delta H°_{fCO_2(g)} + 2\Delta H°_{fNH_3(g)}] - [\Delta H°_{fCO(NH_2)_2(aq)} + \Delta H°_{fH_2O(l)}]$$

$$= \left[1 \text{ mol} \times \left(\frac{-393.5 \text{ kJ}}{\text{mol}} \right) + 2 \text{ mol} \times \left(\frac{-46.19 \text{ kJ}}{\text{mol}} \right) \right]$$
$$- \left[1 \text{ mol} \times \left(\frac{-319.2 \text{ kJ}}{\text{mol}} \right) + 1 \text{ mol} \times \left(\frac{-285.9 \text{ kJ}}{\text{mol}} \right) \right]$$

$$= (-485.9 \text{ kJ}) - (-605.1 \text{ kJ})$$

$$= +119.2 \text{ kJ}$$

In Example 13.3 we found $\Delta S°$ to be $+354.8$ J/K. To calculate $\Delta G°$ we also need the Kelvin temperature, which we will express to four significant figures to match the number of significant figures in $\Delta S°$. Since standard temperature is *exactly* 25 °C, $T = (25.0 + 273.2)$ K $= 298.2$ K. Also, we must be careful to express $\Delta H°$ and $T\Delta S°$ in the same energy units. Substituting into the equation for $\Delta G°$,

$$\Delta G° = +119.2 \text{ kJ} - (298.2 \text{ K})(0.3548 \text{ kJ/K})$$
$$= +119.2 \text{ kJ} - 105.8 \text{ kJ}$$
$$= +13.4 \text{ kJ}$$

Therefore, for this reaction, $\Delta G° = +13.4$ kJ.

■ **Practice Exercise 6** Use the data in Table 5.2 and Table 13.1 to calculate $\Delta G°$ for the formation of iron oxide (hematite, an iron ore). The equation for the reaction is

$$4Fe(s) + 3O_2(g) \longrightarrow 2Fe_2O_3(s)$$

EXAMPLE 13.4
Calculating $\Delta G°$ from $\Delta H°$ and $\Delta S°$

To be exact: $T_K = t_C + 273.15$

354.8 J/K = 0.3548 kJ/K

TABLE 13.2   Standard Free Energies of Formation of Typical Substances (25 °C, 1 atm)

Substance	ΔG_f° (kJ mol^{-1})	Substance	ΔG_f° (kJ mol^{-1})
$Ag(s)$	0.00	$H_2O(g)$	−228.6
$AgCl(s)$	−109.7	$H_2O(l)$	−237.2
$Al(s)$	0.00	$HCl(g)$	−95.27
$Al_2O_3(s)$	−1576.4	$HNO_3(l)$	−79.91
$C(s, \text{graphite})$	0.00	$H_2SO_4(l)$	−689.9
$CO(g)$	−137.3	$HC_2H_3O_2(l)$	−392.5
$CO_2(g)$	−394.4	$Hg(l)$	0.00
$CH_4(g)$	−50.79	$Hg(g)$	+31.8
$CH_3Cl(g)$	−58.6	$K(s)$	0.00
$CH_3OH(l)$	−166.2	$KCl(s)$	−408.3
$CO(NH_2)_2(s)$	−197.2	$K_2SO_4(s)$	−1316.4
$CO(NH_2)_2(aq)$	−203.8	$N_2(g)$	0.00
$C_2H_2(g)$	+209	$NH_3(g)$	−16.7
$C_2H_4(g)$	+68.12	$NH_4Cl(s)$	−203.9
$C_2H_6(g)$	−32.9	$NO(g)$	+86.69
$C_2H_5OH(l)$	−174.8	$NO_2(g)$	+51.84
$C_8H_{18}(l)$	+17.3	$N_2O(g)$	+103.6
$Ca(s)$	0.00	$N_2O_4(g)$	+98.28
$CaCO_3(s)$	−1128.8	$Na(s)$	0.00
$CaCl_2(s)$	−750.2	$Na_2CO_3(s)$	−1048
$CaO(s)$	−604.2	$NaHCO_3(s)$	−851.9
$Ca(OH)_2(s)$	−896.76	$NaCl(s)$	−384.0
$CaSO_4(s)$	−1320.3	$NaOH(s)$	−382
$CaSO_4 \cdot \frac{1}{2}H_2O(s)$	−1435.2	$Na_2SO_4(s)$	−1266.8
$CaSO_4 \cdot 2H_2O(s)$	−1795.7	$O_2(g)$	0.00
$Cl_2(g)$	0.00	$PbO(s)$	−189.3
$Fe(s)$	0.00	$S(s)$	0.00
$Fe_2O_3(s)$	−741.0	$SO_2(g)$	−300.4
$H_2(g)$	0.00	$SO_3(g)$	−370.4

In Section 5.6 you learned that it is useful to tabulate standard heats of formation, ΔH_f°, because they can be used with Hess's law to calculate ΔH° for many different reactions. Standard free energies of formation, ΔG_f°, can be used in similar calculations to obtain ΔG°.

Standard free energies of formation

$$\Delta G^\circ = (\text{sum of } \Delta G_f^\circ \text{ of products}) - (\text{sum of } \Delta G_f^\circ \text{ of reactants}) \qquad (13.9)$$

The ΔG_f° values for some typical substances are found in Table 13.2. Example 13.5 shows how we can use them to calculate ΔG° for a reaction.

EXAMPLE 13.5
Calculating ΔG° from ΔG_f°

What is ΔG° for the combustion of the liquid ethanol, C_2H_5OH, to give $CO_2(g)$ and $H_2O(g)$?

SOLUTION First, we need the balanced equation for the reaction.

$$C_2H_5OH(l) + 3O_2(g) \longrightarrow 2CO_2(g) + 3H_2O(g)$$

Now we use Equation 13.9.

$$\Delta G^\circ = [2\Delta G^\circ_{fCO_2(g)} + 3\Delta G^\circ_{fH_2O(g)}] - [\Delta G^\circ_{fC_2H_5OH(l)} + 3\Delta G^\circ_{fO_2(g)}]$$

As with ΔH°_f, the ΔG°_f for any element in its standard state is zero. Therefore, using the data from Table 13.2,

$$\Delta G^\circ = \left[2\text{ mol} \times \left(\frac{-394.4\text{ kJ}}{\text{mol}}\right) + 3\text{ mol} \times \left(\frac{-228.6\text{ kJ}}{\text{mol}}\right) \right]$$
$$- \left[1\text{ mol} \times \left(\frac{-174.8\text{ kJ}}{\text{mol}}\right) + 1\text{ mol} \times \left(\frac{0\text{ kJ}}{\text{mol}}\right) \right]$$
$$= (-1474.6\text{ kJ}) - (-174.8\text{ kJ})$$
$$= -1299.8\text{ kJ}$$

The standard free energy change for the reaction equals -1299.8 kJ.

■ **Practice Exercise 7** Calculate $\Delta G^\circ_{reaction}$ in kilojoules for the following reactions using the data in Table 13.2.

(a) $2NO(g) + O_2(g) \rightarrow 2NO_2(g)$

(b) $Ca(OH)_2(s) + 2HCl(g) \rightarrow CaCl_2(s) + 2H_2O(g)$

One of the chief uses of spontaneous chemical reactions is the production of useful work. For example, fuels are burned in gasoline or diesel engines to power automobiles and heavy machinery, and chemical reactions in batteries start our autos and run all sorts of modern electronic gadgets, including cellular phones and laptop computers.

When chemical reactions occur, however, their energy is not always harnessed to do work. For instance, if gasoline is burned in an open dish, the energy evolved is lost entirely as heat and no useful work is accomplished. Engineers, therefore, seek ways to capture as much energy as possible in the form of work. One of their primary goals is to maximize the efficiency with which chemical energy is converted to work and to minimize the amount of energy transferred unproductively to the environment as heat.

Scientists have discovered that the maximum conversion of chemical energy to work occurs if a reaction is carried out under conditions that are said to be reversible. A process is defined as **reversible** if its driving force is opposed by another force that is just the slightest bit weaker, so that any increase in the opposing force will cause the direction of the change to be reversed. An example of an almost reversible process would be the use of a 12.00 V battery to charge an 11.99 V battery. The electricity from the 12.00 V battery is opposed by a "force" of 11.99 V.

Although we can obtain the maximum work by carrying out a reaction reversibly, a reversible process proceeds at such an extremely slow speed that the work would be of no use to us. Our goal, then, is to approach reversibility for maximum efficiency, but to carry out the reaction at a pace that will deliver work at acceptable rates.

The relationship of useful work to reversibility was illustrated earlier (Section 13.2) in our discussion of the discharge of an automobile battery. Recall that when the battery is shorted with the heavy wrench, no work is done and all the energy appears as heat. In this case there is nothing opposing the discharge, and it occurs in the most thermodynamically irreversible manner pos-

13.8
FREE ENERGY AND MAXIMUM WORK

A reversible process requires an infinite number of tiny steps. This takes forever to accomplish.

sible. However, when the current is passed through a small electric motor, the motor itself offers resistance to the passage of the electricity and the discharge takes place slowly. In this instance, the discharge occurs in a more nearly reversible manner because of the opposition provided by the motor, and a relatively large amount of the available energy appears in the form of the work accomplished by the motor.

The preceding discussion leads naturally to the question: Is there a limit to the fraction of the available energy in a reaction that can be harnessed as useful work? The answer to this question is to be found in the Gibbs free energy.

The maximum amount of energy produced by a reaction that can be theoretically harnessed as work is equal to ΔG.

This is the energy that need not be lost as heat and is therefore *free* to be used for work. Thus, by determining the value of ΔG, we can find out whether or not a given reaction will be an effective source of useful energy. Also, by comparing the actual amount of work derived from a given system with the ΔG values for the reactions involved, we can measure the efficiency of the system.

EXAMPLE 13.6
Calculating Maximum Work

Calculate the maximum work available, expressed in kilojoules, from the oxidation of 1 mol of octane, $C_8H_{18}(l)$, by oxygen to give $CO_2(g)$ and $H_2O(l)$ at 25 °C and 1 atm.

ANALYSIS The maximum work is equal to ΔG for the reaction. Standard thermodynamic conditions are specified, so we need to calculate $\Delta G°$.

SOLUTION First we need a balanced equation for the reaction. For the combustion of one mole of C_8H_{18} we have

$$C_8H_{18}(l) + 12\tfrac{1}{2} O_2(g) \longrightarrow 8CO_2(g) + 9H_2O(l)$$

Then we apply Equation 13.9

$$\Delta G = [8\Delta G°_{fCO_2(g)} + 9\Delta G°_{fH_2O(l)}] - [\Delta G°_{fC_8H_{18}(l)} + 12.5\Delta G°_{fO_2(g)}]$$

Referring to Table 13.2 and dropping the canceled mol units,

$$\Delta G° = [8 \times (-394.4) \text{ kJ} + 9 \times (-237.2) \text{ kJ}]$$
$$- [1 \times (+17.3) \text{ kJ} + 12.5 \times (0) \text{ kJ}]$$
$$= (-5290 \text{ kJ}) - (+17.3 \text{ kJ})$$
$$= -5307 \text{ kJ}$$

Thus, at 25 °C and 1 atm, we can expect no more than 5307 kJ of work from the oxidation of 1 mol of C_8H_{18}.

■ **Practice Exercise 8** Calculate the maximum work that could be obtained at 25 °C and 1 atm from the oxidation of 1.00 mol of aluminum by $O_2(g)$ to give $Al_2O_3(s)$. (The combustion of aluminum in booster rockets provides part of the energy that lifts the space shuttle off its launching pad.)

We have seen that when the value of ΔG for a given change is negative, the change occurs spontaneously. We have also seen that a change is nonspontaneous when ΔG is positive. However, when ΔG is neither positive nor negative, the change is neither spontaneous nor nonspontaneous—the system is in a state of equilibrium. This occurs when ΔG is equal to zero.

13.9

FREE ENERGY AND EQUILIBRIUM

 Equilibrium criterion, $\Delta G = 0$

When a system is in a state of dynamic equilibrium,

$$G_{products} = G_{reactants} \qquad \text{and} \qquad \Delta G = 0$$

Let's again consider the freezing of water.

$$H_2O(l) \rightleftharpoons H_2O(s)$$

Below 0 °C, ΔG for this change is negative and the freezing is spontaneous. On the other hand, above 0 °C we find that ΔG is positive and freezing is nonspontaneous. When the temperature is exactly 0 °C, $\Delta G = 0$ and an ice–water mixture exists in a condition of equilibrium. As long as heat isn't added or removed from the system, neither freezing nor melting is spontaneous and the ice and liquid water can exist together indefinitely.

Work and Dynamic Equilibrium

We have identified ΔG as a quantity that specifies the amount of work that is available from a system. Since ΔG is zero at equilibrium, the amount of work available is zero also. Therefore, when a system is at equilibrium, no work can be extracted from it. As an example, let's consider again the common lead storage battery that we use to start our car.

When the battery is fully charged, there are virtually no products of the discharge reaction present. The chemical reactants, however, are present in large amounts. Therefore, the total free energy of the reactants far exceeds the total free energy of products and, since $\Delta G = G_{products} - G_{reactants}$, the ΔG of the system has a large negative value. This means that a lot of work is available. As the battery discharges, the reactants are converted to products and $G_{products}$ gets larger while $G_{reactants}$ gets smaller; thus ΔG becomes less negative, and less work is available. Finally, the battery reaches equilibrium. The total free energies of the reactants and the products have become equal, so $G_{products} - G_{reactants} = 0$ and $\Delta G = 0$. No further work can be extracted and we say the battery is dead.

Equilibrium in Phase Changes

When we have equilibrium in any system, we know that $\Delta G = 0$. For a phase change such as $H_2O(l) \longrightarrow H_2O(s)$, equilibrium can only exist at one particular temperature at atmospheric pressure. In this instance, that temperature is 0 °C. Above 0 °C, only liquid water can exist, and below 0 °C all the liquid will freeze to give ice. This yields an interesting relationship between ΔH and ΔS for a phase change. Since $\Delta G = 0$,

$$\Delta G = 0 = \Delta H - T\Delta S$$

Therefore,

$$\Delta H = T\Delta S$$

and

$$\Delta S = \frac{\Delta H}{T} \tag{13.10}$$

Thus, if we know ΔH for the phase change and the temperature at which the two phases coexist, we can calculate ΔS for the phase change. Another interesting relationship that we can obtain is

$$T = \frac{\Delta H}{\Delta S} \tag{13.11}$$

Thus, if we know ΔH and ΔS, we can calculate the temperature at which equilibrium will occur.

EXAMPLE 13.7
Calculating the Equilibrium Temperature for a Phase Change

ΔH and ΔS do not change much with changes in temperature. This is because temperature changes affect the enthalpies and entropies of both the reactants and products by about the same amount, so the differences between reactants and products stay fairly constant.

For the phase change $Br_2(l) \rightarrow Br_2(g)$, $\Delta H° = +31.0$ kJ/mol and $\Delta S° = 92.9$ J/mol K. Assuming that ΔH and ΔS are nearly temperature independent, calculate the approximate temperature at which $Br_2(l)$ will be in equilibrium with $Br_2(g)$ at 1 atm.

SOLUTION The temperature at which equilibrium exists is given by Equation 13.11,

$$T = \frac{\Delta H}{\Delta S}$$

If ΔH and ΔS do not depend on temperature, then we can use $\Delta H°$ and $\Delta S°$ in this equation. That is,

$$T = \frac{\Delta H°}{\Delta S°}$$

Substituting the data given in the problem,

$$T = \frac{3.10 \times 10^4 \text{ J mol}^{-1}}{92.9 \text{ J mol}^{-1} \text{ K}^{-1}}$$

$$= 334 \text{ K}$$

The Celsius temperature is $334 - 273 = 61$ °C. Notice that we were careful to express $\Delta H°$ in joules, not kilojoules, so the units would cancel correctly.

■ **Practice Exercise 9** The heat of vaporization of mercury is 60.7 kJ/mol. For $Hg(l)$, $S° = 76.1$ J/mol K and for $Hg(g)$, $S° = 175$ J/mol K. Calculate the temperature at which there is equilibrium between mercury vapor and liquid at a pressure of 1 atm (the standard state pressure for gaseous mercury).

Free Energy Diagram for a Phase Change

We have said that in a phase change such as $H_2O(l) \rightarrow H_2O(s)$, equilibrium can exist for a given pressure only at one particular temperature; for water at a pressure of 1 atm, this temperature is 0 °C, the freezing point. At other temperatures, the phase change proceeds entirely to completion in one direction or another. One way to gain a better understanding of this is by studying **free energy diagrams,** which depict how the free energy changes as we proceed from the ''reactants'' to the ''products.''

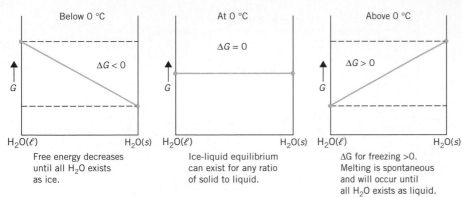

FIGURE 13.13

Free energy diagram for conversion of $H_2O(l)$ to $H_2O(s)$. At the left of each diagram, the system consists entirely of $H_2O(l)$. At the right is $H_2O(s)$. The horizontal axis represents the extent of conversion from $H_2O(l)$ to $H_2O(s)$.

Figure 13.13 illustrates three different free energy diagrams for water–ice mixtures. On the left of each diagram is indicated the free energy of pure liquid water, and on the right the free energy of the pure solid. Points along the horizontal axis represent mixtures of both phases. Thus, going from left to right across the graph we are able to see how the free energy varies as a system that consists entirely of $H_2O(l)$ changes ultimately to one that consists entirely of $H_2O(s)$.

Below 0 °C, we see that the free energy of the liquid is higher than that of the solid. You've learned that a spontaneous change will occur if the free energy can decrease. Therefore, the first diagram tells us that if we start with the liquid phase, or any mixture of liquid or solid, freezing will occur until only the solid is present. This is because the free energy decreases continually until all the liquid has frozen.

Above 0 °C, we have the opposite situation. The free energy decreases in the direction of $H_2O(s) \rightarrow H_2O(l)$, and it continues to drop until all the solid has melted. This means that if we have ice, or any mixture of ice and liquid water, it will continue to melt until only the liquid is present.

Above or below 0 °C the system is unable to establish an equilibrium mixture of liquid and solid. Melting or freezing occurs until only one phase is present. However, at 0 °C there is no change in free energy if either melting or freezing occurs, so there is no driving force for either change. Therefore, as long as a system of ice and liquid water is insulated from warmer or colder surroundings, any particular mixture of the two phases is stable and a state of equilibrium exists.

Free Energy Diagrams for Chemical Reactions

The free energy changes that occur in most chemical reactions are more complex than those in phase changes. For example, Figure 13.14 shows the free energy diagram for the decomposition of N_2O_4 (a substance used to power rockets) into NO_2 (a compound often found in polluted air).

$$N_2O_4(g) \longrightarrow 2NO_2(g)$$

Notice that in going from reactant to product, the free energy has a minimum. It drops below that of either pure N_2O_4 or pure NO_2.

Any system will spontaneously seek the lowest point on its free energy curve. If we begin with pure $N_2O_4(g)$, some $NO_2(g)$ will be formed, because proceeding in the direction of NO_2 leads to a lowering of the free energy. If we be-

FIGURE 13.14

Free energy diagram for the decomposition of $N_2O_4(g)$. The minimum on the curve indicates the composition of the reaction mixture at equilibrium.

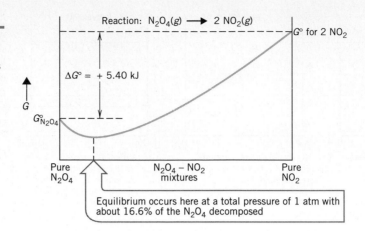

Reaction: $N_2O_4(g) \longrightarrow 2 NO_2(g)$

$G°$ for $2 NO_2$

$\Delta G° = +5.40 \text{ kJ}$

$G°_{N_2O_4}$

G

Pure N_2O_4 · $N_2O_4 - NO_2$ mixtures · Pure NO_2

Equilibrium occurs here at a total pressure of 1 atm with about 16.6% of the N_2O_4 decomposed

gin with pure $NO_2(g)$, a change also will occur. Going downhill on the free energy curve now takes place as the reverse reaction occurs [i.e., $2NO_2(g) \rightarrow N_2O_4(g)$]. Once the bottom of the "valley" is reached, the system has come to equilibrium. The composition of the mixture of N_2O_4 and NO_2 will remain constant because any change would require an uphill climb. Free energy increases are not spontaneous, so this doesn't happen.

An important thing to notice in Figure 13.14 is that some reaction takes place spontaneously in the forward direction even though $\Delta G°$ is positive. However, the reaction doesn't proceed far before equilibrium is reached. For comparison, Figure 13.15 shows the shape of the free energy curve for a reaction with a negative $\Delta G°$. We see here that at equilibrium there has been a much greater conversion of reactants to products. Thus, $\Delta G°$ *tells us where the position of equilibrium lies between pure reactants and pure products.* When $\Delta G°$ is positive, the position of equilibrium lies close to the reactants and little reaction occurs by the time equilibrium is reached. When $\Delta G°$ is negative, the position of equilibrium lies close to the products and a large amount of products will have formed by the time equilibrium is reached.

Using $\Delta G°$ to predict outcome of reaction

Predicting the Outcome of a Chemical Reaction

In general, the value of $\Delta G°$ for most reactions is much larger numerically than the $\Delta G°$ for the $N_2O_4 \rightarrow 2NO_2$ reaction. In addition, the extent to which

FIGURE 13.15

Free energy curve for a reaction having a negative $\Delta G°$. Since $G°_B < G°_A$, $\Delta G° < 0$.

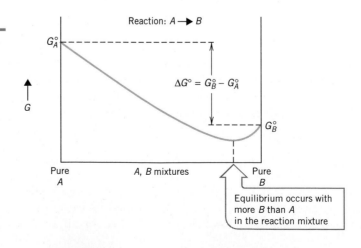

Reaction: $A \longrightarrow B$

$G°_A$

$\Delta G° = G°_B - G°_A$

$G°_B$

G

Pure A · A, B mixtures · Pure B

Equilibrium occurs with more B than A in the reaction mixture

a reaction proceeds is very sensitive to the size of $\Delta G°$. If the $\Delta G°$ value for a reaction is reasonably large—about 20 kJ or more—almost no observable reaction will occur when $\Delta G°$ is positive. On the other hand, the reaction will go almost to completion if $\Delta G°$ is both large and negative.[2] From a practical standpoint, then, *the size and sign of $\Delta G°$ serve as indicators of whether an observable spontaneous reaction will occur.*

Would we expect to observe the following reaction at 25 °C and 1 atm?

$$NH_4Cl(s) \longrightarrow NH_3(g) + HCl(g)$$

EXAMPLE 13.8
Using $\Delta G°$ as a Predictor of the Outcome of a Reaction

ANALYSIS We will need to determine the magnitude and sign of $\Delta G°$ for the reaction. If $\Delta G°$ is reasonably large and positive, the reaction won't be observed. If it is reasonably large and negative, we can expect to see the reaction go nearly to completion.

SOLUTION First let's calculate $\Delta G°$ for the reaction using the data in Table 13.2. The procedure is the same as that discussed earlier.

$$\begin{aligned}
\Delta G° &= [\Delta G°_{fNH_3(g)} + \Delta G°_{fHCl(g)}] - [\Delta G°_{fNH_4Cl(s)}] \\
&= [(-16.7 \text{ kJ}) + (-95.27 \text{ kJ})] - [-203.9 \text{ kJ}] \\
&= +91.9 \text{ kJ}
\end{aligned}$$

Because $\Delta G°$ is large and positive, we shouldn't expect to observe the spontaneous formation of any products at this temperature.

■ **Practice Exercise 10** Use the data in Table 13.2 to determine whether the reaction

$$SO_2(g) + O_2(g) \longrightarrow SO_3(g)$$

should "occur spontaneously" at 25 °C.

■ **Practice Exercise 11** Use the data in Table 13.2 to determine whether we should expect to see the formation of $CaCO_3(s)$ in the following reaction at 25 °C.

$$CaCl_2(s) + H_2O(g) + CO_2(g) \longrightarrow CaCO_3(s) + 2HCl(g)$$

Equilibrium and Temperature

So far, we have confined our discussion of the relationship of free energy and equilibrium to a special case, 25 °C. But what about other temperatures? Equilibria certainly can exist under conditions other than 25 °C, and it would be valuable to know how temperature affects the position of equilibrium.

At 25 °C, we've seen that the position of equilibrium is determined by the differences between the free energy of pure products and the free energy of pure reactants. This difference is given by $\Delta G°_{298 \text{ K}}$, where we have now used the subscript "298 K" to indicate the temperature. $\Delta G°_{298 \text{ K}}$ is defined as

$$\Delta G°_{298 \text{ K}} = (G°_{products})_{298 \text{ K}} - (G°_{reactants})_{298 \text{ K}}$$

298 K = 25 °C

---

[2] To actually see a change take place, the speed of a spontaneous reaction must be reasonably fast. For example, the decomposition of the nitrogen oxides into N_2 and O_2 is thermodynamically spontaneous ($\Delta G°$ is negative) but their rates of decomposition are so slow that these substances appear to be stable and some are obnoxious air pollutants.

At temperatures other than 25 °C, it is still the difference between the free energy of the products and reactants that determines the position of equilibrium. We might write this as ΔG_T°. Thus, at a temperature other than 25 °C (298 K), we have

$$\Delta G_T^\circ = (G_{products}^\circ)_T - (G_{reactants}^\circ)_T$$

where $(G_{products}^\circ)_T$ and $(G_{reactants}^\circ)_T$ are the total free energies of the pure products and reactants, respectively, at this other temperature.

Next, we must find a way to compute ΔG_T°. Earlier we saw that ΔG° can be obtained from the equation

$$\Delta G^\circ = \Delta H^\circ - (298\ \text{K})\Delta S^\circ$$

At a different temperature, T, the equation becomes

$$\Delta G_T^\circ = \Delta H_T^\circ - T\Delta S_T^\circ$$

Now we seem to be getting closer to our goal. If we can compute or estimate the values of ΔH_T° and ΔS_T° for a reaction, we have solved our problem.

The size of ΔG_T° obviously depends very strongly on the temperature—the equation above has temperature as one of its variables. However, the magnitudes of the ΔH and ΔS for a reaction are relatively insensitive to the temperature. This is because the enthalpies and entropies of *both* the reactants and products increase about equally with increasing temperature, so their differences, ΔH and ΔS, remain nearly the same. As a result, we can use $\Delta H_{298\ \text{K}}^\circ$ and $\Delta S_{298\ \text{K}}^\circ$ as reasonable approximations of ΔH_T° and ΔS_T°. This allows us to rewrite the equation for ΔG_T° as

$$\Delta G_T^\circ \cong \Delta H_{298\ \text{K}}^\circ - T\Delta S_{298\ \text{K}}^\circ \qquad (13.12)$$

The following examples illustrate how this equation is useful.

The magnitudes of ΔH and ΔS are relatively insensitive to temperature changes. However, ΔG is very temperature sensitive because $\Delta G = \Delta H - T\Delta S$.

Calculating ΔG_T°

EXAMPLE 13.9
ΔG° at Temperatures Other than 25 °C

Earlier we saw that at 25 °C the value of ΔG° for the reaction

$$N_2O_4(g) \longrightarrow 2NO_2(g)$$

has a value of $+5.40$ kJ. What is the value of $\Delta G_{373\ \text{K}}^\circ$ for this reaction?

SOLUTION To make use of Equation 13.12, we need values of ΔH° and ΔS°. The ΔH° for the reaction can be calculated from the data in Table 5.2 on p. 209. Here we find the following standard heats of formation.

$$N_2O_4(g) \qquad \Delta H_f^\circ = +9.67\ \text{kJ/mol}$$
$$NO_2(g) \qquad \Delta H_f^\circ = +33.8\ \text{kJ/mol}$$

We combine these by a Hess's law calculation to compute ΔH° for the reaction.

$$\Delta H^\circ = [2\Delta H_{f\ NO_2(g)}^\circ] - [\Delta H_{f\ N_2O_4(g)}^\circ]$$
$$= \left[2\ \text{mol} \times \left(\frac{33.8\ \text{kJ}}{\text{mol}} \right) \right] - \left[1\ \text{mol} \times \left(\frac{9.67\ \text{kJ}}{\text{mol}} \right) \right]$$
$$= +57.9\ \text{kJ}$$

Next, we compute ΔS° for the reaction using data from Table 13.1.

$$N_2O_4(g) \qquad S^\circ = 304\ \text{J/mol K}$$
$$NO_2(g) \qquad S^\circ = 240.5\ \text{J/mol K}$$

This is also a Hess's law kind of calculation.

$$\Delta S^\circ = \left[2 \text{ mol} \times \left(\frac{240.5 \text{ J}}{\text{mol K}} \right) \right] - \left[1 \text{ mol} \times \left(\frac{304 \text{ J}}{\text{mol K}} \right) \right]$$

$$= +177 \text{ J/K, or } 0.177 \text{ kJ/K}$$

Now we can substitute into the equation for $\Delta G^\circ_{373 \text{ K}}$, using $T = 373$ K.

$$\Delta G^\circ_{373 \text{ K}} = (+57.9 \text{ kJ}) - (373 \text{ K})(0.177 \text{ kJ/K})$$
$$= -8.1 \text{ kJ}$$

Notice that at this higher temperature, the sign of ΔG°_T has become negative.

Using the results of Example 13.9, sketch a free energy diagram for the decomposition of N_2O_4 into NO_2 at 100 °C. For this reaction, how does the position of equilibrium change as the temperature is increased?

SOLUTION Since ΔG°_T is negative for the reaction at 100 °C, the G° of the reactants must be higher than the G° of the products. Therefore, when we draw the free energy curve, the minimum must lie closer to the products than to the reactants, and the diagram has the appearance shown in the margin.

Comparing this diagram to the one in Figure 13.14, we see that at the higher temperature the minimum lies closer to the products. This means that at the higher temperature the decomposition will proceed farther toward completion when equilibrium has been reached.

■ **Practice Exercise 12** Use the data in Table 13.2 to determine $\Delta G^\circ_{298 \text{ K}}$ for the reaction

$$2NaHCO_3(s) \longrightarrow Na_2CO_3(s) + CO_2(g) + H_2O(g)$$

Then calculate ΔG° for the reaction at 200 °C using data in Tables 5.2 and 13.1. How does the position of equilibrium for this reaction change as the temperature is increased?

EXAMPLE 13.10
Position of Equilibrium as a Function of Temperature

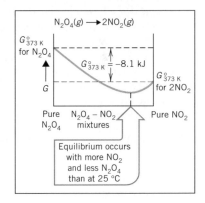

SUMMARY

First Law of Thermodynamics The change in the **internal energy** of a system, ΔE, equals the sum of the heat absorbed by the system, q, and the work done on the system, w. ΔE is a state function, but q and w are not. The values of q and w depend on how the change takes place. For pressure–volume work, $w = -P\Delta V$, where $\Delta V = V_{\text{final}} - V_{\text{initial}}$. The heat of reaction at constant volume, q_V, is equal to ΔE, whereas the heat of reaction at constant pressure is equal to ΔH. The value of ΔH differs from ΔE by the work expended in pushing back the atmosphere when the change occurs at constant atmospheric pressure. In general, the difference between ΔE and ΔH is quite small. For a chemical reaction, $\Delta H = \Delta E + \Delta n_{\text{gas}}RT$, where Δn_{gas} is the change in the number of moles of gas on going from reactants to products.

Spontaneity A spontaneous change occurs without outside assistance. A nonspontaneous change requires contin-

uous help and can occur only if it is accompanied by and linked to some spontaneous event.

The potential energy change is one factor that influences spontaneity. Exothermic changes, with their negative values of ΔH, tend to proceed spontaneously. Events that occur with an increase in randomness also tend to occur by themselves. Randomness is associated with statistical probability—the more random the distribution of a collection of particles, the greater the probability of its occurrence. The thermodynamic quantity associated with randomness is **entropy**. An increase in entropy favors a spontaneous change. In general, gases have much higher entropies than liquids, which have somewhat higher entropies than solids. Entropy increases with volume for a gas and with the temperature. During a chemical reaction, the entropy tends to increase if the complexity of the molecules decreases (e.g., forming diatomic molecules from triatomic molecules).

Second Law of Thermodynamics   This law states that the entropy of the universe increases whenever a spontaneous change occurs. All of our activities, being traced ultimately to spontaneous events, increase the total entropy or disorder in the world.

Third Law of Thermodynamics   The entropy of a pure crystalline substance is equal to zero at absolute zero (0 K). Because we know where the zero point is on the entropy scale, it is possible to measure absolute amounts of entropy possessed by substances. Standard entropies, $S°$, are calculated for 25 °C and 1 atm (Table 13.1) and can be used to calculate $\Delta S°$ for chemical reactions.

$\Delta S°$ = (sum of $S°$ of products) − (sum of $S°$ of reactants)

Gibbs Free Energy   The Gibbs free energy change, ΔG, allows us to determine the combined effects of enthalpy and entropy changes on the spontaneity of a chemical or physical change. A change is spontaneous only if the free energy of the system decreases (ΔG is negative). When ΔH and ΔS have the same algebraic sign, the temperature becomes the critical factor in determining spontaneity.

When ΔG is measured at 25 °C and 1 atm, it is the standard free energy change, $\Delta G°$. As with enthalpy changes, the standard free energies of formation, $\Delta G_f°$ (Table 13.2), can be used to obtain $\Delta G°$ for chemical reactions by a Hess's law type of calculation.

$\Delta G°$ = (sum of $\Delta G_f°$ products) − (sum of $\Delta G_f°$ reactants)

For any system, the value of ΔG is equal to the maximum amount of energy that can be obtained in the form of useful work. This maximum work can be obtained only if the change takes place reversibly. All real changes are irreversible and we always obtain less work than is theoretically available; the rest is lost as heat.

Free Energy and Equilibrium   When a system reaches equilibrium, $\Delta G = 0$ and no useful work can be obtained from it. At any particular pressure, an equilibrium between two phases of a substance (e.g., liquid ⇌ solid, or solid ⇌ vapor) can only occur at one temperature. The entropy change can be computed as $\Delta S = \Delta H/T$. The temperature at which the equilibrium occurs can be calculated from $T = \Delta H/\Delta S$.

In chemical reactions, a minimum on the free energy curve occurs partway between pure reactants and pure products. This minimum can be approached from either the reactants or products, and the composition of the equilibrium mixture is determined by where the minimum lies along the reactant → product axis; when it lies close to the products, the proportion of product to reactants is large and the reaction goes far toward completion.

When a reaction has a value of $\Delta G°$ that is both large and negative, it will appear to occur spontaneously because a lot of products will be formed by the time equilibrium is reached. If $\Delta G°$ is large and positive, no observable reaction will occur because only tiny amounts of products will be formed. The sign and magnitude of $\Delta G°$ can therefore be used to predict the apparent spontaneity of a chemical reaction.

At a temperature other than 25 °C, the value of $\Delta G_T°$ can be calculated using $\Delta H°$ and $\Delta S°$ as approximations of $\Delta H_T°$ and $\Delta S_T°$. The equation is $\Delta G_T° = \Delta H_{298\,K}° - T\Delta S_{298\,K}°$. Computing $\Delta G_T°$ allows us to see how temperature affects the position of equilibrium in a chemical reaction.

TOOLS YOU HAVE LEARNED

Tool	Function
$\Delta H = \Delta E + \Delta n_{gas}RT$ **(page 547)**	To convert between ΔE and ΔH for a reaction.
Predicting the sign of ΔS (page 554)	Enables you to determine whether the entropy change favors spontaneity.
Standard entropies (page 559)	Used to calculate the value of $\Delta S°$ for a reaction.
Sign of ΔG (page 563)	Determines whether or not a change is spontaneous.
Standard free energies of formation (page 564)	Calculate $\Delta G°$ for a reaction from tabulated values of $\Delta G_f°$.

Tool	Function
Equilibrium criterion: $\Delta G = 0$ (page 567)	When this condition is fulfilled, a system is at equilibrium, so this condition is a criterion for equilibrium.
Value of $\Delta G°$ for a reaction (page 570)	The sign of $\Delta G°$ tells us whether or not a reaction can be observed.
$\Delta G_T° \cong \Delta H_{298}° - T\Delta S_{298}°$ (page 572)	To calculate the value of $\Delta G°$ at temperatures other than 25 °C.

THINKING IT THROUGH

Remember, you are not asked to obtain answers for the following problems. Instead, assemble the data necessary to solve the problems and describe how you would use the data to obtain the answers. For numerical problems, set up the calculation using appropriate conversion factors.

The problems are divided into two groups. Those in Level 1 are significantly more challenging than those in Level 1 and provide an opportunity to really hone your problem solving skills.

Level 1 Problems

1. Describe how you would calculate the work done by a gas as it expands at constant temperature from a volume of 3.00 L and a pressure of 5.00 atm to a volume of 8.00 L. The external pressure against which the gas expands is 1.00 atm. Set up the calculation so the answer is in units of joules. (1 atm = 101,325 Pa.)

2. An ideal gas in a cylinder fitted with a piston expands at constant temperature from a pressure of 5 atm and a volume of 12 L to a final volume of 6.0 L against a constant opposing pressure of 2.0 atm. How much heat does the gas absorb, expressed in units of L·atm (liter × atm)? Set up the calculation.

3. The reaction

$$2N_2O(g) \longrightarrow 2N_2(g) + O_2(g)$$

has $\Delta H° = -163.14$ kJ. What is the value of ΔE for the decomposition of 180 g of N_2O at 25 °C? If we assume that ΔH doesn't change appreciably with temperature, what is ΔE for this same reaction at 200 °C? (Set up the calculations.)

4. What factors must you consider to determine the sign of ΔS for the reaction $2N_2O(g) \rightarrow 2N_2(g) + O_2(g)$ if it occurs at constant temperature?

5. For the reaction in Question 4, what additional information do we need to determine whether this reaction should tend to proceed spontaneously in the forward direction?

6. What factors must you consider to determine the sign of ΔS for the forward reaction

$$2HI(g) \rightleftharpoons H_2(g) + I_2(s)$$

7. What is the value of $\Delta S_f°$ for HI? (Set up the calculation.)

8. The reaction

$$2C_4H_{10}(g) + 13O_2(g) \longrightarrow 8CO_2(g) + 10H_2O(g)$$

has $\Delta G° = -5407$ kJ. What is the value of $\Delta G_f°$ for $C_4H_{10}(g)$? (Set up the calculation.)

9. What is the value of $\Delta G_{500\,K}°$ for the reaction in Question 8? (Set up the calculation.)

10. A 10.0 L vessel at 20 °C contains butane, $C_4H_{10}(g)$, at a pressure of 2.00 atm. What is the maximum amount of work that can be obtained by the combustion of this butane if the gas is brought to a pressure of 1 atm and the temperature is brought to 25 °C? Assume the products are also returned to this same temperature and pressure. (Set up the calculation.)

11. Given the following reactions and their values of $\Delta G°$, explain in detail how you would calculate the value of $\Delta G_f°$ for $N_2O_5(g)$. Set up the calculation.

$$2H_2(g) + O_2(g) \longrightarrow 2H_2O(l) \qquad \Delta G° = -474.4 \text{ kJ}$$

$$N_2O_5(g) + H_2O(l) \longrightarrow 2HNO_3(l) \qquad \Delta G° = -37.6 \text{ kJ}$$

$$\tfrac{1}{2}N_2(g) + \tfrac{3}{2}O_2(g) + \tfrac{1}{2}H_2(g) \longrightarrow HNO_3(l) \qquad \Delta G° = -79.91 \text{ kJ}$$

Level 2 Problems

12. In the SI, pressure is expressed in pascals. One pascal is equal to one newton (N, the SI unit of force) per square meter: 1 Pa = 1 N/m². Work is force times distance, and the joule is defined therefore as 1 J = 1 N·m. Show that when the pressure is expressed in pascals and volume in cubic meters, the product $P\Delta V$ yields work in units of joules.

13. When an ideal gas expands or contracts at constant temperature, $\Delta E = 0$. In terms of the definition of an ideal gas and the kinetic theory interpretation of temperature, explain why this is true.

14. The reaction in Question 6 has $\Delta H° = -53.2$ kJ. How can you determine how the sign and magnitude of $\Delta G_T°$ will be affected by a change in temperature?

15. For the reaction in Question 6, how can you tell whether the position of equilibrium lies to the left or the right at 25 °C?

REVIEW EXERCISES

Answers to questions whose numbers are printed in color are given in Appendix D. More challenging questions are marked with asterisks.

First Law of Thermodynamics

13.1 What is the origin of the name *thermodynamics*?

13.2 What kinds of energy contribute to the internal energy of a system? Why can't we measure or calculate a system's internal energy?

13.3 How is a change in the internal energy defined in terms of the initial and final internal energies?

13.4 What is the algebraic sign of ΔE for an endothermic change? Why?

13.5 State the first law of thermodynamics in words. How is a change in the internal energy expressed in terms of heat and work? Define the meaning of the symbols, including the significance of their algebraic signs.

13.6 Which quantities in the statement of the first law are state functions and which are not?

13.7 Explain how the first law of thermodynamics precludes the existence of perpetual motion machines, which would create energy from nothing.

13.8 A certain system absorbs 300 J of heat and has 700 J of work performed on it. What is the value of ΔE for the change? Is the overall change exothermic or endothermic?

13.9 The value of ΔE for a certain change is −1455 J. During the change, the system absorbs 812 J of heat. Did the system do work, or was work done on the system? How much work, expressed in joules, was involved?

13.10 Which thermodynamic quantity corresponds to the heat of reaction at constant volume? Which quantity corresponds to the heat of reaction at constant pressure?

13.11 How is the rate at which energy is withdrawn from a system related to the amount of that energy which can appear as useful work?

13.12 What are the units of $P\Delta V$ if pressure is expressed in pascals and volume is expressed in cubic meters?

*13.13 If pressure is expressed in atmospheres and volume is expressed in liters, $P\Delta V$ has units of L atm (liters × atmospheres). In Chapter 10 you learned that 1 atm = 101,325 Pa, and in Chapter 1 you learned that 1 L = 1 dm³. Use this information to determine the number of joules corresponding to 1 L atm.

*13.14 Suppose that you were pumping an automobile tire with a hand pump that pushed 24.0 in.³ of air into the tire on each stroke, and that during one such stroke the opposing pressure in the tire was 30.0 lb/in.² above the normal atmospheric pressure of 14.7 lb/in.². Calculate the number of joules of work accomplished during this stroke. (1 L atm = 101.325 J)

13.15 How are ΔE and ΔH related to each other?

13.16 If there is a decrease in the number of moles of gas during an exothermic chemical reaction, which is numerically larger, ΔE or ΔH? Why?

13.17 How much work is accomplished by the following chemical reaction, which occurs inside a bomb calorimeter?

$$2C_4H_{10}(g) + 13O_2(g) \longrightarrow 8CO_2(g) + 10H_2O(g)$$

*13.18 Consider the reaction between aqueous solutions of baking soda, $NaHCO_3$, and vinegar, $HC_2H_3O_2$.

$$NaHCO_3(aq) + HC_2H_3O_2(aq) \longrightarrow$$
$$NaC_2H_3O_2(aq) + H_2O(l) + CO_2(g)$$

If this reaction occurs at atmospheric pressure ($P = 1$ atm), how much work is done by the system in pushing back the atmosphere when 1.00 mol $NaHCO_3$ reacts at a temperature of 25 °C? (Hint: Review the gas laws.)

13.19 Calculate $\Delta H°$ and ΔE for the following reactions at 25 °C. (If necessary, refer to the data in Table E.1 in Appendix E.)

(a) $3PbO(s) + 2NH_3(g) \rightarrow 3Pb(s) + N_2(g) + 3H_2O(g)$
(b) $NaOH(s) + HCl(g) \rightarrow NaCl(s) + H_2O(l)$
(c) $C_2H_4(g) + H_2O(g) \rightarrow C_2H_5OH(l)$
(d) $2CH_4(g) \rightarrow C_2H_6(g) + H_2(g)$

13.20 Calculate $\Delta H°$ and ΔE for the following reactions at 25 °C. (If necessary, refer to the data in Table E.1 in Appendix E.)

(a) $2C_2H_2(g) + 5O_2(g) \rightarrow 4CO_2(g) + 2H_2O(g)$

(b) $C_2H_2(g) + 5N_2O(g) \rightarrow 2CO_2(g) + H_2O(g) + 5N_2(g)$
(c) $NH_4Cl(s) \rightarrow NH_3(g) + HCl(g)$
(d) $(CH_3)_2CO(l) + 4O_2(g) \rightarrow 3CO_2(g) + 3H_2O(g)$

Spontaneous Change and Enthalpy

13.21 What is *spontaneous change?*

13.22 List five changes that you have encountered recently that occurred spontaneously. List five changes that are nonspontaneous that you have caused to occur.

13.23 Which of the items that you listed in Review Exercise 13.22 are exothermic and which are endothermic?

13.24 What role does the enthalpy change play in determining the spontaneity of an event?

13.25 Use the data from Table 5.2 to calculate $\Delta H°$ for the following reactions. On the basis of their values of $\Delta H°$, which are favored to occur spontaneously?
(a) $CaO(s) + CO_2(g) \rightarrow CaCO_3(s)$
(b) $C_2H_2(g) + 2H_2(g) \rightarrow C_2H_6(g)$
(c) $3CaO(s) + 2Fe(s) \rightarrow 3Ca(s) + Fe_2O_3(s)$
(d) $Ca(OH)_2(s) \rightarrow CaO(s) + H_2O(l)$
(e) $2NaCl(s) + H_2SO_4(l) \rightarrow Na_2SO_4(s) + 2HCl(g)$

13.26 Use the data from Table 5.2 to calculate $\Delta H°$ for the following reactions. On the basis of their values of $\Delta H°$, which are favored to occur spontaneously?
(a) $2C_2H_2(g) + 5O_2(g) \rightarrow 4CO_2(g) + 2H_2O(g)$
(b) $C_2H_2(g) + 5N_2O(g) \rightarrow 2CO_2(g) + H_2O(g) + 5N_2(g)$
(c) $Fe_2O_3(s) + 2Al(s) \rightarrow Al_2O_3(s) + 2Fe(s)$
(d) $NH_4Cl(s) \rightarrow NH_3(g) + HCl(g)$
(e) $Ag(s) + KCl(s) \rightarrow AgCl(s) + K(s)$

Entropy

13.27 When solid potassium iodide is dissolved in water, a cooling of the mixture occurs because the solution process is endothermic for these substances. Explain, in terms of what happens to the molecules and ions, why this mixing occurs spontaneously.

13.28 What is *entropy?*

13.29 Will the entropy change for each of the following be positive or negative?
(a) Moisture condenses on the outside of a cold glass.
(b) Raindrops form in a cloud.
(c) Gasoline vaporizes in the carburetor of an automobile engine.
(d) Air is pumped into a tire.
(e) Frost forms on the windshield of your car.
(f) Sugar dissolves in coffee.

13.30 On the basis of our definition of entropy, suggest why entropy is a state function.

13.31 State the second law of thermodynamics.

13.32 In animated cartoons, visual effects are often created (for amusement) that show events that ordinarily don't occur in real life because they are accompanied by huge entropy decreases. Can you think of an example of this? Explain why there is an entropy decrease in your example.

13.33 How is entropy related to pollution? What are our chances of eliminating pollution from the environment?

13.34 When a coin is tossed, there is a 50–50 chance of it landing either heads or tails. Suppose that you tossed four coins. On the basis of the different possible outcomes, what is the probability of all four coins coming up heads? What is the probability of an even heads–tails distribution?

*13.35 Suppose that you had two containers of equal volume, sharing a common wall with a hole in it. Suppose there were four molecules in this system. What is the probability that all four would be in one container at the same time? What is the probability of finding an even distribution of molecules between the two containers? (Hint: Try Review Exercise 13.34 first.) What do the results suggest about why gases expand spontaneously?

13.36 Predict the algebraic sign of the entropy change for the following reactions.
(a) $PCl_3(g) + Cl_2(g) \rightarrow PCl_5(g)$
(b) $SO_2(g) + CaO(s) \rightarrow CaSO_3(s)$
(c) $CO_2(g) + H_2O(l) \rightarrow H_2CO_3(aq)$
(d) $Ni(s) + 2HCl(aq) \rightarrow H_2(g) + NiCl_2(aq)$

13.37 Predict the algebraic sign of the entropy change for the following reactions.
(a) $I_2(s) \rightarrow I_2(g)$
(b) $Br_2(g) + 3Cl_2(g) \rightarrow 2BrCl_3(g)$
(c) $NH_3(g) + HCl(g) \rightarrow NH_4Cl(s)$
(d) $CaO(s) + H_2O(l) \rightarrow Ca(OH)_2(s)$

Third Law of Thermodynamics and Standard Entropies

13.38 What is the third law of thermodynamics?

13.39 Would you expect the entropy of an alloy (a solution of two metals) to be zero at 0 K? Explain your answer.

13.40 Why does entropy increase with increasing temperature?

13.41 Calculate $\Delta S°$ for the following reactions in J/K from the data in Table 13.1. On the basis of their values of $\Delta S°$, which of these reactions are favored to occur spontaneously?
(a) $N_2(g) + 3H_2(g) \rightarrow 2NH_3(g)$
(b) $CO(g) + 2H_2(g) \rightarrow CH_3OH(l)$
(c) $2C_2H_6(g) + 7O_2(g) \rightarrow 4CO_2(g) + 6H_2O(g)$
(d) $Ca(OH)_2(s) + H_2SO_4(l) \rightarrow CaSO_4(s) + 2H_2O(l)$
(e) $S(s) + 2N_2O(g) \rightarrow SO_2(g) + 2N_2(g)$

13.42 Calculate $\Delta S°$ for the following reactions in J/K, using the data in Table 13.1.
(a) $Ag(s) + \frac{1}{2}Cl_2(g) \rightarrow AgCl(s)$
(b) $H_2(g) + \frac{1}{2}O_2(g) \rightarrow H_2O(g)$
(c) $H_2(g) + \frac{1}{2}O_2(g) \rightarrow H_2O(l)$
(d) $CaCO_3(s) + H_2SO_4(l) \rightarrow$
$$CaSO_4(s) + H_2O(g) + CO_2(g)$$
(e) $NH_3(g) + HCl(g) \rightarrow NH_4Cl(s)$

13.43 Calculate $\Delta S_f°$ for the following compounds in J/mol K.
(a) $C_2H_4(g)$ (d) $CaSO_4 \cdot 2H_2O(s)$
(b) $N_2O(g)$ (e) $HC_2H_3O_2(l)$
(c) $NaCl(s)$

13.44 Calculate ΔS_f° for the following compounds in J/mol K.
(a) $Al_2O_3(s)$
(b) $CaCO_3(s)$
(c) $N_2O_4(g)$
(d) $NH_4Cl(s)$
(e) $CaSO_4 \cdot \frac{1}{2}H_2O(s)$

13.45 Nitrogen dioxide, NO_2, an air pollutant, dissolves in rainwater to form a dilute solution of nitric acid. The equation for the reaction is

$$3NO_2(g) + H_2O(l) \longrightarrow 2HNO_3(l) + NO(g)$$

Calculate ΔS° for this reaction in J/K.

13.46 Good wine will turn to vinegar if it is left exposed to air because the alcohol is oxidized to acetic acid. The equation for the reaction is

$$C_2H_5OH(l) + O_2(g) \longrightarrow HC_2H_3O_2(l) + H_2O(l)$$

Calculate ΔS° for this reaction in J/K.

Gibbs Free Energy

13.47 What is the equation expressing the change in the Gibbs free energy for a reaction occurring at constant temperature and pressure?

13.48 In what circumstances will a change be spontaneous:
(a) At all temperatures?
(b) At low temperatures but not at high temperatures?
(c) At high temperatures but not at low temperatures?

13.49 In what circumstances will a change be nonspontaneous regardless of the temperature?

13.50 Phosgene, $COCl_2$, was used as a war gas during World War I. It reacts with the moisture in the lungs to produce HCl, which causes the lungs to fill with fluid, leading to the death of the victim. For $COCl_2(g)$, $S^\circ = 284$ J/mol K and $\Delta H_f^\circ = -223$ kJ/mol. Use this information and the data in Table 13.1 to calculate ΔG_f° for $COCl_2(g)$ in kJ/mol.

13.51 Aluminum oxidizes rather easily, but forms a thin protective coating of Al_2O_3 that prevents further oxidation of the aluminum beneath. Use the data for ΔH_f° (Table 5.2) and S° to calculate ΔG_f° for $Al_2O_3(s)$ in kJ/mol.

13.52 Compute ΔG° in kJ for the following reactions, using the data in Table 13.2.
(a) $SO_3(g) + H_2O(l) \rightarrow H_2SO_4(l)$
(b) $2NH_4Cl(s) + CaO(s) \rightarrow$
$$CaCl_2(s) + H_2O(l) + 2NH_3(g)$$
(c) $CaSO_4(s) + 2HCl(g) \rightarrow CaCl_2(s) + H_2SO_4(l)$
(d) $C_2H_4(g) + H_2O(g) \rightarrow C_2H_5OH(l)$
(e) $Ca(s) + 2H_2SO_4(l) \rightarrow$
$$CaSO_4(s) + SO_2(g) + 2H_2O(l)$$

13.53 Compute ΔG° in kJ for the following reactions, using the data in Table 13.2.
(a) $2HCl(g) + CaO(s) \rightarrow CaCl_2(s) + H_2O(g)$
(b) $H_2SO_4(l) + 2NaCl(s) \rightarrow 2HCl(g) + Na_2SO_4(s)$
(c) $3NO_2(g) + H_2O(l) \rightarrow 2HNO_3(l) + NO(g)$
(d) $2AgCl(s) + Ca(s) \rightarrow CaCl_2(s) + 2Ag(s)$
(e) $NH_3(g) + HCl(g) \rightarrow NH_4Cl(s)$

13.54 Plaster of Paris, $CaSO_4 \cdot \frac{1}{2}H_2O(s)$, reacts with liquid water to form gypsum, $CaSO_4 \cdot 2H_2O(s)$. Write a chemical equation for the reaction and calculate ΔG° in kJ, using the data in Table 13.2.

13.55 When phosgene, the war gas described in Review Exercise 13.50, reacts with water vapor, the products are $CO_2(g)$ and $HCl(g)$. Write an equation for the reaction and compute ΔG° in kJ. For $COCl_2(g)$, ΔG_f° is -210 kJ/mol.

13.56 Given the following,
$$4NO(g) \longrightarrow 2N_2O(g) + O_2(g) \qquad \Delta G^\circ = -139.56 \text{ kJ}$$
$$2NO(g) + O_2(g) \longrightarrow 2NO_2(g) \qquad \Delta G^\circ = -69.70 \text{ kJ}$$
calculate ΔG° for the reaction
$$2N_2O(g) + 3O_2(g) \longrightarrow 4NO_2(g)$$

13.57 Given these reactions and their ΔG° values,
$$COCl_2(g) + 4NH_3(g) \longrightarrow CO(NH_2)_2(s) + 2NH_4Cl(s)$$
$$\Delta G^\circ = -332.0 \text{ kJ}$$
$$COCl_2(g) + H_2O(l) \longrightarrow CO_2(g) + 2HCl(g)$$
$$\Delta G^\circ = -141.8 \text{ kJ}$$
$$NH_3(g) + HCl(g) \longrightarrow NH_4Cl(s) \qquad \Delta G^\circ = -91.96 \text{ kJ}$$
Calculate the value of ΔG° for the reaction
$$CO(NH_2)_2(s) + H_2O(l) \longrightarrow CO_2(g) + 2NH_3(g)$$

13.58 Many biochemical reactions have positive values for ΔG° and seemingly should not be expected to be spontaneous. They occur, however, because they are chemically coupled with other reactions that have negative values of ΔG°. An example is the set of reactions that forms the beginning part of the sequence of reactions involved in the metabolism of glucose, a sugar. Given these reactions and their corresponding ΔG° values,

Glucose + phosphate $\longrightarrow$ glucose-6-phosphate + H_2O
$$\Delta G^\circ = +13.13 \text{ kJ}$$

ATP + $H_2O \longrightarrow$ ADP + phosphate $\qquad \Delta G^\circ = -32.22 \text{ kJ}$
calculate ΔG° for the coupled reaction

Glucose + ATP $\longrightarrow$ glucose-6-phosphate + ADP

Free Energy and Work

13.59 How is free energy related to useful work?

13.60 What is a reversible process?

13.61 When glucose is oxidized by the body to generate energy, this energy is stored in molecules of ATP (adenosine triphosphate). However, of the total energy released in the oxidation of glucose, only 38% actually becomes stored in the ATP. What happens to the rest of the energy?

13.62 Why are real, observable changes not considered to be reversible processes?

13.63 Gasohol is a mixture of gasoline and ethanol (grain alcohol), C_2H_5OH. Calculate the maximum work that could be obtained at 25 °C and 1 atm by burning 1 mol of C_2H_5OH.

$$C_2H_5OH(l) + 3O_2(g) \longrightarrow 2CO_2(g) + 3H_2O(g)$$

13.64 What is the maximum amount of useful work that could possibly be obtained at 25 °C and 1 atm from the combustion of 48.0 g of natural gas, $CH_4(g)$, to give $CO_2(g)$ and $H_2O(g)$?

*13.65 Ethyl alcohol, C_2H_5OH, has been suggested as an alternative to gasoline as a fuel. In Example 13.5 we calculated $\Delta G°$ for combustion of 1 mol of C_2H_5OH; in Example 13.6 we calculated $\Delta G°$ for combustion of 1 mol of octane. Let's assume that gasoline has the same properties as octane (one of its major constituents). The density of C_2H_5OH is 0.7893 g/mL; the density of octane, C_8H_{18}, is 0.7025 g/mL. Calculate the maximum work that could be obtained by burning 1 gallon (3.78 liters) of both C_2H_5OH and C_8H_{18}. On a volume basis, which is a better fuel?

Free Energy and Equilibrium

13.66 In what way is free energy related to equilibrium?

13.67 Considering the fact that the formation of a bond between two atoms is exothermic and is accompanied by an entropy decrease, explain why all chemical compounds decompose into individual atoms if heated to a high enough temperature.

13.68 When a warm object is placed in contact with a cold one, they both gradually come to the same temperature. On a molecular level, explain how this is related to entropy and spontaneity.

13.69 Sketch the shape of the free energy curve for a chemical reaction that has a positive $\Delta G°$. Indicate the composition of the reaction mixture corresponding to equilibrium.

13.70 Sketch a graph to show how the free energy changes during a phase change such as the melting of a solid.

13.71 Chloroform, formerly used as an anesthetic and now believed to be a carcinogen (cancer-causing agent), has a heat of vaporization $\Delta H_{vaporization} = 31.4$ kJ/mol. The change, $CHCl_3(l) \rightarrow CHCl_3(g)$ has $\Delta S° = 94.2$ J/mol K. At what temperature do we expect $CHCl_3$ to boil (i.e., at what temperature will liquid and vapor be in equilibrium at 1 atm pressure)?

13.72 For the melting of aluminum, $Al(s) \rightarrow Al(l)$, $\Delta H° = 10.0$ kJ/mol and $\Delta S° = 9.50$ J/mol K. Calculate the melting point of Al. (The actual melting point is 660 °C.)

13.73 Isooctane, an important constituent of gasoline, has a boiling point of 99.3 °C and a heat of vaporization of 37.7 kJ/mol. What is ΔS (in J/mol K) for the vaporization of 1 mol of isooctane?

13.74 Acetone (nail polish remover) has a boiling point of 56.2 °C. The change, $(CH_3)_2CO(l) \rightarrow (CH_3)_2CO(g)$, has $\Delta H° = 31.9$ kJ/mol. What is $\Delta S°$ for this change?

13.75 Determine whether the following reaction will be spontaneous.

$C_2H_4(g) + 2HNO_3(l) \longrightarrow$
$$HC_2H_3O_2(l) + H_2O(l) + NO(g) + NO_2(g)$$

13.76 Which of the following reactions (unbalanced) would be expected to be spontaneous at 25 °C and 1 atm?
(a) $PbO(s) + NH_3(g) \rightarrow Pb(s) + N_2(g) + H_2O(g)$
(b) $NaOH(s) + HCl(g) \rightarrow NaCl(s) + H_2O(l)$

(c) $Al_2O_3(s) + Fe(s) \rightarrow Fe_2O_3(s) + Al(s)$
(d) $2CH_4(g) \rightarrow C_2H_6(g) + H_2(g)$

13.77 Many reactions that have large, negative values of $\Delta G°$ are not actually observed to happen at 25 °C and 1 atm. Why?

Effect of Temperature on the Free Energy Change

13.78 Why is the value of ΔG for a change so dependent on temperature?

13.79 What equation can be used to obtain an approximate value for $\Delta G_T°$ for a reaction at a temperature other than 25 °C?

13.80 Suppose a reaction has a negative $\Delta H°$ and a negative $\Delta S°$. Will more or less product be present at equilibrium as the temperature is raised?

13.81 Calculate the $\Delta G_T°$ in kJ for the following reactions at 100 °C, using data in Tables 5.2 and 13.1.
(a) $C_2H_4(g) + H_2(g) \rightarrow C_2H_6(g)$
(b) $5SO_3(g) + 2NH_3(g) \rightarrow$
$$2NO(g) + 5SO_2(g) + 3H_2O(g)$$
(c) $N_2O(g) + NO_2(g) \rightarrow 3NO(g)$

Additional Exercises

*13.82 When an ideal gas expands at a constant temperature, $\Delta E = 0$ for the change. Why?

*13.83 When a real gas expands at a constant temperature, $\Delta E > 0$ for the change. Why?

*13.84 A cylinder fitted with a piston contains 5.00 L of a gas at a pressure of 4.00 atm. The entire apparatus is contained in a water bath to maintain a constant temperature of 25 °C. The piston is released and the gas expands until the pressure inside the cylinder equals the atmospheric pressure outside, which is 1 atm. Assume ideal gas behavior and calculate the amount of work done by the gas as it expands at constant temperature.

*13.85 The experiment described in the preceding question is repeated, but this time a weight, which exerts a pressure of 2 atm, is placed on the piston. When the gas expands, its pressure drops to this 2 atm pressure. Then the weight is removed and the gas is allowed to expand again to a final pressure of 1 atm. Throughout both expansions the temperature of the apparatus was held at a constant 25 °C. Calculate the amount of work performed by the gas in each step. How does the combined total amount of work in this two-step expansion compare to the amount of work done by the gas in the one-step expansion described in the preceding question?

13.86 What is the algebraic sign of ΔS for the following changes?
(a) Sugar dissolves in water.
(b) $NaCl(s) \rightarrow Na^+(aq) + Cl^-(aq)$
(c) $NH_3(g) \xrightarrow{H_2O} NH_3(aq)$
(d) Naphthalene$(g) \rightarrow$ naphthalene(s)
(e) A gas is cooled from 40 °C to 25 °C. (P constant)
(f) A gas is compressed at constant temperature from 4.0 L to 2.0 L.

13.87 When potassium iodide dissolves in water, the mixture becomes cool. For this change, which is of a larger magnitude, ΔS or ΔH?

13.88 The reaction $Cl_2(g) + Br_2(g) \rightarrow 2BrCl(g)$ has a very small value for $\Delta S°$ ($+11.6$ J/K). Why?

13.89 For a phase change, why is there only one temperature at which there can be an equilibrium between the phases?

*13.90 Cars, trucks, and other machines that use gas or diesel engines for power have cooling systems. In terms of thermodynamics, what makes these cooling systems necessary?

Bond Energies (Special Topic 13.1)

*13.91 Calculate the bond energy in the nitrogen molecule and the oxygen molecule.

13.92 Use the data in Table E.3 to compute the approximate atomization energy of NH_3.

13.93 Approximately how much energy would be released during the formation of the bonds in one mole of acetone molecules? Acetone has the structural formula

$$
\begin{array}{ccc}
H & :O: & H \\
| & \| & | \\
H-C- & C- & C-H \\
| & & | \\
H & & H
\end{array}
$$

13.94 The standard heat of formation of ethylene, $C_2H_4(g)$, is $\Delta H_f° = +52.284$ kJ/mol. Calculate the C$=$C bond energy in this molecule.

13.95 The standard heat of formation of ethanol vapor, $C_2H_5OH(g)$, is $\Delta H_f° = -235.3$ kJ/mol. Use the data in Table E.2 and the average bond energies for C—C, C—H, and O—H bonds to estimate the C—O bond energy in this molecule. The structure of the molecule is

$$
\begin{array}{ccc}
H & H & \\
| & | & \\
H-C- & C- & \ddot{O}-H \\
| & | & \\
H & H &
\end{array}
$$

13.96   Carbon disulfide, CS_2, has the Lewis structure $:\ddot{S}\!=\!C\!=\!\ddot{S}:$ and for $CS_2(g)$, $\Delta H_f^\circ = +115.3$ kJ/mol. Use the data in Table E.2 to calculate the average $C\!=\!S$ bond energy in this molecule.

13.97   For $SF_6(g)$, $\Delta H_f^\circ = -1096$ kJ/mol. Use the data in Table E.2 to calculate the average $S\!-\!F$ bond energy in SF_6.

13.98   Gaseous hydrogen sulfide, H_2S, has $\Delta H_f^\circ = -20.15$ kJ/mol. Use data in Table E.2 to calculate the average $S\!-\!H$ bond energy in this molecule.

13.99   Use the results of Review Exercise 13.97 and the data in Table E.2 to calculate the standard heat of formation of $SF_4(g)$. The measured value of ΔH_f° for $SF_4(g)$ is -718.4 kJ/mol. What is the percentage difference between your calculated value of ΔH_f° and the experimentally determined value?

13.100   For the substance $SO_2F_2(g)$, $\Delta H_f^\circ = -858$ kJ/mol.

The structure of the SO_2F_2 molecule is

$$:\ddot{F}\!-\!\overset{\displaystyle :\ddot{O}}{\underset{\displaystyle \ddot{O}:}{\overset{\displaystyle \|}{\underset{\displaystyle \|}{S}}}}\!-\!\ddot{F}:$$

Use the value of the $S\!-\!F$ bond energy calculated in Review Exercise 13.97 and the data in Table E.2 to determine the average $S\!=\!O$ bond energy in SO_2F_2.

13.101   Use the data in Tables E.2 and E.3 to estimate the standard heat of formation of acetylene, $H\!-\!C\!\equiv\!C\!-\!H$, in the gaseous state.

13.102   What would be the approximate heat of formation of CCl_4 vapor at 25 °C and 1 atm?

13.103   Which substance should have the more exothermic heat of formation, CF_4 or CCl_4?

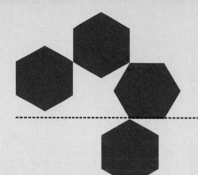

CHEMICALS IN USE 11
Synthetic Diamonds and Diamond Coatings

Ever since it was discovered that cut and polished diamonds sparkle brightly in the light, they have fascinated countless people. Gem-quality diamonds grace crowns and scepters, gowns, lapels, and fingers and have been celebrated in song. Because of the unusual hardness of diamond, diamonds of lesser quality are also valued, and diamond-studded cutting and grinding tools are used extensively by industries of all sorts. Imagine the amazement of scientists when it was discovered in 1797 that diamonds are simply a form of pure carbon. Another form, graphite, is "dirt cheap," so people naturally asked whether graphite could be converted into its more valuable cousin. At the atomic level, graphite consists of covalent *sheets* of carbon atoms that slip over each other without much difficulty, which helps to make graphite a (dry) lubricant. Diamond, in contrast, is a covalent solid with an interlocking, *tetrahedral network* of atoms (see Section 8.9).

As you might expect, many attempted to make diamond from graphite. But try as they might, no one succeeded. In 1938, a careful thermodynamic analysis of the problem was performed, and the results of it are summarized in Figure 11a. Here we have a graph of $\Delta G/T$ versus the absolute temperature for the conversion of graphite to diamond.

$$C(s, \text{graphite}) \longrightarrow C(s, \text{diamond})$$

For this change to occur *spontaneously,* ΔG must be negative, as we learned in Chapter 13. Because T is always positive, the ratio $\Delta G/T$ must therefore be negative for the conversion to happen spontaneously. On the graph of Figure 11a, the regions where the ratio is negative are shaded.

Each line on the graph shows how $\Delta G/T$ varies with temperature for a given pressure. Thus, the line labeled 1 atm shows that at a pressure of 1 atm, $\Delta G/T$ becomes less positive as the temperature increases. However, notice that $\Delta G/T$ never becomes negative at 1 atm. It is thus thermodynamically impossible to convert graphite to diamond at atmospheric pressure.

From the graph, we can also see that $\Delta G/T$ becomes less positive (more negative) with increasing pressure. At 20,000 atm, the ratio is negative at temperatures below 470 K (approximately 200 °C), and at higher pressures $\Delta G/T$ becomes negative at still higher temperatures. From the graph, therefore, scientists were able to establish the conditions of temperature and pressure at which the conversion from graphite to diamond is at least theoretically possible. This was not the end of the problem, however, because suitable materials still had to be found that would allow the transition to occur at a reasonable rate. Success was elusive, but it finally came in the early 1950s. Pressures in excess of 100,000 atm and temperatures above 2800 °C were needed. The diamonds that were made were of industrial quality, and before long, over a million carats of industrial diamonds were being synthesized annually. Some of the synthetic diamonds are of substantial size — as large as 2 carats (see Figure 11b).

DIAMOND COATINGS

When a mixture of 1% methane (CH_4) in hydrogen at a pressure of 50 millibars (38 torr) is passed over a glowing wire at 2200 °C, their molecules crack into atoms of carbon and hydrogen (see Figure 11c). The "substrate" in the figure is a surface 5 to 15 mm above the wire onto which the carbon atoms deposit. There they gradually form a film of carbon *with a tetrahedral network structure.* The film, in other words, is diamond, not graphite.

SOME APPLICATIONS OF DIAMOND-COATED SURFACES

Many instruments, including computers, have components constantly endangered by temperature increases. A semiconductor laser diode, for example, must be protected from a temperature increase (see Figure 11d). Such parts must, therefore, be given ways to lose heat rapidly into a "heat sink." Materials of high *thermal conductivity* are needed for the most effective heat sinks, and diamond has a higher thermal conductivity than all metals, being two to four times higher (depending on the impurities) than copper, silver or gold. For several years, large, gem-quality, yellow diamonds have been manufactured for making heat sinks. Diamond coated surfaces might someday be made to work as well and at lower cost.

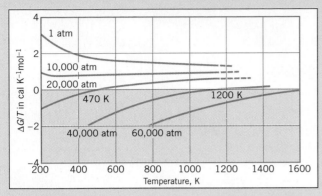

FIGURE 11a

Thermodynamic analysis of the conversion of graphite to diamond. (From *Chemical and Engineering News,* April 5, 1971, p. 51. Used by permission.)

FIGURE 11b

Large, single-crystal synthetic diamonds of about 8 mm in diameter. Their yellow color is caused by traces of nitrogen. Shown also are cut pieces of synthetic diamond that are used to make computer parts. (Courtesy of Sumitomo Electric Industries, Japan.)

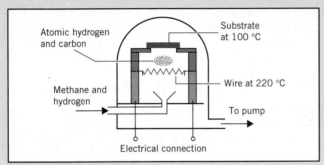

FIGURE 11c

The heated-filament method to prepare a thin film of diamond. A 1% mixture of methane in hydrogen at low pressure encounters a glowing filament. The carbon atoms generated deposit on the substrate. (From P. K. Bachmann and R. Messier, *Chemical and Engineering News,* May 15, 1989. Used by permission.)

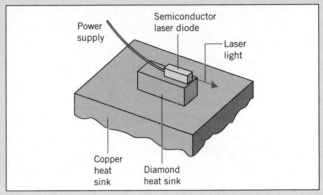

FIGURE 11d

The use of diamond for heat sinks for semiconductors. Heat generated in the semiconductor laser diode moves rapidly into the diamond heat sink and from there into the copper heat sink. (From P. K. Bachmann and R. Messier, *Chemical and Engineering News,* May 15, 1989. Used by permission.)

Diamond point-contact transistors in semiconducting devices are able to withstand operating temperatures up to five times higher than those made of silicon. This opens the way for the use of such devices in environments like jet engines or vehicle motors.

Another potential application of diamond films is to manufacture wear-resistant cutting tools. A diamond coating of only 3 μm in thickness greatly prolongs the life of such tools and so reduces costs.

Speakers for high-frequency sounds—"tweeters" —are improved by using diamond-coated diaphragms. With such a surface, sound distortions in the high-frequency region are largely shifted to frequencies above those the human ear can discern. The reproduction of sound is thus of much higher fidelity.

Other applications being developed for diamond coatings are for scratch-resistant coatings for lenses, sunglasses, and watch glasses.

Questions

1. For the conversion of graphite to diamond, what is the significance of the value of ΔG being *negative?*

2. Ultrahigh pressures tend to favor the conversion of graphite to diamond. Why?

3. What specific *particles* constitute the "building blocks" for the diamond structure when diamond films are made?

4. What physical property of diamond is behind the uses of diamond films for making computer parts?

Butterflies, like these two exploring for nectar among purple lobelia blossoms, have metabolisms that are greatly affected by the temperature of their surroundings. Temperature, as we'll learn in this chapter, is one of many factors that affect reaction rates.

Chapter 14

Kinetics: The Study of Rates of Reaction

In Chapter 13 you learned that the sign of $\Delta G°$ determines whether a reaction is possible (i.e., whether significant amounts of products will be formed when the reaction reaches equilibrium). However, to actually observe something happen in a chemical system, the reaction must not only be possible, it must also occur rapidly enough so that the products are formed within a reasonable time. For this reason we now turn our attention to **kinetics,** the study of the factors that govern how rapidly reactions occur.

When we examine common everyday reactions, we find some are fast and others are slow. For example, when a mixture of gasoline and air in the cylinder of an engine is ignited, a very rapid (almost instantaneous) reaction occurs that can cause the engine to propel a car. Most reactions are much slower than this, however. The cooking of foods, for instance, involves chemical reactions, and as any hungry person waiting for a hamburger to cook knows, these reactions take a while. Fortunately for us, so do the biochemical reactions involved in digestion and metabolism. If biochemical reactions were as rapid as the combustion of gasoline vapor, our lives would be over in an instant.

Some reactions going on around us are quite slow. For instance, the oil paints an artist uses dry slowly, first by evaporation of the solvent, and then by the gradual reaction of oxygen with the linseed oil resins in the paint. This may take weeks or even months. The gradual curing of cement and concrete is another example. Although cement solidifies relatively quickly, chemical reactions within the mixture continue for years as the cement becomes harder and harder.

In scientific terms, the speed at which a reaction occurs is called the **rate of reaction.** It tells us how quickly (or slowly) the reactants disappear and the

14.1
SPEEDS AT WHICH REACTIONS OCCUR

Thermodynamics (Chapter 13) tells us if a reaction can occur. *Kinetics* (this chapter) tells us how fast it will. The study of *chemical equilibria* (next chapter) concerns the relative amounts of reactants and products that remain when nothing further happens.

585

One of the concerns of chemical engineers is how *rapidly* heat evolves in a large-scale exothermic reaction; heat generally must be efficiently "pumped" away as it is released or the system blows up!

Propane is the fuel used for cooking in many rural areas as well as in many campers and pleasure boats.

products form. Such information can be quite valuable. For instance, knowing the rates of reactions and the factors that control them allows the manufacturers of chemical products to adjust the conditions of reactions efficiently so as to improve productivity and hold down operating costs. Of more fundamental importance, however, is the fact that by studying the rates of reactions we can learn a great deal about *how* reactants change into products.

For a chemical reaction, the balanced chemical equation describes the net overall change that's observed. Usually, however, the net change is actually the result of a series of simpler reactions. Consider, for example, the combustion of propane, C_3H_8.

$$C_3H_8(g) + 5O_2(g) \longrightarrow 3CO_2(g) + 4H_2O(g)$$

This reaction simply cannot occur in a single, simultaneous collision between one propane molecule and five oxygen molecules. Anyone who has ever played billiards knows how seldom three balls come together with only one "click" even on a flat, two-dimensional surface. It is easy to understand, then, how unlikely is the *simultaneous* collision in three-dimensional space of six reactant molecules. For the combustion of propane to proceed rapidly, which indeed it does, a series of much more probable collision steps involving chemical species of fleeting existence must take place. *The series of individual steps leading to the overall observed reaction is called the* **mechanism of the reaction.** Information about reaction mechanisms is one of the dividends that the study of reaction rates can give.

14.2 FACTORS THAT AFFECT REACTION RATES

The rates of nearly all chemical reactions are controlled primarily by five factors: (1) the chemical nature of the reactants; (2) the ability of the reactants to come in contact with each other; (3) the concentrations of the reactants; (4) the temperature of the reaction system; and (5) the availability of agents called *catalysts* that affect the rate of the reaction but are not themselves consumed. We'll take up each in turn.

The Nature of the Reactants

Fundamental differences in chemical reactivity, which are controlled by the tendencies toward bond formation, are a major factor in determining the rates of reactions. Some reactions are just naturally fast and others are naturally slow. For example, a freshly exposed surface of metallic sodium tarnishes almost instantly if exposed to air and moisture because sodium loses electrons so easily. Iron also reacts with air and moisture to form rust, but the reaction is much slower under identical conditions because iron doesn't lose electrons as easily as sodium.

The Ability of the Reactants to Meet

Most reactions involve two or more reactants, and they must be able to come in contact with each other for the reaction to occur. This is one of the primary reasons that reactions are so often carried out in liquid solutions or in the gas phase. In these states, the particles of the reactants—molecules or ions—are able to intermingle and collide with each other easily.

A common example of how the ability of the reactants to meet affects the rate of a reaction is the combustion of gasoline. Liquid gasoline burns rapidly,

FIGURE 14.1

An explosion in this grain elevator in New Orleans, Louisiana, killed 35 people in December 1977.

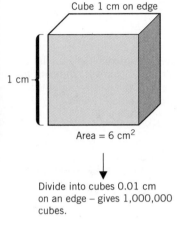

Cube 1 cm on edge

1 cm

Area = 6 cm^2

Divide into cubes 0.01 cm on an edge – gives 1,000,000 cubes.

0.01 cm

Total area = 600 cm^2

but if the gasoline is vaporized and mixed with air before being ignited, the reaction is virtually instantaneous; the mixture explodes. An *explosion* is simply an extremely rapid reaction that quickly generates hot expanding gases. When all of the reactants are in the same phase, the reaction that occurs is called a **homogeneous reaction.** Examples are reactions in solution and gas phase reactions like gasoline vapor combustion.

When the reactants are present in different phases—when one is a gas and the other a liquid or a solid, for example—the reaction is called a **heterogeneous reaction.** In a heterogeneous reaction, the reactants are able to meet only at the interface between the phases, so *the area of contact between the phases determines the rate of the reaction.* This area is controlled by the sizes of the particles of the reactants. For example, lighting a large log with a match is very difficult because there is not enough contact between the wood and the oxygen in the air for combustion to occur quickly. However, small twigs or slivers of wood (kindling) are relatively easy to ignite. Their greater surface area allows them to react faster with oxygen.

When the particle sizes are extremely small, even heterogeneous reactions can be explosive. Figure 14.1 shows the result of an explosion in a grain elevator that was used to store wheat. The explosion occurred when fine particles of grain dust mixed with air were accidentally ignited by a chance spark. Although heterogeneous reactions like this are obviously serious and important, they are difficult to analyze. In this chapter, therefore, we'll focus mostly on homogeneous systems.

The Concentrations of the Reactants

The rates of both homogeneous and heterogeneous reactions are affected by the concentrations of the reactants. For example, wood burns relatively quickly in air but extremely rapidly in pure oxygen. Even red-hot steel wool, which only sputters and glows in air, bursts into flame when thrust into pure oxygen (see Figure 14.2).

FIGURE 14.2

Steel wool, after being heated to redness in a flame, burns spectacularly when dropped into pure oxygen.

The Temperature of the System

Almost all chemical reactions occur faster at higher temperatures than they do at lower temperatures. If you have not already done so, notice that insects move more slowly in cool temperatures. Insects are cold-blooded creatures, which means that their body temperatures are determined by the temperature of their surroundings. As the air cools, the rate of chemical metabolism in insect bodies slows down, and they become sluggish.

Mountain trekkers report that mosquito activity is a reliable "thermometer." Below 55 °C, the mosquitos "go away" for the night.

The Presence of Catalysts

Catalysts are substances that increase the rates of chemical reactions without being used up. They affect every moment of our lives. For example, the enzymes that direct our body chemistry are catalysts. So are many of the substances that the chemical industry uses to make gasoline, plastics, fertilizers, and other products that have become virtual necessities in our lives. We will discuss how catalysts affect reaction rates later in Section 14.9.

Two-part epoxy glues consist of a resin and a catalyst. The catalyst promotes the hardening of the resin into a solid mass.

14.3 ▬▬
MEASURING THE RATE OF REACTION

In general, $\dfrac{1}{x} = x^{-1}$.

A **rate** is always expressed as a ratio in which units of time appear in the denominator. For example, if you have a job, you are probably paid at a certain rate—say, five dollars per hour. Since the word *per* can be translated to mean *divided by*, this pay rate can be written as a fraction:

$$\text{Rate of pay} = \frac{5 \text{ dollars}}{1 \text{ hr}}$$

The fraction $\dfrac{1}{\text{hr}}$ can also be written as hr^{-1}, so your pay rate can be expressed

$$\text{Rate of pay} = 5 \text{ dollars hr}^{-1}$$

We can also look at this as the way your wealth changes with a change in time.

$$\text{Rate of pay} = \text{rate of change of wealth} = \frac{\text{change in wealth}}{\text{change in time}}$$

Because your wealth is increasing, we write the rate as a positive quantity. However, if you hire a tutor for $20 per hour, your wealth would now be decreasing with time, so we now write it with a negative sign.

$$\text{Rate of change of wealth} = \frac{-20 \text{ dollars}}{1 \text{ hr}} = -20 \text{ dollars hr}^{-1}$$

Rates of Chemical Reactions

When a chemical reaction occurs, the concentrations of the reactants and products change with time. The concentrations of the reactants decrease as the reactants are used up, while the concentrations of the products increase as the products form. The rate of a chemical reaction, therefore, is expressed in terms of the rates at which concentrations change.

Once again, we use the symbol Δ to mean a change (final minus initial).

$$\text{Rate of reaction} = \frac{\text{change in concentration}}{\text{change in time}} = \frac{\Delta(\text{concentration})}{\Delta(\text{time})}$$

The concentration unit normally used is molarity, and the unit of time is usually the second (abbreviated s). This means that reaction rates are most frequently expressed with the following units.

$$\frac{\text{mol/L}}{\text{s}}$$

Since $1/L$ and $1/s$ can also be written as L^{-1} and s^{-1}, the units for the rate can be expressed as $\text{mol } L^{-1} s^{-1}$. For instance, if the concentration of a certain product of a reaction increases by 0.50 mol/L each second, the rate of its formation is $0.50 \text{ mol } L^{-1} s^{-1}$. Similarly, if the concentration of a reactant decreases by 0.20 mol/L per second, its rate of reaction is $-0.20 \text{ mol } L^{-1} s^{-1}$. Notice that this rate is expressed with a negative sign because the concentration is decreasing with time.

> Sometimes the units $\text{mol } L^{-1} s^{-1}$ are expressed as $M s^{-1}$. (Remember that molarity, M, means $\text{mol } L^{-1}$.)

In any reaction, the relative rates at which reactants are consumed and the products are formed are related by the coefficients of the balanced equation for the reaction. For example, in the combustion of propane described earlier,

$$C_3H_8(g) + 5O_2(g) \longrightarrow 3CO_2(g) + 4H_2O(g)$$

> The algebraic sign of the rate just indicates whether the substance is being formed or consumed in the reaction.

five moles of O_2 are consumed for each mole of C_3H_8. Therefore, oxygen is reacting five times faster than propane. Similarly, CO_2 forms three times faster and H_2O four times faster than C_3H_8 reacts.

Butane, C_4H_{10}, burns in oxygen to give CO_2 and H_2O according to the equation

$$2C_4H_{10}(g) + 13O_2(g) \longrightarrow 8CO_2(g) + 10H_2O(g)$$

EXAMPLE 14.1
Expressing Rates of Reaction

If at a certain moment the butane is reacting at a rate of $-0.20 \text{ mol } L^{-1} s^{-1}$, what is the rate at which oxygen is being consumed and what are the rates at which the products are being formed?

ANALYSIS The question asks about the rates *relative to the given rate* of the reaction of butane, so it's basically a simple stoichiometry problem. This is because the *magnitudes* of the rates relative to each other are in the same relationship as the coefficients in the balanced equation.

SOLUTION For oxygen

$$\frac{0.20 \text{ mol } C_4H_{10}}{L \text{ s}} \times \frac{13 \text{ mol } O_2}{2 \text{ mol } C_4H_{10}} = \frac{1.3 \text{ mol } O_2}{L \text{ s}}$$

Using a minus sign, because oxygen is being consumed, oxygen is reacting at a rate of $-1.3 \text{ mol } L^{-1} s^{-1}$. For CO_2 and H_2O, we have similar calculations.

$$\frac{0.20 \text{ mol } C_4H_{10}}{L \text{ s}} \times \frac{8 \text{ mol } CO_2}{2 \text{ mol } C_4H_{10}} = \frac{0.80 \text{ mol } CO_2}{L \text{ s}}$$

$$\frac{0.20 \text{ mol } C_4H_{10}}{L \text{ s}} \times \frac{10 \text{ mol } H_2O}{2 \text{ mol } C_4H_{10}} = \frac{1.0 \text{ mol } H_2O}{L \text{ s}}$$

Because the product concentrations are increasing with time, their rates are expressed as positive quantities. Therefore,

$$\text{Rate of formation of } CO_2 = +0.80 \text{ mol } L^{-1} s^{-1}$$

$$\text{Rate of formation of } H_2O = +1.0 \text{ mol } L^{-1} s^{-1}$$

■ **Practice Exercise 1** When heated above 500 °C, potassium nitrate decomposes according to the equation

$$4KNO_3(s) \longrightarrow 2K_2O(s) + 2N_2(g) + 5O_2(g)$$

If oxygen is being formed at a rate of 0.30 mol L^{-1} s^{-1}, what are the rates of formation of the other products? What is the rate at which the KNO_3 is decomposing?

Because the rates of reaction of the reactants and products are all related, it doesn't matter which substance we monitor when we measure concentrations during an experiment to determine the rate. For example, to study the decomposition of hydrogen iodide, HI, into H_2 and I_2

$$2HI(g) \longrightarrow H_2(g) + I_2(g)$$

it is easiest to monitor the I_2 concentration because it is the only colored substance in the reaction. As the reaction proceeds, purple iodine vapor is formed and there are instruments that are able to relate the intensity of the purple color to concentration.

TABLE 14.1   Data, at 508 °C, for the Reaction $2HI(g) \rightarrow H_2(g) + I_2(g)$

Concentration of HI (mol/L)	Time (s)
0.100	0
0.0716	50
0.0558	100
0.0457	150
0.0387	200
0.0336	250
0.0296	300
0.0265	350

Change of Reaction Rate with Time

Commonly, the rate of a reaction changes as the reactants are consumed. This is because the rate usually depends on the concentrations of the reactants, and these change as the reaction proceeds. For example, Table 14.1 contains data for the decomposition of hydrogen iodide at a temperature of 508 °C. These data, which show the way that the HI concentration changes with time, are plotted in Figure 14.3, where the vertical axis gives the molar concentration of HI.

Notice in Figure 14.3 that the HI concentration drops fairly rapidly during the first 50 seconds of the reaction, which means that the initial rate is relatively fast. In the interval between 300 and 350 seconds, however, the concentration changes by only a small amount, so the rate has slowed considerably. Thus, the steepness of the concentration versus time curve reflects the rate of the reaction—the steeper the curve, the higher the rate.

FIGURE   14.3

The change in the concentration of HI with time for the reaction

$$2HI(g) \longrightarrow H_2(g) + I_2(g)$$

at 508 °C. The data that are plotted are in Table 14.1.

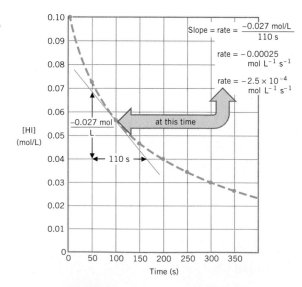

The rate at which the HI is being consumed at any particular moment can be determined from the slope of the concentration *versus* time curve measured at a point corresponding to the time in question. The slope (or tangent), which is equal to the rate, is the ratio of the change in concentration to the change in time as read off the graph. In Figure 14.3, for example, the rate of the decomposition of hydrogen iodide is determined for a time 100 seconds from the start of the reaction. After the tangent to the curve is drawn, we measure a concentration change (-0.027 mol/L) and the time change (110 s) from the graph. The ratio of these quantities,

> If necessary, see Appendix A for a discussion of slope.

$$\text{Rate} = \frac{-0.027 \text{ mol/L}}{110 \text{ s}}$$

gives the rate. At this point in the reaction, the HI is reacting at a rate of -2.5×10^{-4} mol L^{-1} s^{-1}. By working Practice Exercise 2, you will see that the rate later on is smaller.

■ **Practice Exercise 2** Use the graph in Figure 14.3 to estimate the reaction rate 250 s after the start of the reaction.

You learned in the last section that the rates of most reactions change when the concentrations of the reactants change. The relationship between reaction rate and molar concentration can be expressed in a mathematical equation in which molar concentrations are represented by formulas within square brackets. Thus, [H$^+$] stands for the molar concentration of H$^+$ in the system.

In general, the rate of a homogeneous reaction is found to be proportional to the product of the concentrations of the reactants, each raised to a power. For example, if we have a chemical equation such as

$$A + B \longrightarrow \text{products}$$

the rate of the reaction would be

$$\text{Rate} \propto [A]^m [B]^n \tag{14.1}$$

The values of the exponents n and m must be determined by experiment. The proportionality symbol, $\propto$, can be replaced by an equals sign if we introduce a proportionality constant, k, which is called the **rate constant** for the reaction. Equation 14.1 then takes the form

$$\text{Rate} = k[A]^m [B]^n \tag{14.2}$$

Equation 14.2 is called the **rate law** for the reaction of A with B. Once we have found values for k, n, and m, the rate law allows us to calculate the rate of the reaction if the concentrations are known. For example, the reaction

$$H_2SeO_3 + 6I^- + 4H^+ \longrightarrow Se + 2I_3^- + 3H_2O \tag{14.3}$$

should have a rate law of the form

$$\text{Rate} = k[H_2SeO_3]^x [I^-]^y [H^+]^z$$

Experimentally, it's been found that for the initial rate of this reaction (i.e., the rate when the reactants are first combined), $x = 1$, $y = 3$, and $z = 2$. The value of the rate constant k depends on the temperature, and for this reaction

14.4
CONCENTRATION AND RATE

 Rate law concept

> The value of k depends on the particular reaction being studied as well as the temperature at which the reaction occurs.

at 0 °C, k equals 5.0×10^5 L^5 mol^{-5} s^{-1}. Substituting the exponents and the value of k into the previous equation gives the rate law.

$$\text{Rate} = (5.0 \times 10^5 \text{ L}^5 \text{ mol}^{-5} \text{ s}^{-1})[\text{H}_2\text{SeO}_3] \; [\text{I}^-]^3 \; [\text{H}^+]^2 \quad \text{(at 0 °C)} \quad (14.4)$$

Equation 14.4 then allows us to calculate the initial rate of the reaction at 0 °C for any set of concentrations of H_2SeO_3, I^-, and H^+.

The units of the rate constant are generally such that the calculated rate will have the units mol L^{-1} s^{-1}.

EXAMPLE 14.2
Calculating Reaction Rate from the Rate Law

What is the rate of Reaction 14.3 at 0 °C when the reactant concentrations are $[\text{H}_2\text{SeO}_3] = 2.0 \times 10^{-2}$ M, $[\text{I}^-] = 2.0 \times 10^{-3}$ M, and $[\text{H}^+] = 1.0 \times 10^{-3}$ M?

ANALYSIS When we already know the rate law (Equation 14.4), a question like this is merely a matter of substituting the given values of molar concentrations into the rate law.

SOLUTION To see how the units work out as we use Equation 14.4, let's write all of the concentration values as well as the rate constant's units in fraction form.

$$\text{Rate} = \frac{5.0 \times 10^5 \text{ L}^5}{\text{mol}^5 \text{ s}} \times \left(\frac{2.0 \times 10^{-2} \text{ mol}}{\text{L}}\right) \times \left(\frac{2.0 \times 10^{-3} \text{ mol}}{\text{L}}\right)^3 \times \left(\frac{1.0 \times 10^{-3} \text{ mol}}{\text{L}}\right)^2$$

To perform the arithmetic, we first raise the concentrations and their units to the appropriate powers.

$$\text{Rate} = \frac{5.0 \times 10^5 \text{ L}^5}{\text{mol}^5 \text{ s}} \times \left(\frac{2.0 \times 10^{-2} \text{ mol}}{\text{L}}\right) \times \left(\frac{8.0 \times 10^{-9} \text{ mol}^3}{\text{L}^3}\right) \times \left(\frac{1.0 \times 10^{-6} \text{ mol}^2}{\text{L}^2}\right)$$

$$\text{Rate} = \frac{8.0 \times 10^{-11} \text{ mol}}{\text{L s}}$$

The answer can also be written as

$$\text{Rate} = 8.0 \times 10^{-11} \text{ mol L}^{-1} \text{ s}^{-1}$$

CHECK There's no simple check, but at least see that the answer has the correct units for a reaction rate.

■ **Practice Exercise 3** The rate law for the decomposition of HI is rate $= k[\text{HI}]^2$. In Figure 14.3, the rate of the reaction was found to be -2.5×10^{-4} mol L^{-1} s^{-1} when the HI concentration was 0.0558 M. (a) What is the value of the rate constant for the reaction? (b) What are the units of k for this reaction?

Exponents in the Rate Law

If you compare Equations 14.3 and 14.4, you will notice that the exponents in the rate law appear to be unrelated to the coefficients in the overall chemical equation. This often is the case. There is, in fact, no way to know for sure what the exponents in the rate law of an overall reaction will be without doing experiments to determine them. Sometimes, however, the coefficients and the exponents are the same by coincidence, as is the case with the rate law for the decomposition of hydrogen iodide.

$$2\text{HI}(g) \longrightarrow \text{H}_2(g) + \text{I}_2(g)$$

$$\text{Rate} = k[\text{HI}]^2$$

The exponent of [HI] in this rate law, namely 2, happens to match the coefficient of HI in the overall chemical equation, but *there is no way we could have actually predicted this match without experimental data.* Therefore, *never* simply assume the exponents and the coefficients are the same; it's a trap that many students fall into.

The exponents in a rate law are used to describe the **order of the reaction**[1] with respect to each reactant. For instance, the decomposition of gaseous N_2O_5 into NO_2 and O_2,

$$2N_2O_5 \longrightarrow 4NO_2 + O_2$$

has the rate law

$$\text{Rate} = k[N_2O_5]^1$$

The exponent on the N_2O_5 concentration is 1 and the reaction is said to be *first order*. The rate law for the decomposition of HI has an exponent of 2 for the HI concentration, so it is a *second-order* reaction.

When the rate law contains the product of the molar concentrations of two or more reactants raised to powers, the order with respect to each reactant can be described. The rate law in Equation 14.4 (page 592) tells us that Reaction 14.3 is first order with respect to H_2SeO_3, third order with respect to I^-, and second order with respect to H^+. The **overall order of a reaction** is the sum of the orders with respect to each reactant. Reaction 14.3 is a sixth-order reaction, overall.

The exponents in a rate law are usually small whole numbers, but fractional and negative exponents are occasionally found. A negative exponent means that the concentration term really belongs in the denominator; as the concentration of the species increases, the rate of reaction decreases. Zero-order reactions also exist. These are reactions whose rate is independent of the concentration of the reactant. An example of a zero-order process is the elimination of ethyl alcohol by the body. Regardless of the alcohol level in the bloodstream, its rate of expulsion from the body is constant. Another example is the decomposition of gaseous ammonia into H_2 and N_2 on a hot platinum surface. The rate at which ammonia decomposes is the same, regardless of its concentration in the gas.

When the exponent is equal to 1, it is usually omitted. Therefore, if no exponent is written, we take it to be 1.

$1 + 3 + 2 = 6$

■ **Practice Exercise 4** The following reaction

$$BrO_3^- + 3SO_3^{2-} \longrightarrow Br^- + 3SO_4^{2-}$$

has the rate law

$$\text{Rate} = k[BrO_3^-][SO_3^{2-}]$$

What is the order of the reaction with respect to each reactant? What is the overall order of the reaction?

Determining the Exponents in a Rate Law

We've mentioned several times that the exponents in the rate law for an overall reaction must be determined experimentally. *This is the only way for us to know for sure what the exponents are.* One way to determine the exponents is to

[1] The reason for describing the *order* of a reaction is to take advantage of a great convenience: the mathematics involved in the treatment of the data are the same for all reactions having the same order. We will not go into this very deeply, but you should be familiar with this terminology used to describe the effects of concentration on the rates of reactions.

TABLE 14.2   Concentration–Rate Data for the Hypothetical Reaction *A* + *B* → Products

Initial Concentrations		Initial Rate of Formation of Products (mol L^{-1} s^{-1})
[A]	[B]	
0.10	0.10	0.20
0.20	0.10	0.40
0.30	0.10	0.60
0.30	0.20	2.40
0.30	0.30	5.40

study how changes in the concentration of each reactant affect the initial rate of the reaction. For example, suppose we have the reaction

$$A + B \longrightarrow \text{products}$$

for which the data in Table 14.2 have been obtained in a series of five experiments. We know the form of the rate law for the reaction will be

$$\text{Rate} = k[A]^n[B]^m$$

For the first three sets of data, the concentration of *B* is constant. Changes in the rate are therefore due *only* to changes in the concentration of *A*. Now all we have to do is figure out what the order of the reaction must be with respect to *A* to give the observed changes in the rate.

Examining the data of Table 14.2, we find that when [A] is doubled, the rate doubles; when [A] is tripled, the rate triples. What must be the exponent on the concentration of *A* for this to be true? The answer is that the exponent *n* must be 1. We can see this if we try a few values for the concentration. When $[A] = 0.10$, $[A]^1 = 0.10$, and when $[A] = 0.20$, $[A]^1 = 0.20$. Notice that when we double the concentration of *A*, we double the value of $[A]^1$, which means that the rate would double.

In the final three sets of data, the concentration of *B* changes while the concentration of *A* is held constant. This time it is the concentration of *B* that affects the rate. Now we see that when [B] is doubled, the rate increases by a factor of 4 (which equals 2^2), and when [B] is tripled, the rate increases by a factor of 9 (which equals 3^2). The only way that *B* can affect the rate this way is if its concentration is squared in the rate law, so the exponent *m* must equal 2. This also can be shown using some trial values for [B]. If $[B] = 0.3$, then $[B]^2$ equals 0.09; if the concentration is doubled, so that $[B] = 0.6$, then $[B]^2$ equals 0.36. Notice that 0.36 is four times as large as 0.09. Doubling the concentration therefore increases $[B]^2$ by a factor of 4, and that increases the rate by a factor of 4.

Having determined the exponents on the concentration terms, we now know that the rate law for the reaction must be

$$\text{Rate} = k[A]^1[B]^2$$

To calculate the value of *k*, we need only substitute rate and concentration data into the rate law for any one of the sets of data.

$$k = \frac{\text{rate}}{[A]^1[B]^2}$$

TABLE 14.3 Relationship between the Order of a Reaction and Changes in Concentration and Rate

Factor By Which the Concentration Is Changed	Factor By Which the Rate Changes	Exponent on the Concentration Term in the Rate Law
2	Rate	0
3	is	0
4	unchanged	0
2	$2 = 2^1$	1
3	$3 = 3^1$	1
4	$4 = 4^1$	1
2	$4 = 2^2$	2
3	$9 = 3^2$	2
4	$16 = 4^2$	2
2	$8 = 2^3$	3
3	$27 = 3^3$	3
4	$64 = 4^3$	3

Using the data from the first set,

$$k = \frac{0.20 \text{ mol L}^{-1} \text{ s}^{-1}}{(0.10 \text{ mol L}^{-1})(0.10 \text{ mol L}^{-1})^2}$$

$$= \frac{0.20 \text{ mol L}^{-1} \text{ s}^{-1}}{0.001 \text{ mol}^3 \text{ L}^{-3}}$$

Canceling units, we see that $\text{mol}/\text{mol}^3 = \text{mol}^{-2}$, and $\text{L}^{-1}/\text{L}^{-3} = \text{L}^2$. Therefore, the value of k with the correct units is

$$k = 2.0 \times 10^2 \text{ L}^2 \text{ mol}^{-2} \text{ s}^{-1}$$

■ **Practice Exercise 5** Use the data from the other four experiments to calculate k for this reaction. What do you notice about the values of k?

The reasoning used to determine the order with respect to each reactant from experimental data is summarized in Table 14.3.

Sulfuryl chloride, SO_2Cl_2, is used to manufacture the antiseptic chlorophenol. The following data were collected on the decomposition of SO_2Cl_2 at a certain temperature.

$$SO_2Cl_2(g) \longrightarrow SO_2(g) + Cl_2(g)$$

Initial Concentration of SO_2Cl_2 (mol L^{-1})	Initial Rate of Formation of SO_2 (mol L^{-1} s^{-1})
0.100	2.2×10^{-6}
0.200	4.4×10^{-6}
0.300	6.6×10^{-6}

What is the rate law and the value of the rate constant for this reaction?

ANALYSIS The first step is to write the general form of the expected rate law so we can see which exponents have to be determined. Then we study the data to see how the rate changes when the concentration is changed by a certain factor.

EXAMPLE 14.3
Determining the Exponents of a Rate Law

Sulfuryl chloride is a dense liquid (bp 69.3 °C) with a pungent odor and is corrosive to the skin and lungs.

SOLUTION We expect the rate law to have the form

$$\text{Rate} = k[SO_2Cl_2]^x$$

We could also use Experiments 2 and 3, but it requires more thought. From the second to the third, $[SO_2Cl_2]$ increases by a factor of $\frac{0.30}{0.20} = 1.5$ and the rate also increases by a factor of $\frac{6.6}{4.4} = 1.5$. Because the rate increases by the same factor as the concentration, the reaction must be first order.

Let's examine the data from the first two experiments. Notice that when we double the concentration from 0.100 to 0.200, the initial rate doubles (from 2.2×10^{-6} to 4.4×10^{-6}). If we look at the first and third, we see that when the concentration triples (from 0.100 M to 0.300 M), the rate also triples (from 2.2×10^{-6} to 6.6×10^{-6}). This behavior tells us that the reaction must be first order in the SO_2Cl_2 concentration. The rate law is therefore

$$\text{Rate} = k[SO_2Cl_2]^1$$

To evaluate k, we can use any of the three sets of data. Choosing the first,

$$k = \frac{\text{rate}}{[SO_2Cl_2]^1}$$
$$= \frac{2.2 \times 10^{-6} \text{ mol L}^{-1} \text{ s}^{-1}}{0.10 \text{ mol L}^{-1}}$$
$$= 2.2 \times 10^{-5} \text{ s}^{-1}$$

EXAMPLE 14.4
Determining the Exponents of a Rate Law

Isoprene is used industrially to make a product identical with natural rubber.

Isoprene forms a dimer called dipentene. (A *dimer* is a compound formed by joining two simpler molecules.)

What is the rate law for this reaction given the following data for the initial rate of formation of dipentene?

Initial Concentration of Isoprene (mol L^{-1})	Initial Rate of Formation of Dipentene (mol L^{-1} s^{-1})
0.50	1.98
1.50	17.8

ANALYSIS With only one reactant, we expect the rate law to be of the form

$$\text{Rate} = k[\text{isoprene}]^x$$

To discover x, we must compare the results of the two experiments.

SOLUTION From the first to the second experiment, we see that the isoprene concentration increases by a factor of 3. The *rate* of the reaction, however, increases by a factor of $(17.8/1.98) = 8.99$. This is very nearly a factor of 9, which is 3^2. Therefore, the value of x is 2 and the rate law is

$$\text{Rate} = k[\text{isoprene}]^2$$

EXAMPLE 14.5
Determining the Exponents of a Rate Law

PROBLEM The following data were measured for the reduction of nitric oxide with hydrogen.

$$2NO(g) + 2H_2(g) \longrightarrow N_2(g) + 2H_2O(g)$$

Initial Concentrations (mol L^{-1})		Initial Rate of Formation of H$_2$O (mol L^{-1} s^{-1})
[NO]	[H$_2$]	
0.10	0.10	1.23×10^{-3}
0.10	0.20	2.46×10^{-3}
0.20	0.10	4.92×10^{-3}

What is the rate law for the reaction?

ANALYSIS This time we have two reactants. To see how their concentrations affect the rate we must vary only one concentration at a time. Therefore, we choose two experiments in which the concentration of one reactant doesn't change and examine the effect of a change in the concentration of the other reactant. Then we repeat the procedure for the second reactant.

SOLUTION We expect the rate law to have the form

$$\text{Rate} = k[\text{NO}]^n[\text{H}_2]^m$$

Let's look at the first two experiments. Here the concentration of NO remains the same, so the rate is being affected by the change in the H$_2$ concentration. When we double the H$_2$ concentration, the rate doubles, so the reaction is first order with respect to H$_2$. This means $m = 1$.

Next, we need to pick two experiments in which the H$_2$ concentration doesn't change. Working with the first and third, we see that [NO] doubles and the rate increases by a factor of $4.92/1.23 = 4.00$. When doubling the concentration quadruples the rate, the reaction is second order, so $n = 2$.

Therefore, the rate law for the reaction is

$$\text{Rate} = k[\text{NO}]^2[\text{H}_2]$$

■ **Practice Exercise 6** Ordinary sucrose (table sugar) reacts with water in an acidic solution to produce two simpler sugars, glucose and fructose, that have the same molecular formulas.

$$\text{C}_{12}\text{H}_{22}\text{O}_{11} + \text{H}_2\text{O} \longrightarrow \text{C}_6\text{H}_{12}\text{O}_6 + \text{C}_6\text{H}_{12}\text{O}_6$$
sucrose   glucose   fructose

In a series of experiments, the following data were obtained.

Initial Sucrose Concentration (mol L^{-1})	Rate of Formation of Glucose (mol L^{-1} s^{-1})
0.10	6.17×10^{-5}
0.20	1.23×10^{-4}
0.50	3.09×10^{-4}

What is the order of the reaction with respect to sucrose?

■ **Practice Exercise 7** A certain reaction follows the equation $A + B \rightarrow C + D$. Experiments yielded the following results.

Initial Concentrations (mol L^{-1})		Initial Rate of Formation of C (mol L^{-1} s^{-1})
[*A*]	[*B*]	
0.40	0.30	1.0×10^{-4}
0.80	0.30	4.0×10^{-4}
0.80	0.60	1.6×10^{-3}

What is the rate law for the reaction?

The formation of nitric oxide from nitrogen and oxygen (from the air) is promoted in vehicle engine cylinders by the high temperature and pressure.

The NO in vehicle exhaust, when it reaches the outside air, is oxidized almost instantly to nitrogen dioxide, NO$_2$, which is both a poison and largely responsible for the reddish color of smog.

14.5
CONCENTRATION
AND TIME

The rate law tells us how the speed of a reaction varies with the concentrations of the reactants. Often, however, we wish to have more information about the status of a reaction than how fast it is going. For instance, if we were manufacturing some chemical in a large reaction vessel, we might want to know what the concentrations of the reactants and products were at some specified time after the reaction has started, so we could decide whether it was time to harvest the products. We might like to know how long it would take for the reactant concentrations to drop below some minimum optimum values, so we could replenish them. Obtaining information of this kind requires an expression of some sort that relates concentration to time.

A rate law is actually a differential equation, and obtaining from it the relationship between concentration and time involves performing the mathematical operation called integration. The equations we discuss in this section are therefore often called integrated rate equations.

The relationship between the concentration of a reactant and time can be obtained from the rate law of a reaction, but the derivation of the equation requires calculus, so we will only look at the results. Here it is useful to know the order of the reaction, because the mathematics is always the same for a reaction of a given order. Even so, for complex reactions the mathematical expressions become complicated, so we will only consider very simple cases here, just to give you a taste of the subject.

Rate law, First-order reactions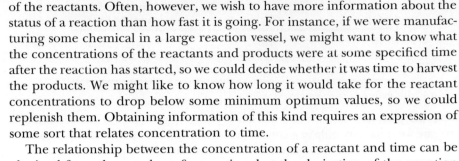

Concentration versus Time for First-Order Reactions

A first order reaction has a rate law of the type

$$\text{Rate} = k[A]$$

The relationship between the concentration of A and time involves natural logarithms, which we discussed in Section 11.7 (page 463).

$$\ln \frac{[A]_0}{[A]_t} = kt \tag{14.5}$$

Logarithms are discussed in Appendix A. Common logarithms are related to natural logarithms by the equation: $\ln x = 2.303 \log x$.

The natural logarithm, ln, of a ratio of concentrations is on the left side of Equation 14.5, where $[A]_t$ is the molar concentration of A at some time t after the start of the reaction and $[A]_0$ is the initial concentration of A ($t = 0$). On the right side we see the product of the rate constant k and time. Equation 14.5 can be used directly in calculations, as illustrated by the following examples.

EXAMPLE 14.6
Concentration–Time
Calculations for First-Order
Reactions

N_2O_5 is the anhydride of nitric acid; it reacts with water as follows: $N_2O_5 + H_2O \longrightarrow 2HNO_3$

Dinitrogen pentaoxide, N_2O_5, is not very stable, and in the gas phase or in solution with a nonaqueous solvent it decomposes by a first-order reaction into N_2O_4 and O_2. The rate law is

$$\text{Rate} = k[N_2O_5]$$

At 45 °C, the rate constant for the reaction in carbon tetrachloride is $6.22 \times 10^{-4}\ \text{s}^{-1}$. If the initial concentration of the N_2O_5 in the solution is 0.100 M, how many minutes will it take for the concentration to drop to 0.0100 M?

ANALYSIS To solve the problem, we simply substitute quantities into Equation 14.5 and solve for t.

SOLUTION First, let's assemble the data.

$$[N_2O_5]_0 = 0.100\ M \qquad\qquad [N_2O_5]_t = 0.0100\ M$$

$$k = 6.22 \times 10^{-4}\ \text{s}^{-1} \qquad\qquad t = ?$$

Substituting,

$$\ln\left(\frac{0.100\ M}{0.0100\ M}\right) = (6.22 \times 10^{-4}\ s^{-1})t$$

$$\ln(10.0) = (6.22 \times 10^{-4}\ s^{-1})t$$

Using a scientific calculator to find the natural logarithm[2] of 10.0

$$2.303 = (6.22 \times 10^{-4}\ s^{-1})t$$

Solving for t,

$$t = \frac{2.303}{6.22 \times 10^{-4}\ s^{-1}}$$

$$= 3.70 \times 10^3\ s$$

This is the time in seconds. Dividing by 60 gives the time in minutes, 61.7 min.

If the initial concentration of N_2O_5 in a carbon tetrachloride solution at 45 °C is 0.500 M, what will its concentration be after exactly 1 hour?

ANALYSIS This time we have to solve for an unknown concentration, which is within the logarithm expression. The easiest way to do this is to first solve for the *ratio* of the concentrations. Once this ratio is known, we can then substitute the known concentration and calculate the unknown one. Remembering that the unit of k involves seconds, not hours, we must convert the given 1 hr into seconds (1 hr = 3600 s).

SOLUTION Let's begin by listing the data.

$$[N_2O_5]_0 = 0.500\ M \qquad [N_2O_5]_t = ?\ M$$

$$k = 6.22 \times 10^{-4}\ s^{-1} \qquad t = 3600\ s$$

Now we solve for the concentration ratio

$$\ln\left(\frac{[N_2O_5]_0}{[N_2O_5]_t}\right) = (6.22 \times 10^{-4}\ s^{-1}) \times 3600\ s$$

$$= 2.24$$

To take the antilogarithm (antiln), we raise e to the 2.24 power.

$$\text{antiln}\left[\ln\left(\frac{[N_2O_5]_0}{[N_2O_5]_t}\right)\right] = \frac{[N_2O_5]_0}{[N_2O_5]_t}$$

$$\text{antiln}(2.24) = e^{2.24}$$

$$= 9.4$$

This means that

$$\frac{[N_2O_5]_0}{[N_2O_5]_t} = 9.4$$

Now we can substitute the known concentration, $[N_2O_5]_0 = 0.500\ M$. This gives

$$\frac{0.500\ M}{[N_2O_5]_t} = 9.4$$

EXAMPLE 14.7
Concentration–Time Calculations for First-Order Reactions

[2] There are special rules for significant figures for logarithms. When taking a logarithm of a quantity, the number of digits written *after the decimal point* equals the number of significant figures in the quantity. Here, 10.0 has three significant figures, so the logarithm of 10.0 has three digits after the decimal.

FIGURE 14.4

(*a*) A graph of concentration versus time for the decomposition of N_2O_5 at 45 °C. (*b*) A straight line is obtained if the logarithm of the concentration is plotted versus time. The slope of the line equals the negative of the rate constant for the reaction.

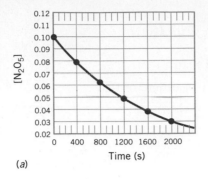

(a)

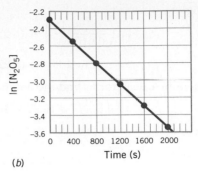

(b)

If you were to do this calculation and find the final concentration to be *larger* than the initial one, you should know you'd made a mistake. Reactants are consumed during a reaction, so their concentrations decrease.

Solving for $[N_2O_5]_t$ gives

$$[N_2O_5]_t = \frac{0.500\ M}{9.4}$$
$$= 0.053\ M$$

Thus, after one hour, the concentration of N_2O_5 will have dropped to 0.053 *M*.

■ **Practice Exercise 8** In Practice Exercise 6, the reaction of sucrose with water in an acidic solution was described.

$$\underset{\text{sucrose}}{C_{12}H_{22}O_{11}} + H_2O \longrightarrow \underset{\text{glucose}}{C_6H_{12}O_6} + \underset{\text{fructose}}{C_6H_{12}O_6}$$

The reaction is first order with a rate constant of $6.2 \times 10^{-5}\ s^{-1}$ at 35 °C, when the H^+ concentration is 0.10 *M*. Suppose, in an experiment, the initial sucrose concentration was 0.40 *M*. (a) What will its concentration be after exactly 2 hours? (b) How many minutes will it take for the concentration of sucrose to drop to 0.30 *M*?

The logarithmic relationship between concentration and time for a first-order reaction (Equation 14.5) provides an elegant way to determine accurately the rate constant by graphical means. Using the properties of logarithms,[3] Equation 14.5 can be rewritten in a form that corresponds to an equation for a straight line.

The equation for a straight line is usually written

$$y = mx + b$$

where *x* and *y* are variables, *m* is the slope, and *b* is the intercept of the line with the *y* axis.

$$\ln[A]_t = -k\ t\ + \ln[A]_0$$
$$\updownarrow \qquad \updownarrow\ \updownarrow \qquad \updownarrow$$
$$y\ =\ m\ x\ +\ \ b$$

Thus, a graph of $\ln[A]_t$ (the vertical axis) versus *t* (the horizontal axis) should give a straight line that has a slope equal to $-k$, as illustrated in Figure 14.4 for the decomposition of N_2O_5 into N_2O_4 and O_2 in the solvent CCl_4.

Concentration versus Time for Second-Order Reactions

Rate law, Second-order reactions

For the sake of simplicity, we will only consider the relationship between concentration and time for a second-order reaction of the kind that has a rate law of the following type.

$$\text{Rate} = k[B]^2$$

---

[3] The logarithm of a quotient, $\ln \dfrac{a}{b}$, can be written as the difference $\ln a - \ln b$.

The equation is rather different from that for a first-order reaction.

$$\frac{1}{[B]_t} - \frac{1}{[B]_0} = kt \qquad (14.6)$$

We have used the same notation as before; $[B]_0$ is the initial concentration of B and $[B]_t$ is the concentration after a time t.

The following example illustrates how Equation 14.6 is applied to calculations.

Nitrosyl chloride, NOCl, is a very corrosive, reddish-yellow gas that is intensely irritating to the eyes and skin. It decomposes slowly to NO and Cl_2 by a second-order reaction with the rate law

$$\text{Rate} = k[\text{NOCl}]^2$$

At a certain temperature, the rate constant for the reaction is $k = 0.020$ L mol^{-1} s^{-1}. If the initial concentration of NOCl in a reaction vessel is 0.050 M, what will the concentration be after 30 minutes?

ANALYSIS We're given a rate law and so can see that it is for a second-order reaction and has the simple form to which our study is limited. Thus, we conclude that we must calculate the molar concentration of NOCl after 30 min (1800 s), $[\text{NOCl}]_t$, by Equation 14.6.

SOLUTION Let's begin by tabulating the data.

$$[\text{NOCl}]_0 = 0.050\ M \qquad [\text{NOCl}]_t = ?\ M$$
$$k = 0.020\ \text{L mol}^{-1}\text{s}^{-1} \qquad t = 1800\ \text{s}$$

The equation we wish to substitute into is

$$\frac{1}{[\text{NOCl}]_t} - \frac{1}{[\text{NOCl}]_0} = kt$$

Making the substitutions gives

$$\frac{1}{[\text{NOCl}]_t} - \frac{1}{0.050\ \text{mol L}^{-1}} = (0.020\ \text{L mol}^{-1}\text{s}^{-1}) \times (1800\ \text{s})$$

Solving for $1/[\text{NOCl}]_t$ gives

$$\frac{1}{[\text{NOCl}]_t} - 20\ \text{L mol}^{-1} = 36\ \text{L mol}^{-1}$$

$$\frac{1}{[\text{NOCl}]_t} = 56\ \text{L mol}^{-1}$$

Taking the reciprocals of both sides gives us the value of $[\text{NOCl}]$.

$$[\text{NOCl}]_t = \frac{1}{56\ \text{L mol}^{-1}}$$
$$= 0.018\ \text{mol L}^{-1}$$
$$= 0.018\ M$$

Thus, the molar concentration of NOCl has decreased from 0.050 M to 0.018 M after 30 minutes.

■ **Practice Exercise 9** For the reaction in the preceding example, determine how many minutes it would take for the NOCl concentration to drop from 0.040 M to 0.010 M.

EXAMPLE 14.8
Concentration–Time Calculations for Second-Order Reactions

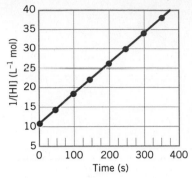

FIGURE 14.5

A graph of $1/[HI]$ versus time for the data in Table 14.1. The slope of the line equals the rate constant for the reaction.

Fast reactions are characterized by large values of k and small values of $t_{1/2}$.

The rate constant k for a second-order reaction, one with a rate following Equation 14.6, can be determined graphically by a method similar to that used for a first-order reaction. Thus, Equation 14.6 can be rearranged as follows to correspond to an equation for a straight line.

$$\frac{1}{[B]_t} = kt + \frac{1}{[B]_0}$$

$$\Updownarrow \quad \Updownarrow\Updownarrow \quad \Updownarrow$$

$$y = mx + b$$

Therefore, when a reaction is second order, a plot of $1/[B]_t$ versus time should yield a straight line with a slope equal to k. Such a graphical solution for a rate constant, using the data in Table 14.1 for the decomposition of HI, is shown in Figure 14.5.

Half-Lives for First- and Second-Order Reactions

A concept that provides a convenient measure of the speed of a reaction, especially for first-order processes, is that of **half-life, $t_{1/2}$,** which is the amount of time required for half of a given reactant to disappear. When the half-life is short, a reaction is rapid because half of the reactant disappears quickly. For any reaction, the half-life is related to the rate constant, but the relationships vary with the order of the reaction. For all first-order reactions there is one equation, and there is a different equation for all second-order reactions.

Half lives

Half-Lives of First-Order Reactions   When a reaction is first order, its half-life can be obtained from Equation 14.5 by setting $[A]_t$ equal to one-half of $[A]_0$.

$$[A]_t = \frac{1}{2}[A]_0$$

Substituting and solving for t, which becomes $t_{1/2}$, gives

Because ln 2 = 0.693, Equation 14.7 is sometimes written

$$t_{1/2} = \frac{0.693}{k}$$

$$t_{1/2} = \frac{\ln 2}{k} \tag{14.7}$$

Because k is a constant for a given reaction, the half-life is also a constant for any particular first-order reaction (at any given temperature). Remarkably, in other words, *the half-life of a first-order reaction is not affected by the initial concentration of the reactant.* This can be illustrated by one of the most common first-order events in nature, the change that radioactive isotopes undergo during radioactive "decay." In fact, you have probably heard the term *half-life* used in reference to the life span of radioactive substances.

Iodine-131, an unstable, radioactive isotope of iodine, undergoes a nuclear reaction that causes it to emit a type of radiation while it is transformed into a stable isotope of xenon.[4] The intensity of the radiation decreases, or *decays,* with time (Figure 14.6). Notice that the time it takes for the first half of the ^{131}I to disappear is 8 days. Then, during the next 8 days half of the remaining ^{131}I disappears, and so on. Regardless of the initial amount, it takes 8 days for half

---

[4] Iodine-131 is used in the diagnosis and treatment of thyroid disorders. The thyroid gland is the only part of the body that uses iodide ions. When a patient is given a dose of $^{131}I^-$ mixed with nonradioactive I^-, both are absorbed by the thyroid. This allows for the testing of thyroid activity and is also used to treat certain types of thyroid cancer.

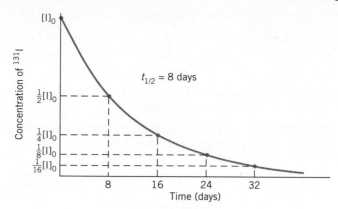

FIGURE 14.6

First-order radioactive decay of iodine-131. The initial concentration of the isotope is represented by $[I]_0$.

of that amount of ^{131}I to disappear, which means that the half-life of ^{131}I is a constant.

Half-Lives of Second-Order Reactions   The half-life of a second-order reaction *does* depend on the initial concentrations of the reactants. We can see this by examining Figure 14.3 (page 590), which follows the decomposition of gaseous HI, a second-order reaction. The reaction begins with a hydrogen iodide concentration of 0.10 *M*. After 125 seconds, the concentration of HI drops to 0.050 *M*, so 125 s is the observed half-life when the initial concentration of HI is 0.050 *M*. If we then take 0.050 *M* as the next "initial" concentration, we find that it takes 250 s (at a *total* elapsed time of 375 seconds) to drop to 0.025 *M*. Thus, halving the initial concentration, from 0.10 *M* to 0.05 *M*, causes a doubling of the half-life, from 125 to 250 seconds. For a second-order reaction like the decomposition of HI, the half-life is inversely proportional to the initial concentration of the reactant and is related to the rate constant by Equation 14.8.

$$t_{1/2} = \frac{1}{k \times \text{(initial concentration of reactant)}}$$ (14.8)

This equation comes from Equation 14.6 by letting $[B]_t = \frac{1}{2}[B]_0$.

The half-life of radioactive iodine-131 is 8.0 days. What fraction of the initial iodine-131 would be present in a patient after 24 days if none of it were eliminated through natural body processes?

ANALYSIS Having learned that iodine-131 is a radioactive isotope and so decays by a first-order process, we know that the half-life is constant and does not depend on the ^{131}I concentration.

SOLUTION A period of 24 days is exactly three half-lives. If we take the fraction initially present to be 1, we can set up a table

Half-life	0	1	2	3
Fraction	1	$\frac{1}{2}$	$\frac{1}{4}$	$\frac{1}{8}$

Half of the iodine-131 is lost in the first half-life, half of that disappears in the second half-life, and so on. Therefore, the fraction remaining after three half-lives is $\frac{1}{8}$.

EXAMPLE 14.9
Half-Life Calculations

The fraction remaining after n half-lives is $\left(\frac{1}{2}\right)^n$, or simply $\frac{1}{2^n}$.

EXAMPLE 14.10
Half-Life Calculations

The reaction $2HI(g) \rightarrow H_2(g) + I_2(g)$ has the rate law, rate = $k[HI]^2$, with $k = 0.079$ L mol^{-1} s^{-1} at 508 °C. What is the half-life for this reaction at this temperature when the initial HI concentration is 0.050 M?

ANALYSIS The rate law tells us that the reaction is second order. To calculate the half-life, we need to use Equation 14.8.

SOLUTION The initial concentration is 0.050 mol L^{-1}; $k = 0.079$ L mol^{-1} s^{-1}. Substituting these values into Equation 14.8 gives

$$t_{1/2} = \frac{1}{(0.079 \text{ L mol}^{-1} \text{ s}^{-1})(0.050 \text{ mol L}^{-1})}$$

$$= 250 \text{ s (rounded)}$$

■ **Practice Exercise 10** In Practice Exercise 6, the reaction of sucrose with water was found to be first order with respect to sucrose. The rate constant for the reaction under the conditions of the experiments was 6.17×10^{-4} s^{-1}. Calculate the value of $t_{1/2}$ for this reaction in minutes. How many minutes would it take for three-quarters of the sucrose to react? (Hint: What fraction of the sucrose remains?)

■ **Practice Exercise 11** Suppose that the value of $t_{1/2}$ for a certain reaction was found to be independent of the initial concentration of the reactants. What could you say about the order of the reaction?

14.6
THEORIES ABOUT REACTION RATES

One reason that a sustained high body temperature is dangerous is because the oxygen-consuming reactions in the body are speeded up, which places demands on the heart to pump harder so as to accelerate the delivery of oxygen to tissues.

In Section 14.2 we mentioned that nearly all reactions proceed faster at higher temperatures. As a rule, the reaction rate increases by a factor of about 2 or 3 for each 10 °C rise in temperature, although the actual amount of increase differs from one reaction to another. To understand why temperature so greatly affects reaction rates we have to imagine what actually happens to the molecules in a reaction system. Stated another way, we need to develop some theoretical models that explain our observations. One of the simplest models is called *collision theory*.

Collision Theory

The basic postulate of **collision theory** is that the rate of a reaction is proportional to the number of effective collisions per second among the reactant molecules. By an *effective collision* we mean one that actually results in the production of product molecules. Anything that can increase the frequency of such collisions should increase the rate. One of the several factors that influences the number of effective collisions per second is the *concentrations* of the reactants. As they increase, the number of collisions per second cannot help but increase. We'll return to the significance of concentration in Section 14.8.

What we wish to focus on in this section is that not *every* collision between reactant molecules actually leads to a chemical change. We know this is true because in a gas or a liquid the reactants undergo an enormous number of collisions per second, and if each were effective at producing a chemical change, all reactions would be over in an instant. Therefore, for most reactions to occur at the nonexplosive rates they do, *only a very small fraction of all the collisions that take place really lead to a net change.* Why is this so?

At the start of the reaction described by Figure 14.2, only about one of every billion billion (10^{18}) collisions leads to a net chemical reaction. In each of the other collisions, the reactant molecules just bounce off each other.

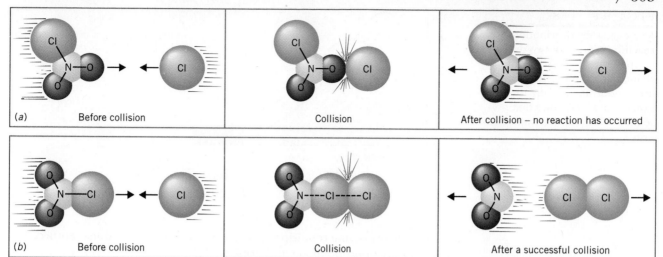

FIGURE 14.7

The importance of molecular orientation during a collision. (*a*) A collision that cannot produce a Cl_2 molecule. (*b*) A collision that can produce NO_2 and Cl_2 molecules.

The Importance of Molecular Orientation In most reactions, when two reactant molecules collide they must be oriented correctly in order for a reaction to occur. For example, the reaction

$$2NO_2Cl \longrightarrow 2NO_2 + Cl_2$$

appears to proceed by a two-step mechanism, one step of which involves the collision of an NO_2Cl molecule with a chlorine atom.

$$NO_2Cl + Cl \longrightarrow NO_2 + Cl_2$$

The orientation of the NO_2Cl molecule when hit by the Cl atom is important (see Figure 14.7). The poor orientation shown in Figure 14.7*a* cannot result in the formation of Cl_2 because the two Cl atoms are not being brought close enough together for a new Cl—Cl bond to form as an N—Cl bond breaks. Figure 14.7*b* shows the necessary orientation if the collision of NO_2Cl and Cl is to be effective (lead to products).

The Importance of Molecular Kinetic Energy The major reason so few collisions actually lead to chemical change is that even in correctly oriented collisions a certain minimum combined kinetic energy of the colliding particles, called the **activation energy, E_a,** must be involved. For a large fraction of all chemical reactions, the activation energy is quite large, and at ordinary temperatures only a small fraction of all well oriented, colliding molecules are traveling fast enough to produce this minimum combined kinetic energy of collision.

The reason for the existence of activation energy can be understood if we examine in detail what actually takes place during a collision. Any chemical change involves a reorganization of chemical bonds; old bonds break as new ones form. For this to happen during a collision, the nuclei of the reacting particles must not only be in the right locations relative to each other but they must also get close enough together. This requires that the molecules on collision course be moving with a combined kinetic energy high enough to overcome the natural repulsions between electron clouds. Otherwise, the molecules simply veer or bounce apart. Only fast-moving molecules with large kinetic energies can collide with enough force to enable their nuclei and

FIGURE 14.8

Kinetic energy distributions for a reaction mixture at two different temperatures. The sizes of the shaded areas under the curves are proportional to the total fractions of the molecules that possess the minimum activation energy.

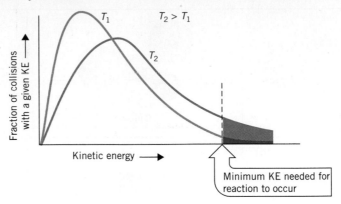

electrons to overcome repulsions and thereby reach the positions required for the necessary bond breaking and bond making.

With the concept of activation energy, we can now explain why the rate of a reaction increases so much with increasing temperature. We'll use the two plots in Figure 14.8, each plot corresponding to a different temperature for the same mixture of reactants. Each curve is a plot of the different *fractions* of all collisions (vertical axis) that have particular values of kinetic energy of collision (horizontal axis). (The total area under a curve represents the total number of collisions because all of the fractions must add up to this total.) Notice what happens to the plots when the temperature is increased. The maximum point shifts to the right and the curve flattens out, which indicates that the average energy of collision increases. However, the minimum kinetic energy needed for an effective collision—the activation energy—is about the same regardless of the temperature.

The tinted areas under the curves in Figure 14.8 represent the sum of all those fractions of the total collisions that have or exceed the activation energy. This sum—we could call it the *reacting fraction*—is relatively much greater at the higher temperature than at the lower temperature. As a result, a much greater fraction of the collisions occurring each second results in a chemical change, so the reactants disappear faster at the higher temperature.

Transition State Theory

Transition state theory is used to explain in detail what happens when reactant molecules come together in a collision. Most often, molecules in a head-on collision slow down, stop, and then fly apart again, unchanged. When a reaction occurs during the collision, however, the particles that separate are chemically different from those that came together. Regardless of what happens to them chemically, however, as the molecules slow down, the total kinetic energy that they possess decreases. Because energy can't disappear, their total potential energy (PE) must increase as their kinetic energy decreases.

The relationship between the activation energy and the total potential energy of the reactants and products can be expressed graphically by a potential-energy diagram (see Figure 14.9). The horizontal axis in Figure 14.9 is called the **reaction coordinate** and represents the extent to which the reactants have changed to the products. It follows the path taken by the reaction as reactant molecules come together and change into product molecules, which separate after the collision. The activation energy appears as a potential energy "hill" between the reactants and products. Only colliding molecules that have com-

The law of conservation of energy requires that the total energy, the sum of KE and PE, be constant in a collision.

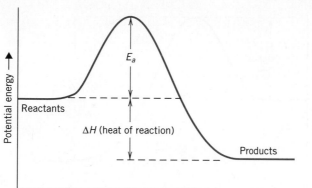

FIGURE 14.9

Potential-energy diagram for an exothermic reaction.

bined kinetic energies at least as large as E_a are able to climb over the hill and produce products.

We can use the potential-energy diagram for a reaction to follow the progress of both an unsuccessful and a successful collision (see Figure 14.10). As two reactant molecules collide, they slow down and their kinetic energy is changed to potential energy; we say that they begin to climb the potential-energy barrier toward the products. If their combined initial kinetic energies are less than E_a, they are unable to reach the top of the hill (Figure 14.10a). Instead, they fall back toward the reactants. They bounce apart chemically unchanged and with their original total kinetic energy; no net reaction has occurred. On the other hand, if the combined kinetic energies of the molecules in collision equal or exceed E_a, and if the molecules are oriented properly, they are able to pass over the activation-energy barrier and form product molecules (Figure 14.10b).

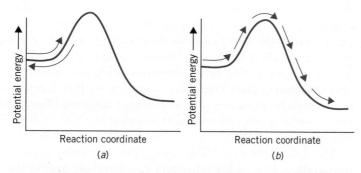

FIGURE 14.10

(*a*) An unsuccessful collision: Molecules separate unchanged. (*b*) A successful collision: The activation-energy barrier is crossed and the products are formed.

Going back to Figure 14.9, we can also see what is meant by the *heat of reaction*—the difference between the potential energy of the products and the potential energy of the reactants. Figure 14.9 is for an *exothermic* reaction because the products have a *lower* potential energy than the reactants. The net decrease in potential energy accompanying a successful collision appears as an increase in the kinetic energies of the emerging product molecules. Thus, the temperature of the system increases during an exothermic reaction because the average molecular kinetic energy of the system increases.

A potential-energy diagram for an endothermic reaction is shown in Figure 14.11. Now the products have a *higher* potential energy than the reactants and, in terms of the heat of reaction, a net input of energy is needed to form the products. Endothermic reactions produce a cooling effect as they proceed because there is a net conversion of molecular kinetic energy to potential energy. As the total molecular kinetic energy decreases, the average molecular kinetic energy decreases as well, and the temperature drops.

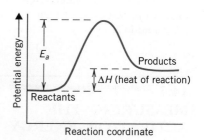

FIGURE 14.11

A potential-energy diagram for an endothermic reaction.

FIGURE 14.12

Activation-energy barrier for the forward and reverse reactions.

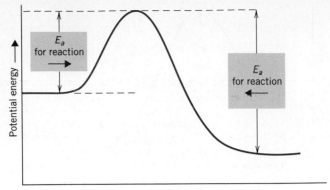

In Chapter 5 we saw that when the direction of a reaction is reversed, the sign given to the enthalpy change, ΔH, is reversed; a reaction that is exothermic in the forward direction is endothermic in the reverse direction, and vice versa. This seems to suggest that reactions are generally reversible. Many are, but if we look again at the energy diagram for a reaction that is exothermic in the forward direction (Figure 14.9), it is obvious that in the opposite direction the reaction is endothermic *and has a significantly higher activation energy* than the forward reaction. What differs most for the forward and reverse directions is the relative height of the activation-energy barrier (Figure 14.12). When too high, a reverse reaction becomes too slow to be of any importance and often cannot be observed.

One of the main reasons for studying activation energies is that they provide information about what actually occurs during an effective collision. For example, in Figure 14.7*b* on page 605, we described a way that NO_2Cl could react successfully with a Cl atom during a collision. During this collision, there is a moment when the N—Cl bond is partially broken and the Cl—Cl bond is partially formed. This brief moment during the reaction is called the reaction's **transition state.** The potential energy of the transition state corresponds to the high point on the potential-energy diagram (Figure 14.13). The unstable chemical species that exists at this instant, $O_2N\cdots Cl\cdots Cl$, with its partially formed and partially broken bonds, is called the **activated complex.**

The size of the activation energy provides information about the relative importance of bond breaking and bond making during the formation of the activated complex. A very high activation energy suggests, for instance, that bond breaking contributes very heavily to the formation of the activated complex because bond breaking is an energy-absorbing process. On the other hand, a small activation energy may mean that bonds of about equal strength are being both broken and formed simultaneously.

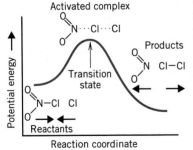

FIGURE 14.13

Formation of an activated complex in the reaction between NO_2Cl and Cl.

14.7 MEASURING THE ACTIVATION ENERGY

As we have said, changing the temperature alters the rate of a reaction and so changes the value of the rate constant, k. The effect can be large, but how much k changes with temperature depends on the magnitude of the activation energy. When the activation energy is very large, for example, even a small increase in the temperature causes a substantial percentage increase in the number of molecules having sufficient energy to react, so a large change in the rate is observed.

Quantitatively, the activation energy is related to the rate constant by a relationship discovered in 1889 by Svante Arrhenius, whose name you may recall from our discussion of electrolytes and acids and bases in Chapter 4. The usual form of the **Arrhenius equation** is

$$k = Ae^{-E_a/RT} \tag{14.9}$$

 Arrhenius equation

where k is the rate constant, A is a proportionality constant sometimes called the **frequency factor,** e is the base of the natural logarithm system, R is the gas constant expressed in energy units ($R = 8.314 \, J \, mol^{-1} \, K^{-1}$), and T is the absolute temperature. When we evaluate the $e^{-E_a/RT}$ term, remembering that E_a/RT appears as a *negative* exponent, we see that an increase in T makes the entire $e^{-E_a/RT}$ term *larger* and so makes the value of k larger. It is largely because T occurs in an *exponential* function in the Arrhenius equation that small changes in T can mean large changes in rate. This is particularly true when the value of E_a is itself large.

Use your scientific calculator to determine some changes in the value of $e^{-E_a/RT}$ at different values of T. You might set E_a at 5.0×10^3 kJ mol^{-1} and see how $e^{-E_a/RT}$ changes between 298 K and 308 K, only a 10 degree difference.

Determining the Activation Energy Graphically

The way Equation 14.9 is normally used is in its logarithmic form. If we take the natural logarithm of both sides, we obtain

$$\ln k = \ln A - E_a/RT$$

Let's rewrite the equation as

$$\ln k = \ln A - (E_a/R) \cdot (1/T) \tag{14.10}$$

We know that the rate constant k varies with the temperature T, which also means that the quantity $\ln k$ varies with the quantity $(1/T)$. These two quantities—$\ln k$ and $1/T$—are variables, and once again we have an equation for a straight line.

Thus, to determine the activation energy, we can make a graph of $\ln k$ versus $1/T$, measure the slope of the line, and then use the relationship

$$\text{Slope} = -E_a/R$$

to calculate E_a. Example 14.11 illustrates how this is done.

The following data were collected for the reaction

$$2NO_2(g) \longrightarrow 2NO(g) + O_2(g)$$

Rate Constant, k (L mol^{-1} s^{-1})	Temperature (°C)
7.8	400
10	410
14	420
18	430
24	440

Determine the activation energy for the reaction in kilojoules per mole.

EXAMPLE 14.11
Determining Energy of Activation Graphically

ANALYSIS Equation 14.10 applies, but the use of rate data to determine the activation energy graphically requires that we plot ln k, not k, versus the *reciprocal* of the *Kelvin* temperature, so we have to convert the given data into ln k and $1/T$ before we can make the graph.

SOLUTION To illustrate, using the first set of data, the conversions are

$$\ln k = \ln(7.8) = 2.05$$

$$\frac{1}{T} = \frac{1}{(400 + 273)\ \text{K}} = \frac{1}{673\ \text{K}}$$

$$= 1.486 \times 10^{-3}\ \text{K}^{-1}$$

We are carrying extra "significant figures" for the purpose of graphing the data. The remaining conversions give the following table.

ln k	$1/T$ (K^{-1})
2.05	1.486×10^{-3}
2.30	1.464×10^{-3}
2.64	1.443×10^{-3}
2.89	1.422×10^{-3}
3.18	1.403×10^{-3}

Then we plot ln k versus $1/T$ as shown in the figure below.

There is a statistical method for computing the slope of the straight line that best fits the data. Ordinarily, this is what chemists would use.

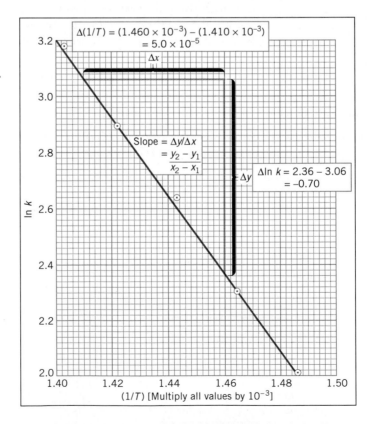

The slope of the curve is obtained as the ratio

$$\text{slope} = \frac{\Delta(\ln k)}{\Delta(1/T)}$$

$$= \frac{-0.70}{5.0 \times 10^{-5} \text{ K}^{-1}}$$

$$= -1.4 \times 10^4 \text{ K} = -E_a/R$$

After changing signs and solving for E_a we have

$$E_a = (8.314 \text{ J mol}^{-1} \text{ K}^{-1})(1.4 \times 10^4 \text{ K})$$
$$= 1.2 \times 10^5 \text{ J mol}^{-1}$$
$$= 1.2 \times 10^2 \text{ kJ mol}^{-1}$$

Calculating the Activation Energy

In obtaining an activation energy, sometimes we may not wish to go to the bother of graphing data, or we may have just two rate constants measured at two temperatures. In such cases the activation energy can be obtained algebraically. From Equation 14.9, the following useful relationship can be derived.

$$\ln\left(\frac{k_2}{k_1}\right) = \frac{-E_a}{R}\left(\frac{1}{T_2} - \frac{1}{T_1}\right) \qquad (14.11)$$

 Arrhenius equation, Alternate form

The following example illustrates the use of this equation.

The decomposition of HI has rate constants $k = 0.079$ L mol^{-1} s^{-1} at 508 °C and $k = 0.24$ L mol^{-1} s^{-1} at 540 °C. What is the activation energy of this reaction in kJ mol^{-1}?

EXAMPLE 14.12
Calculating an Energy of Activation from Two Rate Constants

ANALYSIS There are insufficient data for preparing a plot of ln k versus $1/T$, but we have enough data to use Equation 14.11 and thus to find E_a algebraically.

SOLUTION Let's begin by organizing the data; a small table will be helpful. Choose one of the rate constants as k_1 (it doesn't matter which one) and then fill in the table.

	k (L mol^{-1} s^{-1})	T (K)
1	0.079	508 + 273 = 781
2	0.24	540 + 273 = 813

We also have $R = 8.314$ J mol^{-1} K^{-1}. Substituting into Equation 14.11 gives us

$$\ln\left(\frac{0.24 \text{ L mol}^{-1} \text{ s}^{-1}}{0.079 \text{ L mol}^{-1} \text{ s}^{-1}}\right) = \frac{-E_a}{8.314 \text{ J mol}^{-1} \text{ K}^{-1}}\left(\frac{1}{813 \text{ K}} - \frac{1}{781 \text{ K}}\right)$$

$$\ln(3.0) = \frac{-E_a}{8.314 \text{ J mol}^{-1} \text{ K}^{-1}}(0.001230 \text{ K}^{-1} - 0.001280 \text{ K}^{-1})$$

$$1.10 = \frac{E_a}{8.314 \text{ J mol}^{-1}}(0.000050)$$

Multiplying both sides by 8.314 J mol^{-1} gives

$$9.15 \text{ J mol}^{-1} = E_a(0.000050)$$

Solving for E_a, we have

$$E_a = 1.8 \times 10^5 \text{ J mol}^{-1}$$

Converting to kilojoules

$$E_a = 1.8 \times 10^2 \text{ kJ mol}^{-1}$$

EXAMPLE 14.13
Calculating the Rate Constant at a Particular Temperature

The reaction $2NO_2 \rightarrow 2NO + O_2$ has an activation energy of 111 kJ/mol. At 400 °C, $k = 7.8$ L mol^{-1} s^{-1}. What is the value of k at 430 °C?

ANALYSIS This time we know the activation energy and k at one temperature. We will need to use Equation 14.11 to obtain k at the other temperature. Since the logarithm term contains the ratio of the rate constants, we will solve for the value of this ratio, substitute the known value of k, and then solve for the unknown k.

SOLUTION Let's begin by writing Equation 14.11.

$$\ln \left(\frac{k_2}{k_1} \right) = \frac{-E_a}{R} \left(\frac{1}{T_2} - \frac{1}{T_1} \right)$$

Organizing the data gives us the following table.

	k (L mol^{-1} s^{-1})	T (K)
1	7.8	400 + 273 = 673 K
2	?	430 + 273 = 703 K

We must use $R = 8.314$ J mol^{-1} K^{-1} and express E_a in joules ($E_a = 1.11 \times 10^5$ J mol^{-1}). Next, we substitute values into the right side of the equation and solve for $\ln(k_2/k_1)$.

$$\ln \left(\frac{k_2}{k_1} \right) = \frac{-1.11 \times 10^5 \text{ J mol}^{-1}}{8.314 \text{ J mol}^{-1} \text{ K}^{-1}} \left(\frac{1}{703 \text{ K}} - \frac{1}{673 \text{ K}} \right)$$

$$= (-1.34 \times 10^4 \text{ K})(-6.3 \times 10^{-5} \text{ K}^{-1})$$

Therefore,

$$\ln \left(\frac{k_2}{k_1} \right) = 0.84$$

Taking the antilog gives the ratio of k_2 to k_1.

$$\frac{k_2}{k_1} = e^{0.84} = 2.3$$

As a quick check, we note that the calculated k for the higher temperature is greater than the k for the lower temperature, as it should be.

Solving for k_2,

$$k_2 = 2.3k_1$$

Substituting the value of k_1 from the data table gives

$$k_2 = 2.3(7.8 \text{ L mol}^{-1} \text{ s}^{-1})$$
$$= 18 \text{ L mol}^{-1} \text{ s}^{-1}$$

■ **Practice Exercise 12** The reaction $CH_3I + HI \rightarrow CH_4 + I_2$ was observed to have rate constants $k = 3.2$ L mol^{-1} s^{-1} at 350 °C and $k = 23$ L mol^{-1} s^{-1} at 400 °C. (a) What is the value of E_a for this reaction expressed in kJ/mol? (b) What would the rate constant for this reaction be at 300 °C?

At the beginning of this chapter we mentioned that most reactions do not occur in a single step. Instead, the net overall reaction is the result of a series of simpler, one-step reactions, each of which is called an *elementary process.* An **elementary process** is defined as a reaction whose rate is proportional to the product of the concentrations of the species *that occur as reactants in that step.* The entire series of elementary processes is called the reaction's **mechanism.** Remember that the exponents of the concentration terms for the rate law of an *overall* reaction bear no *necessary* relationship to the coefficients in the overall balanced equation; the exponents in the overall rate law must be determined experimentally. What makes an elementary process "elementary" is that such a necessary relationship between coefficients and exponents does exist, as we are about to see.

For most reactions, the individual elementary processes cannot actually be observed; instead, we only see the net reaction. Therefore, the mechanism a chemist writes is really a *theory* about what occurs step by step as the reactants are changed to the products.

Because the individual steps in a mechanism usually can't be observed, arriving at a chemically sensible set of elementary processes for a reaction is not at all a simple task. The task is disciplined, however, by a major overriding test of a mechanism: *The overall rate law derived from the mechanism must agree with the observed rate law for the overall reaction.* Making reasonable guesses about elementary processes requires a lot of "scientific intuition" and much more chemical knowledge than has presently been provided to you, so you won't have to formulate reaction mechanisms on your own. Nevertheless, to understand the science of chemistry better, it is worthwhile knowing how the study of reaction rates can provide clues to a reaction's mechanism.

In Section 14.7 you learned that the basic postulate of the collision theory is that the rate of a reaction is proportional to the number of effective collisions per second between the reactants. You also learned that, for a given set of conditions, the number of effective collisions is only a small fraction of the total. Nevertheless, if the *total* number of collisions were in some way doubled, the number of effective collisions would be doubled too. With this in mind, let's see how we can predict the rate law from postulated elementary processes. The first step in such a prediction is the writing of rate laws for the elementary processes themselves.

Predicting the Rate Law for an Elementary Process

Let's suppose that we know for a fact that a certain collision process takes place during a particular reaction. In Example 14.8 we introduced the decomposition of nitrosyl chloride into nitric oxide, NO, and chlorine.

$$2NOCl \longrightarrow 2NO + Cl_2$$

Suppose that this reaction involves more than one step, one being a collision between a molecule of NOCl and an *atom* of Cl.

$$NOCl + Cl \longrightarrow NO + Cl_2 \qquad (14.12)$$

Such a step is an example of an elementary process. As with any reaction, we anticipate that its rate law will have the form

$$Rate = k[NOCl]^x [Cl]^y$$

Let's see if we can figure out what x and y are, based on our knowledge that this reaction occurs by collisions between NOCl molecules and Cl atoms. To do this, we will imagine varying the amounts of each reactant and observing how these changes affect the rate.

To start, suppose we were able to double the number of NOCl molecules in the container (thereby doubling the NOCl concentration) while keeping the number of Cl atoms the same. There now would be twice as many NOCl molecules with which each Cl could collide, so the number of NOCl-to-Cl collisions should double and the rate of the reaction should double. Our conclusion then, is that doubling the concentration of NOCl should double the rate of this elementary process, which means that in the rate law for this process the NOCl concentration must appear with an exponent of 1 (i.e., $x = 1$).

A similar argument applies to doubling of the Cl concentration in the container while keeping the number of NOCl molecules the same. Each molecule of NOCl would now have twice as many Cl atoms with which to collide, so the NOCl-to-Cl collision frequency should double and the rate should double. Thus, doubling the concentration of Cl doubles the rate, so in the rate law [Cl] should also have an exponent of 1 (i.e., $y = 1$). The rate law for the elementary process is therefore

$$\text{Rate} = k[\text{NOCl}]^1 [\text{Cl}]^1$$

Ordinarily, of course, we don't write exponents that are equal to one. However, this time we've done so to point out that for the *elementary process,* the exponents in its rate law are the *same* as the coefficients of the reactants in the chemical equation for the process.

Let's do another exercise involving another elementary process, one involving collisions between two identical molecules leading directly to the products shown.

$$2\text{NO}_2 \longrightarrow \text{NO}_3 + \text{NO} \qquad (14.13)$$

What can we say about the rate law of this elementary process? If the NO_2 concentration were doubled, there would be *twice* as many individual NO_2 molecules and *each* would have *twice* as many neighbors with which to collide. The number of NO_2-to-NO_2 collisions per second would therefore be increased by a factor of 4, resulting in an increase in the rate by a factor of 4, which is 2^2. Earlier we saw that when the doubling of a concentration leads to a fourfold increase in the rate, the concentration of that reactant is raised to the second power in the rate law. Thus, if Equation 14.13 represents an elementary process, its rate law should be

$$\text{Rate} = k[\text{NO}_2]^2$$

Once again the exponent in the rate law for the *elementary process* is the same as the coefficient in the chemical equation. Thus, our two exercises have illustrated that *the rate law for an elementary process can be predicted from the chemical equation for the process.*

The exponents in the rate law for an elementary process are equal to the coefficients of the reactants in the chemical equation for that elementary process.

Recall from Section 14.4 that when the rate is doubled by doubling the reactant concentration, the exponent on the concentration term is 1.

There are twice as many NO_2 molecules, each of which collides twice as often, so the total number of collisions per second doubly doubles; i.e., it increases by a factor of 4.

Let us emphasize that this rule applies only to **elementary processes.** If all we know is the balanced equation for the overall reaction and we don't know what elementary processes are involved, then the only way we can find the exponents of the rate law is by doing experiments as described earlier.

Predicting Reaction Mechanisms

How does the ability to predict the rate law of an elementary process help chemists predict reaction mechanisms? To answer this question, let's look at two relatively simple reactions and what are believed to be their mechanisms. (There are many more complex reaction systems, and Special Topic 14.1 describes one type that is particularly important.)

First, let's consider the gaseous reaction

$$2NO_2Cl \longrightarrow 2NO_2 + Cl_2 \qquad (14.14)$$

This is found to be a first-order reaction; its experimentally determined rate law is

$$\text{Rate} = k[NO_2Cl]$$

The first question we might ask is: Could the overall reaction (Equation 14.14) occur in a single step by the collision of two NO_2Cl molecules? The answer is no, because then it would be an elementary process and the rate law predicted for it would include the term $[NO_2Cl]^2$. Since the experimental rate law is first order in NO_2Cl, the predicted and experimental rate laws don't agree, so we must look further to find the mechanism of the reaction.

Based on chemical intuition and other information that we won't discuss here, chemists believe the actual mechanism of the reaction in Equation 14.14 is the following two-step sequence of elementary processes.

$$NO_2Cl \longrightarrow NO_2 + Cl$$

$$NO_2Cl + Cl \longrightarrow NO_2 + Cl_2$$

The Cl formed here is called a *reactive intermediate.* We never actually observe the Cl because it reacts so quickly.

Notice that when the two reactions are added the intermediate, Cl, drops out and we obtain the net overall reaction given in Equation 14.14. *Being able to add the elementary processes and thus to obtain the overall reaction is another major test of a mechanism.*

In a multistep mechanism such as this, it is usually found that one step is much slower than the others. In this mechanism, for example, it is believed that the first step is slow and that once a Cl atom is formed, it reacts very rapidly with another NO_2Cl molecule to give the final products.

The slow step in a mechanism is called the **rate-determining step** or **rate-limiting step.** This is because the final products of the overall reaction cannot appear faster than the products of the slow step. In the mechanism above, then, the first reaction is the rate-determining step because the final products can't be formed faster than the rate at which Cl atoms are formed.

The rate-determining step is similar to a slow worker on an assembly line. Regardless of how fast the other workers are, the production rate depends on how quickly the slow worker does his or her job. The factors that control the speed of the rate-determining step therefore also control the overall rate of the reaction. This means that *the rate law for the rate-determining step is directly related to the rate law for the overall reaction.*

SPECIAL TOPIC 14.1 / FREE RADICALS

A **free radical** is a very reactive species that contains one or more unpaired electrons. Examples are chlorine atoms formed when a Cl_2 molecule absorbs a photon (light) of the appropriate energy:

$$Cl_2 + \text{light energy } (h\nu) \longrightarrow 2Cl\cdot$$

(A dot placed next to the symbol of an atom or molecule represents an unpaired electron and indicates that the particle is a free radical.) The reason free radicals are so reactive is because of the tendency of electrons to become paired through the formation of either ions or covalent bonds.

Free radicals are important in many gaseous reactions, including those responsible for the production of photochemical smog in urban areas. Reactions involving free radicals have useful applications, too. Many plastics are made by polymerization reactions that take place by mechanisms that involve free radicals. In addition, free radicals play a part in one of the most important processes in the petroleum industry, *thermal cracking*. This reaction is used to break C—C and C—H bonds in long-chain hydrocarbons to produce the smaller molecules that give gasoline a higher octane rating. An example is the formation of free radicals in the thermal-cracking reaction of butane. When butane is heated to 700–800 °C, one of the major reactions that occurs is

$$CH_3\text{---}CH_2\text{:}CH_2\text{---}CH_3 \xrightarrow{\text{heat}} CH_3CH_2\cdot + CH_3CH_2\cdot$$

The central C—C bond of butane is shown here as a pair of dots, :, rather than the usual dash. When the bond is broken, the electron pair is divided between the two free radicals that are formed. This reaction produces two ethyl radicals, $CH_3CH_2\cdot$.

Free-radical reactions tend to have high initial activation energies because chemical bonds must be broken to form the radicals. Once the free radicals are formed, however, reactions in which they are involved tend to be very rapid. In many cases, a free radical reacts with a reactant molecule to give a product molecule plus another free radical. Reactions that involve such a step are called **chain reactions.**

Many explosive reactions are chain reactions involving free radical mechanisms. One of the most studied reactions of this type is the formation of water from hydrogen and oxygen. The elementary processes involved can be described according to their roles in the mechanism.

The reaction begins with an **initiation step** that gives free radicals.

$$H_2 + O_2 \xrightarrow{\text{hot surface}} 2OH\cdot \qquad \text{(initiation)}$$

The chain continues with a **propagation step,** which produces the product plus another free radical.

$$OH\cdot + H_2 \longrightarrow H_2O + H\cdot \qquad \text{(propagation)}$$

The reaction of H_2 and O_2 is explosive because the mechanism also contains **branching steps.**

$$\left. \begin{array}{l} H\cdot + O_2 \longrightarrow OH\cdot + O\cdot \\ O\cdot + H_2 \longrightarrow OH\cdot + H\cdot \end{array} \right\} \text{branching}$$

Thus, the reaction of one $H\cdot$ with O_2 leads to the net production of two $OH\cdot$ plus an $O\cdot$. Every time an $H\cdot$ reacts with oxygen, then, there is an increase in the number of free radicals in the

---

Because the rate-determining first step is an elementary process, we can predict its rate law from the coefficients of the reactants. The coefficient of NO_2Cl is 1, so the rate law predicted for the first step is

$$\text{Rate} = k[NO_2Cl]$$

Notice that the predicted rate law derived for this two-step mechanism agrees with the experimentally measured rate law. Although this doesn't *prove* that the mechanism is correct, it does provide considerable support for it. From the standpoint of kinetics, therefore, the mechanism is a reasonable one.

The second reaction we will study is the gas phase reaction

$$2NO + 2H_2 \longrightarrow N_2 + 2H_2O \qquad (14.15)$$

which has the experimentally determined rate law

$$\text{Rate} = k[NO]^2[H_2] \qquad (\textit{experimental})$$

We can quickly see that the overall reaction could *not* be an elementary process because if it were, the exponent on $[H_2]$ would be 2. Obviously, a mechanism involving two or more steps must be involved.

For a proposed mechanism to be acceptable, the rate law predicted by the rate-determining step should be the same as the rate law determined experimentally for the overall reaction.

system. The free-radical concentration grows rapidly, and the reaction rate becomes explosively fast.

Chain mechanisms also contain termination steps, which remove free radicals from the system. In the reaction of H_2 and O_2, the wall of the reaction vessel serves to remove $H\cdot$, which tends to halt the chain process.

$$2H\cdot \xrightarrow{\text{wall}} H_2$$

FREE RADICALS AND AGING

Direct experimental evidence also exists for the presence of free radicals in functioning biological systems. These highly reactive species play many roles, but one of the most interesting is their apparent involvement in the aging process. One theory suggests that free radicals attack protein molecules in collagen. Collagen is composed of long strands of fibers of proteins and is found throughout the body, especially in the flexible tissues of the lungs, skin, muscles, and blood vessels. Attack by free radicals seems to lead to cross-linking between these fibers, which stiffens them and makes them less flexible. The most readily observable result of this is the stiffening and hardening of the skin that accompanies aging or too much sunbathing (see Figure 1).

Free radicals also seem to affect fats (lipids) within the body by promoting the oxidation and deactivation of enzymes that are normally involved in the metabolism of lipids. Over a long period, the accumulated damage gradually reduces the efficiency with which the cell carries out its activities. Interestingly, vitamin E appears to be a natural free-radical inhibitor. Diets that are deficient in vitamin E produce effects resembling radiation damage and aging.

FIGURE 1

People exposed to sunlight over long periods, like this woman from Nepal (a small country between India and Tibet), tend to develop wrinkles because ultraviolet radiation causes changes in their skin. This is particularly evident when compared to the smooth skin of the child that she's holding.

Still another theory of aging suggests that free radicals attack the DNA in the nuclei of cells. DNA is the substance that is responsible for directing the chemical activities required for the successful functioning of the cell. Reactions with free radicals cause a gradual accumulation of errors in the DNA, which reduce the efficiency of the cell and can lead, ultimately, to malfunction and cell death.

A chemically reasonable mechanism that yields the correct form for the rate law is the following.

$$2NO + H_2 \longrightarrow N_2O + H_2O \qquad \text{(slow)}$$
$$N_2O + H_2 \longrightarrow N_2 + H_2O \qquad \text{(fast)}$$

The equations add to give the correct overall equation. Further, the molecule N_2O is known (its name is nitrous oxide; it is used as an anesthetic in medicine and dentistry) and it reacts with H_2 to give N_2 and H_2O. If the first step is slow, the coefficients of NO and H_2 in the rate-determining (slow) step predict the rate law

$$\text{Rate} = k[NO]^2[H_2] \qquad (\textit{predicted})$$

Although this matches the experimental rate law, there is a serious flaw in the proposed mechanism—it has an elementary process that involves the simultaneous collision between three molecules, two of which must be NO and one H_2. This is such an unlikely event that if it were really involved in the mechanism, the overall reaction would be extremely slow. Therefore, when chemists propose mechanisms for reactions, they avoid whenever possible elementary processes involving more than two-body **bimolecular collisions.**

Chemists believe the reaction in Equation 14.15 proceeds by the following three-step sequence of bimolecular elementary processes.

$$2NO \rightleftharpoons N_2O_2 \qquad \text{(fast)}$$

$$N_2O_2 + H_2 \longrightarrow N_2O + H_2O \qquad \text{(slow)}$$

$$N_2O + H_2 \longrightarrow N_2 + H_2O \qquad \text{(fast)}$$

In this mechanism the first step is proposed to be a rapidly established equilibrium in which the unstable intermediate N_2O_2 forms in the forward reaction and then quickly decomposes into NO by the reverse reaction. The rate-determining step is the reaction of N_2O_2 with H_2 to give N_2O and a water molecule. The third step is the reaction mentioned above. Once again, notice that the three steps add to give the net overall change.

Since the second step is rate determining, the rate law for the reaction should match the rate law for this step. We predict this to be

$$\text{Rate} = k[N_2O_2][H_2] \qquad (14.16)$$

However, the experimental rate law does not contain the species N_2O_2. Therefore, we must find a way to express the concentration of N_2O_2 in terms of the reactants in the overall reaction. To do this, let's look closely at the first step of the mechanism, which we view as a reversible reaction.

The rate in the forward direction, in which NO is the reactant, is

$$\text{Rate (forward)} = k_f[NO]^2$$

The rate of the reverse reaction, in which N_2O_2 is the reactant, is

$$\text{Rate (reverse)} = k_r[N_2O_2]$$

If we view this as a dynamic equilibrium, then the rate of the forward and reverse reactions are equal, which means that

$$k_f[NO]^2 = k_r[N_2O_2] \qquad (14.17)$$

Since we would like to eliminate $[N_2O_2]$ from the rate law in Equation 14.16, let's solve Equation 14.17 for $[N_2O_2]$.

$$[N_2O_2] = \frac{k_f}{k_r}[NO]^2$$

Substituting into the rate law in Equation 14.16 yields

$$\text{Rate} = k\left(\frac{k_f}{k_r}\right)[NO]^2[H_2]$$

Combining all the constants into one (k') gives

$$\text{Rate} = k'[NO]^2[H_2]$$

Now the rate law derived from the mechanism matches the rate law obtained experimentally. We are able to say, therefore, that the three-step mechanism does appear to be reasonable on the basis of kinetics.

The procedure we have worked through here applies to many reactions that proceed by mechanisms which involve sequential steps. Steps that precede the rate-determining step are considered to be rapidly established equilibria involving unstable intermediates.

Recall that in a dynamic equilibrium forward and reverse reactions occur at equal rates.

Because N_2O_2 is proposed to be very unstable and therefore quickly decomposes, only tiny amounts of it are present in the reaction mixture. At such low concentrations it is not detectable, so its existence can only be postulated.

Summary Although chemists may devise other experiments to help prove or disprove the correctness of a mechanism, one of the strongest pieces of evidence is the experimentally measured rate law for the overall reaction. No matter how reasonable a particular mechanism may appear, if its elementary processes cannot yield a predicted rate law that matches the experimental one, the mechanism is wrong and must be discarded.

■ **Practice Exercise 13** Ozone, O_3, reacts with nitric oxide, NO, to form nitrogen dioxide and oxygen.

$$NO + O_3 \longrightarrow NO_2 + O_2$$

It's one of the reactions involved in smog. It is believed to occur by a one-step mechanism (the reaction above). If this is so, what is the expected rate law for the reaction?

■ **Practice Exercise 14** The mechanism for the decomposition of NO_2Cl is

$$NO_2Cl \longrightarrow NO_2 + Cl$$
$$NO_2Cl + Cl \longrightarrow NO_2 + Cl_2$$

What would the predicted rate law be if the second step in the mechanism was the rate-determining step?

A **catalyst** is a substance that increases the rate of a chemical reaction without itself being used up. In other words, all of the catalyst added at the start of a reaction is present chemically unchanged after the reaction has gone to completion.

Although the catalyst is not part of the overall reaction, it does participate by changing the mechanism of the reaction. The catalyst provides a path to the products that has a lower activation energy than that of the uncatalyzed reaction, as illustrated in Figure 14.14*a*. In Figure 14.14*b*, we see that when the low-energy pathway is available, a larger total fraction of molecules possess sufficient kinetic energy to react. As a result, the reaction proceeds faster.

14.9 CATALYSTS

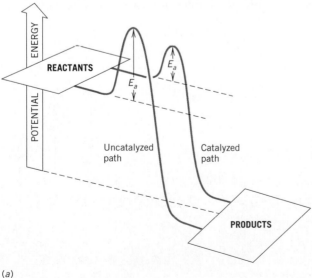

(a)

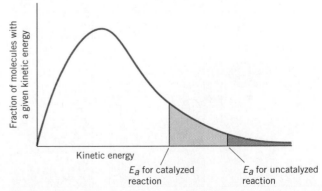

(b)

FIGURE 14.14

Effect of a catalyst on a reaction. (*a*) The catalyst provides an alternative, low-energy path from the reactants to the products. (*b*) A larger fraction of molecules have sufficient energy to react when the catalyzed path is available.

Catalysts can be divided into two groups—**homogeneous catalysts,** which exist in the same phase as the reactants, and **heterogeneous catalysts,** which exist in a separate phase.

Homogeneous Catalysts

Catalysis is the action or effect produced by a catalyst.

An example of homogeneous catalysis is found in the now-outdated lead chamber process used for manufacturing sulfuric acid. To make sulfuric acid by this process, sulfur is burned to give SO_2, which is then oxidized to SO_3. Dissolving SO_3 in water gives H_2SO_4.

$$S + O_2 \longrightarrow SO_2$$

$$SO_2 + \tfrac{1}{2}O_2 \longrightarrow SO_3$$

$$SO_3 + H_2O \longrightarrow H_2SO_4$$

In the modern process for making sulfuric acid, the *contact process,* vanadium pentaoxide, V_2O_5, is the catalyst for the oxidation of sulfur dioxide to sulfur trioxide.

Unassisted, the second reaction, oxidation of SO_2 to SO_3, occurs slowly. In the lead chamber process, the SO_2 is combined with a mixture of NO, NO_2, air, and steam in large lead-lined reaction chambers. The NO_2 readily oxidizes the SO_2 to give NO and SO_3. The NO is then reoxidized to NO_2 by oxygen.

$$NO_2 + SO_2 \longrightarrow NO + SO_3$$

$$NO + \tfrac{1}{2}O_2 \longrightarrow NO_2$$

Because the NO_2 is re-formed in the second reaction and is therefore recycled over and over, only small amounts of it are needed in the reaction mixture to do an effective job.

The NO_2 serves as a catalyst by being an oxygen carrier and by providing a low-energy path for the oxidation of SO_2 to SO_3. (Notice that the NO_2 is regenerated, as must be true for a catalyst.) The steam in the reaction mixture reacts with the SO_3 as it is formed and produces the sulfuric acid.

Heterogeneous Catalysts

A heterogeneous catalyst functions by promoting a reaction on its surface. One or more of the reactant molecules are adsorbed on the surface of the catalyst where an interaction with the surface increases their reactivity. An

FIGURE 14.15

Catalytic formation of ammonia molecules on the surface of a catalyst. Nitrogen and hydrogen are adsorbed and dissociate into atoms that then combine to form ammonia molecules. In the final step, ammonia molecules leave the surface and the whole process can be repeated.

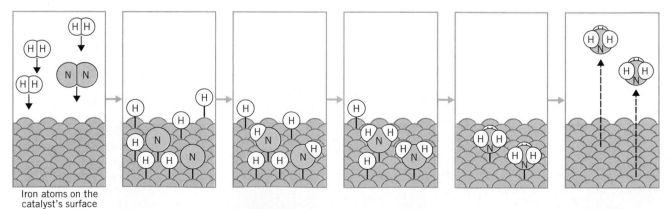

Iron atoms on the catalyst's surface

(b)

FIGURE 14.16

(*a*) Catalytic cracking towers at a Standard Oil refinery. (*b*) A variety of catalysts are available as beads, powders, or in other forms for various refinery operations.

(*a*)

example is the synthesis of ammonia from hydrogen and nitrogen by the Haber process.

$$3H_2 + N_2 \longrightarrow 2NH_3$$

The synthesis of ammonia is described in detail in Chemicals in Use 12.

This is one of the most important industrial reactions in the world because ammonia and nitric acid (which is made from ammonia) are necessary for the production of fertilizers. The reaction takes place on the surface of an iron catalyst that contains traces of aluminum and potassium oxides. It is thought that hydrogen molecules and nitrogen molecules dissociate while being held on the catalytic surface. The hydrogen atoms then combine with the nitrogen atoms to form ammonia. Finally, the completed ammonia molecule breaks away, freeing the surface of the catalyst for further reaction. This sequence of steps is illustrated in Figure 14.15.

Heterogeneous catalysts are used in many important commercial processes. The petroleum industry uses heterogeneous catalysts to crack hydrocarbons into smaller fragments and then re-form them into the useful components of gasoline (Figure 14.16). The availability of such catalysts allows refineries to produce gasoline, jet fuel, or heating oil from crude oil in any ratio necessary to meet the demands of the marketplace.

A vehicle that uses unleaded gasoline is equipped with a catalytic converter (Figure 14.17) designed to lower the concentrations of pollutants in the exhaust. These pollutants are primarily carbon monoxide, unburned hydrocarbons, and nitrogen oxides. Air is mixed with the exhaust stream and passed over a catalyst that adsorbs CO, NO, and O_2. The NO dissociates into N and O atoms, and the O_2 also dissociates into atoms. Pairing of nitrogen atoms then produces N_2, and oxidation of CO by oxygen atoms produces CO_2. Unburned hydrocarbons are also oxidized to CO_2 and H_2O.

In the United States, the use of leaded gasoline has been phased out. While it was still available, however, drivers of cars fitted with catalytic converters had to avoid using the leaded gas because it could destroy the ability of the catalyst to function. Leaded gasoline contains tetraethyllead, $Pb(C_2H_5)_4$, which serves as an octane booster. On combustion, however, lead is released and is deposited on the catalyst in the converter, blocking active sites on the catalyst's

FIGURE 14.17

A modern catalytic converter of the type used in about 80% of new cars. Part of the converter has been cut away to reveal the porous ceramic material that serves as the support for the catalyst.

surface. Thus, the lead "poisons" the surface by destroying its catalytic properties.

The poisoning of catalysts is a major problem in many industrial processes. Methyl alcohol (methanol, CH_3OH), for example, is a promising fuel. It can be made from coal and steam by the reaction

$$C \text{ (from coal)} + H_2O \longrightarrow CO + H_2$$

followed by

$$CO + 2H_2 \longrightarrow CH_3OH$$

A catalyst for the second step is copper(I) ion held in solid solution with zinc oxide. However, traces of sulfur, a contaminant in coal, must be avoided because sulfur reacts with the catalyst and destroys its catalytic activity. Biological poisons often work in a similar way, through catalysts (enzymes). For example, nerve poisons block enzyme sites required for the transport of nerve impulses.

SUMMARY

Reaction Rates. The speeds or **rates** of reaction are controlled by five factors: (1) the nature of the reactants, (2) the ability of reactants to meet, (3) the concentrations of the reactants, (4) the temperature, and (5) the presence of catalysts. The rates of **heterogeneous reactions** are determined largely by the area of contact between the phases; the rates of **homogeneous reactions** are determined by the concentrations of the reactants. The rate is measured by monitoring the change in reactant or product concentrations with time.

$$\text{Rate} = \Delta(\text{concentration})/\Delta(\text{time})$$

Rate is expressed with a negative sign for reactants because the concentration change is negative. In any chemical reaction, the rates of formation of products and disappearance of reactants are related by the coefficients of the balanced overall chemical equation.

Rate Laws The **rate law** for a reaction relates the reaction rate to the molar concentrations of the reactants. The rate is proportional to the product of the molar concentrations of the reactants, each raised to an appropriate power. These exponents must be determined by experiments in which the concentrations are varied and the effects of the variations on the rate are measured. The proportionality constant, k, is called the rate constant. Its value depends on temperature but, of course, not on the concentrations of the reactants. The sum of the exponents in the rate law is the **order** (overall order) of the reaction.

Concentration and Time Equations exists that relate the concentration of a reactant at a given time t to the initial concentration and the rate constant. The time required for half of a reactant to disappear is the **half-life,** $t_{1/2}$. For a first-order reaction, the half-life is a constant that depends only on the rate constant for the reaction; it is independent of the initial concentration. The half-life for a second-order reaction is inversely proportional both to the initial concentration of the reactant and to the rate constant.

Theories of Reaction Rate According to **collision theory,** the rate of a reaction depends on the number of **effective collisions** per second of the reactant particles, which is only an extremely small fraction of the total number of collisions per second. One reason for the relatively small number of successful collisions is that some collisions require special orientations of the reactant molecules. The major reason, however, is that the molecules must jointly possess a minimum molecular kinetic energy called the **activation energy,** E_a. As the temperature increases, a larger fraction of reactant molecules have this necessary energy, so more collisions are effective each second and the reaction is faster.

Transition state theory visualizes how the energies of molecules and the orientations of their nuclei interact as they collide. In this theory, the energy of activation is viewed as an energy barrier on the potential-energy diagram for the reaction. The *heat of reaction* is the net potential-energy difference between the reactants and the products. In reversible reactions, the values of E_a for both the forward and reverse reactions can be identified on an energy diagram. The species at the high point on an energy diagram is the **activated complex** and is said to be in the **transition state.**

Determining the Activation Energy The **Arrhenius equation** lets us see how changes in the activation energy and the temperature affect a rate constant. The Arrhenius equation also lets us determine E_a either graphically or by a calculation using the appropriate form of the equation (Equation 14.9). The calculation requires two rate constants determined at two temperatures. The graphical method uses more values of rate constants at more temperatures and thus usually yields more accurate results. The activation energy and the rate constant at one temperature can be used to calculate the rate constant at another temperature.

Reaction Mechanisms   The detailed sequence of elementary processes that lead to the net chemical change is the **mechanism** of the reaction. Since intermediates usually cannot be detected, the mechanism is a theory. Support for a mechanism comes from matching the predicted rate law for the mechanism with the rate law obtained from experimen-

tal data. For the **rate-determining step** or for any **elementary process** the rate law has exponents equal to the coefficients.

Catalysis   **Catalysts** are substances that increase the rate but are not consumed by the reaction. They function by providing an alternative path for the reaction that has a smaller activation energy than the uncatalyzed reaction. **Homogeneous catalysts** are in the same phase as the reactants. **Heterogeneous catalysts** provide a path of lower activation energy by having a surface on which the reactants are adsorbed and react.

Tools you have learned

The table below lists the tools you have learned in this chapter that are applicable to problem solving. Review them if necessary, and refer to them when working on the Thinking-It-Through problems and the Review Exercises that follow.

Tool	Function
Rate law of a reaction (pages 591, 598, and 600)	Use data on how concentration affects rate to evaluate the exponents of the concentration terms in a rate law, determine the overall order of a reaction from the coefficients of the concentration terms in its rate law, and calculate a rate constant.
Half-lives (page 602)	Calculate a half-life from experimental data and reaction order or calculate the rate of disappearance of a reactant.
Arrhenius equation (pages 609 and 611)	To determine an activation energy graphically and to interrelate rate constants, activation energy, and temperature.

THINKING IT THROUGH

The goal for each of the following problems is to give you practice in thinking your way through problems. The goal is not *to find the answer itself; instead, you are only asked to assemble the available information needed to obtain the answer, state what additional data (if any) are needed, and describe how you would use the data to answer the question. For problems involving unit conversions, list the relationships among the units that are needed to carry out the conversions. Construct the conversion factors that can be formed from these relationships. Then set up the solution to the problem by arranging the conversion factors so the units cancel correctly to give the desired units of the answer.*

The problems are divided into two groups. Those in Level 2 are more challenging than those in Level 1 and provide an opportunity to really hone your problem-solving skills. Both levels, however, may include questions that draw on material studied in earlier chapters.

Level 1 Problems

1. Consider the following reaction

$$C_3H_8(g) + 5O_2(g) \longrightarrow 3CO_2(g) + 4H_2O(g)$$

If at a given moment C_3H_8 is reacting at a rate of 0.400 mol L^{-1} s^{-1}, what are the rates of formation of CO_2 and H_2O? What is the rate at which O_2 is reacting?

2. Suppose the following data were collected for the reaction

$$2A \longrightarrow B + 2C$$

Concentration of A (mol L^{-1})	Time (minutes)
0.2000	0
0.1902	10
0.1721	20
0.1482	30
0.1213	40
0.0945	50
0.0700	60

How would you determine the rate at which A is reacting at $t = 25$ minutes? What are the rates at which B and C are being formed at this time?

3. The following reaction: $I^-(aq) + OCl^-(aq) \rightarrow IO^-(aq) + Cl^-(aq)$ occurs in a basic solution. When the $[OH^-] = 1.00$ M, the reaction is second order with a rate constant equal to 60.0 L mol^{-1} s^{-1}. If $[I^-] = 0.0200$ M and $[OCl^-] = 0.0300$ M at a certain time, what is the rate at which OI^- is being formed?

4. Carbon-14 is a radioactive isotope that decays by the nuclear reaction

$$^{14}_{6}C \longrightarrow {}^{14}_{7}N + {}^{0}_{-1}e$$

It is a first-order process with a half-life of 5770 years. What is the rate constant for the decay? How many years will it take for 1/100 of the ^{14}C in a sample of naturally occurring carbon (a mixture of carbon isotopes, including ^{14}C) to decay?

5. The reaction $2NO_2 \rightarrow 2NO + O_2$ has the rate constant $k = 0.63$ L mol^{-1} s^{-1} at 600 K. If the initial concentration of NO_2 in a reaction vessel is 0.050 M, how long will it take for the concentration to drop to 0.020 M?

6. For the reaction described in the preceding question, how many molecules of NO_2 react each second when the NO_2 concentration is 3.0×10^{-3} M?

Level 2 Problems

7. The following data were collected at a certain temperature for the reaction

$$2N_2O_5(g) \longrightarrow 4NO_2(g) + O_2(g)$$

Experiment	$[N_2O_5]$	Rate of Formation of O_2 (mol L^{-1} s^{-1})
1	0.0325	0.055
2	0.0250	0.042
3	0.0125	0.021

What is the order of the reaction with respect to N_2O_5? What is the value of the rate constant at this temperature?

8. For the reaction

$$2N_2O_5(g) \longrightarrow 4NO_2(g) + O_2(g)$$

it was observed at a particular moment that the pressure in a 1.0 L container of these gases at 25 °C was increasing at the rate of 20.0 torr s^{-1}. What was the rate at which the N_2O_5 was reacting expressed in the units mol L^{-1} s^{-1}?

9. At high temperatures, collisions between argon atoms and oxygen molecules can lead to the formation of oxygen atoms.

$$Ar + O_2 \longrightarrow Ar + O + O$$

At 5000 K, the rate constant for this reaction is 5.49×10^6 L mol^{-1} s^{-1}. At 10,000 K, $k = 9.86 \times 10^8$ L mol^{-1} s^{-1}. Estimate the activation energy for the reaction.

10. A mechanism proposed for the reaction

$$F_2(g) + 2NO_2(g) \longrightarrow 2NO_2F(g)$$

consists of the following elementary processes.

$$F_2 + NO_2 \longrightarrow NO_2F + F \qquad \text{(step 1)}$$

$$F + NO_2 \longrightarrow NO_2F \qquad \text{(step 2)}$$

The rate law for the reaction is Rate $= k[NO_2][F_2]$. If this really *is* the mechanism, which is the rate-determining step?

11. For the reaction described in the preceding question, explain the evidence that shows that it does not occur by the one-step mechanism: $F_2(g) + 2NO_2(g) \longrightarrow 2NO_2F(g)$.

REVIEW EXERCISES

Answers to questions whose numbers are printed in color are given in Appendix D. More challenging questions are marked with asterisks.

Factors That Affect Reaction Rate

14.1 What does *rate of reaction* mean in qualitative terms?

14.2 Give an example from everyday experience of (a) a very fast reaction, (b) a moderately fast reaction, and (c) a slow reaction.

14.3 In terms of reaction rates, what is an explosion?

14.4 From an economic point of view, why would industrial corporations want to know about the factors that affect the rate of a reaction?

14.5 Suppose we compared two reactions, one requiring the simultaneous collision of three molecules and the other requiring a collision between two molecules. From the standpoint of statistics, and all other factors being equal, which reaction should be faster? Explain your answer.

14.6 List the five factors that affect the rates of chemical reactions.

14.7 What is a *homogeneous reaction?* Give an example.

14.8 What is a *heterogeneous reaction?* Give an example.

14.9 Why are chemical reactions usually carried out in solution?

14.10 What is the major factor that affects the rate of a heterogeneous reaction?

14.11 How does particle size affect the rate of a heterogeneous reaction? Why?

14.12 The rate of hardening of epoxy glue depends on the amount of hardener that is mixed into the glue. What factor affecting reaction rates does this illustrate?

14.13 What effect does temperature have on reaction rate?

14.14 What is a catalyst?

14.15 In cool weather, the number of chirps per minute from crickets diminishes. How can this be explained?

14.16 On the basis of what you learned in Chapter 11, why do foods cook faster in a pressure cooker than in an open pot of boiling water?

14.17 Persons who have been submerged in very cold water and who are believed to have drowned sometimes can be revived. On the other hand, persons who have been submerged in warmer water for the same length of time have died. Explain this in terms of factors that affect the rates of chemical reactions.

Measuring Rates of Reaction

14.18 Cindy Crawford makes $20,000 in one day of photo sessions. If 500 photos are taken, what is her rate of pay expressed as dollars per photo? If each photo takes 1/125 of a second and only the time for the shutter to fall is used to compute her pay, what is her pay rate expressed as dollars per second?

14.19 The winners of the Superbowl earn approximately $54,000 for the game. If a player is on the field for the entire game (1.0 hour), what is the player's rate of pay expressed in dollars per second?

14.20 What are the units of reaction rate?

14.21 In the reaction $3H_2 + N_2 \rightarrow 2NH_3$, how does the rate of disappearance of hydrogen compare to the rate of disappearance of nitrogen? How does the rate of appearance of NH_3 compare to the rate of disappearance of nitrogen?

14.22 The following data were collected at a certain temperature for the reaction

$$SO_2Cl_2 \longrightarrow SO_2 + Cl_2$$

Time (min)	$[SO_2Cl_2]$ (mol L^{-1})
0	0.1000
100	0.0876
200	0.0768
300	0.0673
400	0.0590
500	0.0517
600	0.0453
700	0.0397
800	0.0348
900	0.0305
1000	0.0267
1100	0.0234

Make a graph of concentration versus time and determine the rate of formation of SO_2 at $t = 200$ minutes and $t = 600$ minutes.

14.23 The following data were collected for the reaction below at a temperature of 530 °C.

$$CH_3CHO \longrightarrow CH_4 + CO$$

$[CH_3CHO]$ (mol L^{-1})	Time (s)
0.200	0
0.153	20
0.124	40
0.104	60
0.090	80
0.079	100
0.070	120
0.063	140
0.058	160
0.053	180
0.049	200

Make a graph of concentration versus time and determine the rate of reaction of CH_3CHO after 60 s and after 120 s.

14.24 For the reaction $2A + B \longrightarrow 3C$, it was found that the rate of disappearance of B was -0.30 mol L^{-1} s^{-1}. What was the rate of disappearance of A and the rate of appearance of C?

14.25 In the combustion of hexane,

$$2C_6H_{14}(g) + 19O_2(g) \longrightarrow 12CO_2(g) + 14H_2O(g)$$

it was found that the rate of reaction of C_6H_{14} was -1.20 mol L^{-1} s^{-1}.
(a) What was the rate of reaction of O_2?
(b) What was the rate of formation of CO_2?
(c) What was the rate of formation of H_2O?

14.26 At a certain moment in the reaction

$$2N_2O_5 \longrightarrow 4NO_2 + O_2$$

N_2O_5 is decomposing at a rate of 2.5×10^{-6} mol L^{-1} s^{-1}.
(a) Express this rate with the proper algebraic sign.
(b) What are the rates of formation of NO_2 and O_2?

Concentration and Rate; Rate Laws

14.27 What is a *rate law*? What is the proportionality constant called?

14.28 What is meant by the *order* of a reaction?

14.29 What are the units of the rate constant for (a) a first-order reaction, (b) a second-order reaction, and (c) a third-order reaction?

14.30 How must the exponents in a rate law be determined?

14.31 Is there any way of using the coefficients in the balanced overall equation for a reaction to predict with certainty what the exponents are in the rate law?

14.32 If the concentration of a reactant is doubled and the reaction rate doubles, what must be the order of the reaction with respect to that reactant?

14.33 If the concentration of a reactant is doubled, by what factor will the rate increase if the reaction is second order with respect to that reactant?

14.34 In an experiment, the concentration of a reactant was tripled. The rate increased by a factor of 27. What is the order of the reaction with respect to that reactant?

14.35 The rate law for the reaction

$$2NO + O_2 \longrightarrow 2NO_2$$

is Rate $= k[NO]^2[O_2]$. At 25 °C, $k = 7.1 \times 10^9$ L^2 mol^{-2} s^{-1}. What is the rate of reaction when $[NO] = 0.0010$ mol L^{-1} and $[O_2] = 0.034$ mol L^{-1}?

14.36 The rate law for the decomposition of N_2O_5 is Rate $= k[N_2O_5]$. If $k = 1.0 \times 10^{-5}$ s^{-1}, what is the reaction rate when the N_2O_5 concentration is 0.0010 mol L^{-1}?

14.37 The rate law for the reaction

$$2HCrO_4^- + 3HSO_3^- + 5H^+ \rightarrow 2Cr^{3+} + 3SO_4^{2-} + 5H_2O$$

is Rate $= k[HCrO_4^-][HSO_3^-]^2[H^+]$.
(a) What is the order of the reaction with respect to each reactant?
(b) What is the overall order of the reaction?

14.38 The rate law for the reaction

$$NO + O_3 \longrightarrow NO_2 + O_2$$

is Rate $= k[NO][O_3]$. What is the order with respect to each reactant and what is the overall order of the reaction?

14.39 Biological reactions usually involve the interaction of an enzyme with a *substrate*, the substance that actually undergoes the chemical change. In many cases, the rate of reaction depends on the concentration of the enzyme but is independent of the substrate concentration. What is the order of the reaction with respect to the substrate in such instances?

14.40 The following data were collected for the reaction

$$M + N \longrightarrow P + Q$$

Initial Concentrations (mol L^{-1})		Initial Rate of Reaction (mol L^{-1} s^{-1})
[M]	[N]	
0.010	0.010	-2.5×10^{-3}
0.020	0.010	-5.0×10^{-3}
0.020	0.030	-4.5×10^{-2}

What is the rate law for the reaction? What is the value of the rate constant (with correct units)?

14.41 Cyclopropane, C_3H_6, is a gas used as a general anesthetic. It undergoes a slow molecular rearrangement to propylene.

cyclopropane   propylene

At a certain temperature, the following data were obtained relating concentration and rate.

Initial Concentration of C_3H_6 (mol L^{-1})	Rate of Formation of Propylene (mol L^{-1} s^{-1})
0.050	2.95×10^{-5}
0.100	5.90×10^{-5}
0.150	8.85×10^{-5}

What is the rate law for the reaction? What is the value of the rate constant, with correct units?

14.42 The reaction of iodide ion with hypochlorite ion, OCl^- (which is found in liquid bleach), follows the equation

$$OCl^- + I^- \longrightarrow OI^- + Cl^-$$

It is a rapid reaction that gives the following rate data.

Initial Concentrations (mol L^{-1})		Rate of Formation of Cl$^-$ (mol L^{-1} s^{-1})
[OCl$^-$]	[I$^-$]	
1.7×10^{-3}	1.7×10^{-3}	1.75×10^4
3.4×10^{-3}	1.7×10^{-3}	3.50×10^4
1.7×10^{-3}	3.4×10^{-3}	3.50×10^4

What is the rate law for the reaction? Determine the value of the rate constant with its correct units.

14.43 The formation of small amounts of nitric oxide, NO, in automobile engines is the first step in the formation of smog. Nitric oxide is readily oxidized to nitrogen dioxide by the reaction

$$2NO(g) + O_2(g) \longrightarrow 2NO_2(g)$$

The following data were collected in a study of the rate of this reaction.

Initial Concentrations (mol L^{-1})		Rate of Formation of NO$_2$ (mol L^{-1} s^{-1})
[O$_2$]	[NO]	
0.0010	0.0010	7.10
0.0040	0.0010	28.4
0.0040	0.0030	255.6

What is the rate law for the reaction? What is the rate constant with its correct units?

14.44 At a certain temperature the following data were collected for the reaction

$$2ICl + H_2 \longrightarrow I_2 + 2HCl$$

Initial Concentrations (mol L^{-1})		Initial Rate of Formation of I$_2$ (mol L^{-1} s^{-1})
[ICl]	[H$_2$]	
0.10	0.10	0.0015
0.20	0.10	0.0030
0.10	0.050	0.00075

Determine the rate law and the rate constant (with correct units) for the reaction.

14.45 The following data were obtained for the reaction of (CH$_3$)$_3$CBr with hydroxide ion at 55 °C.

$$(CH_3)_3CBr + OH^- \longrightarrow (CH_3)_3COH + Br^-$$

Initial Concentrations (mol L^{-1})		Initial Rate of Formation of (CH$_3$)$_3$COH (mol L^{-1} s^{-1})
[(CH$_3$)$_3$CBr]	[OH$^-$]	
0.10	0.10	1.0×10^{-3}
0.20	0.10	2.0×10^{-3}
0.30	0.10	3.0×10^{-3}
0.10	0.20	1.0×10^{-3}
0.10	0.30	1.0×10^{-3}

What is the rate law for the reaction? What is the value of the rate constant (with correct units) at this temperature?

Concentration and Time

14.46 Give the equations that relate concentration to time for (a) a first-order reaction and (b) a second-order reaction.

14.47 Data for the decomposition of SO$_2$Cl$_2$ according to the equation

$$SO_2Cl_2(g) \longrightarrow SO_2(g) + Cl_2(g)$$

are given in Review Exercise 14.22. Show graphically that these data fit a first-order rate law. Graphically determine the rate constant for the reaction.

*14.48 For data in Review Exercise 14.23, decide graphically whether the reaction is first or second order. Determine the rate constant for the reaction described in that exercise.

14.49 The decomposition of SO$_2$Cl$_2$ described in Review Exercise 14.47 has a first-order rate constant $k = 2.2 \times 10^{-5}$ s^{-1} at 320 °C. If the initial SO$_2$Cl$_2$ concentration in a container is 0.0040 M, what will its concentration be (a) after 1.00 hr, (b) after 1.00 day?

14.50 The decomposition of hydrogen iodide follows the equation

$$2HI(g) \longrightarrow H_2(g) + I_2(g)$$

The reaction is second order and has a rate constant equal to 1.6×10^{-3} L mol^{-1} s^{-1} at 700 °C. If the initial concentration of HI in a container is 3.4×10^{-2} M, how many minutes will it take for the concentration to be reduced to 8.0×10^{-4} M?

14.51 The decomposition of HI follows the equation

$$2HI(g) \longrightarrow H_2(g) + I_2(g)$$

The reaction is second order and has a rate constant equal to 1.6×10^{-3} L mol^{-1} s^{-1} at 700 °C. At 2.5×10^3 min after a particular experiment began, the HI concentration was equal to 4.5×10^{-4} mol L^{-1}. What was the initial molar concentration of HI in the reaction vessel?

14.52 The concentration of a drug in the body is often expressed in units of milligrams per kilogram of body weight. The initial dose of a drug in an animal was 25.0 mg/kg body weight. After 2.00 hr, this concentration had dropped to 15.0 mg/kg body weight. If the drug is eliminated metabolically by a first-order process, what is the rate constant for the process in units of min^{-1}?

14.53 In the preceding question, what must the initial dose of the drug be in order for the drug concentration 3.00 hr afterward to be 5.0 mg/kg body weight?

14.54 If it takes 75.0 min for the concentration of a reactant to drop to 20% of its initial value in a first-order reaction, what is the rate constant for the reaction in the units min^{-1}?

Half-Lives

14.55 What is meant by the term *half-life*?

14.56 How is the half-life of a first-order reaction affected by the initial concentration of the reactant?

14.57 How is the half-life of a second-order reaction affected by the initial reactant concentration?

14.58 Derive the equations for $t_{1/2}$ for first- and second-order reactions from Equations 14.5 and 14.6, respectively.

14.59 A certain first-order reaction has a rate constant $k = 1.6 \times 10^{-3} \text{ s}^{-1}$. What is the half-life for this reaction?

14.60 The decomposition of NOCl,

$$2NOCl \longrightarrow 2NO + Cl_2$$

is a second-order reaction with $k = 6.7 \times 10^{-4} \text{ L mol}^{-1} \text{ s}^{-1}$ at 400 K. What is the half-life of this reaction if the initial concentration of NOCl is 0.20 mol L^{-1}?

14.61 Using the graph from Review Exercise 14.22, determine the time required for the SO_2Cl_2 concentration to drop from 0.100 mol L^{-1} to 0.050 mol L^{-1}. How long does it take for the concentration to drop from 0.050 mol L^{-1} to 0.025 mol L^{-1}? What is the order of this reaction? (Hint: How is the half-life related to concentration?)

14.62 Using the graph from Review Exercise 14.23, determine how long it takes for the CH_3CHO concentration to decrease from 0.200 mol L^{-1} to 0.100 mol L^{-1}. How long does it take the concentration to drop from 0.100 mol L^{-1} to 0.050 mol L^{-1}? What is the order of this reaction? (Hint: How is the half-life related to concentration?)

14.63 The half-life of a certain first-order reaction is 15 min. What fraction of the original reactant concentration will remain after 2.0 hr?

14.64 A particular substance is known to decay by a first-order reaction. The concentration of this substance was found to be exactly one-sixth its original concentration after 106 s. Calculate the half-life of the substance.

14.65 Strontium-90 has a half-life of 28 years. How long will it take for all of the strontium-90 presently on earth to be reduced to $\frac{1}{32}$ of its present amount?

Effect of Temperature on Rate

14.66 What is the basic postulate of collision theory?

14.67 What two factors affect the effectiveness of molecular collisions in producing chemical change?

14.68 In terms of the kinetic theory, why does an increase in temperature increase the reaction rate?

14.69 Draw the potential-energy diagram for an endothermic reaction. Indicate on the diagram the activation energy for both the forward and the reverse reactions. Also indicate the heat of reaction.

14.70 Explain, in terms of the law of conservation of energy, why an endothermic reaction leads to a cooling of the reaction mixture if heat cannot enter from outside the system.

14.71 Define *transition state* and *activated complex*.

14.72 Draw a potential-energy diagram for an exothermic reaction and indicate on the diagram the location of the transition state.

14.73 Suppose a certain slow reaction is found to have a very small activation energy. What does this suggest about

the importance of molecular orientation in the formation of the activated complex?

14.74 The decomposition of carbon dioxide,

$$CO_2 \longrightarrow CO + O$$

has a very large activation energy of approximately 460 kJ/mol. Explain why this is consistent with a mechanism that involves the breaking of a C=O bond.

Calculations Involving the Activation Energy

14.75 State the Arrhenius equation and define the symbols.

14.76 The following data were collected for a reaction.

Rate Constant (L mol^{-1} s^{-1})	Temperature (°C)
2.88×10^{-4}	320
4.87×10^{-4}	340
7.96×10^{-4}	360
1.26×10^{-3}	380
1.94×10^{-3}	400

Determine the activation energy for the reaction in kJ/mol both graphically and by calculation using Equation 14.11. For the calculation of E_a, use the first and last sets of data in the table above.

14.77 Rate constants were measured at various temperatures for the reaction

$$HI(g) + CH_3I(g) \longrightarrow CH_4(g) + I_2(g)$$

The following data were obtained.

Rate Constant (L mol^{-1} s^{-1})	Temperature (°C)
1.91×10^{-2}	205
2.74×10^{-2}	210
3.90×10^{-2}	215
5.51×10^{-2}	220
7.73×10^{-2}	225
1.08×10^{-1}	230

Determine the activation energy in kJ/mol both graphically and by calculation using Equation 14.11. For the calculation of E_a, use the first and last sets of data in the table above.

14.78 The decomposition of NOCl,

$$2NOCl \longrightarrow 2NO + Cl_2$$

has $k = 9.3 \times 10^{-5} \text{ L mol}^{-1} \text{ s}^{-1}$ at 100 °C and $k = 1.0 \times 10^{-3} \text{ L mol}^{-1} \text{ s}^{-1}$ at 130 °C. What is E_a for this reaction in kJ mol^{-1}? Use the data at 100 °C to calculate the frequency factor.

14.79 The conversion of cyclopropane, an anesthetic, to propylene (see Review Exercise 14.41) has a rate constant $k = 1.3 \times 10^{-6} \text{ s}^{-1}$ at 400 °C and $k = 1.1 \times 10^{-5} \text{ s}^{-1}$ at 430 °C.

(a) What is the activation energy in kJ/mol?

(b) What is the value of A for this reaction, calculated using Equation 14.9?

(c) What is the rate constant for the reaction at 350 °C?

14.80 The reaction of CO_2 with water to form carbonic acid,

$$CO_2(aq) + H_2O \longrightarrow H_2CO_3(aq)$$

has $k = 3.75 \times 10^{-2}$ s^{-1} at 25 °C and $k = 2.1 \times 10^{-3}$ s^{-1} at 0 °C. What is the activation energy for this reaction in kJ/mol?

14.81 If a reaction has $k = 3.0 \times 10^{-4}$ s^{-1} at 25 °C and an activation energy of 100.0 kJ/mol, what will the value of k be at 50 °C?

14.82 It was mentioned that the rates of many reactions approximately double for each 10 °C rise in temperature. Assuming a starting temperature of 25 °C, what would the activation energy be, in kJ, if the rate of a reaction were to be twice as large at 35 °C?

14.83 The decomposition of N_2O_5 has an activation energy of 103 kJ/mol and a frequency factor of 4.3×10^{13} s^{-1}. What is the rate constant for this decomposition at (a) 20 °C and (b) 100 °C?

14.84 At 35 °C, the rate constant for the reaction

$$\underset{\text{sucrose}}{C_{12}H_{22}O_{11}} + H_2O \longrightarrow \underset{\text{glucose}}{C_6H_{12}O_6} + \underset{\text{fructose}}{C_6H_{12}O_6}$$

is $k = 6.2 \times 10^{-5}$ s^{-1}. The activation energy for the reaction is 108 kJ mol^{-1}. What is the rate constant for the reaction at 45 °C?

Reaction Mechanisms

14.85 What is the definition of an *elementary process?* How are elementary processes related to the mechanism of a reaction?

14.86 What is a *rate-determining step?*

14.87 In what way is the rate law for a reaction related to the rate-determining step?

14.88 A reaction has the following mechanism.

$$2NO \longrightarrow N_2O_2$$
$$N_2O_2 + H_2 \longrightarrow N_2O + H_2O$$
$$N_2O + H_2 \longrightarrow N_2 + H_2O$$

What is the net overall change that occurs in this reaction?

14.89 If the reaction $NO_2 + CO \longrightarrow NO + CO_2$ occurs by a one-step collision process, what would be the expected rate law for the reaction? The actual rate law is rate = $k[NO_2]^2$. Could the reaction actually occur by a one-step collision between NO_2 and CO? Explain.

14.90 Oxidation of NO to NO_2—one of the reactions in the production of smog—appears to involve carbon monoxide. A possible mechanism is

$$CO + \cdot OH \longrightarrow CO_2 + H\cdot$$
$$H\cdot + O_2 \longrightarrow HOO\cdot$$
$$HOO\cdot + NO \longrightarrow \cdot OH + NO_2$$

(The formulas with dots represent extremely reactive species with unpaired electrons and are called *free radicals.*) Write the net chemical equation for the reaction.

14.91 In the absence of CO, the oxidation of NO to NO_2 by O_2 appears to follow the mechanism

$$2NO \longrightarrow N_2O_2 \qquad \text{(step 1)}$$
$$N_2O_2 + O_2 \longrightarrow 2NO_2 \qquad \text{(step 2)}$$

(a) If step 1 is slow, what is the expected rate law for the reaction?

(b) If step 2 is slow and the first step is rapid, what is the expected rate law for the reaction? How should we properly represent the equation for the first step?

14.92 The reaction $2NO(g) + Br_2(g) \rightarrow 2NOBr(g)$ is believed to follow the mechanism

$$NO(g) + Br_2(g) \longrightarrow NOBr_2(g)$$
$$NOBr_2(g) + NO \longrightarrow 2NOBr(g)$$

The rate law for the reaction is rate = $k[NO]^2[Br_2]$.

(a) Which is the slow step in the mechanism?

(b) Why is it believed that the overall reaction is not an elementary process?

14.93 Show that the following two mechanisms give the same net overall reaction.

Mechanism 1

$$OCl^- + H_2O \longrightarrow HOCl + OH^-$$
$$HOCl + I^- \longrightarrow HOI + Cl^-$$
$$HOI + OH^- \longrightarrow H_2O + OI^-$$

Mechanism 2

$$OCl^- + H_2O \longrightarrow HOCl + OH^-$$
$$I^- + HOCl \longrightarrow ICl + OH^-$$
$$ICl + 2OH^- \longrightarrow OI^- + Cl^- + H_2O$$

14.94 The experimental rate law for the reaction $NO_2 + CO \rightarrow CO_2 + NO$ is rate = $k[NO_2]^2$. If the mechanism is

$$2NO_2 \longrightarrow NO_3 + NO \qquad \text{(slow)}$$
$$NO_3 + CO \longrightarrow NO_2 + CO_2 \qquad \text{(fast)}$$

show that the predicted rate law is the same as the experimental rate law.

***14.95** In the preceding question, what would the rate law for the reaction be if the second step in the mechanism were rate determining?

Catalysts

14.96 How does a catalyst increase the rate of a chemical reaction?

14.97 What is a *homogeneous catalyst?* How does it function, in general terms?

14.98 What is a *heterogeneous catalyst?* How does it function?

14.99 What is the difference in meaning between the terms *adsorption* and *absorption?* (If necessary, use a dictionary.) Which one applies to heterogeneous catalysts?

14.100 What does the catalytic converter do in the exhaust system of an automobile? Why can't leaded gasoline be used in cars equipped with catalytic converters?

Additional Exercises

*14.101   For the reaction and data given in Exercise 14.23, make a graph of concentration versus time for the *formation* of CH_4. What are the rates of formation of CH_4 at $t = 40$ s and $t = 100$ s?

14.102   The age of wine can be determined by measuring the trace amount of radioactive tritium, ^3H, present in a sample. Tritium is formed from hydrogen in water vapor in the upper atmosphere by cosmic bombardment, so all naturally occurring water contains a small amount of this isotope. Once in a bottle of wine, however, the formation of additional tritium from the water is negligible, so the tritium initially present gradually diminishes by a first-order radioactive decay with a half-life of 12.5 years. If a bottle of wine is found to have a tritium concentration that is 0.100 that of freshly bottled wine (i.e., $[^3H]_t = 0.100 [^3H]_0$), what is the age of the wine?

*14.103   Suppose a reaction occurs with the mechanism

(1) $2A \rightleftharpoons A_2$ (fast)

(2) $A_2 + E \longrightarrow B + C$ (slow)

in which the first step is a very rapid reversible reaction that can be considered to be essentially an equilibrium (forward and reverse reactions occurring at the same rate) and the second is a slow step.

(a) Write the rate law for the forward reaction in step (1).

(b) Write the rate law for the reverse reaction in step (1).
(c) Write the rate law for the rate-determining step.
(d) What is the chemical equation for the net reaction that occurs in this chemical change?
(e) Use the results of parts (a) and (b) to rewrite the rate law of the rate-determining step in terms of the concentrations of the reactants in the overall balanced equation for the reaction.

*14.104   A reaction that has the stoichiometry

$$2A + B \longrightarrow products$$

was found to yield the following data.

Initial Concentrations (mol L^{-1})		Initial Rate of Reaction of A
[A]	[B]	(mol L^{-1} s^{-1})
0.020	0.030	-0.0150
0.025	0.030	-0.0188
0.025	0.040	-0.0334

(a) What is the rate law for the reaction?
(b) What is the rate constant for the reaction with its correct units?

*14.105   The decomposition of urea in 0.10 M HCl follows the reaction

$$(NH_2)_2CO + 2H^+ + H_2O \longrightarrow 2NH_4^+ + CO_2(g)$$

At 60 °C, $k = 5.84 \times 10^{-6}$ min^{-1} and at 70 °C, $k = 2.25 \times 10^{-5}$ min^{-1}. If this reaction is run at 80 °C starting with a urea concentration of 0.0020 M, how many minutes will it take for the urea concentration to drop to 0.0012 M?

*14.106 Show that for a reaction that obeys the general rate law

$$\text{Rate} = k[A]^n$$

a graph of log(rate) versus log[A] should yield a straight line with a slope equal to the order of the reaction. For the reaction in Review Exercise 14.22, measure the rate of the reaction at $t = 150$, 300, 450, and 600 s. Then graph log (rate) versus log [SO$_2$Cl$_2$] and determine the order of the reaction with respect to SO$_2$Cl$_2$.

14.107 The molecular structure and bonding in O$_2$ were discussed in Section 8.7 as was the bonding in N$_2$. Molecular nitrogen is very unreactive, whereas molecular oxygen is very reactive. On the basis of what you learned in Section 8.7 and Special Topic 14.1, what fact about O$_2$ is responsible, at least in part, for its reactivity?

*14.108 In the upper atmosphere, a layer of ozone, O$_3$, shields the Earth from harmful ultraviolet radiation. The ozone is generated by the reactions

$$O_2 \xrightarrow{h\nu} 2O$$
$$O + O_2 \longrightarrow O_3$$

Release of chlorofluorocarbon aerosol propellants and refrigerant fluids such as Freon 12 (CCl$_2$F$_2$) into the atmosphere threatens to at least partially destroy the ozone shield by the reactions

$$CCl_2F_2 \longrightarrow CClF_2 + Cl$$
$$Cl + O_3 \longrightarrow ClO + O_2 \qquad (1)$$
$$ClO + O \longrightarrow Cl + O_2 \qquad (2)$$

Read Special Topic 14.1 and then explain how reactions 1 and 2 constitute a chain reaction that can destroy huge numbers of O$_3$ molecules as well as O atoms that might otherwise react with O$_2$ to replenish the ozone.

*14.109 The following question is based on Special Topic 14.1. The reaction of hydrogen and bromine appears to follow the mechanism

$$Br_2 \xrightarrow{h\nu} 2Br\cdot$$
$$Br\cdot + H_2 \longrightarrow HBr + H\cdot$$
$$H\cdot + Br_2 \longrightarrow HBr + Br\cdot$$
$$2Br\cdot \longrightarrow Br_2$$

(a) Identify the initiation step in the mechanism.
(b) Identify any propagation steps.
(c) Identify the termination step.

The mechanism also contains the reaction

$$H\cdot + HBr \longrightarrow H_2 + Br\cdot$$

How does this reaction affect the rate of formation of HBr?

TEST OF FACTS AND CONCEPTS: CHAPTERS 10-14

We pause again to provide an opportunity for you to test your grasp of concepts, your knowledge of scientific terms, and your skills at solving chemistry problems. If you can answer the following questions, you are in a good position to go on to the next chapters.

1. A 15.5 L sample of neon at 25.0 °C and a pressure of 748 torr is kept at 25.0 °C as it is allowed to expand to a final volume of 25.4 L. What is the final pressure?

2. An 8.95 L sample of nitrogen at 25.0 °C and 1.00 atm is compressed to a volume of 0.895 L and a pressure of 5.56 atm. What must its final temperature be?

3. What is the molecular mass of a gaseous element if 6.45 g occupies 1.92 L at 745 torr and 25.0 °C? Which element is it?

4. What is the molecular mass of a gaseous element if at room temperature it effuses through a pinhole 2.16 times as rapidly as xenon? Which element is it?

5. Hydrogen peroxide, H_2O_2, is decomposed by potassium permanganate according to the following equation.

$$5H_2O_2 + 2KMnO_4 + 3H_2SO_4 \longrightarrow$$
$$5O_2 + 2MnSO_4 + K_2SO_4 + 8H_2O$$

What is the minimum number of milliliters of 0.125 M $KMnO_4$ required to prepare 375 mL of dry O_2 when the gas volume is measured at 22.0 °C and 738 torr?

6. A sample of 248 mL of wet nitrogen gas was collected over water at a total gas pressure of 736 torr and a temperature of 21.0 °C. (The vapor pressure of water at 21.0 °C is 18.7 torr.) The nitrogen was produced by the reaction of sulfamic acid, HNH_2SO_3, with 425 mL of a solution of sodium nitrite according to the following equation.

$$NaNO_2 + HNH_2SO_3 \longrightarrow N_2 + NaHSO_4 + H_2O$$

Calculate what must have been the molar concentration of the sodium nitrite.

7. Which has a higher value of the van der Waals constant a, a gas whose molecules are polar or one whose molecules are nonpolar? Explain.

8. What kinds of attractive forces, including chemical bonds, would be present between the particles in
(a) $H_2O(l)$ (c) $CH_3OH(l)$ (e) $NaCl(s)$
(b) $CCl_4(l)$ (d) $BrCl(l)$ (f) $Na_2SO_4(s)$

9. Based on what you've learned in these chapters, explain:
(a) Why a breeze cools you when you're perspiring.

(b) Why droplets of water form on the outside of a glass of cold soda on a warm, humid day.
(c) Why you feel more uncomfortable on a warm, humid day than on a warm, dry day.
(d) The origin of the energy in a violent thunderstorm.
(e) Why clouds form as warm, moist air flows over a mountain range.

10. Trimethylamine, $(CH_3)_3N$, is a substance responsible in part for the smell of fish. It has a boiling point of 3.5 °C and a molecular weight of 59.1. Dimethylamine, $(CH_3)_2NH$, has a similar odor and boils at a slightly higher temperature, 7 °C, even though it has a somewhat lower molecular mass (45.1). How can this be explained in terms of the kinds of attractive forces between their molecules?

11. Methanol, CH_3OH, commonly known as wood alcohol, has a boiling point of 64.7 °C. Methylamine, a fishy-smelling chemical found in herring brine, has a boiling point of -6.3 °C. Ethane, a hydrocarbon present in petroleum, has a boiling point of -88 °C.

methanol   methylamine   ethane
bp 64.7 °C   bp -6.3 °C   bp -88 °C

Each has nearly the same molecular mass. Account for the large differences in their boiling points in terms of the attractive forces between their molecules.

12. Tin tetraiodide (stannic iodide) has the formula SnI_4. It forms soft, yellow to reddish crystals that melt at about 143 °C. What kind of solid does SnI_4 form? What kind of bonding occurs in SnI_4?

13. Sketch the phase diagram for a substance that has a triple point at 25 °C and 100 torr, a normal boiling point of 150 °C, and a melting point at 1 atm of 27 °C. Is the solid more dense or less dense than the liquid? Where on the curve would the critical temperature and critical pressure be? What phase would exist at 30 °C and 10.0 torr?

14. Where in the periodic table are the very reactive metals located? Where are the least reactive ones located?

15. Write molecular equations for the reaction of O_2 with (a) magnesium, (b) aluminum, (c) phosphorus, (d) sulfur.

16. Compound XY is an ionic compound that dissociates as it dissolves in water. The lattice energy of XY is -600 kJ/mol. The hydration energy of its ions is -610 kJ/mol.

(a) Write the thermochemical equations for the two steps in the formation of a solution of *XY* in water.

(b) Write the sum of these two equations in the form of a thermochemical equation, showing the net ΔH.

(c) Draw an enthalpy diagram for the formation of this solution.

17. At 20 °C a 40.00% (v/v) solution of ethyl alcohol, C_2H_5OH, in water, has a density of 0.9369 g/mL. The density of pure ethyl alcohol at this temperature is 0.7907 g/mL and that of water is 0.9982 g/mL.

(a) Calculate the molar concentration and the molal concentration of C_2H_5OH in this solution.

(b) Calculate the concentration of C_2H_5OH in this solution in mole fraction and mole percent.

(c) The vapor pressure of ethyl alcohol at 20 °C is 41.0 torr and that of water is 17.5 torr. If the 40.00% (v/v) solution were ideal, what would be the vapor pressure of each component over the solution?

18. Estimate the boiling point of 1.0 molal $Al(NO_3)_3$, assuming that it dissociates entirely into Al^{3+} and NO_3^- ions in solution.

19. Squalene is an oil found chiefly in shark liver but is also present in low concentrations in olive oil, wheat germ oil, and yeast. A qualitative analysis disclosed that its molecules consist entirely of carbon and hydrogen. When a sample of squalene with a mass of 0.5680 g was burned in pure oxygen, there was obtained 1.8260 g of carbon dioxide and 0.6230 g of water.

(a) Calculate the empirical formula of squalene.

(b) When 0.1268 g of squalene was dissolved in 10.50 g of molten camphor, the freezing point of this solution was 177.3 °C. (The melting point of pure camphor is 178.4 °C, and its molal freezing point depression constant is 37.5 °C m^{-1}.) Calculate the molecular mass of squalene and determine its molecular formula.

20. If a gas in a cylinder pushes back a piston against a constant opposing pressure of 3.0×10^5 pascals and undergoes a volume change of 0.50 m³, how much work will the gas do, expressed in joules?

21. Glycine, one of the important amino acids, has the structure

$$
\begin{array}{ccc}
& H & O \\
& | & \| \\
H-N & -C & -C-O-H \\
& | & | \\
& H & H
\end{array}
$$

Calculate the atomization energy of this molecule from the data in Table E.3.

22. Use bond energies in Table E.3 to calculate the approximate energy that would be absorbed or given off in the formation of 25.0 g of C_2H_6 by the following reaction in the gas phase.

$$
\begin{array}{ccc}
& & H \quad H \\
& & | \quad | \\
H-C\equiv C-H + 2H_2 \longrightarrow & H-C-C-H \\
& & | \quad | \\
& & H \quad H
\end{array}
$$

23. What would be the algebraic signs of ΔS for the following reactions?

(a) $Br_2(l) + Cl_2(g) \rightarrow 2BrCl(g)$

(b) $CaO(s) + CO_2(g) \rightarrow CaCO_3(s)$

24. Which of the following states has the greatest entropy?

(a) $2H_2O(l)$ (d) $2H_2(g) + O_2(g)$

(b) $2H_2O(s)$ (e) $4H(g) + 2O(g)$

(c) $2H_2(l) + O_2(g)$

25. Calculate $\Delta S°$, $\Delta H°$, and $\Delta G_T°$ (at 400 °C) using energy units of joules or kilojoules for these reactions.

(a) $CaSO_4 \cdot 2H_2O(s) \rightarrow CaSO_4 \cdot \frac{1}{2}H_2O(s) + \frac{3}{2}H_2O(g)$

(b) $NaOH(s) + NH_4Cl(s) \rightarrow NaCl(s) + NH_3(g) + H_2O(g)$

(c) $SO_3(g) \rightarrow SO_2(g) + \frac{1}{2}O_2$

26. A reaction has the stoichiometry: $3A + B \rightarrow C + D$. The following data were obtained for the initial rate of formation of *C* at various concentrations of *A* and *B*.

Initial Concentrations		Initial Rate of Formation
[A]	[B]	of C (mol L^{-1} s^{-1})
0.010	0.010	2.0×10^{-4}
0.020	0.020	8.0×10^{-4}
0.020	0.010	8.0×10^{-4}

(a) What is the rate law for the reaction?

(b) What is the value of the rate constant?

(c) What is the rate at which *C* is formed if $[A] = 0.017$ *M* and $[B] = 0.033$ *M*?

27. Organic compounds that contain large proportions of nitrogen and oxygen tend to be unstable and are easily decomposed. Hexanitroethane, $C_2(NO_2)_6$, decomposes according to the equation

$$C_2(NO_2)_6 \longrightarrow 2NO_2 + 4NO + 2CO_2$$

The reaction in CCl_4 as a solvent is first order with respect to $C_2(NO_2)_6$. At 70 °C, $k = 2.41 \times 10^{-6}$ s^{-1} and at 100 °C, $k = 2.22 \times 10^{-4}$ s^{-1}.

(a) What is the half-life of $C_2(NO_2)_6$ at 70 °C? What is the half-life at 100 °C?

(b) If 0.100 mol of $C_2(NO_2)_6$ is dissolved in CCl_4 at 70 °C to give 1.00 L of solution, what will be the $C_2(NO_2)_6$ concentration after 500 minutes?

(c) What is the value of the activation energy of this reaction, expressed in kilojoules?

(d) What is the reaction's rate constant at 120 °C?

A waterfall in Grand Canyon National Park in Arizona pours into a pond that then empties by another waterfall. Because water enters and leaves the pond at the same rate, the amount of water in the pond does not change. The situation is akin to a dynamic equilibrium. In this chapter we will examine how the concept of dynamic equilibrium applies to chemical systems.

Chapter 15

Chemical Equilibrium—General Concepts

You already encountered the concept of a dynamic equilibrium several times earlier in this book. Recall that such an equilibrium is established when two opposing processes occur at equal rates. In a liquid–vapor equilibrium, for example, the processes are evaporation and condensation. In a chemical equilibrium, they are the forward and reverse reactions represented by the chemical equation. For instance, you learned that acetic acid ($HC_2H_3O_2$) rapidly establishes the equilibrium

$$HC_2H_3O_2(aq) + H_2O \rightleftharpoons H_3O^+(aq) + C_2H_3O_2^-(aq)$$

in which $HC_2H_3O_2$ and H_2O are reacting at the same rate as H_3O^+ and $C_2H_3O_2^-$. Let's review how this chemical equilibrium comes about.

When acetic acid is added to the water, the forward reaction (ionization) occurs rapidly at first, but then slows as the number of $HC_2H_3O_2$ molecules decreases. Just the opposite effect is found for the reverse reaction. As the concentrations of the ions increase, the reverse reaction speeds up. Eventually, the rates of the forward and reverse reactions become equal, so the concentrations of the ions cease to change. At this point the system has reached a state of dynamic equilibrium.

The reaction of acetic acid with water is just one example of a general phenomenon. For almost any reaction, the concentrations change rapidly at first and then approach steady equilibrium values as shown in Figure 15.1. If the system is not disturbed, these concentrations will never change. Keep in

15.1
DYNAMIC EQUILIBRIUM IN CHEMICAL SYSTEMS

The reaction read from left to right is the *forward reaction*. The reaction read from right to left is the *reverse reaction*.

It is an *equilibrium* because the concentrations don't change. It is *dynamic* because the opposing reactions never cease.

635

FIGURE 15.1

As a reaction proceeds, the concentrations of the reactants and products approach steady (constant) values.

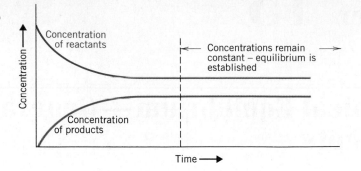

Sometimes an equal sign (=) is used in place of the double arrows.

mind, however, that even after equilibrium has been reached, both the forward and reverse reactions continue to occur. This is what is implied by the double arrows, $\rightleftharpoons$.

Almost all chemical systems eventually reach a state of dynamic equilibrium, although sometimes this equilibrium is extremely difficult (or even impossible) to detect. This is because in some reactions the amounts of either the reactants or products present at equilibrium are virtually zero. For instance, when a strong acid such as HCl is dissolved in water, it reacts so completely to produce H_3O^+ and Cl^- that no detectable amount of HCl molecules remains; we say that the HCl is completely ionized. Similarly, in water vapor at room temperature, there are no detectable amounts of either H_2 or O_2 from the equilibrium

$$2H_2O(g) \rightleftharpoons 2H_2(g) + O_2(g)$$

The water molecules are so stable that we can't detect whether any of them decompose. Even in cases such as these, however, it is often convenient to presume that an equilibrium does exist.

In this chapter, we will study the equilibrium condition both qualitatively and quantitatively. By knowing the factors that affect the composition of an equilibrium system we have some control over the outcome of a reaction. For example, the formation of the pollutant nitric oxide (NO) by reaction of N_2 and O_2 in a gasoline engine can be reduced, in a predictable way, by lowering the combustion temperature within the engine. In this chapter we will learn why. Because many biological substances are weak acids, knowledge of the principles of equilibrium helps biologists to understand how organisms control the acidity of fluids within them. We'll use the basic principles of equilibrium developed in this chapter to learn more about acids and bases in Chapter 16.

15.2 REACTION REVERSIBILITY

The ionization of acetic acid is an example of a reaction that is able to proceed in either direction. Acetic acid molecules are able to react with water molecules to form the ions, and the ions are able to react to form the molecules. For most chemical reactions the composition of the equilibrium mixture is found to be independent of whether we begin the reaction from the "reactant side" or the "product side." To illustrate this, let's consider some experiments that we might perform at 25 °C on the gaseous decomposition of N_2O_4 into NO_2,

$$N_2O_4(g) \rightleftharpoons 2NO_2(g)$$

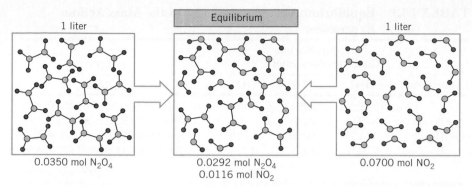

1 liter

0.0350 mol N₂O₄

Equilibrium

0.0292 mol N₂O₄
0.0116 mol NO₂

1 liter

0.0700 mol NO₂

FIGURE 15.2

Reaction reversibility. The same equilibrium composition is reached from either the forward or reverse direction, provided the *overall* system composition is the same.

Suppose we set up the two experiments shown in Figure 15.2. In the first 1-liter flask we place 0.0350 mol N_2O_4. Since no NO_2 is present, some N_2O_4 must decompose for the mixture to reach equilibrium, so the reaction will proceed in the forward direction (i.e., from left to right). When equilibrium is reached, we find the concentration of N_2O_4 has dropped to 0.0292 mol/L and the concentration of NO_2 has become 0.0116 mol/L.

In the second 1-L flask we place 0.0700 mol of NO_2 (*precisely* the amount of NO_2 that would form if 0.0350 mol of N_2O_4 decomposed completely). In this second flask there is no N_2O_4 present initially, so NO_2 molecules must combine, following the reverse reaction (right to left) shown above, to give enough N_2O_4 for equilibrium. If we measure the concentrations at equilibrium in the second flask, we find, once again, 0.0292 mol/L of N_2O_4 and 0.0116 mol/L of NO_2.

0.0700 mol of NO_2 could be formed from 0.0350 mol of N_2O_4.

We see here that the identical composition of the system results whether we begin with pure NO_2 or pure N_2O_4, just as long as the total amount of nitrogen and oxygen to be divided between these two substances is the same. For a given *overall* composition we always reach the same equilibrium concentrations whether equilibrium is approached from the forward or reverse direction. It is in this sense that chemical reactions are said to be reversible.

For any chemical system at equilibrium there exists a simple relationship among the molar concentrations of the reactants and products. To show you this we will use the gaseous reaction of hydrogen with iodine to form hydrogen iodide. The equation for the reaction is

$$H_2(g) + I_2(g) \rightleftharpoons 2HI(g)$$

Let's see what happens if we set up several experiments in which we start with different amounts of the reactants and/or product, as illustrated in Figure 15.3. Notice that when equilibrium is reached the concentrations of H_2, I_2, and HI are different for each experiment. This isn't particularly surprising, but what is amazing is that the relationship among the concentrations is very simple and can actually be predicted from the balanced equation for the reaction.

For each experiment in Figure 15.3, if we square the molar concentration of HI at equilibrium and then divide this by the product of the equilibrium molar concentrations of H_2 and I_2, we obtain the same numerical value. This is shown in Table 15.1 where we have once again used square brackets around formulas as symbols for molar concentrations.

15.3
THE EQUILIBRIUM LAW FOR A REACTION

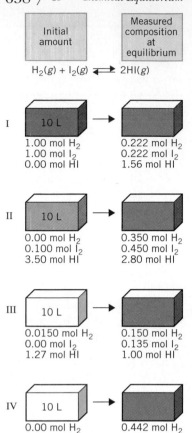

FIGURE 15.3

Four experiments to study the equilibrium among H_2, I_2, and HI gases at 440 °C.

TABLE 15.1 Equilibrium Concentrations and the Mass Action Expression

Experiment	Equilibrium Concentrations (mol L^{-1})			$\dfrac{[HI]^2}{[H_2][I_2]}$
	[H$_2$]	**[I$_2$]**	**[HI]**	
I	0.0222	0.0222	0.156	$(0.156)^2/(0.0222)(0.0222) = 49.4$
II	0.0350	0.0450	0.280	$(0.280)^2/(0.0350)(0.0450) = 49.8$
III	0.0150	0.0135	0.100	$(0.100)^2/(0.0150)(0.0135) = 49.4$
IV	0.0442	0.0442	0.311	$(0.311)^2/(0.0442)(0.0442) = 49.5$
				Average $= 49.5$

The fraction used to calculate the values in the last column of Table 15.1,

$$\frac{[HI]^2}{[H_2][I_2]}$$

is called the **mass action expression.** The origin of this term isn't important; just consider it a name we use to refer to this fraction. The numerical value of the mass action expression is called the **reaction quotient,** and is often symbolized by the letter Q. In Table 15.1, we see that when H_2, I_2, and HI are in dynamic equilibrium at 440 °C, the reaction quotient is equal to essentially a constant value of 49.5. In fact, if we repeated the experiments in Figure 15.3 over and over again starting with different amounts of H_2, I_2, and HI, we would still obtain the same reaction quotient, provided the systems had reached equilibrium and the temperature was 440 °C. Therefore, for this reaction we can write

$$\frac{[HI]^2}{[H_2][I_2]} = 49.5 \quad \text{(at 440 °C)} \qquad (15.1)$$

This is the **equilibrium law** for the system. It tells us that for a mixture of these three gases to be at equilibrium at 440 °C, the value of the mass action expression (the reaction quotient) must equal 49.5. If the reaction quotient has any other value, then the gases are not in equilibrium and they must react further. The constant 49.5 which characterizes this equilibrium system is called the **equilibrium constant.** The equilibrium constant is usually symbolized by K_c (the subscript c because we write the mass action expression using molar concentrations). Thus, we can state the equilibrium law as follows:

$$\frac{[HI]^2}{[H_2][I_2]} = K_c = 49.5 \quad \text{(at 440 °C)} \qquad (15.2)$$

It is often useful to think of an equilibrium law such as Equation 15.2 as a *condition* that must be met for equilibrium to exist.

For chemical equilibrium to exist in a reaction mixture, the reaction quotient Q must be equal to the equilibrium constant, K_c.

As you've probably noticed, we have repeatedly mentioned the temperature when referring to the value of K_c. This is because the value of the equilibrium constant changes when the temperature changes. Thus, if we had performed the experiments in Figure 15.3 at a temperature other than 440 °C, we would have obtained a different value for K_c.

Predicting the Equilibrium Law

An important fact about the mass action expression and the equilibrium law is that it can *always* be predicted from the balanced chemical equation for the reaction. For example, for the general chemical equation

$$dD + eE \rightleftharpoons fF + gG$$

where *D, E, F,* and *G* represent chemical formulas and *d, e, f,* and *g* are their coefficients, the mass action expression is

$$\frac{[F]^f[G]^g}{[D]^d[E]^e}$$

The coefficients in the balanced equation are the same as the exponents in the mass action expression.

The condition for equilibrium in this reaction is given by the equation

$$\frac{[F]^f[G]^g}{[D]^d[E]^e} = K_c$$

where the only concentrations that satisfy the equation are *equilibrium concentrations.*

Notice that in writing the mass action expression the molar concentrations of the products are always placed in the numerator and those of the reactants appear in the denominator. Also note that after being raised to appropriate powers the concentration terms are *multiplied,* not added.

In general, it is necessary to specify the temperature when giving a value of K_c, because K_c changes when the temperature changes. For example, for the reaction

$$CH_4(g) + H_2O(g) \rightleftharpoons$$
$$CO(g) + 3H_2(g)$$

$K_c = 1.78 \times 10^{-3}$ at 800 °C

$K_c = 4.68 \times 10^{-2}$ at 1000 °C

$K_c = 5.67$ at 1500 °C

Notice that even though we cannot predict the *rate law* from the balanced overall equation, we can predict the *equilibrium law.*

 Mass action expression and equilibrium law

Write the equilibrium law for the reaction

$$N_2(g) + 3H_2(g) \rightleftharpoons 2NH_3(g)$$

which is used to synthesize ammonia industrially.

SOLUTION The equilibrium law sets the mass action expression equal to the equilibrium constant. To form the mass action expression, we place the concentrations of the products in the numerator and the concentrations of the reactants in the denominator. The coefficients in the equation become exponents on the concentrations. Therefore,

$$\frac{[NH_3]^2}{[N_2][H_2]^3} = K_c$$

Notice that we omit writing the exponent when it is equal to one.

■ **Practice Exercise 1** Write the equilibrium law for each of the following:
(a) $2H_2(g) + O_2(g) \rightleftharpoons 2H_2O(g)$
(b) $NH_3(aq) + H_2O(l) \rightleftharpoons NH_4^+(aq) + OH^-(aq)$

EXAMPLE 15.1
Writing the Equilibrium Law

The rule that we always write the concentrations of the products in the numerator of the mass action expression and the concentrations of the reactants in the denominator is not required by nature. It is simply a convention chemists have agreed on. Certainly, if the mass action expression is equal to a constant,

$$\frac{[\text{HI}]^2}{[\text{H}_2][\text{I}_2]} = K_c$$

its reciprocal is also equal to a constant (let's call it K_c'),

$$\frac{[\text{H}_2][\text{I}_2]}{[\text{HI}]^2} = \frac{1}{K_c} = K_c'$$

The reason that chemists choose to stick to a set pattern—always placing the concentrations of the products in the numerator—is to avoid having to specify mass action expressions along with tabulated values of equilibrium constants. If we have the chemical equation for the equilibrium, we can *always* construct the correct mass action expression from it. For example, suppose we're told that at a particular temperature $K_c = 10.0$ for the reaction

$$2\text{NO}_2(g) \rightleftharpoons \text{N}_2\text{O}_4(g)$$

From the chemical equation we can write the correct mass action expression and the correct equilibrium law.

$$K_c = \frac{[\text{N}_2\text{O}_4]}{[\text{NO}_2]^2}$$

The balanced chemical equation contains all the information we need to write the equilibrium law.

Manipulating Equations for Chemical Equilibria

Manipulating equilibrium equations

Sometimes it is useful to be able to combine chemical equilibria to obtain the equation for some other reaction of interest. In doing this, we perform various operations such as reversing an equation, multiplying the coefficients by some factor, and adding the equations to give the desired equation. In our discussion of thermochemistry, you learned how such manipulations affect ΔH values. Some different rules apply to changes in the mass action expressions and equilibrium constants.

Changing the Direction of an Equilibrium   *When the direction of an equation is reversed, the new equilibrium constant is the reciprocal of the original.* You have just seen this in the discussion above. As another example, when we reverse the equilibrium

$$\text{PCl}_3 + \text{Cl}_2 \rightleftharpoons \text{PCl}_5 \qquad K_c = \frac{[\text{PCl}_5]}{[\text{PCl}_3][\text{Cl}_2]}$$

we obtain

$$\text{PCl}_5 \rightleftharpoons \text{PCl}_3 + \text{Cl}_2 \qquad K_c' = \frac{[\text{PCl}_3][\text{Cl}_2]}{[\text{PCl}_5]}$$

The mass action expression for the second reaction is the reciprocal of that for the first, so K_c' equals $1/K_c$.

Multiplying the Coefficients by a Factor   *When the coefficients in an equation are multiplied by a factor, the equilibrium constant is raised to a power equal to that* *factor.* For example, suppose we multiply the coefficients of the equation

$$PCl_3 + Cl_2 \rightleftharpoons PCl_5 \qquad K_c = \frac{[PCl_5]}{[PCl_3][Cl_2]}$$

by 2. This gives

$$2PCl_3 + 2Cl_2 \rightleftharpoons 2PCl_5 \qquad K_c'' = \frac{[PCl_5]^2}{[PCl_3]^2[Cl_2]^2}$$

Comparing mass action expressions, we see that $K_c'' = K_c^2$.

Adding Chemical Equilibria   *When chemical equilibria are added, their equilibrium constants are multiplied.* For example, suppose we add the following two equations.

$$2N_2 + O_2 \rightleftharpoons 2N_2O \qquad K_{c1} = \frac{[N_2O]^2}{[N_2]^2[O_2]}$$

$$2N_2O + 3O_2 \rightleftharpoons 4NO_2 \qquad K_{c2} = \frac{[NO_2]^4}{[N_2O]^2[O_2]^3}$$

$$2N_2 + 4O_2 \rightleftharpoons 4NO_2 \qquad K_{c3} = \frac{[NO_2]^4}{[N_2]^2[O_2]^4}$$

We have numbered the equilibrium constants just to distinguish one from the other.

If we multiply the mass action expression for K_{c1} by that for K_{c2}, we obtain the mass action expression for K_{c3}.

$$\frac{[N_2O]^2}{[N_2]^2[O_2]} \times \frac{[NO_2]^4}{[N_2O]^2[O_2]^3} = \frac{[NO_2]^4}{[N_2]^2[O_2]^4}$$

Therefore, $K_{c1} \times K_{c2} = K_{c3}$.

■ **Practice Exercise 2**   At 25 °C, $K_c = 7.0 \times 10^{25}$ for the reaction

$$2SO_2(g) + O_2(g) \rightleftharpoons 2SO_3(g)$$

What is the value of K_c for the reaction: $SO_3(g) \rightleftharpoons SO_2(g) + \frac{1}{2}O_2(g)$?

■ **Practice Exercise 3**   At 25 °C, the following reactions have the equilibrium constants noted to the right of their equations.

$$2CO(g) + O_2(g) \rightleftharpoons 2CO_2(g) \qquad K_c = 3.3 \times 10^{91}$$

$$2H_2(g) + O_2(g) \rightleftharpoons 2H_2O(g) \qquad K_c = 9.1 \times 10^{80}$$

Use these data to calculate K_c for the reaction

$$H_2O(g) + CO(g) \rightleftharpoons CO_2(g) + H_2(g)$$

When all the reactants and products are gases, we can formulate mass action expressions in terms of partial pressures as well as molar concentrations. This is possible because the molar concentration of a gas is proportional to its partial pressure. This comes from the ideal gas law,

$$PV = nRT$$

Solving for P gives

$$P = \left(\frac{n}{V}\right)RT$$

15.4
EQUILIBRIUM LAWS FOR GASEOUS REACTIONS

If you double the molar concentration of a gas without changing its temperature, you double its pressure.

The quantity n/V has units of mol/L and is simply the molar concentration. Therefore, we can write

$$P = (\text{molar concentration}) \times RT \qquad (15.3)$$

This equation applies whether the gas is by itself in a container or part of a mixture. In the case of a gas mixture, P is the partial pressure of the gas.

The relationship expressed in Equation 15.3 lets us write the mass action expression for reactions between gases in terms either of molarities or of partial pressures. However, when we make a switch we can't expect the numerical values of the equilibrium constants to be the same, so we use two different symbols for K. When molar concentrations are used, we use the symbol K_c. When partial pressures are used, then K_p is the symbol. For example, the equilibrium law for the reaction of nitrogen with hydrogen to form ammonia.

$$N_2(g) + 3H_2(g) \rightleftharpoons 2NH_3(g)$$

can be written in either of the following two ways

$$\frac{[NH_3]^2}{[N_2][H_2]^3} = K_c \qquad \left(\begin{array}{c}\text{because molar concentrations are used} \\ \text{in the mass action expression}\end{array}\right)$$

$$\frac{P_{NH_3}^2}{P_{N_2}P_{H_2}^3} = K_p \qquad \left(\begin{array}{c}\text{because partial pressures are used in} \\ \text{the mass action expression}\end{array}\right)$$

The equilibrium molar concentrations can be used to calculate K_c, whereas the equilibrium partial pressures can be used to calculate K_p. We will discuss how to convert between K_c and K_p in Section 15.7.

EXAMPLE 15.2
Writing Expressions for K_p

Write the expression for K_p for the reaction

$$N_2O_4(g) \rightleftharpoons 2NO_2(g)$$

SOLUTION For K_p we use partial pressures in the mass action expression.

$$K_p = \frac{P_{NO_2}^2}{P_{N_2O_4}}$$

■ **Practice Exercise 4** Using partial pressures, write the equilibrium law for the reaction

$$H_2(g) + I_2(g) \rightleftharpoons 2HI(g)$$

15.5
THE SIGNIFICANCE OF THE MAGNITUDE OF K

Whether we work with K_p or K_c, a bonus of always writing the mass action expression with the product concentrations in the numerator is that the size of the equilibrium constant gives us a measure of how far the reaction proceeds toward completion when equilibrium is reached. For example, the reaction

$$2H_2(g) + O_2(g) \rightleftharpoons 2H_2O(g)$$

has $K_c = 9.1 \times 10^{80}$ at 25 °C. This means that when there is an equilibrium between these gases,

$$K_c = \frac{[H_2O]^2}{[H_2]^2[O_2]} = \frac{9.1 \times 10^{80}}{1}$$

By writing K_c as a fraction, $(9.1 \times 10^{80})/1$, we see that the only way for the numerator to be so much larger than the denominator is for the concentration of H_2O to be enormous in comparison to the concentrations of H_2 and O_2. This means that at equilibrium most of the hydrogen and oxygen atoms in the system are found in the H_2O and very few are present in H_2 and O_2. Thus, the large value of K_c tells us that the reaction between H_2 and O_2 goes essentially to completion.

The reaction between N_2 and O_2 to give NO

$$N_2(g) + O_2(g) \rightleftharpoons 2NO(g)$$

has a very small equilibrium constant; $K_c = 4.8 \times 10^{-31}$ at 25 °C. The equilibrium law for this reaction is

$$\frac{[NO]^2}{[N_2][O_2]} = 4.8 \times 10^{-31}$$

Since $10^{-31} = 1/10^{31}$, we can write this as

$$\frac{[NO]^2}{[N_2][O_2]} = \frac{4.8}{10^{31}}$$

Here the denominator is huge compared with the numerator, so the concentrations of N_2 and O_2 must be very much larger than the concentration of NO. This means that in a mixture of N_2 and O_2 at this temperature, very little NO is formed. The reaction hardly proceeds at all toward completion before equilibrium is reached.

The relationship between the equilibrium constant and the position of equilibrium can thus be summarized as follows:

When K is very large	The reaction proceeds far toward completion. The position of equilibrium lies far toward the products.
When $K \approx 1$	The concentrations of reactants and products are nearly the same at equilibrium. The position of equilibrium lies midway between reactants and products.
When K is very small	Extremely small amounts of products are formed. The position of equilibrium lies far toward the reactants.

Notice that we have omitted the subscript for K in this summary. The same qualitative predictions about the extent of reaction apply whether we use K_p or K_c.

One of the ways that we can use equilibrium constants is to compare the extents to which two or more reactions proceed to completion. Take care in making such comparisons, however, because unless the K's are greatly different, the comparison is valid only if both reactions have the same number of reactant and product molecules appearing in their balanced chemical equations.

■ **Practice Exercise 5** Which of the following reactions will tend to proceed farthest toward completion?
(a) $H_2(g) + Br_2(g) \rightleftharpoons 2HBr(g)$ $K_c = 1.4 \times 10^{-21}$
(b) $2NO(g) \rightleftharpoons N_2(g) + O_2(g)$ $K_c = 2.1 \times 10^{30}$
(c) $2BrCl \rightleftharpoons Br_2 + Cl_2$ (in CCl_4 solution) $K_c = 0.145$

Actually, you would need about 200,000 L of water vapor at 25 °C just to find one molecule of O_2 and two molecules of H_2.

In air at 25 °C, the equilibrium concentration of NO *should be* about 10^{-17} mol/L. It is usually higher because NO is formed in various reactions, such as those responsible for air pollution caused by automobiles.

 Significance of the magnitude of K

15.6
CALCULATING EQUILIBRIUM CONSTANTS FROM THERMODYNAMIC DATA

In Chapter 13 you learned qualitatively that the position of equilibrium in a reaction is determined by the sign and magnitude of $\Delta G°$. If $\Delta G°$ is large and negative, the reaction mixture at equilibrium contains lots of products. But if $\Delta G°$ is large and positive, hardly any products are present at equilibrium. You also learned that a reaction is spontaneous if it proceeds in a direction that leads to a lowering of the free energy—that is, when ΔG is negative.

Quantitatively, the relationship between ΔG and $\Delta G°$ is expressed by the following equation, which we will not attempt to justify.

Reaction quotient and spontaneity of reaction

$$\Delta G = \Delta G° + RT \ln Q \qquad (15.4)$$

ΔG, which determines spontaneity, can be either positive or negative depending on where the composition of the system stands relative to equilibrium. $\Delta G°$ is a constant for a reaction; it is the difference between $G°_{products}$ and $G°_{reactants}$.

Here R is the gas constant in appropriate energy units (e.g., $8.314 \, \text{J mol}^{-1} \, \text{K}^{-1}$), T is the Kelvin temperature, and $\ln Q$ is the natural logarithm of the reaction quotient. For gaseous reactions, Q is calculated using partial pressures expressed in atmospheres[1]; for reactions in solution, Q is calculated from molar concentrations. Equation 15.4 allows us to predict the direction of the spontaneous change in a reaction mixture if we know $\Delta G°$ and the composition of the mixture, as illustrated in Example 15.3.

EXAMPLE 15.3
Determining the Direction of a Spontaneous Reaction

The reaction $2NO_2(g) \rightleftharpoons N_2O_4(g)$ has $\Delta G°_{298} = -5.40$ kJ per mole of N_2O_4. In a reaction mixture, the partial pressure of NO_2 is 0.25 atm and the partial pressure of N_2O_4 is 0.60 atm. In which direction must this reaction proceed to reach equilibrium?

ANALYSIS Since we know that reactions proceed spontaneously *toward* equilibrium, we are really being asked to determine whether the reaction will proceed spontaneously in the forward or reverse direction. We can use Equation 15.4 to calculate ΔG for the forward reaction. If ΔG is negative, then the forward reaction is spontaneous. However, if the calculated ΔG is positive, the forward reaction is nonspontaneous and it is really the reverse reaction that is spontaneous.

SOLUTION First, we need the correct form for the mass action expression so we can calculate Q correctly. Expressed in terms of partial pressures, the mass action expression is

$$\frac{P_{N_2O_4}}{P_{NO_2}^2}$$

Therefore, the equation we will use is

$$\Delta G = \Delta G° + RT \ln \left(\frac{P_{N_2O_4}}{P_{NO_2}^2} \right)$$

Next, let's assemble the data:

$$\Delta G°_{298} = -5.40 \text{ kJ mol}^{-1}$$
$$= -5.40 \times 10^3 \text{ J mol}^{-1}$$

$$R = 8.314 \text{ J mol}^{-1} \text{ K}^{-1} \qquad P_{N_2O_4} = 0.60 \text{ atm}$$

$$T = 298 \text{ K} \qquad P_{NO_2} = 0.25 \text{ atm}$$

---

[1] In Chapter 13, we noted in a footnote that the SI has adopted a standard unit of pressure called the bar, which is not quite equal to 1 atm. For simplicity, we will continue to refer to pressures in atmospheres. Thermodynamic quantities at 1 atm and at 1 bar differ by only a very small (virtually insignificant) amount.

Notice we have changed the energy units of $\Delta G°$ to joules so they will be compatible with those calculated using R. Substituting quantities gives[2]

$$\Delta G = -5.40 \times 10^3 \text{ J mol}^{-1} + (8.314 \text{ J mol}^{-1} \text{ K}^{-1})(298 \text{ K}) \ln \left[\frac{0.60 \text{ atm}}{(0.25 \text{ atm})^2} \right]$$

$$= -5.40 \times 10^3 \text{ J mol}^{-1} + (8.314 \text{ J mol}^{-1} \text{ K}^{-1})(298 \text{ K})(2.26)$$
$$= -5.40 \times 10^3 \text{ J mol}^{-1} + 5.60 \times 10^3 \text{ J mol}^{-1}$$
$$= +2.0 \times 10^2 \text{ J mol}^{-1}$$

Since ΔG is positive, the forward reaction is nonspontaneous. The reverse reaction is the one that will occur, and to reach equilibrium, some N_2O_4 will have to decompose.

There is no simple check for reasonableness in this example. However, we can check to be sure the energy units in both terms on the right are the same. Notice that here we have changed the units for $\Delta G°$ to joules to match those of R. Also notice that the temperature is expressed in kelvins, to match the temperature units in R.

■ **Practice Exercise 6** In which direction will the reaction described in the preceding example proceed to reach equilibrium if the partial pressure of NO_2 is 0.60 atm and the partial pressure of N_2O_4 is 0.25 atm?

Thermodynamic Equilibrium Constants

In Chapter 13 you learned that when a system reaches equilibrium the free energy of the products equals the free energy of the reactants and ΔG equals zero. We also know that at equilibrium the reaction quotient equals the equilibrium constant.

$$\text{At equilibrium} \quad \begin{cases} \Delta G = 0 \\ Q = K \end{cases}$$

If we substitute these into Equation 15.4, we obtain

$$0 = \Delta G° + RT \ln K$$

which can be rearranged to give

$$\Delta G° = -RT \ln K \qquad (15.5)$$

 Calculating thermodynamic equilibrium constants

The equilibrium constant K calculated from this equation is often called the **thermodynamic equilibrium constant** and corresponds to K_p for reactions involving gases (with partial pressures expressed in atmospheres) and to K_c for reactions in solution (with concentrations expressed in mol/L).

Equation 15.5 is useful because it permits us to determine equilibrium constants from either measured or calculated values of $\Delta G°$. As you learned in Chapter 13, $\Delta G°$ can be determined by a Hess's law type of calculation from tabulated values of $\Delta G_f°$. Equation 15.5 also allows us to obtain values of $\Delta G°$ from equilibrium constants.

[2] As we noted in the footnote on page 599, when taking the logarithm of a quantity the number of digits *after the decimal point* should equal the number of significant figures in the quantity. Since 0.60 atm and 0.25 atm both have two significant figures, the logarithm of the quantity in square brackets (2.26) is rounded to give two digits after the decimal point.

EXAMPLE 15.4
Thermodynamic Equilibrium
Constants

The brownish haze associated with air pollution is caused by nitrogen dioxide, NO_2, a red-brown gas. Nitric oxide, NO, is oxidized to NO_2 by oxygen.

$$2NO(g) + O_2(g) \rightleftharpoons 2NO_2(g)$$

The value of K_p for this reaction is 1.7×10^{12} at 25.00 °C. What is $\Delta G°$ for the reaction, expressed in joules? In kilojoules?

SOLUTION We must substitute into the equation

$$\Delta G° = -RT \ln K_p$$

First, let's list the data. (For the temperature, we will use the equation $T_K = t_C + 273.15$ to be sure we have enough significant figures.)

$$R = 8.314 \text{ J mol}^{-1} \text{ K}^{-1}$$

$$T = 298.15 \text{ K}$$

$$K_p = 1.7 \times 10^{12}$$

Substituting values,

$$\begin{aligned}
\Delta G° &= -(8.314 \times 298.15) \ln(1.7 \times 10^{12}) \\
&= -(8.314 \times 298.15) \times (28.16) \\
&= -6.980 \times 10^4 \text{ J} \quad \text{(to four significant figures)}
\end{aligned}$$

Expressed in kilojoules, $\Delta G° = -69.80$ kJ.

CHECKING THE ANSWER FOR REASONABLENESS The value of K_p tells us that the position of equilibrium lies far to the right, which, as you learned in Chapter 13, means $\Delta G°$ must be large and negative. Therefore, the answer, -69.82 kJ, seems reasonable.

EXAMPLE 15.5
Thermodynamic Equilibrium
Constants

Sulfur dioxide reacts with oxygen when it passes over the catalyst in automobile catalytic converters. The product is SO_3.

$$2SO_2(g) + O_2(g) \rightleftharpoons 2SO_3(g)$$

For this reaction, $\Delta G° = -1.40 \times 10^2$ kJ at 25 °C. What is the value of K_p?

SOLUTION Once again we use the equation

$$\Delta G° = -RT \ln K_p$$

Our data are

$$R = 8.314 \text{ J mol}^{-1} \text{ K}^{-1}$$

$$T = 298 \text{ K}$$

$$\begin{aligned}
\Delta G° &= -1.40 \times 10^2 \text{ kJ} \\
&= -1.40 \times 10^5 \text{ J}
\end{aligned}$$

To calculate K_p, let's first solve for $\ln K_p$.

$$\ln K_p = \frac{-\Delta G°}{RT}$$

Substituting values gives

$$\begin{aligned}
\ln K_p &= \frac{-(-1.40 \times 10^5)}{(8.314)(298)} \\
&= +56.5
\end{aligned}$$

To calculate K_p, we take the antilogarithm,[3]

$$K_p = e^{56.5}$$
$$= 3 \times 10^{24}$$

If necessary, review the discussion of logarithms and antilogarithms in Appendix A.

Notice that we have expressed the answer to only one significant figure. As discussed earlier, when taking a logarithm, the number of digits written after the decimal place equals the number of significant figures in the number. Conversely, the number of significant figures in the antilogarithm equals the number of digits after the decimal in the logarithm.

CHECKING THE ANSWER FOR REASONABLENESS The value of $\Delta G°$ is large and negative, so the position of equilibrium should favor the products. The large value of K_p is therefore reasonable.

■ **Practice Exercise 7** The reaction $N_2(g) + 3H_2(g) \rightleftharpoons 2NH_3(g)$ has $K_p = 6.9 \times 10^5$ at 25.0 °C. Calculate $\Delta G°$ for this reaction in units of kilojoules.

■ **Practice Exercise 8** The reaction $H_2(g) + I_2(g) \rightleftharpoons 2HI(g)$ has $\Delta G° = +3.3$ kJ at 25.0 °C. What is the value of K_p at this temperature?

Thermodynamic Equilibrium Constants at Temperatures Other than 25 °C

In Section 13.9, we discussed the way temperature affects the position of equilibrium in a reaction. At 25 °C, the relative proportions of reactants and products at equilibrium are determined by $\Delta G°_{298}$, and we have just seen that we can calculate the value of the equilibrium constant at 25 °C from $\Delta G°_{298}$. As the temperature moves away from 25 °C, the position of equilibrium also changes because of changes in the value of $\Delta G°_T$ (the equivalent of $\Delta G°$, but at a temperature other than 25 °C). Therefore, $\Delta G°_T$ can be used to calculate K at temperatures other than 25 °C by the same methods used in Examples 15.4 and 15.5.

The decomposition of nitrous oxide, N_2O, has $K_p = 1.8 \times 10^{36}$ at 25 °C. The equation is

$$2N_2O(g) \rightleftharpoons 2N_2(g) + O_2(g)$$

For this reaction, $\Delta H° = -163$ kJ and $\Delta S° = +148$ J K^{-1}. What is the approximate value for K_p for this reaction at 40 °C?

EXAMPLE 15.6
Calculating K at Temperatures Other than 25 °C

ANALYSIS To apply Equation 15.5, we need to have the value of $\Delta G°$ at 40 °C (313 K). Let's represent this as $\Delta G°_{313 K}$. We can estimate $\Delta G°_{313}$ using the values of $\Delta H°$ and $\Delta S°$ measured at 25 °C.

$$\Delta G°_{313} \approx \Delta H°_{298} - (313 \text{ K})\Delta S°_{298}$$

You learned in Chapter 13 that this is permitted because $\Delta H°$ and $\Delta S°$ do not change much with temperature.

Substituting the values of $\Delta H°$ and $\Delta S°$ provided in the problem gives

$$\Delta G°_{313} \approx -1.63 \times 10^5 \text{ J} - (313 \text{ K})(+148 \text{ J K}^{-1})$$

[3] The rule is that the number of digits after the decimal point in a quantity equals the number of significant figures in its antilogarithm. Since 56.5 has only one digit past the decimal point, its antilogarithm has only one significant figure.

Notice that we've converted kilojoules to joules. Performing the arithmetic gives

$$\Delta G^{\circ}_{313} \approx -2.09 \times 10^5 \, \text{J} \quad \text{(rounded)}$$

The next step is to use this value of ΔG°_{313} to compute K_p with Equation 15.5. First, let's solve for $\ln K_p$.

$$\ln K_p = \frac{-\Delta G^{\circ}_{313}}{RT}$$

Substituting, with $R = 8.314 \, \text{J mol}^{-1} \, \text{K}^{-1}$ and $T = 313$ K, gives

$$\ln K_p = \frac{2.09 \times 10^5}{(8.314)(313)}$$
$$= +80.3$$

Taking the antilogarithm,

$$K_p = e^{80.3}$$
$$= 7 \times 10^{34}$$

Notice that at this higher temperature, N_2O is actually slightly more stable than at 25 °C, as reflected in the slightly smaller value for the equilibrium constant for its decomposition.

■ **Practice Exercise 9** The reaction $N_2(g) + 3H_2(g) \rightleftharpoons 2NH_3(g)$ has a standard heat of reaction of -92.4 kJ and a standard entropy of reaction equal to -198.3 J/K. Estimate the value of K_p for this reaction at 50 °C.

15.7
THE RELATIONSHIP BETWEEN K_p AND K_c

For some reactions K_p is equal to K_c, but for many others the two constants have different values. It is therefore desirable to have a way to calculate one from the other. Converting between K_p and K_c uses the relationship between partial pressure and molarity described in Section 15.4. Equation 15.3 can be used to change K_p to K_c by substituting

$$(\text{molar concentration}) \times RT$$

for the partial pressure of each gas in the mass action expression for K_p. Similarly, K_c can be changed to K_p by solving Equation 15.3 for the molar concentrations, and then substituting the result, P/RT, into the appropriate expression for K_c. This sounds like a lot of work, and it is. Fortunately, there is a general equation that we can use to make these conversions simply.

Relationship between K_p and K_c

$$K_p = K_c(RT)^{\Delta n_g} \tag{15.6}$$

In this equation, the value of Δn_g is equal to the change in the *number of moles of gas* in going from the reactants to the products.

$$\Delta n_g = (\text{moles of } \textit{gaseous} \text{ products}) - (\text{moles of } \textit{gaseous} \text{ reactants})$$

We use the coefficients of the balanced equation for the reaction to calculate the numerical value of Δn_g. For example, the equation

$$N_2(g) + 3H_2(g) \rightleftharpoons 2NH_3(g) \tag{15.7}$$

tells us that 2 mol of NH_3 are formed when 1 mol of N_2 and 3 mol of H_2 react.

In other words, 2 mol of gaseous product are formed from a total of 4 mol of gaseous reactants. That's a decrease of 2 mol of gas, so Δn_g for this reaction equals -2.

For some reactions, the value of Δn_g is equal to zero. An example is the decomposition of HI.

$$2HI(g) \rightleftharpoons H_2(g) + I_2(g)$$

Notice that if we take the coefficients to mean moles, there are 2 mol of gas on each side of the equation. This means that $\Delta n_g = 0$. Since (RT) raised to the zero power is equal to 1, $K_p = K_c$.

Δn_g is calculated from the coefficients of the equation, taking them to stand for moles.

At 500 °C, the reaction between N_2 and H_2 to form ammonia

$$N_2(g) + 3H_2(g) \rightleftharpoons 2NH_3(g)$$

has $K_c = 6.0 \times 10^{-2}$. What is the numerical value of K_p for this reaction?

SOLUTION The equation that we wish to use is

$$K_p = K_c(RT)^{\Delta n_g}$$

In the discussion above, we saw that $\Delta n_g = -2$ for this reaction. All we need now are appropriate values of R and T. The temperature, T, must be expressed in kelvins. (When used to stand for temperature, a capital letter T in an equation always means the absolute temperature.) Next we must choose the appropriate value for R. Referring back to Equation 15.3, we see that partial pressures would be in atm and concentrations would be in mol/L. The only value of R that is consistent with these units is $R = 0.0821$ L atm mol^{-1} K^{-1}, and this is the *only* value of R that can be used in Equation 15.6. Assembling the data, then, we have

$$K_c = 6.0 \times 10^{-2} \qquad \Delta n_g = -2$$
$$T = (500 + 273) \text{ K} = 773 \text{ K} \qquad R = 0.0821 \text{ L atm mol}^{-1} \text{ K}^{-1}$$

Substituting these into the equation for K_p gives

$$K_p = (6.0 \times 10^{-2}) \times [(0.0821) \times (773)]^{-2}$$
$$= (6.0 \times 10^{-2}) \times (63.5)^{-2}$$
$$= 1.5 \times 10^{-5}$$

In this case, K_p has a numerical value quite different from that of K_c.

EXAMPLE 15.7
Converting between K_p and K_c

At 25 °C, K_p for the reaction

$$N_2O_4(g) \rightleftharpoons 2NO_2(g)$$

has a value of 0.140. Calculate the value of K_c.

SOLUTION Once again, the equation we need is

$$K_p = K_c(RT)^{\Delta n_g}$$

This time, $\Delta n_g = 2 - 1 = +1$. Now let's tabulate the data.

$$K_p = 0.140 \qquad \Delta n_g = +1$$
$$T = 298 \text{ K} \qquad R = 0.0821 \text{ L atm mol}^{-1} \text{ K}^{-1}$$

EXAMPLE 15.8
Converting between K_p and K_c

Solving the equation for K_c gives

$$K_c = \frac{K_p}{(RT)^{\Delta n_g}}$$

Substituting values into this equation yields

$$K_c = \frac{0.140}{[(0.0821) \times (298)]^1}$$
$$= 5.72 \times 10^{-3}$$

Once again, there is a substantial difference between the values of K_p and K_c.

THINGS TO CHECK In working these problems, check to be sure you have used the correct value for R and that the temperature is expressed in kelvins.

■ **Practice Exercise 10** Methanol, CH_3OH, is a promising fuel that can be synthesized from carbon monoxide and hydrogen according to the equation

$$CO(g) + 2H_2(g) \rightleftharpoons CH_3OH(g)$$

For this reaction at 200 °C, $K_p = 3.8 \times 10^{-2}$. What is the value of K_c at this temperature?

■ **Practice Exercise 11** Nitrous oxide, N_2O, is a gas used as an anesthetic; it is sometimes called "laughing gas." This compound has a strong tendency to decompose into nitrogen and oxygen following the equation

$$2N_2O(g) \rightleftharpoons 2N_2(g) + O_2(g)$$

but the reaction is so slow that the gas appears to be stable at room temperature (25 °C). The decomposition reaction has $K_c = 7.3 \times 10^{34}$. What is the value of K_p for this reaction at 25 °C?

15.8 HETEROGENEOUS EQUILIBRIA

In a **homogeneous reaction**—or a **homogeneous equilibrium**—all of the reactants and products are in the same phase. Equilibria among gases are homogeneous because all gases mix freely with each other, so a single phase exists. There are also many equilibria in which reactants and products are dissolved in the same liquid phase.

When more than one phase exists in a reaction mixture, we call it a **heterogeneous reaction.** A common example is the combustion of wood, in which a solid fuel combines with gaseous oxygen. Another is the thermal decomposition of sodium bicarbonate (baking soda), which occurs when the compound is sprinkled on a fire.

$$2NaHCO_3(s) \longrightarrow Na_2CO_3(s) + H_2O(g) + CO_2(g)$$

Many cooks keep a box of baking soda nearby because this reaction makes baking soda an excellent fire extinguisher for burning fats or oil. The fire is smothered by the products of the reaction.

Heterogeneous reactions are able to reach equilibrium, just as homogeneous reactions can. If $NaHCO_3$ is placed in a sealed container so that no CO_2 or H_2O can escape, the gases and solids come to a heterogeneous equilibrium.

$$2NaHCO_3(s) \rightleftharpoons Na_2CO_3(s) + H_2O(g) + CO_2(g)$$

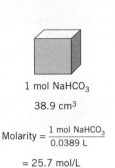

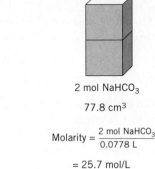

FIGURE 15.4

The concentration of a substance in the solid state is a constant. Doubling the number of moles also doubles the volume, but the *ratio* of moles to volume remains the same.

1 mol NaHCO₃

38.9 cm³

$$\text{Molarity} = \frac{1 \text{ mol NaHCO}_3}{0.0389 \text{ L}}$$

$$= 25.7 \text{ mol/L}$$

2 mol NaHCO₃

77.8 cm³

$$\text{Molarity} = \frac{2 \text{ mol NaHCO}_3}{0.0778 \text{ L}}$$

$$= 25.7 \text{ mol/L}$$

Following our usual procedure, we can write the equilibrium law for this reaction as

$$\frac{[Na_2CO_3(s)]\,[H_2O(g)]\,[CO_2(g)]}{[NaHCO_3(s)]^2} = K$$

However, the equilibrium law for reactions involving pure liquids and solids can be written in an even simpler form. This is because the concentration of a pure liquid or solid is unchangeable; that is, *for any pure liquid or solid, the ratio of amount of substance to volume of substance is a constant.* For example, if we had a 1 mol crystal of $NaHCO_3$, it would occupy a volume of 38.9 cm³. Two moles of $NaHCO_3$ would occupy twice this volume, 77.8 cm³ (Figure 15.4), but the *ratio* of moles to liters (i.e., the molar concentration) remains the same. For $NaHCO_3$, the concentration of the substance in the solid is

$$\frac{1 \text{ mol}}{0.0389 \text{ L}} = \frac{2 \text{ mol}}{0.0778 \text{ L}} = 25.7 \text{ mol/L}$$

This is the concentration of $NaHCO_3$ in the solid, regardless of the size of the solid sample. In other words, the concentration of $NaHCO_3$ is constant, provided that some of it is present in the reaction mixture.

Similar reasoning shows that the concentration of Na_2CO_3 in pure solid Na_2CO_3 is a constant, too. This means that the equilibrium law now has three constants, K plus two of the concentration terms. It makes sense to combine all of the numerical constants together.

$$[H_2O(g)]\,[CO_2(g)] = \frac{K[NaHCO_3(s)]^2}{[Na_2CO_3(s)]} = K_c$$

The equilibrium law for a heterogeneous reaction is written without concentration terms for pure solids or liquids. Equilibrium constants that are given in tables represent all of the constants combined.[4]

---

[4] Thermodynamics handles heterogeneous equilibria in a more elegant way by expressing the mass action expression appropriate for Equations 15.4 and 15.5 in terms of "effective concentrations," or **activities.** In doing this, thermodynamics *defines* the *activity* of any pure liquid or solid as equal to 1, which means terms involving such substances drop out of the mass action expression.

EXAMPLE 15.9
Writing the Equilibrium Law
for a Heterogeneous Reaction

The air pollutant sulfur dioxide can be removed from a gas mixture by passing the mixture over calcium oxide. The equation is

$$CaO(s) + SO_2(g) \rightleftharpoons CaSO_3(s)$$

Write the equilibrium law for this system.

SOLUTION The concentrations of the two solids, CaO and $CaSO_3$, are incorporated into the equilibrium constant K_c for the reaction. The only concentration term that should appear in the mass action expression is that of SO_2. Therefore, the equilibrium law is simply

$$\frac{1}{[SO_2(g)]} = K_c$$

■ **Practice Exercise 12** Write the equilibrium law for the following heterogeneous reactions.
(a) $2Hg(l) + Cl_2(g) \rightleftharpoons Hg_2Cl_2(s)$
(b) $NH_3(g) + HCl(g) \rightleftharpoons NH_4Cl(s)$
(c) Dissolving solid Ag_2CrO_4 in water: $Ag_2CrO_4(s) \rightleftharpoons 2Ag^+(aq) + CrO_4^{2-}(aq)$

15.9 ▬▬▬
LE CHÂTELIER'S PRINCIPLE AND CHEMICAL EQUILIBRIA

Le Châtelier's principle

In Section 15.10 you will see that it is possible to perform calculations that tell us what the composition of an equilibrium system is. However, many times we really don't need to know exactly what the equilibrium concentrations are. Instead, we may want to know what actions we should take to control the relative amounts of the reactants or products at equilibrium. For instance, if we were designing gasoline engines, we would like to know what could be done to minimize the formation of nitrogen oxide pollutants. Or, if we were preparing ammonia, NH_3, by the reaction of N_2 with H_2, we might want to know how to maximize the yield of NH_3.

Le Châtelier's principle, introduced in Chapter 11, provides us with the means for making qualitative predictions about changes in chemical equilibria. It does this in much the same way that it allows us to predict the effects of outside influences on equilibria that involve physical changes, such as liquid–vapor equilibria. Recall that **Le Châtelier's principle** states that *if an outside influence upsets an equilibrium, the system undergoes a change in a direction that counteracts the disturbing influence, and, if possible, returns the system to equilibrium.* Let's examine what kinds of "outside influences" can affect chemical equilibria.

1. **Adding or removing a reactant or product.** If we add or remove a reactant or product, we change one of the concentrations in the system. This changes the value of Q so that it is no longer equal to K_c. The result is that the equilibrium is upset. For the system to return to equilibrium, the concentrations must change in a way that causes Q to equal K_c again. This is brought about by the chemical reaction proceeding either to the right or left, and Le Châtelier's principle lets us predict which way the reaction goes.

A reaction will proceed in a direction that partially consumes a reactant or product that is added, or partially replaces a reactant or product that is removed.

As an example, let's study the equilibrium between two ions of copper.

$$Cu(H_2O)_4^{2+}(aq) + 4Cl^-(aq) \rightleftharpoons CuCl_4^{2-}(aq) + 4H_2O$$

$$\text{blue} \qquad\qquad\qquad\qquad \text{yellow}$$

As noted, $Cu(H_2O)_4^{2+}$ is blue and $CuCl_4^{2-}$ is yellow. Mixtures of the two have an intermediate color and therefore appear blue green, as illustrated in Figure 15.5.

If we add chloride ion to an equilibrium mixture of these copper ions, the system can get rid of some of it by reaction with $Cu(H_2O)_4^{2+}$. This gives more $CuCl_4^{2-}$, and we say that the equilibrium has "shifted to the right." In this new position of equilibrium, there is less $Cu(H_2O)_4^{2+}$ and more $CuCl_4^{2-}$ and uncombined H_2O. There is also more Cl^-, because not all that we add reacts. In other words, in this new position of equilibrium, *all* the concentrations have changed in a way that causes Q to become equal to K_c. Similarly, the position of equilibrium is shifted to the left when we add water to the mixture. The system is able to get rid of some of the H_2O by reaction with $CuCl_4^{2-}$, so more of the blue $Cu(H_2O)_4^{2+}$ is formed.

If we were able to remove a reactant or product, the position of equilibrium would also be changed. For example, if we add Ag^+ to a solution that contains both copper ions in equilibrium, we see an enhancement of the blue color. As the Ag^+ reacts with Cl^- to form insoluble AgCl, the equilibrium shifts to the left to replace some of the Cl^-. In this case, the system replaces part of the substance that is removed.

2. **Changing the volume in gaseous reactions.** Changing the volume of a system composed of gaseous reactants and products changes their molar concentrations. It also changes the pressure, which is how we analyze the effect of this disturbing influence.

   Let's consider the equilibrium

$$3H_2(g) + N_2(g) \rightleftharpoons 2NH_3(g)$$

If we reduce the volume of the reaction mixture, we expect the pressure to increase. The system can oppose the pressure change if it is able to reduce the number of molecules of gas, because fewer molecules of gas exert a lower pressure. If the reaction proceeds to the right, two NH_3 molecules appear when four molecules (one N_2 and three H_2) disappear. Therefore, this equilibrium is shifted to the right when the volume of the reaction mixture is reduced.

   Now let's look at the equilibrium

$$H_2(g) + I_2(g) \rightleftharpoons 2HI(g)$$

If this reaction proceeds in either direction, there is no change in the number of molecules of gas. This reaction, then, cannot respond to pressure changes, so changing the volume of the reaction vessel has virtually no effect on the equilibrium.

   The simplest way to analyze the effects of a volume change on an equilibrium system is to count the number of molecules of gaseous substances on both sides of the equation.

Reducing the volume of a gaseous reaction mixture causes the reaction to decrease the number of molecules of gas, if it can.

As a final note here, moderate pressure changes have essentially no effect on reactions involving only liquids or solids. Substances in these states are

$Cu(H_2O)_4^{2+}$ and $CuCl_4^{2-}$ are called *complex ions*. Some of the interesting properties and applications of complex ions of metals are discussed in Chapter 19.

The reaction shifts in a direction that will remove a substance that's been added or replace a substance that's been removed.

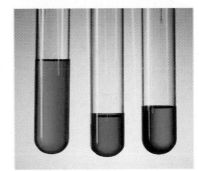

FIGURE   15.5

The solution in the center contains a mixture of blue $Cu(H_2O)_4^{2+}$ and yellow $CuCl_4^{2-}$ and has a blue-green color. At the right is some of the same solution after the addition of concentrated HCl. It has a more pronounced green color because the equilibrium is shifted toward $CuCl_4^{2-}$. At the left is some of the original solution after the addition of water. It is blue because the equilibrium has shifted toward $Cu(H_2O)_4^{2+}$.

virtually incompressible, and reactions involving them have no way to counter-act pressure changes.

You may recall that this same kind of analysis was used in Chapter 12 to predict how solubility changes with temperature.

3. **Changes of temperature.** To determine how a chemical equilibrium responds to a temperature change, we must know how the energy of the system changes when the reaction occurs. For instance, the reaction to produce NH_3 from N_2 and H_2 is exothermic; heat is evolved when NH_3 molecules are formed ($\Delta H_f^\circ = -46.19$ kJ/mol from Table 5.2). If we include heat as a product in the equilibrium equation

$$3H_2(g) + N_2(g) \rightleftharpoons 2NH_3(g) + \text{heat}$$

analyzing the effects of temperature changes becomes simple.

To raise the temperature, we add heat from some external source, such as a Bunsen burner. The equilibrium above can counter the temperature increase by absorbing some of the added heat. It does this by proceeding from right to left because the decomposition of some NH_3 to give more N_2 and H_2 is endothermic and consumes some of the heat that we've added. Thus, raising the temperature shifts this equilibrium to the left. In general,

Increasing the temperature shifts a reaction in a direction that produces an endothermic change.

The effect of temperature on the equilibrium involving the complex ions of copper described earlier is demonstrated in Figure 15.6. Here we have used our observations to determine that the reaction is endothermic in the forward direction.

An important point to note in these examples is that when we change the temperature, the concentrations change *even though the volume stays the same and no chemical substances have been added or removed.* Thus, the system comes to a new position of equilibrium in which the value of the mass action expression has changed, which means that K has changed. We can predict how K changes with temperature from the sign of the heat of reaction. For instance, let's look once again at the equilibrium law for the formation of ammonia, an exothermic reaction.

The only factor that can change K_c or K_p for a given reaction is a change in temperature.

$$\frac{[NH_3]^2}{[N_2][H_2]^3} = K_c$$

As we've noted, when the temperature increases, the concentration of NH_3 decreases while the concentrations of N_2 and H_2 increase. Therefore, the numerator of the mass action expression becomes *smaller* and the denomina-

FIGURE 15.6

The effect of temperature on the equilibrium

$Cu(H_2O)_4{}^{2+} + 4Cl^- \rightleftharpoons$
$\qquad CuCl_4{}^{2-} + 4H_2O$

In the center is an equilibrium mixture of the two complexes. When the solution is cooled in ice (*left*), the equilibrium shifts toward the blue $Cu(H_2O)_4{}^{2+}$. When heated in boiling water (*right*), the equilibrium shifts toward $CuCl_4{}^{2-}$. This behavior indicates that the reaction is endothermic in the forward direction.

tor becomes *larger*. This gives a smaller reaction quotient and therefore a smaller value of K_c. Thus,

---

When the temperature of an exothermic reaction is increased, the value of the equilibrium constant becomes smaller.

Of course, just the opposite occurs for an endothermic reaction.

4. **Effect of a catalyst.** Recall that catalysts are substances that affect the speeds of chemical reactions without actually being used up. However, catalysts do not affect the position of equilibrium in a system. The reason is that a catalyst affects both the forward and reverse reactions equally. Both are speeded up to the same degree, so adding a catalyst to a system has no net effect on the system's equilibrium composition. The catalyst's only effect is to bring the system to equilibrium faster.

5. **Addition of an inert gas at constant volume.** A change in volume is not the only way to change the pressure in an equilibrium system of gaseous reactants and products. The pressure can also be changed by keeping the volume the same and adding another gas. If this gas cannot react with the gases already present (i.e., if the added gas is *inert* toward the substances in equilibrium), the concentrations of the reactants and products won't change. The concentrations will continue to satisfy the equilibrium law and the reaction quotient will continue to equal K_c, so there will be no change in the position of equilibrium.

The reaction $N_2O_4(g) \rightleftharpoons 2NO_2(g)$ is endothermic, with $\Delta H° = +56.9$ kJ. How will the amount of NO_2 at equilibrium be affected by (a) adding N_2O_4, (b) lowering the pressure by increasing the volume of the container, (c) raising the temperature, and (d) adding a catalyst to the system? Which of these changes will alter the value of K_c?

EXAMPLE 15.10
Application of Le Châtelier's Principle

SOLUTION

(a) Adding N_2O_4 will cause the equilibrium to shift to the right—in a direction that will consume some of the added N_2O_4. The amount of NO_2 will increase.

(b) When the pressure in the system drops, the system responds by producing more molecules of gas, which will tend to raise the pressure and partially offset the change. Since more gas molecules are formed if some N_2O_4 decomposes, the amount of NO_2 at equilibrium will increase.

(c) Because the reaction is endothermic, we write the equation showing heat as a reactant

$$\text{heat} + N_2O_4(g) \rightleftharpoons 2NO_2(g)$$

Raising the temperature is accomplished by adding heat, so the system will respond by absorbing heat. This means that the equilibrium will shift to the right. Therefore when equilibrium is reestablished, there will be more NO_2 present.

(d) A catalyst causes a reaction to reach equilibrium more quickly, but it has no effect on the position of chemical equilibrium. Therefore, the amount of NO_2 at equilibrium will not be affected.

Finally, the *only* change that alters K is the temperature change. Raising the temperature (adding heat) will increase K_c for this endothermic reaction.

■ **Practice Exercise 13** Consider the equilibrium $PCl_3(g) + Cl_2(g) \rightleftharpoons PCl_5(g)$, for which $\Delta H° = -88$ kJ. How will the amount of Cl_2 at equilibrium be affected by: (a) adding PCl_3, (b) adding PCl_5, (c) raising the temperature, and (d) decreasing the volume of the container? How (if at all) will each of these changes affect K_p for the reaction?

15.10 EQUILIBRIUM CALCULATIONS

You have seen that the magnitude of an equilibrium constant gives us some feel for the extent to which the reaction proceeds at equilibrium. Sometimes, however, it is necessary to have more than merely a qualitative knowledge of equilibrium concentrations. This requires that we be able to use the equilibrium law for purposes of calculation.

Equilibrium calculations for gaseous reactions can be performed using either K_p or K_c, but for reactions in solution we must use K_c. Whether we deal with concentrations or partial pressures, however, the same basic principles apply.

Overall, we can divide equilibrium calculations into two main categories:

1. Calculating equilibrium constants from known equilibrium concentrations or partial pressures.

2. Calculating one or more equilibrium concentrations or partial pressures using the known value of K_c or K_p.

Calculating K_c from Equilibrium Concentrations

You've seen that we can obtain equilibrium constants from thermodynamic data. However, another way to determine the value of K_c is to carry out the reaction, measure the concentrations of reactants and products after equilibrium has been reached, and then use the equilibrium values in the equilibrium law to compute K_c. As an example, let's look again at the decomposition of N_2O_4.

$$N_2O_4(g) \rightleftharpoons 2NO_2(g)$$

In Section 15.2, we saw that if 0.0350 mol of N_2O_4 is placed into a 1 L flask at 25 °C, the concentrations of N_2O_4 and NO_2 at equilibrium are

$$[N_2O_4] = 0.0292 \text{ mol/L}$$

$$[NO_2] = 0.0116 \text{ mol/L}$$

To calculate K_c for this reaction, we substitute the equilibrium concentrations into the mass action expression of the equilibrium law.

$$\frac{[NO_2]^2}{[N_2O_4]} = K_c$$

$$\frac{(0.0116)^2}{(0.0292)} = K_c$$

Performing the arithmetic gives

$$K_c = 4.61 \times 10^{-3}$$

Although calculating an equilibrium constant this way is easy, sometimes we have to do a little work to figure out what all the concentrations are, as shown in Example 15.11.

EXAMPLE 15.11
Calculating K_c from
Equilibrium Concentrations

At a certain temperature, a mixture of H_2 and I_2 was prepared by placing 0.200 mol of H_2 and 0.200 mol of I_2 into a 2.00-liter flask. After a period of time the equilibrium

$$H_2(g) + I_2(g) \rightleftharpoons 2HI(g)$$

was established. The purple color of the I_2 vapor was used to monitor the reaction, and from the decreased intensity of the purple color it was determined that, at equilibrium, the I_2 concentration had dropped to 0.020 mol/L. What is the value of K_c for this reaction at this temperature?

ANALYSIS The first step in any equilibrium problem is to write the balance chemical equation and the related equilibrium law. The equation is already given, and the equilibrium law corresponding to it is

$$\frac{[HI]^2}{[H_2][I_2]} = K_c$$

To calculate the value of K_c we must substitute the *equilibrium concentrations* of H_2, I_2, and HI into the mass action expression. But what are they? We have been given only one directly, the value of $[I_2]$. To obtain the others, we have to do some reasoning based on the chemical equation.

The system starts with a set of initial concentrations, and to reach equilibrium a chemical change occurs. There is no III present initially, so the reaction above must proceed to the right, because there has to be some HI present for the system to be at equilibrium. This change will increase the HI concentration and decrease the concentrations of H_2 and I_2. If we can determine what the *changes* are, we can figure out the equilibrium concentrations. *In fact, the key to solving almost all equilibrium problems is determining how the concentrations change as the system comes to equilibrium.*

To help us in our analysis we will construct a **concentration table** beneath the chemical equation. We will do this for most of the equilibrium problems we encounter from now on, so let's examine the general form of the table.

In the first row we write the initial *molar concentrations* of the reactants and products. (All entries in the table will be *molar concentrations*, not *moles*. These are the only quantities that satisfy the equilibrium law.) Then, in the second row we enter the changes in concentration, using a positive sign if the concentration is increasing and a negative sign if it is decreasing. Finally, adding the changes to the initial concentrations gives the equilibrium concentrations.

$$\binom{\text{equilibrium}}{\text{concentration}} = \binom{\text{initial}}{\text{concentration}} + \binom{\text{change in}}{\text{concentration}}$$

*To set up the concentration table, it is essential that you realize that the changes in concentration **must be in the same ratio as the coefficients of the balanced equation**.* This is because the changes are caused by the chemical reaction proceeding in one direction or the other. In this problem, the coefficients of H_2 and I_2 are the same, so as the reaction proceeds to the right, the concentrations of H_2 and I_2 must decrease by the *same* amount. Similarly, because the coefficient of HI is twice that of H_2 or I_2, the concentration of HI must increase by *twice* as much. Now let's look at the finished concentration table, how the individual entries are obtained, and the solution of the problem.

SOLUTION To construct the table, we begin by entering the data given in the statement of the problem. This is shown in bold type. The values in regular type are then derived as described below.

	$H_2(g)$ +	$I_2(g)$	$\longrightarrow$	$2HI(g)$
Initial concentrations (M)	**0.100**	**0.100**		**0.000**
Changes in concentration (M)	-0.080	-0.080		$+2(0.080)$
Equilibrium concentrations (M)	0.020	**0.020**		0.160

In any system, the initial concentrations are controlled by the person doing the experiment.

Initial Concentrations Notice that we have calculated the ratio of moles to liters for both the H_2 and I_2, $(0.200 \text{ mol}/2.00 \text{ L}) = 0.100 \ M$. Because no HI was placed into the reaction mixture, its initial concentration has been set to zero.

The changes in concentration are controlled by the stoichiometry of the reaction.

Changes in Concentration We have been given both the initial and equilibrium concentrations of I_2, so by difference we can calculate the change for I_2. The other changes are then calculated from the mole ratios specified in the chemical equation. As noted above, because H_2 and I_2 have the same coefficients (i.e., 1), their changes are equal. The coefficient of HI is 2, so its change must be twice that of I_2. In addition, notice that the changes for both reactants have the same sign, which is opposite that of the product. *This relationship among the signs of the changes is always true and can serve as a useful check when you construct a concentration table.*

Equilibrium Concentrations For H_2 and HI, we just add the change to the initial value.

Now we can substitute the equilibrium concentrations into the mass action expression and calculate K_c.

$$K_c = \frac{(0.160)^2}{(0.020)(0.020)}$$
$$= 64$$

The Concentration Table—A Summary

Concentration table for equilibrium calculations

Let's review some key points that apply not only to this problem, but to others that deal with equilibrium calculations.

1. The only values that we can substitute into the mass action expression in the equilibrium law are equilibrium concentrations—the values that appear in the last row of the table.

2. When we enter initial concentrations into the table, they should be in units of moles per liter (mol/L). The initial concentrations are those present in the reaction mixture when it's prepared; we imagine that no reaction occurs until everything is mixed.

By expressing concentrations in mol/L, we follow what happens to reactants and products in each liter of solution.

3. The changes in concentration always occur in the same ratio as the coefficients in the balanced equation. For example, if we were dealing with the equilibrium

$$3H_2(g) + N_2(g) \rightleftharpoons 2NH_3(g)$$

and found that the $N_2(g)$ concentration decreases by 0.10 M during the approach to equilibrium, the entries in the "change" row would be as follows:

	$3H_2(g)$ +	$N_2(g)$	$\rightleftharpoons$	$2NH_3(g)$
Change in concentration:	$-3 \times (0.10 \ M)$	$-1 \times (0.10 \ M)$		$+2 \times (0.10 \ M)$
	$\downarrow$	$\downarrow$		$\downarrow$
	$-0.30 \ M$	$-0.10 \ M$		$+0.20 \ M$

4. In constructing the "change" row, be sure the reactant concentrations all change in the same direction, and that the product concentrations all change in the opposite direction. If the concentrations of the reactants decrease, all the entries for the reactants in the "change" row should have a minus sign, and all the entries for the products should be positive.

If the initial concentration of some reactant is zero, its change must be positive (an increase) because the final concentration can't be negative.

Keep these ideas in mind as we construct the concentration tables for other equilibrium problems.

■ **Practice Exercise 14** In a particular experiment, it was found that when $O_2(g)$ and $CO(g)$ were mixed and reacted according to the equation

$$2CO(g) + O_2(g) \rightleftharpoons 2CO_2(g)$$

the O_2 concentration decreased by 0.030 mol/L when the reaction reached equilibrium. How had the concentrations of CO and CO_2 changed?

■ **Practice Exercise 15** An equilibrium was established for the reaction

$$CO(g) + H_2O(g) \rightleftharpoons CO_2(g) + H_2(g)$$

at 500 °C. (This is an industrially important reaction for the preparation of hydrogen.) At equilibrium, the following concentrations were found in the reaction vessel: [CO] = 0.180 M, [H_2O] = 0.0411 M, [CO_2] = 0.150 M, and [H_2] = 0.200 M. What is the value of K_c for this reaction?

■ **Practice Exercise 16** A student placed 0.200 mol of $PCl_3(g)$ and 0.100 mol of $Cl_2(g)$ into a 1.00 L container at 250 °C. After the reaction

$$PCl_3(g) + Cl_2(g) \rightleftharpoons PCl_5(g)$$

came to equilibrium it was found that the flask contained 0.120 mol of PCl_3.
(a) What were the initial concentrations of the reactants and product?
(b) By how much had the concentrations changed when the reaction reached equilibrium?
(c) What were the equilibrium concentrations?
(d) What is the value of K_c for this reaction at this temperature?

Calculating Equilibrium Concentrations Using K_c

In the simplest calculation of this type, all but one of the equilibrium concentrations are known, as illustrated in Example 15.12.

EXAMPLE 15.12
Using K_c to Calculate Concentrations at Equilibrium

The reversible reaction

$$CH_4(g) + H_2O(g) \rightleftharpoons CO(g) + 3H_2(g)$$

has been used as a commercial source of hydrogen. At 1500 °C, an equilibrium mixture of these gases was found to have the following concentrations: [CO] = 0.300 M, [H_2] = 0.800 M, and [CH_4] = 0.400 M. At 1500 °C, K_c = 5.67 for this reaction. What was the equilibrium concentration of $H_2O(g)$ in this mixture?

SOLUTION The first step, once we have the chemical equation for the equilibrium, is to write the equilibrium law for the reaction.

$$K_c = \frac{[CO][H_2]^3}{[CH_4][H_2O]}$$

The equilibrium constant and all of the equilibrium concentrations except that for H_2O are known, so we substitute these values into the equilibrium law and solve for the unknown quantity.

$$5.67 = \frac{(0.300)(0.800)^3}{(0.400)[H_2O]}$$

Solving for $[H_2O]$ gives

$$[H_2O] = \frac{(0.300)(0.800)^3}{(0.400)(5.67)}$$

$$= 0.0677\ M$$

■ **Practice Exercise 17** Ethyl acetate, $CH_3CO_2C_2H_5$, is an important solvent used in lacquers, adhesives, the manufacture of plastics, and even as a food flavoring. It is produced from acetic acid and ethanol by the reaction

$$\underset{\text{acetic acid}}{CH_3CO_2H(l)} + \underset{\text{ethanol}}{C_2H_5OH(l)} \rightleftharpoons CH_3CO_2C_2H_5(l) + H_2O(l)$$

At 25 °C, $K_c = 4.10$ for this reaction. In a reaction mixture, the following equilibrium concentrations were observed: $[CH_3CO_2H] = 0.210\ M$, $[H_2O] = 0.00850\ M$, and $[CH_3CO_2C_2H_5] = 0.910\ M$. What was the concentration of C_2H_5OH in the mixture?

Calculating Equilibrium Concentrations Using K_c and Initial Concentrations

A more complex type of calculation involves the use of initial concentrations and K_c to compute equilibrium concentrations. Although some of these problems can be so complicated that a computer is needed to solve them, we can learn the general principles involved by working on simple calculations.

EXAMPLE 15.13
Using K_c to Calculate Equilibrium Concentrations

The reaction

$$CO(g) + H_2O(g) \rightleftharpoons CO_2(g) + H_2(g)$$

has $K_c = 4.06$ at 500 °C. If 0.100 mol of CO and 0.100 mol of $H_2O(g)$ are placed in a 1.00 L reaction vessel at this temperature, what are the concentrations of the reactants and products when the system reaches equilibrium?

ANALYSIS The key to solving this kind of problem is recognizing that at equilibrium the mass action expression must equal K_c.

$$\frac{[CO_2][H_2]}{[CO][H_2O]} = 4.06 \quad \text{(at equilibrium)}$$

We must find values for the concentrations that satisfy this condition.

Because we don't know what these concentrations are, we represent them algebraically as unknowns. This is where the concentration table is very helpful. To build the table, we need quantities to enter into the "initial concentrations," "changes in concentration," and "equilibrium concentrations" rows.

Initial Concentrations The initial concentrations of CO and H_2O are each 0.100 mol/1.00 L = 0.100 *M*. Since no CO_2 or H_2 are initially placed into the reaction vessel, their initial concentrations both are zero.

Changes in Concentration Some CO_2 and H_2 must form for the reaction to reach equilibrium. This also means that some CO and H_2O must react. But how much? If we knew the answer, we could calculate the equilibrium concentrations. Therefore, the changes in concentration are our unknown quantities.

Let us allow x to be equal to the number of moles per liter of CO that react. The change in the concentration of CO is then $-x$ (it is negative because the change decreases the CO concentration). Because CO and H_2 react in a $1:1$ mole ratio, the change in the H_2O concentration is also $-x$. Since 1 mol each of CO_2 and H_2 is formed from 1 mol of CO, the CO_2 and H_2 concentrations each increase by x (their changes are $+x$).

> We could just as easily have chosen x to be the number of mol/L of H_2O that reacts or the number of mol/L of CO_2 or H_2 that forms. There's nothing special about having chosen CO to define x.

Equilibrium Concentrations We obtain the equilibrium concentrations as

$$\begin{pmatrix} \text{equilibrium} \\ \text{concentration} \end{pmatrix} = \begin{pmatrix} \text{initial} \\ \text{concentration} \end{pmatrix} + \begin{pmatrix} \text{change in} \\ \text{concentration} \end{pmatrix}$$

SOLUTION Here is the completed concentration table.

	$CO(g)$	$+$ $H_2O(g)$	$\rightleftharpoons$ $CO_2(g)$	$+$ $H_2(g)$
Initial concentrations (*M*)	0.100	0.100	0.0	0.0
Change in concentrations (*M*)	$-x$	$-x$	$+x$	$+x$
Equilibrium concentrations (*M*)	$0.100 - x$	$0.100 - x$	x	x

Note that the last line in the table tells us that the equilibrium CO and H_2O concentrations are equal to the number of moles per liter that were present initially minus the number of moles per liter that react. The equilibrium concentrations of CO_2 and H_2 equal the number of moles per liter of each that forms, since no CO_2 or H_2 are present initially.

> The coefficients of x can be the same as the coefficients in the balanced chemical equation. This would assure us that the changes are in the same ratio as the coefficients in the equation.

Next, we substitute the quantities from the "equilibrium concentration" row into the mass action expression and solve for x.

> The equilibrium concentration values must satisfy the equation given by the equilibrium law.

$$\frac{(x)\,(x)}{(0.100 - x)\,(0.100 - x)} = 4.06$$

which we can write as

$$\frac{x^2}{(0.100 - x)^2} = 4.06$$

In this problem we can solve the equation for x most easily by taking the square root of both sides.

$$\frac{x}{(0.100 - x)} = \sqrt{4.06} = 2.01$$

Clearing fractions gives

$$x = 2.01\,(0.100 - x)$$

$$x = 0.201 - 2.01x$$

Collecting terms in x gives

$$x + 2.01x = 0.201$$

$$3.01x = 0.201$$

$$x = 0.0668$$

Now that we know the value of x, we can calculate the equilibrium concentrations from the last row of the table.

$$[CO] = 0.100 - x = 0.100 - 0.0668 = 0.033\ M$$

$$[H_2O] = 0.100 - x = 0.100 - 0.0668 = 0.033\ M$$

$$[CO_2] = x = 0.0668\ M$$

$$[H_2] = x = 0.0668\ M$$

CHECKING THE ANSWER One way to check the answers is to substitute the equilibrium concentrations we've found into the mass action expression and evaluate the reaction quotient. If our answers are correct, Q should equal K_c. Let's do this.

$$Q = \frac{(0.0668)^2}{(0.033)^2} = 4.1$$

Rounding K_c to two significant figures gives 4.1, so the calculated concentrations satisfy the equilibrium law.

EXAMPLE 15.14

Using K_c to Calculate Equilibrium Concentrations

In the preceding example it was stated that the reaction

$$CO(g) + H_2O(g) \rightleftharpoons CO_2(g) + H_2(g)$$

has $K_c = 4.06$ at 500 °C. Suppose 0.0600 mol each of CO and H_2O are mixed with 0.100 mol each of CO_2 and H_2 in a 1.00 L reaction vessel. What will be the concentrations of all the substances be when the mixture reaches equilibrium at this temperature?

ANALYSIS We will proceed in much the same way as in the preceding example. However, this time determining the algebraic signs of x will not be quite as simple because none of the initial concentrations is zero. The best way to determine the algebraic signs is to use the initial concentrations to calculate the initial reaction quotient. Then we can compare Q to K_c and, by reasoning, figure out which way the reaction must proceed to make Q equal to K_c.

SOLUTION The equilibrium law for the reaction is

$$\frac{[CO_2]\,[H_2]}{[CO]\,[H_2O]} = K_c$$

Let's use the initial concentrations, shown in the first row of the concentration table below, to determine the initial value of the reaction quotient.

$$Q_{initial} = \frac{(0.100)\,(0.100)}{(0.0600)\,(0.0600)} = 2.78 < K_c$$

As indicated, $Q_{initial}$ is less than K_c, so the system is not at equilibrium. To reach equilibrium Q must become larger, which requires an increase in the concentrations of CO_2 and H_2 as the reaction proceeds. This means that for CO_2 and H_2, x must be positive, and therefore, for CO and H_2, x must be negative.

Here is the completed concentration table.

	CO(g)	+ H₂O(g)	⇌	CO₂(g)	+ H₂(g)
Initial concentrations (*M*)	0.0600	0.0600		0.100	0.100
Change in concentrations (*M*)	$-x$	$-x$		$+x$	$+x$
Equilibrium concentrations (*M*)	$0.0600 - x$	$0.0600 - x$		$0.100 + x$	$0.100 + x$

Substituting equilibrium quantities into the mass action expression in the equilibrium law gives us

$$\frac{(0.100 + x)^2}{(0.0600 - x)^2} = 4.06$$

Taking the square root of both sides yields

$$\frac{0.100 + x}{0.0600 - x} = 2.01$$

To solve for x we first multiply each side by $0.0600 - x$ to obtain

$$0.100 + x = 2.01(0.0600 - x)$$

$$0.100 + x = 0.121 - 2.01x$$

Collecting terms in x to one side and the constants to the other gives

$$x + 2.01x = 0.121 - 0.100$$

$$3.01x = 0.021$$

$$x = 0.0070$$

Now we can calculate the equilibrium concentrations:

$$[CO] = [H_2O] = (0.0600 - x) = 0.0600 - 0.0070 = 0.0530 \; M$$

$$[CO_2] = [H_2] = (0.100 + x) = 0.100 + 0.0070 = 0.107 \; M$$

CHECKING THE ANSWER As a check, let's evaluate the reaction quotient using the calculated equilibrium concentrations.

$$Q = \frac{(0.107)^2}{(0.0530)^2} = 4.08$$

This is acceptably close to the value of K_c. (That it is not *exactly* equal to K_c is because of the rounding of answers during the calculations.)

In each of the preceding two examples, we were able to simplify the solution by taking the square root of both sides of the algebraic equation obtained by substituting equilibrium concentrations into the mass action expression. Such simplifications are not always possible, however, as illustrated in the next example.

At a certain temperature $K_c = 4.50$ for the reaction

$$N_2O_4(g) \rightleftharpoons 2NO_2(g)$$

If 0.300 mol of N_2O_4 is placed into a 2.00 L container at this temperature, what will be the equilibrium concentrations of both gases?

ANALYSIS As in the preceding example, at equilibrium the mass action expression must be equal to K_c.

$$\frac{[NO_2]^2}{[N_2O_4]} = 4.50$$

We will need to find algebraic expressions for the equilibrium concentrations and substitute them into the mass action expression. To obtain these, we set up the concentration table for the reaction.

EXAMPLE 15.15
Using K_c to Calculate
Equilibrium Concentrations

Initial Concentrations The initial concentration of N_2O_4 is 0.300 mol/2.00 L = 0.150 *M*. Since no NO_2 was placed in the reaction vessel, its initial concentration is 0.000 *M*.

Changes in Concentrations There is no NO_2 in the reaction mixture, so we know its concentration must increase. This means the N_2O_4 concentration must decrease as some of the NO_2 is formed. Let's allow x to be the number of moles per liter of N_2O_4 that reacts, so the change in the N_2O_4 concentration is $-x$. Because of the stoichiometry of the reaction, the NO_2 concentration must increase by $2x$, so its change in concentration is $+2x$.

Equilibrium Concentrations As before, we add the change to the initial concentration in each column to obtain expressions for the equilibrium concentrations.

SOLUTION Here is the concentration table.

Notice that we have chosen the coefficients of *x* to be the same as the coefficients in the balanced chemical equation.

	$N_2O_4(g) \rightleftharpoons 2NO_2(g)$	
Initial concentrations (*M*)	0.150	0.000
Changes in concentrations (*M*)	$-x$	$+2x$
Equilibrium concentrations (*M*)	$0.150 - x$	$2x$

Now we substitute the equilibrium quantities into the mass action expression.

$$\frac{(2x)^2}{(0.150 - x)} = 4.50$$

or

$$\frac{4x^2}{(0.150 - x)} = 4.50$$

This time the left side of the equation is not a perfect square, so we cannot just take the square root of both sides as in Example 15.13. However, because the equation involves terms in x^2, x, and a constant, we can use the quadratic formula. Recall that for a quadratic equation of the form

$$ax^2 + bx + c = 0$$

$$x = \frac{-b \pm \sqrt{b^2 - 4ac}}{2a}$$

Expanding our equation above gives

$$4x^2 = 4.50(0.150 - x)$$
$$= 0.675 - 4.50x$$

Arranging terms in the standard order gives

$$4x^2 + 4.50x - 0.675 = 0$$

Therefore, the quantities we will substitute into the quadratic formula are as follows: $a = 4$, $b = 4.50$, and $c = -0.675$. Making these substitutions gives

$$x = \frac{-4.50 \pm \sqrt{(4.50)^2 - 4(4)(-0.675)}}{2(4)}$$

$$= \frac{-4.50 \pm \sqrt{31.05}}{8}$$

$$= \frac{-4.50 \pm 5.57}{8}$$

Because of the ± term, there are two values of x that satisfy the equation, $x = 0.134$, and $x = -1.26$. However, only the first value, $x = 0.134$, makes any sense chemically. Using this value, the equilibrium concentrations are

$$[N_2O_4] = 0.150 - 0.134 = 0.016 \ M$$

$$[NO_2] = 2(0.134) = 0.268 \ M$$

Notice that if we had used the negative root, -1.26, both equilibrium concentrations would be negative. Negative concentrations are impossible, so $x = -1.26$ is not acceptable *for chemical reasons*. In general, whenever you use the quadratic equation in a chemical calculation, one root will be satisfactory and the other will lead to answers that are nonsense.

CHECKING THE ANSWER Once again, we can evaluate the reaction quotient using the calculated equilibrium values. When we do this, we obtain $Q = 4.49$, which is acceptably close to the value of K_c given.

Equilibrium problems can be even more complex than the one we have just discussed; however, you will not have to deal with them. In fact, in most of the problems you encounter there will be ways of simplifying the algebra. Only rarely will it even be necessary to resort to the quadratic equation.

■ **Practice Exercise 18** During an experiment, 0.200 mol of H_2 and 0.200 mol of I_2 were placed into a 1.00-liter vessel where the reaction

$$H_2(g) + I_2(g) \rightleftharpoons 2HI(g)$$

came to equilibrium. For this reaction, $K_c = 49.5$ at the temperature of the experiment. What were the equilibrium concentrations of H_2, I_2, and HI?

Equilibrium Calculations When K_c Is Very Small

Many chemical reactions have equilibrium constants that are either very large or very small. For example, most weak acids have very small values for K_c. Therefore, only very tiny amounts of products form when these weak acids react with water.

When the K_c for a reaction is very small the equilibrium calculations can usually be simplified considerably.

Hydrogen, a potential fuel, is found in great abundance in water. Before the hydrogen can be used as a fuel, however, it must be separated from the oxygen; the water must be split into H_2 and O_2. One possibility is thermal decomposition, but this requires very high temperatures. Even at 1000 °C, $K_c = 7.3 \times 10^{-18}$ for the reaction

$$2H_2O(g) \rightleftharpoons 2H_2(g) + O_2(g)$$

If at 1000 °C the H_2O concentration in a reaction vessel is set initially at 0.100 M, what will the H_2 concentration be when the reaction reaches equilibrium?

SOLUTION We know that at equilibrium

$$\frac{[H_2]^2[O_2]}{[H_2O]^2} = 7.3 \times 10^{-18}$$

We now set up the concentration table.

EXAMPLE 15.16
Simplifying Equilibrium Calculations for Reactions with Small K_c

Initial Concentrations The initial concentration of H_2O is 0.100 *M*; those of H_2 and O_2 are both 0.0 *M*.

Changes in Concentration We know the changes must be in the same ratio as the coefficients in the balanced equation, so we place x's in this row with coefficients equal to those in the chemical equation. Because there are no products present initially, their changes must be positive and the change for the water must be negative. This gives

	$2H_2O(g)$ $\rightleftharpoons$ $2H_2(g)$ + $O_2(g)$		
Initial concentrations (*M*)	0.100	0.0	0.0
Changes in concentration (*M*)	$-2x$	$+2x$	$+x$
Equilibrium concentrations (*M*)	$0.100 - 2x$	$2x$	x

When we substitute the equilibrium quantities into the mass action expression we get

$$\frac{(2x)^2 x}{(0.100 - 2x)^2} = 7.3 \times 10^{-18}$$

or

$(2x)^2 x = (4x^2)x = 4x^3$

$$\frac{4x^3}{(0.100 - 2x)^2} = 7.3 \times 10^{-18}$$

This is a *cubic equation* (one term involves x^3) and can be rather difficult to solve unless we can simplify it. In this instance we are able to do so because the very small value of K_c tells us that hardly any of the H_2O will decompose. Whatever the actual value of x, we know that it is going to be very small. This means that $2x$ will also be small, so when this tiny value is subtracted from 0.100, the result will still be very, very close to 0.100. We will make the assumption, then, that the denominator will be essentially unchanged from 0.100 by subtracting $2x$ from it; that is, we will assume that $0.100 - 2x \approx 0.100$.

Even before we solve the problem, we know that hardly any H_2 and O_2 will be formed, because K_c is so small.

This assumption greatly simplifies the math. We now have

$$\frac{4x^3}{(0.100)^2} = 7.3 \times 10^{-18}$$

$$4x^3 = (0.0100)(7.3 \times 10^{-18}) = 7.3 \times 10^{-20}$$

$$x^3 = 1.8 \times 10^{-20}$$

$$x = \sqrt[3]{1.8 \times 10^{-20}}$$

$$= 2.6 \times 10^{-7}$$

Always be sure to check your assumptions when solving a problem of this kind.

Notice that the value of x that we've obtained is indeed very small. If we double it and subtract the answer from 0.100, we still get 0.100 when we round to the correct number of significant figures (the third decimal place).

$$0.100 - 2x = 0.100 - 2(2.6 \times 10^{-7}) = 0.09999948$$
$$= 0.100 \quad \text{(rounded correctly)}$$

This check verifies that our assumption was valid. Finally, we have to obtain the H_2 concentration. Our table gives

$$[H_2] = 2x$$

Therefore,

$$[H_2] = 2(2.6 \times 10^{-7}) = 5.2 \times 10^{-7} \, M$$

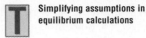

The simplifying assumption made in the preceding example is valid because a very small number is subtracted from a much larger one. We could also have neglected x (or $2x$) if it were a very small number that was being added to a much larger one. Remember that you can only neglect an x that's *added* or *subtracted;* you can never drop an x that occurs as a multiplying or dividing factor. Some examples are

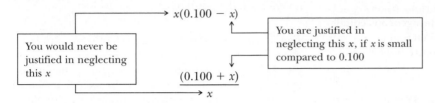

As a rule of thumb, you can expect that these simplifying assumptions will be valid if the concentration from which x is subtracted, or to which x is added, is at least 1000 times greater than K. For instance, in the preceding example, $2x$ was subtracted from 0.100. Since 0.100 is much larger than $1000 \times (7.3 \times 10^{-18})$ we expect the assumption $0.100 - 2x \approx 0.100$ to be valid. However, even though the simplifying assumption is expected to be valid, always check to see if it really is after finishing the calculation. If the assumption proves invalid, then some other way to solve the algebra must be found.

■ **Practice Exercise 19** In air at 25 °C and 1 atm, the N_2 concentration is 0.033 M and the O_2 concentration is 0.00810 M. The reaction

$$N_2(g) + O_2(g) \rightleftharpoons 2NO(g)$$

has $K_c = 4.8 \times 10^{-31}$ at 25 °C. Taking the N_2 and O_2 concentrations given above as initial values, calculate the equilibrium NO concentration that should exist in our atmosphere from this reaction at 25 °C.

SUMMARY

Dynamic Equilibrium When the forward and reverse reactions in a chemical system occur at equal rates, a dynamic equilibrium exists and the concentrations of the reactants and products remain constant. For a given overall chemical composition, the amounts of reactants and products that are present at equilibrium are the same regardless of whether the equilibrium is approached from the direction of pure reactants, of pure products, or of any mixture of reactants and products.

The Equilibrium Law The **mass action expression** is a fraction. The concentrations of the products, raised to powers equal to their coefficients in the chemical equation, are multiplied together in the numerator. The denominator is constructed in the same way from the concentrations of the reactants raised to powers equal to their coefficients. The numerical value of the mass action expression is called the **reaction quotient.** At equilibrium, the reaction quotient is equal to the **equilibrium constant, K_c.** If partial pressures

of gases are used in the mass action expression, K_p is obtained. The magnitude of the equilibrium constant is roughly proportional to the extent to which the reaction proceeds to completion when equilibrium is reached.

Thermodynamic Equilibrium Constants The spontaneity of a reaction is determined by ΔG (how the free energy changes with a change in concentration). This is related to the standard free energy change, $\Delta G°$, by the equation $\Delta G = \Delta G° + RT \ln Q$, where Q is the reaction quotient for the system. At equilibrium, $\Delta G° = -RT \ln K$, where $K = K_p$ for gaseous reactions and $K = K_c$ for reactions in solution. For temperatures other than 25 °C, we can estimate $\Delta G_T°$ by the equation $\Delta G_T° = \Delta H_{298}° - T\Delta S_{298}°$ and then use it to calculate K.

Relating K_p to K_c The values of K_p and K_c are only equal if the same number of moles of gas are represented on both sides of the chemical equation. When the number of moles of gas are different, K_p is related to K_c by the equation $K_p = K_c(RT)^{\Delta n_g}$. Remember to use $R = 0.0821$ L atm mol^{-1} K^{-1} and T = absolute temperature. Also, be careful to calculate Δn_g as the difference between the number of moles of *gaseous* products and the number of moles of *gaseous* reactants in the balanced equation.

Heterogeneous Equilibria   An equilibrium involving substances in more than one phase is a **heterogeneous equilibrium.** The mass action expression for a heterogeneous equilibrium omits concentration terms for pure liquids and/or pure solids.

Le Châtelier's Principle   This principle states that *when an equilibrium is upset, a chemical change occurs in a direction that opposes the disturbing influence and brings the system to equilibrium again.* Adding a reactant or a product causes a reaction to occur that uses up part of what has been added. Removing a reactant or a product causes a reaction that replaces part of what has been removed. Increasing the pressure (by reducing the volume) drives a reaction in the direction of the fewer number of moles of gas. Pressure changes have virtually no effect on equilibria involving only solids and liquids. Raising the temperature causes an equilibrium to shift in an endothermic direction. The value of K increases with increasing temperature for reactions that are endothermic in the forward direction. A change in temperature is the only factor that changes K. Addition of a catalyst has no effect on the position of an equilibrium.

Equilibrium Calculations   The initial concentrations in a chemical system are controlled by the person who combines the chemicals at the start of the reaction. The changes in concentration are determined by the stoichiometry of the reaction. Only equilibrium concentrations satisfy the equilibrium law. When these are used, the mass action expression is equal to K_c. When a change in concentration is expected to be very small compared to the initial concentration, the change may be neglected and the algebraic equation derived from the equilibrium law can be simplified. In general, this simplification is valid if the initial concentration is at least 1000 times larger than K.

Tools You Have Learned

In this chapter you have learned the following tools that we apply to various aspects of solving problems dealing with chemical equilbria.

Tool	Function
Mass action expression and the equilibrium law (page 639)	Constructing the mass action expression and setting it equal to K (either K_c for reactions in solution or K_p for gaseous reactions) provides the fundamental relationship between the concentrations of reactants and products in an equilibrium system.
Manipulation of equilibrium equations (page 640)	Permits you to determine K for a particular reaction by manipulating or combining one or more different chemical equations and their associated equilibrium constants.
$\Delta G = \Delta G° + RT \ln Q$ (page 644)	Determine whether a reaction is at equilibrium from a knowledge of the reaction quotient and $\Delta G°$.
$\Delta G° = -RT \ln K$ (page 645)	Relates $\Delta G°$ to the equilibrium constant; K equals K_p for gaseous reactions and K_c for reactions in solution.
$K_p = K_c(RT)^{\Delta n_g}$ (page 648)	To convert between K_p and K_c. Remember that Δn_g is the change in the number of moles of gas, which is computed using the coefficients of gaseous products and reactants.
Magnitude of the equilibrium constant (page 643)	The magnitude of K is related to the extent to which a reaction proceeds toward completion when equilibrium is reached. The size of K determines whether you are able to apply simplifying assumptions in solving problems.
Le Châtelier's principle (page 652)	Used to analyze the effects of disturbances on the position of equilibrium.

Tool	Function
Concentration table (page 658)	Used in solving problems in chemical equilibria. Incorporates initial concentrations, changes in concentration, and equilibrium concentration expressions.
Simplifying assumptions (page 667)	Used to simplify the algebra when working equilibrium problems for reactions for which K is very small.

THINKING IT THROUGH

Remember, you are not asked to obtain answers for the following problems. Instead, assemble the data necessary to solve the problems and describe how *you would use the data to obtain the answers. For numerical problems, set up the calculation using appropriate conversion factors.*

The problems are divided into two groups. Those in Level 2 are significantly more challenging than those in Level 1 and provide an opportunity to really hone your problem solving skills.

Level 1 Problems

1. Suppose the following data were collected at 25 °C for the system

$$NO_2(g) + NO(g) \rightleftharpoons N_2O_3(g)$$

Equilibrium Concentrations (M)		
[NO$_2$]	[NO]	[N$_2$O$_3$]
0.020	0.033	8.6×10^{-3}
0.040	0.0085	4.4×10^{-3}
0.015	0.012	2.3×10^{-3}

Explain in detail how you would use these data to show that they obey the principles of the equilibrium law.

2. The reaction $NO_2(g) + NO(g) \rightleftharpoons N_2O(g) + O_2(g)$ has $K_c = 0.914$ at a certain temperature. What manipulations are necessary to determine, at this same temperature, the value of K_c for the following reaction?

$$2N_2O(g) + 2O_2(g) \rightleftharpoons 2NO_2(g) + 2NO(g)$$

3. In the preceding question, is there sufficient information to calculate the value of K_p for the reaction

$$NO_2(g) + NO(g) \rightleftharpoons N_2O(g) + O_2(g)$$

4. The reaction

$$NO_2(g) + NO(g) \rightleftharpoons N_2O(g) + O_2(g)$$

has $\Delta H° = -42.9$ kJ. List the kinds of changes that would alter the amount of NO$_2$ present in the system at equilibrium. What could be done to this system *without* changing the amount of NO$_2$ at equilibrium?

5. At 100 °C, $K_c = 0.135$ for the reaction

$$3H_2(g) + N_2(g) \rightleftharpoons 2NH_3(g)$$

In a reaction mixture at equilibrium at this temperature, [NH$_3$] = 0.030 M, and [N$_2$] = 0.50 M. Explain how you can calculate the molar concentration of H$_2(g)$. (Set up the calculation.)

6. For the reaction mixture described in the preceding question, explain how you can calculate the partial pressures of each of the gases in the mixture. (Set up the calculation.)

7. How would you use the results obtained in the preceding equation to calculate the value of K_p for the reaction

$$3H_2(g) + N_2(g) \rightleftharpoons 2NH_3(g)$$

8. The reaction $N_2O_3(g) \rightleftharpoons NO(g) + NO_2(g)$ is to be studied at a certain temperature. A mixture was prepared in which the initial concentrations of NO and NO$_2$ were 0.40 M and 0.60 M, respectively. When equilibrium was reached, the concentration of N$_2$O$_3$ was measured to be 0.13 M. What were the equilibrium concentrations of NO and NO$_2$? (Explain each step in the solution to the problem.)

Level 2 Problems

9. One function of the catalysts in an automotive catalytic converter is to promote the decomposition of nitrogen oxides into nitrogen and oxygen. How is this related to chemical equilibrium and Le Châtelier's principle?

10. Which gas, HI or HBr, is more fully dissociated into its gaseous elements at 150 °C? What are the equilibrium concentrations of the reactants and products in containers of these gases if the initial concentrations of HI and HBr are each 0.0250 M? (Explain how you would obtain the answers to these questions.)

11. The following are equilibrium constants calculated for the reaction

$$3H_2(g) + N_2(g) \rightleftharpoons 2NH_3(g)$$

T (°C)	K_p	K_c
25	7.11×10^5	4.26×10^8
300	4.61×10^{-9}	1.02×10^{-5}
400	2.62×10^{-10}	8.00×10^{-7}

Explain how you can determine in a simple way how the amount of NH_3 at equilibrium varies as the temperature of the reaction mixture is increased.

12. At room temperature (25 °C), the gas ClNO is impure because it decomposes slightly according to the equation

$$2ClNO(g) \rightleftharpoons Cl_2(g) + 2NO(g)$$

The extent of decomposition is about 5%. What is the approximate value of ΔG°_{298} for this reaction at this temperature? (Explain each step in the solution to the problem.)

13. The reaction

$$N_2O(g) + O_2(g) \rightleftharpoons NO_2(g) + NO(g)$$

has $\Delta H^\circ = -42.9$ kJ and $\Delta S^\circ = -26.1$ J/K. Suppose 0.100 mol of N_2O and 0.100 mol of O_2 were placed in a container at 500 °C and this equilibrium was established. Explain how you would determine what percentage of the N_2O would have reacted.

14. Suppose we begin the reaction

$$2H_2(g) + O_2(g) \rightleftharpoons 2H_2O(g) \qquad K_c = 1.4 \times 10^{17}$$

with the following initial concentrations: $[H_2] = 0.020$ M, $[O_2] = 0.030$ M, and $[H_2O] = 0.00$ M. If we let x equal the number of moles per liter of O_2 that react, why are we not justified in neglecting x in expressing the equilibrium concentrations of H_2 and O_2?

15. In the preceding question, how could the initial concentrations be recalculated so as to make the simplifying assumption valid when the calculation is performed? (Hint: The position of equilibrium does not depend on the direction from which equilibrium is approached.)

16. For the equilibrium

$$3NO_2(g) \rightleftharpoons N_2O_5(g) + NO(g)$$

$K_c = 1.0 \times 10^{-11}$. If a 4.00 L container initially holds 0.20 mol of NO_2, how many moles of N_2O_5 will be present when this system reaches equilibrium? (Describe each step in the solution to the problem.)

17. To study the following reaction at 20 °C,

$$NO(g) + NO_2(g) + H_2O(g) \rightleftharpoons 2HNO_2(g)$$

a mixture of $NO(g)$, $NO_2(g)$, and $H_2O(g)$ was prepared. For NO, NO_2, and HNO_2, the initial concentrations were $[NO] = [NO_2] = 2.59 \times 10^{-3}$ M and $[HNO_2] = 0.0$ M. The initial partial pressure of $H_2O(g)$ was 17.5 torr. When equilibrium was reached, the HNO_2 concentration was 4.0×10^{-4} M. Explain in detail how to calculate the equilibrium constant, K_c, for this reaction.

REVIEW EXERCISES

Answers to questions whose numbers are printed in color are given in Appendix D. More challenging questions are marked with asterisks.

General

15.1 Sketch a graph showing how the concentrations of the reactants and products of a typical chemical reaction vary with time during the course of the reaction. Assume no products are present at the start of the reaction.

15.2 What is meant when we say that chemical reactions are *reversible?*

Mass Action Expression, K_p and K_c

15.3 What is an *equilibrium law?*

15.4 How is the term *reaction quotient* defined?

15.5 Under what conditions does the reaction quotient equal K_c?

15.6 Write the equilibrium law for each of the following reactions in terms of molar concentrations.

(a) $2PCl_3(g) + O_2(g) \rightleftharpoons 2POCl_3(g)$
(b) $2SO_3(g) \rightleftharpoons 2SO_2(g) + O_2(g)$
(c) $N_2H_4(g) + 2O_2(g) \rightleftharpoons 2NO(g) + 2H_2O(g)$
(d) $N_2H_4(g) + 6H_2O_2(g) \rightleftharpoons 2NO_2(g) + 8H_2O(g)$
(e) $SOCl_2(g) + H_2O(g) \rightleftharpoons SO_2(g) + 2HCl(g)$

15.7 Write the equilibrium law for each of the following gaseous reactions in terms of molar concentrations.

(a) $3Cl_2(g) + NH_3(g) \rightleftharpoons NCl_3(g) + 3HCl(g)$
(b) $PCl_3(g) + PBr_3(g) \rightleftharpoons PCl_2Br(g) + PClBr_2(g)$
(c) $NO(g) + NO_2(g) + H_2O(g) \rightleftharpoons 2HNO_2(g)$
(d) $H_2O(g) + Cl_2O(g) \rightleftharpoons 2HOCl(g)$
(e) $Br_2(g) + 5F_2(g) \rightleftharpoons 2BrF_5(g)$

15.8 Write the equilibrium law for each of the following reactions in aqueous solution.

(a) $Ag^+(aq) + 2NH_3(aq) \rightleftharpoons Ag(NH_3)_2{}^+(aq)$

(b) $Cd^{2+}(aq) + 4SCN^-(aq) \rightleftharpoons Cd(SCN)_4{}^{2-}(aq)$
(c) $HClO(aq) + H_2O \rightleftharpoons H_3O^+(aq) + ClO^-(aq)$

15.9 The following reaction in aqueous solution has $K_c = 1 \times 10^{-85}$ at 25 °C.

$$7IO_3^- + 9H_2O + 7H^+ \rightleftharpoons I_2 + 5H_5IO_6$$

What is the equilibrium law for this reaction?

15.10 Write the equilibrium law for each of the following reactions in terms of molar concentrations.
(a) $H_2(g) + Cl_2(g) \rightleftharpoons 2HCl(g)$
(b) $\frac{1}{2}H_2(g) + \frac{1}{2}Cl_2(g) \rightleftharpoons HCl(g)$
How does K_c for reaction (a) compare with K_c for reaction (b)?

15.11 At 25 °C, $K_c = 1 \times 10^{-85}$ for the reaction

$$7IO_3^- + 9H_2O + 7H^+ \rightleftharpoons I_2 + 5H_5IO_6$$

What is the value of K_c for the reaction

$$I_2 + 5H_5IO_6 \rightleftharpoons 7IO_3^- + 9H_2O + 7H^+$$

15.12 Write the equilibrium law for the reaction

$$2HCl(g) \rightleftharpoons H_2(g) + Cl_2(g)$$

How does K_c for this reaction compare with K_c for reaction (a) in Review Exercise 15.10?

15.13 Write the equilibrium law for the reactions in Review Exercise 15.6 in terms of partial pressures.

15.14 Write the equilibrium law for the reactions in Review Exercise 15.7 in terms of partial pressures.

15.15 When a chemical equation and its equilibrium constant are given, why is it not necessary to also specify the form of the mass action expression?

15.16 At 225 °C, $K_p = 6.3 \times 10^{-3}$ for the reaction

$$CO(g) + 2H_2(g) \rightleftharpoons CH_3OH(g)$$

Would we expect this reaction to go nearly to completion?

15.17 Here are some reactions and equilibrium constants.
(a) $2CH_4(g) \rightleftharpoons C_2H_6(g) + H_2(g)$ $K_c = 9.5 \times 10^{-13}$
(b) $CH_3OH(g) + H_2(g) \rightleftharpoons CH_4(g) + H_2O(g)$
  $K_c = 3.6 \times 10^{20}$
(c) $H_2(g) + Br_2(g) \rightleftharpoons 2HBr(g)$ $K_c = 2.0 \times 10^9$
Arrange these reactions in order of their increasing tendency to go toward completion.

15.18 Use the following equilibria
$2CH_4(g) \rightleftharpoons C_2H_6(g) + H_2(g)$ $K_c = 9.5 \times 10^{-13}$
$CH_4(g) + H_2O(g) \rightleftharpoons CH_3OH(g) + H_2(g)$
  $K_c = 2.8 \times 10^{-21}$
to calculate K_c for the reaction

$$2CH_3OH(g) + H_2(g) \rightleftharpoons C_2H_6(g) + 2H_2O(g)$$

Thermodynamics and Equilibrium

15.19 Write the equation that relates the free energy change to the value of the reaction quotient for a reaction.

15.20 How is the equilibrium constant related to the standard free energy change for a reaction? (Write the equation.)

15.21 What is a thermodynamic equilibrium constant?

15.22 What is the value of $\Delta G°$ for a reaction for which $K = 1$?

15.23 Calculate the value of the thermodynamic equilibrium constant for the following reactions at 25 °C. (Refer to the data in Appendix E.1.)
(a) $2PCl_3(g) + O_2(g) \rightleftharpoons 2POCl_3(g)$
(b) $2SO_3(g) \rightleftharpoons 2SO_2(g) + O_2(g)$
(c) $N_2H_4(g) + 2O_2(g) \rightleftharpoons 2NO(g) + 2H_2O(g)$
(d) $N_2H_4(g) + 6H_2O_2(g) \rightleftharpoons 2NO_2(g) + 8H_2O(g)$

15.24 For each of the reactions in the preceding exercise, calculate K_p at 300 °C.

15.25 The reaction $NO_2(g) + NO(g) \rightleftharpoons N_2O(g) + O_2(g)$ has $\Delta G°_{1273} = -9.67$ kJ. A 1.00 L reaction vessel at 1000 °C contains 0.0200 mol NO_2, 0.040 mol NO, 0.015 mol N_2O, and 0.0350 mol O_2. Is the reaction at equilibrium? If not, in which direction will the reaction proceed to reach equilibrium?

15.26 The reaction $CO(g) + H_2O(g) \rightleftharpoons HCHO_2(g)$ has $\Delta G°_{673} = +79.8$ kJ. If a mixture at 400 °C contains 0.040 mol CO, 0.022 mol H_2O, and 3.8×10^{-3} mol $HCHO_2$ in a 2.50 L container, is the reaction at equilibrium? If not, in which direction will the reaction proceed spontaneously?

15.27 A reaction that can convert coal to methane (the chief component of natural gas) is

$$C(s) + 2H_2(g) \rightleftharpoons CH_4(g)$$

for which $\Delta G° = -50.79$ kJ. What is the value of K_p for this reaction at 25 °C? Does this value of K_p suggest that studying this reaction as a means of methane production is worth pursuing?

15.28 One of the important reactions in living cells from which the organism draws energy is the reaction of adenosine triphosphate (ATP) with water to give adenosine diphosphate (ADP) and free phosphate ion.

$$ATP + H_2O \rightleftharpoons ADP + PO_4{}^{3-}$$

The value of $\Delta G°_T$ for this reaction at 37 °C (normal human body temperature) is -33 kJ/mol. Calculate the value of the equilibrium constant for the reaction at this temperature.

15.29 What is the value of the equilibrium constant for a reaction for which $\Delta G° = 0$? What will happen to the composition of the system if we begin the reaction with the pure products?

15.30 A certain reaction at 25 °C has $K_p = 10.0$. What is the value of $\Delta G°$ for this reaction?

15.31 Methanol, a potential replacement for gasoline as an automotive fuel, can be made from H_2 and CO by the reaction

$$CO(g) + 2H_2(g) \rightleftharpoons CH_3OH(g)$$

At 500 K, this reaction has $K_p = 6.25 \times 10^{-3}$. Calculate $\Delta G°_{500}$ for this reaction in units of kilojoules.

15.32 Refer to the data in Table 13.2 to determine $\Delta G°$ in kilojoules for the reaction $2NO(g) + 2CO(g) \rightleftharpoons N_2(g) + 2CO_2(g)$. What is the value of K_p for this reaction at 25 °C?

15.33 Use the data in Appendix E.1 to calculate ΔG_T° in kilojoules for the reaction

$$2NO(g) + 2CO(g) \rightleftharpoons N_2(g) + 2CO_2(g)$$

at 500 °C. What is the value of K_p for the reaction at this temperature?

Converting between K_p and K_c

15.34 State the equation relating K_p to K_c and define all terms. Which is the only value of R that can be properly used in this equation?

15.35 Use the ideal gas law to show that the partial pressure of a gas is directly proportional to its molar concentration. What is the proportionality constant?

15.36 A 345 mL container holds NH_3 at a pressure of 745 torr and a temperature of 45 °C. What is the molar concentration of ammonia in the container?

15.37 In a certain container at 145 °C the concentration of water vapor is 0.0200 M. What is the partial pressure of H_2O in the container?

15.38 For which of the following reactions does $K_p = K_c$?
(a) $N_2(g) + O_2(g) \rightleftharpoons 2NO(g)$
(b) $2H_2(g) + C_2H_2(g) \rightleftharpoons C_2H_6(g)$
(c) $2NO(g) + O_2(g) \rightleftharpoons 2NO_2(g)$
(d) $CO_2(g) + H_2(g) \rightleftharpoons CO(g) + H_2O(g)$
(e) $PCl_3(g) + Cl_2(g) \rightleftharpoons PCl_5(g)$

15.39 The reaction $CO(g) + 2H_2(g) \rightleftharpoons CH_3OH(g)$ has $K_p = 6.3 \times 10^{-3}$ at 225 °C. What is the value of K_c at this temperature?

15.40 The reaction $HCHO_2(g) \rightleftharpoons CO(g) + H_2O(g)$ has $K_p = 1.6 \times 10^6$ at 400 °C. What is the value of K_c for this reaction at this temperature?

15.41 The reaction $N_2O(g) + NO_2(g) \rightleftharpoons 3NO(g)$ has $K_c = 4.2 \times 10^{-4}$ at 500 °C. What is the value of K_p at this temperature?

15.42 One possible way of removing NO from the exhaust of a gasoline engine is to cause it to react with CO in the presence of a suitable catalyst.

$$2NO(g) + 2CO(g) \rightleftharpoons N_2(g) + 2CO_2(g)$$

At 300 °C, this reaction has $K_c = 2.2 \times 10^{59}$. What is K_p at 300 °C?

15.43 At 773 °C the reaction $CO(g) + 2H_2(g) \rightleftharpoons CH_3OH(g)$ has $K_c = 0.40$. What is K_p at this temperature?

15.44 The reaction $COCl_2(g) \rightleftharpoons CO(g) + Cl_2(g)$ has $K_p = 4.6 \times 10^{-2}$ at 395 °C. What is K_c at this temperature?

Heterogeneous Equilibria

15.45 What is the difference between a *heterogeneous equilibrium* and a *homogeneous equilibrium*?

15.46 Calculate the molar concentration of water in (a) 18.0 mL of H_2O, (b) 100.0 mL of H_2O, and (c) 1.00 L of H_2O. Assume that the density of water is 1.00 g/mL.

15.47 Refer to the latest edition of the *CRC Handbook of Chemistry and Physics* to find the density of sodium chloride. Calculate the molar concentration of NaCl in pure solid NaCl.

15.48 Why do we omit the concentrations of pure liquids and solids from the mass action expression for heterogeneous reactions?

15.49 Write the equilibrium law corresponding to K_c for each of the following heterogeneous reactions.
(a) $2C(s) + O_2(g) \rightleftharpoons 2CO(g)$
(b) $2NaHSO_3(s) \rightleftharpoons Na_2SO_3(s) + H_2O(g) + SO_2(g)$
(c) $2C(s) + 2H_2O(g) \rightleftharpoons CH_4(g) + CO_2(g)$
(d) $CaCO_3(s) + 2HF(g) \rightleftharpoons CaF_2(s) + H_2O(g) + CO_2(g)$

15.50 Write the equilibrium law corresponding to K_c for each of the following heterogeneous reactions.
(a) $CaCO_3(s) + SO_2(g) \rightleftharpoons CaSO_3(s) + CO_2(g)$
(b) $AgCl(s) + Br^-(aq) \rightleftharpoons AgBr(s) + Cl^-(aq)$
(c) $Cu(OH)_2(s) \rightleftharpoons Cu^{2+}(aq) + 2OH^-(aq)$
(d) $Mg(OH)_2(s) \rightleftharpoons MgO(s) + H_2O(g)$
(e) $CuSO_4 \cdot 5H_2O(s) \rightleftharpoons CuSO_4(s) + 5H_2O(g)$

15.51 The heterogeneous reaction $2HCl(g) + I_2(s) \rightleftharpoons 2HI(g) + Cl_2(g)$ has $K_c = 1.6 \times 10^{-34}$ at 25 °C. Suppose 0.100 mol of HCl and solid I_2 is placed in a 1.00 L container. What will be the equilibrium concentrations of HI and Cl_2 in the container?

15.52 At 25 °C, $K_c = 360$ for the reaction

$$AgCl(s) + Br^-(aq) \rightleftharpoons AgBr(s) + Cl^-(aq)$$

If solid AgCl is added to a solution containing 0.10 M Br$^-$, what will be the equilibrium concentrations of Br$^-$ and Cl$^-$?

Le Châtelier's Principle

15.53 State Le Châtelier's principle in your own words.

15.54 How will the equilibrium

$$\text{heat} + CH_4(g) + 2H_2S(g) \rightleftharpoons CS_2(g) + 4H_2(g)$$

be affected by the following?
(a) The addition of $CH_4(g)$
(b) The addition of $H_2(g)$
(c) The removal of $CS_2(g)$
(d) A decrease in the volume of the container
(e) An increase in temperature

15.55 The reaction $CO(g) + 2H_2(g) \rightleftharpoons CH_3OH(g)$ has $\Delta H° = -18$ kJ. How will the amount of CH_3OH present at equilibrium be affected by the following?
(a) Adding $CO(g)$
(b) Removing $H_2(g)$
(c) Decreasing the volume of the container
(d) Adding a catalyst
(e) Increasing the temperature

15.56 Consider the equilibrium

$$N_2O(g) + NO_2(g) \rightleftharpoons 3NO(g) \qquad \Delta H° = +155.7 \text{ kJ}$$

In which direction will this equilibrium be shifted by the following changes?
(a) Addition of N_2O (c) Addition of NO
(b) Removal of NO_2
(d) Increasing the temperature of the reaction mixture
(e) Addition of helium gas to the reaction mixture at constant volume
(f) Decreasing the volume of the container at constant temperature

15.57 In Review Exercises 15.55 and 15.56, which change(s) will alter the value of K_c?

15.58 Consider the equilibrium $2NO(g) + Cl_2(g) \rightleftharpoons 2NOCl(g)$ for which $\Delta H° = -77.07$ kJ. How will the amount of Cl_2 at equilibrium be affected by the following?
(a) Removal of $NO(g)$
(b) Addition of $NOCl(g)$
(c) Raising the temperature
(d) Decreasing the volume of the container

Equilibrium Calculations

15.59 At 773 °C, a mixture of $CO(g)$, $H_2(g)$, and $CH_3OH(g)$ was allowed to come to equilibrium. The following equilibrium concentrations were then measured: $[CO] = 0.105$ M, $[H_2] = 0.250$ M, and $[CH_3OH] = 0.00261$ M. Calculate K_c for the reaction
$$CO(g) + 2H_2(g) \rightleftharpoons CH_3OH(g)$$

15.60 At a certain temperature, $K_c = 0.18$ for the equilibrium
$$PCl_3(g) + Cl_2(g) \rightleftharpoons PCl_5(g)$$
Suppose a reaction vessel at this temperature contained these three gases at the following concentrations: $[PCl_3] = 0.0420$ M, $[Cl_2] = 0.0240$ M, $[PCl_5] = 0.00500$ M.
(a) Is the system in a state of equilibrium?
(b) If not, which direction will the reaction have to proceed to get to equilibrium?

15.61 At 460 °C, the reaction
$$SO_2(g) + NO_2(g) \rightleftharpoons NO(g) + SO_3(g)$$
has $K_c = 85.0$. A reaction flask at 460 °C contains these gases at the following concentrations: $[SO_2] = 0.00250$ M, $[NO_2] = 0.00350$ M, $[NO] = 0.0250$ M, and $[SO_3] = 0.0400$ M.
(a) Is the reaction at equilibrium?
(b) If not, which way will the reaction have to proceed to arrive at equilibrium?

15.62 The reaction $N_2O_4(g) \rightleftharpoons 2NO_2(g)$ has $K_p = 0.140$ at 25 °C. In a reaction vessel containing these gases in equilibrium at this temperature, the partial pressure of N_2O_4 was 0.250 atm.
(a) What was the partial pressure of the NO_2 in the reaction mixture?
(b) What was the total pressure of the mixture of gases?

15.63 Ethylene, C_2H_4, and water react under appropriate conditions to give ethanol. The reaction is
$$C_2H_4(g) + H_2O(g) \rightleftharpoons C_2H_5OH(g)$$
An equilibrium mixture of these gases at a certain temperature had the following concentrations: $[C_2H_4] = 0.0148$ M, $[H_2O] = 0.0336$ M, $[C_2H_5OH] = 0.180$ M. What is the value of K_c?

15.64 At a certain temperature, the reaction
$$CO(g) + 2H_2(g) \rightleftharpoons CH_3OH(g)$$
has $K_c = 0.500$. If a reaction mixture at equilibrium contains 0.180 M CO and 0.220 M H_2, what is the concentration of CH_3OH?

15.65 At 25 °C, $K_c = 0.145$ for the following reaction in the solvent CCl_4.
$$2BrCl \rightleftharpoons Br_2 + Cl_2$$
If the initial concentration of BrCl in the solution is 0.050 M, what will the equilibrium concentrations of Br_2 and Cl_2 be?

15.66 At 25 °C, $K_c = 0.145$ for the following reaction in the solvent CCl_4.
$$2BrCl \rightleftharpoons Br_2 + Cl_2$$
If the initial concentration of each substance in a solution is 0.0400 M, what will their equilibrium concentrations be?

15.67 At 25 °C, $K_c = 0.145$ for the following reaction in the solvent CCl_4.
$$2BrCl \rightleftharpoons Br_2 + Cl_2$$
If the initial concentrations of Br_2 and Cl_2 are each 0.0250 M, what will their equilibrium concentrations be?

*****15.68** At 25 °C, $K_c = 0.145$ for the following reaction in the solvent CCl_4.
$$2BrCl \rightleftharpoons Br_2 + Cl_2$$
A solution was prepared with the following initial concentrations: $[BrCl] = 0.0400$ M, $[Br_2] = 0.0300$ M, and $[Cl_2] = 0.0200$ M. What will their equilibrium concentrations be?

15.69 $K_c = 64$ for the reaction
$$N_2(g) + 3H_2(g) \rightleftharpoons 2NH_3(g)$$
at a certain temperature. Suppose it was found that an equilibrium mixture of these gases contained 0.360 M NH_3 and 0.0192 M N_2. What would be the concentration of H_2 in the mixture?

15.70 At high temperature, 0.500 mol of HBr was placed in a 1.00 L container where it decomposed to give the equilibrium
$$2HBr(g) \rightleftharpoons H_2(g) + Br_2(g)$$
At equilibrium the concentration of Br_2 was measured to be 0.0955 M. What is K_c for this reaction at this temperature?

15.71 A 0.100 mol sample of formaldehyde vapor, CH_2O, was placed in a heated 1.00 L vessel and some of it decomposed. The reaction is
$$CH_2O(g) \rightleftharpoons H_2(g) + CO(g)$$
At equilibrium, the $CH_2O(g)$ concentration was 0.066 mol/L. Calculate the value of K_c for this reaction.

15.72 The reaction $NO_2(g) + NO(g) \rightleftharpoons N_2O(g) + O_2(g)$ reached equilibrium at a certain high temperature. Originally, the reaction vessel contained the following initial concentrations: $[N_2O] = 0.184$ M, $[O_2] = 0.377$ M, $[NO_2] = 0.0560$ M, and $[NO] = 0.294$ M. The concentration of the NO_2, the only colored gas in the mixture, was monitored by following the intensity of the color. At equilibrium, the NO_2 concentration had become 0.118 M. What is the value of K_c for this reaction at this temperature?

15.73 The equilibrium constant, K_c, for the reaction
$$SO_3(g) + NO(g) \rightleftharpoons NO_2(g) + SO_2(g)$$

was found to be 0.500 at a certain temperature. If 0.240 mol of SO_3 and 0.240 mol of NO are placed in a 2.00 L container and allowed to react, what will be the equilibrium concentration of each gas?

15.74 For the reaction in Exercise 15.73, a reaction mixture is prepared in which 0.300 mol NO_2 and 0.300 mol of SO_2 are placed in a 2.00 L vessel. After the system reaches equilibrium, what will be the equilibrium concentrations of all four gases? How do these equilibrium values compare to those calculated in Exercise 15.73? Account for your observation.

15.75 At a certain temperature the reaction

$$CO(g) + H_2O(g) \rightleftharpoons CO_2(g) + H_2(g)$$

has $K_c = 0.400$. Exactly 1.00 mol of each gas was placed in a 100 L vessel and the mixture underwent reaction. What was the equilibrium concentration of each gas?

15.76 The reaction $2HCl(g) \rightleftharpoons H_2(g) + Cl_2(g)$ has $K_c = 3.2 \times 10^{-34}$ at 25 °C. If a reaction vessel contains initially 0.0500 mol/L of HCl and then reacts to reach equilibrium, what will be the concentrations of H_2 and Cl_2?

15.77 At 400 °C, $K_c = 2.9 \times 10^4$ for the reaction

$$HCHO_2(g) \rightleftharpoons CO(g) + H_2O(g)$$

A mixture was prepared with the following initial concentrations: $[CO] = 0.20\ M$, $[H_2O] = 0.30\ M$. No formic acid, $HCHO_2$, was initially present. What was the equilibrium concentration of $HCHO_2$?

15.78 At 200°C, $K_c = 1.4 \times 10^{-10}$ for the reaction

$$N_2O(g) + 2NO_2(g) \rightleftharpoons 3NO(g)$$

If 0.200 mol of N_2O and 0.400 mol NO_2 are placed in a 4.00 L container, what would the NO concentration be if this equilibrium were established?

15.79 At a certain temperature, $K_c = 0.914$ for the reaction

$$NO_2(g) + NO(g) \rightleftharpoons N_2O(g) + O_2(g)$$

If 0.200 mol of NO_2 and 0.200 mol of NO are placed in a 5.00 L container, what will be the equilibrium concentration of each gas?

15.80 For the reaction in Exercise 15.79, a mixture was prepared containing 0.200 mol of each gas in a 5.00 L container. What will be the equilibrium concentration of each gas?

*15.81 At a certain temperature, $K_c = 0.18$ for the equilibrium

$$PCl_3(g) + Cl_2(g) \rightleftharpoons PCl_5(g)$$

If 0.026 mol of PCl_5 is placed in a 2.00 L vessel at this temperature, what will the concentration of PCl_3 be at equilibrium?

*15.82 At 460 °C, the reaction

$$SO_2(g) + NO_2(g) \rightleftharpoons NO(g) + SO_3(g)$$

has $K_c = 85.0$. Suppose 0.100 mol of SO_2, 0.0600 mol of NO_2, 0.0800 mol of NO, and 0.120 mol of SO_3 are placed in a 10.0 L container at this temperature. What will the concentrations of all the gases be when the system reaches equilibrium?

*15.83 The reaction $H_2(g) + Br_2(g) \rightleftharpoons 2HBr(g)$ has $K_c = 2.0 \times 10^9$ at 25 °C. If 0.100 mol of H_2 and 0.200 mol of

Br$_2$ are placed in a 10.0 L container, what will all the equilibrium concentrations be at 25 °C?

*15.84 At a certain temperature, K_c = 0.500 for the reaction

$$SO_3(g) + NO(g) \rightleftharpoons NO_2(g) + SO_2(g)$$

If 0.100 mol SO$_3$ and 0.200 mol NO are placed in a 2.00 L container and allowed to come to equilibrium, what will the NO$_2$ and SO$_2$ concentrations be?

Additional Exercises

*15.85 At a certain temperature, K_c = 0.914 for the reaction

$$NO_2(g) + NO(g) \rightleftharpoons N_2O(g) + O_2(g)$$

A mixture was prepared containing 0.200 mol NO$_2$, 0.300 mol NO, 0.150 mol N$_2$O, and 0.250 mol O$_2$ in a 4.00 L container. What will be the equilibrium concentration of each gas?

*15.86 At 1000 °C, K_c = 4.3 × 10^4 for the reaction

$$HCHO_2(g) \rightleftharpoons CO(g) + H_2O(g)$$

If 0.200 mol of HCHO$_2$ is placed in a 1.00 L vessel, what will be the concentrations of CO and H$_2$O at equilibrium?

*15.87 At 27 °C, K_p = 1.5 × 10^{18} for the reaction

$$3NO(g) \rightleftharpoons N_2O(g) + NO_2(g)$$

If 0.030 mol of NO were placed in a 1.00 L vessel and this equilibrium were established, what would be the equilibrium concentrations of N$_2$O and NO$_2$?

15.88 Consider the equilibrium

$$2NaHSO_3(s) \rightleftharpoons Na_2SO_3(s) + H_2O(g) + SO_2(g)$$

How will the position of equilibrium be affected by
(a) Adding NaHSO$_3$ to the reaction vessel
(b) Removing Na$_2$SO$_3$ from the reaction vessel
(c) Adding H$_2$O to the reaction vessel
(d) Increasing the volume of the reaction vessel

*15.89 Consider the equilibrium

$$CuSO_4 \cdot 5H_2O(s) \rightleftharpoons CuSO_4(s) + 5H_2O(g)$$

(a) Calculate ΔG_T° for this reaction at 150 °C.
(b) What is the value of K_p for this reaction at 150 °C?
(c) What is the partial pressure of water over solid CuSO$_4 \cdot 5H_2O(s)$ at 150 °C?

*15.90 For the reaction below, K_p = 1.6 × 10^6 at 400 °C

$$HCHO_2(g) \rightleftharpoons CO(g) + H_2O(g)$$

A mixture of CO(g) and H$_2$O(l) was prepared in a 2.00 L reaction vessel at 25 °C in which the pressure of CO was 0.177 atm. The mixture also contained 0.391 g of H$_2$O. The vessel was sealed and heated to 400 °C. When equilibrium was reached, what was the partial pressure of the HCHO$_2$?

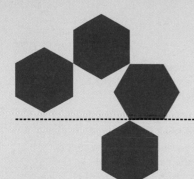

CHEMICALS IN USE 12
Ammonia

Throughout the nineteenth century chemists sought a way to make ammonia directly from its elements. Most succeeded, but with disappointingly low yields. Yet, the efforts did establish important facts. To make ammonia from nitrogen and hydrogen, some kind of metal or metal oxide is needed as a catalyst, high pressures favor the conversion, and high temperatures decompose ammonia into its elements.

Carrying out one of history's earliest major applications of the principles of thermodynamics to a practical problem, Fritz Haber (1868–1934), a German physical chemist, found that the best set of conditions involved pressures and temperatures possibly too high for the materials used to make industrial equipment. Karl Bosch, an engineer for the German company that became interested in Haber's work (Badische Anilin and Soda Fabrik), solved these problems, however. Hence, the process is called the Haber–Bosch method. The first plant to use it went into production in 1913, and Haber won the 1918 Nobel Prize in chemistry for his basic research.

The formation of ammonia involves the following equilibrium:

$$N_2(g) + 3H_2(g) \rightleftharpoons 2NH_3(g) \qquad \Delta H° = -92.2 \text{ kJ}$$

Because the substances are all gases, the coefficients give us not only mole ratios but also volume ratios (Avogadro's principle). The reaction thus changes 4 volumes of reactants into 2 volumes of product. Hence, an *increase* in the pressure favors the formation of ammonia. The equilibrium shifts to the right to absorb this volume-reducing stress in accordance with Le Châtelier's principle (which was known to Haber).

TABLE 12a   Mole Percent of Ammonia at Equilibrium

Temp (°C)	Pressure (atm)						
	10	30	50	100	300	600	1000
200	51	68	74	82	90	95	98
300	15	30	39	52	71	84	93
400	3.9	10	15	25	47	65	80
500	1.2	3.5	5.6	11	26	42	57
600	0.49	1.4	2.3	4.5	14	23	31
700	0.23	0.68	1.1	2.2	7.3	13	13

Because the forward reaction is exothermic, *heating* the mixture shifts the equilibrium to the *left* and lowers the yield of ammonia. Thus, the mixture should be *cooled* to improve the overall yield of ammonia. Many bonds have to be broken among the reactants, however, so that others can form in the product. Thus, the mixture must nevertheless be heated despite the fact that this makes the position of the equilibrium less favorable.

Haber and his students made many measurements of the concentrations of ammonia present at equilibrium under several temperatures and pressures (see Table 12a). Notice in the table how the yield of ammonia increases with pressure (at a given temperature) and how it decreases with increasing temperature (at a given pressure), just as the analysis suggests.

The catalyst consists of very small particles of iron doped with traces of alumina, silica, magnesium oxide, and potassium hydroxide. Without a catalyst, the rate of formation of ammonia is almost zero even at 500 °C, so we can see how vital the catalyst is. With it, most operations are run in the range of 380 °C to 450 °C at a pressure of about 200 atm. Figure 12a shows how a reaction vessel is designed for carrying out the final step.

WATER, AIR, AND METHANE AS THE ACTUAL REACTANTS

The raw materials that enter a Haber–Bosch plant are not the pure nitrogen and hydrogen of the simplified equation. Instead, the starting materials are water, air, and natural gas, the later being almost entirely methane, CH_4. Water and methane supply the hydrogen, and air furnishes the nitrogen. Of course, air also supplies oxygen, but this is removed in a clever and resourceful way.

The methane and water are taken in just those proportions that supply enough hydrogen not only for the ammonia synthesis but also to react with and so remove all of the oxygen in the air. This regenerates some of the water, *and it also supplies the heat energy needed to run the whole operation.* The carbon in methane ends up in carbon dioxide, and this is removed by its reaction with potassium carbonate.

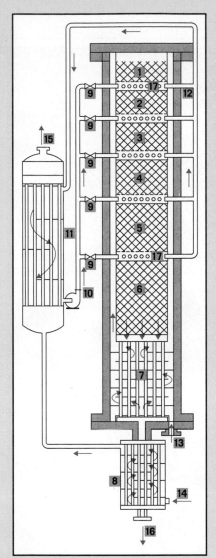

FIGURE 12a

Schematic of an ammonia synthesizer employing the Haber–Bosch process. A mixture of nitrogen and hydrogen enters the system at 13 and then eventually flows through beds of the catalyst (1–6). The product exits at 16. The circulations indicated by the arrows involving pipes in 7–12, 14, 15, and 17 are flows of coolant to carry away the heat produced by this exothermic reaction. (From Supplement I, Volume III, *Mellor's Comprehensive Treatise on Inorganic and Theoretical Chemistry*, page 266, Longmans, Green and Co., London, 1964.)

shift the equilibria to the right). Now air is injected into the flowing gas stream, and some of the hydrogen just made reacts with the oxygen in the air to leave water and nitrogen.

$$2H_2 + [O_2 + 4N_2] \rightleftharpoons 2H_2O + 4N_2$$

This raises the internal temperature to about 1100 °C, and some of the heat generated is used in three ways. First, it helps to force virtually all of the still unchanged CH_4 to react with water, giving more hydrogen.

$$CH_4 + H_2O \rightleftharpoons CO + 3H_2$$

Second, part of the released heat is the energy that operates the gas compressors. Third, a portion of the heat is used to make the high-temperature steam injected along with methane at the very start of the whole operation. The loss of heat for these purposes cools the gas mixture, which now is chiefly CO, CO_2, H_2O, H_2, and N_2. The carbon monoxide must be removed because it inactivates ("poisons") the final Haber–Bosch iron catalyst. The gas mixture, therefore, is next channeled through a series of catalysts that remove the CO by the following reaction. More hydrogen is made, however.

$$CO + H_2O \rightleftharpoons CO_2 + H_2$$

The final mixture of gases that enters the last reaction vessel has only 1 to 2 μg of CO per liter.

Before the gas stream enters the last vessel, it is passed through a concentrated solution of potassium carbonate, which removes both water and carbon dioxide:

$$CO_2 + H_2O + K_2CO_3 \longrightarrow 2KHCO_3$$

The gas stream now consists of hydrogen and nitrogen in essentially the correct mole ratio for the final reaction that makes ammonia. The trace of argon from the original air is still present, but it causes no interference.

Taking the combination $[O_2 + 4N_2]$ to represent air in its almost exactly correct natural mole ratio of oxygen and nitrogen, we can write the following equation for the entire overall operation.

$$7CH_4 + 2[O_2 + 4N_2] + 17H_2O + 7K_2CO_3 \longrightarrow$$
$$16NH_3 + 14KHCO_3$$

This equation represents the net effect of many steps.

STEPS IN THE HABER–BOSCH PROCESS

In the first step, methane and steam are passed under high pressure (30 atm) and temperature (750 °C) over a nickel catalyst where the following equilibria develop, the products being favored.

$$CH_4 + H_2O \rightleftharpoons CO + 3H_2$$

$$CH_4 + 2H_2O \rightleftharpoons CO_2 + 4H_2$$

Together these steps use all but about 9% of the CH_4. (Although the equations are written as equilibria, the subsequent reactions of the products help

Questions

Where appropriate, include equations in your answers.

1. How is natural gas used in the industrial synthesis of NH_3?

2. What becomes of the carbon in CH_4 in the overall process?

3. How is O_2 removed from air in the Haber–Bosch operation?

4. Why does a high temperature lower the yield of ammonia?

5. Why does a high pressure raise the yield of ammonia?

6. Explain how carbon monoxide interferes with the Haber–Bosch reaction.

For some, nothing tastes better on a hot dog than sauerkraut, a product made by the fermentation of cabbage in dilute brine (salt) solution. Sauerkraut owes its tartness to the presence of lactic acid, a weak monoprotic acid formed in the fermentation process. In this chapter we will discuss the equilibria involved in the ionization of weak acids, such as lactic acid, as well as weak bases.

Chapter 16

Acid–Base Equilibria

You learned earlier that weak acids and bases, such as the acetic acid in vinegar and the ammonia in glass cleaners, are incompletely ionized in water. For each of these substances an equilibrium exists between the molecular acid or base and the ions formed by its ionization. In this chapter we will explore these equilibria in depth, but before we do this, we must first discuss a particularly important equilibrium that exists in *all* aqueous solutions—the ionization of water.

In pure water and in any aqueous solution, water molecules participate in the following ionization reaction.

$$H_2O + H_2O \rightleftharpoons H_3O^+ + OH^-$$

This is also called an *autoionization reaction* because the water molecules react with each other. The equilibrium law for the reaction, written following the procedures we developed in Chapter 15, is

$$\frac{[H_3O^+][OH^-]}{[H_2O]^2} = K_c$$

In pure water, its molar concentration, $[H_2O]$, has a constant value (which happens to be 55.6 M), so the quantity in the denominator, $[H_2O]^2$, can be combined with K_c to give a new constant that we will designate as K_w.

$$[H_3O^+][OH^-] = K_c \cdot [H_2O]^2 = K_w$$

Often we omit the water molecule that carries the hydrogen ion and write H^+ in place of H_3O^+. The equation for the ionization of water can then be simplified as

$$H_2O \rightleftharpoons H^+ + OH^-$$

16.1
IONIZATION OF WATER AND THE pH CONCEPT

We have omitted the usual (*aq*) following the symbols for the ions for the sake of simplicity. We will do this throughout the remainder of this chapter, but you should realize that all ions in aqueous solution are hydrated.

Taking the density of water to be 1.00 g mL^{-1} or 1.00×10^3 g L^{-1}, the molar concentration of water (at 18.0 g mol^{-1}) is 55.6 mol L^{-1}.

This yields the equilibrium law

Ion product expression for water

$$[H^+][OH^-] = K_w \qquad (16.1)$$

Because the constant K_w is equal to a product of ion concentrations, it is called the **ion product constant of water.**

In pure water, H^+ and OH^- ions are formed in equal numbers by the autoionization reaction, so their concentrations are equal. At 25 °C in pure water it is found that $[H^+] = [OH^-] = 1.0 \times 10^{-7}$ mol L^{-1}, so at this temperature,[1]

$$K_w = (1.0 \times 10^{-7})(1.0 \times 10^{-7})$$

$$K_w = 1.0 \times 10^{-14} \qquad \text{(at 25 °C)} \qquad (16.2)$$

Throughout this chapter we will often substitute H^+ for H_3O^+. By now, however, you know that a hydrogen ion (proton) cannot exist by itself in water, but always becomes attached to a water molecule. We use H^+ just for simplicity in writing certain expressions.

TABLE 16.1 K_w at Various Temperatures

Temperature (°C)	K_w
0	1.5×10^{-15}
10	3.0×10^{-15}
20	6.8×10^{-15}
25	1.0×10^{-14}
30	1.5×10^{-14}
40	3.0×10^{-14}
50	5.5×10^{-14}
60	9.5×10^{-14}

As with other equilibrium constants, the value of K_w varies with temperature (see Table 16.1). For simplicity, in our discussions in this chapter we will deal with systems at 25 °C (unless noted otherwise), so we will not be constantly specifying the temperature. Be sure to learn the value of K_w at 25 °C, because you will need to use if frequently.

A dilute solution is almost all water, and in such a solution the molar concentration of water is nearly identical to its value in pure water. When expressed to three significant figures, the molarity of water in both pure water and dilute aqueous solutions is 55.6 *M*.

The Autoionization of Water in Aqueous Solutions   In the preceding discussion we focused our attention on pure water, for which Equations 16.1 and 16.2 apply perfectly. For dilute solutions, such as those we will examine later in this chapter, it is also reasonable to consider the molar concentration of water to have effectively the same constant value of 55.6 *M*. Therefore, *it is safe to assume that Equations 16.1 and 16.2, which we derived for pure water, also apply to dilute aqueous solutions.*

The significance of the preceding statement should not be overlooked. **For any aqueous solution, regardless of what other equilibria are taking place, the equilibrium autoionization of water is also present and in the solution, the product of [H$^+$] and [OH$^-$] is equal to K_w.**

Criteria for Acidic, Basic, and Neutral Solutions

A neutral solution is one that is neither acidic nor basic. Another way of saying this is that in a neutral solution, the concentrations of H_3O^+ and OH^- are equal. This is so in pure water, where $[H_3O^+] = [OH^-] = 1.0 \times 10^{-7}$ *M*. In fact, these same concentrations of H_3O^+ and OH^- exist in *any* neutral solution, because the product of their concentrations must equal K_w.

An acidic solution is one in which the H_3O^+ concentration is larger than that of OH^-, and a basic solution exists when the molar concentration of OH^- exceeds that of H_3O^+. The *relative* concentrations of H_3O^+ and OH^- thus enable us to define acidic and basic solutions.

In any aqueous solution there are both H_3O^+ and OH^-, because of the equilibrium ionization of water. Even the most acidic solution has some OH^-, and even the most basic solution has some H_3O^+.

Neutral solution	$[H_3O^+] = [OH^-] = 1.0 \times 10^{-7}$ *M*
Acidic solution	$[H_3O^+] > [OH^-]$
Basic solution	$[H_3O^+] < [OH^-]$

[1] Only the *numerical part* of a molar concentration is used in calculating K_w.

In a sample of blood at 25 °C, $[H^+] = 4.6 \times 10^{-8}$ M. Find the molar concentration of OH^- and decide whether the sample is acidic, basic, or neutral.

ANALYSIS Is there a relationship between $[H^+]$ and $[OH^-]$? Of course, $[H^+]$ and $[OH^-]$ are *always* related to each other and to K_w by Equation 16.1, so this is the equation we need.

SOLUTION We know that $K_w = 1.0 \times 10^{-14} = [H^+][OH^-]$. So we substitute the given value of $[H^+]$ into this equation and solve for $[OH^-]$.

$$1.0 \times 10^{-14} = (4.6 \times 10^{-8})[OH^-]$$

Therefore,

$$[OH^-] = \frac{1.0 \times 10^{-14}}{4.6 \times 10^{-8}}$$
$$= 2.2 \times 10^{-7} \ M$$

When we compare $[H^+] = 4.6 \times 10^{-8}$ M and $[OH^-] = 2.2 \times 10^{-7}$ M, we see that $[OH^-] > [H^+]$, so the blood is basic, although very slightly so.

■ **Practice Exercise 1** An aqueous solution of sodium bicarbonate, $NaHCO_3$, has a molar concentration of hydroxide ion of 7.8×10^{-6} M. What is the molar concentration of hydrogen ion? Is the solution acidic, basic, or neutral?

The pH Concept

In most solutions of weak acids and bases, the concentrations of H^+ and OH^- are very small. Typical values are those in Example 16.1 above. To allow easy comparisons of small values of H^+, a logarithmic system was developed called **pH.** The pH of a solution is defined as follows:

$$pH = -\log[H^+] \tag{16.3}$$

This equation allows us to calculate the pH given $[H^+]$. We can also calculate $[H^+]$ if we know the pH. To accomplish this last operation, Equation 16.3 can be rearranged using the properties of logarithms to give

$$[H^+] = 10^{-pH} \tag{16.4}$$

The pH concept proved so successful for representing and comparing small numbers that it was extended to cover quantities other than $[H^+]$.

For a quantity X, we have the following definition

$$pX = -\log X \tag{16.5}$$

Thus, we can define the pOH of a solution as

$$pOH = -\log[OH^-]$$

and we can define pK_w as

$$pK_w = -\log K_w$$

The numerical value of pK_w is 14.00.

$$pK_w = -\log(1.0 \times 10^{-14})$$
$$= -(-14.00) = 14.00$$

EXAMPLE 16.1
Finding $[H^+]$ from $[OH^-]$ or Finding $[OH^-]$ from $[H^+]$

Appendix A has a review of logarithms. Actual values of logarithms are obtained by using your scientific calculator.

 pH equation

S. P. L. Sørenson (1868–1939) developed the pH concept in 1909.

 General defining equation for pX

Notice that *common logs* (base 10 logs) are used in pH calculations. Be sure that you don't use the *natural log* function on your scientific calculator by mistake.

TABLE 16.2   pH, [H⁺], [OH⁻], and pOH

pH	[H⁺]ᵃ	[OH⁻]ᵃ	pOH	
0	1	1×10^{-14}	14	
1	1×10^{-1}	1×10^{-13}	13	
2	1×10^{-2}	1×10^{-12}	12	acidic
3	1×10^{-3}	1×10^{-11}	11	solutions
4	1×10^{-4}	1×10^{-10}	10	
5	1×10^{-5}	1×10^{-9}	9	
6	1×10^{-6}	1×10^{-8}	8	
7	1×10^{-7}	1×10^{-7}	7	neutral solution
8	1×10^{-8}	1×10^{-6}	6	
9	1×10^{-9}	1×10^{-5}	5	
10	1×10^{-10}	1×10^{-4}	4	
11	1×10^{-11}	1×10^{-3}	3	basic
12	1×10^{-12}	1×10^{-2}	2	solutions
13	1×10^{-13}	1×10^{-1}	1	
14	1×10^{-14}	1	0	

ᵃ Concentrations are in mol L⁻¹ at 25 °C.

 Sum of pH and pOH

A useful relationship among pH, pOH, and pK_w can be derived from the equation for K_w (Equation 16.1).

$$pH + pOH = pK_w = 14.00 \qquad (16.6)$$

This tells us that in an aqueous solution of *any* solute at 25 °C, the sum of pH and pOH is always 14.00.

Definitions of Acidic, Basic, and Neutral Solutions According to pH   In pure water, or in any solution that is neutral, $[H^+] = [OH^-] = 1.0 \times 10^{-7}$ M. The pH of such a solution is 7.00.[2]

$$\begin{aligned} pH &= -\log(1.0 \times 10^{-7}) \\ &= -(-7.00) \\ &= 7.00 \end{aligned}$$

An acidic solution is one in which $[H^+]$ is larger than 10^{-7} M; this gives a pH that is *less than* 7.00. Similarly, a basic solution is one in which the $[H^+]$ is less than 10^{-7} M and the pH is *larger than* 7.00.

At 25 °C:

	pH < 7.00	**Acidic solution**
	pH > 7.00	**Basic solution**
	pH = 7.00	**Neutral solution**

Table 16.2 gives the relationships among pH, [H⁺], [OH⁻], and pOH. Notice how the pH values become larger as the concentration of hydrogen ion becomes smaller. The pH for a number of common substances is given on a pH scale in Figure 16.1.

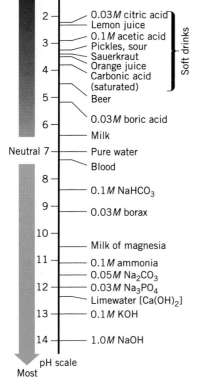

Most acidic

1 —— 0.1 M HCl

2 —— 0.03 M citric acid
—— Lemon juice
—— 0.1 M acetic acid
3 —— Pickles, sour
—— Sauerkraut
—— Orange juice
4 —— Carbonic acid (saturated)
5 —— Beer

6 —— 0.03 M boric acid
—— Milk

Neutral 7 —— Pure water
—— Blood

8 —— 0.1 M NaHCO₃

9 —— 0.03 M borax

10 ——
—— Milk of magnesia
11 —— 0.1 M ammonia
—— 0.05 M Na₂CO₃
12 —— 0.03 M Na₃PO₄
—— Limewater [Ca(OH)₂]
13 —— 0.1 M KOH

14 —— 1.0 M NaOH

pH scale

Most basic

Soft drinks

FIGURE 16.1

The pH scale.

[2] Recall that the rule for significant figures in logarithms is that *the number of decimal places in the logarithm of a number equals the number of significant figures in the number.*

pH Calculations

Let us now work several examples involving pH calculations. The first two calculate a pH from either $[H^+]$ or $[OH^-]$. Another finds the value of $[H^+]$ given a pH.

EXAMPLE 16.2
Calculating pH and pOH from $[H^+]$

Because rain washes pollutants out of the air, the lakes in many parts of the world are undergoing changes in pH. In a New England state, the water in one lake was found to have $[H^+] = 3.2 \times 10^{-5}$ mol L^{-1}. Calculate the pH and pOH of the lake's water and decide whether it is acidic or basic.

ANALYISIS Given a value of $[H^+]$, we use the defining equation for pH directly.

SOLUTION Using Equation 16.3, which defines pH,

$$\begin{aligned} pH &= -\log[H^+] \\ &= -\log(3.2 \times 10^{-5}) \\ &= -(-4.49) = 4.49 \end{aligned}$$

The pH is less than 7.00, so the lake's water is acidic—too acidic, in fact, for game fish to survive. (Young game fish generally cannot withstand water with a pH less than about 5.5.) Because the sum of pH and pOH equals 14.00, the pOH of the lake water is

$$\begin{aligned} pOH &= 14.00 - 4.49 \\ &= 9.51 \end{aligned}$$

■ **Practice Exercise 2** A carbonated beverage was found to have a hydrogen ion concentration of 3.67×10^{-4} mol L^{-1}. What is the pH and pOH of this beverage. Is it acidic, basic, or neutral?

EXAMPLE 16.3
Calculating pH from $[OH^-]$

What is the pH of a sodium hydroxide solution at 25 °C in which $[OH^-] = 0.0026$ *M*?

ANALYSIS The easiest way to find pH when working with a basic solution is to find the pOH first, and then subtract it from 14.00.

SOLUTION Because $[OH^-] = 0.0026$ *M*, we enter this value into the defining equation for pOH:

$$\begin{aligned} pOH &= -\log[OH^-] \\ &= -(\log 0.0026) \\ &= 2.59 \end{aligned}$$

Now we can subtract this from 14.00 to find the pH.

$$\begin{aligned} pH &= 14.00 - 2.59 \\ &= 11.41 \end{aligned}$$

Notice that the pH is well above 7 in this basic solution.

■ **Practice Exercise 3** The water in which a soil sample had soaked overnight was found to have $[OH^-] = 1.47 \times 10^{-9}$ mol L^{-1}. What was the pH?

EXAMPLE 16.4
Calculating [H⁺] from pH

"Calcareous soil" is soil rich in calcium carbonate. The pH of such soil generally ranges from just over 7 to as high as 8.3. One particular sample of this soil was found to have a pH of 8.14. What value of [H⁺] corresponds to a pH of 8.14? Is the soil acidic or basic?

ANALYSIS To calculate [H⁺] from pH, we use Equation 16.4.

$$[H^+] = 10^{-pH}$$

SOLUTION We substitute the pH value, 8.14, into the equation,

$$[H^+] = 10^{-8.14}$$

Because the pH has two digits following the decimal point, we obtain two significant figures in the hydrogen ion concentration. Therefore, the answer correctly rounded is,

$$[H^+] = 7.2 \times 10^{-9} \ M$$

Because $[H^+] < 1 \times 10^{-7} \ M$, the soil is basic.

■ **Practice Exercise 4** Find the values of [H⁺] that correspond to each of the following values of pH. (Remember that the answers should be rounded to two significant figures.) State whether each solution is acidic or basic. (a) 2.90 (the approximate pH of lemon juice), (b) 3.85 (the approximate pH of sauerkraut), (c) 10.81 (the pH of milk of magnesia, a laxative), (d) 4.11 (the pH of orange juice, on the average), (e) 11.61 (the pH of dilute, household ammonia)

Measuring pH

pH is measured either with a pH meter—the most accurate and precise method—or by touching an indicator paper with a drop of the solution and comparing the new color on the paper with a color code. As we learned in Chapter 4, dyes whose colors in an aqueous solution depend on the acidity of the solution are called *acid–base indicators*. Table 16.3 gives several examples of indicator dyes. Litmus is available as "litmus paper," either red or blue. Below pH 4.7, litmus is red and above pH 8.3 it is blue. The transition for the color change occurs over a pH range of 4.7 to 8.3 with the center of this change at about 6.5, very nearly a neutral pH.

TABLE 16.3   Common Acid–Base Indicators

Indicator	Approximate pH Range	Color Change (lower to higher pH)
Methyl green	0.2–1.8	Yellow to blue
Thymol blue	1.2–2.8	Yellow to blue
Methyl orange	3.2–4.4	Red to yellow
Ethyl red	4.0–5.8	Colorless to red
Bromocresol purple	5.2–6.8	Yellow to purple
Bromothymol blue	6.0–7.6	Yellow to blue
Litmus	4.7–8.3	Red to blue
Cresol red	7.0–8.8	Yellow to red
Thymol blue	8.0–9.6	Yellow to blue
Phenolphthalein	8.2–10.0	Colorless to pink
Thymolphthalein	9.4–10.6	Colorless to blue
Alizarin yellow R	10.1–12.0	Yellow to red
Clayton yellow	12.2–13.2	Yellow to amber

FIGURE 16.2

A strip of pH test paper, which contains a mixture of acid–base indicators, is being used here to estimate the pH of lemon juice.

FIGURE 16.3

A pH meter.

Some commercial test papers (e.g., Hydrion) are impregnated with several dyes. The containers of these test papers carry the color code. (See Figure 16.2.) When we discuss acid–base titrations later in this chapter, we will see why it is useful to have a selection of indicators that change colors at widely different ranges of pH values.

Values of pH can be determined to within ± 0.01 pH unit with the better models of pH meters. (See Figure 16.3.) These instruments have to be used when solutions are themselves highly colored. Another advantage of a pH meter is that it can be used to follow the change in the pH of a solution while some reaction is occurring.

Acid–base equilibria are established very quickly, so the measured pH of a solution gives the equilibrium hydrogen ion concentration.

Acids and bases, which are solutes that affect the pH of a solution, can be classified as either strong or weak. As with all strong electrolytes, strong acids and bases are fully dissociated in solution, so it is a simple matter to determine for their solutions the concentrations of H^+ and OH^-, and the pH.

The most common strong monoprotic acids are listed in the margin; be sure you know their names and formulas. Because they are completely ionized in solution, *we do not write their reactions with water as equilibria*. Thus, we can represent the ionization of HCl in either of the two following ways, depending on if we want to represent H_3O^+ as H^+.

$$HCl(g) + H_2O \longrightarrow H_3O^+(aq) + Cl^-(aq)$$

$$HCl(g) \xrightarrow{H_2O} H^+(aq) + Cl^-(aq)$$

The strong bases are largely confined to the Groups IA and IIA metal hydroxides. Those from Group IA are the *soluble* strong bases, and those from Group IIA are the relatively *insoluble* strong bases. Remember that the terms

16.2
SOLUTIONS OF STRONG ACIDS AND BASES

Strong Monoprotic Acids

$HClO_4$	perchloric acid
HNO_3	nitric acid
HCl	hydrochloric acid
HBr	hydrobromic acid
HI	hydriodic acid

Later in this chapter you will find it especially important to recognize the strong acids and bases *and* that they are completely dissociated!

strong and *weak* refer to percentage dissociation, not to solubility in water. Although strong bases like $Ca(OH)_2$ and $Mg(OH)_2$ are not very soluble in water, the little that does dissolve dissociates entirely into ions. This is why we classify them as *strong* bases. It's not possible to use the Group IIA metal hydroxides, however, to obtain a high concentration of hydroxide ion in water. When chemists need such a solution, they turn to the Group IA metal hydroxides, particularly NaOH or KOH. These are both very soluble in water and completely dissociated. We do not use equilibrium arrows for their dissociation. For example,

$$NaOH(s) \xrightarrow{H_2O} Na^+(aq) + OH^-(aq)$$

pH of Dilute Solutions of Strong Acids and Bases

The pH concept is almost never used for solutions of strong acids with concentrations greater than 1 *M*. In 1.00 *M* HCl, for example, we know that $[H^+]$ is also 1.00 *M*, because HCl is a strong acid and fully ionized. Each mole of HCl releases one mole of H^+. The numerical part of 1.00 *M* involves no exponential and so is easy to use by itself; there is no need to reexpress $[H^+]$ as a pH. If we wish, of course, we could calculate the pH when $[H^+]$ = 1.00 *M*; pH = $-\log(1.00)$ = 0.00. Similarly, in a 2.00 *M* HCl solution, the pH = $-\log(2.00)$ = -0.301. There is nothing wrong with negative pH values like this, but they offer no advantage over the actual value of $[H^+]$. The pH concept was invented for convenience in working with small values of $[H^+]$ that involve negative exponents.

When *dilute* solutions (less than 1.00 *M*) of strong acids or bases are used, we again refer to the pH concept. Unless the solution is extremely dilute, like 1×10^{-6} *M* or less, we can ignore the tiny contribution made to the value of $[H^+]$ by the self-ionization of water. In 0.001 *M* HCl, for example, $[H^+]$ = 0.001 *M*. We can rewrite this value exponentially as $[H^+]$ = 1×10^{-3} *M*, and the pH is 3.

Similarly, in dilute solutions of strong bases with molarities greater than 1×10^{-6} *M*, we can ignore the extremely small amount of OH^- contributed by the self-ionization of water. In 0.005 *M* NaOH, we can assume that $[OH^-]$ = 0.005 *M* because for each mole of NaOH put into solution, 1 mol of OH^- has been released. In 0.00045 *M* $Ca(OH)_2$, the concentration of OH^- ion is twice the molarity of the solute, or 0.00090 *M* OH^-, because each mole of this base releases 2 mol of OH^- when it dissociates in water.

EXAMPLE 16.5
Calculation of $[H^+]$ and pH in a Solution of a Strong Acid and a Strong Base

Calculate the values of pH, pOH, and $[OH^-]$ in the following solutions: (a) 0.020 *M* HCl and (b) 0.00035 *M* $Ba(OH)_2$.

ANALYSIS Both HCl and $Ba(OH)_2$ (a Group IIA hydroxide) are strong electrolytes, so we assume 100% ionization for each. From each mole of HCl, we expect 1 mole of H^+, and from each mole of $Ba(OH)_2$, there are 2 moles of OH^- liberated.

SOLUTION (a) In 0.020 *M* HCl, $[H^+]$ = 0.020 *M*. Therefore,

$$pH = -\log(0.020)$$
$$= 1.70$$

Thus in 0.020 M HCl, the pH is 1.70. The pOH is $(14.00 - 1.70) = 12.30$. To find $[OH^-]$, we can use this value of pOH.

$$[OH^-] = 10^{-pOH}$$
$$= 10^{-12.30}$$
$$= 5.0 \times 10^{-13} \ M$$

Notice how much smaller $[OH^-]$ is in this acidic solution than it is in pure water. (b) As noted, 1 mole of $Ba(OH)_2$ dissolved per liter produces 2 mol of OH^- per liter. Therefore,

$$[OH^-] = 2 \times 0.00035 \ M$$
$$= 0.00070 \ M$$

$$pOH = -\log(0.00070)$$
$$= 3.15$$

Thus the pOH of this solution is 3.15 and the pH $= (14.00 - 3.15) = 10.85$.

■ **Practice Exercise 5** Calculate $[H^+]$ and the pH in 0.0050 M NaOH.

Suppression of the Ionization of Water by Acidic or Basic Solutes In a solution of an acid, there are actually two sources of $[H^+]$. One is from the ionization of the acid solute itself and the other is from the autoionization of water. Thus,

$$[H^+]_{total} = [H^+]_{from \ solute} + [H^+]_{from \ H_2O}$$

> A similar equation applies to the OH⁻ concentration in a solution of a base.
>
> $[OH^-]_{total} =$
> $\qquad [OH^-]_{from \ solute} + [OH^-]_{from \ H_2O}$

However, the second term—the contribution from the water—can be neglected because it is so small compared to the contribution from the solute. For instance, in the previous example, we saw that in 0.020 M HCl the $[OH^-] = 5.0 \times 10^{-13}$ M. The only source of OH^- in the solution is from the ionization of water, and the amounts of OH^- and H^+ *formed by the ionization of water* must be equal. Therefore, $[H^+]_{from \ H_2O} = 5.0 \times 10^{-13} \ M$.

If we now look at the total $[H^+]$ for this solution, we have

$$[H^+]_{total} = \underset{\text{(from HCl)}}{0.020 \ M} + \underset{\text{(from H}_2\text{O)}}{5.0 \times 10^{-13} \ M}$$

$$= 0.020 \ M \ \text{(rounded correctly)}$$

In the solution of the acid, the autoionization of the solvent is suppressed by the H^+ produced by the solute. This occurs in all solutions that we will deal with in this chapter, so *in a solution of any* **acid** *it is safe to assume that all the H^+ comes from the solute.* Similarly, *it is also safe to assume that in any solution of a* **base**, *all the OH^- comes from the dissociation of the solute.*

> This is Le Châtelier's principle in action. The H⁺ from the solute shifts the position of equilibrium in the autoionization reaction in the direction of H₂O, so the amount of H⁺ formed per liter is less than 1.0×10^{-7}.

Weak acids and bases are not fully ionized in water and exist in solution in equilibria with the ions formed by their reactions with water. To deal quantitatively with these equilibria it is essential that you be able to write correct chemical equations for the equilibrium reactions, from which you can then obtain the corresponding correct equilibrium laws.

16.3
IONIZATION CONSTANTS FOR WEAK ACIDS AND BASES

Reaction of a Weak Acid with Water

In aqueous solutions, all weak acids behave the same way. They are Brønsted acids and therefore proton donors. Some examples are $HC_2H_3O_2$, HSO_4^-,

Acids do not have to be *molecular.* Many ions are also proton donors.

and NH_4^+. In water, these participate in the following equilibria.

$$HC_2H_3O_2 + H_2O \rightleftharpoons H_3O^+ + C_2H_3O_2^-$$

$$HSO_4^- + H_2O \rightleftharpoons H_3O^+ + SO_4^{2-}$$

$$NH_4^+ + H_2O \rightleftharpoons H_3O^+ + NH_3$$

Notice that in each case, the acid reacts with water to give H_3O^+ and the corresponding conjugate base. We can represent these reactions in a general way using HA to stand for the formula of the acid.

Ionization equation for a weak acid

$$HA + H_2O \rightleftharpoons H_3O^+ + A^-$$

As you can see, HA does not have to be electrically neutral; it can be a molecule such as $HC_2H_3O_2$, but it can also be a negative ion such as HSO_4^- or a positive ion such as NH_4^+. (Of course, the actual charge on the conjugate base will then depend on the charge on the parent acid.)

We can also write a general equation for the equilibrium law for the reaction. Following past practice, we have

$$K_c = \frac{[H_3O^+][A^-]}{[HA][H_2O]}$$

Earlier we said that in dilute aqueous solutions, $[H_2O]$ can be considered a constant, so it can be incorporated in the equilibrium constant. Doing this gives

$$K_c \times [H_2O] = \frac{[H_3O^+][A^-]}{[HA]} = K_a$$

Some references call K_a the acid *dissociation* constant.

The strength of a weak acid is measured by its value of K_a; the larger the K_a, the stronger the acid.

The new constant K_a is called an **acid ionization constant**. Abbreviating H_3O^+ as H^+, the equation for the ionization of the acid can be simplified as

$$HA \rightleftharpoons H^+ + A^-$$

from which the expression for K_a is obtained directly.

K_a expression for weak acids

$$K_a = \frac{[H^+][A^-]}{[HA]} \qquad (16.7)$$

EXAMPLE 16.6
Writing the K_a Expression for a Weak Acid

Nitrous acid, HNO_2, is a weak acid. Write the equation for its equilibrium ionization in water and the appropriate K_a expression.

SOLUTION The conjugate base of HNO_2 is NO_2^-, so the equation for the ionization reaction is

$$HNO_2 + H_2O \rightleftharpoons H_3O^+ + NO_2^-$$

In the K_a expression, we leave out the H_2O that appears on the left and, for simplicity, represent H_3O^+ as H^+.

$$K_a = \frac{[H^+][NO_2^-]}{[HNO_2]}$$

Alternatively, we could have written the simplified equation for the ionization reaction, from which the K_a expression above is obtained directly.

$$HNO_2 \rightleftharpoons H^+ + NO_2^-$$

■ **Practice Exercise 6** For each of the following acids, write the equation for its ionization in water and the appropriate expression for K_a. (a) $HCHO_2$, (b) $(CH_3)_2NH_2^+$, (c) $H_2PO_4^-$

Values of K_a are usually quite small and can be conveniently represented in logarithmic form similar to pH. Thus, we can define the pK_a of an acid as

$$pK_a = -\log K_a$$

Also, $K_a = 10^{-pK_a}$

The strength of a weak acid is determined by its value of K_a; the larger the K_a, the stronger the acid. Because of the negative sign in the defining equation for pK_a, the stronger the acid, the *smaller* its value of pK_a. The values of K_a and pK_a for some typical weak acids are given in Table 16.4.

The values of K_a for strong acids are very large and are not tabulated. For many strong acids, K_a values have not been measured.

Reaction of a Weak Base with Water

As with weak acids, all weak bases behave in a similar manner in water. They are weak Brønsted bases and therefore proton acceptors. Examples are NH_3 and $C_2H_3O_2^-$. Their reactions with water are

$$NH_3 + H_2O \rightleftharpoons NH_4^+ + OH^-$$

$$C_2H_3O_2^- + H_2O \rightleftharpoons HC_2H_3O_2 + OH^-$$

We see that in each instance, the base reacts with water to give OH^- and the corresponding conjugate acid. We can also represent these reactions by a general equation. If we represent the base by the symbol B, the reaction is

$$B + H_2O \rightleftharpoons BH^+ + OH^-$$

 Ionization equation for a weak base

TABLE 16.4 K_a and pK_a Values for Weak Monoprotic Acids at 25 °C

Name of Acid	Formula	K_a	pK_a
Iodic acid	HIO_3	1.7×10^{-1}	0.23
Chloroacetic acid	$HC_2H_2O_2Cl$	1.36×10^{-3}	2.87
Nitrous acid	HNO_2	7.1×10^{-4}	3.15
Hydrofluoric acid	HF	6.8×10^{-4}	3.17
Cyanic acid	$HOCN$	3.5×10^{-4}	3.46
Formic acid	$HCHO_2$	1.8×10^{-4}	3.74
Barbituric acid	$HC_4H_3N_2O_3$	9.8×10^{-5}	4.01
Butanoic acid	$HC_4H_7O_2$	1.52×10^{-5}	4.82
Acetic acid	$HC_2H_3O_2$	1.8×10^{-5}	4.74
Propanoic acid	$HC_3H_5O_2$	1.34×10^{-5}	4.87
Hydrazoic acid	HN_3	1.8×10^{-5}	4.74
Hypochlorous acid	$HOCl$	3.0×10^{-8}	7.52
Hydrogen cyanide (*aq*)	HCN	6.2×10^{-10}	9.21
Phenol	HC_6H_5O	1.3×10^{-10}	9.89
Hydrogen peroxide	H_2O_2	1.8×10^{-12}	11.74

This yields the equilibrium law

$$K_c = \frac{[BH^+][OH^-]}{[B][H_2O]}$$

The quantity $[H_2O]$ in the denominator is effectively a constant that can be combined with K_c to give a new constant we will call the **base ionization constant, K_b.**

$$K_b = \frac{[BH^+][OH^-]}{[B]} \qquad (16.8)$$

EXAMPLE 16.7
Writing the K_b Expression for a Weak Base

Hydrazine is a rocket fuel.

H H
| |
H—N—N—H
hydrazine, N_2H_4

⎡ H H ⎤⁺
⎢ | | ⎥
⎢ H—N—N—H ⎥
⎣ | ⎦
 H
hydrazinium ion, $N_2H_5^+$

Hydrazine, N_2H_4, is a weak base. Write the equation for its reaction with water and write the expression for K_b.

SOLUTION The conjugate acid of N_2H_4 is $N_2H_5^+$. Therefore, the equation for the ionization reaction in water is

$$N_2H_4 + H_2O \rightleftharpoons N_2H_5^+ + OH^-$$

In the expression for K_b we omit the H_2O, so the equilibrium law is

$$K_b = \frac{[N_2H_5^+][OH^-]}{[N_2H_4]}$$

■ **Practice Exercise 7** For each of the following bases, write the equation for its ionization in water and the appropriate expression for K_b. (a) $(CH_3)_3N$, (b) SO_3^{2-}, (c) NH_2OH

Because the K_b values for weak bases are usually small numbers, the same kind of logarithmic notation is often used to represent the equilibrium constants. Thus, pK_b is defined as

$$pK_b = -\log K_b$$

Table 16.5 lists some molecular bases and their corresponding values of K_b and pK_b.

TABLE 16.5 K_b and pK_b Values for Weak Molecular Bases at 25 °C

Name of Base	Formula	K_b	pK_b
Butylamine	$C_4H_9NH_2$	5.9×10^{-4}	3.23
Methylamine	CH_3NH_2	4.4×10^{-4}	3.36
Ammonia	NH_3	1.8×10^{-5}	4.74
Hydrazine	N_2H_4	1.7×10^{-6}	5.77
Strychnine	$C_{21}H_{22}N_2O_2$	1.0×10^{-6}	6.00
Morphine	$C_{17}H_{19}NO_3$	7.5×10^{-7}	6.13
Hydroxylamine	$HONH_2$	6.6×10^{-9}	8.18
Pyridine	C_5H_5N	1.5×10^{-9}	8.82
Aniline	$C_6H_5NH_2$	4.4×10^{-10}	9.36

Conjugate Acid–Base Pairs and Their Values of K_a and K_b

Formic acid, $HCHO_2$, a typical weak acid, ionizes according to the equation

$$HCHO_2 + H_2O \rightleftharpoons H_3O^+ + CHO_2^-$$

As you've seen, we write its K_a expression as

$$K_a = \frac{[H^+][CHO_2^-]}{[HCHO_2]}$$

The conjugate base of formic acid is the formate ion, CHO_2^-, and when a solute that contains this ion (e.g., $NaCHO_2$) is dissolved in water, the solution is slightly basic. In other words, the formate ion is a weak base in water and participates in the equilibrium

$$CHO_2^- + H_2O \rightleftharpoons HCHO_2 + OH^-$$

The K_b expression for this base is

$$K_b = \frac{[HCHO_2][OH^-]}{[CHO_2^-]}$$

There is an interesting relationship between the equilibrium constants for this acid–base pair; the product of K_a times K_b equals K_w. We can see this by multiplying the mass action expressions.

$$K_a \times K_b = \frac{[H^+]\cancel{[CHO_2^-]}}{\cancel{[HCHO_2]}} \times \frac{\cancel{[HCHO_2]}[OH^-]}{\cancel{[CHO_2^-]}} = [H^+][OH^-] = K_w$$

In fact, this same relationship exists for *any* acid–base conjugate pair.

For *any* acid–base conjugate pair:

$$K_a \times K_b = K_w \tag{16.9}$$

 Relationship between K_a and K_b

Another useful relationship that can be derived from Equation 16.9 is

$$pK_a + pK_b = pK_w = 14.00 \tag{16.10}$$

There are some important consequences of the relationship expressed in Equation 16.9. One is that it is not necessary to tabulate both K_a and K_b for the members of an acid–base pair; if one K is known, the other can be calculated. For example, the K_a for $HCHO_2$ and the K_b for NH_3 will be found in most tables of acid–base equilibrium constants, but these tables usually will not contain the equilibrium constants for the ions that are the conjugates. Thus, usually tables will not contain the K_b for CHO_2^- or the K_a for NH_4^+.

Tables of ionization constants usually give values for only the molecular member of an acid–base pair.

■ **Practice Exercise 8** The value of K_a for $HCHO_2$ is 1.8×10^{-4}. What is the value of K_b for the CHO_2^- ion?

A more significant observation is that *there is an inverse relationship between the strengths of the acid and base members of a pair.* This is illustrated graphically in

O
‖
H—C—O—H
formic acid, $HCHO_2$

O
‖
H—C—O$^-$
formate ion, CHO_2^-

FIGURE 16.4

The relative strengths of conjugate acid–base pairs. The stronger the acid is, the weaker is its conjugate base. The weaker the acid is, the stronger is its conjugate base.

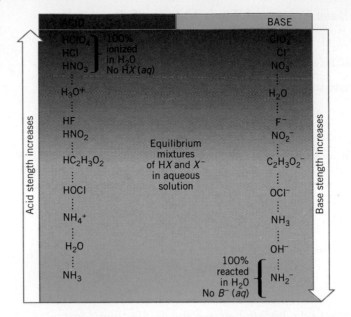

Figure 16.4. Because the product of K_a and K_b is a constant, the larger the value of K_a, the smaller the value of K_b. In other words, *the stronger the conjugate acid, the weaker is the conjugate base*. We will say more about this relationship in Section 16.5.

16.4 EQUILIBRIUM CALCULATIONS

Our goal in this section is to develop a general strategy for dealing quantitatively with the equilibria of weak acids and bases in water. Generally, these calculations fall into two categories: (1) calculating the value of K_a or K_b from the initial concentration of the acid or base and the measured pH of the solution (or some other data about the equilibrium composition of the solution), and (2) calculating equilibrium concentrations given K_a or K_b and initial concentrations.

Calculating K_a or K_b from Initial Concentrations and Equilibrium Data

In problems of this type, our goal is to obtain the equilibrium concentrations that are needed to evaluate the mass action expression in the K_a or K_b expression. (The reason we need *equilibrium* values, of course, is that the value of Q at equilibrium equals the equilibrium constant.) Usually, we are given the molar concentration of the acid or base as it would appear on the label of the bottle containing the solution. Also provided is information from which we can obtain the equilibrium concentrations. This might be the measured pH of the solution, which always provides the equilibrium concentration of H^+. Instead, we might be given the **percentage ionization** of the acid or base, which we can define as follows:

Percentage ionization

$$\text{Percentage ionization} = \frac{\text{amount ionized}}{\text{amount available}} \times 100\% \qquad (16.11)$$

Let's look at some examples that illustrate how to determine K_a and K_b from the kind of data mentioned.

Formic acid, $HCHO_2$, is a monoprotic acid. In a 0.100 *M* solution of formic acid, the pH is 2.38 at 25 °C. Calculate the K_a and pK_a for formic acid at this temperature.

ANALYSIS In any problem like this we begin by writing the equation for the equilibrium so that we can set up the expression for K_a. We know the solute is an acid, and we know the general equation for the ionization of an acid, which we apply to the solute in question.

$$HCHO_2 \rightleftharpoons H^+ + CHO_2^- \qquad K_a = \frac{[H^+][CHO_2^-]}{[HCHO_2]}$$

Keep in mind that the molar concentrations in the K_a expression all refer to *equilibrium* values. Also, in this problem the equilibrium concentration of H^+ must be identical to that of CHO_2^- because the ionization gives these ions in a 1 : 1 ratio. However, we do not have the values of either $[H^+]$ or $[CHO_2^-]$; all we have is the pH of the 0.100 *M* solution. We must start by finding $[H^+]$ from pH. This gives us the *equilibrium* value of $[H^+]$. Then we'll put together a concentration table.

SOLUTION First we convert pH into $[H^+]$.

$$[H^+] = 10^{-2.38}$$
$$= 4.2 \times 10^{-3} \, M$$
$$= 0.0042 \, M \text{ (the form most useful next)}$$

Because $[CHO_2^-] = [H^+]$, we now know that $[CHO_2^-] = 0.0042 \, M$. With these data we can now set up our concentration table.

	$HCHO_2$	$\rightleftharpoons$	H^+	$+$	CHO_2^-
Initial concentrations (*M*)	0.100		0		0 (Note 1)
Changes in concentrations caused by the ionization (*M*)	− 0.0042		+ 0.0042		+ 0.0042 (Note 2)
Final concentrations at equilibrium (*M*)	(0.100 − 0.0042) = 0.096 (correctly rounded)		0.0042		0.0042

Note 1. The *initial* concentration of the acid is what the label tells us, 0.100 *M*; we take it to be the concentration before ionization occurs. The *initial* values of $[H^+]$ and $[CHO_2^-]$ are set equal to zero; these correspond to the concentrations of these two ions from solutes. (As we noted earlier, the ultratrace concentration of H^+ contributed by the self-ionization of water can be safely ignored. We will ignore it in all calculations involving acids in this chapter.)

Note 2. From the pH we have obtained the concentrations of $[H^+]$ and $[CHO_2^-]$ at equilibrium. The *changes* for the ions are obtained by difference. For every H^+ ion that forms by the ionization, there is one less molecule of the initial acid. The minus sign in − 0.0042 in the column for $HCHO_2$ is used because the concentration of this species is *decreased* by the ionization.

The last row of data gives us the equilibrium concentrations that we now use to calculate K_a.

$$K_a = \frac{(4.2 \times 10^{-3})(4.2 \times 10^{-3})}{0.096}$$
$$= 1.8 \times 10^{-4}$$

EXAMPLE 16.8
Calculating K_a and pK_a from pH

To emphasize how we go about thinking about the problem, we're working through the analysis as though we had never discussed formic acid before.

$[H^+]$ means $[H^+]_{\text{at equilibrium}}$.
$[CHO_2^-]$ means $[CHO_2^-]_{\text{at equilibrium}}$.
$[HCHO_2]$ means $[HCHO_2]_{\text{at equilibrium}}$.

The pH of a solution is *always* related to the actual hydrogen ion concentration. For *weak* acids, $[H^+]$ is never the same as—is always less than—the molar concentration of the weak acid itself.

Formica is Latin for "ant." Formic acid is partly responsible for the pain of an ant sting.

Thus the acid ionization constant for formic acid is 1.8×10^{-4}. To find pK_a, we take the negative logarithm of K_a.

$$pK_a = -\log K_a$$
$$= -\log(1.8 \times 10^{-4})$$
$$pK_a = 3.74$$

EXAMPLE 16.9

Calculating K_b and pK_b from pH

H   H
| |
H—C—N—H
|
H
methylamine

Once we have identified the solute as a base, we know how to write the chemical equation for the equilibrium.

Methylamine, CH_3NH_2, is a weak base and one of several substances that give herring brine its pungent odor. In 0.100 M CH_3NH_2, 6.3% of the base has undergone ionization. What is the K_b of methylamine?

ANALYSIS In this problem, we have been given the percentage ionization of the base. We will use this to calculate the equilibrium concentrations of the ions in the solution. After we know these, solving the problem follows the same path as in the preceding example.

SOLUTION The first step is to write the chemical equation for the equilibrium and the equilibrium law. Because CH_3NH_2 is a weak base, we have

$$CH_3NH_2(aq) + H_2O \rightleftharpoons CH_3NH_3^+(aq) + OH^-(aq)$$

$$K_b = \frac{[CH_3NH_3^+][OH^-]}{[CH_3NH_2]}$$

We take 6.3% of 0.100 M to find the moles per liter of the base that has ionized.

The percentage ionization tells us that 6.3% of the CH_3NH_2 has reacted. Therefore, the moles per liter of the base that has ionized at equilibrium in this solution is:

$$\text{Moles per liter of } CH_3NH_2 \text{ ionized} = 0.063 \times 0.100 \ M = 0.0063 \ M$$

This represents the *change* in the concentration of CH_3NH_2, which we can now use to determine the concentrations of the other species at equilibrium. To do this, we can set up the concentration table.

	H_2O +	CH_3NH_2	$\rightleftharpoons$	$CH_3NH_3^+$ +	OH^-
Initial concentrations (M)		0.100		0	0 (Note 1)
Changes in concentrations caused by the ionization (M)		−0.0063		+0.0063	+0.0063 (Note 2)
Final concentrations at equilibrium (M)		(0.100 − 0.0063) = 0.094 (properly rounded)		0.0063	0.0063

Notice that because we know *both* the initial concentration of CH_3NH_2 and the change in this value, we need make no assumptions in computing the final concentration.

Note 1. Once again, we record the concentrations of solute species, assuming no initial ionization.
Note 2. The concentration of CH_3NH_2 decreases by 0.0063 M, so the concentrations of the ions each increase by this amount. (Remember: However the reactants change, the products change in the opposite direction.)

Now we can calculate K_b.

$$K_b = \frac{(0.0063)(0.0063)}{(0.094)}$$
$$= 4.2 \times 10^{-4}$$

and

$$pK_b = -\log(4.2 \times 10^{-4})$$
$$= 3.38$$

■ **Practice Exercise 9** When butter turns rancid, its foul odor is mostly that of butyric acid, a weak acid similar to acetic acid in structure. In a 0.0100 *M* solution of butyric acid the acid is 4.0% ionized at 20 °C. Calculate the K_a and pK_a of butyric acid at this temperature. (You don't need to know the formula for butyric acid; you can use any symbol you wish for the acid and its conjugate base—for example, H*Bu* and *Bu*⁻.)

■ **Practice Exercise 10** Few substances are more effective in relieving intense pain than morphine. Morphine is an alkaloid—an alkali-like compound obtained from plants—and alkaloids are all weak bases. In 0.010 *M* morphine, the pH is 10.10. Calculate the K_b and pK_b for morphine.

Calculating Equilibrium Concentrations from K_a (or K_b) and Initial Concentrations

Acid–base equilibrium problems of this type often present a confusing picture to students. Usually, the difficulty is getting started on the right foot. Therefore, before we begin, let's take an overview of the "landscape" to develop a strategy for selecting the correct attack on the problem.

Almost any problem in which you are given K_a or K_b falls into one of three categories: (1) the solution contains a weak acid as its *only* solute, (2) the solution contains a weak base as its *only* solute, and (3) the solution contains *both* a weak acid and its conjugate base. The following describes the approach we take for each of these conditions.

1. If the solution contains only a weak acid as the solute, then the problem must be solved using K_a. This means that the correct chemical equation for the problem is the ionization of the weak acid. It also means that if you are given the K_b for the acid's conjugate base, you will have to calculate the K_a in order to solve the problem.

 Strategy for approaching equilibrium calculations

2. If the solution contains only a weak base, the problem must be solved using K_b. The correct chemical equation is for the ionization of the weak base. If the statement of the problem provides you with the K_a for the base's conjugate acid, then you must calculate the K_b to solve the problem.

3. Solutions that contain two solutes, one a weak acid and the other its conjugate base, have some special properties that we will discuss later. To work problems for these kinds of mixtures, we can use either K_a or K_b—it doesn't matter which we use, we will obtain the same answers either way. However, if we elect to use K_a, then the chemical equation we use must be for the ionization of the acid. If we decide to use K_b, then the correct chemical equation is for the ionization of the base.

Conditions 1 and 2 apply to solutions of weak molecular acids and bases and to solutions of salts that contain an ion that is an acid or a base. Condition 3 applies to solutions called buffers, which are discussed in Section 16.7.

When you begin the solution of a problem, decide which of the three conditions above applies. This will guide you in selecting the correct path to follow.

Simplifications in Acid–Base Equilibrium Calculations

In Chapter 15, you learned that when the equilibrium constant is small it is frequently possible to make simplifying assumptions that greatly reduce the algebraic effort in obtaining equilibrium concentrations. For most acid–base equilibrium problems such simplifications are particularly useful.

Let's consider a solution of 1.0 *M* acetic acid, $HC_2H_3O_2$, for which $K_a = 1.8 \times 10^{-5}$. What is involved in determining the equilibrium concentrations in the solution?

696 / 16 • Acid–Base Equilibria

To answer this question, we begin with the chemical equation for the equilibrium. Because the only solute in the solution is the weak acid, we must use K_a and therefore write the equation for the ionization of the acid. Let's use the simplified version.

$$HC_2H_3O_2 \rightleftharpoons H^+ + C_2H_3O_2^-$$

The equilibrium law is

$$K_a = \frac{[H^+][C_2H_3O_2^-]}{[HC_2H_3O_2]} = 1.8 \times 10^{-5}$$

We will take the given concentration, 1.0 M, to be the initial concentration of $HC_2H_3O_2$ (i.e., the concentration of the acid before any ionization has occurred). This concentration will drop slightly as the acid ionizes and the ions form. If we let x be the amount of acetic acid that ionizes per liter, we can construct the following concentration table.

acetic acid (structural formula: $H-C-C-O-H$ with H atoms and O)

	$HC_2H_3O_2 \rightleftharpoons$	$H^+ +$	$C_2H_3O_2^-$
Initial concentrations (M)	1.0	0	0
Changes in concentrations caused by the ionization (M)	$-x$	$+x$	$+x$
Concentrations at equilibrium (M)	$1.0 - x$	x	x

The initial concentrations of the ions are set equal to zero because none of them has been supplied by a solute.

The equilibrium constant, 1.8×10^{-5}, is quite small, so we can anticipate that very little of the acetic acid will be ionized at equilibrium. This means x will be very small, so we make the approximation $(1.0 - x) \approx 1.0$. Substituting values into the K_a expression,

$$1.8 \times 10^{-5} = \frac{(x)(x)}{1.0} = \frac{x^2}{1.0}$$

Solving for x gives $x = 0.0042$ M. We see that the value of x is indeed negligible compared to 1.0 M (i.e., if we subtract 0.0042 M from 1.0 M and round correctly, we obtain 1.0 M).

Notice that when we make this approximation, *the initial concentration of the acid is used as if it were the equilibrium concentration*. The approximation is valid when the equilibrium constant is small and the concentrations of the solutes are reasonably high—conditions that will apply to situations you will encounter in all but Section 16.6. In Section 16.6 we will discuss the conditions under which the approximation is not valid.

Let's look at some examples that illustrate typical acid–base equilibrium problems.

As we will discuss later, if the solute concentration is at least 400 times the value of K, we can use initial concentrations as though they were equilibrium values.

EXAMPLE 16.10
Calculating the Values of [H⁺] and pH for a Solution of a Weak Acid from Its K_a Value

A student planned an experiment that would use 0.10 M propionic acid, $HC_3H_5O_2$. Calculate the values of $[H^+]$ and pH of this sample. For this acid, $K_a = 1.34 \times 10^{-5}$.

ANALYSIS First, we note that the only solute in the solution is a weak acid, so we know we will have to use K_a and write the equation for the ionization of the acid.

The initial concentration of the acid is 0.10 *M*, and the initial concentrations of the ions are both 0 *M*. Then we construct the concentration table.

SOLUTION All concentrations in the table are in moles per liter.

	$HC_3H_5O_2 \rightleftharpoons$	H^+ +	$C_3H_5O_2^-$
Initial concentrations (*M*)	0.10	0	0
Changes in concentrations caused by the ionization (*M*)	$-x$	$+x$	$+x$
Final concentrations at equilibrium (*M*)	$(0.10 - x) \approx 0.10$	x	x

Notice that the equilibrium concentrations of H^+ and $C_3H_5O_2^-$ are the same; they are represented by *x*. Using our simplification, the concentration of $HC_3H_5O_2$ is 0.10 *M*. Substituting these quantities into the K_a expression gives

$$K_a = \frac{[H^+][C_3H_5O_2^-]}{[HC_3H_5O_2]} = \frac{(x)(x)}{(0.10)} = 1.34 \times 10^{-5}$$

Solving for *x* gives

$$x = 1.2 \times 10^{-3}$$

Because $x = [H^+]$,

$$[H^+] = 1.2 \times 10^{-3} \, M$$

Finally, we calculate the pH

$$pH = -\log(1.2 \times 10^{-3})$$
$$= 2.92$$

■ **Practice Exercise 11** Nicotinic acid, $HC_2H_4NO_2$, is a B vitamin. It is also a weak acid with $K_a = 1.4 \times 10^{-5}$. What is the $[H^+]$ and the pH of a 0.050 *M* solution of $HC_2H_4NO_2$?

A solution of hydrazine, N_2H_4, has a concentration of 0.25 *M*. What is the pH of the solution and what is the percentage ionization of the hydrazine? Hydrazine has $K_b = 1.7 \times 10^{-6}$.

ANALYSIS Hydrazine must be a weak base, because we have its value of K_b (the "b" in K_b tells us this is a *base* ionization constant). Since hydrazine is the only solute in the solution, we will have to write the equation for the ionization of a weak base and then set up the K_b expression.

SOLUTION Let's write the chemical equation and the K_b expression.

$$N_2H_4 + H_2O \rightleftharpoons N_2H_5^+ + OH^-$$

$$K_b = \frac{[N_2H_5^+][OH^-]}{[N_2H_4]}$$

The only sources of $N_2H_5^+$ and OH^- are the ionization of the N_2H_4, so they will form in equal amounts. Let's call their equilibrium concentrations *x*.

$$[N_2H_5^+] = [OH^-] = x$$

Because K_b is so small, $[N_2H_4] \approx 0.25 \, M$. (In other words, we are using the initial concentration as the equilibrium concentration; that's our approximation, which

EXAMPLE 16.11
Calculating the pH of a Solution and the Percentage Ionization of the Solute

we expect to be valid.) Substituting into the K_b expression,

$$\frac{(x)(x)}{0.25} = 1.7 \times 10^{-6}$$

Solving for x gives: $x = 6.5 \times 10^{-4}$. This value represents the hydroxide ion concentration, from which we can calculate the pOH.

$$\begin{aligned} \text{pOH} &= -\log(6.5 \times 10^{-4}) \\ &= 3.19 \end{aligned}$$

The pH of the solution can then be obtained from the relationship

$$\text{pH} + \text{pOH} = 14.00$$

Thus,

$$\begin{aligned} \text{pH} &= 14.00 - 3.19 \\ &= 10.81 \end{aligned}$$

To calculate the percentage ionization, we need to know the amount of the base that has ionized. This value is also equal to x, because the amount of $N_2H_5^+$ that forms per liter equals the amount of N_2H_4 that ionizes. Therefore,

$$\begin{aligned} \text{Percentage ionization} &= \frac{6.5 \times 10^{-4}}{0.25} \times 100\% \\ &= 0.26\% \end{aligned}$$

The base is 0.26% ionized.

■ **Practice Exercise 12** Pyridine, C_5H_5N, is a bad-smelling liquid for which $K_b = 1.5 \times 10^{-9}$. What is the pH of a 0.010 M aqueous solution of pyridine?

■ **Practice Exercise 13** Phenol is an organic compound that in water has $K_a = 1.3 \times 10^{-10}$. What is the pH of a 0.15 M solution of phenol in water?

16.5
SOLUTIONS OF SALTS: IONS AS WEAK ACIDS AND BASES

In Section 16.3 you saw that weak acids and bases are not limited to molecular substances. For instance, on page 688 NH_4^+ was given as an example of a weak acid, and on page 689 $C_2H_3O_2^-$ was cited as an example of a weak base. To prepare solutions of these ions, however, we cannot simply add them to water. Ions always come to us in compounds in which there is both a cation *and* an anion. Therefore, to place NH_4^+ in water, we need a salt such as NH_4Cl, and to place $C_2H_3O_2^-$ in water, we need a salt such as $NaC_2H_3O_2$.

Because a salt contains two ions, the pH of its solution can potentially be affected by either the cation or the anion, or perhaps even by both. Therefore, we have to consider *both* ions as we assess the effect of the salt on the pH of the solution.

Cations as Acids

Conjugate Acids of *Molecular* Bases Are Weak Acids The conjugate acid of any molecular base is a cation. For example, NH_4^+ is the conjugate acid of the molecular base, NH_3. The equilibrium and K_a expression for NH_4^+ are

$$NH_4^+(aq) \rightleftharpoons NH_3(aq) + H^+(aq) \qquad K_a = \frac{[NH_3][H^+]}{[NH_4^+]}$$

Because the K_a values for ions are seldom tabulated, we would usually expect to calculate the K_a value using the relationship: $K_a \cdot K_b = K_w$. In Table 16.5

the K_b for NH_3 is given as 1.8×10^{-5}. Therefore, for NH_4^+

$$K_a = \frac{K_w}{K_b} = \frac{1.0 \times 10^{-14}}{1.8 \times 10^{-5}} = 5.6 \times 10^{-10}$$

The hydrazinium ion, $N_2H_5^+$, is also a weak acid.

$$N_2H_5^+(aq) \rightleftharpoons N_2H_4(aq) + H^+(aq) \qquad K_a = \frac{[N_2H_4][H^+]}{[N_2H_5^+]}$$

The tabulated K_b for N_2H_4 is 1.7×10^{-6}. From this, the calculated value of K_a for $N_2H_5^+$ equals 5.9×10^{-9}.

These examples illustrate that *the conjugate acids of molecular bases tend to be weak acids that are capable of affecting the pH of a solution.* This leads to the following generalization.

Salts that contain cations of weak molecular bases can affect the pH of a solution. These cations are weak acids.

 Acidity of cations

Hydrates of Metal Ions with High Charge Densities Are Weak Acids

In Section 9.3, you learned that small, highly charged cations such as Al^{3+} are acidic in water because the water molecules that surround the metal ion are able to release H^+ ion rather easily.

$$[Al(H_2O)_6]^{3+}(aq) + H_2O \rightleftharpoons [Al(H_2O)_5(OH)]^{2+}(aq) + H_3O^+(aq)$$

Behind this behavior is the ability of the high positive charge density on Al^{3+} in $[Al(H_2O)_6]^{3+}$ to draw considerable electron density from the O—H bonds of the water molecules. This weakens these O—H bonds enough to make an H_2O molecule in the hydrate a proton donor.

Salts that contain small ions with large charges, such as Al^{3+} and Cr^{3+}, often produce solutions that are acidic. The equilibria in such solutions can be treated by the same procedures that we've used for other weak acids, but we will not discuss them further in this book. (Salts that may be of concern to us in this chapter will not contain these ions.)

It's the charge density (charge per unit volume) that matters, not just the size of the charge, so if the cation is very small, like Be^{2+}, the ratio of its smaller $2+$ charge to its small volume can yet be large enough to make its hydrate a weak acid.

Metal Ions with Small Charges Are "Nonacids"

None of the singly charged cations of the Group IA metals—Li^+, Na^+, K^+, Rb^+, or Cs^+—affects the pH of an aqueous solution. Except for Be^{2+}, neither do any of the doubly charged cations of the Group IIA metals—Mg^{2+}, Ca^{2+}, Sr^{2+}, or Ba^{2+}. In none of these is the ratio of charge to size apparently large enough.

Anions as Bases

When a molecular Brønsted acid loses a proton, its conjugate base is formed. Thus Cl^- is the conjugate base of HCl, and $C_2H_3O_2^-$ is the conjugate base of $HC_2H_3O_2$.

Acid	Base
HCl	Cl^-
$HC_2H_3O_2$	$C_2H_3O_2^-$

Although both Cl^- and $C_2H_3O_2^-$ are bases, only the latter affects the pH of an aqueous solution. Why?

Earlier you learned that there is an inverse relationship between the strength of an acid and its conjugate base—the stronger the acid, the weaker is the conjugate base. Therefore, when an acid is *extremely strong*, as in the case of HCl or other "strong" acids that are 100% ionized, the conjugate base is *extremely weak*—too weak to affect in a measurable way the pH of a solution. Consequently, we have the following generalization.

Anions that don't affect the pH

The anion of a strong acid is too weak to influence the pH of a solution.

Acetic acid is much weaker than HCl, as evidenced by its value of K_a (1.8 × 10^{-5}). Because acetic acid is a weak acid, its conjugate base is much stronger than Cl^-. We can calculate the value of K_b for $C_2H_3O_2^-$ from the K_a for $HC_2H_3O_2$ using the equation $K_a \cdot K_b = K_w$.

$$K_b = \frac{1.0 \times 10^{-14}}{1.8 \times 10^{-5}} = 5.6 \times 10^{-10}$$

This leads to another conclusion:

Anions that are basic

The anion of a weak acid is a weak base and can influence the pH of a solution.

Predicting the Acid–Base Properties of a Salt

To decide whether any given salt will affect the pH of an aqueous solution, we examine each of its ions and see what it alone might do. There are three possibilities:

If neither the cation nor the anion can affect the pH, the salt solution will be neutral (provided no other acidic or basic solutes are present).

1. If only the cation of the salt is acidic, the solution will be acidic.
2. If only the anion of the salt is basic, the solution will be basic.
3. If a salt has a cation that is acidic and an anion that is basic, the pH of the solution is determined by the *relative* strengths of the acid and base.

Let's work some examples to show how to use these generalizations.

EXAMPLE 16.12
Predicting the Effect of a Salt on the pH of a Solution

Sodium hypochlorite, NaOCl, is an ingredient in many common bleaching and disinfecting agents. Will a NaOCl solution be acidic, basic, or neutral?

ANALYSIS This is a salt, so it dissociates in water 100%.

$$NaOCl(s) \xrightarrow{\text{H}_2\text{O}} Na^+(aq) + OCl^-(aq)$$

We take each ion, in turn, and examine its effect of the acidity of the solution.

SOLUTION The Na^+ ion is of a Group IA metal, so it does not form an acidic hydrate. The cation, therefore, is neutral and is *not* acidic.

The OCl^- ion is the conjugate base of HOCl, which must be a weak acid (because it isn't on the list of strong acids you're expected to know). Therefore, OCl^- is a weak base and its presence should tend to make the solution basic.

Of the two ions in the salt, one is basic and the other is neutral. The answer to the question, therefore, is that this salt solution will be basic.

■ **Practice Exercise 14** Is a solution of $NaNO_2$ acidic, basic, or neutral?

■ **Practice Exercise 15** Is a solution of KCl acidic, basic, or neutral?

■ **Practice Exercise 16** Is a solution of NH_4Br acidic, basic, or neutral?

What is the pH of a 0.10 M solution of NaOCl? For HOCl, $K_a = 3.0 \times 10^{-8}$.

ANALYSIS Problems such as this are just like the other acid–base equilibrium problems you have learned to solve. We proceed as follows: (1) We determine the nature of the solute—is it a weak acid, a weak base, or are both a weak acid and weak base present? (2) We write the appropriate chemical equation and equilibrium law. (3) We proceed with the solution.

SOLUTION The solute is a salt, which is dissociated into the ions Na^+ and OCl^-. Only the latter can affect the pH. It is the conjugate *base* of HOCl, so for the purposes of problem solving, the active solute species is a weak base.

On page 695 you learned that when the only solute is a weak base, we must use the K_b expression to solve the problem. Therefore, the next step is to write the appropriate chemical equation and the K_b expression.

$$OCl^- + H_2O \rightleftharpoons HOCl + OH^- \qquad K_b = \frac{[HOCl][OH^-]}{[OCl^-]}$$

The data provided in the problem is the K_a for HOCl. But, we can easily calculate K_b because $K_a \cdot K_b = K_w$. Thus, for OCl^-,

$$K_b = \frac{1.0 \times 10^{-14}}{3.0 \times 10^{-8}} = 3.3 \times 10^{-7}$$

The only sources of HOCl and OH^- are the reaction of the OCl^-, so they will form in equal amounts. Let's call their equilibrium concentrations x.

$$[HOCl] = [OH^-] = x$$

Because K_b is so small, $[OCl^-] \approx 0.10$ M. (Once again, we are using the initial concentration as the equilibrium concentration; that's our approximation, which we expect to be valid.) Substituting into the K_b expression,

$$\frac{(x)(x)}{0.10} = 3.3 \times 10^{-7}$$

$$x = 1.8 \times 10^{-4} \ M$$

This value of x represents the OH^- concentration, from which we can calculate the pOH and then the pH.

$$pOH = -\log(1.8 \times 10^{-4})$$
$$= 3.74$$

$$pH = 14.00 - pOH$$
$$= 14.00 - 3.74$$
$$= 10.26$$

The pH of this solution is 10.26.

EXAMPLE 16.13
Calculating the pH of a Salt Solution

EXAMPLE 16.14

Calculating the pH of a Salt Solution

What is the pH of a 0.20 M solution of hydrazinium chloride, N_2H_5Cl? Hydrazine, N_2H_4, is a weak base with $K_b = 1.7 \times 10^{-6}$.

ANALYSIS This problem is quite similar to the preceding one. Looking over the statement of the problem, we should realize that N_2H_5Cl is a salt composed of $N_2H_5^+$ (the conjugate acid of N_2H_4) and Cl^-. Since N_2H_4 is a weak base, we expect the $N_2H_5^+$ ion to be a weak acid and thereby affect the pH of the solution. On the other hand, Cl^- is the conjugate base of HCl (a strong acid) and is too weak to influence the pH. Therefore, the only active solute species is the acid, $N_2H_5^+$, which means to solve the problem we write the equation for the ionization of the acid and use the K_a expression.

SOLUTION We begin with the chemical equation for the equilibrium and write the K_a expression.

Notice that we calculate K_a from the given value of K_b.

$$N_2H_5^+ \rightleftharpoons H^+ + N_2H_4 \qquad K_a = \frac{[H^+][N_2H_4]}{[N_2H_5^+]} = \frac{1.0 \times 10^{-14}}{1.7 \times 10^{-6}} = 5.9 \times 10^{-9}$$

When the $N_2H_5^+$ reacts, equal amounts of H^+ and N_2H_4 are formed, so we let the equilibrium concentrations of each be equal to x.

$$[H^+] = [N_2H_4] = x$$

We also assume that $[N_2H_5^+] \approx 0.20\ M$ and then substitute quantities into the mass action expression.

$$\frac{(x)(x)}{0.20} = 5.9 \times 10^{-9}$$

$$x = 3.4 \times 10^{-5}\ M$$

Since $x = [H^+]$, the pH of the solution is

$$pH = -\log(3.4 \times 10^{-5})$$
$$= 4.47$$

■ **Practice Exercise 17** What is the pH of a 0.10 M solution of $NaNO_2$?

■ **Practice Exercise 18** What is the pH of a 0.10 M solution of NH_4Br?

Solutions Made with the Salt of a Weak Acid and a Weak Base

Many salts involve ions both of which are capable of affecting the pH of the salt solution. Whether or not the salt has a net effect on the pH now depends on the relative strengths of its ions in functioning one as an acid and the other as a base. If they are matched in their respective strengths, the salt has no effect on pH. In ammonium acetate, for example, the ammonium ion is an acidic cation and the acetate ion is a basic anion. However, the K_a of NH_4^+ is 5.6×10^{-10} and the K_b of $C_2H_3O_2^-$ just happens to be the same, 5.6×10^{-10}. The cation tends to produce H^+ ions to the same extent that the anion tends to produce OH^-. So in aqueous ammonium acetate, $[H^+] = [OH^-]$, and the solution has a pH of 7.

Consider, now, ammonium formate, NH_4CHO_2. The formate ion, CHO_2^-, is the conjugate base of the weak acid, formic acid, so it is a Brønsted base. Its K_b is 5.6×10^{-11}. Comparing this value to the (slightly larger) K_a of the ammonium ion, 5.6×10^{-10}, we see that NH_4^+ is slightly stronger as an acid than the formate ion can be as a base. So a solution of ammonium formate will be slightly acidic. We are not concerned here about calculating a pH, only in predicting whether the solution is acidic, basic, or neutral.

Will an aqueous solution that is 0.20 M NH_4F be acidic, basic, or neutral?

ANALYSIS This is a salt with an acidic cation, and the anion is the conjugate base of a weak acid, HF. So the anion, F^-, is basic. The question, then, is, "How do the two ions compare?"

SOLUTION The K_a of NH_4^+ (calculated from the K_b for NH_3) is 5.6×10^{-10}. Similarly, the K_b of F^- is 1.5×10^{-11} (calculated from the K_a for HF, 6.8×10^{-4}). Comparing the two equilibrium constants, we see that the acid (NH_4^+) is stronger than the base (F^-), so the solution is slightly acidic.

■ **Practice Exercise 19** Will an aqueous solution of ammonium cyanide, NH_4CN, be acidic, basic, or neutral?

EXAMPLE 16.15
Predicting How a Salt Affects the pH of Its Solution

In the preceding two sections we have used initial concentrations of solutes as though they were equilibrium concentrations when we performed calculations. This is only an approximation, as we discussed on page 696, but it is one that works for many situations. Unfortunately, it does not work in all cases, so now that you have learned the basic approach to solving equilibrium problems, we will examine those conditions under which simplifying approximations do and do not work. We will also study how to solve problems when the approximations cannot be used.

16.6
EQUILIBRIUM CALCULATIONS WHEN SIMPLIFICATIONS FAIL

When Simplifying Assumptions Fail

When a weak acid, HA, ionizes in water, its concentration is reduced as the ions form. If we let x represent the amount of acid that ionizes per liter, the equilibrium concentration becomes

$$[HA]_{\text{equilib}} = [HA]_{\text{initial}} - x$$

For reasons beyond the scope of this text,[3] the accuracy of acid–base equilibrium calculations is limited to about $\pm 5\%$, so as long as x does not exceed $\pm 5\%$ of $[HA]_{\text{initial}}$, we assume its value is negligible and say that

$$[HA]_{\text{equilib}} \approx [HA]_{\text{initial}}$$

It is not difficult to show that for x to be less than or equal to $\pm 5\%$ of $[HA]_{\text{initial}}$, $[HA]_{\text{initial}}$ must be greater than or equal to 400 times the value of K_a.

$$[HA]_{\text{initial}} \geq 400 \cdot K_a$$

 Conditions under which simplications are valid

For the equilibrium problems in the preceding sections, this condition was fulfilled, so our simplifications were valid. We now want to study what to do when $[HA]_{\text{initial}} < 400 \cdot K_a$ (i.e., when we cannot justify simplifying the algebra).

---

[3] To properly perform equilibrium calculations, instead of molar concentrations we should use quantities called *activities*. The activity of a substance is its effective concentration, which is affected by its electrical charge and the concentrations of other charged species in the solution. By using molar concentrations we introduce errors in the calculations that limit the accuracy we are able to obtain. Using molar concentrations, however, makes the problems much easier to cope with mathematically.

The Quadratic Solution

When the algebraic equation obtained by substituting quantities into the equilibrium law is a quadratic equation, we can use the quadratic formula to obtain the solution. This is illustrated by the following example.

EXAMPLE 16.16
Using the Quadratic Formula in Equilibrium Problems

Chloroacetic acid, $HC_2H_2O_2Cl$, has $K_a = 1.4 \times 10^{-3}$. What is the pH of a 0.010 M solution of $HC_2H_2O_2Cl$?

ANALYSIS Before we begin the solution, we check to see whether we can use our usual simplifying approximation. We do this by calculating $400 \cdot K_a$ and comparing the result to the initial concentration of the acid.

$$400 \cdot K_a = 400(1.4 \times 10^{-3}) = 0.56$$

The initial concentration of $HC_2H_2O_2Cl$ is *less than* 0.56, so we know the simplification does not work.

SOLUTION Let's begin by writing the chemical equation and the K_a expression. (We know we must use K_a because the only solute is the acid.)

$$HC_2H_2O_2Cl \rightleftharpoons H^+ + C_2H_2O_2Cl^-$$

$$K_a = \frac{[H^+][C_2H_2O_2Cl^-]}{[HC_2H_2O_2Cl]} = 1.4 \times 10^{-3}$$

The initial concentration of the acid will be reduced by an amount x as it undergoes ionization to form the ions. From this, let's build the concentration table.

	$HC_2H_2O_2Cl \rightleftharpoons$	H^+ +	$C_2H_2O_2Cl^-$
Initial concentrations (M)	0.010	0	0
Changes in concentrations (M)	$-x$	$+x$	$+x$
Equilibrium concentrations (M)	$(0.100 - x)$	x	x

Substituting equilibrium concentrations into the equilibrium law gives

$$\frac{(x)(x)}{(0.010 - x)} = 1.4 \times 10^{-3}$$

This time we cannot neglect x. To solve the problem, we first clear fractions by multiplying both sides by $(0.010 - x)$. This gives

$$x^2 = (0.010 - x)1.4 \times 10^{-3}$$

$$x^2 = (1.4 \times 10^{-5}) - (1.4 \times 10^{-3})x$$

The general quadratic equation has the form

$$ax^2 + bx + c = 0$$

The values of x that make this equation true are related to the coefficients, a, b, and c by the quadratic formula:

$$x = \frac{-b \pm \sqrt{b^2 - 4ac}}{2a}$$

Rearranging our equation to follow the general form gives

$$x^2 + (1.4 \times 10^{-3})x - (1.4 \times 10^{-5}) = 0$$

so we make the substitutions $a = 1$, $b = 1.4 \times 10^{-3}$, and $c = -1.4 \times 10^{-5}$. Entering these into the quadratic formula gives

$$x = \frac{-1.4 \times 10^{-3} \pm \sqrt{(1.4 \times 10^{-3})^2 - 4(1)(-1.4 \times 10^{-5})}}{2(1)}$$

$$= \frac{-1.4 \times 10^{-3} \pm \sqrt{2.0 \times 10^{-6} + 5.6 \times 10^{-5}}}{2}$$

$$= \frac{-1.4 \times 10^{-3} \pm \sqrt{5.8 \times 10^{-5}}}{2}$$

$$= \frac{-1.4 \times 10^{-3} \pm 7.6 \times 10^{-3}}{2}$$

Because of the $\pm$ sign we obtain two values for x, but as you saw in Chapter 15, only one of them makes any sense. Here are the two values:

$$x = 3.1 \times 10^{-3} M \quad \text{and} \quad x = -4.5 \times 10^{-3} M$$

We know that x cannot be negative because that would give a negative concentrations for the ions, so we must choose the first value as the correct one. This yields the following equilibrium concentrations.

$$[H^+] = 3.1 \times 10^{-3} M$$

$$[C_2H_2O_2Cl^-] = 3.1 \times 10^{-3} M$$

$$[HC_2H_2O_2Cl] = 0.010 - 0.0031$$
$$= 0.007 M$$

Notice that indeed, x is not negligible compared to the initial concentration, so the simplifying approximation would not have been valid.

Finally, we calculate the pH of the solution.

$$pH = -\log(3.1 \times 10^{-3})$$
$$= 2.51$$

■ **Practice Exercise 20** Calculate the pH of a 0.0010 M solution of dimethylamine, $(CH_3)_2NH$, for which $K_b = 9.6 \times 10^{-4}$.

Solving by Successive Approximations

If you worked your way through the discussion of the use of the quadratic equation, you'll surely be better able to appreciate the *method of successive approximations*. It's not only much faster, particularly with a scientific calculator, but just as accurate.

If we were to attempt to use the simplifying approximation in the preceding example, we would obtain the following:

$$\frac{x^2}{0.010} = 1.4 \times 10^{-3}$$

for which we obtain the solution $x = 3.7 \times 10^{-3}$. We will call this our *first approximation* to a solution to the equation

$$\frac{(x)(x)}{(0.010 - x)} = 1.4 \times 10^{-3}$$

Let's put $x = 3.7 \times 10^{-3}$ into the term in the denominator, $(0.010 - x)$, and recalculate x. This gives us

$$\frac{x^2}{(0.010 - 0.0037)} = 1.4 \times 10^{-3}$$

Or,

$$x^2 = (0.006) \times 1.4 \times 10^{-3}$$

Taking only the positive root,

$$x = 2.9 \times 10^{-3}$$

Notice that this value of x is much closer to the value calculated using the quadratic equation (which gave $x = 3.1 \times 10^{-3}$). Now we'll call $x = 2.9 \times 10^{-3}$ our *second approximation,* and repeat the process. This gives us:

$$\frac{x^2}{(0.010 - 0.0029)} = 1.4 \times 10^{-3}$$

$$x^2 = (0.007) \times 1.4 \times 10^{-3}$$

$$x = 3.1 \times 10^{-3}$$

This is the identical value obtained using the quadratic equation. Notice also, that the *change* in x between the two approximations grew smaller. The second approximation gave a smaller correction than the first. Each succeeding approximation differs from the preceding one by smaller and smaller amounts. We stop the calculation when the difference between two approximations is insignificant. Try this approach by reworking Practice Exercise 20 and solving for $[OH^-]$ in the 0.0010 M solution using the method of successive approximations.

16.7
BUFFERS: THE CONTROL OF pH

Every life form is extremely sensitive to slight changes in pH.

We all know that acids cause things made of metals, like cars or battery terminals, to corrode faster. Chefs know that adding just a little lemon juice to milk makes it curdle. If the pH of your blood were to change from what it should be, within the range of 7.38 to 7.42, either to 7.00 or to 8.00, you would die. Thus, a change in pH can cause chemical reactions, sometimes entirely unwanted. Fortunately, there are ways to protect systems against large changes in pH.

By a careful choice of solutes, a solution can be prepared that will experience no more than a small change in pH even if strong acid or strong base is added or is produced by some reaction. Such mixtures of solutes are called **buffers,** because they do just that—they buffer the system against a change in pH. The solution itself is said to be *buffered* or it is described as a *buffer solution.* There is almost no area of experimental work in chemistry, or in its applications in other fields, where the concept of buffers is not important. Buffers introduce no new concepts to what we have already studied, just a new application.

Components of Buffers

Usually, a buffer consists of two solutes. One provides a weak Brønsted acid and the other provides its conjugate base. If the acid is molecular, then the conjugate base is *supplied by a soluble salt of the acid.* Thus a common buffer system consists of acetic acid plus sodium acetate, with the salt's acetate ion serving as the Brønsted base. Your blood uses carbonic acid (H_2CO_3), a weak diprotic acid, and the bicarbonate ion, its conjugate Brønsted base, to maintain a remarkably constant pH in the face of the body's production of organic acids by metabolism. Another common buffer consists of the weakly acidic cation, NH_4^+, supplied by a salt like NH_4Cl, and its conjugate base, NH_3.

One important point about buffers is the distinction between keeping a solution at a particular pH and keeping it neutral—at a pH of 7. Although it is certainly possible to prepare a buffer to work at pH 7, buffers can be made that will work around any pH value throughout the pH scale. One topic we must study, therefore, is how to pick the weak acid and its salt—and their mole ratio—that would make a buffer good for some chosen pH.

Buffer systems do *not* protect a solution against a tide of strong acid or base, just relatively small amounts. So another topic we have to consider is the *capacity* of a buffer, the amount of strong acid or base it can absorb before the pH changes a lot.

Most of the carbonic acid in blood is actually in the form of hydrated molecules of CO_2, but the net effect is the same with respect to the blood's buffer system. (See *Chemicals in Use 13.*)

How a Buffer Works

For simplicity, we will first confine our discussion to the HA/A^- type of buffer system, like the acetic acid–sodium acetate buffer. Let's first see *how* this system can buffer a solution.

To work, a buffer must be able to neutralize either a strong acid or strong base that is added. This is precisely what the weak base and weak acid components of the buffer do. If we add extra H^+ to the buffer (from a strong acid) the weak conjugate base can react with it as follows.

$$H^+(aq) + A^-(aq) \longrightarrow HA(aq)$$

Thus, the added acid changes some of the buffer's Brønsted base, A^-, to its conjugate (weak) acid, HA. This reaction prevents a large buildup of H^+ that would otherwise be caused by the addition of the strong acid.

A similar response occurs when a strong base is added to the buffer. The OH^- from the strong base will neutralize some HA.

$$HA(aq) + OH^-(aq) \longrightarrow A^-(aq) + H_2O$$

Here the added OH^- changes some of the buffer's Brønsted acid, HA, into its conjugate base, A^-. This prevents a buildup of OH^-, which would otherwise cause a large change in the pH. Thus, one member of a buffer team neutralizes H^+ that might get into the solution, and the other member neutralizes OH^-.

 Reactions when strong acid or base are added to a buffer

Calculating the pH of a Buffer Solution

The Acetic Acid–Acetate Ion Buffer System
The acetate buffer is a solution of both acetic acid and sodium acetate, which provides the acetate ion in solution. The $HC_2H_3O_2$ in the buffer neutralizes OH^- as follows:

$$HC_2H_3O_2(aq) + OH^-(aq) \longrightarrow C_2H_3O_2^-(aq) + H_2O$$

and its acetate ion neutralizes H^+ as follows.

$$C_2H_3O_2^-(aq) + H^+(aq) \longrightarrow HC_2H_3O_2(aq)$$

A mixture of $HC_2H_3O_2$ and $C_2H_3O_2^-$ is called the *acetate buffer.*

Let us now use this buffer system to introduce some important simplifications in buffer calculations that are both legitimate and very useful.[4]

EXAMPLE 16.17
Calculating the pH of a Buffer

To study the effects of a weakly acidic medium on the rate of corrosion of a metal alloy, a student prepared a solution by making it both 0.11 M $NaC_2H_3O_2$ and also 0.090 M $HC_2H_3O_2$. What is the pH of this solution?

ANALYSIS The buffer solution contains both the weak acid $HC_2H_3O_2$ and its conjugate base $C_2H_3O_2^-$. When both solute species are present we can use *either* K_a or K_b to perform calculations. In our tables we find $K_a = 1.8 \times 10^{-5}$ for $HC_2H_3O_2$, so the simplest approach is to use the equation for the ionization of the acid. We will also be able to use the simplifying approximations developed earlier; these always work for buffers, so we will be able to use the initial concentrations as though they were equilibrium values.

SOLUTION We begin with the chemical equation and the expression for K_a.

$$HC_2H_3O_2 \rightleftharpoons H^+ + C_2H_3O_2^- \qquad K_a = \frac{[H^+][C_2H_3O_2^-]}{[HC_2H_3O_2]} = 1.8 \times 10^{-5}$$

Let's set up the concentration table this time so we can proceed carefully. We will take the initial concentrations of $HC_2H_3O_2$ and $C_2H_3O_2^-$ to be the values given in the problem. There's no H^+ present from a strong acid, so we set this concentration equal to zero. If the initial concentration of H^+ is zero, its concentration must increase on the way to equilibrium, so under H^+ in the change row we enter $+x$. The other changes follow from that. Here's the completed table.

	$HC_2H_3O_2$ $\rightleftharpoons$	H^+ +	$C_2H_3O_2^-$
Initial concentrations (M)	0.090	0	0.11
Changes in concentrations (M)	$-x$	$+x$	$+x$
Equilibrium concentrations (M)	$(0.090 - x) \approx 0.090$	x	$(0.11 + x) \approx 0.11$

For buffer solutions the quantity x will always be quite small, so it is safe to make the simplifying approximations. What remains, then, is to substitute the final molar concentrations from the table into the K_a expression.

$$1.8 \times 10^{-5} = \frac{(x)(0.11)}{(0.090)}$$

[4] An equation that can be derived from the K_a expression, and often used by biochemists, is called the *Henderson–Hasselbalch* equation,

$$pH = pK_a + \log \frac{[\text{conjugate base}]}{[\text{conjugate acid}]}$$

For a solution of acetic acid, $HC_2H_3O_2$, which has $K_a = 1.8 \times 10^{-5}$ and $pK_a = 4.74$, this equation takes the form

$$pH = 4.74 + \log \frac{[C_2H_3O_2^-]}{[HC_2H_3O_2]}$$

As an exercise, you should consider deriving the Henderson–Hasselbalch equation from the equilibrium expression corresponding to K_a.

Solving for x gives us

$$x = \frac{(0.090) \times 1.8 \times 10^{-5}}{(0.11)}$$
$$= 1.5 \times 10^{-5}$$

Because x equals $[H^+]$, we now have $[H^+] = 1.5 \times 10^{-5}$ M. Then we calculate pH:

$$pH = -\log(1.5 \times 10^{-5})$$
$$= 4.82$$

Thus the pH of the buffer is 4.82.

> Notice how small x is compared to the initial concentrations. The simplification was valid.

■ **Practice Exercise 21** Calculate the pH of a buffered solution made up as 0.015 M sodium acetate and 0.10 M acetic acid.

Permissible Simplifications in Buffer Calculations There are two useful simplifications that we can use in working buffer calculations. The first is the one we made in Example 16.17. In buffer solutions the initial concentrations of both the acid and its conjugate base are generally large compared to any changes in concentration that occur as a result of the ionization of the acid.

We will be justified in using the *initial* concentrations of both the weak acid and its conjugate base as though they were equilibrium values.

There is a further simplification that can be derived from the units for molar concentration, mol L^{-1}. Let's enter these units for the acid and its conjugate base into the mass action expression. For an acid HA,

$$K_a = \frac{[H^+][A^-]}{[HA]} = \frac{[H^+](\text{mol } A^- \text{ L}^{-1})}{(\text{mol } HA \text{ L}^{-1})} = \frac{[H^+](\text{mol } A^-)}{(\text{mol } HA)} \quad (16.12)$$

Notice that the unit L^{-1} cancels from numerator and denominator. This means that for a given acid–base pair, the $[H^+]$ is determined by the *mole* ratio of conjugate base to conjugate acid; we don't *have* to use molar concentrations.

*For buffer solutions **only**,* we can use either molar concentrations or moles in the K_a (or K_b) expression to express the amounts of the members of the conjugate acid–base pair (but we must use the same units for each member of the pair).

A further consequence of the relationship derived above is that the pH of a buffer should not change if the buffer is diluted. Dilution does not change the number of moles of the solutes, so the mole ratio remains constant and so does $[H^+]$.

The Ammonia–Ammonium Ion Buffer A solution of ammonium chloride in aqueous ammonia is a buffer because it contains the weak acid NH$_4^+$, released by dissociation of the NH$_4$Cl, and its conjugate base, NH$_3$. The buffer is able to neutralize OH$^-$ by the following reaction.

$$NH_4^+(aq) + OH^-(aq) \longrightarrow NH_3(aq) + H_2O$$

The conjugate base can neutralize H$^+$.

$$NH_3(aq) + H^+(aq) \longrightarrow NH_4^+(aq)$$

EXAMPLE 16.18
Calculating the pH of an
Ammonia–Ammonium Ion
Buffer

To study the influence of an alkaline medium on the rate of a reaction, a student prepared a buffer solution by dissolving 0.12 mol of NH_3 and 0.095 mol of NH_4Cl in water. What is the pH of the buffer?

ANALYSIS The pH of the buffer is determined by the mole ratio of the members of the acid–base pair, so to calculate the pH we will be able to use the moles of NH_3 and NH_4^+ directly in the mass action expression. To set up the equilibrium law we can use either the K_a for NH_4^+ or the K_b for NH_3. Since K_b is tabulated, we will use it and write the equation for the ionization of the base.

SOLUTION We begin with the chemical equation and the K_b expression.

$$NH_3 + H_2O \rightleftharpoons NH_4^+ + OH^- \qquad K_b = \frac{[NH_4^+][OH^-]}{[NH_3]} = 1.8 \times 10^{-5}$$

The solution contains 0.12 mol of NH_3 and 0.095 mol of NH_4^+ from the complete dissociation of the salt NH_4Cl. We can enter these quantities into the mass action expression and solve for $[OH^-]$.

$$1.8 \times 10^{-5} = \frac{(0.095)[OH^-]}{0.12}$$

Solving for $[OH^-]$ gives

$$[OH^-] = 2.3 \times 10^{-5}$$

To calculate the pH, we obtain pOH and subtract it from 14.00.

$$pOH = -\log(2.3 \times 10^{-5})$$
$$= 4.64$$

$$pH = 14.00 - 4.64$$
$$= 9.36$$

■ **Practice Exercise 22** Determine the pH of the buffer in the preceding example using the K_a for NH_4^+, which you can calculate from K_b for NH_3. (If you work the problem correctly, you should obtain the same answer as above.)

Preparation of a Buffer with a Given pH

In experimental work, a chemist or biologist usually first decides the pH at which a particular system should be buffered and then chooses the buffer components that will best deliver this pH. This choice requires careful attention to what most affects the pH of a buffer.

Biologists also must be concerned with possible toxic side effects produced by the components of a buffer.

The Factors That Govern the pH of a Buffer Solution We can more clearly see the two factors that dominate the pH of a buffered solution by rearranging Equation 16.12 to solve for $[H^+]$.

$$[H^+] = K_a \frac{[HA]}{[A^-]} \qquad (16.13)$$

or

$$[H^+] = K_a \frac{\text{mol } HA}{\text{mol } A^-} \qquad (16.14)$$

The first factor to affect $[H^+]$ (and therefore pH) is the K_a of the weak acid. The second is the *ratio* of the molarities (in Equation 16.13) or the ratio of moles (in Equation 16.14). The quantities we use in these two equations, of course, can be the initial values—either concentrations or moles. So let's rewrite them as follows

$$[H^+] = K_a \frac{[HA]_{initial}}{[A^-]_{initial}} \qquad (16.15)$$

$$[H^+] = K_a \frac{(mol\ HA)_{initial}}{(mol\ A^-)_{initial}} \qquad (16.16)$$

Notice particularly what happens if we prepare a buffer so as to make the concentrations of the two buffer components identical. Then the ratio $[HA]_{initial}/[A^-]_{initial}$ in Equation 16.15 equals 1, which means that $[H^+] = K_a$ and therefore pH = pK_a.

When $[H^+] = K_a$, then $-\log[H^+] = -\log K_a$. So pH = pK_a.

Selecting the Weak Acid for the Preparation of a Buffer Solution Usually buffers are made so that the ratio $[HA]_{initial}/[A^-]_{initial}$ is not greatly different from one. Consequently, *what mostly determines where, on the pH scale, a buffer can work best is the pK_a of the weak acid.* Thus to prepare a specific buffer for use at a prechosen pH, we first select a weak acid whose pK_a is near the pH we desire. Almost never can an acid be found, however, whose pK_a *exactly* equals the pH we want. So we try for an acid whose pK_a is *close* to this pH. Then, by experimentally adjusting the ratio $[HA]_{initial}/[A^-]_{initial}$, we can make a final adjustment to get the desired pH.

The desirable range for the ratio $[HA]_{initial}/[A^-]_{initial}$ is from 10/1 to 1/10. Outside this range we can run into problems with the solubilities of the solutes, or we can have a buffer of low capacity. (The question of buffer *capacity* will be studied soon.) The log of 10/1 is 1, and the log of 1/10 is -1. Therefore, the range for the ratio $[HA]_{initial}/[A^-]_{initial}$ of 10/1 to 1/10 works out to the desirable range of pH values for a buffer normally being

$$pH = pK_a \pm 1 \qquad (16.17)$$

A solution buffered at pH 5.00 is needed in a chemistry experiment. Can we use acetic acid and sodium acetate to make it? If so, what ratio of acetic acid to acetate ion is needed?

ANALYSIS We first check the pK_a of acetic acid to see if it is in the desired range of pH = $pK_a \pm 1$. If it is, then we proceed to calculate the ratio.

SOLUTION Because we want the pH to be 5.00, the pK_a of the selected acid should be 5.00 ± 1, meaning the range of pK_a between 4.00 and 6.00. Because $K_a = 1.8 \times 10^{-5}$ for acetic acid, $pK_a = 4.74$. So the pK_a of acetic acid falls in the desired range, and acetic acid can be used together with the acetate ion to make the buffer.

We will use Equation 16.16 to answer the second question: What mole ratio of solutes is needed?

$$[H^+] = K_a \frac{(mol\ HC_2H_3O_2)_{initial}}{(mol\ C_2H_3O_2^-)_{initial}} \qquad (Equation\ 16.16)$$

EXAMPLE 16.19
Preparing a Buffer Solution to Have a Predetermined pH

Solving for the mole ratio gives

$$\frac{(\text{mol } HC_2H_3O_2)_{\text{initial}}}{(\text{mol } C_2H_3O_2^-)_{\text{initial}}} = \frac{[H^+]}{K_a}$$

The desired pH = 5.00, so $[H^+] = 1.0 \times 10^{-5}$; also $K_a = 1.8 \times 10^{-5}$. Substituting gives

$$\frac{(\text{mol } HC_2H_3O_2)_{\text{initial}}}{(\text{mol } C_2H_3O_2^-)_{\text{initial}}} = \frac{1.0 \times 10^{-5}}{1.8 \times 10^{-5}} = 0.56$$

This is the *mole* ratio of the buffer components we want. So we have to prepare the solution such that

$$\text{mol } HC_2H_3O_2 = 0.56 \times \text{mol } C_2H_3O_2^-$$

Suppose that we want 1.0 L of the solution, and that we decide to use sodium acetate as the source of the acetate ion. We could, for example, dissolve 0.10 mol of $NaC_2H_3O_2$ in water and then add 0.056 mol of $HC_2H_3O_2$, making the final volume equal to 1.0 L. We could have chosen quantities at one-tenth this, for example, 0.010 mol of $NaC_2H_3O_2$ and 0.0056 mol of $HC_2H_3O_2$. The *ratio* would still be the same. The solution would still be buffered at a pH of 5.00.

■ **Practice Exercise 23** A student needed an aqueous buffer for a pH of 3.90. Would formic acid and its salt, sodium formate, make a good pair for this purpose? If so, what mole ratio of the acid, $HCHO_2$, to the anion of this salt, CHO_2^-, is needed?

Buffer Capacity

Earlier we hinted at the question of the *capacity* of a buffer—how much extra acid or base the buffer solution can absorb before its buffering ability is essentially destroyed. It would always be possible to add so much strong acid to a buffered solution that all the base present would be neutralized. Or all of the buffer's acid could be destroyed by adding too much strong base. The *ratio* of anion to acid influences only the pH of the buffered solution, not its capacity.

A buffer of low capacity might be needed if its ions and molecules could, if too concentrated, interfere with another use of the solution.

The buffer's *capacity* is determined by the sizes of the actual molarities of these components. So the chemist must decide before making the buffer solution what outer limits to the change in pH can be tolerated—0.1 unit, 0.2 unit, or something else. Then, consistent with the necessary *ratio* of acid to anion, the chemist calculates the actual number of moles of each to dissolve in the solution. Let's do a calculation to show how even a relatively small amount of extra acid (or base) can severely tax a buffer's capacity.

EXAMPLE 16.20
The Capacity of a Buffer

A student prepared 1.00 L of an acetic acid–sodium acetate buffer that included $1.00 \, M \, HC_2H_3O_2$ and $1.00 \, M \, NaC_2H_3O_2$. This entire solution was intended for an experiment in which 0.15 mol of H^+ was expected to be generated by a reaction without changing the volume of the solution. Is there enough buffer capacity to neutralize this much strong acid without letting the pH change by more than 0.1 unit?

ANALYSIS To answer the question, we need to know the initial pH of the buffer and the pH after the addition of the H_3O^+. We use the given concentrations of the $HC_2H_3O_2$ and $C_2H_3O_2^-$ to obtain the initial pH. Next, we determine how the concentrations of the $HC_2H_3O_2$ and $C_2H_3O_2^-$ change when the system absorbs the strong acid. We then use these new concentrations to calculate the new pH of the buffer. The difference between the new and original pH's tells us how much the pH changes.

SOLUTION First, we calculate the pH of the buffer before addition of the H_3O^+. For acetic acid, we have seen that

$$K_a = \frac{[H^+][C_2H_3O_2^-]}{[HC_2H_3O_2]} = 1.8 \times 10^{-5}$$

The given data tell us that initially,

$$[HC_2H_3O_2]_{initial} = 1.00 \ M$$

$$[C_2H_3O_2^-]_{initial} = 1.00 \ M$$

Substituting these values lets us calculate the initial pH of the buffer.

$$\frac{[H^+](1.00)}{1.00} = 1.8 \times 10^{-5}$$

$$[H^+] = 1.8 \times 10^{-5}$$

$$pH = 4.74$$

The H^+ formed in the solution reacts with acetate as follows.

$$H^+ + C_2H_3O_2^- \longrightarrow HC_2H_3O_2$$

The reaction is nearly complete, in the sense that virtually all the added H^+ is consumed. As an approximation, we assume complete reaction, so for each mole of H^+ added, a mole of $C_2H_3O_2^-$ is changed to a mole of $HC_2H_3O_2$. Since 0.15 mol of H^+ is added,

$$[HC_2H_3O_2]_{final} = (1.00 + 0.15)M = 1.15 \ M$$

$$[C_2H_3O_2^-]_{final} = (1.00 - 0.15)M = 0.85 \ M$$

We now substitute these values into the mass action expression to calculate the final $[H^+]$ in the solution.

$$\frac{[H^+](0.85)}{(1.15)} = 1.8 \times 10^{-5}$$

$$[H^+] = 2.4 \times 10^{-5} \ mol \ L^{-1}$$

From this we calculate the final pH,

$$pH = 4.62$$

By adding 0.15 mol of HCl to the buffer, its pH has gone from 4.74 to 4.62, a change greater than the limit of ± 0.1 pH unit. The buffer solution, in other words, could not absorb even 0.15 mol of H^+ without exceeding the limit, even though it began with 1.00 mol of the base, $C_2H_3O_2^-$.

■ **Practice Exercise 24** Suppose that the buffer of Example 16.20 is used in an experiment in which 0.11 mol of OH^- ion will be generated (with no volume change). Can the buffer handle this without having the pH change by more than 0.1 unit? Calculate the new pH.

Example 16.20 and Practice Exercise 24 demonstrate an important point. The mole quantities of buffer components have to be considerably greater—upwards of 10 times greater—than the expected influx of strong acid or base if the change in the pH of the system is to be kept within 0.1 pH units of the original. Generally, chemists try for even smaller changes, particularly in chemical research on biological systems, where a change of 0.1 pH units in some internal fluid is usually lethal to a living system.

The Effectiveness of a Buffer System

Although buffers do not have unlimited *capacities,* they still operate remarkably well in preventing wide swings in pH. The question of *capacity* is related to the question of buffer *effectiveness.* How well does a buffer system work? How well does it actually hold the pH? In Example 16.20, we added 0.15 mol of strong acid to 1.00 L of a buffer and the pH changed by 0.12 pH unit. Just think of how much the pH would change had we added this much acid to 1.00 L of pure water. The solution would now have

$$[H^+] = 0.15 \text{ mol L}^{-1}$$

and a pH of

$$\begin{aligned} pH &= -\log(0.15) \\ &= 0.82 \end{aligned}$$

With the buffer, the pH changes from 4.74 to 4.62; without the buffer the pH (starting with pure water) changes from 7.00 to 0.82. This is an enormous difference. Buffers, unless overwhelmed by the addition of excessive amounts of strong acid or base, truly prevent wide swings of pH.

16.8 ACID–BASE TITRATIONS REVISITED

The *titrant* is the solution being slowly added from a buret to a solution in the receiving flask.

In Section 4.11, we studied the overall procedure for an acid–base titration, and we saw how titration data can be used to calculate molar concentrations. The titration is halted at the **end point** when the change in color of the indicator occurs. Ideally, the end point should occur at the **equivalence point,** when stoichiometrically equivalent amounts of acid and base have combined. To obtain this ideal result, the selection of the specific indicator requires foresight. We will understand this better by studying how the pH of the solution being titrated changes with the addition of titrant.

When the pH of a solution at different stages of a titration is plotted against the volume of titrant added, we obtain a *titration curve.* The values of pH in these plots can be calculated by the procedures studied in this chapter, so this discussion will serve as a review.

Titration of a Strong Acid by a Strong Base

The titration of HCl(*aq*) with standard NaOH(*aq*) illustrates the titration of a strong acid by a strong base. The molecular and net ionic equations are

$$HCl(aq) + NaOH(aq) \longrightarrow NaCl(aq) + H_2O$$

$$H^+(aq) + OH^-(aq) \longrightarrow H_2O$$

Let's consider what happens to the pH of a solution, initially 25.00 mL of 0.2000 *M* HCl, as small amounts of the titrant, 0.2000 *M* NaOH, are added. We will calculate the pH of the resulting solution at various stages of the titration, retaining only two significant figures, and plot the values against volume of titrant.

Before any titrant has been run into the receiving flask, the pH of the HCl solution in this flask is found as follows. Because HCl is a strong acid, we know that

$$[H^+] = [HCl] = 0.2000 \ M$$

So

$$pH = -\log(0.2000)$$
$$= 0.70 \text{ (the initial pH)}$$

The amount of HCl initially present in 25.00 mL of 0.2000 M HCl is

$$25.00 \text{ mL HCl soln} \times \frac{0.2000 \text{ mol HCl}}{1000 \text{ mL HCl soln}} = 5.000 \times 10^{-3} \text{ mol HCl}$$

Now suppose we add 10.00 mL of 0.2000 M NaOH from the buret. The amount of NaOH added then is

$$10.00 \text{ mL NaOH soln} \times \frac{0.2000 \text{ mol NaOH}}{1000 \text{ mL NaOH soln}} = 2.000 \times 10^{-3} \text{ mol NaOH}$$

This base neutralizes some of the acid, with one H$^+$ ion removed for each OH$^-$ ion added, as the net ionic equation shows. We calculate the amount of HCl left, therefore, by simple subtraction:

$$(5.000 \times 10^{-3} - 2.000 \times 10^{-3}) \text{ mol HCl} = 3.000 \times 10^{-3} \text{ mol HCl remaining}$$

The total volume of the solution is now $(25.00 + 10.00)$ mL $= 35.00$ mL $= 0.03500$ L, so the $[H^+]$ is calculated as

$$[H^+] = \frac{3.000 \times 10^{-3} \text{ mol}}{0.03500 \text{ L}} = 8.571 \times 10^{-2} \ M$$

The corresponding pH is 1.07.

Table 16.6 shows the results of the calculations of the pH of the solution in the receiving flask after the addition of further small volumes of the titrant. After 25.00 mL of 0.2000 M NaOH is added, all of the HCl is neutralized. At this point the solution is a mixture of Na$^+$ and Cl$^-$ ions, neither of which have an effect on pH, so the solution is neutral with a pH $= 7.00$.

> Despite the high precision assumed for the molarity and the volume of the titrant, we will retain only two significant figures in the calculated pH. Precision higher than this is hard to obtain and seldom sought in actual lab work.

TABLE 16.6   Titration of 25.00 mL of 0.2000 M HCl with 0.2000 M NaOH

Initial Volume of HCl (mL)	Initial Amount of HCl (mol)	Volume of NaOH Added (mL)	Amount of NaOH (mol)	Amount of Excess Reagent (mol)	Total Volume of Solution (mL)	Molarity of Ion in Excess (mol L^{-1})	pH
25.00	5.000×10^{-3}	0	0	5.000×10^{-3}(H$^+$)	25.00	0.2000(H$^+$)	0.70
25.00	5.000×10^{-3}	10.00	2.000×10^{-3}	3.000×10^{-3}(H$^+$)	35.00	8.571×10^{-2}(H$^+$)	1.07
25.00	5.000×10^{-3}	20.00	4.000×10^{-3}	1.000×10^{-3}(H$^+$)	45.00	2.222×10^{-2}(H$^+$)	1.65
25.00	5.000×10^{-3}	24.00	4.800×10^{-3}	2.000×10^{-4}(H$^+$)	49.00	4.082×10^{-3}(H$^+$)	2.39
25.00	5.000×10^{-3}	24.90	4.980×10^{-3}	2.000×10^{-5}(H$^+$)	49.90	4.000×10^{-4}(H$^+$)	3.40
25.00	5.000×10^{-3}	24.99	4.998×10^{-3}	2.000×10^{-6}(H$^+$)	49.99	4.000×10^{-5}(H$^+$)	4.40
25.00	5.000×10^{-3}	25.00	5.000×10^{-3}	0	50.00	0	7.00
25.00	5.000×10^{-3}	25.01	5.002×10^{-3}	2.000×10^{-6}(OH$^-$)	50.01	3.999×10^{-5}(OH$^-$)	9.60
25.00	5.000×10^{-3}	25.10	5.020×10^{-3}	2.000×10^{-5}(OH$^-$)	50.10	3.992×10^{-4}(OH$^-$)	10.60
25.00	5.000×10^{-3}	26.00	5.200×10^{-3}	2.000×10^{-4}(OH$^-$)	51.00	3.922×10^{-3}(OH$^-$)	11.59
25.00	5.000×10^{-3}	50.00	1.000×10^{-2}	5.000×10^{-3}(OH$^-$)	75.00	6.666×10^{-2}(OH$^-$)	12.82

FIGURE 16.5

Titration curve for titrating a strong acid (0.2000 *M* HCl) with a strong base (0.2000 *M* NaOH).

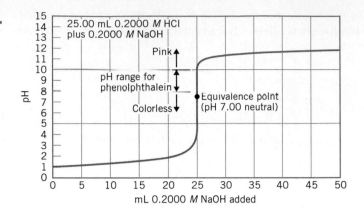

Any further addition of titrant is simply the addition of more and more OH⁻ ion to (an increasing volume of) the solution in the flask. So after 25.00 mL of NaOH solution has been added, the pH calculation reflects only a dilute solution of NaOH.

Figure 16.5 shows a plot of the pH of the solution in the receiving flask versus the milliliters of the 0.2000 *M* NaOH solution added—the *titration curve* for the titration of a strong acid with a strong base. Notice how slowly the pH of the solution increases until we are almost exactly at the equivalence point, pH = 7.00. With the addition of more base, the curve continues to rise very sharply and then, almost as suddenly, bends to reflect a very gradual increase in pH. Notice also that the equivalence point occurs at a pH of 7.00, *which characterizes the titration of any strong monoprotic acid with any strong base.* The only ions at the equivalence point in our example, aside from the traces of those from the self-ionization of water, are Na⁺ and Cl⁻. Na⁺ is not an acidic cation, and Cl⁻ is not a basic anion, so the solution at this point is neutral with a pH of 7.00.

Titration of a Weak Acid by a Strong Base

Let us do the calculations and draw a titration curve for the titration of 0.2000 *M* $HC_2H_3O_2$ with 0.2000 *M* NaOH. The molecular and net ionic equations are

$$HC_2H_3O_2(aq) + NaOH(aq) \longrightarrow NaC_2H_3O_2(aq) + H_2O$$

$$HC_2H_3O_2(aq) + OH^-(aq) \longrightarrow C_2H_3O_2^-(aq) + H_2O$$

Let us start with 25.00 mL of 0.2000 *M* $HC_2H_3O_2$ and use 0.2000 *M* NaOH as the titrant. Because the concentrations of both solutions are the same, we know that we will need exactly 25.00 mL of the base to reach the equivalence point. With this as background, let's look at what is involved in calculating the pH at various points along the titration curve.

1. **Before the titration begins.** The solution at this point is simply a solution of the weak acid, $HC_2H_3O_2$. We must use K_a to calculate the pH.

2. **During the titration but before the equivalence point.** As we add NaOH to the $HC_2H_3O_2$ the neutralization reaction produces $C_2H_3O_2^-$, so the solution contains both $HC_2H_3O_2$ and $C_2H_3O_2^-$ (it is a buffer solution). We can use K_a to calculate the pH here as well.

3. **At the equivalence point.** Here all the $HC_2H_3O_2$ has reacted and the solution contains the salt $NaC_2H_3O_2$ (i.e., we have a mixture of the ions Na^+ and $C_2H_3O_2^-$). The anion is a weak base, so we use K_b to calculate the pOH and then the pH.

4. **After the equivalence point.** The OH^- introduced by the further addition of NaOH has nothing with which to react and causes the solution to become basic. The concentration of OH^- produced by the unreacted base is used to calculate the pOH and ultimately the pH.

We will look at each of these briefly.

Before the Titration Begins   The question here is, What is the pH of $0.2000\ M\ HC_2H_3O_2$? For acetic acid, $K_a = 1.8 \times 10^{-5}$; we let $x = [H^+] = [C_2H_3O_2^-]$, and use $[HC_2H_3O_2] = 0.2000\ M$.

We have worked a similar problem on page 696.

$$K_a = \frac{[H^+][C_2H_3O_2^-]}{[HC_2H_3O_2]} = \frac{(x)(x)}{(0.2000)} = 1.8 \times 10^{-5}$$

$$x = 1.9 \times 10^{-3}$$

So, $[H^+] = 1.9 \times 10^{-3}\ M$, which yields a pH of 2.72. This is the pH before any NaOH is added.

During the Titration but Before the Equivalence Point   Once we add NaOH, the solution contains both $HC_2H_3O_2$ and $C_2H_3O_2^-$, so we now have a buffer system. Let's calculate the pH after 10.00 mL of the base has been added. The amount of NaOH added is

$$10.00\ \text{mL NaOH soln} \times \frac{0.2000\ \text{mol NaOH}}{1000\ \text{mL soln}} = 2.000 \times 10^{-3}\ \text{mol NaOH}$$

We have thereby neutralized 2.000×10^{-3} mol of $HC_2H_3O_2$, and we have produced 2.000×10^{-3} mol of $C_2H_3O_2^-$. The amount of $HC_2H_3O_2$ remaining in the solution (realizing we began with 5.000×10^{-3} mol), is simply

$$(5.000 \times 10^{-3} - 2.000 \times 10^{-3})\ \text{mol } HC_2H_3O_2 = 3.000 \times 10^{-3}\ \text{mol } HC_2H_3O_2$$

The total volume is now $(25.00\ \text{mL} + 10.00\ \text{mL}) = 35.00\ \text{mL} = 0.03500\ \text{L}$. The concentrations of acetate ion and acetic acid now are, therefore,

$$[C_2H_3O_2^-] = \frac{2.000 \times 10^{-3}\ \text{mol}}{0.03500\ \text{L}} = 5.714 \times 10^{-2}\ \text{mol L}^{-1}$$

$$[HC_2H_3O_2] = \frac{3.000 \times 10^{-3}\ \text{mol}}{0.03500\ \text{L}} = 8.571 \times 10^{-2}\ \text{mol L}^{-1}$$

We can now substitute these values into the mass action expression and calculate $[H^+]$.

$$\frac{[H^+](5.714 \times 10^{-2})}{(8.571 \times 10^{-2})} = 1.8 \times 10^{-5}$$

Therefore,

$$[H^+] = 2.7 \times 10^{-5}\ M$$

The pH calculates to be

$$pH = 4.57$$

As long as we have not reached the equivalence point—as long as unreacted acetic acid remains together with acetate ion—we can compute the pH of the solution in the receiving flask by this kind of buffer calculation.

At the Equivalence Point Once we exactly reach the equivalence point, the solution is simply a dilute solution of sodium acetate. It no longer is a buffer. We now have to deal with the reaction of the acetate ion with water and the associated base ionization constant.

$$C_2H_3O_2^-(aq) + H_2O \rightleftharpoons HC_2H_3O_2(aq) + OH^-(aq)$$

$$K_b = \frac{[OH^-][HC_2H_3O_2]}{[C_2H_3O_2^-]} = \frac{K_w}{K_a} = 5.6 \times 10^{-10}$$

At the equivalence point, we have added 25.00 mL of 0.2000 *M* NaOH to the original 25.00 mL of 0.2000 *M* $HC_2H_3O_2$. The volume is now 50.00 mL = 0.05000 L. We began with 5.000×10^{-3} mol of $HC_2H_3O_2$, so we now have 5.000×10^{-3} mol of $C_2H_3O_2^-$ (in 0.05000 L). The concentration of $C_2H_3O_2^-$, therefore, is

$$[C_2H_3O_2^-] = \frac{5.000 \times 10^{-3} \text{ mol}}{0.05000 \text{ L}} = 0.1000 \text{ } M$$

To solve the problem we realize that a small amount of the $C_2H_3O_2^-$ reacts to give equal amounts of OH^- and $HC_2H_3O_2$. We let the concentrations of these equal *x*, so $[OH^-] = [HC_2H_3O_2] = x$. Substituting into the mass action expression gives

$$\frac{(x)(x)}{0.1000} = 5.6 \times 10^{-10}$$

$$x = 7.5 \times 10^{-6}$$

It is interesting that although we describe the overall reaction in the titration as neutralization, at the equivalence point the solution is slightly basic, not neutral.

This means that $[OH^-] = 7.5 \times 10^{-6}$ *M*, so the pOH calculates to be 5.12. Finally, the pH is $(14.00 - 5.12)$ or 8.88. Thus the pH at the equivalence point in this titration is 8.88, on the basic side because the solution contains the Brønsted base, $C_2H_3O_2^-$. *In the titration of any weak acid with a strong base, the pH at the equivalence point will be greater than 7.*

After the Equivalence Point What happens if we continue to add NaOH(*aq*) to the solution beyond the equivalence point? Now the additional OH^- shifts the following equilibrium to the left.

$$C_2H_3O_2^-(aq) + H_2O \rightleftharpoons HC_2H_3O_2(aq) + OH^-(aq)$$

The generation of additional OH^- *by this route* is thus suppressed as we add more and more NaOH solution, so effectively the only source of OH^- to affect the pH is from the base added after the equivalence point. From here on, therefore, the pH calculations become identical to those for the last half of the titration curve for HCl and NaOH. Each point is just a matter of calculating the additional number of moles of NaOH, finding the new final volume, taking the ratio (the molar concentration), and then calculating pOH and pH.

**TABLE 16.7   Titration of 25.00 mL of 0.2000 *M*
HC$_2$H$_3$O$_2$ with 0.2000 *M* NaOH**

Volume of Base Added (mL)	Molar Concentration of Species in Parentheses	pH
none	1.9×10^{-3}(H$^+$)	2.72
10.00	2.7×10^{-5}(H$^+$)	4.57
24.90	7.2×10^{-8}(H$^+$)	7.14
24.99	7.1×10^{-9}(H$^+$)	8.14
25.00	7.5×10^{-6}(OH$^-$)	8.88
25.01	4.0×10^{-5}(OH$^-$)	9.60
25.10	4.0×10^{-4}(OH$^-$)	10.60
26.00	3.9×10^{-3}(OH$^-$)	11.59
35.00	3.3×10^{-2}(OH$^-$)	12.52

The Titration Curve   Table 16.7 summarizes the titration data, and the titration curve is given in Figure 16.6. Once again notice how the change in pH occurs slowly at first, then shoots up through the equivalence point, and finally looks just like the last half of the curve in Figure 16.5.

■ **Practice Exercise 25** Suppose we titrate 20.0 mL of 0.100 *M* HCHO$_2$ with 0.100 *M* NaOH. Calculate the pH (a) before any base is added, (b) when half of the HCHO$_2$ has been neutralized, (c) after a total of 15.0 mL of base has been added, and (d) at the equivalence point.

Titration of a Weak Base by a Strong Acid

The calculations involved here are nearly identical to those of the weak acid–strong base titration. We will review what is required but will not actually perform the calculations.

If we titrate 25.00 mL of 0.2000 *M* NH$_3$ with 0.2000 *M* HCl, the molecular and net ionic equations are

$$NH_3(aq) + HCl(aq) \longrightarrow NH_4Cl(aq)$$

$$NH_3(aq) + H^+(aq) \longrightarrow NH_4^+(aq)$$

1. **Before the titration begins.** The solution at this point is simply a solution of the weak base, NH$_3$. Since the only solute is a base, we must use K_b to calculate the pH.

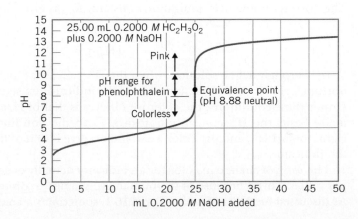

FIGURE 16.6

Titration curve for titrating a weak acid (0.2000 *M* acetic acid) with a strong base (0.2000 *M* NaOH).

FIGURE 16.7

Titration curve for the titration of a weak base (0.2000 M NH$_3$) with a strong acid (0.2000 M HCl).

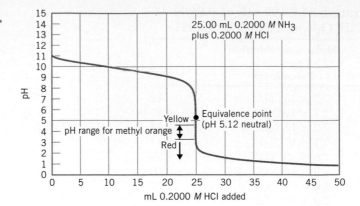

2. **During the titration but before the equivalence point.** As we add HCl to the NH$_3$ the neutralization reaction produces NH$_4^+$, so the solution contains both NH$_3$ and NH$_4^+$ (it is a buffer solution). We can use K_b for NH$_3$ to calculate the pH here as well. (We could also use K_a if it were available.)

3. **At the equivalence point.** Here all the NH$_3$ has reacted and the solution contains the salt NH$_4$Cl. The anion, Cl$^-$, comes from the strong acid HCl and is so weak a base that it cannot affect the pH. However, the cation NH$_4^+$ is a weak acid derived from the weak base NH$_3$, so we use its K_a to calculate the pH.

4. **After the equivalence point.** The H$^+$ introduced by further addition of HCl has nothing with which to react and causes the solution to become acidic. The concentration of H$^+$ produced by the unreacted HCl is used directly to calculate the pH.

Figure 16.7 illustrates the titration curve for this system.

Acid–Base Indicators

Most dyes that work as acid–base indicators are also weak acids. Therefore, let's represent an indicator with the formula H*In*. In its un-ionized state, H*In* has one color. Its conjugate base, *In*$^-$, has a different color, the more strikingly different the better. In solution, the indicator is involved in a typical acid–base equilibrium:

$$\underset{\substack{\text{Acid form} \\ \text{(one color)}}}{\text{H}In(aq)} \rightleftharpoons \text{H}^+(aq) \quad + \quad \underset{\substack{\text{Base form} \\ \text{(another color)}}}{In^-(aq)}$$

The corresponding acid ionization constant, K_{In}, is given by

$$K_{\text{In}} = \frac{[\text{H}^+][In^-]}{[\text{H}In]}$$

In a strongly acidic solution, when the H$^+$ concentration is high, this equilibrium is shifted to the left and most of the indicator exists in its "acid form." Under these conditions, the color we observe is that of H*In*. If the solution is made basic, the H$^+$ concentration drops and the equilibrium shifts to the right, toward *In*$^-$, and the color we would observe is that of the "base form" of the indicator.

The *observed* change in color for an indicator actually occurs over a range of pH values. This is because of the human eye's ability to discern color changes. As discussed further in Special Topic 16.1, sometimes a change of as much as

SPECIAL TOPIC 16.1 / COLOR RANGES FOR INDICATORS

Suppose that the color of H*In* is red and the color of its anion, *In*⁻, is yellow. In terms of the eye's abilities, the ratio of [H*In*]/[*In*⁻] has to be roughly at least as large as 10 to 1 for the eye to see the color as definitely red. And this ratio has to be at least as small as 1 to 10 for the eye to see the color as definitely yellow. Therefore, the color transition that the eye actually sees at the end point occurs as the ratio of [H*In*] to [*In*⁻] changes from 10/1 to 1/10.

We can calculate the relationship of pH to pK_a for the indicator at the start of this range as follows. We know that

$$K_{In} = \frac{[H^+][In^-]}{[HIn]}$$

Therefore, at the start of the range, we must have

$$K_{In} = [H^+] \times \frac{10}{1}$$

$$pK_{In} = pH - \log 10$$

Or, at the start of the transition we should have

$$pH = pK_{In} - 1$$

At the other end of the range, when [H*In*]/[*In*⁻] is 1/10, we have

$$K_{In} = [H^+] \times \frac{1}{10}$$

$$pK_{In} = pH - \log(10^{-1})$$

So at the end of the color transition, the pH should be

$$pH = pK_{In} + 1$$

Therefore, the *range* of pH over which a definite change in color occurs is

$$pH = pK_{In} \pm 1$$

A range of ± 1 pH unit on either side of the value of pK_{In} corresponds to a span of 2 pH units over which the eye can definitely notice the color change.

2 pH units is needed for some indicators—more for litmus—before the eye notices the color change. This is why tables of acid–base indicators, such as Table 16.3, provide approximate pH ranges for the color changes.

How Acid–Base Indicators Work   In a typical acid–base titration, you've seen that as we pass the equivalence point, there is a sudden and large change in the pH. For example, in the titration of HCl with NaOH described earlier, the pH just one-half drop before the equivalence point (when 24.97 mL of the base has been added) is 3.92. Just one-half drop later (when 25.03 mL of base has been added) we have passed the equivalence point and the pH has risen to 10.08. This large swing in pH (from 3.92 to 10.08) causes a sudden shift in the position of equilibrium for the indicator, and we go from a condition where most of the indicator is in its acid form to a condition in which most is in the base form. This is observed visually as a change in color from that of the acid form to that of the base form.

Selecting the Best Indicator for an Acid–Base Titration   The explanation in the preceding paragraph assumes that the midpoint of the color change of an indicator corresponds to the pH at the equivalence point. At this midpoint we expect that there will be equal amounts of both forms of the indicator, which means that [*In*⁻] = [H*In*]. If this condition is true, then

$$K_{In} = \frac{[H^+][\cancel{In^-}]}{[\cancel{HIn}]}$$

so

$$K_{In} = [H^+]$$

FIGURE 16.8

Colors of some common acid–
base indicators.

pH 8.2   pH 10.0
Phenolphthalein

pH 6.0   pH 7.6
Bromothymol blue

pH 3.2   pH 4.4
Methyl orange

pH 9.4   pH 10.6
Thymolphthalein

Taking the negative logarithms of both sides gives us

$$-\log K_{\text{In}} = -\log[\text{H}^+]$$

or

$$pK_{\text{In}} = pH_{\text{(at the equivalence point)}}$$

This important result tells us that once we know the pH that the solution should have at the equivalence point of a titration, we also know the pK_{In} that the indicator should have to function most effectively. Thus, we want an indicator whose pK_{In} equals (or is as close as possible to) the pH at the equivalence point. Phenolphthalein, for example, changes a solution from colorless to pink as the solution's pH changes over a range of 8.2 to 10.0 (Figure 16.8). *Phenolphthalein is therefore the perfect indicator for the titration of a weak acid by a strong base*, where the pH of the equivalence point is on the basic side (refer back to Figure 16.6).

Phenolphthalein also works very well for the titration of a strong acid by a strong base. As discussed earlier, just one drop of titrant can bring the pH from below 7 to nearly 10, which spans the pH range for this indicator.

The Importance of Color Intensity in an Indicator   Another factor involving indicators is that we want to use as little of one as possible. Because they are weak acids, they are also neutralized and consume titrant. We have to accept some loss of titrant this way, but no more than absolutely necessary. So the best indicators are those with the most intense colors. Then, even extremely small amounts can give striking colors and color changes.

SUMMARY

The Autoionization of Water and the pH Concept Water reacts with itself to produce small amounts of H^+ and OH^- ions. The concentrations of these ions both in pure water *and* in dilute aqueous solutions are related by the expression

$$[H^+][OH^-] = K_w = 1.0 \times 10^{-14} \quad \text{(at 25 °C)}$$

In pure water, $[H^+] = [OH^-] = 1.0 \times 10^{-7}$ M. The equilibrium constant K_w is called the **ion product constant of water.**

The **pH** of the solution is useful for describing and comparing low molar concentrations of H^+ ions. Equations you should know are

$$pH = -\log[H^+] \quad \text{and} \quad [H^+] = 10^{-pH}$$

A comparable expression, **pOH,** can be used to describe low OH^- ion concentrations.

$$pOH = -\log[OH^-] \quad \text{and} \quad [OH^-] = 10^{-pOH}$$

At 25 °C, (pH + pOH) = 14.00. A solution is acidic if its pH is less than 7.00 and basic if its pH is greater than 7.00. The smaller the pH, the more concentrated is the solution in hydrogen ions.

Solutions of Strong Acids and Strong Bases In calculating the pH of such solutions it is important to remember that strong acids and bases are 100% dissociated; we never express their dissociation reactions in water as equilibria.

Acid and Base Ionization Constants A weak acid HA ionizes according to the general equation

$$HA \rightleftharpoons H^+ + A^-$$

The equilibrium constant is called the **acid ionization constant, K_a.**

$$K_a = \frac{[H^+][A^-]}{[HA]}$$

A weak base B ionizes by the general equation

$$B + H_2O \rightleftharpoons BH^+ + OH^-$$

The equilibrium constant is called the **base ionization constants, K_b.**

$$K_b = \frac{[OH^-][BH^+]}{[B]}$$

The smaller the values of K_a (or K_b), the weaker are the substances as Brønsted acids (or bases).

Another way to compare the relative strengths of acids or bases is to use the negative logarithms of K_a and K_b, called **pK_a** and **pK_b,** respectively.

$$pK_a = -\log K_a$$
$$pK_b = -\log K_b$$

The *smaller* the pK_a, the *stronger* is the acid. For a conjugate acid-base pair,

$$K_a K_b = K_w$$

and

$$pK_a + pK_b = 14.00 \quad \text{(at 25 °C)}$$

Equilibrium Calculations—Determining K_a and K_b The values of K_a and K_b can be obtained from initial concentrations of the acid or base and either the pH of the solution or the percentage ionization of the acid or base. The measured pH gives the equilibrium value for $[H^+]$. The percentage ionization is defined as

$$\text{Percentage ionization} = \frac{\text{amount ionized}}{\text{amount available}} \times 100\%$$

Equilibrium Calculations—Determining Equilibrium Concentrations When K_a or K_b Is Known Problems fall into one of three categories: (1) the only solute is a weak acid (we must use the K_a expression), (2) the only solute is a weak base (we must use the K_b expression), and (3) the solution contains both a weak acid and its conjugate base (we can use either K_a or K_b). When the initial molar concentration of the acid (or base) is larger than 400 times the value of K_a (or K_b), it is safe to use initial concentrations of acid or base as though they were equilibrium values in the mass action expression. When this approximation cannot be used, we use either the quadratic formula or the method of successive approximations (the latter being faster).

Buffers A solution that contains both a weak acid and its conjugate base is called a **buffer,** because it is able to absorb $[H^+]$ from a strong acid or $[OH^-]$ from a strong base without suffering large changes in pH. For the general acid–base pair, HA and A^-, the following reactions neutralize H^+ and OH^-.

When H^+ is added to the buffer: $\quad A^- + H^+ \longrightarrow HA$

When OH^- is added to the buffer: $\quad HA + OH^- \longrightarrow A^- + H_2O$

The pH of a buffer is controlled by the ratio of weak acid to weak base, expressed in terms of either molarities or moles.

$$[H^+] = K_a \times \frac{[HA]}{[A^-]} = K_a \times \frac{\text{moles } HA}{\text{moles } A^-}$$

Because the $[H^+]$ is determined by the mole ratio of HA to A^-, dilution does not change the pH of a buffer.

In performing buffer calculations, the usually valid simplifications are

$$[HA] = [HA]_{initial}$$
$$[A^-] = [A^-]_{initial}$$

The value of $[A^-]_{initial}$ is found from the molarity of the *salt* of the weak acid in the buffer. The value of $[HA]_{initial}$, the molarity of the weak acid itself in the buffer, can be used for $[HA]$. Once $[H^+]$ is found by this calculation, the pH is calculated.

Buffer calculations can also be performed using K_b for the weak base component. Similar equations apply; the principal difference being that it is the OH^- concentration that is calculated. Thus, using K_b we have

$$[OH^-] = K_b \times \frac{[\text{conjugate base}]}{[\text{conjugate acid}]}$$
$$= K_b \times \frac{\text{moles conjugate base}}{\text{moles conjugate acid}}$$

Buffers are most effective when the pK_a of the acid member of the buffer pair lies within ± 1 unit of the desired pH.

Although the molar *ratio* contributes to the value of the pH at which a buffer works, the individual *concentrations* of acid and conjugate give the buffer its *capacity* for taking on extra acid or base with little change in pH.

Ions as Acids or Bases A solution of a salt is acidic if the cation is acidic but the anion is neutral. Metal ions with high charge density generally are acidic, but those of Groups IA and IIA (except Be^{2+}) are not. Cations such as NH_4^+, which are the conjugate acids of weak *molecular* bases, are themselves proton donors and are acidic.

When the anion of a salt is the conjugate base of a *weak* acid, the anion is a weak base. Anions of strong acids, such as Cl^- and NO_3^-, are such weak Brønsted bases that they cannot affect the pH of a solution.

If the salt is derived from a weak acid *and* a weak base, the net effect has to be determined on a case by case basis by determining which of the two ions is the stronger.

Acid–Base Titrations and Indicators The pH at the equivalence point depends on the ions of the salt made by the titration. In the titration of a strong acid by a strong base, the pH at the equivalence point is 7 because neither of the ions produced in the neutralization affects the pH of an aqueous solution.

For the titration of a weak acid by a strong base, the approach toward calculation of the pH varies depending on which stage of the titration is considered. Before the titration begins, the only solute is the weak acid. After the titration starts but before the equivalence point is reached, the solution is a buffer. At the equivalence point, the active solute is the conjugate base of the weak acid. After the equivalence point, excess base determines the pH of the mixture. At the equivalence point the pH is above 7.

The indicator chosen for a titration should have the center of its color change near the pH at the equivalence point. For the titration of a strong or weak acid with a strong base, phenolphthalein is the best indicator. When a weak base is titrated with a strong acid, the pH at the equivalence point is less than 7. Several indicators, including methyl orange, work satisfactorily.

TOOLS YOU HAVE LEARNED

The following table lists the important tools needed to solve problems dealing with acid–base equilibria. You must have all of them at your fingertips when you begin to analyze a problem. The key to success is in choosing the correct tools to apply to the solution.

Tool	Function
$[H^+][OH^-] = K_w$ (page 680)	Permits calculating either the H^+ or OH^- concentration when the other is known.
$pX = -\log X$ (page 681)	This equation is the basis for calculating pH, pOH, and pK.
$pH + pOH = 14.00$ (page 682)	To calculate pH from pOH and vice versa.
$HA \rightleftharpoons H^+ + A^-$ $K_a = \dfrac{[H^+][A^-]}{[HA]}$ (page 688)	Following this general form, you can write the correct chemical equation for the ionization of any weak acid in water and construct the correct equilibrium law to be used in calculations.

Tool	Function
$B + H_2O \rightleftharpoons BH^+ + OH^-$ $K_b = \dfrac{[BH^+][OH^-]}{[B]}$ (page 689, 690)	Following this general form, you can write the correct chemical equation for the ionization of any weak base in water and construct the correct equilibrium law to be used in calculations.
Percentage ionization = $\dfrac{\text{amount ionized}}{\text{amount available}} \times 100\%$ (page 692)	To calculate the percentage ionization from the initial concentration of acid (or base) and the change in the concentrations of the ions. If the percentage ionization and pH are known, the K_a (or K_b) can be calculated.
$K_a \times K_b = K_w$ (page 691)	Enables you to calculate K_a given K_b or vice versa.
Identification of the solute as (1) only a weak acid, (2) only a weak base, or (3) a mixture of a weak acid and its conjugate base (page 695)	This identification points the way in solving problems by enabling you to write the correct equilibrium chemical equation, and therefore the correct equilibrium law. This is the critical first step necessary to solve any problems involving acid–base equilibria.
Identification of acidic cations and basic anions (page 699, 700)	Enables you to determine whether a salt solution is acidic, basic, or neutral. It is the first step in calculating the pH of a salt solution. Use it in conjunction with the preceding tool to establish the correct direction in analyzing the problem.
Simplifications in calculations (page 703)	To determine whether initial concentrations can be used as though they are equilibrium values in the mass action expression.
Reactions when H^+ or OH^- are added to a buffer (page 707)	Permits you to determine the effect of strong acid or strong base on the pH of a buffer. Adding H^+ lowers $[A^-]$ and raises $[HA]$, adding OH^- lowers $[HA]$ and raises $[A^-]$.

THINKING IT THROUGH

Remember, you are not asked to obtain answers for the following problems. Instead, assemble the data necessary to solve the problems and describe how you would use the data to obtain the answers. For numerical problems, set up the calculation using appropriate conversion factors.

The problems are divided into two groups. Those in Level 2 are significantly more challenging than those in Level 1 and provide an opportunity to really hone your problem solving skills.

Level 1 Problems

1. What experimental information is required to determine the value of K_a for a weak acid? Explain how you would use this information to calculate K_a.

2. Propionic acid, $HC_3H_5O_2$, is a monoprotic acid. You have prepared a 0.100 M solution of this acid. Do you have enough data to calculate the K_a of propionic acid? If you do not, state what additional data are needed.

3. A 0.10 M solution of a weak base is prepared. The value of K_b for the base is 2.0×10^{-8}. Describe how you would calculate the pH of the solution. Explain any simplifications you would make and how you would justify their validity.

4. A 0.10 M solution of a weak acid with $K_a = 2.5 \times 10^{-6}$ has a pH of 3.28. Is the weak acid the only solute in the solution? Describe in detail how you would answer this question.

5. What information do you need to actually prepare, in the laboratory, a buffer that has a pH of 5.00 using acetic acid and potassium acetate? Describe any calculations that would be necessary.

6. How would you calculate the acetate ion concentration in a solution that contains 0.0010 M HCl and 0.10 M $HC_2H_3O_2$? Describe any simplifications you would make in solving the problem.

7. A 0.035 M solution of an acid has a pH of 3.5. Is the solute a strong acid or not? How can you tell?

8. Acetic acid ($HC_2H_3O_2$) and propionic acid ($HC_3H_5O_2$) have nearly identical pK_a values. If you had an aqueous solution that is 0.10 M in acetic acid and also 0.10 M in the sodium salt of the other acid, propionic acid, would this solution function as a *buffer*? Explain. Write net ionic equations that describe its chemical behavior if a small amount of hydrochloric acid is added or if a small amount of sodium hydroxide is added.

Level 2 Problems

9. A simple way to measure the pK_a of an acid is to neutralize half of the acid with a stong base (such as NaOH)

to form an aqueous solution and then measure the pH with a pH meter. Explain why this method avoids having to perform equilibrium calculations.

10. Suppose 25.0 mL of 0.10 M HCl is added to 50.0 mL of 0.10 M KNO_2. Explain how you would calculate the pH of the resulting mixture.

11. The value of K_w increases as the temperature of water is raised.

(a) Explain what effect increasing the temperature of pure water will have on its pH. (Will the pH increase, decrease, or stay the same?)

(b) As the temperature of pure water increases, will the water become more acidic, more basic, or remain neutral? Justify your answer.

12. A solution of dichloroacetic acid, $HC_2HO_2Cl_2$, has a concentration of 0.50 M. Describe how you would calculate the freezing point of this solution. (Since it is a dilute aqueous solution, assume that the molarity equals the molality.)

13. Suppose 250 mL of $NH_3(g)$ measured at 20 °C and 750 torr is dissolved in 100 mL of 0.050 M HCl. Describe how you would calculate the pH of the resulting solution.

14. How many milliliters of 0.100 M HCl would have to be added to 100 mL of a 0.100 M solution of $HC_2H_3O_2$ to make the $C_2H_3O_2^-$ concentration equal 1.0×10^{-4} M? (Explain all the calculations that are necessary to obtain the answer.)

REVIEW EXERCISES

Answers to questions whose numbers are printed in color are given in Appendix D. More challenging questions are marked with asterisks.

Self-Ionization of Water and pH

16.1 Write the chemical equation for the autoionization of water and the equilibrium law for K_w.

16.2 How are acidic, basic, and neutral solutions in water defined (a) in terms of $[H^+]$ and $[OH^-]$ and (b) in terms of pH?

16.3 At the temperature of the human body, 37 °C, the value of K_w is 2.4×10^{-14}. Calculate the $[H^+]$, $[OH^-]$, pH, and pOH of pure water at this temperature. What is the relation between pH, pOH, and K_w at this temperature? Is water neutral at this temperature?

16.4 Deuterium oxide, D_2O, ionizes like water. At 20 °C its K_w, or ion product constant analogous to that of water, is 8.9×10^{-16}. Calculate $[D^+]$ and $[OD^-]$ in deuterium oxide at 20 °C. Calculate also the pD and the pOD.

16.5 Calculate the H^+ concentration in each of the following solutions in which the hydroxide ion concentrations are
(a) 0.0024 M (b) 1.4×10^{-5} M
(c) 5.6×10^{-9} M (d) 4.2×10^{-13} M

16.6 Calculate the OH^- concentration in each of the following solutions in which the hydrogen ion concentrations are
(a) 3.5×10^{-8} M (b) 0.0065 M
(c) 2.5×10^{-13} M (d) 7.5×10^{-5} M

16.7 A certain brand of beer had a hydrogen ion concentration equal to 1.9×10^{-5} mol L^{-1}. What is the pH of the beer?

16.8 A soft drink was put on the market with $[H^+] = 1.4 \times 10^{-5}$ mol L^{-1}. What is its pH?

16.9 Calculate the pH of each of the solutions in Exercises 16.5 and 16.6.

16.10 Calculate the molar concentrations of H^+ and OH^- in solutions that have the following pH values.
(a) 3.14, (b) 2.78, (c) 9.25, (d) 13.24, (e) 5.70

16.11 Calculate the molar concentrations of H^+ and OH^- in solutions that have the following pOH values.
(a) 8.26, (b) 10.25, (c) 4.65, (d) 6.18, (e) 9.70

Solutions of Strong Acids and Bases

16.12 What is the pH of 0.010 *M* HCl?

16.13 What is the pH of a 0.0050 *M* solution of HNO_3?

16.14 A sodium hydroxide solution is prepared by dissolving 6.0 g NaOH in 1.00 L of solution. What is the pOH and the pH of the solution?

16.15 A solution was made by dissolving 0.837 g $Ba(OH)_2$ in 100 mL final volume. What is the pOH and the pH of the solution?

16.16 A solution of $Ca(OH)_2$ has a measured pH of 11.60. What is the molar concentration of the $Ca(OH)_2$ in the solution?

16.17 A solution of HCl has a pH of 2.50. How many grams of HCl are there in 250 mL of this solution?

Acid and Base Ionization Constants: K_a, K_b, pK_a, and pK_b

16.18 Write the chemical equation for the ionization of each of the following weak acids in water. (For any polyprotic acids, write only the equation for the first step in the ionization.)
(a) HNO_2, (b) H_3PO_4, (c) $HAsO_4^{2-}$, (d) $(CH_3)_3NH^+$

16.19 For each of the acids in Exercise 16.18, write the appropriate K_a expression.

16.20 Write the chemical equation for the ionization of each of the following weak bases in water.
(a) $(CH_3)_3N$, (b) AsO_4^{3-}, (c) NO_2^-, (d) $(CH_3)_2N_2H_2$

16.21 For each of the bases in Exercise 16.20, write the appropriate K_b expression.

16.22 Benzoic acid, $C_6H_5CO_2H$, is an organic acid whose sodium salt, $C_6H_5CO_2Na$, has long been used as a safe food additive to protect beverages and many foods against harmful yeasts and bacteria. The acid is monoprotic. Write the equation for its K_a.

16.23 Write the equation for the equilibrium that the benzoate ion, $C_6H_5CO_2^-$ (Review Exercise 16.22), would produce in water as it functions as a Brønsted base. Then write the expression for the K_b of the conjugate base of benzoic acid.

16.24 The pK_a of HCN is 9.21 and that of HF is 3.17. Which is the stronger Brønsted base: CN^- or F^-?

16.25 The K_a for HF is 6.8×10^{-4}. What is the K_b for F^-?

16.26 The barbiturate ion, $C_4H_3N_2O_3^-$, has $K_b = 1.0 \times 10^{-10}$. What is K_a for barbituric acid?

16.27 Hydrogen peroxide, H_2O_2, is a weak acid with $K_a = 1.8 \times 10^{-12}$. What is the value of K_b for the HO_2^- ion?

16.28 Methylamine, CH_3NH_2, resembles ammonia in

odor and basicity. Its K_b is 4.4×10^{-4}. Calculate the K_a of its conjugate acid.

16.29 Lactic acid, $HC_3H_5O_3$, is responsible for the sour taste of sour milk. At 25 °C, its $K_a = 1.4 \times 10^{-4}$. What is the K_b of its conjugate base, the lactate ion, $C_3H_5O_3^-$?

16.30 Iodic acid, HIO_3, has a pK_a of 0.77.
(a) What is the formula and the K_b of its conjugate base?
(b) Is its conjugate base a stronger or a weaker base than the acetate ion?

Equilibrium Calculations

16.31 Periodic acid, HIO_4, is an important oxidizing agent and a moderately strong acid. In a 0.10 *M* solution, $[H^+] = 3.8 \times 10^{-2}$ mol L^{-1}. Calculate the K_a and pK_a for periodic acid.

16.32 Chloroacetic acid, $HC_2H_2ClO_2$, is a stronger monoprotic acid than acetic acid. In a 0.10 *M* solution, this acid is 11.0% ionized. Calculate the K_a and pK_a for chloroacetic acid.

16.33 Ethylamine, $CH_3CH_2NH_2$, has a strong, pungent odor similar to that of ammonia. Like ammonia, it is a Brønsted base. A 0.10 *M* solution has a pH of 11.86. Calculate the K_b and pK_b for ethylamine.

16.34 Hydroxylamine, $HONH_2$, like ammonia, is a Brønsted base. A 0.15 *M* solution has a pH of 10.12. What are K_b and pK_b for hydroxylamine?

16.35 Refer to the data in the preceding question to calculate the percentage ionization of the base in 0.15 *M* $HONH_2$.

16.36 What is the pH of 0.125 *M* pyruvic acid? Its K_a is 3.2×10^{-3}.

16.37 What is the pH of 0.15 *M* HN_3? For HN_3, $K_a = 1.8 \times 10^{-5}$.

16.38 What is the pH of a 1.0 *M* solution of hydrogen peroxide, H_2O_2? For this solute, $K_a = 1.8 \times 10^{-12}$.

16.39 Phenol, also known as carbolic acid, is sometimes used as a disinfectant. What are the concentrations of all of the substances in a 0.050 *M* solution of phenol, HC_6H_5O? What percentage of the phenol is ionized? For this acid, $K_a = 1.3 \times 10^{-10}$.

16.40 Codeine, a cough suppressant extracted from crude opium, is a weak base with a pK_b of 5.79. What will be the pH of a 0.020 *M* solution of codeine? (Use *Cod* as a symbol for codeine.)

16.41 Deuteroammonia, ND_3, is a weak base with a pK_b of 4.96 at 25 °C. What is the pH of a 0.20 *M* solution of this compound?

16.42 A solution of acetic acid has a pH of 2.54. What is the concentration of acetic acid in this solution?

Acid–Base Properties of Salt Solutions

16.43 Aspirin is acetylsalicyclic acid, a monoprotic acid whose K_a value is 3.27×10^{-4}. Does a solution of the sodium salt of aspirin in water test acidic, basic, or neutral? Explain.

16.44 The K_b value of the oxalate ion, $C_2O_4^{2-}$, is 1.9×10^{-10}. Is a solution of $K_2C_2O_4$ acidic, basic, or neutral? Explain.

16.45 Consider the following compounds and suppose that 0.5 M solutions are prepared of each: NaI, KF, $(NH_4)_2SO_4$, KCN, $KC_2H_3O_2$, $CsNO_3$, and KBr. Write the *formulas* of those that have solutions that are (a) acidic, (b) basic, and (c) neutral.

16.46 Will an aqueous solution of $AlCl_3$ turn litmus red or blue? Explain.

16.47 Explain why the beryllium ion is a more acidic cation than the calcium ion.

16.48 Ammonium nitrate is commonly used in fertilizer mixtures as a source of nitrogen for plant growth. What effect, if any, will this compound have on the acidity of the moisture in the ground? Explain.

16.49 Calculate the pH of 0.20 M NaCN.

16.50 Calculate the pH of 0.40 M KNO_2.

16.51 Calculate the pH of 0.15 M CH_3NH_3Cl. For CH_3NH_2, $K_b = 4.4 \times 10^{-4}$.

16.52 A weak base B forms the salt $BHCl$, composed of the ions BH^+ and Cl^-. A 0.15 M solution of the salt has a pH of 4.28. What is the value of K_b for the base B?

16.53 Calculate the number of grams of NH_4Br that have to be dissolved in 1.00 L of water at 25 °C to have a solution with a pH of 5.16.

*16.54** The conjugate acid of a molecular base has the hypothetical formula, BH^+, and has a pK_a of 5.00. A solution of a salt of this cation, BHY, tests slightly basic. Will the conjugate acid of Y^-, HY, have a pK_a greater than 5.00 or less than 5.00? Explain.

16.55 Many drugs that are natural Brønsted bases are put into aqueous solution as their much more soluble salts with strong acids. The powerful painkiller morphine, for example, is very slightly soluble in water, but morphine nitrate is quite soluble. We may represent morphine by the symbol *Mor* and its conjugate acid as H—*Mor*$^+$. The pK_b of morphine is 6.13. What is the calculated pH of a 0.20 M solution of H—*Mor*$^+$?

16.56 Quinine, an important drug in treating malaria, is a weak Brønsted base that we may represent as *Qu*. To make it more soluble in water, it is put into a solution as its conjugate acid, which we may represent as H—*Qu*$^+$. What is the calculated pH of a 0.15 M solution of H—*Qu*$^+$? Its pK_a is 8.52 at 25 °C.

Equilibrium Calculations When Simplifications Fail

16.57 Generally, under what conditions are we *unable* to use the initial concentration of an acid or base as though it were the equilibrium concentration in the mass action expression?

16.58 What is the percentage ionization in a 0.15 M solution of HF? What is the pH of the solution?

16.59 What is the percentage ionization in 0.0010 M acetic acid? What is the pH of the solution?

*16.60** What is the pH of a 1.0×10^{-7} M solution of HCl?

*16.61** The hydrogen sulfate ion, HSO_4^-, is a moderately strong Brønsted acid with a K_a of 1.0×10^{-2}.
(a) Write the chemical equation for the ionization of the acid and give the appropriate K_a expression.

(b) What is the value of $[H^+]$ in 0.010 M HSO_4^- (furnished by the salt, $NaHSO_4$)? Do NOT make simplifying assumptions; solve the quadratic equation.
(c) What is the calculated $[H^+]$ in 0.010 M HSO_4^-, obtained by using the usual simplifying assumption?
(d) How much error is introduced by incorrectly using the simplifying assumption?

16.62 *para*-Aminobenzoic acid (PABA) is a powerful sunscreening agent whose salts were once used widely in sun tanning lotions. The parent acid, which we may symbolize as H–*Paba*, is a weak acid with a pK_a of 4.92 (at 25 °C). What is the $[H^+]$ and pH of a 0.030 M solution of this acid?

16.63 Barbituric acid, $HC_4H_3N_2O_3$ (which we will abbreviate H–*Bar*), was discovered by the Nobel Prize–winning organic chemist Adolph von Baeyer and named after his friend, Barbara. It is the parent compound of widely used sleeping drugs, the barbiturates. Its pK_a is 4.01. What is the $[H^+]$ and pH of a 0.050 M solution of H–*Bar*?

Buffers

16.64 Write ionic equations that illustrate how each pair of compounds can serve as a buffer pair.
(a) H_2CO_3 and $NaHCO_3$ (the "carbonate" buffer in blood)
(b) NaH_2PO_4 and Na_2HPO_4 (the "phosphate" buffer inside body cells)
(c) NH_4Cl and NH_3

16.65 Which buffer would be better able to hold a steady pH on the addition of strong acid, buffer 1 or buffer 2? Explain.
  Buffer 1 is a solution containing 0.10 M NH_4Cl and 1 M NH_3.
  Buffer 2 is a solution containing 1 M NH_4Cl and 0.10 M NH_3.

16.66 What is the pH of a solution that contains 0.15 M $HC_2H_3O_2$ and 0.25 M $C_2H_3O_2^-$? Use $K_a = 1.8 \times 10^{-5}$ for $HC_2H_3O_2$.

16.67 Rework the preceding problem using the K_b for the acetate ion. (Be sure to write the proper chemical equation and equilibrium law.)

16.68 By how much will the pH change if 0.050 mol of HCl is added to 1.00 L of the buffer in Exercise 16.66?

16.69 By how much will the pH change if 50.0 mL of 0.10 M NaOH is added to 500 mL of the buffer in Exercise 16.66?

16.70 A buffer is prepared containing 0.25 M NH_3 and 0.14 M NH_4^+.
(a) Calculate the pH of the buffer using the K_b for NH_3.
(b) Calculate the pH of the buffer using the K_a for NH_4^+.

16.71 By how much will the pH change if 0.020 mol of HCl is added to 1.00 L of the buffer in Exercise 16.70?

16.72 By how much will the pH change if 75 mL of 0.10 M KOH is added to 200 mL of the buffer in Exercise 16.70?

16.73 How many grams of sodium acetate, $NaC_2H_3O_2$, would have to be added to 1.0 L of 0.15 M acetic acid (pK_a 4.74) to make the solution a buffer for pH 5.00?

16.74 How many grams of sodium formate, $NaCHO_2$, would have to be dissolved in 1.0 L of 0.12 M formic acid (pK_a 3.74) to make the solution a buffer for pH 3.80?

16.75 What mole ratio of NH_4Cl to NH_3 would buffer a solution at pH 9.25?

16.76 How many grams of ammonium chloride would have to be dissolved in 500 mL of 0.20 M NH_3 to prepare a solution buffered at pH 10.00?

16.77 How many grams of ammonium chloride have to be dissolved into 125 mL of 0.10 M NH_3 to make it a buffer with a pH of 9.15?

16.78 Suppose 25.00 mL of 0.100 M HCl is added to an acetate buffer prepared by dissolving 0.100 mol of acetic acid and 0.110 mol of sodium acetate in 500 mL of solution. What are the initial and final pH values? What would be the pH if the same amount of HCl solution were added to 125 mL of pure water?

16.79 How many milliliters of 0.15 M HCl would have to be added to 100 mL of the buffer described in Exercise 16.78 to make the pH decrease by 0.05 pH unit? How many milliliters of the same HCl solution would, if added to 100 mL of pure water, make the pH decrease by 0.05 pH unit?

Acid–Base Titrations

16.80 What can make the titrated solution at the equivalence point in an acid–base titration have a pH not equal to 7.00? How does this possibility affect the choice of an indicator?

16.81 Explain why ethyl red is a better indicator than phenolphthalein in the titration of dilute ammonia by dilute hydrochloric acid.

16.82 What is a good indicator for titrating potassium hydroxide with hydrobromic acid? Explain.

16.83 In the titration of an acid with a base, what condition concerning the quantities of reactants ought to be true at the equivalence point?

16.84 When 50 mL of 0.10 M formic acid is titrated with 0.10 M sodium hydroxide, what is the pH at the equivalence point? (Be sure to take into account the change in volume during the titration.) What is a good indicator for this titration?

16.85 When 25 mL of 0.10 M aqueous ammonia is titrated with 0.10 M hydrobromic acid, what is the pH at the equivalence point? What is a good indicator?

*16.86 For the titration of 25.00 mL of 0.1000 M HCl with 0.1000 M NaOH, calculate the pH of the resulting solution after each of the following quantities of base has been added to the original solution (you must take into account the change in total volume). Construct a graph showing the titration curve for this experiment.

(a) 0 mL   (d) 24.99 mL   (g) 25.10 mL
(b) 10.00 mL   (e) 25.00 mL   (h) 26.00 mL
(c) 24.90 mL   (f) 25.01 mL   (i) 50.00 mL

*16.87 For the titration of 25.00 mL of 0.1000 M acetic acid with 0.1000 M NaOH, calculate the pH (a) before the

addition of any NaOH solution, (b) after 10.00 mL of the base has been added, (c) after half of the $HC_2H_3O_2$ has been neutralized, and (d) at the equivalence point.

*16.88 For the titration of 25.00 mL of 0.1000 M ammonia with 0.1000 M HCl, calculate the pH (a) before the addition of any HCl solution, (b) after 10.00 mL of the acid has been added, (c) after half of the NH_3 has been neutralized, and (d) at the equivalence point.

Additional Exercises

16.89 Find among the pH values on the right which value most closely matches the pH of the solutions described in the first column.

A. 2.0×10^{-3} M HCl	11.70
B. 5.0×10^{-4} M NaOH	4.70
C. 0.115 M $HC_2H_3O_2$ that is	2.70
also 0.115 M $NaC_2H_3O_2$	10.70

16.90 What is the pH of a solution that is 0.100 M in HCl and also 0.125 M in $HC_2H_3O_2$? What is the concentration of acetate ion in this solution?

16.91 A solution is prepared by mixing 300 mL of 0.500 M NH_3 and 100 mL of 0.500 M HCl. Assuming that the volumes are additive, what is the pH of the resulting mixture?

16.92 What is the pH of 0.120 M NH_4NO_3?

16.93 A solution is prepared by dissolving 15.0 g of $HC_2H_3O_2$ and 25.0 g of $NaC_2H_3O_2$ in 750 mL of solution (the final volume).
(a) What is the pH of the solution?
(b) What is the pH of the solution after 25.00 mL of 0.25 M NaOH is added?
(c) What is the pH of the original solution after 25.0 mL of 0.40 M HCl is added?

*16.94 For an experiment involving what happens to the growth of a particular fungus in a slightly acidic medium, a biochemist needs 250 mL of an acetate buffer with a pH of 5.12. The buffer solution has to be able to hold the pH to within ± 0.10 pH unit of 5.12 even if 0.0100 mol of NaOH or 0.0100 mol of HCl enters the solution.
(a) What is the minimum number of grams of acetic acid and of sodium acetate dihydrate that must be used to prepare the buffer?
(b) Describe the buffer by giving its molarity in acetic acid and its molarity in sodium acetate.
(c) What is the pH of an unbuffered solution made by adding 0.0100 mol of NaOH to 250 mL of pure water?
(d) What is the pH of an unbuffered solution made by adding 0.0100 mol of HCl to 250 mL of pure water?

16.95 Predict whether the pH of 0.120 M NH_4CN is greater than, less than, or equal to 7.00. Give your reasons.

*16.96 What is the approximate freezing point of a 0.50 M solution of dichloroacetic acid, $HC_2HO_2Cl_2$ ($K_a = 5.0 \times 10^{-2}$)? Assume the density of the solution is 1.0 g/mL.

*16.97 How many milliliters of ammonia gas measured at 25 °C and 740 torr must be dissolved in 250 mL of 0.050 M HNO_3 to give a solution with a pH of 9.26?

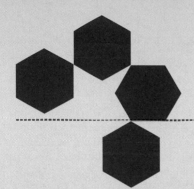

CHEMICALS IN USE 13
Buffers and Breathing

In biochemistry, *respiration* refers to the transport of the blood gases, oxygen and carbon dioxide, and to their reactions in *metabolism*—the manifold of all chemical reactions in the body. Small changes in the pH of the blood have deadly consequences for respiration because the H^+ ion participates in the equilibria of respiration in its earliest phases. Two of them are the following, where H*Hb* represents hemoglobin, a protein in blood, and HbO_2^- is oxyhemoglobin.[1]

$$HHb(aq) + O_2(aq) \rightleftharpoons H^+(aq) + HbO_2^-(aq) \quad (1)$$

$$HCO_3^-(aq) + H^+(aq) \rightleftharpoons H_2O + CO_2(aq) \quad (2)$$

Oxyhemoglobin of Equilibrium 1 is the form in which oxygen is carried in the blood. The bicarbonate ion and carbon dioxide of Equilibrium 2 are the chief forms in which carbon dioxide is transported in the blood. Very little carbonic acid is present in blood. However, $CO_2(aq)$ "stands in" for H_2CO_3 as the chief base neutralizer in blood because it is able to *directly* neutralize hydroxide ion by the following reaction.

$$CO_2(aq) + OH^-(aq) \longrightarrow HCO_3^-(aq) \quad (3)$$

Both Equilibria 1 and 2 shift to the right in the lungs and to the left in actively metabolizing cells. The shifts occur best when the pH of the blood is 7.35 ± 0.05. To hold the blood's pH within this range, the system relies heavily on buffers. Without the blood buffers, we could not live.

THE BLOOD'S CARBONATE BUFFER

Together, $CO_2(aq)$ and $HCO_3^-(aq)$ constitute one of the important buffers in blood, the carbonate buffer. It neutralizes extra base by the reaction of Equation 3. Extra acid is neutralized by the bicarbonate ion.

$$HCO_3^-(aq) + H^+(aq) \longrightarrow CO_2(aq) + H_2O$$

The relevant buffer equation, which can be derived from the defining equation for the first acid ioniza-

[1] We have simplified the first equilibrium by showing only one O_2 instead of the four O_2 molecules known to be taken up by each H*Hb* molecule.

tion constant of carbonic acid (see page 734), is as follows, only we use CO_2 for H_2CO_3.

$$pH = pK_a - \log \frac{[CO_2]}{[HCO_3^-]} \quad (4)$$

The value of pK_a for the first ionization of carbonic acid is 6.4, but *it cannot be used here* because it is for a solution at a temperature of 25 °C, not 37 °C, the temperature of the blood. Specialists in blood chemistry have agreed to an "apparent" pK_a value of 6.1, so we can write the following equation for the CO_2/HCO_3^- buffer in blood.

$$pH = 6.1 - \log \frac{[CO_2]}{[HCO_3^-]} \quad \text{(at 37 °C)}$$

Normally, in arterial blood, $[HCO_3^-]$ is 24 mmol L^{-1}, and $[CO_2]$ is 1.2 mmol L^{-1}. The pH of human arterial blood is then calculated to be 7.4.

$$pH = 6.1 - \log \frac{[1.2 \text{ mmol } L^{-1}]}{[24 \text{ mmol } L^{-1}]}$$
$$= 7.4$$

Statistically, the average human arterial blood in health is 7.35, so the calculated and the observed values agree well.

THE RESPONSE OF THE BLOOD TO EXCESS ACID

To illustrate how the carbonate buffer works, let's suppose that the blood is challenged with a sudden influx of acid in the equivalent of 10 mmol of H^+ ion per liter of blood. (Acids are metabolically produced at higher rates in some conditions, like diabetes.) An extra 10 mmol per liter of H^+ would neutralize 10 mmol L^{-1} of HCO_3^- and so reduce $[HCO_3^-]$ from 24 mmol L^{-1} to 14 mmol L^{-1}. The HCO_3^- would change almost entirely to $CO_2(aq)$, so 10 mmol L^{-1} of *new* $CO_2(aq)$ would appear in the blood. If the new CO_2 could not be removed by breathing, the value of $[CO_2]$ would increase by 10 mmol L^{-1}, from 1.2 mmol L^{-1} to 11.2 mmol L^{-1}. This would be fatal because, by Equation 4, the resulting pH of the blood would be 6.2, much too low to permit life to continue. (Even a drop to a pH of 7.2 is serious.)

$$pH = 6.1 - \log \frac{[11.2 \text{ mmol } L^{-1}]}{[14 \text{ mmol } L^{-1}]}$$
$$= 6.2$$

However, essentially all the $CO_2(aq)$ leaves the blood in the lungs and is breathed out as $CO_2(g)$. Although the level of HCO_3^- stays the same, at 14 mmol L^{-1}, the level of $CO_2(aq)$ quickly drops back to 1.2 mmol L^{-1}. So the pH quickly bounces back up to 7.2.

$$pH = 6.1 - \log \frac{[1.2 \text{ mmol } L^{-1}]}{[14 \text{ mmol } L^{-1}]}$$
$$= 7.2$$

A blood pH of 7.2 is still dangerously low for health, but it doesn't cause immediate death. The breathing mechanism provides further upward adjustment of the blood pH. The body, when healthy, responds quickly to a lowering of the blood pH by increasing the rate of breathing. This works to remove more CO_2. It's quite common for rapid breathing to pull the level of $CO_2(aq)$ in blood from 1.2 mmol L^{-1} down to 0.70 mmol L^{-1}, sometimes a bit lower. This would bring the pH of the blood in the example situation—the sudden influx of 10 mmol L^{-1} of H^+—back to 7.4.

$$pH = 6.1 - \log \frac{[0.7 \text{ mmol } L^{-1}]}{[14 \text{ mmol } L^{-1}]}$$
$$= 7.4$$

Thus two mechanisms have protected the system against the otherwise lethal influx of 10 mmol L^{-1} of strong acid. Neither of these two mechanisms could save the situation alone.

HIGH-ALTITUDE DISEASE

Those who take themselves to high altitude (> 3000 m) without having allowed sufficient time for becoming acclimatized have the opposite problem, a *loss* of "acid" from the body, not a gain. They are in an environment with a partial pressure of oxygen lower than at sea level (Figure 13a). To compensate, a climber breathes at a higher rate. But overbreathing at high altitude may cause the loss of CO_2 to occur too rapidly, *and this has the same effect as a loss of acid.* (Remember that CO_2 is a base neutralizer.) The loss of acid is equivalent to a *gain* of base, so the pH of the blood *increases* with overbreathing at high altitude. One metabolic problem is now an impaired ability to *release* O_2 from HbO_2^- at tissues needing oxygen. This is because a lack of H^+ prevents Equilibrium 1 from shifting to the *left* as needed and where needed. The

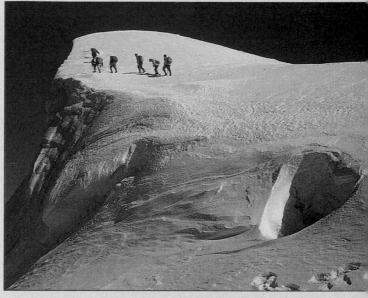

FIGURE 13a

Strenuous exertion at high altitude places a heavy burden on the ability of the blood's carbonate buffer. Overbreathing removes too much base-neutralizing CO_2 from the blood, and the pH of the blood increases.

combination of this problem, the release of oxygen, and the lower availability of oxygen at high altitude makes the situation life threatening. The only remedy (other than supplemental oxygen) is a *prompt* return to a lower elevation.

Suggested Readings

H. Valtin and F. J. Gennari, *Acid–Base Disorders*, Little, Brown and Company, Boston, 1987.

M. P. Ward, J. S. Milledge, and J. B. West, *High Altitude Medicine and Physiology*, University of Pennsylvania Press, Philadelphia, 1989.

Questions

1. Suppose that the blood is suddenly made to accept 12 mmol L^{-1} of hydrochloric acid.
   (a) What is the resulting pH of the blood if no CO_2 is allowed to escape?
   (b) What is the resulting pH of the blood if 12 mmol L^{-1} of CO_2 can be quickly exhaled?
   (c) What additional change will occur to remove more CO_2?

2. In uncontrolled hysterics, the body involuntarily removes CO_2 at an excessive rate. How does this affect the pH of the blood, increase it or decrease it? Explain.

3. By taking an overdose of a narcotic, a person began to breathe in a slow and shallow manner. How does this affect the pH of the blood, increase it or decrease it? Explain.

Brightly colored stalks of coral are formed by marine organisms that extract calcium ions and carbonate ions from sea water to form insoluble calcium carbonate, the chief constituent of the coral. In this chapter we discuss the solubility equilibria of ionic compounds such as $CaCO_3$, as well as the equilibria involved in the ionization of weak polyprotic acids such as H_2CO_3, which provides the carbonate ions.

Chapter 17

Solubility and Simultaneous Equilibria

17.1
Ionization of Polyprotic Acids
17.2
The pH of Solutions of Salts of Polyprotic Acids

17.3
Solubility Equilibria for Salts
17.4
Solubility Equilibria of Metal Oxides and Sulfides

In Chapter 16 our focus was on acids and bases that involve the exchange of a single proton. Examples include acetic acid, which is able to donate one H^+ per molecule, and the acetate ion, which is a base able to accept one H^+. We now turn our attention to more complex situations, and we begin with equilibria involving polyprotic acids—acids such as H_2CO_3 and H_3PO_4 that have the ability to donate two or more protons per molecule. In this chapter we will also discuss solubility equilibria and ways that we can selectively precipitate metal ions by manipulating systems in which two or more equilibria exist simultaneously.

Equilibria Involving Polyprotic Acids

When polyprotic acids ionize, they do so by a series of steps, each of which releases one proton. For weak polyprotic acids, such as H_2CO_3 and H_3PO_4, each step is an equilibrium. Even sulfuric acid, which we consider a strong acid, is not completely ionized. Loss of the first proton to yield the HSO_4^- ion is complete, but the loss of the second proton is incomplete and involves an equilibrium. In this section, we will focus our attention on weak polyprotic acids.

Let's begin with the weak diprotic acid, H_2CO_3. In water, the acid ionizes in two steps each of which is an equilibrium that transfers an H^+ ion to a water molecule.

$$H_2CO_3 + H_2O \rightleftharpoons H_3O^+ + HCO_3^-$$

$$HCO_3^- + H_2O \rightleftharpoons H_3O^+ + CO_3^{2-}$$

As usual, we can use H^+ in place of H_3O^+ and simplify these equations to give

$$H_2CO_3 \rightleftharpoons H^+ + HCO_3^-$$

$$HCO_3^- \rightleftharpoons H^+ + CO_3^{2-}$$

Each step has its own ionization constant, K_a, which we identify as K_{a_1} for the first step and K_{a_2} for the second. For carbonic acid,

$$K_{a_1} = \frac{[H^+][HCO_3^-]}{[H_2CO_3]} = 4.5 \times 10^{-7}$$

$$K_{a_2} = \frac{[H^+][CO_3^{2-}]}{[HCO_3^-]} = 4.7 \times 10^{-11}$$

Notice that each ionization makes a contribution to the total molar concentration of H^+, and one of our goals here is to relate the K_a values and the concentration of the acid to this molarity. At first glance, this seems to be a formidable task, but certain simplifications are justified that make the problem relatively simple to solve.

Simplifications in Calculations Involving Polyprotic Acids

K_{a_1} is nearly 10,000 times greater than K_{a_2}.

The principal factor that simplifies calculations involving polyprotic acids is the large differences between successive ionization constants. Notice that for H_2CO_3 K_{a_1} is much larger than K_{a_2}. Similar differences between K_{a_1} and K_{a_2} are observed for almost all diprotic acids. The reason for this is that an H^+ is lost much more easily from the neutral H_2A molecule than from the HA^- ion. The stronger attraction of the opposite charges inhibits the second ionization. Typically, K_{a_1} is greater than K_{a_2} by a factor of between 10^4 and 10^5, as the data in Table 17.1 show. For a triprotic acid, the second acid ionization constant is

Additional K_a values of polyprotic acids are located in Appendix E.7.

similarly greater than the third, as illustrated in Table 17.1 by phosphoric acid, H_3PO_4.

Because K_{a_1} is so much larger than K_{a_2}, virtually all the H^+ in the solution comes from the first step in the ionization. In other words,

$$[H^+]_{total} = [H^+]_{first step} + [H^+]_{second step}$$

$$[H^+]_{first step} \gg [H^+]_{second step}$$

TABLE 17.1 Acidic Ionization Constants for Polyprotic Acids

Name	Formula	Acid Ionization Constant for Successive Ionizations (25°C)		
		1st	2nd	3rd
Carbonic acid	H_2CO_3	4.5×10^{-7}	4.7×10^{-11}	
Hydrogen sulfide	$H_2S(aq)$	1.1×10^{-7}	1.3×10^{-13}	
Phosphoric acid	H_3PO_4	7.1×10^{-3}	6.3×10^{-8}	4.5×10^{-13}
Sulfuric acid	H_2SO_4	Large	1.0×10^{-2}	
Selenic acid	H_2SeO_4	Large	1.2×10^{-2}	
Telluric acid	H_6TeO_6	2×10^{-8}	1×10^{-11}	
Sulfurous acid	H_2SO_3	1.2×10^{-2}	6.6×10^{-8}	
Selenous acid	H_2SeO_3	3.5×10^{-3}	5×10^{-8}	
Tellurous acid	H_2TeO_3	3×10^{-3}	2×10^{-8}	
Ascorbic acid	$H_2C_6H_6O_6$	6.8×10^{-5}	2.7×10^{-12}	
Oxalic acid	$H_2C_2O_4$	6.5×10^{-2}	6.1×10^{-5}	
Citric acid	$H_3C_6H_5O_7$	7.4×10^{-4}	1.7×10^{-5}	4.0×10^{-7}

Therefore, we make the approximation that

$$[\text{H}^+]_{\text{total}} \approx [\text{H}^+]_{\text{first step}}$$

This means that as far as calculating the H^+ concentration is concerned, we can treat the acid as though it were a monoprotic acid and ignore the second step in the ionization. Example 17.1 illustrates how this is applied.

EXAMPLE 17.1
Calculating the Concentrations of Solute Species in a Solution of a Polyprotic Acid

Calculate the concentrations of all the species produced in the ionization of $0.050 \ M \ \text{H}_2\text{CO}_3$ and the pH of the solution.

ANALYSIS As in any equilibrium problem, we begin with the chemical equations and the K_a expressions.

$$\text{H}_2\text{CO}_3 \rightleftharpoons \text{H}^+ + \text{HCO}_3^- \qquad K_{a_1} = \frac{[\text{H}^+][\text{HCO}_3^-]}{[\text{H}_2\text{CO}_3]} = 4.5 \times 10^{-7}$$

$$\text{HCO}_3^- \rightleftharpoons \text{H}^+ + \text{CO}_3^{2-} \qquad K_{a_2} = \frac{[\text{H}^+][\text{CO}_3^{2-}]}{[\text{HCO}_3^-]} = 4.7 \times 10^{-11}$$

We will want to calculate the concentrations of H^+, HCO_3^-, and CO_3^{2-} and the concentration of H_2CO_3 that remains at equilibrium. As noted in the preceding discussion, the problem is simplified considerably by assuming that the second reaction occurs to a negligible extent compared to the first. This assumption (which we will justify later in the problem) allows us to calculate the H^+ and HCO_3^- concentrations as though the solute were a monoprotic acid. This type of problem is one we've worked on in Chapter 16.

To obtain $[\text{CO}_3^{2-}]$, we will have to use the second step in the ionization. Once again, the large difference between K_{a_1} and K_{a_2} permits some simplifications. The first involves $[\text{H}^+]$, which we have already examined.

$$[\text{H}^+]_{\text{total}} \approx [\text{H}^+]_{\text{first step}}$$

The second involves $[\text{HCO}_3^-]$, for which we can write

$$[\text{HCO}_3^-]_{\text{total}} = [\text{HCO}_3^-]_{\substack{\text{formed in} \\ \text{first step}}} - [\text{HCO}_3^-]_{\substack{\text{lost in} \\ \text{second step}}}$$

However, because $K_{a_1} \gg K_{a_2}$, we expect that

$$[\text{HCO}_3^-]_{\substack{\text{formed in} \\ \text{first step}}} \gg [\text{HCO}_3^-]_{\substack{\text{lost in} \\ \text{second step}}}$$

Therefore, we can make the approximation that

$$[\text{HCO}_3^-]_{\text{total}} \approx [\text{HCO}_3^-]_{\substack{\text{formed in} \\ \text{first step}}}$$

SOLUTION Treating H_2CO_3 as though it were a monoprotic acid yields a problem similar to many others we worked in Chapter 16. We know some of the H_2CO_3 ionizes; let's call this amount x. If x moles per liter of H_2CO_3 ionizes, then x moles per liter of H^+ and HCO_3^- are formed, and the concentration of H_2CO_3 is reduced by x.

$$[\text{H}^+] = [\text{HCO}_3^-] = x$$

$$[\text{H}_2\text{CO}_3] = 0.050 - x$$

Substituting into the expression for K_{a_1} gives

$$\frac{(x)(x)}{(0.050 - x)} = 4.7 \times 10^{-7}$$

Notice that $0.050 < 400 \times K_{a_1}$, so it is safe to use the initial concentration, 0.050 *M*, as though it were the equilibrium concentration.

Because K_{a_1} is so small, we expect that $0.050 - x \approx 0.050$. (This is the usual approximation we made in solving problems of this kind in Chapter 16.) The algebra then reduces to

$$\frac{x^2}{0.050} = 4.5 \times 10^{-7}$$

Solving for x gives $x = 1.5 \times 10^{-4}$, so $[H^+] = [HCO_3^-] = 1.5 \times 10^{-4}$ *M*. From this we can now calculate the pH.

$$pH = -\log(1.5 \times 10^{-4}) = 3.82$$

Now let's calculate the concentration of CO_3^{2-}. We obtain this from the expression for K_{a_2}

$$K_{a_2} = \frac{[H^+][CO_3^{2-}]}{[HCO_3^-]} = 4.7 \times 10^{-11}$$

In our analysis, we concluded that we can use the following approximations (where "eq" stands for equilibrium):

$$[H^+]_{eq} \approx [H^+]_{first\,step}$$

$$[HCO_3^-]_{eq} \approx [HCO_3^-]_{first\,step}$$

Substituting the values obtained above gives us

$$\frac{1.5 \times 10^{-4}\,[CO_3^{2-}]}{1.5 \times 10^{-4}} = 4.7 \times 10^{-11}$$

$$[CO_3^{2-}] = 4.7 \times 10^{-11} = K_{a_2}$$

Let's summarize the results. At equilibrium, we have

$$[H_2CO_3] = 0.050\ M$$

$$[H^+] = [HCO_3^-] = 1.5 \times 10^{-4}\ M \text{ (and pH = 3.82)}$$

$$[CO_3^{2-}] = 4.7 \times 10^{-11}\ M$$

CHECKING THE SIMPLIFYING ASSUMPTIONS In the second step of the ionization, H^+ and CO_3^{2-} are formed in equal amounts from HCO_3^-. This means that the contribution to $[H^+]$ *by the second step* equals $[CO_3^{2-}]$. Therefore, $[H^+]_{second\,step} = 4.7 \times 10^{-11}$ *M*. Comparing this value to the $[H^+]$ computed using K_{a_1}, we see that the second step of the ionization contributes a negligible amount to the total hydrogen ion concentration in the solution.

$$[H^+]_{total} = [H^+]_{first\,step} + [H^+]_{second\,step}$$

$$= (1.5 \times 10^{-4}\ M) + (4.7 \times 10^{-11}\ M)$$

$$= 1.5 \times 10^{-4}\ M \text{ (correctly rounded)}$$

Thus, we were safe in treating H_2CO_3 as though it were a monoprotic acid when we computed the H^+ concentration.

Our second approximation,

$$[HCO_3^-]_{total} \approx [HCO_3^-]_{formed\,in\,first\,step}$$

was also valid. The concentration of HCO_3^- formed in the first step equals 1.5×10^{-4} *M*. This value is decreased by 4.7×10^{-11} mol L^{-1} when the H^+ and CO_3^{2-} form in the second step. Because 4.7×10^{-11} is negligible compared to 1.5×10^{-4}, the equilibrium concentration of HCO_3^- equals the concentration computed using K_{a_1}.

One of the most interesting observations we can derive from the preceding example is that *in a solution that contains a polyprotic acid **as the only solute**, the concentration of the anion formed in the second step of the ionization is numerically the same as K_{a_2}.*

■ **Practice Exercise 1** Ascorbic acid (vitamin C) is a diprotic acid, $H_2C_6H_6O_6$. See Table 17.1. Calculate $[H^+]$, pH, and $[C_6H_6O_6{}^{2-}]$ in a 0.10 M solution of ascorbic acid.

Solutions of Weak Diprotic Acids and a Strong Acid

Later in this chapter we will discuss equilibria involving the solubilities of salts and how it is possible to separate metal ions selectively by adjusting the concentration of an anion so that it precipitates one metal ion but not another. For example, we might wish to separate calcium and magnesium ions by making use of the relative insolubilities of their salts with the oxalate ion, $C_2O_4{}^{2-}$. This can be done by carefully controlling the concentration of oxalate ion in the solution. We are able to do this by making use of Le Châtelier's principle. Consider the equilibria for the ionization of oxalic acid.

$$H_2C_2O_4 \rightleftharpoons H^+ + HC_2O_4{}^-$$

$$HC_2O_4{}^- \rightleftharpoons H^+ + C_2O_4{}^{2-}$$

If we increase the H^+ concentration by adding a strong acid, we will shift both of these equilibria to the left, which will lower the $C_2O_4{}^{2-}$ concentration. If we raise the pH and thereby decrease the H^+ concentration, both equilibria will shift to the right and the concentration of $C_2O_4{}^{2-}$ will increase. Thus, by controlling the H^+ concentration in the solution, we are able to adjust the oxalate ion concentration.

A simple relationship between the H^+ concentration and the oxalate ion concentration is obtained by multiplying the expressions for K_{a_1} and K_{a_2}.

Caution: We can use this combined expression *only* if we know two of the three concentration terms and wish to calculate the third.

$$K_{a_1} \times K_{a_2} = \frac{[H^+][HC_2O_4{}^-]}{[H_2C_2O_4]} \times \frac{[H^+][C_2O_4{}^{2-}]}{[HC_2O_4{}^-]} = \frac{[H^+]^2[C_2O_4{}^{2-}]}{[H_2C_2O_4]}$$

Similar expressions apply for other diprotic acids, as the following example illustrates.

What must the pH of a solution of 0.10 M $H_2C_2O_4$ be to make the $C_2O_4{}^{2-}$ concentration be 1.0×10^{-4} M?

SOLUTION We will use the expression for $K_{a_1} \times K_{a_2}$ obtained in the preceding paragraph, using $[H_2C_2O_4] = 0.10$ M and $[C_2O_4{}^{2-}] = 1.0 \times 10^{-4}$ M. The values of K_{a_1} and K_{a_2} for $H_2C_2O_4$ are given in Table 17.1 as $K_{a_1} = 6.5 \times 10^{-2}$ and $K_{a_2} = 6.1 \times 10^{-5}$. Their product is 4.0×10^{-6}. Therefore, substituting

$$4.0 \times 10^{-6} = \frac{[H^+]^2(1.0 \times 10^{-4})}{(0.10)}$$

Solving for $[H^+]$ gives

$$[H^+] = 0.063 \ M$$

$$pH = -\log(0.063)$$

$$= 1.20$$

To obtain this pH we would probably add a strong acid to the solution.

EXAMPLE 17.2
Controlling the Concentration of the Anion of a Diprotic Acid

The pH of a 0.10 M solution of oxalic acid is 2.11, so a lower pH would have to be obtained by adding additional H^+ from a strong acid.

■ **Practice Exercise 2** What concentration of $C_2O_4^{2-}$ would exist in 0.20 M $H_2C_2O_4$ if the solution is buffered to a pH of 2.00?

17.2
**THE pH OF
SOLUTIONS OF
SALTS OF
POLYPROTIC ACIDS**

The salts of polyprotic acids discussed here are limited to those with *nonacidic* cations, like their sodium salts or others of the metal ions of Group IA. We want to deal, in other words, only with salts in which just the anion affects the pH of a solution. Sodium carbonate, Na_2CO_3, is a typical example. It is the salt of H_2CO_3, and the salt's carbonate ion is a Brønsted base that is responsible for *two* equilibria that furnish OH^- ion and so affect the pH of the solution. It gives the following equilibrium directly, and some OH^- is produced.

$$CO_3^{2-}(aq) + H_2O \rightleftharpoons HCO_3^-(aq) + OH^-(aq) \qquad (17.1)$$

The bicarbonate ion is the other product of the equilibrium, however, *and HCO_3^- is also a Brønsted base.* So it generates more OH^- ion by the following equilibrium.

$$HCO_3^-(aq) + H_2O \rightleftharpoons H_2CO_3(aq) + OH^-(aq) \qquad (17.2)$$

The question now is, "How should we calculate the pH of a solution in which OH^- is produced by more than one equilibrium like this?"

To answer this question, let us first see if there is a legitimate way to simplify matters. Let's compare the K_b values for the successive equilibria. The equation for K_b for the equilibrium in which CO_3^{2-} is the base, Equation 17.1, is the following.

$$K_{b_1} = \frac{[HCO_3^-][OH^-]}{[CO_3^{2-}]} = 2.1 \times 10^{-4}$$

This K_{b_1} value for CO_3^{2-}, 2.1×10^{-4}, is calculated from the relationship that applies to any conjugate acid–base pair, $K_aK_b = K_w$, or

$$K_b = \frac{K_w}{K_a}$$

The conjugate acid of CO_3^{2-} is HCO_3^-, and its *acid* ionization constant is the K_{a_2} for carbonic acid: $K_{a_2} = 4.7 \times 10^{-11}$ (Table 17.1). Therefore,

$$K_{b_1} = \frac{K_w}{K_{a_2}} = \frac{1.0 \times 10^{-14}}{4.7 \times 10^{-11}} = 2.1 \times 10^{-4}$$

For the base ionization constant of HCO_3^-, we use the second equilibrium above, Equation 17.2, and write

$$K_{b_2} = \frac{[H_2CO_3][OH^-]}{[HCO_3^-]} = 2.2 \times 10^{-8}$$

The value of K_{b_2} is also found by using $K_aK_b = K_w$. H_2CO_3 is the conjugate acid of HCO_3^-, so we must use $K_{a_1} = 4.5 \times 10^{-7}$ (Table 17.1). Therefore, the K_{b_2} value is calculated as follows.

$$K_{b_2} = \frac{K_w}{K_{a_1}} = \frac{1.0 \times 10^{-14}}{4.5 \times 10^{-7}} = 2.2 \times 10^{-8}$$

Now we can compare the two K_b values. For CO_3^{2-}, $K_{b_1} = 2.1 \times 10^{-4}$, and for the (much) weaker base, HCO_3^-, K_{b_2} is 2.2×10^{-8}. Thus CO_3^{2-} has a K_b nearly 10,000 times that of HCO_3^-. This means that the reaction of CO_3^{2-} with water (Equation 17.1) generates relatively far more OH^- than the reaction of HCO_3^- (Equation 17.2). The contribution of the latter to the total pool of OH^- is relatively so small that we can safely ignore it. The simplification here is very much the same as in our treatment of the ionization of weak polyprotic acids.

$$CO_3^{2-} + H_2O \rightleftharpoons HCO_3^- + OH^-$$
$$K_{b_1} = 2.1 \times 10^{-4}$$
$$HCO_3^- + H_2O \rightleftharpoons H_2CO_3 + OH^-$$
$$K_{b_2} = 2.2 \times 10^{-8}$$

$$[OH^-]_{\text{total}} = [OH^-]_{\substack{\text{formed in} \\ \text{first step}}} + [OH^-]_{\substack{\text{formed in} \\ \text{second step}}}$$

$$[OH^-]_{\substack{\text{formed in} \\ \text{first step}}} \gg [OH^-]_{\substack{\text{formed in} \\ \text{second step}}}$$

$$[OH^-]_{\text{total}} \approx [OH^-]_{\substack{\text{formed in} \\ \text{first step}}}$$

Thus, when we calculate the pH of a solution of a basic anion of a polyprotic acid, *we may work exclusively with this anion and not with others that can form in additional equilibria.* Let's study an example of how this works.

EXAMPLE 17.3
Calculating the pH of a Solution of a Salt of a Weak Diprotic Acid

What is the pH of a 0.15 M solution of Na_2CO_3?

ANALYSIS We can ignore the reaction of HCO_3^- with water so the only relevant equilibrium is

$$CO_3^{2-} + H_2O \rightleftharpoons HCO_3^- + OH^- \qquad K_{b_1} = \frac{[HCO_3^-][OH^-]}{[CO_3^{2-}]} = 2.1 \times 10^{-4}$$

(The value of K_{b_1} was calculated in the discussion preceding this example.) Also, the initial concentration of CO_3^{2-} (0.15 M) is larger than $400 \times K_{b_1}$, so we can use 0.15 M as the equilibrium concentration of CO_3^{2-}.

$400 \times K_{a_1} = 0.084$. Because $0.15 > 0.084$, we are able to use the initial concentration of CO_3^{2-} as the equilibrium concentration.

SOLUTION Some of the CO_3^{2-} will react; we'll let this be x mol L^{-1}. The concentration table, then, is as follows.

	$CO_3^{2-} + H_2O \rightleftharpoons$	HCO_3^- +	OH^-
Initial concentrations (M)	0.15	0	0
Changes in concentrations caused by the ionization (M)	$-x$	$+x$	$+x$
Final concentrations at equilibrium (M)	$(0.15 - x) \approx 0.15$	x	x

Let's now insert the values in the last row of the table into the equilibrium expression for the carbonate ion.

$$\frac{[HCO_3^-][OH^-]}{[CO_3^{2-}]} = \frac{(x)(x)}{0.15} = 2.1 \times 10^{-4}$$

$$x = 5.6 \times 10^{-3}$$

Notice that the value of x really is sufficiently smaller than 0.15 to justify our using 0.15 as a good approximation of $(0.15 - x)$. Because $x = 5.6 \times 10^{-3}$

$$[OH^-] = 5.6 \times 10^{-3} \text{ mol L}^{-1}$$

$$pOH = -\log(5.6 \times 10^{-3}) = 2.25$$

Hence,

$$pH = 14.00 - 2.25$$
$$= 11.75$$

Thus, the pH of 0.15 M Na$_2$CO$_3$ is calculated to be 11.75. Once again we see that an aqueous solution of a salt composed of a basic anion and a neutral cation is basic.

■ **Practice Exercise 3** What is the pH of a 0.20 M solution of Na$_2$SO$_3$ at 25 °C? For the diprotic acid, H$_2$SO$_3$, $K_{a_1} = 1.2 \times 10^{-2}$ and $K_{a_2} = 6.6 \times 10^{-8}$.

17.3 SOLUBILITY EQUILIBRIA FOR SALTS

Throughout Chapter 16 and so far in Chapter 17 we have been concerned with quantitative ways of dealing with equilibria involving weak acids and bases. In these studies we found that equilibrium equations and equilibrium constants can be used in a variety of useful calculations.

We now turn our attention to sparingly soluble substances and to *quantitative* ways to describe "How soluble?" In Chapter 4 you learned to classify salts as "soluble" or "insoluble" according to a set of solubility rules. The rules tell us, for example, that PbCl$_2$ and AgCl are both insoluble salts. However, if chloride ion is slowly added to a solution containing both Pb^{2+} and Ag$^+$ ions, nearly all of the Ag$^+$ is precipitated as AgCl *before* any Pb^{2+} separates from the solution as PbCl$_2$. This happens because AgCl is much less soluble than PbCl$_2$. To study such differences in solubility, we must examine solubility equilibria quantitatively.

The solubility equilibria for sparingly soluble metal oxides and sulfides are discussed in Section 17.4.

Solubility Product Constant, K_{sp}

Let's begin with the compound AgCl. When placed in water, a small amount of this salt dissolves and the following equilibrium is established once the solution becomes saturated.

$$AgCl(s) \rightleftharpoons Ag^+(aq) + Cl^-(aq)$$
(In a saturated solution of AgCl)

In dealing with solubility equilibria, we are considering saturated solutions of salts.

We can write an equilibrium expression following the procedure developed in Chapter 15.

$$\frac{[Ag^+(aq)][Cl^-(aq)]}{[AgCl(s)]} = K_c$$

This expression can be simplified, however, because the reaction is heterogeneous. Pure solid AgCl rests at the bottom of the solution, and as you learned in Chapter 15, the concentration of a substance within a pure solid is constant. Therefore, the denominator of the mass action expression, [AgCl(s)], is a constant that can be combined with K_c to give a new constant.

$$[Ag^+(aq)][Cl^-(aq)] = K_c [AgCl(s)] = \text{a new constant}$$

Solubility product constant

This new constant is called the **solubility product constant, K_{sp},** because it equals the product of concentration terms for the ions dissolved in a *saturated*

TABLE 17.2 Solubility Product Constants

Type	Salt Ions of Salt	K_{sp} (25°C)
Halides	$CaF_2 \rightleftharpoons Ca^{2+} + 2F^-$	3.9×10^{-11}
	$PbF_2 \rightleftharpoons Pb^{2+} + 2F^-$	3.6×10^{-8}
	$AgCl \rightleftharpoons Ag^+ + Cl^-$	1.8×10^{-10}
	$AgBr \rightleftharpoons Ag^+ + Br^-$	5.0×10^{-13}
	$AgI \rightleftharpoons Ag^+ + I^-$	8.3×10^{-17}
	$PbCl_2 \rightleftharpoons Pb^{2+} + 2Cl^-$	1.7×10^{-5}
	$PbBr_2 \rightleftharpoons Pb^{2+} + 2Br^-$	2.1×10^{-6}
	$PbI_2 \rightleftharpoons Pb^{2+} + 2I^-$	7.9×10^{-9}
Hydroxides	$Al(OH)_3 \rightleftharpoons Al^{3+} + 3OH^-$	$3 \times 10^{-34}(a)$
	$Ca(OH)_2 \rightleftharpoons Ca^{2+} + 2OH^-$	6.5×10^{-6}
	$Fe(OH)_2 \rightleftharpoons Fe^{2+} + 2OH^-$	7.9×10^{-16}
	$Fe(OH)_3 \rightleftharpoons Fe^{3+} + 3OH^-$	1.6×10^{-39}
	$Mg(OH)_2 \rightleftharpoons Mg^{2+} + 2OH^-$	7.1×10^{-12}
	$Zn(OH)_2 \rightleftharpoons Zn^{2+} + 2OH^-$	$3.0 \times 10^{-16}(b)$
Carbonates	$Ag_2CO_3 \rightleftharpoons 2Ag^+ + CO_3^{2-}$	8.1×10^{-12}
	$CaCO_3 \rightleftharpoons Ca^{2+} + CO_3^{2-}$	$4.5 \times 10^{-9}(c)$
	$SrCO_3 \rightleftharpoons Sr^{2+} + CO_3^{2-}$	9.3×10^{-10}
	$BaCO_3 \rightleftharpoons Ba^{2+} + CO_3^{2-}$	5.0×10^{-9}
	$CoCO_3 \rightleftharpoons Co^{2+} + CO_3^{2-}$	1.0×10^{-10}
	$NiCO_3 \rightleftharpoons Ni^{2+} + CO_3^{2-}$	1.3×10^{-7}
	$ZnCO_3 \rightleftharpoons Zn^{2+} + CO_3^{2}$	1.0×10^{-10}
Chromates	$Ag_2CrO_4 \rightleftharpoons 2Ag^+ + CrO_4^{2-}$	1.2×10^{-12}
	$PbCrO_4 \rightleftharpoons Pb^{2+} + CrO_4^{2-}$	2.8×10^{-13}
Sulfates	$CaSO_4 \rightleftharpoons Ca^{2+} + SO_4^{2-}$	2.4×10^{-5}
	$SrSO_4 \rightleftharpoons Sr^{2+} + SO_4^{2-}$	3.2×10^{-7}
	$BaSO_4 \rightleftharpoons Ba^{2+} + SO_4^{2-}$	1.1×10^{-10}
	$PbSO_4 \rightleftharpoons Pb^{2+} + SO_4^{2-}$	6.3×10^{-7}
Oxalates	$CaC_2O_4 \rightleftharpoons Ca^{2+} + C_2O_4^{2-}$	2.3×10^{-9}
	$MgC_2O_4 \rightleftharpoons Mg^{2+} + C_2O_4^{2-}$	8.6×10^{-5}
	$BaC_2O_4 \rightleftharpoons Ba^{2+} + C_2O_4^{2-}$	1.2×10^{-7}
	$FeC_2O_4 \rightleftharpoons Fe^{2+} + C_2O_4^{2-}$	2.1×10^{-7}
	$PbC_2O_4 \rightleftharpoons Pb^{2+} + C_2O_4^{2-}$	2.7×10^{-11}

[a] Alpha form.
[b] Amorphous form.
[c] Calcite form.

solution of a sparingly soluble substance. There is no denominator left in the mass action expression, and the equilibrium condition in a saturated solution of AgCl is

$$[Ag^+][Cl^-] = K_{sp}$$

The mass action expression in this kind of solubility equilibrium is often called the **ion product** because it is a product of ion concentrations.

 Ion product of a salt

The solubilities of salts change with temperature, so a value of K_{sp} applies to solutions only at the temperature at which it was determined. Some typical K_{sp} values are in Table 17.2 and in Appendix E.5 (Insoluble oxides and sulfides make up a special group to be discussed in Section 17.4.)

Silver chromate, Ag_2CrO_4, precipitates when sodium chromate is added to a solution of silver nitrate.

Many salts produce more than one of a given ion when they dissociate. Silver chromate, Ag_2CrO_4, is an example. Each of its formula units gives two Ag^+ ions and one CrO_4^{2-} ion. Its solubility equilibrium is

$$Ag_2CrO_4(s) \rightleftharpoons 2Ag^+(aq) + CrO_4^{2-}(aq)$$

Mass action expressions for salts like this include exponents for the associated concentration terms that equal the coefficients of the formulas of the ions. For example, the solubility product expression for Ag_2CrO_4 is

$$[Ag^+]^2[CrO_4^{2-}] = K_{sp}$$

The exponent of 2 appears in the solubility product because the solubility equilibrium shows 2 Ag^+ ions.

■ **Practice Exercise 4** Write the K_{sp} expressions for these equilibria.
(a) $BaCrO_4(s) \rightleftharpoons Ba^{2+}(aq) + CrO_4^{2-}(aq)$
(b) $Ag_3PO_4(s) \rightleftharpoons 3Ag^+(aq) + PO_4^{3-}(aq)$

Calculating K_{sp} from Molar Solubility Data

One way to obtain the value of K_{sp} for a salt is to measure its **molar solubility** in water—the number of moles of solute dissolved in one liter of its *saturated* solution. The salt is stirred with water for some time, long enough to ensure that equilibrium is established. A portion of the solution is filtered to remove solids and then analyzed to determine the amount of the salt dissolved in it. From such data the molar solubility is calculated. Then the K_{sp} can be computed, as illustrated in the next three examples.

EXAMPLE 17.4
Calculating K_{sp} from Molar Solubility Data

Silver bromide, AgBr, is the light-sensitive compound in nearly all photographic film. At 25 °C, one L of water can dissolve 7.1×10^{-7} mol of AgBr. Calculate the K_{sp} of AgBr at this temperature from these data.

ANALYSIS As usual, we begin with the chemical equation for the equilibrium and the equilibrim law (here, the expression for K_{sp}).

$$AgBr(s) \rightleftharpoons Ag^+(aq) + Br^-(aq)$$

$$K_{sp} = [Ag^+][Br^-]$$

We will construct a concentration table for this and other K_{sp} problems by first listing, as *initial concentrations,* the concentrations of any ions common to the solute that are *already present* in the solvent. In this case, the solvent is water, which contains neither Ag^+ nor Br^-, the two ions in AgBr. In the "change" row, we note how the concentrations of the ions change as the AgBr dissolves. Because dissolving the salt always increases these concentrations, the quantities entered here will be positive. As usual, we obtain the equilibrium values by adding the "initial concentrations" to the "changes."

SOLUTION We have been given the moles of AgBr needed to make one liter of a *saturated* solution, 7.1×10^{-7} mol L^{-1}. We use this information to construct the concentration table below.

	AgBr(s)	$\rightleftharpoons$	$Ag^+(aq)$	+	$Br^-(aq)$
Initial concentrations (M)	No entries in this column		0 (Note 1)		0
Changes in concentrations when AgBr dissolves (M)			$+7.1 \times 10^{-7}$ (Note 2)		$+7.1 \times 10^{-7}$
Equilibrium concentrations (M)			7.1×10^{-7} (Note 3)		7.1×10^{-7}

There are no entries in the column under AgBr(s) because this substance does not appear in the K_{sp} expression.

Note 1. When AgBr is placed in *pure* water, the *initial* concentrations of its ions are zero.

Note 2. When some AgBr dissolves, Ag^+ and Br^- ions are released in a 1:1 ratio. So when 7.1×10^{-7} mol of AgBr dissolves (per liter), 7.1×10^{-7} mol of Ag^+ ion and 7.1×10^{-7} mol of Br^- ion go into solution. The concentrations of these species thus *change* (increase) by these amounts.

Note 3. By adding the initial concentrations and the changes, we obtain the final (equilibrium) concentrations.

We now substitute the equilibrium concentrations into the K_{sp} expression.

$$K_{sp} = [Ag^+][Br^-]$$
$$= (7.1 \times 10^{-7})(7.1 \times 10^{-7})$$
$$= 5.0 \times 10^{-13}$$

The K_{sp} of AgBr is thus 5.0×10^{-13} at 25 °C.

The molar solubility of silver chromate, Ag_2CrO_4, in pure water is 6.7×10^{-5} mol L^{-1} at 25 °C. What is K_{sp} for Ag_2CrO_4?

EXAMPLE 17.5
Calculating K_{sp} from Molar Solubility Data

ANALYSIS The equilibrium equation and K_{sp} expression are

$$Ag_2CrO_4(s) \rightleftharpoons 2Ag^+(aq) + CrO_4^{2-}(aq) \qquad K_{sp} = [Ag^+]^2[CrO_4^{2-}]$$

The solute is pure water, so neither Ag^+ nor CrO_4^{2-} are present before the salt dissolves. The initial concentrations of Ag^+ and CrO_4^{2-} are therefore zero. When one liter of pure water dissolves 6.7×10^{-5} mol of Ag_2CrO_4, we can see by the stoichiometry of the equilibrium equation that we must obtain 6.7×10^{-5} mol of CrO_4^{2-} and $2 \times (6.7 \times 10^{-5}$ mol) of Ag^+.

SOLUTION With this information, we prepare the concentration table.

	$Ag_2CrO_4(s) \rightleftharpoons$	$2Ag^+(aq)$	+	$CrO_4^{2-}(aq)$
Initial concentrations (M)		0		0
Changes in concentrations when Ag_2CrO_4 dissolves (M)		$+2 \times (6.7 \times 10^{-5})$ $= +1.3 \times 10^{-4}$		$+6.7 \times 10^{-5}$
Equilibrium concentrations (M)		1.3×10^{-4}		6.7×10^{-5}

Substituting the equilibrium concentrations into the mass action expression for K_{sp} gives

$$K_{sp} = (1.3 \times 10^{-4})^2(6.7 \times 10^{-5})$$
$$= 1.1 \times 10^{-12}$$

So the K_{sp} of Ag_2CrO_4 is 1.1×10^{-12} at 25 °C.

EXAMPLE 17.6
Calculating K_{sp} from Molar Solubility Data

At 25 °C, the molar solubility of $PbCl_2$ in a 0.10 M NaCl solution is 1.7×10^{-3} mol L^{-1}. Calculate the K_{sp} for $PbCl_2$.

ANALYSIS Again, we begin by writing the equation for the equilibrium and the K_{sp} expression.

$$PbCl_2(s) \rightleftharpoons Pb^{2+}(aq) + 2Cl^-(aq) \qquad K_{sp} = [Pb^{2+}][Cl^-]^2$$

We assume, as usual, that any soluble salt, like NaCl, is fully dissociated in water.

In this problem, the $PbCl_2$ is being dissolved not in pure water but in 0.10 M NaCl, which contains 0.10 M Cl^-. (It also contains 0.10 M Na^+, but that's not important here because Na^+ doesn't affect the equilibrium and so doesn't appear in the K_{sp} expression.) The initial concentration of Pb^{2+} is zero because none is in solution to begin with. The *initial* concentration of Cl^-, however, is 0.10 M. When the $PbCl_2$ dissolves in the NaCl solution, the Pb^{2+} concentration increases by 1.7×10^{-3} M and the Cl^- concentration increases by $2 \times (1.7 \times 10^{-3}$ $M)$. With these data we build the concentration table.

	$PbCl_2(s) \rightleftharpoons Pb^{2+}(aq)$ +	$2Cl^-(aq)$
Initial concentrations (M)	0	0.10
Changes in concentrations when $PbCl_2$ dissolves (M)	$+1.7 \times 10^{-3}$	$+2 \times (1.7 \times 10^{-3})$ = $+3.4 \times 10^{-3}$
Equilibrium concentrations (M)	1.7×10^{-3}	$0.10 + 0.0034 = 0.10$ (rounded to the second decimal place)

For each formula unit of $PbCl_2$ that dissolves, we get one Pb^{2+} ion and two Cl^- ions.

SOLUTION Substituting the equilibrium concentrations into the K_{sp} expression gives

$$K_{sp} = (1.7 \times 10^{-3})(0.10)^2$$
$$= 1.7 \times 10^{-5}$$

The K_{sp} of $PbCl_2$ is 1.7×10^{-5} at 25 °C.

■ **Practice Exercise 5** One liter of water is able to dissolve 2.08×10^{-3} mol of PbF_2 at 25 °C. Calculate the K_{sp} for PbF_2.

■ **Practice Exercise 6** At 25 °C, the molar solubility of $CoCO_3$ in a 0.10 M Na_2CO_3 solution is 1.0×10^{-9} mol L^{-1}. What is the K_{sp} for $CoCO_3$?

■ **Practice Exercise 7** The molar solubility of PbF_2 in a 0.10 M $Pb(NO_3)_2$ solution at 25 °C is 3.0×10^{-4} mol L^{-1}. Calculate K_{sp} for PbF_2.

Calculating Molar Solubility from K_{sp}

Besides calculating K_{sp} from solubility information, we can also compute (rather *estimate*) solubilities from values of K_{sp}. The following examples show how to make such calculations.

EXAMPLE 17.7
Calculating Molar Solubility from K_{sp}

What is the molar solubility of AgCl in pure water at 25 °C?

ANALYSIS To solve this problem, we need three things.

1. The equation for the equilibrium in a saturated solution of AgCl.

2. The K_{sp} equation, which we can figure out for ourselves from the equation for the equilibrium.

3. The value of K_{sp} for AgCl.

The relevant equations are

$$AgCl(s) \rightleftharpoons Ag^+(aq) + Cl^-(aq) \qquad K_{sp} = [Ag^+][Cl^-] = 1.8 \times 10^{-10}$$

The value of K_{sp} is from Table 17.2. Now we can build a concentration table. Neither the Ag^+ nor the Cl^- ion is present in solution at the start, so their *initial* concentrations are zero. Let's define x as the molar solubility of the salt—the number of moles of AgCl that dissolves in one liter. Since 1 mol of AgCl yields 1 mol Ag^+ and 1 mol Cl^-, the concentrations of each of these ions increase by x. Thus, our concentration table is

> We assume that all of the AgCl that dissolves breaks down into Ag^+ and Cl^-.

	$AgCl(s) \rightleftharpoons$	$Ag^+(aq)$ +	$Cl^-(aq)$
Initial concentrations (M)		0	
Changes in concentration when AgCl dissolves (M)		$+x$	$+x$
Equilibrium concentrations (M)		x	x

> Notice that the coefficients of x are the same as the coefficients of Ag^+ and Cl^- in the equation for the equilibrium.

SOLUTION We make the substitutions into the K_{sp} expression:

$$(x)(x) = 1.8 \times 10^{-10}$$

$$x = 1.3 \times 10^{-5}$$

Thus the calculated molar solubility of AgCl in water at 25 °C is 1.3×10^{-5} mol L^{-1}.

Calculate the molar solubility of lead iodide, PbI_2, in water at 25 °C.

ANALYSIS The relevant equations and K_{sp} data (from Table 17.2) are

$$PbI_2(s) \rightleftharpoons Pb^{2+}(aq) + 2I^-(aq) \qquad K_{sp} = [Pb^{2+}][I^-]^2 = 7.9 \times 10^{-9}$$

The solvent is water, so the initial concentrations of the ions are zero. As before, we will define x as the molar solubility of the salt. When we do this, the coefficients of x in the "change" row are made identical to the coefficients of the ions in the chemical equation, thereby assuring that the changes in the concentrations are in the correct mole ratio.

EXAMPLE 17.8
Calculating Molar Solubility from K_{sp}

	$PbI_2(s) \rightleftharpoons$	$Pb^{2+}(aq)$ +	$2I^-(aq)$
Initial concentrations (M)		0	0
Changes in concentrations when PbI_2 dissolves (M)		$+x$	$+2x$
Equilibrium concentrations (M)		x	$2x$

SOLUTION Substituting,

$$K_{sp} = (x)(2x)^2 = 4x^3 = 7.9 \times 10^{-9}$$

$$x^3 = 2.0 \times 10^{-9}$$

$$x = 1.3 \times 10^{-3}$$

Thus, the molar solubility of PbI_2 is 1.3×10^{-3} mol L^{-1}.

The precipitation of yellow PbI_2 occurs when the two colorless solutions, one with sodium iodide and the other of lead(II) nitrate, are mixed.

■ **Practice Exercise 8** What is the calculated molar solubility of AgBr in water?

■ **Practice Exercise 9** Calculate the molar solubility of Ag_2CO_3 in water.

The common ion effect. The test tube initially contained a saturated solution of NaCl in equilibrium with some undissolved salt. The equilibrium is $NaCl(s) \rightleftharpoons Na^+(aq) + Cl^-(aq)$. We see in here the result of adding a few drops of conc. HCl, which contains a high concentration of Cl^-, an ion common to both the acid and the salt. The common ion has shifted the equilibrium to the left, and some crystals of NaCl were caused to precipitate.

The Common Ion Effect and Solubility

Suppose that we stir some calcium carbonate with water long enough to establish the following equilibrium.

$$CaCO_3(s) \rightleftharpoons Ca^{2+}(aq) + CO_3^{2-}(aq)$$

Then we add to the solution a very soluble salt of calcium, like $CaCl_2$. This puts additional Ca^{2+} into solution, *and it upsets the above equilibrium.* The ion product is no longer equal to K_{sp}. The effect that the added Ca^{2+} has on the equilibrium can be predicted using Le Châtelier's principle, which tells us that the equilibrium responds in a way that will absorb as much as possible of the added Ca^{2+}. It can do this only by shifting to the left *causing CO_3^{2-} to precipitate* as $CaCO_3$. Eventually, equilibrium is reestablished, but with a lower concentration of CO_3^{2-} in solution.

In the newly equilibrated system, there are two sources of Ca^{2+}, the added $CaCl_2$ and the $CaCO_3$ still in solution. Because Ca^{2+} is common to both sources, it is called a **common ion.** The addition of the common ion lowers the solubility of $CaCO_3$; it is less soluble in the presence of $CaCl_2$ (or any other soluble calcium salt) than it is in pure water. This lowering of the solubility of an ionic compound by the addition of a common ion is called the **common ion effect.** Figure 17.1 illustrates how even a relatively soluble salt, NaCl, can be forced out of its saturated solution simply by adding concentrated hydrochloric acid, which supplies the common ion, Cl^-.

The common ion effect can dramatically lower the solubility of a salt, as shown in the next example.

EXAMPLE 17.9
Calculation Involving the Common Ion Effect

What is the molar solubility of PbI_2 in a 0.10 M NaI solution?

ANALYSIS As usual, we begin with the chemical equation for the equilibrium and the appropriate K_{sp} expression. The value of K_{sp} is obtained from Table 17.2.

$$PbI_2(s) \rightleftharpoons Pb^{2+}(aq) + 2I^-(aq) \qquad K_{sp} = [Pb^{2+}][I^-]^2 = 7.9 \times 10^{-9}$$

As before, we imagine that we are adding the PbI_2 to the NaI solution, so we need to know the concentrations of any of the ions of PbI_2 present in the NaI solution. These will be our initial concentrations.

The salt NaI does not contain any Pb^{2+}, but it does have I^-. Therefore, initially, no Pb^{2+} is in solution, but there is 0.10 mol of I^- per liter from the 0.10 mol L^{-1} of NaI.

Next, we let x be the molar solubility of PbI_2. When x moles of PbI_2 dissolves per liter, the concentration of Pb^{2+} changes by $+x$ and that of I^- by twice as much, or $+2x$. Finally, the equilibrium concentrations are obtained by summing the initial concentrations and the changes. Here is the concentration table.

	$PbI_2(s) \rightleftharpoons Pb^{2+}(aq) + 2I^-(aq)$	
Initial concentrations (M)	0	0.10
Changes in concentrations when $PbCl_2$ dissolves (M)	$+x$	$+2x$
Equilibrium concentrations (M)	x	$0.10 + 2x$

SOLUTION Substituting equilibrium values into the K_{sp} expression gives

$$K_{sp} = (x)(0.10 + 2x)^2 = 7.9 \times 10^{-9}$$

Just a brief inspection reveals that solving this expression for x will be difficult if we cannot simplify the math. Fortunately, a simplification is possible, because the small value of K_{sp} for PbI_2 tells us that the salt has a very low solubility. This means very little of the salt will dissolve, so x (or even $2x$) will be quite small. Let's assume that $2x$ will be much smaller than 0.10. If this is so, then

$$0.10 + 2x \approx 0.10$$

Substituting 0.10 M for the I^- concentration gives

$$K_{sp} = (x)(0.10)^2 = 7.9 \times 10^{-9}$$

$$x = \frac{7.9 \times 10^{-9}}{(0.10)^2}$$

$$= 7.9 \times 10^{-7} \, M$$

Thus, the molar solubility of PbI_2 in 0.10 M NaI is 7.9×10^{-7} M.

CHECKING THE SIMPLIFICATION Before we leave this problem, we should double check to see that our simplifying assumption was valid. Notice that $2x$, which equals 1.6×10^{-6}, is indeed vastly smaller than 0.10, just as we anticipated.

In Example 17.8 we found that the molar solubility of PbI_2 in *pure* water is 1.3×10^{-3} M. In water that contains 0.10 M NaI, the solubility of PbI_2 is 7.9×10^{-7} M, well over a thousand times smaller. As we said, the common ion effect can cause huge reductions in solubilities of sparingly soluble compounds.

The most common mistake that students make with problems like Example 17.9 is to use the coefficient of a formula in the solubility equilibrium at the wrong moment in the calculation. The coefficient of I^- in the PbI_2 equilibrium is 2. The mistake is to use this 2 to double the *initial* concentration of I^-. However, it can be used only to double the *change* in concentration. The *initial* concentration of I^- was provided not by PbI_2 but by NaI. The *change* in the concentration of I^- comes solely from the small dissociation of PbI_2, so the coefficient of 2 enters into the calculation only to describe this change.

To avoid this mistake, always view the formation of the final solution as a two-step process. You begin with a solvent into which the "insoluble solid" will be placed. In some problems, the solvent may be pure water, in which case the initial concentrations of the ions will be zero. In other problems, like Example 17.9, the solvent will be a *solution* that contains a common ion. When this is so, first decide what the concentration of the common ion is and enter this value into the "initial concentration" row of the table. Next, imagine that the solid is added to the solvent and a little of it dissolves. *The amount that dissolves is what gives the values in the "change" row.* These entries must be in the same ratio as the coefficients in the equilibrium, which is accomplished if we let x be the molar solubility of the salt. Then the coefficients of x will be the same as the coefficients of the ions in the chemical equation for the equilibrium.

■ **Practice Exercise 10** What is the molar solubility of AgI in 0.20 M NaI solution? Compare it to its calculated solubility in pure water.

■ **Practice Exercise 11** What is the molar solubility of $Fe(OH)_3$ in a solution where the OH^- concentration is initially 0.050 M?

Determining Whether a Precipitate Will Form in a Solution

Sometimes in setting up an experiment, the plans tentatively include an aqueous solution in which several ions are to be dissolved (or will form). Before proceeding, the chemist must ask, "Will any combination of the desired ions, at the planned concentrations, be of a salt that is too insoluble to be in solution?" If so, the plans for the experiment are changed.

We can use K_{sp} data and the planned (or expected) values of ion concentrations to predict whether a precipitate will form. We first use the ion concentrations to compute ion products. Then these are compared to K_{sp} values. Remember that K_{sp} equals the ion product *for a saturated solution,* the most concentrated solution that is stable (meaning that no precipitate is forming). A *saturated* solution of AgBr in water, for example, contains $7.1 \times 10^{-7}\,\mathrm{mol\,L^{-1}}$ of AgBr. This yields $[Ag^+] = [Br^-] = 7.1 \times 10^{-7}\,M$ and an ion product, $[Ag^+][Br^-]$, that equals 5.0×10^{-13}, which is the K_{sp} of AgBr. But it is not necessary to have a *saturated* solution to compute an ion product. In a specific experiment, we could plan to have ion concentrations that give a computed ion product that's much less than K_{sp}. For example, in an *unsaturated* solution of AgBr that contains just $7.1 \times 10^{-8}\,\mathrm{mol\,L^{-1}}$ of the solute, the concentrations are $[Ag^+] = [Br^-] = 7.1 \times 10^{-8}\,\mathrm{mol\,L^{-1}}$. The computed ion product now is

$$[Ag^+][Br^-] = (7.1 \times 10^{-8})(7.1 \times 10^{-8}) = 5.0 \times 10^{-15}$$

> A substance will not precipitate from a solution if its concentration is such that the solution is either unsaturated or saturated.

This is less than the K_{sp} of AgBr (5.0×10^{-13}). No precipitate of AgBr will form in this solution because it is actually capable of dissolving more AgBr.

A *supersaturated* solution is one that contains more solute than is necessary for saturation. It is unstable, and there is a tendency for the extra solute to precipitate. In a supersaturated solution of AgBr, the computed ion product would be larger than K_{sp}.

> In a supersaturated solution, the concentrations of the ions are larger than in a saturated solution, so the ion product is larger than K_{sp}.

When comparing a computed ion product with the K_{sp} for the salt, we can tell whether a particular combination of ion concentrations will produce a precipitate or not. Here are the possibilities.

> It is usually difficult to prevent the extra salt from precipitating out of a supersaturated solution. (Sodium acetate is a notable exception. Supersaturated solutions of this salt are easily made.)

Precipitate will form	Ion product $> K_{sp}$ (supersaturated)
No precipitate will form	$\begin{cases} \text{Ion product} = K_{sp} \text{ (saturated)} \\ \text{Ion product} < K_{sp} \text{ (unsaturated)} \end{cases}$

Let's now consider these possibilities in some practical calculations.

EXAMPLE 17.10
Predicting Whether or Not a Precipitate Will Form

A student wished to prepare 1.0 L of a solution containing 0.015 mol of NaCl and 0.15 mol of $Pb(NO_3)_2$. Knowing from the solubility rules that the chloride of Pb^{2+} is "insoluble," there was concern that a precipitate of $PbCl_2$ might form. Will it?

ANALYSIS To answer this question, we will compute the ion product for $PbCl_2$ *using the concentrations of the ions in the solution to be prepared.* If the computed ion product is larger than K_{sp}, then a precipitate will form. To perform the calculation correctly, we need the correct form of the ion product. We can obtain this by writing the solubility equilibrium and the K_{sp} expression that applies to a saturated solution.

$$PbCl_2(s) \rightleftharpoons Pb^{2+}(aq) + 2Cl^-(aq) \qquad K_{sp} = [Pb^{2+}][Cl^-]^2$$

From Table 17.2, $K_{sp} = 1.7 \times 10^{-5}$.

SOLUTION The planned solution would have the following concentrations:

$$[Pb^{2+}] = 0.15 \text{ mol L}^{-1} = 0.15\ M$$

$$[Cl^-] = 0.015 \text{ mol L}^{-1} = 0.015\ M$$

We use these values to compute the ion product for $PbCl_2$.

$$[Pb^{2+}][Cl^-]^2 = (0.15)(0.015)^2 = 3.4 \times 10^{-5}$$

Notice that this value is larger than the K_{sp} of $PbCl_2$, 1.7×10^{-5}. This means a precipitate of $PbCl_2$ will form if the preparation of this solution is attempted.

■ **Practice Exercise 12** Will a precipitate of $CaSO_4$ form in a solution if the Ca^{2+} concentration is 0.0025 M and the SO_4^{2-} concentration is 0.030 M?

■ **Practice Exercise 13** Will a precipitate form in a solution containing $3.4 \times 10^{-4}\ M\ CrO_4^{2-}$ and $4.8 \times 10^{-5}\ M\ Ag^+$?

EXAMPLE 17.11
Predicting Whether or Not a Precipitate Will Form

What possible precipitate might form if we mix 50.0 mL of 0.0010 M $CaCl_2$ with 50.0 mL of 0.010 M Na_2SO_4? Will one form? (Assume that the volumes are additive.)

ANALYSIS In this problem we are being asked, in effect, whether a metathesis reaction will occur between $CaCl_2$ and Na_2SO_4. We should be able to predict whether this *might* occur by using the solubility rules presented in Chapter 4 (page 146). If the solubility rules suggest a precipitate, we can then calculate the ion product for the compound using the concentrations of the ions in the final solution. If this ion product exceeds K_{sp} for the salt, then a precipitate will form.

To calculate the ion product correctly requires that we use the concentrations of the ions *after the solutions have been mixed*. Therefore, before computing the ion product, we must first recall that mixing the solutions dilutes each of the solutes.

SOLUTION Let's begin by writing the equation for the potential metathesis reaction between $CaCl_2$ and Na_2SO_4. We use the solubility rules to determine whether each product is soluble or insoluble.

$$CaCl_2(aq) + Na_2SO_4(aq) \longrightarrow CaSO_4(s) + 2NaCl(aq)$$

The solubility rules indicate that we expect a precipitate of $CaSO_4$. But are the concentrations of Ca^{2+} and SO_4^{2-} actually high enough?

In the original solutions, the 0.0010 M $CaCl_2$ contains 0.0010 M Ca^{2+} and the 0.010 M Na_2SO_4 contains 0.010 M SO_4^{2-}. What are the concentrations of these ions after dilution? To determine these we use the equation that applies to all dilution problems involving molarity.

$$M_i V_i = M_f V_f \qquad (17.3)$$

This equation was developed in Chapter 4 on page 165.

Solving for M_f gives

$$M_f = \frac{M_i V_i}{V_f} \qquad (17.4)$$

The initial volumes of both solutions are 50.0 mL and when the two solutions are combined, the final total volume is 100 mL. Therefore,

$$[Ca^{2+}]_{final} = \frac{(50.0 \text{ mL}) (0.0010 \text{ } M)}{100.0 \text{ mL}} = 0.00050 \text{ } M = 5.0 \times 10^{-4} \text{ } M$$

$$[SO_4^{2-}]_{final} = \frac{(50.0 \text{ mL}) (0.010 \text{ } M)}{100.0 \text{ mL}} = 0.0050 \text{ } M = 5.0 \times 10^{-3} \text{ } M$$

The units for V_i and V_f can be any we please, provided they are the same in both. The units for M, of course, are mol L^{-1}.

Now we use these to calculate the ion product for $CaSO_4$, which we can write from the dissociation reaction of the salt.

$$CaSO_4(s) \rightleftharpoons Ca^{2+} + SO_4^{2-}$$

$$\text{Ion product for } CaSO_4 = [Ca^{2+}] [SO_4^{2-}]$$

Substituting the concentrations computed above gives

$$\text{Ion product} = (5.0 \times 10^{-4}) (5.0 \times 10^{-3}) = 2.5 \times 10^{-6}$$

In Table 17.2, the K_{sp} for $CaSO_4$ is given as 2.4×10^{-5}. Notice that the ion product is *smaller* than K_{sp}, which means that a precipitate will not form when these solutions are poured together.

■ **Practice Exercise 14** What precipitate might be expected if we pour together 100.0 mL of 1.0×10^{-3} M $Pb(NO_3)_2$ and 100.0 mL of 2.0×10^{-3} M $MgSO_4$? Will some form? (Assume that the volumes are additive.)

■ **Practice Exercise 15** What precipitate might be expected if we pour together 50.0 mL of 0.10 M $Pb(NO_3)_2$ and 20.0 mL of 0.040 M NaCl? Will some form? (Assume that the volumes are additive.)

17.4
SOLUBILITY EQUILIBRIA OF METAL OXIDES AND SULFIDES

Most water-insoluble metal oxides, like Fe_2O_3, dissolve in acid. For example,

$$Fe_2O_3(s) + 6H^+(aq) \longrightarrow 2Fe^{3+}(aq) + 3H_2O$$

The acid neutralizes the oxide ion and thus releases the metal ion from the solid. Other metal oxides dissolve in water without the aid of an acid. They do so, however, by reacting with water instead of by a simple dissociation of ions that remain otherwise unchanged. Sodium oxide, for example, consists of Na^+ and O^{2-} ions, and it readily dissolves in water. The solution, however, does not contain the oxide ion, O^{2-}. Instead, the hydroxide ion forms. The equation for the reaction is

$$Na_2O(s) + H_2O \longrightarrow 2NaOH(aq)$$

This actually involves the reaction of oxide ions with water as the crystals of Na_2O break up.

$$O^{2-}(s) + H_2O \longrightarrow 2OH^-(aq)$$

The oxide ion is simply too powerful a base to exist in water at any concentration worthy of experimental note. We can understand why from the extraordinarily high (estimated) value of K_b for O^{2-}, 1×10^{22}. Thus there is no way to supply *oxide ions* to an aqueous solution in order to form an insoluble metal oxide directly. Oxide ions react with water, instead, to generate the hydroxide ion. When an insoluble metal *oxide* instead of an insoluble metal *hydroxide* does precipitate from a solution, it forms because the specific metal ion is able to react with OH$^-$, extract O^{2-}, and leave H$^+$ or a neutralized form in the solution. The silver ion, for example, precipitates as brown silver oxide, Ag$_2$O, when OH$^-$ is added to aqueous silver salts.

$$2Ag^+(aq) + 2OH^-(aq) \longrightarrow Ag_2O(s) + H_2O$$

Hydrogen Sulfide as a Diprotic Acid When we shift from oxygen to sulfur in Group VIA and consider metal sulfides, we find many similarities to oxides. One is that the sulfide ion, S^{2-}, like the oxide ion, does not exist in any ordinary aqueous solution. The sulfide ion has not been detected in an aqueous solution even in the presence of 8 M NaOH, where one might think that the reaction

$$OH^- + HS^- \longrightarrow H_2O + S^{2-}$$

could generate some detectable S^{2-}. An 8 M NaOH solution is a concentration well outside the bounds of the "ordinary." Thus, Na$_2$S, like Na$_2$O, dissolves in water *by reacting with it,* not by releasing an otherwise unchanged divalent anion, S^{2-}.

$$Na_2S(s) + H_2O \longrightarrow 2Na^+(aq) + HS^-(aq) + OH^-(aq)$$

Just as some metal oxides form by a reaction between a metal ion and OH$^-$, many metal sulfides form when their metal ions are exposed to HS$^-$. Some metal ions even react with H$_2$S directly to form sulfides. Simply bubbling hydrogen sulfide gas into an aqueous solution of any one of a number of ions — Cu^{2+}, Pb^{2+}, and Ni^{2+}, for example — causes their sulfides to precipitate. Many of these have distinctive colors (Figure 17.2) that can be used to help identify which metal ion is in solution. A typical reaction is that of Cu^{2+} with H$_2$S.

$$Cu^{2+}(aq) + H_2S(aq) \rightleftharpoons CuS(s) + 2H^+(aq) \qquad K = 1.7 \times 10^{15}$$

The extremely large value of the equilibrium constant tells us that the forward reaction is essentially the only reaction; the reverse reaction is very unimportant. If we turn this equilibrium around, we would have something that looks very much like an equilibrium constant for defining a solubility product. Let's look at this possibility more closely.

K_{spa} Values for Metal Sulfides

CuS is extremely insoluble in aqueous acid judging from the value of K for the reaction of Cu^{2+} with H$_2$S just described. However, if we were to write the

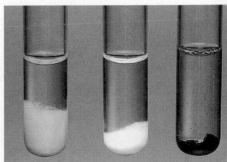

A brown, mudlike precipitate of silver oxide forms as sodium hydroxide solution is added to a solution of silver nitrate.

FIGURE 17.2

The colors of some metal sulfides. From left to right: (*a*) CuS, CdS, As$_2$S$_3$, SnS$_2$, Sb$_2$S$_3$; (*b*) MnS, ZnS, and FeS.

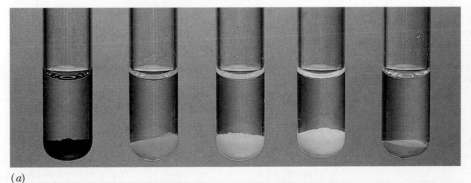

(*a*)

equilibrium that describes the solubility of CuS in water, we would write:

$$CuS(s) \rightleftharpoons Cu^{2+}(aq) + S^{2-}(aq)$$

This assumes that S^{2-} exists in water, and it doesn't, as we said. So we must instead write the equilibrium when CuS(s) dissolves in water as follows in order to show which anions are actually produced.

$$CuS(s) + H_2O \rightleftharpoons Cu^{2+}(aq) + HS^-(aq) + OH^-(aq) \qquad (17.5)$$

The solubility product constant for CuS must therefore be expressed by the following equation.

$$K_{sp} = [Cu^{2+}][HS^-][OH^-]$$

The last column in Table 17.3 gives the K_{sp} values, defined in this new way, for several metal sulfides. Notice particularly how much the K_{sp} values vary—from 2×10^{-53} to 3×10^{-11}, a spread by a factor of 10^{42}. Several metal sulfides, the *acid-insoluble sulfides,* have K_{sp} values so low that they do not dissolve in acid. The cations in this group can be precipitated from the other cations simply by bubbling hydrogen sulfide into a sufficiently acidic solution that contains several metal ions. When the solution is acidic, we have to treat the solubility equilibria differently, however. In acid, HS^- and OH^- would be neutralized leaving their conjugate acids, H_2S and H_2O. When we introduce acid ($2H^+$) into the left side of equilibrium 17.5, for example, to neutralize the bases on the right side (HS^- and OH^-), we obtain

$$CuS(s) + H_2O + 2H^+(aq) \rightleftharpoons Cu^{2+}(aq) + H_2S(aq) + H_2O$$

After we cancel the H_2O from each side, we have the net equation for the CuS(s) solubility equilibrium *in dilute acid.*

$$CuS(s) + 2H^+(aq) \rightleftharpoons Cu^{2+}(aq) + H_2S(aq)$$

TABLE 17.3 Metal Ions Separable by Selective Precipitation of Sulfides[a]

Metal Ion	Sulfide	K_{spa}	K_{sp}
Acid-insoluble Sulfides			
Hg^{2+}	HgS (black form)	2×10^{-32}	2×10^{-53}
Ag^+	Ag_2S	6×10^{-30}	6×10^{-51}
Cu^{2+}	CuS	6×10^{-16}	6×10^{-37}
Cd^{2+}	CdS	3×10^{-7}	3×10^{-28}
Pb^{2+}	PbS	3×10^{-7}	3×10^{-28}
Sn^{2+}	SnS	1×10^{-5}	1×10^{-26}
Base-insoluble Sulfides *(Acid-soluble Sulfides)*			
Zn^{2+}	α–ZnS	3×10^{-4}	3×10^{-25}
	β–ZnS	3×10^{-2}	3×10^{-23}
Co^{2+}	CoS	5×10^{-1}	5×10^{-22}
Ni^{2+}	NiS	4×10^1	4×10^{-20}
Fe^{2+}	FeS	6×10^2	6×10^{-19}
Mn^{2+}	MnS (pink form)	3×10^{10}	3×10^{-11}
	MnS (green form)	3×10^7	3×10^{-14}

[a] Data are for 25°C. See R.J. Meyers, *J. Chem. Ed.*, Vol. 63, 1986, page 689.

This changes the mass action expression for the solubility product equilibrium, which we will now call the **acid solubility product, K_{spa}.** The "a" in the subscript indicates that the medium is acidic.

 Acid solubility product constant

$$K_{spa} = \frac{[Cu^{2+}][H_2S]}{[H^+]^2}$$

Table 17.3 gives K_{spa} values for several metal sulfides. Notice that all K_{spa} values are about 10^{21} larger than the K_{sp} values. Metal sulfides are clearly vastly more soluble in dilute acid than in water. Yet several—the acid-insoluble sulfides— are so insoluble that even the most soluble of them, SnS, barely dissolves, even in moderately concentrated acid.

Selective Precipitation of the Sulfides of Metal Ions

The large differences in K_{spa} values between the acid-insoluble metal sulfides and those that are base insoluble make it possible to separate members of these two families of cations from each other when they are in the same solution. This can be done by selectively precipitating the sulfides of the acid-insoluble cations at a pH that keeps the other cations in solution. A solution saturated in H_2S is used; in it, $[H_2S] = 0.1\ M$. Let's work an example to show how we can calculate the pH conditions necessary to accomplish the selective precipitation of metal sulfides. As we do this, we will see the reason for the classification of the metal sulfides as being "acid insoluble" or "base insoluble." We will use Cu^{2+} and Ni^{2+} ions to represent a cation from each class.

Over what range of hydrogen ion concentrations (and pH) is it possible to separate Cu^{2+} from Ni^{2+} when both metal ions are present in a solution at a concentration of $0.010\ M$ and the solution is made saturated in H_2S (where $[H_2S] = 0.1\ M$)?

ANALYSIS We must work with the chemical equilibria and their associated equilibrium expressions (the equations for their K_{spa}).

$$CuS(s) + 2H^+(aq) \rightleftharpoons Cu^{2+}(aq) + H_2S(aq) \qquad K_{spa} = \frac{[Cu^{2+}][H_2S]}{[H^+]^2} = 6 \times 10^{-16}$$

$$NiS(s) + 2H^+(aq) \rightleftharpoons Ni^{2+}(aq) + H_2S(aq) \qquad K_{spa} = \frac{[Ni^{2+}][H_2S]}{[H^+]^2} = 4 \times 10^{1}$$

The K_{spa} values tell us that the NiS is much more soluble in an acidic solution than CuS. Therefore, we want to make the H^+ concentration large enough to prevent NiS from precipitating, and small enough that CuS does precipitate. The problem reduces to two questions, then. The first is "What hydrogen ion concentration would be needed to keep the Cu^{2+} *in solution?*" (The answer to this question will give us the *upper limit* on $[H^+]$; we would really want a lower H^+ concentration so CuS *will* precipitate.) The second question is "What is the hydrogen ion concentration just before NiS precipitates?" The answer to this will be the *lower limit* on $[H^+]$. At any lower H^+ concentration, NiS will precipitate, so we want an H^+ concentration equal to or larger than this value. Once we know these limits, we know that any hydrogen ion concentration in between them will permit CuS to precipitate but retain Ni^{2+} in solution.

EXAMPLE 17.12
Selective Precipitation of the Sulfides of Acid-Insoluble Cations from Base-Insoluble Cations

SOLUTION We will find the upper limit first. If CuS *does not precipitate*, the Cu^{2+} concentration will be the given value, 0.010 *M*, so we substitute this along with the H_2S concentration (0.1 *M*) into the expression for K_{spa}.

$$K_{spa} = 6 \times 10^{-16} = \frac{(0.010)(0.1)}{[H^+]^2}$$

Now we solve for $[H^+]$.

$$[H^+]^2 = \frac{(0.010)(0.1)}{6 \times 10^{-16}} = 2 \times 10^{12}$$

$$[H^+] = 1 \times 10^6 \ M$$

If we could make $[H^+] = 1 \times 10^6$ *M*, we could prevent CuS from forming. However, it isn't possible to have a million moles of H^+ per liter! What the calculated $[H^+]$ tells us, therefore, is that *no matter how acidic the solution is, we cannot prevent CuS from precipitating when we saturate the solution with H_2S.* (You can now see why CuS is classed as an "acid-insoluble sulfide.") In other words, we don't have to worry about the solution being so acidic that Cu^{2+} is prevented from precipitating as CuS(*s*). There is no upper limit to $[H^+]$. Regardless of how acidic we make the solution, as soon as we start to bubble H_2S into it, CuS will precipitate.

To obtain the lower limit, we calculate the $[H^+]$ required to give an equilibrium concentration of Ni^{2+} equal to 0.010 *M*. If we keep the value of $[H^+]$ *equal to or larger than* this value, then NiS will be prevented from precipitating. The calculation is exactly like the one above. First, we substitute values into the K_{spa} expression.

$$K_{spa} = \frac{(0.010)(0.1)}{[H^+]^2} = 4 \times 10^1$$

Once again, we solve for $[H^+]$.

$$[H^+]^2 = \frac{(0.010)(0.1)}{4 \times 10^1}$$

$$[H^+] = 5 \times 10^{-3} \ M$$

The pH is therefore

$$pH = 2.3$$

If we maintain the pH of the solution of 0.010 *M* Cu^{2+} and 0.010 *M* Ni^{2+} at 2.3 or lower (more acidic), as we make the solution saturated in H_2S, virtually all the Cu^{2+} will precipitate as CuS, but all the Ni^{2+} will stay in solution.

In actual experimental work involving the separation of acid-insoluble cations from base-insoluble cations, a solution more acidic than pH 2.3 (calculated in Example 17.12) is employed. A pH of about 0.5 is used, which corresponds to $[H^+] = 0.3$ *M*. We can also see why NiS can be classified as a *base-insoluble sulfide*. If the solution is *basic* when it is made saturated in H_2S, the pH will surely be larger than 2.3 and NiS will precipitate.

■ **Practice Exercise 16** Consider a solution containing Hg^{2+} and Fe^{2+}, both at molarities of 0.010 *M*. It is to be saturated with H_2S. Calculate the lower limit to the pH of this solution that must be maintained to keep Fe^{2+} in solution while Hg^{2+} precipitates as HgS.

The principles of selective precipitation apply also to other systems, like metal oxalates and metal carbonates, as some Review Exercises will illustrate.

SUMMARY

Equilibria in Solutions of Polyprotic Acids   A polyprotic acid has a K_a value for the ionization of each of its hydrogen ions. Successive values of K_a often differ by a factor of 10^4 to 10^5. This allows us to calculate the pH of a solution of a polyprotic acid by using just the value of K_{a_1}. If the polyprotic acid is the only solute, the anion formed in the second step of the ionization, A^{2-}, has a concentration equal to K_{a_2}. The concentration of A^{2-} can be controlled by adjusting the pH of the solution. The combined expression obtained from $K_{a_1} \times K_{a_2}$ is useful only when two of the three quantities $[H^+]$, $[H_2A]$, and $[A^{2-}]$ are known.

pH of Solutions of Salts of Polyprotic Acids   The anions of weak polyprotic acids are bases that react with water in successive steps, the last of which has the molecular polyprotic acid as a product. For a diprotic acid, $K_{b_1} = K_w / K_{a_2}$ and $K_{b_2} = K_w / K_{a_1}$. Usually, $K_{b_1} \gg K_{b_2}$, so virtually all the OH^- produced in the solution comes from the first step. The pH of the solution can be calculated using just K_{b_1} and the reaction

$$A^{2-} + H_2O \rightleftharpoons HA^- + OH^-$$

Solubility Equilibria for Salts   The **ion product** of a sparingly soluble salt is the product of the molar concentrations of its ions, each concentration raised to the power equal to the subscript of the ion in the salt's formula. At a given temperature, this *ion product* in a *saturated* solution of the salt equals a constant called the **solubility product constant, or K_{sp}.**

The **common ion effect** is the ability of an ion of a soluble salt to suppress the solubility of a sparingly soluble compound that has the same (the "common") ion.

If soluble salts are mixed together in the same solution, a cation of one and an anion of another will precipitate if the ion product exceeds the solubility product constant of the salt formed from them.

When metal oxides and sulfides dissolve in water, neither O^{2-} nor S^{2-} ions exist in the solution. Instead, these anions react with water, and OH^- or HS^- ions form.

Selective Precipitation   Metal sulfides are vastly more soluble in an acidic solution than in water. To express their solubility equilibria in acid, the **acid solubility constant** or K_{spa} is used. Several metal sulfides have such low values of K_{spa} that they are insoluble even at low pH. These acid-insoluble cations can thus be selectively separated from the base-insoluble cations by making the solution both quite acidic as well as saturated in H_2S.

TOOLS YOU HAVE LEARNED

The following tools were introduced in this chapter. You will find them useful for solving certain problems.

Tool	Function
Solubility product constant, K_{sp} (page 740)	Calculate K_{sp} from the molar solubility of the salt. Calculate a solubility from the K_{sp}. Calculate a solubility when a common ion is present.
Ion product of a salt (page 741)	Can be used to predict whether a precipitate of the salt will form in a solution by comparing the ion product with K_{sp}
Acid solubility product constant, K_{spa} (page 753)	Use K_{spa} data to calculate the solubility of a metal sulfide at a given pH. Use K_{spa} data for two or more metal sulfides to calculate the pH at which one will selectively precipitate from a solution saturated in H_2S.

THINKING IT THROUGH

For each of the following, assemble the available information needed to obtain the answer, state what additional data (if any) are needed, and describe how you would use the data to answer the question. Keep in mind that the goal is not to find the answer itself, but instead to find the path that leads to the answer.

The problems are divided into two groups. Those in Level 2 are significantly more challenging than those in Level 1 and provide an opportunity to really hone your problem-solving skills.

Level 1 Problems

1. Describe how you would calculate the value of K_{b_1} for the PO_4^{3-} ion.

2. In a solution of sodium sulfite, there are actually three sources of OH^- ion. Write chemical equations for the three reactions involved. Which source(s) of OH^- can be ignored, and why?

3. Show that the concentration of PO_4^{3-} in a 0.10 *M* solution of H_3PO_4 is not equal to K_{a_3} for this acid.

4. A saturated solution of $CaCO_3$ is prepared in which excess, solid $CaCO_3$ rests at the bottom of the container.

(a) What equilibrium is present involving the solute?
(b) What compound might be added to the solution to *reduce* the molar concentration of Ca^{2+}?
(c) What compound might be added to the solution to *increase* the molar concentration of Ca^{2+}?
(d) How do the actions of parts (b) and (c) illustrate Le Châtelier's principle?
(e) Do any of these actions illustrate the common ion effect?

5. If 50.0 mL each of 0.0100 *M* solutions of NaBr and $Pb(NO_3)_2$ are poured together, will a precipitate form? If so, what is its formula? Describe how you would calculate the concentrations of the ions at equilibrium.

6. Suppose you wished to control the PO_4^{3-} concentration in a solution of H_3PO_4 by controlling the pH of the solution. If you assume you know the H_3PO_4 concentration, what combined equation would be useful for this purpose?

7. Will $PbBr_2$ be less soluble in 0.10 *M* $Pb(C_2H_3O_2)_2$ or 0.10 *M* NaBr? (Describe how you would perform the calculations necessary to answer the question.)

Level 2 Problems

8. What is the concentration of H_2CO_3 in a 0.10 *M* solution of Na_2CO_3? (Explain how you would calculate the answer. In what way is this calculation similar to finding the CO_3^{2-} concentration in a solution of H_2CO_3?)

9. What is the highest concentration of Pb^{2+} that can exist in a solution of 0.10 *M* HCl? (Explain how you would calculate the answer.)

10. If a solution of 0.10 *M* Mn^{2+} and 0.10 *M* Cd^{2+} is gradually made basic, what will the concentration of Cd^{2+} be when $Mn(OH)_2$ just begins to precipitate? Assume no change in the volume of the solution. (Explain how you would approach the problem and what data you would need to answer the question.)

11. A solution of $MgBr_2$ can be changed to a solution of $MgCl_2$ by adding AgCl(*s*) and stirring the mixture well. In terms of the equilibria involved, explain how this happens.

REVIEW EXERCISES

Answers to questions whose numbers are printed in color are given in Appendix D. More challenging questions are marked with asterisks.

Polyprotic Acids

17.1 Arsenic acid, H_3AsO_4, is a triprotic acid.
(a) Write the three equilibrium equations that represent the successive ionizations of this acid.
(b) The negative logarithms of the acid ionization constants for the equilibria are, respectively, 2.2, 6.9, and 11.5 at 25 °C. What are the acid ionization constants?
(c) Which is the stronger acid, $HAsO_4^{2-}$ or HPO_4^{2-}? How can you tell?

17.2 When sulfur dioxide, an air pollutant from the burning of sulfur-containing coal or oil, dissolves in water,

some reacts to form sulfurous acid, usually represented as H_2SO_3.

$$H_2O + SO_2(g) \rightleftharpoons H_2SO_3(aq)$$

Write the expressions for K_{a_1} and K_{a_2}.

17.3 Carbon dioxide, also produced by the burning of coal or oil, will react with water to produce carbonic acid, H_2CO_3.
(a) Write the equilibrium equations and the equations for K_{a_1} and K_{a_2}.
(b) Which is the stronger acid: sulfurous acid or carbonic acid?

17.4 If water is treated as any other weak diprotic Brønsted acid, what is the equation expressing its K_a? How does this expression differ from K_w? Write the expression for K_{a_2} for water.

*17.5 Phosphorous acid, H_3PO_3, is actually a diprotic acid for which $K_{a_1} = 1.0 \times 10^{-2}$ and $K_{a_2} = 2.6 \times 10^{-7}$. What are the values of $[H^+]$, $[H_2PO_3^-]$, and $[HPO_3^{2-}]$ in a 1.0 M solution of H_3PO_3? What is the pH of the solution?

17.6 Tellurium, in the same family as sulfur, forms an acid analogous to sulfuric acid and called telluric acid. It exists, however, as H_6TeO_6 (which looks like the formula $H_2TeO_4 + 2H_2O$). It is a diprotic acid. $K_{a_1} = 2 \times 10^{-8}$ and $K_{a_2} = 1 \times 10^{-11}$. Calculate $[H^+]$, pH, and $[H_4TeO_6^{2-}]$ in a 0.25 M solution of H_6TeO_6.

17.7 Tartaric acid is a diprotic acid for which $K_{a_1} = 9.2 \times 10^{-4}$ and $K_{a_2} = 4.3 \times 10^{-5}$. Calculate the molar concentrations of H^+ and the two anions of tartaric acid in a solution that has a tartaric acid concentration of 0.10 M. (Use the symbol H_2Tar for tartaric acid.)

17.8 Some people who take megadoses of ascorbic acid will drink a solution containing as much as 6.0 g of ascorbic acid dissolved in a glass of water. Assuming the volume to be 250 mL, calculate the pH of this solution.

17.9 What is the molar concentration of HCO_3^- ion in a solution of H_2CO_3 whose pH is 3.5? Calculate also the molar concentration of CO_3^{2-}.

17.10 What molar concentration of HCO_3^- can exist in equilibrium in arterial blood at a pH of 7.35 when the molar concentration of CO_2 is 2.2×10^{-2} mol L^{-1}? Assume that all of the CO_2 is present as H_2CO_3.

17.11 A 0.20 M solution of tartaric acid, $H_2C_4H_4O_6$ ($K_{a_1} = 9.2 \times 10^{-4}$ and $K_{a_2} = 4.3 \times 10^{-5}$), contains enough HCl so that its $[H^+]$ equals 0.030 M. What is the molar concentration of tartrate ion, $C_4H_4O_6^{2-}$, in the solution?

*17.12 Calculate the concentrations of all solute species in 0.30 M H_3PO_4. (Use the method of successive approximations for the first step in the ionization.)

Salts of Polyprotic Acids

17.13 Consider the following salts: Na_3PO_4, $LiHCO_3$, $(NH_4)_2SO_4$, K_2HPO_4, K_2SO_3, and $NaHSO_4$.
(a) Write the formulas of the salts that give an acidic solution in water.
(b) Write the formulas of the salts that give a basic solution in water.
(c) Write the formulas of the salts that give a neutral solution in water.

17.14 Write the equations for the chemical equilibria that exist in solutions of (a) Na_2SO_3, (b) Na_3PO_4, (c) $K_2C_4H_4O_6$.

17.15 Calculate the pH of 0.12 M Na_2SO_3.

17.16 Calculate the pH of 0.15 M K_2CO_3.

17.17 How many grams of $Na_2CO_3 \cdot 10H_2O$ have to be dissolved in 1.00 L of water at 25 °C to have a solution with a pH of 11.62?

17.18 Calculate the number of grams of $Na_2SO_3 \cdot 7H_2O$ that you would have to dissolve in 0.50 L of water at 25 °C to have a solution with a pH of 10.15.

*17.19 The pH of a 0.10 M Na_2CO_3 solution is adjusted to 12.00 using a strong base. What is the concentration of CO_3^{2-} in this solution?

Solubility Products

17.20 What is the difference between an *ion product* and an *ion product constant*?

17.21 The solubility product constant can be described as a modified equilibrium constant.
(a) Write the full equilibrium constant for the system:
$$Ba_3(PO_4)_2(s) \rightleftharpoons 3Ba^{2+}(aq) + 2PO_4^{3-}(aq)$$
(b) How is this equilibrium constant expression modified and how is the modification justified?
(c) Write the solubility product constant expression for the equilibrium in part (a).

17.22 Write the K_{sp} expression for each of the following compounds.
(a) CaF_2 (d) $Fe(OH)_3$
(b) Ag_2CO_3 (e) PbI_2
(c) $PbSO_4$ (f) $Cu(OH)_2$

17.23 Write the K_{sp} expression for each of the following compounds.
(a) AgI (d) $Al(OH)_3$
(b) Ag_3PO_4 (e) $ZnCO_3$
(c) $PbCrO_4$ (f) $Zn(OH)_2$

17.24 What is the common ion effect? How does Le Châtelier's principle explain it? Use the solubility equilibrium for AgCl to illustrate the common ion effect.

17.25 With respect to K_{sp}, what conditions must be met if a precipitate is going to form in a solution?

17.26 Barium sulfate, $BaSO_4$, is so insoluble that it can be swallowed without significant danger, even though Ba^{2+} is toxic. At 25 °C, 1.00 L of water dissolves only 0.00245 g of $BaSO_4$.
(a) How many moles of $BaSO_4$ dissolve per liter?
(b) What are the molar concentrations of Ba^{2+} and SO_4^{2-} in a saturated $BaSO_4$ solution?
(c) Calculate K_{sp} for $BaSO_4$.

17.27 At 25 °C, the molar solubility of Ag_3PO_4 is 1.8×10^{-5} mol L^{-1}. Calculate K_{sp} for this salt.

17.28 A student found that 0.800 g $AgC_2H_3O_2$ is able to dissolve in 100 mL of water. What is the molar solubility of $AgC_2H_3O_2$? What is the value of K_{sp} for this salt?

17.29 The molar solubility of $Ba_3(PO_4)_2$ in water at 25 °C is 1.4×10^{-8} mol L^{-1}. What is the value of K_{sp} for this salt?

17.30 It was found that the molar solubility of $BaSO_3$ in 0.10 M $BaCl_2$ is 8.0×10^{-6} M. What is the value of K_{sp} for $BaSO_3$?

17.31 A student prepared a saturated solution of $CaCrO_4$ and found that when 156 mL of the solution was evaporated, 0.649 g of $CaCrO_4$ was left behind. What is the value of K_{sp} for this salt?

17.32 Copper(I) chloride has $K_{sp} = 1.9 \times 10^{-7}$. Calculate the molar solubility of CuCl in (a) pure water, (b) 0.0200 M HCl solution, (c) 0.200 M HCl solution, and (d) 0.150 M CaCl$_2$ solution.

17.33 Gold(III) chloride, AuCl$_3$, has $K_{sp} = 3.2 \times 10^{-25}$. Calculate the molar solubility of AuCl$_3$ in (a) pure water, (b) 0.010 M HCl solution, (c) 0.010 M MgCl$_2$ solution, and (d) 0.010 M Au(NO$_3$)$_3$ solution.

17.34 At 25 °C, the value of K_{sp} for LiF is 1.7×10^{-3}, and for BaF$_2$ it is 1.7×10^{-6}. Which salt, LiF or BaF$_2$, has the larger molar solubility in water? Calculate the molar solubility of each.

17.35 At 25 °C, the value of K_{sp} for AgCN is 2.2×10^{-16} and that for Zn(CN)$_2$ is 3×10^{-16}. In terms of grams per 100 mL of solution, which salt is the more soluble in water? Calculate the solubility of each in terms of these units.

17.36 A salt whose formula is of the form MX has a value of K_{sp} equal to 3.2×10^{-10}. Another sparingly soluble salt, MX_3, must have what value of K_{sp} if the molar solubilities of the two salts are to be identical?

*17.37 A salt having a formula of the type M_2X_3 has $K_{sp} = 2.2 \times 10^{-20}$. Another salt, M_2X, has to have what K_{sp} value if M_2X has twice the molar solubility of M_2X_3?

17.38 Calcium sulfate is found in plaster. At 25 °C the value of K_{sp} for CaSO$_4$ is 2.4×10^{-5}. What is the calculated molar solubility of CaSO$_4$ in water?

17.39 Chalk is CaCO$_3$, and at 25 °C its $K_{sp} = 4.5 \times 10^{-9}$. What is the molar solubility of CaCO$_3$? How many grams of CaCO$_3$ dissolve in 100 mL of aqueous solution?

17.40 Calculate the molar solubility of PbI$_2$ in water. Its K_{sp} equals 7.9×10^{-9}.

17.41 Calculate the molar solubility of Ag$_2$CO$_3$ in water. At 25 °C, $K_{sp} = 8.1 \times 10^{-12}$. (Ignore the reaction of the CO$_3{}^{2-}$ ion with water.)

17.42 Calculate the molar solubility of Ag$_2$CrO$_4$ in 0.200 M AgNO$_3$ at 25 °C. For Ag$_2$CrO$_4$ at 25 °C, $K_{sp} = 1.2 \times 10^{-12}$.

17.43 What is the pH of a saturated solution of Mg(OH)$_2$?

17.44 What is the molar solubility of Mg(OH)$_2$ in 0.20 M NaOH? For Mg(OH)$_2$, $K_{sp} = 7.1 \times 10^{-12}$.

17.45 Does a precipitate of PbCl$_2$ form when 0.0150 mol of Pb(NO$_3$)$_2$ and 0.0120 mol of NaCl are dissolved in 1.00 L of solution?

17.46 Silver acetate, AgC$_2$H$_3$O$_2$, has $K_{sp} = 2.3 \times 10^{-3}$. Does a precipitate form when 0.015 mol of AgNO$_3$ and 0.25 mol of Ca(C$_2$H$_3$O$_2$)$_2$ are dissolved in a total volume of 1.00 L of solution?

17.47 Does a precipitate of PbBr$_2$ form if 50.0 mL of 0.0100 M Pb(NO$_3$)$_2$ is mixed with (a) 50.0 mL of 0.0100 M KBr and (b) 50.0 mL of 0.100 M NaBr?

17.48 Would a precipitate of silver acetate form if 22.0 mL of 0.100 M AgNO$_3$ were added to 45.0 mL of 0.0260 M NaC$_2$H$_3$O$_2$? For AgC$_2$H$_3$O$_2$, $K_{sp} = 2.3 \times 10^{-3}$.

*17.49 Both AgCl and AgI are very sparingly soluble salts, but the solubility of AgI is much less than that of AgCl, as can be seen by their K_{sp} values. Suppose that a solution contains both Cl$^-$ and I$^-$ with [Cl$^-$] = 0.050 M and [I$^-$] = 0.050 M. If Solid AgNO$_3$ is added to 1.00 L of this mixture (so that no appreciable change in volume occurs), what is the value of [I$^-$] when AgCl first begins to precipitate?

17.50 Calculate the molar solubility of CaSO$_4$ in 0.015 M CaCl$_2$.

17.51 Suppose that Na$_2$SO$_4$ is added gradually to 100 mL of a solution that contains both Ca^{2+} ion (0.15 M) and Sr^{2+} ion (0.15 M).

(a) What will the Sr^{2+} concentration be (in mol L^{-1}) when CaSO$_4$ just begins to precipitate?

(b) What percentage of the strontium ion has precipitated when CaSO$_4$ just begins to precipitate?

Oxides, Sulfides, and Selective Precipitations

17.52 Potassium oxide is readily soluble in water, but the resulting solution contains essentially no oxide ion. Explain, using an equation, what happens to the oxide ion.

17.53 Consider cobalt(II) sulfide.
(a) Write its solubility equilibrium and K_{sp} equation for a saturated solution in water.
(b) Write its solubility equilibrium and K_{spa} equation for a saturated solution in aqueous acid.

17.54 Does iron(II) sulfide dissolve in 8 M HCl? Perform the calculations that prove your answer.

17.55 What value of $[H^+]$ and what pH permit the selective precipitation of the sulfide of just one of the two metal ions in a solution that has a concentration of 0.010 M Pb^{2+} and 0.010 M Co^{2+}?

17.56 Can a selective separation of Mn^{2+} from Sn^{2+} be accomplished by a suitable adjustment of the pH of a solution that is 0.010 M in Mn^{2+}, 0.010 M in Sn^{2+}, and saturated in H_2S? (Assume the green form of MnS in Table 17.3.)

Additional Exercises

17.57 Magnesium hydroxide, $Mg(OH)_2$, found in milk of magnesia, has a solubility of 7.05×10^{-3} g L^{-1} at 25 °C.
(a) What is the solubility of $Mg(OH)_2$ in mol L^{-1}?
(b) What are the molar concentrations of Mg^{2+} and OH^- in a saturated solution of $Mg(OH)_2$?
(c) Calculate K_{sp} for $Mg(OH)_2$.

*17.58 Suppose that 50.0 mL of 0.12 M AgNO$_3$ is added to 50.0 mL of 0.048 M NaCl solution.
(a) What mass of AgCl would form?
(b) Calculate the final concentrations of all of the ions in the solution that is in contact with the precipitate.
(c) What percentage of the Ag^+ ion has precipitated?

*17.59 A sample of hard water was found to have 278 ppm Ca^{2+} ion. Into 1.00 L of this water, 1.00 g of Na_2CO_3 was dissolved. What is the new concentration of Ca^{2+} in parts per million? (Assume that the addition of Na_2CO_3 does not change the volume, and assume that the density of the aqueous solutions involved are all 1.00 g mL^{-1}.)

*17.60 What value of $[H^+]$ and what pH would allow the selective separation of the carbonate of just one of the two metal ions in a solution that has a concentration of 0.010 M La^{3+} and 0.010 M Pb^{2+}? For $La_2(CO_3)_3$, $K_{sp} = 4.0 \times 10^{-34}$; for PbCO$_3$, $K_{sp} = 7.4 \times 10^{-14}$. A saturated solution of CO_2 in water has a concentration of H_2CO_3 equal to 3.3×10^{-2} M.

*17.61 When solid NH$_4$Cl is added to a suspension of $Mg(OH)_2(s)$, some of the $Mg(OH)_2$ dissolves.
(a) Write equations for *all* the chemical equilibria that exist in the solution after the addition of the NH$_4$Cl.
(b) Use Le Châtelier's principle to explain why adding NH$_4$Cl causes $Mg(OH)_2$ to dissolve.
(c) How many moles of NH$_4$Cl must be added to 1.0 L of a suspension of $Mg(OH)_2$ to dissolve 0.10 mol of $Mg(OH)_2$?
(d) What is the pH of the solution after the 0.10 mol of $Mg(OH)_2$ has dissolved in the solution containing the NH$_4$Cl?

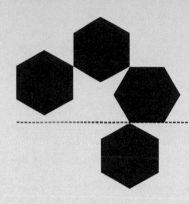

CHEMICALS IN USE 14
Limestone Caverns and Coral Reefs

Of all of the types of stone used in constructing buildings, limestone and the closely related dolomite are employed in greater masses than any other building stone in the United States. *Limestone* consists mostly of calcium carbonate, $CaCO_3$. Sometimes the magnesium ion replaces some of the calcium ion, and when its relative abundance is about the same as the calcium ion the mineral is *dolomite* and has the formula $MgCa(CO_3)_2$.

Limestone forms by the chemical activities of marine organisms like corals, snails, clams, and algae. These use carbon dioxide and calcium ions from the surrounding water to build calcium carbonate into shells and other hardened parts. When the organisms die and decay, the calcium carbonate remains as a deposit.

LIMESTONE CAVERNS AND CHEMICAL EROSION

Carlsbad Caverns (New Mexico), Mammoth Cave (Kentucky), Wind Caves (South Dakota), and Luray Cavern (Virginia) are famous limestone formations that began their formation by an interesting kind of *chemical* erosion rather than by moving wind or water.

The chemical erosion of limestone occurs when it is in contact with water containing acids, whether they are organic acids from the decay of plants, or the acids in "acid rain," or simply carbonic acid (H_2CO_3) produced when carbon dioxide dissolves in water. Combustion, respiration, and decay processes all produce CO_2 naturally, and its maximum concentration in water at 25 °C and 1 atm is 0.033 M. Within the capillary cracks of a limestone formation, however, there is a "bottling effect" that makes the pressure slightly higher than 1 atm. Under the higher pressure, more CO_2 dissolves, more H_2CO_3 forms from it, and more H^+ appears in solution by the ionization of H_2CO_3. The rate of chemical attack on limestone is higher where the concentration of H^+ ions is greater, so the cracks enlarge. Let us study this is more detail using equilibrium expressions and Le Châtelier's principle.

At a higher pressure, a gas is more soluble in water (Henry's law, page 505). As soon as you open a carbonated beverage, for example, the pressure in the container suddenly decreases, becomes the same as that of the atmosphere, and some CO_2 fizzes out. The following equilibrium is involved.

$$CO_2(g) + \text{pressure} \underset{}{\overset{\text{water}}{\rightleftharpoons}} CO_2(aq) \qquad (1)$$

Any change in pressure would put a stress on the equilibrium, and the equilibrium must shift in whichever direction "absorbs" the stress (Le Châtelier's principle). Thus, an increase in pressure makes CO_2 dissolve; a decrease makes CO_2 come out of solution. Now let's look at the *solution*, which contains $CO_2(aq)$, and at the associated equilibria.

A small fraction of the dissolved CO_2 molecules participate in the following equilibrium to give a small concentration of carbonic acid, H_2CO_3.

$$CO_2(aq) + H_2O \rightleftharpoons H_2CO_3(aq) \qquad (2)$$

FIGURE 14*a*

Stalactites and stalagmites are prominent features of most large limestone caverns.

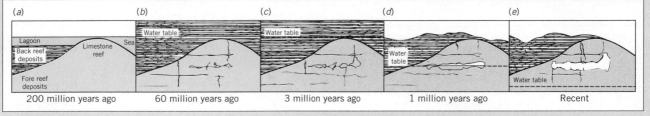

FIGURE 14b

The sequence of events that led to the formation of the Carlsbad Caverns of New Mexico. (a) A limestone reef formed in an ancient sea. (b) The sea level dropped, but the water table remained above the reef. Water rich in CO_2 infiltrated the limestone. (c) The water-filled spaces grew in size. (d) The water table dropped as geological events occurred. (e) As CO_2-rich water percolated through the fissures and as the water table continued to drop, stalactites and stalagmites grew. (Courtesy of John Barnell, N.M.)

If, because of higher pressure, the concentration of $CO_2(aq)$ increases, then the equilibrium *must* shift to the right to use up some of the extra CO_2. Thus, more $H_2CO_3(aq)$ forms when the pressure is higher.

Carbonic acid is a weak acid, so a small percentage of its molecules ionize in the following equilibrium.

$$H_2CO_3(aq) \rightleftharpoons H^+(aq) + HCO_3^-(aq) \qquad (3)$$

Under standard conditions, 1 atm and 25 °C, the concentration of H^+ in water saturated with CO_2 is about 2×10^{-5} M, but when the concentration of H_2CO_3 is higher (because of higher CO_2 pressure), the equilibrium must shift to the right, which makes the level of H^+ greater than normal. In short, a higher pressure makes carbonated water more acidic.

It is the hydrogen ion that attacks limestone; specifically it attacks the carbonate ions present in this rock and changes them to bicarbonate ions, HCO_3^-. Thus the higher the acidity, the more the following equilibrium shifts to the right to give species that are much more soluble in water than $CaCO_3$.

$$CaCO_3(s) + H^+(aq) \rightleftharpoons Ca^{2+}(aq) + HCO_3^-(aq) \qquad (4)$$

It is this change of carbonate ions to the more soluble bicarbonate ions that allows limestone to undergo chemical erosion. And because of the "bottling effect" that leads to a higher concentration of H^+ in the water in fine cracks, the limestone surrounding such cracks dissolves more rapidly than limestone elsewhere. These relatively simple chemical properties account for the onsets of the hollowings of many of the great limestone caverns of the world as well as for the formation of stalactites, stalagmites, columns, walls, travertine terraces, drip curtains, and other beautiful features that fascinate tourists and geologists alike (see Figure 14a).

The sequence that probably happened to create limestone caverns, at least those that formed below

the water table from ancient coral reefs, is as follows (Figure 14b). (The *water table* is the imaginary underground surface below which the ground is saturated with water.) First, water rich in carbon dioxide invades fine cracks and fissures. Then these openings enlarge as the limestone slowly dissolves. The enlarging spaces remain filled with water that, being below ground and "bottled," is still rich in carbon dioxide. Slowly these eroding actions create large spaces.

Eventually, much of the water itself finds outlets, the caves and caverns drain, and the underground "bottles" of space become uncorked, so to speak. However, many fissures in the walls of the caverns remain, and water still rich in carbon dioxide trickles through them carrying away dissolved calcium ions and bicarbonate ions.

When such seepages leave the fissures, the pressure drops slightly and the water loses some of its dissolved carbon dioxide. This loss of CO_2 leads to leftward shifts of Equilibria (1) – (4). When Equilibrium (4) shifts to the left, insoluble $CaCO_3$ precipitates. In other words, after the water leaves a crack in the cavern wall, the $CaCO_3$ comes out of solution. When this happens *before* a drop lets go from a cavern ceiling, a spikelike formation, called a stalactite, grows just a little more downward. Otherwise, the calcium carbonate leaves the solution when the drop hits the cave floor and helps to build up a heavier growth, called a stalagmite. If the water drips from a long, meandering joint crack, the result is a drip curtain.

Questions

1. Write the equilibrium expressions for the following systems.
   (a) The formation of carbonic acid from dissolved carbon dioxide. (b) The ionization of carbonic acid. (c) The action of hydrogen ion on calcium carbonate as it occurs within a fissure of a limestone cavern. (d) The dissolving of carbon dioxide under pressure.

2. Marble is a special form of calcium carbonate. Marble statues corrode when in contact with rainwater rich in dissolved acids. Write an equation that explains how this happens.

3. A drop in pressure shifts several equilibria involved with the formation of stalactites. Writing the changes that occur *as forward reactions only,* not the equilibrium expressions, explain how stalactites form.

Concepts of electrochemistry, studied in this chapter, have made possible the battery packs that power laptop computers, such as the one in use by this businesswoman on an idyllic tropical beach.

Chapter 18

Electrochemistry

18.1 ELECTRICITY AND CHEMICAL CHANGE

Oxidation and reduction (redox) reactions occur in many chemical systems. Examples include the rusting of iron, the action of bleach on stains, and the reactions of photosynthesis in the leaves of green plants. All of these changes involve the transfer of electrons from one chemical species to another. Whether redox reactions spontaneously cause electrons to flow through a wire, or electrons from an outside source force redox reactions to happen, the processes are described as **electrochemical changes.** The study of such changes is called **electrochemistry.**

The applications of electrochemistry are widespread. In industry, many important chemicals, including liquid bleach (sodium hypochlorite) and lye (sodium hydroxide), are manufactured by electrochemical reactions. Batteries, which produce electrical energy by means of chemical reactions, are used to power toys, flashlights, electronic calculators, laptop computers, heart pacemakers, radios, tape recorders, and even some automobiles. In the laboratory, electrical measurements enable us to monitor chemical reactions of all sorts, even those in systems as tiny as a living cell.

In this chapter we will study the factors that affect the outcome of electrochemical changes. Besides the practical applications, fundamental information about chemical reactions is available from electrical measurements—for example, free energy changes and equilibrium constants. We will see that electrochemistry is a very versatile tool for investigating chemical and biological systems.

Clinical chemists routinely use electrochemical reactions to determine the concentrations of substances in blood.

18.2 ELECTROLYSIS

When electricity is passed through a molten (melted) ionic compound or through a solution of an electrolyte, a chemical reaction occurs that we call **electrolysis.** A typical electrolysis apparatus, called an **electrolysis cell** or **elec-**

FIGURE 18.1

An electrolysis cell in which the passage of an electric current decomposes molten sodium chloride into metallic sodium and gaseous chlorine. Unless the products are kept apart, they react on contact to re-form NaCl.

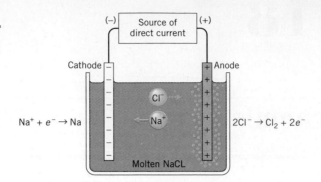

Sodium chloride melts at 801 °C.

To study electrochemical changes we must use direct current, in which electrons move in only one direction, not in the oscillating, back and forth pattern of alternating current.

At the melting point of NaCl, metallic sodium is a liquid.

trolytic cell, is shown in Figure 18.1. This particular cell contains molten sodium chloride. (A substance undergoing electrolysis must be molten or in solution so its ions can move freely and conduction can occur.) *Inert electrodes* —electrodes that won't react with the molten NaCl—are dipped into the cell and then connected to a source of direct current (DC) electricity.

When electricity starts to flow, chemical changes begin to happen. At the positive electrode, the *anode,* oxidation occurs as electrons are pulled from negatively charged chloride ions. The DC source pumps these electrons through the external electrical circuit to the negative electrode, the *cathode.* At the cathode, reduction takes place as the electrons are pushed onto positively charged sodium ions.

The chemical changes that occur at the electrodes can be described by chemical equations.

$$Na^+(l) + e^- \longrightarrow Na(l) \qquad \text{(cathode)}$$

$$2Cl^-(l) \longrightarrow Cl_2(g) + 2e^- \qquad \text{(anode)}$$

You may recognize these as half-reactions of the type we used in Section 4.9 to balance equations by the ion–electron method. This is no accident. The half-reactions generated by the ion–electron method correspond to the chemical changes that occur at the electrodes during electrochemical changes.

The identification of an electrode as the anode or cathode depends on whether oxidation or reduction occurs there. In any apparatus in which an electrochemical reaction is taking place, the following definitions apply:

Identification of electrodes

The **anode** is the electrode at which oxidation occurs.

The **cathode** is the electrode at which reduction occurs.

Conduction of Charge in Electrochemical Cells

The conduction of electricity by a metal like copper occurs by the movement of *electrons.* However, in a molten salt such as sodium chloride, or in a solution of an electrolyte, electrical charge is carried through the liquid by the movement of *ions.* The transport of electrical charge by ions is called **electrolytic conduction,** and it can occur only when chemical reactions take place at the electrodes.

In an electrolytic cell, the electrical current is carried by a metal wire (or bar) in the external circuit.

When charged electrodes are dipped into molten NaCl they become surrounded by a layer of ions of opposite charge. Let's look at what happens at one of them, the anode. Here we find a coating of negative ions on the surface of the electrode, and as these are oxidized their places are quickly taken by others from the surrounding liquid. This leaves the surrounding liquid positively charged, so other negative ions from further away move toward the anode to keep the liquid there electrically neutral. In this way, negative ions gradually migrate toward the anode. By a similar process, positive ions diffuse through the liquid toward the negatively charged cathode, where they become reduced.

Negative ions are called **anions** because they move toward the anode; positive ions are **cations** because they move toward the cathode.

Cell Reactions

The overall reaction that takes place in the electrolysis cell is called the **cell reaction.** To obtain it, we add the individual electrode half-reactions together, but only after we make sure that the number of electrons gained in one half-reaction equals the number lost in the other, a requirement of every redox reaction. The procedure we use is the one described as the *ion–electron method* (Section 4.9). Thus, to obtain the cell reaction for the electrolysis of molten NaCl, we multiply the half-reaction for the reduction of sodium by 2 and then add the two half-reactions to obtain the net reaction. (Notice that $2e^-$ appears on each side, and so they cancel.)

Electrons are *transferred,* so no net gain or loss of electrons occurs.

$$2Na^+(l) + 2e^- \longrightarrow 2Na(l) \qquad \text{(cathode)}$$
$$2Cl^-(l) \longrightarrow Cl_2(g) + 2e^- \qquad \text{(anode)}$$
$$\overline{2Na^+(l) + 2Cl^-(l) + 2e^- \longrightarrow 2Na(l) + Cl_2(g) + 2e^-} \qquad \text{(cell reaction)}$$

As you know, table salt is quite stable. It doesn't normally decompose, because the reaction of sodium and chlorine to form sodium chloride is highly spontaneous. Therefore, we often write the word *electrolysis* above the arrow in the equation to show that electricity is the driving force for this otherwise nonspontaneous reaction.

$$2Na^+(l) + 2Cl^-(l) \xrightarrow{\text{Electrolysis}} 2Na(l) + Cl_2(g)$$

Electrolysis Reactions in Aqueous Solutions

When electrolysis is carried out in an aqueous solution, the electrode reactions are more difficult to predict because the oxidation and reduction of water can also occur. This happens, for example, when electrolysis is carried out in a solution of potassium nitrate (see Figure 18.2). The products are hydrogen and oxygen. At the cathode, water is reduced, not K^+.

$$2H_2O(l) + 2e^- \longrightarrow H_2(g) + 2OH^-(aq) \qquad \text{(cathode)}$$

At the anode, water is oxidized, not the nitrate ion.

$$2H_2O(l) \longrightarrow O_2(g) + 4H^+(aq) + 4e^- \qquad \text{(anode)}$$

When electrons appear as a reactant, the process is reduction; when they appear as a product, it is oxidation.

FIGURE 18.2

Electrolysis of a solution of potassium nitrate gives hydrogen gas and oxygen gas as products.

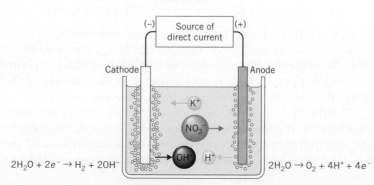

$2H_2O + 2e^- \rightarrow H_2 + 2OH^-$ $2H_2O \rightarrow O_2 + 4H^+ + 4e^-$

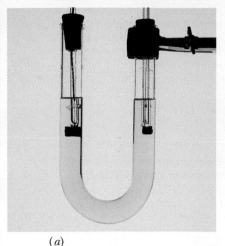

(a)

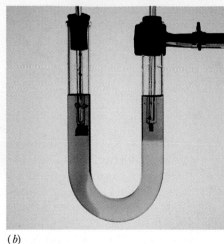

(b)

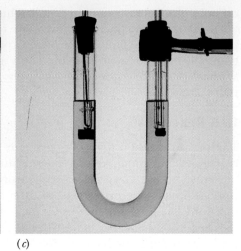

(c)

FIGURE 18.3

Electrolysis of a solution of potassium nitrate, KNO_3, in the presence of indicators. (*a*) The initial yellow color indicates that the solution is neutral (neither acidic nor basic). (*b*) As the electrolysis proceeds, H^+ produced at the anode along with O_2 causes the solution there to become pink. At the cathode, H_2 is evolved and OH^- ions are formed, which turns the solution around that electrode a light tan. (*c*) After the electrolysis is stopped and the solution is stirred, the color becomes yellow again as the H^+ and OH^- ions formed by the electrolysis neutralize each other.

Color changes of an acid–base indicator dissolved in the solution confirm that the solution becomes basic around the cathode and acidic around the anode as the electrolysis proceeds (see Figure 18.3). In addition, the gases H_2 and O_2 can be separately collected.

The overall cell reaction can be obtained as before. Because the number of electrons lost has to equal the number gained, the cathode reaction must occur twice each time the anode reaction occurs once.

$$4H_2O(l) + 4e^- \longrightarrow 2H_2(g) + 4OH^-(aq) \qquad \text{(cathode)}$$

$$2H_2O(l) \longrightarrow O_2(g) + 4H^+(aq) + 4e^- \qquad \text{(anode)}$$

After adding, we combine the coefficients for water and cancel the electrons from both sides to obtain the cell reaction.

$$6H_2O(l) \longrightarrow 2H_2(g) + O_2(g) + 4H^+(aq) + 4OH^-(aq)$$

Notice that hydrogen ions and hydroxide ions are produced in equal numbers. If the solution is stirred, they combine to form water.

$$6H_2O(l) \longrightarrow 2H_2(g) + O_2(g) + \underbrace{4H^+(aq) + 4OH^-(aq)}_{\downarrow}$$
$$4H_2O$$

The net change, then, is

$$2H_2O \xrightarrow{\text{Electrolysis}} 2H_2(g) + O_2(g)$$

At this point you may have begun to wonder whether the potassium nitrate serves any function, since neither K^+ nor NO_3^- ions are changed by the reaction. Nevertheless, if the electrolysis is attempted with just plain distilled water, nothing happens. There is no current flow, and no H_2 or O_2 forms. Apparently, the potassium nitrate must have some function.

The function of KNO_3 (or some other electrolyte) is to maintain electrical neutrality in the vicinity of the electrodes. If the KNO_3 were not present and the electrolysis were to occur anyway, the solution around the anode would become filled with H^+ ions, with no negative ions to balance their charge.

Similarly, the solution surrounding the cathode would become filled with hydroxide ions with no nearby positive ions. Separations like this of oppositely charged ions would demand too much energy to occur.

When KNO_3 is in the solution, K^+ ions move toward the cathode, where they mingle with the OH^- ions as they are formed. The NO_3^- ions move toward the anode and mingle with the H^+ ions as they are produced there. In this way, at any moment, each small region of the solution contains the same number of positive and negative charges.

Predicting the outcome of electrolysis reactions in aqueous solutions can be tricky because the reactions at the electrode surfaces are often complex, especially when they involve the evolution of hydrogen or oxygen. For example, the oxidation of H_2O to give O_2 is thermodynamically easier (in terms of the sign and magnitude of $\Delta G°$) than the oxidation of Cl^- to give Cl_2. The two competing half-reactions at the anode when aqueous sodium chloride is electrolyzed would be

$$2H_2O(l) \longrightarrow O_2(g) + 4H^+(aq) + 4e^-$$

$$2Cl^-(aq) \longrightarrow Cl_2(g) + 2e^-$$

As we said, oxygen should form by the first half-reaction, but because of the complexity of the electrode reactions, chlorine forms instead (provided that the chloride ion concentration is reasonably high). Thus, even though the evolution of O_2 is thermodynamically preferred, we don't see this happen because of complicating factors.

Although it can be hard to anticipate beforehand what will happen in the electrolysis of aqueous solutions, we still can use what we learn experimentally about one electrolysis to predict what will happen in others. For instance, when a solution of $CuBr_2$ is electrolyzed, the cathode becomes coated with a deposit of copper and the solution around the anode acquires an orange-yellow color (see Figure 18.4). These observations tell us that copper ion is being reduced at the cathode and that bromide ion is being oxidized to bromine at the anode. The electrode half-reactions are

$$Cu^{2+}(aq) + 2e^- \longrightarrow Cu(s) \qquad \text{(cathode)}$$

$$2Br^-(aq) \longrightarrow Br_2(aq) + 2e^- \qquad \text{(anode)}$$

The net cell reaction is

$$Cu^{2+}(aq) + 2Br^-(aq) \xrightarrow{\text{Electrolysis}} Cu(s) + Br_2(aq)$$

The behavior of a particular ion in an aqueous solution toward oxidation or reduction by electrolysis is the same regardless of the source of the ion. Therefore, once we know what happens in the electrolysis of $CuBr_2$, we can predict at least partially what will happen in the electrolysis of other salts that contain either of these ions. Solutions of $CuCl_2$, $CuSO_4$, $Cu(NO_3)_2$, and $Cu(C_2H_3O_2)_2$ all contain Cu^{2+} ion, and when they are electrolyzed we can expect that in each case Cu^{2+} will be reduced at the cathode. (However, we can't say at this point what will happen at the anode.) Similarly, solutions of KBr, NaBr, $CaBr_2$, and $FeBr_2$ all contain Br^-, so when solutions of these salts are electrolyzed we can expect that Br_2 will be formed at the anode.

FIGURE 18.4

Electrolysis of a solution of copper(II) bromide. The blue color of the solution is due to copper ion [actually, $Cu(H_2O)_4^{2+}$ ion]. At the anode on the left, bromide ion is oxidized to bromine, which imparts an orange-yellow color to the solution. At the cathode, copper is reduced. Here we see it as a black deposit building up on the electrode.

EXAMPLE 18.1
Predicting the Outcome of an Electrolysis

From the results of the electrolysis of solutions of KNO_3 and of $CuBr_2$, predict the products that will appear if a solution of $Cu(NO_3)_2$ is electrolyzed. Give the net cell reaction.

SOLUTION A solution of $Cu(NO_3)_2$ contains Cu^{2+} and NO_3^- ions. The Cu^{2+} will move toward the cathode and NO_3^- will move toward the anode. The electrolyses described previously in this section tell us that Cu^{2+} is more easily reduced than water at the cathode and that water is more easily oxidized than NO_3^- at the anode. The electrode reactions, therefore, will be

$$Cu^{2+}(aq) + 2e^- \longrightarrow Cu(s) \qquad \text{(cathode: reduction of } Cu^{2+})$$

$$2H_2O(l) \longrightarrow O_2(g) + 4H^+(aq) + 4e^- \quad \text{(anode: oxidation of } H_2O)$$

To obtain the cell reaction, we must multiply the cathode reaction by 2 so that the numbers of electrons gained and lost are equal. After combining half-reactions, we obtain

$$2Cu^{2+}(aq) + 2H_2O(l) \longrightarrow 2Cu(s) + O_2(g) + 4H^+(aq)$$

■ **Practice Exercise 1** From the electrolysis reactions described in this section, predict the products that will form when a solution of KBr undergoes electrolysis. Write the equation for the net cell reaction.

18.3
STOICHIOMETRIC RELATIONSHIPS IN ELECTROLYSIS

Michael Faraday (1791–1867), a British scientist and both a chemist and a physicist, made key discoveries leading to electric motors, generators, and transformers.

Much of the early research in electrochemistry was performed by Michael Faraday. It was he who coined the terms anode, cathode, electrode, electrolyte, and electrolysis. In about 1833, Faraday discovered that the amount of chemical change that occurs during electrolysis is directly proportional to the amount of electrical charge that is passed through an electrolysis cell. For example, the reduction of copper ion at a cathode is given by the equation

$$Cu^{2+}(aq) + 2e^- \longrightarrow Cu(s)$$

The equation tells us that to deposit one mole of metallic copper requires two moles of electrons. Therefore, to deposit two moles of copper requires four moles of electrons, and that takes twice as much electricity. The half-reaction for an oxidation or reduction, therefore, relates the amount of chemical substance consumed or produced to the amount of electrons that the electric current must supply. To use this information, however, we must be able to relate it to electrical measurements that can be made in the laboratory.

The SI unit of electric current is the **ampere (A)** and the SI unit of charge is the **coulomb (C).** A coulomb is the amount of charge that passes by a given point in a wire when an electric current of one ampere flows for one second. This means that coulombs are the product of amperes of current multiplied by seconds. Thus

$$1 \text{ coulomb} = 1 \text{ ampere} \times 1 \text{ second}$$

A·s means amperes × seconds. The dot means "multiplied by."

$$1 \text{ C} = 1 \text{ A·s}$$

For example, if a current of 4 A flows through a wire for 10 s, 40 C pass by a given point in the wire.

$$(4 \text{ A}) \times (10 \text{ s}) = 40 \text{ A} \cdot \text{s}$$
$$= 40 \text{ C}$$

Experimentally, it has been determined that 1 mol of electrons carries a charge of 96,485 C. We will use this number rounded to three significant figures.

$$1 \text{ mol } e^- \Longleftrightarrow 9.65 \times 10^4 \text{ C} \quad \text{(to three significant figures)}$$

In electrochemistry, 1 mol of electrons is called 1 **faraday** ($\mathcal{F}$), in honor of Michael Faraday. Thus,

$$1 \mathcal{F} = 9.65 \times 10^4 \text{ C}$$

 Faraday constant

The number of coulombs per faraday is called the **Faraday constant.**

Now we have a way to relate laboratory measurements to the amount of chemical change that occurs during an electrolysis. Measuring the current in amperes and the time in seconds allows us to calculate the charge sent through the system in coulombs. From this we can get the amount of electrons (in moles), which we can then use to calculate the amount of chemical change produced.

How many grams of copper are deposited on the cathode of an electrolytic cell if an electric current of 2.00 A is run through a solution of $CuSO_4$ for a period of 20.0 min?

EXAMPLE 18.2
Calculations Related to Electrolysis

ANALYSIS The balanced half-reaction serves as our tool for relating chemical change to amounts of electricity. The ion being reduced is Cu^{2+}, so the half-reaction is

$$Cu^{2+} + 2e^- \longrightarrow Cu$$

Therefore,

$$1 \text{ mol Cu} \Longleftrightarrow 2 \text{ mol } e^-$$

The product of current (in amperes) and time (in seconds) will give us charge (in coulombs). We can relate this to the number of moles of electrons by the Faraday constant. Then, from the number of moles of electrons we calculate moles of copper, from which we calculate the mass of copper by using the atomic mass.

SOLUTION First we convert minutes to seconds; $20.0 \text{ min} = 1.20 \times 10^3 \text{ s}$. Then we multiply the current by the time to obtain the number of coulombs ($1 \text{ A} \cdot \text{s} = 1 \text{ C}$).

$$20.0 \text{ min} \times \frac{60 \text{ s}}{1 \text{ min}} = 1.20 \times 10^3 \text{ s}$$

$$(1.20 \times 10^3 \text{ s}) \times (2.00 \text{ A}) = 2.40 \times 10^3 \text{ A} \cdot \text{s}$$
$$= 2.40 \times 10^3 \text{ C}$$

Because $1 \text{ mol } e^- \Longleftrightarrow 9.65 \times 10^4 \text{ C}$,

$$2.40 \times 10^3 \text{ C} \times \frac{1 \text{ mol } e^-}{9.65 \times 10^4 \text{ C}} = 0.0249 \text{ mol } e^-$$

Next, we use the relationship between mol e^- and mol Cu from the balanced half-reaction along with the atomic mass of copper.

$$0.0249 \text{ mol } e^- \times \left(\frac{1 \text{ mol Cu}}{2 \text{ mol } e^-}\right) \times \left(\frac{63.5 \text{ g Cu}}{1 \text{ mol Cu}}\right) = 0.791 \text{ g Cu}$$

The electrolysis will deposit 0.791 g of copper on the cathode.

In this string of conversion factors, we have replaced 9.65 × 10⁴ C by 9.65 × 10⁴ A·s. Remember, 1 C = 1 A·s.

We could have combined all these steps in a single calculation by stringing together the various conversion factors and using the factor–label method to cancel units.

$$2.00 \; A \times 20.0 \; min \times \frac{60 \; s}{1 \; min} \times \frac{1 \; mol \; e^-}{9.65 \times 10^4 \; A \cdot s} \times \frac{1 \; mol \; Cu}{2 \; mol \; e^-} \times \frac{63.5 \; g \; Cu}{1 \; mol \; Cu}$$
$$= 0.790 \; g \; Cu$$

The small difference between the two answers is because of rounding off in the stepwise calculation.

EXAMPLE 18.3
Calculations Related to Electrolysis

Electroplating is an important application of electrolysis. How much time would it take in minutes to deposit 0.500 g of metallic nickel on a metal object using a current of 3.00 A? The nickel is reduced from the +2 oxidation state.

ANALYSIS We need an equation for the reduction. Because the nickel is reduced to the free metal from the +2 state, we can write

$$Ni^{2+} + 2e^- \longrightarrow Ni$$

This gives the relationship

$$1 \; mol \; Ni \Leftrightarrow 2 \; mol \; e^-$$

We wish to deposit 0.500 g of Ni, which we can convert to moles. Then we can calculate the number of moles of electrons required, which in turn is used with the Faraday constant to determine the number of coulombs required. Because this is the product of amperes and seconds, we can calculate the time needed to deposit the metal.

$$\text{Faraday constant} = \frac{9.65 \times 10^4 \; C}{mol \; e^-}$$

SOLUTION First, we calculate the number of moles of electrons required.

$$0.500 \; g \; Ni \times \left(\frac{1 \; mol \; Ni}{58.7 \; g \; Ni} \right) \times \left(\frac{2 \; mol \; e^-}{1 \; mol \; Ni} \right) = 0.0170 \; mol \; e^-$$

Then we calculate the number of coulombs needed.

$$0.0170 \; mol \; e^- \times \left(\frac{9.65 \times 10^4 \; C}{1 \; mol \; e^-} \right) = 1.64 \times 10^3 \; C$$
$$= 1.64 \times 10^3 \; A \cdot s$$

This tells us that the product of current multiplied by time equals $1.64 \times 10^3 \; A \cdot s$. The current is 3.00 A. Dividing $1.64 \times 10^3 \; A \cdot s$ by 3.00 A gives the time required in seconds, which can then be converted to minutes.

$$\left(\frac{1.64 \times 10^3 \; A \cdot s}{3.00 \; A} \right) \times \left(\frac{1 \; min}{60 \; s} \right) = 9.11 \; min$$

We could also have combined these calculations in a single string of conversion factors.

$$0.500 \; g \; Ni \times \frac{1 \; mol \; Ni}{58.7 \; g \; Ni} \times \frac{2 \; mol \; e^-}{1 \; mol \; Ni} \times \frac{9.65 \times 10^4 \; A \cdot s}{1 \; mol \; e^-} \times \frac{1}{3.00 \; A} \times \frac{1 \; min}{60 \; s} = 9.13 \; min$$

As before, the small difference between the two answers is caused by rounding in the stepwise calculation.

What current is needed to deposit 0.500 g of chromium metal from a solution of Cr^{3+} in a period of 1.00 hr?

SOLUTION This problem is quite similar to Example 18.3. First we write the balanced half-reaction.

$$Cr^{3+} + 3e^- \longrightarrow Cr$$

This gives the relationship: 1 mol Cr $\Longleftrightarrow$ 3 mol e^-, which we use as before.

$$0.500 \text{ g Cr} \times \left(\frac{1 \text{ mol Cr}}{52.0 \text{ g Cr}}\right) \times \left(\frac{3 \text{ mol } e^-}{1 \text{ mol Cr}}\right) = 0.0288 \text{ mol } e^-$$

Then we find the number of coulombs.

$$0.0288 \text{ mol } e^- \times \left(\frac{9.65 \times 10^4 \text{ C}}{1 \text{ mol } e^-}\right) = 2.78 \times 10^3 \text{ C}$$

$$= 2.78 \times 10^3 \text{ A} \cdot \text{s}$$

Since we want the metal to be deposited in 1 hr (3600 s), the current is

$$\frac{2.78 \times 10^3 \text{ A} \cdot \text{s}}{3600 \text{ s}} = 0.772 \text{ A}$$

We can also set up the calculation as a string of conversion factors.

$$0.500 \text{ g Cr} \times \frac{1 \text{ mol Cr}}{52.0 \text{ g Cr}} \times \frac{3 \text{ mol } e^-}{1 \text{ mol Cr}} \times \frac{9.65 \times 10^4 \text{ A} \cdot \text{s}}{1 \text{ mol } e^-} \times \frac{1}{3600 \text{ s}} = 0.773 \text{ A}$$

■ **Practice Exercise 2** How many moles of hydroxide ion will be produced at the cathode during the electrolysis of water with a current of 4.00 A for a period of 200 s? The cathode reaction is $2e^- + 2H_2O \rightarrow H_2 + 2OH^-$.

■ **Practice Exercise 3** How many minutes will it take for a current of 10.0 A to deposit 3.00 g of gold from a solution of $AuCl_3$?

■ **Practice Exercise 4** What current must be supplied to deposit 3.00 g of gold from a solution of $AuCl_3$ in 20.0 min?

> **EXAMPLE 18.4**
> Calculations Related to
> Electrolysis

> Once again, rounding in the step-
> wise calculation causes a slight
> discrepancy between the two an-
> swers.

Besides being a useful tool in the chemistry laboratory, electrolysis has many important industrial applications. In this section we will briefly examine the chemistry of electroplating and the production of some of our most common chemicals.

Electroplating

Electroplating is the application by electrolysis of a thin (generally 0.03 to 0.05 mm thick) ornamental or protective coating of one metal over another. It is a common technique for improving the appearance and durability of metal objects. For instance, a thin, shiny coating of metallic chromium is applied over steel automobile bumpers to make them attractive and to prevent rusting. Silver and gold plating are applied to jewelry made from less expensive metals, and silver plate is common on eating utensils (knives, forks, spoons, etc.).

> **18.4**
> **INDUSTRIAL**
> **APPLICATIONS OF**
> **ELECTROLYSIS**

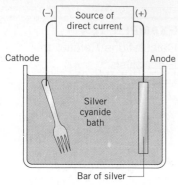

FIGURE 18.5

Apparatus for electroplating silver.

Aluminum is used today as a structural metal, in alloys, and in such products as aluminum foil, electrical wire, window frames, and kitchen utensils.

A large cell can produce as much as 900 lb of aluminum per day.

Figure 18.5 illustrates a typical apparatus used for plating silver. Silver ion in the solution is reduced at the cathode, where it is deposited as metallic silver on the object to be plated. At the anode, silver from the metal bar is oxidized, replenishing the supply of the silver ion in the solution. As time passes, silver is gradually transferred from the bar at the anode onto the object at the cathode.

The exact composition of the electroplating bath varies, depending on the metal to be deposited, and can affect the appearance and durability of the finished surface. For example, silver deposited from a solution of silver nitrate ($AgNO_3$) does not stick to other metal surfaces very well. However, if it is deposited from a solution of silver cyanide containing $Ag(CN)_2^-$, the coating adheres well and is bright and shiny. Other metals that are electroplated from a cyanide bath are gold and cadmium. Nickel, which can also be applied as a protective coating, is plated from a nickel sulfate solution, and chromium is plated from a chromic acid (H_2CrO_4) solution.

Production of Aluminum

Aluminum does not occur as the free metal in nature. It is even so reactive that the traditional methods for obtaining a metal from its ores, like electrolysis, do not work. Thus, until the latter part of the nineteenth century, aluminum was so costly to isolate that only the very wealthy could afford aluminum products. Early efforts to produce aluminum by the electrolysis of a molten form of an aluminum compound were unfeasible because its anhydrous salts (those with no water of hydration) are difficult to prepare and aluminum oxide, Al_2O_3, has such a high melting point (over 2000 °C) that no practical method of melting it could be found.

In 1886, a young student at Oberlin College, Charles M. Hall, discovered that Al_2O_3 dissolves in the molten form of a mineral called cryolite, Na_3AlF_6, to give a conducting mixture with a relatively low melting point from which aluminum could be produced electrolytically. The process was also discovered by Paul Héroult in France at nearly the same time, and today this method for producing aluminum is usually called the **Hall–Héroult process** (see Figure 18.6). Purified aluminum oxide, which is obtained from an ore called *bauxite,* is dissolved in molten cryolite in which the oxide dissociates to give Al^{3+} and O^{2-} ions. At the cathode, aluminum ions are reduced to produce the free metal, which forms as a layer of molten aluminum below the less dense solvent. At the carbon anodes, oxide ion is oxidized to give free O_2.

$$Al^{3+} + 3e^- \longrightarrow Al(l) \qquad \text{(cathode)}$$

$$2O^{2-} \longrightarrow O_2(g) + 4e^- \qquad \text{(anode)}$$

FIGURE 18.6

Diagram of the apparatus used to produce aluminum electrolytically by the Hall–Héroult process.

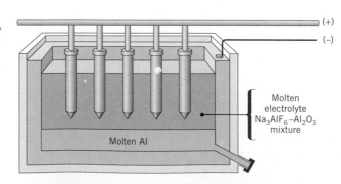

The net cell reaction is

$$4Al^{3+} + 6O^{2-} \longrightarrow 4Al(l) + 3O_2(g)$$

The oxygen formed at the anode attacks the carbon electrodes (producing CO_2), so the electrodes must be replaced frequently.

The production of aluminum consumes enormous amounts of electrical energy and is therefore very costly, not only in terms of dollars but also in terms of energy resources. For this reason, recycling of aluminum has a high priority as we seek to minimize our use of energy.

Production of Magnesium

Metallic magnesium has a number of structural uses because of its low density ("light weight"). Around the home, for example, you may have a magnesium alloy ladder.

The major source of magnesium is seawater in which, on a mole basis, Mg^{2+} is the third most abundant ion, exceeded only by Na^+ and Cl^-. To obtain metallic magnesium, seawater is made basic, which causes Mg^{2+} ions to precipitate as $Mg(OH)_2$. The precipitate is separated by filtration and dissolved in hydrochloric acid.

$$Mg(OH)_2 + 2HCl \longrightarrow MgCl_2 + 2H_2O$$

The resulting solution is evaporated to give solid $MgCl_2$, which is then melted and electrolyzed. Free magnesium is deposited at the cathode and chlorine gas is produced at the anode.

$$MgCl_2(l) \xrightarrow{\text{Electrolysis}} Mg(l) + Cl_2(g)$$

Production of Sodium

Sodium is prepared by the electrolysis of molten sodium chloride (see Section 18.2). The metallic sodium and the chlorine gas that form must be kept apart or they will react violently and re-form NaCl. The **Downs cell** accomplishes this separation (see Figure 18.7).

A row of electrolytic cells fills this aluminum production plant in Ghana (a country located on the west coast of Africa). In the foreground, a crane is about to lift a large ladle filled with molten aluminum.

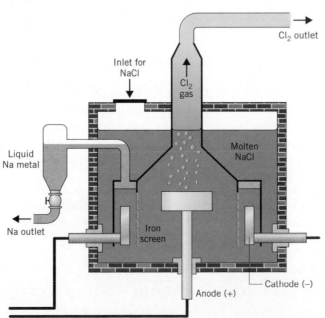

FIGURE   18.7

Cross section of the Downs cell used for the electrolysis of molten sodium chloride. The cathode is a circular ring that surrounds the anode. The electrodes are separated from each other by an iron screen. During the operation of the cell, molten sodium collects at the top of the cathode compartment, from which it is periodically drained. The chlorine gas bubbles out of the anode compartment and is collected.

PVC is widely used to make pipes and conduits for water and sanitary systems.

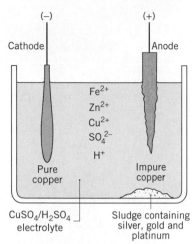

FIGURE 18.8

Purification of copper by electrolysis.

Copper refining is one of the chief sources of gold in the United States.

Copper cathodes, 99.96% pure, are pulled from the electrolytic refining tanks at Kennecott's Utah copper refinery. It takes about 28 days for the impure copper anodes to dissolve and deposit the pure metal on the cathodes.

Both sodium and chlorine are commercially important. Chlorine is used largely to manufacture plastics such as polyvinyl chloride (PVC), many solvents, and industrial chemicals. A small percentage of the annual chlorine production is used to chlorinate drinking water.

Sodium has been used in the manufacture of tetraethyllead, an octane booster for gasoline that has been phased out in the U.S., but which is still used in many other countries. Sodium is also used as a coolant in certain nuclear reactors, and in the production of sodium vapor lamps. These lamps, with their familiar bright yellow color, have the advantage of giving off most of their energy in a portion of the spectrum that humans can see. In terms of useful light output versus energy input, they are about 15 times more efficient than ordinary incandescent light bulbs and about three to four times more efficient than the bluish white mercury vapor lamps sometimes used for street lighting.

Refining of Copper

One of the most interesting and economically attractive applications of electrolysis is the purification or refining of metallic copper. When copper is first obtained from its ore, it is about 99% pure. The impurities—mostly silver, gold, platinum, iron, and zinc—decrease the electrical conductivity of the copper enough that even 99% pure copper must be further refined before it can be used in electrical wire.

The impure copper is used as the anode in an electrolysis cell that contains a solution of copper sulfate and sulfuric acid as the electrolyte (see Figure 18.8). The cathode is a thin sheet of very pure copper. When the cell is operated at the correct voltage, only copper and impurities more easily oxidized than copper (iron and zinc) dissolve at the anode. The less active metals simply fall off the electrode and settle to the bottom of the container. At the cathode, copper ions are reduced, but the zinc ions and iron ions remain in solution because they are more difficult to reduce than copper. Gradually, the impure copper anode dissolves and the copper cathode, about 99.96% pure, grows larger. The accumulating sludge—called anode mud—is removed periodically, and the value of the silver, gold, and platinum recovered from it virtually pays for the entire refining operation.

Electrolysis of Brine

One of the most important commercial electrolysis reactions is the electrolysis of concentrated aqueous sodium chloride solutions called **brine.** An apparatus that could be used for this in the laboratory is shown in Figure 18.9. At the cathode, water is much more easily reduced than sodium ion, so H_2 forms.

$$2H_2O(l) + 2e^- \longrightarrow H_2(g) + 2OH^-(aq) \qquad \text{(cathode)}$$

As we noted earlier, even though water is more easily oxidized than chloride ion, complicating factors at the electrodes actually allow chloride ion to be oxidized instead. At the anode, therefore, we observe the formation of Cl_2.

$$2Cl^-(aq) \longrightarrow Cl_2(g) + 2e^- \qquad \text{(anode)}$$

The net cell reaction is therefore

$$2Cl^-(aq) + 2H_2O(l) \longrightarrow H_2(g) + Cl_2(g) + 2OH^-(aq)$$

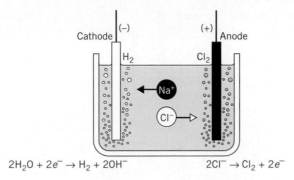

FIGURE 18.9

The electrolysis of brine.

$2H_2O + 2e^- \rightarrow H_2 + 2OH^-$ $2Cl^- \rightarrow Cl_2 + 2e^-$

If we include the sodium ion, already in the solution as a spectator ion and not involved in the electrolysis directly, we can see why this is such an important reaction.

$$\underbrace{2Na^+(aq) + 2Cl^-(aq)}_{2NaCl(aq)} + 2H_2O \xrightarrow{\text{Electrolysis}}$$

$$H_2(g) + Cl_2(g) + \underbrace{2Na^+(aq) + 2\,OH^-(aq)}_{2NaOH(aq)}$$

Thus, the electrolysis converts inexpensive salt to valuable chemicals: H_2, Cl_2, and NaOH. The hydrogen is used to make other chemicals, including hydrogenated vegetable oils. The chlorine is used for the purposes just mentioned. Among the uses of sodium hydroxide, one of industry's most important bases, are the manufacture of soap and paper, the neutralization of acids in industrial reactions, and the purification of aluminum ores.

Sodium hydroxide is commonly known as *lye* or as *caustic soda*.

    In the industrial electrolysis of brine to make pure NaOH, an apparatus as simple as that in Figure 18.9 cannot be used for three chief reasons. First, it is necessary to capture the H_2 and Cl_2 separately to prevent them from mixing and reacting (explosively). Second, the NaOH from the reaction is contaminated with unreacted NaCl. Third, if Cl_2 is left in the presence of NaOH, the solution becomes contaminated by hypochlorite ion (OCl^-), which forms by the reaction of Cl_2 with OH^-.

$$Cl_2 + 2OH^- \longrightarrow Cl^- + OCl^- + H_2O$$

In one manufacturing operation, however, the Cl_2 is not removed as it forms, and its reaction with hydroxide ion is used to manufacture aqueous sodium hypochlorite. For this purpose, the solution is stirred vigorously during the electrolysis so that very little Cl_2 escapes. As a result, a stirred solution of NaCl gradually changes during electrolysis to a solution of NaOCl, a dilute solution of which is sold as liquid laundry bleach (e.g., Clorox).

    Most of the pure NaOH manufactured today is made in an apparatus called a **diaphragm cell.** The design varies somewhat, but Figure 18.10 illustrates its basic features. The cell consists of an iron wire mesh cathode that encloses a porous asbestos shell—the diaphragm. The NaCl solution is added to the top of the cell and seeps slowly through the diaphragm. When it contacts the iron cathode, hydrogen is evolved and is pumped out of the surrounding space. The solution, now containing dilute NaOH, drips off the cell into the reservoir below. Meanwhile, within the cell, chlorine is generated at the anodes dipping

FIGURE 18.10

A *diaphragm cell* used in the production of NaOH by the electrolysis of aqueous NaCl. This is a cross section of a cylindrical cell in which the NaCl solution is surrounded by an asbestos diaphragm supported by an iron mesh cathode. (From J. E. Brady and G. E. Humiston, *General Chemistry: Principles and Structure*, 4th ed. Copyright © 1986, John Wiley & Sons, New York. Used by permission.)

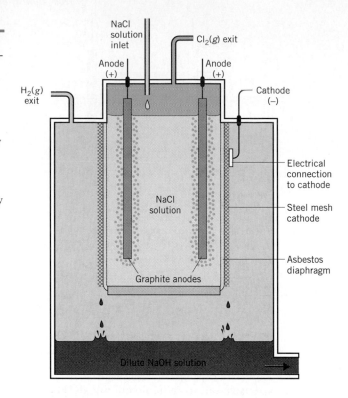

The mercury cell, which gives pure NaOH, poses an environmental hazard because of the potential for mercury pollution. The waste products discharged from plants using mercury cells must be carefully monitored to prevent this.

into the NaCl solution. Because there is no OH^- in this solution, the Cl_2 can't react to form OCl^- ion and simply bubbles out of the solution and is captured.

The NaOH obtained from the diaphragm cell, although free of OCl^-, is still contaminated by small amounts of NaCl. The production of pure NaOH is accomplished using a **mercury cell** (see Figure 18.11). In this apparatus, mercury serves as the cathode, and the cathode reaction actually involves the reduction of sodium ions to sodium atoms, which dissolve in the liquid mercury. The mercury–sodium solution is pumped to another chamber and ex-

FIGURE 18.11

Electrolysis of aqueous sodium chloride (brine) using a *mercury cell*. At the anode, chloride ions are oxidized to chlorine, Cl_2. At the cathode, sodium ions are reduced to sodium atoms, which dissolve in the mercury. This mercury is pumped to a separate compartment, where it is exposed to water. As sodium atoms come to the surface of the mercury, they react with water to give H_2 and NaOH. (From J. E. Brady and G. E. Humiston, *General Chemistry: Principles and Structure*, 4th ed. Copyright © 1986, John Wiley & Sons, New York. Used by permission.)

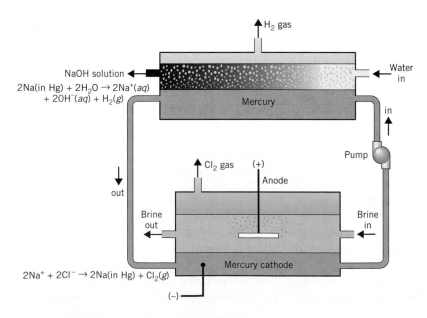

posed to water free of NaCl. Here, the sodium atoms react with water to give H_2 and NaOH. The net overall chemical change is still the conversion of NaCl into H_2, Cl_2, and NaOH, but the NaOH produced in the mercury cell isn't contaminated by NaCl.

If you have silver fillings in your teeth, you may have experienced a strange and perhaps even unpleasant sensation while accidentally biting on a piece of aluminum foil. The sensation is caused by a very mild electric shock produced by a voltage difference between your metal fillings and the aluminum. In effect, you created a battery. Batteries not too much different from this serve to power all sorts of gadgets, as mentioned at the beginning of this chapter. The energy for batteries comes from spontaneous redox reactions in which the electron transfer is forced to take place through a wire. Cells that provide electricity in this way are called **galvanic cells,** after Luigi Galvani (1737–1798), an Italian anatomist who discovered that electricity can cause the contraction of muscles.

If a shiny piece of metallic copper is placed into a solution of silver nitrate, a spontaneous reaction occurs. A grayish white deposit forms on the copper, and the solution itself becomes pale blue because of hydrated Cu^{2+} ions (see Figure 18.12). The equation is

$$2Ag^+(aq) + Cu(s) \longrightarrow Cu^{2+}(aq) + 2Ag(s)$$

No usable energy can be harnessed from this reaction, however, because all of the released energy appears as heat.

To produce *electrical* energy, the two half-reactions involved in the net reaction must be made to occur in separate containers or compartments called **half-cells.** An apparatus to accomplish this—a galvanic cell—is illustrated in

18.5
GALVANIC CELLS

Galvanic cells are also called **voltaic cells,** after Italian scientist Alessandro Volta (1745–1827), the inventor of the battery.

FIGURE 18.12

(*Left*) A coiled piece of copper wire next to a beaker containing a silver nitrate solution. (*Center*) When the copper wire is placed in the solution, copper dissolves, giving the solution its blue color, and metallic silver deposits as glittering crystals on the wire. (*Right*) After a while, much of the copper has dissolved and nearly all of the silver has deposited as the free metal.

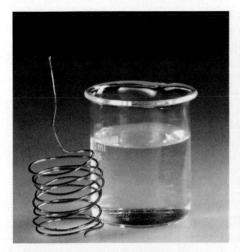

FIGURE 18.13

A galvanic cell.

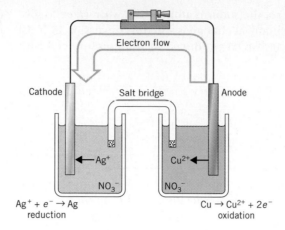

Figure 18.13. On the left, a silver electrode dips into a solution of $AgNO_3$, and, on the right, a copper electrode dips into a $Cu(NO_3)_2$ solution. The two solutions are connected by an external electrical circuit as well as by a *salt bridge,* the function of which will be described shortly. When the circuit is completed by closing the switch, the reduction of Ag^+ to Ag occurs spontaneously in the beaker on the left and oxidation of Cu to Cu^{2+} occurs spontaneously in the beaker on the right. Because of the nature of the reactions taking place, we can identify the silver electrode as the cathode and the copper electrode as the anode.

> **The anode is always the electrode at which oxidation takes place, and the cathode is where reduction takes place.**

$$Ag^+(aq) + e^- \longrightarrow Ag(s) \qquad \text{(reduction; cathode)}$$

$$Cu(s) \longrightarrow Cu^{2+}(aq) + 2e^- \qquad \text{(oxidation; anode)}$$

When these reactions take place, electrons left behind by oxidation of the copper travel as an electric current through the external circuit to the cathode where they are picked up by the silver ions, which are thereby reduced.

For a galvanic cell to work, the solutions in both half-cells must remain electrically neutral. This requires that ions be permitted to enter or leave the solutions. For example, when copper is oxidized, the solution surrounding the electrode becomes filled with Cu^{2+} ions, so negative ions are needed to balance their charge. Similarly, when Ag^+ ions are reduced, NO_3^- ions are left behind in the solution and positive ions are needed to maintain neutrality. The salt bridge shown in Figure 18.13 serves these purposes. A **salt bridge** is a tube filled with an electrolyte solution, commonly KNO_3 or KCl, and fitted with porous plugs at each end. During operation of the cell, negative ions can diffuse from the salt bridge into the copper half-cell, or Cu^{2+} ions can leave the solution and enter the salt bridge. Both processes help keep this half-cell electrically neutral. At the other half-cell, positive ions from the salt bridge can enter or negative NO_3^- ions can leave to keep it electrically neutral.

Without the salt bridge, electrical neutrality couldn't be maintained and no electrical current could be produced by the cell. Therefore, *electrolytic contact* —contact by means of a solution containing ions—must be maintained for the cell to function.

Charges on the Electrodes

In an *electrolysis* cell, the cathode carries a negative charge and the anode carries a positive charge. In a *galvanic* cell, these charges are reversed. At the anode of the galvanic cell in Figure 18.13, copper atoms leave the electrode and enter the solution as Cu^{2+} ions. The electrons that are left behind give the anode a negative charge. At the cathode, electrons are joining Ag^+ ions to produce neutral atoms, but the effect is the same as if Ag^+ ions became part of the electrode, so the cathode acquires a positive charge. The difference in charge between cathode and anode is what causes the electric current to flow when the circuit is complete. Remember, however, that it is the nature of the chemical change, not the electrical charge, that determines whether we label an electrode as a cathode or an anode.

> The difference in charge between the electrodes is forced by the spontaneity of the overall reaction, that is, by the favorable free energy change.

Electrolytic Cell	Galvanic Cell
Cathode is negative (reduction)	Cathode is positive (reduction)
Anode is positive (oxidation)	Anode is negative (oxidation)

 Identification of electrodes

Even though the charges on the cathode and anode differ between electrolytic cells and galvanic cells, the ions in solution always move in the same direction. A cation is a positive ion that always moves away from the anode toward the cathode. In both types of cells, positive ions move toward the cathode. They are attracted there by the negative charge on the cathode in an electrolysis cell; they diffuse toward the cathode in our galvanic cell to balance the charge of negative ions left behind when the Ag^+ ions are reduced. Similarly, anions are negative ions that move away from the cathode and toward the anode. They are attracted to the positive anode in an electrolysis cell, and they diffuse toward the anode in our galvanic cell to balance the charge of the Cu^{2+} ions entering the solution.

> The events in an *electrolytic cell* are forced by electrical energy provided from the outside; those of the *galvanic cell* deliver electrical energy spontaneously to the outside.

Cell Notation

As a matter of convenience, chemists have devised a shorthand way of describing the makeup of a galvanic cell. For example, the copper–silver cell that we have been using in our discussion is described as follows.

$$Cu(s) \,|\, Cu^{2+}(aq) \,\|\, Ag^+(aq) \,|\, Ag(s)$$

By convention, the anode half-cell is specified on the left, with the electrode material of the anode given first. In this case, the anode is copper metal, but in other galvanic cells an electrode could be an inert metal, like platinum, provided that the redox reaction in the half-cell is between two dissolved species (e.g., Fe^{2+} and Fe^{3+}). The single vertical bar represents a phase boundary—here, between the copper electrode and the solution that surrounds it. The double vertical bars represent the salt bridge separating the two half-cells. On the right, the cathode half-cell is described, with the material of the cathode given last. Thus, the electrodes themselves (copper and silver) are specified at opposite ends of the cell description.

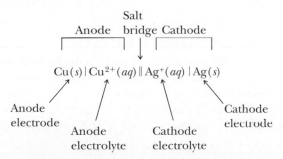

EXAMPLE 18.5
Describing Galvanic Cells

The following spontaneous reaction occurs when metallic zinc is dipped into a solution of copper sulfate.

$$Zn(s) + Cu^{2+}(aq) \longrightarrow Zn^{2+}(aq) + Cu(s)$$

Describe a galvanic cell that could take advantage of this reaction. What are the half-cell reactions? What is the abbreviated cell notation? Make a sketch of the cell and label the cathode and anode, the charges on each electrode, the direction of ion flow, and the direction of electron flow.

ANALYSIS Answering all these questions relies on identifying the anode and cathode from the equation for the cell reaction. The first step, therefore, is to divide the cell reaction into half-reactions. Then, the oxidation takes place at the anode and the reduction at the cathode.

SOLUTION The half-reactions are

$$Zn(s) \longrightarrow Zn^{2+}(aq) + 2e^-$$

$$Cu^{2+}(aq) + 2e^- \longrightarrow Cu(s)$$

Zinc is oxidized, so it is the anode. The anode half-cell is therefore a zinc electrode dipping into a solution that contains Zn^{2+} [e.g., $Zn(NO_3)_2$ or $ZnSO_4$]. Copper ion is reduced, so the cathode half-cell consists of a copper electrode dipping into a solution continuing Cu^{2+} [e.g., $Cu(NO_3)_2$ or $CuSO_4$].

The cell notation places the zinc anode half-cell on the left and the copper cathode half-cell on the right.

$$\underset{\text{anode}}{Zn(s) \,|\, Zn^{2+}(aq)} \,\|\, \underset{\text{cathode}}{Cu^{2+}(aq) \,|\, Cu(s)}$$

A sketch of the cell is shown in the margin. The anode always carries a negative charge in a galvanic cell, so the zinc electrode is negative and the copper electrode is positive. Electrons in the external circuit travel from the negative electrode to the positive electrode (i.e., from the Zn anode to the Cu cathode). Anions move toward the anode and cations move toward the cathode.

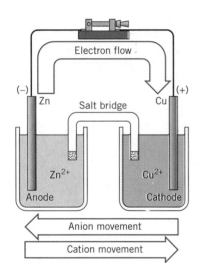

■ **Practice Exercise 5** Sketch and label a galvanic cell that makes use of the following spontaneous redox reaction.

$$Mg(s) + Fe^{2+}(aq) \longrightarrow Mg^{2+}(aq) + Fe(s)$$

Write the half-reactions for the anode and cathode. Give the abbreviated cell notation.

■ **Practice Exercise 6** Write the anode and cathode half-reactions for the following galvanic cell.

$$Al(s) \,|\, Al^{3+}(aq) \,\|\, Pb^{2+}(aq) \,|\, Pb(s)$$

18.6
CELL POTENTIALS AND REDUCTION POTENTIALS

The magnitude of the ability of a galvanic cell to push electrons through the external circuit is expressed as the **potential** or **electromotive force (emf)** of the cell. Electromotive force is the force with which an electric current is pushed through a wire and is described by an electrical unit called the **volt (V)**. Strictly speaking, the number of volts or the *voltage* is a measure of the joules of energy that can be delivered per coulomb of charge as the current moves through the circuit. Thus, a current flowing under an emf of 1 volt can deliver 1 joule of energy per coulomb.

Electrical current and emf are often likened to the rate of flow of water in a hose ("current") and the pressure of the water ("emf").

$$1 \text{ V} = 1 \text{ J/C} \qquad (18.1)$$

We will find this definition important in Section 18.8.

Cell Potentials of Galvanic Cells

The voltage or potential of a galvanic cell varies with the amount of current flowing through the circuit. The *maximum* potential that a given cell can generate is called its **cell potential,** E_{cell}, and it depends on the composition of the electrodes, the concentrations of the ions in the half-cells, and the temperature. Therefore, to compare the potentials for different cells we use the **standard cell potential,** symbolized $E°_{\text{cell}}$. This is the potential of the cell when all of the ion concentrations are 1.00 M, the temperature is 25 °C, and any gases that are involved in the cell reaction are at a pressure of 1 atm.

If current is drawn from a cell, some of the cell's emf is lost overcoming its own internal resistance, and the measured voltage isn't E_{cell}.

Standard cell potentials

Cell potentials are rarely larger than a few volts. For example, the standard cell potential for the galvanic cell constructed from silver and copper electrodes shown in Figure 18.14 is only 0.46 V, and one cell in an automobile battery produces only about 2 V. Batteries that generate higher voltages contain a number of cells arranged in series so that their emfs are additive.

Reduction Potentials

It is useful to imagine that the measured overall cell potential arises from a competition between the two half-cells for electrons. We think of each half-cell reaction as having a certain natural tendency to proceed as a *reduction*. The magnitude of this tendency is expressed by the half-reaction's **reduction potential.** When measured under standard conditions, namely, 25 °C, concentrations of 1.00 M for all solutes, and a pressure of 1 atm, the reduction potential is called the **standard reduction potential.** When two half-cells are connected, the one with the larger reduction potential—the one with the greater tendency to undergo reduction—acquires electrons from the half-cell with the lower reduction potential, which is therefore forced to undergo oxidation. The measured cell potential actually represents the magnitude of the *difference* between the reduction potential of one half-cell and the reduction potential of the other. In general, therefore,

Standard reduction potentials

$$E°_{\text{cell}} = \begin{pmatrix} \text{standard reduction} \\ \text{potential of} \\ \text{substance reduced} \end{pmatrix} - \begin{pmatrix} \text{standard reduction} \\ \text{potential of} \\ \text{substance oxidized} \end{pmatrix} \qquad (18.2)$$

As an example, let's look at the copper–silver cell. From the cell reaction,

$$2\text{Ag}^+(aq) + \text{Cu}(s) \longrightarrow 2\text{Ag}(s) + \text{Cu}^{2+}(aq)$$

we can see that silver ion is reduced and copper is oxidized. If we compare the two possible reduction half-reactions,

$$\text{Ag}^+(aq) + e^- \longrightarrow \text{Ag}(s)$$

$$\text{Cu}^{2+}(aq) + 2e^- \longrightarrow \text{Cu}(s)$$

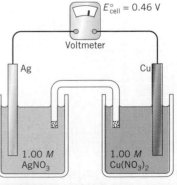

FIGURE 18.14

A cell designed to generate the standard cell potential.

SPECIAL TOPIC 18.1 / CORROSION OF IRON AND CATHODIC PROTECTION

A problem that has plagued humanity ever since the discovery of methods for obtaining iron and other metals from their ores has been corrosion—the reaction of a metal with substances in the environment. The rusting of iron in particular is a serious problem because iron and steel have so many uses.

The rusting of iron is a complex chemical reaction that involves both oxygen and moisture (see Figure 1). Iron won't rust in pure water that's oxygen free, and it won't rust in pure oxygen in the absence of moisture. The corrosion process is apparently electrochemical in nature, as shown in the accompanying diagram. At one place on the surface, iron becomes oxidized in the presence of water and enters solution as Fe^{2+}.

$$Fe(s) \longrightarrow Fe^{2+}(aq) + 2e^-$$

At this location the iron is acting as an anode.

The electrons that are released when the iron is oxidized travel through the metal to some other place where the iron is exposed to oxygen. This is where reduction takes place (it's a cathodic region on the metal surface), and oxygen is reduced to give hydroxide ion.

$$\tfrac{1}{2}O_2(aq) + H_2O + 2e^- \longrightarrow 2OH^-(aq)$$

The iron(II) ions that are formed at the anodic regions gradually diffuse through the water and eventually contact the hydroxide ions. This causes a precipitate of $Fe(OH)_2$ to form, which is very easily oxidized by O_2 to give $Fe(OH)_3$. This hydroxide readily loses water. In fact, complete dehydration gives the oxide,

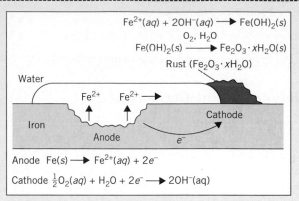

Anode $Fe(s) \longrightarrow Fe^{2+}(aq) + 2e^-$

Cathode $\tfrac{1}{2}O_2(aq) + H_2O + 2e^- \longrightarrow 2OH^-(aq)$

FIGURE 1

Iron dissolves in anodic regions to give Fe^{2+}. Electrons travel through the metal to cathodic sites where oxygen is reduced, forming OH^-. The combination of the Fe^{2+} and OH^-, followed by air oxidation, gives rust. (From J. E. Brady and G. E. Humiston, *General Chemistry: Principles and Structure,* 3rd ed., John Wiley and Sons, New York, 1982. Used by permission.)

$$2Fe(OH)_3 \longrightarrow Fe_2O_3 + 3H_2O$$

When partial dehydration of the $Fe(OH)_3$ occurs, *rust* is formed. It has a composition that lies between that of the hydroxide and that of the oxide, Fe_2O_3, and is usually referred to as a hydrated oxide. Its formula is generally represented as $Fe_2O_3 \cdot xH_2O$.

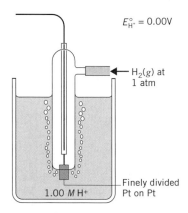

FIGURE 18.15

The hydrogen electrode. The half reaction is $2H^+(aq) + 2e^- \rightleftharpoons H_2(g)$.

the one for Ag^+ must have a greater tendency to proceed than the one for Cu^{2+}, because it is the silver ion that is actually reduced. This means that the standard reduction potential of Ag^+ must be algebraically larger than the standard reduction potential of Cu^{2+}. In other words, if we knew the values of $E°_{Ag^+}$ and $E°_{Cu^{2+}}$, we could calculate $E°_{cell}$ with Equation 18.2 by subtracting the smaller reduction potential from the larger one.

$$E°_{cell} = E°_{Ag^+} - E°_{Cu^{2+}}$$

Assigning Standard Reduction Potentials

Unfortunately there is no way to measure the standard reduction potential of an isolated half-cell. All we can measure is the difference in potential produced when two half-cells are connected. Therefore, to assign values to the various standard reduction potentials, a reference electrode has been arbitrarily chosen and its standard reduction potential has been assigned a value of *exactly* 0 V. This reference electrode is called the **standard hydrogen electrode** (see Figure 18.15). Gaseous hydrogen at a pressure of 1 atm is bubbled over a

This mechanism for the rusting of iron explains one of the more interesting aspects of this damaging process. Perhaps you have noticed that when rusting occurs on the body of a car, the rust appears at and around a break (or a scratch) in the surface of the paint, but the damage extends under the painted surface for some distance. Apparently, the Fe^{2+} ions that are formed at the anode sites are able to diffuse rather long distances to the hole in the paint, where they finally react with air to form the rust.

CATHODIC PROTECTION

One way to prevent the rusting of iron is to coat it with another metal. This is done with "tin" cans, which are actually steel cans that have been coated with a thin layer of tin. However, if the layer of tin is scratched and the iron beneath is exposed, the corrosion is accelerated because iron has a lower reduction potential than tin; the iron becomes the anode in an electrochemical cell and is easily oxidized.

Another way to prevent corrosion is called *cathodic protection.* It involves placing the iron in contact with a metal that is *more easily* oxidized. This causes iron to be a cathode and the other metal to be the anode. If corrosion occurs, iron is protected from oxidation because it is cathodic and the other metal reacts instead.

Zinc is most often used to provide cathodic protection to other metals. For example, zinc sacrificial anodes can be attached to the rudder of a boat (see Figure 2). When submerged, the zinc will gradually corrode but the metal of the rudder will not. Periodically, the anodes are replaced to provide continued protection.

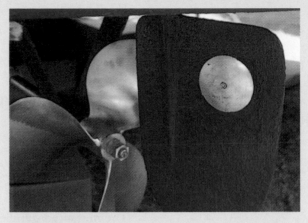

FIGURE 2
Before this boat is launched, a shiny new zinc anode disk is attached to its bronze rudder to provide cathodic protection. Over time, the zinc will corrode instead of the less reactive bronze. (The rudder is painted with a special blue paint to inhibit the growth of barnacles.)

Steel objects that must withstand the weather are often coated with a layer of zinc, a process called galvanizing. You've seen this on chain-link fences and metal garbage pails. Even if the steel is exposed through a scratch, it is prevented from being oxidized because it is in contact with a metal that is more easily oxidized.

platinum electrode coated with very finely divided platinum, which serves as a catalyst for the electrode reaction. This electrode is surrounded by a solution whose temperature is 25 °C and in which the hydrogen ion concentration is 1.00 *M.* The half-cell reaction at the platinum surface, written as a reduction, is

$$2H^+(aq, 1.00\ M) + 2e^- \rightleftharpoons H_2(g, 1\ atm) \qquad E^\circ_{H^+} = 0\ V$$

The double arrows indicate only that the reaction is reversible, not that there is true equilibrium. Whether the half-reaction occurs as reduction or oxidation depends on the reduction potential of the half-cell with which it is paired.

Figure 18.16 illustrates the hydrogen electrode connected to a copper half-cell to form a galvanic cell. To obtain the cell reaction, we have to know what is oxidized and what is reduced, and we need to know which is the cathode and which is the anode. We can determine this by measuring the charges on the electrodes, because we know that in a galvanic cell the cathode is the positive electrode and the anode is the negative electrode. When we use a voltmeter to measure the potential of the cell, we find that the copper electrode carries a positive charge and the hydrogen electrode a negative charge. Therefore, copper must be the cathode, and Cu^{2+} is reduced to Cu when the cell oper-

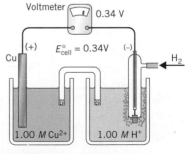

FIGURE 18.16

A galvanic cell composed of copper and hydrogen half-cells. The cell reaction is $Cu^{2+}(aq) + H_2(g) \rightarrow Cu(s) + 2H^+(aq)$.

ates. Similarly, a hydrogen electrode must be the anode, and H_2 is oxidized to H^+. The half-reactions and cell reaction, therefore, are

$$Cu^{2+}(aq) + 2e^- \longrightarrow Cu(s) \qquad \text{(cathode)}$$

$$H_2(g) \longrightarrow 2H^+(aq) + 2e^- \qquad \text{(anode)}$$

$$Cu^{2+}(aq) + H_2(g) \longrightarrow Cu(s) + 2H^+(aq) \qquad \text{(cell reaction)}$$

Using Equation 18.2, we can express E°_{cell} in terms of $E^\circ_{\text{Cu}^{2+}}$ and $E^\circ_{\text{H}^+}$.

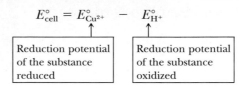

$$E^\circ_{\text{cell}} = E^\circ_{\text{Cu}^{2+}} - E^\circ_{\text{H}^+}$$

| Reduction potential of the substance reduced | Reduction potential of the substance oxidized |

The measured standard cell potential is 0.34 V and $E^\circ_{\text{H}^+}$ equals 0.00 V. Therefore,

$$0.34 \text{ V} = E^\circ_{\text{Cu}^{2+}} - 0.00 \text{ V}$$

Relative to the hydrogen electrode, then, the standard reduction potential of Cu^{2+} is $+0.34$ V. (We have written the value with a plus sign because some reduction potentials are negative, as we will see.)

Now let's look at a galvanic cell set up between a zinc electrode and a hydrogen electrode (see Figure 18.17). This time we find that the hydrogen electrode is positive and the zinc electrode is negative; here, hydrogen is the cathode and zinc is the anode. This means that hydrogen ion is being reduced and zinc is being oxidized. The half-reactions and cell reaction are therefore

$$2H^+(aq) + 2e^- \longrightarrow H_2(g) \qquad \text{(cathode)}$$

$$Zn(s) \longrightarrow Zn^{2+}(aq) + 2e^- \qquad \text{(anode)}$$

$$2H^+(aq) + Zn(s) \longrightarrow H_2(g) + Zn^{2+}(aq) \qquad \text{(cell reaction)}$$

From Equation 18.2, the standard cell potential is

$$E^\circ_{\text{cell}} = E^\circ_{\text{H}^+} - E^\circ_{\text{Zn}^{2+}}$$

Substituting into this the measured standard cell potential of 0.76 V and $E^\circ_{\text{H}^+} = 0.000$ V,

$$0.76 \text{ V} = 0.00 \text{ V} - E^\circ_{\text{Zn}^{2+}}$$

$$E^\circ_{\text{Zn}^{2+}} = -0.76 \text{ V}$$

Notice that the reduction potential of zinc is negative. A negative reduction potential simply means that the substance is not as easily reduced as H^+. In this case, it tells us that Zn is oxidized when it is paired with the hydrogen electrode.

The standard reduction potentials of many half-reactions can be compared to that for the hydrogen electrode in the manner described above. Table 18.1 lists values obtained for some typical half-reactions. They are arranged in decreasing order—the half-reactions at the top have the greatest tendency to occur as reduction, while those at the bottom have the least tendency to occur as reduction.

In a galvanic cell, the measured cell potential is *always* a positive value. This is important to remember.

Remember, in a galvanic cell, the cathode is (+) and the anode is (−).

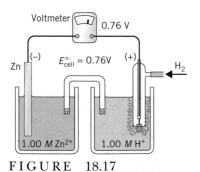

FIGURE 18.17

A galvanic cell composed of zinc and hydrogen half-cells. The cell reaction is $Zn(s) + 2H^+(aq) \rightarrow Zn^{2+}(aq) + H_2(g)$.

TABLE 18.1   Standard Reduction Potentials at 25 °C

Half-Reaction	$E°$ (volts)
$F_2(g) + 2e^- \rightleftharpoons 2F^-(aq)$	+2.87
$PbO_2(s) + SO_4{}^{2-}(aq) + 4H^+(aq) + 2e^- \rightleftharpoons PbSO_4(s) + 2H_2O$	+1.69
$2HOCl(aq) + 2H^+(aq) + 2e^- \rightleftharpoons Cl_2(g) + 2H_2O$	+1.63
$MnO_4{}^-(aq) + 8H^+(aq) + 5e^- \rightleftharpoons Mn^{2+}(aq) + 4H_2O$	+1.51
$PbO_2(s) + 4H^+(aq) + 2e^- \rightleftharpoons Pb^{2+}(aq) + 2H_2O$	+1.46
$BrO_3{}^-(aq) + 6H^+(aq) + 6e^- \rightleftharpoons Br^-(aq) + 3H_2O$	+1.44
$Au^{3+}(aq) + 3e^- \rightleftharpoons Au(s)$	+1.42
$Cl_2(g) + 2e^- \rightleftharpoons 2Cl^-(aq)$	+1.36
$O_2(g) + 4H^+(aq) + 4e^- \rightleftharpoons 2H_2O$	+1.23
$Br_2(aq) + 2e^- \rightleftharpoons 2Br^-(aq)$	+1.07
$NO_3{}^-(aq) + 4H^+(aq) + 3e^- \rightleftharpoons NO(g) + 2H_2O$	+0.96
$Ag^+(aq) + e^- \rightleftharpoons Ag(s)$	+0.80
$Fe^{3+}(aq) + e^- \rightleftharpoons Fe^{2+}(aq)$	+0.77
$I_2(s) + 2e^- \rightleftharpoons 2I^-(aq)$	+0.54
$NiO_2(s) + 2H_2O + 2e^- \rightleftharpoons Ni(OH)_2(s) + 2OH^-(aq)$	+0.49
$Cu^{2+}(aq) + 2e^- \rightleftharpoons Cu(s)$	+0.34
$SO_4{}^{2-} + 4H^+(aq) + 2e^- \rightleftharpoons H_2SO_3(aq) + H_2O$	+0.17
$2H^+(aq) + 2e^- \rightleftharpoons H_2(g)$	0.00
$Sn^{2+}(aq) + 2e^- \rightleftharpoons Sn(s)$	-0.14
$Ni^{2+}(aq) + 2e^- \rightleftharpoons Ni(s)$	-0.25
$Co^{2+}(aq) + 2e^- \rightleftharpoons Co(s)$	-0.28
$PbSO_4(s) + 2e^- \rightleftharpoons Pb(s) + SO_4{}^{2-}(aq)$	-0.36
$Cd^{2+}(aq) + 2e^- \rightleftharpoons Cd(s)$	-0.40
$Fe^{2+}(aq) + 2e^- \rightleftharpoons Fe(s)$	-0.44
$Cr^{3+}(aq) + 3e^- \rightleftharpoons Cr(s)$	-0.74
$Zn^{2+}(aq) + 2e^- \rightleftharpoons Zn(s)$	-0.76
$2H_2O + 2e^- \rightleftharpoons H_2(g) + 2OH^-(aq)$	-0.83
$Al^{3+}(aq) + 3e^- \rightleftharpoons Al(s)$	-1.66
$Mg^{2+}(aq) + 2e^- \rightleftharpoons Mg(s)$	-2.37
$Na^+(aq) + e^- \rightleftharpoons Na(s)$	-2.71
$Ca^{2+}(aq) + 2e^- \rightleftharpoons Ca(s)$	-2.76
$K^+(aq) + e^- \rightleftharpoons K(s)$	-2.92
$Li^+(aq) + e^- \rightleftharpoons Li(s)$	-3.05

Substances located to the left of the double arrows are *oxidizing agents,* because they become reduced when the reactions proceed in the forward direction. The best oxidizing agents are those most easily reduced, and they are located at the top of the table (e.g., F_2).

Substances located to the right of the double arrows are *reducing agents;* they become oxidized when the reactions proceed from right to left. The best reducing agents are those found at the bottom of the table (e.g., Li).

We mentioned earlier that the standard cell potential of the silver–copper galvanic cell has a value of 0.46 V. The cell reaction is

$$2Ag^+(aq) + Cu(s) \longrightarrow 2Ag(s) + Cu^{2+}(aq)$$

and we have seen that the reduction potential of Cu^{2+}, $E°_{Cu^{2+}}$, is +0.34 V. What is the value of $E°_{Ag^+}$, the reduction potential of Ag^+?

ANALYSIS Since we know the potential of the cell and one of the two reduction potentials, we will need to use Equation 18.2 to calculate the unknown reduction potential. This requires that we identify the substance oxidized and the substance reduced.

SOLUTION Silver changes from Ag^+ to Ag, so it is reduced; copper is oxidized from Cu to Cu^{2+}. Therefore, according to Equation 18.2,

$$E°_{cell} = E°_{Ag^+} - E°_{Cu^{2+}}$$

EXAMPLE 18.6
Calculating Half-Cell Potentials

Substituting values for E_{cell}° and $E_{Cu^{2+}}^{\circ}$,

$$0.46 \text{ V} = E_{Ag^+}^{\circ} - 0.34 \text{ V}$$

$$E_{Ag^+}^{\circ} = 0.46 \text{ V} + 0.34 \text{ V}$$

$$= 0.80 \text{ V}$$

The standard reduction potential of silver ion is therefore $+0.80$ V.

■ **Practice Exercise 7** The galvanic cell described in Practice Exercise 5 has a standard cell potential of 1.93 V. The standard reduction potential of Fe^{2+} corresponding to the half-reaction $Fe^{2+}(aq) + 2e^- \rightleftharpoons Fe(s)$ is -0.44 V. Calculate the standard reduction potential of magnesium. Check your answer by referring to Table 18.1.

18.7
USING STANDARD REDUCTION POTENTIALS

One of the goals of chemistry is to be able to predict reactions, and the half-reactions and standard reduction potentials of Table 18.1 serve this purpose particularly well for redox reactions. One question is the feasibility of a reaction. Would a reaction occur, for example, between Cl_2 and Br^- to give Cl^- and Br_2? Or would Br_2 react with Cl^- to give Br^- and Cl_2? What reaction would happen if the oxidized and reduced form of one species, like Fe and a salt of Fe^{2+}, were mixed with a similar pair of another, like Ni and a salt of Ni^{2+}? Would Fe reduce Ni^{2+}, or would Ni reduce Fe^{2+}? How can we use the data in Table 18.1 to calculate the standard cell potential when a given redox reaction is set up as a galvanic cell? Let's look at some examples of these questions.

Predicting Spontaneous Redox Reactions

Standard reduction potentials

It's easy to predict if a spontaneous reaction occurs between the substances in two half-reactions, because we know that *the half-reaction with the more positive reduction potential always takes place as written, namely, as a reduction, while the other half-reaction is forced to run in reverse, namely, as an oxidation.*

EXAMPLE 18.7
Predicting a Spontaneous Reaction

What spontaneous reaction occurs if Cl_2 and Br_2 are added to a solution that contains both Cl^- and Br^-?

ANALYSIS We know that the more easily reduced substance *will* be reduced. If we can find reduction potentials for Cl_2 and Br_2, we can use the E° values to determine this and proceed as before.

SOLUTION There are two possible reduction reactions.

$$Cl_2 + 2e^- \longrightarrow 2Cl^-$$

$$Br_2 + 2e^- \longrightarrow 2Br^-$$

Strictly speaking, the E° values only tell us what to expect under standard conditions. However, only when E_{cell}° is very small can changes in the concentrations change the direction of the spontaneous reaction.

Referring to Table 18.1, we find that Cl_2 has a more positive reduction potential (1.36 V) than does Br_2 (1.07 V). This means Cl_2 will be reduced. Therefore, the reaction that actually occurs has the following half-reactions.

$$Cl_2 + 2e^- \longrightarrow 2Cl^- \qquad \text{(a reduction)}$$

$$2Br^- \longrightarrow Br_2 + 2e^- \qquad \text{(an oxidation)}$$

The net reaction will be

$$Cl_2 + 2Br^- \longrightarrow Br_2 + 2Cl^-$$

Experimentally, chlorine does indeed oxidize bromide ion to bromine, a fact used to recover bromine from seawater and natural brine solutions.

The reactants and products of *spontaneous* redox reactions are easy to spot when reduction potentials are listed in order of most positive to least positive (most negative), as in Table 18.1. This makes predicting possible redox reactions particularly easy. With *any* pair of half-reactions, the one higher up in the table has the more positive reduction potential and occurs as a reduction. The other half-reaction occurs as an oxidation. Therefore, the *reactants* are found on the left side of the equation of the higher half-reaction and on the right side of the lower half-reaction.

The activity series of the metals (Section 9.5), also used to predict reactions, is derived from Table 18.1.

Predict the reaction that will occur when Ni and Fe are added to a solution that contains both Ni^{2+} and Fe^{2+}.

EXAMPLE 18.8
Predicting the Outcome of Redox Reactions

ANALYSIS We are given a situation involving possible changes of ions to atoms or of atoms to ions. In other words, the system involves a possible redox reaction, and so we need Table 18.1. In this table, we see that Ni^{2+} has a more positive (less negative) reduction potential than Fe^{2+}, so Ni^{2+} is reduced and its half-cell reaction will be written just as in Table 18.1. The half-cell reaction for Fe^{2+} in Table 18.1, however, must be reversed; it is Fe that will be oxidized.

SOLUTION The half-cell reactions are

$$Ni^{2+}(aq) + 2e^- \longrightarrow Ni(s)$$
$$\underline{Fe(s) \longrightarrow Fe^{2+}(aq) + 2e^- \qquad \text{(oxidation)}}$$
$$Ni^{2+}(aq) + Fe(s) \longrightarrow Ni(s) + Fe^{2+}(aq) \qquad \text{(net reaction)}$$

■ **Practice Exercise 8** Use the positions of the half-reactions in Table 18.1 to predict the spontaneous reaction when Br^-, SO_4^{2-}, H_2SO_3, and Br_2 are mixed in an acidic solution.

Determining the Cell Reaction and Cell Potential of a Galvanic Cell

In any cell, we know that the substance most easily reduced (with the highest reduction potential) is the one that undergoes reduction. Example 18.9 shows how reduction potentials can be used to predict the standard cell potential and the overall cell reaction in a galvanic cell.

A galvanic cell is constructed using electrodes made of lead and lead(IV) oxide (PbO_2) with sulfuric acid as the electrolyte. The half-reactions and their reduction potentials in this system are

EXAMPLE 18.9
Predicting the Cell Reaction and Cell Potential of a Galvanic Cell

$$PbO_2(s) + 4H^+(aq) + SO_4^{2-}(aq) + 2e^- \rightleftharpoons PbSO_4(s) + 2H_2O \quad E^\circ_{PbO_2} = 1.69 \text{ V}$$
$$PbSO_4(s) + 2e^- \rightleftharpoons Pb(s) + SO_4^{2-}(aq) \quad E^\circ_{PbSO_4} = -0.36 \text{ V}$$

What is the cell reaction and what is the standard potential of the cell?

ANALYSIS In the spontaneous cell reaction, the substance that is more easily reduced will be the one that undergoes reduction. This means that the half-reaction with the larger (more positive) reduction potential will take place as reduction, while the other half-reaction will be reversed and occur as oxidation.

SOLUTION PbO_2 has a larger, more positive reduction potential than $PbSO_4$, so the first half-reaction will occur in the direction written. The second must be reversed to occur as an oxidation. In the cell, therefore, the half-reactions are

$$PbO_2(s) + 4H^+(aq) + SO_4^{2-}(aq) + 2e^- \longrightarrow PbSO_4(s) + 2H_2O$$

$$Pb(s) + SO_4^{2-}(aq) \longrightarrow PbSO_4(s) + 2e^-$$

Adding the two half-reactions and canceling electrons gives the cell reaction,

$$PbO_2(s) + Pb(s) + 4H^+(aq) + 2SO_4^{2-}(aq) \longrightarrow 2PbSO_4(s) + 2H_2O$$

(This is the reaction that takes place in a lead storage battery of the type used to start cars.)

The cell potential is obtained by using Equation 18.2.

$$E°_{cell} = (E° \text{ of substance reduced}) - (E° \text{ of substance oxidized})$$

Since the first half-reaction occurs as a reduction and the second as an oxidation,

$$E°_{cell} = E°_{PbO_2} - E°_{PbSO_4}$$
$$= (1.69 \text{ V}) - (-0.36 \text{ V})$$
$$= 2.05 \text{ V}$$

> Remember that half-reactions are combined following the same procedure used in the ion–electron method of balancing redox reactions (Section 4.9).

EXAMPLE 18.10
Predicting the Cell Reaction and Cell Potential of a Galvanic Cell

What would be the cell reaction and the standard cell potential of a galvanic cell employing the following half-reactions?

$$Al^{3+}(aq) + 3e^- \rightleftharpoons Al(s) \qquad E°_{Al^{3+}} = -1.66 \text{ V}$$

$$Cu^{2+}(aq) + 2e^- \rightleftharpoons Cu(s) \qquad E°_{Cu^{2+}} = +0.34 \text{ V}$$

Which half-cell would be the anode?

SOLUTION The half-reaction with the more positive reduction potential will occur as a reduction; the other will occur as an oxidation. In this cell, then, Cu^{2+} is reduced and Al is oxidized. To obtain the cell reaction, we add the two half-reactions, remembering that the electrons must cancel.

$$3[Cu^{2+}(aq) + 2e^- \longrightarrow Cu(s)] \qquad \text{(reduction)}$$

$$\underline{2[Al(s) \longrightarrow Al^{3+}(aq) + 3e^-]} \qquad \text{(oxidation)}$$

$$3Cu^{2+}(aq) + 2Al(s) \longrightarrow 3Cu(s) + 2Al^{3+}(aq) \qquad \text{(cell reaction)}$$

The anode in the cell is aluminum because that is where oxidation takes place.

To obtain the cell potential, we substitute into Equation 18.2.

$$E°_{cell} = E°_{Cu^{2+}} - E°_{Al^{3+}}$$
$$= (0.34 \text{ V}) - (-1.66 \text{ V})$$
$$= 2.00 \text{ V}$$

An important point to notice here is that although we multiply the half-reactions by factors to make the electrons cancel, *we do not multiply the reduction potentials by these factors.*[1] To obtain the cell potential, we simply subtract one reduction potential from the other.

■ **Practice Exercise 9** What is the overall cell reaction and the standard cell potential of a galvanic cell employing the following half-reactions?

$$NiO_2(s) + 2H_2O + 2e^- \rightleftharpoons Ni(OH)_2(s) + 2OH^-(aq) \qquad E^\circ_{NiO_2} = 0.49 \text{ V}$$

$$Fe(OH)_2(s) + 2e^- \rightleftharpoons Fe(s) + 2OH^-(aq) \qquad E^\circ_{Fe(OH)_2} = -0.88 \text{ V}$$

These are the reactions in an Edison cell, a type of rechargeable storage battery.

■ **Practice Exercise 10** What is the overall cell reaction and the standard cell potential of a galvanic cell employing the following half-reactions?

$$Cr^{3+}(aq) + 3e^- \rightleftharpoons Cr(s) \qquad E^\circ_{Cr^{3+}} = -0.74 \text{ V}$$

$$MnO_4^-(aq) + 8H^+(aq) + 5e^- \rightleftharpoons Mn^{2+}(aq) + 4H_2O \qquad E^\circ_{MnO_4^-} = +1.51 \text{ V}$$

Determining whether a Reaction Is Spontaneous from the Calculated Cell Potential

Since we can predict the spontaneous redox reaction that will take place among a mixture of reactants, it should be possible to predict whether or not a particular reaction, as written, can occur spontaneously. We can do this by calculating the cell potential that corresponds to the reaction in question and seeing whether the potential is *positive*.

In a galvanic cell, the calculated cell potential for the spontaneous reaction is always positive.

For example, to obtain the cell potential for a spontaneous reaction in our previous examples, we subtracted the reduction potentials in a way that gave a positive answer. Therefore, if we compute the cell potential for a particular reaction based on the way the equation is written and the potential comes out positive, we know the reaction is spontaneous. If the calculated cell potential comes out negative, however, the reaction is nonspontaneous. In fact, it is really spontaneous in the opposite direction.

Determine whether the following reactions are spontaneous as written. If they are not, give the reaction that is spontaneous.

(1) $Cu(s) + 2H^+(aq) \rightarrow Cu^{2+}(aq) + H_2(g)$

(2) $3Cu(s) + 2NO_3^-(aq) + 8H^+(aq) \rightarrow 3Cu^{2+}(aq) + 2NO(g) + 4H_2O$

ANALYSIS If we can find the half-reactions for each of these and their corresponding reduction potentials, we should be able to calculate the values of E°_{cell} for the reactions as written. The signs of E°_{cell} will then tell us whether the reactions are spontaneous.

EXAMPLE 18.11
Determining Whether a Reaction Is Spontaneous by Using the Calculated Cell Potential

---

[1] Reduction potentials are intensive quantities; they have the units volts, which are joules *per coulomb*. The same number of joules are available for each coulomb of charge regardless of the total number of electrons shown in the equation. Therefore, reduction potentials are never multiplied by factors before they are subtracted to give the cell potential.

SOLUTION (1) The half-reactions involved in this reaction are

$$Cu(s) \longrightarrow Cu^{2+}(aq) + 2e^-$$

$$2H^+(aq) + 2e^- \longrightarrow H_2(g)$$

The H^+ is reduced and Cu is oxidized. Substituting values from Table 18.1 into Equation 18.2,

$$E_{cell}^\circ = E_{H^+}^\circ - E_{Cu^{2+}}^\circ$$

$$E_{cell}^\circ = (0.00 \text{ V}) - (0.34 \text{ V})$$

$$= -0.34 \text{ V}$$

The calculated cell potential is negative, so reaction (1) is not spontaneous in the forward direction. The spontaneous reaction is actually the reverse of (1).

$$Cu^{2+}(aq) + H_2(g) \longrightarrow Cu(s) + 2H^+(aq)$$

(2) The half-reactions involved in this equation are

$$Cu(s) \longrightarrow Cu^{2+}(aq) + 2e^-$$

$$NO_3^-(aq) + 4H^+(aq) + 3e^- \longrightarrow NO(g) + 2H_2O$$

The Cu is oxidized while the NO_3^- is reduced. According to Equation 18.2,

$$E_{cell}^\circ = E_{NO_3^-}^\circ - E_{Cu^{2+}}^\circ$$

Substituting values from Table 18.1 gives

$$E_{cell}^\circ = (0.96 \text{ V}) - (0.34 \text{ V})$$

$$= +0.62 \text{ V}$$

Because the calculated cell potential is positive, reaction (2) is spontaneous in the forward direction, as written.

Copper dissolves in HNO₃ because the acid contains the oxidizing agent NO₃⁻.

■ **Practice Exercise 11** Which of the following reactions occur spontaneously in the forward direction?
(a) $Br_2(aq) + Cl_2(g) + 2H_2O \rightarrow 2Br^-(aq) + 2HOCl(aq) + 2H^+(aq)$
(b) $3Zn(s) + 2Cr^{3+}(aq) \rightarrow 3Zn^{2+}(aq) + 2Cr(s)$

18.8 CELL POTENTIALS AND THERMODYNAMICS

The fact that cell potentials allow us to predict the spontaneity of redox reactions is no coincidence. There is a relationship between the cell potential and the free energy change for a reaction. In Chapter 13 we saw that ΔG for a reaction is a measure of the maximum useful work that can be obtained from a chemical reaction.

$$-\Delta G = \text{maximum useful work} \qquad (18.3)$$

Determining a Free Energy Change from a Cell Potential

In an electrical system, work is supplied by the electric current that is pushed along by the potential of the cell. It can be calculated from the equation

$$\text{Maximum work} = n\mathscr{F}E_{cell} \qquad (18.4)$$

$$1 \mathscr{F} = \frac{9.65 \times 10^4 \text{ C}}{1 \text{ mol } e^-}$$

where n is the number of moles of electrons transferred, $\mathscr{F}$ is the Faraday constant (9.65×10^4 coulombs per mole of electrons), and E_{cell} is the potential of the cell in volts. To see that Equation 18.4 gives work (which has the units of energy) we can analyze the units. In Equation 18.1 you saw that 1 volt = 1 joule/coulomb. Therefore,

$$\text{Maximum work} = (\text{mol } e^-) \times \left(\frac{\text{coulombs}}{\text{mol } e^-} \right) \times \left(\frac{\text{joule}}{\text{coulomb}} \right) = \text{joule}$$

$$\qquad\qquad\qquad \updownarrow \qquad\qquad\quad \updownarrow \qquad\qquad\quad \updownarrow$$

$$\qquad\qquad\qquad n \qquad\qquad\quad \mathscr{F} \qquad\qquad\quad E_{\text{cell}}$$

Combining Equations 18.3 and 18.4 gives us

$$\Delta G = - n\mathscr{F}E_{\text{cell}} \qquad\qquad (18.5)$$

If we are dealing with the *standard* cell potential, we can calculate the *standard* free energy change.

$$\Delta G^\circ = - n\mathscr{F}E_{\text{cell}}^\circ \qquad\qquad (18.6)$$

Note than when E_{cell}° is positive, ΔG will be negative and the cell reaction will be spontaneous. We used this idea in the preceding section in predicting spontaneous cell reactions.

Calculate ΔG° for the following reaction, given that its standard cell potential is 0.320 V at 25 °C.

$$NiO_2(s) + 2Cl^-(aq) + 4H^+(aq) \longrightarrow Cl_2(g) + Ni^{2+}(aq) + 2H_2O$$

EXAMPLE 18.12
Calculating the Standard
Free Energy Change

ANALYSIS This is a straightforward application of Equation 18.6. Taking the coefficients in the equation to stand for *moles,* two moles of Cl^- are oxidized to Cl_2 and two moles of electrons are transferred ($n = 2$).

SOLUTION Using Equation 18.6, we have

$$\Delta G^\circ = - (2 \text{ mol } e^-) \times \left(\frac{9.65 \times 10^4 \text{ C}}{1 \text{ mol } e^-} \right) \times \left(\frac{0.320 \text{ J}}{\text{C}} \right)$$

$$= - 6.18 \times 10^4 \text{ J}$$

$$= - 61.8 \text{ kJ}$$

■ **Practice Exercise 12** Calculate ΔG° for the reaction that takes place in the galvanic cell studied in Practice Exercise 7.

Determining Equilibrium Constants

One of the useful applications of electrochemistry is in the determination of equilibrium constants. In Chapter 15 you saw that ΔG° is related to the equilibrium constant (K_p for gaseous reactions and K_c for reactions in solution).

$$\Delta G^\circ = - RT \ln K_c$$

Now we have seen that ΔG° is related to E_{cell}°.

$$\Delta G^\circ = - n\mathscr{F}E_{\text{cell}}^\circ$$

Therefore, E_{cell}° and the equilibrium constant are also related. Equating the right sides of the two equations above and solving for E_{cell}° gives

$$E_{\text{cell}}^\circ = \frac{RT}{n\mathscr{F}} \ln K_c$$

Appendix A.2 discusses the relationship between common and natural logarithms.

For historical reasons, this equation is usually expressed in terms of common logs (base 10 logarithms), and for reactions at 25 °C (298 K), all of the constants (R, T, and $\mathscr{F}$) can be combined to give 0.0592 joules/coulomb. Because joules/coulomb equals volts, the final equation that relates a standard cell potential to the corresponding equilibrium constant is

Standard cell potentials

$$E^\circ_{\text{cell}} = \frac{0.0592 \text{ V}}{n} \log K_{\text{c}} \qquad (18.7)$$

As before, n is the number of moles of electrons transferred in the cell reaction as it is written.

EXAMPLE 18.13
Calculating Equilibrium
Constants from E°_{cell}

Be sure to remember that this equation uses common logarithms.

Calculate K_{c} for the reaction in Example 18.12.

SOLUTION The problem simply involves substituting values into Equation 18.7. The reaction in Example 18.12 has $E^\circ_{\text{cell}} = 0.320$ V and $n = 2$. Substituting these values gives

$$0.320 \text{ V} = \frac{0.0592 \text{ V}}{2} \log K_{\text{c}}$$

$$\frac{2(0.320 \text{ V})}{(0.0592 \text{ V})} = \log K_{\text{c}}$$

$$10.8 = \log K_{\text{c}}$$

Taking the antilogarithm,

$$K_{\text{c}} = 10^{10.8} = 6 \times 10^{10}$$

■ **Practice Exercise 13** The calculated standard cell potential for the reaction

$$\text{Cu}^{2+}(aq) + 2\text{Ag}(s) \rightleftharpoons \text{Cu}(s) + 2\text{Ag}^+(aq)$$

is $E^\circ_{\text{cell}} = -0.46$ V. Calculate K_{c} for this reaction. Does this reaction proceed very far toward completion?

18.9

**EFFECT OF
CONCENTRATION
ON CELL POTENTIALS**

When all of the ion concentrations in a cell are 1.00 M and when the partial pressures of any gases involved in the cell reaction are 1 atm, the cell potential is equal to the standard potential. When the concentrations or pressures change, however, so does the potential. For example, in an operating cell or battery, the potential gradually drops as the reactants are used up and as the cell reaction approaches its natural equilibrium status. When it reaches equilibrium, the potential has dropped to zero—the battery is dead.

The Nernst Equation: The Relationship of Cell Potential to Ion Concentrations

The effect of concentration on the cell potential can be obtained from thermodynamics. In Chapter 15, you learned that the free energy change is related to the reaction quotient Q by the equation

$$\Delta G = \Delta G° + RT \ln Q$$

Substituting for ΔG and $\Delta G°$ from Equations 18.5 and 18.6 gives

$$- n\mathscr{F}E_{\text{cell}} = - n\mathscr{F}E°_{\text{cell}} + RT \ln Q$$

Dividing both sides by $- n\mathscr{F}$ gives

$$E_{\text{cell}} = E°_{\text{cell}} - \frac{RT}{n\mathscr{F}} \ln Q \qquad (18.8)$$

Using common instead of natural logarithms and calculating the constants for 25 °C give an equation called the **Nernst equation,** named after Walter Nernst.

Walter Nernst (1864–1941), a German chemist and physicist, was also the discoverer of the third law of thermodynamics.

$$E_{\text{cell}} = E°_{\text{cell}} - \frac{0.0592 \text{ V}}{n} \log Q \qquad (18.9)$$

 Nernst equation

A cell employs the following half-reactions.

$$\text{Ni}^{2+}(aq) + 2e^- \rightleftharpoons \text{Ni}(s) \qquad E°_{\text{Ni}^{2+}} = - 0.25 \text{ V}$$

$$\text{Cr}^{3+}(aq) + 3e^- \rightleftharpoons \text{Cr}(s) \qquad E°_{\text{Cr}^{3+}} = - 0.74 \text{ V}$$

Calculate the potential if $[\text{Ni}^{2+}] = 1.0 \times 10^{-4} \, M$ and $[\text{Cr}^{3+}] = 2.0 \times 10^{-3} \, M$.

EXAMPLE 18.14
Calculating the Effect of Concentration on $E°_{\text{cell}}$

ANALYSIS Because the concentrations are not 1.00 M, we must use the Nernst equation, but first we need the cell reaction. We need it to determine the number of electrons transferred, n, and we need it to determine the correct form of the mass action expression from which we calculate Q. We must also note that the reacting system is heterogeneous; both solid metals and a liquid solution of their dissolved ions are involved. So we have to remember that a mass action expression does not contain concentration terms for solids, like Ni and Cr.

SOLUTION Nickel has the more positive (less negative) reduction potential, so its half-reaction will occur as a reduction. This means that chromium will be oxidized. Making electron gain equal to electron loss, the cell reaction is found as follows.

$$3[\text{Ni}^{2+}(aq) + 2e^- \longrightarrow \text{Ni}(s)] \qquad \text{(reduction)}$$
$$\underline{2[\text{Cr}(s) \longrightarrow \text{Cr}^{3+}(aq) + 3e^-]} \qquad \text{(oxidation)}$$
$$3\text{Ni}^{2+}(aq) + 2\text{Cr}(s) \longrightarrow 3\text{Ni}(s) + 2\text{Cr}^{3+}(aq) \qquad \text{(cell reaction)}$$

The mass action expression, Q, for the cell reaction is

$$Q = \frac{[\text{Cr}^{3+}]^2}{[\text{Ni}^{2+}]^3}$$

A total of six electrons is transferred, which means $n = 6$. Now we can write the Nernst equation for the system.

$$E_{\text{cell}} = E°_{\text{cell}} - \frac{0.0592 \text{ V}}{6} \log \frac{[\text{Cr}^{3+}]^2}{[\text{Ni}^{2+}]^3}$$

Next we need $E°_{\text{cell}}$. Since Ni^{2+} is reduced,

$$E°_{\text{cell}} = E°_{\text{Ni}^{2+}} - E°_{\text{Cr}^{3+}}$$
$$= (- 0.25 \text{ V}) - (- 0.74 \text{ V})$$
$$= 0.49 \text{ V}$$

Substituting values into the equation, we get

$$E_{cell} = 0.49 \text{ V} - \frac{0.0592 \text{ V}}{6} \log \frac{(2.0 \times 10^{-3})^2}{(1.0 \times 10^{-4})^3}$$

$$= 0.49 \text{ V} - (0.00987 \text{ V}) \log(4.0 \times 10^6)$$

$$= 0.49 \text{ V} - (0.00987 \text{ V})(6.60)$$

$$= 0.49 \text{ V} - 0.065 \text{ V}$$

$$= 0.42 \text{ V}$$

The potential of the cell is 0.42 V.

■ **Practice Exercise 14** In a certain zinc–copper cell,

$$\text{Zn}(s) + \text{Cu}^{2+}(aq) \longrightarrow \text{Zn}^{2+}(aq) + \text{Cu}(s)$$

the ion concentrations are $[\text{Cu}^{2+}] = 0.010 \ M$ and $[\text{Zn}^{2+}] = 1.0 \ M$. What is the cell potential at 25 °C?

Determining Concentrations from Experimental Cell Potentials

One of the principal uses of the relationship between concentration and cell potential is in the measurement of concentrations. Experimental determination of cell potentials combined with modern developments in electronics has provided a means of monitoring and analyzing the concentrations of all sorts of substances in solution, even some that are not themselves ionic and that are not involved directly in electrochemical changes. The basic concept behind all these approaches is illustrated in Example 18.15.

EXAMPLE 18.15
Using the Nernst Equation to Determine Concentrations

A chemist needs to measure the concentration of Cu^{2+} in a large number of samples of water in which the copper ion concentration is expected to be quite small. The apparatus consists of a silver electrode, dipping into a 1.00 M solution of AgNO_3, connected by a salt bridge to a second half-cell containing a copper electrode that can be dipped successively into each water sample. In the analysis of one sample, the cell potential was measured to be 0.62 V, the copper electrode serving as the anode. What was the concentration of copper ion in this sample?

ANALYSIS In this problem, we have the cell potential, E_{cell}, and we can calculate $E°_{cell}$. The unknown quantity is one of the concentration terms in the Nernst equation.

SOLUTION The first step is to write the proper equation for the cell reaction, because we need it to compute $E°_{cell}$ and to construct the mass action expression for use in the Nernst equation. Because copper is the anode, it is being oxidized. This also means that Ag^+ is being reduced. Therefore, the equation for the cell reaction is

$$\text{Cu}(s) + 2\text{Ag}^+(aq) \longrightarrow \text{Cu}^{2+}(aq) + 2\text{Ag}(s)$$

Two electrons are transferred, so the Nernst equation is

$$E_{cell} = E°_{cell} - \frac{0.0592 \text{ V}}{2} \log \frac{[\text{Cu}^{2+}]}{[\text{Ag}^+]^2}$$

The mass action expression, Q, for this reaction is

$$Q = \frac{[\text{Cu}^{2+}]}{[\text{Ag}^+]^2}$$

SPECIAL TOPIC 18.2 / MEASUREMENT OF pH

The Nernst equation tells us that the emf of a cell changes with the concentrations of the ions involved in the cell reaction. As noted in the conclusion to Example 18.15, one of the most useful results of this is that cell potential measurements provide a way indirectly to measure and monitor ion concentrations in aqueous solutions.

Scientists have developed a number of specialized electrodes that can be dipped into one solution after another and whose potential is affected in a reproducible way by the concentration of only one species in the solution. An example is the glass electrode of a pH meter for the measurement of pH (see Figure 1). The electrode is constructed from a hollow glass tube sealed with a special thin-walled glass membrane at the bottom. The tube is filled part way with a dilute solution of HCl, and dipping into this HCl solution is a silver wire coated with a layer of silver chloride. The potential of the electrode is controlled by the difference between the hydrogen ion concentrations inside and outside the thin glass membrane at the bottom. Since the H^+ concentration inside the electrode is constant, the electrode's potential varies only with the concentration of H^+ in the solution outside. In fact, this potential is proportional to the logarithm of the H^+ concentration and therefore to the pH of the outside solution.

A glass electrode is always used with another reference electrode whose potential is a constant, and this gives a galvanic cell whose potential depends on the pH of the solution into which the electrodes are immersed. The measurement of the potential of this galvanic cell and the translation of the potential into pH are the job of a pH meter such as the one shown in Figure 16.3.

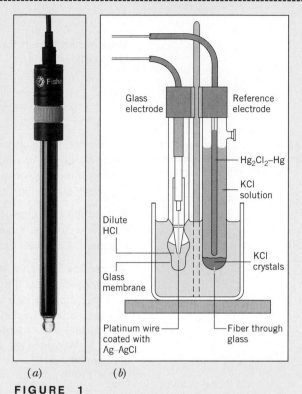

(a)   *(b)*

FIGURE 1

(*a*) Photograph of a typical glass electrode. (*b*) Cutaway view of the construction of a glass electrode and a reference electrode.

The value of E°_{cell} can be obtained from the tabulated standard reduction potentials in Table 18.1. Following our usual procedure and recognizing that silver ion is reduced,

$$
\begin{aligned}
E^\circ_{\text{cell}} &= E^\circ_{\text{Ag}^+} - E^\circ_{\text{Cu}^{2+}} \\
&= (0.80 \text{ V}) - (0.34 \text{ V}) \\
&= 0.46 \text{ V}
\end{aligned}
$$

Now we can substitute values into the Nernst equation and solve for the concentration ratio in the mass action expression.

$$
0.62 \text{ V} = 0.46 \text{ V} - \frac{0.0592 \text{ V}}{2} \log \frac{[\text{Cu}^{2+}]}{[\text{Ag}^+]^2}
$$

Solving for $\log([\text{Cu}^{2+}]/[\text{Ag}^+]^2)$ gives

$$
\log \frac{[\text{Cu}^{2+}]}{[\text{Ag}^+]^2} = -5.4
$$

Taking the antilog gives us the value of the mass action expression.

$$\frac{[Cu^{2+}]}{[Ag^+]^2} = 4 \times 10^{-6}$$

Since we know that the concentration of Ag^+ is 1.00 *M*, we can now solve for the Cu^{2+} concentration.

$$\frac{[Cu^{2+}]}{(1.00)^2} = 4 \times 10^{-6}$$

$$[Cu^{2+}] = 4 \times 10^{-6} \, M$$

The ease of such operations and the fact that they lend themselves well to automation and computer analysis make electrochemical analyses especially attractive to scientists.

Notice that this is indeed a very small concentration, and it can be obtained very easily by simply measuring the potential generated by the electrochemical cell. Determining the concentrations in many samples is also very simple — just change the water sample and measure the potential again.

■ **Practice Exercise 15** In the analysis of two other water samples by the procedure described in Example 18.15, cell potentials (E_{cell}) of 0.57 V and 0.82 V were obtained. Calculate the Cu^{2+} ion concentration in each of these samples.

■ **Practice Exercise 16** A galvanic cell was constructed by connecting a nickel electrode that was dipped into 1.20 *M* $NiSO_4$ solution to a chromium electrode that was dipping into a solution containing Cr^{3+} at an unknown concentration. The potential of the cell was measured to be 0.552 V, with the chromium serving as the anode. The standard cell potential for this system was determined to be 0.487 V. What was the concentration of Cr^{3+} in the solution of unknown concentration?

18.10 PRACTICAL APPLICATIONS OF GALVANIC CELLS

One of the most familiar uses of galvanic cells is the generation of portable electrical energy. In this section we will briefly examine some of the more common present-day uses of galvanic cells, popularly called batteries, and some promising future applications.

The Lead Storage Battery

The common **lead storage battery** used to start an automobile is composed of a number of galvanic cells, each having an emf of about 2 V, that are connected in series so that their voltages are additive. Most automobile batteries contain six such cells and give about 12 V, but 6, 24, and 32 V batteries are also available.

A typical lead storage battery is shown in Figure 18.18. The anode of each cell is composed of a set of lead plates, the cathode consists of another set of plates that hold a coating of PbO_2, and the electrolyte is sulfuric acid. When the battery is discharging the electrode reactions are

$$PbO_2(s) + 4H^+(aq) + SO_4^{2-}(aq) + 2e^- \longrightarrow PbSO_4(s) + 2H_2O \quad \text{(cathode)}$$

$$Pb(s) + SO_4^{2-}(aq) \longrightarrow PbSO_4(s) + 2e^- \quad \text{(anode)}$$

The net reaction taking place in each cell is

$$PbO_2(s) + Pb(s) + \underbrace{4H^+(aq) + 2SO_4^{2-}(aq)}_{2H_2SO_4} \longrightarrow 2PbSO_4(s) + 2H_2O$$

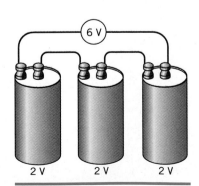

Cells connected in series to give a higher voltage.

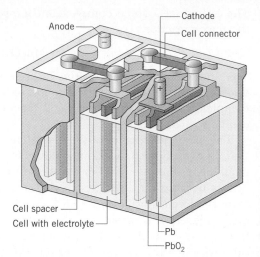

A 6 V lead storage battery is composed of three cells connected in series. Each cell produces about 2 V.

F I G U R E   18.19

A battery hydrometer. Battery acid is drawn into the glass tube. The depth to which the float sinks is inversely proportional to the concentration of the acid and, therefore, to the state of charge of the battery.

As the cell discharges, the sulfuric acid concentration decreases, which provides a convenient means of checking the state of the battery. Because the density of a sulfuric acid solution decreases as its concentration drops, the concentration can be determined very simply by measuring the density with a **hydrometer,** which consists of a rubber bulb that is used to draw the battery fluid into a glass tube containing a float (see Figure 18.19). The depth to which the float sinks is inversely proportional to the density of the liquid—the deeper the float sinks, the lower is the density and the weaker is the charge on the battery. The narrow neck of the float is usually marked to indicate the state of charge of the battery.

The principal advantage of the lead storage battery is that the cell reactions that occur spontaneously during discharge can be reversed by the application of voltage from an external source. In other words, the battery can be recharged by electrolysis. The reaction for battery recharge is

$$2PbSO_4(s) + 2H_2O \xrightarrow{\text{Electrolysis}} PbO_2(s) + Pb(s) + 4H^+(aq) + 2SO_4^{2-}(aq)$$

Disadvantages of the lead storage battery are that it is very heavy and that its corrosive sulfuric acid can spill.

The most modern lead storage batteries use a lead–calcium alloy as the anode. This reduces the need to have the individual cells vented and the battery can be sealed, thereby preventing spillage of the electrolyte.

The Zinc–Carbon Dry Cell

The ordinary, relatively inexpensive, 1.5 V **zinc–carbon dry cell,** or Leclanché cell (after inventor George Leclanché), is used to power flashlights, tape recorders, and the like, but it is not really dry (see Figure 18.20). Its outer shell is made of zinc, which serves as the anode. The exposed outer surface at the bottom of the cell is the negative end of the battery. The cathode—the positive terminal of the battery—consists of a carbon (graphite) rod surrounded by a moist paste of graphite powder, manganese dioxide, and ammonium chloride.

The anode reaction is simply the oxidation of zinc.

$$Zn(s) \longrightarrow Zn^{2+}(aq) + 2e^- \qquad \text{(anode)}$$

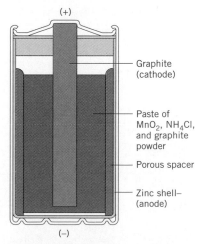

(+)

Graphite (cathode)

Paste of MnO_2, NH_4Cl, and graphite powder

Porous spacer

Zinc shell– (anode)

(–)

F I G U R E   18.20

A cross-section of a zinc–carbon dry cell.

The cathode reaction is complex, and a mixture of products is formed. One of the major reactions is

$$2MnO_2(s) + 2NH_4^+(aq) + 2e^- \longrightarrow Mn_2O_3(s) + 2NH_3(aq) + H_2O \quad \text{(cathode)}$$

The ammonia that forms at the cathode reacts with some of the Zn^{2+} produced from the anode to form a complex ion, $Zn(NH_3)_4^{2+}$. Because of the complexity of the cathode half-cell reaction, no simple overall cell reaction can be written.

The major advantages of the dry cell are its relatively low price and the fact that it normally works without leaking. One disadvantage is that the cell loses its ability to function rather rapidly under heavy current drain, a loss that occurs because the products of the cell reaction cannot easily diffuse away from the electrodes. If unused for a while, however, these batteries spontaneously rejuvenate somewhat as the products diffuse. Another disadvantage of the zinc–carbon dry cell is that it can't be recharged. Devices that claim otherwise simply drive the reaction products away from the electrodes, which allows the cell to function again. This doesn't work very many times, however. Before too long the zinc casing develops holes and the battery must be discarded.

The common alkaline flashlight battery, or **alkaline dry cell,** also uses Zn and MnO_2 as reactants, but under basic conditions (Figure 18.21). The half-cell reactions are

$$Zn(s) + 2OH^-(aq) \longrightarrow ZnO(s) + H_2O + 2e^- \quad \text{(anode)}$$
$$2MnO_2(s) + H_2O + 2e^- \longrightarrow Mn_2O_3(s) + 2OH^-(aq) \quad \text{(cathode)}$$

and the voltage is about 1.54 V. It has a longer shelf-life and is able to deliver higher currents for longer periods than the less expensive zinc–carbon cell.

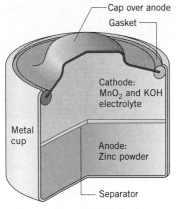

FIGURE 18.21

A simplified diagram of an alkaline dry cell.

The Nickel–Cadmium Storage Cell

The **nickel–cadmium storage cell,** the *nicad battery,* powers rechargeable electronic calculators, electric shavers, and power tools. It produces a potential of about 1.4 V, which is slightly lower than that of the zinc–carbon cell. The electrode reactions in the cell are

$$Cd(s) + 2OH^-(aq) \longrightarrow Cd(OH)_2(s) + 2e^- \quad \text{(anode)}$$
$$NiO_2(s) + 2H_2O + 2e^- \longrightarrow Ni(OH)_2(s) + 2OH^-(aq) \quad \text{(cathode)}$$

The nickel–cadmium battery is rechargeable and can be sealed to prevent leakage, which is particularly important in electronic devices. For the storage of a given amount of electrical energy, however, the nicad battery is considerably more expensive than the lead storage battery.

Specialized Batteries

As everyone knows, the field of electronics has undergone fantastic changes in the past decade or so. Electronic calculators, digital watches, electronically controlled cameras that focus automatically and adjust themselves for the correct exposure, and portable computers have been made possible by microelectronics centered around almost invisible circuits etched into the surfaces of silicon chips. This trend toward miniaturization has also been taken up by the makers of the batteries that power the electronics. Two of the common miniature batteries in use today are the mercury battery and the silver oxide battery.

The Mercury Battery   This battery, one of the earliest small batteries to be developed for commercial use, first appeared in the early 1940s. It consists of a zinc anode and a mercury(II) oxide (HgO) cathode. The electrolyte is a concentrated solution of potassium hydroxide held in pads of absorbent material between the electrodes (see Figure 18.22). The reactions that take place at the electrodes are relatively simple.

$$Zn(s) + 2OH^-(aq) \longrightarrow ZnO(s) + H_2O + 2e^- \text{ (anode)}$$

$$HgO(s) + H_2O + 2e^- \longrightarrow Hg(l) + 2OH^-(aq) \quad \text{(cathode)}$$

The net cell reaction is

$$Zn(s) + HgO(s) \longrightarrow ZnO(s) + Hg(l)$$

The mercury battery develops a potential of about 1.35 V. An especially useful feature of this cell is that its potential remains nearly constant throughout its life. By contrast, the potential of the ordinary dry cell drops off slowly but continuously as it is used.

The Silver Oxide Battery   This battery is a more recent entry into the world of miniature batteries and is now used in many wristwatches, calculators, and autoexposure cameras. The anode material in the cell is zinc and the substance that reacts at the cathode is silver oxide, Ag_2O (see Figure 18.23). The electrolyte is basic and the half-reactions that occur at the electrodes are

$$Zn(s) + 2OH^-(aq) \longrightarrow ZnO(s) + H_2O + 2e^- \text{ (anode)}$$

$$Ag_2O(s) + H_2O + 2e^- \longrightarrow 2Ag(s) + 2OH^-(aq) \quad \text{(cathode)}$$

The overall net cell reaction is

$$Zn(s) + Ag_2O(s) \longrightarrow ZnO(s) + 2Ag(s)$$

As you might expect, these cells are quite expensive because they contain silver. An advantage that they have over the mercury battery, however, is that they generate a potential of about 1.5 V.

Fuel Cells   The production of usable energy by the combustion of fuels is an extremely inefficient process. Largely because of the constraints set by fundamental thermodynamic principles, modern electric power plants are unable to harness more than about 35 to 40% of the potential energy in oil, coal, or natural gas. The gasoline or diesel engine has an efficiency of only about 25 to 30%. The rest of the energy is lost to the surroundings as heat, which is the reason that a vehicle must have an effective cooling system.

Fuel cells are electrochemical cells that "burn" fuel under conditions that are much more nearly thermodynamically reversible than simple combustion. They therefore achieve much greater efficiencies—75% is quite feasible. Figure 18.24 illustrates a hydrogen–oxygen fuel cell. The electrolyte, a hot concentrated solution of potassium hydroxide in the center compartment, is in contact with two porous electrodes, for example, porous carbon and thin

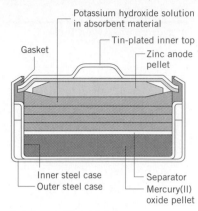

FIGURE 18.22

A mercury battery.

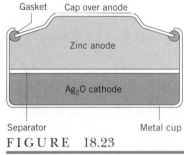

FIGURE 18.23

A silver oxide battery.

If we could "burn" our petroleum in fuel cells to produce energy, fuel consumption could be halved.

FIGURE 18.25

A hydrogen–oxygen fuel cell.

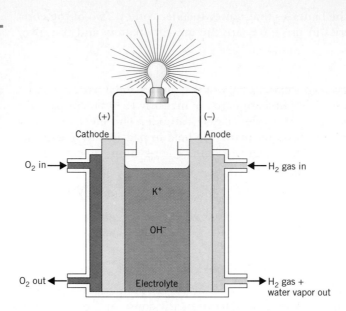

porous nickel. Gaseous hydrogen and oxygen under pressure are circulated so as to come in contact with the electrodes. At the cathode, oxygen is reduced.

$$O_2(g) + 2H_2O + 4e^- \longrightarrow 4OH^-(aq) \qquad \text{(cathode)}$$

At the anode, hydrogen is oxidized to water.

$$H_2(g) + 2OH^-(aq) \longrightarrow 2H_2O + 2e^- \qquad \text{(anode)}$$

Part of the water formed at the anode leaves as steam mixed with the circulating hydrogen gas. The net cell reaction, after making electron loss equal to electron gain, is

$$2H_2(g) + O_2(g) \longrightarrow 2H_2O$$

The advantages of the fuel cell are obvious. There is no electrode material to be replaced, as there is in an ordinary battery. The fuel can be fed in continuously to produce power. In fact, hydrogen–oxygen fuel cells have been used on spacecraft for just this reason. The disadvantages at the present time include their high cost and the rather large size that is necessary to produce useful quantities of electrical power.

In spacecraft, the product of the reaction in the hydrogen–oxygen fuel cell is used as drinking water.

SUMMARY

Electrolysis In an **electrolytic cell,** a flow of electricity causes reduction at a negatively charged **cathode** and oxidation at a positively charged **anode.** Ion movement instead of electron transport occurs in the electrolyte. The electrode reactions are determined by which species is most easily reduced and which is most easily oxidized, but in aqueous solutions complex surface effects at the electrodes can alter the natural order. In the electrolysis of water, an electrolyte must be present to maintain electrical neutrality at the electrodes.

Quantitative Aspects of Electrolysis A **faraday** ($\mathscr{F}$) is the charge carried by 1 mol of electrons, 9.65×10^4 C. The product of current **(amperes)** and time (seconds) gives coulombs. These relationships and the half-reactions that occur at the anode or cathode permit us to relate the amount of chemical change to measurements of current and time.

Applications of Electrolysis **Electroplating,** the production of aluminum and magnesium, the refining of copper, and the electrolysis of molten and aqueous sodium chloride are examples of practical applications of electrolysis.

Galvanic Cells   In a **galvanic cell** a spontaneous redox reaction, in which the individual half-reactions occur in separated **half-cells,** causes the electron transfer to occur through an external electrical circuit. Reduction occurs at the positively charged cathode and oxidation takes place at the negatively charged anode. The half-cells must be connected electrolytically to complete an electrical circuit. A **salt bridge** accomplishes this—it permits electrical neutrality to be maintained. The **emf** or **voltage** produced by a cell is equal to the **standard cell potential** when all ion concentrations are 1 *M* and the partial pressures of any gases involved equal 1 atm. The **cell potential** can be considered to be the difference between the **reduction potentials** of the half-cells. The half-cell with the higher reduction potential undergoes reduction and forces the other to undergo oxidation. The reduction potentials of isolated half-cells can't be measured, but values are assigned by choosing the **hydrogen electrode** as a reference electrode with a reduction potential of 0 V. Species more easily reduced than H^+ have positive reduction potentials; those less easily reduced have negative reduction potentials. Reduction potentials can be used to predict the cell reaction and to calculate E°_{cell}. They can also be used to predict spontaneous redox reactions not occurring in galvanic cells and to predict whether or not a given reaction is spontaneous.

Thermodynamics and Cell Potentials   The values of ΔG° and K_c for a reaction can be calculated from E°_{cell}. The Nernst equation relates the cell potential to the standard cell potential and the reaction quotient. It allows the cell potential to be calculated for ion concentrations other than 1.00 *M*. The important equations in this section are

$$\Delta G^\circ = -n\mathscr{F}E^\circ_{cell}$$

$$E^\circ_{cell} = \frac{0.0592 \text{ V}}{n} \log K_c$$

$$E_{cell} = E^\circ_{cell} - \frac{0.0592 \text{ V}}{n} \log Q$$

Practical Galvanic Cells   The **lead storage battery** and the **nickel–cadmium battery** are rechargeable. The **zinc–carbon cell**—the common dry cell—is not. The common **alkaline battery** uses essentially the same reactions as the less expensive dry cell. The **zinc–mercury(II) oxide cell** and the **zinc–silver oxide cell** are commonly used in small electronic devices. **Fuel cells** consume fuel that can be fed continuously. For the production of usable energy, they offer much higher thermodynamic efficiencies than can be obtained by burning the fuel in conventional power plants or internal combustion engines.

TOOLS YOU HAVE LEARNED

The table below lists the tools you have learned in this chapter that are applicable to problem solving. Review them if necessary, and refer to them when working on the Thinking-It-Through problems and the Review Exercises that follow.

Tool	Function
Identification of electrodes (pages 764 and 779)	To label electrodes and to describe galvanic cells. <div align="center">**Electrolytic Cell**</div> Cathode is negative (reduction). Anode is positive (oxidation). <div align="center">**Galvanic Cell**</div> Cathode is positive (reduction). Anode is negative (oxidation).
Faraday constant (page 769)	To relate lab data [coulombs (amperes times seconds) and moles] to amount of chemical change in an electrolysis.
Standard reduction potentials (pages 781 and 786)	To calculate standard cell potentials.
Standard cell potentials, E°_{cell} (page 781)	To predict redox reactions; to calculate free energy changes (page 791), equilibrium constants (page 792), and concentrations of species at equilibrium (page 794).
Nernst equation (page 793)	To calculate cell voltage or to calculate the concentration of a species in an equilibrium.

THINKING IT THROUGH

The goal for each of the following problems is to give you practice in thinking your way through problems. The goal is not to find the answer itself; instead, you are only asked to assemble the available information needed to obtain the answer, state what additional data (if any) are needed, and describe how you would use the data to answer the question. For problems involving unit conversions, list the relationships among the units that are needed to carry out the conversions. Construct the conversion factors that can be formed from these relationships. Then set up the solution to the problem by arranging the conversion factors so the units cancel correctly to give the desired units of the answer.

The problems are divided into two groups. Those in Level 2 are more challenging than those in Level 1 and provide an opportunity to really hone your problem-solving skills. Both levels, however, may include questions that draw on material studied in earlier chapters.

Level 1 Problems

1. When a solution of $Pb(NO_3)_2$ is electrolyzed using inert electrodes, Pb is deposited on the cathode. On the basis of this information and other facts that you learned in this chapter, answer the following:

(a) What do we expect to observe at the anode?
(b) What is the cathode reaction?
(c) What is the cell reaction?

2. In the balanced cell reaction formed by combining the following half-reactions,

$$MnO_4^- + 8H^+ + 5e^- \longrightarrow Mn^{2+} + 4H_2O$$

$$Cl^- + 3H_2O \longrightarrow ClO_3^- + 6H^+ + 6e^-$$

how many moles of electrons are transferred?

3. How many grams of Pb will be deposited on the cathode of an electrolytic cell if 500 mL of a solution of $Pb(NO_3)_2$ is electrolyzed for a period of 30.00 min at a current of 4.00 A?

4. At a current of 2.00 A, how many minutes would it take to deposit a coating of silver on a metal object by electrolysis of $Ag(CN)_2^-$ if the silver coating is to be 0.020 mm thick and is to have an area of 25 cm²? The density of silver is 10.50 g/cm³.

Level 2 Problems

5. After the electrolysis described in the preceding question, what is the pH of the solution?

6. How many milliliters of H_2 gas will be generated by the electrolysis of water using a KNO_3 electrolyte if a current of 1.25 A is passed through the electrolysis cell for a period of 10.00 minutes?

7. A current was passed through two electrolytic cells connected in series (so the same current passed through both of them). In one cell was a solution of $Cr_2(SO_4)_3$, and in the other, a solution of $AgNO_3$. If 1.45 g of Ag was deposited on the cathode in the second cell, how many grams of Cr were deposited on the cathode in the first cell? The temperature of the cell was 25 °C.

8. Will the reaction $Au + NO_3^- + 4H^+ \rightarrow Au^{3+} + NO + 2H_2O$ occur spontaneously?

9. In a galvanic cell established between Fe(s) and $Fe^{2+}(aq)$ plus Co(s) and $Co^{2+}(aq)$, which electrode will carry the positive charge?

10. What is the value of the equilibrium constant at 25 °C for the following reaction?

$$2Br^-(aq) + I_2(s) \rightleftharpoons Br_2(aq) + 2I^-(aq)$$

11. What is the value of K_c for the reaction in Problem 10 at 35 °C?

12. What is the cell notation for the galvanic cell that uses the following half-reactions?

$$MnO_4^- + 8H^+ + 5e^- \longrightarrow Mn^{2+} + 4H_2O$$
$$E^\circ_{MnO_4^-} = +1.507 \text{ V}$$

$$ClO_3^- + 6H^+ + 6e^- \longrightarrow Cl^- + 3H_2O$$
$$E^\circ_{ClO_3^-} = +1.451$$

13. It was desired to determine the molar concentration of Sn^{2+} in a solution. The following cell was established:

$$Sn(s) \mid Sn^{2+}(? \; M) \parallel Cl^-(1.00 \times 10^{-2} \; M) \mid AgCl(s), Ag(s)$$

The half-reaction

$$AgCl(s) + e^- \rightleftharpoons Ag(s) + Cl^-(aq)$$

has $E^\circ_{AgCl} = 0.2223$ V and $E^\circ_{Sn^{2+}} = -0.1375$ V. The tin electrode was dipped into a solution of Sn^{2+} and the potential of the cell was measured to be 0.3114 V. What is the concentration of Sn^{2+} in the solution?

14. What is the potential of a cell in which the cell reaction is

$$NiO_2(s) + 4H^+(aq) + 2Ag(s) \longrightarrow$$
$$Ni^{2+}(aq) + 2H_2O + 2Ag^+(aq)$$

if the pH is 2.50 and the concentrations of Ni^{2+} and Ag^+ are each 0.020 M?

REVIEW EXERCISES

Answers to questions whose numbers are printed in color are given in Appendix D. More challenging questions are marked with asterisks.

Electrolysis

18.1 Define the following terms.
(a) oxidation   (b) reduction
(c) electrochemical change   (d) anode
(e) cathode   (f) cell reaction

18.2 What electrical charges do the anode and the cathode carry in an electrolytic cell?

18.3 How do *electrolytic conduction* and *metallic conduction* differ?

18.4 Why must electrolysis reactions occur at the electrodes in order for electrolytic conduction to continue?

18.5 What is the difference between a *direct current* and an *alternating current*?

18.6 What is the purpose of writing the word *electrolysis* over the arrow in a chemical equation?

18.7 Why must NaCl be melted before it is electrolyzed to give Na and Cl_2?

18.8 What does the term *inert electrode* mean?

18.9 Write the anode, cathode, and overall cell reactions for the electrolysis of molten NaCl.

18.10 Write half-reactions for the oxidation and the reduction of water.

18.11 What happens to the pH of the solution near the cathode and anode during the electrolysis of KNO_3?

18.12 What function does KNO_3 serve in the electrolysis of a KNO_3 solution?

18.13 Write the anode reaction for the electrolysis of an aqueous solution that contains (a) NO_3^-, (b) Br^-, and (c) NO_3^- and Br^-.

18.14 Write the cathode reaction for the electrolysis of an aqueous solution that contains (a) K^+, (b) Cu^{2+}, and (c) K^+ and Cu^{2+}.

18.15 What products would we expect at the electrodes if a solution containing both KBr and $Cu(NO_3)_2$ were electrolyzed?

Stoichiometric Relationships in Electrolysis

18.16 What is a *faraday*? What relationships relate faradays to current and time measurements?

18.17 How many coulombs are passed through an electrolysis cell by (a) a current of 4.00 A for 600 s? (b) A current of 10.0 A for 20 min? (c) A current of 1.50 A for 6.00 hr?

18.18 How many moles of electrons correspond to the answers to each part of Review Exercise 18.17?

18.19 How many moles of electrons are required to (a) reduce 0.20 mol Fe^{2+} to Fe? (b) Oxidize 0.70 mol Cl^- to Cl_2? (c) Reduce 1.50 mol Cr^{3+} to Cr? (d) Oxidize 1.0×10^{-2} mol Mn^{2+} to MnO_4^-?

18.20 How many moles of electrons are required to (a) produce 5.00 g Mg from molten $MgCl_2$? (b) Form 41.0 g Cu from a $CuSO_4$ solution?

18.21 How many moles of Cr^{3+} would be reduced to Cr by the same amount of electricity that produces 12.0 g Ag from a solution of $AgNO_3$?

18.22 In Exercise 18.21, if a current of 4.00 A were used, how many minutes would the electrolysis take?

18.23 How many grams of $Fe(OH)_2$ are produced at an iron anode when a basic solution undergoes electrolysis at a current of 8.00 A for 12.0 min?

18.24 How many milliliters of gaseous H_2, measured at standard conditions of temperature and pressure, would be produced at the cathode in the electrolysis of water with a current of 0.750 A for 15.00 min?

18.25 A solution of NaCl in water was electrolyzed with a current of 2.50 A for 15.0 min. How many milliliters of 0.100 M HCl would be required to neutralize the resulting solution?

18.26 How many hours would it take to produce 75.0 g of metallic chromium by the electrolytic reduction of Cr^{3+} with a current of 2.25 A?

18.27 How many hours would it take to generate 35.0 g of lead from $PbSO_4$ during the charging of a storage battery using a current of 1.50 A? The half-reaction is

$$PbSO_4 + 2e^- \longrightarrow Pb + SO_4^{2-}$$

18.28 How many amperes would be needed to produce 60.0 g of magnesium during the electrolysis of molten $MgCl_2$ in 2.00 hr?

18.29 A large electrolysis cell that produces metallic aluminum from Al_2O_3 by the Hall–Héroult process is capable of yielding 900 lb (409 kg) of aluminum in 24.0 hr. What current is required?

18.30 How many grams of Cl_2 would be produced in the electrolysis of molten NaCl by a current of 4.25 A for 35.0 min?

*__18.31__ A solution containing vanadium (chemical symbol V) in an unknown oxidation state was electrolyzed with a current of 1.50 A for 30.0 min. It was found that 0.475 g of V was deposited on the cathode.
(a) How many moles of electrons were transferred in the electrolysis reaction?
(b) How many moles of vanadium were deposited?
(c) What was the original oxidation state of the vanadium ion?

Applications of Electrolysis

18.32 What is *electroplating*? Sketch an apparatus to electroplate silver.

18.33 Describe the Hall–Héroult process for producing metallic aluminum. What half-reaction occurs at the anode? What half-reaction occurs at the cathode? What is the overall cell reaction?

18.34 In the Hall–Héroult process, why must the carbon anodes be replaced frequently?

18.35 Describe how magnesium is recovered from seawater. What electrochemical reaction is involved?

18.36 How is metallic sodium produced? What are some uses of metallic sodium? Write equations for the anode and cathode reactions.

18.37 Describe the electrolytic refining of copper. What economic advantages offset the cost of electricity for this process? What chemical reactions occur at (a) the anode and (b) the cathode?

18.38 Describe the electrolysis of aqueous sodium chloride. How do the products of the electrolysis compare for stirred and unstirred reactions? Write chemical equations for the reactions that occur at the electrodes.

18.39 What are the advantages and disadvantages in the electrolysis of aqueous NaCl of (a) the diaphragm cell and (b) the mercury cell?

Galvanic Cells

18.40 What is a galvanic cell? What is a half-cell?

18.41 What is the function of a *salt bridge?*

18.42 In the copper–silver cell, why must the Cu^{2+} and Ag^+ solutions be kept in separate containers?

18.43 Which redox processes take place at the anode and cathode in a galvanic cell? What electrical charges do the anode and cathode carry in a galvanic cell?

18.44 Explain how the movement of the ions relative to the electrodes is the same in both galvanic and electrolytic cells.

18.45 When magnesium metal is placed into a solution of copper sulfate, the magnesium dissolves to give Mg^{2+} and copper metal is formed. Write a net ionic equation for this reaction. Describe how you could use the reaction in a galvanic cell. Which metal, copper or magnesium, is the cathode?

18.46 Aluminum will displace tin from solution according to the equation $2Al(s) + 3Sn^{2+}(aq) \rightarrow 2Al^{3+}(aq) + 3Sn(s)$. What would be the individual half-cell reactions if this were the cell reaction in a galvanic cell? Which metal would be the anode and which the cathode?

18.47 At first glance, the following equation may appear to be balanced: $MnO_4^- + Sn^{2+} \rightarrow SnO_2 + MnO_2$. What is wrong with it?

18.48 Write the half-reactions and the balanced cell reaction for the following galvanic cells.
(a) $Cd(s) \mid Cd^{2+}(aq) \parallel Au^{3+}(aq) \mid Au(s)$
(b) $Pb(s), PbSO_4(s) \mid SO_4{}^{2-}(aq) \parallel H^+(aq),$
$\qquad\qquad SO_4{}^{2-}(aq) \mid PbO_2(s), PbSO_4(s)$
(c) $Cr^{3+}(aq) \mid Cr(s) \parallel Cu^{2+}(aq) \mid Cu(s)$

18.49 Write the half-reactions and the balanced cell reaction for the following galvanic cells.
(a) $Zn(s) \mid Zn^{2+}(aq) \parallel Cr^{3+}(aq) \mid Cr(s)$
(b) $Fe(s) \mid Fe^{2+}(aq) \parallel Br_2(aq), Br^-(aq) \mid Pt(s)$
(c) $Mg(s) \mid Mg^{2+}(aq) \parallel Sn^{2+}(aq) \mid Sn(s)$

18.50 Write the cell notation for the following galvanic cells. For half-reactions in which all the reactants are in solution or are gases, assume the use of inert platinum electrodes.
(a) $Cd^{2+}(aq) + Fe(s) \rightarrow Cd(s) + Fe^{2+}(aq)$
(b) $Br_2(aq) + 2Cl^-(aq) \rightarrow Cl_2(g) + 2Br^-(aq)$
(c) $Au^{3+}(aq) + 3Ag(s) \rightarrow Au(s) + 3Ag^+(aq)$

18.51 Write the cell notation for the following galvanic cells. For half-reactions in which all the reactants are in solution or are gases, assume the use of inert platinum electrodes.
(a) $NO_3^-(aq) + 4H^+(aq) + 3Fe^{2+}(aq) \rightarrow 3Fe^{3+}(aq) + NO(g) + 2H_2O$
(b) $NiO_2(s) + 4H^+(aq) + 2Ag(s) \rightarrow Ni^{2+}(aq) + 2H_2O + 2Ag^+(aq)$
(c) $Mg(s) + Cd^{2+}(aq) \rightarrow Mg^{2+}(aq) + Cd(s)$

Cell Potentials and Reduction Potentials

18.52 What is the difference between a *cell potential* and a *standard cell potential?*

18.53 How are standard reduction potentials combined to give the standard cell potential for a spontaneous reaction?

18.54 What is *emf?* What are its units?

18.55 What ratio of units gives volts?

18.56 What are the units of amperes × volts × seconds?

18.57 Is it possible to measure the emf of an isolated half-cell? Explain your answer.

18.58 Describe the hydrogen electrode. What is the value of its standard reduction potential?

18.59 What do the positive and negative signs of reduction potentials tell us?

18.60 If $E°_{Cu^{2+}}$ had been chosen as the standard reference electrode and had been assigned a potential of 0.00 V, what would the reduction potential of the hydrogen electrode be relative to it?

18.61 Using a voltmeter, how can you tell which is the anode and which is the cathode in a galvanic cell?

Using Standard Reduction Potentials

18.62 For each pair of substances, use Table 18.1 to choose the better reducing agent.
(a) $Sn(s)$ or $Ag(s)$ (b) $Cl^-(aq)$ or $Br^-(aq)$
(c) $Co(s)$ or $Zn(s)$ (d) $I^-(aq)$ or $Au(s)$

18.63 For each pair of substances, use Table 18.1 to choose the better oxidizing agent.
(a) $NO_3^-(aq)$ or $MnO_4^-(aq)$
(b) $Au^{3+}(aq)$ or $Co^{2+}(aq)$
(c) $PbO_2(s)$ or $Cl_2(g)$
(d) $NiO_2(s)$ or $HOCl(aq)$

18.64 For each pair of substances, use Table 18.1 to choose the one that is more easily oxidized.
(a) $Fe^{2+}(aq)$ or $Br^-(aq)$ (b) $F^-(aq)$ or $Ag(s)$
(c) $Ca(s)$ or $Ni(s)$ (d) $H_2(g)$ or $K(s)$

18.65 Use the data in Table 18.1 to calculate the standard cell potential for each of the following reactions:
(a) $Cd^{2+}(aq) + Fe(s) \rightarrow Cd(s) + Fe^{2+}(aq)$
(b) $Br_2(aq) + 2Cl^-(aq) \rightarrow Cl_2(g) + 2Br^-(aq)$
(c) $Au^{3+}(aq) + 3Ag(s) \rightarrow Au(s) + 3Ag^+(aq)$

18.66 Use the data in Table 18.1 to calculate the standard cell potential for each of the following reactions:
(a) $NO_3^-(aq) + 4H^+(aq) + 3Fe^{2+}(aq) \rightarrow 3Fe^{3+}(aq) + NO(g) + 2H_2O$
(b) $NiO_2(s) + 4H^+(aq) + 2Ag(s) \rightarrow Ni^{2+}(aq) + 2H_2O + 2Ag^+(aq)$
(c) $Mg(s) + Cd^{2+}(aq) \rightarrow Mg^{2+}(aq) + Cd(s)$

18.67 From the positions of the half-reactions in Table 18.1, determine whether the following reactions are spontaneous.
(a) $2Au^{3+} + 6I^- \rightarrow 3I_2 + 2Au$
(b) $3Fe^{2+} + 2NO + 4H_2O \rightarrow 3Fe + 2NO_3^- + 8H^+$
(c) $3Ca + 2Cr^{3+} \rightarrow 2Cr + 3Ca^{2+}$

18.68 Use the data in Table 18.1 to determine which of the following reactions should occur spontaneously.
(a) $Br_2 + 2Cl^- \rightarrow Cl_2 + 2Br^-$
(b) $Ni^{2+} + Fe \rightarrow Fe^{2+} + Ni$
(c) $H_2SO_3 + H_2O + Br_2 \rightarrow 4H^+ + SO_4^{2-} + 2Br^-$

18.69 Compare Table 9.3 with Table 18.1. What can you say about the basis for the activity series for metals?

18.70 Use the data in Table 18.1 to calculate the standard cell potential for the reaction
$$MnO_4^- + 8H^+ + 5Fe^{2+} \longrightarrow 5Fe^{3+} + Mn^{2+} + 4H_2O$$

18.71 Use the data in Table 18.1 to calculate the standard cell potential for the reaction
$$2Ag^+ + Fe \longrightarrow 2Ag + Fe^{2+}$$

18.72 Write the cell notation for the galvanic cells formed by using the following pairs of half-reactions. Calculate their standard cell potentials.
(a) $Co^{2+}(aq) + 2e^- \rightleftharpoons Co(s)$
    $Zn^{2+}(aq) + 2e^- \rightleftharpoons Zn(s)$
(b) $Ni^{2+}(aq) + 2e^- \rightleftharpoons Ni(s)$
    $Mg^{2+}(aq) + 2e^- \rightleftharpoons Mg(s)$
(c) $Au^{3+}(aq) + 3e^- \rightleftharpoons Au(s)$
    $Sn^{2+}(aq) + 2e^- \rightleftharpoons Sn(s)$

18.73 Write the cell notation for the galvanic cells formed by using the following pairs of half-reactions. For half-reactions in which all the reactants are in solution, assume the use of inert platinum electrodes. Calculate their standard cell potentials.
(a) $BrO_3^-(aq) + 6H^+(aq) + 6e^- \rightleftharpoons Br^-(aq) + 3H_2O$
    $Cu^{2+}(aq) + 2e^- \rightleftharpoons Cu(s)$
(b) $Fe^{3+}(aq) + e^- \rightleftharpoons Fe^{2+}(aq)$
    $Ag^+(aq) + e^- \rightleftharpoons Ag(s)$
(c) $NO_3^-(aq) + 4H^+(aq) + 3e^- \rightleftharpoons NO(g) + 2H_2O$
    $MnO_4^-(aq) + 8H^+(aq) + 5e^- \rightleftharpoons Mn^{2+}(aq) + 4H_2O$

18.74 Make a sketch of a galvanic cell for which the cell notation is
$$Ag(s) \mid Ag^+(aq) \parallel Fe^{3+}(aq) \mid Fe(s)$$
(a) Label the anode and the cathode.
(b) Indicate the charge on each electrode.

18.75 Make a sketch of a galvanic cell in which inert platinum electrodes are used in the half-cells for the system
$$Pt(s) \mid Fe^{2+}(aq), Fe^{3+}(aq) \parallel Br_2(aq), Br^-(aq) \mid Pt(s)$$
Label the diagram and indicate the composition of the electrolytes in the two cell compartments. Show the signs of the electrodes and label the anode and cathode.

18.76 From the half-reactions below, determine the cell reaction and standard cell potential.
$$BrO_3^- + 6H^+ + 6e^- \rightleftharpoons Br^- + 3H_2O \quad E^\circ_{BrO_3^-} = 1.44\ V$$
$$I_2 + 2e^- \rightleftharpoons 2I^- \quad E^\circ_{I_2} = 0.54\ V$$

18.77 What is the standard cell potential and the net reaction in a galvanic cell that has the following half-reactions?
$$MnO_2 + 4H^+ + 2e^- \rightleftharpoons Mn^{2+} + 2H_2O \quad E^\circ_{MnO_2} = 1.23\ V$$
$$PbCl_2 + 2e^- \rightleftharpoons Pb + 2Cl^- \quad E^\circ_{PbCl_2} = -0.27\ V$$

18.78 What will be the spontaneous reaction among H_2SO_3, $S_2O_3^{2-}$, HOCl, and Cl_2? The half-reactions involved are
$$2H_2SO_3 + 2H^+ + 4e^- \rightleftharpoons S_2O_3^{2-} + 3H_2O \quad E^\circ_{H_2SO_3} = 0.40\ V$$
$$2HOCl + 2H^+ + 2e^- \rightleftharpoons Cl_2 + 2H_2O \quad E^\circ_{HOCl} = 1.63\ V$$

18.79 What will be the spontaneous reaction among Br_2, I_2, Br^-, and I^-?

18.80 Will the following reaction occur spontaneously?
$$SO_4^{2-} + 4H^+ + 2I^- \longrightarrow H_2SO_3 + I_2 + H_2O$$
Use the data in Table 18.1 to answer this question.

18.81 Use the data below to determine whether the reaction
$$S_2O_8^{2-} + Ni(OH)_2 + 2OH^- \longrightarrow 2SO_4^{2-} + NiO_2 + 2H_2O$$
will occur spontaneously.
$$NiO_2 + 2H_2O + 2e^- \rightleftharpoons Ni(OH)_2 + 2OH^- \quad E^\circ_{NiO_2} = 0.49\ V$$
$$S_2O_8^{2-} + 2e^- \rightleftharpoons 2SO_4^{2-} \quad E^\circ_{S_2O_8^{2-}} = 2.01\ V$$

Cell Potentials and Thermodynamics

18.82 Write the equation that relates the standard cell potential to the standard free energy change for a reaction.

18.83 What is the equation that relates the equilibrium constant to the cell potential?

18.84 What is the maximum amount of work, expressed in joules, that can be obtained from the discharge of a silver oxide battery (see p, 799) if 0.500 g of Ag_2O reacts? E° for the cell reaction is 1.50 V.

*18.85 A watt is a unit of electrical power and is equal to one joule per second (1 watt = 1 J s^{-1}). How many hours can a calculator drawing 5×10^{-4} watt be operated by a mercury battery (see p. 799) having a cell potential equal to 1.34 V if a mass of 1.00 g of HgO is available at the cathode?

18.86 Given the following half-reactions and their standard reduction potentials,

$$2ClO_3^- + 12H^+ + 10e^- \rightleftharpoons Cl_2 + 6H_2O \quad E^\circ_{ClO_3^-} = 1.47 \text{ V}$$

$$S_2O_8^{2-} + 2e^- \rightleftharpoons 2SO_4^{2-} \qquad E^\circ_{S_2O_8^{2-}} = 2.01 \text{ V}$$

calculate (a) E°_{cell}, (b) ΔG° for the cell reaction, and (c) the value of K_c for the cell reaction.

18.87 Calculate ΔG° for the reaction

$$2MnO_4^- + 6H^+ + 5HCHO_2 \longrightarrow 2Mn^{2+} + 8H_2O + 5CO_2$$

for which $E^\circ_{cell} = 1.69$ V.

18.88 Calculate ΔG° for the following reaction *as written*

$$2Br^- + I_2 \longrightarrow 2I^- + Br_2$$

18.89 Calculate K_c for the system, $Ni^{2+} + Co \rightleftharpoons Ni + Co^{2+}$. Use the data in Table 18.1. Assume $T = 298$ K.

18.90 The system $2AgI + Sn \rightleftharpoons Sn^{2+} + 2Ag + 2I^-$ has a calculated $E^\circ_{cell} = -0.015$ V. What is the value of K_c for this system?

18.91 Determine the value of K_c at 25 °C for the reaction

$$2H_2O + 2Cl_2 \rightleftharpoons 4H^+ + 4Cl^- + O_2$$

The Effect of Concentration on Cell Potential

18.92 You have learned that the principles of thermodynamics allow the following equation to be derived, $\Delta G = \Delta G^\circ + RT \ln Q$, where Q is the reaction quotient. Without referring to the text, use this equation and the relationship between ΔG and the cell potential to derive the Nernst equation.

18.93 The cell reaction during the discharge of a lead storage battery is

$$Pb(s) + PbO_2(s) + 4H^+(aq) + 2SO_4^{2-}(aq) \longrightarrow 2PbSO_4(s) + 2H_2O$$

The standard cell potential is 2.05 V. What is the correct form of the Nernst equation for this reaction?

18.94 The cell reaction

$$NiO_2(s) + 4H^+(aq) + 2Ag(s) \longrightarrow Ni^{2+}(aq) + 2H_2O + 2Ag^+(aq)$$

has $E^\circ_{cell} = 2.48$ V. What will be the cell potential at a pH of 5.00 when the concentrations of Ni^{2+} and Ag^+ are each 0.010 M?

18.95 $E^\circ_{cell} = 0.135$ V for the reaction

$$3I_2(s) + 5Cr_2O_7^{2-}(aq) + 34H^+(aq) \longrightarrow$$
$$6IO_3^-(aq) + 10Cr^{3+}(aq) + 17H_2O$$

What is E°_{cell} if $[Cr_2O_7^{2-}] = 0.010$ M, $[H^+] = 0.10$ M, $[IO_3^-] = 0.00010$ M, and $[Cr^{3+}] = 0.0010$ M?

*18.96 Suppose that a galvanic cell were set up having the net cell reaction

$$Zn(s) + 2Ag^+(aq) \longrightarrow Zn^{2+}(aq) + 2Ag(s)$$

The Ag^+ and Zn^{2+} concentrations in their respective half-cells initially are 1.00 M, and each half-cell contains 100 mL of electrolyte solution. If this cell delivers current at a constant rate of 0.10 A, what will the cell potential be after 15.00 hr?

18.97 A cell was set up having the following reaction.

$$Mg(s) + Cd^{2+}(aq) \longrightarrow Mg^{2+}(aq) + Cd(s) \quad E^\circ_{cell} = 1.97 \text{ V}$$

The magnesium electrode was dipping into a 1.00 M solution of MgSO$_4$ and the cadmium was dipping into a solution of unknown Cd^{2+} concentration. The potential of the cell was measured to be 1.54 V. What was the unknown Cd^{2+} concentration?

18.98 A silver wire coated with AgCl is sensitive to the presence of chloride ion because of the half-cell reaction

$$AgCl(s) + e^- \rightleftharpoons Ag(s) + Cl^- \qquad E^\circ_{AgCl} = 0.2223 \text{ V}$$

A student, wishing to measure the chloride ion concentration in a number of water samples, constructed a galvanic cell using the AgCl electrode as one half-cell and a copper wire dipping into 1.00 M CuSO$_4$ solution as the other half-cell. In one analysis, the potential of the cell was measured to be 0.0925 V with the copper half-cell serving as the cathode. What was the chloride ion concentration in the water? (Take $E^\circ_{Cu^{2+}} = 0.3419$ V.)

*18.99 A galvanic cell was set up having the following half-reactions.

$$Fe^{2+}(aq) + 2e^- \rightleftharpoons Fe(s) \qquad E^\circ_{Fe^{2+}} = -0.447 \text{ V}$$
$$Cu^{2+}(aq) + 2e^- \rightleftharpoons Cu(s) \qquad E^\circ_{Cu^{2+}} = +0.3419 \text{ V}$$

The copper half-cell contained 100 mL of 1.00 M CuSO$_4$. The iron half-cell contained 50.0 mL of 0.100 M FeSO$_4$. To the iron half-cell was added 50.0 mL of 0.500 M NaOH solution. The mixture was stirred and the cell potential was measured to be 1.175 V. Calculate the value of K_{sp} for Fe(OH)$_2$.

*18.100 Suppose a galvanic cell was constructed using a Cu/Cu^{2+} half-cell (in which the molar concentration of Cu^{2+} was 1.00 M) and a hydrogen electrode having a partial pressure of H$_2$ equal to 1 atm. The hydrogen electrode dips into a solution of unknown hydrogen ion concentration, and the two half-cells are connected by a salt bridge. The precise value of $E^\circ_{Cu^{2+}}$ is $+0.3419$ V.

(a) Derive an equation for the pH of the solution with the unknown hydrogen ion concentration, expressed in terms of E_{cell} and E°_{cell}.

(b) If the pH of the solution were 5.15, what would be the observed emf of the cell?

(c) If the emf of the cell were 0.645 V, what would be the pH of the solution?

Practical Galvanic Cells

18.101 What are the anode and cathode reactions during the discharge of a lead storage battery? How can a battery produce an emf of 12 V if the cell reaction has a standard potential of only 2 V?

18.102 What are the anode and cathode reactions during the charging of a lead storage battery? What are the reactions during discharge of the battery?

18.103 What reactions occur at the electrodes in the ordinary dry cell?

18.104 What chemical reactions take place at the electrodes in an alkaline dry cell?

18.105 Give the half-cell reactions and the cell reaction that take place in a nicad battery during discharge. What are the reactions that take place during the charging of the cell?

18.106 Describe the chemical reactions that take place in (a) a silver oxide battery and (b) a mercury battery.

18.107 What advantages do fuel cells offer over conventional means of obtaining electrical power by the combustion of fuels?

18.108 How is a hydrometer constructed? How does it measure density? Why can a hydrometer be used to check the state of charge of a lead storage battery?

18.109 What role does the element calcium play in the construction of a modern automobile battery?

Additional Exercises

18.110 In the absence of unusual electrode effects, what should be expected to be observed at the cathode when electrolysis is performed on a solution that contains $NiSO_4$ and $CdSO_4$ at equal concentrations?

18.111 Consider the following half-reactions and their standard reduction potentials.

$$S_2O_8^{2-} + 2e^- \rightleftharpoons 2SO_4^{2-} \quad E^\circ_{S_2O_8^{2-}} = +2.01 \text{ V}$$

$$O_2 + 4H^+ + 4e^- \rightleftharpoons 2H_2O \quad E^\circ_{H_2O} = +1.36 \text{ V}$$

What is the most likely product at the anode in the electrolysis of Na_2SO_4?

*18.112 The electrolysis of 250 mL of a brine solution (NaCl) was carried out for a period of 20.00 min with a current of 2.00 A. The resulting solution was titrated with 0.620 M HCl. How many milliliters of the HCl solution was required for the titration?

*18.113 A solution of NaCl is neutral, with an expected pH of 7. If electrolysis was carried out on 500 mL of an NaCl solution with a current of 0.500 A, how many seconds would it take for the pH of the solution to rise to a value of 9.00?

18.114 It was desired to determine the reduction potential of Pt^{2+} ion. A galvanic cell was set up in which one half-cell consisted of a Pt electrode dipping into a 0.0100 M solution of $Pt(NO_3)_2$ and the other was prepared by dipping a silver wire coated with AgCl into a 0.100 M solution of HCl. The emf of the cell was measured to be 0.778 V and it was found that the Pt electrode carried a positive charge. Given the following half-reaction and its reduction potential,

$$AgCl(s) + e^- \rightleftharpoons Ag(s) + Cl^-(aq) \quad E^\circ_{AgCl} = 0.2223 \text{ V}$$

calculate the standard reduction potential for the half-reaction

$$Pt^{2+}(aq) + 2e^- \rightleftharpoons Pt(s)$$

18.115 For Exercise 18.114, what is the value of the standard free energy change for the cell reaction?

18.116 Consider the following half-reactions and their reduction potentials:

$$MnO_4^- + 8H^+ + 5e^- \longrightarrow Mn^{2+} + 4H_2O$$
$$E^\circ_{MnO_4^-} = +1.507 \text{ V}$$

$$ClO_3^- + 6H^+ + 6e^- \longrightarrow Cl^- + 3H_2O$$
$$E^\circ_{ClO_3^-} = +1.451$$

(a) What is the value of E°_{cell}?

(b) What is the value of ΔG° for this reaction?

(c) What is the value of K_c for the reaction at 25 °C?

(d) Write the Nernst equation for this reaction.

(e) What is the emf of the cell when $[MnO_4^-] = 0.20$ M, $[Mn^{2+}] = 0.050$ M, $[Cl^-] = 0.0030$ M, $[ClO_3^-] = 0.110$ M, and the pH of the solution is 4.25?

18.117 Consider the following galvanic cell.

$$Ag(s) \mid Ag^+(3.0 \times 10^{-4} \text{ } M) \parallel Fe^{3+}(1.1 \times 10^{-3} \text{ } M),$$
$$Fe^{2+}(0.040 \text{ } M) \mid Pt(s)$$

Calculate the cell potential. Determine the sign of the electrodes in the cell. Write the equation for the spontaneous cell reaction.

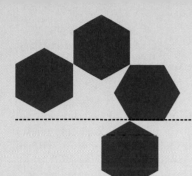

CHEMICALS IN USE 15
Ion Selective Electrodes

Many practical uses of electrochemistry rely on its unique role as an interface between chemical systems and electronic devices that display, record, and manipulate data. The key to using this interface has been the development of electrodes that are *selectively* sensitive to chemicals of interest. Such electrodes are called **ion selective electrodes** because of their ability to respond to certain ions while ignoring others.

In any galvanic cell, there are two electrodes. For the measurement of concentrations, one electrode is a *reference electrode* whose reduction potential is constant. The other is called an *indicating electrode;* it is the ion selective electrode whose potential is sensitive to the concentration of the species of interest. Both electrodes are connected to some appropriate device that displays the emf of the cell or translates that potential into concentration information.

We introduced you to the concept of an ion selective electrode in Special Topic 18.2, where we discussed the measurement of pH by the use of the hydrogen glass electrode (see Figure 15a). This is one of a number of electrodes whose construction is based on the same basic design. Inside is a reference solution, and dipping into it is the device's own internal reference electrode, which provides the electrical contact with the emf readout device. The thin glass membrane is ion sensitive and allows the electrode to detect a difference between the ion concentration inside and outside the membrane. With suitable changes in the composition of the glass membrane, this kind of electrode can be made selectively sensitive to the concentrations of a number of singly charged cations including H^+, Li^+, Na^+, K^+, Ag^+, and NH_4^+.

The glass electrode is called a *solid-state membrane electrode* because of the nature of the ion-sensitive material used in its construction. On the right of Figure 15a is illustrated the construction of a newer type based basically on the same principle. It uses a conducting, ionic, crystalline material such as LaF_3 or polycrystalline Ag_2S or $AgCl$. Although the two types appear different, they function in essentially the same way.

Another important class of ion selective electrode uses a different type of membrane material between

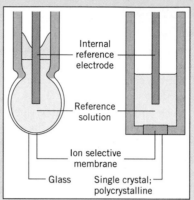

FIGURE 15a

Ion selective electrodes with solid-state membranes.

the reference solution and the solution being tested (see Figure 15b). It is called a *liquid ion-exchange membrane,* which is composed of a water-insoluble organic liquid that's capable of transporting some specific ion (for example, Ca^{2+} or K^+) across the membrane boundary. This ability makes the electrode respond *selectively* to the particular ion being transported.

APPLICATIONS IN MEDICINE AND RESEARCH

Some of the most interesting applications of ion selective electrodes have been in the areas of medicine and biological research. For example, an important cell used to measure the carbon dioxide level in blood is based on the hydrogen glass electrode that we use to measure pH (see Figure 15c). The key to the functioning of the cell is the thin silicone rubber mem-

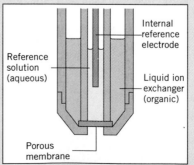

FIGURE 15b

A liquid ion-exchange membrane electrode. The organic liquid ion exchanger seeps into the porous membrane and provides the ion selectivity of the electrode.

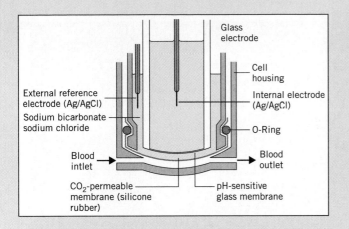

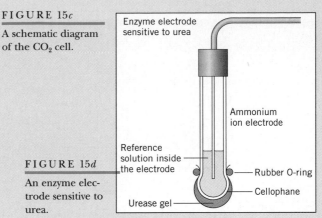

FIGURE 15c

A schematic diagram of the CO_2 cell.

FIGURE 15d

An enzyme electrode sensitive to urea.

brane that separates the blood sample from a dilute $NaHCO_3$ solution in contact with the glass electrode. The rubber membrane is permeable to CO_2 which diffuses into the solution that contains the HCO_3^-. The CO_2 participates in the equilibrium

$$CO_2(aq) + H_2O \rightleftharpoons HCO_3^-(aq) + H^+(aq)$$

and thereby influences the pH of the solution, which is detected by the glass electrode. In an indirect way, then, the potential of the glass electrode reflects the CO_2 concentration just outside the glass membrane, and this in turn is a measure of the CO_2 concentration in the blood that's in contact with the silicone rubber membrane. In practice, the electrode is first calibrated with solutions of known CO_2 partial pressures and then used to measure the CO_2 content of a blood sample.

We mentioned earlier that by suitably modifying the glass membrane, electrodes can be made that respond to ions other than H^+. Sodium-type glass electrodes have been used in clinical laboratories to measure serum sodium levels. A clinical use of the second type of electrode in Figure 15a is in the measurement of chloride ion in sweat. Similarly, ion-exchange membrane electrodes have been used successfully to monitor serum levels of calcium ion and potassium ion.

In the laboratory there have been a number of interesting developments that have made biochemical reactions potentially easier to study. One very clever modification of the ammonium ion glass electrode, for example, is shown in Figure 15d. The glass membrane of the electrode is coated with a gel in which is embedded an enzyme such as urease, a biological catalyst that promotes the hydrolysis of one particular molecule, urea.

$$CO(NH_2)_2 + H_2O \xrightarrow{\text{Urease}} CO_2 + 2NH_3$$

The gel–urease layer is held in place by a cellophane membrane through which the urea is able to diffuse. On contact with the enzyme, hydrolysis occurs and the ammonia that's formed then reacts with water according to the following equilibrium

$$NH_3(aq) + H_2O \rightleftharpoons NH_4^+(aq) + OH^-(aq)$$

The ammonium ion formed in this reaction is then detected by the electrode. Once again, we have an electrode that indirectly senses a particular chemical, and this time the chemical isn't even ionic. Similar electrodes have been developed for ethyl alcohol, cholesterol, glucose ("blood sugar"), and several other substances in body fluids.

In 1991, the Nobel Prize in physiology or medicine was awarded to two German scientists, Erwin Neher and Bert Sakmann, for their use of microelectrodes in the study of ion transport through cell membranes. They found that in a cell membrane there are specific channels through which ions flow, and that a single channel molecule can alter its shape to control the flow of ions within a few millionths of a second. Their work has contributed to a better understanding of the cellular mechanisms of several diseases, including diabetes and cystic fibrosis.

Questions

1. What equation forms the basis for measuring ion concentrations by the use of appropriately designed galvanic cells?

2. What ions can be measured by glass electrodes with suitable glass membranes?

3. What is an ion-exchange membrane electrode? How does it differ from a glass electrode?

4. In your own words, describe how the CO_2 electrode is able to detect varying CO_2 levels in the blood.

5. How does the enzyme electrode work?

Tropical rain forests such as the one shown here on Saint Lucia, the second largest of the Windward Islands in the Caribbean Sea, are responsible in large measure for the Earth's oxygen. The leaves of green plants hold the pigment chlorophyll, which enables the energy of sunlight to convert carbon dioxide and water to glucose plus molecular oxygen. At the heart of chlorophyll is a magnesium ion held in a complex ion. The structure and properties of complex ions is the principal focus of this chapter.

Chapter 19

Metal Complexes and Their Equilibria

In our previous discussions of metal-containing compounds, we left you with the impression that the only kinds of bonds in which metals are ever involved are ionic bonds. For some metals, like the alkali metals of Group IA, this is essentially true. But for many others, especially the transition metals and the post-transition metals, it is not. This is because the ions of many of these metals are able to behave as Lewis acids (i.e., as electron pair acceptors in the formation of coordinate covalent bonds). Thus, by participating in Lewis acid–base reactions they become *covalently* bonded to other atoms. Copper(II) ion is a typical example.

In aqueous solutions of copper(II) salts, like $CuSO_4$ or $Cu(NO_3)_2$, the copper is not present as simple Cu^{2+} ions. Instead, each Cu^{2+} ion becomes bonded to four water molecules to give a pale blue ion with the formula $Cu(H_2O)_4^{2+}$ (see Figure 19.1). We call this species a **complex ion** because it is composed of a number of simpler species (i.e., it is *complex*, not simple). The chemical equation for the formation of the $Cu(H_2O)_4^{2+}$ ion is

$$Cu^{2+} + 4H_2O \longrightarrow Cu(H_2O)_4^{2+}$$

which can be diagrammed using Lewis structures as follows:

FIGURE 19.1

A solution containing copper sulfate has a pale blue color because it contains the complex ion $Cu(H_2O)_4^{2+}$.

811

As you can see in this analysis, the Cu^{2+} ion accepts pairs of electrons from the water molecules, so Cu^{2+} is a Lewis acid and the water molecules are each Lewis bases.

The number of complex ions formed by metals, especially the transition metals, is enormous, and the study of the properties, reactions, structures, and bonding in complex ions like $Cu(H_2O)_4{}^{2+}$ has become an important specialty within chemistry. A part of learning about this specialty is becoming familiar with some of the terminology.

A Lewis base that attaches itself to a metal ion is called a **ligand.** Ligands can be neutral molecules with unshared pairs of electrons (like H_2O), or they can be anions (like Cl^- or OH^-). The atom in the ligand that actually provides the electron pair is called the **donor atom,** and the metal ion is the **acceptor.** The result of combining a metal ion with one or more ligands is a *complex ion,* or simply just a **complex.** The word "complex" avoids problems when the particle formed is electrically neutral, as sometimes happens. Compounds that contain complex ions are generally referred to as **coordination compounds** because the bonds in a complex ion can be viewed as coordinate covalent bonds. Sometimes the complex itself is called a **coordination complex.**

> In the space we have available here our goal will simply be to give you a taste of the subject.

> "Ligand" is from the Latin word *ligare,* meaning "to bind."

> Most of the nutritionally essential trace metal ions that our bodies require, like Cu^{2+}, Zn^{2+}, Mn^{2+}, and many others, are involved as complex ions.

> Recall that a *Lewis base* is an electron pair donor in the formation of a coordinate covalent bond.

Ligands

As we've noted, ligands may be either anions or neutral molecules. In either case, *ligands are Lewis bases* and, therefore, contain at least one atom with one or more lone pairs (unshared pairs) of electrons.

Anions that serve as ligands include many simple monatomic ions, such as the halide ions (F^-, Cl^-, Br^-, and I^-) and the sulfide ion (S^{2-}). Common polyatomic anions that are ligands are nitrite ion ($NO_2{}^-$), cyanide ion (CN^-), hydroxide ion (OH^-), thiocyanate ion (SCN^-), and thiosulfate ion ($S_2O_3{}^{2-}$). (This is only a small sampling, not a complete list.)

The most common neutral molecule that serves as a ligand is water, and most of the reactions of metal ions in aqueous solutions are actually reactions of their complex ions—ions in which the metal is attached to some number of water molecules. This number isn't always the same, however. Copper(II) ion, for example, forms the complex ion $Cu(H_2O)_4{}^{2+}$ (as we've noted earlier), but cobalt(II) combines with water molecules to form $Co(H_2O)_6{}^{2+}$. Another common neutral ligand is ammonia, NH_3, which has one lone pair of electrons on the nitrogen atom. If ammonia is added to an aqueous solution containing the $Cu(H_2O)_4{}^{2+}$ ion, for example, the color deepens dramatically as ammonia molecules displace water molecules (Figure 19.2).

$$Cu(H_2O)_4{}^{2+}(aq) + 4NH_3(aq) \longrightarrow Cu(NH_3)_4{}^{2+}(aq) + 4H_2O$$
$$\text{(pale blue)} \qquad\qquad\qquad\qquad \text{(deep blue)}$$

Each of the ligands that we have discussed so far is able to use just one atom to attach itself to a metal ion. Such ligands are called **monodentate ligands,** indicating that they have only "one tooth" with which to "bite" the metal ion.

There are also many ligands that have two or more donor atoms, and collectively they are referred to as **polydentate ligands.** The most common of these have two donor atoms, so they are called **bidentate ligands.** When they form complexes, *both* donor atoms become attached to the same metal ion. Oxalate ion and ethylenediamine (abbreviated *en* in writing the formula for a complex) are examples of bidentate ligands.

FIGURE 19.2

On the left is a solution containing the pale blue $Cu(H_2O)_4{}^{2+}$ ion. Adding ammonia produces a dramatic deepening of the color as the intensely blue $Cu(NH_3)_4{}^{2+}$ ion is formed, which is seen at the right.

:O: :O:

⊖:Ö—C—C—Ö:⊖

oxalate ion

H$_2$N̈—CH$_2$—CH$_2$—N̈H$_2$

ethylenediamine, en

When these ligands become attached to a metal ion, ring structures are formed as shown. Complexes that contain such ring structures are called **chelates.**[1]

An oxalate complex

An ethylenediamine complex

Structures like these are important in "complex ion chemistry," as we shall see later in this chapter.

One of the most common polydentate ligands is a compound called ethylenediaminetetraacetic acid, mercifully abbreviated EDTA. The H atoms attached to the oxygen atoms are easily removed as protons, which gives an anion with a charge of 4 −. The structures of EDTA and its anion, EDTA^{4-}, are as follows, with the donor atoms shown in color.

EDTA (sometimes, H$_4$EDTA)

EDTA^{4-}

As you can see, the EDTA^{4-} ion has six donor atoms, and this permits it to wrap itself around a metal ion and form very stable complexes.

EDTA is a particularly useful and important ligand. It is relatively nontoxic, which allows it to be used in small amounts in foods to retard spoilage. If you look at the labels on bottles of salad dressings, for example, you often will find that one of the ingredients is CaNa$_2$EDTA (calcium disodium EDTA). The EDTA^{4-} available from this salt forms soluble complex ions with any traces of metal ions that might otherwise promote reactions of the salad oils with oxygen, and thereby lead to spoilage.

Sometimes the ligand EDTA is abbreviated using small letters, i.e., edta.

The structure of an EDTA complex. The nitrogen atoms are pale blue, oxygen is red, carbon is gray, and hydrogen is white. The metal ion is in the center of the complex bonded to the two nitrogens and four oxygens.

The *calcium* salt of EDTA is used because the EDTA^{4-} ion would otherwise extract Ca^{2+} ions from bones, and that would be harmful.

---

[1] The term *chelate* comes from the Greek *chele*, meaning claw. These bidentate ligands grasp the metal ions with two "claws" (donor atoms) somewhat like a crab holds its prey. (Who says scientists have no imagination?)

Salad dressings such as these contain EDTA as a preservative.

Many shampoos contain Na_4EDTA to soften water. The $EDTA^{4-}$ binds to Ca^{2+}, Mg^{2+}, and Fe^{3+} ions, which removes them from the water and prevents them from interfering with the action of soaps in the shampoo.

EDTA is also sometimes added in small amounts to whole blood to prevent clotting. It ties up calcium ions, which the clotting process requires. EDTA has even been used in poison treatment because it can help remove poisonous heavy metal ions, like Pb^{2+}, from the body when they have been accidentally ingested.

Writing Formulas for Complexes

When we write the formula for a complex, we follow two rules:

1. The symbol for the metal ion is always given first, followed by the ligands.

2. The charge on the complex is the algebraic sum of the charge on the metal ion and the charges on the ligands.

For example, the formula of the complex ion of Cu^{2+} and H_2O, which we mentioned earlier, was written $Cu(H_2O)_4{}^{2+}$ with the Cu first followed by the ligands. The charge on the complex is $2+$ because the copper ion has a charge of $2+$ and the water molecules are neutral. Copper(II) ion also forms a complex ion with four cyanide ions, CN^-, $Cu(CN)_4{}^{2-}$. The metal ion contributes two $(+)$ charges and the four ligands contribute a total of four $(-)$ charges, one for each cyanide ion. The algebraic sum is therefore $2-$, so the complex ion has a charge of $2-$.

The formula for a complex ion is often placed within brackets, with the charge *outside*. The two complexes just mentioned would thus be written as $[Cu(H_2O)_4]^{2+}$ and $[Cu(CN)_4]^{2-}$. The brackets emphasize that the ligands are attached to the metal ion and are not free to roam about. One of the many complex ions formed by the chromium(III) ion contains five water molecules and one chloride ion as ligands. To indicate that all are attached to the Cr^{3+}

Square brackets here do not mean molar concentration. It is usually clear from the context of a discussion whether we intend the brackets to mean molarity.

ion, we use brackets and write the complex ion as $[Cr(H_2O)_5Cl]^{2+}$. When isolated as a chloride salt, the formula is written $[Cr(H_2O)_5Cl]Cl_2$, in which $[Cr(H_2O)_5Cl]^{2+}$ is the cation and so is written first. The formula $[Cr(H_2O)_5Cl]Cl_2$ clearly shows that five water molecules and a chloride ion are bonded to the chromium ion, and the other two chloride ions are present to provide electrical neutrality for the salt.

In Chapter 2 you learned about hydrates, and one was the beautiful blue hydrate of copper sulfate, $CuSO_4 \cdot 5H_2O$. It was much too early then to make the distinction, but the formula should have been written $[Cu(H_2O)_4]SO_4 \cdot H_2O$ to show that four of the five water molecules are held in the crystal as part of the complex ion $[Cu(H_2O)_4]^{2+}$. The fifth water molecule is held in the crystal by being hydrogen bonded to the sulfate ion.

Many other hydrates of metal salts actually contain complex ions of the metals in which water is the ligand. Cobalt salts like cobalt(II) chloride, for example, crystallize from aqueous solutions as hexahydrates (meaning they contain six H_2O molecules per formula unit of the salt). The compound $CoCl_2 \cdot 6H_2O$ (Figure 19.3) actually is $[Co(H_2O)_6]Cl_2$ and contains the pink complex $[Co(H_2O)_6]^{2+}$. This ion also gives solutions of cobalt(II) salts a pink color as can also be seen in Figure 19.3. Although most hydrates of metal salts contain complex ions, the distinction is seldom made, and it's acceptable to write the formula for these hydrates in the usual fashion, e.g., $CuSO_4 \cdot 5H_2O$ instead of $[Cu(H_2O)_4]SO_4 \cdot H_2O$.

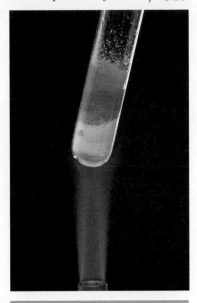

Crystals of $CuSO_4 \cdot 5H_2O$ lose water when heated. This destroys the blue complex and leaves the nearly white anhydrous $CuSO_4$.

Write the formula for the complex ion formed by the metal ion Cr^{3+} and six NO_2^- ions as ligands. Decide whether the complex could be isolated as a chloride salt or a potassium salt, and write the formula for the appropriate salt.

ANALYSIS The net charge on the complex ion must first be determined. If it is a positive ion, it will be isolated as a chloride salt; if it is a negative ion, then a potassium salt forms.

SOLUTION Six NO_2^- ions contribute a total charge of $6-$; the metal contributes a charge of $3+$. The algebraic sum is $(6-) + (3+) = 3-$. The formula of the complex ion is therefore $[Cr(NO_2)_6]^{3-}$.

Because the complex is an anion, it requires a cation to form a neutral salt, so the complex would be isolated as a potassium salt, not as a chloride salt. For the salt to be electrically neutral, three K^+ ions are required for each complex ion, $[Cr(NO_2)_6]^{3-}$. The formula of the salt is therefore $K_3[Cr(NO_2)_6]$.

■ **Practice Exercise 1** Write the formula of the complex ion formed by Ag^+ and two thiosulfate ions, $S_2O_3^{2-}$. If we were able to isolate this complex ion as its ammonium salt, what would be the formula for the salt?

■ **Practice Exercise 2** Aluminum chloride crystallizes from aqueous solutions as a hexahydrate. Write the formula for the salt and suggest a formula for the complex ion formed by aluminum ion and water.

EXAMPLE 19.1
Writing the Formula for a Complex Ion

F I G U R E 19.3
The ion $[Co(H_2O)_6]^{2+}$ is pink and gives its color both to crystals and to an aqueous solution of $CoCl_2 \cdot 6H_2O$.

The Chelate Effect

An interesting aspect of the complexes formed by ligands such as ethylenediamine and oxalate ion is their stabilities compared to similar complexes formed by monodentate ligands. For example, the complex $[Ni(en)_3]^{2+}$ is considerably more stable than $[Ni(NH_3)_6]^{2+}$, even though both complexes have six

Monodentate ligands can supply only one donor atom to a metal.

nitrogen atoms bound to a Ni^{2+} ion. This phenomenon is called the **chelate effect,** so named because it occurs in compounds that have these *chelate ring* structures.

The reason for the chelate effect appears to be related to the probability of the ligand being removed from the vicinity of the metal ion when a donor atom becomes detached. If one end of a bidentate ligand comes loose from the metal, the donor atom cannot wander very far because the other end of the ligand is still attached. There is a high probability that the loose end will become reattached to the metal ion before the other end can let go, so overall the ligand appears to be bound tightly. With a monodentate ligand, however, there is nothing to hold the ligand near the metal ion if it becomes detached. The ligand can easily wander off into the surrounding solution and be lost. As a result, a monodentate ligand doesn't behave as if it is as firmly attached to the metal ion as a polydentate ligand.

19.2 NOMENCLATURE OF METAL COMPLEXES

The naming of chemical compounds was introduced in Chapter 2 where we discussed the nomenclature system for simple inorganic compounds. This system, revised and kept up to date by the International Union of Pure and Applied Chemistry (IUPAC), has been extended to cover metal complexes. Below are some rules that have been developed to name coordination complexes. As you will see, some of the names arrived at by following the rules are difficult to pronounce at first and may even sound a little odd. However, the primary purpose of this and any other system of nomenclature is to provide a method that gives each unique compound its own unique name, and that permits us to write the formula of the compound given the name.

Nomenclature rules for complexes

Rules of Nomenclature for Coordination Complexes

1. **Cationic species are named before anionic species.** This is the same rule that applies to other ionic compounds such as NaCl, where we name the cation first followed by the anion (sodium chloride).

2. **Within a complex, the ligands are named first, in alphabetical order, followed by the metal ion.** This is *opposite* to the sequence in which they appear in the formula of the complex. For example, the complex $[Co(NH_3)_6]^{3+}$ is named by specifying the ammonia first, followed by the cobalt.

3. **The names of anionic ligands always end in the suffix -*o*.**
   (a) Ligands whose names end in *-ide* have this suffix changed to *-o.*

Anion		Ligand
chloride	Cl^-	chloro
bromide	Br^-	bromo
cyanide	CN^-	cyano
oxide	O^{2-}	oxo

(b) Ligands whose names end in *-ite* or *-ate* become *-ito* and *-ato*, respectively.

Anion		Ligand
carbonate	CO_3^{2-}	carbonato
thiosulfate	$S_2O_3^{2-}$	thiosulfato
thiocyanate	SCN^-	thiocyanato (when bonded through sulfur)
		isothiocyanato (when bonded through nitrogen)
oxalate	$C_2O_4^{2-}$	oxalato
nitrite	NO_2^-	nitrito (when bonded through oxygen; written ONO in formula for complex)[a]

[a] An exception to this is when the nitrogen of the NO_2^- ion is bonded to the metal, in which case the ligand is named nitro.

4. **A neutral ligand is given the same name as the neutral molecule.** Thus the molecule ethylenediamine, when serving as a ligand, is called ethylenediamine in the name of the complex. Two very important exceptions to this, however, are water and ammonia. These are named as follows when they serve as ligands.

H_2O aqua NH_3 ammine (note the double *m*)

5. **When there is more than one of a particular ligand, their number is specified by the prefixes di- = 2, tri- = 3, tetra- = 4, penta- = 5, hexa- = 6, and so forth. When confusion might result by using these prefixes, the following are used instead: bis- = 2, tris- = 3, tetrakis- = 4.** Following this rule, the presence of two chloride ligands in a complex would be indicated as *dichloro* (notice, too, the ending on the ligand name). However, if two ethylenediamine ligands are present, use of the prefix *di-* might cause confusion. Someone reading the name might wonder whether diethylenediamine means two ethylenediamine molecules or one molecule of a substance called diethylenediamine. To avoid this problem we place the ligand name in parentheses preceded by *bis;* that is, *bis(ethylenediamine).*

Bis is employed here so that if the name is used in verbal communication it is clear that the meaning is two ethylenediamine molecules.

6. **Negative (anionic) complex ions always end in the suffix *-ate*.** This suffix is appended to the English name of the metal atom in most cases.

Element	Metal as named in anionic complex
Aluminum	aluminate
Chromium	chromate
Manganese	manganate
Nickel	nickelate
Cobalt	cobaltate
Zinc	zincate
Platinum	platinate
Vanadium	vanadate

For metals whose symbols are derived from their Latin names, the suffix *-ate* is appended to the Latin stem. (An exception, however, is mercury.)

Element	Stem	Metal as named in anionic complex
Iron	ferr-	ferrate
Copper	cupr-	cuprate
Lead	plumb-	plumbate
Silver	argent-	argentate
Gold	aur-	aurate
Tin	stann-	stannate

For neutral or positively charged complexes, the metal *always* is specified with the English name for the element, *without any suffix.*

7. **The oxidation state of the metal in the complex is written in Roman numerals within parentheses following the name of the metal.** For example,

$[Co(NH_3)_6]^{3+}$ is the hexaamminecobalt(III) ion
$[CuCl_4]^{2-}$ is the tetrachlorocuprate(II) ion

No space

> As you learned earlier, the charge on the complex is the algebraic sum of the charges on the ligands and the charge on the metal ion.

Note that there are no spaces between the names of the ligands and the name of the metal, and that there is no space between the name of the metal and the parentheses that enclose the oxidation state expressed in Roman numerals.

The following are some additional examples

> Notice that the alphabetical order of the ligands is determined by the first letter in the name of the ligand, not the first letter in the prefix.

$[Ni(CN)_4]^{2-}$	tetracyanonickelate(II) ion
$[CoCl_6]^{3-}$	hexachlorocobaltate(III) ion
$[Co(NH_3)_4Cl_2]^+$	tetraamminedichlorocobalt(III) ion
$Na_3[Co(NO_2)_6]$	sodium hexanitrocobaltate(III)
$[Ag(NH_3)_2]^+$	diamminesilver(I) ion
$[Ag(S_2O_3)_2]^{3-}$	dithiosulfatoargentate(I) ion
$[Mn(en)_3]Cl_2$	tris(ethylenediamine)manganese(II) chloride
$[Pt(NH_3)_2Cl_2]$	diamminedichloroplatinum(II)

■ **Practice Exercise 3** Name the following compounds: (a) $K_3[Fe(CN)_6]$ and (b) $[Cr(en)_2Cl_2]_2SO_4$.

■ **Practice Exercise 4** Write the formula for each of the following: (a) hexachlorostannate(IV) ion and (b) ammonium hexacyanoferrate(II).

19.3 COORDINATION NUMBER AND STRUCTURE

One of the most interesting aspects of the study of complexes is the kinds of structures that are formed. In many ways, this is related to the **coordination number** of the metal ion, which is *the number of donor atoms attached to the metal ion.* For example, in the complex $[Cu(H_2O)_4]^{2+}$ the copper is surrounded by the four oxygen atoms that belong to the water molecules, so the coordination number of Cu^{2+} in this complex is 4. Similarly, the coordination number of

Cr^{3+} in the $[Cr(H_2O)_6]^{3+}$ ion is 6, and the coordination number of Ag^+ in $[Ag(NH_3)_2]^+$ is 2.

Sometimes the coordination number isn't immediately obvious from the formula of the complex. For example, you learned that there are many polydentate ligands which contain more than one donor atom that can bind simultaneously to a metal ion. Often, a metal is able to accommodate two or more polydentate ligands to give complexes with formulas such as $[Cr(en)_2(H_2O)_2]^{3+}$ and $[Cr(en)_3]^{3+}$. In each of these examples, the coordination number of the Cr^{3+} is 6. In the $[Cr(en)_3]^{3+}$ ion, there are three ethylenediamine ligands that each supply two donor atoms, for a total of 6, and in $[Cr(en)_2(H_2O)_2]^{3+}$, the two ethylenediamine ligands supply a total of 4 donor atoms and the two H_2O molecules supply another 2; again the total is 6.

Structures of Complexes

For metal complexes, there are certain geometries that are usually associated with particular coordination numbers.

Coordination Number 2   Examples are complexes such as $[Ag(NH_3)_2]^+$ and $[Ag(CN)_2]^-$. Usually, these complexes have a linear structure such as

$$[H_3N—Ag—NH_3]^+ \qquad \text{and} \qquad [NC—Ag—CN]^-$$

(Since the Ag^+ ion has a filled d subshell, it behaves as any of the representative elements as far as predicting geometry by VSEPR theory, so these structures are exactly what we would expect based on that theoretical model.)

> Ordinarily, VSEPR theory isn't used to predict the structures of transition metal complexes because it can't be relied on to give correct results when the metal has a partially filled d subshell.

Coordination Number 4   Two common geometries occur when four ligand atoms are bonded to a metal ion—tetrahedral and square planar. These are illustrated in Figure 19.4. The tetrahedral geometry is usually found with metal ions that have completely filled d subshells, such as Zn^{2+}. The complexes $[Zn(NH_3)_4]^{2+}$ and $[Zn(OH)_4]^{2-}$ are examples.

Square planar geometries are observed for complexes of Cu^{2+}, Ni^{2+}, and especially Pt^{2+}. Examples are $[Cu(NH_3)_4]^{2+}$, $[Ni(CN)_4]^{2-}$, and $[PtCl_4]^{2-}$. The most well-studied square planar complexes are those of Pt^{2+}, because they are considerably more stable than the others.

Coordination Number 6   The most common coordination number for complex ions is 6. Examples are $[Al(H_2O)_6]^{3+}$, $[Co(C_2O_4)_3]^{3-}$, $[Ni(en)_3]^{2+}$, and $[Co(EDTA)]^-$. With few exceptions, all complexes with a coordination number of 6 are octahedral. This holds true for those formed from both

> Recall that EDTA has six donor atoms.

FIGURE   19.4

Tetrahedral and square planar geometries that occur for complexes in which the metal ion has a coordination number of four. For the copper complex, we are viewing the square planar structure tilted back into the plane of the paper.

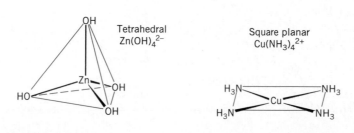

Tetrahedral
$Zn(OH)_4{}^{2-}$

Square planar
$Cu(NH_3)_4{}^{2+}$

FIGURE 19.5

Octahedral complexes can be formed with either monodentate ligands such as water or with polydentate ligands such as ethylenediamine (en). To simplify the drawing of the ethylenediamine complex, the atoms joining the donor nitrogen atoms in the ligand, $-CH_2-CH_2-$, are represented as the curved line between the N atoms. Also notice that the nitrogen atoms of the bidentate ligand span adjacent positions within the octahedron. This is the case for all polydentate ligands that you will encounter in this book.

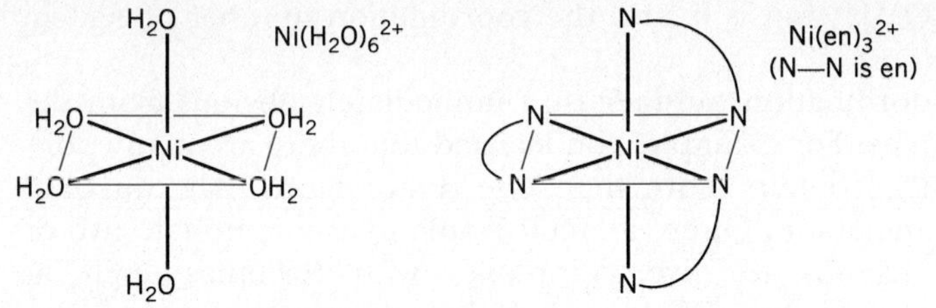

monodentate and bidentate ligands, as illustrated in Figure 19.5. Octahedral coordination is also observed for metal ions in a variety of biologically important complexes. One example is the structure of cyanocobalamin, the form of vitamin B_{12} found in vitamin pills, as shown in Figure 19.6.

The principal ligand structure in many biologically active complexes is the square planar arrangement of nitrogen atoms in a polydentate ligand structure called the *porphyrin structure*.

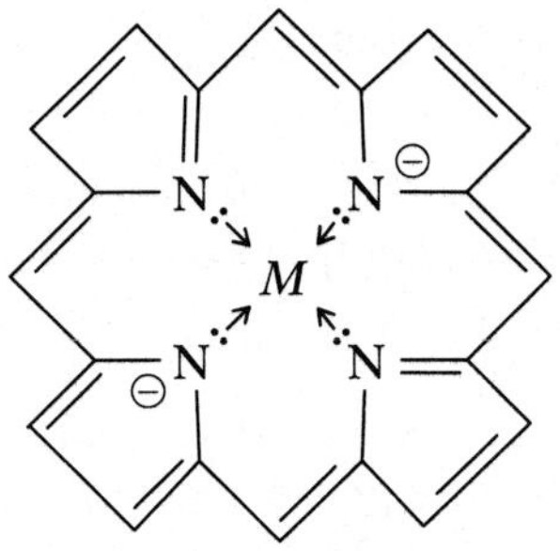

FIGURE 19.6

The structure of cyanocobalamin. Notice the cobalt ion in the center of the square planar arrangement of nitrogen atoms that are part of the ligand structure. Overall, the cobalt is surrounded octahedrally by donor atoms.

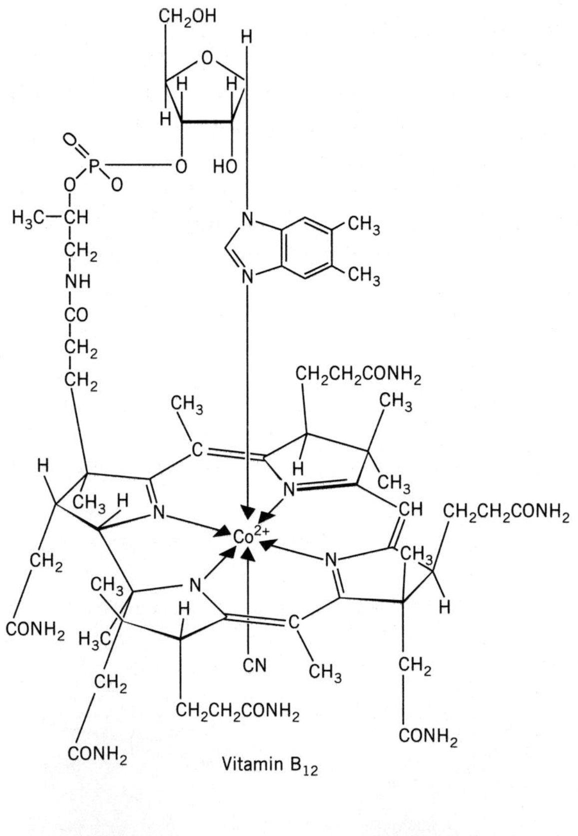

Vitamin B_{12}

(a)

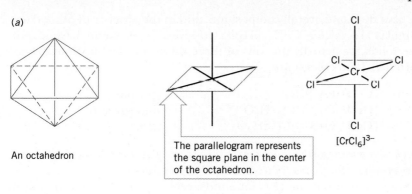

An octahedron

The parallelogram represents the square plane in the center of the octahedron.

$[CrCl_6]^{3-}$

FIGURE   19.7

Two simplified representations of the octahedral complex, $[CrCl_6]^{3-}$. (a) A drawing similar to the one you learned to construct in Chapter 8. (b) An alternative method of representing the octahedron.

(b)

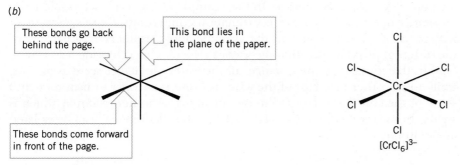

These bonds go back behind the page.

This bond lies in the plane of the paper.

These bonds come forward in front of the page.

$[CrCl_6]^{3-}$

Related ligand structures are found in hemoglobin and myoglobin, which contain a porphyrin ligand that holds Fe^{2+} (see Section 21.10). These bind oxygen as an additional ligand attached to the Fe^{2+} ion. Chlorophyll contains a similar ligand structure in which Mg^{2+} is held.

In describing the shapes of octahedral complexes, most chemists use one of the simplified drawings of the octahedron shown in Figure 19.7.

A description of how to sketch the structure of an octahedral molecule or ion is given in Chapter 8 on page 312.

■ **Practice Exercise 5** What is the coordination number of the metal ion in (a) $[Cr(C_2O_4)_3]^{3-}$, (b) $[Co(C_2O_4)_2Cl_2]^{3-}$, (c) $[Cr(en)(C_2O_4)_2]^-$, and (d) $[Co(EDTA)]^-$?

When you write the chemical formula for a particular compound, you might be tempted to think that you should also be able to predict exactly what the structure of the molecule or ion is. As you may realize by now, this just isn't possible in many cases. Sometimes we can use simple rules to make reasonable structural guesses, as in the discussion of the drawing of Lewis structures in Chapter 7, but these rules apply only to simple molecules and ions. For more complex substances, there usually is no way of knowing for sure what the structure of the molecule or ion is without performing the necessary experiments to determine the structure.

One of the reasons that structures can't be predicted with certainty from chemical formulas alone is that there are usually many different ways that the atoms in the formula can be arranged. In fact, it is frequently possible to isolate two or more compounds that actually have the same chemical formula. For example, three different solids, each with its own characteristic color and other properties, can be isolated from a solution of chromium(III) chloride.

19.4
ISOMERS OF
COORDINATION
COMPOUNDS

Chromium(III) chloride purchased from chemical supply companies is actually $[Cr(H_2O)_4Cl_2]Cl \cdot 2H_2O$. Its green color both in the solid state and in solution is due to the complex ion, $[Cr(H_2O)_4Cl_2]^+$.

Ag_2SO_4 is soluble but $BaSO_4$ is insoluble. $BaBr_2$ is soluble but $AgBr$ is insoluble.

All three have the same overall composition, and in the absence of other data, their formulas are written $CrCl_3 \cdot 6H_2O$. However, experiments have shown that these solids are actually the salts of three different complex ions. Their actual formulas (and colors) are

$[Cr(H_2O)_6]Cl_3$	purple
$[Cr(H_2O)_5Cl]Cl_2 \cdot H_2O$	blue-green
$[Cr(H_2O)_4Cl_2]Cl \cdot 2H_2O$	green

Even though their overall compositions are the same, each of these substances is a distinct chemical compound with its own characteristic set of properties.

The existence of two or more compounds, each having the same chemical formula, is known as **isomerism**. In the example above, each salt is said to be an **isomer** of $CrCl_3 \cdot 6H_2O$. For coordination compounds, there's a variety of ways for isomerism to occur. In the example above, isomers exist because of the different possible ways that the water molecules and chloride ions can be held in the crystals. In one instance, all the water molecules serve as ligands, while in the other two, part of the water is present as water of hydration and some of the chloride is bonded to the metal ion. Another example, which is similar in some respects, is $Cr(NH_3)_5SO_4Br$. This "substance" can be isolated as two isomers.

$$[Cr(NH_3)_5SO_4]Br \qquad and \qquad [Cr(NH_3)_5Br]SO_4$$

They can be distinguished chemically by their differing abilities to react with Ag^+ and Ba^{2+}. The first isomer reacts in aqueous solution with Ag^+ to give a precipitate of AgBr, but it doesn't react with Ba^{2+}. This tells us that Br^- exists as a free ion in the solution. It also suggests the SO_4^{2-} is bound to the chromium and is unavailable to react with Ba^{2+} to give insoluble $BaSO_4$.

The second isomer reacts in solution with Ba^{2+} to give a precipitate of $BaSO_4$, which means there is free SO_4^{2-} in the solution. There is no reaction with Ag^+, however, because Br^- is bonded to the chromium and is not available as free Br^- in the solution. Thus, we see that even though both isomers have the same overall composition, they behave chemically in quite different ways and are therefore distinctly different compounds.

Stereoisomerism

One of the most interesting kinds of isomerism found among coordination compounds is called **stereoisomerism,** which is defined as *differences among isomers that arise as a result of the various possible orientations of their atoms in space.* In other words, when stereoisomerism exists we have compounds in which the same atoms are attached to each other, but they differ in the way those atoms are arranged in space relative to one another.

One form of stereoisomerism is called **geometric isomerism;** it is best understood by considering an example. Consider the square planar complexes having the formula $Pt(NH_3)_2Cl_2$. There are two ways to arrange the ligands around the platinum, as illustrated below. In one isomer, called the **cis isomer,** the chloride ions are *next to each other* and the ammonia molecules are also next to each other. In the other isomer, called the **trans isomer,** identical ligands are *opposite each other*. In identifying (and naming) isomers, *cis* means "on the same side," and *trans* means "on opposite sides."

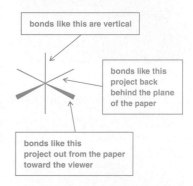

Geometric isomerism also occurs for octahedral complexes. For example, consider the ions $[Cr(H_2O)_4Cl_2]^+$ and $[Cr(en)_2Cl_2]^+$. Both can be isolated as cis and trans isomers.

$[Cr(H_2O)_4Cl_2]^+$

Curved lines connecting nitrogen atoms represent —CH₂—CH₂—, which links the nitrogen atoms in ethylenediamine

$[Cr(en)_2Cl_2]^+$

$\frown$
N   N is en
(ethylenediamine)

In the cis isomer, the chloride ligands are both on the same side of the metal ion; in the trans isomer, the chloride ligands are on opposite ends of a line that passes through the center of the metal ion.

Chirality

There is a second kind of stereoisomerism that is much more subtle than geometric isomerism. This occurs when molecules are exactly the same except for one small difference—they differ from each other in the same way that your left hand differs from your right hand.

Although similar in appearance, your left and right hands are not exactly alike, as you discover if you try to fit your right hand into a left-hand glove (Figure 19.8). There is, however, a special relationship between the two hands, which you can see if you stand in front of a mirror. If you hold your left hand so that it is facing the mirror, and then look at its reflection as illustrated in Figure 19.9, you will see that it looks exactly like your right hand. If it were possible to reach "through the looking glass," your right-hand glove would fit the reflection of your left hand perfectly.

In this analysis, we say our left and right hands are *mirror images of each other.* The two thumbtacks in the margin are also mirror images of each other. The mirror image of one looks just like the other. Yet, there is a special difference between thumbtacks and hands, which is seen in a test for "handedness" that we call **superimposability.**

The isomer on the left, *cis*-$Pt(NH_3)_2Cl_2$, is the anticancer drug known as *cisplatin*. It is interesting that only the cis isomer of this compound is clinically active against tumors. The trans isomer is totally ineffective.

The structures of octahedral complexes are drawn here according to the following key:

bonds like this are vertical

bonds like this project back behind the plane of the paper

bonds like this project out from the paper toward the viewer

FIGURE 19.8

A left-hand glove won't fit the right hand.

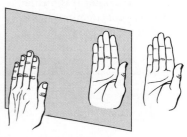

FIGURE 19.9

The image of the left hand reflected in the mirror appears the same as the right hand.

Thumbtacks and their mirror images are identical; they are superimposable.

Superimposability is the ability of two objects to fit "one within the other" with no mismatch of parts. Your left and right hands lack this ability and we say they are *nonsuperimposable*. This is why your right hand doesn't fit into your left-hand glove; your right hand does not match *exactly* the space that corresponds to your left hand. A pair of thumbtacks, on the other hand, are superimposable. If you imagine merging two thumbtacks so one goes inside the other, there is no mismatch of parts.

In the final analysis, *only if two objects pass the test of superimposability may we call them identical.* Your left and right hands are mirror images of each other, but they are not superimposable and are therefore not identical. A pair of thumbtacks are mirror images of each other and *are* superimposable, so they are identical.

The technical term for handedness is **chirality.**[2] Two plain thumbtacks do not have chirality, but your left and right hands do. Two gloves in a pair also have chirality. The test for chirality is the test for superimposability.

> If one object is the mirror image of the other, and if they are not superimposable, then the objects are chiral.

In the realm of molecules and ions, we can apply the same test for handedness as before to determine whether two structures are actually the same (and therefore not really two structures, but one), or whether they are different.

The most common examples of chirality among coordination compounds occur with octahedral complexes that contain two or three bidentate ligands—for instance, $[Co(en)_2Cl_2]^+$ and $[Co(en)_3]^{3+}$. For the complex $[Co(en)_3]^{3+}$, the two nonsuperimposable isomers, called **enantiomers,** are shown in Figure 19.10. For the complex $[Co(en)_2Cl_2]^+$, only the *cis* isomer is chiral, as described in Figure 19.11.

As you can see, chiral isomers differ in only a very minor way from each other. This difference is so subtle that most of their properties are identical. They have identical melting points and boiling points and, in nearly all of their reactions, their behavior is exactly alike. The only way that the difference

FIGURE 19.10

The two isomers of $[Co(en)_3]^{3+}$. Isomer II is constructed as the mirror image of Isomer I. No matter how Isomer II is turned about, it is not superimposable on Isomer I.

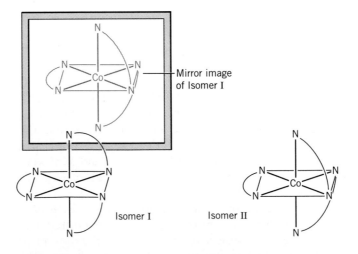

[2] The term **chiral** comes from the Greek *cheir*, meaning "hand."

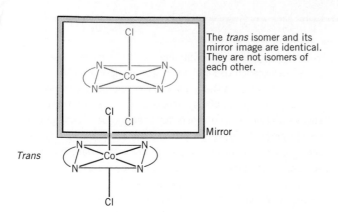

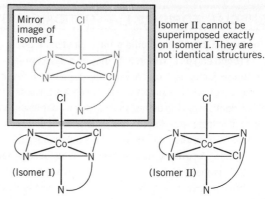

The *trans* isomer and its mirror image are identical. They are not isomers of each other.

Mirror

Trans

Cis

Isomer II cannot be superimposed exactly on Isomer I. They are not identical structures.

(Isomer I)

(Isomer II)

Isomer II has the same structure as the mirror image of isomer I

FIGURE 19.11

Isomers of the $[Co(en)_2Cl_2]^+$ ion. The mirror image of the trans isomer can be superimposed exactly on the original, so the trans isomer is not chiral. The cis isomer (Isomer I) is chiral, however, because its mirror image (Isomer II) cannot be superimposed on the original.

between chiral molecules or ions manifests itself is in the way that they interact with physical or chemical "probes" that also have a handedness about them. For example, if two reactants are both chiral, then a given isomer of one of them will usually behave slightly differently toward each of the two isomers of the other reactant. As you will see in Chapter 21, this has very profound effects in biochemical reactions, in which nearly all of the molecules involved are chiral.

Chiral isomers like those described in Figures 19.10 and 19.11 also have a peculiar effect on polarized light, which is described in Special Topic 19.1. Because of this phenomenon, chiral isomers are said to be **optical isomers.**

19.5 BONDING IN COMPLEXES

Complexes of the transition metals differ from the complexes of other metals in two special ways: (1) they are usually colored, whereas complexes of the representative metals are usually (but not always) colorless, and (2) their magnetic properties are often affected by the ligands attached to the metal ion. For example, it is not unusual for a given metal ion to form complexes with different ligands to give a rainbow of colors, as illustrated in Figure 19.12 for a series of complexes of cobalt. Also, because the transition metal ions often have incompletely filled *d* subshells, we expect to find many of them with unpaired *d* electrons, and compounds that contain them should be paramagnetic. But

FIGURE 19.12

Each of these brightly colored solutions contains a complex ion of Co^{3+}. The variety of colors arises because of the different ligands (molecules or anions) that are bonded to the cobalt ion in the complexes.

SPECIAL TOPIC 19.1 / OPTICAL ACTIVITY

The molecules of enantiomers have identical internal geometries (bond angles and lengths). They also have the same atoms or groups joined to the same central atoms. As a result, enantiomers have identical polarities, which also causes them to have identical boiling points, melting points, densities, and most other physical properties. One way enantiomers do differ, however, is in their interactions with plane polarized light.

Light is *electromagnetic radiation* that possesses both electric and magnetic components which behave like vectors. These vectors oscillate in directions perpendicular to the direction in which the light wave is traveling (Figure 1a). In ordinary light, the oscillations of the electric and magnetic fields of the photons are oriented randomly around the direction of the light beam. In *plane polarized light,* all the vibrations occur in the same plane (Figure 1b). Ordinary light can be polarized in several ways. One is to pass it through a special film of plastic, as in a pair of

Polaroid sunglasses. This has the effect of filtering out all the vibrations except those that are in one plane (Figure 1b).

When plane polarized light is passed through a nonchiral substance, the plane of the vibrations is unaffected. However, chiral substances have a peculiar effect on polarized light. A chiral isomer will rotate the plane of the polarized light as the light passes through it (Figure 2). Looking toward the light source, we find that one enantiomer rotates the plane of polarization clockwise whereas the other enantiomer rotates it counterclockwise. Because of their effects on polarized light, chiral substances are said to be **optically active.**

The phenomenon of optical activity is not difficult to observe. In Figure 3a we see a pair of polarizing filters with a beaker containing a concentrated solution of sugar (a compound with chiral molecules) placed between them. The filters are oriented so that their planes of polarization are at an angle of 90°. As a

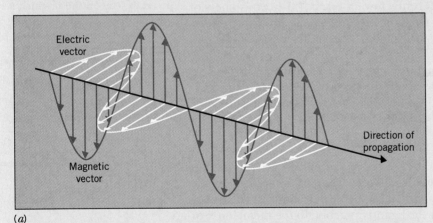

(a)

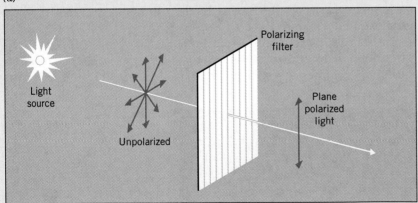

(b)

FIGURE 1
(a) Light possesses electric and magnetic vectors that oscillate perpendicular to the direction of propagation of the light wave. (b) In unpolarized light, the vectors are oriented randomly, but in plane *polarized light,* all the vectors are in the same plane.

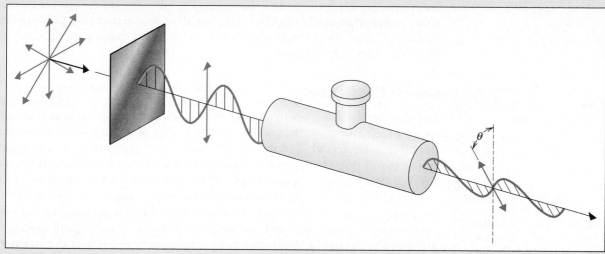

FIGURE 2

When plane polarized light passes through a solution of a chiral substance, the plane of polarization is rotated either to the left or to the right. In this illustration, the plane of polarization of the light is rotated to the left (as seen facing the light source).

result, the vibrations that pass through the filter in back are blocked by one in front. Notice, however, that some of the light that passes through the solution in the beaker is not blocked by the filter in front. This is because the sugar rotates the plane of polarization as the light passes through the solution. Because the light from the first filter has been rotated by the solution, it is not blocked completely by the second.

In Figure 3*b*, we see the same apparatus, but the filter in front has been rotated to the right (clockwise) by about 70°.

Where the light doesn't pass through the solution, it is no longer blocked completely. However, the light passing through the solution *is* blocked. Since we have had to rotate the filter to the right to cause the light coming through the solution to be blocked, the enantiomer in the solution must rotate the plane of polarization to the right. A solution of the other enantiomer at the same concentration would produce a similar rotation, but in the opposite direction.

(a)   *(b)*

FIGURE 3

(*a*) Polarized light rotated by the sugar solution is not blocked completely by crossed polarizing filters. (*b*) Rotating the front polarizer to the right causes polarized light passing through the sugar solution to be blocked.

for a given metal ion, the number of unpaired electrons is not always the same from one complex to another. For example, Fe^{2+} has four of its six $3d$ electrons unpaired in the $Fe(H_2O)_6^{2+}$ ion, but all of its electrons are paired in the $Fe(CN)_6^{4-}$ ion. As a result, the $Fe(H_2O)_6^{2+}$ ion is paramagnetic and the $Fe(CN)_6^{4-}$ ion is diamagnetic.

Crystal Field Theory

Any theory that attempts to explain the bonding in complex ions must also explain their colors and magnetic properties. One of the simplest theories that does this is the **crystal field theory.** The theory gets its name from its original use in explaining the behavior of transition metal ions in crystals. It was discovered later that the theory works well for transition metal complexes, too.

Crystal field theory ignores covalent bonding in complexes. It assumes that the primary stability comes from the electrostatic attractions between the positively charged metal ion and the negative charges of the ligand anions or dipoles. The crystal field theory's unique approach, though, is the way it examines how the negative charges on the ligands affect the energy of the complex by influencing the energies of the d orbitals of the metal ion, and it is this that we focus our attention on here. To understand the theory, therefore, it is essential that you know how the d orbitals are shaped and especially how they are oriented in space relative to each other. The d orbitals were described in Chapter 6, and they are illustrated again in Figure 19.13.

More complete theories consider the covalent nature of metal–ligand bonding, but crystal field theory nevertheless provides a useful model for explaining the colors and magnetic properties of complexes.

FIGURE 19.13

The shapes and directional properties of the five d orbitals of a d subshell.

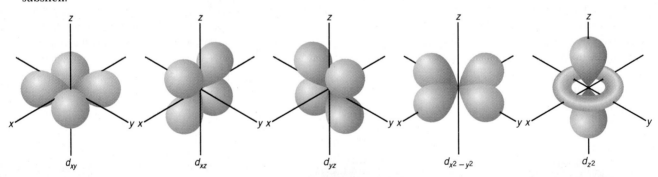

d_{xy} d_{xz} d_{yz} $d_{x^2-y^2}$ d_{z^2}

The labels for the d orbitals come from the mathematics of quantum mechanics.

FIGURE 19.14

An octahedral complex ion with ligands that lie along the x, y, and z axes.

First, notice that four of the d orbitals have the same shape but point in different directions. These are the $d_{x^2-y^2}$, d_{xy}, d_{xz}, and d_{yz}. Each has four lobes of electron density. The fifth d orbital, labeled d_{z^2}, has two lobes that point in opposite directions along the z axis plus a small donut-shaped ring of electron density around the center that is concentrated in the xy plane.

Of prime importance to us are the *directions* in which the lobes of the d orbital point. Notice that three of them—d_{xy}, d_{xz}, and d_{yz}—point *between* the x, y, and z axes. The other two—the d_{z^2} and $d_{x^2-y^2}$ orbitals—have their maximum electron densities *along* the x, y, and z axes.

Now let's consider constructing an octahedral complex within this coordinate system. We can do this by bringing ligands in along each of the axes as shown in Figure 19.14. The question we want to answer is "How do these ligands affect the energies of the d orbitals?"

In an isolated atom or ion, all the d orbitals of a given d subshell have the same energy. Therefore, an electron will have the same energy regardless of which d orbital it occupies. In an octahedral complex, however, this is no longer true. If the electron is in the $d_{x^2-y^2}$ or d_{z^2} orbital, it is forced to be nearer the negative charge of the ligands than if it is in a d_{xy}, d_{xz}, or d_{yz} orbital. Since

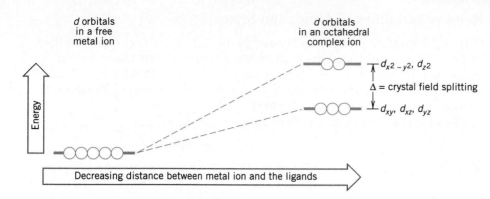

The changes in the energies of the d orbitals of a metal ion as an octahedral complex is formed.

the electron itself is negatively charged and is repelled by the charges of the ligands, the electron's potential energy will be higher in the $d_{x^2-y^2}$ and d_{z^2} orbitals than in a d_{xy}, d_{xz}, or d_{yz} orbital. Therefore, as the complex is formed, the d subshell actually splits into *two* new energy levels as shown in Figure 19.15. Here we see that regardless of which orbital the electron occupies, its energy increases because it is repelled by the negative charges of the approaching ligands. However, the electron is repelled *more* (and has a higher energy) if it is in an orbital that points directly at the ligands than if it occupies an orbital that points between them.

In an octahedral complex, the energy difference between the two sets of d-orbital energy levels is called the **crystal field splitting.** It is usually given the symbol Δ (delta), and its magnitude depends on the following factors:

The nature of the ligand. *Some ligands produce a larger splitting of the energies of the d orbitals than others.* For a given metal ion, for example, cyanide always gives a large value of Δ and F^- always gives a small value. We will have more to say about this later.

The oxidation state of the metal. *For a given metal and ligand, the size of Δ increases with an increase in the oxidation number of the metal.* As electrons are removed from a metal and the charge on the ion becomes more positive, the ion becomes smaller. This means that the ligands are attracted to the metal more strongly and they can approach the center of the complex more closely. As a result, they also approach the d orbitals along the x, y, and z axes more closely, and thereby cause a greater repulsion. This causes a greater splitting of the two d-orbital energy levels and a larger value of Δ.

The row in which the metal occurs in the periodic table. *For a given ligand and oxidation state, the size of Δ increases going down a group.* In other words, an ion of an element in the first row of transition elements has a smaller value of Δ than the ion of a heavier member of the same group. Thus, comparing complexes of Ni^{2+} and Pt^{2+} with the same ligand, we find that the platinum complex has the larger crystal field splitting.

The explanation of this is that for the larger ion (e.g., Pt^{2+}), the d orbitals are larger, are more diffuse, and extend further from the nucleus in the direction of the ligands. This produces a larger repulsion between the ligands and the orbitals that point at them.

The magnitude of Δ is very important in determining the properties of complexes, including the stabilities of oxidation states of the metal ions, the colors of complexes, and their magnetic properties. Let's look at some examples.

Relative Stabilities of Oxidation States

One of the properties of chromium is that the Cr^{2+} ion is very easily oxidized to Cr^{3+}. This is easily explained by crystal field theory. In water, we expect these ions to exist as the complexes, $Cr(H_2O)_6{}^{2+}$ and $Cr(H_2O)_6{}^{3+}$, respectively. Let's examine the energies and electron populations of the d-orbital energy levels in each complex (Figure 19.16).

The element chromium has the electron configuration

$$Cr \qquad\qquad [Ar]\ 3d^5 4s^1$$

Removing two electrons gives the Cr^{2+} ion, and removing three gives Cr^{3+}.

$$Cr^{2+} \qquad\qquad [Ar]\ 3d^4$$
$$Cr^{3+} \qquad\qquad [Ar]\ 3d^3$$

Next, we distribute the d electrons among the various d orbitals, but for Cr^{2+} we have to make a choice. Should they all be forced into the lower of the two energy levels, or should they be spread out? From the diagram we see that the fourth electron does not pair with one of the others in the lower-energy d-orbital level. Instead, three electrons go into the lower level and the fourth is in the upper level. We will discuss *why* this happens later, but for now, let's use the two energy diagrams to explain why Cr^{2+} is so easy to oxidize.

There are actually two factors that favor the oxidation of Cr(II) to Cr(III). First, the electron that is removed from Cr^{2+} to give Cr^{3+} comes from the *higher* energy level, so oxidizing the chromium(II) *removes* a high-energy electron. The second factor is the effect caused by increasing the oxidation state of the chromium. As we've pointed out, this increases the magnitude of Δ, and as you can see, the energy of the three electrons that remain is lowered. Thus, both the removal of a high-energy electron and the lowering of the energy of the electrons that are left behind help make the oxidation occur, and the $Cr(H_2O)_6{}^{2+}$ ion is very easily oxidized to $Cr(H_2O)_6{}^{3+}$.

Colors of Complex Ions

When light is absorbed by an atom, molecule, or ion, the energy of the photon raises an electron from one energy level to another. In many substances, such as sodium chloride, the energy difference between the highest energy populated level and the lowest energy unpopulated level is quite large, so the frequency of a photon that carries the necessary energy lies outside the visible region of the spectrum. The substance appears white because visible light is unaffected; it is reflected unchanged.

Remember, $E = h\nu$. The energy of the photon absorbed determines the frequency (and wavelength) of the absorbed light.

FIGURE 19.16

Energy level diagrams for the $[Cr(H_2O)_6]^{2+}$ and $[Cr(H_2O)_6]^{3+}$ ions.

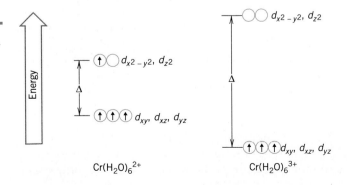

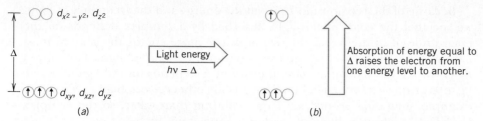

FIGURE 19.17

(*a*) The electron distribution in the ground state of the $Cr(H_2O)_6^{3+}$ ion. (*b*) Light energy raises an electron from the lower energy set of *d* orbitals to the higher energy set.

For complex ions of the transition metals, the energy difference between the *d*-orbital energy levels is not very large, and photons with frequencies in the visible region of the spectrum are able to raise an electron from the lower energy set of *d* orbitals to the higher energy set. This is shown in Figure 19.17 for the $Cr(H_2O)_6^{3+}$ ion.

As you know, white light contains photons of all the frequencies and colors in the visible spectrum. If we shine white light through a solution of a complex, the light that passes through has all colors *except* those that have been absorbed. It is not difficult to determine what will be seen if we know what colors are being absorbed. All we need is a color wheel like the one shown in Figure 19.18. Across from each other on the color wheel are **complementary colors.** Blue-green is the complementary color to red, and yellow is the complementary color to violet. If a substance absorbs a particular color when bathed in white light, the color of the reflected or transmitted light is the complementary color. In the case of the $Cr(H_2O)_6^{3+}$ ion, the light absorbed when the electron is raised from one set of *d* orbitals to the other has a frequency of 5.22×10^{14} Hz, which is the color of yellow light. This is why a solution of this ion appears violet.[3]

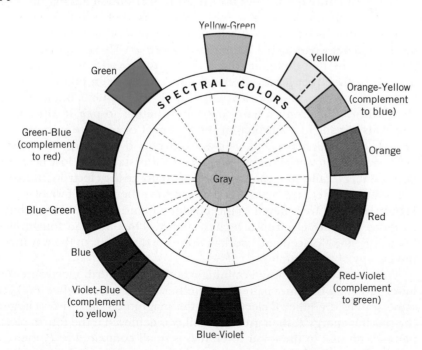

FIGURE 19.18

A color wheel. Colors that are across from each other are called complementary colors. When a substance absorbs a particular color, light that is reflected or transmitted has the color of its complement. Thus, something that absorbs red light appears green-blue, and vice versa.

[3] The perception of color is actually somewhat more complex than this because of the varying sensitivity of the human eye to various wavelengths. For example, the eye is much more sensitive to green than to red. If a compound reflects both of these colors with equal intensity, it will appear greenish simply because the eye sees green better than it sees red.

Because of the relationship between the energy and the frequency of light, we see that the color of the light absorbed by a complex depends on the magnitude of Δ; the larger the size of Δ, the more energy the photon must have and the higher will be the frequency of the absorbed light. For a given metal in a given oxidation state, the size of Δ depends on the ligand. Some ligands give a large crystal field splitting, while others give a small splitting. For example, ammonia produces a larger splitting than water, so the complex $Cr(NH_3)_6^{3+}$ absorbs light of higher energy and higher frequency than $Cr(H_2O)_6^{3+}$. [The ion $Cr(NH_3)_6^{3+}$ absorbs blue light and appears yellow.] Because changing the ligand changes Δ, the same metal ion is able to form a variety of complexes with a large range of colors.

A ligand that produces a large crystal field splitting with one metal ion also produces a large Δ in complexes with other metals. For example, cyanide ion is a very effective ligand and always gives a very large Δ, regardless of the metal to which it is bound. Ammonia is less effective than cyanide ion, but more effective than water. Thus, ligands can be arranged in order of their effectiveness at producing a large crystal field splitting. This sequence is called the **spectrochemical series.** Such a series containing some common ligands arranged in order of their decreasing strength is

$$CN^- > NO_2^- > en > NH_3 > H_2O > C_2O_4^{2-} > OH^- > F^- > Cl^- > Br^- > I^-$$

For a given metal ion, cyanide ion produces the largest Δ and iodide produces the smallest.

The order of the ligands can be determined by measuring the frequencies of the light absorbed by complexes.

Magnetic Properties

Let's return to the question of the electron distribution among the *d* orbitals in Cr^{2+} complexes. As you have seen above, this ion has four *d* electrons, and we noted that in placing these electrons in the *d* orbitals we had to make a decision about where to place the fourth electron. There's no question about the fate of the first three, of course. They just spread out across the three *d* orbitals in the lower level with their spins unpaired. In other words, we just follow Hund's rule, which we learned to apply in Chapter 6. But when we come to the fourth electron we have to decide whether to pair it with one of the electrons already in a *d* orbital of the lower set or to place it in one of the *d* orbitals of the higher set.[4] If we place it in the lower energy level, we give it extra stability (lower energy), but some of this stability is lost because it requires energy, called the **pairing energy, P,** to force the electron into an orbital that's already occupied by an electron. On the other hand, if we place it in the higher level, we are relieved of this burden of pairing the electron, but it also tends to give the electron a higher energy. Thus, for the fourth electron, "pairing" and "placement" work in opposite directions in the way they affect the energy of the complex.

The critical factor in determining whether the fourth electron enters the lower level and becomes paired, or whether it enters the higher level with the same spin as the other *d* electrons, is the magnitude of Δ. If Δ is larger than the pairing energy *P*, then greater stability is achieved if the fourth electron is paired with one in the lower level. If Δ is small compared to *P*, then greater stability is obtained by spreading the electrons out as much as possible. The complexes $[Cr(H_2O)_6]^{2+}$ and $[Cr(CN)_6]^{4-}$ illustrate this well.

[4] We've never had to make this kind of decision before because the energy levels in atoms were always widely spaced. In complex ions, however, the spacing between the two *d*-orbital energy levels is fairly small.

FIGURE 19.21

The splitting pattern of the *d* orbitals changes as the geometry of the complex changes.

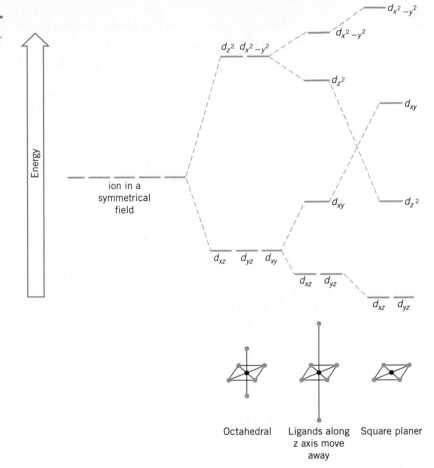

Octahedral   Ligands along z axis move away   Square planer

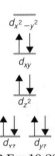

FIGURE 19.22

Distribution of the electrons among the *d* orbitals of nickel in the diamagnetic $[Ni(CN)_4]^{2-}$ ion.

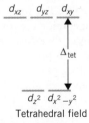

Tetrahedral field

FIGURE 19.23

Splitting pattern of the *d* orbitals for a tetrahedral complex.
$$\Delta_{tet} \approx \tfrac{4}{9}\Delta_{oct}$$

19.21. The repulsions felt by the *d* orbitals that point in the *z* direction are reduced, so we find that the energies of the d_{z^2}, d_{xz}, and d_{yz} orbitals drop. At the same time, the energies of the orbitals in the *xy* plane feel greater repulsions, so the $d_{x^2-y^2}$, and d_{xy} rise in energy.

Nickel(II) ion (which has eight *d* electrons) forms a complex with cyanide ion that is square planar and diamagnetic. In this complex the strong field produced by the cyanide ions yields a large energy separation between the d_{xy} and $d_{x^2-y^2}$, orbitals, so that a low-spin complex results. The electron distribution in this complex is illustrated in Figure 19.22.

Tetrahedral Complexes The splitting pattern for the *d* orbitals in a tetrahedral complex is illustrated in Figure 19.23. Notice that the order of the energy levels is exactly opposite to that in an octahedral complex. In addition, the size of Δ is also much smaller for a tetrahedral complex than for an octahedral one. (Actually, $\Delta_{tet} \approx \tfrac{4}{9}\Delta_{oct}$ for the same metal ion with the same ligand.) This small Δ is always less than the pairing energy, so tetrahedral complexes are always high-spin complexes.

19.6

COMPLEX ION EQUILIBRIA IN AQUEOUS SOLUTIONS

In an aqueous solution, the formation of a complex ion is really a reaction in which water molecules are replaced by other ligands. Thus, when NH_3 is added to a solution of copper ion, the water molecules in the $Cu(H_2O)_4^{2+}$ ion are replaced, one after another, by molecules of NH_3 until the complex $Cu(NH_3)_4^{2+}$ is formed. Each successive reaction is an equilibrium, so the entire chemical system involves many species and is quite complicated. Fortu-

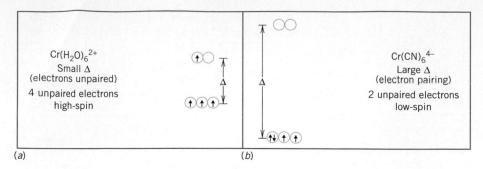

FIGURE 19.19

The effect of Δ on the electron distribution in a complex with four d electrons. (*a*) When Δ is small, the electrons remain unpaired. (*b*) When Δ is large, the lower energy level accepts all four electrons and two electrons become paired.

Water is a ligand that does not produce a large Δ, so $P > \Delta$, and minimum pairing of electrons takes place. This explains the energy level diagram for the $[Cr(H_2O)_6]^{2+}$ complex in Figure 19.19*a*. When cyanide is the ligand, however, a very large Δ is obtained, and this leads to pairing of the fourth electron with one in the lower set of d orbitals. This is shown in Figure 19.19*b*. It is interesting to note that it can be demonstrated experimentally that $[Cr(H_2O)_6]^{2+}$ has 4 unpaired electrons and the $[Cr(CN)_6]^{4-}$ ion has just 2.

For octahedral chromium(II) complexes, there are two possibilities in terms of the number of unpaired electrons. They contain either four or two, depending on the magnitude of Δ. When there is the maximum number of unpaired electrons, the complex is described as being **high spin;** when there is the minimum number of unpaired electrons it is described as being **low spin.** High- and low-spin octahedral complexes are possible when the metal has a d^4, d^5, d^6, or d^7 electron configuration. Let's look at another example—one containing the Fe^{2+} ion, which has the electron configuration

$$Fe^{2+} \ [Ar] \ 3d^6$$

At the beginning of this section we mentioned that the $[Fe(H_2O)_6]^{2+}$ ion is paramagnetic and has four unpaired electrons, while the $[Fe(CN)_6]^{4-}$ ion is diamagnetic, meaning it has no unpaired electrons. Now we can see why, by referring to Figure 19.20. Water produces a weak splitting and a minimum amount of pairing of electrons. When the six d electrons in the Fe^{2+} ion are distributed, one must be paired in the lower level after the upper level is filled. The result is four unpaired d electrons, and a high-spin complex. Cyanide ion, however, produces a large splitting, so $\Delta > P$. This means that maximum pairing of electrons in the lower level takes place, and a low-spin complex is formed. Six electrons are just the right amount to completely fill all three of these d orbitals, and since all the electrons are paired, the complex is diamagnetic.

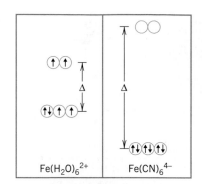

FIGURE 19.20

The distribution of d electrons in (*a*) $[Fe(H_2O)_6]^{2+}$ and (*b*) $[Fe(CN)_6]^{4-}$. The magnitude of Δ for the cyanide complex is much larger than for the water complex. This produces a maximum pairing of electrons in the lower energy set of d orbitals in $[Fe(CN)_6]^{4-}$.

Crystal Field Theory and Other Geometries

The crystal field theory can be extended to geometries other than octahedral. The effect that changing the structure of the complex has on the energies of the d orbitals is to change the splitting pattern.

Square Planar Complexes   We can form a square planar complex from an octahedral one by removing the ligands that lie along the z axis. As this happens, the ligands in the xy plane are able to approach the metal a little closer because are no longer being repelled by ligands along the z axis. The effect of these changes on the energies of the d orbitals is illustrated in Figure

nately, when the ligand concentration is *large* relative to that of the metal ion, the concentrations of the intermediate complexes are very small and we can work only with the *overall* reaction for the formation of the final complex. Our study of complex ion equilibria will be limited to these situations. The equilibrium equation for the formation of $Cu(NH_3)_4^{2+}$, therefore, can be written as though the complex forms in one step:

$$Cu(H_2O)_4^{2+} + 4NH_3 \rightleftharpoons Cu(NH_3)_4^{2+} + 4H_2O$$

We will simplify this equation for the purposes of dealing quantitatively with the equilibrium by omitting the water molecules. (It's safe to do this because the concentration of H_2O in aqueous solutions is taken to be effectively a constant and need not be included in mass action expressions.) In simplified form, we write the equilibrium above as follows:

$$Cu^{2+}(aq) + 4NH_3(aq) \rightleftharpoons Cu(NH_3)_4^{2+}(aq)$$

We have two goals here: to study such equilibria themselves and to learn how they can be used to influence the solubilities of metal ion salts.

Formation Constants of Complex Ions

When the chemical equation for the equilibrium is written so that the complex ion is the product, the equilibrium constant for the reaction is called the **formation constant, K_{form}**. The equilibrium law for the formation of $Cu(NH_3)_4^{2+}$ in the presence of excess NH_3, for example, is

$$\frac{[Cu(NH_3)_4^{2+}]}{[Cu^{2+}][NH_3]^4} = K_{form}$$

Sometimes this equilibrium constant is called the **stability constant.** The larger its value, the greater is the concentration of the complex at equilibrium, and so the more stable is the complex.

Table 19.1 provides several more examples of complex ion equilibria and their associated equilibrium constants. (Additional examples are in Appendix E.6.) Notice that the most stable complex in the table, $Co(NH_3)_6^{3+}$, has the largest K_{form}.

In this section we will omit the brackets when writing the formulas of complexes and use brackets only when we mean molar concentration.

According to Le Châtelier's principle, when the concentration of ammonia is high, the position of equilibrium in this reaction is shifted far to the right, so effectively all the aqua complex is changed to the ammine complex.

 Formation constants for complexes

The brackets here mean molar concentration. Notice that the entire formula for the complex, including its charge, is enclosed within the brackets.

TABLE 19.1   Formation Constants and Instability Constants for Some Complex Ions

Ligand	Equilibrium	K_{form}	K_{inst}
NH_3	$Ag^+ + 2NH_3 \rightleftharpoons Ag(NH_3)_2^+$	1.6×10^7	6.3×10^{-8}
	$Co^{2+} + 6NH_3 \rightleftharpoons Co(NH_3)_6^{2+}$	5.0×10^4	2.0×10^{-5}
	$Co^{3+} + 6NH_3 \rightleftharpoons Co(NH_3)_6^{3+}$	4.6×10^{33}	2.2×10^{-34}
	$Cu^{2+} + 4NH_3 \rightleftharpoons Cu(NH_3)_4^{2+}$	1.1×10^{13}	9.1×10^{-14}
	$Hg^{2+} + 4NH_3 \rightleftharpoons Hg(NH_3)_4^{2+}$	1.8×10^{19}	5.6×10^{-20}
F^-	$Al^{3+} + 6F^- \rightleftharpoons AlF_6^{3-}$	1×10^{20}	1×10^{-20}
	$Sn^{4+} + 6F^- \rightleftharpoons SnF_6^{2-}$	1×10^{25}	1×10^{-25}
Cl^-	$Hg^{2+} + 4Cl^- \rightleftharpoons HgCl_4^{2-}$	5.0×10^{15}	2.0×10^{-16}
Br^-	$Hg^{2+} + 4Br^- \rightleftharpoons HgBr_4^{2-}$	1.0×10^{21}	1.0×10^{-21}
I^-	$Hg^{2+} + 4I^- \rightleftharpoons HgI_4^{2-}$	1.9×10^{30}	5.3×10^{-31}
CN^-	$Fe^{2+} + 6CN^- \rightleftharpoons Fe(CN)_6^{4-}$	1.0×10^{24}	1.0×10^{-24}
	$Fe^{3+} + 6CN^- \rightleftharpoons Fe(CN)_6^{3-}$	1.0×10^{31}	1.0×10^{-31}

Instability Constants

Some chemists prefer to describe the relative stabilities of complex ions differently from our treatment. The *inverses* of formation constants are cited and are called **instability constants,** K_{inst}. This approach focuses attention on the *breakdown* of the complex, not its formation. Therefore, the associated equilibrium equation is written as the reverse of the formation of the complex. The equilibrium for the copper–ammonia complex would be written as follows, for example:

$$Cu(NH_3)_4{}^{2+}(aq) \rightleftharpoons Cu^{2+}(aq) + 4NH_3(aq)$$

The equilibrium constant for the reaction is called the *instability constant*, and the mass action expression associated with it is

$$K_{inst} = \frac{[Cu^{2+}][NH_3]^4}{[Cu(NH_3)_4{}^{2+}]} = \frac{1}{K_{form}}$$

K_{inst} is called an *instability* constant because the larger its value is, the more *unstable* the complex is. The data in the last column of Table 19.1 show this. The least stable complex in the table, $Co(NH_3)_6{}^{2+}$, has the largest value of K_{inst}.

The Effect of Complex Ion Formation on the Solubilities of Salts

The silver halides are extremely insoluble salts; the K_{sp} of AgBr at 25°C, for example, is only 5.0×10^{-13}. In a saturated solution, the concentration of each ion of AgBr is only 7.1×10^{-7} mol L^{-1}. Suppose that we start with a saturated solution in which undissolved AgBr rests on the bottom and equilibrium exists. Now suppose that we start to add aqueous ammonia to the system. NH_3 molecules are strong ligands for silver ions, and they begin to form complexes, $Ag(NH_3)_2{}^+$, with the trace amount of Ag^+ ion initially in solution. The reaction is

Complex ion equilibrium: $Ag^+(aq) + 2NH_3(aq) \rightleftharpoons Ag(NH_3)_2{}^+(aq)$

Because the forward reaction withdraws *uncomplexed* Ag^+ ions from solution, it upsets the solubility equilibrium.

Solubility equilibrium: $AgBr(s) \rightleftharpoons Ag^+(aq) + Br^-(aq)$

Ammonia, by withdrawing uncomplexed Ag^+ ions supplied by this solubility equilibrium, causes it to shift to the right to generate more Ag^+ ions from $AgBr(s)$, another example of Le Châtelier's principle at work. In other words, ammonia induces more $AgBr(s)$ to dissolve. Our example illustrates a general phenomenon:

The solubility of a slightly soluble salt increases when one of its ions can be changed to a soluble complex ion.

To analyze what happens let's put the two equilibria together.

Complex ion
  equilibrium: $Ag^+(aq) + 2NH_3(aq) \rightleftharpoons Ag(NH_3)_2{}^+(aq)$
Solubility equilibrium: $AgBr(s) \rightleftharpoons Ag^+(aq) + Br^-(aq)$
Sum of equilibria: $AgBr(s) + 2NH_3(aq) \rightleftharpoons Ag(NH_3)_2{}^+(aq) + Br^-(aq)$

Ag^+ cancels from both sides when the equations are added.

The equilibrium constant for the net overall reaction is written in the usual way. The term for [AgBr(*s*)] is omitted because it refers to a solid and so has a constant value.

$$K_c = \frac{[\text{Ag(NH}_3)_2{}^+][\text{Br}^-]}{[\text{NH}_3]^2}$$

K_c for this expression is the product of K_{form} and K_{sp}, which we can show as follows. We have the following equations for K_{form} and K_{sp}:

$$K_{\text{form}} = \frac{[\text{Ag(NH}_3)_2{}^+]}{[\text{Ag}^+][\text{NH}_3]^2} \qquad \text{and} \qquad K_{\text{sp}} = [\text{Ag}^+][\text{Br}^-]$$

Then,

$$K_c = K_{\text{form}} \times K_{\text{sp}} = \frac{[\text{Ag(NH}_3)_2{}^+]}{[\text{Ag}^+][\text{NH}_3]^2} \times [\text{Ag}^+][\text{Br}^-] = \frac{[\text{Ag(NH}_3)_2{}^+][\text{Br}^-]}{[\text{NH}_3]^2}$$

Recall that when equilibria are added, their equilibrium constants are multiplied.

We know the values for K_{form} and K_{sp} for the silver bromide–ammonia system, so by multiplying the two, we find the overall value of K_c.

$$K_c = (1.6 \times 10^7)(5.0 \times 10^{-13})$$
$$= 8.0 \times 10^{-6}$$

This approach thus gives us a way to calculate the solubility of a sparingly soluble salt when one of its ligands is put into its solution. The next example shows how this works.

How many moles of AgBr can dissolve in 1.0 L of 1.0 M NH$_3$?

ANALYSIS A few preliminaries have to be done before we can take advantage of a concentration table. We need the overall equation and its associated equation for the equilibrium constant. The overall equilibrium is

$$\text{AgBr}(s) + 2\text{NH}_3(aq) \rightleftharpoons \text{Ag(NH}_3)_2{}^+(aq) + \text{Br}^-(aq)$$

The equation for K_c is

$$K_c = \frac{[\text{Ag(NH}_3)_2{}^+][\text{Br}^-]}{[\text{NH}_3]^2}$$

The value of K_c, the product of K_{form} and K_{sp}, as we learned above, is

$$K_c = 8.0 \times 10^{-6}$$

Now let's prepare the table.

EXAMPLE 19.2
Calculating the Solubility of a Slightly Soluble Salt in the Presence of a Ligand

	AgBr(*s*) + 2NH$_3$(*aq*) $\rightleftharpoons$ Ag(NH$_3$)$_2{}^+$(*aq*) + Br$^-$(*aq*)		
Initial concentrations (*M*)	1.0	0	0
Changes in concentrations caused by NH$_3$ (*M*)	−2*x* (Note 1)	+*x*	+*x*
Equilibrium concentrations (*M*)	(1.0 − 2*x*)	*x*	*x* (Note 2)

Note 1. The 2 in −2*x* signifies that each mole of AgBr that dissolves removes twice as many moles of NH$_3$.

Note 2. Letting the concentration of Br$^-$ equal that of Ag(NH$_3$)$_2{}^+$ is valid because *and only because* K_{form} is such a large number. So essentially *all* Ag$^+$ ions that do dissolve from the insoluble AgBr are changed to the complex ion. There are relatively few uncomplexed Ag$^+$ ions in the solution.

SOLUTION Let's substitute the values in the last row of the concentration table into the equation for K_c.

$$K_c = \frac{(x)(x)}{(1.0 - 2x)^2} = 8.0 \times 10^{-6}$$

This can be simplified by taking the square root of both sides. Then,

$$\frac{x}{(1.0 - 2x)} = \sqrt{8.0 \times 10^{-6}} = 2.8 \times 10^{-3}$$

Solving for x gives

$$x = 2.8 \times 10^{-3}$$

In other words, 2.8×10^{-3} mol of AgBr dissolves in 1.0 L of 1.0 M NH$_3$. This is not very much, of course, but in contrast, only 7.1×10^{-7} mol of AgBr dissolves in 1.0 L of pure water. Thus AgBr is nearly 4000 times more soluble in the 1.0 M NH$_3$ than in pure water.

■ **Practice Exercise 6** Calculate the solubility of silver chloride in 0.10 M NH$_3$ and compare it with its solubility in pure water. (Refer to Table 17.2 for the K_{sp} for AgCl.)

■ **Practice Exercise 7** How many moles of NH$_3$ have to be added to 1.0 L of water to dissolve 0.20 mol of AgCl? The complex ion Ag(NH$_3$)$_2^+$ forms.

SUMMARY

Complex Ions of Metals **Coordination compounds** contain **complex ions** (also called **complexes** or **coordination complexes**), formed from a metal ion and a number of ligands. **Ligands** are Lewis bases and may be **monodentate, bidentate,** or, in general, **polydentate,** depending on the number of **donor atoms** that they contain. Water is the most common monodentate ligand. Polydentate ligands bind to a metal through two or more donor atoms and yield ring structures called **chelates.** Common bidentate ligands are oxalate ion and ethylenediamine (en); a common polydentate ligand is ethylenediaminetetraacetic acid (EDTA), which has six donor atoms.

In the formula of a complex, the metal is written first, followed by the formulas of the ligands. Brackets are often used to enclose the set of atoms that make up the complex, with the charge on the complex written outside the brackets. The charge on the complex is the algebraic sum of the charges on the metal ion and the charges on the ligands.

Complexes of polydentate ligands are more stable than similar complexes formed with monodentate ligands, because a polydentate ligand is less likely to be lost completely if one of its donor atoms becomes detached from the metal ion. This phenomenon is called the **chelate effect.**

Nomenclature of Complexes Complexes are named following a set of rules developed by the IUPAC. These are summarized on pages 816 to 818.

Coordination Number and Structure The **coordination number** of a metal ion in a complex is the number of donor atoms attached to the metal ion. Polydentate ligands supply two or more donor atoms, which must be taken into account when determining the coordination number from the formula of the complex. Geometries associated with common coordination numbers are: for coordination number 2, linear; for coordination number 4, tetrahedral and square planar (especially for Pt^{2+} complexes); and for coordination number 6, octahedral. The *porphyrin structure* is found in many biologically important metal complexes. Be sure you can draw octahedral complexes in the simplified ways described in Figure 19.7.

Isomers of Coordination Compounds When two or more distinct compounds have the same chemical formula, they are **isomers** of each other. **Stereoisomers** have the same atoms attached to each other, but the atoms are arranged differently in space. In a **cis isomer,** attached groups of atoms are on the same side of some reference plane through the molecule. In a **trans isomer,** they are on opposite sides. Cis and trans isomerism is a form of geometrical isomerism. **Chiral** isomers are exactly the same in every way but one—they are not **superimposable** on their mirror images. These kinds of isomers exist for complexes of the type *M(AA)$_3$*, where *M* is a metal ion and *AA* is a bidentate ligand, and also for complexes of the type *cis-M(AA)$_2$a$_2$*, where *a* is a monodentate ligand. Chiral isomers that are related as object to mirror images are said to be **enantiomers** and are **optical isomers.**

Crystal Field Theory   In an octahedral complex, the ligands influence the energies of the *d* orbitals by splitting the *d* subshell into two energy sublevels. The lower one consists of the d_{xz}, d_{yz}, and d_{xy} orbitals; the higher energy level consists of the d_{z^2} and $d_{x^2-y^2}$ orbitals. The energy difference between the two new *d* sublevels is the **crystal field splitting, Δ,** and for a given ligand it increases with an increase in the oxidation state of the metal. For a given metal ion, Δ depends on the ligand, and it depends on the period number in which the metal is found.

In the **spectrochemical series,** ligands are arranged in order of their ability to cause a large Δ. Cyanide ion produces the largest crystal field splitting, iodide ion the smallest. **Low-spin** complexes result when Δ is larger than the **pairing energy**—the energy needed to cause two electrons to become paired in the same orbital. **High-spin** complexes occur when Δ is smaller than the pairing energy. Light of energy equal to Δ is absorbed when an electron is raised from the lower energy set of *d* orbitals to the higher set, and the color of the complex is determined by the colors that remain in the transmitted light. Crystal field theory can also explain the relative stabilities of oxidation states, in many cases.

Different splitting patterns of the *d* orbitals occur for other geometries. The patterns for square planar and tetrahedral geometries are described in Figures 19.21 and 19.23, respectively. The value of Δ for a tetrahedral complex is only about 4/9 that of Δ for an octahedral complex.

Tools you have learned

The following tools were introduced in this chapter and are useful in solving problems.

Tool	Function
Nomenclature rules for complexes (page 816)	To name metal complexes and to write correct formulas for complexes.
Formation constants (stability constants) for complexes (page 835)	To make judgments concerning relative stabilities of complexes. To calculate how the solubility of a sparingly soluble salt changes when its cation is able to form a complex ion with a ligand that is added to the solution.

THINKING IT THROUGH

For the following, identify the information needed to solve the problem and show (or explain) what must be done with it.

1. Explain how you could determine whether the complex $[\text{Co}(\text{EDTA})]^-$ is chiral.

2. Suppose that some dipositive cation, M^{2+}, is able to form a complex ion with a ligand, L, by the following equation.

$$M^{2+} + 2L \rightleftharpoons [M(L)_2]^{2+}$$

The cation also forms a sparingly soluble salt, MCl_2. In which of the following circumstances would a given quantity of ligand be more able to bring larger quantities of the salt into solution?

(a) $K_{\text{form}} = 1 \times 10^2$ and $K_{\text{sp}} = 1 \times 10^{-15}$
(b) $K_{\text{form}} = 1 \times 10^{10}$ and $K_{\text{sp}} = 1 \times 10^{-20}$

3. Silver forms a sparingly soluble iodide salt, AgI, and it also forms a soluble iodide complex, AgI_2^-. How much potassium iodide must be added to 100 mL of a saturated AgI solution in contact with solid AgI to dissolve 0.020 mol of AgI? (Explain in detail how you obtain the answer. Hint: Be sure to include *all* the iodide that's added to the solution.)

REVIEW EXERCISES

Answers to questions whose numbers are printed in color are given in Appendix D. Challenging questions are marked with asterisks.

Complex Ions

19.1 The formation of the complex ion $[Cu(H_2O)_4]^{2+}$ is described as a Lewis acid–base reaction. Explain.
(a) What is the formula of the Lewis acid and the Lewis base in this reaction?
(b) What is the formula of the ligand?
(c) What is the name of the species that provides the donor atom?
(d) What atom is the donor atom and why is it so designated?
(e) What is the name of the species that is the acceptor?

19.2 To be a ligand, a species should also be a Lewis base. Explain.

19.3 Why are substances that contain complex ions often called coordination compounds?

19.4 Give the names of two species we mentioned that are electrically neutral, monodentate ligands.

19.5 Give the formulas of four species that have 1− charges and are monatomic, monodentate ligands.

19.6 Use Lewis structures to diagram the formation of $Cu(NH_3)_4^{2+}$ and $CuCl_4^{2-}$ ions from their respective components.

19.7 What must be true about the structure of a ligand classified as bidentate?

19.8 What is a chelate? Use Lewis structures to diagram the way that the oxalate ion, $C_2O_4^{2-}$, functions as a chelating agent.

19.9 How many donor atoms does $EDTA^{4-}$ have?

19.10 Explain how a salt of $EDTA^{4-}$ can retard the spoilage of salad dressing.

19.11 How does a salt of $EDTA^{4-}$ in shampoo make the shampoo work better in hard water?

19.12 The cobalt(III) ion, Co^{3+}, forms a 1:1 complex with $EDTA^{4-}$. What is the net charge, if any, on this complex, and what would be a suitable formula for it (using the symbol EDTA)?

19.13 The iron(III) ion forms a complex with six cyanide ions, called the ferricyanide ion. What is the net charge on this complex ion, and what is its formula?

19.14 The silver ion forms a complex ion with two ammonia molecules. What is the formula of this ion? Can this complex ion exist as a salt with the sodium ion or with the chloride ion? Write the formula of the possible salt. (Use brackets and parentheses correctly.)

***19.15** In which copper(II) complex, $[Cu(H_2O)_4]^{2+}$ or $[Cu(NH_3)_4]^{2+}$, are the bonds from the acceptor to the donor atoms stronger? What evidence can you cite for this? Describe what one does in the lab and what one sees during a test for aqueous copper(II) ion that involves these species.

19.16 Which complex is more stable, $[Cr(NH_3)_6]^{3+}$ or $[Cr(en)_3]^{3+}$? Why?

Naming Complexes

19.17 How would the following molecules or ions be named as ligands when writing the name of a complex ion? (a) $C_2O_4^{2-}$, (b) S^{2-}, (c) Cl^-, (d) $(CH_3)_2NH$ (dimethylamine), (e) NH_3, (f) N^{3-}, (g) SO_4^{2-}, (h) $C_2H_3O_2^-$

19.18 Give IUPAC names for each of the following.
(a) $[Ni(NH_3)_6]^{2+}$ (f) $[AgI_2]^-$
(b) $[Cr(NH_3)_3Cl_3]^-$ (g) SnS_3^{2-}
(c) $[Co(NO_2)_6]^{3-}$ (h) $[Co(en)_2(H_2O)_2]_2(SO_4)_3$
(d) $[Mn(CN)_4(NH_3)_2]^{2-}$ (i) $[Cr(NH_3)_5Cl]SO_4$
(e) $[Fe(C_2O_4)_3]^{3-}$ (j) $K_3[Co(C_2O_4)_3]$

19.19 Write chemical formulas for each of the following.
(a) tetraaquadicyanoiron(III) ion
(b) tetraammineoxalatonickel(II)
(c) diaquatetracyanoferrate(III) ion
(d) potassium hexathiocyanatomanganate(III)
(e) tetrachlorocuprate(II) ion

19.20 Write chemical formulas for each of the following.
(a) tetrachloroaurate(III) ion
(b) bis(ethylenediamine)dinitroiron(III) ion
(c) sodium tetraamminedicarbonatocobaltate(III)
(d) ethylenediaminetetraacetatoferrate(II) ion
(e) diamminedichloroplatinum(II)

Coordination Number and Structure

19.21 In what ways is the porphyrin structure important in biological systems?

19.22 What is *coordination number*? What structures are generally observed for complexes in which the central metal ion has a coordination number of 4?

19.23 If a metal ion is held in the center of a porphyrin ring structure, what is its coordination number? (Assume the porphyrin is the only ligand.)

19.24 What is the coordination number of Fe in $[Fe(en)(H_2O)_2Cl_2]$? What is the oxidation number of iron in this complex?

19.25 Make a sketch of the structure of an octahedral complex that contains only identical monodentate ligands.

19.26 Make a sketch of the structure of the octahedral $[Co(EDTA)]^-$ ion. Remember that donor atoms in a polydentate ligand span adjacent positions in the octahedron.

19.27 NTA is the abbreviation for nitrilotriacetic acid, a substance that was used at one time in detergents. Its structure is

$$:N \left\langle \begin{array}{l} CH_2-C\!\!-\!\!OH \\[4pt] CH_2-C\!\!-\!\!OH \\[4pt] CH_2-C\!\!-\!\!OH \end{array} \right.$$

The four donor atoms of this ligand are shown in red. Sketch the structure of an octahedral complex containing the anion of this ligand formed by the loss of a hydrogen from each of the OH groups. Assume that two water molecules are also attached to the metal ion and that each oxygen donor atom in the NTA is bonded to a position in the octahedron that is adjacent to the nitrogen of the NTA.

19.28 The compound shown below is called diethylenetriamine and is abbreviated "dien".

$$H_2\ddot{N}-CH_2-CH_2-\ddot{N}H-CH_2-CH_2-\ddot{N}H_2$$

When it bonds to a metal, it is a ligand with three donor atoms.
(a) Which are most likely the donor atoms?
(b) What is the coordination number of cobalt in the complex $Co(dien)_2^{3+}$?
(c) Sketch the structure of the complex $Co(dien)_2^{3+}$.
(d) Which complex would be expected to be more stable in aqueous solution, $Co(dien)_2^{3+}$ or $Co(NH_3)_6^{3+}$?
(e) What would be the structure of triethylenetetraamine?

Isomers of Coordination Compounds

19.29 What are *isomers*?

19.30 Define *stereoisomerism, geometric isomerism, chiral isomers,* and *enantiomers*.

19.31 What are cis and trans isomers?

*19.32 Below are two structures drawn for a complex. Are they actually different isomers, or are they identical? Explain your answer.

Structure I   Structure II

19.33 What condition must be fulfilled in order for a molecule or ion to be chiral?

*19.34 Below is a structure for one of the isomers of the complex $[Co(dien)(H_2O)_3]^{3+}$. Are isomers of this complex chiral? Justify your answer.

$N\frown N\frown N$ represents dien

*19.35 A solution was prepared by dissolving 0.500 g of $CrCl_3\cdot6H_2O$ in 100 mL of water. A silver nitrate solution was added and gave a precipitate of AgCl that was filtered from the mixture, washed, dried, and weighed. The AgCl had a mass of 0.538 g.
(a) What is the formula of the complex ion of chromium in this compound?
(b) What is the correct formula for the compound?
(c) Sketch the structure of the complex ion in this compound.
(d) How many different isomers of the complex can be drawn?

19.36 What is *cisplatin?* Draw its structure.

19.37 Sketch and label the isomers of the square planar complex $Pt(NH_3)_2ClBr$.

19.38 The complex $Pt(NH_3)_2Cl_2$ can be obtained as two distinct isomeric forms. Make a model of a tetrahedron and show that if the complex were tetrahedral, two isomers would be impossible.

19.39 The complex $Co(NH_3)_3Cl_3$ can exist in two isomeric forms. Sketch them.

19.40 Sketch the chiral isomers of $[Cr(en)_2Cl_2]^+$. Is there a nonchiral isomer of this complex?

19.41 Is the complex $[Co(EDTA)]^-$ chiral? Illustrate your answers with sketches.

19.42 Sketch the chiral isomers of $[Co(C_2O_4)_3]^{3-}$.

19.43 What are optical isomers?

Bonding in Complexes

19.44 On appropriate coordinate axes, sketch and label the five *d* orbitals.

19.45 Which *d* orbitals point *between* the *x, y,* and *z* axes? Which point along the coordinate axes?

19.46 Explain why an electron in a $d_{x^2-y^2}$ or d_{z^2} orbital in an octahedral complex will experience greater repulsions because of the presence of the ligands than an electron in a $d_{xy}, d_{yz},$ or d_{yz} orbital.

19.47 Sketch the *d*-orbital energy level diagram for a typical octahedral complex.

19.48 Explain why octahedral cobalt(II) complexes are easily oxidized to cobalt(III) complexes. Sketch the *d*-orbital energy diagram and assume a large value of Δ when placing electrons in the *d* orbitals.

19.49 In which complex do we expect to find the larger Δ? (a) $Cr(H_2O)_6^{2+}$ or $Cr(H_2O)_6^{3+}$, (b) $Cr(en)_3^{3+}$ or $CrCl_6^{3-}$

19.50 In each pair below, which complex is expected to absorb light of the shortest wavelength? Justify your answers.
(a) $[Ru(NH_3)_5Cl]^{2+}$ or $[Fe(NH_3)_5Cl]^{2+}$
(b) $[Ru(NH_3)_6]^{2+}$ or $[Ru(NH_3)_6]^{3+}$

19.51 Explain how the same metal in the same oxidation state is able to form complexes of different colors.

19.52 Arrange the following complexes in order of increasing wavelength of the light absorbed by them: $Cr(H_2O)_6^{3+}$, $CrCl_6^{3-}$, $Cr(en)_3^{3+}$, $Cr(CN)_6^{3-}$, $Cr(NO_2)_6^{3-}$, CrF_6^{3-}, $Cr(NH_3)_6^{3+}$.

19.53 If a complex appears red, what color light does it absorb? What color light is absorbed if the complex appears yellow?

19.54 What does the term *spectrochemical series* mean? How can the order of the ligands in the series be determined?

19.55 A complex CoA_6^{3+} is red. The complex CoB_6^{3+} is green. Which ligand, A or B, produces the larger crystal field splitting, Δ? Explain your answer.

19.56 Referring to the two ligands A and B, described in the preceding question, which complex would be expected to be more easily oxidized, CoA_6^{2+} or CoB_6^{2+}? Explain your answer.

19.57 Referring to the complexes in the preceding two questions, would the color of CoA_6^{2+} more likely be red or blue?

19.58 Which complex should absorb light at the longer wavelength? (a) $Fe(H_2O)_6^{2+}$ or $Fe(CN)_6^{4-}$, (b) $Mn(CN)_6^{3-}$ or $Mn(CN)_6^{4-}$

19.59 Which complex should be expected to absorb light of the highest frequency, $Cr(H_2O)_6^{3+}$, $Cr(en)_3^{3+}$, or $Cr(CN)_6^{3-}$?

19.60 What do the terms *low-spin complex* and *high-spin complex* mean?

19.61 For which d-orbital electron configurations are both high-spin and low-spin complexes possible?

19.62 Would the complex CoF_6^{4-} more likely be low-spin or high-spin? Could it be diamagnetic?

19.63 Sketch the d-orbital energy level diagrams for $Fe(H_2O)_6^{3+}$ and $Fe(CN)_6^{3-}$ and predict the number of unpaired electrons in each.

19.64 Indicate by means of a sketch what happens to the d-orbital electron configuration of the $Fe(CN)_6^{4-}$ ion when it absorbs a photon of visible light.

19.65 The complex $Co(C_2O_4)_3^{3-}$ is diamagnetic. Sketch the d-orbital energy level diagram for this complex and indicate the electron population of the orbitals.

*****19.66** Consider the complex $[M(H_2O)_2Cl_4]^-$ illustrated below on the left. Suppose the structure of this complex is distorted to give the structure on the right, where the water molecules along the z axis have moved away from the metal somewhat and the four chloride ions along the x and y axes have moved closer. What effect will this distortion have on the energy level splitting pattern of the d orbitals? Use a sketch of the splitting pattern to give your answer.

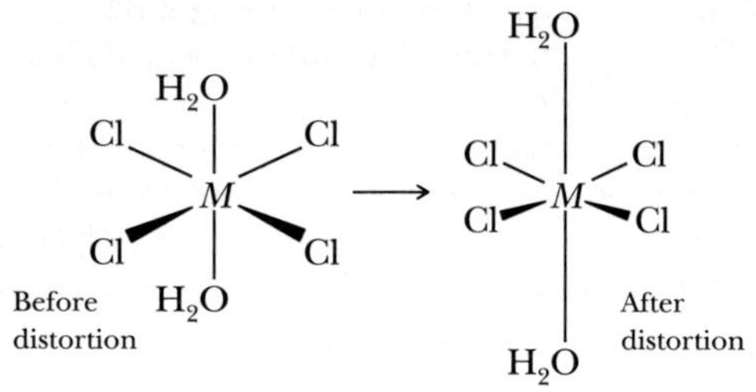

Complex Ion Equilibria

19.67 Write the chemical equilibria and equilibrium laws that correspond to K_{form} for the following complexes. (a) $CuCl_4^-$, (b) AgI_2^-, (c) $Cr(NH_3)_6^{3+}$

19.68 Write the chemical equilibria and equilibrium laws that correspond to K_{inst} for the following complexes. (a) $Ag(S_2O_3)_2^{3-}$, (b) $Zn(NH_3)_4^{2+}$, (c) SnS_3^{2-}

19.69 Using Le Châtelier's principle, explain how the addition of aqueous ammonia dissolves silver chloride. If HNO_3 is added after the AgCl has dissolved in the NH_3 solution, it causes AgCl to reprecipitate. Explain why.

19.70 For $PbCl_3^-$, $K_{form} = 2.5 \times 10^1$. If a solution containing this complex ion is diluted with water, $PbCl_2$ precipitates. Write the equations for the equilibria involved and use them together with Le Châtelier's principle to explain how this happens.

19.71 Write equilibria that correspond to K_{inst} for each of the following complex ions and write the equations for K_{inst}. (a) $Co(NH_3)_6^{3+}$, (b) HgI_4^{2-}, (c) $Fe(CN)_6^{4-}$

19.72 Write the equilibria that are associated with the equations for K_{form} for each of the following complex ions. Write also the equations for the K_{form} of each. (a) $Hg(NH_3)_4^{2+}$, (b) SnF_6^{2-}, (c) $Fe(CN)_6^{3-}$

19.73 The value of K_{inst} for $SnCl_4^{2-}$ is 5.6×10^{-3}.
(a) What is the value of K_{form}?

(b) Is this complex ion more or less stable than those in Table 19.1?

*19.74 How many grams of solid NaCN have to be added to 1.2 L of water to dissolve 0.11 mol of $Fe(OH)_3$ in the form of $Fe(CN)_6^{3-}$? Use data as needed from Tables 17.2 and 19.1.

*19.75 Silver ion forms a complex with thiosulfate ion, $S_2O_3^{2-}$, that has the formula $Ag(S_2O_3)_2^{3-}$. This complex has $K_{form} = 2.0 \times 10^{13}$. How many grams of AgBr ($K_{sp} = 5.0 \times 10^{-13}$) will dissolve in 125 mL of 1.20 M $Na_2S_2O_3$ solution?

Additional Exercises[5]

19.76 Note 2 in Example 19.2 (page 837) says that "there are relatively few uncomplexed Ag^+ ions in the solution." Calculate the molar concentration of Ag^+ ion actually left after the complex forms as described in Example 19.2.[5]

19.77 Why are the following two square planar complexes *not* chiral?[5]

---

[5] Contributed by Professor Mark Benvenuto, University of Detroit.

TEST OF FACTS AND CONCEPTS: CHAPTERS 15–19

Once again we pause so you can test your understanding of concepts, your knowledge of scientific terms, and your skills at solving chemistry problems. Read through the following questions carefully, and answer each as fully as possible. When necessary, review topics you are uncertain of. If you can answer these questions correctly, you are ready to go on to the next group of chapters.

1. Write the appropriate mass action expression, using molar concentrations, for these reactions.
(a) $NO_2(g) + N_2O(g) \rightleftharpoons 3NO(g)$
(b) $CaSO_3(s) \rightleftharpoons CaO(s) + SO_2(g)$
(c) $NiCO_3(s) \rightleftharpoons Ni^{2+}(aq) + CO_3^{2-}(aq)$

2. At a certain temperature, the reaction $2HF(g) \rightleftharpoons H_2(g) + F_2(g)$ has $K_c = 1 \times 10^{-13}$. Does this reaction proceed far toward completion when equilibrium is reached? If 0.010 mol HF were placed in a 1.00 L container and the system were permitted to come to equilibrium, what would be the concentrations of H_2 and F_2 in the container?

3. At 100 °C, the reaction $2NO_2(g) \rightleftharpoons N_2O_4(g)$ has $K_p = 6.5 \times 10^{-2}$. What is the value of K_c at this temperature?

4. For the reaction $3NO(g) \rightleftharpoons NO_2(g) + N_2O(g)$, calculate K_p at 25 °C using values of ΔG_f° from Table 13.2.

5. Calculate the value of K_c at 25 °C for the reaction in Question 4.

6. Consider the reaction

$$CH_4(g) + Cl_2(g) \rightleftharpoons CH_3Cl(g) + HCl(g)$$

(a) Calculate ΔG_{473}° for this reaction at 200 °C.
(b) What is the value of K_p for this reaction at 200 °C?
(c) What is the value of K_c for this reaction at 200 °C?

7. At 1000 °C, the redox reaction $NO_2(g) + SO_2(g) \rightleftharpoons NO(g) + SO_3(g)$ has $K_c = 3.60$. If 0.100 mol NO_2 and 0.100 mol SO_2 are placed in a 5.00 L container and allowed to react, what will all the concentrations be when equilibrium is reached? What will the new equilibrium concentrations be if 0.010 mol NO and 0.010 mol SO_3 are added to this original equilibrium mixture?

8. For the reaction in the preceding question, $\Delta H^\circ = -41.8$ kJ. How will the equilibrium concentration of NO be affected if:
(a) More NO_2 is added to the container?
(b) Some SO_3 is removed from the container?
(c) The temperature of the reaction mixture is raised?
(d) Some SO_2 is removed from the mixture?
(e) The pressure of the gas mixture is lowered by expanding the volume to 10.0 L?

9. At 60 °C, $K_w = 9.5 \times 10^{-14}$. What is the pH of pure water at this temperature? Why can we say that this water is neither acidic nor basic?

10. The pK_b of methylamine, CH_3NH_2, is 3.36. Calculate the pK_a of its conjugate acid, $CH_3NH_3^+$.

11. At 25 °C, the water in a natural pool of water in one of the western states was found to contain hydroxide ions at a concentration of 4.7×10^{-7} g OH^- per liter. Calculate the pH of this water and state if it is acidic, basic, or neutral.

12. The first antiseptic to be used in surgical operating rooms was phenol, C_6H_5OH, a weak acid and a potent bactericide. A 0.550 M solution of phenol in water was found to have a pH of 5.07.
(a) Write the chemical equation for the equilibrium involving C_6H_5OH in the solution.
(b) Write the equilibrium law corresponding to K_a for C_6H_5OH.
(c) Calculate the values of K_a and pK_a for phenol.
(d) Calculate the values of K_b and pK_b for the phenoxide ion, $C_6H_5O^-$.

13. The pK_a of saccharin, $HC_7H_4NSO_3$, a sweetening agent, is 11.68.
(a) What is the pK_b of the saccharinate ion, $C_7H_4NSO_3^-$?
(b) Does a solution of sodium saccharinate in water have a pH of 7, or is the solution acidic or basic? If the pH is not 7, calculate the pH of a 0.010 M solution of sodium saccharinate in water.

14. At 25 °C the value of K_b of codeine, a pain-killing drug, is 1.63×10^{-6}. Calculate the pH of a 0.0115 M solution of codeine in water.

15. Ascorbic acid, $H_2C_6H_6O_6$, is a diprotic acid usually known as vitamin C. For this acid, pK_{a_1} is 4.10 and pK_{a_2} is 11.79. When 125 mL of a solution of ascorbic acid was evaporated to dryness, the residue of pure ascorbic acid had a mass of 3.12 g.
(a) Calculate the molar concentration of ascorbic acid in the solution before it was evaporated.
(b) Calculate the pH of the solution and the molar concentration of the ascorbate ion, $C_6H_6O_6^{2-}$, before the solution was evaporated.

16. What ratio of molar concentrations of sodium acetate to acetic acid can buffer a solution at a pH of 4.50?

17. Write a chemical equation for the reaction that would occur in a buffer composed of $NaC_2H_3O_2$ and $HC_2H_3O_2$ if (a) some HCl were added and (b) some NaOH were added.

18. If 0.020 mol of NaOH were added to 500 mL of a sodium acetate–acetic acid buffer that contains 0.10 M $NaC_2H_3O_2$ and 0.15 M $HC_2H_3O_2$, by how many pH units will the pH of the buffer change?

19. Methylamine, CH_3NH_2, is a weak base. Write the

chemical equation for the equilibrium that occurs in an aqueous solution of this solute. Write the equilibrium law corresponding to K_b for CH_3NH_2.

20. How would each of the following aqueous solutions test, acidic, basic, or neutral? (Assume that each is at least 0.2 M.)

(a) KNO_3
(c) NH_4I
(b) $CrCl_3$
(d) K_2HPO_4

21. When 50.00 mL of an acid with a concentration of 0.115 M (for which $pK_a = 4.87$) is titrated with 0.100 M NaOH, what is the pH at the equivalence point? What would be a good indicator for this titration?

22. Calculate the pH of a 0.050 M solution of sodium ascorbate, $Na_2C_6H_6O_6$. For ascorbic acid, $H_2C_6H_6O_6$, $K_{a_1} = 7.9 \times 10^{-5}$ and $K_{a_2} = 1.6 \times 10^{-12}$.

23. When 25.0 mL of 0.100 M NaOH was added to 50.0 mL of a 0.100 M solution of a weak acid, HX, the pH of the mixture reached a value of 3.56. What is the value of K_a for the weak acid?

24. How many grams of solid NaOH would have to be added to 100 mL of a 0.100 M solution of NH_4Cl to give a mixture with a pH of 9.26?

25. The molar solubility of silver chromate, Ag_2CrO_4, in water is 6.7×10^{-5} M. What is K_{sp} for Ag_2CrO_4?

26. What is the pH of a saturated solution of $Mg(OH)_2$?

27. What is the solubility of $Fe(OH)_2$ in grams per liter if the solution is buffered to a pH of 10.00?

28. Suppose 30.0 mL of 0.100 M $Pb(NO_3)_2$ is added to 20.0 mL of 0.500 M KI.

(a) How many grams of PbI_2 will be formed?
(b) What will the molar concentrations of all the ions be in the mixture after equilibrium has been reached?

29. How many moles of NH_3 must be added to 1.00 L of solution to dissolve 1.00 g of $CuCO_3$? For $CuCO_3$, $K_{sp} = 2.5 \times 10^{-10}$. Ignore the hydrolysis of CO_3^{2-}, but consider the formation of the complex ion, $Cu(NH_3)_4^{2+}$.

30. Over what pH range must a solution be buffered to achieve a selective separation of the carbonates of barium, $BaCO_3$ ($K_{sp} = 5.0 \times 10^{-9}$), and lead, $PbCO_3$ ($K_{sp} = 7.4 \times 10^{-14}$)? The solution is initially 0.010 M in Ba^{2+} and 0.010 M in Pb^{2+}.

31. What current would be required to deposit 0.100 g of nickel in 20.0 minutes from a solution of $NiSO_4$?

32. What is the purpose of a salt bridge in a galvanic cell?

33. Use data from Table 18.1 to calculate the value of K_c at 25 °C for the reaction

$$O_2 + 4Br^- + 4H^+ \longrightarrow 2Br_2 + 2H_2O$$

34. A galvanic cell was assembled as follows. In one compartment, a copper electrode was immersed in a 1.00 M solution of $CuSO_4$. In the other compartment, a manganese electrode was immersed in a 1.00 M solution of $MnSO_4$. The potential of the cell was measured to be 1.52 V, with the Mn electrode as the negative electrode. What is the value of $E°$ for the following half-reaction?

$$Mn^{2+}(aq) + 2e^- \rightleftharpoons Mn(s)$$

35. Calculate $E°_{cell}$ for the reaction

$$3Cu(s) + 2NO_3^-(aq) + 8H^+(aq) \longrightarrow$$
$$2NO(g) + 4H_2O + 3Cu^{2+}(aq)$$

36. What is the value of K_c for the reaction described in Question 35?

37. Consider a galvanic cell formed by using the half-reactions

$$NiO_2(s) + 2H_2O + 2e^- \rightleftharpoons Ni(OH)_2(s) + 2OH^-(aq)$$
$$E° = +0.49 \text{ V}$$

$$PbO_2(s) + H_2O + 2e^- \rightleftharpoons PbO(s) + 2OH^-(aq)$$
$$E° = +0.25 \text{ V}$$

(a) Write the equation for the spontaneous cell reaction.
(b) Calculate the value of $E°_{cell}$ for this system.
(c) Calculate $\Delta G°$ for the spontaneous cell reaction.

38. A galvanic cell was constructed in which one half-cell consists of a silver electrode coated with silver chloride dipping into a solution that contains chloride ion and the second half-cell consists of a nickel electrode dipping into a solution that contains Ni^{2+}. The half-cell reactions and their reduction potentials are

$$AgCl(s) + e^- \rightleftharpoons Ag(s) + Cl^-(aq) \quad E° = +0.222 \text{ V}$$

$$Ni^{2+}(aq) + 2e^- \rightleftharpoons Ni(s) \quad E° = -0.257 \text{ V}$$

(a) Write the equation for the spontaneous cell reaction.
(b) Calculate $E°_{cell}$ for the cell reaction.
(c) Write the Nernst equation for the cell.
(d) Calculate the cell potential if $[Cl^-] = 0.020$ M and $[Ni^{2+}] = 0.10$ M.
(e) The galvanic cell was used to measure an unknown chloride ion concentration. The Ni^{2+} concentration was 0.200 M and the measured cell potential was 0.388 V. What was the chloride ion concentration?

39. Show how ethylenediamine and oxalate ion are able to function as bidentate ligands. What is the *chelate effect*? How does EDTA aid in softening "hard water"?

40. Sketch the stereoisomers of (a) $[Co(en)_2Cl_2]^+$ and (b) $[Cr(en)_3]^{3+}$. Label cis and trans isomers. Which of the structures are chiral?

41. What test is applied to a molecular or ionic structure to determine whether it is chiral?

42. The octahedral complex $[Cr(NH_3)_3Cl_3]$ has two stereoisomers. Sketch their structures. Be sure that the two structures that you draw are indeed different.

43. Sketch an energy level diagram showing the energies of the d orbitals in an octahedral crystal field. Label the diagram by identifying the d orbitals in each energy level and the crystal field splitting, Δ.

44. The complex $Pt(NH_3)_2Cl_2$ can be isolated as two isomers (cis and trans) because it has a square-planar structure. Show that if it were tetrahedral, only one isomer (i.e., only one structure) would be possible. Do this by trying to draw another tetrahedral isomer and then show that it can be rotated to give the first one.

45. The nitrite ion gives a very strong crystal field splitting in the complex ion $[Co(NO_2)_6]^{3-}$. How many unpaired electrons would be present in this complex ion?

An eerie blue glow—Tcherenkoff radiation—accompanies the decay of short-lived radionuclides as spent fuel elements from a nuclear reactor are held submerged in water for at least two years. Fundamental principles concerning atomic radiation and its uses are discussed in this chapter.

Chapter 20

Nuclear Reactions and Their Role in Chemistry

The nuclei of the atoms of most isotopes are exceptionally stable and are completely unaffected by events in the electron energy levels. Many elements, however, have one or more isotopes whose atoms possess unstable nuclei, and these carry properties both extremely useful and (when mishandled) dangerous.

Isotopes with unstable atomic nuclei are **radioactive,** so called because they emit high-energy streams of particles or electromagnetic radiation, often both simultaneously. Radioactive isotopes are sometimes termed **radionuclides,** and they undergo changes new to our study, *nuclear reactions*. Scientists exploit these reactions to study chemical changes, to conduct many kinds of analyses (including the dating of rocks and ancient objects), to diagnose or treat many diseases, and to obtain energy.

Interconvertibility of Mass and Energy

We begin our study of radionuclides by reexamining two physical laws that we have assumed to be separate and independent, the laws of conservation of energy and conservation of mass. For chemical reactions these laws can be treated as independent, but not for nuclear reactions. Albert Einstein, reflecting on the natures of mass, length, and time on his way to the theory of relativity, saw that these two laws are different aspects of a deeper, more general law.

With respect to mass, Einstein realized that if other laws of physics were to be "saved," that is, if they were to hold true in the emerging atomic and nuclear physics, the mass of a particle, *m*, had to be redefined. Einstein said that the mass of a particle is not actually constant. It changes with the particle's

20.1
CONSERVATION OF MASS–ENERGY

Albert Einstein (1879–1955);
Nobel Prize, 1921 (physics)

velocity, v, the velocity relative to the observer. A particle's mass is related to this velocity, and to the velocity of light, c, by the following equation.

$$m = \frac{m_0}{\sqrt{1 - (v/c)^2}} \qquad (20.1)$$

$c = 3.00 \times 10^8$ m s^{-1}, the speed of light.

Notice what happens when the particle has no velocity (relative to the observer) and v is zero. The ratio v/c is then zero, the whole denominator reduces to a value of 1, and Equation 20.1 becomes

$$m = m_0$$

This is why the symbol m_0 stands for the particle's *rest mass*. Rest mass is what we measure in all lab operations, because any object, like a chemical sample, is either at rest (from our viewpoint) or is not moving extraordinarily rapidly. Only as the particle's velocity approaches the speed of light, c, does the v/c term in Equation 20.1 become important. Only then does it measurably affect the denominator and so affect the mass. As v approaches c, the ratio v/c approaches 1, and so $(1 - v/c)$ gets closer and closer to 0. The whole denominator, in other words, approaches a value of 0. If it actually reached 0, then $m = (m_0 \div 0)$ goes to infinity. In other words, the mass, m, of the particle moving at the velocity of light would be infinitely great, a physical impossibility. Thus, the speed of light is an absolute upper limit on the speed that any particle can have.

Even at $v = 1000$ m s^{-1} (about 2250 mph), the denominator (unrounded) is 0.99999333, or within 7×10^{-4}% of 1.

At the velocities of everyday experience, the mass of anything calculated by Equation 20.1 equals the rest mass to several significant figures. The difference cannot be detected by weighing devices. Thus, in all of our normal work, mass appears to be conserved, and the law of conservation of mass functions as if it is an independent law in chemistry.

In a universe where mass changes with velocity, Einstein realized that a new understanding of energy is also necessary if the other law, energy conservation, is to hold. He had to postulate that mass and energy are interconvertible, just like potential and kinetic energy. What is conserved for the energy of a system is *all* its forms, including the system's mass calculated as an equivalent amount of energy. Alternatively, what is conserved about the mass of a system is *all* its forms, including the system's energy expressed as an equivalent amount of mass. The single, deeper law that summarizes these conclusions is now called the **law of conservation of mass–energy.**

Law of Conservation of Mass–Energy

The sum of all the energy in the universe and of all the mass (expressed as an equivalent in energy) is a constant.

Einstein equation

Einstein Equation

Einstein was able to show that when mass converts to energy, the amount of the change in energy, ΔE, is related to the change in rest mass, Δm_0, by the following equation, now called the **Einstein equation.**

$$\Delta E = \Delta m_0 c^2$$

The Einstein equation is given as $E = mc^2$ in the popular press.

Again, c is the velocity of light, 3.00×10^8 m s^{-1}.

Because the velocity of light is very large (and its square, 9×10^{16}, far more so), even an enormous value of ΔE corresponds to an extremely small change in mass, Δm_0. The combustion of methane, for example, releases considerable heat per mole.

$$CH_4(g) + 2O_2(g) \longrightarrow CO_2(g) + 2H_2O(l) \qquad \Delta H° = -890 \text{ kJ}$$

The source of the 890 kJ of heat energy is a loss of mass for the system that, when calculated by the Einstein equation, amounts only to 9.89 ng. (We leave the actual calculation to a Practice Exercise.) The 890 kJ thus originates in the conversion of 9.89 ng of mass into energy. Such a small change in mass, roughly 10 nanograms out of a total mass of 80 grams (16 g of CH_4 plus 64 g of O_2), cannot be detected by weighing balances. It amounts to a loss of about $1 \times 10^{2-7}\%$ of the mass.

The importance of the Einstein equation became clear when atomic fission was first observed in 1939, and since then the equation has become a routine tool among nuclear physicists. We will see how the Einstein equation can help us calculate the energy available by nuclear changes in the next section.

Methane is the chief constituent of natural gas, the fuel for Bunsen burners in most labs.

The actual mass of an atomic nucleus is always a little smaller than the sum of the rest masses of all of its nucleons, its protons and neutrons. The mass difference represents mass that changed into energy as the nucleons gathered to form the nucleus, this energy being released from the system. The absorption of the same amount of energy by the nucleus would be required to break it apart into its nucleons again, so the energy is called the **binding energy** of the nucleus. It is not energy possessed by the nucleus but, instead, the energy the nucleus would have to be given to break it up. Thus, the *higher* the binding energy, the more stable is the nucleus. The nuclear binding energy of a nucleus can be calculated by using the Einstein equation. We can illustrate this calculation by finding the nuclear binding energy of helium-4.

20.2 NUCLEAR BINDING ENERGIES

Nuclear Binding Energy of Helium   Helium has atomic number 2, and the nucleus of one helium-4 atom has 4 nucleons, 2 protons and 2 neutrons. The rest mass of one helium-4 nucleus is known to be 4.001506 u. However, the sum of the rest masses of its four separated nucleons—4.031875 u—is slightly more. This is calculated from the rest mass of a proton, 1.00727252 u, and that of a neutron, 1.008665 u, as follows.

The atomic mass unit, u, equals $1.6605402 \times 10^{-24}$ g.

$$\text{For 2 protons: } 2 \times 1.00727252 \text{ u} = 2.01454504 \text{ u}$$

$$\text{For 2 neutrons: } 2 \times 1.008665 \text{ u} = \underline{2.017330 \text{ u}}$$

$$\text{Total rest mass of nucleons in }{}^4\text{He} = 4.031875 \text{ u}$$

The difference between the calculated and measured rest masses for the helium-4 nucleus is 0.030369 u, found by

$$4.031875 \text{ u} - 4.001506 \text{ u} = 0.030369 \text{ u}$$

Using Einstein's equation, the energy equivalent of the mass difference of 0.030369 u is found as follows, where we have to convert u (the atomic mass unit) to kilograms by the relationships 1 u = 1.6605402 × 10⁻²⁴ g, 1 kg = 1000 g, and 1 J = 1 kg m² s⁻².

$$\Delta E = \Delta mc^2 = (0.030369 \text{ u}) \underbrace{\frac{(1.6605402 \times 10^{-24} \text{ g})}{1 \text{ u}} \frac{1 \text{ kg}}{1000 \text{ g}}}_{\Delta m \text{ (in kg)}} \underbrace{(3.00 \times 10^8 \text{ m s}^{-1})^2}_{c^2}$$

$$= 4.54 \times 10^{-12} \text{ kg m}^2 \text{ s}^{-2}$$
$$= 4.54 \times 10^{-12} \text{ J}$$

There are four nucleons in the helium-4 nucleus, so the binding energy per nucleon is $(4.54 \times 10^{-12} \text{ J})/4$ nucleons or 1.14×10^{-12} J/nucleon.

The formation of just one nucleus of helium-4 releases 4.54×10^{-12} J. If we could make Avogadro's number or 1 mol of helium-4 nuclei—the total mass would be only 4 grams—the net release of energy would be:

$$(6.02 \times 10^{23} \text{ nuclei}) \times (4.54 \times 10^{-12} \text{ J/nucleus}) = 2.73 \times 10^{12} \text{ J}$$

This amount of energy could keep a 100-watt light bulb lit for nearly 900 years! The formation of a nucleus from its nucleons is called *nuclear fusion,* which we will study further later, but you can see by the calculation that the energy potentially available from nuclear fusion is enormous.

Relationship between Binding Energy per Nucleon and Nuclear Stability

Figure 20.1 shows a plot of binding energies per nucleon versus mass numbers for most of the elements. The curve passes through a maximum at iron-56.

FIGURE 20.1

Binding energies per nucleon. The energy unit here is the megaelectron volt or MeV; 1 MeV = 10⁶ eV = 1.602 × 10⁻¹³ J. (From D. Halliday and R. Resnick, *Fundamentals of Physics,* 2nd ed., John Wiley & Sons, Inc., revised 1986. Used by permission.)

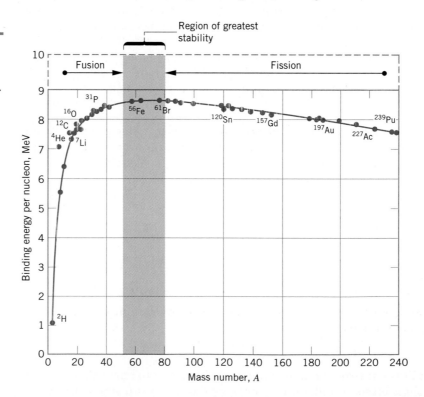

The nuclei of atoms of the iron-56 isotope are thus the most stable of all. The plot in Figure 20.1, however, does not have a sharp maximum, so we find the most stable isotopes in nature among the isotopes of a large number of elements with intermediate mass numbers.

As we follow the plot of Figure 20.1 to the highest mass numbers, the nuclei become less stable. Among the heaviest atoms, therefore, we would expect to find isotopes that could change to more stable forms by undergoing *nuclear fission,* a spontaneous breaking apart into isotopes of intermediate mass number. (We will study fission in more detail later.)

On the other hand, research on *nuclear fusion* is concentrated on fusing the nuclei of the lightest of all isotopes, those of hydrogen. The fusion of such nuclei would give the most gain in nuclear stability as nuclei of higher mass numbers form from them.

Quite often you'll see the terms atomic fusion and atomic fission used for nuclear fusion and nuclear fission.

20.3 RADIOACTIVITY

Except for hydrogen, all atomic nuclei have more than one proton. Each proton bears a positive charge, and we have learned that like charges repel. So we might well ask how can *any* nucleus be stable? What keeps the protons from repelling each other? And what holds neutrons within the nuclear package? Without any electrical charge, why don't the neutrons simply drift away? As physicists addressed such questions, they came to realize that electrostatic forces of attraction and repulsion—the kind present, for example, among the ions in a crystal of sodium chloride—are not the only forces at work in the atomic nucleus. Protons do, indeed, repel each other, but another force, a force of attraction called the *nuclear strong force,* is also acting. The nuclear strong force is able to overcome the electrostatic force of repulsion between protons, and it binds the nucleons into a package. The neutrons, by helping to keep the protons farther apart, also lessen repulsions between protons.

Still another nuclear force is called the *weak force,* and it is involved with beta decay.

Adjacent neutrons experience no electrostatic repulsion between each other, only the strong force (of attraction).

Radioactive Decay One consequence of the difference between the nuclear strong force and the electrostatic force occurs among nuclei carrying large numbers of protons but with too few intermingled neutrons to dilute the electrostatic proton–proton repulsions. Such nuclei are often unstable. They carry more energy than other arrangements of nucleons accessible to them. To achieve less energy and thus more stability, radionuclides undergo **radioactive decay;** they eject small nuclear fragments, and many simultaneously release high-energy electromagnetic radiation.

Alpha Radiation

About 50 of the approximately 350 naturally occurring isotopes are radioactive. Their radiation consists principally of three kinds: alpha, beta, and gamma radiation.

Alpha radiation consists of a stream of the nuclei of helium atoms, called **alpha particles,** symbolized as $_2^4He$, where 4 is the mass number and 2 is the atomic number. The alpha particle bears a charge of 2+, but the charge is omitted from the symbol.

Alpha particles are the most massive of any commonly emitted by radionuclides. When ejected (see Figure 20.2), alpha particles move through the

Alpha-emitting radionuclides are common among the elements near the end of the periodic table.

FIGURE 20.2

Emission of an alpha particle.

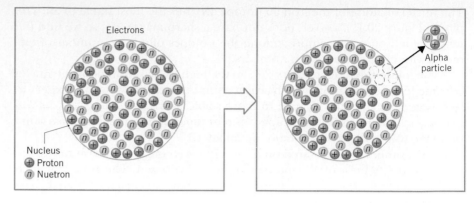

atom's electron orbitals reaching emerging speeds of up to one-tenth the speed of light. Their size, however, prevents them from going far. Within a few centimeters of travel in air, alpha particles collide with air molecules, lose kinetic energy, pick up electrons, and become neutral helium atoms. Alpha particles cannot penetrate the skin, although enough exposure causes a severe skin burn. If carried in air or on food into the soft tissues of the lungs or the intestinal tract, emitters of alpha particles can cause serious harm, including cancer.

Nuclear Equations

Nuclear equations

To symbolize the decay of a nucleus, we construct a **nuclear equation,** which we can illustrate by the alpha decay of uranium-238 to thorium-234.

$$^{238}_{92}\text{U} \longrightarrow {}^{234}_{90}\text{Th} + {}^{4}_{2}\text{He}$$

Unlike changes that occur in chemical reactions, nuclear reactions produce new isotopes, so we need special principles for balancing nuclear equations. A nuclear equation is balanced when

1. The sums of the mass numbers on each side are equal.
2. The sums of the atomic numbers on each side are equal.

In the above nuclear equation, the atomic numbers balance $(90 + 2 = 92)$, and the mass numbers balance $(234 + 4 = 238)$. Notice that electrical charges are not included, even though they are there (initially). The alpha particle, for example, has a charge of $2+$. Initially, therefore, the thorium particle has as much opposite charge, $2-$. These charged particles, however, eventually pick up or lose electrons either from each other or from molecules in the matter through which they travel.

Beta Radiation

Tritium is a synthetic radionuclide.

Naturally occurring **beta radiation** consists of a stream of electrons, which in this context are called **beta particles.** In a nuclear equation, the beta particle has the symbol $_{-1}^{0}e$, because the electron's mass number is 0 and its charge is $1-$. Tritium is a beta emitter that decays by the following equation.

$$^{3}_{1}\text{H} \longrightarrow {}^{3}_{2}\text{He} + {}^{0}_{-1}e + \bar{\nu}$$
$$\text{tritium} \qquad \text{helium-3} \quad \text{beta particle} \quad \text{antineutrino}$$
$$\text{(electron)}$$

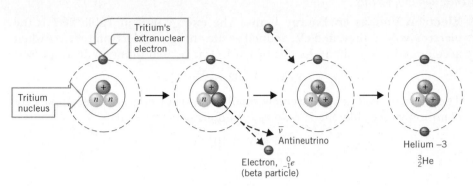

FIGURE 20.3

Emission of a beta particle.

Both the antineutrino (to be described shortly) and the beta particle come from the atom's nucleus, not its electron shells. We do not think of them as having a prior existence in the nucleus, any more than a photon exists before its emission from an excited atom (see Figure 20.3). Both the beta particle and the antineutrino are created during the decay process in which a neutron is transformed into a proton.

$$\underset{\substack{\text{neutron}\\(\text{in nucleus})}}{{}_{0}^{1}n} \longrightarrow \underset{\substack{\text{beta particle}\\(\text{emitted})}}{{}_{-1}^{0}e} + \underset{\substack{\text{proton}\\(\text{remains in}\\\text{nucleus})}}{{}_{1}^{1}p} + \underset{\substack{\text{antineutrino}}}{\bar{\nu}}$$

Unlike alpha particles, which are all emitted with the same discrete energy from a given radionuclide, beta particles emerge from a given beta emitter with a continuous spectrum of energies. Their energies vary from zero up to some characteristic fixed upper limit for each radionuclide. This fact gave nuclear physicists considerable trouble, partly because it was an apparent violation of energy conservation. To solve this problem, Wolfgang Pauli proposed in 1927 that beta emission is accompanied by yet another decay particle, this one electrically neutral and almost massless. Enrico Fermi suggested the name *neutrino* ("little neutral one"), but eventually it was named the *antineutrino*, symbolized by $\bar{\nu}$.

An electron is extremely small, so a beta particle is less likely to collide with the molecules of anything through which it travels. Depending on its initial kinetic energy, a beta particle can travel up to 300 cm in dry air, much farther than alpha particles. Only the highest energy beta particles can penetrate the skin, however.

Gamma Radiation

Gamma radiation, which often accompanies either alpha or beta radiation, consists of high-energy photons given the symbol ${}_{0}^{0}\gamma$ or, often, simply γ in equations.

The emission of gamma radiation involves transitions between energy levels *within* the nucleus. Nuclei have energy levels of their own, much as atoms have orbital energy levels. When a nucleus emits an alpha or beta particle, it sometimes is left in an excited energy state. By the emission of a gamma-ray photon, the nucleus relaxes into a more stable state.

Gamma radiation is an extremely penetrating radiation and is effectively blocked only by very dense materials, like lead. Table 20.1 shows the relative penetrating abilities of alpha, beta, and gamma radiation.

Enrico Fermi (1901–1954); Nobel Prize, 1938 (physics)

Electron Volt as an Energy Unit   The energy unit in Table 20.1 is the **electron volt,** abbreviated **eV,** defined as the energy an electron receives when accelerated under the influence of 1 volt. It is related to the joule as follows.

$$1 \text{ eV} = 1.602 \times 10^{-19} \text{ J}$$

The electron volt is an extremely small amount of energy, so multiples are commonly used, like the kilo-, mega-, and gigaelectron volt.

$$1 \text{ keV} = 10^3 \text{ eV}$$

$$1 \text{ MeV} = 10^6 \text{ eV}$$

$$1 \text{ GeV} = 10^9 \text{ eV}$$

The gamma radiation from cobalt-60, once widely used in cancer therapy, consists of photons with energies of 1.173 MeV and 1.332 MeV.

X Rays

X rays, like gamma rays, consist of high-energy electromagnetic radiation, but their energies are usually less than those of gamma radiation. X rays are emitted by some synthetic radionuclides, but normally they are generated by special machines in which a high-energy electron beam is focused onto a metal target. The beam knocks electrons out of atomic orbitals in the target and creates *holes,* which are actually empty or half-filled orbitals. If a hole is created in an orbital of low energy, then an electron at a higher energy level drops

In the interconversion of mass and energy,

$$1 \text{ eV} = 1.783 \times 10^{-36} \text{ kg}$$
$$1 \text{ GeV} = 1.783 \times 10^{-24} \text{ g}$$

X rays used in diagnosis typically have energies of 100 keV or less.

TABLE 20.1   Penetrating Abilities of Some Common Radiations[a]

Type of Radiation	Common Sources	Approximate Energy When from These Sources	Approximate Depth of Penetration of Radiation into		
			Dry Air	Tissue	Lead
Alpha rays	Radium-226 Radon-222 Polonium-210	5 MeV	4 cm	0.05 mm[c]	0
Beta rays	Tritium Strontium-90 Iodine-131 Carbon-14	0.05 to 1 MeV	6 to 300 cm[b]	0.06 to 4 mm[c]	0.005 to 0.3 mm
			Thickness to Reduce Initial Intensity by 10%		
Gamma rays	Cobalt-60 Cesium-137 Radium-226 decay products	1 MeV	400 cm	50 cm	30 mm
X Rays					
Diagnostic		Up to 90 keV	120 m	15 cm	0.3 mm
Therapeutic		Up to 250 keV	240 m	30 cm	1.5 mm

[a] Data are from J. B. Little, *The New England Journal of Medicine,* Vol. 275, pages 929–938, 1966.

[b] The range of beta particles in air is about 30 cm per MeV. Thus, a 2 MeV beta particle has a range of about 60 cm in air.

[c] The protective layer of skin is about 0.07 mm thick. To penetrate it, alpha particles need about 7.5 MeV of energy and beta particles about 0.07 MeV.

down to fill the hole. This creates a hole higher up, so an electron at a still higher level drops into it, making another hole even higher up. Thus, a cascade of electron transitions occurs. A huge number of frequencies are emitted, and those of the highest energy are in the X ray region of the spectrum. The chief difference, therefore, between X rays and gamma rays is that X rays normally come from transitions involving electron energy levels and gamma rays are from transitions between nuclear energy levels.

Cesium-137, $^{137}_{55}\text{Cs}$, one of the radioactive wastes from a nuclear power plant or an atomic bomb explosion, emits beta and gamma radiation. Write the nuclear equation for the decay of cesium-137.

ANALYSIS We start with an incomplete equation using the given information and then figure out any other data needed to complete the equation.

SOLUTION The incomplete nuclear equation is

$$^{137}_{55}\text{Cs} \longrightarrow ^{0}_{-1}e + ^{0}_{0}\gamma + \underline{}$$

mass number goes here
atomic symbol goes here
atomic number goes here

The atomic symbol can be obtained from the table inside the front cover after we have determined the atomic number, Z. Z is found using the fact that the atomic number (55) on the left side of the equation must equal the sum of the atomic numbers on the right side.

$$55 = -1 + 0 + Z$$

$$Z = 56$$

The periodic table tells us that element 56 is Ba (barium). To determine which isotope of barium forms, we recall that the sums of the mass numbers on either side of the equation must also be equal. Letting A be the mass number of the barium isotope,

$$137 = 0 + 0 + A$$

$$A = 137$$

The balanced nuclear equation, therefore, is

$$^{137}_{55}\text{Cs} \longrightarrow ^{0}_{-1}e + ^{0}_{0}\gamma + ^{137}_{56}\text{Ba}$$

■ **Practice Exercise 1** Radium-226, $^{226}_{88}\text{Ra}$, is an alpha and gamma emitter. Write a balanced nuclear equation for its decay. (Marie Curie earned one of her two Nobel Prizes for isolating the element radium, which soon became widely used to treat cancer.)

■ **Practice Exercise 2** Write the balanced nuclear equation for the decay of strontium-90, a beta emitter. (Strontium-90 is one of the many radionuclides present in the wastes of operating nuclear power plants.)

EXAMPLE 20.1
Writing a Balanced Nuclear Equation

Ions of cesium, which is in the same family as sodium, travel in the body to many of the same sites where sodium ions go.

Marie Curie (1867–1934); Nobel Prizes, 1903 (physics) and 1911 (chemistry)

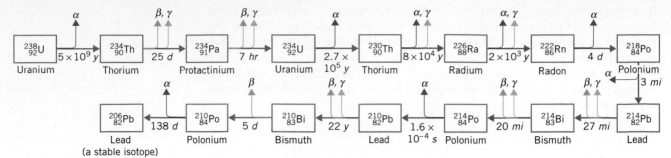

FIGURE 20.4

The uranium-238 radioactive disintegration series. The time given beneath each arrow is the half-life period of the preceding isotope (y = year, m = month, d = day, hr = hour; mi = minute, and s = second).

Radioactive Disintegration Series

Sometimes one radionuclide does not decay to a stable isotope but decays instead to another unstable radionuclide. The decay of one radionuclide after another will continue until a stable isotope forms. The sequence of such successive nuclear reactions is called a **radioactive disintegration series.** Four series occur naturally. Uranium-238 is at the head of one (Figure 20.4).

We studied half-lives in Section 14.5. One half-life period in nuclear science is the time it takes for a given sample of a radionuclide to decay to one-half of its initial amount (Table 20.2). Because radioactive decay is a first-order process, the period of time taken by one half-life period is independent of the initial number of nuclei.

Positron and Neutron Emissions; Electron Capture

Many synthetic isotopes emit **positrons,** particles with the mass of an electron but a positive instead of a negative charge. A positron is a positive beta particle,

TABLE 20.2   Typical Half-Life Periods

Element	Isotope	Half-life	Radiations or Mode of Decay
Naturally occurring radionuclides			
Potassium	$^{40}_{19}K$	1.3×10^9 yr	beta, gamma
Tellurium	$^{123}_{52}Te$	1.2×10^{13} yr	electron capture
Neodymium	$^{144}_{60}Nd$	5×10^{15} yr	alpha
Samarium	$^{149}_{62}Sm$	4×10^{14} yr	alpha
Rhenium	$^{187}_{75}Re$	7×10^{10} yr	beta
Radon	$^{222}_{86}Rn$	3.82 day	alpha
Radium	$^{226}_{88}Ra$	1590 yr	alpha, gamma
Thorium	$^{230}_{90}Th$	8×10^4 yr	alpha, gamma
Uranium	$^{238}_{92}U$	4.51×10^9 yr	alpha
Synthetic radionuclides			
Tritium	$^{3}_{1}T$	12.26 yr	beta
Oxygen	$^{15}_{8}O$	124 s	positron
Phosphorus	$^{32}_{15}P$	14.3 day	beta
Technetium	$^{99m}_{43}Tc$	6.02 hr	gamma
Iodine	$^{131}_{53}I$	8.07 day	beta
Cesium	$^{137}_{55}Cs$	30 yr	beta
Strontium	$^{90}_{38}Sr$	28.1 yr	beta
Americium	$^{243}_{95}Am$	7.37×10^3 yr	alpha

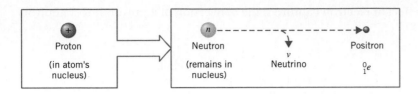

FIGURE 20.5

The emission of a positron replaces a proton by a neutron.

a positive electron, and its symbol is $^{0}_{1}e$. It forms in the nucleus by the conversion of a proton to a neutron (Figure 20.5). Positron emission, like beta emission, is accompanied by a chargeless and virtually massless particle, a *neutrino* (ν), the counterpart of the antineutrino ($\bar{\nu}$) in the realm of antimatter (defined below). Cobalt-54, for example, is a positron emitter and changes to a stable isotope of iron.

$$^{54}_{27}\text{Co} \longrightarrow {}^{54}_{26}\text{Fe} + \underset{\text{positron}}{{}^{0}_{1}e} + \underset{\text{neutrino}}{\nu}$$

The symbol $^{+}\beta$ is sometimes used for the positron.

After a positron is emitted, it eventually collides with an electron and the two annihilate each other (Figure 20.6). Their masses change into the energy of two gamma-ray photons called *annihilation radiation photons*, each with an energy of 511 keV.

$$^{0}_{-1}e + {}^{0}_{1}e \longrightarrow 2\,{}^{0}_{0}\gamma$$

Because a positron destroys a particle of ordinary matter (an electron), it is called a particle of antimatter. To be classified as **antimatter**, a particle must have a counterpart among particles of ordinary matter, and the two must annihilate each other when they collide. For example a neutron destroys an antineutron.

Neutron emission, another kind of nuclear reaction, does not lead to an isotope of a different element. Krypton-87, for example, decays as follows to krypton-86.

$$^{87}_{36}\text{Kr} \longrightarrow {}^{86}_{36}\text{Kr} + \underset{\text{neutron}}{{}^{1}_{0}n}$$

Electron capture, yet another kind of nuclear reaction, is very rare among natural isotopes but common among synthetic radionuclides. Vanadium-50 nuclei, for example, can capture orbital K shell or L shell electrons, change to nuclei of stable atoms of titanium, and emit X rays and neutrinos.

The K shell is the shell with principal quantum number $n = 1$; the L shell is at $n = 2$.

$$^{50}_{23}\text{V} + {}^{0}_{-1}e \xrightarrow{\text{electron capture}} {}^{50}_{22}\text{Ti} + \text{X rays} + \nu$$

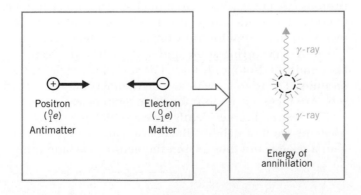

FIGURE 20.6

Gamma radiation is produced when a positron and an electron collide.

The net effect of electron capture is the conversion of a proton into a neutron (Figure 20.7).

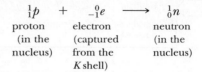

$$\underset{\substack{\text{proton} \\ \text{(in the} \\ \text{nucleus)}}}{{}_{1}^{1}p} + \underset{\substack{\text{electron} \\ \text{(captured} \\ \text{from the} \\ K\text{ shell)}}}{{}_{-1}^{0}e} \longrightarrow \underset{\substack{\text{neutron} \\ \text{(in the} \\ \text{nucleus)}}}{{}_{0}^{1}n}$$

The proton can also be given the symbol ${}_{1}^{1}$H in nuclear equations.

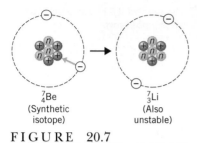

${}_{4}^{7}$Be
(Synthetic isotope)

${}_{3}^{7}$Li
(Also unstable)

FIGURE 20.7

Electron capture is the collapse of an orbital electron into the nucleus, and this changes a proton into a neutron.

Electron capture does not change an atom's mass number, only its atomic number. It also leaves a hole in the K or L shell, and the atom emits photons of X rays as other orbital electrons drop down to fill the hole. Moreover, the nucleus that has just captured an orbital electron can emit a gamma-ray photon.

It is commonly thought that the rate of the radioactive decay of a particular isotope is independent of its oxidation state or any other aspect of its chemical combination or its physical state (pressure and temperature, for example). When decay is by alpha or beta emission, this appears to be true; at least modern instruments have not yet detected any variations. When decay is by electron capture, however, some very small differences in rates of decay have been observed. Beryllium-7, for example, decays about 0.3% faster as the metal than as the oxide, BeO. More orbital electron density exists close to the beryllium nucleus in the beryllium *atom* than in the Be^{2+} ion of the oxide. The ability of the beryllium nucleus to capture an electron is therefore greater when it is in the more electron-dense environment of its atom.

20.4 BAND OF STABILITY

When all known isotopes of each element, both stable and unstable, are arrayed on a plot according to numbers of protons and neutrons, an interesting zone can be defined (Figure 20.8). The two curved lines in the array of Figure 20.8 enclose this zone, called the **band of stability,** within which lie all stable nuclei. (No isotope above element 83, bismuth, is included in Figure 20.8 because none has a *stable* isotope.) Within the band of stability are also some radionuclides, because smooth lines cannot be drawn to exclude them. Any isotope not represented anywhere on the array, inside or outside the band of stability, probably has a half-life too short to permit its detection. An isotope with 50 neutrons and 60 protons, for example, would be too unstable to justify the time and money for an attempt to make it.

Notice that with increasing numbers of protons, the *ratio* of neutrons to protons gradually moves farther and farther away from a simple 1:1 ratio (indicated in the array of Figure 20.8 by the straight line). As more and more protons become packed into a nucleus, still larger numbers of neutrons are needed to provide the compensating nuclear strong force and to dilute proton–proton repulsions by the electrostatic force.

Isotopes occurring above and to the left of the band of stability tend to be beta emitters. Isotopes lying below and to the right of the band are positron emitters. The isotopes with atomic numbers above 83 tend to be alpha emitters. Are there any reasons for these tendencies?

The beta emitters evidently have too high a neutron/proton ratio; they are *above* the band of stability. Beta decay causes a nucleus to lose a neutron and gain a proton and thus *decrease* the neutron/proton ratio.

$$\ _{0}^{1}n \longrightarrow \ _{1}^{1}p + \ _{-1}^{0}e + \bar{\nu}$$

FIGURE 20.8

The band of stability.

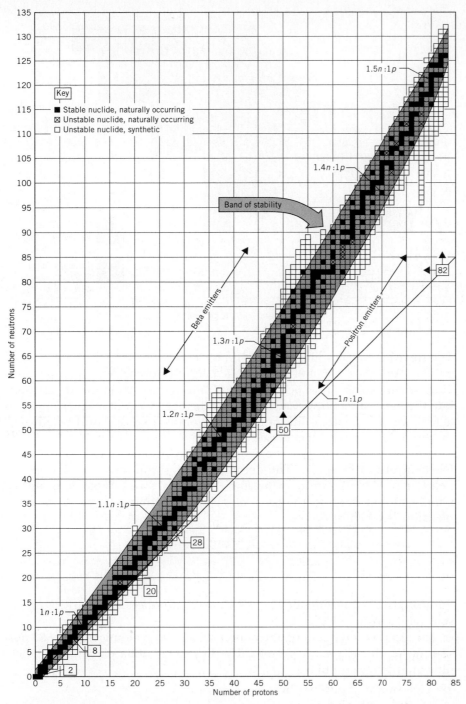

The beta decay of fluorine-20, for example, lets the neutron/proton ratio in its nucleus decrease from 11/9 to 10/10.

$$^{20}_{9}\text{F} \longrightarrow {}^{20}_{10}\text{Ne} + {}^{0}_{-1}e + \bar{\nu}$$

$$\frac{\text{neutron}}{\text{proton}} = \frac{11}{9} \qquad \frac{10}{10}$$

FIGURE 20.9

Beta decay from magnesium-27 and fluorine-20 reduces their neutron/proton ratio and moves them closer to the band of stability. Positron decay from magnesium-23 and fluorine-17 increases this ratio and moves these nuclides closer to the band of stability, too. Shown here is an enlarged section of Figure 20.8.

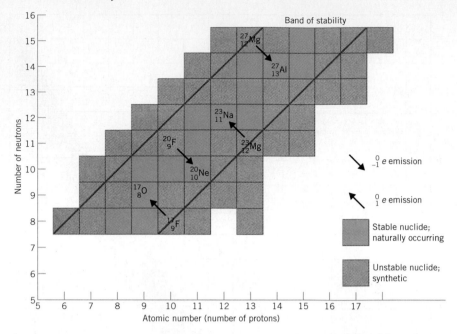

The surviving nucleus, that of neon-10, is closer to the center of the band of stability. An enlargement of the fluorine part of Figure 20.8 is given in Figure 20.9, which explains how beta emission from fluorine-20 moves the nucleus into the band of stability. Notice also how the beta decay of magnesium-27 to aluminum-27 is another illustration of a favorable lowering of the neutron/proton ratio caused by beta decay.

Positron emission *increases* the neutron/proton ratio in nuclei that have too few neutrons to be stable. The net effect of positron emission is to gain a neutron and lose a proton. A fluorine-17 nucleus, for example, improves its stability and moves into the band of stability by emitting a positron and a neutrino (see Figure 20.9).

$$^{17}_{9}F \longrightarrow {}^{17}_{8}O + {}^{0}_{1}e + \nu$$

$$\frac{\text{neutron}}{\text{proton}} = \quad \frac{8}{9} \qquad \frac{9}{8}$$

Positron decay by magnesium-23 to sodium-23 produces a similarly favorable shift (also shown in Figure 20.9).

The alpha emitters among the radionuclides above atomic number 83 possess nuclei with too many protons. The loss of an alpha particle is a quick, direct way to lose protons.

Odd–Even Rule

Nature favors even numbers for protons and neutrons, as summarized by the **odd–even rule.**

Odd–even rule

Odd–Even Rule

When the numbers of neutrons and protons in a nucleus are both even, the isotope is far more likely to be stable than when both numbers are odd.

Of the 264 stable isotopes, only five have odd numbers of *both* protons and neutrons, whereas 157 have even numbers of both. The rest have an odd number of one nucleon and an even number of the other. To see this, notice in Figure 20.8 how the horizontal lines with the largest numbers of dark squares (stable isotopes) most commonly correspond to even numbers of neutrons. Similarly, the vertical lines with the most dark squares most often correspond to even numbers of protons.

The odd–even rule is related to the spins of nucleons. Both protons and neutrons spin, like orbital electrons. When two protons or two neutrons have paired spins, meaning the spins are opposite, their combined energy is less than when the spins are unpaired. Only when there are *even* numbers of protons and of neutrons can all spins be paired and so give the nucleus less energy and more stability. The least stable nuclei tend to be those with both an odd proton and an odd neutron.

> The isotopes ^2_1H, ^6_3Li, $^{10}_5B$, $^{14}_7N$, and $^{138}_{57}La$ all have odd numbers of both protons and neutrons.

Nuclear Energy Levels and "Magic Numbers"

Another rule of thumb about nuclear stability is based on *magic numbers* of nucleons. Isotopes with specific numbers of protons or neutrons, the **magic numbers,** are more stable than the rest (and, according to a theory whose details we cannot discuss, *ought to be more stable*). The magic numbers of nucleons are 2, 8, 20, 28, 50, 82, and 126, and where they fall is shown in Figure 20.8 (except for magic number 126). When the numbers of *both* protons and neutrons are the *same* magic number, as they are in ^4_2He, $^{16}_8O$, and $^{40}_{20}Ca$, the isotope is very stable. The synthesis of $^{100}_{50}Sn$, also having two identical magic numbers, was announced in 1994. Compared to nearby radionuclides (whose half-lives are in milliseconds), tin-100 is much more stable (a half-life of several seconds). Thus, although tin-100 lies well outside the band of stability, it is stable enough to be observed. One stable isotope of lead, $^{208}_{82}Pb$, involves two different magic numbers, 82 protons and 126 neutrons. Another possible lead isotope, lead-164, lies so far outside of the band of stability that no scientist thinks its synthesis is possible.

> Magic numbers do not cancel the need for a favorable neutron/proton ratio. An atom with 82 protons and 82 neutrons lies far outside the band of stability, and yet 82 is a magic number.

The existence of magic numbers supports the hypothesis that nuclei have energy levels analogous to electron energy levels. As you know, electron levels are associated with their own special numbers, those that equal the maximum number of electrons allowed in a principal energy level: 2, 8, 18, 32, 50, 72, and 98 (for principal levels 1, 2, 3, 4, 5, 6, and 7, respectively). The total numbers of electrons in the atoms of the most chemically stable elements—the noble gases—also make up a special set: 2, 10, 18, 36, 54, and 86 electrons. Thus, special sets of numbers are not unique to nuclei.

The change of one isotope into another is called **transmutation,** and radioactive decay is only one cause. Another is the bombardment of nuclei by high-energy particles, which can be alpha particles from natural emitters, neutrons from atomic reactors, or protons made by stripping electrons from hydrogen. To make them better bombarding particles, protons and alpha particles can be accelerated in an electrical field (Figure 20.10). This gives them greater energy and enables them to sweep through the target atom's orbital electrons and become buried in its nucleus. Although beta particles can be accelerated, their disadvantage as bombardment missiles is that they are repelled by a target atom's own electrons.

20.5 TRANSMUTATION

FIGURE 20.10

In this linear accelerator at Brookhaven National Laboratory on Long Island, New York, protons can be accelerated to just under the speed of light and given up to 33 GeV of energy before they strike their target. (1 GeV = 10^9 eV.)

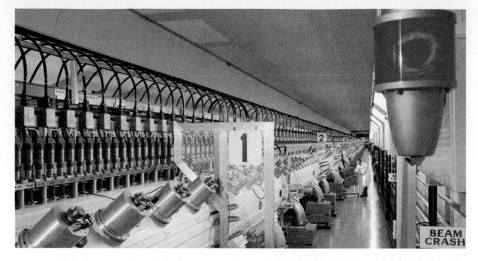

Formation and Decay of Compound Nuclei

Both the energy and the mass of a bombarding particle enter the target nucleus at the moment of capture. The energy of the new nucleus, called a **compound nucleus,** quickly becomes distributed among all of the nucleons, but the nucleus is nevertheless rendered somewhat unstable. To get rid of the excess energy, a compound nucleus generally ejects something, a neutron, proton, or electron, and often emits gamma radiation as well. This leaves a new nucleus of an isotope different than the original target, so overall a transmutation has occurred.

Compound here refers only to the idea of *combination,* not to a chemical.

The first example of this kind of nuclear reaction was observed by Ernest Rutherford. When he let alpha particles pass through a chamber containing nitrogen atoms, an entirely new radiation was generated, one much more penetrating than alpha radiation. It proved to be a stream of protons (Figure 20.11). Rutherford was able to show that the protons came from the decay of the compound nuclei of fluorine-18, produced when nitrogen-14 nuclei captured alpha particles.

$$\underset{\substack{\text{alpha}\\\text{particle}}}{{}^{4}_{2}\text{He}} + \underset{\substack{\text{nitrogen}\\\text{nucleus}}}{{}^{14}_{7}\text{N}} \longrightarrow \underset{\substack{\text{fluorine}\\\text{(compound}\\\text{nucleus)}}}{{}^{18}_{9}\text{F*}} \longrightarrow \underset{\substack{\text{oxygen}\\\text{(a rare but}\\\text{stable isotope)}}}{{}^{17}_{8}\text{O}} + \underset{\substack{\text{proton}\\\text{(high}\\\text{energy)}}}{{}^{1}_{1}p}$$

The asterisk, *, symbolizes a high-energy nucleus, a *compound nucleus.*

FIGURE 20.11

Transmutation of nitrogen into oxygen. When the nucleus of nitrogen-14 captures an alpha particle it becomes a compound nucleus of fluorine-18. This then expels a proton and becomes the nucleus of oxygen-17.

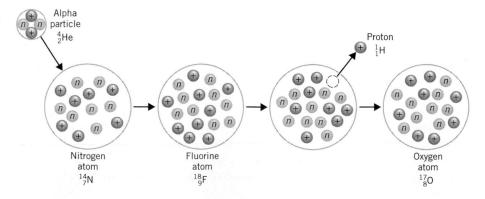

In the synthesis of alpha particles from lithium-7, protons are used as bombardment particles. The resulting compound nucleus, that of beryllium-8, splits in two.

$$\underset{\text{proton}}{{}^{1}_{1}p} + \underset{\text{lithium}}{{}^{7}_{3}\text{Li}} \longrightarrow \underset{\text{beryllium}}{{}^{8}_{4}\text{Be*}} \longrightarrow \underset{\substack{\text{alpha}\\\text{particles}}}{2\,{}^{4}_{2}\text{He}}$$

Decay Modes of Compound Nuclei A given compound nucleus can be made in a variety of ways. Aluminum-27, for example, forms by any of the following routes.

$${}^{4}_{2}\text{He} + {}^{23}_{11}\text{Na} \longrightarrow {}^{27}_{13}\text{Al*}$$

$${}^{1}_{1}p + {}^{26}_{12}\text{Mg} \longrightarrow {}^{27}_{13}\text{Al*}$$

$$\underset{\text{deuteron}}{{}^{2}_{1}\text{H}} + {}^{25}_{12}\text{Mg} \longrightarrow {}^{27}_{13}\text{Al*}$$

Once it forms, the compound nucleus, ${}^{27}_{13}\text{Al*}$, has no memory of how it was made. It only knows how much energy it has. Depending on this energy, different paths of decay are available, and all of the following routes have been observed. They illustrate how the synthesis of such a large number of synthetic isotopes, some stable and others unstable, has been possible.

A deuteron is the nucleus of a deuterium atom, just as a proton is the nucleus of a protium (hydrogen) atom.

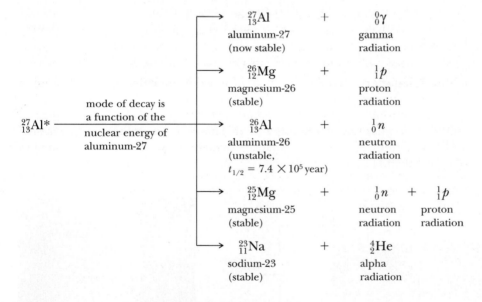

$${}^{27}_{13}\text{Al*} \xrightarrow{\substack{\text{mode of decay is}\\\text{a function of the}\\\text{nuclear energy of}\\\text{aluminum-27}}}$$

$$\underset{\substack{\text{aluminum-27}\\\text{(now stable)}}}{{}^{27}_{13}\text{Al}} + \underset{\substack{\text{gamma}\\\text{radiation}}}{{}^{0}_{0}\gamma}$$

$$\underset{\substack{\text{magnesium-26}\\\text{(stable)}}}{{}^{26}_{12}\text{Mg}} + \underset{\substack{\text{proton}\\\text{radiation}}}{{}^{1}_{1}p}$$

$$\underset{\substack{\text{aluminum-26}\\\text{(unstable,}\\t_{1/2} = 7.4 \times 10^5\,\text{year})}}{{}^{26}_{13}\text{Al}} + \underset{\substack{\text{neutron}\\\text{radiation}}}{{}^{1}_{0}n}$$

$$\underset{\substack{\text{magnesium-25}\\\text{(stable)}}}{{}^{25}_{12}\text{Mg}} + \underset{\substack{\text{neutron}\\\text{radiation}}}{{}^{1}_{0}n} + \underset{\substack{\text{proton}\\\text{radiation}}}{{}^{1}_{1}p}$$

$$\underset{\substack{\text{sodium-23}\\\text{(stable)}}}{{}^{23}_{11}\text{Na}} + \underset{\substack{\text{alpha}\\\text{radiation}}}{{}^{4}_{2}\text{He}}$$

Radionuclides Above Atomic Number 83

Over a thousand isotopes have been made by transmutations. Most do not occur naturally; they appear in the band of stability (see Figure 20.8) as open squares, nearly 900 in number. The naturally occurring isotopes above atomic number 83 all have very long half-lives. Others might have existed, but their half-lives probably were too short to permit them to last into our era. All of the elements from neptunium (atomic number 93) and up, the **transuranium elements,** are synthetic. Elements from atomic numbers 93 to 103 complete the actinide series of the periodic table, which starts with element 90, thorium. Beyond this series, elements 104–111 have also been made. The synthesis of elements 112–114 is expected during the remaining 1990s.

The study of the chemistry of elements 104–111 must be done virtually one atom at a time because so little is available and their half-lives are so short.

To make the heaviest elements, bombarding particles larger than neutrons are used, such as alpha particles or the nuclei of heavier atoms. Element 110, for example, was made in 1994 by the fusion of nickel-26 and lead-208, followed by the emission of a neutron. Some interest is also focused on the synthesis of element 126, an element yet to be made but one with a magic number.

20.6 DETECTING AND MEASURING RADIATION

Atomic radiation is often described as **ionizing radiation** because it creates ions in matter through which it travels by knocking electrons from molecules. When the matter is a gas at low pressure, we have the basis of one kind of radiation detector.

Detection Devices

Geiger Counter   The Geiger–Müller tube (Figure 20.12), one part of a *Geiger counter,* detects beta and gamma radiation having energy high enough to penetrate the tube's window. In operation, a metallized surface inside the tube works as a cathode, a wire electrode is the anode, and a potential difference exists between the two. The tube contains a gas at low pressure, but no current flows until ionizing radiation creates gaseous ions to carry it. Then a brief pulse of electricity activates a current amplifier, a pulse counter, and an audible click.

Atmospheric clouds form in air supersaturated with water vapor and in which condensation is started by dust particles.

Cloud Chambers   When a particle of ionizing radiation moves through a chamber saturated in a vapor, such as that of water, alcohol, or liquid hydrogen, microdroplets of condensed vapor form along its path. These appear as visible tracks that can be photographed. The device is called a *cloud chamber,* and several fundamental atomic particles were initially inferred by the analysis of cloud chamber photographs.

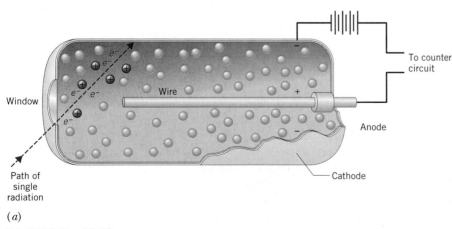

(a)

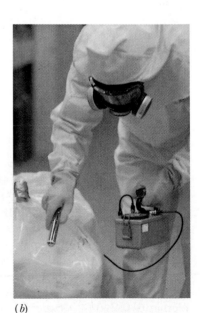

(b)

FIGURE   20.12

The Geiger–Müller tube. (*a*) In this schematic, the bluish spheres within the tube represent atoms of argon at very low pressure. (*b*) A Geiger counter, using a Geiger–Müller tube, is employed to monitor the radioactivity from a bag of waste.

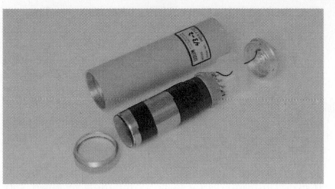

FIGURE 20.13
Scintillation probe. Energy received from radiations striking the phosphor at the end of this probe is processed by the photomultiplier unit and sent to an instrument (not shown) for recording.

Scintillation Counters A *scintillation counter* (Figure 20.13) has a surface coated with a special chemical that emits a tiny flash of light when struck by a particle of radiation. These flashes can be magnified electronically and automatically counted.

Film Dosimeters The darkening of a photographic film exposed to radiation over a period of time is proportional to the total quantity of radiation received. *Film dosimeters* work on this principle (Figure 20.14), and people who work near radiation sources use them. Each person keeps a log of total exposure, and if a predetermined limit is exceeded, the worker must be reassigned to an unexposed workplace.

Units for Radiation

The Curie and the Becquerel, Units of Activity *Activity* in this context refers to the number of nuclear disintegrations per second occurring in a radioactive sample. The SI unit of activity is the **becquerel (Bq),** and it equals one disintegration or other kind of nuclear transformation per second.

The **curie (Ci),** named after Marie Curie, the discoverer of radium, is an older unit of activity.

$$1 \text{ Ci} = 3.7 \times 10^{10} \text{ disintegrations s}^{-1}$$

A 1.0 g sample of radium has this activity. In becquerels,

$$1 \text{ Ci} = 3.7 \times 10^{10} \text{ Bq}$$

If a radiological laboratory has a source rated as 1.5 Ci, it delivers $1.5 \times 3.7 \times 10^{10}$ Bq or 5.6×10^{10} Bq.

The becquerel is named after Henri Becquerel (1852–1908), the discoverer of radioactivity, who won a Nobel Prize in 1903.

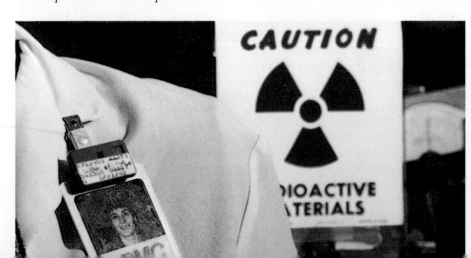

FIGURE 20.14
Badge dosimeter. The doses of various kinds of radiation received by the film in the dosimeter for predetermined periods of time are entered into the worker's log.

The Rad and the Gray, Units of Radiation Dose To describe the quantity of radiation absorbed by some material, units of *absorbed dose* (or simply *dose*) have been defined. The SI unit of absorbed dose is the **gray (Gy)** and 1 gray corresponds to 1 joule of energy absorbed per kilogram of absorbing material. The **rad** is an older unit of absorbed dose. A dose of 1 rad signifies the absorption of 10^{-5} joule per gram of tissue.

$$1 \text{ Gy} = 1 \text{ J kg}^{-1}$$

$$1 \text{ rad} = 10^{-5} \text{ J g}^{-1}$$

$$1 \text{ Gy} = 100 \text{ rad}$$

If every individual in a large population received 450 rad (4.5 Gy), roughly half of the population would die in 60 days.

The Sievert and the Rem, Units of Dose Equivalent A dose of 1 Gy of gamma radiation is not biologically the same as 1 Gy of beta radiation or of neutrons. The gray, therefore, does not serve as a good basis of comparison when working with *biological* effects. The **sievert (Sv)** is the SI unit that satisfies the need for a way to express dose equivalent that is additive for different kinds of radiation and different target tissues. If we let D stand for absorbed dose in grays, Q for quality factor (meaning relative effectiveness for causing harm in tissue), and N for any other modifying factors, then the dose equivalent, H, in sieverts is defined as follows.

$$H = DQN$$

In other words, we multiply the absorbed dose (D) from some radiation by a factor (Q) that takes into account biologically significant properties of the radiation and by any other factor (N) bearing on the net effect. The value of Q for alpha radiation is 20, but is only 1.0 for beta and gamma radiation. The large size and charge of the alpha particle, relative to other radiation particles, enable it to strike molecules with considerable (although undesirable) efficiency and so give this particle its large value of Q.

The older unit of dose equivalent, the **rem,** is the dose that has the effect of one roentgen in a human being. The **roentgen,** an older unit of exposure intensity for X-ray or gamma-ray radiation, equals a number of electrical charges totaling 2.1×10^9 that are created in 1 cm³ of dry air by the radiation. If members of a large population are exposed to 650 roentgens, roughly half would die in 1 to 4 weeks.

Sieverts or rems, unlike grays or rads, are additive. We could not, for example, say that 10 rad to the chest from gamma radiation and 5 rad to the head from neutron radiation add up to 15 rad for the whole body. After these rads are converted to rems, however, the figures can be added to find the whole-body dose in rems. The United States government has set guidelines of 0.3 rem per week as the maximum exposure workers may receive. (For comparison, a chest X ray typically involves about 0.007 rem or 7 mrem.)

Radiation Sickness A whole body dose of 25 rem induces noticeable changes in human blood. A set of symptoms called *radiation sickness* sets in at about 100 rem and is severe at 200 rem. Among the symptoms are nausea, vomiting, a drop in the white cell count, diarrhea, dehydration, prostration, hemorrhaging, and loss of hair. At roughly 400 rem absorbed per person in a large population, half would die in 60 days. A 600-rem dose would kill every-

The gray is named after Harold Gray, a British radiologist.

Rad stands for **r**adiation **a**bsorbed **d**ose.

Rem stands for **r**oentgen **e**quivalent for **m**an.

The results of radiation damage appear earliest in tissues with rapidly dividing cells, like bone marrow and the gastrointestinal tract.

SPECIAL TOPIC 20.1 / RADON-222 IN THE ENVIRONMENT

A few years ago an engineer unintentionally but repeatedly set off radiation alarms at his workplace, a nuclear power plant in Pennsylvania. Such alarms are triggered when radiation levels surpass a predetermined value. The problem eventually was traced to his home, where the radiation level was 2700 picocuries per liter of air, compared with the level in the average house of 1 picocurie per liter. (The picocurie is 10^{-12} curie.) The chief culprit was radon-222, a natural intermediate in the uranium-238 disintegration series (Figure 20.4). (One picocurie per liter is equivalent to the decay of two radon atoms per minute.) The engineer had carried radon-222 and its decay products on his clothing from his home to the power plant.

One result of this episode was to send geologists scurrying to map the general area for concentrations of uranium-238. They found that the bedrock of the Reading Prong, a geological formation that cuts across Pennsylvania, New Jersey, New York, and up into the New England states, is relatively rich in uranium.

Radon-222 is a noble gas and is as *chemically* unreactive as argon, krypton, and the other noble gases. But this decay prod-

uct of uranium-238 has an unstable nucleus. Its half-life is 4 days, and it emits both alpha and gamma radiation. Being a gas, it diffuses through rocks and soil and enters buildings. In fact, buildings can act much like chimneys and draw radon as a chimney draws smoke. Radon-222 in the air is breathed into the lungs. Because of its short half-life, some of it decays in the lungs before it is exhaled. The decay products are not gases, and several have short half-lives. When these are left in the lungs, their radiation can cause lung cancer, and scientists of the National Cancer Institute estimate that about 10% of all lung cancer deaths are caused, at least in part, to indoor exposure to radon.

The Environmental Protection Agency (EPA) has recommended an upper limit of 4 picocuries per liter in the air of homes. Estimates released in the late 1980s by the EPA were that roughly 1% of the homes in the United States have radon levels that generate radiation of more than 10 picocuries per liter. Those who estimate risks present in day-to-day activities say that such a level poses less a risk to life than the risk of driving a car.

one in the exposed group within a week. Many workers received at least 400 rem in the moments following the steam explosion that tore apart one of the nuclear reactors at the Ukraine's energy park near Chornobyl in 1986.

Radiation-Produced Free Radicals   Considering how harmful even small absorbed doses can be, it is somewhat astonishing that a rem involves an extremely small quantity of energy. Obviously, the harm cannot be associated with the *heat* equivalent of the dose. Instead, the danger comes from the ability of ionizing radiation to create unstable ions or neutral species with odd (unpaired) electrons, called *free radicals*. Water, for example, can interact as follows with ionizing radiation.

$$\text{H}-\overset{\cdot\cdot}{\underset{\cdot\cdot}{\text{O}}}-\text{H} \xrightarrow{\text{radiation}} [\text{H}-\overset{\cdot}{\underset{\cdot\cdot}{\text{O}}}-\text{H}]^{+} + \,_{-1}^{0}e$$

The new cation, $[\text{H}-\overset{\cdot}{\underset{\cdot\cdot}{\text{O}}}-\text{H}]^{+}$, is unstable and one breakup path is

$$[\text{H}-\overset{\cdot}{\underset{\cdot\cdot}{\text{O}}}-\text{H}]^{+} \longrightarrow \underset{\text{proton}}{\text{H}^{+}} + \underset{\substack{\text{hydroxyl} \\ \text{radical}}}{:\overset{\cdot}{\underset{\cdot\cdot}{\text{O}}}-\text{H}}$$

The proton might pick up a stray electron to become $\text{H}\cdot$, a hydrogen atom. Both the hydrogen atom and the hydroxyl radical are examples of free radicals (sometimes called simply *radicals*). They lack noble gas configurations and so are chemically very reactive. What they do next depends on the other chemical species nearby, but radicals can set off a series of totally undesirable chemical reactions inside a living cell. This is what makes the injury from absorbed radiation of far greater magnitude than the energy alone could inflict. A dose of 600 rem is a lethal dose in a human, but the same dose absorbed by pure water causes the ionization of only one water molecule in every 36 million.

Background Radiation

It is impossible to be free from all exposure to ionizing radiation. Above us come cosmic ray showers. From below as well as from building stone comes the radiation of crustal radionuclides. Radon gas seeps into basements from underground formations. Diagnostic X rays, both medical and dental, make small contributions. They all add up to the **background radiation,** which averages 360 mrem per person annually in the United States.

As the data in Table 20.3 show, a little over half of the background radiation is from radon-222 and its decay products (see Special Topic 20.1). Some is from radium, because the top 40 cm of soil holds an average of 1 gram of radium per square kilometer. Carbon-14, made by cosmic rays, enters the food chain via the carbon dioxide used in photosynthesis. Thus, by the food we ingest, carbon-14 contributes to radiation released within us. Naturally occurring potassium-40, a beta emitter, adds its radiation wherever potassium ions are found in the body. The presence of carbon-14 and potassium-40 together means that about 5×10^5 nuclear disintegrations occur per minute inside an adult human.

Radium salts were once put into the paint used for the numerals on watch faces to make them glow in the dark.

Radiation Protection

As the data in Table 20.1 show, gamma radiation and X rays are effectively shielded only by very dense materials, like lead. Otherwise, one should stay as far from a source as possible, because the intensity of radiation diminishes with

TABLE 20.3 Average Radiation Doses Received Annually by the United States Population[a]

Type of Radiation	Dose (mrem)	Percentage of Total
Natural radiation—295 mrem, 82%		
Radon[b]	200	55
Cosmic rays[c]	27	8
Rocks and soil	28	8
From inside the body	40	11
Artificial radiation—65 mrem, 18%		
Medical X rays[d]	39	11
Nuclear medicine	14	4
Consumer products	10	3
Others	2	<1
Total[e]	360	

[a] Data from "Ionizing Radiation Exposure of the Population of the United States." Report 93, National Council on Radiation Protection, 1987. These are averages. Individual exposures vary widely.

[b] See Special Topic 20.1.

[c] Travelers in jet airplanes receive about 1 mrem per 1000 miles of travel.

[d] A normal chest X ray entails an exposure of 7–20 mrem.

[e] The U.S. federal standard for maximum safe occupational exposure in the U.S. is roughly 5000 mrem/yr.

the *square* of the distance. This relationship is the **inverse square law,** which can be written mathematically as

$$\text{Radiation intensity} \propto \frac{1}{d^2}$$

Inverse square law

where d is the distance from the source. The inverse square law means that when the intensity, I_1, is known at distance d_1, then the intensity I_2 at distance d_2 can be calculated by the following equation.

$$\frac{I_1}{I_2} = \frac{d_2^2}{d_1^2}$$

When the ratio is taken, the proportionality constant cancels, so we don't have to know what its value is.

This law applies only to a small source that radiates equally in all directions and with no intervening shields.

■ **Practice Exercise 3** If an operator 10 m from a small source is exposed to 1.4 units of radiation, what will be the intensity of the radiation if he moves to 1.2 m from the source?

   The use of a medical X ray is called an *invasive* diagnostic tool because it can cause harm. The only reason X rays are used in medicine is that their potential benefits are judged to outweigh their harm. To minimize this harm, radiologists now use very fast film so as to shorten exposure times significantly.

The fact that the *chemical* properties of both the radioactive and stable isotopes of an element are the same is exploited in several uses of radionuclides. Their ordinary properties, both chemical and physical, enable scientists to get them into place in systems of interest. Then their radiation is exploited for medical or analytical purposes. Tracer analysis is an example.

20.7 APPLICATIONS OF RADIOACTIVITY

Radioactive Tracers

In **tracer analysis,** the chemical form and properties of a radionuclide enable the system to distribute it to a particular location. The intensity of the radiation then tells something about how that site is working. In the form of the iodide ion, for example, iodine-131 is carried by the body to the thyroid gland, the only user of iodide ion in the body. The gland takes up the iodide ion to synthesize the hormone thyroxine. An underactive thyroid gland is unable to concentrate iodide ion normally and will emit less intense radiation under standard test conditions.

   Tracer analyses are also used to pinpoint the locations of brain tumors, which are uniquely able to concentrate the pertechnetate ion, TcO_4^-, made from technetium-99m.[1] This strong gamma emitter, which resembles the chloride ion in some respects, is one of the most widely used radionuclides in medicine.

The tumor appears as a "hot spot" under radiological analysis.

Neutron Activation Analysis

A number of stable nuclei can be changed into emitters of gamma radiation by capturing neutrons, and this make possible a procedure called **neutron activa-**

---

[1] The m in technetium-99m means that the isotope is in a metastable form. Its nucleus is at a higher energy level than the nucleus in technetium-99, to which technetium-99m decays

tion analysis. Neutron capture followed by gamma emission can be represented by the following equation (where A is a mass number and X is a hypothetical atomic symbol).

$$^{A}X \ + \ ^{1}_{0}n \ \longrightarrow \ ^{(A+1)}X^* \ \longrightarrow \ ^{(A+1)}X \ + \ ^{0}_{0}\gamma$$

| isotope
of element
X being
analyzed | neutron | compound
nucleus
(unstable) | more stable
form of a new
isotope of X | gamma
radiation |

A neutron-enriched compound nucleus emits gamma radiation at its own unique frequencies, and these frequencies are now known for each isotope. (Not all isotopes, however, become gamma emitters by neutron capture.) By measuring the *frequencies* emitted, the element can be identified. By measuring the *intensity* of the gamma radiation, the *concentration* of the element can be determined.

Neutron activation analysis is so sensitive that concentrations as low as $10^{-9}\%$ can be determined. A museum might have a lock of hair of some famous but long dead person suspected of having been slowly murdered by arsenic poisoning. If so, some arsenic would be in the hair, and neutron activation analysis could find it without destroying the specimen.

Radiological Dating

The determination of the age of a geological deposit or an archaeological find by the use of the radionuclides that are naturally present is called **radiological dating.** It is based partly on the premise that the half-lives of radionuclides have been constant throughout the entire geological period. This premise is supported by the finding that half-lives are insensitive to all environmental forces such as heat, pressure, magnetism, or electrical stresses.

In geological dating, a pair of isotopes is sought that are related as a "parent" to a "daughter" in a radioactive disintegration series, like the uranium-238 series (see Figure 20.4). Uranium-238 (as "parent") and lead-206 (as "daughter") have thus been used as a radiological dating pair of isotopes. The half-life of uranium-238 is very long, a necessary criterion for geological dating. After the concentrations of uranium-238 and lead-206 are determined in a rock specimen, the *ratio* of the concentrations together with the half-life of uranium-238 is used to calculate the age of the rock.

For ^{238}U, $t_{1/2} = 4.51 \times 10^9$ yr
For ^{40}K, $t_{1/2} = 1/3 \times 10^9$ yr

Probably the most widely used isotopes for dating rock are the potassium-40/argon-40 pair. Potassium-40 is a naturally occurring radionuclide with a half-life nearly as long as that of uranium-238. One of its modes of decay is electron capture, and argon-40 forms.

$$^{40}_{19}K + ^{0}_{-1}e \longrightarrow ^{40}_{18}Ar$$

The argon produced by this reaction remains trapped within the crystal lattices of the rock and is freed only when the rock sample is melted. How much has accumulated is measured with a mass spectrometer (page 53), and the observed ratio of argon-40 to potassium-40, together with the half-life of the parent, permits the age of the specimen to be estimated. Because the half-lives of uranium-238 and potassium-40 are so long, samples have to be at least 300,000 years old for either of the two parent–daughter isotope pairs to work.

Carbon-14 Dating For dating organic remains, like wooden or bone objects from ancient tombs, *carbon-14 analysis* is used. The ratio of carbon-14 (a beta emitter) to stable carbon in the ancient sample is measured and compared to the ratio in recent organic remains. The data can then be used to calculate the age of the sample.

There are two approaches to carbon-14 dating. The older method, introduced by its discoverer, Willard F. Libby (Nobel Prize in chemistry, 1960), measures the *radioactivity* of a sample taken from the specimen. The radioactivity is proportional to the concentration of carbon-14. The newer method relies on a device resembling a mass spectrometer that is able to separate the atoms of carbon-14 from the other isotopes of carbon (as well as from nitrogen-14) *and count all of them,* not just the carbon-14 atoms that decay. This approach permits the use of smaller samples (0.5–5 mg versus 1–20 g for the Libby method); it works at much higher efficiencies; and it gives more precise dates. Objects of up to 70,000 years old can be dated, but the highest accuracy involves systems no older than 7000 years.

Carbon-14 has been produced in the atmosphere for many thousands of years by the action of cosmic rays on atmospheric nitrogen. Primary cosmic rays are streams of particles from the sun, mostly protons but also the nuclei of many elements (as high as atomic number 26). Their collisions with atmospheric atoms and molecules generate secondary cosmic rays, which include all subatomic particles. Secondary cosmic rays thus contribute to our background radiation, and people living at high altitudes receive more of this radiation than other people.

Neutrons of the proper energy in secondary cosmic rays can transmute nitrogen atoms into carbon-14 atoms according to the following equation (Figure 20.15).

$$\underset{\substack{\text{neutron}\\\text{(in cos-}\\\text{mic ray)}}}{^{1}_{0}n} + \underset{\substack{\text{nitrogen}\\\text{atom in}\\\text{the atmo-}\\\text{sphere}}}{^{14}_{7}N} \longrightarrow \underset{\substack{\text{compound}\\\text{nucleus}}}{^{15}_{7}N^{*}} \longrightarrow \underset{\substack{\text{carbon-14}\\\text{atom (a}\\\text{beta}\\\text{emitter)}}}{^{14}_{6}C} + \underset{\substack{\text{high-energy}\\\text{proton}}}{^{1}_{1}p}$$

The newly formed carbon-14 diffuses into the lower atmosphere. It becomes oxidized to carbon dioxide and enters the Earth's biosphere by means of photosynthesis. Carbon-14 thus becomes incorporated into plant substances and into the materials of animals that eat plants. As the carbon-14 decays, more is ingested by the living thing. The net effect is an overall equilibrium involving carbon-14 in the global system. As long as the plant or animal is alive, its ratio of carbon-14 atoms to carbon-12 atoms is constant.

At death, no new carbon-14 can become part of the organism. The organism's remains have as much of the isotope as they can ever have, and they now slowly lose their carbon-14 by decay. The decay is a first-order process with a rate independent of the *number* of original carbon atoms. The ratio of car-

Extraordinary precautions are taken to ensure that specimens are not contaminated by more recent sources of carbon or carbon compounds.

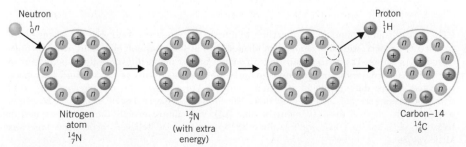

Neutron $^{1}_{0}n$

Nitrogen atom $^{14}_{7}N$

$^{14}_{7}N$ (with extra energy)

Proton $^{1}_{1}H$

Carbon–14 $^{14}_{6}C$

FIGURE 20.15

Carbon-14 forms in the atmosphere when neutrons in secondary cosmic radiation are captured by nitrogen-14 nuclei.

bon-14 to carbon-12, therefore, can be related to the years that have elapsed between the time of death and the time of the measurement. The critical assumption in carbon-14 dating is that the steady-state availability of carbon-14 from the atmosphere has remained largely unchanged since life emerged on Earth.[2]

In contemporary biological samples the ratio $^{14}C/^{12}C$ is about 1.2×10^{-12}. Thus, each fresh 1.0 g sample of biological carbon in equilibrium with the $^{14}CO_2$ of the atmosphere has a ratio of 5.8×10^{10} atoms of carbon-14 to 4.8×10^{22} atoms of carbon-12. The ratio decreases by a factor of 2 for each half-life period of ^{14}C (5730 years). It can be shown that when expressed as a function of time in years, t, the ratio of carbon-14 to carbon-12 is given by the following equation.

$$\frac{^{14}C}{^{12}C} = 1.2 \times 10^{-12}\, e^{-t/8270} \qquad (20.2)$$

Thus, if a sample of some ancient wooden artifact is found to have a ratio of carbon-14 to carbon-12 of 3.0×10^{-13}, the age of the artifact is found as follows, using Equation 20.2.

$$3.0 \times 10^{-13} = 1.2 \times 10^{-12}\, e^{-t/8270}$$

Rearranging,

$$e^{-t/8270} = 0.25$$

Taking the natural logarithms of both sides gives

$$-t/8270 = \ln 0.25 = -1.4$$

Solving for t gives an age of 1.2×10^4 years.

20.8
NUCLEAR FUSION

In the first two sections of this chapter we learned that enormous yields of energy are potentially available from nuclear **fusion,** the fusing of two nuclei. Immensely difficult scientific and engineering problems have to be solved, however, before a working fusion power station can be realized.

Deuterium, an isotope of hydrogen, is a key fuel in all approaches to fusion. It is naturally present as 0.015% of all hydrogen atoms, including those of the molecules of the Earth's waters. Despite this low percentage, the Earth has so much water and the potential energy yield from fusion is so great that the deuterium in just 0.005 cubic kilometers of ocean would supply the energy needs of the United States for 1 year.

The world ocean has a volume of about 1.4×10^9 km³. A volume of 0.005 km³ roughly equals the capacity of 45 petroleum supertankers (700,000 gallons each).

[2] The available atmospheric pool of carbon-14 atoms fluctuates somewhat with the intensities of cosmic ray showers, with slow, long-term changes in Earth's magnetic field, and with the huge injections of carbon-12 into the atmosphere from the large-scale burning of coal and petroleum in the 1900s. To reduce the uncertainties in carbon-14 dating, results of the method have been corrected against dates made by tree-ring counting. For example, an uncorrected carbon-14 dating of a Viking site at L'anse aux Meadows, Newfoundland, gave a date of A.D. 895 ± 30. When corrected, the date of the settlement became A.D. 997, almost exactly the time indicated in Icelandic sagas for Leif Eriksson's landing at "Vinland," now believed to be the L'anse aux Meadows site.

Thermonuclear Fusion

The central scientific problem with fusion is to get the fusing nuclei close enough for a long enough time that the nuclear strong force (of attraction) can overcome the electrostatic force (of repulsion). As we learned in Section 20.3, the strong force acts over a much shorter range than the electrostatic force. Two nuclei on a collision course, therefore, repel each other virtually until they touch and get into the range of the strong force. The energies of two approaching nuclei must therefore be very substantial if they are to overcome this electrostatic barrier. Such energies, moreover, must be achieved by large numbers of nuclei all at once, in batch after batch, if there is to be any practical generation of electrical power by nuclear fusion. Relatively isolated fusion events achieved in huge accelerators will not do. The only practical way to give batch quantities of nuclei enough energy is to transfer *thermal* energy to them, and so the overall process is called *thermonuclear fusion*.

The atoms whose nuclei we want to fuse must first be stripped of their electrons. Thus, a high energy cost is exacted from the start but, overall, the energy yield will more than pay for it. The product is an electrically neutral, gaseous mixture of nuclei and unattached electrons called *plasma*. The plasma must then be made so dense that like-charged nuclei are within 2 fm of each other, meaning a plasma density of roughly 200 g cm^{-3} as compared with 200 mg cm^{-3} under ordinary conditions. To achieve this, the plasma must be confined at a pressure of several billion atmospheres long enough for the separate nuclei to fuse. The temperature needed is several times the temperature at the center of our sun.

2 fm = 2 femtometer
= 2×10^{-15} meter.

Two broad approaches to fusion have been studied, *inertial confinement fusion* and *magnetic confinement fusion*. Both seek the fusion of tritium and deuterium nuclei by the following nuclear reaction.

$$_1^3\text{H} \;+\; _1^2\text{H} \;\longrightarrow\; _2^4\text{He} \;+\; _0^1 n$$

tritium   deuterium   helium   neutron

Using nuclei of hydrogen isotopes keeps electrostatic repulsions to a minimum because they have charges of only 1+. Any other nuclei would have charges double or more in size, and their electrostatic repulsions would make fusion impossible.

Inertial Confinement Fusion

Inertial confinement means enclosing the fuel, tritium and deuterium, within a pellet made of glass, plastic, or metal with a diameter of about 1 mm. The pellet is then given so much energy so rapidly from all angles by the use of focused X rays that the material on its outer surface suddenly vaporizes and itself becomes a plasma. If the explosive action is spherically symmetrical, an implosive reaction to the tritium/deuterium fuel will occur and compress the fuel into a much denser, very hot ball with a radius much smaller than the fuel initially had. Now fusion could occur and more energy would be released than needed to launch these steps. There would be a net gain in energy.

Magnetic Confinement Fusion

One of the most actively studied approaches to controlled nuclear fusion uses a magnetic "bottle" for the confinement of the hot, high-density plasma long enough for fusion. No known material has the strength to withstand the re-

FIGURE 20.16

Fusion by magnetic confinement. The design of a tokamak fusion reactor is seen in this cutaway.

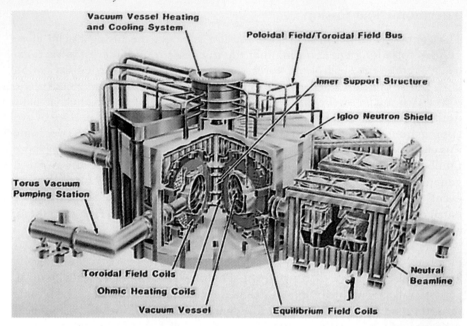

The paths of moving, electrically charged particles are changed by magnetic forces.

Tokamak is a Russian term for a magnetic confinement device.

quired temperatures and pressures of the plasma, so scientists studying magnetic confinement have devised ways to shape and hold magnetic forces in configurations that prevent entrapped plasma from leaking out. Figure 20.16 illustrates a device, called a *tokamak,* which gives the magnetic bottle a torus or doughnut shape. The plasma can be heated well above the requisite temperature by being bombarded with intense beams of electrically neutral deuterium atoms, using special injectors. Of high importance to the success of magnetic confinement fusion is the retention and use of some of the thermal energy of the alpha particles for heating the plasma and additional fuel. Without the retention of this heat, the temperature needed for fusion could not be attained, but the thermal energy of the alpha particles is apparently captured. In late 1994, the Tokamak Fusion Test Reactor (Princeton University) generated a short burst of 10.7×10^6 watts of power, a record high. The goal of a *net* production of usable energy remains far off, however. Further research depends on more advanced tokamaks.

How thermonuclear fusion is believed to occur in the sun is described in Special Topic 20.2. Thermonuclear fusion is also the source of energy released in the explosion of a hydrogen bomb. The energy needed to trigger the fusion is provided by the explosion of a fission bomb based on either uranium or plutonium.

20.9 NUCLEAR FISSION

Because of their electrical neutrality, neutrons penetrate an atom's electron cloud relatively easily and reach the nucleus. Enrico Fermi discovered in the early 1930s that even slow-moving, *thermal neutrons* can be captured. (Thermal neutrons are those whose average kinetic energy puts them in thermal equilibrium with their surroundings at room temperature.)

Fission Reactions

When Fermi directed thermal neutrons at a uranium target, several different species of nuclei were produced, a result that Fermi could not explain. None

SPECIAL TOPIC 20.2 / THERMONUCLEAR FUSION IN THE SUN

At the center of the sun, according to present estimates, the temperature is about 1.5×10^7 K, the pressure is about 2×10^{11} atm, and the density is about 1.5×10^5 kg m^{-3} (roughly 13 times the density of lead). The matter there, all in the gaseous state, consists mostly of a plasma of protons (35% by mass), alpha particles (65% by mass), electrons, and photons. The conditions are suitable for thermonuclear fusion, and the chief process is called the proton–proton cycle.

The positrons produced combine with electrons in the plasma, annihilate each other, and generate gamma radiation. Virtually all the neutrinos escape the sun and move into the solar system (and beyond), carrying with them a little less than 2% of the energy generated by the cycle. Not counting the energy of the neutrinos, each operation of one cycle generates 26.2 MeV or 4.20×10^{-12} J of energy, which is equivalent to 2.53×10^{12} J per *mole* of alpha particles produced. This is the source of the solar energy radiated throughout our system, which can continue in this way for another 5 billion years.

THE PROTON–PROTON CYCLE

$$2[{}_1^1H + {}_1^1H] \longrightarrow 2{}_1^2H + 2{}_1^0e + 2\nu$$

$$2[{}_1^1H + {}_1^2H] \longrightarrow 2{}_2^3He$$

$$\underline{{}_2^3He + {}_2^3He \longrightarrow {}_2^4He + {}_1^1H + {}_1^1H}$$

$$\text{Net:} \quad 4{}_1^1H \longrightarrow {}_2^4He + 2{}_1^0e + 2\nu$$

had an atomic number above 82. He and other physicists did not follow up on a suggestion by Ida Noddack-Tacke, a chemist, that nuclei of much lower atomic number probably formed. The forces holding a nucleus together were considered too strong to permit such a result.

Discovery of Fission Two German chemists, Otto Hahn and Fritz Strassman, finally verified in 1939 that uranium fission produces several isotopes much lighter than uranium, for example, barium-140 (atomic number 56). Shortly after the report of Hahn and Strassman, two German physicists, Lise Meitner and Otto Frisch, were able to show that a nucleus of one of the rare isotopes in natural uranium, uranium-235, splits into roughly two equal parts after it captures a slow neutron. They named this breakup nuclear **fission.** The general reaction can be represented as follows.

$$\text{Otto Hahn received the 1944 Nobel Prize in chemistry.}$$

$${}_{92}^{235}U + {}_0^1n \longrightarrow X + Y + b{}_0^1n$$

X and Y can be a large variety of nuclei with intermediate atomic numbers. Over 30 have been identified. The coefficient b has an average value of 2.47, the average number of neutrons produced by fission events. A typical specific fission is

$${}_{92}^{235}U + {}_0^1n \longrightarrow {}_{92}^{236}U^* \longrightarrow {}_{36}^{94}Kr + {}_{56}^{139}Ba + 3{}_0^1n$$

What actually undergoes fission is the compound nucleus of uranium-236. It has 144 neutrons and 92 protons for a neutron/proton ratio of roughly 1.6. Initially, the emerging krypton and barium isotopes have the same ratio, and this is much too high for them. The neutron/proton ratio for stable isotopes with 36 to 56 protons is nearer 1.2 to 1.3 (see Figure 20.8). Therefore, the initially formed, neutron-rich krypton and barium nuclides promptly eject neutrons, called *secondary neutrons,* that generally have much higher energies than thermal neutrons.

An isotope that can undergo fission after neutron capture is called a **fissile isotope.** The naturally occurring fissile isotope of uranium used in reactors is uranium-235, whose abundance among the uranium isotopes today is only 0.72%. Two other fissile isotopes, uranium-233 and plutonium-239, can be made in nuclear reactors.

Nuclear Chain Reactions   The secondary neutrons released by fission become thermal neutrons as they are slowed by collisions with surrounding materials. They can now be captured by unchanged uranium-235 atoms. Because each fission event produces, on the average, more than two new neutrons, the potential exists for a **nuclear chain reaction** (Figure 20.17). A *chain reaction* is a self-sustaining process whereby products from one event cause one or more repetitions. If the sample of uranium-235 is small enough, the loss of neutrons to the surroundings is sufficiently rapid to prevent a chain reaction. However, at a certain *critical mass* of uranium-235, a few kilograms, this loss of neutrons is insufficient to prevent a sustained reaction. A virtually instantaneous fission of the sample ensues, in other words, an atomic bomb explosion. To trigger an atomic bomb, therefore, two subcritical masses of uranium-235 (or plutonium-239) are forced together to form a critical mass.

$$1 \text{ MeV} = 1.602 \times 10^{-13} \text{ J}$$

Energy Yield from Fission   The binding energy per nucleon (Figure 20.1) in uranium-235 (about 7.6 MeV) is lower than the binding energies of the new nuclides (about 8.5 MeV). The net change for a single fission event can be calculated as follows (where we intend only *two* significant figures in each result).

FIGURE 20.17

Nuclear chain reaction. Whenever the concentration of a fissile isotope is high enough (at or above the critical mass), the neutrons released by one fission can be captured by enough unchanged nuclei to cause more than one additional fission event. In civilian nuclear reactors, the fissile isotope is too dilute for this to get out of control. In addition, control rods of nonfissile materials that are able to capture excess neutrons can be inserted or withdrawn from the reactor core to make sure that the heat generated can be removed as fast as it forms.

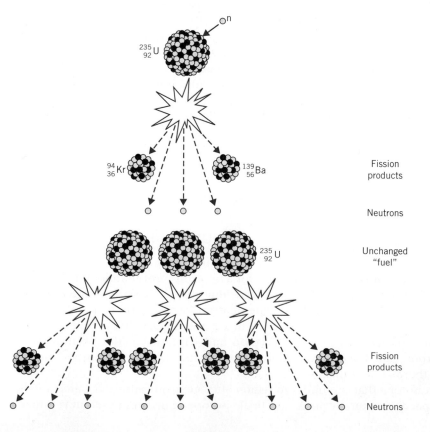

Binding energy in krypton-94:
$$(8.5 \text{ Me V/nucleon}) \times 94 \text{ nucleons} = 800 \text{ MeV}$$
Binding energy in barium-139:
$$(8.5 \text{ MeV/nucleon}) \times 139 \text{ nucleons} = \underline{1200 \text{ MeV}}$$
$$\text{Total binding energy of products} = 2000 \text{ MeV}$$
Binding energy in uranium-235:
$$(7.6 \text{ MeV/nucleon}) \times 235 \text{ nucleons} = 1800 \text{ MeV}$$

The difference in total binding energy is (2000 MeV − 1800 MeV), or 200 MeV, which has to be taken as just a rough calculation. This is the energy released by each fission event going by the equation given. The energy produced by the fission of 1 kg of uranium-235 calculates to be roughly 8×10^{13} J, enough to keep a 100-watt light bulb in energy for 3000 years.

Electrical Energy from Fission

Virtually all civilian nuclear power plants throughout the world operate on the same general principles. The heat of fission is used either directly or indirectly to increase the pressure of some gas that then drives an electrical generator.

The energy available from 1 kg of uranium-235 is equivalent to the energy of 3000 tons of soft coal or 13,200 barrels of oil.

The heart of a nuclear power plant is the *reactor,* where fission takes place in the *fuel core.* The nuclear fuel is generally uranium oxide, enriched to 2–4% in uranium-235 and formed into glasslike pellets. These are housed in long, sealed metal tubes called *cladding.* Bunches of tubes are assembled in spacers that permit a coolant to circulate around the tubes. A reactor has several such assemblies in its fuel core. The coolant carries away the heat of fission.

To convert secondary neutrons to thermal neutrons, the fuel core has a *moderator,* which is the coolant water itself in nearly all civilian reactors. Collisions between secondary neutrons and moderator molecules heat up the moderator. This heat energy eventually generates steam that enables an electric turbine to run. Ordinary light water (H_2O) is a good moderator, but so are the heavy water (D_2O) and graphite.

Two main types of reactors dominate civilian nuclear power, the *boiling water reactor* and the *pressurized water reactor.* Both use light water as the moderator and so are sometimes called *light water reactors.* Roughly two-thirds of the reactors in the United States are the pressurized water type (Figure 20.18).

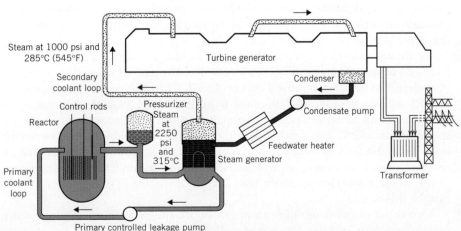

FIGURE 20.18

Pressurized water reactor, the type used in most of the nuclear power plants in the United States. Water in the primary coolant loop is pumped around and through the fuel elements in the core, and it carries away the heat of the nuclear chain reactions. The hot water delivers its heat to the cooler water in the secondary coolant loop, where steam is generated to drive the turbines. (Drawing from WASH-1261, U.S. Atomic Energy Commission, 1973.)

Such a reactor has two loops, and water circulates in both. The primary loop moves water through the reactor core, where it picks up thermal energy from fission. The water is kept in the *liquid* state by being under high pressure (hence the name *pressurized* water reactor).

The hot water in the primary loop transfers thermal energy to the secondary loop at the steam generator (Figure 20.18). This makes steam at high temperature and pressure, which is piped to the turbine. As the steam drives the turbine, the steam pressure drops. The condenser at the end of the turbine, cooled by water circulating from a river or lake or from huge cooling towers, forces a maximum pressure drop within the turbine by condensing the steam to liquid water. The returned water is then recycled to high-pressure steam. (In the boiling water reactor, there is only one coolant loop. The water heated in the reactor itself is changed to the steam that drives the turbine.)

Safety Features of Nuclear Reactors

There is no danger of a reactor undergoing an atomic bomb explosion. An atomic bomb requires pure uranium-235 or plutonium-239. The concentration of fissile isotopes in a reactor is in the range of 2 to 4%, and much of the remainder is the common, nonfissile uranium-238. It is not possible, therefore, for a critical mass to form, even if all the fuel melts into one pool.

The safe operation of a reactor requires that the multiplication of slow neutrons be controlled throughout the *fission cycle.* This cycle begins with the production of fast neutrons (with some leakage away from the fissile material), then their moderation (also with some leakage), and next the use of the surviving neutrons to cause additional fission events. These produce more fast neutrons, and additional rounds of the cycle occur.

The ratio of neutrons at the end of the cycle to those that started it is called the *multiplication factor, k.* During smooth, safe, continuous operation, the value of k in the reactor should be exactly 1. When it is, the reactor is said to be *critical.* When $k > 1$, fission is accelerating and the reactor is *supercritical.* When $k < 1$, fission is slowing down, and the reactor is *subcritical.* For safe startup, energy production, and shutdown, the reactor is designed with the built-in capacity to go supercritical for brief periods, then be returned to its critical mode, and finally made to go subcritical for shutting down. To make this possible, the reactor core incorporates *control rods,* typically made of cadmium or boron. When fully inserted among the cladding tubes, the control rods capture enough neutrons to keep the reactor subcritical. If removed long enough, the reactor goes supercritical. By moving the control rods in and out, the reactor itself is controlled.

As fission continues, wastes build up in the cladding. They dilute unchanged atoms of uranium-235 and thus slow the rate of fission. In time, the fuel elements must be removed and replaced. They still contain uranium-235 as well as some plutonium-239, made from uranium-238 as the reactor operates (to be discussed in Special Topic 20.3). Both are fissile isotopes, so unspent uranium-235 and newly made plutonium-239 ideally should be recovered from the fuel elements for recycling as fuel. A major issue with civilian nuclear power, however, is the diversion of plutonium-239 to military purposes. Some countries no doubt have acquired the fissile material for atomic bombs this way.

Two other major issues of nuclear power are the prevention of a loss of the primary coolant and the disposal of radioactive wastes.

Loss of Coolant Emergencies

No single aspect of a reactor's operation is more important than the removal of the heat generated by fission as fast as it is produced. The primary coolant must do this or the temperature of the core will increase very rapidly. The conversion of any water remaining in the primary loop into steam could quickly develop enough pressure to blow the reactor apart. This happened to one of the reactors of the Chornobyl power plant in the Ukraine on April 26, 1986.

Even if insufficient steam is present from which to generate enough pressure for a steam explosion, the temperature in the core can make the cladding melt. The molten mass, in which fission energy continues to be generated, could melt its way through the containment structure and on into the ground.

To prevent a core meltdown and a breach of the containment vessel, engineers have installed emergency core cooling systems designed to flood the reactor with water at the first sign of impending danger, before the failure of cladding. While this emergency water temporarily halts a steep increase in temperature, other action is taken to shut the reactor down.

The reactor that blew held 3×10^9 Ci at the time of the accident. About 3 to 4% of this was released. ("Chornobyl" is the *National Geographic* spelling; others use "Chernobyl.")

Radioactive Wastes

Radioactive wastes from nuclear power plants occur as gases, liquids, and solids. The gases are mostly radionuclides of krypton and xenon but, with the exception of xenon-85 ($t_{1/2}$ 10.4 years), the gases have short half-lives and decay quickly. During decay, they must be contained, and this is one function of the cladding. Other dangerous radionuclides produced by fission include iodine-131, strontium-90, and cesium-137.

Milk from dairy cows feeding on pastures exposed to iodine-131 fallout will contain this isotope. Iodine-131 in milk consumed by people is a problem because, as we said earlier, the human thyroid gland concentrates iodide ion to make the hormone thyroxine. Once in the gland, beta radiation from iodine-131 could cause harm, possibly thyroid cancer, possibly impaired thyroid function. An effective countermeasure to iodione-131 poisoning is to take doses of ordinary iodine (as sodium iodide). Statistically, this increases the likelihood that the thyroid will take up stable iodide ion rather than the unstable anion of iodine-131.

Cesium-137 and strontium-90 also pose problems to humans. Cesium is in Group IA together with sodium, so radiating cesium-137 cations travel to some of the same places in the body where sodium ions go. Strontium is in Group IIA with calcium, so strontium-90 cations can replace calcium ions in bone tissue, sending radiation into bone marrow and possibly causing leukemia. The half-lives of cesium-137 and strontium-90, however, are relatively short. Some radionuclides in wastes are so long-lived that solid reactor wastes must be kept away from all human contact for dozens of centuries, longer than any nation has ever yet endured.

Cesium-137 has a half-life of 30 years; that of strontium-90 is 28.1 years. Both are beta emitters.

Critics of nuclear power argue that until the problem of safely storing radioactive wastes is solved, no new power plants should be authorized. Proponents believe that the problem will safely be solved. In the meanwhile, in the early 1990s no new nuclear power plants were being planned in the United States, almost entirely for economic reasons. Research to make light water reactors safer has continued, however, and reactors of quite different design are being studied.

The entire reactor of a power plant becomes a solid waste problem when its working life is over.

Another approach to nuclear power involves the conversion of nonfissile radionuclides into fissile nuclides in *breeder reactors* (see Special Topic 20.3).

SPECIAL TOPIC 20.3 / THE BREEDER REACTOR

Naturally occurring uranium is 99.275% uranium-238, 0.720% uranium-235, and 0.005% uranium-234. When a reactor uses uranium fuel, naturally occurring uranium is processed to increase the concentration of uranium-235 to 2 to 4%, which means that 96 to 98% of the uranium atoms in the fuel are the nonfissile uranium-238. Therefore, in any operating nuclear reactor fueled with uranium, the uranium-238 atoms are mostly "neutral stuff." We say "mostly" because the following nuclear reaction of uranium-238 occurs to a small extent, and it produces a fissile isotope, plutonium-239.

$$ {}_0^1n + {}_{92}^{238}\text{U} \longrightarrow {}_{92}^{239}\text{U}^* \longrightarrow {}_{94}^{239}\text{Pu} + 2\,{}_{-1}^{0}e + 2\bar{\nu} $$

Uranium-fueled nuclear reactors thus manufacture plutonium fuel out of nonfissile uranium-238. The plutonium-239 can be extracted and used as fuel, too.

If plutonium-239 were mixed with uranium-238 in a reactor, in principle it would be possible to exploit all of the uranium-238 as a source of fuel. The plutonium-239 would serve as a catalyst because it would not be consumed. As seen in Figure 1, for every atom of plutonium-239 that undergoes fission and generates energy, another atom of plutonium-239 is made from an atom of uranium-238. The system thus breeds its own fuel from what initially was not a fuel, uranium-238, and a reactor capable of this is called a **breeder reactor.** Enough uranium-238 is on hand today, stockpiled from the processing of spent nuclear fuel elements, to supply the entire energy needs of the United States for centuries using breeder reactor technology. One of the major problems with the breeder is that it produces plutonium-239, which can be diverted to make atomic bombs. The machinery for effective international control of weapons-grade fissile material is not working to the satisfaction of many.

Although a breeder reactor breeds fuel, the system is not "free." Many expenses are involved, enough so that in relation to the cost of fuel for ordinary reactors the cost per megawatt of power from breeders today is judged too high compared with any other source of power. The successful experimental breeder reactor in Great Britain was shut down permanently in 1994. France is still using the technology. Japan's experimental breeder reactor has succeeded, but commercial developments probably will not be seen for three or more decades. The United States (as of 1996) has no fast breeder reactor development program.

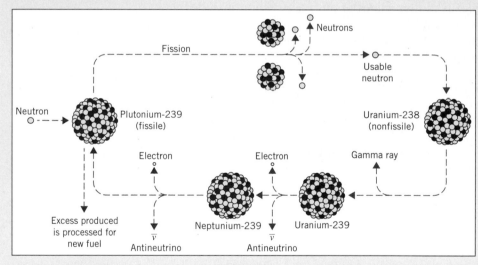

FIGURE 1

The breeding cycle whereby uranium-238, which is nonfissile, is changed into plutonium-239, which is fissile.

SUMMARY

The Einstein Equation Einstein realized that mass and energy are interconvertible. The **Einstein equation,** $\Delta E = \Delta mc^2$ (where c is the speed of light), lets us calculate one from the other. The total of the energy in the universe and all the mass calculated as an equivalent of energy is a constant, which is the **law of conservation of mass–energy.**

Nuclear Binding Energies When a nucleus forms from its nucleons, some mass changes into energy. This amount of energy, the **nuclear binding energy,** would be required to break up the nucleus again. The higher the binding energy per nucleon, the more stable is the nucleus.

Radioactivity The *electrostatic force* by which protons repel each other is overcome in the nucleus by the *nuclear strong force.* The ratio of neutrons to protons is a factor in nuclear stability. By radioactivity, the naturally occurring **radionuclides** adjust their neutron/proton ratios, lower their energies, and so become more stable by emitting **alpha** or **beta radiation,** sometimes gamma radiation as well.

The loss of an **alpha particle** leaves a nucleus with four fewer units of mass number and two fewer of atomic number. Loss of a **beta particle** leaves a nucleus with the same mass number and an atomic number one unit higher. **Gamma radiation** lets a nucleus lose some energy without a change in mass or atomic number.

Depending on the specific isotope, synthetic radionuclides emit alpha, beta, and gamma radiation. Some emit **positrons** (positive electrons) that produce gamma radiation by annihilation collisions with electrons. Other synthetic radionuclides decay by **electron capture** and emit X rays. Some radionuclides emit neutrons.

Nuclear equations are balanced when the mass numbers and atomic numbers on either side of the arrow respectively balance. The energies of emission are usually described in **electron volts** or multiples thereof.

A few very long-lived radionuclides in nature, like ^{238}U, are at the heads of **radioactive disintegration series,** which represent the successive decays of "daughter" radionuclides until a stable isotope forms.

Nuclear Stability Stable nuclides generally fall within a curving band, called the **band of stability,** when all known nuclides are plotted according to their numbers of neutrons and protons. Radionuclides with too high neutron/proton ratios eject beta particles to adjust their ratios downward. Those with neutron/proton ratios too low generally emit positrons to change their ratios upward.

Isotopes whose nuclei consist of even numbers of both neutrons and protons are generally much more stable than those with odd numbers of both; this is the **odd–even rule.** Having all neutrons paired and all protons paired is energetically better (more stable) than having any nucleon unpaired.

Isotopes with specific numbers of protons or neutrons, the **magic numbers** of 2, 8, 20, 28, 50, 82, and 126, are generally more stable than others.

Transmutation When a bombardment particle—a proton, deuteron, alpha particle, or a neutron—is captured, the resulting **compound nucleus** contains the energy of both the captured particle and its nucleons. The mode of decay of the compound nucleus is a function of its extra energy, not its extra mass. Many radionuclides have been made by these nuclear reactions, including all of the elements from 93 and higher.

Detecting and Measuring Radiations Instruments to detect and measure radiation—Geiger counters, for example —take advantage of the ability of radiation to generate ions in air or other matter.

The electron volt describes how much energy a radiation carries; 1 eV = 1.602×10^{-19} J.

The **curie (Ci)** and the **becquerel (Bq)** describe how active a source is; 1 Ci = 3.7×10^{10} disintegrations/s, and 1 Bq = 1 disintegration/s.

The SI unit of absorbed dose, the **gray (Gy),** is used to describe how much energy is absorbed by a unit mass of absorber; 1 Gy = 1 J/kg. An older unit, the **rad,** is equal to 0.01 Gy.

The SI unit of the **sievert** and the older unit of the **rem** are used to compare doses absorbed by different tissues and caused by different kinds of radiation. A 600 rem whole-body dose is lethal.

The normal **background radiation** causes millirem exposures per year. Naturally occurring radon, cosmic rays, radionuclides in soil and stone building materials, medical X rays, and releases from nuclear tests or from nuclear power plants all contribute to this background.

Protection against radiation can be achieved by using dense shields (e.g., lead or thick concrete), by avoiding overuse of radionuclides or X rays in medicine, and by taking advantage of the **inverse square law.** This law tells us that the intensity of radiation decreases with the square of the distance from its source.

Applications Radiological dating uses the known half-lives of naturally occurring radionuclides to date geological and archeological objects. During life, living systems take up carbon-14. Following the organism's death, the age of the remains can be determined from the ratio of carbon-12 to carbon-14.

Nuclear Fusion To achieve the **fusion** of the nuclei of isotopes of hydrogen, their atoms must be heated to a temperature high enough to form a plasma—a neutral, gaseous mixture of nuclei and electrons. The plasma pressure must make the plasma so dense that bare nuclei can get sufficiently near each other that the nuclear strong force overcomes the repulsive electrostatic force between protons. *Inertial confinement* of the fuel in pellets during laser bombardment is one approach to fusion, but *magnetic confinement* of the plasma is the technology being most studied.

Fission Uranium-235, which occurs naturally, and plutonium-239, which can be made from uranium-238, are **fissile isotopes** that serve as the fuel in present-day reactors. When either isotope captures a thermal neutron, the isotope splits in one of several ways to give two smaller isotopes plus energy and more neutrons. The neutrons can generate additional fission events, enabling a nuclear chain reaction. If a critical mass of a fissile isotope is allowed to form, the **nuclear chain reaction** proceeds out of control, and the material detonates as an atomic bomb explosion.

Fission Reactors The goal of controlling the chain reaction and keeping it from becoming supercritical has resulted in various designs for commercial reactors. *Pressurized water reactors* are most commonly used, and these have two loops of circulating fluids. In the primary loop, water circulates around the reactor core and absorbs the heat of fission. In the secondary loop, water accepts the heat and changes to high-pressure steam, which drives the electrical generator. If the circulation of fluid in the primary loop fails, the reactor must be shut down immediately to prevent a core meltdown and a possible breach of the reactor's containment vessel.

The major problems with nuclear energy are the storage of radioactive wastes, the potential for loss-of-coolant accidents, the diversion of plutonium-239 to weapons (nuclear proliferation), and current high economic costs of building safe power plants.

Tools You Have Learned

The table below lists the tools you have learned in this chapter that are applicable to problem solving. Review them if necessary, and refer to them when working on the Thinking-It-Through problems and the Review Exercises that follow.

Tool	Function
Einstein equation (page 848)	To calculate the energy equivalent of a given amount of mass, or the mass that is equivalent to a given amount of energy.
Rules for balancing nuclear equations (page 852)	The sums of the mass numbers and of the atomic numbers on each side of the equals sign must equal.
Odd–even rule (page 860)	To make judgments concerning the relative stabilities of isotopes.
Inverse square law (page 869)	To calculate how a given radiation intensity will change by changing the distance from the source.

THINKING IT THROUGH

The goal for each of the following problems is to give you practice in thinking your way through problems. The goal is not *to find the answer itself; instead, you are only asked to assemble the available information needed to obtain the answer, state what additional data (if any) are needed, and describe how you would use the data to answer the question.*

1. What specific information must be looked up in a table in order to balance the nuclear equation for the alpha decay of americium-241?

2. If it is known that lead-206 is the end product of a radioactive disintegration series, what do we know about the lead-206 atom (besides its mass)?

3. Without using a table, suggest the likeliest mass number for the most stable isotope of the element of atomic number 20.

4. If you want to reduce your exposure to the radiation of some radioactive source by one-half, what can you do without changing the source or interposing shielding devices?

REVIEW EXERCISES

Answers to questions whose numbers are printed in color are given in Appendix D. More challenging questions are marked with asterisks.

Conservation of Mass – Energy

20.1 When a substance is described as *radioactive*, what do we know about it?

20.2 In chemical calculations involving chemical reactions we can regard the law of conservation of mass as a law independent of the law of conservation of energy despite Einstein's union of the two. What fact(s) make this possible?

20.3 According to Einstein, how can we know that the absolute upper limit on the speed of any object is the speed of light?

20.4 Calculate the mass in kilograms of a 1.00 kg object when its velocity, relative to us, is (a) 3.00×10^7 m s^{-1}, (b) 2.90×10^8 m s^{-1}, and (c) 2.99×10^8 m s^{-1}. (Notice the progression of these numbers toward the velocity of light, 3.00×10^8 m s^{-1}.)

20.5 State the following.
(a) law of conservation of mass-energy
(b) Einstein equation

20.6 Calculate the mass equivalent in grams of 1.00 kJ.

20.7 Show that the mass equivalent to the energy released by the combustion of 1 mol of methane (890 kJ) is 9.89 ng.

20.8 Calculate the quantity of mass in nanograms that is changed into energy when one mole of liquid water forms by the combustion of hydrogen, all measurements being made at 1 atm and 25 °C.

Nuclear Binding Energies

20.9 Why isn't the sum of the masses of all nucleons in one nucleus equal to the mass of the actual nucleus?

20.10 Calculate the binding energy in J/nucleon of the deuterium nucleus whose mass is 2.0135 u.

20.11 Calculate the binding energy in J/nucleon of the tritium nucleus whose mass is 3.01550 u.

Radioactivity

20.12 Three kinds of radiation make up nearly all of the radiation observed from naturally occurring radionuclides. What are they?

20.13 Give the composition of each of the following.
(a) alpha particle   (c) positron
(b) beta particle   (d) deuteron

20.14 Why is the penetrating ability of alpha radiation less than that of beta or gamma radiation?

20.15 Complete the following nuclear equations by writing the symbols of the missing particles.
(a) $^{211}_{82}\text{Pb} \rightarrow ^{0}_{-1}e + \underline{\quad}$
(b) $^{177}_{73}\text{Ta} \xrightarrow{\text{electron capture}} \underline{\quad}$
(c) $^{220}_{86}\text{Rn} \rightarrow ^{4}_{2}\text{He} + \underline{\quad}$
(d) $^{19}_{10}\text{Ne} \rightarrow ^{0}_{1}e + \underline{\quad}$

20.16 Write the symbols of the missing particles to complete the following nuclear equations.
(a) $^{245}_{96}\text{Cm} \rightarrow ^{4}_{2}\text{He} + \underline{\quad}$

(b) $^{146}_{56}\text{Ba} \rightarrow ^{0}_{-1}e + \underline{\quad}$ (d) $^{68}_{32}\text{Ge} \xrightarrow{\text{electron capture}} \underline{\quad}$
(c) $^{58}_{29}\text{Cu} \rightarrow ^{0}_{1}e + \underline{\quad}$

20.17 Write a balanced nuclear equation for each of the following changes.
(a) alpha emission from plutonium-242
(b) beta emission from magnesium-28
(c) positron emission from silicon-26
(d) electron capture by argon-37

20.18 Write the balanced nuclear equation for each of the following nuclear reactions.
(a) electron capture by iron-55
(b) beta emission by potassium-42
(c) positron emission by ruthenium-93
(d) alpha emission by californium-251

20.19 Write the symbols, including the atomic and mass numbers, for the radionuclides that would give each of the following products.
(a) fermium-257 by alpha emission
(b) bismuth-211 by beta emission
(c) neodymium-141 by positron emission
(d) tantalum-179 by electron capture

20.20 Each of the following nuclides forms by the decay mode described. Write the symbols of the parents, giving both atomic and mass numbers.
(a) rubidium-80 formed by electron capture
(b) antimony-121 formed by beta emission
(c) chromium-50 formed by positron emission
(d) californium-253 formed by alpha emission

20.21 Write the symbol of the nuclide that forms from cobalt-58 when it decays by electron capture.

20.22 With respect to their modes of formation, how do gamma rays and X rays differ?

20.23 What happens in electron capture to generate X rays?

Nuclear Stability

20.24 What data are plotted and what criterion is used to identify the actual band in the band of stability?

20.25 Both barium-123 and barium-140 are radioactive, but which is more likely to have the *longer* half-life? Explain your answer.

20.26 Tin-112 is a stable nuclide but indium-112 is radioactive and has a very short half life ($t_{1/2} = 14$ min). What does tin-112 have that indium-112 does not to account for this difference in stability?

20.27 Lanthanum-139 is a stable nuclide but lanthanum-140 is unstable ($t_{1/2} = 40$ hr). What rule of thumb concerning nuclear stability is involved?

20.28 As the atomic number increases, the neutron/proton ratio increases. What does this suggest is a factor in nuclear stability?

20.29 Radionuclides of high atomic number are far more likely to be alpha emitters than those of low atomic number. Offer an explanation for this phenomenon.

20.30 Although lead-164 has two magic numbers, 82 protons and 82 neutrons, it is unknown. Lead-208, however, is known and stable. What problem accounts for the nonexistence of lead-164?

20.31 What decay particle is emitted from a nucleus of low to intermediate atomic number but a relatively high neutron/proton ratio? How does the emission of this particle benefit the nucleus?

20.32 What decay particle is emitted from a nucleus of low to intermediate atomic number but a relatively low neutron/proton ratio? How does the emission of this particle benefit the nucleus?

20.33 What does electron capture do to the neutron/proton ratio in a nucleus, increase it, decrease it, or leave it alone? Which kinds of radionuclides are more likely to undergo this change, those above or those below the band of stability?

20.34 If an atom of potassium-38 had the option of decaying by positron emission or beta emission, which route would it likely take, and why? Write the nuclear equation.

20.35 Suppose that an atom of argon-37 could decay by either beta emission or electron capture. Which route would it likely take, and why? Write the nuclear equation.

20.36 If we begin with 3.00 mg of iodine-131 ($t_{1/2} =$ 8.07 hr), how much remains after 6 half-life periods?

20.37 A sample of technetium-99m with a mass of 9.00 ng will have decayed to how much of this radionuclide after 4 half-life periods (about 1 day)?

Transmutations

20.38 When vanadium-51 captures a deuteron ($^2_1$H), what compound nucleus forms? (Write the symbol.) This particle expels a proton (^1_1p). Write the nuclear equation for the overall change from vanadium-51.

20.39 The alpha-particle bombardment of fluorine-19 generates sodium-22 and neutrons. Write the nuclear equation, including the intermediate compound nucleus.

20.40 Gamma-ray bombardment of bromine-81 causes a transmutation in which neutrons are one product. Write the symbol of the other product.

20.41 Neutron bombardment of cadmium-115 results in neutron capture and the release of gamma radiation. Write the nuclear equation.

20.42 When manganese-55 is bombarded by protons, neutrons are released. What else forms? Write the nuclear equation.

Detecting and Measuring Radiations

20.43 What specific property of radiation is used by the Geiger counter?

20.44 Dangerous doses of radiation can actually involve very small quantities of energy. Explain.

20.45 What units, SI and common, are used to describe each of the following?

(a) the *activity* of a radioactive sample
(b) the *energy* of a particle or of a photon of radiation
(c) the quantity of energy absorbed by a given mass
(d) dose equivalents for comparing biological effects

20.46 A sample giving 3.7×10^{10} disintegrations/s has what activity in Ci and in Bq?

20.47 In general terms, how is the rem related to the rad? Why was the rem invented?

Applications of Radionuclides

20.48 Why should a radionuclide used in diagnostic work have a short half-life? If the half-life is too short, what problem arises?

20.49 An alpha emitter is not used in diagnostic work. Why?

20.50 In general terms, explain how neutron activation analysis is used and how it works.

20.51 What is one assumption in the use of the uranium–lead ratio for dating ancient geologic formations?

20.52 List some of the radiation that make up our background radiation.

20.53 A sample of waste has a radioactivity, caused solely by strontium-90 (beta emitter, $t_{1/2}$ 28.1 year), of 0.245 Ci g^{-1}. How many years will it take for its activity to decrease to 1.00×10^{-6} Ci g^{-1}?

20.54 A worker in a laboratory unknowingly became exposed to a sample of radiolabeled sodium iodide made from iodine-131 (beta emitter, $t_{1/2} = 8.07$ day). The mistake was realized 28.0 days after the accidental exposure at which time the activity of the sample was 25.6×10^{-5} Ci/g. The safety officer needed to know how active the sample was at the time of the exposure. Calculate this value in Ci/g.

20.55 Technetium-99m (gamma emitter, $t_{1/2} = 6.02$ hr) is widely used for diagnosis in medicine. A sample prepared in the early morning for use that day had an activity of 4.52×10^{-6} Ci. What will its activity be at the end of the day, that is, after 8.00 hr?

20.56 If exposure from a distance of 1.60 m gave a worker a dose of 8.4 rem, how far should the worker move away from the source to reduce the dose to 0.50 rem for the same period?

20.57 During work with a radioactive source, a worker was told that he would receive 50 mrem at a distance of 4.0 m during 30 min of work. What would be the received dose if the worker moved closer, to 0.50 m for the same period?

20.58 A 500 mg sample of rock was found to have 2.45×10^{-6} mol of potassium-40 ($t_{1/2} = 1.3 \times 10^9$ yr) and 2.45×10^{-6} mol of argon-40. How old was the rock? (What assumption is made about the origin of the argon-40?)

20.59 If a rock sample was found to contain 1.16×10^{-7} mol of argon-40, how much potassium-40 would also have to be present for the rock to be 1.3×10^9 years old?

20.60 A tree killed by being buried under volcanic ash was found to have a ratio of carbon-14 atoms to carbon-12 atoms of 4.8×10^{-14}. How long ago did the eruption occur?

*20.61 If a sample used for carbon-14 dating is contaminated by air, there is a potentially serious problem with the method. What is it?

Nuclear Fusion

20.62 The word *plasma* is used in discussions of nuclear fusion. What does this word mean?

20.63 Write the nuclear equation for the fusion of tritium and deuterium.

20.64 What is a tokamak?

Nuclear Fission

20.65 Why is it easier for a nucleus to capture a neutron than a proton?

20.66 What do each of the following terms mean?
(a) thermal neutron
(b) nuclear fission
(c) fissile isotope

20.67 Which fissile isotope occurs in nature?

20.68 Which fissile isotope is commonly a by-product in the operation of a nuclear reactor using uranium-235 fuel (mixed with uranium-238)?

20.69 What fact about the fission of uranium-235 makes it possible for a *chain reaction* to occur?

20.70 Explain in general terms why fission generates more neutrons than needed to initiate it.

20.71 Complete the following nuclear equation by supplying the symbol for the other product of the fission.

$$^{235}_{92}U + ^{1}_{0}n \longrightarrow ^{94}_{38}Sr + \underline{\hspace{1cm}} + 2^{1}_{0}n$$

20.72 Both products of the fission in Review Exercise 20.71 are unstable. According to Figures 20.8 and 20.9, what is the most likely way for each of them to decay, by alpha emission, beta emission, or by positron emission? Explain.

20.73 What are some of the possible fates of the extra neutrons produced by the fission shown in Review Exercise 20.71?

20.74 Why would there be a *subcritical mass* of a fissile isotope? (Why isn't *any* mass of uranium-235 critical?)

20.75 What purpose is served by a *moderator* in a nuclear reactor?

20.76 What does the term *multiplication factor* mean in connection with a nuclear reactor?

20.77 How would the multiplication factors compare between two reactors when one is operating continuously at full power and the other is operating continuously at half power?

20.78 When a reactor is said to be *critical,* what is true about the multiplication factor within the reactor?

20.79 What do power plant engineers do to keep a reactor from going supercritical?

20.80 If in a pressurized water reactor a break occurs in a primary coolant line, what is one of the first things to happen within the reactor?

20.81 How might hydrogen gas be generated following a loss-of-coolant accident in a reactor?

20.82 Besides retaining the actual fuel in a reactor, what important function does a cladding tube have?

20.83 Why is there no possibility of an atomic bomb explosion from a nuclear power plant?

Radon-222 in the Environment (Special Topic 20.1)

20.84 How is radon-222 produced in the environment?

20.85 What radiation is emitted by radon-222?

20.86 Besides its own radiation what other factors make radon-222 particularly dangerous in the lungs?

Solar Fusion (Special Topic 20.2)

20.87 What is the overall equation for the fusion of protons that is believed to occur in the sun?

20.88 Of all the particles produced by solar fusion, which one has the highest penetrating ability?

Breeder Reactor (Special Topic 20.3)

20.89 What does it mean when uranium-238 is described as a *fertile* isotope?

20.90 Write the nuclear equation for the breeding of plutonium-239 from uranium-238.

Additional Exercises

20.91 Calculate the binding energy in J/nucleon of the nucleus of an atom of iron-56. The observed mass of one *atom* is 55.9349 u. What information lets us know that no isotope has a *larger* binding energy per nucleon?

20.92 Calculate the binding energy in J/nucleon of uranium-235. The observed mass of one *atom* is 235.0439 u.

20.93 Give the nuclear equation for each of these changes.
(a) positron emission by carbon-10
(b) alpha emission by curium-243
(c) electron capture by vanadium-49
(d) beta emission by oxygen-20

*20.94 If a positron is to be emitted spontaneously, how much more *mass* (as a minimum) must an *atom* of the parent have than an *atom* of the daughter nuclide? Explain.

20.95 There is a gain in binding energy per nucleon when light nuclei fuse to form heavier nuclei. Yet, a tritium atom and a deuterium atom, in a mixture of these isotopes, does not spontaneously fuse to give helium (and energy). Explain why not.

20.96 ^{214}Bi decays to isotope A by alpha emission; A then decays to B by beta emission, which decays to C by another beta emission. Element C decays to D by still another beta emission, and D decays by alpha emission to a stable isotope, E. What is the proper symbol of element E? (*Contributed by Prof. W. J. Wysochansky, Georgia Southwestern College.*)

20.97 ^{15}O decays by positron emission with a half-life of 124 s. (a) Give the proper symbol of the product of the decay. (b) How much of a 750-mg sample of ^{15}O remains after 5.0 min of decay? (*Contributed by Prof. W. J. Wysochansky, Georgia Southwestern College.*)

20.98 Alpha decay of ^{238}U forms ^{234}Th. What kind of decay from ^{234}Th produces ^{234}Ac? (*Contributed by Prof. Mark Benvennto, University of Detroit–Mercy.*)

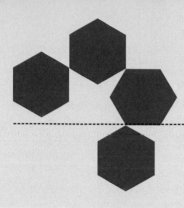

CHEMICALS IN USE 16
Positron Emitters and the PET Scan

POSITRON EMISSION TOMOGRAPHY

As described in Chapter 20, a number of synthetic radionuclides emit positrons, particles with the same mass as an electron but carrying one unit of *positive* charge. A positron forms by the conversion of a nuclear proton into a neutron.

$$\underset{\text{Proton}}{{}^{1}_{1}p} \longrightarrow \underset{\text{Neutron}}{{}^{1}_{0}n} + \underset{\text{Positron}}{{}^{0}_{1}e}$$

Following emission, a positron lasts only briefly before it collides with an electron. Both are destroyed and replaced by an equivalent quantity of gamma radiation.

To make a medically useful technology out of the gamma radiation, a positron-emitting radionuclide must be incorporated into a molecule with chemical properties that enable the natural reactions of the body to carry it into the tissue to be studied. The tissue then has a gamma radiator *on the inside*. Thus, instead of X rays being sent through the body from the outside, the radiation originates within the site. The overall procedure is called *positron emission tomography*, or **PET** for short (Figure 16*a*).

Three positron emitters are often used: oxygen-15, nitrogen-13, and carbon-11. Glucose, for example, can be made in which one carbon atom is carbon-11 instead of the usual carbon-12. Glucose is one of a relatively small number of organic species whose molecules are able to cross what medical specialists call the blood–brain barrier and so get inside brain cells. If some part of the brain is experiencing abnormal glucose metabolism, this will be reflected in the way in which positron-emitting glucose is handled, and gamma radiation detectors on the outside can pick up the differences (Figure 16*a*).

The use of the PET scan has led to the discovery that glucose metabolism in the brain is altered in schizophrenia, manic depression, and Alzheimer's disease. Because of nicotine, the brains of smokers show a general reduction in glucose use (Figure 16*b*).

MAGNETIC RESONANCE IMAGING

Because gamma radiation of any source is *ionizing* radiation, the PET scan is an *invasive* procedure, meaning that something harmful is allowed to invade the patient. The procedure is therefore used only when the possible benefits are believed to outweigh the risks of not using this diagnostic tool. *Magnetic resonance imaging* or **MRI,** is a noninvasive procedure that uses the magnetic properties of certain atomic nuclei for diagnosis.

Some atomic nuclei spin much as the earth or an electron spins about its axis. For example, the nucleus of a hydrogen atom, the proton, spins on an axis. When either an electron or a proton spins, you actually have a tiny bit of electrical charge in circulation. *Anytime electrical charge is in circulation, there is an associated magnetism.* (Electromagnets work on this principle. The current flowing in the wire wrapped around an iron core is current in circulation and it sets up a magnetic field.) A spinning proton is thus a spinning nuclear magnet.

Unpaired spins are needed for magnetic properties.

FIGURE 16*a*

The patient is shown undergoing a brain scan performed by means of PET technology.

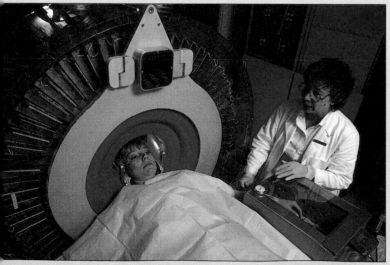

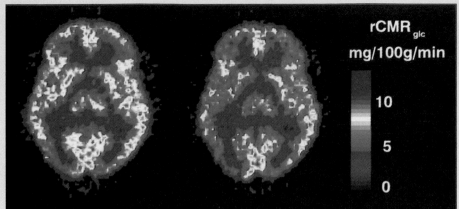

FIGURE 16b

Left: The control, a normal brain scan using PET technology. *Center:* The PET scan of the brain of a volunteer injected with nicotine. *Right:* The color code indicates the rates of glucose metabolism. Notice how widespread is the reduction in the rate of glucose metabolism over the control. (Photos courtesy of E. D. London.)

Thus, only atomic nuclei that have odd numbers of protons or neutrons behave as though they were tiny magnets, hence, the M or magnetic part of MRI. In nuclei with even numbers of protons and neutrons (like carbon-12 and oxygen-16) all of the nucleons are in *paired* spins, so the nuclei of these abundant elements have no magnetic properties. The nucleus of ordinary hydrogen is the most abundant nuclear magnet in living systems, because hydrogen atoms are parts of water molecules and all biochemicals.

When molecules with spinning nuclear magnets are in a strong, externally applied magnetic field and are simultaneously bathed with properly tuned radio-frequency radiation (which is of very low energy), the nuclear magnets flip their spins. (This is the R or resonance part of MRI.) As the nuclei resonate, electromagnetic radiation of very low frequency and so of low energy radiates, and the radiation is picked up by detectors.

A tumor has a different concentration of water than normal tissue nearby, so the tumorous region of a patient in an MRI scanner will send out low-energy electromagnetic radiation in a novel way. Such MRI data are fed into computers, which produce an image much like that of an X ray or a PET scan, but without having subjected the patient to ultrahigh-energy electromagnetic radiation such as X rays or gamma rays.

MRI scanning has proved to be especially useful for studying soft tissue, the sort of tissue least well studied by X rays. Different soft tissues have different population densities of water molecules or of fat molecules (which are loaded with hydrogen atoms). And tumors and cancerous tissue have their own water inventories, as we said. Calcium ions do not produce any signals to confuse MRI scanning, so bone, which is rich in calcium, is transparent to MRI.

MRI technology has developed during the 1980s into a method superior to other techniques for diagnosing tumors at the rear and base of the skull and equal to other methods for finding other brain tumors. MRI is now the preferred technology for assessing problems in joints (particularly the knees) and in the spinal cord, such as ruptured (herniated or "slipped") disks (see Figure 16c).

Questions

1. Compare the positron and the electron in mass and charge.

2. How does a positron form in a positron-emitting radionuclide?

3. What property makes the life of a positron very short?

4. In a PET scan what radiation is converted into a result resembling an X ray?

5. Why is the MRI less harmful a technique than the PET scan?

6. How does MRI complement the use of X rays?

FIGURE 16c

An MRI scan of a 7-month-old child revealed a malignant tumor pushing its way into the spinal cord. (The tumor was treated in time.)

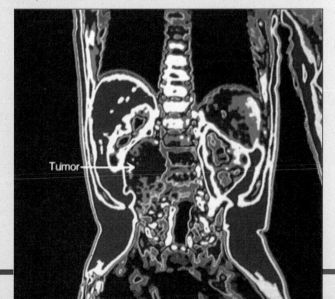

The abilities of chemists to turn coal, oil, air, and water into strong, lightweight plastics for fabrics, hulls, and other equipment have changed all sports and created others, like sailboarding. Plastics and other organic materials, including the most important kinds of biochemicals, are introduced in this chapter.

Chapter 21

Organic Compounds, Polymers, and Biochemicals

Organic chemistry is the study of the preparation, properties, identification, and reactions of those compounds of carbon not classified as inorganic. Inorganic carbon compounds include the oxides of carbon, the bicarbonates and carbonates of metal ions, the metal cyanides, and a handful of other compounds. Nearly all of the several million carbon compounds are thus organic.

Virtually all plastics, synthetic and natural fibers, dyes and drugs, insecticides and herbicides, ingredients in perfumes and flavoring agents, and all petroleum products are organic compounds. All the foods you eat consist chiefly of organic compounds in the families of the carbohydrates, fats and oils, proteins, and vitamins. The substances that make up furs and feathers, hides and skins, and all cell membranes are also organic.

The Uniqueness of the Element Carbon What makes the existence of so many organic compounds possible is not just the multivalency of carbon—its atoms always have four bonds in organic compounds. Rather, carbon atoms are unique in their ability to form strong covalent bonds to each other *while also holding the atoms of other nonmetals strongly*. The molecules in polyethylene, for example, have *carbon chains* that are thousands of carbon atoms long, each carbon holding hydrogen atoms.

21.1
NATURE OF ORGANIC CHEMISTRY

Nearly all medications, and all enzymes, hormones, neurotransmitters, and genes are organic compounds.

Sulfur atoms can also form long chains, but they are unable to hold the atoms of any other element strongly at the same time.

polyethylene (a small segment of one molecule)

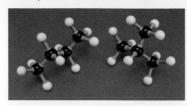

(a)

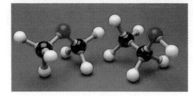

(b)

FIGURE 21.1

(a) The isomers of C_4H_{10}. Butane is on the left and 2-methylpropane (isobutane) is on the right. (b) The isomers of C_2H_6O. Dimethyl ether is on the left and ethanol is on the right.

The longest known sequence of atoms for other members of the carbon family, each also holding hydrogens, is eight for silicon, five for germanium, two for tin, and none for lead.

Isomerism among Organic Compounds Another reason for the huge number of organic compounds is *isomerism*, introduced in Section 7.4 (page 280) where the isomers of C_4H_{10}, butane and isobutane, were discussed. *Isomers,* recall, are compounds with identical molecular formulas but whose molecules have different structures. Two other examples of isomers are ethyl alcohol and dimethyl ether (Figure 21.1).

butane
BP −0.5 °C

isobutane
BP −11.7 °C

ethyl alcohol
BP 78.5 °C

dimethyl ether
BP −23 °C

As the number of carbons per molecule increases, the number of possible isomers for any given formula becomes astronomic.

Formula	Number of Isomers
C_8H_{18}	18
$C_{10}H_{22}$	75
$C_{20}H_{42}$	366,319
$C_{40}H_{82}$	6.25×10^{13} (estimated)

Few of the possible isomers of the compounds with large numbers of carbon atoms have actually been made, but nothing except too much crowding within their molecules prevents their existence.

Organic Families and Their Functional Groups

Functional groups

The study of the huge number of organic compounds is manageable because they can be sorted into *organic families* defined by *functional groups.* **Functional groups** are small structural units within molecules at which most of the compound's chemical reactions occur (Table 21.1). For example, as we learned in Section 7.4, all *alcohols* have the *alcohol group,* and a molecule of the simplest member of the alcohol family, methyl alcohol, has only one carbon. All members of the family of *carboxylic acids* have the *carboxyl group,* and the member of this family having two carbons per molecule is the very familiar weak acid, acetic acid.

alcohol group

methyl alcohol

carboxyl group

acetic acid

TABLE 21.1 Some Important Families of Organic Compounds

Family	Characteristic Structural Feature[a]	Example
Hydrocarbons	Only C and H present **Families of Hydrocarbons:**	
	Alkanes: only single bonds	CH_3CH_3
	Alkenes: C=C	CH_2=CH_2
	Alkynes: C≡C	HC≡CH
	Aromatic: Benzene ring	⬡
Alcohols	ROH	CH_3CH_2OH
Ethers	ROR′	CH_3OCH_3
Aldehydes	$\overset{\displaystyle O}{\overset{\|}{R}}CH$	$\overset{\displaystyle O}{\overset{\|}{CH_3}}CH$
Ketones	$\overset{\displaystyle O}{\overset{\|}{R}}CR'$	$\overset{\displaystyle O}{\overset{\|}{CH_3}}CCH_3$
Carboxylic acids	$\overset{\displaystyle O}{\overset{\|}{R}}COH$	$\overset{\displaystyle O}{\overset{\|}{CH_3}}COH$
Esters	$\overset{\displaystyle O}{\overset{\|}{R}}COR'$	$\overset{\displaystyle O}{\overset{\|}{CH_3}}COCH_3$
Amines	RNH_2, RNHR′, RNR′R″	CH_3NH_2 CH_3NHCH_3 $\overset{\displaystyle CH_3}{\overset{\|}{CH_3}}NCH_3$
Amides	$\overset{\displaystyle O \quad R''(H)}{\overset{\| \quad \|}{RC}}$—NR′(H)	$\overset{\displaystyle O}{\overset{\|}{CH_3}}CNH_2$

[a] R, R′, and R″ represent hydrocarbon groups—*alkyl groups*—defined in the text.

One family in Table 21.1, the *alkane* family, has no functional group, just C—C and C—H single bonds. These bonds are virtually nonpolar, because C and H are so alike in electronegativity. Therefore, alkane molecules are the least able of all organic molecules to attract ions or polar molecules, so they are unable to react with polar or ionic reactants such as strong acids and bases and common oxidizing agents, like the dichromate or the permanganate ion.

The significance of functional groups will become clearer when we contrast the relative lack of chemical properties of alkanes with those of other organic families. A *major goal* of this chapter, in fact, is to demonstrate how the presence of functional groups enables us to organize and understand the chemical and physical properties of organic compounds.

Condensed Structures The structural formulas in Table 21.1 are "condensed" because this saves both space and time in writing structures *without sacrificing any structural information*. In condensed structures, C—H bonds are usually "understood." When three H atoms are attached to carbon, they are set along side of the C, as in CH_3, but sometimes H_3C. When a carbon holds two other H atoms, the condensed symbol is usually CH_2, sometimes H_2C. Thus the condensed structure of ethanol is CH_3CH_2OH, and that of dimethyl ether is CH_3OCH_3 or H_3COCH_3.

Alkanes do react with fluorine, chlorine, bromine, and hot nitric acid. They also burn, a reaction with oxygen.

Ethanol is the alcohol in beverages.

Functional Groups and Polar Reactants

Amines are organic derivatives of ammonia and are weakly basic compounds.

When a polar group of atoms, like the OH group or even a halogen atom, is attached to carbon, the molecule has a polar site. It now can attract polar and ionic reactants and undergo chemical changes, but generally at or near just this functional group. This is why compounds of widely varying size but with the same functional group display very similar kinds of reactions. The reactions of *amines*, for example, are similar from compound to compound, so we can learn the handful of kinds of reactions exhibited by all amines and then adapt this knowledge to a specific example. For example, because the amines are compounds with the NH_2 group, like ammonia ($H-NH_2$), they are Brønsted bases, like ammonia. The simplest amine is methylamine, CH_3NH_2, and it reacts with hydrochloric acid as follows. (We will show the parallel reaction of ammonia first.)

$$NH_3(aq) + HCl(aq) \longrightarrow NH_4^+(aq) + Cl^-(aq)$$

$$\underset{\text{methylamine}}{CH_3NH_2(aq)} + HCl(aq) \longrightarrow \underset{\substack{\text{methylammonium} \\ \text{ion}}}{CH_3NH_3^+(aq)} + Cl^-(aq)$$

Ethylamine, $CH_3CH_2NH_2$, another member of the amine family, gives the same kind of reaction.

$$\underset{\text{ethylamine}}{CH_3CH_2NH_2(l)} + HCl(aq) \longrightarrow \underset{\text{ethylammonium ion}}{CH_3CH_2NH_3^+(aq)} + Cl^-(aq)$$

Only the NH_2 group of the amine changes in these reactions, and it changes in the identical way in both methylamine and ethylamine as well as in all amines having longer alkane-like chains.

The Symbol R in Structural Formulas

The symbol R is from the German *radikal*.

We could write equations for the reactions of hundreds of other amines with hydrochloric acid, but the equations would all be alike. Therefore we can summarize in just one equation the reactions of hydrochloric acid with all amines. To write this summarizing equation, we use the symbol R to represent any and all purely alkane-like groups, like the CH_3 group of methylamine or the CH_3CH_2 group of ethylamine.

$$R-NH_2 + HCl \longrightarrow R-NH_3^+ + Cl^-$$

The study of organic chemistry is thus not the study of individual compounds, taking them one at a time. It is the study of the common properties of functional groups, how to change one into another, and how each group affects physical properties. The symbol R makes this study easier.

Open-Chain and Ring Compounds

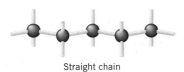

Straight chain

Branched chain

The continuous sequence of carbon atoms in polyethylene is called a **straight chain,** which means *only that no carbon atom holds more than two other carbons. It means nothing concerning the conformation* of the molecule, the particular shape in which it might exist as the result of twisting or contorting the chain. Even if we made a molecular model of polyethylene and coiled it into a spiral, we would still call its carbon skeleton a straight chain.

The term *straight* signifies only the idea of *unbranched.* **Branched chains** are also very common. Isooctane, for example, has a *main chain* of five carbon atoms carrying three CH$_3$ *branches*.

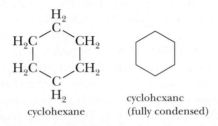

isooctane

Isooctane, one of the many alkanes in gasoline, is the standard for the octane ratings of various types of gasoline. Pure isooctane is assigned an octane rating of 100.

Multivalent atoms, like C, N, and O, can be incorporated into **rings** as well as into open chains. Cyclohexane, for example, has a ring of six carbon atoms.

cyclohexane   cyclohexane
(fully condensed)

Carbon ring

Just about everything is "understood" in the very convenient, fully condensed structure of cyclohexane. By convention, polygons like the hexagon for cyclohexane can be used to represent rings provided that we understand the following rules.

1. C occurs at each corner unless O or N (or another multivalent atom) is explicitly written at a corner.

2. A line connecting two corners is a covalent bond between adjacent ring atoms.

3. Remaining bonds, as required by the covalence of the atom at a corner, are understood to hold H atoms.

4. Double bonds are always explicitly shown.

We can illustrate these rules with the following cyclic compounds.

The rule about double bonds will be modified with benzene and compounds like it.

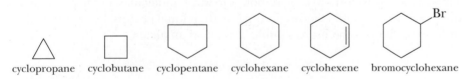

cyclopropane   cyclobutane   cyclopentane   cyclohexane   cyclohexene   bromocyclohexane

There is no theoretical upper limit to the size of a ring.

Compounds like pyrrole, piperidine, and tetrahydropyran, whose molecules carry an atom other than carbon in a ring position, are called **heterocyclic compounds.**

The tetrahydropyran ring occurs in molecules of sugar.

pyrrole   piperidine   tetrahydropyran

The atom other than carbon is called the *heteroatom,* and its requisite number of covalent bonds must be shown or clearly understood. Nitrogen, for example, normally forms three bonds in uncharged species, so three bonds must be indicated for it in nitrogen heterocycles. Oxygen forms two bonds (at neutral sites), so its heterocyclic compounds must show two bonds from oxygen. Rings can have more than one heteroatom.

Structure and Physical Properties: General Principles

One goal of organic chemistry is to enable the prediction of properties from molecular structure. Functional groups, heteroatoms, and molecular size all influence the physical properties in ways that permit some broad generalizations. Functional groups and molecular size determine, for example, whether a compound is soluble in water.

Effects of Functional Groups on Solubility in Water   Piperidine and tetrahydropyran (shown above) are both freely soluble in water, whereas cyclohexane is insoluble. The cyclohexane molecule has only C—C and C—H single bonds. The C—H bond is too nonpolar to enable its H atom to form hydrogen bonds (page 444). Hydrogen bonds occur largely when H is attached to O or N atoms—two of the three most electronegative atoms. These atoms are able to induce a sizable δ+ charge on the H atoms of an O—H or an N—H group. This δ+ charge is attracted only to a *sizable δ−*, one located only on the O or N of another molecule, like a water molecule. Piperidine and tetrahydropyran, for example, are able to form hydrogen bonds (⋯) to water molecules as follows.

Strong hydrogen bonds form between molecules of HF, but HF is not an organic compound so it is not treated here.

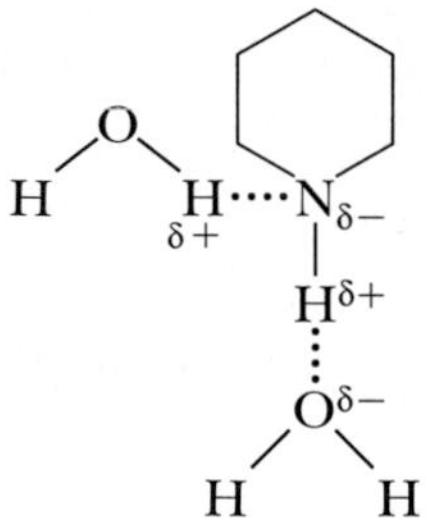

Piperidine hydrogen bonded to two water molecules

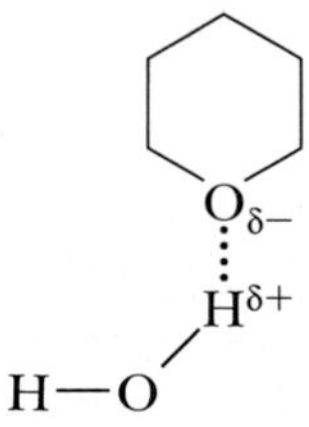

Tetrahydropyran hydrogen bonded to a water molecule

Thus the ability to form hydrogen bonds to water molecules, made possible by the presence of N and O atoms in organic functional groups, helps molecules with these atoms to be more soluble in water than alkanes.

Effects of Molecular Size on Physical Properties   As the alkane-like portion of a molecule becomes larger and larger *in comparison to a functional group involving N or O atoms,* the substance's solubility in water decreases. The cholesterol molecule, for example, has an OH group, but the group is overwhelmed by the huge hydrocarbon group. So cholesterol is practically insoluble in water.

The insolubility of cholesterol in water accounts for its separation from the blood stream as heart disease slowly develops.

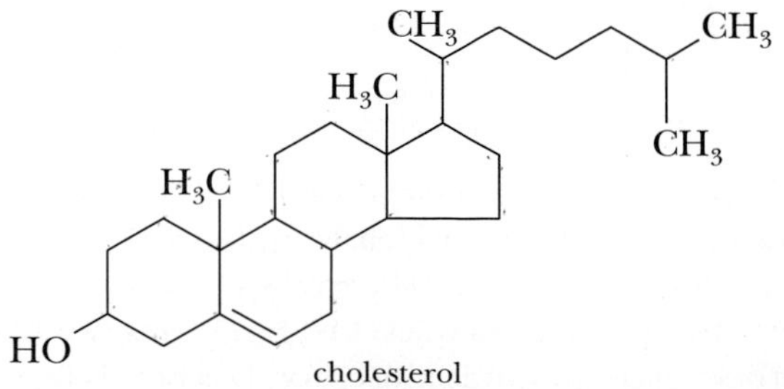

cholesterol

Cholesterol, however, is correspondingly more soluble in such nonpolar organic solvents as benzene and chloroform. (Remember, "like dissolves like.")

The alkanes make up just one family of a multifamily group of compounds called the **hydrocarbons,** whose molecules consist only of C and H atoms. Besides the alkanes, the hydrocarbons include *alkenes, alkynes,* and *aromatic hydrocarbons* (Table 21.1). All hydrocarbons are insoluble in water. All hydrocarbons burn and, in sufficient oxygen, carbon dioxide and water are the sole products.

The chief structural difference among the hydrocarbon families is the presence of double or triple bonds. The alkanes are examples of **saturated organic compounds,** compounds with only single bonds regardless of family. Regardless of the family, any compound with double or triple bonds is unsaturated, so the alkenes and alkynes are **unsaturated organic compounds.** The aromatic hydrocarbons are also unsaturated because the carbon atoms of their rings, when represented by simple Lewis structures, also have double bonds.

Benzene:
Lewis structure

Benzene:
alternative structure

21.2 HYDROCARBONS

Saturated implies here, as elsewhere, that there is no more room for something, such as no room for holding more atoms or groups.

The special nature of the bonds in benzene is discussed in Section 8.8.

Sources of Hydrocarbons

Virtually all of the usable supplies of hydrocarbons are obtained from the *fossil fuels*—coal, petroleum, and natural gas. One of the operations in petroleum refining is to boil crude oil (petroleum freed of natural gas) and then selectively condense the vapors between prechosen ranges of temperatures. The liquid collected at each range is called a *fraction,* and the operation is *fractional distillation.* Each fraction is used for a particular purpose, and each consists of a rather complex mixture of compounds—almost entirely hydrocarbons and mostly alkanes. *Gasoline,* for example, is a fraction boiling roughly between 40 and 200 °C. The *kerosene* and *jet fuel* fraction overlaps this range, going from 175 to 325 °C. The molecules of the alkanes in gasoline generally have from 5 to 10 carbon atoms; those in kerosene, from 12 to 18. *Paraffin wax* is part of the nonvolatile residue of petroleum refining, and it consists of alkanes with over 20 carbons per molecule. Low-boiling fractions of crude oil, those having molecules of 4 to 8 carbons, are used as nonpolar solvents.

Fuels are surveyed in *Chemicals in Use 2,* page 86.

Diesel fuel is another fraction of petroleum.

Alkanes

All open-chain alkanes (those without rings) have the general formula C_nH_{2n+2}, where n equals the number of carbon atoms. Table 21.2 gives the structures, names, and some properties of the first 10 unbranched open-chain alkanes. Their boiling points steadily increase with molecular mass, illustrating how London forces become greater with molecular size. The alkanes are generally much less dense than water, so oil spills float on water.

London forces are discussed in Section 11.2.

TABLE 21.2   Straight-Chain Alkanes

IUPAC Name	Molecular Formula	Structure	Boiling Point (°C)	Melting Point (°C)	Density (g mL^{-1}, 20°C)
Methane	CH_4	CH_4	-161.5	-182.5	
Ethane	C_2H_6	CH_3CH_3	-88.6	-183.3	
Propane	C_3H_8	$CH_3CH_2CH_3$	-42.1	-189.7	
Butane	C_4H_{10}	$CH_3(CH_2)_2CH_3$	-0.5	-138.4	
Pentane	C_5H_{12}	$CH_3(CH_2)_3CH_3$	36.1	-129.7	0.626
Hexane	C_6H_{14}	$CH_3(CH_2)_4CH_3$	68.7	-95.3	0.659
Heptane	C_7H_{16}	$CH_3(CH_2)_5CH_3$	98.4	-90.6	0.684
Octane	C_8H_{18}	$CH_3(CH_2)_6CH_3$	125.7	-56.8	0.703
Nonane	C_9H_{20}	$CH_3(CH_2)_7CH_3$	150.8	-53.5	0.718
Decane	$C_{10}H_{22}$	$CH_3(CH_2)_8CH_3$	174.1	-29.7	0.730

IUPAC rules for naming compounds

IUPAC System of Nomenclature   Chemists representing the national chemical societies of several countries form the International Union of Pure and Applied Chemistry or IUPAC. One committee of the IUPAC, the Commission on Nomenclature of Organic Chemistry, has devised rules for creating names of organic compounds. The intent of the rules is to permit only one name for each compound and to permit the drawing of only one structure from a name.

We cannot go into the IUPAC rules in any detailed manner, but we'll illustrate their chief principles. Thus, the last syllable in an IUPAC name designates the family to which the compound belongs. All saturated hydrocarbons, for example, have names ending in *-ane*. The names of hydrocarbons with double bonds end in *-ene,* and those with triple bonds end in *-yne*. Each compound in a given family is regarded as having been made by attaching *substituents* to a *parent chain* or *parent ring*.

IUPAC Rules for Naming the Alkanes

1. The name ending for all alkanes (and cycloalkanes) is *-ane*.

2. The *parent chain* is the longest continuous chain of carbons in the structure. For example, the branched-chain alkane

$$CH_3$$
$$|$$
$$CH_3CH_2CHCH_2CH_2CH_3$$

is regarded as being "made" from the following parent

$$CH_3CH_2CH_2CH_2CH_2CH_3$$

by replacing a hydrogen atom on the third carbon from the left with CH_3.

$$CH_3 \searrow \overset{\nearrow}{H}$$
$$CH_3CH_2CHCH_2CH_2CH_3 \longrightarrow CH_3CH_2CHCH_2CH_2CH_3$$
$$\hspace{7.5cm} |$$
$$\hspace{7cm} CH_3$$

3. A prefix is attached to the name ending *ane* to specify the number of carbon atoms *in the parent chain*. The prefixes through chain lengths of 10 carbons are as follows. The names in Table 21.2 show their use.

meth-	1 C	hex-	6 C
eth-	2 C	hept-	7 C
prop-	3 C	oct-	8 C
but-	4 C	non-	9 C
pent-	5 C	dec-	10 C

Because the parent chain of our example has six carbons, the parent chain is named hexane—*hex* for six carbons and *ane* for being in the alkane family. Therefore, the alkane whose name we are devising is regarded as a derivative of this parent, *hexane*.

4. The carbon atoms of the parent chain are numbered starting from whichever end of the chain gives the location of the first branch the lower of two possible numbers. Thus the correct direction for numbering our example is from left to right, not right to left, because this locates the branch (CH_3) at position 3, not position 4.

$$\underset{\substack{1 \quad 2 \quad 3 \quad 4 \quad 5 \quad 6}}{CH_3CH_2\overset{\displaystyle CH_3}{\overset{|}{C}}HCH_2CH_2CH_3} \qquad \underset{\substack{6 \quad 5 \quad 4 \quad 3 \quad 2 \quad 1}}{CH_3CH_2\overset{\displaystyle CH_3}{\overset{|}{C}}HCH_2CH_2CH_3}$$

(correct direction of numbering)   (incorrect direction of numbering)

5. Each branch attached to the parent chain is named. Therefore, we must pause and learn the names of some of the alkane-like branches.

The Alkyl Groups   Any branch that consists only of carbon and hydrogen and has only single bonds is called an **alkyl group,** and the names of all alkyl groups end in *-yl*. Think of an alkyl group as an alkane minus one of its hydrogen atoms. For example,

methane $\xrightarrow{\text{Remove one H}}$ methyl   or CH_3-

ethane $\xrightarrow{\text{Remove one H}}$ ethyl   or CH_3CH_2-

Two alkyl groups can be obtained from propane because the middle position in its chain of three is not equivalent to either of the end positions.

propane $\xrightarrow{\text{Remove one H}}$ propyl   or $CH_3CH_2CH_2-$

propane $\xrightarrow{\text{Remove one H}}$ isopropyl   or CH_3CHCH_3

We will not need to know the IUPAC names for any alkyl groups with four or more carbon atoms.

6. The name of each alkyl group is attached to the name of the parent as a prefix, placing its chain location number in front and separating the number from the name by a hyphen. For instance, the name of our original example is 3-methylhexane.

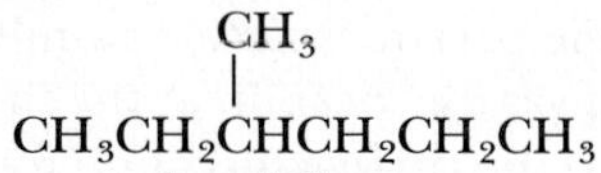

$$CH_3CH_2CHCH_2CH_2CH_3$$
with CH_3 branch

3-methylhexane

7. When two or more groups are attached to the parent, name each and locate each with a number. The names of alkyl substituents are assembled in their alphabetical order. Always use *hyphens* to separate numbers from words. Here is an application.

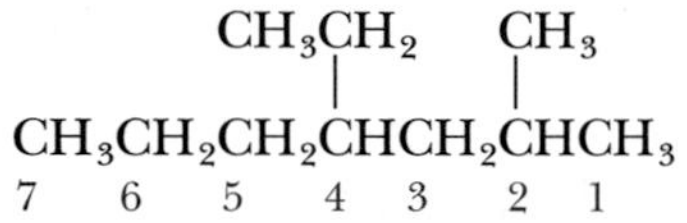

$$CH_3CH_2CH_2CHCH_2CHCH_3$$
with CH_3CH_2 and CH_3 branches

7 6 5 4 3 2 1

4-ethyl-2-methylheptane

8. When two or more substituents are identical, use such multiplier prefixes as di- (for 2), tri- (for 3), tetra- (for 4), and so forth; specify the location number of every group. Always separate a number from another number in a name by a *comma*. For example,

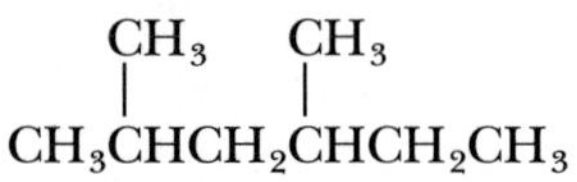

$$CH_3CHCH_2CHCH_2CH_3$$
with two CH_3 branches

Correct name:	2,4-dimethylhexane
Incorrect names:	2,4-methylhexane
	3,5-dimethylhexane
	2-methyl-4-methylhexane
	2-4-dimethylhexane

9. When identical groups are on the *same* carbon, repeat the number of this carbon in the name. For example,

$$CH_3CCH_2CH_2CH_3$$
with two CH_3 branches

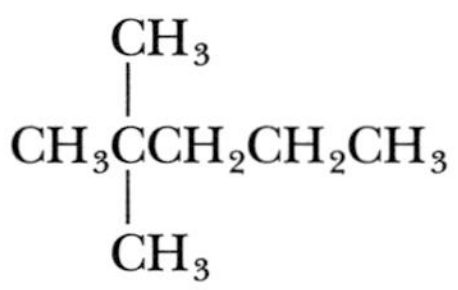

Correct name:	2,2-dimethylpentane
Incorrect names:	2-dimethylpentane
	2,2-methylpentane
	4,4-dimethylpentane

These are not all of the IUPAC rules for alkanes, but they will handle all of our needs.

Unsystematic or *common names* are still widely used for many compounds. The common name of 2-methylpropane is *isobutane,* for example.

EXAMPLE 21.1
Using the IUPAC Rules to Name an Alkane

What is the IUPAC name for the following compound?

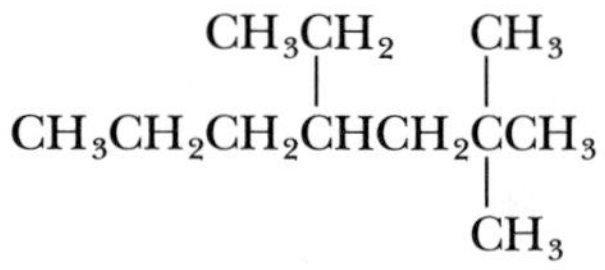

$$CH_3CH_2CH_2CHCH_2CCH_3$$
with CH_3CH_2, CH_3, and CH_3 branches

ANALYSIS The compound is an alkane because it is a hydrocarbon with only single bonds. We must therefore use the IUPAC rules for alkanes.

SOLUTION The ending to the name must be *-ane.* The next step is to find the longest chain. This chain is seven carbons long, so the name of the parent alkane is *heptane*. We have to number the chain from right to left, as follows, in order to reach the first branch with the lower number.

$$\underset{7\quad6\quad5\quad4\quad3}{\text{CH}_3\text{CH}_2\text{CH}_2\text{CHCH}_2\text{CCH}_3} \overset{\overset{\displaystyle\text{CH}_3\text{CH}_2}{|}}{}\overset{\overset{\displaystyle\text{CH}_3}{2|}}{}\underset{|1}{}$$
$$\underset{\text{CH}_3}{}$$

At carbon 2 there are two one-carbon methyl groups. At carbon 4, there is a two carbon ethyl group. Alphabetically, *ethyl* comes before *methyl*, so we must assemble these names as follows to make the final name. (Names of alkyl groups are alphabetized *before* any prefixes such as di- or tri- are affixed.)

4-ethyl-2,2-dimethylheptane

| hyphen separates a number from a word | comma separates two numbers | no hyphen, no comma, no space, |

■ **Practice Exercise 1** Write the IUPAC names of the following compounds. In searching for the parent chain, be sure to look for the longest continuous chain of carbons *even if the chain twists and goes around corners.*

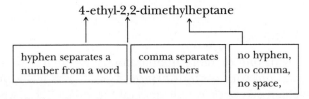

(a)
$$\text{CH}_3\text{CH}_2$$
$$\qquad\qquad\text{CHCH}_3$$
$$\text{CH}_2\text{CH}_2$$
$$|$$
$$\text{CH}_3$$

(b)
$$\text{CH}_3\qquad\text{CH}_2\text{CH}_2\text{CH}_3$$
$$\qquad\qquad|$$
$$\qquad\quad\text{CHCHCH}_2\text{CH}_3$$
$$\text{CH}_3\text{CH}$$
$$|$$
$$\text{CH}_3$$

(c)
$$\qquad\quad\text{CH}_3\quad\text{CH}_3\quad\text{CH}_3$$
$$\qquad\quad|\qquad\;|\qquad\;|$$
$$\text{CH}_3\text{CH}_2\text{CHCHCHCH}_2\text{CHCH}_3$$
$$\qquad\qquad\quad|$$
$$\qquad\qquad\text{CH}_3\text{CH}_2$$

Chemical Properties of Alkanes   Alkanes, as we have indicated, are generally stable at room temperature toward such different reactants as concentrated sulfuric acid (or any other common acid), concentrated aqueous bases (like NaOH), and even the most reactive metals. Fluorine attacks virtually all organic compounds, including the alkanes, to give mixtures of products. Like all hydrocarbons, the alkanes burn in air to give carbon dioxide and water. *Hot* nitric acid, chlorine, and bromine also react with alkanes. The chlorination of methane, for example, can be made to give the following compounds depending on the relative quantities of reactants.

| CH_3Cl | CH_2Cl_2 | CHCl_3 | CCl_4 |
| methyl chloride | methylene chloride | chloroform | carbon tetrachloride |

These are the common names for the chlorinated derivatives of methane, not the IUPAC names.

When heated at high temperatures in the absence of air, alkanes "crack," meaning that they break up into smaller molecules. The cracking of methane, for example, yields finely powdered carbon and hydrogen.

$$\text{CH}_4 \xrightarrow{\text{High temperatures}} \text{C} + 2\text{H}_2$$

The carbon is used as a filler in auto tires, and the hydrogen is used either as a fuel or as a raw material for the chemical industry. The cracking of methane is also the source of carbon atoms for the deposition of diamond films, as described in *Chemicals in Use 11*. The controlled cracking of ethane gives ethene, commonly called ethylene.

$$CH_3CH_3 \xrightarrow{\text{High temperatures}} CH_2{=}CH_2 + H_2$$

ethane · ethene

HOCH$_2$CH$_2$OH

ethylene glycol

Ethylene, one of the most important raw materials in the organic chemicals industry, is used to make polyethylene plastic items as well as ethyl alcohol and ethylene glycol (an antifreeze).

Alkenes and Alkynes

The molecular orbital structures of double and triple bonds are discussed in Section 8.6.

Hydrocarbons with one or more double bonds are members of the **alkene family.** Hydrocarbons with carbon–carbon triple bonds make up the **alkyne family.** Open-chain alkenes have the general formula, C_nH_{2n}, where n equals the number of carbon atoms. Open-chain alkynes have the general formula, C_nH_{2n-2}.

Alkenes and alkynes, like all hydrocarbons, are insoluble in water and are flammable, as we've said. The most familiar alkenes are ethene and propene (commonly called propylene), the raw materials for the manufacture of polyethylene and polypropylene, respectively. Ethyne ("acetylene"), an important alkyne, is widely used in oxyacetylene torches, which can quickly generate enough heat of combustion to cut steel.

The IUPAC accepts both *ethene* and *ethylene* as the name of CH$_2$=CH$_2$. The common name of propene is *propylene*, and other simple alkenes have common names as well.

$$CH_2{=}CH_2 \qquad CH_3CH{=}CH_2 \qquad HC{\equiv}CH$$

ethene · · · · · · · propene · · · · · · · · · ethyne
(ethylene) · · · · · (propylene) · · · · · · · (acetylene)

IUPAC rules for naming compounds **T**

Naming the Alkenes The IUPAC rules for the names of alkenes are adaptations of those for alkanes, but with two important differences. The first is that the parent chain *must include the double bond* even if this means that the parent chain is shorter than another that occurs in the structure. The second difference is that the parent *alkene* chain must be *numbered from whichever end gives the first carbon of the double bond the lower of two possible numbers*. This (lower) number and a hyphen precedes the name of the parent, when ambiguity about the double bond's location would otherwise exist. The locations of branches are not a factor. Otherwise, alkyl groups are named and located as before. Some examples of correctly named alkenes are as follows.

cyclohexene

$$CH_3CH_2CH{=}CH_2 \qquad CH_3CH{=}CHCH_3$$

1-butene · · · · · · · · · · · · · · · · 2-butene

$$\overset{\displaystyle CH_3}{\overset{|}{}} \qquad \overset{\displaystyle CH_3}{\overset{|}{}}$$
$$CH_3CH_2CHCH_2CH{=}CCH_3$$

2,5-dimethyl-2-heptene

No number is needed to name CH$_2$=CH$_2$, ethene, or CH$_3$CH=CH$_2$, propene.

Notice that only one number is needed to locate the double bond, the number of the first carbon of the double bond to be reached as the chain is numbered.

Some alkenes have two double bonds and are called *dienes*. Some have three double bonds and are *trienes*, and so forth. Each double bond has to be located by a number.

$$CH_2{=}CHCH{=}CHCH_3 \qquad CH_2{=}CHCH_2CH{=}CH_2 \qquad CH_2{=}CHCH{=}CHCH{=}CH_2$$

1,3-pentadiene · · · · · · · · · · · 1,4-pentadiene · · · · · · · · · · 1,3,5-hexatriene

Geometric Isomerism Among the Alkenes   As explained in Section 8.6, there is no free rotation at a double bond (see page 342). Many alkenes, therefore, exhibit **geometric isomerism.** Thus *cis*-2-butene and *trans*-2-butene are **geometric isomers** of each other. They not only have the same molecular formula, C_4H_8, but also the same skeletons and the same organization of atoms and bonds: $CH_3CH{=}CHCH_3$. They differ in the *directions* taken by the two CH_3 groups attached at the double bond.

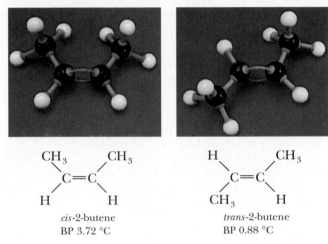

cis-2-butene
BP 3.72 °C

trans-2-butene
BP 0.88 °C

Because ring structures also lock out free rotation, geometric isomers of ring compounds are also possible. These two isomers of 1,2-dimethylcyclopropane are examples.

trans isomer
BP 28 °C

cis isomer
BP 37 °C

Cis means "on the same side;" *trans* means "on opposite sides." This difference in orientation gives the two geometric isomers of 2-butene measurable differences in physical properties, as their boiling points show. Because each has a double bond, however, the *chemical* properties of *cis-* and *trans*-2-butene are very similar.

Addition Reactions of Alkenes   Electron-seeking species are naturally attracted to the electron density at the pi bond of the double bond. Thus, alkenes react readily with protons provided by strong proton donors, making possible several **addition reactions,** reactions in which the pieces of a reactant become separately attached to the carbons at the ends of a double bond. Ethene, for example, readily reacts with hydrogen chloride as follows:

$$CH_2{=}CH_2 + H{-}Cl(g) \longrightarrow Cl{-}CH_2{-}CH_3$$

We say that the hydrogen chloride molecule *adds* across the double bond, one piece of the HCl molecule going to the carbon at one end and the other piece going to the other end. The pair of electrons of the pi bond move out and take H^+ from the HCl molecule, which releases Cl^-. A (+) charge is left at one end of the original carbon–carbon double bond as a bond to H forms at the other end. The result of this step is a very unstable cation, called a *carbocation,* an ion with a positive charge on carbon. The positively charged carbon in the ethyl carbocation then quickly attracts the Cl^- ion to give the product, 1-chloroethane. We may visualize the relocations of electron pairs between the reactants as follows.

A carbocation, having a carbon with only a sextet of valence electrons, not an octet, generally has only a fleeting existence.

$$CH_2{=}CH_2 + H{-}Cl \longrightarrow \overset{+}{C}H_2{-}CH_2 \longrightarrow Cl{-}CH_2{-}CH_3$$

1-chloroethane

+ Cl^-

ethyl carbocation

2-Butene also adds HCl according to the same pattern as ethene.

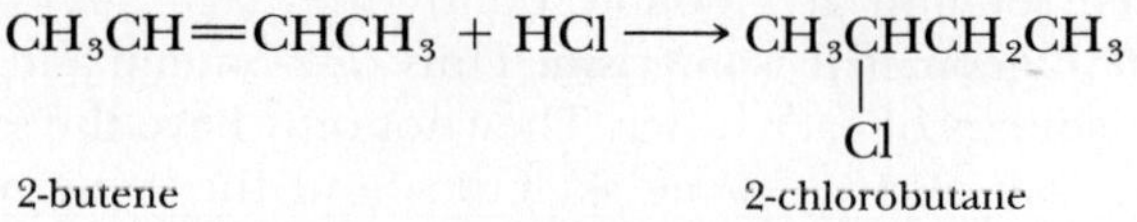

$$CH_3CH{=}CHCH_3 + HCl \longrightarrow CH_3\underset{\underset{Cl}{|}}{C}HCH_2CH_3$$

2-butene   2-chlorobutane

In the same manner, the alkene bond also adds hydrogen bromide, hydrogen iodide, and sulfuric acid ("hydrogen sulfate").

A water molecule will add to a double bond provided that an acid catalyst, like sulfuric acid, is present. The catalyst participates in the first step by donating a proton to one end of the double bond to create a carbocation. This unstable species preferentially attracts what is the most abundant electron-rich species present, namely, a water molecule.

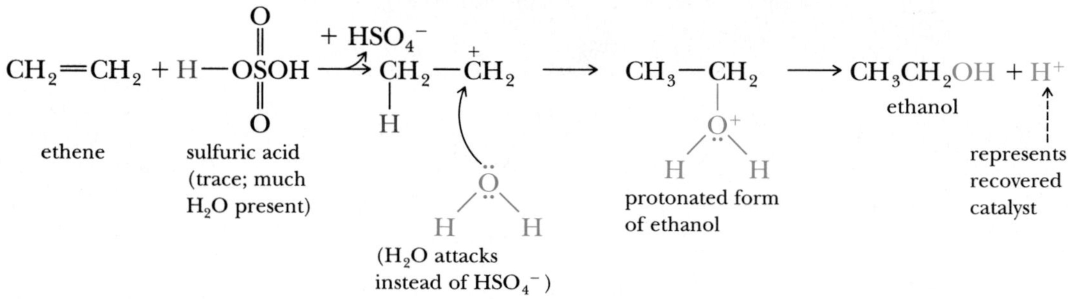

$$CH_2{=}CH_2 + H{-}OSOH \longrightarrow CH_2{-}CH_2 \longrightarrow CH_3{-}CH_2 \longrightarrow CH_3CH_2OH + H^+$$

ethene

sulfuric acid (trace; much H_2O present)

(H_2O attacks instead of HSO_4^-)

protonated form of ethanol

ethanol

represents recovered catalyst

Note that in the *overall* result, the pieces of the water molecule, H and OH, become attached at the different ends of the double bond. 2-Butene (either cis or trans) gives a similar reaction.

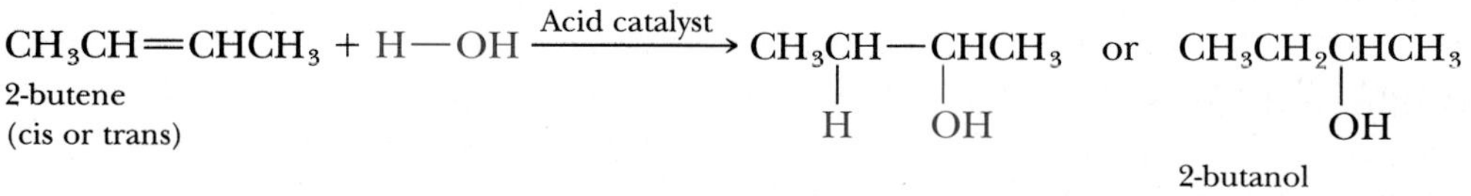

$$CH_3CH{=}CHCH_3 + H{-}OH \xrightarrow{\text{Acid catalyst}} CH_3\underset{\underset{H}{|}}{C}H{-}\underset{\underset{OH}{|}}{C}HCH_3 \text{ or } CH_3CH_2\underset{\underset{OH}{|}}{C}HCH_3$$

2-butene (cis or trans)

2-butanol

Alkynes give similar addition reactions.

Other chemicals that add to the double bond are chlorine, bromine, and hydrogen. Chlorine and bromine react rapidly at room temperature. Ethene, for example, reacts with bromine to give 1,2-dibromoethane.

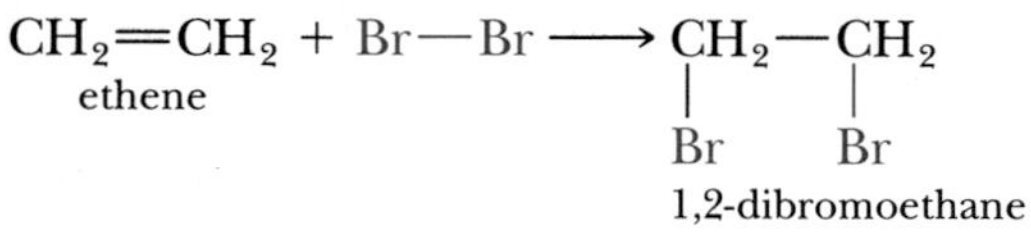

$$CH_2{=}CH_2 + Br{-}Br \longrightarrow \underset{\underset{Br}{|}}{C}H_2{-}\underset{\underset{Br}{|}}{C}H_2$$

ethene

1,2-dibromoethane

The Addition of Hydrogen—Hydrogenation The product of the addition of hydrogen to an alkene is an alkane and the reaction is called *hydrogenation*. Hydrogenation requires a catalyst—powdered platinum, for example—and sometimes a higher temperature and pressure than available under the room atmosphere. The hydrogenation of 2-butene (cis or trans) gives butane.

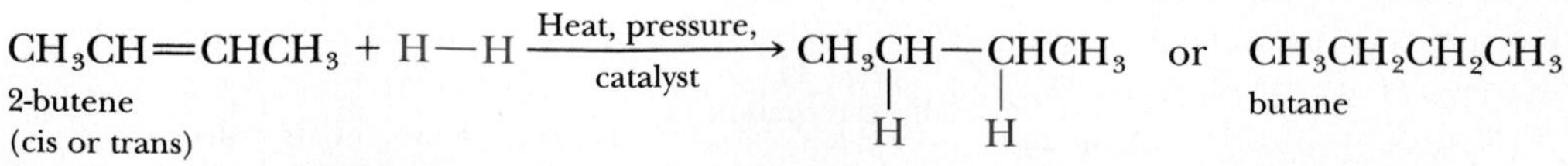

$$CH_3CH{=}CHCH_3 + H{-}H \xrightarrow[\text{catalyst}]{\text{Heat, pressure,}} CH_3\underset{\underset{H}{|}}{C}H{-}\underset{\underset{H}{|}}{C}HCH_3 \text{ or } CH_3CH_2CH_2CH_3$$

2-butene (cis or trans)

butane

Oxidation of a Double Bond   Ozone reacts with anything that has carbon–carbon double or triple bonds, and the reaction breaks the molecules into fragments at each such site, giving a variety of products. Because many important compounds in living systems have alkene double bonds, ozone is a very dangerous material in the wrong places.

Ozone, by attacking the double bonds in the chlorophyll of green plants, is able to prevent photosynthesis and so kill the plants.

Aromatic Hydrocarbons

The Benzene Ring   The most common **aromatic compounds** contain the *benzene ring*, a ring of six carbon atoms, each holding an insufficient number of hydrogen atoms to let us describe the system as saturated. The benzene ring can be represented in either of two ways, by an older symbol using alternating single and double bonds or by a circle. For example,

Hydrocarbons and their oxygen or nitrogen derivatives that are not aromatic are called *aliphatic compounds.*

benzene   or

toluene   or
(methylbenzene)

ethylbenzene   or

Except when writing a mechanism of a reaction involving the benzene ring, most chemists prefer to use the hexagon with the circle. This symbol does not signify what the presence of a double bond normally means, namely the ability to give addition reactions with considerable ease. Unlike an alkene, the benzene ring strongly resists addition reactions.

Historically, the term "aromatic" arose because many of the compounds with pleasant fragrances found long ago had benzene rings, and chemists thought that perhaps it was the ring that imparted this property. However, aromatic compounds like aspirin are entirely without any odor. Thus, any compound with a benzene ring is classified today as an aromatic compound, regardless of its odor or its functional groups. Benzene itself is the simplest aromatic hydrocarbon, and its delocalized bonds are described in Section 8.8 (see page 352).

The theory of resonance pictures the benzene molecule as a hybrid structure (see page 296). The chief contributors or resonance structures are the following.

aspirin

benzene (resonance structures)

Thus, resonance theory tells us that each carbon–carbon bond in benzene is alike, and each such bond is only a *partial* double bond. The theory suggests, in other words, not to expect benzene to behave like ordinary alkenes because it does not have ordinary alkene double bonds.

In the molecular orbital view of benzene, discussed in Section 8.8, the delocalization of the ring's pi electrons strongly stabilizes the ring, thus explaining why benzene does not easily give those reactions that interfere with the delocalization of electron density. Thus, the benzene ring most commonly undergoes **substitution reactions,** reactions in which one of the ring H atoms is substituted by another atom or group. For example, benzene reacts with chlorine in the presence of iron(III) chloride to give chlorobenzene, not a 1,2-dichloro compound. (To dramatize this point, we must use a resonance structure for benzene.)

$$\text{benzene} + Cl_2 \xrightarrow[HCl]{FeCl_3} \text{chlorobenzene (Cl)} \qquad Not \qquad \text{(Cl, Cl)}$$

chlorobenzene   (This would form if chlorine *added* to the double bond.)

You can infer that *substitution,* but not addition, leaves intact the closed-circuit, delocalized, and very stable pi electron network of the benzene ring.

Provided that a suitable catalyst is present, benzene reacts by substitution with chlorine, bromine, and nitric acid as well as with sulfuric acid. (Recall that Cl_2 and Br_2 readily *add* to alkene double bonds.)

$$C_6H_6 + Br_2 \xrightarrow{FeBr_3 \text{ catalyst}} C_6H_5{-}Br + HBr$$
bromobenzene

$$C_6H_6 + HNO_3 \xrightarrow{H_2SO_4 \text{ catalyst}} C_6H_5{-}NO_2 + H_2O$$
nitrobenzene

The sulfonic acids are roughly as strong acids as sulfuric acid.

$$C_6H_6 + H_2SO_4 \longrightarrow C_6H_5{-}SO_3H + H_2O$$
benzenesulfonic acid

21.3

ALCOHOLS AND ETHERS, ORGANIC DERIVATIVES OF WATER

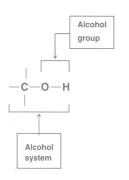

Alcohol group

Alcohol system

IUPAC rules for naming compounds

An **alcohol** is any compound with an OH group attached to a carbon with three other groups also attached by *single bonds.* Using the symbol R to represent any alkyl group, alcohols have ROH as their general structure. The four simplest (and most common) alcohols are the following. (Their common names are in parentheses below their IUPAC names.)

CH_3OH	CH_3CH_2OH	$CH_3CH_2CH_2OH$	CH_3CHCH_3
methanol	ethanol	1-propanol	$\overset{|}{OH}$
(methyl alcohol)	(ethyl alcohol)	(propyl alcohol)	2-propanol
BP 65 °C	BP 78.5 °C	BP 97 °C	(isopropyl alcohol)
			BP 82 °C

Ethanol is the alcohol in beverages and is also added to gasoline to make "gasohol."

IUPAC Names of Alcohols   The parent chain of an alcohol must be the longest *that includes the carbon holding the OH group.* The name ending is *-ol,* which replaces the *-e* ending of the name of the hydrocarbon that corresponds to the parent. The chain is numbered to give the site of the OH group the lower number regardless of where alkyl substituents occur.

Ethers   Molecules of **ethers** contain two alkyl groups joined to one oxygen, the two R groups being alike or different. We give only the common names for the following examples.

CH_3OCH_3 $CH_3CH_2OCH_2CH_3$ $CH_3OCH_2CH_3$ R—O—R′

dimethyl ether   diethyl ether   methyl ethyl ether   ethers
BP − 23 °C   BP 34.5 °C   BP 11 °C   (general structure)

Diethyl ether was once the "ether" most widely used as an anesthetic in surgery.

The simple ethers are very low-boiling compounds because hydrogen bonds cannot exist between neighboring molecules. Consequently, the boiling points of ethers are lower than those of alcohols of comparable molecular masses. For example, 1-butanol, $CH_3CH_2CH_2CH_2OH$ (BP 117 °C) boils 83 degrees higher than its isomer, diethyl ether (BP 34.5 °C).

Major Reactions of Alcohols and Ethers

Ethers are almost as chemically inert as alkanes. They burn (as do alkanes), and they are split apart when boiled in concentrated acids. The alcohols, in contrast, have a rich chemistry. We'll look at their oxidation and dehydration reactions as well as some substitution reactions.

Oxidation Reactions of Alcohols   When the carbon atom of the alcohol system also holds at least one H atom, this H atom and its bonding pair of electrons can be removed by an oxidizing agent. The number of H atoms remaining at this carbon then determines the family to which the product belongs. When an alcohol is oxidized, an H atom attached to the alcohol carbon is transferred away as $H{:}^-$ to an acceptor, and the H atom of the OH group leaves as H^+. These two pieces become part of a water molecule, with the oxidizing agent providing the O atom for H_2O.

The oxidation of an alcohol of the RCH_2OH type produces first an aldehyde and then a carboxylic acid.

$$RCH_2OH \xrightarrow{\text{Oxidation}} \underset{\text{aldehyde}}{\overset{\overset{\textstyle O}{\|}}{RCH}} \xrightarrow{\substack{\text{Further}\\\text{oxidation}}} \underset{\text{carboxylic acid}}{\overset{\overset{\textstyle O}{\|}}{RCOH}}$$

The net ionic equation for the formation of the aldehyde when dichromate ion is used is

$$3RCH_2OH + Cr_2O_7{}^{2-} + 8H^+ \longrightarrow 3RCH{=}O + 2Cr^{3+} + 7H_2O$$

For example, the oxidation of 1-propanol to propanal occurs as follows.

$$3CH_3CH_2CH_2OH + Cr_2O_7{}^{2-} + 8H^+ \longrightarrow 3CH_3CH_2CH{=}O + 2Cr^{3+} + 7H_2O$$
1-propanol (BP 97 °C)   propanal (BP 49 °C)

Aldehydes are much more easily oxidized than alcohols. Therefore, unless propanal is boiled out of the solution as it forms (the boiling points indicate that this is possible), it and not 1-propanol will consume still unreacted oxidizing agent and be changed to propanoic acid, $CH_3CH_2CO_2H$.

The oxidation of an alcohol of the R_2CHOH type produces a ketone. For example, the oxidation of 2-propanol gives propanone.

$$3CH_3\overset{\overset{\textstyle OH}{|}}{C}HCH_3 + Cr_2O_7{}^{2-} + 8H^+ \longrightarrow 3CH_3\overset{\overset{\textstyle O}{\|}}{C}CH_3 + 2Cr^{3+} + 7H_2O$$
2-propanol   propanone
                                                (acetone)

Pentane (BP 36.1 °C), like diethyl ether in having no OH group but having nearly the identical molecular mass, has about the same boiling point as diethyl ether.

In the body, methanol causes blindness and often death, and ethanol causes loss of coordination and inhibitions. When abused, ethanol ruins the liver and the brain.

The aldehyde group, CH=O, is one of the most easily oxidized, and the carboxyl group, CO_2H, is one of the most oxidation resistant of the functional groups.

Propanone is nearly always called by its common name, *acetone*.

Ketones strongly resist oxidation, so the ketone does not have to be removed from the oxidizing agent as it forms.

Alcohols of the type R_3COH have no removable H atom on the alcohol carbon, so they cannot be oxidized in the same manner as those of the type RCH_2OH or R_2CHOH.

EXAMPLE 21.2
Alcohol Oxidation Products

What organic product can be made by the oxidation of 2-butanol with dichromate ion? If no oxidation can occur, state so.

ANALYSIS We first have to look at the structure of 2-butanol.

$$\overset{\displaystyle OH}{\overset{|}{CH_3CHCH_2CH_3}}$$
2-butanol

2-Butanol can be oxidized because it has an H atom on the alcohol carbon (carbon-2). The oxidation results in the detachment of both this H atom and the one joined to the O atom, leaving a double bond to O.

SOLUTION We carry out the changes that the analysis found.

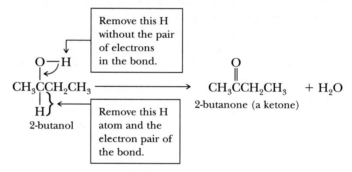

The product is a ketone, 2-butanone. The starting material had two alkyl groups, CH_3 and CH_2CH_3. It doesn't matter what these alkyl groups are, however, because the reaction takes the same course in all cases. All alcohols of the R_2CHOH type can be oxidized to ketones in this way.

■ **Practice Exercise 2** What are the structures of the products that form by the oxidation of the following alcohols.
(a) ethanol   (b) 3-pentanol

Dehydration Reactions of Alcohols   In the presence of a strong acid, like concentrated sulfuric acid, an alcohol molecule can undergo the loss of a water molecule leaving behind a carbon–carbon double bond. For example,

$$\underset{\underset{H}{|} \quad \underset{OH}{|}}{CH_2 - CH_2} \xrightarrow[\text{heat}]{\text{Acid catalyst,}} CH_2{=}CH_2 + H_2O$$
$$\text{ethanol} \qquad\qquad\qquad\qquad \text{ethene}$$

$$\underset{\underset{H}{|} \quad \underset{OH}{|}}{CH_3CH - CH_2} \xrightarrow[\text{heat}]{\text{Acid catalyst,}} CH_3CH{=}CH_2 + H_2O$$
$$\text{1-propanol} \qquad\qquad\qquad\qquad \text{propene}$$

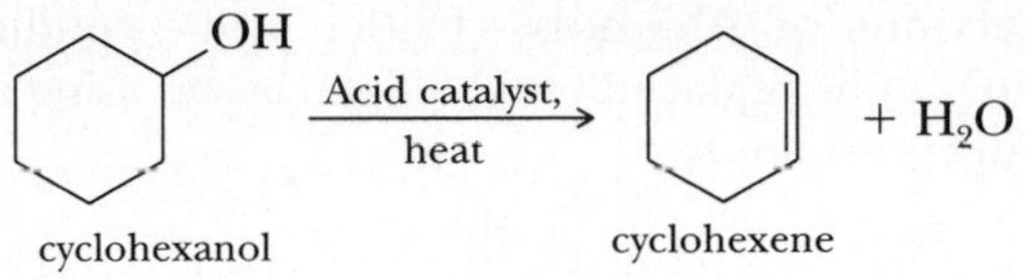

cyclohexanol $\xrightarrow[\text{heat}]{\text{Acid catalyst,}}$ cyclohexene $+ H_2O$

These are examples of another kind of organic reaction, the **elimination reaction.** What makes alcohol dehydration possible is the proton-accepting ability of the oxygen atom of the OH group. Alcohols, therefore, react with concentrated, strong acids to give an equilibrium mixture involving the protonated form of the alcohol. Ethanol, for example, reacts with (and dissolves in) concentrated sulfuric acid (H_2SO_4, but written next as $H\!-\!OSO_3H$) by the following reaction.

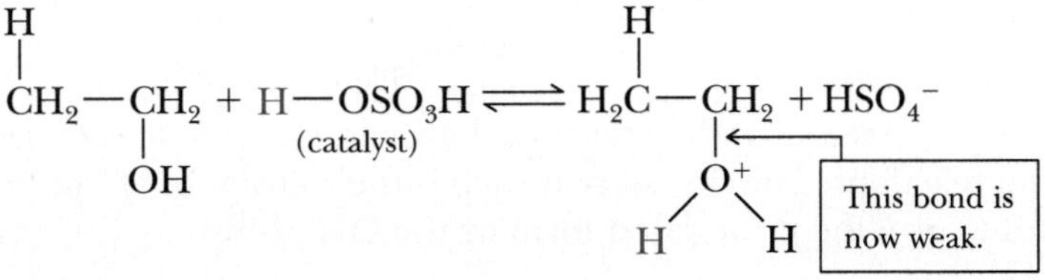

This is like the reaction of H_2SO_4 with H_2O to give the H_3O^+ and HSO_4^- ions.

The organic cation is nothing more than the ethyl derivative of the hydronium ion, $CH_3CH_2OH_2^+$, and *all three bonds to oxygen in this cation are weak,* just like all three bonds to oxygen in H_3O^+.

The next step in the dehydration of ethanol is the separation of a water molecule, which takes with it the electron pair that held it to the CH_2 group and leaves behind a positive charge on carbon.

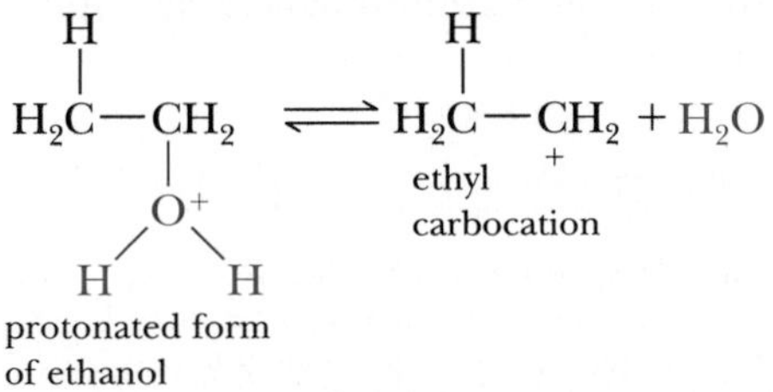

The ethyl carbocation, $CH_3CH_2^+$, to become more stable, loses a proton, donating it to some proton acceptor in the medium. The electron pair holding the departing proton to carbon stays behind to become the second bond of the new double bond in the product. All carbon atoms now have outer octets. We'll use the HSO_4^- ion (written here as $^-OSO_3H$) as the proton acceptor.

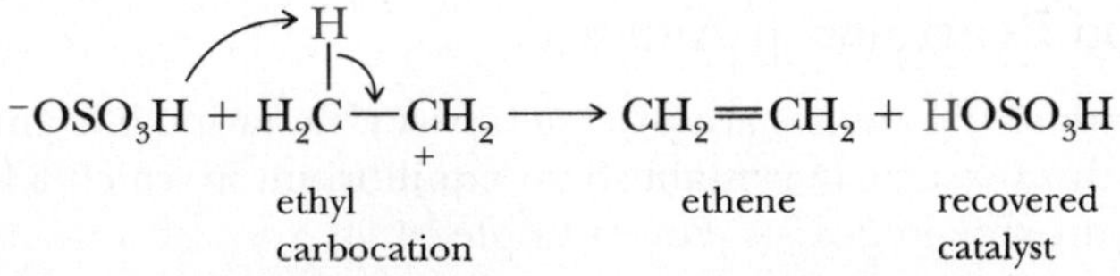

The last few steps—the separation of H_2O, loss of the proton, formation of the double bond, and recovery of the catalyst—probably occur simultaneously. Notice two things about the acid catalyst, H_2SO_4. First, it works to convert a species with strong bonds, the alcohol, to one with weak bonds strategically located, the protonated alcohol. Second, the catalyst is recovered.

Substitution Reactions of Alcohols Under acidic conditions, the OH group of an alcohol can be replaced by a halogen atom, using a concentrated hydrohalogen acid. For example,

$$\underset{\text{ethanol}}{CH_3CH_2OH} + HI(\text{concd}) \xrightarrow{\text{Heat}} \underset{\substack{\text{iodoethane}\\ \text{(ethyl iodide)}}}{CH_3CH_2I} + H_2O$$

$$\underset{\text{1-propanol}}{CH_3CH_2CH_2OH} + HBr(\text{concd}) \xrightarrow{\text{Heat}} \underset{\substack{\text{1-bromopropane}\\ \text{(propyl bromide)}}}{CH_3CH_2CH_2Br} + H_2O$$

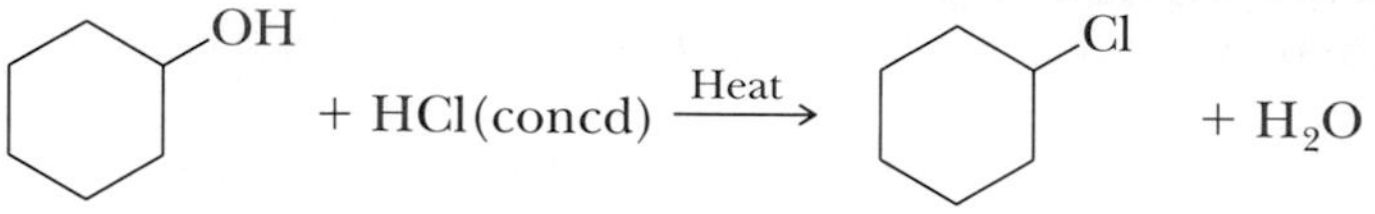

cyclohexanol chlorocyclohexane

These reactions, like the earlier reaction between chlorine and benzene, are **substitution reactions.** The first step in each is the transfer of H^+ to the OH of the alcohol to give the protonated form of the OH group.

$$R{-}OH + H^+ \longrightarrow R{-}OH_2^{\;+}$$

Once again, the acid catalyst works to weaken an important bond. Given the high concentration of halide ion, X^-, it is this species that successfully interacts with $R{-}OH_2^{\;+}$ to displace OH_2 and give $R{-}X$.

21.4 AMINES, ORGANIC DERIVATIVES OF AMMONIA

The hydrocarbon groups, which are all methyl groups in these structures, do not have to be identical.

CH_3CH_2OH, BP 78.5 °C

$CH_3CH_2NH_2$, BP 17 °C

$CH_3CH_2CH_3$, BP − 42 °C

Amines are organic derivatives of ammonia in which one, two, or three hydrocarbon groups have replaced hydrogens. Examples of amines together with their common (not IUPAC) names are

$$\underset{\substack{\text{ammonia}\\ \text{BP } -33.4\ ^\circ\text{C}}}{H{-}\ddot{N}(H){-}H} \qquad \underset{\substack{\text{methylamine}\\ \text{BP } -8\ ^\circ\text{C}}}{H_3C{-}\ddot{N}(H){-}H} \qquad \underset{\substack{\text{dimethylamine}\\ \text{BP } 8\ ^\circ\text{C}}}{H_3C{-}\ddot{N}(H){-}CH_3} \qquad \underset{\substack{\text{trimethylamine}\\ \text{BP } 3\ ^\circ\text{C}}}{H_3C{-}\ddot{N}(CH_3){-}CH_3}$$

The N—H bond is not as polar as the O—H bond, so amines boil at lower temperatures than alcohols of comparable molecular masses. (Why does trimethylamine boil lower than dimethylamine, however, despite its larger size?) Amines of low molecular mass are soluble in water. Hydrogen bonding between molecules of water and the amine facilitates this.

Basicity and Reactions of Amines

As we said earlier, the amines are Brønsted bases. Behaving like ammonia, the amines that dissolve in water establish an equilibrium in which a low concentration of hydroxide ion exists. For example,

$$\underset{\text{ethylmethylamine}}{CH_3CH_2{-}\ddot{N}(H){-}CH_3(aq)} + H_2O \rightleftharpoons \underset{\text{ethylmethylammonium ion}}{CH_3CH_2{-}\overset{+}{N}(H)_2{-}CH_3(aq)} + OH^-(aq)$$

As a result of this kind of equilibrium and its resulting hydroxide ion, aqueous solutions of amines test basic to litmus and have pH values above 7.

When an amine is mixed with a stronger proton donor than water, for example, the hydronium ion in hydrochloric acid, the amine and acid react almost quantitatively. The amine accepts a proton and changes almost 100% into its protonated form. For example,

$$CH_3CH_2-\overset{\overset{\displaystyle H}{|}}{\underset{\displaystyle ..}{N}}-CH_3(aq) + HCl(aq) \longrightarrow CH_3CH_2-\overset{\overset{\displaystyle H}{|}}{\underset{\underset{\displaystyle H}{|}}{N^+}}-CH_3(aq) + Cl^-(aq)$$

ethylmethylamine

ethylmethylammonium ion
(a protonated amine)

Even water-insoluble amines give this reaction. By thus changing into ions that can be hydrated by water, amines otherwise insoluble in water become much more soluble. Many important medicinal chemicals, like quinine, are amines, and they are usually supplied in their protonated forms so that they can be administered as aqueous solutions, not as solids.

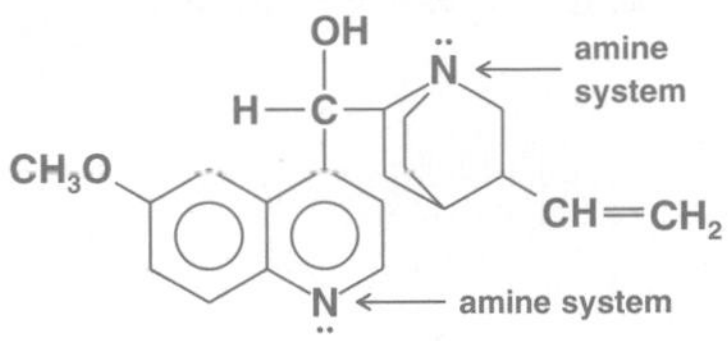

Acidity of Protonated Amines   A protonated amine is a substituted ammonium ion. Like the ammonium ion itself, protonated amines are weak Brønsted acids. They can neutralize strong base. For example,

$$CH_3NH_3^+(aq) + OH^-(aq) \longrightarrow CH_3NH_2(aq) + H_2O$$

methyl
ammonium ion

methylamine

This reverses the protonation of an amine and releases the uncharged amine molecule.

quinine
(an antimalarial drug)

We have introduced the carbon–oxygen double bond, C=O, or the **carbonyl group,** in Section 7.4 (see page 280). The group occurs in several organic families, however, so it uniquely defines none. What is attached to the carbon atom in C=O determines the specific family.

**21.5
ORGANIC
COMPOUNDS WITH
CARBONYL GROUPS**

Types and Names of Carbonyl Compounds

When the carbonyl group holds an H atom plus a hydrocarbon group (or a second H), the compound is an **aldehyde.** When C=O holds two hydrocarbon groups at C, the compound is a **ketone.** A **carboxylic acid** carries an OH on the carbon of the carbonyl group.

AL-de-hide
KEY-tone

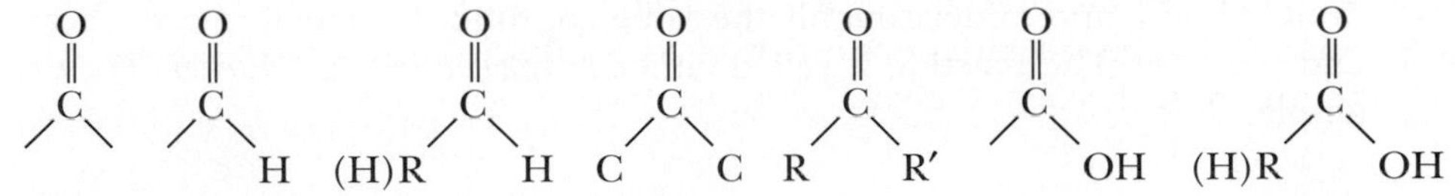

carbonyl group | aldehyde group | aldehydes | keto group | ketones | carboxyl group | carboxylic acids

The aldehyde group is often condensed to CHO, and the double bond of the carbonyl group is "understood." The ketone group is sometimes (but not often) condensed to CO. The carboxyl group is often written as CO_2H and sometimes as COOH (but be sure to remember that this group does not have the peroxide system, —O—O—).

The carbonyl group occurs widely. As an aldehyde group, it is present in the molecules of most sugars, like glucose. Another common sugar, fructose, has the keto group. The carboxyl group is present in all of the building blocks of proteins, the amino acids.

The carbonyl group is a polar group, and it helps to make compounds having it much more soluble in water than hydrocarbons of roughly the same molecular mass.

Aldehydes and Ketones

IUPAC rules for naming compounds

Nomenclature For the IUPAC names of aldehydes, the parent chain is the longest chain *that includes the aldehyde group*. The name of the parent aldehyde is made by replacing the ending *-e* of the name of the corresponding hydrocarbon with *-al*. Thus, the three-carbon aldehyde is named *propanal*, because "propane" is the name of the three-carbon alkane and the *-e* in propan*e* is replaced by *-al*. The numbering of the chain always starts by assigning the carbon of the aldehyde group position 1. This rule, therefore, makes it unnecessary to include the number locating the aldehyde group in the name, as illustrated by the name 2-methylpropanal.

$$
\begin{array}{cccc}
\overset{O}{\overset{\|}{HCH}} & \overset{O}{\overset{\|}{CH_3CH}} & \overset{O}{\overset{\|}{CH_3CH_2CH}} & \overset{CH_3}{\overset{|}{CH_3CH}}\overset{O}{\overset{\|}{-CH}} \\
\text{methanal} & \text{ethanal} & \text{propanal} & \text{2-methylpropanal} \\
\text{BP} -21\ ^\circ\text{C} & \text{BP } 21\ ^\circ\text{C} & \text{BP } 49\ ^\circ\text{C} & \text{BP } 64\ ^\circ\text{C}
\end{array}
$$

CH_3OH (molecular mass, 32), BP 65 °C
$CH_2{=}O$ (molecular mass, 30), BP −21 °C

Notice that aldehydes, which cannot form hydrogen bonds between their own molecules, boil lower than alcohols of comparable molecular masses.

For the IUPAC names of ketones, the parent chain must include the carbonyl group and be numbered from whichever end reaches the carbonyl carbon first. The number of the ketone group's location must be part of the name whenever there would otherwise be uncertainty.

We need not write 2-propanone, because if the carbonyl carbon is anywhere else in a three-carbon chain, the compound is the *aldehyde*, propanal.

$$
\begin{array}{ccc}
\overset{O}{\overset{\|}{CH_3CCH_3}} & \overset{O}{\overset{\|}{CH_3CH_2CCH_2CH_3}} & \overset{CH_3}{\overset{|}{CH_3CHCH_2CH_2}}\overset{O}{\overset{\|}{CCH_3}} \\
\text{propanone} & \text{3-pentanone} & \text{5-methyl-2-hexanone} \\
\text{(acetone)} & \text{BP } 101.5\ ^\circ\text{C} & \text{BP } 145\ ^\circ\text{C} \\
\text{BP } 56.5\ ^\circ\text{C} & &
\end{array}
$$

Hydrogenation of Aldehydes and Ketones The double bond in the carbonyl group of both aldehydes and ketones adds hydrogen just like the alkene double bond, and under roughly the same conditions—metal catalyst, heat, and pressure. The reaction is called either *hydrogenation* or *reduction*. For example,

$$
\underset{\text{ethanal}}{\overset{O}{\overset{\|}{CH_3CH}}} + H{-}H \xrightarrow[\text{catalyst}]{\text{Heat, pressure,}} \underset{\underset{H}{|}}{\overset{O{-}H}{\overset{|}{CH_3CH}}} \quad \text{or} \quad \underset{\text{ethanol}}{CH_3CH_2OH}
$$

$$\underset{\text{propanone}}{\overset{\displaystyle\overset{\text{O}}{\|}}{\text{CH}_3\text{CCH}_3}} + \text{H}-\text{H} \xrightarrow[\text{catalyst}]{\text{Heat, pressure,}} \underset{\text{H}}{\overset{\text{O}-\text{H}}{\text{CH}_3\text{CCH}_3}} \quad \text{or} \quad \underset{\text{2-propanol}}{\overset{\text{OH}}{\text{CH}_3\text{CHCH}_3}}$$

> Propanone (acetone) is commonly used as a fingernail polish remover.

Thus the H atoms take up positions at opposite ends of the carbonyl group's double bond, which then becomes a single bond holding an OH group.

Oxidation of Aldehydes Aldehydes and ketones are in separate families because of their remarkably different behavior toward oxidizing agents. As we noted on page 905, aldehydes are easily oxidized, but ketones strongly resist oxidation. Even in storage in a bottle, aldehydes are slowly oxidized by the oxygen of the air trapped in the bottle.

Carboxylic Acids and Their Derivatives

IUPAC Names of Carboxylic Acids In the IUPAC names for carboxylic acids, the parent chain must include the carboxyl carbon, which is given position number 1. The name of the hydrocarbon with the same number of carbons as the parent is changed by replacing the terminal *-e* with *-oic acid*. For example,

> **T** IUPAC rules for naming compounds

> Because carboxylic acids have both a lone oxygen and an OH group, their molecules strongly hydrogen bond to each other. Their high boiling points, relative to alcohols of comparable molecular mass, reflect this.

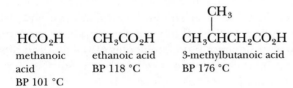

HCO_2H	$\text{CH}_3\text{CO}_2\text{H}$	$\text{CH}_3\text{CHCH}_2\text{CO}_2\text{H}$
methanoic acid	ethanoic acid	3-methylbutanoic acid
BP 101 °C	BP 118 °C	BP 176 °C

The carboxyl group is a weakly acidic group, but all carboxylic acids neutralize such bases as the hydroxide, bicarbonate, and carbonate ions. The general equation for the reaction with OH^- is

$$\text{RCO}_2\text{H} + \text{NaOH} \xrightarrow{\text{H}_2\text{O}} \text{RCO}_2\text{Na} + \text{H}_2\text{O}$$

> RCO_2Na is a salt consisting of Na^+ and RCO_2^-, the *carboxylate ion*.

Esters of Carboxylic Acids Carboxylic acids are used to synthesize two important derivatives of the acids, *esters* and *amides*. In **esters,** the OH of the carboxyl group is replaced by OR.

> A *derivative* of a carboxylic acid is a compound that can be made from the acid, or which can be changed to the acid by hydrolysis.

$$\underset{\text{esters}}{\overset{\displaystyle\overset{\text{O}}{\|}}{(\text{H})\text{RCOR}'}} \quad \text{or} \quad (\text{H})\text{RCO}_2\text{R}' \qquad \text{for example:} \quad \underset{\substack{\text{octyl ethanoate}\\ \text{(fragrance of oranges)}}}{\overset{\displaystyle\overset{\text{O}}{\|}}{\text{CH}_3\text{CO}(\text{CH}_2)_7\text{CH}_3}}$$

The IUPAC name of an ester begins with the name of the alkyl group attached to the O atom. This is followed by a separate word, one taken from the name of the parent carboxylic acid but altered by changing *-ic acid* to *-ate*. For example,

> Esters are responsible for many pleasant fragrances in nature.
>
> | Ester | Source of Aroma |
> |---|---|
> | $\text{HCO}_2\text{CH}_2\text{CH}_3$ | rum |
> | $\text{HCO}_2\text{CH}_2\text{CH}(\text{CH}_3)_2$ | raspberries |
> | $\text{CH}_3\text{CO}_2(\text{CH}_2)_4\text{CH}_3$ | bananas |
> | $\text{CH}_3\text{CO}_2(\text{CH}_2)_2\text{CH}(\text{CH}_3)_2$ | pears |
> | $\text{CH}_3\text{CO}_2(\text{CH}_2)_7\text{CH}_3$ | oranges |
> | $\text{CH}_3(\text{CH}_2)_2\text{CO}_2\text{CH}_2\text{CH}_3$ | pineapples |
> | $\text{CH}_3(\text{CH}_2)_2\text{CO}_2(\text{CH}_2)_4\text{CH}_3$ | apricots |

HCO_2CH_3	$\text{CH}_3\text{CO}_2\text{CH}_2\text{CH}_3$	$\text{CH}_3\text{CHCH}_2\text{CO}_2\text{CHCH}_3$
methyl methanoate	ethyl ethanoate	isopropyl 3-methylbutanoate
BP 31.5 °C	BP 77 °C	BP 142 °C

Formation of Esters One way to prepare an ester is to heat a solution of the parent carboxylic acid and alcohol in the presence of an acid catalyst. (We'll not go into the mechanism of what happens.) The following kind of equilibrium forms, but a substantial stoichiometric excess of the alcohol (which is usually the less expensive reactant) is commonly used to drive the position of the equilibrium toward the ester. (This works when the ester is easily separated from the other substances, as it often is.)

$$\underset{\substack{\text{carboxylic} \\ \text{acid}}}{\text{RCOH}} + \underset{\text{alcohol}}{\text{HOR}'} \underset{\text{heat}}{\overset{\text{H}^+ \text{catalyst,}}{\rightleftharpoons}} \underset{\text{ester}}{\text{RCOR}'} + \text{H}_2\text{O}$$

For example,

$$\underset{\substack{\text{butanoic acid} \\ \text{(BP 166 °C)}}}{\text{CH}_3\text{CH}_2\text{CH}_2\text{COH}} + \underset{\substack{\text{ethanol} \\ \text{(BP 78.5 °C)}}}{\text{HOCH}_2\text{CH}_3} \underset{\text{heat}}{\overset{\text{H}^+ \text{catalyst,}}{\rightleftharpoons}} \underset{\substack{\text{ethyl butanoate (BP 120 °C)} \\ \text{(fragrance of pineapple)}}}{\text{CH}_3\text{CH}_2\text{CH}_2\text{COCH}_2\text{CH}_3} + \text{H}_2\text{O}$$

Hydrolysis of Esters An ester is hydrolyzed to its parent acid and alcohol when the ester is heated together with a stoichiometric excess of water (plus an acid catalyst). The identical equilibrium as shown above forms, but the excess of water shifts it to the left to favor the carboxylic acid and alcohol.

Ester groups abound among the molecules of the fats and oils in our diets. The hydrolysis of their ester groups occurs when we digest them, except that an enzyme is the catalyst, not a strong acid. In fact, this digestion occurs in a region of the intestinal tract where the fluids are slightly basic, so carboxylic acids form not as free acids but as their *anions*. We may illustrate this by the action of aqueous sodium hydroxide on a simple ester, ethyl acetate.

The conversion of an ester into a *salt* of its acid portion plus the alcohol is called *saponification*.

$$\underset{\substack{\text{ethyl ethanoate} \\ \text{(ethyl acetate)}}}{\text{CH}_3\text{COCH}_2\text{CH}_3} + \text{NaOH}(aq) \overset{\text{Heat}}{\longrightarrow} \underset{\substack{\text{ethanoate ion} \\ \text{(acetate ion)}}}{\text{CH}_3\text{CO}^-(aq)} + \text{Na}^+(aq) + \underset{\text{ethanol}}{\text{HOCH}_2\text{CH}_3}$$

Amides of Carboxylic Acids Carboxylic acids can also be converted to *amides*, a functional group found in proteins. In **amides,** the OH of the carboxyl group is replaced by trivalent nitrogen, which may also hold any combination of H atoms or hydrocarbon groups.

$$\underset{\text{simple amides}}{(\text{H})\text{R}-\overset{\text{O}}{\overset{\|}{\text{C}}}-\text{NH}_2} \quad \text{or} \quad (\text{H})\text{RCONH}_2 \quad \text{for example} \quad \underset{\substack{\text{ethanamide} \\ \text{(acetamide)}}}{\text{CH}_3\overset{\text{O}}{\overset{\|}{\text{C}}}\text{NH}_2}$$

The *simple amides* are those in which the nitrogen bears no hydrocarbon groups, only 2 H atoms. Either or both of these H atoms can be replaced by hydrocarbon groups, however, and the resulting substance is still in the amide family.

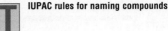
IUPAC rules for naming compounds

The IUPAC names of the simple amides are devised by first writing the name of the parent carboxylic acid. Then its ending, *-ic acid,* is replaced by *-amide.* For example.

$$CH_3CH_2CONH_2$$
propanamide

$$CH_3CH_2CH_2CH_2CONH_2$$
pentanamide

$$\overset{\overset{\displaystyle CH_3}{|}}{CH_3CHCH_2CH_2CONH_2}$$
4-methylpentanamide

One of the ways to prepare simple amides parallels that of the synthesis of esters, by heating a mixture of the carboxylic acid and an excess of ammonia.

In general:

$$\underset{\substack{\text{carboxylic}\\\text{acid}}}{\overset{\overset{\displaystyle O}{\|}}{RCOH}} + \underset{\text{ammonia}}{H-NH_2} \xrightarrow{\text{Heat}} \underset{\substack{\text{simple}\\\text{amide}}}{\overset{\overset{\displaystyle O}{\|}}{RCNH_2}} + H_2O$$

An example:

$$\underset{\text{ethanoic acid}}{\overset{\overset{\displaystyle O}{\|}}{CH_3COH}} + \underset{\text{ammonia}}{H-NH_2} \xrightarrow{\text{Heat}} \underset{\text{ethanamide}}{\overset{\overset{\displaystyle O}{\|}}{CH_3CNH_2}} + H_2O$$

Amides, like esters, can be hydrolyzed. When simple amides are heated with water, they change back to their parent carboxylic acids and ammonia. Both strong acids and strong bases promote the reaction. As the following equations show, the reaction is the reverse of the formation of an amide.

In general:

$$\underset{\substack{\text{simple}\\\text{amide}}}{\overset{\overset{\displaystyle O}{\|}}{RCNH_2}} + H-OH \xrightarrow{\text{Heat}} \underset{\substack{\text{carboxylic}\\\text{acid}}}{\overset{\overset{\displaystyle O}{\|}}{RCOH}} + NH_3$$

An example:

$$\underset{\text{ethanamide}}{\overset{\overset{\displaystyle O}{\|}}{CH_3CNH_2}} + H-OH \xrightarrow{\text{Heat}} \underset{\text{ethanoic acid}}{\overset{\overset{\displaystyle O}{\|}}{CH_3COH}} + NH_3$$

Nonbasicity of Amides Despite the presence of the NH_2 group in simple amides, the amides are not Brønsted bases like the amines or ammonia. The carbonyl group is the cause. Because of its O atom, the entire carbonyl group is electronegative. It acts to draw electron density toward itself and so tightens the unshared pair of electrons on N enough to render the N atom of an amide unable to accept a proton in dilute acid. Amides, therefore, are neutral compounds in an acid–base sense.

21.6
ORGANIC POLYMERS

FIGURE 21.2

Light weight synthetic fabrics have reduced the work of packing strong tents into remote areas (here, the Karakoram Range in Pakistan).

$CH_3CH \!=\! CH_2$
propylene

From the Greek: *poly-*, many, + *meros*, parts.

Types of Polymers

Nearly all of the organic compounds that we have studied so far have relatively low molecular masses. Both in nature and in the world of synthetics, however, many substances consist of **macromolecules,** molecules made up of hundreds or even thousands of atoms.

Macromolecular substances are almost everywhere you look. Trees and anything made of wood derive their strength from lignins and cellulose, both consisting of enormous molecules that overlap and intertwine. Anything useful for making thread and cloth consists of macromolecules. As paints cure, some of their molecules change into other molecules of enormous sizes. Recording tapes, skis, composites in recreational vehicles and their tires, backpacking gear, and all sorts of other recreational equipment derive strength from macromolecules (Figure 21.2). Synthetic macromolecules are examples of how chemists have been able to take very ordinary substances in nature, like coal, oil, air, and water, and make new materials, never seen before, with useful applications.

Some macromolecular substances have more structural order than others. A **polymer,** for example, is a macromolecular substance all of whose molecules have a small characteristic structural feature that repeats itself over and over. The molecules of polypropylene, for example, have the following system.

$$\underset{\text{polypropylene}}{\text{etc.}\!-\!CH_2\underset{|}{\overset{CH_3}{C}}HCH_2\underset{|}{\overset{CH_3}{C}}HCH_2\underset{|}{\overset{CH_3}{C}}HCH_2\underset{|}{\overset{CH_3}{C}}HCH_2\underset{|}{\overset{CH_3}{C}}HCH_2\underset{|}{\overset{CH_3}{C}}HCH_2\underset{|}{\overset{CH_3}{C}}HCH_2\underset{|}{\overset{CH_3}{C}}H\!-\!\text{etc.}}$$

If you study this structure, you can see that one structural unit occurs repeatedly (actually thousands of times). In fact, the structure of a polymer is usually represented by the use of only its repeating unit, enclosed in parentheses, with a subscript *n* standing for several thousand units.

$$\underset{\substack{\text{repeating unit} \\ \text{in polypropylene}}}{-CH_2\underset{|}{\overset{CH_3}{C}}H-} \qquad \underset{\text{polypropylene}}{(\!CH_2\underset{|}{\overset{CH_3}{C}}H\!)_n}$$

The value of *n* in the structure is not a constant for every molecule in a given sample. Thus, a polymer does not consist of molecules identical *size,* just identical in *kind,* having the same repeating unit. Notice that despite the *-ene* ending to "polypropylene," the substance has no double bonds. The polymer is named after its starting material.

The repeating unit of a polymer is contributed by a chemical raw material called a **monomer.** Thus propylene is the monomer for polypropylene. The reaction that makes a polymer out of a monomer is called **polymerization,** and the verb is "to polymerize."

There are several variations in polymer structure. If we let *a* represent a repeating unit, then the polymer in some instances is simply.

a-a-a-a-a-a-a-a-a-a-a-a-a-a-a-a-a-a-a-a-a-etc.

Thus, when *a* equals CH_2CH_2, the molecule is polyethylene.

etc.—CH_2CH_2—CH_2CH_2—CH_2CH_2—CH_2CH_2—etc.   or $+CH_2CH_2+_{\overline{n}}$

polyethylene, portion of structure

During polymerization, a long chain might develop branches that themselves are long. For example:

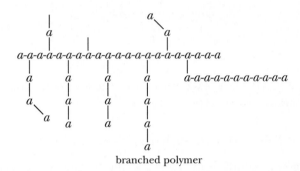

branched polymer

Copolymers   Sometimes two monomers are mixed together and polymerized; one provides a unit *a* and the other unit *b*. The operation is called *copolymerization*, and the product is a *copolymer*. Saran, for example, is a copolymer of $CH_2=CCl_2$ and $CH_2=CHCl$, and its molecules have the following units in their structure.

—CH_2CCl_2—CH_2CHCl—

Some features of the structure of Saran

The Saran structure is actually not this uniform; the units provided by the monomers do not always alternate.

Several possibilities exist when copolymers form. A copolymer might have its units *a* and *b* joined in a very regular alternating manner; for example,

a-b-a-b-a-b-a-b-a-b-a-b-a-b-a-b-a-b-a-b-a-b-a-b-etc.   or $(a\text{-}b)_n$

an alternate copolymer

In *block* copolymers, larger groups of units appear more or less repeatedly.

a-a-b-b-b-a-a-b-b-b-a-a-b-b-b-a-a-b-b-b-a-a-b-b-b-etc.

a block copolymer

Or there might be a more random organization, such as

a-a-a-b-b-a-a-b-b-b-b-a-a-a-a-a-b-b-a-b-b-b-a-a-a-b-

a random copolymer

In some polymers, units from one monomer have been grafted during polymerization onto the back bone of another polymer.

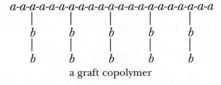

a graft copolymer

In some polymers, *cross-linkages* form during polymerization.

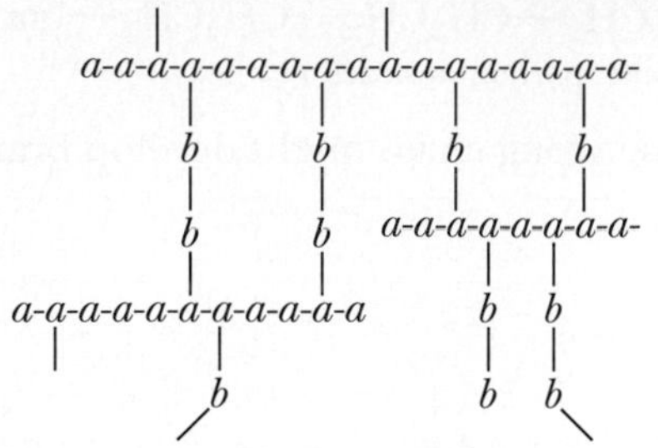

cross-linked copolymer

For these copolymers, except those of the alternate type, it isn't possible to specify one particular repeating unit.

General Principles of Polymer Formation

Before going further into a few specific polymers, let us survey the general principles that apply to the formation of any of them. Broadly speaking, there are two types of polymers, *addition polymers* and *condensation polymers*.

Olefin (OH-la-fin) is an old name for alkenes.

Although promoters serve the *functions* of catalysts, they are not called catalysts because they are not recovered unchanged.

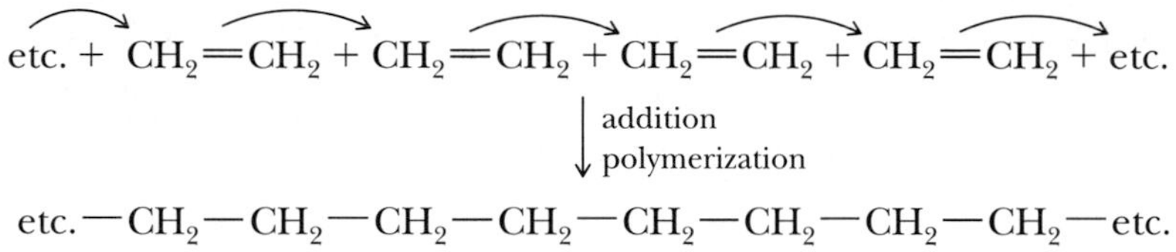

Addition Polymers   Addition polymers are those that form when their monomer molecules simply reorganize their bonds and link end to end. The *polyolefins,* which include polypropylene and polyethylene, are common examples. "Catalysts" are necessary for polymerization reactions, but because they are used up, they are called *promoters* instead. The addition polymerization of ethylene illustrates how bonds reorganize. In the following equation, we can think of molecular portions of the promoter molecules as launching the electron shifts indicated by the curved arrows. Pieces of the promoter molecules generally terminate the chains (and might be considered as the providers of the "etc." units).

$$\text{etc.} + \text{CH}_2{=}\text{CH}_2 + \text{CH}_2{=}\text{CH}_2 + \text{CH}_2{=}\text{CH}_2 + \text{CH}_2{=}\text{CH}_2 + \text{etc.}$$

addition polymerization

$$\text{etc.} - \text{CH}_2 - \text{CH}_2 - \text{CH}_2 - \text{CH}_2 - \text{CH}_2 - \text{CH}_2 - \text{CH}_2 - \text{CH}_2 - \text{etc.}$$

polyethylene

When propylene polymerizes, the same kind of reorganization of bonds occurs. The methyl groups do not appear at random along the chain; they are at every other position.

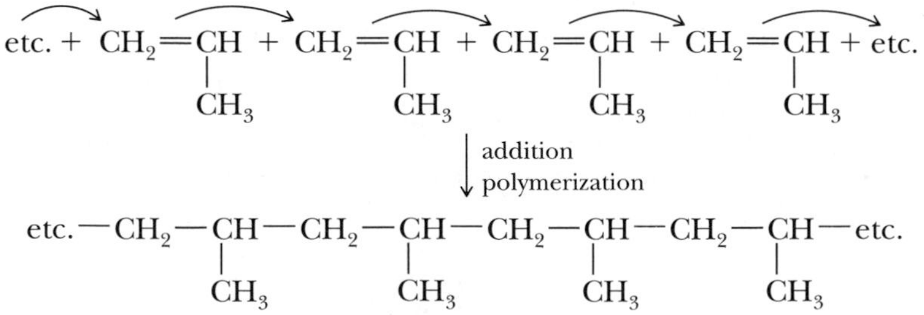

$$\text{etc.} + \text{CH}_2{=}\underset{\underset{\text{CH}_3}{|}}{\text{CH}} + \text{CH}_2{=}\underset{\underset{\text{CH}_3}{|}}{\text{CH}} + \text{CH}_2{=}\underset{\underset{\text{CH}_3}{|}}{\text{CH}} + \text{CH}_2{=}\underset{\underset{\text{CH}_3}{|}}{\text{CH}} + \text{etc.}$$

addition polymerization

$$\text{etc.} - \text{CH}_2 - \underset{\underset{\text{CH}_3}{|}}{\text{CH}} - \text{CH}_2 - \underset{\underset{\text{CH}_3}{|}}{\text{CH}} - \text{CH}_2 - \underset{\underset{\text{CH}_3}{|}}{\text{CH}} - \text{CH}_2 - \underset{\underset{\text{CH}_3}{|}}{\text{CH}} - \text{etc.}$$

polypropylene

Condensation Polymers   Condensation polymers are generally copoly-mers. Their monomers have functional groups that can react with each other, causing a small molecule to split out, usually water. We have already seen reactions in which water splits out. The formation of an ester, for example, involves the splitting out of water between a carboxylic acid and an alcohol.

$$
\underset{\text{R—C—OH}}{\overset{\text{O}}{\parallel}} + \text{H—OR}' \longrightarrow \underset{\text{RC—OR}'}{\overset{\text{O}}{\parallel}} + \text{H}_2\text{O}
$$

Similarly, the formation of an amide is the splitting out of water between a carboxylic acid and an amine.

$$
\underset{\text{R—C—OH}}{\overset{\text{O}}{\parallel}} + \underset{\text{H—N—R}'}{\overset{\text{H}}{|}} \longrightarrow \underset{\text{RC—N—R}'}{\overset{\text{O}\quad\text{H}}{\parallel\quad|}} + \text{H}_2\text{O}
$$

To exploit such reactions to make polymer molecules, all that is needed is to put *two* functional groups on each of the two starting materials. Let's represent two monomers as *aAa* and *bBb,* where *a* and *b* are pieces of functional groups that are able to split out together as *a-b.* The reaction then links *A* and *B* units together as follows.

$$n\ aAa + n\ bBb \rightarrow \text{A-B-A-B-A-B-A-B-A-B-A-B-etc.} + n\ a\text{-}b$$

The monomer *aAa* might, for example, be a dicarboxylic acid, and *bBb* might be a diol (di-alcohol). Their copolymerization would give a *polyester,* like Dacron. If *bBb* is a diamine, on the other hand, its copolymerization with a dicarboxylic acid, *aAa,* would give a *polyamide,* like nylon. We'll return to the specific examples after surveying the effects of molecular size and geometry on some of the chemical and physical properties of polymers.

General Effects of Molecular Size and Geometry on Properties of Polymers

For polymers to be useful, they must be as chemically inert as possible under the conditions in which they are employed. They must not be readily attacked by air or any of its pollutants; they must be stable toward water and microorganisms; and they must sometimes be stable even at elevated temperatures, like Teflon on a nonstick frying pan.

Polyolefins like polyethylene and polypropylene, which are entirely alkane-like, are as chemically inert as almost any polymer can be. Their unreactivity toward water (and the oxygen of the air) makes them particularly useful for storing food material. Other polymers, like polyesters and polyamides, have functional groups that water can attack, because water hydrolyzes esters and amides. For hydrolysis to happen, however, the water molecules have to be able to get at the functional groups, and this is inhibited by the sheer sizes of polymer molecules. Most polymers are so insoluble in water, and water is so insoluble in them, that no reaction with water can occur rapidly. At the same time, their stability toward water (and soil microorganisms) makes most polymers a major problem in landfill operations. Such waste-disposal problems have led fast-food restaurants and grocery stores to change how they package food items.

Teflon-coated frying pan.

$\text{—(CF}_2\text{—CF}_2\text{)}_n$
Teflon

Landfill problems have moved increasing numbers of communities to using incineration and recycling for waste disposal.

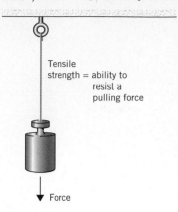

Tensile strength = ability to resist a pulling force

↓ Force

Polymers and Physical Properties Beyond chemical stability, physical properties are the most sought-after features of polymers. What is nice about Teflon are its chemical inertness and its slipperiness toward just about anything. Nylon (a polyamide) isn't eaten by moths—a chemical property, in fact—but its superior tensile strength as well as its ability to be made into fibers and fabrics of great beauty is what makes nylon such a valuable product. Dacron (a polyester) does not mildew, like cotton, and when made into fibers for a sail, it is superior to cotton (Figure 21.3). Dacron has greater strength with lower mass than cotton, and Dacron fibers do not stretch as much.

FIGURE 21.3

The exceptional wet strength and lightness of Dacron is put to good use in windsurfing (here, at the World Cup of windsurfing in Guadeloupe).

The best fiber-forming polymers have molecules that are not only very long but also have shapes that let the molecules align side by side, overlap end to end, and twist into cables of molecules, which further overlap and twist into fibers (Figure 21.4). Because the polymer molecules are large, substantial London forces of attraction occur between them. Hydrogen bonds are also present in some fiber-forming polymers, like the polyamides.

FIGURE 21.4

When molecules of polymers are very long and symmetrical about their axes, they can overlap and twist into cablelike systems that, in turn, can intertwine to give strong fibers.

Polymer molecules overlap

Overlapping molecules twist together to form cables

Cables intertwine to form fibers

Polypropylene is a polymer with methyl groups on every other carbon. The *geometry* of the projection of these groups can be varied according to the promoter used for polymerization, and the projections of the methyl groups are an additional factor in fiber-forming ability. When propene is polymerized so that all of its methyl groups are on the same side of the chain (Figure 21.5), particularly strong fibers can be made. This property plus chemical inertness make polypropylene fibers widely used to make flexible tubing, valves, and mesh for use in medicine. Polypropylene fiber is also used to make indoor–outdoor carpeting (Figure 21.6).

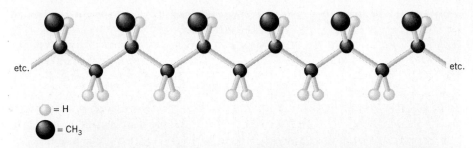

FIGURE 21.5

Propene can be polymerized so that all of the side-chain groups, CH_3, project on the same side of the main chain. With this geometry, the polymer molecules are able to assemble into strong fibers.

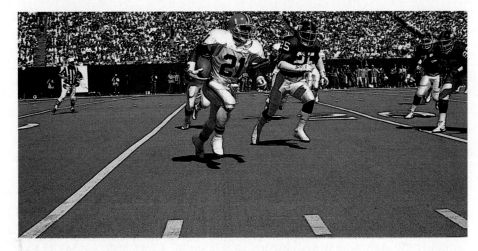

FIGURE 21.6

Artificial turf on a football field.

Polyolefins

We have used polypropylene and polyethylene, two polyolefins, to introduce various aspects of polymers and polymerization. Hundreds of alkenes and their halogen derivatives have been tested as monomers, and Table 21.3 gives some examples of those that have been particularly successful.

TABLE 21.3   Polymers of Substituted Alkenes

Polymer	Monomer	Uses
Polyvinyl chloride (PVC)	$CH_2{=}CHCl$	Insulation, credit cards, bottles, plastic pipe
Saran	$CH_2{=}CCl_2$ and $CH_2{=}CHCl$	Packaging film, fibers, tubing
Teflon	$F_2C{=}CF_2$	Nonstick surfaces, valves
Orlon	$CH_2{=}CHC{\equiv}N$	Fabrics
Polystyrene	⬡$-CH{=}CH_2$	Foamed items, insulation
Lucite	$CH_2{=}\overset{\overset{\textstyle CH_3}{\vert}}{C}CO_2CH_3$	Windows, coatings, molded items

Natural Polymer

Rubber	$CH_2{=}\overset{\overset{\textstyle CH_3}{\vert}}{C}CH{=}CH_2$	Tires, hoses, boots

FIGURE 21.7

Styrofoam, made of polystyrene, is widely used as an insulation.

$$\begin{array}{c} C_6H_5 \\ | \\ \text{-(-CH}_2\text{CH-)}_n \end{array}$$
Polystyrene

Teflon, a polymer of $F_2C{=}CF_2$, contains only carbon and fluorine atoms. Chemically, it is so inert that only molten sodium and potassium attack it.

Polystyrene is a polymer of styrene, $C_6H_5{-}CH{=}CH_2$, and its structure is like that of polypropylene but with phenyl groups, $C_6H_5{-}$, instead of methyl groups appended to every other carbon. Sometimes a gas, like carbon dioxide, is blown through molten polystyrene as it is molded into articles. As the hot liquid congeals, tiny pockets of gas are trapped, and the product is a foamed plastic, like the familiar polystyrene cups or insulation materials (Figure 21.7).

Polyacrylates

The *acrylates* are a family of polymers somewhat like the polyolefins, but made from various derivatives of acrylic acid. The alkene group in these monomers makes it possible to polymerize them much as ethylene and propylene are polymerized.

$$CH_2{=}CHCO_2H$$
acrylic acid

$$\begin{array}{c} CH_3 \\ | \\ CH_2{=}CCO_2H \end{array}$$
methacrylic acid

$$\begin{array}{c} C{\equiv}N \\ | \\ CH_2{=}CCO_2CH_3 \end{array}$$
methyl α-cyanoacrylate
(monomer for superglue)

$$\begin{array}{c} CH_3 \\ | \\ CH_2{=}CCO_2CH_3 \end{array}$$
methyl methacrylate
(the monomer for Lucite)

FIGURE 21.8

The man's shoes are glued to the top of the frame by superglue.

It's a good idea to have some acetone (fingernail polish remover) on hand when you use superglue in case you glue your fingers together.

Methyl α-cyanoacrylate is an unusual monomer in that its polymerization is promoted by water. When you buy ''superglue,'' you are buying a small tube of this monomer. A thin, invisible film of moisture is on almost any surface wherever superglue would be used, and this water promotes the polymerization. Superglue forms an exceptionally strong adhesive (see Figure 21.8).

One of the skills you should learn as you study addition polymers is the ability to convert a monomer structure into that of the polymer. Example 21.3 shows one way to do this.

EXAMPLE 21.3
Writing the Structure of an
Addition Polymer

Write a portion of the structure of the polymer that can form from methyl α-cyano-acrylate, the monomer of superglue.

ANALYSIS In the addition polymerization of alkenes—and the acrylate monomers are just substituted alkenes—a double bond disappears, and the remaining units are joined end to end in a regular array.

SOLUTION We start with the structure of the monomer.

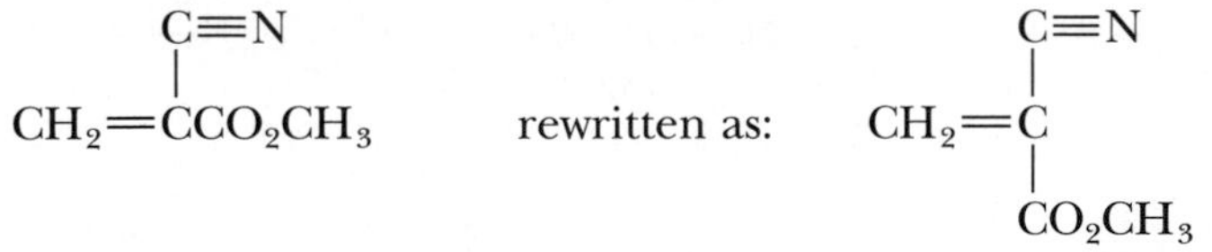

$$CH_2=CCO_2CH_3 \qquad \text{rewritten as:} \qquad CH_2=C$$

with C≡N above and CO₂CH₃ below.

Let's rewrite four of these, omitting the double bond and joining them by single bonds as we go.

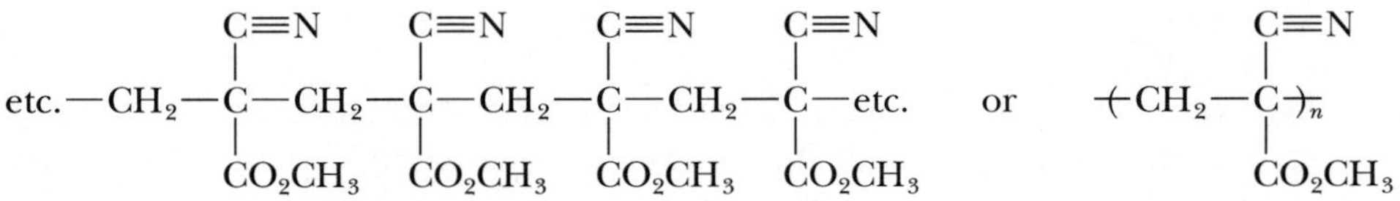

$$\text{etc.}-CH_2-C-CH_2-C-CH_2-C-CH_2-C-\text{etc.} \qquad \text{or} \qquad +CH_2-C+_n$$

poly(methyl α-cyanoacrylate)

■ **Practice Exercise 3** Write the structure of the polymer of cyanoethene, $CH_2=CH-CN$. The fiber made from this polymer is called Orlon and finds use in making lustrous sweaters and other fabrics. Show four repeating units in one representation, and also show the condensed structure.

Natural Rubber

Dienes, alkenes with two double bonds, are also important monomers. We will note just one diene polymer, naturally occurring rubber. Rubber is a member of a large family of polymers called **elastomers,** so named because of their ability to recover their shapes after being stretched or otherwise distorted.

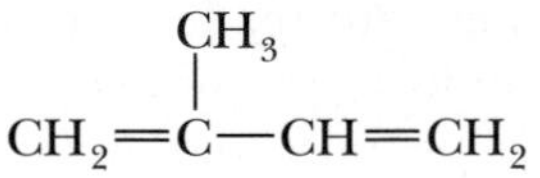

$$CH_2=C-CH=CH_2$$

with CH₃ above.

isoprene
(2-methyl-1,3-
butadiene)

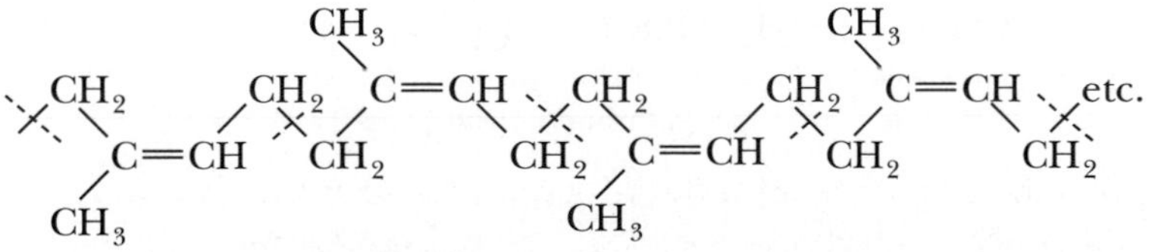

natural rubber
(note the repeating units, furnished by
isoprene, between the dashed lines)

Natural rubber retains several alkene groups, which is why it is easily attacked by ozone. Chemists have largely solved this problem by finding antioxidant compounds to add to the rubber when rubber products are made.

A copolymer of isoprene and
2-methylpropene (isobutylene) is
a synthetic elastomer called
butyl rubber.

Polyesters and Polyamides

Polyesters and polyamides are *condensation polymers*. To make a polyester, one monomer used is a dicarboxylic acid and the other is a diol (an organic compound with two alcohol groups). Water molecules split out during the copolymerization, and a succession of ester groups forms as the polymer chain grows.

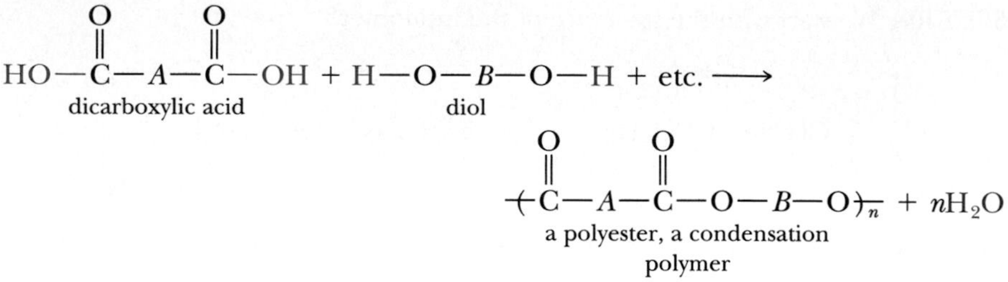

$$
\underset{\text{dicarboxylic acid}}{HO-\overset{O}{\overset{\|}{C}}-A-\overset{O}{\overset{\|}{C}}-OH} + \underset{\text{diol}}{H-O-B-O-H} + etc. \longrightarrow
$$

$$
\underset{\substack{\text{a polyester, a condensation}\\ \text{polymer}}}{\left(\overset{O}{\overset{\|}{C}}-A-\overset{O}{\overset{\|}{C}}-O-B-O\right)_n} + nH_2O
$$

Dacron   The polyester made from terephthalic acid and ethylene glycol is called *Dacron* when it is made into fibers and *Mylar* when it is cast as a thin film to make the backings for cassette tapes.

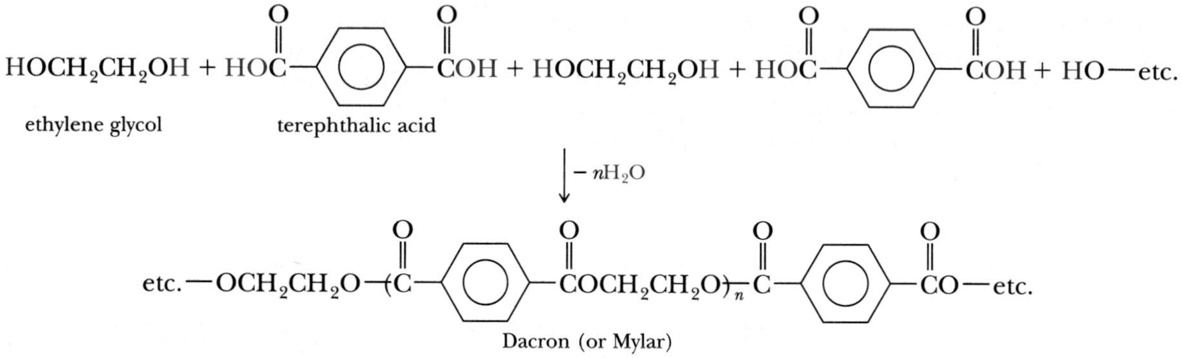

$$
\underset{\text{ethylene glycol}}{HOCH_2CH_2OH} + \underset{\text{terephthalic acid}}{HO\overset{O}{\overset{\|}{C}}-\underset{\bigcirc}{}-\overset{O}{\overset{\|}{C}}OH} + HOCH_2CH_2OH + HO\overset{O}{\overset{\|}{C}}-\underset{\bigcirc}{}-\overset{O}{\overset{\|}{C}}OH + HO-etc.
$$

$$\downarrow -nH_2O$$

$$
etc.-OCH_2CH_2O-\left(\overset{O}{\overset{\|}{C}}-\underset{\bigcirc}{}-\overset{O}{\overset{\|}{C}}OCH_2CH_2O\right)_n-\overset{O}{\overset{\|}{C}}-\underset{\bigcirc}{}-\overset{O}{\overset{\|}{C}}O-etc.
$$

Dacron (or Mylar)

Nylon   A number of polymers are made from diamines and dicarboxylic acids, and they make up a family of polyamides called the *nylon family*. One example is nylon-66, probably the most common member. We can visualize its relationship to its monomers by the following equation.

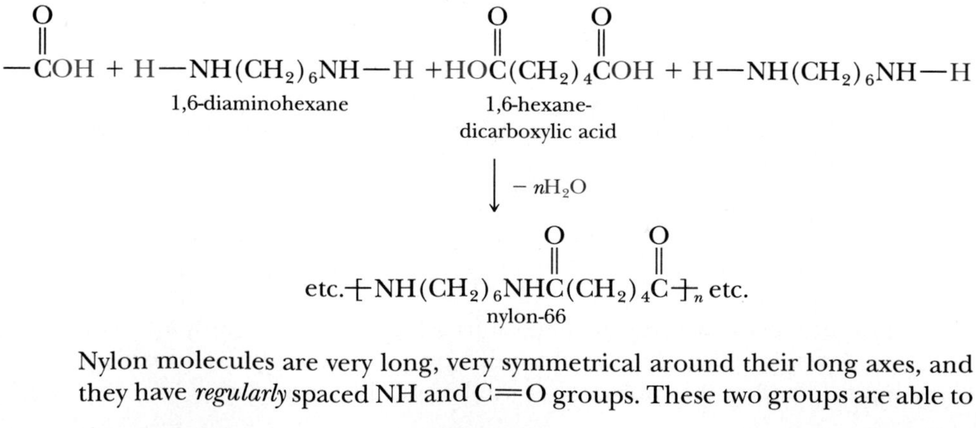

$$
-\overset{O}{\overset{\|}{C}}OH + \underset{\text{1,6-diaminohexane}}{H-NH(CH_2)_6NH-H} + \underset{\substack{\text{1,6-hexane-}\\\text{dicarboxylic acid}}}{HO\overset{O}{\overset{\|}{C}}(CH_2)_4\overset{O}{\overset{\|}{C}}OH} + H-NH(CH_2)_6NH-H
$$

$$\downarrow -nH_2O$$

$$
etc.\left(NH(CH_2)_6NH\overset{O}{\overset{\|}{C}}(CH_2)_4\overset{O}{\overset{\|}{C}}\right)_n etc.
$$

nylon-66

Nylon molecules are very long, very symmetrical around their long axes, and they have *regularly* spaced NH and C=O groups. These two groups are able to

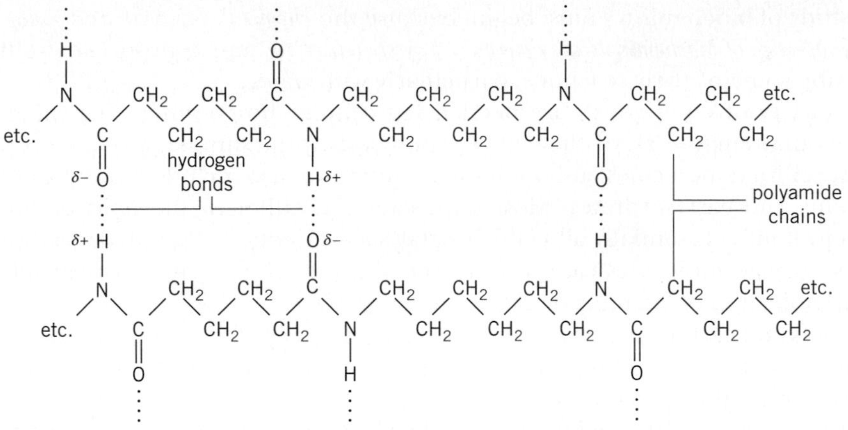

FIGURE 21.9

Hydrogen bonds occur at regular intervals between the long chains of nylon molecules in a nylon fiber.

attract each other by a hydrogen bond

$$N—H \cdots O=C$$

hydrogen bond ($\cdots$)

As nylon molecules aggregate side by side, innumerable hydrogen bonds are set up *between* the chains, which makes possible very strong fibers (see Figure 21.9).

The world of living things includes natural polymers of major importance, substances such as proteins, starch, cellulose, and the chemicals of genes. Our study of the principles of polymer structure just concluded will therefore serve us well as we shift our attention to those substances at the heart of the molecular basis of life.

The functional groups selected for study in the preceding sections were chosen for their importance to biochemical compounds. **Biochemistry** is the systematic study of the chemicals of living systems, their organization into cells, and their chemical interactions. Biochemicals have no life in themselves as isolated compounds, yet life has a molecular basis. Only when chemicals are organized into cells in tissues can interactions occur that enable tissue repair, cell reproduction, the generation of energy, the removal of wastes, and the management of a number of other functions. The reactions that sustain life obey all of the known laws of chemistry. So also do the reactions by which a disease sets in. The final, deepest level of understanding a disease comes only when we know its chemistry. This is true whether the disease is caused by bacteria, viruses, a vitamin deficiency, or genetic faults. Thus, for many years the Nobel Prizes in physiology or medicine have nearly always gone to scientists who are basically chemists, being specialists in biochemistry or molecular biology.

Three Important Needs for Life   A living system needs materials, energy, and information or "blueprints." Our focus in this chapter will be on the substances that supply them—carbohydrates, lipids, proteins, and nucleic acids—the basic materials whose molecules, together with water and a few kinds of ions, make up cells and tissues. Our brief survey of biochemistry will focus mainly on the *structures* of selected biochemical materials. This is where

21.7
MAJOR TYPES OF BIOCHEMICALS

A major focus of *molecular biology* is the chemical reactions by which genes work.

Starch, table sugar, and cotton are all carbohydrates.

any study of biochemistry must begin, because the *chemical, physical, and biological properties of biochemicals are determined by structure.* Where appropriate, we'll describe some of their reactions, particularly with water.

A variety of compounds are needed for cellular functioning. The membranes that enclose all of the cells of your body, for example, are made up mostly of lipid molecules, although the membranes also include molecules of proteins and carbohydrates. Most hormones are either in the lipid or the protein family. Essentially all cellular catalysts—enzymes—are proteins, but many enzyme molecules cannot function without the presence of relatively small molecules of vitamins or certain metal ions.

Lipids and carbohydrates are also our major sources of the chemical energy we need to function. In times of fasting or starvation, however, the body is also able to draw on its proteins for energy.

The information needed to operate a living system is borne by molecules of nucleic acids. Their structures carry the *genetic code* that instructs cells how to make its proteins, including its enzymes. Hundreds of diseases, like cystic fibrosis and sickle-cell anemia, are caused by defects in the molecular structures of nucleic acids. Viruses, like the AIDS virus, work by taking over the genetic machinery of a cell.

Lipids include the fats and oils in our diet, like butter, margarine, salad oils, and baking shortenings (e.g., lard).

Meat and egg albumin are particularly rich in proteins.

Individual genes are actually sections of molecules of a nucleic acid called DNA for short.

21.8 CARBOHYDRATES

Carbohydrates are naturally occurring polyhydroxyaldehydes or polyhydroxyketones, or else they are compounds that react with water to give these compounds. The carbohydrates include common sugar—sucrose—as well as starch and cellulose.

Monosaccharides   Carbohydrates unable to react with water are called **monosaccharides.** The most common is glucose, a pentahydroxyaldehyde, and probably the most widely occurring structural unit in the entire living world. Glucose is the chief carbohydrate in blood, and it provides the building unit for such important polysaccharides as cellulose and starch. Fructose, a pentahydroxyketone, is produced together with glucose when we digest sugar, and honey is also rich in fructose.

Glucose is sometimes called corn sugar because it is manufactured from cornstarch by hydrolysis.

$$CH_2CH-CH-CH-CH-CH=O$$
$$\quad|\quad|\quad\quad|\quad\quad|\quad\quad|$$
$$OH\ OH\quad OH\quad OH\quad OH$$

glucose (open-chain form)
a polyhydroxyaldehyde

$$CH_2CH-CH-CH-CHCCH_2$$
$$\quad|\quad|\quad\quad|\quad\quad|\quad\quad|\quad(=O)$$
$$OH\ OH\quad OH\quad OH\ OH$$

fructose (open-chain form)
a polyhydroxyketone

Cyclic Forms of Monosaccharides   When dissolved in water, the molecules of most carbohydrates exist in a mobile equilibrium involving more than one structure. Glucose, for example, exists as two cyclic forms and one open-chain form in equilibrium in water (Figure 21.10). The open-chain form, the only one with a free aldehyde group, is represented by less than 0.1% of the solute molecules. Yet, the glucose solution is able to give the chemical reactions of a polyhydroxyaldehyde, because the equilibrium shifts to supply more of any of its members when a specific reaction occurs to one (in accordance with Le Châtelier's principle). For example, *all* of the glucose molecules dissolved in water, regardless of which form they are in, can undergo the oxida-

α-Glucose Open form (polyhydroxyaldehyde) β-Glucose

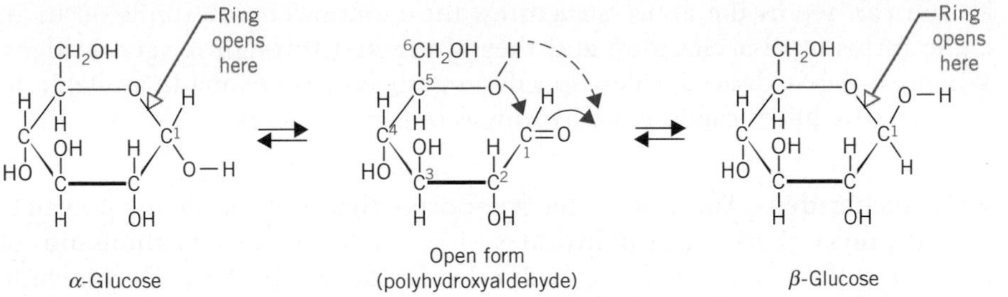

FIGURE 21.10

The three forms of glucose in equilibrium in an aqueous solution. The α- and β-forms differ only in the direction at which the OH group at carbon-1 projects. The curved arrows in the open form show how bonds become reoriented as it closes into a cyclic form. Depending on how the CH=O group is turned at the moment of ring closure, the new OH group at C-1 takes up one of two possible orientations. Galactose, another nutritionally important monosaccharide, differs from glucose only in the projection of the OH group at carbon-4.

tion of the aldehyde group of the open-chain form of glucose to a carboxyl group. The cyclic forms simply open up as the equilibrium shifts during oxidation.

Disaccharides   Carbohydrates whose molecules are split into two monosaccharide molecules by reacting with water are called **disaccharides.** *Sucrose* (table sugar, cane sugar, or beet sugar) is an example, and its hydrolysis gives glucose and fructose.

The six-membered rings of the cyclic forms of monosaccharides are not actually flat rings, but this is a complication that will not concern us. The essential structural features are shown by the flat rings widely used in discussions of carbohydrate chemistry.

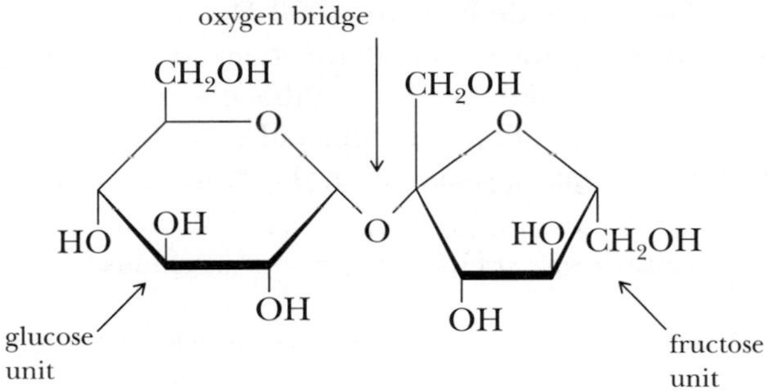

sucrose

We could represent sucrose as Glu—O—Fru, where Glu is a glucose unit and Fru is a fructose unit, both joined by an oxygen bridge, —O—. The hydrolysis of sucrose, the chemical reaction by which we digest this disaccharide, can thus be represented as follows.

The digestion of carbohydrates is the reaction of disaccharides and starch with water, catalyzed by special enzymes present in the digestive juices.

$$\text{Glu—O—Fru} + H_2O \xrightarrow[\text{(hydrolysis)}]{\text{Digestion}} \text{glucose + fructose}$$
sucrose

Lactose (milk sugar) hydrolyzes to glucose and galactose (Gal), an isomer of glucose, and this is the reaction by which we digest lactose.

Many adults lack the enzyme *lactase* needed to digest the lactose in ordinary milk products, and they suffer intestinal distress after drinking milk or eating ice cream.

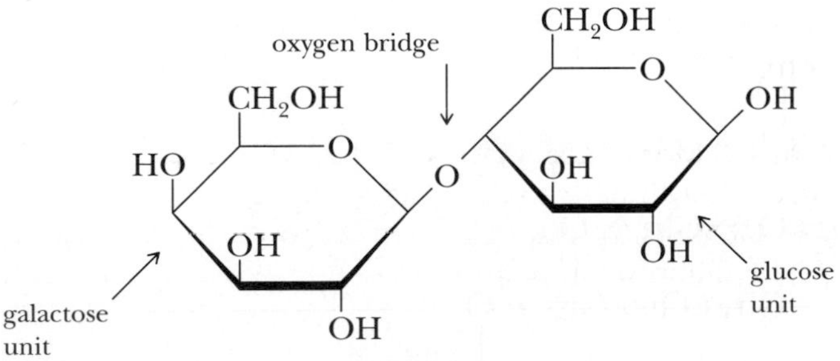

lactose

$$\text{Gal—O—Glu} + H_2O \xrightarrow[\text{(hydrolysis)}]{\text{Digestion}} \text{galactose + glucose}$$
lactose

As you can see in the above structures, the monosaccharide units occur in cyclic forms in disaccharides, and they are linked through oxygen bridges. When we digest disaccharides, specific digestive enzymes made available in our digestive juices catalyze the hydrolysis of these bridges.

Polysaccharides   Some of the most important naturally occurring polymers are the **polysaccharides,** carbohydrates whose molecules involve thousands of monosaccharide units linked to each other by oxygen bridges. They include starch, glycogen, and cellulose. The complete hydrolyses of all three yield only glucose, so their structural differences involve details of the oxygen bridges.

Plants store glucose units for energy needs in molecules of *starch,* found often in seeds and tubers (e.g., potatoes). Starch consists of two kinds of glucose polymers. The structurally simpler kind is *amylose,* which makes up roughly 20% of starch. We may represent its structure as follows, where O is the oxygen bridge linking glucose units.

$$\text{Glu}\text{(O}-\text{Glu)}_n\text{OH}$$
amylose; n is very large

The average amylose molecule has over 1000 glucose units tied together by oxygen bridges. These are the sites that are attacked and broken when water reacts with amylose during the digestion of this material.

When starch is eaten and digested, molecules of glucose are released and are eventually delivered into circulation in the bloodstream.

$$\text{Amylose} + n\text{H}_2\text{O} \xrightarrow[\text{(hydrolysis)}]{\text{Digestion}} n \text{ glucose}$$

The bulk of starch is made up of *amylopectin,* whose molecules are even larger than those of amylose. The amylopectin molecule consists of several amylose molecules linked by oxygen bridges from the end of one amylose unit to a site somewhere along the "chain" of another amylose unit (Figure 21.11).

Animals store glucose units for energy as *glycogen,* a polysaccharide with a molecular structure very similar to that of amylopectin. When we eat starchy foods and deliver glucose molecules into the bloodstream, any excess glucose not needed to maintain a healthy concentration in the blood is removed from circulation by particular tissues, like the liver and muscles. Liver and muscle cells convert glucose to glycogen. Later, during periods of high energy de-

Human blood has roughly 100 mg of glucose per 100 mL. The brain normally relies on the chemical energy of glucose for its function.

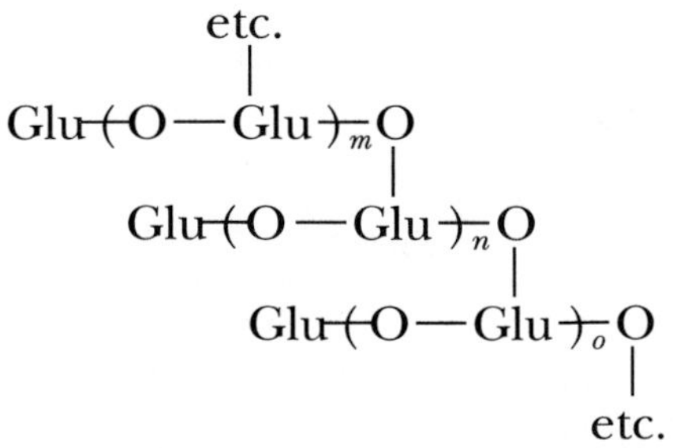

FIGURE 21.11

Amylopectin, a major constituent of starch. Molecular masses of starch samples from different plants range from 50,000 to several million. (A molecular mass of 1 million corresponds to about 6000 glucose units.)

Amylopectin (*m*, *n*, and *o* are large numbers)

mand or fasting, glucose units are released from the glycogen reserves so that the concentration of glucose in the blood stays high enough for the needs of the brain and other tissues.

Cellulose is a polymer of glucose, much like amylose, but with the oxygen bridges oriented with different geometries. The difference has been called one of nature's "tremendous trifles," because we can digest starch but not cellulose. Cellulose is a major constituent of the walls of plant cells, and it makes up nearly 100% of cotton. We lack the enzyme needed to hydrolyze cellulose, so we are unable to use cellulosic materials like lettuce for food, only for fiber. Animals that eat grass and leaves, however, have bacteria living in their digestive tracts that convert cellulose into small molecules, which the host organism then appropriates for its own use.

21.9 LIPIDS

Lipids are natural products that tend to dissolve in such nonpolar solvents as diethyl ether and benzene but do not dissolve in water. Molecules of lipids are relatively nonpolar with large segments that are entirely hydrocarbon-like.

The lipid family is huge and diverse, because the only structural requirement is that large portions of lipid molecules be so hydrocarbon-like that they have poor solubility in water. Thus, the lipid family includes cholesterol as well as sex hormones, like estradiol and testosterone. You can see from their structures how largely hydrocarbon-like they are.

We see here another illustration of the "like dissolves like" rule (see Section 12.1).

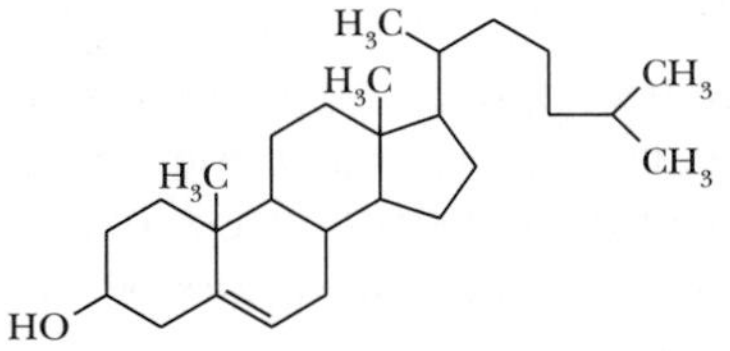

cholesterol

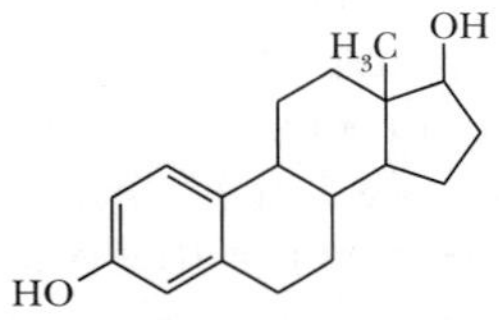

estradiol (a female sex hormone)

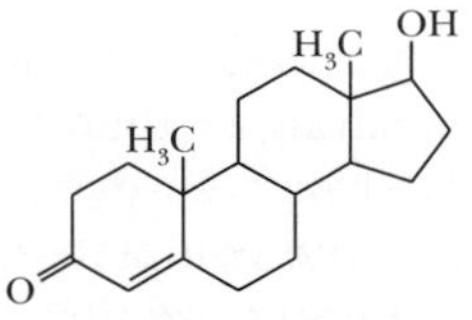

testosterone (male sex hormone)

Triacylglycerols

The lipid family also includes the edible fats and oils in our diets—substances such as olive oil, corn oil, peanut oil, butterfat, lard, and tallow. These are **triacylglycerols** (commonly *triglycerides*)—esters between glycerol, an alcohol with three OH groups, and any three of several long-chain carboxylic acids.

CH_2OH
$CHOH$
CH_2OH
glycerol

$$CH_2OCR$$
$$CHOCR'$$
$$CH_2OCR''$$

triacylglycerol (general structural features)

$$CH_2OC(CH_2)_7CH=CH(CH_2)_7CH_3 \leftarrow \text{oleic acid unit}$$
$$CHOC(CH_2)_{16}CH_3 \leftarrow \text{stearic acid unit}$$
$$CH_2OC(CH_2)_7CH=CHCH_2CH=CH(CH_2)_4CH_3 \leftarrow \text{linoleic acid unit}$$

triacylglycerol—typical molecule present in a vegetable oil

TABLE 21.4   Common Fatty Acids

Fatty Acid	Number of Carbon Atoms	Structure	Melting Point (°C)
Myristic acid	14	$CH_3(CH_2)_{12}CO_2H$	54
Palmitic acid	16	$CH_3(CH_2)_{14}CO_2H$	63
Stearic acid	18	$CH_3(CH_2)_{16}CO_2H$	70
Oleic acid	18	$CH_3(CH_2)_7CH{=}CH(CH_2)_7CO_2H$	4
Linoleic acid	18	$CH_3(CH_2)_4CH{=}CHCH_2CH{=}CH(CH_2)_7CO_2H$	−5
Linolenic acid	18	$CH_3CH_2CH{=}CHCH_2CH{=}CHCH_2CH{=}CH(CH_2)_7CO_2H$	−11

Fatty acids have an *even* number of C atoms because they are made in living systems from two-carbon acetate ions.

Fatty Acids   The carboxylic acids used to make triacylglycerols, the **fatty acids,** generally have just one carboxyl group on an *unbranched* chain of an *even* number of carbon atoms (Table 21.4). Their long hydrocarbon chains make triacylglycerols mostly like hydrocarbons in physical properties, including insolubility in water. Many fatty acids have alkene groups.

Lard is hog fat. Tallow is beef fat.

Lipids obtained from vegetable sources, like olive oil, corn oil, and peanut oil, are called *vegetable oils* and are liquids at room temperature. Lipids from animal sources, like lard and tallow, are called *animal fats* and are solids at room temperature. The vegetable oils generally have more alkene double bonds per molecule than animal fats and so are said to be *polyunsaturated*. The double bonds are usually cis and so the molecules are kinked, making it more difficult for them to nestle close together, experience London forces, and so be in the solid state.

Cooking oils are advertised as *polyunsaturated* because of their several alkene groups per molecule.

Digestion of Triacylglycerols   We digest the triacylglycerols by hydrolysis, meaning by their reaction with water. Our digestive juices in the upper intestinal tract have enzymes called *lipases* that catalyze these reactions. For example, the complete digestion of the triacylglycerol shown above occurs by the following reaction.

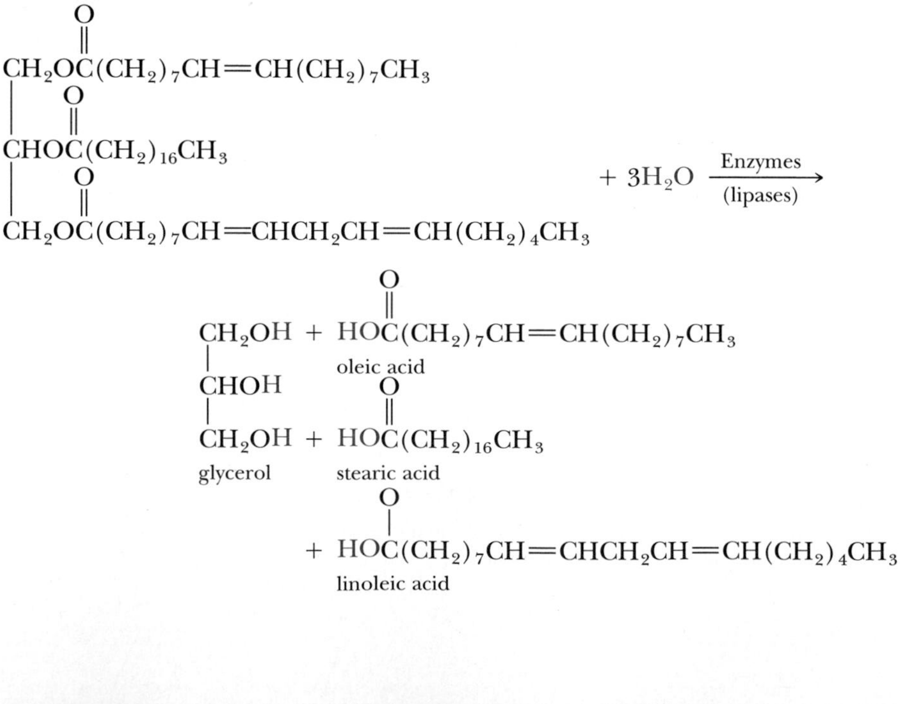

$$
\begin{array}{l}
CH_2O\overset{\displaystyle O}{\overset{\|}{C}}(CH_2)_7CH{=}CH(CH_2)_7CH_3 \\
CHO\overset{\displaystyle O}{\overset{\|}{C}}(CH_2)_{16}CH_3 \\
CH_2O\overset{\displaystyle O}{\overset{\|}{C}}(CH_2)_7CH{=}CHCH_2CH{=}CH(CH_2)_4CH_3
\end{array}
\quad + \; 3H_2O \xrightarrow[\text{(lipases)}]{\text{Enzymes}}
$$

$$
\begin{array}{ll}
CH_2OH & + \; HO\overset{\displaystyle O}{\overset{\|}{C}}(CH_2)_7CH{=}CH(CH_2)_7CH_3 \\
& \qquad\qquad \text{oleic acid} \\
CHOH & \\
CH_2OH & + \; HO\overset{\displaystyle O}{\overset{\|}{C}}(CH_2)_{16}CH_3 \\
\text{glycerol} & \quad\;\; \text{stearic acid} \\
& + \; HO\overset{\displaystyle O}{\overset{\|}{C}}(CH_2)_7CH{=}CHCH_2CH{=}CH(CH_2)_4CH_3 \\
& \qquad\qquad \text{linoleic acid}
\end{array}
$$

Triacylglycerol digestion occurs where the medium has a pH above 7, so the fatty acids actually form as their anions, $RCO_2{}^-$.

Powerful surface-active compounds in the bile, a fluid that empties into the digestive tract from the gallbladder, emulsify triacylglycerols as they move through the digestive tract and so aid in their digestion.

Saponification of Triacylglycerols   When the hydrolysis of a triacylglycerol is carried out in the presence of sufficient base, like sodium hydroxide or sodium carbonate, the fatty acids are released as their anions, not as the free acids. The reaction is called *saponification,* and a mixture of the salts of long-chain fatty acids is ordinary *soap.* We discussed how soaps and detergents work in Section 12.10, page 530.

Surface-active agents, like detergents and soap, lower the surface tension of water and so help to make oil-in-water emulsions.

Hydrogenation of Vegetable Oils   Vegetable oils are generally less expensive to produce than butterfat, but because the oils are liquids, few people care to use them as bread spreads. Remember that animal fats, like butterfat, are solids at room temperature, and vegetable oils differ from animal fats only in the number of carbon–carbon double bonds per molecule. Simply adding hydrogen to the double bonds of a vegetable oil, therefore, changes the lipid from a liquid to a solid. If only one of the three double bonds in the triacylglycerol molecule shown above were hydrogenated, the product (with two double bonds) would melt above room temperature.

Hydrogenated vegetable oils are chemically identical to the triacylglycerols in animal fats.

Chemists have learned how to control the addition of hydrogen so that not all double bonds in a vegetable oil are hydrogenated. If they were, the product would be like lard and, for most people, would be as unpleasant as lard as a bread spread. Thus, the *limited* hydrogenation of a vegetable oil changes the oil into a solid used as the chief ingredient in margarine. The better margarines melt on the tongue, like butter, which is one reason for butter's popularity. The peanut oil in "old-fashioned" peanut butter is similarly given limited hydrogenation to make a homogeneous spread.

The Lipids of Cell Membranes   The lipids involved in the structures of cell membranes in animals are not the relatively simple triacylglycerols. Some are diacylglycerols with the third site on the glycerol unit taken up by an attachment to a phosphate unit. This, in turn, is joined to an amino alcohol unit by an ester-like network. The phosphate unit carries one negative charge, and the amino unit has a positive charge. These lipids are called *glycerophospholipids.* Lecithin is one example.

The amino alcohol unit in lecithin is contributed by choline, a cation:
$$HOCH_2CH_2\overset{+}{N}(CH_3)_3$$
Ethanolamine (in its protonated form), $HOCH_2CH_2NH_3^+$, is another amino alcohol that occurs in phospholipids.

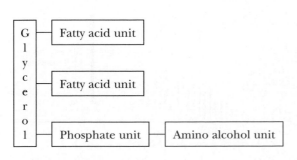

glycerophospholipid, general structure

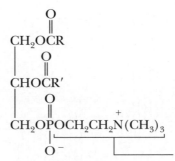

a glycerophospholipid
(phosphatidylcholine, "lecithin")

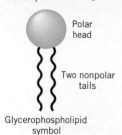

Polar
head

Two nonpolar
tails

Glycerophospholipid
symbol

Hydrophobic and *hydrophilic* have been introduced on page 530.

The glycerophospholipids, members of a family of *phospholipids*, illustrate that it is possible for lipid molecules to carry polar sites, even ionic units, and still not be very soluble in water. The combination of both nonpolar and polar or ionic units within the same molecule enable the glycerophospholipids and similar phospholipids to be major building units for the membranes of animal cells.

The Lipid Bilayer of Cell Membranes   The purely hydrocarbon-like portions of a glycerophospholipid molecule, the long R groups contributed by the fatty acid units, are *hydrophobic*. The portions bearing the electrical charges are *hydrophilic*. In an aqueous medium, therefore, the molecules of a glycerophospholipid aggregate in a way that minimizes the exposure of the hydrophobic side chains to water and maximizes contact between the hydrophilic sites and water. These interactions are roughly what take place when glycerophospholipid molecules aggregate to form the *lipid bilayer* membrane of an animal cell (Figure 21.12). The hydrophobic side chains intermingle in the center of the layer where water molecules do not occur. The hydrophilic groups are exposed to the aqueous medium inside and outside of the cell. Not shown in Figure 21.12 are cholesterol and cholesterol ester molecules, which help to stiffen the membranes. Thus cholesterol is essential to the cell membranes of animals.

If a pin were stuck through a cell membrane and then withdrawn, the membrane would automatically close up. It is remarkable that a unit as vital to life as a cell depends for its structural integrity on the sum of many weak forces, the forces involved in water-attracting and water-avoiding dipole–dipole interactions, hydrogen bonds, and London forces.

Neurotransmitters are small molecules that travel between the end of one nerve cell and the surface of the next to transmit the nerve impulse.

Cell membranes also include protein units, which provide several services. Some are molecular recognition sites for molecules of hormones and neurotransmitters. Others provide channels for the movements of ions, like Na^+, K^+, Ca^{2+}, Cl^-, HCO_3^-, and others. Some are channels for the transfer of small organic molecules, like glucose.

FIGURE   21.12

Animal cell membrane, showing its bilayer organization.

Hydrophilic
areas

Hydrophobic
areas

Hydrophilic
areas

Peripheral
protein

Hydrophilic
heads

Hydrophobic
tails

Glycerophosholipid molecules

Integral
protein

The proteins are a huge family of substances that make up about half of the human body's dry weight. They are found in all cells and in virtually all parts of cells. They are the stress-bearing constituents of skin, muscles, and tendons. In teeth and bones, proteins function like steel rods in concrete, namely, as reinforcing members without which the minerals would crumble. Proteins serve as enzymes, hormones, and neurotransmitters. They carry oxygen in the blood stream as well as some of the waste products of metabolism. No other group of compounds has such a variety of functions in living systems.

The dominant structural units of **proteins** are macromolecules called **polypeptides,** which are made from a set of monomers called **amino acids.** Most protein molecules include, besides their polypeptide units, small organic molecules or metal ions, and the whole protein lacks its characteristic biological function without these species (Figure 21.13).

21.10
PROTEINS

Proteins truly merit the name, taken from the Greek *proteios,* meaning "of the first rank."

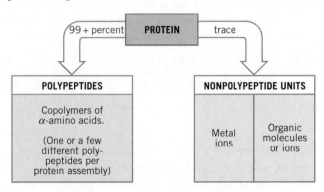

FIGURE 21.13

Components of proteins. Some proteins consist exclusively of polypeptide molecules, but most also have nonpolypeptide units such as small organic molecules or metal ions, or both.

Amino Acids   The monomer units for polypeptides are a group of about 20 α-amino acids all of which share the following structural features. R stands for a structural group, an *amino acid side chain.* Some examples of the set of 20 amino acids used to make proteins are given below. All are known by their common names. Each also has a three-letter symbol.

The alpha (α) position is the carbon of the chain right next to the carbonyl carbon.

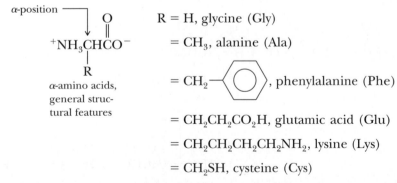

The simplest amino acid is aminoacetic acid, or glycine, for which the "side chain" is H. Glycine, like all of the amino acids in their pure states, exists as a *dipolar ion.* Such an ion results from an internal self-neutralization, by the transfer of a proton from the proton-donating carboxyl group to the proton-accepting amino group.

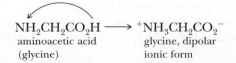

Polypeptides   Polypeptides are copolymers of the amino acids. The carboxyl group of one amino acid becomes joined to the amino group of another by means of the same kind of carbonyl–nitrogen bond found in amides, but here called the **peptide bond.** Let's see how two amino acids, glycine and alanine, can become linked by a (multistep) splitting out of water.

$$^+NH_3CH_2C-O^- + H-\overset{H}{\underset{H}{\overset{+}{N}}}CHCO^- \xrightarrow[\text{by several steps}]{\text{In the body,}} {}^+NH_3CH_2C-NHCHCO^- + H_2O$$

glycine (Gly)   alanine (Ala)   peptide bond   glycylalanine (Gly-Ala)

The product of this reaction, glycylalanine, is an example of a *dipeptide*. (Notice how the three-letter symbols for the amino acids make up what biochemists sometimes use as the *structural* formulas of such products.)

We could have taken glycine and alanine in different roles and written the equation for the formation of a different dipeptide, Ala-Gly.

$$^+NH_3CHC-O^- + H-\overset{H}{\underset{H}{\overset{+}{N}}}CH_2CO^- \xrightarrow[\text{by several steps}]{\text{In the body,}} {}^+NH_3CHC-NHCH_2CO^- + H_2O$$

alanine (Ala)   glycine (Gly)   alanylglycine (Ala-Gly)   peptide bond

The artificial sweetener aspartame (NutraSweet) is the methyl ester of a dipeptide.

$$\begin{array}{l}
\overset{O}{\overset{\|}{C}}OCH_3 \quad \leftarrow \text{methyl ester group}\\
CHCH_2C_6H_5\\
NH \quad \leftarrow \text{peptide bond}\\
C=O\\
CHCH_2CO_2^-\\
NH_3^+
\end{array}$$

aspartame

We can think of the formation of these two dipeptides as being like the formation of two-letter words from the letters N and O. Taken in one order we get NO; in the other, ON. They have entirely different meanings, yet are made of the same pieces.

Note that each dipeptide has one CO_2^- at one end of the chain and one NH_3^+ group at the other end. Each end of a dipeptide molecule, therefore, could become involved in the formation of yet another peptide bond involving any of the 20 amino acids. For example, if glycylalanine were to combine with phenylalanine as follows, a *tripeptide* would form with the sequence Gly-Ala-Phe.

$$^+NH_3CH_2C-NHCHC-O^- + H-\overset{H}{\underset{H}{\overset{+}{N}}}CHCO^- \longrightarrow$$

glycylalanine (Gly-Ala)   phenylalanine (Phe)

peptide bonds

$$^+NH_3CH_2C-NHCHC-NHCHCO^- + H_2O$$

glycylalanylphenylalanine (Gly-Ala-Phe)

The tripeptide still has one CO_2^- and one NH_3^+ group at opposite ends of the chain, so it could react at either end with still another amino acid to make a tetrapeptide. Each additional unit still leaves the product with one CO_2^- and one NH_3^+ group. You can see how a very long sequence of amino acid units might be joined together. Note how *the sequence of side chains could be in any order*. We took only one sequence for purposes of illustration, but there are actually six different ways of putting together three different amino acids. Using the three letter symbols only, they are

Gly-Ala-Phe	Ala-Gly-Phe	Phe-Gly-Ala
Gly-Phe-Ala	Ala-Phe-Gly	Phe-Ala-Gly

All six tripeptides have identical "backbones."

$$-N-C-\overset{O}{\overset{\|}{C}}-N-C-\overset{O}{\overset{\|}{C}}-N-C-\overset{O}{\overset{\|}{C}}-$$

The six tripeptides differ only in the sequence in which the three side chains, H, CH_3, and $CH_2C_6H_5$, are located at the α-carbon atoms.

To write the general structure of the backbone of a polypeptide, all we need to add are parentheses to indicate a very long macromolecule.

α-positions

$$-N-C-\overset{O}{\overset{\|}{C}}(N-C-\overset{O}{\overset{\|}{C}})_n N-C-\overset{O}{\overset{\|}{C}}-$$

polypeptide "backbone" (n can equal several thousand)

The value of n varies by several orders of magnitude, but all polypeptides have backbones of the same general structure and all have regularly spaced side chains.

Proteins   Many proteins consist of single polypeptides. Most proteins, however, involve two or more aggregated polypeptides. These are identical in some proteins, but in others the aggregating polypeptides are different. Moreover, a relatively small organic molecule may be included in the aggregation, and a metal ion is sometimes present, as well. Thus, the terms "protein" and "polypeptide" are not synonyms. Hemoglobin has all of the features just described (Figure 21.14). It is made of four polypeptides—two similar pairs— and one molecule of heme, the organic compound that causes the red color of blood. Heme, in turn, holds an iron(II) ion. The *entire* package is the protein, hemoglobin. If one piece is missing or altered in any way—for example, if Fe^{3+} instead of Fe^{2+} is present—the substance is not hemoglobin and does not transport oxygen in the blood.

Hemoglobin is the oxygen carrier in blood.

The Importance of Protein Shape   Notice in Figure 21.14 how the strands of each polypeptide unit in hemoglobin are coiled and that the coils are kinked and twisted. Such shapes of polypeptides are determined by the amino acid sequence, because the side chains are of different sizes and some are hydrophilic and others are hydrophobic. Polypeptide molecules become twisted and coiled in whatever way minimizes the contact of hydrophobic groups with the surrounding water and maximizes the contacts of hydrophilic groups with water molecules. The final shape of a protein, called its *native form,* is as critical to its ability to function as anything else about its molecular architecture. For example, just the exchange of one side-chain R group by another

FIGURE 21.14

Hemoglobin. Its four polypeptide chains occupy the twisted tubelike forms. Two units are alike, the *α*-chains; the two *β*-chains are also alike. The flat disks are the heme molecules. (From R. E. Dickerson and I. E. Geis, *The Structure and Action of Proteins.* © 1969, W. A. Benjamin, Inc., Menlow Park, CA. All rights reserved. Used by permission.)

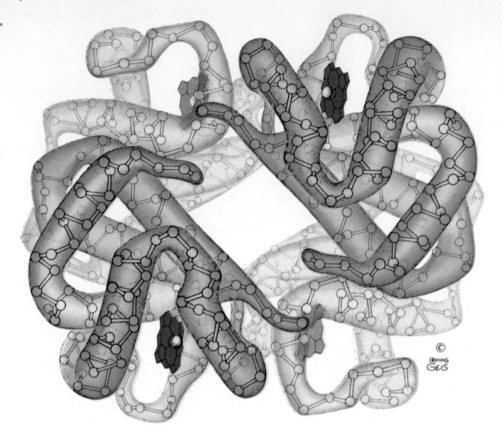

Heme

In one of the subunits of the hemoglobin in those with sickle-cell anemia, an isopropyl group is a side chain where a $CH_2CH_2CO_2H$ side chain should be.

changes the shape of hemoglobin and causes a debilitating condition known as sickle-cell anemia. Physical agents, like heat, and chemicals, like poisons and certain solvents, can distort the native form of a protein and render it biologically useless without breaking any covalent bond. When this happens, we say that the protein has been *denatured,* and the change is nearly always irreversible.

The shapes of polypeptides change even when H^+ ions are donated to accepting sites or are released from donating sites. Thus any change in the pH of the protein's surrounding medium can have disastrous consequences for the ability of the protein to function. This is why a living system must exercise extremely tight control over the pH of its fluids by means of its buffers (and, incidentally, why the study of medicine requires extensive grounding in acid–base chemistry).

Glutamic acid has a proton-donating side chain.

$$—CH_2CH_2CO_2H$$

Lysine has a proton-accepting side chain.

$$—CH_2CH_2CH_2CH_2NH_2$$

Lock-and-Key Theory of Enzyme Action   The catalysts in living cells are called **enzymes,** and virtually all are proteins. Some enzymes require metal ions, such as Mn^{2+}, Co^{2+}, Cu^{2+}, and Zn^{2+}, all of which are on the list of the *trace elements* that must be in a good diet. Some enzymes also require molecules of the B vitamins to be complete enzymes.

The molecule whose reaction an enzyme catalyzes is called the *substrate,* and enzymes are specific in the substrates they accept. An enzyme that catalyzes the hydrolysis of an ester bond, for example, is inefficient or useless for the hydrolysis of a peptide bond. Everything that we discussed earlier about the importance of protein shape comes to bear on enzyme specificity. We think of the interaction of an enzyme and its substrate as a series of equilibria:

$$E + S \rightleftharpoons E—S \rightleftharpoons E—S^* \rightleftharpoons E—P \rightleftharpoons E + P$$

enzyme   substrate   enzyme–substrate complex   substrate-activated $E—S$ complex   enzyme–product complex   enzyme (recovered)   product

The formation of the *enzyme–substrate complex* is the key step. The shape of the enzyme molecule and that of the substrate are complementary, meaning that the two molecules are able to fit to each other somewhat as a key fits to its tumbler lock. According to this **lock-and-key mechanism,** if the enzyme and substrate molecules cannot fit, there is no catalysis.

Enzymes enhance the rates of reactions remarkably. Carbonic anhydrase, for example, catalyzes the reestablishment of the following equilibrium after it has experienced a stress, like the removal of HCO_3^- ion.

$$CO_2 + H_2O \underset{\text{Carbonic anhydrase}}{\rightleftharpoons} HCO_3^- + H^+$$

It is important to realize that whenever an enzyme is involved with an *equilibrium,* it does *not* shift the equilibrium one way or another. It only quickly reestablishes equilibrium if other events cause a shift. For example, if another reaction continuously removes HCO_3^- and upsets the equilibrium shown above, each molecule of carbonic anhydrase is able to catalyze the conversion of CO_2 to HCO_3^- at a rate of 600,000 molecules per second! Nothing like this rate enhancement exists among reactions affected by ordinary catalysts, like strong acids.

Some of our most dangerous poisons work by deactivating enzymes, often those needed for the transmission of nerve signals. For example, the botulinus toxin that causes botulism, a deadly form of food poisoning, deactivates an enzyme in the nervous system. Heavy metal ions, like Hg^{2+} or Pb^{2+}, are poisons because they deactivate enzymes.

Heavy metal ions bond to the HS groups of cysteine side chains in polypeptides.

21.11
NUCLEIC ACIDS

The enzymes of an organism are made under the chemical direction of a family of compounds called the *nucleic acids.* Both the similarities and the uniqueness of every species as well as every individual member of a species depend on structural features of these compounds.

DNA and RNA   The **nucleic acids** occur as two broad types: **RNA,** or ribonucleic acids; and **DNA,** or deoxyribonucleic acids. DNA is the actual chemical of a gene, the individual unit of heredity and the chemical basis through which we inherit all of our characteristics. Our purpose in this section is to study the general structural features of the nucleic acids, how DNA molecules direct their own duplications, and how they guide the assembling of amino acid units into a unique sequence in a polypeptide.

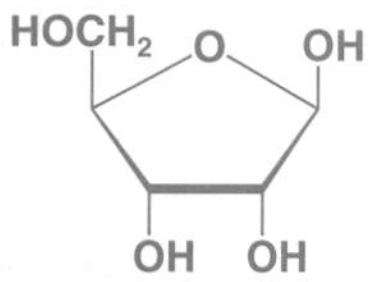

Figure 21.15 gives a short segment of a DNA molecule together with some simpler representations. The main chains or "backbones" of DNA molecules consist of alternating units contributed by phosphoric acid and a simple monosaccharide. One difference between DNA and RNA is in the monosaccharide, being ribose in RNA (hence the R in the symbol RNA) and it is deoxyribose in DNA. These two sugars differ by just one OH group (and *deoxy* means "lacking an oxygen unit"). Thus, both DNA and RNA have the following systems, where *G* stands for *group,* each *G* unit representing a unique nucleic acid side chain.

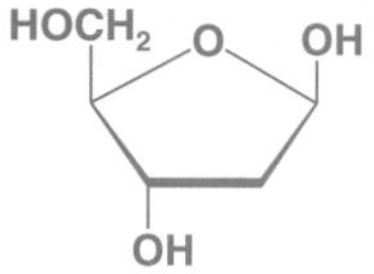

ribose

deoxyribose

$$\begin{matrix} G^1 && G^2 && G^3 \\ | && | && | \\ \text{phosphate—sugar—phosphate—sugar—phosphate—sugar—etc.} \end{matrix}$$

Backbone system in all nucleic acids—many thousands of repeating units long
In DNA, the sugar is deoxyribose.
In RNA, the sugar is ribose.

The side chains are all heterocyclic amines whose molecular shapes have much to do with their function. Being amines, these side chains are often referred to as the *bases* of the nucleic acids and are represented by single letters—A for adenine, T for thymine, U for uracil, G for guanine, and C for cytosine.

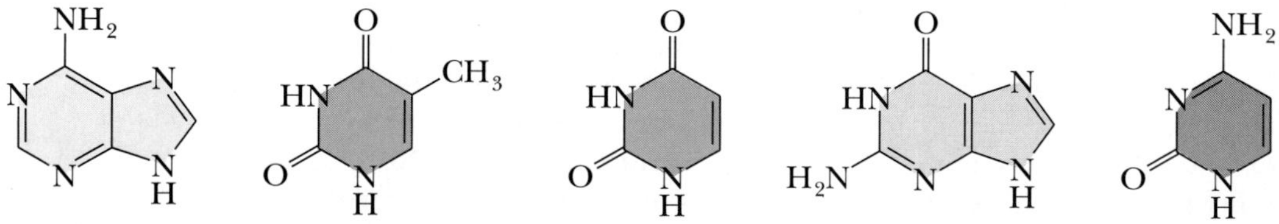

adenine
A

thymine
T

uracil
U

guanine
G

cytosine
C

A, T, G, and C occur in DNA. A, U, G, and C are in RNA. These few bases are the "letters" of the genetic alphabet.

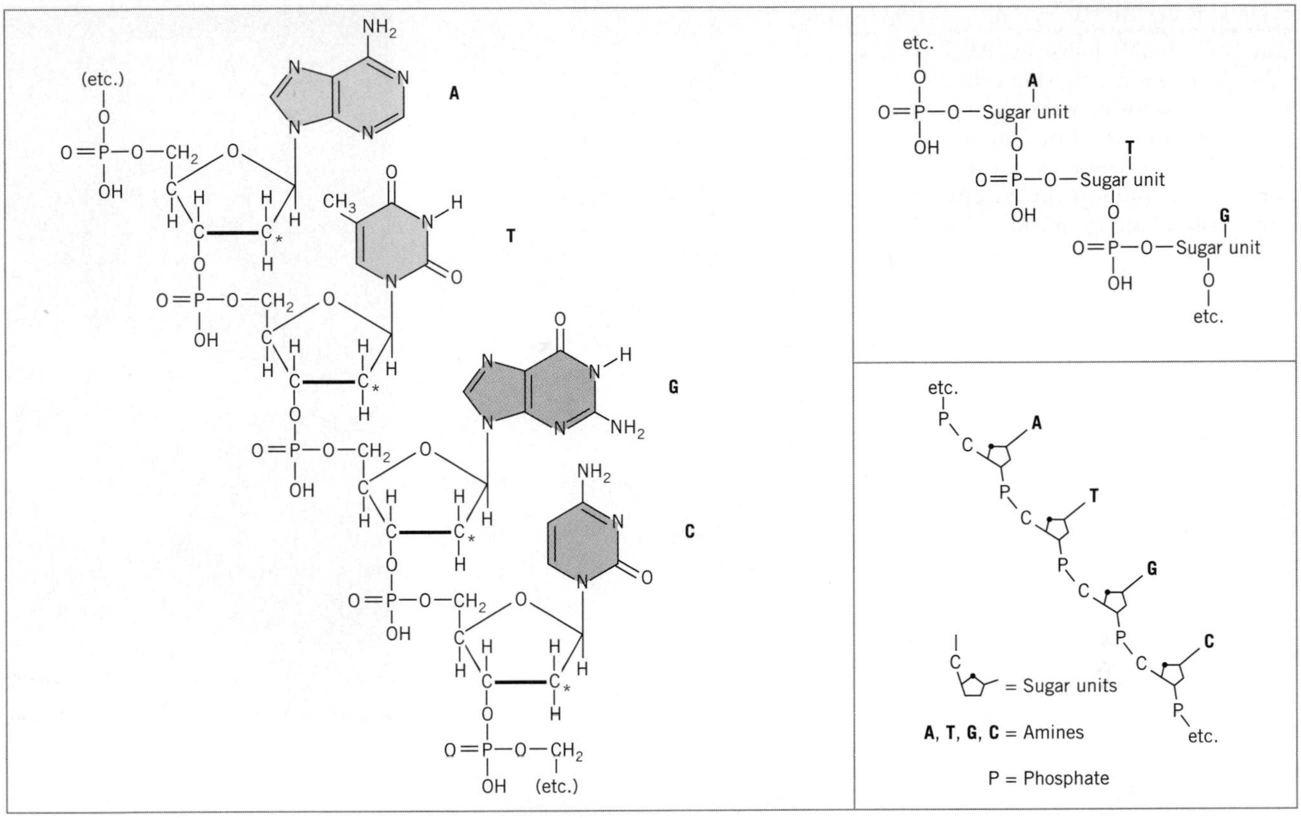

FIGURE 21.15

Nucleic acids. A segment of a DNA chain featuring each of the four DNA bases. When the sites marked by asterisks each carry an OH group, the main "backbone" would be that of RNA. In RNA U would replace T. The insets show how simplified versions of a DNA strand can be drawn.

The uniqueness of an individual gene lies in two features, namely, the length of the backbone and the sequence in which the bases occur. All of an individual's hereditary information resides in these structural aspects of its DNA molecules.

The DNA Double Helix   In 1953, F. H. C. Crick of England and J. D. Watson, an American, deduced that DNA occurs in cells as two intertwined, oppositely running strands of molecules coiled like a spiral staircase and called the **DNA double helix** (Figure 21.16). Hydrogen bonds hold the two strands side by side.

The side-chain bases have both N—H groups and O=C units between which hydrogen bonds ($\cdots$) can exist.

$$N—H \cdots O=C$$

However, the best molecular geometries for maximum hydrogen bonding occur in only particular choices of *pairs* of bases. The members of each pair have complementary structures; their functional groups are in exactly the right locations to allow hydrogen bonds between pair members. Figure 21.17

Crick and Watson shared the 1962 Nobel Prize in physiology and medicine with Maurice Wilkins, having used X-ray data from Rosalind Franklin in deducing the helical structure of DNA.

The DNA in one human cell nucleus has an estimated 3 to 5 million base pairs.

FIGURE 21.16

The DNA double helix. (*a*) A schematic drawing in which the hydrogen bonds between the two strands are indicated by dotted lines. (*b*) A model of a short section of a DNA double helix. The "backbones" are in purple.

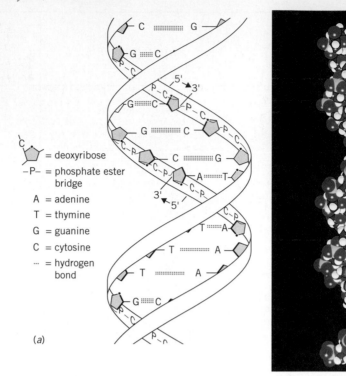

(*a*)

(*b*)

FIGURE 21.17

Base pairing in DNA. The hydrogen bonds are indicated by dotted lines.

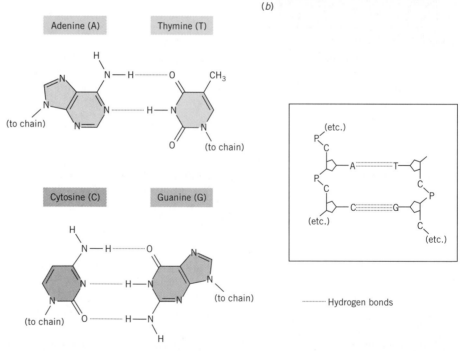

The actual mole ratios of G to C and of A to T in all DNA samples from different species, determined in the late 1940s by Erwin Chargaff, was known to be 1 to 1, and the Crick–Watson structure of DNA fits this fact.

shows how adenine (A) pairs with thymine (T), and how cytosine (C) pairs with guanine (G). A pairs only with T in DNA, never with G or C.

Adenine (A) can also pair with uracil (U), but U occurs in RNA and the A-to-U pairing is an important factor in the work of RNA, as we will see. In DNA, only G can pair with C; only A can pair with T. Thus opposite every G on one strand in a DNA double helix, a C occurs on the other strand. Opposite every A on one strand in DNA, a T is found on the other.

The Replication of DNA

Prior to cell division, the cell produces duplicate copies of its DNA so that each daughter cell will have a complete set. Such reproductive duplication is called DNA **replication.**

The accuracy of replication results from the limitations of the base pairings: A only with T and C only with G (Figure 21.18). The unattached letters A, T, G, and C represented in Figure 21.18 are the monomers of DNA, called *nucleotides,* molecules made of a phosphate–sugar unit holding one particular base. The nucleotides are made by the cell and are present in the cellular "soup."

An enzyme catalyzes each step of replication. As it occurs, the two strands of the parent DNA double helix separate, and the monomers for new strands assemble along the exposed single strands. Their order of assembling is determined altogether by the specificity of base pairing. For example, a base such as T on a parent strand can accept only a nucleotide with the base A. Two daughter double helixes result that are identical to the parent, and each carries one strand from the parent. One new double helix goes to one daughter cell and the second goes to the other.

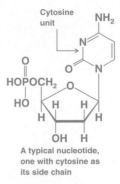

Cytosine unit

A typical nucleotide, one with cytosine as its side chain

FIGURE 21.18

Base pairing and the replication of DNA.

Two "daughter" DNA double helices

"Parent" DNA double helix

Genes A single human gene has between 1000 and 3000 bases, but they do not occur continuously on a DNA molecule. In multicelled organisms, a gene is neither a whole DNA molecule nor a continuous sequence in one. A single gene consists of the totality of particular *segments* of a DNA chain that, taken together, carry the necessary genetic instruction for a particular polypeptide. The separated sections making up a gene are called **exons**—a unit that helps to *ex*press a message. The sections of a DNA strand between exons are called **introns**—units that *in*terrupt the gene.

Something like 10,000 to 100,000 genes are needed to account for all of the kinds of proteins in a human body.

DNA-Directed Synthesis of Polypeptides

Each polypeptide in a cell is made under the direction of its own gene. In broad outline, the steps between a gene and an enzyme occur as follows. (We momentarily ignore both the introns as well as how the cell ignores them.)

$$\text{DNA} \xrightarrow[\substack{\text{(The genetic message is read} \\ \text{off in the cell nucleus and} \\ \text{transferred to RNA.)}}]{\text{Transcription}} \text{RNA} \xrightarrow[\substack{\text{(The genetic message, now} \\ \text{on RNA outside the nucleus, is} \\ \text{used to direct the synthesis of} \\ \text{a polypeptide.)}}]{\text{Translation}} \text{Polypeptide}$$

gene

The step labeled **transcription** accomplishes just that—the genetic message, represented by a sequence of bases on DNA, is transcribed into a *complementary* sequence of bases on RNA, but U, not T, is on the RNA. **Translation** means the conversion of the base sequence on RNA into a side-chain sequence on a new polypeptide. It is like translating from one language (the DNA/RNA base sequences) to another language (the polypeptide side-chain sequence).

The cell tells which amino acid unit must come next in a growing polypeptide by means of a code that relates unique groups of nucleic acid bases to specific amino acids. The code, in other words, enables the cell to translate from the four-letter RNA alphabet (A, U, G, and C) to the 20-letter alphabet of amino acids (the amino acid side chains). To see how the cell does this, we have to learn more about RNA.

The RNA of a Cell   Four types of RNA are involved in the connection between gene and enzyme. One is called *ribosomal* RNA, or rRNA, and it is packaged together with enzymes in small granules called *ribosomes*. Ribosomes are manufacturing stations for polypeptides.

Another RNA is called *messenger* RNA, or mRNA, and it brings the blueprints for a specific polypeptide from the cell nucleus to the manufacturing station (a ribosome). mRNA is the carrier of the genetic message to the polypeptide assembly site.

mRNA is made from another kind of RNA called *heterogeneous nuclear* RNA, or hnRNA. This is the RNA first made at the direction of a DNA unit, so it contains sections that were specified by both exons and introns (hence, *heterogeneous*).

The last type of RNA is called *transfer* RNA, or tRNA. It has responsibility for picking up the prefabricated parts, the amino acids, and getting them from the cell "soup" to the ribosome.

> Little agreement exists on the name *heterogeneous nuclear RNA* among biochemistry references. Some call it pre-RNA; others call it *primary transcript RNA,* or ptRNA.

Polypeptide Synthesis   When some chemical signal tells a cell to make a particular enzyme, the cell nucleus makes the hnRNA that corresponds to the gene (Figure 21.19). As we indicated, the base sequence in hnRNA is an exact complement to a base sequence in the DNA being transcribed—both the exon sections of the DNA and the introns. Reactions and enzymes in the cell nucleus then remove those parts of the new hnRNA that were caused by introns. It is as if the message, now on hnRNA, is edited to remove nonsense sections. The result of this editing is messenger RNA.

The mRNA now moves outside the cell nucleus and becomes joined to a ribosome where the enzymes for polypeptide synthesis are found. The polypeptide manufacturing site now awaits the arrival of tRNA molecules bearing amino acid units. All is in readiness for polypeptide synthesis—for *translation*. Figure 21.20 shows how a tRNA molecule carries an amino acid unit in one place and at another site has three base pairs with an important role in what follows.

The Genetic Code   To continue with our use of the language of coding, we can describe the genetic message as being written in code words made of three letters—three bases. For example, when the bases G, G, and U occur side by side on an mRNA chain, they together specify the amino acid glycine. Although three other triplets also mean glycine, one of the genetic codes for

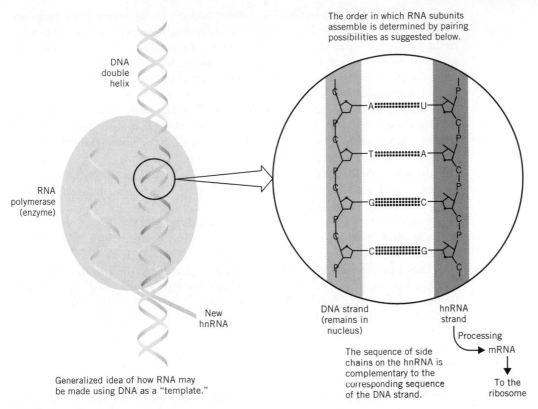

The order in which RNA subunits assemble is determined by pairing possibilities as suggested below.

DNA double helix

RNA polymerase (enzyme)

New hnRNA

Generalized idea of how RNA may be made using DNA as a "template."

DNA strand (remains in nucleus)

hnRNA strand

Processing

mRNA

To the ribosome

The sequence of side chains on the hnRNA is complementary to the corresponding sequence of the DNA strand.

FIGURE 21.19

The DNA-directed synthesis of *heterogeneous nuclear* RNA (hnRNA). Once made, hnRNA is processed to make *messenger* RNA, mRNA. This carries the genetic message from the cell nucleus to a protein assembly site.

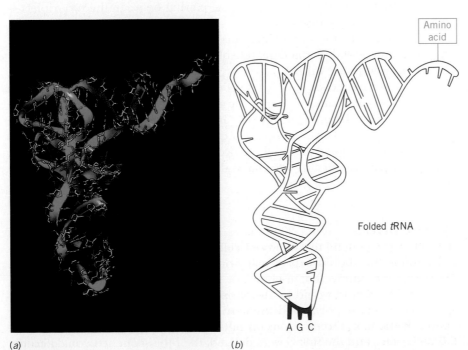

Amino acid

Folded *t*RNA

A G C

(a)   (b)

FIGURE 21.20

Transfer RNA, tRNA. (*a*) A molecular model of the tRNA for phenylalanine. Its anticodon occurs at the tip of the base, and the site of attachment of the phenylalanine molecule is on the upper right tip. (*b*) A highly schematic representation of a tRNA molecule holding an amino acid unit. The anticodon, AGC, is seen at its bottom.

glycine is GGU. Similarly, GCU is a code "word" for alanine. *Three* bases are used for each code word to enable a connection between the four-letter genetic "alphabet" and the 20-letter amino acid alphabet. When four bases are taken in *triplets,* 4^3 or 64 combinations are possible, more than enough for 20 amino acids. If doublets of bases are used, only 4^2 or 16 combinations would be available as code words, not enough.

Each triplet of bases as it occurs on mRNA is called a **codon,** and each codon corresponds to a specific amino acid. Which amino acid goes with which codon constitutes the **genetic code.** One of the most remarkable features of the code is that it is essentially universal. The codon specifying alanine in humans, for example, also specifies alanine in the genetic machinery of bacteria, aardvarks, camels, rabbits, and stinkbugs. Our chemical kinship with the entire living world is profound (even humbling).

A triplet of bases complementary to a codon occurs on tRNA and is called an **anticodon.** In Figure 21.20, the triplet AGC has been arbitrarily selected to illustrate an anticodon.

A tRNA molecule, bearing the amino acid corresponding to its anticodon, can line up at a strand of mRNA only where the anticodon "fits" by hydrogen bonding to the matching codon. If the anticodon is ACC, then the A on the anticodon must find a U on the mRNA at the same time that its neighboring two C bases find neighboring G bases on the mRNA strand. Thus the anticodon and codon line up as shown in the margin, where the dotted lines indicate hydrogen bonds.

mRNA codons determine the sequence in which tRNA units can line up, and this sets the sequence in which amino acid units become joined in a polypeptide. Figure 21.21 carries this description further, where we enter the process soon after special steps (not shown) have launched it.

In Figure 21.21*a*, a tRNA–glycine unit arrives at mRNA where it finds two situations—a dipeptide already started, and a codon to which the tRNA–glycine can fit. In Figure 21.21*b* the tRNA–glycine unit is in place. It is the only such unit that could be in this place. Base pairing between the two triplets—codon to anticodon—ensures this.

In Figure 21.21*c,* the whole dipeptide unit moves over and is joined to the glycine unit by a newly formed peptide bond. This leaves a tripeptide attached via a tRNA unit to the mRNA.

Because the next codon, GCU, is coded for the amino acid alanine, only a tRNA–alanine unit can come into the next site. So alanine, with its methyl group as the side chain, can be the only amino acid that enters the polypeptide chain at this point. These steps repeat over and over until a special codon is reached that is designed to stop the process. The polypeptide is now complete. It has a *unique* sequence of side chains that has been initially specified by the molecular structure of a gene.

A::::U
C:::G
C:::G
anticodon   codon on
on tRNA   mRNA

Knowledge of the locations and structures of genes will help to cure at least some genetic defects, and the *Human Genome Project* is a major scientific effort to "map" all of the genes of the body by locating where they are on the various chromosomes.

Genetic Defects About 2000 diseases are attributed to various kinds of defects in the genetic machinery of cells. With so many steps from gene to polypeptide, we should expect many opportunities for things to go wrong. Suppose, for example, that just one base in a gene were wrong. Then one base on an mRNA strand would be the wrong base—"wrong" meaning with respect to getting the polypeptide we want. This would change every remaining codon. If the next three codons on mRNA, for example, were UCU–GGU–GCU–U–etc., and the first G were deleted, then the sequence would become

Methionine
$CH_3SCH_2CH_2-CH$
    |
    NH_2
    |
    $C=O$
    |
    NH
Serine $HOCH_2-CH$
    |
    $O=C$

Glycine $C=O$ CH_2 NH_2

*t*RNA for glycine

A G A

A U G U C U G G U G C U

Codons —————— *mRNA*

(*a*) A growing polypeptide is shown here just two amino acid units long. A third amino acid, glycine, carried by its own tRNA is moving into place at the codon GGU. (Behind the assemblage shown here is a large particle called a ribosome, not shown, that stabilizes the assemblage and catalyzes the reactions.)

$CH_3SCH_2CH_2-CH$
    |
    NH_2
    |
    $C=O$
    |
    NH NH_2
    | |
$HOCH_2-CH$ CH_2
    | |
    $O=C$ $C=O$

A G A C C A

A U G U C U G G U G C U

(*b*) The tRNA- glycine unit is in place, and now the dipeptide will transfer to the glycine, form another peptide bond, and make a tripeptide unit.

$CH_3SCH_2CH_2-CH$
    |
    NH_2
    |
    $C=O$
    |
    NH
$HOCH_2-CH$
    |
    $O=C$
    |
    NH
    CH_2
    $C=O$

A G A C C A

A U G U C U G G U G C U

(*c*) The tripeptide unit has formed, and the next series of reactions that will add another amino acid unit can now start.

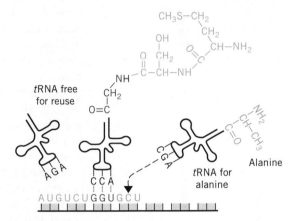

*t*RNA free for reuse

A G A

C C A

*t*RNA for alanine

Alanine
$CH-CH_3$
    |
    NH_2
    $C=O$

A U G U C U G G U G C U

(*d*) One tRNA leaves to be reused. Another tRNA carrying a fourth amino acid unit, alanine, is moving into place along the mRNA strand to give an assemblage like that shown in part (*b*). Then the tripeptide unit will transfer to alanine as another peptide bond is made and a tetrapeptide unit forms. A cycle of events thus occurs as the polypeptide forms.

FIGURE 21.21

Polypeptide synthesis at an mRNA strand.

UCU–GUG–CUU–etc. All remaining triplets change! You can imagine how this could lead to an entirely different polypeptide than what we want.

Or suppose that the GGU codon were replaced by, say, a GCU codon so that the UCU–GGU–GCU–etc. becomes UCU–GCU–GCU–etc. One amino acid in the resulting polypeptide would not be what we want. Something like this makes the difference, for example, between normal hemoglobin and sickle-cell hemoglobin and the difference between health and a tragic genetic disease.

Atomic radiation, like gamma rays, beta rays, or even X rays, as well as chemicals that mimic radiation can cause genetic defects, too. A stray radiation hitting a gene might cause two opposite bases to fuse together chemically and so make the cell reproductively dead. When it came time for it to divide, the cell could not replicate its genes, could not divide, and that would be the end of it. If this happens to enough cells in a tissue, the organism might be fatally affected.

Viruses   Viruses are packages of chemicals usually consisting of nucleic acid and protein. Their nucleic acids are able to take over the genetic machinery of the cells of particular host tissues, manufacture more virus particles, and multiply enough to burst the host cell. Because cancer cells divide irregularly, and usually more rapidly than normal cells, some viruses are thought to be among the agents that can cause cancer.

The protein of a virus is a protective overcoat for its nucleic acid, and it provides an enzyme to catalyze a breakthrough by the virus into a host cell.

Genetic Engineering

Insulin is a polypeptide hormone made by the pancreas, and we require insulin to manage the concentration of glucose in the blood. People who do not make insulin have diabetes. In severe diabetes, the individual must take insulin by injection (to avoid the protein-digesting enzymes of the digestive tract). People who once relied on the insulin isolated from the pancreases of animals—cows, pigs, and sheep, for example—sometimes experienced allergic responses because the insulin molecules from these animals are not identical to those of human insulin. Diabetics were also vulnerable to the continuing availability of animal pancreases. *Human* insulin, however, is now being manufactured by a process that involves the production of *recombinant DNA*. It represents one of the important advances in scientific technology of this century.

Figure 21.22 shows how recombinant DNA technology works. It starts with bacteria, which make their own polypeptides using the same genetic code as humans. Bacteria, unlike humans, possess DNA not only in their chromosomes but also in large, circular, supercoiled DNA molecules called *plasmids*. Each plasmid carries just a few genes, but several copies of a plasmid can exist in one bacterial cell. Each plasmid can replicate independently of the chromosome.

Plasmids can be isolated from bacteria. By the use of special enzymes, the plasmids can be given new DNA material, a new gene. The new gene might have the base triplets for directing the synthesis of a polypeptide completely foreign to the bacteria, like the subunits of human insulin. The plasmids are snipped open by special enzymes, the new DNA combines with the open ends, and then the plasmid loops are reclosed. The DNA in these altered plasmids is called **recombinant DNA.** The technology associated with this is often referred to as **genetic engineering.**

Many Nobel Prizes have been won for the fundamental research that made genetic engineering possible.

The remarkable feature of bacteria with altered plasmids made of recombinant DNA is that when the bacteria multiply, the new bacteria have within them the *altered* plasmids. When these multiply, still more altered plasmids are made. Between their cell divisions, the bacteria manufacture the proteins for which they are genetically programmed, including the proteins specified by the recombinant DNA. In this way, bacteria can be induced to make *human* insulin. The technology is not limited to bacteria; yeast cells can also be used for genetic engineering.

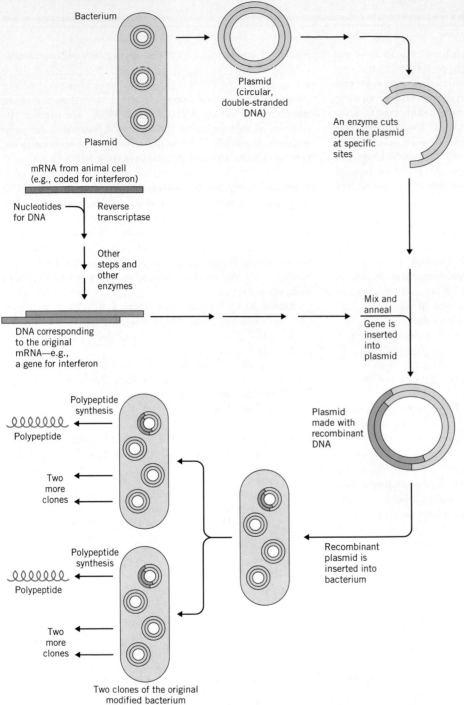

FIGURE 21.22

Recombinant DNA is made by inserting a DNA strand, coded for some protein not made by the bacteria, into the circular DNA of the bacterial plasmid.

One of the hopes of specialists in genetic engineering is to find ways to correct genetic defects. Considerable research is being done, for example, to alter the nucleic acids of viruses to enable them to "piggyback" into human cells the DNA that will cure cystic fibrosis. Hemophiliacs, those who bleed extensively from even the slightest cut because they lack a factor for blood clotting, may be helped by the genetic engineering of this clotting factor. Many other applications of genetic engineering are being investigated.

SUMMARY

Organic Compounds **Organic chemistry** is the study of the covalent compounds of carbon, except for its oxides, carbonates, bicarbonates, and a few others. **Functional groups** attached to nonfunctional and hydrocarbon-like groups are the basis for the classification of organic compounds. Members of a family have the same kinds of chemical properties. Their physical properties are a function of the relative proportions of functional and nonfunctional groups. Opportunities for hydrogen bonding strongly influence the boiling points and water solubilities of compounds with OH or NH groups.

Saturated Hydrocarbons The **alkanes**—saturated compounds found chiefly in petroleum—are the least polar and the least reactive of the organic families. Their carbon skeletons can exist as **straight chains, branched chains,** or **rings.** When they are open chain, there is free rotation about single bonds. Rings can include heteroatoms, but **heterocyclic compounds** are not hydrocarbons.

By the IUPAC rules, the names of alkanes end in *-ane*. A prefix denotes the carbon content of the parent chain, the longest continuous chain, which is numbered from whichever end is nearest a branch. The **alkyl groups** are alkanes minus a hydrogen.

Alkanes in general do not react with strong acids or bases or strong redox reactants. At high temperatures, their molecules crack to give H_2 and unsaturated compounds.

Unsaturated Hydrocarbons The lack of free rotation at a carbon–carbon double bond makes **geometric isomerism** possible. The pi bond electrons make alkenes act as Brønsted bases, enabling alkenes to undergo **addition reactions** with hydrogen chloride and water (in the presence of an acid catalyst). Alkenes also add hydrogen (in the presence of a metal catalyst). Bromine and chlorine add to alkenes under mild, uncatalyzed conditions. Strong oxidizing agents, like ozone, readily attack alkenes and break their molecules apart.

Aromatic hydrocarbons, like benzene, do not give the addition reactions shown by alkenes. Instead, they undergo **substitution reactions** that leave the energy lowering, pi electron network intact. In the presence of suitable catalysts, benzene reacts with chlorine, bromine, nitric acid, and sulfuric acid.

Alcohols and Ethers The **alcohols,** ROH, and the **ethers,** ROR′, are alkyl derivatives of water. Ethers have almost as few chemical properties as alkanes. Alcohols undergo an **elimination reaction** that splits out H_2O and changes them to alkenes. Concentrated hydrohalogen acids, like HI, react with alcohols by a **substitution reaction,** which replaces the OH group by halogen. Oxidizing agents convert alcohols of the type RCH_2OH first into aldehydes and then into carboxylic acids. Oxidizing agents convert alcohols of the type

R_2CHOH into ketones. Alcohols form esters with carboxylic acids.

Amines The simple **amines,** RNH_2, as well as the more substituted relatives, RNHR′ and RNR′R″, are organic derivatives of ammonia and thus are weak bases that can neutralize strong acids. The conjugate acids of amines are good proton donors and can neutralize strong bases.

Carbonyl Compounds Aldehydes, $RCH{=}O$, among the most easily oxidized organic compounds, are oxidized to carboxylic acids. **Ketones,** RCOR′, strongly resist oxidation. Aldehydes and ketones add hydrogen to make alcohols.

The **carboxylic acids,** RCO_2H, are weak acids but neutralize strong bases. Their anions are Brønsted bases. Carboxylic acids can also be changed to esters by heating them with alcohols.

Esters, $RCO_2R′$, react with water to give back their parent acids and alcohols. When aqueous base is used for hydrolysis, the products are the salts of the parent carboxylic acids as well as alcohols.

The simple **amides,** $RCONH_2$, can be made from acids and ammonia, and they can be hydrolyzed back to their parent acids and trivalent nitrogen compounds. The amides are not basic.

IUPAC Nomenclature of Compounds with Functional Groups Each family of compounds has a characteristic name ending:

-ane	alkanes	-al	aldehydes
-ene	alkenes	-one	ketones
-yne	alkynes	-oic acid	carboxylic acids
-ol	alcohols		

Each member of a family has a defined "parent," which is the longest chain that includes the functional group. How the chain is numbered varies with the family, but generally the numbering starts from whichever end of the parent chain reaches the functional group with the lower number. The names of side chains, such as the alkyl groups, are included in the name ahead of the name of the parent, and their locations are designated by the numbers of the carbons at which they are attached to the parent.

Polymers **Macromolecules** with more or less repeating structural units are called **polymers.** They are made from one or more **monomers** by polymerization, either by **addition polymerization** or by **condensation polymerization.**

Monomers for the **polyolefins** include ethylene, propylene, and styrene. When side chains are possible, the alkenes polymerize so that the side chains take up regular positions on the main chain. Monomers for the **acrylates** also contain alkene groups, but generally have carbonyl-containing groups as well.

The condensation of a dicarboxylic acid with a diol produces a **polyester**, like Dacron. The condensation of a dicarboxylic acid with a diamine gives a member of the nylon family, which are all **polyamides.**

The polymers most useful for making filaments and fibers are those with very long, symmetrical molecules. If regularly spaced polar groups occur that can attract each other, either by ordinary dipole–dipole attractions or by hydrogen bonds, the fibers are particularly strong.

Carbohydrates   The glucose unit is present in starch, glycogen, cellulose, sucrose, lactose, and simply as a **monosaccharide** in honey. It's the chief "sugar" in blood. As a monosaccharide, glucose is a pentahydroxyaldehyde in one form but it also exists in two cyclic forms. The latter are joined by oxygen bridges in the **disaccharides** and **polysaccharides.** In lactose (milk sugar), a glucose and a galactose unit are joined. In sucrose (table sugar), glucose and fructose units are linked. The major polysaccharides—starch (amylose and amylopectin), glycogen, and cellulose—are all polymers of glucose but with different patterns and geometries of branching. The digestion of the disaccharides and starch requires the hydrolyses of the oxygen bridges to give monosaccharides.

Lipids   Natural products in living systems that are relatively soluble in nonpolar solvents, but not in water, are all in a large group of compounds called **lipids.** They include sex hormones and cholesterol, the triacylglycerols of nutritional importance, and the glycerophospholipids and other phospholipids needed for cell membranes.

The **triacylglycerols** are triesters between glycerol and three of a number of long-chain, often unsaturated **fatty acids.** In the digestion of the triacylglycerols, the hydrolysis of the ester groups occurs, and anions of fatty acids form (together with glycerol).

Molecules of glycerophospholipids have large segments that are hydrophobic and others that are hydrophilic. Glycerophospholipid molecules are the major components of cell membranes, where they are arranged in a lipid bilayer.

Amino Acids, Polypeptides, and Proteins   The **amino acids** in nature are mostly compounds with NH_3^+ and CO_2^- groups joined to the same carbon atom (called the alpha position of the amino acid). About 20 amino acids are important to the structures of polypeptides and proteins.

When two amino acids are linked by a **peptide bond** (an amide bond), the compound is a dipeptide. In **polypeptides,** several (sometimes thousands) of regularly spaced peptide bonds occur, involving as many amino acid units. The uniqueness of a polypeptide lies in its chain length and in the sequence of the side chain groups.

Some **proteins** consist of only a polypeptide, but most proteins involve two or more (sometimes different) polypeptides and often a nonpolypeptide (an organic molecule) and a metal ion.

Nearly all **enzymes** are proteins. Many involve multiple polypeptide strands, nonprotein molecules like vitamins, and trace metal ions. Enzymes are unusually selective in their **substrates** and exceptionally efficient at rate enhancements. They work by fitting to their substrates in a **lock-and-key mechanism** to form an **enzyme–substrate complex,** where the chemical change occurs. Major poisons work by deactivating enzymes or by preventing the binding of substrate to enzyme.

Nucleic Acids   The formation of a polypeptide with the correct amino acid sequence needed for a given polypeptide is under the control of **nucleic acids.** These are polymers whose backbones are made of a repeating sequence of pentose sugar units. On each sugar unit is a **base,** a heterocyclic amine such as adenine (A), thymine (T), guanine (G), cytosine (C), or uracil (U). In **DNA,** the sugar is deoxyribose, and the bases are A, T, G, and C. In **RNA,** the sugar is ribose and the bases are A, U, G, and C.

DNA exists in cell nuclei as **double helices.** The two strands are associated side by side, with hydrogen bonds occurring between particular pairs of bases. A is always paired to T; C is always paired to G. The bases, A, U, G, and C, are the four letters of the genetic alphabet, and the specific combination that codes for one amino acid in polypeptide synthesis are three bases, side by side, on a strand. An individual gene in higher organisms consists of sections of a DNA chain, called exons, separated by other sections called introns. The **replication** of DNA is the synthesis by the cell of exact copies of the original two strands of a DNA double helix.

Polypeptide Synthesis   The DNA that carries the gene for a particular polypeptide is used to direct the synthesis of a molecule of hnRNA. The hnRNA is then modified by the removal of those segments corresponding to the introns of the DNA. This leaves messenger RNA, the carrier of the message of the gene to the ribosomes. The overall change from gene to mRNA is called **translation.** mRNA is like an assembly line at a factory (ribosomes) awaiting the arrival of parts—the amino acids. These are borne on transfer RNA molecules, tRNA.

A unit of three bases on mRNA constitutes a **codon** for one amino acid. A tRNA–amino acid unit has an **anticodon** that can pair to a codon on mRNA. In this way a tRNA–amino acid unit "recognizes" only one place on mRNA for delivering the amino acid to the polypeptide developing at the mRNA assembly line. The key that relates codons to specific amino acids is the **genetic code** and it is essentially the same code for all organisms in nature. The involvement of mRNA in directing the synthesis of a polypeptide is called **translation.**

Even small changes or deletions in the sequences of bases on DNA can have large consequences in the forms of genetic diseases. Viruses are able to use their own nucleic acids to take over the genetic apparatus of the host tissue. In **genetic engineering,** the genetic machinery of microorganisms is taken over, forced to take **recombinant DNA,** and made to manufacture polypeptides not otherwise synthesized by the organism.

TOOLS YOU HAVE LEARNED

The table below lists the tools you have learned in this chapter that are applicable to problem solving. Review them, if necessary, and refer to them when working on the Review Exercises that follow.

Tool	Function
Functional group concept (page 890)	To place the structure of an organic compound into an organic family and so predict the kinds of reactions the compound can give.
IUPAC rules of nomenclature (pages 896, 900, 904, 910, 911, 913)	To write a unique name for a given structure, or only one structure for a given name.

REVIEW EXERCISES

Answers to questions whose numbers are printed in color are given in Appendix D. More challenging questions are marked with asterisks.

Structural Formulas

21.1 Which of the following compounds are inorganic?
(a) CH_3OH (b) CO (c) CCl_4 (d) $NaHCO_3$ (e) K_2CO_3

21.2 In general terms, what makes possible so many organic compounds?

21.3 What is the condensed structure of R if R-CH_3 is ethane?

21.4 Which of the following structures are possible, given the numbers of bonds that various atoms can form?
(a) $CH_2CH_2CH_3$
(b) $CH_3{=}CHCH_2CH_3$
(c) $CH_3CH{=}CH_2CH_2CH_3$

21.5 Write full (expanded) structures for each of the following molecular formulas. Remember how many bonds the various kinds of atoms must have. In some you will have to use double or triple bonds. (Hint: A trial-and-error approach will have to be used.)
(a) CH_5N (g) NH_3O
(b) CH_2Br_2 (h) C_2H_2
(c) $CHCl_3$ (i) N_2H_4
(d) C_2H_6 (j) HCN
(e) CH_2O_2 (k) C_2H_3N
(f) CH_2O (l) CH_4O

21.6 Write neat, condensed structures of the following.
(a)

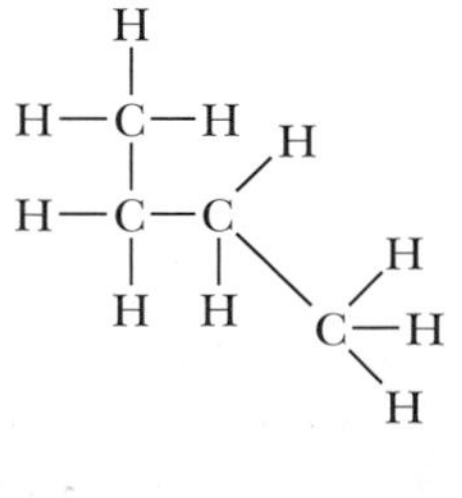

(b)

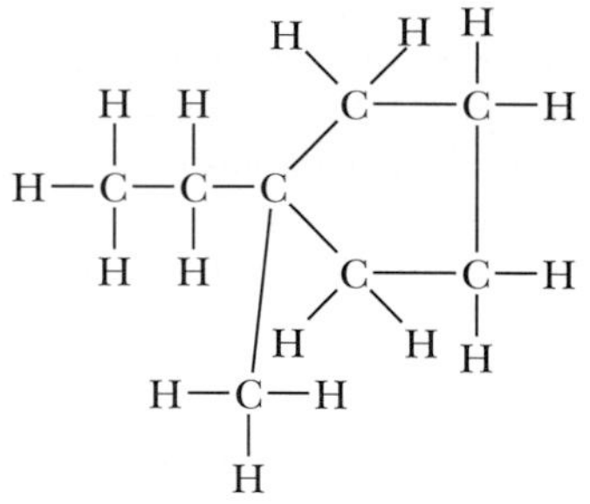

(c)

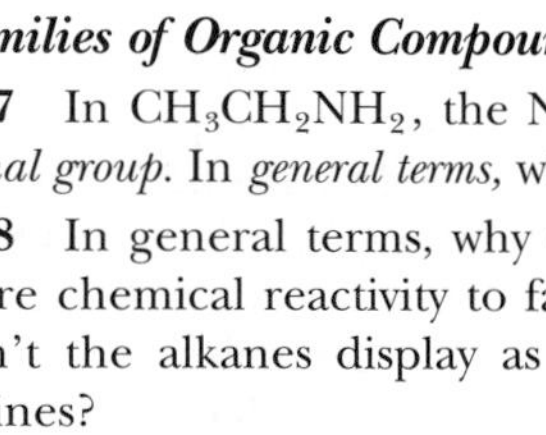

Families of Organic Compounds

21.7 In $CH_3CH_2NH_2$, the NH_2 group is called the *functional group*. In *general terms*, why is it called this?

21.8 In general terms, why do functional groups impart more chemical reactivity to families that have them? Why don't the alkanes display as many reactions as, say, the amines?

21.9 Name the family to which each compound belongs.
(a) $CH_3CH{=}CH_2$
(b) CH_3CH_2OH
(c)

$$CH_3CH_2\overset{\displaystyle O}{\overset{\displaystyle \|}{C}}OCH_3$$

(d)

$$CH_3CH_2CH_2\overset{\overset{\displaystyle O}{\|}}{C}OH$$

(e) $CH_3CH_2CH_2NH_2$
(f) $HOCH_2CH_2CH_3$
(g) $CH_3C{\equiv}CH$
(h)

$$CH_3CH_2\overset{\overset{\displaystyle O}{\|}}{C}H$$

(i)

$$CH_3\overset{\overset{\displaystyle O}{\|}}{C}CH_2CH_2CH_3$$

(j) $CH_3{-}O{-}CH_2CH_3$
(k) $CH_3CH_2NH_2$
(l)

$$CH_3CH_2\overset{\overset{\displaystyle O}{\|}}{C}NH_2$$

21.10 Identify by letter the compounds in Review Exercise 21.9 that are saturated compounds.

Isomers

21.11 Decide whether the members of each pair are identical, are isomers, or are unrelated.

(a) $CH_3{-}CH_3$ and
$$CH_3 \\ \ \ | \\ CH_3$$

(b)
$$CH_3 \\ \ \ | \\ CH_2 \\ \ \ | \\ CH_3$$
and
$$CH_2 \\ / \ \ \backslash \\ CH_3 \ \ CH_3$$

(c) CH_3CH_2OH and $CH_3CH_2CH_2OH$

(d) $CH_3CH{=}CH_2$ and
$$CH_2{-}CH_2 \\ \ \backslash \ \ / \\ CH_2$$

(e)
$$H{-}\overset{\overset{\displaystyle O}{\|}}{C}{-}CH_3 \quad \text{and} \quad CH_3{-}\overset{\overset{\displaystyle O}{\|}}{C}{-}H$$

(f)
$$\overset{\displaystyle CH_3}{\underset{\displaystyle |}{}} \\ CH_3CHCH_3$$
and
$$\overset{\displaystyle CH_3}{\underset{\displaystyle |}{}} \\ CH_3CH \\ \ \ \ | \\ \ \ \ CH_3$$

(g) $CH_3CH_2CH_2NH_2$ and $CH_3CH_2{-}NH{-}CH_3$

(h)
$$CH_3CH_2\overset{\overset{\displaystyle O}{\|}}{C}OH \quad \text{and} \quad HO\overset{\overset{\displaystyle O}{\|}}{C}CH_2CH_3$$

(i)
$$H\overset{\overset{\displaystyle O}{\|}}{C}OCH_2CH_3 \quad \text{and} \quad CH_3CH_2\overset{\overset{\displaystyle O}{\|}}{C}OH$$

(j)
$$H\overset{\overset{\displaystyle O}{\|}}{C}OCH_2CH_2OH \quad \text{and} \quad HOCH_2CH_2\overset{\overset{\displaystyle O}{\|}}{C}OH$$

(k)
$$CH_3\overset{\overset{\displaystyle O}{\|}}{C}CH_2CH_3 \quad \text{and} \quad CH_3CH_2\overset{\overset{\displaystyle O}{\|}}{C}CH_3$$

(l) $CH_3{-}CH{-}CH_3$
$$CH_2{-}CH_2 \quad CH_3 \\ \qquad\qquad\quad | \quad\quad \backslash \ CH_3 \\ \qquad CH_2{-}C{-}CH \\ \qquad\qquad | \quad\quad \backslash CH_3 \\ \qquad\qquad CH_3$$

and

$$\overset{\displaystyle CH_3}{\underset{\displaystyle |}{}} \qquad\qquad\qquad \overset{\displaystyle CH_3}{\underset{\displaystyle |}{}} \ \overset{\displaystyle CH_3}{\underset{\displaystyle |}{}} \\ CH_3{-}CH{-}CH_2{-}CH_2{-}CH_2{-}C{-}{-}CH{-}CH_3 \\ \qquad\qquad\qquad\qquad\qquad\qquad | \\ \qquad\qquad\qquad\qquad\qquad\qquad CH_3$$

(m)
$$CH_3{-}NH{-}\overset{\overset{\displaystyle O}{\|}}{C}{-}CH_3 \quad \text{and} \quad CH_3CH_2\overset{\overset{\displaystyle O}{\|}}{C}NH_2$$

(n) $H{-}O{-}O{-}H$ and $H{-}O{-}H$

Physical Properties and Structure

21.12 Explain why $CH_3CH_2CH_2OH$ is more soluble in water than $CH_3CH_2CH_2CH_3$.

21.13 Of the two compounds in Exercise 21.12, which has the higher boiling point? Explain.

21.14 Examine the structures of the following compounds.

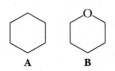

(a) Which has the highest boiling point?
(b) Which has the lowest boiling point?
(c) Explain your answers to parts a and b.

21.15 Which of the following compounds is more soluble in water? Explain.

$$\text{A} \qquad\qquad \text{B}$$

21.16 Which of the following compounds has the higher boiling point? Explain.

$$\overset{\displaystyle CH_3}{\underset{\displaystyle |}{}} \\ CH_3CH_2CHCH_2CHCH_3 \qquad CH_3CH_2CH_2CH_2CH_3 \\ \qquad\qquad\quad | \\ \qquad\qquad\quad CH_3$$
$$\quad\text{A} \qquad\qquad\qquad\qquad \text{B}$$

Names of Hydrocarbons

21.17 Write the IUPAC names of the following hydrocarbons.

(a) $CH_3CH_2CH_2CH_2CH_3$
(b) $CH_3CH_2CH_2CHCH_3$
$$\qquad\qquad\qquad\quad | \\ \qquad\qquad\qquad\quad CH_3$$

(c)
$$\overset{\displaystyle CH_3}{\underset{\displaystyle |}{}} \\ CH_3CHCH_2CHCH_2CH_3 \\ \qquad\qquad\quad | \\ \qquad\qquad\quad CH_3$$

(d)
$$\overset{\displaystyle CH_3}{\underset{\displaystyle |}{}} \\ CH_3CH_2CHCH_2CHCH_3 \\ \qquad\qquad\quad | \\ \qquad\qquad\quad CH_3$$

(e) $CH_3CH_2CH{=}CHCH_2CH_3$
(f)
$$\overset{\displaystyle CH_3}{\underset{\displaystyle |}{}} \\ CH_3CHCH{=}CHCH_3$$

21.18 Write the condensed structures of the following compounds.
(a) 2,2-dimethyloctane
(b) 1,3-dimethylcyclopentane
(c) 1,1-diethylcyclohexane
(d) 6-ethyl-5-isopropyl-7-methyl-1-octene
(e) *cis*-2-pentene

Properties of Hydrocarbons

21.19 What prevents free rotation about a double bond?

*21.20 If at either end of a carbon–carbon double bond the two atoms or groups are identical, can the alkene exist as a pair of cis and trans isomers? Write the structures of the cis and trans isomers, if any, for the following.

(a) $CH_2{=}CH_2$
(b) $CH_3CH{=}CH_2$
(c)
$$CH_3\overset{\displaystyle CH_3}{\underset{|}{C}}{=}CH_2$$
(d)
$$CH_3\overset{\displaystyle CH_3}{\underset{|}{C}}{=}CHCH_3$$
(e)
$$CH_3CH{=}\overset{\displaystyle CH_3}{\underset{|}{C}}CH_2CH_3$$
(f)
$$CH_3CH_2\overset{\displaystyle Cl}{\underset{|}{C}}{=}\overset{\displaystyle Cl}{\underset{|}{C}}CH_2CH_3$$

*21.21 Propene is known to react with concentrated sulfuric acid as follows.

$$CH_3CH{=}CH_2 + H_2SO_4 \rightarrow CH_3\underset{\underset{OSO_3H}{|}}{C}HCH_3$$

The reaction occurs in two steps. Write the equation for the first.

21.22 Write the structures of the products that form when ethylene reacts with each of the following substances by an addition reaction. (Assume that needed catalysts or other conditions are provided.) (a) H_2, (b) Cl_2, (c) Br_2, (d) $HCl(g)$, (e) $HBr(g)$, (f) H_2O (in acid).

21.23 Repeat Exercise 21.22 using 2-butene.

21.24 Repeat Exercise 21.22 using cyclohexene.

21.25 In general terms, why doesn't benzene give the same kinds of addition reactions shown by cyclohexene?

21.26 If benzene were to *add* one mole of Br_2, what would form? What forms, instead, when Br_2, benzene, and an $FeBr_3$ catalyst are heated together? (Write the structures.)

Alcohols and Ethers

21.27 Write condensed structures and the IUPAC names for all of the saturated alcohols with three carbon atoms or fewer per molecule.

21.28 Write the condensed structures and the IUPAC names for all of the possible alcohols with the general formula $C_4H_{10}O$.

21.29 Write the condensed structures of all of the possible ethers with the general formula $C_4H_{10}O$. Following the pattern for the common names of ethers given in the chapter, what are the likely common names of these ethers?

21.30 "3-Butanol" is not a proper name, but a structure could still be written for it. What is this structure, and what IUPAC name should be used?

21.31 Write the structures of the isomeric alcohols of the formula $C_4H_{10}O$ that could be oxidized to aldehydes. Write the structure of the isomer that could be oxidized to a ketone.

21.32 Write the structures of the aldehydes or ketones that can be prepared from the following compounds by oxidation.

21.33 Briefly explain how the C—O bond in isopropyl alcohol is weakened when a strong acid is present.

*21.34 Write the structure of the product of the acid-catalyzed dehydration of 2-propanol. Write the mechanism of the reaction.

*21.35 Write the structures of the products of the acid-catalyzed dehydrations of the compounds given in Review Exercise 21.32.

21.36 When ethanol is heated in the presence of an acid catalyst, ethene and water form; an elimination reaction occurs. When 2-butanol is heated under similar conditions, *two* alkenes form (although not in equal quantities). Write the structures of these two alkenes.

*21.37 If the formation of the two alkenes in Review Exercise 21.36 were determined purely by statistics, then the mole ratio of 1-butene to 2-butene should be in the ratio of what whole numbers? Why?

*21.38 The major product of the elimination reaction described in Exercise 21.36 is 2-butene. In the light of Exercise 21.37, is the ratio of alkenes that form determined simply by statistics? Suggest a possible thermodynamic reason for the observed major product.

21.39 Write the structures of the substitution products that form when the alcohols of Review Exercise 21.32 are heated with concentrated hydriodic acid.

Amines

21.40 Amines, RNH_2, do not have as high boiling points as alcohols of comparable molecular masses. Why?

21.41 Write the equilibrium that is present when $CH_3CH_2CH_2NH_2$ is in an aqueous solution.

21.42 Write the products that can be expected to form in the following situations. If no reaction occurs, write "no reaction."

(a) $CH_3CH_2CH_2NH_2 + HBr(aq) \rightarrow$
(b) $CH_3CH_2CH_2CH_3 + HI(aq) \rightarrow$
(c) $CH_3CH_2CH_2NH_3^+ + H_3O^+ \rightarrow$
(d) $CH_3CH_2CH_2NH_3^+ + OH^- \rightarrow$

Carbonyl Compounds

21.43 Why do aldehydes and ketones have boiling points that are lower than those of their corresponding alcohols?

21.44 Acetic acid boils at 118 °C, higher even than 1-propanol, which boils at 98 °C. How can the higher boiling point of acetic acid be explained?

21.45 Methyl ethanoate has a higher molecular mass than its parent acid, ethanoic acid. Yet methyl ethanoate (BP 59 °C) boils much lower than ethanoic acid (BP 118 °C). How can this be explained?

21.46 Ethanamide is a solid at room temperature. 1-Aminobutane, which has about the same molecular mass, is a liquid. Explain.

21.47 Write condensed structures of the following compounds.
(a) 3-methylbutanal
(b) 4-methyl-2-octanone
(c) 2-chloropropanoic acid
(d) isopropyl ethanoate
(e) 2-methylbutanamide

21.48 Write condensed structures of the following compounds.
(a) 2,3-butanedione
(b) butanedicarboxylic acid
(c) 2-aminopropanal
(d) cyclohexyl 2-methylpropanoate
(e) sodium 2,3-dimethylbutanoate

21.49 Which of the following two compounds could be easily oxidized? Write the structure of the product of the oxidation.

$$\underset{\textbf{A}}{CH_3CH_2\overset{\displaystyle O}{\overset{\|}{C}}H} \qquad \underset{\textbf{B}}{CH_3CH_2\overset{\displaystyle O}{\overset{\|}{C}}\underset{\underset{CH_3}{|}}{C}HCH_3}$$

21.50 Which of the following two compounds has the chemical property given below? Write the structure of the product of the reaction.

$$\underset{\textbf{A}}{CH_3CH_2CH_2OH} \qquad \underset{\textbf{B}}{CH_3CH_2\overset{\displaystyle O}{\overset{\|}{C}}OH} \qquad \underset{\textbf{C}}{CH_3CH_2\overset{\displaystyle O}{\overset{\|}{C}}H}$$

(a) is easily oxidized
(b) neutralizes NaOH
(c) forms an ester with methyl alcohol
(d) can be oxidized to an aldehyde
(e) can be dehydrated to an alkene

21.51 Write the equation for the equilibrium that is present in a solution of propanoic acid and methanol with a trace of strong acid.

21.52 Write the structures of the products that form in each of the following situations. If no reaction occurs, write "no reaction."
(a) $CH_3CH_2CO_2^- + HCl(aq) \rightarrow$

(b) $CH_3CH_2CO_2CH_3 + H_2O \xrightarrow{\text{Heat}}$
(c) $CH_3CH_2CH_2CO_2H + NaOH(aq) \rightarrow$
(d) $CH_3CH_2CO_2H + NH_3 \xrightarrow{\text{Heat}}$
(e) $CH_3CH_2CH_2CONH_2 + H_2O \xrightarrow{\text{Heat}}$
(f) $CH_3CH_2CO_2CH_3 + NaOH(aq) \xrightarrow{\text{Heat}}$

21.53 Classify the following compounds as amines or amides. If another functional group occurs, name it too.
(a) $CH_3CH_2NH_2$
(b) $CH_3CH_2NHCH_3$
(c) $CH_3CH_2\overset{\displaystyle O}{\overset{\|}{C}}NH_2$
(d) $CH_3\overset{\displaystyle O}{\overset{\|}{C}}CH_2NH_2$

21.54 What organic products with smaller molecular sizes form when the following compound is heated with water?

$$CH_3CH_2\underset{\underset{CH_3CH_2}{|}}{N}\overset{\displaystyle O}{\overset{\|}{C}}CH_3$$

21.55 Which of the following species give the reaction specified? Write the structures of the products that form.

$$CH_3CH_2\overset{\displaystyle O}{\overset{\|}{C}}NH_2 \qquad CH_3CH_2CH_2NH_2 \qquad CH_3CH_2NH_3^+$$

(a) neutralizes dilute hydrochloric acid
(b) hydrolyzes
(c) neutralizes dilute sodium hydroxide

21.56 Why are amides neutral but amines are basic?

Polymers

21.57 How are *macromolecules* and *polymers* alike? How are they different?

21.58 Polyvinyl chloride, or PVC, is widely used to manufacture drainage pipes. It is a polymer of vinyl chloride, $CH_2{=}CHCl$. Write the structure of PVC, first showing a molecular portion of four successive monomer units and then the condensed structure.

*21.59 If the following two compounds polymerized in the same way that Dacron forms, what would be the repeating unit of their polymer?

$$HO\overset{\displaystyle O}{\overset{\|}{C}}CH_2CH_2\overset{\displaystyle O}{\overset{\|}{C}}OH \qquad HOCH_2CH_2OH$$

*21.60 Kodel, a polyester made from the following monomers, is used to make fibers for weaving crease-resistant fabrics.

$$HOCH_2{-}\langle\bigcirc\rangle{-}CH_2OH \qquad HO\overset{\displaystyle O}{\overset{\|}{C}}{-}\langle\bigcirc\rangle{-}\overset{\displaystyle O}{\overset{\|}{C}}OH$$

monomers for Kodel

What is the structure of the repeating unit in Kodel?

*21.61 If the following two compounds polymerized in the same way that nylon forms, what would be the repeating unit in the polymer?

$$HOC\text{—}\bigcirc\text{—}COH \qquad NH_2CH_2CH_2CH_2CH_2NH_2$$

with O double bonded above each C.

Biochemistry

21.62 What is *biochemistry*?

21.63 What are the three fundamental needs for sustaining life and what are the general names for the substances that supply these needs?

Carbohydrates

21.64 How are carbohydrates defined?

21.65 What monosaccharide forms when the following polysaccharides are completely hydrolyzed? (a) starch, (b) glycogen, and (c) cellulose.

21.66 Glucose exists in three forms only one of which is a polyhydroxyaldehyde. How then can an aqueous solution of any form of glucose give the reactions of an aldehyde?

21.67 What compounds form when sucrose is digested?

21.68 The digestion of lactose gives what compounds?

21.69 The complete hydrolysis of the following compounds gives what product(s)? (Give the names only.)
(a) amylose   (b) amylopectin

21.70 Describe the relationships among amylose, amylopectin, and starch.

21.71 Why are humans unable to use cellulose as a source of glucose?

21.72 What function is served by glycogen in the body?

Lipids

21.73 How are lipids defined?

21.74 Why are lipids more soluble than carbohydrates in nonpolar solvents?

21.75 Cholesterol is not an ester, yet it is defined as a lipid. Why?

21.76 A product such as corn oil is advertised as "polyunsaturated," but all nutritionally important lipids have unsaturated C=O groups. What does "polyunsaturated" refer to?

21.77 Is it likely that the following compound could be obtained by the hydrolysis of a naturally occurring lipid? Explain.

$$CH_3(CH_2)_2CH\text{=}CHCH_2CH\text{=}CHCH_2CH\text{=}CH(CH_2)_7CO_2H$$

*21.78 Write the structure of a triacylglycerol that could be made from palmitic acid, oleic acid, and linoleic acid.

*21.79 Write the structures of the products of the complete hydrolysis of the following triacylglycerol.

$$CH_2OC(CH_2)_7CH\text{=}CHCH_2CH\text{=}CH(CH_2)_4CH_3$$
$$\overset{O}{\underset{|}{CHOC(CH_2)_{12}CH_3}}$$
$$CH_2OC(CH_2)_7CH\text{=}CH(CH_2)_7CH_3$$

with O double bonded above each OC.

21.80 Write the structure of the triacylglycerol that would result from the complete hydrogenation of the lipid whose structure is given in Exercise 21.79.

21.81 If the compound in Exercise 21.79 is saponified, what are the products? Give their names and structures.

21.82 What parts of glycerophospholipid molecules provide hydrophobic sites? Hydrophilic sites?

21.83 In general terms, describe the structure of the membrane of an animal cell. What kinds of molecules are present? How are they arranged? What forces keep the membrane intact? What functions are served by the protein components?

Amino Acids, Polypeptides, and Proteins

21.84 Describe the specific ways in which the monomers for all polypeptides are (a) alike and (b) different.

21.85 What is the peptide bond?

21.86 Write the structure of the dipeptide that could be hydrolyzed to give two molecules of glycine.

21.87 What is the structure of the tripeptide that could be hydrolyzed to give three molecules of alanine?

21.88 What are the structures of the two dipeptides that could be hydrolyzed to give glycine and phenylalanine?

21.89 Describe the structural way in which two *isomeric* polypeptides would be different.

21.90 Describe the structural ways in which two different polypeptides can differ.

21.91 Why is a distinction made between the terms *polypeptide* and *protein*?

21.92 In general terms, what forces are at work that determine the final *shape* of a polypeptide strand?

21.93 Explain in general terms how a change in pH might alter the shape of a polypeptide.

1.94 Define the following terms in a narrative that describes their relationships in living systems.
(a) enzyme
(b) substrate
(c) enzyme–substrate complex
(d) lock-and-key mechanism

21.95 What is meant by enzyme *specificity,* and what causes it?

21.96 For an enzyme involved in an equilibrium, what does *rate enhancement* refer to?

21.97 Write the equilibrium equation in which carbonic anhydrase participates.

21.98 The most lethal chemical poisons act on what kinds of substances? Give examples.

Nucleic Acids and Heredity

21.99 In general terms only, how does the body solve the problem of getting a particular amino acid sequence in a polypeptide rather than a mixture of randomly organized sequences?

21.100 In what specific structural way does DNA carry genetic messages?

21.101 What kind of force occurs between the two DNA strands in a double helix?

21.102 How are the two DNA strands in a double helix structurally related?

21.103 In what ways do DNA and RNA differ structurally?

21.104 Which base pairs with (a) A in DNA, (b) A in RNA, and (c) C in DNA or RNA?

21.105 A specific gene in the DNA of a higher organism is said to be *segmented* or *interrupted*. Explain.

21.106 On what kind of RNA are codons found?

21.107 On what kind of RNA are anticodons found?

21.108 What function does each of the following have in polypeptide synthesis?
(a) a ribosome   (b) mRNA   (c) tRNA

21.109 What kind of RNA forms directly when a cell uses DNA at the start of the synthesis of a particular enzyme? What kind of RNA is made from it?

21.110 The process of *transcription* begins with which nucleic acid and ends with which one?

21.111 The process of *translation* begins with which nucleic acid? What is the end result of translation?

21.112 What is a virus, and (in general terms only) how does it infect a tissue?

21.113 Genetic engineering is able (in general terms) to accomplish what?

21.114 What is *recombinant DNA?*

Additional Problems

*21.115 The H_2SO_4-catalyzed addition of water to 2-methylpropene could give two isomeric alcohols.
(a) Write their condensed structures.
(b) Actually, only one forms. It is not possible to oxidize this alcohol to an aldehyde or to a ketone having four carbons. Which alcohol, therefore, is the product?
(c) Write the structures of the two possible carbocations that could form when sulfuric acid donates a proton to 2-methylpropene.
(d) One of the two carbocations of part (c) is more stable than the other. Which one is it, and how can you tell?

21.116 An unknown alcohol was either **A** or **B**. When the unknown was oxidized with an *excess* of strong oxidizing agent, sodium dichromate in acid, the *organic* product isolated was able to neutralize sodium hydroxide.

(a) Which was the alcohol? Give the reasons for this choice.
(b) Write the balanced net ionic equation for the oxidation.

21.117 Write the structures of the organic products that can be expected to form in each situation. If no reaction occurs, write "no reaction."

*21.118 A synthetic polymer that competes with natural rubber for making vehicle tires is called SBR for styrene-butyl-rubber. It is made by copolymerizing styrene (25%) and butadiene (75%). Write the structure of one repeating unit of SBR if it has originated as an alternate copolymer from one molecule of each monomer.

*21.119 A 0.5574 mg sample of an organic acid, when burned in pure oxygen, gave 1.181 mg of CO_2 and 0.3653 mg H_2O. Calculate the empirical formula of the acid.

*21.120 When 0.2081 g of an organic acid was dissolved in 50.00 mL of 0.1016 M NaOH, it took 21.78 mL of 0.1182 M HCl to neutralize the NaOH that was not used up by the sample of the organic acid. Calculate the molecular mass that is equivalent to the data given. Is the answer necessarily the *molecular mass* of the organic acid? Explain.

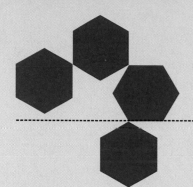

CHEMICALS IN USE 17
Genetic Fingerprinting

We have learned in Chapter 21 that genes are made up of deoxyribonucleic acid, or DNA. Because all people are *humans,* most of our DNA is common to us all. Parts of DNA molecules differ from person to person, however, so, unless you have an identical sibling, your DNA is also special. No one else has ever had DNA *identical* to yours; no one else ever will. However, every cell in your body has the entire set of the DNA molecules unique to you.

The "building blocks" of DNA are four nucleotides, each bearing one of four DNA "bases," A, T, G, or C. If one each were joined together, one possible structure would be ATGC; a few others are TGCA, GCAT, and CATG. DNA molecules are normally much longer, of course, involving millions of bases.

In the entire set of human genes, the human *genome,* there are many regions of DNA molecules that consist of nucleotide sequences repeated in tandem. These regions all have a common core sequence, but the *number* and the *lengths* of the repeated sequences vary widely from person to person. The variations are so considerable between individuals that they are the basis of the most important advance in forensic science (the science of crime evidence) since the development of fingerprints. The new method is called DNA profiling, DNA typing, or DNA fingerprinting.

We'll first study the technique used when the sample is relatively large; for example, a spot of blood the size of a dime. The sample to be studied is isolated and its DNA is extracted. Using Figure 17a, let's see how a DNA profile analysis now proceeds. In Step 1, the DNA is cut into pieces called *restriction fragments,* that is, fragments produced by the use of special enzymes called *restriction enzymes.* Restriction enzymes are isolated from bacteria, which use them to protect themselves against viruses that attack bacteria. (Yes, bacteria get "diseases," too!) Restriction enzymes catalyze the hydrolysis of DNA but do so only at places on a DNA strand that are adjacent to particular, short sequences of side chain bases 4 to 8 base pairs in length. For example, one commonly used restriction enzyme (about 100 are known) cleaves a DNA double helix at sites indicated by the arrows below, namely, adjacent to a series of four bases, AATT paired on the opposite strand of the double helix to TTAA.

$$\text{etc.}-G \overset{\downarrow}{\vert} A-A-T-T-C-\text{etc.}$$
$$\text{etc.}-C-T-T-A-A \overset{\vert}{\underset{\uparrow}{}} G-\text{etc.}$$

Section of a DNA double helix (arrows are sites of hydrolysis, catalyzed by a restriction enzyme)

Notice that one restriction fragment from the top DNA strand will end on its left with AATTC. A fragment from the bottom ends on its right with CTTAA. The remainders of each of the restriction fragments vary widely in length from individual to individual. In other words, humans have restriction fragments that, in *length,* are *polymorphic* (from the Greek *polymorphism* for "multiform"). In short, humans display *restriction-fragment length polymorphism,* which, mercifully, we'll return to calling RFLPs. The variations in length and relative amounts of an individual's set of RFLPs constitute that person's genetic profile or "DNA fingerprint." After chopping up the person's DNA, the technique continues with separating the RFLPs and displaying them.

Being polymorphic in length means that the RFLPs are polymorphic in *molecular mass.* This makes possible a technique called *gel electrophoresis,* which sorts and separates the RFLPs by mass. "Gel" refers to the supporting medium designed to carry the sample of RFLPs. The gel is a semisolid, semiliquid material that lets ions move through it. *Electrophoresis,* from the Greek *phoresis,* "being carried," refers to the movement of ions in an electrical field caused by an electrical force. In Step 2 (Figure 17a), a solution of RFLPs is spread in a narrow band at one end of the gel-coated surface, a current is turned on, and the ions move out, *lighter ions moving more rapidly than heavier ions,* assuming that the charges on the ions are equal. Several samples can be run side by side at the same time on the same gel-coated plate and so under identical conditions. The RFLPs are thus separated and now exist in narrow, parallel zones in the gel; what remains is to locate the zones and make them visible.

A blotting technique transfers the (invisible) RFLPs to a nylon membrane (Step 3), which is then immersed in a solution containing a *DNA probe.* The DNA probe consists of radiolabeled molecules of short lengths of DNA (Step 4). One or more atoms in the probe molecules are from a beta-emitting isotope.

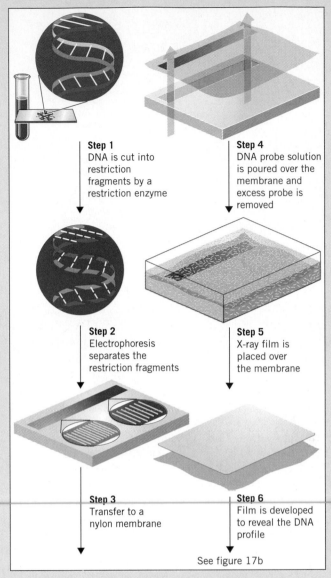

Step 1
DNA is cut into restriction fragments by a restriction enzyme

Step 2
Electrophoresis separates the restriction fragments

Step 3
Transfer to a nylon membrane

Step 4
DNA probe solution is poured over the membrane and excess probe is removed

Step 5
X-ray film is placed over the membrane

Step 6
Film is developed to reveal the DNA profile

See figure 17b

FIGURE 17a

Several steps (discussed in the text) are used to obtain a profile of the RFLPs from the DNA obtained from a crime scene sample or a defendant. (Courtesy of Cellmark Diagnostics, Germantown, Maryland.)

By base pairing, the short DNA probe molecules bind to the RFLPs wherever the RFLPs are on the nylon surface, and any excess probe is then washed off. The binding of probe to restriction fragments makes each zone of the restriction fragments a beta-emitting zone, but the zones are still invisible. To find the zones, a sheet of X ray film, which is also affected by beta radiation, is placed over the nylon surface (Step 5), and the beta radiation acts on the film. After several days, the X-ray film is developed (Step 6), and the dark bands appear wherever bands of separated RFLPs have been present.

The final result is called a *DNA profile,* and it looks much like a section of a grocery store's bar code, a series of parallel lines on a piece of paper that differ in number, thickness, and spatial separation from person to person (see Figure 17*b*). When the profile of a suspect matches that of a crime scene sample of blood or semen, it's similar to having a match of fingerprints. Without a match, the suspect is innocent.

Usually, to increase the precision of the method, four or more different restriction enzymes are allowed to work on samples, each enzyme chopping up the DNA into a unique set of RFLPs to give several "bar codes." It's like using the prints of several fingers, each one unique.

If two DNA profiles do not match, another suspect is sought. In London, England, about 20% of rape suspects have been cleared by failures of DNA profile matches. When a match occurs, however, the suspect is almost certainly guilty (many experts say, *certainly*), assuming that all the standards of handling evidence and doing the profiling have been followed.

When the crime scene specimen is very small, then the DNA must first be "amplified" or cloned so that the amount of DNA fragments is large enough for the rest of the analysis. A reaction called the *polymerase chain reaction,* catalyzed by a polymerase enzyme, catalyzes the cloning.

Questions

1. What fact about cells makes it possible to use cells from any part of the body of a suspect for DNA fingerprinting in a rape case for which a semen sample has been obtained?

2. What is meant by RFLP? What does each letter mean?

3. How are RFLPs used to give the genetic "bar code"?

V	E	S		V	E	S
No Match				Match		

FIGURE 17b

DNA analysis in crime lab testing. The DNA profiles of the victim (V) and the suspect (S) are compared to the evidence (E) obtained at the crime scene. On the left, there is no match between S and E, but on the right, the match occurs. In neither can the evidence be attributed to the victim. (Courtesy of Cellmark Diagnostics, Germantown, Maryland.)

Appendix A

Review of Mathematics

A.1 EXPONENTIALS

Very large and very small numbers are often expressed as powers of ten. This is often called **exponential notation.** When the quantity is expressed as a number between 1 and 10 multiplied by 10 raised to a power (e.g., 3.2×10^5), we also call it **scientific notation.** Some examples are given in Table A.1.

If you have a scientific calculator, it almost certainly is able to express numbers in scientific notation and perform arithmetic on numbers expressed this way. In fact, scientific calculators are discussed later in this Appendix. However, even though you have a calculator that works with numbers in scientific notation, it is important that you understand how to write these numbers and how to perform arithmetic with them. If you are unsure of yourself, work through the following review.

Positive Exponents   Suppose you come across a number such as 6.4×10^4. This is just an alternative way of writing 64,000. In other words, the number 6.4 is multiplied by 10 four times.

$$6.4 \times 10^4 = 6.4 \times 10 \times 10 \times 10 \times 10 = 64,000$$

Instead of actually writing out all of these 10s, notice that the decimal point is simply moved four places to the right in going from 6.4×10^4 to 64,000. Here are a few other examples. Study them to see how the decimal point changes.

TABLE A.1

Number	Exponential Form
1	1×10^0
10	1×10^1
100	1×10^2
1000	1×10^3
10,000	1×10^4
100,000	1×10^5
1,000,000	1×10^6
0.1	1×10^{-1}
0.01	1×10^{-2}
0.001	1×10^{-3}
0.0001	1×10^{-4}
0.00001	1×10^{-5}
0.000001	1×10^{-6}
0.0000001	1×10^{-7}

$$53.476 \times 10^2 = 5347.6$$
$$0.0016 \times 10^5 = 160$$
$$0.000056 \times 10^3 = 0.056$$

In some calculations it is helpful to reexpress large numbers in scientific notation. To do this, count the number of places you would have to move the decimal point to the left to put it just after the first digit in the number. For example, you would have to move the decimal point four places left in the following number.

$$6\ 0\ 5\ 3\ 0$$

Therefore, we can rewrite 60530 as 6.0530×10^4. Thus the number of moves equals the exponent.

Sometimes it is necessary to change a number from one exponential form to another. For example, suppose we wish to reexpress the number 6.0530×10^4 as something times 10^2.

$$6.0530 \times 10^4 = \boxed{} \times 10^2$$

What number goes in the box? Although there are rules that you could learn, it is better to think it through as follows. The number 6.0530×10^4 is a product of two parts, 6.0530 and 10^4. Together they equal 60530. We want to change the exponential part to 10^2, which means the exponential part of the product is becoming *smaller*. The only way the product can remain equal to 60530 is if the decimal part becomes *larger*. Since we are decreasing the exponential part by a factor of 100 (i.e., 10^2), we must increase the decimal part by a factor of 100, which means we change it from 6.0530 to 605.30. Therefore,

$$6.0530 \times 10^4 = \boxed{605.30} \times 10^2$$

Negative Exponents   A number such as 1.4×10^{-4} is an alternative way of writing

$$\frac{1.4}{10 \times 10 \times 10 \times 10}$$

Thus a negative exponent tells us how many times to divide by ten. Dividing by ten, of course, can be done by simply moving the decimal point. Thus, 1.4×10^{-4} can be reexpressed as 0.00014. Notice that the decimal point is four places to the *left* of the 1. Here are some examples of numbers expressed as negative exponentials and their equivalents.

$$3567.9 \times 10^{-3} = 3.5679$$

$$0.01456 \times 10^{-2} = 0.0001456$$

$$45691 \times 10^{-3} = 45.691$$

To express a number smaller than 1 in scientific notation, we count the number of places the decimal has to be moved to put it to the right of the first digit. The number that we count equals the negative exponent. For example, you would have to move the decimal point five places to the right to put it to the right of the first digit in

$$0.000068901$$

This number can be reexpressed as 6.8901×10^{-5}.

Multiplying Numbers Written in Scientific Notation

The real value of exponential forms of numbers comes in carrying out multiplications and divisions of large and small numbers. Suppose we want to multiply 1000 by 10000. Doing this by a long process fills the page with zeros (and some 1s). But notice the following relationships.

Numbers	1000	×	10000		
	$[10 \times 10 \times 10]$	×	$[10 \times 10 \times 10 \times 10]$	=	10,000,000
Exponential form	10^3	×	10^4	=	10^7
Exponents	3	+	4	=	7

If we *add* 3 and 4, the exponents of the exponential forms of 1000 and 10000, we get the exponent for the answer. Suppose we want to multiply 4160 by 20000. If we reexpress each number in scientific notation we have

$$(4.16 \times 10^3) \times (2 \times 10^4)$$

This now simplifies to

$$(4.16) \times (2) \times [10^3 \times 10^4]$$

The answer can be easily worked without losing the location of the decimal point. The exponents are added to give the exponent on 10 in the answer, and the numbers 4.16 and 2 are multiplied. The product is 8.32×10^7. Thus there are 2 steps in multiplying numbers that are written in scientific notation.

1. Multiply the decimal parts of the numbers (the parts that precede the 10s).
2. Add the exponents of the 10s algebraically to obtain the exponent on the 10 in the answer.

Dividing Numbers Written in Scientific Notation

Suppose we want to divide 10,000,000 by 1000. Doing this by long division would be primitive to say the least. Notice the following relationships.

$$\frac{10,000,000}{1000} = \frac{10^7}{10^3} = 10^4 = 10,000$$

The exponent in the result is $(7 - 3)$. So, to divide numbers in exponential form, we subtract exponents.

To extend this operation one step, let's divide 0.00468 by 0.0000400. The answer can be obtained as follows, expressing the numbers in scientific notation.

$$\frac{4.68 \times 10^{-3}}{4.00 \times 10^{-5}} = \frac{4.68}{4.00} \times 10^{[(-3) - (-5)]}$$
$$= 1.17 \times 10^2$$

Notice that the division proceeds in two steps.

1. Carry out the division of the numbers standing before the 10s.
2. Algebraically subtract the exponent in the denominator from the exponent in the numerator.

The Pocket Calculator and Exponentials

In scientific work the concept of an exponential is very important because it is much easier to handle either very large or very small numbers in exponential form. The review we've provided in this appendix is intended to help you refresh your memory from secondary school about these kinds of numbers. As noted earlier, almost any scientific calculator gives you the ability to multiply and divide large or small numbers when they are reexpressed as exponentials. We encourage you to study the *Instruction Manual* for your pocket calculator and read Section A.4 of this appendix so that you can do these kinds of calculations easily. However, we caution you that doing this without a basic understanding of exponentials will leave many traps in your path to skill in performing chemistry calculations.

A.2 LOGARITHMS

A **logarithm** is simply an exponent. For example, if we have an expression

$$N = a^b$$

we see that to obtain N we must raise a to the b power. The exponent b is said to be the logarithm of N; it is the exponent to which we must raise a, which we call the *base*, to obtain the number N.

$$\log_a N = b$$

We can define a logarithm system for any base we want. For example, if we choose the base 2, then

$$\log_2 8 = 3$$

which means that to obtain the number 8, we must raise the base (2) to the third power.

$$8 = 2^3$$

The most frequently encountered logarithm systems in the sciences are *common logarithms* and *natural logarithms,* which are discussed separately below. However, for any system of logarithms there are some useful relationships that evolve from the behavior of exponents in arithmetic operations. We encounter these at various times throughout the sciences, so it is important that you are familiar with them.

For Multiplication:

> If $N = A \times B$
>
> $\log N = \log A + \log B$

For Division:

> If $N = \dfrac{A}{B}$
>
> $\log N = \log A - \log B$

For Exponentials:

> If $N = A^b$
>
> $\log N = b \log A$

Common Logarithms

The **common logarithm** or **log** of a number N is the exponent to which 10 must be raised to equal the number N. Thus, if

$$N = 10^x$$

then the common logarithm of N, or **log N,** is simply x.

$$\log N = x$$

As mentioned earlier, there is another kind of logarithm called a **natural logarithm** that we'll take up later. However, when we use the terms "logarithm" or "log N" without specifying the base, we always mean common logarithms related to the base 10.

Some examples of common logarithms are as follows.

$N = 10$	$\log N = 1$	because	$10 = 10^1$
$= 100$	$= 2$		$100 = 10^2$
$= 1000$	$= 3$		$1000 = 10^3$
$= 0.1$	$= -1$		$0.1 = 10^{-1}$
$= 0.00001$	$= -5$		$0.00001 = 10^{-5}$

The logarithms of these numbers were easy to figure out by just using the definition of logarithms, but usually it isn't this simple. For example, what is the value of x in the following equations?

$$4.23 = 10^x$$

$$\log 4.23 = x$$

In other words, to what power x must we raise 10 to obtain 4.23? Obviously, the exponent is not a whole number. Therefore, to determine the value of x we have to use a table of logarithms or a hand-held calculator with the capability of using logs. If you are using a calculator for calculations involving logarithms, you should refer to the instruction booklet for directions.

Antilogarithms

There are times when we know what the logarithm of a number is and we seek to find the number. The procedure is called taking the **antilogarithm** or **antilog.** Thus, if

$$N = 10^x$$

and

$$\log N = x$$

then,

$$N = \text{antilog of } x$$

Finding an antilog is also discussed in Section A.4.

Natural Logarithms

In the sciences there are many phenomena in which observed quantities are related to each other logarithmically, which means that there is an exponential relationship between them. However, this exponential relationship does not involve 10 raised to a power. Instead it involves powers of a number symbolized by the letter e, which is said to be the base of the system of natural logarithms.

The quantity e is an irrational number, just like another quantity you are familiar with, π. This means that e is a nonrepeating decimal. To eight significant figures its value is

$$e = 2.7182818 \ldots$$

To truly appreciate the significance of this number and the system of logarithms based on it, you must have a good foundation in calculus. A discussion of calculus is beyond the scope of this text, so we will not attempt to explain the origin of the value of e further. Nevertheless, we can still use the results and discuss the system of **natural logarithms** that have the number e as its base.

Natural logarithms use the symbol **ln** instead of log. Thus, if the number N equals e^x,

$$N = e^x$$

then

$$\log_e N = \ln N = x$$

Scientific calculators have the ability to calculate natural logarithms and their corresponding antilogarithms. This is discussed in Section A.4 of this appendix. Sometimes it is useful to know the simple relationship between common and natural logarithms. To four significant figures,

$$\ln N = 2.303 \log N$$

In other words, to find ln N, simply find the common log of N and multiply it by 2.303.

A.3 GRAPHING

One of the most useful ways of describing the relationship between two quantities is by means of a graph. It allows us to obtain an overall view of how one quantity changes when the other changes. Constructing graphs from experimental data, or from data calculated from an equation, is therefore a common operation in the sciences, so it is important that you understand how to present data graphically. Let's look at an example.

Suppose that we want to know how the volume of a given amount of gas varies as we change its pressure. In Chapter

10 we find an equation that gives the pressure–volume relationship for a fixed amount of gas at a constant temperature,

$$PV = \text{constant} \qquad (A.1)$$

where P = pressure and V = volume.

Suppose, now, that we want to see graphically how the pressure and volume of a gas are related as the pressure is raised from 1.0 to 10.0 atm in steps of 1.0 atm. Let's also suppose that for this sample of gas the value of the constant in our equation is 0.25 L atm. To construct the graph we first must decide how to label and number the axes.

Usually, the horizontal axis, called the **abscissa,** is chosen to correspond to the **independent variable**—the variable whose values are chosen first and from which the values of the **dependent variable** are determined. In our example, we are choosing values of pressure (1.0 atm, 2.0 atm, etc.), so we will label the abscissa "pressure." We will also mark off the axis evenly from 1.0 to 10.0 atm, as shown in Figure A.1. Next we put the label "volume" on the vertical axis, called the **ordinate.** Before we can number this axis, however, we need to know over what range the volume will vary. Using Equation A.1 we can calculate the data in Table A.2, and we see that the volume ranges from a low of 0.025 L to a high of 0.25 L. We can therefore mark off the ordinate evenly in increments of 0.025 L, starting at the bottom with 0.025 L. Notice that in labeling the axes we have indicated the units of P and V in parentheses.

Next we plot the data as shown in Figure A.2. To clearly show each plotted point, we use a small dot that has its center located at the coordinates of the point. Then we draw a *smooth* curve through the points.

Sometimes, when plotting experimentally measured data, the points do not fall exactly on a smooth line (Figure A.3). However, nature generally is not irregular even though that's what the data appear to suggest. The fluctua-

TABLE A.2

Pressure (atm)	Volume (L)
1.0	0.25
2.0	0.125
3.0	0.083
4.0	0.062
5.0	0.050
6.0	0.042
7.0	0.036
8.0	0.031
9.0	0.028
10.0	0.025

tions are usually due to experimental error of some sort. Therefore, rather than draw an irregular line connecting all the data points (the dashed line), we draw a smooth curve that passes as close as possible to all of the points, even though it may not actually pass through any of them.

Slope One of the properties of curves and lines on a graph is their *slope* or steepness. Consider Figure A.4, which is a straight line drawn on a set of xy coordinate axes. The slope of the line is defined as the change in y, Δy, divided by the change in x, Δx.

$$\text{slope} = \frac{\Delta y}{\Delta x} = \frac{y_2 - y_1}{x_2 - x_1}$$

Curved lines have slopes too, but the slope changes from point to point on the curve. To obtain the slope graphically, we can draw a line that is tangent to the curve at the point where we want to know the slope. The slope of this tangent is the same as the slope of the curve at this point. In Figure A.5 we see a curve for which the slope is determined at two different points. At point M the curve is rising steeply and

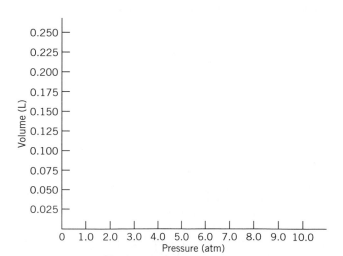

FIGURE A.1

Choosing and labeling the axes of a graph.

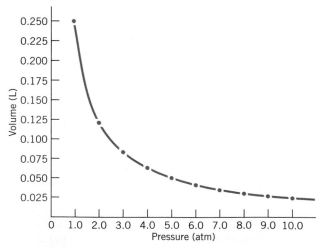

FIGURE A.2

Plotting points and drawing the curve.

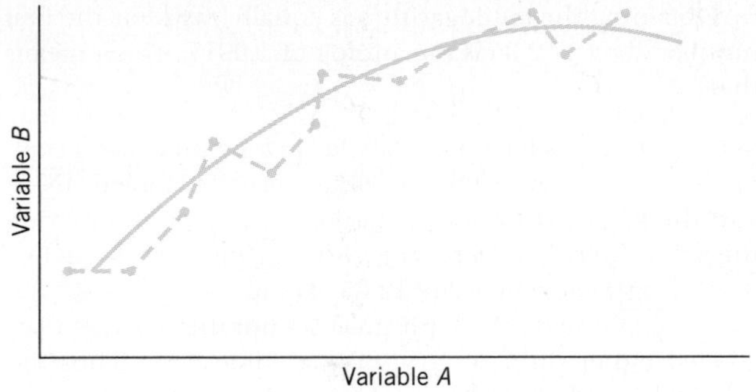

FIGURE A.3

Experimental data do not all fall on a smooth curve when they are plotted because they contain experimental error. A smooth curve is drawn as close to all of the data points as possible, rather than the jerky dashed line.

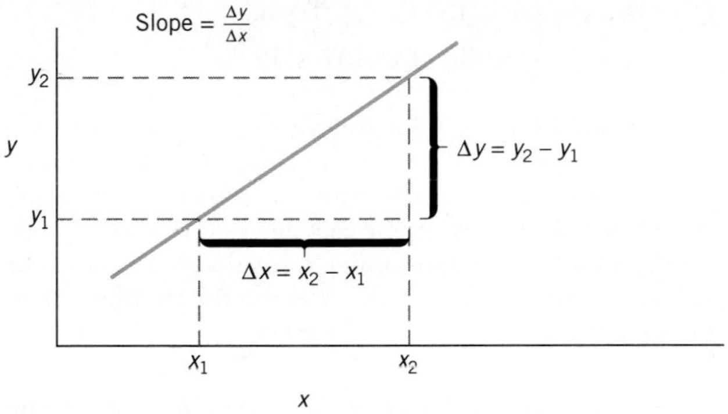

FIGURE A.4

Determining the slope of a straight line.

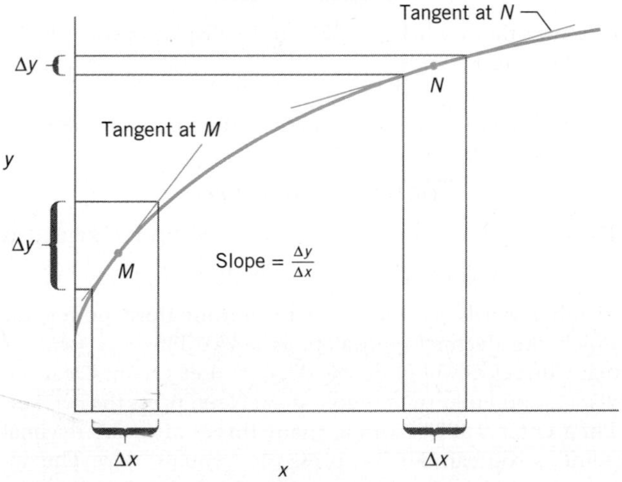

FIGURE A.5

For a given Δx, Δy is much larger at M than it is at N. Therefore, the slope at M is larger than at N.

$\Delta y/\Delta x$ for the tangent is large. At point N the curve is rising less steeply, and $\Delta y/\Delta x$ for the tangent at N is small. Therefore, the slope at M is larger than the slope at N.

A.4 ELECTRONIC CALCULATORS

Performing arithmetic calculations is commonplace in all of the sciences, and electronic calculators have made these operations simple and routine. So-called **scientific calculators** are inexpensive (most cost less than this book) and are able to perform complex functions at the touch of a button. But these electronic wizards are only as versatile as the person who presses the keys.

This section is not intended to teach you how to use your particular calculator—each calculator comes with its own instruction booklet. Instead, our goal is to point out common errors that are easy to make with some kinds of calculators, to explain how to speed up certain types of calculations, and to explain how certain kinds of calculations are done on the calculator.

General Operation

Most scientific calculators use an *algebraic entry system,* in which keys are pressed in the order that the quantities and mathematical operations appear in the computation. Thus

$$2 \times 3 = 6$$

is entered by pressing keys in this order: 2, ×, 3, = . We will indicate this by using boxes to represent the keys.

2	×	3	=

When the keys are pressed in this sequence, the answer, 6, appears on the display. In explaining how to use the calculator for certain calculations, we will assume an algebraic entry system.

Scientific Notation

Earlier in this appendix and in Section 1.3 we saw that numbers expressed as

$$2.3 \times 10^5$$
$$6.7 \times 10^{-12}$$

are said to be in **exponential notation** or **scientific notation.** Scientific calculators are able to do arithmetic with numbers such as these, provided they are entered correctly.

Many calculators have a key labeled EXP or EE that is used to enter the exponential portion of numbers that are expressed in scientific notation. If we read the number 2.3×10^5 as "two point three times ten to the fifth," the EXP (or EE) key should be read as ". . . times ten to the. . . ." Thus the sequence of keys that enter the number 2.3×10^5 is

2	.	3	EXP	5

Try this on your own calculator and check to be sure that the number is correctly displayed after you've entered it. If not, check the instruction booklet that came with the calculator or ask your instructor for help.

When entering a number such as 6.7×10^{-12}, be sure to press the "change sign" or $\boxed{+/-}$ key after pressing the $\boxed{\text{EXP}}$ key. Calculators will not give the correct entry if the $\boxed{-}$ key is pressed instead of the $\boxed{+/-}$ key. For example, clear the display and then try the following sequence of keystrokes on your calculator.

$$\boxed{6}\,\boxed{.}\,\boxed{7}\,\boxed{\text{EXP}}\,\boxed{+/-}\,\boxed{1}\,\boxed{2}$$

If you wish to multiply 2.3×10^5 and 6.7×10^{-12}, enter the first one, press the $\boxed{\times}$ key, then enter the second number, and finally press the $\boxed{=}$ key. The answer displayed on the calculator should be 1.541×10^{-6}.

$$(2.3 \times 10^5) \times (6.7 \times 10^{-12}) = 1.541 \times 10^{-6}$$

If your display does not show an exponential, your calculator's instruction manual will tell you how to proceed.

Chain Calculations   It is not uncommon to have to calculate the value of a fraction in which there is a string of numbers that must be multiplied together in the numerator and another string of numbers to be multiplied together in the denominator. For example,

$$\frac{5.0 \times 7.3 \times 8.5}{6.2 \times 2.5 \times 3.9}$$

Many students will compute the value of the numerator and write down the answer, then compute the value of the denominator and write it down, and finally divide the value of the numerator by the value of the denominator (whew!). Although this gives the correct answer, there is a simpler way to do the arithmetic that doesn't require writing down any intermediate values. The procedure is as follows:

1. Enter the first value from the numerator (5.0).
2. Any of the other values from the numerator are entered by *multiplication* and any of the values from the denominator are entered by *division*.

The following is one sequence that gives the answer.

$$5.0 \times 7.3 \times 8.5 \div 6.2 \div 2.5 \div 3.9 = 5.13234 \ldots$$

Notice that each value from the denominator is entered by division. It also doesn't matter in what sequence the numbers are entered. The following gives the same answer.

$$5.0 \div 6.2 \times 7.3 \div 2.5 \div 3.9 \times 8.5 = 5.13234 \ldots$$

Logarithms and Antilogarithms   Your scientific calculator works with both common logarithms and natural logarithms. To find the common log of a number, enter the number in the calculator and press the $\boxed{\log}$ or $\boxed{\text{LOG}}$ button. Try this for the following and check your results. (The values here are rounded so they have the same number of digits after the decimal point as there are significant figures in the number. This is the rule for significant figures in logarithms.)

$$\log 12.35 = 1.0917 \text{ (rounded)}$$

$$\log(3.70 \times 10^{-4}) = -3.432 \text{ (rounded)}$$

Obtaining the antilogarithm is equally easy. For the first number above, 12.35 is the antilog of 1.0917, which means that

$$12.35 = 10^{1.0917}$$

Some calculators have a key labeled $\boxed{10^x}$; others use a combination of an inverse function key, usually labeled $\boxed{\text{INV}}$, and the key used to obtain the logarithm. Thus, entering the value 1.0917 and pressing either $\boxed{10^x}$ or the sequence $\boxed{\text{INV}}\,\boxed{\log}$ yields the value 12.35. Try it.

To obtain natural logarithms of a number we use a key labeled either $\boxed{\ln x}$ or $\boxed{\text{LN}}$ or sometimes $\boxed{\ln}$. Thus the natural logarithm of 12.35 is obtained by entering the number and pressing the $\boxed{\ln x}$ key. Try it.

$$\ln 12.35 = 2.5137 \text{ (rounded)}$$

The value 12.35 is the antiln of 2.5137, which means

$$12.35 = e^{2.5137}$$

Therefore, to obtain the antiln of 2.5137, we enter the number into the calculator and press either the key labeled $\boxed{e^x}$ or the sequence $\boxed{\text{INV}}\,\boxed{\ln x}$. Try it.

$$\text{antiln of } 2.5137 = 12.35$$

Powers and Roots   Squares and square roots are handled easily with keys labeled $\boxed{x^2}$ and $\boxed{\sqrt{x}}$. For higher powers and higher roots (e.g., cube roots or fourth roots) we use the key that on most scientific calculators is labeled $\boxed{x^y}$. To use this function, we must enter two values—the number and its exponent. For example, to raise the number 2.6 to the 5th power,

$$(2.6)^5$$

we enter the number 2.6, press the $\boxed{x^y}$ key, enter the value of the exponent (5), and then press either the "equals" key, or the next function key desired if we are performing a chain calculation. Thus, the sequence

$$\boxed{2}\,\boxed{.}\,\boxed{6}\,\boxed{x^y}\,\boxed{5}\,\boxed{=}$$

produces the result 118.81376 on the display of the calculator. Therefore,

$$(2.6)^5 = 118.81376$$

To take a root, the procedure is quite similar, because we make use of the relationship

$$\sqrt[n]{(\text{number})} = (\text{number})^{1/n}$$

Thus, the cube root of a number such as 56.4 is obtained as

$$\sqrt[3]{56.4} = (56.4)^{1/3} = (56.4)^{0.333333 \ldots}$$

In other words, we raise 56.4 to the one-third power, for which the decimal equivalent is 0.3333333 . . . (with as many threes as will fit on the display). Let's try it. Clear the display and enter the number 56.4. Next, press the $\boxed{x^y}$ key. Then enter 0.33333 with as many threes after the decimal point as you can. Finally, press the "equals" key. The answer should be 3.834949. . . . If we had wanted to take the 5th root of 56.4, we would have raised the number to the 1/5 power, or 0.20. Try it. You should get 2.24004 . . . as the answer. (Your calculator may handle x^y functions differently. If so, consult the manual.)

Appendix B

Electron Configurations of the Elements

Atomic Number			Atomic Number			Atomic Number		
1	H	$1s^1$	38	Sr	[Kr] $5s^2$	75	Re	[Xe] $6s^2 4f^{14} 5d^5$
2	He	$1s^2$	39	Y	[Kr] $5s^2 4d^1$	76	Os	[Xe] $6s^2 4f^{14} 5d^6$
3	Li	[He] $2s^1$	40	Zr	[Kr] $5s^2 4d^2$	77	Ir	[Xe] $6s^2 4f^{14} 5d^7$
4	Be	[He] $2s^2$	41	Nb	[Kr] $5s^1 4d^4$	78	Pt	[Xe] $6s^1 4f^{14} 5d^9$
5	B	[He] $2s^2 2p^1$	42	Mo	[Kr] $5s^1 4d^5$	79	Au	[Xe] $6s^1 4f^{14} 5d^{10}$
6	C	[He] $2s^2 2p^2$	43	Tc	[Kr] $5s^2 4d^5$	80	Hg	[Xe] $6s^2 4f^{14} 5d^{10}$
7	N	[He] $2s^2 2p^3$	44	Ru	[Kr] $5s^1 4d^7$	81	Tl	[Xe] $6s^2 4f^{14} 5d^{10} 6p^1$
8	O	[He] $2s^2 2p^4$	45	Rh	[Kr] $5s^1 4d^8$	82	Pb	[Xe] $6s^2 4f^{14} 5d^{10} 6p^2$
9	F	[He] $2s^2 2p^5$	46	Pd	[Kr] $4d^{10}$	83	Bi	[Xe] $6s^2 4f^{14} 5d^{10} 6p^3$
10	Ne	[He] $2s^2 2p^6$	47	Ag	[Kr] $5s^1 4d^{10}$	84	Po	[Xe] $6s^2 4f^{14} 5d^{10} 6p^4$
11	Na	[Ne] $3s^1$	48	Cd	[Kr] $5s^2 4d^{10}$	85	At	[Xe] $6s^2 4f^{14} 5d^{10} 6p^5$
12	Mg	[Ne] $3s^2$	49	In	[Kr] $5s^2 4d^{10} 5p^1$	86	Rn	[Xe] $6s^2 4f^{14} 5d^{10} 6p^6$
13	Al	[Ne] $3s^2 3p^1$	50	Sn	[Kr] $5s^2 4d^{10} 5p^2$	87	Fr	[Rn] $7s^1$
14	Si	[Ne] $3s^2 3p^2$	51	Sb	[Kr] $5s^2 4d^{10} 5p^3$	88	Ra	[Rn] $7s^2$
15	P	[Ne] $3s^2 3p^3$	52	Te	[Kr] $5s^2 4d^{10} 5p^4$	89	Ac	[Rn] $7s^2 6d^1$
16	S	[Ne] $3s^2 3p^4$	53	I	[Kr] $5s^2 4d^{10} 5p^5$	90	Th	[Rn] $7s^2 6d^2$
17	Cl	[Ne] $3s^2 3p^5$	54	Xe	[Kr] $5s^2 4d^{10} 5p^6$	91	Pa	[Rn] $7s^2 5f^2 6d^1$
18	Ar	[Ne] $3s^2 3p^6$	55	Cs	[Xe] $6s^1$	92	U	[Rn] $7s^2 5f^3 6d^1$
19	K	[Ar] $4s^1$	56	Ba	[Xe] $6s^2$	93	Np	[Rn] $7s^2 5f^4 6d^1$
20	Ca	[Ar] $4s^2$	57	La	[Xe] $6s^2 5d^1$	94	Pu	[Rn] $7s^2 5f^6$
21	Sc	[Ar] $4s^2 3d^1$	58	Ce	[Xe] $6s^2 4f^1 5d^1$	95	Am	[Rn] $7s^2 5f^7$
22	Ti	[Ar] $4s^2 3d^2$	59	Pr	[Xe] $6s^2 4f^3$	96	Cm	[Rn] $7s^2 5f^7 6d^1$
23	V	[Ar] $4s^2 3d^3$	60	Nd	[Xe] $6s^2 4f^4$	97	Bk	[Rn] $7s^2 5f^9$
24	Cr	[Ar] $4s^1 3d^5$	61	Pm	[Xe] $6s^2 4f^5$	98	Cf	[Rn] $7s^2 5f^{10}$
25	Mn	[Ar] $4s^2 3d^5$	62	Sm	[Xe] $6s^2 4f^6$	99	Es	[Rn] $7s^2 5f^{11}$
26	Fe	[Ar] $4s^2 3d^6$	63	Eu	[Xe] $6s^2 4f^7$	100	Fm	[Rn] $7s^2 5f^{12}$
27	Co	[Ar] $4s^2 3d^7$	64	Gd	[Xe] $6s^2 4f^7 5d^1$	101	Md	[Rn] $7s^2 5f^{13}$
28	Ni	[Ar] $4s^2 3d^8$	65	Tb	[Xe] $6s^2 4f^9$	102	No	[Rn] $7s^2 5f^{14}$
29	Cu	[Ar] $4s^1 3d^{10}$	66	Dy	[Xe] $6s^2 4f^{10}$	103	Lr	[Rn] $7s^2 5f^{14} 6d^1$
30	Zn	[Ar] $4s^2 3d^{10}$	67	Ho	[Xe] $6s^2 4f^{11}$	104	Rf	[Rn] $7s^2 5f^{14} 6d^2$
31	Ga	[Ar] $4s^2 3d^{10} 4p^1$	68	Er	[Xe] $6s^2 4f^{12}$	105	Ha	[Rn] $7s^2 5f^{14} 6d^3$
32	Ge	[Ar] $4s^2 3d^{10} 4p^2$	69	Tm	[Xe] $6s^2 4f^{13}$	106	Sg	[Rn] $7s^2 5f^{14} 6d^4$
33	As	[Ar] $4s^2 3d^{10} 4p^3$	70	Yb	[Xe] $6s^2 4f^{14}$	107	Ns	[Rn] $7s^2 5f^{14} 6d^5$
34	Se	[Ar] $4s^2 3d^{10} 4p^4$	71	Lu	[Xe] $6s^2 4f^{14} 5d^1$	108	Hs	[Rn] $7s^2 5f^{14} 6d^6$
35	Br	[Ar] $4s^2 3d^{10} 4p^5$	72	Hf	[Xe] $6s^2 4f^{14} 5d^2$	109	Mt	[Rn] $7s^2 5f^{14} 6d^7$
36	Kr	[Ar] $4s^2 3d^{10} 4p^6$	73	Ta	[Xe] $6s^2 4f^{14} 5d^3$	110	Uun	[Rn] $7s^2 5f^{14} 6d^8$
37	Rb	[Kr] $5s^1$	74	W	[Xe] $6s^2 4f^{14} 5d^4$	111	Uuu	[Rn] $7s^2 5f^{14} 6d^9$

Appendix C

Equivalent Weights and Normality

C.1 THE EQUIVALENT IN ACID–BASE REACTIONS

For certain kinds of reactions, it is possible to define chemical quantities that always combine in a one-to-one ratio, regardless of the coefficients in the balanced equation. One is acid–base reactions and the other is redox reactions.

In an acid–base neutralization or proton-transfer reaction, the quantity of acid that provides one mole of H^+ will *always* neutralize the quantity of base that provides one mole of OH^-. These two always combine in a $1:1$ ratio.

$$H^+ + OH^- \longrightarrow H_2O$$

Suppose that 36 g of some acid is able to furnish exactly one mole of H^+, and 40 g of a base is able to furnish exactly 1 mol of OH^-. Because 1 mol of H^+ reacts with 1 mol of OH^-, we know that the 36 g of the acid is just enough to react with the 40 g of base. We can make this statement without even knowing what the acid or base is; we only have to know how much of the acid gives 1 mol H^+ and how much of the base gives 1 mol OH^-.

This reasoning forms the basis of the definition of another kind of chemical quantity called the **equivalent,** abbreviated **eq.** The exact definition depends on the kind of reaction, acid–base or redox but, in both, the *equivalent* is always defined so that equivalents of reactants always react in a one-to-one ratio. This is the key concept to remember. If there are two reactants, *A* and *B*,

1 eq of reactant *A* reacts with exactly 1 eq of reactant *B*.

For acids and bases, the definitions of the equivalent that meet this standard are the following.

One equivalent of an acid is the amount of the acid that is able to furnish 1 mol of hydrogen ion.
One equivalent of a base is the amount of the base that is able to neutralize 1 mol of hydrogen ion.

From the chemical equation for the neutralization reaction, we see that these definitions assure a one-to-one combining ratio for equivalents of acids and bases.

C.2 THE EQUIVALENT IN REDOX REACTIONS

The definitions of equivalents for a redox reaction parallel those for acid–base reactions. They retain the central principle of a $1:1$ ratio between an equivalent of an oxidizing agent and an equivalent of a reducing agent. If one particle of an oxidizing agent accepts one electron, something else must donate that electron. Thus the *equivalents* for redox reactions are defined as follows.

One equivalent of an oxidizing agent is the amount of the substance that is able to accept 1 mol of electrons.
One equivalent of a reducing agent is the amount of the substance that is able to donate 1 mol of electrons.

C.3 EQUIVALENT WEIGHTS

Let's now apply these principles to real chemicals and see how the equivalent is related to the more familiar chemical unit, the *mole.* The *equivalent* is not an SI unit. Many chemists have moved away from using it. In certain fields, however, including many scientific fields outside of chemistry, the concept of the equivalent is widely employed.

Equivalents of Acids and Bases Consider the acids HCl and H_2SO_4. One mole of HCl is enough acid to supply 1 mol of H^+, so 1 mol of HCl must be the same as 1 eq HCl.

$$1 \text{ mol } HCl \Leftrightarrow 1 \text{ mol } H^+$$

So,

$$1 \text{ mol } HCl = 1 \text{ eq } HCl$$

On the other hand, 1 mol of H_2SO_4 is enough acid to supply 2 mol of H^+, provided the H_2SO_4 is completely neutralized. For complete neutralization, then, 1 mol of H_2SO_4 must be equal to 2 eq H_2SO_4.

$$1 \text{ mol } H_2SO_4 \Leftrightarrow 2 \text{ mol } H^+$$

So,

$$1 \text{ mol } H_2SO_4 = 2 \text{ eq } H_2SO_4 \text{ (for complete reaction)}$$

Thus, there is a simple relationship between moles and

equivalents for acids. *The number of equivalents in 1 mol of an acid is equal to the number of hydrogen ions that are neutralized when one molecule of the acid reacts.*

Phosphoric acid, H_3PO_4, is a triprotic acid when its *complete* neutralization entails the reaction of three H^+ ions per molecule. Thus, 1 mol of H_3PO_4 equals 3 equivalents of this acid

$$1 \text{ mol } H_3PO_4 \Leftrightarrow 3 \text{ mol } H^+$$

1 mol H_3PO_4 = 3 eq H_3PO_4 (for complete neutralization)

You can see the pattern, which we can summarize as follows.

If an acid is monoprotic
$$1 \text{ mol of acid} = 1 \text{ eq of acid}$$
If an acid is diprotic
$$1 \text{ mol of acid} = 2 \text{ eq of acid}$$
If an acid is triprotic
$$1 \text{ mol of acid} = 3 \text{ eq of acid}$$

The size of an equivalent does depend on the reaction. We emphasized for the case of phosphoric acid that we were referring to its *complete* neutralization. But suppose that the following reaction were being used.

$$H_3PO_4 + 2NaOH \longrightarrow Na_2HPO_4 + 2H_2O$$

Here the H_3PO_4 molecule is donating only two H^+. In this reaction, we have to write:

$$1 \text{ mol } H_3PO_4 \Leftrightarrow 2 \text{ mol } H^+$$

So for this partial neutralization,

$$1 \text{ mol } H_3PO_4 = 2 \text{ eq } H_3PO_4$$

This must caution us about the use of the concept of the equivalent.

Determining the number of equivalents per mole of base is as simple as for an acid. Consider the two bases NaOH and $Ba(OH)_2$. One mole of NaOH is enough base to supply 1 mol of OH^-, so an equivalent of NaOH must be the same as one mole of NaOH.

$$1 \text{ mol NaOH} \Leftrightarrow 1 \text{ mol } OH^-$$
$$1 \text{ mol NaOH} = 1 \text{ eq NaOH}$$

One mole of $Ba(OH)_2$ is able to supply 2 mol of OH^- when completely neutralized, so 1 mol of $Ba(OH)_2$ must then be 2 eq base.

$$1 \text{ mol } Ba(OH)_2 \Leftrightarrow 2 \text{ mol } OH^-$$
$$1 \text{ mol } Ba(OH)_2 = 2 \text{ eq } Ba(OH)_2$$

Thus, for a metal hydroxide, the number of equivalents in 1 mol of the base is equal to the number of hydroxides in one formula unit of the base. Thus, for $Al(OH)_3$,

$$1 \text{ mol } Al(OH)_3 = 3 \text{ eq } Al(OH)_3$$

When the base is not one that supplies OH^- ions, the same principles apply. For example, one HCO_3^- ion can neutralize one H^+ ion.

$$HCO_3^- + H^+ \longrightarrow CO_2 + H_2O$$

Thus,

$$1 \text{ mol NaHCO}_3 \Leftrightarrow 1 \text{ mol of neutralized } H^+$$
$$1 \text{ mol NaHCO}_3 = 1 \text{ eq NaHCO}_3$$

The carbonate ion, however, can neutralize two H^+ ions.

$$CO_3^{2-} + 2H^+ \longrightarrow CO_2 + H_2O$$

When this is the stoichiometry of the neutralization reaction, then we can write

$$1 \text{ mol Na}_2CO_3 \Leftrightarrow 2 \text{ mol of neutralized } H^+$$
$$1 \text{ mol Na}_2CO_3 = 2 \text{ eq Na}_2CO_3$$

Equivalent Weights of Acids and Bases In working problems using equivalents, it is often useful to know the mass of an equivalent of each of the reactants. Earlier, for example, we described a case in which 36 g of an acid gave 1 mol of H^+ and 40 g of a base gave 1 mol of OH^-. These quantities provide 1 eq each of the acid and the base. Knowing how much an equivalent of each of them weighs establishes a mass relationship between the two reactants that can be used in stoichiometric calculations. Thus, if we had a sample of 20 g of the base (0.50 eq of base), we know that we would only need 18 g of the acid (0.50 eq of acid).

The mass in grams of one equivalent is called the **equivalent weight.**[1] An equivalent weight, in other words, is the number of grams per equivalent, the ratio of the number of grams in a mole to the number of equivalents in a mole.

$$\text{Equivalent weight} = \frac{\text{no. g per mol}}{\text{no. eq per mol}}$$

We know the following two facts about sulfuric acid, for example. First, when it is neutralized completely,

$$1 \text{ mol } H_2SO_4 = 2 \text{ eq } H_2SO_4$$

Second,

$$1 \text{ mol } H_2SO_4 = 98.07 \text{ g } H_2SO_4$$

Therefore, we can write

$$98.07 \text{ g } H_2SO_4 = 2 \text{ eq } H_2SO_4$$

The equivalent weight, therefore, is found by taking the ratio of grams to equivalents:

$$\frac{98.07 \text{ g } H_2SO_4}{2 \text{ eq } H_2SO_4} = 49.04 \text{ g eq}^{-1}$$

The same kind of calculations apply to any base.

---

[1] Equivalent *weight* is the term almost universally used, although equivalent *mass* would be correct. We will follow here the usual custom.

Equivalents of Oxidizing Agents and Reducing Agents

Careful attention to stoichiometry is also essential in calculating the equivalent weights of oxidizing and reducing agents. Consider, for example, the reaction of potassium permanganate as an oxidizing agent. We have learned in Chapter 4 that when a reaction using $KMnO_4$ is carried out in an *acidic* solution, the permanganate ion is reduced to the manganese(II) ion by the following half-reaction.

$$\underset{+7}{MnO_4^-}(aq) + 8H^+(aq) + 5e^- \longrightarrow \underset{+2}{Mn^{2+}}(aq) + 4H_2O$$

Mn goes from the $+7$ to the $+2$ state by a gain of five electrons. We do not, in fact, need the half-reaction to know this. All we need is just the change in oxidation numbers themselves to figure out how many electrons are gained (or lost) by some reactant. Because each mole of MnO_4^- gains 5 mol of electrons, each mole of this oxidizing agent must correspond to 5 eq. In other words,

$$1 \text{ mol } KMnO_4 \Leftrightarrow 5 \text{ mol } e^- \quad \text{(acidic medium)}$$

$$1 \text{ mol } KMnO_4 = 5 \text{ eq } KMnO_4 \quad \text{(acidic medium)}$$

In a neutral or slightly basic medium, however, MnO_4^- is reduced to MnO_2.

$$\underset{+7}{MnO_4^-} \longrightarrow \underset{+4}{MnO_2}$$

Now, Mn changes from its $+7$ oxidation state to the $+4$ state, which means a three-electron change. Thus a gain of only 3 mol of electrons by each mole of MnO_4^- occurs as MnO_2 forms, so there are 3 eq/mol of $KMnO_4$ when it is used in a neutral or basic medium.

$$1 \text{ mol } KMnO_4 \Leftrightarrow 3 \text{ mol of } e^- \quad \text{(neutral or basic medium)}$$

$$1 \text{ mol } KMnO_4 = 3 \text{ eq } KMnO_4 \quad \text{(neutral or basic medium)}$$

Thus *the number of equivalents per mole of an oxidizing agent equals the number of moles of electrons accepted by 1 mol during a redox reaction.* A similar statement applies to reducing agents. *The number of equivalents per mole of a reducing agent equals the number of moles of electrons donated by 1 mol during a redox reaction.* To determine the equivalents per mole for an oxidizing agent or a reducing agent, we use the oxidation states of the reactants and products.

Consider sodium dichromate, $Na_2Cr_2O_7$, which contains $Cr_2O_7^{2-}$, a strong oxidizing agent. In an acidic medium, the chromium in the dichromate ion changes from a $+6$ oxidation state to a $+3$ state.

$$\underset{+6}{Cr_2O_7^{2-}} \longrightarrow \underset{+3}{2Cr^{3+}}$$

This means that *each* Cr in $Cr_2O_7^{2-}$ gains three electrons, but there are two Cr in $Cr_2O_7^{2-}$. So each whole $Cr_2O_7^{2-}$ ion gains six electrons. In other words,

$$1 \text{ mol } Cr_2O_7^{2-} \Leftrightarrow 6 \text{ mol of } e^-$$

$$1 \text{ mol } Na_2Cr_2O_7 = 6 \text{ eq } Na_2Cr_2O_7$$

Thus, we must consider both the change in the oxidation number *and* the number of atoms in the actual reactant that undergo the change.

Consider another example, that of oxalic acid, $H_2C_2O_4$. When it acts as a reducing agent, it becomes oxidized to CO_2.

$$\underset{+3}{H_2C_2O_4} \longrightarrow \underset{+4}{2CO_2}$$

Carbon is oxidized from a $+3$ state in $H_2C_2O_4$ to a $+4$ state in CO_2, but *two carbons* in each molecule of the reactant, $H_2C_2O_4$, make this change. It is a $1e^-$ change *per carbon atom* in $H_2C_2O_4$. Each whole molecule of $H_2C_2O_4$ must, therefore, give up two electrons. Therefore,

$$1 \text{ mol } H_2C_2O_4 \Leftrightarrow 2 \text{ mol of } e^-$$

$$1 \text{ mol } H_2C_2O_4 = 2 \text{ eq } H_2C_2O_4$$

The reduction of sodium chromate to chromium(III) oxide provides another example that relates moles to equivalents. The change is

$$\underset{+6}{Na_2CrO_4} \longrightarrow \underset{+3}{Cr_2O_3}$$

Thus each Cr in Na_2CrO_4 changes from the $+6$ to the $+3$ state. This corresponds to a three-electron change for each Cr in Na_2CrO_4. We are concerned about equivalents of *reactants,* so only the number of Cr atoms in the *reactant* concerns us, and each one changes by three electrons. Therefore,

$$1 \text{ mol } Na_2CrO_4 \Leftrightarrow 3 \text{ mol of } e^-$$

$$1 \text{ mol } Na_2CrO_4 = 3 \text{ eq } Na_2CrO_4$$

Equivalent Weights of Oxidizing Agents and Reducing Agents

Calculating the equivalent weight of an oxidizing agent or a reducing agent is exactly like that for acids and bases. For $Na_2Cr_2O_7$, for example, we know the following.

$$1 \text{ mol } Na_2Cr_2O_7 = 6 \text{ eq } Na_2Cr_2O_7$$

$$1 \text{ mol } Na_2Cr_2O_7 = 261.97 \text{ g } Na_2Cr_2O_7$$

Therefore,

$$261.97 \text{ g } Na_2Cr_2O_7 = 6 \text{ eq } Na_2Cr_2O_7$$

Because *an equivalent weight is simply the ratio of grams per equivalent,* we set up the ratio and carry through the calculation.

$$\frac{261.97 \text{ g } Na_2Cr_2O_7}{6 \text{ eq } Na_2Cr_2O_7} = 43.66 \text{ g eq}^{-1} \text{ for } Na_2Cr_2O_7$$

Retaining four significant figures, the equivalent weight of $Na_2Cr_2O_7$ is 43.66 g; 43.66 g of $Na_2Cr_2O_7$ takes up 1 mol of electrons in any redox reaction in which Cr changes from the $+6$ to the $+3$ state.

You must always remember that the number of equivalents per mole varies not only from compound to compound but also according to the experimental conditions (acidic, basic or neutral solution) and the nature of the other reactant. The equivalent weight of a compound is not a constant under all conditions.

C.4 NORMALITY

In working with equivalents when the reactants are in solution, it is convenient to define a quantity similar to molarity, but in terms of equivalents instead of moles. This quantity is called the *normality* of a solution. The **normality** of a solution is a concentration unit that expresses the number of *equivalents* of solute per liter of solution.

$$\text{Normality} = \frac{\text{equivalents of solute}}{1 \text{ L of solution}}$$

Thus, a solution that contains one equivalent of solute per liter of solution has a concentration that we specify as "one normal," abbreviated 1 *N*. Like molarity, it provides a conversion factor that relates equivalents and volume. For example, a solution labeled 2.00 *N* H_2SO_4 contains 2.00 eq of H_2SO_4 per liter (1000 mL), and provides the two conversion factors

$$\frac{2.00 \text{ eq } H_2SO_4}{1 \text{ L soln}} \quad \text{and} \quad \frac{1 \text{ L soln}}{2.00 \text{ eq } H_2SO_4}$$

One of the kinds of problems that often comes up in the lab when a stock solution's concentration is given in the units of molarity is to convert this to the units of normality. There is a useful rule of thumb for this calculation.

The normality of a solution equals its molarity multiplied by the number of equivalents per mole.

$$\text{normality} = \text{molarity} \times \text{no. of eq per mol}$$

This holds true both for acid–base systems and redox systems.

Titrations Using Normality as a Concentration Unit One of the benefits of defining equivalents and normality is that it makes calculations for titrations very simple. Consider the units that we obtain if we multiply the normality of a solution by the volume used in a particular experiment.

$$\underbrace{\frac{\text{eq solute}}{\text{L soln}}}_{N} \times \underbrace{\text{L soln}}_{V} = \text{eq solute}$$

Thus, normality (*N*) times volume in liters (*V*) gives equivalents. In any reaction, the number of equivalents of one reactant exactly equals the number of equivalents of another; it's the way equivalents are defined. So we can write the following relationship that applies to any reaction between two substances, *A* and *B*, in which the concentrations are in normality.

$$N_A V_A = N_B V_B \qquad (C.1)[2]$$

Equation C.1 applies both to acid–base reactions and to redox reactions. This is the beauty of the concept of equivalents. *By definition*, we can rely on the 1:1 ratio of equivalents that makes Equation C.1 possible, regardless of the system.

---

[2] Don't try to use this equation when the concentrations are in units of molarity. It won't necessarily work because we cannot say that there is *always* a 1:1 ratio of the *moles* of one reactant to the *moles* of another.

Answers to Practice Exercises and Selected Review Exercises

CHAPTER 1

Practice Exercises

1. (a) mg   (b) μm   (c) ps
   (a) 1×10^{-9} (b) 1×10^{-2} (c) 1×10^{-12}
   (a) cg   (b) Mm   (c) ns
2. 50 °F = 10 °C, 68 °F = 293 K
3. (a) 108 in.   (b) 1.25×10^5 cm   (c) 0.0107 ft
   (d) 8.58 km/L
4. 2.70 g/mL
5. (a) 0.272 cm³   (b) 171 g
6. sp.gr. = 2.70, d = 168 lb/ft³
7. 0.902 g/mL, 7.52 lb/gal

Selected Review Exercises

1.16   (a) 0.01 10^{-2}
        (b) 0.001 10^{-3}
        (c) 1000 10^3
        (d) 0.000001 10^{-6}
        (e) 0.000000001 10^{-9}
        (f) 0.000000000001 10^{-12}
        (g) 1,000,000 10^6

1.18   (a) 0.01   (b) 1000   (c) 10^{12} (d) 0.1
      (e) 0.001   (f) 0.001

1.24   (a) 140 °F   (b) 68 °F   (c) 7.5 °C   (d) 15 °C
      (e) 313 K   (f) 253 K

1.26   99.43 °F

1.28 -269 °C, -452 °F

1.30   90 K

1.33   (a) four   (b) five   (c) four   (d) two   (e) four
      (f) one

1.35   (a) 0.69   (b) 83.24   (c) 0.006   (d) 22.84
      (e) 775.4

1.37   (a) 2.34×10^3 (b) 3.10×10^7 (c) 2.87×10^{-4}
      (d) 4.50×10^4 (e) 4.00×10^{-6} (f) 3.24×10^5

1.39   (a) 210,000   (b) 0.00000335   (c) 3800
      (d) 0.0000000000046   (e) 0.00000346
      (f) 850,000,000

1.41   (a) 2.0×10^4 (b) 8.0×10^7 (c) 1.0×10^3
      (d) 2.4×10^5 (e) 2.0×10^{18}

1.46   (a) 3.20×10^{-3} km   (b) 8.2×10^3 μg
      (c) 7.53×10^{-5} kg   (d) 0.1375 L   (e) 25 mL
      (f) 3.42×10^{-9} dm

1.48   (a) 91 cm   (b) 2.3 kg   (c) 2800 mL
      (d) 200 mL   (e) 88 km/hr   (f) 80.5 km

1.50   360 mL

1.52   2205 lb

1.54   190 cm

1.56   (a) 5.2 m²   (b) 3100 mm²   (c) 1.0×10^2 L

1.58   2452 km/hr

1.60 1.0×10^2 km/hr

1.69   The carbon dioxide and water vapor have less potential energy, because the reaction of oxygen with gasoline (combustion) releases potential energy when the products are formed. The excess potential energy of the gasoline and oxygen is transformed to heat.

1.75   44 kg, 97 lb

1.77   31.6 mL

1.79   11 g/mL

1.81   1.04

1.83 1.20×10^3 lb

1.85   Since the density closely matches the known value, we conclude that this is an authentic sample of ethylene glycol.

1.89   (a) 9.556 mL   (b) 1.477 g mL^{-1}

1.91   (a) $2.50 \Leftrightarrow 30$ min   (b) $8.75   (c) 88.2 min

1.94   (a) 1.040×10^3 g L^{-1} (b) 1.040×10^3 kg m³

1.96   4.74 miles/hr

1.98   (a) The average kinetic energy of the system must decrease because the system becomes "cool," i.e., the temperature decreases.
      (b) The total heat of this system decreases as the system cools. Therefore, the total kinetic energy must decrease.

(c) Since energy must be conserved in this system (an insulated container), the potential energy must increase as a result of the decrease in kinetic energy.

1.100 (a) The only temperature at which any absolute scale overlaps any other absolute scale is zero.
(b) 703 R

CHAPTER 2

Practice Exercises

1. (a) 1 Ni, 2 Cl   (b) 1 Fe, 1 S, 4 O
   (c) 3 Ca, 2 P, 8 O   (d) 1 Co, 2 N, 12 O, 12 H

2. 1 Mg, 2 O, 4 H, 2 Cl

3. $Mg(OH)_2(s) + 2HCl(aq) \rightarrow MgCl_2(aq) + 2H_2O(l)$

4. 26.9814 u

5. 5.2955 times as heavy as ^{12}C

6. 10.8 u

7. $^{240}_{94}Pu$

8. 17 protons, 17 electrons, 18 neutrons

9. (a) K, Ar, Al   (b) Cl   (c) Ba   (d) Ne   (e) Li
   (f) Ce

10. (a) NaF   (b) Na_2O (c) MgF_2 (d) Al_4C_3

11. (a) $CrCl_3$ and $CrCl_2$, Cr_2O_3 and CrO
    (b) CuCl, $CuCl_2$, Cu_2O, CuO

12. (a) Na_2CO_3 (b) $(NH_4)_2SO_4$ (c) $KC_2H_3O_2$
    (d) $Sr(NO_3)_2$ (e) $Fe(C_2H_3O_2)_3$

13. (a) potassium sulfide   (b) magnesium phosphide
    (c) nickel(II) chloride   (d) iron(III) oxide

14. (a) Al_2S_3 (b) SrF_2 (c) TiO_2 (d) $CrBr_2$

15. phosphorus trichloride, sulfur dioxide, dichlorine heptaoxide

16. hydrofluoric acid and hydrobromic acid, sodium fluoride and sodium bromide

17. sodium arsenate

18. $NaHSO_3$, sodium hydrogen sulfite

Selected Review Exercises

2.8 (a) 2 K, 2 C, 4 O   (b) 2 H, 1 S, 3 O
     (c) 12 C, 26 H   (d) 4 H, 2 C, 2 O
     (e) 9 H, 2 N, 1 P, 4 O   (f) 4 C, 10 H, 1 O

2.10 (a) 1 Ni, 2 Cl, 8 O   (b) 1 Cu, 1 C, 3 O
      (c) 2 K, 2 Cr, 7 O   (d) 2 C, 4 H, 2 O
      (e) 2 N, 9 H, 1 P, 4 O   (f) 3 C, 8 H, 3 O

2.15 (a) 6 N, 3 O   (b) 14 C, 28 H, 14 O
      (c) 4 Na, 4 H, 4 C, 12 O   (d) 4 N, 8 H, 2 C, 2 O
      (e) 2 Cu, 2 S, 18 O, 20 H   (f) 10 K, 10 Cr, 35 O

2.16 (a) 6   (b) 3   (c) 27

2.24 (c)

2.26 (a) 2/1   (b) 1.19 g Cl

2.28 $1.998079824 \times 10^{-23}$ g

2.30 2.0158 u

2.32 24.31 u

2.41 (a) $^{131}_{53}I$ (b) $^{90}_{38}Sr$ (c) $^{137}_{55}Ce$ (d) $^{18}_{9}F$

2.42

	Neutrons	Protons	Electrons
(a)	138	88	88
(b)	8	6	6
(c)	124	82	82
(d)	12	11	11

2.44

	Neutrons	Protons	Electrons
(a)	46	35	36
(b)	32	26	23
(c)	34	29	27
(d)	50	37	36

2.51 Strontium and calcium are in the same group of the periodic table, so they are expected to have similar chemical properties. Strontium should therefore form compounds that are similar to those of calcium, including the sorts of compounds found in bone.

2.53 Cadmium is in the same periodic table group as zinc, but silver is not. Therefore, cadmium would be expected to have properties similar to those of zinc, whereas silver would not.

2.69 copper and gold

2.78 (a) Cl_2 (b) S_8 (c) P_4 (d) N_2 (e) O_2 (f) H_2

2.80 HAt

2.82 Bi_2O_3 and Bi_2O_5

2.84 $C_{10}H_{22}$

2.94 Ca^{2+} and Cl^-

2.97 This requires a gain of three electrons. There are thus 7 protons and 10 electrons.

2.99 $RbCl_3$ does not represent an electrically neutral formula unit. SNa_2 has the anion first and the cation second.

2.106 (a) NaBr   (b) KI   (c) BaO   (d) $MgBr_2$
      (e) BaF_2

2.111 (a) KNO_3 (b) $Ca(C_2H_3O_2)_2$ (c) NH_4Cl
      (d) $Fe_2(CO_3)_3$ (e) $Mg_3(PO_4)_2$

2.113 (a) PbO and PbO_2 (b) SnO and SnO_2
      (c) MnO and Mn_2O_3 (d) FeO and Fe_2O_3
      (e) Cu_2O and CuO

2.115 (a) $Ca(s) + Cl_2(g) \rightarrow CaCl_2(s)$
      (b) $2Mg(s) + O_2(g) \rightarrow 2MgO(s)$
      (c) $4Al(s) + 3O_2(g) \rightarrow 2Al_2O_3(s)$
      (d) $S(s) + 2Na(s) \rightarrow Na_2S(s)$ or
      $S_8(s) + 16Na(s) \rightarrow 8Na_2S(s)$

2.122 (a) calcium sulfide   (b) sodium fluoride
      (c) aluminum bromide   (d) magnesium carbide

(e) sodium phosphide (f) lithium nitride
(g) barium arsenide (h) aluminum oxide

2.124 (a) silicon dioxide (b) chlorine trifluoride
(c) xenon tetrafluoride (d) disulfur dichloride
(e) tetraphosphorus decaoxide (f) dinitrogen
pentaoxide (g) dichlorine heptaoxide
(h) arsenic pentachloride

2.126 (a) periodic acid (b) iodic acid (c) iodous
acid (d) hypoiodous acid (e) hydroiodic acid

2.128 (a) sodium nitrite (b) potassium phosphate
(c) potassium permanganate (d) ammonium
acetate (e) barium sulfate (f) ferric carbonate;
iron(III) carbonate (g) potassium thiocyanate
(h) sodium thiosulfate

2.130 (a) chromous chloride; chromium(II) chloride
(b) ammonium acetate (c) potassium iodate
(d) chlorous acid (e) calcium sulfite (f) silver
cyanide (g) zinc(II) bromide (h) hydrogen selenide (i) hydroselenic acid (j) vanadium(III)
nitrate (k) cobaltous acetate; cobalt(II) acetate
(l) auric sulfide; gold(III) sulfide (m) aurous
sulfide; gold(I) sulfide (n) germanium tetrabro-
mide (o) potassium chromate (p) ferrous hydroxide; iron(II) hydroxide (q) diiodine tetra-
oxide (r) tetraiodine nonaoxide (s) tetraphosphorus triselenide

2.131 (a) Na_2HPO_4 (b) Li_2Se (c) NaH
(d) $Cr(C_2H_3O_2)_3$ (e) $Ni(CN)_2$ (f) Fe_2O_3
(g) SnS_2 (h) SbF_5 (i) Al_2Cl_6 (j) As_4O_{10}
(k) $Mg(OH)_2$ (l) $Cu(HSO_4)_2$ (m) NH_4SCN
(n) $K_2S_2O_3$

2.133 (a) $Au(NO_3)_3$ (b) $CuSO_4$ (c) NH_4BrO_3
(d) $Ni(IO_3)_2$ (e) PbO_2 (f) Sb_2S_3 (g) $PtCl_2$
(h) CdS (i) $Ba(C_2H_3O_2)_2$ (j) Hg_2Cl_2
(k) $HgCl_2$ (l) $Sr_3(PO_4)_2$ (m) Ba_3As_2
(n) $Co(OH)_2$ (o) $Al_2(SO_3)_3$ (p) $(NH_4)_2Cr_2O_7$
(q) I_2O_5 (r) P_4S_7 (s) S_2F_{10}

2.135 (a) sodium bicarbonate (b) potassium dihydrogen phosphate (c) ammonium hydrogen phosphate

2.137 (a) 28.1 (b) 35.7

2.139 (a) $3H_2SO_4 + 2Al(OH)_3 \rightarrow Al_2(SO_4)_3 + 6H_2O$
(b) $H_2SO_4 + Ca(OH)_2 \rightarrow CaSO_4 + 2H_2O$

2.141 (a) hypochlorous acid and sodium hypochlorite;
NaOCl
(b) iodous acid and sodium iodite; $NaIO_2$
(c) bromic acid and sodium bromate; $NaBrO_3$
(d) perchloric acid and sodium perchlorate;
$NaClO_4$

CHAPTER 3

Practice Exercises

1. 3.44 mol N
2. 1.11 mol S

3. 28.4 g Ag
4. 2.906×10^{12} atoms Pb
5. 106.01 g/mol
6. 13.2 g
7. 0.467 mol H_2SO_4
8. 59.5 g Fe
9. 25.94% N, 74.06% O; no other elements are present.
10. 36.86% N, 63.14% O
11. NO
12. N_2O_5
13. Na_2SO_4
14. CH_2O
15. N_2H_4
16. $3CaCl_2(aq) + 2K_3PO_4(aq) \rightarrow$
$$Ca_3(PO_4)_2(s) + 6KCl(aq)$$
17. $2Fe(s) + 3Cl_2(g) \rightarrow 2FeCl_3(s)$
18. 0.183 mol H_2SO_4
19. 0.958 mol O_2
20. 78.5 g Al_2O_3
21. 30.01 g NO
22. percent yield = 86.0%

Selected Review Exercises

3.4 9.03×10^{23} atoms, 18 g
3.5 (a) $1 S \Leftrightarrow 2 O$ (b) $2 As \Leftrightarrow 3 O$ (c) $1 Pb \Leftrightarrow 6 N$
(d) $2 Al \Leftrightarrow 6 Cl$
3.8 0.226 mol V
3.12 2.906×10^9 atoms Pb
3.14 (a) 84.01 g/mole (b) 294.20 g/mole
(c) 96.11 g/mole (d) 342.17 g/mole
(e) 249.72 g/mole
3.17 (a) 50.1 g Ca (b) 34.9 g Fe (c) 34.9 g C_4H_{10}
(d) 139 g $(NH_4)_2CO_3$ (e) 0.340 g $KMnO_4$
3.19 (a) 0.215 mol $CaCO_3$ (b) 9.16×10^{-2} mol NH_3
(c) 7.94×10^{-2} mol $Sr(NO_3)_2$
(d) 4.31×10^{-8} mol Na_2CrO_4
3.21 (a) 2.40 g (b) 7.21 g (c) 3.09 g (d) 3.96 g
(e) 20.9 g
3.24 0.0750 mol Ca, 3.01 g Ca
3.26 1.30 mol N, 62.5 g $(NH_4)_2CO_3$
3.28 3.0 to 1.0 NaOH to tallow
3.30 9.1×10^{11} mol NH_3
3.32 (a) 23.95 g O (b) 35.93 g O
3.34 25.23 g O, 0.7947 g H
3.36 (a) 72.71% O (b) 13.65 g C
3.38 NH_3 is 82.4% N, $CO(NH_2)_2$ is 46.6% N
3.40 yes
3.44 (a) SCl (b) CH_2O (c) C_2H_5 (d) AsO_3 (e) HO
3.46 P_2O_3

3.48 Sb_2S_5, Sb_2S_5

3.50 $HgC_2H_3O_2$, $Hg_2C_4H_6O_4$

3.52 (a) NaO
  (b) 1.437 g Na
  (c) Any integer multiple or any whole number fraction of the value 1.437 g will be a possible amount of Na to combine with 1.000 g of O to make a new compound of sodium and oxygen; 2(1.437 g Na) = 2.874 g Na, 3(1.437 g Na) = 4.311 g Na, (1/2)(1.437 g Na) = 0.7185 g Na, (1/3)(1.437 g Na) = 0.4790 g Na, etc.
  (d) 2.875 g Na
  (e) 1:2, Law of Multiple Proportions
  (f) Na_2O

3.54 $CrCl_3$

3.56 $C_{19}H_{30}O_2$

3.59 14 moles of Fe

3.64 (a) $Ca(OH)_2 + 2HCl \rightarrow CaCl_2 + 2H_2O$
  (b) $2AgNO_3 + CaCl_2 \rightarrow Ca(NO_3)_2 + 2AgCl$
  (c) $2Fe_2O_3 + 3C \rightarrow 4Fe + 3CO_2$
  (d) $2NaHCO_3 + H_2SO_4 \rightarrow$
      $Na_2SO_4 + 2H_2O + 2CO_2$
  (e) $2C_4H_{10} + 13O_2 \rightarrow 8CO_2 + 10H_2O$

3.69 (a) 0.030 mol $Na_2S_2O_3$ (b) 0.24 mol HCl
  (c) 0.15 mol H_2O

3.71 (a) $4P + 5O_2 \rightarrow P_4O_{10}$ (b) 0.276 mol O_2
  (c) 0.0500 mol P_4O_{10} (d) 0.0912 mol P

3.73 (a) $Fe_2O_3 + 3H_2 \rightarrow 2Fe + 3H_2O$ (b) 44 mol Fe
  (c) 36 mol H_2 (d) 5.1×10^3 g Fe_2O_3

3.75 0.4805 mol HNO_3

3.77 (a) Zn is the limiting reactant. (b) 44.7 g ZnS
  (c) 21.3 g S

3.79 (a) The limiting reactant is PCl_5. (b) 0.360 mol H_3PO_4, 1.80 mol HCl

3.82 (a) 27.58 g H_2SO_4 (b) 88.43%

3.84 9.2 g toluene

3.86 12.9 g $(NH_2)_2CO$

3.88 (a) $Mg(OH)_2 + 2HBr \rightarrow MgBr_2 + 2H_2O$
  (b) $2HCl + Ca(OH)_2 \rightarrow CaCl_2 + 2H_2O$
  (c) $Al_2O_3 + 3H_2SO_4 \rightarrow Al_2(SO_4)_3 + 3H_2O$
  (d) $2KHCO_3 + H_3PO_4 \rightarrow$
      $K_2HPO_4 + 2H_2O + 2CO_2$
  (e) $C_9H_{20} + 14O_2 \rightarrow 9CO_2 + 10H_2O$

3.90 The empirical formula is CHNO, and the molecular formula is $C_3H_3N_3O_3$.

CHAPTER 4

Practice Exercises

----

  1. (a) $MgCl_2(s) \rightarrow Mg^{2+}(aq) + 2Cl^-(aq)$
  (b) $Al(NO_3)_3(s) \rightarrow Al^{3+}(aq) + 3NO_3^-(aq)$
  (c) $Na_2CO_3(s) \rightarrow 2Na^+(aq) + CO_3^{2-}(aq)$
  (d) $(NH_4)_2SO_4(s) \rightarrow 2NH_4^+(aq) + SO_4^{2-}(aq)$

  2. (a) $HCHO_2 + H_2O \rightarrow H_3O^+ + CHO_2^-$
  (b) $H_3PO_4 + H_2O \rightarrow H_3O^+ + H_2PO_4^-$
      $H_2PO_4^- + H_2O \rightarrow H_3O^+ + HPO_4^{2-}$
      $HPO_4^{2-} + H_2O \rightarrow H_3O^+ + PO_4^{3-}$

  3. $HNO_2(aq) + H_2O \rightleftharpoons H_3O^+(aq) + NO_2^-(aq)$

  4. (a) molecular: $HCl(aq) + KOH(aq) \rightarrow$
      $H_2O + KCl(aq)$
    ionic: $H^+(aq) + Cl^-(aq) + K^+(aq) +$
      $OH^-(aq) \rightarrow H_2O + K^+(aq) + Cl^-(aq)$
    net ionic: $H^+(aq) + OH^-(aq) \rightarrow H_2O$
  (b) molecular: $HCHO_2(aq) + LiOH(aq) \rightarrow$
      $H_2O + LiCHO_2(aq)$
    ionic: $HCHO_2(aq) + Li^+(aq) + OH^-(aq) \rightarrow$
      $H_2O + Li^+(aq) + CHO_2^-(aq)$
    net ionic: $HCHO_2(aq) + OH^-(aq) \rightarrow$
      $H_2O + CHO_2^-(aq)$
  (c) molecular: $N_2H_4(aq) + HCl(aq) \rightarrow N_2H_5Cl(aq)$
    ionic: $N_2H_4(aq) + H^+(aq) + Cl^-(aq) \rightarrow$
      $N_2H_5^+(aq) + Cl^-(aq)$
    net ionic: $N_2H_4(aq) + H^+(aq) \rightarrow N_2H_5^+(aq)$

  5. molecular: $Al(OH)_3(s) + 3HCl(aq) \rightarrow$
      $AlCl_3(aq) + 3H_2O$
    ionic: $Al(OH)_3(s) + 3H^+(aq) + 3Cl^-(aq) \rightarrow$
      $Al^{3+}(aq) + 3Cl^-(aq) + 3H_2O$
    net ionic: $Al(OH)_3(s) + 3H^+(aq) \rightarrow$
      $Al^{3+}(aq) + 3H_2O$

  6. molecular: $MgCl_2(aq) + 2NaOH(aq) \rightarrow$
      $Mg(OH)_2(s) + 2NaCl(aq)$
    ionic: $Mg^{2+}(aq) + 2Cl^-(aq) + 2Na^+(aq) +$
      $2OH^-(aq) \rightarrow Mg(OH)_2(s) + 2Cl^-(aq) + 2Na^+(aq)$
    net ionic: $Mg^{2+}(aq) + 2OH^-(aq) \rightarrow Mg(OH)_2(s)$

  7. (a) $Ag^+(aq) + Cl^-(aq) \rightarrow AgCl(s)$
  (b) $Pb^{2+}(aq) + S^{2-}(aq) \rightarrow PbS(s)$
  (c) no reaction

  8. Ca is oxidized and is the reducing agent; Cl_2 is reduced and is the oxidizing agent.

  9. (a) Ni +2, Cl −1 (b) Mg +2, Ti +4, O −2
  (c) K +1, Cr +6, O −2 (d) H +1, P +5, O −2
  (e) V +3, C 0, H +1, O −2

 10. +8/3

 11. $SnCl_3^- + 2HgCl_2 + Cl^- \rightarrow SnCl_6^{2-} + Hg_2Cl_2$

 12. $3Cu + 2NO_3^- + 8H^+ \rightarrow 3Cu^{2+} + 2NO + 4H_2O$

 13. $2MnO_4^- + 3C_2O_4^{2-} + 4OH^- \rightarrow$
      $2MnO_2 + 6CO_3^{2-} + 2H_2O$

 14. 0.1999 M

 15. 333 mL

 16. 3.47 g $CaCl_2$

 17. Dilute 25.0 mL of 0.500 M H_2SO_4 to a total volume of 100 mL.

 18. 26.8 mL NaOH soln

 19. 0.40 M Fe^{3+}, 1.2 M Cl^-

 20. 0.750 M PO_4^{3-}

 21. 60.0 mL KOH

22. 1.20×10^{-2} mol $BaSO_4$, 0.0 M Ba^{2+}, 0.480 M Cl^-, 0.300 M Mg^{2+}, 6.0×10^{-2} M SO_4^{2-}

23. (a) 5.41×10^{-3} mol Ca^{2+} (b) 5.41×10^{-3} mol Ca^{2+} (c) 5.41×10^{-3} mol $CaCl_2$ (d) 0.600 g $CaCl_2$ (e) 30.0% $CaCl_2$

24. 0.178 M H_2SO_4

25. 0.0220 M HCl

26. (a) $5Sn^{2+} + 2MnO_4^- + 16H^+ \rightarrow 5Sn^{4+} + 2Mn^{2+} + 8H_2O$ (b) 0.1199 g Sn (c) 39.97% Sn (d) 50.73% SnO_2

Selected Review Exercises

4.6 (a) $LiCl(s) \rightarrow Li^+(aq) + Cl^-(aq)$
(b) $BaCl_2(s) \rightarrow Ba^{2+}(aq) + 2Cl^-(aq)$
(c) $Al(C_2H_3O_2)_3(s) \rightarrow Al^{3+}(aq) + 3C_2H_3O_2^-(aq)$
(d) $(NH_4)_2CO_3(s) \rightarrow 2NH_4^+(aq) + CO_3^{2-}(aq)$
(e) $FeCl_3(s) \rightarrow Fe^{3+}(aq) + 3Cl^-(aq)$

4.11 $HClO_4(l) + H_2O(l) \rightarrow H_3O^+(aq) + ClO_4^-(aq)$

4.13 (a), (c), and (d)

4.16 $HNO_2(aq) + H_2O(l) \rightleftharpoons H_3O^+(aq) + NO_2^-(aq)$

4.19 $HClO_3(aq) + H_2O(l) \rightarrow H_3O^+(aq) + ClO_3^-(aq)$

4.21 $H_3PO_4(aq) + H_2O(l) \rightleftharpoons H_3O^+(aq) + H_2PO_4^-(aq)$
$H_2PO_4^-(aq) + H_2O(l) \rightleftharpoons$
$\qquad\qquad\qquad H_3O^+(aq) + HPO_4^{2-}(aq)$
$HPO_4^{2-}(aq) + H_2O(l) \rightleftharpoons H_3O^+(aq) + PO_4^{3-}(aq)$

4.25 (a) molecular: $Ca(OH)_2(aq) + 2HNO_3(aq) \rightarrow$
$\qquad\qquad\qquad Ca(NO_3)_2(aq) + 2H_2O$
ionic: $Ca^{2+}(aq) + 2OH^-(aq) + 2H^+(aq) +$
$\quad 2NO_3^-(aq) \rightarrow Ca^{2+}(aq) + 2NO_3^-(aq) +$
$\qquad\qquad\qquad\qquad\qquad 2H_2O$
net: $H^+(aq) + OH^-(aq) \rightarrow H_2O$
(b) molecular: $Al_2O_3(s) + 6HCl(aq) \rightarrow$
$\qquad\qquad\qquad 2AlCl_3(aq) + 3H_2O$
ionic: $Al_2O_3(s) + 6H^+(aq) + 6Cl^-(aq) \rightarrow$
$\qquad\qquad\qquad 2Al^{3+}(aq) + 6Cl^-(aq) + 3H_2O$
net: $Al_2O_3(s) + 6H^+(aq) \rightarrow$
$\qquad\qquad\qquad 2Al^{3+}(aq) + 3H_2O$
(c) molecular: $Zn(OH)_2(s) + H_2SO_4(aq) \rightarrow$
$\qquad\qquad\qquad ZnSO_4(aq) + 2H_2O$
ionic: $Zn(OH)_2(s) + 2H^+(aq) + SO_4^{2-}(aq) \rightarrow$
$\qquad\qquad Zn^{2+}(aq) + SO_4^{2-}(aq) + 2H_2O$
net: $Zn(OH)_2(s) + 2H^+(aq) \rightarrow$
$\qquad\qquad\qquad Zn^{2+}(aq) + 2H_2O$

4.28 (a) ionic: $2NH_4^+(aq) + CO_3^{2-}(aq) +$
$\quad Mg^{2+}(aq) + 2Cl^-(aq) \rightarrow$
$\qquad 2NH_4^+(aq) + 2Cl^-(aq) + MgCO_3(s)$
net: $Mg^{2+}(aq) + CO_3^{2-}(aq) \rightarrow MgCO_3(s)$
(b) ionic: $Cu^{2+}(aq) + 2Cl^-(aq) + 2Na^+(aq) +$
$\quad 2OH^-(aq) \rightarrow Cu(OH)_2(s) + 2Na^+(aq) +$
$\qquad\qquad\qquad\qquad\qquad 2Cl^-(aq)$
net: $Cu^{2+}(aq) + 2OH^-(aq) \rightarrow Cu(OH)_2(s)$
(c) ionic: $3Fe^{2+}(aq) + 3SO_4^{2-}(aq) + 6Na^+(aq) +$
$\quad 2PO_4^{3-}(aq) \rightarrow Fe_3(PO_4)_2(s) + 6Na^+(aq) +$
$\qquad\qquad\qquad\qquad\qquad 3SO_4^{2-}(aq)$
net: $3Fe^{2+}(aq) + 2PO_4^{3-}(aq) \rightarrow Fe_3(PO_4)_2(s)$

(d) ionic: $2Ag^+(aq) + 2C_2H_3O_2^-(aq) +$
$\quad Ni^{2+}(aq) + 2Cl^-(aq) \rightarrow 2AgCl(s) +$
$\qquad\qquad\qquad Ni^{2+}(aq) + 2C_2H_3O_2^-(aq)$
net: $2Ag^+(aq) + 2Cl^-(aq) \rightarrow 2AgCl(s)$

4.30 $Cu^{2+}(aq) + S^{2-}(aq) \rightarrow CuS(s)$

4.32 molecular: $AgNO_3(aq) + NaBr(aq) \rightarrow$
$\qquad\qquad\qquad AgBr(s) + NaNO_3(aq)$
ionic: $Ag^+(aq) + NO_3^-(aq) + Na^+(aq) +$
$\quad Br^-(aq) \rightarrow AgBr(s) + Na^+(aq) + NO_3^-(aq)$
net: $Ag^+(aq) + Br^-(aq) \rightarrow AgBr(s)$

4.34 (a) $CO_2(g)$ (b) $H_2S(g)$ (c) $SO_2(g)$

4.36 (a) $2H^+(aq) + CO_3^{2-}(aq) \rightarrow H_2O(l) + CO_2(g)$
(b) $NH_4^+(aq) + OH^-(aq) \rightarrow NH_3(g) + H_2O(l)$
(c) $H^+(aq) + HSO_3^-(aq) \rightarrow H_2O(l) + SO_2(g)$

4.37 The soluble ones are (a), (b), and (d).

4.38 The insoluble ones are (a), (d), and (f).

4.41 (a) $Na_2SO_3(aq) + Ba(NO_3)_2(aq) \rightarrow$
$\qquad\qquad\qquad BaSO_3(s) + 2NaNO_3(aq)$
ionic: $2Na^+(aq) + SO_3^{2-}(aq) + Ba^{2+}(aq) +$
$\quad 2NO_3^-(aq) \rightarrow BaSO_3(s) + 2Na^+(aq) +$
$\qquad\qquad\qquad\qquad\qquad 2NO_3^-(aq)$
net: $Ba^{2+}(aq) + SO_3^{2-}(aq) \rightarrow BaSO_3(s)$
(b) $K_2S(aq) + ZnCl_2(aq) \rightarrow ZnS(s) + 2KCl(aq)$
ionic: $2K^+(aq) + S^{2-}(aq) + Zn^{2+}(aq) +$
$\quad 2Cl^-(aq) \rightarrow ZnS(s) + 2K^+(aq) + 2Cl^-(aq)$
net: $Zn^{2+}(aq) + S^{2-}(aq) \rightarrow ZnS(s)$
(c) $2NH_4Br(aq) + Pb(C_2H_3O_2)_2(aq) \rightarrow$
$\qquad 2NH_4C_2H_3O_2(aq) + PbBr_2(s)$
ionic: $2NH_4^+(aq) + 2Br^-(aq) + Pb^{2+}(aq) +$
$\quad 2C_2H_3O_2^-(aq) \rightarrow$
$\qquad 2NH_4^+(aq) + 2C_2H_3O_2^-(aq) + PbBr_2(s)$
net: $Pb^{2+}(aq) + 2Br^-(aq) \rightarrow PbBr_2(s)$
(d) $2NH_4ClO_4(aq) + Cu(NO_3)_2(aq) \rightarrow$
$\qquad Cu(ClO_4)_2(aq) + 2NH_4NO_3(aq)$
ionic: $2NH_4^+(aq) + 2ClO_4^-(aq) +$
$\quad Cu^{2+}(aq) + 2NO_3^-(aq) \rightarrow Cu^{2+}(aq) +$
$\qquad 2ClO_4^-(aq) + 2NO_3^-(aq) + 2NH_4^+(aq)$
net: No reaction

4.43 (a) $3HNO_3(aq) + Cr(OH)_3(s) \rightarrow$
$\qquad\qquad\qquad Cr(NO_3)_3(aq) + 3H_2O(l)$
ionic: $3H^+(aq) + 3NO_3^-(aq) +$
$\quad Cr(OH)_3(s) \rightarrow Cr^{3+}(aq) + 3NO_3^-(aq) +$
$\qquad\qquad\qquad\qquad\qquad 3H_2O(l)$
net: $3H^+(aq) + Cr(OH)_3(s) \rightarrow$
$\qquad\qquad\qquad Cr^{3+}(aq) + 3H_2O(l)$
(b) $HClO_4(aq) + NaOH(aq) \rightarrow$
$\qquad\qquad\qquad NaClO_4(aq) + H_2O(l)$
ionic: $H^+(aq) + ClO_4^-(aq) + Na^+(aq) +$
$\quad OH^-(aq) \rightarrow Na^+(aq) + ClO_4^-(aq) + H_2O(l)$
net: $H^+(aq) + OH^-(aq) \rightarrow H_2O(l)$
(c) $Cu(OH)_2(s) + 2HC_2H_3O_2(aq) \rightarrow$
$\qquad\qquad\qquad Cu(C_2H_3O_2)_2(aq) + 2H_2O(l)$
ionic: $Cu(OH)_2(s) + 2HC_2H_3O_2(aq) \rightarrow$
$\qquad\qquad Cu^{2+}(aq) + 2C_2H_3O_2^-(aq) + 2H_2O(l)$

net: same as ionic.

(d) $ZnO(s) + 2HBr(aq) \rightarrow ZnBr_2(aq) + H_2O(l)$
ionic: $ZnO(s) + 2H^+(aq) + 2Br^-(aq) \rightarrow$
$$Zn^{2+}(aq) + 2Br^-(aq) + H_2O(l)$$
net: $ZnO(s) + 2H^+(aq) \rightarrow$
$$Zn^{2+}(aq) + H_2O(l)$$

4.45 The electrical conductivity would decrease regularly, until one solution had neutralized the other, forming a nonelectrolyte:

$$Ba^{2+}(aq) + 2OH^-(aq) + 2H^+(aq) +$$
$$SO_4^{2-}(aq) \longrightarrow BaSO_4(s) + 2H_2O(l)$$

Once the point of neutralization had been reached, the addition of excess sulfuric acid would cause the conductivity to increase, because sulfuric acid is a strong electrolyte itself.

4.49 (b) and (d)

4.50 These reactions have the following "driving forces": (a) formation of insoluble $Cr(OH)_3$; (b) formation of water, a weak electrolyte; (c) formation of a gas, CO_2; (d) formation of a weak electrolyte, $H_2C_2O_4$.

4.52 The magnesium is oxidized and the oxygen is reduced. Consequently, Mg is the reducing agent and the O_2 is the oxidizing agent.

4.55 (a) -2 (b) $+4$ (c) 0 (d) -3

4.57 The sum of the oxidation numbers should be zero: (a) $O-2$, $Na+1$, $H+1$, $P+5$; (b) $Ba+2$, $O-2$, $Mn+6$; (c) $Na+1$, $O-2$, $S+2.5$; (d) $F-1$ (Note: Except in F_2, fluorine is always in a -1 oxidation state), $Cl+3$

4.59 (a) $+2$ (b) $+4$ (c) $+3$ (d) $+5$ (e) -2
(f) 0 (g) -1 (h) -3 (i) $-\frac{1}{3}$

4.62 The sum of the oxidation numbers should be zero: (a) $O-2$, $Na+1$, $Cl+1$; (b) $O-2$, $Na+1$, $Cl+3$; (c) $O-2$, $Na+1$, $Cl+5$; (d) $O-2$, $Na+1$, $Cl+7$

4.64 The sum of the oxidation numbers should be zero: (a) $S-2$, $Pb+2$; (b) $Cl-1$, $Ti+4$; (c) $O-2$, $Sr+2$, $I+5$; (d) $S-2$, $Cr+3$

4.66 (a) -2 (b) 0 (c) $+4$ (d) $+4$

4.68 (a) substance reduced (and oxidizing agent): HNO_3
substance oxidized (and reducing agent): H_3AsO_3
(b) substance reduced (and oxidizing agent): $HOCl$
substance oxidized (and reducing agent): NaI
(c) substance reduced (and oxidizing agent): $KMnO_4$
substance oxidized (and reducing agent): $H_2C_2O_4$
(d) substance reduced (and oxidizing agent): H_2SO_4
substance oxidized (and reducing agent): Al

4.71 (a) $BiO_3^- + 6H^+ + 2e^- \rightarrow Bi^{3+} + 3H_2O$ This is reduction of BiO_3^-.
(b) $Pb^{2+} + 2H_2O \rightarrow PbO_2 + 4H^+ + 2e^-$ This is oxidation of Pb^{2+}.
(c) $NO_3^- + 10H^+ + 8e^- \rightarrow NH_4^+ + 3H_2O$ This constitutes reduction of NO_3^-.
(d) $6H_2O + Cl_2 \rightarrow 2ClO_3^- + 12H^+ + 10e^-$ This constitutes oxidation of Cl_2.

4.73 (a) $OCl^- + 2S_2O_3^{2-} + 2H^+ \rightarrow$
$$S_4O_6^{2-} + Cl^- + H_2O$$
(b) $2NO_3^- + Cu + 4H^+ \rightarrow$
$$2NO_2 + Cu^{2+} + 2H_2O$$
(c) $3AsO_3^{3-} + IO_3^- \rightarrow I^- + 3AsO_4^{3-}$
(d) $Zn + SO_4^{2-} + 4H^+ \rightarrow Zn^{2+} + SO_2 + 2H_2O$
(e) $NO_3^- + 4Zn + 10H^+ \rightarrow$
$$4Zn^{2+} + NH_4^+ + 3H_2O$$
(f) $2Cr^{3+} + 3BiO_3^- + 4H^+ \rightarrow$
$$Cr_2O_7^{2-} + 3Bi^{3+} + 2H_2O$$
(g) $I_2 + 5OCl^- + H_2O \rightarrow 2IO_3^- + 5Cl^- + 2H^+$
(h) $2Mn^{2+} + 5BiO_3^- + 14H^+ \rightarrow$
$$2MnO_4^- + 5Bi^{3+} + 7H_2O$$
(i) $3H_3AsO_3 + Cr_2O_7^{2-} + 8H^+ \rightarrow$
$$3H_3AsO_4 + 2Cr^{3+} + 4H_2O$$
(j) $2I^- + HSO_4^- + 3H^+ \rightarrow I_2 + SO_2 + 2H_2O$

4.75 (a) $2CrO_4^{2-} + 3S^{2-} + 4H_2O \rightarrow$
$$2CrO_2^- + 3S + 8OH^-$$
(b) $3C_2O_4^{2-} + 2MnO_4^- + 4H_2O \rightarrow$
$$6CO_2 + 2MnO_2 + 8OH^-$$
(c) $4ClO_3^- + 3N_2H_4 \rightarrow 4Cl^- + 6NO + 6H_2O$
(d) $NiO_2 + 2Mn(OH)_2 \rightarrow$
$$Ni(OH)_2 + Mn_2O_3 + H_2O$$
(e) $3SO_3^{2-} + 2MnO_4^- + H_2O \rightarrow$
$$3SO_4^{2-} + 2MnO_2 + 2OH^-$$
(f) $3S_2O_8^{2-} + 2CrO_2^- + 8OH^- \rightarrow$
$$2CrO_4^{2-} + 6SO_4^{2-} + 4H_2O$$
(g) $3SO_3^{2-} + 2CrO_4^{2-} + H_2O \rightarrow$
$$3SO_4^{2-} + 2CrO_2^- + 2OH^-$$
(h) $2O_2 + N_2H_4 \rightarrow 2H_2O_2 + N_2$
(i) $4Fe(OH)_2 + O_2 + 2H_2O \rightarrow 4Fe(OH)_3$
(j) $4Au + 16CN^- + 3O_2 + 6H_2O \rightarrow$
$$4Au(CN)_4^- + 12OH^-$$

4.77 $4OCl^- + S_2O_3^{2-} + H_2O \rightarrow 4Cl^- + 2SO_4^{2-} + 2H^+$

4.81 (a) $1.00\ M$ NaOH (b) $0.577\ M$ $CaCl_2$
(c) $3.33\ M$ KOH (d) $0.2\ M$ $H_2C_2O_4$

4.83 (a) 1.46 g NaCl (b) 16 g $C_6H_{12}O_6$
(c) 6.1 g H_2SO_4

4.85 (a) 0.0438 mol K^+, 0.0438 mol OH^-
(b) 0.015 mol Ca^{2+}, 0.030 mol Cl^-
(c) 0.15 mol NH_4^+, 0.074 mol CO_4^{2-}
(d) 0.022 mol Al^{3+}, 0.033 mol SO_4^{2-}

4.86 (a) $0.25\ M$ Cr^{2+}, $0.50\ M$ NO_3^- (b) $0.10\ M$ Cu^{2+}, $0.10\ M$ SO_4^{2-} (c) $0.48\ M$ Na^+, $0.16\ M$ PO_4^{3-}
(d) $0.15\ M$ Al^{3+}, $0.23\ M$ SO_4^{2-}

4.88 $0.070\ M$ Na_3PO_4

4.90 $0.11\ M$ H_2SO_4

4.92 3.00×10^2 mL

4.94 5.0×10^1 mL

4.96 0.1035 M H_2SO_4

4.98 1.54 g $MgSO_4 \cdot 7H_2O$

4.100 0.71 g $NaHCO_3$

4.102 1.00×10^2 mL of $(NH_4)_2SO_4$ solution

4.104 (a) 8.00×10^{-3} mol AgCl (b) 0.220 M Na^+, 0.060 M Cl^-, 0.160 M NO_3^-, the Ag^+ is 0 M

4.106 17.4 mL $KMnO_4$

4.108 50.6 g PbO_2

4.110 (a) $2Mn^{2+}(aq) + 5BiO_3^-(aq) + 14H^+(aq) \rightarrow$ $2MnO_4^-(aq) + 5Bi^{3+}(aq) + 7H_2O(l)$ (b) 72.3 g $NaBiO_3$

4.113 0.114 M HCl

4.115 2.67×10^{-3} mol $HC_3H_5O_3$

4.117 (a) 0.168 g Cl (b) $FeCl_2$

4.119 63.6% aspirin

4.121 (a) $2CrO_4^{2-} + 3SO_3^{2-} + H_2O \rightarrow 2CrO_2^- +$ $3SO_4^{2-} + 2OH^-$ (b) 1.68×10^{-2} mol CrO_4^{2-} (c) 25.4% Cr

4.123 2.096% $NaNO_2$

4.125 (a) 4.388×10^{-3} mol I_3^- (b) 8.776×10^{-3} mol NO_2^- (c) 54.85% $NaNO_2$

4.127 (a) 9.463% Cu (b) 18.40% $CuCO_3$

4.128 (a) strong electrolyte (b) nonelectrolyte
(c) strong electrolyte (d) nonelectrolyte
(e) weak electrolyte (f) nonelectrolyte
(g) strong electrolyte (h) weak electrolyte

4.130 (a) molecular: $Na_2S(aq) + H_2SO_4(aq) \rightarrow$ $H_2S(g) + Na_2SO_4(aq)$
net ionic: $S^{2-}(aq) + 2H^+(aq) \rightarrow H_2S(g)$
(b) molecular: $LiHCO_3(aq) + HNO_3(aq) \rightarrow$ $LiNO_3(aq) + H_2O(l) + CO_2(g)$
net ionic: $H^+(aq) + HCO_3^-(aq) \rightarrow$ $H_2O(l) + CO_2(g)$
(c) molecular: $(NH_4)_3PO_4(aq) + 3KOH(aq) \rightarrow$ $K_3PO_4(aq) + 3NH_3(g) + 3H_2O(l)$
net ionic: $NH_4^+(aq) + OH^-(aq) \rightarrow$ $NH_3(g) + H_2O(l)$
(d) molecular: $K_2SO_3(aq) + 2HCl(aq) \rightarrow$ $H_2SO_3(aq) + 2KCl(aq)$
net ionic: $SO_3^{2-}(aq) + 2H^+(aq) \rightarrow SO_2(g) +$ H_2O
(e) molecular: $BaCO_3(s) + 2HBr(aq) \rightarrow$ $BaBr_2(aq) + CO_2(g) + H_2O(l)$
net ionic: $BaCO_3(s) + 2H^+(aq) \rightarrow$ $Ba^{2+}(aq) + CO_2(g) + H_2O(l)$
(f) no reaction

4.132 82.2% Na_2CO_3

4.134 Ce^{2+}

CHAPTER 5

Practice Exercises

1. 5200 J, 5.2 kJ, 1200 cal, 1.2 kcal

2. 3.7 kJ, 74 kJ/mol

3. $2.5H_2(g) + 1.25O_2(g) \rightarrow$ $2.5H_2O(g)$ $\Delta H = -647.3$ kJ

4.

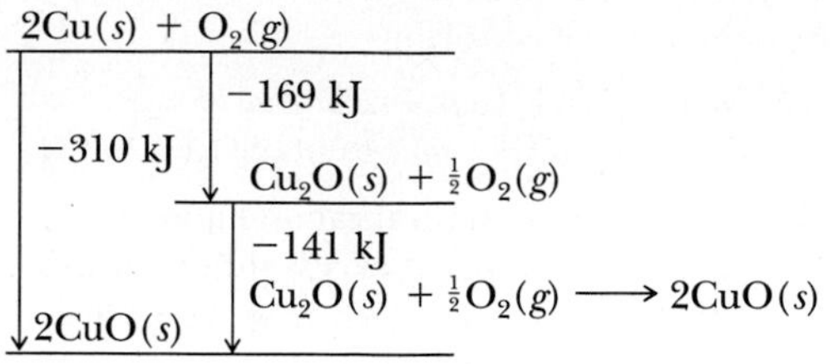

5. $\Delta H° = -44.0$ kJ

6. $Na(s) + \frac{1}{2}H_2(g) + C(s) + \frac{3}{2}O_2(g) \rightarrow$ $NaHCO_3(s)$ $\Delta H_f° = -947.7$ kJ/mol

7. (a) -113.1 kJ (b) -177.8 kJ

Selected Review Exercises

5.12 7.32 kJ

5.14 135 J

5.16 silver

5.17 $t_{final} = 26.4$ °C

5.19 (a) 1.26×10^3 J (b) 1.26×10^3 J (c) 34.7 g

5.21 1.5 days

5.23 113 J mol^{-1} °C^{-1}

5.25 3.8×10^3 J, -53 kJ mol^{-1}

5.27 -58.72 kJ mol^{-1}

5.36 (a) $2CO(g) + O_2(g) \rightarrow 2CO_2(g)$ $\Delta H° =$ -566 kJ (b) -283 kJ mol^{-1}

5.38 $4Al(s) + 2Fe_2O_3(s) \rightarrow 2Al_2O_3(s) + 4Fe(s)$ $\Delta H° = -1708$ kJ

5.40 $10CaO(s) + 10H_2O(l) \rightarrow 10Ca(OH)_2(s)$ $\Delta H° = -653$ kJ

5.42

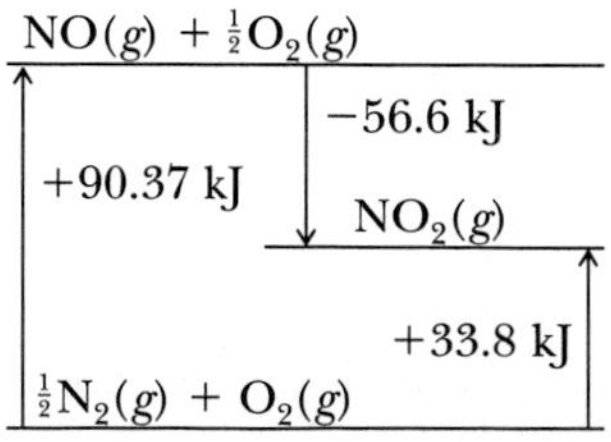

The enthalpy change for the reaction $NO(g) + \frac{1}{2}O_2(g) \rightarrow NO_2(g)$ is -56.6 kJ as seen in the figure above.

5.45 $NO(g) + \frac{1}{2}O_2(g) \rightarrow NO_2(g)$ $\Delta H° = -56.570$ kJ

5.47 $HCl(g) + NaNO_2(s) \rightarrow NaCl(s) + HNO_2(l)$ $\Delta H° = -78.61$ kJ

5.49 $\Delta H° = 364$ kJ. Since $\Delta H°$ has a positive value, it is the reverse reaction that is likely spontaneous.

5.51 $Ca(OH)_2(aq) + 2HCl(aq) \rightarrow CaCl_2(aq) + 2H_2O(l)$ $\quad \Delta H° = -111$ kJ

5.53 Only (b)

5.56 (a) $\frac{1}{2}H_2(g) + \frac{1}{2}Cl_2(g) \rightarrow HCl(g)$
$\Delta H_f° = -92.30$ kJ/mol
(b) $\frac{1}{2}N_2(g) + 2H_2(g) + \frac{1}{2}Cl_2(g) \rightarrow NH_4Cl(s)$
$\Delta H_f° = -315.4$ kJ/mol
(c) $C(s) + \frac{1}{2}O_2(g) + 2H_2(g) + N_2(g) \rightarrow$
$CO(NH_2)_2(s)$ $\quad \Delta H_f° = -333.19$ kJ/mol
(d) $2Na(s) + C(s) + \frac{3}{2}O_2(g) \rightarrow Na_2CO_3(s)$
$\Delta H_f° = -1131$ kJ/mol

5.57 $+463.9$ kJ. This reaction is endothermic.

5.59 $P(s) + \frac{3}{2}H_2(g) + 2O_2(g) \rightarrow H_3PO_4(l)$
$\Delta H_f° = -1259$ kJ/mol

5.61 $\frac{1}{2}H_2(g) + \frac{1}{2}Br_2(l) \rightarrow HBr(g)$ $\quad \Delta H_f° = -43$ kJ/mol

CHAPTER 6

Practice Exercises

1. 5.45×10^{14} Hz
2. 2.874 m
3. 656.5 nm, red
4. $3s, 3p, 3d; 4s, 4p, 4d, 4f$
5. nine
6. (a) Mg $1s^22s^22p^63s^2$
   (b) Ge $1s^22s^22p^63s^23p^63d^{10}4s^24p^2$
   (c) Cd $1s^22s^22p^63s^23p^63d^{10}4s^24p^64d^{10}5s^2$
   (d) Gd
   $1s^22s^22p^63s^23p^63d^{10}4s^24p^64d^{10}4f^75s^25p^65d^16s^2$
7. (a) Na [orbital diagram: $1s$ ⇅ $2s$ ⇅ $2p$ ⇅⇅⇅ $3s$ ↑]
   (b) S [orbital diagram: $1s$ ⇅ $2s$ ⇅ $2p$ ⇅⇅⇅ $3s$ ⇅ $3p$ ⇅↑↑]
   (c) Fe [orbital diagram: $1s$ ⇅ $2s$ ⇅ $2p$ ⇅⇅⇅ $3s$ ⇅ $3p$ ⇅⇅⇅ $3d$ ⇅↑↑↑↑ $4s$ ⇅]
8. (a) [Ne] $3s^23p^3$ (b) [Kr] $4d^{10}5s^25p^2$
9. (a) $4s^24p^4$ (b) $5s^25p^2$ (c) $5s^25p^5$
10. (a) Sn (b) Ga (c) Fe (d) S^{2-}
11. (a) Be (b) C

Selected Review Exercises

6.9 6.98×10^{14} Hz
6.11 1.02×10^{15} Hz
6.13 2.98 m
6.15 5.0×10^6 m, 5.0×10^3 km
6.22 2.7×10^{-19} J, 1.6×10^5 J mol^{-1}
6.24 6.32×10^{22} photons

6.27 (a) violet (b) 7.31×10^{14} s^{-1} (c) 4.85×10^{-19} J

6.28 No, the wavelength is 1090 nm, which is in the infrared region of the spectrum.

6.32 2.04×10^{-18} J, 97.3 nm, ultraviolet

6.43 (a) $n = 1$ (b) $n = 3$

6.45 (a) $n = 3, \ell = 0$ (b) $n = 5, \ell = 2$ (c) $n = 4, \ell = 3$

6.48 (a) $m_\ell = 1, 0,$ or -1 (b) $m_\ell = 3, 2, 1, 0, -1, -2,$ or -3

6.50 No, the maximum value of ℓ in the fourth shell ($n = 4$) is 3.

6.52 When $m_\ell = -4$, the minimum value of ℓ is 4 and the minimum value of n is 5.

6.54 There are eleven values for m_ℓ: $-5, -4, -3, -2, -1, 0, 1, 2, 3, 4,$ and 5. Thus there are eleven orbitals.

6.65 (a) S $1s^22s^22p^63s^23p^4$
(b) K $1s^22s^22p^63s^23p^64s^1$
(c) Ti $1s^22s^22p^63s^23p^63d^24s^2$
(d) Sn $1s^22s^22p^63s^23p^63d^{10}4s^24p^64d^{10}5s^25p^2$

6.69 (a) Mg [orbital diagram: $1s$ ⇅ $2s$ ⇅ $2p$ ⇅⇅⇅ $3s$ ⇅]
(b) Ti [orbital diagram: $1s$ ⇅ $2s$ ⇅ $2p$ ⇅⇅⇅ $3s$ ⇅ $3p$ ⇅⇅⇅ $3d$ ↑↑○○○ $4s$ ⇅]

6.71 (a) zero (b) three (c) three

6.72 (a) [Ar] $3d^84s^2$ (b) [Xe] $6s^1$
(c) [Ar] $3d^{10}4s^24p^2$ (d) [Ar] $3d^{10}4s^24p^5$

6.73 (a) Ni [Ar] [orbital diagram: $3d$ ⇅⇅⇅↑↑ $4s$ ⇅]
(b) Cs [Xe] [orbital diagram: $6s$ ↑]
(c) Ge [Ar] [orbital diagram: $3d$ ⇅⇅⇅⇅⇅ $4s$ ⇅ $4p$ ↑↑○]
(d) Br [Ar] [orbital diagram: $3d$ ⇅⇅⇅⇅⇅ $4s$ ⇅ $4p$ ⇅⇅↑]

6.78 (a) 5 (b) 4 (c) 4 (d) 6

6.79 (a) $3s^1$ (b) $3s^23p^1$ (c) $4s^24p^2$ (d) $3s^23p^3$

6.80 (a) Na [orbital diagram: $3s$ ↑]
(b) Al [orbital diagram: $3s$ ⇅ $3p$ ↑○○]
(c) Ge [orbital diagram: $4s$ ⇅ $4p$ ↑↑○]
(d) P [orbital diagram: $3s$ ⇅ $3p$ ↑↑↑]

6.83 (a) 5 (b) 3 (c) 2 (d) 0 (e) 0

6.92 (a) 1 (b) 6 (c) 7

6.93 (a) Na (b) Sb

6.95 Sb is a bit larger than Sn, according to Figure 6.24.

6.97 $Mg^{2+} < Na^+ < Ne < F^- < O^{2-} < N^{3-}$

6.99 (a) Na (b) Co^{2+} (c) Cl^-

6.102 (a) C (b) O (c) Cl

6.111 (a) Cl (b) Br (c) Si

6.114 (a) 9.57×10^3 g (b) 1.91×10^4 g

6.116 (a) 434.2 nm, in the violet region of the visible spectrum (b) 97.25 nm, in the ultraviolet region of the spectrum (c) 2628 nm, in the infrared region of the spectrum

6.118 (a) This diagram violates the aufbau principle. Specifically, the *s* orbital should be filled before filling the higher energy *p* orbitals.
(b) This diagram violates the aufbau principle. Specifically, the lower energy *s* orbital should be filled completely before filling the *p* orbitals.
(c) This diagram violates the aufbau principle. Specifically, the lower energy *s* orbital should be filled completely before filling the *p* orbitals.

6.120 $n = 4$, $\ell = 0$, $m_\ell = 0$, $m_s = \pm\frac{1}{2}$

6.124 (a) 328 kJ/mol (b) 141 kJ/mol (c) -844 kJ/mol (The last of these is exothermic.)

CHAPTER 7

Practice Exercise

1. Cr [Ar] $3d^5 4s^1$ (a) Cr^{2+} [Ar] $3d^4$
(b) Cr^{3+} [Ar] $3d^3$ (c) Cr^{6+} [Ar]

2. (a) :S̈e (b) :Ï· (c) ·Ca·

3. ·Mg⤵Ö: → Mg^{2+} $\left[:\ddot{O}:\right]^{2-}$

4. (a) Br (b) C (c) Cl

5. (a) OSO (b) ONO (c) HOClO
(d) HOPOH with O above and H below

6. SO_2, 18 e^-; PO_4^{3-}, 32 e^-; NO^+, 10 e^-

7. :F̈—Ö—F̈: $\left[\begin{array}{c}H\\|\\H-N-H\\|\\H\end{array}\right]^+$:Ö—S̈=Ö:

$\left[:\ddot{O}-\overset{\cdot\cdot}{N}=O:\right]^-$:F̈—Cl̈—F̈: H—Ö—Cl—Ö: with :Ö: above

8. (a) ²⁻:N̈—N≡O:⊕⊕ (b) $\left[\ddot{S}=C=\overset{\cdot\cdot}{N}\ominus\right]^-$

9. (a) :Ö=S=Ö: (b) :Ö=Cl—Ö—H
(c) H—Ö—P—Ö II with :O: above (double bond) and :Ö—H below

10. $\left[:\ddot{O}-N=\ddot{O}:\right]^- \longleftrightarrow \left[\ddot{O}=N-\ddot{O}:\right]^-$

$\left[:\ddot{O}-\overset{:O:}{P}-\ddot{O}:\right]^{3-} \longleftrightarrow \left[:\ddot{O}-\overset{:\ddot{O}:}{P}=\ddot{O}\right]^{3-} \longleftrightarrow$

$\left[:\ddot{O}-\overset{:\ddot{O}:}{P}-\ddot{O}:\right]^{3-} \longleftrightarrow \left[\ddot{O}=\overset{:\ddot{O}:}{P}-\ddot{O}:\right]^{3-}$

11. (a) F^- is a Lewis base, because it has lone pairs of electrons and cannot accept any additional electrons.
(b) $BeCl_2$ is a Lewis acid, because the valence shell of Be has less than an octet of electrons.

Selected Review Exercises

7.5 Magnesium loses two electrons; bromine gains an electron.

7.9 Pb^{2+} [Xe] $4f^{14} 5d^{10} 6s^2$ Pb^{4+} [Xe] $4f^{14} 5d^{10}$

7.11 (a) [Ar] $3d^8$ (b) [Ar] $3d^2$ (c) [Ar]
(d) [Ar] $3d^9$ (e) [Xe] $5d^8$

7.13 (a) ·S̈i· (b) ·S̈b· (c) ·Ba· (d) ·Äl· (e) :S̈·

7.15 (a) $[K]^+$ (b) $[Al]^{3+}$ (c) $\left[:\ddot{S}:\right]^{2-}$
(d) $\left[:\ddot{S}i:\right]^{4-}$ (e) $[Mg]^{2+}$

7.17 (a) :B̈r· :B̈r· Ca ⟶ $2\left[:\ddot{B}r:\right]^- + [Ca]^{2+}$

(b) Al :Ö: :Ö: Al :Ö: ⟶ $2[Al]^{3+} + 3\left[:\ddot{O}:\right]^{2-}$

(c) K· K· S̈: ⟶ $2K^+ + \left[:\ddot{S}:\right]^{2-}$

7.26 7.23×10^{-19} J

7.28 (a) :B̈r· + ·B̈r: → :B̈r—B̈r:
(b) 2H· + ·Ö· → H—Ö: with H below

(c) $3H\cdot\ +\ \cdot\ddot{N}\cdot\ \rightarrow\ H-\underset{\underset{H}{|}}{\overset{\overset{H}{|}}{N}}-H$

7.30 (a) one (b) four (c) two (d) three (e) one

7.32 (a) H_2Se (b) H_3As (c) SiH_4

7.41 $H-\underset{\underset{H}{|}}{\overset{\overset{H}{|}}{C}}-\underset{\underset{H}{|}}{\overset{\overset{H}{|}}{C}}-\underset{\underset{H}{|}}{\overset{\overset{H}{|}}{C}}-\underset{\underset{H}{|}}{\overset{\overset{H}{|}}{C}}-\underset{\underset{H}{|}}{\overset{\overset{H}{|}}{C}}-H$

There are 12 H atoms, and the formula is C_5H_{12}.

7.44 $CH_3CH_2CO_2H + H_2O \rightleftharpoons H_3O^+ + CH_3CH_2CO_2^-$
It is a weak acid.

7.46 $CH_3NHCH_2CH_3(aq) + CH_3CH_3CO_2H(aq) \rightarrow$
$CH_3NH_2CH_2CH_3^+(aq) + CH_3CH_2CO_2^-(aq)$

7.51 The noble gases are assigned electronegativity values of zero because they have a complete shell of electrons.

7.52 (a) S (b) Si (c) Br

7.53 N—S

7.57 (a) 4 (b) 6 (c) 2

7.59 (a) Cl / Cl Si Cl / Cl (b) F P F / F (c) H P H / H (d) Cl S Cl

7.61 (a) 32 (b) 26 (c) 8 (d) 20

7.63 (a) $\ddot{\underset{..}{Cl}}-\underset{\underset{:\ddot{Cl}:}{|}}{\overset{\overset{:\ddot{Cl}:}{|}}{Si}}-\ddot{\underset{..}{Cl}}:$ (b) $:\ddot{F}-\underset{\underset{:\ddot{F}:}{|}}{\overset{\overset{:\ddot{F}:}{|}}{P}}-\ddot{F}:$

(c) $H-\underset{\underset{H}{|}}{\overset{\overset{..}{P}}{}}-H$ (d) $:\ddot{\underset{..}{Cl}}-\ddot{\underset{..}{S}}-\ddot{\underset{..}{Cl}}:$

7.65 (a) $\ddot{\underset{..}{S}}=C=\ddot{\underset{..}{S}}$ (b) $[:C\equiv N:]^-$

(c) $:\ddot{\underset{..}{O}}-\underset{\underset{:\ddot{O}:}{\|}}{Se}-\ddot{\underset{..}{O}}:$ (d) $:\ddot{\underset{..}{O}}-Se=\ddot{\underset{..}{O}}$

7.66 (a) $H-\ddot{\underset{..}{O}}-N=\ddot{\underset{..}{O}}$ (b) $H-\ddot{\underset{..}{O}}-\underset{\underset{:\ddot{O}:}{|}}{\overset{\overset{:\ddot{O}:}{|}}{Cl}}:$

(c) $H-\ddot{\underset{..}{O}}-\underset{\underset{:O:}{|}}{\overset{\overset{..}{Se}}{}}-\ddot{\underset{..}{O}}-H$

7.68 (a) Lewis structure of TeF_5 with F atoms and Te center (b) Lewis structure of ClF_5

(c) $:\ddot{F}:$ (d) $:\ddot{F}$ Xe $\ddot{F}:$ / $Xe:$ / $:\ddot{F}:$ and $:\ddot{F}$ Xe $\ddot{F}:$ / $:\ddot{F}$ $\ddot{F}:$

7.77 Atoms with no formal charge indicated have zero formal charge.

(a) $H-\overset{..}{\underset{..}{O}}-\overset{\oplus}{\underset{..}{Cl}}-\ddot{\underset{..}{O}}:^{\ominus}$ (b) $\overset{:\ddot{O}:^{\ominus}}{\underset{\underset{:\ddot{O}:^{\ominus}}{}}{|}}$ $:\ddot{O}=S-\ddot{\underset{..}{O}}:$ with $S^{2\oplus}$

(c) $\ddot{\underset{..}{O}}=\overset{\oplus}{S}-\ddot{\underset{..}{O}}:^{\ominus}$ (d) $:\ddot{\underset{..}{Cl}}-\overset{\oplus}{\underset{\underset{:\ddot{O}:}{|}}{N}}-\ddot{\underset{..}{O}}:^{\ominus}$ with O above

(e) $:\ddot{F}-\overset{\oplus}{\underset{\underset{:\ddot{F}:}{|}}{\overset{\overset{:\ddot{F}:}{|}}{N}}}-\ddot{\underset{..}{O}}:^{\ominus}$

7.79 The formal charges on all of the atoms of the left structure are zero, therefore, the potential energy of this molecule is lower and it is more stable.

7.81 $H-\overset{\oplus}{N}\equiv\overset{\oplus}{N}-\ddot{\underset{..}{N}}:^{2\ominus}$

7.85 Resonance causes the average number of bonds in each N—O linkage of NO_3^- to be 1.33. Resonance causes the average number of electron pair bonds in each linkage of NO_2 to be 1.5. We conclude that the N—O bond in NO_2^- should be shorter than that in NO_3^-.

7.88 $:O\equiv C-\ddot{\underset{..}{O}}: \longleftrightarrow :\ddot{\underset{..}{O}}-C\equiv O:$

These are not preferred structures because, in each Lewis diagram, one oxygen bears a formal charge of $+1$ whereas the other bears a formal charge of -1. The structure with the formal charges of zero has a lower potential energy and is more stable.

7.90 The Lewis structure obtained using the rules of Figure 7.7 is

$$\left[\ \overset{\ominus\ \ddot{O}:}{\underset{\underset{:O:^{\ominus}}{|}}{\overset{\overset{|}{}}{:\ddot{O}-\underset{}{S}^{2\oplus}-\ddot{O}:^{\ominus}}}}\ \right]^{2-}$$

$$\left[\ \overset{:\ddot{O}:^{\ominus}}{\underset{\underset{:O:^{\ominus}}{|}}{\overset{\overset{|}{}}{:\ddot{O}=S=\ddot{O}:}}}\ \right]^{2-} \longleftrightarrow \left[\ \overset{\overset{O}{\|}}{\underset{\underset{..O..}{\|}}{:\ddot{O}^{\ominus}-S-\ddot{O}^{\ominus}:}}\ \right]^{2-}$$

7.93

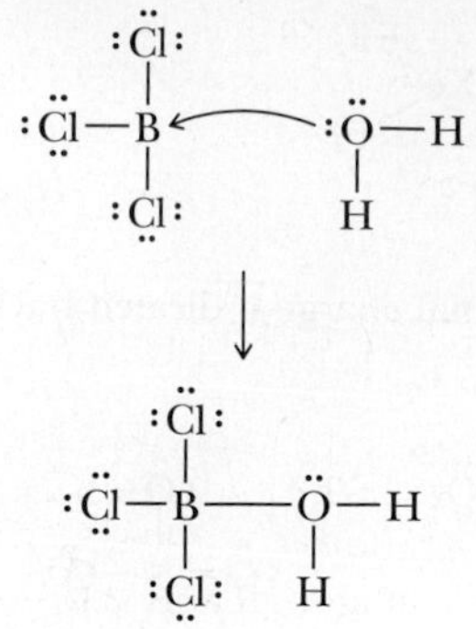

7.95

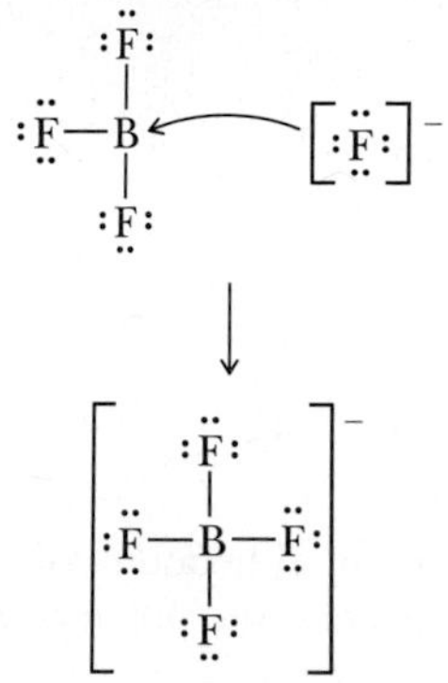

7.98

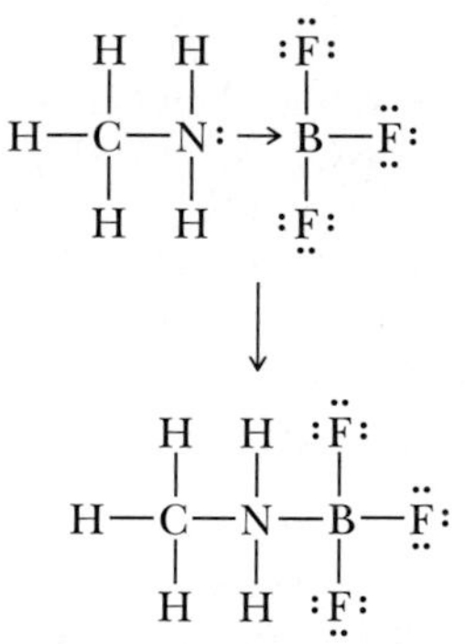

7.100

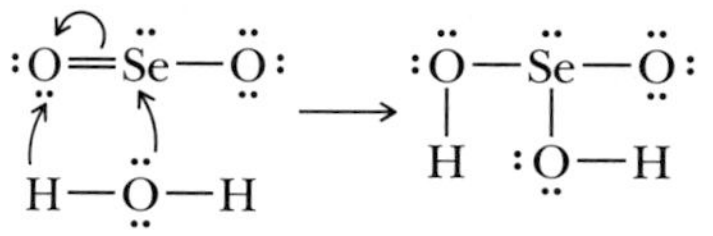

7.102

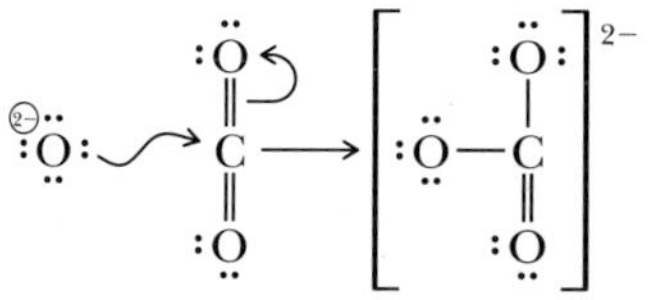

7.109 Only the top two are acceptable. The others violate the octet rule.

7.111

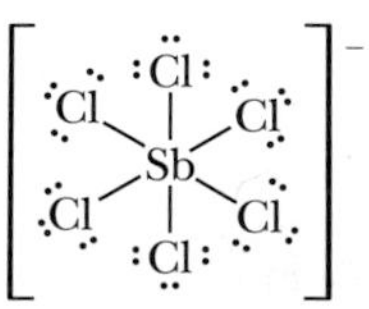

CHAPTER 8

Practice Exercises

1. trigonal bipyramidal

2. SO_3^{2-}, trigonal pyramidal; XeO_4, tetrahedral; OF_2, nonlinear (bent)

3. planar triangular

4. SO_2, $BrCl$, AsH_3, CF_2Cl_2

5. Bond formed by overlap of hydrogen $1s$ orbital with the half-filled $3p$ orbital of chlorine. Orbitals overlap as shown in Figure 8.12 for HF. Orbital diagram:
   Cl (in HCl) 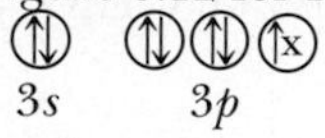
   (x = H electron)

6. Phosphorus uses its three half-filled $3p$ orbitals to overlap with the three hydrogen $1s$ orbitals. The arrangement is similar to that for H_2S (Figure 8.13). Orbital diagram: P (in PH_3) 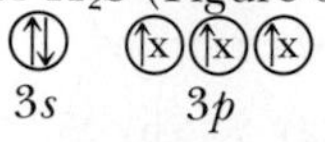
   (x = H electrons)

7. As (in $AsCl_5$) 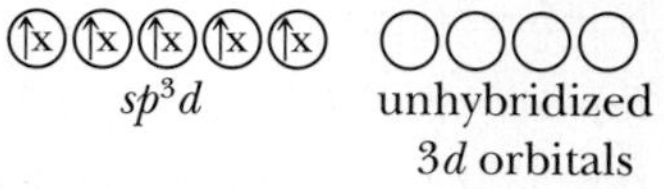
   (x = Cl electrons)

8. (a) sp^3 (b) sp^3d

9. (a) sp^3 (b) sp^3d

10. sp^3d^2; the ion is octahedral.
    P (in PCl_6^-) 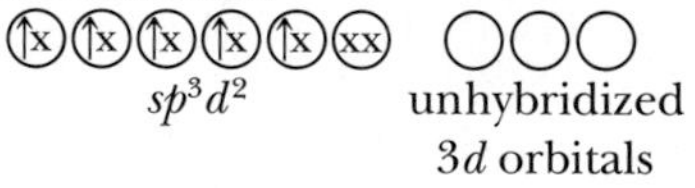
    (x = Cl electrons)

11. Electron population of molecular orbitals is
    $(\sigma_{2s})^2(\sigma_{2s}^*)^2(\pi_{2p_y})^2(\pi_{2p_z})^2(\sigma_{2p_x})^2(\pi_{2p_y}^*)^1(\pi_{2p_z}^*)^0(\sigma_{2p_x}^*)^0$
    The net bond order is 2.5.

Selected Review Exercises

8.6 (a) nonlinear (b) trigonal bipyramidal (c) pyramidal (d) nonlinear

8.8 (a) tetrahedral (b) square planar (c) octahedral (d) linear

8.10 (a) pyramidal (b) tetrahedral (c) pyramidal (d) tetrahedral

8.12 all angles 120°

8.18 (a), (b), and (c)

8.27 The $1s$ atomic orbitals of the hydrogen atoms overlap with the mutually perpendicular p atomic orbitals of the selenium atom. Se (in H_2Se) 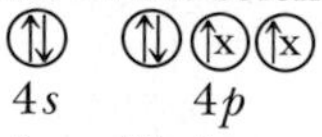
(x = H electrons)

8.32 Elements in period 2 do not have a d subshell in the valence level.

8.35 (a)

$$\left[\begin{array}{c} :\ddot{O}: \\ | \\ :\ddot{O}-Cl-\ddot{O}: \\ \ddot{}\quad\ddot{} \end{array}\right]^{-}$$

tetrahedral orientation of electron pairs on Cl, sp^3 hybridization

(b)

$$\begin{array}{c} :\ddot{O}: \\ | \\ \ddot{O}=S-\ddot{O}: \end{array}$$

planar triangular orientation of electron pairs on S, sp^2 hybridization

(c) $:\ddot{F}-\ddot{O}-\ddot{F}:$

tetrahedral orientation of electron pairs on O, sp^3 hybridization

(d)

$$\left[\begin{array}{c} :\ddot{C}l: \quad :\ddot{C}l: \\ \diagdown \quad | \\ :\ddot{C}l-Sb-\ddot{C}l: \\ | \quad \diagup \\ :\ddot{C}l: \quad :\ddot{C}l: \end{array}\right]^{-}$$

octahedral orientation of electron pairs on Sb, sp^3d^2 hybridization

(e)

$$\begin{array}{c} :\ddot{C}l: \\ | \\ :\ddot{B}r-\ddot{C}l: \\ | \\ :\ddot{C}l: \end{array}$$

trigonal bipyramidal orientation of electron pairs on Br, sp^3d hybridization

(f)

$$\begin{array}{c} :\ddot{F}: \\ | \\ :\ddot{F}-Xe-\ddot{F}: \\ | \\ :\ddot{F}: \end{array}$$

octahedral orientation of electron pairs on Xe, sp^3d^2 hybridization

8.37 (a) Lewis diagram:
$$\begin{array}{c} :\ddot{C}l-\ddot{A}s-\ddot{C}l: \\ | \\ :\ddot{C}l: \end{array}$$

Hybrid orbital diagram for As: ⊛⊛⊛⊛
sp^3

(x = Cl electrons)

(b) Lewis diagram: $:\ddot{F}-\ddot{C}l-\ddot{F}:$
$$\begin{array}{c} | \\ :\ddot{F}: \end{array}$$

Hybrid orbital diagram for Cl:
⊛⊛⊛⊛⊛ ○○○○
sp^3d $3d$

(x = F electrons)

(c) Lewis diagram:
$$\begin{array}{c} :\ddot{C}l: \\ | \quad :\ddot{C}l: \\ :\ddot{C}l-Sb \diagup \\ | \quad \diagdown :\ddot{C}l: \\ :\ddot{C}l: \end{array}$$

Hybrid orbital diagram for Sb:
⊛⊛⊛⊛⊛ ○○○○○
sp^3d $5d$

(x = Cl electrons)

(d) Lewis diagram:
$$\begin{array}{c} \ddot{S}e \\ \diagup \quad \diagdown \\ :\ddot{C}l \qquad \ddot{C}l: \end{array}$$

Hybrid orbital diagram for Se: ⊛⊛⊛⊛
sp^3

(x = Cl electrons)

8.41 The 60° bond angle in cyclopropane is much less than this optimum bond angle. This means that the bonding within the ring cannot be accomplished through the desirable "head on" overlap of hybrid orbitals from each C atom. As a result, the overlap of the hybrid orbitals in cyclopropane is less effec-

tive than that in the more normal, noncyclic propane molecule, and this makes the C—C bonds in cyclopropane comparatively weaker than those in the noncyclic molecule.

8.45 90°

8.48 Sn (in $SnCl_6^{2-}$) ⊗⊗⊗⊗⊗⊗ ○○○
          sp^3d^2 $4d$

8.54 Whereas the arrangement around an sp^2-hybridized atom may have 120° bond angles (which well accommodates a six-membered ring), the geometrical arrangement around an atom that is sp-hybridized must be linear, as in C—C≡C—C systems.

8.56 (a) sp-hybridized N atom: ⊛⊛ ○○
          sp $2p$

(b)

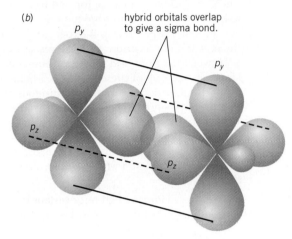

hybrid orbitals overlap to give a sigma bond.

p_y and p_z orbitals overlap to give two pi bonds

(c) Sigma bond framework

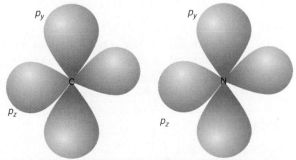

pi bonds formed by overlap of p orbitals

(d) The HCN bond angle should be 180°.

8.58 Each carbon atom is sp^2 hybridized, and each C—Cl bond is formed by the overlap of an sp^2 hybrid of carbon with a p atomic orbital of a chlorine atom. The C=C double bond consists first of a C—C σ bond formed by "head on" overlap of sp^2 hybrids from each C atom. Second, the C=C double bond consists of a side-to-side overlap of unhybridized p orbitals of each C atom, to give one π bond. The molecule is planar, and the expected bond angles are all 120°.

8.64 As shown in Table 8.1, the bond order of Li$_2$ is 1.0. The bond order of Be$_2$ would be zero. Yes, Be$_2{}^+$ could exist since the bond order would be $\frac{1}{2}$.

8.65 (a) O$_2{}^+$ (b) O$_2$ (c) N$_2$

8.71 (a) The C—C single bonds are formed from head-to-head overlap of C atom sp^2 hybrids.
(b) Sideways or π type overlap is expected between the first and the second carbon atoms, as well as between the third and fourth carbon atoms. However, since all of these atomic p orbitals are properly aligned, there can be continuous π type overlap between all four carbon atoms.
(c) We expect completely delocalized π type bonding among the carbon atoms.
(d) The bond is shorter because of the extra electron density associated with the delocalized bond.

8.81 The $4s$ level in calcium is filled but overlaps the vacant $4p$ conduction band and provides for conductivity.

8.83 Since germanium is in Group IV, an element from Group III must be added to make a p-type semiconductor. Here we could use boron, aluminum, or gallium.

8.88 The double bonds are predicted to be between S and O atoms. Hence, the Cl—S—Cl angle diminishes under the influence of the S=O double bonds.

8.90 1. sp^3 2. sp 3. sp^2 4. sp^2

8.91 1. one σ bond. 2. one σ bond and two π bonds 3. one σ bond 4. one σ bond and one π bond

8.93 Only the p_y orbital can form a π bond with d_{xy}, if the internuclear axis is the x axis.

CHAPTER 9

Practice Exercises

1. In each case the conjugate base is obtained by removing a proton from the acid: (a) OH$^-$ (b) I$^-$
(c) NO$_2{}^-$ (d) H$_2$PO$_4{}^-$ (e) HPO$_4{}^{2-}$ (f) PO$_4{}^{3-}$ (g) H$^-$ (h) NH$_3$

2. In each case the conjugate acid is obtained by adding a proton to the base: (a) H$_2$O$_2$ (b) HSO$_4{}^-$ (c) HCO$_3{}^-$ (d) HCN (e) NH$_3$ (f) NH$_4{}^+$ (g) H$_3$PO$_4$ (h) H$_2$PO$_4{}^-$

3. The Brønsted acids are H$_2$PO$_4{}^-(aq)$ and H$_2$CO$_3(aq)$. The Brønsted bases are HCO$_3{}^-(aq)$ and HPO$_4{}^{2-}(aq)$.

$$\text{HCO}_3{}^-(aq) + \text{H}_2\text{PO}_4{}^-(aq) \rightleftharpoons \text{H}_2\text{CO}_3(aq) + \text{HPO}_4{}^{2-}(aq)$$
base / acid / acid / base

4.
$$\text{PO}_4{}^{3-}(aq) + \text{HC}_2\text{H}_3\text{O}_2(aq) \rightleftharpoons \text{HPO}_4{}^{2-}(aq) + \text{C}_2\text{H}_3\text{O}_2{}^-(aq)$$
base / acid / acid / base

5. HPO$_4{}^{2-}(aq)$ + OH$^-(aq)$ → PO$_4{}^{3-}(aq)$ + H$_2$O, HPO$_4{}^{2-}$ acting as an acid
HPO$_4{}^{2-}(aq)$ + H$_3$O$^+(aq)$ → H$_2$PO$_4{}^-(aq)$ + H$_2$O, HPO$_4{}^{2-}$ acting as a base

6. (a) HBr (b) H$_2$Te (c) CH$_3$SH

7. (a) HClO$_4$ (b) H$_2$SeO$_4$

8. HClO < HClO$_2$ < HClO$_3$ < HClO$_4$

9. (a) molecular: Mg(s) + 2HCl(aq) → MgCl$_2(aq)$ + H$_2(g)$
ionic: Mg(s) + 2H$^+(aq)$ + 2Cl$^-(aq)$ → Mg$^{2+}(aq)$ + 2Cl$^-(aq)$ + H$_2(g)$
net ionic: Mg(s) + 2H$^+(aq)$ → Mg$^{2+}(aq)$ + H$_2(g)$
(b) molecular: 2Al(s) + 6HCl(aq) → 2AlCl$_3(aq)$ + 3H$_2(g)$
ionic: 2Al(s) + 6H$^+(aq)$ + 6Cl$^-(aq)$ → 2Al$^{3+}(aq)$ + 6Cl$^-(aq)$ + 3H$_2(g)$
net ionic: 2Al(s) + 6H$^+(aq)$ → 2Al$^{3+}(aq)$ + 3H$_2(g)$

10. (a) 2Al(s) + 3Cu$^{2+}(aq)$ → 2Al$^{3+}(aq)$ + 3Cu(s)
(b) No reaction

11. 2C$_4$H$_{10}(l)$ + 13O$_2(g)$ → 8CO$_2(g)$ + 10H$_2$O(g)

12. C$_2$H$_5$OH(l) + 3O$_2(g)$ → 2CO$_2(g)$ + 3H$_2$O(g)

13. 4Fe(s) + 3O$_2(g)$ → 2Fe$_2$O$_3(s)$

14. P$_4(s)$ + 5O$_2(g)$ → P$_4$O$_{10}(s)$

Selected Review Exercises

9.2 (a) HF (b) N$_2$H$_5{}^+$ (c) C$_5$H$_5$NH$^+$ (d) HO$_2{}^-$ (e) H$_2$CrO$_4$

9.4 (a)
$$\text{HNO}_3 + \text{N}_2\text{H}_4 \rightleftharpoons \text{N}_2\text{H}_5{}^+ + \text{NO}_3{}^-$$
acid / base / acid / base

(b)

conjugate pair

$$N_2H_5^+ + NH_3 \rightleftharpoons NH_4^+ + N_2H_4$$

acid   base   acid   base

conjugate pair

(c)

conjugate pair

$$H_2PO_4^- + CO_3^{2-} \rightleftharpoons HCO_3^- + HPO_4^{2-}$$

acid   base   acid   base

conjugate pair

(d)

conjugate pair

$$HIO_3 + HC_2O_4^- \rightleftharpoons H_2C_2O_4 + IO_3^-$$

acid   base   acid   base

conjugate pair

9.7

conjugate pair

$$HOCl(aq) + H_2O \rightleftharpoons H_3O^+(aq) + OCl^-(aq)$$

acid   base   acid   base

conjugate pair

OCl^- is a stronger base than water, and H_3O^+ is a stronger acid than HOCl.

9.9 $C_2H_3O_2^-$ is a stronger base than NO_2^-.

9.11   (a) H_2Se (b) HI   (c) HIO_4 (d) $HClO_4$

9.14   (a) The relative strength of the binary acids increases from top to bottom in a group of the periodic table, so we expect HAt to be a stronger acid than HI.   (b) $HAtO_4$

9.19   (a) ionic: $Mn(s) + 2H^+(aq) + SO_4^{2-}(aq) \rightarrow$
$$Mn^{2+}(aq) + H_2(g) + SO_4^{2-}(aq)$$
net ionic: $Mn(s) + 2H^+(aq) \rightarrow$
$$Mn^{2+}(aq) + H_2(g)$$
(b) ionic: $Cd(s) + 2H^+(aq) + SO_4^{2-}(aq) \rightarrow$
$$Cd^{2+}(aq) + H_2(g) + SO_4^{2-}(aq)$$
net ionic: $Cd(s) + 2H^+(aq) \rightarrow$
$$Cd^{2+}(aq) + H_2(g)$$
(c) ionic: $Sn(s) + 2H^+(aq) + SO_4^{2-}(aq) \rightarrow$
$$Sn^{2+}(aq) + H_2(g) + SO_4^{2-}(aq)$$
net ionic: $Sn(s) + 2H^+(aq) \rightarrow$
$$Sn^{2+}(aq) + H_2(g)$$
(d) ionic: $Ni(s) + 2H^+(aq) + SO_4^{2-}(aq) \rightarrow$
$$Ni^{2+}(aq) + H_2(g) + SO_4^{2-}(aq)$$
net ionic: $Ni(s) + 2H^+(aq) \rightarrow$
$$Ni^{2+}(aq) + H_2(g)$$
(e) ionic: $2Cr(s) + 6H^+(aq) + 3SO_4^{2-}(aq) \rightarrow$
$$2Cr^{3+}(aq) + 3H_2(g) + 3SO_4^{2-}(aq)$$
net ionic: $2Cr(s) + 6H^+(aq) \rightarrow$
$$2Cr^{3+}(aq) + 3H_2(g)$$

9.21   (a) $3Ag(s) + 4HNO_3(aq) \rightarrow$
$$3AgNO_3(aq) + 2H_2O(l) + NO(g)$$
(b) $Ag(s) + 2HNO_3(aq) \rightarrow$
$$AgNO_3(aq) + H_2O(l) + NO_2(aq)$$

9.23 $Cu(s) \rightarrow Cu^{2+}(aq) + 2e^-$
$H_2SO_4(aq) + 2H^+(aq) + 2e^- \rightarrow SO_2(g) + 2H_2O(l)$

$Cu(s) + H_2SO_4(aq) + 2H^+(aq) \rightarrow$
$$Cu^{2+}(aq) + SO_2(g) + 2H_2O(l)$$
The molecular equation is
$Cu(s) + 2H_2SO_4(aq) \longrightarrow$
$$CuSO_4(aq) + SO_2(g) + 2H_2O(l)$$

9.24   Cu < Pb < Fe < Al

9.26   (c) zinc and   (d) magnesium

9.28   (a) No reaction
(b) $2Cr(s) + 3Pb^{2+}(aq) \rightarrow 2Cr^{3+}(aq) + 3Pb(s)$
(c) $2Ag^+(aq) + Fe(s) \rightarrow 2Ag(s) + Fe^{2+}(aq)$
(d) $3Ag(s) + Au^{3+}(aq) \rightarrow Au(s) + 3Ag^+(aq)$

9.30   (a) $Zn(s) + Sn^{2+}(aq) \rightarrow Zn^{2+}(aq) + Sn(s)$
(b) $2Cr(s) + 6H^+(aq) \rightarrow 2Cr^{3+}(aq) + 3H_2(g)$
(c) No reaction
(d) $Mn(s) + Pb^{2+}(aq) \rightarrow Mn^{2+}(aq) + Pb(s)$
(e) $Zn(s) + Co^{2+}(aq) \rightarrow Zn^{2+}(aq) + Co(s)$

9.36   calcium > iron > silver > iridium

9.40   (a) $2C_6H_6(l) + 15O_2(g) \rightarrow 12CO_2(g) + 6H_2O(g)$
(b) $C_3H_8(g) + 5O_2(g) \rightarrow 3CO_2(g) + 4H_2O(g)$
(c) $C_{21}H_{44}(s) + 32O_2(g) \rightarrow 21CO_2(g) + 22H_2O(g)$
(d) $2C_{12}H_{26}(l) + 37O_2(g) \rightarrow 24CO_2(g) + 26H_2O(g)$
(e) $C_{18}H_{36}(l) + 27O_2(g) \rightarrow 18CO_2(g) + 18H_2O(g)$

9.42 $2CH_3OH(l) + 3O_2(g) \rightarrow 2CO_2(g) + 4H_2O(g)$

9.44 $C_{12}H_{22}O_{11}(s) + 12O_2(g) \rightarrow 12CO_2(g) + 11H_2O(g)$

9.46   The products will be CO_2, H_2O, and SO_2:
$2C_2H_6S(l) + 9O_2(g) \longrightarrow$
$$4CO_2(g) + 2SO_2(g) + 6H_2O(g)$$

9.48   (a) $2Zn(s) + O_2(g) \rightarrow 2ZnO(s)$
(b) $4Al(s) + 3O_2(g) \rightarrow 2Al_2O_3(s)$
(c) $2Mg(s) + O_2(g) \rightarrow 2MgO(s)$
(d) $2Fe(s) + O_2(g) \rightarrow 2FeO(s)$
Alternatively we have
$4Fe(s) + 3O_2(g) \rightarrow 2Fe_2O_3(s)$
(e) $2Ca(s) + O_2(g) \rightarrow 2CaO(s)$

9.50   (a) F_2 (b) P_4 (c) Br_2 (d) S_8 (e) Cl_2 (f) S_8

9.51 192 g mol^{-1}

9.52 -427 kJ

CHAPTER 10

Practice Exercises

1. 1.03×10^5 mm, 33.9 ft

2.   688 torr

3.   750 torr

4.   214 mL

5. 132 g mol^{-1}, xenon

6. 138 g mol^{-1}, P_2F_4

7. $9.00 \text{ L } O_2$

8. $1.06 \text{ g } Na_2CO_3$

9. $98.1 \text{ g } O_2$

10.   732 torr, 283 mL at STP

11. $X_{O_2} = 0.153$, 15.3 mol %

12. 128.9 g mol^{-1}, HI

Selected Review Exercises

10.4 Since the density of water is approximately 13 times lower than that of mercury, a barometer constructed with water as the movable liquid would have to be some 13 times longer than one constructed using mercury. Also, the vapor pressure of water is large enough that the closed end of the barometer may fill with sufficient water vapor so as to affect atmospheric pressure readings. In fact, the measurement of atmospheric pressure would be about 18 torr too low, due to the presence of water vapor in the closed end of the barometer.

10.6 (a) 735 torr (b) 0.974 atm (c) 738 mm Hg (d) 10.9 torr

10.8 (a) 79.3 kPa (b) 21.3 kPa (c) 4.00×10^{-2} kPa

10.9 The best conceivable vacuum pump (one capable of producing a perfect vacuum on the column of water being pulled up) could cause the water in the pipe attached to the pump to rise only 33.9 ft above the surface of the water in the pit. The water at the bottom of a 35 ft pit could not be removed by use of a vacuum pump.

10.12 826 torr

10.17 796 torr

10.19 5.73 L

10.21 $-53\,°C$

10.23 715 torr

10.25 1060 torr

10.27 (a) 210 torr (b) 204 torr

10.29 645 K, 372 °C

10.34 24.1 L

10.36 4.16 atm

10.38 195 atm

10.40 151 moles

10.42 (a) 4 moles, 128 g (b) 2 moles

10.44 (a) 1.34 g L^{-1} (b) 1.25 g L^{-1} (c) 3.17 g L^{-1} (d) 1.78 g L^{-1}

10.46 1.28 g L^{-1}

10.48 27.6 g mol^{-1}

10.50 88.4 g mol^{-1}

10.52 13.5 mg

10.54 15.0 L

10.56 10.7 L

10.58 (a) 0.780 mol, 19.4 L (b) 4.02×10^4 mol, 2.58×10^3 kg

10.59 (a) 85.74% C, 14.28% H (b) CH_2 (c) C_3H_6

10.60 79.287 kPa N_2, 21.332 kPa O_2. These two gases are normally the predominant constituents in air. Other gases must account for the remaining pressure.

10.62 739 torr

10.64 718 torr, 272 mL

10.66 285 mL

10.70 146 g mol^{-1}

10.71 The rate of effusion of the ^{235}U isotope is 1.0043 times faster than the ^{238}U isotope.

10.85 50 lb in.$^{-2}$

10.87 472 °C

10.89 (a) $Zn(s) + 2HCl(aq) \rightarrow H_2(g) + ZnCl_2(aq)$ (b) 0.499 mol, 32.6 g (c) 0.998 mol (d) 125 mL

10.91 1.85 g $CaCO_3$, 4.63 mL HCl

10.93 (a) 40.00% S, 60.00% O (b) SO_3 (c) SO_3

10.96 (a) 2.7×10^{-5} L (b) 1.2×10^{-2} mL

10.97 652 torr

10.98 4.7×10^{-2} g Zn

CHAPTER 11

Practice Exercises

1. Number of molecules in the vapor will increase; number of molecules in the liquid will decrease.

2. approximately 75 °C

3. 477 torr

4. Adding heat will shift the equilibrium to the right, which produces more vapor and increases the vapor pressure.

5. solid $\rightarrow$ gas (sublimation)

6. liquid

7. covalent (network) solid

8. molecular solid

Selected Review Exercises

11.9 The larger molecule has the greater London force of attraction and hence the higher boiling point: C_8H_{18}.

11.14 (a) London, dipole–dipole, hydrogen bonding (b) London (c) London, dipole–dipole (d) London (e) London, dipole–dipole

11.23 Water should have the greater surface tension because it has the stronger intermolecular force (i.e., hydrogen bonding).

11.26 Glycerol ought to wet the surface of glass quite nicely, because the oxygen atoms at the surface of glass can form effective hydrogen bonds to the O—H groups of glycerol.

11.30 diethyl ether

11.31 The snow dissipates by sublimation.

11.43 diethyl ether

11.45 Air with 100% humidity is saturated with water vapor, meaning that the partial pressure of water in the air has become equal to the vapor pressure of water at that temperature. Since vapor pressure increases with increasing temperature, so too should the total amount of water vapor in humid (saturated) air increase with increasing temperature.

11.47 At the temperature of the cool glass, the equilibrium vapor pressure of the water is lower than the partial pressure of water in the air. The air in contact with the cool glass is induced to relinquish some of its water, and condensation occurs.

11.49 Although the cold air outside the building may be nearly saturated at the low temperature, the same air inside at a higher temperature is not now saturated. In fact, the water content of the air at the higher temperature may be only a fraction of the maximum % humidity for that temperature. The % humidity of the air at the indoor temperature is, therefore, comparatively low.

11.53 about 73 °C

11.55 butanol

11.59 Inside the lighter, the liquid butane is in equilibrium with its vapor, which exerts a pressure somewhat above normal atmospheric pressure. If the liquid butane were spilled on a desktop, it would vaporize (boil) because it would be at a pressure of only about 1 atm, which is somewhat less than the vapor pressure of butane. The intermolecular forces in butane are weak.

11.61 The hydrogen bond network in HF is less extensive than in water, because it is a monohydride not a dihydride.

11.64 305 kJ

11.68 water

11.70 ethanol

11.72 (a) 0 °C (b) 47.9 g

11.74 $CH_4 < CF_4 < HCl < HF$

11.76 252 torr

11.79 This is an endothermic system, and adding heat to the system will shift the position of the equilibrium to the right.

11.86 Carbon dioxide does not have a normal boiling point because its triple point lies above one atmosphere.

11.88 The solid's temperature increases until the solid–liquid line is reached, at which point the solid sublimes. After it has completely vaporized, the vapor's temperature increases to 0 °C.

11.92 The critical temperature of hydrogen is below room temperature because, at room temperature, it cannot be liquified by the application of pressure. The critical temperature of butane is above room temperature because butane can be liquified by the application of pressure.

11.101 one

11.102 four

11.104 407 pm

11.109 molecular solid

11.111 covalent solid

11.113 covalent solid

11.114 metallic

11.120 The principal attractive forces are ion–ion forces and ion–induced dipole attractions. These are of overwhelming strength compared to London forces, which do technically exist.

11.122 The comparatively weak intermolecular forces in acetone allow for very *rapid* evaporization. This provides for significantly faster cooling than does the slow evaporation of ethylene glycol. The intermolecular forces in ethylene glycol cause it to evaporate slowly.

11.124 At the higher temperature, all of the carbon dioxide exists in the gas phase, because the critical temperature has been exceeded.

11.126 86 °C

11.128 656 pm

CHAPTER 12

Practice Exercises

1. 8.49×10^{-4} g O_2, 1.48×10^{-3} g N_2
2. 2.50 g NaOH, 248 g H_2O, 248 mL H_2O
3. 20 g solution
4. 16.0 g CH_3OH
5. 0.400 *m*
6. 16.0 *m*
7. 6.82 *M*
8. 9.02 torr
9. 44.0 torr
10. 100.16 °C
11. 157 g mol^{-1}
12. 214 torr
13. 5.4×10^2 g mol^{-1}
14. For 100% dissociation, $T_f = -0.882$ °C; if 0% dissociated, $T_f = -0.441$ °C

Selected Review Exercises

12.17 (a) $KCl(s) \rightarrow K^+(g) + Cl^-(g)$ $\Delta H = +690$ kJ
$K^+(g) + Cl^-(g) \rightarrow K^+(aq) + Cl^-(aq)$
$\Delta H = -686$ kJ
(b) $KCl(s) \rightarrow K^+(aq) + Cl^-(aq)$ $\Delta H = +4$ kJ

12.18 $+18$ kJ/mol, 450% larger

12.24 350 g

12.27 0.038 g/L

12.32 0.197 molal, 0.197 *M*. A solvent must have a density close to 1 g/mL for this to happen. Also, the volume of the solvent must not change appreciably on addition of the solute (i.e., the solution must be dilute).

12.34 (a) 3.00% $NaNO_3$ (b) 0.359 *M*

12.37 (a) 2.872 *M* (b) 3.089 *m*

12.42 23.7 torr

12.44 130 torr

12.46 (a) $X_{solute} = 0.1600$, $X_{solvent} = 0.8400$
(b) 0.0581 moles (c) 314 g/mol

12.50 (a) 22 moles (b) 1.2×10^3 mL (c) The proper ratio of ethylene glycol to water is 1.2 qt to 1 qt.

12.52 (a) 101 °C (b) -3.72 °C (c) 23.0 torr

12.54 24 g/mol

12.56 (a) 35.4 g (b) 59.9% (c) 70.54% C, 13.78% H, 15.68% O, $C_6H_{14}O$ (d) 102 g/mol, $C_6H_{14}O$

12.63 glucose

12.68 17.4 torr

12.70 Na_2CO_3

12.74 (a) -0.415 °C (b) This is due to the smaller size of the lithium cation, which is strongly hydrated.

12.76 12%

12.88 (a) 6.109 *m* (b) 21.97% (c) 26.26%

CHAPTER 13

Practice Exercises

1. $\Delta E° = -214.6$ kJ, % difference = 1.14%

2. (a) negative (b) positive

3. (a) negative (b) negative

4. (a) negative (b) negative (c) positive

5. (a) -229 J/K (b) -120.9 J/K

6. $\Delta G° = -1482$ kJ

7. (a) -69.7 kJ (b) 120.1 kJ

8. Maximum work obtainable is 788 kJ

9. 614 K or 341 °C

10. $\Delta G° = -70.0$ kJ; the reaction should be spontaneous.

11. $\Delta G° = +53.9$ kJ; no products should be observed in the reaction.

12. $\Delta G°_{473} = -29$ kJ; the position of equilibrium is shifted toward the products at the higher temperature.

Selected Review Exercises

13.8 $\Delta E = 1000$ J; endothermic

13.14 121 J

13.18 2.47×10^3 J

13.19 (a) $\Delta H° = +18.5$ kJ, $\Delta E = 13.5$ kJ
(b) $\Delta H° = -180$ kJ, $\Delta E = -178$ kJ
(c) $\Delta H° = -88$ kJ, $\Delta E = -83$ kJ
(d) $\Delta H° = 65.3$ kJ, $\Delta E = 65.3$ kJ

13.25 (a) $\Delta H° = -178$ kJ, favored (b) $\Delta H° = -311.42$ kJ, favored (c) $\Delta H° = +1084$ kJ, not favorable from the standpoint of enthalpy alone (d) $\Delta H° = +65.2$ kJ, not favored from the standpoint of enthalpy alone (e) $\Delta H° = +64.2$ kJ, not favored from the standpoint of enthalpy alone

13.29 (a) negative (b) negative (c) positive (d) negative (e) negative (f) positive

13.34 The probability of all heads is 1 in 16 or 1/16 = 0.0625. The probability of two heads and two tails is 6 in 16 or 6/16 = 0.375.

13.36 (a) negative (b) negative (c) negative (d) positive

13.41 (a) $\Delta S° = -198.3$ J/K, not spontaneous from the standpoint of entropy (b) $\Delta S° = -332.3$ J/K, not favored from the standpoint of entropy alone (c) $\Delta S° = +92.6$ J/K, favorable from the standpoint of entropy alone (d) $\Delta S° = +14$ J/K, favorable from the standpoint of entropy alone (e) $\Delta S° = +159$ J/K, favorable from the standpoint of entropy alone

13.43 (a) $\Delta S° = -52.8$ J mol^{-1} K^{-1} (b) $\Delta S° = -74.0$ J mol^{-1} K^{-1} (c) $\Delta S° = -90.1$ J mol^{-1} K^{-1} (d) $\Delta S° = -755.4$ J mol^{-1} K^{-1} (e) $\Delta S° = -318$ J mol^{-1} K^{-1}

13.45 $\Delta S° = -269.7$ J/K

13.50 $\Delta G°_f = -209$ kJ/mol

13.52 (a) $\Delta G° = -83$ kJ (b) $\Delta G° = -8.8$ kJ (c) $\Delta G° = +70.7$ kJ (d) $\Delta G° = -14.3$ kJ (e) $\Delta G° = -715$ kJ

13.54 $\Delta G° = -4.7$ kJ

13.56 $\Delta G° = +0.16$ kJ

13.58 $\Delta G = -19.09$ kJ

13.63 $\Delta G° = -1299.8$ kJ

13.65 Ethanol: 8.42×10^4 kJ/gallon; octane: 1.23×10^5 kJ/gallon. Octane is the better fuel.

13.71 $T_{eq} = 333$ K

13.73 $\Delta S = 101.2$ J mol^{-1} K^{-1}

13.75 $\Delta G° = -399.5$ kJ. Yes, the reaction is spontaneous.

13.80 Less product will be present at equilibrium.

13.84 $w = -15.0$ L atm

13.85 $w_1 = -10.0$ L atm, $w_2 = -10.0$ L atm, $w_{total} = -20.0$ L atm. The two-step expansion does more work than the one-step expansion from Review Exercise 13.84.

13.87 ΔH must be a larger positive quantity than $T\Delta S$, in order for ΔG to be positive.

13.92 1.16×10^3 kJ/mol

13.94 605 kJ/mol

13.96 577.7 kJ/mol

13.98 366.46 kJ/mol

13.100 588 kJ/mol

13.102 -149 kJ/mol

CHAPTER 14

Practice Exercises

1. Rate$(K_2O) = 0.12$ mol L^{-1} s^{-1}
    Rate$(N_2) = 0.12$ mol L^{-1} s^{-1}
    Rate$(KNO_3) = -0.24$ mol L^{-1} s^{-1}

2. You should obtain a value close to 9.4×10^{-5} mol L^{-1} s^{-1}.

3. (a) 8.0×10^{-2} (b) L mol^{-1} s^{-1}

4. Order with respect to [BrO$_3^-$] = 1; order with respect to [SO$_3^{2-}$] = 1; overall order = 2.

5. k has a constant value of 2.0×10^2 L^2 mol^{-2} s^{-1}.

6. first order

7. Rate = $k[A]^2[B]^2$

8. (a) 0.26 M (b) 77 min

9. 63 min

10. $t_{1/2} = 18.7$ min; 37.4 min

11. The reaction is first order.

12. (a) 1.4×10^2 kJ mol^{-1} (b) 0.30 L mol^{-1} s^{-1}

13. Rate = $k[NO]^1[O_3]^1$

14. Rate = $k[NO_2Cl]^1[Cl]^1$

Selected Review Exercises

14.15 In cool weather, the rates of metabolic reactions of cold-blooded insects decrease, because of the effect of temperature on rate.

14.17 The low temperature causes the rate of metabolism to be very low.

14.22 Rate at $t = 200$ min is 1.0×10^{-4} mol L^{-1} min^{-1}; rate at $t = 600$ min is 5.9×10^{-5} mol L^{-1} min^{-1}.

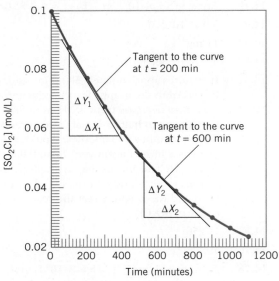

14.24 Rate of disappearance of $A = -0.60$ mol L^{-1} s^{-1}; rate of appearance of $C = 0.90$ mol L^{-1} s^{-1}.

14.26 (a) -2.5×10^{-6} mol L^{-1} s^{-1} (b) Rate of formation of NO$_2$ = 5.0×10^{-6} mol L^{-1} s^{-1}; rate of formation of O$_2$ = 1.3×10^{-6} mol L^{-1} s^{-1}.

14.35 2.4×10^2 mol L^{-1} s^{-1}

14.37 (a) For HCrO$_4^-$, the order is 1. For HSO$_3^-$, the order is 2. For H$^+$, the order is 1. (b) The overall order is $1 + 2 + 1 = 4$.

14.40 Rate = $k[M][N]^2$, $k = 2.5 \times 10^3$ L^2 mol^{-2} s^{-1}

14.42 Rate = $k[OCl^-][I^-]$, $k = 6.1 \times 10^9$ L mol^{-1} s^{-1}

14.44 $k[ICl][H_2]$, $k = 1.5 \times 10^{-1}$ L mol^{-1} s^{-1}

14.47

The data do yield a straight line when $\ln[SO_2Cl_2]_t$ is plotted against the time, t. The slope of this line equals $- k$. Plotting the data yields a value of 1.32×10^{-3} min^{-1} for k.

14.49 (a) $x = 3.7 \times 10^{-3}\ M$ (b) $x = 6.0 \times 10^{-4}\ M$

14.51 $5.0 \times 10^{-4}\ M$

14.53 11 mg/kg

14.59 4.3×10^{2} seconds

14.61 It requires approximately 500 min (as determined from the graph) for the concentration of SO_2Cl_2 to decrease from 0.100 M to 0.050 M (i.e., to decrease to half its initial concentration). Likewise, in another 500 minutes, the concentration decreases by half again (i.e., from 0.050 M to 0.025 M). This means that the half-life of the reaction is independent of the initial concentration, and we conclude that the reaction is first order in SO_2Cl_2.

14.63 1/256

14.76 79.3 kJ/mol

14.78 $E_a = 98.9$ kJ/mol, $A = 6.6 \times 10^9$ L mol^{-1} s^{-1}

14.80 78 kJ/mol

14.82 52.9 kJ/mol

14.84 2.3×10^{-4} s^{-1}

14.88 $2NO + 2H_2 \rightarrow N_2 + 2H_2O$

14.94 The predicted rate law is based on the rate-determining step: Rate $= k[NO_2]^2$

14.102 41.6 yr

14.104 (a) Rate $= k[A]^1[B]^2$ (b) $k = 8.33 \times 10^2$ L^2 mol^{-2} s^{-1}

CHAPTER 15

Practice Exercises

1. (a) $K_c = \dfrac{[H_2O]^2}{[H_2]^2[O_2]}$ (b) $K_c = \dfrac{[NH_4^+][OH^-]}{[NH_3][H_2O]}$

2. $K_c = 1.2 \times 10^{-13}$

3. 1.9×10^5

4. $K_p = \dfrac{P_{HI}^2}{P_{H_2}P_{I_2}}$

5. reaction (b)

6. The reaction will proceed to the right.

7. -33.3 kJ

8. 0.26

9. 3.84×10^4

10. $K_c = 57$

11. $K_p = 1.8 \times 10^{36}$

12. (a) $K_c = \dfrac{1}{[Cl_2(g)]}$ (b) $K_c = \dfrac{1}{[NH_3(g)][HCl(g)]}$
  (c) $K_c = [Ag^+]^2[CrO_4^{2-}]$

13. (a) shift to the right (b) shift to the left (c) shift to the left (d) shift to the right

14. $[O_2]$ decreases by 0.030 mol L^{-1}, $[CO]$ decreases by 0.060 mol L^{-1}, $[CO_2]$ increases by 0.060 mol L^{-1}.

15. $K_c - 4.06$

16. (a) $[PCl_3] = 0.200\ M$, $[Cl_2] = 0.100\ M$, $[PCl_5] = 0.000\ M$ (b) $[PCl_3]$ and $[Cl_2]$ decrease by 0.080 M, $\{PCl_5\}$ increases by 0.080 M (c) $[PCl_3] = 0.120\ M$, $[Cl_2] = 0.020\ M$, $[PCl_5] = 0.080\ M$ (d) $K_c = 33$

17. $8.98 \times 10^{-3}\ M$

18. $[H_2] = [I_2] = 0.044\ M$, $[HI] = 0.311\ M$

19. $1.1 \times 10^{-17}\ M$

Selected Review Exercises

15.6 (a) $K_c = \dfrac{[POCl_3]^2}{[PCl_3]^2[O_2]}$ (b) $K_c = \dfrac{[SO_2]^2[O_2]}{[SO_3]^2}$
  (c) $K_c = \dfrac{[NO]^2[H_2O]^2}{[N_2H_4][O_2]^2}$
  (d) $K_c = \dfrac{[NO_2]^2[H_2O]^8}{[N_2H_4][H_2O_2]^6}$
  (e) $K_c = \dfrac{[SO_2][HCl]^2}{[SOCl_2][H_2O]}$

15.8 (a) $K_c = \dfrac{[Ag(NH_3)_2^+]}{[Ag^+][NH_3]}$
  (b) $K_c = \dfrac{[Cd(SCN)_4^{2-}]}{[Cd^{2+}][SCN^-]^4}$
  (c) $K_c = \dfrac{[H_3O^+][ClO^-]}{[HClO]}$

15.10 (a) $K_c = \dfrac{[HCl]^2}{[H_2][Cl_2]}$ (b) $K_c = \dfrac{[HCl]}{[H_2]^{1/2}[Cl_2]^{1/2}}$
  K_c for reaction (b) is the square root of K_c for reaction (a).

15.13 (a) $K_p = \dfrac{(P_{POCl_3})^2}{(P_{PCl_3})^2(P_{O_2})}$ (b) $K_p = \dfrac{(P_{SO_2})^2(P_{O_2})}{(P_{SO_3})^2}$
  (c) $K_p = \dfrac{(P_{NO})^2(P_{H_2O})^2}{(P_{N_2H_4})(P_{O_2})^2}$
  (d) $K_p = \dfrac{(P_{NO_2})^2(P_{H_2O})^8}{(P_{N_2H_4})(P_{H_2O_2})^6}$
  (e) $K_p = \dfrac{(P_{SO_2})(P_{HCl})^2}{(P_{SOCl_2})(P_{H_2O})}$

15.17 (a) < (c) < (b)

15.23 (a) $K_p = 10^{263}$ (b) $K_p = 2.90 \times 10^{-25}$
  (c) $K_p = 4.96 \times 10^{77}$ (d) $K_p = 10^{219}$

15.25 Since the value of Q is less than the value of K, the system is not at equilibrium and must shift to the right to reach equilibrium.

15.27 $K_p = 8.000 \times 10^8$. This is a favorable reaction, since the equilibrium lies far to the side favoring products, and is worth studying as a method for methane production.

15.31 $\Delta G^\circ_{500} = 21.1$ kJ

15.33 $K_p = 1.82 \times 10^{40}$

15.36 $0.0375\ M$

15.38 (a) and (d)

15.39 $K_c = 11$

15.41 $K_p = 2.7 \times 10^{-2}$

15.43 $K_p = 5.4 \times 10^{-5}$

15.46 (a) $55.5\ M$ (b) $55.49\ M$ (c) $55.5\ M$

15.49 (a) $K_c = \dfrac{[CO]^2}{[O_2]}$ (b) $K_c = [H_2O][SO_2]$

(c) $K_c = \dfrac{[CH_4][CO_2]}{[H_2O]^2}$ (d) $K_c = \dfrac{[H_2O][CO_2]}{[HF]^2}$

15.51 $[HI] = 1.47 \times 10^{-12}\ M$, $[Cl_2] = 7.37 \times 10^{-13}\ M$

15.54 (a) The system shifts to the right to consume some of the added methane. (b) The system shifts to the left to consume some of the added hydrogen. (c) The system shifts to the right to make some more carbon disulfide. (d) The system shifts to the left to decrease the amount of gas. (e) The system shifts to the right to absorb some of the added heat.

15.56 (a) right (b) left (c) left (d) right (e) no effect (f) left

15.58 (a) increase (b) increase (c) increase (d) decrease

15.59 $K_c = 0.398$

15.60 (a) This is not the value of the equilibrium constant, and we conclude that the system is not at equilibrium. (b) Since the value of the reaction quotient for this system is larger than that of the equilibrium constant, the system must shift to the left to reach equilibrium.

15.62 (a) $P_{NO_2} = 0.187$ atm (b) $P_{total} = 0.437$ atm

15.64 $[CH_3OH] = 4.36 \times 10^{-3}\ M$

15.66 $[Cl_2] = [Br_2] = 0.0259\ M$, $[BrCl] = 0.0682\ M$

15.68 $[Cl_2] = 0.0147\ M$, $[Br_2] = 0.0247\ M$, $[BrCl] = 0.0504\ M$

15.70 $K_c = 0.0955$

15.72 $K_c = 0.915$

15.74 $[NO] = [SO_3] = 0.0878$ mol/L, $[NO_2] = [SO_2] = 0.062$ mol/L. Although these two systems reach different equilibrium positions, the equilibrium concentrations of the four substances in each experiment give the same value for the equilibrium constant when substituted into the mass action expression.

15.76 $[H_2] = [Cl_2] = 8.9 \times 10^{-19}\ M$

15.78 $[NO] = 8.9 \times 10^{-5}\ M$

15.80 $[N_2O] = [O_2] = 0.0391\ M$, $[NO_2] = [NO] = 0.0409\ M$

15.82 $[SO_2] = 0.0046\ M$, $[NO_2] = 0.00058\ M$, $[NO] = 0.01342\ M$, $[SO_3] = 0.0174\ M$

15.84 $[NO_2] = [SO_2] = 0.0281\ M$

15.86 $[CO] = [H_2O] = 0.200\ M$

15.88 (a) Equilibrium is unaffected by the addition of a solid. (b) Equilibrium is unaffected by the removal of a solid. (c) The equilibrium will shift to the left. (d) The equilibrium will shift to the right.

CHAPTER 16

Practice Exercises

1. $1.3 \times 10^{-9}\ M$, basic

2. pH = 3.44, pOH = 10.56, acidic

3. pH = 5.17

4. (a) $[H^+] = 1.3 \times 10^{-3}\ M$, acidic (b) $[H^+] = 1.4 \times 10^{-4}\ M$, acidic (c) $[H^+] = 1.5 \times 10^{-11}\ M$, basic (d) $[H^+] = 7.8 \times 10^{-5}\ M$, acidic (e) $[H^+] = 2.5 \times 10^{-12}\ M$, basic

5. pH = 11.70, $[H^+] = 2.0 \times 10^{-12}\ M$

6. (a) $HCHO_2 \rightleftharpoons H^+ + CHO_2^-$ $K_a = \dfrac{[H^+][CHO_2^-]}{[HCHO_2]}$

(b) $(CH_3)_2NH_2^+ \rightleftharpoons H^+ + (CH_3)_2NH$
$K_a = \dfrac{[H^+][(CH_3)_2NH]}{[(CH_3)_2NH_2^+]}$

(c) $H_2PO_4^- \rightleftharpoons H^+ + HPO_4^{2-}$ $K_a = \dfrac{[H^+][HPO_4^{2-}]}{[H_2PO_4^-]}$

7. (a) $(CH_3)_3N + H_2O \rightleftharpoons (CH_3)_3NH^+ + OH^-$
$K_b = \dfrac{[(CH_3)_3NH^+][OH^-]}{[(CH_3)_3N]}$

(b) $SO_3^{2-} + H_2O \rightleftharpoons HSO_3^- + OH^-$
$K_b = \dfrac{[HSO_3^-][OH^-]}{[SO_3^{2-}]}$

(c) $NH_2OH + H_2O \rightleftharpoons NH_3OH^+ + OH^-$
$K_b = \dfrac{[NH_3OH^+][OH^-]}{[NH_2OH]}$

8. 5.6×10^{-11}

9. $K_a = 1.7 \times 10^{-5}$, $pK_a = 4.78$

10. $K_b = 1.6 \times 10^{-6}$, $pK_b = 5.79$

11. $[H^+] = 8.4 \times 10^{-4}\ M$, pH = 3.08

12. pH = 8.59

13. pH = 5.36

14. basic

15. neutral

16. acidic

17. pH = 8.07

18. pH = 5.13

19. basic

20. pH = 10.79

21. pH = 3.92

22. pH = 9.36

23. ratio = $\dfrac{0.70}{1.00}$

24. pH = 4.84

25. (a) 2.37 (b) 3.74 (c) 4.22 (d) 8.22

Selected Review Exercises

16.3 $[H^+] = [OH^-] = 1.5 \times 10^{-7} M$, pH = pOH = 6.82, pK_w = pH + pOH. Water is neutral at this temperature because the concentration of the hydrogen ion is the same as the concentration of the hydroxide ion.

16.5 (a) $4.2 \times 10^{-12} M$ (b) $7.1 \times 10^{-10} M$
(c) $1.8 \times 10^{-6} M$ (d) $2.4 \times 10^{-2} M$

16.6 (a) $2.9 \times 10^{-7} M$ (b) $1.5 \times 10^{-12} M$
(c) $4.0 \times 10^{-2} M$ (d) $1.3 \times 10^{-10} M$

16.8 4.85

16.10 (a) $[H^+] = 7.2 \times 10^{-4} M$, $[OH^-] = 1.4 \times 10^{-11} M$ (b) $[H^+] = 1.7 \times 10^{-3} M$, $[OH^-] = 6.0 \times 10^{-12} M$ (c) $[H^+] = 5.6 \times 10^{-10} M$, $[OH^-] = 1.8 \times 10^{-5} M$ (d) $[H^+] = 5.8 \times 10^{-14} M$, $[OH^-] = 1.7 \times 10^{-1} M$ (e) $[H^+] = 2.0 \times 10^{-6} M$, $[OH^-] = 5.0 \times 10^{-9} M$

16.12 2.00

16.14 pOH = 0.83, pH = 13.17

16.16 $2.0 \times 10^{-3} M$

16.18 (a) $HNO_2 \rightleftharpoons H^+ + NO_2^-$
(b) $H_3PO_4 \rightleftharpoons H^+ + H_2PO_4^-$
(c) $HAsO_4^{2-} \rightleftharpoons H^+ + AsO_4^{3-}$
(d) $(CH_3)_3NH^+ \rightleftharpoons H^+ + (CH_3)_3N$

16.20 (a) $(CH_3)_3N + H_2O \rightleftharpoons (CH_3)_3NH^+ + OH^-$
(b) $AsO_4^{3-} + H_2O \rightleftharpoons HAsO_4^{2-} + OH^-$
(c) $NO_2^- + H_2O \rightleftharpoons HNO_2 + OH^-$
(d) $(CH_3)_2N_2H_2 + H_2O \rightleftharpoons (CH_3)_2N_2H_3^+ + OH^-$

16.22 $K_a = \dfrac{[H^+][C_6H_5CO_2^-]}{[C_6H_5CO_2H]}$

16.24 CN^-

16.26 1.0×10^{-4}

16.28 2.3×10^{-11}

16.30 (a) IO_3^-, 5.9×10^{-14} (b) weaker

16.32 1×10^{-3}, 3.0

16.34 1.1×10^{-7}, 6.96

16.36 1.74

16.38 5.89

16.40 10.26

16.42 $0.47 M$

16.44 The oxalate ion reacts with water to give a basic solution according to the equation $C_2O_4^{2-} + H_2O \rightleftharpoons HC_2O_4^- + OH^-$.

16.49 11.26

16.51 5.72

16.53 8.2 g

16.55 4.29

16.58 6.7%, 2.00

16.60 6.70

16.62 $6.0 \times 10^{-4} M$, 3.22

16.66 4.97

16.68 -0.23 pH units

16.70 (a) 9.51 (b) 9.50

16.72 0.19 pH units

16.74 8.8 g

16.76 0.96 g

16.78 4.786, 4.766, 2.32

16.84 8.23, Cresol red

16.87 (a) 2.8724 (b) 4.5686 (c) 4.7447 (d) 8.7236

16.90 1.000, $2.25 \times 10^{-5} M$

16.92 5.08

16.94 (a) 3.8 g $HC_2H_3O_2$, 18 g $NaC_2H_3O_2 \cdot 2H_2O$
(b) $0.25 M HC_2H_3O_2$, $0.60 M NaC_2H_3O_2$
(c) 12.602 (d) 1.398

16.95 Since K_b for CN^- is larger than K_a for NH_4^+, the NH_4CN solution will be basic.

16.97 630 mL NH_3

CHAPTER 17

Practice Exercises

1. $[H^+] = 2.6 \times 10^{-3} M$, pH = 2.58, $[C_6H_6O_6^{2-}] = 2.7 \times 10^{-12} M$

2. $8.0 \times 10^{-3} M$

3. pH = 10.24

4. (a) $K_{sp} = [Ba^{2+}][CrO_4^{2-}]$
(b) $K_{sp} = [Ag^+]^3[PO_4^{3-}]$

5. 3.60×10^{-8}

6. 1.0×10^{-10}

7. 3.8×10^{-8}

8. $7.1 \times 10^{-7} M$

9. $1.3 \times 10^{-4} M$

10. $4.2 \times 10^{-16} M$

11. $1.3 \times 10^{-35} M$. This is less than one formula unit per liter, so $Fe(OH)_3$ is essentially completely insoluble in the basic solution.

12. Ion product $= 7.5 \times 10^{-5}$, which is larger than K_{sp}; a precipitate forms.

13. Ion product $= 7.8 \times 10^{-13}$, which is smaller than K_{sp}; a precipitate will not form.

14. Ion product for $PbSO_4 = 5.0 \times 10^{-7} < K_{sp}$, no precipitate

15. Ion product for $PbCl_2 = 8.6 \times 10^{-6} < K_{sp}$, no precipitate

16. pH ≤ 2.9

Selected Review Exercises

17.1 (a) $H_3AsO_4 \rightleftharpoons H_2AsO_4^- + H^+$
$H_2AsO_4^- \rightleftharpoons HAsO_4^{2-} + H^+$
$HAsO_4^{2-} \rightleftharpoons AsO_4^{3-} + H^+$
(b) $K_1 = 6 \times 10^{-3}$, $K_2 = 1 \times 10^{-7}$, $K_3 = 3 \times 10^{-12}$
(c) $HAsO_4^{2-}$ is the stronger acid because it has the larger K_a (see Table 17.1).

17.3 (a) $H_2CO_3 \rightleftharpoons HCO_3^- + H^+$ $K_{a_1} = \dfrac{[HCO_3^-][H^+]}{[H_2CO_3]}$

$HCO_3^- \rightleftharpoons CO_3^{2-} + H^+$ $K_{a_2} = \dfrac{[CO_3^{2-}][H^+]}{[HCO_3^-]}$

(b) sulfurous acid

17.5 $0.095\ M$, $0.095\ M$, $2.6 \times 10^{-7}\ M$, 1.022

17.7 $0.0091\ M$, $0.0091\ M$, $4.3 \times 10^{-5}\ M$

17.9 $3 \times 10^{-4}\ M$, $4.7 \times 10^{-11}\ M$

17.11 $4.3 \times 10^{-5}\ M$

17.13 (a) $(NH_4)_2SO_4$, $NaHSO_4$ (b) Na_3PO_4, $LiHCO_3$, K_2HPO_4, K_2SO_3
(c) none

17.15 10.13

17.17 25 g

17.19 $0.10\ M$

17.22 (a) $K_{sp} = [Ca^{2+}][F^-]^2$ (b) $K_{sp} = [Ag^+]^2[CO_3^{2-}]$
(c) $K_{sp} = [Pb^{2+}][SO_4^{2-}]$
(d) $K_{sp} = [Fe^{3+}][OH^-]^3$
(e) $K_{sp} = [Pb^{2+}][I^-]^2$
(f) $K_{sp} = [Cu^{2+}][OH^-]^2$

17.26 (a) 1.05×10^{-5} moles (b) $1.05 \times 10^{-5}\ M$
(c) 1.10×10^{-10}

17.28 $4.79 \times 10^{-2}\ M$, 2.29×10^{-3}

17.30 8.0×10^{-7}

17.32 (a) $4.4 \times 10^{-4}\ M$ (b) $9.5 \times 10^{-6}\ M$
(c) $9.5 \times 10^{-7}\ M$
(d) $6.3 \times 10^{-7}\ M$

17.34 LiF is more soluble.

17.36 2.8×10^{-18}

17.38 $4.9 \times 10^{-3}\ M$

17.40 $1.3 \times 10^{-3}\ M$

17.42 $3.0 \times 10^{-11}\ M$

17.44 $1.8 \times 10^{-10}\ M$

17.46 Yes, since $Q > K_{sp}$, a precipitate will form.

17.48 Since $Q < K_{sp}$, no precipitate will form.

17.50 $1.5 \times 10^{-3}\ M$

17.55 At a $[H^+]$ greater than $0.045\ M$, PbS will precipitate and CoS will not. At a pH lower than 1.35, PbS will precipitate and CoS will not.

17.57 (a) $1.21 \times 10^{-4}\ M$ (b) $[Mg^{2+}] = 1.21 \times 10^{-4}\ M$, $[OH^-] = 2.42 \times 10^{-4}\ M$ (c) $K_{sp} = 7.09 \times 10^{-12}$

17.58 (a) 0.35 g AgCl (b) $[Ag^+] = 3.6 \times 10^{-2}\ M$, $[NO_3^-] = 6.0 \times 10^{-2}\ M$, $[Na^+] = 2.4 \times 10^{-2}\ M$, $[Cl^-] = 5.0 \times 10^{-9}\ M$ (c) 40%

17.60 If $[H^+] = 6.6 \times 10^{-5}\ M$ (pH = 4.18), $La_2(CO_3)_3$ will not precipitate but $PbCO_3$ will precipitate. At pH = 3.51 and below, neither carbonate will precipitate.

CHAPTER 18

Practice Exercises

1. Products are bromine and hydrogen gas, and the net cell reaction is
$$2H_2O + 2Br^-(aq) \rightarrow Br_2(aq) + H_2(g) + 2OH^-(aq)$$

2. 8.29×10^{-3} mol OH^-

3. 7.33 min

4. 3.67 A

5. anode: $Mg(s) \rightarrow Mg^{2+}(aq) + 2e^-$
cathode: $Fe^{2+}(aq) + 2e^- \rightarrow Fe(s)$
cell notation: $Mg|Mg^{2+}\|Fe^{2+}|Fe$

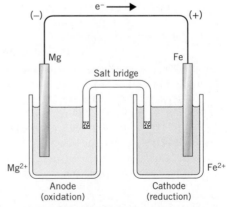

6. anode: $Al \rightarrow Al^{3+} + 3e^-$
cathode: $Pb^{2+} + 2e^- \rightarrow Pb$

7. $E^\circ_{Mg^{2+}} = -2.37$ V

8. $Br_2(aq) + H_2SO_3(aq) + H_2O \rightarrow 2Br^-(aq) + SO_4^{2-}(aq) + 4H^+(aq)$

9. $NiO_2(s) + Fe(s) + 2H_2O \rightarrow Ni(OH)_2(s) + Fe(OH)_2(s)$ $E^\circ_{cell} = 1.37$ V

10. $3MnO_4^-(aq) + 24H^+(aq) + 5Cr(s) \rightarrow$
$5Cr^{3+}(aq) + 3Mn^{2+}(aq) + 12H_2O \quad E°_{cell} = 2.25\ V$

11. Reaction (b); $E°_{cell} = +0.02\ V$

12. $\Delta G° = -372\ kJ$

13. $K_c = 2.9 \times 10^{-16}$

14. $E_{cell} = 1.04\ V$

15. Cu^{2+} concentrations: $1.9 \times 10^{-4}\ M$ and $6.9 \times 10^{-13}\ M$

16. $[Cr^{3+}] = 6.7 \times 10^{-4}\ M$

Selected Review Exercises

18.13   (a) $2H_2O(l) \rightarrow 4H^+(aq) + 4e^- + O_2(g)$
(b) $2Br^-(aq) \rightarrow Br_2(aq) + 2e^-$
(c) $2Br^-(aq) \rightarrow Br_2(aq) + 2e^-$

18.15 $Cu(s)$ and $Br_2(aq)$

18.17   (a) $2.40 \times 10^3\ C$ (b) $1.2 \times 10^4\ C$
(c) $3.24 \times 10^4\ C$

18.19   (a) $0.40\ mol\ e^-$ (b) $0.70\ mol\ e^-$
(c) $4.50\ mol\ e^-$ (d) $5.0 \times 10^{-2}\ mol\ e^-$

18.21 $0.0371\ mol$

18.23 $2.68\ g\ Fe(OH)_2$

18.25 $233\ mL\ HCl$

18.27 $6.04\ hr$

18.29 $5.08 \times 10^4\ A$

18.31   (a) $0.0280\ mol\ e^-$ (b) $9.33 \times 10^{-3}\ mol\ V$
(c) V^{3+}

18.45 $Mg(s) + Cu^{2+}(aq) \rightarrow Mg^{2+}(aq) + Cu(s)$, copper is the cathode.

18.48   (a) anode: $Cd(s) \rightarrow Cd^{2+}(aq) + 2e^-$
cathode: $Au^{3+}(aq) + 3e^- \rightarrow Au(s)$
cell: $3Cd(s) + 2Au^{3+}(aq) \rightarrow$
$3Cd^{2+}(aq) + 2Au(s)$
(b) anode: $Pb(s) + SO_4^{2-}(aq) \rightarrow PbSO_4(s) + 2e^-$
cathode: $PbO_2(s) + SO_4^{2-}(aq) +$
$4H^+(aq) + 2e^- \rightarrow Pb\,SO_4(s) + 2H_2O(l)$
cell: $Pb(s) + PbO_2(s) + 2SO_4^{2-}(aq) +$
$4H^+(aq) \rightarrow 2PbSO_4(s) + 2H_2O(l)$
(c) anode: $Cr(s) \rightarrow Cr^{3+}(aq) + 3e^-$
cathode: $Cu^{2+}(aq) + 2e^- \rightarrow Cu(s)$
cell: $2Cr(s) + 3Cu^{2+}(aq) \rightarrow$
$2Cr^{3+}(aq) + 3Cu(s)$

18.50   (a) $Fe(s)\,|\,Fe^{2+}(aq)\,\|\,Cd^{2+}(aq)\,|\,Cd(s)$
(b) $Pt(s), Cl^-(aq)\,|\,Cl_2(g)\,\|\,Br_2(aq)\,|\,Br^-(aq), Pt(s)$
(c) $Ag(s)\,|\,Ag^+(aq)\,\|\,Au^{3+}(aq)\,|\,Au(s)$

18.60 $-0.34\ V$

18.62   (a) Sn (b) Br^- (c) Zn (d) I^-

18.64   (a) Fe^{2+} (b) Ag (c) Ca (d) K

18.65   (a) $0.04\ V$ (b) $-0.29\ V$ (c) $0.62\ V$

18.67   (a) spontaneous   (b) not spontaneous
(c) spontaneous

18.70 $0.74\ V$

18.72   (a) $Zn(s)\,|\,Zn^{2+}(aq)\,\|\,Co^{2+}(aq)\,|\,Co(s)$
$E°_{cell} = 0.48\ V$
(b) $Mg(s)\,|\,Mg^{2+}(aq)\,\|\,Ni^{2+}(aq)\,|\,Ni(s)$
$E°_{cell} = 2.12\ V$
(c) $Sn(s)\,|\,Sn^{2+}(aq)\,\|\,Au^{3+}(aq)\,|\,Au(s)$
$E°_{cell} = 1.56\ V$

18.74

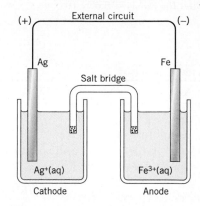

18.76 $BrO_3^-(aq) + 6I^-(aq) + 6H^+(aq) \rightarrow$
$3I_2(s) + Br^-(aq) + 3H_2O \quad E°_{cell} = 0.90\ V$

18.78 $4HOCl + 2H^+ + S_2O_3^{2-} \rightarrow$
$2Cl_2 + H_2O + 2H_2SO_3$

18.80   Since the overall cell potential is negative, we conclude that the reaction is not spontaneous in the direction written.

18.84 $626\ J$

18.86   (a) $E°_{cell} = 0.54\ V$ (b) $\Delta G° = -5.2 \times 10^2\ kJ$
(c) $K_c = 1.7 \times 10^{91}$

18.88 $\Delta G° = 1.0 \times 10^2\ kJ$

18.90 $K_c = 0.31$

18.94 $E_{cell} = 2.07\ V$

18.96 $E_{cell} = 1.54\ V$

18.98 $[Cl^-] = 0.348\ M$

18.100   (a) $\dfrac{(E_{cell} - E°_{cell})}{0.0592\ V} = -\log[H^+] = pH$
(b) $E_{cell} = 0.647\ V$ (c) $pH = 5.12$

18.110   Reduction of these two metal ions should take place at different applied voltages. First, Ni should "plate out" at $-0.25\ V$, followed by Cd at $-0.40\ V$.

18.112 $40.2\ mL\ HCl$

18.114 $E° = 1.118\ V$

18.115 $\Delta G° = -1.73 \times 10^2\ kJ$

CHAPTER 19

Practice Exercises

1. $[Ag(S_2O_3)_2]^{3-}$, $(NH_4)_3[Ag(S_2O_3)_2]$

2. $[Al(H_2O)_6]^{3+}$

3. (a) potassium hexacyanoferrate(III)
   (b) dichlorobis(ethylenediamine)chromium(III) sulfate

4. (a) $[SnCl_6]^{2-}$ (b) $(NH_4)_4[Fe(CN)_6]$

5. The coordination number is 6 in each of them.

6. In the NH_3 solution, molar solubility = 1.4×10^{-2} M; in pure water, molar solubility = 1.3×10^{-5} M.

7. 14 mol NH_3

Selected Review Exercises

19.1 The metal ion is a Lewis acid, accepting a pair of electrons from the ligand, which serves as a Lewis base.

    (a) The Lewis acid is Cu^{2+}, and the Lewis base is H_2O.
    (b) The ligand is H_2O.
    (c) Water provides the donor atom, oxygen.
    (d) Oxygen is the donor atom, because it is attached to the copper ion.
    (e) The copper ion is the acceptor.

19.12 The net charge is -1, and the formula is $[Co(EDTA)]^-$.

19.14 This is the ion $[Ag(NH_3)_2]^+$, whih can exist as the chloride salt $[Ag(NH_3)_2]Cl$.

19.15 The bonds to NH_3 are known to be stronger, because the test for copper(II) ion requires the ready displacement of water ligands by ammonia ligands to give the ion $[Cu(NH_3)_4]^{2+}$, which has a recognizable deep blue color.

19.18 (a) hexaamminenickel(II) ion (b) triammine-trichlorochromate(II) ion (c) hexanitrocobaltate(III) ion (d) diamminetetracyanomanganate(II) ion (e) tris(oxalato)ferrate(III) ion (f) diiodoargentate(I) ion (g) trisulfidostannate(IV) ion (h) diaquabis(ethylenediamine)cobalt(III) sulfate (i) pentaamminechlorochromium(III) sulfate (j) potassium tris(oxalato)cobaltate(III)

19.19 (a) $[Fe(H_2O)_4(CN)_2]^+$ (b) $Ni(NH_3)_4(C_2O_4)$
   (c) $[Fe(H_2O)_2(CN)_4]^-$ (d) $K_3[Mn(SCN)_6]$
   (e) $[CuCl_4]^{2-}$

19.24 The coordination number is 6, and the oxidation number of the iron atom is +2.

19.28 (a) The nitrogen atoms are the donor atoms.
   (b) This is $2 \times 3 = 6$.
   (c)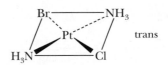

   (d) $Co(dien)_2^{3+}$, due to the chelate effect
   (e)
$$H_2\ddot{N}-CH_2CH_2-\ddot{N}H-CH_2CH_2-\ddot{N}H-CH_2CH_2-\ddot{N}H_2$$

19.32 Since both are the cis isomer, they are identical.

One can be superimposed on the other after simple rotation.

19.34 No. Neither isomer is chiral, because the mirror image may be superimposed on the original.

19.35 (a) $[Cr(Cl)(H_2O)_5]^{2+}$
   (b) $[Cr(Cl)(H_2O)_5]Cl_2 \cdot H_2O$
   (c)

$$\left[\begin{array}{c} OH_2 \\ H_2O \underset{H_2O}{\overset{\displaystyle |}{\underset{\displaystyle |}{Cr}}} \overset{OH_2}{\underset{Cl}{}} \end{array}\right]^{2+}$$

   (d) There is only one isomer.

19.37

Br———NH₃   cis
Cl———NH₃ (Pt)

Br———NH₃   trans
H₃N———Cl (Pt)

19.39

Cl——|—NH₃   Cl——|—NH₃
| Co | H₃N Co
Cl——|—NH₃   H₃N——|—Cl
   NH₃   Cl

19.42

$$\left[\begin{array}{c} O \\ O—|—O \\ Co \\ O———O \\ | \\ O \end{array}\right]^{3-} \quad \left[\begin{array}{c} O \\ O—|—O \\ Co \\ O———O \\ | \\ O \end{array}\right]^{3-}$$

(The curved lines represent C_2O_2 parts of the $C_2O_4^{2-}$ ligand.)

19.48

The oxidation of Co^{2+} to give Co^{3+} removes a high-energy electron. The energy separation, Δ, also increases as the oxidation state increases. This stabilizes the compound and explains the ease of oxidation.

19.49 (a) $Cr(H_2O)_6^{3+}$ (b) $Cr(en)_3^{3+}$

19.50 (a) The value of Δ increases down a group. Therefore, we choose $[Ru(NH_3)_5Cl]^{2+}$.
   (b) The value of Δ increases with oxidation state of the metal. Therefore, we choose $[Ru(NH_3)_6]^{3+}$.

19.52 $Cr(CN)_6^{3-} < Cr(NO_2)_6^{3-} < Cr(en)_3^{3+} < Cr(NH_3)_6^{3+} < Cr(H_2O)_6^{3+} < CrF_6^{3-} < CrCl_6^{3-}$

19.55 Ligand *A* produces the larger splitting. The absorbed colors are the complements of the perceived colors: CoA_6^{3+}, absorbed color is green; CoB_6^{3+}, absorbed color is red. The absorbed color with the highest energy (shortest wavelength) is green. We conclude that ligand *A* is higher in the spectrochemical series.

19.57 The color of CoA_6^{3+} is red. The complex with the lower oxidation state should have a smaller value for Δ. Therefore, the 2+ ion should absorb in the red, and thus appear blue.

19.59 $Cr(CN)_6^{3-}$

19.62 This is a weak-field complex of Co^{2+}, and it should be a high-spin d^7 case. It cannot be diamagnetic; even if it were low spin, we would still have one unpaired electron.

19.67 (a) $Cu^{2+} + 4Cl^- \rightleftharpoons CuCl_4^{2-}$ $K_{form} = \dfrac{[CuCl_4^{2-}]}{[Cu^{2+}][Cl^-]^4}$

(b) $Ag^+ + 2I^- \rightleftharpoons AgI_2^-$ $K_{form} = \dfrac{[AgI_2^-]}{[Ag^+][I^-]^2}$

(c) $Cr^{3+} + 6NH_3 \rightleftharpoons Cr(NH_3)_6^{3+}$

$K_{form} = \dfrac{[Cr(NH_3)_6^{3+}]}{[Cr^{3+}][NH_3]^6}$

19.69 $AgCl(s) \rightleftharpoons Ag^+ + Cl^-$, $Ag^+ + 2NH_3 \rightleftharpoons$ $Ag(NH_3)_2^+$. Silver chloride is an insoluble solid. However, any Ag^+ ions present react with added NH_3 to form the $Ag(NH_3)_2^+$ complex ion. According to Le Châtelier's principle, as NH_3 is added to a solution containing Ag^+ ions, the complex ion forms, using up the Ag^+ ions. This disrupts the equilibrium and forces AgCl to dissolve.

19.71 (a) $Co(NH_3)_6^{3+} \rightleftharpoons Co^{3+} + 6NH_3$

$K_{inst} = \dfrac{[Co^{3+}][NH_3]^6}{[Co(NH_3)_6^{3+}]}$

(b) $HgI_4^{2-} \rightleftharpoons Hg^{2+} + 4I^-$ $K_{inst} = \dfrac{[Hg^{2+}][I^-]^4}{[HgI_4^{2-}]}$

(c) $Fe(CN)_6^{4-} \rightleftharpoons Fe^{2+} + 6CN^-$

$K_{inst} = \dfrac{[Fe^{2+}][CN^-]^6}{[Fe(CN)_6^{4-}]}$

19.73 (a) $K_{form} = 18$ (b) Less stable

19.75 12.2 g AgBr

CHAPTER 20

Practice Exercises

1. $^{226}_{88}Ra \rightarrow {}^{222}_{86}Rn + {}^4_2He + {}^0_0\gamma$

2. $^{90}_{38}Sr \rightarrow {}^{90}_{39}Y + {}^{0}_{-1}e$

3. 100 units

Selected Review Exercises

20.4 (a) 1.01 kg (b) 3.91 kg (c) 12.3 kg

20.6 1.11×10^{-11} g

20.8 3.18 ng

20.10 1.8×10^{-13} J per nucleon

20.15 (a) $^{211}_{83}Bi$ (b) $^{177}_{72}Hf$ (c) $^{216}_{84}Po$ (d) $^{19}_{9}F$

20.17 (a) $^{242}_{94}Pu \rightarrow {}^4_2He + {}^{238}_{92}U$

(b) $^{28}_{12}Mg \rightarrow {}^{0}_{-1}e + {}^{28}_{13}Al$

(c) $^{26}_{14}Si \rightarrow {}^{0}_{1}e + {}^{26}_{13}Al$

(d) $^{37}_{18}Ar + {}^{0}_{-1}e \rightarrow {}^{37}_{17}Cl$

20.21 $^{58}_{26}Fe$

20.25 Barium-140 should have the longer half-life because it has an even number of protons and neutrons.

20.30 The neutron-to-proton ratio of lead-164 is too low.

20.34 The more likely process is positron emission, because this produces a product having a higher neutron-to-proton ratio: $^{38}_{19}K \rightarrow {}^{0}_{1}e + {}^{38}_{18}Ar$.

20.36 0.0469 mg

20.38 $^{53}_{24}Cr^*$; $^{51}_{23}V + {}^2_1H \rightarrow {}^{53}_{24}Cr^* \rightarrow {}^1_1p + {}^{52}_{23}V$

20.40 $^{80}_{35}Br$

20.42 $^{55}_{26}Fe$; $^{55}_{25}Mn + {}^1_1p \rightarrow {}^1_0n + {}^{55}_{26}Fe$

20.46 1.0 Ci, 3.7×10^{10} Bq

20.53 502 yr

20.55 1.80×10^{-6} Ci

20.58 1.3×10^9 yr. Assume that all of the argon-40 that is found in the rock must have come from the potassium-40 (i.e., that the rock contains no other source of argon-40).

20.60 2.7×10^4

20.63 $^3_1H + {}^2_1H \rightarrow {}^4_2He + {}^1_0n$

20.71 $^{140}_{54}Xe$

20.72 Both will be beta emitters. Both lie above the band of stability, and they can move closer to it by emitting beta particles.

CHAPTER 21

Practice Exercises

1. (a) 3-methylhexane (b) 4-ethyl-2,3-dimethylheptane (c) 5-ethyl-2,4,6-trimethyloctane

2. (a) $CH_3CH{=}O \xrightarrow[\text{oxidation}]{\text{further}} CH_3{-}\overset{\displaystyle O}{\overset{\|}{C}}{-}OH$

(b) $CH_3CH_2{-}\overset{\displaystyle O}{\overset{\|}{C}}{-}CH_2CH_3$

3.

$$-CH_2-CH-CH_2-CH-CH_2-CH-CH_2-CH-$$

(with $C≡N$ groups above each CH, and a bracketed repeating unit $-(CH_2-CH)_n-$ with $C≡N$ above)

Selected Review Exercises

21.1 (b), (d), and (e)

21.4 none are possible

21.5

(a)
$$H-\underset{\overset{|}{H}}{\overset{\overset{H}{|}}{C}}-\underset{\overset{|}{H}}{\overset{\overset{H}{|}}{N}}-H$$

(b)
$$Br-\underset{\overset{|}{H}}{\overset{\overset{H}{|}}{C}}-Br$$

(c)
$$H-\underset{\overset{|}{Cl}}{\overset{\overset{Cl}{|}}{C}}-Cl$$

(d)
$$H-\underset{\overset{|}{H}}{\overset{\overset{H}{|}}{C}}-\underset{\overset{|}{H}}{\overset{\overset{H}{|}}{C}}-H$$

(e)
$$H-\overset{\overset{O}{\|}}{C}-O-H$$

(f)
$$H-\overset{\overset{O}{\|}}{C}-H$$

(g)
$$H-\underset{\overset{|}{H}}{N}-O-H$$

(h) $H-C≡C-H$

(i)
$$H-\underset{\overset{|}{H}}{N}-\underset{\overset{|}{H}}{N}-H$$

(j) $H-C≡N$

(k)
$$H-\underset{\overset{|}{H}}{\overset{\overset{H}{|}}{C}}-C≡N$$

(l)
$$H-\underset{\overset{|}{H}}{\overset{\overset{H}{|}}{C}}-O-H$$

21.9 (a) alkene (b) alcohol (c) ester (d) carboxylic acid (e) amine (f) alcohol (g) alkyne (h) aldehyde (i) ketone (j) ether (k) amine (l) amide

21.10 The saturated compounds are (b), (e), (f), (j), and (k).

21.11 (a) identical (b) identical (c) unrelated (d) isomers (e) identical (f) identical (g) isomers (h) identical (i) isomers (j) isomers (k) identical (l) identical (m) isomers (n) unrelated

21.14 (a) C (b) A (c) Compound C has two OH groups, and it will have the greatest capacity for hydrogen bonding. Compound A has no polar groups and has only the very weak London forces.

21.15 Compound B is more soluble in water because of its capacity for hydrogen bonding.

21.17 (a) pentane (b) 2-methylpentane (c) 2,4-dimethylhexane (d) 2,4-dimethylhexane (e) 3-hexene (f) 4-methyl-2-pentene

21.20 No, the groups must be different in order to have cis and trans isomers. Isomerism is possible only for (e) and (f).

(e)

cis
$$\underset{H}{\overset{H_3C}{}}C=C\underset{CH_2CH_3}{\overset{CH_3}{}}$$

trans
$$\underset{H_3C}{\overset{H}{}}C=C\underset{CH_2CH_3}{\overset{CH_3}{}}$$

(f)

cis
$$\underset{CH_3CH_2}{\overset{Cl}{}}C=C\underset{CH_2CH_3}{\overset{Cl}{}}$$

trans
$$\underset{CH_3CH_2}{\overset{Cl}{}}C=C\underset{Cl}{\overset{CH_2CH_3}{}}$$

21.23 (a) $CH_3CH_2CH_2CH_3$

(b)
$$CH_3\underset{\overset{|}{Cl}}{CH}-\underset{\overset{|}{Cl}}{CH}CH_3$$

(c)
$$CH_3\underset{\overset{|}{Br}}{CH}-\underset{\overset{|}{Br}}{CH}CH_3$$

(d)
$$CH_3\underset{\overset{|}{H}}{CH}-\underset{\overset{|}{Cl}}{CH}CH_3$$

(e)
$$CH_3\underset{\overset{|}{H}}{CH}-\underset{\overset{|}{Br}}{CH}CH_3$$

(f)
$$CH_3\underset{\overset{|}{H}}{CH}-\underset{\overset{|}{OH}}{CH}CH_3$$

21.28
$CH_3CH_2CH_2CH_2OH$ 1-butanol
$CH_3CH_2CHOHCH_3$ 2-butanol
$(CH_3)_2CHCH_2OH$ 2-methyl-1-propanol
$(CH_3)_3COH$ 2-methyl-2-propanol

21.31 Those that can be oxidized to aldehydes are $CH_3CH_2CH_2CH_2OH$ and $(CH_3)_2CHCH_2OH$. One can be oxidized to a ketone: $CH_3CH_2CHOHCH_3$.

21.32 (a) (cyclopentanone) (b) (phenyl $-\overset{\overset{O}{\|}}{C}CH_3$)

(c) (phenyl $-CH_2\overset{\overset{H}{|}}{C}=O$)

21.36 $CH_2=CHCH_2CH_3$, $CH_3CH=CHCH_3$

21.39 (a) (cyclopentane with I) (b) (phenyl $-\underset{\overset{|}{I}}{CH}CH_3$)

(c) (phenyl $-CH_2CH_2I$)

21.42 (a) $CH_3CH_2CH_2NH_3{}^+Br^-$ (b) no reaction (c) no reaction (d) $CH_3CH_2CH_2NH_2 + H_2O$

21.47 (a)

$$CH_3CHCH_2CH$$ with CH_3 and O groups

(b)

$$CH_3CCH_2CHCH_2CH_2CH_2CH_3$$ with O and CH_3 groups

(c)

$$CH_3CHClC-OH$$ with O group

(d)

$$CH_3C-O-CH(CH_3)_2$$ with O group

(e)

$$CH_3CH_2CHC-NH_2$$ with O and CH_3 groups

21.50 (a) This is C, giving

$$CH_3CH_2C-OH$$

(b) This is B, giving

$$CH_3CH_2C-O^-$$

(c) This is B, giving

$$CH_3CH_2C-OCH_3$$

(d) This is A, giving

$$CH_3CH_2C-H$$

(e) This is A, giving $CH_3CH=CH_2$

21.51 $CH_3CH_2CO_2H + CH_3OH \rightleftharpoons$
$$CH_3CH_2CO_2CH_3 + H_2O$$

21.52 (a) $CH_3CH_2CO_2H$ (b) $CH_3CH_2CO_2H +$
CH_3OH (c) $Na^+ + CH_3CH_2CH_2CO_2^- + H_2O$
(d) $CH_3CH_2CONH_2 + H_2O$
(e) $CH_3CH_2CH_2CO_2H + NH_3$
(f) $Na^+ + CH_3CH_2CO_2^- + CH_3OH$

21.54 $CH_3CO_2H + CH_3CH_2NHCH_2CH_3$

21.55 (a) The amine neutralizes HCl:
$CH_3CH_2CH_2NH_2 + HCl \longrightarrow CH_3CH_2CH_2NH_3^+ + Cl^-$
(b) The amide is hydrolyzed:
$$CH_3CH_2CNH_2 + H_2O \longrightarrow CH_3CH_2COH + NH_3$$
(c) The alkylammonium cation neutralizes sodium hydroxide:
$CH_3CH_2NH_3^+ + OH^- \longrightarrow CH_3CH_2NH_2 + H_2O$

21.58
$$-CH_2-CH-CH_2-CH-CH_2-CH-CH_2-CH-$$ each with Cl
$$-(-CH_2CHCl-)_n$$

21.59
$$-OCCH_2CH_2COCH_2CH_2-$$ with O groups

21.61
$$-C-\bigcirc-C-HN(CH_2)_4NH-$$ with O groups

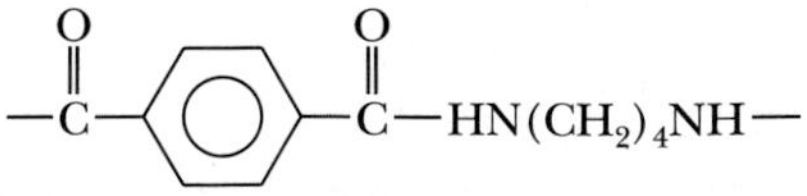

21.65 Each yields glucose on complete hydrolysis.

21.67 The digestion of sucrose yields glucose and fructose.

21.69 They each give glucose.

21.75 It is soluble in nonpolar solvents, and it is a substance found in living things.

21.77 No, because it has an odd number of carbon atoms.

21.79 These are the trialcohol $CH_2-CH-CH_2$ with OH OH OH and the three carboxylic acids:
$$HOC(CH_2)_7CH=CHCH_2CH=CH(CH_2)_4CH_3$$
$$HOC(CH_2)_{12}CH_3$$
$$HOC(CH_2)_7CH=CH(CH_2)_7CH_3$$

21.81 First, we have glycerol, $CH_2-CH-CH_2$ with OH OH OH, and then three anions, as follows.

(1) linoleate anion:
$$^-O-C(CH_2)_7CH=CHCH_2CH=CH(CH_2)_4CH_3$$
(2) myristate anion:
$$^-O-C(CH_2)_{12}CH_3$$
(3) oleate anion:
$$^-O-C(CH_2)_7CH=CH(CH_2)_7CH_3$$

21.86
$$NH_2CH_2C-NHCH_2COH$$ with O groups

21.88
$$NH_2CHC-NHCH_2COH$$ with $CH_2C_6H_5$ $NH_2CH_2C-NH-CHCOH$ with $CH_2C_6H_5$

21.100 The genetic messages are carried as a sequence of side-chain bases on a DNA segment.

21.102 This is shown in Figures 21.16 and 21.17. Only one particular base can be found opposite another in the double helix. Both structures also have sugar–phosphate backbones.

21.104 (a) In DNA, A pairs with T. (b) In RNA, A pairs with U. (c) C pairs with G.

21.106 Codons are found on mRNA.

21.107 Anticodons are found on tRNA.

21.109 This is hnRNA (also called ptRNA), which is used in making mRNA.

21.110 Transcription begins with DNA and ends with the synthesis of mRNA.

21.111 Translation begins with mRNA, uses tRNA, and ends with a specific polypeptide.

21.115 (a) $(CH_3)_2CHCH_2OH$ and $(CH_3)_3COH$
(b) The one that cannot be oxidized is $(CH_3)_3COH$.
(c)

$$CH_3—_+C—CH_3 \quad CH_3—CH—_+CH_2$$
(with CH$_3$ substituents above each)

(d) The first one is more stable, since it is the one that leads to the observed product.

21.117 (a) $^-O_2C—C_6H_4—CO_2^-$
(b) $CH_3CHOHCH_2CH_2CH_3$
(c) $CH_3NHCH_2CH_2CH_3$
(d) $CH_3CH_2OCH_2CH_2CO_2H + CH_3OH$
(e) C_5H_8OH (f) $C_6H_5—CH_2—CHBr—C_6H_5$

21.119 $C_{20}H_{30}O_9$

Appendix E

Tables of Selected Data

TABLE E.1   Thermodynamic Data for Selected Elements, Compounds, and Ions (at 25 °C)

Substance	ΔH_f° (kJ/mol)	S° (J/mol K)	ΔG_f° (kJ/mol)
Aluminum			
Al(s)	0	28.3	0
Al^{3+}(aq)	−524.7		−481.2
$AlCl_3$(s)	−704	110.7	−629
Al_2O_3(s)	−1676	51.0	−1576.4
$Al_2(SO_4)_3$(s)	−3441	239	−3100
Arsenic			
As(s)	0	35.1	0
AsH_3(g)	+66.4	223	+68.9
As_4O_6(s)	−1314	214	−1153
As_2O_5(s)	−925	105	−782
H_3AsO_3(aq)	−742.2		
H_3AsO_4(aq)	−902.5		
Barium			
Ba(s)	0	66.9	0
Ba^{2+}(aq)	−537.6	9.6	−560.8
$BaCO_3$(s)	−1219	112	−1139
$BaCrO_4$(s)	−1428.0		
$BaCl_2$(s)	−860.2	125	−810.8
BaO(s)	−553.5	70.4	−525.1
$Ba(OH)_2$(s)	−998.22	112	−875.3
$Ba(NO_3)_2$(s)	−992	214	−795
$BaSO_4$(s)	−1465	132	−1353
Beryllium			
Be(s)	0	9.50	0
$BeCl_2$(s)	−468.6	89.9	−426.3
BeO(s)	−611	14	−582
Bismuth			
Bi(s)	0	56.9	0
$BiCl_3$(s)	−379	177	−315
Bi_2O_3(s)	−576	151	−497
Boron			
B(s)	0	5.87	0
BCl_3(g)	−404	290	−389

TABLE E.1 **Thermodynamic Data for Selected Elements, Compounds, and Ions (at 25°C)** *(Continued)*

Substance	ΔH_f° (kJ/mol)	S° (J/mol K)	ΔG_f° (kJ/mol)
Boron			
$B_2H_6(g)$	+36	232	+87
$B_2O_3(s)$	−1273	53.8	−1194
$B(OH)_3(s)$	−1094	88.8	−969
Bromine			
$Br_2(l)$	0	152.2	0
$Br_2(g)$	+30.9	245.4	+3.11
$HBr(g)$	−36	198.5	+53.1
$Br^-(aq)$	−121.55	82.4	−103.96
Cadmium			
$Cd(s)$	0	51.8	0
$Cd^{2+}(aq)$	−75.90	−73.2	−77.61
$CdCl_2(s)$	−392	115	−344
$CdO(s)$	−258.2	54.8	−228.4
$CdS(s)$	−162	64.9	−156
$CdSO_4(s)$	−933.5	123	−822.6
Calcium			
$Ca(s)$	0	41.4	0
$Ca^{2+}(aq)$	−542.83	−53.1	−553.58
$CaCO_3(s)$	−1207	92.9	−1128.8
$CaF_2(s)$	−741	80.3	−1166
$CaCl_2(s)$	−795.8	114	−750.2
$CaBr_2(s)$	682.8	130	−663.6
$CaI_2(s)$	−535.9	143	
$CaO(s)$	−635.5	40	−604.2
$Ca(OH)_2(s)$	−986.6	76.1	−896.76
$Ca_3(PO_4)_2(s)$	−4119	241	−3852
$CaSO_3(s)$	−1156		
$CaSO_4(s)$	−1433	107	−1320.3
$CaSO_4 \cdot \frac{1}{2}H_2O(s)$	−1573	131	−1435.2
$CaSO_4 \cdot 2H_2O(s)$	−2020	194.0	−1795.7
Carbon			
$C(s, graphite)$	0	5.69	0
$C(s, diamond)$	+1.88	2.4	+2.9
$CCl_4(l)$	−134	214.4	−65.3
$CO(g)$	−110	197.9	−137.3
$CO_2(g)$	−394	213.6	−394.4
$CO_2(aq)$	−413.8	117.6	−385.98
$H_2CO_3(aq)$	−699.65	187.4	−623.08
$HCO_3^-(aq)$	−691.99	91.2	−586.77
$CO_3^{2-}(aq)$	−677.14	−56.9	−527.81
$CS_2(l)$	+89.5	151.3	+65.3
$CS_2(g)$	+117	237.7	+67.2
$HCN(g)$	+135.1	201.7	+124.7
$CN^-(aq)$	+150.6	94.1	+172.4
$CH_4(g)$	−74.9	186.2	−50.79
$C_2H_2(g)$	+227	200.8	+209
$C_2H_4(g)$	+51.9	219.8	+68.12

(continued)

TABLE E.1   Thermodynamic Data for Selected Elements, Compounds, and Ions (at 25°C)   (Continued)

Substance	ΔH_f° (kJ/mol)	S° (J/mol K)	ΔG_f° (kJ/mol)
Carbon			
$C_2H_6(g)$	−84.5	229.5	−32.9
$C_3H_8(g)$	−104	269.9	−23
$C_4H_{10}(g)$	−126	310.2	−17.0
$C_6H_6(l)$	+49.0	173.3	+124.3
$CH_3OH(l)$	−238	126.8	−166.2
$C_2H_5OH(l)$	−278	161	−174.8
$HCHO_2(g)$	−363	251	+335
$HC_2H_3O_2(l)$	−487.0	160	−392.5
$CH_2O(g)$	−108.6	218.8	−102.5
$CH_3CHO(g)$	−167	250	−129
$(CH_3)_2CO(l)$	−248.1	200.4	−155.4
$C_6H_5CO_2H(s)$	−385.1	167.6	−245.3
$CO(NH_2)_2(s)$	−333.5	104.6	−197.2
$CO(NH_2)_2(aq)$	−319.2	173.8	−203.8
$CH_2(NH_2)CO_2H(s)$	−532.9	103.5	−373.4
Chlorine			
$Cl_2(g)$	0	223.0	0
$Cl^-(aq)$	−167.2	56.5	−131.2
$HCl(g)$	−92.5	186.7	−95.27
$HCl(aq)$	−167.2	56.5	−131.2
$HClO(aq)$	−131.3	106.8	−80.21
Chromium			
$Cr(s)$	0	23.8	0
$Cr^{3+}(aq)$	−232		
$CrCl_2(s)$	−326	115	−282
$CrCl_3(s)$	−563.2	126	−493.7
$Cr_2O_3(s)$	−1141	81.2	−1059
$CrO_3(s)$	−585.8	72.0	−506.2
$(NH_4)_2Cr_2O_7(s)$	−1807		
$K_2Cr_2O_7(s)$	−2033.01		
Cobalt			
$Co(s)$	0	30.0	0
$Co^{2+}(aq)$	−59.4	−110	−53.6
$CoCl_2(s)$	−325.5	106	−282.4
$Co(NO_3)_2(s)$	−422.2	192	−230.5
$CoO(s)$	−237.9	53.0	−214.2
$CoS(s)$	−80.8	67.4	−82.8
Copper			
$Cu(s)$	0	33.15	0
$Cu^{2+}(aq)$	+64.77	−99.6	+65.49
$CuCl(s)$	−137.2	86.2	−119.87
$CuCl_2(s)$	−172	119	−131
$Cu_2O(s)$	−168.6	93.1	−146.0
$CuO(s)$	−155	42.6	−127
$Cu_2S(s)$	−79.5	121	−86.2
$CuS(s)$	−53.1	66.5	−53.6

TABLE E.1 **Thermodynamic Data for Selected Elements, Compounds, and Ions (at 25°C)** *(Continued)*

Substance	ΔH_f° (kJ/mol)	S° (J/mol K)	ΔG_f° (kJ/mol)
Copper			
$CuSO_4(s)$	-771.4	109	-661.8
$CuSO_4 \cdot 5H_2O(s)$	-2279.7	300.4	-1879.7
Fluorine			
$F_2(g)$	0	202.7	0
$F^-(aq)$	-332.6	-13.8	-278.8
$HF(g)$	-271	173.5	-273
Gold			
$Au(s)$	0	47.7	0
$Au_2O_3(s)$	$+80.8$	125	$+163$
$AuCl_3(s)$	-118	148	-48.5
Hydrogen			
$H_2(g)$	0	130.6	0
$H_2O(l)$	-285.9	69.96	-237.2
$H_2O(g)$	-241.8	188.7	-228.6
$H_2O_2(l)$	-187.8	109.6	-120.3
$H_2O_2(g)$	-136.3	233	-105.6
$H_2Se(g)$	$+76$	219	$+62.3$
$H_2Te(g)$	$+154$	234	$+138$
Iodine			
$I_2(s)$	0	116.1	0
$I_2(g)$	$+62.4$	260.7	$+19.3$
$HI(g)$	$+26$	206	$+1.30$
Iron			
$Fe(s)$	0	27	0
$Fe^{2+}(aq)$	-89.1	-137.7	-78.9
$Fe^{3+}(aq)$	-48.5	-315.9	-4.7
$Fe_2O_3(s)$	-822.2	90.0	-741.0
$Fe_3O_4(s)$	-1118.4	146.4	-1015.4
$FeS(s)$	-100.0	60.3	-100.4
$FeS_2(s)$	-178.2	52.9	-166.9
Lead			
$Pb(s)$	0	64.8	0
$Pb^{2+}(aq)$	-1.7	10.5	-24.4
$PbCl_2(s)$	-359.4	136	-314.1
$PbO(s)$	-217.3	68.7	-187.9
$PbO_2(s)$	-277	68.6	-219
$Pb(OH)_2(s)$	-515.9	88	-420.9
$PbS(s)$	-100	91.2	-98.7
$PbSO_4(s)$	-920.1	149	-811.3
Lithium			
$Li(s)$	0	28.4	0
$Li^+(aq)$	-278.6	10.3	
$LiF(s)$	-611.7	35.7	-583.3

(continued)

TABLE E.1 Thermodynamic Data for Selected Elements, Compounds, and Ions (at 25°C) *(Continued)*

Substance	ΔH_f° (kJ/mol)	S° (J/mol K)	ΔG_f° (kJ/mol)
Lithium			
LiCl(s)	−408	59.29	−383.7
LiBr(s)	−350.3	66.9	−338.87
Li$_2$O(s)	−596.5	37.9	−560.5
Li$_3$N(s)	−199	37.7	−155.4
Magnesium			
Mg(s)	0	32.5	0
Mg^{2+}(aq)	−466.9	−138.1	−454.8
MgCO$_3$(s)	−1113	65.7	−1029
MgF$_2$(s)	−1124	79.9	−1056
MgCl$_2$(s)	−641.8	89.5	−592.5
MgCl$_2 \cdot$ 2H$_2$O(s)	−1280	180	−1118
Mg$_3$N$_2$(s)	−463.2	87.9	−411
MgO(s)	−601.7	26.9	−569.4
Mg(OH)$_2$(s)	−924.7	63.1	−833.9
Manganese			
Mn(s)	0	32.0	0
Mn^{2+}(aq)	−223	−74.9	−228
MnO$_4^-$(aq)	−542.7	191	−449.4
KMnO$_4$(s)	−813.4	171.71	−713.8
MnO(s)	−385	60.2	−363
Mn$_2$O$_3$(s)	−959.8	110	−882.0
MnO$_2$(s)	−520.9	53.1	−466.1
Mn$_3$O$_4$(s)	−1387	149	−1280
MnSO$_4$(s)	−1064	112	−956
Mercury			
Hg(l)	0	76.1	0
Hg(g)	+61.32	175	+31.8
Hg$_2$Cl$_2$(s)	−265.2	192.5	−210.8
HgCl$_2$(s)	−224.3	146.0	−178.6
HgO(s)	−90.83	70.3	−58.54
HgS(s, red)	−58.2	82.4	−50.6
Nickel			
Ni(s)	0	30	0
NiCl$_2$(s)	−305	97.5	−259
NiO(s)	−244	38	−216
NiO$_2$(s)			−199
NiSO$_4$(s)	−891.2	77.8	−773.6
NiCO$_3$(s)	−664.0	91.6	−615.0
Ni(CO)$_4$(g)	−220	399	−567.4
Nitrogen			
N$_2$(g)	0	191.5	0
NH$_3$(g)	−46.0	192.5	−16.7
NH$_4^+$(aq)	−132.5	113	−79.37
N$_2$H$_4$(g)	+95.40	238.4	+159.3
N$_2$H$_4$(l)	+50.6	121.2	+149.4
NH$_4$Cl(s)	−314.4	94.6	−203.9

TABLE E.1 Thermodynamic Data for Selected Elements, Compounds, and Ions (at 25°C) *(Continued)*

Substance	ΔH_f° (kJ/mol)	S° (J/mol K)	ΔG_f° (kJ/mol)
Nitrogen			
$NO(g)$	+90.4	210.6	+86.69
$NO_2(g)$	+34	240.5	+51.84
$N_2O(g)$	+81.5	220.0	+103.6
$N_2O_4(g)$	+9.16	304	+98.28
$N_2O_5(g)$	+11	356	+115
$HNO_3(l)$	−174.1	155.6	−79.9
$NO_3^-(aq)$	−205.0	146.4	−108.74
Oxygen			
$O_2(g)$	0	205.0	0
$O_3(g)$	+143	238.8	+163
$OH^-(aq)$	−230.0	−10.75	−157.24
Phosphorus			
$P(s, \text{white})$	0	41.09	0
$P_4(g)$	+314.6	163.2	+278.3
$PCl_3(g)$	−287.0	311.8	−267.8
$PCl_5(g)$	−374.9	364.6	−305.0
$PH_3(g)$	+5.4	210.2	+12.9
$P_4O_6(s)$	−1640		
$POCl_3(g)$	−1109.7	646.5	−1019
$POCl_3(l)$	−1186	26.36	−1035
$P_4O_{10}(s)$	−2984	228.9	−2698
$H_3PO_4(s)$	−1279	110.5	−1119
Potassium			
$K(s)$	0	64.18	0
$K^+(aq)$	−252.4	102.5	−283.3
$KF(s)$	−567.3	66.6	−537.8
$KCl(s)$	−436.8	82.59	−408.3
$KBr(s)$	−393.8	95.9	−380.7
$KI(s)$	−327.9	106.3	−324.9
$KOH(s)$	−424.8	78.9	−379.1
$K_2O(s)$	−361	98.3	−322
$K_2SO_4(s)$	−1433.7	176	−1316.4
Silicon			
$Si(s)$	0	19	0
$SiH_4(g)$	+33	205	+52.3
$SiO_2(s, \text{alpha})$	−910.0	41.8	−856
Silver			
$Ag(s)$	0	42.55	0
$Ag^+(aq)$	+105.58	72.68	+77.11
$AgCl(s)$	−127.1	96.2	−109.8
$AgBr(s)$	−100.4	107.1	−96.9
$AgNO_3(s)$	−124	141	−32
$Ag_2O(s)$	−31.1	121.3	−11.2

(continued)

TABLE E.1 **Thermodynamic Data for Selected Elements, Compounds, and Ions (at 25°C)** *(Continued)*

Substance	ΔH_f° (kJ/mol)	S° (J/mol K)	ΔG_f° (kJ/mol)
Sodium			
Na(s)	0	51.0	0
Na$^+$(aq)	−240.12	59.0	−261.91
NaF(s)	−571	51.5	−545
NaCl(s)	−413	72.38	−384.0
NaBr(s)	−360	83.7	−349
NaI(s)	−288	91.2	−286
NaHCO$_3$(s)	−947.7	102	−851.9
Na$_2$CO$_3$(s)	−1131	136	−1048
Na$_2$O$_2$(s)	−510.9	94.6	−447.7
Na$_2$O(s)	−510	72.8	−376
NaOH(s)	−426.8	64.18	−382
Na$_2$SO$_4$(s)	−1384.49	149.49	−1266.83
Sulfur			
S(s, rhombic)	0	31.8	0
SO$_2$(g)	−297	248	−300
SO$_3$(g)	−396	256	−370
H$_2$S(g)	−20.6	206	−33.6
H$_2$SO$_4$(l)	−813.8	157	−689.9
H$_2$SO$_4$(aq)	−909.3	20.1	−744.5
SF$_6$(g)	−1209	292	−1105
Tin			
Sn(s, white)	0	51.6	0
Sn^{2+}(aq)	−8.8	−17	−27.2
SnCl$_4$(l)	−511.3	258.6	−440.2
SnO(s)	−285.8	56.5	−256.9
SnO$_2$(s)	−580.7	52.3	−519.6
Zinc			
Zn(s)	0	41.6	0
Zn^{2+}(aq)	−153.9	−112.1	−147.06
ZnCl$_2$(s)	−415.1	111	−369.4
ZnO(s)	−348.3	43.6	−318.3
ZnS(s)	−205.6	57.7	−201.3
ZnSO$_4$(s)	−982.8	120	−874.5

TABLE E.2 Heats of Formation of Gaseous Atoms from Elements in Their Standard States

Element	ΔH_f° (kJ mol^{-1})^a	Element	ΔH_f° (kJ mol^{-1})^a
Group IA		**Group IVA**	
H	217.89	C	716.67
Li	161.5	Si	450
Na	108.2	**Group VA**	
K	89.62	N	472.68
Rb	82.0	P	332.2
Cs	78.2		
Group IIA		**Group VIA**	
Be	324.3	O	249.17
Mg	146.4	S	276.98
Ca	178.2		
Sr	163.6	**Group VIIA**	
Ba	177.8	F	79.14
		Cl	121.47
Group IIIA		Br	112.38
		I	107.48
B	560		
Al	329.7		

^{a} All values in this table are positive because forming the gaseous atoms from the elements is endothermic; it involves bond breaking.

TABLE E.3 Average Bond Energies

Bond	Bond Energy (kJ mol^{-1})
C—C	348
C=C	612
C≡C	960
C—H	412
C—N	305
C=N	613
C≡N	890
C—O	360
C=O	743
C—F	484
C—Cl	338
C—Br	276
C—I	238
H—H	436
H—F	565
H—Cl	431
H—Br	366
H—I	299
H—N	388
H—O	463
H—S	338
H—Si	376

TABLE E.4 Vapor Pressure of Water as a Function of Temperature

Temp (°C)	Vapor pressure (torr)	Temp (°C)	Vapor pressure (torr)	Temp (°C)	Vapor pressure (torr)	Temp (°C)	Vapor pressure (torr)
0	4.58	26	25.2	52	102.1	78	327.3
1	4.93	27	26.7	53	107.2	79	341.0
2	5.29	28	28.3	54	112.5	80	355.1
3	5.68	29	30.0	55	118.0	81	369.7
4	6.10	30	31.8	56	123.8	82	384.9
5	6.54	31	33.7	57	129.8	83	400.6
6	7.01	32	35.7	58	136.1	84	416.8
7	7.51	33	37.7	59	142.6	85	433.6
8	8.04	34	39.9	60	149.4	86	450.9
9	8.61	35	41.2	61	156.4	87	468.7
10	9.21	36	44.6	62	163.8	88	487.1
11	9.84	37	47.1	63	171.4	89	506.1
12	10.5	38	49.7	64	179.3	90	525.8
13	11.2	39	52.4	65	187.5	91	546.0
14	12.0	40	55.3	66	196.1	92	567.0
15	12.8	41	58.3	67	205.0	93	588.6
16	13.6	42	61.5	68	214.2	94	610.9
17	14.5	43	64.8	69	223.7	95	633.9
18	15.5	44	68.3	70	233.7	96	657.6
19	16.5	45	71.9	71	243.9	97	682.1
20	17.5	46	75.6	72	254.6	98	707.3
21	18.6	47	79.6	73	265.7	99	733.2
22	19.8	48	83.7	74	277.2	100	760.0
23	21.1	49	88.0	75	289.1		
24	22.4	50	92.5	76	301.4		
25	23.8	51	97.2	77	314.1		

TABLE E.5 Solubility Product Constants

Salt	Solubility Equilibrium	K_{sp}
Fluorides		
MgF_2	$MgF_2(s) \rightleftharpoons Mg^{2+}(aq) + 2F^-(aq)$	6.6×10^{-9}
CaF_2	$CaF_2(s) \rightleftharpoons Ca^{2+}(aq) + 2F^-(aq)$	3.9×10^{-11}
SrF_2	$SrF_2(s) \rightleftharpoons Sr^{2+}(aq) + 2F^-(aq)$	2.9×10^{-9}
BaF_2	$BaF_2(s) \rightleftharpoons Ba^{2+}(aq) + 2F^-(aq)$	1.7×10^{-6}
LiF	$LiF(s) \rightleftharpoons Li^+(aq) + F^-(aq)$	1.7×10^{-3}
PbF_2	$PbF_2(s) \rightleftharpoons Pb^{2+}(aq) + 2F^-(aq)$	3.6×10^{-8}
Chlorides		
$CuCl$	$CuCl(s) \rightleftharpoons Cu^+(aq) + Cl^-(aq)$	1.9×10^{-7}
$AgCl$	$AgCl(s) \rightleftharpoons Ag^+(aq) + Cl^-(aq)$	1.8×10^{-10}
Hg_2Cl_2	$Hg_2Cl_2(s) \rightleftharpoons Hg_2^{2+}(aq) + 2Cl^-(aq)$	1.2×10^{-18}
$TlCl$	$TlCl(s) \rightleftharpoons Tl^+(aq) + Cl^-(aq)$	1.8×10^{-4}
$PbCl_2$	$PbCl_2(s) \rightleftharpoons Pb^{2+}(aq) + 2Cl^-(aq)$	1.7×10^{-5}
$AuCl_3$	$AuCl_3(s) \rightleftharpoons Au^{3+}(aq) + 3Cl^-(aq)$	3.2×10^{-25}

TABLE E.5   Solubility Product Constants   *(Continued)*

Salt	Solubility Equilibrium	K_{sp}

Bromides

CuBr	$CuBr(s) \rightleftharpoons Cu^+(aq) + Br^-(aq)$	5×10^{-9}
AgBr	$AgBr(s) \rightleftharpoons Ag^+(aq) + Br^-(aq)$	5.0×10^{-13}
Hg_2Br_2	$Hg_2Br_2(s) \rightleftharpoons Hg_2^{2+}(aq) + 2Br^-(aq)$	5.6×10^{-23}
$HgBr_2$	$HgBr_2(s) \rightleftharpoons Hg^{2+}(aq) + 2Br^-(aq)$	1.3×10^{-19}
$PbBr_2$	$PbBr_2(s) \rightleftharpoons Pb^{2+}(aq) + 2Br^-(aq)$	2.1×10^{-6}

Iodides

CuI	$CuI(s) \rightleftharpoons Cu^+(aq) + I^-(aq)$	1×10^{-12}
AgI	$AgI(s) \rightleftharpoons Ag^+(aq) + I^-(aq)$	8.3×10^{-17}
Hg_2I_2	$Hg_2I_2(s) \rightleftharpoons Hg_2^{2+}(aq) + 2I^-(aq)$	4.7×10^{-29}
HgI_2	$HgI_2(s) \rightleftharpoons Hg^{2+}(aq) + 2I^-(aq)$	1.1×10^{-28}
PbI_2	$PbI_2(s) \rightleftharpoons Pb^{2+}(aq) + 2I^-(aq)$	7.9×10^{-9}

Hydroxides

$Mg(OH)_2$	$Mg(OH)_2(s) \rightleftharpoons Mg^{2+}(aq) + 2OH^-(aq)$	7.1×10^{-12}
$Ca(OH)_2$	$Ca(OH)_2(s) \rightleftharpoons Ca^{2+}(aq) + 2OH^-(aq)$	6.5×10^{-6}
$Mn(OH)_2$	$Mn(OH)_2(s) \rightleftharpoons Mn^{2+}(aq) + 2OH^-(aq)$	1.6×10^{-13}
$Fe(OH)_2$	$Fe(OH)_2(s) \rightleftharpoons Fe^{2+}(aq) + 2OH^-(aq)$	7.9×10^{-16}
$Fe(OH)_3$	$Fe(OH)_3(s) \rightleftharpoons Fe^{3+}(aq) + 3OH^-(aq)$	1.6×10^{-39}
$Co(OH)_2$	$Co(OH)_2(s) \rightleftharpoons Co^{2+}(aq) + 2OH^-(aq)$	1×10^{-15}
$Co(OH)_3$	$Co(OH)_3(s) \rightleftharpoons Co^{3+}(aq) + 3OH^-(aq)$	3×10^{-45}
$Ni(OH)_2$	$Ni(OH)_2(s) \rightleftharpoons Ni^{2+}(aq) + 2OH^-(aq)$	6×10^{-16}
$Cu(OH)_2$	$Cu(OH)_2(s) \rightleftharpoons Cu^{2+}(aq) + 2OH^-(aq)$	4.8×10^{-20}
$V(OH)_3$	$V(OH)_3(s) \rightleftharpoons V^{3+}(aq) + 3OH^-(aq)$	4×10^{-35}
$Cr(OH)_3$	$Cr(OH)_3(s) \rightleftharpoons Cr^{3+}(aq) + 3OH^-(aq)$	2×10^{-30}
Ag_2O	$Ag_2O(s) + H_2O \rightleftharpoons 2Ag^+(aq) + 2OH^-(aq)$	1.9×10^{-8}
$Zn(OH)_2$	$Zn(OH)_2(s) \rightleftharpoons Zn^{2+}(aq) + 2OH^-(aq)$	3.0×10^{-16}
$Cd(OH)_2$	$Cd(OH)_2(s) \rightleftharpoons Cd^{2+}(aq) + 2OH^-(aq)$	5.0×10^{-15}
$Al(OH)_3$ (alpha form)	$Al(OH)_3(s) \rightleftharpoons Al^{3+}(aq) + 3OH^-(aq)$	3×10^{-34}

Cyanides

AgCN	$AgCN(s) \rightleftharpoons Ag^+(aq) + CN^-(aq)$	2.2×10^{-16}
$Zn(CN)_2$	$Zn(CN)_2(s) \rightleftharpoons Zn^{2+}(aq) + 2CN^-(aq)$	3×10^{-16}

Sulfites

$CaSO_3$	$CaSO_3(s) \rightleftharpoons Ca^{2+}(aq) + SO_3^{2-}(aq)$	3×10^{-7}
Ag_2SO_3	$Ag_2SO_3(s) \rightleftharpoons 2Ag^+(aq) + SO_3^{2-}(aq)$	1.5×10^{-14}
$BaSO_3$	$BaSO_3(s) \rightleftharpoons Ba^{2+}(aq) + SO_3^{2-}(aq)$	8×10^{-7}

Sulfates

$CaSO_4$	$CaSO_4(s) \rightleftharpoons Ca^{2+}(aq) + SO_4^{2-}(aq)$	2.4×10^{-5}
$SrSO_4$	$SrSO_4(s) \rightleftharpoons Sr^{2+}(aq) + SO_4^{2-}(aq)$	3.2×10^{-7}
$BaSO_4$	$BaSO_4(s) \rightleftharpoons Ba^{2+}(aq) + SO_4^{2-}(aq)$	1.1×10^{-10}
$RaSO_4$	$RaSO_4(s) \rightleftharpoons Ra^{2+}(aq) + SO_4^{2-}(aq)$	4.3×10^{-11}
Ag_2SO_4	$Ag_2SO_4(s) \rightleftharpoons 2Ag^+(aq) + SO_4^{2-}(aq)$	1.5×10^{-5}
Hg_2SO_4	$Hg_2SO_4(s) \rightleftharpoons Hg_2^{2+}(aq) + SO_4^{2-}(aq)$	7.4×10^{-7}
$PbSO_4$	$PbSO_4(s) \rightleftharpoons Pb^{2+}(aq) + SO_4^{2-}(aq)$	6.3×10^{-7}

(continued)

TABLE E.5   Solubility Product Constants   *(Continued)*

Salt	Solubility Equilibrium	K_{sp}
Chromates		
$BaCrO_4$	$BaCrO_4(s) \rightleftharpoons Ba^{2+}(aq) + CrO_4{}^{2-}(aq)$	2.1×10^{-10}
$CuCrO_4$	$CuCrO_4(s) \rightleftharpoons Cu^{2+}(aq) + CrO_4{}^{2-}(aq)$	3.6×10^{-6}
Ag_2CrO_4	$Ag_2CrO_4(s) \rightleftharpoons 2Ag^+(aq) + CrO_4{}^{2-}(aq)$	1.2×10^{-12}
Hg_2CrO_4	$Hg_2CrO_4(s) \rightleftharpoons Hg_2{}^{2+}(aq) + CrO_4{}^{2-}(aq)$	2.0×10^{-9}
$CaCrO_4$	$CaCrO_4(s) \rightleftharpoons Ca^{2+}(aq) + CrO_4{}^{2-}(aq)$	7.1×10^{-4}
$PbCrO_4$	$PbCrO_4(s) \rightleftharpoons Pb^{2+}(aq) + CrO_4{}^{2-}(aq)$	1.8×10^{-14}
Carbonates		
$MgCO_3$	$MgCO_3(s) \rightleftharpoons Mg^{2+}(aq) + CO_3{}^{2-}(aq)$	3.5×10^{-8}
$CaCO_3$	$CaCO_3(s) \rightleftharpoons Ca^{2+}(aq) + CO_3{}^{2-}(aq)$	4.5×10^{-9}
$SrCO_3$	$SrCO_3(s) \rightleftharpoons Sr^{2+}(aq) + CO_3{}^{2-}(aq)$	9.3×10^{-10}
$BaCO_3$	$BaCO_3(s) \rightleftharpoons Ba^{2+}(aq) + CO_3{}^{2-}(aq)$	5.0×10^{-9}
$MnCO_3$	$MnCO_3(s) \rightleftharpoons Mn^{2+}(aq) + CO_3{}^{2-}(aq)$	5.0×10^{-10}
$FeCO_3$	$FeCO_3(s) \rightleftharpoons Fe^{2+}(aq) + CO_3{}^{2-}(aq)$	2.1×10^{-11}
$CoCO_3$	$CoCO_3(s) \rightleftharpoons Co^{2+}(aq) + CO_3{}^{2-}(aq)$	1.0×10^{-10}
$NiCO_3$	$NiCO_3(s) \rightleftharpoons Ni^{2+}(aq) + CO_3{}^{2-}(aq)$	1.3×10^{-7}
$CuCO_3$	$CuCO_3(s) \rightleftharpoons Cu^{2+}(aq) + CO_3{}^{2-}(aq)$	2.3×10^{-10}
Ag_2CO_3	$Ag_2CO_3(s) \rightleftharpoons 2Ag^+(aq) + CO_3{}^{2-}(aq)$	8.1×10^{-12}
Hg_2CO_3	$Hg_2CO_3(s) \rightleftharpoons Hg_2{}^{2+}(aq) + CO_3{}^{2-}(aq)$	8.9×10^{-17}
$ZnCO_3$	$ZnCO_3(s) \rightleftharpoons Zn^{2+}(aq) + CO_3{}^{2-}(aq)$	1.0×10^{-10}
$CdCO_3$	$CdCO_3(s) \rightleftharpoons Cd^{2+}(aq) + CO_3{}^{2-}(aq)$	1.8×10^{-14}
$PbCO_3$	$PbCO_3(s) \rightleftharpoons Pb^{2+}(aq) + CO_3{}^{2-}(aq)$	7.4×10^{-14}
Phosphates		
$Mg_3(PO_4)_2$	$Mg_3(PO_4)_2(s) \rightleftharpoons 3Mg^{2+}(aq) + 2PO_4{}^{3-}(aq)$	6.3×10^{-26}
$SrHPO_4$	$SrHPO_4(s) \rightleftharpoons Sr^{2+}(aq) + HPO_4{}^{2-}(aq)$	1.2×10^{-7}
$BaHPO_4$	$BaHPO_4(s) \rightleftharpoons Ba^{2+}(aq) + HPO_4{}^{2-}(aq)$	4.0×10^{-8}
$LaPO_4$	$LaPO_4(s) \rightleftharpoons La^{3+}(aq) + PO_4{}^{3-}(aq)$	3.7×10^{-23}
$Fe_3(PO_4)_2$	$Fe_3(PO_4)_2(s) \rightleftharpoons 3Fe^{2+}(aq) + 2PO_4{}^{3-}(aq)$	1×10^{-36}
Ag_3PO_4	$Ag_3PO_4(s) \rightleftharpoons 3Ag^+(aq) + PO_4{}^{3-}(aq)$	2.8×10^{-18}
$FePO_4$	$FePO_4(s) \rightleftharpoons Fe^{3+}(aq) + PO_4{}^{3-}(aq)$	4.0×10^{-27}
$Zn_3(PO_4)_2$	$Zn_3(PO_4)_2(s) \rightleftharpoons 3Zn^{2+}(aq) + 2PO_4{}^{3-}(aq)$	5×10^{-36}
$Pb_3(PO_4)_2$	$Pb_3(PO_4)_2(s) \rightleftharpoons 3Pb^{2+}(aq) + 2PO_4{}^{3-}(aq)$	3.0×10^{-44}
$Ba_3(PO_4)_2$	$Ba_3(PO_4)_2(s) \rightleftharpoons 3Ba^{2+}(aq) + 2PO_4{}^{3-}(aq)$	5.8×10^{-38}
Ferrocyanides		
$Zn_2[Fe(CN)_6]$	$Zn_2[Fe(CN)_6](s) \rightleftharpoons 2Zn^{2+}(aq) + Fe(CN)_6{}^{4-}(aq)$	2.1×10^{-16}
$Cd_2[Fe(CN)_6]$	$Cd_2[Fe(CN)_6](s) \rightleftharpoons 2Cd^{2+}(aq) + Fe(CN)_6{}^{4-}(aq)$	4.2×10^{-18}
$Pb_2[Fe(CN)_6]$	$Pb_2[Fe(CN)_6](s) \rightleftharpoons 2Pb^{2+}(aq) + Fe(CN)_6{}^{4-}(aq)$	9.5×10^{-19}

TABLE E.6 Formation Constants of Complexes (25° C)

Halide Complexes

Complex Ion Equilibrium	K_f
$Al^{3+} + 6F^- \rightleftharpoons [AlF_6]^{3-}$	2.5×10^4
$Al^{3+} + 4F^- \rightleftharpoons [AlF_4]^-$	2.0×10^8
$Be^{2+} + 4F^- \rightleftharpoons [BeF_4]^{2-}$	1.3×10^{13}
$Sn^{4+} + 6F^- \rightleftharpoons [SnF_6]^{2-}$	1×10^{25}
$Cu^+ + 2Cl^- \rightleftharpoons [CuCl_2]^-$	3×10^5
$Ag^+ + 2Cl^- \rightleftharpoons [AgCl_2]^-$	1.8×10^5
$Pb^{2+} + 4Cl^- \rightleftharpoons [PbCl_4]^{2-}$	2.5×10^{15}
$Zn^{2+} + 4Cl^- \rightleftharpoons [ZnCl_4]^{2-}$	1.6
$Hg^{2+} + 4Cl^- \rightleftharpoons [HgCl_4]^{2-}$	5.0×10^{15}
$Cu^+ + 2Br^- \rightleftharpoons [CuBr_2]^-$	8×10^5
$Ag^+ + 2Br^- \rightleftharpoons [AgBr_2]^-$	1.7×10^7
$Hg^{2+} + 4Br^- \rightleftharpoons [HgBr_4]^{2-}$	1×10^{21}
$Cu^+ + 2I^- \rightleftharpoons [CuI_2]^-$	8×10^8
$Ag^+ + 2I^- \rightleftharpoons [AgI_2]^-$	1×10^{11}
$Pb^{2+} + 4I^- \rightleftharpoons [PbI_4]^{2-}$	3×10^4
$Hg^{2+} + 4I^- \rightleftharpoons [HgI_4]^{2-}$	1.9×10^{30}

Ammonia Complexes

Complex Ion Equilibrium	K_f
$Ag^+ + 2NH_3 \rightleftharpoons [Ag(NH_3)_2]^+$	1.6×10^7
$Zn^{2+} + 4NH_3 \rightleftharpoons [Zn(NH_3)_4]^{2+}$	7.8×10^8
$Cu^{2+} + 4NH_3 \rightleftharpoons [Cu(NH_3)_4]^{2+}$	1.1×10^{13}
$Hg^{2+} + 4NH_3 \rightleftharpoons [Hg(NH_3)_4]^{2+}$	1.8×10^{19}
$Co^{2+} + 6NH_3 \rightleftharpoons [Co(NH_3)_6]^{2+}$	5.0×10^4
$Co^{3+} + 6NH_3 \rightleftharpoons [Co(NH_3)_6]^{3+}$	4.6×10^{33}
$Cd^{2+} + 6NH_3 \rightleftharpoons [Cd(NH_3)_6]^{2+}$	2.6×10^5
$Ni^{2+} + 6NH_3 \rightleftharpoons [Ni(NH_3)_6]^{2+}$	2.0×10^8

Cyanide Complexes

Complex Ion Equilibrium	K_f
$Fe^{2+} + 6CN^- \rightleftharpoons [Fe(CN)_6]^{4-}$	1.0×10^{24}
$Fe^{3+} + 6CN^- \rightleftharpoons [Fe(CN)_6]^{3-}$	1.0×10^{31}
$Ag^+ + 2CN^- \rightleftharpoons [Ag(CN)_2]^-$	5.3×10^{18}
$Cu^+ + 2CN^- \rightleftharpoons [Cu(CN)_2]^-$	1.0×10^{16}
$Cd^{2+} + 4CN^- \rightleftharpoons [Cd(CN)_4]^{2-}$	7.7×10^{16}
$Au^+ + 2CN^- \rightleftharpoons [Au(CN)_2]^-$	2×10^{38}

Complexes with Other Monodentate Ligands

Complex Ion Equilibrium	K_f
Methylamine (CH_3NH_2)	
$Ag^+ + 2CH_3NH_2 \rightleftharpoons [Ag(CH_3NH_2)_2]^+$	7.8×10^6
Thiocyanate ion (SCN^-)	
$Cd^{2+} + 4SCN^- \rightleftharpoons [Cd(SCN)_4]^{2-}$	1×10^3
$Cu^{2+} + 2SCN^- \rightleftharpoons [Cu(SCN)_2]$	5.6×10^3
$Fe^{3+} + 3SCN^- \rightleftharpoons [Fe(SCN)_3]$	2×10^6
$Hg^{2+} + 4SCN^- \rightleftharpoons [Hg(SCN)_4]^{2-}$	5.0×10^{21}

Hydroxide Ion

Complex Ion Equilibrium	K_f
$Cu^{2+} + 4OH^- \rightleftharpoons [Cu(OH)_4]^{2-}$	1.3×10^{16}
$Zn^{2+} + 4OH^- \rightleftharpoons [Zn(OH)_4]^{2-}$	2×10^{20}

Complexes with Bidentate Ligands[a]

Complex Ion Equilibrium	K_f
$Mn^{2+} + 3\,en \rightleftharpoons [Mn(en)_3]^{2+}$	6.5×10^5
$Fe^{2+} + 3\,en \rightleftharpoons [Fe(en)_3]^{2+}$	5.2×10^9
$Co^{2+} + 3\,en \rightleftharpoons [Co(en)_3]^{2+}$	1.3×10^{14}
$Co^{3+} + 3\,en \rightleftharpoons [Co(en)_3]^{3+}$	4.8×10^{48}
$Ni^{2+} + 3\,en \rightleftharpoons [Ni(en)_3]^{2+}$	4.1×10^{17}
$Cu^{2+} + 2\,en \rightleftharpoons [Cu(en)_2]^{2+}$	3.5×10^{19}
$Mn^{2+} + 3\,bipy \rightleftharpoons [Mn(bipy)_3]^{2+}$	1×10^6
$Fe^{2+} + 3\,bipy \rightleftharpoons [Fe(bipy)_3]^{2+}$	1.6×10^{17}
$Ni^{2+} + 3\,bipy \rightleftharpoons [Ni(bipy)_3]^{2+}$	3.0×10^{20}
$Co^{2+} + 3\,bipy \rightleftharpoons [Co(bipy)_3]^{2+}$	8×10^{15}
$Mn^{2+} + 3\,phen \rightleftharpoons [Mn(phen)_3]^{2+}$	2×10^{10}
$Fe^{2+} + 3\,phen \rightleftharpoons [Fe(phen)_3]^{2+}$	1×10^{21}
$Co^{2+} + 3\,phen \rightleftharpoons [Co(phen)_3]^{2+}$	6×10^{19}
$Ni^{2+} + 3\,phen \rightleftharpoons [Ni(phen)_3]^{2+}$	2×10^{24}
$Co^{2+} + 3C_2O_4^{2-} \rightleftharpoons [Co(C_2O_4)_3]^{4-}$	4.5×10^6
$Fe^{3+} + 3C_2O_4^{2-} \rightleftharpoons [Fe(C_2O_4)_3]^{3-}$	3.3×10^{20}

Complexes of Other Ligands[a]

Complex Ion Equilibrium	K_f
$Zn^{2+} + EDTA^{4-} \rightleftharpoons [Zn(EDTA)]^{2-}$	3.8×10^{16}
$Mg^{2+} + 2NTA^{3-} \rightleftharpoons [Mg(NTA)_2]^{4-}$	1.6×10^{10}
$Ca^{2+} + 2NTA^{3-} \rightleftharpoons [Ca(NTA)_2]^{4-}$	3.2×10^{11}

[a] en, ethylenediamine; bipy, bipyridyl; phen, 1,10-phenanthroline; $EDTA^{4-}$, ethylendiaminetetraacetate ion; and NTA^{3-}, nitrilotriacetate ion.

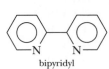

bipyridyl

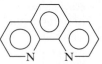

1,10-phenanthroline

TABLE E.7 Ionization Constants of Weak Acids and Bases (Alternative Formulas in Brackets)

Monoprotic Acids	Name	K_a
$HC_2O_2Cl_3$ [Cl_3CCO_2H]	trichloroacetic acid	2.2×10^{-1}
HIO_3	iodic acid	1.69×10^{-1}
$HC_2HO_2Cl_2$ [Cl_2CHCO_2H]	dichloroacetic acid	5.0×10^{-2}
$HC_2H_2O_2Cl$ [ClH_2CCO_2H]	chloroacetic acid	1.36×10^{-3}
HNO_2	nitrous acid	7.1×10^{-4}
HF	hydrofluoric acid	6.8×10^{-4}
$HOCN$	cyanic acid	3.5×10^{-4}
$HCHO_2$ [HCO_2H]	formic acid	1.8×10^{-4}
$HC_3H_5O_3$ [$CH_3CH(OH)CO_2H$]	lactic acid	1.38×10^{-4}
$HC_4H_3N_2O_3$	barbituric acid	9.8×10^{-5}
$HC_7H_5O_2$ [$C_6H_5CO_2H$]	benzoic acid	6.28×10^{-5}
$HC_4H_7O_2$ [$CH_3CH_2CH_2CO_2H$]	butanoic acid	1.52×10^{-5}
HN_3	hydrazoic acid	1.8×10^{-5}
$HC_2H_3O_2$ [CH_3CO_2H]	acetic acid	1.8×10^{-5}
$HC_3H_5O_2$ [$CH_3CH_2CO_2H$]	propanoic acid	1.34×10^{-5}
$HOCl$	hypochlorous acid	3.0×10^{-8}
$HOBr$	hypobromous acid	2.1×10^{-9}
HCN	hydrocyanic acid	6.2×10^{-10}
HC_6H_5O	phenol	1.3×10^{-10}
HOI	hypoiodous acid	2.3×10^{-11}
H_2O_2	hydrogen peroxide	1.8×10^{-12}

Polyprotic Acids	Name	K_{a_1}	K_{a_2}	K_{a_3}
H_2SO_4	sulfuric acid	Large	1.0×10^{-2}	
H_2CrO_4	chromic acid	5.0	1.5×10^{-6}	
$H_2C_2O_4$	oxalic acid	5.6×10^{-2}	5.4×10^{-5}	
H_3PO_3	phosphorous acid	3×10^{-2}	1.6×10^{-7}	
H_2SO_3	sulfurous acid	1.2×10^{-2}	6.6×10^{-8}	
H_2SeO_3	selenous acid	4.5×10^{-3}	1.1×10^{-8}	
H_2TeO_3	tellurous acid	3.3×10^{-3}	2.0×10^{-8}	
$H_2C_3H_2O_4$ [$HO_2CCH_2CO_2H$]	malonic acid	1.4×10^{-3}	2.0×10^{-6}	
$H_2C_8H_4O_4$	phthalic acid	1.1×10^{-3}	3.9×10^{-6}	
$H_2C_4H_4O_6$	tartaric acid	9.2×10^{-4}	4.3×10^{-5}	
$H_2C_6H_6O_6$	ascorbic acid	6.8×10^{-5}	2.7×10^{-12}	
H_2CO_3	carbonic acid	4.5×10^{-7}	4.7×10^{-11}	
H_3PO_4	phosphoric acid	7.1×10^{-3}	6.3×10^{-8}	4.5×10^{-13}
H_3AsO_4	arsenic acid	5.6×10^{-3}	1.7×10^{-7}	4.0×10^{-12}
$H_3C_6H_5O_7$	citric acid	7.1×10^{-4}	1.7×10^{-5}	6.3×10^{-6}

Weak Bases	Name	K_b
$(CH_3)_2NH$	dimethylamine	9.6×10^{-4}
CH_3NH_2	methylamine	4.4×10^{-4}
$CH_3CH_2NH_2$	ethylamine	4.3×10^{-4}
$(CH_3)_3N$	trimethylamine	7.4×10^{-5}
NH_3	ammonia	1.8×10^{-5}
N_2H_4	hydrazine	9.6×10^{-7}
NH_2OH	hydroxylamine	6.6×10^{-9}
C_5H_5N	pyridine	1.5×10^{-9}
$C_6H_5NH_2$	aniline	4.1×10^{-10}
PH_3	phosphine	1×10^{-28}

TABLE E.8 Standard Reduction Potentials (25 °C)

$E°$ (Volts)	Half-Cell Reaction
+ 2.87	$F_2(g) + 2e^- \rightleftharpoons 2F^-(aq)$
+ 2.08	$O_3(g) + 2H^+(aq) + 2e^- \rightleftharpoons O_2(g) + H_2O$
+ 2.05	$S_2O_8^{2-}(aq) + 2e^- \rightleftharpoons 2SO_4^{2-}(aq)$
+ 1.82	$Co^{3+}(aq) + e^- \rightleftharpoons Co^{2+}(aq)$
+ 1.77	$H_2O_2(aq) + 2H^+(aq) + 2e^- \rightleftharpoons 2H_2O$
+ 1.69	$PbO_2(s) + SO_4^{2-}(aq) + 4H^+(aq) + 2e^- \rightleftharpoons PbSO_4(s) + 2H_2O$
+ 1.63	$2HOCl(aq) + 2H^+(aq) + 2e^- \rightleftharpoons Cl_2(g) + 2H_2O$
+ 1.51	$Mn^{3+}(aq) + e^- \rightleftharpoons Mn^{2+}(aq)$
+ 1.507	$MnO_4^-(aq) + 4H^+(aq) + 3e^- \rightleftharpoons MnO_2(s) + 2H_2O$
+ 1.49	$MnO_4^-(aq) + 8H^+(aq) + 5e^- \rightleftharpoons Mn^{2+}(aq) + 4H_2O$
+ 1.46	$PbO_2(s) + 4H^+(aq) + 2e^- \rightleftharpoons Pb^{2+}(aq) + 2H_2O$
+ 1.44	$BrO_3^-(aq) + 6H^+(aq) + 6e^- \rightleftharpoons Br^-(aq) + 3H_2O$
+ 1.42	$Au^{3+}(aq) + 3e^- \rightleftharpoons Au(s)$
+ 1.36	$Cl_2(g) + 2e^- \rightleftharpoons 2Cl^-(aq)$
+ 1.33	$Cr_2O_7^{2-}(aq) + 14H^+(aq) + 6e^- \rightleftharpoons 2Cr^{3+}(aq) + 7H_2O$
+ 1.24	$O_3(g) + H_2O + 2e^- \rightleftharpoons O_2(g) + 2OH^-(aq)$
+ 1.23	$MnO_2(s) + 4H^+(aq) + 2e^- \rightleftharpoons Mn^{2+}(aq) + 2H_2O$
+ 1.23	$O_2(g) + 4H^+(aq) + 4e^- \rightleftharpoons 2H_2O$
+ 1.20	$Pt^{2+}(aq) + 2e^- \rightleftharpoons Pt(s)$
+ 1.07	$Br_2(aq) + 2e^- \rightleftharpoons 2Br^-(aq)$
+ 0.96	$NO_3^-(aq) + 4H^+(aq) + 3e^- \rightleftharpoons NO(g) + 2H_2O$
+ 0.94	$NO_3^-(aq) + 3H^+(aq) + 2e^- \rightleftharpoons HNO_2(aq) + H_2O$
+ 0.91	$2Hg^{2+}(aq) + 2e^- \rightleftharpoons Hg_2^{2+}(aq)$
+ 0.87	$HO_2^-(aq) + H_2O + 2e^- \rightleftharpoons 3OH^-(aq)$
+ 0.80	$NO_3^-(aq) + 4H^+(aq) + 2e^- \rightleftharpoons 2NO_2(g) + 2H_2O$
+ 0.80	$Ag^+(aq) + e^- \rightleftharpoons Ag(s)$
+ 0.77	$Fe^{3+}(aq) + e^- \rightleftharpoons Fe^{2+}(aq)$
+ 0.69	$O_2(g) + 2H^+(aq) + 2e^- \rightleftharpoons H_2O_2(aq)$
+ 0.54	$I_2(s) + 2e^- \rightleftharpoons 2I^-(aq)$
+ 0.49	$NiO_2(s) + 2H_2O + 2e^- \rightleftharpoons Ni(OH)_2(s) + 2OH^-(aq)$
+ 0.45	$SO_2(aq) + 4H^+(aq) + 4e^- \rightleftharpoons S(s) + 2H_2O$
+ 0.401	$O_2(g) + 2H_2O + 4e^- \rightleftharpoons 4OH^-(aq)$
+ 0.34	$Cu^{2+}(aq) + 2e^- \rightleftharpoons Cu(s)$
+ 0.27	$Hg_2Cl_2(s) + 2e^- \rightleftharpoons 2Hg(l) + 2Cl^-(aq)$
+ 0.25	$PbO_2(s) + H_2O + 2e^- \rightleftharpoons PbO(s) + 2OH^-(aq)$
+ 0.2223	$AgCl(s) + e^- \rightleftharpoons Ag(s) + Cl^-(aq)$
+ 0.172	$SO_4^{2-}(aq) + 4H^+(aq) + 2e^- \rightleftharpoons H_2SO_3(aq) + H_2O$
+ 0.169	$S_4O_6^{2-}(aq) + 2e^- \rightleftharpoons 2S_2O_3^{2-}(aq)$
+ 0.16	$Cu^{2+}(aq) + e^- \rightleftharpoons Cu^+(aq)$
+ 0.15	$Sn^{4+}(aq) + 2e^- \rightleftharpoons Sn^{2+}(aq)$
+ 0.14	$S(s) + 2H^+(aq) + 2e^- \rightleftharpoons H_2S(g)$
+ 0.07	$AgBr(s) + e^- \rightleftharpoons Ag(s) + Br^-(aq)$
0.00	$2H^+(aq) + 2e^- \rightleftharpoons H_2(g)$
− 0.13	$Pb^{2+}(aq) + 2e^- \rightleftharpoons Pb(s)$
− 0.14	$Sn^{2+}(aq) + 2e^- \rightleftharpoons Sn(s)$
− 0.15	$AgI(s) + e^- \rightleftharpoons Ag(s) + I^-(aq)$
− 0.25	$Ni^{2+}(aq) + 2e^- \rightleftharpoons Ni(s)$
− 0.28	$Co^{2+}(aq) + 2e^- \rightleftharpoons Co(s)$
− 0.34	$In^{3+}(aq) + 3e^- \rightleftharpoons In(s)$
− 0.34	$Tl^+(aq) + e^- \rightleftharpoons Tl(s)$
− 0.36	$PbSO_4(s) + 2e^- \rightleftharpoons Pb(s) + SO_4^{2-}(aq)$
− 0.40	$Cd^{2+}(aq) + 2e^- \rightleftharpoons Cd(s)$
− 0.44	$Fe^{2+}(aq) + 2e^- \rightleftharpoons Fe(s)$
− 0.56	$Ga^{3+}(aq) + 3e^- \rightleftharpoons Ga(s)$

(continued)

TABLE E.8   Standard Reduction Potentials (25 °C)   *(Continued)*

$E°$ (Volts)	Half-Cell Reaction
-0.58	$PbO(s) + H_2O + 2e^- \rightleftharpoons Pb(s) + 2OH^-(aq)$
-0.74	$Cr^{3+}(aq) + 3e^- \rightleftharpoons Cr(s)$
-0.76	$Zn^{2+}(aq) + 2e^- \rightleftharpoons Zn(s)$
-0.81	$Cd(OH)_2(s) + 2e^- \rightleftharpoons Cd(s) + 2OH^-(aq)$
-0.83	$2H_2O + 2e^- \rightleftharpoons H_2(g) + 2OH^-(aq)$
-0.88	$Fe(OH)_2(s) + 2e^- \rightleftharpoons Fe(s) + 2OH^-(aq)$
-0.91	$Cr^{2+}(aq) + 2e^- \rightleftharpoons Cr(s)$
-1.16	$N_2(g) + 4H_2O + 4e^- \rightleftharpoons N_2H_4(aq) + 4OH^-(aq)$
-1.18	$V^{2+}(aq) + 2e^- \rightleftharpoons V(s)$
-1.216	$ZnO_2^{-2}(aq) + 2H_2O + 2e^- \rightleftharpoons Zn(s) + 4OH^-(aq)$
-1.63	$Ti^{2+}(aq) + 2e^- \rightleftharpoons Ti(s)$
-1.66	$Al^{3+}(aq) + 3e^- \rightleftharpoons Al(s)$
-1.79	$U^{3+}(aq) + 3e^- \rightleftharpoons U(s)$
-2.02	$Sc^{3+}(aq) + 3e^- \rightleftharpoons Sc(s)$
-2.36	$La^{3+}(aq) + 3e^- \rightleftharpoons La(s)$
-2.37	$Y^{3+}(aq) + 3e^- \rightleftharpoons Y(s)$
-2.37	$Mg^{2+}(aq) + 2e^- \rightleftharpoons Mg(s)$
-2.71	$Na^+(aq) + e^- \rightleftharpoons Na(s)$
-2.76	$Ca^{2+}(aq) + 2e^- \rightleftharpoons Ca(s)$
-2.89	$Sr^{2+}(aq) + 2e^- \rightleftharpoons Sr(s)$
-2.90	$Ba^{2+}(aq) + 2e^- \rightleftharpoons Ba(s)$
-2.92	$Cs^+(aq) + e^- \rightleftharpoons Cs(s)$
-2.92	$K^+(aq) + e^- \rightleftharpoons K(s)$
-2.93	$Rb^+(aq) + e^- \rightleftharpoons Rb(s)$
-3.05	$Li^+(aq) + e^- \rightleftharpoons Li(s)$

Glossary

This Glossary has the definitions of the Key Terms that were marked in boldface throughout the chapters plus a few additional terms. The numbers in parentheses that follow the definitions are the numbers of the sections in which the glossary entries received their principal discussions.

A

Abscissa The horizontal axis. (Appendix A.3)

Absolute Zero 0 K. Nature's lowest temperature. (10.3)

Acceptor Ion The central metal ion in a complex ion. (19.1)

Accuracy Freedom from error. The closeness of a measurement to the true value. (1.4)

Acid *Arrhenius theory*—a substance that produces hydronium ions (hydrogen ions) in water. (2.9, 4.3)
Brønsted theory—a proton donor. (9.1)
Lewis theory—an electron-pair acceptor. (7.9)

Acid–Base Indicator A dye with one color in acid and another color in base. (4.11, 16.1, 16.8)

Acid–Base Neutralization The reaction of an acid with a base. (2.9, 7.9)

Acid Deposition The settling of acidic materials, whether they are gases, liquids or solids, onto surfaces; acid rain. (*Chemicals in Use* 10)

Acidic Anhydride An oxide that reacts with water to make the solution acidic. (4.3, 9.3)

Acidic Solution An aqueous solution in which $[H^+] > [OH^-]$. (16.1)

Acid Ionization Constant (K_a) $K_a = \dfrac{[H^+][A^-]}{[HA]}$ for the equilibrium,

$$HA \rightleftharpoons H^+ + A^- \quad (16.3)$$

Acid Rain Rain made acidic by dissolved sulfur and nitrogen oxides. (*Chemicals in Use* 10)

Acid Salt A salt of a partially neutralized polyprotic acid, for example, $NaHSO_4$ or $NaHCO_3$. (2.9)

Acid Solubility Product The special solubility product expression for metal sulfides in dilute acid and related to the equation for their dissolving. For a divalent metal sulfide, MS,

$$MS(s) + 2H^+(aq) \longrightarrow M^{2+}(aq) + H_2S(aq)$$

$$K_{spa} = \frac{[M^{2+}][H_2S]}{[H^+]^2} \quad (17.4)$$

Actinide Elements (Actinide Series) Elements 90–103. (2.4)

Activated Complex The chemical species that exists with partly broken and partly formed bonds in the transition state. (14.6)

Activation Energy The minimum kinetic energy that must be possessed by the reactants in order to give an effective collision (one that produces products). (14.6)

Activity Series A list of metals in their order of reactivity as reducing agents. (9.5)

Addition Reaction The addition of a molecule to a double or triple bond. (21.2)

Alcohol An organic compound whose molecules have the OH group attached to tetrahedral carbon. (21.3)

Aldehyde An organic compound whose molecules have the group —CH=O. (21.5)

Alkali Metals The Group IA elements (except hydrogen)—lithium, sodium, potassium, rubidium, and cesium. (2.4)

Alkaline Dry Cell A zinc–manganese dioxide galvanic cell of 1.54 V used commonly in flashlight batteries. (18.10)

Alkaline Earth Metals The Group IIA elements—beryllium, magnesium, calcium, strontium, barium, and radium. (2.4)

Alkalis The alkali metals. (2.4)

Alkane A hydrocarbon whose molecules have only single bonds. (2.6, 21.2)

Alkene A hydrocarbon whose molecules have one or more double bonds. (21.2)

Alkyl Group An organic group of carbon and hydrogen atoms related to an alkane but with one less hydrogen atom (e.g., CH_3—, methyl; CH_3CH_2—, ethyl). (21.2)

Alkyne A hydrocarbon whose molecules have one or more triple bonds. (21.2)

Alpha Particle (^4_2He) The nucleus of a helium atom. (20.3)

Alpha Radiation A high-velocity stream of alpha particles produced by radioactive decay. (20.3)

Amide An organic compound whose molecules have any one of the following groups. (21.5)

$$\overset{O}{\underset{\|}{}}\ \overset{O}{\underset{\|}{}}\ \overset{O}{\underset{\|}{}}$$
$$—CNH_2 \quad —CNHR \quad —CNR_2$$

Amine An organic compound whose molecules contain the group NH_2, NHR, or NR_2. (21.4)

α-Amino Acid One of 20 monomers of polypeptides. (21.10)

Amorphous Solid A noncrystalline solid. A glass. (11.13)

Ampere (A) The SI unit for electric current; one coulomb per second. (18.3)

Amphiprotic Compound A compound that can act either as a proton donor or a proton acceptor; an amphoteric compound. (9.1)

Amphoteric Compound A compound that can react as either an acid or a base. (9.1)

Angstrom (Å) $1 \text{ Å} = 10^{-10} \text{ m} = 100$ pm = 0.1 nm. (6.9)

Anion A negatively charged ion. (2.7)

Anode The positive electrode in a gas discharge tube. The electrode at which oxidation occurs during an electrochemical change. (18.2)

Antibonding Molecular Orbital A molecular orbital that denies electron density to the space between nuclei and destabilizes a molecule when occupied by electrons. (8.8)

Anticodon A triplet of bases on a tRNA molecule that pairs to a matching triplet—a codon—on an mRNA molecule during mRNA-directed polypeptide synthesis. (21.1)

Antilogarithm When $N = a^b$, N is the antilog of b. (Appendix A.2)

Antimatter Any particle annihilated by a particle of ordinary matter. (21.3)

Aqueous Solution A solution that has water as the solvent. (4.1)

Aromatic Compound An organic compound whose molecules have the benzene ring system. (21.2)

Arrhenius Acid See *Acid*.

Arrhenius Base See *Base*.

Arrhenius Equation An equation that relates the rate constant of a reaction to the reaction's activation energy. (14.7)

Association The joining together of molecules by hydrogen bonds. (12.9)

Atmosphere, Standard (Atm) 101,325 Pa. The pressure that supports a column of mercury 760 mm high at 0 °C; 760 torr. (5.5, 10.2)

Atmospheric Pressure The weight per unit area of the Earth's surface that is caused by the gases of the atmosphere. (10.2)

Atom A neutral particle having one nucleus; the smallest representative sample of an element. (2.1)

Atomic Mass The average mass (in u) of the atoms of the isotopes of a given element as they occur naturally. (2.2)

Atomic Mass Unit (u) $1.6605402 \times 10^{-24}$ g; 1/12 the mass of one atom of carbon-12. (2.2)

Atomic Number The number of protons in a nucleus. (2.3)

Atomic Spectrum The line spectrum produced when energized or excited atoms emit electromagnetic radiation. (6.2)

Atomic Weight See *Atomic Mass*.

Atomization Energy (ΔH_{atom}) The energy needed to rupture all of the bonds in one mole of a substance in the gas state and produce its atoms, also in the gas state. (Special Topic 13.1)

Aufbau Principle A set of rules enabling the construction of an electron structure of an atom from its atomic number. (6.5)

Avogadro's Number 6.0221367×10^{23}; the number of particles or formula units in one mole. (3.1)

Avogadro's Principle Equal volumes of gases contain equal numbers of molecules when they are at identical temperatures and pressures. (10.4)

Axial Bonds Covalent bonds oriented parallel to the vertical axis in a trigonal bipyramidal molecule. (8.1)

B

Background Radiation The atomic radiation from the natural radionuclides in the environment and from cosmic radiation. (20.6)

Balance An apparatus for measuring mass. (1.3)

Balanced Equation A chemical equation that has on opposite sides of the arrow the same number of each atom and the same net charge. (2.1, 3.5)

Band Gap The energy separation between the filled valence shell and the nearest conduction band (empty band). (8.9)

Band of Stability The envelope that encloses just the stable nuclides in a plot of all nuclides constructed according to their numbers of neutrons versus their numbers of protons. (20.4)

Band Theory of Solids An energy band of closely spaced energy levels exists in solids, some of which levels make up a valence band and others of which make up the (partially filled or empty) conduction band. (8.9)

Bar The standard pressure for thermodynamic quantities; 1 bar = 10^5 pascals. (13.6)

Barometer An apparatus for measuring atmospheric pressure. (10.2)

Base *Arrhenius theory*—a substance that releases OH^- ions in water. (2.9, 4.3)
Brønsted theory—a proton acceptor. (9.1)
Lewis theory—an electron pair donor. (7.9)

Base Ionization Constant, K_b $K_b = \dfrac{[BH^+][OH^-]}{[B]}$ for the equilibrium,

$$B + H_2O \rightleftharpoons BH^+ + OH^- \quad (16.3)$$

Base Units The units of the fundamental measurements of the SI. (1.3)

Basic Anhydride An oxide that can neutralize acid or that reacts with water to give OH^-. (4.3, 9.3)

Basic Solution An aqueous solution in which $[H^+] < [OH^-]$. (16.1)

Battery One or more galvanic cells arranged to serve as a practical source of electricity. (18.10)

Becquerel (Bq) 1 disintegration s^{-1}. The SI unit for the activity of a radioactive source. (20.6)

Bent Molecule A molecule that is nonlinear. (8.2)

Beta Particle ($_{-1}^{0}e$) An electron emitted by radioactive decay. (20.3)

Beta Radiation A stream of electrons produced by radioactive decay. (20.3)

Bidentate Ligand A ligand that has two atoms that can become simultaneously attached to the same metal ion. (19.1)

Binary Acid An acid with the general formula H_nX, where X is a nonmetal. (2.9, 9.2)

Binary Compound A compound

composed of two different elements. (2.7)

Binding Energy, Nuclear The energy equivalent of the difference in mass between an atomic nucleus and the sum of the masses of its nucleons. (20.2)

Biochemistry The study of the organic substances in organisms. (21.7)

Body-Centered Cubic (bcc) Unit Cell A unit cell having identical atoms, molecules, or ions at the corners of a cube plus one more particle in the center of the cube. (11.10)

Boiling Point The temperature at which the vapor pressure of the liquid equals the atmospheric pressure. (11.6)

Boiling Point Elevation A colligative property of a solution by which the solution's boiling point is higher than that of the pure solvent. (12.7)

Bond Angle The angle formed by two bonds that extend from the same atom. (8.1)

Bond Dipole A dipole within a molecule associated with a specific bond. (8.3)

Bond Dissociation Energy See *Bond Energy.*

Bond Distance See *Bond Length.*

Bond Energy The energy needed to break one mole of a particular bond to give electrically neutral fragments. (7.3, 7.7, 8.7)

Bonding Molecular Orbital A molecular orbital that introduces a buildup of electron density between nuclei and stabilizes a molecule when occupied by electrons. (8.8)

Bond Length The distance between two nuclei that are held by a chemical bond. (7.3, 7.7)

Bond Order The *net* number of pairs of bonding electrons.
Bond order = 1/2 × (No. of bonding e^- − No. of antibonding e^-). (7.7, 8.8)

Boundary That which divides a system from its surroundings. (5.3)

Boyle's Law See *Pressure–Volume Law.*

Bragg Equation $n\lambda = 2d \sin \theta$. The equation used to convert X-ray diffraction data into crystal structure. (11.11)

Branched-Chain Compounds An organic compound in whose molecules the carbon atoms do not all occur one after another in a continuous sequence. (21.2)

Branching Step A step in a chain reaction that produces more chain-propagating species than it consumes. (Special Topic 14.1)

Breeder Reactor A device in which a fissile isotope is continuously made from a fertile isotope. (Special Topic 20.3)

Brine An aqueous solution of sodium chloride, often with other salts. (18.4)

Brønsted Acid See *Acid.*

Brønsted Base See *Base.*

Brownian Movement The random, erratic motions of colloidally dispersed particles in a fluid. (12.10)

Buffer (a) A pair of solutes that can keep the pH of a solution almost constant if either acid or base is added. (b) A solution containing such a pair of solutes. (16.7)

Buret A long tube of glass usually marked in mL and 0.1 mL units and equipped with a stopcock for the controlled addition of a liquid to a receiving flask. (4.11)

By-product The substances formed by side reactions. (3.8)

C

Calorie (cal) 4.184 J. The energy that will raise the temperature of 1.00 g of water from 14.5 to 15.5 °C. (In popular books on foods, the term *Calorie,* with a capital C, means 1000 cal or 1 kcal.) (1.5)

Calorimeter An apparatus used in the determination of the heat of a reaction. (5.4)

Calorimetry The science of measuring the quantities of heat that are in-

volved in chemical or physical changes. (5.4)

Carbohydrates Polyhydroxyaldehydes or polyhydroxyketones or substances that yield these by hydrolysis and that are obtained from plants or animals. (21.8)

Carbonyl Group The carbon–oxygen double bond, C=O. (21.5)

Carboxylic Acids An organic compound whose molecules have the carboxyl group, CO_2H. (21.5)

Catalysis Rate enhancement caused by a catalyst. (14.2, 14.9)

Catalyst A substance that in relatively small proportion accelerates the rate of a reaction without being permanently chemically changed. (14.2, 14.9)

Cathode The negative electrode in a gas discharge tube. The electrode at which reduction occurs during an electrochemical change. (18.2)

Cathode Ray A stream of electrons ejected from a hot metal and accelerated toward a positively charged site in a vacuum tube. (Special Topic 2.1)

Cation A positively charged ion. (2.7)

Cell Potential, E_{cell} The emf of a galvanic cell when no current is drawn from the cell. (18.6)

Cell Reaction The overall chemical change that takes place in an electrolytic cell or a galvanic cell. (18.2)

Celsius Scale A temperature scale on which water freezes at 0 °C and boils at 100 °C (at 1 atm) and that has 100 divisions called Celsius degrees between these two points. (1.3)

Centimeter (cm) 0.01 m. (1.3)

Chain Reaction A self-sustaining change in which the products of one event cause one or more new events. (Special Topic 14.1, 20.9)

Change of State Transformation of matter from one physical state to another. In thermochemistry, any change in a variable used to define the state of a particular system—a change in composition, pressure, volume, or temperature. (11.4)

Charles' Law See *Temperature–Volume Law.*

Chelate A complex ion containing rings formed by polydentate ligands. (19.1)

Chelate Effect The extra stability found in complexes that contain chelate rings. (19.1)

Chemical Bond The force of electrical attraction that holds atoms together in compounds. (2.6, 7.1)

Chemical Change A change that converts substances into other substances; a chemical reaction. (1.2)

Chemical Energy The potential energy of chemicals that is transferred during chemical reactions. (1.5, 5.1)

Chemical Equation A before-and-after description that uses formulas and coefficients to represent a chemical reaction. (2.1, 3.5)

Chemical Equilibrium See *Dynamic Equilibrium.*

Chemical Property The ability of a substance, either by itself or with other substances, to undergo a change into new substances. (1.5)

Chemical Reaction A change in which new substances (products) form from starting materials (reactants). (1.2)

Chemical Symbol A formula for a substance. (2.1)

Chemistry The study of the compositions of substances and the ways by which their properties are related to their compositions. (1.2)

Chirality The "handedness" of an object; the property of an object (like a molecule) that makes it unable to be superimposed onto a model of its own mirror image. (19.4)

Cis Isomer A stereoisomer whose uniqueness is in having two groups on the same side of some reference plane. (19.4, 21.2)

Clausius–Clapeyron Equation The relationship between the vapor pressure, the temperature, and the molar heat of vaporization of a substance (where C is a constant).

$$\ln P = \frac{\Delta H_{vap}}{RT} + C \qquad (11.7)$$

Closed End Manometer See *Manometer.*

Codon An individual unit of hereditary instruction that consists of three side-by-side side chains on a molecule of mRNA. (21.11)

Coefficients Numbers in front of formulas in chemical equations. (2.1)

Colligative Property A property such as vapor pressure lowering, boiling point elevation, freezing point depression, and osmotic pressure whose physical value depends only on the ratio of the numbers of moles of solute and solvent particles and not on their chemical identities. (12.6)

Collision Theory The rate of a reaction is proportional to the number of collisions that occur each second between the reactants. (14.6)

Colloidal Dispersion (Colloid) A homogeneous mixture in which the particles of one of more components have at least one dimension in the range of 1 to 1000 nm—larger than those in a solution but smaller than those in a suspension. (12.10)

Combined Gas Law See *Gas Law, Combined.*

Combustion A rapid reaction with oxygen accompanied by a flame and the evolution of heat and light. (19.8)

Common Ion The ion in a mixture of ionic substances that is common to the formulas of at least two. (17.3)

Common Ion Effect The solubility of one salt is reduced by the presence of another having a common ion. (17.3)

Competing Reaction A reaction that reduces the yield of the main product by forming by-products. (3.8)

Complementary Color The color of the reflected or transmitted light when one component of white light is removed by absorption. (19.5)

Complex Ion (or simply a *Complex*) The combination of one or more anions or neutral molecules (ligands) with a metal ion. (19.1)

Compound A substance consisting of chemically combined atoms from two or more elements and present in a definite ratio. (2.1)

Compound Nucleus An atomic nucleus carrying excess energy following its capture of some bombarding particle. (20.5)

Compressibility Capability to undergo a reduction in volume under increasing pressure. (11.3)

Concentrated Solution A solution that has a large ratio of the amounts of solute to solvent. (4.1)

Concentration The ratio of the quantity of solute to the quantity of solution (or the quantity of solvent). (See *Molal Concentration, Molar Concentration, Mole Fraction, Normality, Percentage Concentration.*) (4.1)

Concentration Table A part of the strategy for organizing data needed to make certain calculations, particularly any involving equilibria. (15.10)

Condensation The change of a vapor to its liquid state. (11.4)

Conduction Band A vacant or partially filled but uninterrupted band in the valence band region of a solid. (8.10)

Conformation A particular relative orientation or geometric form of a flexible molecule. (8.5)

Conjugate Acid The species in a conjugate acid–base pair that has the greater number of H^+ units. (9.1, 16.3)

Conjugate Acid–Base Pair Two substances (ions or molecules) whose formulas differ by only one H^+ unit. (9.1, 16.3)

Conjugate Base The species in a conjugate acid–base pair that has the fewer number of H^+ units. (9.1, 16.3)

Conservation of Energy, Law of See *Law of Conservation of Energy.*

Conservation of Mass–Energy, Law of See *Law of Conservation of Mass–Energy.*

Continuous Spectrum The electromagnetic spectrum corresponding to the mixture of frequencies present in white light. (6.2)

Contributing Structure   One of a set of two or more Lewis structures used in applying the theory of resonance to the structure of a compound. A resonance structure. (7.8)

Conversion Factor   A ratio constructed from the relationship between two units such as 2.54 cm/1 in., from 1 in. = 2.54 cm. (1.4)

Cooling Curve   A graph showing how the temperature of a substance changes as heat is removed from it at a constant rate as the substance undergoes changes in its physical state. (11.7)

Coordinate Covalent Bond   A covalent bond in which both electrons originated from one of the joined atoms, but otherwise like a covalent bond in all respects. (7.9)

Coordination Complex   See *Coordination Compound.*

Coordination Compound   A complex or its salt. (19.1)

Coordination Number   The number of donor atoms that surround a metal ion. (19.3)

Copolymer   A polymer made from two or more different monomers. (21.6)

Core Electrons   The inner electrons of an atom that are not exposed to the electrons of other atoms when chemical bonds form. (6.5)

Corrosion   The slow oxidation of metals exposed to air or water. (Special Topic 18.1)

Coulomb (C)   The SI unit of electrical charge; the charge on 6.25×10^{18} electrons; the amount of charge that passes a fixed point of a wire conductor when a current of 1 A flows for 1 s. (18.3)

Covalent Bond   A chemical bond that results when atoms share electron pairs. (7.3)

Covalent Crystal   A crystal in which the lattice positions are occupied by atoms that are covalently bonded to the atoms at adjacent lattice sites. (11.12)

Critical Mass   The mass of a fissile isotope above which a self-sustaining chain reaction occurs. (20.9)

Critical Point   The point at the end of a vapor pressure versus temperature curve for a liquid and that corresponds to the critical pressure and the critical temperature. (11.9)

Critical Pressure (P_c)   The vapor pressure of a substance at its critical temperature. (11.9)

Critical Temperature (T_c)   The temperature above which a substance cannot exist as a liquid regardless of the pressure. (11.9)

Crystal Field Splitting (Δ)   The difference in energy between sets of d orbitals in a complex ion. (19.5)

Crystal Field Theory   A theory that considers the effects of the polarities or the charges of the ligands in a complex ion on the energies of the d orbitals of the central metal ion. (19.5)

Crystal Lattice   The repeating symmetrical pattern of atoms, molecules, or ions that occurs in a crystal. (11.10)

Cubic Meter (m^3)   The SI derived unit of volume. (1.3)

Curie (Ci)   A unit of activity for radioactive samples, equal to 3.7×10^{10} disintegrations s^{-1}. (20.6)

D

Dalton   One atomic mass unit, u. (2.2)

Dalton's Atomic Theory   Matter consists of tiny, indestructible particles called atoms. All atoms of one element are identical. The atoms of different elements have different masses. Atoms combine in definite ratios by atoms when they form compounds. (2.2)

Dalton's Law of Partial Pressures   See *Partial Pressures, Law of.*

Data   The information (often in the form of physical quantities) obtained in an experiment or other experience or from references. (1.2)

Decimal Multipliers   Factors—exponentials of 10 or decimals—that are

used to define larger or smaller SI units. (1.3)

Decomposition   A chemical reaction that changes one substance into two or more other substances. (2.1)

Delocalization Energy   The difference between the energy a substance would have if its molecules had no delocalized molecular orbitals and the energy it has because of such orbitals. (8.9)

Delocalized Molecular Orbital   A molecular orbital that spreads over more than two nuclei. (8.9)

Density   The ratio of an object's mass to its volume. (1.6)

Derived Unit   Any unit defined solely in terms of base units. (1.3)

Dialysis   The passage of small molecules and ions, but not species of a colloidal size, through a semipermeable membrane. (12.8)

Diamagnetism   The condition of not capable of being attracted to a magnet. (6.4)

Diaphragm Cell   An electrolytic cell used to manufacture sodium hydroxide by the electrolysis of aqueous sodium chloride. (18.4)

Diatomic Substance   A substance made from the atoms of only two elements. (2.7)

Diffraction   Constructive and destructive interference by waves. (6.3)

Diffraction Pattern   The image formed on a screen or a photographic film caused by the diffraction of electromagnetic radiation such as visible light or X rays. (11.11)

Diffusion   The spontaneous intermingling of one substance with another. (11.3)

Dilute Solution   A solution in which the ratio of the quantities of solute to solvent is small. (4.1)

Dimensional Analysis   See *Factor-Label Method.*

Dipole   Partial positive and partial negative charges separated by a distance. (7.5)

Dipole–Dipole Attractions   Attractions between molecules that are dipoles. (11.2)

Dipole Moment The product of the sizes of the partial charges in a dipole multiplied by the distance between them; a measure of the polarity of a molecule. (7.5)

Diprotic Acid An acid that can furnish two H^+ per molecule. (2.9)

Disaccharide A carbohydrate whose molecules can be hydrolyzed to two monosaccharides. (21.8)

Dissociation The separation of pre-existing ions when an ionic compound dissolves or melts. (4.2)

DNA Deoxyribonucleic acid; a nucleic acid that hydrolyzes to deoxyribose, phosphate ion, adenine, thymine, guanine, and cytosine, and that is the carrier of genes. (21.11)

DNA Double Helix Two oppositely running strands of DNA held in a helical configuration by interstrand hydrogen bonds. (21.11)

Donor Atom The atom on a ligand that makes an electron pair available in the formation of a complex. (19.1)

Doped Semiconductor A semiconductor chip (silicon) to which a trace amount of an element (like boron or arsenic) has been added to alter the conducting properties of the semiconductor. (8.10)

Double Bond A covalent bond consisting of one sigma bond and one pi bond. (7.3, 8.6)

Double Replacement Reaction (Metathesis Reaction) A reaction of two salts in which cations and anions exchange partners (e.g., $AgNO_3$ + $NaCl \rightarrow AgCl$ + $NaNO_3$). (4.5)

Downs Cell An electrolytic cell for the industrial production of sodium. (18.4)

Ductility A metal's ability to be drawn (or stretched) into wire. (2.4)

Dynamic Equilibrium A condition in which two opposing processes are occurring at equal rates. (4.3)

E

Effective Nuclear Charge The net positive charge an outer electron experiences as a result of the partial screening of the full nuclear charge by core electrons. (6.9)

Effusion The movement of a gas through a very tiny opening into a region of lower pressure. (10.7)

Effusion, Law of (Graham's Law) The rates of effusion of gases are inversely proportional to the square roots of their densities when compared at identical pressures and temperatures.

$$\text{Effusion rate} \propto \frac{1}{\sqrt{d}} \ (\text{constant } P, T)$$

where d is the gas density. (10.7)

Einstein Equation $\Delta E = \Delta m_0 c^2$ where ΔE is the energy obtained when a quantity of rest mass, Δm_0, is destroyed, or the energy lost when this quantity of mass is created. (20.1)

Elastomers Polymers with elastic properties. (21.6)

Electrochemical Change A chemical change that is caused by or that produces electricity. (18.1)

Electrochemistry The study of electrochemical changes. (18.1)

Electrolysis The production of a chemical change by the passage of electricity through a solution that contains ions or through a molten ionic compound. (18.1)

Electrolysis Cell An apparatus for electrolysis. (18.2, 18.5)

Electrolyte A compound that conducts electricity either in solution or in the molten state. (4.2)

Electrolytic Cell See *Electrolysis Cell.*

Electrolytic Conduction The transport of electrical charge by ions. (18.2)

Electrolyze To pass electricity through an electrolyte and cause a chemical change. (18.2)

Electromagnetic Energy Energy transmitted by wavelike oscillations in the strengths of electrical and magnetic fields; light energy. (6.1)

Electromagnetic Radiation The successive series of oscillations in the strengths of electrical and magnetic fields associated with light, microwaves, gamma rays, ultraviolet rays, infrared rays, and the like. (6.1)

Electromagnetic Spectrum The distribution of frequencies of electromagnetic radiation among various types of such radiation—microwave, infrared, visible, ultraviolet, X, and gamma rays. (6.1)

Electromotive Force (emf) The voltage produced by a galvanic cell and that can make electrons move in a conductor. (18.6)

Electron (e^- or $_{-1}^{0}e$) (a) A subatomic particle with a charge of $1-$ and mass of 0.0005485712 u ($9.1093897 \times 10^{-28}$ g) and that occurs outside an atomic nucleus. The particle that moves when an electric current flows. (2.3) (b) A beta particle. (20.3)

Electron Affinity The energy change (usually expressed in kJ mol^{-1}) that occurs when an electron adds to an isolated gaseous atom or ion. (6.9)

Electron Capture The capture by a nucleus of an orbital electron and that changes a proton into a neutron in the nucleus. (20.3)

Electron Cloud Because of its wave properties, an electron's influence spreads out like a cloud around the nucleus. (6.8)

Electron Configuration The distribution of electrons in an atom's orbitals. (6.5)

Electron Density The concentration of the electron's charge within a given volume. (6.8)

Electronegativity The relative ability of an atom to attract electron density toward itself when joined to another atom by a covalent bond. (7.5)

Electronic Structure The distribution of electrons in an atom's orbitals. (6.1, 6.5)

Electron-Pair Bond A covalent bond. (7.3)

Electron Spin The spinning of an electron about its axis that is believed to occur because the electron behaves as a tiny magnet. (6.4)

Electron Volt (eV) The energy an

electron receives when it is accelerated under the influence of 1 V and equal to 1.6×10^{-19} J. (20.3)

Electroplating Depositing a thin metallic coating on an object by electrolysis. (18.4)

Element A substance in which all of the atoms have the same atomic number. A substance that cannot be broken down by chemical reactions into anything that is both stable and simpler. (2.1)

Elementary Process One of the individual steps in the mechanism of a reaction. (14.8)

Elimination Reaction The loss of a small molecule from a larger molecule as in the elimination of water from an alcohol. (21.3)

Emission Spectrum See *Atomic Spectrum.*

Empirical Facts Facts discovered by performing experiments. (1.2)

Empirical Formula A chemical formula that uses the smallest whole-number subscripts to give the proportions by atoms of the different elements present. (3.4)

Emulsifying Agent A substance that stabilizes an emulsion. (12.10)

Emulsion A colloidal dispersion of one liquid in another. (12.10)

Enantiomers Stereoisomers whose molecular structures are related as an object to its mirror image but that cannot be superimposed. (19.4)

Endergonic Descriptive of a change in which a system's free energy increases. (13.6)

Endothermic Descriptive of a change in which a system's internal energy increases. (5.3, 5.5)

End Point The moment in a titration when the indicator changes color and the titration is ended. (4.11, 16.8)

Energy Something that matter possesses by virtue of an ability to do work. (1.5, 5.1)

Energy Band A large number of closely spaced energy levels in a solid formed by combining atomic orbitals of similar energy from each of the atoms in the solid. (8.10)

Energy Level A particular energy an electron can have in an atom or a molecule. (6.2)

Enthalpy (H) The heat content of a system. (5.5, 13.2)

Enthalpy Change (ΔH) The difference in enthalpy between the initial state and the final state for some change. (5.5, 13.2)

Enthalpy of Solution See *Heat of Solution.*

Entropy (S) The thermodynamic quantity that describes the degree of randomness of a system. The greater the disorder or randomness, the higher is the statistical probability of the state and the higher is the entropy. (13.4)

Entropy Change (ΔS) For a change, the sum of the values of S for the products minus the sum of the values of S for the reactants. (13.5)

Enzyme A catalyst in a living system. (21.10)

Equation of State of an Ideal Gas See *Gas Law, Ideal.*

Equatorial Bond A covalent bond located in the plane perpendicular to the long axis of a trigonal bipyramidal molecule. (8.1)

Equilibrium See *Dynamic Equilibrium.*

Equilibrium Constant The value that the mass action expression has when the system is at equilibrium. (15.3)

Equilibrium Law The mathematical equation for a particular equilibrium system that sets the mass action expression equal to the equilibrium constant. (15.3)

Equilibrium Vapor Pressure of a Liquid The pressure exerted by a vapor in equilibrium with its liquid state. (11.5)

Equilibrium Vapor Pressure of a Solid The pressure exerted by a vapor in equilibrium with its solid state. (11.5)

Equivalence See *Stoichiometric Equivalence.* (3.1)

Equivalence Point The moment in a titration when the number of equivalents of the reactant added from a buret equals the number of equivalents of another reactant in the receiving flask. (16.8)

Equivalent (eq) (a) The amount of an acid (or base) that provides 1 mol of H^+ (or 1 mole of OH^-). (b) The amount of an oxidizing agent (or a reducing agent) that can accept (or provide) 1 mole of electrons in a particular redox reaction. (Appendix C)

Equivalent Weight The mass in grams of one equivalent. (Appendix C)

Ester An organic compound whose molecules have the ester group. (21.5)

$$\underset{\text{Ester group}}{-\overset{\overset{\displaystyle O}{\|}}{C}-O-C}$$

Ether An organic compound in whose molecules two hydrocarbon groups are joined to an oxygen. (21.3)

Evaporate To change from a liquid to a vapor. (11.3)

Exact Number A number obtained by a direct count or that results by a definition and that is considered to have an infinite number of significant figures. (1.4)

Exergonic Descriptive of a change in a system in which there is a free energy decrease. (13.6)

Exon One of a set of sections of a DNA molecule (separated by introns) that, taken together, constitute a gene. (21.11)

Exothermic Descriptive of a change in which energy leaves a system and enters the surroundings. (5.3, 5.5)

Exponential Notation See *Scientific Notation.*

Extensive Property A property of an object that is described by a physical quantity whose magnitude is proportional to the size or amount of the object (e.g., mass or volume). (1.5)

F

Face-Centered Cubic (fcc) Unit Cell A unit cell having identical atoms, molecules, or ions at the corners of a cube and also in the center of each face of the cube. (11.10)

Factor-Label Method A problem-solving technique that uses the correct cancellation of the units of physical quantities as a guide for the correct setting-up of the solution to the problem. (1.4)

Fahrenheit Scale A temperature scale on which water freezes at 32 °F and boils at 212 °F (at 1 atm) and between which points there are 180 degree divisions called Fahrenheit degrees. (1.3)

Family of Elements See *Group.*

Faraday (𝓕) One mole of electrons; 9.65×10^4 coulombs. (18.3)

Faraday Constant (𝓕) 9.65×10^4 coulomb/mol e^-. (18.3)

Fatty Acid One of several long-chain carboxylic acids produced by the hydrolysis (digestion) of a lipid. (21.9)

First Law of Thermodynamics A formal statement of the law of conservation of energy. See *Law of Conservation of Energy.* (13.2)

First-Order Reaction A reaction with a rate law in which rate = $k[A]^1$, where A is a reactant. (14.4)

Fissile Isotope An isotope capable of undergoing fission following neutron capture. (20.9)

Fission The breaking apart of atomic nuclei into smaller nuclei accompanied by the release of energy, and the source of energy in nuclear reactors. (20.9)

Force Anything that can cause an object to change its motion or direction. (10.2)

Formal Charge The apparent charge on an atom in a molecule or polyatomic ion as calculated by a set of rules that generally assign a bonding pair of electrons to the more electronegative of the two atoms held by the bond. (7.7)

Formation Constant (K_{form}) The equilibrium constant for an equilibrium involving the formation of a complex ion. Also called the stability constant. (19.6)

Formula A representation of the composition of a substance that uses the chemical symbols of the elements present and subscripts that describe the proportions by atoms. (2.1, 2.6, 2.7)

Formula Mass The sum of the atomic masses (in u) of all of the atoms represented in the chemical formula of an ionic compound. (3.2)

Formula Unit A particle that has the composition given by the chemical formula. (2.7)

Fossil Fuels Coal, oil, and natural gas. (*Chemicals in Use* 2)

Free Energy See *Gibbs Free Energy.*

Free Energy Diagram A plot of the changes in free energy for a multi-component system versus the composition. (13.9)

Free Radical An atom, molecule, or ion that has one or more unpaired electrons. (Special Topic 14.1, *Chemicals in Use* 8, 20.6)

Freezing Point Depression A colligative property of a liquid solution by which the freezing point of the solution is lower than that of the pure solvent. (12.7)

Frequency (ν) The number of cycles per second of electromagnetic radiation. (6.1)

Frequency Factor The proportionality constant, A, in the Arrhenius equation. (14.7)

Fuel Cell An electrochemical cell in which electricity is generated from the reactions that accompany the burning of a fuel. (18.10)

Functional Group The group of atoms of an organic molecule that enters into a characteristic set of reactions that are independent of the rest of the molecule. (20.1)

Fusion (a) Melting. (11.7) (b) The formation of atomic nuclei by the joining together of the nuclei of lighter atoms. (20.8)

G

G See *Gibbs Free Energy.*

ΔG See *Gibbs Free Energy Change.*

ΔG° See *Standard Free Energy Change.*

Galvanic Cell An electrochemical cell in which a spontaneous redox reaction produces electricity. (18.5)

Gamma Radiation Electromagnetic radiation with wavelengths in the range of 1 Å or less (the shortest wavelengths of the spectrum). (20.3)

Gas Constant, Universal (R) $R = 0.0821$ liter atm mol^{-1} K^{-1} (10.4)

Gas Law, Combined For a given mass of gas, the product of its pressure and volume divided by its Kelvin temperature is a constant.

$$PV/T = \text{a constant} \qquad (10.3)$$

Gas Law, Ideal $PV = nRT$ (10.4)

Gay-Lussac's Law See *Pressure–Temperature Law.*

Genetic Code The correlation of codons with amino acids. (21.11)

Geometric Isomer One of a set of isomers that differ only in geometry. (19.4, 21.2)

Geometric Isomerism The existence of isomers whose molecules have identical atomic organizations but different geometries; cis/trans isomers. (19.4, 21.2)

Gibbs Free Energy (G) A thermodynamic quantity that relates enthalpy (H), entropy (S), and temperature (T) by the equation

$$G = H - TS \qquad (13.6)$$

Gibbs Free Energy Change (ΔG) The difference in free energy between the initial and final states for some reaction or physical change. (13.6)

Glass Any amorphous solid. (11.13, *Chemicals in Use* 9)

Glass Electrode A hollow glass tube containing the chemicals needed to complete an electrode and fitted with a thin-walled membrane at its bottom across which a potential difference can be created whose magnitude depends on the concentration of some ion in solution (e.g.,

H^+) and the kind of reference electrode used. (*Chemicals in Use* 15)

Graham's Law See *Effusion, Law of.*

Gram (g) 0.001 kg. (1.3)

Gray (Gy) The SI unit of radiation absorbed dose.

$$1 \text{ Gy} = 1 \text{ J kg}^{-1} \qquad (20.6)$$

Greenhouse Effect The retention of solar energy made possible by the ability of the greenhouse gases (e.g., CO_2, CH_4, H_2O, and the chlorofluorocarbons) to absorb outgoing radiation and re-radiate some of it back to Earth. (*Chemicals in Use* 4)

Ground State The lowest energy state of an atom or molecule. (6.2, 6.3)

Group A vertical column of elements in the periodic table. (2.4)

H

ΔH See *Enthalpy Change.*

ΔH_{atom} See *Atomization Energy.*

$\Delta H°$ See *Standard Heat of Reaction.*

$\Delta H_f°$ See *Standard Heat of Formation.*

ΔH_{fusion} See *Molar Heat of Fusion.*

$\Delta H_{sublimation}$ See *Molar Heat of Sublimation.*

$\Delta H_{vaporization}$ See *Molar Heat of Vaporization.*

Haber–Bosch Process An industrial synthesis of ammonia from nitrogen and hydrogen under pressure and heat in the presence of a catalyst. (*Chemicals in Use* 12)

Half-Cell That part of a galvanic cell in which either oxidation or reduction takes place. (18.5)

Half-Life ($t_{1/2}$) The time required for a reactant concentration or the mass of a radionuclide to be reduced by half. (14.5, 20.3)

Half-Reaction A hypothetical reaction that constitutes exclusively either the oxidation or the reduction half of a redox reaction and in whose equation the correct formulas for all species taking part in the change are given together with enough electrons to give the correct electrical balance. (12.3)

Halogen Family Group VIIA in the periodic table—fluorine, chlorine, bromine, iodine, and astatine. (2.4)

Hall–Héroult Process A method for manufacturing aluminum by the electrolysis of aluminum oxide in molten cryolite. (18.4)

Hard Water Water with dissolved Mg^{2+}, Ca^{2+}, Fe^{2+}, or Fe^{3+} ions at a concentration high enough (above 25 mg L^{-1}) to interfere with the use of soap. (Special Topic 4.1)

Heat A form of energy. (1.5, 5.1)

Heat Capacity The quantity of heat needed to raise the temperature of an object by 1 °C. (5.4)

Heating Curve A plot of the temperature change in a sample versus the quantity of heat added. (11.7)

Heat of Formation, Standard See *Standard Heat of Formation.*

Heat of Reaction The heat exchanged between a system and its surroundings when a chemical change occurs in the system. (5.5)

Heat of Reaction at Constant Pressure The heat of a reaction in an open system. (5.5)

Heat of Reaction, Standard See *Standard Heat of Reaction.*

Heat of Solution (ΔH_{soln}) The energy exchanged between the system and its surroundings when one mole of a solute dissolves in a solvent to make a dilute solution. (12.2)

Henry's Law See *Pressure–Solubility Law.*

Hertz (Hz) 1 cycle s^{-1}; the SI unit of frequency. (6.1)

Hess's Law For any reaction that can be written in steps, the standard heat of reaction is the same as the sum of the standard heats of reaction for the steps. (5.5)

Hess's Law Equation For the change,
$$aA + bB + \ldots \longrightarrow nN + mM + \ldots$$
$$\Delta H° = \left\{ \begin{array}{l} \text{sum of } \Delta H_f° \text{ of all} \\ \text{of the reactants.} \end{array} \right\} - \left\{ \begin{array}{l} \text{sum of } \Delta H_f° \text{ of all} \\ \text{of the products.} \end{array} \right\} \quad (5.5)$$

Heterocyclic Compound A compound whose molecules have rings that include one or more multivalent atoms other than carbon. (21.1)

Heterogeneous Catalyst A catalyst that is in a different phase than the reactants and onto whose surface the reactant molecules are adsorbed and where they react. (14.9)

Heterogeneous Equilibrium An equilibrium involving more than one phase. (15.8)

Heterogeneous Mixture A mixture that has two or more phases with different properties. (2.1)

Heterogeneous Reaction A reaction in which not all of the chemical species are in the same phase. (14.2, 15.8)

High-Spin Complex A complex ion or coordination compound in which there is the maximum number of unpaired electrons. (19.5)

Homogeneous Catalyst A catalyst that is in the same phase as the reactants. (14.9)

Homogeneous Equilibrium An equilibrium system in which all components are in the same phase. (15.8)

Homogeneous Mixture A mixture that has only one phase and that has uniform properties throughout. (2.1)

Homogeneous Reaction A reaction in which all of the chemical species are in the same phase. (14.2, 15.8)

Hund's Rule Electrons that occupy orbitals of equal energy are distributed with unpaired spins as much as possible among all such orbitals. (6.5)

Hybrid Atomic Orbitals Orbitals formed by mixing two or more of the basic atomic orbitals of an atom and that make possible more effective overlaps with the orbitals of adjacent atoms than do ordinary atomic orbitals. (8.5)

Hydrate A compound that contains molecules of water in a definite ratio to other components. (2.1)

Hydration The development in an aqueous solution of a cage of water molecules about ions or polar molecules of the solute. (12.1)

Hydration Energy The enthalpy change associated with the hydration of gaseous ions or molecules as they dissolve in water. (12.2)

Hydrocarbon An organic compound whose molecules consist entirely of carbon and hydrogen atoms. (2.6, 21.2)

Hydrogen Bond An extra strong dipole–dipole attraction between a hydrogen bound covalently to nitrogen, oxygen, or fluorine, and another nitrogen, oxygen, or fluorine atom. (11.2)

Hydrogen Electrode The standard of comparison for reduction potentials and for which $E_{H^+}^\circ$ has a value of 0.00 V (25 °C, 1 atm) when [H^+] = 1 M in the reversible half-cell reaction:

$$2H^+(aq) + 2e^- \rightleftharpoons H_2(g) \quad (18.6)$$

Hydrometer A device for measuring specific gravity. (18.10)

Hydrophilic Group A polar molecular unit capable of having dipole–dipole attractions or hydrogen bonds with water molecules. (12.10)

Hydrophobic Group A nonpolar molecular unit with no affinity for the molecules of a polar solvent, like water. (12.10)

Hypothesis A tentative explanation of the results of experiments. (1.2)

I

Ideal Gas A hypothetical gas that obeys the gas laws exactly. (10.3)

Ideal Gas Law $PV = nRT$ (10.4)

Ideal Solution A hypothetical solution that would obey the vapor pressure–concentration law (Raoult's law) exactly. (12.2)

Immiscible Insoluble. (12.1)

Incompressible Incapable of losing volume under increasing pressure. (11.3)

Indicator A chemical put in a solution being titrated and whose change in color signals the end point. (4.11, 16.1, 16.8)

Induced Dipole A dipole created when the electron cloud of an atom or a molecule is distorted by a neighboring dipole or by an ion. (11.2)

Inert Gases See *Noble Gases.*

Initiation Step The step in a chain reaction that produces reactive species that can start chain propagation steps. (Special Topic 14.1)

Inner Transition Elements Members of the two long rows of elements below the main body of the periodic table—elements 58–71 and elements 90–103. (2.4)

Inorganic Compound A compound made from any elements except those compounds of carbon classified as organic compounds. (2.9)

Instability Constant (K_{inst}) The reciprocal of the formation constant for an equilibrium in which a complex ion forms. (19.6)

Instantaneous Dipole A momentary dipole in an atom, ion, or molecule caused by the erratic movement of electrons. (11.2)

Insulated System A system with boundaries that permit no energy to pass through. (5.3)

Intensive Property A property whose physical quantity is independent of the size of the sample, such as density or temperature. (1.5)

Intermolecular Attractions Attractions *between* neighboring molecules. (11.2)

Internal Energy (E) The sum of all of the kinetic energies and potential energies of the particles within a system. (13.2)

International System of Units (SI) The successor to the metric system of measurements that retains most of the units of the metric system and their decimal relationships but employs new reference standards. (1.3)

Intron One of a set of sections of a DNA molecule that separate the exon sections of a gene from each other. (21.11)

Inverse Square Law The intensity of a radiation is inversely proportional to the square of the distance from its source. (20.6)

Ion An electrically charged particle on the atomic or molecular scale of size. (2.7)

Ion–Electron Method A method for balancing redox reactions that uses half-reactions. (4.9)

Ionic Bond The attractions between ions that hold them together in ionic compounds. (7.1)

Ionic Character The extent to which a covalent bond has a dipole moment and is polarized. (7.5)

Ionic Compound A compound consisting of positive and negative ions. (2.7)

Ionic Crystal A crystal that has ions located at the lattice points. (11.12)

Ionic Equation A chemical equation in which soluble strong electrolytes are written in dissociated or ionized form. (4.4)

Ionic Reaction A chemical reaction in which ions are involved. (4.3)

Ion–Induced Dipole Attraction An electrostatic attraction to a dipole induced in a nearby molecule (or ion) by an ion. (11.2)

Ionization Energy (IE) The energy needed to remove an electron from an isolated, gaseous atom, ion, or molecule (usually given in units of $kJ\ mol^{-1}$). (6.9)

Ionization Reaction A reaction of chemical particles that produces ions. (4.3)

Ionizing Radiation Any high-energy radiation—X rays, gamma rays, or radiation from radionuclides—that generates ions as it passes through matter. (20.6)

Ion Product The mass action expression for the solubility equilibrium involving the ions of a salt and equal to the product of the molar concentrations of the ions, each concentration raised to a power that equals the number of ions obtained from one formula unit of the salt. (17.3)

Ion Product Constant of Water (K_w) $K_w = [H^+][OH^-]$ (16.1)

Ion Product of Water $[H^+][OH^-]$ (16.1)

Ion Selective Electrode An electrode sensitive only to one kind of ion. (*Chemicals in Use* 15)

Isomer One of a set of compounds that have identical molecular formulas but different structures. (7.4, 19.4, 21.2)

Isomerism The existence of sets of isomers. (7.4, 19.4, 21.2)

Isotopes Constituents of the same element but whose atoms have different mass numbers. (2.2)

IUPAC Rules The formal rules for naming substances as developed by the International Union of Pure and Applied Chemistry. (21.2)

J

Joule (J) The SI unit of energy.

$$1\,J = 1\ kg\ m^2\ s^{-2}$$
$$1\,J = 4.184\ cal\ (exactly) \quad (1.5)$$

K

K See *Kelvin*.

K_a See *Acid Ionization Constant*.

K_b See *Base Ionization Constant*.

K_{sp} See *Solubility Product Constant*.

K_{spa} See *Acid Solubility Product*.

K_w See *Ion Product Constant of Water*.

K-Capture See *Electron Capture*.

Kelvin (K) One degree on the Kelvin scale of temperature and identical in size to the Celsius degree. (1.3)

Kelvin Scale The temperature scale on which water freezes at 273.15 K and boils at 373.15 K and that has 100 degree divisions called kelvins between these points. K = °C + 273.15. (1.3)

Ketone An organic compound whose molecules have the carbonyl group (C=O) flanked by hydrocarbon groups. (21.5)

Kilogram (kg) The base unit for mass in the SI and equal to the mass of a cylinder of platinum–iridium alloy kept by the International Bureau of Weights and Measures at Sèvres, France. 1 kg = 1000 g. (1.3)

Kinetic Energy (KE) Energy of motion. KE = $(1/2)\,mv^2$. (1.5, 5.1)

Kinetics The study of the factors that govern how rapidly reactions occur. (14.1)

Kinetic Theory of Gases A set of postulates used to explain the gas laws. A gas consists of an extremely large number of very tiny, very hard particles in constant, random motion. They have negligible volume and, between collisions, experience no forces between themselves. (10.8)

Kinetic Theory of Matter The particles of substances (atoms, ions, molecules) are in a state of constant agitation or motion, except at absolute zero (0 K). (5.2)

L

Lanthanide Elements Elements 58–71. (2.4)

Lattice Energy Energy released by the imaginary process in which isolated ions come together to form a crystal of an ionic compound. (7.1)

Law A description of behavior (and not an *explanation* of behavior) based on the results of many experiments. (1.2)

Law of Conservation of Energy The energy of the universe is constant; it can be neither created nor destroyed but only transferred and transformed. (1.5)

Law of Conservation of Mass No detectable gain or loss in mass occurs in chemical reactions. Mass is conserved. (2.2)

Law of Conservation of Mass–Energy The sum of all the mass in the universe and of all of the energy, expressed as an equivalent in mass (calculated by the Einstein equation), is a constant. (20.1)

Law of Definite Proportions In a given chemical compound, the elements are always combined in the same proportion by mass. (2.2)

Law of Gas Effusion See *Effusion, Law of*.

Law of Heat Summation (Hess's Law) For any reaction that can be written in steps, the standard heat of

reaction is the sum of the standard heats of reaction for the steps. (5.5)

Law of Multiple Proportions Whenever two elements form more than one compound, the different masses of one element that combine with the same mass of the other are in a ratio of small whole numbers. (2.2)

Law of Partial Pressures See *Partial Pressures, Dalton's Law of*.

Lead Storage Battery A galvanic cell of about 2 V involving lead and lead(IV) oxide in sulfuric acid. (18.10)

Le Châtelier's Principle When a system that is in dynamic equilibrium is subjected to a disturbance that upsets the equilibrium, the system undergoes a change that counteracts the disturbance and, if possible, restores the equilibrium. (11.8, 15.9)

Lewis Acid An electron-pair acceptor. (7.9)

Lewis Base An electron-pair donor. (7.9)

Lewis Structure (Lewis Formula) A structural formula drawn with Lewis symbols and that uses dots and dashes to show the valence electrons and shared pairs of electrons. (7.2, 7.3, 7.6)

Lewis Symbol The symbol of an element that includes dots to represent the valence electrons of an atom of the element. (7.2, 7.3, 7.6)

Ligand A molecule or an anion that can bind to a metal ion to form a complex. (19.1)

Like Dissolves Like Rule Strongly polar and ionic solutes tend to dissolve in polar solvents, and nonpolar solutes tend to dissolve in nonpolar solvents. (12.1)

Limiting Reactant The reactant that determines how much product can form when nonstoichiometric amounts of reactants are used. (3.7)

Linear Molecule A molecule all of whose atoms lie on a straight line. (8.1)

Lipid Any substance found in plants

or animals that can be dissolved in nonpolar solvents. (21.7)

Liter (L) 1 dm³. 1 L = 1000 mL = 1000 cm³ (1.3)

Ln *x* The natural logarithm (to the base *e*) of *x*. (Appendix A.2)

Lock-and-Key Mechanism A mechanism of enzyme action that postulates the combining of an enzyme and its substrate as allowed by the complementary shapes of their molecules. (21.10)

Logarithm In general, the exponent, *b*, in $N = a^b$. In *common* logarithms, the exponent, *x*, in $N = 10^x$. In *natural* logarithms, the exponent, *x*, in $N = e^x$, where $e = 2.7182818 . . .$ (Appendix A.2)

London Forces Weak attractive forces caused by instantaneous dipole–induced dipole attractions. (11.2)

Lone Pair A pair of electrons in the valence shell of an atom that is not shared with another atom. An unshared pair of electrons. (8.2)

Low-Spin Complex A coordination compound or a complex ion with electrons paired as much as possible in the lower energy set of *d* orbitals. (19.5)

M

Macromolecule A molecule whose molecular mass is very large. (21.6)

Magic Numbers The numbers 2, 8, 20, 28, 50, 82, and 126, numbers whose significance in nuclear science is that a nuclide in which the number of protons or neutrons equals a magic number has nuclei that are relatively more stable than those of other nuclides nearby in the band of stability. (20.4)

Magnetic Quantum Number (m_ℓ) A quantum number that can have values from $-\ell$ to $+\ell$. (6.3)

Main Reaction The desired reaction between the reactants as opposed to competing reactions that give by-products. (3.8)

Malleability A metal's ability to be hammered or rolled into thin sheets. (2.4)

Manometer A device for measuring the pressure within a closed system. The two types—*closed end* and *open end*—differ according to whether the operating fluid (e.g., mercury) is exposed at one end to the atmosphere or not. (10.2)

Mass A measure of the amount of matter that there is in a given sample. (1.5)

Mass Action Expression A fraction in which the numerator is the product of the molar concentrations of the products, each raised to a power equal to its coefficient in the equilibrium equation, and the denominator is the product of the molar concentrations of the reactants, each also raised to the power that equals its coefficient in the equation. (For gaseous reactions, partial pressures can be used in place of molar concentrations.) (15.3)

Mass Number The numerical sum of the protons and neutrons in an atom of a given isotope. (2.3)

Matter Anything that has mass and occupies space. (1.5)

Maximum Work The maximum amount of energy that can be harnessed as work (instead of released as heat) and is equal to the standard Gibbs free energy change for the reaction. (13.9)

Mechanism of a Reaction The series of individual steps (called elementary processes) in a chemical reaction that gives the net, overall change. (14.1, 14.8)

Melting Point The temperature at which a substance melts; the temperature at which a solid is in equilibrium with its liquid state. (11.4)

Mercury Battery A galvanic cell of about 1.35 V involving zinc and mercury(II) oxide. (18.10)

Mercury Cell A device for the manufacture of sodium hydroxide. (18.4)

Metal An element or an alloy that is a good conductor of electricity, that has a shiny surface, and that is malleable and ductile; an element that normally forms positive ions and has an oxide that is basic. (2.5)

Metallic Crystal A solid having positive ions at the lattice positions that are attracted to a "sea of electrons" that extends throughout the entire crystal. (11.12)

Metalloids Elements with properties that lie between those of metals and nonmetals, and that are found in the periodic table around the diagonal line running from boron (B) to astatine (At). (2.4)

Metathesis Reaction See *Double Replacement Reaction.*

Meter (m) The SI base unit for length. (1.3)

Metric System A decimal system of units for physical quantities taken over by the SI. (1.3; See also *International System of Units*)

Micelle A colloidal-sized group of ions, such as the anions of a detergent, that have clustered to maximize the contact between the ions' hydrophilic heads and water and to minimize the contact between the ions' hydrophobic parts and water. (12.10)

Milliliter (mL) 0.001 L.

1000 mL = 1 L (1.3)

Millimeter (mm) 0.001 m.

1000 mm = 1 m (1.3)

Millimeter of Mercury (mm Hg) A unit of pressure equal to 1/760 atm. 760 mm Hg = 1 atm. 1 mm Hg = 1 torr. (10.2)

Miscible Mutually soluble. (12.1)

Mixture Any matter consisting of two or more substances physically combined in no particular proportion by mass. (2.1)

Model, Scientific A picture or a mental construction derived from a set of ideas and assumptions that are imagined to be true because they can be used to explain certain observations and measurements, for example, the model of an ideal gas. (10.8)

Molal Boiling Point Elevation Constant (k_b) The number of degrees (°C) per unit of molal concentration that a boiling point of a solution is higher than that of the pure solvent. (12.7)

Molal Concentration (*m*) The number of moles of solute in 1000 g of solvent. (12.5)

Molal Freezing Point Depression Constant (*k_f*) The number of degrees (°C) per unit of molal concentration that a freezing point of a solution is lower than that of the pure solvent. (12.7)

Molality The molal concentration. (12.5)

Molar Concentration (*M*) The number of moles of solute per liter of solution. The molarity of a solution. (4.10)

Molar Heat Capacity The heat that can raise the temperature of 1 mol of a substance by 1 °C; the heat capacity per mole. (5.4)

Molar Heat of Fusion The heat absorbed when 1 mol of a solid melts to give 1 mol of the liquid at constant temperature and pressure. (11.7)

Molar Heat of Sublimation The heat absorbed when 1 mol of a solid sublimes to give 1 mol of its vapor at constant temperature and pressure. (11.7)

Molar Heat of Vaporization The heat absorbed when 1 mol of a liquid changes to 1 mol of its vapor at constant temperature and pressure. (11.7)

Molarity See *Molar Concentration.*

Molar Mass See *Formula Mass.*

Molar Solubility The number of moles of solute required to give 1 L of a saturated solution of the solute. (17.3)

Molar Volume, Standard The volume of 1 mol of a gas under standard conditions of temperature and pressure; 22.4 L mol^{-1}. (10.4)

Mole (mol) The SI unit for amount of substance; the formula mass in grams of an element or compound; a quantity of chemical substance that contains Avogadro's number of formula units (6.02 × 10^{23}). (3.1)

Molecular Compound A compound consisting of neutral (but often polar) molecules. (2.6)

Molecular Crystal A crystal that has molecules or individual atoms at the lattice points. (11.12)

Molecular Equation A chemical equation that gives the full formulas of all of the reactants and products and that is used to plan an actual experiment. (4.4)

Molecular Formula A chemical formula that gives the actual composition of one molecule. (2.6, 3.4)

Molecular Mass The sum of the atomic masses of all of the atoms present in one molecule of a molecular compound. (3.2)

Molecular Orbital (MO) An orbital that extends over two or more atomic nuclei. (8.8)

Molecular Orbital Theory (MO Theory) A theory about covalent bonds that views a molecule as a collection of positive nuclei surrounded by electrons distributed among a set of bonding and antibonding orbitals of different energies. (8.4, 8.8)

Molecular Weight See *Molecular Mass.*

Molecule A neutral particle composed of two or more atoms combined in a definite ratio of whole numbers. (2.6, 7.3)

Mole Fraction The ratio of the number of moles of one component of a mixture to the total number of moles of all components. (10.6)

Mole Percent The mole fraction of a component expressed as a percentage. (10.6)

Monodentate Ligand A ligand that can attach itself to a metal ion by only one atom. (19.1)

Monomer A substance of relatively low formula mass that is used to make a polymer. (21.6)

Monoprotic Acid An acid that can furnish one H$^+$ per molecule. (2.9)

Monosaccharide A carbohydrate that cannot be hydrolyzed. (21.8)

N

Nernst Equation (18.9)

$$E_{cell} = E°_{cell} - \frac{0.0592}{n} \log Q$$

Net Ionic Equation An ionic equation from which spectator ions have been omitted. It is balanced when both atoms and electrical charge balance. (4.4)

Neutralization, Acid–Base The destruction of an acid by a base or of a base by an acid. (2.9, 7.9)

Neutral Solution A solution in which [H$^+$] = [OH$^-$]. (16.1)

Neutron (*n*, ^1_0n) A subatomic particle with a charge of zero, a mass of 1.008665 u (1.674954 × 10^{-24} g), and that exists in all atomic nuclei except those of the hydrogen-1 isotope. (2.3)

Neutron Activation Analysis A technique to analyze for trace impurities in a sample by studying the frequencies and intensities of the gamma radiation they emit after they have been rendered radioactive by neutron bombardment of the sample. (20.7)

Neutron Emission A nuclear reaction in which a neutron is ejected. (20.3)

Nickel–Cadmium Battery A galvanic cell of about 1.4 V involving cadmium and nickel(IV) oxide. (18.10)

Noble Gases Group 0 in the periodic table—helium, neon, argon, krypton, xenon, and radon. (2.4)

Node A place where the amplitude or intensity of a wave is zero. (6.3)

Nomenclature The names of substances and the rules for devising names. (2.9, 21.2)

Nonelectrolyte A compound that in its molten state or in solution cannot conduct electricity. (4.2)

Nonlinear Molecule A molecule in which the atoms do not lie in a straight line. (8.2)

Nonmetal (Nonmetallic Element) A nonductile, nonmalleable, nonconducting element that tends to form negative ions (if it forms them at all) far more readily than positive ions and whose oxide is likely to show acidic properties. (2.4)

Nonoxidizing Acid An acid in which the anion is a poorer oxidizing agent than the hydrogen ion (e.g., HCl, H_2SO_4, H_3PO_4). (9.4)

Nonpolar Covalent Bond An electron pair bond at the ends of which are atoms of equal or very nearly equal electronegativity. (7.5)

Nonvolatile Descriptive of a substance with a high boiling point, a low vapor pressure, and that does not evaporate. (12.6)

Normal Boiling Point The temperature at which the vapor pressure of a liquid equals 1 atm. (11.6)

Normality (*N*) The number of equivalents of solute per liter of solution. (Appendix C)

n-Type Semiconductor A semiconductor doped with an impurity that causes the moving charge to consist of negatively charged electrons. (8.9)

Nuclear Equation A description of a nuclear reaction that uses the special symbols of isotopes, that describes some kind of nuclear transformation or disintegration, and that is balanced when the sums the atomic numbers on either side of the arrow are equal and the sums of the mass numbers are also equal. (20.3)

Nuclear Reaction A change in the composition or energy of the nuclei of isotopes accompanied by one or more events such as the radiation of nuclear particles or electromagnetic energy, transmutation, fission, or fusion. (20.1)

Nucleic Acids Polymers in living cells that store and translate genetic information and whose molecules hydrolyze to give a sugar unit (ribose from ribonucleic acid, RNA, or deoxyribose from deoxyribonucleic acid, DNA), a phosphate, and a set of four or five nitrogen-containing, heterocyclic bases (adenine, thymine, guanine, cytosine, and uracil). (21.11)

Nucleon A proton or a neutron. (2.3)

Nucleus The hard, dense core of an atom that holds the atom's protons and neutrons. (2.3)

O

Octahedral Molecule A molecule in which the planar surfaces of an octahedron are created by connecting adjacent nuclei with imaginary lines. (8.1)

Octahedron An eight-sided figure that can be envisioned as two square pyramids sharing the common square base. (8.1)

Octet (of electrons) Eight electrons in the valence shell of an atom. (7.1, 7.3)

Octet Rule An atom tends to gain or lose electrons until its outer shell has eight electrons. (7.1, 7.3)

Odd–Even Rule When the numbers of protons and neutrons in an atomic nucleus are both even, the isotope is more likely to be stable than when both numbers are odd. (20.4)

Open-End Manometer See *Manometer*.

Optical Isomers Stereoisomers other than geometric (cis/trans) isomers and that include substances that can rotate the plane of plane polarized light. (19.5)

Orbital An electron waveform with a particular energy and a unique set of values for the quantum numbers n, ℓ, and m_ℓ. (6.3)

Orbital Diagram A diagram showing an atom's orbitals in which the electrons are represented by arrows to indicate paired and unpaired spins. (6.5)

Order (of a Reaction) The sum of the exponents in the rate law is the *overall* order. Each exponent gives the order of the reaction with respect to a specific reactant. (14.4)

Ordinate The vertical axis. (Appendix A.3)

Organic Chemistry The study of the compounds of carbon that are not classified as inorganic. (2.6, 21.1)

Organic Compound Any compound of carbon other than a carbonate, bicarbonate, cyanide, cyanate, carbide, or gaseous oxide. (2.6, 21.1)

Osmosis The passage of solvent molecules, but not those of solutes, through a semipermeable membrane; the limiting case of dialysis. (12.8)

Osmotic Pressure The back pressure that would have to be applied to prevent osmosis; one of the colligative properties. (12.8)

Outer Shell The occupied shell in an atom having the highest principal quantum number (*n*). (6.6)

Overlap A portion of two orbitals from different atoms that share the same space in a molecule. (8.4)

Oxidation A change in which an oxidation number increases (becomes more positive). A loss of electrons. (9.4)

Oxidation Number The charge that an atom in a molecule or ion would have if all of the electrons in its bonds belonged entirely to the more electronegative atoms; the oxidation state of an atom. (4.8)

Oxidation–Reduction Reaction A chemical reaction in which changes in oxidation numbers occur. (4.8, 9.4)

Oxidation State See *Oxidation Number*.

Oxidizing Acid An acid in which the anion is a stronger oxidizing agent than H^+ (e.g., $HClO_4$, HNO_3). (9.4)

Oxidizing Agent The substance that causes oxidation and that is itself reduced. (9.4)

Oxoacid An acid that contains oxygen besides hydrogen and another element (e.g., HNO_3, H_3PO_4, H_2SO_4). (2.9, 9.2)

P

Pairing Energy The energy required to force two electrons to become paired and occupy the same orbital. (19.5)

Paramagnetism The weak magnetism of a substance whose atoms, molecules, or ions have unpaired electrons. (6.4)

Partial Charge Charges at opposite ends of a dipole that are fractions of full 1+ or 1− charges. (7.5)

Partial Pressure The pressure contributed by an individual gas to the total pressure of a gas mixture. (10.6)

Partial Pressures, Law of (Dalton's Law of Partial Pressures) The total pressure of a mixture of gases equals the sum of their partial pressures. (10.6)

Parts per Billion (ppb) The number of parts of one component of a mixture to 10^9 total parts. (Special Topic 12.2)

Parts per Million (ppm) The number of parts of one component of a mixture to 10^6 total parts. (Special Topic 12.2)

Pascal (Pa) The SI unit of pressure equal to 1 newton m^{-2};

133.3224 Pa = 1 torr. (10.2)

Pauli Exclusion Principle No two electrons in an atom can have the same values for all four of their quantum numbers. (6.4)

Peptide Bond The amide linkage in molecules of polypeptides. (21.10)

Percentage by Weight (Percentage by Mass) (a) The number of grams of an element combined in 100 g of a compound. (3.3) (b) The number of grams of a substance in 100 g of a mixture or solution. (12.5, Special Topic 12.2)

Percentage Composition A list of the percentages by weight of the elements in a compound. (3.3)

Percentage Concentration A ratio of the amount of solute to the amount of solution expressed as a percent.

Weight/weight The grams of solute in 100 g of solution. (12.5)
Volume/volume The volumes of solute in 100 volumes of solution. (Special Topic 12.2)

Percentage Ionization

$$\frac{[\text{Amt of species ionized}]}{[\text{Initial amt of species}]} \times 100\%$$

(16.4)

Percentage Yield The ratio (taken as a percent) of the mass of product obtained to the mass calculated from the reaction's stoichiometry. (3.8)

Period A horizontal row of elements in the periodic table. (2.4)

Periodic Law The properties of the elements are a periodic function of their atomic numbers. (2.4)

Periodic Table A display of the symbols of all of the elements, in order of increasing atomic number, in rows (periods) and columns (groups) that organizes the families of the elements into columns. (2.4, inside front cover)

PET See *Positron Emission Tomography.*

pH −log[H$^+$]. (16.1)

Phase A homogeneous region within a sample. (2.1)

Phase Diagram A pressure–temperature graph on which are plotted the temperatures and the pressures at which equilibrium exists between the states of a substance. It defines regions of T and P in which the solid, liquid, and gaseous states of the substance can exist. (11.9)

Photon A unit of energy in electromagnetic radiation equal to $h\nu$, where ν is the frequency of the radiation and h is Planck's constant. (6.1)

Physical Change Any change in which substances do not change into other substances. (1.5)

Physical Law A relationship between two or more physical properties of a system, usually expressed as a mathematical equation, that describes how a change in one property affects the others. (1.2)

Physical Property A property that can be specified without reference to another substance and that can be measured without causing a chemical change. (1.5)

Physical State The condition of aggregation of a substance's formula units, whether as a solid, a liquid, or a gas. (1.5)

Pi Bond (π Bond) A bond formed by the sideways overlap of a pair of p orbitals and that concentrates electron density into two separate regions that lie on opposite sides of a plane that contains an imaginary line joining the nuclei. (8.6)

pK_a − log K_a (16.3)

pK_b − log K_b (16.3)

pK_w − log K_w (16.1)

Planar Triangular Molecule A molecule in which a central atom holds three other atoms located at the corners of an equilateral triangle and that includes the central atom at its center. (8.1)

Planck's Constant (h) The ratio of the energy of a photon to its frequency; $6.6260755 \times 10^{-34}$ J Hz^{-1}. (6.1)

p–n Junction The interface between n- and p-type semiconductors in a transistor. (8.9)

pOH −log[OH$^-$] (16.1)

Polar Covalent Bond (Polar Bond) A covalent bond in which more than half of the bond's negative charge is concentrated around one of the two atoms. (7.5)

Polar Molecule A molecule in which individual bond polarities do not cancel and in which, therefore, the centers of density of negative and positive charges do not coincide. (7.5)

Polyatomic Ion An ion composed of two or more atoms. (2.7)

Polydentate Ligand A ligand that has two or more atoms that can become simultaneously attached to a metal ion. (19.1)

Polymer A substance consisting of macromolecules that have repeating structural units. (21.6)

Polymerization A chemical reaction that converts a monomer into a polymer. (21.6)

Polypeptide A polymer of α-amino acids that makes up all or most of a protein. (21.10)

Polyprotic Acid An acid that can furnish more than one H$^+$ per molecule. (2.9)

Polysaccharide A carbohydrate whose molecules can be hydrolyzed to hundreds of monosaccharide molecules. (21.8)

Position of Equilibrium The relative amounts of the substances on both sides of the double arrows in the equation for an equilibrium. (4.3, 11.8)

Positron ($_0^1e$) A positively charged particle with the mass of an electron. (20.3)

Positron Emission Tomography (PET) A technique to obtain an X-ray-like image of part of the body by letting the body take in positron-emitting isotopes and then studying the gamma emission. (*Chemicals in Use* 16)

Post-transition Metal A metal that occurs in the periodic table immediately to the right of a row of transition elements. (2.7)

Potential See *Electromotive Force.*

Potential Energy Stored energy. (1.5, 5.1)

Precipitate A solid that separates from a solution usually as the result of a chemical reaction. (4.5)

Precipitation In chemistry, the formation of a precipitate. (4.5)

Precision How reproducible measurements are; the fineness of a measurement as indicated by the number of significant figures reported in the physical quantity. (1.4)

Pressure Force per unit area. (10.2)

Pressure–Concentration Law See *Vapor Pressure–Concentration Law.*

Pressure–Solubility Law (Henry's Law) The concentration of a gas dissolved in a liquid at any given temperature is directly proportional to the partial pressure of this gas above the solution. (12.4)

Pressure–Temperature Law (Gay-Lussac's Law) The pressure of a given mass of gas is directly proportional to its Kelvin temperature if the volume is kept constant. $P \propto T$. (10.3)

Pressure–Volume Law (Boyle's Law) The volume of a given mass of a gas

is inversely proportional to its pressure if the temperature is kept constant. $V \propto 1/P$. (10.3)

Pressure–Volume Work The energy transferred as work when a system expands or contracts against the pressure exerted by the surroundings. At constant pressure, work = $-P \Delta V$. (13.2)

Principal Quantum Number (n) The quantum number that defines the principal energy levels and that can have values of 1, 2, 3, . . . , ∞. (6.3)

Products The substances produced by a chemical reaction and whose formulas follow the arrows in chemical equations. (2.1)

Propagation Step A step in a chain reaction for which one product must serve in a succeeding propagation step as a reactant and for which another (final) product accumulates with each repetition of the step. (Special Topic 14.1)

Property A characteristic of matter. (1.4)

Protein A macromolecular substance found in cells that consists wholly or mostly of one or more polypeptides that often are combined with an organic molecule or a metal ion. (21.10)

Proton ($_1^1p$ or $_1^1H$) (a) A subatomic particle, with a charge of 1+ and a mass of 1.00727252 u (1.6726430 × 10^{-24} g), that is found in atomic nuclei. (2.4) (2) The name often used for the hydrogen ion and symbolized as H^+. (2.3)

Proton Acceptor A Brønsted base. (9.1)

Proton Donor A Brønsted acid. (9.1)

p-Type Semiconductor A semiconductor doped with an impurity that enables the charge to move as positively charged holes. (8.9)

Pure Substance An element or a compound. (2.1)

Q

Qualitative Analysis The use of experimental procedures to determine what elements are present in a substance. (3.3)

Qualitative Observation Observations that do not involve numerical information. (1.3)

Quantitative Analysis The use of experimental procedures to determine the percentage composition of a compound or the percentage of a component of a mixture. (3.3)

Quantitative Observation An observation involving a measurement and numerical information. (1.3)

Quantized Descriptive of a discrete, definite amount as of *quantized energy*. (6.2)

Quantum The energy of one photon. (6.1)

Quantum Mechanics See *Wave Mechanics.* (6.3)

Quantum Number A number related to the energy, shape, or orientation of an orbital, or to the spin of an electron. (6.2)

R

R See *Gas Constant, Universal.*

Rad A unit of radiation absorbed dose equal to 10^{-5} J g^{-1} or 10^{-2} Gy. (20.6)

Radiation The emission of electromagnetic energy or nuclear particles. (20.1)

Radioactive The ability to emit various types of atomic radiation or gamma rays. (20.1)

Radioactive Decay The change of a nucleus into another nucleus (or into a more stable form of the same nucleus) by the loss of a small particle or a gamma ray photon. (20.3)

Radioactive Disintegration Series A sequence of nuclear reactions beginning with a very long-lived radionuclide and ending with a stable isotope of lower atomic number. (20.3)

Radioactivity The emission of one or more kinds of radiation from an isotope with unstable nuclei. (20.3)

Radiological Dating A technique for measuring the age of a geologic formation or an ancient artifact by determining the ratio of the concentrations of two isotopes, one radioactive and the other a stable decay product. (20.7)

Radionuclide A radioactive isotope. (20.1)

Raoult's Law The vapor pressure of one component above a mixture of molecular compounds equals the product of its vapor pressure when pure and its mole fraction. (12.6)

Rare Earth Metals The lanthanides. (2.4)

Rate A ratio in which a unit of time appears in the denominator, for example, 40 mile hr^{-1} or 3.0 mol L^{-1} s^{-1}. (14.3)

Rate Constant The proportionality constant in the rate law; the rate of reaction when all reactant concentrations are 1 M. (14.4)

Rate-Determining Step (Rate-Limiting Step) The slowest step in a reaction mechanism. (14.8)

Rate Law An equation that relates the rate of a reaction to the molar concentrations of the reactants raised to powers. (14.4)

Rate of Reaction How quickly the reactants disappear and the products form and usually expressed in units of mol L^{-1} s^{-1}. (14.1)

Reactant, Limiting See *Limiting Reactant.*

Reactants The substances brought together to react and whose formulas appear before the arrow in a chemical equation. (2.1)

Reaction Coordinate The horizontal axis of a potential energy diagram of a reaction. (14.6)

Reaction Quotient (Q) The numerical value of the mass action expression. See *Mass Action Expression.* (15.3)

Recombinant DNA DNA in a bacterial plasmid altered by the insertion of DNA from another organism. (21.11)

Redox Reaction An oxidation–reduction reaction. (9.4)

Reducing Agent A substance that causes reduction and is itself oxidized. (9.4)

Reduction A change in which an oxidation number decreases (becomes less positive and more negative). A gain of electrons. (9.4)

Reduction Potential A measure of the tendency of a given half-reaction to occur as a reduction. (18.6)

Rem A dose in rads multiplied by a factor that takes into account the variations that different kinds of radiation have in their damage-causing abilities in tissue. (20.6)

Replication In nucleic acid chemistry, the reproductive duplication of DNA double helixes prior to cell division. (21.11)

Representative Element An element in one of the A groups in the periodic table. (2.4)

Resonance A concept in which the actual structure of a molecule or polyatomic ion is represented as a composite or average of two or more Lewis structures, which are called the resonance or contributing structures (and none of which has real existence). (7.8)

Resonance Energy The difference in energy between a substance and its principal resonance (contributing) structure. (7.8)

Resonance Hybrid The actual structure of a molecule or polyatomic ion taken as a composite or average of the resonance or contributing structures. (7.8)

Resonance Structure A Lewis structure that contributes to the hybrid structure in resonance-stabilized systems; a contributing structure. (7.8)

Reverse Osmosis The use of pressure to force the migration of solvent from a solution through a semipermeable membrane. (Special Topic 12.3)

Reversible Process A process that occurs by an infinite number of steps during which the driving force for the change is just barely greater than the force that resists the change. (13.8)

Ring, Carbon A closed-chain sequence of carbon atoms. (21.1)

RNA Ribonucleic acid; a nucleic acid that gives ribose, phosphate ion, adenine, uracil, guanine, and cytosine when hydrolyzed. (21.11)

Roentgen Unit of exposure intensity for X rays or gamma radiation. (20.6)

Rydberg Equation An equation used to calculate the wavelengths of all the spectral lines of hydrogen. (6.2)

S

Salt An ionic compound in which the anion is not OH$^-$ or O^{2-} and the cation is not H$^+$. (2.9, 4.4)

Salt Bridge A tube that contains an electrolyte that connects the two half-cells of a galvanic cell. (18.5)

Saponification The reaction of an organic ester with a strong base to give an alcohol and the salt of the organic acid. (21.9)

Saturated Organic Compound A compound whose molecules have only single bonds. (21.2)

Saturated Solution A solution that holds as much solute as it can at a given temperature. A solution in which there is an equilibrium between the dissolved and the undissolved states of the solute. (4.1)

Scientific Law See *Law.*

Scientific Method The observation, explanation, and testing of an explanation by additional experiments. (1.2)

Scientific Notation The representation of a quantity as a decimal number between 1 and 10 multiplied by 10 raised to a power (e.g., 6.02×10^{23}). (Appendices A.1, A.4)

Secondary Quantum Number (ℓ) The quantum number whose values can be 0, 1, 2, . . . ,$(n-1)$, where n is the principal quantum number. (6.3)

Second Law of Thermodynamics Whenever a spontaneous event takes place, it is accompanied by an increase in the entropy of the universe. (13.4)

Second-Order Reaction A reaction with a rate law of the type rate =

$k[A]^2$ or rate $= k[A][B]$, where A and B are reactants. (14.4)

Semiconductor A substance that conducts electricity weakly. (2.4)

Shell All of the orbitals associated with a given value of n (the principal quantum number). (6.3)

SI (International System of Units) The modified metric system adopted in 1960 by the General Conference on Weights and Measures. (1.3)

Side Reaction A reaction the occurs simultaneously with another reaction (the main reaction) in the same mixture to produce by-products. (3.8)

Sievert (Sv) The SI unit for dose equivalent. (20.6)

Sigma Bond (σ Bond) A bond formed by the head-to-head overlap of two atomic orbitals and in which electron density becomes concentrated along and around the imaginary line joining the two nuclei. (8.6)

Significant Figures The number of digits in a physical quantity that are known to be certain plus one more. (1.4)

Silver Oxide Battery A galvanic cell of about 1.5 V involving zinc and silver oxide and used when miniature batteries are needed. (18.10)

Simple Cubic Unit Cell A cell or unit of crystal structure that has atoms, molecules, or ions only at the corners of a cube. (11.10)

Single Bond A covalent bond in which a single pair of electrons is shared. (7.3)

Single Replacement Reaction A reaction in which one element replaces another in a compound; usually a redox reaction. (9.5)

Skeleton Equation An unbalanced equation showing only the formulas of reactants and products. (4.9)

Soap A salt of a fatty acid whose anions form micelles and which is able to hold oil and grease particles in suspension in water. (12.10)

Sol The colloidal dispersion of a solid in a fluid. (12.10)

Solar Battery (Solar Cell) A silicon wafer doped with arsenic and placed over a silicon wafer doped with boron to give a system that conducts electricity when light falls on it. (8.9)

Solubility The ratio of the quantity of solute to the quantity of solvent in a saturated solution and that is usually expressed in units of (g solute)/ (100 g solvent) at a specified temperature. (4.1)

Solubility Product Constant (K_{sp}) The equilibrium constant for the solubility of a salt and that, for a saturated solution, is equal to the product of the molar concentrations of the ions, each raised to a power equal to the number of its ions in one formula unit of the salt. (17.3)

Solute Something dissolved in a solvent to make a solution. (4.1)

Solution A homogeneous mixture in which all particles are of the size of atoms, small molecules, or small ions. (3.1, 4.1, 12.1)

Solvation The development of a cage-like network of a solution's solvent molecules about a molecule or ion of the solute. (12.1)

Solvation Energy The enthalpy of the interaction of gaseous molecules or ions of solute with solvent molecules during the formation of a solution. (12.2)

Solvent A medium, usually a liquid, into which something (a solute) is dissolved to make a solution. (4.1)

sp Hybrid Orbital A hybrid orbital formed by mixing one s and one p atomic orbital. The angle between a pair of sp hybrid orbitals is 180°. (8.5)

sp^2 Hybrid Orbital A hybrid orbital formed by mixing one s and two p atomic orbitals. The angle between two sp^2 hybrid orbitals is 120°. (8.5)

sp^3 Hybrid Orbital A hybrid orbital formed by mixing one s and three p atomic orbitals. The angle between two sp^3 hybrid orbitals is 109.5° (8.5)

sp^3d Hybrid Orbital A hybrid orbital formed by mixing one s, three p, and one d atomic orbital. sp^3d hybrids point to the corners of a trigonal bipyramid. (8.5)

sp^3d^2 Hybrid Orbital A hybrid orbital formed by mixing one s, three p, and two d atomic orbitals. sp^3d^2 hybrids point to the corners of an octahedron. (8.5)

Specific Gravity The ratio of the density of a substance to the density of water. (1.6)

Specific Heat The quantity of heat that will raise the temperature of 1 g of a substance by 1 °C, usually in units of cal g^{-1} °C^{-1} or J g^{-1} °C^{-1}. (5.4)

Spectator Ion An ion whose formula appears in an ionic equation identically on both sides of the arrow, that does not participate in the reaction, and that is excluded from the net ionic equation. (4.4)

Spectrochemical Series A listing of ligands in order of their ability to produce a large crystal field splitting. (19.5)

Spin Quantum Number (m_s) The quantum number associated with the spin of a subatomic particle and for the electron can have a value of $+(\frac{1}{2})$ or $+(\frac{1}{2})$. (6.4)

Spontaneous Change A change that occurs by itself without outside assistance. (13.3)

Square Planar Molecule A molecule with a central atom having four bonds that point to the corners of a square. (8.2)

Square Pyramid A pyramid with four triangular sides and a square base. (8.2)

Stability Constant See *Formation Constant.*

Stabilization Energy See *Resonance Energy.*

Standard Atmosphere See *Atmosphere, Standard.*

Standard Cell Potential (E°_{cell}) The potential of a galvanic cell at 25 °C and when all ion concentrations are exactly 1 *M* and the partial pressures of all gases are 1 atm. (18.6)

Standard Conditions of Temperature and Pressure (STP) 273 K (0 °C) and 1 atm (760 torr). (10.4)

Standard Enthalpy Change ($\Delta H°$) See *Standard Heat of Reaction.*

Standard Enthalpy of Formation ($\Delta H_f°$) See *Standard Heat of Formation.*

Standard Entropy ($S°$) The entropy of 1 mol of a substance at 25 °C and 1 atm. (13.5)

Standard Entropy Change ($\Delta S°$) The entropy change of a reaction when determined with reactants and products at 25 °C and 1 atm and on the scale of the mole quantities given by the coefficients of the balanced equation. (13.5)

Standard Entropy of Formation ($\Delta S_f°$) The value of $\Delta S°$ for the formation of one mole of a substance from its elements in their standard states. (13.5)

Standard Free Energy Change ($\Delta G°$) $\Delta G° = \Delta H° - T\Delta S°$ (13.8)

Standard Free Energy of Formation ($\Delta G_f°$) The value of $\Delta G°$ for the formation of *one* mole of a compound from its elements in their standard states. (13.7)

Standard Heat of Formation ($\Delta H_f°$) The amount of heat absorbed or evolved when one mole of the compound is formed from its elements in their standard states. (5.6)

Standard Heat of Reaction ($\Delta H°$) The enthalpy change of a reaction when determined with reactants and products at 25 °C and 1 atm and on the scale of the mole quantities given by the coefficients of the balanced equation. (5.5)

Standard Molar Volume See *Molar Volume, Standard.*

Standard Reduction Potential The reduction potential of a half-reaction at 25 °C when all ion concentrations are 1 *M* and the partial pressures of all gases are 1 atm. (18.6)

Standard Solution Any solution whose concentration is accurately known. (4.11)

Standard State The condition in which a substance is in its most stable form at 25 °C and 1 atm. (5.6)

Standing Wave A wave whose peaks and nodes do not change position. (6.3)

State Function A function or variable whose value depends only on the initial and final states of the system and not on the path taken by the system to get from the initial to the final state. (*P, V, T, H, S,* and *G* are all state functions.) (5.5)

State of Matter A physical state of a substance: solid, liquid, or gas. (1.5. See also *Standard State.*)

State of a System The set of specific values of the physical properties of a system—its composition, physical form, concentration, temperature, pressure, and volume. (5.5)

Stereoisomerism The existence of isomers whose structures differ only in spatial orientations (e.g., geometric isomers and optical isomers). (19.4)

Stock System A system of nomenclature that uses Roman numerals to specify oxidation states. (2.9)

Stoichiometric Equivalence The ratio by moles between two elements in a formula or two substances in a chemical reaction. (3.1)

Stoichiometry A description of the relative quantities by moles of the reactants and products in a reaction as given by the coefficients in the balanced equation. (3.1)

Stopcock A device on a buret that is used to control the flow of titrant. (4.11)

Stored Energy See *Potential Energy.*

STP See *Standard Conditions of Temperature and Pressure.*

Straight-Chain Compound An organic compound in whose molecules the carbon atoms are joined in one continuous open-chain sequence. (21.1)

Strong Acid An acid that is essentially 100% ionized in water. A good proton donor. An acid with a large value of K_a. (4.3, 16.2)

Strong Base Any powerful proton acceptor. A base with a large value of K_b. A metal hydroxide that dissociates essentially 100% in water. (4.3, 16.2)

Strong Electrolyte Any substance that ionizes or dissociates in water to essentially 100%. (4.3)

Structural Formula A chemical formula that shows how the atoms of a molecule or polyatomic ion are arranged, to which other atoms they are bonded, and the kinds of bonds (single, double, or triple). (7.3)

Subatomic Particles Electrons, protons, neutrons, and atomic nuclei. (2.3)

Sublimation The conversion of a solid directly into a gas without passing through the liquid state. (11.3)

Subshell All of the orbitals of a given shell that have the same value of their secondary quantum number, ℓ. (6.3)

Substances See *Pure Substance.*

Substitution Reaction The replacement of an atom or group on a molecule by another atom or group. (21.2, 21.3)

Supercooled The condition of a substance in its liquid state below its freezing point. (11.7)

Supercooled Liquid A liquid at a temperature below its freezing point. An amorphous solid. (11.7)

Supercritical Fluid A substance at a temperature above its critical temperature. (11.9)

Superimposability A test of structural chirality in which a model of one structure and a model of its mirror image are compared to see if the two could be made to blend perfectly, with every part of one coinciding simultaneously with the parts of the other. (19.4)

Supersaturated Solution A solution whose concentration of solute exceeds the equilibrium concentration. (4.1)

Surface Tension A measure of the amount of energy needed to expand the surface area of a liquid. (11.3)

Surfactant A substance that lowers the surface tension of a liquid and promotes wetting. (11.3)

Surroundings That part of the universe other than the system being studied and separated from the system by a real or an imaginary boundary. (2.3)

Suspension A homogeneous mixture in which the particles of at least one component are larger than colloidal particles (> 1000 nm in at least one dimension). (12.10)

System That part of the universe under study and separated from the surroundings by a real or an imaginary boundary. (5.3)

T

$t_{1/2}$ See *Half-Life.*

Temperature A measure of the hotness or coldness of something and proportional to the average molecular kinetic energy of the atoms, molecules, or ions present. (1.5)

Temperature–Volume Law (Charles' Law) The volume of a given mass of a gas is directly proportional to its Kelvin temperature if the pressure is kept constant. $V \propto T$. (10.3)

Tetrahedral Molecule A molecule with a central atom bonded to four other atoms located at the corners of an imaginary tetrahedron. (8.1)

Tetrahedron A four-sided figure with four triangular faces and shaped like a pyramid. (8.1)

Theoretical Yield The yield of a product calculated from the reaction's stoichiometry. (3.8)

Theory A tested explanation of the results of many experiments. (1.2)

Thermal Property A physical property, like heat capacity or heat of fusion, that concerns a substance's ability to absorb heat without changing chemically. (5.4)

Thermochemical Equation A balanced chemical equation accompanied by the value of $\Delta H°$ that corresponds to the mole quantities specified by the coefficients. (5.5)

Thermochemistry The study of the energy changes of chemical reactions. (5.1)

Thermodynamic Equilibrium Constant (K) The equilibrium constant that is calculated from $\Delta G°$ (the standard free energy change) for a reaction at T (in K) by the equation, $\Delta G° = RT \ln K$. (15.6)

Thermodynamics The study of the laws that govern the energy and entropy changes of physical and chemical events. (13.1)

Third Law of Thermodynamics For a pure crystalline substance at 0 K, $S = 0$. (13.5)

Titrant The solution added from a buret during a titration. (4.11)

Titration An analytical procedure in which a solution of unknown concentration is combined slowly and carefully with a standard solution until a color change of some indicator or some other signal shows that equivalent quantities have reacted. Either solution can be the titrant in a buret with the other solution being in a receiving flask. (4.11)

Torr A unit of pressure equal to 1/760 atm. 1 mm Hg. (10.2)

Total Energy The sum of the kinetic and potential energies of a system. (5.1)

Tracer Analysis The use of radioactivity to trace the movement of something or to locate the site of the radioactivity. (20.7)

Transcription The synthesis of mRNA at the direction of DNA. (21.11)

Trans Isomer A stereoisomer whose uniqueness lies in having two groups that project on opposite sides of a reference plane. (19.4, 21.2)

Transition Elements The elements located between Groups IIA and IIIA in the periodic table. (2.4)

Transition Metals The transition elements. (2.4)

Transition State The brief moment during an elementary process in a reaction mechanism when the species involved have acquired the minimum amount of potential energy needed for a successful reaction, an amount of energy that corresponds to the high point on a potential energy diagram of the reaction. (14.6)

Transition State Theory A theory about the formation and breakup of activated complexes. (14.6)

Translation The synthesis of a polypeptide at the direction of a molecule of mRNA. (21.11)

Transmutation The conversion of one isotope into another. (20.5)

Transuranium Elements Elements 93–103. (20.5)

Traveling Wave A wave whose peaks and nodes move. (6.3)

Triacylglycerol An ester of glycerol and three fatty acids. (21.9)

Trigonal Bipyramid A six-sided figure made of two three-sided pyramids that share a common face. (8.1)

Trigonal Bipyramidal Molecule A molecule with a central atom holding five other atoms that are located at the corners of a trigonal bipyramid. (8.1)

Triple Bond A covalent bond in which three pairs of electrons are shared. (7.3, 8.6)

Triple Point The temperature and pressure at which the liquid, solid, and vapor states of a substance can coexist in equilibrium. (11.9)

Triprotic Acid An acid that can furnish three H^+ ions per molecule. (2.9)

Tyndall Effect The scattering of light by colloidally dispersed particles that gives a milky appearance to the mixture. (12.10)

U

Uncertainty Principle There is a limit to our ability to measure a particle's speed and position simultaneously. (6.8)

Unit Cell The smallest portion of a crystal that can be repeated over and over in all directions to give the crystal lattice. (11.10)

Universal Gas Constant (*R*) The ratio of *PV* to *nT* for gases,

$R = 0.0821$ L atm mol^{-1} K^{-1}. (10.4)

Unsaturated Compound A compound whose molecules have one or more double or triple bonds. (21.2)

Unsaturated Solution Any solution with a concentration less than that of a saturated solution of the same solute and solvent. (4.1)

V

Vacuum An enclosed space containing no matter whatsoever. A *partial vacuum* is an enclosed space containing a gas at a very low pressure. (10.2)

Valence Band The vacant or partially filled band of outer-shell orbitals in a solid. (8.9)

Valence Bond Theory A theory of covalent bonding that views a bond as being formed by the sharing of one pair of electrons between two overlapping atomic or hybrid orbitals. (8.4)

Valence Electrons The electrons of an atom in its valence shell that participate in the formation of chemical bonds. (6.6)

Valence Shell The electron shell with the highest principal quantum number, *n*, that is occupied by electrons. (6.6)

Valence Shell Electron Pair Repulsion Theory (VSEPR Theory) The geometry of molecule at any given central atom is determined by the repulsions of both bonding and nonbonding (lone pair) electrons in the valence shell. (8.2)

Van der Waals Constants Empirical constants that make the van der Waals equation conform to the gas law behavior of a real gas. (10.9)

Van der Waals Equation An equation of state for a real gas that corrects the ideal gas law for the excluded volume of the gas and for intermolecular attractions. (10.9)

Van't Hoff Factor The ratio of the observed freezing point depression to the value calculated on the assumption that the solute dissolves as un-ionized molecules. (12.9)

Vapor Pressure The pressure exerted by the vapor above a liquid (usually referring to the *equilibrium* vapor pressure when the vapor and liquid are in equilibrium with each other). (10.6, 11.5)

Vapor Pressure–Concentration Law (Raoult's Law) The vapor pressure of one component above a mixture of molecular compounds equals the product of its vapor pressure when pure and its mole fraction. (12.6)

Viscosity The resistance of a fluid to a change in its form. (11.3)

Visible Spectrum That region of the electromagnetic spectrum whose frequencies can be detected by the human eye. (6.1)

Volatile Descriptive of a liquid that has a low boiling point, a high vapor pressure at room temperature, and therefore evaporates easily. (12.6)

Volt (V) The SI unit of electromotive force or emf in joules per coulomb.

1 V $= 1$ J C^{-1} (18.6)

Voltaic Cell See *Galvanic Cell.*

VSEPR Theory See *Valence Shell Electron Pair Repulsion Theory.*

V-Shaped Molecule See *Nonlinear Molecule.*

W

Wavelength (λ) The distance between crests in the wavelike oscillations of electromagnetic radiations. (6.1)

Wave Mechanics A theory of atomic structure based on the wave properties of matter. (6.3)

Weak Acid An acid with a low percentage ionization in solution; a poor proton donor; an acid with a low value of K_a. (4.3, 16.2)

Weak Base A base with a low percentage ionization in solution; a poor proton acceptor; a base with a low value of K_b. (4.3, 16.2)

Weak Electrolyte A substance that has a low percentage ionization or dissociation in solution. (4.3)

Weighing The operation of measuring the mass of something. (1.5)

Weight The force with which something is attracted to the Earth by gravity. (1.5)

Weight Fraction The ratio of the mass of one component of a mixture to the total mass. (12.5)

Wetting The spreading of a liquid across a solid surface. (11.3)

X

X Ray A stream of very high energy photons emitted by substances when they are bombarded by high-energy beams of electrons or emitted by radionuclides that have undergone *K*-electron capture. (20.3)

Y

Yield, Actual The quantity of product isolated from a chemical reaction. (3.8)

Yield, Percentage The ratio, given as a percent, of the quantity of product actually obtained in a reaction to the theoretical yield. (3.8)

Yield, Theoretical The quantity of a product calculated by the stoichiometry of the reaction. (3.8)

Z

Zinc–Carbon Dry Cell A galvanic cell of about 1.5 V involving zinc and manganese dioxide under mildly acidic conditions. (18.10)

Photo Credits

CHAPTER 1

Opener: Scott Rutherford/Black Star. *Figure 1.1:* Michael Watson. *Page 3:* Stephen Frisch. *Figure 1.2:* Michael Watson. *Figure 1.3:* National Bureau of Standards. *Figures 1.4, 1.5, and 1.6a):* Michael Watson. *Figure 1.6b):* Courtesy Central Scientific Co. *Figure 1.6c):* Cole-Parmer Instrument Co. *Figure 1.7:* Michael Watson. *Page 19:* Chris Noble/Tony Stone Images/New York, Inc. *Page 21 (left):* West Stock Photography. *Page 21 (right):* Michael Schumann/SABA. *Figure 1.12:* Sinclair Stammers/Science Photo Library/ Photo Researchers.

CHEMICALS IN USE 1

Figure 1: Bettmann Archive

CHAPTER 2

Opener: John Reader/Science Photo Library/Photo Researchers. *Figures 2.2 and 2.3:* Michael Watson. *Figure 2.4:* Breck Kent. *Figure 2.6:* Michael Watson. *Figure 2.7:* Peter Lerman. *Figure 2.8:* Robert Capece. *Page 56:* Chemical Club Library. *Figure 2.15:* George Haling/Photo Researchers. *Figure 2.16:* Courtesy General Cable Corp. *Figure 2.17:* OPC, Inc. *Page 61 (top):* Russ Kinne/Photo Researchers. *Figure 2.18:* Michael Watson. *Figure 2.19:* Dan McCoy/Rainbow. *Page 75:* Michael Watson. *Page 78 (top):* Robert Capece.

CHEMICALS IN USE 2

Figure 2a): Kent & Donna Dannen. *Figure 2b):* Bob Anderson/Masterfile.

CHAPTER 3

Opener: R. H. Richard Tomkins/ Gamma Liaison. *Page 91:* Photri, Inc.

Figure 3.1: Michael Watson. *Page 93 (top):* David Parker/Science Photo Library/Photo Researchers. *Page 93 (bottom):* E.R. Degginger/Animals Animals. *Figure 3.2:* Michael Watson. *Page 95:* Steven Mangold/West Light. *Page 105:* Michael Watson. *Figure 3.4:* Orgo-Thermit, Inc. *Page 112:* Peter Garfield/Tony Stone Images/New York, Inc.

CHEMICALS IN USE 3

Figure 3a): Luiz Claudio Marigo/Peter Arnold, Inc. *Figure 3b):* George Hall/ Woodfin Camp & Associates.

CHAPTER 4

Opener: Nick Nicholson/The Image Bank. *Figure 4.1:* Courtesy Miles Laboratories. *Figures 4.2, 4.3, and 4.4:* Michael Watson. *Page 131 (top):* Robert Capece. *Page 131 (bottom):* Ken Karp. *Figure 4.4:* Michael Watson. *Page 137:* Bill Stormont/The Stock Market. *Figure 4.5:* Andy Washnik. *Figure 4.6:* Michael Watson. *Figure 4.7:* OPC, Inc. *Figure 4.9:* Michael Watson. *Page 144 (top):* The Denver Post. *Figure 4.10:* Andy Washnik. *Page 147:* Courtesy of Betz Company. *Page 151:* OPC, Inc. *Page 157:* Peter Lerman. *Figures 4.11 and 4.12:* Michael Watson. *Figure 4.13:* OPC, Inc. *Page 167:* Michael Watson. *Page 170:* Robert Capece. *Figure 4.14:* Michael Watson. *Figure 4.15:* Andy Washnik.

CHAPTER 5

Opener: John Kelly/The Image Bank.

CHAPTER 6

Opener: Otto Rogge/The Stock Market. *Figure 6.1a):* Peter Arnold/Peter

Arnold, Inc. *Figure 6.1b):* IPA/The Image Works. *Figure 6.1c):* Edward Degginger/Bruch Coleman, Inc. *Figure 6.5:* From "The Gift of Color," Eastman Kodak Company. *Page 229:* Henley & Savage/The Stock Market. *Figure 6.10:* Phil Savoie/Bruch Coleman, Inc. *Figure 6.12:* Jody Dole/The Image Ban. *Page 235:* Courtesy Perkin Elmer Corporation. *Page 240:* Ray Pfortner/Peter Arnold, Inc.

CHEMICALS IN USE 5

Figure 5b): Any Sacks/Tony Stone Images/New York, Inc. *Figure 5c):* Larry Keenan/The Image Bank.

CHAPTER 7

Opener: Gary Newkirk/Allsport. *Page 280:* Andy Washnik.

CHEMICALS IN USE 6

Figure 2: David Perker/IMI/University of Birmingham TC Consortium/ Science Photo Library/Photo Researchers

CHAPTER 8

Opener: Bruch Stoddard/FPG International. *Page 355:* Andy Washnik.

CHEMICALS IN USE 7

Figure 7a): Used by permission from M.R. Schoeberl and D.L. Hartmann, *Science,* January 4, 1991, p. 47.

CHAPTER 9

Opener: Bruch Hands/Tony Stone Images/New York, Inc. *Figure 9.1:* Andy Washnik. *Figures 9.2, 9.3, 9.4 and 9.5:* Michael Watson. *Figures 9.6 and*

9.7: OPC, Inc. *Figure 9.9:* Robert Capece. *Figure 9.10:* Lynn Johnson.

CHAPTER 10

Opener: Novosti Press Agency/Science Photo Library/Photo Researchers. *Figures 10.1 and 10.2:* OPC, Inc. *Page 421:* OPC, Inc.

CHEMICALS IN USE 8

Figure 8a) and b): Jim Mendenhall.

CHAPTER 11

Opener: Jeff Gnass. *Page 444:* Comstock, Inc. *Page 449:* IPA/Peter Arnold, Inc. *Figures 11.8 and 11.10:* Michael Watson. *Page 452:* Michael Watson. *Page 459:* Courtesy Corning. *Page 462:* Michael Watson. *Page 469:* Courtesy Mountain House Foods. *Figure 11.26:* Robert Capece. *Page 472:* Eric Grave/ Science Source/Photo Researchers. *Page 476 (left):* Courtesy Professor M.H.F. Wilkins, Biophysics Dept., Kings College, London. *Page 476 (right):* Peter Menzel/Stock, Boston. *Figure 11.38:* Robert Capece. *Page 480:* Courtesy Allied Signal, Inc.

CHEMICALS IN USE 9

Figure 9a): Sylvain Grandadam/Photo Researchers. *Figure 9b):* Kimble Glass Company.

CHAPTER 12

Opener: David Muench Photography. *Page 519:* Dan McCoy/Black Star. *Page 521:* Courtesy Recovery Engineering. *Page 528:* John Clare duBois/Photo Researchers. *Page 529:* Lawrence Livermore Laboratory. *Figure 12.27:* OPC, Inc.

CHEMICALS IN USE 10

Figure 10: Simon Fraser/Science Photo Library/Photo Researchers.

CHAPTER 13

Opener: Chromosohm/Sohm/Photo Researchers. *Figure 13.5a):* Bob Torrez/Tony Stone Images/New York, Inc. *Figure 13.5b):* Raphael Koskas/Tony Stone Images/New York, Inc. *Figure 13.5c):* Lowell J. Georgia/Photo Researchers. *Figure 13.11:* Ira Wyman/Sygma. *Page 557:* Peter Menzel/Stock Boston.

CHEMICALS IN USE 11

Figure 11b): Courtesy Sumitomo Electric, USA.

CHAPTER 14

Opener: Gay Bumgarner/Tony Stone Images/New York, Inc. *Figure 14.1:* Courtesy USDA. *Figure 14.2:* OPC, Inc. *Page 617:* Leonard Lee Rue III/Photo Researchers. *Figure 14.16a):* Courtesy American Petroleum Institute. *Figure 14.16b):* Courtesy Englehard Corporation. *Figure 14.17:* Courtesy AC Spark Plug.

CHAPTER 15

Opener: Alex Stewart/The Image Bank. *Figures 15.5 and 15.6:* Michael Watson.

CHAPTER 16

Opener: Paul Simcock/The Image Bank. *Figure 16.2:* Andy Washnik. *Figure 16.3:* Courtesy Fisher Scientific. *Figure 16.8:* OPC, Inc.

CHEMICALS IN USE 13

Figure 13a): E.A. Heininger/Rapho/ Photo Researchers.

CHAPTER 17

Opener: Mike Severns/Tony Stone Images/New York, Inc. *Pages 742, 745 and 746:* Michael Watson. *Figure 17.1:* Michael Watson. *Figure 17.2:* OPC, Inc. *Page 751 (top):* Michael Watson.

CHEMICALS IN USE 14

Figure 14a): David Muench Photography.

CHAPTER 18

Opener: Duka/Photo Network. *Figures 18.3 and 18.4:* Michael Watson. *Page 773:* Kaiser Aluminum. *Figure 18.12:* Michael Watson. *Page 774:* ASARCO Inc. *Page 783:* James Brady. *Page 795:* Courtesy Fisher Scientific. *Figure 18.19:* OPC, Inc.

CHAPTER 19

Opener: Bob Krist/Tony Stone Images/ New York, Inc. *Figures 19.1 and 19.2:* Andy Washnik. *Page 814:* Andy Washnik. *Pages 815 (top), 822, and 824:* Michael Watson. *Figures 19.3, 19.8, and 19.12:* Michael Watson. *Page 827:* Michael Watson.

CHAPTER 20

Opener: Roger Ressmeyer/Starlight. *Page 847:* Philippe Halsman Studio. *Page 853:* Courtesy Los Alamos Scientific Scientific Laboratory, University of California. *Page 855:* AIP Neils Bohr Library. *Figure 20.10:* Courtesy Brookhaven National Laboratory. *Figure 20.12:* William Rivelli/The Image Bank. *Figure 20.13:* Courtesy Ludlum Measurements, Inc. *Figure 20.14:* Yoav Levy/Phototake.

CHEMICALS IN USE 16

Figure 16a): Hank Morgan/Rainbow. *Figure 16b):* Courtesy of E. D. London, National Institute on Drug Abuse. *Figure 16c):* Howard Sochurek/Woodfin Camp & Associates.

CHAPTER 21

Opener: Darrell Wong/Tony Stone Images/New York, Inc. *Figure 21.1a):* Robert Capece. *Figure 21.1b):* Michael Watson. *Pages 897 and 901:* Robert Ca-pece. *Figure 21.2:* Ned Gillette/The Stock Market. *Page 917:* Courtesy Regal Ware. *Figure 21.3:* Presse-Sports/Picture Group. *Figure 21.6:* Robert Tringali Jr./Sports Chrome Inc. *Figure 21.7:* Michael Ventura/Bruce Coleman, Inc. *Figure 21.8:* Courtesy Loctite Corporation. *Figure 21.16b):* Nelson Max/Peter Arnold, Inc. *Figure 21.20a):* Tripos Associates.

Index

Page references preceded by *S-* are for items in the *Descriptive Chemistry of the Elements* text. Numbers set in italics refer to tables.